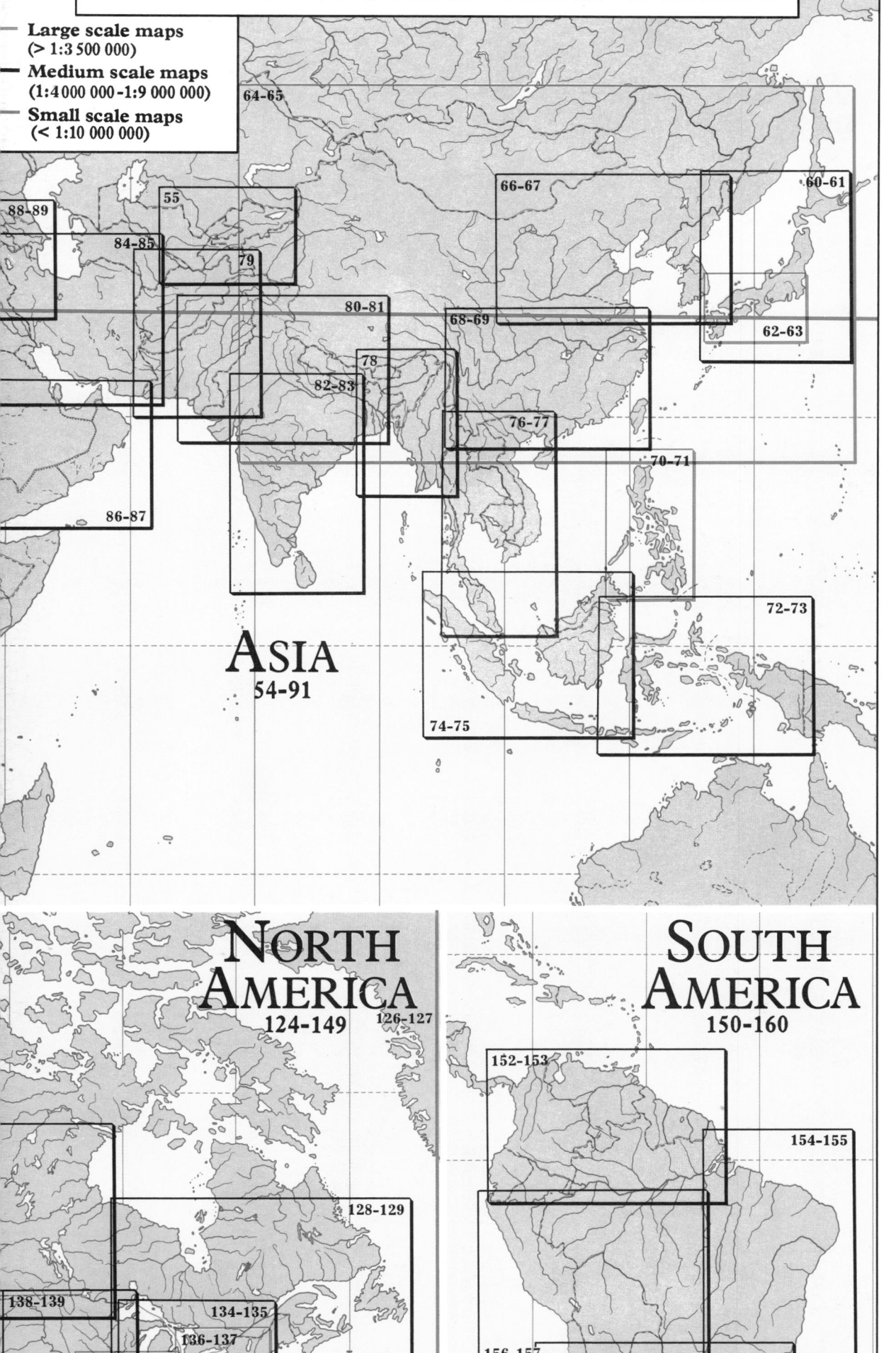

KEY TO WORLD MAP PAGES

- **Large scale maps**
  (> 1:3 500 000)
- **Medium scale maps**
  (1:4 000 000–1:9 000 000)
- **Small scale maps**
  (< 1:10 000 000)

ASIA
54-91

NORTH AMERICA
124-149   126-127

SOUTH AMERICA
150-160

ATLAS

OF THE

WORLD

OXFORD

OXFORD

ATLAS
OF THE
WORLD

THIRD EDITION

© 1995 Reed International Books Limited

George Philip Limited,
an imprint of Reed Books,
Michelin House, 81 Fulham Road,
London SW3 6RB, England

Cartography by Philip's

Published in North America by
Oxford University Press, Inc.,
198 Madison Avenue,
New York, N.Y. 10016, U.S.A.

Oxford is a registered trademark of
Oxford University Press

Library of Congress Cataloging-in-Publication Data

Atlas of the world / [cartography by Philip's].—3rd ed.
    p. cm.
    © 1995 Reed International Books Limited.
    George Philip Limited, an imprint of Reed
    Books, London. Published in North America
    by Oxford University Press—CIP t.p. verso
    Audience: For ages 12 through college.
    Includes index.
    ISBN 0–19–521172–3
    1. Atlases. 2. Children's atlases. [1. Atlases.]
    I. George Philip Limited. II. Oxford University
    Press.
G1021.A7545 1995 <G&M>
912—dc20   95–8192
                                        CIP
                                      MAP AC

ISBN 0–19–521172–3

Printing (last digit): 9 8 7 6 5 4 3 2 1

Printed in Spain

WORLD MAPS

The reference maps which form the main body of this atlas have been prepared in accordance with the highest standards of international cartography to provide an accurate and detailed representation of the Earth. The scales and projections used have been carefully chosen to give balanced coverage of the world, while emphasizing the most densely populated and economically significant regions. A hallmark of Philip's mapping is the use of hill shading and relief coloring to create a graphic impression of landforms: this makes the maps exceptionally easy to read. However, knowledge of the key features employed in the construction and presentation of the maps will enable the reader to derive the fullest benefit from the atlas.

Map sequence

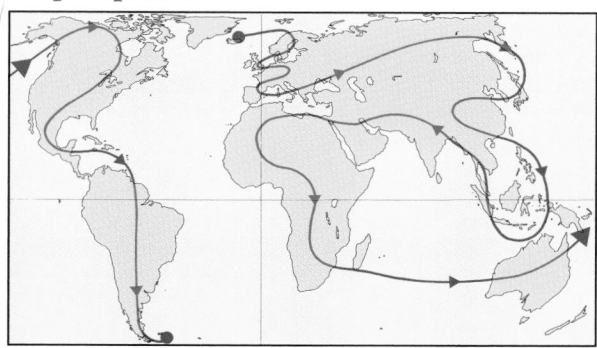

The atlas covers the Earth continent by continent: first Europe; then its land neighbor Asia (mapped north before south, in a clockwise sequence), then Africa, Australia and Oceania, North America and South America. This is the classic arrangement adopted by most cartographers since the 16th century. For each continent, there are maps at a variety of scales. First, physical relief and political maps of the whole continent; then a series of larger-scale maps of the regions within the continent, each followed, where required, by still larger-scale maps of the most important or densely populated areas. The governing principle is that by turning the pages of the atlas, the reader moves steadily from north to south through each continent, with each map overlapping its neighbors. A key map showing this sequence, and the area covered by each map, can be found on the endpapers of the atlas.

Map presentation

With very few exceptions (e.g. for the Arctic and Antarctic), the maps are drawn with north at the top, regardless of whether they are presented upright or sideways on the page. In the borders will be found the map title; a locator diagram showing the area covered and the page numbers for maps of adjacent areas; the scale; the projection used; the degrees of latitude and longitude; and the letters and figures used in the index for locating place names and geographical features. Physical relief maps also have a height reference panel identifying the colors used for each layer of contouring.

Map symbols

Each map contains a vast amount of detail which can only be conveyed clearly and accurately by the use of symbols. Points and circles of varying sizes locate and identify the relative importance of towns and cities; different styles of type are employed for administrative, geographical and regional place names. A variety of pictorial symbols denote landscape features such as glaciers, marshes and reefs, and man-made structures including roads, railroads, airports, canals and dams. International borders are shown by red lines. Where neighboring countries are in dispute, for example in the Middle East, the maps show the *de facto* boundary between nations, regardless of the legal or historical situation. The symbols are explained on the first page of the World Maps section of the atlas.

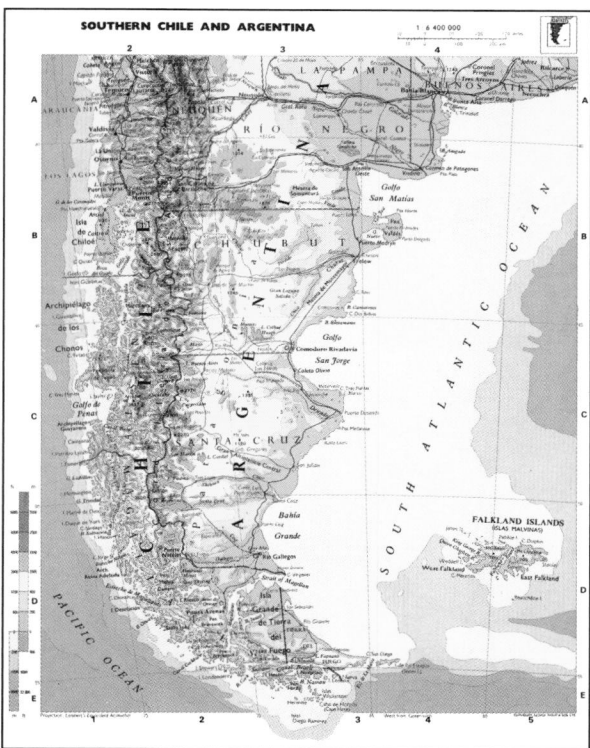

Map scales

1: 16 000 000
1 inch = 252 statute miles

The scale of each map is given in the numerical form known as the 'representative fraction'. The first figure is always one, signifying one unit of distance on the map; the second figure, usually in millions, is the number by which the map unit must be multiplied to give the equivalent distance on the Earth's surface. Calculations can easily be made in centimeters and kilometers, by dividing the Earth units figure by 100 000 (i.e. deleting the last five 0s). Thus 1:1 000 000 means 1 cm = 10 km. The calculation for inches and miles is more laborious, but 1 000 000 divided by 63 360 (the number of inches in a mile) shows that 1:1 000 000 means approximately 1 inch = 16 miles. The table below provides distance equivalents for scales down to 1:50 000 000.

LARGE SCALE		
1: 1 000 000	1 cm = 10 km	1 inch = 16 miles
1: 2 500 000	1 cm = 25 km	1 inch = 39.5 miles
1: 5 000 000	1 cm = 50 km	1 inch = 79 miles
1: 6 000 000	1 cm = 60 km	1 inch = 95 miles
1: 8 000 000	1 cm = 80 km	1 inch = 126 miles
1: 10 000 000	1 cm = 100 km	1 inch = 158 miles
1: 15 000 000	1 cm = 150 km	1 inch = 237 miles
1: 20 000 000	1 cm = 200 km	1 inch = 316 miles
1: 50 000 000	1 cm = 500 km	1 inch = 790 miles
SMALL SCALE		

Measuring distances

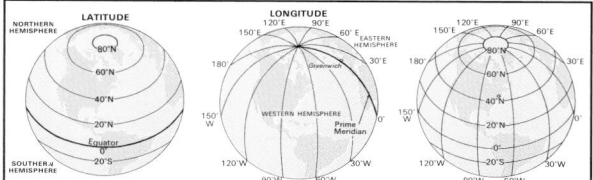

Although each map is accompanied by a scale bar, distances cannot always be measured with confidence because of the distortions involved in portraying the curved surface of the Earth on a flat page. As a general rule, the larger the map scale (i.e. the lower the number of Earth units in the representative fraction), the more accurate and reliable will be the distance measured. On small-scale maps such as those of the world and of entire continents, measurement may only be accurate along the 'standard parallels', or central axes, and should not be attempted without considering the map projection.

Map projections

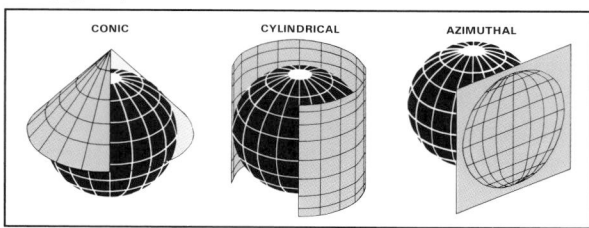

Unlike a globe, no flat map can give a true scale representation of the world in terms of area, shape and position of every region. Each of the numerous systems that have been devised for projecting the curved surface of the Earth on to a flat page involves the sacrifice of accuracy in one or more of these elements. The variations in shape and position of landmasses such as Alaska, Greenland and Australia, for example, can be quite dramatic when different projections are compared.

For this atlas, the guiding principle has been to select projections that involve the least distortion of size and distance. The projection used for each map is noted in the border. Most fall into one of three categories – conic, cylindrical or azimuthal – whose basic concepts are shown above. Each involves plotting the forms of the Earth's surface on a grid of latitude and longitude lines, which may be shown as parallels, curves or radiating spokes.

Latitude and longitude

Accurate positioning of individual points on the Earth's surface is made possible by reference to the geometrical system of latitude and longitude. Latitude *parallels* are drawn west–east around the Earth and numbered by degrees north and south of the Equator, which is designated 0° of latitude. Longitude *meridians* are drawn north–south and numbered by degrees east and west of the *prime meridian*, 0° of longitude, which passes through Greenwich in England. By referring to these co-ordinates and their subdivisions of minutes (¹⁄₆₀th of a degree) and seconds (¹⁄₆₀th of a minute), any place on Earth can be located to within a few hundred yards. Latitude and longitude are indicated by blue lines on the maps; they are straight or curved according to the projection employed. Reference to these lines is the easiest way of determining the relative positions of places on different maps, and for plotting compass directions.

Name forms

For ease of reference, both English and local name forms appear in the atlas. Oceans, seas and countries are shown in English throughout the atlas; country names may be abbreviated to their commonly accepted form (e.g. Germany, not The Federal Republic of Germany). Conventional English forms are also used for place names on the smaller-scale maps of the continents. However, local name forms are used on all large-scale and regional maps, with the English form given in brackets only for important cities – the large-scale map of Russia and Central Asia thus shows Moskva (Moscow). For countries which do not use a Roman script, place names have been transcribed according to the systems adopted by the British and US Geographic Names Authorities. For China, the Pin Yin system has been used, with some more widely known forms appearing in brackets, as with Beijing (Peking). Both English and local names appear in the index, the English form being cross-referenced to the local form.

CONTENTS

NOTE
The titles to the World Maps
list the main countries, states
and provinces covered by
each map. A name given in
italics indicates that only part
of the country is shown on
the map.

Netherlands, Belgium and Luxembourg 1:1 000 000

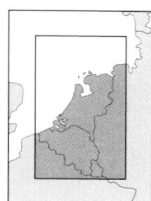

20–21

Northern France 1:2 000 000

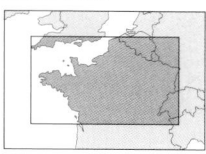

22–23

Southern France 1:2 000 000
Corsica, Monaco

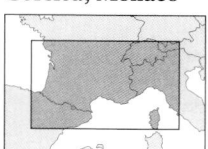

24–25

Germany 1:2 000 000

26–27

Switzerland 1:800 000
Liechtenstein

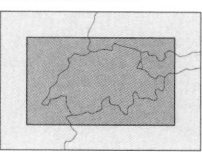

28–29

Austria, Czech Republic, Slovak Republic and Hungary 1:2 000 000
Poland

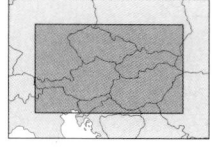

30–31

Malta, Crete, Corfu, Rhodes and Cyprus 1:800 000 / 1:1 040 000

32

Balearics, Canaries and Madeira 1:800 000 / 1:1 600 000
Mallorca, Menorca, Ibiza

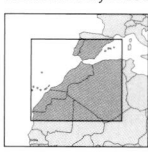

33

Eastern Spain 1:2 000 000
Andorra

34–35

Western Spain and Portugal 1:2 000 000

36–37

Northern Italy, Slovenia and Croatia 1:2 000 000
San Marino, Slovenia, *Croatia*

38–39

Southern Italy 1:2 000 000
Sardinia, Sicily

40–41

The Lower Danube 1:2 000 000
Bosnia-Herzegovina, Yugoslavia, Macedonia, Bulgaria

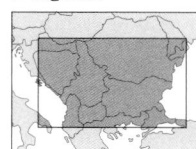

42–43

Greece and Albania 1:2 000 000

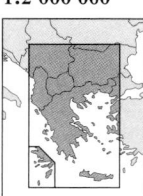

44–45

Romania 1:2 000 000

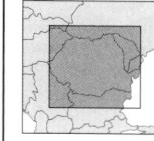

46

Poland 1:2 000 000

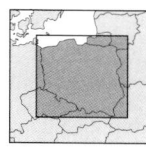

47

Eastern Europe and Turkey 1:8 000 000

48–49

Baltic States, Belarus and Ukraine 1:4 000 000
Russia, Estonia, Latvia, Lithuania, Belarus, Moldova, Ukraine

50–51

Volga Basin and the Caucasus 1:4 000 000
Russia, Georgia, *Armenia, Azerbaijan*

52–53

ASIA

Southern Urals 1:4 000 000
Russia

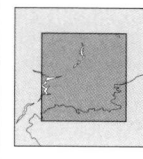

54

Central Asia 1:4 000 000
Kazakhstan, Kyrgyzstan, Tajikistan, *Uzbekistan*

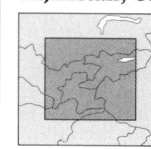

55

Russia and Central Asia 1:16 000 000
Russia, Kazakhstan, Turkmenistan, Uzbekistan

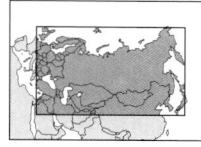

56–57

Asia: Physical 1:40 000 000

58

Asia: Political 1:40 000 000

59

Japan 1:4 000 000
Ryukyu Islands

60–61

Southern Japan 1:2 000 000

62–63

China 1:12 000 000
Mongolia

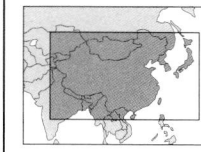

64–65

Northern China and Korea 1:4 800 000
North Korea, South Korea

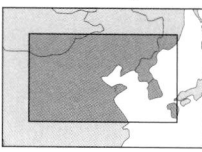

66–67

Southern China 1:4 800 000
Hong Kong, Taiwan, Macau

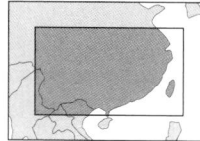

68–69

Philippines 1:3 200 000

70–71

Eastern Indonesia 1:5 600 000

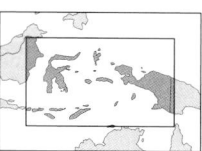

72–73

Western Indonesia 1:5 600 000
Malaysia, Singapore, Brunei

74–75

Mainland South-East Asia 1:4 800 000
Thailand, Vietnam, Cambodia, Laos

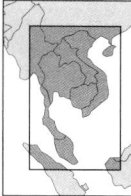

76–77

Bangladesh, North-Eastern India and Burma 1:4 800 000
Bhutan

78

Afghanistan and Pakistan
1:5 600 000

 79

The Indo-Gangetic Plain
1:4 800 000
India, Nepal, *Pakistan*, Kashmir

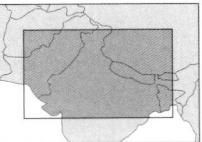 80–81

**Southern India and
Sri Lanka** 1:4 800 000

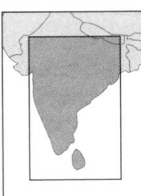 82–83

The Middle East 1:5 600 000
Iran, Iraq, *Saudi Arabia*, Kuwait

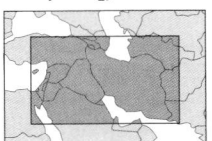 84–85

**Southern Arabian
Peninsula** 1:5 600 000
Saudi Arabia, Yemen, United
Arab Emirates, Oman, Qatar

 86–87

**Turkey and
Transcaucasia** 1:4 000 000
Turkey, Syria, Georgia,
Armenia, Azerbaijan, *Iraq*

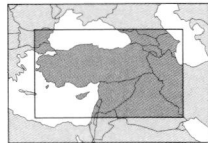 88–89

**Arabia and the Horn of
Africa** 1:12 000 000
Saudi Arabia, Oman, Yemen,
Somalia, Ethiopia, Eritrea,
Djibouti

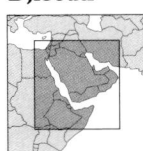 90

The Near East 1:2 000 000
Israel, Lebanon, *Jordan*

 91

AFRICA

Africa: Physical
1:33 600 000
92

Africa: Political
1:33 600 000
93

The Nile Valley 1:6 400 000
Egypt, Sudan, Eritrea, *Ethiopia*
The Nile Delta 1:3 200 000

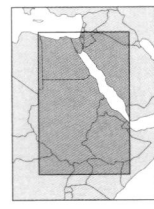 94–95

Central North Africa
1:6 400 000
Libya, Chad, *Niger*

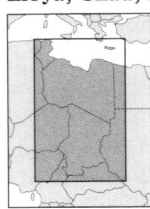 96–97

North-West Africa
1:6 400 000
Algeria, Morocco, Tunisia,
Mauritania, *Niger*, *Mali*

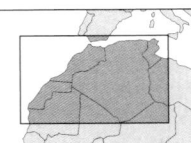 98–99

West Africa 1:6 400 000
Nigeria, Ivory Coast, Ghana,
Senegal, Guinea, Burkina Faso

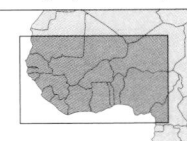 100–101

Central Africa 1:6 400 000
Zaïre, Angola, Cameroon, Congo,
Gabon, Central African Republic

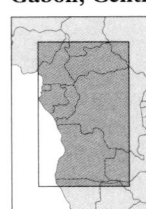 102–103

Southern Africa 1:6 400 000
South Africa, Zimbabwe,
Madagascar, *Mozambique*,
Botswana, Namibia

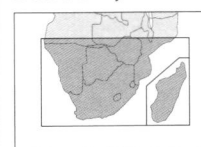 104–105

East Africa 1:6 400 000
Kenya, Tanzania, Zambia,
Uganda, Malawi

 106–107

Horn of Africa 1:6 400 000
Somalia, *Eritrea*, *Ethiopia*,
Djibouti

 108

Indian Ocean 1:40 000 000
Mauritius, Réunion, Seychelles,
Maldives
109

AUSTRALIA AND OCEANIA

**Australia and Oceania:
Physical and Political**
1:16 000 000
110–111

Western Australia
1:6 400 000
Northern Territory

 112–113

Eastern Australia
1:6 400 000
Queensland, Tasmania,
New South Wales

 114–115

South-East Australia
1:3 200 000
New South Wales, Victoria,
South Australia

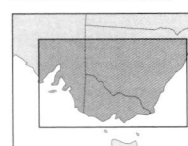 116–117

**New Zealand – North
Island** 1:2 800 000

 118

**New Zealand – South
Island** 1:2 800 000

 119

Papua New Guinea
1:5 200 000

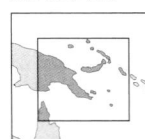 120

VIII

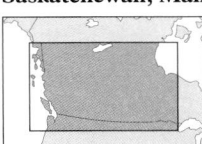

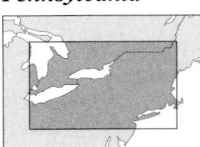
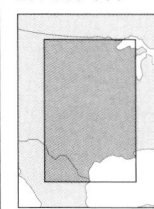
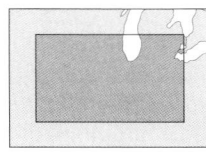

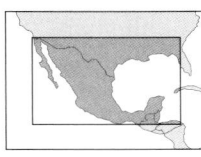

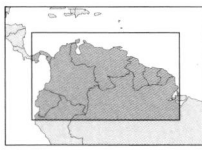

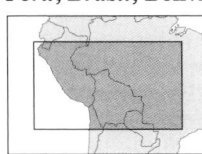
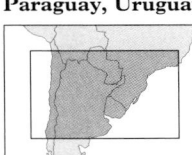

IX

WORLD STATISTICS: COUNTRIES

This alphabetical list includes all the countries and territories of the world. If a territory is not completely independent, then the country it is associated with is named. The area figures give the total area of land, inland water and ice. Units for areas and populations are thousands. The annual income is the Gross National Product per capita in US dollars. The figures are the latest available, usually 1994.

Country/Territory	Area km² Thousands	Area miles² Thousands	Population Thousands	Capital	Annual Income US $
Adélie Land (Fr.)	432	167	0.03	–	–
Afghanistan	652	252	18,879	Kabul	450
Albania	28.8	11.1	3,414	Tirana	820
Algeria	2,382	920	27,325	Algiers	1,840
American Samoa (US)	0.20	0.08	53	Pago Pago	6,000
Amsterdam Is. (Fr.)	0.05	0.02	0.03	–	–
Andorra	0.45	0.17	65	Andorra la Vella	14,000
Angola	1,247	481	10,674	Luanda	620
Anguilla (UK)	0.1	0.04	8	The Valley	6,800
Antigua & Barbuda	0.44	0.17	65	St John's	4,770
Argentina	2,767	1,068	34,182	Buenos Aires	2,780
Armenia	29.8	11.5	3,548	Yerevan	780
Aruba (Neths)	0.19	0.07	69	Oranjestad	6,000
Ascension Is. (UK)	0.09	0.03	1.5	Georgetown	–
Australia	7,687	2,968	17,847	Canberra	17,260
Australian Antarctic Territory	6,120	2,363	0	–	–
Austria	83.9	32.4	7,918	Vienna	22,380
Azerbaijan	86.6	33.4	7,472	Baku	740
Azores (Port.)	2.2	0.87	238	Ponta Delgada	–
Bahamas	13.9	5.4	272	Nassau	12,070
Bahrain	0.68	0.26	549	Manama	6,910
Bangladesh	144	56	117,787	Dhaka	220
Barbados	0.43	0.17	261	Bridgetown	6,540
Belarus	207.6	80.1	10,163	Minsk	2,930
Belgium	30.5	11.8	10,080	Brussels	20,880
Belize	23	8.9	210	Belmopan	2,220
Benin	113	43	5,246	Porto-Novo	410
Bermuda (UK)	0.05	0.02	63	Hamilton	27,800
Bhutan	47	18.1	1,614	Thimphu	180
Bolivia	1,099	424	7,237	La Paz/Sucre	680
Bosnia-Herzegovina	51	20	3,527	Sarajevo	2,454
Botswana	582	225	1,443	Gaborone	2,790
Bouvet Is. (Nor.)	0.05	0.02	0.02	–	–
Brazil	8,512	3,286	159,143	Brasília	2,770
British Antarctic Terr. (UK)	1,709	660	0.3	Stanley	–
British Indian Ocean Terr. (UK)	0.08	0.03	0	–	–
Brunei	5.8	2.2	280	Bandar Seri Begawan	14,120
Bulgaria	111	43	8,818	Sofia	1,330
Burkina Faso	274	106	10,046	Ouagadougou	300
Burma (Myanmar)	677	261	45,555	Rangoon	500
Burundi	27.8	10.7	6,209	Bujumbura	210
Cambodia	181	70	9,968	Phnom Penh	202
Cameroon	475	184	12,871	Yaoundé	820
Canada	9,976	3,852	29,141	Ottawa	20,710
Canary Is. (Spain)	7.3	2.8	1,494	Las Palmas/Santa Cruz	–
Cape Verde Is.	4	1.6	381	Praia	850
Cayman Is. (UK)	0.26	0.10	30	George Town	20,000
Central African Republic	623	241	3,235	Bangui	410
Chad	1,284	496	6,183	Ndjamena	222
Chatham Is. (NZ)	0.96	0.37	0.05	Waitangi	–
Chile	757	292	14,044	Santiago	2,730
China	9,597	3,705	1,208,841	Beijing (Peking)	470
Christmas Is. (Aus.)	0.14	0.05	2	The Settlement	–
Cocos (Keeling) Is. (Aus.)	0.01	0.005	0.6	West Island	–
Colombia	1,139	440	34,545	Bogotá	1,330
Comoros	2.2	0.86	630	Moroni	500
Congo	342	132	2,516	Brazzaville	1,030
Cook Is. (NZ)	0.24	0.09	19	Avarua	900
Costa Rica	51.1	19.7	3,347	San José	1,960
Croatia	56.5	21.8	4,504	Zagreb	1,800
Crozet Is. (Fr.)	0.51	0.19	35	–	–
Cuba	111	43	10,960	Havana	1,580
Cyprus	9.3	3.6	734	Nicosia	9,820
Czech Republic	78.9	30.4	10,295	Prague	2,450
Denmark	43.1	16.6	5,173	Copenhagen	26,000
Djibouti	23.2	9	566	Djibouti	1,000
Dominica	0.75	0.29	71	Roseau	2,520
Dominican Republic	48.7	18.8	7,684	Santo Domingo	1,050
Ecuador	284	109	11,220	Quito	1,070
Egypt	1,001	387	61,636	Cairo	640
El Salvador	21	8.1	5,641	San Salvador	1,170
Equatorial Guinea	28.1	10.8	389	Malabo	330
Eritrea	94	36	3,437	Asmara	150
Estonia	44.7	17.3	1,541	Tallinn	2,760
Ethiopia	1,128	436	53,435	Addis Ababa	110
Falkland Is. (UK)	12.2	4.7	2	Stanley	–
Faroe Is. (Den.)	1.4	0.54	47	Tórshavn	23,660
Fiji	18.3	7.1	771	Suva	2,010
Finland	338	131	5,083	Helsinki	21,970
France	552	213	57,747	Paris	22,260
French Guiana (Fr.)	90	34.7	141	Cayenne	2,500
French Polynesia (Fr.)	4	1.5	215	Papeete	6,000
Gabon	268	103	1,283	Libreville	4,450
Gambia, The	11.3	4.4	1,081	Banjul	370
Georgia	69.7	26.9	5,450	Tbilisi	850
Germany	357	138	80,278	Berlin/Bonn	23,030
Ghana	239	92	16,944	Accra	450
Gibraltar (UK)	0.007	0.003	28	Gibraltar Town	15,080
Greece	132	51	10,416	Athens	7,290
Greenland (Den.)	2,176	840	58	Godthåb (Nuuk)	9,000
Grenada	0.34	0.13	92	St George's	2,310
Guadeloupe (Fr.)	1.7	0.66	421	Basse-Terre	7,000
Guam (US)	0.55	0.21	147	Agana	6,000
Guatemala	109	42	10,322	Guatemala City	980
Guinea	246	95	6,501	Conakry	570
Guinea-Bissau	36.1	13.9	1,050	Bissau	220
Guyana	215	83	825	Georgetown	330
Haiti	27.8	10.7	7,035	Port-au-Prince	380
Honduras	112	43	5,493	Tegucigalpa	580
Hong Kong (UK)	1.1	0.40	5,838	Victoria	15,360
Hungary	93	35.9	10,161	Budapest	2,970
Iceland	103	40	266	Reykjavik	23,880
India	3,288	1,269	918,570	New Delhi	310
Indonesia	1,905	735	194,615	Jakarta	670
Iran	1,648	636	65,758	Tehran	2,200
Iraq	438	169	19,925	Baghdad	2,000
Ireland	70.3	27.1	3,539	Dublin	12,210
Israel	27	10.3	5,458	Jerusalem	13,220
Italy	301	116	57,157	Rome	20,460
Ivory Coast	322	125	13,780	Yamoussoukro	670
Jamaica	11	4.2	2,429	Kingston	1,340
Jan Mayen Is. (Nor.)	0.38	0.15	0.06	–	–
Japan	378	146	124,815	Tokyo	28,190
Johnston Is. (US)	0.002	0.0009	1	–	–
Jordan	89.2	34.4	5,198	Amman	1,120
Kazakhstan	2,717	1,049	17,027	Alma-Ata	1,680
Kenya	580	224	27,343	Nairobi	310
Kerguelen Is. (Fr.)	7.2	2.8	0.7	–	–
Kermadec Is. (NZ)	0.03	0.01	0.1	–	–
Kiribati	0.72	0.28	77	Tarawa	750
Korea, North	121	47	23,483	Pyongyang	1,040
Korea, South	99	38.2	44,563	Seoul	6,790
Kuwait	17.8	6.9	1,633	Kuwait City	16,150
Kyrgyzstan	198.5	76.6	4,667	Bishkek	820
Laos	237	91	4,742	Vientiane	250
Latvia	65	25	2,583	Riga	1,930
Lebanon	10.4	4	2,915	Beirut	1,400
Lesotho	30.4	11.7	1,996	Maseru	590
Liberia	111	43	2,941	Monrovia	400
Libya	1,760	679	5,225	Tripoli	5,800
Liechtenstein	0.16	0.06	30	Vaduz	33,000
Lithuania	65.2	25.2	3,706	Vilnius	1,310
Luxembourg	2.6	1	401	Luxembourg	35,160
Macau (Port.)	0.02	0.006	398	Macau	2,000
Macedonia	25.3	9.8	2,142	Skopje	1,812
Madagascar	587	227	14,303	Antananarivo	230
Madeira (Port.)	0.81	0.31	253	Funchal	–
Malawi	118	46	10,843	Lilongwe	210
Malaysia	330	127	19,695	Kuala Lumpur	2,790
Maldives	0.30	0.12	246	Malé	460
Mali	1,240	479	10,462	Bamako	254
Malta	0.32	0.12	364	Valletta	7,300
Marshall Is.	0.18	0.07	52	Dalap-Uliga-Darrit	1,500
Martinique (Fr.)	1.1	0.42	375	Fort-de-France	4,000
Mauritania	1,025	396	2,217	Nouakchott	530
Mauritius	2.0	0.72	1,104	Port Louis	2,700
Mayotte (Fr.)	0.37	0.14	101	Mamoundzou	–
Mexico	1,958	756	91,858	Mexico City	3,470
Micronesia, Fed. States	0.70	0.27	121	Palikir	1,500
Midway Is. (US)	0.005	0.002	2	–	–
Moldova	33.7	13	4,420	Kishinev	1,300
Monaco	0.002	0.0001	31	Monaco	16,000
Mongolia	1,567	605	2,363	Ulan Bator	112
Montserrat (UK)	0.10	0.04	11	Plymouth	5,800
Morocco	447	172	26,488	Rabat	1,030
Mozambique	802	309	15,527	Maputo	60
Namibia	825	318	1,500	Windhoek	1,610
Nauru	0.02	0.008	11	Yaren District	–
Nepal	141	54	21,360	Katmandu	170
Netherlands	41.5	16	15,397	Amsterdam/The Hague	20,480
Neths Antilles (Neths)	0.99	0.38	197	Willemstad	6,000
New Caledonia (Fr.)	19	7.3	178	Nouméa	4,000
New Zealand	269	104	3,531	Wellington	12,300
Nicaragua	130	50	4,275	Managua	340
Niger	1,267	489	8,846	Niamey	284
Nigeria	924	357	88,515	Abuja	320
Niue (NZ)	0.26	0.10	2	Alofi	–
Norfolk Is. (Aus.)	0.03	0.01	2	Kingston	–
Northern Mariana Is. (US)	0.48	0.18	47	Saipan	11,500
Norway	324	125	4,318	Oslo	25,820
Oman	212	82	2,077	Muscat	6,480
Pakistan	796	307	136,645	Islamabad	420
Palau	0.46	0.18	17	Koror	2,260
Panama	77.1	29.8	2,585	Panama City	2,420
Papua New Guinea	463	179	4,205	Port Moresby	950
Paraguay	407	157	4,830	Asunción	1,380
Peru	1,285	496	23,331	Lima	950
Peter 1st Is. (Nor.)	0.18	0.07	0	–	–
Philippines	300	116	66,188	Manila	770
Pitcairn Is. (UK)	0.03	0.01	0.07	Adamstown	–
Poland	313	121	38,341	Warsaw	1,910
Portugal	92.4	35.7	9,830	Lisbon	7,450
Puerto Rico (US)	9	3.5	3,646	San Juan	6,330
Qatar	11	4.2	540	Doha	15,860
Queen Maud Land (Nor.)	2,800	1,081	0	–	–
Réunion (Fr.)	2.5	0.97	644	Saint-Denis	4,000
Romania	238	92	22,922	Bucharest	1,130
Ross Dependency (NZ)	435	168	0	–	–
Russia	17,075	6,592	147,370	Moscow	2,510
Rwanda	26.3	10.2	7,750	Kigali	250
St Christopher & Nevis	0.36	0.14	41	Basseterre	3,960
St Helena (UK)	0.12	0.05	6	Jamestown	–
St Lucia	0.62	0.24	141	Castries	2,500
St Paul Is. (Fr.)	0.007	0.003	0	–	–
St Pierre & Miquelon (Fr.)	0.24	0.09	6	Saint Pierre	–
St Vincent & Grenadines	0.39	0.15	111	Kingstown	1,730
San Marino	0.06	0.02	25	San Marino	20,000
São Tomé & Príncipe	0.96	0.37	130	São Tomé	350
Saudi Arabia	2,150	830	17,451	Riyadh	7,510
Senegal	197	76	8,102	Dakar	780
Seychelles	0.46	0.18	73	Victoria	5,110
Sierra Leone	71.7	27.7	4,402	Freetown	160
Singapore	0.62	0.24	2,821	Singapore	15,730
Slovak Republic	49	18.9	5,333	Bratislava	1,930
Slovenia	20.3	7.8	1,942	Ljubljana	7,150
Solomon Is.	28.9	11.2	366	Honiara	690
Somalia	638	246	9,077	Mogadishu	150
South Africa	1,220	471	40,555	Pretoria/Cape Town	2,670
South Georgia (UK)	3.8	1.4	0.05	–	–
South Sandwich Is. (UK)	0.38	0.15	0	–	–
Spain	505	195	39,568	Madrid	13,970
Sri Lanka	65.6	25.3	18,125	Colombo	540
Sudan	2,506	967	27,361	Khartoum	277
Surinam	163	63	418	Paramaribo	4,280
Svalbard (Nor.)	62.9	24.3	4	Longyearbyen	–
Swaziland	17.4	6.7	832	Mbabane	1,090
Sweden	450	174	8,738	Stockholm	27,010
Switzerland	41.3	15.9	7,131	Bern	36,080
Syria	185	71	14,171	Damascus	1,170
Taiwan	36	13.9	20,659	Taipei	8,780
Tajikistan	143.1	55.2	5,933	Dushanbe	490
Tanzania	945	365	28,846	Dodoma	110
Thailand	513	198	58,183	Bangkok	1,840
Togo	56.8	21.9	4,010	Lomé	390
Tokelau (NZ)	0.01	0.005	2	Nukunonu	–
Tonga	0.75	0.29	98	Nuku'alofa	1,100
Trinidad & Tobago	5.1	2	1,292	Port of Spain	3,940
Tristan da Cunha (UK)	0.11	0.04	0.33	Edinburgh	–
Tunisia	164	63	8,733	Tunis	1,720
Turkey	779	301	60,771	Ankara	1,980
Turkmenistan	488.1	188.5	4,010	Ashkhabad	1,230
Turks & Caicos Is. (UK)	0.43	0.17	14	Cockburn Town	5,000
Tuvalu	0.03	0.01	9	Fongafale	600
Uganda	236	91	20,621	Kampala	170
Ukraine	603.7	233.1	51,465	Kiev	1,820
United Arab Emirates	83.6	32.3	1,861	Abu Dhabi	20,020
United Kingdom	243.3	94	58,091	London	17,790
United States of America	9,373	3,619	260,631	Washington, DC	23,240
Uruguay	177	68	3,167	Montevideo	3,340
Uzbekistan	447.4	172.7	22,349	Tashkent	850
Vanuatu	12.2	4.7	165	Port-Vila	1,120
Vatican City	0.0004	0.0002	1	–	–
Venezuela	912	352	21,378	Caracas	2,910
Vietnam	332	127	72,931	Hanoi	220
Virgin Is. (UK)	0.15	0.06	18	Road Town	–
Virgin Is. (US)	0.34	0.13	104	Charlotte Amalie	12,000
Wake Is.	0.008	0.003	0.30	–	–
Wallis & Futuna Is. (Fr.)	0.20	0.08	19	Mata-Utu	–
Western Sahara	266	103	272	El Aaiún	–
Western Samoa	2.8	1.1	169	Apia	940
Yemen	528	204	13,873	Sana	520
Yugoslavia	102.3	39.5	10,763	Belgrade	3,000
Zaïre	2,345	905	42,552	Kinshasa	220
Zambia	753	291	9,196	Lusaka	460
Zimbabwe	391	151	11,002	Harare	570

WORLD STATISTICS: CITIES

This list shows the principal cities with more than 500,000 inhabitants (for China and Japan, only cities with more than 1 million inhabitants are included). The figures are taken from the most recent census or estimate available, and as far as possible are the population of the metropolitan area, e.g. greater New York, Mexico or London. All the figures are in thousands. Local name forms have been used for the smaller cities (e.g. Kraków).

Afghanistan
Kabul 1,424
Algeria
Algiers 1,722
Oran 664
Angola
Luanda 1,544
Argentina
Buenos Aires 11,256
Córdoba 1,198
Rosario 1,096
Mendoza 775
La Plata 640
San Miguel de Tucumán 622
Mar del Plata 520
Armenia
Yerevan 1,254
Australia
Sydney 3,657
Melbourne 3,081
Brisbane 1,302
Perth 1,193
Adelaide 1,050
Austria
Vienna 1,560
Azerbaijan
Baku 1,149
Bangladesh
Dhaka 6,105
Chittagong 2,041
Khulna 877
Rajshahi 517
Belarus
Minsk 1,613
Gomel 506
Belgium
Brussels 954
Bolivia
La Paz 1,126
Santa Cruz 696
Bosnia-Herzegovina
Sarajevo 526
Brazil
São Paulo 9,480
Rio de Janeiro 5,336
Salvador 2,056
Belo Horizonte 2,049
Fortaleza 1,758
Brasília 1,596
Curitiba 1,290
Recife 1,290
Nova Iguaçu 1,286
Pôrto Alegre 1,263
Belém 1,246
Manaus 1,011
Goiânia 921
Campinas 846
Guarulhos 781
São Gonçalo 748
São Luis 695
Duque de Caxias 665
Maceió 628
Santo André 614
Natal 607
Teresina 598
São Bernado de Campo 565
Osasco 563
Campo Grande 526
São João de Meriti 508
Bulgaria
Sofia 1,141
Burkina Faso
Ouagadougou 634
Burma (Myanmar)
Rangoon 2,513
Mandalay 533
Cambodia
Phnom Penh 800
Cameroon
Douala 884
Yaoundé 750
Canada
Toronto 3,893
Montréal 3,127
Vancouver 1,603
Ottawa-Hull 921
Edmonton 840
Calgary 754
Winnipeg 652
Québec 646
Hamilton 600
Central African Rep.
Bangui 597

Chad
Ndjamena 530
Chile
Santiago 5,343
China
Shanghai 12,320
Beijing (Peking) 9,750
Tianjin 7,790
Chongqing 6,511
Wenzhou 5,948
Guangzhou 5,669
Hangzhou 5,234
Shenyang 5,055
Dalian 4,619
Jinzhou 4,448
Wuhan 4,273
Qingdao 4,205
Chengdu 4,025
Jilin 3,974
Nanjing 3,682
Jinan 3,376
Xi'an 2,911
Harbin 2,830
Yingkou 2,789
Dandong 2,574
Anshan 2,517
Nanchang 2,471
Zibo 2,460
Lanzhou 2,340
Lupanshui 2,247
Fushun 2,045
Taiyuan 2,177
Changchun 2,110
Kunming 1,976
Tianshui 1,967
Zhengzhou 1,943
Fuxin 1,693
Zigong 1,673
Fuzhou 1,652
Liaoyang 1,612
Zhaozhuang 1,612
Botou 1,593
Hepei 1,541
Guiyang 1,530
Huainan 1,519
Tangshan 1,500
Linyi 1,385
Qiqihar 1,380
Tai'an 1,370
Changsha 1,330
Shijiazhuang 1,320
Huaibei 1,308
Pingxiang 1,305
Xintao 1,272
Yangcheng 1,265
Yulin 1,255
Dongguang 1,230
Chao'an 1,227
Hohhot 1,206
Baotou 1,200
Suining 1,195
Luoyang 1,190
Macheng 1,190
Xintai 1,167
Yichun 1,167
Ürümqi 1,160
Puyang 1,125
Datong 1,110
Handan 1,110
Shaoxing 1,091
Ningbo 1,090
Zhongshan 1,073
Nanning 1,070
Huangshi 1,069
Laiwu 1,054
Leshan 1,039
Heze 1,017
Linhai 1,012
Changshu 1,004
Colombia
Bogotá 4,921
Cali 1,624
Medellin 1,581
Barranquilla 1,019
Cartagena 688
Congo
Brazzaville 938
Pointe-Noire 576
Croatia
Zagreb 931
Cuba
Havana 2,119
Czech Republic
Prague 1,216
Denmark
Copenhagen 1,337

Dominican Rep.
Santo Domingo 1,601
Ecuador
Guayaquil 1,508
Quito 1,101
Egypt
Cairo 6,800
Alexandria 3,380
El Giza 2,144
Shubra el Kheima 834
El Salvador
San Salvador 1,522
Ethiopia
Addis Ababa 2,213
Finland
Helsinki 977
France
Paris 9,319
Lyons 1,262
Marseilles 1,087
Lille 959
Bordeaux 696
Toulouse 650
Nice 516
Georgia
Tbilisi 1,279
Germany
Berlin 3,446
Hamburg 1,669
Munich 1,229
Cologne 957
Frankfurt 654
Essen 627
Dortmund 601
Stuttgart 592
Düsseldorf 578
Bremen 553
Duisburg 537
Hanover 517
Leipzig 503
Ghana
Accra 965
Greece
Athens 3,097
Guatemala
Guatemala 2,000
Guinea
Conakry 705
Haiti
Port-au-Prince 1,144
Honduras
Tegucigalpa 679
Hong Kong
Kowloon 2,031
Hong Kong 1,251
Tsuen Wan 690
Hungary
Budapest 2,016
India
Bombay 12,572
Calcutta 10,916
Delhi 8,081
Madras 5,361
Hyderabad 4,280
Bangalore 4,087
Ahmadabad 3,298
Pune 2,485
Kanpur 2,111
Nagpur 1,661
Lucknow 1,642
Surat 1,517
Jaipur 1,515
Kochi 1,140
Coimbatore 1,136
Vadodara 1,115
Indore 1,104
Patna 1,099
Madurai 1,094
Bhopal 1,064
Vishakhapatnam 1,052
Varanasi 1,026
Ludhiana 1,012
Agra 956
Jabalpur 887
Allahabad 858
Meerut 847
Vijayawada 845
Jamshedpur 834
Trivandrum 826
Dhanbad 818
Kozhikode 801
Asansol 764
Nasik 722
Gwalior 720
Tiruchchirappalli 711
Amritsar 709

Durg-Bhilai 689
Mysore 652
Jodhpur 649
Hubli-Dharwad 648
Solapur 621
Faridabad 614
Ranchi 614
Bareilly 608
Srinagar 595
Aurangabad 592
Guwahati 578
Chandigarh 575
Salem 574
Cochin 564
Kota 536
Ghaziabad 520
Jullundur 520
Indonesia
Jakarta 7,886
Surabaya 2,224
Medan 1,806
Bandung 1,567
Semarang 1,027
Palembang 787
Ujung Pandang 709
Malang 512
Iran
Tehran 6,476
Mashhad 1,759
Esfahan 1,127
Tabriz 1,089
Shiraz 965
Ahvaz 725
Qom 681
Bakhtaran 624
Iraq
Baghdad 3,841
Diyala 961
As Sulaymaniyah 952
Arbil 770
Mosul 644
Kadhimain 521
Ireland
Dublin 1,024
Israel
Tel Aviv-Jaffa 1,844
Jerusalem 544
Italy
Rome 2,791
Milan 1,432
Naples 1,206
Turin 992
Palermo 734
Genoa 701
Ivory Coast
Abidjan 1,929
Jamaica
Kingston 588
Japan
Tokyo 11,936
Yokohama 3,220
Osaka 2,624
Nagoya 2,155
Sapporo 1,672
Kobe 1,477
Kyoto 1,461
Fukuoka 1,237
Kawasaki 1,174
Hiroshima 1,086
Kitakyushu 1,026
Jordan
Amman 1,272
Az-Zarqa 605
Kazakhstan
Alma-Ata 1,147
Karaganda 613
Kenya
Nairobi 1,429
Korea, North
Pyongyang 2,639
Hamhung 775
Chongjin 754
Chinnamp'o 691
Sinuiju 500
Korea, South
Seoul 10,628
Pusan 3,798
Taegu 2,229
Inchon 1,818
Kwangju 1,145
Taejon 1,062
Ulsan 683
Puch'on 668
Suwon 645
Songnam 541
Chonju 517

Kyrgyzstan
Bishkek 641
Latvia
Riga 917
Lebanon
Beirut 1,500
Tripoli 500
Libya
Tripoli 990
Lithuania
Vilnius 593
Macedonia
Skopje 563
Madagascar
Antananarivo 802
Malaysia
Kuala Lumpur 938
Mali
Bamako 746
Mexico
Mexico City 15,048
Guadalajara 2,847
Monterrey 2,522
Puebla 1,055
León 872
Ciudad Juárez 798
Tijuana 743
Culiacán Rosales 602
Mexicali 602
Acapulco de Juárez 592
Mérida 557
Chihuahua 530
San Luis Potosí 526
Aguascalientés 506
Moldova
Kishinev 667
Mongolia
Ulan Bator 575
Morocco
Casablanca 2,409
Rabat-Salé 893
Fès 562
Marrakesh 549
Mozambique
Maputo 1,070
Netherlands
Amsterdam 1,091
Rotterdam 1,069
The Hague 694
Utrecht 543
New Zealand
Auckland 896
Nicaragua
Managua 682
Nigeria
Lagos 1,347
Ibadan 1,295
Kano 700
Ogbomosho 661
Norway
Oslo 746
Pakistan
Karachi 5,181
Lahore 2,953
Faisalabad 1,104
Rawalpindi 795
Hyderabad 752
Multan 722
Gujranwala 659
Peshawar 556
Paraguay
Asunción 945
Peru
Lima-Callao 6,415
Arequipa 635
Trujillo 532
Callao 515
Philippines
Manila 6,720
Quezon City 1,667
Davao 868
Cebu 641
Caloocan 629
Poland
Warsaw 1,655
Lódz 847
Kraków 751
Wroclaw 643
Poznan 590
Portugal
Lisbon 2,561
Oporto 1,174
Romania
Bucharest 2,064
Russia
Moscow 8,957

St Petersburg 5,004
Novosibirsk 1,472
Nizhniy Novgorod 1,451
Yekaterinburg 1,413
Samara 1,271
Omsk 1,193
Chelyabinsk 1,170
Perm 1,108
Kazan 1,107
Ufa 1,100
Volgograd 1,031
Rostov 1,027
Voronezh 958
Krasnoyarsk 925
Saratov 909
Togliatti 677
Vladivostok 675
Krasnodar 671
Barnaul 665
Izhevsk 651
Irkutsk 644
Simbirsk 638
Yaroslavl 637
Khaborovsk 626
Novokuznetsk 614
Tula 591
Orenburg 574
Kemerovo 559
Penza 553
Tyumen 550
Ryazan 533
Kirov 525
Naberezhnyye-Chelny 517
Astrakhan 512
Tomsk 506
Lipetsk 504
Saudi Arabia
Riyadh 2,000
Jedda 1,400
Mecca 618
Medina 500
Senegal
Dakar 1,730
Singapore
Singapore 2,818
Somali Republic
Mogadishu 1,000
South Africa
Cape Town 1,912
Johannesburg 1,726
East Rand 1,038
Durban 982
Pretoria 823
Port Elizabeth 652
West Rand 647
Vereeniging 540
Spain
Madrid 3,121
Barcelona 1,707
Valencia 753
Seville 659
Zaragoza 586
Málaga 512
Sri Lanka
Colombo 1,863
Sudan
Khartoum 561
Omdurman 526
Sweden
Stockholm 1,503
Gothenburg 734
Switzerland
Zürich 840
Syria
Damascus 1,451
Aleppo 1,445
Holms 518
Taiwan
Taipei 2,718
Kaohsiung 1,396
Taichung 774
Tainan 690
Panchiao 543
Tajikistan
Dushanbe 602
Tanzania
Dar es Salaam 1,361
Thailand
Bangkok 5,876
Togo
Lomé 500
Tunisia
Tunis 1,395
Turkey
Istanbul 7,309

Ankara 3,022
Izmir 2,665
Adana 1,430
Bursa 1,031
Konya 1,015
Gaziantep 760
Içel 701
Kayseri 588
Uganda
Kampala 773
Ukraine
Kiev 2,643
Kharkiv 1,622
Dnipropetrovsk 1,190
Donetsk 1,121
Odessa 1,096
Zaporizhya 898
Lviv 807
Kryvyy Rih 729
Mariupol 523
Mykolayiv 515
Luhansk 505
United Kingdom
London 6,933
Birmingham 1,012
Leeds 725
Glasgow 681
Sheffield 532
United States
New York 18,087
Los Angeles 14,532
Chicago 8,066
San Francisco 6,253
Philadelphia 5,899
Detroit 4,665
Boston 4,172
Washington, DC 3,924
Dallas 3,885
Houston 3,711
Miami 3,193
Atlanta 2,834
Cleveland 2,760
Seattle 2,559
San Diego 2,498
Minneapolis-SP. 2,464
St Louis 2,444
Baltimore 2,382
Pittsburgh 2,243
Phoenix 2,122
Tampa 2,098
Denver 1,848
Cincinnati 1,744
Milwaukee 1,607
Kansas City 1,566
Sacramento 1,481
Portland 1,478
Norfolk 1,396
Columbus 1,377
San Antonio 1,303
Indianapolis 1,250
New Orleans 1,239
Buffalo 1,189
Charlotte 1,162
Hartford 1,086
Salt Lake City 1,072
Albany 861
San Jose 782
Jacksonville 672
Memphis 610
Uruguay
Montevideo 1,384
Uzbekistan
Tashkent 2,094
Venezuela
Caracas 2,784
Maracaibo 1,364
Valencia 1,032
Maracay 800
Barquisimeto 745
Ciudad Guayana 524
Vietnam
Ho Chi Minh 3,924
Hanoi 3,056
Haiphong 1,448
Yugoslavia
Belgrade 1,137
Zaïre
Kinshasa 3,804
Lubumbashi 739
Mbuji-Mayi 613
Kolwezi 544
Zambia
Lusaka 982
Zimbabwe
Harare 1,189
Bulawayo 622

WORLD STATISTICS: DISTANCES

The table shows air distances in miles and kilometers between thirty major cities. Known as 'Great Circle' distances, these measure the shortest routes between the cities, which aircraft use where possible. The maps show the world centered on six individual cities, and illustrate, for example, why direct flights from Japan to northern America and Europe are across the Arctic regions, and Singapore is on the direct line route from Europe to Australia. The maps have been constructed on an Azimuthal Equidistant projection, on which all distances measured through the center point are true to scale. The circular lines are drawn at 5,000, 10,000 and 15,000 km (3,100, 6,200 and 9,300 miles) from the central city.

Distance chart — values above the diagonal are in Kms; values below the diagonal are in Miles. (The "Kms" label is printed at upper-left and "Miles" at lower-right of the chart.)

	Berlin	Bombay	Buenos Aires	Cairo	Calcutta	Caracas	Chicago	Copenhagen	Darwin	Hong Kong	Honolulu	Johannesburg	Lagos	Lisbon	London	Los Angeles	Mexico City	Moscow	Nairobi	New York	Paris	Peking	Reykjavik	Rio de Janeiro	Rome	Singapore	Sydney	Tokyo	Toronto	Wellington
Berlin	Berlin	3907	7400	1795	4370	5241	4402	222	8044	5440	7310	5511	3230	1436	557	5785	6047	1000	3958	3967	545	4860	1482	6230	734	6179	10002	5545	4037	11272
Bombay	6288	Bombay	9275	2706	1034	9024	8048	3990	4510	2683	8024	4334	4730	4982	4467	8700	9728	3126	2816	7793	4356	2956	5179	8332	3837	2432	6313	4189	7760	7686
Buenos Aires	11909	14925	Buenos Aires	7341	10268	3167	5599	7498	9130	11481	7558	5025	4919	5964	6917	6122	4591	8374	6463	5298	6867	11972	7106	1214	6929	9867	7332	11410	5650	6202
Cairo	2890	4355	11814	Cairo	3541	6340	6127	1992	7216	5064	8838	3894	2432	2358	2180	7580	7687	1803	2197	5605	1994	4688	3272	6149	1325	5137	8959	5737	10268
Calcutta	7033	1664	16524	5699	Calcutta	9609	7978	4395	3758	1653	7048	5256	5727	5639	4946	8152	9494	3438	3839	7921	4883	2031	5398	9366	4486	1800	5678	3195	7805	7055
Caracas	8435	14522	5096	10203	15464	Caracas	2502	5215	11221	10166	6009	6847	4810	4044	4664	3612	2228	6175	7173	2131	4738	8947	4297	2825	5196	11407	9534	8801	2406	8154
Chicago	7084	12953	9011	3206	12839	4027	Chicago	4250	9361	7783	4247	8689	5973	3992	3949	1742	1694	4971	8005	711	4132	6588	2956	5311	4809	9369	9243	6299	435	8358
Copenhagen	357	6422	12067	9860	7072	8392	6840	Copenhagen	8017	5388	7088	5732	3436	1540	592	5594	5912	970	4167	3845	638	4475	1306	6345	951	6195	9968	5403	3892	11160
Darwin	12946	7257	14693	11612	6047	18059	15065	12903	Darwin	2654	5369	6611	8837	9391	8605	7888	9091	7053	6472	9971	8582	3735	8632	9948	8243	2081	1957	3375	9630	3309
Hong Kong	8754	4317	18478	8150	2659	16360	12526	8671	4271	Hong Kong	5543	6669	7360	6853	5980	7232	8775	4439	5453	8047	5984	1220	6015	11001	5769	1615	4582	1786	7810	5857
Honolulu	11764	12914	12164	14223	11343	9670	6836	11407	8640	8921	Honolulu	11934	10133	7821	7228	2558	3781	7036	10739	4958	7437	5070	6081	8290	8026	6721	5075	3854	4638	4669
Johannesburg	8870	6974	8088	6267	8459	11019	13984	9225	10639	10732	19206	Johannesburg	2799	5089	5637	10362	9063	5692	1818	7979	5426	7276	6797	4420	4811	5381	6860	8418	8310	7308
Lagos	5198	7612	7916	3915	9216	7741	9612	5530	14222	11845	16308	4505	Lagos	2360	3118	7713	6879	3886	2366	5268	2929	7119	4175	3750	2510	6925	9643	8376	5560	9973
Lisbon	2311	8018	9600	3794	9075	6501	6424	2478	15114	11028	12587	8191	3799	Lisbon	987	5668	5391	2427	4015	3369	903	6007	1832	4805	1157	7385	11295	6928	3565	12163
London	928	7190	11131	3508	7961	7507	6356	952	13848	9623	11632	9071	5017	1588	London	5442	5552	1552	4237	3463	212	5057	1172	5778	889	6743	10558	5942	3545	11691
Los Angeles	9311	14000	9852	12200	13120	5812	2804	9003	12695	11639	4117	16676	12414	9122	8758	Los Angeles	1549	6070	9659	2446	5645	6251	4310	6310	6331	8776	7502	5475	2170	6719
Mexico City	9732	15656	7389	12372	15280	3586	2726	9514	14631	14122	6085	14585	11071	8676	8936	2493	Mexico City	6664	9207	2090	5717	7742	4635	4780	6365	10321	8058	7024	2018	6897
Moscow	1610	5031	13477	2902	5534	9938	8000	1561	11350	7144	11323	9161	6254	3906	2498	9769	10724	Moscow	3942	4666	1545	3600	2053	7184	1477	5237	9008	4651	4637	10283
Nairobi	6370	4532	10402	3536	6179	11544	12883	6706	10415	8776	17282	2927	3807	6461	6819	15544	14818	6344	Nairobi	7358	4029	5727	5395	5548	3350	4635	7552	6996	7570	8490
New York	6385	12541	8526	9020	12747	3430	1145	6188	16047	12950	7980	12841	8477	5422	5572	3936	3364	7510	11842	New York	3626	6828	2613	4832	4280	9531	9935	6741	356	8951
Paris	876	7010	11051	3210	7858	7625	6650	1026	13812	9630	11968	8732	4714	1454	342	9085	9200	2486	6485	5836	Paris	5106	1384	5708	687	6671	10539	6038	3738	11798
Peking	7822	4757	19268	7544	3269	14399	10603	7202	6011	1963	8160	11710	11457	9668	8138	10060	12460	5794	9216	10988	8217	Peking	4897	10773	5049	2783	5561	1304	6557	6700
Reykjavik	2385	8335	11437	5266	8687	6915	4757	2103	13892	9681	9787	10938	6718	2948	1887	6936	7460	3304	8683	4206	2228	7882	Reykjavik	6135	2048	7155	10325	5469	2600	10725
Rio de Janeiro	10025	13409	1953	9896	15073	4546	8547	10211	16011	17704	13342	7113	6035	7734	9299	10155	7693	11562	8928	7777	9187	17338	9874	Rio de Janeiro	5725	9763	8389	11551	5180	7367
Rome	1180	6175	11151	2133	7219	8363	7739	1531	13265	9284	12916	7743	4039	1861	1431	10188	10243	2376	5391	6888	1105	8126	3297	9214	Rome	6229	10143	6127	4399	11523
Singapore	9944	3914	15879	8267	2897	18359	16046	9969	3349	2599	10816	8660	11145	11186	10852	14123	16810	8428	7460	15390	10737	8428	11514	15712	10025	Singapore	3915	3306	9350	5298
Sydney	16096	10160	11800	14418	9138	15343	14875	16042	3150	7374	8168	11040	15519	18178	16992	12073	12969	14497	12153	15989	16962	16617	13501	16324	6300	Sydney	4861	9800	1383
Tokyo	8924	6742	18362	9571	5141	14164	10137	8696	5431	2874	6202	13547	13480	11149	9562	8811	11304	7485	11260	10849	9718	2099	8802	18589	9861	5321	7823	Tokyo	6410	5762
Toronto	6497	12488	9093	9233	12561	3873	700	6265	15498	12569	7465	13374	8948	5737	5704	3492	3247	7462	12183	574	6015	10552	4184	8336	7080	15047	15772	10316	Toronto	8820
Wellington	18140	12370	9981	16524	11354	13122	13451	17961	5325	9427	7513	11761	16050	19575	18814	10814	11100	16549	13664	14405	18987	10782	17260	11855	18545	8526	2226	9273	14194	Wellington

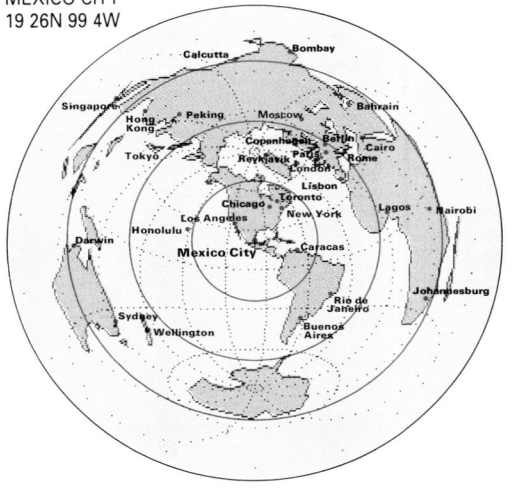

MEXICO CITY
19 26N 99 4W

LONDON
51 28N 0 27W

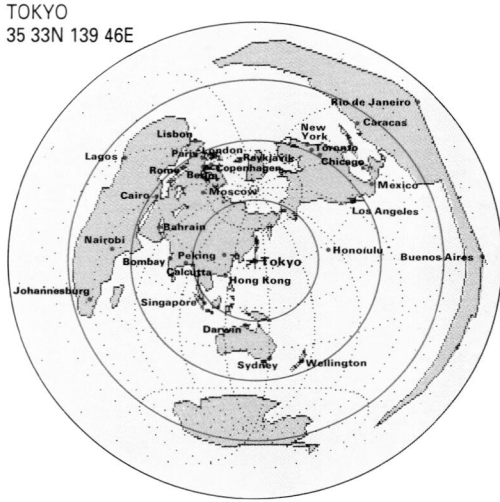

TOKYO
35 33N 139 46E

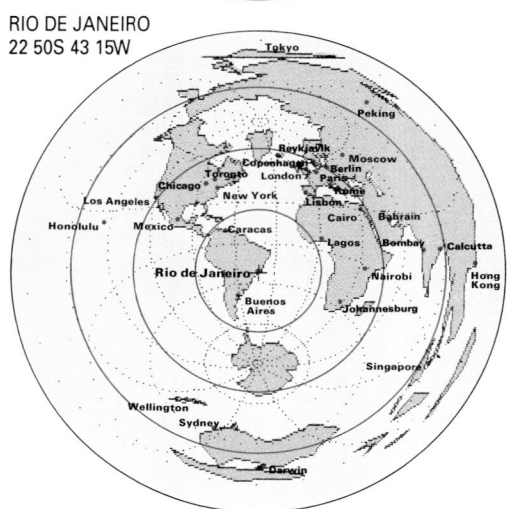

RIO DE JANEIRO
22 50S 43 15W

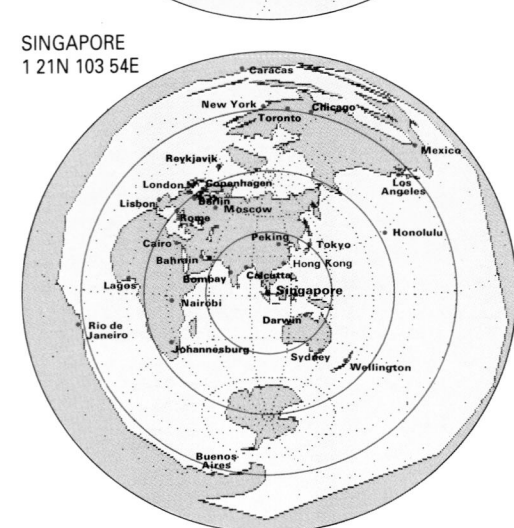

SINGAPORE
1 21N 103 54E

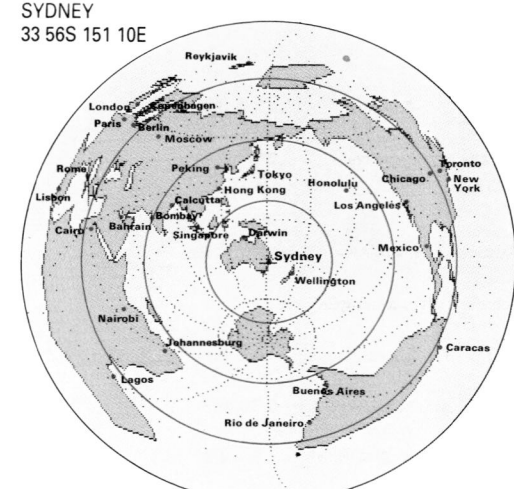

SYDNEY
33 56S 151 10E

World Statistics: Climate

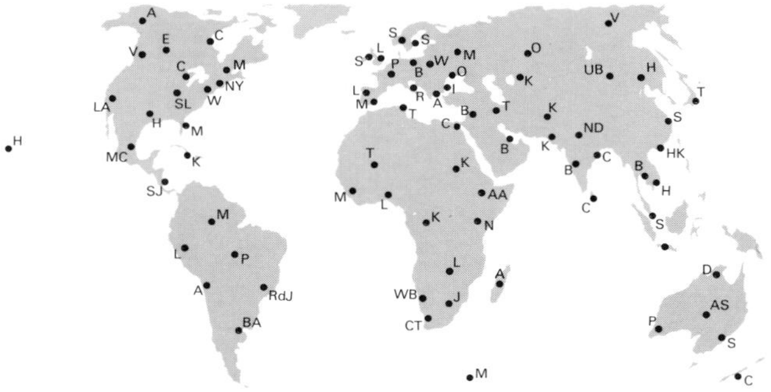

Rainfall and temperature figures are provided for more than 70 cities around the world. As climate is affected by altitude, the height of each city is shown in feet beneath its name. For each month, the figures in red show average temperature in degrees Fahrenheit, and in blue the total rainfall or snow in inches; the average annual temperature and total annual rainfall are at the end of the rows.

EUROPE	Jan.	Feb.	Mar.	Apr.	May	June	July	Aug.	Sept.	Oct.	Nov.	Dec.	Year
Athens, Greece	2.4	1.5	1.5	0.9	0.9	0.6	0.2	0.3	0.6	2	2.2	2.8	15.8
351 ft	50	50	54	61	68	77	82	82	75	68	59	52	64
Berlin, Germany	1.8	1.6	1.3	1.7	1.9	2.6	2.9	2.7	1.9	1.9	1.8	1.7	23.7
180 ft	30	32	39	48	57	63	66	64	59	48	41	34	48
Istanbul, Turkey	4.3	3.6	2.8	1.8	1.5	1.3	1.3	1.2	2.3	3.2	4.1	4.7	32.1
374 ft	41	43	45	52	61	68	73	73	68	61	54	46	57
Lisbon, Portugal	4.4	3	4.3	2.1	1.7	0.6	0.1	0.2	1.3	2.4	3.7	4.1	27.9
253 ft	52	54	57	61	63	68	72	73	70	64	57	54	63
London, UK	2.1	1.6	1.5	1.5	1.8	1.8	2.2	2.3	1.9	2.2	2.5	1.9	23.3
16 ft	39	41	45	48	54	61	64	63	59	52	46	41	52
Málaga, Spain	2.4	2	2.4	1.8	1	0.2	0	0.1	1.1	2.5	2.5	2.4	18.7
108 ft	54	55	61	63	66	84	77	79	73	68	61	55	66
Moscow, Russia	1.5	1.5	1.4	1.5	2.1	2.3	3.5	2.8	2.3	1.8	1.9	2.1	24.6
512 ft	9	14	25	43	55	61	64	63	54	43	30	19	39
Odessa, Ukraine	2.2	2.4	1.2	0.8	1.3	1.3	1.7	1.5	1.5	0.5	1.4	2.8	18.6
210 ft	27	30	36	48	59	68	72	72	64	54	48	34	51
Paris, France	2.2	1.8	1.4	1.7	2.2	2.1	2.3	2.5	2.2	2	2	2	24.4
246 ft	37	39	46	52	59	64	68	66	63	54	45	39	53
Rome, Italy	2.8	2.4	2.2	2	1.8	1.5	0.6	0.8	2.5	3.9	5.1	3.7	29.3
56 ft	46	48	52	57	64	72	77	77	72	63	55	50	61
Shannon, Irish Republic	3.7	2.6	2.2	2.1	2.4	2.2	3	3.1	3.4	3.4	3.8	4.6	36.6
7 ft	41	41	45	48	54	57	61	61	57	52	46	43	50
Stockholm, Sweden	1.7	1.2	1	1.2	1.3	1.8	2.4	3	2.4	1.9	2.1	1.9	21.8
144 ft	27	27	30	41	50	59	64	63	54	45	37	32	44

ASIA	Jan.	Feb.	Mar.	Apr.	May	June	July	Aug.	Sept.	Oct.	Nov.	Dec.	Year
Bahrain	0.3	0.7	0.5	0.3	<0.1	0	0	0	0	0	0.7	0.7	3.3
16 ft	63	64	70	77	84	90	91	93	88	82	75	66	79
Bangkok, Thailand	0.3	0.8	1.4	2.3	7.8	6.3	6.3	6.9	12	8.1	2.6	0.2	55
7 ft	79	82	84	86	84	84	82	82	82	82	79	77	82
Beirut, Lebanon	7.5	6.2	3.7	2.1	0.7	0.1	<0.1	<0.1	0.2	2	5.2	7.3	35
112 ft	79	82	84	86	84	84	82	82	82	82	79	77	82
Bombay, India	0.1	0.1	0.1	<0.1	0.7	19.1	24.3	13.4	10.4	2.5	0.5	0.1	71.4
36 ft	75	75	79	82	86	84	81	81	81	82	81	79	80
Calcutta, India	0.4	1.2	1.4	1.7	5.5	11.7	12.8	12.9	9.9	4.5	0.8	0.2	63
20 ft	68	72	81	86	86	86	84	84	84	82	73	66	79
Colombo, Sri Lanka	3.5	2.7	5.8	9.1	14.6	8.8	5.3	4.3	6.3	13.7	12.4	5.8	92.3
23 ft	79	79	81	82	82	81	81	81	81	81	79	79	80
Harbin, China	0.2	0.2	0.4	0.9	1.7	3.7	4.4	4.1	1.8	1.3	0.3	0.2	19.3
525 ft	0	5	23	43	55	66	72	70	57	39	21	3	38
Ho Chi Minh, Vietnam	0.6	0.1	0.5	1.7	8.7	13	12.4	10.6	13.2	10.6	4.5	2.2	78.1
30 ft	79	81	84	86	84	82	82	82	81	81	81	79	82
Hong Kong	1.3	1.8	2.9	5.4	11.5	15.5	15	14.2	10.1	4.5	1.7	1.2	85.2
108 ft	61	59	64	72	79	82	82	82	81	77	70	64	73
Jakarta, Indonesia	11.8	11.8	8.3	5.8	4.5	3.8	2.5	1.7	2.6	4.4	5.6	8	70.8
26 ft	79	79	81	81	81	81	81	81	81	81	81	79	80
Kabul, Afghanistan	1.2	1.4	3.7	4	0.8	0.2	0.1	0.1	<0.1	0.6	0.8	0.4	13.3
5,953 ft	27	30	43	55	64	72	77	75	68	57	45	37	54
Karachi, Pakistan	0.5	0.4	0.3	0.1	0.1	0.7	3.2	1.6	0.5	<0.1	0.1	0.2	7.8
13 ft	66	68	75	82	86	88	86	84	82	82	75	68	79
Kazalinsk, Kazakhstan	0.4	0.4	0.5	0.5	0.6	0.2	0.2	0.3	0.3	0.4	0.5	0.6	4.9
207 ft	10	12	27	43	64	73	77	73	61	46	30	19	45
New Delhi, India	0.9	0.7	0.5	0.3	0.5	2.9	7.1	6.8	4.6	0.4	0.1	0.4	25.2
715 ft	57	63	73	82	91	93	88	86	84	79	68	59	77
Omsk, Russia	0.6	0.3	0.3	0.5	1.2	2	2	2	1.1	1	0.7	0.8	12.6
279 ft	-7	-1	10	30	50	61	64	61	50	34	12	0	30
Shanghai, China	1.9	2.3	3.3	3.7	3.7	7.1	5.8	5.6	5.1	2.8	2	1.4	44.7
23 ft	39	41	48	57	68	75	82	82	73	66	54	45	61
Singapore	9.9	6.8	7.6	7.4	6.8	6.8	6.7	7.7	7	8.2	10	10.1	95.1
33 ft	79	81	82	82	82	82	82	81	81	81	81	81	81
Tehran, Iran	1.8	1.5	1.8	1.4	0.5	0.1	0.1	0.1	0.1	0.3	0.8	1.2	9.8
4,002 ft	36	41	48	61	70	79	86	84	77	64	54	43	62
Tokyo, Japan	1.9	2.9	4.2	5.3	5.8	6.5	5.6	6	9.2	8.2	3.8	2.2	61.6
20 ft	37	39	45	55	63	70	77	79	73	63	52	43	58
Ulan Bator, Mongolia	<0.1	<0.1	0.1	0.2	0.4	1.1	3	2	0.9	0.2	0.2	0.1	8.2
4,346 ft	-14	-5	9	30	43	57	61	57	46	30	9	-7	26
Verkhoyansk, Russia	0.2	0.2	0.1	0.2	0.3	0.9	1.1	1	0.5	0.3	0.3	0.2	5.4
328 ft	-57	-48	-25	5	32	54	57	48	36	5	-35	-53	1

AFRICA	Jan.	Feb.	Mar.	Apr.	May	June	July	Aug.	Sept.	Oct.	Nov.	Dec.	Year
Addis Ababa, Ethiopia	<0.1	0.1	1	5.3	8.4	7.9	8.1	9.4	4	1.1	<0.1	0	45.4
8,036 ft	66	68	68	68	66	64	64	66	70	72	70	68	68
Antananarivo, Madagas.	11.8	11	7	2.1	0.7	0.3	0.3	0.4	0.7	2.4	5.3	11.3	53.3
4,500 ft	70	70	70	66	64	59	57	59	63	66	70	70	65
Cairo, Egypt	0.2	0.2	0.2	0.1	0.1	<0.1	0	0	<0.1	<0.1	0.1	0.2	1.1
380 ft	55	59	64	70	77	82	82	82	79	75	68	59	71
Cape Town, South Africa	0.6	0.3	0.7	1.9	3.1	3.3	3.5	2.6	1.7	1.2	0.7	0.4	20
56 ft	70	70	68	63	57	55	54	55	57	61	64	66	62
Johannesburg, S. Africa	4.5	4.3	3.5	1.5	1	0.3	0.3	0.3	0.9	2.2	4.2	4.9	28
5,461 ft	68	68	64	61	55	50	52	55	61	64	66	68	61
Khartoum, Sudan	<0.1	<0.1	<0.1	<0.1	0.1	0.3	2.1	2.8	0.7	0.2	<0.1	0	6.2
1,279 ft	75	77	82	88	91	93	90	88	90	90	82	77	85
Kinshasa, Zaïre	5.3	5.7	7.7	7.7	6.2	0.3	0.1	0.1	1.2	4.7	8.7	5.6	53.4
1,066 ft	79	79	81	81	79	75	73	75	77	79	79	79	78
Lagos, Nigeria	1.1	1.8	4	5.9	10.6	18.1	11	2.5	5.5	8.1	2.7	1	72.4
10 ft	81	82	84	82	82	79	79	77	79	79	82	82	81
Lusaka, Zambia	9.1	7.5	5.6	0.7	0.1	<0.1	<0.1	0	<0.1	0.4	3.6	5.9	32.9
4,189 ft	70	72	70	70	66	61	61	64	72	75	73	72	69
Monrovia, Liberia	1.2	2.2	3.8	8.5	20.3	38.3	39.2	14.7	29.3	30.4	9.3	5.1	202.4
75 ft	79	79	81	81	79	77	75	77	77	77	79	79	78
Nairobi, Kenya	1.5	2.5	4.9	8.3	6.2	1.8	0.6	0.9	1.2	2.1	4.3	3.4	37.8
5,970 ft	66	66	66	66	64	61	61	61	64	66	64	64	64
Timbuktu, Mali	<0.1	<0.1	0.1	<0.1	0.2	0.9	3.1	3.2	1.5	0.1	<0.1	<0.1	9.1
987 ft	72	75	82	90	93	95	90	86	90	88	82	73	85
Tunis, Tunisia	2.5	2	1.6	1.4	0.7	0.3	0.1	0.3	1.3	2	1.9	2.4	16.6
216 ft	50	52	55	61	66	73	79	81	77	68	61	52	65
Walvis Bay, Namibia	<0.1	0.2	0.3	0.1	0.1	<0.1	<0.1	0.1	<0.1	<0.1	<0.1	<0.1	0.9
23 ft	66	66	66	64	63	61	59	57	57	59	63	64	62

AUSTRALIA, NEW ZEALAND AND ANTARCTICA	Jan.	Feb.	Mar.	Apr.	May	June	July	Aug.	Sept.	Oct.	Nov.	Dec.	Year
Alice Springs, Australia	1.7	1.3	1.1	0.4	0.6	0.5	0.3	0.3	0.3	0.7	1.2	1.5	10
1,899 ft	84	82	77	68	59	54	54	57	64	73	79	82	70
Christchurch, N. Zealand	2.2	1.7	1.9	1.9	2.6	2.6	2.7	1.9	1.8	1.7	1.9	2.2	25.1
33 ft	61	61	57	54	48	43	43	45	48	54	57	61	53
Darwin, Australia	15.2	12.3	10	3.8	0.6	0.1	<0.1	0.1	0.5	2	4.7	9.4	58.7
98 ft	84	84	84	84	82	79	77	79	82	84	86	84	83
Mawson, Antarctica	0.4	1.2	0.8	0.4	1.7	7.1	0.2	1.6	0.1	0.8	0	0	14.3
46 ft	32	23	14	7	5	3	0	0	-1	9	23	30	12
Perth, Australia	0.3	0.4	0.8	1.7	5.1	7.1	6.7	5.9	3.4	2.2	0.8	0.5	34.8
197 ft	73	73	72	66	61	57	55	55	59	61	66	72	64
Sydney, Australia	3.5	4	5	5.3	5	4.6	4.6	3	2.9	2.8	2.9	2.9	46.5
138 ft	72	72	70	64	59	55	54	55	59	64	66	70	63

NORTH AMERICA	Jan.	Feb.	Mar.	Apr.	May	June	July	Aug.	Sept.	Oct.	Nov.	Dec.	Year
Anchorage, Alaska, USA	0.8	0.7	0.6	0.4	0.5	0.7	1.6	2.6	2.6	2.2	1	0.9	14.6
131 ft	12	18	23	36	45	54	57	55	48	36	23	12	35
Chicago, Ill., USA	2	2	2.6	2.8	3.4	3.5	3.3	3.2	3.1	2.6	2.4	2	32.9
823 ft	25	27	36	48	57	68	73	72	66	54	41	30	50
Churchill, Man., Canada	0.6	0.5	0.7	0.9	1.3	1.7	1.8	2.3	2	1.7	1.5	0.8	15.9
43 ft	-17	-14	-3	14	28	43	54	52	41	28	10	-7	19
Edmonton, Alta., Canada	1	0.7	0.7	0.9	1.7	3	3.5	3.1	1.5	0.7	0.6	1	18.5
2,217 ft	5	14	23	39	52	59	63	61	52	43	25	14	37
Honolulu, Hawaii, USA	4.1	2.6	3.1	1.9	1	0.7	0.9	1.1	1.4	1.9	2.5	4.1	25.3
39 ft	73	64	66	68	72	75	77	79	79	75	72	66	72
Houston, Tex., USA	3.5	3	3.3	3.6	4.7	4.6	3.9	3.9	4.1	3.7	3.5	4.3	46.1
39 ft	54	55	63	70	75	81	82	84	79	72	61	54	69
Kingston, Jamaica	0.9	0.6	0.9	1.2	4	3.5	1.5	3.6	3.9	7.1	2.9	1.4	31.5
112 ft	77	77	77	79	79	82	82	82	81	81	79	79	80
Los Angeles, Calif., USA	3.1	3	2.8	1	0.4	0.1	<0.1	<0.1	0.2	0.6	1.2	2.6	15
312 ft	55	57	57	61	63	66	70	72	70	64	61	57	63
Mexico City, Mexico	0.5	0.2	0.4	0.8	2.1	4.7	6.7	6	5.1	2	0.7	0.3	29.5
7,574 ft	54	55	61	64	66	66	63	64	64	61	57	55	61
Miami, Fla., USA	2.8	2.1	2.5	3.2	6.8	7	6.1	6.3	8	9.2	2.8	2	58.8
26 ft	68	68	72	73	77	81	82	82	81	77	72	70	75
Montréal, Que., Canada	2.8	2.6	2.9	2.9	2.6	3.2	3.5	3.6	3.5	3	3.2	3.4	37.3
187 ft	14	16	27	21	55	64	70	68	59	48	36	19	41
New York, N.Y., USA	3.7	3.8	3.6	3.2	3.2	3.3	4.2	4.3	3.4	3.5	3	3.6	42.8
315 ft	30	30	37	50	61	68	73	73	70	59	45	36	53
St Louis, Mo., USA	2.3	2.5	3.5	3.8	4.5	4.5	3.5	3.4	3.2	2.9	2.8	2.5	39.4
567 ft	32	34	45	55	66	75	79	79	72	59	46	36	56
San José, Costa Rica	0.6	0.2	0.8	1.8	9	9.5	8.3	9.5	12	11.8	5.7	1.6	70.8
3,759 ft	66	66	70	70	72	70	70	70	70	68	68	66	69
Vancouver, B.C., Canada	6.1	4.5	4	2.4	2	1.8	1.3	1.6	2.6	4.5	5.9	7.2	43.8
46 ft	37	41	43	48	54	59	63	63	57	50	43	39	50
Washington, D.C., USA	3.4	3	3.6	3.3	3.7	3.9	4.4	4.3	3.7	2.9	2.6	3.1	41.9
72 ft	34	36	45	54	64	73	77	75	68	57	46	37	56

SOUTH AMERICA	Jan.	Feb.	Mar.	Apr.	May	June	July	Aug.	Sept.	Oct.	Nov.	Dec.	Year
Antofagasta, Chile	0	0	0	<0.1	<0.1	0.1	0.2	0.1	<0.1	0.1	<0.1	0	0.6
308 ft	70	70	68	64	61	59	57	57	59	61	64	66	63
Buenos Aires, Argentina	3.1	2.8	4.3	3.5	3	2.4	2.2	2.4	3.1	3.4	3.3	3.9	37.4
89 ft	73	73	70	63	55	48	50	52	55	59	66	72	61
Lima, Peru	0.1	<0.1	0.1	<0.1	0.2	0.2	0.3	0.3	0.3	0.1	0.1	<0.1	1.7
394 ft	73	75	75	72	66	63	63	61	63	64	66	70	68
Manaus, Brazil	9.8	9.1	10.3	8.7	6.7	3.3	2.3	1.5	1.8	4.2	5.6	8	71.3
144 ft	82	82	82	81	82	82	82	82	84	84	84	82	83
Paraná, Brazil	11.3	9.3	9.4	4	0.5	<0.1	0.1	0.2	1.1	5	9.1	12.2	62.2
853 ft	73	73	73	73	73	70	70	72	75	75	75	73	73
Rio de Janeiro, Brazil	4.9	4.8	5.1	4.2	3.1	2.1	1.6	1.7	2.6	3.1	4.1	5.4	42.8
200 ft	79	79	77	75	72	70	70	70	70	72	73	77	74

WORLD STATISTICS: PHYSICAL DIMENSIONS

Each topic list is divided into continents and within a continent the items are listed in order of size. The order of the continents is as in the atlas, Europe through to South America. Certain lists down to this mark > are complete; below they are selective. The world top ten are shown in square brackets; in the case of mountains this has not been done because the world top 30 are all in Asia. The figures are rounded as appropriate.

WORLD, CONTINENTS, OCEANS

	km²	miles²	%
The World	509,450,000	196,672,000	–
Land	149,450,000	57,688,000	29.3
Water	360,000,000	138,984,000	70.7
Asia	44,500,000	17,177,000	29.8
Africa	30,302,000	11,697,000	20.3
North America	24,241,000	9,357,000	16.2
South America	17,793,000	6,868,000	11.9
Antarctica	14,100,000	5,443,000	9.4
Europe	9,957,000	3,843,000	6.7
Australia & Oceania	8,557,000	3,303,000	5.7
Pacific Ocean	179,679,000	69,356,000	49.9
Atlantic Ocean	92,373,000	35,657,000	25.7
Indian Ocean	73,917,000	28,532,000	20.5
Arctic Ocean	14,090,000	5,439,000	3.9

SEAS

Pacific	km²	miles²
South China Sea	2,974,600	1,148,500
Bering Sea	2,268,000	875,000
Sea of Okhotsk	1,528,000	590,000
East China & Yellow	1,249,000	482,000
Sea of Japan	1,008,000	389,000
Gulf of California	162,000	62,500
Bass Strait	75,000	29,000

Atlantic	km²	miles²
Caribbean Sea	2,766,000	1,068,000
Mediterranean Sea	2,516,000	971,000
Gulf of Mexico	1,543,000	596,000
Hudson Bay	1,232,000	476,000
North Sea	575,000	223,000
Black Sea	462,000	178,000
Baltic Sea	422,170	163,000
Gulf of St Lawrence	238,000	92,000

Indian	km²	miles²
Red Sea	438,000	169,000
The Gulf	239,000	92,000

MOUNTAINS

Europe		m	ft
Mont Blanc	France/Italy	4,807	15,771
Monte Rosa	Italy/Switzerland	4,634	15,203
Dom	Switzerland	4,545	14,911
Weisshorn	Switzerland	4,505	14,780
Matterhorn/Cervino	Italy/Switzerland	4,478	14,691
Mt Maudit	France/Italy	4,465	14,649
Finsteraarhorn	Switzerland	4,274	14,022
Aletschhorn	Switzerland	4,182	13,720
Jungfrau	Switzerland	4,158	13,642
Barre des Ecrins	France	4,103	13,461
Schreckhorn	Switzerland	4,078	13,380
Gran Paradiso	Italy	4,061	13,323
Piz Bernina	Italy/Switzerland	4,049	13,284
Ortles	Italy	3,899	12,792
Monte Viso	Italy	3,841	12,602
Grossglockner	Austria	3,797	12,457
Wildspitze	Austria	3,774	12,382
Weisskügel	Austria/Italy	3,736	12,257
Balmhorn	Switzerland	3,709	12,169
Dammastock	Switzerland	3,630	11,909
Tödi	Switzerland	3,620	11,877
Presanella	Italy	3,556	11,667
Monte Adamello	Italy	3,554	11,660
Mulhacén	Spain	3,478	11,411
Pico de Aneto	Spain	3,404	11,168
Posets	Spain	3,375	11,073
Marmolada	Italy	3,342	10,964
> Etna	Italy	3,340	10,958
Musala	Bulgaria	2,925	9,596
Olympus	Greece	2,917	9,570
Gerlachovka	Slovak Republic	2,655	8,711
Galdhöpiggen	Norway	2,469	8,100
Pietrosul	Romania	2,305	7,562
Hvannadalshnúkur	Iceland	2,119	6,952
Narodnaya	Russia	1,894	6,214
Ben Nevis	UK	1,343	4,406

Asia		m	ft
Everest	China/Nepal	8,848	29,029
Godwin Austen (K2)	China/Kashmir	8,611	28,251
Kanchenjunga	India/Nepal	8,598	28,208
Lhotse	China/Nepal	8,516	27,939
Makalu	China/Nepal	8,481	27,824
Cho Oyu	China/Nepal	8,201	26,906
Dhaulagiri	Nepal	8,172	26,811
Manaslu	Nepal	8,156	26,758
Nanga Parbat	Kashmir	8,126	26,660
Annapurna	Nepal	8,078	26,502
Gasherbrum	China/Kashmir	8,068	26,469
Broad Peak	China/Kashmir	8,051	26,414
Gosainthan	China	8,012	26,286
Disteghil Sar	Kashmir	7,885	25,869
Nuptse	Nepal	7,879	25,849
Masherbrum	Kashmir	7,821	25,659
Nanda Devi	India	7,817	25,646
Rakaposhi	Kashmir	7,788	25,551
Kanjut Sar	India	7,760	25,459
Kamet	India	7,756	25,446
Namcha Barwa	China	7,756	25,446
Gurla Mandhata	China	7,728	25,354
Muztag	China	7,723	25,338
Kongur Shan	China	7,719	25,324
Tirich Mir	Pakistan	7,690	25,229
> Saser	Kashmir	7,672	25,170
K'ula Shan	Bhutan/China	7,543	24,747
Pik Kommunizma	Tajikistan	7,495	24,590
Aling Gangri	China	7,314	23,996
Elbrus	Russia	5,642	18,510
Demavend	Iran	5,604	18,386
Ararat	Turkey	5,165	16,945
Gunong Kinabalu	Malaysia (Borneo)	4,101	13,455
Yu Shan	Taiwan	3,997	13,113
Fuji-san	Japan	3,776	12,388
Rinjani	Indonesia	3,726	12,224
Mt Rajang	Philippines	3,364	11,037
Pidurutalagala	Sri Lanka	2,524	8,281

Africa		m	ft
Kilimanjaro	Tanzania	5,895	19,340
Mt Kenya	Kenya	5,199	17,057
Ruwenzori	Uganda/Zaïre	5,109	16,762
Ras Dashan	Ethiopia	4,620	15,157
Meru	Tanzania	4,565	14,977
Karisimbi	Rwanda/Zaïre	4,507	14,787
Mt Elgon	Kenya/Uganda	4,321	14,176
Batu	Ethiopia	4,307	14,130
Guna	Ethiopia	4,231	13,882
Toubkal	Morocco	4,165	13,665
Irhil Mgoun	Morocco	4,071	13,356
Mt Cameroon	Cameroon	4,070	13,353
Amba Ferit	Ethiopia	3,875	13,042
Teide	Spain (Tenerife)	3,718	12,198
Thabana Ntlenyana	Lesotho	3,482	11,424
> Emi Koussi	Chad	3,415	11,204
Mt aux Sources	Lesotho/S. Africa	3,282	10,768
Mt Piton	Réunion	3,069	10,069

Oceania		m	ft
Puncak Jaya	Indonesia	5,029	16,499
Puncak Trikora	Indonesia	4,750	15,584
Puncak Mandala	Indonesia	4,702	15,427
> Mt Wilhelm	Papua New Guinea	4,508	14,790
Mauna Kea	USA (Hawaii)	4,205	13,796
Mauna Loa	USA (Hawaii)	4,170	13,681
Mt Cook	New Zealand	3,753	12,313
Mt Balbi	Solomon Is.	2,439	8,002
Orohena	Tahiti	2,241	7,352
Mt Kosciusko	Australia	2,237	7,339

North America		m	ft
Mt McKinley	USA (Alaska)	6,194	20,321
Mt Logan	Canada	5,959	19,551
Citlaltepetl	Mexico	5,700	18,701
Mt St Elias	USA/Canada	5,489	18,008
Popocatepetl	Mexico	5,452	17,887
Mt Foraker	USA (Alaska)	5,304	17,401
Ixtaccihuatl	Mexico	5,286	17,342
Lucania	Canada	5,227	17,149
Mt Steele	Canada	5,073	16,644
Mt Bona	USA (Alaska)	5,005	16,420
Mt Blackburn	USA (Alaska)	4,996	16,391
Mt Sanford	USA (Alaska)	4,940	16,207
Mt Wood	Canada	4,848	15,905
Nevado de Toluca	Mexico	4,670	15,321
Mt Fairweather	USA (Alaska)	4,663	15,298
Mt Whitney	USA	4,418	14,495
Mt Elbert	USA	4,399	14,432
Mt Harvard	USA	4,395	14,419
Mt Rainier	USA	4,392	14,409
Blanca Peak	USA	4,372	14,344
Long's Peak	USA	4,345	14,255
Nevado de Colima	Mexico	4,339	14,235
Mt Shasta	USA	4,317	14,163
Tajumulco	Guatemala	4,220	13,845
> Gannett Peak	USA	4,202	13,786
Mt Waddington	Canada	3,994	13,104
Mt Robson	Canada	3,954	12,972
Chirripó Grande	Costa Rica	3,837	12,589
Pico Duarte	Dominican Rep.	3,175	10,417

South America		m	ft
Aconcagua	Argentina	6,960	22,834
Illimani	Bolivia	6,882	22,578
Bonete	Argentina	6,872	22,546
Ojos del Salado	Argentina/Chile	6,863	22,516
Tupungato	Argentina/Chile	6,800	22,309
Pissis	Argentina	6,779	22,241
Mercedario	Argentina/Chile	6,770	22,211
Huascaran	Peru	6,768	22,204
Llullaillaco	Argentina/Chile	6,723	22,057
Nudo de Cachi	Argentina	6,720	22,047
Yerupaja	Peru	6,632	21,758
N. de Tres Cruces	Argentina/Chile	6,620	21,719
Incahuasi	Argentina/Chile	6,600	21,654
Ancohuma	Bolivia	6,550	21,489
Sajama	Bolivia	6,542	21,463
Coropuna	Peru	6,425	21,079
Ausangate	Peru	6,384	20,945
Cerro del Toro	Argentina	6,380	20,932
Ampato	Peru	6,310	20,702
> Chimborasso	Ecuador	6,267	20,561
Cotopaxi	Ecuador	5,896	19,344
S. Nev. de S. Marta	Colombia	5,800	19,029
Cayambe	Ecuador	5,796	19,016
Pico Bolivar	Venezuela	5,007	16,427

Antarctica		m	ft
Vinson Massif		4,897	16,066
Mt Kirkpatrick		4,528	14,855
Mt Markham		4,349	14,268

OCEAN DEPTHS

Atlantic Ocean	m	ft	
Puerto Rico (Milwaukee) Deep	9,220	30,249	[7]
Cayman Trench	7,680	25,197	[10]
Gulf of Mexico	5,203	17,070	
Mediterranean Sea	5,121	16,801	
Black Sea	2,211	7,254	
North Sea	660	2,165	
Baltic Sea	463	1,519	
Hudson Bay	258	846	

Indian Ocean	m	ft
Java Trench	7,450	24,442
Red Sea	2,635	8,454
Persian Gulf	73	239

Pacific Ocean	m	ft	
Mariana Trench	11,022	36,161	[1]
Tonga Trench	10,882	35,702	[2]
Japan Trench	10,554	34,626	[3]
Kuril Trench	10,542	34,587	[4]
Mindanao Trench	10,497	34,439	[5]
Kermadec Trench	10,047	32,962	[6]
Peru-Chile Trench	8,050	26,410	[8]
Aleutian Trench	7,822	25,662	[9]
Middle American Trench	6,662	21,857	

Arctic Ocean	m	ft
Molloy Deep	5,608	18,399

LAND LOWS

		m	ft
Caspian Sea	Europe	–28	–92
Dead Sea	Asia	–403	–1,322
Lake Assal	Africa	–156	–512
Lake Eyre North	Oceania	–16	–52
Death Valley	N. America	–86	–282
Valdés Peninsula	S. America	–40	–131

Rivers

XV

WORLD : REGIONS IN THE NEWS

Maps show the situation in May 1995

THE BREAK UP OF YUGOSLAVIA

The former country of Yugoslavia comprised six republics. In 1991 Slovenia and Croatia declared independence. Bosnia-Herzegovina followed in 1992 and Macedonia in 1993. Yugoslavia now comprises the remaining two republics, Serbia and Montenegro.

YUGOSLAVIA
Population : 10,763,000 (Serb 62.6%, Albanian 16.5%, Montenegrin 5%, Hungarian 3.3%, Muslim 3.2%)
Serbia
Population : 5,824,211 (Serb 87.7%) excluding the former autonomous provinces of Kosovo and Vojvodina
Kosovo
Population : 1,956,196 (Albanian 81.6%, Serb 9.9%)
Vojvodina
Population : 2,014,000 (Serb 56.8%, Hungarian 16.9%)
Montenegro Population : 615,035 (Montenegrin 61.9%, Muslim 14.6%, Albanian 6.6%)
CROATIA
Population : 4,504,000 (Croat 78.1%, Serb 12.2%)
SLOVENIA
Population : 1,942,000 (Slovene 88%, Croat 3%, Serb 2%)
MACEDONIA (F.Y.R.O.M.)
Population : 2,142,000 (Macedonian 64%, Albanian 21.7%, Turkish 5%, Romanian 3%, Serb 2%)
BOSNIA - HERZEGOVINA
Population : 3,527,000 (Muslim 49%, Serb 31.2%, Croat 17.2%)

Civil war between Serbs and other ethnic groups continues in Bosnia-Herzegovina. The large scale map on the left shows the situation in early 1995.

FORMER YUGOSLAVIA
0   50   100 miles

— · — · — International boundaries
— · · — · · Republic boundaries
— — — Province boundaries
◎ Capital cities

BOSNIA-HERZEGOVINA
0   25   50 miles

☐ Under Croatian control
▨ Under Serbian control
▨ Under Muslim control

THE NEAR EAST
0   10   20   30 miles

ISRAEL Population : 5,458,000 (inc. East Jerusalem and Jewish settlers in the areas under Israeli administration. (Jewish 82%, Arab Muslim 13.8%, Arab Christian 2.5%, Druze 1.7%)
West Bank Population : 973,500 (Palestinian Arabs 97% [of whom Arab Muslim 85%, Jewish 7%, Christian 8%])
Gaza Strip Population : 658,200 (Arab Muslim 98%)
JORDAN Population : 5,198,000 (Arab 99% [of whom about 50% are Palestinian Arab])

— · · — · · 1949 Armistice Line
— — — 1974 Cease-fire Lines
Efrata ● Main Jewish settlements in the West Bank and Gaza Strip
Halhul ☐ Main Palestinian Arab towns in the West Bank and Gaza Strip

THE CAUCASUS
0   50   100 miles

— · — · — International boundaries
— · · — · · Republic boundaries

Georgia, Armenia and Azerbaijan achieved independence in 1991. Abkhazia, Ajaria and South Ossetia seek independence from Georgia. Chechenia has been trying to break away from Russia since 1991, but Russia has resisted with military force. Hostility also continues between Armenia and Azerbaijan over the enclave of Nagorno-Karabakh.

RUSSIA
North Ossetia
Population : 695,000 (Ossetian 53%, Russian 29%, Chechen 5.2% Ingush 5% [expelled in 1992])
Chechenia
Population : 1,308,000 (Chechen and Ingush 70.7%, Russian 23.1%
Neighboring **Ingushetia** (now split from Chechenia)
Population : 250,000 (mainly Ingush)
GEORGIA
Population : 5,450,000 (Georgian 70.1%, Armenian 8.1%, Russian 6.3%, Azerbaijani 5.7%, Ossetian 3%, Greek 2%, Abkhazian 2%)
Abkhazia
Population : 537,500 (Georgian 45.7%, Abkhazian 17.8%, Armenia 14.6%, Russian 14.3%)
Ajaria
Population : 382,000 (Georgian 82.8%, Russian 7.7%, Armenian 4
South Ossetia
Population : 99,800 (Ossetian 66.2 Georgian 29%)
ARMENIA
Population : 3,548,000 (Armenian 93.3%, Azerbaijani 2.6%)
Nagorno-Karabakh
Population : 192,400 (Armenian 76.9%, Azerbaijani 21.5%)
AZERBAIJAN
Population : 7,472,000 (Azerbaijan 82.7%, Russian 5.6%, Armenian 5.6%, Lezgin 2.4%)
Naxçivan
Population : 300,400 (Azerbaijani 95.9%)

MOLDOVA
0   25   50   75 miles

▨ Separatist regions

Population : 4,420,000 (Moldovan 64.5%, Ukrainian 13.9%, Russian 14%, Gagauzi 3.5%, Jewish 2%, Bulgarian 2%)

ECUADOR AND PERU
0   50   100 miles

— · · — · · 1995 disputed bord
▨ Disputed territory allocated to Peru in 1942

CARTOGRAPHY BY PHILIP'S. COPYRIGHT REED INTERNATIONAL BOO

INTRODUCTION TO WORLD GEOGRAPHY

THE UNIVERSE

About 15,000 million years ago, time and space began with the most colossal explosion in cosmic history: the 'Big Bang' that initiated the universe. According to current theory, in the first millionth of a second of its existence it expanded from a dimensionless point of infinite mass and density into a fireball about 18,000 million miles across; and it has been expanding ever since.

It took almost a million years for the primal fireball to cool enough for atoms to form. They were mostly hydrogen, still the most abundant material in the universe. But the new matter was not evenly distributed around the young universe, and a few 1,000 million years later atoms in relatively dense regions began to cling together under the influence of gravity, forming distinct masses of gas separated by vast expanses of empty space. To begin with, these first protogalaxies were dark places: the universe had cooled. But gravitational attraction continued, condensing matter into coherent lumps inside the galactic gas clouds. About 3,000 million years later, some of these masses had contracted so much that internal pressure produced the high temperatures necessary to bring about nuclear fusion: the first stars were born.

There were several generations of stars, each feeding on the wreckage of its extinct predecessors as well as the original galactic gas swirls. With each new generation, progressively larger atoms were forged in stellar furnaces and the galaxy's range of elements, once restricted to hydrogen, grew larger. About 10,000 million years after the Big Bang, a star formed on the outskirts of our galaxy with enough matter left over to create a retinue of planets. Nearly 5,000 million years after that, a few planetary atoms had evolved into structures of complex molecules that lived, breathed and eventually pointed telescopes at the sky.

They found that their Sun is just one of more than 100,000 million stars in the home galaxy alone. Our galaxy, in turn, forms part of a local group of 25 or so similar structures, some much larger than our own; there are at least 100 million other galaxies in the universe as a whole. The most distant ever observed, a highly energetic galactic core known only as Quasar PKS 2000–330, lies about 15,000 million light-years away.

LIFE OF A STAR

For most of its existence, a star produces energy by the nuclear fusion of hydrogen into helium at its core. The duration of this hydrogen-burning period – known as the main sequence – depends on the star's mass; the greater the mass, the higher the core temperatures and the sooner the star's supply of hydrogen is exhausted. Dim, dwarf stars consume their hydrogen slowly, eking it out over 1,000 billion years or more. The Sun, like other stars of its mass, should spend about 10,000 million years on the main sequence; since it was formed less than 5,000 million years ago, it still has half its life left.

Once all a star's core hydrogen has been fused into helium, nuclear activity moves outward into layers of unconsumed hydrogen. For a time, energy production sharply increases: the star grows hotter and expands enormously, turning into a so-called red giant. Its energy output will increase a thousandfold, and it will swell to a hundred times its present diameter.

After a few hundred million years, helium in the core will become sufficiently compressed to initiate a new cycle of nuclear fusion: from helium to carbon. The star will contract somewhat, before beginning its last expansion, in the Sun's case engulfing the Earth and perhaps Mars. In this bloated condition, the Sun's outer layers will break off into space, leaving a tiny inner core, mainly of carbon, that shrinks progressively under the force of its own gravity: dwarf stars can attain a density more than 10,000 times that of normal matter, with crushing surface gravities to match. Gradually, the nuclear fires will die down, and the Sun will reach its terminal stage: a black dwarf, emitting insignificant amounts of energy.

However, stars more massive than the Sun may undergo another transformation. The additional mass allows gravitational collapse to continue indefinitely: eventually, all the star's remaining matter shrinks to a point, and its density approaches infinity – a state that will not permit even subatomic structures to survive.

The star has become a black hole: an anomalous 'singularity' in the fabric of space and time. Although vast coruscations of radiation will be emitted by any matter falling into its grasp, the singularity itself has an escape velocity that exceeds the speed of light, and nothing can ever be released from it. Within the boundaries of the black hole, the laws of physics are suspended, but no physicist can ever observe the extraordinary events that may occur.

THE END OF THE UNIVERSE

The likely fate of the universe is disputed. One theory (top right) dictates that the expansion begun at the time of the Big Bang will continue 'indefinitely', with aging galaxies moving further and further apart in an immense, dark graveyard. Alternatively, gravity may overcome the expansion (bottom right). Galaxies will fall back together until everything is again concentrated at a single point, followed by a new Big Bang and a new expansion, in an endlessly repeated cycle. The first theory is supported by the amount of visible matter in the universe; the second assumes there is enough dark material to bring about the gravitational collapse.

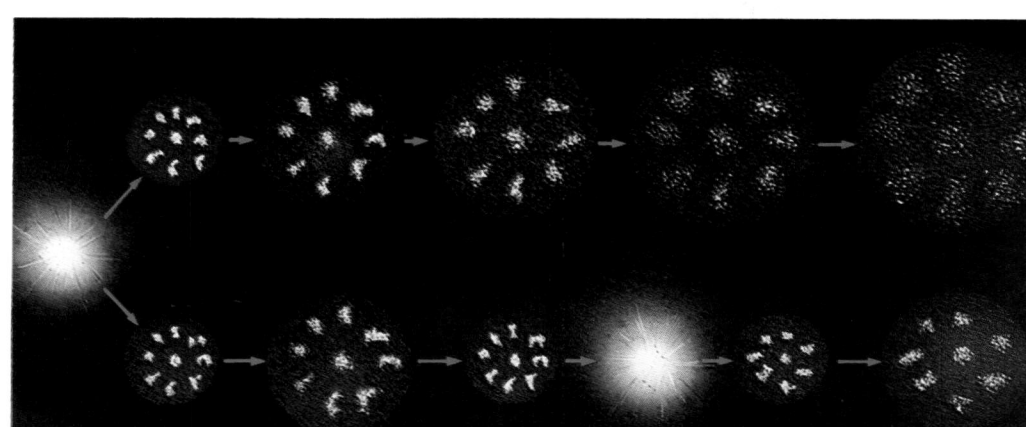

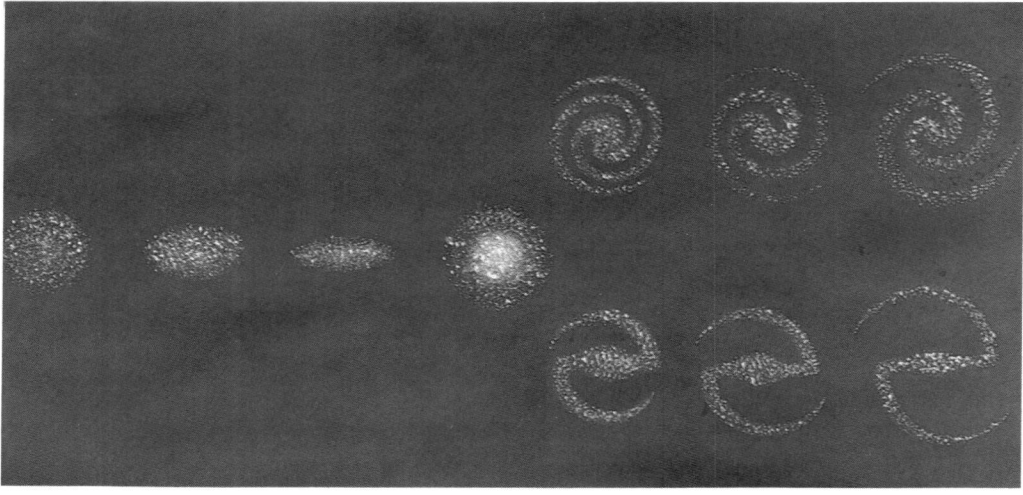

GALACTIC STRUCTURES

The universe's 100 million galaxies show clear structural patterns, originally classified by the American astronomer Edwin Hubble in 1925. Spiral galaxies like our own (top row) have a central, almost spherical bulge and a surrounding disk composed of spiral arms. Barred spirals (bottom row) have a central bar of stars across the nucleus, with spiral arms trailing from the ends of the bar. Elliptical galaxies (far left) have a uniform appearance, ranging from a flattened disk to a near sphere. So-called SO galaxies (left row, right) have a central bulge, but no spiral arms. A few have no discernible structure at all. Galaxies also vary enormously in size, from dwarfs only 2,000 light-years across to great assemblies of stars 80 or more times larger.

THE HOME GALAXY

The Sun and its planets are located in one of the spiral arms, a little less than 30,000 light-years from the galactic center and orbiting around it in a period of more than 200 million years. The center is invisible from the Earth, masked by vast, light-absorbing clouds of interstellar dust. The galaxy is probably around 12 billion years old and, like other spiral galaxies, has three distinct regions. The central bulge is about 30,000 light-years in diameter. The disk in which the Sun is located is not much more than 1,000 light-years thick but 100,000 light-years from end to end. Around the galaxy is the halo, a spherical zone 150,000 light-years across, studded with globular star-clusters and sprinkled with individual suns.

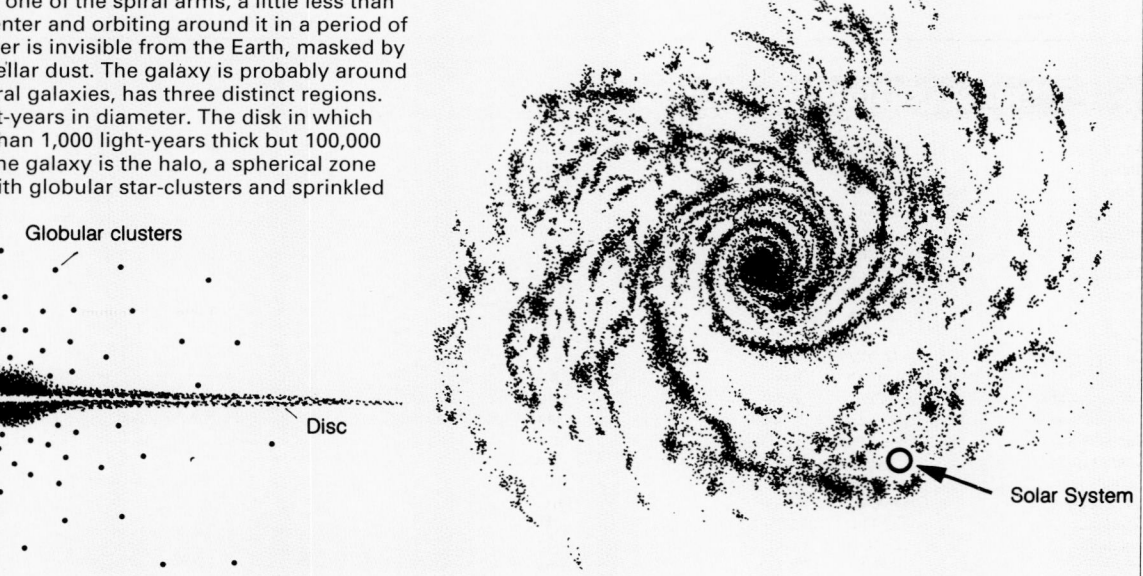

Globular clusters

Bulge

Disc

Solar System

CARTOGRAPHY BY PHILIP'S. COPYRIGHT REED INTERNATIONAL BOOKS LTD

STAR CHARTS

Star charts are drawn as projections of a vast, hollow sphere with the observer in the middle. Each circle below represents one hemisphere, centered on the north and south celestial poles respectively – projections of the Earth's poles in the heavens. At the present era, the north pole is marked by the star Polaris; the south pole has no such convenient reference point. The rectangular map shows the stars immediately above and below the celestial equator.

Astronomical coordinates are normally given in terms of 'Right Ascension' for longitude and 'Declination' for latitude or altitude. Since the stars appear to rotate around the Earth once every 24 hours, Right Ascension is measured eastward – counterclockwise – in hours and minutes. One hour is equivalent to 15 angular degrees; zero on the scale is the point at which the Sun crosses the celestial equator at the spring equinox, known to astronomers as the First Point in Aries. Unlike the Sun, stars always rise and set at the same point on the horizon. Declination measures (in degrees) a star's angular distance above or below the celestial equator.

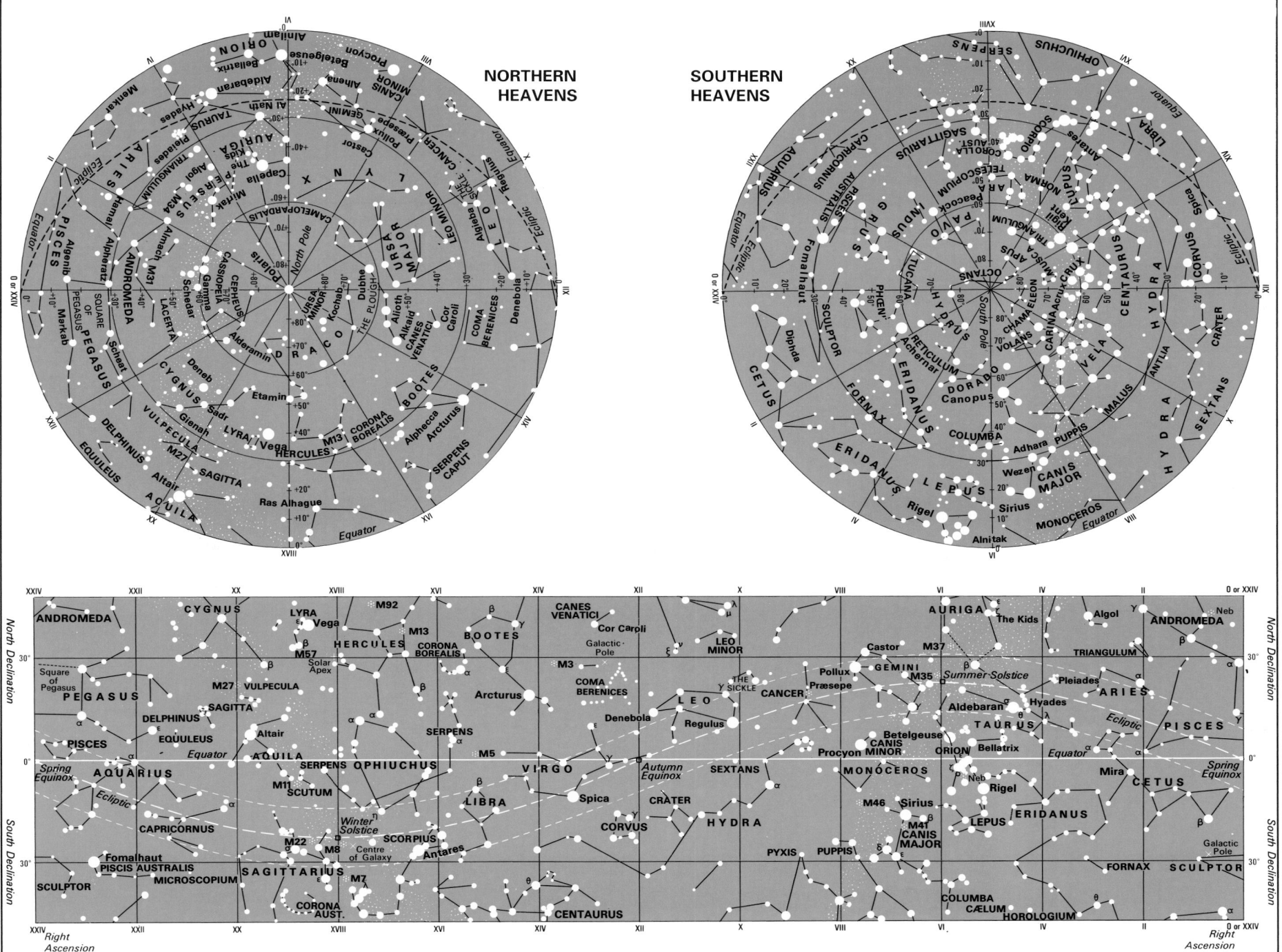

NORTHERN HEAVENS

SOUTHERN HEAVENS

THE CONSTELLATIONS

The constellations and their English names

Andromeda	Andromeda	Circinus	Compasses	Lacerta	Lizard	Piscis Austrinus	Southern Fish
Antila	Air Pump	Columba	Dove	Leo	Lion	Puppis	Ship's Stern
Apus	Bird of Paradise	Coma Berenices	Berenice's Hair	Leo Minor	Little Lion	Pyxis	Mariner's Compass
Aquarius	Water Carrier	Corona Australis	Southern Crown	Lepus	Hare	Reticulum	Net
Aquila	Eagle	Corona Borealis	Northern Crown	Libra	Scales	Sagitta	Arrow
Ara	Altar	Corvus	Crow	Lupus	Wolf	Sagittarius	Archer
Aries	Ram	Crater	Cup	Lynx	Lynx	Scorpius	Scorpion
Auriga	Charioteer	Crux	Southern Cross	Lyra	Harp	Sculptor	Sculptor
Boötes	Herdsman	Cygnus	Swan	Mensa	Table	Scutum	Shield
Caelum	Chisel	Delphinus	Dolphin	Microscopium	Microscope	Serpens	Serpent
Camelopardalis	Giraffe	Dorado	Swordfish	Monoceros	Unicorn	Sextans	Sextant
Cancer	Crab	Draco	Dragon	Musca	Fly	Taurus	Bull
Canes Venatici	Hunting Dogs	Equuleus	Little House	Norma	Level	Telescopium	Telescope
Canis Major	Great Dog	Eridanus	Eridanus	Octans	Octant	Triangulum	Triangle
Canis Minor	Little Dog	Fornax	Furnace	Ophiuchus	Serpent Bearer	Triangulum Australe	Southern Triangle
Capricornus	Goat	Gemini	Twins	Orion	Orion	Tucana	Toucan
Carina	Keel	Grus	Crane	Pavo	Peacock	Ursa Major	Great Bear
Cassiopeia	Cassiopeia	Hercules	Hercules	Pegasus	Winged Horse	Ursa Minor	Little Bear
Centaurus	Centaur	Horologium	Clock	Perseus	Perseus	Vela	Sails
Cepheus	Cepheus	Hydra	Water Snake	Phoenix	Phoenix	Virgo	Virgin
Cetus	Whale	Hydrus	Sea Serpent	Pictor	Easel	Volans	Flying Fish
Chamaeleon	Chameleon	Indus	Indian	Pisces	Fishes	Vulpecula	Fox

THE NEAREST STARS

The 20 nearest stars, excluding the Sun, with their distance from Earth in light-years*

Proxima Centauri	4.3
Alpha Centauri A	4.3
Alpha Centauri B	4.3
Barnard's Star	6.0
Wolf 359	8.1
Lal 21185	8.2
Sirius A	8.7
Sirius B	8.7
UV Ceti A	9.0
UV Ceti B	9.0
Ross 154	9.3
Ross 248	10.3
Epsilon Eridani	10.8
L 789-6	11.1
Ross 128	11.1
61 Cygni A	11.2
61 Cygni B	11.2
Procyon A	11.3
Procyon B	11.3
Epsilon Indi	11.4

Many of the nearest stars, like Alpha Centauri A and B, are doubles, orbiting about their common center of gravity and to all intents and purposes equidistant from Earth. Many of them are dim objects, with no name other than the designation given by the astronomers who investigated them. However, they include Sirius, the brightest star in the sky, and Procyon, the seventh brightest. Both are far larger than the Sun: of the nearest stars, only Epsilon Eridani is similar in size and luminosity.

* A light-year equals approx. 5,870,000,000,000 miles

THE SOLAR SYSTEM

Lying 27,000 light-years from the center of one of billions of galaxies that comprise the observable universe, our Solar System contains nine planets and their moons, innumerable asteroids and comets, and a miscellany of dust and gas, all tethered by the immense gravitational field of the Sun, the middling-sized star whose thermonuclear furnaces provide them all with heat and light. The Solar System was formed about 4,600 million years ago, when a spinning cloud of gas, mostly hydrogen but seeded with other, heavier elements, condensed enough to ignite a nuclear reaction and create a star. The Sun still accounts for almost 99.9% of the system's total mass; one planet, Jupiter, contains most of the remainder.

By composition as well as distance, the planetary array divides quite neatly in two: an inner system of four small, solid planets, including the Earth, and an outer system, from Jupiter to Neptune, of four huge gas giants. Between the two groups lies a scattering of asteroids, perhaps as many as 40,000; possibly the remains of a planet destroyed by some unexplained catastrophe, they are more likely to be debris left over from the Solar System's formation, prevented by the gravity of massive Jupiter from coalescing into a larger body. The ninth planet, Pluto, seems to be a world of the inner system type: small, rocky and something of an anomaly.

By the 1990s, however, the Solar System also included some newer anomalies: several thousand spacecraft. Most were in orbit around the Earth, but some had probed far and wide around the system. The valuable information beamed back by these robotic investigators has transformed our knowledge of our celestial environment.

Much of the early history of science is the story of people trying to make sense of the errant points of light that were all they knew of the planets. Now, people have themselves stood on the Earth's Moon; probes have landed on Mars and Venus, and orbiting radars have mapped far distant landscapes with astonishing accuracy. In the 1980s, the US *Voyagers* skimmed all four major planets of the outer system, bringing new revelations with each close approach. Only Pluto, inscrutably distant in an orbit that takes it 50 times the Earth's distance from the Sun, remains unvisited by our messengers.

ORBITS OF THE PLANETS

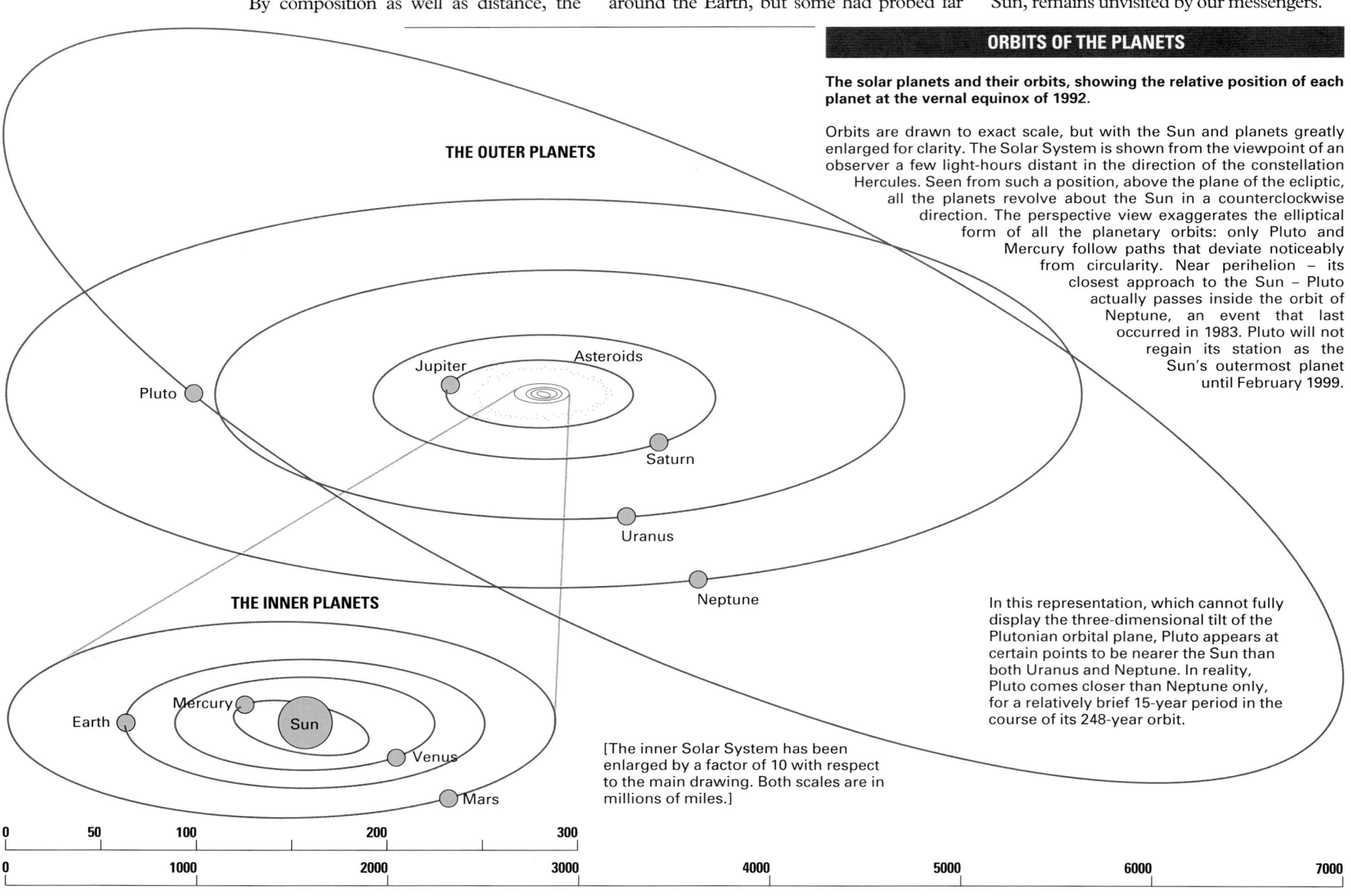

The solar planets and their orbits, showing the relative position of each planet at the vernal equinox of 1992.

Orbits are drawn to exact scale, but with the Sun and planets greatly enlarged for clarity. The Solar System is shown from the viewpoint of an observer a few light-hours distant in the direction of the constellation Hercules. Seen from such a position, above the plane of the ecliptic, all the planets revolve about the Sun in a counterclockwise direction. The perspective view exaggerates the elliptical form of all the planetary orbits: only Pluto and Mercury follow paths that deviate noticeably from circularity. Near perihelion – its closest approach to the Sun – Pluto actually passes inside the orbit of Neptune, an event that last occurred in 1983. Pluto will not regain its station as the Sun's outermost planet until February 1999.

THE OUTER PLANETS

Pluto — Jupiter — Asteroids — Saturn — Uranus — Neptune

THE INNER PLANETS

Mercury — Earth — Sun — Venus — Mars

[The inner Solar System has been enlarged by a factor of 10 with respect to the main drawing. Both scales are in millions of miles.]

In this representation, which cannot fully display the three-dimensional tilt of the Plutonian orbital plane, Pluto appears at certain points to be nearer the Sun than both Uranus and Neptune. In reality, Pluto comes closer than Neptune only, for a relatively brief 15-year period in the course of its 248-year orbit.

0	50	100	200	300

0	1000	2000	3000	4000	5000	6000	7000

PLANETARY DATA

	Mean distance from Sun (million mi)	Mass (Earth = 1)	Period of orbit (Earth years)	Period of rotation (Earth days)	Equatorial diameter (mi)	Average density (water = 1)	Surface gravity (Earth = 1)	Escape velocity (mi/sec)	Number of known satellites
Sun	–	332,946	–	25.38	870,000	1.41	27.9	385.6	–
Mercury	36.4	0.06	0.241	58.67	3,049	5.5	0.38	2.67	0
Venus	67.3	0.8	0.615	243	7,565	5.25	0.90	6.48	0
Earth	93.5	1.0	1.00	0.99	7,973	5.52	1.00	6.99	1
Mars	142.1	0.1	1.88	1.02	4,246	3.94	0.38	3.14	2
Jupiter	486.2	317.8	11.86	0.41	89,250	1.33	2.64	37.64	16
Saturn	891.9	95.2	29.63	0.42	75,000	0.706	1.16	22.66	17
Uranus	1795.2	14.5	83.97	0.45	32,500	1.70	1.11	14.0	15
Neptune	2814.2	17.2	164.8	0.67	30,250	1.77	1.21	14.94	8
Pluto	3683.9	0.002	248.63	6.38	1,500	5.50	0.47	3.19	1

Planetary days are given in sidereal time – that is, with respect to the stars rather than the Sun. Most of the information in the table was confirmed by spacecraft and often obtained from photographs and other data transmitted back to the Earth. In the case of Pluto, however, only earthbound observations have been made, and no spacecraft can hope to encounter it until well into the next century. Given the planet's small size and great distance, figures for its diameter and rotation period cannot be definitive.

Since Pluto does not appear to be massive enough to account for the perturbations in the orbits of Uranus and Neptune that led to its 1930 discovery, it is quite possible that a tenth and even more distant planet may exist. Once Pluto's own 248-year orbit has been observed for long enough, further discrepancies may give a clue as to any tenth planet's whereabouts. Even so, distance alone would make it very difficult to locate, especially since telescopes powerful enough to find it are normally engaged in galactic study.

THE PLANETS

Mercury is the closest planet to the Sun and hence the fastest-moving. It has no significant atmosphere and a cratered, wrinkled surface very similar to that of Earth's moon.

Venus has much the same physical dimensions as Earth. However, its carbon dioxide atmosphere is 90 times as dense, creating a runaway greenhouse effect that makes the Venusian surface, at 890°F, the hottest of all the planets in the Solar System. Radar mapping shows relatively level land with volcanic regions whose sulfurous discharges explain the sulfuric acid rains reported by soft-landing space probes before they succumbed to Venus's fierce climate.

Earth seen from space is easily the most beautiful of the inner planets; it is also, and more objectively, the largest, as well the only home of known life. Living things are the main reason why the Earth is able to retain a substantial proportion of corrosive and highly reactive oxygen in its atmosphere, a state of affairs that contradicts the laws of chemical equilibrium; the oxygen in turn supports the life that constantly regenerates it.

Mars was once considered the likeliest of the other planets to share Earth's cargo of life: the seasonal expansion of dark patches strongly suggested vegetation and the planet's apparent icecaps indicated the vital presence of water. But close inspection by spacecraft brought disappointment: chemical reactions account for the seeming vegetation, the icecaps are mainly frozen carbon dioxide, and whatever oxygen the planet once possessed is now locked up in the iron-bearing rock that covers its cratered surface and gives it its characteristic red hue.

Jupiter masses almost three times as much as all the other planets together; had it scooped up a little more matter during its formation, it might have evolved into a small companion star for the Sun. The planet is mostly gas, under intense pressure in the lower atmosphere above a core of fiercely compressed hydrogen and helium. The upper layers form strikingly-colored rotating belts, the outward sign of the intense storms created by Jupiter's rapid diurnal rotation. Close approaches by spacecraft have shown an orbiting ring system, and discovered several previously unknown moons: Jupiter has at least 16 moons.

Saturn is structurally similar to Jupiter, rotating fast enough to produce an obvious bulge at its equator. Ever since the invention of the telescope, however, Saturn's rings have been the feature that has attracted most observers. *Voyager* probes in 1980 and 1981 sent back detailed pictures that showed them to be composed of thousands of separate ringlets, each in turn made up of tiny icy particles, interacting in a complex dance that may serve as a model for the study of galactic and even larger structures.

Uranus was unknown to the ancients. Although it is faintly visible to the naked eye, it was not discovered until 1781. Its composition is broadly similar to Jupiter and Saturn, though its distance from the Sun ensures an even colder surface temperature. Observations in 1977 suggested the presence of a faint ring system, amply confirmed when *Voyager 2* swung past the planet in 1986.

Neptune is always more than 2,500 million miles from Earth, and despite its diameter of over 30,000 miles, it can only be seen by telescope. Its 1846 discovery was the result of mathematical predictions by astronomers seeking to explain irregularities in the orbit of Uranus, but until *Voyager 2* closed with the planet in 1989, little was known of it. Like Uranus, it has a ring system; *Voyager's* photographs revealed a total of eight moons.

Pluto is the most mysterious of the solar planets, if only because even the most powerful telescopes can scarcely resolve it from a point of light to a disk. It was discovered as recently as 1930, like Neptune as the result of perturbations in the orbits of the two then outermost planets. Its small size, as well as its eccentric and highly tilted orbit, has led to suggestions that it is a former satellite of Neptune, somehow liberated from its primary. In 1978 Pluto was found to have a moon of its own, Charon, apparently half the size of Pluto itself.

Mean distance from Sun in million miles

Mercury	36.4
Venus	67.3
Earth	93.5
Mars	142.1
Jupiter	486.2
Saturn	891.9
Uranus	1795.2
Neptune	2814.2
Pluto	3683.9

THE EARTH: TIME AND MOTION

The basic unit of time measurement is the day, that is, one rotation of the Earth on its axis. The subdivision of the day into hours, minutes and seconds is arbitrary and simply for our convenience. Our present calendar is based on the solar year of 365.24 days, the time taken by the Earth to orbit the Sun. As the Earth rotates from west to east, the Sun appears to rise in the east and set in the west. When the Sun is setting in Shanghai, on the opposite side of the world New York is just emerging into sunlight. Noon, when the Sun is directly overhead, is coincident at all places on the same meridian, with shadows pointing directly toward the poles.

Calendars based on the movements of the Sun and Moon have been used since ancient times. The Julian Calendar, with its leap year, introduced by Julius Caesar, fixed the average length of the year at 365.25 days, which was about 11 minutes too long (the Earth completes its orbit in 365 days, 5 hours, 48 minutes and 46 seconds of mean solar time). The cumulative error was rectified by the Gregorian Calendar, introduced by Pope Gregory XIII in 1582, when he decreed that the day following 4 October was 15 October , and that century years did not count as leap years unless divisible by 400. England did not adopt the reformed calendar until 1752, when the country found itself 11 days behind the continent.

Britain imposed the Gregorian Calendar on all its possessions, including the American colonies. All dates preceding 2 September were marked 'OS', for 'Old Style'.

EARTH DATA

Maximum distance from Sun (Aphelion): 94,452,780 miles
Minimum distance from Sun (Perihelion): 91,342,080 miles
Obliquity of the ecliptic: 23° 27' 08"
Length of year – solar tropical (equinox to equinox): 365.24 days
Length of year – sidereal (fixed star to fixed star): 365.26 days
Length of day – mean solar day: 24h, 03m, 56s.
Length of day – mean sidereal day: 23h, 56m, 04s

Superficial area: 197,000,000 sq mi
Land surface: 57,500,000 sq mi (29.2%)
Water surface: 139,500,000 sq mi (70.8%)
Equatorial circumference: 24,903 mi
Polar circumference: 24,860 mi
Equatorial diameter: 7,926.7 mi
Polar diameter: 7,900.0 mi
Equatorial radius: 3,963.4 mi
Polar radius: 3,950.0 mi
Volume of the Earth: 260,000 x 10^6 cu mi
Mass of the Earth: 6.5 x 10^21 tons

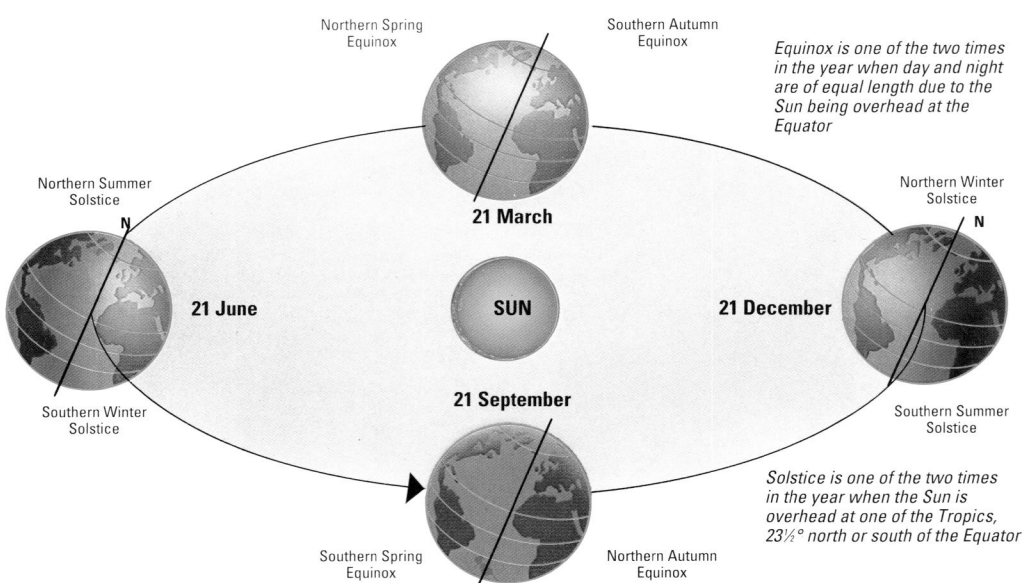

Equinox is one of the two times in the year when day and night are of equal length due to the Sun being overhead at the Equator

Solstice is one of the two times in the year when the Sun is overhead at one of the Tropics, 23½° north or south of the Equator

THE SEASONS

The Earth revolves around the Sun once a year in a 'counterclockwise' direction, tilted at a constant angle of 66½°. In June, the northern hemisphere is tilted toward the Sun: as a result it receives more hours of sunshine in a day and therefore has its warmest season, summer. By December, the Earth has rotated halfway round the Sun so that the southern hemisphere is tilted toward the Sun and has its summer; the hemisphere that is tilted away from the Sun has winter. On 21 June the Sun is directly overhead at the Tropic of Cancer (23½° N), and this is midsummer in the northern hemisphere. Midsummer in the southern hemisphere occurs on 21 December, when the Sun is overhead at the Tropic of Capricorn (23½° S).

DAY AND NIGHT

The Sun appears to rise in the east, reach its highest point at noon, and then set in the west, to be followed by night. In reality, it is not the Sun that is moving but the Earth revolving from west to east.

At the summer solstice in the northern hemisphere (21 June), the Arctic has total daylight and the Antarctic total darkness. The opposite occurs at the winter solstice (21 December). At the Equator, the length of day and night are almost equal all year.

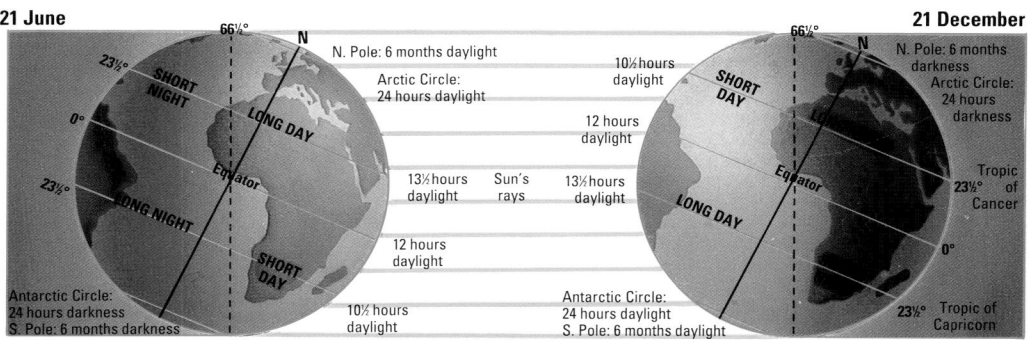

THE SUN'S PATH

The diagrams on the left illustrate the apparent path of the Sun at (A) the Equator, (B) in mid-latitude (45°), (C) at the Arctic Circle (66½°), and (D) at the North Pole, where there are six months of continuous daylight and six months of continuous night.

MEASUREMENTS OF TIME

Astronomers distinguish between solar time and sidereal time. Solar time derives from the period taken by the Earth to rotate on its axis: one rotation defines a solar day. But the speed of the Earth along its orbit around the Sun is not constant. The length of day – or 'apparent solar day', as defined by the apparent successive transits of the Sun – is irregular because the Earth must complete more than one rotation before the Sun returns to the same meridian. The constant sidereal day is defined as the interval between two successive apparent transits of a star, or the first point of Aries, across the same meridian. If the Sun is at the equinox and overhead at a meridian one day, then the next day it will be to the east by approximately 1°. Thus, the Sun will not cross the meridian until four minutes after the sidereal noon.

From the diagrams on the right it is possible to discover the time of sunrise or sunset on a given date and for latitudes between 60°N and 60°S.

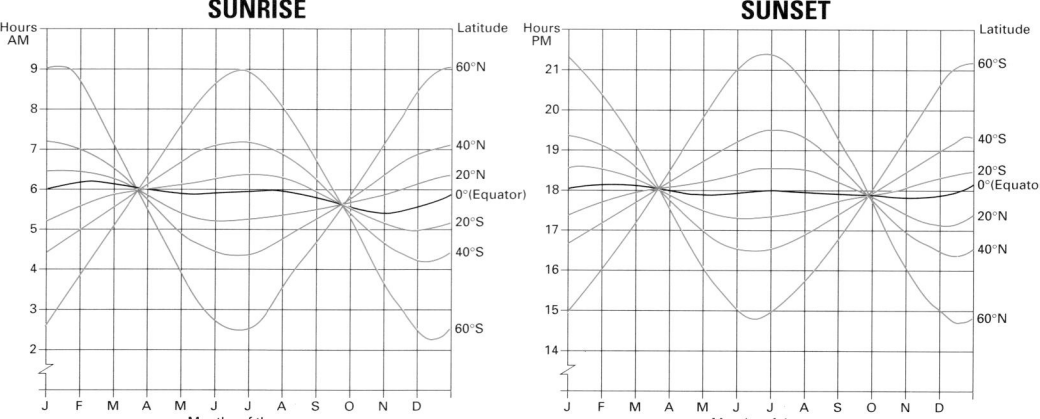

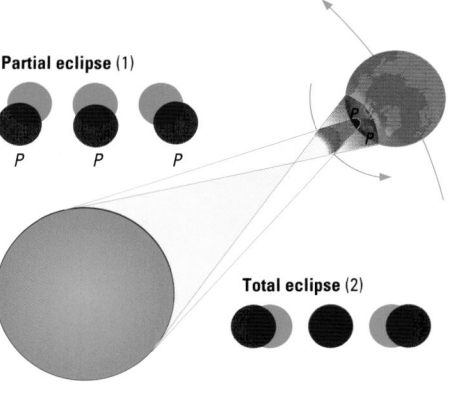

THE MOON

PHASES OF THE MOON

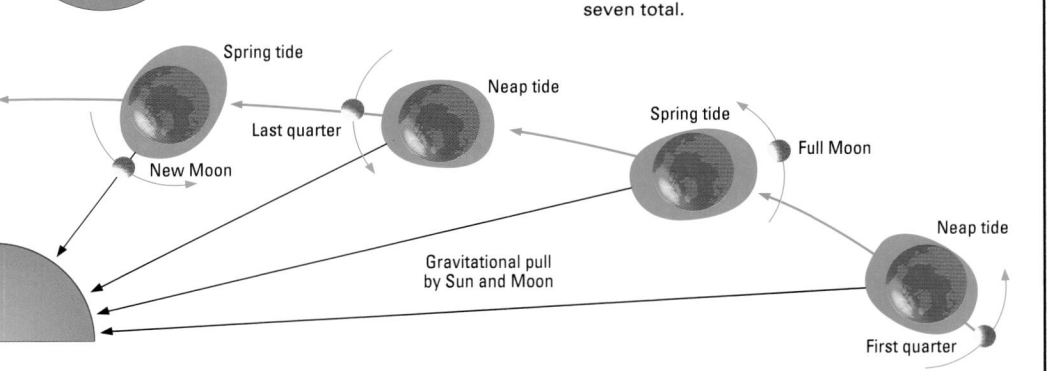

New Moon — Crescent — First quarter — Gibbous — Full Moon — Gibbous — Last quarter — Crescent — New Moon

MOON DATA

Distance from Earth
The Moon orbits at a mean distance of 238,731 mi, at an average speed of 2,289 mph in relation to the Earth.

Size and mass
The average diameter of the Moon is 2,159.3 mi. It is 400 times smaller than the Sun but is about 400 times closer to the Earth, so we see them as the same size. The Moon has a mass of 7.975×10^{19} tons, with a density 3.344 times that of water.

Visibility
Only 59% of the Moon's surface is directly visible from Earth. Reflected light takes 1.25 seconds to reach Earth – compared to 8 minutes 27.3 seconds for light to reach us from the Sun.

Temperature
With the Sun overhead, the temperature on the lunar equator can reach 243°F [117.2°C]. At night it can sink to –261°F [–162.7°C].

The Moon rotates more slowly than the Earth, making one complete turn on its axis in just over 27 days. Since this corresponds to its period of revolution around the Earth, the Moon always presents the same hemisphere or face to us, and we never see 'the dark side'. The interval between one Full Moon and the next (and between new Moons) is about 29½ days – a lunar month. The apparent changes in the shape of the Moon are caused by its changing position in relation to the Earth; like the planets, it produces no light of its own and shines only by reflecting the rays of the Sun.

Partial eclipse (1)

P   P   P

Total eclipse (2)

Lunar eclipse

ECLIPSES

When the Moon passes between the Sun and the Earth, it causes a partial eclipse of the Sun (1) if the Earth passes through the Moon's outer shadow (P), or a total eclipse (2) if the inner cone shadow crosses the Earth's surface. In a lunar eclipse, the Earth's shadow crosses the Moon and, again, provides either a partial or total eclipse. Eclipses of the Sun and the Moon do not occur every month because of the 5° difference between the plane of the Moon's orbit and the plane in which the Earth moves. In the 1990s only 14 lunar eclipses are possible, for example, seven partial and seven total; each is visible only from certain, and variable, parts of the world. The same period witnesses 13 solar eclipses – six partial (or annular) and seven total.

TIDES

The daily rise and fall of the ocean's tides are the result of the gravitational pull of the Moon and that of the Sun, though the effect of the latter is only 46.6% as strong as that of the Moon. This effect is greatest on the hemisphere facing the Moon and causes a tidal 'bulge'. When lunar and solar forces pull together, with Sun, Earth and Moon in line (near New and Full Moons), higher 'spring tides' (and lower low tides) occur; when lunar and solar forces are least coincidental with the Sun and Moon at an angle (near the Moon's first and third quarters), 'neap tides' occur, which have a small tidal range.

Spring tide — Neap tide — Last quarter — New Moon — Spring tide — Full Moon — Neap tide — First quarter

Gravitational pull by Sun and Moon

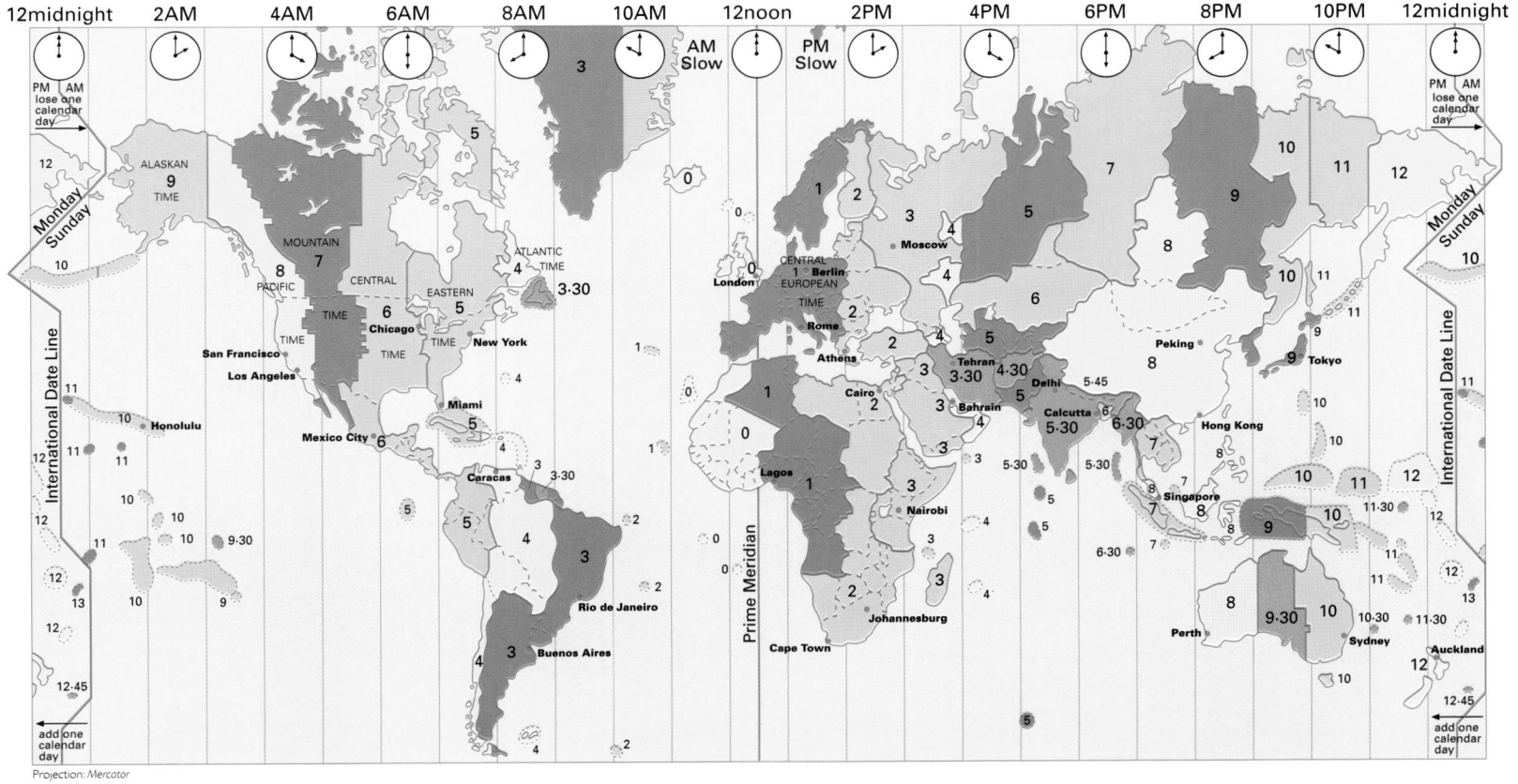

12midnight 2AM 4AM 6AM 8AM 10AM 12noon 2PM 4PM 6PM 8PM 10PM 12midnight

Projection: Mercator

TIME ZONES

The Earth rotates through 360° in 24 hours, and so moves 15° every hour. The world is divided into 24 standard time zones, each centered on lines of longitude at 15° intervals. The Greenwich meridian lies at the center of the first zone. All places to the west of Greenwich are one hour behind for every 15° of longitude; places to the east are ahead by one hour for every 15°. When it is 12 noon at the Greenwich meridian, 180° east it is midnight of the same day – while 180° west the day is just beginning. To overcome this, the International Date Line was established, approximately following the 180° meridian. Thus, if you traveled eastward from Japan (140° East) to Samoa (170° West), you would pass from Sunday night into Sunday morning.

 Zones slow or fast of Greenwich Mean Time

 Half-hour zones

 The time when it is 12 noon at Greenwich

THE EARTH: GEOLOGY

The origin of the Earth is still open to conjecture, although the most widely accepted theory is that it was formed from a solar cloud consisting mainly of hydrogen about 4,600 million years ago. The cloud condensed, forming the planets. The lighter elements floated to the surface of the Earth, where they cooled to form a crust; the inner material remained hot and molten. The first rocks were formed over 3,500 million years ago, but the Earth's surface has since been constantly altered.

The crust consists of a brittle, low-density material, varying from 3 miles to 30 miles thick beneath the continents, which is predominantly made up of silica and aluminum: hence its name, 'sial'. Below the sial is a basaltic layer known as 'sima', comprising mainly silica and magnesium. The crust accounts for only 1.5% of the Earth's volume.

The mantle lies immediately below the crust, with a distinct change in density and chemical properties. The rock here is rich in iron and magnesium silicates, with temperatures reaching 2,900°F. The rigid upper mantle extends down to a depth of about 600 miles, below which is a more viscous lower mantle measuring about 1,200 miles thick.

The outer core, measuring about 1,300 miles thick, consists of molten iron and nickel at temperatures ranging from 3,800°F to 9,000°F, possibly separated from the less dense mantle by an oxidized shell. About 3,100 miles below the planetary surface is a liquid transition zone, below which is the solid inner core, a sphere of about 1,650 miles diameter, where rock is three times as dense as in the crust. The temperature at the center of the Earth is probably about 9,000°F.

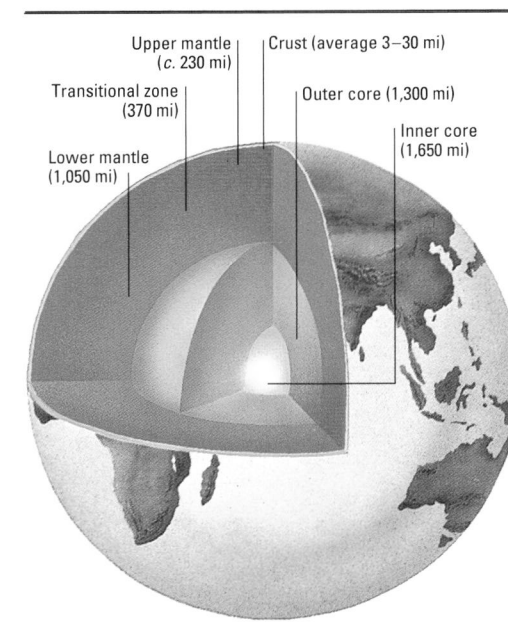

Upper mantle (c. 230 mi)
Crust (average 3–30 mi)
Transitional zone (370 mi)
Outer core (1,300 mi)
Inner core (1,650 mi)
Lower mantle (1,050 mi)

The complementary, almost jigsaw-puzzle fit of the Atlantic coasts led to Alfred Wegener's proposition of continental drift in Germany (1915). His theory suggested that an ancient super-continent, which he called Pangaea, incorporating all the Earth's land masses, gradually split up to form the continents we know today.

By 180 million years ago, Pangaea had divided into two major groups and the southern part, Gondwanaland, had itself begun to break up with India and Antarctica-Australia becoming isolated.

By 135 million years ago, the widening of the splits in the North Atlantic and Indian Oceans persisted, a South Atlantic gap had appeared, and India continued to move 'north' toward Asia.

By 65 million years ago, South America had completely split from Africa.

To form today's pattern, India 'collided' with Asia (crumpling up sediments to form the Himalayas); South America rotated and moved west to connect with North America; Australia separated from Antarctica and moved north; and the familiar gap developed between Greenland and Europe.

CONTINENTAL DRIFT

About 200 million years ago the original Pangaea land mass began to split into two continental groups, which further separated over time to produce the present-day configuration.

Laurasia

Gondwana

180 million years ago

135 million years ago

Present day

~~~~~~ Trench
————— Rift
New ocean floor
Zones of slippage

PLATE TECTONICS

The original debate about the drift theory of Wegener and others formed a long prelude to a more radical idea: plate tectonics. The discovery that the continents are carried along on the top of slowly-moving crustal plates (which float on heavier liquid material – the lower mantle – much as icebergs do on water) provided the mechanism for the drift theories to work. The plates converge and diverge along margins marked by seismic and volcanic activity. Plates diverge from mid-ocean ridges where molten lava pushes up and forces the plates apart at a rate of up to 1.5 inches a year; converging plates form either a trench (where the oceanic plates sink below the lighter continental rock) or mountain ranges (where two continents collide).

The debate about plate tectonics is not over, however. In addition to abiding questions such as what force actually moves the plates (massive convection currents in the Earth's interior is the most popular explanation), and why so many volcanoes and earthquakes occur in mid-plate (such as Hawaii and central China), evidence began to emerge in the early 1990s that, with more sophisticated equipment and models, the whole theory might be in doubt.

VOLCANOES

Of some 850 volcanoes that have produced recorded eruptions, nearly three-quarters lie in the 'Ring of Fire' that surrounds the Pacific Ocean. The 1980s was a bad decade for loss of life here, with three major eruptions – Mount St Helens, USA, in 1980; El Chichon, Mexico, in 1982; and Nevado del Ruiz, Colombia, in 1985 – killing 25,000 people. This is not because the world is becoming less geologically stable: it is simply that populations are growing fast, with over 350 million people now living in areas vulnerable to seismic activity.

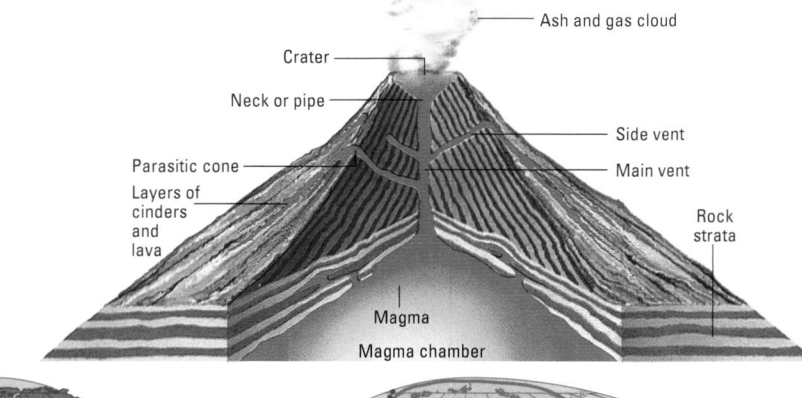

Ash and gas cloud
Crater
Neck or pipe
Side vent
Parasitic cone
Main vent
Layers of cinders and lava
Rock strata
Magma
Magma chamber

Shield cone

Hornit cone

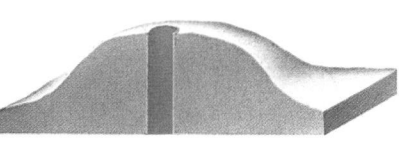

Cinder cone

Caldera

DISTRIBUTION

▲ Land volcanoes active since 1700

• Submarine volcanoes

✛ Geysers

— Boundaries of tectonic plates

↙ Direction of movement along plate boundaries

Volcanoes can suddenly erupt after lying dormant for centuries: in 1991 Mount Pinatubo, Philippines, burst into life after sleeping for more than 600 years.

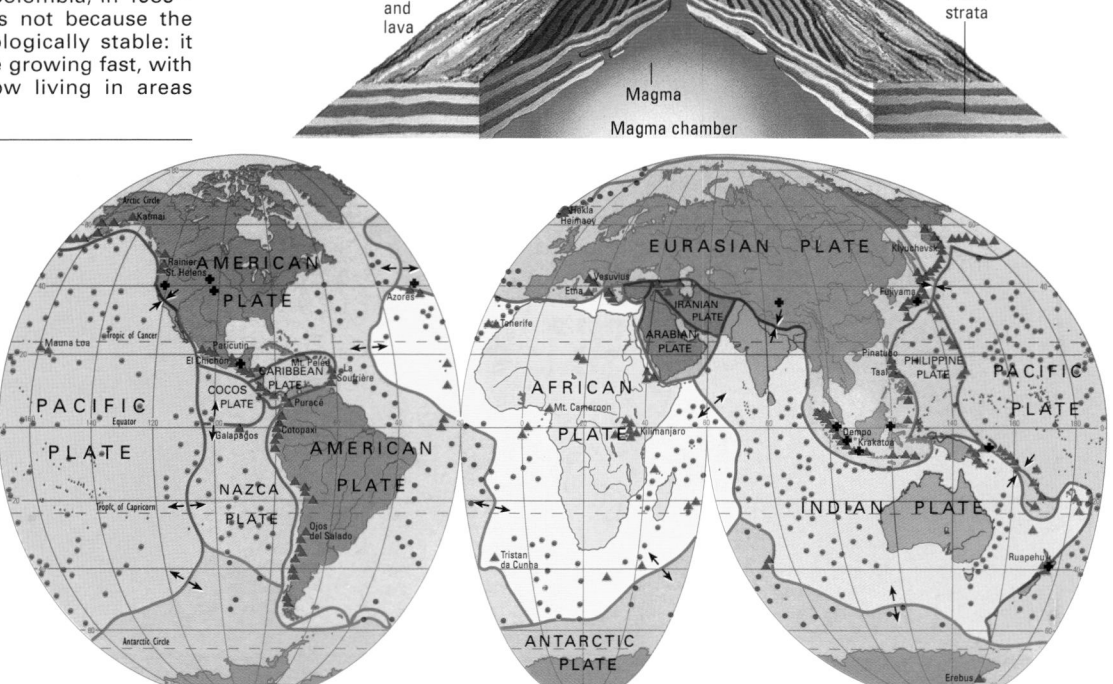

EURASIAN PLATE
AMERICAN PLATE
IRANIAN PLATE
ARABIAN PLATE
PACIFIC PLATE
AFRICAN PLATE
PHILIPPINE PLATE
CARIBBEAN PLATE
COCOS PLATE
PACIFIC PLATE
AMERICAN PLATE
NAZCA PLATE
INDIAN PLATE
ANTARCTIC PLATE

4600	4600
2000	
1000	
	PRE-CAMBRIAN

Time, in millions of years before the present, is shown on a sliding scale, greatly compressed in the distant past.

570	Cambrian
500	Ordovician
430	Silurian
395	Devonian
345	
	Carboniferous
280	Permian
225	Triassic
190	Jurassic
135	Cretaceous
65	Paleocene
53	Eocene
37	Oligocene
26	Miocene
12	Pliocene
2	Pleistocene

Eras: PALEOZOIC, MESOZOIC, CAINOZOIC

Tertiary, Quaternary

Holocene 12,000 BP to present

ERA **PERIOD** **EPOCH**

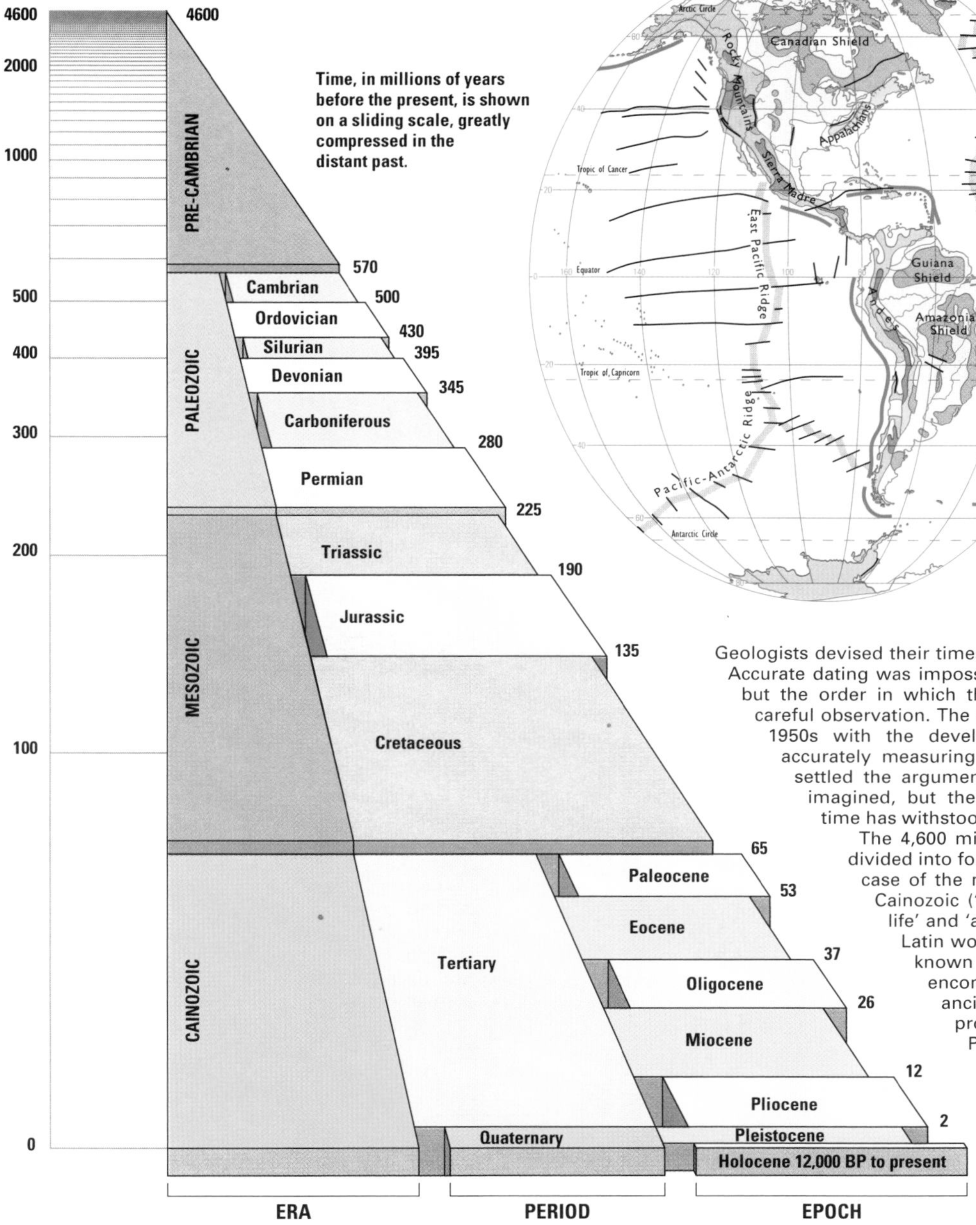

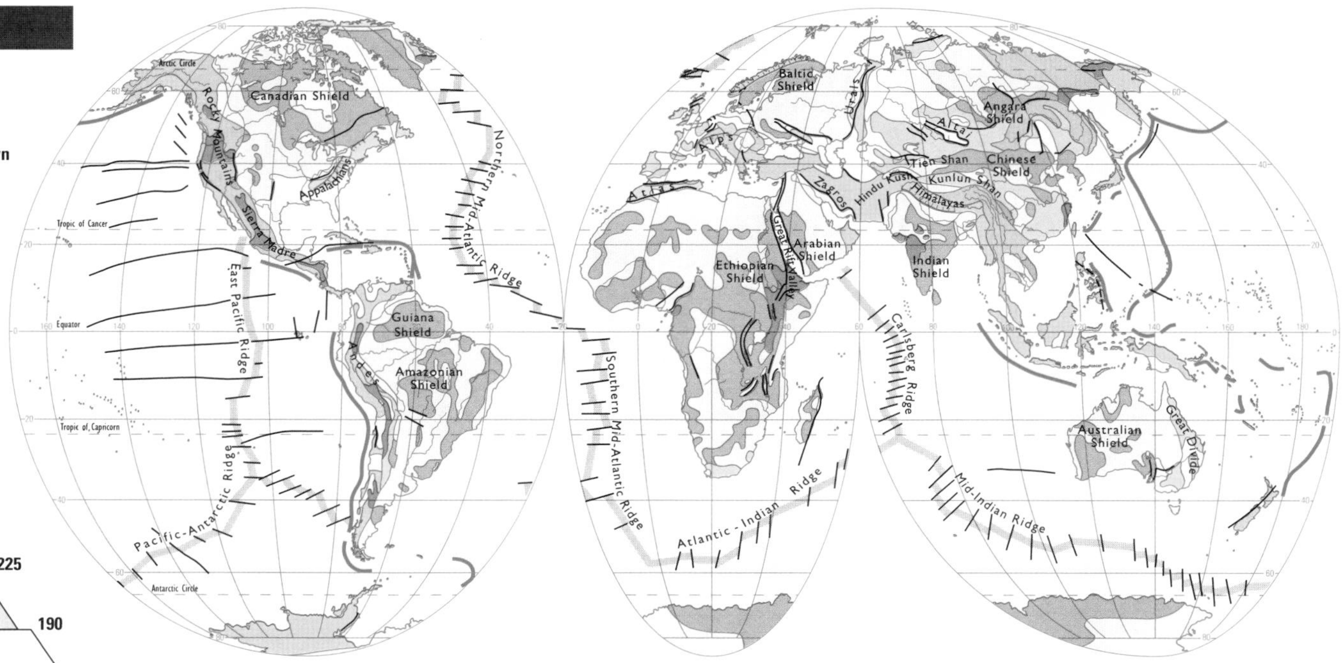

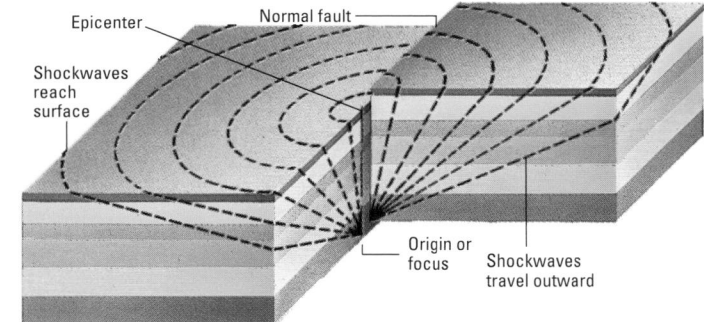

Geologists devised their timescale on the basis of relative, not calendar, ages. Accurate dating was impossible and estimates were often bitterly disputed, but the order in which the rocks were formed could be deduced from careful observation. The advent of radioactive dating – culminating in the 1950s with the development of a mass spectrometer capable of accurately measuring tiny quantities of isotopes – appears to have settled the arguments. The Earth is far older than geologists first imagined, but their painstakingly-created structure of geological time has withstood the advent of high technology.

The 4,600 million years since the formation of the Earth are divided into four great eras, further split into periods and, in the case of the most recent era, epochs. The present era is the Cainozoic ('new life'), extending backward through 'middle life' and 'ancient life' to the Pre-Cambrian, named after the Latin word for Wales, the location of some of the earliest known fossils. Most of the Earth's geological history is encompassed by the Pre-Cambrian: though traces of ancient life have since been found, it was largely the proliferation of fossils from the beginning of the Paleozoic era onward, some 570 million years ago, which first allowed precise subdivisions to be made.

Like the Cambrian, most are named after regions exemplifying a period's geology. Others – such as the Carboniferous ('coal-bearing') or the Cretaceous ('chalk-bearing') – are more directly descriptive.

- Pre-Cambrian shields
- Sedimentary cover on Pre-Cambrian shields
- Paleozoic (Caledonian and Hercynian) folding
- Sedimentary cover on Paleozoic folding
- Mesozoic folding
- Sedimentary cover on Mesozoic folding
- Cainozoic (Alpine) folding
- Sedimentary cover on Cainozoic folding
- Intensive Mesozoic and Cainozoic vulcanism
- Principal faults
- Oceanic marginal troughs
- Mid-oceanic ridges
- Overthrust faults

EARTHQUAKES

Earthquake magnitude is usually rated according to either the Richter or the Modified Mercalli scale, both devised by seismologists in the 1930s. The Richter scale measures absolute earthquake power with mathematical precision: each step upward represents a tenfold increase in shockwave amplitude. Theoretically, there is no upper limit, but the largest earthquakes measured have been rated at between 8.8 and 8.9. The 12–point Mercalli scale, based on observed effects, is often more meaningful, ranging from I (earthquakes noticed only by seismographs) to XII (total destruction); intermediate points include V (people awakened at night; unstable objects overturned), VII (collapse of ordinary buildings; chimneys and monuments fall) and IX (conspicuous cracks in ground; serious damage to reservoirs).

NOTABLE EARTHQUAKES SINCE 1900

Year	Location	Mag.	Deaths
1906	San Francisco, USA	8.3	503
1906	Valparaiso, Chile	8.6	22,000
1908	Messina, Italy	7.5	83,000
1915	Avezzano, Italy	7.5	30,000
1920	Gansu (Kansu), China	8.6	180,000
1923	Yokohama, Japan	8.3	143,000
1927	Nan Shan, China	8.3	200,000
1932	Gansu (Kansu), China	7.6	70,000
1934	Bihar, India/Nepal	8.4	10,700
1935	Quetta, India*	7.5	60,000
1939	Chillan, Chile	8.3	28,000
1939	Erzincan, Turkey	7.9	30,000
1960	Agadir, Morocco	5.8	12,000
1962	Khorasan, Iran	7.1	12,230
1963	Skopje, Yugoslavia**	6.0	1,000
1964	Anchorage, Alaska	8.4	131
1968	N.E. Iran	7.4	12,000
1970	N. Peru	7.7	66,794
1972	Managua, Nicaragua	6.2	5,000
1974	N. Pakistan	6.3	5,200
1976	Guatemala	7.5	22,778
1976	Tangshan, China	8.2	650,000
1978	Tabas, Iran	7.7	25,000
1980	El Asnam, Algeria	7.3	20,000
1980	S. Italy	7.2	4,800
1985	Mexico City, Mexico	8.1	4,200
1988	N.W. Armenia	6.8	55,000
1990	N. Iran	7.7	36,000
1993	Maharashtra, India	6.4	30,000
1994	Los Angeles, USA	6.6	57
1995	Kobe, Japan	7.2	5,000

The highest magnitude recorded on the Richter scale is 8.9, in Japan on 2 March 1933 (2,990 deaths). The most devastating quake ever was at Shaanxi (Shensi) province, central China, on 24 January 1566, when an estimated 830,000 people were killed.

* now Pakistan
** now Macedonia

Epicenter — Normal fault

Shockwaves reach surface

Origin or focus — Shockwaves travel outward

DISTRIBUTION

1976 ●	Principal earthquakes and dates
	Oceanic marginal troughs
	Mobile land areas
	Submarine zones of mobile land areas
	Stable land platforms
	Submarine extensions of stable land platforms
	Mid-oceanic volcanic ridges
	Oceanic platforms

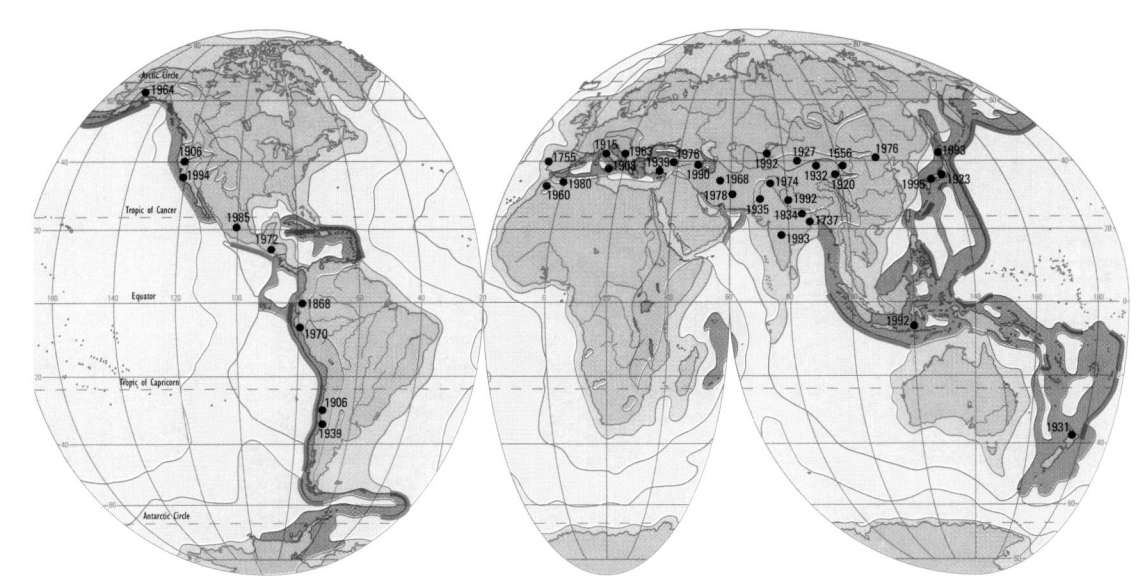

Earthquakes are a series of rapid vibrations originating from the slipping or faulting of parts of the Earth's crust when stresses within build to breaking point, and usually occur at depths between 5 and 20 miles.

THE EARTH: OCEANS

The Earth is a misnamed planet: more than 70% of its total surface area – 139,079,000 square miles – is covered by its oceans and seas. This great cloak of liquid water gives the planet its characteristic blue appearance from space, and is one of two obvious differences between the Earth and its near-neighbors in space, Mars and Venus. The other difference is the presence of life, and the two are closely linked.

In a strict geographical sense, the Earth has only three oceans: the Atlantic, Pacific and Indian Oceans. Subdivided vertically instead of horizontally, however, there are many more. The most active is the sunlit upper layer, home of most sea life and the vital interface between air and water. In this surface zone, huge energies are exchanged between the oceans and the atmosphere above; it is also a kind of membrane through which the ocean breathes, absorbing great quantities of carbon dioxide and partially exchanging them for oxygen, largely through the phytoplankton, tiny plants that photo-synthesize solar energy and provide the food base for all other marine life.

As depth increases, so light and color gradually fade away, the longer wavelengths dying first. At 150 feet, the ocean is a world of green, blue and violet; at 300 feet, only blue remains; by 650 feet, there is only a dim twilight. The temperature falls away with the light, until just before 3,500 feet – the precise depth varies – there occurs a temperature change almost as abrupt as the transition between air and water far above.

Below this thermocline, at a near-stable 37°F, the waters are forever unmoved by the winds of the upper world and are stirred only by the slow action of deep ocean currents. The pressure is crushing, reaching 1,000 atmospheres in the deepest trenches: a force of more than 7 tons bearing down on every square inch.

Yet even here the oceans support life, and not only the handful of strange, deep-sea creatures that find a living in the near-empty abyss. The deep ocean serves as a gigantic storehouse both for heat and for assorted atmospheric chemicals, regulating and balancing the proportions of various trace compounds and elements, and ensuring a large measure of stability for both the climate and the ecology that depend on it.

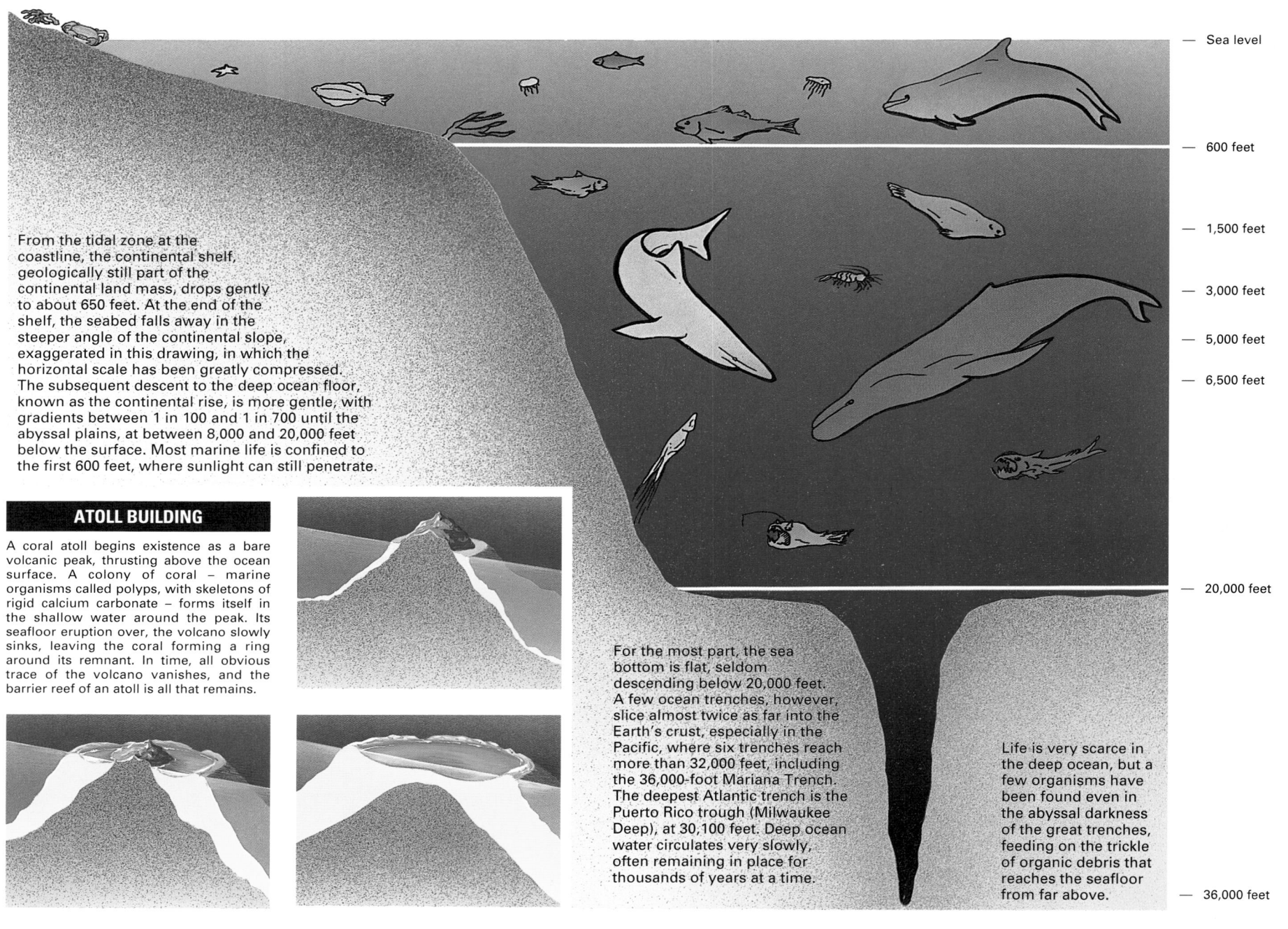

From the tidal zone at the coastline, the continental shelf, geologically still part of the continental land mass, drops gently to about 650 feet. At the end of the shelf, the seabed falls away in the steeper angle of the continental slope, exaggerated in this drawing, in which the horizontal scale has been greatly compressed. The subsequent descent to the deep ocean floor, known as the continental rise, is more gentle, with gradients between 1 in 100 and 1 in 700 until the abyssal plains, at between 8,000 and 20,000 feet below the surface. Most marine life is confined to the first 600 feet, where sunlight can still penetrate.

— Sea level
— 600 feet
— 1,500 feet
— 3,000 feet
— 5,000 feet
— 6,500 feet
— 20,000 feet
— 36,000 feet

ATOLL BUILDING

A coral atoll begins existence as a bare volcanic peak, thrusting above the ocean surface. A colony of coral – marine organisms called polyps, with skeletons of rigid calcium carbonate – forms itself in the shallow water around the peak. Its seafloor eruption over, the volcano slowly sinks, leaving the coral forming a ring around its remnant. In time, all obvious trace of the volcano vanishes, and the barrier reef of an atoll is all that remains.

For the most part, the sea bottom is flat, seldom descending below 20,000 feet. A few ocean trenches, however, slice almost twice as far into the Earth's crust, especially in the Pacific, where six trenches reach more than 32,000 feet, including the 36,000-foot Mariana Trench. The deepest Atlantic trench is the Puerto Rico trough (Milwaukee Deep), at 30,100 feet. Deep ocean water circulates very slowly, often remaining in place for thousands of years at a time.

Life is very scarce in the deep ocean, but a few organisms have been found even in the abyssal darkness of the great trenches, feeding on the trickle of organic debris that reaches the seafloor from far above.

PROFILE OF AN OCEAN

The deep ocean floor is no more uniform than the surface of the continents, although it was not until the development of effective sonar equipment that it was possible to examine submarine contours in detail. The Atlantic (right) and the Pacific show similar patterns. Offshore comes the continental shelf, sliding downward to the continental slope and the steeper continental rise, after which the seabed rolls onward into the abyssal plains. In the wide Pacific, these are interrupted by gently-rising abyssal hills; in both oceans, the plains extend all the way to the mid-oceanic ridges, where the upwelling of new crustal material is constantly forcing the oceans wider. Volcanic activity is responsible for the formation of seamounts and tablemounts, or guyots, their flat-topped equivalents. In this cross-section, only the Azores are high enough to break the surface and become islands.

Massachusetts (Nantucket Sound)

Kelvin seamounts

Co[...]
seam[...]

6,000 feet
12,000 feet

Abyssal plain

OCEAN CURRENTS

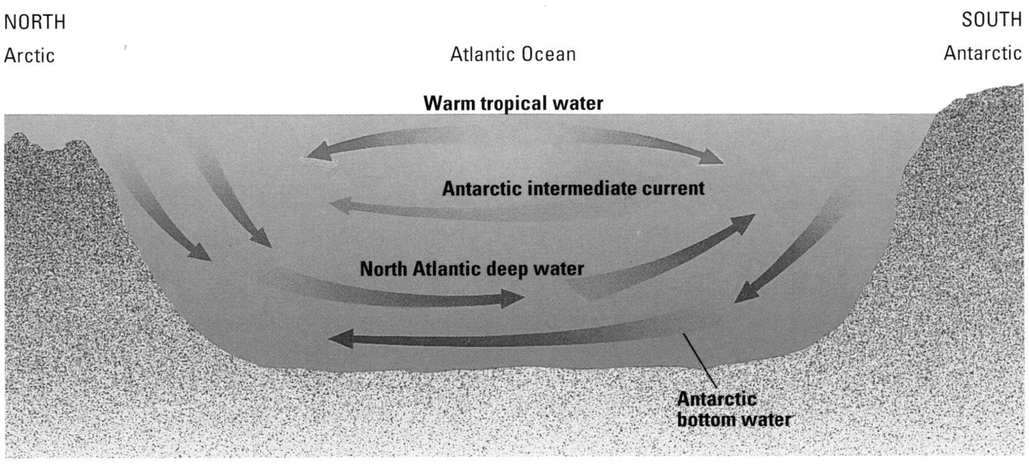

NORTH
Arctic

Atlantic Ocean

SOUTH
Antarctic

Warm tropical water

Antarctic intermediate current

North Atlantic deep water

Antarctic bottom water

Moving immense quantities of energy as well as billions of tons of water every hour, the ocean currents are a vital part of the great heat engine that drives the Earth's climate. They themselves are produced by a twofold mechanism. At the surface, winds push huge masses of water before them; in the deep ocean, below an abrupt temperature gradient that separates the churning surface waters from the still depths, density variations cause slow vertical movements.

The pattern of circulation of the great surface currents is determined by the displacement known as the Coriolis effect. As the Earth turns beneath a moving object – whether it is a tennis ball or a vast mass of water – it appears to be deflected to one side. The deflection is most obvious near the Equator, where the Earth's surface is spinning eastward at over 1,000 mph; currents moving poleward are curved clockwise in the northern hemisphere and counterclockwise in the southern.

The result is a system of spinning circles known as gyres. The Coriolis effect piles up water on the left of each gyre, creating a narrow, fast-moving stream that is matched by a slower, broader returning current on the right. North and south of the Equator, the fastest currents are located in the west and in the east respectively. In each case, warm water moves from the Equator and cold water returns to it. Cold currents often bring an upwelling of nutrients with them, supporting the world's most economically important fisheries.

Depending on the prevailing winds, some currents on or near the Equator may reverse their direction in the course of the year – a seasonal variation on which Asian monsoon rains depend, and whose occasional failure can bring disaster to millions of people.

CURRENTS AND TEMPERATURES

(Northern Hemisphere: winter)

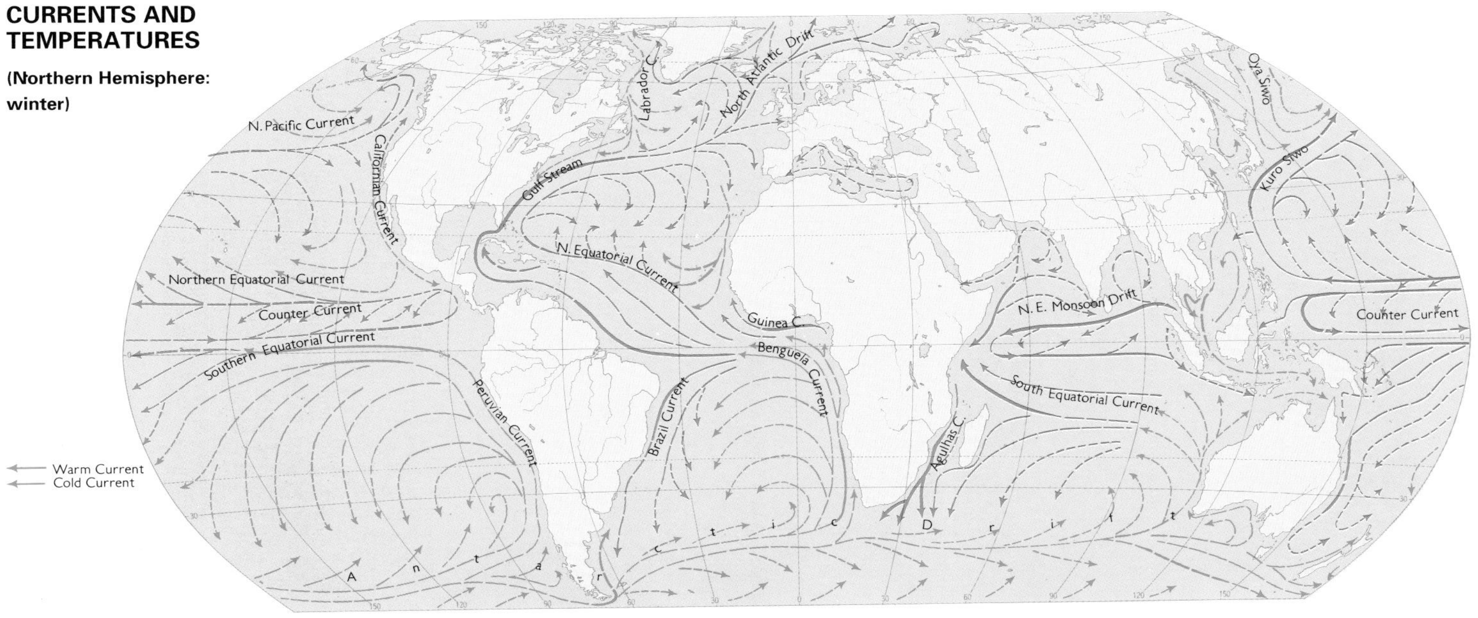

Warm Current
Cold Current

CURRENTS AND TEMPERATURES

(Northern Hemisphere: summer)

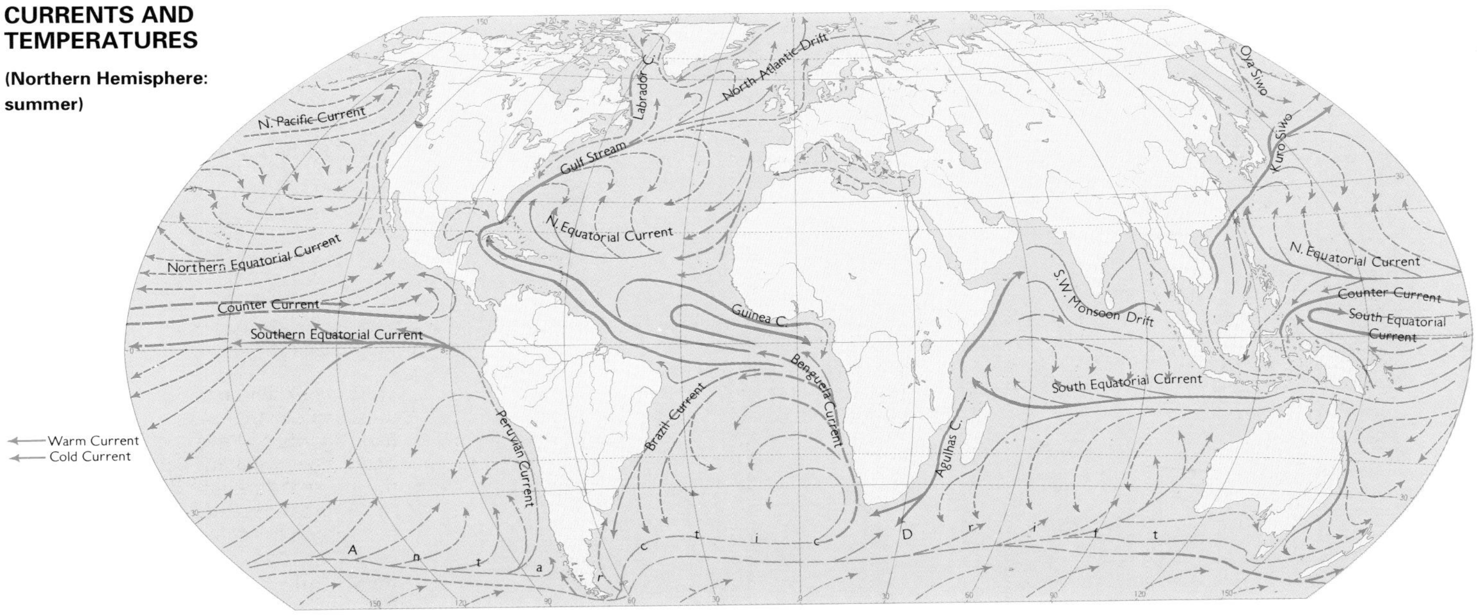

Warm Current
Cold Current

SEAWATER

The chemical composition of the sea, in parts per million, excluding the elements of water itself

Chlorine	19,400
Sodium	10,800
Magnesium	1,290
Sulfur	904
Calcium	411
Potassium	392
Bromine	67
Strontium	8.1
Boron	4.5
Fluorine	1.3
Lithium	0.17
Rubidium	0.12
Phosphorus	0.09
Iodine	0.06
Barium	0.02
Arsenic	0.003
Cesium	0.0003

Seawater also contains virtually every other element, although the quantities involved are too small for reliable measurement. In natural conditions, its composition is broadly consistent across the world's seas and oceans; but in coastal areas especially, variations, sometimes substantial, may be caused by the presence of industrial waste and sewage sludge.

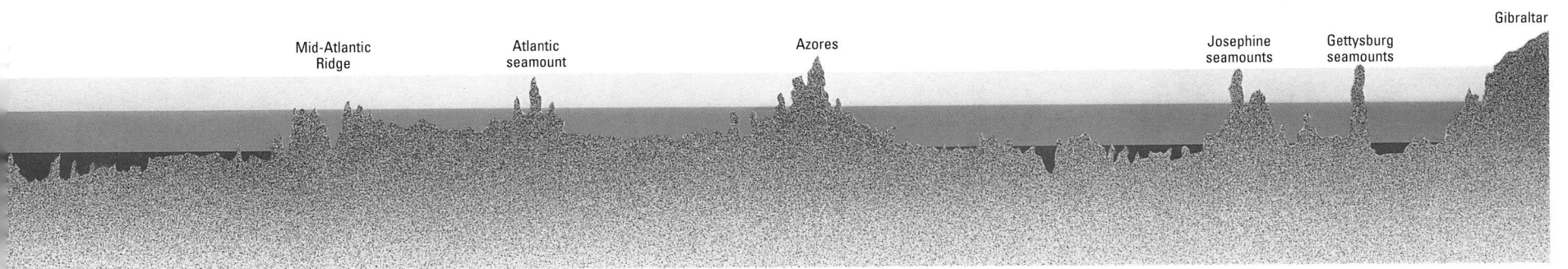

Gibraltar
Mid-Atlantic Ridge
Atlantic seamount
Azores
Josephine seamounts
Gettysburg seamounts

THE EARTH: ATMOSPHERE

Extending from the surface far into space, the atmosphere is a meteor shield, a radiation deflector, a thermal blanket and a source of chemical energy for the Earth's diverse inhabitants. Five-sixths of its mass is found in the first 10 miles, the troposphere, no thicker in relative terms than the skin of an onion. Clouds, cyclonic winds, precipitation and virtually all the phenomena we call weather occur in this narrow layer. Above, a thin layer of ozone blocks ultraviolet radiation. Beyond 60 miles, atmospheric density is lower than most laboratory vacuums, yet these tenuous outer reaches, composed largely of hydrogen and helium, trap cosmic debris and incoming high-energy particles alike.

CIRCULATION OF THE AIR

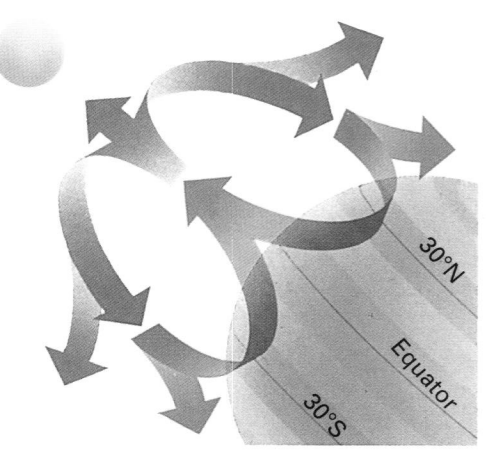

STRUCTURE OF ATMOSPHERE TEMPERATURE PRESSURE

CHEMICAL STRUCTURE

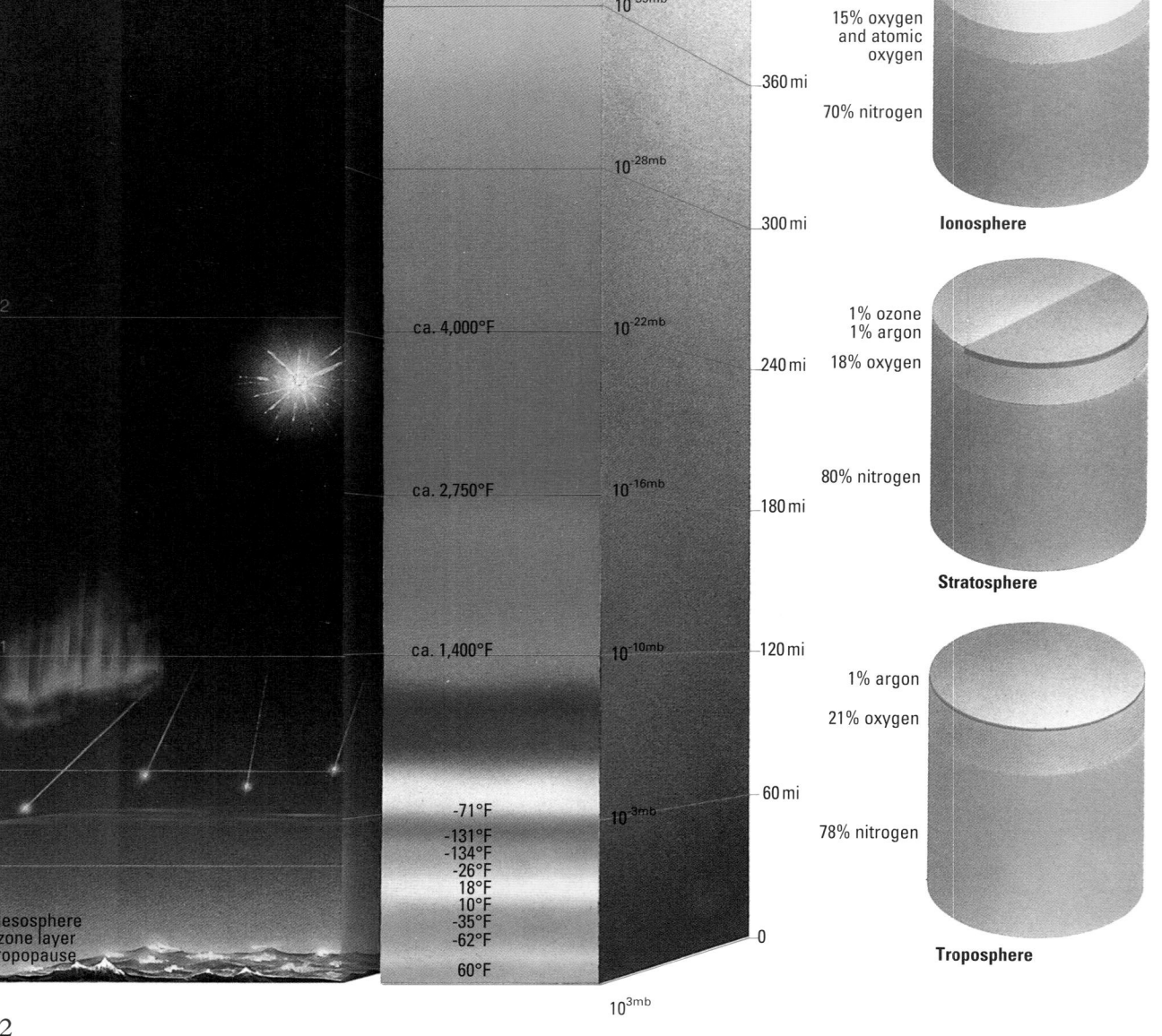

Inner:
50% helium
50% hydrogen

Middle:
25% helium
75% hydrogen

Outer:
100% hydrogen

Exosphere

15% helium

15% oxygen
and atomic
oxygen

70% nitrogen

Ionosphere

1% ozone
1% argon

18% oxygen

80% nitrogen

Stratosphere

1% argon

21% oxygen

78% nitrogen

Troposphere

Temperature / Pressure / Altitude scale

10^-53mb

540 mi

10^-47mb

480 mi

10^-41mb

420 mi

10^-35mb

360 mi

10^-28mb

300 mi

ca. 4,000°F   10^-22mb   240 mi

ca. 2,750°F   10^-16mb   180 mi

ca. 1,400°F   10^-10mb   120 mi

-71°F
-131°F
-134°F   10^-3mb   60 mi
-26°F
18°F
10°F
-35°F
-62°F   0
60°F

10^3mb

F2

F1

E

D

Mesosphere
Ozone layer
Tropopause

Exosphere
The atmosphere's upper layer has no clear outer boundary, merging imperceptibly with interplanetary space. Its lower boundary, at an altitude of approximately 370 miles, is almost equally vague. The exosphere is mainly composed of hydrogen and helium in changing proportions, with a small quantity of atomic oxygen up to 370 miles. Helium vanishes with increasing altitude, and above 1,500 miles the exosphere is almost entirely composed of hydrogen.

Ionosphere
Gas molecules in the ionosphere, mainly helium, oxygen and nitrogen, are electrically charged – ionized – by the Sun's radiation. Within the ionosphere's range of 30 to 350 miles in altitude, they group themselves into four layers, known conventionally as D, E, F1 and F2, all of which can reflect radio waves of differing frequencies. The high energy of ionospheric gas gives it a notional temperature of more than 3,600°F, although its density is negligible. The auroras – *aurora borealis* and its southern counterpart, *aurora australis* – occur in the ionosphere when charged particles from the Sun interact with the Earth's magnetic fields, at their strongest near the poles.

Stratosphere
Separated at its upper and lower limits by the distinct thresholds of the stratopause and the tropopause, the stratosphere is a remarkably stable layer between 30 miles and about 10 miles. Its temperature rises from –65°F at its lower extent to approximately 32°F near the stratopause, where a thin layer of ozone absorbs ultraviolet radiation. 'Mother-of-pearl' or nacreous cloud occurs at about 15 miles' altitude. Stratospheric air contains enough ozone to make it poisonous, although it is in any case far too rarefied to breathe.

Troposphere
The narrowest of all the atmospheric layers, the troposphere extends up to 10 miles at the Equator but only 5 miles at the poles. Since this thin region contains about 85% of the atmosphere's total mass and almost all of its water vapor, it is also the realm of the Earth's weather. Temperatures fall steadily with increasing height by about 0.5°F for every 100 feet above sea level.

Heated by the relatively high surface temperatures near the Earth's Equator, air expands and rises to create a belt of low pressure. Moving northward toward the poles, it gradually cools, sinking once more and producing high-pressure belts at about latitudes 30° North and South. Water vapor carried with the air falls as rain, releasing vast quantities of energy as well as liquid water when it condenses.

The high- and low-pressure belts are both areas of comparative calm, but between them, blowing from high-pressure to low-pressure areas, are the prevailing winds. The atmospheric circulatory system is enormously complicated by the Coriolis effect brought about by the spinning Earth: winds are deflected to the right in the northern hemisphere and to the left in the southern, giving rise to the typically cyclonic pattern of swirling clouds carried by the moving masses of air.

Although clouds appear in an almost infinite variety of shapes and sizes, there are recognizable features that form the basis of a classification first put forward by Luke Howard, a London chemist, in 1803 and later modified by the World Meteorological Organization. The system is derived from the altitude of clouds and whether they form hairlike filaments ('cirrus'), heaps or piles ('cumulus'), or layers ('stratus'). Each characteristic carries some kind of message – not always a clear one – to forecasters about the weather to come.

CLASSIFICATION OF CLOUDS

Altitude at which clouds are formed

Cirrus

Cirrocumulus

Altostratus

Altocumulus

Cumulus

Cumulonimbus

Stratocumulus

Stratus

High clouds — 35,000 / 30,000 / 25,000

Middle clouds — 20,000 / 15,000

Low clouds — 10,000 / 5,000

feet

Clouds form when damp, usually rising, air is cooled. Thus they form when a wind rises to cross hills or mountains; when a mass of air rises over, or is pushed up by, another mass of denser air; or when local heating of the ground causes convection currents.

The types of clouds are classified according to altitude as high, middle or low. The high ones, composed of ice crystals, are cirrus, cirrostratus and cirrocumulus. The middle clouds are altostratus, a grey or bluish striated, fibrous, or uniform sheet producing light drizzle, and altocumulus, a thicker and fluffier version of cirrocumulus.

The low clouds include nimbostratus, a dark grey layer that brings almost continuous rain or snow; cumulus, a detached 'heap' – brilliant white in sunlight but dark and flat at the base; and stratus, which forms dull, overcast skies at low altitudes.

Cumulonimbus, associated with storms and rains, heavy and dense with a flat base and a high, fluffy outline, can be tall enough to occupy middle as well as low altitudes.

PRESSURE AND WINDS

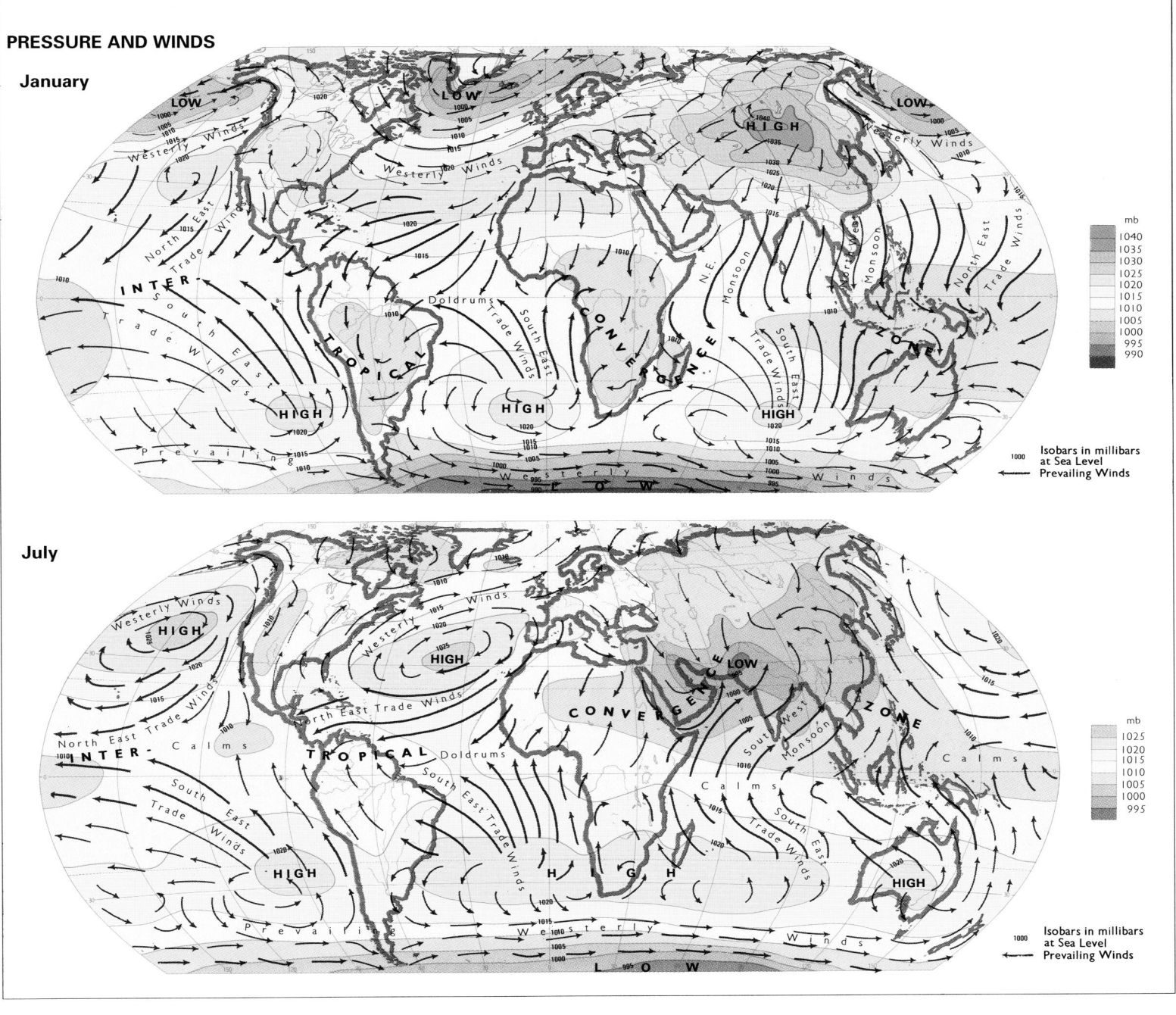

January

July

Isobars in millibars at Sea Level
Prevailing Winds

mb
1040
1035
1030
1025
1020
1015
1010
1005
1000
995
990

mb
1025
1020
1015
1010
1005
1000
995

CLIMATE RECORDS

Pressure and winds

Highest barometric pressure: Agata, Siberia, 1,083.8 mb [32 in] at altitude 862 ft, 31 December 1968.

Lowest barometric pressure: Typhoon Tip, 300 miles west of Guam, Pacific Ocean, 870 mb [25.69 in], 12 October 1979.

Highest recorded wind speed: Mt Washington, New Hampshire, USA, 231 mph, 12 April 1934. This is three times as strong as hurricane force on the Beaufort Scale.

Windiest place: Commonwealth Bay, George V Coast, Antarctica, where gales frequently reach over 200 mph.

Worst recorded storm: Bangladesh (then East Pakistan) cyclone*, 13 November 1970 – over 300,000 dead or missing. The 1991 cyclone, Bangladesh's and the world's second worst in terms of loss of life, killed an estimated 138,000 people.

Worst recorded tornado: Missouri/Illinois/Indiana, USA, 18 March 1925 – 792 deaths. The tornado was only 300 yds wide.

* Tropical cyclones are known as hurricanes in Central and North America and as typhoons in the Far East

THE EARTH: CLIMATE

Climate is weather in the long term: the seasonal pattern of hot and cold, wet and dry, averaged over time. At the simplest level, it is caused by the uneven heating of the Earth. Surplus heat at the Equator passes toward the poles, leveling out the energy differential. Its passage is marked by a ceaseless churning of the atmosphere and the oceans, further agitated by the Earth's diurnal spin and the motion it imparts to moving air and water. The heat's means of transport – by winds and ocean currents, by the continual evaporation and recondensation of water molecules – is the weather itself.

There are four basic types of climate, each of which is open to considerable subdivision: tropical, desert, temperate and polar. But although latitude is obviously a critical factor,

it is not the only determinant. The differential heating of land and sea, the funneling and interruption of winds and ocean currents by land masses and mountain ranges, and the transpiration of vegetation: all the factors combine to add complexity. New York, Naples and the Gobi Desert share almost the same latitude, for example, but their climates are very different. And although the sheer intricacy of the weather system often defies day-to-day prediction in these or any other places – despite the many satellites and number-crunching supercomputers with which present-day meteorologists are now equipped – their climatic patterns retain a year-on-year stability.

They are not indefinitely stable, however. The planet regularly passes through long,

cool periods lasting about 100,000 years: these are the Ice Ages, probably caused by recurring long-term oscillations in the Earth's orbital path and fluctuations in the Sun's energy output. In the present era, the Earth is nearest to the Sun in the middle of the northern hemisphere's winter; 11,000 years ago, at the end of the last Ice Age, the northern winter fell with the Sun at its most distant.

Left to its own devices, the climate even now should be drifting toward another glacial period. But global warming caused by increasing carbon dioxide levels in the atmosphere, largely the result of 20th-century fuel-burning and deforestation, may well precipitate change far faster than the great, slow cycles of the Solar System.

Legend

Tropical rainy climates
All mean monthly temperatures above 64°F.

Af	Rain forest climate
Am	Monsoon climate
Aw	Savanna climate

Dry climates
Low rainfall combined with a wide range of temperatures

| BS | Steppe climate |
| BW | Desert climate |

Warm temperate rainy climates
The mean temperature is below 64°F but above 26°F and that of the warmest month is over 50°F.

Cw	Dry winter climate
Cs	Dry summer climate
Cf	Climate with no dry season

Cold temperate rainy climates
The mean temperature of the coldest month is below 37°F but that of the warmest month is still over 50°F.

| Dw | Dry winter climate |
| Df | Climate with no dry season |

Polar climates
The mean temperature of the warmest month is below 50°F, giving permanently frozen subsoil.

| ET | Tundra climate |

The mean temperature of the warmest month is below 32°F, giving permanent ice and snow.

| EF | Polar climate |

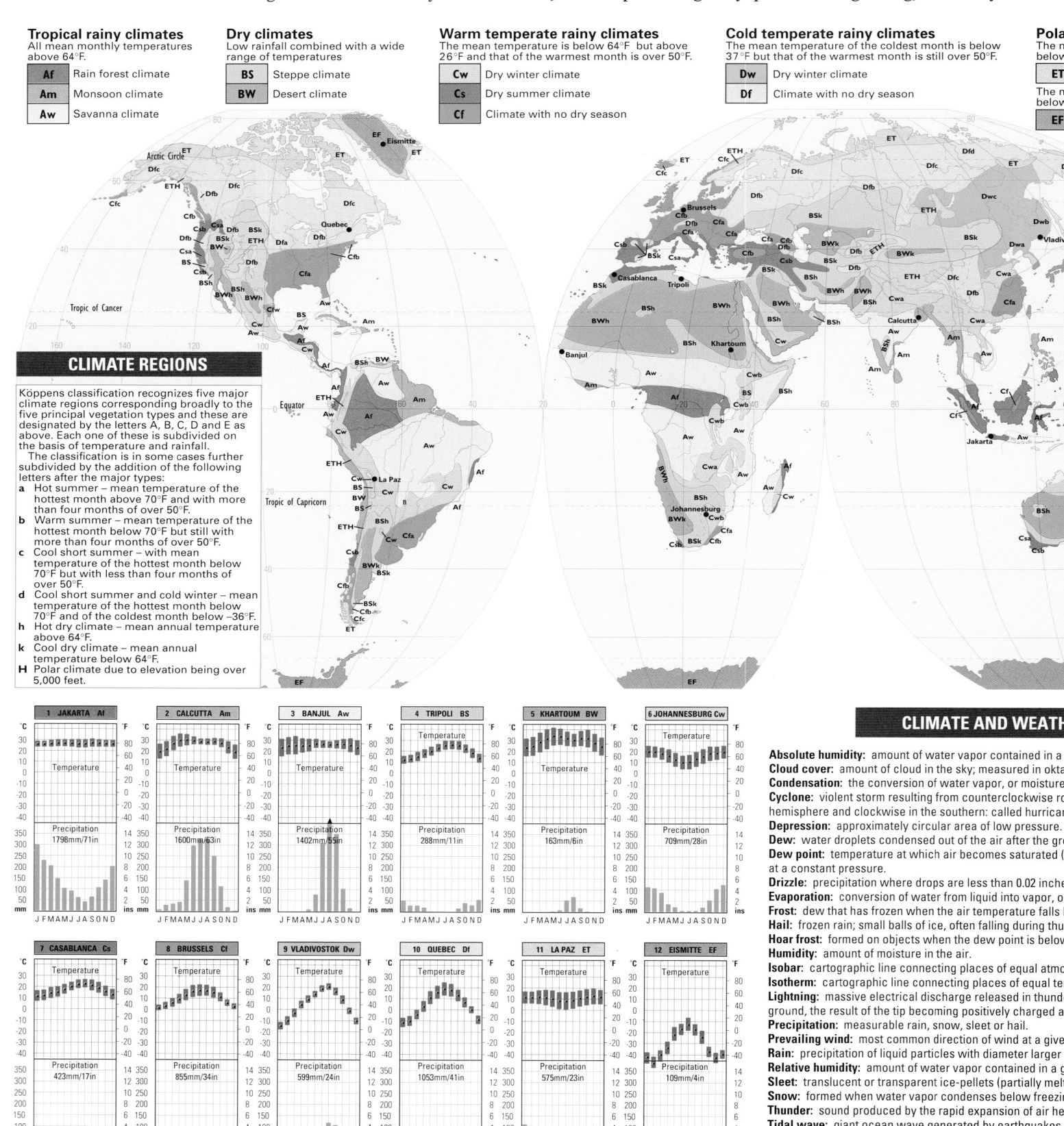

CLIMATE REGIONS

Köppens classification recognizes five major climate regions corresponding broadly to the five principal vegetation types and these are designated by the letters A, B, C, D and E as above. Each one of these is subdivided on the basis of temperature and rainfall.

The classification is in some cases further subdivided by the addition of the following letters after the major types:
a Hot summer – mean temperature of the hottest month above 70°F and with more than four months of over 50°F.
b Warm summer – mean temperature of the hottest month below 70°F but still with more than four months of over 50°F.
c Cool short summer – with mean temperature of the hottest month below 70°F but with less than four months of over 50°F.
d Cool short summer and cold winter – mean temperature of the hottest month below 70°F and of the coldest month below –36°F.
h Hot dry climate – mean annual temperature above 64°F.
k Cool dry climate – mean annual temperature below 64°F.
H Polar climate due to elevation being over 5,000 feet.

Climate graphs

1 JAKARTA Af — Temperature — Precipitation 1798mm/71in
2 CALCUTTA Am — Temperature — Precipitation 1600mm/63in
3 BANJUL Aw — Temperature — Precipitation 1402mm/55in
4 TRIPOLI BS — Temperature — Precipitation 288mm/11in
5 KHARTOUM BW — Temperature — Precipitation 163mm/6in
6 JOHANNESBURG Cw — Temperature — Precipitation 709mm/28in
7 CASABLANCA Cs — Temperature — Precipitation 423mm/17in
8 BRUSSELS Cf — Temperature — Precipitation 855mm/34in
9 VLADIVOSTOK Dw — Temperature — Precipitation 599mm/24in
10 QUEBEC Df — Temperature — Precipitation 1053mm/41in
11 LA PAZ ET — Temperature — Precipitation 575mm/23in
12 EISMITTE EF — Temperature — Precipitation 109mm/4in

CLIMATE AND WEATHER TERMS

Absolute humidity: amount of water vapor contained in a given volume of air.
Cloud cover: amount of cloud in the sky; measured in oktas (from 1 – 8), with 0 clear, and 8 total cover.
Condensation: the conversion of water vapor, or moisture in the air, into liquid.
Cyclone: violent storm resulting from counterclockwise rotation of winds in the northern hemisphere and clockwise in the southern: called hurricane in N. America, typhoon in the Far East.
Depression: approximately circular area of low pressure.
Dew: water droplets condensed out of the air after the ground has cooled at night.
Dew point: temperature at which air becomes saturated (reaches a relative humidity of 100%) at a constant pressure.
Drizzle: precipitation where drops are less than 0.02 inches in diameter.
Evaporation: conversion of water from liquid into vapor, or moisture in the air.
Frost: dew that has frozen when the air temperature falls below freezing point.
Hail: frozen rain; small balls of ice, often falling during thunderstorms.
Hoar frost: formed on objects when the dew point is below freezing point.
Humidity: amount of moisture in the air.
Isobar: cartographic line connecting places of equal atmospheric pressure.
Isotherm: cartographic line connecting places of equal temperature.
Lightning: massive electrical discharge released in thunderstorm from cloud to cloud or cloud to ground, the result of the tip becoming positively charged and the bottom negatively charged.
Precipitation: measurable rain, snow, sleet or hail.
Prevailing wind: most common direction of wind at a given location.
Rain: precipitation of liquid particles with diameter larger than 0.02 inches.
Relative humidity: amount of water vapor contained in a given volume of air at a given temperature.
Sleet: translucent or transparent ice-pellets (partially melted snow).
Snow: formed when water vapor condenses below freezing point.
Thunder: sound produced by the rapid expansion of air heated by lightning.
Tidal wave: giant ocean wave generated by earthquakes (tsunami) or cyclonic winds.
Tornado: severe funnel-shaped storm that twists as hot air spins vertically (waterspout at sea).
Whirlwind: rapidly rotating column of air, only a few yards across, made visible by dust.

WINDCHILL FACTOR

In sub-zero weather, even moderate winds significantly reduce effective temperatures. The chart below shows the windchill effect across a range of speeds. Figures in the pink zone are not dangerous to well-clad people; in the blue zone, the risk of serious frostbite is acute.

	Wind speed (mph)				
	5	15	25	35	45
30°F	27	9	1	-4	-6
25°F	21	2	-7	-12	-14
20°F	16	-5	-15	-20	-22
15°F	12	-11	-22	-27	-30
10°F	7	-18	-29	-35	-38
5°F	0	-25	-36	-43	-46
0°F	-5	-31	-44	-52	-54
-5°F	-10	-38	-51	-58	-62
-10°F	-15	-45	-59	-67	-70
-15°F	-21	-51	-66	-74	-78
-20°F	-26	-58	-74	-82	-85

BEAUFORT WIND SCALE

Named for the 19th-century British naval officer who devised it, the Beaufort Scale assesses wind speed according to its effects. It was originally designed as an aid for sailors, but has since been adapted for use on land.

Scale	Wind speed km/h	mph	Effect
0	0-1	0-1	**Calm** Smoke rises vertically
1	1-5	1-3	**Light air** Wind direction shown only by smoke drift
2	6-11	4-7	**Light breeze** Wind felt on face; leaves rustle; vanes moved by wind
3	12-19	8-12	**Gentle breeze** Leaves and small twigs in constant motion; wind extends small flag.
4	20-28	13-18	**Moderate** Raises dust and loose paper; small branches move
5	29-38	19-24	**Fresh** Small trees in leaf sway; crested wavelets on inland waters
6	39-49	25-31	**Strong** Large branches move; difficult to use umbrellas; overhead wires whistle
7	50-61	32-38	**Near gale** Whole trees in motion; difficult to walk against wind
8	62-74	39-46	**Gale** Twigs break from trees; walking very difficult
9	75-88	47-54	**Strong gale** Slight structural damage
10	89-102	55-63	**Storm** Trees uprooted; serious structural damage
11	103-117	64-72	**Violent storm** Widespread damage
12	118+	73+	**Hurricane**

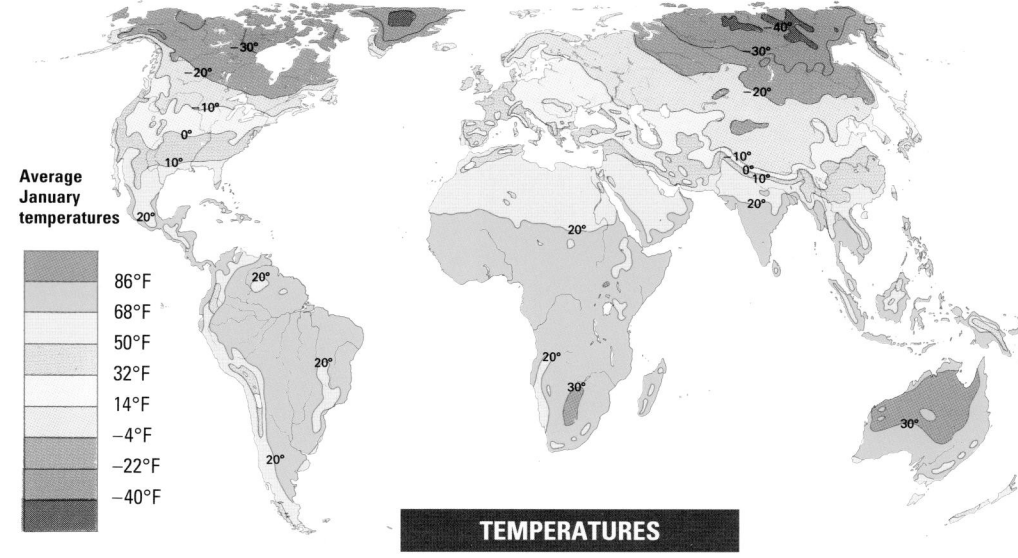

Average January temperatures

- 86°F
- 68°F
- 50°F
- 32°F
- 14°F
- −4°F
- −22°F
- −40°F

TEMPERATURES

Average July temperatures

- 86°F
- 68°F
- 50°F
- 32°F
- 14°F

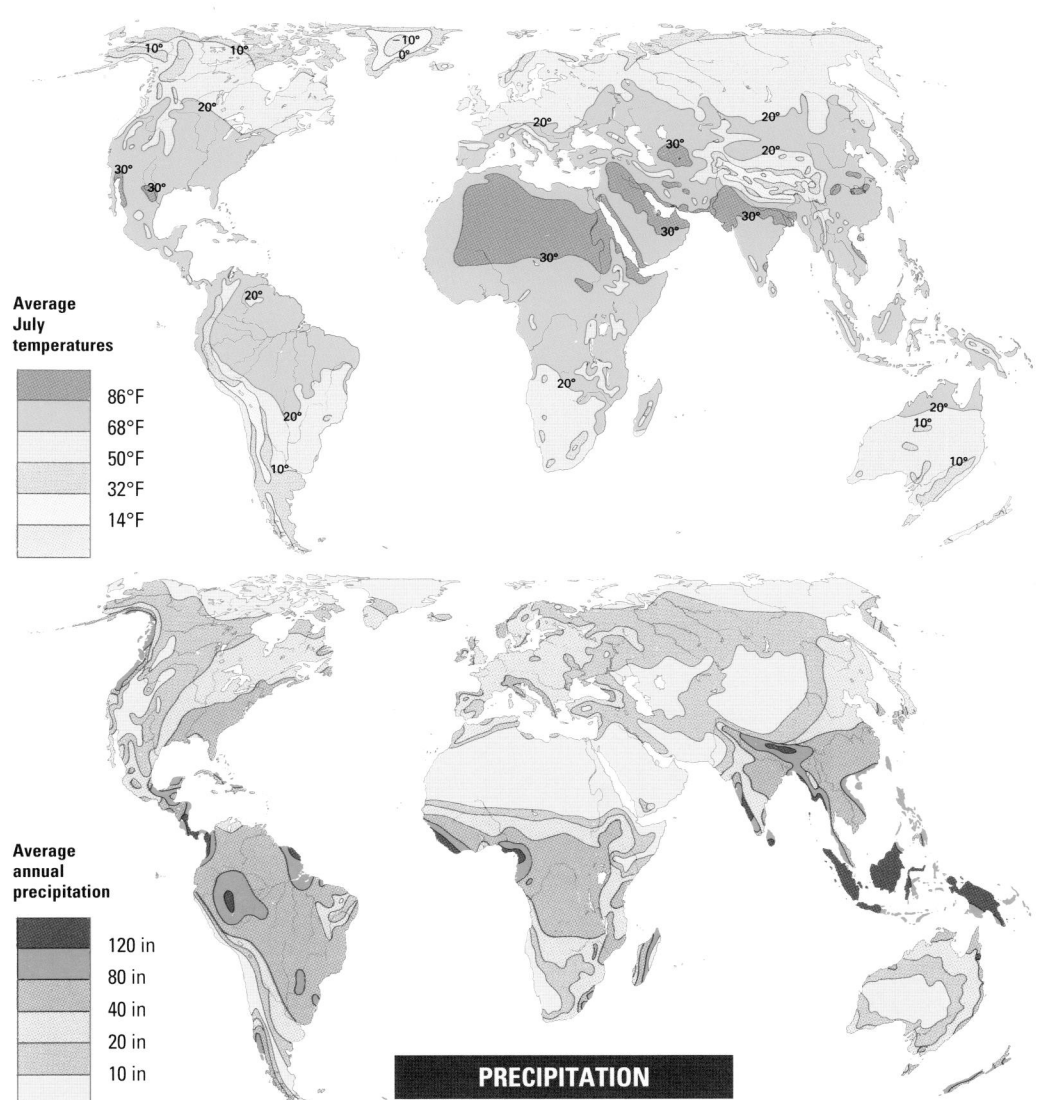

Average annual precipitation

- 120 in
- 80 in
- 40 in
- 20 in
- 10 in

PRECIPITATION

CLIMATE RECORDS

Temperature

Highest recorded temperature: Al Aziziyah, Libya, 136.4°F, 13 September 1922.

Highest mean annual temperature: Dallol, Ethiopia, 94°F, 1960–66.

Longest heatwave: Marble Bar, W. Australia, 162 days over 100°F, 23 October 1923 to 7 April 1924.

Lowest recorded temperature (outside poles): Verkhoyansk, Siberia, –90°F, 6 February 1933. Verkhoyansk also registered the greatest annual range of temperature: –94°F to 98°F.

Lowest mean annual temperature: Polus Nedostupnosti, Pole of Cold, Antarctica, –72°F.

Precipitation

Driest place: Arica, N. Chile, 0.03 in per year (60-year average).

Longest drought: Calama, N. Chile: no recorded rainfall in 400 years to 1971.

Wettest place (average): Tututendo, Colombia: mean annual rainfall 463.4 in.

Wettest place (12 months): Cherrapunji, Meghalaya, N.E. India, 1,040 in, August 1860 to August 1861. Cherrapunji also holds the record for rainfall in one month: 37 in, July 1861.

Wettest place (24 hours): Cilaos, Réunion, Indian Ocean, 73.6 in, 15–16 March 1952.

Heaviest hailstones: Gopalganj, Bangladesh, up to 2.25 lb, 14 April 1986 (killed 92 people).

Heaviest snowfall (continuous): Bessans, Savoie, France, 68 in, over 19 hours, 5–6 April 1969.

Heaviest snowfall (season/year): Paradise Ranger Station, Mt Rainier, Washington, USA, 1,224.5 in, 19 February 1971 to 18 February 1972.

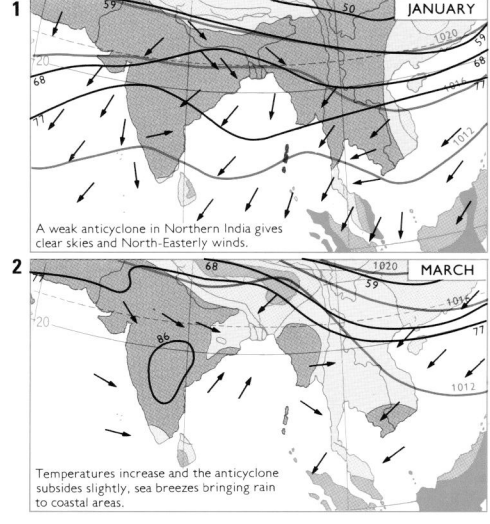

1 JANUARY A weak anticyclone in Northern India gives clear skies and North-Easterly winds.

2 MARCH Temperatures increase and the anticyclone subsides slightly, sea breezes bringing rain to coastal areas.

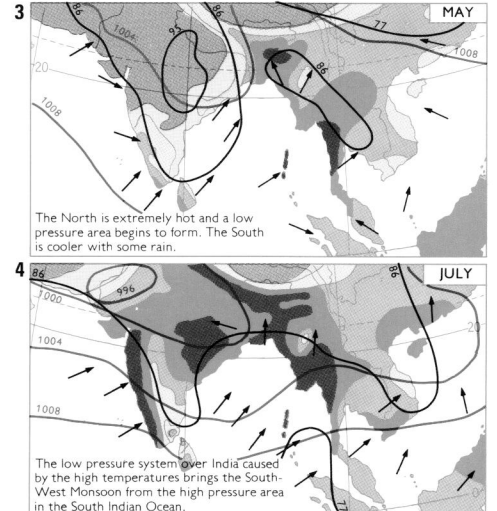

3 MAY The North is extremely hot and a low pressure area begins to form. The South is cooler with some rain.

4 JULY The low pressure system over India caused by the high temperatures brings the South-West Monsoon from the high pressure area in the South Indian Ocean.

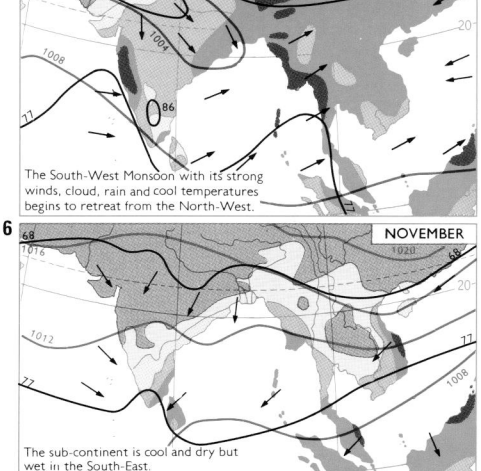

5 SEPTEMBER The South-West Monsoon with its strong winds, cloud, rain and cool temperatures begins to retreat from the North-West.

6 NOVEMBER The sub-continent is cool and dry but wet in the South-East.

THE MONSOON

While it is crucial to the agriculture of South Asia, the monsoon that follows the dry months is unpredictable – in duration as well as intensity. A season of very heavy rainfall, causing disastrous floods, can be succeeded by years of low precipitation, leading to serious drought.

Monthly rainfall

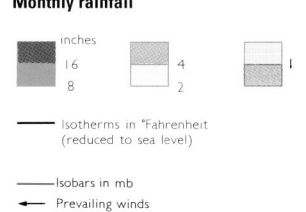

inches
- 16
- 8
- 4
- 2
- 1

— Isotherms in °Fahrenheit (reduced to sea level)

— Isobars in mb

← Prevailing winds

THE EARTH: WATER AND LAND USE

Fresh water is essential to all terrestrial life, from the humblest bacterium to the most advanced technological society. Yet freshwater resources form a minute fraction of the Earth's 340 billion cubic miles of water: most human needs must be met from the 480 cubic miles circulating in rivers at any one time. Agriculture accounts for huge quantities: without large-scale irrigation, most of the world's people would starve. And since fresh water is just as essential for most industrial processes – the smelting of a ton of nickel, for example, requires about 4,000 tons of water – the combination of a rapidly growing population and advancing industry have put water supplies under strain.

Fortunately, water is seldom used up: the planet's hydrological cycle circulates it with benign efficiency, at least on a global scale. More locally, though, human activity can cause severe shortages: water for industry and agriculture is being withdrawn from many river basins and underground aquifers faster than natural recirculation can replace it.

THE HYDROLOGICAL CYCLE

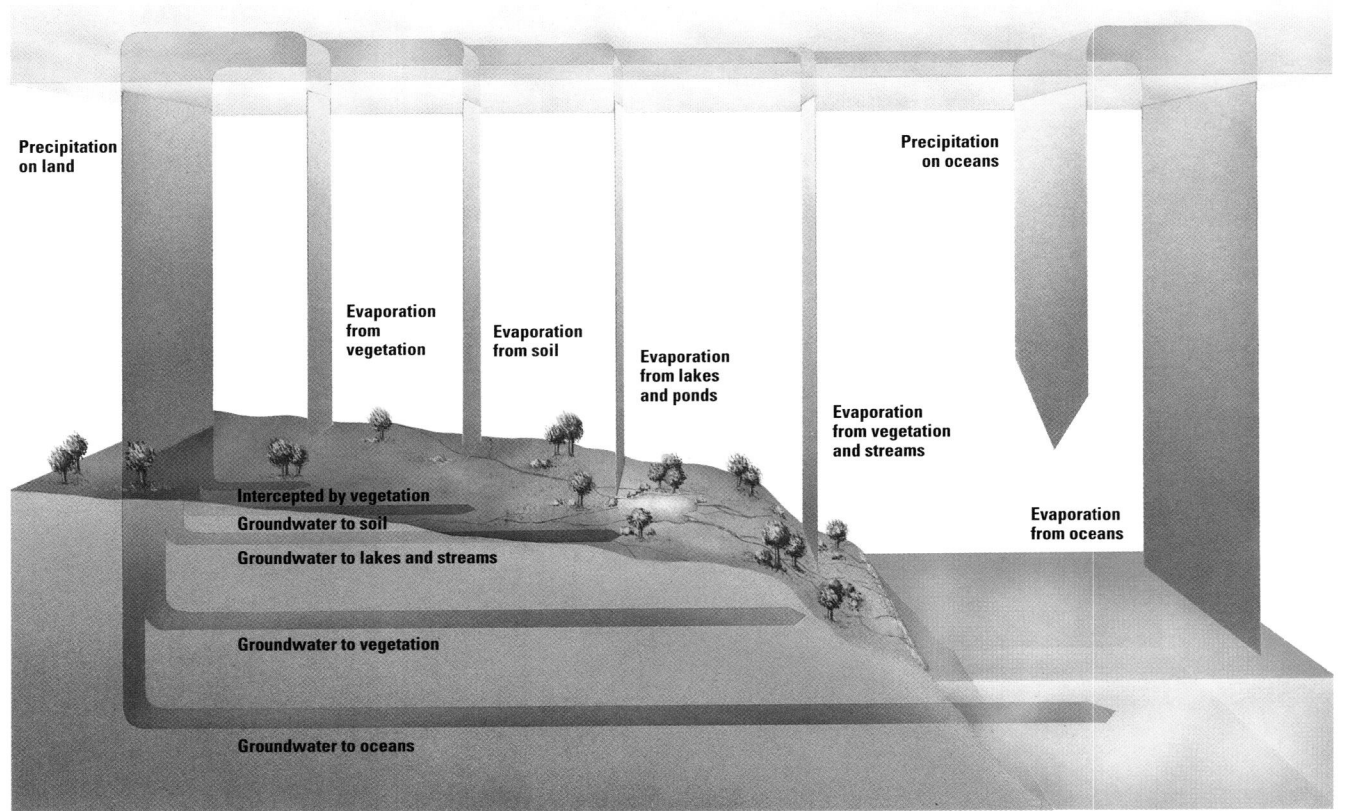

Precipitation on land

Evaporation from vegetation

Evaporation from soil

Evaporation from lakes and ponds

Precipitation on oceans

Evaporation from vegetation and streams

Evaporation from oceans

Intercepted by vegetation
Groundwater to soil
Groundwater to lakes and streams
Groundwater to vegetation
Groundwater to oceans

WATER DISTRIBUTION

The distribution of planetary water, by percentage. Oceans and icecaps together account for more than 99% of the total; the breakdown of the remainder is estimated.

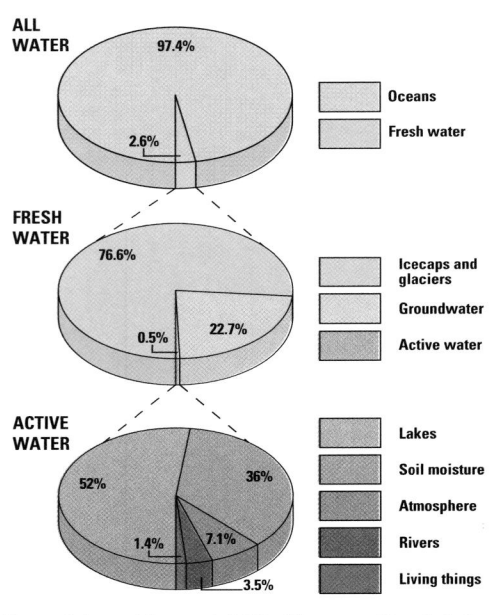

ALL WATER — 97.4%, 2.6%
- Oceans
- Fresh water

FRESH WATER — 76.6%, 0.5%, 22.7%
- Icecaps and glaciers
- Groundwater
- Active water

ACTIVE WATER — 52%, 36%, 1.4%, 7.1%, 3.5%
- Lakes
- Soil moisture
- Atmosphere
- Rivers
- Living things

Almost all the world's water is 3,000 million years old, and all of it cycles endlessly through the hydrosphere, though at different rates. Water vapor circulates over days, even hours, deep ocean water circulates over millennia, and icecap water remains solid for millions of years.

Water vapor is constantly drawn into the air from the Earth's rivers, lakes, seas and plant transpiration. In the atmosphere, it circulates around the planet, transporting energy as well as water itself. When the vapor cools it falls as rain or snow, and returns to the surface to evaporate once more. The whole cycle is driven by the Sun.

WATER RUNOFF

Annual freshwater runoff by continent in cubic miles

- Asia
- North America
- South America
- Australasia
- Europe
- Africa

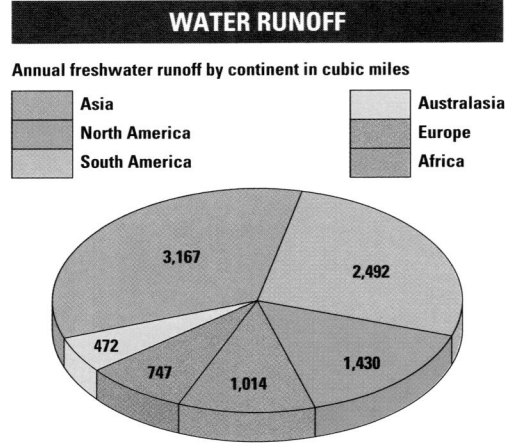

3,167 | 2,492 | 472 | 747 | 1,014 | 1,430

WATER UTILIZATION

The percentage breakdown of water usage by sector, selected countries (latest available year)

Domestic | Industrial | Agriculture

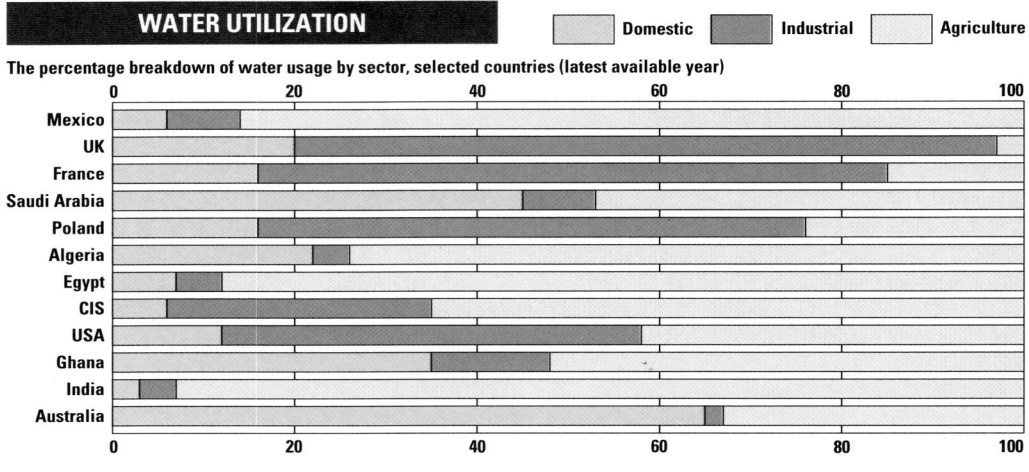

Mexico
UK
France
Saudi Arabia
Poland
Algeria
Egypt
CIS
USA
Ghana
India
Australia

0   20   40   60   80   100

WATER SUPPLY

Percentage of total population with access to safe drinking water (latest available year)

- Over 90% with safe water
- 75 – 90% with safe water
- 60 – 75% with safe water
- 45 – 60% with safe water
- 30 – 45% with safe water
- Under 30% with safe water

Least well-provided countries

Mozambique	22%	Afghanistan	29%
Madagascar	23%	Burma	32%
Central African Rep.	24%	Papua New Guinea	33%
Vietnam	24%	Uganda	33%
Ethiopia	25%	Bhutan	34%

WATERSHEDS

The world's major rivers; the world's 20 longest are shown in square brackets, led by the Nile and the Amazon.

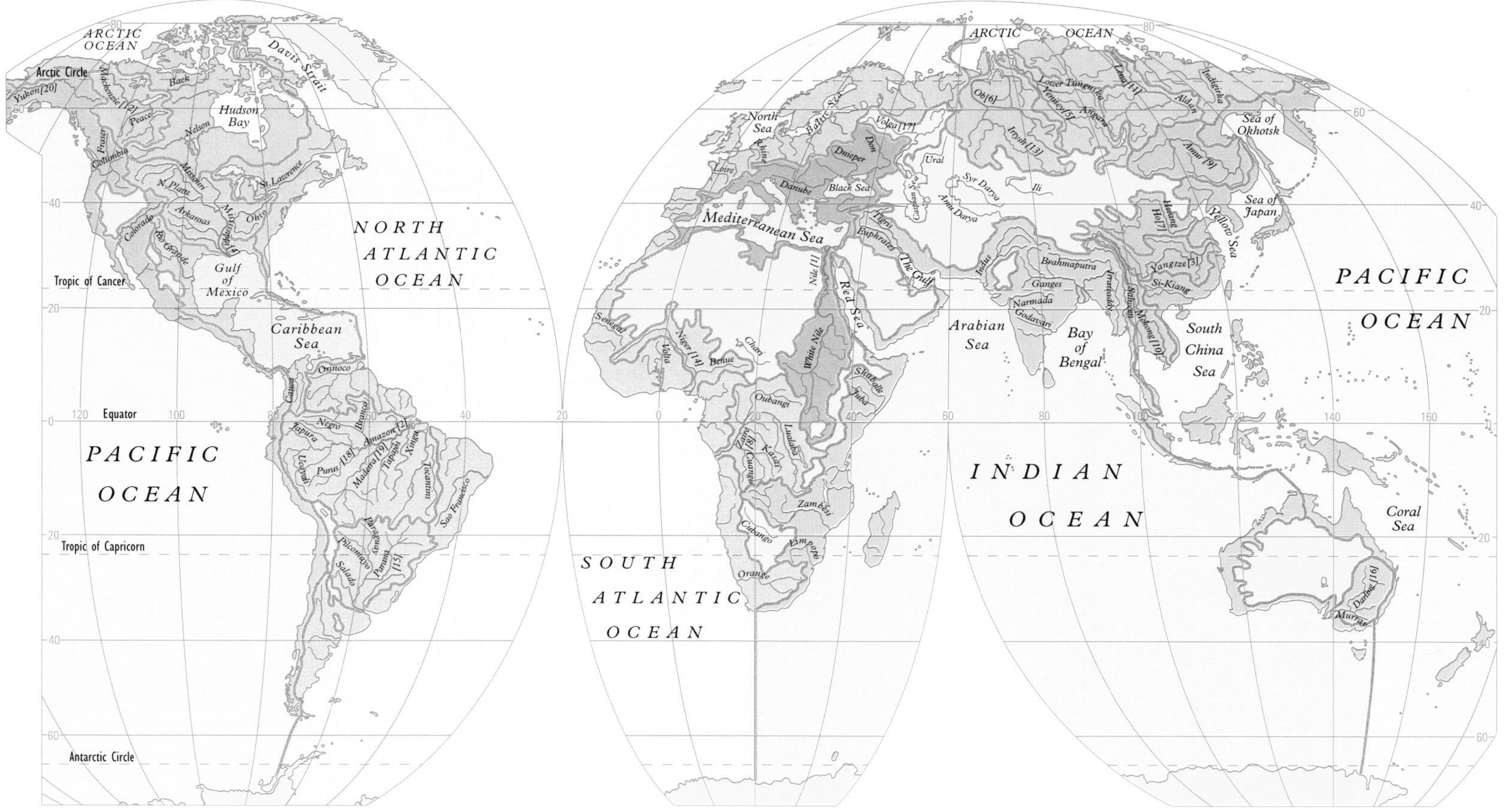

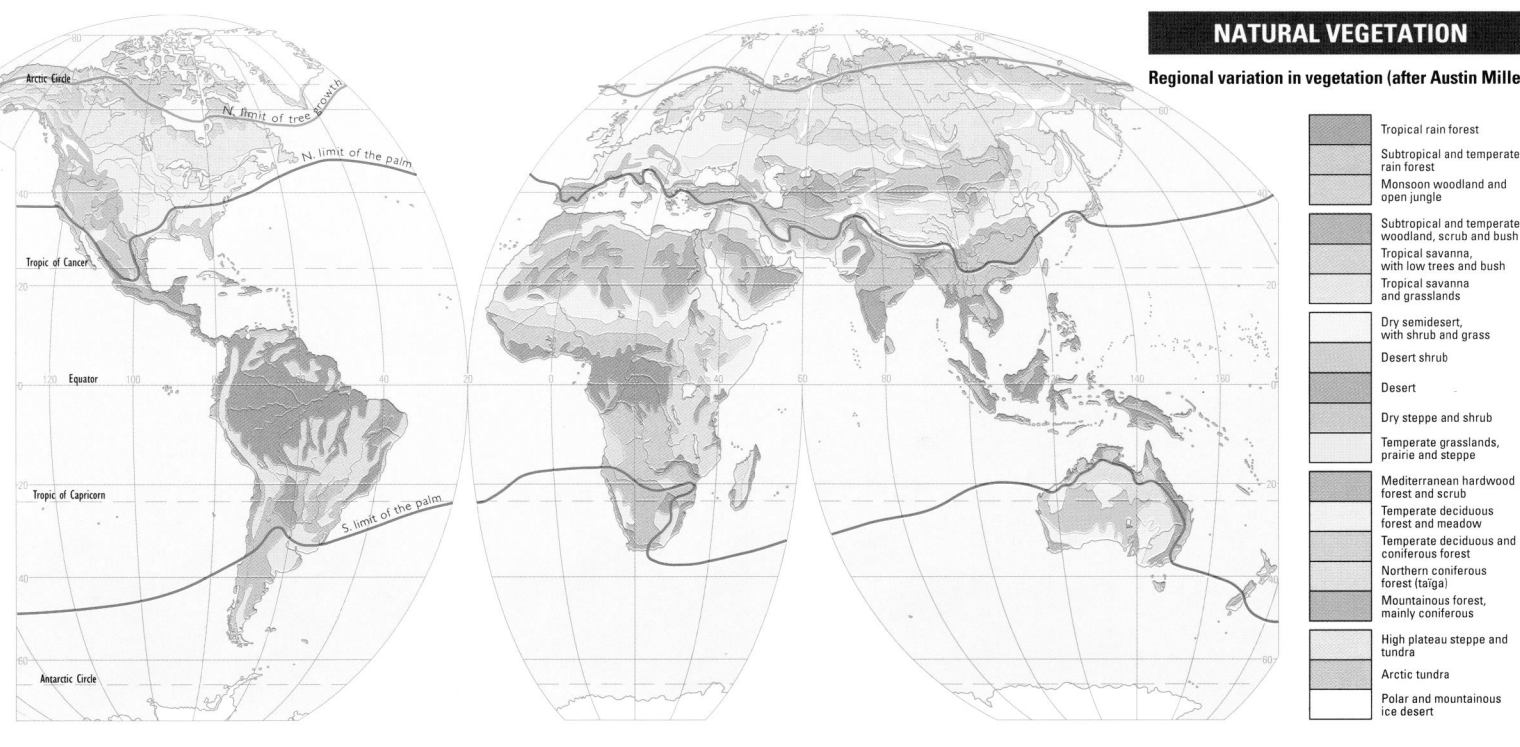

WHERE THE RIVERS RUN

- Pacific Ocean
- Indian Ocean
- Arctic Ocean
- Atlantic Ocean
- Caribbean Sea-Gulf of Mexico
- Mediterranean Sea
- Inland basins, icecaps and deserts

The map shows the direction of freshwater flow on a continental scale; the chart opposite indicates the quantities involved. The rate of runoff varies seasonally and is affected by the surface vegetation.

LAND USE BY CONTINENT

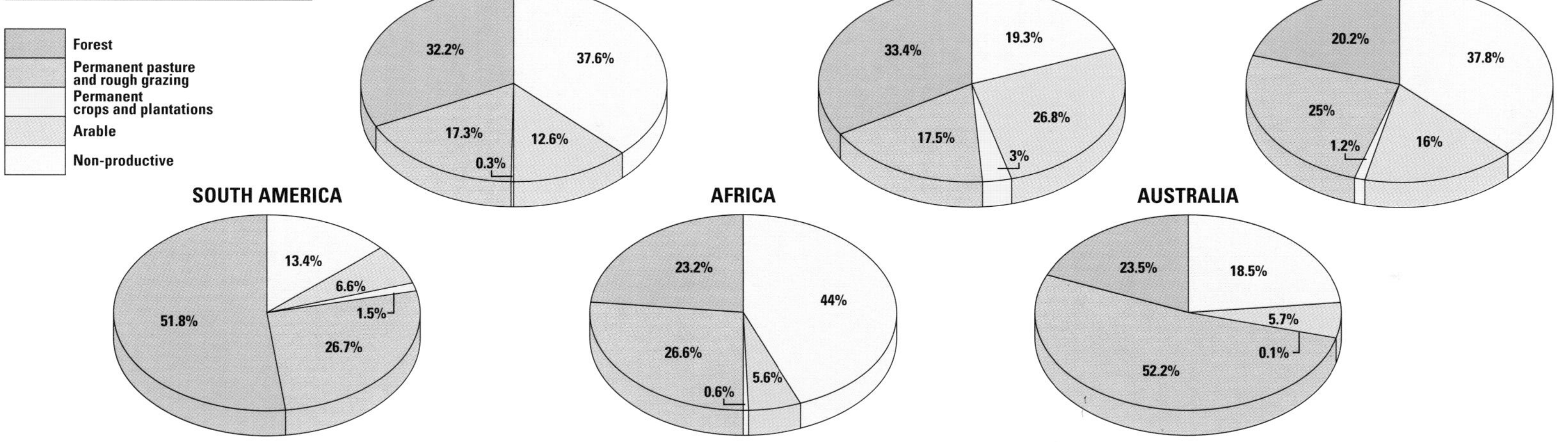

- Forest
- Permanent pasture and rough grazing
- Permanent crops and plantations
- Arable
- Non-productive

NORTH AMERICA
32.2% 37.6% 17.3% 0.3% 12.6%

EUROPE
33.4% 19.3% 17.5% 3% 26.8%

ASIA
20.2% 37.8% 25% 1.2% 16%

SOUTH AMERICA
13.4% 6.6% 1.5% 51.8% 26.7%

AFRICA
23.2% 44% 26.6% 0.6% 5.6%

AUSTRALIA
23.5% 18.5% 5.7% 52.2% 0.1%

The proportion of productive land has reached its upper limit in Europe, and in Asia more than 80% of potential cropland is already under cultivation. Elsewhere, any increase is often matched by corresponding losses due to desertification and erosion; projections for 2025 show a decline in cropland per capita for all continents, most notably in Africa.

NATURAL VEGETATION

Regional variation in vegetation (after Austin Miller)

- Tropical rain forest
- Subtropical and temperate rain forest
- Monsoon woodland and open jungle
- Subtropical and temperate woodland, scrub and bush
- Tropical savanna, with low trees and bush
- Tropical savanna and grasslands
- Dry semidesert, with shrub and grass
- Desert shrub
- Desert
- Dry steppe and shrub
- Temperate grasslands, prairie and steppe
- Mediterranean hardwood forest and scrub
- Temperate deciduous forest and meadow
- Temperate deciduous and coniferous forest
- Northern coniferous forest (taiga)
- Mountainous forest, mainly coniferous
- High plateau steppe and tundra
- Arctic tundra
- Polar and mountainous ice desert

The map illustrates the natural 'climax vegetation' of a region, as dictated by its climate and topography. In most cases, human agricultural activity has drastically altered the vegetation pattern. Western Europe, for example, lost most of its broadleaf forest many centuries ago, while irrigation has turned some natural semidesert into productive land.

THE EARTH: LANDSCAPE

Above and below the surface of the oceans, the features of the Earth's crust are constantly changing. The phenomenal forces generated by convection currents in the molten core of our planet carry the vast segments, or 'plates', of the crust across the globe in an endless cycle of creation and destruction. New crust emerges along the central depths of the oceans, where molten magma flows from the margins of neighboring plates to form the massive mid-ocean ridges. The sea floor spreads, and where ocean plates meet continental plates, they dip back into the Earth's core to melt once again into magma.

Less dense, the continental plates 'float' among the oceans, drifting into and apart from each other at a rate which is almost imperceptibly slow. A continent may travel little more than an inch each year – in an average lifetime, America will move no more than a man's height – yet in the vast span of geological time, this process throws up giant mountain ranges and opens massive rifts in the land's surface.

The world's greatest mountain ranges have been formed in this way: the Himalayas by the collision of the Indo-Australian and Eurasian plates; the Andes by the meeting of the Nazca and South American plates. The Himalayas are a classic example of 'fold mountains', formed by the crumpling of the Earth's surface where two land masses have been driven together. The coastal range of the Andes, by contrast, was formed by the upsurge of molten volcanic rock created by the friction of the continent 'overriding' the ocean plate.

However, the destruction of the landscape begins as soon as it is formed. Wind, water, ice and sea, the main agents of erosion, mount a constant assault that even the hardest rocks cannot withstand. Mountain peaks may dwindle by only a fraction of an inch each year, but if they are not uplifted by further movements of the crust they will eventually be reduced to rubble. Water is the most powerful destroyer – it has been estimated that 100 billion tons of rock is washed into the oceans every year.

When water freezes, its volume increases by about 9%, and no rock is strong enough to resist this pressure. Where water has penetrated tiny fissures or seeped into softer rock, a severe freeze followed by a thaw may result in rockfalls or earthslides, creating major destruction in a few minutes. Over much longer periods, acidity in rainwater breaks down the chemical composition of porous rocks, such as limestone, eating away the rock to form deep caves and tunnels. Chemical decomposition also occurs in riverbeds and glacier valleys, hastening the process of mechanical erosion.

Rivers and glaciers, like the sea itself, generate much of their effect through abrasion – pounding the landscape with the debris they carry with them. But, as well as destroying, they also create new landscapes, many of them spectacular: vast deltas, as seen at the mouth of the Mississippi or the Nile; cliffs, rock arches and stacks, as found along the south coast of Australia; and the fjords cut by long-melted glaciers in British Columbia, Norway and New Zealand.

The vast ridges that divide the Earth's crust beneath each of the world's major oceans mark the boundaries between tectonic plates which are moving very gradually in opposite directions. As the plates shift apart, molten magma rises from the Earth's core to seal the rift and the sea floor slowly spreads toward the continental land masses. The rate of sea floor spreading has been calculated by magnetic analysis of the rock – at about 1.5 inches a year in the North Atlantic. Near the ocean shore, underwater volcanoes mark the line where the continental rise begins. As the plates meet, much of the denser ocean crust dips beneath the continental plate and melts back into the magma.

THE SPREADING EARTH

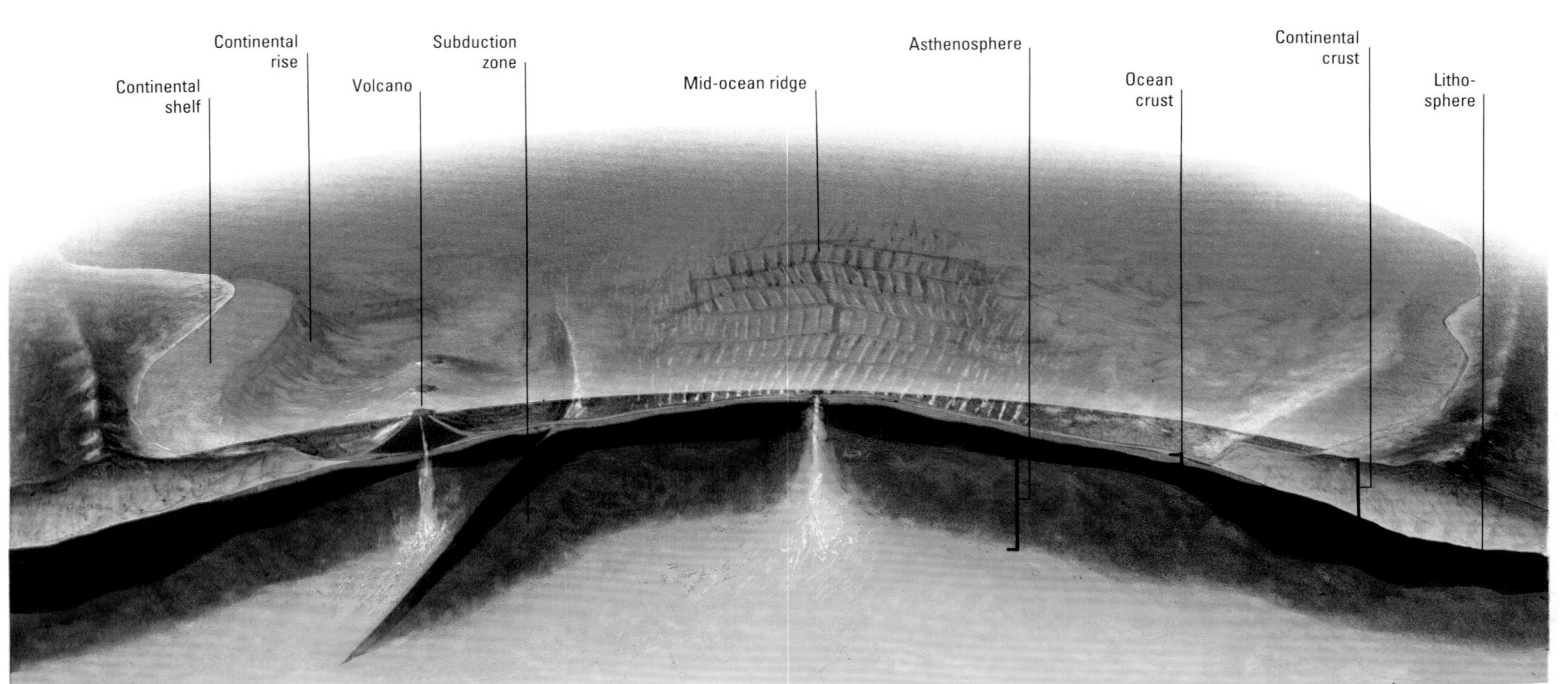

Continental shelf · Continental rise · Volcano · Subduction zone · Mid-ocean ridge · Asthenosphere · Ocean crust · Continental crust · Lithosphere

TYPES OF ROCK

Rocks are divided into three types, according to the way in which they are formed:

Igneous rocks, including granite and basalt, are formed by the cooling of magma from within the Earth's crust.

Metamorphic rocks, such as slate, marble and quartzite, are formed below the Earth's surface by the compression or baking of existing rocks.

Sedimentary rocks, like sandstone and limestone, are formed on the surface of the Earth from the remains of living organisms and eroded fragments of older rocks.

MOUNTAIN BUILDING

Mountains are formed when pressures on the Earth's crust caused by continental drift become so intense that the surface buckles or cracks. This happens most dramatically where two tectonic plates collide: the Rockies, Andes, Alps, Urals and Himalayas resulted from such impacts. These are all known as fold mountains, because they were formed by the compression of the rocks, forcing the surface to bend and fold like a crumpled rug.

The other main building process is when the crust fractures to create faults, allowing rock to be forced upward in large blocks; or when the pressure of magma within the crust forces the surface to bulge into a dome, or erupts to form a volcano. Large mountain ranges may reveal a combination of those features; the Alps, for example, have been compressed so violently that the folds are fragmented by numerous faults and intrusions of molten rock.

Over millions of years, even the greatest mountain ranges can be reduced by erosion to a rugged landscape known as a peneplain.

Types of fold: Geographers give different names to the degrees of fold that result from continuing pressure on the rock strata. A simple fold may be symmetric, with even slopes on either side, but as the pressure builds up, one slope becomes steeper and the fold becomes asymmetric. Later, the ridge or 'anticline' at the top of the fold may slide over the lower ground or 'syncline' to form a recumbent fold. Eventually, the rock strata may break under the pressure to form an overthrust and finally a nappe fold.

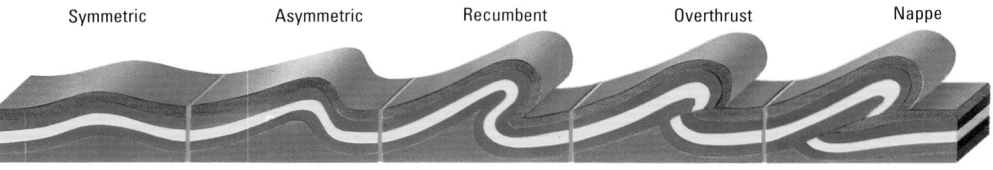

Symmetric · Asymmetric · Recumbent · Overthrust · Nappe

Types of faults: Faults are classified by the direction in which the blocks of rock have moved. A normal fault results when a vertical movement causes the surface to break apart; compression causes a reverse fault. Sideways movement causes shearing, known as a strike-slip fault. When the rock breaks in two places, the central block may be pushed up in a horst fault, or sink in a graben fault.

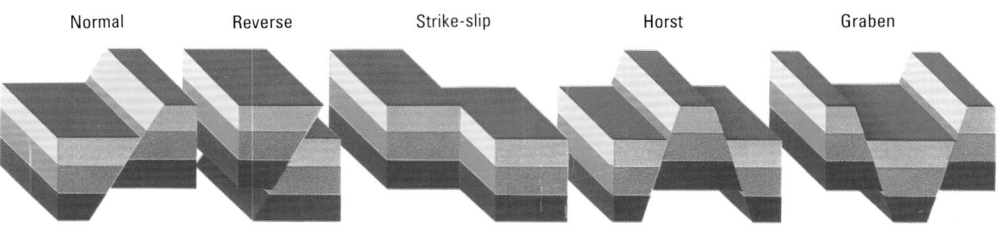

Normal · Reverse · Strike-slip · Horst · Graben

MOLDING THE LAND

While hidden forces of extraordinary power are moving the continents from below the Earth's crust, the more familiar elements of wind, water, heat and cold combine to sculpt the land surface. Erosion by weathering is seen in desert regions, where rocks degrade into sand through the effects of changing temperatures and strong winds.

The power of water is fiercer still. In severe storms, giant waves pound the shoreline with rocks and boulders, and often destroy concrete coastal defenses; but even in quieter conditions, the sea steadily erodes cliffs and headlands and creates new land in the form of sand dunes, spits and salt marshes.

Rivers, too, are incessantly at work shaping the landscape on their way to join the sea. In highland regions, where the flow is rapid, they cut deep gorges and V-shaped valleys. As they reach more gentle slopes, rivers release some of the debris they have carried downstream, broadening out and raising levees along their banks by depositing mud and sand. In the lowland plains, they may drift into meanders, depositing more sediment and even building deltas when they finally approach the sea.

Ice has created some of the world's dramatic landscapes. As glaciers move slowly downhill, they scrape away rock from the mountains and valley sides, creating spectacular features.

SHAPING FORCES: THE SEA

In areas of hard rock, waves cut steep cliffs and form underwater platforms; debris is deposited as a terrace. Bays are formed when sections of soft rock are carved away between headlands of harder rock; these are then battered until the headlands are reduced to rock arches and stacks.

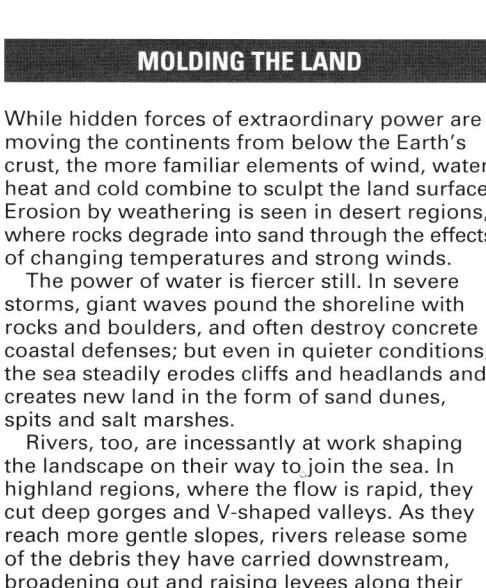

Headland

Cliff

Wave-cut platform

Wave-built terrace

Arch

Stack

Cove

SHAPING FORCES: RIVERS

Rivers shape the landscape according to the speed of their flow. In their youthful, upland stage they erode soft rocks quickly, cutting steep narrow valleys and tumbling in waterfalls over harder rock. As they mature, they deposit some debris and erode outward to widen the valley. In their old age, where the gradient is minimal, they meander across wide plains, depositing deep layers of sediment.

Waterfall

Gorge

V-shaped valley

Tree line

Natural levee

Meanders

Floodplain

Sediment

Man-made levee

YOUTH

MATURITY

OLD AGE

SHAPING FORCES: GLACIERS

Glaciers are formed from compressed snow accumulating in a valley head or cirque. They move downhill at a rate of a few inches to several yards each day, eroding large quantities of rocks, debris or moraine, that are caught up by the glacier and add to the abrasive power of the ice. Glaciers create numerous distinctive landscape features: among the most easily recognized are hanging valleys, cut by tributary glaciers; terminal moraine and drumlins formed by rock debris deposited when a glacier retreats; and the broad U-shape that distinguishes a glacial valley from one cut by a river.

Col

Lateral moraine

Ice-dammed lake

U-shaped valley

Truncated spur

Hanging valley

Arête

Crevasse

Medial moraine

Drumlins

Snout

Outwash plain

Terminal moraine

THE EARTH: ENVIRONMENT

Unique among the planets, the Earth has been the home of living creatures for most of its existence. Precisely how these improbable assemblies of self-replicating chemicals ever began remains a matter of conjecture, but the planet and its passengers have matured together for a very long time. Over 3,000 million years, life has not only adapted to its environment, but it has also slowly changed that environment to suit itself.

The planet and its biosphere – the entirety of its living things – function like a single organism. The British scientist James Lovelock, who first stated this 'Gaia hypothesis' in the 1970s, went further: the planet, he declared, actually was a living organism, equipped on a colossal scale with the same sort of stability-seeking mechanisms

used by lesser lifeforms like bacteria and humans to keep themselves running at optimum efficiency.

Lovelock's theory was inspired by a study of the Earth's atmosphere whose constituents he noted were very far from the state of chemical equilibrium observed elsewhere in the Solar System. The atmosphere has contained a substantial amount of free oxygen for the last 2,000 million years; yet without constant renewal, the oxygen molecules would soon be locked permanently in oxides. The nitrogen, too, would find chemical stability, probably in nitrates (accounting for some of the oxygen). Without living plants and algae to remove it, carbon dioxide would steadily increase from its present-day 0.03%; in a few million years, it

would form a thick blanket similar to the atmosphere of lifeless Venus, where surface temperatures reach 900°F.

It is not enough, however, for the biosphere simply to produce oxygen. While falling concentrations would first be uncomfortable and ultimately prove fatal for most contemporary life, at levels above the current 21% even moist vegetation is highly inflammable, and a massive conflagration becomes almost inevitable – a violent form of negative feedback to set the atmosphere on the path back to sterile equilibrium.

Fortunately, the biosphere has evolved over aeons into a subtle and complex control system, sensing changes and reacting to them quickly but gently, tending always to maintain the balance it has achieved.

Air-sea interface

The ocean surface is the location of most of the great systems of heat exchange that keep the Earth functioning properly. In addition, the ocean absorbs and circulates critical atmospheric gases.

The high atmosphere

On the edge of space, the ionized outer atmosphere shields the Earth from meteors and high-energy solar particles. Below, a layer of ozone traps ultraviolet radiation.

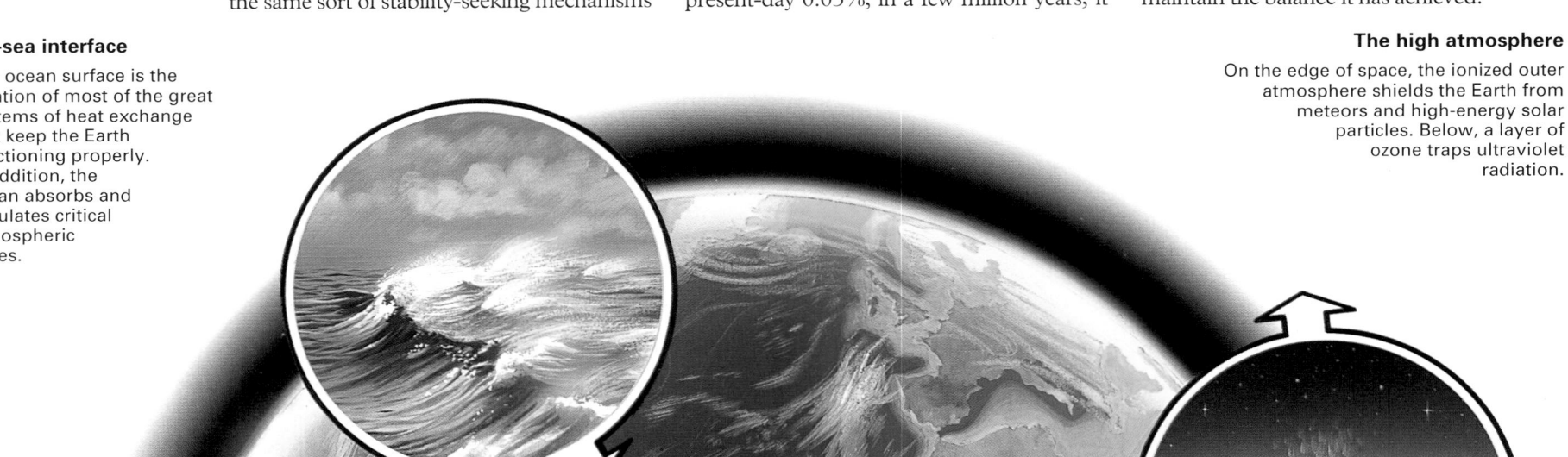

Tropical vegetation

The lush growth of rain forest and other vegetation in the Earth's tropical zones is one of the most important oxygen generators on the planet. Large-scale transpiration influences rainfall and climate patterns both locally and far afield.

Continental shelves

The warm, shallow fringes amount to 21% of the Earth's total ocean area but contain a far higher proportion of its plant and animal life. Vulnerable to coastal and marine pollution, plankton and other plants in these waters are key elements in the carbon and oxygen cycles upon which all life depends.

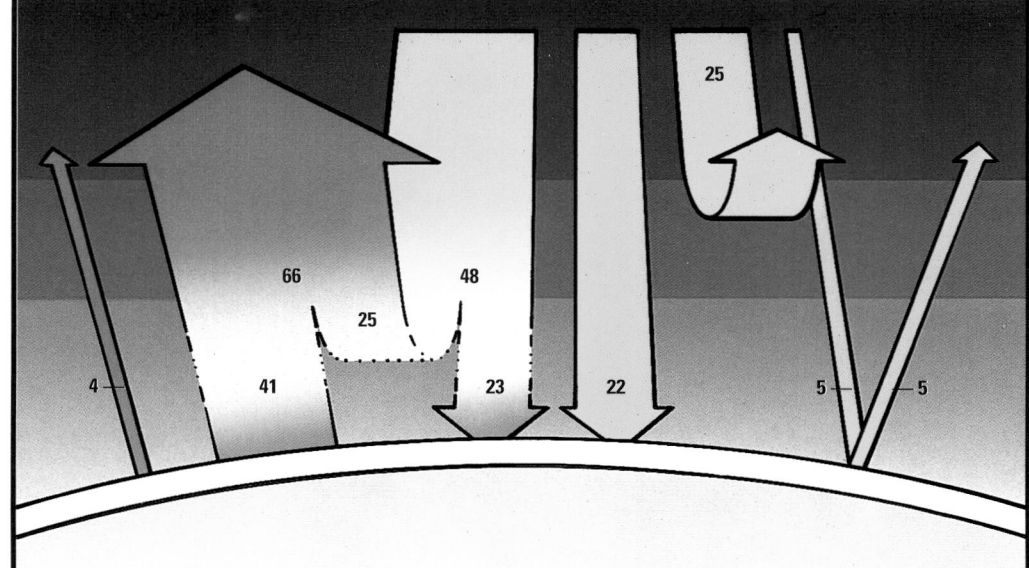

THE EARTH'S ENERGY BALANCE

Apart from a modest quantity of internal heat from its molten core, the Earth receives all of its energy from the Sun. If the planet is to remain at a constant temperature, it must reradiate exactly as much energy as it receives. Even a minute surplus would lead to a warmer Earth, a deficit to a cooler one; because the planetary energy budget is constantly audited by the laws of physics, which do not permit juggling, it must balance with absolute precision. The temperature at which thermal equilibrium is reached depends on a multitude of interconnected factors. Two of the most important are the relative brightness of the Earth – its index of reflectivity, called the 'albedo' – and the heat-trapping capacity of the atmosphere – the celebrated 'greenhouse effect'.

Because the Sun is very hot, most of its energy arrives in the form of relatively short-wave radiation: the shorter the waves, the more energy they carry. Some of the incoming energy is reflected straight back into space, exactly as it arrived; some is absorbed by the atmosphere on its way toward the surface; some is absorbed by the Earth itself. Absorbed energy heats the Earth and its atmosphere alike. But since its temperature is very much lower than that of the Sun, outgoing energy is emitted at much longer infra-red wavelengths. Some of the outgoing radiation escapes directly into outer space; some of it is reabsorbed by the atmosphere. Atmospheric energy eventually finds its way back into space, too, after a complex series of interactions. These include the air movements we call the weather and, almost incidentally, the maintenance of life on Earth.

This diagram does not attempt to illustrate the actual mechanisms of heat exchange, but gives a reasonable account (in percentages) of what happens to 100 energy 'units'. Short-wave radiation is shown in yellow, long-wave in red.

THE CARBON CYCLE

Most of the constituents of the atmosphere are kept in constant balance by complex cycles in which life plays an essential and indeed a dominant part. The control of carbon dioxide, which left to its own devices would be the dominant atmospheric gas, is possibly the most important, although since all the Earth's biological and geophysical cycles interact and interlock, it is hard to separate them even in theory and quite impossible in practice.

The Earth has a huge supply of carbon, only a small quantity of which is in the form of carbon dioxide. Of that, around 98% is dissolved in the sea; the fraction circulating in the air amounts to only 340 parts per million of the atmosphere, where its capacity as a greenhouse gas is the key regulator of the planetary temperature. In turn, life regulates the regulator, keeping carbon dioxide concentrations below danger level.

If all life were to vanish tomorrow from the Earth, the atmosphere would begin the process of change immediately, although it might take several million years to achieve a new, inorganic stability. First, the oxygen content would begin to fall away; with no more assistance than a little solar radiation, a few electrical storms and its own high chemical potential, oxygen would steadily combine with atmospheric nitrogen and volcanic outgassing. In doing so, it would yield sufficient acid to react with carbonaceous rocks such as limestone, releasing carbon dioxide. Once carbon dioxide levels exceeded about 1%, its greenhouse power would increase disproportionately. Rising temperatures – well above the boiling point of water – would speed chemical reactions; in time, the Earth's atmosphere would consist of little more than carbon dioxide and superheated water vapor.

Living things, however, circulate carbon. They do so first by simply existing: after all, the carbon atom is the basic building block of living matter. During life, plants absorb atmospheric carbon dioxide, incorporating the carbon itself into their structure – leaves and trunks in the case of land plants, shells in the case of plankton and the tiny creatures that feed on it. The oxygen thereby freed is added to the atmosphere, at least for a time. Most plant carbon is returned to circulation when the plants die and decay, combining once more with the oxygen released during life. However, a small proportion – about one part in 1,000 – is removed almost permanently, buried beneath mud on land, or at sea sinking as dead matter to the ocean floor. In time, it is slowly compressed into sedimentary rocks such as limestone and chalk.

But in the evolution of the Earth, nothing is quite permanent. On an even longer timescale, the planet's crustal movements force new rock upward in mid-ocean ridges. Limestone deposits are moved, and sea levels change; ancient limestone is exposed to weathering, and a little of its carbon is released to be fixed in turn by the current generation of plants.

The carbon cycle has continued quietly for an immensely long time, and without gross disturbance there is no reason why it would not continue almost indefinitely in the future. However, human beings have found a way to release fixed carbon at a rate far faster than existing global systems can recirculate it. Oil and coal deposits represent the work of millions of years of carbon accumulation; but it has taken only a few human generations of high-energy scavenging to endanger the entire complex regulatory cycle.

AIR

Organic decay, animal respiration and burning

Plankton photosynthesis

Absorbtion by living plants

Plankton respiration

LAND

Mineral washout

SEA

Sea shells to sedimentary rock

[98% of existing carbon dioxide held in solution in the sea]

THE GREENHOUSE EFFECT

Constituting barely 0.03% of the atmosphere, carbon dioxide has a hugely disproportionate effect on the Earth's climate and even its habitability. Like the glass panes in a greenhouse, it is transparent to most incoming short-wave radiation, which passes freely to heat the planet beneath. But when the warmed Earth retransmits that energy, in the form of longer-wave infra-red radiation, the carbon dioxide functions as an opaque shield, so that the planetary surface (like the interior of a greenhouse) stays relatively hot.

The recent increases in CO_2 levels are causing alarm: global warming associated with a runaway greenhouse effect could bring disaster. But a serious reduction would be just as damaging, with surface temperatures falling dramatically; during the last Ice Age, for example, the carbon dioxide concentration was around 180 parts per million, and a total absence of the gas would likely leave the planet a ball of ice, or at best frozen tundra.

The diagram shows incoming sunlight as yellow; high-energy ultraviolet (blue) is trapped by the ozone layer, while outgoing heat from the warmed Earth (red) is partially retained by carbon dioxide.

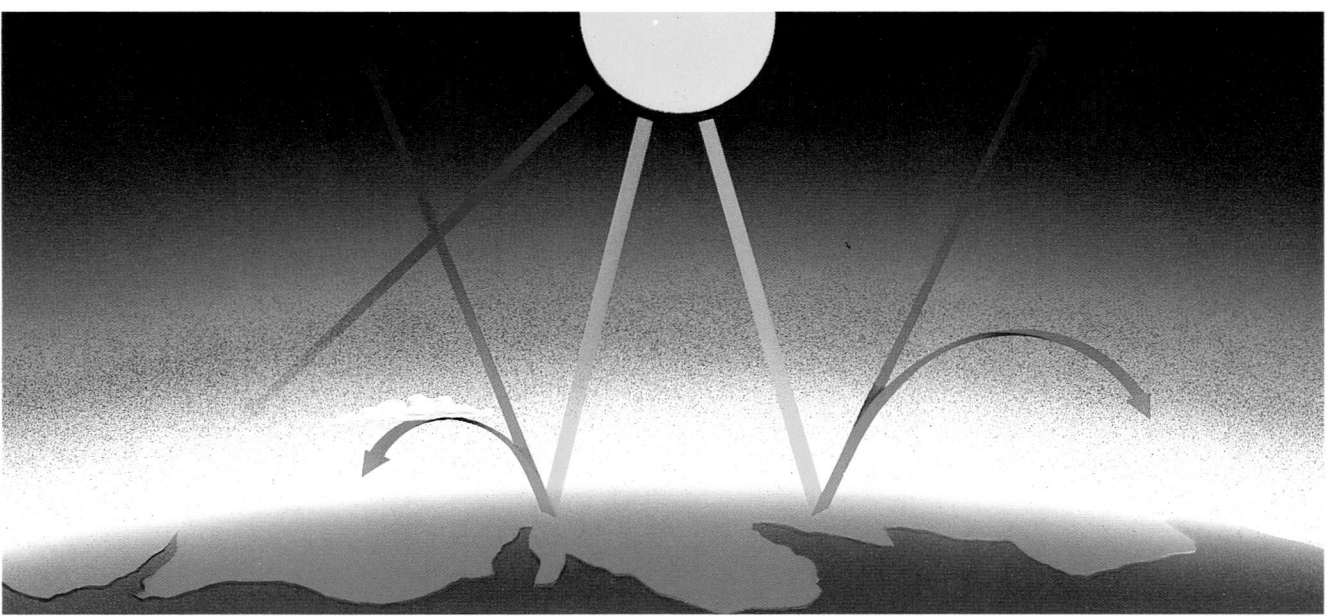

PEOPLE: DEMOGRAPHY

As the 20th century draws to its close, the Earth's population increases by nearly 10,000 every hour – enough to fill a new major city every week. The growth is almost entirely confined to the developing world, which accounted for 67% of total population in 1950 and is set to reach 84% by 2025. In developed countries, populations are almost static, and in some places, such as Germany, are actually falling. In fact, there is a clear correlation between wealth and low fertility: as incomes rise, reproduction rates drop.

The decline is already apparent. With the exception of Africa, the actual rates of increase are falling nearly everywhere. The population structure, however, ensures that human numbers will continue to rise even as fertility diminishes. Developed nations, like the USA or UK, have an even spread across ages, and usually a growing proportion of elderly people: the over-75s often outnumber the under-5s, and women of child-bearing age form only a small part of the total. Developing nations fall into a pattern somewhere between that of Kenya and Brazil: the great majority of their people are in the younger age groups, about to enter their most fertile years. In time, even Kenya's population profile should resemble the developed model, but the transition will come about only after a few more generations' growth.

It remains to be seen whether the planet will tolerate the population growth that seems inevitable before stability is reached. More people consume more resources, increasing the strain on an already troubled environment. However, more people should mean a greater supply of human ingenuity – the only commodity likely to resolve the crisis.

LARGEST NATIONS

The world's most populous nations, in millions (1994)

1.	China	1,209
2.	India	919
3.	USA	261
4.	Indonesia	195
5.	Brazil	159
6.	Russia	147
7.	Pakistan	137
8.	Japan	125
9.	Bangladesh	118
10.	Mexico	92
11.	Nigeria	89
12.	Germany	81
13.	Vietnam	73
14.	Philippines	66
15.	Iran	66
16.	Egypt	62
17.	Turkey	61
18.	Thailand	58
19.	UK	58
20.	France	58
21.	Italy	57
22.	Ethiopia	53
23.	Ukraine	51
24.	Burma	46

CROWDED NATIONS

Population per square mile (1994), exc. nations of less than one million.

1.	Hong Kong	2,049.1
2.	Singapore	1,756.7
3.	Bangladesh	315.8
4.	Mauritius	224.4
5.	Taiwan	221.6
6.	South Korea	173.8
7.	Puerto Rico	156.4
8.	Netherlands	143.2
9.	Belgium	127.6
10.	Japan	127.5
11.	Rwanda	113.8
12.	Lebanon	108.2
13.	India	107.9
14.	Sri Lanka	106.7
15.	El Salvador	103.7
16.	Trinidad & Tobago	97.8
17.	UK	92.2
18.	Germany	86.8
19.	Jamaica	85.3
20.	Israel	78.0

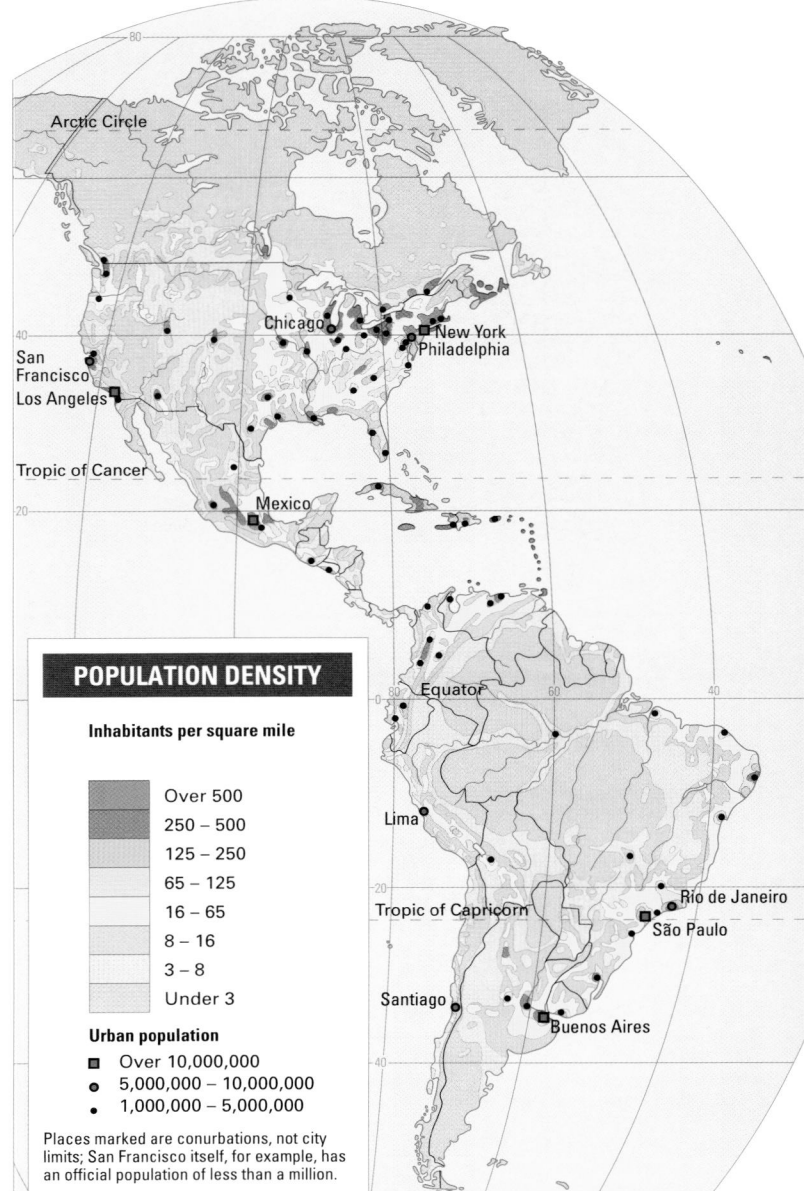

POPULATION DENSITY

Inhabitants per square mile

- Over 500
- 250 – 500
- 125 – 250
- 65 – 125
- 16 – 65
- 8 – 16
- 3 – 8
- Under 3

Urban population

- ■ Over 10,000,000
- ⬤ 5,000,000 – 10,000,000
- • 1,000,000 – 5,000,000

Places marked are conurbations, not city limits; San Francisco itself, for example, has an official population of less than a million.

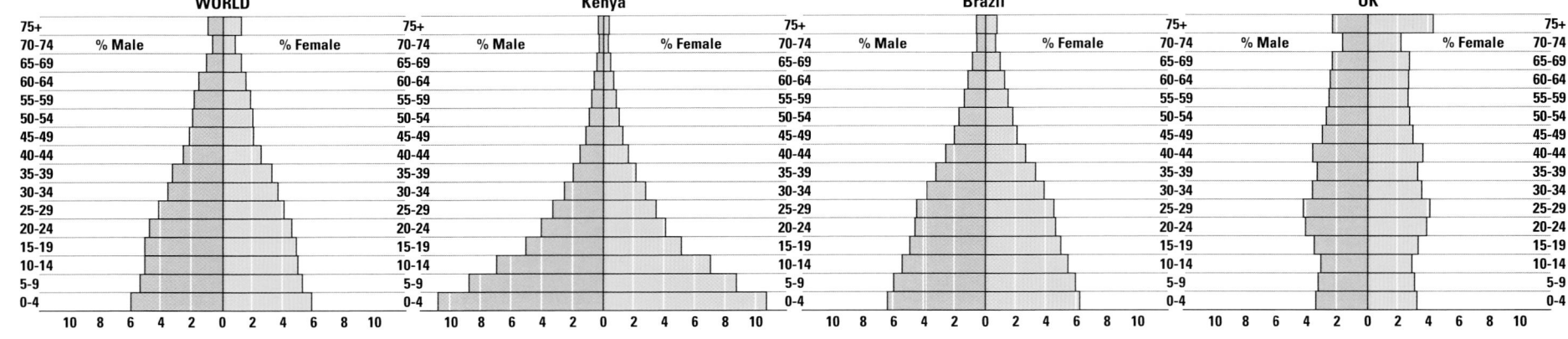

WORLD — Kenya — Brazil — UK

(Age pyramids: age groups from 0–4 up to 75+, % Male and % Female, scale 10 8 6 4 2 0 2 4 6 8 10)

RATES OF GROWTH

Apparently small rates of population growth lead to dramatic increases over two or three generations. The table below translates annual percentage growth into the number of years required to double a population.

% change	Doubling time
0.5	139.0
1.0	69.7
1.5	46.6
2.0	35.0
2.5	28.1
3.0	23.4
3.5	20.1
4.0	17.7

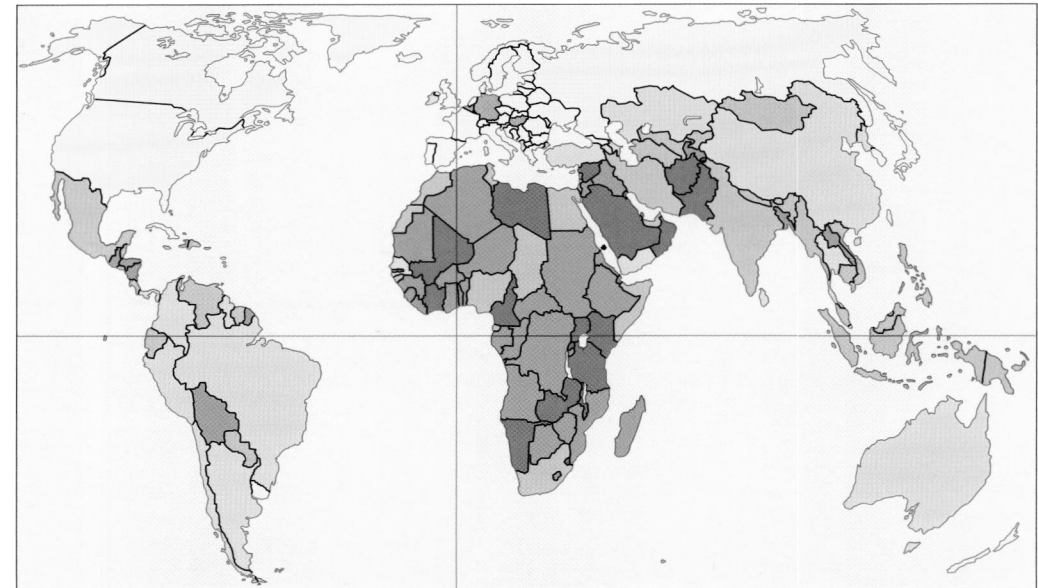

POPULATION CHANGE 1990–2000

The predicted population change for the years 1990–2000

- Over 40% population gain
- 30 – 40% population gain
- 20 – 30% population gain
- 10 – 20% population gain
- 0 – 10% population gain
- No change or population loss

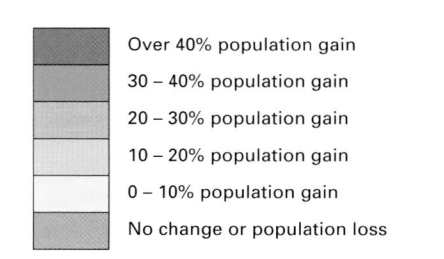

Top 5 countries		Bottom 5 countries	
Kuwait	+75.9%	Belgium	−0.1%
Namibia	+62.5%	Hungary	−0.2%
Afghanistan	+60.1%	Grenada	−2.4%
Mali	+55.5%	Germany	−3.2%
Tanzania	+54.6%	Tonga	−3.2%

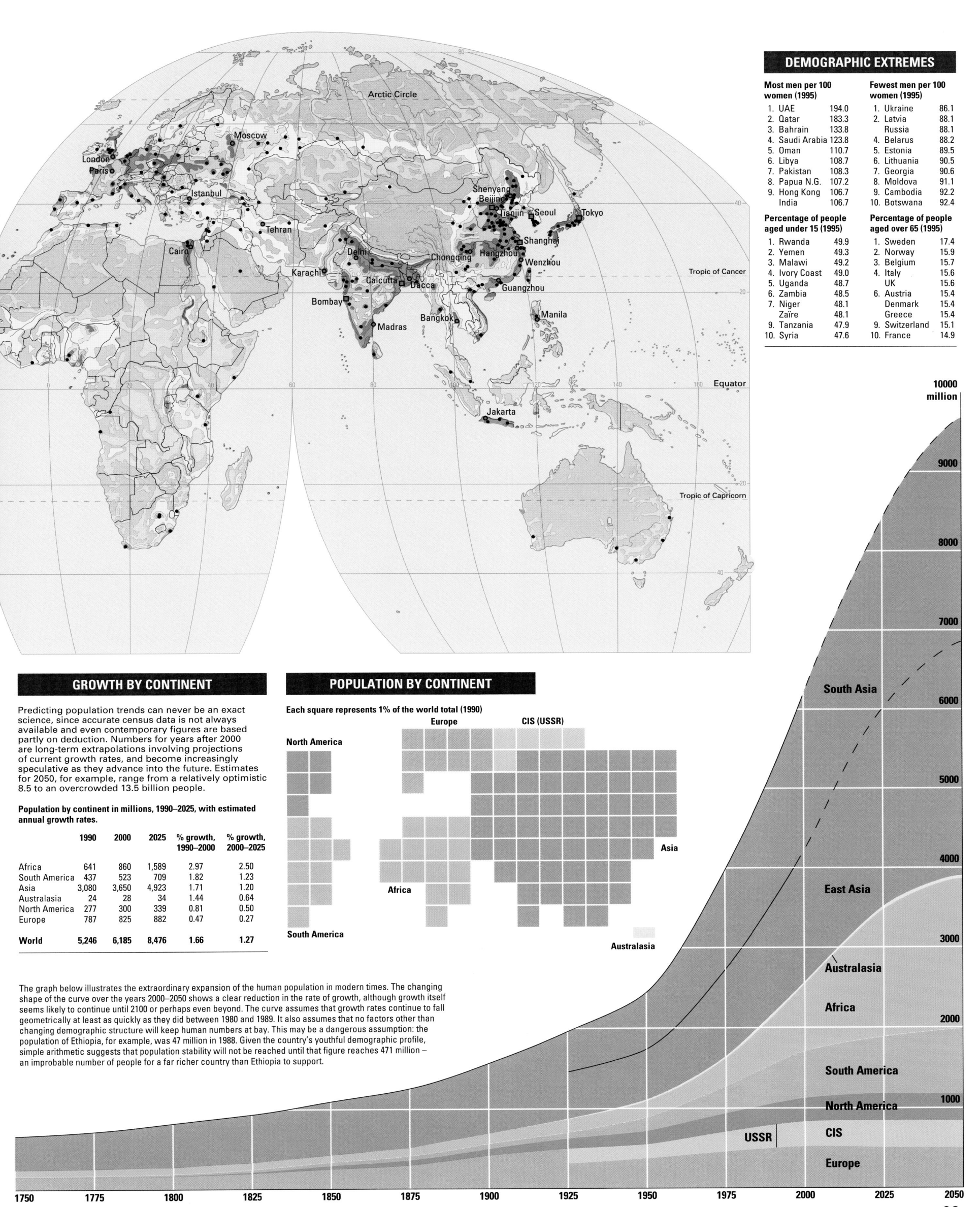

Arctic Circle

London
Paris
Moscow
Istanbul
Tehran
Cairo
Karachi
Delhi
Bombay
Calcutta
Dacca
Madras
Bangkok
Shenyang
Beijing
Tianjin
Seoul
Tokyo
Shanghai
Chongqing
Hangzhou
Wenzhou
Guangzhou
Manila
Jakarta

Tropic of Cancer

Equator

Tropic of Capricorn

10000 million
9000
8000
7000
6000
5000
4000
3000
2000
1000

South Asia
East Asia
Australasia
Africa
South America
North America
USSR
CIS
Europe

GROWTH BY CONTINENT

Predicting population trends can never be an exact science, since accurate census data is not always available and even contemporary figures are based partly on deduction. Numbers for years after 2000 are long-term extrapolations involving projections of current growth rates, and become increasingly speculative as they advance into the future. Estimates for 2050, for example, range from a relatively optimistic 8.5 to an overcrowded 13.5 billion people.

Population by continent in millions, 1990–2025, with estimated annual growth rates.

	1990	2000	2025	% growth, 1990–2000	% growth, 2000–2025
Africa	641	860	1,589	2.97	2.50
South America	437	523	709	1.82	1.23
Asia	3,080	3,650	4,923	1.71	1.20
Australasia	24	28	34	1.44	0.64
North America	277	300	339	0.81	0.50
Europe	787	825	882	0.47	0.27
World	**5,246**	**6,185**	**8,476**	**1.66**	**1.27**

POPULATION BY CONTINENT

Each square represents 1% of the world total (1990)

North America
Europe
CIS (USSR)
Asia
Africa
South America
Australasia

The graph below illustrates the extraordinary expansion of the human population in modern times. The changing shape of the curve over the years 2000–2050 shows a clear reduction in the rate of growth, although growth itself seems likely to continue until 2100 or perhaps even beyond. The curve assumes that growth rates continue to fall geometrically at least as quickly as they did between 1980 and 1989. It also assumes that no factors other than changing demographic structure will keep human numbers at bay. This may be a dangerous assumption: the population of Ethiopia, for example, was 47 million in 1988. Given the country's youthful demographic profile, simple arithmetic suggests that population stability will not be reached until that figure reaches 471 million – an improbable number of people for a far richer country than Ethiopia to support.

1750　1775　1800　1825　1850　1875　1900　1925　1950　1975　2000　2025　2050

PEOPLE: CITIES

In 1750, barely three humans in every hundred lived in a city; by 2000, more than half the world's population will find a home in some kind of urban area. In 1850, only London and Paris had more than a million inhabitants; by 2000, at least 24 cities will each contain over 10 million people. The increase is concentrated in the Third World, if only because levels of urbanization in most developed countries – more than 90% in the UK and Belgium, and almost 75% in the USA, despite that country's great open spaces – have already reached practical limits.

Such large-scale concentration is relatively new to the human race. Although city life has always attracted country dwellers in search of trade, employment or simply human contact, until modern times they paid a high price. Crowding and poor sanitation ensured high death rates, and until about 1850, most cities needed a steady flow of incomers simply to maintain their population levels: for example, there were 600,000 more deaths than births in 18th-century London, and some other large cities showed an even worse imbalance.

With improved public health, cities could grow from their own human resources, and large-scale urban living became commonplace in the developed world. Since about 1950, the pattern has been global. Like their counterparts in 19th-century Europe and the USA, the great new cities are driven into rapid growth by a kind of push-pull mechanism. The push is generated by agricultural overcrowding: only so many people can live from a single plot of land and population pressure drives many into towns. The pull comes from the possibilities of economic improvement – an irresistible lure to the world's rural hopefuls.

Such improvement is not always obvious: the typical Third World city, with millions of people living (often illegally) in shanty towns and many thousands existing homelessly on the ill-made streets, does not present a great image of prosperity. Yet modern shanty towns are healthier than industrializing Pittsburgh or Manchester in the last century, and these human ant-hills teem with industry as well as squalor: throughout the world, above-average rates of urbanization have gone hand-in-hand with above-average rates of economic growth. Surveys demonstrate that Third World city dwellers are generally better off than their rural counterparts, whose poverty is less concentrated but often more desperate. This only serves to increase the attraction of the city for the rural poor.

However, the sheer speed of the urbanization process threatens to overwhelm the limited abilities of city authorities to provide even rudimentary services. The 24 million people expected to live in Mexico City by 2000, for example, would swamp a more efficient local government than Mexico can provide. Improvements are often swallowed up by the relentless rise in urban population: although safe drinking water should reach 75% of Third World city dwellers by the end of the century – a considerable achievement – population growth will add 100 million to the list of those without it.

THE URBANIZATION OF THE EARTH

City-building, 1850–2000; each white spot represents a city of at least 1 million inhabitants.

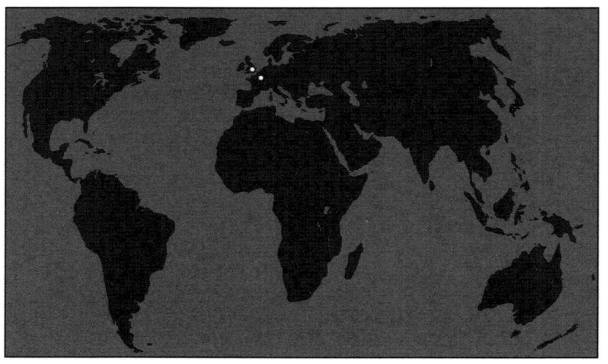

1850

1900

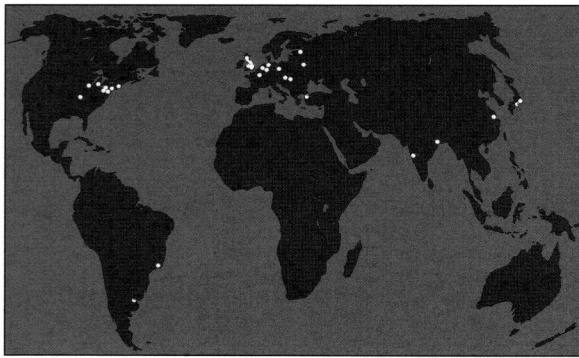

1925

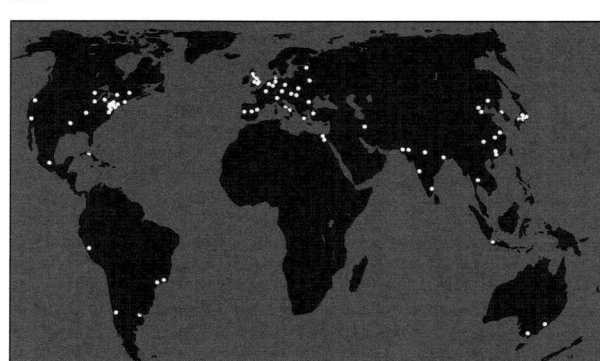

1950

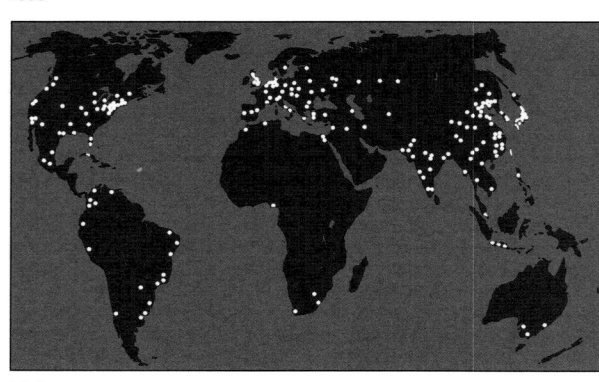

1975

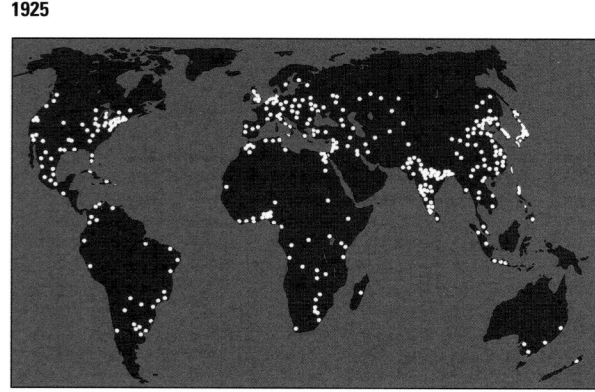

2000

URBAN POPULATION

Percentage of total population living in towns and cities (1990)

- Over 75%
- 50 – 75%
- 25 – 50%
- 10 – 25%
- Under 10%

Most urbanized		Least urbanized	
Singapore	100%	Bhutan	5%
Belgium	97%	Burundi	7%
Kuwait	96%	Rwanda	8%
Hong Kong	93%	Burkina Faso	9%
UK	93%	Nepal	10%

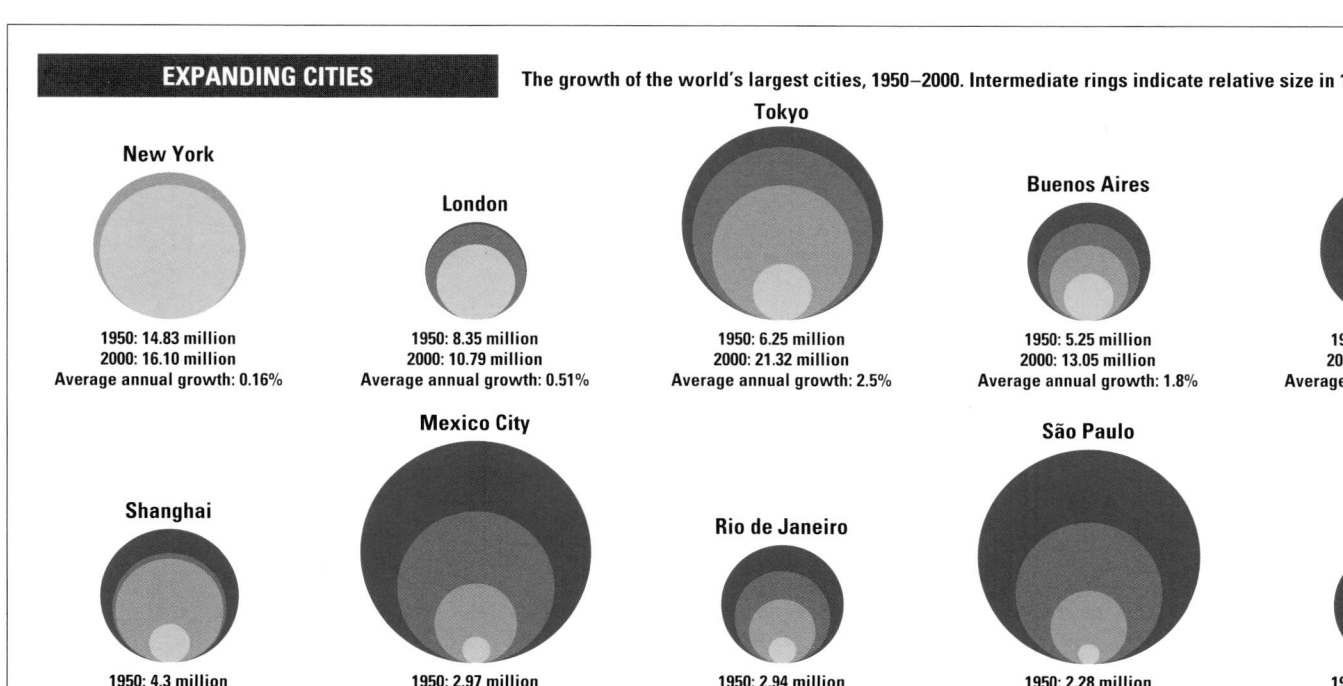

EXPANDING CITIES

The growth of the world's largest cities, 1950–2000. Intermediate rings indicate relative size in 1970 and 1985.

New York
1950: 14.83 million
2000: 16.10 million
Average annual growth: 0.16%

London
1950: 8.35 million
2000: 10.79 million
Average annual growth: 0.51%

Tokyo
1950: 6.25 million
2000: 21.32 million
Average annual growth: 2.5%

Buenos Aires
1950: 5.25 million
2000: 13.05 million
Average annual growth: 1.8%

Calcutta
1950: 4.45 million
2000: 15.94 million
Average annual growth: 2.6%

Shanghai
1950: 4.3 million
2000: 14.69 million
Average annual growth: 2.5%

Mexico City
1950: 2.97 million
2000: 24.44 million
Average annual growth: 4.3%

Rio de Janeiro
1950: 2.94 million
2000: 13.0 million
Average annual growth: 3.0%

São Paulo
1950: 2.28 million
2000: 23.6 million
Average annual growth: 4.8%

Seoul
1950: 1.45 million
2000: 12.97 million
Average annual growth: 4.5%

Each set of circles illustrates a city's size in 1950, 1970, 1985 and 2000. In most cases, expansion has been steady and, often, explosive. New York and London, however, went through patches of negative growth during the period. In New York, the world's largest city in 1950, population reached a peak around 1970. London shrank slightly between 1970 and 1985 before resuming a very modest rate of increase. In both cases, the divergence from world trends can be explained in part by counting methods: each is at the center of a great agglomeration, and definitions of where 'city limits' lie may vary over time. But their relative decline also matches a pattern often seen in mature cities in the developed world, where urbanization, already at a very high level, has reached a plateau.

CITIES IN DANGER

As the decade of the 1980s advanced, most industrial countries, alarmed by acid rain and urban smog, took significant steps to limit air pollution. These controls, however, are expensive to install and difficult to enforce, and clean air remains a luxury most developed as well as developing cities must live without.

Those taking part in the United Nations' Global Environment Monitoring System (see right) frequently show dangerous levels of pollutants ranging from soot to sulfur dioxide and photochemical smog; air in the majority of cities without such sampling equipment is likely to be at least as bad.

URBAN AIR POLLUTION

The world's most polluted cities: number of days each year when sulfur dioxide levels exceeded the WHO threshold of 150 micrograms per cubic meter (about 0.00015 oz per cubic foot). Figures are averaged over 4 to 15 years, 1970s – 1980s

Sulfur dioxide is the main pollutant associated with industrial cities. According to the World Health Organization, more than seven days in a year above 150 µg per cubic meter bring a serious risk of respiratory disease: at least 600 million people live in urban areas where SO_2 concentrations regularly reach damaging levels.

Manila, Philippines
Calcutta, India
Milan, Italy
Zagreb, Croatia
Guangzhou, China
Madrid, Spain
Peking (Beijing), China
Xian, China
Seoul, South Korea
Tehran, Iran
Shenyang, China

120   90   60   30

LARGEST CITIES

By early next century for the first time in history, the majority of the world's population will live in cities. Below is a list of the world's largest cities, in millions of inhabitants, based on estimates for the year 2000.*

1.	Mexico City	25.6
2.	Tokyo-Yokohama	24.2
3.	São Paulo	22.1
4.	Shanghai	17.0
5.	New York	16.8
6.	Calcutta	15.7
7.	Bombay	15.4
8.	Beijing	14.0
9.	Los Angeles	13.9
10.	Jakarta	13.7
11.	Delhi	13.2
12.	Buenos Aires	12.9
	Lagos	12.9
14.	Seoul	12.7
	Tianjin	12.7
16.	Rio de Janeiro	12.5
17.	Dhaka	12.2
18.	Manila	11.8
	Cairo	11.8
20.	Karachi	11.7
21.	London	10.5
22.	Bangkok	10.3
23.	Istanbul	9.5
24.	Moscow	9.0
25.	Osaka	8.6
	Paris	8.6
27.	Tehran	8.5
28.	Bangalore	8.2
	Lima-Callao	8.2
30.	Madras	7.8
31.	Chicago	7.3
32.	Bogotá	6.4
33.	Shenyang	6.3
34.	Hong Kong	6.1
35.	Lahore	5.9
	Madrid	5.9
37.	Santiago	5.6
38.	Milan	5.4
	St Petersburg	5.4
40.	Philadelphia	4.5

[City populations are based on urban agglomerations rather than legal city limits. In some cases where two adjacent cities have merged into one concentration, such as Tokyo-Yokohama, they have been regarded as a single unit.]

* For a list of current city estimates, see page XI.

INFORMAL CITIZENS

Proportion of population living in squatter settlements, selected cities in the developing world (1980s)

Urbanization in most Third World countries has been coming about far faster than local governments can provide services and accommodation for the new city dwellers. Many – in some cities, most – find their homes in improvised squatter settlements, often unconnected to power, water and sanitation networks. Yet despite their ramshackle housing and marginal legality, these communities are often the most dynamic part of a city economy. They are also growing in size; and given the squatters' reluctance to be counted by tax-demanding authorities, the percentages shown here are likely to be underestimates.

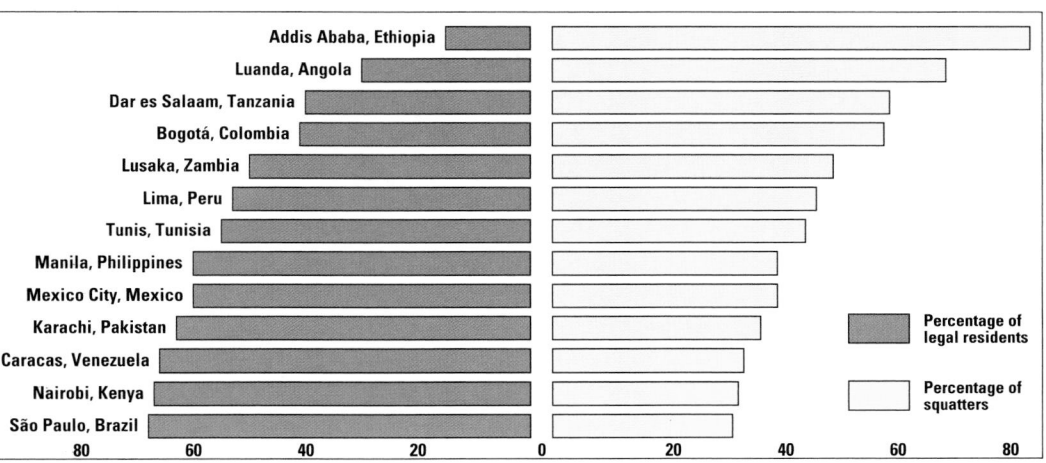

Addis Ababa, Ethiopia
Luanda, Angola
Dar es Salaam, Tanzania
Bogotá, Colombia
Lusaka, Zambia
Lima, Peru
Tunis, Tunisia
Manila, Philippines
Mexico City, Mexico
Karachi, Pakistan
Caracas, Venezuela
Nairobi, Kenya
São Paulo, Brazil

Percentage of legal residents

Percentage of squatters

80   60   40   20   0   20   40   60   80

URBAN ADVANTAGES

Despite overcrowding and poor housing, living standards in the developing world's cities are almost invariably better than in the surrounding countryside. Resources – financial, material and administrative – are concentrated in the towns, which are usually also the centers of political activity and pressure. Governments – frequently unstable, and rarely established on a solid democratic base – are usually more responsive to urban discontent than rural misery.

In many countries, especially in Africa, food prices are often kept artificially low, appeasing underemployed urban masses at the expense of agricultural development. The imbalance encourages further cityward migration, helping to account for the astonishing rate of post-1950 urbanization and putting great strain on the ability of many nations to provide even modest improvements for their people.

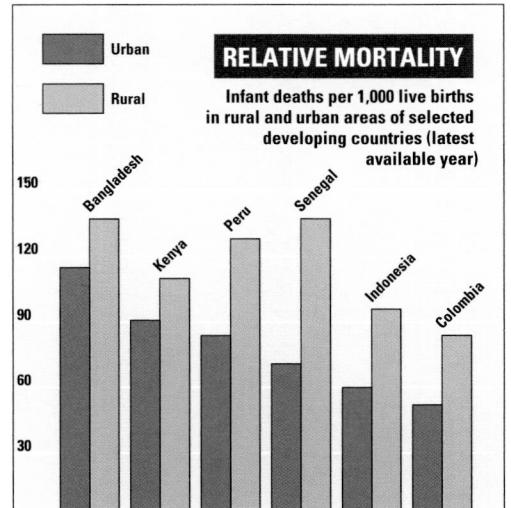

Urban
Rural

RELATIVE MORTALITY
Infant deaths per 1,000 live births in rural and urban areas of selected developing countries (latest available year)

150
120
90
60
30

Bangladesh
Kenya
Peru
Senegal
Indonesia
Colombia

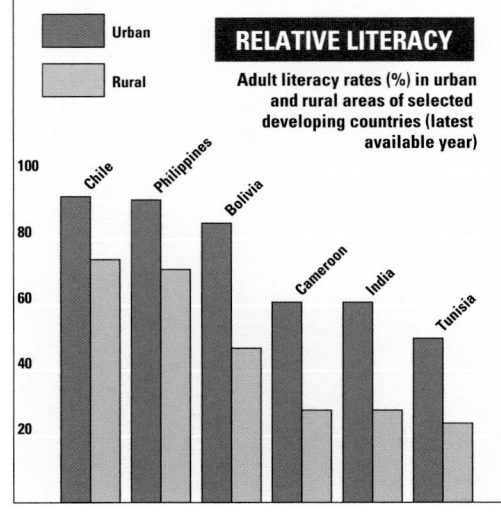

Urban
Rural

RELATIVE LITERACY
Adult literacy rates (%) in urban and rural areas of selected developing countries (latest available year)

100
80
60
40
20

Chile
Philippines
Bolivia
Cameroon
India
Tunisia

PEOPLE: THE HUMAN FAMILY

Strictly speaking, all human beings belong to a single race – *Homo sapiens* has no subspecies. But although all humans are interfertile, anthropologists and geneticists distinguish three main racial types: Caucasoid, Negroid and Mongoloid. Racial differences reflect not so much evolutionary origin as long periods of separation.

Racial affinities are not always obvious. The Caucasoid group stems from Europe, North Africa and India, but still includes Australian aboriginals within its broad type; Mongoloid peoples comprise American Indians and Eskimos as well as most Chinese, central Asians and Malays; Negroids are mostly of African origin, but also include the Papuan peoples of New Guinea.

Migration in modern times has mingled racial groups to an unprecedented extent, and most nations now have some degree of racially mixed population.

Language is almost the definition of a particular human culture; the world has well over 5,000, most of them with only a few hundred thousand speakers. In one important sense, all languages are equal; although different vocabularies and linguistic structures greatly influence patterns of thought, all true human languages can carry virtually unlimited information. But even if, for example, there is no theoretical difference in the communicative power of English and one of the 500 or more tribal languages of Papua New Guinea, an English speaker has access to much more of the global culture than a Papuan who knows no other tongue.

Like language, religion encourages the internal cohesion of a single human group at the expense of creating gulfs of incomprehension between different groups. All religions satisfy a deep-seated human need, assigning men and women to a comprehensible place in what most of them still consider a divinely ordered world. But religion is also a means by which a culture can assert its individuality; the startling rise of Islam in the late 20th century is partly a response by large sections of the developing world to the secular, Western-inspired world order from which many non-Western peoples feel excluded. Like uncounted millions of human beings before them, they find in their religion not only a personal faith but also a powerful group identity.

WORLD MIGRATION

The greatest voluntary migration was the colonization of North America by 30–35 million European settlers during the 19th century. The greatest forced migration involved 9–11 million Africans taken as slaves to America 1550–1860. The migrations shown on the map are mostly international as population movements within borders are not usually recorded. Many of the statistics are necessarily estimates as so many refugees and migrant workers enter countries illegally and unrecorded. Emigrants may have a variety of motives for leaving, thus making it difficult to distinguish between voluntary and involuntary migrations.

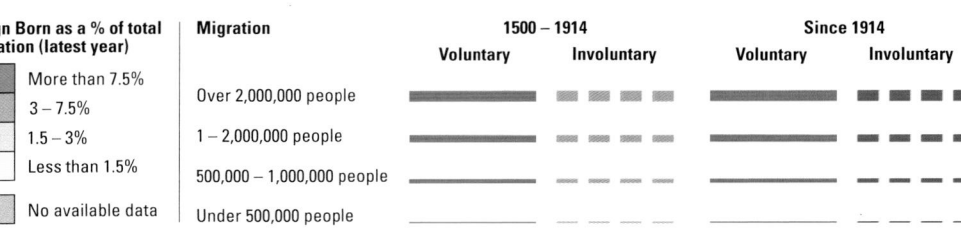

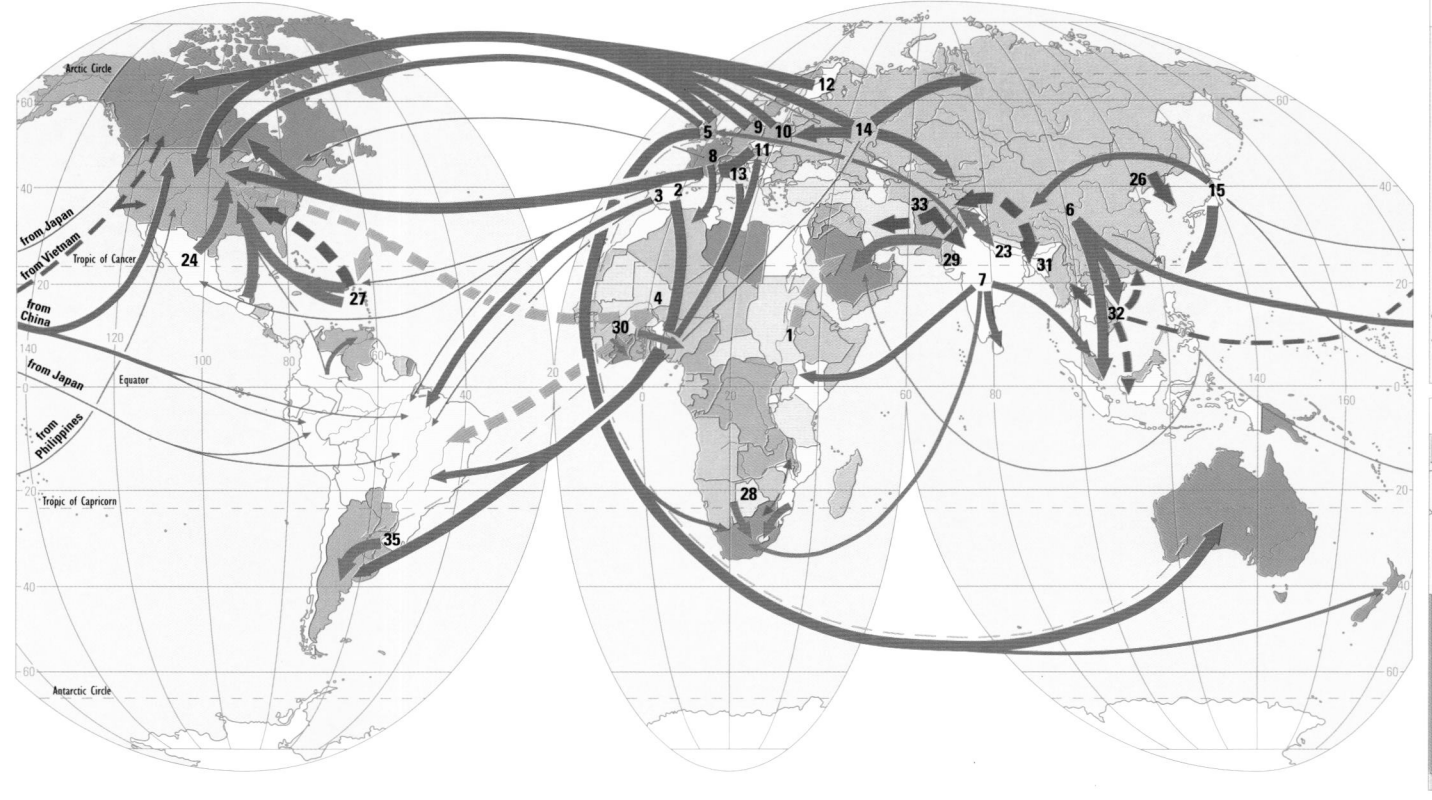

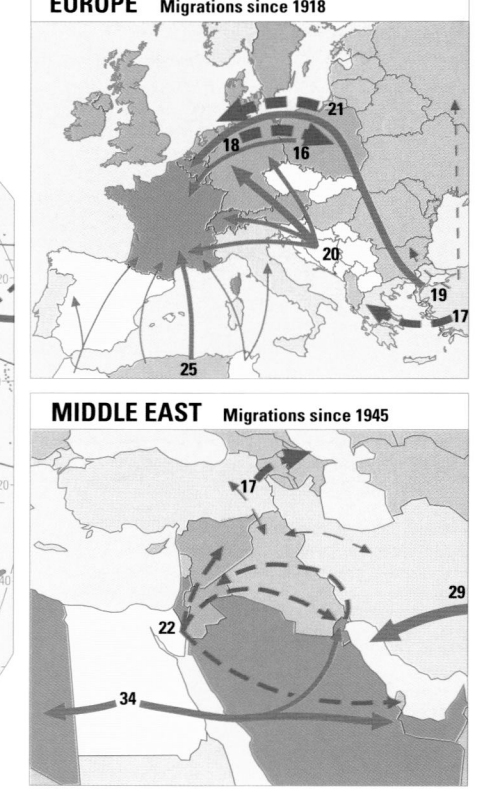

BUILDING THE USA

US Immigration 1820–1990

'Give me your tired, your poor/Your huddled masses yearning to breathe free....'

So starts Emma Lazarus's poem *The New Colossus*, inscribed on the Statue of Liberty. For decades the USA was the magnet that attracted millions of immigrants, notably from Central and Eastern Europe, the flow peaking in the early years of this century.

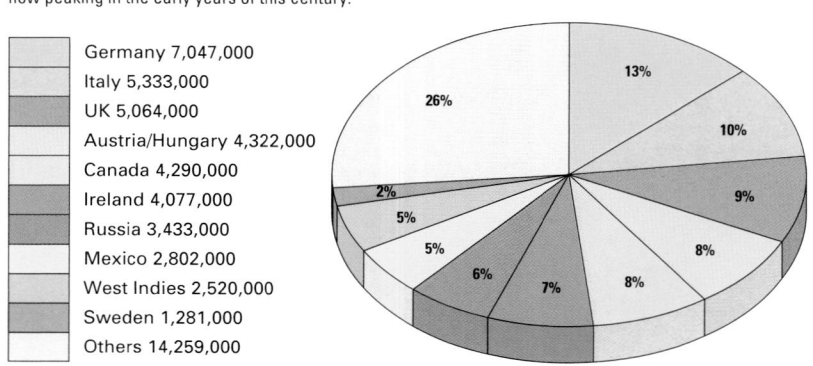

Germany	7,047,000
Italy	5,333,000
UK	5,064,000
Austria/Hungary	4,322,000
Canada	4,290,000
Ireland	4,077,000
Russia	3,433,000
Mexico	2,802,000
West Indies	2,520,000
Sweden	1,281,000
Others	14,259,000

Major world migrations since 1500 (over 1,000,000 people)

1. North African and East African slaves to Arabia (4.3m) 1500–1900
2. Spanish to South and Central America (2.3m) 1530–1914
3. Portuguese to Brazil (1.4m) 1530–1914
4. West African slaves to South America (4.6m) 1550–1860
   to Caribbean (4m) 1580–1860
   to North and Central America (1m) 1650–1820
5. British and Irish to North America (13.5m) 1620–1914
   to Australasia and South Africa (3m) 1790–1914
6. Chinese to South-east Asia (22m) 1820–1914
   to North America (1m) 1880–1914
7. Indian migrant workers (3m) 1850–1914
8. French to North Africa (1.5m) 1850–1914
9. Germans to North America (5m) 1850–1914
10. Poles to North America (3.6m) 1850–1914
11. Austro-Hungarians to North America (3.2m) 1850–1914
    to Western Europe (3.4m) 1850–1914
    to South America (1.8m) 1850–1914
12. Scandinavians to North America (2.7m) 1850–1914
13. Italians to North America (5m) 1860–1914
    to South America (3.7m) 1860–1914
14. Russians to North America (2.2m) 1880–1914
    to Western Europe (2.2m) 1880–1914
    to Siberia (6m) 1880–1914
    to Central Asia (4m) 1880–1914
15. Japanese to Eastern Asia, South-east Asia and America (8m) 1900–1914
16. Poles to Western Europe (1m) 1920–1940
17. Greeks and Armenians from Turkey (1.6m) 1922–1923
18. European Jews to extermination camps (5m) 1940–1944
19. Turks to Western Europe (1.9m) 1940–
20. Yugoslavs to Western Europe (2m) 1940–
21. Germans to Western Europe (9.8m) 1945–1947
22. Palestinian refugees (2m) 1947–
23. Indian and Pakistani refugees (15m) 1947
24. Mexicans to North America (9m) 1950–
25. North Africans to Western Europe (1.1m) 1950–
26. Korean refugees (5m) 1950–1954
27. Latin Americans and West Indians to North America (4.7m) 1960–
28. Migrant workers to South Africa (1.5m) 1960–
29. Indians and Pakistanis to The Gulf (2.4m) 1970–
30. Migrant workers to Nigeria and Ivory Coast (3m) 1970–
31. Bangladeshi and Pakistani refugees (2m) 1972
32. Vietnamese and Cambodian refugees (1.5m) 1975–
33. Afghan refugees (6.1m) 1979–
34. Egyptians to The Gulf and Libya (2.9m) 1980–
35. Migrant workers to Argentina (2m) 1980–

INDO-EUROPEAN FAMILY

1 Balto-Slavic group (incl. Russian, Ukrainian)
2 Germanic group (incl. English, German)
3 Celtic group
4 Greek
5 Albanian
6 Iranian group
7 Armenian
8 Romance group (incl. Spanish, Portuguese, French, Italian)
9 Indo-Aryan group (incl. Hindi, Bengali, Urdu, Punjabi, Marathi)
10 CAUCASIAN FAMILY

AFRO-ASIATIC FAMILY

11 Semitic group (incl. Arabic)
12 Kushitic group
13 Berber group
14 KHOISAN FAMILY
15 NIGER-CONGO FAMILY
16 NILO-SAHARAN FAMILY
17 URALIC FAMILY

ALTAIC FAMILY

18 Turkic group
19 Mongolian group
20 Tungus-Manchu group
21 Japanese and Korean

SINO-TIBETAN FAMILY

22 Sinitic (Chinese) languages
23 Tibetic-Burmic languages
24 TAI FAMILY

AUSTRO-ASIATIC FAMILY

25 Mon-Khmer group
26 Munda group
27 Vietnamese
28 DRAVIDIAN FAMILY (incl. Telugu, Tamil)
29 AUSTRONESIAN FAMILY (incl. Malay-Indonesian)
30 OTHER LANGUAGES

OFFICIAL LANGUAGES

Language	Total population	World %
English	1,400m	27.0%
Chinese	1,070m	19.1%
Hindi	700m	13.5%
Spanish	280m	5.4%
Russian	270m	5.2%
French	220m	4.2%
Arabic	170m	3.3%
Portuguese	160m	3.0%
Malay	160m	3.0%
Bengali	150m	2.9%
Japanese	120m	2.3%

Languages form a kind of tree of development, splitting from a few ancient proto-tongues into branches that have grown apart and further divided with the passage of time. English and Hindi, for example, both belong to the great Indo-European family, although the relationship is only apparent after much analysis and comparison with non-Indo-European languages such as Chinese or Arabic; Hindi is part of the Indo-Aryan subgroup, whereas English is a member of Indo-European's Germanic branch; French, another Indo-European tongue, traces its descent through the Latin, or Romance, branch. A few languages – Basque is one example – have no apparent links with any other, living or dead. Most modern languages, of course, have acquired enormous quantities of vocabulary from each other.

MOTHER TONGUES

Native speakers of the major languages, in millions (1989)

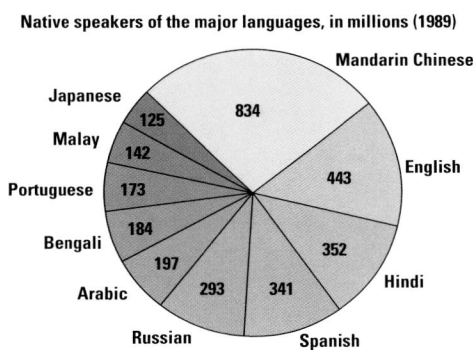

- Mandarin Chinese 834
- English 443
- Hindi 352
- Spanish 341
- Russian 293
- Arabic 197
- Bengali 184
- Portuguese 173
- Malay 142
- Japanese 125

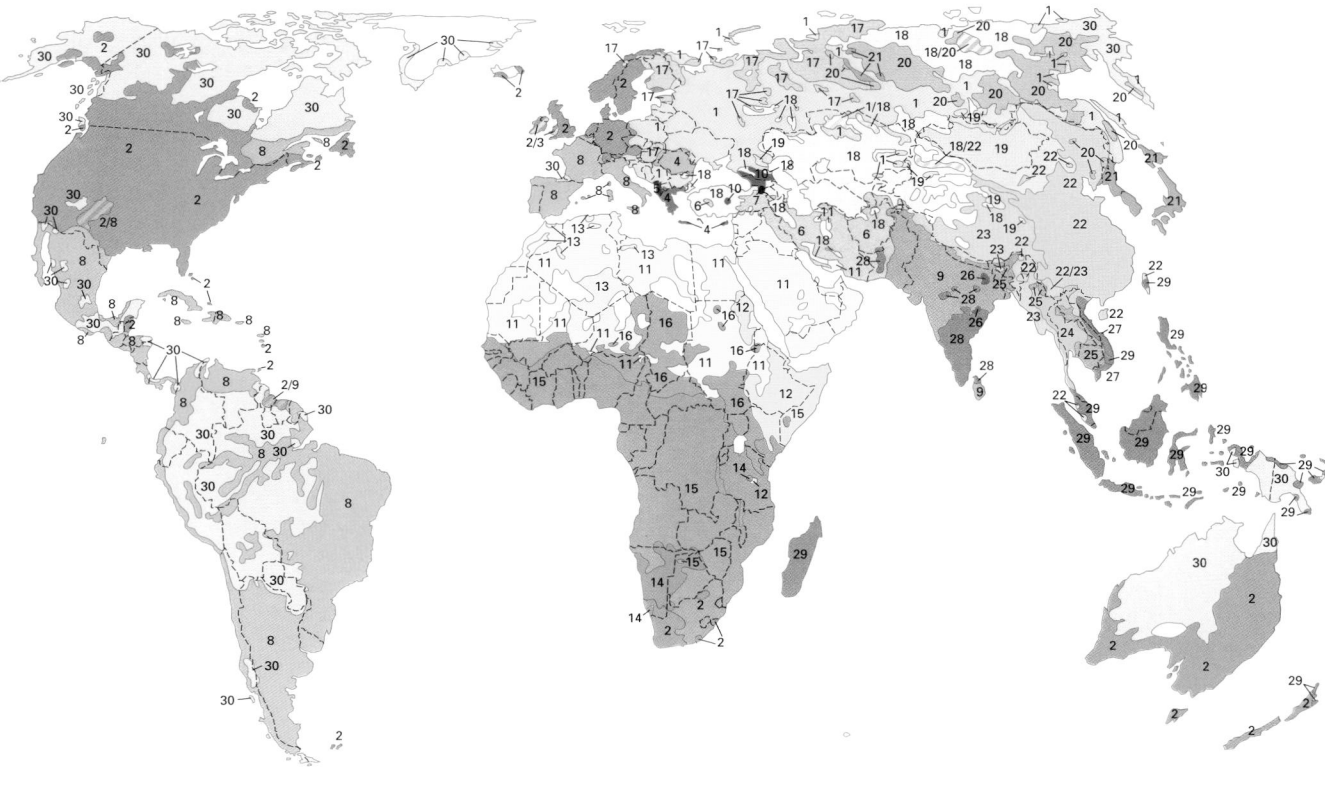

- Roman Catholicism
- Orthodox and other Eastern Churches
- Protestantism
- Sunni Islam
- Shia Islam
- Buddhism
- Hinduism
- Confucianism
- Judaism
- Shintoism
- Primitive Religions

Religions are not as easily mapped as the physical contours of landscape. Divisions are often blurred and frequently overlapping: most nations include people of many different faiths – or no faith at all. Some religions, like Islam and Christianity, have proselytes worldwide; others, like Hinduism and Confucianism, are restricted to a particular area, though modern migrations have taken some Indians and Chinese very far from their cultural origins. It is also difficult to show the degree to which religion exercises control over daily life: Christian Western Europe, for example, is nowadays far less dominated by its religion than are the Islamic nations of the Middle East. Similarly, figures for the major faiths' adherents make no distinction between nominal believers enrolled at birth and those for whom religion is a vital part of existence.

RELIGIOUS ADHERENTS

Christian	1,667m
Roman Catholic	952m
Protestant	337m
Orthodox	162m
Anglican	70m
Other Christian	148m
Muslim	881m
Sunni	841m
Shia	40m
Hindu	663m
Buddhist	312m
Chinese Folk	172m
Tribal	92m
Jewish	18m
Sikhs	17m

PEOPLE: CONFLICT & COOPERATION

Humans are social animals, rarely functioning well except in groups. Evolution has made them so: hunter-gatherers in cooperative bands were far more effective than animals that prowled alone. Agriculture, the building of cities and industrialization are all developments that depended on human cooperative ability – and in turn increased the need for it.

Unfortunately, human groups do not always cooperate so well with other human groups, and friction between them sometimes leads to cooperatively organized violence. War is itself a very human activity, with no real equivalent in any other species. Always murderous, it is sometimes purposeful and

may even be very effective. The colonization of the Americas and Australia, for example, was in effect the waging of aggressive war by well-armed Europeans against indigenous peoples incapable of offering a serious defense.

Most often, war achieves little but death and ruin. The great 20th-century wars accomplished nothing for the nations involved in them, although the world paid a price of between 50 and 100 million dead as well as immense material damage. The relative peace in the postwar developed world is at least partly due to the nuclear weapons with which rival powers have armed themselves – weapons so powerful that their

use would leave a scarcely habitable planet with no meaningful distinction between victor and vanquished.

Yet warfare remains endemic: the second half of the 20th century was one of the bloodiest periods in history, and death by organized violence remains unhappily common. The map below attempts to show the serious conflicts that have scarred the Earth since 1945. Most are civil wars in poor countries, rather than international conflicts between rich ones; some of them are still unresolved, while others, like apparently extinct volcanoes, may erupt again at intervals, adding to the world's miserable population of refugees.

THE WORLD'S REFUGEES

Refugees and their national origin; the host nations and the relative size of their refugee populations (1991)

■ Refugees in millions

◔ Refugees as a proportion of host country's population

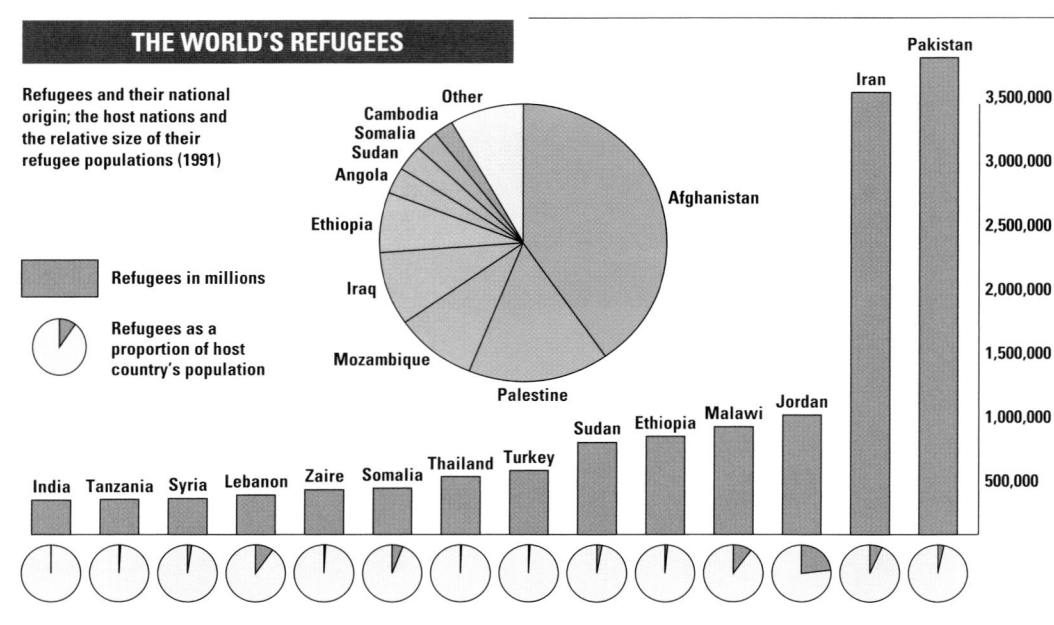

The pie-chart shows the origins of the world's refugees, while the bar-chart shows their destinations. According to the United Nations High Commissioner for Refugees, in 1990 there were almost 15 million refugees, a number that has continued to increase and is almost certain to be amplified during the decade. Some have fled from climatic change, some from economic disaster and others from political persecution; the great majority, however, are the victims of war.

All but a few who make it overseas seek asylum in neighboring countries, which are often the least equipped to deal with them and where they are rarely welcome. Lacking any rights or power, they frequently become an embarrassment and a burden to their reluctant hosts.

Usually, the best any refugee can hope for is rudimentary food and shelter in temporary camps that all too often become semipermanent, with little prospect of assimilation by host populations: many Palestinians, for example, have been forced to live in camps since 1948.

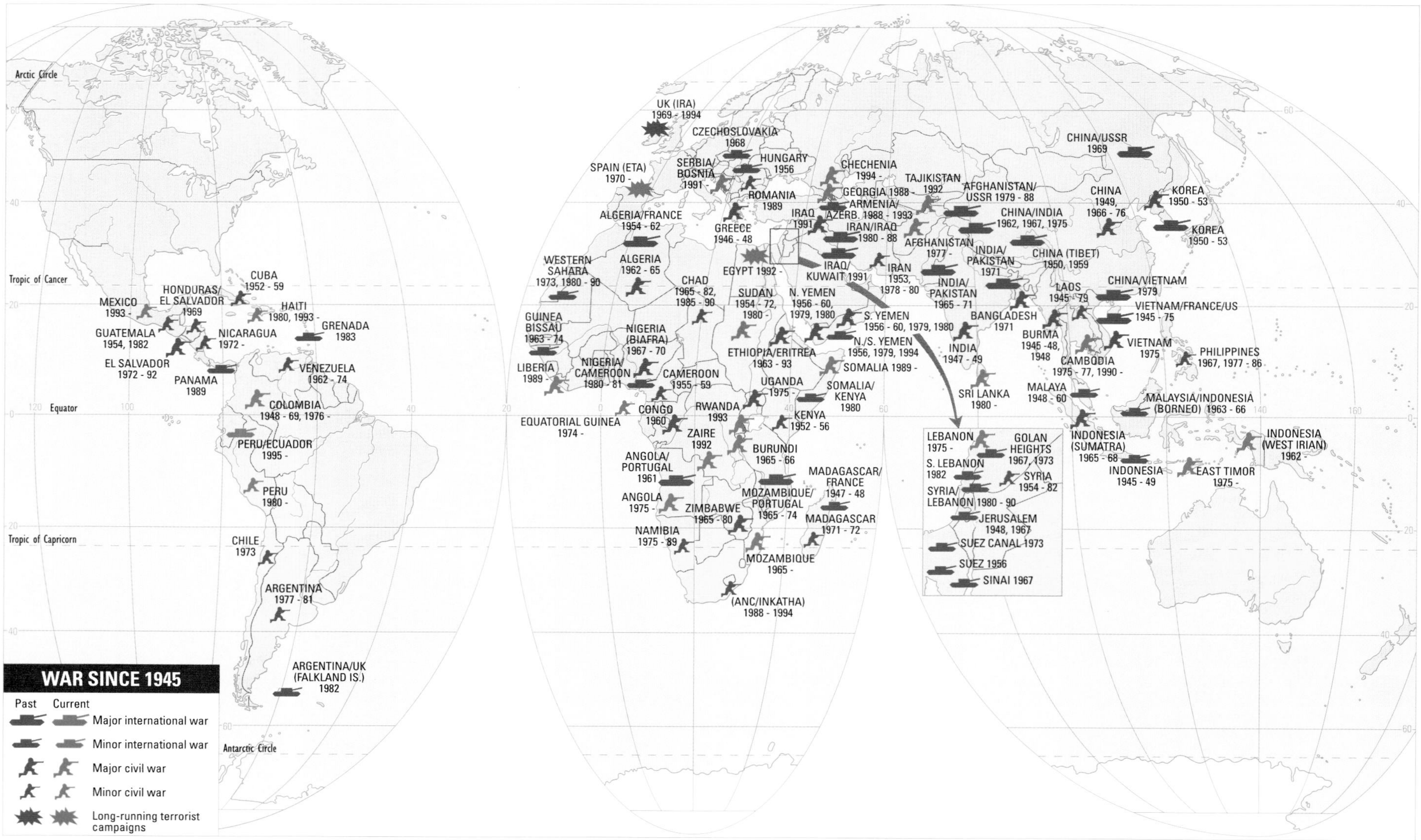

WAR SINCE 1945

Past	Current	
		Major international war
		Minor international war
		Major civil war
		Minor civil war
		Long-running terrorist campaigns

CARTOGRAPHY BY PHILIP'S. COPYRIGHT REED INTERNATIONAL BOOKS LTD

UNITED NATIONS

The United Nations Organization was born as World War II drew to its conclusion. Six years of strife had strengthened the world's desire for peace, but an effective international organization was needed to help achieve it. That body would replace the League of Nations which, since its inception in 1920, had signally failed to curb the aggression of at least some of its member nations. At the United Nations Conference on International Organization held in San Francisco, the United Nations Charter was drawn up. Ratified by the Security Council and signed by the 51 original members, it came into effect on 24 October 1945.

The Charter set out the aims of the organization: to maintain peace and security, and develop friendly relations between nations; to achieve international cooperation in solving economic, social, cultural and humanitarian problems; to promote respect for human rights and fundamental freedoms; and to harmonize the activities of nations in order to achieve these common goals.

By 1995, the UN had expanded to 185 member countries; it is the largest international political organization, employing over 25,000 people worldwide; its headquarters in New York accounts for 7,000 staff and it also has major offices in Rome, Geneva and Vienna.

The United Nations has six principal organs:

The General Assembly
The forum at which member nations discuss moral and political issues affecting world development, peace and security meets annually in September, under a newly-elected President whose tenure lasts one year. Any member can bring business to the agenda, and each member nation has one vote. Decisions are made by simple majority, save for matters of very great importance, when a two-thirds majority is required.

The Security Council
A legislative and executive body, the Security Council is the primary instrument for establishing and maintaining international peace by attempting to settle disputes between nations. It has the power to dispatch UN forces to stop aggression, and member nations undertake to make armed forces, assistance and facilities available as required. The Security Council has ten temporary members elected by the General Assembly for two-year terms, and five permanent members – China, France, Russia, UK and USA.

The Economic and Social Council
By far the largest United Nations executive, the Council operates as a conduit between the General Assembly and the many United Nations agencies it instructs to implement Assembly decisions, and whose work it coordinates. The Council also sets up commissions to examine economic conditions, collects data and issues studies and reports, and may make recommendations to the Assembly.

The Secretariat
This is the staff of the United Nations, and its task is to administer the policies and programs of the UN and its organs, and assist and advise the Head of the Secretariat, the Secretary-General – a full-time, non-political, appointment made by the General Assembly.

The Trusteeship Council
The Council administers trust territories with the aim of promoting their advancement. Only one remains – the Trust Territory of the Pacific Is. (Palau).

The International Court of Justice (the World Court)
The World Court is the judicial organ of the United Nations. It deals only with United Nations disputes and all members are subject to its jurisdiction. There are 15 judges, elected for nine-year terms by the General Assembly and the Security Council. The Court sits in The Hague.

United Nations agencies and programs, and intergovernmental agencies coordinated by the UN, contribute to harmonious world development. Social and humanitarian operations include:

United Nations Development Program (UNDP) Plans and funds projects to help developing countries make better use of resources.
United Nations International Childrens' Fund (UNICEF) Created at the General Assembly's first session in 1945 to help children in the aftermath of World War II, it now provides basic health care and aid worldwide.
United Nations Fund for Population Activities (UNFPA) Promotes awareness of population issues and family planning, providing appropriate assistance.
Food and Agriculture Organization (FAO) Aims to raise living standards and nutrition levels in rural areas by improving food production and distribution.
United Nations Educational, Scientific and Cultural Organization (UNESCO) Promotes international cooperation through broader and better education.
World Health Organization (WHO) Promotes and provides for better health care, public and environmental health and medical research.

Membership There are seven independent states which are not members of the UN – Kiribati, Nauru, Switzerland, Taiwan, Tonga, Tuvalu and Vatican City. Official languages are Chinese, English, French, Russian, Spanish and Arabic.
Funding The UN budget for 1994–95 is US $2.6 billion. Contributions are assessed by the members' ability to pay, with the maximum 25% of the total, the minimum 0.01%. Contributions for 1992–94 were: USA 25%, Japan 12.45%, Germany 8.93%, Russia 6.71%, France 6%, UK 5.02%, Italy 4.29%, Canada 3.11% (others 28.49%).
Peacekeeping The UN has been involved in 33 peacekeeping operations worldwide since 1948 and there are currently 17 areas of UN patrol. In July 1993 there were 80,146 'blue berets' from 74 countries.

United Nations agencies are involved in many aspects of international trade, safety and security:

General Agreement on Tariffs and Trade (GATT) Sponsors international trade negotiations and advocates a common code of conduct.
International Maritime Organization (IMO) Promotes unity amongst merchant shipping, especially in regard to safety, marine pollution and standardization.
International Labor Organization (ILO) Seeks to improve labor conditions and promote productive employment to raise living standards.
World Meteorological Organization (WMO) Promotes co-operation in weather observation, reporting and forecasting.
World Intellectual Property Organization (WIPO) Seeks to protect intellectual property such as artistic copyright, scientific patents and trademarks.
Disarmament Commission Considers and makes recommendations to the General Assembly on disarmament issues.
International Atomic Energy Agency (IAEA) Fosters development of peaceful uses for nuclear energy, establishes safety standards and monitors the destruction of nuclear material designed for military use.

The World Bank comprises three United Nations agencies:

International Monetary Fund (IMF) Cultivates international monetary cooperation and expansion of trade.
International Bank for Reconstruction and Development (IBRD) Provides funds and technical assistance to developing countries.
International Finance Corporation (IFC) Encourages the growth of productive private enterprise in less developed countries.

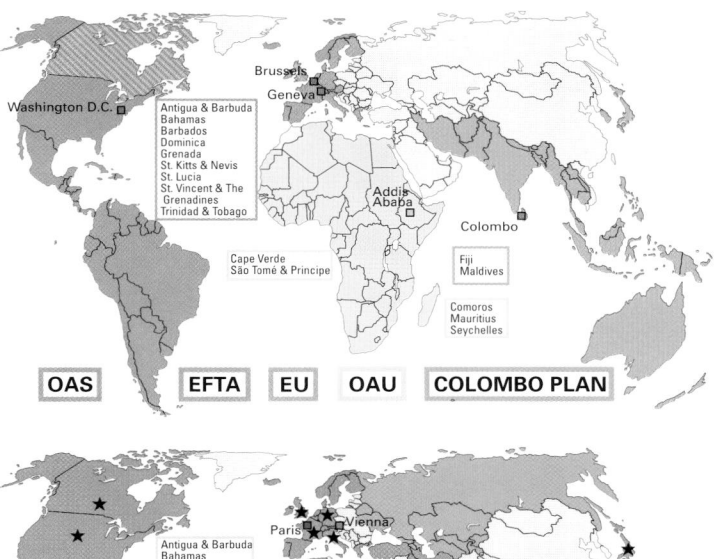

Antigua & Barbuda
Bahamas
Barbados
Dominica
Grenada
St. Kitts & Nevis
St. Lucia
St. Vincent & The Grenadines
Trinidad & Tobago

Cape Verde
São Tomé & Principe

Fiji
Maldives

Comoros
Mauritius
Seychelles

| OAS | EFTA | EU | OAU | COLOMBO PLAN |

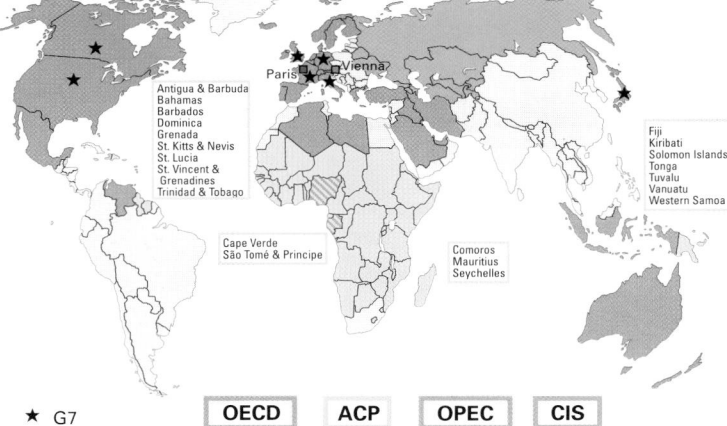

Antigua & Barbuda
Bahamas
Barbados
Dominica
Grenada
St. Kitts & Nevis
St. Lucia
St. Vincent & Grenadines
Trinidad & Tobago

Cape Verde
São Tomé & Principe

Comoros
Mauritius
Seychelles

Fiji
Kiribati
Solomon Islands
Tonga
Tuvalu
Vanuatu
Western Samoa

★ G7 | OECD | ACP | OPEC | CIS |

Cyprus
Malta

Antigua & Barbuda
Bahamas
Barbados
Dominica
Grenada
St. Kitts-Nevis
St. Lucia
St. Vincent & Grenadines
Trinidad & Tobago

Bahrain
Comoros
Palestine

Brunei
Maldives
Mauritius
Seychelles
Singapore

W. Samoa
Nauru
Tonga
Solomon Is.
Tuvalu
Kiribati
Vanuatu

| NATO | LAIA | ARAB LEAGUE | COMMONWEALTH | ASEAN |

EU As from December 1993 the European Union (EU) refers to matters of foreign policy, security and justice. The European Community (EC) refers to all other matters. The 15 members – Austria, Belgium, Denmark, Finland, France, Germany, Greece, Ireland, Italy, Luxembourg, Netherlands, Portugal, Spain, Sweden and the UK – aim to integrate economies, coordinate social developments and bring about political union. These members of what is now the world's biggest market share agricultural and industrial policies and tariffs on trade.
EFTA European Free Trade Association (formed in 1960). Portugal left the original 'Seven' in 1989 to join the EC, followed by Austria, Finland and Sweden in 1995. There are now only four members: Norway, Iceland, Liechtenstein and Switzerland.
ACP African-Caribbean-Pacific countries associated with the EC (1963).
NATO North Atlantic Treaty Organization (formed in 1949). It continues after 1991 despite the winding up of the Warsaw Pact.
OAS Organization of American States (1948). It aims to promote social and economic cooperation between developed countries of North America and developing nations of Latin America.
ASEAN Association of Southeast Asian Nations (1967).
OAU Organization of African Unity (1963). Its 53 members represent over 94% of Africa's population.
LAIA Latin American Integration Association (1980) superceded the Latin American Free Trade Association formed in 1961.
OECD Organization for Economic Cooperation and Development (1961). The 25 major Western free-market economies.* 'G7' is its 'inner group' of USA, Canada, Japan, UK, Germany, Italy and France. *Mexico joined in May 1994.
COMMONWEALTH The Commonwealth of Nations evolved from the British Empire; it comprises 19 nations recognizing the British monarch as head of state and 32 with their own heads of state.
CIS The Commonwealth of Independent States (1991) comprises the countries of the former Soviet Union except for Estonia, Latvia and Lithuania.
OPEC Organization of Petroleum Exporting Countries (1960). It controls about three-quarters of the world's oil supply. Ecuador withdrew formally on 1 January 1993.
ARAB LEAGUE (1945) The League's aim is to promote economic, social, political and military cooperation.
COLOMBO PLAN (1951) Its 26 members aim to promote economic and social development in Asia and the Pacific.

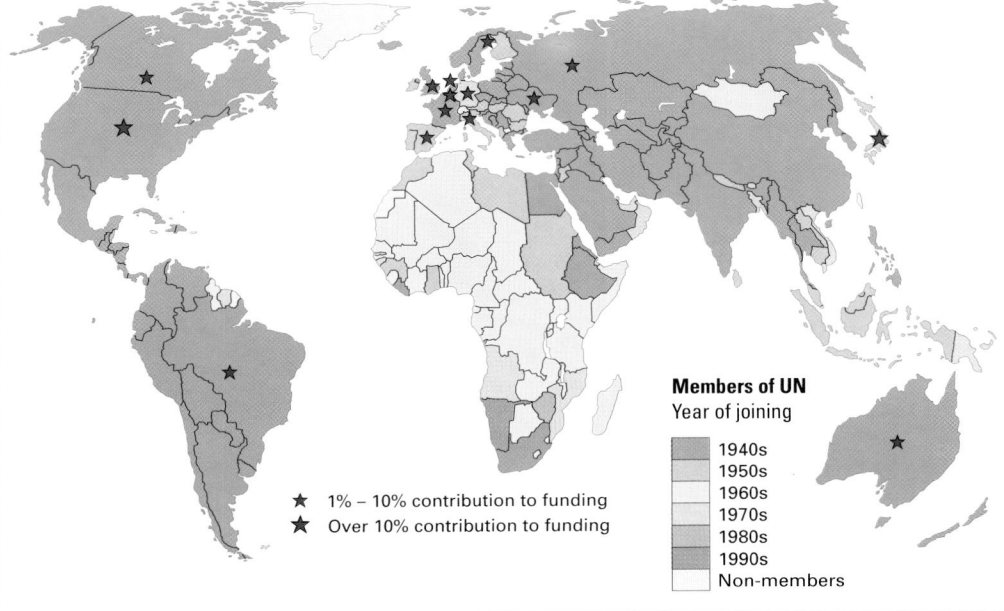

Members of UN
Year of joining

- 1940s
- 1950s
- 1960s
- 1970s
- 1980s
- 1990s
- Non-members

★ 1% – 10% contribution to funding
★ Over 10% contribution to funding

PRODUCTION: AGRICULTURE

The invention of agriculture transformed human existence more than any other development, though it may not have seemed much of an improvement to its first practitioners. Primitive farming required brutally hard work, and it tied men and women to a patch of land, highly vulnerable to local weather patterns and to predators, especially human predators – drawbacks still apparent in much of the world today. It is difficult to imagine early humans being interested in such an existence while there were still animals around to hunt and wild seeds and berries to gather. Probably the spur was population pressure, with consequent overhunting and scarcity.

Despite its difficulties, the new life style had a few overwhelming advantages. It supported far larger populations, eventually including substantial cities, with all the varied cultural and economic activities they allowed. Later still, it furnished the surpluses that allowed industrialization – another enormous step in the course of human development.

Machines relieved many farmers of their burden of endless toil, and made it possible for relatively small numbers to provide food for more than 5,000 million people.

Now, as in the past, the whole business of farming involves the creation of a severely simplified ecology, under the tutelage and for the benefit of the farmer. Natural plant life is divided into crops, to be protected and nurtured, and weeds, the rest, to be destroyed. From the earliest days, crops were selectively bred to increase their food yield, usually at the expense of their ability to survive, which became the farmer's responsibility; 20th-century plant geneticists have carried the technique to highly productive extremes. Due mainly to new varieties of rice and wheat, world grain production has increased by 70% since 1965, more than doubling in the developing countries, although such high yields demand equally high consumption of fertilizers and pesticides to maintain them. Mechanized farmers in North America and Europe continue to turn out huge surpluses, although not without environmental costs.

Where production is inadequate, the reasons are as likely to be political as agricultural. Africa, the only continent where food production per capita is actually falling, suffers acutely from economic mismanagement, as well as from the perennial problems of war and banditry. Dismal harvests in the USSR, despite its excellent farmland, helped bring about the collapse of the Soviet system.

There are other limits to progress too. Increasing population puts relentless pressure on farmers not only to maintain high yields but also to increase them. Most of the world's potential cropland is already under the plow. The overworking of marginal land is one of the prime causes of desertification; new farmlands burned out of former rain forests are seldom fertile for long. Human numbers may yet outrun the land's ability to feed them, as they did almost 10,000 years ago.

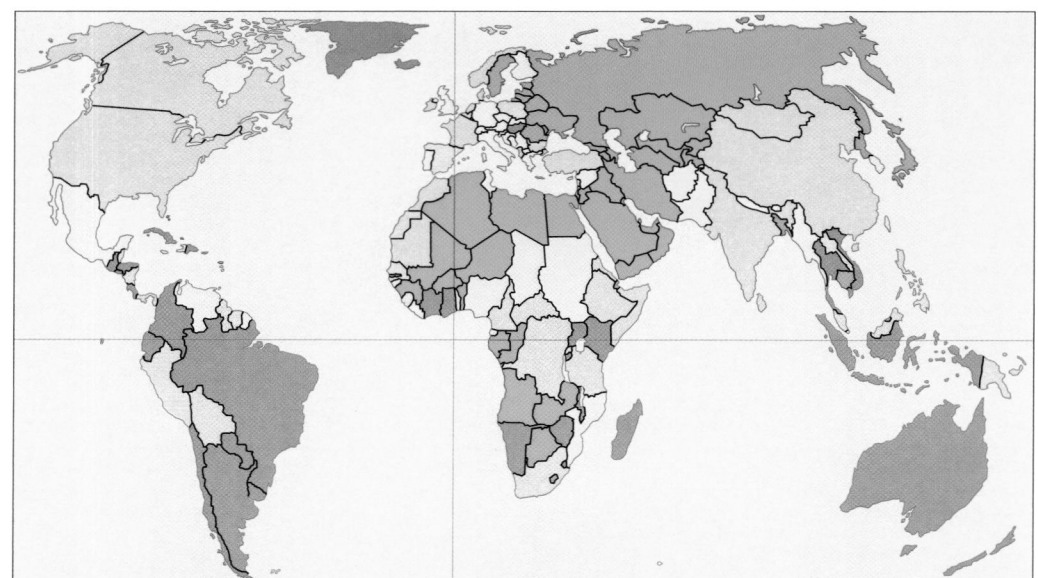

SELF-SUFFICIENCY IN FOOD

Balance of trade in food products as a percentage of total trade in food products (latest available year)

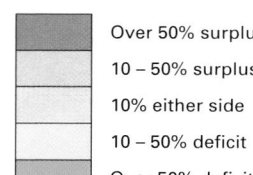

	Over 50% surplus
	10 – 50% surplus
	10% either side
	10 – 50% deficit
	Over 50% deficit

Most self-sufficient		Least self-sufficient	
Argentina	95%	Algeria	–98%
Zimbabwe	87%	Djibouti	–97%
Honduras	81%	Yemen	–95%
Malawi	81%	Zambia	–95%
Costa Rica	79%	Japan	–91%
Iceland	78%	Gabon	–90%
Chile	75%	Kuwait	–75%
Uruguay	75%	Brunei	–89%
Ecuador	74%	Burkina Faso	–82%

LAND USE

	Arable
	Arable and pasture
	Market gardening
	Woods and forests
	Rough grazing
	Non-productive
	Pasture
	Savanna
	Fishing
	Industrial areas

STAPLE CROPS

Wheat: Grown in a range of climates, with most varieties – including the highest-quality bread wheats – requiring temperate conditions. Mainly used in baking, it is also used for pasta and breakfast cereals.

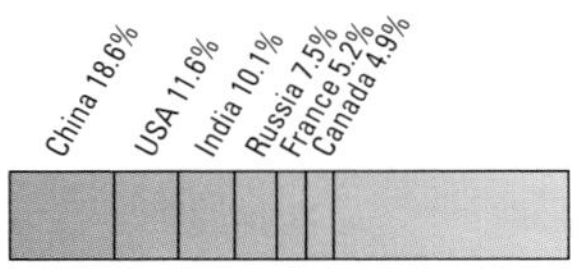

China 18.6%
USA 11.6%
India 10.1%
Russia 7.5%
France 5.2%
Canada 4.9%

World total (1993): 620,902,700 tons

Maize: Originating in the New World and still an important human food in Africa and Latin America, in the developed world it is processed into breakfast cereals, oil, starches and adhesives. It is also used for animal feed.

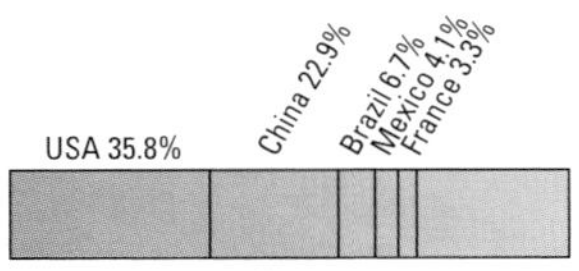

USA 35.8%
China 22.9%
Brazil 6.7%
Mexico 4.1%
France 3.3%

World total (1993): 495,627,000 tons

Oats: Most widely used to feed livestock, but eaten by humans as oatmeal or porridge. Oats have a beneficial effect on the cardio-vascular system, and human consumption is likely to increase.

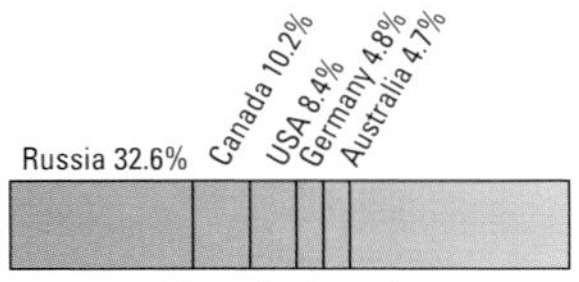

Russia 32.6%
Canada 10.2%
USA 8.4%
Germany 4.8%
Australia 4.7%

World total (1993): 38,987,300 tons

Millet: The name covers a number of small grained cereals, members of the grass family with a short growing season. Used to produce flour, meal and animal feed, and fermented to make beer, especially in Africa.

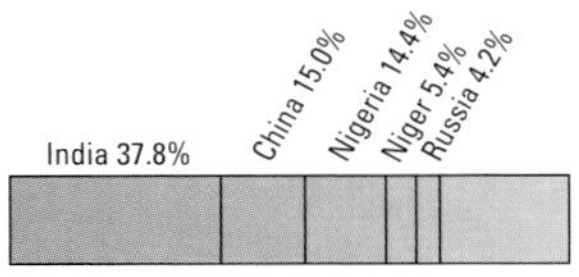

India 37.8%
China 15.0%
Nigeria 14.4%
Niger 5.4%
Russia 4.2%

World total (1993): 29,086,200 tons

Cassava: A tropical shrub that needs high rainfall (over 125 inches annually) and a 10–30 month growing season to produce its large, edible tubers. Used as flour by humans, as cattle feed and in industrial starches.

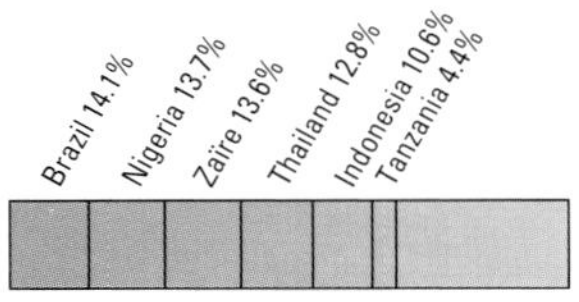

Brazil 14.1%
Nigeria 13.7%
Zaïre 13.6%
Thailand 12.8%
Indonesia 10.6%
Tanzania 4.4%

World total (1993): 168,990,800 tons

Rice: Thrives on the high humidity and temperatures of the Far East, where it is the traditional staple food of half the human race. Usually grown standing in water, rice responds well to continuous cultivation, with three or four crops annually.

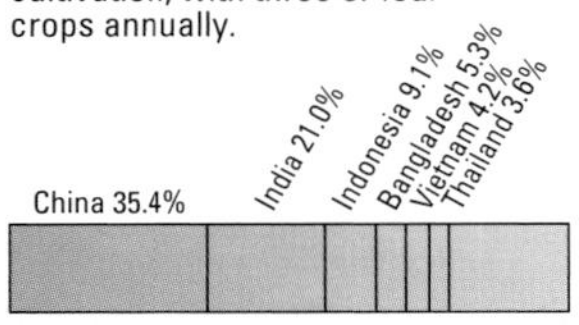

China 35.4%
India 21.0%
Indonesia 9.1%
Bangladesh 5.3%
Vietnam 4.2%
Thailand 3.6%

World total (1993): 580,154,300 tons

Barley: Primarily used as animal feed, but widely eaten by humans in Africa and Asia. Elsewhere, malted barley furnishes beer and spirits. Able to withstand the dry heat of subarid tropics, its growing season is only 80 days.

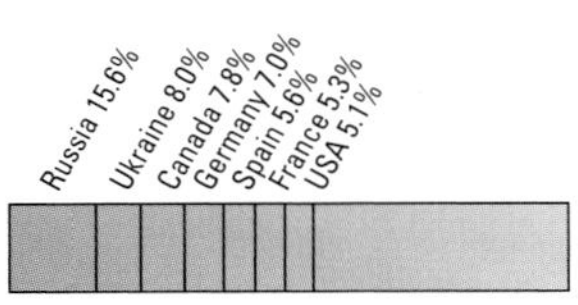

Russia 15.6%
Ukraine 8.0%
Canada 7.8%
Germany 7.0%
Spain 5.6%
France 5.2%
USA 5.1%

World total (1993): 187,400,400 tons

Rye: Hardy and tolerant of poor and sandy soils, it is an important foodstuff and animal feed in Central and Eastern Europe. Rye produces a dark, heavy bread as well as alcoholic drinks.

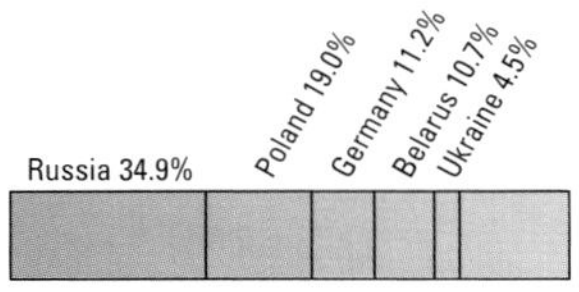

Russia 34.9%
Poland 19.0%
Germany 11.2%
Belarus 10.7%
Ukraine 4.5%

World total (1993): 28,820,000 tons

Potatoes: The most important of the edible tubers, potatoes grow in well-watered, temperate areas. Weight for weight less nutritious than grain, they are a human staple as well as an important animal feed.

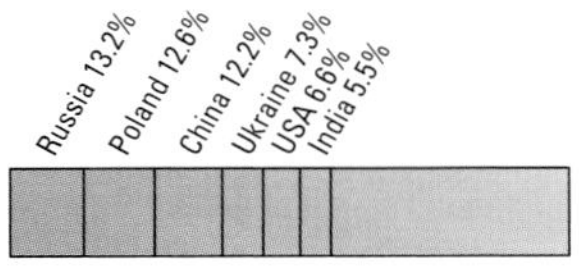

Russia 13.2%
Poland 12.6%
China 12.2%
Ukraine 7.3%
USA 6.6%
India 5.5%

World total (1993): 317,001,300 tons

Soya: Beans from soya bushes are very high (30–40%) in protein. Most are processed into oil and proprietary protein foods. Consumption since 1950 has tripled, mainly due to the health-conscious developed world.

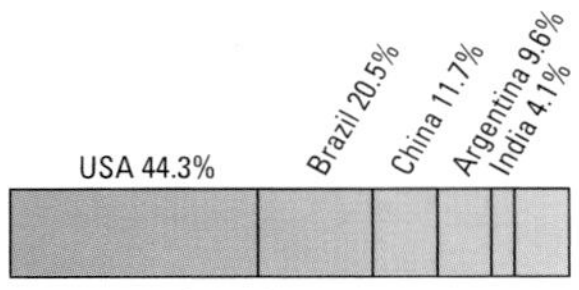

USA 44.3%
Brazil 20.5%
China 11.7%
Argentina 9.6%
India 4.1%

World total (1993): 122,112,100 tons

Cereals are grasses with starchy, edible seeds; every important civilization has depended on them as a source of food. The major cereal grains contain about 10% protein and 75% carbohydrate; grain is easy to store, handle and transport, and contributes more than any other group of foods to the energy and protein content of human diet. If all the cereals were consumed directly by man, there would be no shortage of food in the world, but a considerable proportion of the total output is used as animal feed.

Starchy tuber crops or root crops, represented here by potatoes and cassava, are second in importance only to cereals as staple foods; easily cultivated, they provide high yields for little effort and store well – potatoes for up to six months, cassava for up to a year in the ground. Protein content is low (2% or less) and starch content high; some minerals and vitamins are present, but populations that rely heavily on these crops may suffer from malnutrition.

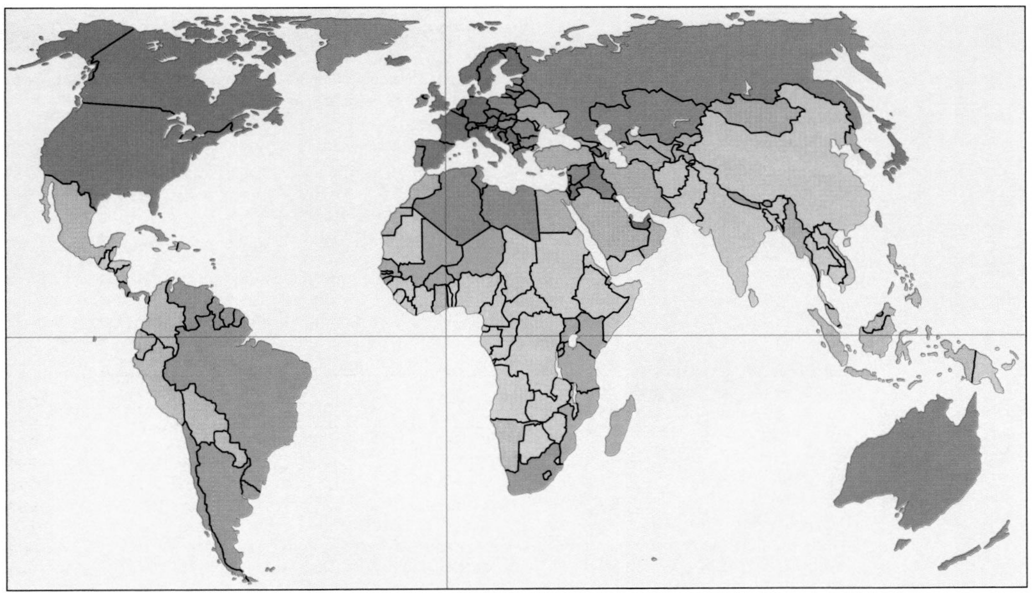

SELF-SUFFICIENCY IN FOOD

Percentage of the total population dependent on agriculture (1991)

- Over 75% dependent
- 50 – 75% dependent
- 25 – 50% dependent
- 10 – 25% dependent
- Under 10% dependent

Top 5 countries		Bottom 5 countries	
Nepal	92%	Singapore	0.9%
Rwanda	91%	Hong Kong	1.2%
Burundi	91%	Bahrain	1.7%
Bhutan	91%	Belgium	1.7%
Niger	87%	UK	1.9%

FOOD & POPULATION

Comparison of food production and population by continent (latest available year). The left column indicates percentage shares of total world food production; the right shows population in proportion.

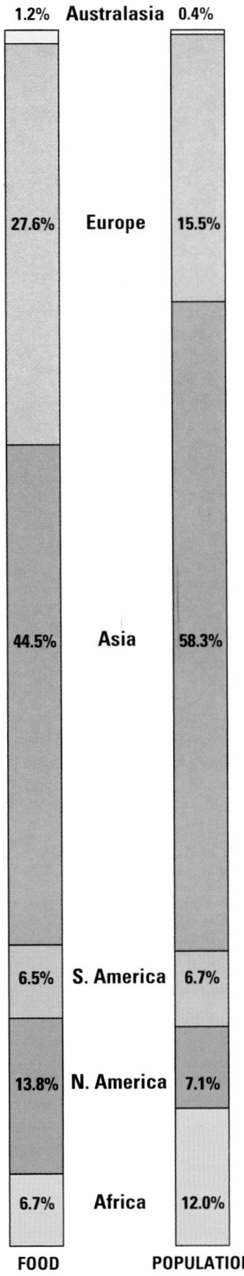

1.2% Australasia 0.4%
27.6% Europe 15.5%
44.5% Asia 58.3%
6.5% S. America 6.7%
13.8% N. America 7.1%
6.7% Africa 12.0%
FOOD
POPULATION

ANIMAL PRODUCTS

Traditionally, food animals subsisted on land unsuitable for cultivation, supporting agricultural production with their fertilizing dung. But free-ranging animals grow slowly and yield less meat than those more intensively reared; the demands of urban markets in the developed world have encouraged the growth of factory-like production methods.

A large proportion of staple crops, especially cereals, are fed to animals, an inefficient way to produce protein but one likely to continue as long as people value meat and dairy products in their diet.

Cheese: Least perishable of all dairy products, cheese is milk fermented with selected bacterial strains to produce a foodstuff with a potentially immense range of flavors and textures. The vast majority of cheeses are made from cow's milk, although sheep and goat cheeses are highly prized.

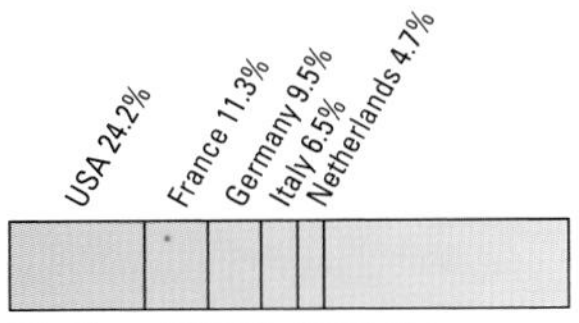

USA 24.2%
France 11.3%
Germany 9.5%
Italy 6.5%
Netherlands 4.7%

World total (1993): 14,886,300 tons

Lamb and Mutton: Sheep are the least demanding of domestic animals. Although unsuited to intensive rearing, they can thrive on marginal pastureland incapable of supporting beef cattle on a commercial scale. Sheep are raised as much for their valuable wool as for the meat that they provide, with Australia the world leader.

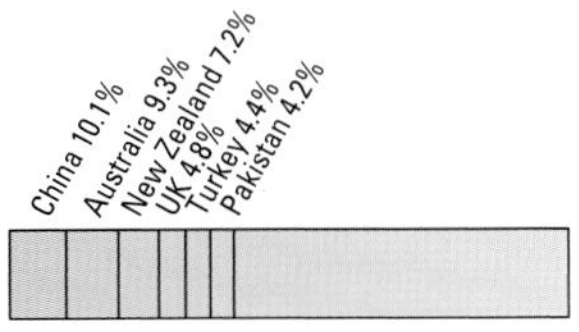

China 10.1%
Australia 9.3%
New Zealand 7.2%
UK 4.8%
Turkey 4.4%
Pakistan 4.2%

World total (1993): 7,605,400 tons

Beef and Veal: Most beef and veal is reared for home markets, and the top five producers are also the biggest consumers. The USA produces nearly a quarter of the world's beef and eats even more.

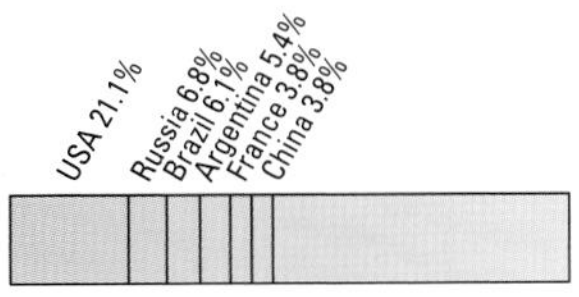

USA 21.1%
Russia 6.8%
Brazil 6.1%
Argentina 5.4%
France 3.8%
China 3.8%

World total (1993): 55,262,900 tons

SUGARS

Sugarcane: Confined to tropical regions, cane sugar accounts for the bulk of international trade in the commodity. Most is produced as a foodstuff, but some countries, notably Brazil and South Africa, distill sugarcane and use the resulting ethyl alcohol to make motor fuels.

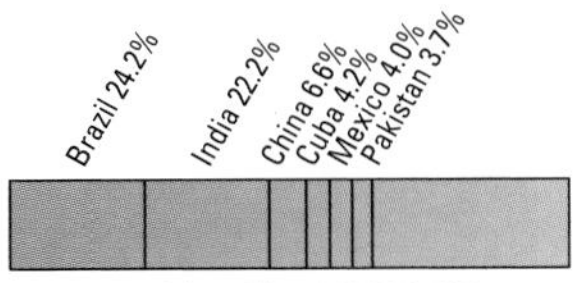

Brazil 24.2%
India 22.2%
China 6.8%
Cuba 4.2%
Mexico 4.0%
Pakistan 3.7%

World total (1993): 1,144,660,000 tons

Milk: Many human groups, including most Asians, find raw milk indigestible after infancy, and it is often only the starting point for other dairy products such as butter, cheese and yogurt. Most world production comes from cows, but sheep's milk and goats' milk are also important.

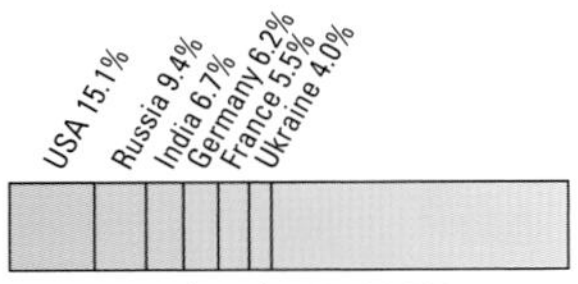

USA 15.1%
Russia 9.4%
India 6.7%
Germany 6.2%
France 5.5%
Ukraine 4.0%

World total (1993): 499,106,300 tons

Butter: A traditional source of vitamin A as well as calories, butter has lost much popularity in the developed world for health reasons, although it remains a valuable food. Most butter from India, the world's largest producer, is clarified into ghee, which has religious as well as nutritional importance.

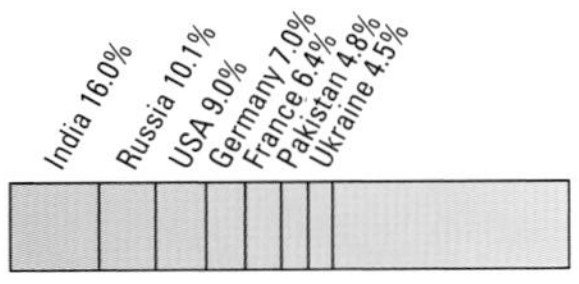

India 16.0%
Russia 10.1%
USA 9.0%
Germany 7.0%
France 6.4%
Pakistan 4.8%
Ukraine 4.5%

World total (1993): 7,651,600 tons

Pork: Although pork is forbidden to many millions, notably Muslims, on religious grounds, more is produced than any other meat in the world, mainly because it is the cheapest. It accounts for about 90% of China's meat output, although per capita meat consumption is relatively low.

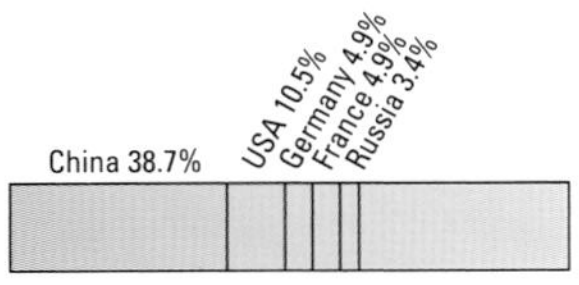

China 38.7%
USA 10.5%
Germany 4.9%
France 4.9%
Russia 3.4%

World total (1993): 81,280,100 tons

Fish: Commercial fishing requires large shoals of fish, often of only one species, within easy reach of markets. Although the great majority are caught wild in the sea, fish-farming of both marine and freshwater species is assuming increasing importance, especially as natural stocks become depleted.

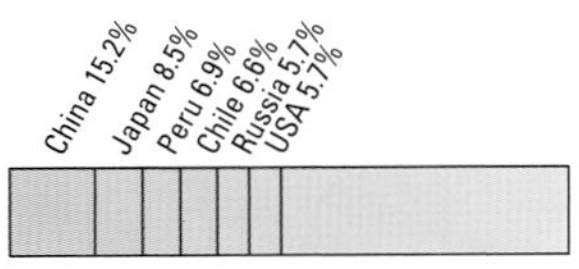

China 15.2%
Japan 8.5%
Peru 6.9%
Chile 6.6%
Russia 5.7%
USA 5.7%

World total (1993): 107,924,300 tons

Sugar beet: A temperate crop closely related to the humble beet, the root's yield after processing is indistinguishable from cane sugar. Sugar beet is steadily replacing sugarcane imports in Europe, to the detriment of the developing countries that rely on it as a major cash crop.

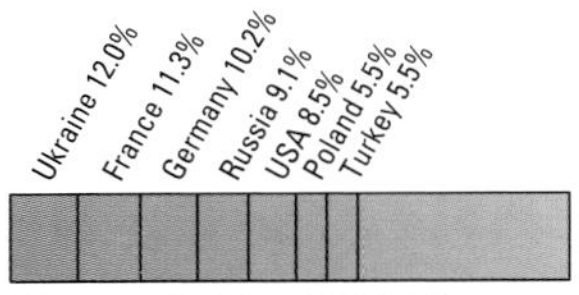

Ukraine 12.0%
France 11.3%
Germany 10.2%
Russia 9.1%
USA 8.5%
Poland 5.5%
Turkey 5.5%

World total (1993): 309,850,200 tons

PRODUCTION: ENERGY

We live in a high-energy civilization. While vast discrepancies exist between rich and poor – a North American consumes 13 times as much energy as a Chinese, for example – even developing nations have more power at their disposal than was imaginable a century ago. Abundant energy supplies keep us warm or cool, fuel our industries and our transport systems, and even feed us: high-intensity agriculture, with its fertilizers, pesticides and machinery, is heavily energy-dependent.

Unfortunately, most of the world's energy comes from fossil fuels: coal, oil and gas deposits laid down over many millions of years. These are the Earth's capital, not its income, and we are consuming that capital at an alarming rate. New discoveries have persistently extended the known reserves: in 1989, the reserves-to-production ratio for oil assured over 45 years' supply, an improvement of almost a decade on the 1970 situation. But despite the effort and ingenuity of prospectors, stocks are clearly limited. They are also very unequally distributed, with the Middle East accounting for most oil reserves, and the CIS, especially Russia, possessing an even higher proportion of the world's natural gas. Coal reserves are more evenly shared, and also more plentiful: coal will outlast oil and gas by a very wide margin.

It is possible to reduce energy demand by improving efficiency: most industrial nations have dramatically increased output since the 1970s without a matching rise in energy consumption. But as fossil stocks continue to diminish, renewable energy sources – solar, wave and wind power, as well as hydro-electricity – must take on greater importance.

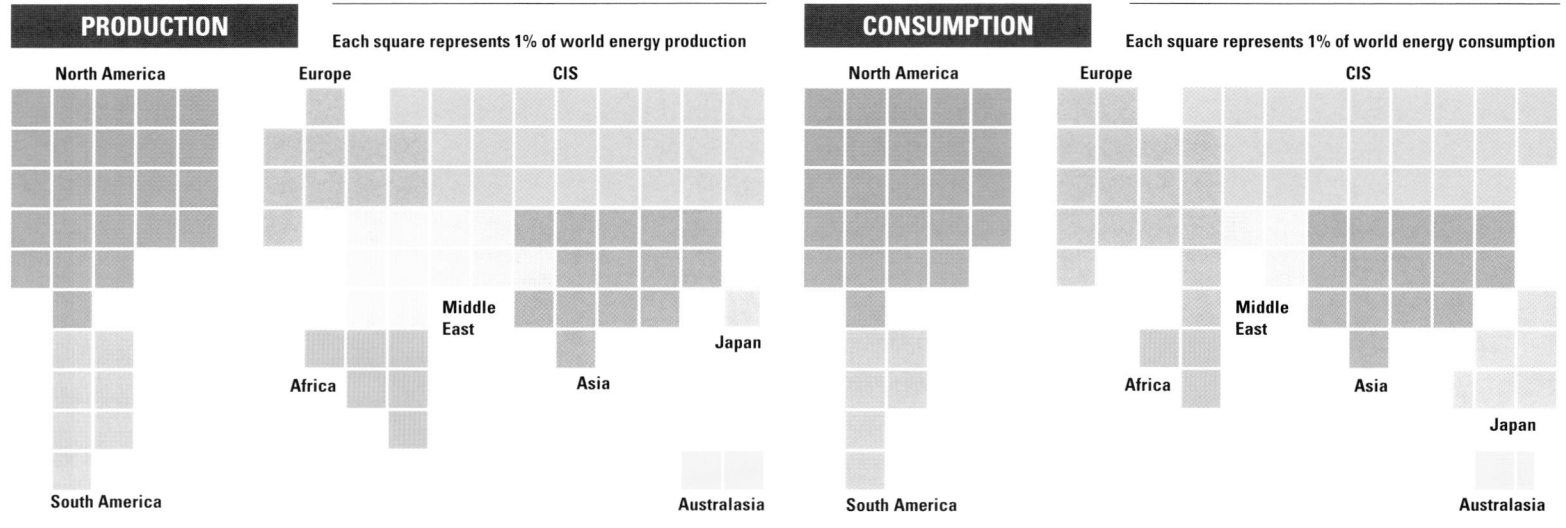

PRODUCTION — Each square represents 1% of world energy production

North America · Europe · CIS · Middle East · Japan · Africa · Asia · South America · Australasia

CONSUMPTION — Each square represents 1% of world energy consumption

North America · Europe · CIS · Middle East · Japan · Africa · Asia · South America · Australasia

CONVERSIONS

For historical reasons, oil is still traded in barrels. The weight and volume equivalents shown below are all based on average density 'Arabian light' crude oil, and should be considered approximate.

The energy equivalents given for a ton of oil are also somewhat imprecise: oil and coal of different qualities will have varying energy contents, a fact usually reflected in their price on world markets.

1 barrel:

0.15 tons
42 gallons
0.136 metric tonnes
159 liters

1 ton:

6.67 barrels
280 gallons
1,075 liters

1 ton oil:

1.5 tons hard coal
3.0 tons lignite
10,900 kWh

(1 ton = 2,000lb; 0.893 long tons; 0.907 metric tonnes)

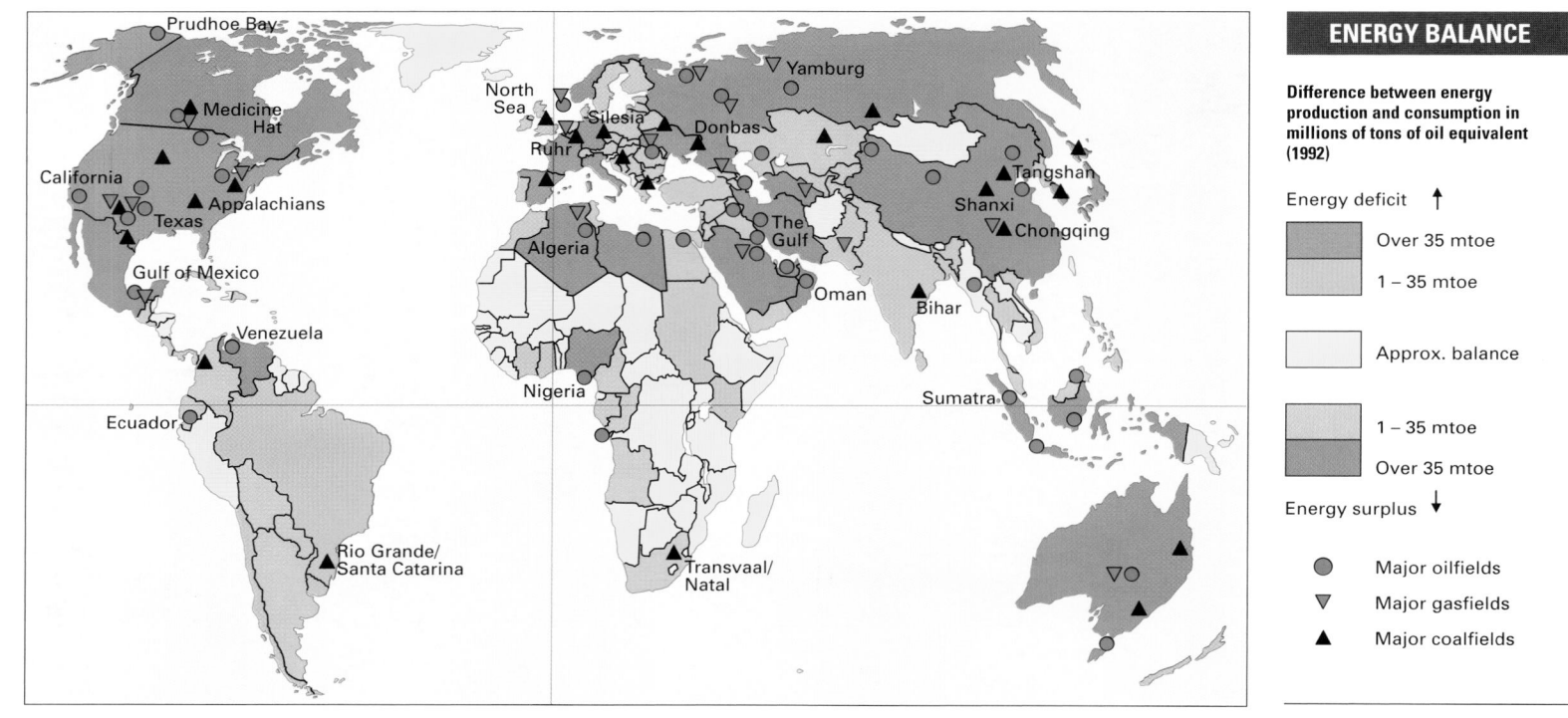

ENERGY BALANCE

Difference between energy production and consumption in millions of tons of oil equivalent (1992)

Energy deficit ↑

Over 35 mtoe
1 – 35 mtoe

Approx. balance

1 – 35 mtoe
Over 35 mtoe

Energy surplus ↓

● Major oilfields
▽ Major gasfields
▲ Major coalfields

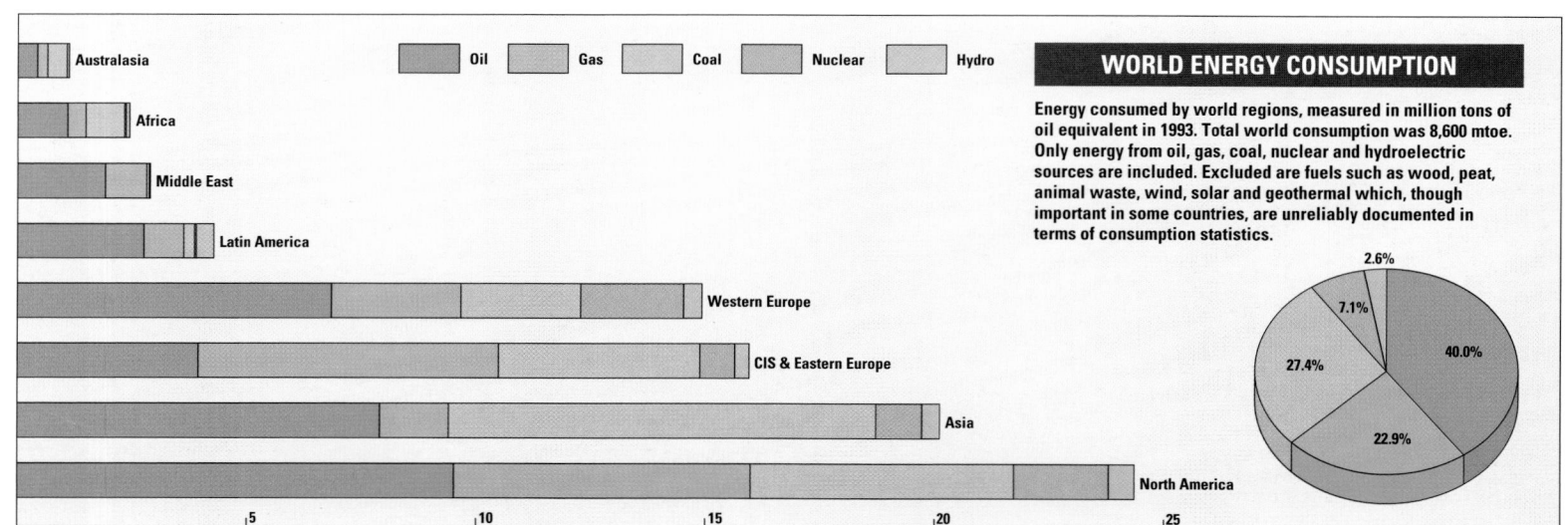

Australasia · Africa · Middle East · Latin America · Western Europe · CIS & Eastern Europe · Asia · North America

Oil · Gas · Coal · Nuclear · Hydro

WORLD ENERGY CONSUMPTION

Energy consumed by world regions, measured in million tons of oil equivalent in 1993. Total world consumption was 8,600 mtoe. Only energy from oil, gas, coal, nuclear and hydroelectric sources are included. Excluded are fuels such as wood, peat, animal waste, wind, solar and geothermal which, though important in some countries, are unreliably documented in terms of consumption statistics.

2.6% · 7.1% · 27.4% · 40.0% · 22.9%

FOSSIL FUEL RESERVES

Known world reserves in years as a multiple of annual production, 1970, 1980 and 1989

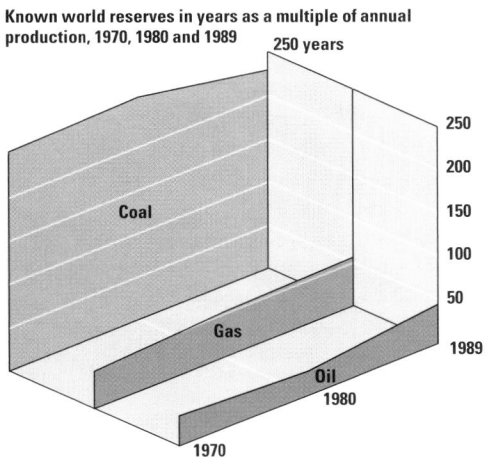

ENERGY AND OUTPUT

Tons of oil equivalent consumed to produce US $1,000 of GDP, four industrial nations (1973–89)

Intensity of energy use is a rough indicator of efficiency: the 1973–4 oil crisis caused a dramatic improvement in each of the countries illustrated, although the USA remains relatively profligate. Reliable figures for Russia and the other republics of the former USSR are hard to obtain, but estimates suggest that for equivalent production they use up to four times as much energy as the USA.

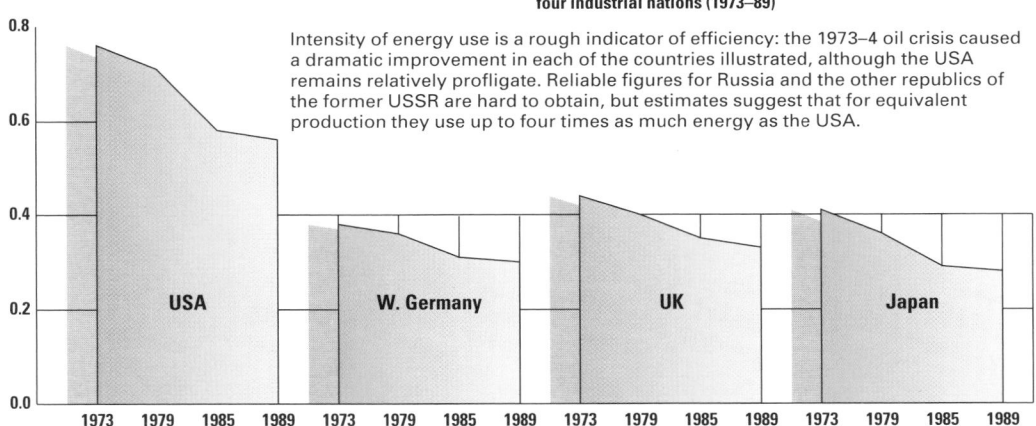

COAL RESERVES

World coal reserves by region and country, thousand million tons (latest available year)

GAS RESERVES

World natural gas reserves by region and country, thousand million tons (latest available year)

Ca: Canada
In: Indonesia
Ma: Malaysia
AD: Abu Dhabi
SA: Saudi Arabia
Qa: Qatar
Iq: Iraq
No: Norway
Ne: Netherlands
Ve: Venezuela
Mx: Mexico
Al: Algeria
Ni: Nigeria

OIL RESERVES

World oil reserves by region and country, thousand million tons (latest available year)

A: Abu Dhabi
Ve: Venezuela
M: Mexico

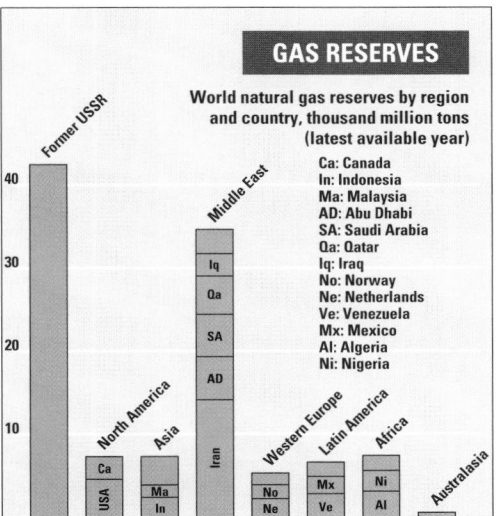
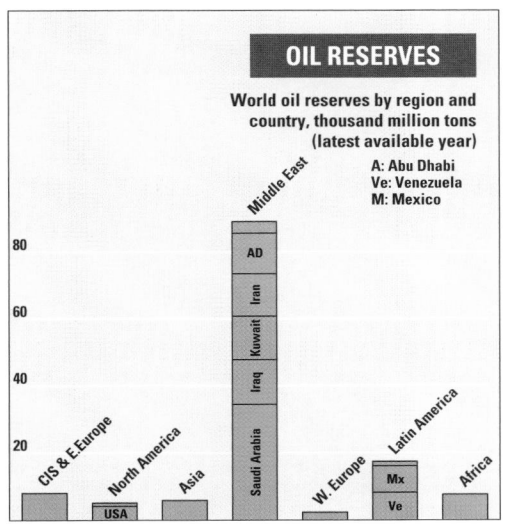

OIL MOVEMENTS

Major world movements of oil in millions of tons (1989)

Middle East to Western Europe	215.5
Middle East to Japan	165.3
Middle East to Asia (exc. Japan and China)	140.5
Latin America to USA	139.0
Middle East to USA	103.7
USSR to Western Europe	86.1
North Africa to Western Europe	103.1
West Africa to Western Europe	43.7
West Africa to USA	65.9
Canada to USA	49.6
South-east Asia to Japan	46.5
Latin America to Western Europe	31.6
Western Europe to USA	31.6
Middle East to Latin America	22.6

Total world movements: 1,738 million tons

Only inter-regional movements in excess of 20 million tons are shown. Other Middle Eastern oil shipments throughout the world totalled 52.2 million tons; miscellaneous oil exports of the then USSR amounted to 97.9 million tons.

FUEL EXPORTS

Fuels as a percentage of total value of exports (latest available year)

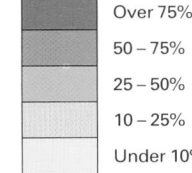

Over 75%
50 – 75%
25 – 50%
10 – 25%
Under 10%

Direction of trade

Coal
Oil

Arrows show the major trade direction of selected fuels, and are proportional to export value.

NUCLEAR POWER

Percentage of electricity generated by nuclear power stations, leading nations (1990)

1.	France	73%	11.	Finland	34%
2.	Belgium	60%	12.	Czechoslovakia	29%
3.	Sweden	52%	13.	Germany	28%
4.	Hungary	46%	14.	Japan	24%
5.	South Korea	43%	15.	UK	22%
6.	Switzerland	40%	16.	USA	20%
7.	Taiwan	38%	17.	Canada	17%
8.	Slovenia	37%	18.	Argentina	14%
9.	Spain	36%	19.	Russia	11%
10.	Bulgaria	34%	20.	Netherlands	5%

The decade 1980–90 was a bad time for the nuclear power industry. Major projects regularly ran vastly overbudget, and fears of long-term environmental damage were heavily reinforced by the 1986 Soviet disaster at Chernobyl. Although the number of reactors in service continued to increase throughout the period, orders for new plants shrank dramatically, and most countries cut back on their nuclear programs.

HYDROELECTRICITY

Percentage of electricity generated by hydroelectrical power stations, leading nations (1990)

1.	Paraguay	99.9%	11.	Zaïre	97.4%
2.	Bhutan	99.6%	12.	Cameroon	97.3%
3.	Norway	99.6%	13.	Nepal	97.0%
4.	Zambia	99.5%	14.	Laos	95.2%
5.	Congo	99.4%	15.	Costa Rica	94.2%
6.	Ghana	99.3%	16.	Iceland	93.5%
7.	Uganda	99.1%	17.	Brazil	92.9%
8.	Rwanda	98.9%	18.	Sri Lanka	92.3%
9.	Burundi	98.5%	19.	Albania	91.1%
10.	Malawi	98.1%	20.	Guatemala	89.7%

Countries heavily reliant on hydroelectricity are usually small and non-industrial: a high proportion of hydroelectric power more often reflects a modest energy budget than vast hydroelectric resources. The USA, for instance, produces only 9% of power requirements from hydroelectricity; yet that 9% amounts to more than three times the HEP generated by all of Africa.

ALTERNATIVE ENERGY SOURCES

Solar: Each year the Sun bestows upon the Earth almost a million times as much energy as is locked up in all the planet's oil reserves, but only an insignificant fraction is trapped and used commercially. In some experimental installations, mirrors focus the Sun's rays on to boilers, whose steam generates electricity by spinning turbines. Solar cells turn the sunlight into electricity directly, and although efficiencies are still low, advancing technology offers some prospect of using the Sun as the main world electricity source by 2100.

Wind: Caused by uneven heating of the Earth, winds are themselves a form of solar energy. Windmills have been used for centuries to turn windpower into mechanical work; recent models, often arranged in banks on gust-swept high ground, usually generate electricity.

Tidal: The energy from tides is potentially enormous, although only a few installations have been built to exploit it. In theory, at least, waves and currents could also provide almost unimaginable power, and the thermal differences in the ocean depths are another huge well of potential energy. But work on extracting it is still in the experimental stage.

Geothermal: The Earth's temperature rises by 1F° for every 50 feet descent, with much steeper temperature gradients in geologically active areas. El Salvador, for example, produces 39% of its electricity from geothermal power stations. More than 130 are operating worldwide.

Biomass: The oldest of human fuels ranges from animal dung, still burned in cooking fires in much of North Africa and elsewhere, to sugarcane plantations feeding high-technology distilleries to produce ethanol for motor vehicle engines. In Brazil and South Africa, plant ethanol provides up to 25% of motor fuel. Throughout the developing world, most biomass energy comes from firewood: although accurate figures are impossible to obtain, it may yield as much as 10% of the world's total energy consumption.

PRODUCTION: MINERALS

Even during the Stone Age, when humans often settled near the outcrops of flint on which their technology depended, mineral resources have attracted human exploiters. Their descendants have learned how to make use of almost every known element. These elements can be found, in one form or another, somewhere in the Earth's bountiful crust. Iron remains the most important, but modern industrial civilization has a voracious appetite for virtually all of them.

Mineral deposits once dictated the site of new industries; today, most industrial countries are heavily dependent on imports for many of their key materials. Most mining, and much refining of raw ores, is done in developing countries, where labor is cheap.

The main map below shows the richest sources of the most important minerals at present; some reserves – lead and mercury, for example – are running very low. The map takes no account of undersea deposits, most of which are considered inaccessible. Growing shortages, though, may encourage submarine mining: plans have already been made to recover the nodules of manganese found widely scattered on ocean floors.

MINERAL EXPORTS

Minerals and metals as a percentage of total exports (latest available year)

- Over 50%
- 10 – 50%
- 5 – 10%
- Under 5%

Direction of trade

- Copper
- Iron
- Bauxite (Aluminum)

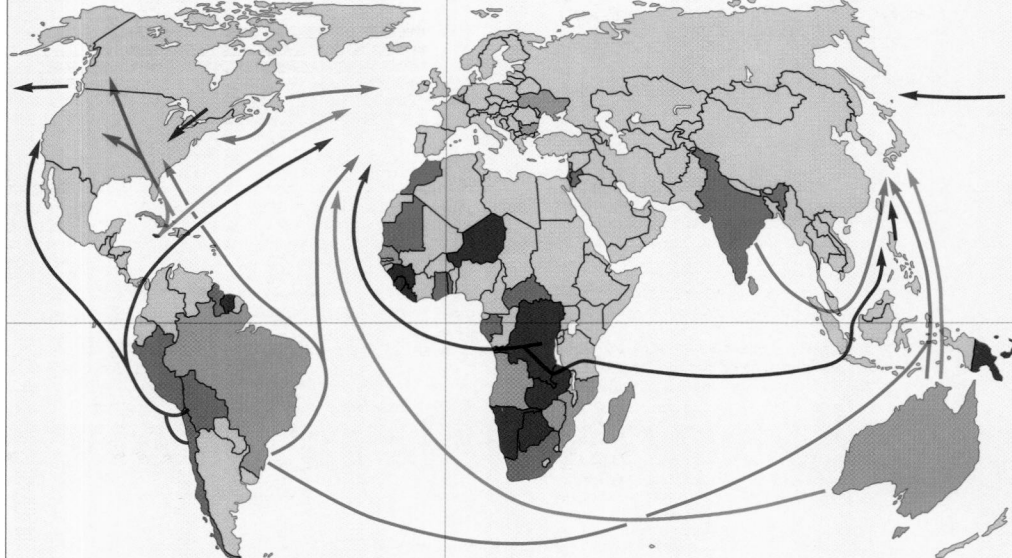

URANIUM

In its pure state, uranium is an immensely heavy, white metal; but although spent uranium is employed as projectiles in anti-missile cannons, where its mass ensures a lethal punch, its main use is as a fuel in nuclear reactors, and in nuclear weaponry. Uranium is very scarce: the main source is the rare ore pitchblende, which itself contains only 0.2% uranium oxide. Only a minute fraction of that is the radioactive U^{235} isotope, though so-called breeder reactors can transmute the more common U^{238} into highly radioactive plutonium.

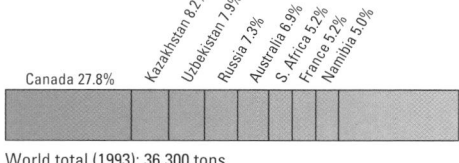

Canada 27.8% | Kazakhstan 8.2% | Uzbekistan 7.9% | Russia 7.3% | Australia 6.9% | S. Africa 5.52% | France 5.2% | Namibia 5.0%

World total (1993): 36,300 tons

METALS

*Figures for aluminum are for refined metal; all other figures refer to ore production

Aluminum: Produced mainly from its oxide, bauxite, which yields 25% of its weight in aluminum. The cost of refining and production is often too high for producer-countries to bear, so bauxite is largely exported. Lightweight and corrosion resistant, aluminum alloys are widely used in aircraft, vehicles, cans and packaging.

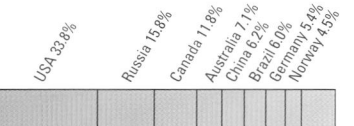

USA 33.8% | Russia 15.8% | Canada 11.8% | Australia 7.1% | China 6.2% | Brazil 6.0% | Germany 5.4% | Norway 4.5%

World total (1993): 21,569,900 tons *

Copper: Derived from low-yielding sulphide ores, copper is an important export for several developing countries. An excellent conductor of heat and electricity, it forms part of most electrical items, and is used in the manufacture of brass and bronze. Major importers include Japan and Germany.

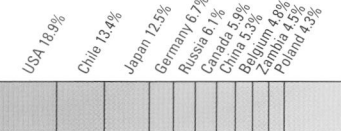

USA 18.9% | Chile 13.4% | Japan 12.5% | Germany 6.7% | Russia 6.1% | Canada 5.9% | China 5.3% | Belgium 4.8% | Zambia 4.5% | Poland 4.3%

World total (1993): 10,450,000 tons *

Lead: A soft metal, obtained mainly from galena (lead sulfide), which occurs in veins associated with iron, zinc and silver sulfides. Its use in vehicle batteries accounts for the USA's prime consumer status; lead is also made into sheeting and piping. Its use as an additive to paints and petrol is decreasing.

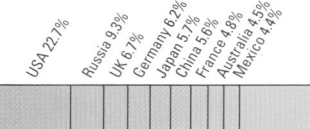

USA 22.7% | Russia 9.3% | UK 6.7% | Germany 6.2% | Japan 5.7% | China 5.6% | Australia 4.6% | France 4.5% | Mexico 4.4%

World total (1993): 5,940,000 tons *

Mercury: The only metal that is liquid at normal temperatures, most is derived from its sulfide, cinnabar, found only in small quantities in volcanic areas. Apart from its value in thermometers and other instruments, most mercury production is used in anti-fungal and anti-fouling preparations, and to make detonators.

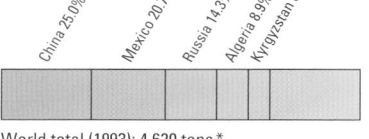

China 25.0% | Mexico 20.7% | Russia 14.3% | Algeria 8.9% | Kyrgyzstan 6.0%

World total (1993): 4,620 tons *

DIAMOND

Most diamond is found in kimberlite, or 'blue ground', a basic peridotite rock; erosion may wash the diamond from its kimberlite matrix and deposit it with sand or gravel on river beds. Only a small proportion of the world's diamond, the most flawless, is cut into gemstones – 'diamonds'; most is used in industry, where the material's remarkable hardness and abrasion resistance finds a use in cutting tools, drills and dies, as well as in styluses. Australia, not among the top 12 producers at the beginning of the 1980s, had by 1986 become world leader and by 1993 was the source of 40.6% of world production. The other main producers were Zaïre (16.3%), Botswana (14.6%), Russia (11.4%) and South Africa (9.7%). Between them, these five nations accounted for over 82% of the world total of 100,850,000 carats.

Tin: Soft, pliable and non-toxic, used to coat 'tin' (tin-plated steel) cans, in the manufacture of foils and in alloys. The principal tin-bearing mineral is cassiterite (SnO_2), found in ore formed from molten rock. Producers and refiners were hit by a price collapse in 1991.

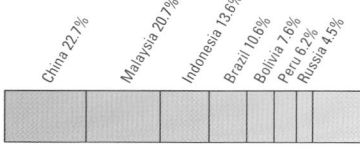

China 22.7% | Malaysia 20.7% | Indonesia 13.6% | Brazil 10.6% | Bolivia 7.6% | Peru 4.5% | Russia 4.5%

World total (1993): 242,000 tons *

Zinc: Often found in association with lead ores, zinc is highly resistant to corrosion, and about 40% of the refined metal is used to plate sheet steel, particularly vehicle bodies – a process known as galvanizing. Zinc is also used in dry batteries, paints and dyes.

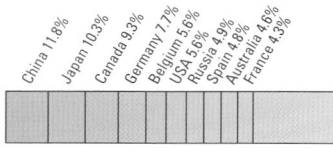

China 11.8% | Japan 10.3% | Canada 9.3% | Germany 7.7% | Belgium 5.6% | USA 5.6% | Russia 4.9% | Spain 4.8% | Australia 4.6% | France 4.3%

World total (1993): 7,839,700 tons *

Gold: Regarded for centuries as the most valuable metal in the world and used to make coins, gold is still recognized as the monetary standard. A soft metal, it is alloyed to make jewelry; the electronics industry values its corrosion resistance and conductivity.

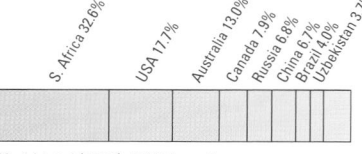

S. Africa 32.6% | USA 17.7% | Australia 13.0% | Canada 7.9% | Russia 6.8% | China 6.7% | Brazil 4.0% | Uzbekistan 3.7%

World total (1993): 2,090 tons *

Silver: Most silver comes from ores mined and processed for other metals (including lead and copper). Pure or alloyed with harder metals, it is used for jewelry and ornaments. Industrial use includes dentistry, electronics, photography and as a chemical catalyst.

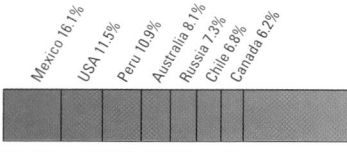

Mexico 16.1% | USA 11% | Peru 10.9% | Australia 8.1% | Russia 7.3% | China 6.8% | Chile 6.2% | Canada 6.2%

World total (1993): 14,300 tons *

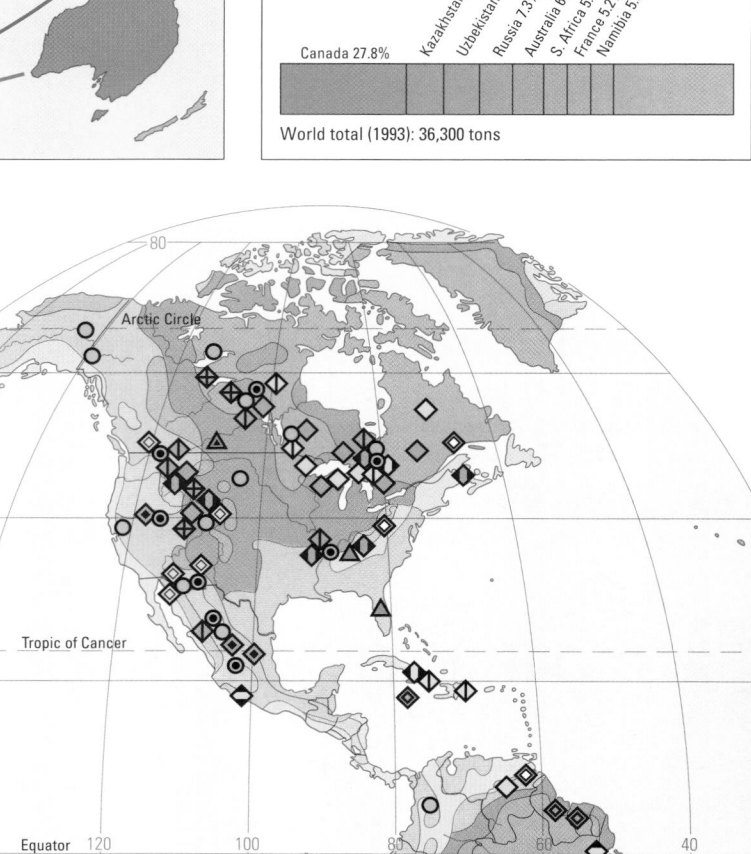

IRON AND FERRO-ALLOYS

Ever since the art of high-temperature smelting was discovered, some time in the second millennium BC, iron has been by far the most important metal known to man. The earliest iron plows transformed primitive agriculture and led to the first human population explosion, while iron weapons – or the lack of them – ensured the rise or fall of entire cultures.

Widely distributed around the world, iron ores usually contain 25–60% iron; blast furnaces process the raw product into pig-iron, which is then alloyed with carbon and other minerals to produce steels of various qualities. From the time of the Industrial Revolution steel has been almost literally the backbone of modern civilization, the prime structural material on which all else is built.

Iron-smelting usually developed close to sources of ore and, later, to the coalfields that fueled the furnaces. Today, most ore comes from a few richly-endowed locations where large-scale mining is possible. Iron and steel plants are generally built at coastal sites so that giant ore carriers, which account for a sizable proportion of the world's merchant fleet, can easily discharge their cargoes.

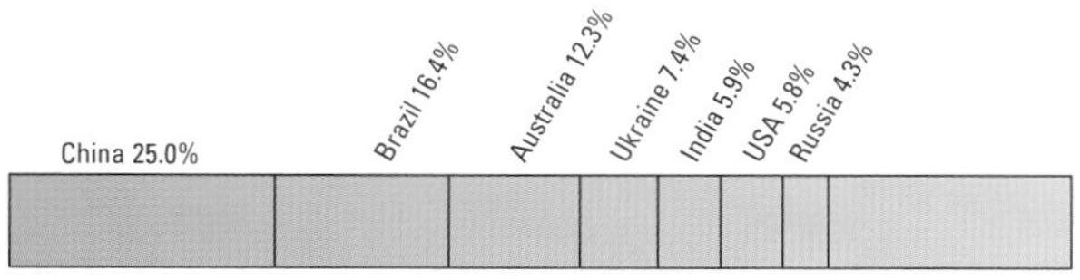

World production of pig-iron and ferro-alloys (1993). All countries with an annual output of more than one million tons are shown

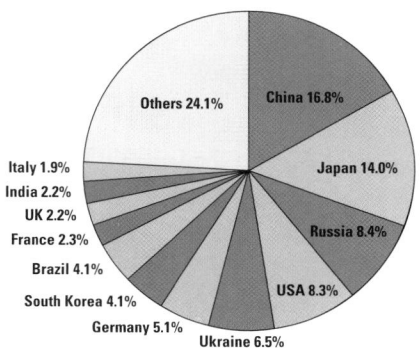

- China 16.8%
- Japan 14.0%
- Russia 8.4%
- USA 8.3%
- Ukraine 6.5%
- Germany 5.1%
- South Korea 4.1%
- Brazil 4.1%
- France 2.3%
- UK 2.2%
- India 2.2%
- Italy 1.9%
- Others 24.1%

Total world production: 590 million tons

Development of world production of pig-iron and ferro-alloys (1945–93) in million tons

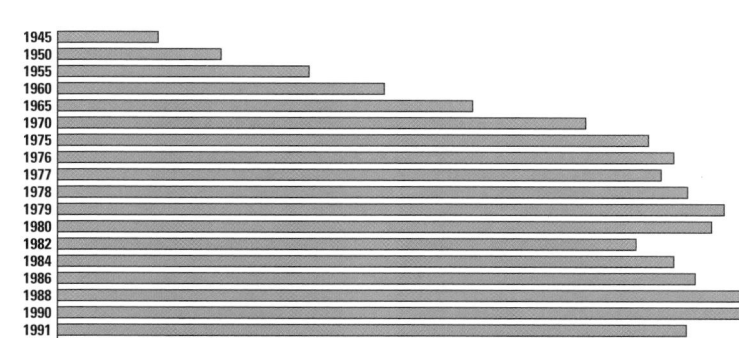

1945, 1950, 1955, 1960, 1965, 1970, 1975, 1976, 1977, 1978, 1979, 1980, 1982, 1984, 1986, 1988, 1990, 1991, 1992, 1993

Chromium: Most of the world's chromium production is alloyed with iron and other metals to produce steels with various different properties. Combined with iron, nickel, cobalt and tungsten, chromium produces an exceptionally hard steel, resistant to heat; chrome steels are used for many household items where utility must be matched with appearance – cutlery, for example. Chromium is also used in production of refractory bricks, and its salts for tanning and dyeing leather and cloth.

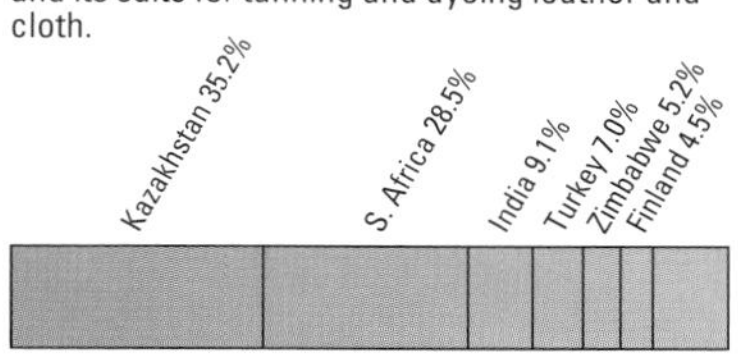

Manganese: In its pure state, manganese is a hard, brittle metal. Alloyed with chrome, iron and nickel, it produces abrasion-resistant steels; manganese-aluminum alloys are light but tough. Found in batteries and inks, manganese is also used in glass production. Manganese ores are frequently found in the same location as sedimentary iron ores. Pyrolusite (MnO_2) and psilomelane are the main economically-exploitable sources.

Nickel: Combined with chrome and iron, nickel produces stainless and high-strength steels; similar alloys go to make magnets and electrical heating elements. Nickel combined with copper is widely used to make coins; cupro-nickel alloy is very resistant to corrosion. Its ores yield only modest quantities of nickel – 0.5 to 3.0% – but also contain copper, iron and small amounts of precious metals. Japan, USA, UK, Germany and France are the principal importers.

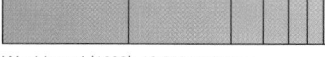

- China 25.0%
- Brazil 16.4%
- Australia 12.3%
- Ukraine 7.4%
- India 5.9%
- USA 5.8%
- Russia 4.5%

World total production of iron ore (1993): 1,034,000,000 tons

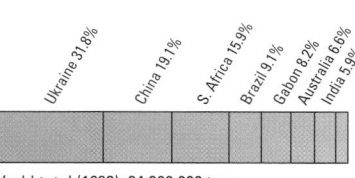

- Kazakhstan 35.2%
- S. Africa 28.5%
- India 9.1%
- Turkey 7.0%
- Zimbabwe 5.2%
- Finland 4.5%

World total (1993): 10,923,000 tons

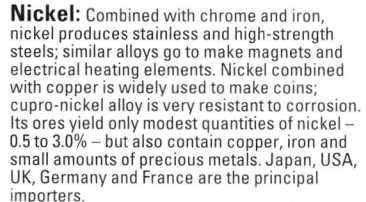

- Ukraine 31.8%
- China 19.1%
- S. Africa 15.9%
- Brazil 9.1%
- Gabon 8.2%
- Australia 6.6%
- India 5.9%

World total (1993): 24,200,000 tons

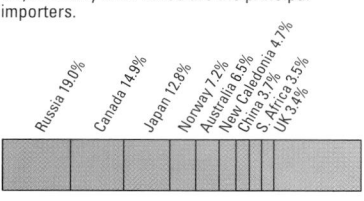

- Russia 19.0%
- Canada 14.9%
- Japan 12.8%
- Norway 7.2%
- Australia 6.5%
- New Caledonia 4.7%
- China 3.7%
- S. Africa 3.5%
- UK 3.4%

World total (1993): 869,000 tons

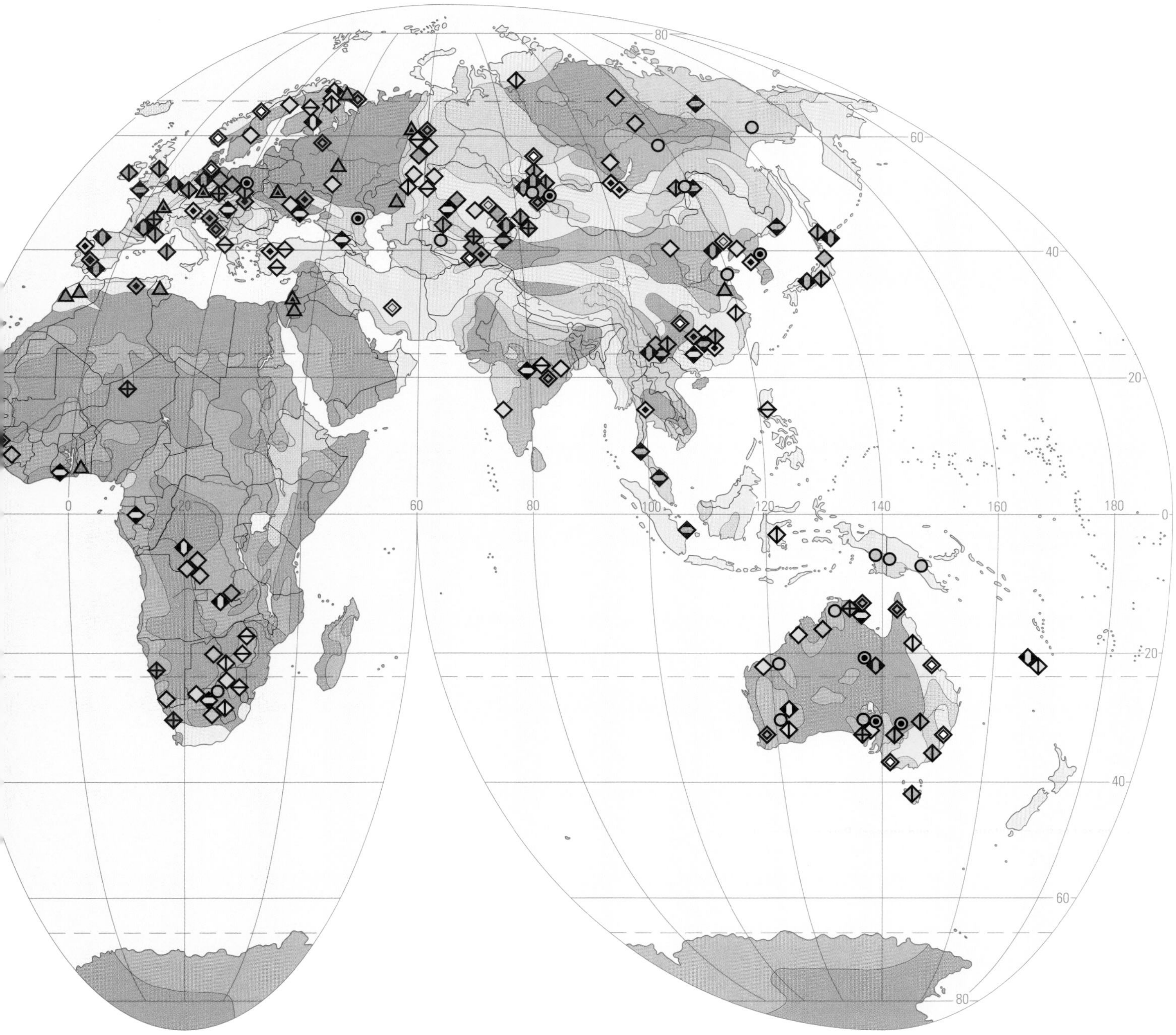

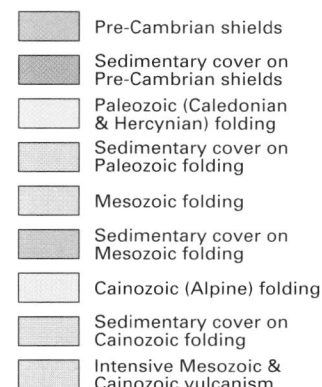

STRUCTURAL REGIONS

- Pre-Cambrian shields
- Sedimentary cover on Pre-Cambrian shields
- Paleozoic (Caledonian & Hercynian) folding
- Sedimentary cover on Paleozoic folding
- Mesozoic folding
- Sedimentary cover on Mesozoic folding
- Cainozoic (Alpine) folding
- Sedimentary cover on Cainozoic folding
- Intensive Mesozoic & Cainozoic vulcanism

DISTRIBUTION

Iron & ferro-alloys

- Chrome
- Cobalt
- Iron Ore
- Manganese
- Molybdenum
- Nickel Ore
- Tungsten

Non-ferrous metals

- Aluminum
- Bauxite
- Copper
- Lead
- Mercury
- Tin
- Zinc
- Uranium

Precious metals & stones

- Diamonds
- Gold
- Silver

Fertilizers

- Phosphates
- Potash

CARTOGRAPHY BY PHILIP'S. COPYRIGHT REED INTERNATIONAL BOOKS LTD

PRODUCTION: MANUFACTURING

In its broadest sense, manufacturing is the application of energy, labor and skill to raw materials in order to transform them into finished goods with a higher value than the various elements used in production.

Since the early days of the Industrial Revolution, manufacturing has implied the use of an organized work force harnessed to some form of machine. The tendency has consistently been for increasingly expensive human labor to be replaced by increasingly complex machinery, which has evolved over time from water-powered looms to fully-integrated robotic plants.

Obviously, not all the world's industries – or manufacturing countries – have reached the same level. Textiles, for example, the foundation of the early Industrial Revolution in the West, can be mass-produced with

fairly modest technology; today, they are usually produced in developing countries, mostly in Asia, where the low labor costs compensate for the large work force that the relatively simple machinery requires. Nevertheless, the trend toward high-technology production, however uneven, seems inexorable. Gains in efficiency make up for the staggering cost of the equipment itself, and the outcome is that fewer and fewer people are employed to produce more and more goods.

One paradoxical result of the increase in industrial efficiency is a relative decline in the importance of the industrial sector of a nation's economy. The economy has already passed through one transition, generations past, when workers were drawn from the land into factories. The second transition releases

labor into what is called the service sector of the economy: a diffuse but vital concept that includes not only such obvious services as transport and administration, but also finance, insurance and activities as diverse as fashion design or the writing of computer software.

The process is far advanced in the mature economies of the West, with Japan not far behind. Almost two-thirds of US wealth, for example, is now generated in the service sector, and less than half of Japan's Gross National Product comes from industry. The shrinkage, though, is only relative: between them, these two industrial giants produce almost twice the amount of manufactured goods as the rest of the world put together. And it is on the solid base of production that their general prosperity is founded.

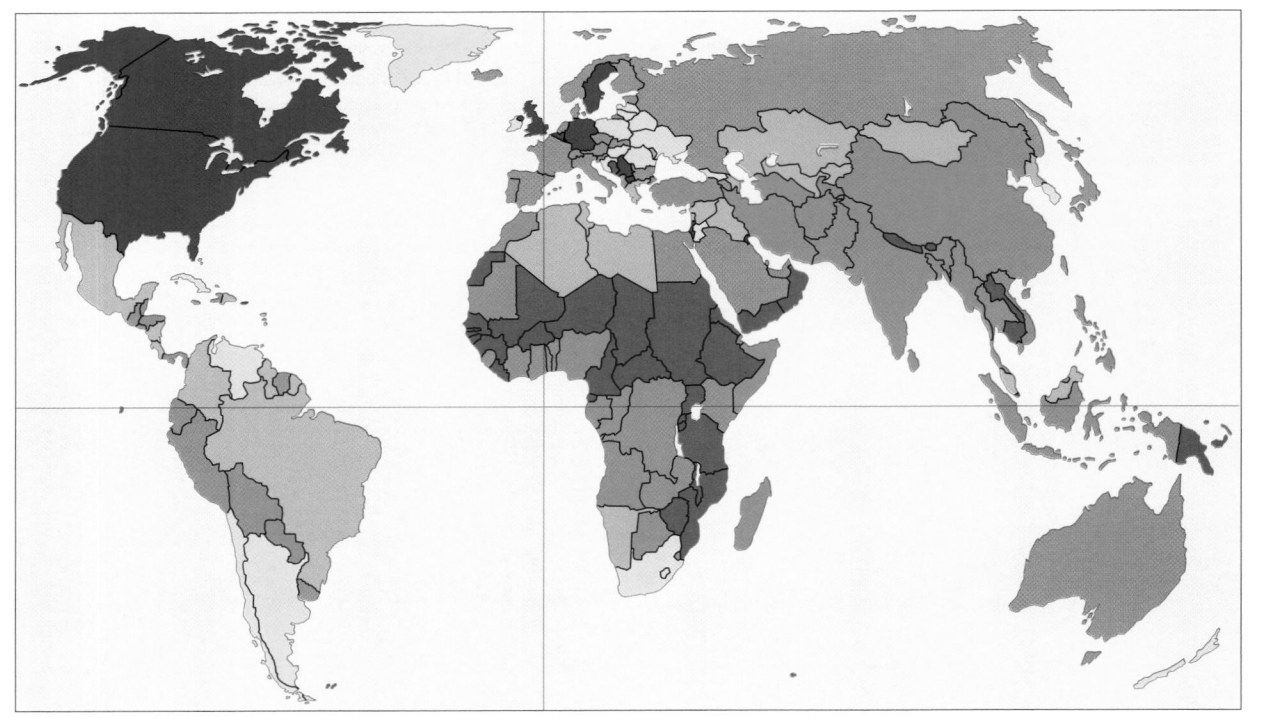

EMPLOYMENT

The number of workers employed in manufacturing for every 100 workers engaged in agriculture

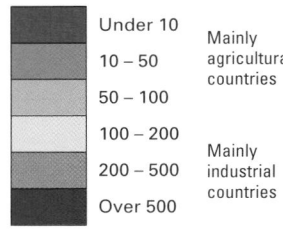

Under 10	Mainly agricultural countries
10 – 50	
50 – 100	
100 – 200	
200 – 500	Mainly industrial countries
Over 500	

Selected countries (latest available year)

Singapore	8,860
Hong Kong	3,532
UK	1,270
Belgium	820
Ex-Yugoslavia	809
Germany	800
Kuwait	767
Bahrain	660
USA	657
Israel	633

DIVISION OF EMPLOYMENT

Distribution of workers between agriculture, industry and services, selected countries (latest available year)

The six countries selected illustrate the usual stages of economic development, from dependence on agriculture through industrial growth to the expansion of the services sector.

- Agriculture
- Industry
- Services

Nepal	Nigeria	Pakistan	Brazil	Hong Kong	USA
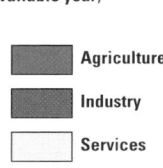	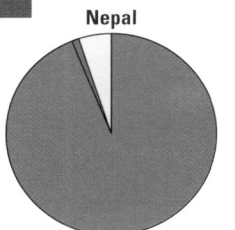	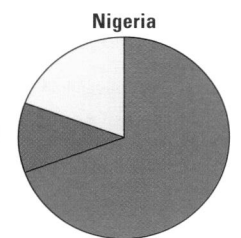	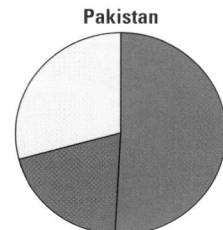	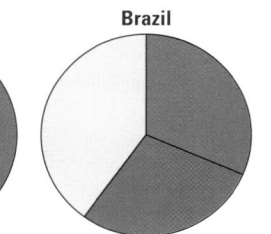	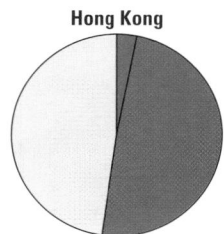

THE WORK FORCE

Percentages of men and women between 15 and 64 in employment, selected countries (latest available year)

The figures include employees and self-employed, who in developing countries are often subsistence farmers. People in full-time education are excluded. Because of the population age structure in developing countries, the employed population has to support a far larger number of non-workers than its industrial equivalent. For example, more than 52% of Kenya's people are under 15, an age group that makes up less than a tenth of the UK population.

- Men
- Women

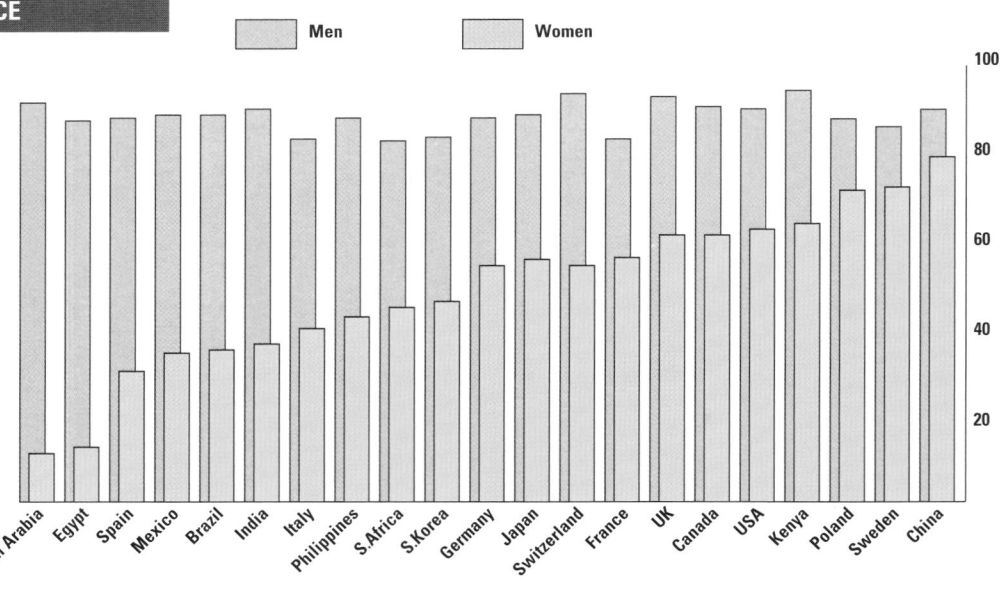

WEALTH CREATION

The Gross National Product (GNP) of the world's largest economies, US $ billion (1991)

1.	USA	5,686,038	21.	Austria	157,538
2.	Japan	3,337,191	22.	Iran	127,366
3.	Germany	1,516,785	23.	Finland	121,982
4.	France	1,167,749	24.	Denmark	121,695
5.	Italy	1,072,198	25.	Ukraine	121,458
6.	UK	963,696	26.	Indonesia	111,409
7.	Canada	568,765	27.	Saudi Arabia	105,133
8.	Spain	486,614	28.	Turkey	103,388
9.	Russia	479,546	29.	Norway	102,885
10.	Brazil	447,324	30.	Argentina	91,211
11.	China	424,012	31.	South Africa	90,953
12.	Australia	287,765	32.	Thailand	89,548
13.	India	284,668	33.	Hong Kong	77,302
14.	Netherlands	278,839	34.	Poland	70,640
15.	South Korea	274,464	35.	Greece	65,504
16.	Mexico	252,381	36.	Israel	59,128
17.	Switzerland	225,890	37.	Portugal	58,451
18.	Sweden	218,934	38.	Venezuela	52,775
19.	Belgium	192,370	39.	Algeria	52,239
20.	Taiwan	161,000	40.	Pakistan	46,725

PATTERNS OF PRODUCTION

Breakdown of industrial output by value, selected countries (latest available year)

	Food & agriculture	Textiles & clothing	Machinery & transport	Chemicals	Other
Algeria	26%	20%	11%	1%	41%
Argentina	24%	10%	16%	12%	37%
Australia	18%	7%	21%	8%	45%
Austria	17%	8%	25%	6%	43%
Belgium	19%	8%	23%	13%	36%
Brazil	15%	12%	24%	9%	40%
Burkina Faso	62%	18%	2%	1%	17%
Canada	15%	7%	25%	9%	44%
Denmark	22%	6%	23%	10%	39%
Egypt	20%	27%	13%	10%	31%
Finland	13%	6%	24%	7%	50%
France	18%	7%	33%	9%	33%
Germany	12%	5%	38%	10%	36%
Greece	20%	22%	14%	7%	38%
Hong Kong	6%	40%	20%	2%	33%
Hungary	6%	11%	37%	11%	35%
India	11%	16%	26%	15%	32%
Indonesia	23%	11%	10%	10%	47%
Iran	13%	22%	22%	7%	36%
Israel	13%	10%	28%	8%	42%
Ireland	28%	7%	20%	15%	28%
Italy	7%	13%	32%	10%	38%
Japan	10%	6%	38%	10%	37%
Kenya	35%	12%	14%	9%	29%
Malaysia	21%	5%	23%	14%	37%
Mexico	24%	12%	14%	12%	39%
Netherlands	19%	4%	28%	11%	38%
New Zealand	26%	10%	16%	6%	43%
Norway	21%	3%	26%	7%	44%
Pakistan	34%	21%	8%	12%	25%
Philippines	40%	7%	7%	10%	35%
Poland	15%	16%	30%	6%	33%
Portugal	17%	22%	16%	8%	38%
Singapore	6%	5%	46%	8%	36%
South Africa	14%	8%	17%	11%	49%
South Korea	15%	17%	24%	9%	35%
Spain	17%	9%	22%	9%	43%
Sweden	10%	2%	35%	8%	44%
Thailand	30%	17%	14%	6%	33%
Turkey	20%	14%	15%	8%	43%
UK	14%	6%	32%	11%	36%
USA	12%	5%	35%	10%	38%
Venezuela	23%	8%	9%	11%	49%

INDUSTRY AND TRADE

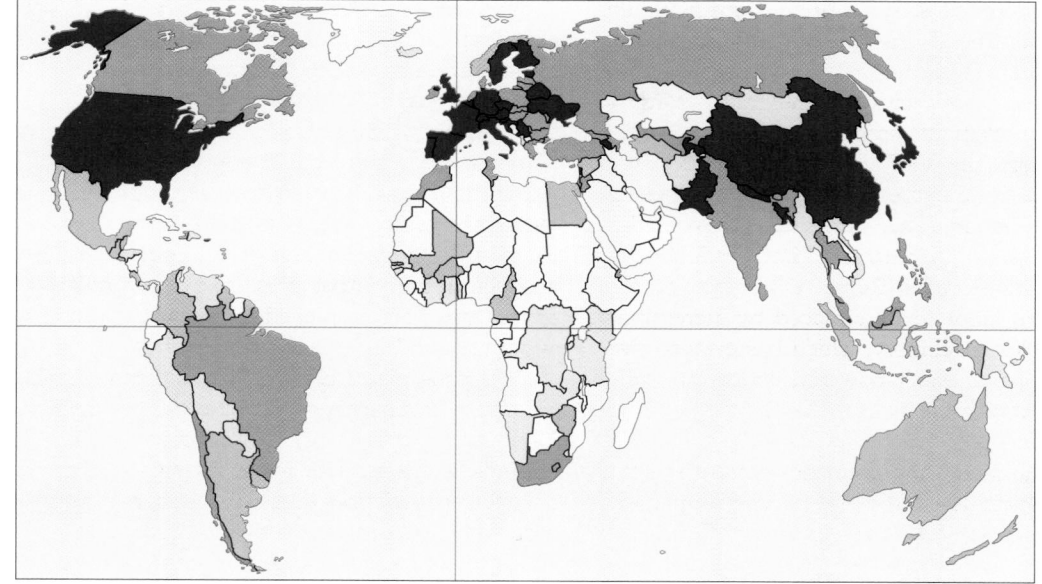

Manufactured goods (including machinery and transport) as a percentage of total exports (latest available year)

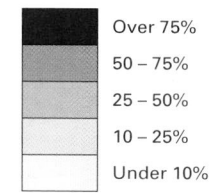

- Over 75%
- 50 – 75%
- 25 – 50%
- 10 – 25%
- Under 10%

The Far East and South-east Asia (Japan 98.3%, Macau 97.8%, Taiwan 92.7%, Hong Kong 93.0%, South Korea 93.4%) are most dominant, but many countries in Europe (e.g. Slovenia 92.4%) are also heavily dependent on manufactured goods.

AUTOMOBILES

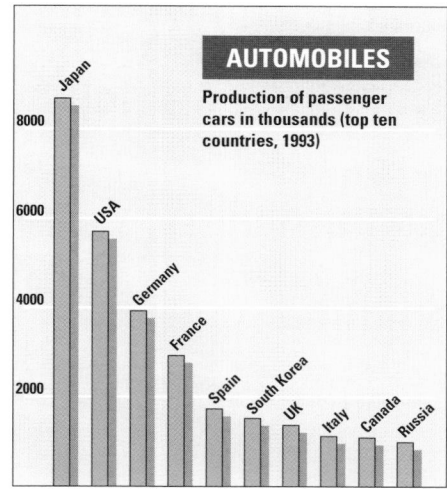

Production of passenger cars in thousands (top ten countries, 1993)

COMMERCIAL VEHICLES

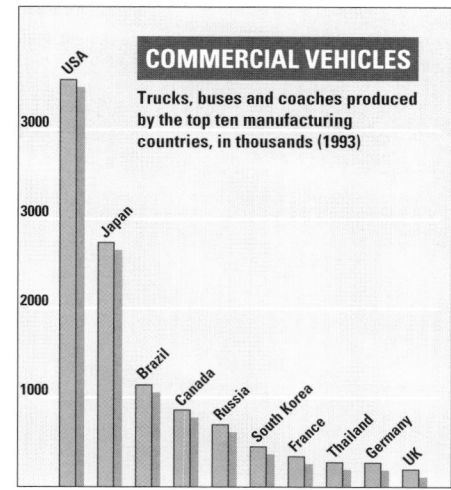

Trucks, buses and coaches produced by the top ten manufacturing countries, in thousands (1993)

TELEVISION SETS

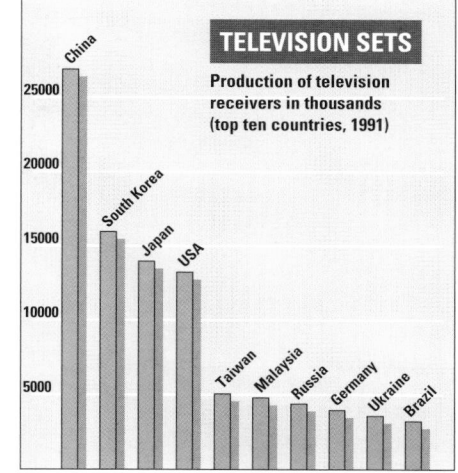

Production of television receivers in thousands (top ten countries, 1991)

STEEL PRODUCTION

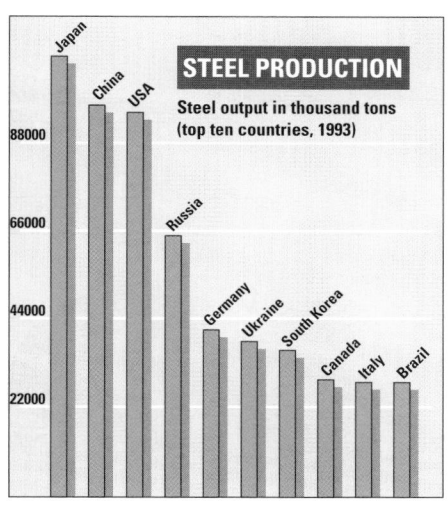

Steel output in thousand tons (top ten countries, 1993)

SHIPBUILDING

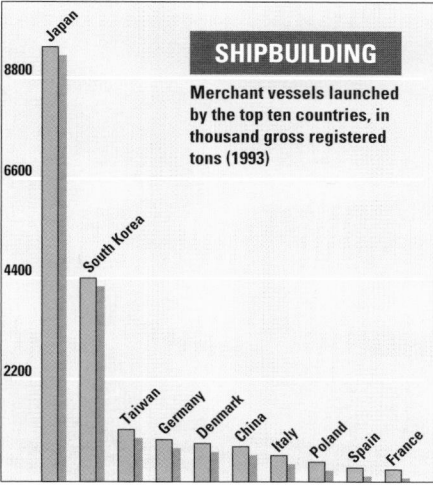

Merchant vessels launched by the top ten countries, in thousand gross registered tons (1993)

NATURAL & SYNTHETIC RUBBER

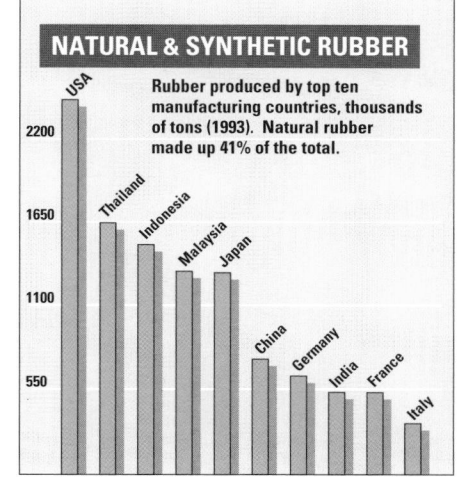

Rubber produced by top ten manufacturing countries, thousands of tons (1993). Natural rubber made up 41% of the total.

RADIO RECEIVERS

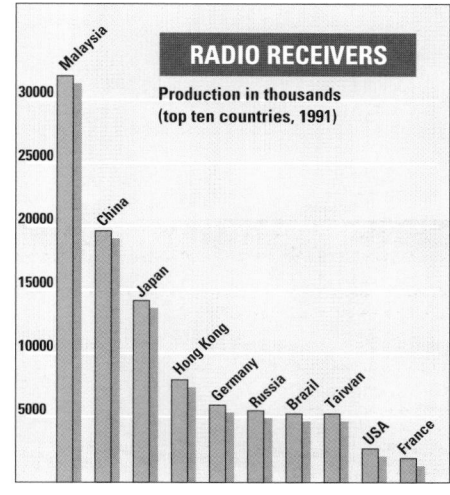

Production in thousands (top ten countries, 1991)

INDUSTRIAL POWER

Industrial output (mining, manufacturing, construction, energy and water production), top 40 nations, US $ billion (1991)

1. USA	1,627	21. Saudi Arabia	56
2. Japan	1,412	22. Indonesia	48
3. Germany	614	23. Spain	47
4. Italy	380	24. Argentina	46
5. France	348	25. Poland	39
6. UK	324	26. Norway	38
7. Ex-Soviet Union	250	27. Finland	37
8. Brazil	161	28. Thailand	36
9. China	155	29. Turkey	33
10. South Korea	127	30. Denmark	31
11. Canada	117	31. Israel	23
12. Australia	93	32. Iran	20
Netherlands	93	33. Ex-Czechoslovakia	19
14. Taiwan	86	34. Hong Kong	17
15. Mexico	85	Portugal (1989)	17
16. Sweden	70	36. Algeria	16
17. Switzerland (1989)	61	Greece	16
18. India	60	38. Iraq	15
19. Austria	59	Philippines	15
Belgium	59	Singapore	15

EXPORTS PER CAPITA

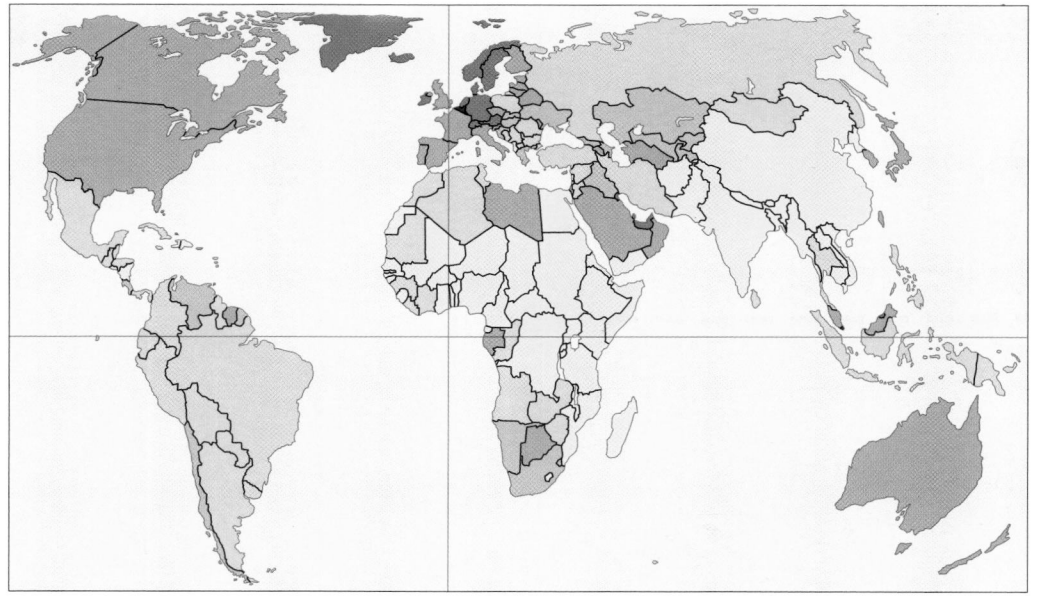

Value of exports in US $, divided by total population (latest available year)

- Over 10,000
- 5,000 – 10,000
- 1,000 – 5,000   [USA 1,733]
- 500 – 1,000
- 100 – 500
- Under 100

Highest per capita

Singapore	22,070
Hong Kong	20,004
Luxembourg	16,013
Belgium	12,258
Switzerland	9,428
Netherlands	9,145

PRODUCTION: TRADE

Thriving international trade is the outward sign of a healthy world economy – the obvious indicator that some countries have goods to sell and others the wherewithal to buy them. Despite local fluctuations, trade throughout the 1980s grew consistently faster than output, increasing in value by almost 50% between 1979-89. It remains dominated by the wealthy, industrialized countries of the Organization for Economic Development:

between them, the 24 OECD members account for almost 75% of world imports and exports in most years. OECD dominance is just as marked in the trade in 'invisibles' – a column in the balance sheet that includes, among other headings, the export of services, interest payments on overseas investments, tourism, and even remittances from migrant workers abroad. In the UK, 'invisibles' account for more than half all trading income.

However, the size of these great trading

economies means that imports and exports usually comprise a fraction of their total wealth: in the case of the export-conscious Japanese, trade in goods and services amounts to less than 18% of GDP. In poorer countries, trade – often in a single commodity – may amount to 50% of GDP or more. And there are oddities: import-export figures for the entrepôt economy of Singapore, the transit point for much Asian trade, are almost double that small nation's total earnings.

WORLD TRADE

Percentage share of total world exports by value (1990)

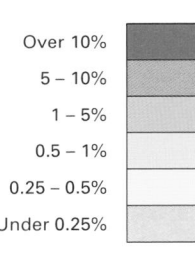

- Over 10%
- 5 – 10%
- 1 – 5%
- 0.5 – 1%
- 0.25 – 0.5%
- Under 0.25%

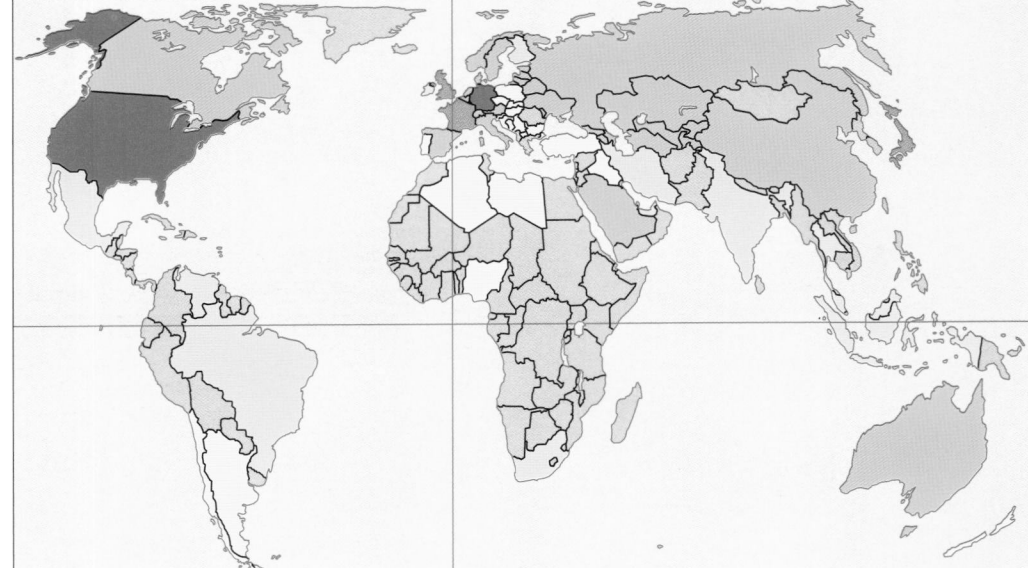

THE GREAT TRADING NATIONS

The imports and exports of the top ten trading nations as a percentage of world trade (latest available year). Each country's trade in manufactured goods is shown in orange.

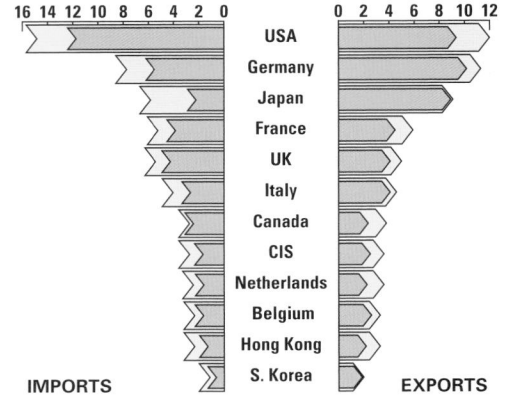

USA
Germany
Japan
France
UK
Italy
Canada
CIS
Netherlands
Belgium
Hong Kong
S. Korea

IMPORTS   EXPORTS

MAJOR EXPORTS

Leading manufactured items and their exporters, by percentage of world total in US $ (latest available year)

AIRCRAFT
Italy 3%, Canada 5%, France 8%, Germany 9%, UK 13%, USA 51%, Other 11%

TELECOMMUNICATIONS GEAR
Italy 3%, Canada 4%, Hong Kong 4%, Sweden 4%, UK 5%, France 5%, Germany 9%, Other 19%, Japan 33%, USA 14%

DATA PROCESSING EQUIPMENT
Singapore 4%, Italy 4%, Canada 4%, Ireland 5%, France 6%, UK 6%, Other 14%, USA 24%, Japan 22%, Germany 11%

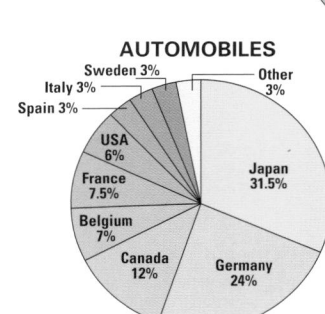

AUTOMOBILES
Sweden 3%, Italy 3%, Spain 3%, USA 6%, France 7.5%, Belgium 7%, Canada 12%, Germany 24%, Japan 31.5%, Other 3%

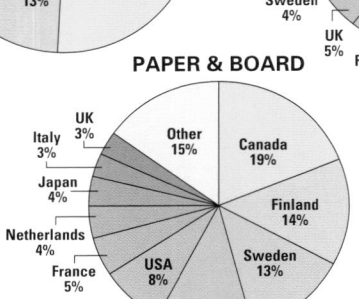

PAPER & BOARD
UK 3%, Italy 3%, Japan 4%, Netherlands 4%, France 5%, USA 8%, Germany 12%, Sweden 13%, Finland 14%, Canada 19%, Other 15%

ELECTRICAL MACHINERY
Belgium 4%, Switzerland 4%, Italy 4%, Netherlands 6%, France 7%, UK 8%, USA 14%, Germany 19%, Japan 22%, Other

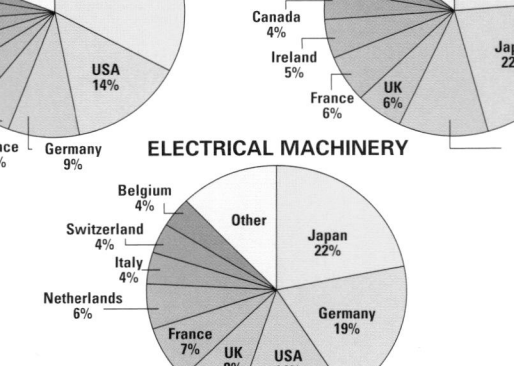

TRADED PRODUCTS

Top ten manufactures traded, by value in billions of US $ (latest available year)

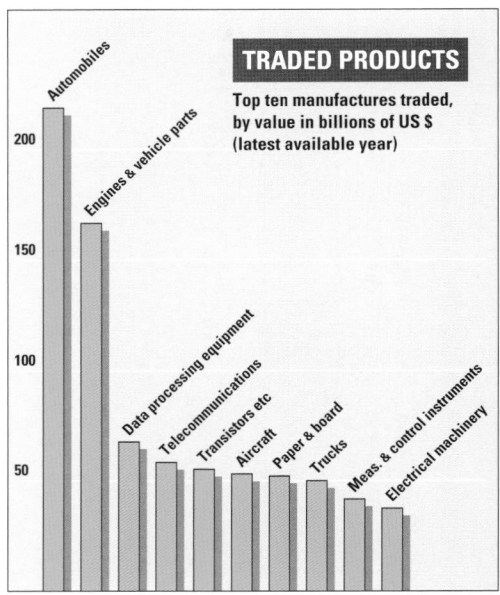

Automobiles
Engines & vehicle parts
Data processing equipment
Telecommunications
Transistors etc
Aircraft
Paper & board
Trucks
Meas & control instruments
Electrical machinery

DEPENDENCE ON TRADE

Value of exports as a percentage of Gross Domestic Product (1991)

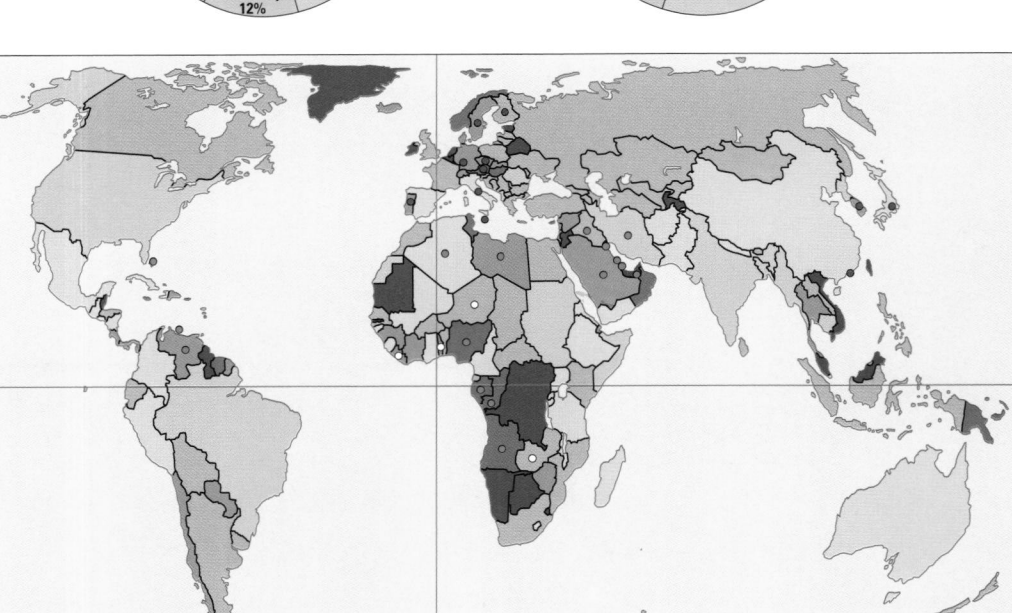

- Over 50% GDP
- 40 – 50% GDP
- 30 – 40% GDP
- 20 – 30% GDP
- 10 – 20% GDP
- Under 10% GDP

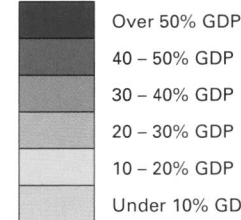

- Most dependent on industrial exports (over 75% of total exports)
- Most dependent on fuel exports (over 75% of total exports)
- Most dependent on mineral and metal exports (over 75% of total exports)

WORLD SHIPPING

While ocean passenger traffic is nowadays relatively modest, sea transport still carries most of the world's trade. Oil and bulk carriers make up the majority of the world fleet, although the general cargo category was the fastest growing in 1989, a year in which total tonnage increased by 1.5%.

Almost 30% of world shipping sails under a 'flag of convenience', whereby owners take advantage of low taxes by registering their vessels in a foreign country the ships will never see, notably Panama and Liberia.

MERCHANT FLEETS

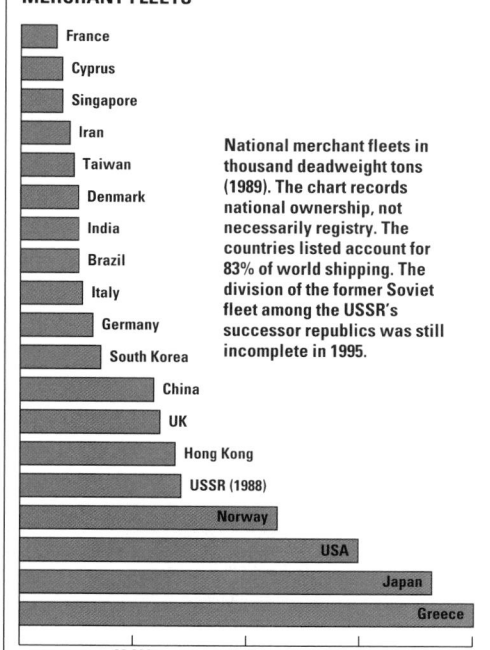

National merchant fleets in thousand deadweight tons (1989). The chart records national ownership, not necessarily registry. The countries listed account for 83% of world shipping. The division of the former Soviet fleet among the USSR's successor republics was still incomplete in 1995.

France
Cyprus
Singapore
Iran
Taiwan
Denmark
India
Brazil
Italy
Germany
South Korea
China
UK
Hong Kong
USSR (1988)
Norway
USA
Japan
Greece

20,000   40,000   60,000   80,000

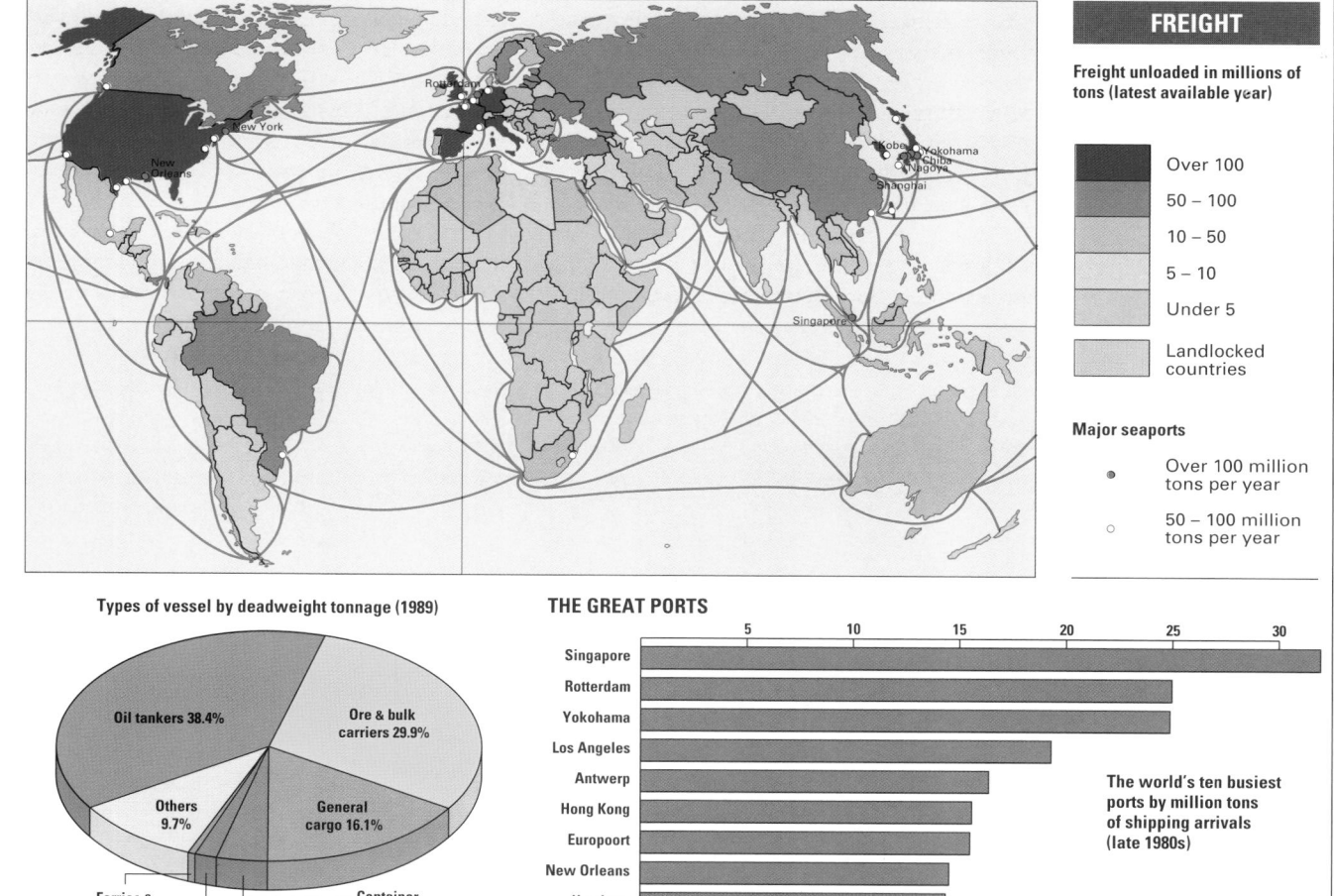

<!-- map labels -->
New Orleans, New York, Rotterdam, Kobe, Yokohama, Chiba, Nagoya, Shanghai, Singapore

FREIGHT

Freight unloaded in millions of tons (latest available year)

- Over 100
- 50 – 100
- 10 – 50
- 5 – 10
- Under 5
- Landlocked countries

Major seaports
- ● Over 100 million tons per year
- ○ 50 – 100 million tons per year

Types of vessel by deadweight tonnage (1989)

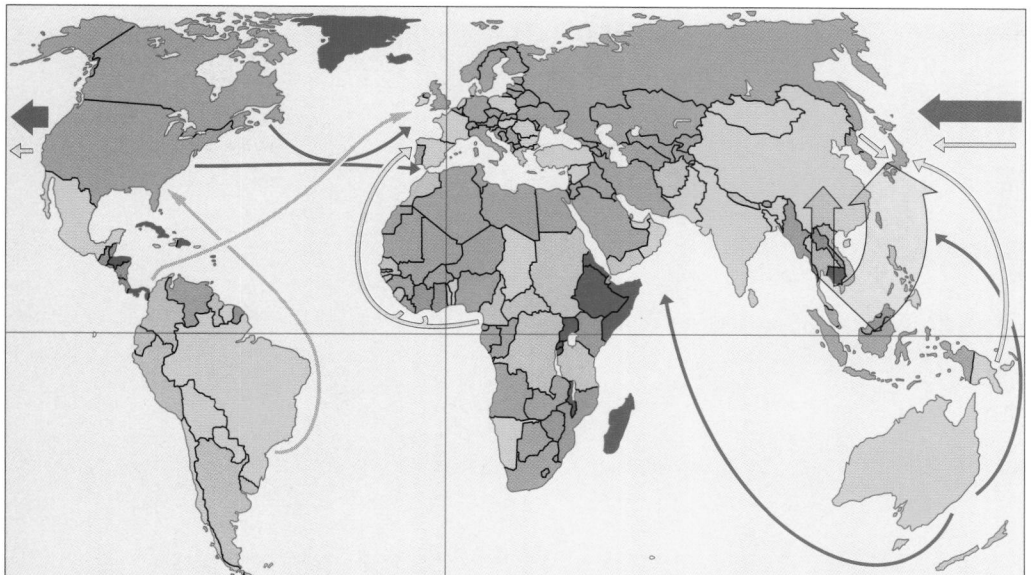

- Oil tankers 38.4%
- Ore & bulk carriers 29.9%
- General cargo 16.1%
- Others 9.7%
- Container ships 3.8%
- Liquid gas carriers 1.6%
- Ferries & passenger ships 0.5%

THE GREAT PORTS

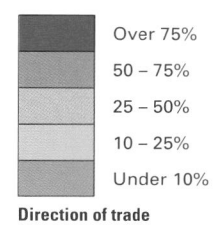

5   10   15   20   25   30

Singapore
Rotterdam
Yokohama
Los Angeles
Antwerp
Hong Kong
Europoort
New Orleans
Hamburg
Kobe

The world's ten busiest ports by million tons of shipping arrivals (late 1980s)

TRADE IN PRIMARY PRODUCTS

Primary products (excluding fuels, minerals and metals) as a percentage of total export value (latest available year)

- Over 75%
- 50 – 75%
- 25 – 50%
- 10 – 25%
- Under 10%

Direction of trade
- ➤ Major movements of cereals
- ➤ Major movements of coffee
- ➤ Major movements of hardwoods

Arrows show the major trade directions of selected primary products, and are proportional to export value.

BALANCE OF TRADE

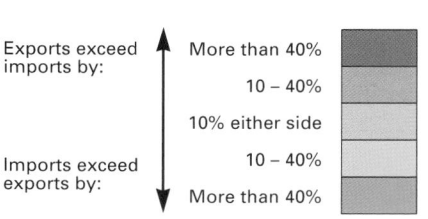

Value of exports in proportion to the value of imports (latest available year)

Exports exceed imports by:
- More than 40%
- 10 – 40%

10% either side

Imports exceed exports by:
- 10 – 40%
- More than 40%

The total world trade balance should amount to zero, since exports must equal imports on a global scale. In practice, at least $100 billion in exports go unrecorded, leaving the world with an apparent deficit and many countries in a better position than public accounting reveals. However, a favorable trade balance is not necessarily a sign of prosperity: many poorer countries must maintain a high surplus in order to service debts, and do so by restricting imports below the levels needed to sustain successful economies.

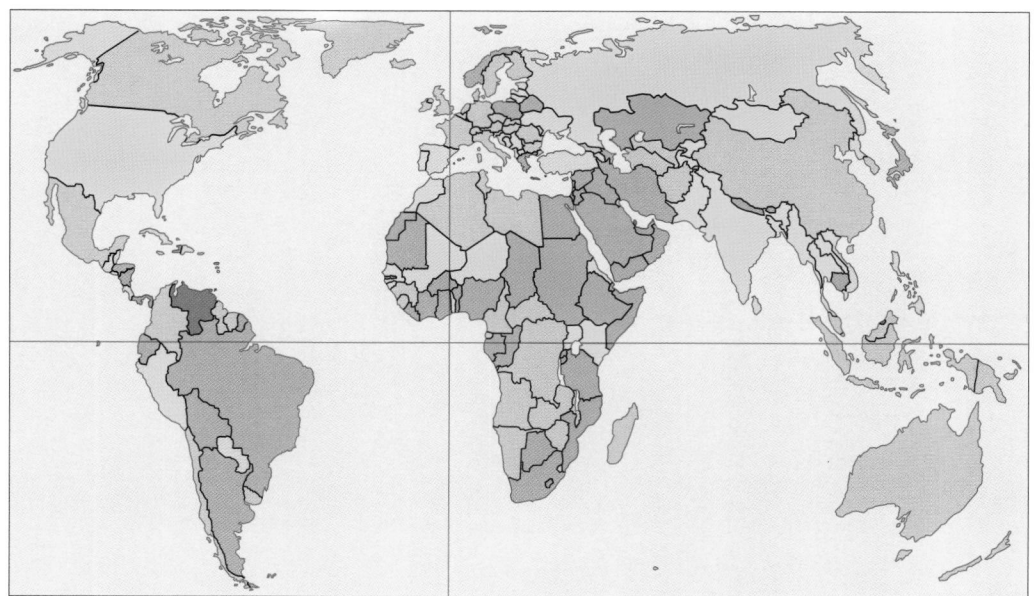

QUALITY OF LIFE: WEALTH

Throughout the 1980s, most of the world became at least slightly richer. There were exceptions: in Africa, the poorest of the continents, many incomes actually fell, and the upheavals in Eastern Europe in 1989 left whole populations awash with political freedom but worse off financially in economies still teetering toward capitalism.

Most of the improvements, however, came to those who were already, in world terms, extremely affluent: the gap between rich and poor grew steadily wider. And in those developing countries that showed significant statistical progress, advances were often confined to a few favored areas, while conditions in other, usually rural, districts went from bad to worse.

The pattern of world poverty varies from region to region. In most of Asia, the process of recognized development is generally under way, with production increases outpacing population growth. By 2000, less than 10% of the Chinese population should be officially rated 'poor': without the means to buy either adequate food or the basic necessities required to take a full part in everyday life. Even India's lower growth rate should be enough to reduce the burden of poverty for at least some of its people. In Latin America, average per capita production is high enough for most countries to be considered 'middle income' in world rankings. But although adequate resources exist, Latin American wealth is distributed with startling inequality. According to a 1990 World Bank report, a tax of only 2% on the richest fifth would raise enough money to pull every one of the continent's 437 million people above the poverty line.

In Africa, solutions will be much harder to find. The bane of high population growth has often been aggravated by incompetent administration, war and a succession of natural disasters. Population is the crux of the problem: numbers are growing anything up to twice as fast as the economies that try to support them. Aid from the developed world is only a partial solution; although Africa receives more aid than any other continent, much has been wasted on overambitious projects or lost in webs of inexperienced or corrupt bureaucracy. Yet without aid, Africa seems doomed to permanent crisis.

The rich countries can afford to increase their spending. The 24 members of the Organization for Economic Cooperation and Development comprise only 16% of the world's population, yet between them the nations accounted for almost 80% of total world production in 1988, a share that is likely to increase as the year 2000 approaches.

CONTINENTAL SHARES

Shares of population and of wealth (GNP) by continent

Generalized continental figures show the startling difference between rich and poor, but mask the successes or failures of individual countries. Japan, for example, with less than 4% of Asia's population, produces almost 70% of the continent's output.

POPULATION

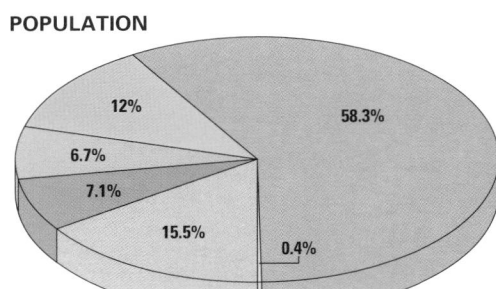

GNP

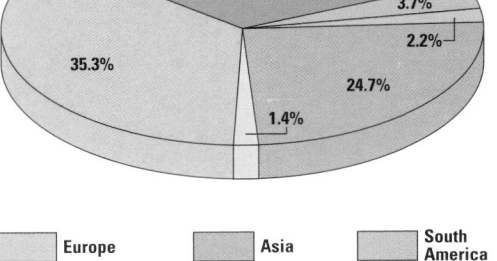

Europe | Asia | South America
Australia | Africa | North America

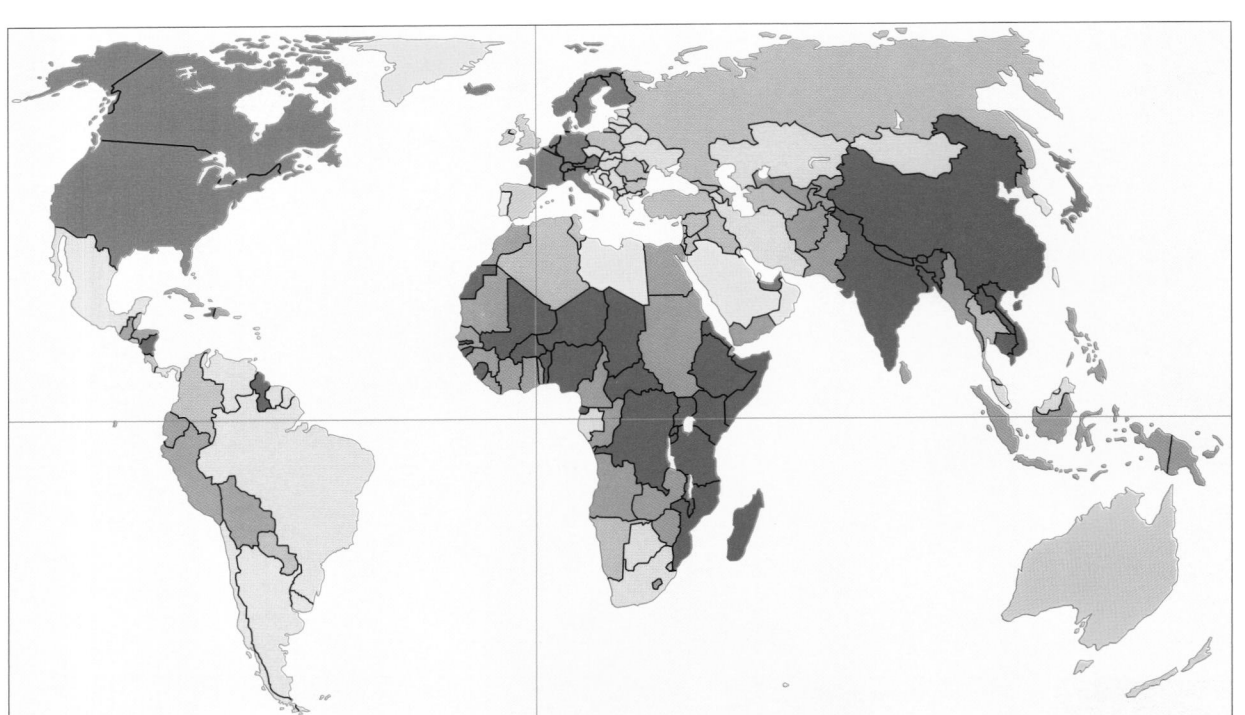

LEVELS OF INCOME

Gross National Product per capita: the value of total production divided by the population (1991)

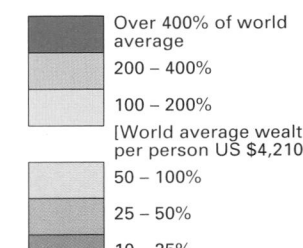

Over 400% of world average

200 – 400%

100 – 200%

[World average wealth per person US $4,210]

50 – 100%

25 – 50%

10 – 25%

Under 10%

Richest countries

Switzerland $33,510
Luxembourg $31,080
Japan $26,920
Sweden $25,490

Poorest countries

Mozambique.......................... $70
Tanzania.............................. $100
Ethiopia.............................. $120
Somalia............................... $150

INDICATORS

The gap between the world's rich and poor is now so great that it is difficult to illustrate it on a single graph. Car ownership in the USA, for example, is almost 2,000 times as common as it is in Bangladesh. Within each income group, however, comparisons have some meaning: the affluent Japanese on their overcrowded island have far fewer cars than the Americans; the Chinese, perhaps because of propaganda value, have more television sets than people in India, whose per capita income is similar, while Nigerians prefer to spend their money on vehicles.

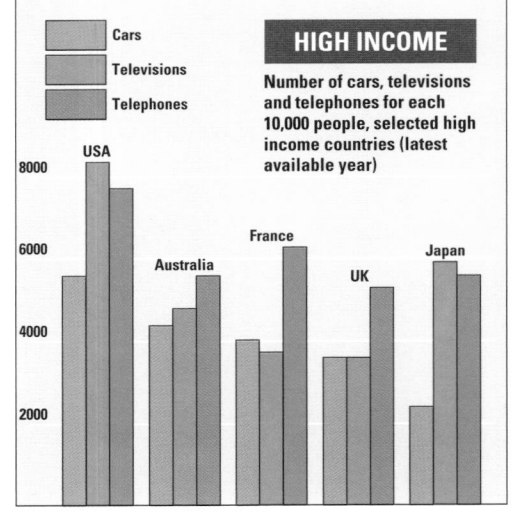

HIGH INCOME

Cars
Televisions
Telephones

Number of cars, televisions and telephones for each 10,000 people, selected high income countries (latest available year)

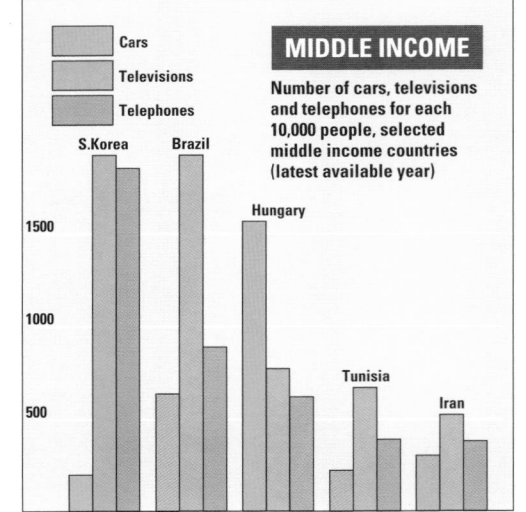

MIDDLE INCOME

Cars
Televisions
Telephones

Number of cars, televisions and telephones for each 10,000 people, selected middle income countries (latest available year)

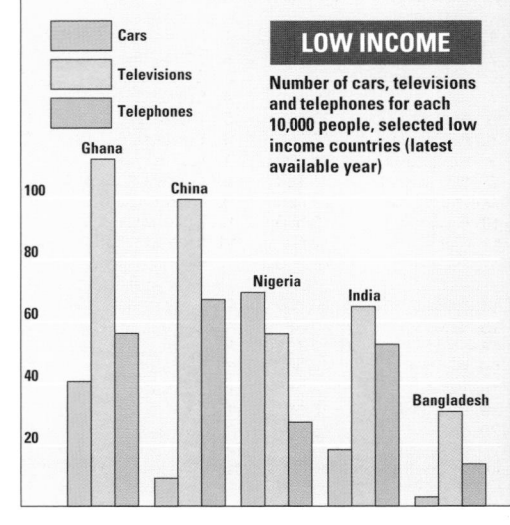

LOW INCOME

Cars
Televisions
Telephones

Number of cars, televisions and telephones for each 10,000 people, selected low income countries (latest available year)

DEBT AND AID

International debtors and the development aid they receive (1989)

■ Debt, $ per capita

■ Aid, $ per capita

Although aid grants make a vital contribution to many of the world's poorer countries, they are usually dwarfed by the burden of debt that developing economies are expected to repay. In the case of Mozambique, aid amounted to more than 70% of GNP. In 1990, the World Bank rated Mozambique as the world's poorest country, yet debt interest payments came to almost 75 times its entire export earnings.

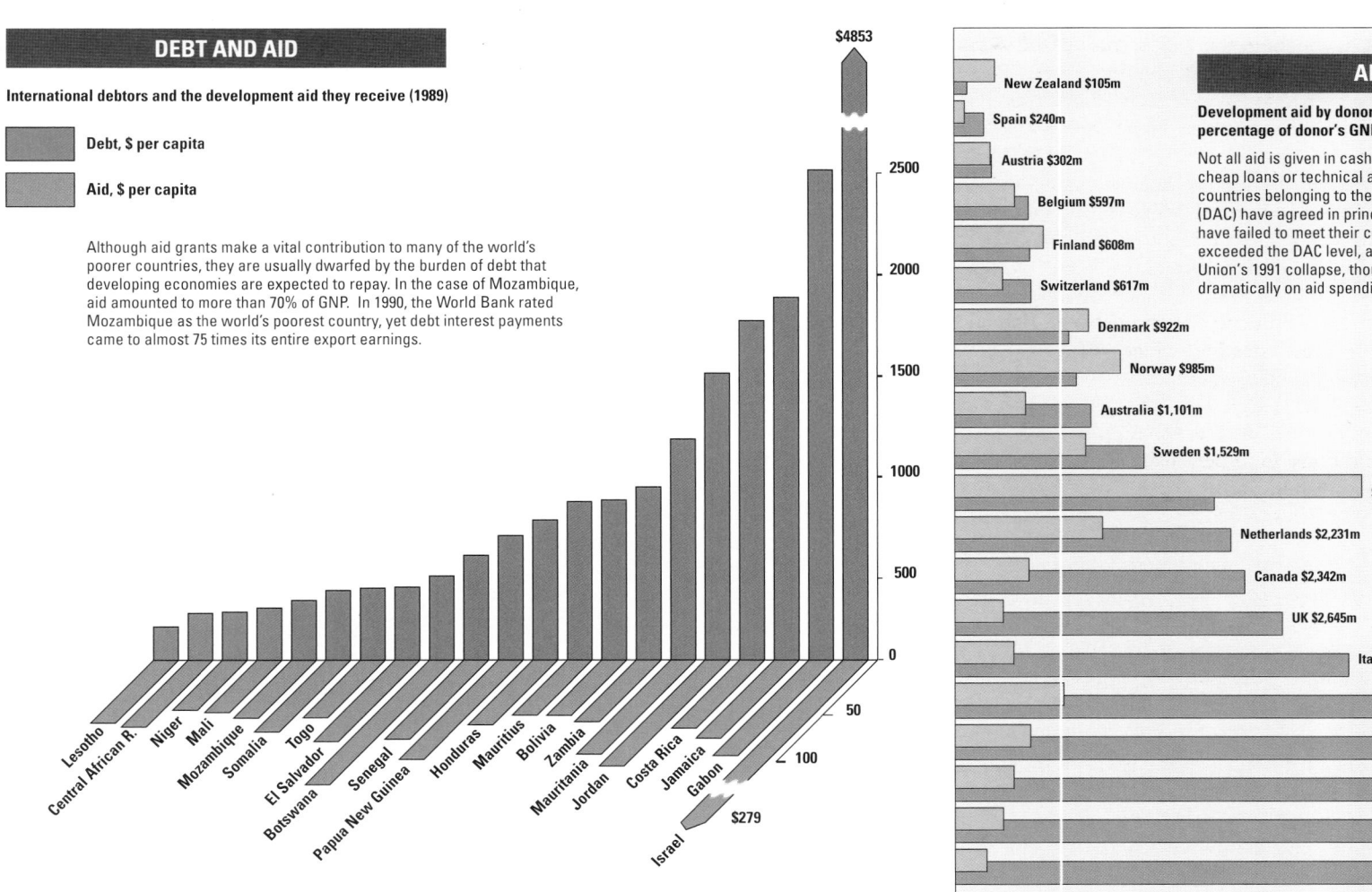

$4853

$279

Lesotho, Central African R., Niger, Mali, Mozambique, Somalia, Togo, El Salvador, Botswana, Senegal, Papua New Guinea, Honduras, Mauritius, Bolivia, Zambia, Mauritania, Jordan, Costa Rica, Jamaica, Gabon, Israel

AID DONORS

Development aid by donor country, in millions of US $ and as a percentage of donor's GNP (1988)

Not all aid is given in cash grants: much is delivered in the form of cheap loans or technical assistance. Since the 1970s, OECD countries belonging to the Development Assistance Committee (DAC) have agreed in principle to give 0.7% of their GNP. Most have failed to meet their commitment. The USSR usually exceeded the DAC level, at least until 1988. After the Soviet Union's 1991 collapse, though, its impoverished heirs cut back dramatically on aid spending.

■ AID in US $

■ % GNP

DAC threshold

New Zealand $105m
Spain $240m
Austria $302m
Belgium $597m
Finland $608m
Switzerland $617m
Denmark $922m
Norway $985m
Australia $1,101m
Sweden $1,529m
Saudi Arabia $2,098m
Netherlands $2,231m
Canada $2,342m
UK $2,645m
Italy $3,183m
USSR $4,212m
France $4,777m
Germany $4,911m
Japan $9,134m
USA $10,141m

.5% 1% 1.5% 2% 2.5%

Inflation (right) is an excellent index of a country's financial stability, and usually its prosperity or at least its prospects. Inflation rates above 20% are generally matched by slow or even negative growth; above 50%, an economy is left reeling. Most advanced countries during the 1980s had to wrestle with inflation that occasionally touched or even exceeded 10%; in Japan, the growth leader, price increases averaged only 1.8% between 1980 and 1988.

Government spending (below right) is more difficult to interpret. Obviously, very low levels indicate a weak state, and high levels a strong one; but in poor countries, the 10–20% absorbed by the government may well amount to most of the liquid cash available, whereas in rich countries most of the 35–50% typically in government hands is returned in services.

GNP per capita figures (below) should also be compared with caution. They do not reveal the vast differences in living costs between different countries: the equivalent of US $100 is worth considerably more in poorer nations than it is in the USA itself.

INFLATION

Average annual rate of inflation (1980–91)

■ Over 50%
■ 20 – 50%
■ 7.5 – 20%
□ 1 – 7.5%
□ Negative inflation
□ No data available

Highest average inflation
Nicaragua 584%
Argentina 417%
Brazil 328%

Lowest average inflation
Oman .... –3.1%
Kuwait............................ –2.7%
Saudi Arabia .... –2.4%

THE WEALTH GAP

The world's richest and poorest countries, by Gross National Product per capita in US $ (1991)

1.	Switzerland	33,510	1.	Mozambique	70
2.	Liechtenstein	33,000	2.	Tanzania	100
3.	Luxembourg	31,080	3.	Ethiopia	120
4.	Japan	26,920	4.	Somalia	150
5.	Sweden	25,490	5.	Uganda	160
6.	Bermuda	25,000	6.	Bhutan	180
7.	Finland	24,400	7.	Nepal	180
8.	Norway	24,160	8.	Guinea-Bissau	190
9.	Denmark	23,660	9.	Cambodia	200
10.	Germany	23,650	10.	Burundi	210
11.	Iceland	22,580	11.	Madagascar	210
12.	USA	22,550	12.	Sierra Leone	210
13.	Canada	21,260	13.	Bangladesh	220
14.	France	20,600	14.	Chad	220
15.	Austria	20,380	15.	Zaïre	220
16.	UAE	19,500	16.	Laos	230
17.	Belgium	19,300	17.	Malawi	230
18.	Italy	18,580	18.	Rwanda	260
19.	Netherlands	18,560	19.	Mali	280
20.	UK	16,750	20.	Guyana	290

GNP per capita is calculated by dividing a country's Gross National Product by its population.

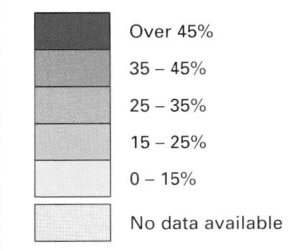

STATE SPENDING

Central government expenditure as a percentage of GNP (latest available year) (‡ estimate)

■ Over 45%
■ 35 – 45%
■ 25 – 35%
■ 15 – 25%
□ 0 – 15%
□ No data available

Top 5 countries
Bulgaria 77.3%
Guinea-Bissau 63.0%
Greece 60.0%
Czechoslovakia 55.6%
Hungary 54.7%

QUALITY OF LIFE: STANDARDS

At first sight, most international contrasts are swamped by differences in wealth. The rich not only have more money, they have more of everything, including years of life. Those with only a little money are obliged to spend most of it on food and clothing, the basic maintenance costs of existence; air travel and tourism are unlikely to feature on the lists of their expenditure. However, poverty and wealth are both relative: slum dwellers living on social security payments in an affluent industrial country have far more resources at their disposal than an average African peasant, but feel their own poverty none the less acutely. A middle-class Indian lawyer cannot command a fraction of the earnings of a counterpart in New York, London or Rome; nevertheless, he rightly sees himself as prosperous.

In 1990, the United Nations Development Program published its first Human Development Index, an attempt to construct a comparative scale by which at least a simplified form of well-being might be measured. The index, running from 1 to 100, combined figures for life expectancy and literacy with a wealth scale that matched incomes against the official poverty lines of a group of industrialized nations. National scores ranged from a startling 98.7 for

Sweden to a miserable 11.6 for Niger, reflecting the all-too-familiar gap between rich and poor.

Comparisons between nations with similar incomes are more interesting, showing the effect of government policies. For example, Sri Lanka was awarded 78.9 against 43.9 for its only slightly poorer neighbor, India; Zimbabwe, at 57.6, had more than double the score of Senegal, despite no apparent disparities in average income. Some development indicators may be interpreted in two ways. There is a very clear correlation, for example, between the wealth of a nation and the level of education that its people enjoy. Education helps create wealth, of course; but are rich countries wealthy because they are educated, or well-educated because they are rich? Women's fertility rates appear to fall almost in direct proportion to the amount of secondary education they receive; but high levels of female education are associated with rich countries, where fertility is already low.

Not everything, though, is married to wealth. The countries cited on these pages have been chosen to give a range covering different cultures as well as different economic power, revealing disparities among rich and among poor as well as between the two obvious groups. Income distribution, for

example, shows that in Brazil (following the general pattern of Latin America) most national wealth is concentrated in a few hands; Bangladesh is much poorer, but what little wealth there is, is more evenly spread.

Among the developed countries the USA, with its poorest 20% sharing less than 5% of the national cake, has a noticeably less even distribution than Japan where, despite massive industrialization, traditional values act as a brake against poverty. Hungary, still enmeshed in Communism when these statistics were compiled, shows the most even distribution of all, which certainly matches with Socialist theory. However, the inequalities in Communist societies, a contributing factor in the demise of most of them in the late 1980s, are not easily measured in money terms. Communist élites are less often rewarded with cash than with power and privilege, commodities not easily expressed statistically.

There are other limits to statistical analysis. Even without taking account of such imponderables as personal satisfaction, it will always be more difficult to measure a reasonable standard of living than a nation's income or its productivity. Lack of money certainly brings misery, but its presence does not guarantee contentment.

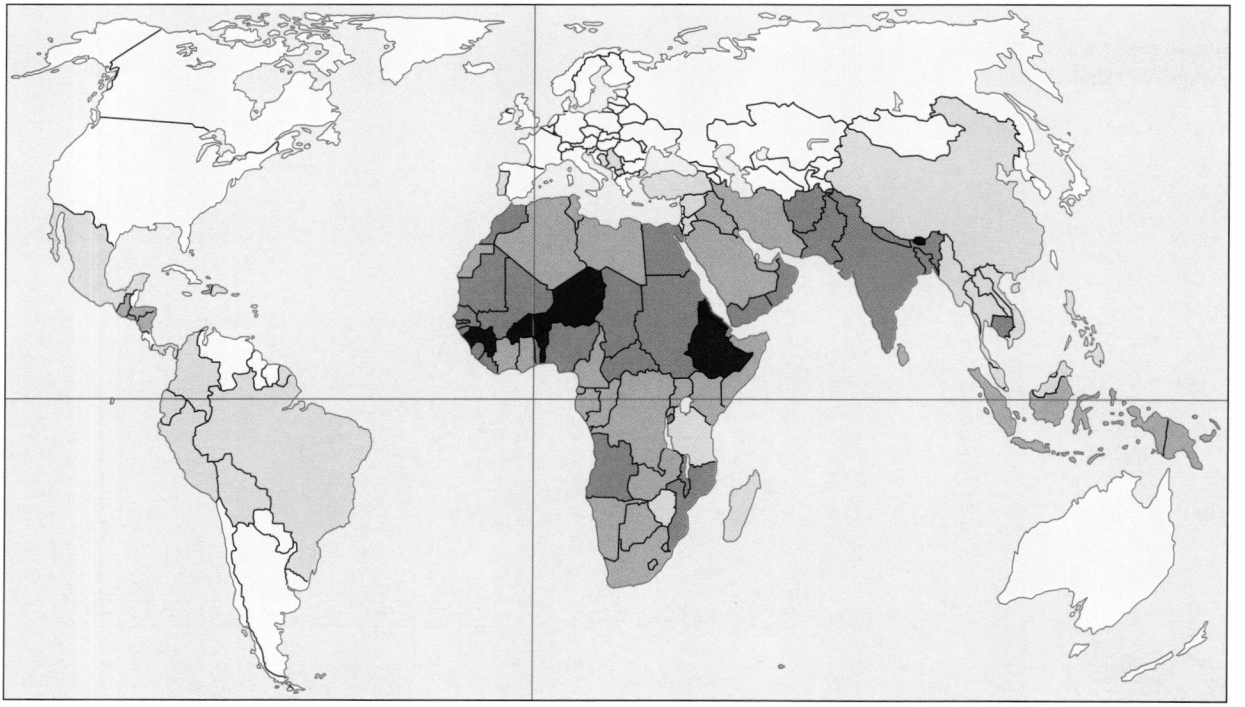

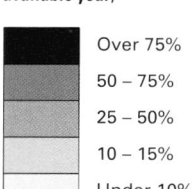

ILLITERACY

Percentage of the total population unable to read or write (latest available year)

- Over 75%
- 50 – 75%
- 25 – 50%
- 10 – 15%
- Under 10%

Educational expenditure per person (latest available year)

Top 5 countries

Sweden	$997
Qatar	$989
Canada	$983
Norway	$971
Switzerland	$796

Bottom 5 countries

Chad	$2
Bangladesh	$3
Ethiopia	$3
Nepal	$4
Somalia	$4

EDUCATION

The developing countries made great efforts in the 1970s and 1980s to bring at least a basic education to their people. Primary school enrolments rose above 60% in all but the poorest nations. Figures often include teenagers or young adults, however, and there are still an estimated 300 million children worldwide who receive no schooling at all. Secondary and higher education are expanding far more slowly, and the gap between rich and poor is probably even larger than it appears from the charts here, while the bare statistics provide no real reflection of educational quality.

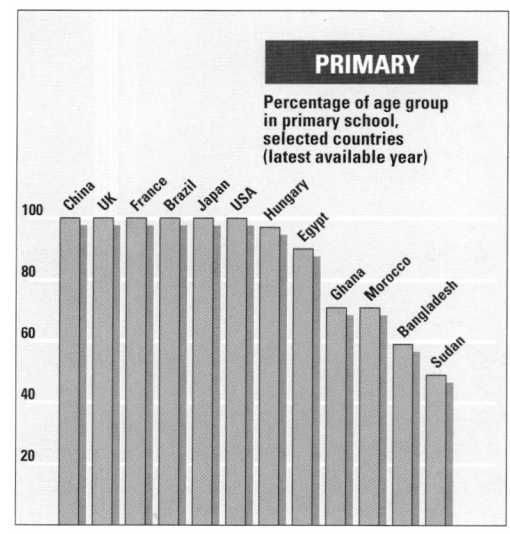

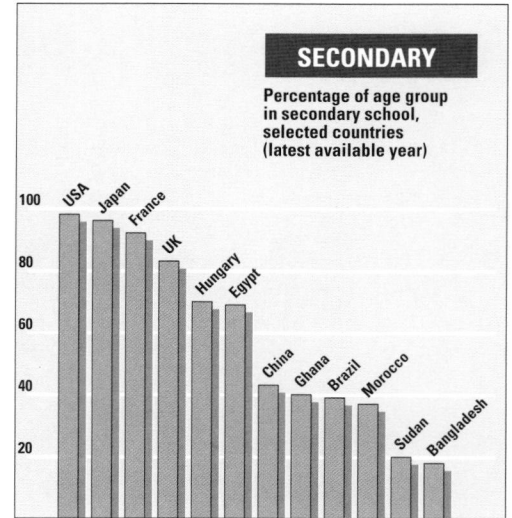

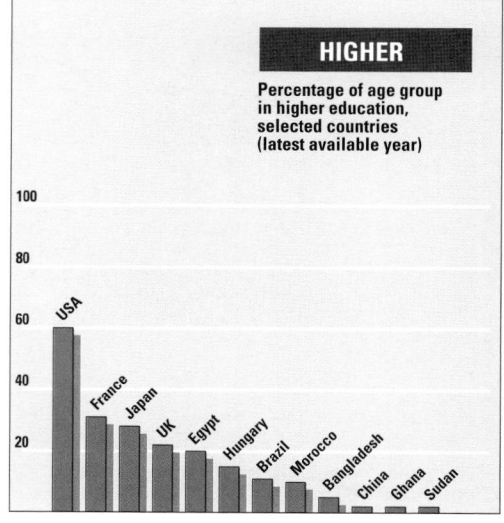

DISTRIBUTION OF SPENDING

Percentage share of household spending (latest available year)

- Food
- Clothing
- Energy & Housing
- Medicine & Education
- Transport
- Other

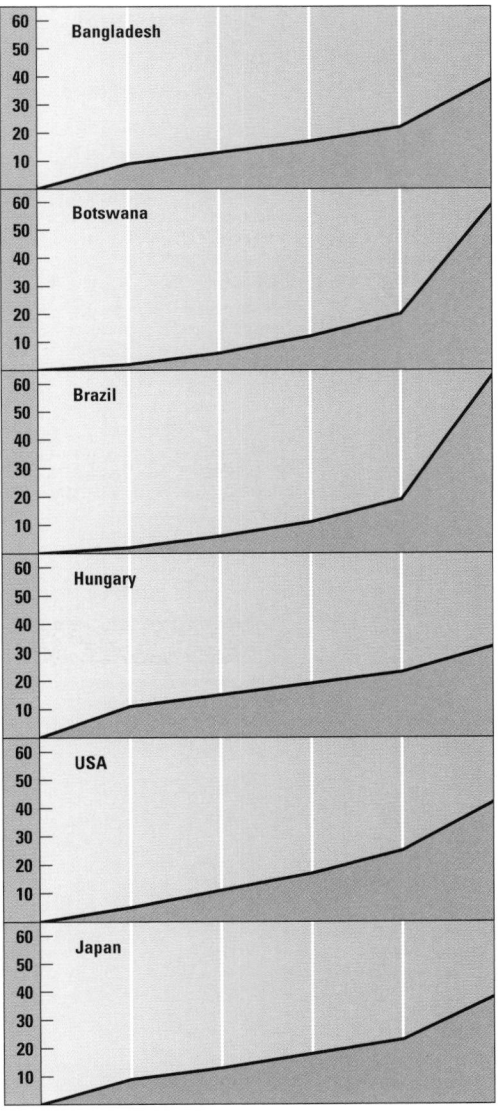

Countries: UK, USA, Japan, Hungary, Brazil, Egypt, Nigeria, B'desh

DISTRIBUTION OF INCOME

Percentage share of household income from poorest fifth to richest fifth, selected countries (latest available year)

Bangladesh
Botswana
Brazil
Hungary
USA
Japan

FERTILITY AND EDUCATION

- Fertility rate: average number of children borne per woman
- Percentage of female age group in secondary education

Fertility rates compared with female education, selected countries (latest available year)

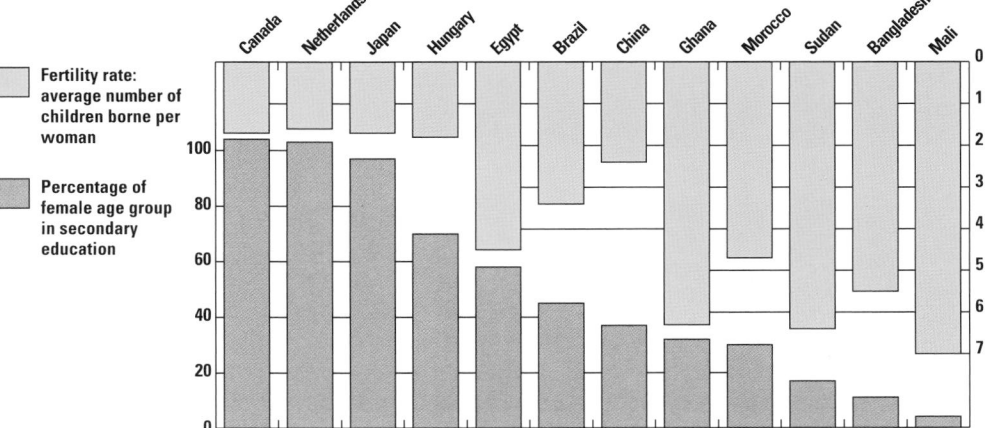

Countries: Canada, Netherlands, Japan, Hungary, Egypt, Brazil, China, Ghana, Morocco, Sudan, Bangladesh, Mali

Since the age group for secondary schooling is usually defined as 12–17 years, percentages for countries with a significant number of 11- or 18-year-olds in secondary school may actually exceed 100. A high proportion of employed women may indicate either an advanced, industrial economy where female opportunities are high, or a poor country where many women's lives are dominated by agricultural toil. The lowest rates are found in Islamic nations, whose religious precepts often exclude women even from fieldwork.

WOMEN AT WORK

Women in paid employment as a percentage of the total workforce (latest available year)

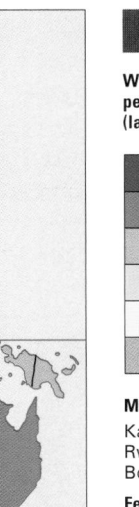

- Over 50%
- 40 – 50%
- 30 – 40%
- 20 – 30%
- 10 – 20%
- Under 10%

Most women in work
Kazakhstan 54%
Rwanda 54%
Botswana 53%

Fewest women in work
Guinea-Bissau.................... 3%
Oman 6%
Afghanistan 8%

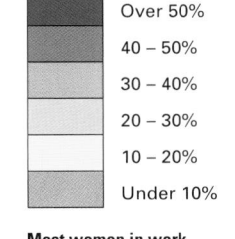

TOURIST SPENDING

Nations spending the most on overseas tourism, US $ million (latest available year)

Germany, USA, UK, Japan, France, Netherlands, Canada, Italy, Austria, Switzerland, Belgium, Sweden

TOURIST EARNING

Nations receiving the most from overseas tourism, US $ million (latest available year)

USA, Spain, Italy, France, UK, Germany, Austria, Switzerland, Canada, Mexico, Belgium, Netherlands

Small economies in attractive areas are often completely dominated by tourism: in some West Indian islands, tourist spending provides over 90% of the total income. In cash terms the USA is the world leader: its 1987 earnings exceeded $15 billion, though that sum amounted to only 0.4% of its GDP.

AIR TRAVEL

Millions of passenger miles [number carried, multiplied by distance flown by each from airport of origin] (latest year)

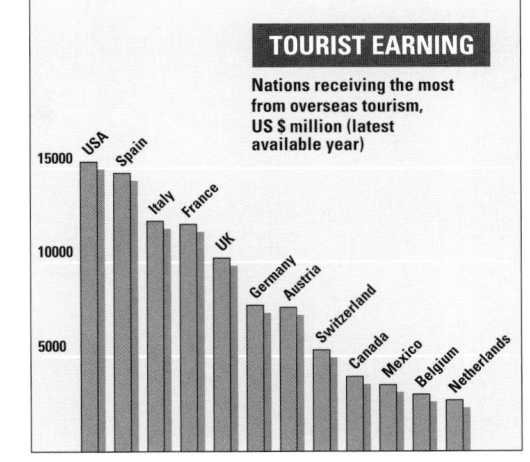

- Over 60,000
- 30,000 – 60,000
- 6,000 – 30,000
- 600 – 6,000
- 300 – 600
- Under 300
- ○ Major airports (over 20 million passengers in 1991)

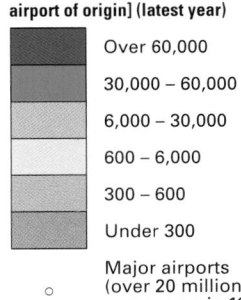

The world's busiest airport in terms of total passengers is Chicago's O'Hare; the busiest international airport is Heathrow, the largest of London's airports

QUALITY OF LIFE: HEALTH

According to statistics gathered in the late 1980s and early 1990s, a third of the world's population has no access to safe drinking water: malaria is on the increase; cholera, thought vanquished, is reappearing in South America; an epidemic of the AIDS virus is gathering force in Africa; and few developing countries can stretch their health care budgets beyond US $2 per person per year.

Yet human beings, by every statistical index, have never been healthier. In the richest nations, where food is plentiful, the demands of daily work are rarely onerous and medical care is both readily available and highly advanced, the average life expectancy is often more than 75 years – approaching the perceived limits for human longevity. In middle-income nations, such as Brazil and the Philippines, life expectancy usually extends at least to the mid-60s; in China, it has already reached 70 years. Even in poverty-stricken Ethiopia and Chad, lifespans are close to 50 years. Despite economic crisis, drought, famine and even war, every country in the world reported an increase between 1965 and 1990.

It was not always so, even in countries then considered rich. By comparison, in 1880 the life expectancy of an average Berliner was under 30 years and infant mortality in the United Kingdom, then the wealthiest nation, stood at 144 per thousand births – a grim toll exceeded today only by three of the poorest African countries (Mali, Sierra Leone and Guinea). Even by 1910, European death rates were almost twice as high as the world average less than 80 years later; infant mortality in Norway, Europe's healthiest country, was then higher than in present-day Indonesia. In far less than a century, human prospects have improved beyond recognition.

In global terms, the transformation is less the result of high-technology medicine – still too expensive for all but a minority, even in rich countries – than of improvements in agriculture and hence nutrition, matched by the widespread diffusion of the basic concepts of disease and public health. One obvious consequence, as death rates everywhere continue to fall, is sustained population growth. Another is the rising expectation of continued improvement felt by both rich and poor nations alike.

In some ways, the task is easier for developing countries, striving with limited resources to attain health levels to which the industrialized world has only recently become accustomed. As the tables below illustrate, infectious disease is rare among the richer nations, while ailments such as cancer, which tend to kill in advanced years, do not seriously impinge on populations with shorter lifespans.

Yet infectious disease is relatively cheap to eliminate, or at least reduce, and it is likely to be easier to raise life expectancy from 60 to 70 years than from 75 to 85 years. The ills of the developed world and its aging population are more expensive to treat – though most poor countries would be happy to suffer from the problems of the affluent. Western nations regularly spend more money on campaigns to educate their citizens out of overeating and other bad habits than many developing countries can devote to an entire health budget – an irony that marks the dimensions of the rich-poor divide.

Indeed, wealth itself may be the most reliable indicator of longevity. Harmful habits are usually the province of the rich; yet curiously, though the dangerous effects of tobacco have been proved beyond doubt, the affluent Japanese combine very high cigarette consumption with the longest life expectancy of all the major nations. Similarly, heavy alcohol consumption seems to have no effect on longevity: the French, world leaders in 1988 and in most previous surveys, outlive the more moderate British by a year, and the abstemious Indians by almost two decades.

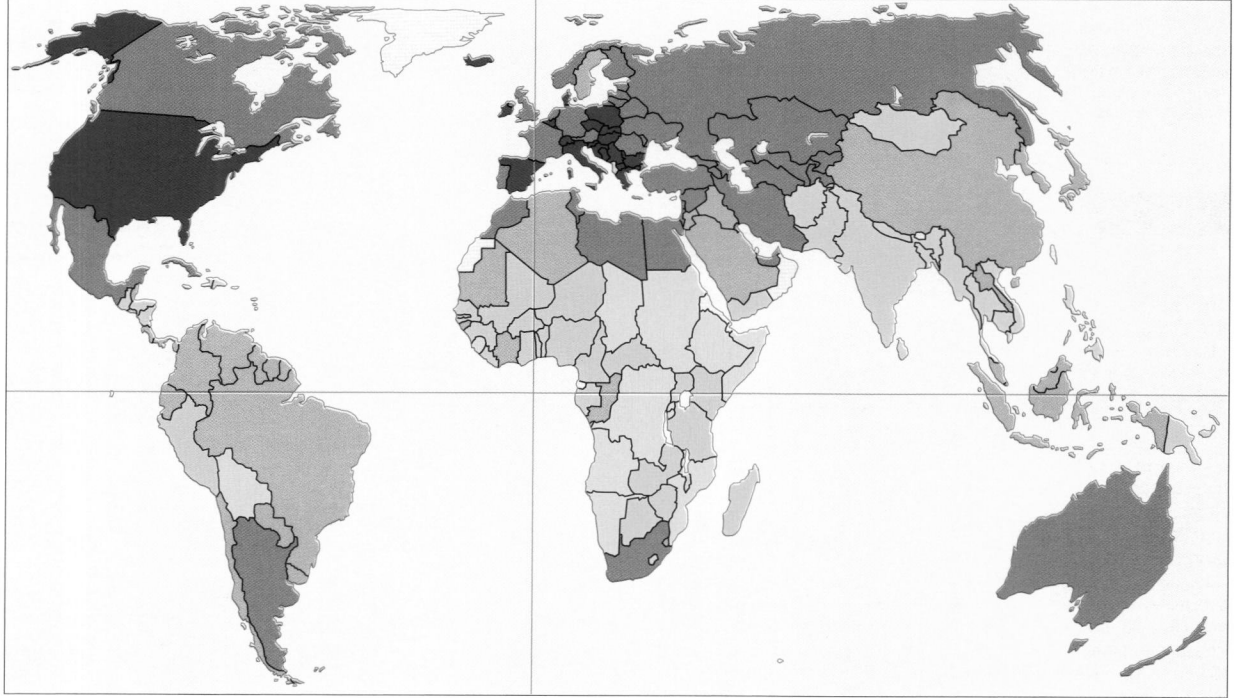

FOOD CONSUMPTION

Average daily food intake in calories per person (1989)

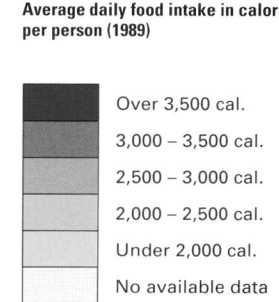

- Over 3,500 cal.
- 3,000 – 3,500 cal.
- 2,500 – 3,000 cal.
- 2,000 – 2,500 cal.
- Under 2,000 cal.
- No available data

Top 5 countries

Belgium	3,902 cal.
Greece	3,825 cal.
Ireland	3,778 cal.
Bulgaria	3,707 cal.
USA	3,650 cal.

Bottom 5 countries

Ethiopia	1,666 cal.
Mozambique	1,679 cal.
Chad	1,742 cal.
Sierra Leone	1,799 cal.
Angola	1,806 cal.

CAUSES OF DEATH

The rich not only live longer, on average, than the poor; they also die from different causes. Infectious and parasitic diseases, all but eliminated in the developed world, remain a scourge in poorer countries. On the other hand, more than two-thirds of the populations of OECD nations eventually succumb to cancer or circulatory disease; the proportion in Latin America is only about 45%. In addition to the three major diseases shown here, respiratory infection and injury also claim more lives in developing nations, which lack the drugs and medical skills required to treat them.

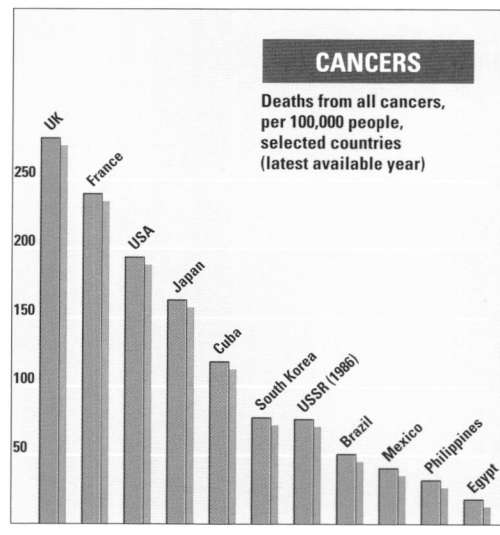

CANCERS

Deaths from all cancers, per 100,000 people, selected countries (latest available year)

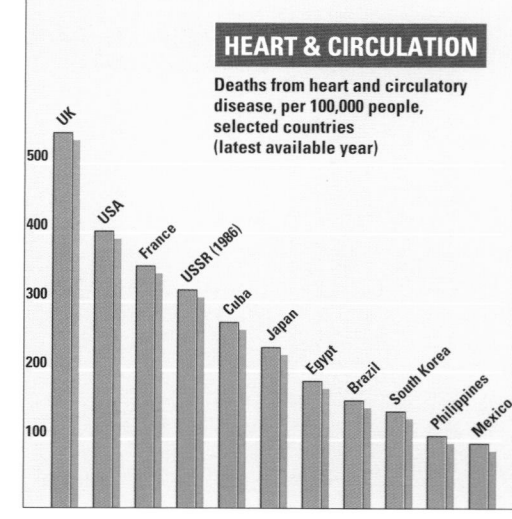

HEART & CIRCULATION

Deaths from heart and circulatory disease, per 100,000 people, selected countries (latest available year)

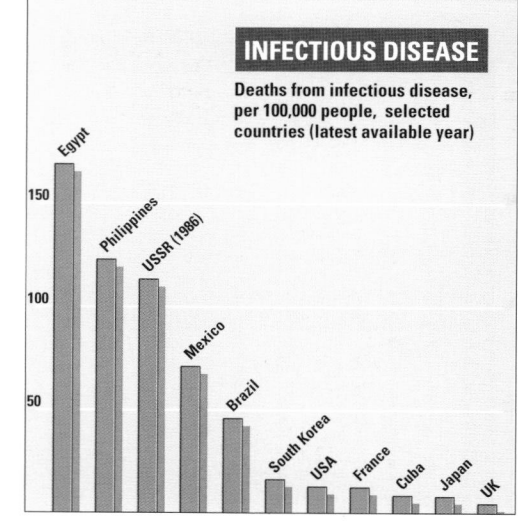

INFECTIOUS DISEASE

Deaths from infectious disease, per 100,000 people, selected countries (latest available year)

LIFE EXPECTANCY

Years of life expectancy at birth, selected countries (1990–95)

The chart shows combined data for both sexes. On average, women live longer than men worldwide, even in developing countries with high maternal mortality rates. Overall, life expectancy is steadily rising, though the difference between rich and poor nations remains dramatic.

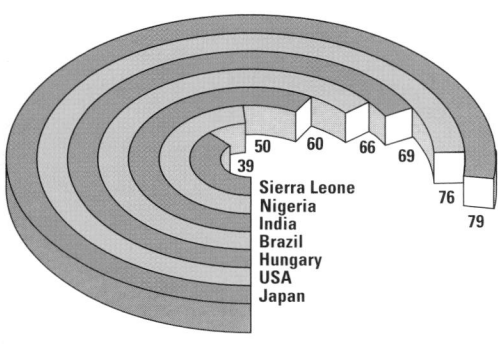

39 Sierra Leone
50 Nigeria
60 India
66 Brazil
69 Hungary
76 USA
79 Japan

CHILD MORTALITY

Number of babies who will die under the age of one, per 1,000 births (average 1990–95)

Over 150 deaths
100 – 150 deaths
50 – 100 deaths
20 – 50 deaths
10 – 20 deaths
Under 10 deaths

Highest child mortality

Afghanistan...................... 162
Mali................................... 159

Lowest child mortality

Iceland 5
Finland.................................. 5

[USA 9]

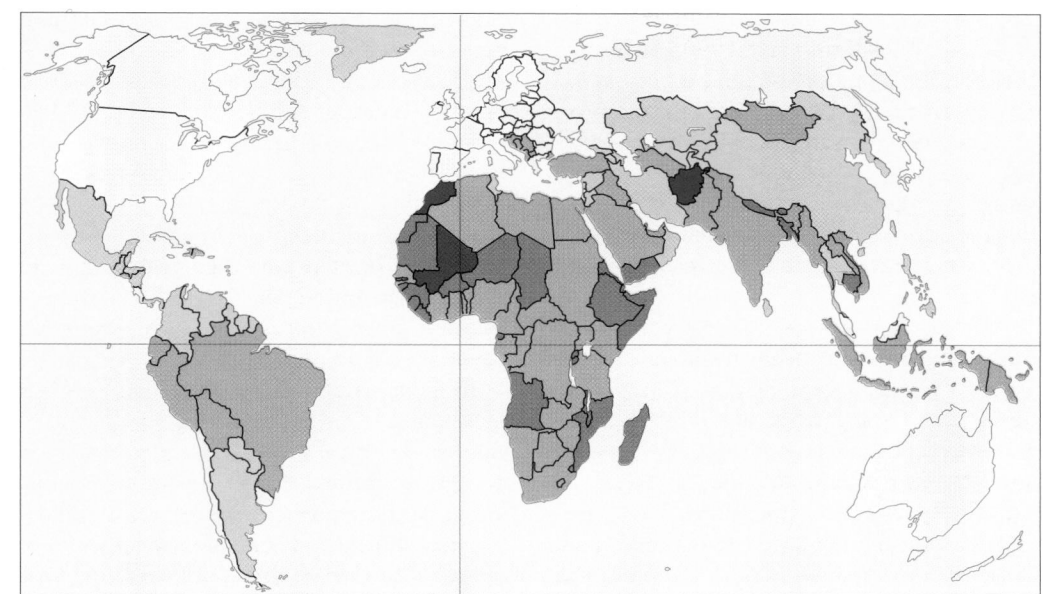
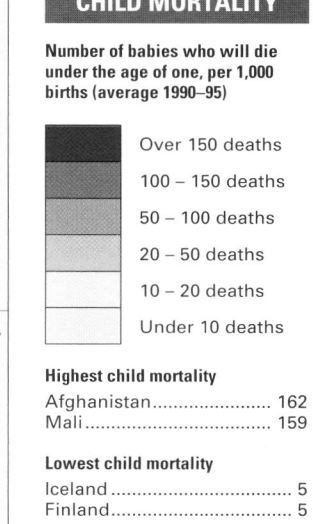

HOSPITAL CAPACITY

Hospital beds available for each 1,000 people (latest available year)

Highest capacity		Lowest capacity	
Finland	14.9	Bangladesh	0.2
Sweden	13.2	Nepal	0.2
France	12.9	Ethiopia	0.3
USSR (1986)	12.8	Mauritania	0.4
Netherlands	12.0	Mali	0.5
North Korea	11.7	Burkina Faso	0.6
Switzerland	11.3	Pakistan	0.6
Austria	10.4	Niger	0.7
Czechoslovakia	10.1	Haiti	0.8
Hungary	9.1	Chad	0.8

[USA 5.9]

The availability of a bed can mean anything from a private room in a well-equipped Californian teaching hospital to a place in the overcrowded annex of a rural African clinic. In the Third World especially, quality of treatment can vary enormously from place to place within the same country.

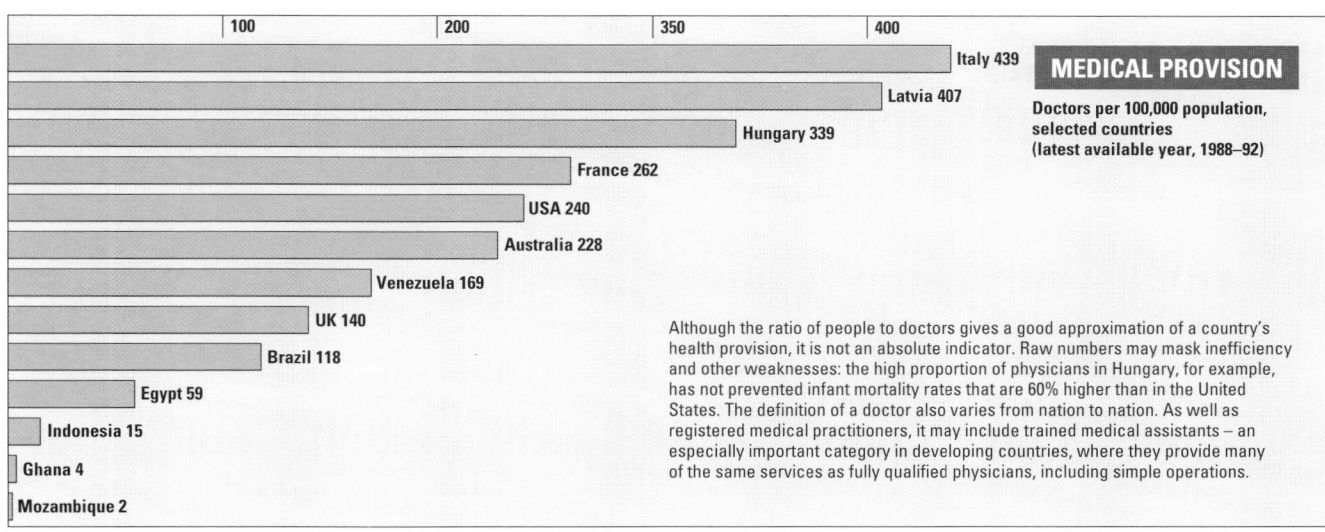

100	200	350	400	

Italy 439
Latvia 407
Hungary 339
France 262
USA 240
Australia 228
Venezuela 169
UK 140
Brazil 118
Egypt 59
Indonesia 15
Ghana 4
Mozambique 2

MEDICAL PROVISION

Doctors per 100,000 population, selected countries (latest available year, 1988–92)

Although the ratio of people to doctors gives a good approximation of a country's health provision, it is not an absolute indicator. Raw numbers may mask inefficiency and other weaknesses: the high proportion of physicians in Hungary, for example, has not prevented infant mortality rates that are 60% higher than in the United States. The definition of a doctor also varies from nation to nation. As well as registered medical practitioners, it may include trained medical assistants – an especially important category in developing countries, where they provide many of the same services as fully qualified physicians, including simple operations.

THE AIDS CRISIS

The Acquired Immune Deficiency Syndrome was first identified in 1981, when American doctors found otherwise healthy young men succumbing to rare infections. By 1984, the cause had been traced to the Human Immunodeficiency Virus (HIV), which can remain dormant for many years and perhaps indefinitely: only half of those known to carry the virus in 1981 had developed AIDS ten years later.

By 1991 the World Health Organization knew of more than 250,000 AIDS cases worldwide and suspected the true number to be at least four times as high. In Western countries in the early 1990s, most AIDS deaths were among male homosexuals or needle-sharing drug-users. However, the disease is spreading fastest among heterosexual men and women, which is its usual vector in the Third World, where most of its victims live. Africa is the most severely hit: a 1992 UN report estimated that 2 million African children will die of AIDS before the year 2000 – and some 10 million will be orphaned.

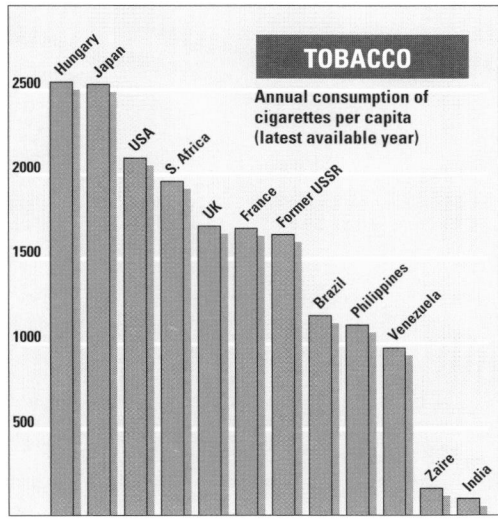

TOBACCO

Annual consumption of cigarettes per capita (latest available year)

Hungary
Japan
USA
S. Africa
UK
France
Former USSR
Brazil
Philippines
Venezuela
Zaire
India

CRIME AND PUNISHMENT

MURDER RATES

Murders per 100,000 population, selected countries (latest available year)

Philippines 38.7
Botswana
USA
Chile
Sudan
Australia
France
Canada
Hungary
UK
Japan
China

Crime rates are difficult to compare internationally. Standards of reporting and detection vary greatly, as do the definitions of many types of crime. Murder is probably the best detected as well as the most heinous, but different legal systems make different distinctions between murder and manslaughter or other forms of culpable homicide. By any reckoning, however, the USA's high murder rate stands out against otherwise similar Western countries, although it is dwarfed by the killings recorded in the very different culture of the Philippines.

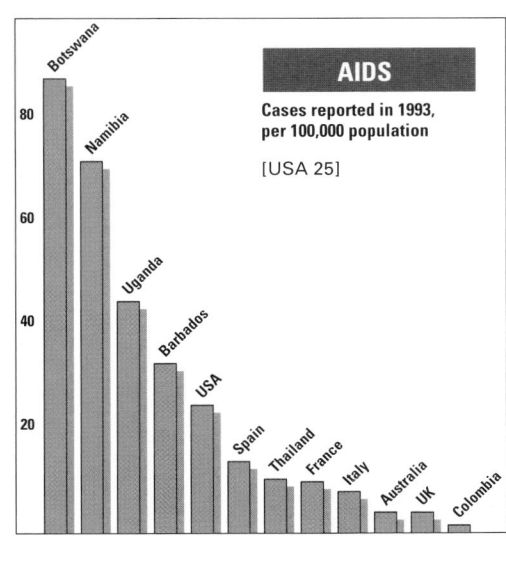

AIDS

Cases reported in 1993, per 100,000 population

[USA 25]

Botswana
Namibia
Uganda
Barbados
USA
Spain
Thailand
France
Italy
Australia
UK
Colombia

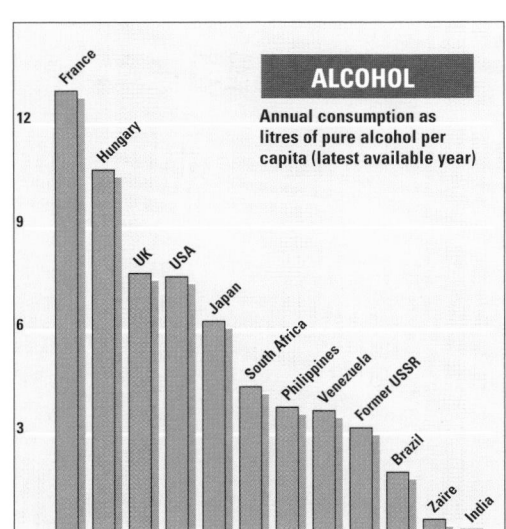

ALCOHOL

Annual consumption as litres of pure alcohol per capita (latest available year)

France
Hungary
UK
USA
Japan
South Africa
Philippines
Venezuela
Former USSR
Brazil
Zaire
India

PRISON POPULATIONS

Prisoners per 100,000 population, selected developed countries (latest available year)

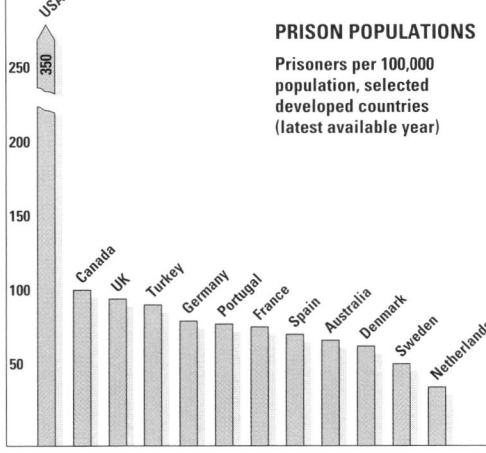

USA 350
Canada
UK
Turkey
Germany
Portugal
France
Spain
Australia
Denmark
Sweden
Netherlands

Differences in prison population reflect penal policies as much as the relative honesty or otherwise of different nations, and by no means all governments publish accurate figures. In more than 50 countries, people are still regularly imprisoned without trial, in 60 torture is a normal part of interrogation, and some 130 retain the death penalty, often administered for political crimes and in secret. Over 2,000 executions were recorded in 1990 by the civil rights organization Amnesty International; the real figure, as Amnesty itself maintains, was almost certainly much higher.

QUALITY OF LIFE: ENVIRONMENT

Humans have always had a dramatic effect on their environment, at least since the invention of agriculture almost 10,000 years ago. Generally, the Earth has accepted human interference without any obvious ill effects: the complex systems that regulate the global environment have managed to absorb substantial damage while maintaining a stable and comfortable home for the planet's trillions of lifeforms. But advancing human technology and the rapidly expanding populations it supports are now threatening to overwhelm the Earth's ability to cope.

Industrial wastes, acid rainfall, expanding deserts and large-scale deforestation all combine to create environmental change at a rate far faster than the Earth can easily accommodate. Equipped with chainsaws

and flamethrowers, humans can now destroy more forest in a day than their ancestors could in a century, upsetting the balance between plant and animal, carbon dioxide and oxygen, on which all life ultimately depends. The fossil fuels that power industrial civilization have pumped enough carbon dioxide and other greenhouse gases into the atmosphere to make climatic change a near-certainty. Chlorofluorocarbons (CFCs) and other man-made chemicals are rapidly eroding the ozone layer, the planet's screen against ultraviolet radiation.

As a result, the Earth's average temperature has risen by almost 1°F since the beginning of this century. Further rises seem inevitable, with 1990 marked as the hottest year worldwide since records began. A warmer Earth probably means a wetter Earth,

with melting icecaps raising sea levels and causing severe flooding in some of the world's most densely populated regions. Other climatic models suggest an alternative doom: rising temperatures could increase cloud cover, reflecting more solar energy back into space and causing a new Ice Age.

Either way, the consequences for humans could be disastrous – perhaps the Earth's own way of restoring the ecological balance over the next few thousand years. Fortunately, there is a far faster mechanism available. Humans have provoked the present crisis, but human ingenuity can respond to it. CFC production is already almost at a standstill, and the first faltering steps toward stabilization and the reduction of carbon dioxide have been taken, with Denmark pioneering the way by taxing emissions in 1991.

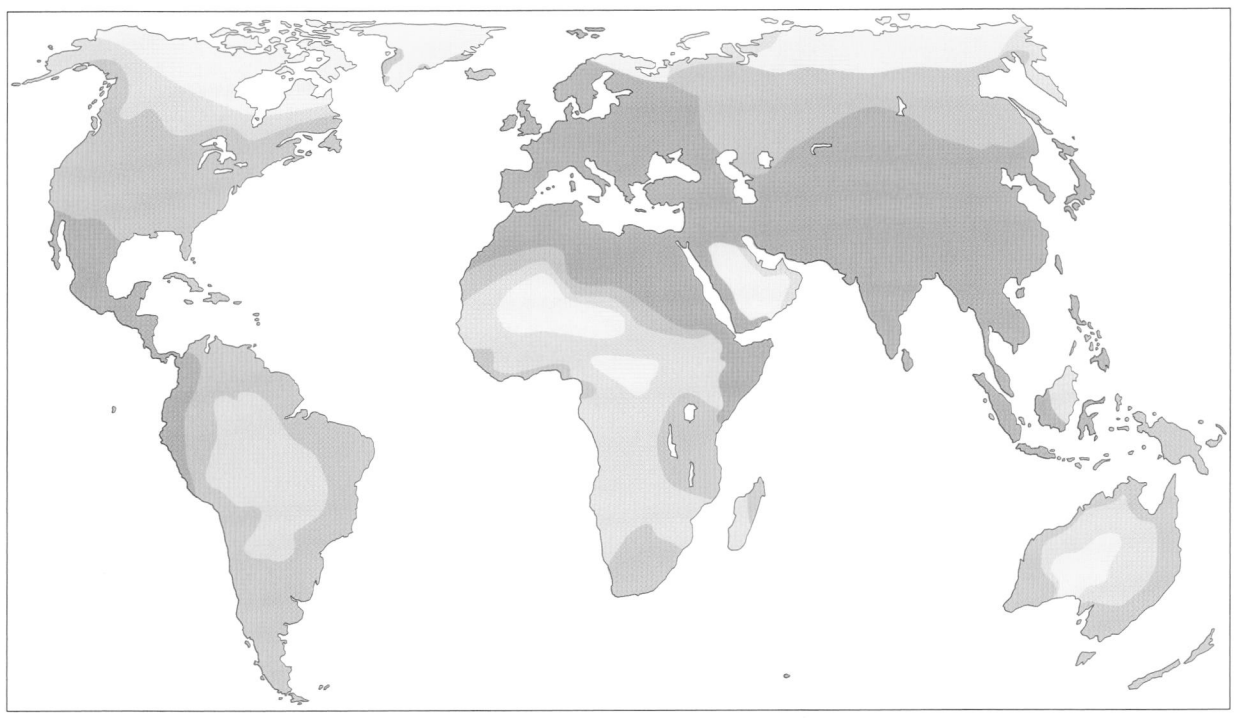

THE HISTORY OF HUMAN EXPANSION

The growth of ecological control: areas where human activity dominates the environment, from primitive times to the year 2000

- By AD 1500
- By AD 1900
- By AD 2000
- Areas not dominated by human activity

THE RISE IN CARBON DIOXIDE

Emissions of carbon dioxide in millions of tons, 1950–91

Atmospheric concentration of carbon dioxide, parts per million, 1750–2000. Pre-1950 data were obtained from air samples trapped in Antarctic ice.

Since the beginning of the Industrial Revolution, human activity has pumped steadily more and more carbon dioxide into the atmosphere. Most of it was quietly absorbed by the oceans, whose immense 'sink' capacity meant that 170 years were needed for levels to increase from the pre-industrial 280 parts per million to 300 (inset graph). But the vast increase in fuel-burning since 1950 (main graph) has overwhelmed even the oceanic sink. Atmospheric concentrations are now rising almost as steeply as carbon dioxide emissions themselves.

GREENHOUSE POWER

Relative contributions to the Greenhouse Effect by the major heat-absorbing gases in the atmosphere

The chart combines greenhouse potency and volume. Carbon dioxide has a greenhouse potential of only 1, but its concentration of 350 parts per million makes it predominant. CFC 12, with 25,000 times the absorption capacity of CO_2, is present only as 0.00044 ppm.

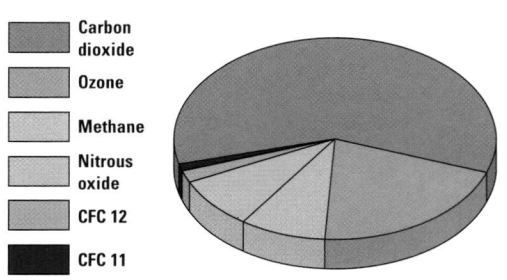

- Carbon dioxide
- Ozone
- Methane
- Nitrous oxide
- CFC 12
- CFC 11

CARBON DIOXIDE

Carbon dioxide released in millions of tons (1980s)

Although most of the net increase in atmospheric carbon dioxide comes from fossil fuel combustion, deforestation and changing land use also contribute.

- Fuel burning
- Deforestation

GLOBAL WARMING

The rise in average temperatures caused by carbon dioxide and other greenhouse gases (1960–2020)

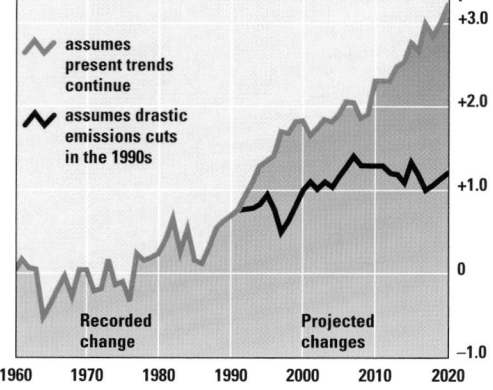

- assumes present trends continue
- assumes drastic emissions cuts in the 1990s

Recorded change   Projected changes

1960 1970 1980 1990 2000 2010 2020

ACID RAIN

Acid rainfall and sources of acidic emissions (1980s)
Acid rain is caused when sulfur and nitrogen oxides in the air combine with water vapor to form sulfuric, nitric and other acids.

 Regions where sulfur and nitrogen oxides are released in high concentrations, mainly from fossil fuel combustion.

• Major cities with high levels of air pollution (including nitrogen and sulfur emissions)

Areas of heavy acid deposition
pH numbers indicate acidity, decreasing from a neutral 7. Normal rain, slightly acid from dissolved carbon dioxide, never exceeds a pH of 5.6.

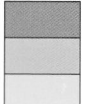

pH less than 4.0 (most acidic)

pH 4.0 to 4.5

pH 4.5 to 5.0

Areas where acid rain is a potential problem

ANTARCTICA

The vast Antarctic ice-sheet, containing some 70% of the Earth's fresh water, plays a crucial role in the circulation of atmosphere and oceans and hence in determining the planetary climate. The frozen southern continent is also the last remaining wilderness – the largest area to remain free from human colonization.

Ever since Amundsen and Scott raced for the South Pole in 1911, various countries have pressed territorial claims over sections of Antarctica, spurred in recent years by its known and suspected mineral wealth: enough iron ore to supply the world at present levels for 200 years, large oil reserves and, probably, the biggest coal deposits on Earth.

However, the 1961 Antarctic Treaty set aside the area for peaceful uses only, guaranteeing freedom of scientific investigation, banning waste disposal and nuclear testing, and suspending the issue of territorial rights. By 1990, the original 12 signatories had grown to 25, with a further 15 nations granted observer status in subsequent deliberations. However, the Treaty itself was threatened by wrangles between different countries, government agencies and international pressure groups.

Finally, in July, 1991, the belated agreement of the UK and the US assured unanimity on a new accord to ban all mineral exploration for a further 50 years. The ban can only be rescinded if all the present signatories, plus a majority of any future adherents, agree. While the treaty has always lacked a formal mechanism for enforcement, it is firmly underwritten by public concern generated by the efforts of environmental pressure groups such as Greenpeace, which has been foremost in the campaign to have Antarctica declared a 'World Park'.

It now seems likely that the virtually uninhabited continent will remain untouched by tourism, staying nuclear-free and dedicated to peaceful scientific research.

DESERTIFICATION

Existing deserts
Areas with a high risk of desertification
Areas with a moderate risk of desertification
Former areas of rain forest
Existing rain forest

DEFORESTATION

Millions of acres of forest cleared annually, tropical countries surveyed 1981–85 and 1987–90. Loss as a percentage of remaining stocks is shown in figures on each column.

1987–90   1981–85

Country	1987–90	1981–85
Brazil	—	0.4
India	4.1	0.3
Indonesia	0.8	0.5
Burma	2.1	0.3
Thailand	2.5	2.4
Vietnam	2.0	0.7
Philippines	1.5	1.0
Costa Rica	7.6	4.0
Cameroon	0.6	0.4

1.5

WATER POLLUTION

Severely polluted sea areas and lakes
Less polluted sea areas and lakes
Areas of frequent oil pollution by shipping

Major oil tanker spills ▸
Major oil rig blow-outs ▲
Offshore dumpsites for industrial and municipal waste ▾
Severely polluted rivers and estuaries ——

Poisoned rivers, domestic sewage and oil spillage have combined in recent years to reduce the world's oceans to a sorry state of contamination, notably near the crowded coasts of industrialized nations. Shipping routes, too, are constantly affected by tanker discharges. Oil spills of all kinds, however, declined significantly during the 1980s, from a peak of over 750,000 tons in 1979 to less than 50,000 tons in 1990. The most notorious tanker spill of that period – when the *Exxon Valdez* (94,999 grt) ran aground in Prince William Sound, Alaska, in March 1989 – released only 267,000 barrels, a relatively small amount compared to the results of blow-outs and war damage. Over 2,500,000 barrels were spilled during the Gulf War of 1991. The worst tanker accident in history occurred in July 1979, when the *Atlantic Empress* and the *Aegean Captain* collided off Trinidad, polluting the Caribbean with 1,890,000 barrels of crude oil.

CITY MAPS

Oslo, Copenhagen 2, Helsinki, Stockholm 3, London 4, Paris 5, The Ruhr 6, Berlin,
Hamburg, Munich 7, Madrid, Barcelona, Lisbon, Athens 8, Turin, Milan, Rome,
Naples 9, Prague, Warsaw, Vienna, Budapest 10, Moscow, St Petersburg 11,
Osaka, Hong Kong, Seoul 12, Tokyo 13, Peking, Shanghai, Tientsin, Canton 14,
Bangkok, Manila, Singapore, Jakarta 15, Delhi, Bombay, Calcutta 16,
Istanbul, Tehran, Baghdad, Karachi 17, Lagos, Cairo, Johannesburg 18, Sydney,
Melbourne 19, Montréal, Toronto 20, Boston 21, New York 22, Philadelphia 24,
Washington, Baltimore 25, Chicago 26, San Francisco 27, Los Angeles 28,
Mexico City 29, Havana, Caracas, Lima, Santiago 30, Rio de Janeiro,
São Paulo 31, Buenos Aires 32

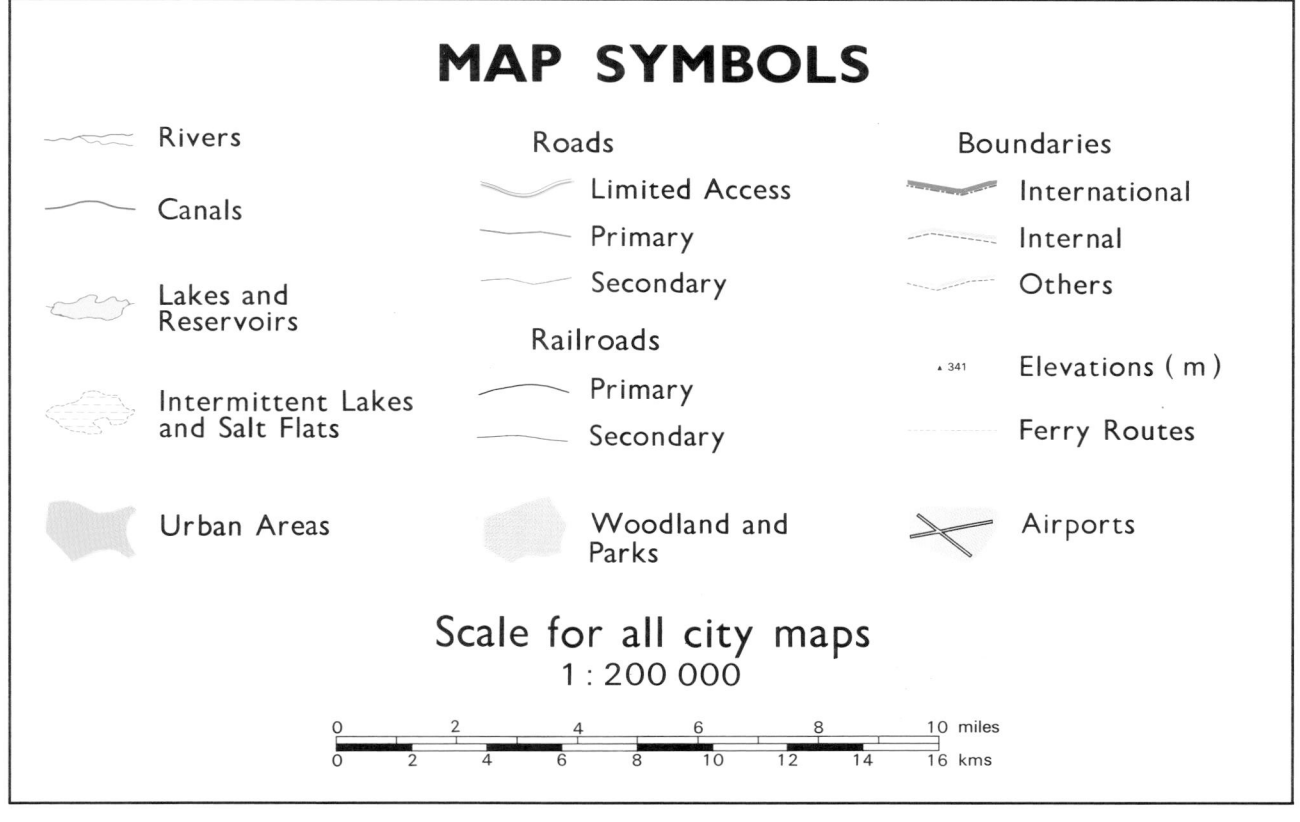

MAP SYMBOLS

Rivers	Roads	Boundaries
Canals	Limited Access	International
	Primary	Internal
Lakes and Reservoirs	Secondary	Others
	Railroads	
Intermittent Lakes and Salt Flats	Primary	• 341 Elevations (m)
	Secondary	Ferry Routes
Urban Areas	Woodland and Parks	Airports

Scale for all city maps
1 : 200 000

0   2   4   6   8   10 miles
0   2   4   6   8   10   12   14   16 kms

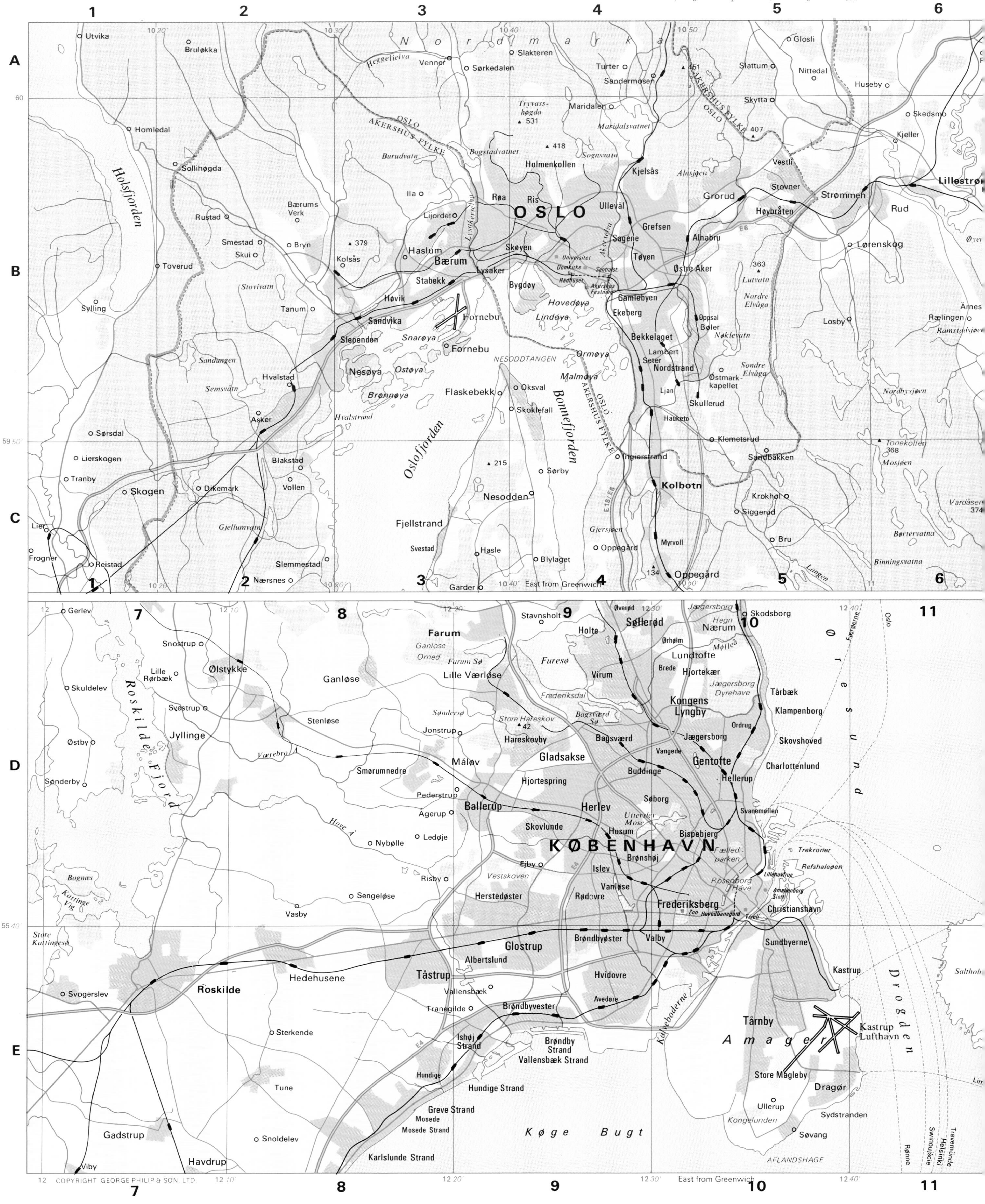

1: 200 000

5 miles
8 km

1 2 3 4 5 6

A

N o r d m a r k a

Utvika
Bruløkka
Heggelielva
Venner
Sørkedalen
Slakteren
Turter
Sandermosen
451
Slattum
Nittedal
Glosli
Huseby

60

Homledal
OSLO
AKERSHUS FYLKE
Tryvass-høgda
531
Maridalen
Maridalsvatnet
Akershus Fylke
Skytta
Skedsmo
Kjeller
407

Sollihøgda
Burudvatn
Bogstadvatnet
418
Sognsvatn
Holmenkollen
Kjelsås
Alnsjøen
Vestli
Strømmen
Lillestrø

B

Rustad
Bærums Verk
Ila
Røa
Ris
Ullevål
Grorud
Høybråten
Rud

Smestad
Skui
Bryn
Kolsås
379
Haslum
OSLO
Skøyen
Sagene
Alnabru
Østre Aker
363
Løvenskog

Toverud
Lijordet
Bærum
Lysaker
Universitet
Domkirke
Sentralst
Rådhuset
Tøyen
Østre Aker
Lutvatn
Nordre Elvåga

Sylling
Stabekk
Hovik
Bygdøy
Akershus Festning
Gamlebyen
Ekeberg
Oppsal
Bøler
Nøklevatn
Ærnes
Rælingen
Ramstadsjø

Holsjfjorden
Sandvika
Snarøya
Førnebu
Hovedøya
Lindøya
Bekkelaget
Lambert Seter
Nordstrand
Sondre Elvåga
Losby
Nordbysjøen

Slependen
Sandangen
Nesøya
Ostøya
Ormøya
Malmøya
Ljan
Skullerud
Østmark-kapellet

Hvalstad
Brønnøya
Flaskebekk
Oksval
Skoklefall
Haukelø
Klemetsrud
Tonekollen

59 50'
Sørsdal
Semsvatn
Asker
Hvalstrand
NESODDTANGEN
Bonnefjorden
AKERSHUS FYLKE OSLO
Ingierstrand
Sandbakken
368
Mosjøen

Lierskogen
215
Sørby
Kolbotn
Krokhol
Vardåsen 374

Tranby
Skogen
Dikemark
Blakstad
Vollen
Nesodden
Gjersjøen
Myrvoll
Siggerud
Bru
Børtervatna

C
Lier
Gjellumvatn
Fjellstrand
Hasle
Oppegård
134
Oppegård
Binningsvatna

Frogner
Reistad
Svestad
Blylaget
Langen

Nærsnes
Slemmestad
Garder
East from Greenwich
Oslofjorden

1 2 3 4 5 6

7 8 9 10 11

Gerlev
Snostrup
Stavnsholt
Øverød
Jægersborg
Skodsborg
Oslo

Farum
Søllerød
Hegn
Nærum

Skuldelev
Lille Rørbæk
Ølstykke
Ganløse Orned
Holte
Ørholm
Lundtofte
Tårbæk

D
Østby
Svestrup
Ganløse
Furum Sø
Lille Værløse
Virum
Brede
Hjortekær
Klampenborg

Roskilde Fjord
Jyllinge
Stenløse
Frederiksdal
Store Hareşkov
42
Bagsværd Sø
Jægersborg Dyrehave
Ordrup
Skovshoved

Sønderby
Værebro Å
Jonstrup
Hareskovby
Bagsværd
Kongens Lyngby
Jægersborg
Charlottenlund

Smørumnedre
Måløv
Vangede
Gentofte
Hellerup

Pederstrup
Hjortespring
Søborg
Svanemøllen

Bognæs
Hove Å
Ågerup
Ballerup
Herlev
Buddinge

Kattinge Vig
Nybølle
Ledøje
Skovlunde
Husum
Utterslev Mose
Bispebjerg

KØBENHAVN
Brønshøj
Fælled-parken
Trekroner
Refshaleøen

Svogerslev
Risby
Vestskoven
Ejby
Islev
Vanløse
Rosenborg Have
Amalienborg Slot
Christianshavn

55 40'
Sengeløse
Herstedøster
Rødovre
Frederiksberg
Zoo
Hovedbanegård
Tivoli

Vasby
Glostrup
Albertslund
Brøndbyøster
Valby
Sundbyerne

Roskilde
Hedehusene
Tåstrup
Hvidovre
Kastrup

Sterkende
Vallensbæk
Tranegilde
Brøndbyvester
Avedøre
Kilviebåderne
Tårnby
Kastrup Lufthavn

Amager
Drogden

Ishøj Strand
Brøndby Strand
Vallensbæk Strand
Store Magleby
Dragør

Gadstrup
Tune
Hundige
Mosede
Mosede Strand
Greve Strand
Ullerup
Sydstranden

E
Viby
Havdrup
Snoldelev
Karlslunde Strand
Hundige Strand
Køge Bugt
Kongelunden
Søvang
AFLANDSHAGE

East from Greenwich

7 8 9 10 11

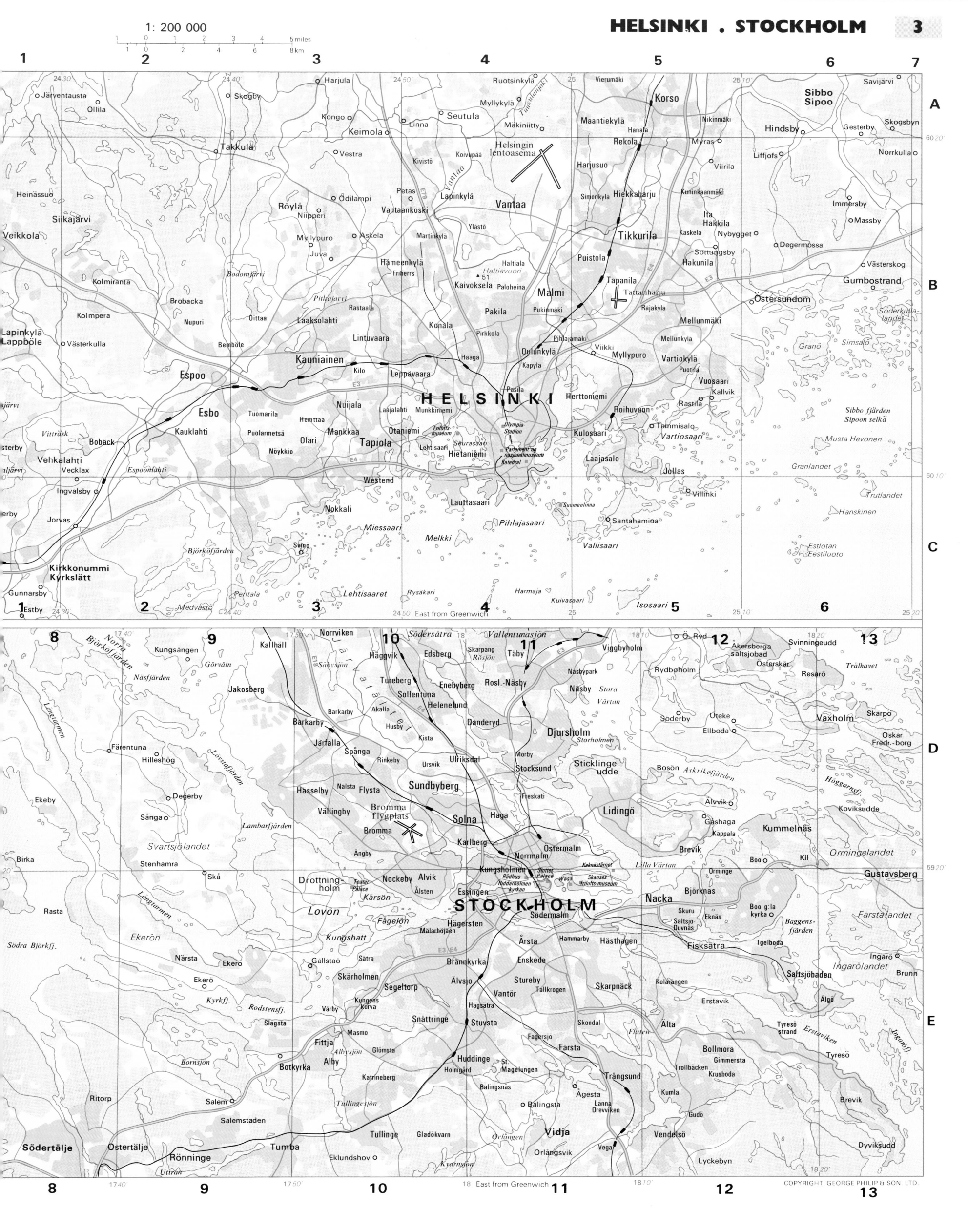

1: 200 000

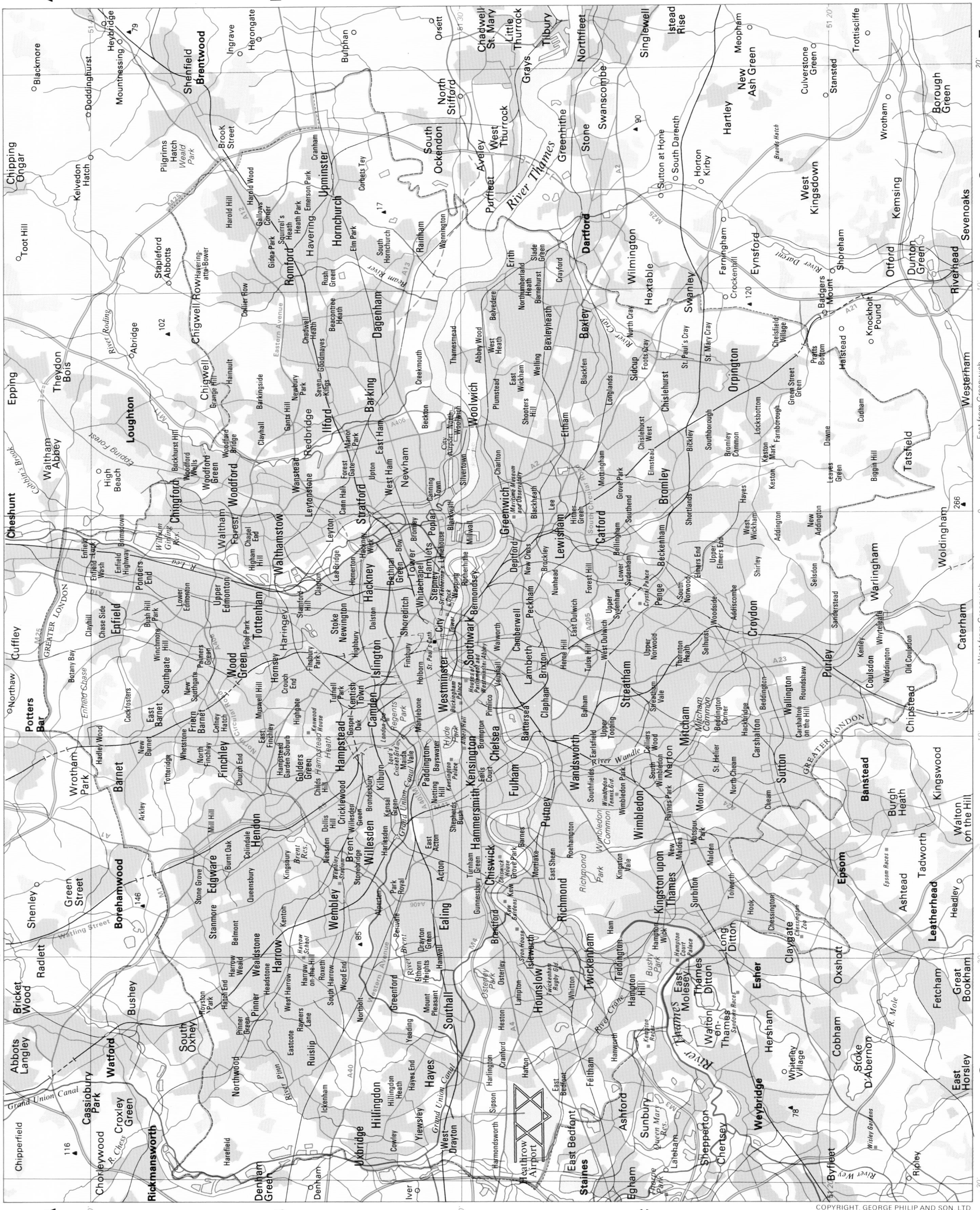

1 : 200 000

5 miles
8 km

East from Greenwich

West from Greenwich

GREATER LONDON

River Thames

River Thames

Epping Forest

Hampstead Heath

Wimbledon Common

Richmond Park

Heathrow Airport

London: Blackmore, Heybridge, Doddinghurst, Mountnessing, Shenfield, Brentwood, Ingrave, Herongate, Chadwell St. Mary, Little Thurrock, Tilbury, Northfleet, Singlewell, Istead Rise, New Meopham, Culverstone Green, Stansted, Trottiscliffe, Borough Green, Wrotham.

Chipping Ongar, Toot Hill, Kelvedon Hatch, Stapleford Abbotts, Pilgrims Hatch, Brook Street, Harold Hill, Harold Wood, Gallows Corner, Cranham, Upminster, Corbets Tey, North Stifford, South Ockendon, Aveley, West Thurrock, Grays, Swanscombe, Greenhithe, Stone, Swanley, Wilmington, Dartford, West Kingsdown, Kemsing, Sevenoaks, Riverhead.

Theydon Bois, Epping, Cheshunt, Waltham Abbey, Loughton, Chigwell, Chingford, Woodford Green, Hainault, Collier Row, Romford, Gidea Park, Emerson Park, Havering, Hornchurch, Elm Park, South Hornchurch, Rainham, Wennington, Purfleet, Erith, Belvedere, Abbey Wood, Slade Green, Crayford, Bexley, Bexleyheath, Hextable, Orpington, Chelsfield, Westerham, Tatsfield.

Northaw, Cuffley, Enfield, Ponders End, Edmonton, Tottenham, Walthamstow, Leytonstone, Wanstead, Ilford, Barking, Dagenham, East Ham, West Ham, Newham, Beckton, Woolwich, Plumstead, Welling, Eltham, Sidcup, Chislehurst, Petts Wood, Keston, Biggin Hill.

Potters Bar, Barnet, New Barnet, East Barnet, Southgate, Wood Green, Haringey, Hornsey, Finsbury Park, Highbury, Islington, Hackney, Stoke Newington, Dalston, Shoreditch, Stepney, Bethnal Green, Tower Hamlets, Poplar, Millwall, Greenwich, Deptford, Lewisham, Catford, Beckenham, Bromley, Shirley, Addington, Croydon, Purley, Coulsdon, Caterham, Warlingham, Woldingham.

Borehamwood, Elstree, Edgware, Hendon, Finchley, Golders Green, Hampstead, Camden, Holborn, City, Westminster, Southwark, Bermondsey, Camberwell, Peckham, Dulwich, Sydenham, Crystal Palace, Penge, Norwood, Streatham, Mitcham, Sutton, Carshalton, Wallington, Banstead, Burgh Heath, Kingswood, Walton on the Hill, Tadworth, Epsom, Ashtead, Leatherhead, Headley.

Watford, Bushey, Stanmore, Harrow, Wealdstone, Pinner, Northwood, Ruislip, Wembley, Willesden, Kilburn, Paddington, Kensington, Chelsea, Hammersmith, Fulham, Putney, Wandsworth, Battersea, Clapham, Balham, Tooting, Wimbledon, Merton, Morden, New Malden, Kingston upon Thames, Surbiton, Tolworth, Chessington, Claygate, Esher, Oxshott, Cobham, Great Bookham, Fetcham.

Chipperfield, Abbots Langley, Kings Langley, Cassiobury Park, Croxley Green, Rickmansworth, Chorleywood, Denham, Uxbridge, Yiewsley, West Drayton, Hillingdon, Hayes, Southall, Greenford, Ealing, Acton, Brentford, Chiswick, Isleworth, Hounslow, Twickenham, Teddington, Hampton, Feltham, Ashford, Staines, Egham, Chertsey, Shepperton, Sunbury, Walton-on-Thames, Weybridge, Byfleet, Ripley.

1: 200 000

0 1 2 3 4 5 miles
0 1 2 3 4 6 8 km

PARIS

Aéroport Charles-de-Gaulle

Aéroport de Paris Le Bourget

MARNE-LA-VALLÉE

Roissy-en-Brie

Ozoir-la-Ferrière

Gretz-Armainvilliers

Évry-les-Châteaux

Aérodrome de Meln-Villaroche

MELUN-SÉNART

Corbeil-Essonnes

EVRY

Brétigny-sur-Orge

Longjumeau

Versailles

St-Cyr-l'École

ST-QUENTIN-EN-YVELINES

Trappes

Les Clayes-sous-Bois

Le Mesnil-St-Denis

Les Mureaux

Vernouillet

Verneuil-sur-Seine

Triel-sur-Seine

Achères

Poissy

St-Germain-en-Laye

Conflans-Ste-Honorine

Maisons-Laffitte

CERGY-PONTOISE

Pontoise

St-Ouen-l'Aumône

Écouen

Goussainville

Sarcelles

Montmorency

Deuil-la-Barre

Enghien-les-Bains

Argenteuil

Colombes

La Garenne-Colombes

Courbevoie

Puteaux

La Défense

Nanterre

Rueil-Malmaison

Suresnes

Boulogne-Billancourt

Sèvres

Meudon

Clamart

Montrouge

Chatillon

Vanves

Issy-les-Moulineaux

Neuilly-sur-Seine

Levallois-Perret

Clichy

St-Ouen

St-Denis

La Courneuve

Aubervilliers

Pantin

Le Pré-St-Gervais

Les Lilas

Bagnolet

Montreuil

Vincennes

St-Mandé

Charenton-le-Pont

St-Maurice

Maisons-Alfort

Alfortville

Créteil

Joinville-le-Pont

Nogent-sur-Marne

Le Perreux-sur-Marne

Champigny-sur-Marne

Chennevières-sur-Marne

St-Maur-des-Fossés

Bonneuil-sur-Marne

Villeneuve-St-Georges

Valenton

Vitry-sur-Seine

Ivry-sur-Seine

Le Kremlin-Bicêtre

Gentilly

Arcueil

Cachan

L'Haÿ-les-Roses

Chevilly-Larue

Thiais

Choisy-le-Roi

Orly

Aéroport de Paris-Orly

Villeneuve-le-Roi

Athis-Mons

Draveil

Vigneux-sur-Seine

Montgeron

Yerres

Brunoy

Villejuif

Rungis

Fresnes

Antony

Bourg-la-Reine

Sceaux

Fontenay-aux-Roses

Le Plessis-Robinson

Châtenay-Malabry

Massy

Palaiseau

Wissous

Noisy-le-Grand

Neuilly-Plaisance

Rosny-sous-Bois

Villemomble

Gagny

Le Raincy

Bondy

Pavillons-sous-Bois

Livry-Gargan

Aulnay-sous-Bois

Sevran

Villepinte

Tremblay-en-France

Drancy

Le Blanc-Mesnil

Bobigny

Noisy-le-Sec

Romainville

Le Bourget

Dugny

Stains

Pierrefitte

Gonesse

Villiers-le-Bel

Garges-lès-Gonesse

Arnouville-lès-Gonesse

VAL-D'OISE

SEINE-ST-DENIS

VAL-DE-MARNE

HAUTS-DE-SEINE

Mitry-Mory

Lagny

Chelles

Gournay-sur-Marne

Noiseau

Ferrières-en-Brie

Pontault-Combault

Brie-Comte-Robert

Combs-la-Ville

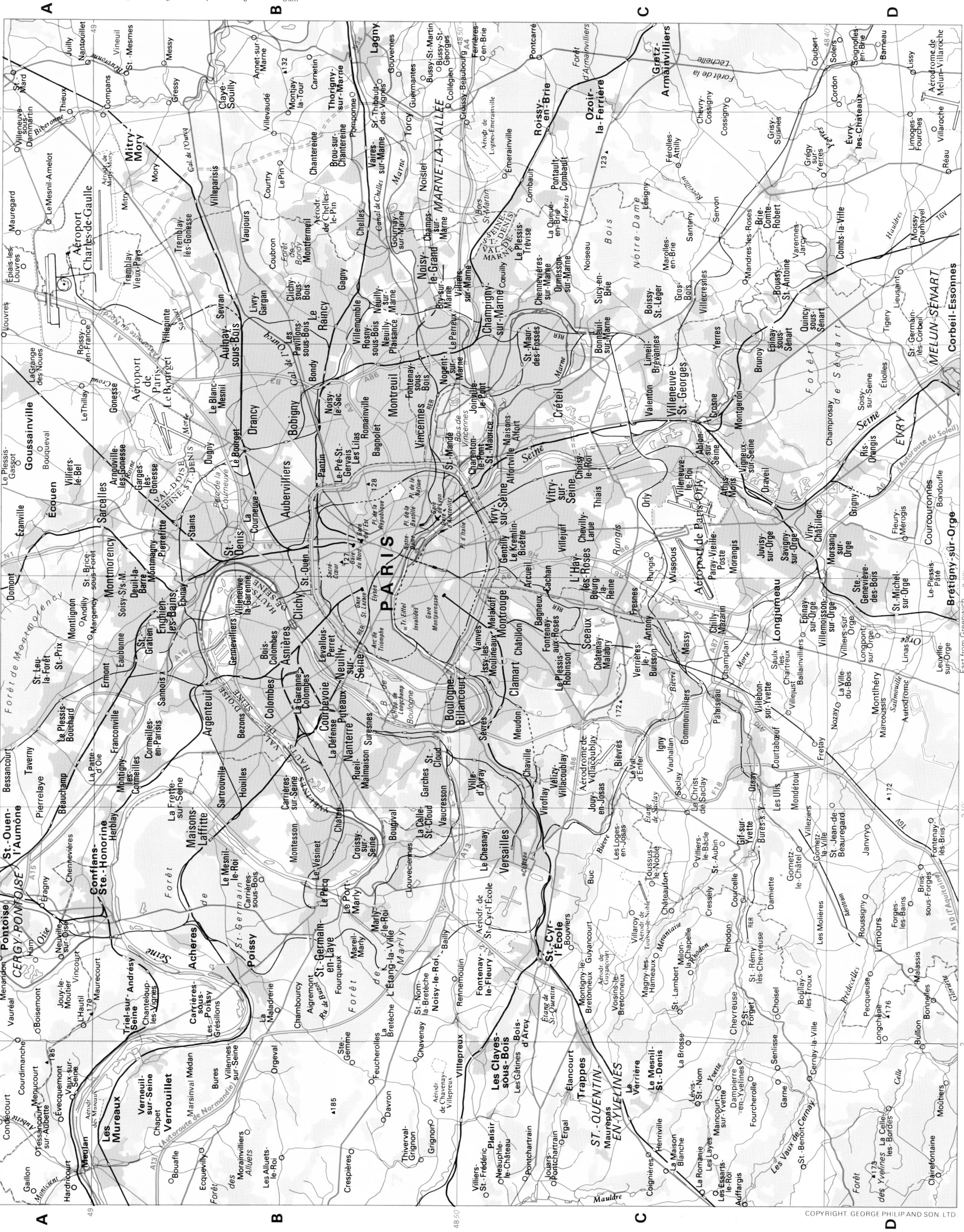

COPYRIGHT GEORGE PHILIP AND SON LTD.

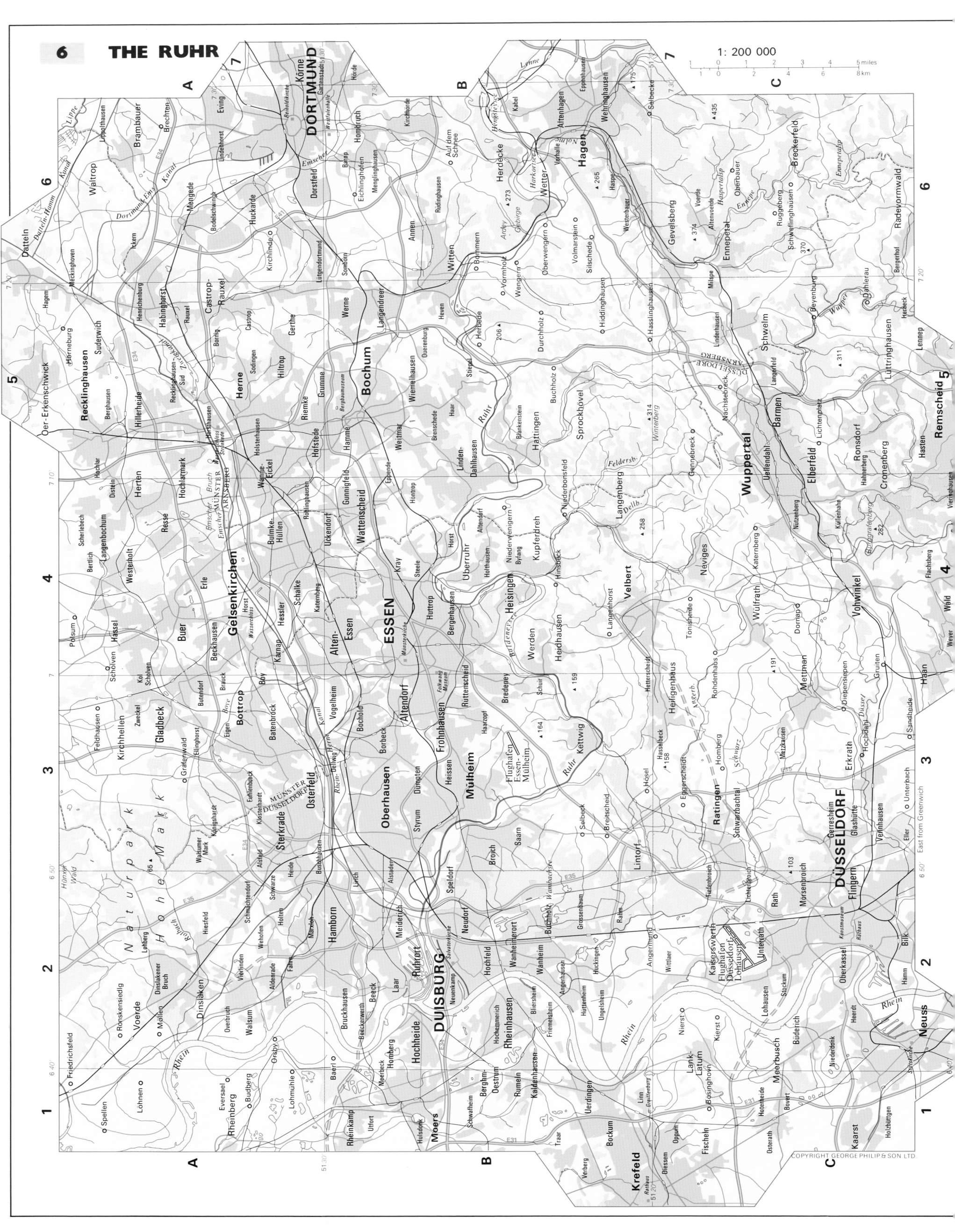

1: 200 000

5 miles
8 km

Berlin map (top)

Pausin, Wansdorf, Botzow, Zepernick, Birkenhöhe, Elisenau, Steinitzsee, Amselhain, Werneuchen

Schönwalde, Hennigsdorf, Frohnau, Glienicke, Schildow, Buch, Schwanebeck, Birkholz, Birkholzaue, Löhme, Rudolfshöhe

Stolpe-Süd, Hermsdorf, Lübars, Blankenfelde, Karow, Neu Lindenberg, Seefeld

Nieder Neuendorf, Heiligensee, Schulzendorf, Waidmannslust, Buchholz, Lindenberg, Blumberg, Krummensee, Wegendorf

Alter Finkenkrug, Siedlung Schönwalde, Heiligensee, Konradshöhe, Tegel, Wittenau, Rosenthal, Niederschönhausen, Blankenburg, BRANDENBURG BERLIN, Ahrensfelde, Trappenfelde, Paulshof, Altlandsberg Nord, Neuhagen

A Waldheim, Falkensee, Johannesstift, Scharfenberg, Flughafen Tegel, Reinickendorf, Pankow, Heinersdorf, Malchow, Wartenberg, Falkenburg, Mehrow, Eiche, Altlandsberg **A**

Finkenkrug, Falkenhagen, Seegefeld, Spandau, Haselhorst, Volkspark Jungfernheide, Wedding, Prenzlauerberg, Weissensee, Hohenschönhausen, Marzahn, Eiche Süd, Honow, Seeberg, Fredersdorf Nord, Friedrichslust

Döberitz, Dallgow, Staaken, Charlottenburg, Siemensstadt, Mitte, Lichtenburg, Hellersdorf, Neuenhagen, Fredersdorf

Seeburg, Olympic Stadium, Deutsche Oper, University, Tiergarten, Zoo Station, Brandenburg Gate, **BERLIN**, Kreuzberg, Friedrichshain, Biesdorf, Kaulsdorf, Mahlsdorf, Dahlwitz-Hoppegarten, Vogelsdorf

B Gross Glienicke, Gatow, Grunewald, Teufelsberg, Rathaus, Schöneberg, Neukölln, Treptow, Karlshorst, Friedrichsfelde, Münchehof, Kleinschönebeck, Schöneiche **B**

Krampnitz, Neu Fahrland, Sacrower See, Kladow, Schwanenwerder, Dahlem, Schmargendorf, Friedenau, Flughafen Tempelhof, Tempelhof, Oberschöneweide, Fichtenau, Schönblick

Nedlitz, Sacrow, Pfaueninsel, Wannsee, Nikolassee, Zehlendorf, Steglitz, Britz, Teltow Kanal, Johannisthal, Aldershof, Köpenick, Grosse Müggelsee, Wilhelmshagen, Springeberg, Erkner

Cecilienhof, Grosser Wannsee, Schlachtensee, Lichterfelde, Mariendorf, Lankwitz, Buckow, Altglienicke, Grünau, Müggelberge, Rahnsdorf, Dämeritzsee

Tiefersee, Potsdam, Dreilinden, Kleinmachnow, Seehof, Osdorf, Grossziethen, Rudow, Wendenschloss, Müggelheim, Gr. Krampe, Neu Buchhorst

Potsdam, Klein Gleinicke, Steinstücken, Stahnsdorf, Teltow, Heinersdorf, Lichtenrade, Schönefeld, Bohnsdorf, Flughafen Schönefeld, Eichwalde, Karolinenhof, Gosen, Neu Zittau, Schmöckwitz

Babelsberg, Kienwerder, Ruhlsdorf, Friederikenhof, East from Greenwich, Kleinziethen

Hamburg map (bottom left)

Rantzau, Quickborn, Harksheide, Tangstedter Forst, Duvenstedter Brook

Renzel, Norderstedt, Glasmoor, Duvenstedt, Wulksfelde, Wohldorf-Ohlstedt

C Hohenraden, Hasloh, Haslohfeld, Moorbek, Glashütte, Lemsahl, Mellingstedt, Bergstedt **C**

Wulfsmühle, Tangstedt, Winzeldorf, Garstedt, Ochsenzoll, Volksdorf

Pinneberg, Bönningstedt, Poppenbüttel, Sasel, Berne

Rellingen, Ellerbek, Egenbüttel, Langenhorn, Hummelsbüttel, Wellingsbüttel, Meiendorf

Halstenbek, Neuegenbüttel, Schnelsen, Flughafen Hamburg, Niendorf, Fuhlsbüttel, Ohlsdorf, Brämfeld, Farmsen

Brande, Friedrichshulde, Eidelstedt, Gross Borstel, Alsterdorf, Steilshoop, Rahlstedt

D Schenefeld, Lurup, Lokstedt, Winterhude, Hinschenfelde, Tonndorf, Wandsbek, Jenfeld **D**

Sulldorf, Osdorf, Stellingen, Harvestehude, Barmbek, Eimsbüttel, Marienthal

Iserbrook, Bahrenfeld, Rotherbaum, Uhlenhorst, Eilbek, **HAMBURG**

Blankenese, Gross-Flottbek, St. Pauli, Hohenfelde, Horn, Billstedt

Nienstedten, Othmarschen, Ottensen, Altona, Hamm, Hammerbrook, Billbrook

Finkenwerder, Waltershof, Steinwerder, Kl. Grasbrook, Rothenburgsort, Kirchsteinbek, Boberg

Rosengarten, Neuenfelde, Veddel, Georgswerder, Moorfleet, Billwerder

Nincop, Francop, Altenwerder, Wilhelmsburg, Tatenberg, Spadenland, Allermöhe

E Neugraben-Fischbek, Moorburg, Kirchof, Hohe Schaar, Moorwerder, Ochsenwerder **E**

Neu Wulmstorf, Hausbruch, Heimfeld, Harburg, Neuland, Fünfhausen

Munich map (bottom right)

Etzenhausen, Riedmoos, Unterschleissheim, Dachau, Mittenheim, Oberschleissheim

Udlding, Dachau Ost, Badersfeld, Hochbrück, Garching, Carlshof

F Gröbenried, Eschenried, Karlsfeld, Lustheim, Dirnismaning, Ismaning **F**

Obermoos Schwaige, Rothschwaige, Neuherberg, Speicher-See

Ludwigsfeld, Feldmoching, Am Hasenbergl, Freimann, Unterföhring

Langwald, Allach, Fasanerie-Nord, Gross-Lappen, Mittl. Isarkanal

Lockhausen, Untermenzing, Moosach, Milbertshofen, Aschheim

D Aubing, Obermenzing, Bern, Nymphenburg, Schwabing, Oberföhring, Johanneskirchen, Dornach, Feldkirchen **D**

Neu Aubing, Blutenburg, Neuhausen, Bogenhausen, Zamdorf, Daglfing, Riem

Freiham, Freimann, Pasing, Laim, Haidhausen, Flughafen München-Riem, Kirchtrudering, Salmdorf

G Gräfelfing, Klein-Hadern, Sendling, Ramersdorf, Berg am Laim, Strasstrudering, Haar **G**

Planegg, Martinsried, Gross-Hadern, **MÜNCHEN**, Thalkirchen, Giesing, Neuperlach, Perlach, Waldtrudering, Keferloh

Krailling, Neuried, Fürstenried, Fasangarten, Solalinden, Gronsdorf

Warnberg, Solln, Harlaching, Unterbiberg, Waldperlach, Öden-Stockach, Putzbrunn

Maxhof, Forstenried, Grosshesselohe, Unterhaching, Westerham, Neubiberg

Pullach, Geiselgasteig, Am Wald, Taufkirchen, Winning, Ottobrunn, Hohenbrunn

Höllriegelskreuth, Grünwald, Furth, Bergham, Wachterhof, Kirchstockbach

Buchenhain, Baierbrunn, Forst, Oberhaching, Höhenkirchen

Laufzorn, Strasslach, Deisenhofen, Brunnthal

1: 200 000

5 miles
8 km

MEDITERRANEAN SEA

BARCELONA
Badalona
Mongat
San Adrián de Besós
Sta. Coloma de Gramanet
La Puntúgala
San Joan
C'an
303 Poyo
151
Llano de Can Gineu
327 Andrés
Beós
La Sagrera
La Vacuna
San Martín
Pueblo Nuevo
Gràcia
La Taxonera
Guinardó
Sagrada Familia
Temple de la Sagrada Familia
Catedral
Barceloneta
Museo Marítimo
Parque de la Ciudadela
La Fransa
Las Corts
Putxet
Sarrià
Pedralbes
Univers
Campo F.C. Barcelona
Sants
Hostafranchs
Hospitalet
Sant Cugat
Valldoreix
La Floresta
512
387
Tibidabo
Vallvidrera
San Pedro Martir
389
S. Just Desvern
Esplugas
Cornellà
435
La Ribera
Santa Cruz de Olorde
Santa Coloma de Cervelló
Colonia Guell
Beri
Sant Boi de Llobregat
S. Feliu de Llobregat
Joan Despí
Mollins de Rey
336 Madrona
Papiol
Castelljbisbal
Pallejà
S. Vicenç dels Horts
Torrellas del Llobregat
S. Clemente del Llobregat
Viladecans
La Pineda
Gavà
Gavamar
Prat de Llobregat
Aeropuerto de Barcelona-Prat
Laguna de la Ricarda
Laguna del Remolo
Río Llobregat

Mahón
Palma
Ibiza
Malaga
Cadiz
Islas Canarias

East from Greenwich

MADRID
Paracuellos del Jarama
Guardias 703
680
Aeropuerto Transoceánico de Barajas
Ciudad Fin de Semana
Barrio de La Estación
San Fernando de Henares
La Moraleja
Valdeveba
Manoteras
Barajas
Canillas
Canillejas
Hortaleza
San 674 Cristóbal
Rivas de Jarama
Rivas-Vaciamadrid
655
Fuencarral
Almanara
Chamartín
Chamartín Station
Tetuán
Colonia Puerta de Hierro
University
Universidad
Chamberí
Ventas
Buenavista
Ciudad Lineal
Pueblo Nuevo
Moratalaz
Vicálvaro
Vallecas
Cumbres de Vallecas
633
Canteras de Vallecas
581
Aravaca
La Estación
El Plantío
Casa de Campo
Latina
Carabanchel Bajo
Carabanchel Alto
Campamento
Cuatro Vientos
Pozuelo de Alarcón
Boadilla del Monte
Majadahonda
Las Rozas de Madrid
703
Portilleros de las Rozas
Carretera de la Coruña
Ventorro 705 del Cano
Humera
Retiro
Aganzuela
Usera
Leganitos
Villaverde
Villaverde Bajo
Mediodía
Palomeras
Entrevías
Mercamadrid
Los Angeles
Perales del Río
Cerro de los Ángeles
Avda de Andalucía
Getafe
Leganés
Móstoles
Alcorcón
Manzanares
Canal de Manzanares
West from Greenwich

ATHINAI
Drafi
Pallini
Hristoupoli
Spata
Markopoulo
Kalivia Thorikou
Gerakas
Karellas
Koropi
Peania
Kitsi
Giika Nera
Barako
230
Vari
765
Néa Ionia
Kholargós
Khalándrion
Filothei
Psikhikón
Kholargós
Zografos
Kaisariani
Patisia
Galátsion
Nea Liósia
Efladhnéfhai
Attiki
University
National Archaeological Museum
Kipseli
Neapolis
Ay. Paraskevi
Koupónia
Ampelokipi
Pangrati
Viron
Imittós
N. Alexandhria
Peristerion
Sepolia
Kolonós
Akropolis
Ambelokipi
Dhafni
Ay. Dhimitrios
Iliopolis
Athinai Ellinikón Airport
Petroúpolis
Verdi
Ltaumi
Koladhoú
Aiyáleo
Távros
Kallithéa
N. Smirni
P. Fáliron
Kalamákion
Glifada
Khaidhárion
Dhafni
Dhamaraküi
Ay. I. Rendis
Moskhatón
Ellinikón
Ayiroupolis
Voula
Vouliagmeni
Skara-mangas
Diflistiria
468
Neapolis
Nikaia
Dhrapetsón
N. Fáliron
Piraévs
Aig. Gheórghios
Koridhallós
Saronikós Kólpos
Voullaagmenis
Vouliagmeni
Ormos Fáliron
Kithnos
Silnos
Oros Imittós
1026
Évzonos
Éjzonos
Oros

East from Greenwich

LISBOA
Atcochete
Montijo
Lagoa da Pedra
Sarilhos Pequenos
Sarilhos Grandes
Moita
Base Aérea
Samoueo
Rosarinho
Alhos Vedros
Loviadio
Santo André
Barra da Bambeira
Santo António da Charneca
S. João da Talha
Sacavém
Beirolas
Matinha
Beato
Xabregas
Moscavide
Olivais
Aeroporto de Portela
108
Apelação
Unhos
Frielas
163 Boavista
Camarate
Povoa de Santo Adrião
S. João de S. Jorge
Sta. Apolónia Station
Praça do Comércio
Barreiro
Coina
Amadora
Amoreira
Loures
Máximenor 357
Damaia
Odivelas
Paia
Pontinha
Lumiar
Campo Grande
Carnide
University
228 Monsanto
Benfica
Estádio Benfica
Alcântara
Alto do Pina
B. Lopes
Campo Pequeno
Calhau
Amadora
Camões
Santo Amaro
Ajuda
Belém
Torre de Belém
Mosteiro dos Jerónimos
Basílica da Estrela
Rato
Campolide
Caparica
Cova da Piedade
Almada
Cacilhas
Trafaria
Porto Brandão
Caxias
Cruz de Pau
Amora
Seixal
Arrentela
Feijó
Sobreda
Costa da Caparica
Paco de Arcos
Oeiras
Linda-a-Pastora
Caxias
Barcarena
Rio de Mouro
Colão
222
Agualva-Cacem
Massamá
Queluz
Belas
283
Carnaxide
210
Carenguejeira
Cañeças
Ada Beja
Camarões
Sabugo
Tapada
320 Piedade
Telhal
Venda Seca
Talaide
Leião
Terrugem
Oeiras
Bugio
Charneca
Cota da Sto. António
ATLANTIC OCEAN
Río de
LISBOA SETÚBAL
Tejo
Rib. de Barcarena

West from Greenwich

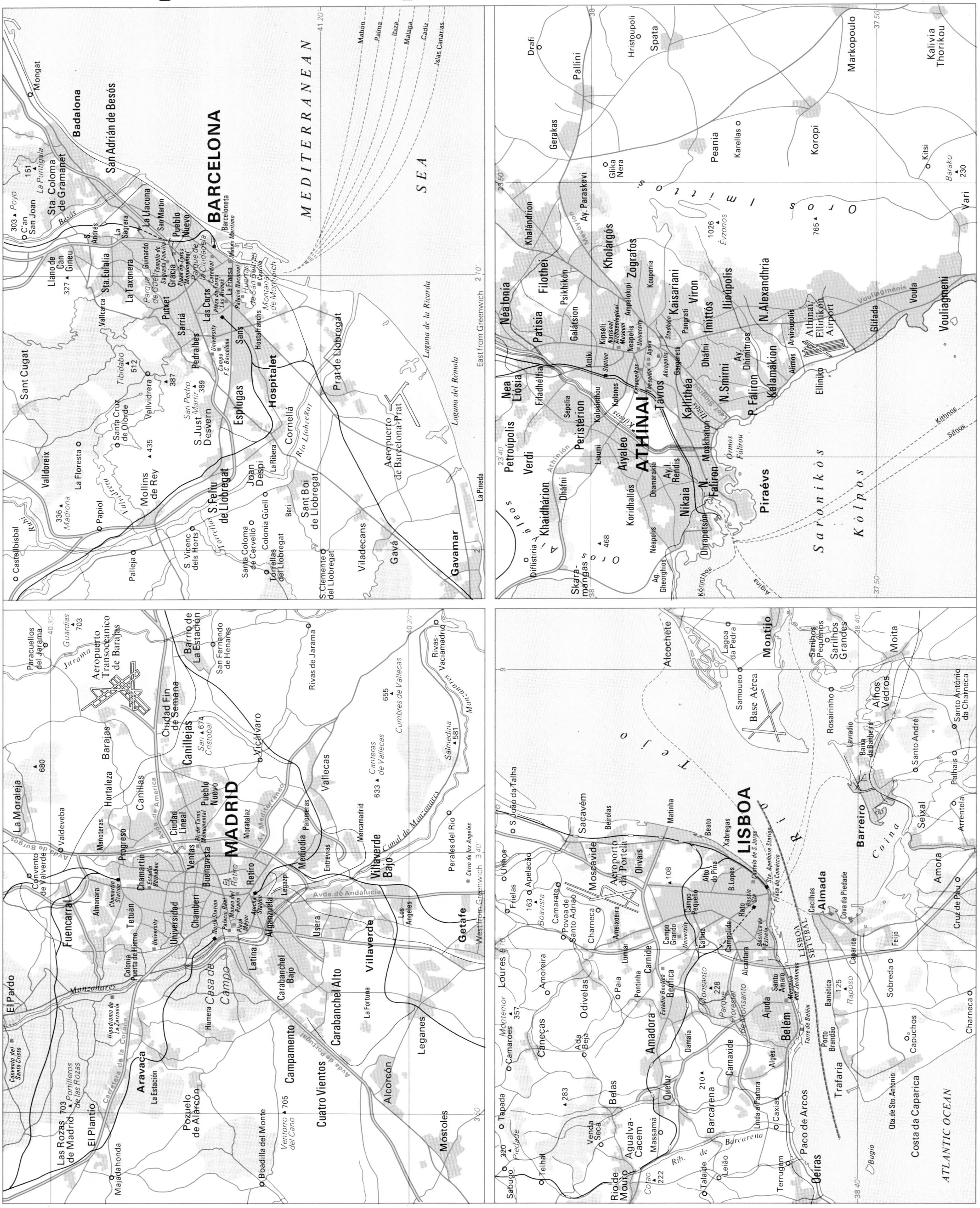

COPYRIGHT GEORGE PHILIP AND SON. LTD.

1 : 200 000

1 0 1 2 3 4 5 miles
1 0 1 2 3 4 5 6 7 8 km

TORINO (Turin)

Monza, Lissone, Desio, Muggiò, Cinisello Bálsamo, Sesto S. Giovanni, Brugherio, Concorezzo, Villasanta, Cologno Monz., Cernusco s. Nav., Segrate, Pioltello, Peschiera Borromeo, Linate, Aeroporto di Linate, S. Donato Mil., MILANO, Lambrate, Crescenzago, Bresso, Cusano Mil., Paderno, Dugnano, Novate Mil., Bollate, Rho, Cornaredo, Córsico, Baggio, Settimo Mil., Trezzano s. Nav., Buccinasco, Assago, Pero, Bareggio, Pregnana Mil., Nerviano, Lainate, Caronno Pert., Saronno, Garbagnate Mil., Solaro, Bovisio-Masciago, Seveso, Varedo, Limbiate, Senago

Volpiano, Settimo Tor., Caselle Tor., Leini, Borgaro Torinese, Venaria, Mathi, Caselette, Alpignano, Pianezza, Collegno, Grugliasco, TORINO, Lucento, Mirafiori, Nichelino, Moncalieri, Lingotto, Beinasco, Orbassano, Rivoli, Rivalta di Torino, Chieri, Gassino Torinese, Castiglione Torinese, Baldissero Tor., Pino Torinese, Pecetto, Cambiano

ROMA (Rome)

ROMA, CITTÀ DEL VATICANO, Trastevere, E.U.R., Ostiense, Centocelle, Cinecittà, Ciampino, Aeroporto di Ciampino, Aeroporto d. Urbe, Acilia, Ponte Galéria, Aeroporto Intercontinentale Leonardo da Vinci, Magliana, Corviale, Monteverde Nuovo, Gianicolense, Appio, Tuscolano, Prenestino, Labicano, Quadraro, Torricola, Cecchignola, Valleranello, Vitinia, Casalotti, Primavalle, Torrevecchia, Ottavia, La Giustiniana, Fidene, Tufello, Mte. Sacro, Pietralata, Tiburtino

NAPOLI (Naples)

NAPOLI, Pórtici, Torre del Greco, Ercolano, S. Giorgio a Crem., Barra, Ponticelli, Poggioreale, Afragola, Casoria, Arzano, Grumo Nevano, Frattamaggiore, Caivano, Acerra, Cardito, Casalnuovo di Náp., Pomigliano d'Arco, Marigliano, Somma Ves., S. Anastasia, Ottaviano, S. Giuseppe, Vesuvio, Mte. Somma 1132, 1277, Boscoreale, Boscotrecase, Torre Annunziata, Pompei, Giugliano in Camp., Marano di Náp., Quarto, Campi Flegrei, Pozzuoli, Bagnoli, Posillipo, Vómero, Soccavo, Pianura, Golfo di Nápoli

COPYRIGHT GEORGE PHILIP & SON. LTD.

1 : 200 000

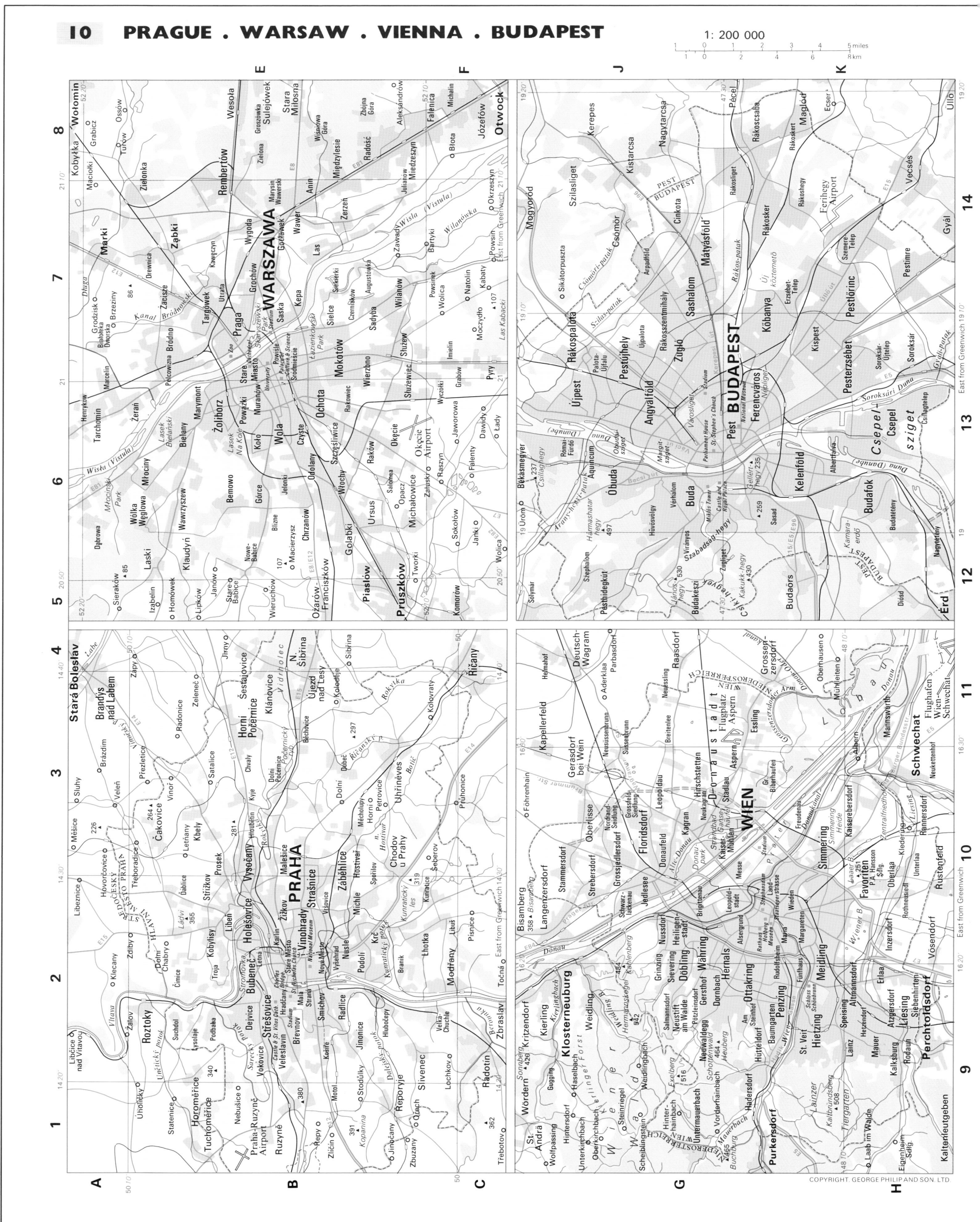

1:200 000

1 0 1 2 3 4 5 miles
1 0 2 4 6 8 km

SANKT-PETERBURG

Gulf of Finland

Lisiy Nos
O. Verperluda
Olgino
Primorskoye Prospekt
Bobylyskaya
Lakhtinskiy
Oz. Lakhtinskiy Razliv
Kolomyagi
Novaya Derevnya
Udelnaya
Ruchyi
Gorelyy
Berngardovka
Lubya
Vsevolozhsk
Lesnoy
Grazhdanka
Rybatskaya
Kalytino
Staraya Derevnya
Bolshaya Nevka
Trudyashchikhsya
Kirov Stadium
Ostrova Kirovskiye
Apterkarskiy Ostrov
Petrogradskaya Storona
Vyborgskaya Storona
Rzhevka
Noyoye Kovalyova
Krasnaya Gorka
O. Volynnyy
Malaya Neva
O. Dekabristov
Fortress of St. Peter & St. Paul
Finland Station
Selytsy
Polyustrovo
Zanevka
Khirvosti
Koltushi
Neva
Admiralteyskaya Storona
Bolshaya-Okhta
Pavlovo
Ostrov Vasilyevskiy
Hermitage & Winter Palace
Old Admiralty
St. Isaac's Cathedral
Alexander Nevsky Abbey
Moskva Station
Staraya
Yanino
Oz. Korkinskoye
Tavry
Okkervil
Kudrovo
Malaya-Okhta
Vitebsk Station
Novosergiyevka
Razmitelevo
Ozerki
Ostrov Kanonerskiy
Ostrov Gutuyevskiy
Baltic Station
Warszawa Station
Volynkina-Derevnya
Obvodnyy kanal
Fontanka
Volodarskoye
Vesolyy Posolok
Myaglovo
Khaboye
Cornaya
LENINGRAD OBLAST
GOROD ST. PETERBURG
Obukhovo
Avtovo
Volkovka
Farforovskaya
Lesnozavodskaya
Strelyna
Kikenka
Posolok Lenina
Uritsk
Ulyanka
Moskovskiy Prospekt
Aleksandrovskoye
Novosaratovka
Sosnovaya
Ligovo
Dakhnoye
Airport
Srednyaya Rogatka
Kupchino
Novoaleksandrovskoye
Rybatskoye
Ust-Slavyanka
East from Greenwich

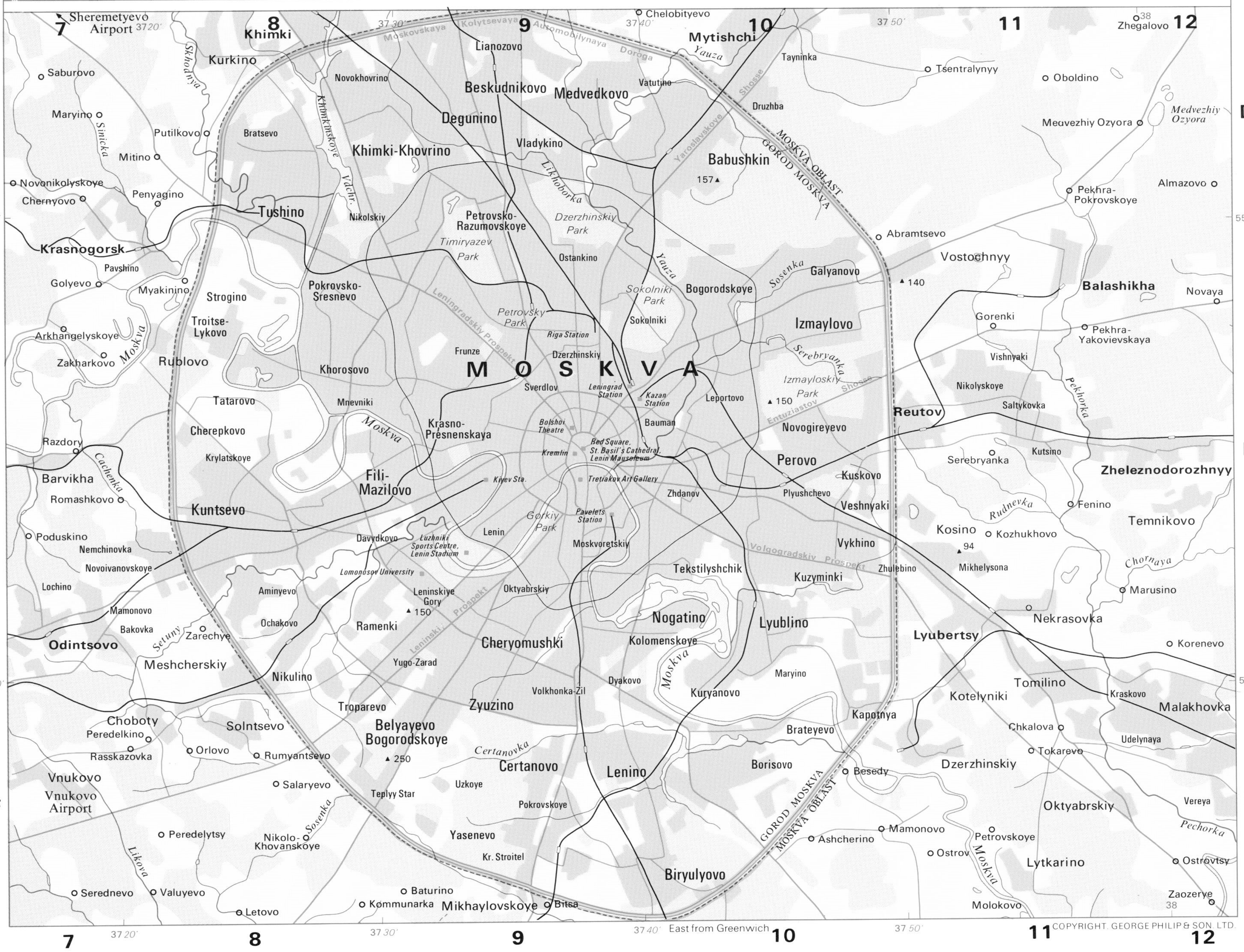

MOSKVA

Sheremetyevo Airport
Khimki
Chelobityevo
Mytishchi
Zhegalovo
Kurkino
Lianozovo
Avtomobilnaya Doroga
Tayninka
Tsentralynyy
Oboldino
Saburovo
Moskovskaya
Koltsevaya
Novokhovrino
Beskudnikovo
Medvedkovo
Vatutino
Yauza
Maryino
Sinichka
Putilkovo
Bratsevo
Khimkinskoye
Degunino
Vladykino
Druzhba
Medvezhiy Ozyora
Meavezhiy Ozyora
Mitino
Khimki-Khovrino
Vachi
Babushkin
Pekhra-Pokrovskoye
Almazovo
Novonikolyskoye
Chernyovo
Penyagino
Nikolskiy
Petrovsko-Razumovskoye
L. Khoborka
Dzerzhinskiy Park
157
Abramtsevo
Tushino
Timiryazev Park
Ostankino
Sosenka
Galyanovo
Vostochnyy
Krasnogorsk
Pavshino
Myakinino
Strogino
Pokrovsko-Sresnevo
Petrovsky Park
Sokolniki Park
Bogorodskoye
140
Balashikha
Novaya
Golyevo
Leningradskiy Prospekt
Riga Station
Sokolniki
Izmaylovo
Gorenki
Pekhra-Yakovievskaya
Troitse-Lykovo
Arkhangelskoye
Frunze
Dzerzhinskiy
Izmayloskiy Park
Serebryanka
Vishnyaki
Zakharkovo
Rublovo
Moskva
Khorosovo
Sverdlov
Leningrad Station
Kazan Station
Leportovo
150
Entuziastov Shosse
Nikolyskoye
Saltykovka
Razdory
Tatarovo
Mnevniki
Krasno-Presnenskaya
Bolshoi Theatre
Bauman
Reutov
Kutsino
Barvikha
Cherepkovo
Krylatskoye
Red Square, St. Basil's Cathedral
Kremlin
Lenin Mausoleum
Novogireyevo
Perovo
Serebryanka
Zheleznodorozhnyy
Romashkovo
Fili-Mazilovo
Kiev Sta.
Tretiakov Art Gallery
Zhdanov
Kuskovo
Plyushchevo
Veshnyaki
Fenino
Kuntsevo
Davydkovo
Luzhniki Sports Centre, Lenin Stadium
Pavelets Station
Gorkiy Park
Moskvoretskiy
Vykhino
Kosino
Kozhukhovo
Temnikovo
Poduskino
Nemchinovka
Novoivanovskoye
Lomonosov University
Lenin
94
Mikhelysona
Lochino
Aminyevo
Ochakovo
Leninskiye Gory
150
Oktyabrskiy
Tekstilyshchik
Kuzyminki
Zhulebino
Lyublino
Lyubertsy
Odintsovo
Meshcherskiy
Ramenki
Nogatino
Kolomenskoye
Maryino
Nekrasovka
Koreneve
Setuny
Bakovka
Zarechye
Nikulino
Cheryomushki
Dyakovo
Kuryanovo
Tomilino
Kraskovo
Malakhovka
Choboty
Peredelkino
Solntsevo
Orlovo
Rumyantsevo
Troparevo
Zyuzino
Volkhonka-Zil
Lenino
Borisovo
Kotelyniki
Chkalova
Rasskazovka
Belyayevo Bogorodskoye
250
Certanovo
Certanovka
Besedy
Tokareva
Udelynaya
Dzerzhinskiy
Vnukovo
Salaryevo
Teplyy Star
Uzkoye
Pokrovskoye
GOROD MOSKVA
MOSKVA OBLAST
Mamonovo
Ashcherino
Petrovskoye
Oktyabrskiy
Vereya
Pechorka
Vnukovo Airport
Peredelytsy
Nikolo-Khovanskoye
Sosenka
Yasenevo
Kr. Stroitel
Biryulyovo
Ostrov
Moskva
Lytkarino
Ostrovtsy
Serednevo
Valuyevo
Letovo
Baturino
Kommunarka
Mikhaylovskoye
Bitsa
Molokovo
Zaozerye

East from Greenwich

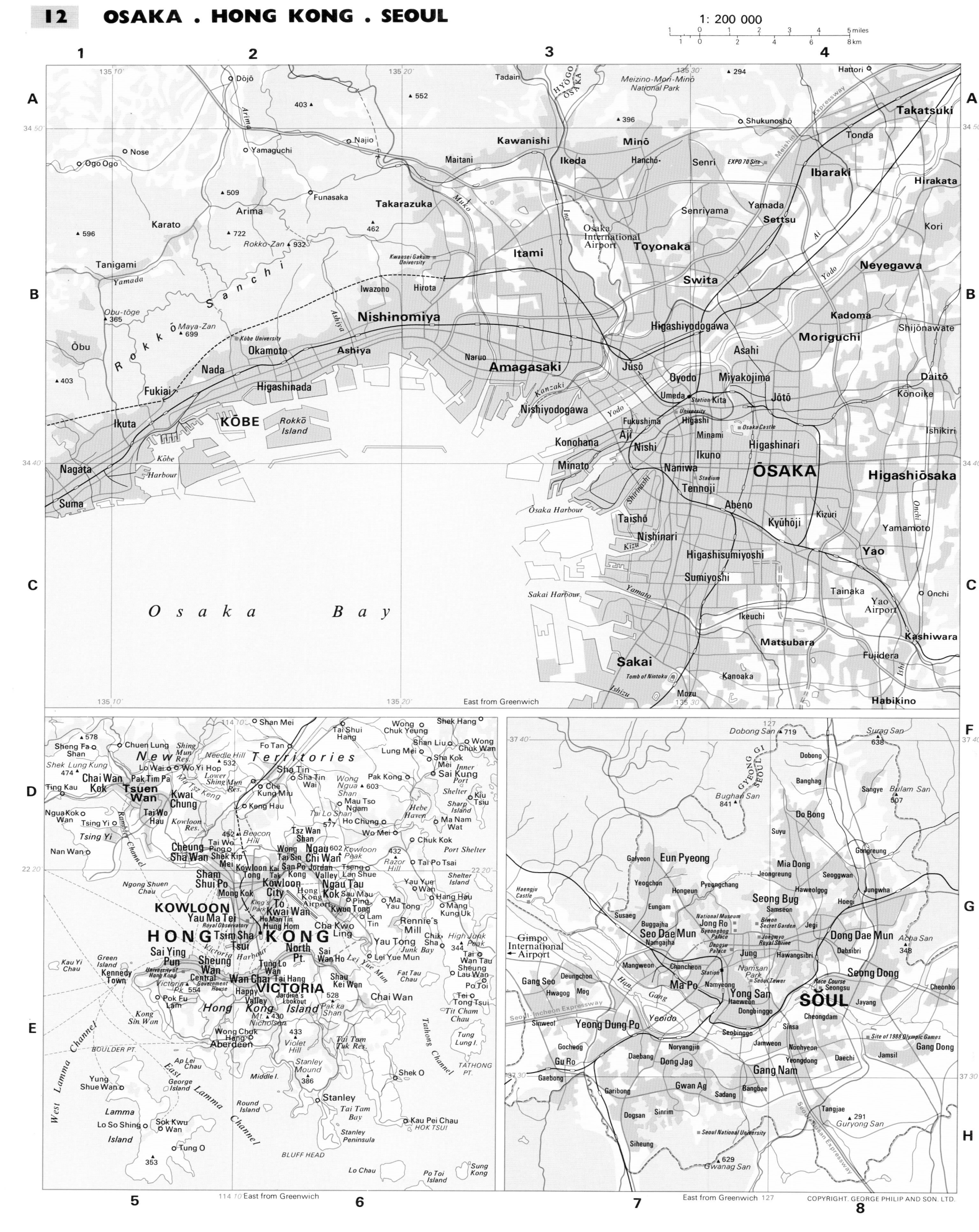

1: 200 000

5 miles
8 km

1 **2** **3** **4**

A

135 10'
135 20'
135 30'

Hattori

○ Dōjō
403 ▲
▲ 552
Tadain
HYOGO ŌSAKA
Meizino-Mori-Minō
National Park
▲ 294
○ Shukunoshō
Takatsuki

34 50'
○ Nose
○ Yamaguchi
○ Najio
Tonda
Maitani
Kawanishi
Minō
Hanchō
Ibaraki
Hirakata

Ogo Ogo
○ Nose
○ Funasaka
Maitani
Ikeda
Senri
EXPO 70 Site
Kori

A

▲ 509
Arima
Takarazuka
Senriyama
Yamada
Settsu
Neyegawa

Karato
▲ 722
462
Osaka
International
Airport
Toyonaka

▲ 596
Rokko-Zan 932
Muko
Itami
Swita
Yodo
Kadoma

B
Tanigami
Yamada
Kwasei Gakuin
University
Iwazono
Hirota
Higashiyodogawa
Moriguchi
Shijōnawate

Obu-tōge
▲ 365
Maya-Zan
▲ 699
Nishinomiya
Naruo
Asahi

B

Ōbu
Kōbe University
Okamoto
Ashiya
Amagasaki
Jūsō
Ōyodo
Miyakojima
Dáitō

▲ 403
Nada
Kanzaki
Umeda Station Kita
Kōnoike

Fukiai
Higashinada
Naniwa
University
Higashi
Osaka Castle
Jōtō

Ikuta
Fukushima
Minami
Ishikiri

KŌBE
Rokkō
Island
Yodo
Konohana
Aji
Nishi
Ikuno
Higashinari

34 40'
Nagata
Kōbe
Harbour
ŌSAKA
Higashiōsaka
34 40'

Suma
Minato
Naniwa
Stadium
Tennoji
Kizuri
Yamamoto

Taishō
Abeno
Kyūhōji

C
Nishinari
Kizu
Yao

Ōsaka Harbour
Higashisumiyoshi
Onchi
C

Shiriuchi
Tainaka
○ Onchi

Osaka Bay
Sakai Harbour
Sumiyoshi
Yao
Airport

Yamato
Matsubara
Kashiwara

Ikeuchi
Fujidera

Tomb of Nintoku
Kanoaka
Ishi

Sakai
Mozu
East from Greenwich
Habikino

135 10'
135 20'
135 30'

5 **6** **7** **8**

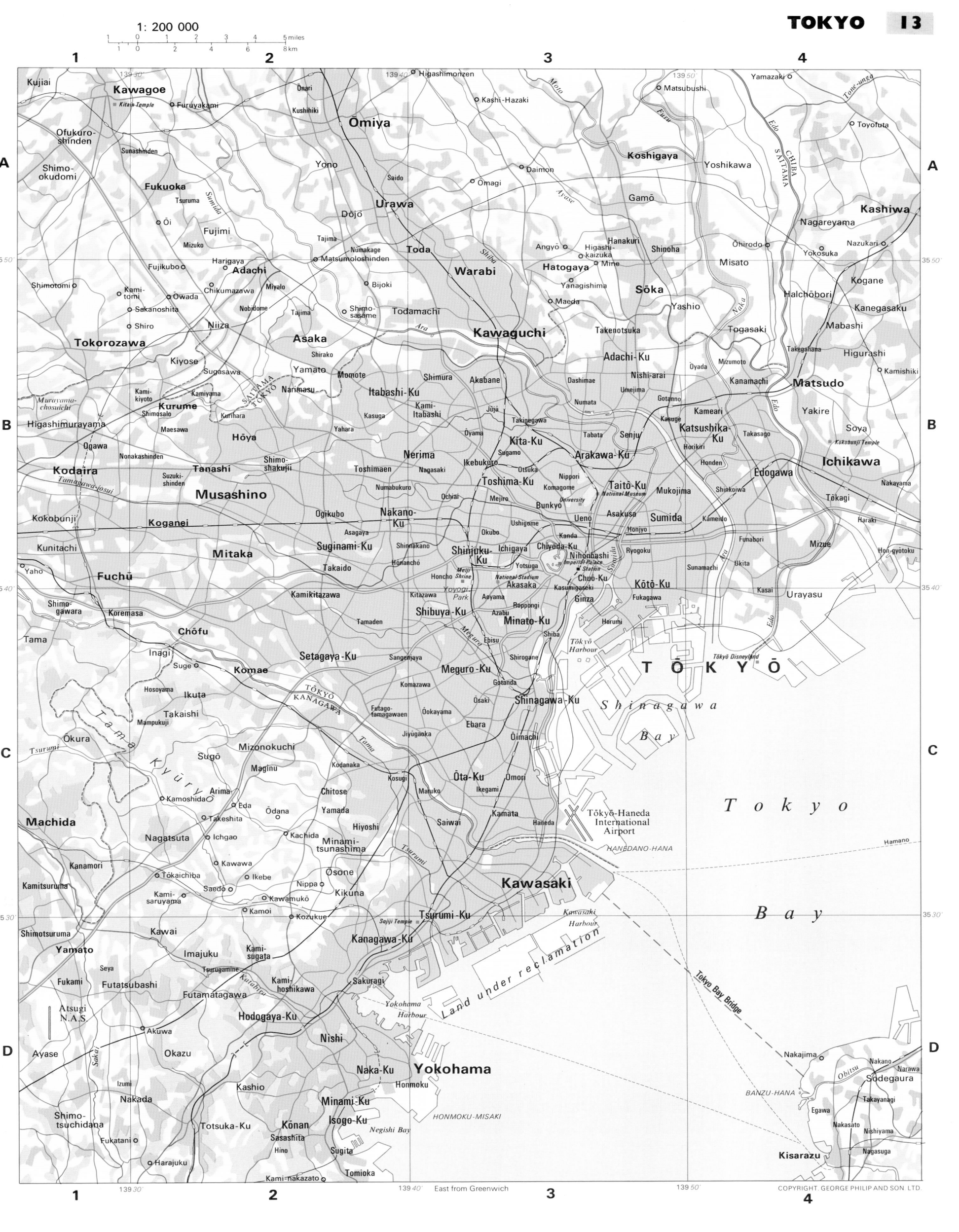

1 0 1 2 3 4 5 miles
1 0 2 4 6 8 km

1 **2** **3** **4**

A

Kujiai
Kawagoe
Ofukuro-shinden
Shimo-okudomi
Kitain Temple
Furuyakami
Ōnari
Higashimonzen
Kushihiki
Kashi-Hazaki
Matsubushi
Yamazaki
Toyofuta
Tone-ung
CHIBA
SAITAMA
Ōmiya
Yono
Saido
Ōmagi
Daimon
Koshigaya
Yoshikawa
Gamō
Nagareyama
Nazukari
Kashiwa

Fukuoka
Tsuruma
Sumida
Urawa
Dōjō
Tajima
Numakage
Angyō
Higashi-kaizuka
Mine
Shinoha
Ōhirodo
Yokosuka
Misato
Hatchōbori
Kogane
Ōi
Fujimi
Mizuko
Toda
Warabi
Hatogaya
Yanagishima
Sōka
Yashio
Takenotsuka
Togasaki
Takegahana
Kanegasaku
Mabashi

Shimotomi
Fujikubo
Harigaya
Matsumoloshinden
Bijoki
Shimosasame
Kawaguchi
Maeda
Ōyada
Mizumoto
Kamishiki
Higurashi

5.50 Adachi Chikumazawa Miyalo Nobidome Todamachī Adachi-Ku Nishi-arai Kanamachi Matsudo **35.50**
Kami-tomi
Ōwada
Sakanoshita
Shiro
Niiza
Yamato
Shirako
Momote
Shimura
Akabane
Dashimae
Umejima
Gotanno
Yakire

B Higashimurayama Kiyose Sugasawa Itabashi-Ku Kasuga Kami-Itabashi Jūjō Takinegawa Tabata Senju Kasuge Kameari Soya Kokobunji Temple **B**
Murayama-chosuichi
Kami-kiyoto
Shimosato
Kamiyama
Kurihara
Yahara
Ōyama
Nerima
Nagasaki
Ikebukuro
Sugamo
Otsuka
Nippori
Komagome
Arakawa-Ku
Horikiri
Honden
Edogawa
Takasago
Ichikawa

Kodaira
Tamagawa-josui
Ogawa
Nonakashinden
Hōya
Shimo-shakujii
Toshimaen
Numabukuro
Ochiai
Mejiro
Bunkyō
Ueno
Asakusa
Sumida
Kameido
Haraki
Tōkagi
Nakayama

Kunitachi
Musashino
Koganei
Ōgikubo
Nakano-Ku
Asagaya
Shinnakano
Okubo
Ushigome
University
Taitō-Ku
National Museum
Mukojima
Shinkoiwa
Hon-gyōtoku

5.40 Yaho Mitaka Suginami-Ku Honanchō Shinjuku-Ku Ichigaya Chiyoda-Ku Nihonbashi Ryogoku Funabori Mizue **35.40**
Fuchū
Takaido
Yotsuga
Imperial Palace
Station
Honcho
Honjyo
Sunamachi
Ūkita
Kasai
Urayasu

Shimo-gawara
Koremasa
Kamikitazawa
Kitazawa
Anyama
Meiji
Shrine
National Stadium
Akasaka
Kasumigaseki
Chūō-Ku
Kōtō-Ku
Fukagawa
Edo

Tama
Chōfu
Inagi
Suge
Komae
Yoyogi
Park
Shibuya-Ku
Roppongi
Azabu
Ginza
Harumi
Tōkyō Disneyland

Hosoyama
Ikuta
Tamaden
Minato-Ku
Ebisu
Shiba
Tōkyō
Harbour
TŌKYŌ

C Okura Takaishi Setagaya-Ku Sangenjaya Meguro-Ku Shirogane Shinagawa **C**
Tsurumi
Mampukuji
Komazawa
Gotanda
Ōsaki
Bay

Sugō
Mizonokuchi
Futago-tamagawaen
Ōokayama
Shinagawa-Ku
Ōimachi

Arima
Maginu
Kōdanaka
Ebara
Jiyūgaoka

Kamoshida
Eda
Ōdana
Kosugi
Matuko
Ōta-Ku
Ōmori
Tokyo

Machida
Takeshita
Chitose
Yamada
Ikegami

Nagatsuta
Ichgao
Kachida
Hiyoshi
Saiwai
Kamata
Haneda
Tōkyō-Haneda
International
Airport
Bay
Hamano

Kanamori
Kawawa
Minami-tsunashima
HANEDANO-HANA

45.30 Kamitsuruma Tōkaichiba Saedō Ikebe Nippa Ōsone Kikuna **Kawasaki** **35.30**
Kami-saruyama
Kawamukō
Kamoi
Kozukue
Seiji Temple
Tsurumi-Ku
Kawasaki
Harbour

Shimotsuruma
Kawai
Kanagawa-Ku
Tokyo Bay Bridge
Nakajima
Nakano
Narawa

Yamato
Imajuku
Kami-sugata
Sakuragi
Nakano
Sōdegaura

Fukami
Seya
Tsurugamine
Kami-hoshikawa
Yokohama
Harbour
Egawa
Takayanagi

Futatsubashi
Futamatagawa
Katabira
Land under reclamation
Obitsu

Atsugi
N.A.S
Hodogaya-Ku
BANZU-HANA

D Akuwa Nishi Nakajima Egawa **D**
Ayase
Okazu
Naka-Ku
Yokohama
Honmoku
Nakasato
Nishiyama

Izumi
Kashio
Honmoku
Nagasuga

Nakada
Minami-Ku
Kōnan
Isogo-Ku
HONMOKU-MISAKI
Negishi Bay
Kisarazu

Shimo-tsuchidana
Totsuka-Ku
Sasashita
Hino

Fukatani
Harajuku
Sugita
Tomioka
Kami-nakazato

1 **2** East from Greenwich **3** **4**

COPYRIGHT. GEORGE PHILIP AND SON. LTD.

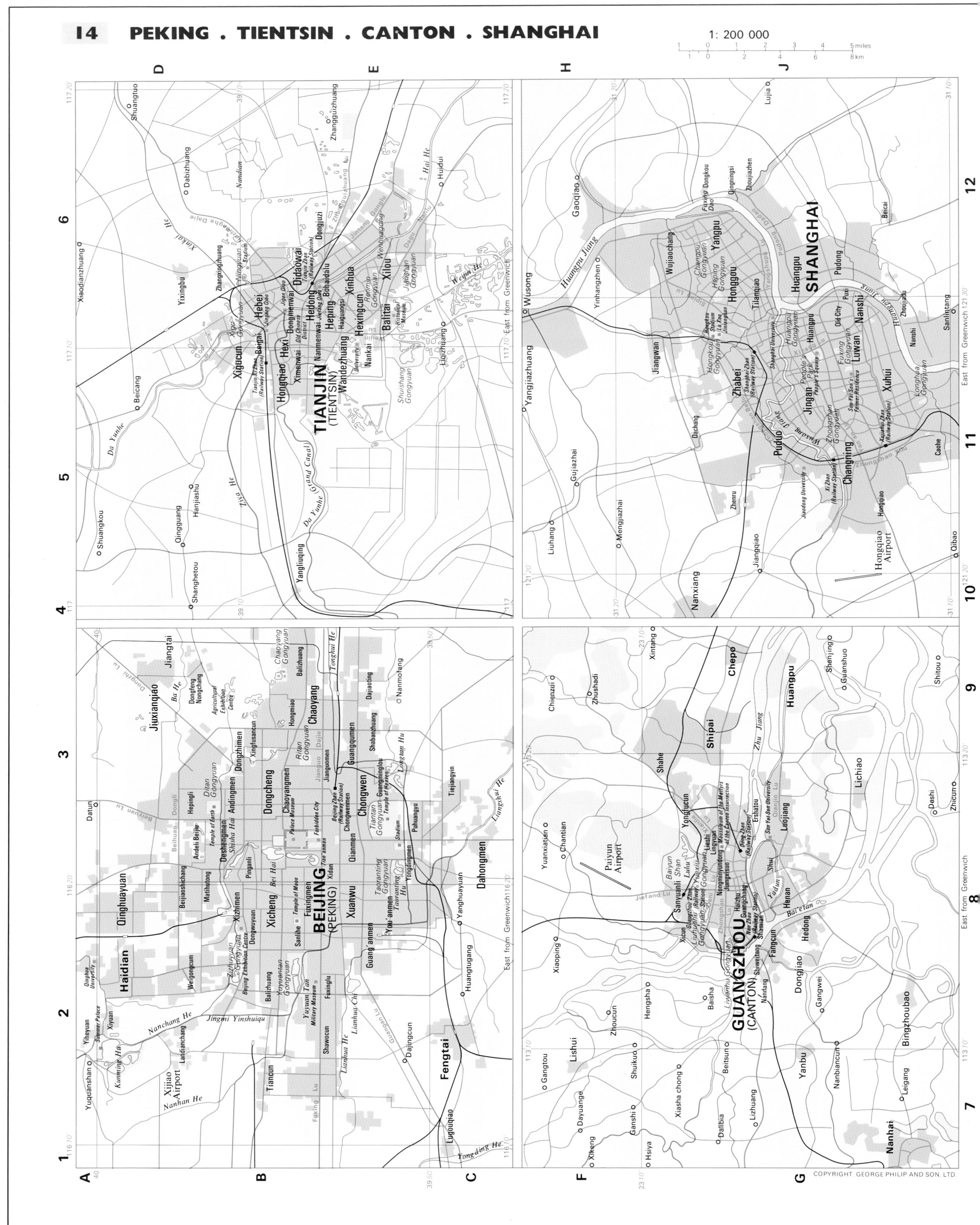

1: 200 000

COPYRIGHT. GEORGE PHILIP AND SON, LTD.

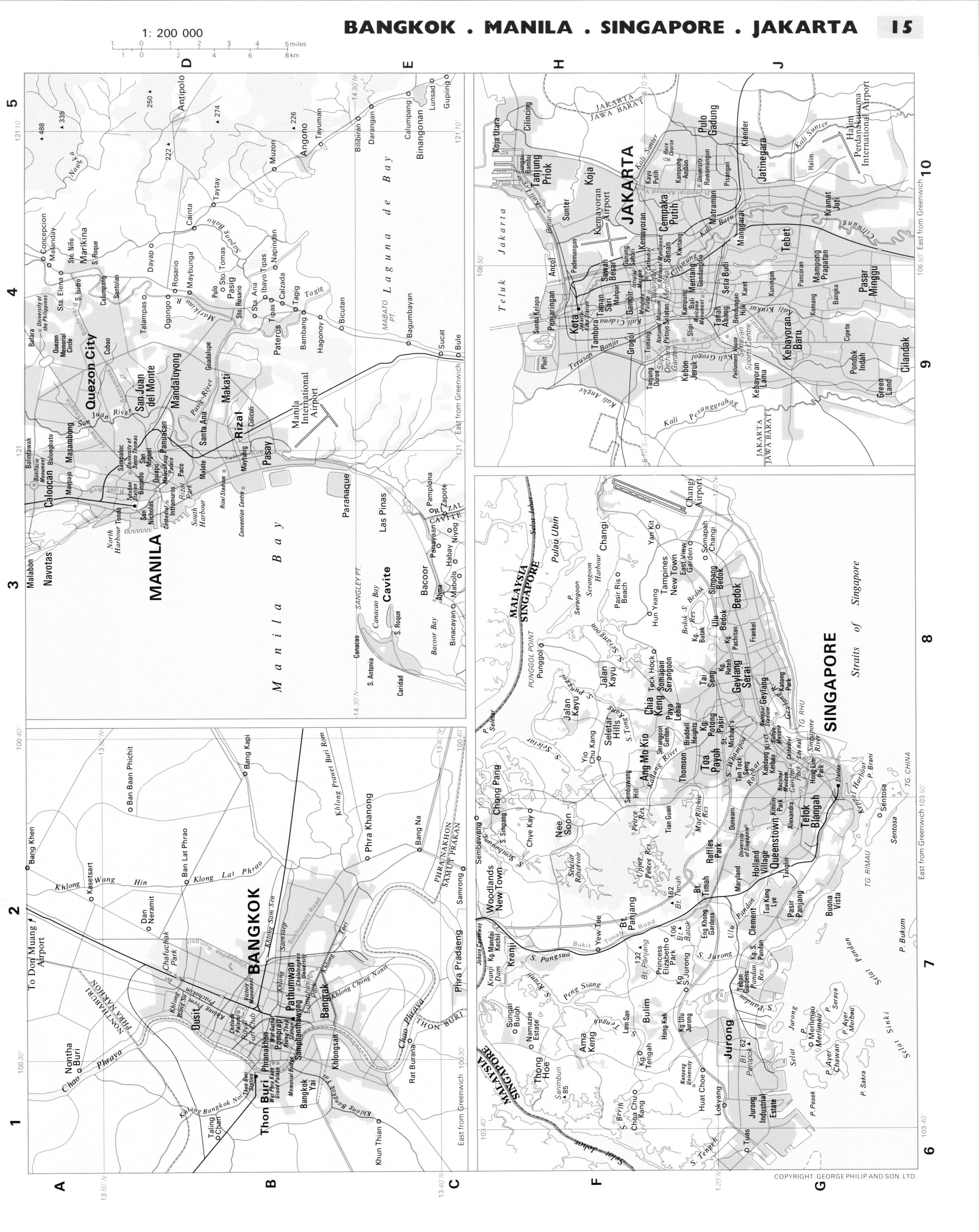

1: 200 000

1: 200 000

5 miles
8 km

Calcutta panel (top):

Madatpur, Naihati, Bhatpara, Panpur, Gauripur, Kankinara, Bidyadharpur, Basudebpur, Mirzapur, Hansia Bariti Bil, Niiganj, Balagarh, Beraberi, Bandipur, Dum Dum Int. Airport, Gopalpur, Aghara, Hatiara, Salt Water, Nabadabad

Bhadreswar, Chunchura, Chandernagore, Alpur, Garulia, Ichapur, Barrackpore Airport, Titagarh, Khardah, Sukchar, Panihati, Kamarhati, Nimta, Phinga, Sodpur, Bhatpur, Madhyamgram, DumDum, Setgachi, Baghati, Banstala, Madhudaha, Gariya

Bhadreswar, Champdani, Baidyabati, Shrirampur, Satghara, Ballabhpur, Rishra, Konnagar, Kotrang, Uttarpara, Bally, Belur, Salkhia, Ghushuri, Lituah, Howrah Bridge, CALCUTTA, Tollygunge, Ballygunge, Bhawanipore, Russa, Alipore

Haora, Garden Reach, Panchur, Behala, Nangi, Baj Baj, Bauria, Batanagar, Maheshtala, Satsuna

Delhi panel (lower left):

Daultpur, Jauli, Rampur, Arzalpur, Bhopura, Shahdara, Nithari, Atta, Aganpur, Loni, Subhepur, Saboli, Maharajpur, Ghaziabad, Ghazipur, Kondli, Khichripur, Chilla Saroda, Agra Canal, Okhla, Kilokri, Khajuri Khas

DELHI, NEW DELHI, Red Fort, Delhi Station, New Delhi Station, India Gate, Safdar Jang Airport, Mujahidpur, Kalkaji, Shakarpur Khas, Yamuna, Shahdara

University, Malakpur, Sabzi Mandi, Sadar Bazar, Rajpur, Coronation Memorial, Wazirabad, Mukandpur, Dahirpur, Shastrinagar, Wazirpur, Azadpur, Shakurpur, Shalimar Bagh, Madipur, Tatarpur, Naraina, Der Sarai, Munirka, Mehram Nagar, Palam Int. Airport

Basai Darapur, Tehar, Delhi Cantonment, Nangal Dewat, Palam, Bagrula, Daulatpur, Nithari, Nangloi Jat, Sahibabad, Rithala, Puth Kalan, Nangloi, Pira, Asalatpur, Mehpalpur

UTTAR PRADESH, DELHI, Grand Trunk Rd, Ring Road, Rohtak Road, The Ridge

Bombay panel (lower right):

Koparkhairna, Kharirna, Juhu, Turambhe, Sampada, Shiraone, Bonsari, Sarsol, Derave, Vahal, Shahabad, Darave, Bamondongri, Selghar, Dhutumkhar

Thana Creek, Ghatkopar, Vashi, Nerul, Man Khurd, Yadaul, Man Budrukh, Gharapuri, Nanole, Sheva Nhava, Karave, Belapurpada, Gavanpada, Jasai, Chirle, Panje, Sonari, Jaskhar

Kurla, Chembur, Trombay, Elephanta Island (Gharapuri) 169, Elephanta Caves, Shet Bandar, Salsette Island, 305, BOMBAY, Bombay Harbour, Mora, Sheva, Panje, Dongri, Punde, Papote

Bandra, Santa Cruz Int. Airport, Mahim, Worli, Parel, Byculla, Mazagaon, Girgaum, Malabar Hill, Back Bay, Colaba, Fort, Victoria Terminus, Gateway of India, ARABIAN SEA, BANDRA POINT, COLABA POINT

East from Greenwich

COPYRIGHT GEORGE PHILIP AND SON. LTD.

1 : 200 000

miles
1 0 1 2 3 4 5 miles
1 0 2 4 6 8 km

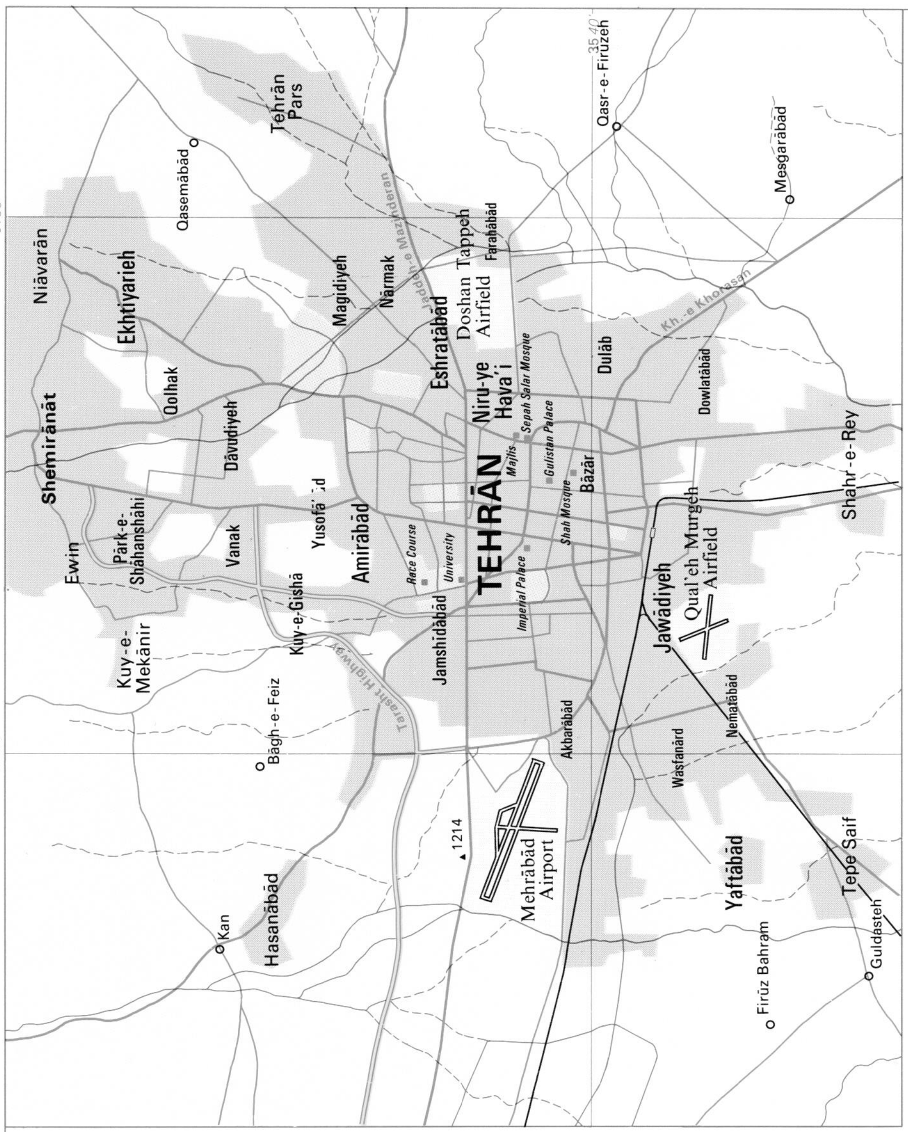

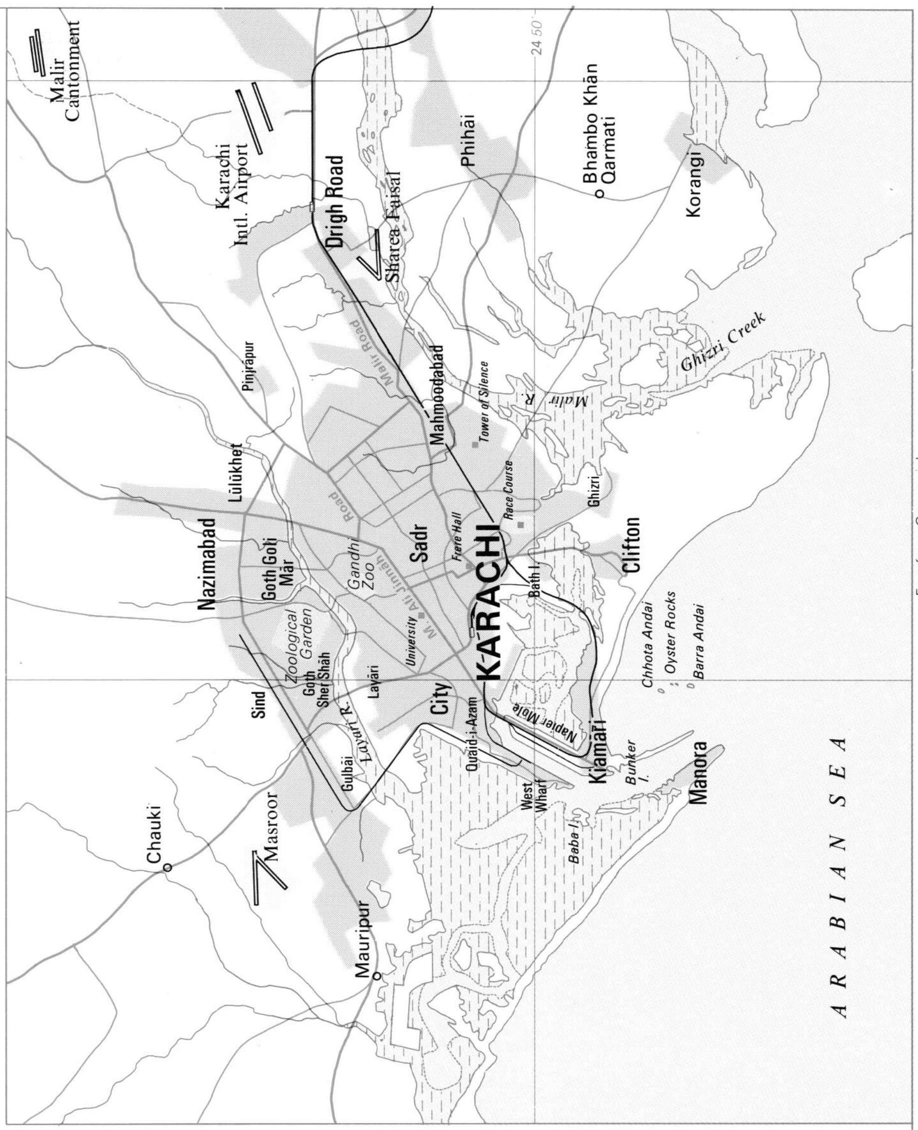

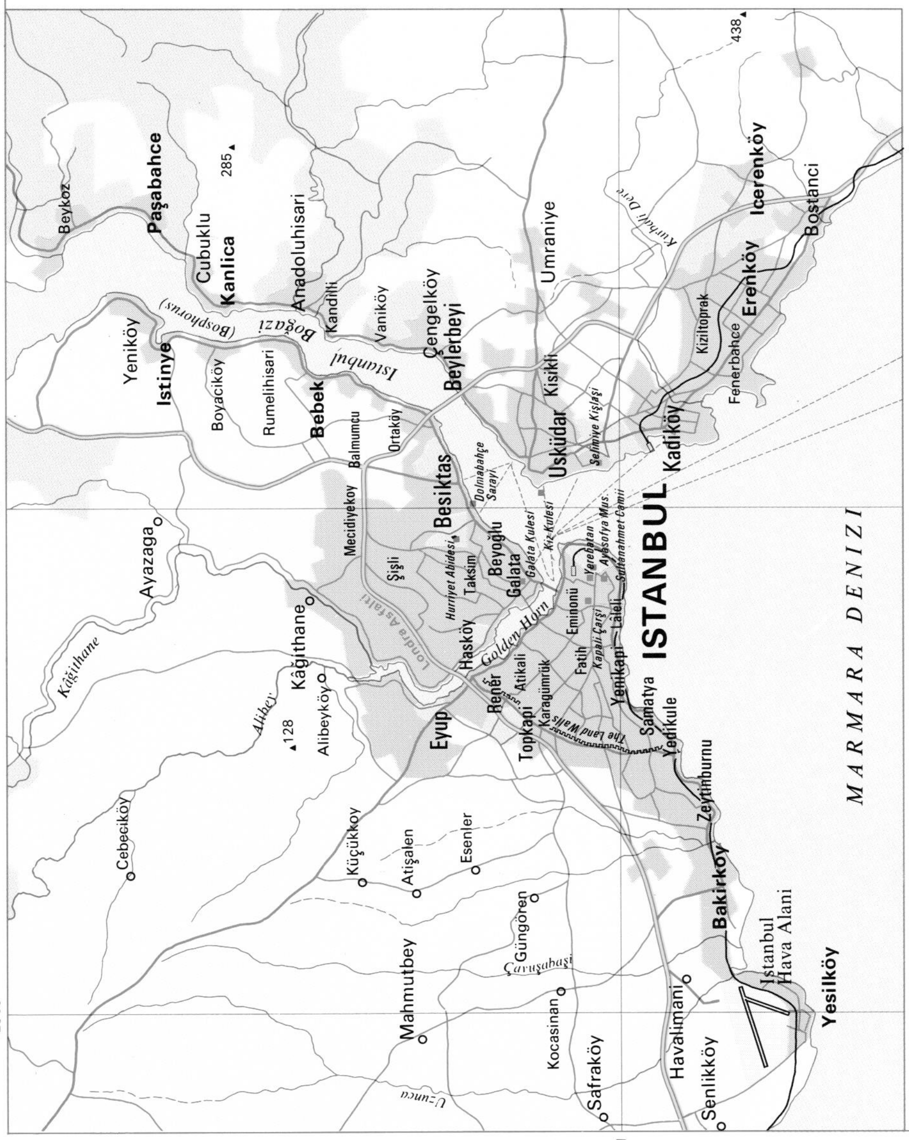

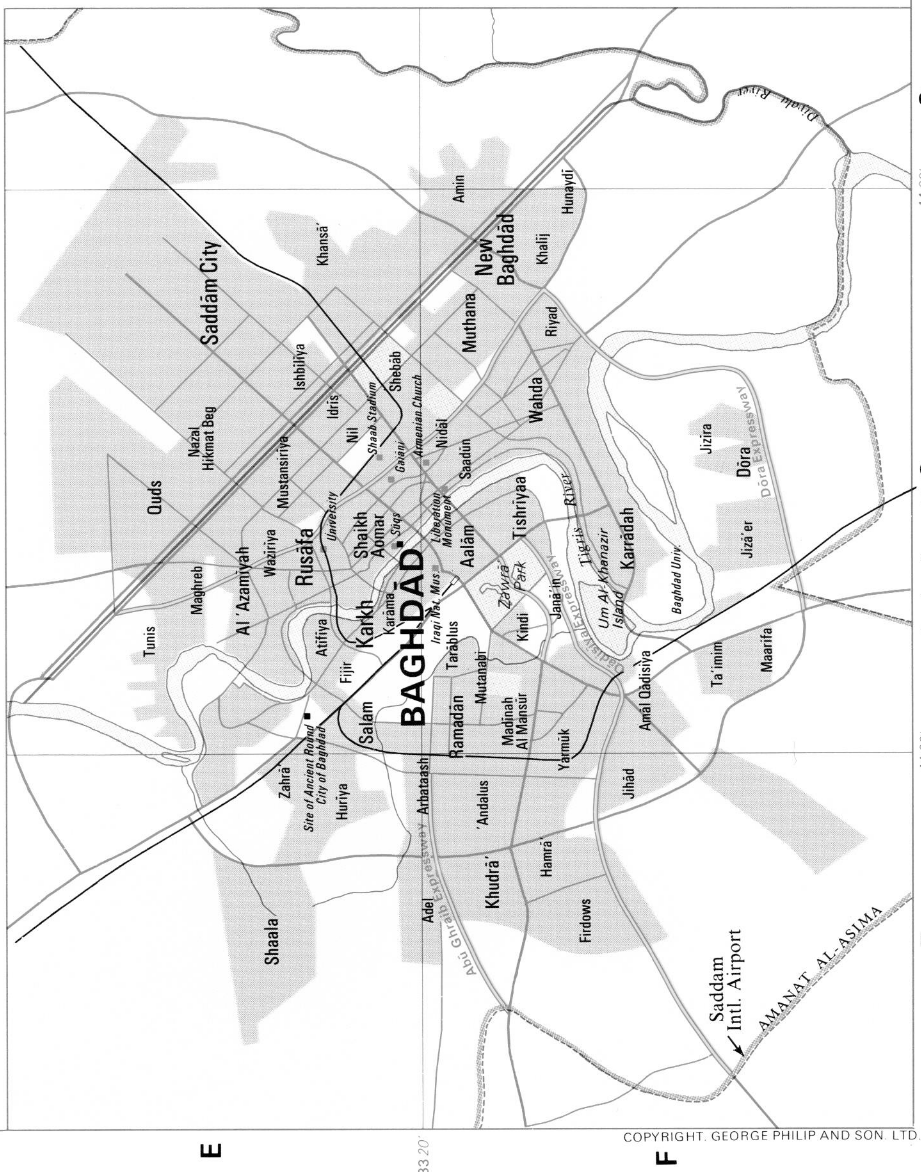

Tehran (C–D, 4–6)

Niāvarān
Tehrān Pars
Qasemābād
Qasr-e-Firūzeh
Ekhtiyarieh
Shemirānāt
Qolhak
Magidiyeh
Narmak
Eshratābād
Doshan Tappeh Airfield
Farahābād
Mesgarābād
Ewin
Pārk-e-Shāhanshāhi
Dāvudiyeh
Niru-ye Havā'i
Sepah Salar Mosque
Dulāb
Dowlatābād
Kuy-e-Mekānir
Vanak
Yusofā'bād
Majlis
Shah Mosque
Bāzār
Shahr-e-Rey
Kuy-e-Gishā
Amirābād
University
Gulistan Palace
TEHRĀN
Jamshidābād
Imperial Palace
Jawādiyeh
Qual'eh Murgeh Airfield
Kan
Hasanābād
Bāgh-e-Feiz
Akbarābād
Wastanārd
Nematābād
Mehrābād Airport
▲1214
Yaftābād
Tepe Saif
Firūz Bahrām
Guldasteh

Istanbul (A–B, 1–3)

Beykoz
Paşabahçe
Çubuklu
Kanlica
Anadoluhisari
Kandilli
Yeniköy
Istinye
Boyacıköy
Rumelihisari
Bebek
Bahçeköy
Ortaköy
Çengelköy
Beylerbeyi
Vanıköy
Üsküdar
Kısıklı
Ümraniye
İçerenköy
Erenköy
Bostancı
Kızıltoprak
Fenerbahçe
Kadıköy
Avazaga
Mecidiyeköy
Şişli
Kağithane
Beyoğlu
Taksim
Galata
Beşiktaş
Dolmabahçe Sarayı
Yenikapı
Hasköy
Eminönü
Fatih
İSTANBUL
Eyüp
Topkapı
Karagümrük
Zeytinburnu
Yedikule
Sahatya
Bakırköy
Istanbul Hava Alani
Yeşilköy
Mahmutbey
Küçükköy
Esenler
Atişalen
Havalimanı
Cebeciköy
Kocasinan
Safraköy
Senlikköy

BOĞAZİÇİ (Bosphorus)
İstanbul Boğazı
MARMARA DENIZI
Golden Horn
The Sea Wall

Karachi (G–H, 10–12)

Malir Cantonment
Karachi Intl. Airport
Dright Road
Sharea Faisal
Phihāi
Bhambo Khān Qarmati
Korangi
Tower of Silence
Mahmoodabad
Pinrāpur
Liāukhet
Sadr
Ghizri Creek
Nazimabad
Goth Goli Mār
Goth Sher Shah
Race Course
Fort Hall
Gandhi Zoo
Ghizri
Clifton
Zoological Garden
Bath I.
University
Lyāri
Layāri
KARACHI
Chhota Andai
Oyster Rocks
Barra Andai
Sind
Quáid-i-Azam
City
Napier Mole
Kiamāri
Bunker I.
Gulbai
Chauki
West Wharf
Baba I.
Manora
Masroor
Mauripur
Baba Is.

ARABIAN SEA

Baghdad (E–F, 7–9)

Saddām City
Khansā'
Amin
Khalij
New Baghdad
Hunaydi
Ishbilīya
Idrīs
Shebāb
Riyad
Wahda
Jizīra
Nazāl
Hikmat Beg
Mustansirīya
Nil
Shaab Stadium
Amanean Church
Aalām
Tishrīya
Karrādah
Dōra
Quds
Umm Ibn
Gikau
Shaikh Omar
Saadun
Mutanabi
Tarabus
Nuidal
Jana'in
Baghdad Univ.
Maghreb
Al 'Azamiyah
Wazīriya
Zawra Park
Um Al Khanazir Island
Jizī'er
Tunis
Atifiya
Karkh
Shaikh
BAGHDĀD
Fijir
Madinah Al Mansūr
Kindi
Tigris River
Ta'imim
Maarifa
Salam
Ramadān
Yarmük
Salām
Huriya
Andalus
Jihād
Hamrā
Khudrā
Hamrā
Firdows
Shaala
Saddam Intl. Airport
Site of Ancient Ruins City of Baghdad

AMANAT AL-ASIMA

East from Greenwich

COPYRIGHT. GEORGE PHILIP AND SON. LTD.

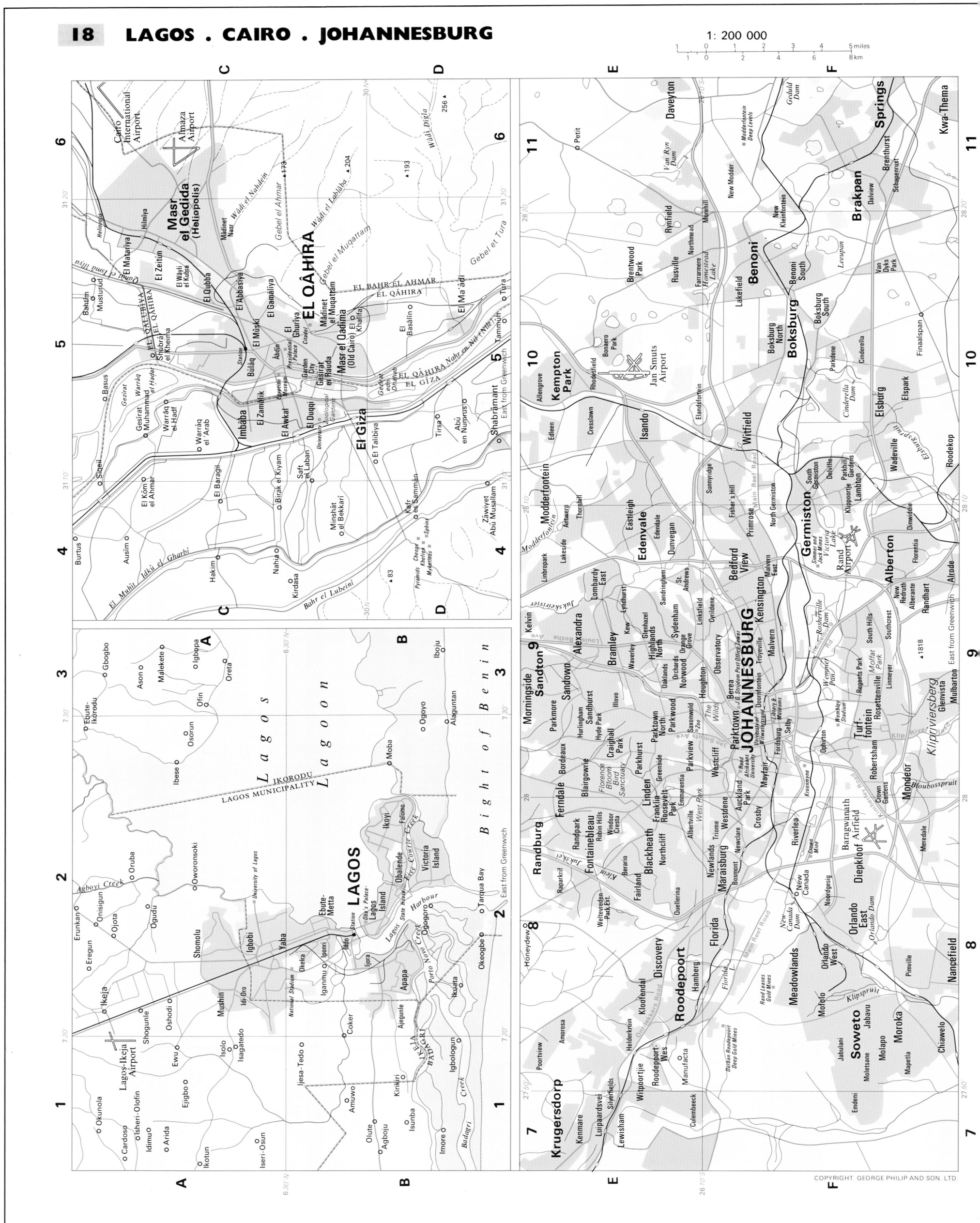

1 : 200 000

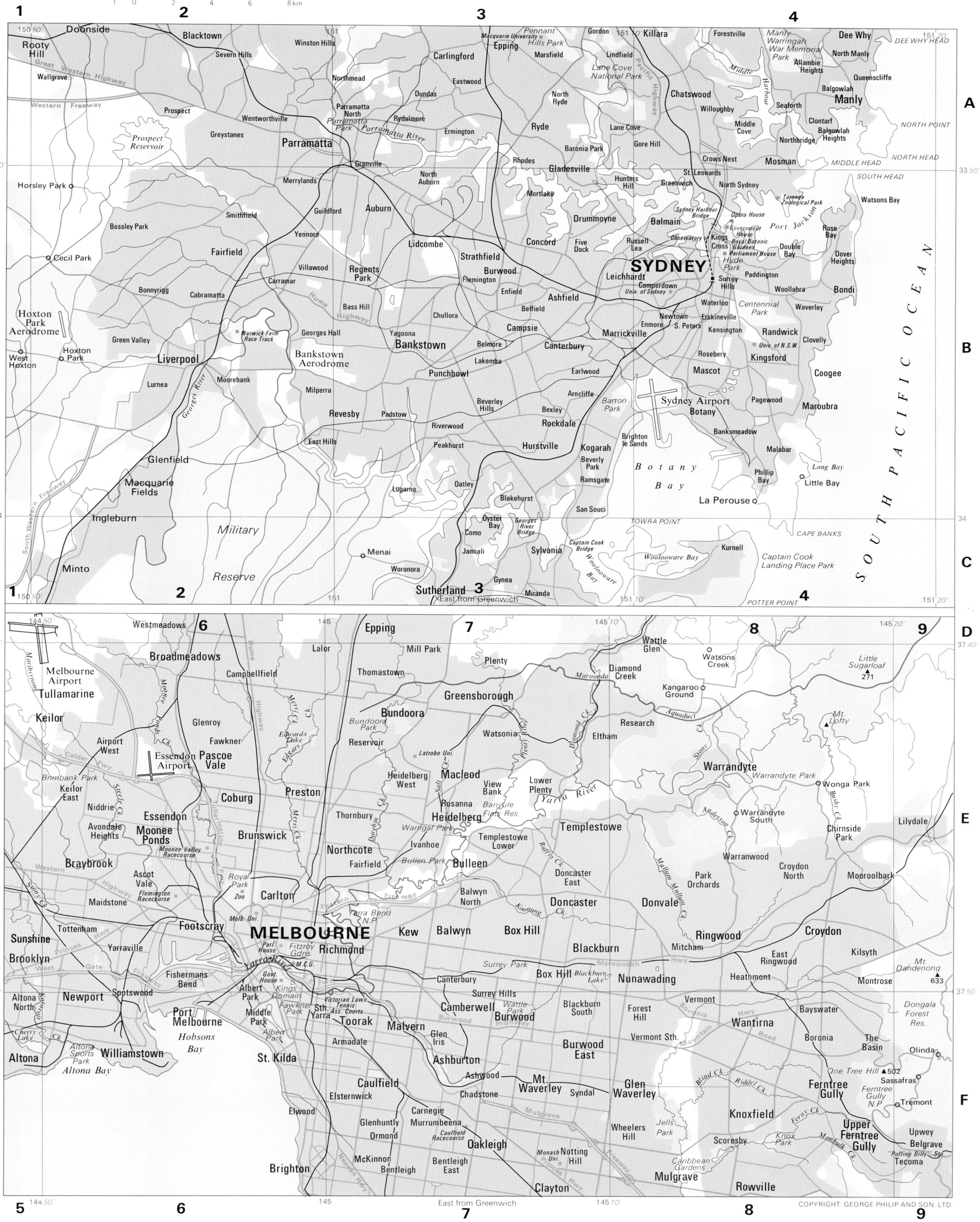

1 0 1 2 3 4 5 miles
1 0 2 4 6 8 km

SYDNEY (upper map)

Doonside, Blacktown, Winston Hills, Carlingford, Epping, Marsfield, Macquarie University, Pennant Hills Park, Gordon, Killara, Forestville, Manly Warringah War Memorial Park, Dee Why, DEE WHY HEAD

Rooty Hill, Severn Hills, Northmead, Eastwood, Dundas, North Ryde, Lindfield, Lane Cove National Park, North Manly, Queenscliffe

Wallgrove, Prospect, Wentworthville, Parramatta North, Parramatta Park, Rydalmere, Ryde, Lane Cove, Chatswood, Willoughby, Seaforth, Clontarf, Balgowlah Heights, Manly, NORTH POINT

Great Western Highway, Western Freeway, Greystanes, Parramatta, Parramatta River, Ermington, Gore Hill, Middle Cove, Northbridge, Balgowlah Heights

Prospect Reservoir, Granville, North Auburn, Rhodes, Gladesville, Baronia Park, Crows Nest, Mosman, MIDDLE HEAD, NORTH HEAD

Horsley Park, Merrylands, Guildford, Auburn, Mortlake, Hunters Hill, Tarongo Zoological Park, Watsons Bay, SOUTH HEAD

Smithfield, Yennora, Villawood, Lidcombe, Strathfield, Concord, Five Dock, Drummoyne, Balmain, Sydney Harbour Bridge, Opera House, Port Jackson, Rose Bay

Bossley Park, Fairfield, Carramar, Regents Park, Burwood, Flemington, Ashfield, Russell Lea, Observatory, Kings Cross, Government House, Royal Botanic Gardens, Parliament House, Double Bay, Dover Heights

Cecil Park, Bonnyrigg, Cabramatta, Bass Hill, Enfield, Belfield, **SYDNEY**, Leichhardt, Camperdown, Univ. of Sydney, Hyde Park, Surry Hills, Paddington, Woollahra, Bondi

Hoxton Park Aerodrome, Green Valley, Warwick Farm Race Track, Georges Hall, Yagoona, Belmore, Campsie, Canterbury, Marrickville, Newtown, S. Peters, Enmore, Waterloo, Erskineville, Kensington, Centennial Park, Waverley, Randwick, Clovelly

West Hoxton, Liverpool, Hoxton Park, Bankstown Aerodrome, **Bankstown**, Lakemba, Roseberry, Univ. of N.S.W., Kingsford, Coogee

Lurnea, Moorebank, Milperra, Punchbowl, Earlwood, Mascot, Maroubra

Georges River, Revesby, Padstow, Beverley Hills, Arncliffe, Barton Park, Sydney Airport, Botany, Pagewood, Banksmeadow

Glenfield, East Hills, Riverwood, Bexley, Rockdale, Brighton le Sands, Malabar

Macquarie Fields, Peakhurst, Hurstville, Kogarah, Beverly Park, Phillip Bay, Long Bay, Little Bay

Ingleburn, Lugarno, Oatley, Blakehurst, Ramsgate, *Botany Bay*, La Perouse

Military Reserve, Oyster Bay, Como, Jannali, San Souci, TOWRA POINT, CAPE BANKS

Minto, Menai, Woronora, Georges River Bridge, Sylvania, Captain Cook Bridge, Woolooware Bay, Kurnell, Captain Cook Landing Place Park

Sutherland, Gynea, Miranda, POTTER POINT, *SOUTH PACIFIC OCEAN*

X East from Greenwich

MELBOURNE (lower map)

Westmeadows, Epping, Wattle Glen, Watsons Creek, Little Sugarloaf 271

Broadmeadows, Lalor, Mill Park, Plenty, Diamond Creek, Kangaroo Ground

Melbourne Airport, Tullamarine, Campbellfield, Thomastown, Greensborough, Research, Mt. Lofty

Keilor, Glenroy, Fawkner, Edwards Lake, Bundoora, Bundoora Park, Reservoir, Watsonia, Eltham, Warrandyte, Warrandyte Park, Wonga Park

Airport West, Pascoe Vale, Preston, Heidelberg West, Macleod, Rosanna, Latrobe Uni., Lower Plenty, Yarra River, Warrandyte South, Chirnside Park, Lilydale

Essendon Airport, Coburg, View Bank, Banyule Flats Res., Templestowe, Warranwood, Croydon North, Mooroolbark

Keilor East, Niddrie, Avondale Heights, Essendon, Moonee Ponds, Brunswick, Thornbury, Heidelberg, Warringal Park, Templestowe Lower, Doncaster East, Park Orchards, Donvale

Braybrook, Ascot Vale, Moonee Valley Racecourse, Northcote, Fairfield, Ivanhoe, Bullen Park, Bulleen, Doncaster, Ringwood, Croydon

Maidstone, Royal Park Zoo, Flemington Racecourse, Carlton, Melb. Uni., Yarra Bend N.P., Balwyn North, Kooling, Mitcham, East Ringwood, Kilsyth

Sunshine, Tottenham, Footscray, **MELBOURNE**, Kew, Balwyn, Box Hill, Blackburn, Mt. Dandenong 633

Brooklyn, Yarraville, Parl. House, Fitzroy Gdns, M.C.G., Richmond, Box Hill, Blackburn Lake, Nunawading, Heathmont, Montrose

Newport, Spotswood, Fishermans Bend, Albert Park, Govt. House, Kings Domain, Canterbury, Surrey Park, Surrey Hills, Vermont, Bayswater

Altona North, Middle Park, Port Melbourne, Fawkner Park, Toorak, Camberwell, Burwood, Blackburn South, Forest Hill, Vermont Sth., Wantirna, Boronia, Dongala Forest Res.

Altona, Williamstown, Hobsons Bay, St. Kilda, Malvern, Glen Iris, Ashburton, Ashwood, Burwood East, Mt. Waverley, Syndal, Glen Waverley, The Basin, Olinda

Altona Sports Park, Altona Bay, Elwood, Caulfield, Elsternwick, Armadale, Carnegie, Chadstone, Mt. Waverley, One Tree Hill 502, Sassafras, Ferntree Gully, Ferntree Gully N.P., Tremont

Brighton, Elwood, Glenhuntly, Murrumbeena, Ormond, Oakleigh, Notting Hill, Monash Uni., Wheelers Hill, Jells Park, Scoresby, Knoxfield, Upper Ferntree Gully, Belgrave, Upwey, Puffing Billy Stn, Tecoma

McKinnon, Bentleigh, Bentleigh East, Clayton, Mulgrave, Rowville, Caribbean Gardens, Caulfield Racecourse

East from Greenwich

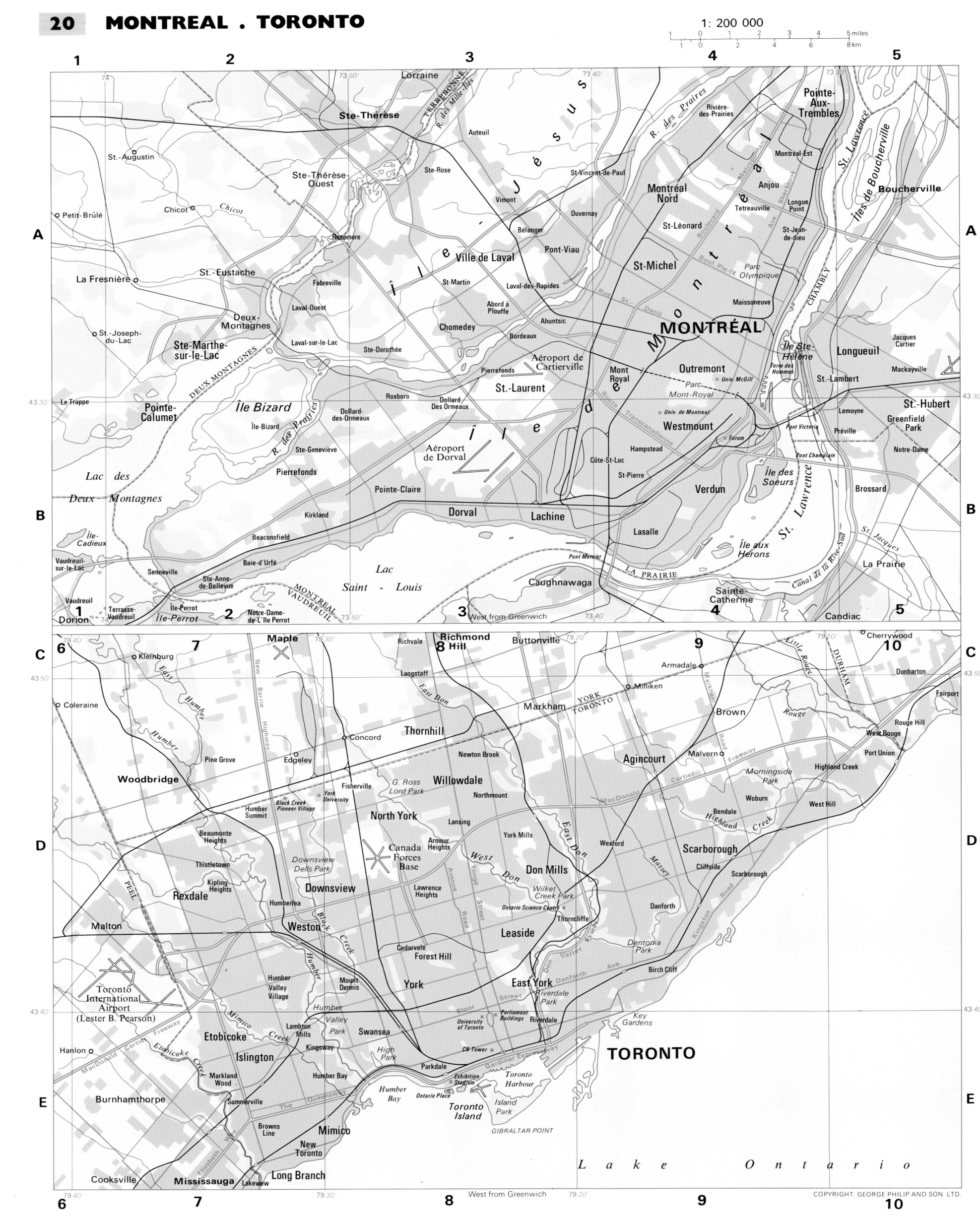

1 : 200 000

Lorraine
Ste-Thérèse
Auteuil
Ste-Rose
St-Vincent-de-Paul
St.-Augustin
Ste-Thérèse-Ouest
Vimont
Montréal-Est
Pointe-Aux-Trembles
Petit-Brûlé
Chicot
Chicot
Montréal Nord
Anjou
Îles de Boucherville
St-Léonard
Longue Point
Tetreauville
Boucherville
La Fresnière
St.-Eustache
Ville de Laval
Bélanger
St-Michel
St-Jean-de-dieu
Pont-Viau
Maissoneuve
Fabreville
St-Martin
Laval-des-Rapides
Deux-Montagnes
Laval-Ouest
Abord à Plouffe
MONTRÉAL
St.-Joseph-du-Lac
Ste-Marthe-sur-le-Lac
Laval-sur-le-Lac
Chomedey
Ahuntsic
Bordeaux
Parc Olympique
Jacques Cartier
Ste-Dorothée
Aéroport de Cartierville
Outremont
Mackayville
Mont Royal
Parc Mont-Royal
Univ. McGill
Longueuil
Le Trappe
Pointe-Calumet
Île Bizard
Roxboro
St.-Laurent
Terre des Hommes
St-Lambert
St.-Hubert
Dollard Des Ormeaux
Lemoyne
Greenfield Park
Île-Bizard
Dollard-des-Ormeaux
Univ. de Montréal
Westmount
Préville
Notre-Dame
Ste-Geneviève
Pont Victoria
Lac des
Deux-Montagnes
Pierrefonds
Aéroport de Dorval
Pointe-Claire
Hampstead
Côte-St-Luc
St-Pierre
Île des Soeurs
Pont Champlain
Brossard
Île-Cadieux
Kirkland
Dorval
Lachine
Verdun
Beaconsfield
Lasalle
Vaudreuil-sur-le-Lac
Baie-d'Urfé
Île aux Herons
La Prairie
St. Lawrence
La Prairie
St. Jacques
Senneville
Ste-Anne-de-Bellevue
Lac
Saint - Louis
Pont Mercier
Canal de la Rive-Sud
Vaudreuil
Terrasse-Vaudreuil
Île-Perrot
Notre-Dame-de-L'Île Perrot
Caughnawaga
Sainte Catherine
Dorion
Île-Perrot
MONTREAL VAUDREUIL
West from Greenwich
Candiac

Maple
Richvale
Richmond Hill
Buttonville
Cherrywood
Kleinburg
Langstaff
Armadale
DURHAM
Dunbarton
Coleraine
Markham
YORK TORONTO
Milliken
Fairport
Thornhill
Brown
Rouge
Rouge Hill
Woodbridge
Pine Grove
Concord
Newton Brook
West Rouge
Edgeley
Fisherville
Willowdale
Agincourt
Malvern
Port Union
York University
G. Ross Lord Park
Northmount
Highland Creek
Humber Summit
Black Creek Pioneer Village
Morningside Park
Beaumonte Heights
North York
Lansing
Woburn
West Hill
Thistletown
Downsview Dells Park
Canada Forces Base
Armour Heights
York Mills
Wexford
Bendale
Kipling Heights
Don Mills
Scarborough
Rexdale
Downsview
Lawrence Heights
Wilket Creek Park
Cliffside
Scarborough
Malton
Humberlea
Ontario Science Centre
Danforth
Weston
Thorncliffe
Leaside
Dentonia Park
Toronto International Airport (Lester B. Pearson)
Cedarvale Forest Hill
East York
Birch Cliff
Hanlon
Humber Valley Village
Mount Dennis
York
Riverdale Park
Key Gardens
Etobicoke
Lambton Mills
Valley Park
Swansea
University of Toronto
Parliament Buildings
Riverdale
Islington
Kingsway
High Park
CN Tower
TORONTO
Markland Wood
Humber Bay
Parkdale
Gardiner Expressway
Toronto Harbour
Burnhamthorpe
Summerville
The Queensway
Humber Bay
Exhibition Stadium
Ontario Place
Island Park
Mimico
Browns Line
New Toronto
Toronto Island
GIBRALTAR POINT
Cooksville
Mississauga
Lakeview
Long Branch
Lake Ontario

West from Greenwich

COPYRIGHT. GEORGE PHILIP AND SON. LTD.

1: 200 000

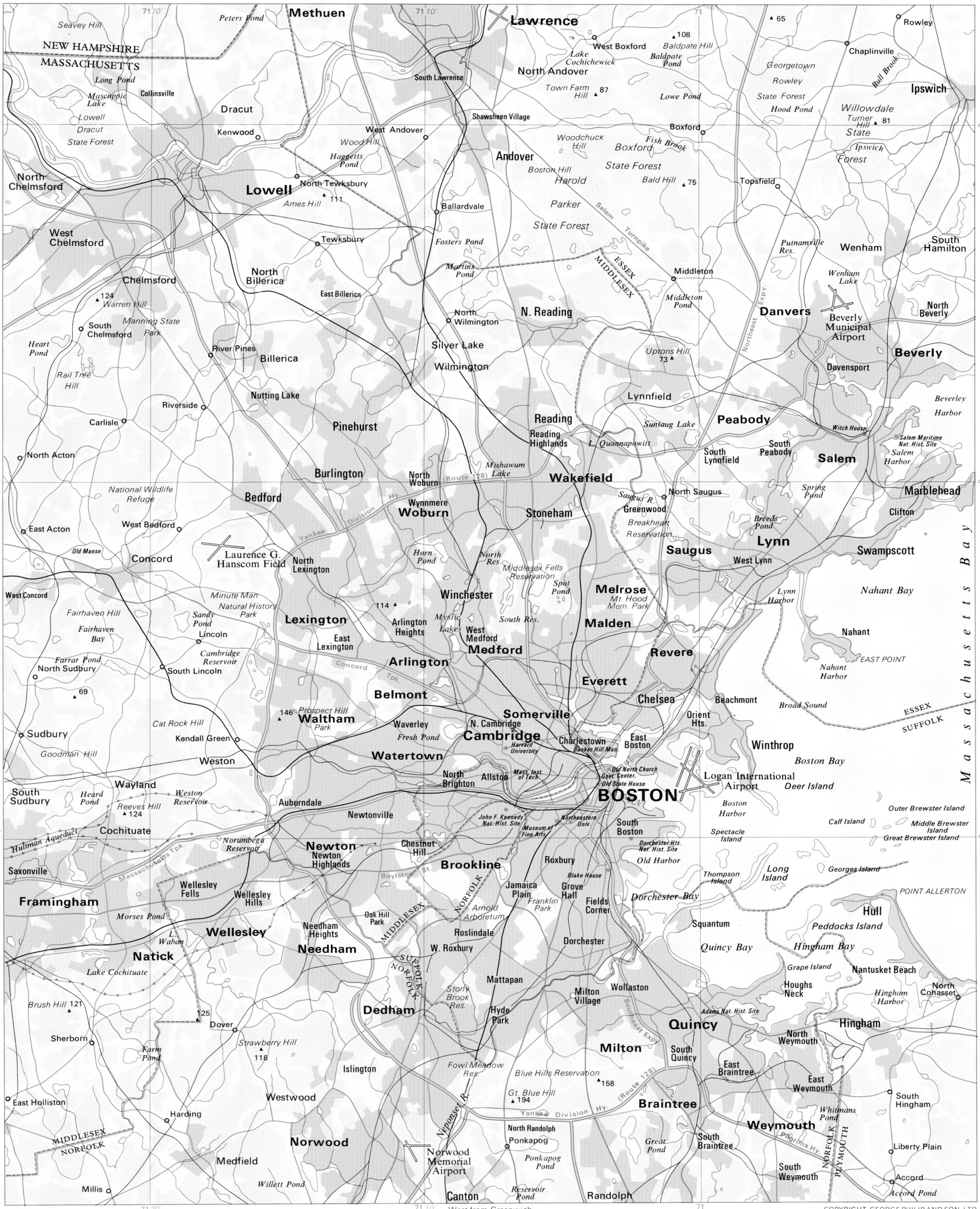

1 0 1 2 3 4 5 miles
1 0 2 4 6 8 km

1 2 3 4

A

NEW HAMPSHIRE
MASSACHUSETTS

Seavey Hill
Peters Pond
Methuen
Lawrence
West Boxford
▲ 65
Rowley
108 ▲
Baldpate Hill
Baldpate Pond
Chaplinville
Lake Cochichewick
Long Pond
Collinsville
Mascuppic Lake
North Andover
Georgetown
Rowley
State Forest
Hood Pond
Willowdale State
Ipswich
Dracut
South Lawrence
Town Farm Hill
87 ▲
Lowe Pond
Boxford
Turner ▲ 81
Ipswich Forest
Lowell Dracut State Forest
Kenwood
West Andover
Shawsheen Village
Woodchuck Hill
Boston Hill
Boxford State Forest
Fish Brook
Putnamsville Res.
Wenham
South Hamilton

B

North Chelmsford
Wood Hill
Haggetts Pond
Andover
Harold
Bald Hill
75 ▲
Topsfield
Wenham Lake
Beverly
West Chelmsford
Lowell
North Tewksbury
111 ▲
Ames Hill
Ballardvale
Parker State Forest
Salem Turnpike
Middleton
ESSEX MIDDLESEX
Middleton Pond
Danvers
Beverly Municipal Airport
North Beverly
Warren Hill ▲ 124
Chelmsford
North Billerica
East Billerica
Tewksbury
Fosters Pond
Martins Pond
N. Reading
Uptons Hill 73 ▲
Beverley Harbor
South Chelmsford
Manning State Park
Heart Pond
Rail Tree Hill
River Pines
Billerica
North Wilmington
Silver Lake
Wilmington
Lynnfield
Suntaug Lake
Peabody
Davensport
Witch House
Salem Maritime Nat. Hist. Site
Salem Harbor

C

Carlisle
Nutting Lake
Pinehurst
Reading
Reading Highlands
L. Quannapowitt
South Lynnfield
Salem
North Acton
Mishawum Lake
North Woburn
(Route 28)
Wakefield
North Saugus
Saugus R.
Breeds Pond
Marblehead
National Wildlife Refuge
Burlington
Division Hy.
Woburn
Wynnmere
Stoneham
Greenwood
Breakheart Reservation
Spring Pond
Clifton
Bedford
West Bedford
Yankee
Horn Pond
North Res.
Middlesex Fells Reservation
Saugus
Lynn
Swampscott
Laurence G. Hanscom Field
North Lexington
Winchester
South Res.
Spot Pond
West Lynn
Old Manse
Concord
Minute Man Natural History Park
Sandy Pond
Lincoln
114 ▲
Arlington Heights
Mystic Lakes
West Medford
Melrose
Mt. Hood Mem. Park
Lynn Harbor
Nahant Bay
West Concord
Fairhaven Hill
Fairhaven Bay
Lexington
East Lexington
Medford
Malden
Nahant
Farrar Pond
North Sudbury
Cambridge Reservoir
South Lincoln
Arlington
Concord Tpk.
Everett
Revere
Beachmont
East Point
Sudbury
69 ▲
Cat Rock Hill
146 ▲
Prospect Hill Park
Waltham
Belmont
Waverley
Fresh Pond
N. Cambridge
Somerville
Chelsea
Orient Hts.
Broad Sound
ESSEX SUFFOLK
Nahant Harbor

D

Goodman Hill
Weston
Kendall Green
Cambridge
Harvard University
Charlestown
Bunker Hill Mon.
East Boston
Winthrop
Boston Bay
Weston
Watertown
North Brighton
Allston
Mass. Inst. of Tech.
Old North Church
Logan International Airport
Deer Island
Wayland
Heard Pond
Reeves Hill ▲ 124
Weston Reservoir
Auburndale
Govt. Center
Old State House
BOSTON
South Boston
Boston Harbor
Outer Brewster Island
South Sudbury
Hultman Aqueduct
Cochituate
Norumbega Reservoir
Newtonville
John F. Kennedy Nat. Hist. Site
Northeastern Univ.
Museum of Fine Arts
Spectacle Island
Calf Island
Middle Brewster Island
Great Brewster Island
Saxonville
Massachusetts Tpke.
Newton
Newton Highlands
Chestnut Hill
Dorchester Hts. Nat. Hist. Site
Old Harbor
Thompson Island
Long Island
Georges Island
POINT ALLERTON
Framingham
Wellesley Fells
Wellesley Hills
Brookline
Boylston St.
Roxbury
Blake House
Jamaica Plain
Grove Hall
Fields Corner
Dorchester Bay
Hull
Peddocks Island
Morses Pond
Oak Hill Park
Arnold Arboretum
Franklin Park
Squantum
Quincy Bay
Hingham Bay
Wellesley
Needham Heights
Roslindale
W. Roxbury
Dorchester
Grape Island
Nantasket Beach
Natick
L. Waban
Needham
MIDDLESEX NORFOLK
SUFFOLK NORFOLK
Stony Brook Res.
Mattapan
Wollaston
Houghs Neck
Hingham Harbor
North Cohasset
Lake Cochituate
Brush Hill ▲ 121
125
Dover
Strawberry Hill
118 ▲
Islington
Hyde Park
Milton Village
Milton
Adams Nat. Hist. Site
Quincy
North Weymouth
Hingham
Sherborn
Farm Pond
Fowl Meadow Res.
Blue Hills Reservation
▲ 158
South Quincy
East Braintree
East Weymouth
Whitmans Pond
East Holliston
Westwood
Gt. Blue Hill ▲ 194
Braintree
Weymouth
South Hingham
Harding
Yankee Division Hy. (Route 128)
Southeast Expy.
MIDDLESEX NORFOLK
Medfield
Norwood
North Randolph
Ponkapog
Great Pond
South Braintree
South Weymouth
Liberty Plain
Millis
Willett Pond
Canton
Reservoir Pond
Ponkapog Pond
Randolph
Pilgrims Hy.
NORFOLK PLYMOUTH
Accord
Accord Pond
Norwood Memorial Airport

West from Greenwich

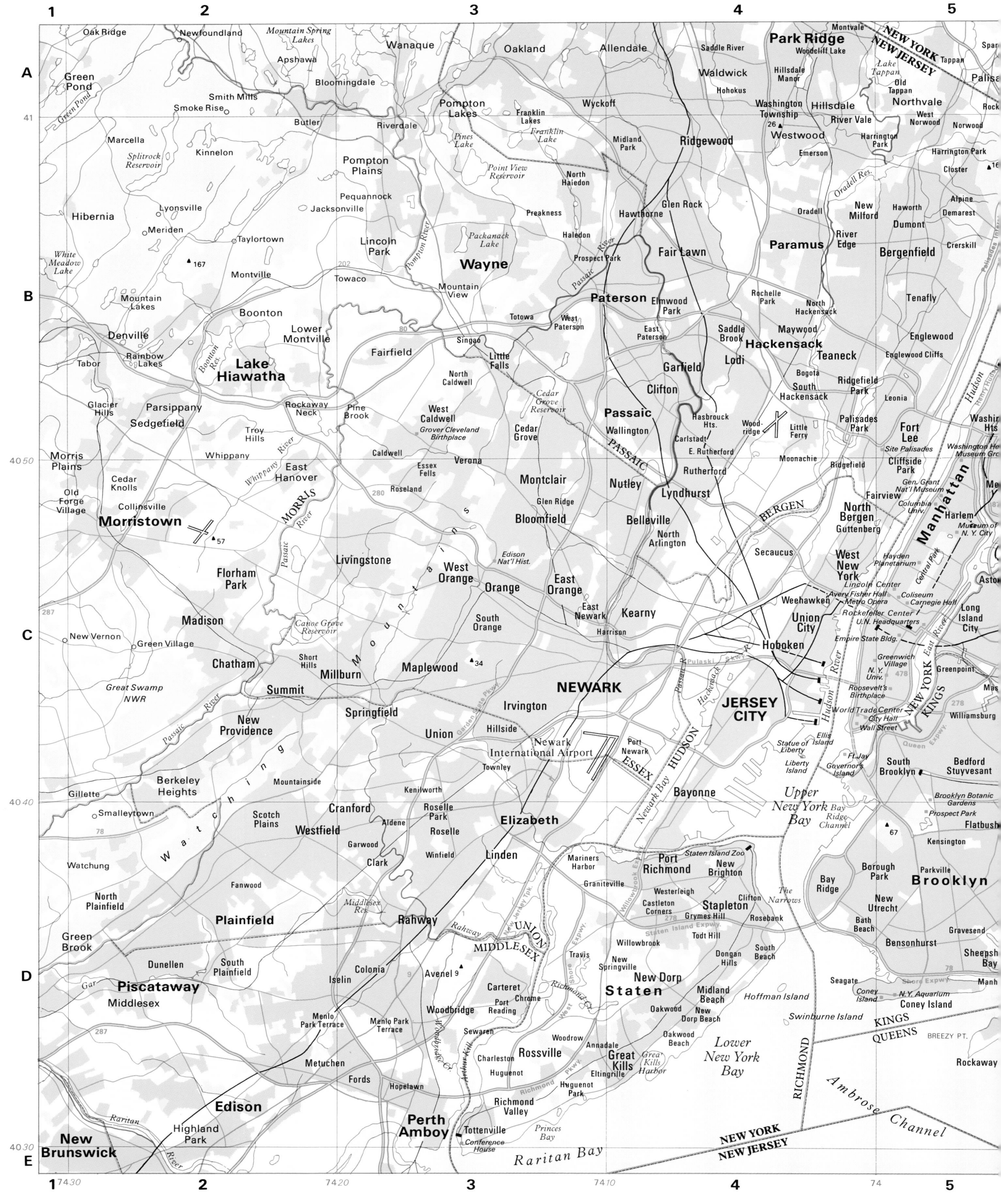

Oak Ridge   Newfoundland   *Mountain Spring Lakes*   Wanaque   Oakland   Allendale   Saddle River   **Park Ridge**   Montvale   **NEW YORK NEW JERSEY**   Tappan

Green Pond   Apshawa   Woodcliff Lake   Palis

Bloomingdale   Waldwick   Hillsdale Manor   *Lake Tappan*

Smith Mills   Pompton Lakes   Wyckoff   Washington Township   Hillsdale   Old Tappan   Northvale

Smoke Rise   Butler   Riverdale   Franklin Lakes   26   Westwood   River Vale   West Norwood   Norwood

Marcella   Kinnelon   *Pines Lake*   *Franklin Lake*   Ridgewood   Harrington Park   Harrington Park

Splitrock Reservoir   Pompton Plains   *Point View Reservoir*   Midland Park   Glen Rock   Emerson   Closter

Hibernia   Lyonsville   Jacksonville   Pequannock   North Haledon   Hawthorne   Oradell   New Milford   Alpine

Meriden   Preakness   Fair Lawn   Demarest

White Meadow Lake   Taylortown   Lincoln Park   Haledon   Prospect Park   **Paramus**   River Edge   **Bergenfield**   Crerskill

167   Montville   202   *Packanack Lake*   **Wayne**   Rochelle Park   North Hackensack

Towaco   Mountain View   **Paterson**   Elmwood Park   Tenafly

Mountain Lakes   **Lake Hiawatha**   Totowa   West Paterson   East Paterson   Maywood   Englewood

Boonton   Fairfield   Singao   **Hackensack**   Englewood Cliffs

Denville   Lower Montville   Little Falls   **Garfield**   Saddle Brook   Teaneck

Rainbow Lakes   *Boonton Res.*   North Caldwell   **Clifton**   Lodi   Bogota

Tabor   Glacier Hills   Parsippany   Rockaway Neck   Pine Brook   West Caldwell   *Cedar Grove Reservoir*   **Passaic**   Wallington   South Hackensack   Ridgefield Park

Sedgefield   Troy Hills   Grover Cleveland Birthplace   Hasbrouck Hts.   Leonia

Morris Plains   Caldwell   Cedar Grove   Carlstadt   Wood-ridge   Little Ferry   Palisades Park

Cedar Knolls   Whippany   East Hanover   Essex Fells   Verona   E. Rutherford   Moonachie   **Fort Lee**

Old Forge Village   Roseland   **Montclair**   Nutley   Rutherford   Ridgefield   Cliffside Park

Collinsville   280   Glen Ridge   **Lyndhurst**   Fairview   Columbia Univ.

Morristown   57   **MORRIS**   **Bloomfield**   **Belleville**   **BERGEN**   **North Bergen**   **Manhattan**

New Vernon   Green Village   *Passaic River*   North Arlington   Secaucus   Guttenberg   Harlem

Florham Park   **Livingstone**   **West Orange**   **Orange**   **East Orange**   **West New York**   Hayden Planetarium

Madison   *Canoe Grove Reservoir*   South Orange   East Newark   **Kearny**   Weehawken   Lincoln Center   Avery Fisher Hall   Metro Opera

Chatham   Short Hills   **Maplewood**   34   Harrison   **Union City**   Rockefeller Center   U.N. Headquarters

287   **Millburn**   Maplewood   **Hoboken**   Empire State Bldg.   **Long Island City**

Summit   *Great Swamp NWR*   Irvington   **NEWARK**   Pulaski Skwy   Greenwich Village   N.Y. Univ.   Astor

New Providence   Springfield   Hillside   **JERSEY CITY**   478   Greenpoint

Gillette   Union   Newark International Airport   Port Newark   Roosevelt's Birthplace   Williamsburg

Berkeley Heights   Townley   **ESSEX**   **HUDSON**   World Trade Center   City Hall   278

Smalleytown   Mountainside   Kenilworth   *Newark Bay*   Wall Street   Ellis Island   **South Brooklyn**

Watchung   Scotch Plains   Cranford   Roselle Park   **Elizabeth**   Bayonne   Statue of Liberty   Ft. Jay   Governor's Island   *Brooklyn Botanic Gardens*

North Plainfield   Westfield   Garwood   Roselle   Linden   Liberty Island   *Upper New York Bay*   Prospect Park   67

78   Clark   Winfield   Mariners Harbor   **Port Richmond**   New Brighton   *Bay Ridge Channel*   Kensington   **Brooklyn**

Green Brook   Fanwood   Graniteville   Westerleigh   Clifton   Borough Park   Parkville   Flatbush

Plainfield   *Middlesex Res.*   **Rahway**   Castleton Corners   **Stapleton**   Rosebank   *The Narrows*   **Bay Ridge**   New Utrecht

Dunellen   South Plainfield   Aldene   *Rahway River*   **UNION**   Willowbrook   Grymes Hill   Todt Hill   Bath Beach   Bensonhurst   Gravesend

Piscataway   Iselin   Colonia   Avenel 9   **MIDDLESEX**   Travis   New Springville   Dongan Hills   South Beach   Seagate   *Shore Expwy*   **Sheepshead Bay**

Middlesex   9   Carteret   Port Reading   Woodbridge   **New Dorp**   Midland Beach   *Hoffman Island*   Coney Island   N.Y. Aquarium   Manh

Menlo Park Terrace   Chrome   **Staten**   Oakwood   New Dorp Beach   *Swinburne Island*   Coney Island

287   Menlo Park Terrace   Sewaren   Woodrow   Oakwood Beach   *Lower New York Bay*   **KINGS QUEENS**   BREEZY PT.

Metuchen   Fords   Charleston   **Rossville**   Annadale   Great Kills   *Great Kills Harbor*   **RICHMOND**   *Ambrose Channel*   Rockaway

Edison   Hopelawn   Eltingville   Huguenot   Richmond Valley   *Richmond Pkwy*   Huguenot Park

Highland Park   *Raritan River*   *Arthur Kill*   Tottenville   Conference House   *Princes Bay*   **NEW YORK NEW JERSEY**

New Brunswick   **Perth Amboy**   *Raritan Bay*

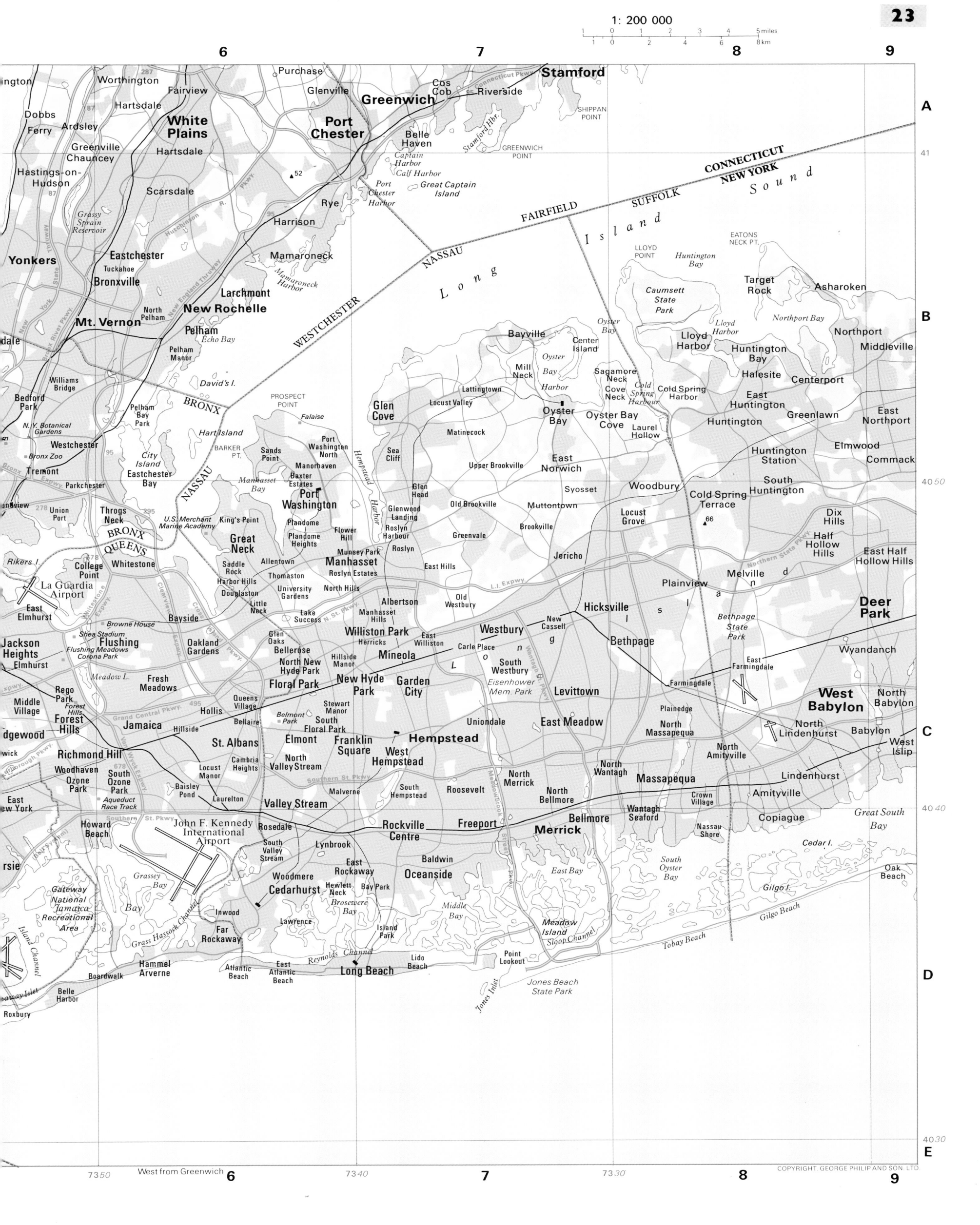

1: 200 000

5 miles
8 km

PHILADELPHIA

Penndel · Bristol · Burlington · Willingboro · Kresson · Atco · Florence

Hulmeville · Croydon · Edgewater Park · Beverly · Delanco · Holiday Lake Amphitheatre Park · Centerton · Hartford · Evesboro · Marlton · West Berlin · Berlin · Albion

Newportville · Eddington · Cornwells Hts. · Delran · Riverside · Rancocas · Ramblewood · Williamstown Junction

Nottingham · Feasterville · Somerton · Holmesburg · Mayfair · Cinnaminson · Maple Shade · Fellowship · Woodstream · Gibbsboro · Lindenwold · Pine Hill · Erial

Trevose · Bustleton · Andalusia · Torresdale · Riverton · Moorestown · Cherry Hill · Willowdale · Clementon · Laurel Springs · Stratford · Somerdale

Feasterville · Bethayres · Bryn Athyn · Fox Chase · Tacony · Palmyra · Merchantville · Collingswood · Westmont · Magnolia · Lawnside · Blackwood · Bells Lake

MONTGOMERY · PHILADELPHIA · BUCKS · Frankford · Wissinoming · Bridesburg · Pennsauken · Camden · Haddonfield · Haddon Heights · Barrington · Blenheim · Turnersville · Hurffville

Ambler · Willow Grove · Abington · Jenkintown · Kensington · Audubon · Mount Ephraim · Bellmawr · Almonesson

Fort Washington · Roslyn · Wyncote · Oak Lane · Logan · Woodlynne · Runnemede · Glendora

Broad Axe · Edge Hill · Glenside · Elkins Park · Olney · Feltonville · Gloucester City · Westville · Colonial Manor · Barnsboro · Mullica Hill

Plymouth Meeting · Enfield · Germantown · Mt. Airy · Chestnut Hill · PHILADELPHIA · Woodbury · Oak Valley · Centre City

Harmonville · Lafayette Hill · Erdenheim · Roxborough · Manayunk · Univ. of Pennsylvania · Drexel Inst. of Tech. · Point Breeze · Thorofare · Mount Royal · Clarksboro · Mickleton · Jefferson

Conshohocken · Belmont Hills · Bala-Cynwyd · Merion Station · Fairmount Park · Camden · Paulsboro · Repaupo

Norristown · Swedesburg · Gulph Mills · Ardmore · Narberth · Wynnewood · Penn Wynne · Upper Darby · Lansdowne · Yeadon · Darby · Gibbstown · Bridgeport

Bridgeport · Haverford · Rosemont · Haverton · Clifton Heights · Aldan · Collingdale · Glenolden · Norwood · Prospect Park

King of Prussia · Radnor · Villanova · Ithan · Broomall · Mania · Pilgrim Corner · Drexel Hill · Springfield · Folcroft · Holmes · Folsom · Woodlyn

Jeffersonville · St. Davids · Wayne · Ethan · Lawrence Park · Marple · Morton · Swarthmore · Ridley Park · Eddystone · Center Square · Swedesboro

Oaks · Audubon · Port Kennedy · Berwyn · Leopard · Newtown Square · Rose Tree · Media · Brookhaven · Woodlyn · Marcus Hook · Bridgeport

Valley Forge · Paoli · Duffryn Mawr · Whitehorse · Edgemont · Sycamore Mills · Lima · Chester · Linwood · Twin Oaks · Pedricktown · Penns Grove

Kimberton · Devault · Malvern · Sugartown · Plumsock · Glen Riddle · Chester Heights · Aston · Village Green · Booth · Marcus Hook · Ogden · Holly Oak

Phoenixville · Goshenville · Milltown · Cheyney · Glen Mills · Thornton · Markham · Concordville · Elam · Brandywine · Arden · Claymont

West Chester · Westtown · Darlington Corners · Dulworthtown · Ward · Chadds Ford · Booth Corner · Boothwyn · Winterthur · Montchanin · Telleyville · Fairfax

Rockland · Westover Hills · Elsmere · Wilmington · Bellefonte

DELAWARE · NEW JERSEY · PENNSYLVANIA · Delaware River · Schuylkill River · Cooper River · Darby Creek · Ridley Creek · Chester Creek · Brandywine Creek · Raccoon Creek · Oldmans Creek · Salem

Philadelphia Airport · Philadelphia International Airport · Independence Natl. Historical Park · Benjamin Franklin Bridge · Walt Whitman Bridge · Vet's Stadium · Spectrum · J.F.K. Stadium · Franklin Roosevelt Park · Red Bank Battle Monument · Valley Forge Historical State Park · Washington Memorial Museum · Ridley Creek State Park · Fort Washington Historical State Park

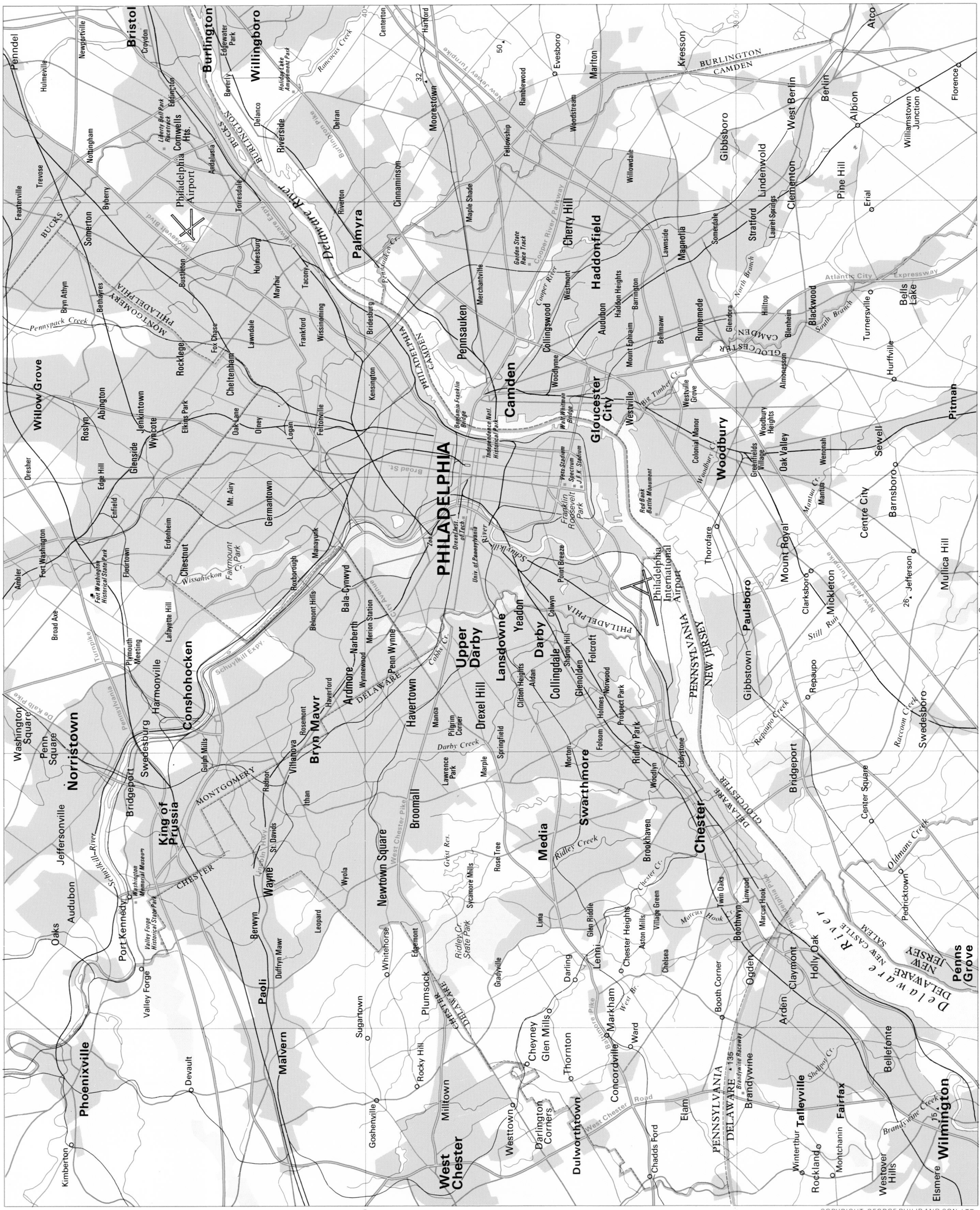

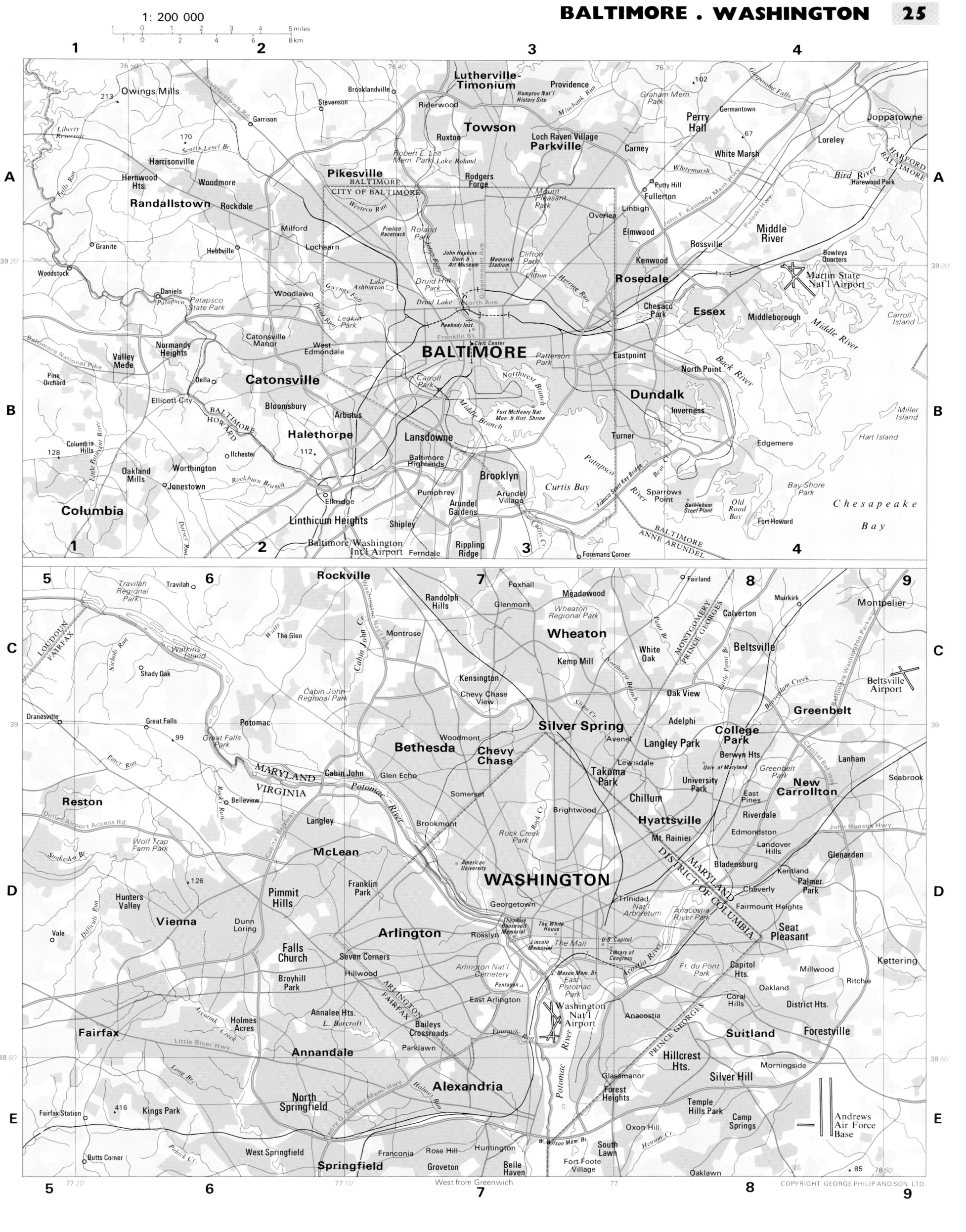

1: 200 000

5 miles
8km

1: 200 000

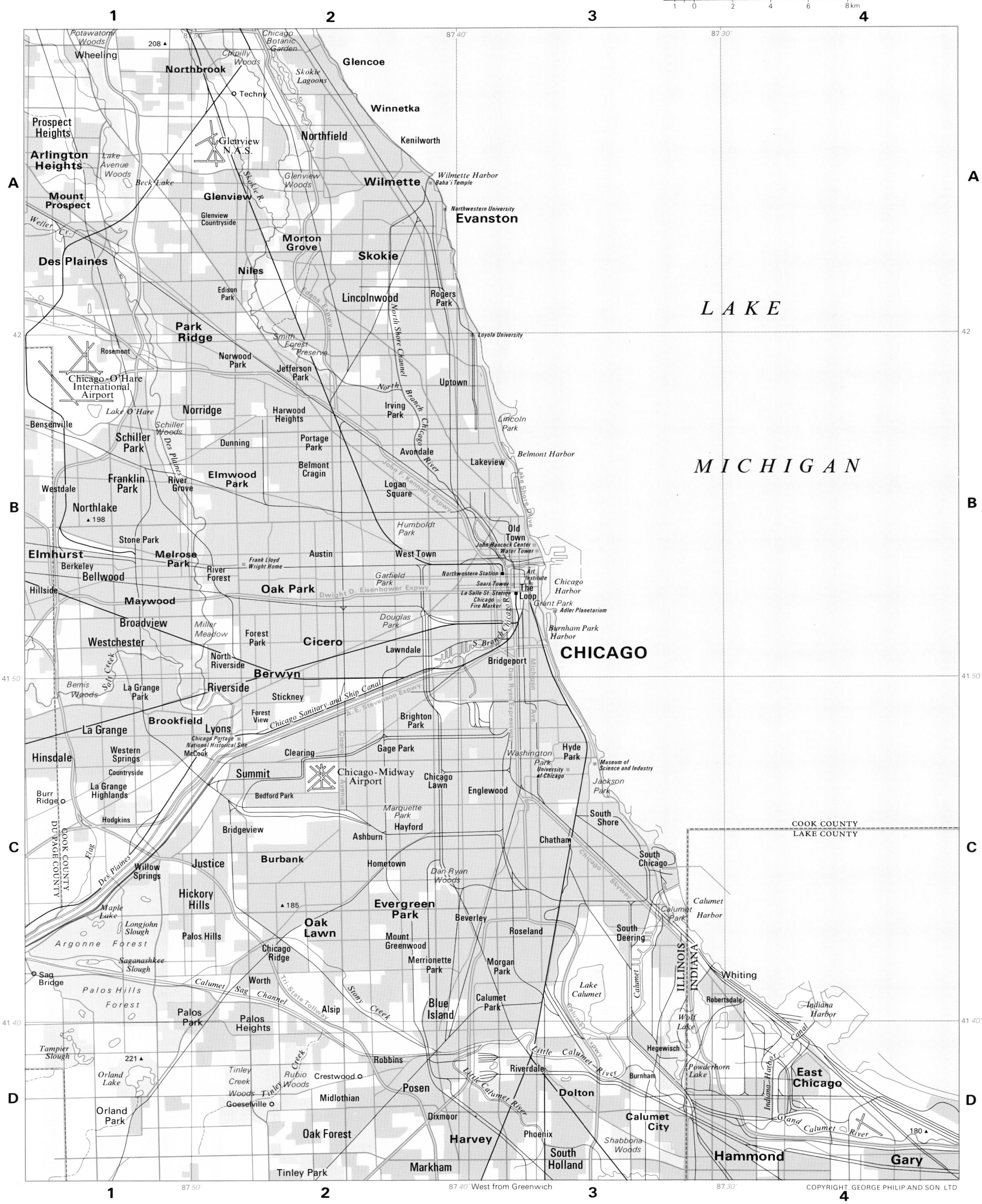

5 miles
8 km

1 **2** **3** **4**

Potawatomi Woods
208 ▲
87 50'
87 40'
87 30'
Wheeling
Chipilly Woods
Chicago Botanic Garden
Skokie Lagoons
Northbrook
Techny
Glencoe

Winnetka

A
Prospect Heights
Arlington Heights
Glenview N.A.S.
Northfield
Kenilworth
Lake Avenue Woods
Beck Lake
Skokie R.
Glenview Woods
Wilmette
Wilmette Harbor
Baha'i Temple
Mount Prospect
Glenview
Northwestern University
Evanston
Glenview Countryside
Morton Grove
A
Des Plaines
Niles
Skokie
Edison Park
Lincolnwood
Rogers Park

Park Ridge
Smith Forest Preserve
North Shore Channel
Loyola University
42
Rosemont
Norwood Park
Jefferson Park
North
Uptown
42
Chicago-O'Hare International Airport
Lake O'Hare
Norridge
Harwood Heights
Portage Park
Irving Park
Lincoln Park
Bensenville
Schiller Woods
Dunning
Des Plaines R.
B
Schiller Park
Belmont Cragin
Avondale
North Branch Chicago River
Lakeview
Belmont Harbor
Franklin Park
Elmwood Park
River Grove
Logan Square
John F. Kennedy Expwy
MICHIGAN
Westdale
Northlake
198 ▲
Stone Park
Humboldt Park
B
Elmhurst
Berkeley
Melrose Park
Frank Lloyd Wright Home
Austin
West Town
Garfield Park
Old Town
John Hancock Center
Water Tower
LAKE
Bellwood
River Forest
Northwestern Station
Art Institute
Chicago Harbor
Maywood
Oak Park
Sears Tower
La Salle St. Station
Chicago Fire Marker
The Loop
Grant Park
Hillside
Dwight D. Eisenhower Expwy
Adler Planetarium
Broadview
Miller Meadow
Forest Park
Douglas Park
Burnham Park Harbor
Westchester
Cicero
Lawndale
S. Branch Chicago R.
CHICAGO
41 50
North Riverside
Berwyn
Bridgeport
41 50
Bemis Woods
Salt Creek
Riverside
La Grange Park
Stickney
Michigan Ave.
Brighton Park
Brookfield
Lyons
Forest View
Chicago Sanitary and Ship Canal
La Grange
Chicago Portage National Historical Site
McCook
Clearing
Gage Park
Washington Park
Hyde Park
Hinsdale
Western Springs
Countryside
Summit
Chicago-Midway Airport
Chicago Lawn
University of Chicago
Museum of Science and Industry
Burr Ridge
La Grange Highlands
Bedford Park
Englewood
Jackson Park
Hodgkins
Bridgeview
Marquette Park
Hayford
Ashburn
C
Flag Cr.
Des Plaines R.
Willow Springs
Justice
Burbank
Hometown
Chatham
South Shore
C
COOK COUNTY
DU PAGE COUNTY
Dan Ryan Woods
South Chicago
COOK COUNTY
LAKE COUNTY
Maple Lake
Hickory Hills
185 ▲
Beverley
Roseland
Calumet Park
Calumet Harbor
Longjohn Slough
Palos Hills
Oak Lawn
Mount Greenwood
ILLINOIS
INDIANA
Argonne Forest
Saganashkee Slough
Evergreen Park
Merrionette Park
Morgan Park
South Deering
Calumet
Whiting
Sag Bridge
Calumet
Chicago Ridge
Worth
Lake Calumet
Robertsdale
Indiana Harbor
Palos Hills Forest
Palos Hills
Alsip
Stony Creek
Blue Island
Calumet Park
Wolf Lake
Robbins
Tri-State Tollway
41 40
Palos Park
Palos Heights
Calumet Sag Channel
Posen
Riverdale
Burnham
Powderhorn Lake
41 40
Tampier Slough
221 ▲
Orland Lake
Tinley Creek Woods
Rubio Woods
Crestwood
Little Calumet River
Dolton
East Chicago
D
Orland Park
Goselville
Midlothian
Dixmoor
Calumet City
Phoenix
Shabbona Woods
Indiana Harbor Canal
180 ▲
D
Oak Forest
Harvey
Hammond
Gary
Tinley Park
Markham
South Holland
West from Greenwich
87 50'
87 40'
87 30'
COPYRIGHT. GEORGE PHILIP AND SON LTD.

1 **2** **3** **4**

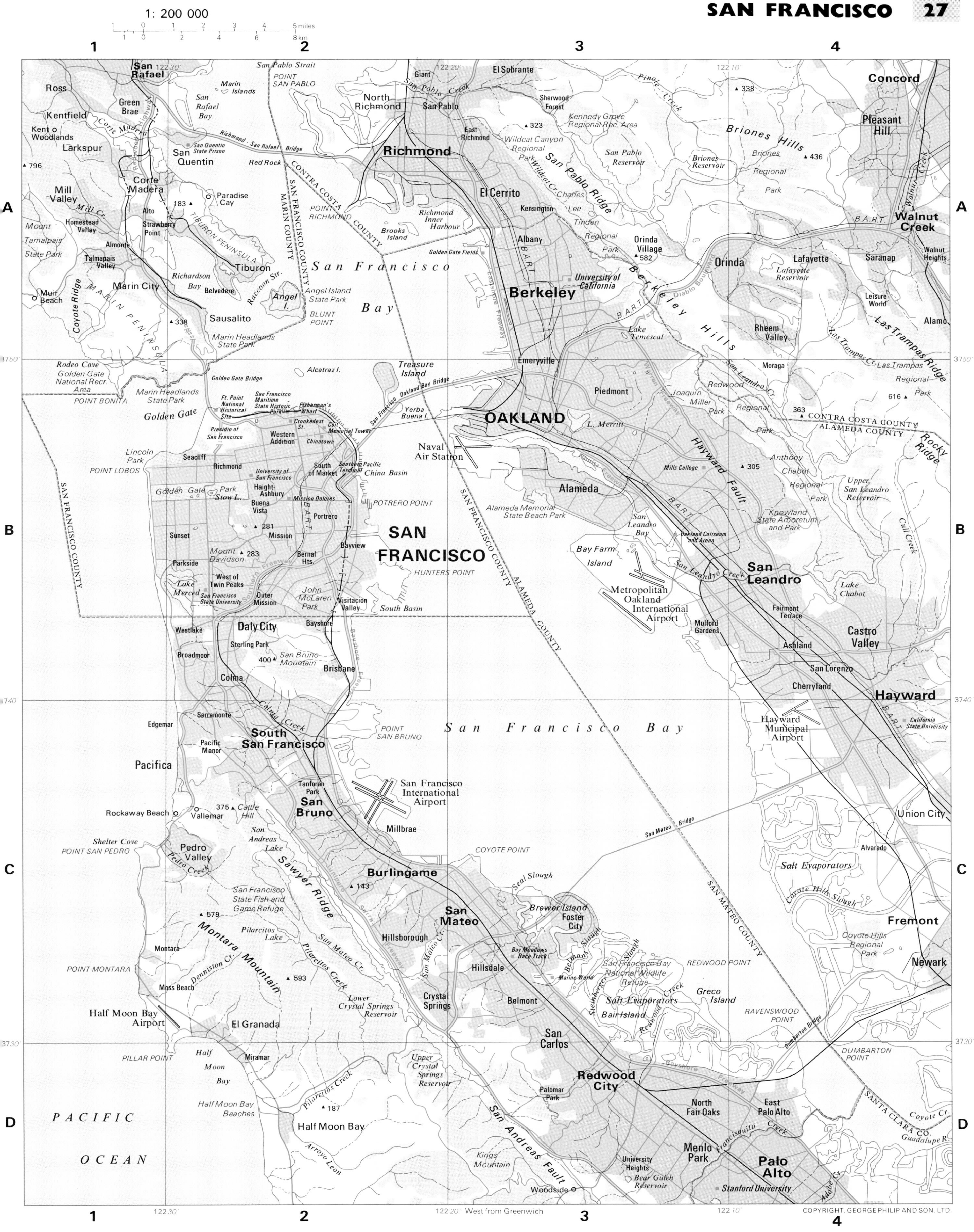

1: 200 000

5 miles
8 km

1 **2** **3** **4**

San Rafael
Ross
Green Brae
Kentfield
Kent o Woodlands
Larkspur
San Quentin
San Quentin State Prison
Red Rock
San Pablo Strait
POINT SAN PABLO
Marin Islands
San Rafael Bay
Richmond - San Rafael Bridge
North Richmond
San Pablo Creek
Giant
El Sobrante
San Pablo
East Richmond
Sherwood Forest
Kennedy Grove Regional Rec. Area
Wildcat Canyon Regional Park
San Pablo Reservoir
Pinole Creek
338
Concord
Pleasant Hill
Briones Hills
Briones Reservoir
Briones Regional Park
436

A
▲796
Mill Valley
Homestead Valley
Almonte
Corte Madera
Alto
183
Paradise Cay
Strawberry Point
CONTRA COSTA COUNTY
MARIN COUNTY
SAN FRANCISCO COUNTY
Brooks Island
Richmond Inner Harbour
Richmond
El Cerrito
Kensington
Tilden Regional Park
Charles Lee Tilden Regional Park
Wildcat Cr.
San Pablo Ridge
Orinda Village
582
Orinda
Lafayette
Lafayette Reservoir
Saranap
Walnut Heights
BART
Walnut Creek
A

Mount Tamalpais State Park
Talmapais Valley
Muir Beach
Coyote Ridge
Marin City
Richardson Bay
Tiburon
Belvedere
Sausalito
338
Raccoon Str.
Angel I.
Angel Island State Park
BLUNT POINT
TIBURON PENINSULA
MARIN PENINSULA
San Francisco Bay
Golden Gate Fields
Albany
Berkeley
University of California
Berkeley Hills
Diablo Boulevard
Moraga
Rheem Valley
Las Trampas Cr.
Las Trampas
Alamo
Leisure World
Las Trampas Ridge

3750
Rodeo Cove
Golden Gate National Recr. Area
POINT BONITA
Marin Headlands State Park
Golden Gate
Golden Gate Bridge
Alcatraz I.
Treasure Island
Oakland Bay Bridge
Yerba Buena I.
Emeryville
Lake Temescal
Piedmont
Joaquin Miller Park
Redwood Regional Park
San Leandro Cr.
363
616
Las Trampas Regional Park
CONTRA COSTA COUNTY
ALAMEDA COUNTY
Rocky Ridge
3750

B
SAN FRANCISCO COUNTY
POINT LOBOS
Lincoln Park
Seacliff
Richmond
Golden Gate Park
Stow L.
Sunset
Parkside
Lake Merced
West of Twin Peaks
Westlake
Broadmoor
Colma
Ft. Point National Historical Site
Presidio of San Francisco
Western Addition
University of San Francisco
Haight-Ashbury
Buena Vista
Mount Davidson
283
San Francisco State University
Outer Mission
San Francisco Maritime State Historic Park
Fisherman's Wharf
Crookedest St.
Chinatown
Coit Memorial Tower
South of Market
Mission Dolores
281
Mission
BART
John McLaren Park
Bayshore
Daly City
Sterling Park
400
San Bruno Mountain
Colma
San Francisco
Southern Pacific Terminal
China Basin
POTRERO POINT
Portrero
Bayview
SAN FRANCISCO
Bernal Hts.
Visitacion Valley
South Basin
HUNTERS POINT
Hunters Point
Brisbane
Naval Air Station
OAKLAND
L. Merritt
Mills College
305
Alameda Memorial State Beach Park
Alameda
San Leandro Bay
Bay Farm Island
Oakland Coliseum and Arena
San Leandro Creek
Metropolitan Oakland International Airport
Mulford Gardens
Hayward Fault
Knowland State Arboretum and Park
Anthony Chabot Regional Park
Upper San Leandro Reservoir
San Leandro
Lake Chabot
Fairmont Terrace
Castro Valley
Ashland
San Lorenzo
Cull Creek
B

3740
Serramonte
Edgemar
Pacific Manor
South San Francisco
Pacifica
Tanforan Park
San Bruno
Millbrae
Colma Creek
POINT SAN BRUNO
San Francisco International Airport
San Mateo Bridge
S a n   F r a n c i s c o   B a y
Cherryland
Hayward Municipal Airport
California State University
Hayward
BART
3740

C
Rockaway Beach
375
Vallemar
Cattle Hill
Shelter Cove
POINT SAN PEDRO
Pedro Valley
Pedro Creek
579
San Andreas Lake
Sawyer Ridge
Millbrae
COYOTE POINT
Coyote Point
Seal Slough
Burlingame
143
Brewer Island
Foster City
Union City
Alvarado
Salt Evaporators
Coyote Hills Slough
Fremont
C

Montara
POINT MONTARA
Moss Beach
Montara Mountain
593
Pilarcitos Lake
San Francisco State Fish and Game Refuge
Pilarcitos Creek
San Mateo Cr.
San Mateo
Hillsborough
Hillsdale
Crystal Springs
Belmont
SAN MATEO COUNTY
Bay Meadows Race Track
Marine World
Belmont Slough
San Francisco Bay National Wildlife Refuge
Steinberger Slough
Salt Evaporators
Bair Island
Greco Island
REDWOOD POINT
RAVENSWOOD POINT
Redwood Creek
Coyote Hills Regional Park
Newark

D
Half Moon Bay Airport
El Granada
PILLAR POINT
Miramar
Half Moon Bay
Half Moon Bay Beaches
Half Moon Bay
187
Lower Crystal Springs Reservoir
Pilarcitos Creek
Upper Crystal Springs Reservoir
Kings Mountain
Woodside
Crystal Springs
San Carlos
Palomar Park
San Andreas Fault
University Heights
Bear Gulch Reservoir
Redwood City
North Fair Oaks
Menlo Park
East Palo Alto
Stanford University
Palo Alto
Dumbarton Bridge
DUMBARTON POINT
Bayshore Freeway
San Francisquito Creek
SANTA CLARA CO.
Coyote Cr.
Guadalupe R.
Adobe Cr.
D

P A C I F I C

O C E A N

122 30'   122 20' West from Greenwich   122 10'   COPYRIGHT GEORGE PHILIP AND SON LTD.

1 **2** **3** **4**

1 : 200 000

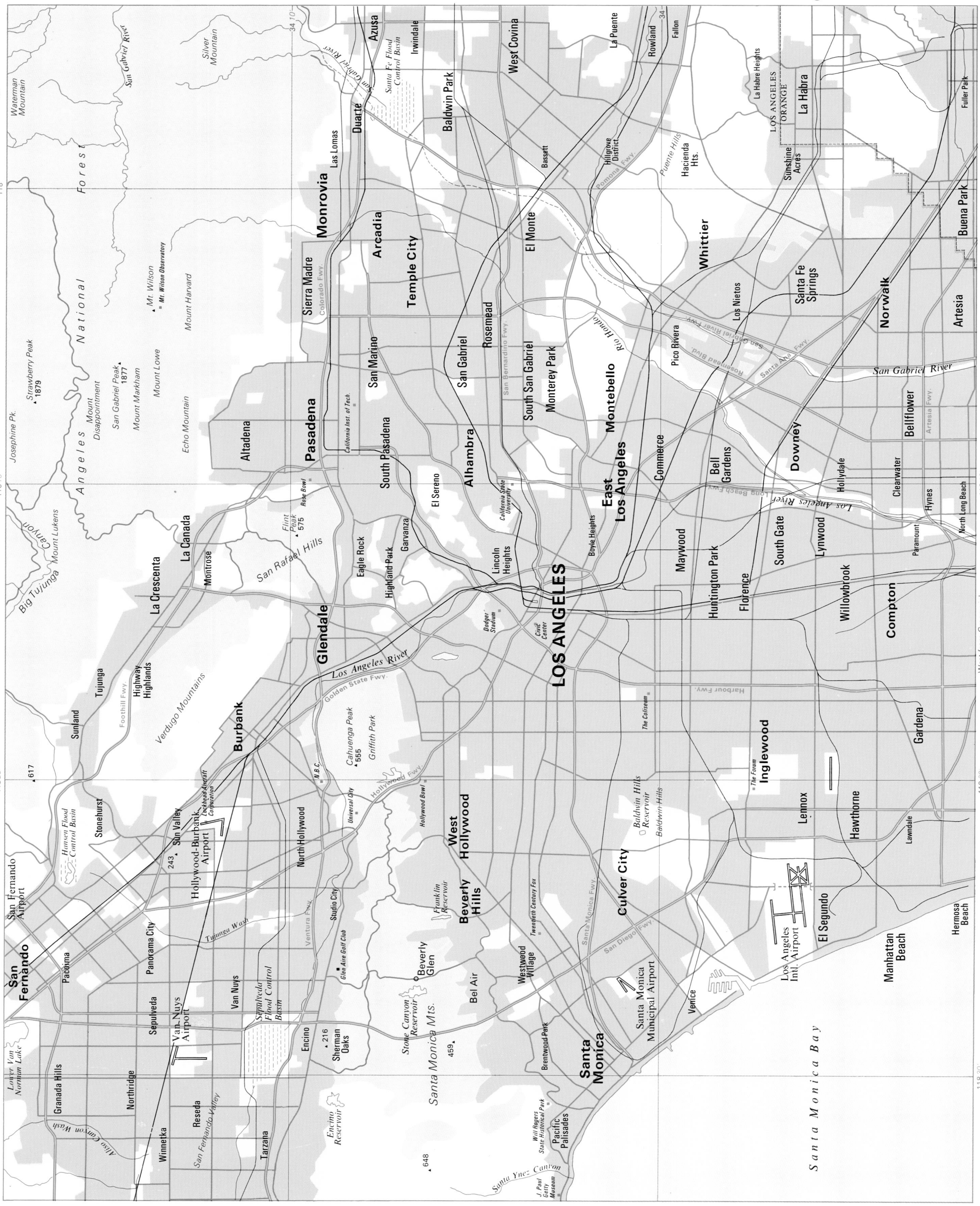

5 miles
8 km

A B C

LOS ANGELES

Waterman Mountain
Silver Mountain
San Gabriel River
Angeles National Forest
Strawberry Peak 1879
Josephine Pk.
San Gabriel Peak 1877
Mount Markham
Mount Lowe
Echo Mountain
Mount Harvard
Mt. Wilson
Mt. Wilson Observatory
Mount Disappointment
Big Tujunga Canyon
Mount Lukens
617

Azusa
Irwindale
Duarte
Las Lomas
Santa Fe Flood Control Basin
West Covina
La Puente
Rowland
Fallon
Baldwin Park
Bassett
La Habra Heights
La Habra Heights
Sunshine Acres
LOS ANGELES
ORANGE
Fuller Park

Monrovia
Sierra Madre
Arcadia
Temple City
El Monte
Hillgrove District
Puente Hills
Hacienda Hts.
Pomona Fwy.
Whittier
Buena Park

Altadena
Pasadena
San Marino
San Gabriel
Rosemead
South San Gabriel
Monterey Park
Montebello
Santa Fe Springs
Los Nietos
Norwalk
Artesia
California Inst. of Tech
Rose Bowl
South Pasadena
San Bernardino Fwy.
Pico Rivera
Rio Hondo
Rosemead Blvd.
Santa Ana Fwy.
San Gabriel River Fwy.
San Gabriel River

La Canada
Montrose
La Crescenta
San Rafael Hills
Eagle Rock
Highland Park
Garvanza
El Sereno
Alhambra
California State University
Boyle Heights
Commerce
East Los Angeles
Bell Gardens
Downey
Bellflower
Clearwater
Hynes
North Long Beach
Artesia Fwy.
Paramount

Tujunga
Highway Highlands
Foothill Fwy.
Verdugo Mountains
Sunland
Glendale
Los Angeles River
Golden State Fwy.
Lincoln Heights
Dodger Stadium
Civic Center
Maywood
Huntington Park
Florence
South Gate
Lynwood
Willowbrook
Compton
Gardena

Stonehurst
Sun Valley
Hansen Flood Control Basin
Burbank
Lockheed Aircraft Corporation
Hollywood-Burbank Airport
243
N.B.C.
Cahuenga Peak 555
Griffith Park
Universal City
Hollywood Fwy.
Hollywood Bowl
West Hollywood
Harbour Fwy.
The Coliseum
Inglewood
The Forum
Lennox
Hawthorne
Lawndale

San Fernando Airport
Panorama City
North Hollywood
Studio City
Ventura Park
Franklin Reservoir
Beverly Hills
Twentieth Century Fox
Culver City
Baldwin Hills Reservoir
Baldwin Hills
Santa Monica Fwy.
Lawndale

San Fernando
Pacoima
Sepulveda
Van Nuys
Sepulveda Flood Control Basin
Bel Air
Westwood Village
San Diego Fwy.
Los Angeles Intl. Airport
El Segundo

Granada Hills
Northridge
Reseda
Van Nuys Airport
Encino
216
Sherman Oaks
459
Stone Canyon Reservoir
Beverly Glen
Brentwood Park
Santa Monica Municipal Airport
Venice
Manhattan Beach
Hermosa Beach

Winnetka
Tarzana
San Fernando Valley
Encino Reservoir
648
Santa Monica Mts.
Will Rogers State Historical Park
Pacific Palisades
Santa Ynez Canyon
J. Paul Getty Museum
Santa Monica
Santa Monica Bay

Aliso Canyon Wash
Lower Van Norman Lake
Tujunga Wash
Glen Aire Golf Club

West from Greenwich

COPYRIGHT. GEORGE PHILIP AND SON. LTD

1: 200 000

1   0   1   2   3   4   5 miles
1   0   2   4   6   8 km

1   **2**   **3**   **4**

Hila   La Colmena
                                   Cerro el Picacho   Ecatepec
San Mateo Tecoloapan   2968   de Morelos   Santa Isabel
                                                                                 Ixtapan
                    Barrientos   Santa María
                                   Cuautepec   Tulpetlac   Planta de
Ciudad   El Alto   Santa   *Evaporación*
López Mateos   Clara
              San Andrés   Cuautepec   *Río Nextipayac*
              Atenco   *Santa   de Madero   Ciudad Azteca
A   Cecilia*   A
                    Tlalnepantla   *Pirámide   Ticomán
San Nicolás Viejo   *de Tenayuca*   San Pedro
                                                          Zacatenco
              Río Tlalnepantla La   Progreso
                              Loma   Nacional
                    San Juan   Juan González
              Ciudad   Ixtacala   *Indios*   Nueva   Romero   *Lago de Texcoco*
19 30'   Satélite   Reynosa   **Villa**   *Verdes*   Aztacoalco   19 30'
Presa de   Tamaulipas   **Gustavo A.** Villa de Guadalupe
Rancho Colorado   **Madero**   *Basílica de Guadalupe*
              Río Los Remedios
Santiago   **Azcapotzalco**   *Zoológica* San Juan
Tepatlaxco   San Juan Toltotepec   de Aragón
              Presa Tenantongo   Nueva
Santa Cruz   **Naucalpan**   Tenochtitlán *Parque San Juan*
Ayotusco   **de Juárez**   *de Aragón*
              Parque Nacional   *Av. Río Consulado*
              de los Remedios **CUAJIMALPA** **CIUDAD DE**   **VENUSTIANO**
San Francisco *Río Sn. Lorenzo*   **MÉXICO**   **CARRANZA**
Chimalpa   San Rafael Champa El Toreo
                    130   **Tacuba**   *Av. Central*   *Aeropuerto*
              San José Río Hondo   *U Central Station*   *Internacional*
B   *Hipódromo de las Americas* Tlatelolco   Chimalhuacán San Pablo   B
La Magdalena   *Catedral* *Tenochtitlán*   Xochitenco   San Pedro
Chichicaspa   Lomas   *Bosque*   *Bellas Artes* *Palacio Nacional*   Xochiaca
              Chapultepec   *de Chapultepec* Ciudadela   *Aeropuerto*   San Lorenzo
              Tecamachaleo   *Castillo de* Tlaxcoaque *Internacional* **Netzahualcóyotl** Chimalco
              Presa   *Chapultepec*   Pantitlán
San Bartolomé *Los Jazmines*   *Viaducto Presidente Miguel Alemán*   San Agustín
Coatepec   Lomas   **Tacubaya** *Palacio* Ciudad   Atlapulco
              Reforma   *de los* Deportiva **Agrícola**
Santa Cruz   *Unidad Santa Fe*   *Deportes* Juan   **Oriental** La Magdalena
Ayotusco   **Iztacalco**   Escutia Tepalcates Atlapac
              Olivar del   Los Piriules **ESTADO DE MÉXICO**
Dos Ríos   Conde   **Benito Juárez**   **DISTRITO FEDERAL**
              Molino de Rosas   **IZTACALCO**   Los
Huixquilucan Mixcoac   **IZTAPALAPA**   Reyes Tecamachalco
Chimalpa   *Presa* Olivar de los   Héroes de Churubusco
Cuajimalpa *de Mixcoac* Padres *Presa*   Santa Martha Acatitla
              Olivar de los *Tarango*   Universidad
General   Padres   **Iztapalapa** *Ibero-Americana* Santa María
Ignacio Allende *Río Cañada de las Helechos* Lomas de   Prado   Aztahuacán
              Villa San Angel Inn Coyoacán Churubusco Los Reyes Santiago
Tlaltenango   **Obregón** San Angel **Coyoacán**   *Parque* Acahualtepec
19 20'   San Bartolo   *Rosedal* *Nacional* Santa Cruz Meyehualco 19 20'
San Lorenzo   Ameyalco Tizapán *Estadio Olímpico* La Candelaria 2460 *Cerro de la*
Acopilco   Santa Rosa San Jerónimo Lidice *Ciudad Universitaria* *Estrella*
              Xochiac   San Francisco San Lorenzo Tezonco **IZTAPALAPA**
Parque   Jardines del   Culhuacán   **TLAHUAC** Tlalpitzáhuac
Nacional   *Pedregal de*   El Vergel
Desierto de   *San Angel*   El Reloj
los Leones   La Nopalera Zapotitlán
              La Magdalena   Pirámide
La Marquesa   Contreras   *de Cuicuilco* *Estadio Azteca*   Tlaltenco
                              San Nicolás   **CUAJIMALPA**
Parque Nacional del Totolapan   **TLALPAN** *Lago de*
Insurgente Miguel Hidalgo   Santa Ursula Xitla **Tlalpan** *Xochimilco* Tlahuac
                              San Pedro Martir   Tepepan *Jardines Flotantes* *Gran Canal* *Cerro Xico*
C   Xitle   *Las Fuentes*   2346   C
                              Brotantes   Xochitepec San Luis
Cerro Xitle San Andrés Totoltepec   San Lucas **Xochimilco** Tlaxialtemalco
3128   La Magdalena   Xochimanca San Gregorio Tulyehualco
              Petlalco   Santiago Atlapulco San Juan Ixtayopan
                    San Miguel   Tepalcatlalpan Santa Cruz **XOCHIMILCO**
San Miguel   Xicalco San Mateo Nativitas Alcapixca **TLAHUAC**
Ajusco   San Andrés Xalpa   **Mixquic**
                    Ahuayucan   San Antonio
              Santa Cecilia   Tecómitl   Tetelco
Parque Nacional   Tepetlapa   San Pedro San Francisco
de Ajusco   Topilejo   Actopan Tecoxpa San Jerónimo
Cerro Ajusco   Miacatlán San Juan y
3937   San Francisco San Salvador San Augustín San Pedro
              Tlalnepantla Cuauhtenco Ohtenco Tezompa
                              San Pablo **Milpa Alta**
                              Ostotepec
                    San Lorenzo   **Santa Ana Tlacotenco**
                    Tlacoyucan
19 10'   Aserradero   19 10'
              Cerro Pelado *Cerro Cuautzin*
              3620   3497

El Guarda   *Cerro Tláloc*
Parres   3690
                    DISTRITO FEDERAL
D   **ESTADO DE MORELOS**   **DISTRITO FEDERAL**   D
                                                 ESTADO DE MORELOS
Parque Nacional *Cerro*
              Chichinautzin
de las Lagunas 3476
de Zempoala
              Tres Marias   *Parque Nacional del Tepozteco*

1   **2**   99 10' West from Greenwich **3**   99 *COPYRIGHT GEORGE PHILIP & SON LTD* **4**

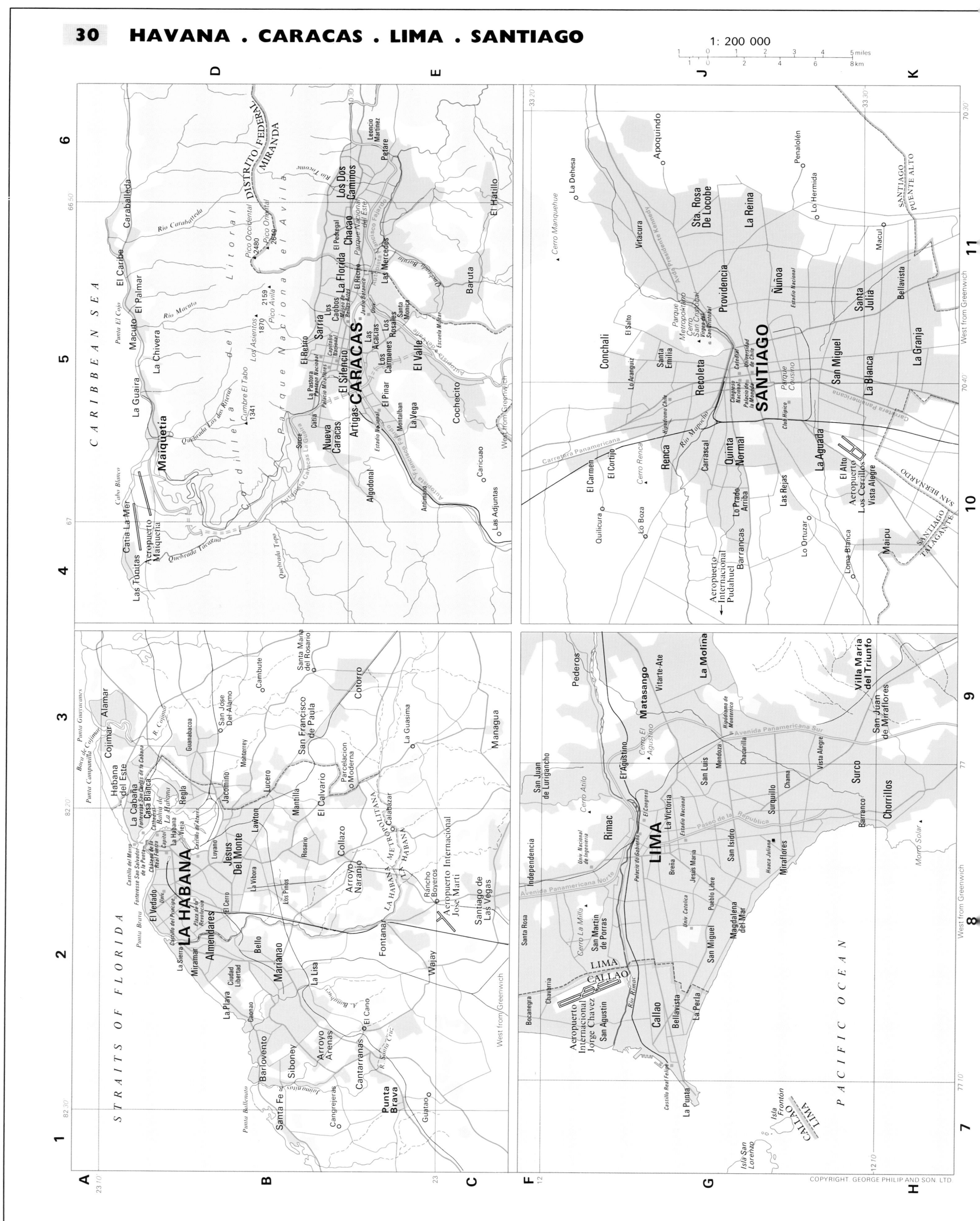

1 : 200 000

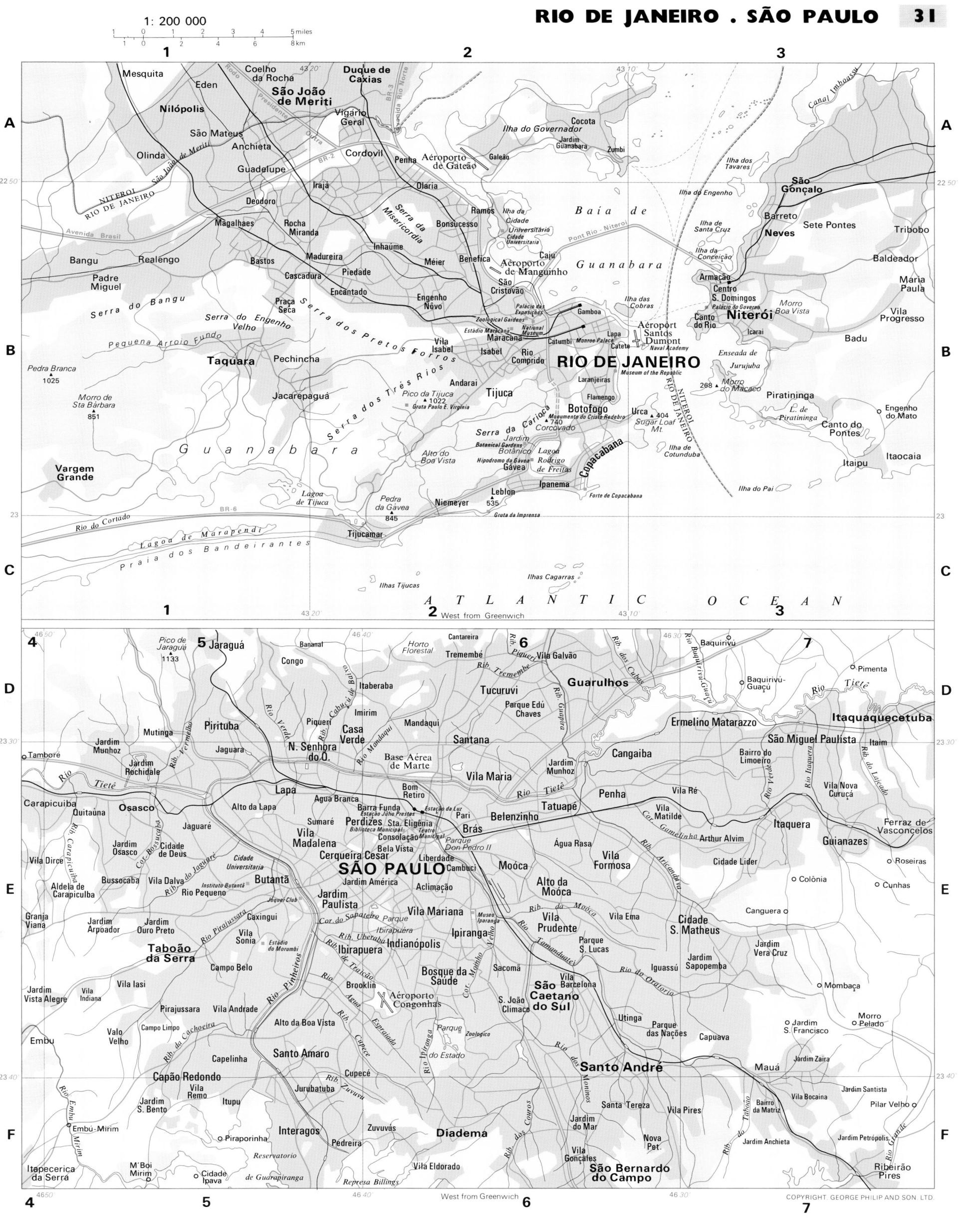

1: 200 000

RIO DE JANEIRO

Mesquita
Eden
São João de Meriti
Nilópolis
São Mateus
Anchieta
Olinda
Guadelupe
Coelho da Rocha
Duque de Caxias
Vigário Geral
Cordovil
Penha
Aéroporto de Gateão
Galeão
Ilha do Governador
Cocota
Jardim Guanabara
Zumbi
Ilha do Engenho
São Gonçalo
Ilha dos Tavares

Irajá
Olaria
Ramos
Ilha da Cidade
Ilha de Santa Cruz
Barreto
Neves
Sete Pontes
Tribobo

Deodoro
Bonsucesso
Universitário Cidade Universitaria
Baia de Guanabara
Ilha da Conceição
Armação
Centro S. Domingos
Palácio do Governo
Baldeador

Magalhães
Rocha Miranda
Madureira
Inhaúme
Méier
Benefica
Caju
Aéroporto de Manguinho
São Cristovão
Ilha das Cobras
Niterói
Canto do Rio
Icaraí
Maria Paula

Bangu
Realengo
Bastos
Cascadura
Piedade
Encantado
Engenho Nôvo
Palacio das Exposições
Zoological Gardens
Estádio Maracana
Gamboa
Lapa
Aéroport Santos Dumont
Naval Academy
Morro Boa Vista
Vila Progresso

Padre Miguel
Serra do Bangu
Praça Seca
Serra do Engenho Velho
Vila Isabel
Isabel
Maracana
National Museum
Catumbi
Monroe Palace
Badu

Pedra Branca 1025
Morro de Sta Bárbara 851
Taquara
Pechincha
Serra dos Pretos Forros
Rio Comprido
RIO DE JANEIRO
Laranjeiras
Museum of the Republic
Enseada de Jurujuba
268 Morro do Macaeo
Piratininga
É. de Piratininga
Engenho do Mato

Jacarepaguá
Serra dos Três Rios
Andarai
Tijuca
Flamengo
Morro do 268
Canto do Pontes

Pico da Tijuca 1022
Gruta Paulo E. Virgínia
Botafogo
Monumento do Cristo Redentor
Urca 404
Sugar Loaf Mt.
Itaocaia

Guanabara
Serra da Carioca
740 Corcovado
Jardim Botânico
Botanical Gardens
Alto do Boa Vista
Hipodromo da Gávea
Lagoa Rodrigo de Freitas
Copacabana
Ilha de Cotunduba

Vargem Grande
Leblon
Ipanema
Forte de Copacabana
Ilha do Pai

Lagoa de Tijuca
Pedra da Gávea 845
Niemeyer 535
Gruta da Imprensa

Rio do Cortado
BR-6
Tijucamar
ATLANTIC OCEAN
Ilhas Cagarras

Lagoa de Marapendi
Praia dos Bandeirantes
Ilhas Tijucas

West from Greenwich

SÃO PAULO

Pico de Jaraguá 1133
Jaraguá
Bananal
Congo
Horto Florestal
Cantareira
Tremembé
Vila Galvão
Baquirivú
Pimenta

Itaberaba
Piqueri
Imirim
Tucuruvi
Parque Edú Chaves
Guarulhos
Baquirivú-Guaçú
Baquirivú-Guaçú

Jardim Munhoz
Mutinga
Piruíba
Casa Verde
Mandaqui
Santana
Cangaiba
Ermelino Matarazzo
Itaquaquecetuba

Tamboré
Jardim Rochidale
Jaguara
N. Senhora do Ó.
Base Aérea de Marte
Vila Maria
Jardim Munhoz
São Miguel Paulista
Itaim

Carapicuiba
Osasco
Quitaúna
Lapa
Agua Branca
Bom Retiro
Belenzinho
Tatuapé
Penha
Vila Ré
Vila Matilde
Vila Nova Curuçá

Vila Dirce
Jardim Osasco
Cidade de Deus
Alto da Lapa
Barra Funda
Estação Julio Prestes
Estação da Luz
Pari
Brás
Água Rasa
Arthur Alvim
Itaquera
Ferraz de Vasconcelos

Aldeia de Carapiculba
Bussocaba
Vila Dalva
Sumaré
Perdizes
Sta. Eligênia
Biblioteca Municipal
Consolação
Teatro Municipal
Vila Formosa
Cidade Líder
Guianazes

Granja Viana
Jardim Arpoador
Jardim Ouro Preto
Cidade Universitária
Instituto Butantã
Vila Madalena
Cerqueira Cesar
SÃO PAULO
Liberdade
Cambuci
Moóca
Alto da Moóca
Vila Prudente
Cidade S. Matheus
Roseiras
Cunhas

Butantã
Jardim América
Aclimação
Vila Ema
Canguera
Colônia

Taboão da Serra
Caxingui
Jardim Paulista
Vila Mariana
Museu Iparanga
Parque S. Lucas
Jardim Vera Cruz

Campo Belo
Ibirapuéra
Indianópolis
Ipiranga
Vila Barcelona
Jardim Sapopemba
Iguassú
Mombaça

Jardim Vista Alegre
Vila Indiana
Vila Iasi
Estádio do Morumbi
Bosque da Saúde
Sacomã
São Caetano do Sul
Utinga
Morro Pelado

Embu
Valo Velho
Pirajussara
Vila Andrade
Brooklin
Aéroporto Congonhas
S. João Climaco
Parque das Nações
Capuava
Jardim S. Francisco

Capão Redondo
Vila Remo
Alto da Boa Vista
Santo Amaro
Parque Zoologico do Estado
Santo André
Jardim Zaira

Jardim S. Bento
Capelinha
Cupecé
Mauá
Jardim Santista
Pilar Velho

Embú-Mirim
Itupu
Jurubatuba
Zuvuvás
Santa Tereza
Vila Pires
Vila Bocaina
Bairro da Matriz

Itapecerica da Serra
Piraporinha
Interagos
Pédreira
Diadema
Jardim do Mar
Nova Pet.
Jardim Anchieta
Jardim Petrópolis

M'Boi Mirim
Cidade Ipava
Reservatorio de Guarapiranga
Represa Billings
Vila Eldorado
Vila Gonçales
São Bernardo do Campo
Ribeirão Pires

West from Greenwich

1 : 200 000

1 2 3 4 5 miles
1 0 2 4 6 8 km

Quilmes
Espeleta
Berazátegui
Villa Augusta
Villa D. Sobral
San Francisco
España
Ranelagh
Villa Giambruno
Bosques
Gdor. Monteverde
Don Bosco
Bernal
San Francisco Solano
Florencio Varela
Villa Dominico
Wilde
Villa Barilari
Rafael Calzada
Claypole
Sarandi
Villa C. Colón
Ministro Rivadavia
Avellaneda
Gerli
Monte Chingolo
José Mármol
Temperley
Almirante Brown
Burzaco
La Boca
Barracas
Villa Alsina
Lanús
Remedios de Escalada
Banfield
Lomas de Zamora
Diamante
Caraza
Santa Catalina
Turdera
Llavallol
Luis Guillón
Monte Grande
Almagro
Nueva Pompeya
Villa Lugano
Fiorito
La Salada
Villa Hogar Alemán
Esteban Echeverria
Ezeiza
San Telmo
Once
Caballito
Flores
Floresta
Parques
Almirante Brown
G. Brown
Villa Madero
Tapiales
Aldo Bonzi
Ciudad General Belgrano
Ezeiza

BUENOS AIRES

Aeroparque de la Ciudad de Buenos Aires
Porto Nuevo
Retiro
Palermo
Belgrano
Nuñez
Saavedra
General Urquiza
La Paternal
Villa Lynch
Villa Devoto
Versailles
Liniers
Ciudadela
Sáenz Peña
Villa Bosch
Lourdes
Villa Alianza
Villa D. F. Sarmiento
Ramos Mejía
Villa Basso
M. J. Haedo
Villa Luzuriaga
Tablada
San Justo
Aeropuerto Ezeiza
West from Greenwich

DISTRITO FEDERAL BUENOS AIRES

R i o   d e   l a   P l a t a

Las Barrancas
I. Anchorena
Olivos
Florida
La Lucila
Munro
Vicente López
Martinez
Acassuso
Beccar
San Isidro
Victoria
Virreyes
Carupá
Tigre
Las Conchas
General Pacheco
Boulogne
Villa Adelina
Carapachay
José L. Suárez
Villa Ballester
San Andres
General San Martin
Santos Lugares
Caseros
El Palomar
Hurlingham
Billinghurst
Moron
Ituzaingo
Villa Ariza
Castelar
San Antonio de Padua
Libertad
Villa Leloir
Villa Reichembach
Villa Leon
Merlo
Moreno
Paso del Rey
San Miguel
Bella Vista
Muniz
General Sarmiento
José C. Paz
Villa de Mayo
Los Polvorines
Don Torcuato
El Talar de Pacheco
Campo de Mayo
Igr. P. Nogues
Grand Bourg
Tortuguitas
Benavidez
Garin
Del Viso
Villa Iglesias
Villa Altube
Piñero
Presidente Derqui
Toro
Villa Rosa

Francisco Alvarez
La Reja
Mariano Acosta
Puente Cascallares
Pontevedra
Isidro Casanova
Rafael Castillo
Laferrere
González Catán
Marcos Paz
20 de Junio

R. Luján
R. Reconquista
R. José L. Suárez
A. Morales
A. La Horqueta
Pinazo

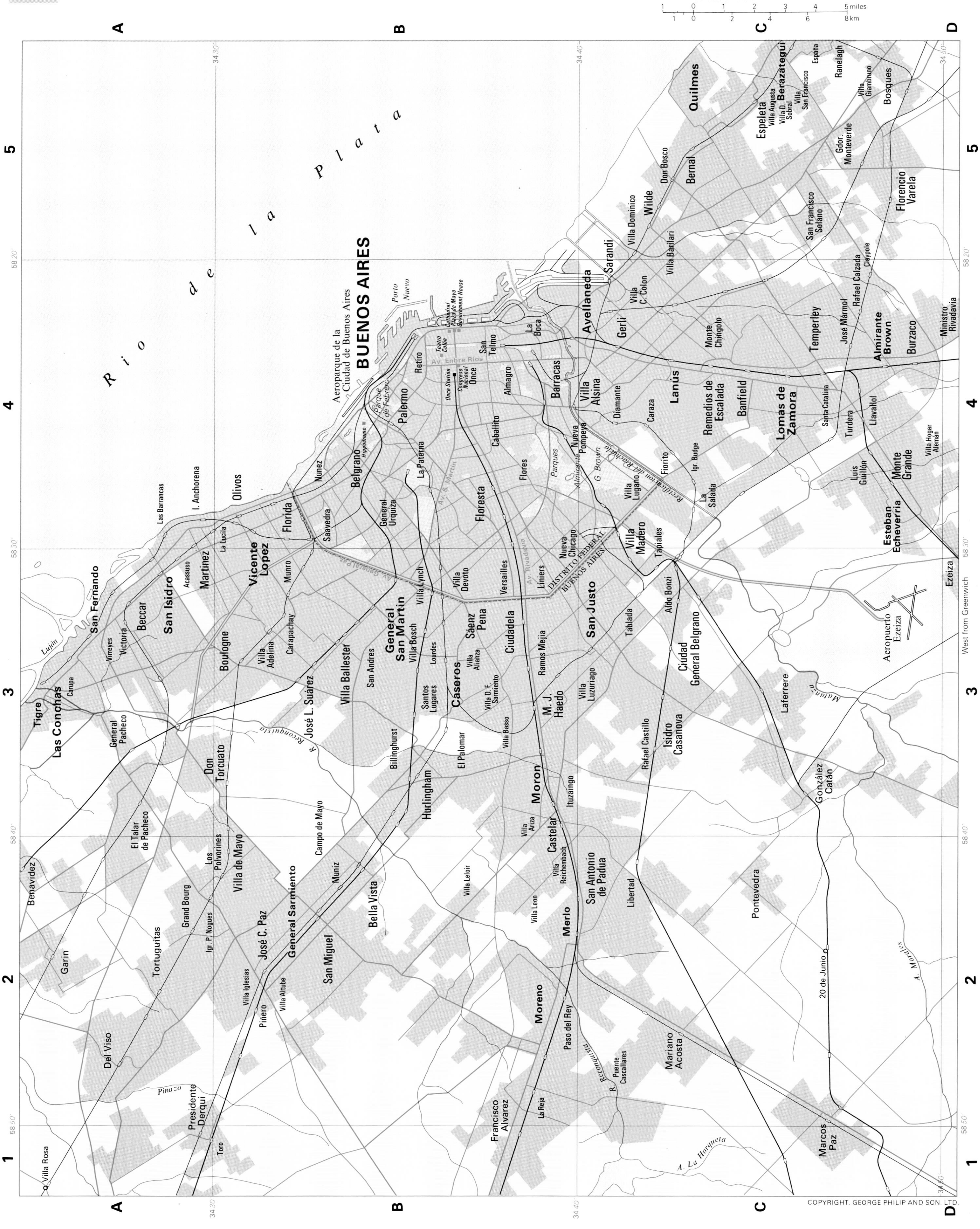

COPYRIGHT. GEORGE PHILIP AND SON. LTD.

INDEX TO CITY MAPS

Place names in this index are given a letter-figure reference to a map square made from the lines of latitude and longitude that appear on the city maps. The full geographic reference is provided in the border of each map. The letter-figure reference will take the reader directly to the square, and by using the geographical coordinates the place sought can be pinpointed within that square.

The location given is the city or suburban center, and not necessarily the name. Lakes, airports and other features having a large area are given coordinates for their centers. Rivers that enter the sea, lake or main stream within the map area have the coordinates of that entrance.

If the river flows through the map, then the coordinates are given to the name. The same rule applies to canals. A river carries the symbol → after its name.

As an aid to identification, every place name is followed by the city map name or its abbreviation; for example, Oakland in California will be followed by S.F. Some of the place names so described will be completely independent of the main city.

An explanation of the alphabetical order rules is to be found at the beginning of the World Map Index.

ABBREVIATIONS USED IN THE INDEX

Ath. – Athinai (Athens)
B. – Baie, Bahía, Bay, Bucht
B.A. – Buenos Aires
Bagd. – Baghdad
Balt. – Baltimore
Bangk. – Bangkok
Barc. – Barcelona
Beij. – Beijing (Peking)
Berl. – Berlin
Bomb. – Bombay
Bost. – Boston
Bud. – Budapest
C. – Cabo, Cap, Cape
Calc. – Calcutta
Car. – Caracas
Chan. – Channel
Chic. – Chicago
Cr. – Creek
E. – East
El Qâ. – El Qâhira (Cairo)
G. – Golfe, Golfo, Gulf, Guba
Gzh. – Guangzhou (Canton)
H.K. – Hong Kong
Hbg. – Hamburg
Hd. – Head
Hels. – Helsinki
Hts. – Heights
I.(s) – Île, Ilha, Insel, Isla, Island, Isle
Ist. – Istanbul
J. – Jabal, Jebel
Jak. – Jakarta
Jobg. – Johannesburg
K. – Kap, Kapp
Kar. – Karachi
Kep. – Kepulauan
Købn. – København (Copenhagen)
L. – Lac, Lacul, Lago, Lagoa, Lake
L.A. – Los Angeles
La Hab. – La Habana (Havana)
Lisb. – Lisboa (Lisbon)
Lon. – London
Mdrd. – Madrid
Melb. – Melbourne
Méx. – México
Mil. – Milano
Mos. – Moskva (Moscow)
Mt. (e) – Mont, Monte, Monti, Montaña, Mountain
Mtrl. – Montréal
Mün. – München (Munich)
N. – Nord, Norte, North, Northern, Nouveau
Nápl. – Nápoli (Naples)
N.Y. – New York City
Os. – Ostrov
Oz. – Ozero
Pen. – Peninsula, Peninsule
Phil. – Philadelphia
Ph. – Park, Peak
Pra. – Praha (Prague)
Pt. – Point
Pta. – Ponta, Punta
Pte. – Pointe
R. – Rio, River
Ra. (s) – Range(s)
Res. – Reserve, Reservoir
Rio J. – Rio de Janeiro
S. – San, South
S.F. – San Francisco
S. Pau. – São Paulo
Sa. – Serra, Sierra
Sd. – Sound
Shang. – Shanghai
Sing. – Singapore
St. – Saint, Sankt, Sint
St-Pet. – St-Peterburg
Sta. – Santa, Station
Ste. – Sainte
Stgo. – Santiago
Sto. – Santo
Stock. – Stockholm
Str. – Strait, Stretto
Syd. – Sydney
Tehr. – Tehran
Tianj. – Tianjin (Tientsin)
Tori. – Torino (Turin)
Trto. – Toronto
W. – West
Wash. – Washington
Wsaw. – Warszawa (Warsaw)

A

Aalām, *Bagd.*	**17 F8**	33 19N	44 23 E
Abada, *Calc.*	**16 E5**	22 32N	88 13 E
Abbadia di Stura, *Tori.*	**9 B3**	45 7N	7 44 E
Abbey Wood, *Lon.*	**4 C5**	51 29N	0 7 E
Abbots Langley, *Lon.*	**4 A2**	51 42N	0 25W
Abeno, *Ōsaka*	**12 C4**	34 38N	135 31 E
Aberdeen, *H.K.*	**12 E6**	22 14N	114 8 E
Abfanggraben, *Mün.*	**7 F11**	48 10N	11 41 E
Abington, *Phil.*	**24 A4**	40 7N	75 7W
Ablon-sur-Seine, *Paris*	**5 C4**	48 43N	2 25 E
Abord à Plouffe, *Mtrl.*	**20 A3**	43 32N	73 43W
Abramtsevo, *Mos.*	**11 E10**	55 49N	37 49 E
Abridge, *Lon.*	**4 B5**	51 38N	0 7 E
Abû en Numrus, *El Qâ.*	**18 D5**	29 57N	31 12 E
Acassuso, *B.A.*	**32 A3**	34 29 S	58 30W
Accord, *Bost.*	**21 D4**	42 10N	70 52W
Accord Pond, *Bost.*	**21 D4**	42 10N	70 53W
Accotink Cr. →, *Wash.*	**25 D6**	38 51N	77 15W
Acerra, *Nápl.*	**9 H13**	40 56N	14 22 E
Acha San, *Sŏul*	**12 G8**	37 33N	127 5 E
Acheres, *Paris*	**5 B2**	48 57N	2 3 E
Acília, *Rome*	**9 G9**	41 47N	12 21 E
Aclimação, *S. Pau.*	**31 E6**	23 34 S	46 37W
Acostia →, *Wash.*	**25 D8**	38 51N	77 1W
Acton, *Lon.*	**4 B3**	51 30N	0 16W
Açúcar, Pão de, *Rio J.*	**31 B3**	22 56 S	43 9W
Ada Beja, *Lisb.*	**8 F7**	38 47N	9 13W
Adabe Cr. →, *S.F.*	**27 D4**	37 26N	122 6W
Adachi, *Tōkyō*	**13 B2**	35 49N	139 34 E
Adachi-Ku, *Tōkyō*	**13 B3**	35 47N	139 47 E
Adams Nat. Hist. Site, *Bost.*	**21 D4**	42 15N	71 0W
Addington, *Lon.*	**4 C4**	51 21N	0 1W
Addiscombe, *Lon.*	**4 C4**	51 22N	0 4W
Adel, *Bagd.*	**17 E7**	33 20N	44 17 E
Adelphi, *Wash.*	**25 C8**	39 0N	76 58W
Aderklaa, *Wien*	**10 G11**	48 17N	16 32 E
Admiralteyskaya Storona, *St-Pet.*	**11 B4**	59 56N	30 20 E
Áffori, *Mil.*	**9 D6**	45 31N	9 10 E
Aflandshage, *Købn.*	**2 E10**	55 33N	12 35 E
Afragola, *Nápl.*	**9 H12**	40 55N	14 18 E
Aganpur, *Delhi*	**16 B3**	28 33N	77 20 E
Agboju, *Lagos*	**18 B1**	6 27N	7 16 E
Agboyi Cr. →, *Lagos*	**18 A2**	6 33N	7 24 E
Ågerup, *Købn.*	**2 D8**	55 43N	12 19 E
Ågesta, *Stock.*	**3 E11**	59 12N	18 6 E
Agincourt, *Trto.*	**20 D9**	43 47N	79 16W
Agnano Terme, *Nápl.*	**9 J12**	40 49N	14 10 E
Agora, *Ath.*	**8 J11**	37 57N	23 43 E
Agra Canal, *Delhi*	**16 B2**	28 33N	77 17 E
Agricola Oriental, *Méx.*	**29 B3**	19 23N	99 4W
Agro Romano, *Rome*	**9 F8**	41 56N	12 17 E
Agua Branca, *S. Pau.*	**31 E5**	23 31 S	46 40W
Agua Espraiada →, *S. Pau.*	**31 E6**	23 36 S	46 41W
Água Rasa, *S. Pau.*	**31 E6**	23 32 S	46 33W
Agualva-Cacem, *Lisb.*	**8 F7**	38 46N	9 15W
Agustino, Cerro El, *Lima*	**30 G8**	12 3 S	76 59W
Ahrensfelde, *Berl.*	**7 A4**	52 34N	13 34 E
Ahuntsic, *Mtrl.*	**20 A3**	43 32N	73 41W
Ai →, *Ōsaka*	**12 B4**	34 46N	135 35 E
Aigremont, *Paris*	**5 B2**	48 54N	2 1 E
Airport West, *Melb.*	**19 E6**	37 42 S	144 52 E
Aiyaleo, *Ath.*	**8 J11**	37 59N	23 40 E
Ajegunle, *Lagos*	**18 B2**	6 26N	7 20 E
Aji, *Ōsaka*	**12 B3**	34 40N	135 27 E
Ajuda, *Lisb.*	**8 F7**	38 42N	9 12W
Ajusco, Parque Nacional de, *Méx.*	**29 C2**	19 12N	99 15W
Akabane, *Tōkyō*	**13 B3**	35 46N	139 42 E
Akalla, *Stock.*	**3 D10**	59 24N	17 55 E
Akasaka, *Tōkyō*	**13 B3**	35 40N	139 43 E
Akbarābād, *Tehr.*	**17 C5**	35 40N	51 20 E
Åkersberga Saltsjobad, *Stock.*	**3 D12**	59 26N	18 15 E
Akerselva →, *Oslo*	**2 B4**	59 54N	10 45 E
Akrópolis, *Ath.*	**8 J11**	37 57N	23 43 E
Akuwa, *Tōkyō*	**13 D2**	35 26N	139 30 E
Al 'Azamiyah, *Bagd.*	**17 E8**	33 22N	44 22 E
Alaguntan, *Lagos*	**18 B2**	6 25N	7 29 E
Alamar, *La Hab.*	**30 B3**	23 9N	82 16W
Alameda, *S.F.*	**27 B3**	37 46N	122 15W
Alameda Memorial State Beach Park, *S.F.*	**27 B3**	37 45N	122 16W
Alamo, *S.F.*	**27 A4**	37 51N	122 2W
Albany, *S.F.*	**27 A3**	37 53N	122 17W
Alberante, *Jobg.*	**18 F9**	26 16 S	28 7 E
Albern, *Wien*	**10 H10**	48 9N	16 29 E
Albert Hall, *Lon.*	**4 C3**	51 29N	0 10W
Albert Park, *Melb.*	**19 F6**	37 51 S	144 58 E
Albertfalva, *Bud.*	**10 K13**	47 26N	19 3 E
Alberton, *Jobg.*	**18 F9**	26 15 S	28 7 E
Albertslund, *Købn.*	**2 E9**	55 39N	12 21 E
Albertson, *N.Y.*	**23 C7**	40 46N	73 38W
Albertville, *Jobg.*	**18 E8**	26 9 S	27 58 E
Albion, *Phil.*	**24 C5**	39 46N	74 57W
Alby, *Stock.*	**3 E10**	59 14N	17 51 E
Albysjön, *Stock.*	**3 E10**	59 14N	17 52 E
Alcantara, *Lisb.*	**8 F7**	38 43N	9 10W
Alcatraz I., *S.F.*	**27 B2**	37 49N	122 25W
Alcochete, *Lisb.*	**8 F9**	38 45N	8 58W
Alcorcón, *Mdrd.*	**8 B2**	40 20N	3 48W
Aldan, *Phil.*	**24 B3**	39 55N	75 17W
Aldela de Carapiculba, *S. Pau.*	**31 E5**	23 34 S	46 49W
Aldene, *N.Y.*	**22 D3**	40 39N	74 17W
Aldenrade, *Ruhr*	**6 A2**	51 31N	6 44 E
Alder Planetarium, *Chic.*	**26 B3**	41 5N	87 36W
Aldershof, *Berl.*	**7 B4**	52 26N	13 33 E
Aldo Bonzi, *B.A.*	**32 C3**	34 42 S	58 31W
Aleksandrovskoye, *St-Pet.*	**11 B4**	59 51N	30 20 E
Aleksandrów, *Wsaw.*	**10 E8**	52 10N	21 14 E
Alexander Nevsky Abbey, *St-Pet.*	**11 B4**	59 54N	30 23 E
Alexandra, *Jobg.*	**18 E9**	26 6 S	28 5 E
Alexandra, *Sing.*	**15 G7**	1 17N	103 49 E
Alexandria, *Wash.*	**25 E7**	38 49N	77 5W
Alfortville, *Paris*	**5 C4**	48 48N	2 24 E
Algés, *Lisb.*	**8 F7**	38 42N	9 13W
Älgo, *Stock.*	**3 E13**	59 16N	18 20 E
Algodonal, *Car.*	**30 E5**	10 29N	66 58W
Alhambra, *L.A.*	**28 B4**	34 5N	118 7W
Alhos Vedros, *Lisb.*	**8 G8**	38 39N	9 1W
Alibey →, *Ist.*	**17 A2**	41 3N	28 56 E
Alibeyköy, *Ist.*	**17 A2**	41 4N	28 56 E
Alima, *Manila*	**15 E3**	14 27N	120 55 E
Alimos, *Ath.*	**8 J11**	37 52N	23 43 E
Aliperti, *Nápl.*	**9 H13**	40 53N	14 28 E
Alipore, *Calc.*	**16 E6**	22 31N	88 20 E
Alipur, *Calc.*	**16 D5**	22 43N	88 12 E
Aliso Canyon Wash →, *L.A.*	**28 A1**	34 15N	118 31W
Allach, *Mün.*	**7 F9**	48 11N	11 27 E
Allambie Heights, *Syd.*	**19 A4**	33 46 S	151 15 E
Allendale, *N.Y.*	**22 A4**	41 1N	74 9W
Allengrove, *Jobg.*	**18 E10**	26 5 S	28 14 E
Allentown, *N.Y.*	**23 C6**	40 47N	73 43W
Allermohe, *Hbg.*	**7 E8**	53 29N	10 7 E
Allerton, Pt., *Bost.*	**21 D4**	42 18N	70 52W
Allston, *Bost.*	**21 C3**	42 21N	71 7W
Alluets, Forêt des, *Paris*	**5 B1**	48 56N	1 55 E
Almada, *Lisb.*	**8 F8**	38 41N	9 8W
Almagro, *B.A.*	**32 B4**	34 38 S	58 24W
Almanara, *Mdrd.*	**8 B2**	40 28N	3 41W
Almaza Airport, *El Qâ.*	**18 C6**	30 5N	31 21 E
Almazovo, *Mos.*	**11 D12**	55 50N	38 3 E
Almendares, *La Hab.*	**30 B2**	23 6N	82 23W
Almendares →, *La Hab.*	**30 B2**	23 7N	82 24W
Almirante Brown, *B.A.*	**32 C4**	34 48 S	58 23W
Almirante G. Brown, Parques, *B.A.*	**32 C4**	34 40 S	58 28W
Almonesson, *Phil.*	**24 C4**	39 48N	75 5W
Almonte, *S.F.*	**27 A1**	37 53N	122 31W
Alnabru, *Oslo*	**2 B5**	59 55N	10 50 E
Alnsjøen, *Oslo*	**2 B5**	59 57N	10 51 E
Alperton, *Lon.*	**4 B3**	51 32N	0 17W
Alpignano, *Tori.*	**9 B1**	45 6N	7 31 E
Alpine, *N.Y.*	**22 B5**	40 57N	73 57W
Alpur, *Calc.*	**16 C6**	22 50N	88 23 E
Alrode, *Jobg.*	**18 F9**	26 17 S	28 7 E
Alsergrund, *Wien*	**10 G10**	48 13N	16 21 E
Alsfeld, *Ruhr*	**6 A3**	51 31N	6 50 E
Alsip, *Chic.*	**26 C2**	41 40N	87 44W
Alstaden, *Ruhr*	**6 B2**	51 28N	6 49 E
Ålsten, *Stock.*	**3 E10**	59 19N	17 57 E
Alster →, *Hbg.*	**7 D8**	53 38N	10 4 E
Alsterdorf, *Hbg.*	**7 D8**	53 36N	10 0 E
Älta, *Stock.*	**3 E12**	59 15N	18 11 E
Altadena, *L.A.*	**28 A4**	34 11N	118 8W
Alte-Donau →, *Wien*	**10 G10**	48 14N	16 25 E
Alte Süderelbe, *Hbg.*	**7 D7**	53 31N	9 52 E
Alten-Essen, *Ruhr*	**6 B4**	51 29N	7 1 E
Altendorf, *Ruhr*	**6 B3**	51 27N	6 58 E
Altenhagen, *Ruhr*	**6 B6**	51 22N	7 27 E
Altenvoerde, *Ruhr*	**6 C6**	51 18N	7 22 E
Altenwerder, *Hbg.*	**7 D7**	53 30N	9 55 E
Alter Finkenkrug, *Berl.*	**7 A1**	52 35N	13 3 E
Altglenicke, *Berl.*	**7 B4**	52 25N	13 32 E
Altlandsberg Nord, *Berl.*	**7 A5**	52 34N	13 43 E
Altmannsdorf, *Wien*	**10 H9**	48 9N	16 18 E
Alto, *S.F.*	**27 A1**	37 54N	122 30W
Alto da Boa Vista, *S. Pau.*	**31 E5**	23 38 S	46 42W
Alto da Lapa, *S. Pau.*	**31 E5**	23 31 S	46 43W
Alto da Moóca, *S. Pau.*	**31 E6**	23 34 S	46 33W
Alto do Pina, *Lisb.*	**8 F8**	38 44N	9 7W
Altona, *Hbg.*	**7 D7**	53 32N	9 56 E
Altona, *Melb.*	**19 F5**	37 51 S	144 49 E
Altona B., *Melb.*	**19 F6**	37 52 S	144 51 E
Altona North, *Melb.*	**19 F5**	37 50 S	144 49 E
Altona Sports Park, *Melb.*	**19 F6**	37 51 S	144 51 E
Altstadt, *Hbg.*	**7 D8**	53 32N	10 0 E
Alvarado, *S.F.*	**27 C4**	37 35N	122 4W
Alvik, *Stock.*	**3 E10**	59 19N	17 58 E
Älvsjo, *Stock.*	**3 E11**	59 16N	18 0 E
Älvvik, *Stock.*	**3 D12**	59 21N	18 15 E
Am Hasenbergl, *Mün.*	**7 F10**	48 12N	11 33 E
Am Steinhof, *Wien*	**10 G9**	48 12N	16 17 E
Am Wald, *Mün.*	**7 G10**	48 3N	11 36 E
Ama Keng, *Sing.*	**15 F7**	1 23N	103 41 E
Amadora, *Lisb.*	**8 F7**	38 45N	9 13W
Amagasaki, *Ōsaka*	**12 B3**	34 42N	135 23 E
Amager, *Købn.*	**2 E10**	55 36N	12 35 E
Amāl Qādisiya, *Bagd.*	**17 F8**	33 16N	44 20 E
Amalienborg Slott, *Købn.*	**2 D10**	55 41N	12 35 E
Amata, *Mil.*	**9 D5**	45 34N	9 8 E
Ambler, *Phil.*	**24 A3**	40 9N	75 13W
Ambrose Channel, *N.Y.*	**22 D5**	40 31N	73 50W
Ameixoeira, *Lisb.*	**8 F8**	38 46N	9 8W
Ames Hill, *Bost.*	**21 B2**	42 38N	71 13W
Amin, *Bagd.*	**17 F8**	33 19N	44 29 E
Aminyevo, *Mos.*	**11 E8**	55 41N	37 25 E
Amirābād, *Tehr.*	**17 C5**	35 43N	51 24 E
Amityville, *N.Y.*	**23 C8**	40 40N	73 23W
Ammersbek →, *Hbg.*	**7 C8**	53 42N	10 7 E
Amora, *Lisb.*	**8 G8**	38 37N	9 6W
Amoreira, *Lisb.*	**8 F7**	38 48N	9 11W
Amorosa, *Jobg.*	**18 E8**	26 5 S	27 52 E
Ampelokipi, *Ath.*	**8 J11**	37 58N	23 47 E
Amper →, *Mün.*	**7 F9**	48 14N	11 25 E
Amselhain, *Berl.*	**7 A5**	52 38N	13 43 E
Amuwo, *Lagos*	**18 B1**	6 28N	7 18 E
Anacostia, *Wash.*	**25 D8**	38 51N	76 59W
Anacostia River Park, *Wash.*	**25 D8**	38 54N	76 57W
Anadoluhisari, *Ist.*	**17 A3**	41 4N	29 3 E
Anandanagar, *Calc.*	**16 C5**	22 51N	88 16 E
Anchieta, *Rio J.*	**31 A1**	22 48 S	43 21W
Ancol, *Jak.*	**15 H9**	6 7 S	106 49 E
Andalus, *Bagd.*	**17 F7**	33 19N	44 18 E
Andalusia, *Phil.*	**24 A5**	40 4N	74 58W
Andarai, *Rio J.*	**31 B2**	22 56 S	43 14W
Andeli Beijie, *Beij.*	**14 B3**	39 57N	116 21 E
Anderson Cr. →, *Melb.*	**19 E8**	37 44 S	145 12 E
Andilly, *Paris*	**5 A3**	49 0N	2 17 E
Andingmen, *Beij.*	**14 B3**	39 55N	116 23 E
Andover, *Bost.*	**21 B3**	42 39N	71 7W
Andrésy, *Paris*	**5 B2**	48 58N	2 3 E
Andrews Air Force Base, *Wash.*	**25 E8**	38 48N	76 52W
Ang Mo Kio, *Sing.*	**15 F8**	1 22N	103 50 E
Ängby, *Stock.*	**3 D10**	59 20N	17 53 E
Angel I., *S.F.*	**27 A2**	37 52N	122 25W
Angel Island State Park, *S.F.*	**27 A2**	37 52N	122 25W
Angerbruch →, *Ruhr*	**6 C3**	51 18N	6 59 E
Angerhausen, *Ruhr*	**6 B2**	51 22N	6 43 E
Angermund, *Ruhr*	**6 C2**	51 19N	6 46 E
Angke, Kali →, *Jak.*	**15 H9**	6 6 S	106 46 E
Angono, *Manila*	**15 D4**	14 31N	121 8 E
Angyalföld, *Bud.*	**10 J13**	47 32N	19 5 E
Angyō, *Tōkyō*	**13 A3**	35 50N	139 45 E
Aniene →, *Rome*	**9 F10**	41 56N	12 35 E
Anik, *Bomb.*	**16 G8**	19 1N	72 53 E
Anin, *Wsaw.*	**10 E7**	52 13N	21 9 E
Anjou, *Mtrl.*	**20 A4**	43 36N	73 33W
Annadale, *N.Y.*	**22 D3**	40 32N	74 10W
Annalee Heights, *Wash.*	**25 D6**	38 51N	77 10W
Annandale, *Wash.*	**25 D6**	38 50N	77 11W
Annen, *Ruhr*	**6 B6**	51 27N	7 22 E
Annet-sur-Marne, *Paris*	**5 B6**	48 55N	2 43 E
Anthony Chabot Regional Park, *S.F.*	**27 B4**	37 46N	122 7W
Antignano, *Nápl.*	**9 H12**	40 51N	14 13 E
Antimano, *Car.*	**30 E5**	10 27N	66 59W
Antipolo, *Manila*	**15 D5**	14 35N	121 10 E
Antony, *Paris*	**5 C3**	48 44N	2 17 E
Antwerp, *Jobg.*	**18 E9**	26 5 S	28 9 E
Aoyama, *Tōkyō*	**13 C3**	35 39N	139 42 E
Ap Lei Chau, *H.K.*	**12 E5**	22 14N	114 9 E
Apapa, *Lagos*	**18 B2**	6 26N	7 21 E
Apelacão, *Lisb.*	**8 F8**	38 48N	9 7W
Apoquindo, *Stgo*	**30 J11**	33 23 S	70 30W
Apshawa, *N.Y.*	**22 A2**	41 1N	74 22W
Apterkarskiy Os., *St-Pet.*	**11 B4**	59 57N	30 20 E
Aquincum, *Bud.*	**10 J13**	47 33N	19 3 E
Ara →, *Tōkyō*	**13 B4**	35 41N	139 50 E
Arakawa-Ku, *Tōkyō*	**13 B3**	35 44N	139 48 E
Arakpur, *Delhi*	**16 B2**	28 35N	77 11 E
Arany-hegyi-patak →, *Bud.*	**10 J13**	47 34N	19 4 E
Aravaca, *Mdrd.*	**8 B2**	40 27N	3 47W
Arbataash, *Bagd.*	**17 E7**	33 20N	44 19 E
Arbutus, *Balt.*	**25 B2**	39 15N	76 41W
Arc de Triomphe, *Paris*	**5 B3**	48 52N	2 17 E
Arcadia, *L.A.*	**28 B4**	34 7N	118 1W
Arceuil, *Paris*	**5 C3**	48 48N	2 19 E
Arden, *Phil.*	**24 C2**	39 48N	75 29W
Ardey Gebirge, *Ruhr*	**6 B6**	51 24N	7 23 E
Ardmore, *Phil.*	**24 A3**	40 0N	75 17W
Ardsley, *N.Y.*	**23 A5**	41 0N	73 50W
Arese, *Mil.*	**9 D5**	45 32N	9 4 E
Arganzuela, *Mdrd.*	**8 B2**	40 23N	3 42W
Argenteuil, *Paris*	**5 B3**	48 56N	2 15 E
Argonne Forest, *Chic.*	**26 C1**	41 42N	87 53W
Ariadana, *Calc.*	**16 E6**	22 39N	88 22 E
Aricanduva →, *S. Pau.*	**31 E6**	23 31 S	46 33W
Arida, *Lagos*	**18 A1**	6 33N	7 16 E
Arima, *Ōsaka*	**12 B2**	34 47N	135 15 E
Arima, *Tōkyō*	**13 C2**	35 33N	139 33 E
Arima →, *Ōsaka*	**12 A2**	34 50N	135 14 E
Arkhangelyskoye, *Mos.*	**11 E7**	55 47N	37 17 E
Arkley, *Lon.*	**4 B3**	51 38N	0 13W
Arlington, *Bost.*	**21 C2**	42 24N	71 10W
Arlington, *Wash.*	**25 D7**	38 53N	77 7W
Arlington Heights, *Bost.*	**21 C2**	42 25N	71 10W
Arlington Heights, *Chic.*	**26 A1**	42 5N	87 55W
Arlington Nat. Cemetery, *Wash.*	**25 D7**	38 52N	77 4W
Armação, *Rio J.*	**31 B3**	22 52 S	43 6W
Armadale, *Melb.*	**19 F7**	37 51 S	145 0 E
Armadale, *Trto.*	**20 C9**	43 50N	79 14W
Armainvilliers, Forêt d', *Paris*	**5 C6**	48 46N	2 42 E
Armour Heights, *Trto.*	**20 D8**	43 45N	79 25W
Arncliffe, *Syd.*	**19 B3**	33 56 S	151 8 E
Arnold Arboretum, *Bost.*	**21 D3**	42 18N	71 8W
Arnouville-les-Gonesse, *Paris*	**5 B4**	48 59N	2 24 E
Arrentela, *Lisb.*	**8 G8**	38 37N	9 6W
Arrone →, *Rome*	**9 F8**	41 55N	12 16 E
Arroyo Arenas, *La Hab.*	**30 B2**	23 3N	82 27W
Arroyo Cr. →, *S.F.*	**26 D2**	37 27N	122 25W
Arroyo Naranjo, *La Hab.*	**30 B2**	23 2N	82 21W
Årsta, *Stock.*	**3 E11**	59 17N	18 3 E
Artesia, *L.A.*	**28 C4**	33 51N	118 4W
Arthur Alvim, *S. Pau.*	**31 E7**	23 32 S	46 28W
Arthur Kill →, *N.Y.*	**22 D3**	40 32N	74 15W
Artigas, *Car.*	**30 E5**	10 29N	66 56W
Arundel Gardens, *Balt.*	**25 B3**	39 12N	76 37W
Arundel Village, *Balt.*	**25 B3**	39 13N	76 35W
Aryiróupolis, *Ath.*	**8 J11**	37 52N	23 44 E
Arzano, *Nápl.*	**9 H12**	40 54N	14 16 E
Asagaya, *Tōkyō*	**13 B2**	35 41N	139 38 E
Asahi, *Ōsaka*	**12 B4**	34 43N	135 31 E
Asaka, *Tōkyō*	**13 B2**	35 47N	139 35 E
Asakusa, *Tōkyō*	**13 B3**	35 42N	139 47 E
Asalatpur, *Delhi*	**16 B1**	28 37N	77 4 E
Asati, *Calc.*	**16 F5**	22 28N	88 15 E
Aschheim, *Mün.*	**7 F11**	48 10N	11 42 E
Ascot Vale, *Melb.*	**19 E6**	37 46 S	144 53 E
Aserradero, *Méx.*	**29 D2**	19 10N	99 16W
Asharoken, *N.Y.*	**23 B8**	40 55N	73 21W
Ashburn, *Chic.*	**26 C2**	41 45N	87 43W
Ashburton, *Melb.*	**19 F7**	37 51 S	145 4 E

Choisy-le-Roi 35

Boullay-les-Troux, *Paris* **5 C2** 48 40N 2 2 E
Boulogne, *B.A.* **32 B3** 34 30 S 58 33W
Boulogne, Bois de, *Paris* **5 B3** 48 51N 2 14 E
Boulogne-Billancourt, *Paris* **5 B3** 48 50N 2 14 E
Bouqueval, *Paris* **5 A4** 49 1N 2 25 E
Bourg-la-Reine, *Paris* **5 C3** 48 46N 2 19 E
Boussy-St.-Antoine, *Paris* **5 C5** 48 41N 2 33 E
Bouviers, *Paris* **5 C2** 48 46N 2 4 E
Bovert, *Ruhr* **6 C1** 51 16N 6 37 E
Bovisa, *Mil.* **9 D6** 45 30N 9 10 E
Bovísio-Masciago, *Mil.* **9 D5** 45 36N 9 8 E
Bow, *Lon.* **4 B4** 51 31N 0 1W
Bowleys Quarters, *Balt.* **25 A4** 39 20N 76 24W
Box Hill, *Melb.* **19 E7** 37 48 S 145 6 E
Boxford State Forest, *Bost.* **21 B3** 42 39N 71 2W
Boy, *Ruhr* **6 A3** 51 31N 7 0 E
Boyacıköy, *Ist.* **17 A3** 41 5N 29 2 E
Boye →, *Ruhr* **6 A3** 51 30N 6 59 E
Boyle Heights, *L.A.* **28 B3** 34 1N 118 12 E
Braddell Heights, *Sing.* **15 F8** 1 20N 103 51 E
Brahmanpur, *Bomb.* **16 G8** 19 5N 72 52 E
Braintree, *Bost.* **21 D3** 42 12N 71 0W
Brakpan, *Jobg.* **18 F11** 26 14 S 28 20 E
Brambauer, *Ruhr* **6 A6** 51 35N 7 26 E
Bramfeld, *Hbg.* **7 D8** 53 36N 10 5 E
Bramley, *Jobg.* **18 E9** 26 7 S 28 4 E
Brande, *Hbg.* **7 D6** 53 37N 9 49 E
Brandenburg Gate, *Berl.* **7 A3** 52 30N 13 21 E
Brandizzo, *Tori.* **9 A3** 45 10N 7 49 E
Brands Hatch, *Lon.* **4 C6** 51 21N 0 15 E
Brandýs nad Labem, *Pra.* **10 A3** 50 10N 14 39 E
Brandywine, *Phil.* **24 C1** 39 49N 75 32W
Brandywine Cr. →, *Phil.* **24 C1** 39 43N 75 31W
Brani, P., *Sing.* **15 G8** 1 15N 103 50 E
Branik, *Pra.* **10 B2** 50 1N 14 25 E
Brännkyrka, *Stock.* **3 E11** 59 17N 18 0 E
Brás, *S. Pau.* **31 E6** 23 32 S 46 36W
Brateyevo, *Mos.* **11 F10** 55 39N 37 45 E
Bratsevo, *Mos.* **11 D8** 55 51N 37 24 E
Brauck, *Ruhr* **6 A3** 51 32N 7 0 E
Brava, Pta., *La Hab.* **30 B2** 23 8N 82 23W
Braybrook, *Melb.* **19 E6** 37 46 S 144 51 E
Brázdim, *Pra.* **10 A3** 50 10N 14 35 E
Breakheart Reservation, *Bost.* **21 C3** 42 28N 71 1W
Brechten, *Ruhr* **6 A6** 51 34N 7 27 E
Breckerfeld, *Ruhr* **6 C6** 51 15N 7 28 E
Brede, *Købn.* **2 D10** 55 47N 12 30 E
Bredeney, *Ruhr* **6 B3** 51 24N 6 59 E
Breeds Pond, *Bost.* **21 C4** 42 29N 70 59W
Breezy Pt., *N.Y.* **22 D5** 40 33N 73 56W
Breitenlee, *Wien* **10 G11** 48 15N 16 30 E
Breitscheid, *Ruhr* **6 B3** 51 21N 6 51 E
Breña, *Lima* **30 G8** 12 3 S 77 3W
Brenschede, *Ruhr* **6 B5** 51 26N 7 12 E
Brent, *Lon.* **4 B3** 51 33N 0 15W
Brent →, *Lon.* **4 B2** 51 30N 0 20W
Brent Res., *Lon.* **4 B3** 51 34N 0 14W
Brentford, *Lon.* **4 C3** 51 28N 0 18W
Brenthurst, *Jobg.* **18 F11** 26 15 S 28 21 E
Brentwood, *Lon.* **4 B6** 51 36N 0 18 E
Brentwood Park, *Jobg.* **18 E10** 26 7 S 28 17 E
Brentwood Park, *L.A.* **28 B2** 34 3N 118 29W
Brera, *Mil.* **9 E6** 45 28N 9 11 E
Bresso, *Mil.* **9 D6** 45 32N 9 11 E
Brétigny-sur-Orge, *Paris* **5 D3** 48 36N 2 18 E
Brevik, *Stock.* **3 D12** 59 20N 18 12 E
Břevnov, *Pra.* **10 B2** 50 4N 14 22 E
Brewer I., *S.F.* **27 C3** 37 33N 122 16W
Bricket Wood, *Lon.* **4 A2** 51 42N 0 21W
Bridesburg, *Phil.* **24 B4** 39 59N 75 4W
Bridgeport, *Chic.* **26 B3** 41 50N 87 38W
Bridgeport, *Phil.* **24 A2** 40 6N 75 21W
Bridgeview, *Chic.* **26 C2** 41 45N 87 48W
Brie-Comte-Robert, *Paris* **5 C5** 48 41N 2 36 E
Brighton, *Melb.* **19 F6** 37 55 S 144 59 E
Brighton le Sands, *Syd.* **19 B3** 33 57 S 151 9 E
Brighton Park, *Chic.* **26 C2** 41 48N 87 41W
Brightwood, *Wash.* **25 D7** 38 57N 77 1W
Brigittenau, *Wien* **10 G10** 48 14N 16 22 E
Briis-sous-Forges, *Paris* **5 D2** 48 37N 2 7 E
Brimbank Park, *Melb.* **19 E6** 37 43 S 144 50 E
Brimsdown, *Lon.* **4 B4** 51 39N 0 0 E
Brione, *Tori.* **9 B1** 45 8N 7 28 E
Briones Hills, *S.F.* **27 A4** 37 56N 122 8W
Briones Regional Park, *S.F.* **27 A4** 37 55N 122 8W
Briones Res., *S.F.* **27 A3** 37 55N 122 11W
Brisbane, *S.F.* **27 B2** 37 40N 122 23W
Bristol, *Phil.* **24 A5** 40 6N 74 53W
Britz, *Berl.* **7 B3** 52 26N 13 27 E
Brixton, *Lon.* **4 C4** 51 27N 0 8W
Broad Axe, *Phil.* **24 A3** 40 8N 75 14W
Broad Sd., *Bost.* **21 C4** 42 23N 70 56W
Broadmeadows, *Melb.* **19 E6** 37 40 S 144 55 E
Broadmoor, *S.F.* **27 B2** 37 41N 122 29W
Broadview, *Chic.* **26 B1** 41 51N 87 52W
Brobacka, *Hels.* **3 B2** 60 15N 24 36 E
Brockley, *Lon.* **4 C4** 51 27N 0 2W
Bródno, *Wsaw.* **10 E7** 52 17N 21 1 E
Bródnowski, Kanal, *Wsaw.* **10 E7** 52 17N 21 3 E
Broich, *Ruhr* **6 B3** 51 25N 6 50 E
Bromley, *Lon.* **4 C5** 51 24N 0 0 E
Bromley-by-Bow, *Lon.* **4 B4** 51 31N 0 0 E
Bromley Common, *Lon.* **4 C5** 51 22N 0 2 E
Bromma, *Stock.* **3 D10** 59 21N 17 55 E
Bromma flygplats, *Stock.* **3 D10** 59 21N 17 56 E
Brompton, *Lon.* **4 C3** 51 29N 0 10W
Brøndby Strand, *Købn.* **2 E9** 55 36N 12 25 E
Brøndbyøster, *Købn.* **2 E9** 55 39N 12 26 E
Brøndbyvester, *Købn.* **2 E9** 55 37N 12 23 E
Brondesbury, *Lon.* **4 B3** 51 32N 0 12W
Brønnøya, *Oslo* **2 B3** 59 51N 10 32 E
Brønshøj, *Købn.* **2 D9** 55 41N 12 29 E
Bronx Zoo, *N.Y.* **23 B5** 40 50N 73 51W
Bronxville, *N.Y.* **23 B6** 40 56N 73 49W
Brook Street, *Lon.* **4 B6** 51 36N 0 17 E
Brookfield, *Chic.* **26 C1** 41 48N 87 50W
Brookhaven, *Phil.* **24 B2** 39 52N 75 23W
Brooklandville, *Balt.* **25 A2** 39 25N 76 40W
Brooklin, *S. Pau.* **31 E6** 23 37 S 46 39W
Brookline, *Bost.* **21 D3** 42 19N 71 8W
Brooklyn, *Balt.* **25 B3** 39 13N 76 36W
Brooklyn, *Melb.* **19 E5** 37 49 S 144 49 E
Brooklyn, *N.Y.* **22 D5** 40 37N 73 57W
Brookmont, *Wash.* **25 D7** 38 57N 77 7W
Brooks I., *S.F.* **27 A2** 37 53N 122 21W
Brookville, *N.Y.* **23 C7** 40 48N 73 33W
Broomall, *Phil.* **24 B2** 39 58N 75 22W
Brosewere B., *N.Y.* **23 D6** 40 37N 73 40W
Brossard, *Mtrl.* **20 B5** 43 27N 73 28W
Brou-sur-Chantereine, *Paris* **5 B5** 48 53N 2 37 E
Brown, *Trto.* **20 D9** 43 48N 79 14W
Browns Line, *Trto.* **20 E7** 43 36N 79 32W
Broyhill Park, *Wash.* **25 D6** 38 52N 77 12W
Bru, *Oslo* **2 C5** 59 47N 10 54 E
Bruckhausen, *Ruhr* **6 B2** 51 29N 6 43 E
Brughério, *Mil.* **9 D6** 45 33N 9 17 E
Bruino, *Tori.* **9 B1** 45 1N 7 27 E
Bruløkka, *Oslo* **2 A2** 60 1N 10 22 E
Brunn, *Stock.* **3 E13** 59 17N 18 25 E
Brunnthal, *Mün.* **7 G11** 48 0N 11 41 E
Brunoy, *Paris* **5 C4** 48 41N 2 29 E
Brunswick, *Melb.* **19 E6** 37 45 S 144 57 E
Brusciano, *Nápl.* **9 H13** 40 55N 14 25 E
Brush Hill, *Bost.* **21 D1** 42 15N 71 22W
Bruzzano, *Mil.* **9 D6** 45 31N 9 10 E
Bry-sur-Marne, *Paris* **5 B5** 48 50N 2 32 E
Bryn, *Oslo* **2 B2** 59 55N 10 27 E
Bryn Athyn, *Phil.* **24 A4** 40 8N 75 3W
Bryn Mawr, *Phil.* **24 A3** 40 1N 75 19W
Brzeziny, *Wsaw.* **10 E7** 52 19N 21 2 E
Bubeneč, *Pra.* **10 B2** 50 6N 14 24 E
Buc, *Paris* **5 C2** 48 46N 2 7 E
Buch, *Berl.* **7 A3** 52 38N 13 29 E
Buchburg, *Wien* **10 G9** 48 13N 16 11 E
Buchenhain, *Mün.* **7 G9** 48 1N 11 29 E
Buchholz, *Berl.* **7 A3** 52 36N 13 25 E
Buchholz, *Ruhr* **6 B2** 51 23N 6 46 E
Buckhurst Hill, *Lon.* **4 B5** 51 37N 0 2 E
Buckingham Palace, *Lon.* **4 B4** 51 30N 0 8W
Buckow, *Berl.* **7 B3** 52 25N 13 26 E
Buda, *Bud.* **10 J13** 47 30N 19 2 E
Budafok, *Bud.* **10 K13** 47 25N 19 2 E
Budakeszi, *Bud.* **10 J12** 47 30N 18 56 E
Budaörs, *Bud.* **10 K12** 47 27N 18 57 E
Budapest, *Bud.* **10 K13** 47 29N 19 3 E
Budatétény, *Bud.* **10 K13** 47 25N 19 1 E
Budberg, *Ruhr* **6 A1** 51 32N 6 38 E
Buddinge, *Købn.* **2 D10** 55 44N 12 30 E
Büderich, *Ruhr* **6 C2** 51 15N 6 41 E
Buena Park, *L.A.* **28 C4** 33 51N 118 1W
Buena Vista, *S.F.* **27 B2** 37 45N 122 26W
Buenavista, *Mdrd.* **8 B2** 40 25N 3 40W
Buenos Aires, *B.A.* **32 B4** 34 36 S 58 22W
Buenos Aires, Aeroparque de la Ciudad de, *B.A.* **32 B4** 34 34 S 58 25W
Buer, *Ruhr* **6 A4** 51 34N 7 2 E
Bufalotta, *Rome* **9 F10** 41 59N 12 33 E
Buggajha, *Sŏul* **12 G7** 37 34N 126 55 E
Bughan San, *Sŏul* **12 G7** 37 38N 126 58 E
Bugio, *Lisb.* **8 G7** 38 39N 9 18W
Bukit Panjang, *Sing.* **15 F7** 1 22N 103 45 E
Bukit Timah, *Sing.* **15 F7** 1 20N 103 47 E
Bulam San, *Sŏul* **12 G8** 37 38N 127 4 E
Bûlâq, *El Qâ.* **18 C5** 30 3N 31 14 E
Bule, *Manila* **15 E4** 14 26N 121 2 E
Bulim, *Sing.* **15 F7** 1 22N 103 43 E
Bull Brook →, *Bost.* **21 A4** 42 41N 70 52W
Bulleen, *Melb.* **19 E7** 37 46 S 145 4 E
Bullen Park, *Melb.* **19 E7** 37 46 S 145 4 E
Bullion, *Paris* **5 D1** 48 37N 1 59 E
Bulmke-Hüllen, *Ruhr* **6 A4** 51 31N 7 7 E
Bulphan, *Lon.* **4 B7** 51 32N 0 21 E
Bundoora, *Melb.* **19 E7** 37 41 S 145 2 E
Bundoora Park, *Melb.* **19 E7** 37 42 S 145 2 E
Bunker I., *Kar.* **17 H10** 24 48N 66 57 E
Bunkyo, *Tōkyō* **13 B3** 35 42N 139 45 E
Bunnefjorden, *Oslo* **2 B4** 59 50N 10 44 E
Buona Vista, *Sing.* **15 G7** 1 16N 103 47 E
Buquirivú-Guaçu →, *S. Pau.* **31 D7** 23 28 S 46 28W
Burbank, *Chic.* **26 C2** 41 44N 87 46W
Burbank, *L.A.* **28 A3** 34 12N 118 18W
Bures, *Paris* **5 C2** 48 56N 1 57 E
Bures-sur-Yvette, *Paris* **5 C2** 48 41N 2 9 E
Burggrafenberg, *Ruhr* **6 C4** 51 13N 7 7 E
Burgh Heath, *Lon.* **4 D3** 51 18N 0 13W
Burlingame, *S.F.* **27 C2** 37 34N 122 20W
Burlington, *Bost.* **21 B2** 42 30N 71 13W
Burlington, *Phil.* **24 A5** 40 4N 74 53W
Burnham, *Chic.* **26 D3** 41 38N 87 33W
Burnham Park Harbor, *Chic.* **26 B3** 41 51N 87 36W
Burnhamthorpe, *Trto.* **20 E7** 43 37N 79 35W
Burnt Oak, *Lon.* **4 B3** 51 36N 0 15W
Burr Ridge, *Chic.* **26 C1** 41 46N 87 54W
Burtus, *El Qâ.* **18 C4** 30 8N 31 8 E
Burudvatn, *Oslo* **2 B3** 59 58N 10 35 E
Burwood, *Melb.* **19 F7** 37 50 S 145 6 E
Burwood, *Syd.* **19 B3** 33 52 S 151 6 E
Burwood East, *Melb.* **19 F7** 37 51 S 145 8 E
Burzaco, *B.A.* **32 C4** 34 49 S 58 23W
Buschhausen, *Ruhr* **6 A3** 51 30N 6 50 E
Bush Hill Park, *Lon.* **4 B4** 51 38N 0 4W
Bushey, *Lon.* **4 B2** 51 38N 0 22W
Bushwick, *N.Y.* **23 C5** 40 41N 73 54W
Bushy Cr. →, *Melb.* **19 E8** 37 42 S 145 17 E
Bushy Park, *Lon.* **4 C2** 51 24N 0 20W
Bussocaba, *S. Pau.* **31 E5** 23 34 S 46 47W
Bussy-St.-Georges, *Paris* **5 B6** 48 50N 2 41 E
Bussy-St.-Martin, *Paris* **5 B6** 48 50N 2 41 E
Bustleton, *Phil.* **24 A4** 40 4N 75 2W
Butantã, *S. Pau.* **31 E5** 23 34 S 46 42W
Butcher I., *Bomb.* **16 H8** 18 57N 72 53 E
Butendorf, *Ruhr* **6 A3** 51 32N 6 59 E
Butler, *N.Y.* **22 B2** 41 0N 74 20W
Buttonville, *Trto.* **20 C8** 43 51N 79 20W
Butts Corner, *Wash.* **25 E6** 38 46N 77 19W
Byailla, *Bomb.* **16 H8** 18 58N 72 50 E
Byberry, *Phil.* **24 A5** 40 6N 74 59W
Byfang, *Ruhr* **6 B4** 51 24N 7 5 E
Byfleet, *Lon.* **4 D2** 51 19N 0 28W
Bygdøy, *Oslo* **2 B4** 59 54N 10 40 E

C

C.N. Tower, *Trto.* **20 E8** 43 38N 79 23W
Caballito, *B.A.* **32 B4** 34 37 S 58 25W
Cabin John, *Wash.* **25 D6** 38 58N 77 10W
Cabin John Cr. →, *Wash.* **25 C7** 39 2N 77 8W
Cabin John Regional Park, *Wash.* **25 C6** 39 0N 77 10W
Cabramatta, *Syd.* **19 B2** 33 53 S 150 56 E
Cabuçú de Baixo →, *S. Pau.* **31 D5** 23 30 S 46 40W
Cachan, *Paris* **5 C3** 48 47N 2 19 E
Cachenka →, *Mos.* **11 E7** 55 46N 37 17 E
Cachoeira →, *S. Pau.* **31 E5** 23 38 S 46 43W
Cacilhas, *Lisb.* **8 F8** 38 41N 9 8W
Cadieux, Î., *Mtrl.* **20 B1** 43 25N 74 1W
Cagarras, Is., *Rio J.* **31 C2** 23 1 S 43 12W
Cahuenga Pk., *L.A.* **28 B3** 34 8N 118 19 E
Cainta, *Manila* **15 D4** 14 34N 121 6 E
Cairo = El Qâhira, *El Qâ.* **18 C5** 30 2N 31 13 E
Cairo Int. Airport, *El Qâ.* **18 C6** 30 7N 31 23 E
Caivano, *Nápl.* **9 H12** 40 57N 14 18 E
Caju, *Rio J.* **31 B2** 22 52 S 43 12W
Čakovice, *Pra.* **10 B3** 50 9N 14 31 E
Calabazar, *La Hab.* **30 B2** 23 1N 82 20W
Calcutta, *Calc.* **16 E6** 22 34N 88 21 E
Caldwell, *N.Y.* **22 B3** 40 50N 74 19W
Calf Harbour, *N.Y.* **23 B7** 40 59N 73 37W
Calf I., *Bost.* **21 C4** 42 20N 70 53W
Calhua, *Lisb.* **8 F8** 38 44N 9 9W
California, Univ. of, *S.F.* **27 A3** 37 52N 122 16W
California Inst. of Tech., *L.A.* **28 B4** 34 8N 118 8W
California State Univ., *L.A.* **28 B3** 34 4N 118 10W
California State Univ., *S.F.* **27 C4** 37 39N 122 6W
Callao, *Lima* **30 G8** 12 3 S 77 8W
Caloocan, *Manila* **15 D3** 14 39N 120 58 E
Calumet →, *Chic.* **26 C3** 41 43N 87 31W
Calumet, L., *Chic.* **26 C3** 41 40N 87 35W
Calumet City, *Chic.* **26 D3** 41 36N 87 32W
Calumet Harbor, *Chic.* **26 C3** 41 43N 87 30W
Calumet Park, *Chic.* **26 C3** 41 40N 87 39W
Calumet Sag Channel →, *Chic.* **26 C2** 41 40N 87 47W
Calumpang, *Manila* **15 D4** 14 37N 121 5 E
Calvairate, *Mil.* **9 E6** 45 27N 9 13 E
Calverton, *Wash.* **25 C8** 39 3N 76 56W
Calvizzano, *Nápl.* **9 H12** 40 54N 14 11 E
Calzada, *Manila* **15 D4** 14 32N 121 4 E
Camarate, *Lisb.* **8 F8** 38 48N 9 7W
Camaroes, *Lisb.* **8 F7** 38 49N 9 14W
Camberwell, *Lon.* **4 C4** 51 28N 0 5W
Camberwell, *Melb.* **19 F7** 37 50 S 145 5 E
Cambria Heights, *N.Y.* **23 C6** 40 41N 73 44W
Cambridge, *Bost.* **21 C3** 42 23N 71 7W
Cambridge Res., *Bost.* **21 C2** 42 24N 71 16W
Cambuci, *S. Pau.* **31 E6** 23 33 S 46 47W
Cambute, *La Hab.* **30 B3** 23 5N 82 16W
Camden, *Lon.* **4 B4** 51 32N 0 8W
Camden, *Phil.* **24 B4** 39 56N 75 7W
Camp Springs, *Wash.* **25 E8** 38 48N 76 55W
Campamento, *Mdrd.* **8 B2** 40 23N 3 46W
Campanilla, Pta., *La Hab.* **30 A3** 23 10N 82 18W
Campbellfield, *Melb.* **19 E6** 37 40 S 144 57 E
Camperdown, *Syd.* **19 B4** 33 53 S 151 11 E
Campi Flegrei, *Nápl.* **9 H11** 40 50N 14 9 E
Campo, Casa de, *Mdrd.* **8 B2** 40 25N 3 45W
Campo Belo, *S. Pau.* **31 E5** 23 36 S 46 44W
Campo de Mayo, *B.A.* **32 B2** 34 32 S 58 40W
Campo Grande, *Lisb.* **8 F8** 38 45N 9 9W
Campo Limpo, *S. Pau.* **31 E5** 23 38 S 46 46W
Campo Pequeño, *Lisb.* **8 F8** 38 44N 9 8W
Campolide, *Lisb.* **8 F8** 38 43N 9 9W
Campsie, *Syd.* **19 B3** 33 54 S 151 6 E
C'an San Joan, *Barc.* **8 D6** 41 28N 2 11 E
Canacao, *Manila* **15 E3** 14 29N 120 54 E
Canacao B., *Manila* **15 E3** 14 29N 120 54 E
Cañada de los Helechos →, *Méx.* **29 B2** 19 21N 99 15W
Canarsie, *N.Y.* **23 D5** 40 38N 73 53W
Candiac, *Mtrl.* **20 B5** 43 23N 73 29W
Caneças, *Lisb.* **8 F7** 38 48N 9 13W
Cangaiba, *S. Pau.* **31 E6** 23 30 S 46 31W
Cangrejeras, *La Hab.* **30 B1** 23 3N 82 30W
Canguera, *S. Pau.* **31 E5** 23 34 S 46 26W
Canillas, *Mdrd.* **8 B3** 40 27N 3 38W
Canillejas, *Mdrd.* **8 B3** 40 26N 3 36W
Cann Hall, *Lon.* **4 B5** 51 33N 0 0 E
Canning Town, *Lon.* **4 B5** 51 30N 0 1 E
Canoe Grove Res., *N.Y.* **22 C2** 40 45N 74 21W
Cantalupo, *Mil.* **9 D4** 45 34N 8 58 E
Cantareira, *S. Pau.* **31 D6** 23 26 S 46 36W
Cantarranas, *La Hab.* **30 B2** 23 2N 82 28W
Canteras de Vallecas, *Mdrd.* **8 B3** 40 20N 3 37W
Canterbury, *Melb.* **19 E7** 37 49 S 145 4 E
Canterbury, *Syd.* **19 B3** 33 55 S 151 7 E
Canto do Rio, *Rio J.* **31 B3** 22 54 S 43 7W
Canton, *Bost.* **21 D3** 42 10N 71 8W
Caohe, *Shang.* **14 J11** 31 10N 121 24 E
Caonao, *La Hab.* **30 B2** 23 5N 82 24W
Capão Redondo, *S. Pau.* **31 E5** 23 39 S 46 45W
Caparica, *Lisb.* **8 F8** 38 40N 9 12W
Caparica, Costa da, *Lisb.* **8 G7** 38 38N 9 15W
Capelinha, *S. Pau.* **31 E5** 23 39 S 46 44W
Capitol Heights, *Wash.* **25 D8** 38 52N 76 55W
Capodichino, Aeroporto di, *Nápl.* **9 H12** 40 52N 14 17 E
Capodimonte, *Nápl.* **9 H12** 40 52N 14 14 E
Capodimonte, Bosco di, *Nápl.* **9 H12** 40 52N 14 15 E
Captain Cook Bridge, *Syd.* **19 C3** 34 0 S 151 7 E
Captain Cook Landing Place Park, *Syd.* **19 C4** 34 1 S 151 14 E
Captain Harbour, *N.Y.* **23 B7** 40 59N 73 37W
Capuava, *S. Pau.* **31 E7** 23 38 S 46 28W
Capuchos, *Lisb.* **8 G7** 38 38N 9 16W
Caraballeda, *Car.* **30 D5** 10 36N 66 50W
Carabanchel Alto, *Mdrd.* **8 B2** 40 22N 3 44W
Carabanchel Bajo, *Mdrd.* **8 B2** 40 23N 3 44W
Carabatteda →, *Car.* **30 D5** 10 37N 66 51W
Caracas, *Car.* **30 E5** 10 30N 66 54W
Carapachay, *B.A.* **32 B3** 34 31 S 58 32W
Carapicuiba, *S. Pau.* **31 E5** 23 31 S 46 49W
Carapicuiba →, *S. Pau.* **31 E5** 23 31 S 46 49W
Caravita, *Nápl.* **9 H13** 40 52N 14 21 E
Caraza, *B.A.* **32 C4** 34 41 S 58 25W
Cardito, *Nápl.* **9 H12** 40 56N 14 17 E
Cardoso, *Lagos* **18 A1** 6 34N 7 16 E
Caribbean Gardens, *Melb.* **19 F8** 37 54 S 145 12 E
Caricuao, *Car.* **30 E5** 10 25N 66 58W
Caridad, *Manila* **15 E3** 14 28N 120 53 E
Carioca, Sa. da, *Rio J.* **31 B2** 22 57 S 43 13W
Carle Place, *N.Y.* **23 C7** 40 44N 73 35W
Carlingford, *Syd.* **19 A3** 33 46 S 151 3 E
Carlisle, *Bost.* **21 B1** 42 31N 71 20W
Carlshof, *Mün.* **7 F11** 48 15N 11 41 E
Carlstadt, *N.Y.* **22 B4** 40 50N 74 6W
Carlton, *Melb.* **19 E6** 37 47 S 144 57 E
Carnaxide, *Lisb.* **8 F7** 38 43N 9 14W
Carnegie, *Melb.* **19 F7** 37 53 S 145 3 E
Carnegie Hall, *N.Y.* **22 C5** 40 45N 73 59W
Carnetin, *Paris* **5 B6** 48 54N 2 42 E
Carney, *Balt.* **25 A3** 39 23N 76 31W
Carnide, *Lisb.* **8 F7** 38 45N 9 10W
Caronno Pert, *Mil.* **9 D5** 45 35N 9 2 E
Carramar, *Syd.* **19 B2** 33 51 S 150 58 E
Carrascal, *Stgo* **30 J10** 33 25 S 70 42W
Carrières-sous-Bois, *Paris* **5 B2** 48 55N 2 6 E
Carrières-sous-Poissy, *Paris* **5 B2** 48 56N 2 2 E
Carrières-sur-Seine, *Paris* **5 B3** 48 54N 2 11 E
Carroll I., *Balt.* **25 B4** 39 19N 76 20W
Carroll Park, *Balt.* **25 B3** 39 16N 76 38W
Carshalton, *Lon.* **4 C3** 51 22N 0 10W
Carshalton on the Hill, *Lon.* **4 C4** 51 20N 0 9W
Carteret, *N.Y.* **22 D3** 40 34N 74 13W
Cartierville, Aéroport de, *Mtrl.* **20 A3** 43 31N 73 42 E
Carugate, *Mil.* **9 D6** 45 32N 9 20 E
Carupa, *B.A.* **32 A3** 34 25 S 58 33W
Casa Blanca, *La Hab.* **30 B3** 23 8N 82 19W
Casa Verde, *S. Pau.* **31 D5** 23 29 S 46 40W
Casalnuovo di Nápoli, *Nápl.* **9 H12** 40 54N 14 20 E
Casalotti, *Rome* **9 F9** 41 54N 12 22 E
Casandrino, *Nápl.* **9 H12** 40 56N 14 15 E
Casavatore, *Nápl.* **9 H12** 40 53N 14 15 E
Cascadura, *Rio J.* **31 B2** 22 52 S 43 19W
Caselette, *Tori.* **9 B1** 45 6N 7 28 E
Caselette, Laghi di, *Tori.* **9 B1** 45 7N 7 29 E
Caselle Torinese, *Tori.* **9 A3** 45 10N 7 38 E
Caseros, *B.A.* **32 B3** 34 36 S 58 34W
Casória, *Nápl.* **9 H12** 40 54N 14 17 E
Cassignanica, *Mil.* **9 E7** 45 27N 9 20 E
Cassiobury Park, *Lon.* **4 B2** 51 39N 0 25W
Castel di Camerletto, *Tori.* **9 B1** 45 6N 7 27 E
Castel di Guido, *Rome* **9 F8** 41 53N 12 17 E
Castel Malnome, *Rome* **9 F8** 41 50N 12 19 E
Castel San Cristina, *Tori.* **9 B3** 45 8N 7 40 E
Castel Sant'Angelo, *Rome* **9 F9** 41 54N 12 27 E
Castelar, *B.A.* **32 B3** 34 39 S 58 39W
Castellbisbal, *Barc.* **8 D4** 41 28N 1 58 E
Castello di Cisterna, *Nápl.* **9 H13** 40 54N 14 24 E
Castelvécchio, *Tori.* **9 B3** 45 1N 7 46 E
Castiglione Torinese, *Tori.* **9 B3** 45 6N 7 48 E
Castleton Corners, *N.Y.* **22 D4** 40 36N 74 8W
Castro Valley, *S.F.* **27 B4** 37 41N 122 5W
Castrop, *Ruhr* **6 A5** 51 32N 7 18 E
Castrop-Rauxel, *Ruhr* **6 A5** 51 33N 7 18 E
Cat Rock Hill, *Bost.* **21 C2** 42 23N 71 18W
Caterham, *Lon.* **4 D4** 51 16N 0 5W
Catete, *Rio J.* **31 B2** 22 54 S 43 10W
Catford, *Lon.* **4 C4** 51 26N 0 1W
Catia, *Car.* **30 D5** 10 31N 66 56W
Catia La Mer, *Car.* **30 D4** 10 36N 67 0W
Catonsville, *Balt.* **25 B2** 39 16N 76 43W
Catonsville Manor, *Balt.* **25 B2** 39 17N 76 44W
Cattle Hill, *S.F.* **27 C2** 37 36N 122 27W
Catumbi, *Rio J.* **31 B2** 22 54 S 43 12W
Caughnawaga, *Mtrl.* **20 B3** 43 24N 73 40W
Caulfield, *Melb.* **19 F7** 37 52 S 145 1 E
Caulfield Racecourse, *Melb.* **19 F7** 37 53 S 145 4 E
Caumsett State Park, *N.Y.* **23 B8** 40 55N 73 27W
Cavite, *Manila* **15 E3** 14 29N 120 54 E
Cavoretto, *Tori.* **9 B3** 45 1N 7 41 E
Caxias, *Lisb.* **8 F7** 38 42N 9 16W
Caxingui, *S. Pau.* **31 E5** 23 35 S 46 43W
Cebeciköy, *Ist.* **17 A2** 41 7N 28 53 E
Cecchignola, *Rome* **9 G10** 41 48N 12 29 E
Cecil Park, *Syd.* **19 B2** 33 52 S 150 51 E
Cecilienhof, *Berl.* **7 B1** 52 25N 13 5 E
Cedar Grove, *N.Y.* **22 B3** 40 50N 74 13W
Cedar Grove Res., *N.Y.* **22 B3** 40 51N 74 12W
Cedar I., *N.Y.* **23 D8** 40 38N 73 22W
Cedar Knolls, *N.Y.* **22 C2** 40 49N 74 27W
Cedarhurst, *N.Y.* **23 D6** 40 37N 73 43W
Cedarvale, *Trto.* **20 D8** 43 41N 79 26W
Celle →, *Paris* **5 D1** 48 36N 1 59 E
Cempaka Putih, *Jak.* **15 J10** 6 10 S 106 51 E
Çengelköy, *Ist.* **17 A3** 41 2N 29 3 E
Centennial Park, *Syd.* **19 B4** 33 53 S 151 14 E
Center Square, *Phil.* **24 C2** 39 46N 75 22W
Centerport, *N.Y.* **23 B8** 40 53N 73 22W
Centerton, *Phil.* **24 B5** 39 59N 74 53W
Centocelle, *Rome* **9 F10** 41 52N 12 34 E
Central Park, *N.Y.* **22 C5** 40 47N 73 58W
Central Park, *Sing.* **15 G8** 1 17N 103 50 E
Centre City, *Phil.* **24 C3** 39 46N 75 11W
Centre I., *N.Y.* **23 B7** 40 54N 73 31W
Cércola, *Nápl.* **9 H13** 40 51N 14 21 E
Cergy-Pontoise, *Paris* **5 A2** 49 1N 2 4 E
Cernay-la-Ville, *Paris* **5 C1** 48 40N 1 58 E
Cernusco sul Navíglio, *Mil.* **9 D6** 45 31N 9 19 E
Cerqueira Cesar, *S. Pau.* **31 E5** 23 33 S 46 40W
Cerro Ajusco, *Méx.* **29 C2** 19 12N 99 15W
Cerro de la Estrella, *Méx.* **29 B3** 19 20N 99 5W
Cerro de los Angeles, *Mdrd.* **8 C2** 40 18N 3 41W
Cerro el Picacho, *Méx.* **29 A3** 19 35N 99 6W
Cerro Maggiore, *Mil.* **9 D4** 45 35N 8 57 E
Certanovka →, *Mos.* **11 F9** 55 38N 37 48 E
Certanovo, *Mos.* **11 F9** 55 38N 37 36 E
Cesano Boscone, *Mil.* **9 E5** 45 26N 9 5 E
Cesate, *Mil.* **9 D5** 45 35N 9 4 E
Cha Kwo Ling, *H.K.* **12 E6** 22 18N 114 13 E
Chabot, L., *S.F.* **27 B4** 37 43N 122 6W
Chacarilla, *Lima* **30 G9** 12 6 S 76 59W
Chadds Ford, *Phil.* **24 B1** 39 52N 75 35W
Chadstone, *Melb.* **19 F7** 37 52 S 145 5 E
Chadwell Heath, *Lon.* **4 B5** 51 34N 0 8 E
Chadwell St. Mary, *Lon.* **4 C7** 51 29N 0 22 E
Chai Wan, *H.K.* **12 E6** 22 16N 114 14 E
Chai Wan Kok, *H.K.* **12 D5** 22 22N 114 6 E
Chakdaha, *Calc.* **16 F5** 22 28N 88 19 E
Chama, *Lima* **30 G8** 12 7 S 77 0W
Chamartin, *Mdrd.* **8 B2** 40 27N 3 40W
Chamberi, *Mdrd.* **8 B2** 40 26N 3 42W
Chambourcy, *Paris* **5 B2** 48 54N 2 2 E
Champdani, *Calc.* **16 D5** 22 48N 88 19 E
Champigny-sur-Marne, *Paris* **5 C5** 48 49N 2 30 E
Champlain, Pont, *Mtrl.* **20 B4** 43 28N 73 31W
Champlan, *Paris* **5 C3** 48 42N 2 16 E
Champrosay, *Paris* **5 D4** 48 39N 2 25 E
Champs-sur-Marne, *Paris* **5 B5** 48 50N 2 34 E
Chamrail, *Calc.* **16 E5** 22 38N 88 17 E
Chancheon, *Sŏul* **12 G7** 37 33N 126 56 E
Chandernagore, *Calc.* **16 C6** 22 51N 88 21 E
Chanditala, *Calc.* **16 D5** 22 43N 88 15 E
Changi, *Sing.* **15 F8** 1 23N 103 59 E
Changi Airport, *Sing.* **15 F8** 1 21N 103 58 E
Changning, *Shang.* **14 J11** 31 13N 121 24 E
Changpu Gongyuan, *Shang.* **14 J12** 31 17N 121 31 E
Chanteloup-les-Vignes, *Paris* **5 B2** 48 58N 2 1 E
Chantereine, *Paris* **5 B5** 48 53N 2 37 E
Chantian, *Gzh.* **14 F8** 23 12N 113 16 E
Chao Phraya →, *Bangk.* **15 B2** 13 40N 100 31 E
Chaoyang, *Beij.* **14 B3** 39 53N 116 28 E
Chaoyang Gongyuan, *Beij.* **14 B3** 39 54N 116 28 E
Chaoyangmen, *Beij.* **14 B3** 39 54N 116 23 E
Chapel End, *Lon.* **4 B4** 51 35N 0 1W
Chapet, *Paris* **5 B1** 48 58N 1 55 E
Chaplinville, *Bost.* **21 A4** 42 42N 70 54W
Chapultepec, Bosque de, *Méx.* **29 B2** 19 25N 99 11W
Chapultepec, Castillo de, *Méx.* **29 B2** 19 25N 99 10W
Charenton-le-Pont, *Paris* **5 C4** 48 49N 2 25 E
Charles-de-Gaulle, Aéroport, *Paris* **5 A5** 49 0N 2 33 E
Charles Lee Tinden Regional Park, *S.F.* **27 A3** 37 53N 122 14W
Charleston, *N.Y.* **22 D3** 40 32N 74 14W
Charlestown, *Bost.* **21 C3** 42 22N 71 4W
Charlottenburg, *Berl.* **7 A2** 52 31N 13 14 E
Charlottenburg, Schloss, *Berl.* **7 A2** 52 31N 13 18 E
Charlottenlund, *Købn.* **2 D10** 55 44N 12 35 E
Charlton, *Lon.* **4 C5** 51 29N 0 1 E
Charneca, *Lisb.* **8 F8** 38 47N 9 8W
Charneca, *Lisb.* **8 G7** 38 37N 9 12W
Chase Side, *Lon.* **4 B4** 51 39N 0 4W
Châteaufort, *Paris* **5 C2** 48 44N 2 5 E
Châtenay-Malabry, *Paris* **5 C3** 48 46N 2 16 E
Chatham, *Chic.* **26 C3** 41 45N 87 36W
Chatham, *N.Y.* **22 C2** 40 44N 74 23W
Châtillon, *Paris* **5 C3** 48 48N 2 17 E
Chatou, *Paris* **5 B2** 48 53N 2 9 E
Chatpur, *Calc.* **16 E6** 22 36N 88 22 E
Chatra, *Calc.* **16 D5** 22 45N 88 19 E
Chatswood, *Syd.* **19 A4** 33 47 S 151 11 E
Chauki, *Kar.* **17 G10** 24 55N 66 56 E
Chavarria, *Lima* **30 G8** 12 0 S 77 7W
Chavenay, *Paris* **5 B1** 48 51N 1 59 E
Chavenay-Villepreux, Aérodrôme de, *Paris* **5 B1** 48 50N 1 58 E
Chaville, *Paris* **5 C3** 48 48N 2 11 E
Che Kung Miu, *H.K.* **12 D6** 22 22N 114 10 E
Cheam, *Lon.* **4 C3** 51 21N 0 12W
Chelles, *Paris* **5 B5** 48 53N 2 35 E
Chelles, Canal de, *Paris* **5 B5** 48 51N 2 35 E
Chells-le-Pin, Aérodrome, *Paris* **5 B5** 48 53N 2 36 E
Chelmsford, *Bost.* **21 B1** 42 35N 71 20W
Chelobityevo, *Mos.* **11 D10** 55 54N 37 40 E
Chelsea, *Bost.* **21 C3** 42 23N 71 1W
Chelsea, *Lon.* **4 C3** 51 29N 0 10W
Chelsea, *Phil.* **24 B2** 39 51N 75 27W
Chelsfield Village, *Lon.* **4 C5** 51 21N 0 8 E
Cheltenham, *Phil.* **24 A4** 40 3N 75 6W
Chembur, *Bomb.* **16 G8** 19 3N 72 53 E
Chennevières, *Paris* **5 A2** 49 0N 2 6 E
Chennevières-sur-Marne, *Paris* **5 C5** 48 47N 2 31 E
Cheongdam, *Sŏul* **12 G8** 37 31N 127 2 E
Cheonho, *Sŏul* **12 G8** 37 32N 127 6 E
Cheops, *El Qâ.* **18 D4** 29 58N 31 8 E
Chepo, *Gzh.* **14 G9** 23 7N 113 23 E
Cherepkovo, *Mos.* **11 E8** 55 45N 37 21 E
Chernyovo, *Mos.* **11 D7** 55 50N 37 17 E
Cherry Hill, *Phil.* **24 B4** 39 54N 75 1W
Cherry L., *Melb.* **19 F5** 37 51 S 144 49 E
Cherryland, *S.F.* **27 B4** 37 40N 122 7W
Cherrywood, *Trto.* **20 C10** 43 51N 79 8W
Chertsey, *Lon.* **4 C2** 51 23N 0 29W
Cheryomushki, *Mos.* **11 E9** 55 40N 37 35 E
Chesaco Park, *Balt.* **25 B3** 39 18N 76 30W
Chesapeake B., *Balt.* **25 B4** 39 12N 76 22W
Cheshunt, *Lon.* **4 A4** 51 42N 0 0 E
Chess →, *Lon.* **4 B2** 51 38N 0 27W
Chessington, *Lon.* **4 C3** 51 21N 0 18W
Chessington Zoo, *Lon.* **4 C3** 51 20N 0 19W
Chester, *Phil.* **24 B2** 39 50N 75 23W
Chester Cr. →, *Phil.* **24 B2** 39 50N 75 21W
Chester Heights, *Phil.* **24 B2** 39 53N 75 27W
Chestnut, *Phil.* **24 A3** 40 4N 75 13W
Chestnut Hill, *Bost.* **21 D2** 42 19N 71 10W
Cheung Sha Wan, *H.K.* **12 D5** 22 20N 114 8 E
Cheverly, *Wash.* **25 D8** 38 55N 76 54W
Chevilly-Larue, *Paris* **5 C4** 48 46N 2 21 E
Chevreuse, *Paris* **5 C2** 48 42N 2 2 E
Chevry-Cossigny, *Paris* **5 C5** 48 43N 2 39 E
Chevy Chase, *Wash.* **25 D7** 38 59N 77 4W
Chevy Chase View, *Wash.* **25 C7** 39 0N 77 4W
Cheyney, *Phil.* **24 B1** 39 55N 75 31W
Chhalera Bangar, *Delhi* **16 B2** 28 33N 77 17 E
Chhinamor, *Calc.* **16 D5** 22 48N 88 17 E
Chhota Andai, *Kar.* **17 H11** 24 48N 66 59 E
Chia Keng, *Sing.* **15 F8** 1 21N 103 52 E
Chiaiano, *Nápl.* **9 H12** 40 53N 14 13 E
Chiaravalle Milanese, *Mil.* **9 E6** 45 24N 9 16 E
Chiawelo, *Jobg.* **18 F8** 26 17 S 27 51 E
Chicago, *Chic.* **26 B3** 41 52N 87 38W
Chicago, Univ. of, *Chic.* **26 C3** 41 47N 87 35W
Chicago Harbor, *Chic.* **26 B3** 41 52N 87 36W
Chicago Lawn, *Chic.* **26 C2** 41 47N 87 40W
Chicago-Midway Airport, *Chic.* **26 C2** 41 47N 87 45W
Chicago-O'Hare Int. Airport, *Chic.* **26 B1** 41 58N 87 53W
Chicago Ridge, *Chic.* **26 C2** 41 41N 87 46W
Chicago Sanitary and Ship Canal, *Chic.* **26 C2** 41 49N 87 45W
Chichinautzin, Cerro, *Méx.* **29 D3** 19 6N 99 8W
Chicot, *Mtrl.* **20 A2** 43 35N 73 56W
Chicot →, *Mtrl.* **20 A2** 43 34N 73 51W
Chienzui, *Gzh.* **14 F9** 23 12N 113 22 E
Chieri, *Tori.* **9 B3** 45 0N 7 49 E
Chigwell, *Lon.* **4 B5** 51 36N 0 5 E
Chigwell Row, *Lon.* **4 B5** 51 37N 0 7 E
Chik Sha, *H.K.* **12 E6** 22 17N 114 16 E
Chikumazawa, *Tōkyō* **13 B2** 35 49N 139 32 E
Childs Hill, *Lon.* **4 B3** 51 33N 0 12W
Chilla Saroda, *Delhi* **16 B2** 28 35N 77 18 E
Chillum, *Wash.* **25 D8** 38 57N 76 58W
Chilly-Mazarin, *Paris* **5 C3** 48 42N 2 17 E
Chimalhuacán, *Méx.* **29 B4** 19 25N 89 57 E
Chimalpa, *Méx.* **29 B1** 19 22N 99 19W
China, Tg., *Sing.* **15 G8** 1 14N 103 50 E
China Basin, *S.F.* **27 B2** 37 46N 122 22W
Chingford, *Lon.* **4 B4** 51 37N 0 1 E
Chingupota, *Calc.* **16 F5** 22 29N 88 14 E
Chipilly Woods, *Chic.* **26 A2** 42 8N 87 48W
Chipperfield, *Lon.* **4 A2** 51 42N 0 29W
Chipping Ongar, *Lon.* **4 A6** 51 41N 0 15 E
Chipstead, *Lon.* **4 D4** 51 17N 0 8W
Chirle, *Bomb.* **16 H9** 18 55N 73 2 E
Chirnside Park, *Melb.* **19 E8** 37 45 S 145 18 E
Chislehurst, *Lon.* **4 C5** 51 24N 0 3 E
Chislehurst West, *Lon.* **4 C5** 51 25N 0 3 E
Chiswick, *Lon.* **4 C3** 51 29N 0 16W
Chiswick House, *Lon.* **4 C3** 51 28N 0 15W
Chitlade Palace, *Bangk.* **15 B2** 13 45N 100 31 E
Chitose, *Tōkyō* **13 C2** 35 33N 139 37 E
Chiyoda-Ku, *Tōkyō* **13 B3** 35 41N 139 44 E
Chkalova, *Mos.* **11 F11** 55 39N 37 56 E
Choa Chu Kang, *Sing.* **15 F7** 1 22N 103 41 E
Choboty, *Mos.* **11 F8** 55 38N 37 20 E
Chodov u Prahy, *Pra.* **10 B3** 50 1N 14 30 E
Chōfu, *Tōkyō* **13 C2** 35 38N 139 32 E
Choisel, *Paris* **5 C2** 48 41N 2 1 E
Choisy-le-Roi, *Paris* **5 C4** 48 46N 2 24 E

Gerthe 37

Ebute-Metta, *Lagos* **18 B2** 6 28N 7 23 E
Ecatepec de Morelos, *Méx.* **29 A3** 19 35N 99 2W
Echo B., *N.Y.* **23 B6** 40 54N 73 45W
Echo Mt., *L.A.* **28 A4** 34 12N 118 8W
Écouen, *Paris* **5 A4** 49 1N 2 22 E
Ecquevilly, *Paris* **5 B1** 48 57N 1 55 E
Ecser, *Bud.* **10 K14** 47 26N 19 19 E
Eda, *Tōkyō* **13 C2** 35 33N 139 33 E
Eddington, *Phil.* **24 A5** 40 5N 74 55W
Eddystone, *Phil.* **24 B2** 39 51N 75 20W
Eden, *Rio J.* **31 A1** 22 47 S 43 23W
Edendale, *Jobg.* **18 E9** 26 8 S 28 9 E
Edenvale, *Jobg.* **18 E9** 26 8 S 28 9 E
Edgars Cr. →, *Melb.* **19 E6** 37 43 S 144 58 E
Edge Hill, *Phil.* **24 A4** 40 7N 75 9W
Edgeley, *Trto.* **20 D7** 43 47N 79 31W
Edgemar, *S.F.* **27 C2** 37 39N 122 29W
Edgemere, *Balt.* **25 B4** 39 14N 76 26W
Edgemont, *Phil.* **24 B2** 39 58N 75 26W
Edgewater Park, *Phil.* **24 A5** 40 3N 74 54W
Edgware, *Lon.* **4 B3** 51 36N 0 15W
Edison, *N.Y.* **22 D2** 40 31N 74 23W
Edison Park, *Chic.* **26 A2** 42 1N 87 48W
Edleen, *Jobg.* **18 E10** 26 5 S 28 12 E
Edmondston, *Wash.* **25 D8** 38 56N 76 54W
Edo →, *Tōkyō* **13 C4** 35 38N 139 52 E
Edogawa, *Tōkyō* **13 B4** 35 43N 139 52 E
Edsberg, *Stock.* **3 D10** 59 26N 17 57 E
Edwards L., *Melb.* **19 E6** 37 42 S 144 59 E
Eestiluoto, *Hels.* **3 C6** 60 7N 25 13 E
Egawa, *Tōkyō* **13 D4** 35 22N 139 54 E
Egenbüttel, *Hbg.* **7 D7** 53 39N 9 51 E
Eggerscheidt, *Ruhr* **6 C3** 51 19N 6 53 E
Egham, *Lon.* **4 C1** 51 25N 0 30W
Eiche, *Berl.* **7 A4** 52 33N 13 35 E
Eiche Sud, *Berl.* **7 A4** 52 33N 13 35 E
Eichlinghofen, *Ruhr* **6 B6** 51 29N 7 24 E
Eichwalde, *Berl.* **7 B4** 52 22N 13 37 E
Eidelstedt, *Hbg.* **7 D7** 53 36N 9 54 E
Eiffel, Tour, *Paris* **5 B3** 48 51N 2 17 E
Eigen, *Ruhr* **6 A3** 51 32N 6 56 E
Eilbek, *Hbg.* **7 D8** 53 34N 10 2 E
Eimsbüttel, *Hbg.* **7 D7** 53 34N 9 57 E
Eissendorf, *Hbg.* **7 E7** 53 27N 9 57 E
Ejby, *Købn.* **2 D9** 55 41N 12 24 E
Ejigbo, *Lagos* **18 A1** 6 33N 7 18 E
Ekeberg, *Oslo* **2 B4** 59 53N 10 46 E
Ekeby, *Stock.* **3 D8** 59 21N 17 35 E
Ekerö, *Stock.* **3 E9** 59 17N 17 46 E
Ekerön, *Stock.* **3 E9** 59 18N 17 41 E
Ekhtiyarieh, *Tehr.* **17 C5** 35 46N 51 28 E
Eklundshov, *Stock.* **3 E10** 59 11N 17 54 E
Eknäs, *Stock.* **3 E12** 59 18N 18 13 E
El 'Abbasiya, *El Qâ.* **18 C5** 30 3N 31 16 E
El Agustino, *Lima* **30 G8** 12 2 S 77 0W
El Alto, *Stgo* **30 J10** 33 29 S 70 42W
El Awkal, *El Qâ.* **18 C5** 30 2N 31 12 E
El Baragil, *El Qâ.* **18 C4** 30 4N 31 9 E
El Basâlîn, *El Qâ.* **18 D5** 29 58N 31 16 E
El Calvario, *La Hab.* **30 B3** 23 3N 82 19W
El Cano, *La Hab.* **30 B2** 23 2N 82 27W
El Caribe, *Car.* **30 D5** 10 36N 66 52W
El Carmen, *Stgo* **30 J10** 33 22 S 70 43W
El Cerrito, *S.F.* **27 A3** 37 54N 122 18W
El Cerro, *La Hab.* **30 B2** 23 6N 82 23W
El Cojo, Pta., *Car.* **30 D5** 10 36N 66 53W
El Cortijo, *Stgo* **30 J10** 33 22 S 70 42W
El Duqqi, *El Qâ.* **18 C5** 30 1N 31 12 E
El Gamâlîya, *El Qâ.* **18 C5** 30 2N 31 15 E
El Ghurîya, *El Qâ.* **18 C5** 30 2N 31 15 E
El Gîza, *El Qâ.* **18 C5** 30 0N 31 12 E
El Granada, *S.F.* **27 C2** 37 30N 122 27W
El Guarda Parres, *Méx.* **29 D2** 19 9N 99 11W
El Hatillo, *Car.* **30 E6** 10 25N 66 49W
El Khalîfa, *El Qâ.* **18 C5** 30 0N 31 15 E
El Kôm el Ahmar, *El Qâ.* **18 C5** 30 6N 31 10 E
El Ma'âdi, *El Qâ.* **18 D5** 29 57N 31 15 E
El Matarîya, *El Qâ.* **18 C5** 30 7N 31 18 E
El Monte, *L.A.* **28 B4** 34 3N 118 1W
El Muhît Idkû el Gharbî →, *El Qâ.* **18 C4** 30 6N 31 6 E
El Mûskî, *El Qâ.* **18 C5** 30 3N 31 15 E
El Palmar, *Car.* **30 D5** 10 36N 66 53W
El Palomar, *B.A.* **32 B3** 34 36 S 58 37W
El Pardo, *Mdrd.* **8 A2** 40 30N 3 46W
El Pedregal, *Car.* **30 D5** 10 30N 66 51W
El Pinar, *Car.* **30 E5** 10 28N 66 56W
El Plantio, *Mdrd.* **8 B1** 40 28N 3 51W
El Qâhira, *El Qâ.* **18 C5** 30 2N 31 13 E
El Qubba, *El Qâ.* **18 C5** 30 4N 31 16 E
El Recreo, *Car.* **30 E5** 10 29N 66 52W
El Reloj, *Méx.* **29 C3** 19 19N 99 9W
El Retiro, *Car.* **30 D5** 10 31N 66 54W
El Salto, *Stgo* **30 J11** 33 22 S 70 38W
El Segundo, *L.A.* **28 C2** 33 55N 118 24W
El Sereno, *L.A.* **28 B3** 34 6N 118 10 E
El Silencio, *Car.* **30 D5** 10 30N 66 55W
El Sobrante, *S.F.* **27 A3** 37 58N 122 17W
El Talar de Pacheco, *B.A.* **32 A3** 34 27 S 58 38W
El Talibîya, *El Qâ.* **18 D5** 29 59N 31 10 E
El Valle, *Car.* **30 E5** 10 29N 66 54W
El Vedado, *La Hab.* **30 B2** 23 8N 82 23W
El Vergel, *Méx.* **29 C3** 19 18N 99 5W
El Wâyli el Kubra, *El Qâ.* **18 C5** 30 5N 31 17 E
El Zamalik, *El Qâ.* **18 C5** 30 3N 31 12 E
Elam, *Phil.* **24 B1** 39 51N 75 32W
Élancourt, *Paris* **5 C1** 48 47N 1 57 E
Elandsfortein, *Jobg.* **18 E10** 26 9 S 28 13 E
Elbe →, *Hbg.* **7 D6** 53 32N 9 49 E
Elberfeld, *Ruhr* **6 C4** 51 15N 7 9 E
Elephanta Caves, *Bomb.* **16 H8** 18 57N 72 57 E
Elephanta I., *Bomb.* **16 H8** 18 57N 72 56 E
Elisenau, *Berl.* **7 A4** 52 38N 13 37 E
Elizabeth, *N.Y.* **22 D3** 40 39N 74 13W
Elkins Park, *Phil.* **24 A4** 40 4N 75 8W
Elkridge, *Balt.* **25 B2** 39 13N 76 42W
Ellboda, *Stock.* **3 D12** 59 24N 18 15 E
Eller, *Ruhr* **6 C3** 51 12N 6 51 E
Ellerbek, *Hbg.* **7 D7** 53 39N 9 52 E
Ellicott City, *Balt.* **25 B2** 39 15N 76 49W
Ellinghorst, *Ruhr* **6 A3** 51 33N 6 57 E
Ellinikón, *Ath.* **8 J11** 37 50N 23 45 E
Ellis I., *N.Y.* **22 C4** 40 41N 74 2W
Elm Park, *Lon.* **4 B6** 51 32N 0 12 E
Elmers End, *Lon.* **4 C4** 51 23N 0 2W
Elmhurst, *Chic.* **26 B1** 41 53N 87 55W
Elmhurst, *N.Y.* **23 C5** 40 44N 73 53W
Elmont, *N.Y.* **23 C6** 40 42N 73 42W
Elmstead, *Lon.* **4 C5** 51 24N 0 2 E
Elmwood, *Balt.* **25 A3** 39 20N 76 31W
Elmwood, *N.Y.* **23 B8** 40 51N 73 20W
Elmwood Park, *Chic.* **26 B2** 41 55N 87 48W
Elmwood Park, *N.Y.* **22 B4** 40 54N 74 7W
Elsburg, *Jobg.* **18 F10** 26 15 S 28 13 E
Elsburgspruit →, *Jobg.* **18 F10** 26 16 S 28 12 E
Elsmere, *Phil.* **24 C1** 39 44N 75 35W
Elspark, *Jobg.* **18 F10** 26 15 S 28 13 E
Elsternwick, *Melb.* **19 F7** 37 52 S 145 0 E
Eltham, *Lon.* **4 C5** 51 27N 0 3 E
Eltham, *Melb.* **19 E7** 37 42 S 145 9 E
Elthorn Heights, *Lon.* **4 B2** 51 31N 0 20W
Eltingrille, *N.Y.* **22 D4** 40 32N 74 9W
Elwood, *Melb.* **19 F6** 37 53 S 144 59 E
Élysée, *Paris* **5 B3** 48 52N 2 19 E
Embu, *S. Pau.* **31 E4** 23 38 S 46 50W
Embu-Mirim, *S. Pau.* **31 F5** 23 41 S 46 49W
Embu Mirim →, *S. Pau.* **31 F5** 23 43 S 46 47W
Emdeni, *Jobg.* **18 F7** 26 14 S 27 49 E
Émerainville, *Paris* **5 C5** 48 48N 2 37 E
Emerson, *N.Y.* **22 B4** 40 57N 74 2W
Emerson Park, *Lon.* **4 B6** 51 34N 0 13 E
Emeryville, *S.F.* **27 B3** 37 49N 122 17W
Eminonu, *Ist.* **17 A2** 41 0N 28 57 E
Emmarentia, *Jobg.* **18 E9** 26 9 S 28 0 E
Emperor's Palace, *Tōkyō* **13 B3** 35 40N 139 45 E
Empire State Building, *N.Y.* **22 C5** 40 44N 73 59W
Emscher →, *Ruhr* **6 A6** 51 30N 7 26 E
Emscher Bruch, *Ruhr* **6 A4** 51 33N 7 8 E
Emscher Zweigkanal, *Ruhr* **6 A4** 51 33N 7 9 E
Encantado, *Rio J.* **31 B2** 22 53 S 43 19W
Encino, *L.A.* **28 B2** 34 9N 118 28W
Encino Res., *L.A.* **28 B1** 34 8N 118 30W
Enebyberg, *Stock.* **3 D10** 59 25N 17 59 E
Enfield, *Lon.* **4 B4** 51 39N 0 4W
Enfield, *Phil.* **24 A3** 40 6N 75 11W
Enfield, *Syd.* **19 B3** 33 53 S 151 6 E
Enfield Chase, *Lon.* **4 A4** 51 40N 0 8W
Enfield Highway, *Lon.* **4 B4** 51 39N 0 2W
Enfield Lock, *Lon.* **4 A4** 51 40N 0 1W
Enfield Wash, *Lon.* **4 B4** 51 39N 0 2W
Eng Khong Gardens, *Sing.* **15 F7** 1 20N 103 46 E
Engenho, I. do, *Rio J.* **31 B3** 22 50 S 43 6W
Engenho Nôvo, *Rio J.* **31 B2** 22 53 S 43 17W
Engenho Velho, Sa. do, *Rio J.* **31 B1** 22 54 S 43 21W
Engenno do Mato, *Rio J.* **31 B3** 22 56 S 43 2W
Enghein-les-Bains, *Paris* **5 B3** 48 58N 2 18 E
Englewood, *Chic.* **26 C3** 41 46N 87 38W
Englewood, *N.Y.* **22 B5** 40 53N 73 58W
Englewood Cliffs, *N.Y.* **22 B5** 40 53N 73 59W
Englischer Garten, *Mün.* **7 G10** 48 9N 11 35 E
Enmore, *Syd.* **19 B4** 33 54 S 151 10 E
Ennepe →, *Ruhr* **6 C6** 51 17N 7 23 E
Ennepetal, *Ruhr* **6 C6** 51 17N 7 21 E
Ennepetalsp →, *Ruhr* **6 C6** 51 14N 7 24 E
Enskede, *Stock.* **3 E11** 59 17N 18 4 E
Entrevias, *Mdrd.* **8 B2** 40 22N 3 40W
Épiais-les-Louvres, *Paris* **5 A5** 49 1N 2 33 E
Epinay, *Paris* **5 B3** 48 57N 2 19 E
Epinay-sous-Sénart, *Paris* **5 C5** 48 41N 2 30 E
Epinay-sur-Orge, *Paris* **5 C3** 48 40N 2 19 E
Eppende, *Ruhr* **6 B4** 51 28N 7 9 E
Eppenhausen, *Ruhr* **6 B6** 51 22N 7 29 E
Epping, *Lon.* **4 A5** 51 41N 0 6 E
Epping, *Melb.* **19 D7** 37 39 S 145 1 E
Epping, *Syd.* **19 A3** 33 46 S 151 5 E
Epping Forest, *Lon.* **4 B5** 51 39N 0 2 E
Epsom, *Lon.* **4 D3** 51 19N 0 15W
Epsom Racecourse, *Lon.* **4 D3** 51 18N 0 15W
Éragny, *Paris* **5 A2** 49 1N 2 5 E
Ercolano, *Nápl.* **9 J13** 40 48N 14 21 E
Érd, *Bud.* **10 K12** 47 23N 18 56 E
Erdenheim, *Phil.* **24 A3** 40 5N 75 12W
Eregun, *Lagos* **18 A2** 6 35N 7 22 E
Erenköy, *Ist.* **17 B3** 40 58N 29 3 E
Ergal, *Paris* **5 C1** 48 47N 1 55 E
Erial, *Phil.* **24 C4** 39 46N 75 0W
Erith, *Lon.* **4 C6** 51 28N 0 11 E
Erkner, *Berl.* **7 B5** 52 25N 13 44 E
Erkrath, *Ruhr* **6 C3** 51 13N 6 54 E
Erlaa, *Wien* **10 H9** 48 9N 16 19 E
Erle, *Ruhr* **6 A4** 51 33N 7 4 E
Ermelino Matarazzo, *S. Pau.* **31 D7** 23 29 S 46 28W
Ermington, *Syd.* **19 A3** 33 48 S 151 4 E
Ermont, *Paris* **5 B3** 48 59N 2 15 E
Ersébet-Telep, *Bud.* **10 K14** 47 27N 19 10 E
Ershatou, *Gzh.* **14 G8** 23 6N 113 18 E
Erskineville, *Syd.* **19 B4** 33 54 S 151 12 E
Erstavik, *Stock.* **3 E12** 59 16N 18 14 E
Erstaviken, *Stock.* **3 E12** 59 16N 18 20 E
Erunkan, *Lagos* **18 A2** 6 36N 7 23 E
Eschenried, *Mün.* **7 F9** 48 13N 11 24 E
Esenler, *Ist.* **17 A2** 41 1N 28 52 E
Esher, *Lon.* **4 C2** 51 22N 0 20W
Eshratābād, *Tehr.* **17 C5** 35 42N 51 27 E
España, *B.A.* **32 C5** 34 46 S 58 14W
Espeleta, *B.A.* **32 C5** 34 45 S 58 15W
Esplugas, *Barc.* **8 D5** 41 22N 2 5 E
Espoo, *Hels.* **3 B2** 60 13N 24 38 E
Espoonlahti, *Hels.* **3 B2** 60 9N 24 31 E
Esposizione Univ. di Roma (E.U.R.), *Rome* **9 G9** 41 49N 12 28 E
Essen, *Ruhr* **6 B4** 51 27N 7 0 E
Essen-Mülheim, Flughafen, *Ruhr* **6 B3** 51 24N 6 56 E
Essendon, *Melb.* **19 E6** 37 44 S 144 55 E
Essendon Airport, *Melb.* **19 E6** 37 43 S 144 54 E
Essex, *Balt.* **25 B4** 39 18N 76 28W
Essex Falls, *N.Y.* **22 C3** 40 49N 74 16W
Essingen, *Stock.* **3 E10** 59 19N 17 59 E
Essling, *Wien* **10 G11** 48 12N 16 30 E
Est, Gare de l', *Paris* **5 B4** 48 52N 2 21 E
Estado, Parque do, *S. Pau.* **31 E6** 23 38 S 46 38W
Estby, *Hels.* **3 C1** 60 5N 24 27W
Este, Parque Nacional del, *Car.* **30 E5** 10 29N 66 50W
Esteban Echeverria, *B.A.* **32 C4** 34 48 S 58 29W
Estlotan, *Hels.* **3 C6** 60 7N 25 13 E
Estrela, Basilica da, *Lisb.* **8 F8** 38 42N 9 9W
Étiolles, *Paris* **5 D4** 48 38N 2 28 E
Etobicoke, *Trto.* **20 E7** 43 39N 79 34W
Etobicoke Cr. →, *Trto.* **20 E7** 43 35N 79 32W
Etzenhauzen, *Mün.* **7 F9** 48 16N 11 27 E
Eun Pyeong, *Sŏul* **12 G7** 37 36N 126 56 E
Eungam, *Sŏul* **12 G7** 37 34N 126 55 E
Evanston, *Chic.* **26 A2** 42 3N 87 40W
Évecquemont, *Paris* **5 A1** 49 0N 1 56 E
Everett, *Bost.* **21 C3** 42 24N 71 3W
Evergreen Park, *Chic.* **26 C2** 41 43N 87 42W
Eversael, *Ruhr* **6 A1** 51 32N 6 39 E
Evesboro, *Phil.* **24 B5** 39 54N 74 55W
Eving, *Ruhr* **6 A6** 51 33N 7 28 E
Evry, *Paris* **5 D4** 48 38N 2 26 E
Évry-les-Châteaux, *Paris* **5 D5** 48 39N 2 38 E
Évzonos, *Ath.* **8 J11** 37 55N 23 49 E
Ewin, *Tehr.* **17 C5** 35 47N 51 23 E
Ewu, *Lagos* **18 A1** 6 33N 7 19 E
Exelberg, *Wien* **10 G9** 48 14N 16 15 E
Eynsford, *Lon.* **4 C6** 51 21N 0 12 E
Eyup, *Ist.* **17 A2** 41 2N 28 55 E
Ez Zeitûn, *El Qâ.* **18 C5** 30 6N 31 18 E
Ézanville, *Paris* **5 A4** 49 1N 2 21 E
Ezeiza, *B.A.* **32 D3** 34 50 S 58 31W
Ezeiza, Aeropuerto, *B.A.* **32 C3** 34 48 S 58 32W

F

Fabreville, *Mtrl.* **20 A2** 43 33N 73 51W
Fælledparken, *Købn.* **2 D10** 55 42N 12 34 E
Fågelön, *Stock.* **3 E10** 59 18N 17 55 E
Fagersjo, *Stock.* **3 E11** 59 14N 18 4 E
Fagnano, *Mil.* **9 E4** 45 24N 8 59 E
Fahrn, *Ruhr* **6 A2** 51 30N 6 45 E
Faibano, *Nápl.* **9 H13** 40 55N 14 27 E
Fair Lawn, *N.Y.* **22 B4** 40 55N 74 7W
Fairfax, *Phil.* **24 C1** 39 47N 75 33W
Fairfax, *Wash.* **25 D6** 38 50N 77 19W
Fairfax Station, *Wash.* **25 E6** 38 48N 77 19W
Fairfield, *Melb.* **19 E7** 37 46 S 145 2 E
Fairfield, *N.Y.* **22 B3** 40 53N 74 17W
Fairfield, *Syd.* **19 B2** 33 52 S 150 56 E
Fairhaven B., *Bost.* **21 C1** 42 25N 71 21W
Fairhaven Hill, *Bost.* **21 C1** 42 26N 71 21W
Fairland, *Jobg.* **18 E8** 26 8 S 27 57 E
Fairland, *Wash.* **25 C8** 39 4N 76 57W
Fairmont Terrace, *S.F.* **27 B4** 37 42N 122 7W
Fairmount Heights, *Wash.* **25 D8** 38 54N 76 54W
Fairmount Park, *Phil.* **24 A3** 40 3N 75 13W
Fairport, *Trto.* **20 D10** 43 49N 79 4W
Fairview, *N.Y.* **22 C5** 40 48N 73 59W
Fairview, *N.Y.* **23 A6** 41 1N 73 46W
Falenica, *Wsaw.* **10 F8** 52 9N 21 12 E
Falenty, *Wsaw.* **10 F6** 52 8N 20 55 E
Falkenburg, *Berl.* **7 A4** 52 34N 13 32 E
Falkenhagen, *Berl.* **7 A1** 52 34N 13 5 E
Falkensee, *Berl.* **7 A1** 52 34N 13 4 E
Fallon, *L.A.* **28 C5** 33 59N 117 54W
Falls Church, *Wash.* **25 D6** 38 53N 77 12W
Falls Run →, *Balt.* **25 A1** 39 21N 76 52W
Falomo, *Lagos* **18 B2** 6 26N 7 26 E
Fangcun, *Gzh.* **14 G8** 23 6N 113 13 E
Fanwood, *N.Y.* **22 D2** 40 37N 74 23W
Far Rockaway, *N.Y.* **23 D6** 40 36N 73 45W
Farahābād, *Tehr.* **17 C5** 35 41N 51 29 E
Färentuna, *Stock.* **3 D8** 59 23N 17 39 E
Farforovskaya, *St-Pet.* **11 B4** 59 52N 30 27 E
Farm Pond, *Bost.* **21 D2** 42 13N 71 20W
Farmingdale, *N.Y.* **23 C8** 40 43N 73 27W
Farmsen, *Hbg.* **7 D8** 53 36N 10 8 E
Farnborough, *Lon.* **4 C5** 51 21N 0 3 E
Farningham, *Lon.* **4 C6** 51 23N 0 12 E
Farrar Pond, *Bost.* **21 C1** 42 24N 71 21W
Farrarmere, *Jobg.* **18 E10** 26 9 S 28 18 E
Farsta, *Stock.* **3 E11** 59 14N 18 5 E
Farstalandet, *Stock.* **3 E13** 59 18N 18 23 E
Farum, *Købn.* **2 D8** 55 48N 12 21 E
Farum Sø, *Købn.* **2 D9** 55 48N 12 21 E
Fasanerie-Nord, *Mün.* **7 F10** 48 11N 11 32 E
Fasangarten, *Mün.* **7 G10** 48 5N 11 36 E
Fat Tau Chau, *H.K.* **12 E6** 22 16N 114 16 E
Fatih, *Ist.* **17 A2** 41 0N 28 56 E
Favoriten, *Wien* **10 H10** 48 9N 16 23 E
Fawkner, *Melb.* **19 E6** 37 42 S 144 56 E
Fawkner Park, *Melb.* **19 F6** 37 50 S 144 58 E
Feasterville, *Phil.* **24 A4** 40 9N 75 0W
Febrero, Parque de, *B.A.* **32 B4** 34 36 S 58 25W
Feijó, *Lisb.* **8 G8** 38 39N 9 9W
Feldersbruch →, *Ruhr* **6 B5** 51 23N 7 4 E
Feldhausen, *Ruhr* **6 A3** 51 36N 6 58 E
Feldkirchen, *Mün.* **7 G11** 48 8N 11 43 E
Feldmoching, *Mün.* **7 F10** 48 14N 11 32 E
Fellowship, *Phil.* **24 B5** 39 56N 74 57W
Feltham, *Lon.* **4 C2** 51 26N 0 24W
Feltonville, *Phil.* **24 A4** 40 1N 75 8W
Fenerbahce, *Ist.* **17 B3** 40 58N 29 2 E
Fengtai, *Beij.* **14 C2** 39 49N 116 14 E
Fenino, *Mos.* **11 E11** 55 43N 37 56 E
Ferencváros, *Bud.* **10 K13** 47 29N 19 5 E
Ferihegyi Airport, *Bud.* **10 K14** 47 26N 19 14 E
Ferndale, *Balt.* **25 B3** 39 11N 76 38W
Ferndale, *Jobg.* **18 E9** 26 5 S 28 0 E
Ferntree Gully, *Melb.* **19 F8** 37 52 S 145 17 E
Ferntree Gully Nat. Park, *Melb.* **19 F8** 37 52 S 145 19 E
Ferny Cr. →, *Melb.* **19 F8** 37 54 S 145 16 E
Férolles-Attilly, *Paris* **5 C5** 48 44N 2 37 E
Ferraz de Vasconcelos, *S. Pau.* **31 E7** 23 32 S 46 22W
Ferrières-en-Brie, *Paris* **5 C6** 48 49N 2 42 E
Ferry, *N.Y.* **23 A5** 41 0N 73 52W
Fetcham, *Lon.* **4 D2** 51 17N 0 21W
Feucherolles, *Paris* **5 B1** 48 52N 1 58 E
Fichtenau, *Berl.* **7 B5** 52 27N 13 42 E
Fields Corner, *Bost.* **21 D3** 42 18N 71 3W
Fiera Camp, *Mil.* **9 E5** 45 28N 9 9 E
Figino, *Mil.* **9 E5** 45 29N 9 4 E
Fijir, *Bagd.* **17 E8** 33 21N 44 21 E
Filadhélfia, *Ath.* **8 H11** 38 2N 23 43 E
Fili-Masilovo, *Mos.* **11 E8** 55 44N 37 29 E
Filothei, *Ath.* **8 H11** 38 2N 23 46 E
Finaalspan, *Jobg.* **18 F10** 26 16 S 28 16 E
Finchley, *Lon.* **4 B3** 51 36N 0 11W
Finkenkrug, *Berl.* **7 A1** 52 33N 13 3 E
Finkenwerder, *Hbg.* **7 D7** 53 32N 9 51 E
Finsbury, *Lon.* **4 B4** 51 31N 0 6W
Finsbury Park, *Lon.* **4 B4** 51 34N 0 6W
Fiorito, *B.A.* **32 C4** 34 42 S 58 26W
Firdows, *Bagd.* **17 F7** 33 17N 44 17 E
Firōz Bahram, *Tehr.* **17 D4** 35 37N 51 14 E
Fischeln, *Ruhr* **6 C1** 51 18N 6 35 E
Fish Brook →, *Bost.* **21 B3** 42 39N 71 1W
Fishermans Bend, *Melb.* **19 E6** 37 49 S 144 55 E
Fisher's Hill, *Jobg.* **18 F10** 26 10 S 28 10 E
Fisherville, *Trto.* **20 D8** 43 46N 79 28W
Fisksätra, *Stock.* **3 E12** 59 17N 18 13 E
Fittja, *Stock.* **3 E10** 59 14N 17 51 E
Fitzroy Gardens, *Melb.* **19 E6** 37 48 S 144 58 E
Five Cowrie Cr. →, *Lagos* **18 B2** 6 26N 7 25 E
Five Dock, *Syd.* **19 B3** 33 52 S 151 8 E
Fjellstrand, *Oslo* **2 C3** 59 47N 10 36 E
Flachsberg, *Ruhr* **6 C4** 51 11N 7 4 E
Flag →, *Chic.* **26 C1** 41 43N 87 55W
Flamengo, *Rio J.* **31 B2** 22 56 S 43 11W
Flaminio, *Rome* **9 F9** 41 55N 12 28 E
Flaskebekk, *Oslo* **2 B3** 59 51N 10 39 E
Flatbush, *N.Y.* **22 D5** 40 39N 73 56W
Flaten, *Stock.* **3 E11** 59 15N 18 9 E
Flemington, *Syd.* **19 B3** 33 53 S 151 4 E
Flemington Racecourse, *Melb.* **19 E6** 37 47 S 144 54 E
Fleury-Mérogis, *Paris* **5 D4** 48 37N 2 21 E
Flingern, *Ruhr* **6 C2** 51 13N 6 48 E
Flint Pk., *L.A.* **28 B3** 34 9N 118 11 E
Floral Park, *N.Y.* **23 C6** 40 43N 73 42W
Florence, *L.A.* **28 C3** 33 57N 118 13W
Florence, *Phil.* **24 C5** 39 44N 74 55W
Florence Bloom Bird Sanctuary, *Jobg.* **18 E9** 26 7 S 28 0 E
Florencio Varela, *B.A.* **32 C5** 34 49 S 58 18W
Florentia, *Jobg.* **18 F9** 26 16 S 28 8 E
Flores, *B.A.* **32 B4** 34 38 S 58 27W
Floresta, *B.A.* **32 B4** 34 37 S 58 27W
Florham Park, *N.Y.* **22 C2** 40 46N 74 23W
Florida, *B.A.* **32 B4** 34 31 S 58 28W
Florida, *Jobg.* **18 F8** 26 10 S 27 55 E
Florida L., *Jobg.* **18 F8** 26 10 S 27 54 E
Floridsdorf, *Wien* **10 G10** 48 15N 16 26 E
Flourtown, *Phil.* **24 A3** 40 6N 75 13W
Flower Hill, *N.Y.* **23 C6** 40 48N 73 40W
Flushing, *N.Y.* **23 C6** 40 45N 73 49W
Flushing Meadows Corona Park, *N.Y.* **23 C5** 40 44N 73 50W
Flysta, *Stock.* **3 D10** 59 22N 17 54 E
Fo Tan, *H.K.* **12 D6** 22 23N 114 11 E
Föhrenhain, *Wien* **10 G10** 48 19N 16 26 E
Folcroft, *Phil.* **24 B3** 39 53N 75 16W
Folsom, *Phil.* **24 B3** 39 53N 75 19W
Fontainebleau, *Jobg.* **18 E8** 26 6 S 27 57 E
Fontana, *La Hab.* **30 B2** 23 1N 82 24W
Fontanka, *St-Pet.* **11 B3** 59 54N 30 16 E
Fontenay-aux-Roses, *Paris* **5 C3** 48 47N 2 17 E
Fontenay-le-Fleury, *Paris* **5 C2** 48 48N 2 2 E
Fontenay-lès-Briis, *Paris* **5 D2** 48 37N 2 9 E
Fontenay-sous-Bois, *Paris* **5 B4** 48 51N 2 28 E
Foots Cray, *Lon.* **4 C5** 51 24N 0 7 E
Footscray, *Melb.* **19 E6** 37 48 S 144 56 E
Forbidden City, *Beij.* **14 B3** 39 53N 116 21 E
Fordham Univ., *N.Y.* **23 B5** 40 51N 73 51W
Fords, *N.Y.* **22 D3** 40 31N 74 19W
Fordsburg, *Jobg.* **18 F9** 26 12 S 28 2 E
Foremans Corner, *Balt.* **25 B3** 39 11N 76 33W
Forest Gate, *Lon.* **4 B5** 51 32N 0 1 E
Forest Heights, *Wash.* **25 E7** 38 48N 77 0W
Forest Hill, *Lon.* **4 C4** 51 26N 0 2W
Forest Hill, *Melb.* **19 F8** 37 50 S 145 10 E
Forest Hill, *Trto.* **20 D8** 43 41N 79 25W
Forest Hills, *N.Y.* **23 C5** 40 42N 73 51W
Forest Park, *Chic.* **26 B2** 41 51N 87 47W
Forest View, *Chic.* **26 C2** 41 48N 87 47W
Forestville, *Syd.* **19 A4** 33 46 S 151 12 E
Forestville, *Wash.* **25 D8** 38 50N 76 52W
Forges-les-Bains, *Paris* **5 D2** 48 37N 2 5 E
Fornacino, *Tori.* **9 B3** 45 9N 7 44 E
Fornebu, *Oslo* **2 B3** 59 53N 10 36 E
Fornebu Airport, *Oslo* **2 B3** 59 56N 10 37 E
Foro Italico, *Rome* **9 F9** 41 56N 12 26 E
Foro Romano, *Rome* **9 F9** 41 53N 12 29 E
Forst Rantzau, *Hbg.* **7 C6** 53 43N 9 49 E
Forstenried, *Mün.* **7 G9** 48 5N 11 29 E
Forstenrieder Park, *Mün.* **7 G9** 48 3N 11 27 E
Fort du Pont Park, *Wash.* **25 D8** 38 52N 76 56W
Fort Foote Village, *Wash.* **25 E7** 38 46N 77 1W
Fort Howard, *Balt.* **25 B4** 39 12N 76 26W
Fort Lee, *N.Y.* **22 B5** 40 50N 73 58W
Fort McHenry Nat. Mon., *Balt.* **25 B3** 35 15N 76 35W
Fort Washington, *Phil.* **24 A3** 40 8N 75 13W
Fort William, *Calc.* **16 E6** 22 33N 88 20 E
Foster City, *S.F.* **27 C3** 37 33N 122 15W
Fosters Pond, *Bost.* **21 B3** 42 36N 71 8W
Fourcherolle, *Paris* **5 C1** 48 42N 1 58 E
Fourmile Run →, *Wash.* **25 D7** 38 50N 77 2W
Fourqueux, *Paris* **5 B2** 48 53N 2 3 E
Fowl Meadow Res., *Bost.* **21 D3** 42 13N 71 8W
Fox Chase, *Phil.* **24 A4** 40 4N 75 5W
Foxhall, *Wash.* **25 C7** 39 4N 77 3W
Framingham, *Bost.* **21 D1** 42 18N 71 24W
Francisco Alvarez, *B.A.* **32 B1** 34 38 S 58 50W
Francisquito Cr. →, *S.F.* **27 D4** 37 27N 122 9W
Franconia, *Wash.* **25 E7** 38 47N 77 7W
Franconville, *Paris* **5 B3** 48 59N 2 13 E
Francop, *Hbg.* **7 D7** 53 30N 9 51 E
Frankel, *Sing.* **15 G8** 1 18N 103 55 E
Frankford, *Phil.* **24 A4** 40 1N 75 5W
Franklin L., *N.Y.* **22 B3** 40 59N 74 13W
Franklin Lakes, *N.Y.* **22 B3** 40 59N 74 13W
Franklin Park, *Bost.* **21 D3** 42 18N 71 5W
Franklin Park, *Chic.* **26 B1** 41 55N 87 52W
Franklin Park, *Wash.* **25 D7** 38 55N 77 9W
Franklin Res., *L.A.* **28 B2** 34 5N 118 24W
Franklin Roosevelt Park, *Jobg.* **18 E8** 26 8 S 27 59 E
Franklin Roosevelt Park, *Phil.* **24 B3** 39 54N 75 10W
Franklin Square, *N.Y.* **23 C6** 40 42N 73 40W
Frattamaggiore, *Nápl.* **9 H12** 40 56N 14 16 E
Frauenkirche, *Mün.* **7 G10** 48 8N 11 34 E
Frederiksberg, *Købn.* **2 D10** 55 40N 12 33 E
Frederiksdal, *Købn.* **2 D9** 55 46N 12 25 E
Fredersdorf, *Berl.* **7 A5** 52 31N 13 45 E
Fredersdorf Nord, *Berl.* **7 A5** 52 32N 13 45 E
Freeport, *N.Y.* **23 D7** 40 39N 73 35W
Freidrichshain, Volkspark, *Berl.* **7 A3** 52 31N 13 25 E
Freiham, *Mün.* **7 G9** 48 8N 11 25 E
Freimann, *Mün.* **7 F10** 48 11N 11 37 E
Fremont, *S.F.* **27 C4** 37 33N 122 2W
Fresh Meadows, *N.Y.* **23 C6** 40 43N 73 47W
Fresh Pond, *Bost.* **21 C3** 42 22N 71 8W
Freskati, *Stock.* **3 D11** 59 22N 18 3 E
Fresnes, *Paris* **5 C3** 48 45N 2 18 E
Fretay, *Paris* **5 C3** 48 40N 2 12 E
Freudenau, *Wien* **10 G10** 48 11N 16 25 E
Friedenau, *Berl.* **7 B3** 52 28N 31 20 E
Friederikenhof, *Berl.* **7 B3** 52 23N 13 21 E
Friedrichsfeld, *Ruhr* **6 A1** 51 37N 6 39 E
Friedrichsfelde, *Berl.* **7 B4** 52 29N 13 33 E
Friedrichshagen, *Berl.* **7 B4** 52 27N 13 37 E
Friedrichshain, *Berl.* **7 A3** 52 31N 13 25 E
Friedrichshulde, *Hbg.* **7 D7** 53 36N 9 51 E
Friedrichslust, *Berl.* **7 A5** 52 33N 13 43 E
Frielas, *Lisb.* **8 F8** 38 49N 9 6W
Friemersheim, *Ruhr* **6 B2** 51 23N 6 42 E
Friern Barnet, *Lon.* **4 B4** 51 37N 0 9W
Friherrs, *Hels.* **3 B3** 60 16N 24 49 E
Frogner, *Oslo* **2 A6** 60 1N 11 6 E
Frohnau, *Berl.* **7 A2** 52 38N 13 16 E
Frohnhausen, *Ruhr* **6 B3** 51 26N 6 56 E
Frontón, I., *Lima* **30 G7** 12 7 S 77 11W
Frunze, *Mos.* **11 E9** 55 47N 37 33 E
Fuchū, *Tōkyō* **13 B1** 35 40N 139 29 E
Fuencarral, *Mdrd.* **8 B2** 40 29N 3 42W
Fuhlenbrock, *Ruhr* **6 A3** 51 32N 6 54 E
Fuhlsbüttel, *Hbg.* **7 D8** 53 37N 10 1 E
Fujidera, *Ōsaka* **12 C4** 34 34N 135 36 E
Fujikubo, *Tōkyō* **13 B2** 35 49N 139 31 E
Fujimi, *Tōkyō* **13 A2** 35 50N 139 33 E
Fukagawa, *Tōkyō* **13 C3** 35 39N 139 48 E
Fukami, *Tōkyō* **13 D1** 35 28N 139 27 E
Fukiai, *Ōsaka* **12 B2** 34 42N 135 12 E
Fukuoka, *Tōkyō* **13 A2** 35 52N 139 31 E
Fukushima, *Ōsaka* **12 B3** 34 41N 135 28 E
Fulatani, *Tōkyō* **13 D1** 35 22N 139 30 E
Fulham, *Lon.* **4 C3** 51 28N 0 12W
Fuller Park, *L.A.* **28 C5** 33 51N 117 56W
Fullerton, *Balt.* **25 A3** 39 22N 76 30W
Funabori, *Tōkyō* **13 B4** 35 41N 139 52 E
Funasaka, *Ōsaka* **12 B2** 34 48N 135 16 E
Fünfhaus, *Wien* **10 G10** 48 11N 16 20 E
Fünfhausen, *Hbg.* **7 E8** 53 27N 10 2 E
Furesø, *Købn.* **2 D9** 55 47N 12 25 E
Fürstenried, *Mün.* **7 G9** 48 5N 11 28 E
Furth, *Mün.* **7 G10** 48 2N 11 35 E
Furu →, *Tōkyō* **13 A3** 35 54N 139 49 E
Furuyakami, *Tōkyō* **13 A2** 35 54N 139 31 E
Futago-tamagawaen, *Tōkyō* **13 C2** 35 36N 139 39 E
Futamatagawa, *Tōkyō* **13 D2** 35 27N 139 31 E
Futatsubashi, *Tōkyō* **13 D1** 35 28N 139 29 E
Fuxing Dao, *Shang.* **14 J12** 31 16N 121 33 E
Fuxing Gongyuan, *Shang.* **14 J11** 31 13N 121 27 E
Fuxinglu, *Beij.* **14 B2** 39 52N 116 16 E
Fuxingmen, *Beij.* **14 B2** 39 53N 116 19 E

G

Gadstrup, *Købn.* **2 E7** 55 34N 12 5 E
Gaebong, *Sŏul* **12 H7** 37 29N 126 52 E
Gage Park, *Chic.* **26 C2** 41 47N 87 42W
Gagny, *Paris* **5 B5** 48 53N 2 32 E
Gaillon, *Paris* **5 A1** 49 1N 1 53 E
Galata, *Ist.* **17 A2** 41 1N 28 58 E
Galátsion, *Ath.* **8 H11** 38 1N 23 45 E
Galeâo, *Rio J.* **31 A2** 22 49 S 43 14W
Galéria →, *Rome* **9 F9** 41 57N 12 20 E
Gallows Corner, *Lon.* **4 B6** 51 35N 0 13 E
Gällstaö, *Stock.* **3 E10** 59 17N 17 51 E
Galyanovo, *Mos.* **11 E10** 55 48N 37 47 E
Galyeon, *Sŏul* **12 G7** 37 36N 126 55 E
Gambir, *Jak.* **15 H9** 6 9 S 106 48 E
Gamboa, *Rio J.* **31 B2** 22 53 S 43 11W
Gambolóita, *Mil.* **9 E6** 45 26N 9 13 E
Gamelinha →, *S. Pau.* **31 E6** 23 31 S 46 31W
Gamlebyen, *Oslo* **2 B4** 59 54N 10 46 E
Gamlebyen, *Shang.* **14 J11** 31 13N 121 29 E
Gamō, *Tōkyō* **13 A3** 35 52N 139 48 E
Gang Dong, *Sŏul* **12 G8** 37 30N 127 5 E
Gang Nam, *Sŏul* **12 G7** 37 30N 126 59 E
Gang Sea, *Sŏul* **12 G7** 37 32N 126 51 E
Gangadharpur, *Calc.* **16 E5** 22 35N 88 11 E
Gangtou, *Gzh.* **14 F7** 23 12N 113 8 E
Gangwei, *Gzh.* **14 G8** 23 4N 113 11 E
Ganløse, *Købn.* **2 D8** 55 47N 12 15 E
Ganløse Orned, *Købn.* **2 D8** 55 48N 12 18 E
Ganshi, *Gzh.* **14 F7** 23 10N 113 8 E
Gants Hill, *Lon.* **4 B5** 51 34N 0 4 E
Gaoqiao, *Shang.* **14 H12** 31 21N 121 34 E
Garbagnate Milanese, *Mil.* **9 D5** 45 34N 9 4 E
Garbatella, *Rome* **9 F10** 41 51N 12 30 E
Garches, *Paris* **5 B3** 48 50N 2 11 E
Garching, *Mün.* **7 F11** 48 14N 11 39 E
Garden City, *El Qâ.* **18 C5** 30 2N 31 14 E
Garden City, *N.Y.* **23 C7** 40 43N 73 37W
Garden Reach, *Calc.* **16 E5** 22 33N 88 15 E
Gardena, *L.A.* **28 C3** 33 53N 118 18W
Garder, *Oslo* **2 C3** 59 45N 10 38 E
Garfield, *N.Y.* **22 B4** 40 52N 74 7W
Garfield Park, *Chic.* **26 B2** 41 52N 87 42W
Gargareta, *Ath.* **8 J11** 37 57N 23 43 E
Garges-lès-Gonesse, *Paris* **5 B4** 48 58N 2 25 E
Garhi Naraina, *Delhi* **16 B1** 28 37N 77 8 E
Garibong, *Sŏul* **12 H7** 37 29N 126 54 E
Garin, *B.A.* **32 A2** 34 25 S 58 44W
Gariya, *Calc.* **16 F6** 22 28N 88 23 E
Garji, *Calc.* **16 C5** 22 50N 88 19 E
Garne, *Paris* **5 C1** 48 41N 1 58 E
Garrison, *Balt.* **25 A2** 39 24N 76 45W
Garstedt, *Hbg.* **7 C7** 53 40N 9 59 E
Gartenstadt, *Ruhr* **6 B6** 51 30N 7 30 E
Garulia, *Calc.* **16 D6** 22 48N 88 22 E
Garvanza, *L.A.* **28 B3** 34 6N 118 11 E
Garwood, *N.Y.* **22 D3** 40 38N 74 18W
Gary, *Chic.* **26 D4** 41 35N 87 23W
Gåshaga, *Stock.* **3 D12** 59 21N 18 13 E
Gássino Torinese, *Tori.* **9 B3** 45 7N 7 49 E
Gästerby, *Hels.* **3 C1** 60 8N 24 27 E
Gateão, Aéroporto de, *Rio J.* **31 A2** 22 49 S 43 15W
Gateway of India, *Bomb.* **16 H8** 18 55N 72 50 E
Gatow, *Berl.* **7 B1** 52 29N 13 11 E
Gaurhati, *Calc.* **16 D6** 22 48N 88 21 E
Gauripur, *Calc.* **16 C6** 22 53N 88 25 E
Gavá, *Barc.* **8 E5** 41 18N 2 0 E
Gavamar, *Barc.* **8 E4** 41 16N 1 58 E
Gavanpada, *Bomb.* **16 H9** 18 57N 73 0 E
Gávea, *Rio J.* **31 B2** 22 58 S 43 13W
Gávea, Pedra da, *Rio J.* **31 B2** 22 59 S 43 18W
Gbogbo, *Lagos* **18 A3** 6 35N 7 31 E
Gebel el Ahmar, *El Qâ.* **18 C5** 30 2N 31 19 E
Gebel el Muqattam, *El Qâ.* **18 C5** 30 1N 31 17 E
Gebel et Tura, *El Qâ.* **18 D5** 29 56N 31 15 E
Geduld Dam, *Jobg.* **18 F11** 26 12 S 28 24 E
Geiselgasteig, *Mün.* **7 G10** 48 3N 11 33 E
Geist Res., *Phil.* **24 B2** 39 57N 75 24W
Gellért hegy, *Bud.* **10 K13** 47 29N 19 3 E
Gelsenkirchen, *Ruhr* **6 A4** 51 32N 7 2 E
General Ignacio Allende, *Méx.* **29 B1** 19 20N 99 21W
General Pacheco, *B.A.* **32 A3** 34 27 S 58 36W
General San Martin, *B.A.* **32 B3** 34 35 S 58 32W
General Sarmiento, *B.A.* **32 B2** 34 32 S 58 43W
General Urquiza, *B.A.* **32 B4** 34 34 S 58 28W
Gennebreck, *Ruhr* **6 C5** 51 18N 7 12 E
Gennevilliers, *Paris* **5 B3** 48 56N 2 17 E
Gentilly, *Paris* **5 C4** 48 48N 2 21 E
Gentofte, *Købn.* **2 D10** 55 44N 12 32 E
Georges →, *Syd.* **19 B2** 33 56 S 150 55 E
Georges Hall, *Syd.* **19 B2** 33 54 S 150 59 E
Georges I., *Bost.* **21 D4** 42 19N 70 55W
Georges River Bridge, *Syd.* **19 C3** 34 0 S 151 6 E
Georgetown, *Wash.* **25 D7** 38 54N 77 3W
Georgetown Rowley State Forest, *Bost.* **21 A4** 42 41N 70 56W
Georgsweide, *Hbg.* **7 D8** 53 30N 10 1 E
Gerasdorf bei Wein, *Wien* **10 G10** 48 17N 16 28 E
Gerberau, *Mün.* **7 F9** 48 12N 11 27 E
Gérbido, *Tori.* **9 B2** 45 2N 7 36 E
Gerlev, *Købn.* **2 D7** 55 49N 12 0 E
Gerli, *B.A.* **32 C4** 34 41 S 58 22W
Germantown, *Balt.* **25 A4** 39 24N 76 28W
Germantown, *Phil.* **24 A3** 40 2N 75 11W
Germiston, *Jobg.* **18 F9** 26 13 S 28 10 E
Gerresheim, *Ruhr* **6 C3** 51 14N 6 51 E
Gersthof, *Wien* **10 G9** 48 14N 16 18 E
Gerthe, *Ruhr* **6 A5** 51 31N 7 16 E

38 Gesîrat el Rauda

Gesîrat el Rauda, El Qâ. 18 C5 30 1N 31 13 E
Gesîrat Muhammad, El Qâ. 18 C5 30 6N 31 11 E
Gesterby, Hels. 3 A6 60 20N 25 17 E
Getafe, Mdrd. 8 C2 40 18N 3 43W
Gevelsberg, Ruhr 6 C6 51 19N 7 21 E
Geylang, Sing. 15 G8 1 18N 103 53 E
Geylang →, Sing. 15 G8 1 18N 103 52 E
Geylang Serai, Sing. 15 G8 1 19N 103 53 E
Gezîrat edn Dhahab, El Qâ. 18 D5 29 59N 31 13 E
Gezîrat Warrâq el Hadar, El Qâ. 18 C5 30 6N 31 13 E
Gharapuri, Bomb. 16 H8 18 57N 72 57 E
Ghatkopar, Bomb. 16 G8 19 4N 72 54 E
Ghazipur, Delhi 16 B2 28 37N 77 19 E
Ghizri, Kar. 17 H11 24 49N 67 2 E
Ghizri Cr. →, Kar. 17 H11 24 47N 67 5 E
Ghonda, Delhi 16 A2 28 41N 77 16 E
Ghushuri, Calc. 16 E6 22 37N 88 21 E
Gianicolense, Rome 9 F9 51 53N 12 28 E
Giant, S.F. 27 A2 37 58N 122 20W
Gibbsboro, Phil. 24 B5 39 50N 74 57W
Gibbstown, Phil. 24 C3 39 49N 75 17W
Gibraltar Pt., Trto. 20 E8 43 36N 79 23W
Gidea Park, Lon. 4 B6 51 35N 0 11 E
Giesing, Mün. 7 G10 48 6N 11 35 E
Gif-sur-Yvette, Paris 5 C2 48 42N 2 8 E
Gilgo Beach, N.Y. 23 D8 40 36N 73 24W
Gilgo I., N.Y. 23 D8 40 37N 73 23W
Gillette, N.Y. 22 C2 40 40N 74 29W
Gimmersta, Stock. 3 E12 59 14N 18 14 E
Ginza, Tōkyō 13 C3 35 39N 139 46 E
Girgaum, Bomb. 16 H8 18 57N 72 50 E
Giugliano in Campània, Nápl. 9 H12 40 55N 14 12 E
Givoletto, Tori. 9 B1 45 9N 7 29 E
Gjellumvatn, Oslo 2 C2 59 47N 10 26 E
Gjersjøen, Oslo 2 C4 59 47N 10 47 E
Glacier Hills, N.Y. 22 B2 40 51N 74 28W
Gladbeck, Ruhr 6 A3 51 34N 6 58 E
Gladesville, Syd. 19 B3 33 50 S 151 8 E
Gladökvarn, Stock. 3 E10 59 11N 17 59 E
Gladsakse, Købn. 2 D9 55 45N 12 25 E
Glashutte, Hbg. 7 C8 53 41N 10 2 E
Glashütte, Ruhr 6 C3 51 13N 6 51 E
Glasmoor, Hbg. 7 C8 53 42N 10 1 E
Glassmanor, Wash. 25 E7 38 49N 77 0W
Glen Cove, N.Y. 23 B7 40 52N 73 38W
Glen Echo, Wash. 25 D7 38 58N 77 8W
Glen Hd., N.Y. 23 C7 40 49N 73 37W
Glen Iris, Melb. 19 F7 37 51 S 145 3 E
Glen Mills, Phil. 24 B2 39 55N 75 29W
Glen Oaks, N.Y. 23 C6 40 45N 73 43W
Glen Riddle, Phil. 24 B2 39 53N 75 26W
Glen Ridge, N.Y. 22 C3 40 48N 74 12W
Glen Rock, N.Y. 22 B4 40 57N 74 7W
Glen Waverley, Melb. 19 F8 37 52 S 145 10 E
Glenardon, Wash. 25 D8 38 56N 76 51W
Glencoe, Chic. 26 A2 42 7N 87 44W
Glendale, L.A. 28 B3 34 9N 118 15 E
Glendora, Phil. 24 B4 39 50N 75 4W
Glenfield, Syd. 19 B2 33 58 S 150 53 E
Glenhazel, Jobg. 18 E9 26 8 S 28 6 E
Glenhuntly, Melb. 19 F7 37 52 S 145 1 E
Glenmont, Wash. 25 C7 39 3N 77 4W
Glenolden, Phil. 24 B3 39 54N 75 17W
Glenroy, Melb. 19 E6 37 42 S 144 55 E
Glenside, Phil. 24 A4 40 6N 75 9W
Glenview, Chic. 26 A2 42 3N 87 48W
Glenview Countryside, Chic. 26 A2 42 3N 87 49W
Glenview Woods, Chic. 26 A2 42 4N 87 46W
Glenville, N.Y. 23 A6 41 1N 73 41W
Glenvista, Jobg. 18 F9 26 17 S 28 3 E
Glenwood Landing, N.Y. 23 C7 40 48N 73 38W
Glienicke, Berl. 7 A2 52 38N 13 18 E
Glömsta, Stock. 3 E10 59 14N 17 55 E
Glosli, Oslo 2 A5 60 1N 10 55 E
Glostrup, Købn. 2 E9 55 39N 12 23 E
Gloucester City, Phil. 24 B4 39 53N 75 7W
Gocheog, Sŏul 12 G7 37 30N 126 52 E
Gocławek, Wsaw. 10 E7 52 14N 21 7 E
Goeselville, Chic. 26 D2 41 37N 87 46W
Goetjensort, Hbg. 7 E8 53 29N 10 2 E
Golabari, Calc. 16 E6 22 35N 88 20 E
Golabki, Wsaw. 10 E6 52 12N 20 52 E
Golden Gate, S.F. 27 B2 37 48N 122 29W
Golden Gate Bridge, S.F. 27 B2 37 49N 122 28W
Golden Gate National Recreation Area, S.F. 27 B1 37 49N 122 31W
Golden Gate Park, S.F. 27 B2 37 46N 122 28W
Golden Horn, Ist. 17 A2 41 1N 28 57 E
Golders Green, Lon. 4 B3 51 34N 0 11W
Golyevo, Mos. 11 E7 55 48N 37 18 E
Gometz-la-Ville, Paris 5 C2 48 40N 2 7 E
Gometz-le-Châtel, Paris 5 C2 48 40N 2 8 E
Gondangdra, Jak. 15 J9 6 11 S 106 49 E
Gonesse, Paris 5 B4 48 59N 2 26 E
Gongreung, Sŏul 12 G8 37 36N 127 3 E
González Catán, B.A. 32 C3 34 46 S 58 38W
Goodman Hill, Bost. 21 C1 42 22N 71 23W
Goodmayes, Lon. 4 B5 51 33N 0 6 E
Gopalnagar, Calc. 16 C5 22 50N 88 13 E
Gopalpur, Calc. 16 E6 22 38N 88 26 E
Górce, Wsaw. 10 E6 52 15N 20 55 E
Gordon, Syd. 19 A3 33 46 S 151 8 E
Gore Hill, Syd. 19 A4 33 49 S 151 10 E
Gorelyy →, St-Pet. 11 A5 60 1N 30 30 E
Gorenki, Mos. 11 E11 55 47N 37 53 E
Gorkiy Park, Mos. 11 E9 55 43N 37 36 E
Görväln, Stock. 3 D9 59 26N 17 45 E
Gose Elbe →, Hbg. 7 E8 53 28N 10 6 E
Gosen, Berl. 7 B5 52 23N 13 43 E
Gosener kanal, Berl. 7 B5 52 24N 13 42 E
Goshenville, Phil. 24 B1 39 59N 75 32W
Gospel Oak, Lon. 4 B4 51 32N 0 9W
Gotanda, Tōkyō 13 C3 35 37N 139 43 E
Gotanno, Tōkyō 13 B3 35 45N 139 49 E
Goth Goli Mār, Kar. 17 G11 24 53N 67 1 E
Goth Sher Shāh, Kar. 17 G10 24 53N 66 59 E
Gournay-sur-Marne, Paris 5 B5 48 51N 2 34 E
Goussainville, Paris 5 A4 49 1N 2 27 E
Gouvernes, Paris 5 B6 48 51N 2 41 E
Governador, I. do, Rio J. 31 A2 22 48 S 43 13W
Governor's I., N.Y. 22 C4 40 41N 74 1W
Grabicz, Wsaw. 10 E8 52 19N 21 12 E
Grabów, Wsaw. 10 F6 52 8N 20 59 E
Gracia, Barc. 8 D6 41 24N 2 10 E
Gradyville, Phil. 24 B2 39 56N 75 27W
Gräfelfing, Mün. 7 G9 48 7N 11 25 E
Grafenwald, Ruhr 6 A3 51 33N 6 54 E
Graham Memorial Park, Balt. 25 A4 39 25N 76 29W
Gran Canal, Méx. 29 A3 19 33N 99 1W
Granada Hills, L.A. 28 A1 34 16N 118 30W
Grand Bourg, B.A. 32 A2 34 29 S 58 42W
Grand Calumet →, Chic. 26 D4 41 37N 87 28W
Grand Union Canal, Lon. 4 A2 51 42N 0 26W
Grande →, S. Pau. 31 F7 23 43 S 46 24W
Grange, Tori. 9 B1 45 7N 7 29 E
Grange Hill, Lon. 4 B5 51 36N 0 5 E
Granite, Balt. 25 A1 39 20N 76 51W
Graniteville, N.Y. 22 D3 40 37N 74 10W
Granja Viana, S. Pau. 31 E4 23 35 S 46 50W
Granlandet, Hels. 3 B6 60 10N 25 15 E
Granö, Hels. 3 B6 60 13N 25 14 E
Grant Park, Chic. 26 B3 41 52N 87 37W
Granville, Syd. 19 A3 33 49 S 151 1 E
Grape I., Bost. 21 D4 42 16N 70 55W
Grass Hassock Channel, N.Y. 23 D6 40 36N 73 47W
Grassey B., N.Y. 23 D6 40 37N 73 47W
Grassy Sprain Res., N.Y. 23 B5 40 58N 73 50W
Gratosóglio, Mil. 9 E6 45 24N 9 1 E
Gratzwalde, Berl. 7 B5 52 28N 13 42 E
Gravesend, N.Y. 22 D5 40 36N 73 56W
Grays, Lon. 4 C6 51 28N 0 19 E
Grazhdanka, St-Pet. 11 B4 59 59N 30 24 E
Great Blue Hill, Bost. 21 D3 42 12N 71 4W
Great Bookham, Lon. 4 D2 51 16N 0 21W
Great Brewster I., Bost. 21 C4 42 19N 70 53W
Great Captain I., N.Y. 23 B7 40 59N 73 37W
Great Falls, Wash. 25 D6 38 59N 77 17W
Great Falls Park, Wash. 25 D6 38 59N 77 14W
Great Kills, N.Y. 22 D4 40 32N 74 9W
Great Kills Harbour, N.Y. 22 D4 40 32N 74 8W
Great Neck, N.Y. 23 C6 40 48N 73 44W
Great Pond, Bost. 21 D3 42 11N 71 2W
Great South B., N.Y. 23 D9 40 39N 73 19W
Greco, Mil. 9 D6 45 30N 9 12 E
Greco I., S.F. 27 C3 37 30N 122 10W
Green Brae, S.F. 27 A1 37 57N 122 31W
Green Brook, N.Y. 22 D2 40 35N 74 26W
Green I., H.K. 12 E5 22 17N 114 6 E
Green Land, Jak. 15 J9 6 17 S 106 46 E
Green Pond, N.Y. 22 A2 41 1N 74 29W
Green Street, Lon. 4 A3 51 40N 0 16W
Green Street Green, Lon. 4 C5 51 21N 0 5 E
Green Valley, Syd. 19 B2 33 54 S 150 53 E
Green Village, N.Y. 22 C2 40 44N 74 27W
Greenbelt, Wash. 25 C8 39 0N 76 52W
Greenbelt Park, Wash. 25 D8 38 58N 76 53W
Greenfield Park, Mtrl. 20 B5 43 29N 73 28W
Greenfields Village, Phil. 24 C4 39 49N 75 9W
Greenford, Lon. 4 B2 51 31N 0 21W
Greenhithe, Lon. 4 C6 51 27N 0 17 E
Greenlawn, N.Y. 23 B8 40 52N 73 22W
Greenpoint, N.Y. 22 C5 40 43N 73 57W
Greensborough, Melb. 19 E7 37 41 S 145 5 E
Greenside, Jobg. 18 E9 26 8 S 28 1 E
Greenvale, N.Y. 23 C7 40 48N 73 35W
Greenville Chauncey, N.Y. 23 B5 40 59N 73 50W
Greenwich, Lon. 4 C4 51 28N 0 0 E
Greenwich, N.Y. 23 A7 41 1N 73 37W
Greenwich, Syd. 19 B4 33 50 S 151 11 E
Greenwich Observatory, Lon. 4 C4 51 28N 0 0 E
Greenwich Pt., N.Y. 23 A7 41 0N 73 34W
Greenwich Village, N.Y. 22 C5 40 44N 73 59W
Greenwood, Bost. 21 C3 42 29N 71 2W
Grefsen, Oslo 2 B4 59 56N 10 47 E
Grégy-sur-Yerres, Paris 5 C5 48 40N 2 37 E
Greiffenburg, Ruhr 6 B1 51 20N 6 37 E
Gressy, Paris 5 B6 48 58N 2 40 E
Greve Strand, Købn. 2 E8 55 34N 12 18 E
Greystanes, Syd. 19 A2 33 49 S 150 58 E
Griebnitzsee, Berl. 7 B1 52 23N 13 8 E
Griffith Park, L.A. 28 B3 34 7N 118 18 E
Grignon, Paris 5 B1 48 50N 1 56 E
Grigny, Paris 5 D4 48 39N 2 23 E
Grinzing, Wien 10 G10 48 15N 16 20 E
Grisy-Suisnes, Paris 5 C6 48 41N 2 40 E
Gröbenried, Mün. 7 F9 48 13N 11 25 E
Grochów, Wsaw. 10 E7 52 15N 21 5 E
Grodzisk, Wsaw. 10 E7 52 19N 21 4 E
Grogol, Jak. 15 H9 6 9 S 106 47 E
Grogol, Kali →, Jak. 15 J9 6 11 S 106 47 E
Gronsdorf, Mün. 7 G11 48 7N 11 42 E
Grorud, Oslo 2 B5 59 57N 10 52 E
Gross Borstel, Hbg. 7 D7 53 36N 9 58 E
Gross Flottbek, Hbg. 7 D7 53 33N 9 53 E
Gross Glienicke, Berl. 7 B1 52 28N 13 6 E
Gross-Hadern, Mün. 7 G9 48 6N 11 29 E
Gross-Lappen, Mün. 7 F10 48 11N 11 35 E
Grosse Krampe, Berl. 7 B5 52 23N 13 40 E
Grosse Müggelsee, Berl. 7 B4 52 26N 13 38 E
Grossenbaum, Ruhr 6 B2 51 22N 6 44 E
Grossenzersdorf, Wien 10 G11 48 12N 16 33 E
Grossenzersdorfer Arm →, Wien 10 G11 48 12N 16 31 E
Grosser Biberhaufen, Wien 10 G10 48 12N 16 28 E
Grosser Wannsee, Berl. 7 B2 52 25N 13 10 E
Grossfeld-Siedlung, Wien 10 G10 48 16N 16 26 E
Grosshesselohe, Mün. 7 G10 48 3N 11 32 E
Grossjedlersdorf, Wien 10 G10 48 16N 16 23 E
Grossziethen, Berl. 7 B3 52 23N 13 26 E
Groszówka, Wsaw. 10 E8 52 14N 21 13 E
Grove Hall, Bost. 21 D3 42 18N 71 4W
Grove Park, Lon. 4 C5 51 25N 0 1 E
Grove Park, Lon. 4 C3 51 28N 0 15W
Groveton, Wash. 25 E7 38 46N 77 6W
Grugliasco, Tori. 9 B2 45 5N 7 34 E
Gruiten, Ruhr 6 C4 51 12N 7 0 E
Grumme, Ruhr 6 B5 51 29N 7 13 E
Grumo Nevano, Nápl. 9 H12 40 56N 14 15 E
Grünau, Berl. 7 B4 52 25N 13 34 E
Grunewald, Berl. 7 B2 52 29N 13 16 E
Grünwald, Mün. 7 G10 48 2N 11 31 E
Grünwalder Forst, Mün. 7 G10 48 2N 11 33 E
Grymes Hill, N.Y. 22 D4 40 36N 74 6W
Gu Ro, Sŏul 12 G7 37 30N 126 51 E
Guadalupe, Manila 15 D4 14 34N 121 2 E
Guadalupe →, S.F. 27 D4 37 28N 122 4W
Guadalupe, Basilica de, Méx. 29 B3 19 29N 99 7W
Guadelupe, Rio J. 31 A2 22 49 S 43 20W
Guanabacoa, La Hab. 30 B3 23 7N 82 18W
Guanabara, Rio J. 31 B1 22 57 S 43 20W
Guanabara, B. de, Rio J. 31 B2 22 52 S 43 10W
Guanabara, Jardim, Rio J. 31 B2 22 48 S 43 11W
Guang'anmen, Beij. 14 B2 39 51N 116 18 E
Guangminglou, Beij. 14 B3 39 51N 116 23 E
Guangqumen, Beij. 14 B3 39 52N 116 25 E
Guangzhou, Gzh. 14 G8 23 6N 113 13 E
Guanshuo, Gzh. 14 G8 23 5N 113 22 E
Guantai, Nápl. 9 H12 40 52N 14 11 E
Guapira →, S. Pau. 31 D6 23 30 S 46 33W
Guarapiranga, Res. de, S. Pau. 31 F5 23 42 S 46 43W
Guardias, Mdrd. 8 B3 40 29N 3 31W
Guarulhos, S. Pau. 31 D6 23 28 S 46 32W
Guasabara, La Hab. 30 B2 23 0N 82 29W
Guayacanes, Pta., La Hab. 30 A3 23 10N 82 16W
Gubernador Monteverde, B.A. 32 C5 34 47 S 58 16W
Gudö, Stock. 3 E12 59 12N 18 12 E
Güell, Parque de, Barc. 8 D6 41 24N 2 10 E
Guermantes, Paris 5 B6 48 51N 2 42 E
Gugging, Wien 10 G9 48 18N 16 15 E
Guianazes, S. Pau. 31 E7 23 32 S 46 24W
Guildford, Syd. 19 B2 33 51 S 150 59 E
Guinardó, Barc. 8 D6 41 24N 2 10 E
Gujiazhai, Shang. 14 H11 31 21N 121 23 E
Gulbāi, Kar. 17 G10 24 52N 66 58 E
Guldasteh, Tehr. 17 D4 35 36N 51 15 E
Gulistan Palace, Tehr. 17 C5 35 40N 51 24 E
Gulph Mills, Phil. 24 A2 40 4N 75 20W
Gumbostrand, Hels. 3 B6 60 15N 25 17 E
Güngören, Ist. 17 A2 41 1N 28 52 E
Gunnarsby, Hels. 3 C1 60 6N 24 28W
Gunnersbury, Lon. 4 C3 51 29N 0 17W
Gunnigfeld, Ruhr 6 B4 51 29N 7 8 E
Gunpowder Falls →, Balt. 25 A4 39 23N 76 36W
Gunung Sahari, Jak. 15 H9 6 9 S 106 49 E
Gupiing, Manila 15 E5 14 27N 121 11 E
Guryong San, Sŏul 12 H8 37 28N 127 3 E
Gustavsberg, Stock. 3 E13 59 19N 18 23 E
Guttenberg, N.Y. 22 C4 40 48N 74 0W
Gutuyevskiy, Os., St-Pet. 11 B3 59 53N 30 15 E
Guyancourt, Paris 5 C2 48 46N 2 4 E
Guyancourt, Aérodrome de, Paris 5 C2 48 45N 2 3 E
Gvali-patak →, Bud. 10 K13 47 23N 19 7 E
Gwan Ag, Sŏul 12 H7 37 29N 126 57 E
Gwanag San, Sŏul 12 H7 37 27N 126 58 E
Gwynns Falls →, Balt. 25 B2 39 19N 76 42W
Gyál, Bud. 10 K14 47 23N 19 13 E
Gyeongbong Palace, Sŏul 12 G7 37 34N 126 58 E
Gynea, Syd. 19 C3 34 1 S 151 5 E

H

Haaga, Hels. 3 B4 60 13N 24 53 E
Haan, Ruhr 6 C3 51 11N 6 59 E
Haar, Mün. 7 G11 48 6N 11 43 E
Haar, Ruhr 6 B3 51 26N 7 13 E
Haarzopf, Ruhr 6 B3 51 25N 6 57 E
Habana del Este, La Hab. 30 B3 23 9N 82 19W
Habay, Manila 15 E3 14 27N 120 56 E
Habikino, Ōsaka 12 C4 34 33N 135 36 E
Habinghorst, Ruhr 6 A5 51 34N 7 18 E
Hacienda Heights, L.A. 28 C5 33 59N 117 59W
Hackbridge, Lon. 4 C4 51 23N 0 9W
Hackensack, N.Y. 22 B4 40 52N 74 4W
Hackney, Lon. 4 B4 51 32N 0 3W
Hackney Wick, Lon. 4 B4 51 32N 0 1W
Haddon Heights, Phil. 24 B4 39 53N 75 3W
Haddonfield, Phil. 24 B4 39 53N 75 2W
Hadersdorf, Wien 10 G9 48 12N 16 14 E
Hadley Wood, Lon. 4 B3 51 39N 0 10W
Haga, Stock. 3 D11 59 21N 18 1 E
Hagem, Ruhr 6 A5 51 38N 7 19 E
Hagen, Ruhr 6 B6 51 21N 7 27 E
Hägersten, Stock. 3 E10 59 18N 17 59 E
Haggetts Pond, Bost. 21 B2 42 39N 71 11W
Häggvik, Stock. 3 D10 59 26N 17 56 E
Hagonoy, Manila 15 D4 14 30N 121 4 E
Hagsätra, Stock. 3 E11 59 15N 18 0 E
Hahipur, Calc. 16 E5 22 47N 88 10 E
Hahnerberg, Ruhr 6 C4 51 12N 7 9 E
Hai He →, Tianj. 14 E6 39 4N 117 17 E
Haidarpur, Delhi 16 A1 28 43N 77 8 E
Haidhausen, Mün. 7 G10 48 7N 11 36 E
Haidian, Beij. 14 B2 39 59N 116 16 E
Haight-Ashbury, S.F. 27 B2 37 46N 122 26W
Haiguangsi, Tianj. 14 E6 39 7N 117 11 E
Hainault, Lon. 4 B5 51 36N 0 6 E
Haizhu Guangchang, Gzh. 14 G8 23 6N 113 14 E
Hakim, El Qâ. 18 C4 30 4N 31 7 E
Hakunila, Hels. 3 B5 60 16N 25 6 E
Halchōbori, Tōkyō 13 B4 35 48N 139 55 E
Haledon, N.Y. 22 B3 40 57N 74 11W
Halesite, N.Y. 23 B8 40 53N 73 24W
Halethorpe, Balt. 25 B2 39 14N 76 41W
Half Hollow Hills, N.Y. 23 C8 40 48N 73 21W
Half Moon B., S.F. 26 D2 37 27N 122 25W
Half Moon Bay Airport, S.F. 27 C1 37 31N 122 30W
Half Moon Bay Beaches, S.F. 26 D2 37 28N 122 28W
Halim, Jak. 15 J10 6 15 S 106 53 E
Halim Perdanakusuma Airport, Jak. 15 J10 6 16 S 106 53 E
Halstead, Lon. 4 D5 51 19N 0 8 E
Halstenbeck, Hbg. 7 D7 53 38N 9 50 E
Haltiala, Hels. 3 B4 60 16N 24 57 E
Haltiavuori, Hels. 3 B4 60 16N 24 56 E
Ham, Lon. 4 C3 51 25N 0 18W
Ham, Paris 5 A2 49 1N 2 3 E
Hamberg, Jobg. 18 E8 26 9 S 27 54 E
Hamborn, Ruhr 6 B2 51 29N 6 46 E
Hamburg, Hbg. 7 D8 53 33N 10 0 E
Hamburg Flughafen, Hbg. 7 D7 53 38N 9 59 E
Hämeenkylä, Hels. 3 B3 60 16N 24 48 E
Hamm, Hbg. 7 D8 53 33N 10 2 E
Hamm, Ruhr 6 C2 51 12N 6 44 E
Hammarby, Stock. 3 E11 59 17N 18 5 E
Hamme, Ruhr 6 B5 51 29N 7 12 E
Hammel Arverne, N.Y. 23 D6 40 35N 73 48W
Hammerbrook, Hbg. 7 D8 53 32N 10 1 E
Hammersmith, Lon. 4 C3 51 29N 0 14W
Hammond, Chic. 26 D4 41 36N 87 29W
Hampstead, Lon. 4 B3 51 33N 0 10W
Hampstead, Mtrl. 20 B4 43 28N 73 37W
Hampstead Garden Suburb, Lon. 4 B3 51 34N 0 11W
Hampstead Heath, Lon. 4 B3 51 33N 0 9W
Hampton Court Palace, Lon. 4 C2 51 24N 0 20W
Hampton Hill, Lon. 4 C2 51 25N 0 21W
Hampton Wick, Lon. 4 C3 51 24N 0 18W
Hamrā, Bagd. 17 F7 33 18N 44 22 E
Han Gang →, Sŏul 12 G7 37 32N 126 55 E
Hanakuri, Tōkyō 13 A3 35 50N 139 47 E
Hanala, Hels. 3 A5 60 20N 25 4 E
Hancho, Ōsaka 12 B3 34 48N 135 28 E
Haneda, Tōkyō 13 C3 35 32N 139 44 E
Hang Hau, H.K. 12 E6 22 19N 114 16 E
Hanjiashu, Tianj. 14 D5 39 11N 117 4 E
Hanlon, Trto. 20 E7 43 38N 79 37W
Hansen Flood Control Basin, L.A. 28 A2 34 15N 118 23W
Hansia, Calc. 16 D6 22 48N 88 24 E
Hanskinen, Hels. 3 C6 60 8N 25 18 E
Hanwell, Lon. 4 B2 51 30N 0 20W
Hanworth, Lon. 4 C2 51 26N 0 23W
Haora, Calc. 16 E5 22 34N 88 18 E
Happy Valley, H.K. 12 E6 22 16N 114 10 E
Harajuku, Tōkyō 13 D2 35 22N 139 30 E
Haraki, Tōkyō 13 B4 35 42N 139 56 E
Harat, Calc. 16 C5 22 52N 88 11 E
Harbor Hills, N.Y. 23 C6 40 46N 73 44W
Harburg, Hbg. 7 E7 53 27N 9 59 E
Harding, Bost. 21 D2 42 12N 71 19W
Hardricourt, Paris 5 A1 49 0N 1 53 E
Harefield, Lon. 4 B2 51 36N 0 28W
Hareskovby, Købn. 2 D9 55 45N 12 23 E
Harewood Park, Balt. 25 A4 39 22N 76 21W
Harigaya, Tōkyō 13 B2 35 49N 139 33 E
Haringey, Lon. 4 B4 51 34N 0 4W
Haripur, Calc. 16 D5 22 42N 88 10 E
Harjula, Hels. 3 A3 60 21N 24 45 E
Harjusuo, Hels. 3 B5 60 19N 25 0 E
Harkortsee, Ruhr 6 B6 51 23N 7 24 E
Harksheide, Hbg. 7 C8 53 43N 10 0 E
Harlaching, Mün. 7 G10 48 5N 11 33 E
Harlem, N.Y. 22 C5 40 48N 73 56W
Harlesden, Lon. 4 B3 51 32N 0 14W
Harlington, Lon. 4 C2 51 29N 0 25W
Harmaja, Hels. 3 C4 60 6N 24 58 E
Harmashatar hegy, Bud. 10 J13 47 33N 19 0 E
Harmondsworth, Lon. 4 C2 51 29N 0 29W
Harmonville, Phil. 24 A3 40 5N 75 18W
Harold Hill, Lon. 4 B6 51 36N 0 14 E
Harold Parker State Forest, Bost. 21 B3 42 37N 71 4W
Harold Wood, Lon. 4 B6 51 35N 0 14 E
Harrington Park, N.Y. 22 B5 40 59N 73 59W
Harrison, N.Y. 22 C4 40 44N 74 9W
Harrison, N.Y. 23 B6 40 57N 73 42W
Harrisonville, Balt. 25 A2 39 22N 76 49W
Harrow, Lon. 4 B2 51 34N 0 20W
Harrow on the Hill, Lon. 4 B2 51 34N 0 21W
Harrow School, Lon. 4 B2 51 34N 0 20W
Harrow Weald, Lon. 4 B2 51 36N 0 20W
Hart I., Balt. 25 B4 39 14N 76 23W
Hart I., N.Y. 23 B6 40 51N 73 46W
Hartford, Phil. 24 B5 39 58N 74 53W
Hartley, Lon. 4 C6 51 22N 0 18 E
Hartsdale, N.Y. 23 A6 41 1N 73 48W
Harumi, Tōkyō 13 C3 35 38N 139 47 E
Harvard, Mt., L.A. 28 A4 34 12N 118 4W
Harvard Univ., Bost. 21 C3 42 23N 71 7W
Harvestehude, Hbg. 7 D7 53 34N 9 58 E
Harvey, Chic. 26 D3 41 36N 87 39W
Harwood Heights, Chic. 26 B2 41 57N 87 46W
Hasanābād, Tehr. 17 C4 35 44N 51 16 E
Hasbrouck Heights, N.Y. 22 B4 40 51N 74 6W
Haselbach, Wien 10 G9 48 18N 16 14 E
Haselhorst, Berl. 7 A2 52 33N 13 14 E
Hasköy, Ist. 17 A2 41 2N 28 57 E
Hasle, Oslo 2 C3 59 46N 10 38 E
Hasloh, Hbg. 7 C7 53 41N 9 54 E
Haslohfeld, Hbg. 7 C7 53 41N 9 56 E
Haslum, Oslo 2 B3 59 55N 10 34 E
Haspe, Ruhr 6 B5 51 21N 7 25 E
Haspertalsp, Ruhr 6 C6 51 17N 7 24 E
Hasselbeck, Ruhr 6 C3 51 19N 6 56 E
Hässelby, Stock. 3 D10 59 22N 17 50 E
Hasslinghausen, Ruhr 6 C5 51 20N 7 16 E
Hasten, Ruhr 6 C5 51 11N 7 11 E
Hästhagen, Stock. 3 E11 59 18N 18 9 E
Hastings-on-Hudson, N.Y. 23 B5 40 59N 73 51W
Hatch End, Lon. 4 B2 51 36N 0 22W
Hatiara, Calc. 16 E6 22 36N 88 26 E
Hatogaya, Tōkyō 13 B3 35 49N 139 44 E
Hattingen, Ruhr 6 B5 51 23N 7 11 E
Hatton, Lon. 4 C2 51 28N 0 25W
Hattori, Ōsaka 12 A4 34 51N 135 36 E
Hauketo, Oslo 2 B4 59 50N 10 48 E
Hauldres →, Paris 5 D5 48 37N 2 37 E
Hausbruch, Hbg. 7 E7 53 28N 9 53 E
Havalimani, Ist. 17 B2 40 59N 28 50 E
Havana = La Habana, La Hab. 30 B2 23 7N 82 21W
Havdrup, Købn. 2 E7 55 33N 12 7 E
Havel →, Berl. 7 A2 52 37N 13 11 E
Havelkanal, Berl. 7 A2 52 36N 13 10 E
Haverford, Phil. 24 A3 40 0N 75 18W
Havering, Lon. 4 B6 51 33N 0 12 E
Havering-atte-Bower, Lon. 4 B6 51 37N 0 11 E
Havertown, Phil. 24 B3 39 58N 75 18W
Hawangsibri, Sŏul 12 G8 37 33N 127 1 E
Haweolgog, Sŏul 12 G8 37 35N 127 1 E
Haworth, N.Y. 22 B5 40 57N 73 59W
Hawthorne, L.A. 28 C2 33 54N 118 21W
Hawthorne, N.Y. 22 B4 40 57N 74 8W
Hayes, Lon. 4 C5 51 22N 0 0 E
Hayes, Lon. 4 B2 51 30N 0 25W
Hayes End, Lon. 4 B2 51 31N 0 25W
Hayford, Chic. 26 C2 41 45N 87 42W
Hayward, S.F. 27 B4 37 40N 122 4W
Hayward Fault, S.F. 27 B3 37 46N 122 10W
Haywood Municipal Airport, S.F. 27 C4 32 39N 122 9W
Headley, Lon. 4 D3 51 16N 0 16W
Headstone, Lon. 4 B2 51 35N 0 21W
Heard Pond, Bost. 21 C1 42 20N 71 23W
Heart Pond, Bost. 21 B1 42 33N 71 20W
Heath Park, Lon. 4 B6 51 34N 0 12 E
Heathmont, Melb. 19 E8 37 49 S 145 14 E
Heathrow Airport, Lon. 4 C2 51 28N 0 27W
Hebbville, Balt. 25 A2 39 20N 76 45W
Hebe Haven, H.K. 12 D6 22 21N 114 15 E
Hebei, Tianj. 14 E6 39 9N 117 11 E
Hedehusene, Købn. 2 E8 55 39N 12 11 E
Hedong, Gzh. 14 G8 23 5N 113 14 E
Hedong, Tianj. 14 E6 39 7N 117 11 E
Heerdt, Ruhr 6 C2 51 13N 6 42 E
Hegewisch, Chic. 26 D3 41 39N 87 32W
Heggelielva →, Oslo 2 A3 60 1N 10 36 E
Heide, Ruhr 6 A3 51 31N 6 50 E
Heidelberg, Melb. 19 E7 37 45 S 145 4 E
Heidelberg West, Melb. 19 E7 37 43 S 145 2 E
Heidemühle, Berl. 7 B5 52 29N 13 40 E
Heidhausen, Ruhr 6 B4 51 22N 7 1 E
Heiligenhaus, Ruhr 6 C3 51 19N 6 57 E
Heiligensee, Berl. 7 A2 52 36N 13 13 E
Heiligenstadt, Wien 10 G10 48 14N 16 21 E
Heimfeld, Hbg. 7 E7 53 28N 9 57 E
Heinässuo, Hels. 3 B1 60 18N 24 27 E
Heinersdorf, Berl. 7 A3 52 34N 13 25 E
Heisingen, Ruhr 6 B4 51 24N 7 3 E
Heissen, Ruhr 6 B3 51 26N 6 54 E
Helderkruin, Jobg. 18 E8 26 7 S 27 51 E
Helenelund, Stock. 3 D10 59 24N 17 58 E
Heliopolis, El Qâ. 18 C5 30 5N 31 19 E
Hellersdorf, Berl. 7 A4 52 32N 13 35 E
Hellerup, Købn. 2 D10 55 44N 12 33 E
Helmahof, Wien 10 G11 48 18N 16 34 E
Helsingfors = Helsinki, Hels. 3 B4 60 10N 24 55 E
Helsinki, Hels. 3 B4 60 10N 24 55 E
Helsinki Airport, Hels. 3 B4 60 18N 24 58 E
Hempstead, N.Y. 23 C7 40 42N 73 37W
Hempstead Harbor, N.Y. 23 B7 40 50N 73 39W
Henan, Gzh. 14 G8 23 5N 113 14 E
Hendon, Lon. 4 B3 51 35N 0 14W
Hengsha, Gzh. 14 G8 23 9N 113 12 E
Hengsteysee, Ruhr 6 B6 51 24N 7 27 E
Hennigsdorf, Berl. 7 A1 52 38N 13 12 E
Henrichenburg, Ruhr 6 A5 51 35N 7 19 E
Henriville, Paris 5 C1 48 44N 1 56 E
Henrykow, Wsaw. 10 E6 52 19N 20 58 E
Henson Cr. →, Wash. 25 E8 38 47N 76 58W
Henttaa, Hels. 3 B3 60 11N 24 45 E
Heping, Tianj. 14 E6 39 7N 117 11 E
Heping Gongyuan, Shang. 14 J12 31 16N 121 30 E
Hepingli, Beij. 14 B3 39 57N 116 23 E
Herbeck, Ruhr 6 C5 51 12N 7 18 E
Herbede, Ruhr 6 B5 51 25N 7 16 E
Herblay, Paris 5 A2 48 59N 2 9 E
Herdecke, Ruhr 6 B6 51 24N 7 25 E
Herlev, Købn. 2 D9 55 43N 12 27 E
Hermannskogel, Wien 10 G9 48 16N 16 17 E
Hermitage and Winter Palace, St-Pet. 11 B3 59 55N 30 19 E
Hermosa Beach, L.A. 28 C2 33 51N 118 23W
Hermsdorf, Berl. 7 A2 52 37N 13 18 E
Hernals, Wien 10 G10 48 13N 16 20 E
Herne, Ruhr 6 A5 51 32N 7 13 E
Herne Hill, Lon. 4 C4 51 27N 0 6W
Hernwood Heights, Balt. 25 A2 39 22N 76 49W
Héroes de Churubusco, Méx. 29 B3 19 21N 99 6W
Herongate, Lon. 4 B7 51 35N 0 21 E
Herons, Î. aux, Mtrl. 20 B4 43 25N 73 34W
Herricks, N.Y. 23 C7 40 45N 73 39W
Herring Run →, Balt. 25 B3 39 18N 76 30W
Hersham, Lon. 4 C2 51 21N 0 22W
Herstedøster, Købn. 2 D9 55 40N 12 22 E
Herten, Ruhr 6 A4 51 35N 7 8 E
Herttoniemi, Hels. 3 B5 60 12N 25 2 E
Hessler, Ruhr 6 A4 51 31N 7 3 E
Heston, Lon. 4 C2 51 28N 0 22W
Hetterscheidt, Ruhr 6 B3 51 20N 6 59 E
Hetzendorf, Wien 10 H9 48 9N 16 17 E
Heuberg, Wien 10 G9 48 13N 16 16 E
Heven, Ruhr 6 B5 51 26N 7 18 E
Hewlett Neck, N.Y. 23 D6 40 37N 73 41W
Hexi, Tianj. 14 E5 39 8N 117 9 E
Hexingcun, Tianj. 14 E6 39 6N 117 10 E
Hextable, Lon. 4 C6 51 24N 0 10 E
Heybridge, Lon. 4 B7 51 39N 0 21 E
Hibernia, N.Y. 22 A2 40 57N 74 29W
Hickory Hills, Chic. 26 C2 41 43N 87 49W
Hicksville, N.Y. 23 C7 40 46N 73 30W
Hiddinghausen, Ruhr 6 B6 51 21N 7 17 E
Hiekkaharju, Hels. 3 B5 60 18N 25 2 E
Hiesfeld, Ruhr 6 A2 51 33N 6 46 E
Hietaniemi, Hels. 3 B4 60 10N 24 54 E
Hietzing, Wien 10 G9 48 10N 16 17 E
Higashi, Ōsaka 12 B4 34 41N 135 30 E
Higashi-kaizuka, Tōkyō 13 B4 35 41N 139 56 E
Higashimonzen, Tōkyō 13 A3 35 55N 139 40 E
Higashimurayama, Tōkyō 13 B1 35 45N 139 26 E
Higashinada, Ōsaka 12 B2 34 42N 135 15 E
Higashinari, Ōsaka 12 B4 34 40N 135 32 E
Higashiōsaka, Ōsaka 12 C4 34 39N 135 37 E
Higashisumiyoshi, Ōsaka 12 C4 34 37N 135 31 E
Higashiyodogawa, Ōsaka 12 B3 34 44N 135 28 E
High Beach, Lon. 4 B5 51 39N 0 1 E
High Junk Pk., H.K. 12 E6 22 17N 114 17 E
High Park, Trto. 20 E8 43 38N 79 27W
Higham Hill, Lon. 4 B4 51 35N 0 2W
Highbury, Lon. 4 B4 51 33N 0 6W
Highgate, Lon. 4 B4 51 34N 0 8W
Highland Cr. →, Trto. 20 D9 43 45N 79 13W
Highland Creek, Trto. 20 D9 43 46N 79 8W
Highland Park, L.A. 28 B3 34 7N 118 13 E
Highland Park, N.Y. 22 D2 40 30N 74 25W
Highlands North, Jobg. 18 E9 26 8 S 28 5 E
Highway Highlands, L.A. 28 A3 34 14N 118 16W
Higurashi, Tōkyō 13 B4 35 47N 139 55 E
Hila, Méx. 29 A2 19 35N 99 17W
Hillcrest Heights, Wash. 25 E8 38 49N 76 57W
Hillerheide, Ruhr 6 A5 51 35N 7 12 E
Hilleshög, Stock. 3 D9 59 23N 17 42 E
Hillgrove District, L.A. 28 B5 34 1N 117 58W
Hillingdon, Lon. 4 B2 51 32N 0 27W
Hillingdon Heath, Lon. 4 B2 51 31N 0 26W
Hillsborough, S.F. 27 C2 37 32N 122 21W
Hillsdale, N.Y. 22 A4 41 0N 74 1W
Hillsdale, S.F. 27 C3 37 32N 122 18W
Hillsdale Manor, N.Y. 22 A4 41 1N 74 3W
Hillside, Chic. 26 B1 41 52N 87 55W
Hillside, N.Y. 23 C6 40 42N 73 46W
Hillside Manor, N.Y. 23 C6 40 44N 73 40W
Hilltop, Phil. 24 C4 39 49N 75 4W
Hillwood, Wash. 25 D7 38 52N 77 9W
Hilmîya, El Qâ. 18 C5 30 6N 31 19 E
Hiltrop, Ruhr 6 A5 51 31N 7 15 E
Hindsby, Hels. 3 A6 60 20N 25 13 E
Hingham, Bost. 21 D4 42 14N 70 54W
Hingham B., Bost. 21 D4 42 17N 70 56W
Hingham Harbor, Bost. 21 D4 42 15N 70 53W
Hino, Tōkyō 13 D2 35 23N 139 35 E
Hinsbeck, Ruhr 6 B4 51 22N 7 4 E
Hinschenfelde, Hbg. 7 D8 53 35N 10 4 E
Hinsdale, Chic. 26 C1 41 47N 87 55W
Hinterhainbach, Wien 10 G8 48 14N 16 13 E
Hintersdorf, Wien 10 G9 48 18N 16 13 E
Hirakata, Ōsaka 12 B4 34 48N 135 38 E
Hirota, Ōsaka 12 B3 34 45N 135 20 E
Hirschstetten, Wien 10 G10 48 14N 16 27 E
Hither Green, Lon. 4 C4 51 26N 0 0 E
Hiyoshi, Tōkyō 13 C2 35 32N 139 38 E
Hjortekær, Købn. 2 D10 55 47N 12 32 E
Hjortespring, Købn. 2 D9 55 44N 12 25 E
Hlubočepy, Pra. 10 B2 50 2N 14 23 E
Ho Chung, H.K. 12 D6 22 21N 114 14 E
Ho Man Tin, H.K. 12 E6 22 19N 114 11 E
Hoboken, N.Y. 22 C4 40 44N 74 3W
Hobsons B., Melb. 19 F6 37 51 S 144 55 E
Hochbrück, Mün. 7 F10 48 14N 11 37 E
Hochdahl, Ruhr 6 C3 51 13N 6 55 E
Hochemmerich, Ruhr 6 A2 51 24N 6 41 E
Hochfeld, Ruhr 6 B2 51 25N 6 45 E
Hochheide, Ruhr 6 B2 51 26N 6 40 E
Hochlar, Ruhr 6 A5 51 36N 7 11 E
Hochlarmark, Ruhr 6 A4 51 34N 7 10 E
Hodgkins, Chic. 26 C1 41 46N 87 53W
Hodogaya-Ku, Tōkyō 13 D2 35 26N 139 35 E
Hoegi, Sŏul 12 G8 37 35N 127 3 E
Hofberg, Wien 10 G10 48 12N 16 21 E
Hoffman I., N.Y. 22 D4 40 34N 74 3W
Hofstede, Ruhr 6 A5 51 29N 7 11 E
Höggarnsfjärden, Stock. 3 D13 59 22N 18 22 E
Hohe Mark, Naturpark, Ruhr 6 A2 51 35N 6 49 E
Hohe Schaar, Hbg. 7 E7 53 29N 9 58 E
Hohenbrunn, Mün. 7 G11 48 2N 11 42 E
Hohenfelde, Hbg. 7 D8 53 33N 10 1 E
Höhenkirchen, Mün. 7 G11 48 1N 11 42 E
Hohenraden, Hbg. 7 C6 53 41N 9 39 E
Hohenschönhausen, Berl. 7 A4 52 33N 13 30 E
Hohenwisch, Hbg. 7 E7 53 29N 9 53 E

Kennedy Grove Regional Rec. Area 39

Hohokus, N.Y. 22 A4 41 0N 74 5W
Hok Tsui, H.K. 12 E6 22 12N 114 15 E
Holborn, Lon. 4 B4 51 31N 0 7W
Holečovice, Pra. 10 B2 50 6N 14 28 E
Holland Village, Sing. 15 G7 1 18N 103 47 E
Hollis, N.Y. 23 C6 40 42N 73 45W
Höllriegelskreuth, Mün. 7 G9 48 2N 11 30 E
Holly Oak, Phil. 24 C2 39 47N 75 27W
Hollydale, L.A. 28 C4 33 55N 118 10W
Hollywood Bowl, L.A. 28 B2 34 6N 118 21W
Hollywood-Burbank Airport, L.A. 28 A2 34 11N 118 21W
Holmenkollen, Oslo 2 B4 59 57N 10 41 E
Holmes, Phil. 24 B3 39 53N 75 18W
Holmes Acres, Wash. 25 D6 38 51N 77 13W
Holmes Run →, Wash. 25 E7 38 48N 77 6W
Holmesburg, Phil. 24 A4 40 2N 75 2W
Holmgård, Stock. 3 E10 59 14N 18 0 E
Holsfjorden, Oslo 2 B1 59 58N 10 17 E
Holsterhausen, Ruhr 6 A5 51 32N 7 11 E
Holte, Købn. 2 D9 55 48N 12 27 E
Holten, Ruhr 6 A2 51 31N 6 47 E
Holthausen, Ruhr 6 B4 51 25N 7 5 E
Holzbüttgen, Ruhr 6 C1 51 13N 6 37 E
Homberg, Ruhr 6 B2 51 27N 6 41 E
Hombruch, Ruhr 6 B6 51 28N 7 27 E
Homerton, Lon. 4 B4 51 32N 0 2W
Homestead Lake, Jobg. 18 F10 26 10 S 28 17 E
Homestead Valley, S.F. 27 A1 37 53N 122 32W
Hometown, Chic. 26 C2 41 44N 87 42W
Homledal, Oslo 2 B1 59 59N 10 18 E
Homówek, Wsaw. 10 E5 52 17N 20 48 E
Hon-gyōtoku, Tōkyō 13 B4 35 41N 139 57 E
Hōnanchō, Tōkyō 13 B2 35 40N 139 39 E
Honcho, Tōkyō 13 B3 35 40N 139 41 E
Honden, Tōkyō 13 B4 35 43N 139 50 E
Honeydew, Jobg. 18 E8 26 4 S 27 55 E
Hong Kah, Sing. 15 F7 1 21N 103 43 E
Hong Kong, H.K. 12 E5 22 17N 114 11 E
Hong Kong, Univ. of, H.K. 12 E5 22 16N 114 8 E
Hong Kong Airport, H.K. 12 E6 22 19N 114 11 E
Hong Kong I., H.K. 12 E6 22 16N 114 11 E
Hong Lim Park, Sing. 15 G8 1 17N 103 50 E
Hongeun, Sŏul 12 G7 37 35N 126 56 E
Hongqiao, Shang. 14 J11 31 12N 121 22 E
Hongqou, Shang. 14 J11 31 16N 121 29 E
Hongkou Gongyuan, Shang. 14 J11 31 17N 121 28 E
Hongmiao, Beij. 14 B3 39 54N 116 26 E
Hongqiao, Tianj. 14 E5 39 8N 117 9 E
Hongqiao Airport, Shang. 14 J10 31 12N 121 19 E
Honjyo, Tōkyō 13 B3 35 41N 139 48 E
Honmoku, Tōkyō 13 D2 35 24N 139 39 E
Hönow, Berl. 7 A4 52 32N 13 38 E
Höntrop, Ruhr 6 B4 51 27N 7 9 E
Hood Pond, Bost. 21 A4 42 40N 70 57W
Hooghly →, Calc. 16 D6 22 41N 88 21 E
Hook, Lon. 4 C3 51 22N 0 17W
Hopelawn, N.Y. 22 D3 40 31N 74 17W
Hörde, Ruhr 6 B7 51 29N 7 30 E
Horikiri, Tōkyō 13 B4 35 44N 139 50 E
Horn, Hbg. 7 D8 53 33N 10 5 E
Horn Pond, Bost. 21 C2 42 28N 71 9W
Hornchurch, Lon. 4 B6 51 33N 0 14 E
Horneburg, Ruhr 6 A5 51 37N 7 17 E
Horni, Pra. 10 B3 50 2N 14 33 E
Horni Počernice, Pra. 10 B3 50 6N 14 36 E
Hornsey, Lon. 4 B4 51 35N 0 7W
Horoměřice, Pra. 10 B1 50 8N 14 20 E
Horsley Park, Syd. 19 B2 33 50 S 150 51 E
Horst, Ruhr 6 B4 51 26N 7 6 E
Horsthausen, Ruhr 6 A5 51 33N 7 12 E
Hortaleza, Mdrd. 8 B3 40 28N 3 38W
Horto Florestal, S. Pau. 31 D6 23 27 S 46 38W
Horton Kirby, Lon. 4 C6 51 23N 0 14 E
Hösel, Ruhr 6 B3 51 20N 6 53 E
Hosoyama, Tōkyō 13 C2 35 36N 139 31 E
Hospitalet, Barc. 8 D5 41 21N 2 6 E
Hostafrancs, Barc. 8 D5 41 21N 2 8 E
Hoterheide, Ruhr 6 C1 51 16N 6 37 E
Houbetin, Pra. 10 B3 50 6N 14 33 E
Houghs Neck, Bost. 21 D4 42 15N 70 57W
Houghton, Jobg. 18 F9 26 10 S 28 3 E
Houilles, Paris 5 B3 48 56N 2 11 E
Hounslow, Lon. 4 C2 51 28N 0 21W
Houses of Parliament, Lon. 4 C4 51 29N 0 7W
Hove Å →, Købn. 2 D8 55 43N 12 7 E
Hovedøya, Oslo 2 B4 59 53N 10 43 E
Høvik, Oslo 2 B3 59 54N 10 34 E
Hovorčovice, Pra. 10 A3 50 10N 14 31 E
Howard Beach, N.Y. 23 D5 40 39N 73 50W
Hoxton Park, Syd. 19 B2 33 55 S 150 51 E
Hoxton Park Aerodrome, Syd. 19 B2 33 54 S 150 50 E
Hōya, Tōkyō 13 B2 35 44N 139 34 E
Høybråten, Oslo 2 B5 59 56N 10 55 E
Hradčany, Pra. 10 B2 50 5N 14 24 E
Hsiya, Gzh. 14 G7 23 9N 113 6 E
Huangpu, Gzh. 14 G9 23 5N 113 23 E
Huangpu, Shang. 14 J12 31 14N 121 30 E
Huangpu Gongyuan, Shang. 14 J11 31 14N 121 29 E
Huangpu Jiang →, Shang. 14 J11 31 11N 121 29 E
Huangtugang, Beij. 14 C2 39 49N 116 15 E
Huat Choe, Sing. 15 F7 1 20N 103 41 E
Huckarde, Ruhr 6 A6 51 32N 7 24 E
Huckingen, Ruhr 6 B2 51 21N 6 44 E
Huddinge, Stock. 3 E11 59 14N 18 0 E
Hudson →, N.Y. 22 B5 40 43N 73 6W
Huertas de San Beltran, Barc. 8 D5 41 22N 2 9 E
Huguenot, N.Y. 22 D3 40 32N 74 13W
Huguenot Park, N.Y. 22 D3 40 31N 74 12W
Huidui, Tianj. 14 E6 39 4N 117 16 E
Huisquilucan →, Méx. 29 B2 19 24N 99 17W
Huixquilucan, Méx. 29 B1 19 21N 99 21W
Hull, Bost. 21 D4 42 18N 70 54W
Hulman Aqueduct, Bost. 21 C1 42 20N 71 23W
Hulmeville, Phil. 24 A5 40 8N 74 55W
Hulsdonk, Ruhr 6 B1 51 27N 6 36 E
Humaljärvi, Hels. 3 B1 60 10N 24 26 E
Humber →, Trto. 20 D7 43 47N 79 38W
Humber B., Trto. 20 E8 43 37N 79 27W
Humber Bay, Trto. 20 E8 43 38N 79 29W
Humber Summit, Trto. 20 D7 43 45N 79 32W
Humber Valley Park, Trto. 20 E8 43 39N 79 29W
Humber Valley Village, Trto. 20 D7 43 40N 79 31W
Humberlea, Trto. 20 D7 43 43N 79 31W
Humboldt Park, Chic. 26 B2 41 54N 87 42W
Humera, Mdrd. 8 B2 40 25N 3 46W
Hummelsbüttel, Hbg. 7 D8 53 39N 10 4 E
Hun Yeang, Sing. 15 F8 1 21N 103 55 E
Hunaydī, Bagd. 17 F8 33 18N 44 29 E
Hundige, Købn. 2 E8 55 35N 12 18 E
Hundige Strand, Købn. 2 E9 55 35N 12 20 E
Hung Hom, H.K. 12 E6 22 18N 114 11 E
Hunters Hill, Syd. 19 B3 33 50 S 151 9 E
Hunters Pt., S.F. 27 B2 37 43N 122 21W
Hunters Valley, Wash. 25 D6 38 54N 77 17W
Huntington, N.Y. 23 B8 40 51N 73 25W
Huntington, Wash. 25 E7 38 47N 77 4W
Huntington B., N.Y. 23 B8 40 54N 73 24W
Huntington Bay, N.Y. 23 B8 40 56N 73 26W
Huntington Park, L.A. 28 C3 33 58N 118 13W
Huntington Station, N.Y. 23 B8 40 50N 73 23W
Hünxer Wald, Ruhr 6 A2 51 37N 6 49 E
Hurffville, Phil. 24 C4 39 45N 75 6W
Hurīya, Bagd. 17 E7 33 21N 44 19 E
Hurlingham, B.A. 32 B3 34 35 S 58 37W
Hurlingham, Jobg. 18 E9 26 6 S 28 2 E
Hurstville, Syd. 19 B3 33 57 S 151 6 E
Husby, Stock. 3 D10 59 24N 17 56 E
Huseby, Oslo 2 A6 60 0N 11 1 E
Hustivař, Pra. 10 B3 50 3N 14 31 E
Husum, Købn. 2 D9 55 42N 12 27 E
Hütteldorf, Wien 10 G9 48 12N 16 15 E
Hüttenheim, Ruhr 6 B2 51 21N 6 43 E
Huttrop, Ruhr 6 B4 51 26N 7 3 E
Hüvösvölgy, Bud. 10 J13 47 32N 19 0 E
Hvalstad, Oslo 2 B2 59 51N 10 27 E
Hvalstrand, Oslo 2 B3 59 50N 10 30 E
Hvidovre, Købn. 2 E9 55 38N 12 27 E
Hwagog, Sŏul 12 G7 37 32N 126 51 E
Hyattsville, Wash. 25 D8 38 57N 76 57W
Hyde Park, Bost. 21 D3 42 15N 71 7W
Hyde Park, Chic. 26 C3 41 47N 87 35W
Hyde Park, Jobg. 18 E9 26 6 S 28 2 E
Hyde Park, Lon. 4 B3 51 30N 0 10W
Hyde Park, Syd. 19 B4 33 52 S 151 12 E
Hynes, L.A. 28 C3 33 52N 118 10W

I

Ibaraki, Ōsaka 12 B4 34 48N 135 34 E
Ibayo Tipas, Manila 15 D4 14 32N 121 4 E
Ibese, Lagos 18 A2 6 33N 7 28 E
Ibirapuera, S. Pau. 31 E5 23 36 S 46 40W
Ibirapuera, Parque, S. Pau. 31 E6 23 35 S 46 38W
Iboju, Lagos 18 B3 6 25N 7 31 E
Icarai, Rio J. 31 B3 22 54 S 43 6W
Icerenköy, Ist. 17 B3 40 58N 29 6 E
Ichapur, Calc. 16 D6 22 48N 88 22 E
Ichgao, Tōkyō 13 C2 35 32N 139 32 E
Ichigaya, Tōkyō 13 B3 35 41N 139 43 E
Ichikawa, Tōkyō 13 B4 35 43N 139 54 E
Ickenham, Lon. 4 B2 51 33N 0 26W
Ickern, Ruhr 6 A6 51 35N 7 21 E
Iddo, Lagos 18 B2 6 28N 7 22 E
Idi-Oro, Lagos 18 A2 6 31N 7 21 E
Idimu, Lagos 18 A1 6 34N 7 17 E
Idris, Bagd. 17 E8 33 22N 44 27 E
Iganmu, Lagos 18 B2 6 28N 7 22 E
Igbobi, Lagos 18 A2 6 31N 7 22 E
Igbologun, Lagos 18 B1 6 24N 7 19 E
Igbopa, Lagos 18 A3 6 32N 7 31 E
Igelboda, Stock. 3 E12 59 17N 18 17 E
Igny, Paris 5 C3 48 44N 2 13 E
Iguassú, S. Pau. 31 E6 23 36 S 46 30W
Ijesa-Tedo, Lagos 18 B1 6 29N 7 19 E
Ijora, Lagos 18 B2 6 27N 7 22 E
Ikebe, Tōkyō 13 C2 35 31N 139 34 E
Ikebukuro, Tōkyō 13 B3 35 43N 139 42 E
Ikeda, Ōsaka 12 B3 34 48N 135 25 E
Ikegami, Tōkyō 13 C3 35 33N 139 42 E
Ikeja, Lagos 18 A2 6 35N 7 20 E
Ikeuchi, Ōsaka 12 C4 34 35N 135 32 E
Ikotun, Lagos 18 A1 6 32N 7 16 E
Ikoyi, Lagos 18 B2 6 26N 7 22 E
Ikuata, Lagos 18 B2 6 27N 7 21 E
Ikuno, Ōsaka 12 B4 34 40N 135 30 E
Ikuta, Ōsaka 12 B2 34 41N 135 10 E
Ikuta, Tōkyō 13 B2 35 36N 139 32 E
Ila, Oslo 2 B3 59 57N 10 35 E
Ilchester, Balt. 25 B2 39 14N 76 46W
Ilford, Lon. 4 B5 51 33N 0 4 E
Ilioúpolis, Ath. 8 J11 37 54N 23 47 E
Illovo, Jobg. 18 E9 26 7 S 28 3 E
Ilsós →, Ath. 8 J11 37 55N 23 41 E
Imajuku, Tōkyō 13 D2 35 28N 139 32 E
Imbâba, El Qâ. 18 C5 30 3N 31 12 E
Imielin, Wsaw. 10 F7 52 9N 21 0 E
Imirim, S. Pau. 31 D6 23 29 S 46 39W
Imittós, Ath. 8 J11 37 55N 23 45 E
Immersby, Hels. 3 B6 60 18N 25 16 E
Imore, Lagos 18 B1 6 25N 7 17 E
Imperial Palace, Tōkyō 13 B3 35 41N 139 45 E
Ina →, Ōsaka 12 B3 34 48N 135 27 E
Inagi, Tōkyō 13 C2 35 38N 139 31 E
Incirano, Mil. 9 D5 45 34N 9 9 E
Independencia, Lima 30 F8 11 59 S 77 3W
Indian Gabe, Delhi 16 B2 28 36N 77 13 E
Indian Museum, Calc. 16 E6 22 33N 88 21 E
Indiana Harbor, Chic. 26 C4 41 40N 87 26W
Indiana Harbor Canal, Chic. 26 D4 41 39N 87 26W
Indianópolis, S. Pau. 31 E6 23 35 S 46 38W
Indios Verdes, Méx. 29 B3 19 29N 99 6W
Ingarö, Stock. 3 E13 59 15N 18 26 E
Ingaröfjärden, Stock. 3 E13 59 17N 18 22 E
Ingarölandet, Stock. 3 E13 59 17N 18 26 E
Ingenieur Budge, B.A. 32 C4 34 43 S 58 27W
Ingierstrand, Oslo 2 C4 59 49N 10 46 E
Ingleburn, Syd. 19 C2 34 0 S 150 52 E
Inglewood, L.A. 28 C3 33 57N 118 19W
Ingrave, Lon. 4 B7 51 35N 0 20 E
Ingvalsby, Hels. 3 C2 60 9N 24 32 E
Inhaúme, Rio J. 31 B2 22 51 S 43 17W
Inner Port Shelter, H.K. 12 D6 22 22N 114 17 E
Interagos, S. Pau. 31 F5 23 43 S 46 40W
Intramuros, Manila 15 D3 14 35N 120 57 E
Invalides, Paris 5 B3 48 51N 2 18 E
Inverness, Balt. 25 B4 39 15N 76 29W
Inwood, N.Y. 23 D6 40 36N 73 45W
Inzersdorf, Wien 10 H10 48 8N 16 21 E
Ipanema, Rio J. 31 B2 22 59 S 43 11W
Ipiranga, S. Pau. 31 E6 23 35 S 46 36W
Ipiranga →, S. Pau. 31 E6 23 35 S 46 37W
Iponri, Lagos 18 B2 6 28N 7 22 E
Ipswich, Bost. 21 A4 42 41N 70 50W
Ipswich →, Bost. 21 A4 42 39N 70 53W
Irajá, Rio J. 31 B2 22 49 S 43 19W
Irving Park, Chic. 26 B2 41 57N 87 42W
Irvington, N.Y. 23 A5 41 2N 73 52W
Irwindale, L.A. 28 B5 34 6N 117 54W
Isabel, Rio J. 31 B2 22 55 S 43 14W
Isagatedo, Lagos 18 A1 6 31N 7 13 E
Isando, Jobg. 18 E10 26 8 S 28 12 E
Isar →, Mün. 7 F11 48 15N 11 41 E
Iselin, N.Y. 22 D3 40 34N 74 19W
Iserbrook, Hbg. 7 D6 53 34N 9 48 E
Iseri-Osun, Lagos 18 A1 6 30N 7 16 E
Ishbīlīya, Bagd. 17 E8 33 22N 44 27 E
Isheri-Olofin, Lagos 18 A1 6 34N 7 16 E
Ishi →, Ōsaka 12 C4 34 34N 135 36 E
Ishikiri, Ōsaka 12 B4 34 40N 135 39 E
Ishizu →, Ōsaka 12 C3 34 33N 135 28 E
Ishøj Strand, Købn. 2 E9 55 36N 12 20 E
Isidro Casanova, B.A. 32 C3 34 42 S 58 36W
Island Channel, N.Y. 23 D5 40 35N 73 52W
Island Park, N.Y. 23 D7 40 36N 73 38W
Island Park, Trto. 20 E8 43 37N 79 22W
Islev, Købn. 2 D9 55 41N 12 27 E
Isleworth, Lon. 4 C3 51 28N 0 19W
Islington, Bost. 21 D2 42 13N 71 13W
Islington, Lon. 4 B4 51 32N 0 6W
Islington, Trto. 20 E7 43 38N 79 30W
Ismaning, Mün. 7 F11 48 13N 11 40 E
Ismayloskiypark, Mos. 11 E10 55 46N 37 46 E
Isogo-Ku, Tōkyō 13 D2 35 23N 139 37 E
Isolo, Lagos 18 A1 6 31N 7 19 E
Isosaari, Hels. 3 C5 60 6N 25 3 E
Issy-les-Moulineaux, Paris 5 C3 48 49N 2 15 E
Istanbul, Ist. 17 B2 41 0N 28 58 E
Istanbul Boğazi, Ist. 17 A3 41 5N 29 3 E
Istanbul Hava Alani, Ist. 17 B2 40 58N 28 50 E
Istead Rise, Lon. 4 C7 51 24N 0 21 E
Istinye, Ist. 17 A3 41 6N 29 3 E
Isunba, Lagos 18 B1 6 25N 7 17 E
Itä Hakkila, Hels. 3 B5 60 17N 25 7 E
Itabashi-Ku, Tōkyō 13 B3 35 46N 139 38 E
Itaberaba, S. Pau. 31 D6 23 28 S 46 39W
Itaeweon, Sŏul 12 G7 37 32N 126 58 E
Itaim, S. Pau. 31 D7 23 29 S 46 23W
Itaipu, Rio J. 31 B3 22 58 S 43 2W
Italie, Place d', Paris 5 C4 48 49N 2 22 E
Itami, Ōsaka 12 B3 34 46N 135 24 E
Itaoca, Rio J. 31 B3 22 58 S 43 2W
Itapecerica da Serra, S. Pau. 31 F5 23 42 S 46 50W
Itaquaquecetuba, S. Pau. 31 D7 23 29 S 46 23W
Itaquera, S. Pau. 31 E7 23 32 S 46 27W
Itaquera →, S. Pau. 31 E7 23 28 S 46 26W
Ithan, Phil. 24 A2 40 1N 75 21W
Itupu, S. Pau. 31 F5 23 40 S 46 43W
Ituzaingo, B.A. 32 B3 34 39 S 58 38W
Ivanhoe, Melb. 19 E7 37 45 S 145 3 E
Iver, Lon. 4 B1 51 32N 0 30W
Ivry-sur-Seine, Paris 5 C4 48 49N 2 22 E
Iwazono, Ōsaka 12 B2 34 45N 135 18 E
Izabelin, Wsaw. 10 E5 52 17N 20 48 E
Izmaylovo, Mos. 11 E10 55 47N 37 47 E
Iztacalco, Méx. 29 B3 19 23N 99 6W
Iztapalapa, Méx. 29 B3 19 21N 99 6W
Izumi, Tōkyō 13 D1 35 25N 139 29 E

J

J. G. Strijdom Post Office Tower, Jobg. 18 F9 26 11 S 28 2 E
J. Paul Getty Museum, L.A. 28 B1 34 2N 118 33W
Jabavu, Jobg. 18 F8 26 14 S 27 52 E
Jabulani, Jobg. 18 F8 26 14 S 27 51 E
Jacarepaguá, Rio J. 31 B1 22 56 S 43 20W
Jackson Heights, N.Y. 23 C5 40 44N 73 53W
Jackson Park, Chic. 26 C3 41 46N 87 34W
Jacksonville, N.Y. 22 B3 40 57N 74 18W
Jacomino, La Hab. 30 B3 23 6N 82 19W
Jacques Cartier, Mtrl. 20 A5 43 31N 73 27W
Jægersborg, Købn. 2 D10 55 45N 12 31 E
Jægersborg Dyrehave, Købn. 2 D10 55 46N 12 33 E
Jægersborg Hegn, Købn. 2 D10 55 49N 12 33 E
Jafarpur, Calc. 16 D6 22 45N 88 22 E
Jagacha, Calc. 16 E5 22 35N 88 17 E
Jagannathpur, Calc. 16 E5 22 43N 88 18 E
Jagatdal, Calc. 16 C6 22 51N 88 23 E
Jagatgagar, Calc. 16 D5 22 48N 88 15 E
Jagatpur, Delhi 16 A2 28 44N 77 13 E
Jagdispur, Calc. 16 E5 22 39N 88 17 E
Jaguara, S. Pau. 31 E5 23 30 S 46 45W
Jaguaré, S. Pau. 31 E5 23 32 S 46 45W
Jaguaré →, S. Pau. 31 E5 23 32 S 46 45W
Jahangirpur, Delhi 16 A2 28 43N 77 12 E
Jaimanitas →, La Hab. 30 B2 23 5N 82 29W
Jakarta, Jak. 15 H10 6 9 S 106 52 E
Jakarta, Teluk, Jak. 15 H9 6 5 S 106 50 E
Jakosberg, Stock. 3 D9 59 25N 17 47 E
Jalan Kayu, Sing. 15 F8 1 24N 103 52 E
Jamaica, N.Y. 23 C6 40 42N 73 48W
Jamaica B., N.Y. 23 D6 40 36N 73 49W
Jamaica Plain, Bost. 21 D3 42 18N 71 6W
Jamshīdābād, Tehr. 17 C5 35 42N 51 22 E
Jamsil, Sŏul 12 G8 37 30N 127 4 E
Jamweon, Sŏul 12 G8 37 30N 127 0 E
Jan Smuts Airport, Jobg. 18 E10 26 7 S 28 14 E
Janai, Calc. 16 D5 22 43N 88 15 E
Janā'in, Bagd. 17 F8 33 18N 44 22 E
Janki, Wsaw. 10 F6 52 8N 20 52 E
Jannali, Syd. 19 C3 34 0 S 151 4 E
Jánoshegy, Bud. 10 J12 47 31N 18 57 E
Janów, Wsaw. 10 E6 52 16N 20 50 E
Janvry, Paris 5 D2 48 38N 2 9 E
Jaraguá, S. Pau. 31 D5 23 27 S 46 44W
Jaraguá, Pico de, S. Pau. 31 D5 23 27 S 46 46W
Jarama →, Mdrd. 8 B3 40 29N 3 32W
Jardim América, S. Pau. 31 E6 23 34 S 46 39W
Jardim Anchieta, S. Pau. 31 F7 23 41 S 46 27W
Jardim Arpoador, S. Pau. 31 E5 23 35 S 46 48W
Jardim do Mar, S. Pau. 31 F6 23 41 S 46 33W
Jardim Munhoz, S. Pau. 31 E6 23 30 S 46 33W
Jardim Osasco, S. Pau. 31 E5 23 33 S 46 47W
Jardim Ouro Preto, S. Pau. 31 E5 23 35 S 46 46W
Jardim Paulista, S. Pau. 31 E5 23 34 S 46 41W
Jardim Petrópolis, S. Pau. 31 F7 23 41 S 46 23W
Jardim Rochidale, S. Pau. 31 E5 23 30 S 46 46W
Jardim Santista, S. Pau. 31 F7 23 40 S 46 24W
Jardim São Bento, S. Pau. 31 F5 23 40 S 46 46W
Jardim São Francisco, S. Pau. 31 E7 23 38 S 46 26W
Jardim Sapopemba, S. Pau. 31 E7 23 36 S 46 29W
Jardim Vera Cruz, S. Pau. 31 E7 23 35 S 46 27W
Jardim Vista Alegre, S. Pau. 31 E5 23 37 S 46 49W
Jardim Zaira, S. Pau. 31 E7 23 39 S 46 25W
Jardines Flotantes, Méx. 29 C3 19 16N 99 6W
Jardine's Lookout, H.K. 12 E6 22 16N 114 11 E
Järfälla, Stock. 3 D10 59 23N 17 51 E
Järvafältet, Stock. 3 D10 59 25N 17 54 E
Järventausta, Hels. 3 A1 60 21N 24 28 E
Jasai, Bomb. 16 H9 18 56N 73 1 E
Jaskhar, Bomb. 16 H8 18 54N 72 58 E
Jatinegara, Jak. 15 J10 6 13 S 106 52 E
Jauli, Delhi 16 A3 28 44N 77 20 E
Jawādiyeh, Tehr. 17 D5 35 39N 51 22 E
Jaworowa, Wsaw. 10 F6 52 9N 20 56 E
Jayang, Sŏul 12 G8 37 32N 127 3 E
Jedlesee, Wien 10 G10 48 15N 16 23 E
Jefferson, Phil. 24 C3 39 45N 75 12W
Jefferson Park, Chic. 26 B2 41 58N 87 46W
Jeffersonville, Phil. 24 A2 40 8N 75 23W
Jegi, Sŏul 12 G8 37 34N 127 1 E
Jells Park, Melb. 19 F8 37 53 S 145 11 E
Jelonki, Wsaw. 10 E6 52 14N 20 54 E
Jenfeld, Hbg. 7 D8 53 34N 10 8 E
Jenkintown, Phil. 24 A4 40 6N 75 8W
Jeongreung, Sŏul 12 G8 37 35N 127 0 E
Jericho, N.Y. 23 C7 40 47N 73 32W
Jerónimes, Mosteiro dos, Lisb. 8 F7 38 41N 9 11W
Jersey City, N.Y. 22 C4 40 42N 74 4W
Jésus, Î., Mtrl. 20 A3 43 36N 73 44W
Jesus Del Monte, La Hab. 30 B2 23 6N 82 21W
Jesús Maria, Lima 30 G8 12 4 S 77 3W
Jhenkari, Calc. 16 D5 22 45N 88 18 E
Jhil Kuranga, Delhi 16 B2 28 39N 77 14 E
Jiangqiao, Shang. 14 J11 31 15N 121 20 E
Jiangtai, Beij. 14 B3 39 57N 116 28 E
Jianguomen, Beij. 14 B3 39 53N 116 24 E
Jiangwan, Shang. 14 J11 31 18N 121 28 E
Jianshan Gongyuan, Tianj. 14 E6 39 5N 117 12 E
Jihād, Bagd. 17 F7 33 17N 44 19 E
Jingan, Shang. 14 J11 31 14N 121 25 E
Jinočany, Pra. 10 B1 50 2N 14 16 E
Jinonice, Pra. 10 B2 50 3N 14 22 E
Jirny, Pra. 10 B4 50 7N 14 41 E
Jiuxianqiao, Beij. 14 B3 39 58N 116 28 E
Jiyźgaoka, Tōkyō 13 C3 35 35N 139 40 E
Jizā'er, Bagd. 17 F8 33 15N 44 23 E
Jizīra, Bagd. 17 F8 33 15N 44 25 E
Joan Despi, Barc. 8 D5 41 22N 2 2 E
Joaquin Miller Park, S.F. 27 B3 37 48N 122 11W
Johannesburg, Jobg. 18 F9 26 11 S 28 2 E
Johanneskirchen, Mün. 7 F10 48 10N 11 38 E
Johannesstift, Berl. 7 A2 52 34N 13 12 E
Johannisthal, Berl. 7 B4 52 26N 13 30 E
John F. Kennedy Int. Airport, N.Y. 23 D6 40 39N 73 45W
John F. Kennedy Nat. Hist. Site, Bost. 21 C3 42 20N 71 7W
John Hancock Center, Chic. 26 B3 41 53N 87 37W
John Hopkins Univ., Balt. 25 B3 39 19N 76 37W
John McLaren Park, S.F. 27 B2 37 43N 122 24W
Joinville-le-Pont, Paris 5 C4 48 49N 2 27 E
Jollas, Hels. 3 B5 60 10N 25 5 E
Jones Beach State Park, N.Y. 23 D7 40 35N 73 32W
Jones Falls →, Balt. 25 B3 39 20N 76 36W
Jones Inlet, N.Y. 23 D7 40 34N 73 34W
Jonestown, Balt. 25 B2 39 13N 76 48W
Jong Ro, Sŏul 12 G7 37 34N 126 58 E
Jongmyo Royal Shrine, Sŏul 12 G7 37 34N 126 59 E
Jonstrup, Købn. 2 D9 55 45N 12 20 E
Joppatowne, Balt. 25 A4 39 24N 76 20W
Jordan Valley, H.K. 12 D6 22 20N 114 12 E
Jorge Chavez, Aeropuerto Int., Lima 30 G8 12 2 S 77 8W
Jorvas, Hels. 3 C2 60 8N 24 30 E
José C. Paz, B.A. 32 B2 34 31 S 58 44W
José L. Suárez, B.A. 32 B3 34 32 S 58 34W
José Mármol, B.A. 32 C4 34 47 S 58 22W
Jose Marti, Aeropuerto Int., La Hab. 30 C2 22 59N 82 22W
Josephine Pk., L.A. 28 A4 34 17N 118 7W
Jōsō, Ōsaka 12 B3 34 42N 135 27 E
Jōtō, Ōsaka 12 B4 34 42N 135 33 E
Jouars-Pontchartrain, Paris 5 C1 48 47N 1 53 E
Jouy-en-Josas, Paris 5 C3 48 46N 2 10 E
Jouy-le-Moutier, Paris 5 A2 49 0N 2 2 E
Józefów, Wsaw. 10 F8 52 8N 21 13 E
Juan Escutia, Méx. 29 B3 19 23N 99 3W
Juan González Romero, Méx. 29 A3 19 30N 99 3W
Juhu, Bomb. 16 G9 19 5N 73 0 E
Juilly, Paris 5 A6 49 0N 2 42 E
Jūjā, Tōkyō 13 B3 35 45N 139 43 E
Jukskeirivier →, Jobg. 18 E9 26 5 S 28 6 E
Julianów, Wsaw. 10 E7 52 10N 21 9 E
Jung, Sŏul 12 G7 37 33N 126 59 E
Jungfernheide, Volkspark, Berl. 7 A2 52 32N 13 18 E
Jungfernsee, Berl. 7 B1 52 25N 13 6 E
Jungwha, Sŏul 12 G8 37 35N 127 3 E
Junk B., H.K. 12 E6 22 17N 114 15 E
Jurong, Sing. 15 G7 1 19N 103 40 E
Jurong, Selat, Sing. 15 G7 1 17N 103 42 E
Jurong, Sungei →, Sing. 15 G7 1 17N 103 45 E
Jurubatuba, S. Pau. 31 F5 23 40 S 46 41W
Jurujuba, Enseada de, Rio J. 31 B3 22 54 S 43 6W
Justice, Chic. 26 C2 41 44N 87 49W
Juusjärvi, Hels. 3 B1 60 12N 24 26 E
Juva, Hels. 3 B3 60 16N 24 45 E
Juvisy-sur-Orge, Paris 5 C4 48 41N 2 21 E
Jwalahari, Delhi 16 B1 28 40N 77 6 E
Jyllinge, Købn. 2 D7 55 45N 12 6 E

K

Kaarst, Ruhr 6 C1 51 13N 6 36 E
Kabaty, Wsaw. 10 F7 52 8N 21 4 E
Kabel, Ruhr 6 B6 51 24N 7 28 E
Kadiköy, Ist. 17 B3 40 59N 29 1 E
Kadoma, Ōsaka 12 B4 34 44N 135 35 E
Kafr es Sammân, El Qâ. 18 D4 29 58N 31 8 E
Kâğithane, Ist. 17 A2 41 4N 28 58 E
Kâğithane →, Ist. 17 A2 41 9N 28 56 E
Kagran, Wien 10 G10 48 14N 16 26 E
Kahlenberg, Wien 10 G9 48 16N 16 19 E
Kai Tak, H.K. 12 D6 22 20N 114 11 E
Kaisariani, Ath. 8 J11 37 57N 23 46 E
Kaiser-Mühlen, Wien 10 G10 48 13N 16 25 E
Kaiserebersdorf, Wien 10 H10 48 9N 16 26 E
Kaiserswerth, Ruhr 6 C2 51 18N 6 44 E
Kaivoksela, Hels. 3 B4 60 15N 24 53 E
Kakukk-hegy, Bud. 10 K12 47 29N 18 57 E
Kalachhara, Calc. 16 D5 22 42N 88 15 E
Kalamákion, Ath. 8 J11 37 52N 23 43 E
Kaldenhausen, Ruhr 6 B1 51 23N 6 39 E
Kalipur, Calc. 16 D5 22 40N 88 17 E
Kalkaji, Delhi 16 B2 28 32N 77 12 E
Kalksburg, Wien 10 H9 48 8N 16 15 E
Kallang →, Sing. 15 F8 1 21N 103 51 E
Kallhäll, Stock. 3 D9 59 26N 17 48 E
Kallithéa, Ath. 8 J11 37 56N 23 43 E
Kallvik, Hels. 3 B5 60 12N 25 8 E
Kaltbründsberg, Wien 10 G9 48 10N 16 13 E
Kaltenleutgeben, Wien 10 H9 48 7N 16 11 E
Kalveboderne, Købn. 2 E10 55 37N 12 31 E
Kalytino, St-Pet. 11 B5 59 59N 30 39 E
Kamaraerdö, Bud. 10 K12 47 26N 18 59 E
Kamarhati, Calc. 16 D6 22 40N 88 23 E
Kamarkunda, Calc. 16 D5 22 49N 88 12 E
Kamata, Tōkyō 13 C3 35 33N 139 43 E
Kamdebpur, Calc. 16 C5 22 53N 88 19 E
Kameari, Tōkyō 13 B4 35 45N 139 50 E
Kameido, Tōkyō 13 B4 35 42N 139 50 E
Kami-hoshikawa, Tōkyō 13 D2 35 27N 139 35 E
Kami-Itabashi, Tōkyō 13 B3 35 45N 139 40 E
Kami-nakazato, Tōkyō 13 D2 35 22N 139 36 E
Kami-saruyama, Tōkyō 13 C2 35 30N 139 31 E
Kami-sugata, Tōkyō 13 D2 35 28N 139 34 E
Kami-tomi, Tōkyō 13 B1 35 48N 139 29 E
Kamikitazawa, Tōkyō 13 C2 35 39N 139 36 E
Kamikiyoto, Tōkyō 13 B2 35 45N 139 30 E
Kamishiki, Tōkyō 13 B4 35 46N 139 57 E
Kamitsuruma, Tōkyō 13 C1 35 30N 139 26 E
Kamiyama, Tōkyō 13 B2 35 46N 139 32 E
Kamoi, Tōkyō 13 C2 35 30N 139 34 E
Kamoshida, Tōkyō 13 C2 35 33N 139 31 E
Kampong Batak, Sing. 15 F8 1 20N 103 54 E
Kampong Mandai Kechil, Sing. 15 F7 1 26N 103 46 E
Kampong Pachitan, Sing. 15 G8 1 19N 103 54 E
Kampong Potong Pasir, Sing. 15 F8 1 20N 103 52 E
Kampong Reteh, Sing. 15 G8 1 19N 103 53 E
Kampong Tengah, Sing. 15 F7 1 22N 103 42 E
Kampong Ulu Jurong, Sing. 15 F7 1 20N 103 42 E
Kampung Ambon, Jak. 15 J10 6 11 S 106 53 E
Kampung Bali, Jak. 15 J9 6 11 S 106 48 E
Kan, Tehr. 17 C4 35 45N 51 16 E
Kanagawa-Ku, Tōkyō 13 D2 35 29N 139 38 E
Kanamachi, Tōkyō 13 B4 35 46N 139 52 E
Kanamori, Tōkyō 13 C1 35 31N 139 27 E
Kanda, Tōkyō 13 B3 35 41N 139 45 E
Kandang Kerbau, Sing. 15 G8 1 18N 103 51 E
Kandilli, Ist. 17 A3 41 4N 29 3 E
Kanegasaku, Tōkyō 13 B4 35 48N 139 56 E
Kangaroo Ground, Melb. 19 E8 37 41 S 145 13 E
Kankinara, Calc. 16 C6 22 51N 88 24 E
Kankurgachi, Calc. 16 E6 22 34N 88 23 E
Kanlica, Ist. 17 A3 41 5N 29 3 E
Kanoaka, Ōsaka 12 C4 34 33N 135 31 E
Kanonerskiy, Os., St-Pet. 11 B3 59 53N 30 13 E
Kanzaki →, Ōsaka 12 B3 34 41N 135 24 E
Kapellerfeld, Wien 10 G10 48 18N 16 29 E
Kapotnya, Mos. 11 F10 55 39N 37 48 E
Käppala, Stock. 3 D12 59 21N 18 13 E
Käpylä, Hels. 3 B4 60 13N 24 57 E
Karachi, Kar. 17 G11 24 50N 67 0 E
Karachi Int. Airport, Kar. 17 G11 24 5N 67 9 E
Karachi Univ., Kar. 17 G11 24 51N 67 0 E
Karagümrük, Ist. 17 A2 41 1N 28 56 E
Karāma, Bagd. 17 E8 33 20N 44 22 E
Karato, Ōsaka 12 B2 34 46N 135 12 E
Karave, Bomb. 16 G9 19 0N 73 0 E
Karet, Jak. 15 J9 6 12 S 106 49 E
Karkar Duman, Delhi 16 B2 28 39N 77 18 E
Karkh, Bagd. 17 E8 33 20N 44 22 E
Karlberg, Stock. 3 D11 59 20N 18 1 E
Karlin, Pra. 10 B2 50 5N 14 26 E
Karlsfeld, Mün. 7 F9 48 13N 11 28 E
Karlshorst, Berl. 7 B4 52 29N 13 31 E
Karlslunde Strand, Købn. 2 E8 55 33N 12 15 E
Karnap, Ruhr 6 A4 51 31N 7 0 E
Karolinenhof, Berl. 7 B4 52 23N 13 38 E
Karow, Berl. 7 A3 52 36N 13 29 E
Karrādah, Bagd. 17 F8 33 17N 44 23 E
Kärsön, Stock. 3 E10 59 19N 17 54 E
Kasai, Tōkyō 13 C4 35 39N 139 52 E
Kasetsart, Bangk. 15 A2 13 50N 100 34 E
Kashi-Hazaki, Tōkyō 13 A3 35 54N 139 42 E
Kashio, Tōkyō 13 D2 35 24N 139 34 E
Kashiwa, Tōkyō 13 A4 35 51N 139 57 E
Kashiwara, Ōsaka 12 C4 34 34N 135 37 E
Kaskela, Hels. 3 B5 60 17N 25 6 E
Kastrup, Købn. 2 E10 55 38N 12 39 E
Kastrup Lufthavn, Købn. 2 E11 55 37N 12 14 E
Kasuga, Tōkyō 13 B2 35 45N 139 38 E
Kasuge, Tōkyō 13 B3 35 45N 139 49 E
Kasumigasek, Tōkyō 13 C3 35 40N 139 46 E
Katabira →, Tōkyō 13 D2 35 27N 139 38 E
Katernberg, Ruhr 6 A4 51 30N 7 4 E
Katong Park, Sing. 15 G8 1 18N 103 53 E
Katrineberg, Stock. 3 E10 59 13N 17 54 E
Katsushika-Ku, Tōkyō 13 B4 35 44N 139 51 E
Kattinge Vig, Købn. 2 D7 55 40N 12 1 E
Kau Pei Chau, H.K. 12 E6 22 12N 114 15 E
Kau Yi Chau, H.K. 12 E5 22 17N 114 4 E
Kauklahti, Hels. 3 B2 60 11N 24 36 E
Kaulsdorf, Berl. 7 B4 52 29N 13 34 E
Kauniainen, Hels. 3 B3 60 13N 24 44 E
Kawagoe, Tōkyō 13 A1 35 54N 139 29 E
Kawaguchi, Tōkyō 13 B3 35 47N 139 43 E
Kawai, Tōkyō 13 D2 35 29N 139 31 E
Kawamukō, Tōkyō 13 C2 35 30N 139 34 E
Kawanishi, Ōsaka 12 B3 34 49N 135 24 E
Kawasaki, Tōkyō 13 C3 35 31N 139 43 E
Kawasaki Harbour, Tōkyō 13 D3 35 30N 139 47 E
Kawawa, Tōkyō 13 C2 35 31N 139 33 E
Kawęczyn, Wsaw. 10 E7 52 16N 21 5 E
Kayu Putih, Jak. 15 J10 6 10 S 106 53 E
Kbely, Pra. 10 B3 50 8N 14 32 E
Kearny, N.Y. 22 C4 40 45N 74 8W
Kebayoran Baru, Jak. 15 J9 6 14 S 106 47 E
Kebayoran Lama, Jak. 15 J9 6 13 S 106 46 E
Kebon Jeruk, Jak. 15 J9 6 11 S 106 46 E
Keferloh, Mün. 7 G11 48 5N 11 43 E
Keilor, Melb. 19 E6 37 42 S 144 50 E
Keilor East, Melb. 19 E6 37 44 S 144 51 E
Keimola, Hels. 3 A3 60 20N 24 49 E
Kelenföld, Bud. 10 K13 47 27N 19 2 E
Kelvedon Hatch, Lon. 4 A6 51 40N 0 16 E
Kelvin, Jobg. 18 E9 26 4 S 28 5 E
Kemang, Jak. 15 J9 6 15 S 106 48 E
Kemayoran, Jak. 15 J10 6 10 S 106 51 E
Kemayoran Airport, Jak. 15 H10 6 8 S 106 50 E
Kemp Mill, Wash. 25 C7 39 0N 77 1W
Kempton Park, Jobg. 18 E10 26 5 S 28 14 E
Kempton Racecourse, Lon. 4 C2 51 24N 0 23W
Kemsing, Lon. 4 D6 51 18N 0 12 E
Kendall Green, Bost. 21 C2 42 22N 71 16W
Kendua, Calc. 16 E5 22 34N 88 10 E
Keng Hau, H.K. 12 D6 22 22N 114 10 E
Kenilworth, Chic. 26 A2 42 5N 87 42W
Kenilworth, N.Y. 22 C3 40 40N 74 16W
Kenley, Lon. 4 D4 51 19N 0 6W
Kenmare, Jobg. 18 E7 26 6 S 27 48 E
Kennedy Grove Regional Rec. Area, S.F. 27 A3 37 56N 122 14W

40 Kennedy Town

Kennedy Town, *H.K.* **12 E5** 22 16N 114 6 E
Kensal Green, *Lon.* **4 B3** 51 32N 0 13W
Kensington, *Jobg.* **18 F9** 26 11 S 28 6 E
Kensington, *Lon.* **4 C3** 51 29N 0 10W
Kensington, *N.Y.* **22 D5** 40 38N 73 57W
Kensington, *Phil.* **24 B4** 39 59N 75 6W
Kensington, *S.F.* **27 A3** 37 54N 122 17W
Kensington, *Syd.* **19 B4** 33 54 S 151 13 E
Kensington, *Wash.* **25 C7** 39 1N 77 4W
Kensington Palace, *Lon.* **4 B3** 51 30N 0 11W
Kent Woodlands, *S.F.* **27 A1** 37 56N 122 34W
Kentfield, *S.F.* **27 A1** 37 57N 122 33W
Kentish Town, *Lon.* **4 B4** 51 32N 0 8W
Kentland, *Wash.* **25 D8** 38 55N 76 53W
Kenton, *Lon.* **4 B3** 51 34N 0 17W
Kenwood, *Balt.* **25 A4** 39 20N 76 30W
Kenwood, *Bost.* **21 B2** 42 40N 71 14W
Kenwood House, *Lon.* **4 B4** 51 34N 0 9W
Kepa, *Wsaw.* **10 E7** 52 13N 21 3 E
Keppel Harbour, *Sing.* **15 G7** 1 15N 103 49 E
Kerameikos, *Ath.* **8 J11** 37 58N 23 42 E
Kerepes, *Bud.* **10 J14** 47 33N 19 17 E
Keston, *Lon.* **4 C5** 51 21N 0 1 E
Keston Mark, *Lon.* **4 C5** 51 21N 0 2 E
Keth Wara, *Delhi* **16 A2** 28 40N 77 13 E
Kettering, *Wash.* **25 D9** 38 53N 76 49W
Kettwig, *Ruhr* **6 B3** 51 22N 6 56 E
Kew, *Jobg.* **18 E9** 26 7 S 28 5 E
Kew, *Lon.* **4 C3** 51 28N 0 17W
Kew, *Melb.* **19 E7** 37 48 S 145 2 E
Kew Gardens, *Lon.* **4 C3** 51 28N 0 17W
Key Gardens, *Trto.* **20 E9** 43 39N 79 18W
Khaboye, *St-Pet.* **11 B6** 59 53N 30 44 E
Khaidhárion, *Ath.* **8 H10** 38 2N 23 38 E
Khairna, *Bomb.* **16 G9** 19 5N 73 0 E
Khalándrion, *Ath.* **8 H11** 38 3N 23 48 E
Khalīj, *Bagd.* **17 F8** 33 18N 44 28 E
Khansā, *Bagd.* **17 E8** 33 21N 44 28 E
Kharavli, *Bomb.* **16 H8** 18 54N 72 55 E
Khardah, *Calc.* **16 D6** 22 43N 88 22 E
Khayala, *Delhi* **16 B1** 28 39N 77 6 E
Khefren, *El Qâ.* **18 D4** 29 58N 31 8 E
Khichripur, *Delhi* **16 B2** 28 37N 77 18 E
Khimki, *Mos.* **11 D8** 55 53N 37 24 E
Khimki-Khovrino, *Mos.* **11 D9** 55 51N 37 31 E
Khimkinskoye Vdkr., *Mos.* **11 D8** 55 51N 37 27 E
Khirvosti, *St-Pet.* **11 B5** 59 56N 30 37 E
Khlongsan, *Bangk.* **15 B1** 13 43N 100 29 E
Kholargós, *Ath.* **8 J11** 37 59N 23 48 E
Khorel, *Calc.* **16 D5** 22 41N 88 18 E
Khorosovo, *Mos.* **11 E8** 55 46N 37 27 E
Khudrā, *Bagd.* **17 F7** 33 19N 44 17 E
Khun Thian, *Bangk.* **15 B1** 13 41N 100 27 E
Khuraiji Khas, *Delhi* **16 B2** 28 38N 77 16 E
Khurigachi, *Calc.* **16 D5** 22 48N 88 21 E
Kiamari, *Kar.* **17 H10** 24 49N 66 58 E
Kidderpore, *Calc.* **16 E5** 22 32N 88 19 E
Kienwerder, *Berl.* **7 B2** 52 22N 13 11 E
Kierling, *Wien* **10 G9** 48 18N 16 16 E
Kierlingbach →, *Wien* **10 G9** 48 18N 16 19 E
Kierlinger Forst, *Wien* **10 G9** 48 17N 16 14 E
Kierst, *Ruhr* **6 C2** 51 18N 6 42 E
Kifisós →, *Ath.* **8 J11** 37 58N 23 42 E
Kikenka →, *St-Pet.* **11 B2** 59 50N 30 3 E
Kikuna, *Tōkyō* **13 C2** 35 30N 139 37 E
Kil, *Stock.* **3 D12** 59 20N 18 19 E
Kilburn, *Lon.* **4 B3** 51 32N 0 11W
Killara, *Syd.* **19 A4** 33 46 S 151 10 E
Kilo, *Hels.* **3 B3** 60 13N 24 47 E
Kilokri, *Delhi* **16 B2** 28 34N 77 15 E
Kilsyth, *Melb.* **19 E8** 37 48 S 145 18 E
Kimberton, *Phil.* **24 A1** 40 7N 75 34W
Kimlin Park, *Sing.* **15 G7** 1 18N 103 49 E
Kindi, *Bagd.* **17 F8** 33 18N 44 22 E
King of Prussia, *Phil.* **24 A2** 40 5N 75 22W
Kings Cross, *Syd.* **19 B4** 33 52 S 151 12 E
Kings Domain, *Melb.* **19 E6** 37 49 S 144 58 E
Kings Mt., *S.F.* **27 D3** 37 27N 122 19W
King's Park, *H.K.* **12 E6** 22 18N 114 10 E
Kings Park, *Wash.* **25 E6** 38 48N 77 17W
King's Point, *N.Y.* **23 C6** 40 48N 73 45W
Kingsbury, *Lon.* **4 B3** 51 34N 0 15W
Kingsford, *Syd.* **19 B4** 33 55 S 151 14 E
Kingston upon Thames, *Lon.* **4 C3** 51 24N 0 17W
Kingston Vale, *Lon.* **4 C3** 51 25N 0 25W
Kingsway, *Trto.* **20 E7** 43 38N 79 32W
Kingswood, *Lon.* **4 D3** 51 17N 0 12W
Kinnelon, *N.Y.* **22 B2** 40 59N 74 23W
Kipling Heights, *Trto.* **20 D7** 43 43N 79 34W
Kipséli, *Ath.* **8 J11** 37 59N 23 45 E
Kirchhellen, *Ruhr* **6 A3** 51 36N 6 56 E
Kirchhörde, *Ruhr* **6 B6** 51 27N 7 27 E
Kirchlinde, *Ruhr* **6 A6** 51 31N 7 22 E
Kirchof, *Hbg.* **7 E8** 53 29N 10 1 E
Kirchsteinbek, *Hbg.* **7 D8** 53 31N 10 7 E
Kirchstockbach, *Mün.* **7 G11** 48 1N 11 40 E
Kirchtrudering, *Mün.* **7 G11** 48 7N 11 40 E
Kirdasa, *El Qâ.* **18 C4** 30 2N 31 6 E
Kirikiri, *Lagos* **18 B1** 6 26N 7 18 E
Kirkkonummi, *Hels.* **3 C1** 60 6N 24 28W
Kirkland, *Mtrl.* **20 B2** 43 26N 73 51W
Kirovskiye, Os., *St-Pet.* **11 B3** 59 57N 30 14 E
Kisarazu, *Tōkyō* **13 D4** 35 21N 139 54 E
Kisikli, *Ist.* **17 A3** 41 1N 29 2 E
Kispest, *Bud.* **10 K13** 47 27N 19 8 E
Kista, *Stock.* **3 D10** 59 24N 17 57 E
Kistarcsa, *Bud.* **10 J14** 47 32N 19 16 E
Kita, *Ōsaka* **12 B4** 34 41N 135 30 E
Kita-Ku, *Tōkyō* **13 B3** 35 44N 139 44 E
Kitain-Temple, *Tōkyō* **13 A1** 35 54N 139 29 E
Kitazawa, *Tōkyō* **13 C3** 35 39N 139 40 E
Kiu Tsiu, *H.K.* **12 D6** 22 22N 114 17 E
Kivistö, *Hels.* **3 B4** 60 18N 24 50 E
Kiyose, *Tōkyō* **13 B2** 35 46N 139 31 E
Kiziltoprak, *Ist.* **17 B3** 40 58N 29 3 E
Kizu →, *Ōsaka* **12 C3** 34 37N 135 27 E
Kizuri, *Ōsaka* **12 C4** 34 38N 135 34 E
Kjeller, *Oslo* **2 B6** 59 58N 11 1 E
Kjelsås, *Oslo* **2 B4** 59 57N 10 47 E
Kladow, *Berl.* **7 B1** 52 27N 13 9 E
Klampenborg, *Købn.* **2 D10** 55 46N 12 35 E
Klánovice, *Pra.* **10 B3** 50 5N 14 40 E
Klaudyń, *Wsaw.* **10 E6** 52 17N 20 50 E
Klecany, *Pra.* **10 A2** 50 10N 14 24 E
Kledering, *Wien* **10 H10** 48 8N 16 26 E
Klein Gleinicke, *Berl.* **7 B1** 52 23N 13 5 E
Klein-Hadern, *Mün.* **7 G9** 48 7N 11 28 E
Klein Jukskei →, *Jobg.* **18 E8** 26 6 S 27 57 E
Kleinburg, *Trto.* **20 C7** 43 51N 79 37W
Kleine Grasbrook, *Hbg.* **7 D7** 53 31N 9 59 E
Kleinmachnow, *Berl.* **7 B2** 52 24N 13 14 E
Kleinschönebeck, *Berl.* **7 B5** 52 29N 13 42 E
Kleinziethen, *Berl.* **7 B3** 52 22N 13 26 E
Klemetsrud, *Oslo* **2 B5** 59 50N 10 51 E
Klender, *Jak.* **15 J10** 6 12 S 106 53 E
Klippoortje, *Jobg.* **18 F10** 26 14 S 28 10 E
Klipriviersberg, *Jobg.* **18 F9** 26 16 S 28 2 E
Klipspruit →, *Jobg.* **18 F8** 26 14 S 27 53 E
Kloofendal, *Jobg.* **18 E8** 26 8 S 27 52 E
Klosterhardt, *Ruhr* **6 A3** 51 32N 6 52 E
Klosterneuburg, *Wien* **10 G9** 48 18N 16 19 E
Knockholt Pound, *Lon.* **4 D5** 51 18N 0 7 E
Knowland State Arboretum and Park, *S.F.* **27 B4** 37 45N 122 7W
Knox Park, *Melb.* **19 F8** 37 54 S 145 15 E
Knoxville, *Melb.* **19 F8** 37 53 S 145 14 E
Köbanya, *Bud.* **10 K13** 47 28N 19 9 E
Kobe, *Ōsaka* **12 B2** 34 41N 135 13 E
Kōbe Harbour, *Ōsaka* **12 C2** 34 39N 135 11 E
København, *Købn.* **2 D9** 55 40N 12 26 E
Kobylisy, *Pra.* **10 B2** 50 7N 14 26 E
Kobyłka, *Wsaw.* **10 D8** 52 20N 21 10 E
Kocasinan, *Ist.* **17 A2** 41 1N 28 50 E
Kočife, *Pra.* **10 B2** 50 3N 14 21 E
Kodaira, *Tōkyō* **13 B1** 35 43N 139 28 E
Kodanaka, *Tōkyō* **13 C2** 35 34N 139 37 E
Kogane, *Tōkyō* **13 B4** 35 49N 139 55 E
Koganei, *Tōkyō* **13 B2** 35 42N 139 31 E
Kogarah, *Syd.* **19 B3** 33 57 S 151 8 E
Køge Bugt, *Købn.* **2 E9** 55 34N 12 24 E
Köhlbrand Rethe, *Hbg.* **7 D7** 53 31N 9 56 E
Köhlfleet, *Hbg.* **7 D7** 53 32N 9 53 E
Koivupää, *Hels.* **3 B4** 60 18N 24 53 E
Koja, *Jak.* **15 H10** 6 8 S 106 52 E
Koja Utara, *Jak.* **15 H10** 6 5 S 106 53 E
Kokobunji, *Tōkyō* **13 B1** 35 42N 139 27 E
Kokobunji-Temple, *Tōkyō* **13 B4** 35 44N 139 55 E
Kol Scholven, *Ruhr* **6 A3** 51 35N 6 59 E
Kolarängen, *Stock.* **3 E12** 59 16N 18 10 E
Kolbotn, *Oslo* **2 C4** 59 48N 10 48 E
Kole Kalyan, *Bomb.* **16 G8** 19 5N 72 50 E
Kolmiranta, *Hels.* **3 B2** 60 15N 24 31 E
Kolmperä, *Hels.* **3 B2** 60 15N 24 32 E
Koło, *Wsaw.* **10 E6** 52 14N 20 56 E
Kolodeje, *Pra.* **10 B3** 50 3N 14 38 E
Kolokinthou, *Ath.* **8 J11** 38 0N 23 42 E
Kolomenskoye, *Mos.* **11 E10** 55 40N 37 40 E
Kolomyagi, *St-Pet.* **11 A3** 60 0N 30 19 E
Kolónos, *Ath.* **8 J11** 37 59N 23 43 E
Kolovraty, *Pra.* **10 B3** 50 0N 14 37 E
Kolsås, *Oslo* **2 B3** 59 55N 10 30 E
Koltushi, *St-Pet.* **11 B5** 59 55N 30 38 E
Komae, *Tōkyō* **13 C2** 35 37N 139 34 E
Komagome, *Tōkyō* **13 B3** 35 43N 139 45 E
Komazawa, *Tōkyō* **13 C3** 35 37N 139 40 E
Komdhara, *Calc.* **16 C5** 22 52N 88 14 E
Kommunarka, *Mos.* **11 F8** 55 35N 37 29 E
Komorów, *Wsaw.* **10 F5** 52 9N 20 48 E
Kona, *Calc.* **16 E5** 22 37N 88 18 E
Konala, *Hels.* **3 B4** 60 14N 24 52 E
Kōnan, *Tōkyō* **13 D2** 35 23N 139 35 E
Kondli, *Delhi* **16 B2** 28 36N 77 19 E
Kong Sin Wan, *H.K.* **12 E5** 22 15N 114 7 E
Kongelunden, *Købn.* **2 E10** 55 34N 12 34 E
Kongens Lyngby, *Købn.* **2 D10** 55 46N 12 30 E
Kongo, *Hels.* **3 A3** 60 20N 24 47 E
Königshardt, *Ruhr* **6 A3** 51 33N 6 51 E
Konnagar, *Calc.* **16 D6** 22 42N 88 21 E
Konohana, *Ōsaka* **12 B3** 34 40N 135 26 E
Kōnoike, *Ōsaka* **12 B4** 34 42N 135 37 E
Konradshöhe, *Berl.* **7 A2** 52 35N 13 13 E
Koonung Cr. →, *Melb.* **19 E7** 37 46 S 145 4 E
Kopanina, *Pra.* **10 B1** 50 3N 14 17 E
Koparkhairna, *Bomb.* **16 G8** 19 6N 72 59 E
Köpenick, *Berl.* **7 B4** 52 26N 13 35 E
Korangi, *Kar.* **17 H11** 24 47N 67 8 E
Koremasa, *Tōkyō* **13 C1** 35 39N 139 29 E
Korenevo, *Mos.* **11 E12** 55 40N 38 0 E
Kori, *Ōsaka* **12 B4** 34 47N 135 38 E
Koridhallós, *Ath.* **8 J10** 37 59N 23 39 E
Korkinskoye, Oz., *St-Pet.* **11 B6** 59 55N 30 42 E
Körne, *Ruhr* **6 A7** 51 30N 7 30 E
Korso, *Hels.* **3 A5** 60 21N 25 5 E
Koshigaya, *Tōkyō* **13 A3** 35 53N 139 47 E
Kosino, *Mos.* **11 E11** 55 43N 37 50 E
Kosugi, *Tōkyō* **13 C2** 35 34N 139 39 E
Kota, *Jak.* **15 H9** 6 7 S 106 48 E
Kotelyniki, *Mos.* **11 F11** 55 39N 37 52 E
Kōtō-Ku, *Tōkyō* **13 B3** 35 40N 139 48 E
Kotrang, *Calc.* **16 D6** 22 41N 88 20 E
Kouponia, *Ath.* **8 J11** 37 57N 23 47 E
Koviksudde, *Stock.* **3 D13** 59 21N 18 21 E
Kowloon, *H.K.* **12 E5** 22 18N 114 10 E
Kowloon City, *H.K.* **12 E6** 22 19N 114 11 E
Kowloon Pk., *H.K.* **12 D6** 22 20N 114 13 E
Kowloon Res., *H.K.* **12 D5** 22 21N 114 9 E
Kowloon Tong, *H.K.* **12 D6** 22 20N 114 10 E
Kozhukhovo, *Mos.* **11 E11** 55 43N 37 53 E
Kozukue, *Tōkyō* **13 C2** 35 30N 139 35 E
Krailling, *Mün.* **7 G9** 48 5N 11 25 E
Kramat Jati, *Jak.* **15 J10** 6 15 S 106 51 E
Krampnitz, *Berl.* **7 B1** 52 27N 13 3 E
Krampnitzsee, *Berl.* **7 B1** 52 27N 13 3 E
Kranji, *Sing.* **15 F7** 1 26N 103 45 E
Kranji, Sungei →, *Sing.* **15 F7** 1 26N 103 44 E
Kranji Dam, *Sing.* **15 F7** 1 26N 103 44 E
Kraskovo, *Mos.* **11 F11** 55 39N 37 58 E
Krasnaya Gorka, *St-Pet.* **11 B5** 59 58N 30 38 E
Krasno-Presnenskaya, *Mos.* **11 E9** 55 45N 37 32 E
Krasnogorsk, *Mos.* **11 E7** 55 49N 37 18 E
Krasnyj Stroitel, *Mos.* **11 F9** 55 36N 37 35 E
Kray, *Ruhr* **6 B4** 51 27N 7 4 E
Krč, *Pra.* **10 B2** 50 2N 14 26 E
Krefeld, *Ruhr* **6 B1** 51 20N 6 33 E
Kremlin, *Mos.* **11 E9** 55 45N 37 38 E
Kresson, *Phil.* **24 B5** 39 51N 74 54W
Kreuzberg, *Berl.* **7 A3** 52 30N 13 24 E
Krishnarampur, *Calc.* **16 D5** 22 43N 88 13 E
Kritzendorf, *Wien* **10 G9** 48 19N 16 18 E
Krokhol, *Oslo* **2 C5** 59 48N 10 55 E
Krugersdorp, *Jobg.* **18 E7** 26 6 S 27 48 E
Krukut, Kali →, *Jak.* **15 J9** 6 13 S 106 48 E
Krumme Lanke, *Berl.* **7 B2** 52 27N 13 14 E
Krummensee, *Berl.* **7 A5** 52 35N 13 41 E
Krupunder, *Hbg.* **7 D7** 53 37N 9 53 E
Krusboda, *Stock.* **3 E12** 59 13N 18 14 E
Krylatskoye, *Mos.* **11 E8** 55 44N 37 23 E
Küçükkoy, *Ist.* **17 A2** 41 3N 28 52 E
Kudrovo, *St-Pet.* **11 B5** 59 54N 30 30 E
Kuivasaari, *Hels.* **3 C5** 60 6N 25 0 E
Kujiai, *Tōkyō* **13 A1** 35 57N 139 26 E
Küllenhahn, *Ruhr* **6 C4** 51 14N 7 8 E
Kulosaari, *Hels.* **3 B5** 60 11N 25 0 E
Kulturpalasset, *Wsaw.* **10 E7** 52 14N 21 0 E
Kumla, *Stock.* **3 E12** 59 13N 18 11 E
Kummelnäs, *Stock.* **3 D12** 59 21N 18 16 E
Kungens kurva, *Stock.* **3 E10** 59 15N 17 53 E
Kungsängen, *Stock.* **3 D9** 59 26N 17 43 E
Kungshatt, *Stock.* **3 E10** 59 17N 17 53 E
Kungsholmen, *Stock.* **3 D11** 59 20N 18 2 E
Kuningan, *Jak.* **15 J9** 6 13 S 106 49 E
Kuninkaanmäki, *Hels.* **3 B5** 60 18N 25 7 E
Kunitachi, *Tōkyō* **13 B1** 35 41N 139 27 E
Kunming Hu, *Beij.* **14 B2** 39 59N 116 13 E
Kunratice, *Pra.* **10 B2** 50 0N 14 28 E
Kunratický →, *Pra.* **10 B2** 50 2N 14 24 E
Kunsthalle, *Hbg.* **7 D8** 53 33N 10 0 E
Kuntsevo, *Mos.* **11 E8** 55 43N 37 23 E
Kupchino, *St-Pet.* **11 B4** 59 50N 30 23 E
Kupferdreh, *Ruhr* **6 B4** 51 23N 7 5 E
Kurbali Dere →, *Ist.* **17 B3** 40 58N 29 1 E
Kurihara, *Tōkyō* **13 B2** 35 45N 139 34 E
Kurkino, *Mos.* **11 D8** 55 53N 37 22 E
Kurla, *Bomb.* **16 G8** 19 4N 72 52 E
Kurmuri, *Bomb.* **16 G8** 19 4N 72 53 E
Kurnell, *Syd.* **19 C4** 34 0 S 151 10 E
Kurume, *Tōkyō* **13 B2** 35 45N 139 31 E
Kuryanovo, *Mos.* **11 F10** 55 39N 37 42 E
Kushihiki, *Tōkyō* **13 A2** 35 54N 139 36 E
Kushtia, *Calc.* **16 E6** 22 31N 88 23 E
Kuskovo, *Mos.* **11 E10** 55 44N 37 48 E
Kutsino, *Mos.* **11 E11** 55 44N 37 55 E
Kuy-e-Gishā, *Tehr.* **17 C5** 35 44N 51 23 E
Kuy-e-Mekānir, *Tehr.* **17 C5** 35 46N 51 22 E
Kuzyminki, *Mos.* **11 E10** 55 42N 37 46 E
Kvarnsjön, *Stock.* **3 E10** 59 11N 17 58 E
Kwa-Thema, *Jobg.* **18 F11** 26 17 S 28 23 E
Kwai Chung, *H.K.* **12 D5** 22 22N 114 7 E
Kwitang, *Jak.* **15 J10** 6 11 S 106 50 E
Kwun Tong, *H.K.* **12 E6** 22 18N 114 13 E
Kyje, *Pra.* **10 B3** 50 6N 14 33 E
Kyōhōji, *Ōsaka* **12 C4** 34 38N 135 33 E
Kyrkfjärden, *Stock.* **3 E9** 59 16N 17 45 E
Kyrkslätt, *Hels.* **3 C1** 60 6N 24 28W

L

La Aguada, *Stgo* **30 J10** 33 28 S 70 40W
La Blanca, *Stgo* **30 K11** 33 30 S 70 40W
La Boca, *B.A.* **32 B4** 34 38 S 58 22W
La Bottàccia, *Rome* **9 F8** 41 54N 12 18 E
La Bretèche, *Paris* **5 B2** 48 51N 2 1 E
La Brosse, *Paris* **5 C1** 48 43N 1 20 E
La Cabana, *La Hab.* **30 B3** 23 8N 82 20W
La Canada, *L.A.* **28 A3** 34 12N 118 12W
La Cassa, *Tori.* **9 A2** 45 11N 7 30 E
La Celle-les-Bordes, *Paris* **5 D1** 48 38N 1 57 E
La Celle-St.-Cloud, *Paris* **5 B2** 48 50N 2 9 E
La Chivera, *Car.* **30 D5** 10 35N 66 54W
La Colmena, *Méx.* **29 A2** 19 35N 99 16W
La Courneuve, *Paris* **5 B4** 48 55N 2 22 E
La Crescenta, *L.A.* **28 A3** 34 13N 118 14W
La Défense, *Paris* **5 B3** 48 53N 2 12 E
La Dehesa, *Stgo* **30 J11** 33 21 S 70 33W
La Estación, *Mdrd.* **8 B2** 40 27N 3 48W
La Floresta, *Barc.* **8 D5** 41 26N 2 3 E
La Florida, *Car.* **30 D5** 10 30N 66 52W
La Fortuna, *Mdrd.* **8 B2** 40 21N 3 46W
La Fransa, *Barc.* **8 D5** 41 22N 2 9 E
La Fresnière, *Mtrl.* **20 A2** 43 33N 73 58W
La Frette-sur-Seine, *Paris* **5 B3** 48 58N 2 11 E
La Garenne-Colombes, *Paris* **5 B3** 48 54N 2 15 E
La Giustiniana, *Rome* **9 F9** 41 59N 12 24 E
La Grange, *Chic.* **26 C1** 41 48N 87 53W
La Grange des Noues, *Paris* **5 A4** 49 1N 2 28 E
La Grange Highlands, *Chic.* **26 C1** 41 46N 87 53W
La Grange Park, *Chic.* **26 C1** 41 49N 87 51W
La Granja, *Stgo* **30 K11** 33 31 S 70 38W
La Guaira, *Car.* **30 D5** 10 36N 66 55W
La Guardia Airport, *N.Y.* **23 C5** 40 46N 73 52W
La Guasima, *La Hab.* **30 B3** 23 0N 82 17W
La Habana, *La Hab.* **30 B2** 23 7N 82 21W
La habana, B. de, *La Hab.* **30 B3** 23 7N 82 20W
La Habana Vieia, *La Hab.* **30 B2** 23 7N 82 20W
La Habra, *L.A.* **28 C5** 33 56N 117 57W
La Habre Heights, *L.A.* **28 C5** 33 59N 117 56W
La Horqueta →, *B.A.* **32 C1** 34 43 S 58 51W
La Lisa, *La Hab.* **30 B2** 23 3N 82 25W
La Llacuna, *Barc.* **8 D6** 41 24N 2 12 E
La Loma, *Méx.* **29 A2** 19 31N 99 11W
La Lucila, *B.A.* **32 B4** 34 30 S 58 29W
La Magdalena Atlipac, *Méx.* **29 B4** 19 22N 89 56 E
La Magdalena Chichicaspa, *Méx.* **29 B2** 19 24N 99 18W
La Magdalena Contreras, *Méx.* **29 C2** 19 17N 99 13W
La Magdalena Petlalco, *Méx.* **29 C2** 19 13N 99 10W
La Maison Blanche, *Paris* **5 C1** 48 44N 1 54 E
La Maladrerie, *Paris* **5 B2** 48 54N 2 1 E
La Marquesa, *Méx.* **29 C1** 19 18N 99 22W
La Milla, Cerro, *Lima* **30 G8** 12 2 S 77 5W
La Molina, *Lima* **30 G9** 12 4 S 76 56W
La Monachina, *Rome* **9 F9** 41 53N 12 21 E
La Moraleja, *Mdrd.* **8 A3** 40 30N 30 38W
La Nopalera, *Méx.* **29 C3** 19 18N 99 5W
La Pastora, *Car.* **30 D5** 10 31N 66 55W
La Paterna, *B.A.* **32 B4** 34 35 S 58 26W
La Patte-d'Oie, *Paris* **5 A3** 49 0N 2 10 E
La Perla, *Lima* **30 G8** 12 4 S 77 7W
La Perouse, *Syd.* **19 B4** 33 59 S 151 14 E
La Pineda, *Barc.* **8 E5** 41 15N 2 1 E
La Pisana, *Rome* **9 F9** 41 51N 12 23 E
La Playa, *La Hab.* **30 B2** 23 6N 82 26W
La Prairie, *Mtrl.* **20 B5** 43 25N 73 29W
La Puente, *L.A.* **28 B5** 34 1N 117 54W
La Punta, *Lima* **30 G7** 12 4 S 77 10W
La Puntigala, *Barc.* **8 D6** 41 27N 2 13 E
La Queue-en-Brie, *Paris* **5 C5** 48 47N 2 34 E
La Reina, *Stgo* **30 J11** 33 26 S 70 33W
La Reja, *B.A.* **32 B2** 34 38 S 58 48W
La Ribera, *Barc.* **8 D5** 41 21N 2 4 E
La Romanie, *Paris* **5 C1** 48 43N 1 53 E
La Rústica, *Rome* **9 F10** 41 54N 12 36 E
La Sagrera, *Barc.* **8 D6** 41 25N 2 11 E
La Salada, *B.A.* **32 C4** 34 43 S 58 28W
La Scala, *Mil.* **9 E6** 45 28N 9 11 E
La Selce, *Rome* **9 F9** 41 53N 12 20 E
La Sierra, *La Hab.* **30 B2** 23 7N 82 24W
La Taxonera, *Barc.* **8 D6** 41 25N 2 10 E
La Vega, *Car.* **30 E5** 10 28N 66 56W
La Verrière, *Paris* **5 C1** 48 45N 1 57 E
La Vibora, *La Hab.* **30 B2** 23 5N 82 22W
La Victoria, *Lima* **30 G8** 12 3 S 77 2W
La Ville-du-Bois, *Paris* **5 D3** 48 39N 2 16 E
Laab im Walde, *Wien* **10 H9** 48 9N 16 10 E
Laaer Berg, *Wien* **10 H10** 48 9N 16 24 E
Laajalahti, *Hels.* **3 B3** 60 11N 24 48 E
Laajasalo, *Hels.* **3 B5** 60 10N 25 1 E
Laaksolahti, *Hels.* **3 B3** 60 14N 24 45 E
Laar, *Ruhr* **6 B2** 51 27N 6 44 E
Lablâba, W. el →, *El Qâ.* **18 C5** 30 1N 31 19 E
Lachine, *Mtrl.* **20 B3** 43 26N 73 42W
Ládvi, *Pra.* **10 B2** 50 8N 14 27 E
Łady, *Wsaw.* **10 F6** 52 8N 20 57 E
Lafayette, *S.F.* **27 A4** 37 52N 122 6W
Lafayette Hill, *Phil.* **24 A3** 40 5N 75 15W
Lafayette Res., *S.F.* **27 A4** 37 52N 122 8W
Laferrere, *B.A.* **32 C3** 34 45 S 58 35W
Lagny, *Paris* **5 B6** 48 52N 2 42 E
Lagoa da Pedra, *Lisb.* **8 F9** 38 43N 8 58W
Lagos, *Lagos* **18 B2** 6 27N 7 23 E
Lagos Harbour, *Lagos* **18 B2** 6 26N 7 23 E
Lagos-Ikeja Airport, *Lagos* **18 A1** 6 34N 7 19 E
Lagos Island, *Lagos* **18 B2** 6 26N 7 23 E
Lagos Lagoon, *Lagos* **18 B2** 6 30N 7 28 E
Laguna de B., *Manila* **15 E4** 14 29N 121 6 E
Laim, *Mün.* **7 G10** 48 7N 11 30 E
Lainate, *Mil.* **9 D5** 45 34N 9 1 E
Lainz, *Wien* **10 H9** 48 10N 16 16 E
Lainzer Tiergarten, *Wien* **10 G9** 48 10N 16 13 E
Lajeado →, *S. Pau.* **31 E7** 23 28 S 46 24W
Lake Avenue Woods, *Chic.* **26 A1** 42 4N 87 53W
Lake Hiawatha, *N.Y.* **22 B2** 40 52N 74 23W
Lakefield, *Jobg.* **18 F10** 26 11 S 28 17 E
Lakemba, *Syd.* **19 B3** 33 55 S 151 5 E
Lakeside, *Jobg.* **18 E9** 26 5 S 28 8 E
Lakeview, *Chic.* **26 B3** 41 56N 87 38W
Lakeview, *Trto.* **20 E7** 43 35N 79 32W
Lakhtinskiy, *St-Pet.* **11 B2** 59 59N 30 9 E
Lakhtinskiy Razliv, Oz., *St-Pet.* **11 B3** 59 59N 30 12 E
Lakshmanpur, *Calc.* **16 E5** 22 38N 88 16 E
Laleham, *Lon.* **4 C2** 51 24N 0 29W
Lâleli, *Ist.* **17 A2** 41 0N 28 57 E
Lalor, *Melb.* **19 E6** 37 40 S 144 59 E
Lam San, *Sing.* **15 F7** 1 22N 103 43 E
Lam Tin, *H.K.* **12 E6** 22 18N 114 14 E
Lambarfjärden, *Stock.* **3 D9** 59 21N 17 48 E
Lambert, *Oslo* **2 B4** 59 52N 10 48 E
Lambeth, *Lon.* **4 C4** 51 28N 0 6W
Lambrate, *Mil.* **9 E6** 45 28N 9 16 E
Lambro →, *Mil.* **9 E6** 45 24N 9 17 E
Lambro, Parco, *Mil.* **9 E6** 45 29N 9 14 E
Lambton, *Jobg.* **18 F10** 26 14 S 28 10 E
Lambton Hills, *Trto.* **20 E7** 43 39N 79 30W
Lamma I., *H.K.* **12 E5** 22 12N 114 7 E
Lampton, *Lon.* **4 C2** 51 28N 0 21W
Landianchang, *Beij.* **14 B2** 39 57N 116 13 E
Landover Hills, *Wash.* **25 D8** 38 56N 76 54W
Landstrasse, *Wien* **10 G10** 48 12N 16 23 E
Landwehr kanal, *Berl.* **7 B3** 52 29N 13 24 E
Lane Cove, *Syd.* **19 A3** 33 48 S 151 9 E
Lane Cove National Park, *Syd.* **19 A3** 33 47 S 151 8 E
Langen, *Oslo* **2 C5** 59 44N 10 57 E
Langenberg, *Ruhr* **6 B4** 51 20N 7 7 E
Langenbochum, *Ruhr* **6 A4** 51 36N 7 7 E
Langendreer, *Ruhr* **6 B5** 51 28N 7 18 E
Langenhorn, *Hbg.* **7 D7** 53 39N 9 59 E
Langenhorst, *Ruhr* **6 B4** 51 21N 7 1 E
Langenzersdorf, *Wien* **10 G10** 48 18N 16 21 E
Langer See, *Berl.* **7 B4** 52 24N 13 37 E
Langerfeld, *Ruhr* **6 C5** 51 16N 7 14 E
Langley, *Wash.* **25 D6** 38 57N 77 11W
Langley Park, *Wash.* **25 D8** 38 59N 76 58W
Langstaff, *Trto.* **20 C8** 43 50N 79 26W
Längtarmen, *Stock.* **3 D8** 59 24N 17 36 E
Langwald, *Mün.* **7 F9** 48 10N 11 25 E
Lanham, *Wash.* **25 D8** 38 59N 76 51W
Lank-Latum, *Ruhr* **6 C2** 51 18N 6 40 E
Lankwitz, *Berl.* **7 B3** 52 25N 13 21 E
Länna Drevviken, *Stock.* **3 E11** 59 12N 18 8 E
L'Annunziatella, *Rome* **9 G10** 41 49N 12 33 E
Lansdowne, *Balt.* **25 B3** 39 14N 76 38W
Lansdowne, *Phil.* **24 B3** 39 56N 75 16W
Lansing, *Trto.* **20 D8** 43 45N 79 24W
Lanús, *B.A.* **32 C4** 34 42 S 58 23W
Lapa, *Rio J.* **31 B2** 22 54 S 43 10W
Lapa, *S. Pau.* **31 E5** 23 31 S 46 42W
Lapangan Merdeka, *Jak.* **15 J9** 6 10 S 106 49 E
Lapinkylä, *Hels.* **3 B4** 60 18N 24 51 E
Lapinkylä, *Hels.* **3 B1** 60 13N 24 27 E
Lappböle, *Hels.* **3 B1** 60 13N 24 27 E
Laranjeiras, *Rio J.* **31 B2** 22 55 S 43 10W
Larchmont, *N.Y.* **23 B6** 40 55N 73 44W
Larkspur, *S.F.* **27 A1** 37 55N 122 31W
Las, *Wsaw.* **10 E7** 52 13N 21 6 E
Las Acacias, *Car.* **30 E5** 10 29N 66 54W
Las Adjuntas, *Car.* **30 E4** 10 25N 67 0W
Las Barrancas, *B.A.* **32 A4** 34 28 S 58 29W
Las Conchas, *B.A.* **32 A3** 34 25 S 58 34W
Las Corts, *Barc.* **8 D5** 41 23N 2 9 E
Las Fuentes Brotantes, *Méx.* **29 C2** 19 16N 99 11W
Las Kabacki, *Wsaw.* **10 F7** 52 7N 21 2 E
Las Lomas, *L.A.* **28 B5** 34 8N 117 59W
Las Mercedes, *Car.* **30 E5** 10 28N 66 51W
Las Pinas, *Manila* **15 E3** 14 29N 120 58 E
Las Rejas, *Stgo* **30 J10** 33 27 S 70 42W
Las Rozas de Madrid, *Mdrd.* **8 B1** 40 29N 3 52W
Las Trampas Cr. →, *S.F.* **27 A4** 37 53N 122 6W
Las Trampas Regional Park, *S.F.* **27 B4** 37 49N 122 3W
Las Trampas Ridge, *S.F.* **27 A4** 37 50N 122 3W
Las Tunitas, *Car.* **30 D4** 10 36N 67 1W
Lasalle, *Mtrl.* **20 B4** 43 26N 73 37W
Lasek Bielański, *Wsaw.* **10 E6** 52 17N 20 57 E
Lasek Na Kole, *Wsaw.* **10 E6** 52 15N 20 56 E
Laski, *Wsaw.* **10 E6** 52 18N 20 50 E
Latina, *Mdrd.* **8 B2** 40 24N 3 44W
Latrobe Univ., *Melb.* **19 E7** 37 43 S 145 3 E
Lattingtown, *N.Y.* **23 B7** 40 52N 73 34W
Laufzorn, *Mün.* **7 G10** 48 0N 11 33 E
Laurel Hollow, *N.Y.* **23 B8** 40 51N 73 28W
Laurel Springs, *Phil.* **24 C4** 39 49N 75 0W
Laurelton, *N.Y.* **23 C6** 40 40N 73 45W
Laurence Hanscom Field, *Bost.* **21 C2** 42 28N 71 16W
Lausdomini, *Nápl.* **9 H13** 40 55N 14 26 E
Lauttasaari, *Hels.* **3 C4** 60 9N 24 53 E
Lava Nuova, *Nápl.* **9 J13** 40 47N 14 23 E
Laval-des-Rapides, *Mtrl.* **20 A3** 43 33N 73 41W
Laval-Ouest, *Mtrl.* **20 A2** 43 32N 73 52W
Laval-sur-le-Lac, *Mtrl.* **20 A2** 43 31N 73 52W
Lavradio, *Lisb.* **8 F8** 38 40N 9 2W
Lawndale, *Chic.* **26 B2** 41 50N 87 42W
Lawndale, *L.A.* **28 C2** 33 52N 118 22W
Lawndale, *Phil.* **24 A4** 40 3N 75 5W
Lawnside, *Phil.* **24 B4** 39 51N 75 1W
Lawrence, *Bost.* **21 A3** 42 43N 71 7W
Lawrence, *N.Y.* **23 D6** 40 36N 73 43W
Lawrence Heights, *Trto.* **20 D8** 43 43N 79 25W
Lawrence Park, *Phil.* **24 B2** 39 57N 75 20W
Lawton, *La Hab.* **30 B2** 23 5N 82 20W
Layāri, *Kar.* **17 G11** 24 52N 67 0 E
Layāri →, *Kar.* **17 G10** 24 52N 66 58 E
Lazienkowski Park, *Wsaw.* **10 E7** 52 13N 21 1 E
Le Blanc-Mesnil, *Paris* **5 B4** 48 56N 2 27 E
Le Bourget, *Paris* **5 B4** 48 56N 2 25 E
Le Chesnay, *Paris* **5 C2** 48 49N 2 8 E
Le Christ de Saclay, *Paris* **5 C3** 48 43N 1 59 E
Le Kremlin-Bicêtre, *Paris* **5 C4** 48 48N 2 21 E
Le Mesnil-Amelot, *Paris* **5 A5** 49 1N 2 35 E
Le Mesnil-le-Roi, *Paris* **5 B2** 48 56N 2 7 E
Le Mesnil-St.-Denis, *Paris* **5 C1** 48 44N 1 57 E
Le Pecq, *Paris* **5 B2** 48 53N 2 6 E
Le Perreux, *Paris* **5 B4** 48 50N 2 29 E
Le Pin, *Paris* **5 B5** 48 54N 2 37 E
Le Plessis-Bouchard, *Paris* **5 A3** 49 0N 2 14 E
Le Plessis-Gassot, *Paris* **5 A4** 49 2N 2 24 E
Le Plessis-Pâté, *Paris* **5 D3** 48 36N 2 19 E
Le Plessis-Robinson, *Paris* **5 C3** 48 47N 2 15 E
Le Plessis-Trévise, *Paris* **5 C5** 48 48N 2 34 E
Le Port-Marly, *Paris* **5 B2** 48 52N 2 6 E
Le Pré-St.-Gervais, *Paris* **5 B4** 48 53N 2 23 E
Le Raincy, *Paris* **5 B5** 48 53N 2 31 E
Le Thillay, *Paris* **5 A4** 49 0N 2 28 E
Le Trappe, *Mtrl.* **20 B1** 43 30N 74 1W
Le Val d'Enfer, *Paris* **5 C3** 48 45N 2 11 E
Le Vésinet, *Paris* **5 B2** 48 54N 2 8 E
Lea →, *Lon.* **4 B4** 51 30N 0 2W
Lea Bridge, *Lon.* **4 B4** 51 33N 0 2W
Leakin Park, *Balt.* **25 B2** 39 18N 76 41W
Leaside, *Trto.* **20 D8** 43 42N 79 22W
Leatherhead, *Lon.* **4 D3** 51 17N 0 19W
Leaves Green, *Lon.* **4 D5** 51 19N 0 1 E
Leblon, *Rio J.* **31 B2** 22 59 S 43 14W
Léchelle, Forêt de la, *Paris* **5 C6** 48 43N 2 41 E
Ledøje, *Købn.* **2 D8** 55 42N 12 18 E
Lee, *Lon.* **4 C5** 51 27N 0 0 E
Leeupan, *Jobg.* **18 F10** 26 13 S 28 18 E
Leganes, *Mdrd.* **8 C2** 40 19N 3 45W
Legazpi, *Mdrd.* **8 B2** 40 23N 3 41W
Legoa, Kali →, *Jak.* **15 H10** 6 6 S 106 52 E
Lehtisaaret, *Hels.* **3 C3** 60 6N 24 46 E
Lehtisaari, *Hels.* **3 B4** 60 10N 24 51 E
Lei Yue Mun, *H.K.* **12 E6** 22 17N 114 14 E
Leião, *Lisb.* **8 F7** 38 43N 9 17W
Leichhardt, *Syd.* **19 B3** 33 53 S 151 9 E
Leigang, *Gzh.* **14 G7** 23 2N 113 6 E
Léini, *Tori.* **9 A3** 45 11N 7 42 E
Leisure World, *S.F.* **27 A4** 37 51N 122 4W
Lemoyne, *Mtrl.* **20 B5** 43 29N 73 29W
Lemsahl, *Hbg.* **7 C8** 53 41N 10 5 E
Lenin, *Mos.* **11 E9** 55 43N 37 34 E
Leningrad = St. Petersburg, *St-Pet.* **11 B3** 59 55N 30 15 E
Lenino, *Mos.* **11 F9** 55 38N 37 39 E
Leninskiye Gory, *Mos.* **11 E9** 55 41N 37 32 E
Lenne →, *Ruhr* **6 B7** 51 25N 7 30 E
Lennep, *Ruhr* **6 C5** 51 11N 7 15 E
Lenni, *Phil.* **24 B2** 39 53N 75 27W
Lennox, *L.A.* **28 C2** 33 56N 118 20W
Leonardo da Vinci, Aeroporto Int., *Rome* **9 G8** 41 47N 12 15 E
Leoncio Martinez, *Car.* **30 E6** 10 29N 66 48W
Leonia, *N.Y.* **22 B5** 40 51N 73 59W
Leopard, *Phil.* **24 A2** 40 1N 75 26W
Leopardi, *Nápl.* **9 J13** 40 45N 14 24 E
Leopoldau, *Wien* **10 G10** 48 15N 16 26 E
Leopoldstadt, *Wien* **10 G10** 48 13N 16 22 E
Leportovo, *Mos.* **11 E10** 55 46N 37 43 E
Leppävaara, *Hels.* **3 B3** 60 13N 24 49 E
Lera, Mte., *Tori.* **9 A1** 45 10N 7 27 E
L'Éremo, *Tori.* **9 B3** 45 2N 7 44 E
Les Alluets-le-Roi, *Paris* **5 B1** 48 54N 1 55 E
Les Clayes-sous-Bois, *Paris* **5 C1** 48 49N 1 59 E
Les Essarts-le-Roi, *Paris* **5 C1** 48 42N 1 53 E
Les Gâtines, *Paris* **5 C1** 48 48N 1 58 E
Les Grésillons, *Paris* **5 B2** 48 56N 2 1 E
Les Layes, *Paris* **5 C1** 48 43N 1 55 E
Les Lilas, *Paris* **5 B4** 48 52N 2 25 E
Les Loges-en-Josas, *Paris* **5 C2** 48 45N 2 8 E
Les Molières, *Paris* **5 C2** 48 40N 2 4 E
Les Mureaux, *Paris* **5 B1** 48 59N 1 54 E
Les Pavillons-sous-Bois, *Paris* **5 B5** 48 54N 2 30 E
Les Vaux de Cernay →, *Paris* **5 C1** 48 41N 1 59 E
Lésigny, *Paris* **5 C5** 48 44N 2 37 E
Lesnosavodskaya, *St-Pet.* **11 B4** 59 51N 30 29 E
Lesnoy, *St-Pet.* **11 B4** 59 59N 30 22 E
Lester B. Pearson Int. Airport, *Trto.* **20 D7** 43 40N 79 38 E
L'Étang-la-Ville, *Paris* **5 B2** 48 52N 2 4 E
Letná, *Pra.* **10 B2** 50 0 S 14 26 E
Letňany, *Pra.* **10 B3** 50 8N 14 30 E
Letovo, *Mos.* **11 F8** 55 34N 37 24 E
Leuville-sur-Orge, *Paris* **5 D3** 48 36N 2 15 E
Levallois-Perret, *Paris* **5 B3** 48 53N 2 16 E
Lévis St.-Nom, *Paris* **5 C1** 48 43N 1 57 E
Levittown, *N.Y.* **23 C7** 40 43N 73 31W
Lewisdale, *Wash.* **25 D8** 38 58N 76 59W
Lewisham, *Jobg.* **18 E7** 26 7 S 27 49 E
Lewisham, *Lon.* **4 C4** 51 27N 0 1W
Lexington, *Bost.* **21 C2** 42 25N 71 12W
Leyton, *Lon.* **4 B4** 51 33N 0 1W
Leytonstone, *Lon.* **4 B5** 51 34N 0 0 E
L'Hautil, *Paris* **5 A2** 49 0N 2 0 E
L'Hay-les-Roses, *Paris* **5 C4** 48 46N 2 20 E
Lhotka, *Pra.* **10 B2** 50 1N 14 26 E
Liangshui He →, *Beij.* **14 C3** 39 48N 116 23 E
Lianhua Chi, *Beij.* **14 B2** 39 52N 116 16 E
Lianhua He →, *Beij.* **14 B2** 39 52N 116 13 E
Lianozovo, *Mos.* **11 D9** 55 53N 37 34 E
Libčice nad Vitavou, *Pra.* **10 A2** 50 11N 14 22 E
Liben, *Pra.* **10 B2** 50 6N 14 27 E
Liberdade, *S. Pau.* **31 E6** 23 33 S 46 37W
Libertad, *B.A.* **32 C2** 34 41 S 58 41W
Liberty I., *N.Y.* **22 C4** 40 41N 74 2W
Liberty Plain, *Bost.* **21 D4** 42 11N 70 52W
Liberty Res., *Balt.* **25 A1** 39 23N 76 52W
Libeznice, *Pra.* **10 A2** 50 11N 14 29 E
Library of Congress, *Wash.* **25 D7** 38 53N 77 0W
Libuč, *Pra.* **10 B2** 50 0N 14 27 E
Lichiao, *Gzh.* **14 G8** 23 3N 113 18 E
Lichtenbroich, *Ruhr* **6 C2** 51 17N 6 49 E
Lichtenburg, *Berl.* **7 A4** 52 31N 13 30 E
Lichtenplatz, *Ruhr* **6 C5** 51 14N 7 11 E
Lichtenrade, *Berl.* **7 B3** 52 23N 13 24 E
Lichterfelde, *Berl.* **7 B2** 52 25N 13 19 E
Licignano di Nápoli, *Nápl.* **9 H13** 40 54N 14 21 E
Lidcombe, *Syd.* **19 B3** 33 52 S 151 3 E
Lidingö, *Stock.* **3 D11** 59 21N 18 8 E
Lido Beach, *N.Y.* **23 D7** 40 35N 73 37W
Lier, *Oslo* **2 C1** 59 47N 10 13 E
Lierskogen, *Oslo* **2 C1** 59 49N 10 15 E
Lieshi Lingyuan, *Gzh.* **14 G8** 23 7N 113 16 E
Liesing, *Wien* **10 H9** 48 8N 16 17 E
Liesing →, *Wien* **10 H10** 48 8N 16 28 E
Lieusant, *Paris* **5 D5** 48 38N 2 33 E
Liffjofs, *Hels.* **3 B6** 60 19N 25 12 E
Ligovo, *St-Pet.* **11 C3** 59 49N 30 10 E
Lijordet, *Oslo* **2 B3** 59 56N 10 36 E
Likhoborka →, *Mos.* **11 D9** 55 50N 37 37 E
Likova →, *Mos.* **11 F8** 55 36N 37 20 E
Lilla Värtan, *Stock.* **3 D12** 59 20N 18 11 E
Lille Rørbæk, *Købn.* **2 D7** 55 47N 12 6 E

Midland Park 41

Lille Værløse, Købn. 2 D9 55 47N 12 22 E
Lillehavfrue, Købn. 2 D10 55 42N 12 35 E
Lillestrøm, Oslo 2 B6 59 57N 11 3 E
Liluah, Calc. 16 E5 22 37N 88 19 E
Lilydale, Melb. 19 E9 37 45 S 145 21 E
Lima, Lima 30 G8 12 3 S 77 2W
Lima, Phil. 24 B2 39 55N 75 26W
Limbiate, Mil. 9 D5 45 35N 9 7 E
Limehouse, Lon. 4 B4 51 30N 0 1W
Limeil-Brévannes, Paris 5 C4 48 44N 2 29 E
Limito, Mil. 9 E6 45 28N 9 19 E
Limoges-Fourches, Paris 5 D5 48 37N 2 39 E
Limours, Paris 5 D2 48 38N 2 4 E
Linas, Paris 5 D3 48 37N 2 16 E
Linate, Mil. 9 E6 45 26N 9 16 E
Linate, Aeroporto Internazionale di, Mil. 9 E6 45 26N 9 16 E
Linbigh, Balt. 25 A3 39 21N 76 31W
Linbropark, Jobg. 18 E9 26 5 S 28 7 E
Lincoln, Bost. 21 C2 42 25N 71 18W
Lincoln Center, N.Y. 22 C5 40 46N 43 59W
Lincoln Heights, L.A. 28 B3 34 4N 118 12 E
Lincoln Memorial, Wash. 25 D7 38 53N 77 2W
Lincoln Park, Chic. 26 B3 41 57N 87 38W
Lincoln Park, N.Y. 22 B3 40 56N 74 18W
Lincoln Park, S.F. 27 B1 37 47N 122 30W
Lincolnwood, Chic. 26 A2 42 1N 87 43W
Linda-a-Pastora, Lisb. 8 F7 38 42N 9 15W
Linden, Jobg. 18 E9 26 8 S 28 0 E
Linden, N.Y. 22 D3 40 38N 74 14W
Linden-Dahlhausen, Ruhr 6 B5 51 25N 7 10 E
Lindenberg, Berl. 7 A4 52 36N 13 31 E
Lindenhorst, Ruhr 6 A6 51 33N 7 27 E
Lindenhurst, N.Y. 23 C8 40 40N 73 22W
Lindenwold, Phil. 24 C5 39 49N 74 59W
Linderhausen, Ruhr 6 C5 51 17N 7 17 E
Lindfield, Syd. 19 A3 33 46 S 151 9 E
Lindøya, Oslo 2 B4 59 53N 10 42 E
Lingotto, Tori. 9 B2 45 1N 7 39 E
Liniers, B.A. 32 B3 34 39 S 58 30W
Linksfield, Jobg. 18 E9 26 9 S 28 6 E
Linmeyer, Jobg. 18 F9 26 15 S 28 4 E
Linn, Ruhr 6 B1 51 20N 6 38 E
Linna, Hels. 3 A4 60 20N 24 50 E
Linthicum Heights, Balt. 25 B2 39 12N 76 47W
Lintorf, Ruhr 6 B3 51 20N 6 50 E
Lintuvaara, Hels. 3 B3 60 14N 24 49 E
Linwood, Phil. 24 C2 39 49N 75 25W
Lioúmi, Ath. 8 J11 38 0N 23 40 E
Lipków, Wsaw. 10 E5 52 16N 20 48 E
Lippaltshausen, Ruhr 6 A6 51 36N 7 26 E
Liqizhuang, Tianj. 14 E6 39 4N 117 10 E
Lirich, Ruhr 6 B2 51 29N 6 49 E
Lisboa, Lisb. 8 F8 38 42N 9 8W
Lisbon = Lisboa, Lisb. 8 F8 38 42N 9 8W
Lishui, Gzh. 14 F7 23 12N 113 9 E
Lisiy Nos, St-Pet. 11 A2 60 1N 30 0 E
Lissone, Mil. 9 D6 45 36N 9 14 E
Lissy, Paris 5 D6 48 38N 2 42 E
Litoral, Cord. del, Car. 30 D5 10 33N 66 54W
Little B., Syd. 19 B4 33 58 S 151 15 E
Little Calumet →, Chic. 26 D3 41 39N 87 34W
Little Falls, N.Y. 22 B3 40 52N 74 14W
Little Ferry, N.Y. 22 B4 40 50N 74 2W
Little Neck, N.Y. 23 C6 40 46N 73 43W
Little Paint Br. →, Wash. 25 C8 39 0N 76 55W
Little Patuxent →, Balt. 25 B1 39 13N 76 51W
Little Rouge →, Trto. 20 C9 43 45N 79 11W
Little Sugarloaf, Melb. 19 E8 37 40 S 145 18 E
Little Thurrock, Lon. 4 C7 51 29N 0 20 E
Liuhang, Shang. 14 H11 31 21N 121 21 E
Liuhuahu Gongyuan, Gzh. 14 G8 23 8N 113 14 E
Liverpool, Syd. 19 B2 33 55 S 150 55 E
Livingstone, N.Y. 22 C3 40 47N 74 19W
Livry-Gargan, Paris 5 B5 48 55N 2 31 E
Liwanhu Gongyuan, Gzh. 14 G8 23 7N 113 13 E
Lizhuang, Gzh. 14 G7 23 6N 113 7 E
Ljan, Oslo 2 B4 59 51N 10 48 E
Llano de Can Gineu, Barc. 8 D6 41 27N 2 10 E
Llavallol, B.A. 32 C4 34 48 S 58 25W
Llobregat →, Barc. 8 D5 41 19N 2 5 E
Lloyd Harbor, N.Y. 23 B8 40 54N 73 26W
Lloyd Pt., N.Y. 23 B8 40 56N 73 29W
Lo Aranguiz, Stgo 30 J11 33 23 S 70 40W
Lo Boza, Stgo 30 J10 33 23 S 70 43W
Lo Chau, H.K. 12 E6 22 11N 114 15 E
Lo Hermida, Stgo 30 J11 33 28 S 70 33W
Lo Ortuzar, Stgo 30 J10 33 28 S 70 43W
Lo Prado Arriba, Stgo 30 J10 33 26 S 70 42W
Lo So Shing, H.K. 12 E5 22 12N 114 7 E
Lo Wai, H.K. 12 D5 22 22N 114 8 E
Lobau, Wien 10 G11 48 10N 16 31 E
Lobos, Pt., S.F. 27 B1 37 46N 122 30W
Loch Raven Village, Balt. 25 A3 39 23N 76 34W
Locham, Mün. 7 G9 48 7N 11 26 E
Lochearn, Balt. 25 A2 39 20N 76 43W
Lochino, Mos. 11 E7 55 41N 37 17 E
Lochkov, Pra. 10 B2 50 0N 14 21 E
Lockhausen, Mün. 7 F9 48 10N 11 24 E
Locksbottom, Lon. 4 C5 51 21N 0 3 E
Locust Grove, N.Y. 23 C8 40 48N 73 29W
Locust Manor, N.Y. 23 C6 40 41N 73 46W
Locust Valley, N.Y. 23 B7 40 52N 73 36W
Lodi, N.Y. 22 B4 40 52N 74 5W
Lofty, Mt., Melb. 19 E8 37 42 S 145 17 E
Logan, Phil. 24 A4 40 2N 75 8W
Logan Int. Airport, Bost. 21 C4 42 22N 71 0W
Logan Square, Chic. 26 B2 41 55N 87 42W
Lognes-Émerainville, Aérodrome de, Paris 5 C5 48 49N 2 37 E
Lohausen, Ruhr 6 C2 51 16N 6 44 E
Lohberg, Ruhr 6 A2 51 34N 6 45 E
Lohme, Berl. 7 A5 52 37N 13 40 E
Löhmühle, Ruhr 6 A1 51 30N 6 39 E
Löhnen, Ruhr 6 A1 51 35N 6 39 E
Lokstedt, Hbg. 7 D7 53 36N 9 56 E
Lokyang, Sing. 15 G7 1 19N 103 40 E
Lölökhet, Kar. 17 G11 24 54N 67 2 E
Loma Blanca, Stgo 30 J10 33 29 S 70 43W
Lomas Chapultepec, Méx. 29 B2 19 25N 99 12W
Lomas de San Angel Inn, Méx. 29 B2 19 20N 99 13W
Lomas de Zamora, B.A. 32 C4 34 45 S 58 24W
Lombardy East, Jobg. 18 E9 26 6 S 28 7 E
Lomonosov Univ., Mos. 11 E9 55 42N 37 31 E
Lomus Reforma, Méx. 29 B2 19 24N 99 14W
London, Lon. 4 B4 51 30N 0 6W
London, City of, Lon. 4 B4 51 30N 0 6W
London, Tower of, Lon. 4 B4 51 30N 0 4W
London Zoo, Lon. 4 B4 51 31N 0 9W
Long B., Syd. 19 B4 33 59 S 151 15 E
Long Beach, N.Y. 23 D7 40 35N 73 39W
Long Branch, Trto. 20 E7 43 35N 79 31W
Long Brook →, Wash. 25 E6 38 49N 77 15W
Long Ditton, Lon. 4 C3 51 22N 0 19W
Long I., Bost. 21 D4 42 19N 70 59W
Long I., N.Y. 23 C7 40 45N 73 30W
Long Island City, N.Y. 22 C5 40 45N 73 56W
Long Island Sd., N.Y. 23 B7 40 57N 73 30W
Long Pond, Bost. 21 A1 42 41N 71 22W
Longchamp, Hippodrôme de, Paris 5 B3 48 51N 2 13 E
Longchêne, Paris 5 D2 48 38N 2 0 E
Longhua Gongyuan, Shang. 14 J11 31 10N 121 26 E
Longjohn Slough, Chic. 26 C1 41 42N 87 52W
Longjumeau, Paris 5 C3 48 41N 2 17 E
Longlands, Lon. 4 C5 51 25N 0 5 E
Longpont-sur-Orge, Paris 5 D3 48 38N 2 17 E
Longtan Hu →, Beij. 14 B3 39 51N 116 24 E
Longue Point, Mtrl. 20 A4 43 35N 73 31W
Longueuil, Mtrl. 20 A5 43 31N 73 29W
Loni, Delhi 16 A2 28 45N 77 17 E
Lord's Cricket Ground, Lon. 4 B3 51 31N 0 10W
Loreley, Balt. 25 A4 39 23N 76 24W
Lørenskog, Oslo 2 B5 59 55N 10 59 E
Loreto, Mil. 9 E6 45 29N 9 12 E
Lorraine, Mtrl. 20 A3 43 39N 73 46W
Los Angeles, L.A. 28 B3 34 3N 118 14 E
Los Angeles, Mdrd. 8 B2 40 20N 3 41W
Los Angeles →, L.A. 28 C3 33 55N 118 10W
Los Angeles Int. Airport, L.A. 28 C2 33 56N 118 23W
Los Asientos, Car. 30 D5 10 32N 66 53W
Los Caobos, Car. 30 D5 10 30N 66 53W
Los Cermenes, Car. 30 E5 10 28N 66 54W
Los Cerrillos, Aeroporto, Stgo 30 J10 33 29 S 70 42W
Los Dos Caminos, Car. 30 D6 10 30N 66 49W
Los dos Riteras →, Car. 30 D5 10 35N 66 57W
Los Jazmines, Presa, Méx. 29 B2 19 25N 99 15W
Los Nietos, L.A. 28 C4 33 57N 118 4W
Los Pinos, La Hab. 30 B2 23 4N 82 22W
Los Pirules, Méx. 29 B3 19 24N 99 2W
Los Polvorines, B.A. 32 B2 34 30 S 58 41W
Los Remedios →, Méx. 29 B2 19 28N 99 13W
Los Remedios, Parque Nacional de, Méx. 29 B2 19 27N 99 15W
Los Reyes, Méx. 29 B4 19 21N 99 6W
Los Rosales, Car. 30 E5 10 28N 66 54W
Losby, Oslo 2 B5 59 53N 10 59 E
Loughton, Lon. 4 B5 51 38N 0 4 E
Loures, Lisb. 8 F7 38 49N 9 10W
Louveciennes, Paris 5 B2 48 51N 2 8 E
Louvres, Paris 5 A5 49 2N 2 30 E
Lovön, Stock. 3 E10 59 18N 17 51 E
Lövstafjärden, Stock. 3 D9 59 23N 17 46 E
Lowe, Mt., L.A. 28 A4 34 13N 118 5W
Lowe Pond, Bost. 21 A3 42 41N 71 0W
Lowell, Bost. 21 B2 42 38N 71 16W
Lowell Dracut State Forest, Bost. 21 B1 42 39N 71 22W
Lower Crystal Springs Res., S.F. 27 C2 37 31N 122 21W
Lower Edmonton, Lon. 4 B4 51 37N 0 3W
Lower Montville, N.Y. 22 B2 40 53N 74 21W
Lower New York B., N.Y. 22 D4 40 32N 74 5W
Lower Plenty, Melb. 19 E7 37 44 S 145 7 E
Lower Shing Mun Res., H.K. 12 D5 22 22N 114 9 E
Lower Sydenham, Lon. 4 C4 51 25N 0 2W
Lower Van Norman L., L.A. 28 A2 34 17N 118 28W
Lübars, Berl. 7 A3 52 37N 13 21 E
Lubeinī, Bahr el →, El Qâ. 18 C4 30 1N 31 5 E
Lubya →, St-Pet. 11 A5 60 1N 30 39 E
Lucento, Tori. 9 B2 45 5N 7 39 E
Lucero, La Hab. 30 B3 23 5N 82 19W
Ludwigsfeld, Mün. 7 F9 48 12N 11 27 E
Lugarno, Syd. 19 B3 33 59 S 151 2 E
Luguqiao, Beij. 14 C2 39 49N 116 10 E
Luhu, Gzh. 14 G8 23 9N 113 16 E
Luipaardsvlei, Jobg. 18 F7 26 6 S 27 49 E
Luis Guillón, B.A. 32 C4 34 48 S 58 26W
Lujia, Shang. 14 J12 31 15N 121 37 E
Lukens, Mt., L.A. 28 A3 34 16N 118 12W
Lumiar, Lisb. 8 F8 38 4N 9 10W
Lundtofte, Købn. 2 D10 55 47N 12 32 E
Lung Mei, H.K. 12 D6 22 23N 114 15 E
Lunsad, Manila 15 E4 14 27N 121 11 E
Luojiang, Gzh. 14 G8 23 5N 113 17 E
Lura →, Mil. 9 D5 45 34N 9 5 E
Lurnea, Syd. 19 B2 33 56 S 150 54 E
Lurup, Hbg. 7 D7 53 35N 9 54 E
Lustheim, Mün. 7 F10 48 14N 11 34 E
Lütgendortmund, Ruhr 6 A6 51 30N 7 20 E
Lutherville-Timonium, Balt. 25 A3 39 25N 76 36W
Lütringhausen, Ruhr 6 C5 51 12N 7 14 E
Lutvatn, Oslo 2 B5 59 54N 10 52 E
Luwan, Shang. 14 J11 31 12N 121 27 E
Luyano, La Hab. 30 B2 23 6N 82 21W
Luzhniki Sports Centre, Mos. 11 E9 55 43N 37 31 E
Lyckebyn, Stock. 3 E12 59 11N 18 13 E
Lynbrook, N.Y. 23 D6 40 38N 73 41W
Lyndhurst, Jobg. 18 E9 26 7 S 28 6 E
Lyndhurst, N.Y. 22 C4 40 49N 74 7W
Lynn, Bost. 21 C4 42 28N 70 57W
Lynn Harbor, Bost. 21 C4 42 26N 70 56W
Lynnfield, Bost. 21 B3 42 32N 71 2W
Lynwood, L.A. 28 C3 33 55N 118 11W
Lyon, Gare de, Paris 5 B4 48 50N 2 22 E
Lyons, Chic. 26 C2 41 48N 87 49W
Lyonsville, N.Y. 22 B2 40 57N 74 26W
Lysaker, Oslo 2 B3 59 54N 10 38 E
Lysakerselva →, Oslo 2 B3 59 55N 10 38 E
Lysolaje, Pra. 10 B2 50 8N 14 22 E
Lytkarino, Mos. 11 F11 55 35N 37 55 E
Lyubertsy, Mos. 11 E11 55 40N 37 52 E
Lyublino, Mos. 11 E10 55 41N 37 44 E

M

Ma Nam Wat, H.K. 12 D6 22 21N 114 16 E
Ma Po, Sŏul 12 G7 37 32N 126 56 E
Ma Tsz Keng, H.K. 12 D5 22 22N 114 7 E
Ma Yau Tong, H.K. 12 E6 22 19N 114 14 E
Maantiekylä, Hels. 3 A5 60 20N 25 3 E
Maarifa, Bagd. 17 F8 33 15N 44 21 E
Mabashi, Tōkyō 13 B4 35 48N 139 55 E
Mabato Pt., Manila 15 E4 14 29N 121 3 E
Mabolo, Manila 15 E3 14 26N 120 56 E
Macaco, Morro do, Rio J. 31 B3 22 56 S 43 6W
McCook, Chic. 26 C2 41 47N 87 49W
McGill Univ., Mtrl. 20 A4 43 30N 73 35W
Machida, Tōkyō 13 C1 35 32N 139 26 E
Macierzysz, Wsaw. 10 E6 52 13N 20 50 E
Maciołki, Wsaw. 10 E7 52 19N 21 9 E
Mackayville, Mtrl. 20 A5 43 30N 73 26W
McKinnon, Melb. 19 F7 37 54 S 145 1 E
Mclean, Wash. 25 D6 38 56N 77 10W
Macleod, Melb. 19 E7 37 43 S 145 4 E
Macopocho →, Stgo 30 J10 33 24 S 70 40W
Macquarie Fields, Syd. 19 B2 33 59 S 150 53 E
Macquarie Univ., Syd. 19 A3 33 46 S 151 7 E
MacRitchie Res., Sing. 15 F7 1 20N 103 49 E
Macul, Stgo 30 K11 33 30 S 70 35W
Macuto, Car. 30 D5 10 36N 66 53W
Macuto →, Car. 30 D5 10 36N 66 53W
Madatpur, Calc. 16 C6 22 53N 88 27 E
Maddalena, Colle della, Tori. 9 B3 45 2N 7 43 E
Madhudaha, Calc. 16 E6 22 30N 88 24 E
Madhyamgram, Calc. 16 D6 22 41N 88 26 E
Madīnah Al Mansōr, Bagd. 17 F8 33 18N 44 20 E
Mâdinet el Muqattam, El Qâ. 18 C5 30 1N 31 15 E
Mâdinet Nasr, El Qâ. 18 C5 30 4N 31 18 E
Madipur, Delhi 16 B1 28 40N 77 8 E
Madison, N.Y. 22 C2 40 45N 74 24W
Madonna della Scala, Tori. 9 B3 44 59N 7 46 E
Madonna dell'Arco, Nápl. 9 H13 40 52N 14 23 E
Madrid, Mdrd. 8 B2 40 24N 3 42W
Madrona, Barc. 8 D5 41 27N 2 1 E
Madureira, Rio J. 31 B2 22 52 S 43 19W
Maeda, Tōkyō 13 B3 35 48N 139 45 E
Maesawa, Tōkyō 13 B2 35 44N 139 31 E
Magalhaes, Rio J. 31 B1 22 51 S 43 22W
Magdalena del Mar, Lima 30 G8 12 5 S 77 5W
Magholpur, Delhi 16 A1 28 41N 77 6 E
Maghreb, Bagd. 17 E8 33 23N 44 22 E
Magidiyeh, Tehr. 17 C5 35 43N 51 28 E
Maginu, Tōkyō 13 C2 35 34N 139 34 E
Magliana, Rome 9 F9 41 50N 12 26 E
Maglód, Bud. 10 K14 47 27N 19 18 E
Magnolia, Phil. 24 B4 39 51N 75 1W
Magny-les-Hameaux, Paris 5 C2 48 44N 2 3 E
Maharajpur, Delhi 16 B2 28 39N 77 19 E
Maheshtala, Calc. 16 E5 22 29N 88 15 E
Mahiari, Calc. 16 E5 22 35N 88 14 E
Mahikpur, Calc. 16 E5 22 32N 88 13 E
Mahim, Bomb. 16 G8 19 2N 72 50 E
Mahim B., Bomb. 16 G7 19 2N 72 49 E
Mahishdanga, Calc. 16 C5 22 53N 88 11 E
Mahlsdorf, Berl. 7 A4 52 30N 13 37 E
Mahmoodabad, Kar. 17 G11 24 51N 67 4 E
Mahmutbey, Ist. 17 A1 41 2N 28 49 E
Mahpar, Jak. 15 H9 6 9 S 106 49 E
Mahul, Bomb. 16 G8 19 0N 72 53 E
Maida Vale, Lon. 4 B3 51 31N 0 11W
Maidstone, Melb. 19 E6 37 47 S 144 52 E
Maincourt-sur-Yvette, Paris 5 C1 48 42N 1 58 E
Maipu, Stgo 30 K10 33 30 S 70 45W
Maiquetia, Car. 30 D5 10 35N 66 57W
Maiquetia Aeropuerto, Car. 30 D4 10 36N 67 0W
Maisons-Alfort, Paris 5 C4 48 48N 2 26 E
Maisons-Laffitte, Paris 5 B2 48 57N 2 8 E
Maissoneuve, Mtrl. 20 A4 43 33N 73 33W
Maitani, Ōsaka 12 B3 34 48N 135 22 E
Majadahonda, Mdrd. 8 B1 40 28N 3 52W
Majlis, Tehr. 17 C5 35 41N 51 25 E
Makati, Manila 15 D4 14 33N 121 7 E
Mäkiniitty, Hels. 3 A4 60 20N 24 58 E
Mala Straná, Pra. 10 B2 50 4N 14 24 E
Malabar, Syd. 19 B4 33 58 S 151 14 E
Malabar Hill, Bomb. 16 H7 18 57N 72 48 E
Malabon, Manila 15 D3 14 39N 120 56 E
Malacanang Palace, Manila 15 D3 14 35N 120 59 E
Malagrotta, Rome 9 F9 41 52N 12 20 E
Malakhovka, Mos. 11 F12 55 39N 38 0 E
Malakoff, Paris 5 C3 48 49N 2 18 E
Malakpur, Delhi 16 A2 28 42N 77 12 E
Malanday, Manila 15 D4 14 38N 121 5 E
Malanghero, Tori. 9 A2 45 12N 7 39 E
Mälarhöjaen, Stock. 3 E10 59 18N 17 58 E
Malaspina, L., Mil. 9 E6 45 28N 9 18 E
Malassis, Paris 5 D2 48 37N 2 3 E
Malate, Manila 15 D3 14 34N 120 59 E
Malaya Neva, St-Pet. 11 B3 59 56N 30 16 E
Malaya-Okhta, St-Pet. 11 B4 59 55N 30 25 E
Malchow, Berl. 7 A3 52 34N 13 29 E
Malden, Bost. 21 C3 42 26N 71 3W
Malden, Lon. 4 C3 51 23N 0 15W
Malešice, Pra. 10 B3 50 5N 14 30 E
Malekete, Lagos 18 A3 6 33N 7 32 E
Malir →, Kar. 17 H11 24 49N 67 4 E
Malir Cantonment, Kar. 17 G12 24 58N 67 10 E
Malmi, Hels. 3 B4 60 15N 25 0 E
Malmøya, Oslo 2 B4 59 52N 10 45 E
Måløv, Købn. 2 D9 55 44N 12 20 E
Malton, Trto. 20 D7 43 42N 79 38W
Malvern, Jobg. 18 F9 26 11 S 28 5 E
Malvern, Melb. 19 F7 37 50 S 145 2 E
Malvern, Phil. 24 A1 40 2N 75 31W
Malvern, Trto. 20 D9 43 47N 79 13W
Malvern East, Jobg. 18 F9 26 11 S 28 7 E
Malverne, N.Y. 23 C6 40 40N 73 40W
Mamaroneck, N.Y. 23 B6 40 56N 73 41W
Mamaroneck Harbour, N.Y. 23 B6 40 56N 73 42W
Mamonovo, Mos. 11 F10 55 36N 37 49 E
Mamonovo, Mos. 11 E8 55 41N 37 18 E
Mampong Prapatan, Jak. 15 J9 6 15 S 106 49 E
Mampukuji, Tōkyō 13 C2 35 36N 139 31 E
Man Budrukh, Bomb. 16 G8 19 2N 72 55 E
Man Khurd, Bomb. 16 G8 19 3N 72 55 E
Managua, La Hab. 30 C3 22 58N 82 17W
Manayunk, Phil. 24 A3 40 1N 75 12W
Mandaluyong, Manila 15 D4 14 35N 121 1 E
Mandaoli, Delhi 16 B2 28 37N 77 17 E
Mandaqui, S. Pau. 31 D6 23 29 S 46 37W
Mandaqui →, S. Pau. 31 D6 23 30 S 46 40W
Mandres-les-Roses, Paris 5 C5 48 42N 2 32 E
Mandvi, Bomb. 16 H8 18 56N 72 50 E
Mang Kung Uk, H.K. 12 E6 22 18N 114 16 E
Manggarai, Jak. 15 J10 6 12 S 106 50 E
Manguinho, Aéroporto de, Rio J. 31 B2 22 52 S 43 14W
Mangwon, Sŏul 12 G7 37 33N 126 55 E
Manhasset, N.Y. 23 C6 40 47N 73 41W
Manhasset B., N.Y. 23 C6 40 49N 73 43W
Manhasset Hills, N.Y. 23 C7 40 45N 73 40W
Manhattan, N.Y. 22 C5 40 48N 73 57W
Manhattan Beach, L.A. 28 C2 33 53N 118 24W
Manhatten, N.Y. 22 D5 40 34N 73 56W
Manila, Manila 15 D3 14 35N 120 58 E
Manila B., Manila 15 D3 14 32N 120 53 E
Manila Int. Airport, Manila 15 D4 14 31N 121 0 E
Mankkaa, Hels. 3 B3 60 11N 24 47 E
Mankunda, Calc. 16 C6 22 50N 88 22 E
Manly, Syd. 19 A4 33 47 S 151 17 E
Manly Warringah War Memorial Park, Syd. 19 A4 33 46 S 151 15 E
Manning State Park, Bost. 21 B1 42 34N 71 20W
Mannsworth, Wien 10 H11 48 8N 16 30 E
Manoa, Phil. 24 B3 39 58N 75 18W
Manor Park, Lon. 4 B5 51 32N 0 1 E
Manora, Kar. 17 H10 24 47N 66 58 E
Manorhaven, N.Y. 23 B6 40 50N 73 41W
Manoteras, Mdrd. 8 B3 40 28N 3 39W
Manquehue, Cerro, Stgo 30 J11 33 21 S 70 35W
Mantegazza, Mil. 9 D4 45 30N 8 58 E
Mantilla, La Hab. 30 B3 23 4N 82 20W
Mantua, Phil. 24 C3 39 47N 75 10W
Mantua Cr. →, Phil. 24 C3 39 47N 75 13W
Manufacta, Jobg. 18 E8 26 9 S 27 51 E
Manzanares, Canal de, Mdrd. 8 C3 40 19N 3 38W
Mapetla, Jobg. 18 F8 26 16 S 27 51 E
Maple, Trto. 20 C7 43 51N 79 30W
Maple L., Chic. 26 C1 41 43N 87 53W
Maple Shade, Phil. 24 B4 39 57N 75 0W
Maplewood, N.Y. 22 C3 40 43N 74 16W
Maracana, Rio J. 31 B2 22 54 S 43 13W
Maraisburg, Jobg. 18 F8 26 10 S 27 57 E
Marano di Nápoli, Nápl. 9 H12 40 53N 14 11 E
Maraoli, Bomb. 16 G8 19 2N 72 53 E
Marapendi, L. de, Rio J. 31 C1 23 0 S 43 23W
Marblehead, Bost. 21 C4 42 29N 70 51W
Marcelin, Wsaw. 10 E6 52 19N 20 59 E
Marcella, N.Y. 22 B2 40 59N 74 29W
Marcos Paz, B.A. 32 C2 34 46 S 58 49W
Marcoussis, Paris 5 D3 48 38N 2 13 E
Marcus Hook, Phil. 24 C2 39 49N 75 25W
Marcus Hook Cr. →, Phil. 24 B2 39 49N 75 24W
Marechiaro, Nápl. 9 J12 40 48N 14 12 E
Mareil-Marly, Paris 5 B2 48 52N 2 4 E
Margareten, Wien 10 G10 48 11N 16 20 E
Margency, Paris 5 A3 49 0N 2 17 E
Margitsziget, Bud. 10 J13 47 31N 19 2 E
Maria, Wien 10 G10 48 11N 16 21 E
Maria Paula, Rio J. 31 B3 22 53 S 43 1W
Marianao, La Hab. 30 B2 23 4N 82 25W
Marianella, Nápl. 9 H12 40 53N 14 13 E
Mariano Acosta, B.A. 32 C2 34 42 S 58 47W
Mariano J. Haedo, B.A. 32 B3 34 39 S 58 35W
Maridalen, Oslo 2 B4 59 59N 10 45 E
Maridalsvatnet, Oslo 2 B4 59 58N 10 46 E
Mariendorf, Berl. 7 B3 52 26N 13 23 E
Marienfelde, Berl. 7 B3 52 24N 13 23 E
Marienthal, Hbg. 7 D8 53 34N 10 4 E
Mariglianella, Nápl. 9 H13 40 55N 14 26 E
Marigliano, Nápl. 9 H13 40 55N 14 27 E
Marikina, Manila 15 D4 14 38N 121 5 E
Marikina →, Manila 15 D4 14 33N 121 3 E
Marin City, S.F. 27 A1 37 52N 122 30W
Marin Headlands State Park, S.F. 27 A2 37 50N 122 28W
Marin Is., S.F. 27 A2 37 57N 122 27W
Marin Pen., S.F. 27 A1 37 50N 122 30W
Marine World, S.F. 27 C3 37 32N 122 16W
Mariners Harbour, N.Y. 22 D3 40 38N 74 10W
Mario, Mt., Rome 9 F9 41 55N 12 27 E
Markham, Chic. 26 D2 41 35N 87 40W
Markham, Phil. 24 B1 39 53N 75 30W
Markham, Trto. 20 C8 43 49N 79 22W
Markham, Mt., L.A. 28 A4 34 14N 118 6W
Marki, Wsaw. 10 E7 52 19N 21 6 E
Markland Wood, Trto. 20 E7 43 38N 79 34W
Marlton, Phil. 24 B5 39 53N 74 55W
Marly, Forêt de, Paris 5 B2 48 52N 2 2 E
Marly-le-Roi, Paris 5 B2 48 52N 2 5 E
Marne →, Paris 5 C4 48 47N 2 29 E
Marne-la-Vallée, Paris 5 B5 48 50N 2 37 E
Marolles-en-Brie, Paris 5 C5 48 44N 2 33 E
Maroonda Aquaduct, Melb. 19 E7 37 40 S 145 9 E
Maroubra, Syd. 19 B4 33 56 S 151 16 E
Marple, Phil. 24 B2 39 56N 75 20W
Marquette Park, Chic. 26 C2 41 46N 87 42W
Marrickville, Syd. 19 B3 33 54 S 151 9 E
Marschlande, Hbg. 7 E8 53 27N 10 6 E
Marsfield, Syd. 19 A3 33 46 S 151 7 E
Marte, Base Aérea de, S. Pau. 31 E6 23 30 S 46 38W
Martesana, Navíglio della, Mil. 9 D6 45 31N 9 17 E
Martin State Nat. Airport, Balt. 25 B4 39 19N 76 24W
Martinez, B.A. 32 A3 34 29 S 58 31W
Martinkylä, Hels. 3 B4 60 17N 24 51 E
Martins Pond, Bost. 21 B3 42 35N 71 7W
Martinsried, Mün. 7 G9 48 6N 11 27 E
Maruko, Tōkyō 13 C3 35 33N 139 40 E
Marusino, Mos. 11 E11 55 41N 37 58 E
Marxloh, Ruhr 6 A2 51 30N 6 47 E
Maryino, Mos. 11 E10 55 40N 37 45 E
Maryland, Sing. 15 G7 1 19N 103 47 E
Maryland, Univ. of, Wash. 25 D8 38 59N 76 56W
Marylebone, Lon. 4 B4 51 31N 0 9W
Marymont, Wsaw. 10 E6 52 16N 20 58 E
Marysin Wawerski, Wsaw. 10 E7 52 14N 21 9 E
Marzahn, Berl. 7 A4 52 32N 13 34 E
Masambong, Manila 15 D4 14 38N 121 0 E
Mascot, Syd. 19 B4 33 55 S 151 12 E
Mascuppic L., Bost. 21 A1 42 40N 71 23W
Masmo, Stock. 3 E10 59 15N 17 53 E
Maspeth, N.Y. 22 C5 40 43N 73 55W
Masr el Gedida, El Qâ. 18 C5 30 5N 31 19 E
Masr el Qadîma, El Qâ. 18 C5 30 0N 31 14 E
Masroor Airport, Kar. 17 G10 24 53N 66 56 E
Massa di Somma, Nápl. 9 H13 40 50N 14 22 E
Massachusetts B., Bost. 21 C4 42 25N 70 50W
Massachusett's Inst. of Tech., Bost. 21 C3 42 22N 71 6W
Massamá, Lisb. 8 F7 38 45N 9 16W
Massapequa, N.Y. 23 C8 40 40N 73 28W
Massby, Hels. 3 B6 60 17N 25 16 E
Massey →, Trto. 20 D9 43 42N 79 19W
Massy, Paris 5 C3 48 43N 2 16 E
Matarza →, B.A. 32 C3 34 47 S 58 25W
Matasango, Lima 30 G9 12 3 S 76 58W
Mathle, Calc. 16 E5 22 34N 88 13 E
Mathitong, Beij. 14 B2 39 56N 116 19 E
Matinecock, N.Y. 23 B7 40 51N 73 36W
Matinha, Lisb. 8 F8 38 45N 9 5W
Matramam, Jak. 15 J10 6 12 S 106 51 E
Matsubara, Ōsaka 12 C4 34 34N 135 33 E
Matsubushi, Tōkyō 13 A3 35 55N 139 49 E
Matsudo, Tōkyō 13 B4 35 46N 139 54 E
Matsumoloshinden, Tōkyō 13 B2 35 50N 139 36 E
Mattapan, Bost. 21 D3 42 16N 71 6W
Mátyásföld, Bud. 10 J14 47 30N 19 12 E
Mau Tso Ngam, H.K. 12 D6 22 22N 114 13 E
Mauá, S. Pau. 31 E7 23 39 S 46 27W
Mauer, Wien 10 H9 48 9N 16 16 E
Mauerbach →, Wien 10 G9 48 12N 16 13 E
Mauldre, Paris 5 C1 48 56N 1 53 E
Maurecourt, Paris 5 B2 48 59N 2 3 E
Mauregard, Paris 5 A5 49 2N 2 34 E
Maurepas, Paris 5 C1 48 46N 1 55 E
Mauripur, Kar. 17 G10 24 52N 66 55 E
Maxhof, Mün. 7 G9 48 4N 11 29 E
Maya-Zan, Ōsaka 12 B2 34 43N 135 12 E
Maybunga, Manila 15 D4 14 34N 121 4 E
Mayfair, Jobg. 18 F9 26 11 S 28 0 E
Mayfair, Phil. 24 A4 40 2N 75 3W
Maypajo, Manila 15 D3 14 38N 120 58 E
Maytubig, Manila 15 D3 14 33N 120 59 E
Maywood, Chic. 26 B1 41 52N 87 51W
Maywood, L.A. 28 C3 33 59N 118 12W
Maywood, N.Y. 22 B4 40 53N 74 3W
Mazagaon, Bomb. 16 H8 18 57N 72 50 E
M'Boi Mirim, S. Pau. 31 F5 23 42 S 46 46W
Meadow I., N.Y. 23 D7 40 36N 73 32W
Meadow L., N.Y. 23 C5 40 44N 73 50W
Meadowlands, Jobg. 18 F8 26 12 S 27 53 E
Meadowood, Wash. 25 C7 39 4N 77 0W
Měcholupy, Pra. 10 B3 50 3N 14 32 E
Měčice, Pra. 10 A3 50 11N 14 31 E
Mecidiyekoy, Ist. 17 A3 41 4N 29 0 E
Meckinghoven, Ruhr 6 A5 51 37N 7 19 E
Médan, Paris 5 B1 48 57N 1 59 E
Medfield, Bost. 21 D2 42 11N 71 18W
Medford, Bost. 21 C3 42 25N 71 7W
Media, Phil. 24 B2 39 55N 75 23W
Mediodia, Mdrd. 8 B3 40 22N 3 39W
Medvastö, Hels. 3 C2 60 5N 24 38 E
Medvedkovo, Mos. 11 D9 55 52N 37 37 E
Medvezhiy Ozyora, Mos. 11 D11 55 52N 37 59 E
Meerbeck, Ruhr 6 B1 51 28N 6 38 E
Meerbusch, Ruhr 6 C2 51 15N 6 40 E
Meguro →, Tōkyō 13 C3 35 37N 139 45 E
Meguro-Ku, Tōkyō 13 C3 35 37N 139 42 E
Mehpalpur, Delhi 16 B1 28 32N 77 7 E
Mehrābād Airport, Tehr. 17 C4 35 41N 51 18 E
Mehram Nagar, Delhi 16 B1 28 34N 77 8 E
Mehrow, Berl. 7 A4 52 34N 13 37 E
Meiderich, Ruhr 6 B2 51 27N 6 47 E
Meidling, Wien 10 G10 48 10N 16 20 E
Meiendorf, Hbg. 7 D8 53 37N 10 8 E
Méier, Rio J. 31 B2 22 52 S 43 15W
Meiji Shrine, Tōkyō 13 B3 35 41N 139 41 E
Meizino-Mori-Minō National Park, Ōsaka 12 A3 34 51N 135 28 E
Mejiro, Tōkyō 13 B3 35 43N 139 43 E
Melbourne, Melb. 19 E6 37 48 S 144 58 E
Melbourne Airport, Melb. 19 E6 37 40 S 144 50 E
Melbourne Univ., Melb. 19 E6 37 47 S 144 57 E
Melito di Nápoli, Nápl. 9 H12 40 55N 14 13 E
Melkki, Hels. 3 C4 60 8N 24 53 E
Mellingstedt, Hbg. 7 C8 53 40N 10 6 E
Mellunkylä, Hels. 3 B5 60 14N 25 5 E
Mellunmäki, Hels. 3 B5 60 14N 25 6 E
Melrose, Bost. 21 C3 42 27N 71 2W
Melrose, N.Y. 22 C5 40 49N 73 55W
Melrose Park, Chic. 26 B1 41 53N 87 50W
Melun-Sénart, Paris 5 D5 48 37N 2 31 E
Melun-Villaroche, Aérodrome de, Paris 5 D6 48 37N 2 41 E
Melville, N.Y. 23 C8 40 47N 73 24W
Menai, Syd. 19 C3 34 1 S 151 1 E
Menandon, Paris 5 A2 49 2N 2 3 E
Mendoza, Lima 30 G9 12 5 S 76 59W
Mengede, Ruhr 6 A6 51 34N 7 23 E
Mengjiazhai, Shang. 14 J11 31 19N 121 21 E
Menglinghausen, Ruhr 6 B6 51 28N 7 26 E
Menlo Park, S.F. 27 D3 37 27N 122 11W
Menlo Park Terrace, N.Y. 22 D3 40 34N 74 18W
Mentang, Jak. 15 J9 6 11 S 106 49 E
Menucourt, Paris 5 A1 49 1N 1 59 E
Meopham, Lon. 4 C7 51 22N 0 21 E
Mérantaise →, Paris 5 C2 48 42N 2 8 E
Mercamadrid, Mdrd. 8 B3 40 21N 3 39W
Merced, L., S.F. 27 B2 37 43N 122 29W
Merchantville, Phil. 24 B4 39 56N 75 3W
Mercier, Pont, Mtrl. 20 B3 43 24N 73 39W
Merdeka Palace, Jak. 15 J9 6 10 S 106 49 E
Meredale, Jobg. 18 F8 26 16 S 27 58 E
Mergellina, Nápl. 9 J12 40 49N 14 12 E
Meriden, N.Y. 22 B2 40 56N 74 29W
Merion Station, Phil. 24 B3 39 59N 75 15W
Merlimau, Sing. 15 G7 1 17N 103 42 E
Merlimau, P., Sing. 15 G7 1 17N 103 42 E
Merlo, B.A. 32 B2 34 39 S 58 43W
Merri Cr. →, Melb. 19 E6 37 47 S 144 59 E
Merrick, N.Y. 23 D7 40 39N 73 32W
Merrionette Park, Chic. 26 C2 41 41N 87 40W
Merritt, L., S.F. 27 B3 37 48N 122 15W
Merrylands, Syd. 19 B2 33 50 S 150 59 E
Merton, Lon. 4 C3 51 24N 0 11W
Mesgarābād, Tehr. 17 D6 35 27N 51 30 E
Meshcherskiy, Mos. 11 E8 55 40N 37 23 E
Mesquita, Rio J. 31 A1 22 46 S 43 25W
Messe, Wien 10 G10 48 13N 16 24 E
Messy, Paris 5 B6 48 58N 2 41 E
Metanópoli, Mil. 9 E6 45 24N 9 15 E
Methuen, Bost. 21 A2 42 42N 71 12W
Metropolitan Opera, N.Y. 22 C5 40 46N 74 59W
Mettman, Ruhr 6 C3 51 15N 6 58 E
Metuchen, N.Y. 22 D2 40 32N 74 21W
Metzkausen, Ruhr 6 C3 51 16N 6 57 E
Meudon, Paris 5 C3 48 48N 2 14 E
Meulan, Paris 5 A1 49 0N 1 54 E
México, Aeropuerto Int. de, Méx. 29 B3 19 25N 99 4W
México, Ciudad de, Méx. 29 B2 19 25N 99 7W
Mezzate, Mil. 9 E6 45 26N 9 17 E
Mia Dong, Sŏul 12 G8 37 36N 127 0 E
Miano, Nápl. 9 H12 40 53N 14 15 E
Miasto, Wsaw. 10 E7 52 15N 21 0 E
Michalin, Wsaw. 10 F8 52 9N 21 13 E
Michałowice, Wsaw. 10 E6 52 10N 20 52 E
Michle, Pra. 10 B2 50 3N 14 28 E
Mickleton, Phil. 24 C3 39 47N 75 14W
Middle →, Balt. 25 B4 39 18N 76 24W
Middle B., N.Y. 23 D7 40 36N 73 36W
Middle Branch →, Balt. 25 B3 39 15N 76 37W
Middle Brewster I., Bost. 21 C4 42 20N 70 51W
Middle Cove, Syd. 19 A4 33 48 S 151 13 E
Middle Harbour, Syd. 19 A4 33 47 S 151 14 E
Middle Hd., Syd. 19 A4 33 49 S 151 16 E
Middle I., H.K. 12 E6 22 16N 114 10 E
Middle Park, Melb. 19 F6 37 50 S 144 57 E
Middle River, Balt. 25 A4 39 20N 76 26W
Middle Village, N.Y. 23 C5 40 43N 73 52W
Middleborough, Balt. 25 B4 39 18N 76 26W
Middlesex, N.Y. 22 D2 40 34N 74 27W
Middlesex Fells Reservation, Bost. 21 C3 42 27N 71 6W
Middlesex Res., N.Y. 22 D3 40 37N 74 19W
Middleton, Bost. 21 B3 42 35N 71 0W
Middleton Pond, Bost. 21 B3 42 35N 71 1W
Middleville, N.Y. 23 B9 40 53N 73 19W
Midland Beach, N.Y. 22 D4 40 34N 74 6W
Midland Park, N.Y. 22 B4 40 59N 74 9W

42 Midlothian

Midlothian, Chic. 26 D2 41 37N 87 43W
Miedzeszyn, Wsaw. 10 E8 52 10N 21 11 E
Międzylesie, Wsaw. 10 E8 52 12N 21 10 E
Miessaari, Hels. 3 C3 60 8N 24 47 E
Mikhaylovskoye, Mos. 11 F9 55 35N 37 35 E
Mikhelysona, Mos. 11 E11 55 42N 37 52 E
Milano, Mil. 9 E5 45 28N 9 10 E
Milano Due, Mil. 9 E6 45 29N 9 16 E
Milano San Felice, Mil. 9 E6 45 28N 9 18 E
Milanolago, Mil. 9 E6 45 27N 9 17 E
Milbertshofen, Mün. 7 F10 48 10N 11 34 E
Milburn, N.Y. 22 C3 40 43N 74 19W
Milford, Balt. 25 A2 39 21N 76 43W
Mill Cr. →, S.F. 27 A1 37 53N 122 31W
Mill Hill, Lon. 4 B3 51 37N 0 14W
Mill Neck, N.Y. 23 B7 40 53N 73 33W
Mill Park, Melb. 19 E7 37 40 S 145 3 E
Mill Valley, S.F. 27 A1 37 54N 122 33W
Millbrae, S.F. 27 C2 37 35N 122 22W
Mille-Iles, R. des →, Mtrl. 20 A3 43 39N 73 46W
Miller L., Balt. 25 B4 39 15N 76 21W
Miller Meadow, Chic. 26 B2 41 51N 87 49W
Milliken, Trto. 20 D9 43 49N 79 17W
Millis, Bost. 21 D1 42 10N 71 21W
Mills College, S.F. 27 B3 37 46N 122 10W
Milltown, Phil. 24 B1 39 57N 75 32W
Millwall, Lon. 4 C4 51 29N 0 0 E
Millwood, Wash. 25 D8 38 52N 76 52W
Milon-la-Chapelle, Paris 5 C2 48 43N 2 3 E
Milpa Alta, Méx. 29 C3 19 11N 99 0W
Milperra, Syd. 19 B2 33 56 S 150 59 E
Milspe, Ruhr 6 C5 51 18N 7 19 E
Milton, Bost. 21 D3 42 14N 71 2W
Milton Village, Bost. 21 D3 42 15N 71 4W
Mimico, Trto. 20 E8 43 36N 79 29W
Mimico Cr. →, Trto. 20 E7 43 37N 79 33W
Minami, Ōsaka 12 B4 34 43N 135 30 E
Minami-Ku, Tōkyō 13 D2 35 24N 139 37 E
Minami-tsunashima, Tōkyō 13 C2 35 32N 139 37 E
Minato, Ōsaka 12 C3 34 39N 135 25 E
Minato-Ku, Tōkyō 13 C3 35 39N 139 44 E
Mine, Tōkyō 13 B3 35 49N 139 46 E
Minebank Run →, Balt. 25 A3 39 24N 76 33W
Mineola, N.Y. 23 C7 40 44N 73 38W
Ministro Rivadavia, B.A. 32 D4 34 50 S 58 22W
Minō, Ōsaka 12 B3 34 49N 135 28 E
Minshât el Bekkarî, El Qâ. 18 C4 30 0N 31 8 E
Minto, Syd. 19 C2 34 1 S 150 51 E
Minute Man Nat. Hist. Park, Bost. 21 C2 42 25N 71 16W
Mirafiori, Tori. 9 B2 45 1N 7 36 E
Miraflores, Lima 30 G8 12 7 S 77 2W
Miramar, La Hab. 30 B2 23 7N 82 25W
Miramar, S.F. 27 D2 37 29N 122 27W
Miranda, Syd. 19 C3 34 2 S 151 6 E
Mirzapur, Calc. 16 D6 22 49N 88 24 E
Misato, Tōkyō 13 B4 35 49N 139 51 E
Misericordia, Sa. da, Rio J. 31 B2 22 51 S 43 17W
Mishawum L., Bost. 21 B3 42 30N 71 8W
Mission, S.F. 27 B2 37 44N 122 25W
Mississauga, Trto. 20 E7 43 35N 79 34W
Mitaka, Tōkyō 13 B2 35 41N 139 34 E
Mitcham, Lon. 4 C3 51 23N 0 10W
Mitcham, Melb. 19 E8 37 48 S 145 12 E
Mitcham Common, Lon. 4 C4 51 23N 0 8W
Mitino, Mos. 11 D8 55 51N 37 20 E
Mitry, Paris 5 B5 48 59N 2 36 E
Mitry-Mory, Paris 5 B5 48 59N 2 38 E
Mitry-Mory, Aérodrome de, Paris 5 B5 48 59N 2 37 E
Mitte, Berl. 7 A3 52 32N 13 24 E
Mittel Isarkanal, Mün. 7 F11 48 12N 11 40 E
Mittenheim, Mün. 7 F10 48 15N 11 33 E
Mixcoac, Presa de, Méx. 29 B2 19 21N 99 14W
Mixquic, Méx. 29 C4 19 13N 98 58W
Miyakojima, Ōsaka 12 B4 34 42N 135 31 E
Miyalo, Tōkyō 13 B3 35 49N 139 35 E
Mizonokuchi, Tōkyō 13 C2 35 35N 139 34 E
Mizue, Tōkyō 13 B4 35 41N 139 54 E
Mizuko, Tōkyō 13 A2 35 50N 139 32 E
Mizumoto, Tōkyō 13 B4 35 46N 139 52 E
Mlocinski Park, Wsaw. 10 E6 52 19N 20 57 E
Mlociny, Wsaw. 10 E6 52 18N 20 55 E
Mnevniki, Mos. 11 E8 55 45N 37 28 E
Moba, Lagos 18 B2 6 26N 7 28 E
Moczydło, Wsaw. 10 F7 52 8N 21 2 E
Modderfontein, Jobg. 18 E10 26 5 S 28 10 E
Modderfontein →, Jobg. 18 E9 26 5 S 28 10 E
Modřany, Pra. 10 B2 50 0N 14 24 E
Moers, Ruhr 6 B1 51 26N 6 37 E
Moffat Park, Jobg. 18 F9 26 15 S 28 4 E
Mofolo, Jobg. 18 F8 26 13 S 27 53 E
Mog, Sŏul 12 G7 37 32N 126 52 E
Mogyorod, Bud. 10 J14 47 35N 19 14 E
Mohili, Bomb. 16 G8 19 5N 72 52 E
Moinho Velho →, S. Pau. 31 E6 23 35 S 46 35W
Moissy-Cramayel, Paris 5 D5 48 37N 2 35 E
Moita, Lisb. 8 G9 38 39N 8 59W
Mokotów, Wsaw. 10 E7 52 12N 21 0 E
Molapo, Jobg. 18 F8 26 15 S 27 51 E
Mole →, Lon. 4 D2 51 14N 0 20W
Moletsane, Jobg. 18 F8 26 14 S 27 50 E
Molino de Rosas, Méx. 29 B2 19 21N 99 14W
Mølleå →, Købn. 2 D10 55 48N 12 35 E
Möllen, Ruhr 6 A2 51 35N 6 41 E
Mollins de Rey, Barc. 8 D5 41 24N 2 1 E
Molokovo, Mos. 11 F11 55 35N 37 53 E
Mombaça, S. Pau. 31 E7 23 37 S 46 25W
Mombello, Mil. 9 D5 45 36N 9 7 E
Momote, Tōkyō 13 B3 35 46N 139 37 E
Monash Univ., Melb. 19 F7 37 54 S 145 8 E
Monbulk Cr. →, Melb. 19 F8 37 55 S 145 12 E
Moncalieri, Tori. 9 B3 45 0N 7 41 E
Moncolombone, Tori. 9 A1 45 12N 7 28 E
Mondeor, Jobg. 18 F9 26 16 S 28 0 E
Moneda, Palacio de la, Stgo 30 J11 33 26 S 70 39W
Mong Kok, H.K. 12 E6 22 19N 114 10 E
Mongat, Barc. 8 D6 41 27N 2 16 E
Mongreno, Tori. 9 B3 45 3N 7 45 E
Moninos →, S. Pau. 31 F6 23 40 S 46 33W
Monrovia, L.A. 28 B4 34 9N 118 1W
Monsanto, Lisb. 8 F7 38 44N 9 12W
Monsanto, Parque Florestal de, Lisb. 8 F7 38 43N 9 11W
Mont Royal, Mtrl. 20 A4 43 30N 73 38W
Mont-Royal, Parc, Mtrl. 20 A4 43 30N 73 36W
Montalban, Car. 30 E5 10 28N 66 56W
Montana de Montjuich, Barc. 8 D5 41 21N 2 9 E
Montara, S.F. 27 C2 37 32N 122 30W
Montara, Pt., S.F. 27 C1 37 32N 122 31W
Montara Mt., S.F. 27 C2 37 32N 122 27W
Montchanin, Phil. 24 C1 39 47N 75 35W
Montclair, N.Y. 22 C3 40 49N 74 12W
Monte Chingolo, B.A. 32 C4 34 43 S 58 22W
Monte Grande, B.A. 32 C4 34 48 S 58 27W
Monte Sacro, Rome 9 F10 41 56N 12 32 E
Montebello, L.A. 28 B4 34 1N 118 8W
Montelera, Tori. 9 B1 45 9N 7 26 E
Montemor, Lisb. 8 F7 38 49N 9 12W
Monterey Park, L.A. 28 B4 34 3N 118 7W
Monterrey, La Hab. 30 B3 23 5N 82 18W
Montespaccato, Rome 9 F9 41 54N 12 23 E
Montesson, Paris 5 B2 48 54N 2 8 E
Monteverde Nuovo, Rome 9 F9 41 52N 12 26 E
Montfermeil, Paris 5 B5 48 54N 2 33 E
Montgeron, Paris 5 C4 48 42N 2 27 E
Montigny-le-Bretonneux, Paris 5 C2 48 46N 2 1 E
Montigny-les-Cormeilles, Paris 5 B3 48 59N 2 11 E
Montijo, Lisb. 8 F9 38 42N 8 58W
Montjay-la-Tour, Paris 5 B6 48 54N 2 40 E
Montlhéry, Paris 5 D3 48 38N 2 16 E
Montlignon, Paris 5 A3 49 0N 2 16 E
Montmagny, Paris 5 B4 48 58N 2 21 E
Montmorency, Paris 5 B3 48 59N 2 19 E
Montmorency, Forêt de, Paris 5 A3 49 2N 2 16 E
Montparnasse, Gare, Paris 5 B3 48 50N 2 19 E
Montpelier, Wash. 25 C8 39 3N 76 50W
Montréal, Mtrl. 20 A4 43 30N 73 33W
Montréal, Î. de, Mtrl. 20 A4 43 30N 73 40W
Montréal, Univ. de, Mtrl. 20 B4 43 29N 73 37W
Montréal-Est, Mtrl. 20 A4 43 37N 73 31W
Montréal Nord, Mtrl. 20 A4 43 36N 73 36W
Montreuil, Paris 5 B4 48 51N 2 27 E
Montrose, L.A. 28 A3 34 12N 118 12W
Montrose, Melb. 19 E8 37 49 S 145 19 E
Montrose, Wash. 25 C7 39 2N 77 7W
Montrouge, Paris 5 C3 48 48N 2 18 E
Montvale, N.Y. 22 A4 41 2N 74 1W
Montville, N.Y. 22 B2 40 55N 74 23W
Monza, Mil. 9 D6 45 35N 9 16 E
Monzoro, Mil. 9 E5 45 27N 9 2 E
Moóca, S. Pau. 31 E6 23 33 S 46 35W
Moóca →, S. Pau. 31 E6 23 35 S 46 35W
Moonachie, N.Y. 22 C4 40 50N 74 2W
Moonee Ponds, Melb. 19 E6 37 45 S 144 53 E
Moonee Valley Racecourse, Melb. 19 E6 37 45 S 144 55 E
Moorbek, Hbg. 7 C7 53 41N 9 58 E
Moorburg, Hbg. 7 E7 53 29N 9 55 E
Moorebank, Syd. 19 B2 33 56 S 150 56 E
Moorestown, Phil. 24 B5 39 58N 74 56W
Moorfleet, Hbg. 7 D8 53 30N 10 4 E
Mooroolbark, Melb. 19 E8 37 46 S 145 19 E
Moorwerder, Hbg. 7 E8 53 28N 10 3 E
Moosach, Mün. 7 F10 48 10N 11 30 E
Mora, Bomb. 16 H8 18 54N 72 55 E
Moraga, S.F. 27 B4 37 49N 122 7W
Morainvilliers, Paris 5 B1 48 55N 1 56 E
Morales →, B.A. 32 C3 34 47 S 58 35W
Morangis, Paris 5 C4 48 42N 2 20 E
Moratalaz, Mdrd. 8 B3 40 24N 3 39W
Morbras →, Paris 5 C5 48 46N 2 30 E
Mörby, Stock. 3 D11 59 23N 18 3 E
Morce →, Paris 5 B4 48 57N 2 25 E
Morden, Lon. 4 C3 51 24N 0 13W
Morehill, Jobg. 18 F11 26 10 S 28 20 E
Moreno, B.A. 32 B2 34 38 S 58 45W
Moreno, Rome 9 G10 41 48N 12 37 E
Morgan Park, Chic. 26 C3 41 41N 87 38W
Moriguchi, Ōsaka 12 B4 34 43N 135 34 E
Morivione, Mil. 9 E6 45 26N 9 12 E
Morningside, Jobg. 18 E9 26 4 S 28 3 E
Morningside, Wash. 25 E8 38 49N 76 53W
Morningside Park, Trto. 20 D9 43 46N 79 12W
Moroka, Jobg. 18 F8 26 15 S 27 52 E
Moron, B.A. 32 B3 34 39 S 58 37W
Morris Plains, N.Y. 22 C2 40 49N 74 29W
Morristown, N.Y. 22 C2 40 47N 74 28W
Morro, Castillo del, La Hab. 30 B2 23 8N 82 21W
Morro Pelado, S. Pau. 31 E7 23 38 S 46 24W
Morro Solar, Lima 30 H8 12 11 S 77 1W
Morsang-sur-Orge, Paris 5 D4 48 39N 2 21 E
Mörsenbroich, Ruhr 6 C2 51 15N 6 48 E
Morses Pond, Bost. 21 D2 42 17N 71 19W
Morte →, Paris 5 C3 48 40N 2 16 E
Mortlake, Lon. 4 C3 51 27N 0 15W
Mortlake, Syd. 19 B3 33 50 S 151 6 E
Morton, Phil. 24 B2 39 54N 75 20W
Morton Grove, Chic. 26 A2 42 2N 87 46W
Mory, Paris 5 B5 48 58N 2 37 E
Moscavide, Lisb. 8 F8 38 47N 9 6W
Moscow = Moskva, Mos. 11 E9 55 45N 37 37 E
Mosede, Købn. 2 E8 55 34N 12 17 E
Mosede Strand, Købn. 2 E8 55 34N 12 17 E
Mosjøen, Oslo 2 C6 50 49N 11 0 E
Moskhaton, Ath. 8 J11 37 55N 23 40 E
Moskva, Mos. 11 E9 55 45N 37 37 E
Moskvoretskiy, Mos. 11 E9 55 42N 37 37 E
Mosman, Syd. 19 A4 33 49 S 151 15 E
Moss Beach, S.F. 27 C2 37 31N 122 30W
Móstoles, Mdrd. 8 C1 40 18N 3 51W
Moto →, Tōkyō 13 A3 35 53N 139 45 E
Motol, Pra. 10 B1 50 3N 14 19 E
Motspur Park, Lon. 4 C3 51 23N 0 14W
Mount Airy, Phil. 24 A3 40 3N 75 10W
Mount Dennis, Trto. 20 D8 43 40N 79 28W
Mount Ephraim, Phil. 24 B4 39 52N 75 5W
Mount Greenwood, Chic. 26 C2 41 42N 87 42W
Mount Hood Memorial Park, Bost. 21 C3 42 26N 71 1W
Mount Pleasant, Lon. 4 B2 51 30N 0 22W
Mount Pleasant Park, Balt. 25 A3 39 22N 76 34W
Mount Prospect, Chic. 26 A1 42 3N 87 54W
Mount Royal, Phil. 24 C3 39 48N 75 12W
Mount Tamalpais State Park, S.F. 27 A1 37 53N 122 34W
Mount Vernon, N.Y. 23 B6 40 54N 73 49W
Mount Waverley, Melb. 19 F7 37 52 S 145 7 E
Mount Wilson Observatory, L.A. 28 A4 34 13N 118 4W
Mountain Lakes, N.Y. 22 B2 40 54N 74 27W
Mountain Spring Ls., N.Y. 22 A2 41 2N 74 21W
Mountain View, N.Y. 22 B3 40 55N 74 15W
Mountainside, N.Y. 22 C2 40 40N 74 21W
Mountnessing, Lon. 4 B7 51 39N 0 21 E
Moûtiers, Paris 5 D1 48 36N 1 58 E
Mozu, Ōsaka 12 C3 34 33N 135 29 E
Müggelberge, Berl. 7 B4 52 25N 13 37 E
Müggelheim, Berl. 7 B5 52 24N 13 40 E
Muggiò, Mil. 9 D6 45 35N 9 13 E
Mugnano di Nápoli, Nápl. 9 H12 40 54N 14 12 E
Mühleiten, Wien 10 G11 48 10N 16 33 E
Mühlenau →, Hbg. 7 C7 53 41N 9 56 E
Mühlenfliess →, Berl. 7 A5 52 32N 13 42 E
Muir Beach, S.F. 27 A1 37 51N 122 34W
Muirkirk, Wash. 25 C8 39 3N 76 53W
Mujahidpur, Delhi 16 B2 28 33N 77 14 E
Mukandpur, Delhi 16 A2 28 44N 77 10 E
Muko →, Ōsaka 12 B3 34 48N 135 22 E
Mukojima, Tōkyō 13 B3 35 43N 139 49 E
Mulbarton, Jobg. 18 F9 26 17 S 28 3 E
Mulford Gardens, S.F. 27 B3 37 42N 122 10W
Mulgrave, Melb. 19 F8 37 55 S 145 12 E
Mülheim, Ruhr 6 B3 51 25N 6 53 E
Mullica Hill, Phil. 24 C3 39 44N 75 13W
Mullum Mullum Cr. →, Melb. 19 E8 37 44 S 145 10 E
Münchehofe, Berl. 7 B5 52 29N 13 40 E
München, Mün. 7 G10 48 8N 11 34 E
München-Riem, Flughafen, Mün. 7 G11 48 7N 11 42 E
Munich = München, Mün. 7 G10 48 8N 11 34 E
Munirka, Delhi 16 B2 28 33N 77 10 E
Muniz, B.A. 32 B2 34 33 S 58 42W
Munkkiniemi, Hels. 3 B4 60 11N 24 52 E
Munro, B.A. 32 B3 34 31 S 58 31W
Munsey Park, N.Y. 23 C6 40 47N 73 40W
Münsterkirche, Ruhr 6 B4 51 27N 7 0 E
Muranów, Wsaw. 10 E6 52 14N 20 58 E
Murayama-chosuichi, Tōkyō 13 B1 35 45N 139 26 E
Murrumbeena, Melb. 19 F7 37 53 S 145 4 E
Musashino, Tōkyō 13 B2 35 42N 139 33 E
Mushin, Lagos 18 A2 6 31N 7 21 E
Musinè, Mte., Tori. 9 B1 45 7N 7 27 E
Musocco, Mil. 9 E5 45 29N 9 8 E
Musta Hevonen, Hels. 3 B6 60 11N 25 14 E
Mustafabad, Delhi 16 A2 28 43N 77 13 E
Mustansiriya, Bagd. 17 E8 33 22N 44 24 E
Musturud, El Qâ. 18 C5 30 8N 31 17 E
Muswell Hill, Lon. 4 B4 51 35N 0 8W
Mutanabi, Bagd. 17 F8 33 19N 44 21 E
Muthana, Bagd. 17 F8 33 19N 44 27 E
Mutinga, S. Pau. 31 D5 23 29 S 46 46W
Muttontown, N.Y. 23 C7 40 49N 73 32W
Muzon, Manila 15 D4 14 32N 121 8 E
Myaglovo, St-Pet. 11 B5 59 53N 30 39 E
Myakinino, Mos. 11 E8 55 48N 37 22 E
Mykerinos, El Qâ. 18 D4 29 28N 31 8 E
Myllykylä, Hels. 3 A4 60 21N 24 57 E
Myllypuro, Hels. 3 B5 60 13N 25 3 E
Myras, Hels. 3 B5 60 19N 25 6 E
Myrvoll, Oslo 2 C4 59 47N 10 48 E
Mystic Lakes, Bost. 21 C3 42 26N 71 8W
Mytishchi, Mos. 11 D10 55 53N 37 44 E

N

Nababpur, Calc. 16 D5 22 42N 88 12 E
Nações, Parque das, S. Pau. 31 E6 23 38 S 46 30W
Nachstebreck, Ruhr 6 C5 51 17N 7 14 E
Nacka, Stock. 3 E12 59 19N 18 10 E
Nada, Ōsaka 12 B2 34 42N 135 13 E
Nærsnes, Oslo 2 C2 59 45N 10 27 E
Nærum, Købn. 2 D10 55 48N 12 33 E
Nagareyama, Tōkyō 13 A4 35 51N 139 54 E
Nagasaki, Tōkyō 13 B3 35 43N 139 40 E
Nagasuga, Tōkyō 13 D4 35 21N 139 57 E
Nagata, Ōsaka 12 C1 34 39N 135 8 E
Nagatsuta, Tōkyō 13 C2 35 32N 139 31 E
Nagytarcsa, Bud. 10 J14 47 31N 19 17 E
Nagytétény, Bud. 10 K12 47 23N 18 59 E
Nahant, Bost. 21 C4 42 25N 70 54W
Nahant B., Bost. 21 C4 42 26N 70 54W
Nahant Harbor, Bost. 21 C4 42 25N 70 55W
Nahdein, W. el →, El Qâ. 18 C5 30 3N 31 19 E
Nahia, El Qâ. 18 C4 30 2N 31 7 E
Naihati, Calc. 16 C6 22 53N 88 25 E
Najafgarh Drain →, Delhi 16 B1 28 39N 77 4 E
Najio, Ōsaka 12 B2 34 49N 135 18 E
Naka →, Tōkyō 13 B4 35 49N 139 52 E
Naka-Ku, Tōkyō 13 D2 35 25N 139 38 E
Nakada, Tōkyō 13 D2 35 24N 139 30 E
Nakajima, Tōkyō 13 D4 35 25N 139 54 E
Nakano, Tōkyō 13 D4 35 25N 139 56 E
Nakano-Ku, Tōkyō 13 B2 35 42N 139 39 E
Nakasato, Tōkyō 13 D4 35 22N 139 55 E
Nakayama, Tōkyō 13 B4 35 43N 139 57 E
Nalikul, Calc. 16 D5 22 49N 88 10 E
Nalpur, Calc. 16 E5 22 31N 88 10 E
Namazie Estate, Sing. 15 F7 1 25N 103 42 E
Namgajha, Sŏul 12 G7 37 33N 126 55 E
Namsan Park, Sŏul 12 G7 37 32N 126 59 E
Namyeong, Sŏul 12 G7 37 32N 126 57 E
Nan Wan, H.K. 12 D5 22 20N 114 5 E
Nanbiancun, Gzh. 14 G7 23 4N 113 10 E
Nancefield, Jobg. 18 F8 26 17 S 27 54 E
Nanchang He →, Beij. 14 B2 39 58N 116 14 E
Nandaha, Calc. 16 D5 22 49N 88 18 E
Nandang, Gzh. 14 G8 23 6N 113 12 E
Nandian, Tianj. 14 D6 39 10N 117 16 E
Nangal Dewat, Delhi 16 B1 28 33N 77 5 E
Nangi, Calc. 16 E5 22 30N 88 13 E
Nangka →, Manila 15 D4 14 38N 121 8 E
Nangloi, Delhi 16 A1 28 41N 77 4 E
Nangloi Jat, Delhi 16 A1 28 41N 77 3 E
Nanhai, Gzh. 14 G7 23 2N 113 6 E
Nanhan He →, Beij. 14 B2 39 57N 116 11 E
Naniwa, Ōsaka 12 C3 34 39N 135 29 E
Nankai, Tianj. 14 E5 39 7N 117 10 E
Nanmenwai, Tianj. 14 E6 39 8N 117 10 E
Nanole, Bomb. 16 G8 19 0N 72 55 E
Nanshi, Shang. 14 J11 31 12N 121 29 E
Nanterre, Paris 5 B3 48 53N 2 12 E
Nantouillet, Paris 5 A6 49 0N 2 42 E
Nantusket Beach, Bost. 21 D4 42 16N 70 52W
Nanxiang, Shang. 14 J10 31 17N 121 18 E
Naoabad, Calc. 16 F6 22 28N 88 26 E
Napara, Calc. 16 E6 22 38N 88 23 E
Napier Mole, Kar. 17 H10 24 49N 66 58 E
Napindan, Manila 15 D4 14 32N 121 5 E
Naples = Nápoli, Nápl. 9 J12 40 50N 14 14 E
Nápoli, Nápl. 9 J12 40 50N 14 14 E
Nápoli, G. di, Nápl. 9 J12 40 46N 14 15 E
Naraina, Delhi 16 B1 28 36N 77 8 E
Narawa, Tōkyō 13 D4 35 25N 139 58 E
Narayanpara, Calc. 16 C5 22 53N 88 18 E
Narberth, Phil. 24 A3 40 0N 75 16W
Narimasu, Tōkyō 13 B2 35 46N 139 35 E
Nārmak, Tehr. 17 C5 35 42N 51 28 E
Närsta, Stock. 3 E9 59 17N 17 43 E
Naruo, Ōsaka 12 B3 34 43N 135 22 E
Näsby, Stock. 3 D11 59 25N 18 5 E
Näsbypark, Stock. 3 D11 59 25N 18 5 E
Näsfjärden, Stock. 3 D9 59 25N 17 41 E
Nassau Shore, N.Y. 23 D8 40 39N 73 26W
Natick, Bost. 21 D2 42 16N 71 19W
Nation, Place de la, Paris 5 B4 48 51N 2 23 E
National Arboretum, Wash. 25 D8 39 54N 76 59W
Nativitas, Méx. 29 C3 19 15N 99 5W
Natolin, Wsaw. 10 F7 52 8N 21 4 E
Naucalpan de Juárez, Méx. 29 B2 19 28N 99 14W
Naupada, Bomb. 16 G8 19 3N 72 50 E
Navíglio di Pavia, Mil. 9 E5 45 24N 9 9 E
Navíglio Grande, Mil. 9 E5 45 25N 9 5 E
Navotas, Manila 15 D3 14 39N 120 56 E
Nazal Hikmat Beg, Bagd. 17 E8 33 23N 44 25 E
Nazimabad, Kar. 17 G11 24 54N 67 1 E
Nazukari, Tōkyō 13 A4 35 50N 139 57 E
Néa Alexandhria, Ath. 8 J11 37 52N 23 46 E
Néa Faliron, Ath. 8 J10 37 55N 23 39 E
Néa Ionía, Ath. 8 H11 38 3N 23 45 E
Néa Liósia, Ath. 8 H11 38 3N 23 43 E
Néa Smirni, Ath. 8 J11 37 54N 23 43 E
Neapolis, Ath. 8 J11 37 58N 23 45 E
Neasden, Lon. 4 B3 51 33N 0 16W
Neauphle-le-Château, Paris 5 C1 48 48N 1 53 E
Nebušice, Pra. 10 B1 50 6N 14 19 E
Nedlitz, Berl. 7 B1 52 25N 13 3 E
Nee Soon, Sing. 15 F7 1 24N 103 49 E
Needham, Bost. 21 D2 42 16N 71 13W
Needham Heights, Bost. 21 D2 42 17N 71 14W
Needle Hill, H.K. 12 D5 22 23N 114 9 E
Negishi B., Tōkyō 13 D2 35 23N 139 38 E
Nehiti, Calc. 16 D5 22 42N 88 16 E
Nekrasovka, Mos. 11 E11 55 41N 37 55 E
Nemaṭābād, Tehr. 17 D5 35 38N 51 21 E
Nemchinovka, Mos. 11 E7 55 42N 37 19 E
Népliget, Bud. 10 K13 47 29N 19 7 E
Neponset →, Bost. 21 D3 42 12N 71 9W
Nerima, Tōkyō 13 B3 35 44N 139 40 E
Nerul, Bomb. 16 G9 19 2N 73 0 E
Nerviano, Mil. 9 D4 45 32N 8 58 E
Nesodden, Oslo 2 C4 59 48N 10 41 E
Nesoddtangen, Oslo 2 B4 59 52N 10 41 E
Nesøya, Oslo 2 B3 59 52N 10 31 E
Nestipayac →, Méx. 29 A4 19 33N 89 57W
Netzahualcóyotl, Méx. 29 B3 19 24N 99 2W
Neu Aubing, Mün. 7 G9 48 8N 11 25 E
Neu Buch, Berl. 7 A4 52 37N 13 31 E
Neu Buchhorst, Berl. 7 B5 52 24N 13 44 E
Neu Fahrland, Berl. 7 B1 52 26N 13 3 E
Neu Lindenberg, Berl. 7 A4 52 36N 13 33 E
Neu Wulmstorf, Hbg. 7 E6 53 27N 9 48 E
Neu Zittau, Berl. 7 B5 52 23N 13 44 E
Neubiberg, Mün. 7 G11 48 4N 11 40 E
Neudorf, Hbg. 7 E8 53 27N 10 4 E
Neudorf, Ruhr 6 B2 51 25N 6 47 E
Neuegenbüttel, Hbg. 7 D7 53 38N 9 54 E
Neuenfelde, Hbg. 7 D6 53 31N 9 48 E
Neuenhagen, Berl. 7 A4 52 32N 13 38 E
Neuenkamp, Ruhr 6 B2 51 26N 6 43 E
Neuessling, Wien 10 G11 48 15N 16 32 E
Neugraben-Fischbek, Hbg. 7 E6 53 28N 9 49 E
Neuhausen, Mün. 7 G10 48 9N 11 32 E
Neuherberg, Mün. 7 F10 48 13N 11 35 E
Neuhönow, Berl. 7 A5 52 34N 13 44 E
Neuilly-Plaisance, Paris 5 B5 48 51N 2 30 E
Neuilly-sur-Marne, Paris 5 B5 48 51N 2 31 E
Neuilly-sur-Seine, Paris 5 B3 48 53N 2 15 E
Neukagran, Wien 10 G10 48 14N 16 27 E
Neukettenhof, Wien 10 H10 48 7N 16 28 E
Neukölln, Berl. 7 B3 52 28N 13 25 E
Neuland, Hbg. 7 E8 53 27N 10 0 E
Neuperlach, Mün. 7 G10 48 6N 11 37 E
Neuried, Mün. 7 G9 48 5N 11 27 E
Neuss, Ruhr 6 C2 51 12N 6 42 E
Neustift am Walde, Wien 10 G9 48 14N 16 17 E
Neusüssenbrunn, Wien 10 G10 48 16N 16 29 E
Neuville-sur-Oise, Paris 5 A2 49 0N 2 3 E
Neuwaldegg, Wien 10 G9 48 14N 16 17 E
Neuwiedenthal, Hbg. 7 E7 53 28N 9 52 E
Neva →, St-Pet. 11 B4 59 56N 30 20 E
Neves, Rio J. 31 B3 22 51 S 43 5W
Neviges, Ruhr 6 C4 51 18N 7 6 E
New Addington, Lon. 4 C4 51 20N 0 0 E
New Ash Green, Lon. 4 C6 51 22N 0 18 E
New Baghdād, Bagd. 17 F8 33 18N 44 28 E
New Barnet, Lon. 4 B3 51 38N 0 10W
New Brighton, N.Y. 22 D4 40 38N 74 5W
New Brunswick, N.Y. 22 D2 40 30N 74 29W
New Canada, Jobg. 18 F8 26 12 S 27 56 E
New Canada Dam, Jobg. 18 F8 26 12 S 27 56 E
New Canal →, Calc. 16 E6 22 33N 88 25 E
New Carrollton, Wash. 25 D8 38 58N 76 52W
New Cassell, N.Y. 23 C7 40 45N 73 32W
New Cross, Lon. 4 C4 51 28N 0 1W
New Delhi, Delhi 16 B2 28 36N 77 11 E
New Dorp, N.Y. 22 D4 40 34N 74 7W
New Dorp Beach, N.Y. 22 D4 40 33N 74 6W
New Hyde Park, N.Y. 23 C7 40 43N 73 39W
New Kleinfontein, Jobg. 18 F11 26 11 S 28 20 E
New Malden, Lon. 4 C3 51 24N 0 15W
New Milford, N.Y. 22 B4 40 56N 74 0W
New Modder, Jobg. 18 F11 26 10 S 28 21 E
New Providence, N.Y. 22 C2 40 42N 74 23W
New Redruth, Jobg. 18 F9 26 15 S 28 7 E
New Rochelle, N.Y. 23 B6 40 55N 73 45W
New South Wales, Univ. of, Syd. 19 B4 33 55 S 151 14 E
New Southgate, Lon. 4 B4 51 37N 0 7W
New Springville, N.Y. 22 D4 40 35N 74 9W
New Territories, H.K. 12 D5 22 23N 114 10 E
New Toronto, Trto. 20 E7 43 35N 79 30W
New Utrecht, N.Y. 22 D5 40 36N 73 59W
New Vernon, N.Y. 22 C2 40 44N 74 30W
New York Aquarium, N.Y. 22 D5 40 33N 73 59W
New York Botanical Gdns., N.Y. 23 B5 40 51N 73 51W
New York Univ., N.Y. 22 C5 40 43N 74 59W
Newabgarj, Calc. 16 D6 22 47N 88 23 E
Newark, N.Y. 22 C3 40 43N 74 10W
Newark, S.F. 27 C4 37 32N 122 2W
Newark B., N.Y. 22 C4 40 40N 74 8W
Newark Int. Airport, N.Y. 22 C3 40 41N 74 10W
Newbury Park, Lon. 4 B5 51 34N 0 5 E
Newclare, Jobg. 18 F8 26 11 S 27 58 E
Newfoundland, N.Y. 22 A2 41 2N 74 25W
Newham, Lon. 4 B5 51 31N 0 2 E
Newlands, Jobg. 18 F8 26 10 S 27 57 E
Newport, Melb. 19 F6 37 50 S 144 51 E
Newportville, Phil. 24 A5 40 7N 74 53W
Newton, Bost. 21 D2 42 19N 71 12W
Newton Brook, Trto. 20 D8 43 47N 79 24W
Newton Highlands, Bost. 21 D2 42 19N 71 13W
Newtonville, Bost. 21 C2 42 20N 71 11W
Newtown, Syd. 19 B4 33 54 S 151 11 E
Newtown Square, Phil. 24 B2 39 59N 75 24W
Neyegawa, Ōsaka 12 B4 34 45N 135 36 E
Ngau Chi Wan, H.K. 12 D6 22 20N 114 12 E
Ngau Tau Kok, H.K. 12 E6 22 19N 114 12 E
Ngong Shuen Chau, H.K. 12 E5 22 19N 114 8 E
Ngua Kok Wan, H.K. 12 D5 22 21N 114 5 E
Niāvarān, Tehr. 17 C5 35 47N 51 29 E
Nibria, Calc. 16 E5 22 36N 88 15 E
Nichelino, Tori. 9 C2 44 59N 7 38 E
Nichols Run →, Wash. 25 C6 39 2N 77 17W
Nicholson, Mt., H.K. 12 E6 22 15N 114 11 E
Nidāl, Bagd. 17 F8 33 19N 44 25 E
Niddrie, Melb. 19 E6 37 44 S 144 51 E
Nieder Neuendorf, Berl. 7 A2 52 36N 13 12 E
Niederbonsfeld, Ruhr 6 B4 51 22N 7 8 E
Niederdonk, Ruhr 6 C2 51 14N 6 41 E
Niederschöneweide, Berl. 7 B3 52 27N 13 30 E
Niederschönhausen, Berl. 7 A3 52 35N 13 25 E
Niederwenigern, Ruhr 6 B4 51 24N 7 8 E
Niemeyer, Rio J. 31 B2 22 59 S 43 16W
Niendorf, Hbg. 7 D7 53 37N 9 57 E
Nienstedten, Hbg. 7 D7 53 33N 9 51 E
Nierst, Ruhr 6 C2 51 19N 6 43 E
Nihonbashi, Tōkyō 13 B3 35 41N 139 46 E
Niipperi, Hels. 3 B3 60 18N 24 45 E
Niiza, Tōkyō 13 B2 35 48N 139 33 E
Nikaia, Ath. 8 J10 37 57N 23 38 E
Nikinmäki, Hels. 3 A5 60 20N 25 8 E
Nikolassee, Berl. 7 B2 52 25N 13 12 E
Nikolo-Khovanskoye, Mos. 11 F8 55 36N 37 27 E
Nikolskiy, Mos. 11 E8 55 49N 37 29 E
Nikolyskoye, Mos. 11 E11 55 46N 37 53 E
Nikulino, Mos. 11 E8 55 40N 37 25 E
Nil, Bagd. 17 E8 33 21N 44 25 E
Nil, Nahr en →, El Qâ. 18 D5 29 57N 31 14 E
Nile = Nil, Nahr en →, El Qâ. 18 D5 29 57N 31 14 E
Niles, Chic. 26 A2 42 1N 87 48W
Nilganj, Calc. 16 D6 22 45N 88 25 E
Nilópolis, Rio J. 31 A1 22 47 S 43 23W
Nimta, Calc. 16 D6 22 40N 88 24 E
Nincop, Hbg. 7 D6 53 30N 9 48 E
Ningyuan, Tianj. 14 E6 39 9N 117 12 E
Nippa, Tōkyō 13 C2 35 31N 139 36 E
Nippori, Tōkyō 13 B3 35 43N 139 45 E
Niru-ye-Hava'i, Tehr. 17 C5 35 41N 51 26 E
Nishi, Ōsaka 12 B3 34 40N 135 28 E
Nishi, Tōkyō 13 D2 35 26N 139 37 E
Nishi-arai, Tōkyō 13 B3 35 46N 139 48 E
Nishinari, Ōsaka 12 C3 34 37N 135 28 E
Nishinomiya, Ōsaka 12 B2 34 44N 135 18 E
Nishiyama, Tōkyō 13 D4 35 22N 139 57 E
Nishiyodogawa, Ōsaka 12 B3 34 41N 135 24 E
Nisida, I. di, Nápl. 9 J11 40 47N 14 10 E
Niterói, Rio J. 31 B3 22 53 S 43 7W
Nithari, Delhi 16 B3 28 34N 77 20 E
Nittedal, Oslo 2 A5 60 0N 10 57 E
Niyog, Manila 15 E3 14 27N 120 57 E
Noapara, Calc. 16 D6 22 49N 88 22 E
Nobidome, Tōkyō 13 B2 35 48N 139 34 E
Nockeby, Stock. 3 E10 59 19N 17 56 E
Noel Park, Lon. 4 B4 51 35N 0 5W
Nogatino, Mos. 11 E10 55 41N 37 41 E
Nogent-sur-Marne, Paris 5 B4 48 50N 2 28 E
Noiseau, Paris 5 C5 48 46N 2 32 E
Noisiel, Paris 5 B5 48 51N 2 37 E
Noisy-le-Grand, Paris 5 B5 48 50N 2 33 E
Noisy-le-Roi, Paris 5 B2 48 50N 2 3 E
Noisy-le-Sec, Paris 5 B4 48 53N 2 27 E
Nokkala, Hels. 3 C3 60 8N 24 45 E
Nøklevatn, Oslo 2 B5 59 52N 10 52 E
Nolme →, Ruhr 6 B6 51 23N 7 26 E
Nomentano, Rome 9 F10 41 55N 12 30 E
Nonakashinden, Tōkyō 13 B2 35 44N 139 30 E
Nongminyundong Jiangxisuo, Gzh. 14 G8 23 7N 113 15 E
Nonhyeon, Sŏul 12 G8 37 30N 127 1 E
Nontha Buri, Bangk. 15 A1 13 50N 100 29 E
Noordgesig, Jobg. 18 F8 26 13 S 27 56 E
Nord, Gare du, Paris 5 B4 48 52N 2 21 E
Nordbysjøen, Oslo 2 B6 59 51N 11 1 E
Nordereble, Hbg. 7 D7 53 32N 9 59 E
Norderelbe →, Hbg. 7 E8 53 29N 10 3 E
Norderstedt, Hbg. 7 C7 53 42N 9 59 E
Nordmarka, Oslo 2 A4 60 1N 10 38 E
Nordrand-Seidlung, Wien 10 G10 48 16N 16 26 E
Nordre Elvåga, Oslo 2 B5 59 53N 10 54 E
Nordstrand, Oslo 2 B4 59 52N 10 48 E
Normandy Heights, Balt. 25 B2 39 17N 76 48W
Norra Björköfjärden, Stock. 3 D8 59 26N 17 39 E
Norridge, Chic. 26 B2 41 57N 87 49W
Norristown, Phil. 24 A2 40 7N 75 20W
Norrkula, Hels. 3 B6 60 19N 25 20 E
Norrmalm, Stock. 3 D11 59 20N 18 3 E
Norrviken, Stock. 3 D10 59 27N 17 52 E
North Acton, Bost. 21 B1 42 30N 71 23W
North Amityville, N.Y. 23 C8 40 41N 73 25W
North Andover, Bost. 21 A3 42 41N 71 7W
North Arlington, N.Y. 22 C4 40 47N 74 7W
North Auburn, Syd. 19 B3 33 50 S 151 3 E
North Babylon, N.Y. 23 C9 40 43N 73 19W
North Bellmore, N.Y. 23 C7 40 40N 73 32W
North Bergen, N.Y. 22 C4 40 48N 74 0W
North Beverly, Bost. 21 B4 42 34N 70 51W
North Billerica, Bost. 21 B2 42 35N 71 16W
North Branch →, Phil. 24 C4 39 50N 75 5W
North Branch Chicago River →, Chic. 26 B2 41 53N 87 42W
North Brighton, Bost. 21 C3 42 21N 71 8W
North Caldwell, N.Y. 22 B3 40 52N 74 15W
North Cambridge, Bost. 21 C3 42 23N 71 7W
North Cheam, Lon. 4 C3 51 22N 0 11W
North Chelmsford, Bost. 21 B1 42 38N 71 23W
North Cohasset, Bost. 21 D4 42 15N 70 50W
North Cray, Lon. 4 C5 51 25N 0 8 E
North Fair Oaks, S.F. 27 D3 37 28N 122 11W
North Finchley, Lon. 4 B3 51 36N 0 10W
North Germiston, Jobg. 18 F9 26 12 S 28 9 E
North Hackensack, N.Y. 22 B4 40 54N 74 2W
North Haledon, N.Y. 22 B3 40 58N 74 11W
North Harbour, Manila 15 D3 14 37N 120 57 E
North Hd., Syd. 19 A4 33 49 S 151 18 E
North Hills, N.Y. 23 C6 40 46N 73 40W
North Hollywood, L.A. 28 B2 34 9N 118 22W
North Lexington, Bost. 21 C2 42 27N 71 14W
North Lindenhurst, N.Y. 23 C8 40 42N 73 22W
North Long Beach, L.A. 28 C3 33 53N 118 10W
North Manly, Syd. 19 A4 33 46 S 151 17 E
North Massapequa, N.Y. 23 C8 40 42N 73 27W
North Merrick, N.Y. 23 C7 40 41N 73 33W
North New Hyde Park, N.Y. 23 C6 40 44N 73 42W
North Pelham, N.Y. 23 B6 40 54N 73 46W
North Plainfield, N.Y. 22 D2 40 37N 74 27W
North Point, Balt. 25 B4 39 16N 76 29W
North Pt., H.K. 12 E6 22 17N 114 12 E
North Pt., Syd. 19 A4 33 48 S 151 18 E
North Randolph, Bost. 21 D3 42 11N 71 5W
North Reading, Bost. 21 B3 42 34N 71 5W
North Res., Bost. 21 C3 42 27N 71 6W
North Richmond, S.F. 27 A2 37 57N 122 22W
North Riverside, Chic. 26 B2 41 50N 87 48W
North Ryde, Syd. 19 A3 33 47 S 151 7 E
North Saugus, Bost. 21 C3 42 29N 71 0W
North Shore Channel →, Chic. 26 B2 41 58N 87 42W

Pitampura Kalan 43

North Springfield, *Wash.* 25 E6 38 48N 77 11W
North Stifford, *Lon.* 4 B6 51 30N 0 18 E
North Sudbury, *Bost.* 21 C1 42 24N 71 24W
North Sydney, *Syd.* 19 B4 33 50 S 151 13 E
North Tewksbury, *Bost.* 21 B2 42 38N 71 14W
North Valley Stream, *N.Y.* 23 C6 40 41N 73 42W
North Wantagh, *N.Y.* 23 C7 40 41N 73 30W
North Weymouth, *Bost.* 21 D4 42 14N 70 56W
North Wilmington, *Bost.* 21 B3 42 34N 71 9W
North Woburn, *Bost.* 21 B2 42 30N 71 10W
North Woolwich, *Lon.* 4 B5 51 30N 0 3 E
North York, *Trto.* 20 D8 43 45N 79 27W
Northaw, *Lon.* 4 A4 51 42N 0 8W
Northbridge, *Syd.* 19 A4 33 49 S 151 15 E
Northbrook, *Chic.* 26 A1 42 7N 87 50W
Northcliff, *Jobg.* 18 E8 26 8 S 27 58 E
Northcote, *Melb.* 19 E7 37 46 S 145 0 E
Northeastern Univ., *Bost.* 21 C3 42 20N 71 4W
Northfield, *Chic.* 26 A2 42 5N 87 45W
Northfleet, *Lon.* 4 C7 51 26N 0 21 E
Northlake, *Chic.* 26 B1 41 54N 87 53W
Northmead, *Jobg.* 18 E10 26 9 S 28 19 E
Northmead, *Syd.* 19 A3 33 47 S 151 0 E
Northmount, *Trto.* 20 D8 43 46N 79 23W
Northolt, *Lon.* 4 B2 51 32N 0 22W
Northport, *N.Y.* 23 B8 40 54N 73 20W
Northport B., *N.Y.* 23 B8 40 54N 73 22W
Northridge, *L.A.* 28 A1 34 14N 118 30W
Northumberland Heath, *Lon.* 4 C6 51 28N 0 10 E
Northvale, *N.Y.* 22 A5 41 0N 73 59W
Northwest Branch ➛, *Balt.* 25 B3 39 16N 76 35W
Northwest Branch ➛, *Wash.* 25 C8 39 2N 76 56W
Northwestern Univ., *Chic.* 26 A2 42 3N 87 40W
Northwood, *Lon.* 4 B2 51 36N 0 25W
Norumbega Res., *Bost.* 21 D2 42 19N 71 17W
Norwalk, *L.A.* 28 C4 33 53N 118 4W
Norwood, *Bost.* 21 D2 42 11N 71 13W
Norwood, *Jobg.* 18 E9 26 9 S 28 4 E
Norwood, *N.Y.* 22 B5 40 59N 73 57W
Norwood, *Phil.* 24 B3 39 53N 75 17W
Norwood Memorial Airport, *Bost.* 21 D3 42 11N 71 9W
Norwood Park, *Chic.* 26 B2 41 59N 87 48W
Noryangjin, *Sŏul* 12 G7 37 30N 126 56 E
Nose, *Ōsaka* 12 B2 34 49N 135 10 E
Nossa Senhora do Ó, *S. Pau.* 31 E5 23 30 S 46 41W
Notre-Dame, *Mtrl.* 20 B5 43 28N 73 28W
Notre-Dame, *Paris* 5 B4 48 51N 2 21 E
Notre-Dame, Bois, *Paris* 5 C5 48 45N 2 34 E
Notre Dame de L'Île Perrot, *Mtrl.* 20 B2 43 23N 73 53W
Notting Hill, *Lon.* 4 B3 51 30N 0 12W
Notting Hill, *Melb.* 19 F7 37 54 S 145 9 E
Nottingham, *Phil.* 24 A5 40 7N 74 58W
Nova Milanese, *Mil.* 9 D6 45 35N 9 12 E
Novate Milanese, *Mil.* 9 D5 45 30N 9 8 E
Novaya Derevnya, *St-Pet.* 11 A3 60 0N 30 19 E
Nové Mesto, *Pra.* 10 B2 50 4N 14 25 E
Novoaleksandrovskoye, *St-Pet.* 11 B4 59 50N 30 31 E
Novogireyevo, *Mos.* 11 E10 55 45N 37 46 E
Novoivanovskoye, *Mos.* 11 E7 55 42N 37 21 E
Novokhovrino, *Mos.* 11 D8 55 53N 37 27 E
Novonikolyskoye, *Mos.* 11 D7 55 50N 37 14 E
Novosaratovka, *St-Pet.* 11 B5 59 50N 30 32 E
Novosergiyevka, *St-Pet.* 11 B5 59 54N 30 34 E
Nowe-Babice, *Wsaw.* 10 E6 52 15N 20 51 E
Nöykkiö, *Hels.* 3 B3 60 10N 24 42 E
Noyoye Kovalyova, *St-Pet.* 11 B5 59 58N 30 34 E
Nozay, *Paris* 5 D3 48 39N 2 14 E
Nueva Atzacoalco, *Méx.* 29 B3 19 29N 99 4W
Nueva Caracas, *Car.* 30 D5 10 30N 66 57W
Nueva Chicago, *B.A.* 32 B4 34 39 S 58 29W
Nueva Pompeya, *B.A.* 32 C4 34 40 S 58 25W
Nueva Tenochtitlán, *Méx.* 29 B3 19 27N 99 5W
Nuijala, *Hels.* 3 B3 60 12N 24 46 E
Numabukuro, *Tōkyō* 13 B2 35 43N 139 39 E
Numakage, *Tōkyō* 13 A2 35 50N 139 37 E
Numata, *Tōkyō* 13 B3 35 45N 139 46 E
Nunawading, *Melb.* 19 E8 37 49 S 145 10 E
Nunez, *B.A.* 32 B4 34 32 S 58 27W
Nunhead, *Lon.* 4 C4 51 27N 0 3W
Ñuñoa, *Stgo* 30 J11 33 27 S 70 35W
Nupuri, *Hels.* 3 B2 60 14N 24 36 E
Nusle, *Pra.* 10 B2 50 3N 14 26 E
Nussdorf, *Wien* 10 G10 48 15N 16 21 E
Nuthe ➛, *Berl.* 7 B1 52 22N 13 5 E
Nutley, *N.Y.* 22 C4 40 49N 74 9W
Nutting L., *Bost.* 21 B2 42 32N 71 16W
Nützenberg, *Ruhr* 6 C4 51 15N 7 8 E
Nybølle, *Købn.* 2 D8 55 42N 12 15 E
Nybygget, *Hels.* 3 B6 60 17N 25 11 E
Nymphenburg, *Mün.* 7 G10 48 9N 11 30 E
Nymphenburg, Schloss, *Mün.* 7 G10 48 9N 11 30 E

O

Oak Beach, *N.Y.* 23 D9 40 38N 73 19W
Oak Forest, *Chic.* 26 D2 41 36N 87 44W
Oak Hill Park, *Bost.* 21 D2 42 17N 71 11W
Oak Lane, *Phil.* 24 A4 40 3N 75 8W
Oak Lawn, *Chic.* 26 C2 41 42N 87 45W
Oak Park, *Chic.* 26 B2 41 52N 87 47W
Oak Ridge, *N.Y.* 22 A2 41 2N 74 28W
Oak Valley, *Phil.* 24 C4 39 48N 75 9W
Oak View, *Wash.* 25 C8 39 1N 76 58W
Oakland, *N.Y.* 22 A3 41 1N 74 13W
Oakland, *S.F.* 27 B3 37 48N 122 18W
Oakland, *Wash.* 25 D8 38 52N 76 54W
Oakland Coliseum, *S.F.* 27 B3 37 44N 122 11W
Oakland Gardens, *N.Y.* 23 C6 40 45N 73 46W
Oakland Int. Airport, *S.F.* 27 B3 37 43N 122 12W
Oakland Mills, *Balt.* 25 B2 39 13N 76 49W
Oakland Naval Air Station, *S.F.* 27 B3 37 47N 122 19W
Oaklands, *Jobg.* 18 E9 26 8 S 28 4 E
Oaklawn, *Wash.* 25 E8 38 46N 76 56W
Oakleigh, *Melb.* 19 F7 37 54 S 145 5 E
Oaks, *Phil.* 24 A2 40 8N 75 26W
Oakwood, *N.Y.* 22 D4 40 34N 74 7W
Oakwood Beach, *N.Y.* 22 D4 40 33N 74 7W
Oatley, *Syd.* 19 B3 33 59 S 151 4 E
Obalende, *Lagos* 18 B2 6 26N 7 25 E
Oba's Palace, *Lagos* 18 B2 6 26N 7 22 E
Oberbauer, *Ruhr* 6 C6 51 17N 7 25 E
Oberföhring, *Mün.* 7 G10 48 10N 11 37 E
Oberhaching, *Mün.* 7 G10 48 1N 11 35 E
Oberhausen, *Ruhr* 6 B3 51 28N 6 54 E
Oberhausen, *Wien* 10 G11 48 10N 16 34 E
Oberkassel, *Ruhr* 6 C2 51 14N 6 45 E
Oberkirchbach, *Wien* 10 G9 48 17N 16 12 E
Oberlaa, *Wien* 10 H10 48 8N 16 24 E
Oberlisse, *Wien* 10 G10 48 17N 16 26 E
Obermenzing, *Mün.* 7 F9 48 10N 11 28 E
Obermoos Schwaige, *Mün.* 7 F9 48 14N 11 27 E
Oberschleissheim, *Mün.* 7 F10 48 15N 11 33 E
Oberschöneweide, *Berl.* 7 B4 52 27N 13 31 E
Oberwengern, *Ruhr* 6 B6 51 23N 7 22 E
Obitsu ➛, *Tōkyō* 13 D4 35 25N 139 56 E
Oboldino, *Mos.* 11 D11 55 53N 37 56 E
Observatory, *Jobg.* 18 F9 26 10 S 28 4 E
Ōbu, *Ōsaka* 12 B1 34 43N 135 8 E
Obu-tōge, *Ōsaka* 12 B1 34 44N 135 9 E
Óbuda, *Bud.* 10 J13 47 33N 19 2 E
Obudaisziget, *Bud.* 10 J13 47 33N 19 3 E
Obukhovo, *St-Pet.* 11 B4 59 53N 30 22 E
Occidental, Pico, *Car.* 30 D5 10 32N 66 51W
Oceanside, *N.Y.* 23 D7 40 38N 73 37W
Ochakovo, *Mos.* 11 E8 55 41N 37 26 E
Ochiai, *Tōkyō* 13 B3 35 43N 139 42 E
Ochota, *Wsaw.* 10 E6 52 13N 20 58 E
Ochsenwerder, *Hbg.* 7 E8 53 28N 10 4 E
Ochsenzoll, *Hbg.* 7 C8 53 41N 10 0 E
Odana, *Tōkyō* 13 C2 35 33N 139 35 E
Oden-Stockach, *Mün.* 7 G11 48 5N 11 41 E
Odilampi, *Hels.* 3 B3 60 18N 24 45 E
Odintsovo, *Mos.* 11 E7 55 40N 37 16 E
Odivelas, *Lisb.* 8 F7 38 47N 9 10W
Odolany, *Wsaw.* 10 E6 52 13N 20 55 E
Oeiras, *Lisb.* 8 F7 38 41N 9 18W
Oella, *Balt.* 25 B2 39 16N 76 46W
Oer-Erkenschwick, *Ruhr* 6 A5 51 38N 7 15 E
Oern, *Mün.* 7 G10 48 10N 11 32 E
Ofin, *Lagos* 18 A3 6 32N 7 30 E
Ofukuro-shinden, *Tōkyō* 13 A1 35 53N 139 28 E
Ogawa, *Tōkyō* 13 B1 35 44N 139 28 E
Ogden, *Phil.* 24 C2 39 49N 75 27W
Ogikubo, *Tōkyō* 13 B2 35 42N 139 37 E
Ogo Ogo, *Ōsaka* 12 B1 34 49N 135 8 E
Ogogoro, *Lagos* 18 B2 6 25N 7 24 E
Ogongo, *Manila* 15 D4 14 35N 121 4 E
Ogoyo, *Lagos* 18 B2 6 25N 7 29 E
Ogudu, *Lagos* 18 A2 6 34N 7 24 E
O'Hare, L., *Chic.* 26 B1 41 57N 87 53W
Ōhirodo, *Tōkyō* 13 A4 35 50N 139 53 E
Ohlsdorf, *Hbg.* 7 D8 53 37N 10 3 E
Ōi, *Tōkyō* 13 A2 35 51N 139 31 E
Ōimachi, *Tōkyō* 13 C3 35 35N 139 43 E
Oise ➛, *Paris* 5 A2 49 2N 2 5 E
Oittaa, *Hels.* 3 B3 60 15N 24 42 E
Ojota, *Lagos* 18 A2 6 35N 7 23 E
Okamoto, *Ōsaka* 12 B2 34 43N 135 15 E
Okazu, *Tōkyō* 13 D2 35 25N 139 31 E
Okęcie, *Wsaw.* 10 E6 52 11N 20 56 E
Okęcie Airport, *Wsaw.* 10 E6 52 10N 20 57 E
Okelra, *Lagos* 18 B2 6 29N 7 22 E
Okeogbe, *Lagos* 18 B2 6 24N 7 23 E
Okhla, *Delhi* 16 B2 28 33N 77 16 E
Okhta ➛, *St-Pet.* 11 B4 59 56N 30 25 E
Okkervil ➛, *St-Pet.* 11 B4 59 56N 30 30 E
Okrzeszyn, *Wsaw.* 10 F7 52 8N 21 8 E
Oksval, *Oslo* 2 B4 59 51N 10 40 E
Oktyabrskiy, *Mos.* 11 F11 55 37N 37 58 E
Oktyabrskiy, *Mos.* 11 E9 55 41N 37 35 E
Okubo, *Tōkyō* 13 B3 35 41N 139 42 E
Okunola, *Lagos* 18 A1 6 35N 7 17 E
Ōkura, *Tōkyō* 13 C1 35 35N 139 27 E
Olari, *Hels.* 3 B3 60 10N 24 44 E
Olaria, *Rio J.* 31 B2 22 50 S 43 16W
Old Brookville, *N.Y.* 23 C7 40 49N 73 35W
Old Cairo, *El Qâ.* 18 C5 30 0N 31 14 E
Old Coulsdon, *Lon.* 4 D4 51 17N 0 6W
Old Forge Village, *N.Y.* 22 C2 40 48N 74 29W
Old Harbor, *Bost.* 21 D3 42 19N 71 1W
Old Road B., *Balt.* 25 B4 39 12N 76 27W
Old Tappan, *N.Y.* 22 A5 41 0N 73 59W
Old Town, *Chic.* 26 B3 41 54N 87 37W
Old Westbury, *N.Y.* 23 C7 40 46N 73 35W
Oldmans Cr. ➛, *Phil.* 24 C2 39 47N 75 26W
Olgino, *St-Pet.* 11 A3 60 0N 30 10 E
Olimpico, Estadio, *Méx.* 29 C2 19 19N 99 11W
Olinda, *Melb.* 19 F9 37 51 S 145 21 E
Olinda, *Rio J.* 31 A1 22 49 S 43 25W
Olivais, *Lisb.* 8 F8 38 45N 9 7W
Olivar de los Padres, *Méx.* 29 B2 19 21N 99 14W
Olivar del Conde, *Méx.* 29 B2 19 22N 99 12W
Olivos, *B.A.* 32 B4 34 30 S 58 28W
Ollila, *Hels.* 3 A2 60 20N 24 32 E
Olney, *Phil.* 24 A4 40 2N 75 8W
Olona ➛, *Mil.* 9 E5 45 29N 9 6 E
Ølstykke, *Købn.* 2 D7 55 47N 12 8 E
Olute, *Lagos* 18 B1 6 27N 7 17 E
Olympia-Stadion, *Hels.* 3 B4 60 11N 24 55 E
Olympique Parc, *Mtrl.* 20 A4 43 33N 73 33W
Ōmagi, *Tōkyō* 13 A3 35 52N 139 43 E
Ōmiya, *Tōkyō* 13 A2 35 54N 139 37 E
Ōmori, *Tōkyō* 13 C3 35 34N 139 43 E
Ōnari, *Tōkyō* 13 A2 35 55N 139 36 E
Once, *B.A.* 32 B4 34 37 S 58 24W
Onchi, *Ōsaka* 12 C4 34 36N 135 38 E
Onchi ➛, *Ōsaka* 12 C4 34 38N 135 37 E
One Tree Hill, *Melb.* 19 F8 37 52 S 145 19 E
Onisigun, *Lagos* 18 A2 6 35N 7 24 E
Ōokayama, *Tōkyō* 13 C3 35 36N 139 40 E
Opacz, *Wsaw.* 10 E6 52 10N 20 53 E
Ophirton, *Jobg.* 18 F9 26 13 S 28 1 E
Oppegård, *Oslo* 2 C4 59 45N 10 49 E
Oppsal, *Oslo* 2 B5 59 53N 10 50 E
Oppum, *Ruhr* 6 C1 51 19N 6 36 E
Oradell, *N.Y.* 22 B4 40 57N 74 2W
Oradell Res., *N.Y.* 22 B4 40 58N 74 0W
Orange, *N.Y.* 22 C3 40 46N 74 15W
Orange Grove, *Jobg.* 18 E9 26 9 S 28 5 E
Oratorio ➛, *S. Pau.* 31 E6 23 36 S 46 32W
Orbassano, *Tori.* 9 B2 45 0N 7 31 E
Orchards, *Jobg.* 18 E9 26 9 S 28 4 E
Ordrup, *Købn.* 2 D10 55 45N 12 34 E
Orech, *Pra.* 10 B1 50 1N 14 17 E
Øresund, *Købn.* 2 D11 55 45N 12 40 E
Oreta, *Lagos* 18 A3 6 31N 7 31 E
Orge ➛, *Paris* 5 D3 48 36N 2 17 E
Orgeval, *Paris* 5 B1 48 55N 1 58 E
Orhølm, *Købn.* 2 D10 55 48N 12 30 E
Orient Heights, *Bost.* 21 C4 42 23N 70 59W
Oriental, Pico, *Car.* 30 D5 10 32N 66 51W
Origgio, *Mil.* 9 D5 45 35N 9 1 E
Orinda, *S.F.* 27 A3 37 52N 122 10W
Orinda Village, *S.F.* 27 A3 37 53N 122 12W
Orland L., *Chic.* 26 D1 41 38N 87 52W
Orland Park, *Chic.* 26 D1 41 37N 87 52W
Orlando Dam, *Jobg.* 18 F8 26 15 S 27 55 E
Orlando East, *Jobg.* 18 F8 26 14 S 27 55 E
Orlando West, *Jobg.* 18 F8 26 13 S 27 54 E
Orlången, *Stock.* 3 E11 59 11N 18 2 E
Orlångsvik, *Stock.* 3 E11 59 11N 18 3 E
Orlovo, *Mos.* 11 F8 55 38N 37 22 E
Orly, *Paris* 5 C4 48 45N 2 23 E
Ormesson-sur-Marne, *Paris* 5 C5 48 47N 2 32 E
Orminge, *Stock.* 3 E12 59 19N 18 14 E
Ormingelandet, *Stock.* 3 D13 59 20N 18 22 E
Ormond, *Melb.* 19 F7 37 54 S 145 1 E
Órmos Fálirou, *Ath.* 8 J11 37 54N 23 40 E
Ormøya, *Oslo* 2 B4 59 52N 10 45 E
Oros Aiyáleos, *Ath.* 8 J10 38 0N 23 36 E
Oros Imittoś, *Ath.* 8 J11 37 53N 23 48 E
Örpadfold, *Bud.* 10 J14 47 32N 19 12 E
Orpington, *Lon.* 4 C5 51 22N 0 6 E
Orsay, *Paris* 5 C3 48 41N 2 11 E
Orsby, *Ruhr* 6 A2 51 31N 6 41 E
Orsett, *Lon.* 4 B7 51 30N 0 22 E
Ortaköy, *Ist.* 17 A3 41 3N 29 1 E
Ortica, *Mil.* 9 E6 45 28N 9 16 E
Oruba, *Lagos* 18 A2 6 34N 7 24 E
Ōsaka, *Ōsaka* 12 C4 34 42N 135 30 E
Ōsaka B., *Ōsaka* 12 C2 34 35N 135 18 E
Ōsaka Castle, *Ōsaka* 12 B4 34 41N 135 30 E
Ōsaka Harbour, *Ōsaka* 12 C3 34 38N 135 25 E
Ōsaka Univ., *Ōsaka* 12 B3 34 41N 135 29 E
Ōsaki, *Tōkyō* 13 C3 35 36N 139 42 E
Osasco, *S. Pau.* 31 E5 23 31 S 46 46W
Osdorf, *Berl.* 7 B3 52 24N 13 20 E
Osdorf, *Hbg.* 7 D7 53 34N 9 50 E
Oshodi, *Lagos* 18 A2 6 33N 7 21 E
Oskar Frederikborg, *Stock.* 3 D13 59 24N 18 24 E
Oslo, *Oslo* 2 B4 59 54N 10 43 E
Oslofjorden, *Oslo* 2 C3 59 40N 10 35 E
Ōsone, *Tōkyō* 13 C2 35 31N 139 37 E
Osorun, *Lagos* 18 A2 6 33N 7 29 E
Ospiate, *Mil.* 9 D5 45 32N 9 6 E
Ossów, *Wsaw.* 10 E8 52 18N 21 12 E
Ostankino, *Mos.* 11 E9 55 49N 37 37 E
Østby, *Købn.* 2 D7 55 45N 12 2 E
Osterath, *Ruhr* 6 C1 51 16N 6 36 E
Österby, *Hels.* 3 B1 60 10N 24 25 E
Osterfeld, *Ruhr* 6 A3 51 30N 6 53 E
Osterley, *Lon.* 4 C2 51 28N 0 21W
Osterley Park, *Lon.* 4 C2 51 29N 0 21W
Östermalm, *Stock.* 3 D11 59 20N 18 4 E
Österskär, *Stock.* 3 D12 59 26N 18 16 E
Östersundom, *Hels.* 3 B6 60 15N 25 10 E
Ostertälje, *Stock.* 3 E8 59 11N 17 39 E
Ostiense, *Rome* 9 F9 41 51N 12 29 E
Østmarkkapellet, *Oslo* 2 B5 59 52N 10 51 E
Ostøya, *Oslo* 2 B3 59 52N 10 34 E
Östra Ryd, *Stock.* 3 D12 59 27N 18 11 E
Østre Aker, *Oslo* 2 B4 59 54N 10 49 E
Ostrov, *Mos.* 11 F11 55 36N 37 50 E
Ostrovtsy, *Mos.* 11 F12 55 36N 38 0 E
Ōta-Ku, *Tōkyō* 13 C3 35 34N 139 41 E
Otaniemi, *Hels.* 3 B3 60 11N 24 49 E
Otford, *Lon.* 4 D6 51 18N 0 11 E
Othmarschen, *Hbg.* 7 D7 53 33N 9 53 E
Otsuka, *Tōkyō* 13 B3 35 43N 139 44 E
Ottakring, *Wien* 10 G9 48 12N 16 18 E
Ottávia, *Rome* 9 F9 41 57N 12 24 E
Ottaviano, *Nápl.* 9 H13 40 50N 14 28 E
Ottensen, *Hbg.* 7 D7 53 33N 9 55 E
Ottobrunn, *Mün.* 7 G11 48 3N 11 40 E
Ottocalli, *Nápl.* 9 H12 40 52N 14 17 E
Otwock, *Wsaw.* 10 F8 52 8N 21 13 E
Ouerenburg, *Ruhr* 6 B5 51 27N 7 16 E
Ouiapo, *Manila* 15 D3 14 35N 120 59 E
Oulunkylä, *Hels.* 3 B4 60 13N 24 58 E
Ourcq, Canal de l', *Paris* 5 B4 48 54N 2 28 E
Ousit, *Bangk.* 15 B2 13 47N 100 31 E
Outer Brewster I., *Bost.* 21 C4 42 20N 70 52W
Outer Mission, *S.F.* 27 B2 37 43N 122 26W
Outremont, *Mtrl.* 20 A4 43 31N 73 36W
Overbruch, *Ruhr* 6 A2 51 32N 6 43 E
Overlea, *Balt.* 25 A3 39 21N 76 32W
Øverød, *Købn.* 2 D9 55 48N 12 28 E
Ōwada, *Tōkyō* 13 B2 35 48N 139 31 E
Owings Mills, *Balt.* 25 A2 39 25N 76 47W
Oworonsoki, *Lagos* 18 A2 6 32N 7 24 E
Oxon Hill, *Wash.* 25 E8 38 48N 76 59W
Oxshott, *Lon.* 4 D2 51 19N 0 21W
Oyada, *Tōkyō* 13 B4 35 46N 139 50 E
Ōyama, *Tōkyō* 13 B3 35 44N 139 42 E
Øyeren, *Oslo* 2 B6 59 55N 11 6 E
Oyodo, *Ōsaka* 12 B3 34 42N 135 29 E
Oyster B., *N.Y.* 23 B7 40 52N 73 31W
Oyster B., *Syd.* 19 C3 34 0 S 151 5 E
Oyster Bay Cove, *N.Y.* 23 B8 40 51N 73 29W
Oyster Bay Harbour, *N.Y.* 23 B7 40 53N 73 32W
Oyster Rock, *Bomb.* 16 H7 18 54N 72 49 E
Oyster Rocks, *Kar.* 17 H11 24 48N 66 59 E
Ozarów-Franciszków, *Wsaw.* 10 E5 52 13N 20 48 E
Ozerki, *St-Pet.* 11 B6 59 53N 30 42 E
Ozoir-la-Ferrière, *Paris* 5 C6 48 46N 2 40 E
Ozone Park, *N.Y.* 23 C5 40 40N 73 50W

P

Pacific Manor, *S.F.* 27 C2 37 38N 122 27W
Pacific Palisades, *L.A.* 28 B1 34 2N 118 32W
Pacifica, *S.F.* 27 C2 37 37N 122 27W
Packanack L., *N.Y.* 22 B3 40 56N 74 15W
Paco, *Manila* 15 D3 14 35N 120 59 E
Paco de Arcos, *Lisb.* 8 F7 38 41N 9 17W
Paddington, *Lon.* 4 B3 51 30N 0 10W
Paddington, *Syd.* 19 B4 33 53 S 151 14 E
Pademangan, *Jak.* 15 H9 6 7 S 106 49 E
Paderno, *Mil.* 9 D5 45 33N 9 9 E
Padre Miguel, *Rio J.* 31 B1 22 52 S 43 25W
Padstow, *Syd.* 19 B3 33 57 S 151 2 E
Pagewood, *Syd.* 19 B4 33 56 S 151 14 E
Pagote, *Bomb.* 16 H8 18 53N 72 59 E
Pai, I. do, *Rio J.* 31 B3 22 59 S 43 5W
Paia, *Lisb.* 8 F7 38 46N 9 11W
Paikpara, *Calc.* 16 E6 22 36N 88 23 E
Paint Br. ➛, *Wash.* 25 C8 38 57N 76 55W
Paiyun Airport, *Gzh.* 14 F8 23 10N 113 15 E
Pak ka Shan, *H.K.* 12 E6 22 16N 114 13 E
Pak Kong, *H.K.* 12 D6 22 22N 114 15 E
Pak Tim Pa, *H.K.* 12 D5 22 22N 114 7 E
Pakila, *Hels.* 3 B4 60 15N 24 55 E
Palace Museum, *Beij.* 14 B3 39 54N 116 21 E
Palaión Fáliron, *Ath.* 8 J11 37 53N 23 42 E
Palaiseau, *Paris* 5 C3 48 42N 2 14 E
Palam, *Delhi* 16 B1 28 35N 77 4 E
Palam Int. Airport, *Delhi* 16 B1 28 32N 77 8 E
Palazzo Reale, *Nápl.* 9 H12 40 50N 14 15 E
Palazzo Reale, *Tori.* 9 B3 45 4N 7 40 E
Palazzolo, *Mil.* 9 D5 45 34N 9 9 E
Palazzuolo, *Nápl.* 9 H13 40 54N 14 28 E
Palermo, *B.A.* 32 B4 34 35 S 58 24W
Palhais, *Lisb.* 8 G8 38 37N 9 2W
Palisades, *N.Y.* 22 A5 41 1N 73 57W
Palisades Park, *N.Y.* 22 B4 40 50N 74 1W
Palleja, *Barc.* 8 D5 41 25N 2 0 E
Palmer Park, *Wash.* 25 D8 38 55N 76 52W
Palmers Green, *Lon.* 4 B4 51 36N 0 6W
Palmyra, *Phil.* 24 A4 40 0N 75 1W
Palo Alto, *S.F.* 27 D4 37 27N 122 8W
Paloheinä, *Hels.* 3 B4 60 15N 24 56 E
Palomar Park, *S.F.* 27 D3 37 29N 122 16W
Palomeras, *Mdrd.* 8 B3 40 22N 3 39W
Palos Heights, *Chic.* 26 D2 41 39N 87 47W
Palos Hills, *Chic.* 26 C2 41 42N 87 49W
Palos Hills Forest, *Chic.* 26 C1 41 40N 87 52W
Palos Park, *Chic.* 26 C1 41 40N 87 50W
Palota-Üjfalu, *Bud.* 10 J13 47 33N 19 7 E
Palpara, *Calc.* 16 E6 22 38N 88 22 E
Palta, *Calc.* 16 D6 22 46N 88 23 E
Pamplona, *Manila* 15 E3 14 27N 120 58 E
Panayaan, *Manila* 15 E3 14 27N 120 57 E
Panchghara, *Calc.* 16 D5 22 44N 88 16 E
Panchur, *Calc.* 16 E5 22 32N 88 16 E
Pancoran, *Jak.* 15 J9 6 14 S 106 49 E
Pandan, Selat, *Sing.* 15 G7 1 16N 103 45 E
Pandan, Sungei ➛, *Sing.* 15 G7 1 18N 103 43 E
Pandan Res., *Sing.* 15 G7 1 18N 103 44 E
Panehpara, *Calc.* 16 E5 22 34N 88 15 E
Pangrati, *Ath.* 8 J11 37 56N 23 45 E
Pangsua, Sungei ➛, *Sing.* 15 F7 1 25N 103 45 E
Panihati, *Calc.* 16 D6 22 41N 88 22 E
Panjang, Bukit, *Sing.* 15 F7 1 22N 103 45 E
Panje, *Bomb.* 16 H8 18 54N 72 57 E
Panke ➛, *Berl.* 7 A3 52 31N 13 22 E
Pankow, *Berl.* 7 A3 52 34N 13 23 E
Panorama City, *L.A.* 28 A2 34 13N 118 26W
Panpur, *Calc.* 16 C6 22 51N 88 26 E
Pantheon, *Rome* 9 F9 41 53N 12 28 E
Pantin, *Paris* 5 B4 48 53N 2 24 E
Pantitlán, *Méx.* 29 B3 19 24N 99 4W
Panuacan, *Manila* 15 D4 14 35N 121 0 E
Panvel Cr. ➛, *Bomb.* 16 H9 18 59N 73 0 E
Paoli, *Phil.* 24 A2 40 2N 75 28W
Papiol, *Barc.* 8 D5 41 25N 2 0 E
Paracuellos del Jarama, *Mdrd.* 8 A3 40 30N 3 31W
Paradise Cay, *S.F.* 27 A2 37 54N 122 28W
Paramount, *L.A.* 28 C3 33 53N 118 11W
Paramus, *N.Y.* 22 B4 40 56N 74 2W
Paranaque, *Manila* 15 D3 14 30N 120 59 E
Paray-Vieille-Poste, *Paris* 5 C4 48 42N 2 20 E
Parbasdorf, *Wien* 10 G11 48 16N 16 35 E
Parbatipur, *Calc.* 16 E5 22 39N 88 13 E
Parcelacion Moderna, *La Hab.* 30 B3 23 2N 82 19W
Parco Regionale, *Mil.* 9 D5 45 35N 9 5 E
Parel, *Bomb.* 16 H7 18 59N 72 49 E
Pari, *S. Pau.* 31 E6 23 32 S 46 36W
Parioli, *Rome* 9 F9 41 55N 12 29 E
Paris, *Paris* 5 B4 48 53N 2 20 E
Paris-Le Bourget, Aéroport de, *Paris* 5 B4 48 58N 2 26 E
Paris-Orly, Aéroport de, *Paris* 5 C4 48 43N 2 22 E
Pārk-e-Shahānshāh, *Tehr.* 17 C5 35 46N 51 24 E
Park Orchards, *Melb.* 19 E8 37 46 S 145 13 E
Park Ridge, *Chic.* 26 A1 42 0N 87 50W
Park Ridge, *N.Y.* 22 A4 41 2N 74 2W
Park Royal, *Lon.* 4 B3 51 31N 0 16W
Parkchester, *N.Y.* 23 C5 40 49N 73 50W
Parkdale, *Trto.* 20 E8 43 38N 79 25W
Parkdene, *Jobg.* 18 F10 26 11 S 28 15 E
Parkhafen, *Hbg.* 7 D7 53 32N 9 54 E
Parkhill Gardens, *Jobg.* 18 F10 26 14 S 28 11 E
Parkhurst, *Jobg.* 18 E9 26 8 S 28 1 E
Parklawn, *Wash.* 25 D7 38 50N 77 7W
Parkmore, *Jobg.* 18 E9 26 5 S 28 2 E
Parkside, *S.F.* 27 B2 37 44N 122 29W
Parktown, *Jobg.* 18 F9 26 10 S 28 2 E
Parktown North, *Jobg.* 18 E9 26 8 S 28 2 E
Parkview, *Jobg.* 18 E9 26 9 S 28 1 E
Parkville, *Balt.* 25 A3 39 23N 76 34W
Parkville, *N.Y.* 22 D5 40 38N 73 57W
Parkwood, *Jobg.* 18 E9 26 9 S 28 2 E
Parque Edú Chaves, *S. Pau.* 31 D6 23 29 S 46 34W
Parramatta, *Syd.* 19 A2 33 49 S 150 59 E
Parramatta ➛, *Syd.* 19 A3 33 49 S 151 3 E
Parramatta North, *Syd.* 19 A3 33 48 S 151 0 E
Parramatta Park, *Syd.* 19 A3 33 48 S 151 0 E
Parsippany, *N.Y.* 22 B2 40 51N 74 26W
Paşabahce, *Ist.* 17 A3 41 6N 29 5 E
Pasadena, *L.A.* 28 B4 34 9N 118 8W
Pasar Minggu, *Jak.* 15 J9 6 16 S 106 49 E
Pasay, *Manila* 15 D3 14 32N 120 59 E
Pascoe Vale, *Melb.* 19 E6 37 43 S 144 55 E
Pasig, *Manila* 15 D4 14 33N 121 4 E
Pasig ➛, *Manila* 15 D4 14 31N 121 6 E
Pasila, *Hels.* 3 B4 60 12N 24 56 E
Pasing, *Mün.* 7 G9 48 8N 11 28 E
Pasir Panjang, *Sing.* 15 G7 1 17N 103 46 E
Pasir Ris Beach, *Sing.* 15 F8 1 22N 103 56 E
Paso del Rey, *B.A.* 32 B2 34 39 S 58 45W
Passaic, *N.Y.* 22 B4 40 51N 74 9W
Passaic ➛, *N.Y.* 22 B3 40 42N 74 10W
Passirana, *Mil.* 9 D5 45 32N 9 2 E
Patapsco ➛, *Balt.* 25 B2 39 19N 76 49W
Patapsco State Park, *Balt.* 25 B2 39 18N 76 47W
Pateres, *Manila* 15 D4 14 32N 121 3 E
Paterson, *N.Y.* 22 B4 40 54N 74 9W
Pathumwan, *Bangk.* 15 B2 13 44N 100 31 E
Patipukun, *Calc.* 16 E6 22 36N 88 24 E
Patisia, *Ath.* 8 H11 38 2N 23 45 E
Patterson Park, *Balt.* 25 B3 39 17N 76 34W
Patul, *Calc.* 16 D4 22 45N 88 9 E
Paulo E. Virginia, Gruta, *Rio J.* 31 B2 22 56 S 43 16W
Paulsboro, *Phil.* 24 C3 39 49N 75 14W
Paulshof, *Berl.* 7 A5 52 34N 13 42 E
Pausin, *Berl.* 7 A1 52 38N 13 2 E
Pavarolo, *Tori.* 9 B3 45 4N 7 49 E
Pavlovo, *St-Pet.* 11 B5 59 55N 30 38 E
Pavne, *Bomb.* 16 G9 19 5N 73 1 E
Pavshino, *Mos.* 11 E7 55 49N 37 20 E
Paya Lebar, *Sing.* 15 F8 1 21N 103 52 E
Paylampur, *Calc.* 16 D5 22 46N 88 15 E
Peabody, *Bost.* 21 B4 42 32N 70 57W
Peabody Inst., *Balt.* 25 B3 39 17N 76 37W
Peakhurst, *Syd.* 19 B3 33 57 S 151 3 E
Pécel, *Bud.* 10 K14 47 29N 19 20 E
Pecetto Torinese, *Tori.* 9 B3 45 1N 7 44 E
Pechincha, *Rio J.* 31 B1 22 55 S 43 20W
Pechorka ➛, *Mos.* 11 F12 55 37N 38 2 E
Peckham, *Lon.* 4 C4 51 28N 0 3W
Pecqueuse, *Paris* 5 D2 48 38N 2 3 E
Peddocks I., *Bost.* 21 D4 42 17N 70 56W
Pederstrup, *Købn.* 2 D9 55 44N 12 20 E
Pedra Branca, *Rio J.* 31 B1 22 55 S 43 26W
Pedralbes, *Barc.* 8 D5 41 23N 2 7 E
Pedregal de San Angel, Jardines del, *Méx.* 29 C2 19 19N 99 12W
Pedreira, *S. Pau.* 31 F5 23 41 S 46 40W
Pedreros, *Lima* 30 G8 12 2 S 76 57W
Pedricktown, *Phil.* 24 C2 39 45N 75 24W
Pedro Cr. ➛, *S.F.* 27 C2 37 35N 122 31W
Pedro Valley, *S.F.* 27 C2 37 35N 122 29W
Peirce Res., *Sing.* 15 F7 1 22N 103 49 E
Pekhra-Pokrovskoye, *Mos.* 11 D11 55 50N 37 56 E
Pekhra-Yakovievskaya, *Mos.* 11 E11 55 47N 37 57 E
Peking = Beijing, *Beij.* 14 B2 39 53N 116 21 E
Pelado, Cerro, *Méx.* 29 D2 19 10N 99 14W
Pelcowizna, *Wsaw.* 10 E7 52 17N 21 0 E
Pelham, *N.Y.* 23 B6 40 54N 73 46W
Pelham B. Park, *N.Y.* 23 B6 40 52N 73 48W
Pelham Manor, *N.Y.* 23 B6 40 53N 73 46W
Penalolén, *Stgo* 30 J11 33 28 S 70 30W
Peng Siang ➛, *Sing.* 15 F7 1 24N 103 43 E
Penge, *Lon.* 4 C4 51 24N 0 3W
Penha, *Rio J.* 31 A2 22 49 S 43 17W
Penha, *S. Pau.* 31 E6 23 31 S 46 32W
Penjaringan, *Jak.* 15 H9 6 7 S 106 48 E
Penn Square, *Phil.* 24 A3 40 8N 75 19W
Penn Wynne, *Phil.* 24 B3 39 59N 75 16W
Pennant Hills Park, *Syd.* 19 A3 33 46 S 151 6 E
Penndel, *Phil.* 24 A5 40 9N 74 54W
Penns Grove, *Phil.* 24 C2 39 44N 75 27W
Pennsauken, *Phil.* 24 B4 39 57N 75 5W
Pennsauken Cr. ➛, *Phil.* 24 B4 39 59N 75 3W
Pennsylvania, Univ. of, *Phil.* 24 B3 39 51N 75 11W
Pennypack Cr. ➛, *Phil.* 24 A4 40 0N 75 3W
Pentala, *Hels.* 3 C3 60 6N 24 40 E
Penyagino, *Mos.* 11 D8 55 50N 37 20 E
Penzing, *Wien* 10 G9 48 11N 16 18 E
Pequannock, *N.Y.* 22 B3 40 57N 74 17W
Pequena Arroio Fundo ➛, *Rio J.* 31 B1 22 58 S 43 21W
Perales del Rio, *Mdrd.* 8 C3 40 18N 3 38W
Perchtoldsdorf, *Wien* 10 H9 48 7N 16 17 E
Perdizes, *S. Pau.* 31 E6 23 32 S 46 39W
Peredelkino, *Mos.* 11 F8 55 38N 37 20 E
Peredelytsy, *Mos.* 11 F8 55 36N 37 21 E
Peristérion, *Ath.* 8 H11 38 1N 23 42 E
Perivale, *Lon.* 4 B3 51 31N 0 18W
Perlach, *Mün.* 7 G10 48 5N 11 37 E
Perlacher Forst, *Mün.* 7 G10 48 4N 11 34 E
Pero, *Mil.* 9 D5 45 30N 9 5 E
Peropok, Bukit, *Sing.* 15 G7 1 19N 103 42 E
Perovo, *Mos.* 11 E10 55 44N 37 45 E
Perrot, Î., *Mtrl.* 20 B2 43 23N 73 56W
Perry Hall, *Balt.* 25 A4 39 24N 76 28W
Perth Amboy, *N.Y.* 22 D3 40 30N 74 16W
Pertusella, *Mil.* 9 D5 45 35N 9 3 E
Pesanggrahag, Kali ➛, *Jak.* 15 J9 6 10 S 106 44 E
Peschiera Borromeo, *Mil.* 9 E6 45 26N 9 19 E
Pesek, P., *Sing.* 15 G7 1 17N 103 41 E
Pest, *Bud.* 10 K13 47 29N 19 4 E
Pesterzsébet, *Bud.* 10 K13 47 26N 19 6 E
Pesthidegkút, *Bud.* 10 J12 47 33N 18 57 E
Pestimre, *Bud.* 10 K14 47 24N 19 11 E
Pestlörinc, *Bud.* 10 K14 47 26N 19 11 E
Pestujhely, *Bud.* 10 J13 47 32N 19 7 E
Petare, *Car.* 30 E6 10 29N 66 48W
Petas, *Hels.* 3 B4 60 15N 24 50 E
Peters Pond, *Bost.* 21 A2 42 43N 71 15W
Petit, *Jobg.* 18 E11 26 6 S 28 22 E
Petit-Brûlé, *Mtrl.* 20 A1 43 35N 74 2W
Petojo Selatan, *Jak.* 15 J9 6 10 S 106 48 E
Petrograd = St. Petersburg, *St-Pet.* 11 B3 59 55N 30 15 E
Petrogradskaya Storona, *St-Pet.* 11 B4 59 56N 30 20 E
Petroúpolis, *Ath.* 8 H11 38 3N 23 40 E
Petrovice, *Pra.* 10 B3 50 2N 14 33 E
Petrovsko-Rasumovskoye, *Mos.* 11 E9 55 49N 37 34 E
Petrovskoye, *Mos.* 11 F11 55 36N 37 53 E
Petrovsky Park, *Mos.* 11 E9 55 47N 37 34 E
Pfaueninsel, *Berl.* 7 B1 52 26N 13 7 E
Phihāi, *Kar.* 17 G11 24 50N 67 8 E
Philadelphia, *Phil.* 24 B3 39 57N 75 11W
Philadelphia Airport, *Phil.* 24 A5 40 4N 75 0W
Philadelphia Int. Airport, *Phil.* 24 B3 39 52N 75 16W
Phillip B., *Syd.* 19 B4 33 58 S 151 14 E
Phinga, *Calc.* 16 D6 22 41N 88 25 E
Phoenix, *Chic.* 26 D3 41 36N 87 37W
Phoenixville, *Phil.* 24 A1 40 7N 75 31W
Phra Khanong, *Bangk.* 15 B2 13 40N 100 36 E
Phra Pradaeng, *Bangk.* 15 C2 13 39N 100 33 E
Phranakhon, *Bangk.* 15 B1 13 44N 100 29 E
Pianezza, *Tori.* 9 B2 45 6N 7 32 E
Pianura, *Nápl.* 9 H11 40 51N 14 10 E
Piaslów, *Wsaw.* 10 E5 52 11N 20 49 E
Pico Rivera, *L.A.* 28 C4 33 59N 118 5W
Piedade, *Lisb.* 8 F7 38 49N 9 16W
Piedade, *Rio J.* 31 B2 22 52 S 43 18W
Piedade, Cova da, *Lisb.* 8 F8 38 40N 9 8W
Piedmont, *S.F.* 27 B3 37 49N 122 14W
Pierrefitte, *Paris* 5 B4 48 58N 2 21 E
Pierrefonds, *Mtrl.* 20 B2 43 27N 73 52W
Pierrelaye, *Paris* 5 A2 49 1N 2 8 E
Pietralata, *Rome* 9 F10 41 55N 12 33 E
Pihlajamäki, *Hels.* 3 B4 60 14N 24 58 E
Pihlajasaari, *Hels.* 3 C4 60 8N 24 55 E
Pikesville, *Balt.* 25 A2 39 22N 76 42W
Pilar Velho, *S. Pau.* 31 F7 23 40 S 46 22W
Pilarcitos Cr. ➛, *S.F.* 27 C2 37 33N 122 24W
Pilarcitos L., *S.F.* 27 C2 37 33N 122 25W
Pilgrim Corner, *Phil.* 24 B3 39 57N 75 19W
Pilgrims Hatch, *Lon.* 4 B6 51 37N 0 17 E
Pillar Pt., *S.F.* 26 D2 37 29N 122 30W
Pimenta, *S. Pau.* 31 D7 23 27 S 46 24W
Pimlico, *Lon.* 4 C4 51 29N 0 8W
Pimmit Hills, *Wash.* 25 D6 38 54N 77 12W
Pimville, *Jobg.* 18 F8 26 16 S 27 54 E
Pinazo ➛, *B.A.* 32 A2 34 29 S 58 49W
Pine Brook, *N.Y.* 22 B3 40 51N 74 18W
Pine Grove, *Trto.* 20 D7 43 47N 79 34W
Pine Hill, *Phil.* 24 C5 39 47N 74 59W
Pine Orchard, *Balt.* 25 B1 39 16N 76 52W
Pinehurst, *Bost.* 21 B2 42 31N 71 12W
Piñero, *B.A.* 32 B2 34 31 S 58 46W
Pines Lake, *N.Y.* 22 B3 40 59N 74 15W
Piney Run ➛, *Wash.* 25 D6 38 58N 77 14W
Pinganli, *Beij.* 14 B3 39 56N 116 20 E
Pinheiros ➛, *S. Pau.* 31 E5 23 37 S 46 44W
Pinjrāpur, *Kar.* 17 G11 24 53N 67 4 E
Pinn ➛, *Lon.* 4 B2 51 30N 0 28W
Pinnau ➛, *Hbg.* 7 C6 53 40N 9 49 E
Pinneberg, *Hbg.* 7 C6 53 40N 9 48 E
Pinner, *Lon.* 4 B2 51 35N 0 22W
Pinner Green, *Lon.* 4 B2 51 35N 0 23W
Pino Torinese, *Tori.* 9 B3 45 2N 7 46 E
Pinole Cr. ➛, *S.F.* 27 A3 37 58N 122 12W
Pioltello, *Mil.* 9 D6 45 30N 9 19 E
Piossasco, *Tori.* 9 C1 44 59N 7 27 E
Piqueri, *S. Pau.* 31 D5 23 29 S 46 41W
Piqueri ➛, *S. Pau.* 31 D6 23 27 S 46 34W
Pira, *Delhi* 16 A1 28 40N 77 7 E
Piraévs, *Ath.* 8 J10 37 54N 23 39 E
Pirajussara, *S. Pau.* 31 E5 23 37 S 46 46W
Pirajussara ➛, *S. Pau.* 31 E5 23 33 S 46 42W
Piraporinha, *S. Pau.* 31 F5 23 41 S 46 44W
Piratininga, *Rio J.* 31 B3 22 56 S 43 4W
Piratininga, L. de, *Rio J.* 31 B3 22 56 S 43 4W
Pirituba, *S. Pau.* 31 D5 23 29 S 46 44W
Pirkkola, *Hels.* 3 B4 60 14N 24 55 E
Pisangan, *Jak.* 15 J10 6 12 S 106 52 E
Piscataway, *N.Y.* 22 D2 40 34N 74 27W
Pisnice, *Pra.* 10 C2 49 59N 14 28 E
Pitampura Kalan, *Delhi* 16 A1 28 41N 77 7 E

44 Pitkäjärvi

Pitkäjärvi, *Hels.* 3 B3 60 15N 24 45 E
Pitman, *Phil.* 24 C4 39 44N 75 7W
Plainedge, *N.Y.* 23 C8 40 43N 73 27W
Plainfield, *N.Y.* 22 D2 40 36N 74 23W
Plainview, *N.Y.* 23 C8 40 46N 73 27W
Plaisir, *Paris* 5 C1 48 49N 1 56 E
Plandome, *N.Y.* 23 C6 40 48N 73 42W
Plandome Heights, *N.Y.* 23 C6 40 48N 73 42W
Planegg, *Mün.* 7 G9 48 6N 11 25 E
Plazo Mayor, *Mdrd.* 8 B2 40 25N 3 43W
Pleasant Hill, *S.F.* 27 A4 37 56N 122 4W
Plenty, *Melb.* 19 E7 37 40 S 145 5 E
Pluit, *Jak.* 15 H9 6 7 S 106 47 E
Plumsock, *Phil.* 24 B2 39 58N 75 28W
Plumstead, *Lon.* 4 C5 51 29N 0 5 E
Plymouth Meeting, *Phil.* 24 A3 40 6N 75 16W
Plyushchevo, *Mos.* 11 E10 55 44N 37 45 E
Po →, *Tori.* 9 B3 45 7N 7 46 E
Po Toi, *H.K.* 12 E6 22 16N 114 17 E
Po Toi I., *H.K.* 12 E6 22 10N 114 15 E
Podbaba, *Pra.* 10 B2 50 7N 14 22 E
Podoli, *Pra.* 10 B2 50 2N 14 25 E
Podra, *Calc.* 16 E5 22 33N 88 16 E
Poduskino, *Mos.* 11 E7 55 43N 37 15 E
Poggioreale, *Nápl.* 9 H12 40 51N 14 17 E
Pogliano Milanese, *Mil.* 9 D4 45 32N 8 59 E
Pohick Cr. →, *Wash.* 25 E6 38 47N 77 16W
Point Breeze, *Phil.* 24 B3 39 54N 75 13W
Point Lookout, *N.Y.* 23 D7 40 35N 73 34W
Point View Res., *N.Y.* 22 B3 40 58N 74 14W
Pointe-Aux-Trembles, *Mtrl.* 20 A4 43 38N 73 30W
Pointe-Calumet, *Mtrl.* 20 B2 43 29N 73 58W
Pointe-Claire, *Mtrl.* 20 B3 43 27N 73 48W
Poissy, *Paris* 5 B2 48 55N 2 2 E
Pok Fu Lam, *H.K.* 12 E5 22 16N 114 7 E
Pokrovsko-Sresnevo, *Mos.* 11 E8 55 48N 37 27 E
Pokrovskoye, *Mos.* 11 F9 55 37N 37 36 E
Póllena, *Nápl.* 9 H13 40 51N 14 22 E
Polsum, *Ruhr* 6 A4 51 37N 7 2 E
Polyustrovo, *St-Pet.* 11 B4 59 57N 30 25 E
Pomigliano d'Arco, *Nápl.* 9 H13 40 54N 14 23 E
Pompei, *Nápl.* 9 J13 40 45N 14 29 E
Pomponne, *Paris* 5 B6 48 52N 2 40 E
Pomprap, *Bangk.* 15 B2 13 44N 100 30 E
Pompton →, *N.Y.* 22 B3 40 53N 74 16W
Pompton Lakes, *N.Y.* 22 A3 41 0N 74 15W
Pompton Plains, *N.Y.* 22 B3 40 58N 74 18W
Ponders End, *Lon.* 4 B4 51 38N 0 2W
Pondok Indah, *Jak.* 15 J9 6 16 S 106 46 E
Ponkapog, *Bost.* 21 D3 42 11N 71 4W
Ponkapog Pond, *Bost.* 21 D3 42 11N 71 5W
Pont-Viau, *Mtrl.* 20 A3 43 34N 73 41W
Pontault-Combault, *Paris* 5 C5 48 47N 2 36 E
Pontcarré, *Paris* 5 C6 48 47N 2 42 E
Pontchartrain, *Paris* 5 C1 48 48N 1 54 E
Ponte Galéria, *Rome* 9 G8 41 48N 12 19 E
Pontes, Canto do, *Rio J.* 31 B3 22 56 S 43 3W
Pontevedra, *B.A.* 32 C2 34 44 S 58 41W
Ponticelli, *Nápl.* 9 H12 40 51N 14 19 E
Pontinha, *Lisb.* 8 F7 38 45N 9 11W
Pontoise, *Paris* 5 A2 49 2N 2 4 E
Poortview, *Jobg.* 18 E8 26 5 S 27 51 E
Poplar, *Lon.* 4 B4 51 30N 0 0 E
Poppenbüttel, *Hbg.* 7 D8 53 39N 10 4 E
Port Chester, *N.Y.* 23 A6 41 0N 73 40W
Port Chester Harbour, *N.Y.* 23 B7 40 58N 73 38W
Port Jackson, *Syd.* 19 B4 33 51 S 151 14 E
Port Kennedy, *Phil.* 24 A2 40 6N 75 25W
Port Melbourne, *Melb.* 19 F6 37 50 S 144 54 E
Port Newark, *N.Y.* 22 C4 40 41N 74 9W
Port Reading, *N.Y.* 22 D3 40 34N 74 13W
Port Richmond, *N.Y.* 22 D4 40 38N 74 7W
Port Shelter, *H.K.* 12 D6 22 20N 114 17 E
Port Union, *Trto.* 20 D10 43 47N 79 7W
Port Washington, *N.Y.* 23 C6 40 49N 73 42W
Port Washington North, *N.Y.* 23 B6 40 50N 73 41W
Portage Park, *Chic.* 26 B2 41 56N 87 45W
Portela, Aeroporto da, *Lisb.* 8 F8 38 46N 9 7W
Pórtici, *Nápl.* 9 J12 40 48N 14 19 E
Porto Brandão, *Lisb.* 8 F7 38 40N 9 12W
Porto Novo Cr. →, *Lagos* 18 B2 6 25N 7 22 E
Porto Nuevo, *B.A.* 32 B4 34 35 S 58 22W
Portrero, *S.F.* 27 B3 37 46N 122 25W
Posen, *Chic.* 26 D2 41 38N 87 41W
Posíllipo, *Nápl.* 9 J12 40 49N 14 13 E
Posíllipo, C. di, *Nápl.* 9 J12 40 48N 14 12 E
Posolok Lenina, *St-Pet.* 11 C2 59 50N 30 5 E
Potawatomi Woods, *Chic.* 26 A1 42 8N 87 53W
Potomac, *Wash.* 25 D6 38 59N 77 13W
Potomac →, *Wash.* 25 D7 38 58N 77 9W
Potrero Pt., *S.F.* 27 B2 37 45N 122 22W
Potsdam, *Berl.* 7 B1 52 23N 13 3 E
Potter Pt., *Syd.* 19 C4 34 2 S 151 13 E
Potters Bar, *Lon.* 4 A4 51 41N 0 10W
Potzham, *Mün.* 7 G10 48 1N 11 36 E
Pötzleinsdorf, *Wien* 10 G9 48 14N 16 17 E
Povoa de Santo Adriao, *Lisb.* 8 F8 38 47N 9 9W
Powderhorn L., *Chic.* 26 D3 41 38N 87 31W
Powičle, *Wsaw.* 10 E7 52 14N 21 1 E
Powõzki, *Wsaw.* 10 E6 52 15N 20 58 E
Powsin, *Wsaw.* 10 F7 52 8N 21 6 E
Powsinek, *Wsaw.* 10 F7 52 9N 21 6 E
Poyo, *Barc.* 8 D6 41 28N 2 12 E
Pozuelo de Alarcón, *Mdrd.* 8 B2 40 25N 3 48W
Praça Seca, *Rio J.* 31 B1 22 53 S 43 20W
Prado, Museo del, *Mdrd.* 8 B2 40 25N 3 42W
Prado Churubusco, *Méx.* 29 B3 19 20N 99 8W
Praga, *Wsaw.* 10 E7 52 15N 21 2 E
Prague = Praha, *Pra.* 10 B2 50 4N 14 25 E
Praha, *Pra.* 10 B2 50 4N 14 25 E
Praha-Ruzyně Airport, *Pra.* 10 B1 50 6N 14 16 E
Praires, R. des →, *Mtrl.* 20 A4 43 38N 73 36W
Prat de Llobregat, *Barc.* 8 E5 41 19N 2 5 E
Prater, *Wien* 10 G10 48 12N 16 25 E
Pratts Bottom, *Lon.* 4 C5 51 20N 0 7 E
Prawet Buri Rom, Khlong →, *Bangk.* 15 B2 13 43N 100 38 E
Preakness, *N.Y.* 22 B3 40 56N 74 12W
Precotto, *Mil.* 9 D6 45 30N 9 13 E
Prédecelles →, *Paris* 5 D2 48 36N 2 7 E
Pregnana Milanese, *Mil.* 9 D4 45 30N 9 0 E
Prem Prachakan, Khlong →, *Bangk.* 15 B2 13 46N 100 35 E
Prenestino Labicano, *Rome* 9 F10 41 53N 12 33 E
Prenzlauerberg, *Berl.* 7 A3 52 32N 13 24 E
Presidente Derqui, *B.A.* 32 A1 34 29 S 58 50W
Presidente Outra, Rodo, *Rio J.* 31 A1 22 47 S 43 21W
Preston, *Melb.* 19 E6 37 44 S 144 59 E
Pretos Forros, Sa. dos, *Rio J.* 31 B2 22 54 S 43 17W
Préville, *Mtrl.* 20 B5 43 28N 73 29W
Přezletice, *Pra.* 10 B3 50 9N 14 34 E
Primavalle, *Rome* 9 F9 41 55N 12 25 E
Primrose, *Jobg.* 18 F9 26 11 S 28 9 E
Princes B., *N.Y.* 22 D3 40 30N 74 12W
Princess Elizabeth Park, *Sing.* 15 F7 1 21N 103 45 E
Progreso, *Mdrd.* 8 B3 40 27N 3 39W
Progreso Nacional, *Méx.* 29 A3 19 30N 99 9W
Prosek, *Pra.* 10 B3 50 7N 14 30 E
Prospect, *Syd.* 19 A2 33 48 S 150 55 E
Prospect Heights, *Chic.* 26 A1 42 5N 87 55W
Prospect Hill Park, *Bost.* 21 C2 42 23N 71 13W
Prospect Park, *N.Y.* 22 B3 40 55N 74 10W
Prospect Park, *Phil.* 24 B3 39 53N 75 18W
Prospect Pt., *N.Y.* 23 B6 40 52N 73 42W
Prospect Res., *Syd.* 19 A2 33 49 S 150 53 E
Providence, *Balt.* 25 A3 39 25N 76 34W
Providencia, *Stgo* 30 J11 33 25 S 70 36W
Průhonice, *Pra.* 10 C3 50 0N 14 33 E
Pruszków, *Wsaw.* 10 E5 52 10N 20 48 E
Psikhikón, *Ath.* 8 H11 38 1N 23 46 E
Pudong, *Shang.* 14 J12 31 13N 121 30 E
Puduo, *Shang.* 14 J11 31 15N 121 24 E
Pueblo Libre, *Lima* 30 G8 12 5 S 77 4W
Pueblo Nuevo, *Barc.* 8 D6 41 23N 2 11 E
Pueblo Nuevo, *Mdrd.* 8 B3 40 25N 3 37W
Puente Cascallares, *B.A.* 32 C2 34 41 S 58 48W
Puente Hills, *L.A.* 28 C5 33 59N 117 59W
Puffing Billy Station, *Melb.* 19 F9 37 54 S 145 20 E
Puhuangyu, *Beij.* 14 B3 39 50N 116 22 E
Puistola, *Hels.* 3 B5 60 16N 25 2 E
Pukinmäki, *Hels.* 3 B4 60 15N 24 57 E
Pullach, *Mün.* 7 G9 48 3N 11 31 E
Pulo, *Manila* 15 D4 14 34N 121 4 E
Pulo Gadung, *Jak.* 15 J10 6 11 S 106 54 E
Pumphrey, *Balt.* 25 B3 39 13N 76 39W
Punchbowl, *Syd.* 19 B3 33 55 S 151 3 E
Punde, *Bomb.* 16 H8 18 53N 72 57 E
Punggol, *Sing.* 15 F8 1 23N 103 54 E
Punggol, Sungei →, *Sing.* 15 F8 1 24N 103 54 E
Punggol Pt., *Sing.* 15 F8 1 24N 103 54 E
Punta Brava, *La Hab.* 30 B2 23 1N 82 29W
Puolarmetsä, *Hels.* 3 B3 60 11N 24 41 E
Puotila, *Hels.* 3 B5 60 13N 25 6 E
Purchase, *N.Y.* 23 A6 41 2N 73 43W
Purfleet, *Lon.* 4 C6 51 29N 0 14 E
Purkersdorf, *Wien* 10 G9 48 12N 16 11 E
Purley, *Lon.* 4 C4 51 20N 0 6W
Puteaux, *Paris* 5 B3 48 53N 2 14 E
Puth Kalan, *Delhi* 16 A1 28 42N 77 4 E
Putilkovo, *Mos.* 11 D8 55 51N 37 22 E
Putnamville Res., *Bost.* 21 B4 42 36N 70 56W
Putney, *Lon.* 4 C3 51 27N 0 13W
Putty Hill, *Balt.* 25 A3 39 22N 76 30W
Putxet, *Barc.* 8 D5 41 24N 2 8 E
Putzbrunn, *Mün.* 7 G11 48 4N 11 42 E
Pyeongchang, *Sŏul* 12 G7 37 35N 126 57 E
Pyramids, *El Qâ.* 18 D4 29 58N 31 7 E
Pyry, *Wsaw.* 10 F6 52 8N 21 0 E

Q

Qanât el Ismâilîya, *El Qâ.* 18 C5 30 7N 31 17 E
Qasemābād, *Tehr.* 17 C6 35 4N 51 3 E
Qasr-e-Firōzeh, *Tehr.* 17 D6 35 29N 51 31 E
Qianmen, *Beij.* 14 B3 39 51N 116 21 E
Qibao, *Shang.* 14 K11 31 9N 121 20 E
Qingguang, *Tianj.* 14 D5 39 11N 117 2 E
Qinghua Univ., *Beij.* 14 A2 40 0N 116 17 E
Qinghuayuan, *Beij.* 14 B2 39 59N 116 19 E
Qingningsi, *Shang.* 14 J12 31 16N 121 33 E
Qolhak, *Tehr.* 17 C5 35 45N 51 26 E
Quadraro, *Rome* 9 F10 41 51N 12 33 E
Quaid-i-Azam, *Kar.* 17 G10 24 50N 66 59 E
Qual'eh Murgeh Airport, *Tehr.* 17 D5 35 38N 51 22 E
Qualiano, *Nápl.* 9 H11 40 55N 14 9 E
Quannapowitt, L., *Bost.* 21 B3 42 30N 71 4W
Quartiere Zingone, *Mil.* 9 E5 45 25N 9 3 E
Quarto, *Nápl.* 9 H11 40 52N 14 8 E
Quds, *Bagd.* 17 E8 33 23N 44 24 E
Quebrada Baruta →, *Car.* 30 E5 10 29N 66 53W
Quebrada Tácagua →, *Car.* 30 D4 10 36N 67 1W
Quebrada Topo →, *Car.* 30 D4 10 32N 67 0W
Queen Mary Res., *Lon.* 4 C2 51 24N 0 27W
Queens Village, *N.Y.* 23 C6 40 43N 73 44W
Queensbury, *Lon.* 4 B3 51 35N 0 16W
Queenscliffe, *Syd.* 19 A4 33 47 S 151 17 E
Queenstown, *Sing.* 15 G7 1 18N 103 48 E
Quellerina, *Jobg.* 18 E8 26 9 S 27 56 E
Queluz, *Lisb.* 8 F7 38 45N 9 14W
Quezon City, *Manila* 15 D4 14 37N 121 2 E
Quickborn, *Hbg.* 7 C7 53 43N 9 54 E
Quilicura, *Stgo* 30 J10 33 22 S 70 43W
Quilmes, *B.A.* 32 C5 34 43 S 58 15W
Quincy, *Bost.* 21 D3 42 14N 71 0W
Quincy B., *Bost.* 21 D4 42 16N 70 59W
Quincy-sous-Sénart, *Paris* 5 C5 48 40N 2 32 E
Quinta Normal, *Stgo* 30 J10 33 26 S 70 40W
Quinto Romano, *Mil.* 9 E5 45 28N 9 7 E
Quirinale, *Rome* 9 F9 41 53N 12 29 E
Quitaúna, *S. Pau.* 31 E5 23 31 S 46 48W

R

Raasdorf, *Wien* 10 G11 48 14N 16 33 E
Raccoon Cr. →, *Phil.* 24 C3 39 48N 75 21W
Raccoon Str., *S.F.* 27 A2 37 52N 122 26W
Radevormwald, *Ruhr* 6 C6 51 12N 7 22 E
Radlett, *Lon.* 4 A3 51 41N 0 19W
Radlice, *Pra.* 10 B2 50 3N 14 23 E
Radnor, *Phil.* 24 A2 40 2N 75 21W
Radonice, *Pra.* 10 B3 50 8N 14 36 E
Radotin, *Pra.* 10 C2 49 59N 14 21 E
Rælingen, *Oslo* 2 B6 59 53N 11 5 E
Rafael Calzada, *B.A.* 32 C4 34 47 S 58 21W
Rafael Castillo, *B.A.* 32 C3 34 42 S 58 36W
Raffles Park, *Sing.* 15 G7 1 19N 103 48 E
Raghunathpur, *Calc.* 16 D5 22 41N 88 16 E
Rahlstedt, *Hbg.* 7 D8 53 35N 10 7 E
Rahm, *Ruhr* 6 B2 51 21N 6 47 E
Rahnsdorf, *Berl.* 7 B5 52 25N 13 41 E
Rahway, *N.Y.* 22 D3 40 36N 74 17W
Rail Tree Hill, *Bost.* 21 B1 42 32N 71 22W
Rainbow Lakes, *N.Y.* 22 B2 40 53N 74 27W
Rainham, *Lon.* 4 B6 51 31N 0 11 E
Rainier, Mt., *Wash.* 25 D8 38 56N 76 57W
Raj Bhawan, *Calc.* 16 E6 22 33N 88 20 E
Rajakylä, *Hels.* 3 B5 60 15N 25 5 E
Rajapur, *Calc.* 16 E5 22 39N 88 11 E
Rajganj, *Calc.* 16 E5 22 34N 88 14 E
Rajpur, *Delhi* 16 A2 28 41N 77 12 E
Rákos-patak →, *Bud.* 10 K14 47 29N 19 12 E
Rákoscsaba, *Bud.* 10 K14 47 28N 19 17 E
Rákoshegy, *Bud.* 10 K14 47 27N 19 14 E
Rákosker, *Bud.* 10 K14 47 28N 19 14 E
Rákoskert, *Bud.* 10 K14 47 27N 19 18 E
Rákosliget, *Bud.* 10 K14 47 29N 19 16 E
Rákospalota, *Bud.* 10 J13 47 34N 19 7 E
Rákosszentmihály, *Bud.* 10 J13 47 31N 19 8 E
Raków, *Wsaw.* 10 E6 52 11N 20 56 E
Rakowiec, *Wsaw.* 10 E6 52 12N 20 58 E
Ramadān, *Bagd.* 17 F8 33 19N 44 20 E
Ramanathpur, *Calc.* 16 D5 22 41N 88 14 E
Rambler Channel, *H.K.* 12 D5 22 21N 114 6 E
Ramblewood, *Phil.* 24 B5 39 55N 74 56W
Ramenki, *Mos.* 11 E8 55 41N 37 28 E
Ramersdorf, *Mün.* 7 G10 48 6N 11 35 E
Ramnathpur, *Calc.* 16 E5 22 35N 88 18 E
Ramos, *Rio J.* 31 B2 22 50 S 43 14W
Ramos Mejia, *B.A.* 32 B3 34 39 S 58 33W
Rampur, *Delhi* 16 A2 28 44N 77 18 E
Ramsgate, *Syd.* 19 B3 33 58 S 151 8 E
Ramstadjøen, *Oslo* 2 B6 59 53N 11 3 E
Rancho Boyeros, *La Hab.* 30 C2 22 59N 82 22W
Rancho Colorado, Presa de, *Méx.* 29 B2 19 29N 99 16W
Rancocas Cr. →, *Phil.* 24 A5 40 2N 74 58W
Rand Afrikaans Univ., *Jobg.* 18 F9 26 11 S 28 0 E
Rand Airport, *Jobg.* 18 F9 26 14 S 28 8 E
Randallstown, *Balt.* 25 A2 39 21N 76 46W
Randburg, *Jobg.* 18 E8 26 5 S 27 57 E
Randhart, *Jobg.* 18 F9 26 16 S 28 7 E
Randolph, *Bost.* 21 D3 42 10N 71 3W
Randolph Hills, *Wash.* 25 C7 39 3N 77 6W
Randpark, *Jobg.* 18 E8 26 6 S 27 58 E
Randwick, *Syd.* 19 B4 33 54 S 151 14 E
Ranelagh, *B.A.* 32 C5 34 47 S 58 14W
Rannersdorf, *Wien* 10 H10 48 7N 16 27 E
Raparkrif, *Jobg.* 18 E8 26 5 S 27 57 E
Raposo, *Lisb.* 8 F7 38 40N 9 11W
Raritan →, *N.Y.* 22 D2 40 30N 74 27W
Raritan B., *N.Y.* 22 E3 40 29N 74 12W
Rasskazovka, *Mos.* 11 F8 55 38N 37 20 E
Rasta, *Stock.* 3 E8 59 18N 17 37 E
Rastaala, *Hels.* 3 B3 60 15N 24 47 E
Rastila, *Hels.* 3 B5 60 12N 25 7 E
Raszyn, *Wsaw.* 10 F6 52 9N 20 54 E
Rat Burana, *Bangk.* 15 B2 13 40N 100 30 E
Ratanpur, *Calc.* 16 D5 22 49N 88 14 E
Rath, *Ruhr* 6 C2 51 16N 6 49 E
Ratingen, *Ruhr* 6 C3 51 18N 6 52 E
Rato, *Lisb.* 8 F8 38 43N 9 8W
Rauxel, *Ruhr* 6 A5 51 34N 7 18 E
Ravenswood Pt., *S.F.* 27 C4 37 30N 122 8W
Rawamangun, *Jak.* 15 J10 6 11 S 106 52 E
Rayners Lane, *Lon.* 4 B2 51 34N 0 23W
Raynes Park, *Lon.* 4 C3 51 24N 0 12W
Raypur, *Calc.* 16 F6 22 28N 88 22 E
Razdory, *Mos.* 11 E7 55 44N 37 17 E
Razmitelevo, *St-Pet.* 11 B5 59 54N 30 39 E
Razor Hill, *H.K.* 12 D6 22 20N 114 15 E
Reading, *Bost.* 21 B3 42 31N 71 5W
Reading Highlands, *Bost.* 21 B3 42 31N 71 5W
Reáglie, *Tori.* 9 B3 45 3N 7 44 E
Real, Palacio, *Mdrd.* 8 B2 40 25N 3 43W
Real Felipe, Castillo, *Lima* 30 G8 12 4 S 77 9W
Real Fuerta, Château de la, *La Hab.* 30 B2 23 8N 82 20W
Realengo, *Rio J.* 31 B1 22 52 S 43 24W
Réau, *Paris* 5 D5 48 36N 2 37 E
Recklinghausen, *Ruhr* 6 A5 51 37N 7 12 E
Recklinghausen-Süd, *Ruhr* 6 A5 51 34N 7 14 E
Recoleta, *Stgo* 30 J11 33 25 S 70 40W
Reconquista →, *B.A.* 32 B3 34 35 S 58 35W
Red Bank Battle Mon., *Phil.* 24 B3 39 52N 75 11W
Red Fort, *Delhi* 16 B2 28 39N 77 14 E
Red Rock, *S.F.* 27 A2 37 55N 122 25W
Red Square, *Mos.* 11 E9 55 45N 37 37 E
Redbridge, *Lon.* 4 B5 51 34N 0 3 E
Redwood City, *S.F.* 27 D3 37 29N 122 14W
Redwood Cr. →, *S.F.* 27 C3 37 31N 122 11W
Redwood Pt., *S.F.* 27 C3 37 32N 122 11W
Redwood Regional Park, *S.F.* 27 B4 37 48N 122 8W
Reeves Hill, *Bost.* 21 C1 42 20N 71 20W
Refshaleøen, *Købn.* 2 D10 55 41N 12 36 E
Regents Park, *Jobg.* 18 F9 26 14 S 28 3 E
Regents Park, *Lon.* 4 B4 51 31N 0 9W
Regents Park, *Syd.* 19 B3 33 52 S 151 1 E
Regi Lagni →, *Nápl.* 9 H13 40 56N 14 23 E
Regina Margherita, *Tori.* 9 B2 45 4N 7 34 E
Regla, *La Hab.* 30 B3 23 7N 82 19W
Rego Park, *N.Y.* 23 C5 40 43N 73 51W
Reiherstieg, *Hbg.* 7 D7 53 30N 9 58 E
Reinickendorf, *Berl.* 7 A3 52 34N 13 22 E
Reinoldikirche, *Ruhr* 6 A6 51 30N 7 28 E
Reistad, *Oslo* 2 C1 59 46N 10 16 E
Reitbrook, *Hbg.* 7 E8 53 28N 10 8 E
Rekola, *Hels.* 3 B5 60 19N 25 4 E
Rellingen, *Hbg.* 7 D7 53 39N 9 50 E
Rembertów, *Wsaw.* 10 E7 52 15N 21 9 E
Remedios de Escalada, *B.A.* 32 C4 34 43 S 58 24W
Rémola, Laguna del, *Barc.* 8 E5 41 16N 2 4 E
Remscheid, *Ruhr* 6 C5 51 11N 7 11 E
Renca, *Stgo* 30 J10 33 24 S 70 42W
Renca, Cerro, *Stgo* 30 J10 33 23 S 70 40W
Rener, *Ist.* 17 A2 41 1N 28 56 E
Renmin Gongyuan, *Tianj.* 14 E6 39 6N 117 12 E
Rennemoulin, *Paris* 5 B2 48 50N 2 2 E
Rennie's Mill, *H.K.* 12 E6 22 18N 114 15 E
Renzel, *Hbg.* 7 C7 53 43N 9 52 E
Repaupo, *Phil.* 24 C3 39 48N 75 18W
Repaupo Cr. →, *Phil.* 24 C3 39 49N 75 20W
Řeporyje, *Pra.* 10 B1 50 1N 14 18 E
République, Place de la, *Paris* 5 B4 48 52N 2 22 E
Repy, *Pra.* 10 B1 50 4N 14 17 E
Resarö, *Stock.* 3 D13 59 25N 18 20 E
Rescaldna, *Mil.* 9 D4 45 36N 8 57 E
Research, *Melb.* 19 E8 37 42 S 145 10 E
Reseda, *L.A.* 28 A1 34 12N 118 31W
Reservoir, *Melb.* 19 E7 37 42 S 145 1 E
Reservoir Pond, *Bost.* 21 D3 42 10N 71 7W
Residenz, *Mün.* 7 G10 48 8N 11 34 E
Resse, *Ruhr* 6 A4 51 34N 7 6 E
Reston, *Wash.* 25 D6 38 57N 77 20W
Retiro, *B.A.* 32 B4 34 35 S 58 23W
Retiro, *Mdrd.* 8 B2 40 24N 3 40W
Reutov, *Mos.* 11 E11 55 45N 37 50 E
Réveillon →, *Paris* 5 C5 48 42N 2 30 E
Revere, *Bost.* 21 C3 42 25N 71 0W
Revesby, *Syd.* 19 B3 33 57 S 151 0 E
Revolucion, Plaza de la, *La Hab.* 30 B2 23 7N 82 23W
Rexdale, *Trto.* 20 D7 43 43N 79 35W
Reynolds Channel, *N.Y.* 23 D6 40 35N 73 41W
Reynosa Tamaulipas, *Méx.* 29 A2 19 30N 99 10W
Rheem Valley, *S.F.* 27 A4 37 50N 122 8W
Rhein-Herne Kanal, *Ruhr* 6 B3 51 29N 6 59 E
Rheinberg, *Ruhr* 6 A1 51 32N 6 37 E
Rheinhausen, *Ruhr* 6 B2 51 24N 6 43 E
Rheinkamp, *Ruhr* 6 B1 51 29N 6 36 E
Rho, *Mil.* 9 D5 45 31N 9 2 E
Rhodes, *Syd.* 19 A3 33 49 S 151 6 E
Rhodesfield, *Jobg.* 18 E10 26 6 S 28 14 E
Rhodon, *Paris* 5 C2 48 42N 2 3 E
Rhodon →, *Paris* 5 C2 48 42N 2 4 E
Rhu, Tg., *Sing.* 15 G8 1 17N 103 51 E
Ribeirão Pires, *S. Pau.* 31 F7 23 42 S 46 23W
Říčanský →, *Pra.* 10 B3 50 5N 14 36 E
Říčany, *Pra.* 10 C3 49 59N 14 39 E
Ricarda, Laguna de la, *Barc.* 8 E5 41 17N 2 6 E
Richardson B., *S.F.* 27 A2 37 52N 122 29W
Richmond, *Lon.* 4 C3 51 27N 0 17W
Richmond, *Melb.* 19 E7 37 48 S 145 0 E
Richmond, *S.F.* 27 B2 37 46N 122 27W
Richmond, *S.F.* 27 A2 37 56N 122 21W
Richmond →, *N.Y.* 22 D3 40 34N 74 11W
Richmond, Pt., *S.F.* 27 A2 37 54N 122 23W
Richmond Hill, *N.Y.* 23 C5 40 41N 73 51W
Richmond Hill, *Trto.* 20 C8 43 51N 79 24W
Richmond Inner Harbour, *S.F.* 27 A2 37 54N 122 20W
Richmond Park, *Lon.* 4 C3 51 26N 0 16W
Richmond Valley, *N.Y.* 22 D3 40 31N 74 13W
Richvale, *Trto.* 20 C8 43 51N 79 26W
Rickers I., *N.Y.* 23 C5 40 47N 73 53W
Rickmansworth, *Lon.* 4 B2 51 38N 0 28W
Riddel Cr. →, *Melb.* 19 F8 37 52 S 145 13 E
Riderwood, *Balt.* 25 A3 39 24N 76 37W
Ridgefield, *N.Y.* 22 C4 40 49N 74 0W
Ridgefield Park, *N.Y.* 22 B4 40 52N 74 1W
Ridgewood, *N.Y.* 23 C5 40 42N 73 53W
Ridley Cr. →, *Phil.* 24 B2 39 51N 75 20W
Ridley Creek State Park, *Phil.* 24 B2 39 57N 75 26W
Ridley Park, *Phil.* 24 B3 39 52N 75 19W
Riedmoos, *Mün.* 7 F10 48 16N 11 32 E
Riem, *Mün.* 7 G11 48 8N 11 41 E
Riemke, *Ruhr* 6 A5 51 30N 7 12 E
Rimac, *Lima* 30 G8 12 2 S 77 2W
Rimau, Tg., *Sing.* 15 G7 1 15N 103 48 E
Ringwood, *Melb.* 19 E8 37 48 S 145 4 E
Rinkeby, *Stock.* 3 D10 59 23N 17 55 E
Rio Comprido, *Rio J.* 31 B2 22 55 S 43 12W
Rio de Janeiro, *Rio J.* 31 B2 22 54 S 43 12W
Rio de Mouro, *Lisb.* 8 F7 38 46N 9 15W
Rio Hondo →, *L.A.* 28 B4 34 2N 118 15W
Rio Pequeno, *S. Pau.* 31 E5 23 34 S 46 44W
Rione Trieste, *Nápl.* 9 H13 40 52N 14 27 E
Ripley, *Lon.* 4 D2 51 17N 0 29W
Rippling Ridge, *Balt.* 25 B3 39 11N 76 37W
Ris, *Oslo* 2 B4 59 56N 10 41 E
Ris-Orangis, *Paris* 5 D4 48 38N 2 24 E
Risby, *Købn.* 2 D8 55 41N 12 19 E
Rishra, *Calc.* 16 D6 22 42N 88 20 E
Ritan Gongyuan, *Beij.* 14 B3 39 53N 116 24 E
Ritchie, *Wash.* 25 D8 38 52N 76 51W
Rithala, *Delhi* 16 A1 28 43N 77 6 E
Ritorp, *Stock.* 3 E8 59 12N 17 38 E
Rivalta di Torino, *Tori.* 9 B1 45 2N 7 31 E
Rivas de Jarama, *Mdrd.* 8 B3 40 22N 3 31W
Rivas-Vaciamadrio, *Mdrd.* 8 C3 40 19N 3 30W
Rivasacco, *Tori.* 9 A1 45 10N 7 29 E
Rive Sud, Canal de la, *Mtrl.* 20 B4 43 24N 73 31W
River Edge, *N.Y.* 22 B4 40 56N 74 1W
River Forest, *Chic.* 26 B2 41 53N 87 49W
River Grove, *Chic.* 26 B1 41 55N 87 50W
River Pines, *Bost.* 21 B2 42 33N 71 17W
River Vale, *N.Y.* 22 B4 40 59N 74 1W
Riverdale, *Chic.* 26 D3 41 38N 87 37W
Riverdale, *N.Y.* 22 B3 40 59N 74 17W
Riverdale, *Trto.* 20 E8 43 39N 79 21W
Riverdale, *Wash.* 25 D8 38 57N 76 54W
Riverdale Park, *Trto.* 20 D8 43 40N 79 21W
Riverhead, *Lon.* 4 D6 51 16N 0 10 E
Riverlea, *Jobg.* 18 F8 26 12 S 27 58 E
Riverside, *Bost.* 21 B2 42 32N 71 18W
Riverside, *Chic.* 26 C2 41 49N 87 49W
Riverside, *N.Y.* 23 B5 40 54N 73 54W
Riverside, *N.Y.* 23 A7 41 1N 73 34W
Riverside, *Phil.* 24 A5 40 2N 74 57W
Riverton, *Phil.* 24 A4 40 0N 75 0W
Riverwood, *Syd.* 19 B3 33 57 S 151 3 E
Rivière-des-Praires, *Mtrl.* 20 A4 43 38N 73 34W
Rivodora, *Tori.* 9 B3 45 5N 7 47 E
Rivoli, *Tori.* 9 B1 45 4N 7 30 E
Riyad, *Bagd.* 17 F8 33 18N 44 27 E
Rizal, *Manila* 15 D4 14 33N 121 0 E
Rizal Park, *Manila* 15 D3 14 35N 120 58 E
Rizel Stadium, *Manila* 15 D3 14 34N 120 59 E
Røa, *Oslo* 2 B3 59 57N 10 39 E
Robassomero, *Tori.* 9 A2 45 11N 7 34 E
Robbins, *Chic.* 26 D2 41 38N 87 42W
Robert E. Lee Memorial Park, *Balt.* 25 A3 39 23N 76 40 E
Robertsdale, *Chic.* 26 C3 41 40N 87 30W
Robertsham, *Jobg.* 18 F9 26 15 S 28 1 E
Robin Hills, *Jobg.* 18 E8 26 6 S 27 58 E
Rocha Miranda, *Rio J.* 31 B1 22 51 S 43 20W
Rochar →, *Sing.* 15 G8 1 18N 103 52 E
Rochelle Park, *N.Y.* 22 B4 40 54N 74 4W
Rock Cr. →, *Wash.* 25 D7 38 54N 77 3W
Rock Creek Park, *Wash.* 25 D7 38 56N 77 2W
Rockaway, *N.Y.* 22 D5 40 32N 73 56W
Rockaway Beach, *S.F.* 27 C2 37 36N 122 29W
Rockaway Islet, *N.Y.* 23 D5 40 34N 73 53W
Rockaway Neck, *N.Y.* 22 B2 40 51N 74 21W
Rockaway Point, *N.Y.* 23 D5 40 33N 73 54W
Rockburn Branch →, *Balt.* 25 B2 39 13N 76 43W
Rockdale, *Balt.* 25 A2 39 21N 76 46W
Rockdale, *Syd.* 19 B3 33 57 S 151 8 E
Rockland, *Phil.* 24 C1 39 47N 75 34W
Rocklege, *Phil.* 24 A4 40 4N 75 5W
Rockleigh, *N.Y.* 22 A5 41 0N 73 55W
Rockville, *Wash.* 25 C6 39 4N 77 10W
Rockville Centre, *N.Y.* 23 D7 40 39N 73 38W
Rocky Hill, *Phil.* 24 B1 39 58N 75 32W
Rocky Ridge, *S.F.* 27 B4 37 47N 122 2W
Rocky Run →, *Wash.* 25 D6 38 58N 77 14W
Rodaon, *Wien* 10 H9 48 8N 16 16 E
Rodeo Cove, *S.F.* 27 B1 37 49N 122 32W
Rodgers Forge, *Balt.* 25 A3 39 22N 76 37W
Roding →, *Lon.* 4 B5 51 30N 0 5 E
Rodość, *Wsaw.* 10 E8 52 11N 21 11 E
Rødovre, *Købn.* 2 D9 55 40N 12 26 E
Rodrigo de Freitas, L., *Rio J.* 31 B2 22 58 S 43 12W
Rodstensfjärden, *Stock.* 3 E9 59 16N 17 48 E
Roehampton, *Lon.* 4 C3 51 27N 0 15W
Rogers Park, *Chic.* 26 A2 42 0N 87 40W
Rohdenhaus, *Ruhr* 6 C4 51 18N 7 0 E
Röhlinghausen, *Ruhr* 6 A4 51 30N 7 9 E
Roihuvuori, *Hels.* 3 B5 60 11N 25 2 E
Roissy, *Paris* 5 C5 48 47N 2 39 E
Roissy-en-France, *Paris* 5 A5 49 0N 2 30 E
Rokkō Sanchi, *Ōsaka* 12 B2 34 44N 135 13 E
Rokko-Zan, *Ōsaka* 12 B2 34 46N 135 16 E
Rokytka →, *Pra.* 10 B3 50 6N 14 27 E
Roland Lake, *Balt.* 25 A3 39 23N 76 38W
Roland Park, *Balt.* 25 A3 39 20N 76 39W
Roma, *Rome* 9 F9 41 54N 12 28 E
Római-Fürdö, *Bud.* 10 J13 47 34N 19 4 E
Romainville, *Paris* 5 B4 48 52N 2 26 E
Romani, *Nápl.* 9 H13 40 52N 14 22 E
Romano Banco, *Mil.* 9 E5 45 25N 9 6 E
Romashkovo, *Mos.* 11 E7 55 43N 37 19 E
Rome = Roma, *Rome* 9 F9 41 54N 12 28 E
Romford, *Lon.* 4 B6 51 34N 0 11 E
Roncáglia, *Tori.* 9 B1 45 2N 7 29 E
Rönninge, *Stock.* 3 E9 59 12N 17 45 E
Ronsdorf, *Ruhr* 6 C5 51 13N 7 11 E
Ronskensiedig, *Ruhr* 6 A2 51 36N 6 41 E
Rontgental, *Berl.* 7 A4 52 38N 13 31 E
Roodekop, *Jobg.* 18 F10 26 17 S 28 11 E
Roodepoort, *Jobg.* 18 E8 26 9 S 27 53 E
Roodepoort-Wes, *Jobg.* 18 E8 26 8 S 27 51 E
Roosevelt, *N.Y.* 23 C7 40 40N 73 35W
Rooty Hill, *Syd.* 19 A2 33 46 S 150 50 E
Roppongi, *Tōkyō* 13 C3 35 39N 139 44 E
Rosairinho, *Lisb.* 8 F8 38 40N 9 0W
Rosanna, *Melb.* 19 E7 37 44 S 145 4 E
Rosario, *La Hab.* 30 B2 23 3N 82 21W
Rosario, *Manila* 15 D4 14 35N 121 4 E
Rose B., *Syd.* 19 B4 33 51 S 151 16 E
Rose Hill, *Wash.* 25 E7 38 47N 77 6W
Rose Tree, *Phil.* 24 B2 39 56N 75 23W
Rosebank, *N.Y.* 22 D4 40 36N 74 4W
Rosebery, *Syd.* 19 B4 33 55 S 151 12 E
Rosedal La Candelaria, *Méx.* 29 B3 19 20N 99 10W
Rosedale, *Balt.* 25 B3 39 19N 76 31W
Rosedale, *N.Y.* 23 D6 40 39N 73 43W
Roseiras, *S. Pau.* 31 E7 23 33 S 46 23W
Roseland, *Chic.* 26 C3 41 42N 87 37W
Roseland, *N.Y.* 22 C3 40 49N 74 17W
Roselle, *N.Y.* 22 D3 40 39N 74 15W
Roselle Park, *N.Y.* 22 D3 40 39N 74 16W
Rosemead, *L.A.* 28 B4 34 4N 118 4W
Rosemere, *Mtrl.* 20 A2 43 34N 73 50W
Rosemont, *Chic.* 26 B1 41 59N 87 52W
Rosemont, *Phil.* 24 A3 40 1N 75 19W
Rosenborg Have, *Købn.* 2 D10 55 41N 12 33 E
Rosengarten, *Hbg.* 7 D6 53 31N 9 49 E
Rosenthal, *Berl.* 7 A3 52 35N 13 22 E
Rosettenville, *Jobg.* 18 F9 26 15 S 28 3 E
Rosherville Dam, *Jobg.* 18 F9 26 13 S 28 6 E
Rósio, *Mil.* 9 E4 45 25N 8 57 E
Rösjön, *Stock.* 3 D11 59 26N 18 0 E
Roskilde, *Købn.* 2 E7 55 38N 12 5 E
Roskilde Fjord, *Købn.* 2 D7 55 45N 12 4 E
Roslags-Näsby, *Stock.* 3 D11 59 25N 18 2 E
Roslindale, *Bost.* 21 D3 42 17N 71 7W
Roslyn, *N.Y.* 23 C7 40 47N 73 38W
Roslyn, *Phil.* 24 A4 40 7N 75 8W
Roslyn Estates, *N.Y.* 23 C6 40 47N 73 40W
Roslyn Harbour, *N.Y.* 23 C7 40 48N 73 38W
Rosne →, *Paris* 5 B4 48 58N 2 25 E
Rosny-sous-Bois, *Paris* 5 B5 48 52N 2 30 E
Ross, *S.F.* 27 A1 37 57N 122 33W
Rosslyn, *Wash.* 25 D7 38 53N 77 4W
Rossville, *Balt.* 25 A4 39 20N 76 28W
Rossville, *N.Y.* 22 D3 40 32N 74 12W
Rosta, *Tori.* 9 B1 45 4N 7 27 E
Rotbach →, *Ruhr* 6 A2 51 34N 6 41 E
Rothenburgsort, *Hbg.* 7 D8 53 32N 10 2 E
Rotherbaum, *Hbg.* 7 D7 53 33N 9 58 E
Rotherhithe, *Lon.* 4 C4 51 29N 0 2W
Rothneusiedl, *Wien* 10 H10 48 8N 16 23 E
Rothschmaige, *Mün.* 7 F9 48 14N 11 27 E
Rouge →, *Trto.* 20 D9 43 47N 79 12W
Rouge Hill, *Trto.* 20 D10 43 48N 79 6W
Round I., *H.K.* 12 E6 22 13N 114 11 E
Roundshaw, *Lon.* 4 C4 51 20N 0 7W
Roussigny, *Paris* 5 D2 48 38N 2 6 E
Rowland, *L.A.* 28 B5 34 0N 117 55W
Rowley, *Bost.* 21 A4 42 43N 70 52W
Rowville, *Melb.* 19 F8 37 55 S 145 14 E
Roxboro, *Mtrl.* 20 A3 43 30N 73 48W
Roxborough, *Phil.* 24 A3 40 1N 75 13W
Roxbury, *Bost.* 21 D3 42 19N 71 5W
Roxbury, *N.Y.* 23 D5 40 33N 73 53W
Roxeth, *Lon.* 4 B2 51 33N 0 20W
Royal Observatory, *H.K.* 12 E6 22 18N 114 10 E
Royal Park, *Melb.* 19 E6 37 46 S 144 57 E
Röylä, *Hels.* 3 B3 60 18N 24 42 E
Royston Park, *Lon.* 4 B2 51 36N 0 22W
Rozas, Portilleros de las, *Mdrd.* 8 B2 40 29N 3 49W
Roztoky, *Pra.* 10 B2 50 9N 14 23 E
Rubbianetta, *Tori.* 9 B2 45 9N 7 34 E
Rubi →, *Barc.* 8 D5 41 26N 2 0 E
Rubio Woods, *Chic.* 26 D2 41 38N 87 45W
Rublovo, *Mos.* 11 E8 55 47N 37 21 E
Ruchyi, *St-Pet.* 11 A4 60 0N 30 25 E
Rud, *Oslo* 2 B6 59 56N 11 0 E
Rüdinghausen, *Ruhr* 6 B6 51 26N 7 22 E
Rudnevka →, *Mos.* 11 E11 55 43N 37 56 E
Rudolfsheim, *Wien* 10 G10 48 12N 16 20 E
Rudolfshöhe, *Berl.* 7 A5 52 37N 13 44 E
Rudow, *Berl.* 7 B3 52 24N 13 29 E
Rueil-Malmaison, *Paris* 5 B3 48 52N 2 11 E
Ruffys Cr. →, *Melb.* 19 E7 37 45 S 145 7 E
Ruggeberg, *Ruhr* 6 C6 51 16N 7 23 E
Ruhlsdorf, *Berl.* 7 B2 52 22N 13 15 E
Ruhr →, *Ruhr* 6 B3 51 27N 6 56 E
Ruhrort, *Ruhr* 6 B2 51 27N 6 44 E
Ruislip, *Lon.* 4 B2 51 34N 0 24W
Rumelihisari, *Ist.* 17 A3 41 4N 29 2 E
Rumeln, *Ruhr* 6 B1 51 24N 6 39 E
Rumyantsevo, *Mos.* 11 F8 55 38N 37 25 E
Rungis, *Paris* 5 C4 48 44N 2 20 E
Runnemede, *Phil.* 24 B4 39 50N 75 4W
Ruotsinkylä, *Hels.* 3 A4 60 21N 24 57 E
Rusāfa, *Bagd.* 17 E8 33 21N 44 23 E
Rush Green, *Lon.* 4 B6 51 33N 0 10 E
Russa, *Calc.* 16 F6 22 29N 88 21 E
Russell Lea, *Syd.* 19 B4 33 52 S 151 10 E
Rustad, *Oslo* 2 B2 59 56N 10 24 E
Rustenfeld, *Wien* 10 H10 48 7N 16 23 E
Rusville, *Jobg.* 18 E10 26 9 S 28 18 E
Rutherford, *N.Y.* 22 C4 40 49N 74 6W
Rüttenscheid, *Ruhr* 6 B3 51 25N 6 58 E
Ruxton, *Balt.* 25 A3 39 23N 76 38W
Ruzyně, *Pra.* 10 B1 50 5N 14 17 E
Rybatskaya, *St-Pet.* 11 B4 59 59N 30 29 E
Rybatskoye, *St-Pet.* 11 B4 59 50N 30 29 E
Rydalmere, *Syd.* 19 A3 33 48 S 151 2 E
Rydboholm, *Stock.* 3 D12 59 26N 18 12 E
Ryde, *Syd.* 19 A3 33 48 S 151 6 E
Rye, *N.Y.* 23 B6 40 58N 73 40W
Rynfield, *Jobg.* 18 E10 26 9 S 28 19 E
Ryogoku, *Tōkyō* 13 B3 35 40N 139 48 E
Rysäkari, *Hels.* 3 C4 60 6N 24 50 E
Rzhevka, *St-Pet.* 11 B5 59 59N 30 31 E

Serangoon Garden 45

S

Saadōn, *Bagd.* 17 F8 33 19N 44 25 E
Saarn, *Ruhr* 6 B3 51 24N 6 51 E
Saavedra, *B.A.* 32 B4 34 33 S 58 29W
Saboli, *Delhi* 16 A2 28 42N 77 18 E
Sabugo, *Lisb.* 8 F7 38 49N 9 17W
Saburovo, *Mos.* 11 D7 55 53N 37 15 E
Säbysjön, *Stock.* 3 D10 59 26N 17 52 E
Sabzi Mandi, *Delhi* 16 A2 28 40N 77 12 E
Sacavém, *Lisb.* 8 F8 38 47N 9 5W
Saclay, *Paris* 5 C3 48 43N 2 10 E
Saclay, Étang de, *Paris* 5 C2 48 44N 2 9 E
Sacoma, *S. Pau.* 31 E6 23 36 S 46 35W
Sacré-Coeur, *Paris* 5 B4 48 53N 2 20 E
Sacrow, *Berl.* 7 B1 52 25N 13 6 E
Sacrower See, *Berl.* 7 B1 52 26N 13 6 E
Sadang, *Sŏul* 12 H7 37 29N 126 58 E
Sadar Bazar, *Delhi* 16 B2 28 39N 77 11 E
Saddām City, *Bagd.* 17 E8 33 23N 44 27 E
Saddle Brook, *N.Y.* 22 B4 40 53N 74 5W
Saddle River, *N.Y.* 22 A4 41 1N 74 6W
Saddle Rock, *N.Y.* 23 C6 40 47N 73 45W
Sadr, *Kar.* 17 G11 24 51N 67 2 E
Sadyba, *Wsaw.* 10 E7 52 11N 21 3 E
Saedo, *Tōkyō* 13 C2 35 30N 139 33 E
Saensaep, Khlong →, *Bangk.* 15 B2 13 44N 100 32 E
Sáenz Pena, *B.A.* 32 B3 34 37 S 58 32W
Safdar Jang Airport, *Delhi* 16 B2 28 35N 77 12 E
Safdar Jangs Tomb, *Delhi* 16 B2 28 35N 77 12 E
Safraköy, *Ist.* 17 A1 41 0N 28 48 E
Saft el Laban, *El Qâ.* 18 C5 30 1N 31 10 E
Sag Bridge, *Chic.* 26 C1 41 41N 87 55W
Sagamore Neck, *N.Y.* 23 B8 40 53N 73 29W
Saganashkee Slough, *Chic.* 26 C1 41 41N 87 53W
Sagene, *Oslo* 2 B4 59 55N 10 46 E
Sagrada Familia, Temple de, *Barc.* 8 D6 41 24N 2 10 E
Sahapur, *Calc.* 16 E5 22 31N 88 11 E
Sahibabad, *Delhi* 16 A1 28 45N 77 4 E
Sai Kung, *H.K.* 12 D6 22 22N 114 16 E
Sai Wan Ho, *H.K.* 12 E6 22 17N 114 12 E
Sai Ying Pun, *H.K.* 12 E5 22 17N 114 8 E
Saido, *Tōkyō* 13 A2 35 52N 139 39 E
Sailmouille →, *Paris* 5 D3 48 37N 2 17 E
St. Albans, *N.Y.* 23 C6 40 42N 73 44W
St. Andrä, *Wien* 10 G9 48 19N 16 12 E
St. Andrews, *Jobg.* 18 E9 26 9 S 28 7 E
St. Aubin, *Paris* 5 C2 48 44N 2 8 E
St. Augustin, *Mtrl.* 20 A2 43 37N 73 58W
St. Basil's Cathedral, *Mos.* 11 E9 55 45N 37 38 E
St.-Benoit, *Paris* 5 C1 48 40N 1 54 E
St.-Brice-sous-Forêt, *Paris* 5 A4 49 0N 2 21 E
St.-Cloud, *Paris* 5 B3 48 50N 2 12 E
St.-Cyr-l'École, *Paris* 5 C2 48 47N 2 4 E
St.-Cyr-l'École, Aérodrome de, *Paris* 5 C2 48 48N 2 4 E
St. Davids, *Phil.* 24 A2 40 2N 75 23W
St.-Denis, *Paris* 5 B4 48 56N 2 20 E
St. Eustache, *Mtrl.* 20 A2 43 33N 73 54W
St.-Forget, *Paris* 5 C2 48 42N 2 0 E
St. Georg, *Hbg.* 7 D8 53 33N 10 1 E
St.-Germain, Forêt de, *Paris* 5 B2 48 57N 2 5 E
St. Germain-en-Laye, *Paris* 5 B2 48 53N 2 4 E
St.-Germain-lès-Corbeil, *Paris* 5 D4 48 37N 2 29 E
St.-Gratien, *Paris* 5 B3 48 58N 2 17 E
St. Helier, *Lon.* 4 C3 51 23N 0 11W
St.-Hubert, *Mtrl.* 20 B5 43 29N 73 25W
St. Isaac's Cathedral, *St-Pet.* 11 B3 59 55N 30 19 E
St. Jacques →, *Mtrl.* 20 B5 43 26N 73 29W
St.-Jean-de-Beauregard, *Paris* 5 D3 48 39N 2 10 E
St.-Jean-de-dieu, *Mtrl.* 20 A4 43 34N 73 31W
St. Joseph-du-Lac, *Mtrl.* 20 A1 43 32N 74 0W
St. Katherine's Dock, *Lon.* 4 B4 51 30N 0 5W
St. Kilda, *Melb.* 19 F6 37 51 S 144 58 E
St. Lambert, *Mtrl.* 20 A5 43 30N 73 29W
St.-Lambert, *Paris* 5 C2 48 43N 2 1 E
St.-Laurent, *Mtrl.* 20 A3 43 30N 73 43W
St. Lawrence, *Mtrl.* 20 A5 43 26N 73 29W
St.-Lazare, Gare, *Paris* 5 B3 48 52N 2 19 E
St.-Léonard, *Mtrl.* 20 A4 43 35N 73 34W
St. Leonards, *Syd.* 19 B4 33 50 S 151 12 E
St. Leu-la-Forêt, *Paris* 5 A3 49 1N 2 14 E
St.-Louis, L., *Mtrl.* 20 B3 43 24N 73 48W
St. Magelungen, *Stock.* 3 E11 59 13N 18 4 E
St.-Mandé, *Paris* 5 B4 48 50N 2 24 E
St.-Mard, *Paris* 5 A6 49 2N 2 41 E
St.-Martin, *Mtrl.* 20 A3 43 33N 73 45W
St.-Martin, Bois, *Paris* 5 C5 48 48N 2 35 E
St. Mary Cray, *Lon.* 4 C5 51 23N 0 7 E
St.-Maur-des-Fossés, *Paris* 5 C4 48 48N 2 29 E
St.-Maurice, *Paris* 5 C4 48 49N 2 24 E
St.-Mesmes, *Paris* 5 B6 48 59N 2 41 E
St. Michaeliskirche, *Hbg.* 7 D7 53 32N 9 59 E
St. Michael's, *Sing.* 15 G8 1 19N 103 51 E
St.-Michel, *Mtrl.* 20 A4 43 34N 73 37W
St.-Michel-sur-Orge, *Paris* 5 D3 48 38N 2 18 E
St. Nikolaus-Kirken, *Pra.* 10 B2 50 5N 14 23 E
St. Nom-la-Bretèche, *Paris* 5 B2 48 51N 2 1 E
St.-Ouen, *Paris* 5 B4 48 56N 2 20 E
St. Ouen-l'Aumône, *Paris* 5 A2 49 2N 2 6 E
St. Pauli, *Hbg.* 7 D7 53 33N 9 57 E
St. Pauls Cathedral, *Lon.* 4 B4 51 30N 0 5W
St. Paul's Cray, *Lon.* 4 C5 51 23N 0 6 E
St. Petersburg, *St-Pet.* 11 B3 59 55N 30 15 E
St.-Pierre, *Mtrl.* 20 B4 43 27N 73 38W
St. Prix, *Paris* 5 A3 49 0N 2 15 E
St.-Quentin, Étang de, *Paris* 5 C2 48 47N 2 0 E
St. Quentin-en-Yvelines, *Paris* 5 C1 48 46N 1 57 E
St.-Rémy-lès-Chevreuse, *Paris* 5 C2 48 42N 2 4 E
St.-Thibault-des-Vignes, *Paris* 5 B6 48 52N 2 41 E
St. Veit, *Wien* 10 G9 48 11N 16 16 E
St.-Vincent-de-Paul, *Mtrl.* 20 A4 43 36N 73 39W
Ste.-Anne-de-Bellevue, *Mtrl.* 20 B2 43 24N 73 55W
Ste.-Catherine, *Mtrl.* 20 B4 43 24N 73 34W
Ste.-Dorothée, *Mtrl.* 20 A3 43 31N 73 48W
Ste.-Gemme, *Paris* 5 B1 48 52N 1 59 E
Ste.-Geneviève, *Mtrl.* 20 B2 43 28N 73 51W
Ste.-Geneviève-des-Bois, *Paris* 5 D3 48 38N 2 19 E
Ste.-Hélène, Î., *Mtrl.* 20 A4 43 31N 73 32W
Ste. Marthe-sur-le-Lac, *Mtrl.* 20 A2 43 31N 73 56W
Ste.-Rose, *Mtrl.* 20 A3 43 37N 73 46W
Ste. Thérèse, *Mtrl.* 20 A3 43 38N 73 49W
Ste. Thérèse-Ouest, *Mtrl.* 20 A2 43 36N 73 50W
Saiwai, *Tōkyō* 13 C3 35 32N 139 41 E
Sakai, *Ōsaka* 12 C3 34 34N 135 27 E
Sakai →, *Tōkyō* 13 D1 35 27N 139 29 E
Sakai Harbour, *Ōsaka* 12 C3 34 36N 135 26 E
Sakanoshita, *Tōkyō* 13 B2 35 48N 139 30 E
Sakra, P., *Sing.* 15 G7 1 15N 103 41 E
Sakuragi, *Tōkyō* 13 D2 35 28N 139 38 E
Salam, *Bagd.* 17 E8 33 20N 44 20 E
Salaryevo, *Mos.* 11 F8 55 37N 37 25 E
Salem, *Bost.* 21 B4 42 30N 70 54W
Salem, *Stock.* 3 E9 59 13N 17 46 E
Salem Harbor, *Bost.* 21 B4 42 30N 70 52W
Salem Maritime Nat. Hist. Site, *Bost.* 21 B4 42 31N 70 52W
Salemstaden, *Stock.* 3 E9 59 13N 17 46 E
Salkhia, *Calc.* 16 E6 22 36N 88 21 E
Salmannsdorf, *Wien* 10 G9 48 14N 16 14 E
Salmdorf, *Mün.* 7 G11 48 7N 11 43 E
Salmedina, *Mdrd.* 8 C3 40 18N 3 35W
Salomea, *Wsaw.* 10 E6 52 11N 20 55 E
Salsette I., *Bomb.* 16 G8 19 2N 72 53 E
Salt Cr. →, *Chic.* 26 C1 41 51N 87 54W
Salt Cr. →, *Melb.* 19 E7 37 45 S 145 4 E
Salt Water L., *Calc.* 16 E6 22 33N 88 26 E
Saltholm, *Købn.* 2 E11 55 38N 12 46 E
Saltsjö-Duvnäs, *Stock.* 3 E12 59 18N 18 12 E
Saltsjöbaden, *Stock.* 3 E12 59 16N 18 18 E
Saltykovka, *Mos.* 11 E11 55 45N 37 54 E
Salvatorkirche, *Ruhr* 6 B2 51 26N 6 45 E
Sam Sen, Khlong →, *Bangk.* 15 B2 13 45N 100 33 E
Samatya, *Ist.* 17 B2 40 59N 28 55 E
Samoueo, *Lisb.* 8 F8 38 43N 8 59W
Sampaloc, *Manila* 15 D3 14 36N 120 59 E
Samphanthawong, *Bangk.* 15 B2 13 44N 100 31 E
Samrong, *Bangk.* 15 C2 13 39N 100 35 E
Samseon, *Sŏul* 12 G8 37 34N 127 0 E
San Agustin, *Lima* 30 G8 12 1 S 77 9W
San Agustin Atlapulco, *Méx.* 29 B4 19 23N 89 57 E
San Andreas Fault, *S.F.* 27 D3 37 27N 122 18W
San Andreas L., *S.F.* 27 C2 37 35N 122 25W
San Andres, *B.A.* 32 B3 34 34 S 58 33W
San Andrés, *Barc.* 8 D6 41 26N 2 11 E
San Andrés Ahuayucan, *Méx.* 29 C3 19 13N 99 7W
San Andrés Atenco, *Méx.* 29 A2 19 32N 99 13W
San Andrés Totoltepec, *Méx.* 29 C2 19 15N 99 10W
San Andrián de Besós, *Barc.* 8 D6 41 25N 2 13 E
San Angel, *Méx.* 29 B2 19 20N 99 11W
San Antonia, *Manila* 15 E3 14 29N 120 53 E
San Antonio de Padua, *B.A.* 32 C2 34 40 S 58 42W
San Augustin Ohtenco, *Méx.* 29 C3 19 12N 99 0W
San Bartolo Ameyalco, *Méx.* 29 C2 19 19N 99 16W
San Bartolomé Coatepec, *Méx.* 29 B2 19 23N 99 18W
San Basilio, *Rome* 9 F10 41 56N 12 35 E
San Bóvio, *Mil.* 9 E6 45 27N 9 18 E
San Bruno, *S.F.* 27 C2 37 36N 122 24W
San Bruno, Pt., *S.F.* 27 C2 37 39N 122 22W
San Bruno Mt., *S.F.* 27 B2 37 41N 122 26W
San Carlos, *S.F.* 27 C3 37 30N 122 16W
San Carlos de la Cabana, Forteresse, *La Hab.* 30 B2 23 8N 82 20W
San Clemente del Llobregat, *Barc.* 8 E4 41 19N 1 59 E
San Cristobal, *Mdrd.* 8 B3 40 25N 3 35W
San Cristobal, Cerro, *Stgo* 30 J11 33 25 S 70 38W
San Cristoforo, *Mil.* 9 E5 45 26N 9 9 E
San Donato Milanese, *Mil.* 9 E6 45 24N 9 16 E
San Felice, *Tori.* 9 B3 45 1N 7 46 E
San Feliu de Llobregat, *Barc.* 8 D5 41 22N 2 2 E
San Fernando, *B.A.* 32 A3 34 26 S 58 32W
San Fernando, *L.A.* 28 A2 34 17N 118 26W
San Fernando Airport, *L.A.* 28 A2 34 17N 118 25W
San Fernando de Henares, *Mdrd.* 8 B3 40 25N 3 31W
San Fernando Valley, *L.A.* 28 A1 34 12N 118 31W
San Francisco, *S.F.* 27 B2 37 46N 122 23W
San Francisco, Univ. of, *S.F.* 27 B2 37 47N 122 27W
San Francisco B., *S.F.* 27 C3 37 39N 122 14W
San Francisco Chimalpa, *Méx.* 29 B1 19 26N 99 20W
San Francisco Culhuacán, *Méx.* 29 C3 19 19N 99 8W
San Francisco de Paula, *La Hab.* 30 B3 23 3N 82 17W
San Francisco Int. Airport, *S.F.* 27 C2 37 37N 122 22W
San Francisco Solano, *B.A.* 32 C5 34 46 S 58 19W
San Francisco State Univ., *S.F.* 27 B2 37 43N 122 28W
San Francisco Tecoxpa, *Méx.* 29 C3 19 12N 99 0W
San Francisco Tlalnepantla, *Méx.* 29 C2 19 12N 99 8W
San Fruttuoso, *Mil.* 9 D6 45 34N 9 14 E
San Gabriel, *L.A.* 28 B4 34 5N 118 5W
San Gabriel →, *L.A.* 28 C4 33 53N 118 6W
San Gabriel Pk., *L.A.* 28 A4 34 14N 118 5W
San Giacomo, *Tori.* 9 A2 45 11N 7 36 E
San Gillio, *Tori.* 9 B2 45 8N 7 32 E
San Giórgio a Crem, *Nápl.* 9 J13 40 50N 14 20 E
San Giovanni a Teduccio, *Nápl.* 9 J12 40 49N 14 18 E
San Giuseppe Vesuviano, *Nápl.* 9 H13 40 50N 14 30 E
San Gregorio Atlapulco, *Méx.* 29 C3 19 15N 99 4W
San Isidro, *B.A.* 32 A3 34 28 S 58 30W
San Isidro, *Lima* 30 G8 12 5 S 77 2W
San Isidro, *Manila* 15 D4 14 38N 121 5 E
San Jerónimo Lídice, *Méx.* 29 C2 19 19N 99 14W
San Jerónimo Miacatlán, *Méx.* 29 C4 19 12N 98 59W
San Jorge, Castelo de, *Lisb.* 8 F8 38 42N 9 8W
San Jose Del Alamo, *La Hab.* 30 B3 23 6N 82 17W
San José Rio Hondo, *Méx.* 29 B2 19 26N 99 14W
San Juan →, *Manila* 15 D4 14 35N 121 0 E
San Juan de Aragón, *Méx.* 29 B3 19 28N 99 4W
San Juan de Aragón, Parque, *Méx.* 29 B3 19 27N 99 4W
San Juan de Lurigancho, *Lima* 30 F8 11 59 S 77 0W
San Juan de Miraflores, *Lima* 30 H9 12 10 S 76 58W
San Juan del Monte, *Manila* 15 D4 14 36N 121 1 E
San Juan Ixtacala, *Méx.* 29 A2 19 31N 99 10W
San Juan Ixtayopan, *Méx.* 29 C4 19 14N 98 59W
San Juan Toltotepec, *Méx.* 29 B2 19 28N 99 15W
San Juan y San Pedro Tezompa, *Méx.* 29 C4 19 12N 98 57W
San Just Desvern, *Barc.* 8 D5 41 22N 2 4 E
San Justo, *B.A.* 32 C3 34 40 S 58 33W
San Leandro, *S.F.* 27 B4 37 43N 122 9W
San Leandro B., *S.F.* 27 B3 37 45N 122 13W
San Leandro Cr. →, *S.F.* 27 B3 37 44N 122 12W
San Lorenzo, *Mil.* 9 D4 45 34N 8 57 E
San Lorenzo, *S.F.* 27 B4 37 41N 122 6W
San Lorenzo →, *Méx.* 29 B2 19 28N 99 17W
San Lorenzo I., *Lima* 30 G7 12 6 S 77 12W
San Lorenzo Acopilco, *Méx.* 29 C1 19 19N 99 20 E
San Lorenzo Chimalco, *Méx.* 29 B4 19 24N 89 58 E
San Lorenzo Tezonco, *Méx.* 29 C3 19 19N 99 3W
San Lorenzo Tlacoyucan, *Méx.* 29 C3 19 10N 99 2W
San Lucas Xochimanca, *Méx.* 29 C3 19 15N 99 6W
San Luis, *Lima* 30 G8 12 4 S 77 0W
San Luis Tlaxialtemalco, *Méx.* 29 C3 19 16N 99 2W
San Marino, *L.A.* 28 B4 34 7N 118 5W
San Martin, *Barc.* 8 D6 41 24N 2 11 E
San Martin de Porras, *Lima* 30 G8 12 1 S 77 5W
San Martino, *Tori.* 9 B3 45 6N 7 47 E
San Mateo, *S.F.* 27 C3 37 33N 122 19W
San Mateo Cr. →, *S.F.* 27 C2 37 31N 122 22W
San Mateo Tecoloapan, *Méx.* 29 A2 19 35N 99 14W
San Mateo Xalpa, *Méx.* 29 C3 19 13N 99 8W
San Máuro Torinese, *Tori.* 9 B3 45 6N 7 45 E
San Miguel, *B.A.* 32 B2 34 32 S 58 43W
San Miguel, *Lima* 30 G8 12 5 S 77 6W
San Miguel, *Manila* 15 D3 14 36N 120 59 E
San Miguel, *Stgo* 30 J11 33 29 S 70 39W
San Miguel Ajusco, *Méx.* 29 C2 19 13N 99 11W
San Miguel Xicalco, *Méx.* 29 C3 19 13N 99 9W
San Nicholas, *Manila* 15 D3 14 36N 120 57 E
San Nicola, *Rome* 9 F9 41 58N 12 21 E
San Nicolás Totolapan, *Méx.* 29 C2 19 16N 99 16W
San Nicolás Viejo, *Méx.* 29 A1 19 31N 99 21W
San Onófrio, *Rome* 9 F9 41 57N 12 25 E
San Pablo, *Méx.* 29 B4 19 25N 89 57 E
San Pablo, *S.F.* 27 A2 37 57N 122 20W
San Pablo, Pt., *S.F.* 27 A2 37 57N 122 25W
San Pablo Cr. →, *S.F.* 27 A2 37 58N 122 22W
San Pablo Ostotepec, *Méx.* 29 C3 19 11N 99 5W
San Pablo Res., *S.F.* 27 A3 37 55N 122 14W
San Pablo Ridge, *S.F.* 27 A3 37 55N 122 15W
San Pablo Str., *S.F.* 27 A2 37 58N 122 25W
San Pancrázio, *Tori.* 9 B2 45 6N 7 32 E
San Pedro, *Méx.* 29 B4 19 24N 89 56 E
San Pedro, Pt., *S.F.* 27 C1 37 35N 122 31W
San Pedro Actopan, *Méx.* 29 C3 19 12N 99 4W
San Pedro Martir, *Barc.* 8 D5 41 23N 2 6 E
San Pedro Martir, *Méx.* 29 C2 19 16N 99 10W
San Pedro Zacatenco, *Méx.* 29 A3 19 30N 99 6W
San Pietro, *Rome* 9 F9 41 53N 12 27 E
San Pietro, *Tori.* 9 B3 45 1N 7 45 E
San Pietro a Patierno, *Nápl.* 9 H12 40 53N 14 17 E
San Pietro all'Olmo, *Mil.* 9 E5 45 29N 9 0 E
San Po Kong, *H.K.* 12 D6 22 20N 114 11 E
San Quentin, *S.F.* 27 A2 37 56N 122 27W
San Rafael, *S.F.* 27 A1 37 58N 122 30W
San Rafael B., *S.F.* 27 A1 37 57N 122 28W
San Rafael Chamapa, *Méx.* 29 B2 19 27N 99 15W
San Rafael Hills, *L.A.* 28 A3 34 10N 118 12W
San Roque, *Manila* 15 D4 14 37N 121 5 E
San Salvador Cuauhtenco, *Méx.* 29 C3 19 11N 99 8W
San Salvador de la Punta, Forteresse, *La Hab.* 30 B2 23 8N 82 21W
San Sebastiano al Vesúvio, *Nápl.* 9 H13 40 50N 14 22 E
San Siro, *Mil.* 9 E5 45 28N 9 7 E
San Souci, *Syd.* 19 B3 33 59 S 151 8 E
San Telmo, *B.A.* 32 B4 34 37 S 58 23W
San Vicenc dels Horts, *Barc.* 8 D5 41 23N 2 0 E
San Vitaliano, *Nápl.* 9 H13 40 55N 14 28 E
San Vito, *Mil.* 9 E5 45 24N 9 0 E
San Vito, *Nápl.* 9 J13 40 49N 14 22 E
San Vito, *Tori.* 9 B3 45 2N 7 41 E
Sandbakken, *Oslo* 2 C5 59 49N 10 54 E
Sandermosen, *Oslo* 2 A4 60 0N 10 48 E
Sanderstead, *Lon.* 4 D4 51 19N 0 4W
Sandheide, *Ruhr* 6 C3 51 12N 6 56 E
Sandhurst, *Jobg.* 18 E9 26 6 S 28 3 E
Sandown, *Jobg.* 18 E9 26 5 S 28 4 E
Sandown Racecourse, *Lon.* 4 C2 51 22N 0 21W
Sandringham, *Jobg.* 18 E9 26 8 S 28 6 E
Sands Point, *N.Y.* 23 B6 40 50N 73 43W
Sandton, *Jobg.* 18 E9 26 5 S 28 4 E
Sandungen, *Oslo* 2 B2 59 52N 10 21 E
Sandvika, *Oslo* 2 B2 59 53N 10 32 E
Sandy Pond, *Bost.* 21 C2 42 26N 71 18W
Sånga, *Stock.* 3 D9 59 21N 17 42 E
Sangano, *Tori.* 9 B1 45 1N 7 26 E
Sangenjaya, *Tōkyō* 13 C2 35 37N 139 39 E
Sangley Pt., *Manila* 15 E3 14 29N 120 54 E
Sangone →, *Tori.* 9 B2 45 1N 7 32 E
Sangye, *Sŏul* 12 G8 37 38N 127 3 E
Sankrail, *Calc.* 16 E5 22 33N 88 13 E
Sanlihe, *Beij.* 14 B2 39 53N 116 18 E
Sanlintang, *Shang.* 14 K11 31 9N 121 29 E
Sannois, *Paris* 5 B3 48 58N 2 15 E
Sanpada, *Bomb.* 16 G9 19 3N 73 0 E
Sans, *Barc.* 8 D5 41 22N 2 7 E
Sant Ambrogio, Basilica di, *Mil.* 9 E6 45 27N 9 10 E
Sant Boi de Llobregat, *Barc.* 8 D5 41 20N 2 2 E
Sant Cugat, *Barc.* 8 D5 41 28N 2 5 E
Santa Ana, *Manila* 15 D4 14 34N 121 0 E
Santa Ana Tlacotenco, *Méx.* 29 C4 19 11N 98 58W
Santa Bárbara, Morro de, *Rio J.* 31 B1 22 56 S 43 26W
Santa Catalina, *B.A.* 32 C4 34 47 S 58 24W
Santa Cecilia Tepetlapa, *Méx.* 29 C3 19 13N 99 5W
Santa Clara, *Méx.* 29 A3 19 33N 99 3W
Santa Coloma de Cervelló, *Barc.* 8 D5 41 21N 2 0 E
Santa Coloma de Gramanet, *Barc.* 8 D6 41 27N 2 12 E
Santa Cruz, *Bomb.* 16 G7 19 4N 72 51 E
Santa Cruz →, *La Hab.* 30 B2 23 4N 82 29W
Santa Cruz, Ilhe de, *Rio J.* 31 B3 22 51 S 43 7W
Santa Cruz Alcapixca, *Méx.* 29 C3 19 14N 99 4W
Santa Cruz Ayotusco, *Méx.* 29 B1 19 22N 99 21W
Santa Cruz de Olorde, *Barc.* 8 D5 41 25N 2 3 E
Santa Cruz Int. Airport, *Bomb.* 16 G8 19 5N 72 51 E
Santa Cruz Meyehualco, *Méx.* 29 B3 19 20N 99 2W
Santa Elena, *Manila* 15 D4 14 38N 121 5 E
Santa Eligênia Consolação, *S. Pau.* 31 E6 23 32 S 46 38W
Santa Emilia, *Stgo* 30 J11 33 23 S 70 39W
Santa Eulalia, *Barc.* 8 D6 41 25N 2 10 E
Santa Fe, *La Hab.* 30 B2 23 4N 82 30W
Santa Fe Flood Control Basin, *L.A.* 28 B5 34 7N 117 57W
Santa Fe Springs, *L.A.* 28 C5 33 56N 118 3W
Santa Isabel Ixtapan, *Méx.* 29 A4 19 35N 89 57W
Santa Julia, *Stgo* 30 K11 33 30 S 70 35W
Santa Lucia, *Nápl.* 9 J12 40 49N 14 15 E
Santa Margherita, *Tori.* 9 B3 45 3N 7 43 E
Santa Maria Aztahuacán, *Méx.* 29 B3 19 21N 99 2W
Santa Maria del Rosario, *La Hab.* 30 B3 23 3N 82 15W
Santa Maria Tulpetlac, *Méx.* 29 A3 19 34N 99 3W
Santa Martha Acatitla, *Méx.* 29 B3 19 21N 99 2W
Santa Monica, *Car.* 30 E5 10 28N 66 53W
Santa Monica, *L.A.* 28 B2 34 1N 118 29W
Santa Monica B., *L.A.* 28 C1 33 58N 118 30W
Santa Monica Mt., *L.A.* 28 B2 34 6N 118 29W
Santa Rosa, *Lima* 30 F8 11 59 S 77 5W
Santa Rosa De Locobe, *Stgo* 30 J11 33 25 S 70 33W
Santa Rosa Xochiac, *Méx.* 29 C2 19 19N 99 17W
Santa Tereza, *S. Pau.* 31 F6 23 40 S 46 33W
Santa Ursula Xitla, *Méx.* 29 C2 19 16N 99 11W
Santa Ynez Canyon →, *L.A.* 28 B1 34 2N 118 33W
Santahamina, *Hels.* 3 C5 60 8N 25 2 E
Santana, *S. Pau.* 31 D6 23 29 S 46 36W
Sant'Anastasia, *Nápl.* 9 H13 40 51N 14 24 E
Sant'Ántimo, *Nápl.* 9 H12 40 56N 14 14 E
Santeny, *Paris* 5 C5 48 43N 2 34 E
Santiago, *Stgo* 30 J11 33 26 S 70 40W
Santiago Acahualtepec, *Méx.* 29 B3 19 20N 99 0W
Santiago de Las Vegas, *La Hab.* 30 C2 22 58N 82 22W
Santiago Tepalcatlalpan, *Méx.* 29 C3 19 14N 99 8W
Santiago Tepatlaxco, *Méx.* 29 B1 19 28N 99 20W
Sant'Ilário, *Mil.* 9 D4 45 34N 8 59 E
Santo Amaro, *Lisb.* 8 F7 38 42N 9 11W
Santo Amaro, *S. Pau.* 31 E5 23 39 S 46 42W
Santo Andre, *Lisb.* 8 G8 38 38N 9 3W
Santo André, *S. Pau.* 31 E6 23 39 S 46 41W
Santo António, Qta. de, *Lisb.* 8 G7 38 39N 9 15W
Santo António da Charneca, *Lisb.* 8 G8 38 37N 9 1W
Santo Niño, *Manila* 15 D4 14 38N 121 5 E
Santo Rosario, *Manila* 15 D4 14 33N 121 4 E
Santo Thomas, Univ. of, *Manila* 15 D3 14 36N 120 59 E
Santo Tomas, *Manila* 15 D4 14 33N 121 4 E
Santolan, *Manila* 15 D4 14 36N 121 4 E
Santos Dumont, Aéroport, *Rio J.* 31 B3 22 54 S 43 9W
Santos Lugares, *B.A.* 32 B3 34 35 S 58 35W
Santoshpur, *Calc.* 16 E5 22 31N 88 16 E
Santragachi, *Calc.* 16 E5 22 35N 88 17 E
Sanyuanli, *Gzh.* 14 G8 23 8N 113 14 E
São Bernardo do Campo, *S. Pau.* 31 F6 23 42 S 46 32W
São Caetano do Sul, *S. Pau.* 31 E6 23 37 S 46 34W
São Cristovão, *Rio J.* 31 B2 22 53 S 43 13W
São Domingos, Centro, *Rio J.* 31 B2 22 53 S 43 6W
São Gonçalo, *Rio J.* 31 A3 22 49 S 43 4W
São João Clímaco, *S. Pau.* 31 E6 23 37 S 46 35W
São João da Talha, *Lisb.* 8 F8 38 49N 9 5W
São João de Meriti, *Rio J.* 31 A1 22 47 S 43 18W
São Lucas, Parque, *S. Pau.* 31 E6 23 35 S 45 32W
São Mateus, *Rio J.* 31 A1 22 48 S 43 22W
São Miguel Paulista, *S. Pau.* 31 D7 23 29 S 46 26W
São Paulo, *S. Pau.* 31 E6 23 32 S 46 37W
Sapa, *Calc.* 16 E5 22 30N 88 15 E
Sapang Baho →, *Manila* 15 D4 14 33N 121 4 E
Sapateiro →, *S. Pau.* 31 E5 23 35 S 46 41W
Sapanap, *S.F.* 27 A4 37 57N 122 4W
Sarandi, *B.A.* 32 C4 34 40 S 58 20W
Saraswati →, *Calc.* 16 E5 22 46N 88 15 E
Sarcelles, *Paris* 5 B4 48 59N 2 22 E
Sarecky →, *Pra.* 10 B3 50 7N 14 23 E
Sarenga, *Calc.* 16 E5 22 31N 88 12 E
Sarilhos Grandes, *Lisb.* 8 F9 38 40N 8 58W
Sarilhos Pequenos, *Lisb.* 8 F9 38 40N 8 58W
Sarimbun, *Sing.* 15 F7 1 24N 103 40 E
Saronikòs Kòlpos, *Ath.* 8 J10 37 52N 23 38 E
Sarriá, *Barc.* 8 D5 41 22N 2 8 E
Sarria, *Car.* 30 E5 10 30N 66 53W
Sarsol, *Bomb.* 16 G9 19 2N 73 0 E
Sartrouville, *Paris* 5 B3 48 56N 2 10 E
Sasad, *Bud.* 10 K13 47 28N 19 0 E
Sasashita, *Tōkyō* 13 D2 35 23N 139 35 E
Sasel, *Hbg.* 7 D8 53 39N 10 6 E
Sashalom, *Bud.* 10 J14 47 30N 19 10 E
Saska, *Wsaw.* 10 E7 52 14N 21 3 E
Sassafras, *Melb.* 19 F9 37 52 S 145 20 E
Satalice, *Pra.* 10 B3 50 7N 14 34 E
Satgachi, *Calc.* 16 E6 22 37N 88 25 E
Satghara, *Calc.* 16 E6 22 43N 88 21 E
Satpukur, *Calc.* 16 E6 22 37N 88 24 E
Sätra, *Stock.* 3 E10 59 17N 17 54 E
Satsuna, *Calc.* 16 F5 22 28N 88 17 E
Sau Mau Ping, *H.K.* 12 E6 22 19N 114 13 E
Saugus, *Bost.* 21 C3 42 28N 71 0W
Saugus →, *Bost.* 21 C3 42 27N 70 58W
Saulx-les-Chartreux, *Paris* 5 C3 48 41N 2 16 E
Sausalito, *S.F.* 27 A2 37 51N 122 28W
Sausset →, *Paris* 5 B5 48 56N 2 28 E
Savigny-sur-Orge, *Paris* 5 C4 48 40N 2 21 E
Savijärvi, *Hels.* 3 A6 60 21N 25 19 E
Savonera, *Tori.* 9 B2 45 7N 7 36 E
Sawah Besar, *Jak.* 15 H9 6 8 S 106 49 E
Sawyer Ridge, *S.F.* 27 C2 37 34N 122 24W
Saxonville, *Bost.* 21 D1 42 19N 71 24W
Saxonwold, *Jobg.* 18 E9 26 9 S 28 2 E
Scarborough, *Trto.* 20 D9 43 44N 79 14W
Scarsdale, *N.Y.* 23 B6 40 58N 73 47W
Sceaux, *Paris* 5 C3 48 46N 2 17 E
Schalke, *Ruhr* 6 A4 51 33N 7 4 E
Schapenrust, *Jobg.* 18 F11 26 15 S 28 21 E
Scharfenberg, *Berl.* 7 A2 52 35N 13 15 E
Scheiblingstein, *Wien* 10 G9 48 16N 16 13 E
Schenefeld, *Hbg.* 7 D7 53 36N 9 52 E
Scherlebech, *Ruhr* 6 A4 51 37N 7 8 E
Schildow, *Berl.* 7 A3 52 38N 13 22 E
Schiller Park, *Chic.* 26 B1 41 56N 87 52W
Schiller Woods, *Chic.* 26 B1 41 57N 87 51W
Schlachtensee, *Berl.* 7 B2 52 26N 13 13 E
Schlossgarten, *Berl.* 7 A2 52 31N 13 18 E
Schmachtendorf, *Ruhr* 6 A2 51 32N 6 48 E
Schmargendorf, *Berl.* 7 B2 52 28N 13 17 E
Schmöckwitz, *Berl.* 7 B5 52 22N 13 38 E
Schnelsen, *Hbg.* 7 D7 53 38N 9 54 E
Scholven, *Ruhr* 6 A4 51 36N 7 0 E
Schönblick, *Berl.* 7 B5 52 27N 13 43 E
Schönbrunn, Schloss, *Wien* 10 G9 48 10N 16 19 E
Schöneberg, *Berl.* 7 B3 52 28N 13 20 E
Schönefeld, *Berl.* 7 B4 52 23N 13 30 E
Schöneiche, *Berl.* 7 B5 52 28N 13 41 E
Schönwalde, *Berl.* 7 A1 52 37N 13 7 E
Schottenwald, *Wien* 10 G9 48 13N 16 16 E
Schuir, *Ruhr* 6 B3 51 23N 6 59 E
Schulzendorf, *Berl.* 7 A2 52 36N 13 15 E
Schuylkill →, *Phil.* 24 B3 39 53N 75 11W
Schwabing, *Mün.* 7 G10 48 10N 11 35 E
Schwafheim, *Ruhr* 6 B1 51 25N 6 36 E
Schwanebeck, *Berl.* 7 A4 52 37N 13 32 E
Schwanenwerder, *Berl.* 7 B2 52 26N 13 10 E
Schwarz →, *Ruhr* 6 C3 51 19N 6 44 E
Schwarzbachtal, *Ruhr* 6 C3 51 17N 6 51 E
Schwarze, *Ruhr* 6 A2 51 31N 6 48 E
Schwarze Berge, *Hbg.* 7 E7 53 27N 9 54 E
Schwarzlackenau, *Wien* 10 G10 48 16N 16 23 E
Schwechat, *Wien* 10 H10 48 8N 16 28 E
Schwefinghaüsen, *Ruhr* 6 C6 51 15N 7 24 E
Schwelm, *Ruhr* 6 C5 51 16N 7 16 E
Scisciano, *Nápl.* 9 H13 40 54N 14 28 E
Scoresby, *Melb.* 19 F8 37 54 S 145 14 E
Scotch Plains, *N.Y.* 22 D2 40 39N 74 22W
Scotts Level Br. →, *Balt.* 25 A2 39 23N 76 45W
Sea Cliff, *N.Y.* 23 B7 40 50N 73 38W
Seabrook, *Wash.* 25 D8 38 58N 76 49W
Seacliff, *S.F.* 27 B2 37 47N 122 28W
Seaforth, *Syd.* 19 A4 33 48 S 151 15 E
Seagate, *N.Y.* 22 D4 40 34N 74 0W
Seal Slough, *S.F.* 27 C3 37 34N 122 17W
Sears Tower, *Chic.* 26 B3 41 52N 87 38W
Seat Pleasant, *Wash.* 25 D8 38 53N 76 53W
Seavey Hill, *Bost.* 21 A1 42 42N 71 23W
Šeberov, *Pra.* 10 B3 50 0N 14 30 E
Secaucus, *N.Y.* 22 C4 40 47N 74 3W
Secondigliano, *Nápl.* 9 H12 40 53N 14 15 E
Seddinsee, *Berl.* 7 B5 52 23N 13 41 E
Sedgefield, *N.Y.* 22 B2 40 51N 74 26W
Sedriano, *Mil.* 9 E4 45 29N 8 58 E
Seeberg, *Berl.* 7 A5 52 33N 13 41 E
Seeburg, *Berl.* 7 A1 52 30N 13 7 E
Seefeld, *Berl.* 7 A5 52 37N 13 40 E
Segefeld, *Berl.* 7 A1 52 33N 13 5 E
Seehof, *Berl.* 7 B2 52 24N 13 17 E
Segeltorp, *Stock.* 3 E10 59 16N 17 56 E
Segrate, *Mil.* 9 E6 45 29N 9 17 E
Seguro, *Mil.* 9 E5 45 29N 9 3 E
Seine →, *Paris* 5 C4 48 48N 2 25 E
Seixal, *Lisb.* 8 G8 38 38N 9 5W
Selbeck, *Ruhr* 6 B3 51 22N 6 51 E
Selbecke, *Ruhr* 6 C6 51 20N 7 28 E
Selby, *Jobg.* 18 F9 26 12 S 28 2 E
Seletar, P., *Sing.* 15 F8 1 26N 103 51 E
Seletar, Sungei →, *Sing.* 15 F8 1 25N 103 51 E
Seletar Hills, *Sing.* 15 F8 1 23N 103 52 E
Seletar Res., *Sing.* 15 F7 1 24N 103 48 E
Selghar, *Bomb.* 16 H9 18 57N 73 1 E
Selhurst, *Lon.* 4 C4 51 23N 0 5W
Selsdon, *Lon.* 4 C4 51 20N 0 3W
Selytsy, *St-Pet.* 11 B6 59 56N 30 52 E
Sembawang, *Sing.* 15 F7 1 26N 103 49 E
Sembawang, Sungei →, *Sing.* 15 F7 1 26N 103 48 E
Sembawang Hill, *Sing.* 15 F7 1 22N 103 49 E
Semsvatn, *Oslo* 2 B2 59 51N 10 25 E
Senago, *Mil.* 9 D5 45 34N 9 7 E
Senan, *Jak.* 15 J10 6 10 S 106 50 E
Sénart, Forêt de, *Paris* 5 D4 48 40N 2 28 E
Senayan Sports Centre, *Jak.* 15 J9 6 12 S 106 47 E
Sendling, *Mün.* 7 G10 48 7N 11 31 E
Sengeløse, *Købn.* 2 D8 55 40N 12 14 E
Senju, *Tōkyō* 13 B3 35 44N 139 48 E
Senlikköy, *Ist.* 17 B1 40 58N 28 47 E
Senlisse, *Paris* 5 C1 48 41N 1 59 E
Senneville, *Mtrl.* 20 B2 43 24N 73 57W
Senri, *Ōsaka* 12 B4 34 49N 135 30 E
Senriyama, *Ōsaka* 12 B4 34 47N 135 30 E
Sentosa, *Sing.* 15 G7 1 15N 103 49 E
Sentosa, P., *Sing.* 15 G7 1 15N 103 49 E
Seo Dae Mun, *Sŏul* 12 G7 37 34N 126 55 E
Seobinngo, *Sŏul* 12 G7 37 31N 126 58 E
Seoggwan, *Sŏul* 12 G8 37 35N 127 2 E
Seong Bug, *Sŏul* 12 G8 37 35N 127 0 E
Seong Dong, *Sŏul* 12 G8 37 33N 127 3 E
Seongsu, *Sŏul* 12 G8 37 32N 127 2 E
Seoul = Sŏul, *Sŏul* 12 G8 37 34N 127 51 E
Seoul National Univ., *Sŏul* 12 H7 37 28N 126 57 E
Seoul Tower, *Sŏul* 12 G7 37 32N 126 59 E
Sepah Salar Mosque, *Tehr.* 17 C5 35 40N 51 25 E
Sepolia, *Ath.* 8 H11 38 1N 23 42 E
Sepulveda, *L.A.* 28 A2 34 13N 118 27W
Sepulveda Flood Control Basin, *L.A.* 28 A2 34 10N 118 28W
Serangoon, P., *Sing.* 15 F8 1 23N 103 55 E
Serangoon, Sungei →, *Sing.* 15 F8 1 23N 103 55 E
Serangoon Garden, *Sing.* 15 F8 1 21N 103 51 E

46 Serangoon Harbour

Serangoon Harbour, *Sing.* 15 F8 1 23N 103 57 E
Seraya, P., *Sing.* 15 G7 1 16N 103 43 E
Serebryanka, *Mos.* 11 E11 55 44N 37 53 E
Serebryanka →, *Mos.* 11 E10 55 47N 37 44 E
Serednevo, *Mos.* 11 F7 55 35N 37 18 E
Serramonte, *S.F.* 27 C2 37 39N 122 28W
Servon, *Paris* 5 C5 48 43N 2 35 E
Šestajovice, *Pra.* 10 B3 50 6N 14 40 E
Sesto San Giovanni, *Mil.* 9 D6 45 31N 9 13 E
Seta Budi, *Jak.* 15 J9 6 12 S 106 49 E
Setagaya-Ku, *Tōkyō* 13 C2 35 37N 139 36 E
Sete Pontes, *Rio J.* 31 B3 22 50 S 43 4W
Seter, *Oslo* 2 B4 59 52N 10 47 E
Séttimo Milanese, *Mil.* 9 E5 45 28N 9 3 E
Séttimo Torinese, *Tori.* 9 B3 45 8N 7 46 E
Settsu, *Ōsaka* 12 B4 34 47N 135 33 E
Setuny →, *Mos.* 11 E8 55 43N 37 21 E
Seurasaari, *Hels.* 3 B4 60 11N 24 53 E
Seutula; *Hels.* 3 A4 60 20N 24 52 E
Seven Corners, *Wash.* 25 D7 38 53N 77 9W
Seven Kings, *Lon.* 4 B5 51 33N 0 5 E
Sevenoaks, *Lon.* 4 D6 51 16N 0 11 E
Severn Hills, *Syd.* 19 A2 33 46 S 150 57 E
Sévesco →, *Mil.* 9 D5 45 35N 9 9 E
Sevran, *Paris* 5 B5 48 56N 2 31 E
Sèvres, *Paris* 5 C3 48 49N 2 13 E
Sewaren, *N.Y.* 22 D3 40 33N 74 15W
Sewell, *Phil.* 24 C4 39 46N 75 8W
Sewri, *Bomb.* 16 H8 18 59N 72 50 E
Seya, *Tōkyō* 13 D1 35 28N 139 28 E
Sforzesco, Castello, *Mil.* 9 E6 45 28N 9 10 E
Sha Kok Mei, *H.K.* 12 D6 22 23N 114 15 E
Sha Tin, *H.K.* 12 D6 22 23N 114 11 E
Sha Tin Wai, *H.K.* 12 D6 22 22N 114 11 E
Shaala, *Bagd.* 17 E7 33 22N 44 16 E
Shabanzhuang, *Beij.* 14 B3 39 51N 116 25 E
Shabbona Woods, *Chic.* 26 D3 41 36N 87 33W
Shabrâmant, *El Qâ.* 18 D5 29 56N 31 11 E
Shadipur, *Delhi* 16 B2 28 38N 77 11 E
Shady Oak, *Wash.* 25 C6 39 1N 77 17W
Shahabad, *Bomb.* 16 G9 19 0N 73 2 E
Shahar, *Bomb.* 16 G8 19 5N 72 52 E
Shahdara, *Delhi* 16 A2 28 40N 77 18 E
Shahe, *Gzh.* 14 G8 23 9N 113 19 E
Shahpur Jel, *Delhi* 16 B2 28 33N 77 12 E
Shahr-e-Rey, *Tehr.* 17 D5 35 36N 51 25 E
Shaikh Aomar, *Bagd.* 17 E8 33 20N 44 23 E
Shakarpur Khas, *Delhi* 16 B2 28 37N 77 14 E
Shakurpur, *Delhi* 16 A1 28 40N 77 8 E
Sham Shui Po, *H.K.* 12 E5 22 19N 114 9 E
Shamepur, *Delhi* 16 A1 28 44N 77 8 E
Shamian, *Gzh.* 14 G8 23 6N 113 13 E
Shamspur, *Delhi* 16 B2 28 36N 77 17 E
Shan Liu, *H.K.* 12 D6 22 23N 114 16 E
Shan Mei, *H.K.* 12 D6 22 24N 114 10 E
Shanghai, *Shang.* 14 J12 31 14N 121 28 E
Shanghetou, *Tianj.* 14 D5 39 11N 117 0 E
Shanjing, *Gzh.* 14 G9 23 4N 113 23 E
Sharea Faisal, *Kar.* 17 G11 24 52N 67 8 E
Sharon Hill, *Phil.* 24 B3 39 54N 75 16W
Sharp I., *H.K.* 12 D6 22 21N 114 17 E
Sharp Park, *S.F.* 27 C2 37 38N 122 29W
Shau Kei Wan, *H.K.* 12 E6 22 16N 114 13 E
Shawocun, *Beij.* 14 B2 39 53N 116 13 E
Shawsheen Village, *Bost.* 21 A3 42 40N 71 7W
Shea Stadium, *N.Y.* 23 C5 40 45N 73 50W
Sheakhala, *Calc.* 16 D5 22 45N 88 10 E
Shebāb, *Bagd.* 17 E8 33 20N 44 26 E
Sheepshead B., *N.Y.* 22 D5 40 35N 73 55W
Shek Hang, *H.K.* 12 D6 22 24N 114 17 E
Shek Kip Mei, *H.K.* 12 D5 22 20N 114 9 E
Shek Lung Kung, *H.K.* 12 D5 22 23N 114 5 E
Shek O, *H.K.* 12 E6 22 13N 114 15 E
Shellpot Cr. →, *Phil.* 24 C1 39 44N 75 30W
Shelter Cove, *S.F.* 27 C1 37 35N 122 30W
Shelter I., *H.K.* 12 E6 22 19N 114 17 E
Shemirānāt, *Tehr.* 17 C5 35 47N 51 25 E
Shenfield, *Lon.* 4 B6 51 37N 0 19 E
Sheng Fa Shan, *H.K.* 12 D5 22 23N 114 5 E
Shenley, *Lon.* 4 A3 51 41N 0 16W
Shepherds Bush, *Lon.* 4 B3 51 30N 0 13W
Shepperton, *Lon.* 4 C2 51 23N 0 26W
Sherborn, *Bost.* 21 D1 42 14N 71 22W
Sherman Oaks, *L.A.* 28 B2 34 8N 118 29W
Sherwood Forest, *S.F.* 27 A3 37 57N 122 16W
Shet Bandar, *Bomb.* 16 H8 18 57N 72 55 E
Sheung Lau Wan, *H.K.* 12 E6 22 16N 114 16 E
Sheung Wan, *H.K.* 12 E5 22 16N 114 9 E
Sheva, *Bomb.* 16 H8 18 56N 72 57 E
Sheva Nhava, *Bomb.* 16 H8 18 57N 72 57 E
Shiba, *Tōkyō* 13 C3 35 38N 139 45 E
Shiba →, *Tōkyō* 13 A3 35 50N 139 44 E
Shibuya-Ku, *Tōkyō* 13 C3 35 39N 139 41 E
Shijōnawate, *Ōsaka* 12 B4 34 44N 135 37 E
Shimo-okudomi, *Tōkyō* 13 A1 35 52N 139 27 E
Shimo-tsuchidana, *Tōkyō* 13 D1 35 23N 139 27 E
Shimogawara, *Tōkyō* 13 C1 35 39N 139 27 E
Shimosalo, *Tōkyō* 13 B2 35 45N 139 31 E
Shimosasame, *Tōkyō* 13 B2 35 48N 139 37 E
Shimoshakujii, *Tōkyō* 13 B2 35 43N 139 35 E
Shimotomi, *Tōkyō* 13 B1 35 48N 139 28 E
Shimotsuruma, *Tōkyō* 13 D1 35 29N 139 26 E
Shimura, *Tōkyō* 13 B2 35 46N 139 41 E
Shinagawa B., *Tōkyō* 13 C3 35 37N 139 45 E
Shinagawa-Ku, *Tōkyō* 13 C3 35 36N 139 44 E
Shing Mun Res., *H.K.* 12 D5 22 23N 114 8 E
Shinjuku-Ku, *Tōkyō* 13 B3 35 41N 139 42 E
Shinkoiwa, *Tōkyō* 13 B4 35 43N 139 51 E
Shinnakano, *Tōkyō* 13 B3 35 41N 139 40 E
Shinoha, *Tōkyō* 13 D2 35 26N 139 49 E
Shipai, *Gzh.* 14 G9 23 8N 113 20 E
Shipley, *Balt.* 25 B3 39 12N 76 39W
Shippan Pt., *N.Y.* 23 A7 41 1N 73 31W
Shirako, *Tōkyō* 13 B2 35 47N 139 36 E
Shiraone, *Bomb.* 16 G9 19 2N 73 1 E
Shirinashi →, *Ōsaka* 12 C3 34 38N 135 27 E
Shirley, *Lon.* 4 C4 51 22N 0 2W
Shiro, *Tōkyō* 13 B2 35 48N 139 30 E
Shirogane, *Tōkyō* 13 C3 35 37N 139 44 E
Shisha Hai, *Beij.* 14 B3 39 55N 116 21 E
Shitou, *Gzh.* 14 G9 23 1N 113 23 E
Shiweitang, *Gzh.* 14 G8 23 6N 113 12 E
Shogunle, *Lagos* 18 A2 6 34N 7 20 E
Shomolu, *Lagos* 18 A2 6 32N 7 22 E
Shooters Hill, *Lon.* 4 C5 51 28N 0 4 E
Shoreditch, *Lon.* 4 B4 51 31N 0 4W
Shoreham, *Lon.* 4 D6 51 19N 0 10 E
Short Hills, *N.Y.* 22 C2 40 44N 74 21W
Shortlands, *Lon.* 4 C5 51 23N 0 0 E
Shrirampur, *Calc.* 16 D5 22 45N 88 21 E
Shuangkou, *Tianj.* 14 D5 39 14N 117 2 E
Shuangtuo, *Tianj.* 14 D6 39 13N 117 19 E
Shubrâ el Kheima, *El Qâ.* 18 C5 30 6N 31 14 E
Shuikuo, *Gzh.* 14 F8 23 10N 113 10 E
Shuishang Gongyuan, *Tianj.* 14 E5 39 5N 117 9 E
Shukunoshō, *Ōsaka* 12 A4 34 50N 135 31 E
Sibbo, *Hels.* 3 A6 60 21N 25 14 E
Sibbo fjärden, *Hels.* 3 B6 60 11N 25 17 E
Siboney, *La Hab.* 30 B2 23 4N 82 28W
Sibpur, *Calc.* 16 E5 22 34N 88 19 E
Sibřina, *Pra.* 10 B4 50 3N 14 40 E
Sidcup, *Lon.* 4 C5 51 25N 0 6 E
Siebenhirten, *Wien* 10 H9 48 8N 16 17 E
Siedlung, *Berl.* 7 A1 52 35N 13 7 E
Siekierki, *Wsaw.* 10 E7 52 12N 21 4 E
Sielce, *Wsaw.* 10 E7 52 12N 21 2 E
Siemensstadt, *Berl.* 7 A2 52 32N 13 16 E
Sieraków, *Wsaw.* 10 E5 52 19N 20 48 E
Sierra Madre, *L.A.* 28 B4 34 9N 118 3W
Sievering, *Wien* 10 G10 48 15N 16 20 E
Siggerud, *Oslo* 2 C5 59 47N 10 52 E
Siheung, *Sŏul* 12 H7 37 28N 126 54 E
Siikajärvi, *Hels.* 3 B2 60 17N 24 31 E
Sikátorpuszta, *Bud.* 10 J14 47 34N 19 10 E
Silampur, *Delhi* 16 B2 28 39N 77 16 E
Silschede, *Ruhr* 6 B6 51 21N 7 22 E
Silver Hill, *Wash.* 25 E8 38 49N 76 55W
Silver L., *Bost.* 21 B3 42 33N 71 W
Silver Mt., *L.A.* 28 A5 34 12N 117 55W
Silver Spring, *Wash.* 25 D8 38 59N 77 2W
Silverfields, *Jobg.* 18 E7 26 7 S 27 49 E
Silvertown, *Lon.* 4 C5 51 29N 0 1 E
Simla, *Calc.* 16 E6 22 35N 88 22 E
Simmer and Jack Mines, *Jobg.* 18 F9 26 12 S 28 8 E
Simmering, *Wien* 10 G10 48 10N 16 24 E
Simmering Heide, *Wien* 10 G10 48 10N 16 26 E
Simonkylä, *Hels.* 3 B5 60 18N 25 1 E
Simpang Bedok, *Sing.* 15 G8 1 19N 103 56 E
Simsalö, *Hels.* 3 B6 60 14N 25 17 E
Singao, *N.Y.* 22 B3 40 53N 74 14W
Singapore, *Sing.* 15 G8 1 17N 103 51 E
Singapore →, *Sing.* 15 G8 1 17N 103 51 E
Singapore, Univ. of, *Sing.* 15 G7 1 19N 103 49 E
Singapore Airport, *Sing.* 15 F8 1 21N 103 54 E
Singlewell, *Lon.* 4 C7 51 25N 0 21 E
Singur, *Calc.* 16 D5 22 48N 88 13 E
Sinicka →, *Mos.* 11 D7 55 52N 37 18 E
Sinki, Selat, *Sing.* 15 G7 1 15N 103 42 E
Sinrim, *Sŏul* 12 H7 37 28N 126 56 E
Sinsa, *Sŏul* 12 G8 37 31N 127 0 E
Sinthi, *Calc.* 16 E6 22 37N 88 23 E
Sinweol, *Sŏul* 12 G7 37 31N 126 51 E
Sipoo, *Hels.* 3 A6 60 21N 25 14 E
Sipoon selkä, *Hels.* 3 B6 60 11N 25 17 E
Sipson, *Lon.* 4 C2 51 29N 0 26W
Siqeil, *El Qâ.* 18 C4 30 7N 31 10 E
Şişli, *Ist.* 17 A2 41 3N 28 58 E
Skå, *Stock.* 3 E9 59 19N 17 44 E
Skärholmen, *Stock.* 3 E10 59 16N 17 53 E
Skarpäng, *Stock.* 3 D11 59 26N 18 0 E
Skarpnäck, *Stock.* 3 E11 59 16N 18 7 E
Skarpö, *Stock.* 3 D13 59 24N 18 22 E
Skedsmo, *Oslo* 2 B6 59 59N 11 2 E
Skhodnya →, *Mos.* 11 D8 55 53N 37 23 E
Skodsborg, *Købn.* 2 D10 55 49N 12 34 E
Skogby, *Hels.* 3 A2 60 21N 24 40 E
Skogen, *Oslo* 2 C1 59 48N 10 18 E
Skogsbyn, *Hels.* 3 A6 60 20N 25 18 E
Skokie, *Chic.* 26 A2 42 2N 87 43W
Skokie →, *Chic.* 26 A2 42 7N 87 46W
Skokie Lagoons, *Chic.* 26 A2 42 7N 87 46W
Skoklefall, *Oslo* 2 B4 59 50N 10 40 E
Sköndal, *Stock.* 3 E11 59 15N 18 6 E
Skovlunde, *Købn.* 2 D9 55 42N 12 25 E
Skovshoved, *Købn.* 2 D10 55 45N 12 35 E
Skøyen, *Oslo* 2 B4 59 55N 10 40 E
Skui, *Oslo* 2 B2 59 55N 10 25 E
Skuldelev, *Købn.* 2 D7 55 56N 12 1 E
Skullerud, *Oslo* 2 B5 59 51N 10 50 E
Skuru, *Stock.* 3 E12 59 18N 18 12 E
Skytta, *Oslo* 2 B5 59 59N 10 54 E
Slade Green, *Lon.* 4 C6 51 27N 0 11 E
Slagsta, *Stock.* 3 E9 59 15N 17 48 E
Slakteren, *Oslo* 2 A4 60 1N 10 40 E
Slattum, *Oslo* 2 A5 60 0N 10 55 E
Slemmestad, *Oslo* 2 C2 59 46N 10 29 E
Slependen, *Oslo* 2 B3 59 52N 10 30 E
Sligo Cr. →, *Wash.* 25 C7 39 0N 77 1W
Slipi, *Jak.* 15 J9 6 11 S 106 47 E
Slipi Orchard Garden, *Jak.* 15 J9 6 10 S 106 46 E
Slivenec, *Pra.* 10 B2 50 1N 14 21 E
Slone Canyon Res., *L.A.* 28 B2 34 6N 118 27W
Sloop Channel, *N.Y.* 23 D7 40 36N 73 31W
Sluhy, *Pra.* 10 A3 50 11N 14 33 E
Służew, *Wsaw.* 10 E7 52 10N 21 1 E
Służewiec, *Wsaw.* 10 E7 52 10N 21 0 E
Smalleytown, *N.Y.* 22 D2 40 39N 74 28W
Smestad, *Oslo* 2 B2 59 55N 10 25 E
Smíchov, *Pra.* 10 B2 50 4N 14 23 E
Smith Forest Preserve, *Chic.* 26 B2 41 59N 87 45W
Smith Mills, *N.Y.* 22 A2 41 0N 74 23W
Smithfield, *Syd.* 19 B2 33 51 S 150 56 E
Smoke Rise, *N.Y.* 22 A2 41 0N 74 24W
Smørumnedre, *Købn.* 2 D8 55 44N 12 17 E
Snakeden Br. →, *Wash.* 25 D6 38 58N 77 17W
Snarøya, *Oslo* 2 B3 59 52N 10 33 E
Snättringe, *Stock.* 3 E10 59 15N 17 58 E
Snoldelev, *Købn.* 2 E7 55 33N 12 10 E
Snostrup, *Købn.* 2 D7 55 48N 12 1 E
Søborg, *Købn.* 2 D9 55 44N 12 30 E
Sobreda, *Lisb.* 8 G7 38 39N 9 11W
Soccavo, *Nápl.* 9 H12 40 50N 14 11 E
Sodegaura, *Tōkyō* 13 D4 35 24N 139 57 E
Söderby, *Stock.* 3 D12 59 24N 18 12 E
Söderkullalandet, *Hels.* 3 B6 60 14N 25 19 E
Södermalm, *Stock.* 3 E11 59 18N 18 5 E
Södersätra, *Stock.* 3 D10 59 27N 17 56 E
Södertälje, *Stock.* 3 E8 59 11N 17 38 E
Sodingen, *Ruhr* 6 A5 51 32N 7 15 E
Sodpur, *Calc.* 16 D6 22 42N 88 24 E
Södra Björkfjärden, *Stock.* 3 E8 59 17N 17 34 E
Soeurs, Î. des, *Mtrl.* 20 B4 43 27N 73 32W
Sofienlund, *Oslo* 2 B4 59 55N 10 44 E
Sognsvatn, *Oslo* 2 B4 59 58N 10 43 E
Soignolles-en-Brie, *Paris* 5 D6 48 39N 2 43 E
Soisy-sous-Montmorency, *Paris* 5 B3 48 59N 2 17 E
Soisy-sur-Seine, *Paris* 5 D4 48 39N 2 27 E
Sōji Temple, *Tōkyō* 13 D3 35 29N 139 40 E
Sok Kwu Wan, *H.K.* 12 E5 22 12N 114 7 E
Sōka, *Tōkyō* 13 B3 35 49N 139 48 E
Sokolniki, *Mos.* 11 E10 55 47N 37 40 E
Sokolniki Park, *Mos.* 11 E10 55 47N 37 40 E
Sokołów, *Wsaw.* 10 F6 52 9N 20 51 E
Solalinden, *Mün.* 7 G11 48 5N 11 46 E
Solaro, *Mil.* 9 D5 45 36N 9 6 E
Solers, *Paris* 5 D6 48 39N 2 43 E
Solingen, *Ruhr* 6 C4 51 10N 7 5 E
Sollentuna, *Stock.* 3 D10 59 25N 17 56 E
Søllerød, *Købn.* 2 D9 55 48N 12 29 E
Sollihøgda, *Oslo* 2 B2 59 58N 10 20 E
Solln, *Mün.* 7 G10 48 4N 11 31 E
Solna, *Stock.* 3 D10 59 21N 17 59 E
Solntsevo, *Mos.* 11 F8 55 39N 37 23 E
Solymár, *Bud.* 10 J12 47 35N 18 56 E
Somapah Changi, *Sing.* 15 F8 1 20N 103 57 E
Somapan Serangoon, *Sing.* 15 F8 1 21N 103 53 E
Somborn, *Ruhr* 6 B6 51 29N 7 20 E
Somerdale, *Phil.* 24 B4 39 50N 75 1W
Somerset, *Wash.* 25 D7 38 57N 77 5W
Somerton, *Phil.* 24 A4 40 7N 75 1W
Somerville, *Bost.* 21 C3 42 22N 71 5W
Somma, Mte., *Nápl.* 9 H13 40 50N 14 25 E
Somma Vesuviana, *Nápl.* 9 H13 40 52N 14 26 E
Sonari, *Bomb.* 16 H8 18 54N 72 59 E
Sønderby, *Købn.* 2 D7 55 44N 12 2 E
Søndersø, *Købn.* 2 D9 55 46N 12 21 E
Sondre Elvåga, *Oslo* 2 B5 59 51N 10 54 E
Sonnberg, *Wien* 10 G9 48 19N 16 15 E
Sørby, *Oslo* 2 C4 59 49N 10 41 E
Sørkedalen, *Oslo* 2 A3 60 1N 10 37 E
Soroksár, *Bud.* 10 K13 47 24N 19 7 E
Soroksár-Újtelep, *Bud.* 10 K13 47 25N 19 7 E
Soroksari Duna →, *Bud.* 10 K13 47 25N 19 5 E
Sørsdal, *Oslo* 2 B1 59 50N 10 16 E
Sosenka →, *Mos.* 11 E10 55 46N 37 42 E
Sosnovaya, *St-Pet.* 11 C2 59 49N 30 8 E
Sottungsby, *Hels.* 3 B5 60 16N 25 8 E
Sŏul, *Sŏul* 12 G8 37 34N 127 51 E
Soundview, *N.Y.* 23 C5 40 49N 73 53W
South Basin, *S.F.* 27 B2 37 42N 122 22W
South Beach, *N.Y.* 22 D4 40 35N 74 4W
South Boston, *Bost.* 21 C3 42 20N 71 2W
South Braintree, *Bost.* 21 D4 42 11N 70 59W
South Branch →, *Phil.* 24 C4 39 50N 75 5W
South Brooklyn, *N.Y.* 22 C5 40 41N 73 59W
South Chelmsford, *Bost.* 21 B1 42 34N 71 22W
South Chicago, *Chic.* 26 C3 41 44N 87 32W
South Darenth, *Lon.* 4 C6 51 23N 0 15 E
South Deering, *Chic.* 26 C3 41 42N 87 33W
South Floral Park, *N.Y.* 23 C6 40 42N 73 41W
South Gate, *L.A.* 28 C3 33 56N 118 12W
South Germiston, *Jobg.* 18 F10 26 11 S 28 13 E
South Hackensack, *N.Y.* 22 B4 40 51N 74 2W
South Hamilton, *Bost.* 21 B4 42 36N 70 52W
South Harbour, *Manila* 15 D3 14 34N 120 58 E
South Harrow, *Lon.* 4 B2 51 33N 0 21W
South Hd., *Syd.* 19 B4 33 50 S 151 16 E
South Hempstead, *N.Y.* 23 C7 40 40N 73 37W
South Hills, *Jobg.* 18 F9 26 14 S 28 5 E
South Hingham, *Bost.* 21 D4 42 12N 70 53W
South Holland, *Chic.* 26 D3 41 36N 87 35W
South Hornchurch, *Lon.* 4 B6 51 32N 0 11 E
South Huntington, *N.Y.* 23 C8 40 49N 73 23W
South Lawn, *Wash.* 25 E7 38 47N 77 0W
South Lawrence, *Bost.* 21 A3 42 41N 71 9W
South Lincoln, *Bost.* 21 C2 42 24N 71 19W
South Lynnfield, *Bost.* 21 B4 42 30N 70 59W
South Norwood, *Lon.* 4 C4 51 23N 0 3W
South Ockendon, *Lon.* 4 B6 51 30N 0 16 E
South of Market, *S.F.* 27 B2 37 46N 122 24W
South Orange, *N.Y.* 22 C3 40 45N 74 14W
South Oxley, *Lon.* 4 B2 51 37N 0 23W
South Oyster B., *N.Y.* 23 D8 40 38N 73 27W
South Ozone Park, *N.Y.* 23 C6 40 41N 73 49W
South Pasadena, *L.A.* 28 B4 34 7N 118 8W
South Peabody, *Bost.* 21 B4 42 30N 70 57W
South Peters, *Syd.* 19 B4 33 54 S 151 11 E
South Plainfield, *N.Y.* 22 D2 40 35N 74 24W
South Quincy, *Bost.* 21 D3 42 13N 71 0W
South Res., *Bost.* 21 C3 42 26N 71 6W
South San Francisco, *S.F.* 27 C2 37 38N 122 26W
South San Gabriel, *L.A.* 28 B4 34 3N 118 6W
South Shore, *Chic.* 26 C3 41 45N 87 34W
South Sudbury, *Bost.* 21 C1 42 21N 71 24W
South Valley Stream, *N.Y.* 23 D6 40 38N 73 43W
South Westbury, *N.Y.* 23 C7 40 44N 73 34W
South Weymouth, *Bost.* 21 D4 42 10N 70 56W
South Wimbledon, *Lon.* 4 C3 51 24N 0 11W
South Yarra, *Melb.* 19 F6 37 50 S 144 59 E
Southall, *Lon.* 4 B2 51 30N 0 22W
Southborough, *Lon.* 4 C5 51 23N 0 3 E
Southcrest, *Jobg.* 18 F9 26 15 S 28 5 E
Southend, *Lon.* 4 C4 51 25N 0 0 E
Southfields, *Lon.* 4 C3 51 26N 0 11W
Southgate, *Lon.* 4 B4 51 38N 0 7W
Southwark, *Lon.* 4 C4 51 29N 0 5W
Søvang, *Købn.* 2 E10 55 34N 12 37 E
Soweto, *Jobg.* 18 F8 26 14 S 27 52 E
Soya, *Tōkyō* 13 B4 35 44N 139 55 E
Spadenland, *Hbg.* 7 E8 53 28N 10 3 E
Spandau, *Berl.* 7 A1 52 33N 13 9 E
Spånga, *Stock.* 3 D10 59 23N 17 53 E
Sparkhill, *N.Y.* 22 A5 41 1N 73 55W
Sparrows Point, *Balt.* 25 B4 39 13N 76 29W
Spectacle I., *Bost.* 21 C4 42 19N 70 59W
Speicher-See, *Mün.* 7 F11 48 12N 11 42 E
Speising, *Wien* 10 H9 48 10N 16 17 E
Speldorf, *Ruhr* 6 B2 51 26N 6 49 E
Spellen, *Ruhr* 6 A1 51 36N 6 36 E
Sphinx, *El Qâ.* 18 D4 29 58N 31 8 E
Spinaceto, *Rome* 9 G9 41 47N 12 27 E
Splitrock Res., *N.Y.* 22 B2 40 58N 74 26W
Spořilov, *Pra.* 10 B3 50 2N 14 29 E
Spot Pond, *Bost.* 21 C3 42 26N 71 4W
Spotswood, *Melb.* 19 F6 37 50 S 144 52 E
Spree →, *Berl.* 7 A2 52 32N 13 12 E
Spreehafen, *Hbg.* 7 D7 53 31N 9 59 E
Spring Pond, *Bost.* 21 C4 42 31N 70 55W
Springeberg, *Berl.* 7 B5 52 26N 13 43 E
Springfield, *N.Y.* 22 C3 40 42N 74 19W
Springfield, *Phil.* 24 B3 39 56N 75 19W
Springfield, *Wash.* 25 E6 38 46N 77 10W
Springs, *Jobg.* 18 F11 26 15 S 28 23 E
Sprockhövel, *Ruhr* 6 B5 51 22N 7 14 E
Squantum, *Bost.* 21 D3 42 17N 71 0W
Squirrel's Heath, *Lon.* 4 B6 51 34N 0 12 E
Srednaya Rogatka, *St-Pet.* 11 C4 59 49N 30 22 E
Śródmieście, *Wsaw.* 10 E7 52 13N 21 0 E
Staaken, *Berl.* 7 A1 52 31N 13 8 E
Staatsoper, *Wien* 10 G10 48 12N 16 21 E
Stabekk, *Oslo* 2 B3 59 54N 10 36 E
Stadlau, *Wien* 10 G10 48 13N 16 27 E
Stahnsdorf, *Berl.* 7 B2 52 23N 13 13 E
Staines, *Lon.* 4 C2 51 26N 0 30W
Stains, *Paris* 5 B4 48 57N 2 22 E
Stamford, *N.Y.* 23 A7 41 2N 73 32W
Stamford Harbor, *N.Y.* 23 A7 41 0N 73 34W
Stamford Hill, *Lon.* 4 B4 51 34N 0 4W
Stammersdorf, *Wien* 10 G10 48 18N 16 24 E
Stamford Univ., *S.F.* 27 D4 37 26N 122 10W
Stanley, *H.K.* 12 E6 22 13N 114 12 E
Stanley Mound, *H.K.* 12 E6 22 13N 114 12 E
Stanley Pen., *H.K.* 12 E6 22 12N 114 12 E
Stanmore, *Lon.* 4 B3 51 36N 0 18W
Stansted, *Lon.* 4 C6 51 20N 0 18 E
Stapleford Abbotts, *Lon.* 4 B6 51 37N 0 10 E
Stapleton, *N.Y.* 22 D4 40 36N 74 5W
Stará Boleslav, *Pra.* 10 A4 50 11N 14 39 E
Stara Milosna, *Wsaw.* 10 E8 52 13N 21 13 E
Staraya, *St-Pet.* 11 B5 59 55N 30 37 E
Staraya Derevnya, *St-Pet.* 11 B3 59 58N 30 15 E
Stare, *Wsaw.* 10 E7 52 15N 21 0 E
Stare Babice, *Wsaw.* 10 E6 52 15N 20 49 E
Staré Mesto, *Pra.* 10 B2 50 5N 14 25 E
State House, *Lagos* 18 B2 6 26N 7 24 E
Staten, *N.Y.* 22 D4 40 34N 74 7W
Staten Island Zoo, *N.Y.* 22 D4 40 38N 74 6W
Statenice, *Pra.* 10 B1 50 9N 14 19 E
Stavnsholt, *Købn.* 2 D9 55 48N 12 24 E
Steele, *Ruhr* 6 B4 51 27N 7 4 E
Steele Creek, *Melb.* 19 E6 37 44 S 144 52 E
Steglitz, *Berl.* 7 B2 52 27N 13 19 E
Stehstücken, *Berl.* 7 B1 52 23N 13 7 E
Steilshoop, *Hbg.* 7 D8 53 36N 10 2 E
Steinberger Slough, *S.F.* 27 C3 37 32N 122 13W
Steinriegel, *Wien* 10 G9 48 16N 16 12 E
Steinstücken, *Berl.* 7 B1 52 21N 13 7 E
Steinwerder, *Hbg.* 7 D7 53 32N 9 57 E
Stellingen, *Hbg.* 7 D7 53 35N 9 56 E
Stenhamra, *Stock.* 3 D9 59 20N 17 40 E
Stenløse, *Købn.* 2 D8 55 46N 12 11 E
Stephansdom, *Wien* 10 G10 48 12N 16 21 E
Stepney, *Lon.* 4 B4 51 30N 0 3W
Sterkende, *Købn.* 2 E8 55 36N 12 10 E
Sterkrade, *Ruhr* 6 A3 51 31N 6 52 E
Sterling Park, *S.F.* 27 B2 37 41N 122 27W
Stevenson, *Balt.* 25 A2 39 24N 76 42W
Stewart Manor, *N.Y.* 23 C6 40 43N 73 40W
Sticklinge udde, *Stock.* 3 D11 59 23N 18 6 E
Stickney, *Chic.* 26 C2 41 49N 87 46W
Stienitzaue, *Berl.* 7 A5 52 38N 13 44 E
Stiepel, *Ruhr* 6 B5 51 25N 7 14 E
Stiftskirche, *Ruhr* 6 C2 51 12N 6 41 E
Still Run →, *Phil.* 24 C3 39 47N 75 16W
Stockholm, *Stock.* 3 E11 59 19N 18 4 E
Stocksund, *Stock.* 3 D11 59 23N 18 3 E
Stockum, *Ruhr* 6 C2 51 16N 6 44 E
Stodůlky, *Pra.* 10 B1 50 3N 14 19 E
Stoke D'Abernon, *Lon.* 4 D2 51 19N 0 23W
Stoke Newington, *Lon.* 4 B4 51 33N 0 4W
Stolpe-Süd, *Berl.* 7 A2 52 37N 13 14 E
Stone, *Lon.* 4 C6 51 26N 0 16 E
Stone Grove, *Lon.* 4 B3 51 37N 0 16W
Stone Park, *Chic.* 26 B1 41 53N 87 52W
Stonebridge, *Lon.* 4 B3 51 32N 0 16W
Stoneham, *Bost.* 21 C3 42 29N 71 5W
Stonehurst, *L.A.* 28 A2 34 15N 118 21W
Stony Brook Res., *Bost.* 21 D3 42 15N 71 8W
Stony Cr. →, *Chic.* 26 C2 41 40N 87 45W
Stony Cr. →, *Melb.* 19 E6 37 49 S 144 53 E
Stora Värtan, *Stock.* 3 D11 59 25N 18 7 E
Store Hareskov, *Købn.* 2 D9 55 46N 12 23 E
Store Kattingesø, *Købn.* 2 E7 55 39N 12 0 E
Store Magleby, *Købn.* 2 E10 55 35N 12 35 E
Storholmen, *Stock.* 3 D11 59 23N 18 8 E
Stovivatn, *Oslo* 2 B2 59 54N 10 26 E
Stovner, *Oslo* 2 B5 59 57N 10 55 E
Stow L., *S.F.* 27 B2 37 46N 122 28W
Strašnice, *Pra.* 10 B2 50 4N 14 28 E
Strandbad Gansehäufe, *Wien* 10 G10 48 13N 16 26 E
Strasslach, *Mün.* 7 G10 48 0N 11 30 E
Strasstrudering, *Mün.* 7 G11 48 6N 11 41 E
Stratford, *Lon.* 4 B5 51 33N 0 0 E
Stratford, *Phil.* 24 C4 39 49N 75 0W
Strathfield, *Syd.* 19 B3 33 52 S 151 5 E
Strawberry Hill, *Bost.* 21 D2 42 14N 71 15W
Strawberry Pk., *L.A.* 28 A4 34 16N 118 7W
Strawberry Pt., *S.F.* 27 A1 37 53N 122 30W
Streatham, *Lon.* 4 C4 51 25N 0 7W
Streatham Vale, *Lon.* 4 C4 51 24N 0 8W
Strebersdorf, *Wien* 10 G10 48 17N 16 23 E
Střešovice, *Pra.* 10 B2 50 5N 14 23 E
Strelyna, *St-Pet.* 11 C1 59 49N 30 0 E
Střížkov, *Pra.* 10 B2 50 7N 14 28 E
Strogino, *Mos.* 11 E8 55 48N 37 24 E
Strømmen, *Oslo* 2 B5 59 56N 10 59 E
Stromovka, *Pra.* 10 B2 50 6N 14 25 E
Strunkede Wasserschloss, *Ruhr* 6 A5 51 33N 7 12 E
Studio City, *L.A.* 28 B2 34 8N 118 24W
Stupinigi, *Tori.* 9 C2 44 59N 7 36 E
Stura di Lanzo →, *Tori.* 9 A2 45 11N 7 47 E
Stureby, *Stock.* 3 E11 59 15N 18 4 E
Stuvsta, *Stock.* 3 E11 59 15N 18 0 E
Styrum, *Ruhr* 6 B3 51 27N 6 52 E
Subhepur, *Delhi* 16 A2 28 44N 77 15 E
Sucat, *Manila* 15 E4 14 27N 121 2 E
Success, L., *N.Y.* 23 C6 40 45N 73 42W
Suchdol, *Pra.* 10 B2 50 8N 14 22 E
Sucre, *Car.* 30 D5 10 31N 66 57W
Sucy-en-Brie, *Paris* 5 C5 48 46N 2 31 E
Sudberg, *Ruhr* 6 C4 51 10N 7 6 E
Sudbury, *Bost.* 21 C1 42 22N 71 24W
Suderelbe →, *Hbg.* 7 E7 53 28N 9 58 E
Suderwich, *Ruhr* 6 A5 51 36N 7 15 E
Sugamo, *Tōkyō* 13 B3 35 44N 139 43 E
Sugar Loaf Mt. = Açúcar, Pão de, *Rio J.* 31 B3 22 56 S 43 9W
Sugartown, *Phil.* 24 B1 39 59N 75 30W
Sugasawa, *Tōkyō* 13 B2 35 46N 139 32 E
Suge, *Tōkyō* 13 C2 35 37N 139 32 E
Suginami-Ku, *Tōkyō* 13 B2 35 41N 139 37 E
Sugita, *Tōkyō* 13 D2 35 22N 139 37 E
Sugō, *Tōkyō* 13 C2 35 34N 139 33 E
Suitland, *Wash.* 25 D8 38 50N 76 55W
Sukchar, *Calc.* 16 D6 22 42N 88 22 E
Sulejówek, *Wsaw.* 10 E8 52 14N 21 14 E
Sulldorf, *Hbg.* 7 D6 53 34N 9 49 E
Sultan Mosque, *Sing.* 15 G8 1 18N 103 51 E
Suma, *Ōsaka* 12 C1 34 38N 135 8 E
Sumaré, *S. Pau.* 31 E5 23 32 S 46 41W
Sumida, *Tōkyō* 13 B3 35 43N 139 48 E
Sumida →, *Tōkyō* 13 B3 35 39N 139 46 E
Sumiyoshi, *Ōsaka* 12 C4 34 36N 135 30 E
Summer Palace, *Beij.* 14 B2 39 59N 116 13 E
Summerville, *Trto.* 20 E7 43 37N 79 33W
Summit, *Chic.* 26 C2 41 47N 87 47W
Summit, *N.Y.* 22 C2 40 43N 74 21W
Sun Valley, *L.A.* 28 A2 34 13N 118 21W
Sunamachi, *Tōkyō* 13 B4 35 40N 139 50 E
Sunashinden, *Tōkyō* 13 A2 35 53N 139 30 E
Sunbury, *Lon.* 4 C2 51 25N 0 26W
Sunda Kelapa, *Jak.* 15 H9 6 6 S 106 48 E
Sundbyberg, *Stock.* 3 D10 59 22N 17 57 E
Sundbyerne, *Købn.* 2 E10 55 39N 12 36 E
Sung Kong, *H.K.* 12 E6 22 11N 114 17 E
Sungai Bambu, *Jak.* 15 H10 6 6 S 106 53 E
Sungei Buloh, *Sing.* 15 F7 1 25N 103 42 E
Sungei Simpang, *Sing.* 15 F8 1 26N 103 49 E
Sunland, *L.A.* 28 A3 34 15N 118 18W
Sunnyridge, *Jobg.* 18 F10 26 10 S 28 10 E
Sunset, *S.F.* 27 B2 37 44N 122 29W
Sunshine, *Melb.* 19 E5 37 48 S 144 49 E
Sunshine Acres, *L.A.* 28 C5 33 56N 117 59W
Suntaug L., *Bost.* 21 B3 42 31N 71 0W
Sunter, *Jak.* 15 H10 6 7 S 106 51 E
Sunter, Kali, *Jak.* 15 J10 6 10 S 106 53 E
Suomenlinna, *Hels.* 3 C4 60 9N 24 59 E
Superga, *Tori.* 9 B3 45 4N 7 46 E
Superga, Basilica di, *Tori.* 9 B3 45 5N 7 45 E
Sura, *Calc.* 16 E6 22 33N 88 24 E
Surag San, *Sŏul* 12 F8 37 40N 127 4 E
Surbiton, *Lon.* 4 C3 51 23N 0 17W
Surco, *Lima* 30 G8 12 9 S 77 0W
Suresnes, *Paris* 5 B3 48 52N 2 13 E
Surquillo, *Lima* 30 G8 12 6 S 77 1W
Surrey Hills, *Melb.* 19 F7 37 50 S 145 5 E
Surrey Hills, *Syd.* 19 B4 33 53 S 151 13 E
Surrey Park, *Melb.* 19 E7 37 49 S 145 6 E
Susaeg, *Sŏul* 12 G7 37 34N 126 54 E
Süssenbrunn, *Wien* 10 G10 48 16N 16 29 E
Sutherland, *Syd.* 19 C3 34 2 S 151 3 E
Sutton, *Lon.* 4 C3 51 21N 0 11W
Sutton at Hone, *Lon.* 4 C6 51 24N 0 14 E
Suyu, *Sŏul* 12 G8 37 37N 127 0 E
Suzukishinden, *Tōkyō* 13 B2 35 43N 139 31 E
Svanemøllen, *Købn.* 2 D10 55 43N 12 34 E
Svartsjölandet, *Stock.* 3 D9 59 20N 17 43 E
Sverdlov, *Mos.* 11 E9 55 46N 37 36 E
Svestad, *Oslo* 2 C3 59 46N 10 36 E
Svestrup, *Købn.* 2 D7 55 46N 12 8 E
Svinningeudd, *Stock.* 3 D12 59 26N 18 17 E
Svinö, *Hels.* 3 C3 60 7N 24 44 E
Svogerslev, *Købn.* 2 E7 55 38N 12 0 E
Swampscott, *Bost.* 21 C4 42 28N 70 53W
Swanley, *Lon.* 4 C6 51 23N 0 9 E
Swanscombe, *Lon.* 4 C6 51 26N 0 18 E
Swansea, *Trto.* 20 E8 43 39N 79 27W
Swarthmore, *Phil.* 24 B2 39 54N 75 22W
Swedesboro, *Phil.* 24 C3 39 45N 75 17W
Swedesburg, *Phil.* 24 A3 40 5N 75 19W
Swinburne I., *N.Y.* 22 D4 40 33N 74 3W
Swita, *Ōsaka* 12 B4 34 45N 135 30 E
Syampur, *Calc.* 16 F5 22 28N 88 12 E
Sycamore Mills, *Phil.* 24 B2 39 57N 75 25W
Sydenham, *Jobg.* 18 E9 26 9 S 28 5 E
Sydney, *Syd.* 19 B4 33 52 S 151 12 E
Sydney, Univ. of, *Syd.* 19 B4 33 54 S 151 11 E
Sydney Airport, *Syd.* 19 B4 33 56 S 151 10 E
Sydney Harbour Bridge, *Syd.* 19 B4 33 51 S 151 12 E
Sydstranden, *Købn.* 2 E10 55 34N 12 38 E
Sylling, *Oslo* 2 B1 59 54N 10 16 E
Sylvania, *Syd.* 19 C3 34 0 S 151 7 E
Syndal, *Melb.* 19 F7 37 52 S 145 9 E
Syon House, *Lon.* 4 C3 51 28N 0 18W
Syosset, *N.Y.* 23 C7 40 49N 73 30W
Szabadság-hegy, *Bud.* 10 J12 47 30N 18 59 E
Szczęśliwice, *Wsaw.* 10 E6 52 12N 20 57 E
Szemere-Telep, *Bud.* 10 K14 47 26N 19 13 E
Széphalom, *Bud.* 10 J12 47 34N 18 57 E
Szilasliget, *Bud.* 10 J14 47 34N 19 16 E

T

Tabata, *Tōkyō* 13 B3 35 44N 139 46 E
Tablada, *B.A.* 32 C3 34 41 S 58 32W
Taboão →, *S. Pau.* 31 F7 23 40 S 46 27W
Taboão da Serra, *S. Pau.* 31 E5 23 36 S 46 45W
Tabor, *N.Y.* 22 B2 40 52N 74 28W
Täby, *Stock.* 3 D11 59 26N 18 2 E
Tacony, *Phil.* 24 A4 40 1N 75 2W
Tacuba, *Méx.* 29 B2 19 26N 99 11W
Tacubaya, *Méx.* 29 B2 19 24N 99 10W
Tadain, *Ōsaka* 12 A3 34 51N 135 24 E
Tadworth, *Lon.* 4 D3 51 17N 0 14W
Tagig, *Manila* 15 D4 14 31N 121 4 E
Tagig →, *Manila* 15 D4 14 31N 121 5 E
Tai Hang, *H.K.* 12 E6 22 16N 114 11 E
Tai Lo Shan, *H.K.* 12 D6 22 21N 114 13 E
Tai Po Tsai, *H.K.* 12 D6 22 20N 114 15 E
Tai Seng, *Sing.* 15 F8 1 20N 103 53 E
Tai Shui Hang, *H.K.* 12 D6 22 24N 114 13 E
Tai Tam B., *H.K.* 12 E6 22 13N 114 13 E
Tai Tam Tuk Res., *H.K.* 12 E6 22 14N 114 13 E
Tai Wan Tau, *H.K.* 12 E6 22 17N 114 17 E
Tai Wo Hau, *H.K.* 12 D5 22 21N 114 7 E
Tai Wo Ping, *H.K.* 12 D5 22 20N 114 9 E
Ta'imim, *Bagd.* 17 F8 33 15N 44 21 E
Tainaka, *Ōsaka* 12 C4 34 36N 135 35 E
Taishō, *Ōsaka* 12 C3 34 38N 135 27 E
Taitō-Ku, *Tōkyō* 13 B3 35 43N 139 47 E
Tajima, *Tōkyō* 13 B2 35 48N 139 36 E
Tajpur, *Calc.* 16 D5 22 44N 88 15 E
Takaido, *Tōkyō* 13 B2 35 40N 139 37 E
Takaishi, *Tōkyō* 13 C2 35 36N 139 34 E
Takarazuka, *Ōsaka* 12 B2 34 47N 135 20 E
Takasago, *Tōkyō* 13 B4 35 44N 139 52 E
Takatsuki, *Ōsaka* 12 A4 34 50N 135 37 E
Takayanagi, *Tōkyō* 13 D4 35 23N 139 57 E
Takegahana, *Tōkyō* 13 B4 35 47N 139 54 E
Takenotsuka, *Tōkyō* 13 B3 35 47N 139 47 E
Takeshita, *Tōkyō* 13 C3 35 33N 139 42 E
Takinogawa, *Tōkyō* 13 B3 35 45N 139 44 E
Takkula, *Hels.* 3 B2 60 19N 24 42 E
Takoma Park, *Wash.* 25 D7 38 58N 77 0W
Taksim, *Ist.* 17 A2 41 2N 28 58 E
Talaide, *Lisb.* 8 F7 38 44N 9 18W
Talampas, *Manila* 15 D4 14 36N 121 4 E
Taling Chan, *Bangk.* 15 B1 13 46N 100 27 E
Talleyville, *Phil.* 24 C1 39 48N 75 32W
Tallkrogen, *Stock.* 3 E11 59 16N 18 4 E
Talmapais Valley, *S.F.* 27 A1 37 52N 122 32W
Tama, *Tōkyō* 13 C1 35 38N 139 26 E
Tama →, *Tōkyō* 13 C2 35 34N 139 30 E
Tama Kyźryō, *Tōkyō* 13 C2 35 36N 139 30 E
Tamaden, *Tōkyō* 13 C2 35 39N 139 38 E
Tamagawa-josui →, *Tōkyō* 13 B1 35 41N 139 47 E
Taman Sari, *Jak.* 15 H9 6 8 S 106 48 E
Tamanduatei →, *S. Pau.* 31 E6 23 37 S 46 38W
Tambora, *Jak.* 15 H9 6 8 S 106 47 E
Tamboré, *S. Pau.* 31 E4 23 30 S 46 50W
Tammisalo, *Hels.* 3 B5 60 11N 25 5 E
Tammûh, *El Qâ.* 18 D5 29 55N 31 15 E
Tampier Slough, *Chic.* 26 D1 41 39N 87 54W
Tan Tock Seng, *Sing.* 15 G8 1 19N 103 50 E
Tanah Abang, *Jak.* 15 J9 6 12 S 106 48 E
Tanashi, *Tōkyō* 13 B2 35 43N 139 32 E
Tanforan Park, *S.F.* 27 C2 37 37N 122 24W
Tangjae, *Sŏul* 12 H8 37 29N 127 2 E
Tanglin, *Sing.* 15 G7 1 18N 103 47 E
Tangstedt, *Hbg.* 7 C7 53 40N 9 51 E
Tangstedter Forst, *Hbg.* 7 C8 53 43N 10 3 E
Tanigami, *Ōsaka* 12 B2 34 45N 135 10 E
Tanjung Duren, *Jak.* 15 J9 6 10 S 106 47 E
Tanjung Priok, *Jak.* 15 H10 6 6 S 106 52 E
Tanum, *Oslo* 2 B2 59 53N 10 28 E
Taoranting Gongyuan, *Beij.* 14 B3 39 51N 116 20 E
Taoranting Hu, *Beij.* 14 B3 39 50N 116 20 E
Tapada, *Lisb.* 8 F7 38 42N 9 10W
Tapanila, *Hels.* 3 B5 60 15N 25 2 E
Tapiales, *B.A.* 32 C3 34 42 S 58 30W
Tapiola, *Hels.* 3 B3 60 10N 24 48 E
Tappan, *N.Y.* 22 A5 41 1N 73 57W
Tappan, L., *N.Y.* 22 A4 41 1N 73 59W
Tapsel, *Tehr.* 17 C5 35 48N 51 23 E
Tapsia, *Calc.* 16 E6 22 32N 88 23 E
Taquara, *Rio J.* 31 B1 22 55 S 43 21W
Tara, *Bomb.* 16 G7 19 5N 72 49 E
Tarâblus, *Bagd.* 17 F8 33 19N 44 23 E
Tarango, Presa, *Méx.* 29 B2 19 21N 99 12W
Tårbæk, *Købn.* 2 D10 55 46N 12 35 E
Tarchomin, *Wsaw.* 10 E6 52 19N 20 58 E
Tardeo, *Bomb.* 16 H7 18 57N 72 48 E
Target Rock, *N.Y.* 23 B8 40 55N 73 26W
Targówek, *Wsaw.* 10 E7 52 16N 21 3 E
Tårnby, *Købn.* 2 E10 55 37N 12 35 E

Villecresnes 47

Taronga Zoo. Park, *Syd.* 19 B4 33 50 S 151 14 E
Tarqua B., *Lagos* 18 B2 6 24N 7 23 E
Tarzana, *L.A.* 28 A1 34 10N 118 32W
Tåstrup, *Købn.* 2 E8 55 39N 12 18 E
Tatarovo, *Mos.* 11 E8 55 45N 37 24 E
Tatarpur, *Delhi* 16 B1 28 38N 77 9 E
Tatenberg, *Hbg.* 7 E8 53 29N 10 3 E
Tathong Channel, *H.K.* 12 E6 22 15N 114 16 E
Tathong Pt., *H.K.* 12 E6 22 14N 114 17 E
Tatsfield, *Lon.* 4 D5 51 17N 0 1 E
Tattariharju, *Hels.* 3 B5 60 15N 25 2 E
Tatuapé, *S. Pau.* 31 E6 23 31 S 46 33W
Taufkirchen, *Mün.* 7 G10 48 2N 11 36 E
Tavares, I. dos, *Rio J.* 31 A3 22 49 S 43 6W
Tavernanova, *Nápl.* 9 H13 40 54N 14 21 E
Taverny, *Paris* 5 A3 49 1N 2 13 E
Távros, *Ath.* 8 J11 37 57N 23 43 E
Tavry, *St-Pet.* 11 B6 59 54N 30 40 E
Taylortown, *N.Y.* 22 B2 40 56N 74 23W
Tayninka, *Mos.* 11 D10 55 53N 37 45 E
Taytay, *Manila* 15 D4 14 34N 121 7 E
Tayuman, *Manila* 15 D4 14 31N 121 9 E
Teaneck, *N.Y.* 22 B4 40 52N 74 1W
Teatro Colón, *B.A.* 32 B4 34 36 S 58 23 E
Teban Gardens, *Sing.* 15 G7 1 19N 103 44 E
Tebet, *Jak.* 15 J10 6 14 S 106 50 E
Tecamachaleo, *Méx.* 29 B2 19 25N 99 14W
Techny, *Chic.* 26 A2 42 6N 87 48W
Teck Hock, *Sing.* 15 F8 1 21N 103 54 E
Tecoma, *Melb.* 19 F9 37 54 S 145 20 E
Teddington, *Lon.* 4 C2 51 25N 0 20W
Tegel, *Berl.* 7 A2 52 34N 13 16 E
Tegel, Flughafen, *Berl.* 7 A2 52 35N 13 15 E
Tegeler Fliess →, *Berl.* 7 A3 52 37N 13 21 E
Tegeler See, *Berl.* 7 A2 52 34N 13 15 E
Tegelort, *Berl.* 7 A2 52 34N 13 13 E
Tehar, *Delhi* 16 B1 28 37N 77 7 E
Tehrān, *Tehr.* 17 C5 35 41N 51 25 E
Tehrān Pars, *Tehr.* 17 C6 35 44N 51 32 E
Tei Tong Tsui, *H.K.* 12 E6 22 16N 114 17 E
Tejo →, *Lisb.* 8 F8 38 45N 9 3W
Tekstilyshchik, *Mos.* 11 E10 55 42N 37 41 E
Tela, *Delhi* 16 A2 28 43N 77 19 E
Telhal, *Lisb.* 8 F7 38 48N 9 18W
Telinipara, *Calc.* 16 D6 22 46N 88 23 E
Telok Blangah, *Sing.* 15 G7 1 17N 103 49 E
Teltow, *Berl.* 7 B2 52 23N 13 17 E
Teltow kanal, *Berl.* 7 B3 52 26N 13 29 E
Temescal, L., *S.F.* 27 A3 37 50N 122 13W
Temnikovo, *Mos.* 11 E12 55 43N 38 1 E
Tempelhof, *Berl.* 7 B3 52 27N 13 23 E
Tempelhof, Flughafen, *Berl.* 7 B3 52 28N 13 27 E
Temperley, *B.A.* 32 C4 34 46 S 58 22W
Temple City, *L.A.* 28 B4 34 6N 118 3W
Temple Hills Park, *Wash.* 25 E8 38 48N 76 56W
Templestowe, *Melb.* 19 E7 37 45 S 145 8 E
Templestowe Lower, *Melb.* 19 E7 37 45 S 145 6 E
Tenafly, *N.Y.* 22 B5 40 54N 73 58W
Tenantongo, Presa, *Méx.* 29 B2 19 28N 99 15W
Tengah →, *Sing.* 15 F7 1 23N 103 43 E
Tengeh, Sungei →, *Sing.* 15 F6 1 20N 103 39 E
Tennoji, *Ōsaka* 12 C4 34 39N 135 30 E
Tenochtitlán, *Méx.* 29 B3 19 26N 99 7W
Tepalcates, *Méx.* 29 B3 19 23N 99 3W
Tepe Saif, *Tehr.* 17 D4 35 36N 51 17 E
Tepepan, *Méx.* 29 C3 19 16N 99 9W
Teplyy Star, *Mos.* 11 F9 55 37N 37 30 E
Tepozteco, Parque Nac. del, *Méx.* 29 D3 19 3N 99 5W
Terrasse Vaudreuil, *Mtrl.* 20 B2 43 23N 73 59W
Terrazzano, *Mil.* 9 D5 45 32N 9 4 E
Terrugem, *Lisb.* 8 F7 38 41N 9 17W
Terusan Banjir, *Jak.* 15 H9 6 7 S 106 46 E
Terzigno, *Nápl.* 9 J13 40 48N 14 29 E
Tessancourt-sur-Aubette, *Paris* 5 A1 49 1N 1 55 E
Testona, *Tori.* 9 C3 44 59N 7 42 E
Tetelco, *Méx.* 29 C4 19 12N 98 57W
Tetreauville, *Mtrl.* 20 A4 43 35N 73 32W
Tetti Neirotti, *Tori.* 9 B2 45 3N 7 32 E
Tetuán, *Mdrd.* 8 B2 40 27N 3 42W
Teufelsberg, *Berl.* 7 B2 52 29N 13 14 E
Tévere →, *Rome* 9 F9 41 56N 12 27 E
Tewksbury, *Bost.* 21 B2 42 37N 71 12W
Texcoco, L. de, *Méx.* 29 B4 19 30N 89 58 E
Thalkirchen, *Mün.* 7 G10 48 6N 11 32 E
Thames Ditton, *Lon.* 4 C2 51 23N 0 20W
Thamesmead, *Lon.* 4 B5 51 30N 0 7 E
Thana Cr. →, *Bomb.* 16 G8 19 4N 72 54 E
The Basin, *Melb.* 19 F8 37 51 S 145 19 E
The Glen, *Wash.* 25 C6 39 2N 77 12W
The Loop, *Chic.* 26 B3 41 52N 87 37W
The Narrows, *N.Y.* 22 D4 40 37N 74 3W
The Ridge, *Delhi* 16 B2 28 37N 77 10 E
The White House, *Wash.* 25 D7 38 53N 77 1W
The Wilds, *Jobg.* 18 F9 26 10 S 28 2 E
Theseión, *Ath.* 8 J11 37 57N 23 43 E
Theydon Bois, *Lon.* 4 A5 51 40N 0 6 E
Thiais, *Paris* 5 C4 48 46N 2 23 E
Thieux, *Paris* 5 A6 49 0N 2 40 E
Thistletown, *Trto.* 20 D7 43 44N 79 34W
Thiverval-Grignon, *Paris* 5 B1 48 51N 1 55 E
Thomaston, *N.Y.* 23 C6 40 47N 73 43W
Thomastown, *Melb.* 19 E7 37 40 S 145 2 E
Thompson I., *Bost.* 21 D4 42 19N 70 59W
Thomson, *Sing.* 15 F8 1 20N 103 50 E
Thon Buri, *Bangk.* 15 B1 13 45N 100 29 E
Thong Hoe, *Sing.* 15 F7 1 25N 103 42 E
Thorigny-sur-Marne, *Paris* 5 B6 48 53N 2 41 E
Thornbury, *Melb.* 19 E7 37 44 S 145 1 E
Thorncliffe, *Trto.* 20 D8 43 42N 79 20W
Thornhill, *Jobg.* 18 E9 26 6 S 28 9 E
Thornhill, *Trto.* 20 D8 43 48N 79 25W
Thornton, *Phil.* 24 B1 39 54N 75 31W
Thornton Heath, *Lon.* 4 C4 51 23N 0 6W
Thorofare, *Phil.* 24 B3 39 50N 75 11W
Throgs Neck, *N.Y.* 23 C6 40 48N 73 49W
Tian Guan, *Sing.* 15 F7 1 21N 103 49 E
Tian'anmen, *Beij.* 14 B3 39 53N 116 21 E
Tiancun, *Beij.* 14 B2 39 54N 116 12 E
Tianjin, *Tianj.* 14 E5 39 7N 117 12 E
Tiantan Gongyuan, *Beij.* 14 B3 39 51N 116 22 E
Tiateloco, *Méx.* 29 B3 19 27N 99 8W
Tibidabo, *Barc.* 8 D5 41 25N 2 6 E
Tiburon, *S.F.* 27 A2 37 52N 122 27W
Tiburon Pen., *S.F.* 27 A2 37 53N 122 28W
Tiburtino, *Rome* 9 F10 41 53N 12 30 E
Ticomán, *Méx.* 29 A3 19 31N 99 8W
Tiefenbroich, *Ruhr* 6 C2 51 18N 6 49 E
Tiefersee, *Berl.* 7 B1 52 24N 13 5 E
Tiejiangyin, *Beij.* 14 C3 39 49N 116 23 E
Tientsin = Tianjin, *Tianj.* 14 E5 39 7N 117 12 E
Tiergarten, *Berl.* 7 A3 52 31N 13 20 E
Tietê →, *S. Pau.* 31 D7 23 28 S 46 24W
Tigery, *Paris* 5 D5 48 38N 2 30 E
Tigre, *B.A.* 32 A3 34 25 S 58 34W
Tigris →, *Bagd.* 17 F8 33 17N 44 23 E
Tijuca, *Rio J.* 31 B2 22 56 S 43 13W
Tijuca, L. de, *Rio J.* 31 B2 22 59 S 43 20W
Tijuca, Pico da, *Rio J.* 31 B2 22 56 S 43 15W
Tijucamar, *Rio J.* 31 C2 23 0 S 43 18W
Tijucas, Is., *Rio J.* 31 C2 23 1 S 43 17W
Tikkurila, *Hels.* 3 B5 60 17N 25 2 E
Tilanqiao, *Shang.* 14 J11 31 15N 121 29 E
Tilbury, *Lon.* 4 C7 51 27N 0 21 E
Timah, Bukit, *Sing.* 15 F7 1 21N 103 46 E
Timiryazev Park, *Mos.* 11 E9 55 49N 37 33 E
Ting Kau, *H.K.* 12 D5 22 22N 114 4 E
Tinley Cr. →, *Chic.* 26 D2 41 39N 87 45W
Tinley Creek Woods, *Chic.* 26 D2 41 38N 87 48W
Tinley Park, *Chic.* 26 D2 41 35N 87 46W
Tipas, *Manila* 15 D4 14 32N 121 4 E
Tirsa, *El Qâ.* 18 D5 29 57N 31 12 E
Tishrīyaa, *Bagd.* 17 F8 33 18N 44 24 E
Tit Cham Chau, *H.K.* 12 E6 22 15N 114 17 E
Titagarh, *Calc.* 16 D6 22 44N 88 22 E
Tivoli, *Købn.* 2 D10 55 40N 12 35 E
Tizapán, *Méx.* 29 C2 19 19N 99 13W
Tlalnepantla, *Méx.* 29 A2 19 32N 99 11W
Tlalnepantla →, *Méx.* 29 A2 19 30N 99 18W
Tláloc, Cerro, *Méx.* 29 D3 19 7N 99 3W
Tlalpan, *Méx.* 29 C2 19 17N 99 10W
Tlalpitzáhuac, *Méx.* 29 C4 19 19N 98 56W
Tlaltenango, *Méx.* 29 B2 19 20N 99 17W
Tlaltenco, *Méx.* 29 C3 19 19N 99 0W
Tlaxcoaque, *Méx.* 29 B3 19 25N 99 8W
To Kwai Wan, *H.K.* 12 E6 22 18N 114 11 E
Toa Payoh, *Sing.* 15 F8 1 20N 103 50 E
Tobay Beach, *N.Y.* 23 D8 40 36N 73 26W
Točná, *Pra.* 10 C2 49 58N 14 25 E
Tocome →, *Car.* 30 D6 10 28N 66 49W
Toda, *Tōkyō* 13 A3 35 50N 139 40 E
Todamachī, *Tōkyō* 13 B2 35 48N 139 39 E
Todt Hill, *N.Y.* 22 D4 40 36N 74 6W
Toei, Khlong →, *Bangk.* 15 B2 13 43N 100 32 E
Togasaki, *Tōkyō* 13 B4 35 47N 139 51 E
Tōkagi, *Tōkyō* 13 B4 35 42N 139 55 E
Tōkaichiba, *Tōkyō* 13 C2 35 31N 139 30 E
Tokarevo, *Mos.* 11 F11 55 38N 37 54 E
Tokorozawa, *Tōkyō* 13 B1 35 47N 139 28 E
Tōkyō, *Tōkyō* 13 C3 35 43N 139 45 E
Tōkyō B., *Tōkyō* 13 C4 35 33N 139 53 E
Tōkyō-Haneda Int. Airport, *Tōkyō* 13 C3 35 33N 139 45 E
Tōkyō Harbour, *Tōkyō* 13 C3 35 38N 139 46 E
Tokyo Univ., *Tōkyō* 13 B3 35 42N 139 46 E
Tollygunge, *Calc.* 16 F6 22 29N 88 21 E
Tolly's Nala, *Calc.* 16 E6 22 33N 88 19 E
Tolworth, *Lon.* 4 C3 51 22N 0 17W
Tomang, *Jak.* 15 J9 6 10 S 106 47 E
Tomba di Nerone, *Rome* 9 F9 41 58N 12 26 E
Tomilino, *Mos.* 11 F11 55 39N 37 55 E
Tomioka, *Tōkyō* 13 D2 35 22N 139 37 E
Tonda, *Ōsaka* 12 B4 34 49N 135 35 E
Tondo, *Manila* 15 D3 14 36N 120 57 E
Tone-unga →, *Tōkyō* 13 A4 35 55N 139 56 E
Tonekollen, *Oslo* 2 C6 50 49N 11 0 E
Tong Kang, Sungei →, *Sing.* 15 F8 1 23N 103 53 E
Tonghui He →, *Beij.* 14 B3 39 53N 116 28 E
Tönisheide, *Ruhr* 6 C4 51 18N 7 3 E
Tonndorf, *Hbg.* 7 D8 53 35N 10 8 E
Toorak, *Melb.* 19 F7 37 50 S 145 1 E
Toot Hill, *Lon.* 4 A6 51 41N 0 11 E
Topilejo, *Méx.* 29 C3 19 12N 99 9W
Topkapi, *Ist.* 17 A2 41 1N 28 55 E
Topsfield, *Bost.* 21 B4 42 38N 70 57W
Tor di Quinto, *Rome* 9 F9 41 56N 12 27 E
Tor Pignattara, *Rome* 9 F10 41 52N 12 31 E
Tor Sapienza, *Rome* 9 F10 41 53N 12 35 E
Torcy, *Paris* 5 B5 48 51N 2 39 E
Torino, *Tori.* 9 B2 45 5N 7 39 E
Toro, *B.A.* 32 B1 34 30 S 58 50W
Toronto, *Trto.* 20 E9 43 39N 79 13W
Toronto, Univ. of, *Trto.* 20 E8 43 39N 79 23W
Toronto Harbour, *Trto.* 20 E8 43 38N 79 21W
Toronto I., *Trto.* 20 E8 43 37N 79 23W
Toronto Int. Airport, *Trto.* 20 D7 43 40N 79 38 E
Torre Annunziata, *Nápl.* 9 J13 40 45N 14 26 E
Torre Cervara, *Rome* 9 F10 41 55N 12 35 E
Torre del Greco, *Nápl.* 9 J13 40 47N 14 21 E
Torre Novo, *Rome* 9 F10 41 51N 12 36 E
Torrellas →, *Barc.* 8 D5 41 23N 2 1 E
Torrellas del Llobregat, *Barc.* 8 D4 41 20N 1 59 E
Torresdale, *Phil.* 24 A5 40 3N 74 59W
Torrevécchia, *Rome* 9 F9 41 55N 12 25 E
Tortuguitas, *B.A.* 32 A2 34 28 S 58 44W
Toshima-Ku, *Tōkyō* 13 B3 35 43N 139 43 E
Toshimaen, *Tōkyō* 13 B2 35 43N 139 38 E
Totowa, *N.Y.* 22 B3 40 54N 74 13W
Totsuka-Ku, *Tōkyō* 13 D2 35 23N 139 32 E
Tottenham, *Lon.* 4 B4 51 35N 0 4W
Tottenham, *Melb.* 19 E6 37 48 S 144 51 E
Tottenville, *N.Y.* 22 D3 40 30N 74 14W
Totteridge, *Lon.* 4 B3 51 37N 0 11W
Toussus-le-Noble, *Paris* 5 C2 48 44N 2 6 E
Toussus-le-Noble, Aérodrome de, *Paris* 5 C2 48 44N 2 6 E
Toverud, *Oslo* 2 B2 59 55N 10 20 E
Towaco, *N.Y.* 22 B3 40 55N 74 18W
Tower Hamlets, *Lon.* 4 B4 51 31N 0 2W
Town Farm Hill, *Bost.* 21 A3 42 40N 71 3W
Townley, *N.Y.* 22 C3 40 41N 74 14W
Towra Pt., *Syd.* 19 C4 34 0 S 151 10 E
Towson, *Balt.* 25 A3 39 24N 76 36W
Tøyen, *Oslo* 2 B4 59 55N 10 47 E
Toyofuta, *Tōkyō* 13 A4 35 54N 139 55 E
Toyonaka, *Ōsaka* 12 B3 34 46N 135 28 E
Traar, *Ruhr* 6 B1 51 22N 6 36 E
Trafaria, *Lisb.* 8 F7 38 40N 9 13W
Tragliata, *Rome* 9 F8 41 58N 12 14 E
Traicão →, *S. Pau.* 31 E6 23 35 S 46 41W
Trälhavet, *Stock.* 3 D13 59 26N 18 22 E
Tranby, *Oslo* 2 C1 59 49N 10 14 E
Tranegilde, *Købn.* 2 E9 55 37N 12 20 E
Trångsund, *Stock.* 3 E11 59 13N 18 8 E
Trappenfelde, *Berl.* 7 A4 52 34N 13 39 E
Trappes, *Paris* 5 C1 48 46N 1 59 E
Trastévere, *Rome* 9 F9 51 53N 12 28 E
Travilah, *Wash.* 25 C6 39 4N 77 15W
Travilah Regional Park, *Wash.* 25 C6 39 4N 77 17W
Travis, *N.Y.* 22 D3 40 35N 74 11W
Treasure I., *S.F.* 27 B2 37 49N 122 22W
Třeboradice, *Pra.* 10 B3 50 9N 14 31 E
Třebotov, *Pra.* 10 C1 49 58N 14 17 E
Trecase, *Nápl.* 9 J13 40 46N 14 26 E
Trekroner, *Købn.* 2 D10 55 42N 12 36 E
Tremblay-lès-Gonesse, *Paris* 5 B5 48 58N 2 30 E
Tremembé, *S. Pau.* 31 D6 23 27 S 46 36W
Tremembe →, *S. Pau.* 31 D6 23 27 S 46 34W
Tremont, *Melb.* 19 F9 37 53 S 145 20 E
Tremont, *N.Y.* 23 B5 40 50N 73 52W
Trenno, *Mil.* 9 E5 45 29N 9 6 E
Treptow, *Berl.* 7 B3 52 29N 13 27 E
Tres Marias, *Méx.* 29 D2 19 3N 99 15W
Trés Rios, Sa. dos, *Rio J.* 31 B2 22 56 S 43 17W
Tretiakov Art Gallery, *Mos.* 11 E9 55 44N 37 38 E
Trevose, *Phil.* 24 A5 40 8N 74 59W
Trezzano sul Navíglio, *Mil.* 9 E5 45 24N 9 4 E
Tribobo, *Rio J.* 31 B3 22 50 S 43 0W
Triel-sur-Seine, *Paris* 5 B2 48 58N 2 0 E
Trieste, *Rome* 9 F10 41 55N 12 30 E
Trinidad, *Wash.* 25 D8 38 54N 76 59W
Triome, *Jobg.* 18 F8 26 10 S 27 58 E
Trionfale, *Rome* 9 F9 41 54N 12 26 E
Triulzo, *Mil.* 9 E6 45 25N 9 16 E
Trócchia, *Nápl.* 9 H13 40 51N 14 23 E
Troitse-Lykovo, *Mos.* 11 E8 55 47N 37 23 E
Troja, *Pra.* 10 B2 50 7N 14 25 E
Trollbäcken, *Stock.* 3 E12 59 14N 18 12 E
Trombay, *Bomb.* 16 G8 19 2N 72 56 E
Troparevo, *Mos.* 11 F8 55 39N 37 29 E
Trottiscliffe, *Lon.* 4 D7 51 18N 0 21 E
Troy Hills, *N.Y.* 22 B2 40 50N 74 23W
Troyeville, *Jobg.* 18 F9 26 11 S 28 4 E
Truc di Miola, *Tori.* 9 A2 45 11N 7 30 E
Trudyashchikhsya, Os., *St-Pet.* 11 B3 59 58N 30 18 E
Trutlandet, *Hels.* 3 C6 60 9N 25 17 E
Tryvasshøgda, *Oslo* 2 B4 59 59N 10 40 E
Tseng Lan Shue, *H.K.* 12 D6 22 20N 114 14 E
Tsentralynyy, *Mos.* 11 D11 55 53N 37 51 E
Tsim Sha Tsui, *H.K.* 12 E6 22 17N 114 10 E
Tsing Yi, *H.K.* 12 D5 22 21N 114 6 E
Tsuen Wan, *H.K.* 12 D5 22 22N 114 7 E
Tsurugamine, *Tōkyō* 13 D2 35 28N 139 33 E
Tsuruma, *Tōkyō* 13 A2 35 52N 139 31 E
Tsurumi →, *Tōkyō* 13 C3 35 32N 139 40 E
Tsurumi-Ku, *Tōkyō* 13 D3 35 30N 139 41 E
Tsz Wan Shan, *H.K.* 12 D6 22 20N 114 11 E
Tua Kang Lye, *Sing.* 15 G7 1 18N 103 46 E
Tuas, *Sing.* 15 G6 1 19N 103 39 E
Tuchoměřice, *Pra.* 10 B1 50 7N 14 16 E
Tuckahoe, *N.Y.* 23 B6 40 56N 73 49W
Tucuruvi, *S. Pau.* 31 D6 23 28 S 46 35W
Tufello, *Rome* 9 F10 41 56N 12 32 E
Tufnell Park, *Lon.* 4 B4 51 33N 0 8W
Tujunga, *L.A.* 28 A3 34 15N 118 16W
Tujunga Wash →, *L.A.* 28 A2 34 12N 118 23W
Tullamarine, *Melb.* 19 E6 37 41 S 144 50 E
Tullinge, *Stock.* 3 E10 59 12N 17 54 E
Tullingesjön, *Stock.* 3 E10 59 12N 17 52 E
Tulse Hill, *Lon.* 4 C4 51 26N 0 6W
Tulyehualco, *Méx.* 29 C3 19 15N 99 0W
Tumba, *Stock.* 3 E9 59 12N 17 49 E
Tune, *Købn.* 2 E8 55 35N 12 10 E
Tung Lo Wan, *H.K.* 12 E6 22 17N 114 11 E
Tung Lung I., *H.K.* 12 E6 22 15N 114 17 E
Tung O, *H.K.* 12 E5 22 11N 114 8 E
Tunis, *Bagd.* 17 E8 33 23N 44 21 E
Tuomarila, *Hels.* 3 B3 60 11N 24 41 E
Tura, *El Qâ.* 18 D5 29 55N 31 16 E
Turambhe, *Bomb.* 16 G9 19 4N 73 0 E
Turdera, *B.A.* 32 C4 34 48 S 58 26W
Tureberg, *Stock.* 3 D10 59 25N 17 55 E
Turffontein, *Jobg.* 18 F9 26 14 S 28 2 E
Turin = Torino, *Tori.* 9 B2 45 5N 7 39 E
Turner, *Balt.* 25 B3 39 14N 76 31W
Turner Hill, *Bost.* 21 A4 42 40N 70 53W
Turnersville, *Phil.* 24 C4 39 46N 75 3W
Turnham Green, *Lon.* 4 C3 51 29N 0 16W
Turów, *Wsaw.* 10 E8 52 19N 21 11 E
Turter, *Oslo* 2 A4 60 0N 10 46 E
Tuscolano, *Rome* 9 F10 41 52N 12 31 E
Tushino, *Mos.* 11 D8 55 50N 37 24 E
Tuusulanjoki →, *Hels.* 3 A4 60 20N 24 54 E
Twickenham, *Lon.* 4 C2 51 26N 0 20W
Twickenham Rugby Ground, *Lon.* 4 C2 51 27N 0 20W
Twin Oaks, *Phil.* 24 B2 39 50N 75 25W
Twórki, *Wsaw.* 10 E5 52 10N 20 49 E
Tyresö, *Stock.* 3 E13 59 14N 18 20 E
Tyresö strand, *Stock.* 3 E12 59 15N 18 17 E

U

Uberaba →, *S. Pau.* 31 E6 23 35 S 46 41W
Uberruhr, *Ruhr* 6 B4 51 25N 7 4 E
Ubin, P., *Sing.* 15 F8 1 24N 103 57 E
Uboldo, *Mil.* 9 D5 45 36N 9 0 E
Uckendorf, *Ruhr* 6 B4 51 29N 7 7 E
Udelnaya, *St-Pet.* 11 A4 60 0N 30 21 E
Udelynaya, *Mos.* 11 F11 55 38N 37 59 E
Udlding, *Mün.* 7 F9 48 15N 11 25 E
Uellendahl, *Ruhr* 6 C5 51 16N 7 10 E
Ueno, *Tōkyō* 13 B3 35 42N 139 46 E
Uerdingen, *Ruhr* 6 B1 51 21N 6 38 E
Uhlenhorst, *Hbg.* 7 D8 53 34N 10 1 E
Üholičky, *Pra.* 10 B1 50 9N 14 19 E
Uhříněves, *Pra.* 10 B3 50 2N 14 35 E
Újezd nad Lesy, *Pra.* 10 B3 50 4N 14 39 E
Újpalota, *Bud.* 10 J13 47 32N 19 8 E
Újpest, *Bud.* 10 J13 47 35N 19 4 E
Ukita, *Tōkyō* 13 B4 35 40N 139 51 E
Ullerup, *Købn.* 2 E10 55 34N 12 36 E
Ullevål, *Oslo* 2 B4 59 56N 10 46 E
Üllo, *Bud.* 10 K14 47 23N 19 20 E
Ulriksdal, *Stock.* 3 D10 59 23N 17 59 E
Ulu Bedok, *Sing.* 15 G8 1 19N 103 55 E
Ulu Pandan →, *Sing.* 15 G7 1 19N 103 45 E
Ulyanka, *St-Pet.* 11 B3 59 50N 30 14 E
Um Al-Khanazir, *Bagd.* 17 F8 33 17N 44 22 E
Umeda, *Ōsaka* 12 B3 34 41N 135 29 E
Umejima, *Tōkyō* 13 B3 35 46N 139 48 E
Umraniye, *Ist.* 17 A3 41 1N 29 4 E
Unětický →, *Pra.* 10 B2 50 9N 14 24 E
Ungelsheim, *Ruhr* 6 B2 51 21N 6 43 E
Unhos, *Lisb.* 8 F8 38 49N 9 7W
Unidad Santa Fe, *Méx.* 29 B2 19 23N 99 13W
Union, *N.Y.* 22 C3 40 42N 74 16W
Union City, *N.Y.* 22 C4 40 45N 74 2W
Union City, *S.F.* 27 C4 37 36N 122 2W
Union Port, *N.Y.* 23 C5 40 48N 73 51W
Uniondale, *N.Y.* 23 C7 40 42N 73 35W
United Nations H.Q., *N.Y.* 22 C5 40 45N 73 59W
Universal City, *L.A.* 28 B2 34 8N 118 21W
Universidad de Chila, *Stgo* 30 J11 33 26 S 70 39W
University Gardens, *N.Y.* 23 C6 40 46N 73 42W
University Heights, *S.F.* 27 D3 37 26N 122 13W
University Park, *Wash.* 25 D8 38 58N 76 56W
Unsani, *Calc.* 16 E5 22 35N 88 15 E
Unterbach, *Ruhr* 6 C3 51 12N 6 53 E
Unterbiberg, *Mün.* 7 G10 48 6N 11 38 E
Unterföhring, *Mün.* 7 F11 48 11N 11 38 E
Unterhaching, *Mün.* 7 G10 48 3N 11 37 E
Unterkirchbach, *Wien* 10 G9 48 17N 16 12 E
Unterlaa, *Wien* 10 H10 48 8N 16 24 E
Untermauerbach, *Wien* 10 G9 48 14N 16 11 E
Untermenzing, *Mün.* 7 F9 48 10N 11 28 E
Unterrath, *Ruhr* 6 C2 51 16N 6 45 E
Unterschleissheim, *Mün.* 7 F10 48 16N 11 35 E
Upminster, *Lon.* 4 B6 51 33N 0 14 E
Upper Brookville, *N.Y.* 23 B7 40 50N 73 35W
Upper Crystal Springs Res., *S.F.* 26 D2 37 28N 122 20W
Upper Darby, *Phil.* 24 B3 39 57N 75 16W
Upper Edmonton, *Lon.* 4 B4 51 36N 0 3W
Upper Elmers End, *Lon.* 4 C4 51 23N 0 1W
Upper Fern Tree Gully, *Melb.* 19 F8 37 53 S 145 18 E
Upper New York B., *N.Y.* 22 D4 40 39N 74 3W
Upper Norwood, *Lon.* 4 C4 51 24N 0 6W
Upper Peirce Res., *Sing.* 15 F7 1 22N 103 47 E
Upper San Leandro Res., *S.F.* 27 B4 37 46N 122 6W
Upper Sydenham, *Lon.* 4 C4 51 26N 0 4W
Upper Tooting, *Lon.* 4 C4 51 25N 0 9W
Upton, *Lon.* 4 B5 51 32N 0 1 E
Uptons Hill, *Bost.* 21 B3 42 33N 71 0W
Uptown, *Chic.* 26 B2 41 58N 87 40W
Upwey, *Melb.* 19 F9 37 53 S 145 20 E
Urawa, *Tōkyō* 13 A2 35 51N 139 39 E
Urayasu, *Tōkyō* 13 C4 35 39N 139 53 E
Urbe, Aeroporto d', *Rome* 9 F10 41 57N 12 30 E
Urca, *Rio J.* 31 B3 22 56 S 43 9W
Uritsk, *St-Pet.* 11 C3 59 49N 30 10 E
Üröm, *Bud.* 10 J13 47 35N 19 1 E
Ursus, *Wsaw.* 10 E6 52 11N 20 52 E
Ursvik, *Stock.* 3 D10 59 23N 17 57 E
Usera, *Mdrd.* 8 B2 40 22N 3 42W
Ushigome, *Tōkyō* 13 B3 35 42N 139 44 E
Usküdar, *Ist.* 17 A3 41 1N 29 0 E
Ust-Slavyanka, *St-Pet.* 11 C5 59 51N 30 32 E
Uteke, *Stock.* 3 D12 59 24N 18 15 E
Utfort, *Ruhr* 6 B1 51 28N 6 37 E
Utinga, *S. Pau.* 31 E6 23 38 S 46 31W
Utrata, *Wsaw.* 10 E7 52 15N 21 4 E
Uttarpara, *Calc.* 16 E5 22 39N 88 21 E
Utterslev Mose, *Købn.* 2 D9 55 42N 12 29 E
Uttran, *Stock.* 3 E9 59 12N 17 43 E
Utvika, *Oslo* 2 A1 60 2N 10 15 E
Uxbridge, *Lon.* 4 B2 51 32N 0 28W
Uzkoye, *Mos.* 11 F9 55 37N 37 32 E
Uzunca →, *Ist.* 17 A1 41 54N 28 50 E

V

Vadaul, *Bomb.* 16 G8 19 2N 72 55 E
Værebro Å →, *Købn.* 2 D8 55 47N 12 7 E
Vahal, *Bomb.* 16 H9 18 58N 73 2 E
Vaires-sur-Marne, *Paris* 5 B5 48 52N 2 38 E
Val della Torre, *Tori.* 9 B1 45 8N 7 27 E
Valby, *Købn.* 2 E9 55 39N 12 29 E
Valcannuta, *Rome* 9 F9 41 52N 12 25 E
Valdeveba, *Mdrd.* 8 B3 40 29N 3 39W
Vale, *Wash.* 25 D5 38 55N 77 20W
Valentino, Parco del, *Tori.* 9 B3 45 3N 7 41 E
Valenton, *Paris* 5 C4 48 44N 2 27 E
Valera, *Mil.* 9 D5 45 33N 9 3 E
Vallcarca, *Barc.* 8 D5 41 25N 2 9 E
Valldoreix, *Barc.* 8 D5 41 27N 2 3 E
Vallecas, *Mdrd.* 8 B3 40 22N 3 37W
Vallemar, *S.F.* 27 C2 37 36N 122 28W
Vallensbæk, *Købn.* 2 E9 55 38N 12 21 E
Vallensbæk Strand, *Købn.* 2 E9 55 36N 12 23 E
Vallentunasjön, *Stock.* 3 D11 59 27N 18 1 E
Valleranello, *Rome* 9 G9 41 46N 12 29 E
Valley Forge, *Phil.* 24 A2 40 5N 75 27W
Valley Forge Hist. State Park, *Phil.* 24 A2 40 5N 75 27W
Valley Mede, *Balt.* 25 B1 39 16N 76 50W
Valley Stream, *N.Y.* 23 C6 40 40N 73 43W
Vällingby, *Stock.* 3 D10 59 21N 17 52 E
Vallisaari, *Hels.* 3 C5 60 7N 25 0 E
Vallvidrera, *Barc.* 8 D5 41 24N 2 6 E
Valo Velho, *S. Pau.* 31 E5 23 38 S 46 47W
Valuyevo, *Mos.* 11 F8 55 30N 37 21 E
Valvidrera →, *Barc.* 8 D5 41 25N 2 0 E
Van Dyks Park, *Jobg.* 18 F10 26 15 S 28 18 E
Van Nuys, *L.A.* 28 A2 34 11N 118 27W
Van Nuys Airport, *L.A.* 28 A2 34 12N 118 29W
Van Ryn Dam, *Jobg.* 18 E11 26 8 S 28 21 E
Vanak, *Tehr.* 17 C5 35 45N 51 23 E
Vangede, *Købn.* 2 D10 55 45N 12 30 E
Vaniköy, *Ist.* 17 A3 41 3N 29 3 E
Vanløse, *Købn.* 2 D9 55 41N 12 28 E
Vantaa, *Hels.* 3 B4 60 18N 24 56 E
Vantaa →, *Hels.* 3 B4 60 13N 24 53 E
Vantaankoski, *Hels.* 3 B4 60 18N 24 50 E
Vantör, *Stock.* 3 E11 59 16N 18 2 E
Vanves, *Paris* 5 C3 48 49N 2 17 E
Vanzago, *Mil.* 9 D4 45 31N 8 59 E
Värby, *Stock.* 3 E10 59 15N 17 52 E
Vardåsen, *Oslo* 2 C6 50 48N 11 6 E
Varedo, *Mil.* 9 D5 45 35N 9 8 E
Varennes-Jarcy, *Paris* 5 C5 48 40N 2 33 E
Vargem Grande, *Rio J.* 31 B1 22 58 S 43 27W
Városliget, *Bud.* 10 J13 47 30N 19 5 E
Vartiokylä, *Hels.* 3 B5 60 13N 25 6 E
Vartiosaari, *Hels.* 3 B5 60 11N 25 5 E
Vasby, *Købn.* 2 D8 55 40N 12 12 E
Vashi, *Bomb.* 16 G8 19 4N 72 59 E
Vasilyevskiy, Os., *St-Pet.* 11 B3 59 55N 30 16 E
Västerkulla, *Hels.* 3 B2 60 16N 24 31 E
Västerskog, *Hels.* 3 B6 60 16N 25 17 E
Vasto, *Nápl.* 9 H12 40 51N 14 16 E
Vatutino, *Mos.* 11 D10 55 52N 37 40 E
Vaucresson, *Paris* 5 B2 48 50N 2 9 E
Vaudreuil, *Mtrl.* 20 B1 43 24N 74 1W
Vaudreuil-sur-le Lac, *Mtrl.* 20 B1 43 25N 74 1W
Vauhallan, *Paris* 5 C3 48 44N 2 12 E
Vaujours, *Paris* 5 B5 48 56N 2 34 E
Vauréal, *Paris* 5 A2 49 2N 2 1 E
Vaux-sur-Seine, *Paris* 5 A1 49 0N 1 57 E
Vauxhall, *Lon.* 4 C4 51 29N 0 7W
Vaxholm, *Stock.* 3 D13 59 24N 18 20 E
Vecklax, *Hels.* 3 B2 60 10N 24 31 E
Vecsés, *Bud.* 10 K14 47 24N 19 16 E
Vedano al Lissone, *Mil.* 9 D6 45 36N 9 16 E
Veddel, *Hbg.* 7 D8 53 31N 10 2 E
Vega, *Stock.* 3 E11 59 11N 18 8 E
Vehkalahti, *Hels.* 3 B2 60 11N 24 30 E
Veikkola, *Hels.* 3 B1 60 17N 24 26 E
Velbert, *Ruhr* 6 B4 51 21N 7 1 E
Veleň, *Pra.* 10 A3 50 10N 14 33 E
Veleslavin, *Pra.* 10 B2 50 5N 14 21 E
Vélizy-Villacoublay, *Paris* 5 C3 48 47N 2 11 E
Velka-Chuchle, *Pra.* 10 B2 50 0N 14 23 E
Venaria, *Tori.* 9 B2 45 8N 7 37 E
Venda Seca, *Lisb.* 8 F7 38 46N 9 15W
Vendelsö, *Stock.* 3 E12 59 12N 18 11 E
Venice, *L.A.* 28 C2 33 59N 118 27W
Venner, *Oslo* 2 A3 60 1N 10 36 E
Vennhausen, *Ruhr* 6 C3 51 13N 6 51 E
Ventas, *Mdrd.* 8 B2 40 26N 3 40W
Ventorro del Cano, *Mdrd.* 8 B2 40 23N 3 49W
Verberg, *Ruhr* 6 B1 51 21N 6 34 E
Verde →, *S. Pau.* 31 E7 23 29 S 46 27W
Verdi, *Ath.* 8 H11 38 2N 23 40 E
Verdugo Mt., *L.A.* 28 A3 34 12N 118 17W
Verdun, *Mtrl.* 20 B4 43 27N 73 35W
Vereya, *Mos.* 11 F12 55 37N 38 2 E
Vérhalom, *Bud.* 10 J13 47 31N 19 1 E
Vermelho →, *S. Pau.* 31 E5 23 30 S 46 46W
Vermont, *Melb.* 19 F8 37 50 S 145 12 E
Vermont South, *Melb.* 19 F8 37 51 S 145 11 E
Verneuil-sur-Seine, *Paris* 5 B1 48 58N 1 59 E
Vernouillet, *Paris* 5 B1 48 58N 1 56 E
Verona, *N.Y.* 22 C3 40 49N 74 15W
Verperluda, Os., *St-Pet.* 11 B2 59 59N 30 0 E
Verrières-le-Buisson, *Paris* 5 C3 48 44N 2 16 E
Versailles, *B.A.* 32 B3 34 38 S 58 31W
Versailles, *Paris* 5 C2 48 48N 2 7 E
Veshnyaki, *Mos.* 11 E10 55 43N 37 48 E
Vesolyy Posolok, *St-Pet.* 11 B4 59 53N 30 28 E
Vestli, *Oslo* 2 B5 59 58N 10 55 E
Vestra, *Hels.* 3 B3 60 19N 24 46 E
Vestskoven, *Købn.* 2 D9 55 41N 12 23 E
Vesuvio, *Nápl.* 9 J13 40 49N 14 25 E
Vets Stadium, *Phil.* 24 B3 39 54N 75 10W
Viby, *Købn.* 2 E7 55 33N 12 1 E
Vicálvaro, *Mdrd.* 8 B3 40 24N 3 36W
Vicente Lopez, *B.A.* 32 B4 34 31 S 58 30W
Victoria, *B.A.* 32 A3 34 27 S 58 32W
Victoria, *H.K.* 12 E6 22 17N 114 11 E
Victoria, Pont, *Mtrl.* 20 B4 43 29N 73 32W
Victoria Gardens, *Bomb.* 16 H8 18 58N 72 50 E
Victoria Harbour, *H.K.* 12 E5 22 17N 114 10 E
Victoria Island, *Lagos* 18 B2 6 25N 7 25 E
Victoria L., *Jobg.* 18 F9 26 13 S 28 9 E
Victoria Lawn Tennis Courts, *Melb.* 19 F7 37 50 S 145 1 E
Victoria Park, *H.K.* 12 E5 22 16N 114 8 E
Vidja, *Stock.* 3 E11 59 12N 18 4 E
Vidrholec, *Pra.* 10 B3 50 5N 14 39 E
Vienna = Wien, *Wien* 10 G10 48 12N 16 22 E
Vienna, *Wash.* 25 D6 38 54N 77 16W
Vieringhausen, *Ruhr* 6 C4 51 10N 7 9 E
Vierlinden, *Ruhr* 6 A2 51 32N 6 45 E
Vierumäki, *Hels.* 3 A5 60 21N 25 2 E
Vierzigstücken, *Hbg.* 7 D6 53 30N 9 49 E
View Bank, *Melb.* 19 E7 37 43 S 145 6 E
Vigário Geral, *Rio J.* 31 A2 22 48 S 43 18W
Vigentino, *Mil.* 9 E6 45 26N 9 13 E
Viggbyholm, *Stock.* 3 D11 59 26N 18 7 E
Vighignolo, *Mil.* 9 E5 45 29N 9 2 E
Vigneux-sur-Seine, *Paris* 5 C4 48 42N 2 24 E
Viikki, *Hels.* 3 B5 60 13N 25 1 E
Viirilä, *Hels.* 3 B5 60 19N 25 8 E
Vila Andrade, *S. Pau.* 31 E5 23 37 S 46 44W
Vila Barcelona, *S. Pau.* 31 E6 23 37 S 46 33W
Vila Bocaina, *S. Pau.* 31 F7 23 40 S 46 26W
Vila Dalva, *S. Pau.* 31 E5 23 34 S 46 46W
Vila Dirce, *S. Pau.* 31 E4 23 33 S 46 50W
Vila Eldorado, *S. Pau.* 31 F6 23 42 S 46 38W
Vila Ema, *S. Pau.* 31 E6 23 35 S 46 31W
Vila Formosa, *S. Pau.* 31 E6 23 36 S 46 44W
Vila Galvão, *S. Pau.* 31 D6 23 27 S 46 34W
Vila Gonçales, *S. Pau.* 31 F6 23 42 S 46 33W
Vila Iasi, *S. Pau.* 31 E5 23 37 S 46 47W
Vila Indiana, *S. Pau.* 31 E5 23 37 S 46 48W
Vila Isabel, *Rio J.* 31 B2 22 54 S 43 15W
Vila Madalena, *S. Pau.* 31 E5 23 32 S 46 41W
Vila Maria, *S. Pau.* 31 E6 23 31 S 46 36W
Vila Mariana, *S. Pau.* 31 E6 23 34 S 46 36W
Vila Matilde, *S. Pau.* 31 E6 23 31 S 46 30W
Vila Nova Curuçá, *S. Pau.* 31 E7 23 31 S 46 25W
Vila Pires, *S. Pau.* 31 F6 23 41 S 46 30W
Vila Progresso, *Rio J.* 31 B3 22 53 S 43 1W
Vila Prudente, *S. Pau.* 31 E6 23 35 S 46 33W
Vila Ré, *S. Pau.* 31 E7 23 31 S 46 29W
Vila Remo, *S. Pau.* 31 F5 23 40 S 46 45W
Vila Sonia, *S. Pau.* 31 E5 23 35 S 46 43W
Viladecans, *Barc.* 8 E5 41 18N 2 1 E
Villa Ada, *Rome* 9 F10 41 55N 12 30 E
Villa Adelina, *B.A.* 32 B3 34 31 S 58 33W
Villa Alianza, *B.A.* 32 B3 34 37 S 58 33W
Villa Alsina, *B.A.* 32 C4 34 40 S 58 24W
Villa Altube, *B.A.* 32 B2 34 31 S 58 45W
Villa Ariza, *B.A.* 32 B3 34 38 S 58 39W
Villa Augusta, *B.A.* 32 C5 34 45 S 58 15W
Villa Ballester, *B.A.* 32 B3 34 33 S 58 33W
Villa Barilari, *B.A.* 32 C4 34 42 S 58 20W
Villa Basso, *B.A.* 32 B3 34 38 S 58 36W
Villa Bosch, *B.A.* 32 B3 34 35 S 58 33W
Villa C. Colon, *B.A.* 32 C4 34 41 S 58 21W
Villa D. F. Sarmiento, *B.A.* 32 B3 34 38 S 58 35W
Villa D. Sobral, *B.A.* 32 C5 34 45 S 58 15W
Villa de Guadalupe, *Méx.* 29 B3 19 29N 99 6W
Villa de Mayo, *B.A.* 32 B2 34 30 S 58 40W
Villa Devoto, *B.A.* 32 B3 34 36 S 58 31W
Villa Dominico, *B.A.* 32 C5 34 41 S 58 19W
Villa Giambruno, *B.A.* 32 C5 34 48 S 58 15W
Villa Gustavo A. Madero, *Méx.* 29 B3 19 29N 99 8W
Villa Hogar Alemán, *B.A.* 32 C4 34 49 S 58 26W
Villa Iglesias, *B.A.* 32 B2 34 30 S 58 45W
Villa Leloir, *B.A.* 32 B2 34 36 S 58 41W
Villa Leon, *B.A.* 32 B2 34 38 S 58 41W
Villa Lugano, *B.A.* 32 C4 34 41 S 58 27W
Villa Luzuriago, *B.A.* 32 C3 34 40 S 58 34W
Villa Lynch, *B.A.* 32 B3 34 35 S 58 31W
Villa Madero, *B.A.* 32 C3 34 41 S 58 30W
Villa Maria del Triunfo, *Lima* 30 G9 12 9 S 76 57W
Villa Obregon, *Méx.* 29 B2 19 20N 99 12W
Villa Reichembach, *B.A.* 32 B2 34 39 S 58 40W
Villa Rosa, *B.A.* 32 A1 34 25 S 58 5W
Villa San Francisco, *B.A.* 32 C5 34 46 S 58 15W
Villacoublay, Aérodrome de, *Paris* 5 C3 48 46N 2 12 E
Village Green, *Phil.* 24 B2 39 52N 75 26W
Villanova, *Phil.* 24 A2 40 1N 75 20W
Villarbasse, *Tori.* 9 B1 45 2N 7 27 E
Villaretto, *Tori.* 9 B3 45 8N 7 41 E
Villaricca, *Nápl.* 9 H12 40 55N 14 11 E
Villaroche, *Paris* 5 D5 48 37N 2 39 E
Villaroy, *Paris* 5 C2 48 46N 2 5 E
Villasanta, *Mil.* 9 D6 45 36N 9 18 E
Villastanza, *Mil.* 9 D4 45 32N 8 57 E
Villaverde, *Mdrd.* 8 B2 40 21N 3 42W
Villaverde Bajo, *Mdrd.* 8 B2 40 20N 3 40W
Villawood, *Syd.* 19 B2 33 52 S 150 58 E
Ville-d'Avray, *Paris* 5 C3 48 49N 2 11 E
Ville de Laval, *Mtrl.* 20 A3 43 34N 73 43W
Villebon-sur-Yvette, *Paris* 5 C3 48 41N 2 14 E
Villecresnes, *Paris* 5 C5 48 43N 2 31 E

48 Villejuif

Villejuif, *Paris* **5 C4** 48 47N 2 21 E
Villejust, *Paris* **5 C3** 48 41N 2 15 E
Villemoisson-sur-Orge, *Paris* **5 C3** 48 40N 2 19 E
Villemomble, *Paris* **5 B5** 48 52N 2 30 E
Villeneuve-la-Garenne, *Paris* **5 B3** 48 56N 2 19 E
Villeneuve-le-Roi, *Paris* **5 C4** 48 43N 2 24 E
Villeneuve-St.-Georges, *Paris* **5 C4** 48 43N 2 27 E
Villeneuve-sous-Dammartin, *Paris* **5 A5** 49 2N 2 38 E
Villennes-sur-Seine, *Paris* **5 B1** 48 56N 2 0 E
Villeparisis, *Paris* **5 B5** 48 56N 2 36 E
Villepinte, *Paris* **5 B5** 48 57N 2 30 E
Villepreux, *Paris* **5 C1** 48 49N 1 59 E
Villevaudé, *Paris* **5 B5** 48 55N 2 39 E
Villeziers, *Paris* **5 C3** 48 40N 2 10 E
Villiers-le-Bâcle, *Paris* **5 C2** 48 44N 2 8 E
Villiers-le-Bel, *Paris* **5 A4** 49 0N 2 23 E
Villiers-St. Frédéric, *Paris* **5 C1** 48 49N 1 53 E
Villiers-sur-Marne, *Paris* **5 C5** 48 49N 2 32 E
Villiers-sur-Orge, *Paris* **5 D3** 48 39N 2 18 E
Villinki, *Hels.* **3 C5** 60 9N 25 6 E
Villoresi, Canale, *Mil.* **9 D4** 45 33N 8 59 E
Vimodrone, *Mil.* **9 D6** 45 30N 9 16 E
Vimont, *Mtrl.* **20 A3** 43 36N 73 43W
Vincennes, *Paris* **5 B4** 48 51N 2 26 E
Vincennes, Bois de, *Paris* **5 C4** 48 49N 2 26 E
Vinohrady, *Pra.* **10 B2** 50 4N 14 26 E
Vinoř, *Pra.* **10 B3** 50 8N 14 34 E
Vinořský →, *Pra.* **10 A3** 50 11N 14 39 E
Violet Hill, *H.K.* **12 E6** 22 15N 114 11 E
Virányos, *Bud.* **10 J12** 47 31N 18 59 E
Virgeo del San Cristóbal, *Stgo* **30 J11** 33 25 S 70 38W
Viroflay, *Paris* **5 C3** 48 48N 2 10 E
Viron, *Ath.* **8 J11** 37 55N 23 46 E
Virreyes, *B.A.* **32 A3** 34 27 S 58 33W
Virum, *Købn.* **2 D9** 55 47N 12 27 E
Viry-Châtillon, *Paris* **5 C4** 48 40N 2 21 E
Vishnyaki, *Mos.* **11 E11** 55 46N 37 53 E
Visitacion Valley, *S.F.* **27 B2** 37 42N 122 23W
Vista Alegre, *Lima* **30 G9** 12 8 S 76 59W
Vista Alegre, *Stgo* **30 K10** 33 30 S 70 43W
Vitacura, *Stgo* **30 J11** 33 23 S 70 35W
Vitarte-Ate, *Lima* **30 G9** 12 3 S 76 57W
Vitinia, *Rome* **9 G9** 41 47N 12 24 E
Vitry-sur-Seine, *Paris* **5 C4** 48 47N 2 23 E
Vitträsk, *Hels.* **3 B1** 60 11N 24 29 E
Vittuone, *Mil.* **9 E4** 45 28N 8 57 E
Vladykino, *Mos.* **11 D9** 55 51N 37 35 E
Vltava →, *Pra.* **10 A2** 50 10N 14 2 E
Vnukovo, *Mos.* **11 F7** 55 37N 37 17 E
Voerde, *Ruhr* **6 C6** 51 18N 7 23 E
Voerde, *Ruhr* **6 A2** 51 35N 6 42 E
Vogelheim, *Ruhr* **6 B3** 51 29N 6 59 E
Vohwinkel, *Ruhr* **6 C4** 51 13N 7 4 E
Voisins-le-Bretonneux, *Paris* **5 C2** 48 45N 2 3 E
Vokovice, *Pra.* **10 B2** 50 5N 14 21 E
Volgelsdorf, *Berl.* **7 B5** 52 30N 13 44 E
Volkhonka-Zil, *Mos.* **11 F9** 55 39N 37 37 E
Volkovka →, *St-Pet.* **11 B4** 59 54N 30 25 E
Volksdorf, *Hbg.* **7 D8** 53 39N 10 8 E
Volla, *Nápl.* **9 H13** 40 52N 14 20 E
Vollen, *Oslo* **2 C2** 59 48N 10 27 E
Volmarstein, *Ruhr* **6 B6** 51 22N 7 22 E
Volodarskoye, *St-Pet.* **11 B4** 59 54N 30 23 E
Volpiano, *Tori.* **9 A3** 45 12N 7 46 E
Volynkina-Derevnya, *St-Pet.* **11 B3** 59 53N 30 18 E
Volynyy, Os., *St-Pet.* **11 B3** 59 57N 30 14 E
Vōmero, *Nápl.* **9 H12** 40 50N 14 13 E
Vorderhainbach, *Wien* **10 G9** 48 13N 16 12 E
Vorhalle, *Ruhr* **6 B6** 51 23N 7 26 E
Vormholz, *Ruhr* **6 B5** 51 24N 7 19 E
Vösendorf, *Wien* **10 H10** 48 7N 16 20 E
Vostochnyy, *Mos.* **11 E11** 55 49N 37 51 E
Vouliagmeni, *Ath.* **8 K11** 37 50N 23 46 E
Vrčovice, *Pra.* **10 B2** 50 4N 14 28 E
Vsevolozhsk, *St-Pet.* **11 A5** 60 0N 30 39 E
Vuosaari, *Hels.* **3 B5** 60 13N 25 8 E
Vyborgskaya Storona, *St-Pet.* **11 B4** 59 57N 30 22 E
Vyčehrad, *Pra.* **10 B2** 50 3N 14 25 E
Vykhino, *Mos.* **11 E10** 55 42N 37 48 E
Vysočany, *Pra.* **10 B2** 50 6N 14 29 E

W

Waban, L., *Bost.* **21 D2** 42 17N 71 18W
Wachterhof, *Mün.* **7 G11** 48 2N 11 42 E
Waddington, *Lon.* **4 D4** 51 18N 0 7W
Wadeville, *Jobg.* **18 F10** 26 15 S 28 11 E
Wahda, *Bagd.* **17 F8** 33 18N 44 26 E
Währing, *Wien* **10 G10** 48 14N 16 20 E
Waidmannslust, *Berl.* **7 A3** 52 36N 13 20 E
Wajay, *La Hab.* **30 B2** 23 0N 82 25W
Wakefield, *Bost.* **21 B3** 42 30N 71 5W
Wald, *Ruhr* **6 C4** 51 11N 7 3 E
Waldesruh, *Berl.* **7 B4** 52 28N 13 37 E
Waldheim, *Berl.* **7 A1** 52 34N 13 3 E
Waldperlach, *Mün.* **7 G11** 48 4N 11 40 E
Waldtrudering, *Mün.* **7 G11** 48 6N 11 42 E
Waldwick, *N.Y.* **22 A4** 41 1N 74 5W
Wall Street, *N.Y.* **22 C4** 40 42N 74 0W
Wallgrove, *Syd.* **19 A2** 33 47 S 150 51 E
Wallington, *Lon.* **4 C4** 51 21N 0 8W
Wallington, *N.Y.* **22 B4** 40 50N 74 8W
Walnut Cr. →, *S.F.* **27 A4** 37 55N 122 3W
Walnut Creek, *S.F.* **27 A4** 37 53N 122 3W
Walnut Heights, *S.F.* **27 A4** 37 52N 122 2W
Walsum, *Ruhr* **6 A2** 51 32N 6 42 E
Walsumer Mark, *Ruhr* **6 A3** 51 33N 6 50 E
Walt Whitman Br., *Phil.* **24 B4** 39 4N 75 9W
Waltershof, *Hbg.* **7 D7** 53 31N 9 54 E
Waltham, *Bost.* **21 C2** 42 23N 71 14W
Waltham Abbey, *Lon.* **4 A5** 51 41N 0 1 E
Waltham Forest, *Lon.* **4 B4** 51 36N 0 0 E
Walthamstow, *Lon.* **4 B4** 51 34N 0 1W
Walton on Thames, *Lon.* **4 C2** 51 22N 0 23W
Walton on the Hill, *Lon.* **4 D3** 51 16N 0 14W
Waltrop, *Ruhr* **6 A6** 51 36N 7 25 E
Walworth, *Lon.* **4 C4** 51 29N 0 5W
Wambachsee, *Ruhr* **6 B2** 51 23N 6 47 E
Wan Chai, *H.K.* **12 E6** 22 16N 114 10 E
Wanaque, *N.Y.* **22 A3** 41 1N 74 17W
Wandezhuang, *Tianj.* **14 E5** 39 6N 117 10 E
Wandle →, *Lon.* **4 C3** 51 27N 0 11W
Wandsbek, *Hbg.* **7 D8** 53 34N 10 4 E
Wandsworth, *Lon.* **4 C3** 51 27N 0 11W
Wang Hin, Khlong →, *Bangk.* **15 A2** 13 50N 100 35 E
Wanheim, *Ruhr* **6 B2** 51 23N 6 45 E
Wanheimerort, *Ruhr* **6 B2** 51 24N 6 45 E
Wanne-Eickel, *Ruhr* **6 A4** 51 31N 7 9 E
Wannsee, *Berl.* **7 B1** 52 25N 13 9 E
Wansdorf, *Berl.* **7 A1** 52 38N 13 5 E
Wanstead, *Lon.* **4 B5** 51 34N 0 1 E
Wantagh Seaford, *N.Y.* **23 D8** 40 39N 73 28W
Wantirna, *Melb.* **19 F8** 37 50 S 145 14 E
Wapping, *Lon.* **4 B4** 51 30N 0 3W
Warabi, *Tōkyō* **13 B3** 35 49N 139 42 E
Ward, *Phil.* **24 B1** 39 52N 75 30W
Warlingham, *Lon.* **4 D4** 51 18N 0 2W
Warnberg, *Mün.* **7 G10** 48 4N 11 31 E
Warngal Park, *Melb.* **19 E7** 37 45 S 145 4 E
Warrandyte, *Melb.* **19 E8** 37 43 S 145 13 E
Warrandyte Park, *Melb.* **19 E8** 37 44 S 145 14 E
Warrandyte South, *Melb.* **19 E8** 37 44 S 145 14 E
Warranwood, *Melb.* **19 E8** 37 46 S 145 14 E
Warrâq el 'Arab, *El Qâ.* **18 C5** 30 4N 31 11 E
Warrâq el Hadf, *El Qâ.* **18 C5** 30 5N 31 12 E
Warren Hill, *Bost.* **21 B1** 42 35N 71 21W
Warsaw = Warszawa, *Wsaw.* **10 E7** 52 14N 21 0 E
Warszawa, *Wsaw.* **10 E7** 52 14N 21 0 E
Wartenberg, *Berl.* **7 A4** 52 34N 13 31 E
Warwick Farm Racetrack, *Syd.* **19 B2** 33 54 S 150 56 E
Wasa, *Stock.* **3 E11** 59 19N 18 5 E
Wasfanārd, *Tehr.* **17 D5** 35 38N 51 20 E
Washington, *Wash.* **25 D7** 38 53N 77 2W
Washington Heights, *N.Y.* **22 B5** 40 51N 73 56W
Washington Memorial Museum, *Phil.* **24 A2** 40 5N 75 26W
Washington Nat. Airport, *Wash.* **25 D7** 38 51N 77 2W
Washington Park, *Chic.* **26 C3** 41 47N 87 36W
Washington Square, *Phil.* **24 A3** 40 9N 75 19W
Washington Township, *N.Y.* **22 A4** 41 0N 74 3W
Wasserschloss, *Ruhr* **6 A4** 51 32N 7 1 E
Watching Mts., *N.Y.* **22 C2** 40 43N 74 20W
Watchung, *N.Y.* **22 D2** 40 38N 74 29W
Waterloo, *Syd.* **19 B4** 33 53 S 151 12 E
Waterman Mt., *L.A.* **28 A5** 34 14N 117 56W
Watertown, *Bost.* **21 C2** 42 22N 71 10W
Watford, *Lon.* **4 A2** 51 40N 0 27W
Watkins Island, *Wash.* **25 C6** 39 2N 77 15W
Watsonia, *Melb.* **19 E7** 37 43 S 145 6 E
Watsons B., *Syd.* **19 B4** 33 50 S 151 18 E
Watsons Creek, *Melb.* **19 E8** 37 40 S 145 13 E
Wattenscheid, *Ruhr* **6 B4** 51 28N 7 8 E
Wattle Glen, *Melb.* **19 D8** 37 39 S 145 11 E
Wattle Park, *Melb.* **19 F7** 37 50 S 145 6 E
Watts →, *Wash.* **25 C6** 39 2N 77 15W
Waverley, *Bost.* **21 C2** 42 23N 71 10W
Waverley, *Jobg.* **18 E9** 26 7 S 28 4 E
Waverley, *Syd.* **19 B4** 33 53 S 151 15 E
Wawer, *Wsaw.* **10 E7** 52 13N 21 8 E
Wawrzyszew, *Wsaw.* **10 E6** 52 17N 20 53 E
Wayland, *Bost.* **21 C1** 42 21N 71 20W
Wayne, *N.Y.* **22 B3** 40 55N 74 15W
Wayne, *Phil.* **24 A2** 40 2N 75 24W
Wazirabad, *Delhi* **16 A2** 28 43N 77 13 E
Wazīrīya, *Bagd.* **17 E8** 33 22N 44 23 E
Wazirpur, *Delhi* **16 A2** 28 41N 77 10 E
Weald Park, *Lon.* **4 B6** 51 37N 0 16 E
Wedding, *Berl.* **7 A3** 52 32N 13 21 E
Weehawken, *N.Y.* **22 C4** 40 45N 74 2W
Wegendorf, *Berl.* **7 A5** 52 36N 13 45 E
Wehofen, *Ruhr* **6 A2** 51 31N 6 46 E
Wehringhausen, *Ruhr* **6 B6** 51 21N 7 28 E
Weidling, *Wien* **10 G9** 48 17N 16 18 E
Weidling →, *Wien* **10 G9** 48 17N 16 19 E
Weidlingbach, *Wien* **10 G9** 48 16N 16 15 E
Weigongcum, *Beij.* **14 B2** 39 57N 116 16 E
Weijin He →, *Tianj.* **14 E6** 39 3N 117 12 E
Weissensee, *Berl.* **7 A3** 52 33N 13 27 E
Weitmar, *Ruhr* **6 B5** 51 27N 7 11 E
Welcome Monument, *Jak.* **15 J9** 6 12N 106 49 E
Weller Creek, *Chic.* **26 A1** 42 2N 87 52W
Wellesley, *Bost.* **21 D2** 42 17N 71 17W
Wellesley Fells, *Bost.* **21 D2** 42 18N 71 18W
Wellesley Hills, *Bost.* **21 D2** 42 18N 71 16W
Welling, *Lon.* **4 C5** 51 27N 0 6 E
Wellingsbüttel, *Hbg.* **7 D8** 53 38N 10 6 E
Weltevreden Park Extension, *Jobg.* **18 E8** 26 7 S 27 56 E
Wembley, *Lon.* **4 B3** 51 33N 0 17W
Wembley Stadium, *Jobg.* **18 F9** 26 13 S 28 1 E
Wembley Stadium, *Lon.* **4 B3** 51 33N 0 16W
Wemmer Pan, *Jobg.* **18 F9** 26 13 S 28 3 E
Wendenschloss, *Berl.* **7 B4** 52 24N 13 35 E
Wengern, *Ruhr* **6 B6** 51 24N 7 20 E
Wenham, *Bost.* **21 B4** 42 36N 70 53W
Wenham L., *Bost.* **21 B4** 42 35N 70 53W
Wenhuagong, *Tianj.* **14 E6** 39 5N 117 14 E
Wennington, *Lon.* **4 B6** 51 30N 0 12 E
Wenonah, *Phil.* **24 C4** 39 47N 75 9W
Wentworthville, *Syd.* **19 A2** 33 48 S 150 58 E
Werden, *Ruhr* **6 B4** 51 23N 7 1 E
Werne, *Ruhr* **6 B5** 51 29N 7 18 E
Werneuchen, *Berl.* **7 A5** 52 38N 13 44 E
Wesoła, *Wsaw.* **10 E8** 52 15N 21 13 E
West Andover, *Bost.* **21 B2** 42 39N 71 10W
West Babylon, *N.Y.* **23 C8** 40 43N 73 21W
West Bedford, *Bost.* **21 C2** 42 28N 71 18W
West Berlin, *Phil.* **24 C5** 39 48N 74 56W
West Boxford, *Bost.* **21 A3** 42 42N 71 3W
West Caldwell, *N.Y.* **22 B3** 40 51N 74 16W
West Chelmsford, *Bost.* **21 B1** 42 36N 71 23W
West Chester, *Phil.* **24 B1** 39 57N 75 35W
West Concord, *Bost.* **21 C1** 42 27N 71 24W
West Covina, *L.A.* **28 B5** 34 4N 117 55W
West Don →, *Trto.* **20 D8** 43 44N 79 24W
West Drayton, *Lon.* **4 B2** 51 30N 0 28W
West Dulwich, *Lon.* **4 C4** 51 26N 0 5W
West Edmondale, *Balt.* **25 B2** 39 17N 76 42W
West Ham, *Lon.* **4 B5** 51 31N 0 1 E
West Harrow, *Lon.* **4 B2** 51 34N 0 21W
West Heath, *Lon.* **4 C5** 51 29N 0 7 E
West Hempstead, *N.Y.* **23 C7** 40 41N 73 38W
West Hill, *Trto.* **20 D9** 43 46N 79 10W
West Hollywood, *L.A.* **28 B2** 34 5N 118 21W
West Hoxton, *Syd.* **19 B1** 33 55 S 150 49 E
West Islip, *N.Y.* **23 C9** 40 41N 73 18W
West Kingsdown, *Lon.* **4 C6** 51 20N 0 15 E
West Lamma Channel, *H.K.* **12 E5** 22 14N 114 4 E
West Lynn, *Bost.* **21 C4** 42 27N 70 58W
West Medford, *Bost.* **21 C3** 42 25N 71 7W
West New York, *N.Y.* **22 C4** 40 46N 74 1W
West Norwood, *N.Y.* **22 B5** 40 59N 73 58W
West of Twin Peaks, *S.F.* **27 B2** 37 43N 122 27W
West Orange, *N.Y.* **22 C3** 40 46N 74 15W
West Park, *Jobg.* **18 E8** 26 9 S 27 59 E
West Paterson, *N.Y.* **22 B3** 40 54N 74 11W
West Rouge, *Trto.* **20 D10** 43 48N 79 7W
West Roxbury, *Bost.* **21 D3** 42 16N 71 9W
West Springfield, *Wash.* **25 E6** 38 47N 77 13W
West Thurrock, *Lon.* **4 C6** 51 28N 0 16 E
West Town, *Chic.* **26 B2** 41 53N 87 42W
West Wharf, *Kar.* **17 H10** 24 49N 66 58 E
West Wickham, *Lon.* **4 C4** 51 22N 0 0 E
Westbury, *N.Y.* **23 C7** 40 45N 73 34W
Westchester, *Chic.* **26 B1** 41 51N 87 53W
Westchester, *N.Y.* **23 B5** 40 51N 73 51W
Westcliff, *Jobg.* **18 F9** 26 10 S 28 1 E
Westdale, *Chic.* **26 B1** 41 55N 87 54W
Westdene, *Jobg.* **18 F8** 26 10 S 27 59 E
Westend, *Hels.* **3 C3** 60 9N 24 48 E
Westerbauer, *Ruhr* **6 B6** 51 20N 7 23 E
Westerham, *Lon.* **4 D5** 51 16N 0 4 E
Westerham, *Mün.* **7 G10** 48 3N 11 36 E
Westerholt, *Ruhr* **6 A4** 51 36N 7 5 E
Westerleigh, *N.Y.* **22 D4** 40 37N 74 7W
Western Addition, *S.F.* **27 B2** 37 47N 122 25W
Western Run →, *Balt.* **25 A2** 39 22N 76 39W
Western Springs, *Chic.* **26 C1** 41 47N 87 52W
Westfalenhalle, *Ruhr* **6 B6** 51 29N 7 27 E
Westfield, *N.Y.* **22 D2** 40 39N 74 21W
Westlake, *S.F.* **27 B2** 37 42N 122 29W
Westmeadows, *Melb.* **19 D6** 37 39 S 144 55 E
Westminster, *Lon.* **4 B4** 51 30N 0 7W
Westminster Abbey, *Lon.* **4 C4** 51 29N 0 7W
Westmont, *Phil.* **24 B4** 39 54N 75 3W
Westmount, *Mtrl.* **20 B4** 43 29N 73 35W
Weston, *Bost.* **21 C2** 42 22N 71 16W
Weston, *Trto.* **20 D7** 43 42N 79 30W
Weston Res., *Bost.* **21 C2** 42 20N 71 11W
Westover Hills, *Phil.* **24 C1** 39 45N 75 35W
Westtown, *Phil.* **24 B1** 39 56N 75 32W
Westville, *Phil.* **24 B4** 39 52N 75 7W
Westville Grove, *Phil.* **24 B4** 39 51N 75 7W
Westwood, *Bost.* **21 D2** 42 12N 71 14W
Westwood, *N.Y.* **22 B4** 40 59N 74 3W
Westwood Village, *L.A.* **28 B2** 34 3N 118 26W
Wetter, *Ruhr* **6 B6** 51 23N 7 23 E
Wexford, *Trto.* **20 D9** 43 45N 79 18W
Wey →, *Lon.* **4 D2** 51 18N 0 27W
Weybridge, *Lon.* **4 C2** 51 22N 0 27W
Weyer, *Ruhr* **6 C4** 51 10N 7 1 E
Weymouth, *Bost.* **21 D4** 42 12N 70 57W
Whampoa, Sungei →, *Sing.* **15 G8** 1 18N 103 52 E
Wheaton, *Wash.* **25 C7** 39 2N 77 2W
Wheaton Regional Park, *Wash.* **25 C7** 39 3N 77 1W
Wheelers Hill, *Melb.* **19 F8** 37 53 S 145 10 E
Wheeling, *Chic.* **26 A1** 42 8N 87 54W
Whetstone, *Lon.* **4 B3** 51 37N 0 10W
Whippany, *N.Y.* **22 B2** 40 49N 74 24W
Whippany →, *N.Y.* **22 B2** 40 50N 74 20W
White Marsh, *Balt.* **25 A4** 39 23N 76 28W
White Meadow L., *N.Y.* **22 B1** 40 55N 74 30W
White Oak, *Wash.* **25 C8** 39 2N 76 59W
White Plains, *N.Y.* **23 A6** 41 0N 73 46W
Whitechapel, *Lon.* **4 B4** 51 31N 0 3W
Whitehorse, *Phil.* **24 B2** 39 59N 75 28W
Whiteley Village, *Lon.* **4 C2** 51 21N 0 25W
Whitemarsh →, *Balt.* **25 A4** 39 22N 76 29W
Whitestone, *N.Y.* **23 C6** 40 47N 73 48W
Whiting, *Chic.* **26 C4** 41 41N 87 30W
Whitmans Pond, *Bost.* **21 D4** 42 12N 70 55W
Whittier, *L.A.* **28 C4** 33 58N 118 2W
Whitton, *Lon.* **4 C2** 51 27N 0 21W
Whyteleafe, *Lon.* **4 D4** 51 18N 0 5W
Wieden, *Wien* **10 G10** 48 11N 16 22 E
Wiemelhausen, *Ruhr* **6 B5** 51 27N 7 13 E
Wien, *Wien* **10 G10** 48 12N 16 22 E
Wien-Schwechat, Flughafen, *Wien* **10 H11** 48 6N 16 34 E
Wiener Berg, *Wien* **10 H10** 48 9N 16 21 E
Wiener Wald, *Wien* **10 G9** 48 16N 16 14 E
Wieruchów, *Wsaw.* **10 E5** 52 14N 20 49 E
Wierzbno, *Wsaw.* **10 E7** 52 11N 21 1 E
Wilanów, *Wsaw.* **10 E7** 52 10N 21 4 E
Wilanówka →, *Wsaw.* **10 F7** 52 13N 21 6 E
Wildcat Canyon Regional Park, *S.F.* **27 A3** 37 56N 122 17W
Wildcat Cr. →, *S.F.* **27 A3** 37 57N 122 15W
Wilde, *B.A.* **32 C5** 34 24 S 58 18W
Wilhelmsburg, *Hbg.* **7 E7** 53 29N 9 59 E
Wilhelmshagen, *Berl.* **7 B5** 52 26N 13 42 E
Wilket Creek Park, *Trto.* **20 D8** 43 43N 79 21W
Willesden, *Lon.* **4 B3** 51 32N 0 15W
Willesden Green, *Lon.* **4 B3** 51 32N 0 13W
Willett Pond, *Bost.* **21 D2** 42 10N 71 14W
William Girling Res., *Lon.* **4 B4** 51 38N 0 1W
Williams Bridge, *N.Y.* **23 B5** 40 52N 73 51W
Williamsburg, *N.Y.* **22 C5** 40 42N 73 56W
Williamstown, *Melb.* **19 F6** 37 51 S 144 52 E
Williamstown Junction, *Phil.* **24 C5** 39 45N 74 56W
Willingboro, *Phil.* **24 A5** 40 2N 74 53W
Williston Park, *N.Y.* **23 C7** 40 45N 73 38W
Willoughby, *Syd.* **19 A4** 33 48 S 151 12 E
Willow Grove, *Phil.* **24 A4** 40 8N 75 7W
Willow Springs, *Chic.* **26 C1** 41 44N 87 51W
Willowbrook, *L.A.* **28 C3** 33 54N 118 13W
Willowbrook, *N.Y.* **22 D4** 40 35N 74 8W
Willowdale, *Phil.* **24 B5** 39 52N 74 58W
Willowdale, *Trto.* **20 D8** 43 46N 79 25W
Willowdale State Forest, *Bost.* **21 B4** 42 39N 70 54W
Wilmette, *Chic.* **26 A2** 42 4N 87 42W
Wilmette Harbor, *Chic.* **26 A2** 42 4N 87 41W
Wilmington, *Bost.* **21 B3** 42 33N 71 9W
Wilmington, *Lon.* **4 C6** 51 25N 0 12 E
Wilmington, *Phil.* **24 C1** 39 44N 75 33W
Wilson, Mt., *L.A.* **28 A4** 34 13N 118 4W
Wimbledon, *Lon.* **4 C3** 51 25N 0 13W
Wimbledon Common, *Lon.* **4 C3** 51 26N 0 14W
Wimbledon Park, *Lon.* **4 C3** 51 25N 0 11W
Wimbledon Tennis Ground, *Lon.* **4 C3** 51 25N 0 12W
Winchester, *Bost.* **21 C3** 42 26N 71 8W
Winchmore Hill, *Lon.* **4 B4** 51 38N 0 5W
Windsor Cresta, *Jobg.* **18 E8** 26 7 S 27 59 E
Winfield, *N.Y.* **22 D3** 40 38N 74 16W
Winnetka, *Chic.* **26 A2** 42 6N 87 43W
Winnetka, *L.A.* **28 A1** 34 10N 118 32W
Winning, *Mün.* **7 G10** 48 2N 11 37 E
Winston Hills, *Syd.* **19 A2** 33 46 S 150 57 E
Winterberg, *Ruhr* **6 C5** 51 19N 7 12 E
Winterhude, *Hbg.* **7 D8** 53 35N 10 0 E
Winterthur, *Phil.* **24 C1** 39 48N 75 35W
Winthrop, *Bost.* **21 C4** 42 22N 70 58W
Winzeldorf, *Hbg.* **7 C7** 53 40N 9 54 E
Wisley Gardens, *Lon.* **4 D2** 51 18N 0 28W
Wiśniowa Góra, *Wsaw.* **10 E8** 52 13N 21 12 E
Wissahickon Cr. →, *Phil.* **24 A3** 40 0N 75 12W
Wissinoming, *Phil.* **24 A4** 40 1N 75 4W
Wissous, *Paris* **5 C3** 48 44N 2 19 E
Witch House, *Bost.* **21 B4** 42 31N 70 54W
Witfield, *Jobg.* **18 F10** 26 11 S 28 12 E
Witpoortjie, *Jobg.* **18 E8** 26 8 S 27 50 E
Witten, *Ruhr* **6 B6** 51 26N 7 20 E
Wittenau, *Berl.* **7 A2** 52 35N 13 19 E
Wittlaer, *Ruhr* **6 C2** 51 19N 6 44 E
Witwatersrand, Univ. of, *Jobg.* **18 F9** 26 11 S 28 1 E
Włochy, *Wsaw.* **10 E6** 52 12N 20 54 E
Wo Mei, *H.K.* **12 D6** 22 21N 114 15 E
Wo Yi Hop, *H.K.* **12 D5** 22 23N 114 8 E
Woburn, *Bost.* **21 C3** 42 29N 71 9W
Woburn, *Trto.* **20 D9** 43 46N 79 12W
Wohldorf-Ohlstedt, *Hbg.* **7 C8** 53 41N 10 7 E
Wola, *Wsaw.* **10 E6** 52 14N 20 57 E
Woldingham, *Lon.* **4 D4** 51 16N 0 1W
Wolf Lake, *Chic.* **26 D3** 41 39N 87 31W
Wolf Trap Farm Park, *Wash.* **25 D6** 38 56N 77 17W
Wolfpassing, *Wien* **10 G9** 48 18N 16 10 E
Wolica, *Wsaw.* **10 F7** 52 9N 21 3 E
Wolica, *Wsaw.* **10 F6** 52 7N 20 51 E
Wólka Węglowa, *Wsaw.* **10 E6** 52 18N 20 52 E
Wollaston, *Bost.* **21 D3** 42 15N 71 2W
Wołomin, *Wsaw.* **10 D8** 52 20N 21 12 E
Woltersdorf, *Berl.* **7 B5** 52 27N 13 44 E
Wong Chuk Hang, *H.K.* **12 E6** 22 15N 114 10 E
Wong Chuk Wan, *H.K.* **12 D6** 22 23N 114 17 E
Wong Chuk Yeung, *H.K.* **12 D6** 22 24N 114 15 E
Wong Ngua Shan, *H.K.* **12 D6** 22 22N 114 14 E
Wong Tai Sin, *H.K.* **12 D6** 22 20N 114 11 E
Wonga Park, *Melb.* **19 E8** 37 44 S 145 17 E
Wood End, *Lon.* **4 B2** 51 33N 0 21W
Wood Green, *Lon.* **4 B4** 51 36N 0 6W
Wood Hill, *Bost.* **21 B2** 42 39N 71 11W
Woodbridge, *N.Y.* **22 D3** 40 33N 74 16W
Woodbridge, *Trto.* **20 D7** 43 47N 79 35W
Woodbridge Cr. →, *N.Y.* **22 D3** 40 32N 74 15W
Woodbury, *N.Y.* **23 C8** 40 49N 73 28W
Woodbury, *Phil.* **24 B4** 39 50N 75 9W
Woodbury Cr. →, *Phil.* **24 B4** 39 51N 75 11W
Woodbury Heights, *Phil.* **24 C4** 39 49N 75 7W
Woodchuck Hill, *Bost.* **21 B3** 42 39N 71 4W
Woodcliff Lake, *N.Y.* **22 A4** 41 1N 74 2W
Woodford, *Lon.* **4 B5** 51 36N 0 1 E
Woodford Bridge, *Lon.* **4 B5** 51 36N 0 3 E
Woodford Green, *Lon.* **4 B5** 51 36N 0 1 E
Woodford Wells, *Lon.* **4 B5** 51 37N 0 1 E
Woodhaven, *N.Y.* **23 C5** 40 41N 73 51W
Woodlands, *Sing.* **15 F7** 1 26N 103 46 E
Woodlawn, *Balt.* **25 B2** 39 19N 76 44W
Woodlyn, *Phil.* **24 B2** 39 52N 75 21W
Woodlynne, *Phil.* **24 B4** 39 54N 75 6W
Woodmere, *N.Y.* **23 D6** 40 38N 73 43W
Woodmont, *Wash.* **25 D7** 38 59N 77 5W
Woodmore, *Balt.* **25 A2** 39 22N 76 47W
Woodridge, *N.Y.* **22 B4** 40 50N 74 4W
Woodrow, *N.Y.* **22 D3** 40 32N 74 11W
Woodside, *Lon.* **4 C4** 51 23N 0 4W
Woodside, *N.Y.* **23 C5** 40 44N 73 54W
Woodside, *S.F.* **27 D3** 37 26N 122 16W
Woodstock, *Balt.* **25 B1** 39 19N 76 52W
Woodstream, *Phil.* **24 B5** 39 54N 74 57W
Woollahra, *Syd.* **19 B4** 33 53 S 151 15 E
Woolooware B., *Syd.* **19 C3** 34 1 S 151 8 E
Woolwich, *Lon.* **4 C5** 51 29N 0 4 E
Wördern, *Wien* **10 G9** 48 19N 16 12 E
World Trade Center, *N.Y.* **22 C4** 40 42N 74 0W
Worli, *Bomb.* **16 G7** 19 1N 72 49 E
Woronora, *Syd.* **19 C3** 34 1 S 151 2 E
Worth, *Chic.* **26 C2** 41 41N 87 47W
Worthington, *Balt.* **25 B2** 39 14N 76 47W
Worthington, *N.Y.* **23 A6** 41 2N 73 49W
Wrotham, *Lon.* **4 D6** 51 18N 0 18 E
Wrotham Park, *Lon.* **4 A3** 51 40N 0 10W
Wuhlgarten, *Berl.* **7 A4** 52 31N 13 34 E
Wujiaochang, *Shang.* **14 J12** 31 18N 121 31 E
Wülfrath, *Ruhr* **6 C4** 51 16N 7 2 E
Wulfsmühle, *Hbg.* **7 C7** 53 41N 9 51 E
Wulksfelde, *Hbg.* **7 C8** 53 42N 10 6 E
Wupper →, *Ruhr* **6 C5** 51 14N 7 18 E
Wuppertal, *Ruhr* **6 C5** 51 17N 7 10 E
Würm →, *Mün.* **7 G9** 48 8N 11 27 E
Würm-kanal, *Mün.* **7 F9** 48 13N 11 29 E
Wusong, *Shang.* **14 H11** 31 22N 121 29 E
Wusong Jiang →, *Shang.* **14 J11** 31 15N 121 29 E
Wyandanch, *N.Y.* **23 C8** 40 44N 73 20W
Wyckoff, *N.Y.* **22 A3** 41 0N 74 10W
Wyczółki, *Wsaw.* **10 F6** 52 9N 20 59 E
Wygoda, *Wsaw.* **10 E7** 52 15N 21 7 E
Wyncote, *Phil.* **24 A4** 40 5N 75 8W
Wynnewood, *Phil.* **24 A3** 40 0N 75 17W
Wynnmere, *Bost.* **21 C3** 42 29N 71 9W
Wyola, *Phil.* **24 A2** 40 0N 75 24W

X

Xabregas, *Lisb.* **8 F8** 38 43N 9 6W
Xiaodianzhuang, *Tianj.* **14 D6** 39 14N 117 14 E
Xiaoping, *Gzh.* **14 F8** 23 12N 113 13 E
Xiasha chong, *Gzh.* **14 G7** 23 8N 113 9 E
Xicheng, *Beij.* **14 B2** 39 54N 116 19 E
Xico, Cerro, *Méx.* **29 C4** 19 15N 98 56W
Xicun, *Gzh.* **14 G8** 23 8N 113 13 E
Xidan, *Beij.* **14 B3** 39 52N 116 20 E
Xigu Gongyuan, *Tianj.* **14 D6** 39 10N 117 10 E
Xigucun, *Tianj.* **14 D5** 39 10N 117 9 E
Xijiao Airport, *Beij.* **14 B2** 39 57N 116 12 E
Xikeng, *Gzh.* **14 F7** 23 11N 113 6 E
Xilou, *Tianj.* **14 E6** 39 5N 117 12 E
Ximenwai, *Tianj.* **14 E5** 39 8N 117 9 E
Xingfusancun, *Beij.* **14 B3** 39 55N 116 25 E
Xinhua, *Tianj.* **14 E6** 39 6N 117 12 E
Xinkai He →, *Tianj.* **14 D6** 39 12N 117 15 E
Xintang, *Gzh.* **14 G9** 23 9N 113 24 E
Xitle, *Méx.* **29 C2** 19 15N 99 12W
Xitle, Cerro, *Méx.* **29 C2** 19 14N 99 12W
Xiyuan, *Beij.* **14 B2** 39 59N 116 14 E
Xizhimen, *Beij.* **14 B2** 39 55N 116 19 E
Xochiaca, *Méx.* **29 B4** 19 24N 89 58 E
Xochimilco, *Méx.* **29 C3** 19 15N 99 5W
Xochimilco, L. de, *Méx.* **29 C3** 19 16N 99 7W
Xochitenco, *Méx.* **29 B4** 19 24N 89 58 E
Xochitepec, *Méx.* **29 C3** 19 15N 99 9W
Xuanwu, *Beij.* **14 B2** 39 52N 116 19 E
Xuhui, *Shang.* **14 J11** 31 11N 121 26 E

Y

Yaba, *Lagos* **18 A2** 6 30N 7 22 E
Yadun Shui, *Gzh.* **14 G8** 23 5N 113 15 E
Yaftābād, *Tehr.* **17 D4** 35 37N 51 17 E
Yagoona, *Syd.* **19 B3** 33 54 S 151 2 E
Yahara, *Tōkyō* **13 B2** 35 44N 139 37 E
Yaho, *Tōkyō* **13 B1** 35 40N 139 26 E
Yakire, *Tōkyō* **13 B4** 35 45N 139 54 E
Yamada, *Ōsaka* **12 B4** 34 47N 135 32 E
Yamada, *Tōkyō* **13 C2** 35 33N 139 37 E
Yamada →, *Ōsaka* **12 B2** 34 45N 135 10 E
Yamaguchi, *Ōsaka* **12 B2** 34 49N 135 14 E
Yamamoto, *Ōsaka* **12 C4** 34 38N 135 37 E
Yamato, *Tōkyō* **13 B2** 35 46N 139 36 E
Yamato, *Tōkyō* **13 D1** 35 29N 139 27 E
Yamato →, *Ōsaka* **12 C3** 34 36N 135 26 E
Yamazaki, *Tōkyō* **13 A4** 35 55N 139 53 E
Yamuna →, *Delhi* **16 B2** 28 37N 77 15 E
Yan Kit, *Sing.* **15 F8** 1 21N 103 58 E
Yanagishima, *Tōkyō* **13 B3** 35 49N 139 45 E
Yanbu, *Gzh.* **14 G7** 23 5N 113 9 E
Yanghuayuan, *Beij.* **14 C2** 39 49N 116 18 E
Yangjiazhuang, *Shang.* **14 H11** 31 22N 121 25 E
Yangliuqing, *Tianj.* **14 E5** 39 8N 117 0 E
Yangpu, *Shang.* **14 J12** 31 16N 121 32 E
Yanino, *St-Pet.* **11 B5** 59 55N 30 36 E
Yao, *Ōsaka* **12 C4** 34 37N 135 36 E
Yao Airport, *Ōsaka* **12 C4** 34 35N 135 36 E
Yarmōk, *Bagd.* **17 F7** 33 18N 44 19 E
Yarra →, *Melb.* **19 E6** 37 51 S 144 53 E
Yarra Bend Nat. Park, *Melb.* **19 E7** 37 47 S 145 0 E
Yarraville, *Melb.* **19 E6** 37 48 S 144 53 E
Yasenevo, *Mos.* **11 F9** 55 36N 37 21 E
Yashio, *Tōkyō* **13 B3** 35 48N 139 49 E
Yau Ma Tei, *H.K.* **12 E5** 22 18N 114 10 E
Yau Tong, *H.K.* **12 E6** 22 17N 114 14 E
Yau Yue Wan, *H.K.* **12 E6** 22 19N 114 15 E
Yauza →, *Mos.* **11 D10** 55 54N 37 43 E
Yeading, *Lon.* **4 B2** 51 31N 0 23W
Yeadon, *Phil.* **24 B3** 39 55N 75 15W
Yedikule, *Ist.* **17 B2** 40 59N 28 55 E
Yenikapi, *Ist.* **17 A2** 41 0N 28 56 E
Yeniköy, *Ist.* **17 A3** 41 6N 29 3 E
Yennora, *Syd.* **19 B2** 33 51 S 150 58 E
Yeogchon, *Sŏul* **12 G7** 37 35N 126 55 E
Yeoido, *Sŏul* **12 G7** 37 31N 126 55 E
Yeong Dung Po, *Sŏul* **12 G7** 37 31N 126 54 E
Yeongdong, *Sŏul* **12 G8** 37 30N 127 1 E
Yerba Buena I., *S.F.* **27 B2** 37 48N 122 21W
Yerres, *Paris* **5 C5** 48 43N 2 30 E
Yerres →, *Paris* **5 C5** 48 43N 2 26 E
Yesilköy, *Ist.* **17 B2** 40 57N 28 50 E
Yew Tee, *Sing.* **15 F7** 1 23N 103 45 E
Yiewsley, *Lon.* **4 B2** 51 31N 0 27W
Yiheyuan, *Beij.* **14 A2** 40 0N 116 14 E
Yinhangzhen, *Shang.* **14 H12** 31 20N 121 31 E
Yio Chu Kang, *Sing.* **15 F8** 1 23N 103 51 E
Yixingbu, *Tianj.* **14 D6** 39 11N 117 12 E
Ylästö, *Hels.* **3 B4** 60 17N 24 35 E
Yodo →, *Ōsaka* **12 B4** 34 45N 135 35 E
Yokohama, *Tōkyō* **13 D3** 35 26N 139 41 E
Yokohama Harbour, *Tōkyō* **13 D2** 35 27N 139 39 E
Yokosuka, *Tōkyō* **13 A4** 35 50N 139 54 E
Yong San, *Sŏul* **12 G7** 37 32N 126 58 E
Yongding He →, *Beij.* **14 C1** 39 49N 116 10 E
Yongdingmen, *Beij.* **14 B3** 39 50N 116 20 E
Yongfucun, *Gzh.* **14 G8** 23 8N 113 17 E
Yonkers, *N.Y.* **23 B5** 40 56N 73 52W
Yono, *Tōkyō* **13 A2** 35 52N 139 37 E
York, *Trto.* **20 D8** 43 40N 79 26W
York Mills, *Trto.* **20 D8** 43 45N 79 22W
Yoshikawa, *Tōkyō* **13 A4** 35 53N 139 50 E
Yotsuga, *Tōkyō* **13 B3** 35 40N 139 44 E
You'anmen, *Beij.* **14 B2** 39 51N 116 19 E
Yoyogi Park, *Tōkyō* **13 C3** 35 40N 139 41 E
Yuanxiatian, *Gzh.* **14 F8** 23 12N 113 17 E
Yuexiu Gongyuan, *Gzh.* **14 G8** 23 8N 113 16 E
Yugo-Zarad, *Mos.* **11 E9** 55 40N 37 30 E
Yung Shue Wan, *H.K.* **12 E5** 22 13N 114 6 E
Yuquanshan, *Beij.* **14 A2** 40 0N 116 13 E
Yusofābād, *Tehr.* **17 C5** 35 43N 51 24 E
Yuyuan Tan, *Beij.* **14 B2** 39 53N 116 16 E
Yuyuantan Gongyuan, *Beij.* **14 B2** 39 54N 116 16 E
Yvelines, Forêt des, *Paris* **5 D1** 48 38N 1 53 E
Yvette →, *Paris* **5 C1** 48 43N 1 57 E

Z

Záběhlice, *Pra.* **10 B2** 50 3N 14 29 E
Żacisze, *Wsaw.* **10 E7** 52 17N 21 4 E
Zahrā, *Bagd.* **17 E7** 33 22N 44 19 E
Zakharkovo, *Mos.* **11 E7** 55 46N 37 18 E
Żalov, *Pra.* **10 A2** 50 10N 14 22 E
Załuski, *Wsaw.* **10 F6** 52 9N 20 55 E
Zamdorf, *Mün.* **7 G10** 48 8N 11 35 E
Zanevka, *St-Pet.* **11 B5** 59 55N 30 31 E
Zaozerye, *Mos.* **11 F12** 55 35N 38 1 E
Zapote, *Manila* **15 E3** 14 27N 120 56 E
Zapotitlán, *Méx.* **29 C3** 19 18N 99 2W
Zápy, *Pra.* **10 B4** 50 9N 14 40 E
Zarechye, *Mos.* **11 E8** 55 41N 37 22 E
Zawady, *Wsaw.* **10 E7** 52 10N 21 6 E
Zâwiyet Abû Musallam, *El Qâ.* **18 D4** 29 56N 31 9 E
Zawrā Park, *Bagd.* **17 F8** 33 18N 44 23 E
Zbójna Góra, *Wsaw.* **10 E8** 52 11N 21 13 E
Zbraslav, *Pra.* **10 C2** 49 58N 14 23 E
Zbuzany, *Pra.* **10 B1** 50 1N 14 17 E
Zdiby, *Pra.* **10 B2** 50 9N 14 27 E
Zehlendorf, *Berl.* **7 B2** 52 26N 13 16 E
Zeleneč, *Pra.* **10 B3** 50 8N 14 39 E
Zempoala, Parque Nac. de las Lagunas de, *Méx.* **29 D2** 19 5N 99 18W
Zepernick, *Berl.* **7 A4** 52 38N 13 33 E
Żerań, *Wsaw.* **10 E6** 52 18N 20 58 E
Zerzeń, *Wsaw.* **10 E7** 52 12N 21 7 E
Zeytinburnu, *Ist.* **17 B2** 40 58N 28 53 E
Zhabei, *Shang.* **14 J11** 31 16N 121 26 E
Zhangguizhuang, *Tianj.* **14 E6** 39 7N 117 19 E
Zhangxingzhuang, *Tianj.* **14 D6** 39 10N 117 12 E
Zhdanov, *Mos.* **11 E10** 55 44N 37 41 E
Zhegalovo, *Mos.* **11 D11** 55 54N 37 59 E
Zheleznodorozhnyy, *Mos.* **11 E12** 55 45N 38 0 E
Zhenru, *Shang.* **14 J11** 31 16N 121 22 E
Zhicun, *Gzh.* **14 G8** 23 0N 113 18 E
Zhongshan Gongyuan, *Shang.* **14 J11** 31 13N 121 24 E
Zhoucun, *Gzh.* **14 F8** 23 11N 113 11 E
Zhoujiadu, *Shang.* **14 J11** 31 11N 121 29 E
Zhoujiazhen, *Shang.* **14 J12** 31 16N 121 33 E
Zhu Jiang →, *Gzh.* **14 G9** 23 6N 113 20 E
Zhulebino, *Mos.* **11 E11** 55 42N 37 50 E
Zhushadi, *Gzh.* **14 F9** 23 11N 113 22 E
Zielona, *Wsaw.* **10 E8** 52 14N 21 11 E
Zielonka, *Wsaw.* **10 E7** 52 18N 21 9 E
Zitadella, *Berl.* **7 A2** 52 31N 13 11 E
Zizhuyuan Gongyuan, *Beij.* **14 B2** 39 55N 116 17 E
Żiżkov, *Pra.* **10 B2** 50 5N 14 28 E
Zličin, *Pra.* **10 B1** 50 3N 14 17 E
Zōbki, *Wsaw.* **10 E7** 52 17N 21 6 E
Zografos, *Ath.* **8 J11** 37 58N 23 47 E
Żoliborz, *Wsaw.* **10 E6** 52 16N 20 57 E
Zugliget, *Bud.* **10 K12** 47 30N 18 59 E
Zugló, *Bud.* **10 J13** 47 31N 19 8 E
Zumbi, *Rio J.* **31 A2** 22 49 S 43 10W
Zuvuvu →, *S. Pau.* **31 F5** 23 40 S 46 42W
Zuvuvus, *S. Pau.* **31 F6** 23 41 S 46 39W
Zweckel, *Ruhr* **6 A3** 51 35N 6 57 E
Zyuzino, *Mos.* **11 F9** 55 39N 37 34 E

WORLD
MAPS

MAP SYMBOLS

SETTLEMENTS

◆ PARIS   ■ Berne   ◉ Livorno   ⦿ Brugge   ◎ Algeciras   ○ Fréjus   ○ Oberammergau   ○ Thira

Settlement symbols and type styles vary according to the scale of each map and indicate the importance of towns on the map rather than specific population figures

∴   Ruins or Archæological Sites   ᵛ   Wells in Desert

ADMINISTRATION

Boundaries

_____ International

_ _ _ International (Undefined or Disputed)

.... Internal

National Parks

International boundaries show the de facto situation where there are rival claims to territory.

Country Names
NICARAGUA

Administrative Areas
KENT
CALABRIA

COMMUNICATIONS

Roads

_____ Primary

⌇ Secondary

.·-·-· Trails and Seasonal

Railroads

⌒ Primary

⌒ Secondary

--·---· Under Construction

✧   Airfields

⋈   Passes

⊣--⊢   Railroad Tunnels

....   Principal Canals

PHYSICAL FEATURES

Perennial Streams

Intermittent Streams

Perennial Lakes

Intermittent Lakes

Swamps and Marshes

Permanent Ice and Glaciers

▲ 2259   Elevations (m)

▾ 2604   Sea Depths (m)

_408_   Elevation of Lake Surface Above Sea Level (m)

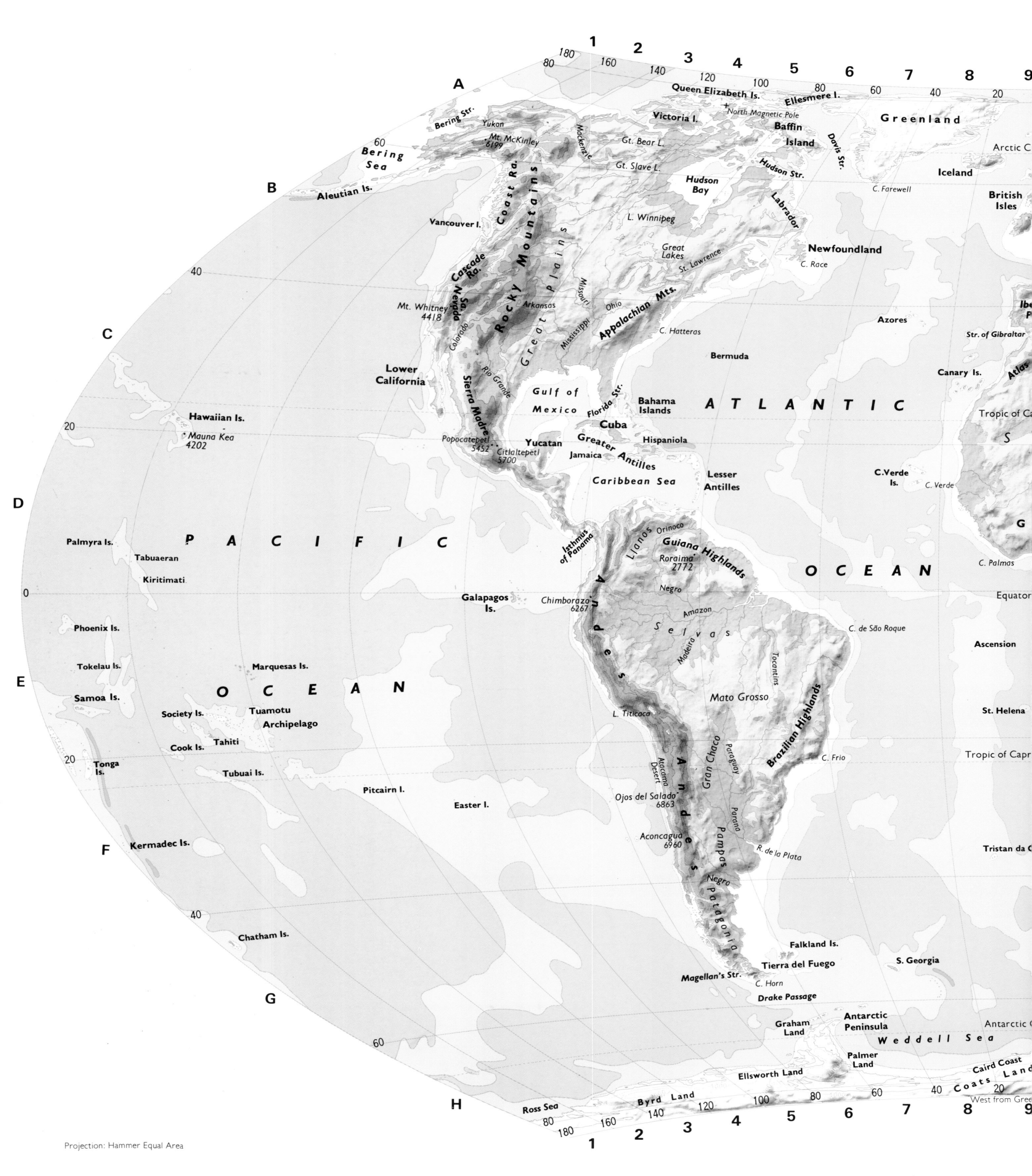

Projection: Hammer Equal Area

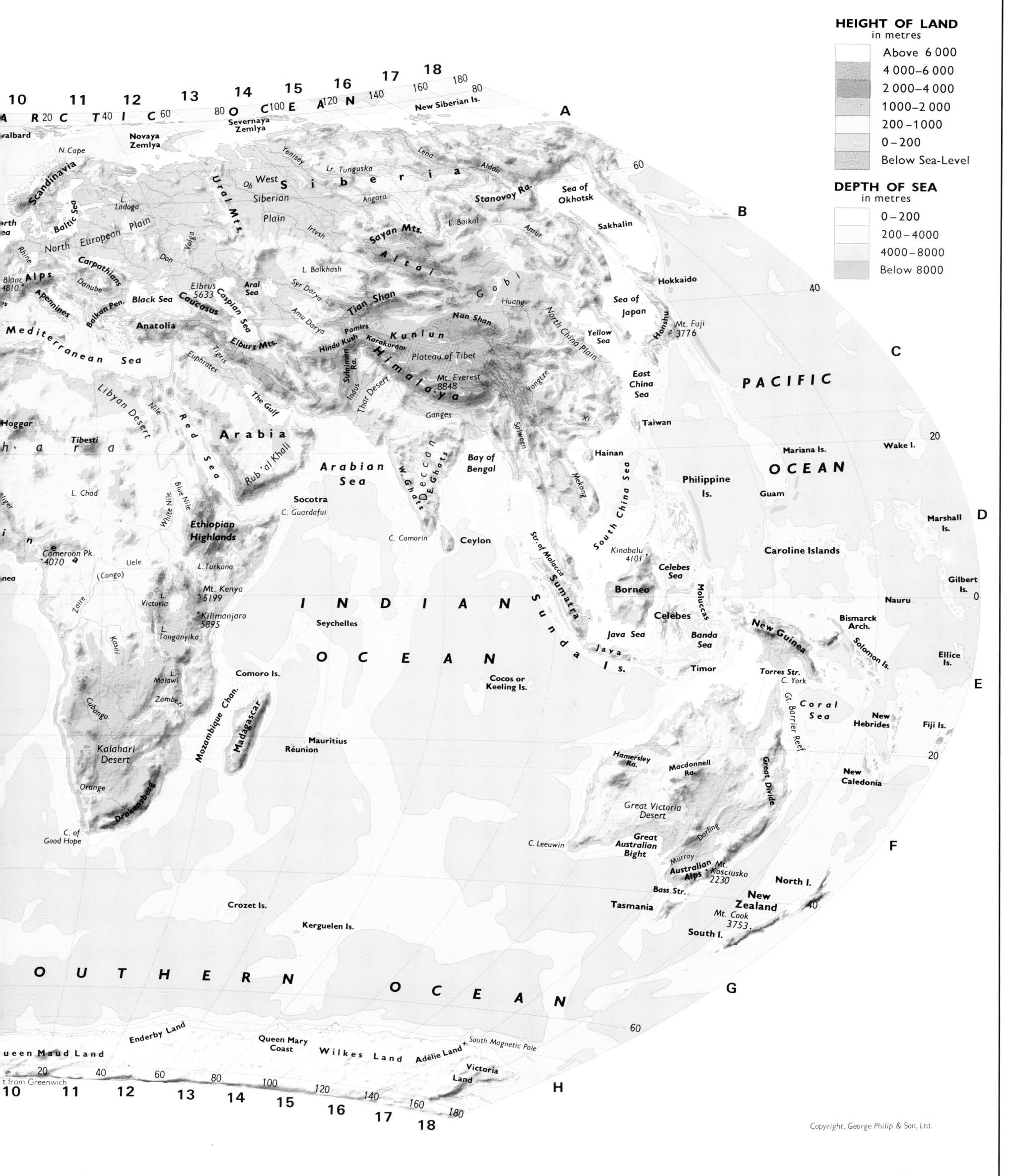

HEIGHT OF LAND
in metres

Above 6 000
4 000–6 000
2 000–4 000
1000–2 000
200–1000
0–200
Below Sea-Level

DEPTH OF SEA
in metres

0–200
200–4000
4000–8000
Below 8000

A R C T I C   O C E A N

10   11   12   13   14   15   16   17   18
180   80

20   40   60   80   100   120   140   160   80

New Siberian Is.

Svalbard   N.Cape   Novaya Zemlya   Severnaya Zemlya

Scandinavia   Yenisey   Lr. Tunguska   Lena   Aldan

Baltic Sea   L. Ladoga   Ob   West Siberian   Siberia   Stanovoy Ra.   Sea of Okhotsk

North Sea   North European Plain   Ural Mts.   Plain   Angara   Sakhalin

Rhine   Volga   Irtysh   L. Baikal   Amur   Hokkaido

Carpathians   Don   Sayan Mts.   Sea of Japan

Mont Blanc   Alps   Danube   Elbrus 5633   Aral Sea   Altai   Gobi   Honshu   Mt. Fuji 3776
4810

Apennines   Balkan Pen.   Black Sea   Caucasus   Caspian Sea   Syr Darya   L. Balkhash   Huang   North China Plain

Anatolia   Elburz Mts.   Amu Darya   Tian Shan   Nan Shan   Yellow Sea   East China Sea

Mediterranean Sea   Euphrates   Pamirs   Hindu Kush   Kunlun   Plateau of Tibet   Yangtze   Taiwan

Tigris   Suleiman Ra.   Karakoram   Himalaya   Mt. Everest 8848   Xi

Libyan Desert   Nile   Thar Desert   Indus   Ganges   Schwen

Hoggar   Tibesti   Red Sea   Arabia   W. Ghats   Bay of Bengal   Hainan

Sahara   Rub 'al Khali   Arabian Sea   Deccan   E. Ghats   Mekong   Mariana Is.   Wake I.

L. Chad   Socotra   South China Sea   Philippine Is.   Guam

Niger   White Nile   Blue Nile   C. Guardafui   Str. of Malacca   Kinabalu 4101   Caroline Islands   Marshall Is.

Ethiopian Highlands   C. Comorin   Ceylon   Sumatra   Borneo   Celebes Sea   Gilbert Is.

Cameroon Pk. 4070   Uele   Sunda Is.   Celebes   Moluccas   Nauru

Zaire (Congo)   L. Turkana   Mt. Kenya 5199   INDIAN   Seychelles   Java Sea   Banda Sea   New Guinea   Bismarck Arch.

L. Victoria   Kilimanjaro 5895   Java   Timor   Solomon Is.   Ellice Is.

L. Tanganyika   Comoro Is.   OCEAN   Torres Str.   New Hebrides

Kasai   L. Malawi   Cocos or Keeling Is.   C. York   Coral Sea   Fiji Is.

Cubango   Zambezi   Mozambique Chan.   Madagascar   Mauritius   Réunion   Hamersley Ra.   Macdonnell Ra.   New Caledonia

Kalahari Desert   Great Barrier Reef   Great Divide

Orange   Drakensberg   Great Victoria Desert   Darling

C. of Good Hope   Crozet Is.   C. Leeuwin   Great Australian Bight   Murray   Australian Alps   Mt. Kosciusko 2230   North I.

Kerguelen Is.   Bass Str.   New Zealand

Tasmania   Mt. Cook 3753   South I.

PACIFIC OCEAN

A   B   C   D   E   F   G   H

60   40   20   0   20   40   60

S O U T H E R N   O C E A N

Queen Maud Land   Enderby Land   Queen Mary Coast   Wilkes Land   Adélie Land   South Magnetic Pole   Victoria Land

st from Greenwich

20   40   60   80   100   120   140   160   180

10   11   12   13   14   15   16   17   18

Projection : Hammer Equal Area

ARCTIC   OCEAN

10   11   12   13   14   15   16   17   18

Svalbard
(Nor.)
Barents   Novaya   Kara
Sea   Zemlya   Sea
Severnaya
Zemlya
New Siberian Is.   East Siberian
Laptev Sea   Sea   Wrangel I.

Murmansk   Norilsk   Verkhoyansk   Arctic Circle   A

NORWAY   Arkhangelsk   Salekhard   Yenisey   Lena   Yakutsk
Oslo   SWEDEN   FINLAND   Ob   Magadan
Helsinki   RUSSIA   Okhotsk   Bering
Stockholm   EST.   ST.PETERSBURG   Sea of   Sea   B
Copenhagen   LATVIA   Perm   Yekaterinburg   Tomsk   Krasnoyarsk   L. Baikal   Okhotsk   Petropavlovsk-
DENMARK   LITH.   MOSCOW   Kazan   Chelyabinsk   Omsk   Novosibirsk   Irkutsk   Ulan Ude   Khabarovsk   Kamchatskiy   International
Amsterdam   Berlin   BELARUS   Volga   Samara   Barnaul   Komsomolsk   Date Line
NETH.   POLAND   Minsk   Saratov   Amur   Kuril Is.
Brussels   GERMANY   Kiev   Vladivostok
LUX.   Prague   Warsaw   UKRAINE   Volgograd   KAZAKHSTAN   Ulan Bator   Harbin   Sapporo
PARIS   Vienna   CZECH   Odessa   Astrakhan   Aral   L. Balkhash   MONGOLIA   Changchun   NORTH
Brussels   AUSTRIA   SLOVAK   Bucharest   Sea   Karaganda   SHENYANG   KOREA   Pyongyang
Milan   HUNG.   ROMANIA   Black   GEORGIA   Alma Ata   Ürümqi   BEIJING   TIANJIN   SEOUL   JAPAN   C
ITALY   CROATIA   YUG.   BULGARIA   Sea   Tbilisi   Bishkek   KYRGYZSTAN   Dalian   SOUTH   TOKYO
Rome   Belgrade   Sofia   Baku   Tashkent   CHINA   Lanzhou   Taiyuan   KOREA   Osaka
Naples   ALB.   GREECE   ISTANBUL   Ankara   ARM.   AZER.   UZBEKISTAN   Samarkand   Xi'an   Hwang Ho   Kitakyushu
Sardinia   Athens   TURKEY   Yerevan   TURKMENISTAN   Dushanbe   Nanjing
Mediterranean   Izmir   Tabriz   Ashkhabad   TAJIKISTAN   Chengdu   Wuhan   SHANGHAI   PACIFIC
Tunis   MALTA   Crete   CYPRUS   SYRIA   TEHRAN   Mashhad   TIBET   Lhasa   CHONGQING   East China   OCEAN
TUNISIA   Tripoli   Beirut   Damascus   Baghdad   Esfahan   Islamabad   Kathmandu   Fuzhou   Sea
Benghazi   Jerusalem   ISR.   AMM   IRAQ   IRAN   Kabul   AFGHANISTAN   Lahore   Kunming   Taipei   Ryukyus
Alexandria   JORDAN   KUWAIT   Shiraz   PAKISTAN   DELHI   NEPAL   GUANGZHOU   TAIWAN   Tropic of Cancer
LIBYA   CAIRO   QATAR   Riyadh   Abu Dhabi   New Delhi   BANGLA-   BURMA   HONG KONG (U.K.)
EGYPT   BAHRAIN   U.A.E.   Muscat   KARACHI   Kanpur   CALCUTTA   DESH   DACCA   Hanoi
Aswan   Mecca   SAUDI   OMAN   Ahmadabad   INDIA   Nagpur   MYANMAR   Hainan   South

NIGER   CHAD   Omdurman   ARABIA   Arabian   BOMBAY   Bay of   Rangoon   VIET-   China   MANILA   D
Niamey   Khartoum   Sana   YEMEN   Sea   Hyderabad   Bengal   BANGKOK   NAM   Sea   PHILIPPINES
Kano   L. Chad   Asmara   Aden   Andaman Is.   THAILAND   MARSHALL IS.
NIGERIA   SUDAN   ERITREA   G. of Aden   Socotra   Bangalore   MADRAS   (India)   Phnom   CAMBODIA
Abuja   DJIBOUTI   (Yemen)   Lakshadweep Is.   Penh   Ho Chi Minh
Ibadan   CENTRAL   Addis Ababa   (India)   Nicobar Is.   City   FEDERATED STATES
Lagos   CAMEROON   AFRICAN   ETHIOPIA   SOMALI   Colombo   (India)   Yap   OF MICRONESIA
Douala   REP.   REP.   SRI LANKA   MALAYSIA   BELAU   Truk   Pohnpei
EQUATORIAL   Bangui   L. Turkana   MALDIVES   Medan   Kuala Lumpur   SABAH   Caroline Is.
GUINEA   Yaoundé   UGANDA   Mogadishu   PEN. MALAYSIA   BRUNEI   Gilbert Is.
SAO   GABON   Kisangani   Kampala   KENYA   SINGAPORE   Borneo
TOME   Libreville   ZAIRE   Victoria   Nairobi   Equator   Palembang   Banjarmasin   IRIAN   NAURU   KIRIBATI
PRINCIPE   CONGO   Kigali   RWANDA   INDIAN   Sumatra   INDONESIA   JAYA
Brazzaville   Bujumbura   BURUNDI   Mombasa   SEYCHELLES   JAKARTA   Ujung Pandang   PAPUA
CABINDA   Kinshasa   Kananga   Zanzibar   Chagos Arch.   OCEAN   Bandung   Surabaya   NEW   New   E
(Angola)   Dar es Salaam   Amirante   (U.K.)   Java   Timor   GUINEA   Ireland
Luanda   Kasai   TANZANIA   Is.   Diego Garcia   Port   New
Lubumbashi   COMOROS   Moresby   Britain   SOLOMON
ANGOLA   Mayotte   Arafura Sea   C. York   IS.   Santa Cruz I.
Benguela   Malawi   (Fr.)   Aldabra Is.   Cocos Is.   Christmas I.   Darwin   TUVALU
ZAMBIA   Lilongwe   Agalega Is.   (Austral.)   (Austral.)   Cairns
Lusaka   MALAWI   Cargados Carajos   VANUATU
NAMIBIA   ZIMBABWE   MOZAMBIQUE   MADAGASCAR   Rodriguez   Port Hedland   Townsville   FIJI
Windhoek   Bulawayo   Antananarivo   MAURITIUS   Tropic of Capricorn   Alice Springs   NEW   Suva
BOTSWANA   RÉUNION   (Fr.)   CALEDONIA
Gaborone   Pretoria   (Fr.)   AUSTRALIA   Rockhampton   (Fr.)   F
Johannesburg   Maputo   Geraldton   Brisbane
SOUTH   SWAZILAND   Amsterdam I.   Kalgoorlie-   Newcastle   Lord Howe
NAMIBIA   LESOTHO   Durban   (Fr.)   Perth   Boulder   Adelaide   Darling   (Austral.)
Cape Town   AFRICA   St.Paul (Fr.)   Fremantle   Great   Sydney   Norfolk I.
C. of Good Hope   Port Elizabeth   Australian   Canberra   (Austral.)
Bight   Melbourne   Tasman   Auckland   NEW
Prince Edward Is.   Crozet Is.   Tasmania   Sea   North I.   ZEALAND
(S.Africa)   (Fr.)   Kerguelen   Hobart   Wellington
(Fr.)   Christchurch
McDonald Is.   Heard I.   Stewart I.   South I.   G
(Austral.)   (Austral.)   Dunedin
ANTARCTICA   Campbell I.   Bounty Is.
(N.Z.)   Auckland Is.   Antipodes Is.
Macquarie I.   (N.Z.)   (N.Z.)
(Austral.)
SOUTHERN   OCEAN   H

East from Greenwich
10   11   12   13   14   15   16   17   18

Antarctic Circle

Ross Sea

Hanoi ◉ Capital Cities

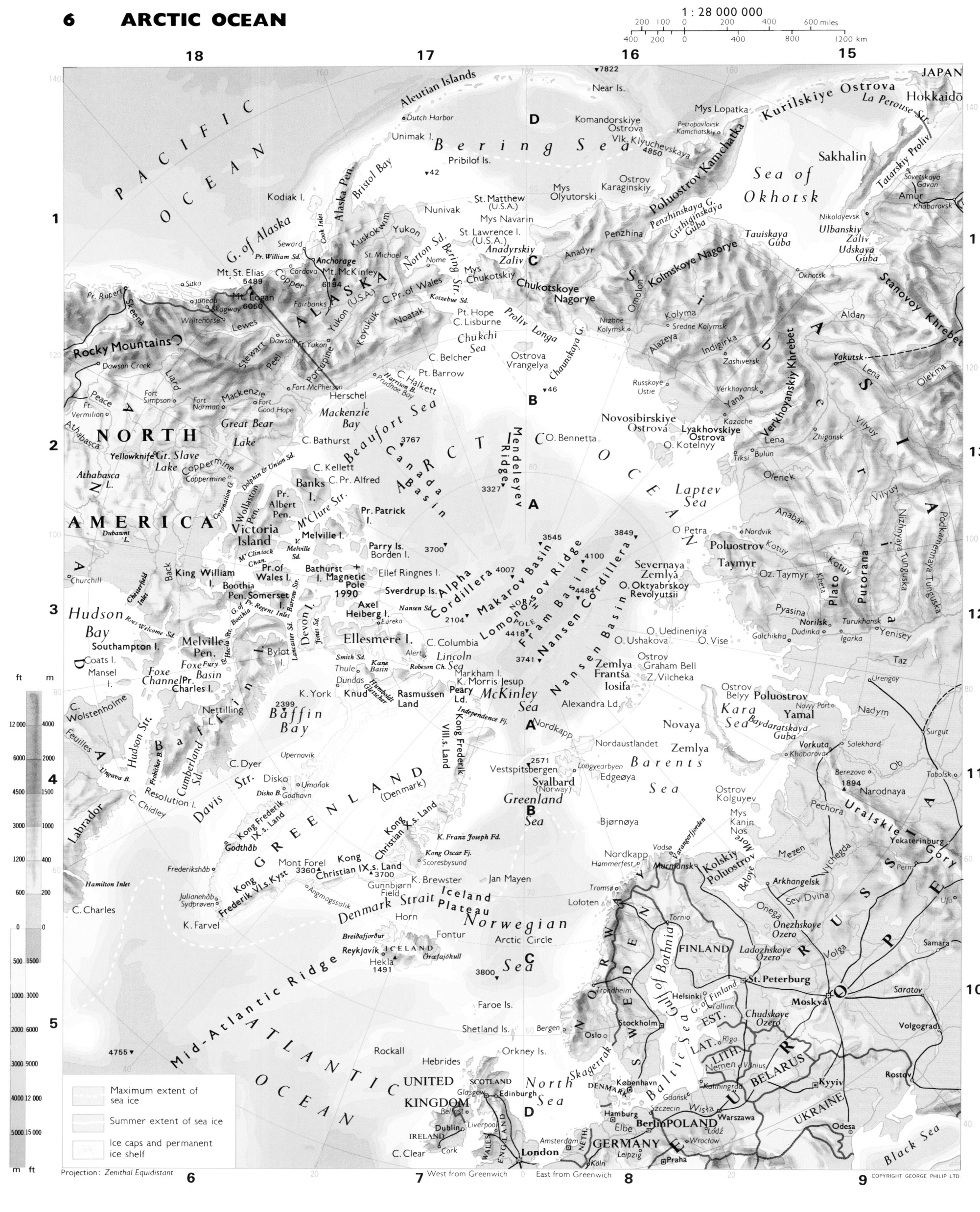

1 : 28 000 000

Projection: Zenithal Equidistant

West from Greenwich   East from Greenwich

COPYRIGHT GEORGE PHILIP LTD.

Maximum extent of sea ice

Summer extent of sea ice

Ice caps and permanent ice shelf

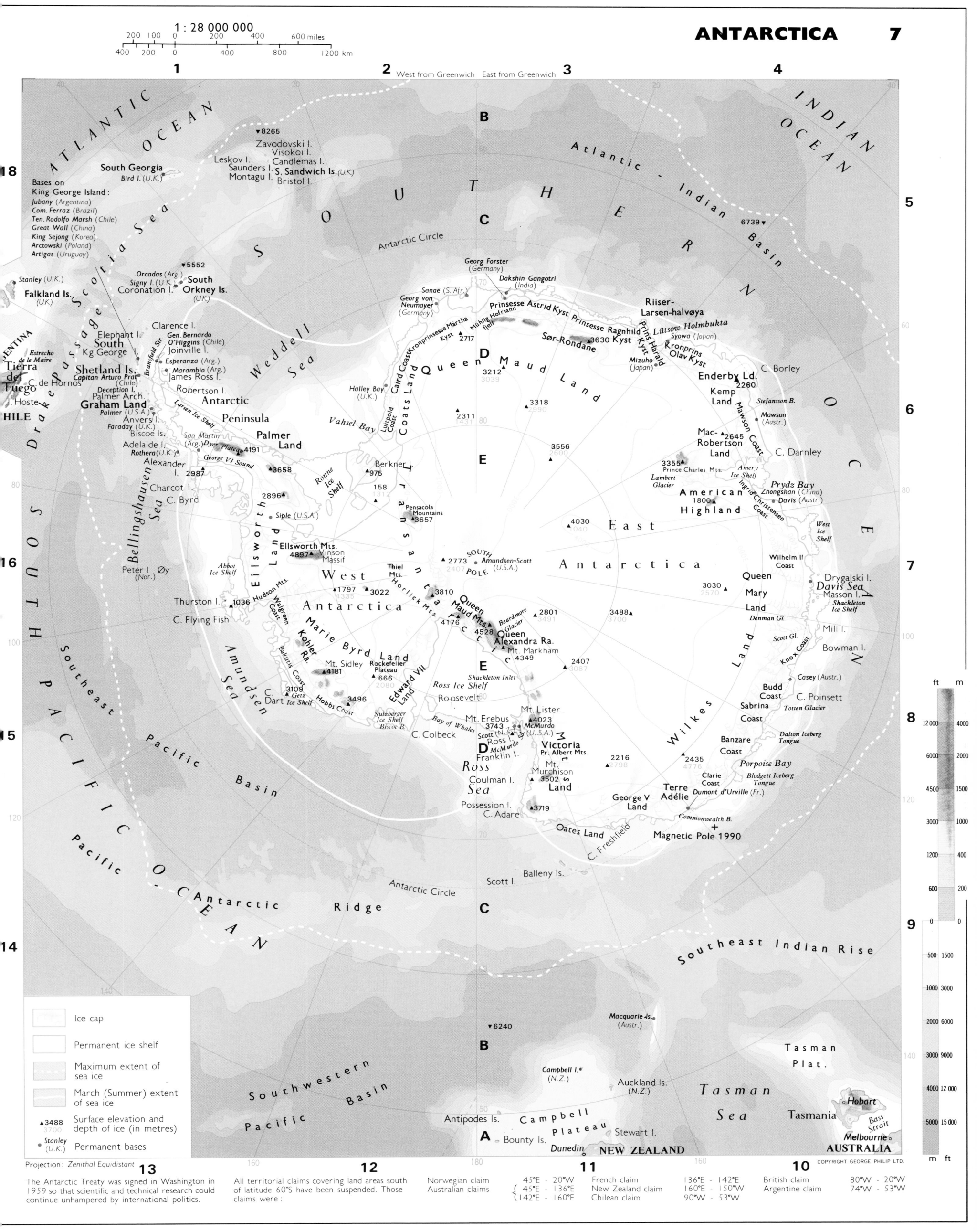

1 : 28 000 000

200 100 0 200 400 600 miles
400 200 0 400 800 1200 km

West from Greenwich East from Greenwich

ATLANTIC OCEAN

SOUTHERN

South Georgia
Bird I. (U.K.)

Bases on
King George Island :
Jubany (Argentina)
Com. Ferraz (Brazil)
Ten. Rodolfo Marsh (Chile)
Great Wall (China)
King Sejong (Korea)
Arctowski (Poland)
Artigas (Uruguay)

Leskov I.
Visokoi I.
Zavodovski I.
Saunders I.
Montagu I.
Candlemas I.
S. Sandwich Is. (U.K.)
Bristol I.

▼8265

Scotia Sea

▼5552

Stanley (U.K.)
Falkland Is.

Orcadas (Arg.)
Signy I. (U.K.)
Coronation I.
South Orkney Is. (U.K.)

Clarence I.
Elephant I.
Gen. Bernardo O'Higgins (Chile)
Joinville I.

ATLANTIC – INDIAN

INDIAN OCEAN

Antarctic Circle

6739 ▼

Georg Forster (Germany)
Dakshin Gangotri (India)
Sanae (S. Afr.)
Georg von Neumayer (Germany)

Riiser-Larsen-halvøya
Prinsesse Astrid Kyst Prinsesse Ragnhild Kyst
Prinsesse Märtha Kyst 2717
Mühlig Hofmann fjell
Sør-Rondane
Lützow Holmbukta
Syowa (Japan)
Prins Harald Kyst
3630 Kyst
Kronprins Olav Kyst
Mizuho (Japan)

ATLANTIC – INDIAN BASIN

OCEAN

Tierra del Fuego
Estrecho de le Maire
C. de Hornos
I. Hoste
CHILE

ARGENTINA

Drake Passage

Shetland Is.
Capitan Arturo Prat (Chile)
Deception I.
Palmer Arch.
Graham Land
Palmer (U.S.A.)
Anvers I.
Faraday (U.K.)
Biscoe Is.
Adelaide I.
Rothera (U.K.)
Alexander I.
Charcot I.
C. Byrd
2987

Esperanza (Arg.)
Marambio (Arg.)
James Ross I.
Robertson I.
Larsen Ice Shelf
San Martin (Arg.)
Dyer Plateau 4191
George VI Sound
▲3658

Weddell Sea
Antarctic Peninsula
Palmer Land
Ronne Ice Shelf
2896▲
Berkner I.
975
158

Bellingshausen Sea

SOUTH

PACIFIC OCEAN

Peter I Øy (Nor.)

Thurston I.
1036
C. Flying Fish

Abbot Ice Shelf

Ellsworth Land
Hudson Mts.
1797 3022
Walgreen Coast
Siple (U.S.A.)
Ellsworth Mts. 4897▲
Vinson Massif
Thiel Mts. 3810
Horlick Mts.
4335

Halley Bay (U.K.)
Vahsel Bay
Caird Coast
Coats Land
Luitpold Coast
Pensacola Mountains
3657

Queen Maud Land
3212
3039
3318
2311
1431
80
3556
4030
4040

East Antarctica

Enderby Ld.
2260
Kemp Land
Stefansson B.
Mawson (Austr.)
Mac-Robertson Land 2645
Prince Charles Mts. 3355
Lambert Glacier
Amery Ice Shelf
C. Borley
C. Darnley
Prydz Bay
Zhongshan (China)
Davis (Austr.)
American Highland
1800

West Antarctica
Marie Byrd Land
Kohler Ra.
Mt. Sidley 4181
Rockefeller Plateau
666
Bakutis Coast
C. Dart
Getz Ice Shelf
3109
3496
Hobbs Coast
Edward VII Land
2080
Sulzberger Ice Shelf
Biscoe B.

2773▲
2407
SOUTH POLE
Amundsen-Scott (U.S.A.)
4176
Queen Maud Mts.
4528
Beardmore Glacier
Queen Alexandra Ra.
Mt. Markham
4349
2801
3491
3488▲
3700
3030
2570
Queen Mary Land
Wilhelm II Coast
West Ice Shelf
Denman Gl.
Scott Gl.
Mill I.
Bowman I.

Transantarctic Mts.

2407
4087

Amundsen Sea
Pacific Basin

C. Colbeck
Bay of Whales
Roosevelt I.
Ross Ice Shelf
Shackleton Inlet
Mt. Erebus 3743
McMurdo (U.S.A.)
Scott (N.Z.)
Ross
4023 McMurdo
Franklin I.
Mt. Lister 4176
Victoria
Pr. Albert Mts.
2216
2798
Mt. Murchison 3502
Land
2435
4776
Wilkes Land
Budd Coast
Casey (Austr.)
Sabrina Coast
Totten Glacier
Banzare Coast
Dalton Iceberg Tongue
Porpoise Bay
Clarie Coast
Blodgett Iceberg Tongue
Dumont d'Urville (Fr.)
Terre Adélie
George V Land
Commonwealth B.
+ Magnetic Pole 1990

Ross Sea
Coulman I.
Possession I.
C. Adare 3719
Oates Land
C. Freshfield

Scott I.
Balleny Is.

Antarctic Circle

Antarctic Ridge

Southeast Indian Rise

Southwestern Pacific Basin

Macquarie Is. (Austr.)
▼6240

Tasman Plat.

Campbell I. (N.Z.)
Auckland Is. (N.Z.)

Tasman Sea

Tasmania
Hobart
Bass Strait
Melbourne
AUSTRALIA

Antipodes Is.
Bounty Is.
Campbell Plateau
Stewart I.
Dunedin
NEW ZEALAND

COPYRIGHT GEORGE PHILIP LTD.

Legend:

	Ice cap
	Permanent ice shelf
	Maximum extent of sea ice
	March (Summer) extent of sea ice
▲3488 / 3700	Surface elevation and depth of ice (in metres)
• Stanley (U.K.)	Permanent bases

Projection: *Zenithal Equidistant*

The Antarctic Treaty was signed in Washington in 1959 so that scientific and technical research could continue unhampered by international politics.

All territorial claims covering land areas south of latitude 60°S have been suspended. Those claims were :

Norwegian claim	45°E – 20°W
Australian claims	{ 45°E – 136°E / 142°E – 160°E }
French claim	136°E – 142°E
New Zealand claim	160°E – 150°W
Chilean claim	90°W – 53°W
British claim	80°W – 20°W
Argentine claim	74°W – 53°W

ft m
12 000 4000
6000 2000
4500 1500
3000 1000
1200 400
600 200
0 0
500 1500
1000 3000
2000 6000
3000 9000
4000 12 000
5000 15 000
m ft

PACIFIC OCEAN

SOUTH ATLANTIC OCEAN

SOUTHERN OCEAN

BRAZIL

ARGENTINA

BOLIVIA

PERU

ECUADOR

PARAGUAY

URUGUAY

CHILE

ANGOLA

NAMIBIA

SOUTH AFRICA

Andes

Mato Grosso

Pampas

Patagonia

Gran Chaco

Amazon

Marañón
Ucayali
Putumayo
Japurá
Negro
Madeira
Purús
Aripuanã
Tapajós
Xingu
Tocantins
Araguaia
Paraná
Paraguay
Pilcomayo
Salado
Uruguay
Colorado

Lima
La Paz
Santiago
Buenos Aires
Montevideo
Asunción
Córdoba
Rosario
Concepción
Valparaíso
Antofagasta
Iquique
Arica
Guayaquil
Iquitos
Leticia
Manaus
Belém
Fortaleza
Recife
Salvador
Belo Horizonte
Rio de Janeiro
São Paulo
Santos
Brasília
Goiânia
Pôrto Alegre
Bahía Blanca
Cape Town

Lago Titicaca
L. Mirim

Río de la Plata

Golfo San Matías
Golfo San Jorge
Pen. Valdés

Tierra del Fuego
Estrecho de Magallanes
Drake Passage
Cape Horn
Falkland Is. (Islas Malvinas)
South Georgia
South Orkney Is.
South Sandwich Is.
South Shetland Is.

Scotia Sea
Weddell Sea
Argentine Basin
Brazil Basin
Angola Basin
Cape Basin
Agulhas Basin
Pacific Basin
South East Pacific Basin
Chile Rise
Mid-Atlantic Ridge
Southern Mid-Atlantic Ridge
Atlantic Indian Ridge
Walvis Ridge

Antarctic Peninsula
Graham Land
Palmer Land
Ellsworth Land
Byrd Land
Queen Maud Land
Enderby Land
Coats Land
Ross Sea

BENGUELA COLD CURRENT
PERUVIAN COLD CURRENT
FALKLAND COLD CURRENT
SOUTH EQUATORIAL CURRENT
BRAZIL CURRENT
CAPE HORN COLD CURRENT
WEST WIND DRIFT

Tropic of Capricorn
Antarctic Circle
Equatorial Limit of Icebergs

St. Helena
Ascension
Tristan da Cunha
Gough I.
Bouvetøya

Richards Deep 8050
6960
6863
6723
6866
6550
6369
6267
6013
6537
6027
5755
638
892
5457
411
302
6739
8428
5552
6212
5385

Projection: Mollweide

Direction of Currents

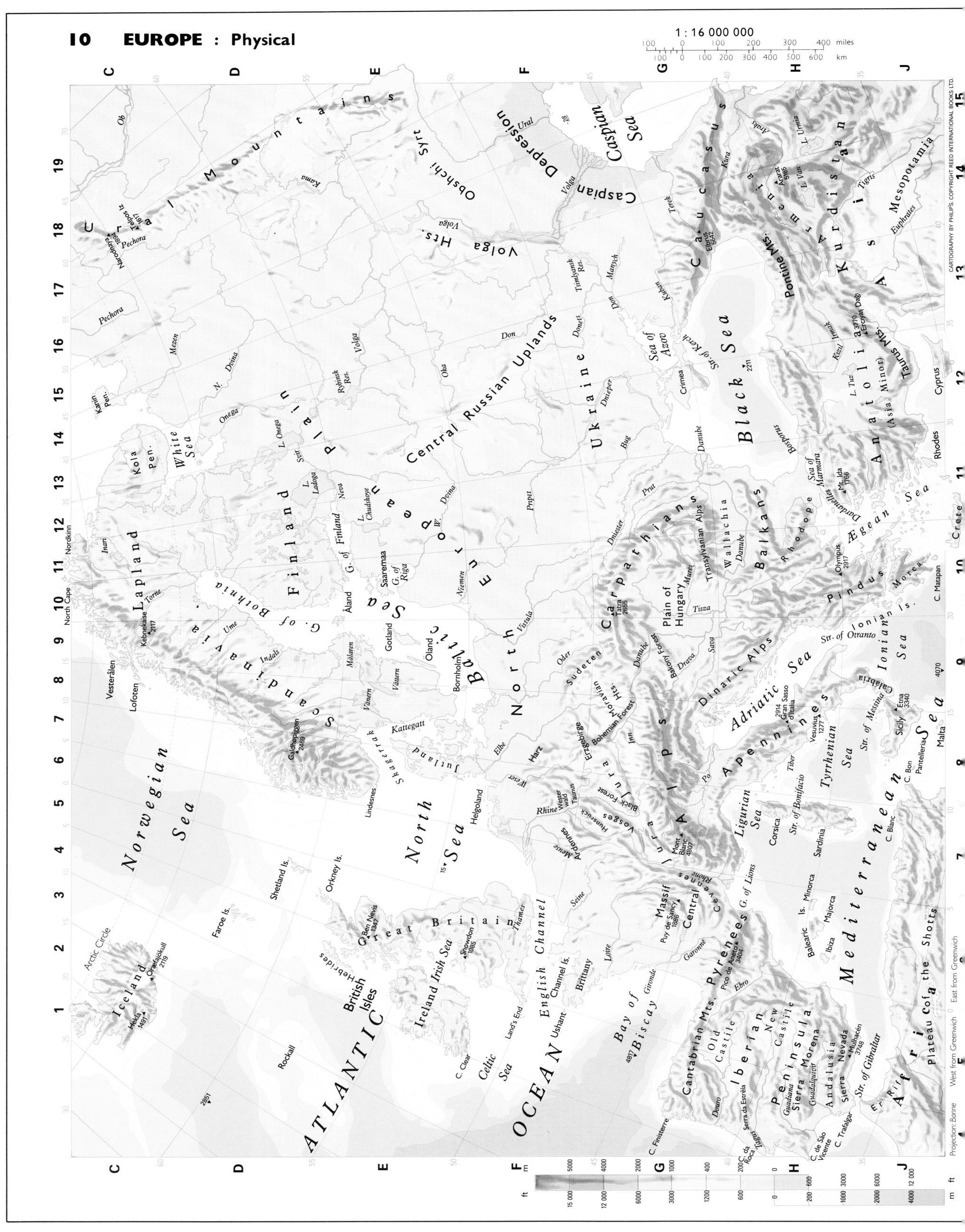

1 : 16 000 000

| 100 | | 0 | 100 | 200 | 300 | 400 | miles |

| 100 | 0 | 100 | 200 | 300 | 400 | 500 | 600 | km |

CARTOGRAPHY BY PHILIP'S. COPYRIGHT REED INTERNATIONAL BOOKS LTD.

Projection: Bonne

West from Greenwich   0   East from Greenwich

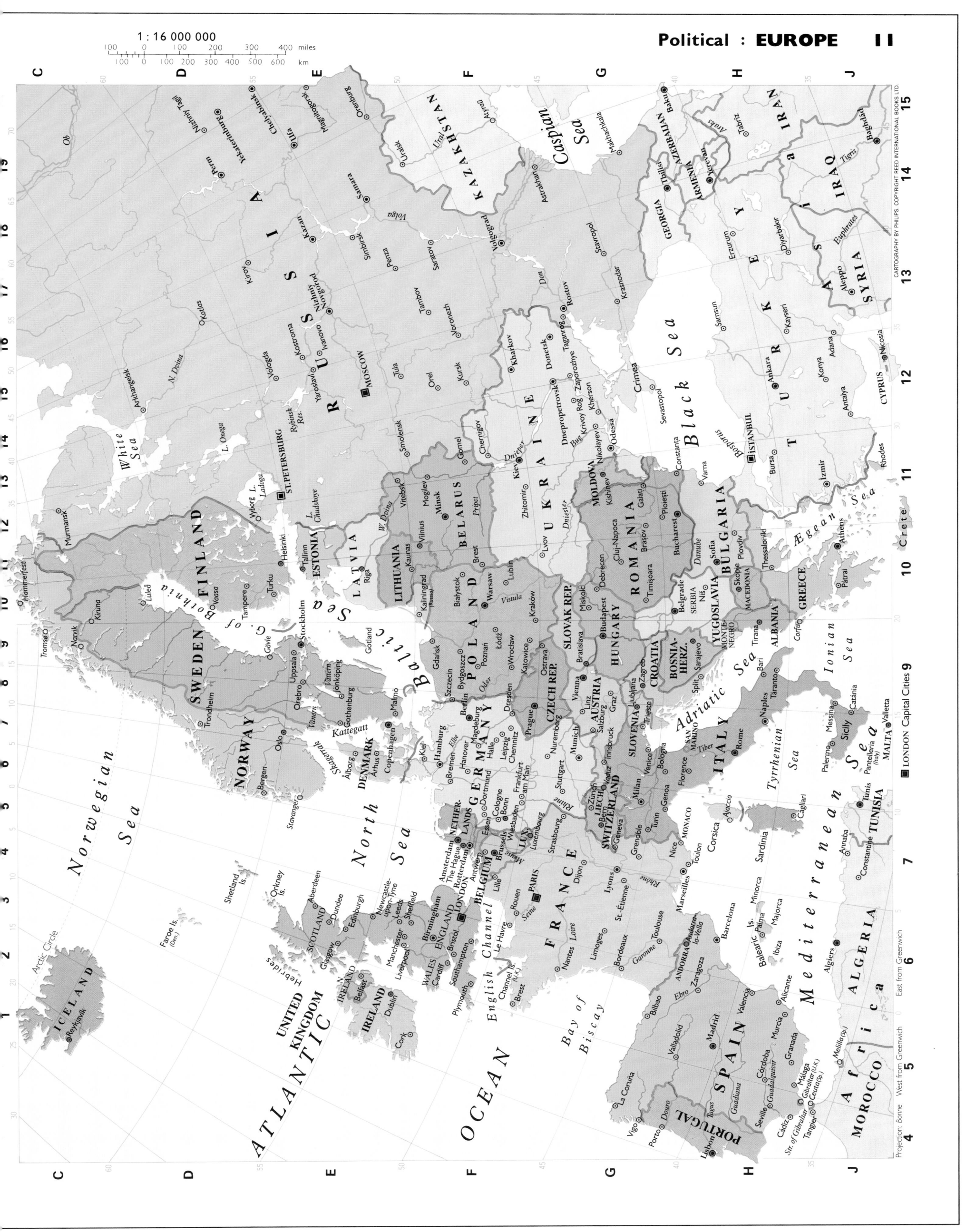

1 : 16 000 000

100   0   100   200   300   400 miles
100   0   100   200   300   400   500   600 km

CARTOGRAPHY BY PHILIP'S. COPYRIGHT REED INTERNATIONAL BOOKS LTD.

Projection: Bonne   West from Greenwich   East from Greenwich

■ LONDON Capital Cities

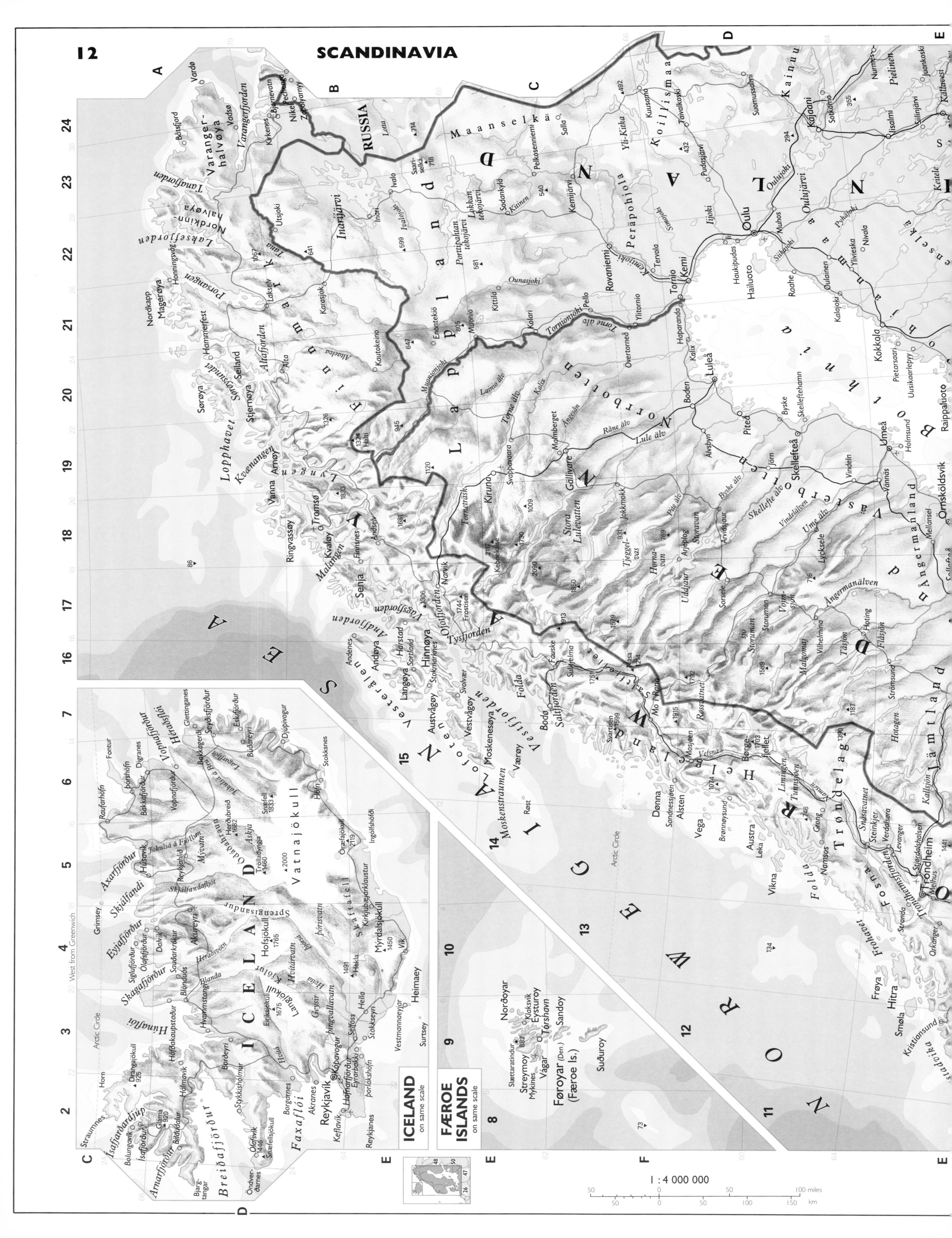

ICELAND
on same scale

FÆROE ISLANDS
on same scale

1 : 4 000 000

RUSSIA

ICELAND

FØROYAR (Den.)
(Færoe Is.)

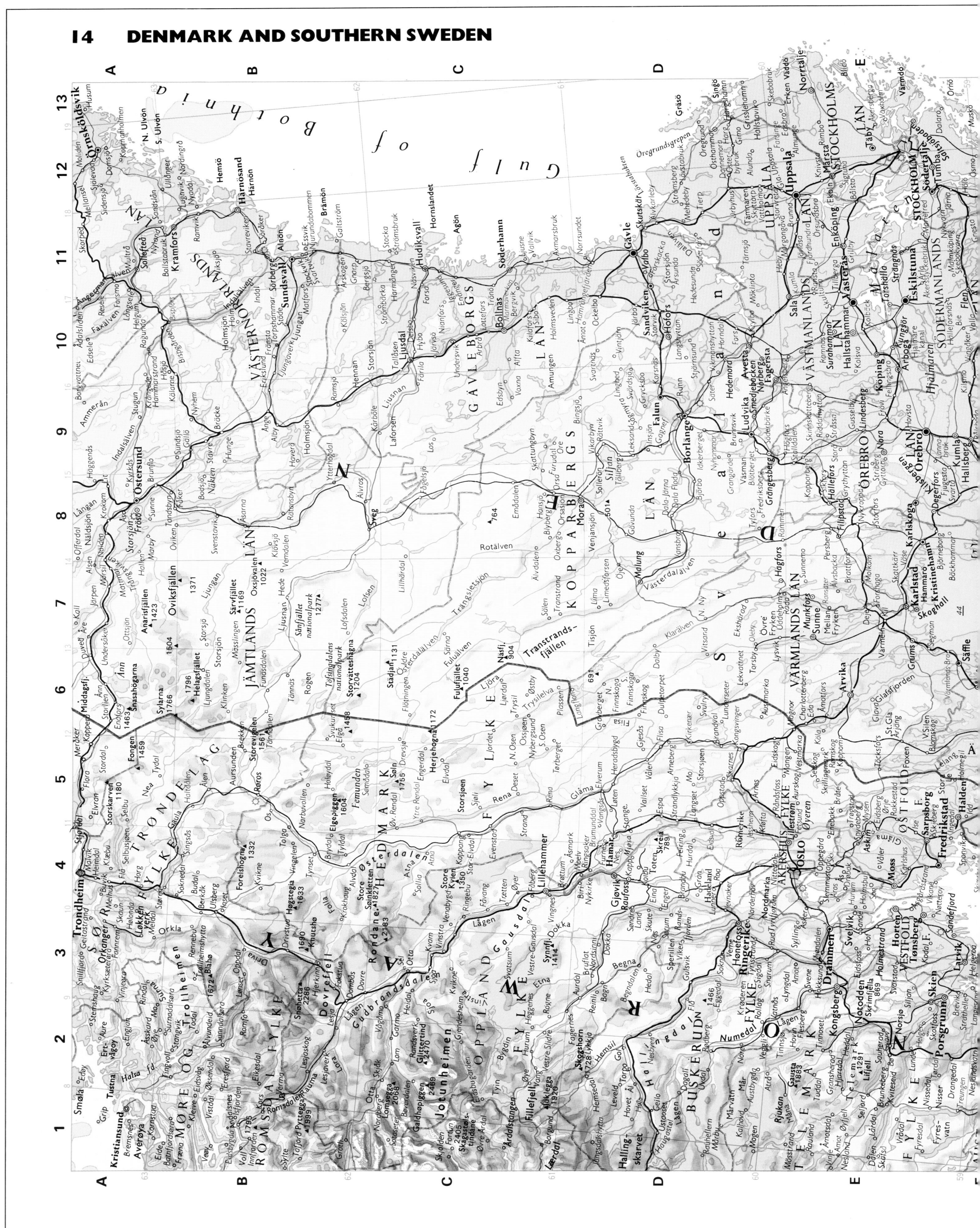

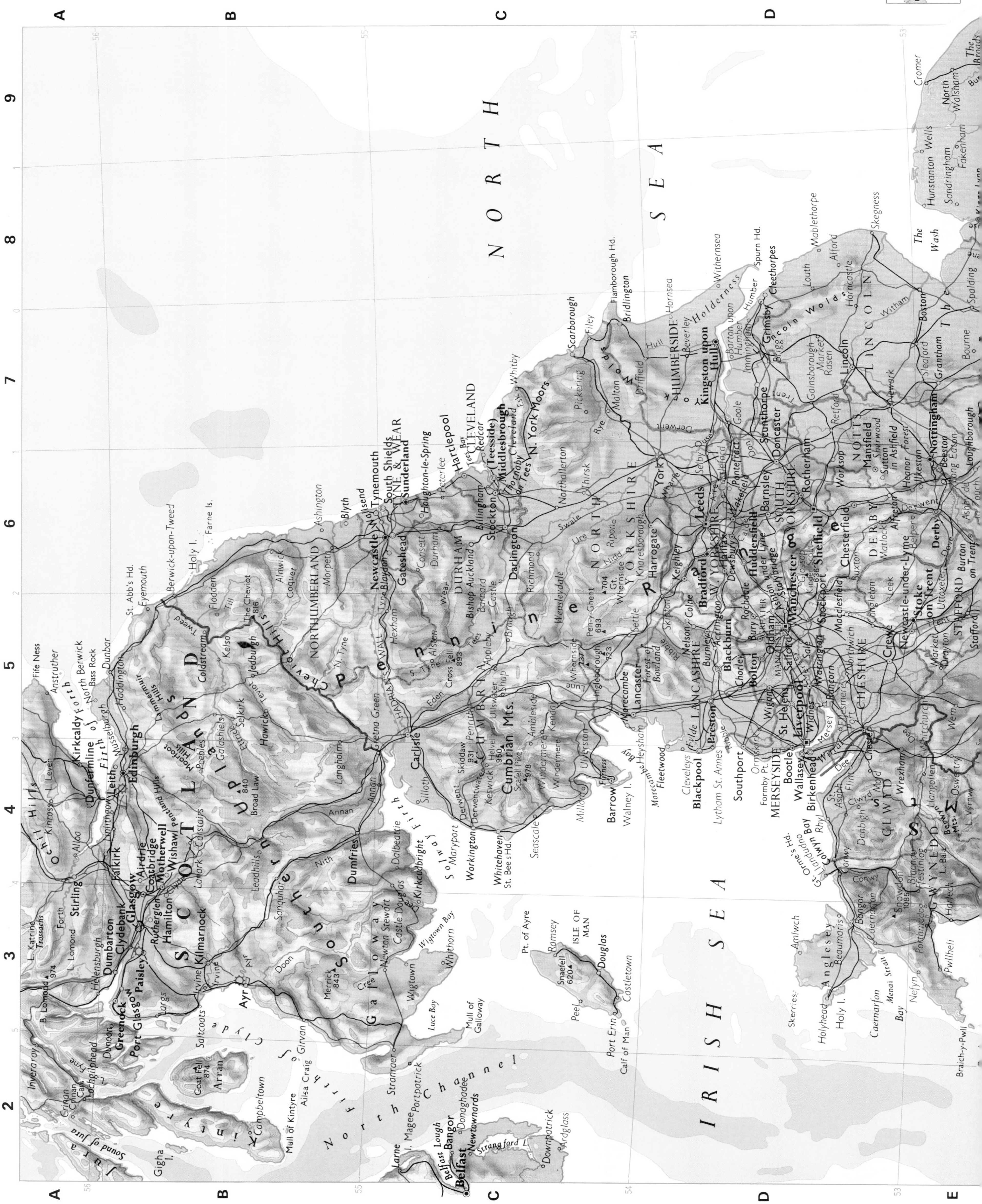
NORTH SEA
IRISH SEA
North Channel
Firth of Clyde
Solway Firth
Firth of Forth
SCOTLAND
Southern Uplands
Cheviot Hills
Pennines
Cumbrian Mts.
N. York Moors
NORTHUMBERLAND
DURHAM
CUMBRIA
TYNE & WEAR
CLEVELAND
NORTH YORKSHIRE
W. YORKSHIRE
SOUTH YORKSHIRE
HUMBERSIDE
LANCASHIRE
MERSEYSIDE
CHESHIRE
DERBY
NOTTS
LINCOLN
STAFFORD
CLWYD
GWYNEDD
ISLE OF MAN
Edinburgh
Glasgow
Stirling
Dundee
Kirkcaldy
Dunfermline
Falkirk
Paisley
Greenock
Kilmarnock
Ayr
Dumfries
Carlisle
Berwick-upon-Tweed
Newcastle
Gateshead
Sunderland
South Shields
Tynemouth
Hartlepool
Middlesbrough
Stockton
Darlington
Scarborough
Bridlington
Kingston upon Hull
York
Leeds
Bradford
Harrogate
Sheffield
Manchester
Liverpool
Preston
Blackpool
Lancaster
Barrow
Douglas
Doncaster
Lincoln
Grimsby
Nottingham
Derby
Stoke on Trent
Chester
Belfast
Bangor

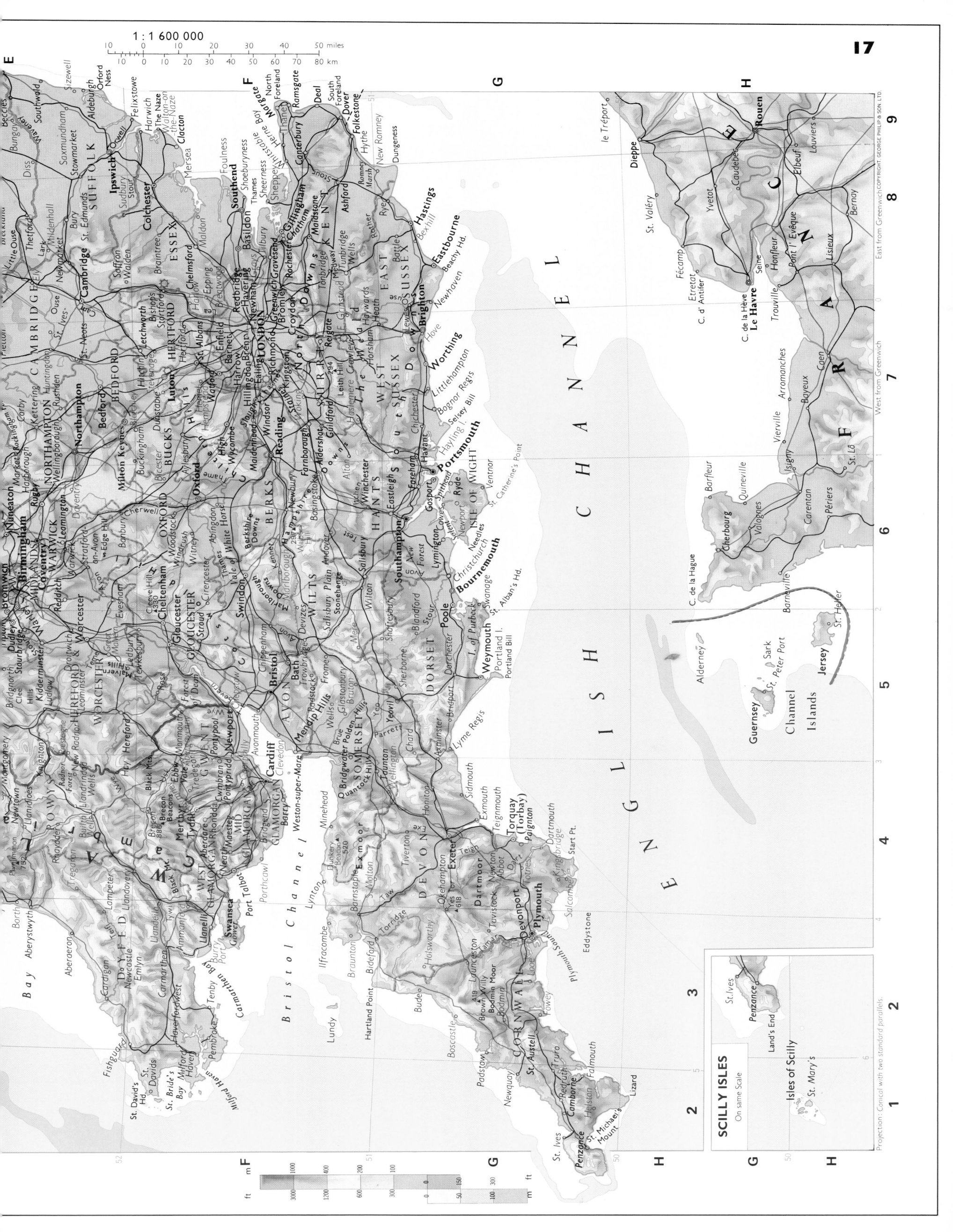

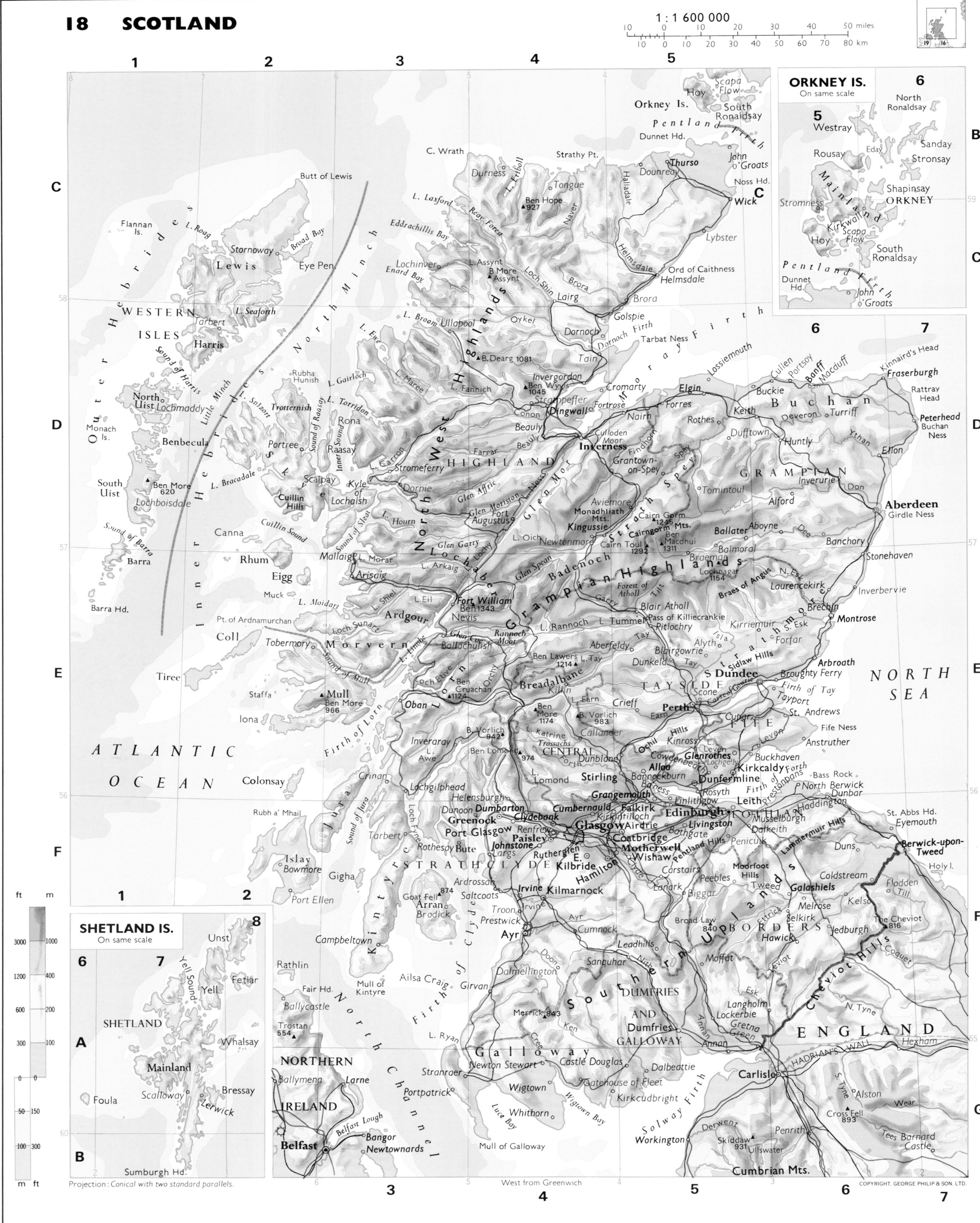

18 SCOTLAND

1 : 1 600 000

ORKNEY IS.
On same scale

SHETLAND IS.
On same scale

Projection: Conical with two standard parallels.

West from Greenwich

COPYRIGHT. GEORGE PHILIP & SON LTD.

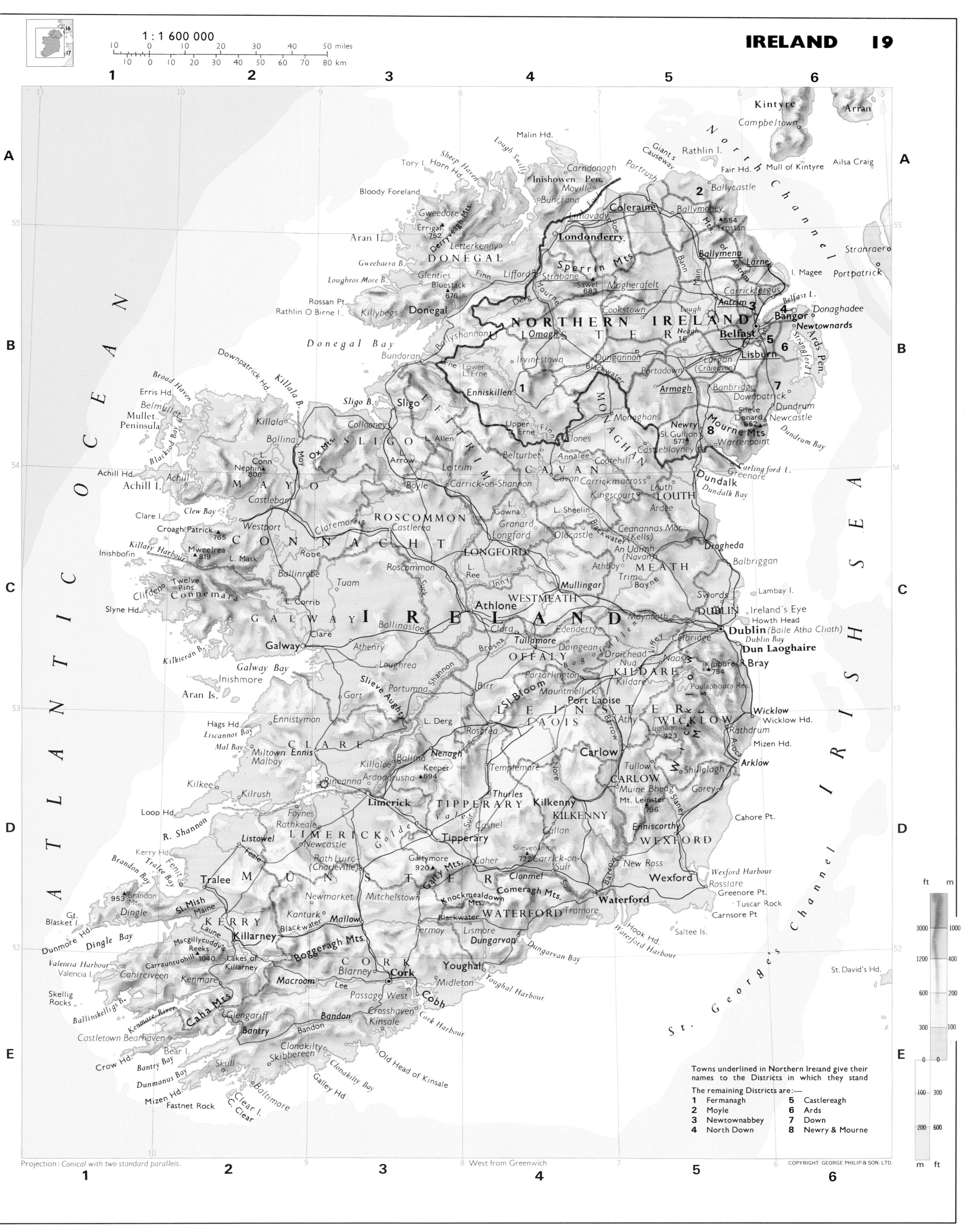

1 : 1 600 000

10   0   10   20   30   40   50 miles
10   0   10   20   30   40   50   60   70   80 km

IRELAND

NORTH CHANNEL

ATLANTIC OCEAN

DONEGAL

NORTHERN IRELAND

ULSTER

MAYO

SLIGO

LEITRIM

CAVAN

MONAGHAN

LOUTH

CONNACHT

ROSCOMMON

LONGFORD

WESTMEATH

MEATH

GALWAY

IRELAND

OFFALY

KILDARE

DUBLIN

LEINSTER

CLARE

TIPPERARY

LAOIS

CARLOW

WICKLOW

KILKENNY

WEXFORD

MUNSTER

LIMERICK

WATERFORD

KERRY

CORK

IRISH SEA

St. George's Channel

Towns underlined in Northern Ireland give their
names to the Districts in which they stand

The remaining Districts are:—

1	Fermanagh	5	Castlereagh
2	Moyle	6	Ards
3	Newtownabbey	7	Down
4	North Down	8	Newry & Mourne

Projection: Conical with two standard parallels.

West from Greenwich

COPYRIGHT. GEORGE PHILIP & SON. LTD.

ft   m
3000 1000
1200 400
600   200
300   100
0   0
100   300
200   600
m   ft

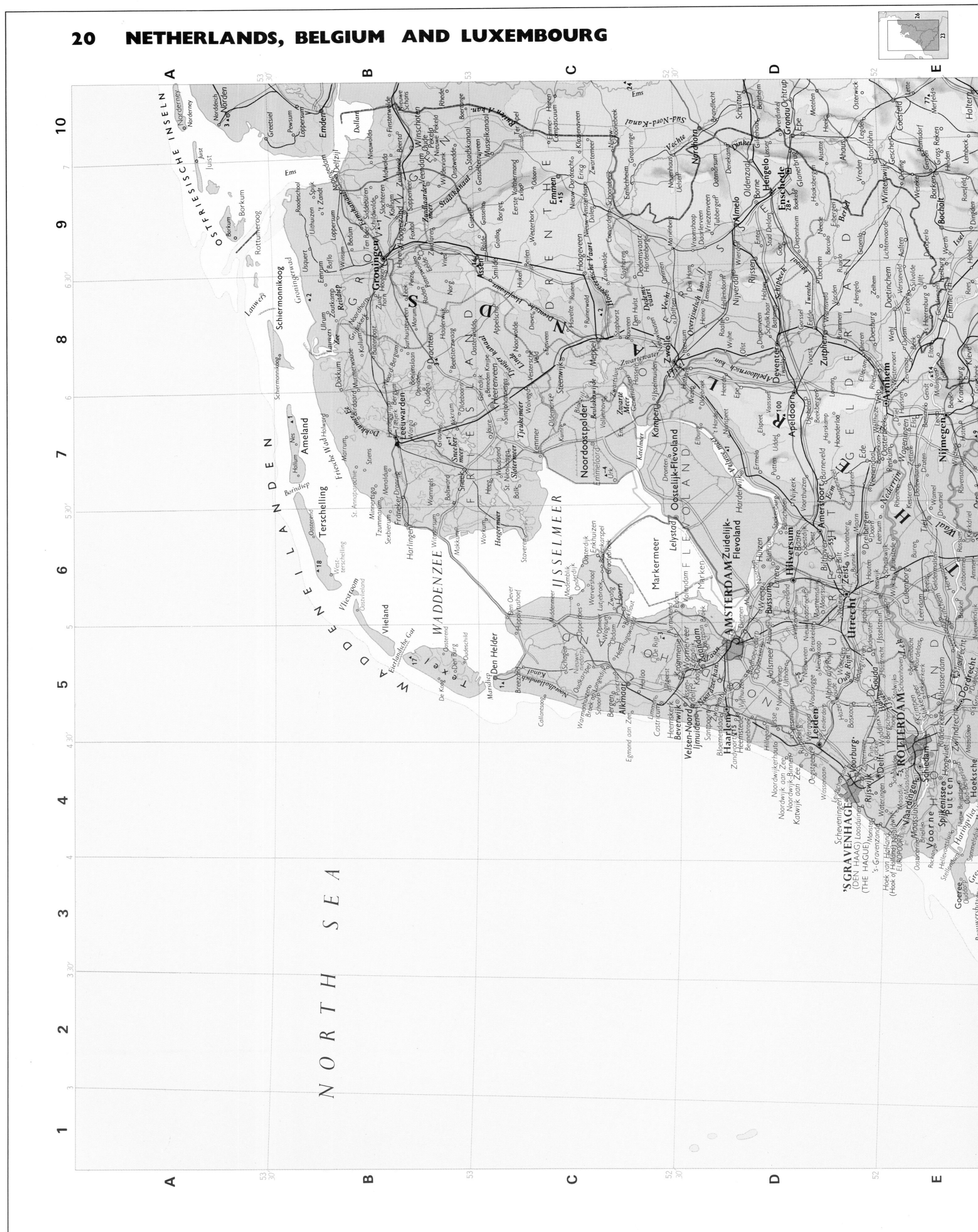

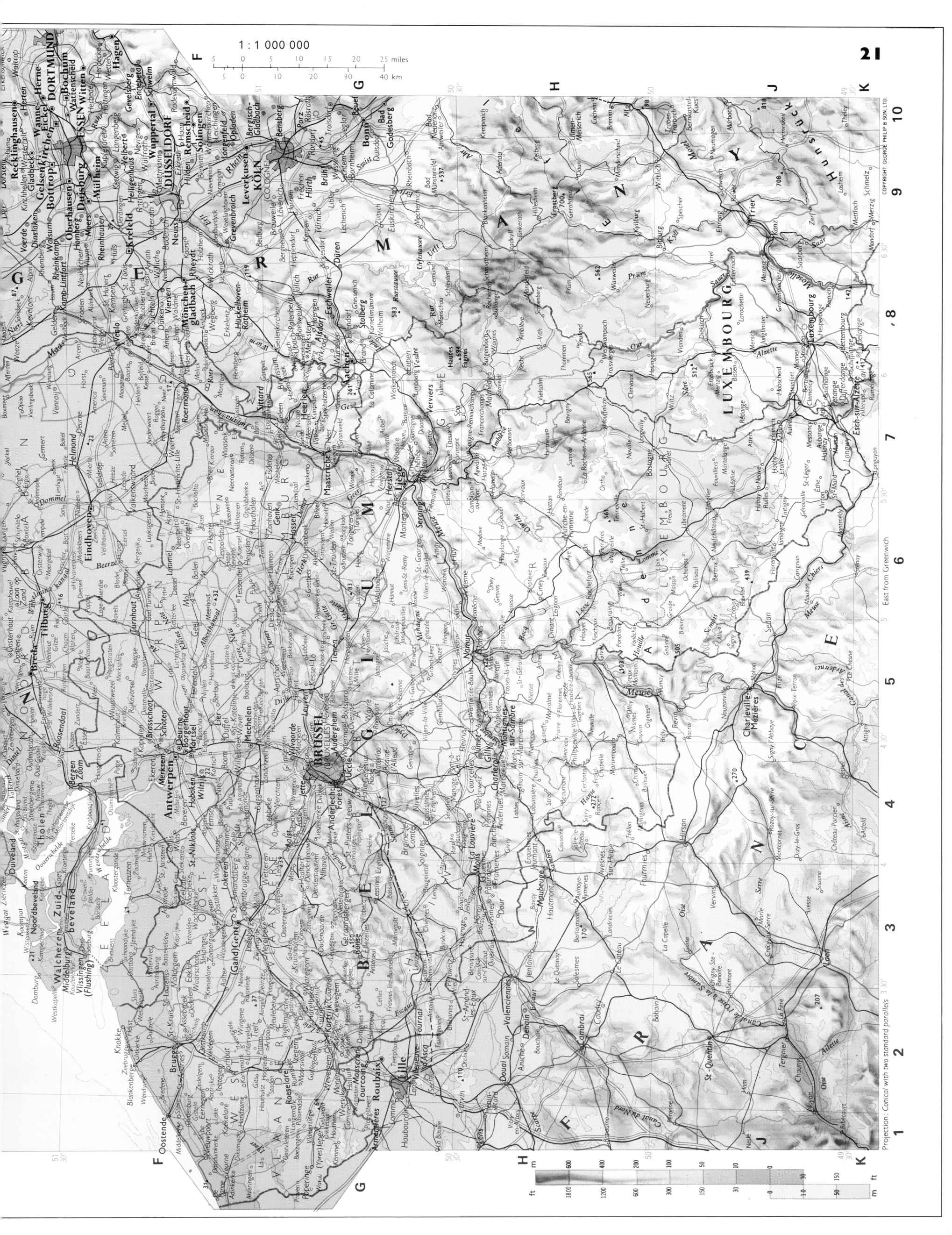

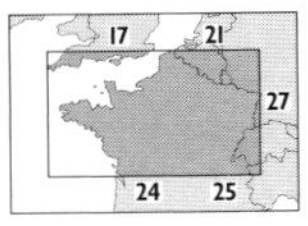
17
21
27
24
25

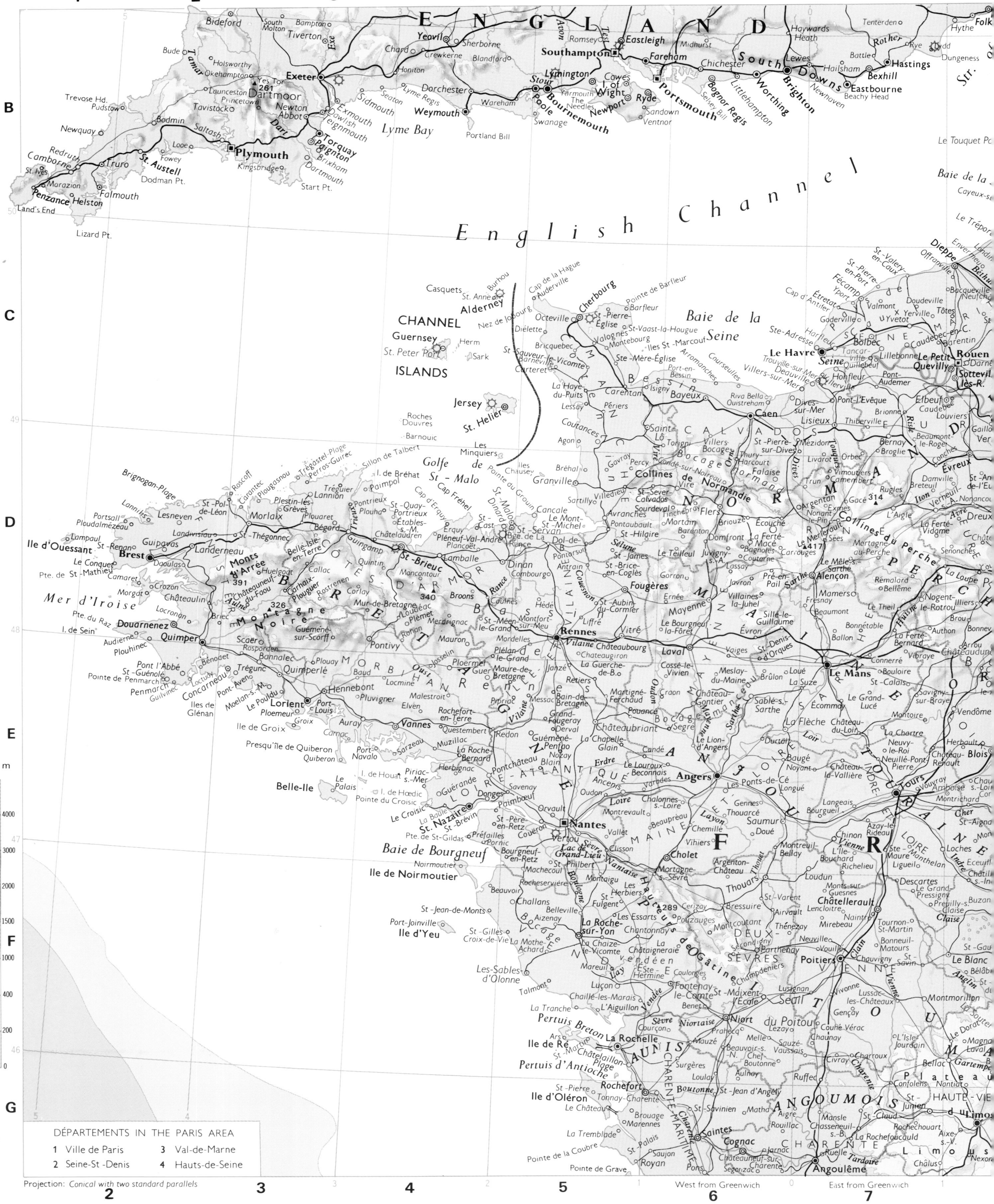
ENGLAND
English Channel
CHANNEL ISLANDS
Guernsey
St. Peter Port
Alderney
Sark
Herm
Jersey
St. Helier
Exeter
Plymouth
Torquay
Paignton
Penzance
Truro
St. Austell
Falmouth
Helston
Land's End
Lizard Pt.
Dartmoor
Lyme Bay
Weymouth
Portland Bill
Poole
Bournemouth
Southampton
Eastleigh
Fareham
Portsmouth
Chichester
Worthing
Brighton
South Downs
Lewes
Eastbourne
Hastings
Bexhill
Beachy Head
Dungeness
Newport
Ryde
Cowes
I. of Wight
Yeovil
Sherborne
Dorchester
Tiverton
Bideford
Baie de la Seine
Cherbourg
Le Havre
Rouen
Dieppe
Caen
Bayeux
Lisieux
Évreux
Alençon
Le Mans
Laval
Rennes
St.-Malo
Golfe de St.-Malo
St.-Brieuc
Brest
Quimper
Lorient
Vannes
Nantes
St.-Nazaire
Angers
Tours
Blois
Saumur
Cholet
Poitiers
Niort
La Rochelle
Rochefort
Saintes
Cognac
Angoulême
Châtellerault
La Roche-sur-Yon
Les Sables-d'Olonne
Ile d'Ouessant
Mer d'Iroise
Belle-Ile
Ile de Noirmoutier
Baie de Bourgneuf
Ile d'Yeu
Ile de Ré
Pertuis Breton
Pertuis d'Antioche
Ile d'Oléron
Pointe de Grave
Royan
BRETAGNE
NORMANDIE
MAINE
ANJOU
TOURAINE
POITOU
AUNIS
ANGOUMOIS
PERCHE
Collines de Normandie
Monts d'Arrée
Montagne Noire
Gâtine
Plateau du Limousin
Loire
Vilaine
Seine
ft m
12 000 4000
9000 3000
6000 2000
4500 1500
3000 1000
1200 400
600 200
0 0
200 600
2000 6000
m ft
DÉPARTEMENTS IN THE PARIS AREA
1 Ville de Paris
2 Seine-St-Denis
3 Val-de-Marne
4 Hauts-de-Seine
Projection: Conical with two standard parallels
West from Greenwich
East from Greenwich

1 : 2 000 000

10 0 10 20 30 40 50 miles
10 0 10 20 30 40 50 60 70 80 km

COPYRIGHT GEORGE PHILIP & SON. LTD.

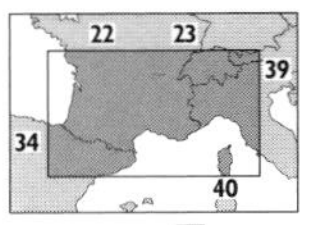

Projection: Conical with two standard parallels

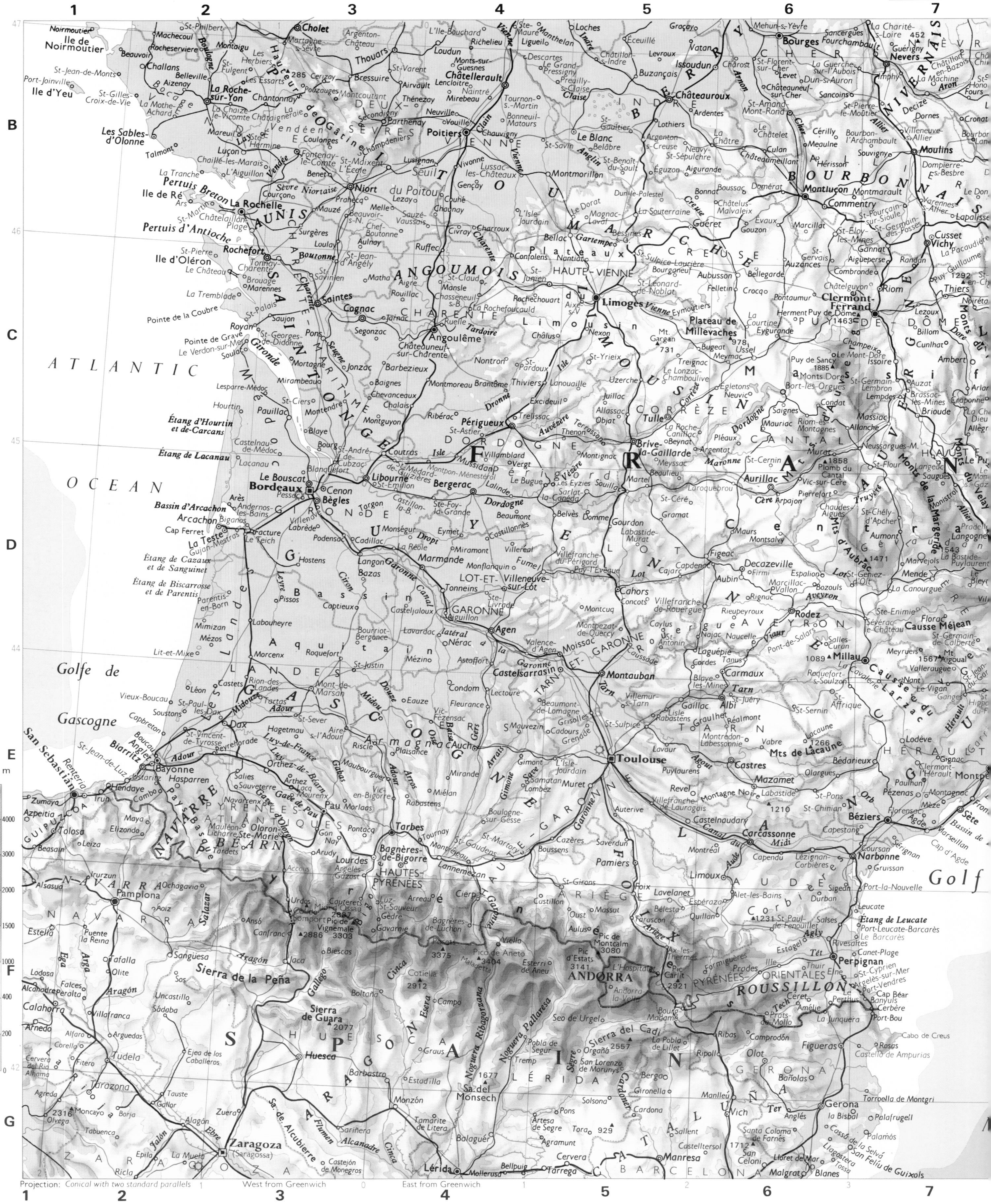

West from Greenwich   East from Greenwich

ATLANTIC

OCEAN

Golfe de Gascogne

SPAIN

Bordeaux

Toulouse

Limoges

Clermont-Ferrand

Bourges

Poitiers

La Rochelle

Zaragoza (Saragossa)

ANDORRA

Pamplona

PYRÉNÉES

PYRENEES

1 : 2 000 000

10 0 10 20 30 40 50 miles
10 0 10 20 30 40 50 60 70 80 km

Grid references: 8 · 9 · 10 · 11 · 12 · 13 · 14

Countries and Regions: SWITZERLAND · FRANCE · ITALY · JURA · VAUD · FRIBOURG · BERN · VALAIS · GRAUBÜNDEN · UNTERWALDEN · HAUTE-SAVOIE · SAVOIE · DAUPHINÉ · PROVENCE · HAUTE-PROVENCE · ALPES-DE-HAUTE-PROVENCE · ALPES-MARITIMES · VAUCLUSE · BOUCHES-DU-RHÔNE · VAR · DRÔME · PIEMONTE · VALLE D'AOSTA · LOMBARDIA · LIGURIA · HAUTE-CORSE · CORSE (CORSICA) · CORSE DU SUD

Seas and water: LIGURIAN SEA · MEDITERRANEAN SEA · Golfe du Lion · Golfo di Génova · Lac Léman (Lac de Genève) · Lac de Neuchâtel · Thuner See · Brienzer See · Vierwaldstätter See · Zuger See · Zuger See · Lac du Bourget · Lac d'Annecy · Lago Maggiore · L. di Como · L. di Lugano · L. d'Iseo · L. d'Orta · Étang de Berre · Étang de Biguglia

Cities (France): Lyon · Genève · Lausanne · Neuchâtel · Bern · Fribourg · Grenoble · Chambéry · Annecy · Valence · Montélimar · Avignon · Nîmes · Arles · Marseille · Aix-en-Provence · Toulon · Cannes · Nice · Antibes · Fréjus · Draguignan · Gap · Digne-les-Bains · Briançon · St-Étienne · Vienne · Mâcon · Bourg-en-Bresse · Beaune · Chalon-sur-Saône · Le Creusot · Autun

Cities (Switzerland): Luzern · Schwyz · Interlaken · Montreux · Vevey · Martigny · Sion · Brig · Zermatt

Cities (Italy): TORINO (Turin) · MILANO (Milan) · Bergamo · Brescia · Novara · Pavia · Cremona · Parma · Piacenza · Alessandria · Asti · Cuneo · Savona · GENOVA (Genoa) · La Spezia · Massa · Carrara · Aosta · Varese · Como · Lecco · Vercelli · Ivrea · Pinerolo · Mondovi · Imperia (Maurizio-Oneglia) · San Remo · Ventimiglia · Monaco · Monte-Carlo · Menton · Livorno

Corsica: Bastia · St-Florent · L'Ile-Rousse · Calvi · Corte · Ajaccio · Sartène · Propriano · Porto-Vecchio · Bonifacio · Monte Cinto 2710 · C. Corse · G. de St-Florent · G. de Porto · G. de Sagone · G. d'Ajaccio · G. de Valinco

Islands: Îles d'Hyères · Île du Levant · Île de Porquerolles · Elba · Capraia · Gorgona · Pianosa · I. de Cavallo

Mountains / spot heights: Mont Blanc 4807 · Matterhorn 4478 · Monte Rosa 4634 · Finsteraarhorn 4274 · Piz Bernina 4049 · Gran Paradiso 4061 · Monte Viso 3841 · Barre des Écrins · Ortles 2899

COPYRIGHT. GEORGE PHILIP & SON. LTD.

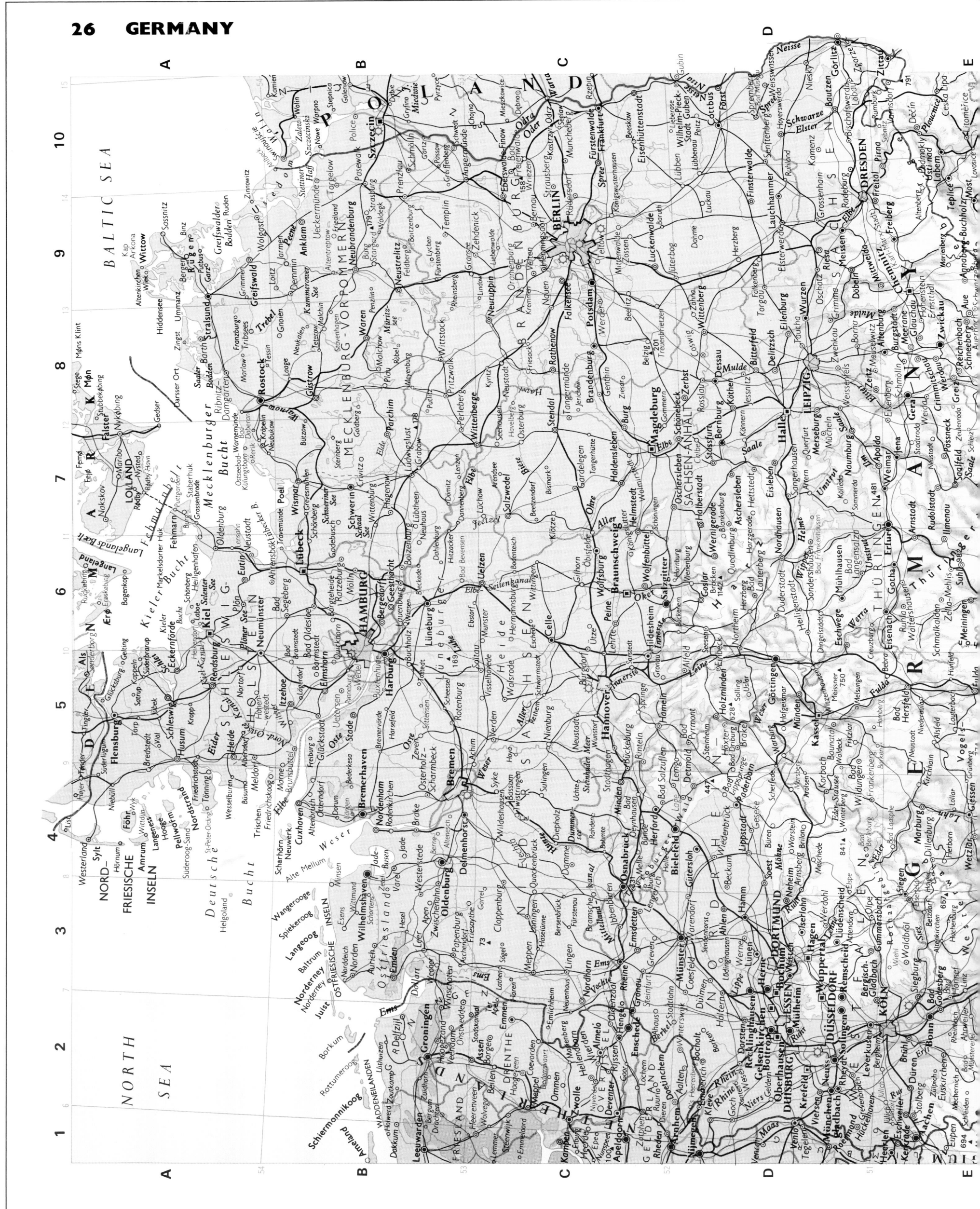

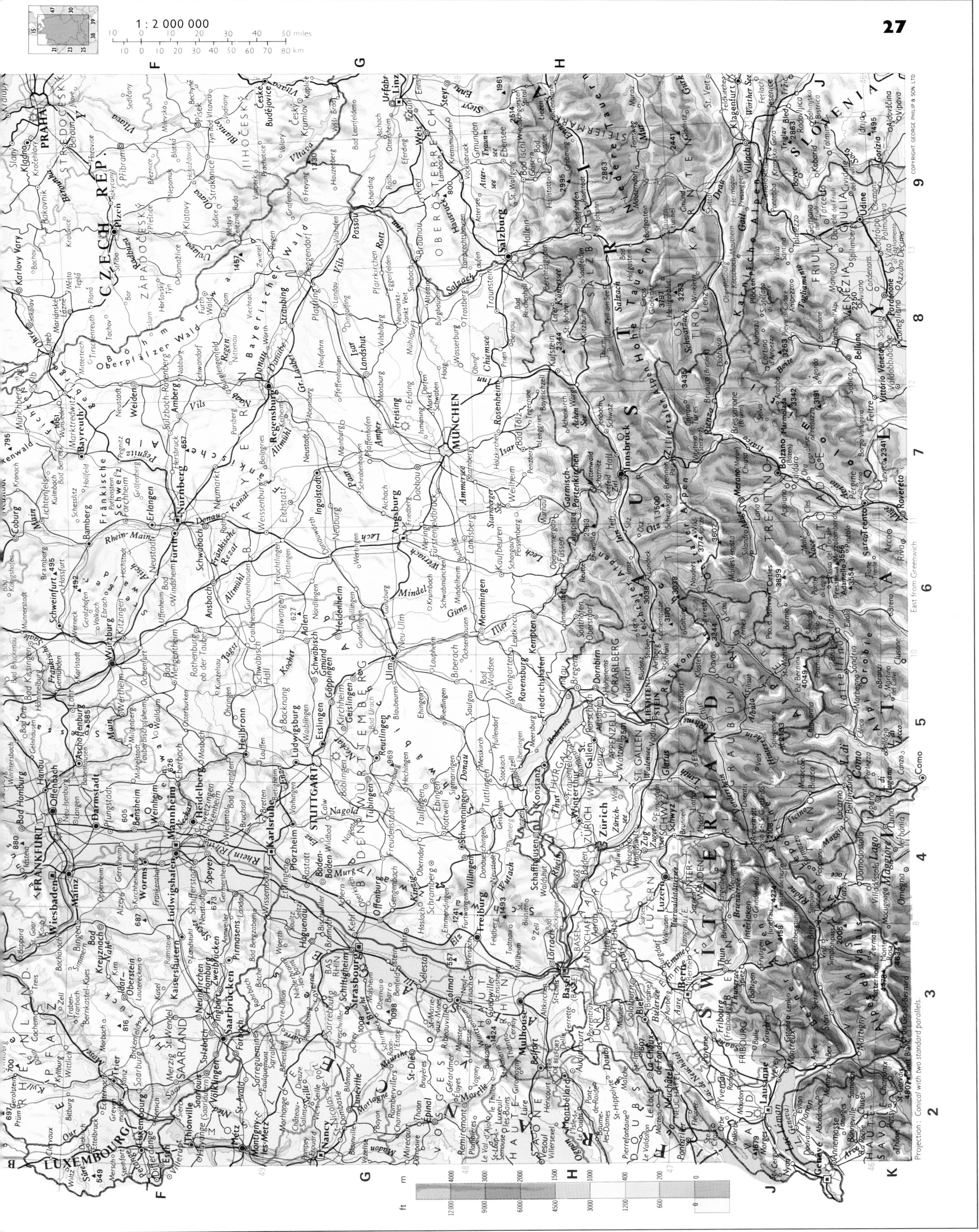

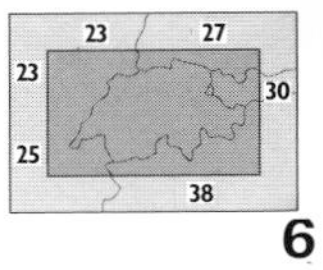
23
27
30
25
38

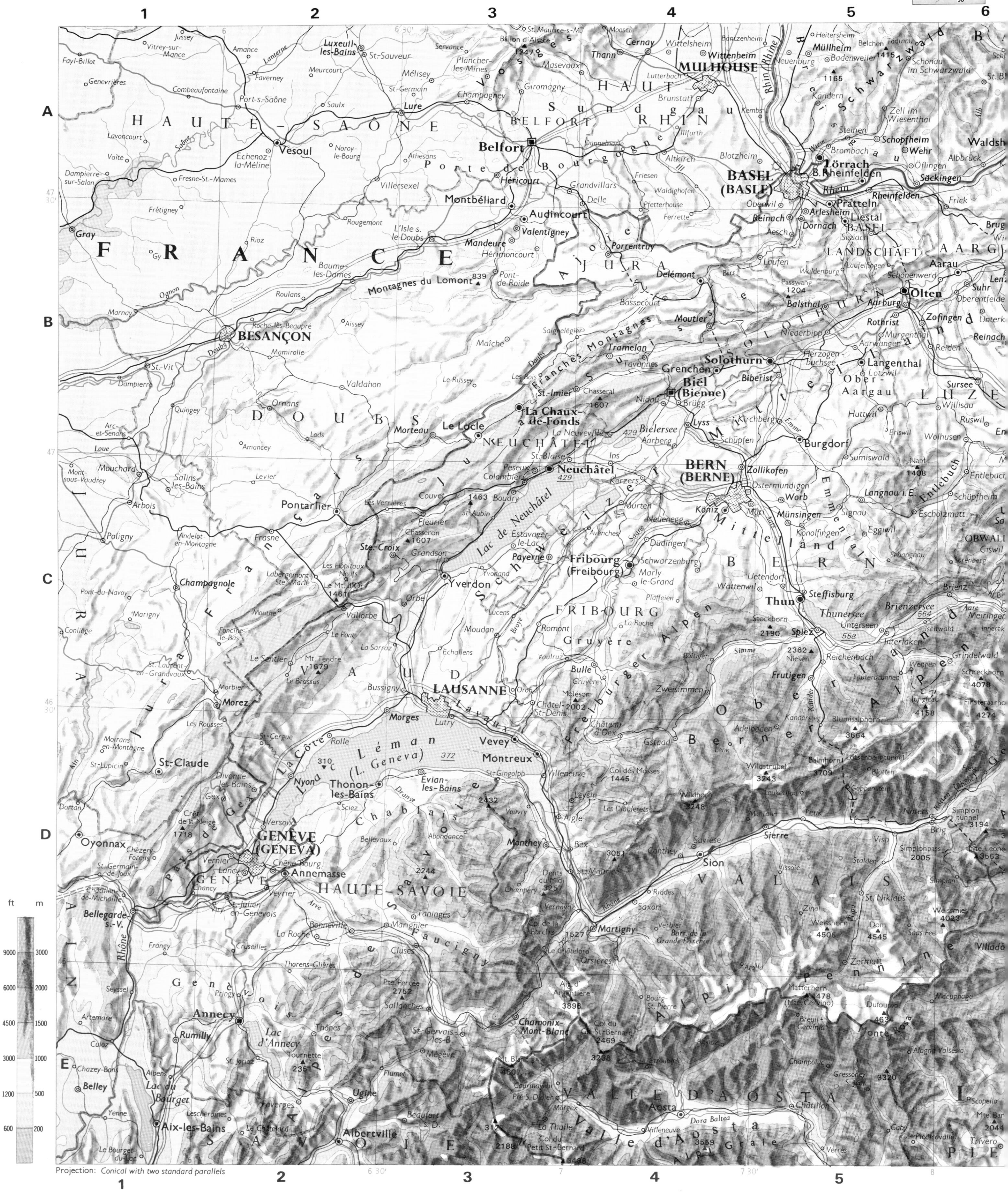
1
2
3
4
5
6
A
B
C
D
E
HAUTE-SAÔNE
FRANCE
DOUBS
JURA
HAUT-RHIN
BELFORT
Sundgau
Porte de Bourgogne
Vosges
Schwarzwald
BASEL (BASLE)
MULHOUSE
Belfort
Vesoul
Lure
Luxeuil-les-Bains
Gray
Montbéliard
Audincourt
Valentigney
Mandeure
Héricourt
L'Isle-s.-le-Doubs
Baume-les-Dames
BESANÇON
Montagnes du Lomont
839
Porrentruy
Delémont
Moutier
Franches Montagnes
Tramelan
St.-Imier
Chasseral
1607
La Chaux-de-Fonds
Le Locle
Morteau
Pontarlier
Neuchâtel
Lac de Neuchâtel
NEUCHÂTEL
429
Biel (Bienne)
Bielersee
Grenchen
Solothurn
SOLOTHURN
Olten
Aarau
AARGAU
Zofingen
Langenthal
Lörrach
Müllheim
Rheinfelden
Säckingen
Liestal
LANDSCHAFT
Pratteln
Arlesheim
Reinach
Dornach
Lyss
Burgdorf
BERN (BERNE)
Köniz
Worb
Münsingen
Thun
Steffisburg
Thunersee
Spiez
Brienzersee
Interlaken
Meiringen
Grindelwald
Schweizer Mittelland
Emmental
Entlebuch
Napf
1408
LUZERN
OBWALDEN
BERN
FRIBOURG
Fribourg (Freiburg)
Gruyère
Bulle
Romont
Payerne
Murten
Estavayer-le-Lac
Yverdon
Grandson
Ste.-Croix
Chasseron
1607
Orbe
Vallorbe
Le Mt. d'Or
1461
Champagnole
Jura
VAUD
Mt. Tendre
1679
LAUSANNE
Morges
Lutry
Lavaux
Vevey
Montreux
Lac Léman (L. Geneva)
372
Nyon
La Côte
Thonon-les-Bains
Évian-les-Bains
Chablais
GENÈVE (GENEVA)
GENÈVE
Annemasse
Pays de Gex
Gex
Divonne-les-Bains
Crêt de la Neige
1718
Oyonnax
St.-Claude
Morez
Bellegarde-s.-V.
Rhône
HAUTE-SAVOIE
Faucigny
Annecy
Lac d'Annecy
Rumilly
Tournette
2351
Thônes
Sallanches
Chamonix-Mont-Blanc
Mt. Blanc
4807
Aiguille d'Argentière
3896
Lac du Bourget
Aix-les-Bains
Belley
Albertville
SAVOIE
Ugine
Megève
Monthey
Aigle
Martigny
Sion
Sierre
Brig
Visp
VALAIS
Zermatt
Matterhorn (Mte. Cervino)
4478
Dom
4545
Weisshorn
4505
Dufourspitze
4634
Monte Rosa
Alpi Pennine
Simplon tunnel
Simplonpass
2005
Gstaad
Zweisimmen
Frutigen
Adelboden
Wildstrubel
3243
Wildhorn
3248
Balmhorn
3709
Lötschberg tunnel
Blümlisalphorn
3664
Jungfrau
4158
Finsteraarhorn
4274
Schreckhorn
4078
Berner Oberland
Berner Alpen
Freiburger Alpen
Col des Mosses
1445
Leysin
Dents du Midi
3257
Col du Gd. St. Bernard
2469
VALLE D'AOSTA
Aosta
Valle d'Aosta
Alpi Graie
Dora Baltea
Chatillon
3559
Col du Petit St.-Bernard
2188
3486
Projection: Conical with two standard parallels
ft
m
9000
3000
6000
2000
4500
1500
3000
1000
1200
500
600
200

Projection: Conical with two standard parallels

ft m
12 000 4000
9000 3000
6000 2000
4500 1500
3000 1000
1200 400
600 200
0 0

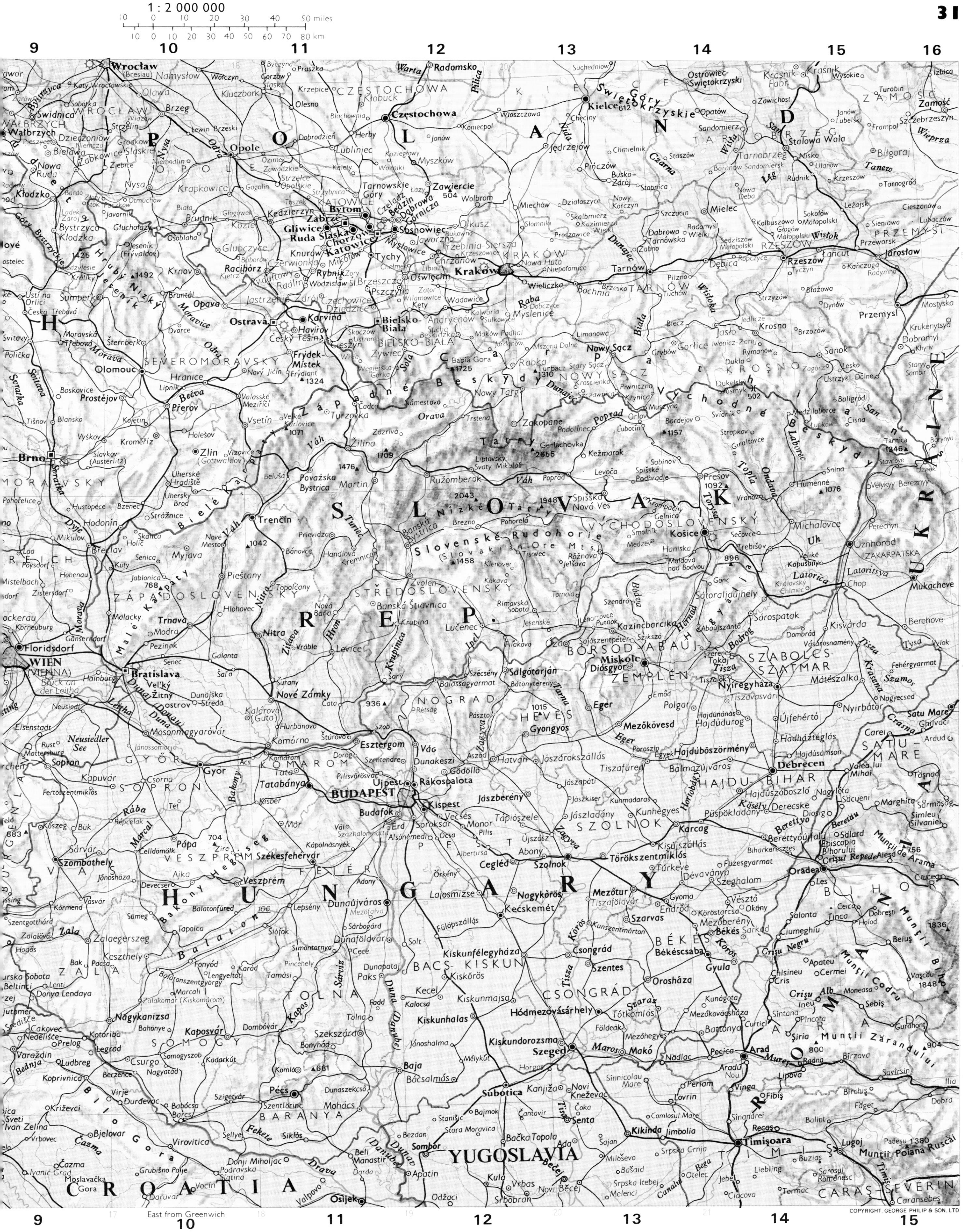

1 : 2 000 000

POLAND

SLOVAK REP.

HUNGARY

YUGOSLAVIA

CROATIA

UKRAINE

Wrocław (Breslau)
Opole
Częstochowa
Kielce
Katowice
Kraków
Rzeszów
Przemyśl
Ostrava
Brno
WIEN (VIENNA)
Bratislava
Nitra
Budapest
Debrecen
Szeged
Pécs
Timişoara
Oradea
Arad

East from Greenwich

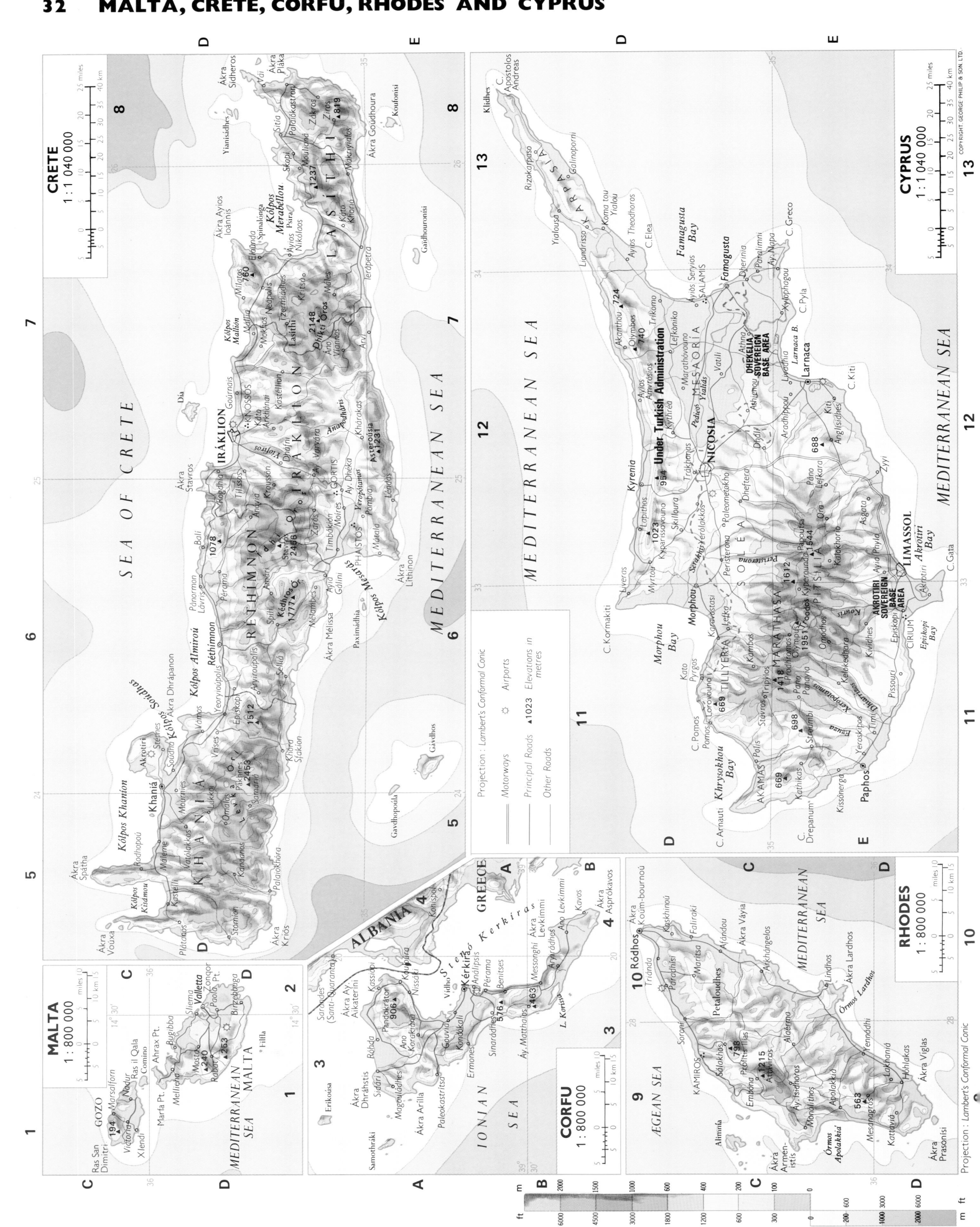

CRETE
1:1 040 000

25 miles
40 km

CYPRUS
1:1 040 000

25 miles
40 km

COPYRIGHT. GEORGE PHILIP & SON LTD.

SEA OF CRETE

MEDITERRANEAN SEA

MEDITERRANEAN SEA

MEDITERRANEAN SEA

Projection : Lambert's Conformal Conic

Motorways
Principal Roads
Other Roads

Airports

▲1023 Elevations in metres

MALTA
1:800 000

10 miles
km

CORFU
1:800 000

10 miles
km

RHODES
1:800 000

miles
km

Projection : Lambert's Conformal Conic

IONIAN SEA

AEGEAN SEA

MEDITERRANEAN SEA

MEDITERRANEAN SEA

ALBANIA

GREECE

m
ft
6000
4500
3000
1800
1200
600
300
0

2000
1500
1000
600
400
200
100
0

200
600
1000
3000
6000

600
1800
3000
6000

Under Turkish Administration

DHEKELIA SOVEREIGN BASE AREA

AKROTIRI SOVEREIGN BASE AREA

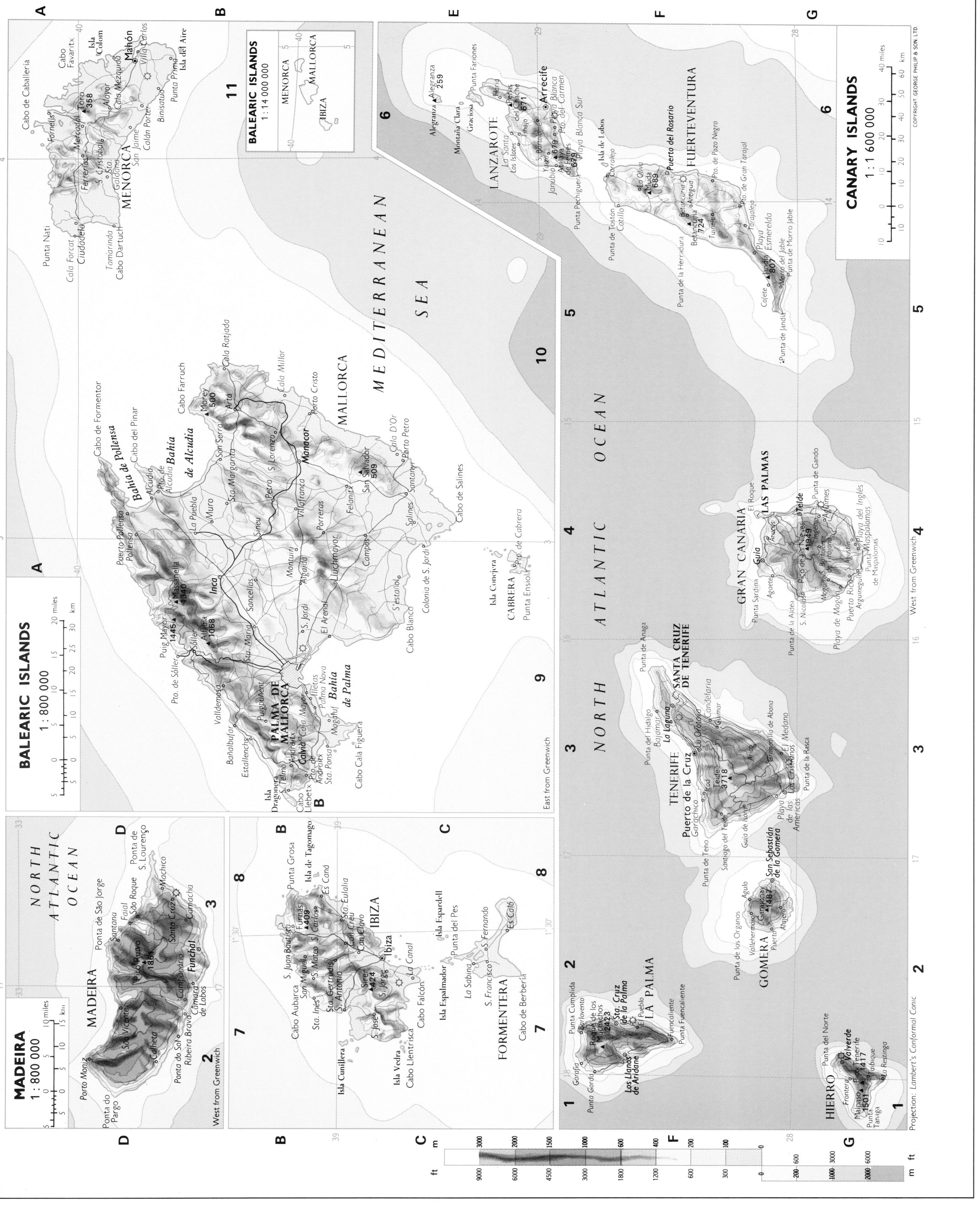

BALEARIC ISLANDS
1 : 800 000

20 miles
km

BALEARIC ISLANDS
1 : 14 000 000

MENORCA
MALLORCA
IBIZA

CANARY ISLANDS
1 : 1 600 000

40 miles
km

MADEIRA
1 : 800 000

10 miles
km

MENORCA
Cabo de Caballería
Fornells
Punta Nati
Cala Forcat
Ciudadela
Cabo Dartuch
Tamarinda
Punta Grosa
Cabo Favaritx
Isla Colom
Mahón
Mercadal
Toro ▲358
Alayor
Cala Mezquida
Villa Carlos
Isla del Aire
Punta Prima
Binisatuet
Calán Porter
San Jaime
San Cristóbal
Ferrerias
Sta. Galdana

MALLORCA
Cabo de Formentor
Puerto Pollensa
Pollensa
Cabo del Pinar
La Puebla
Alcudia
Bahía de Alcudia
Pto. de Alcudia
Cabo Farruch
Cala Ratjada
Artá
Morey ▲500
Cala Millor
Porto Cristo
Manacor
Cala D'Or
Porto Petro
San Salvador ▲509
Felanitx
Santany
Cabo de Salines
Salines
Campos
Llucmayor
Montuiri
Porreras
Villafranca
Petra
Sta. Margarita
S. Lorenzo
San Serra
Inca
Sineu
Sancellas
Sta. Maria
Muro
Puig Mayor ▲1445
Alfabia ▲1068
Massanella ▲1340
Puerto de Sóller
Sóller
Pto. de Sóller
Valldemosa
Bañalbufar
Estallenchs
Puigpuñent
PALMA DE MALLORCA
Calviá
Andraitx
Pto. de Andraitx
Sta. Ponsa
Cabo Cala Figuera
Isla Dragonera
Cabo Llebetx
Bahía de Palma
Moghuf
Palma Nova
Illetas
Bahía de Palma
El Arenal
S. Jordi
S'Estanol

MEDITERRANEAN SEA

CABRERA
Isla Conejera
Pto. de Cabrera
Punta Ensiola

IBIZA
Punta Grosa
Isla de Tagomago
Es Caná
Sta. Eulalia
S. Carlos
Cabo Aubarca
Furnás ▲409
S. Juan Bautista
San Miguel
S. Mateo
S. Gertrudis
Sta. Inés
San Antonio
Sirer
S. Jorge
Cana ▲424
Ibiza
Cabo Falcón
Isla Cunillera
Isla Vedra

FORMENTERA
Isla Espardell
Punta del Pes
Isla Espalmador
S. Francisco
S. Fernando
La Sabina
La Canal
Es Caló
Cabo de Berbería

MADEIRA
NORTH ATLANTIC OCEAN
Porto Moniz
Ponta do Pargo
S. Vicente
Ponta de São Jorge
Santana
Faial
São Roque
Machico
Ponta de S. Lourenço
Pico Ruivo ▲1861
Camacha
Funchal
Santa Cruz
Câmara de Lobos
Ribeira Brava
Ponta do Sol
Calheta

CANARY ISLANDS

LANZAROTE
Alegranza ▲259
Montaña Clara
Graciosa
La Santa
Los Islotes
Arrecife
Punta Fariones
Puerto del Carmen
Playa Blanca Sur
Isla de Lobos

FUERTEVENTURA
Corralejo
La Oliva ▲689
Puerto del Rosario
Betancuria ▲724
Tuineje
Pájara
Playa de Esmerelda ▲807
Morro Jable
Punta de Jandía
Punta de Morro Jable
Punta de Tostón
Cotillo

GRAN CANARIA
Las Palmas
El Roque
Teide
Guía
Pico de las Nieves ▲1949
Agaete
Punta Sardina
Telde
Punta de Gando
Playa del Inglés
Maspalomas
Punta de Maspalomas
Puerto Rico
Arguineguín
Mogán
San Nicolás
Puerto de Mogán

TENERIFE
Punta de Anaga
SANTA CRUZ DE TENERIFE
La Laguna
Candelaria
Güímar
Puerto de la Cruz
Icod
Teide ▲3718
Garachico
Buenavista
Punta de Teno
Adeje
Playa de las Américas
El Médano
Los Cristianos
Punta de la Rasca
Guía de Isora
Santiago del Teide

GOMERA
San Sebastián de la Gomera
Garajonay ▲1482
Agulo
Hermigua
Vallehermoso
Chipude
Punta de los Órganos
Valle Gran Rey
Playa de Santiago

LA PALMA
Punta Cumplida
Barlovento
Santo Domingo
Roque de los Muchachos ▲2423
Sta. Cruz de la Palma
El Pueblo
Los Llanos de Aridane
Fuencaliente
Punta Fuencaliente
Punta Gorda
Punta de la Herradura

HIERRO
Punta del Norte
Valverde ▲1501
Frontera
Pico de Malpaso ▲1417
La Restinga
Punta Orchilla

NORTH ATLANTIC OCEAN

MEDITERRANEAN SEA

East from Greenwich
West from Greenwich

Projection: Lambert's Conformal Conic

m — ft
3000 — 9000
2000 — 6000
1500 — 4500
1000 — 3000
600 — 1800
400 — 1200
200 — 600
100 — 300
0

m — ft
6000 — 2000
3000 — 1000
600 — 200
0
200

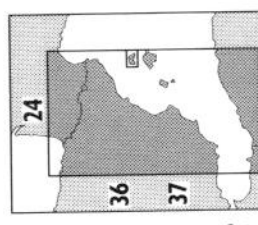

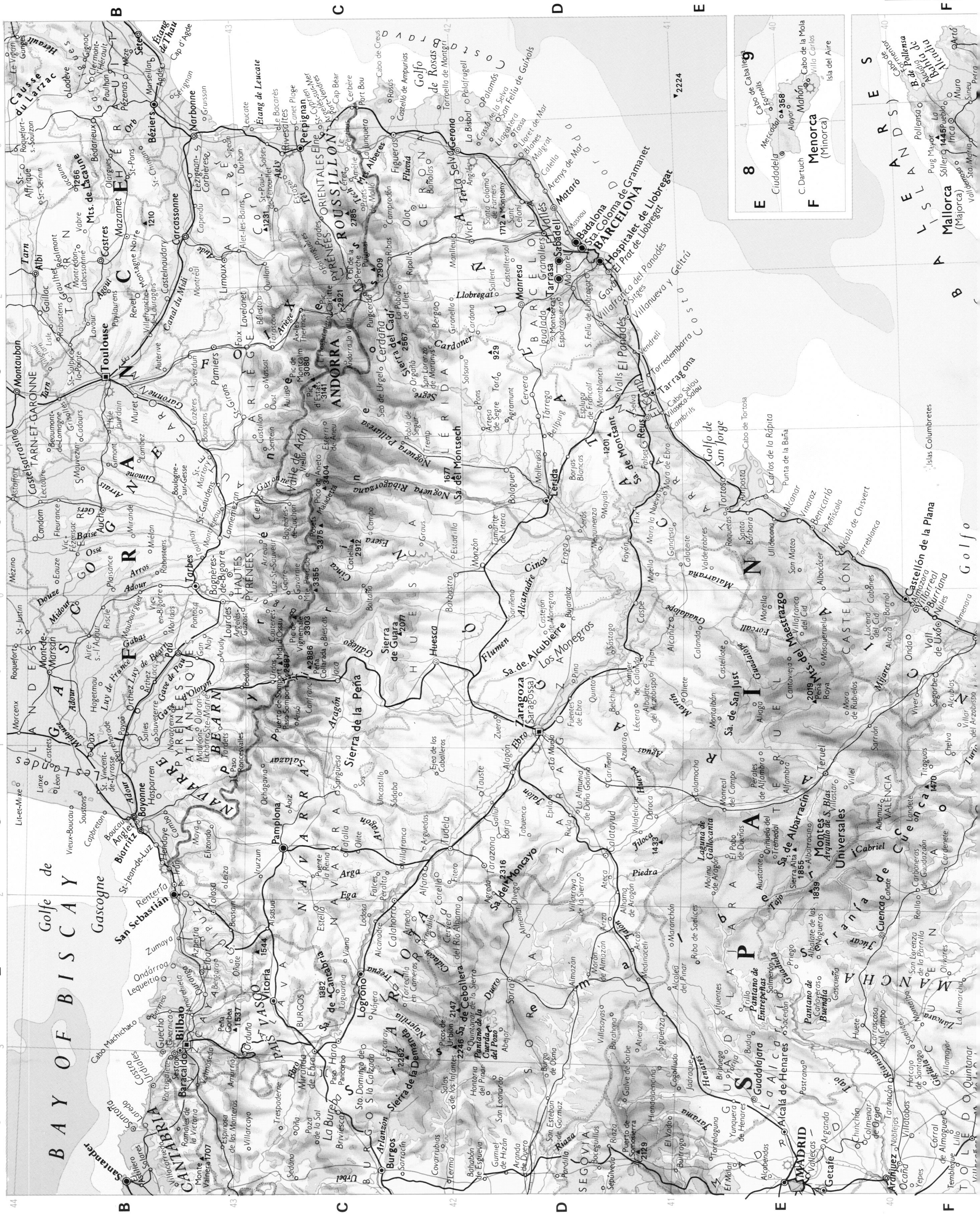

BAY OF BISCAY

Golfe de Gascogne

PYRÉNÉES

ANDORRA

ROUSSILLON

BARCELONA

Hospitalet de Llobregat

Badalona

Sta Coloma de Gramanet

Sabadell

Tarrasa

Toulouse

Carcassonne

Narbonne

Béziers

Perpignan

Gerona

Figueras

Golfo de Rosas

Cabo de Creus

Tarragona

Reus

Lérida

Zaragoza (Saragossa)

Huesca

Pamplona

San Sebastián

Bilbao

Vitoria

Logroño

Burgos

Soria

Teruel

VALENCIA

Castellón de la Plana

MADRID

Getafe

Aranjuez

Guadalajara

Alcalá de Henares

CANTABRIA

NAVARRA

PAIS VASCO

RIOJA

ARAGON

CATALUÑA

Mallorca (Majorca)

Menorca (Minorca)

ISLAS BALEARES

Ciudadela

Golfo de Valencia

Golfo de San Jorge

Sierra de Guara

Sierra de la Demanda

Montes Universales

Serranía de Cuenca

LA MANCHA

Ebro

Duero

Tajo

Peña de Oroel 1765

Aneto 3404

Maladeta 3375

Pic d'Estats 3141

Pic de Montcalm 3080

Monte Perdido 3355

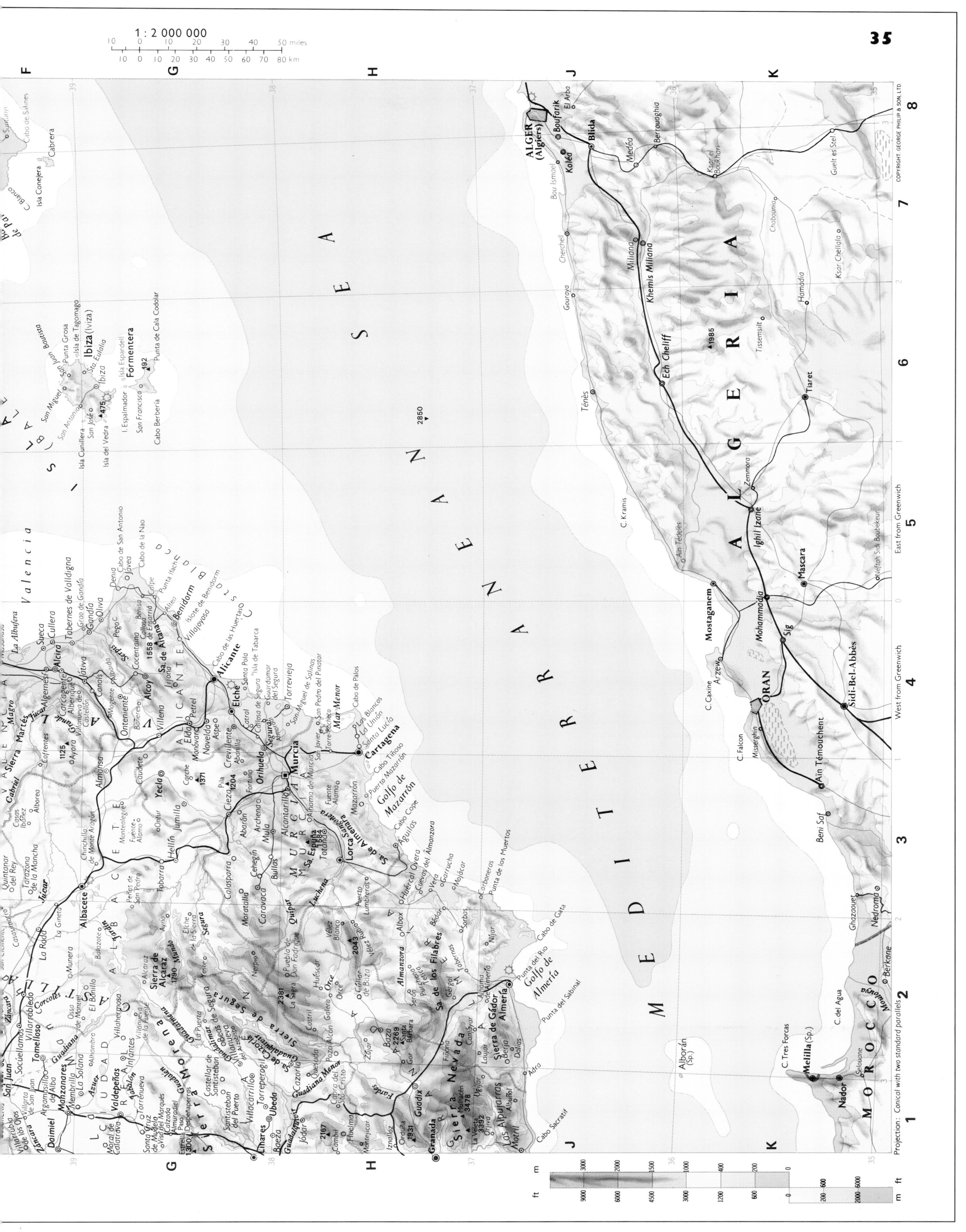

1 : 2 000 000

10   10   20   30   40   50 miles
10 0 10 20 30 40 50 60 70 80 km

Projection: Conical with two standard parallels

COPYRIGHT GEORGE PHILIP & SON, LTD.

BALEARIC ISLANDS

San Miguel
San Antonio
San José
Ibiza (Iviza)
Isla Conejera
Cabrera
Cabo de Salinas
Isla Grosa
Isla de Tagomago
Ibiza
Isla Espardell
San Francisco
Isla Espardell
Formentera
192
Punta de Cala Codolar
Cabo Berbería

MEDITERRANEAN   SEA

2850

MEDITERRANEAN   SEA

ALGERIA

ALGER (Algiers)
Boufarik
El Arba
Koléa
Blida
Medéa
Berrouaghia
Bou Ismael
Ksar el Boukhari
Guelt es Stel
Miliana
Khemis Miliana
Ech Cheliff
1985
Tassemalt
Chaboumia
Hamadia
Ksar Chellala
Cherchell
Gouraya
Ténès
C. Kramis
Aïn Tedeles
Ighil Izone
Tiaret
Mostaganem
Mohammadia
Sig
Mascara
Arzew
Zemmora
Oued Sidi Boubékeur
ORAN
C. Caxine
C. Falcon
Misserghin
Sidi-Bel-Abbès
Aïn Témouchent
Beni Saf
Ghazaouet
Nedroma
Berkane
Selouane
MOROCCO
Nador
Melilla (Sp.)
C. Tres Forcas
C. del Agua
Alborán (Sp.)
Cabo Sacratif

East from Greenwich
West from Greenwich

Valencia
La Albufera
Sueca
Cullera
Alcira
Tabernes de Valldigna
Játiva
Gandía
Oliva
Denia
Cabo de San Antonio
Jávea
Cabo de la Nao
Cocentaina
Benidorm
Benisa
Calpe
Altea
Villajoyosa
Alcoy
1558 Sa. de Aitana
Alicante
Cabo de las Huertas
Santa Pola
Elche
Crevillente
Orihuela
Murcia
Elda Petrel
Novelda
Aspe
Cabo de Palos
Mar Menor
Cartagena
Cabo Tiñoso
Golfo de Mazarrón
Cabo Cope
Águilas
Lorca
Sa. de Almenara
Totana
Sa. Espuña 1884
Cartagena
MURCIA
Alcantarilla
Cieza
1204 Pila
Jumilla
Yecla
1371
Villena
Ontenient
Onteniente
Albacete
Hellín
Cabriel
Sierra Martés
1125
Cofrentes
Ayora
Almansa
Montealegre
Chinchilla de Monte Aragón
Sierra Nevada
3478 Mulhacén
3392 La Veleta
Granada
Guadix
Sierra de Gádor
Almería
Golfo de Almería
Sierra de los Filabres 2168
Baza
Huércal Overa
Vélez Rubio
Cuevas del Almanzora
Punta del Río
Punta del Sabinal

CÁDIZ
JAÉN
Linares
Úbeda
Baeza
Jódar
Sierra de Cazorla
Sierra de Segura
1790 Mundo
Sierra de Alcaraz
Alcaraz

ft m
9000 3000
6000 2000
4500 1500
3000 1000
1500 400
600   200
0   0
m ft
200 600
2000 6000

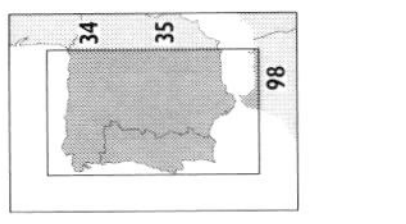

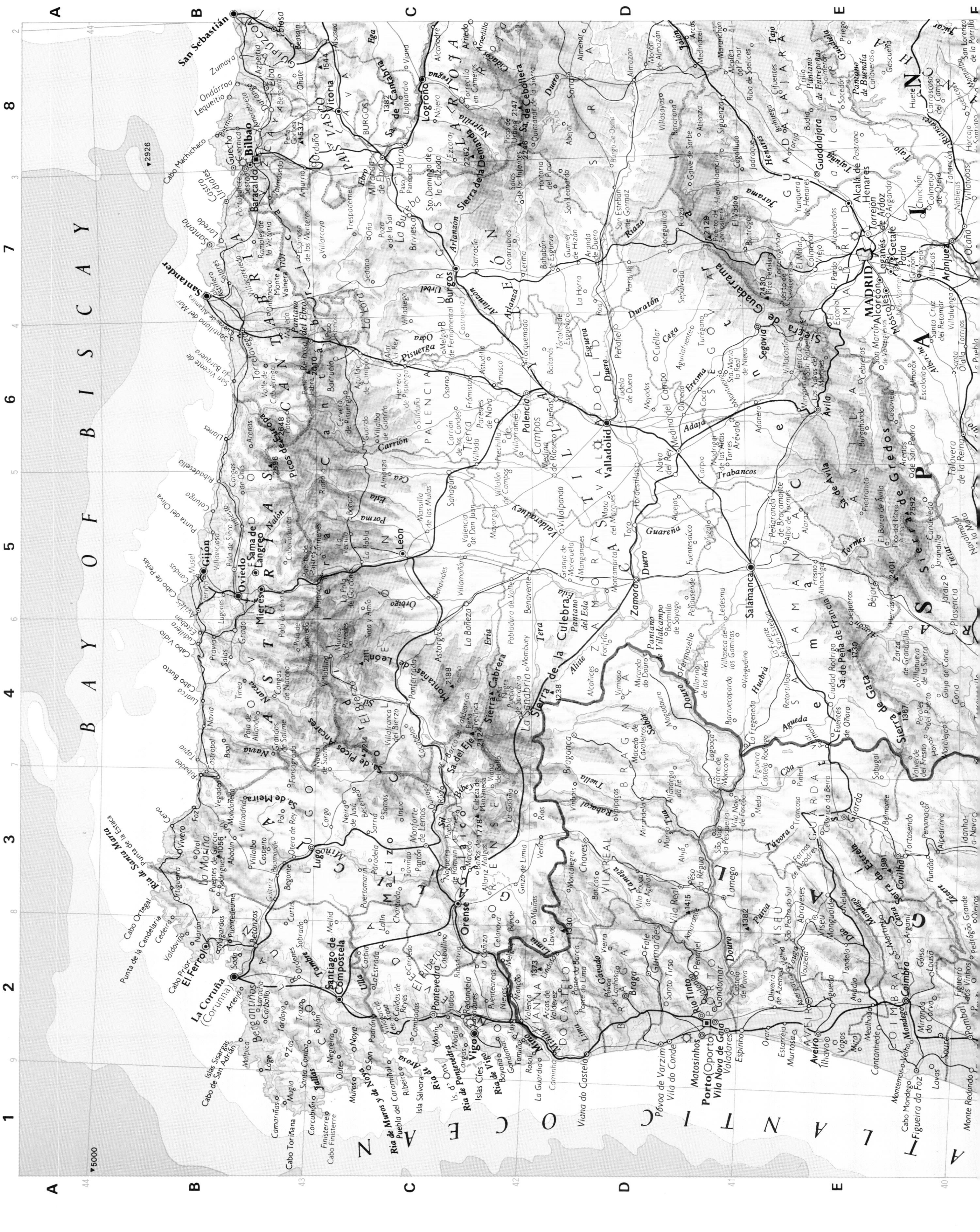

BAY OF BISCAY

ATLANTIC OCEAN

San Sebastián
Bilbao
Baracaldo
Santander
Gijón
Oviedo
Mieres
La Coruña (Coruña)
El Ferrol
Santiago de Compostela
Pontevedra
Vigo
Orense
Lugo
Vitoria
Logroño
BURGOS
Burgos
PALENCIA
Palencia
Valladolid
León
Zamora
Salamanca
Ávila
Segovia
MADRID
Guadalajara
Alcalá de Henares

PAÍS VASCO
RIOJA
CANTABRIA
ASTURIAS
CASTILLA
LEÓN
GALICIA
SALAMANCA
GUADALAJARA

Duero
Ebro
Pisuerga
Carrión
Esla
Tormes
Tajo
Jarama
Henares

PORTO (Oporto)
Vila Nova de Gaia
Matosinhos
Aveiro
Coimbra
Braga
Guimarães
Vila Real
Lamego
Viseu
Chaves
Bragança

BRAGANÇA
VILA REAL
VISEU
COIMBRA
AVEIRO
PORTO

Douro
Mondego
Mondego

Sierra de Guadarrama
Sierra de Gredos
Picos de Europa
Sierra de la Demanda
Sierra de la Culebra

1 : 2 000 000

10   0   10   20   30   40   50 miles
10   0 10 20 30 40 50 60 70 80 km

MEDITERRANEAN

SEA

MOROCCO

Strait of Gibraltar

Golfo de Cádiz

Golfo de Almería

Alborán (Sp.)

PORTUGAL

Montes de Toledo

Sierra Nevada   3478

Sierra de Gádor

Sierra de los Filabres

Sierra de Segura

Sierra de Cazorla

Sierra de Aracena

Sierra Morena

LISBOA (LISBON)

Setúbal

Évora

Faro

Badajoz

Mérida

Cáceres

Sevilla (Seville)

Córdoba

Jaén

Granada

Almería

Málaga

Antequera

Cádiz

Jerez de la Frontera

Algeciras

Gibraltar (Br.)

La Línea de la Concepción

Ceuta (Sp.)

Tetouan

Tánger (Tanger)

Larache

Melilla (Sp.)

Ciudad Real

Valdepeñas

Linares

Úbeda

Huelva

West from Greenwich

Projection: Conical with two standard parallels

ft   m
9000   3000
6000   2000
4500   1500
3000   1000
1200   400
600   200
0   0
200   600
2000   6000
ft   m

FRANCHE-COMTÉ
SWITZERLAND
BERN
VAUD
FRIBOURG
VALAIS
GRAUBÜNDEN
TICINO
VORARLBERG
LIECHTENSTEIN
ST GALLEN

Lyon (Lyons)
Genève (GENEVA)
Lausanne
Bern (BERNE)
Luzern
Grenoble
Chambéry
Annecy
Aosta
TORINO (Turin)
Moncalier
Asti
Alessandria
Novi Ligure
Cúneo
Mondovì
Savona
GENOVA (Genoa)
La Spezia
Carrara
Massa

PIEMONTE
LOMBARDIA
VALLE D'AOSTA
DAUPHINÉ
PROVENCE
HAUTE-PROVENCE
ALPES MARITIMES
VAUCLUSE
BOUCHES-DU-RHÔNE

Milano (Milan)
Como
Varese
Bergamo
Brescia
Pavia
Cremona
Piacenza
Parma
Reggio
Módena
Novara
Vercelli
Vigévano
Lodi
Monza

Lago Maggiore
L. di Como
L. di Lugano
L. d'Orta
L. d'Iseo
Lago di Garda
Golfo di Génova
Riviera di Ponente
Riviera di Levante

LIGURIAN SEA

Marseille (Marseilles)
Toulon
Aix-en-Provence
Avignon
Arles
Salon
Nice
Cannes
Antibes
Monaco
Monte-Carlo
Menton
Ventimiglia
San Remo
Imperia (Maurizio-Oneglia)
Draguignan
Grasse
Fréjus
St-Raphaël
Ste-Maxime
St-Tropez
ILES D'HYÈRES
I. du Levant
I. de Porquerolles
Côte d'Azur

Livorno (Leghorn)
Pisa
Lucca
Péscia
Pistóia
Cécina
Piombino
Elba
Portoferráio
Gorgona
Capraia
TOSCANA
Arcipelago Toscano
Pianosa
Montecristo
Giglio

CORSE (CORSICA)
HAUTE-CORSE
CORSE-DU-SUD
Bastia
St-Florent
Calvi
L'Île Rousse
Ajaccio
Monte Cinto 2710
Bonifacio
G. de Porto
G. de St-Florent
G. de Sagone
G. d'Ajaccio
Iles Sanguinaires

Projection: Conical with two standard parallels
East from Greenwich

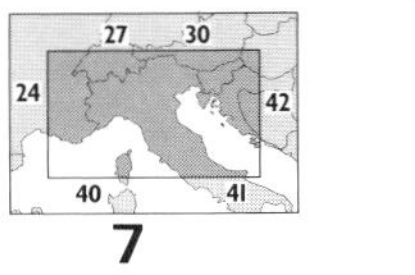

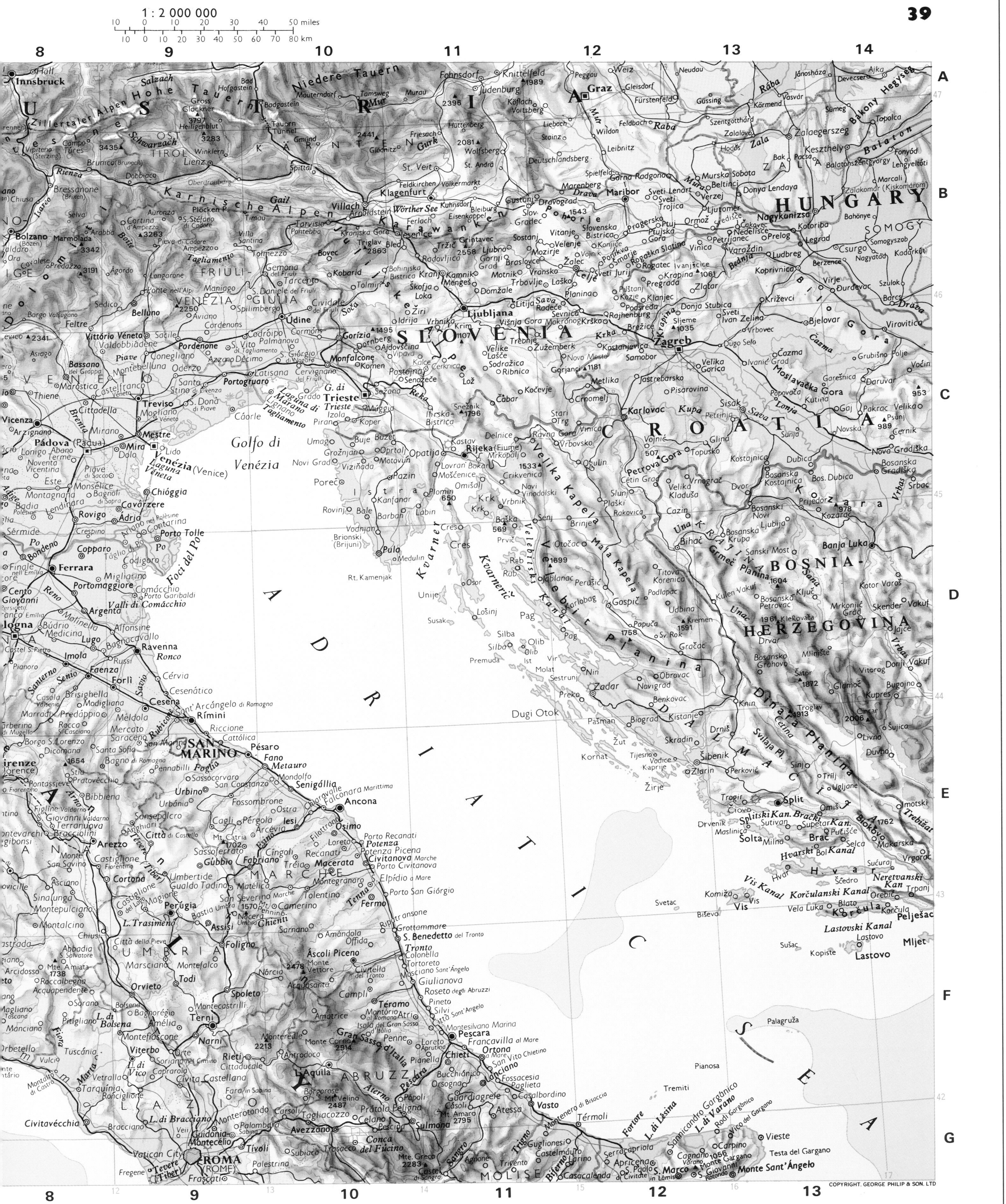

1 : 2 000 000

10 0 10 20 30 40 50 miles

10 0 10 20 30 40 50 60 70 80 km

8   9   10   11   12   13   14

HUNGARY

SLOVENIA

CROATIA

BOSNIA-HERZEGOVINA

SOMOGY

Innsbruck · Hall · Salzach · Hohe Tauern · Niedere Tauern · Fohnsdorf · Knittelfeld · Peggau · Graz · Weiz · Neudau · Devecser · Ajka · Bikony Hegység

Zillertaler Alpen · Gross Glockner · Heiligenblut · Mauterndorf · Tamsweg · Murau · Judenburg · Köflach · Voitsberg · Gleisdorf · Fürstenfeld · Güssing · Rába · Jánosháza · Körmend · Vasvár · Sümeg · Topolca

Osti Tirol · Lienz · Spittal · Gmünd · Friesach · Gurk · Wolfsberg · Deutschlandsberg · Leibnitz · Feldbach · Körmend · Zalaegerszeg · Keszthely · Fonyód · Balaton

Kärnten · Villach · Klagenfurt · Feldkirchen · Völkermarkt · St. Veit · St. Andrä · Wildon · Spielfeld · Radgona · Murska Sobota · Lengyeltóti · Marcali

Karnische Alpen · Gail · Tarvisio · Arnoldstein · Karawanken · Bleiburg · Slov. Bistrica · Maribor · Sveti Lenart · Verzej · Donya Lendava · Nagykanizsa · Bahónye

Cortina d'Ampezzo · Tolmezzo · Kranjska Gora · Jesenice · Bled · Radovljica · Kamnik · Celje · Sveti Jurij · Rogaska Slatina · Vinica · Varaždin · Ludbreg · Csurgo · Somogyszob · Kadárkút

Bolzano (Bözen) · Marmolada · Belluno · Udine · Gorizia · Ljubljana · Škofja Loka · Domžale · Litija · Sava · Krško · Zagreb · Bjelovar · Virovitica

FRIULI-VENEZIA GIULIA · Pordenone · Monfalcone · Trieste · Koper · Postojna · Novo Mesto · Karlovac · Sisak · Kutina · Daruvar

VENETO · Treviso · Mestre · Venézia (Venice) · Golfo di Venézia · Laguna Véneta · Istra · Rijeka (Fiume) · Opatija · Petrova Gora · Banja Luka

Padova (Padua) · Mira · Chióggia · Rovigo · Adria · Porto Tolle · Po · Pula · Cres · Losinj · Rab · Bihać · Sanski Most

Ferrara · Ravenna · Rímini · SAN MARINO · Pésaro · Fano · Ancona · Zadar · Šibenik · Split · Knin · Sinj · Livno

Firenze (Florence) · Arezzo · Perúgia · Assisi · Foligno · Spoleto · Terni · Ascoli Piceno · Téramo · Pescara · Chieti

UMBRIA · MARCHE · ABRUZZI · L'Aquila · Viterbo · Rieti · Tívoli · ROMA (ROME) · Vatican City · Frascati

MOLISE · Vasto · Térmoli · Monte Sant'Ángelo · Vieste · Testa del Gargano

ADRIATIC SEA

Golfo di Venézia

8   9   10   11   12   13

COPYRIGHT GEORGE PHILIP & SON. LTD

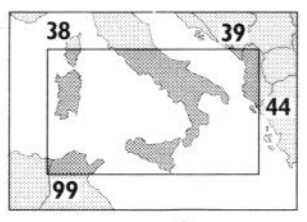

CORSE
CORSICA

Iles Sanguinaires
G. d'Ajaccio
C. di Muro
Petretto
Tarravo
Ulacudine
2136 Zonza
Levie
Solenzara
Favone
CORSE-DU-SUD
G. de Valinco
Propriano
Sartène
Porto-Vecchio
Iles Cerbicales
Bonifacio
I. de Cavallo

Bouches de Bonifacio
Santa Teresa Gallura
La Maddalena
Maddalena
Caprera
Costa
Smeralda

Asinara
Punta dello Scorno
Golfo dell'
Asinara
Coghinas
Aggius
Tempio Pausania
Calangianus
Pto. Cervo
Arzachena
G. di Olbia
Golfo Aranci
Olbia
Tavolara

Porto Tórres
Sorso
Sennori
1362
M. Limbara
Sássari
Osilo
Oschiri
C. dell'Argentiera
L. di Coghinas
Ozieri
Posada
Fertília
Alghero
Pattada
Buddusò
Siniscola
Villanova
Monteleone
1259
Bonorva
Bitti
C. Comino
Temo
Bosa
Macomer
Nuoro
Orune
Dorgali
Oliena
Golfo di
Orosei
C. Mannu
Ghilarza
L. del Tirso
Fonni
Monti del
Gennargentu
Baunei
C. di Monte Santu
Golfo di
Oristano
Cábras
Oristano
Sorgona
1834
Laconi
Arbatax
Lanusei
SARDEGNA
M. Arci
812
Terralba
Arborea
Muravera
Mandas
Tertenia
SARDINIA
Gúspini
Montevale
Sanluri
Murri
Villacidro
S. Vito
Villaputzu
C. Pécora
Arbus
1236
Gonnostanadiga
Serramanna
Fluminimaggiore
M. Línas
Sestu
Sinnai
C. Ferrato
Iglésias
Siliqua
Selárgius
 Pta. Serpeddi
Muravera
Portoscuso
Gonnesa
Assemini
1069
San Pietro
Carbonia
Quartu Sant'Elena
Serpentara
Sant'Antioco
1116
Cágliari
Sant'
Antíoco
Santadi
Golfo di
Cágliari
C. Carbonara
Porto Botte
Pula
Teulada
G. di Pálmas
C. Spartivento

TYRRHENIAN
SEA
3719
3589

ROMA
(Rome)
Vatican City
Tívoli
Subiaco
Trosacco
Conca
del Fucino
Fregene
Palestrina
Valmontone
Lido di Óstia
(Lido di Roma)
Velletri
Anagni
Alatri
Véroli
Sora
Pomézia
Lazio
Cori
Ferentino
Monte S. Gio
Arpino
Albano
Aprília
Cisterna di Latina
Ceccano
Ceprano
Cassino
Latina
Sezze
Priverno
Sannino
Ánzio
Nettuno
Pontinia
Fondi
1533
Sabáudia
Monte Circeo
541
Terracina
Gaeta
Fórmia
Minturno
Gariglian
Golfo di
Gaeta
Zannone
Palmarola
Ísole
Ponziane
Ponza
283
Ventotene
788
Ischia

PALERMO
C. San Vito
Castellammare del Golfo
G. di Castellammare
Terrasini
Favarotta
C. Gallo
Bagheria
Levanzo
Trápani
Érice
1110
Alcamo
Montreale
Partinico
Misilmeri
Términ
Ísole Égadi
Paceco
S. Giuseppe
Iato
Marineo
Maréttimo
Calatafimi
Camporeale
Corleone
Favignana
Salemi
Gibellina
1613
Belsito
Stagnone
Partanna
Bisacquino
Prizzi
Lercara
Friddi
Leo M
Marsala
Castelvetrano
Sambuca
di Sicília
SICI
Mazara
del Vallo
Menfi
Búrgio
Mussomeli
C
Campobello di Mazara
Belice
Ribera
Platani
San Cato
Calt
Sciacca
Caltabellotta
Racalmuto
Cani
Cattólica Eraclea
Siculiana
Raffadali
Aragona
Nar
Porto Empédocle
Siciliana
Agrigento
Favara
Palma di Montechiaro
Campobello di Lica
Li

Sicilian Channel

Iles de la
Galite

Ustica

C. Blanc
Cani
Bizerte
(Binzert)
Plane
C. Serrat
Menzel-Bourguiba
Zembra
El Kala
Tabarka
Mateur
Golfe de Tunis
C. Bon
ALGERIA
TUNIS
Halq el Oued
Kelibia
Béja
Tébourba
Menzel-
Temime
Bou Salem
Soliman
TUNISIA
Nabeul
Téboursouk
Zaghouan
Hammamet
Pantelleria
Pantelleria
836
(It.)
Medjerda
MEDITE
1319
Me

Projection: Conical with two standard parallels
East from Greenwich

ft / m scale:
9000 / 3000
6000 / 2000
4500 / 1500
3000 / 1000
1200 / 400
600 / 200
0 / 0
200 / 600
2000 / 6000
4000 / 12.000
m / ft

1 : 2 000 000

10 0 10 20 30 40 50 miles
10 0 10 20 30 40 50 60 70 80 km

7   8   9   10   11   12

ADRIATIC SEA

Drini

K. iMyzhllit
te Skënderbeut

Shëngjini   Lezha Rrësheni
Rubiku
Mati

Bishti i Pallës   TIRANA
Shijaku   DURRËSI
Kavaja   Ishmi   Kruja
Durrësi   TIRANA
(Durazzo)   (Tirana)
Kalaja e Turrës   Rrogozhina
Çerriku

Shkumbini   Lushnja   ELBASANI
Semani   Kuçova
Fieri   Devolli
BERATI
Levani   Berati

Selenica
Laguna e Nartës   I. Sazani
Gjiri i Vlorës   Kanina   VLORA
Kep i Gjuhes   Karaburuni   (Valona)
Orika   M. e Kendervices
Dukati   2130   Gribes

Himara

Erikoúsa   Sarandë

Othonoí   Karousádhes   Kassiópi

Samothráki   Korakiádha   Liapádhes

Kérkira   **Kérkira**
(Corfu)   Gastoúri

Áyios Matthaíos
Argyrádhes   Levkímmi

A

B

C

40

ADRIATIC SEA

BRUZZI   Trento   Montenero   L. di Lésina   Sannicandro   Rodi Gargánnico
Guglionesi   di Bisaccia   Apricena   Vico del Gargano   Vieste
Agnone   Trivento   Serracapriola   Cagnano   Monte Gargano
OBISE   Casacalenda   Santa Croce   Carpino   Varano   Testa del Gargano
Biferno   S. Paolo di Civitate   Rotondo   1056   Monte Sant'Ángelo
Frosolone   di Magliano   S. Severo   S. Giovanni
ISerniа   S. Marco   Manfredónia
Iafro   Campobasso   Lucera   G. di Manfredónia
Boiano   Cercemaggiore   Fóggia   Zapponeta
Piedimonte   Riccia   Cervaro   Cerignola
d'Alife   S. Bartolomeo   Trinitápoli   Margherita di Savoia
in Galdo   Biccari   Tróia   Barletta
Colle Sannita   Orta Nova   Trani   Biscéglie
Cereto   Pontelandolfo   Cerignola   Canosa   Andria   Molfetta
Caserta   Benevento   Ariano   Candela   di Púglia   Giovinazzo
Maddaloni   Grottaminarda   Irpino   Ruvo   Terlizzi   Bari
anise   Mirabella   Lavello   Corato   Bitonto   Mola di Bari
Afrágola   Avellino   Ofanto   Minervino   Palo   del Colle   Rutigliano
Nola   Somma   Melfi   Múrge 686   Grumo   Casamássima   Polignano a Mare
OLI   1277 Vesuviana   Montella   Spinazzola   Acquaviva   Castellana Grotte   Monópoli
Nápoli   Boscotrecase   Calitri   Rionero   Palazzo   delle Fonti   Fasano
Annunziata   Sarno   Nocera   in Vulture   S. Gervásio   Gióia   Putignano   Ostuni
ammare di Stabia   Gragnano   Inferiore   Forenza   del Colle   Noci   Cisternino   Cegle Messápico
Sorrento   Battipaglia   Lioni   Acerenza   Santéramo   Martina   Carovigno
Capri   Éboli   Muro   Irsina   in Colle   Franca   S. Vito dei Normanni   Bríndisi
G. di Salerno   Lucano   Matera   Laterza   Móttola   Francavilla Fontana
Sele   Potenza   Tricárico   Grassano   Massafra   Oria   Mesagne
M. Alburno   BASILICATA   Basento   Ginosa   Taranto   Latiano   Squinzano
1742   Marsico Nuovo   Pomárico   Is. Coradi   Sava   Mandúria   Pietro Vernótico
Picerno   1836   Ferrandina   S. Giórgio Iónico   Lizzano   Campi   Lecce
Rocca d'Aspide   Stigliano   Pisticci   Salentina   Leverano
Teggiano   Brádano   Bernalda   Marúggio   Nardò   Martano
Agrópoli   Sala Consilina   Corleto   Perticara   Tursi   Golfo di   Galátone   Galatina   Máglie
Capáccio   Pádola   Agri   Senise   Montalbano Iónico   Táranto   Casarano   Otranto
Castellabate   Vallo della Lucánia   Moliterno   S. Arcángelo   Sinni   Gallípoli   Paràbita   Poggiardo
Punta Licosa   M. Sirino   Latrónico   Rócale   Ugento   Tricase
Pisciotta   1225   Lagónegro   2005   Rotondella   Presicce   Gagliano del Capo
Póllica   Camerota   Láuria   C. Santa Maria di Leuca   C. d'Otranto
C. Palinuro   Murolucano   Monte Pollino   Amendolaro   C. Santa Maria di Leuca

Strait of Otranto

ALBANIA

IONIAN SEA

G. di Policastro   Mormanno   Castrovíllari   Trebisacce
Scalea   2271   Morano   Cassano Iónio
Lao   Cálabro   Cordata   Crati
Belvedere   Verbicaro   Spezzano   Ascrati
Marittimo   Roggiano   Grávina   C. Trionto
Fagnano   Castello   Demétrio   Rossano
Cetraro   Montalto Uffugo   Bisignano   Acri   Còrone
Páola   Fuscaldo   Luzzi   Longobucca   Ciró   Pta. dell'Alice
S. Lúcido   Rende   Cosenza   Aprigliano   Marina di Ciró
Fiumefreddo Brúzio   Lago   1929   Sila   Neto   Stróngoli
Amantea   Rogliano   Cotronei   Petilia   Policastro   Crotone
Aiello Cálabro   Decollatura   Mesoraca   Sersale   C. delle Colonne
Nocera Terinese   Nicastro   Gimigliano   Cutro   Ísola di Capo Rizzuto
Gizzeria   Tiriolo   Tacina   C. Rizzuto
Sambiase   Maida   Bórgia   Catanzaro
Golfo di   Filadélfia   Girifalco
Sant'Eufémia   Pizzo   S. Onofrio   Golfo di Squillace
926 Strómboli   Tropea   Filadélfia   Chiaravalle   Centrale
Isole Eólie o Lípari (Æolian Is.)   Capo Vaticano   Vibo Valéntia   Serra S. Bruno   3065
Filicudi   Salina   Nicótera   1423   Guardavalle
962   Malfa   Panarea   G. di Gióia   Laureana   Capa Stilo
Alicudi   602   Lípari   Rosarno   di Borrella   Roccella Iónica
499   Lípari   Gióia Táuro   Mámmola   Gioiosa Iónica
Vulcano   Palmi   Taurianova   Siderno Marina
Bagnara   Oppido   Locri
S. Ágata di Militello   C. Calavà   Barcellona   Cittanova   Mamertino   Ardore Marina
Sant'Ágata di Militello   Pirzino   Bagnara   Villa S. Giovanni   Bovalino Marina
San Fratello   Pozza di Gotto   Mi. Peloritani   1956   Locri
Milazzo   Náso   Aspromonte
Monti   Montalbano   Elicona   Messina   S. Stéfano   Santa Teresa
Nébrodi   Castroreale   Réggio   di Riva   Bova Marina
1847   Patti   di Cálabra
telbuono Mistretta Tortorici   Péllaro   C. Spartivento
Monti   Cesarò   Randazzo   Stir. di Messina   Palizzi
Petralia   Capizzi   Taormina   Mélito   4116
Nicosia   Troina   Bronte   di Porto Salvo
Alimena   Agira   Etna   Riposto
IA   Leonforte   3340   Giarre
Nicosia   Regalbuto   Adrano   Santo Stéfano
916 Valguarnera Caropepe   Biancavilla   Acireale
Enna   Paternò   Belpasso
Pietraperzia   Centuripe   Misterbianco
Aidone   Ramacca   Catánia
SICILIA   Palagonia   Golfo di
Barrafranca   Scordia   Simeto   Catánia
Mazzarino   Militello   Lentini   Carlentini
Mont.   in Val di Catánia   Augusta
Caltagirone   Francofonte
Grammichele   Vizzini   986   Siracusa
Butera   Monti Iblei   Florídia
Niscemi   Sortino   Siracusa
Chiaramonte   Canicattini   G. di
Gela   Gulfi   Palazzolo   Avola
Vittória   Cómiso   Acréide   Noto
Santa Croceo   Ragusa   Módica   Noto
Camerina   Scicli   Rosolini
Pozzallo   Íspica   Pachino
Channel   C. Passero

RANEAN SEA

MEDITERRANEAN SEA

39

D

38

37

E

F

7   8   9   10   11   12

15   16   17   18   19

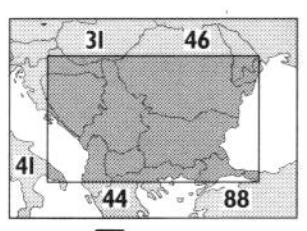

Grid columns: 1 2 3 4 5 6 7

Grid rows: A B C D E F G

Countries / Regions:
HUNGARY
CROATIA
BOSNIA-
HERZEGOVINA
YUGOSLAVIA
VOIVODINA
SERBIA
MONTENEGRO
KOSOVO
ALBANIA
MACEDONIA
SOMOGY
BARANYA
TOLNA
CSONGRÁD
ARAD
BIHOR
CARAŞ-SEVERIN
TIMIŞ

Seas:
ADRIATIC SEA

Selected place names:
Nagykanizsa, Kotoriba, Prelog, Legrad, Ludbreg, Koprivnica, Križevci, Bjelovar, Čazma, Kaposvár, Dombóvár, Bonyhád, Szekszárd, Pécs, Mohács, Siklós, Baja, Kiskunhalas, Kiskunmajsa, Hódmezővásárhely, Orosháza, Gyula, Szeged, Makó, Arad, Timişoara, Lugoj, Hunedoara, Reşiţa, Caransebeş

Subotica, Sombor, Senta, Kikinda, Bečej, Novi Bečej, Zrenjanin, Vršac, Osijek, Vinkovci, Vukovar, Novi Sad, Ruma, Sremska Mitrovica, Šabac, Beograd (Belgrade), Zemun, Pančevo, Smederevo, Požarevac

Banja Luka, Doboj, Tuzla, Bijeljina, Zvornik, Valjevo, Kragujevac, Jagodina, Ćuprija, Paraćin, Zaječar, Vidin, Negotin, Bor

Zenica, Sarajevo, Višegrad, Užice (Titovo Užice), Čačak, Kraljevo, Kruševac, Niš, Leskovac, Pirot

Mostar, Foča, Pljevlja, Nova Varoš, Prijepolje, Novi Pazar, Priština, Vranje

Dubrovnik, Cavtat, Nikšić, Podgorica, Cetinje, Kotor, Bar, Ulcinj, Shkodra (Shkodër), Prizren, Skopje, Kumanovo

Tirana (Tiranë), Durrësi (Durazzo), Elbasani, Ohrid, Bitola, Prilep, Veles (Titov Veles), Štip

Korčula, Lastovo, Mljet, Hvar

Rivers / water:
Drava, Sava, Danube (Dunav), Tisa, Drina, Bosna, Vrbas, Morava, Zapadna Morava, Velika Morava, Timok, Vardar, Neretva, Lim, Tara, Skadarsko Jezero, Ohridsko Jezero (L. Ohri), Prepansko Jezero (L. Prespa), Dojransko Jezero (L. Doiran)

Spot heights (selected):
681, 953, 989, 459, 539, 916, 1328, 1943, 2112, 2006, 2067, 2228, 2155, 1762, 1969, 2522, 2253, 1833, 1734, 2017, 1492, 1560, 1808, 2483, 2656, 2694, 2604, 2764, 2540, 2242, 2259, 2181, 2600, 1754, 2252, 1593, 1224, 1226, 1848, 1380, 1441, 1409, 1706, 1922, 2168

Elevation scale:
ft   m
9000 3000
6000 2000
4500 1500
3000 1000
1200 400
600   200
0   0
200   600
2000 6000
m   ft

Projection: Conical with two standard parallels

East from Greenwich

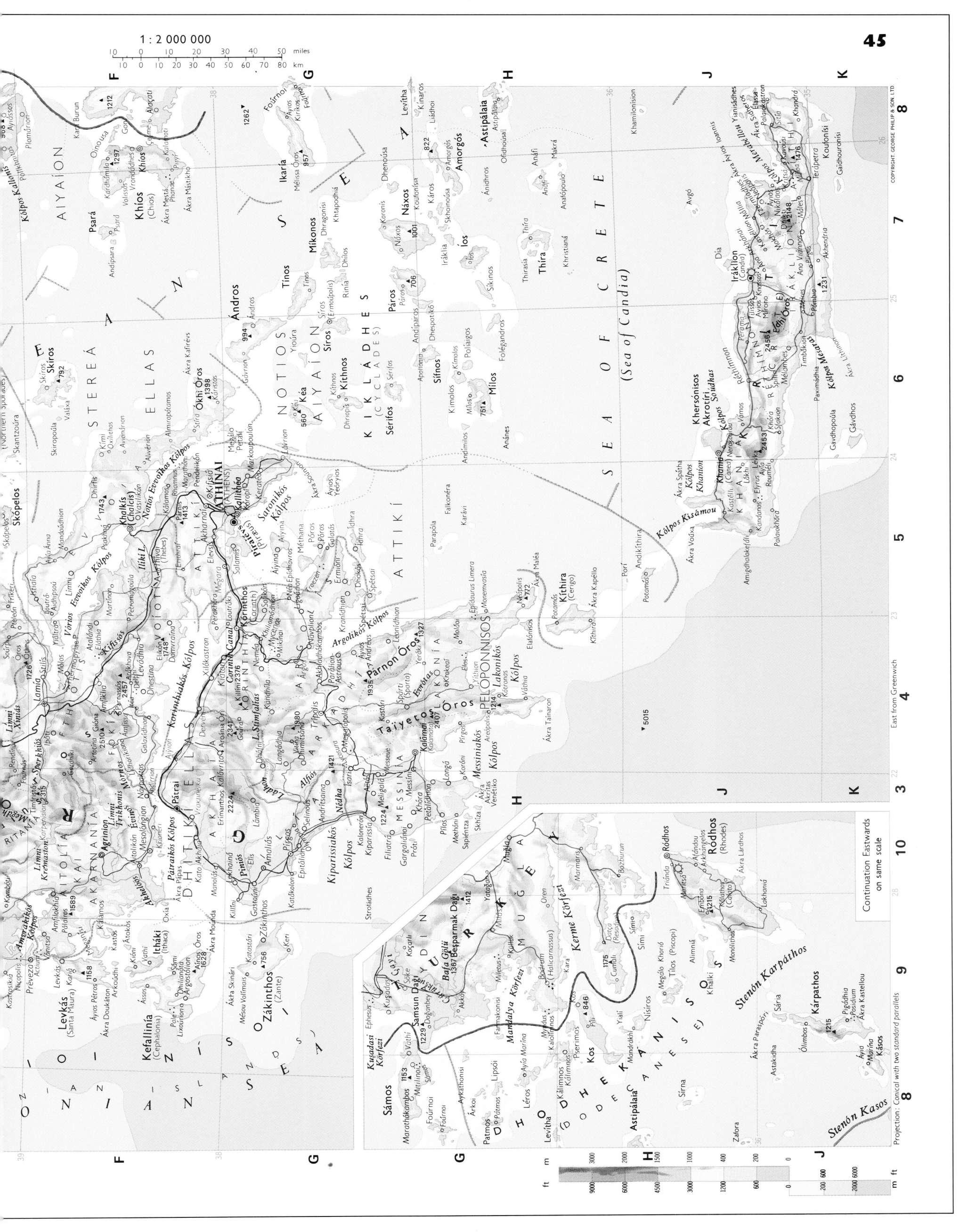

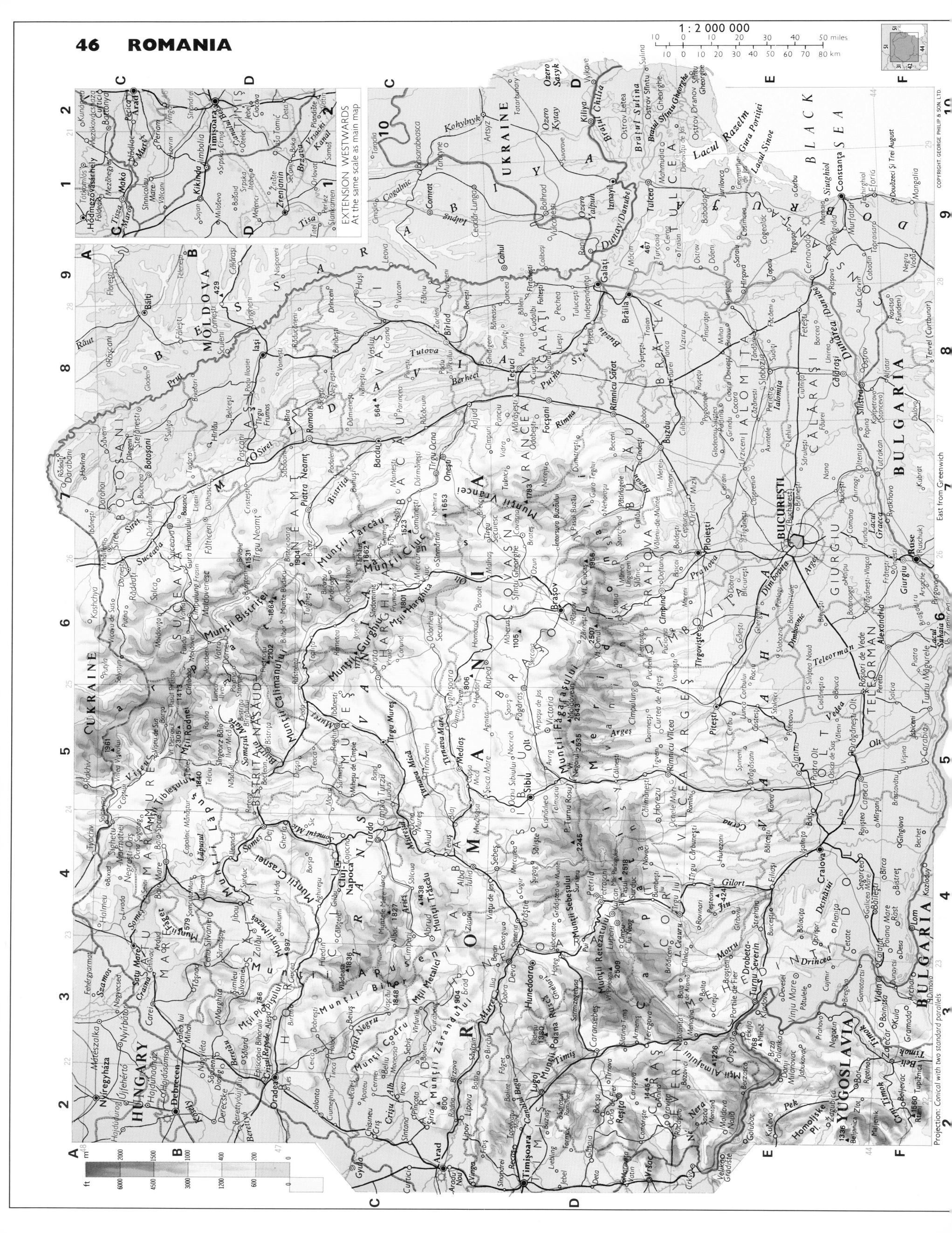

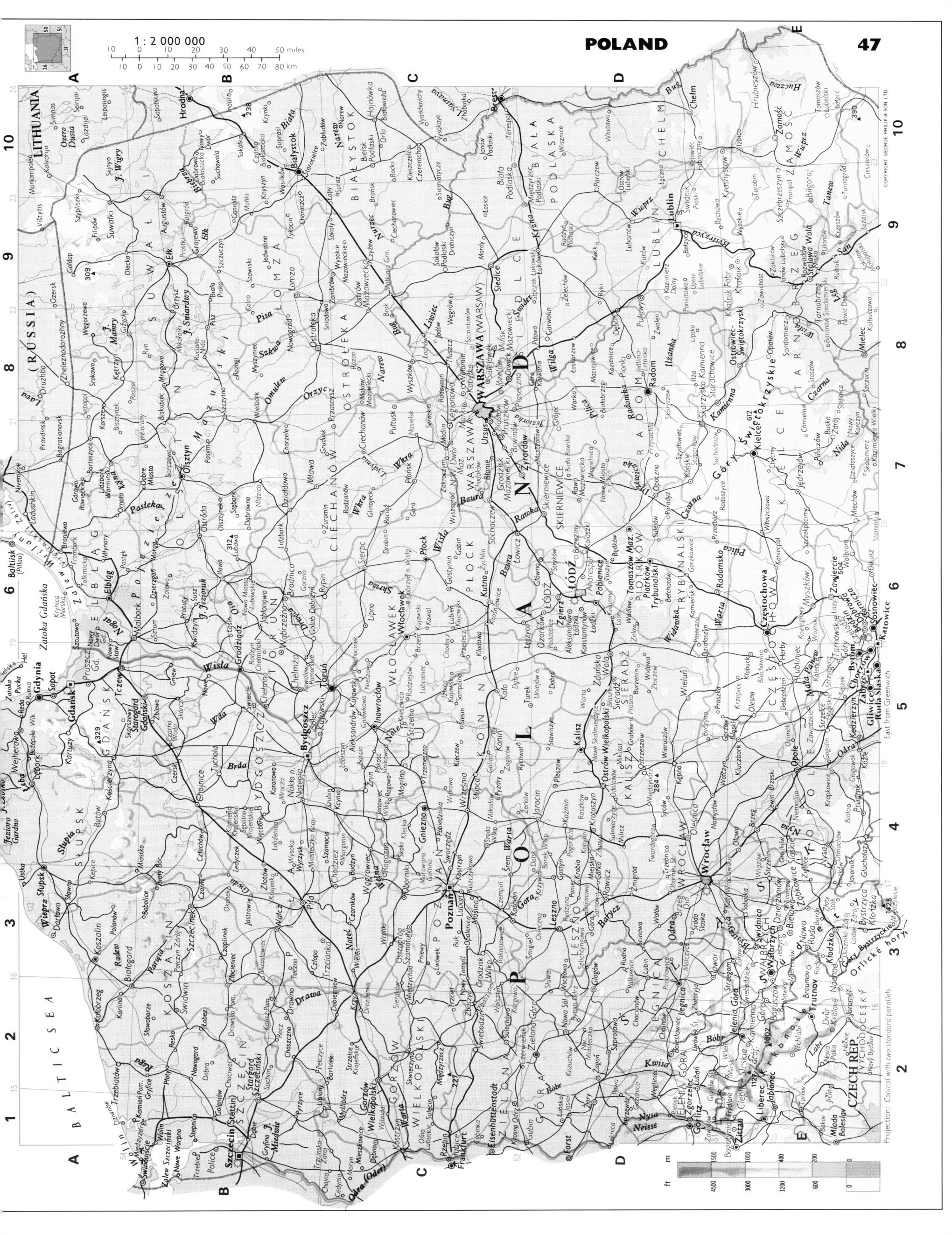

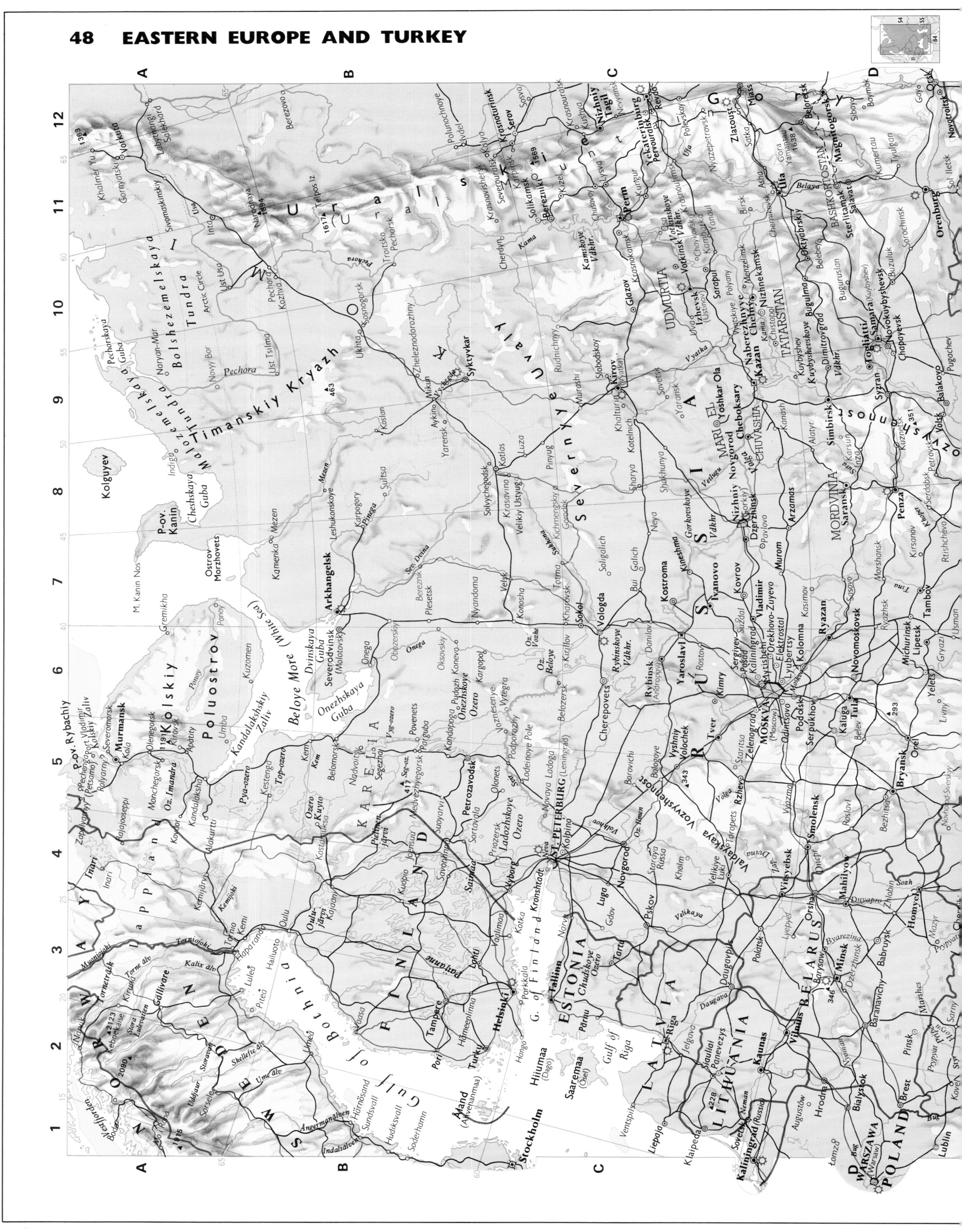

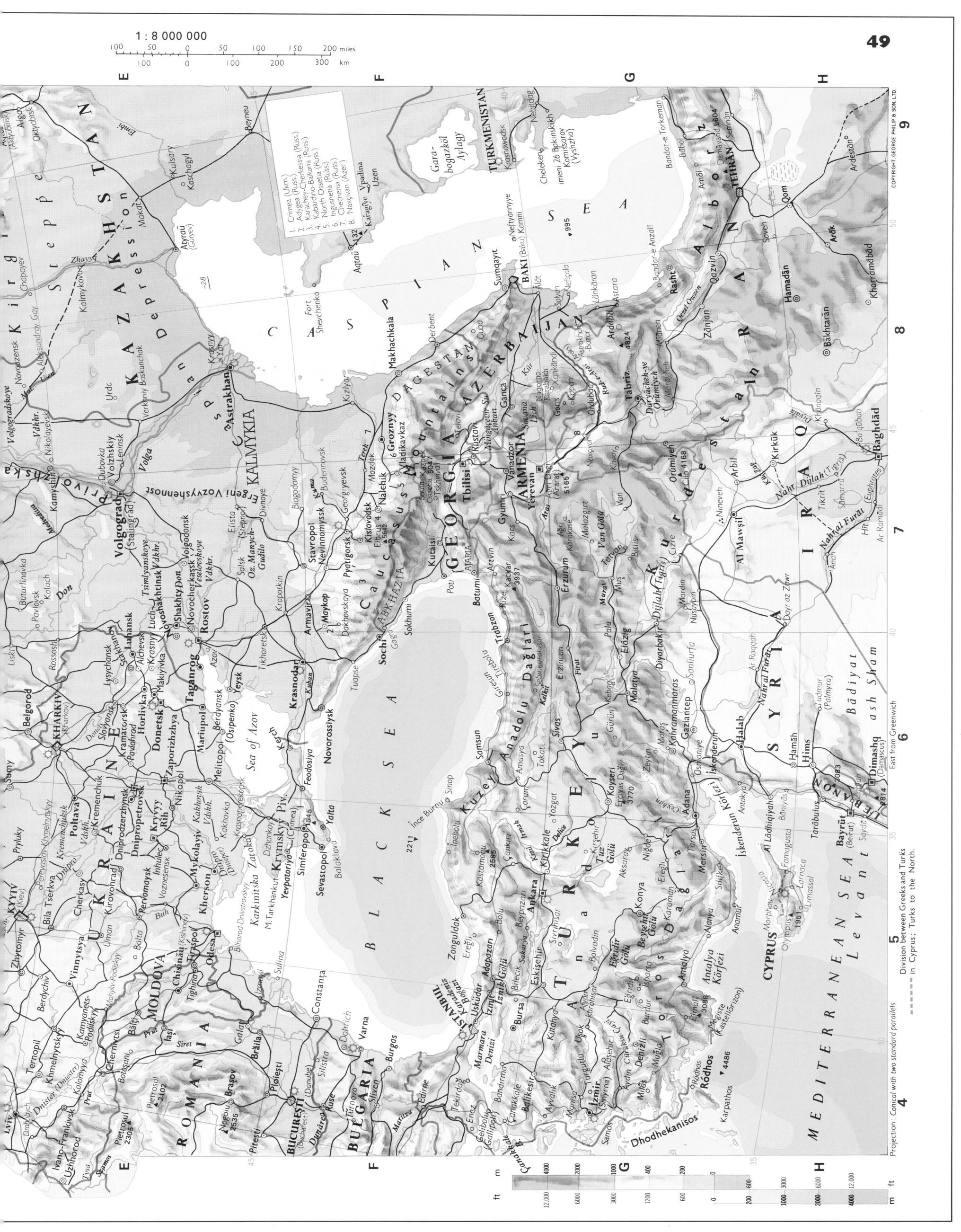

1 : 4 000 000

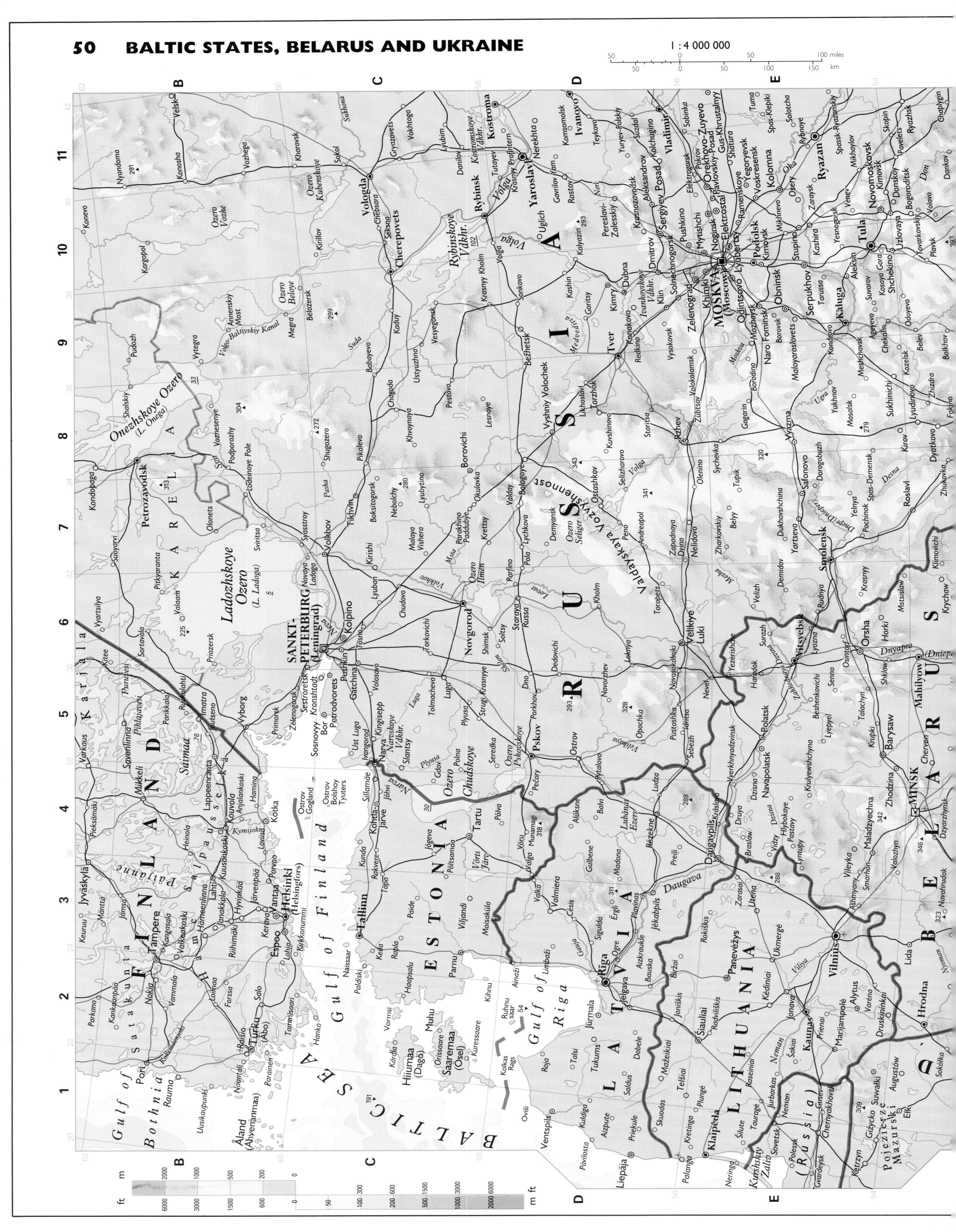

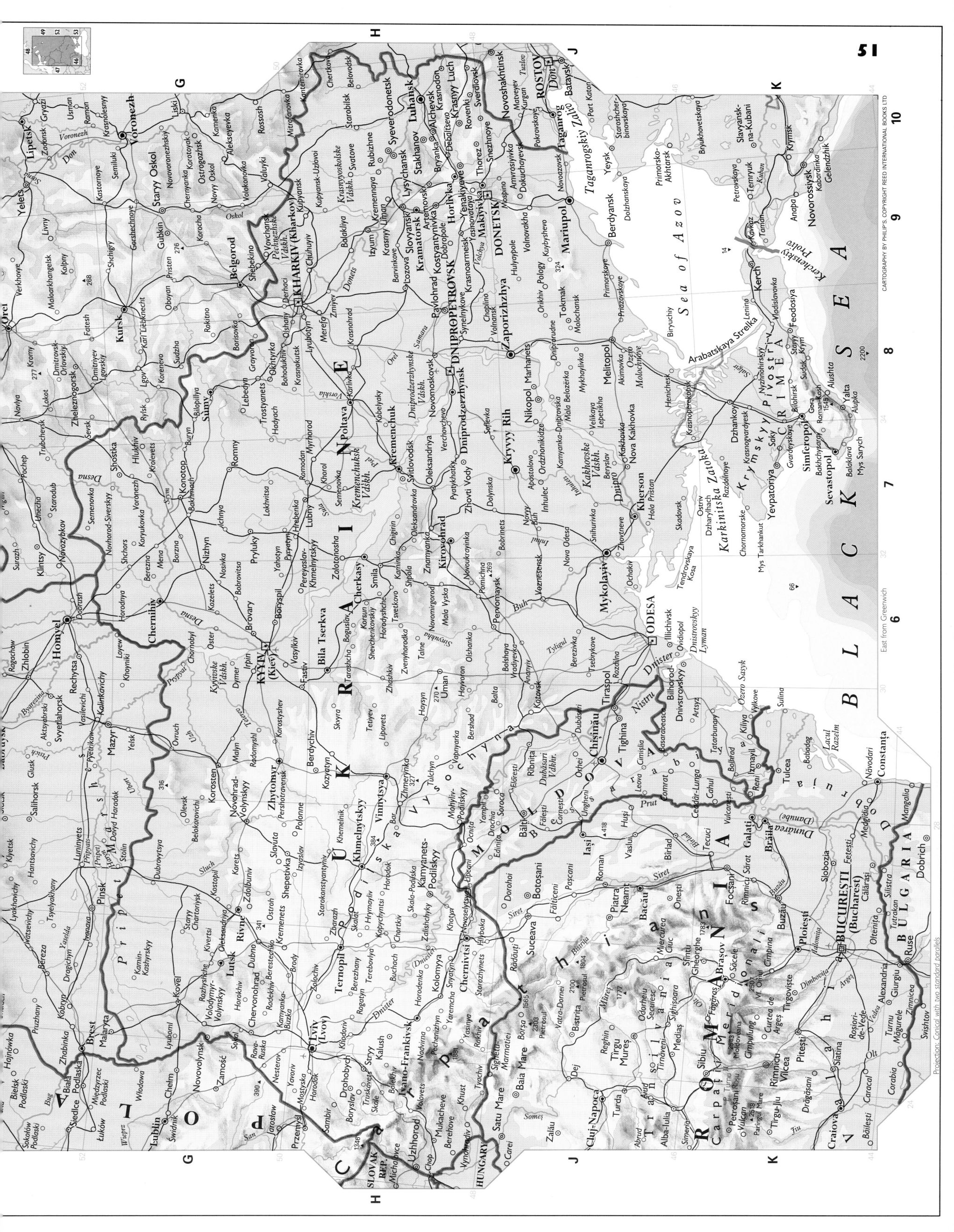

Yelets
Lipetsk
Orel
Voronezh
VORONEZH

ROSTOV

Kursk
Belgorod
KHARKOV (Kharkiv)
Luhansk
Sumy
Poltava
DNIPROPETROVSK
DONETSK

Mariupol

Sea of Azov

Novorossiysk

Kremenchuk

Kyyvy Rih
Zaporizhzhya

Melitopol

Kherson

Simferopol
C R I M E A
Sevastopol
Yalta

B L A C K   S E A

KYIV (Kiev)

Cherkasy

Kirovohrad

Mykolayiv

ODESA

Chernihiv
Homyel

U K R A I N E

Zhytomyr
Vinnytsya

Chişinău
Tighina

MOLDOVA

Tiraspol

Galaţi
Brăila

BUCUREŞTI (Bucharest)

R O M A N I A

Lviv (Lvov)
Ivano-Frankivsk
Chernivtsi

Ternopil
Khmelnytskyy

C a r p a t h i a n s

Cluj-Napoca
Braşov

B U L G A R I A
Ruse

Constanţa

Pinsk
Brest
Lutsk
Rivne

SLOVAK REP.
HUNGARY

Satu Mare

P o l i s y a

East from Greenwich

CARTOGRAPHY BY PHILIP'S COPYRIGHT REED INTERNATIONAL BOOKS LTD

Projection: Conical with two standard parallels

UDMURTIA

TATARSTAN

MARI EL

CHUVASHIA

NIZHNIY NOVGOROD

MORDVINIA

SAMARA

RUSSIA

UKRAINE

Glazov · Bolezino · Yar · Igra · Kumeny · Nolinsk · Medvedok · Arkul · Uni · Uva · Mozhga · Malmyzh · Sosnovka · Vyatskiye Polyany · Kukmor · Arsk · Zelenodolsk

Nizhnekamskoye Vdkhr. · Naberezhnye Chelny · Nizhnekamsk · Zainsk · Aktash · Almetyevsk · Chistopol · Bugulma

KAZAN · Volga · Kamskoye Ustye · Bulgar · Buinsk · Tetyushi · Novodevichye

Kotelnich · Leninskoye · Sovetsk · Sorvizhi · Yaransk · Sernur · Yoshkar Ola · Medvedevo · Kozmodemyansk · Cheboksary · Novocheboksarsk · Volzhsk

Togliatti · Zhigulevsk · Novokuybyshevsk · SAMARA · Chapayevsk · Privolzhye · Oktyabrsk · Syzran

Simbirsk · Dimitrovgrad · Ulyanovsk · Surskoye · Alatyr · Shumerlya · Kanash · Kirya

Balakovo · Volsk · Marks · Engels · SARATOV · Pugachev · Yershov

Kuznetsk · Penza · Serdobsk · Petrovsk · Atkarsk · Kalininsk · Balashov · Rtishchevo · Arkadak · Kamyshin

Saransk · Ruzayevka · Nizhniy Lomov · Mokshan · Gorodishche · Inza · Nikolsk · Komsomolsky

Arzamas · Murom · Vyksa · Kasimov · Kulebaki · Vladimir

Dzerzhinsk · NIZHNIY NOVGOROD · Bor · Balakhna · Gorodets · Pavlovo · Bogorodsk

Oral · Zhayyq

Rybinsk · Yaroslavl · Kostroma · Ivanovo · Kineshma · Shuya · Furmanov

MOSKVA (Moscow) · Khimki · Mytishchi · Noginsk · Elektrostal · Orekhovo-Zuyevo · Podolsk · Obninsk · Serpukhov · Kolomna · Ryazan · Kaluga · Tula · Novomoskovsk · Michurinsk · Tambov · Kirsanov · Uvarovo

Tver · Rzhev · Vyshniy Volochek · Vyazma · Bryansk · Orel · Kursk · Belgorod · KHARKIV (Kharkov)

Lipetsk · Voronezh · Yelets · Gryazi · Staryy Oskol · Gubkin

Vladimir · Tambov · Morshansk · Rasskazovo · Kotovsk

Sumy · Poltava

Volgogradskoye Vdkhr. · Kuybyshevskoye Vdkhr. · Saratovskoye Vdkhr. · Rybinskoye Vdkhr. · Gorkovskoye Vdkhr.

Don · Volga · Oka · Sura · Moksha

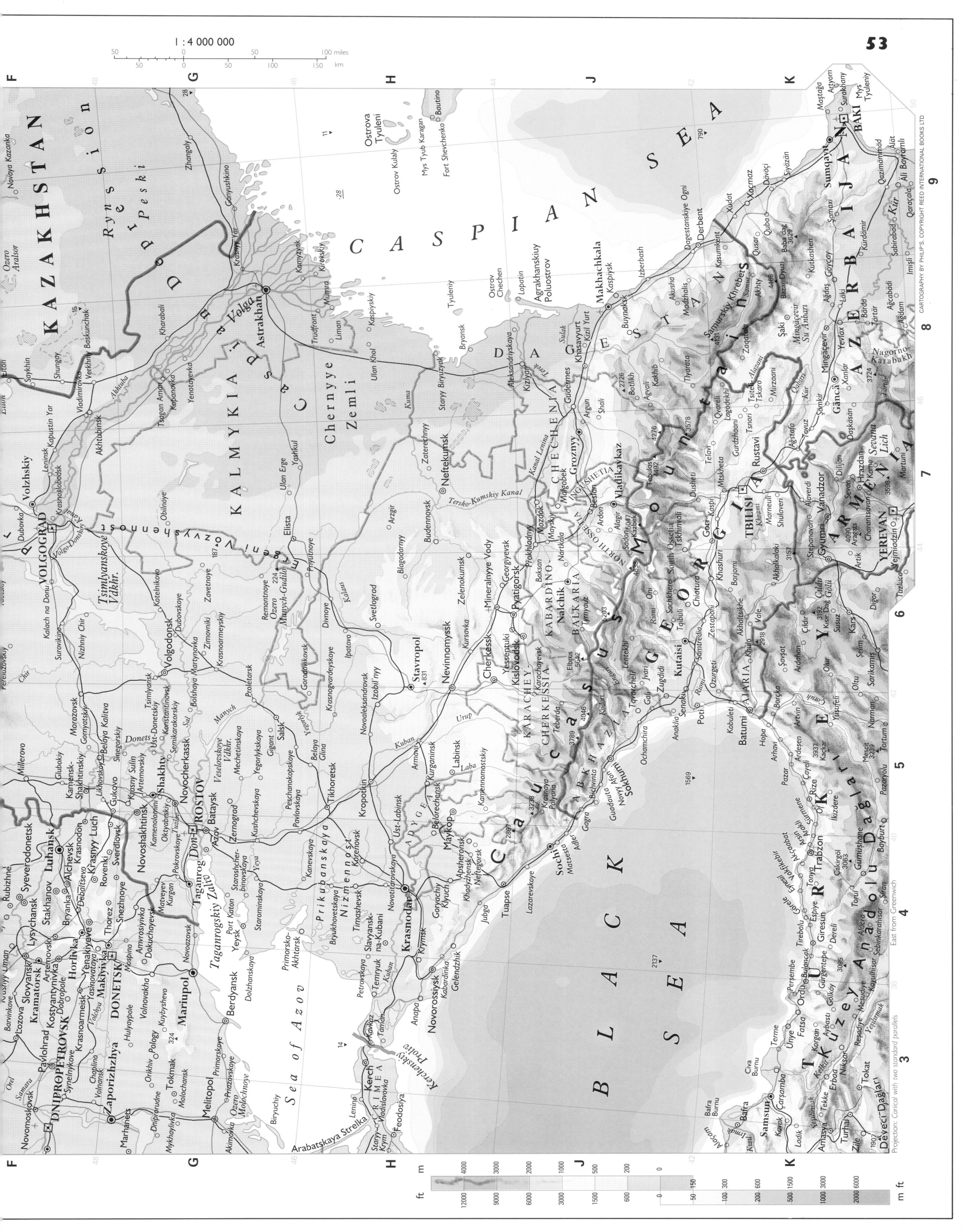

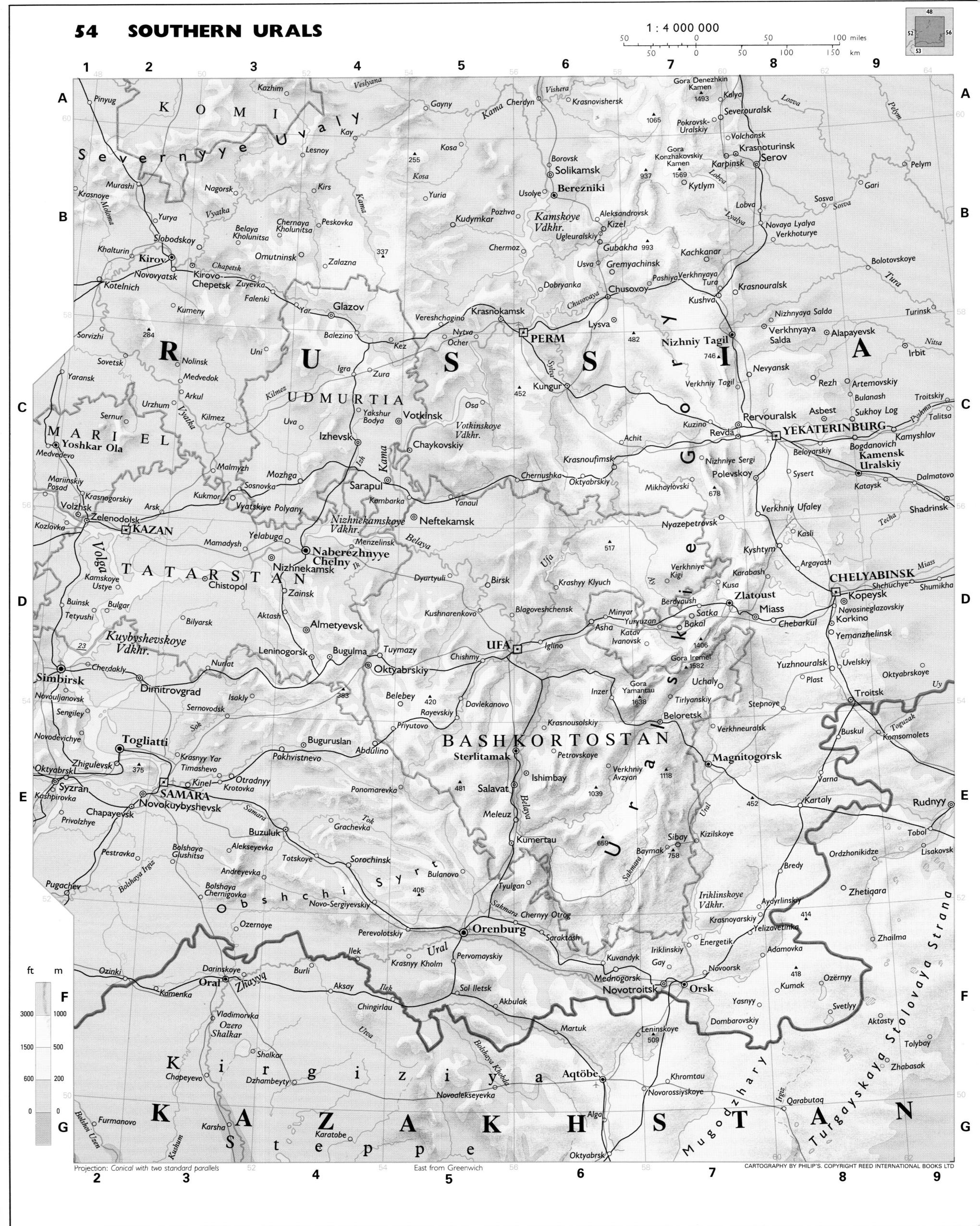

1 : 4 000 000

Projection: Conical with two standard parallels

East from Greenwich

CARTOGRAPHY BY PHILIP'S. COPYRIGHT REED INTERNATIONAL BOOKS LTD

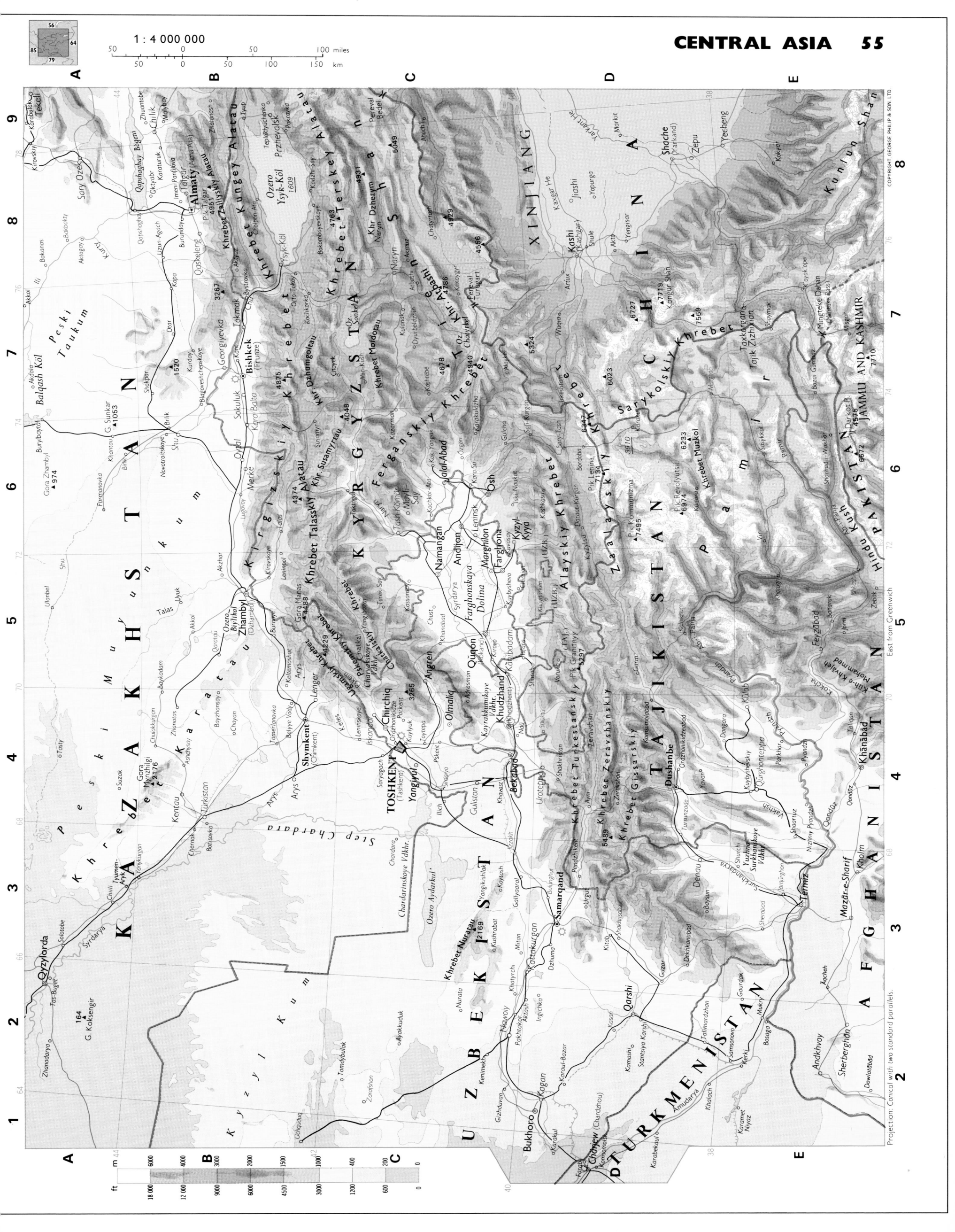

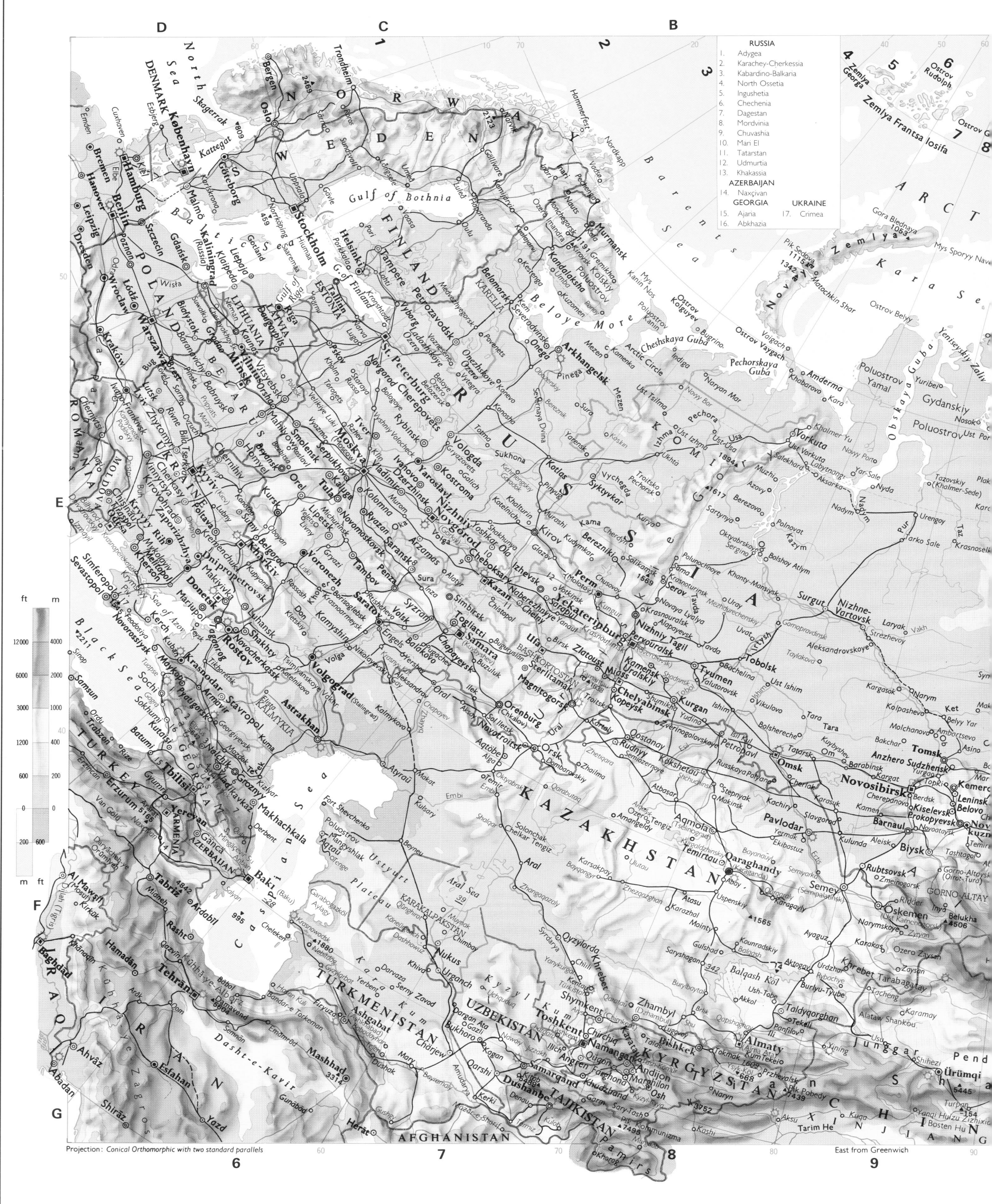

	RUSSIA
1.	Adygea
2.	Karachey-Cherkessia
3.	Kabardino-Balkana
4.	North Ossetia
5.	Ingushetia
6.	Chechenia
7.	Dagestan
8.	Mordvinia
9.	Chuvashia
10.	Mari El
11.	Tatarstan
12.	Udmurtia
13.	Khakassia
	AZERBAIJAN
14.	Naxçivan
	GEORGIA
15.	Ajaria
16.	Abkhazia
	UKRAINE
17.	Crimea

Projection: *Conical Orthomorphic with two standard parallels*

East from Greenwich

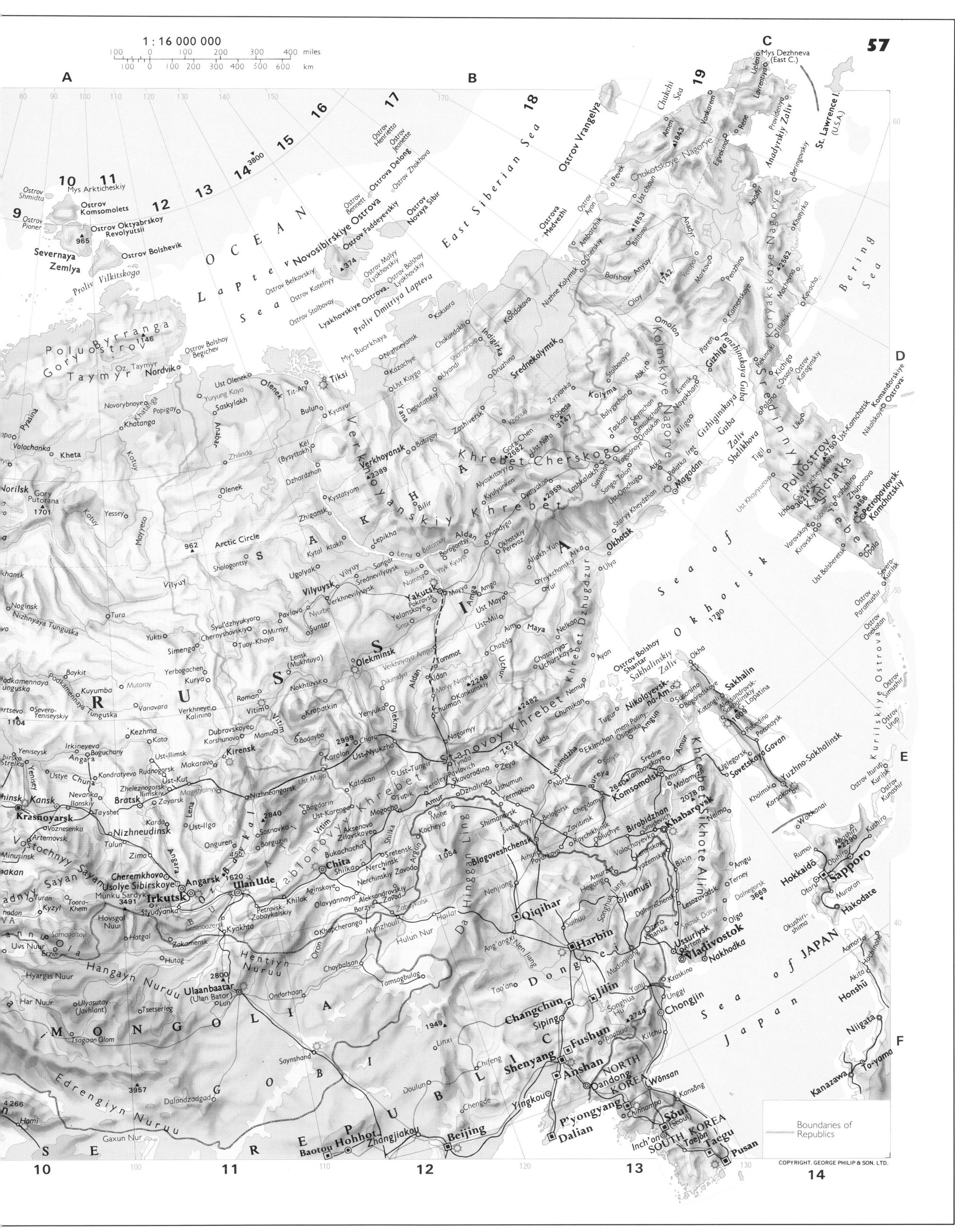

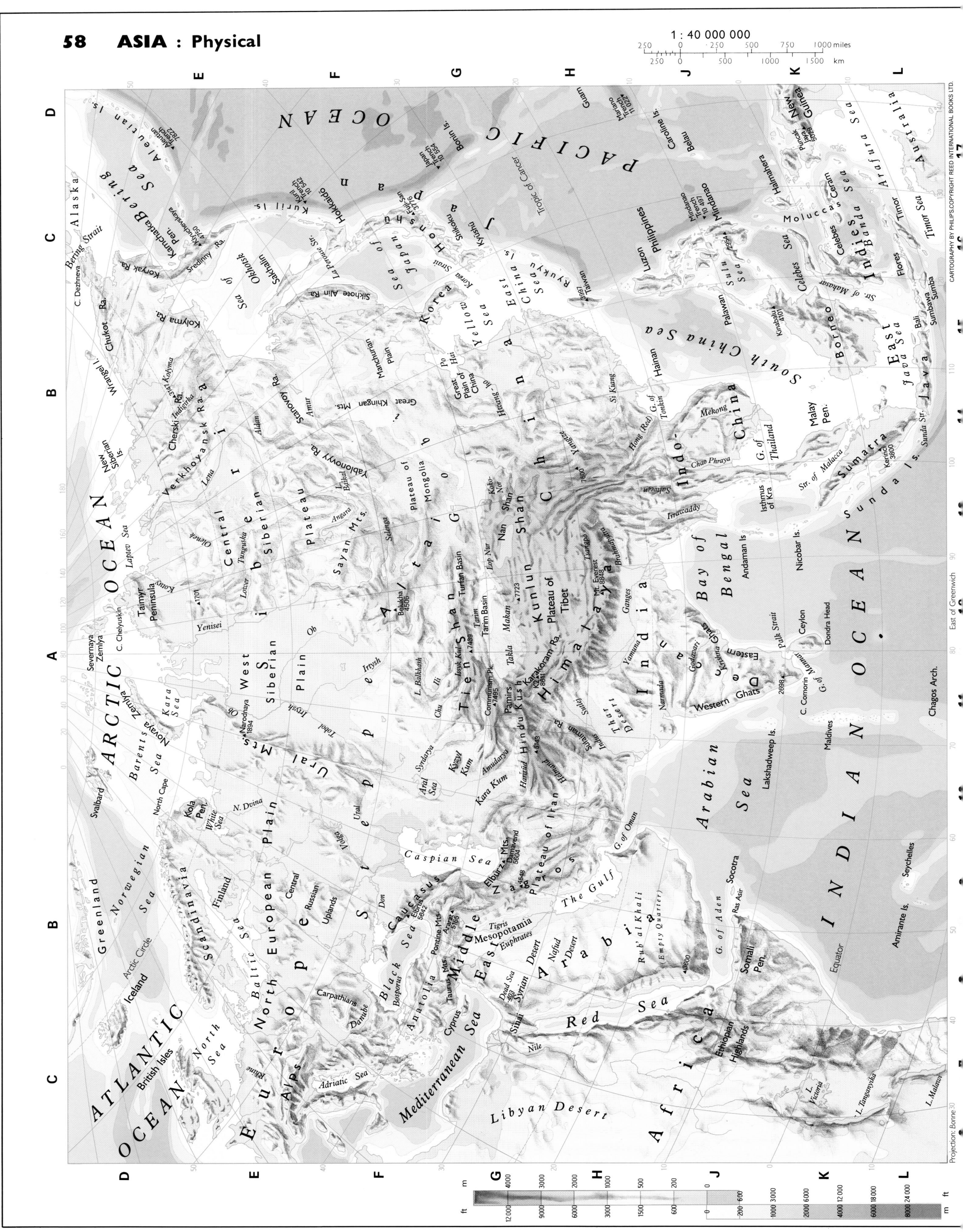

1 : 40 000 000

250   0   250   500   750   1000 miles

250   0   500   1000   1500 km

CARTOGRAPHY BY PHILIP'S.
COPYRIGHT REED INTERNATIONAL BOOKS LTD.

Projection: Bonne

East of Greenwich

ARCTIC OCEAN

ATLANTIC OCEAN

PACIFIC OCEAN

INDIAN OCEAN

Europe

Asia

Africa

Greenland

Iceland

Bering Sea

Alaska

Scandinavia

North European Plain

Steppes

Caucasus

Black Sea

Middle East

Mediterranean Sea

Red Sea

Arabia

Rub' al Khali (Empty Quarter)

Arabian Sea

Caspian Sea

Ural Mts.

West Siberian Plain

Central Siberian Plateau

Plateau of Mongolia

Altai

Tien Shan

Plateau of Tibet

Himalaya

Hindu Kush

Karakoram Ra.

Kunlun Shan

Takla Makan

Tarim Basin

Turfan Basin

China

Great Plain of China

Manchurian Plain

Great Khingan Mts.

Korea

Japan

Sea of Japan

Yellow Sea

East China Sea

South China Sea

Indo-China

Bay of Bengal

India

Deccan

Western Ghats

Eastern Ghats

Ceylon

Maldives

Borneo

Sumatra

Java

Philippines

New Guinea

Australia

ft   m
12000   4000
9000   3000
6000   2000
3000   1000
1500   500
600   200
0
-200 -600
1000 3000   2000 6000
4000 12000   6000 18000
8000 24000   m ft

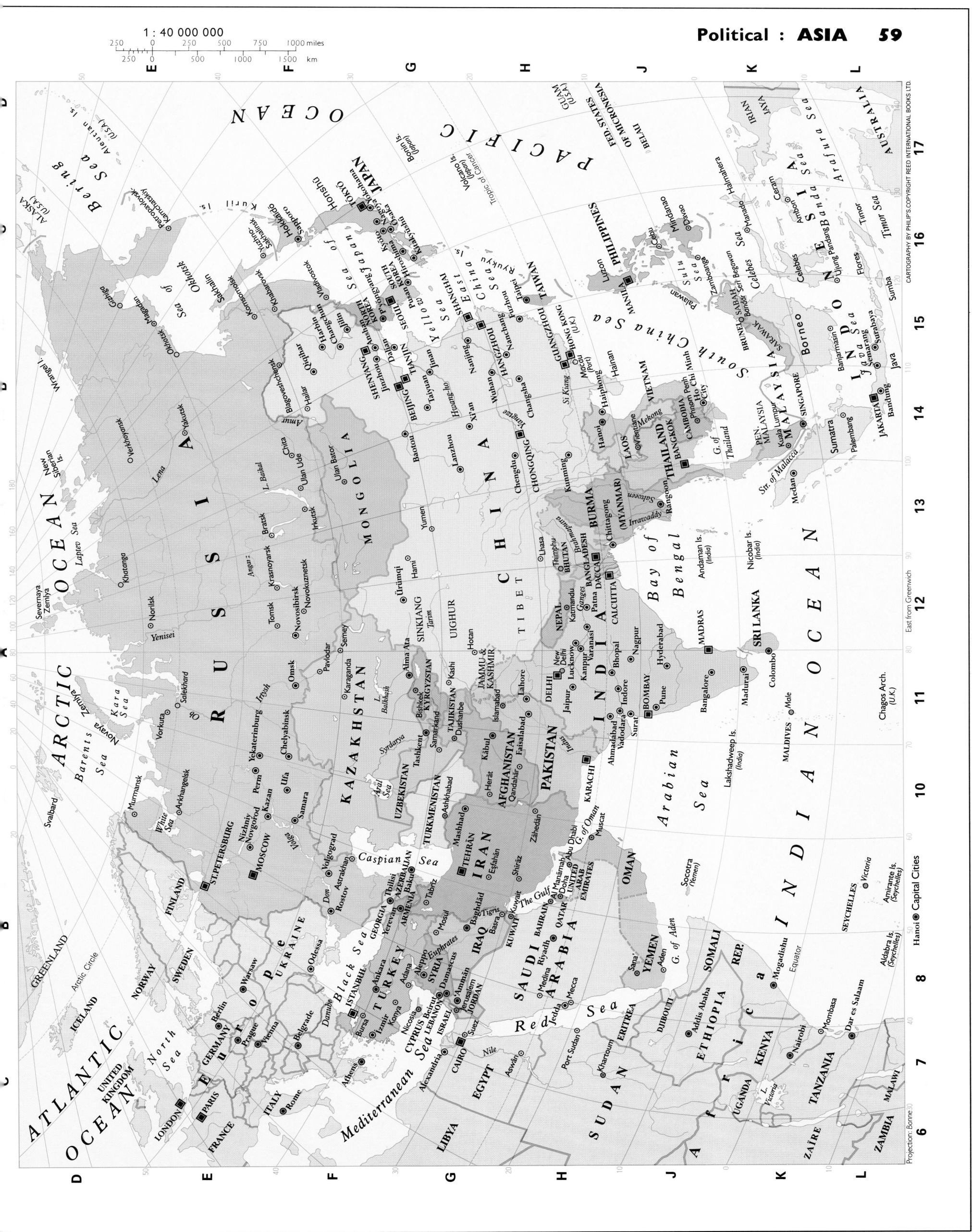

1 : 40 000 000

250 0 250 500 750 1000 miles

250 0 500 1000 1500 km

CARTOGRAPHY BY PHILIP'S. COPYRIGHT REED INTERNATIONAL BOOKS LTD.

East from Greenwich

Hanoi ⊙ Capital Cities

Projection: Bonne

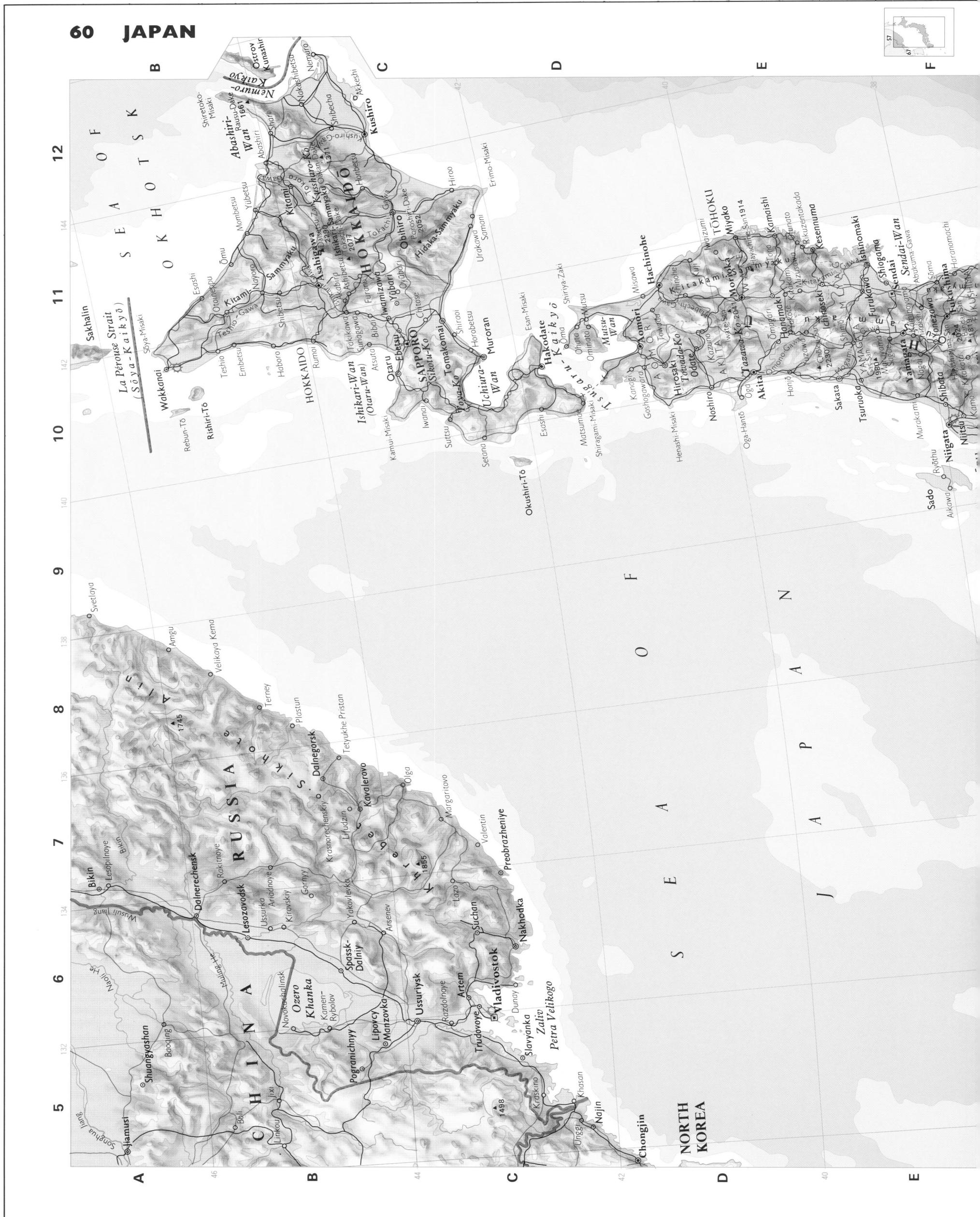

A B C D E F

12 11 10 9 8 7 6 5

SEA OF OKHOTSK

Sakhalin

Ostrov Kunashir

Nemuro-Kaikyō

HOKKAIDO

Shiretoko-Misaki

Abashiri-Wan

Rausu-Dake 1661

Nakashibetsu

Nemuro

Abashiri

Shari

Shibecha

Akkeshi

Kushiro

Ōsoro-ko

Ōmu

Mombetsu

Yūbetsu

Esashi

Otoineppu

Teshio-Gawa

Kushiro-Ko

Okan-Dake 1371

San-Dake 2077

Daisetsu-Zan 2290

Sammyaku

Asahigawa

Kitami

Nayoro

Kitami

Sammyaku

Tokachi-Dake 2077

Obihiro-Ko

Obihiro

Tokachi-Gawa

Hiroo

Erimo-Misaki

Urakawa

Samani

Hidaka-Sammyaku 2052

Horoshiri-Dake 2052

Ōmu

Bibai

Bibaushi

Shibetsu

Takikawa

Sunagawa

Iwamizawa

Furano

Yūbari

Chitose

Shiraoi

Horobetsu

Muroran

Uchiura-Wan

Tomakomai

Toya-Ko

Shikotsu-Ko

Noboribetsu

Ishikari-Wan

(Otaru-Wan)

Ishikari

Otaru

Ebetsu

SAPPORO

Esan-Misaki

Setana

Okushiri-Tō

Kamui-Misaki

Suttsu

Iwanai

Rebun-Tō

Rishiri-Tō

Wakkanai

Sōya-Misaki

La Pérouse Strait
(Sōya-Kaikyō)

HOKKAIDO

Haboro

Rumoi

Teshio

Embetsu

TŌHOKU

Ōminato

Shiranuka-Misaki

Shiragami-Misaki

Matsumae

Esashi

Hakodate

Tsugaru-Kaikyō

Tsugaru-Hantō

Ōma

Ōhata

Mutsu

Mutsu-Wan

Aomori

Misawa

Noheji

Hachinohe

Misawa

AOMORI

Kanagi

Goshogawara

Henashi-Misaki

Ajigasawa

Fukaura

Ōwani

Hirosaki

Hanamaki

Ōdate

Kazuno

Kosaka

Towada-Ko

Towada-Ko

Kitakami-Gawa

San 1914

Iwaizumi

Miyako

Morioka

MORIOKA

Kitakami

Tōno

Rikuzentakada

Kamaishi

Kesennuma

Ōfunato

Kitakami-Sammyaku

Ichinoseki

Furukawa

Ishinomaki

Sendai

Shiogama

Sendai-Wan

Haranomachi

Abukuma-Gawa

Sōma

Shinchi

Fukushima

Takekoma

YAMAGATA

Yamagata

Murayama

Shinjō

Mogami-Gawa

Oga-Hantō

Oga

AKITA

Akita

Honjō

Yuzawa

Yokote

Ōmagari

Kakunodate

Tazawa-Ko

Hachimantai 1614

Chōkai-San 2230

Sakata

Tsuruoka

Shibata

Niitsu

Niigata

Ryōtsu

Sado

Akkawa

Murakami

Sendai-Wan

Shinjō

Honjō

Noshiro

Takanosu

Ōdate

Mogami

SEA OF JAPAN

RUSSIA

CHINA

NORTH KOREA

Svetlaya

Amgu

Velikaya Kema

Terney

Plastun

Tetyukhe Pristan

Dalnegorsk

Krasnorechensky

Lifudzin

Kavalerovo

Olga

Margaritovo

Valentin

Preobrazheniye

Sikhotē Alin

1745

Bikin

Bikin

Lesopilnoye

Dalnerechensk

Rokitnoye

1855

Krasny Yar

Gornyy

Lazo

Kirovskiy

Sŭchan

Arseñev

Spassk-Dalniy

Yakovlevka

Lesozavodsk

Ussurka

Aradnoye

Novosselishskinsk

Ozero Khanka

Kamen-Rybolov

Spassk-Dalniy

Artem

Ussuriysk

Razdolnoye

Trudovoye

Vladivostok

Nakhodka

Dunay

Slavyanka

Zaliv Petra Velikogo

Kraskino

Khasan

1498

Ungi

Najin

Chongjin

Pogranichnyy

Liporcy

Monzovka

Jiamusi

Baoqing

Shuangyashan

Linkou

Boli

Jixi

Sōnghua Jiāng

Wusuli Jiang

Naoli He

Muling He

46 44 42 40 38

144 142 140 138 136 134 132

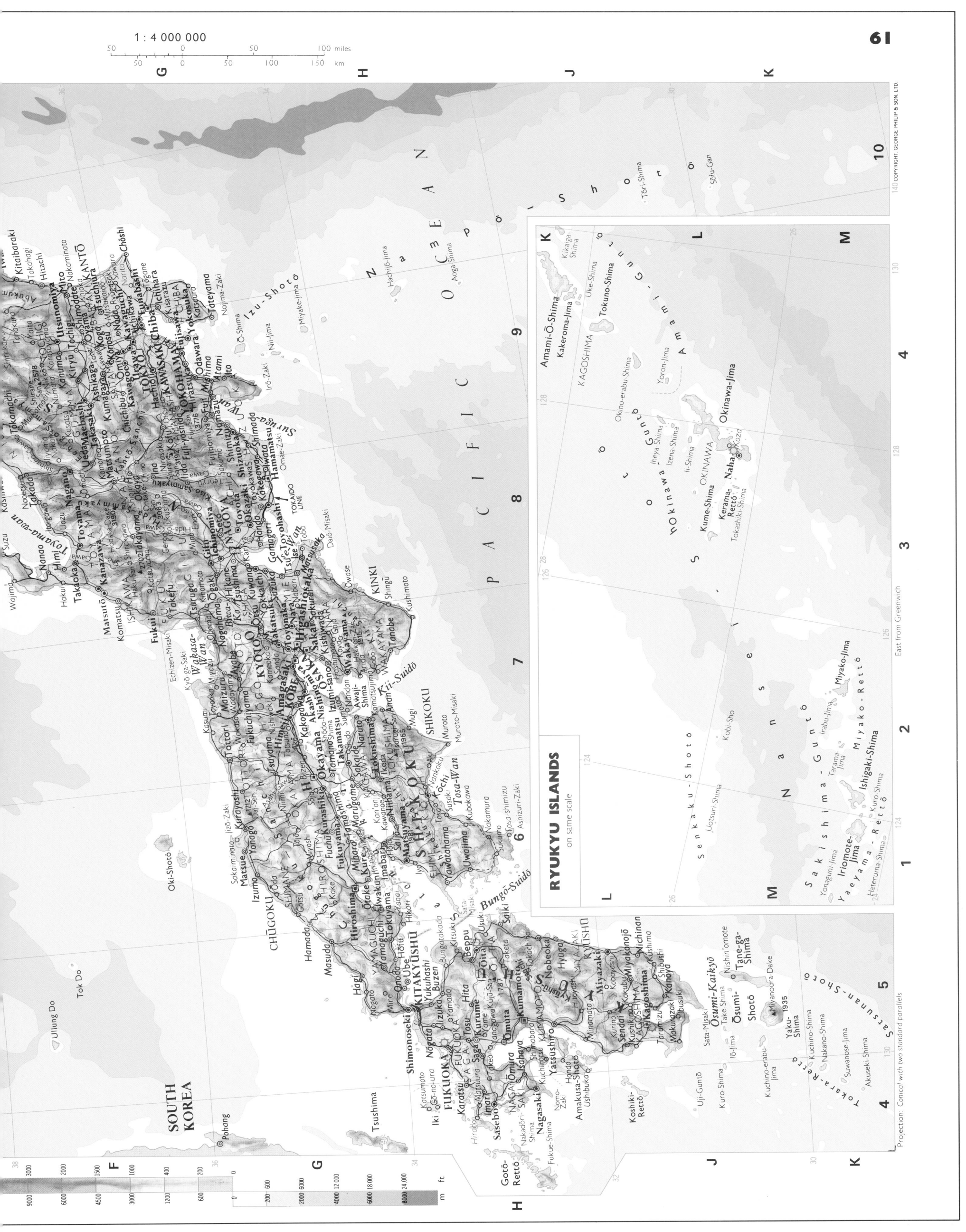

140 COPYRIGHT GEORGE PHILIP & SON LTD

1 : 4 000 000

50   0   50   100 miles

50   0   50   100   150 km

G   H   J   K

10

9

8

7

6

PACIFIC OCEAN

Izu-Shotō

TOKYO
YOKOHAMA
KAWASAKI
CHIBA
KANTŌ

NAGOYA
KYOTO
KOBE
OSAKA
KINKI

SHIKOKU
CHŪGOKU

HIROSHIMA

KITAKYŪSHŪ
FUKUOKA
KUMAMOTO
KYŪSHŪ
KAGOSHIMA

Shimonoseki
Nagasaki
Sasebo
Gotō-Rettō

SOUTH
KOREA

Tsushima
Tok Do
Ullung Do
Pohang

Bungo-Suidō
Ōsumi-Kaikyō
Tane-ga-Shima
Yaku-Shima
Tokara-Rettō
Satsunan-Shotō

RYUKYU ISLANDS
on same scale

Amami-Ō-Shima
Kakeroma-Jima
Tokuno-Shima
KAGOSHIMA
Okino-erabu-Shima
Yoron-Jima

OKINAWA
Okinawa-Jima
Naha
Kume-Shima
Kerama-Rettō

Nansei
Sakishima-Guntō
Senkaku-Shotō
Miyako-Rettō
Miyako-Jima
Ishigaki-Shima
Iriomote-Jima
Yaeyama-Rettō
Yonaguni-Jima
Hateruma-Shima

K   L   M

4

3

2

1

East from Greenwich

Projection: Conical with two standard parallels

ft
24 000
18 000
12 000
6000
4000
2000
1500
600
200
0

m
8000
6000
4000
3000
2000
1000
600
400
200
0

F   G   H

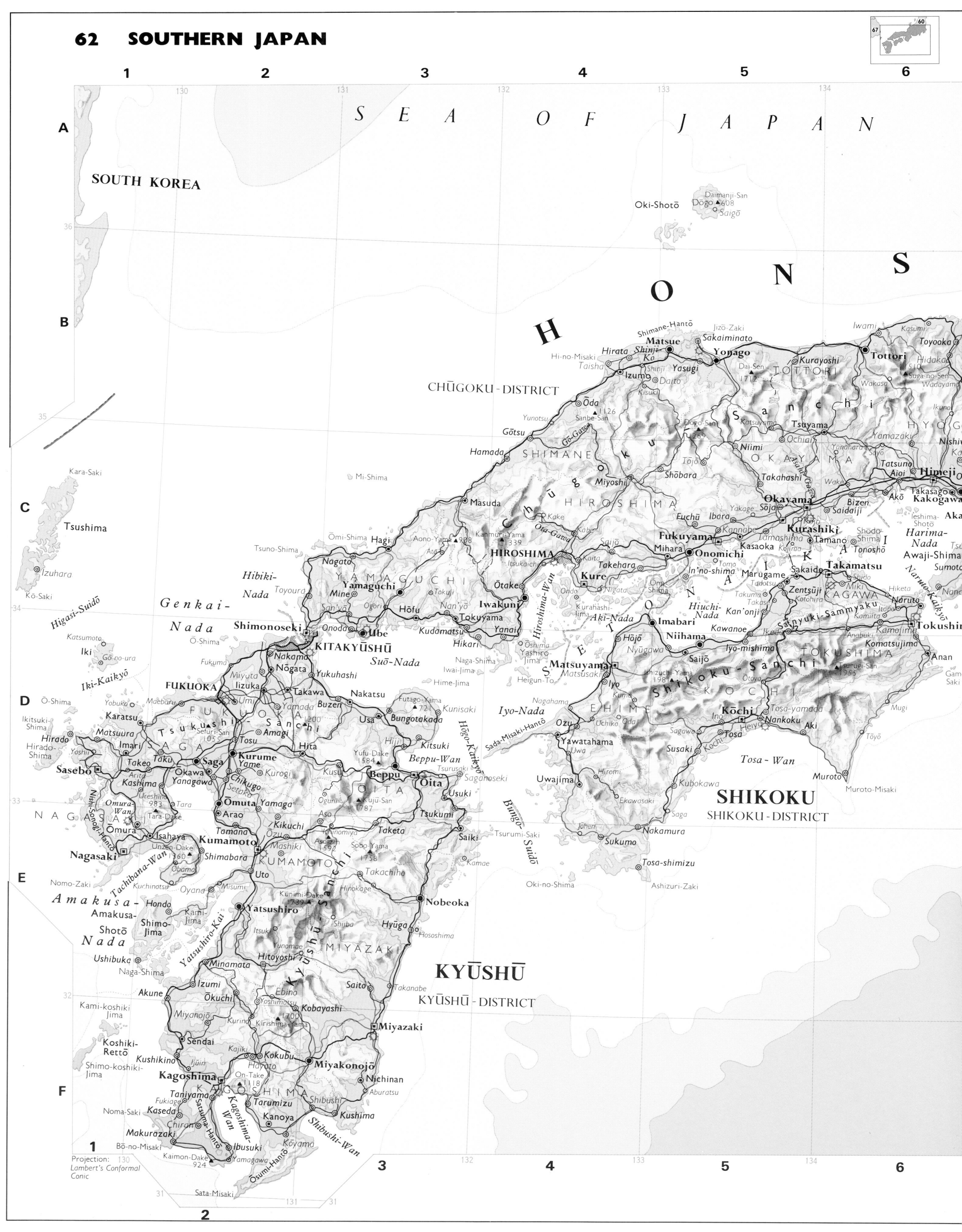

SEA OF JAPAN

SOUTH KOREA

HONS$\bar{\text{U}}$

CHŪGOKU-DISTRICT

Oki-Shotō
Dōgo ▲608
Saigō
Daimanji-San

Shimane-Hantō
Jizō-Zaki
Iwami
Kasumi
Toyooka
Hi-no-Misaki
Hirata
Shinji
Matsue
Yonago
Kurayoshi
Tottori
510
Hidaka
Taisha
Shinji-Ko
Yasugi
Dai-Sen
Suga-no-Sen
Wadayama
Izumo
Daitō
1712
TOTTORI
Wakasa
Sanbe-San
Kisuki
Dōgo-San
Katsuyama
Tsuyama
Yamazaki
HYŌG
Ōda
1126
Sanbe-San
1264
Sayō
Yanahara
Ochiai
Nishi
Yunotsu
Go-Gatsu
Tōjō
Ikuno
Gōtsu
SHIMANE
Miyoshi
Shōbara
OKAYAMA
Yamazaki
Tatsuno
Hamada
CHŪGOKU
Kake
Takahashi
Wake
Aioi
HimeJi
Masuda
Kanmuri-Yama
1339
Osa-Gawa
Euchū
Ibara
Sōja
Bizen
Takasago
Akō
Kakogawa
Mi-Shima
Aono-Yama ▲908
Kabi
Kannabe
Kasaoka
Tamano
Saidaiji
Ieshima-Shotō
Ōmi-Shima
Hagi
Saijō
Fukuyama
Kurashiki
Shōdo-Shima
Harima-Nada
Nagato
Tsuno-Shima
Itsukaichi
Mihara
Onomichi
Tamashima
Tonoshō
Awaji-Shima
Sumoto
Hibiki-Nada
YAMAGUCHI
HIROSHIMA
Takehara
Toma
In'no-shima
Marugame
Takamatsu
Hiketa
Mine
San'yō
Ōtake
Kure
Sakaide
Zentsūji
KAGAWA
Naruto-Kaikyō
Genkai-Nada
Toyoura
Yamaguchi
Tokuji
Nan'yō
Kurahashi-Jima
Ōmi
Takuma
Kotohira
Miki
Tokushi
Kara-Saki
Tsushima
Shimonoseki
Onoda
Ube
Kudamatsu
Hiroshima-Wan
Aki-Nada
Hōjō
Imabari
Kawanoe
Sanuki-Sammyaku
Kamita
Komatsujima
Higasi-Suidō
Ō-Shima
Hōfu
Tokuyama
Yanai
Ōshima
Yashiro-Jima
Niigata
Hiuchi-Nada
Niihama
Iyo-mishima
TOKUSHIMA
Katsumoto
Izuhara
Hikari
Naga-Shima
Iwai-Jima
Matsuyama
Nyūgawa
Saijō
Anan
Iki
Gō-no-ura
Iki-Kaikyō
Kō-Saki
FUKUOKA
Heigun-Tō
Matsusaki
Iyo
Shikoku-Sammyaku
Ishizuchi-Yama
1981
Tosa-yamada
1955
Ame
Ikitsuki-Shima
Hirado
KITAKYŪSHŪ
Suō-Nada
Nakama
Nōgata
Yukuhashi
Buzen
Kunisaki
Iyo-Nada
EHIME
KŌCHI
Kōchi
Nankoku
Mugi
Ō-Shima
Miyuta
Iizuka
Takawa
Futago-Yama ▲72
Usa
Sada-Misaki-Hantō
Ōzu
Aki
Tōyō
Yobuko
Mdeburu
Umi
Yamada
1200
Amagi
Hita
Hiji
Kitsuki
Sagawa
Kochi
Tosa
Karatsu
Tsukushi-Sanchi
Tosu
Beppu-Wan
Yawatahama
Inō
Heiyo
Matsuura
Imari
Sefuri-San
1055
Kurume
Yufu-Dake
1584
Beppu
Tsurusaki
Uchiko
Oda
Hiyo
Uwa
Susaki
Kubokawa
Tōyō
Takeo
Taku
Yame
Kurogi
Kusu
Beppu
Ōita
Sagane
Uwajima
Tosa-Wan
Muroto
Sasebo
SAGA
Saga
Okawa
Chikugo
Setaka
Oita
Usuki
Hiromi
Muroto-Misaki
Arita
Kashima
Yanagawa
OITA
Kuju-San
1787
Ekawasaki
Saga
Nakamura
SHIKOKU
Ōmura-Wan
983
Tara
Arao
Yamaga
Oguni
Aso
Tsukumi
SHIKOKU-DISTRICT
NAGASA
Tara-Dake
Ōmuta
Kikuchi
Kinomiya
Sobo-Yama
1756
Kamae
Jōhen
Sukumo
Isahaya
Kumamoto
Mashiki
Asagiri
Taketa
Saiki
Tsurumi-Saki
Ōmura
360
Kikuchi
Ozu
1592
Kyūshū-Sanchi
Ashizuri-Zaki
Nagasaki
Shimabara
KUMAMOTO
Takachiho
Bungo-Suidō
Nomo-Zaki
Obama
Uto
Misumi
Hinokage
Oki-no-Shima
Tosa-shimizu
Tachibana-Wan
Kuchinotsu
Oyana
Kunimi-Dake
739
Hinokage
Nobeoka
Tosa-shimizu
Amakusa-Shotō
Hondo
Kami-Jima
Shiiba
Hyūga
Amakusa-Shimo-Jima
Yatsushiro
Itsuki
Yunomae
MIYAZAKI
Nada
Yatsushiro-Kai
Hitoyoshi
Kyūshū-Sanchi
Hososhima
Ushibuka
Minamata
KYŪSHŪ
Naga-Shima
Saito
Takanabe
KYŪSHŪ-DISTRICT
Koshiki-Rettō
Izumi
Akune
Ōkuchi
Ebino
Yoshimatsu
Kobayashi
Kami-koshiki-Jima
Miyanojō
Kurino
Kirishima-Yama
1700
Miyazaki
Shimo-koshiki-Jima
Sendai
Kajiki
Kokubu
Nichinan
Kushikino
Ijūin
Hayato
Miyakonojō
Aburatsu
Kagoshima
On-Take
KAGOSHIMA
Shibushi
Taniyama
1118
Kanoya
Kushima
Fukiage
Tarumizu
Kagoshima-Wan
Shibushi-Wan
Noma-Saki
Kaseda
Chiran
Ibusuki
Kōyama
Makurazaki
Satsuma-Hantō
Kanoya
Ōsumi-Hantō
Bō-no-Misaki
Kaimon-Dake
924
Yamagawa
Sata-Misaki

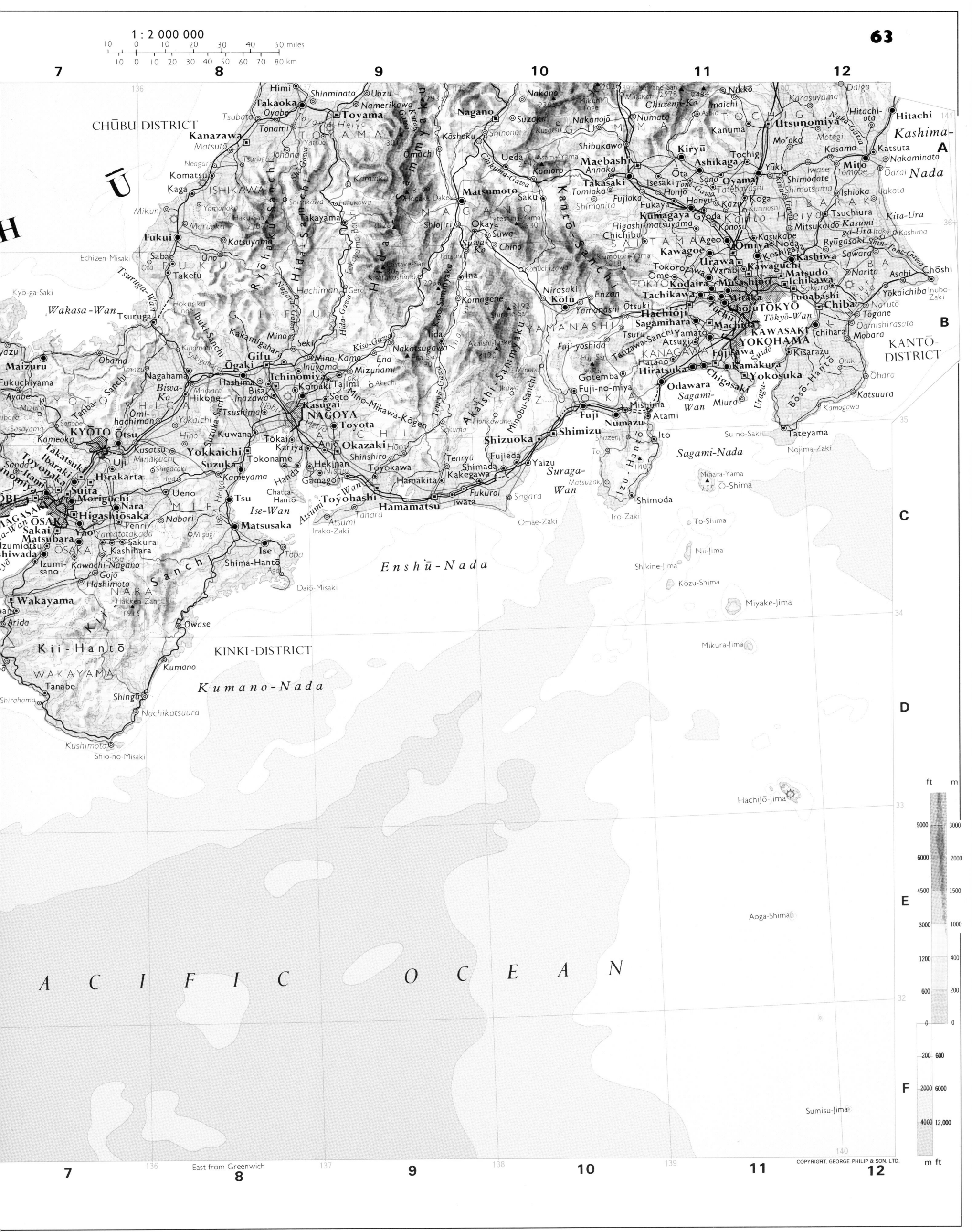

1 : 2 000 000

10 0 10 20 30 40 50 miles
10 0 10 20 30 40 50 60 70 80 km

CHŪBU-DISTRICT

Ū

H

7   8   9   10   11   12

A

Kashima-
Nada

B

KANTŌ-
DISTRICT

C

KINKI-DISTRICT

Kii-Hantō

Kumano-Nada

Enshū-Nada

Suruga-Wan

Sagami-Nada

D

E

F

ACIFIC OCEAN

East from Greenwich

7   8   9   10   11   12

COPYRIGHT. GEORGE PHILIP & SON. LTD.

ft m

9000 3000
6000 2000
4500 1500
3000 1000
1200 400
600 200
0 0
200 600
2000 6000
4000 12,000

m ft

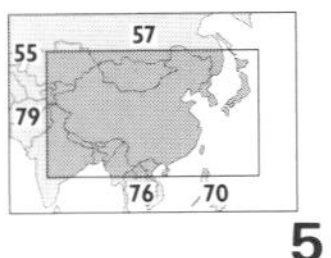

1 2 3 4 5

B

RUSSIA

Cheremkhovo
Angarsk Irkutsk

Qaraghandy

KAZAKHSTAN

Karsakpay
Zhezqazghan

1565

Mointy
Kounradski
Balqash

342 Balqash Köl

Shu

Zhambyl
Bishkek
Ysyk-Köl
Almaty
1609

KYRGYZSTAN

Namangan
Andijon
Artux
Kashi
Shule
Shache

Taxkorgan Tajik
Zizhixian

Karakoram
8126
8611

JAMMU &
KASHMIR

Srinagar
Leh
Rutog
Gar

Nanda
Devi
7817
Burang

Dehra Dun
Meerut
DELHI
Moradabad
Bareilly
Aligarh
Agra
KANPUR
Gwalior
Lucknow
Jhansi
Allahabad
Sagar

I N D I A

Jabalpur

Tropic of Cancer

Nagpur
Chanda
Raipur
Bilaspur

Warangal
Vizianagaram
Berhampur
Vishakhapatnam

BAY OF

BENGAL

Semey
Ridder
Öskeman
Zyryan

Belukha
4506

Qarqaraly

Qarghaly

Ayaguz

Taldyqorghan

Qapshaghay

Qapshaghay Bögeni

Ala Tau

Bole

Yining
Ili

Naryn
Wensu
Aksu
Kuqa

Pik Pobedy
7439

Yecheng
Pishan

1635

Hotan
(Khotan)
Yutian

Rubtsovsk

Gorno-Altaysk

Zapadnyy Sayan

Tannu Ola

Altay
Fuyun

Fuhai

Tacheng

Khrebet
Tarabagatay

Ozero
Zaysan

Ozero Alakol

Karamay

Dzhungarskiye Vorota

Usu
Manas

Junggar Pendi

Ürümqi 5445

Qitai

Turpan
-154
Aydingkol
Hu

Yanqi
Korla

Bosten (Bagrax) Hu

ZIZHIQU

Kuruktag

XINJIANG

UYGUR

Tarim He

Tarim Pendi

Qarqan He
Qiemo

Ruoqiang

Munku Sardyk 3491

Khuabem

Hatgal

Hövsgöl
Nuur

Nauski

Selenge Mörön
Babushk

455

Ues
Nuur

Ulaangom

Har Us
Nuur
Hov

Döröö
Nuur

Hyargas
Nuur

Ulyasutay

Hangayn Nuruu

MONGO

4362

Aerhtai

(Altai) Shan

Barkol Kazak Zizhixian

Hami

4925

Gaxun Nur

Lop Nor

Dunhuang

Anxi

Altun Shan

Yumen
Jiayuguan

6348

Bugun

Shara

Buyanhongor

Orhon Gol

Tsetserleg

Ulan Bate

Dzu

Dalandzadgad

G

Qilian Shan

Zhangye

Shandan

Alxa Zuoqi

Wuhai

Pingluo

Yinchuan

NINGXIA
Wuzh
HUIZU
ZIZHIQU

C

Ayakkum Hu

Huh Xil Shan

Tart

Mangnai

Da Qaidam

Qaidam Pendi

Har Hu

Tianjun

Golmud

Dulan
Gonghe

Qinghai Hu 3205

Minhe

Xining

LANZHOU

Baiyin

Qingyang

Dehra Dun

Wuluk omushih
Ling 7723

K u n l u n   S h a n

Q I N G H A I

Ngoring
Hu
4237

Gyaring Hu

Maqen

6094

Yushu

Bayan Har Shan

Huang He

Songpan

Min Xian

Sengpan

Tianshui

Baoji

Guyuan

Pingliang

Dingxi

X I Z A N G

Mapam Yumco

Zhongba

Tanggula (Dangla) Shan

(T I

B E

T)

Siling Co
4495

Nam Co
4627

Amdo

Nagqu

Nyainqêntanglha Shan

Qamdo

Bomi

Lancang Jiang

Lhasa

Jinsha Jiang

Yalong Jiang

Garze

Datue Shan

Ninglang Shan

Shaluli

Amdo

Wudu

4113

Hanzhong

Jialing He

Mianyang

Daxian

S I C H U A N

Santai

Nanchong

Hechua

Nanda
Devi

Burang

H i m

Dhaulagiri
8221

Ngamring

NEPAL

Katmandu

Everest
8848

Gorakhpur

Darbhanga

Zhongba

Lhozê

Xigaze

Xainza

Nyainqêntanglha Shan

Yarlung Zangbo Jiang

Namcha Barwa
7756

Yamzho Yumco

Pagri

Punakha

BHUTAN

Brahmaputra

Tezpur

Dibrugarh

Sadiya

5887

Zayu

Nu Jiang

Zhongdian

Xichang

Goggu
Shan
7600

CHENGDU

Leshan

Wutongqia

Yibin

Zhaotong

Zunyi

Zigong

Luzhou

Neijiang

Chang Jiang

G U I Z H

Guiya

Anshun

Shuicheng

Daliang Shan

Lijiang

Dukou

Huize

Zhanyi

Dali

Dongchuan

KUNMING

Xingyi

GU

Meerut

Ganga

Ghaghra

Patna

Varanasi

Gaya

Rajshahi

Asansol

Berhampore

Barddhaman

Bhatpara

Haora

CALCUTTA

Khulna

Ranchi

Jamshedpur

Kharagpur

Baleshwar

Cuttack

Indravati

Makanadi

BANGLADESH

DHAKA
Narayanganj

Khasi Hills

Silchar

Imphal 3824

Chittagong

Koch Bihar

Gauhati

B U R M A

(M Y A N M A R)

Victoria
3053

Mandalay

Monywa

Shwebo

Akyab

Arakan Yoma

Pegu Yoma

Irrawaddy

Salween

Taunggyi

Yamethin

Toungoo

THAILAND
(SIAM)

2163

Mekong

Luang
Prabang

2711

LAOS

Myitkyina

Patkai Bum

3411

Tengchong

Bhamo

Lashio

Baoshan

Luxi

Y U N N A N

Anning

Yuxi

Chengjiang

Shiping

Jinggu

Jiangcheng

Gejiu

Mengzi

Hekou

Wenshan

Hongshu He

Bose

ZHU

Nanning

Pingxiang

3143

VIETNAM

Hanoi Haiph

Hoa Binh

Gulf

Tonki

Projection: Bonne

East from Greenwich

ft m

18 000 6000

12 000 4000

9000 3000

6000 2000

4500 1500

3000 1000

1200 400

600 200

0 0

200 600

m ft

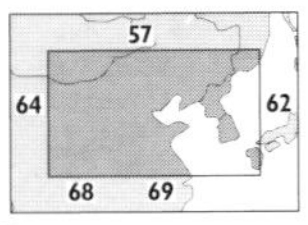

ÖVÖR HANGAY
▲3582
Arts Bogd Uul

M O N G O L I A

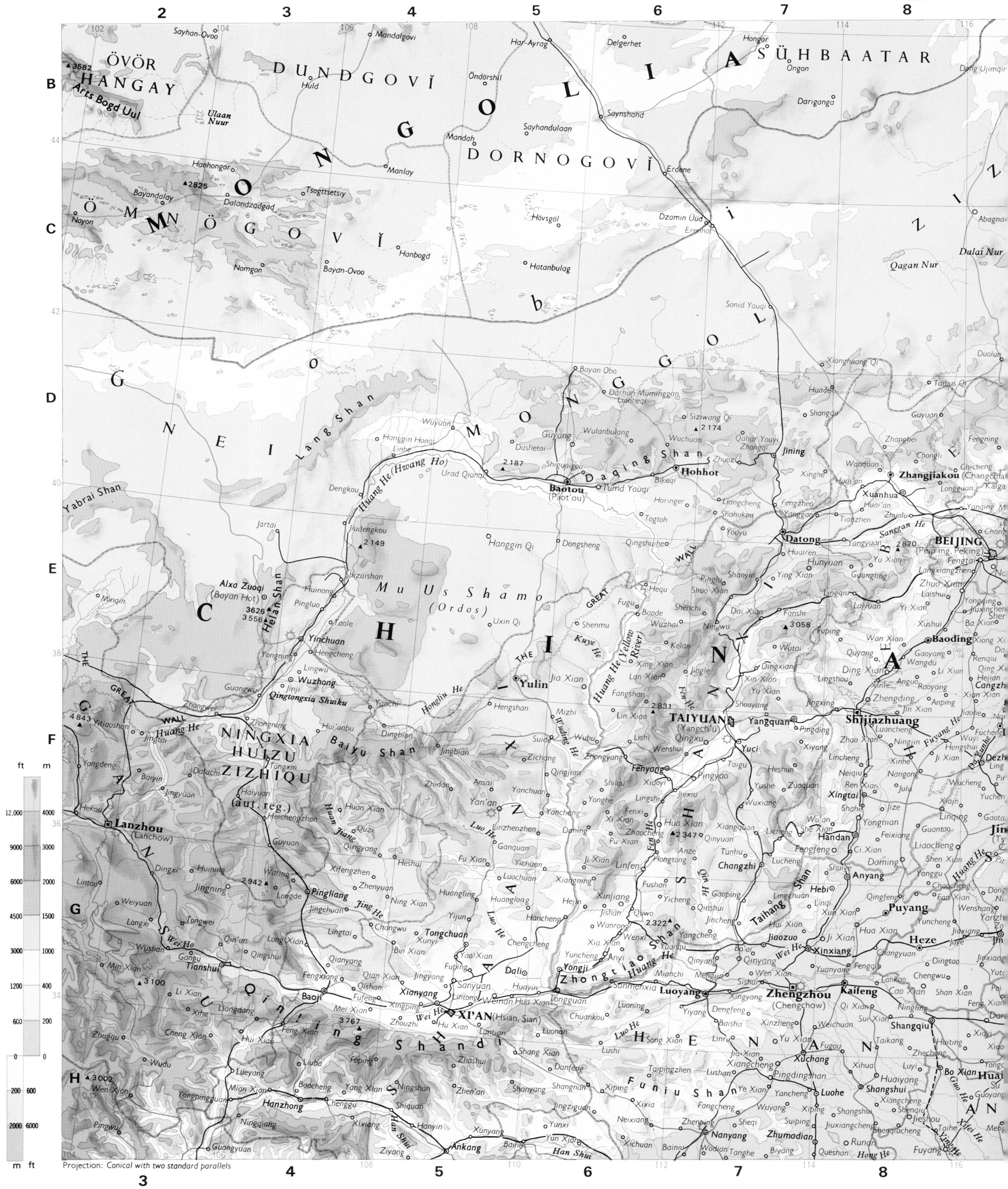

DUNDGOVĬ

SÜHBAATAR

Sayhan-Ovoo
Mandalgovi
Har-Ayrag
Delgerhet
Hongor
Ongon
Dang Ujimqin

Huld
Öndörshil
Saynshand
Dariganga

DORNOGOVĬ

Hanhongor
▲2825
Bayandalay
Dalandzadgad
Tsogttsetsiy
Manlay
Sayhandulaan
Mandah
Erdene
Linhe

ÖMNÖGOVĬ
Noyon
Nomgon
Hanbogd
Bayan-Ovoo
Hövsgöl
Hatanbulag
Dzamin Üüd
Erenhot
Qagan Nur
Dalai Nur
Abagnar

Sonid Youqi
Xianghuang Qi
Duolun

Bayan Obo
Darhan Muminggan Lianheqi
Siziwang Qi
Shangdu
Tatbus Qi

NEI MONGOL
Lang Shan
Wuyuan
Guyang
Wulanbulang
▲2174
Qahar Youyi Zhongqi
Jining
Zhangbei
Fengning
Chongli
Chicheng

Huang He (Hwang Ho)
Hanggin Hongi
Linhe
Urad Qianqi
Dashetai
Shiguaigou
Wuchuan
Zhuozi
Hohhot
Xinghe
Wanquan
Zhangjiakou
Changcheng
Kalga
Longguan
Yanqing

▲2187
Baotou (Paot'ou)
Tumd Youqi
Horinger
Liangcheng
Fengzhen
Yanggao
Huai'an
Xuanhua
Xuanhua
Tianzhen
Zhuolu
Yanqing

Daqing Shan
Dengkou
Jartai
Jiudengkou
▲2149
Hanggin Qi
Dongsheng
Qingshuihe
Togtoh
Youyu
Shahukou
Datong
Yangyuan
▲2870
BEIJING (Peip'ing, Peking)

Alxa Zuoqi (Bayan Hot)
▲3626
3556
Helan Shan
Huinong
Shizuishan
Mu Us Shamo (Ordos)
Uxin Qi
Hequ
Fugu
Shenmu
Wuzhai
Ninwu
▲3058
Fanshi
Wutai
Fuping
Lingqiu

Mingin
Yabrai Shan
Pingluo
Taole
Yinchuan
Hengcheng
Lingwu
Wuzhong
Qingtongxia Shuiku
Yanchi
Honglu He
Hengshan
Yulin
Mizhi
Lin Xian
Baode
Shenchi
Kelan
Kuye He
Dai Xian
Xing Xian
Lan Xian
Jingle
Xin Xian
Dingxiang
Shouyang
Baoding

Zhongning
Zhongwei
Hui'anbu
Dingbian
Jingbian
Suide
Wubu
Zhongyang
▲2831
Fen He
TAIYUAN (Yangch'u) Qingxu
Yangquan
Shijiazhuang
Zhengding

NINGXIA HUIZU ZIZHIQU (aut. reg.)
Haiyuan
Tongxin
Baiyu Shan
Zichang
Ansai
Yanchuan
Yonghe
Lishi
Xiaoyi
Pingyao
Taigu
Yuci
Yuci
Heshun
Zuoquan
Xingtai

Lanzhou (Lanchow)
Hekou
Heichengzhen
Guyuan
Huan Xian
Zhidan
Yan'an
Yanchang
Xi Xian
Fenxi
Jiexiu
Yushe
Wuxiang
Ren Xian
Handan

Yongdeng
Baiyin
Jingyuan
Dalachi
Huan Jiang
Quzi
Fu Xian
Yichuan
Luochuan
Xiangning
Linfen
Fushan
Anze
Tunliu
Gaoping
Fengfeng
Ci Xian
Daming
Anyang

Dingxi
Huining
Wating
▲2942
Longde
Pingliang
Zhenyuan
Heshui
Huangling
Hejin
Xinjiang
Yicheng
Qinshui
Jincheng
Changzhi
Hui Xian
Linqi
Puyang

Weiyuan
Jingning
Jinghuan
Jing He
Ning Xian
Changwu
Huanglong
Yijun
Qinyuan
▲2347
Taihang
Hua Xian
Xiangyuan
Jiaozuo
Ji Xian
Heze

Longxi
Tongwei
Qin'an
Long Xian
Lingtai
Chengcheng
Xia Xian
Wenxi
▲2322
Yangcheng
Yuanqu
Bo'ai
Qinyang
Xinxiang
Changyuan
Dingtao

Min Xian
Gangu
Tianshui
Qingshui
Qianyang
Bin Xian
Yao Xian
Yongji
Wanrong
Huang He
Wei He
Mianchi
Mengjin
Yuanyang
Lankao
Jinxiang

▲3100
Li Xian
Liangdang
Qishan
Fufeng
Fengxiang
Qian Xian
Jingyang
Sanyuan
Dali
Yuncheng
Zhongtiao
Luoyang
Zhengzhou (Chengchow)
Kaifeng
Cao Xian
Shan Xian
Xiayi

Zhugqu
Cheng Xian
Xihe
Baoji
Xianyang
Lintong
Huayin
Tongguan
Sanmenxia
Xingyang
Qi Xian
Ningling
Shangqiu

XI'AN (Hsian, Sian)
▲3767
Wei He
Hu Xian
Lantian
Chuankou
Luoning
Luo He
Yiyang
Dengfeng
Baisha
Xinzheng
Weichuan
Sui Xian

Wudu
Qin Ling Shandi
Mei Xian
Zhouzhi
Foping
Zhashui
Shang Xian
Danfeng
Lushi
Song Xian
Yu Xian
Fugou
Taikang

▲3002
Tangpingguan
Mian Xian
Yang Xian
Ningshan
Zhen'an
Shangnan
Xichuan
Xiping
Xiangcheng
Xuchang
Xihua
Luyi
Bo Xian

Funiu Shan
Ningqiang
Hanzhong
Chenggu
Shiquan
Xixiang
Hanyin
Xunyang
Yun Xian
Neixiang
Zhenping
Fangcheng
Pingdingshan
Yancheng
Luohe
Huaiyang

Guangyuan
Pingwu
Ziyang
Ankang
Baihe
Han Shui
Hong He
Zhenghe
Sheqi
Nanyang
Zhumadian
Suiping
Biyang
Queshan

Projection: Conical with two standard parallels

ft m / 12,000 4000 / 9000 3000 / 6000 2000 / 4500 1500 / 3000 1000 / 1200 400 / 600 200 / 0 0 / 200 600 / 2000 6000
m ft

1 : 4 800 000

50   0   50   100   150 miles

50   0   50   100   150   200   km

9   10   11   12   13   14   15   16

IQU

Horqin Youyi Qianqi   Zhenlai   Nen Jiang   HARBIN (Haerhpin)   Bin Xian   Yanshou

B

Baicheng   Tao'er He   Maoxing   Zhaoyuan   Acheng   Turiy Rog

Hulin He   Tuquan   Songhua Jiang   Shuangcheng   Yimianpo   Mudan Jiang   Ozero Khanka

Tao'an   Anguang   Fuyu   Changchunling   Sanchahe   Wuchang   Hengdaohezi   Mudanjiang   Maqiaohe   RUSSIA

JILIN   Qian Gorlos   Beitaabaizhan   Songhua Jiang   Yusha   Shanhetun   Hailin   Xiachengzi   Suiyang   Suifenhe   Pogranichnyy   Golenki

Jarud Qi   Zhanyu   Shenjingzi   Kuoshan   Dehui   Gangyao   Jingpo Hu   Ning'an   Muling   Dongning   Pokrovka   Ussuriysk (Voroshilov) Razdolnoye

1949   Chongling   Fulongquan   Jiutai   Changchun   Jilin (Kirin, Chilin)   Jiaohe   Dunhua   Daxinggou   Wangqing   Mingyuegou   Shixian   Hunchun   Kraskino   Artem   Tavrichanka

Bairin Zuoqi   Tongliao   Kailu   Huaidezhen   Maolin   Songhua Hu   Yitong   Panshi   Huadian   Emu   Chunyang   Yanji   Tumen   Vladivostok   Slavyanka   Posyet

Xar Moron He   Xinkai He   Shuangliao   Lishu   Huaide   Shuangyang   Liaoyuan   Dongfeng   Huinan   Hailong   Jingyu   Fusong   Helong   Antu   Puryang   Pugodong   Najin

C

Hexigten Qi   Laoha He   Jargalang   Siping   Xifeng   Hunjiang   1677   Paektu-san   Musan   Sosura   Unggi

Linxi   Bairin Youqi   Wutonghaolai   Kangping   Zhangwu   Faku   Kaiyuan   Shanchengzhen   Jinyu   Changbai   Hyesan   Hoeryang   Simpungdong   Ondaejin

Ongniud Qi   Wanfu   Xiawa   Liu He   Zhangwu   WALL   Tieling   Qingyuan   Liuhe   Xinbin   2541   Nanam   Kyongsong   Chuuronjin

2020   Chifeng   Beipiao   Qinghemen   Fuxin   Xinlitun   Xinmin   Heishui   FUSHUN   WILLOW   Hailong   Tonghua   Linjiang   Chunggang-up   Hachon   Chongjin

D

1885   Chaoyang   Liaoyang   Benxi   Qingchengzi   Huajianzi   1845   Ji'an   Manpojin   Inpundong   Kasan-dong   Pungsan   Iwon   Kosongni   Tanchon

Chengde   Lingyuan   Jinzhou   Jinxi   Niuzhuang   Panshan   Anshan   Anping   Lianshanguan   Kuandian   Chosan   Kanggye   Kuup-tong   2522   Pujon-chosuji   Kwangchon   Hapsu   Orhyangdong   Kimchaek (Songjin)   Musudan

LIAONING   Don bei (Manchuria)   Changbai Shan

SHENYANG (Mukden)

Hun Jiang   Yalu Jiang   Chosan   Koin-dong   Changjin-chosuji   Sinhung   Simpo   Sinpo

Huanren   Taegwan   Pyoktong   Kujang   Changjin   Sori   Hungnam

Liaodong Wan   Gai Xian   Yingkou   Fengcheng   Cao He   Kou Ji   Supung Sk.   Sonchon   Chongju   Oro   Hamhung   Hungnam

E

Yingkou   Xiuyan   Sakchu   Anju   Yongbyon   Tokchon   SEA   OF

Liaodong Bandao   Xiongyuecheng   Wanfu   Dandong   Sinuiju   Yongampo   Sonchon   Sunchon   Unsan   Sinchangni   Songchon   Tongjoson Man

Suizhong   Qinhuangdao   Fu Xian   Gushan   Donggou   Yalu Jiang   P'YONGYANG   Chunghwa   Kangdong   Tongyang   Anbyon   Wonsan

Tangshan   Jin Xian   Xinjin   Zhuanghe   Changyon   Chinnampo   Songnim   Pyongsan   Hoeyang   Kumhwa   Kosong

TIANJIN (Tientsin, T'ienching)   Lushun   DALIAN (Luda)   Korea Bay   Cho-do   Sariwon   Suan   Chiha-ri   Changdo-ri   Kangsong   Yangyang

Tanggu   Dagu   Bo Hai (Gulf of Chihli)   Haeju   Kumchon   Sinmak   Nam-chon   Chorwon   Kumhwa   Hwachon-chosuji   1578   Chumunjin

F

Qixou   Changli   Leting   Huimin   Penglai   Longkou   Yantai   Ongjin   Yonan   Munsan   Uijongbu   Chunchon   Hongchon   Kangnung   Ullung-do

Paengnyong-do   Cease Fire Line   Kaesong   Panmunjom   Kanghwa   SOUL (Seoul)   Hoengsong   Samchok

Huang He   Laizhou Wan   Weihai   Yongdungpo   Ichon   Yoju   Wonju   Yongwol

INCH'ON   Suwon   Chungju   Chechon   Ulchin

Huang Xian   Muping   Wendeng   Pyongtaek   Chonan   Chungju   Yongju

SOUTH KOREA   Chochiwon   Yongdong   Andong   Yongdok

923   Shandong Bandao   Rushan   Yesan   Chongju   Mungyong   Uisong   Chongha

Zibo   Yidu   Fangzi   Pingdu   Gaomi   Jimo   Hongsong   Taejon   Sangju   Pohang

Weifang   Jiao Xian   Chengyang   Nonsan   Kimchon   Yongchon

G

Xintai   Laiwu   1108   QINGDAO (Ch'ingtao)   Kanggyong   Iri   Yongdong   Waegwan   Kyongju

Kunsan   Chonju   Kochang   TAEGU   Chongdo   Ulsan

HUANG HAI (Yellow Sea)   Puan   Imje   Hamyang   Koryong   Miryang   Tongne

Linyi   Haizhou Wan   Chongup   Namwon   1915   Chinju   Masan   PUSAN

Sago-ri   KWANGJU   Hadong   Chinhae   Chungmu   Sasuna

Ganyu   Lianyungang   Suncheon   Samchonpo   Korea   Izuhara

Lanshan   Tancheng   Mokpo   Changhung   Posong   Polgyo   Yosu   Chindo   Tsushima   Iki

H

Linyi   Lianyungang (Hsinhailien)   Haenam   Strait   Tsushima-kaikyo   JAPAN   Karatsu   Imari

JIANGSU   Guannan   Chenjiagang   Xiangshui   Cheju   Cheju-do   Onpyong-ni   Nakadori-jima   Sasebo   Kashima

Qingjiang   Huai'an   Da Yunhe   Binhai   Hallim   1950   Sogwi-po   Omura   Isahaya

Bengbu   Fengyang   Hongze Hu   Baoying   Liuzhuang   Funing   Mosulpo   Nagasaki   Kuchinotsu

Gaoyou Hu   Yancheng   Xinghua   Dongtai   Fukue-jima

9   118   East from Greenwich   120   11   122   12   124   13   126   14   128   15

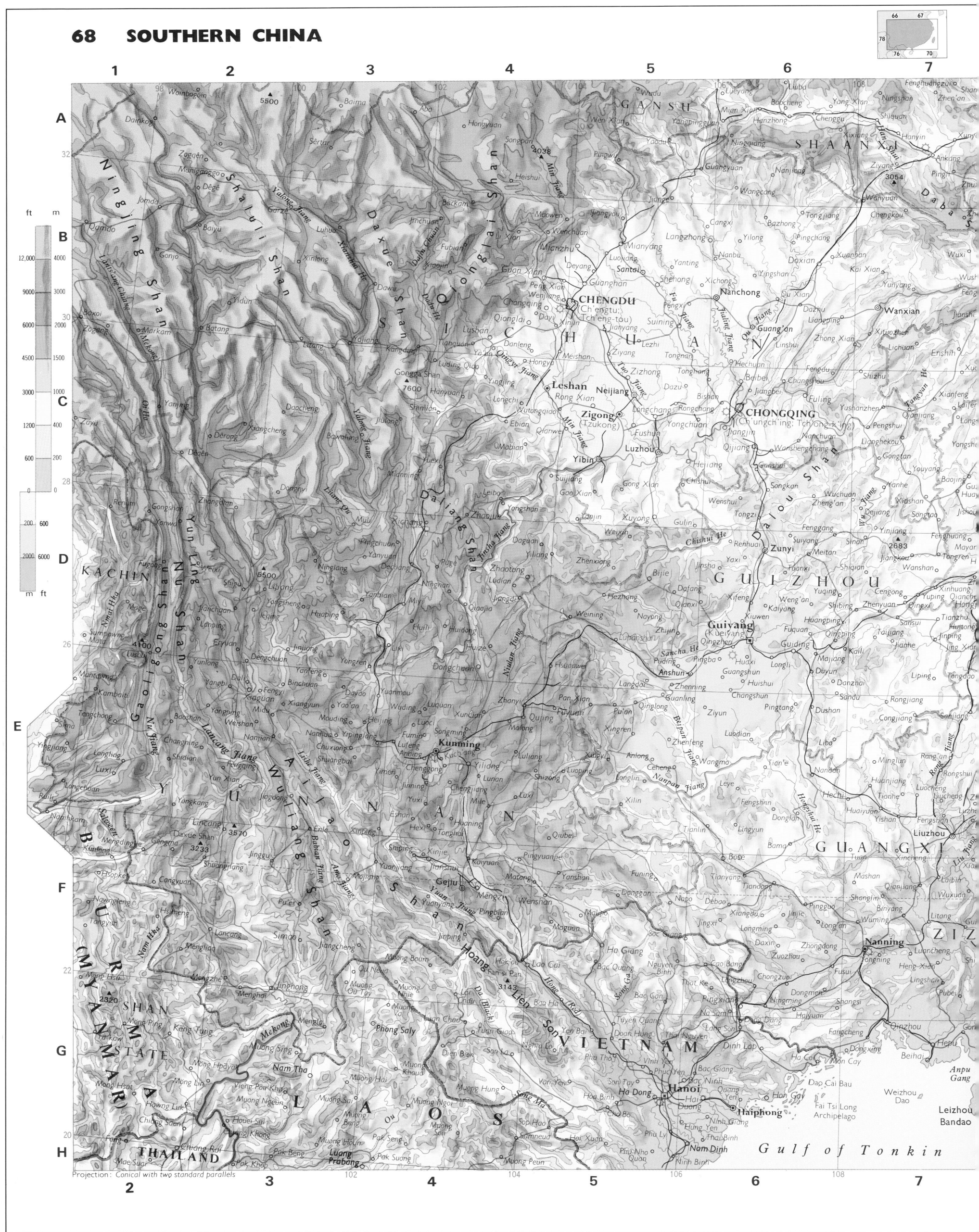

1 : 4 800 000

50   0   50   100   150 miles

50 0 50 100 150 200 km

8   9   10   11   12   13   14

A

B

C

D

E

F

G

H

HENAN

Shangnan · Xiping · Wuyang · Xiangcheng · Shenqiu · Jieshou · Xifei He · Guzhen · Mengcheng · Wuhe · Hongze · Xinghua · Dongtai

Jingziguan · Xixia · Xichuan · Fangcheng · Suiping · Jixiangcheng · Linquan · Yingshang · Yingshang · Taihe · Madian · Fengyang · Gaoyou Hu · Gaoyou

Yunxi · Yun Xian · Neixiang · Sheqi · Zhumadian · Xincai · Shangcai · Runan · Fuyang · Fengtai · Bengbu · **JIANGSU**

Shiyan · Danjiangkou · Zhenping · **Nanyang** · Wadian · Tanghe · Biyang · Queshan · Zhengyang · Mingang · Shou Xian · Huainan · Dingyuan · Lai'an · Yangzhou · Guazhou · Taizhou · Hai'an · Rugao

Hanshui · Guanghua · Xinye · Baokang · Nanzhang · Dongjinwan · Luoshan · Huangchuan · Gushi · Chengxi Hu · Chengdong Hu · Hefei · Chao Xian · He Xian · Zhenjiang (Chenchiang) · Danyang · Jiangyin · Changshu · Chongming Dao

Xiangfan · Yicheng · Maping · Yingshan · Sui Xian · Xinyang · Shangcheng · Jinzhai · Lu'an · **Hefei** · Chao Hu · Ma'anshan · **Changzhou** (Ch'angchou) · Wuxi (Wuhsi) · Kunshan · Jiading · Baoshan · **SHANGHAI** (Changhai)

Jing Shan · Zhongxiang · Fenghe · Hong'an · Macheng 1834 · Yuexi · Qianshan · Tongcheng · Nanling · Jing Xian · Xuanzheng · Guangde · Wuxing · Suzhou (Suchow) · Wujiang · Qingpu · Songjiang · Fengxian

Yichang (Ich'ang) · Dongyang · Jingmen · **Wuhan** (Wou-han) · Hankou · Huangpi · Luotian · Taihu · Wangjiang · Donglu · Jiuhua · Ningguo · Qingyang · Lin'an · **Hangzhou** (Hangchou; Hangchow) · Haining · Pinghu · **Zhoushan Dao**

Changyang · Yuan'an · Tianmen · Hanchuan · Hanyang · Huangguang · Qichun · Susong · Dongzhi · Shitai · Jixi · Yi Xian 1810 · Jixi · **Ningbo** (Ningpo) · Fenghua · Daqu Shan · Taohua Dao

Zigui · Yidu · Changyang · Songzi · Shashi · Qianjiang · Mianyang · Paizhou · Daye · Meichuan · Guangli · Dongzhi · Pengze · Hukou · Yi Xian · Tunxi · Chun'an · Meicheng · Tonglu · Zhuji · Sheng Xian · Xinchang · Fenghua · Liuheng Dao

HUBEI · Gong'an · Haoxue · Jiayu · Yangxin · Xianning · Ruichang · Jiujiang · Shimenjie · She Xian · Fuyang · Shaoxing · Ninghai · Niutou Shan

Shimen · Linli · Jinshi · Nan Xian · Linxiang · Puqi · De'an · Duchang · Jingdezhen · Wuyuan · Kaihua · Lanxi · **Jinhua** · Yongkang · Tiantai · Linhai

Dayong · Cili · Li Xian · Anxiang · Huarong · **Poyang Hu** · Yongxiu · Leping · Dexing · Yushan · Changshan · Jiangshan · Wuyi · Xianju · Huangyan

Changde · Hanshou · Yuanjiang · **Dongting Hu** · Pingjiang · Tongcheng · Xiushui · Jing'an · Anyi · Yugan · Wannian · Hengfeng · Yiyang · Shangrao · Guangfeng 1725 · Yunhe · Qingtian · Wenling · Taizhou Liedao

Taoyuan · Yiyang · Ningxiang · Xiangyin · Xinjian · **Nanchang** · Jin Jiang · Gao'an · Guixi · Yanshan · Chong'an · Songxi · Taishun · Fuding · Yuhuan Dao

Yuanling · Chenxi · Anhua · **Xiang Jiang** · Wugang · Xiangyin · Tonggu · Shanggao · Fengxin · Dongxiang · Linchuan · Jinxi · Zixi · Pucheng · **Wenzhou** (Wenchow) · Rui'an · Nanji Shan

Changsha · Liuyang · Wanzai · Yichun · Xinyu · Xingan · Jin Jiang · Guixi · **Wuyi Shan** · Zhenghe · Pingnan · Zherong · Sansha · Xiapu

Meichengzhen · Xupu · Xiangtan · Zhuzhou · Liling · **JIANGXI** · Fengcheng · Yujiang · Guangchang · Jianyang · 1629 · Fu'an

Lianyuan · Shuangfeng · Lou Di · Xiangxiang · Pingxiang · **Wugong Shan** · Gu'an · Ji'an · Xinfeng · Nancheng · Lichuan · Shunchang · Jian'ou · Ningde · Luoyuan · Sansha

HUNAN · **Shaoyang** · Hengshan · You Xian · Lianhua · Anfu · Jishui · Le'an · Nanfeng · Taining · Jiangle · Nanping · Gutian · Mingxi · Shaxian · Minqing · Mazu Dao

Xinhua · Hengyang · Chaling · Gujiang · Taihe · Yihuang · Nancheng · Mingxi · Sha Xian · Youxi · Fuzhou (Fuchou; Fuchou)

Xuefeng Shan · Qidong · Anren · Ling Xian · Xincheng · Wan'an · Ningdu · Shicheng · Qin Jiang · Ninghua · Qingliu · Yong'an · Datian · Yongtai · Fuqing

Dong'an · **Hengyang** · Huangyangsi · Qiyang · Changning · Yongxing · Suichuan · 2164 · Xingguo · Kudu · Changting · Lianchen · Maiyuan · Hanjiang · Pingtan · Xinghua Wan

Shaoyang · Xinning · Lingling · Leiyang · Chaling · Zixing · Chongyi · Shangyou · **Ganzhou** · Huichang · Ruijin · Liancheng · Xianyou · **Quanzhou** (Ch'uanchou) · Nanri Dao

Quanzhou · Guidong · Shadi · Nankang · Xinfeng · Yudu · Longkou · Xinfeng · Datian · Hua'an · Nan'an · Jinjiang

Xing'an · Guanyang · Ningyuan · Jiahe · Yizhang · Renhua · Nanxiong · Anyuan · Wuping · Jiaoling · Yongchun · Anxi · Tong'an · **Xiamen** (Hsiamen; Amoy)

Guilin · Jiangyong · Dao Xian · Lanshan · Linwu · Lechang · Dayu · Shixing · Longnan · Zhenping · Yongding · Changtai · **Zhangzhou** · Nanjing · Longhai · Jinmen Dao

Gongcheng · Zizhixian · Lian Xian · Lianshan · Ruyuan · Quannan · Dinghan · Pingyuan · Jiaoling · Zhangping · Zhao'an · Dongshan

Lipu · Zhongshan · **Shaoguan** · Wengcheng · Heping · Xingning · Longchuan · Huachuan · Mei Xian · Han Jiang · Yunxiao · Zhangpu

Xiuren · Fuchuan · Pingle · Yangshan · Qingyuan · Fogang · Longmen · Heyuan · Zijin · Puning · Chaoyang · **Chao'an** · Nan'ao

GUANGZU · He Xian · Zhongshan · Qingyuan · Zengcheng · Boluo · Huizhou · Haifeng · Lufeng · Huilai · Jieyang · **Shantou** (Swatow) · Haimen

Wuzhou · Hexian · Sihui · Hekou · Shilong · Dongguan · Haifeng · Honghai Wan · **Tropic of Cancer**

Teng Xian · Fengkai · Xi Jiang · **Guangzhou** (Kwangchou; Canton) · Panyu · Xiaolan · Shenzhen

Xun Jiang · Deqing · Zhaoqing · Sanshui · **Foshan** · Guohe · Shunde · Zhongshan · **HONG KONG** (U.K.) · Kowloon

Beiliu · Luoding · Yunfu · Yunan · Xinxing · Kaiping · Xinhui · Taishan · Zhuhai · **Macau** (Macao) (Port.) · Dangan Liedao

Luchuan · Xinyi · 1703 · Enping · Yangchun · Doumen · Chixi · Gaolan Dao · Shangchuan Dao · Xiachuan Dao

Huazhou · Dianbai · **Maoming** · Yangjiang · Yangdong · Hailing Dao

Wuchuan · Zhanjiang · Donghai Dao · Naozhou Dao

SOUTH CHINA SEA

Strait   For   For   **Formosa**

TAIWAN (FORMOSA)

Danshui · **Jilong** · Taoyuan · Tooyuan · **TAIBEI** (Taipei) · Yilan

Xinzhu · Dayi · Migoli · Xue Shan 3931 · Dongshi · Luodong

Yuanli · Dajia · Taizhong (T'aichung) · Hualian

Zhanghua · Lugang · Erlin · Yunlin · Jiayi (Chiayi) · Yu Shan 3997 · Taidong

Tainan · Jiali · Pozi · Qishan · Qishan · Huoshao Dao

Gaoxiong (Kaohsiung) · Fangliao · Pingdong · Lan Yu · Eluanbi

Luzon Strait

Tungsha Tao

East from Greenwich

COPYRIGHT. GEORGE PHILIP & SON. LTD.

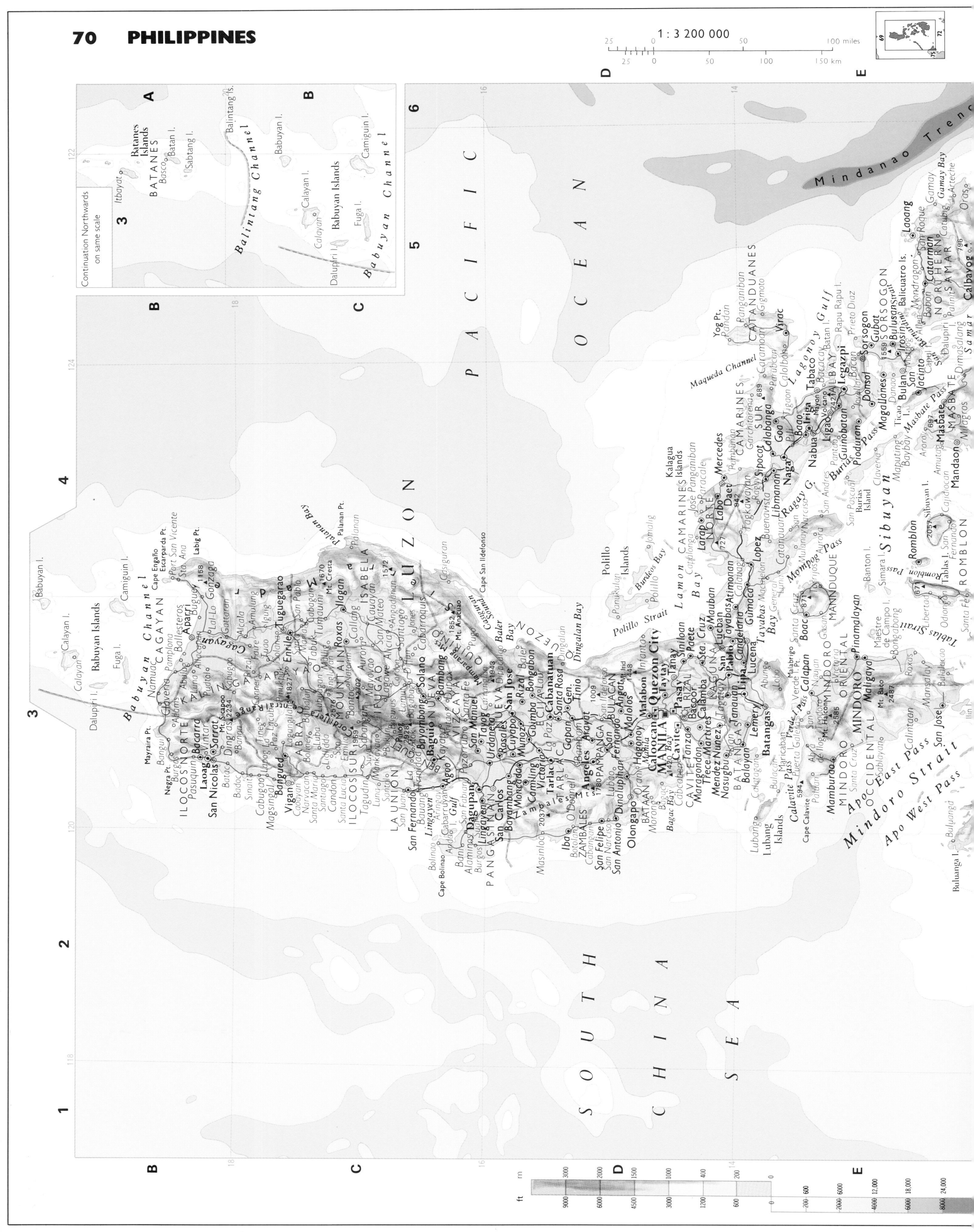

1 : 3 200 000

25   0   50   100 miles

25   0   50   100   150 km

A   B

Batanes
Islands

BATANES

Bosco
Batan I.
Sabtang I.
Itbayat
Balintang Is.

Continuation Northwards
on same scale

3

Babuyan Is.

Calayan

Camiguin I.

Babuyan Islands

Fuga I.

Dalupiri I.

Balintang Channel

Babuyan Channel

P A C I F I C

O C E A N

Mindanao Trench

L U Z O N

S O U T H

C H I N A

S E A

MANILA

Quezon City

Mindoro Strait

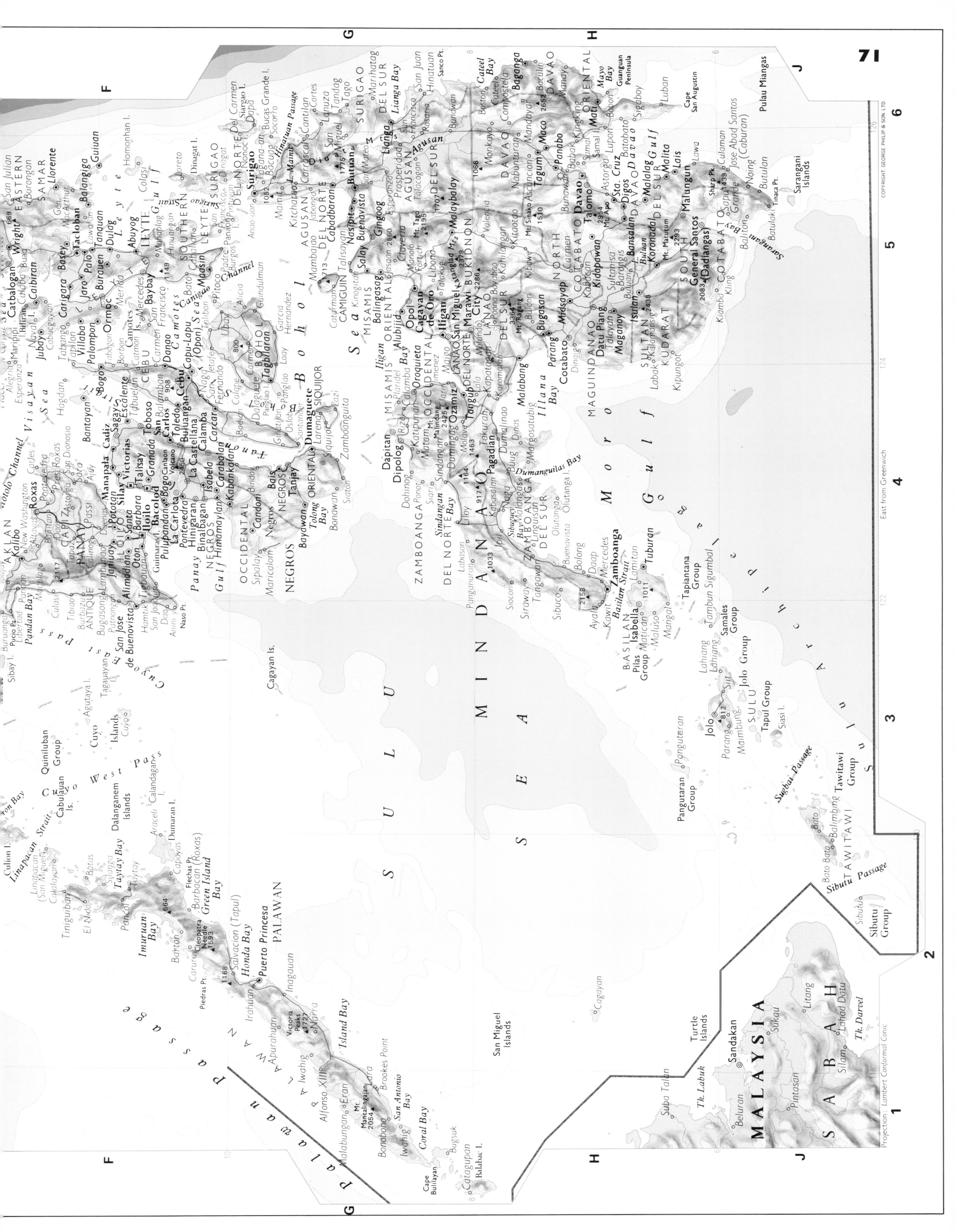

Projection: Lambert Conformal Conic

East from Greenwich

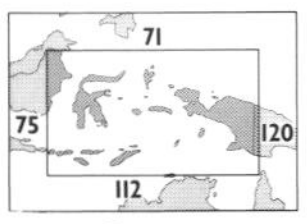

1

2

3

MALAYSIA 1346

Semporna 120

Tawau

Teluk Sebuku Mandul

Malinau Bunyu

Sesayap Tarakan

A

Nomeh Tanjungselor

Berau Tanjungbatu Maratua

Kongkemul 2053 Tanjungredeb Dumaring

B O R N E O

TIMUR Equator Dumaring

Muarakaman Bontang

Tenggarong **Samarinda**

Sangasanga

Sungaitiram

Tanahgrogot

Selat Makasar

Langeru **Balikpapan**

0

Kepulauan Balabalangan

Kotabaru

Sebuku

Karambu

Pulau Laut

Kepulauan Masalima

C E L E B E S

S E A

▼5315

Karakelong Bulu Beo

Kepulauan Talaud

Kaburuang

Tahuna

Sangihe

Kepulauan Sangihe

Siau

Tahulandang

Biaro

Bangka Rau Bebereb

Morotai

Doi

Tolitoli Buol Paleleh

Teluk Dondo Malino Sumalata

Ogoamas 2300 Tentolomatihan Kuandang

2913 2707 Tomini Gambuta 1954

Manado 2022 Kema

Amurang Tondano

Halmahera

Ternate **Ternate** Soasiu

Tidore

Teluk Weda

UTARA Tilamuta Tg. Flesko

Makian Weda

Kayoa **Wosi**

A

Donggala Toboli Parigi

Palu 3127 Poso

Lariang *Danau Poso* Tojo

T E N G A H

SULAWESI (CELEBES) Kolonodale

Mamuju 3074 Masamba Malili

Manasa Palopo

Onang Balease 3016

SELATAN *Danau Matana*

Makale *Danau Towuti*

Polewali 3455 Tangkelebuke 1782

Majene Suli Malili

Teluk Mandar Enrekang Mondeodo

Parepare Pinrang Mekongga 2790

Palanro Rappang **TENGGARA** Kolaka

Watansoppeng Singkang Kendari

Sumpangbinangae Watampone Manui

Pangkajene Marek

Ujung Pandang Maros

Sungguminasa Lompobatang

Patalasang 2871 Bulukumba **Kabaena**

Bantaeng

Bontosunggu

Salayar Benteng

Teluk Tomini

Kepulauan Togian

Muotong Poh Maliku

Luwuk

Peleng

Toili Banggai

Tokala 2630 *Kepulauan Banggai*

Taliabu

Mangole

Auponhia Sanana

Kepulauan Sula

Sanana

Kaupalatmada Namlea

Buru 2429

Wamulan Tifu Kayeli

Leksula Namrole

Teluk Tolo

Buapinang Raha

Pising **Muna**

Lawele

Buton (Butung)

Baubau

Binongko

Batuata (Watuata I.)

M O L U C C A

S E A

Mayu

Kepulauan Bacan

Bisa

Bacan

Mandioli

Kepulauan Obilatu Obi

Loji Sesepe Fluk

Obi

M O L U C C K

S E R A

Piru

Ambon

Lisa Amb

Monse

Wowoni

I N D O N E S

Wangiwangi

Kepulauan Tukangbesi

B A N D A

Gunungapi

▼5888

Dama

5

Kepulauan Bonerate

Tanahjampea Kalao

Bonerate

Kalaotoa

Moyo Sangeang

Lombok

Rinjani 3726 Tambora 2821

Selong **Sumbawa Besar** Raba

Mataram Taliwang Dompu Sape

Parado Rinca

Sumbawa

Selat Sumba

NUSA TENGGARA BARAT

Memboro

Waikabubak **Sumba**

Waingapu

Melolo

Baing

F L O R E S   S E A

L e s s e r   S u n d a   I s l a n d s

Labuhanbajo

Komodo Ruteng **Flores** Maumere

Aimere Ende

Larantuka Adonara Pantar Kalabahi

Solor Lomblen

NUSA TENGGARA TIMUR

S A W U   S E A

Semau

Raijua Sawu

Dana Roti

Baa

Wetar Wesiri

Iwaki *Selat Wetar* Kisar

Alor Ataúro

Selat Ombai Manatuto

Dili

TIMOR TIMUR

Atapupu

Pante Makasar Atambua

Naikliu Kefamenanu

Pariti Soe **Timor**

Kupang Nikiniki

Romang

Kepulau

Leti Moa

Kisar **Kepulauan Leti**

Baukau Tutuala

Viqeke

T I M O R   S E A

East from Greenwich

D

ft m

12,000 4000

9000 3000

6000 2000

4500 1500

3000 1000

1200 400

600 200

0 0

200 600

2000 6000

4000 12,000

6000 18,000

8000 24,000

m ft

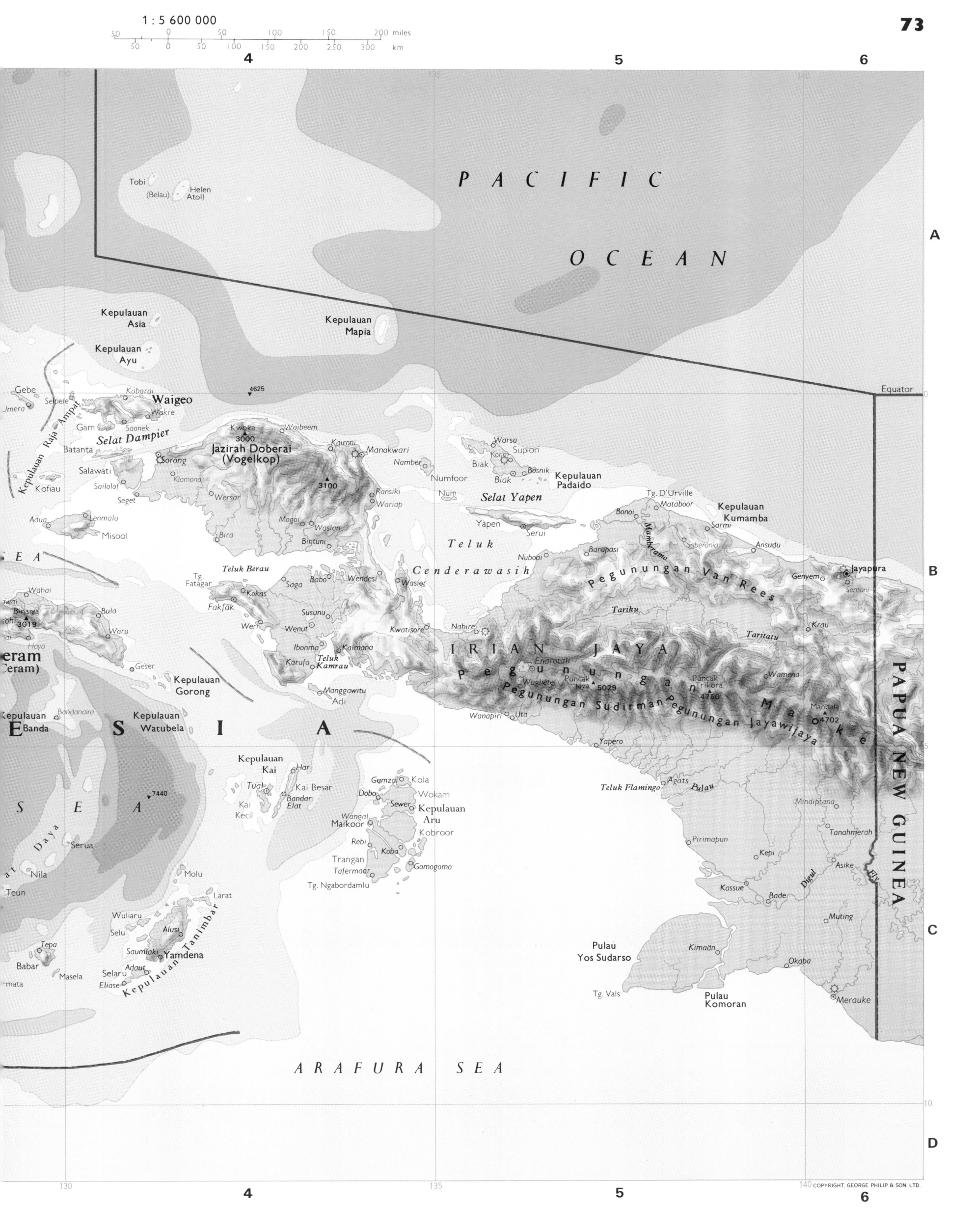

1 : 5 600 000

50   0   50   100   150   200 miles

50   0   50   100   150   200   250   300 km

4   **5**   **6**

P A C I F I C

O C E A N

Tobi
(Belau)   Helen
Atoll

Kepulauan
Asia

Kepulauan
Mapia

Kepulauan
Ayu

▼4625

A

Equator

Gebe
Imera   Selpele   Kabarai   Waigeo
Wakre

Kwoka
3000   Waibeem

Kairani   Warsa
Supion

Gam   Saonek   Jazirah Doberai
(Vogelkop)   Manokwari   Biak   Bosnik   Kepulauan
Padaido

Raja Ampat   Batanta   Sorong   Nomber   Numfoor   Biak

Salawati   Klamono   ▲3100   Ransiki   Num   Selat Yapen   Kepulauan
Kumamba

Kofiau   Sailolof   Seget   Wersar   Wariap   Yapen   Serui   Tg. D'Urville   Sarmi   Ansudu

Adua   Lenmalu   Mogoi   Wasian   Teluk   Mataboor   Bonoi   Barapasi   Srberania

Misool   Bira   Bintuni   Cenderawasih   Nuboai   Pegunungan Van Rees   Krau   Jayapura

SEA   Teluk Berau   Saga   Babo   Wasior   Nabire   Tariku   Genyem   Sentani

Wahai   Tg.
Fatagar   Kokas   Wendesi   Kwatisore   Taritatu

Bula   Fakfak   Susunu   IRIAN   JAYA   Mamberamo   B

Binaiya
3019   Waru   Weri   Wenut   Ibonma   Kaimana   Enarotali   Pegunungan   Wamena

Haya   Geser   Karufa   Teluk
Kamrau   Waghete   Puncak
Jaya   Puncak
Trikora   Mandala
▲4702

eram
(Ceram)   Kepulauan
Gorong   Manggawitu
Adi   Wanapiri   Uta   Pegunungan Sudirman Pegunungan Jayawijaya

Kepulauan
Banda   Bandanaira   Kepulauan
Watubela   E   S   I   A   Yapero

Banda   Wanggar   Teluk Flamingo   Agats   Pulau   Mindiptana

S E A   Kepulauan
Kai   Har   Pirimapun   Kepi   Tanahmerah

▼7440   Tual   Kai Besar   Gumzai   Kola   Kassue   Bade   Asike

Daya   Bandar
Elat   Dobo   Wokam   PAPUA NEW GUINEA

Serua   Kai
Kecil   Sewer   Kepulauan
Aru   Muting

Nila   Wongai
Maikoor   Kobroor   Okaba

Teun   Rebi   Koba   Merauke

Molu   Trangan   Gomogomo

Wuliaru   Larat   Tafermaar   C

Selu   Alusi   Tg. Ngabordamlu   Pulau
Yos Sudarso   Kimaän

Tepa   Saumlaki   Kepulauan Tanimbar   Pulau
Komoran

Babar   Selaru   Yamdena

rmata   Masela   Adaut
Eliase   Tg. Vals

A R A F U R A   S E A

D

4   **5**   **6**

COPYRIGHT. GEORGE PHILIP & SON. LTD.

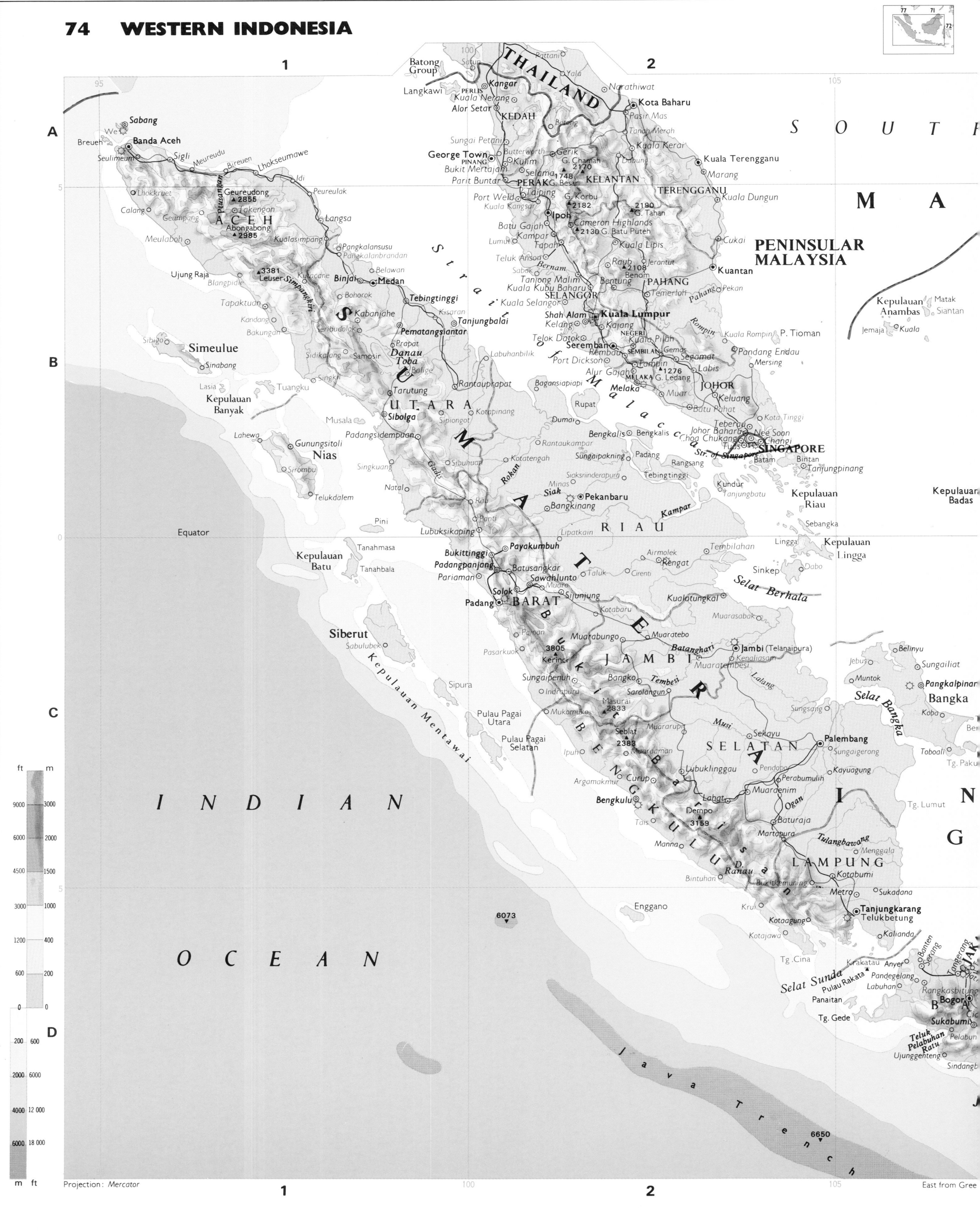

SOUTH

MA

THAILAND

Batong Group
Langkawi
Kuala Nerang
Alor Setar
Kangar
PERLIS
KEDAH
George Town
PINANG
Butterworth
Kulim
Bukit Mertajam
Parit Buntar
Sungai Petani
Batong
Gerik
G. Chamah 2170
Selama 748
Taiping
PERAK
G. Korbu 2182
Ipoh
Batu Gajah
Kampar
Tapah
Cameron Highlands
2130 G. Batu Puteh
Teluk Anson
Bernam
Tanjong Malim
Lumut
Sabak
Kuala Kubu Baharu
SELANGOR
Kuala Selangor
Kelang
Shah Alam
Kajang
Telok Datok
Port Dickson
Seremban
Rembau
NEGERI
SEMBILAN
2108
Bentung
Benom
PAHANG
Temerloh
Rompin
Kuala Rompin
Kuala Lipis
Jerantut
Rayb
Gemas
Tampin
1276
MELAKA
G. Ledang
Melaka
Muar
Batu Pahat
JOHOR
Keluang
Labis
Segamat
Mersing
Pandang Endau
P. Tioman
Kepulauan Anambas
Matak
Siantan
Jemaja
Kuala
Kota Baharu
Pasir Mas
Tanah Merah
Kuala Kerar
Narathiwat
Pattani
Yala
Satun
Kuala Terengganu
Marang
TERENGGANU
Kuala Dungun
Cukai
Kuantan
Pekan
PENINSULAR MALAYSIA

KELANTAN

Sabang
We
Breuen
Seulimeum
Banda Aceh
Sigli
Meureudu
Bireuen
Lhokseumawe
Idi
Peureulak
Langsa
Kualasimpang
Pangkalansusu
Pangkalanbrandan
Belawan
Binjai
Medan
Tebingtinggi
Kisaran
Tanjungbalai
Labuhanbilik
ACEH
Geureudong 2855
Takengon
Abongabong
2985
Leuser 3381
Pinusungkit
Simpangkiri
Lhokkruet
Geumpang
Calang
Meulaboh
Tapaktuan
Kandang
Bakungan
Sidikalang
Kabanjahe
Pematangsiantar
Prapat
Danau Toba
Samosir
Bolge
Tarutung
UTARA
Sibolga
SUMATERA
Sibigo
Sinabang
Simeulue
Sinabang
Kepulauan Banyak
Lasia
Tuangku
Nias
Gunungsitoli
Sirombu
Telukdelam
Padangsidempuan
Sibuhuan
Gadis
Singkuang
Natal
Rao
Panti
Lubuksikaping
Payakumbuh
Bukittinggi
Padangpanjang
Batusangkar
Pariaman
Sawahlunto
Solok
Padang
BARAT
3805
Kerinci
Sungaipenuh
Muarabungo
Bangko
Tembesi
Batanghari
Jambi (Telanaipura)
Muaratembesi
JAMBI
Sarolangun
Masurai 2833
Mukomuko
Seblat
2383
Muaraaman
Muararupit
Sekayu
Musi
Palembang
SELATAN
Curup
Lubuklinggau
Argamakmur
Bengkulu
BENGKULU
Tais
Dempo 3159
Manna
Bintuhan
D. Ranau
Krui
Kotabumi
Metro
Kotaagung
Kotaadung
Tanjungkarang
Telukbetung
Kalianda
Tg. Cina
LAMPUNG
Bukitkemuning
Sukadana
Enggano
6073

Str. of Singapore
Dumai
Bengkalis
Bengkalis
Rupat
Siaksriinderapura
Sungaipakning
Padang
Rangsang
Tebingtinggi
Kundur
Tanjungbatu
Batam
Bintan
Tanjungpinang
Kepulauan Riau
Kepulauan Badas
Johor Baharu
Nee Soon
Teberau
Kota Tinggi
Changi
SINGAPORE
Tudong
Choa Chukang

RIAU
Kotapinang
Rantauprapat
Rantaukampar
Kotatengah
Minas
Pekanbaru
Bangkinang
Lipatkain
Taluk
Cirenti
Airmolek
Rengat
Tembilahan
Bangka
Selat Berhata
Lingga
Sinkep
Dabo
Kepulauan Lingga
Bansiapiapi
Alur Gajah
Kualatungkal
Muarasabak
Batanghari
Lalang
Sungsang
Jebus
Muntok
Selat Bangka
Koba
Toboali
Sungailiat
Pangkalpinang
Bangka
BANGKA
Tg. Pakur
Tg. Lumut

Straits of Malacca

Kuala Kangsar
Lumut

Sungaigerong
Perabumulih
Muaraenim
Ogan
Lahat
Baturaja
Martapura
Kayuagung
Tulangbawang
Menggala
Pendopo
Kotabumi

Kotajawa

Indian Ocean
I N D I A N
O C E A N

Siberut
Sabulubek
Kepulauan Mentawai
Sipura
Pulau Pagai Utara
Pulau Pagai Selatan
Ipuh
Tanahmasa
Tanahbala
Kepulauan Batu
Pini
Musala
Equator
Lahewa

Selat Sunda
Krakatau
Pulau Rakata
Anyer
Pandegelang
Labuhan
Panaitan
Tg. Gede
Teluk Pelabuhan Ratu
Ujunggenteng
Sindangb
Pelabuh
Sukabumi
Bogor
Rangkasbitung
Tangerang
JAKARTA
Banten
Serang
B
Cic

Java Trench
6650

East from Gree

ft   m
9000 3000
6000 2000
4500 1500
3000 1000
1200 400
600   200
0   0
200   600
2000 6000
4000 12 000
6000 18 000
m   ft

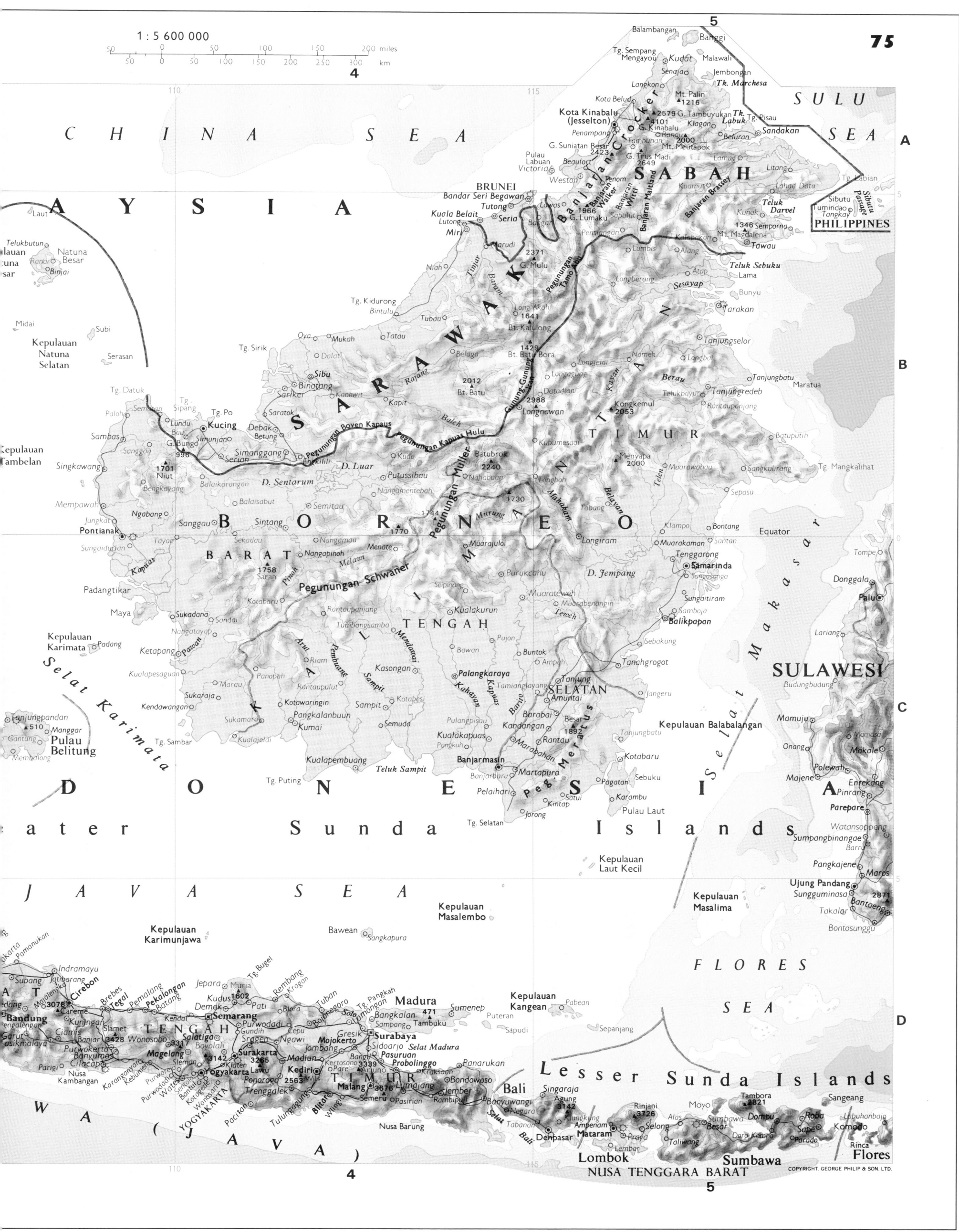

1 : 5 600 000

50 · 0 · 50 · 100 · 150 · 200 miles
50 · 0 · 50 · 100 · 150 · 200 · 250 · 300 km

4

C H I N A   S E A

A Y S I A

SULU   SEA

Balambangan
Banggi
Tg. Sempang
Mengayou
Kudat
Malawali
Langkon
Jembongan
Tk. Marchesa
Kota Belud
Mt. Palin
1216
2579 G.
Tambuyukan
Tk. Labuk
Tg. Pisau
Kota Kinabalu
(Jesselton)
4101
G. Kinabalu
Klagan
Beluran
Sandakan
Penampang
Mt. Meutapok
2000
Lamag
G. Suniaten Besar
2423
Mt. Trus Madi
2649
Kuamut
Litang
Lahad Datu
Pulau
Labuan
Victoria
Beaufort
Weston
Penom
G. Lumaku
1966
Sapulut
Kalabakang
Mt. Magdalena
Teluk
Darvel
BRUNEI
Bandar Seri Begawan
Tutong
Lawas
Walker
1346 Semporna
Tawau
Kuala Belait
Seria
Bagan
Pemangan
Tumindao
Tangkay
Sibutu Passage
Lutong
Miri
PHILIPPINES
Marudi
Lumbis
Alang
Teluk Sebuku
Lama
Niah
2371
G. Mulu
Pegunungan Tama Abu
Longberang
Sesayap
Bunyu
Tg. Kidurong
Barum
Long Akah
1641
Atap
Longnawan
Bintulu
Tubau
Nomeh
Tarakan
Tatau
Belaga
Bt. Kalulong
1429
Bt. Batu Bora
Longgelai
Longbal
Mukah
Berau
Tanjungbatu
Oya
Dalat
Rajang
Kanowit
Bt. Batu
2012
Gunung Gunung
Longanggung
Datadian
Tanjungredeb
Maratua
Sibu
Sarikei
Kapit
2988
Longnawan
Menyapa
2053
Kongkemul
Binatang
Saratok
Baleh
Kubumesaai
2000
Tg. Datuk
Tg.
Sipang
Pegunungan Boven Kapuas
Pegunungan Kapuas Hulu
Kuda
Kewan
Telukbayur
Tg.
Sambas
Lundu
Bau
Kucing
Debak
Betung
998
Putussibau
Longboh
Muarawahau
Sangkulirang
Tg. Mangkalihat
Singkawang
G. Bungo
Simunjan
Serian
D. Luar
Nangamentebah
Batubrok
2240
Mahakam
Tabang
Sepasu
Palph
Niut
1701
Balaikarangan
Balaisabut
Nahabuan
1730
Belayan
Batuputih
Mempawah
Ngabang
Sintang
D. Sentarum
Semitau
Murung
1744
Klampo
Bontang
Equator
Ngkat
Pontianak
Sekadau
Nangamau
1770
Muarajuloi
Longiram
Muarakaman
Saritan
Tompe
B O R N E O
B A R A T
Tayam
Nangapinoh
Menate
Seipinang
Muaratewen
Tenggarong
Samarinda
Donggala
Padangtikar
1758
Saran
Melawi
Purukcahu
Muarabengkang
Sungaitiram
Sagasanga
Kapuas
Pinuh
Pembuang
Muarateweh
Samboja
Palu
Maya
Sukadana
Sandai
Rantauparanjang
Tumbangsamba
Mendawai
D. Jempang
Teweh
Balikpapan
Lariang
Kepulauan
Karimata
Padang
Arut
Riam
K A L I   T E N G A H
Bawan
Buntok
Sebakung
Tanahgrogot
Ketapang
Pawan
Panopah
Kasongan
Pujon
Ampah
SULAWESI
Kualapesaguan
Kotawaringin
Rantaupulut
Sampit
Kotabesi
Pulangpisau
Budungbudung
Sukaraja
Pangkalanbuun
Semuda
Palangkaraya
Kahayan
Tamianglayang
Jangeru
Mamuju
Kumai
Tg. Sambar
Kualajelai
Kualakapuas
Barito
Tanjung
Amuntai
SELATAN
Tanjungbatu
Mamosh
Onang
Makale
Kualapembuang
Teluk Sampit
Pongkuh
Barabai
Marabahan
Kepulauan Balabalangan
Polewah
Enrekang
Tg. Puting
Banjarmasin
Kandangan
Rantau
Kotabaru
Majene
Pinrang
Banjarbaru
Martapura
Pagatan
Sebuku
Parepare
Pelaihari
Karamb
Watansoppeng
Satui
Pulau Laut
Jorong
Kintap
Tg. Selatan
Peg. Meratus
1892
Kepulauan
Laut Kecil
Sumpangbinangae
Barru
Pangkajene
Maros
D O N E S I A
Ujung Pandang
Sungguminasa
2871
Bantaeng
ater
S u n d a
I s l a n d s
Kepulauan
Masalima
Takalar
Bontosunggu

J A V A   S E A

Kepulauan
Masalembo
Bawean
Sangkapura
F L O R E S

Kepulauan
Karimunjawa

Kepulauan
Kangean
Pabean
Puteran
Sapudi
S E A

J A V A

Jakarta
Pamanukan
Indramayu
Subang
Jatibarang
Cirebon
Brebes
Tegal
Pemalang
Pekalongan
Batang
Kudus
1602
Pati
Jepara
Muria
Rembang
Tg. Bugel
Tuban
Tg. Pangkah
Madura
Sumenep
Kepulauan Kangean
3078
Careme
Kuningan
Demak
Blora
Bojonegoro
Bangkalan
Tambuku
471
Bandung
Majalengka
Semarang
Gundih
Cepu
Sampang
Garut
Cianjur
Slamet
Wonosobo
T E N G A H
Purwodadi
Ngawi
Gresik
Surabaya
3331
Salatiga
Boyolali
Jombang
Sidoarjo
Selat Madura
Tasikmalaya
Parigi
Banjar
Purwokerto
3142
Surakarta
3265
Mojokerto
Pasuruan
Probolinggo
Cilacap
Banyumas
Magelang
Klaten
Lawu
Kediri
Panarukan
Kebumen
Yogyakarta
Ponorogo
3339
Bondowoso
Bali
Singaraja
Nusa
Kambangan
Karangayar
Wonogiri
2563
T I M U R
3676
Lumajang
Banyuwangi
Agung
Rinjani
Moyo
Tambora
2821
Sangeang
YOGYAKARTA
Pacitan
Blitar
Wlingi
3142
Negara
Tjangkuk
3726
Sumbawa
Dompu
Roba
Tulungagung
Blitar
Semeru
Pasirian
Tabanan
Alas
Sape
Trenggalek
Rambipuji
Jember
Besar
Parado
W A (JAVA)
Nusa Barung
Denpasar
Mataram
Ampenam
Praya
Selong
Taliwang
Dara Kompa
Labuhanbajo
Rinca
L e s s e r   S u n d a   I s l a n d s
Selat
Bali
Lombok
Sumbawa
Flores
Komodo
NUSA TENGGARA BARAT

SABAH
SARAWAK
KALIMANTAN
TIMUR
Selat Makasar
Selat Karimata
Pulau
Belitung
Tanjungpandan
510
Manggar
Gantung
Membalong

Natuna
Besar
Kepulauan
Natuna
Selatan
Midai
Subi
Serasan
Kepulauan
Tambelan

Pegunungan Muller
Pegunungan Schwaner
Banjaran Crocker

GUANGXI ZHUANGZU ZIZHIQU AUTONOMOUS REGION

Nanning

YUNNAN

Gulf of Tonkin

HAINAN

Zhanjiang (Tsamkong)

Leizhou Bandao

Haikou

Haiphong

HANOI

Red River Delta

Nam Dinh

Da Nang

Hue

Central Highlands

TONKIN

BAC PHAN

Son La

Cao Nguyen Tran Ninh

Luang Prabang

Vientiane

LAOS

Plaine des Jarres

VIETNAM

Chaine Annamitique

Mekong

Pakse

Boloven

Phnom Dangrek

CAMBODIA

Battambang

Tonle Sap

THAILAND

Khorat

Nakhon Ratchasima (Khorat)

Udon Thani

Ubon Ratchathani

Thiu Khao Phetchabun

Khorat

Chao Phraya

BANGKOK (Krung Thep)

Thon Buri

Phitsanulok

Lampang

Chiang Mai

Nan

Yom

Ping

BURMA (MYANMAR)

SHAN STATE

Mandalay

Rangoon

Pegu

Sittang

Salween

Dawna Range

Bilauk Taung

TENASSERIM

Gulf of Martaban

KAYAH

KAREN

Mekong

Lancang Jiang

Chindwin

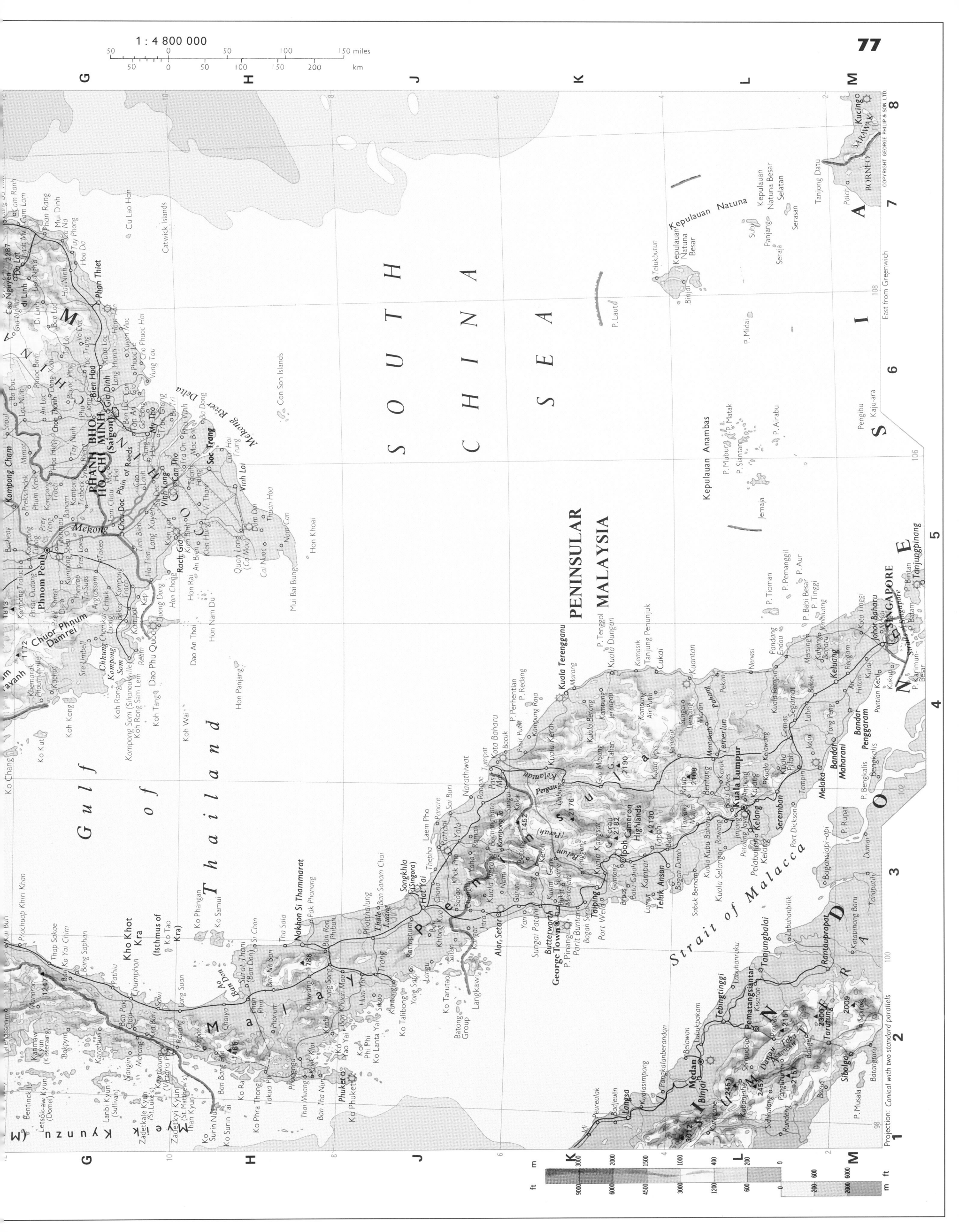

1 : 4 800 000

50   0   50   100   150 miles
50   0   50   100   150   200   km

G   H   J   K   L   M

SOUTH

CHINA

SEA

Kepulauan Natuna

Kepulauan
Natuna
Besar

Natuna Besar
Selatan

Serasan

Tanjong Datu

Kucing
SARAWAK
BORNEO

East from Greenwich

PENINSULAR
MALAYSIA

Kepulauan Anambas

P. Mubur
P. Matak
P. Siantan

Jemaja

Pengibu
Kaju-ara

SINGAPORE

INDONESIA

Strait of Malacca

Projection: Conical with two standard parallels

COPYRIGHT GEORGE PHILIP & SON LTD

Gulf

of

Thailand

MINH

HO CHI MINH
(Saigon)

Mekong River Delta

Mekong

Phnom Penh

Chuor Phnum
Damrei

PHANH BHO

Con Son Islands

Kho Khot
Kra
(Isthmus of
Kra)

Nakhon Si Thammarat

Songkhla
(Singora)
Hat Yai

Phuket
Ko Phuket

M a l a y

George Town
P. Pinang
Butterworth

Alor Setar

Kuala Terengganu

Kuala Lumpur

Seremban

Melaka

Bandar
Maharani

Johor Baharu

Medan
Binjai
Langsa

Pematangsiantar

Rantauprapat

Tebingtinggi

Tanjungbalai

ft   m
9000   3000
6000   2000
        1500
4500   1000
3000   600
        400
1200   200
600   0
0   -200
        -600
        -2000
6000
2000   m
ft   K   L   M

1 : 4 800 000

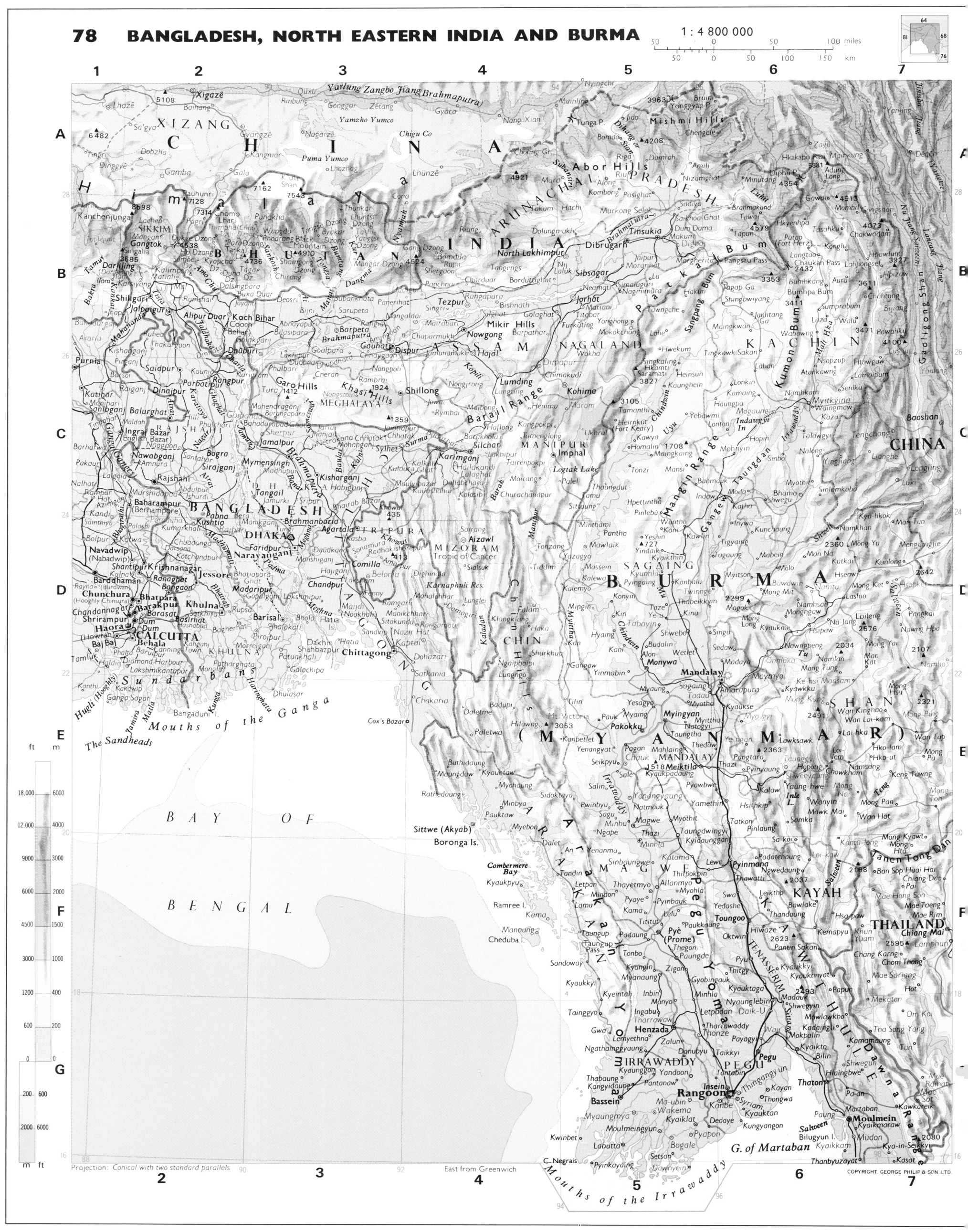

Projection: Conical with two standard parallels · East from Greenwich

COPYRIGHT. GEORGE PHILIP & SON. LTD.

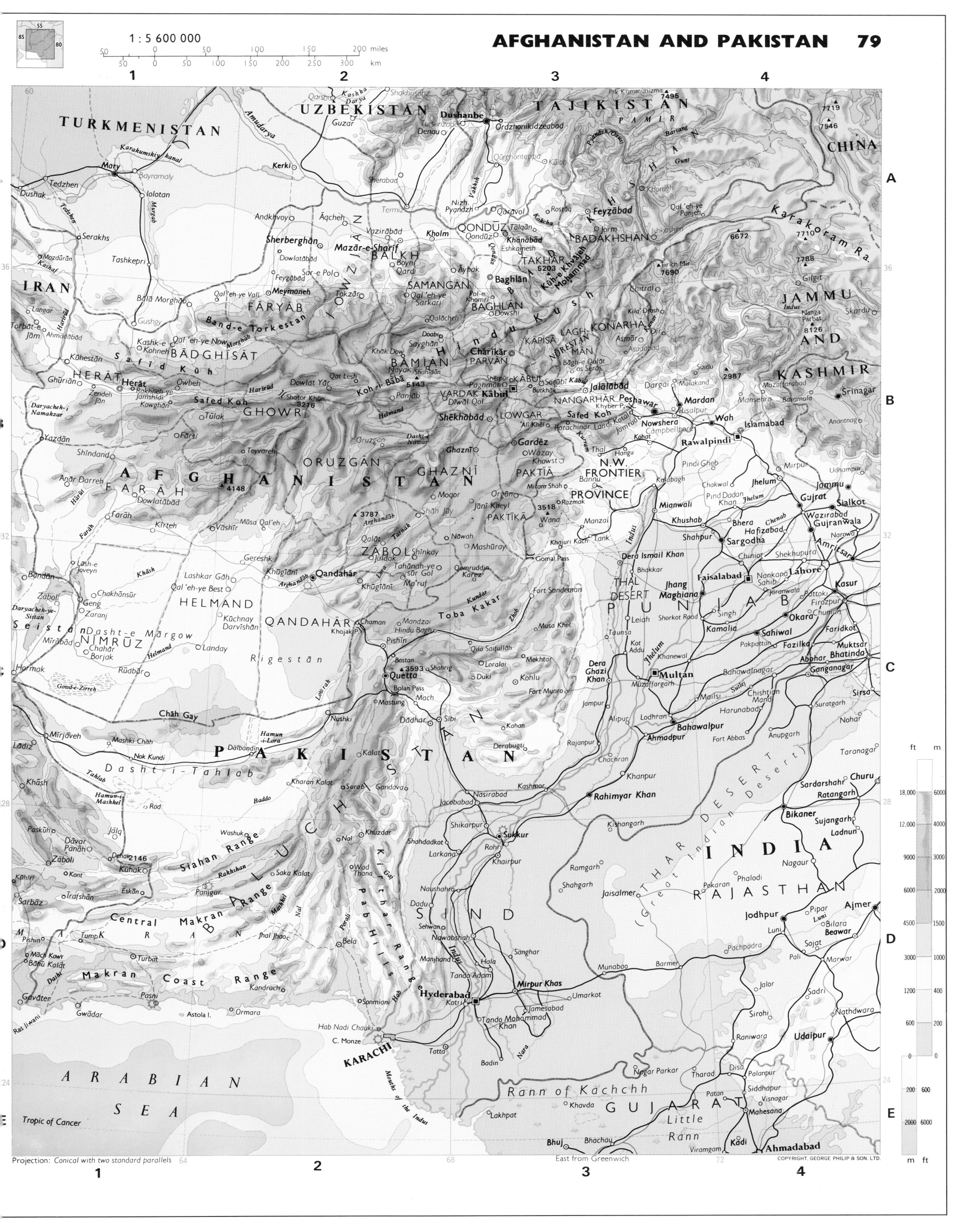

1 : 5 600 000

East from Greenwich

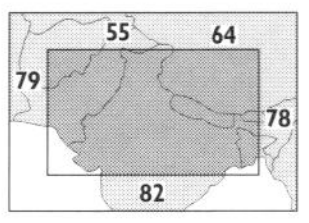

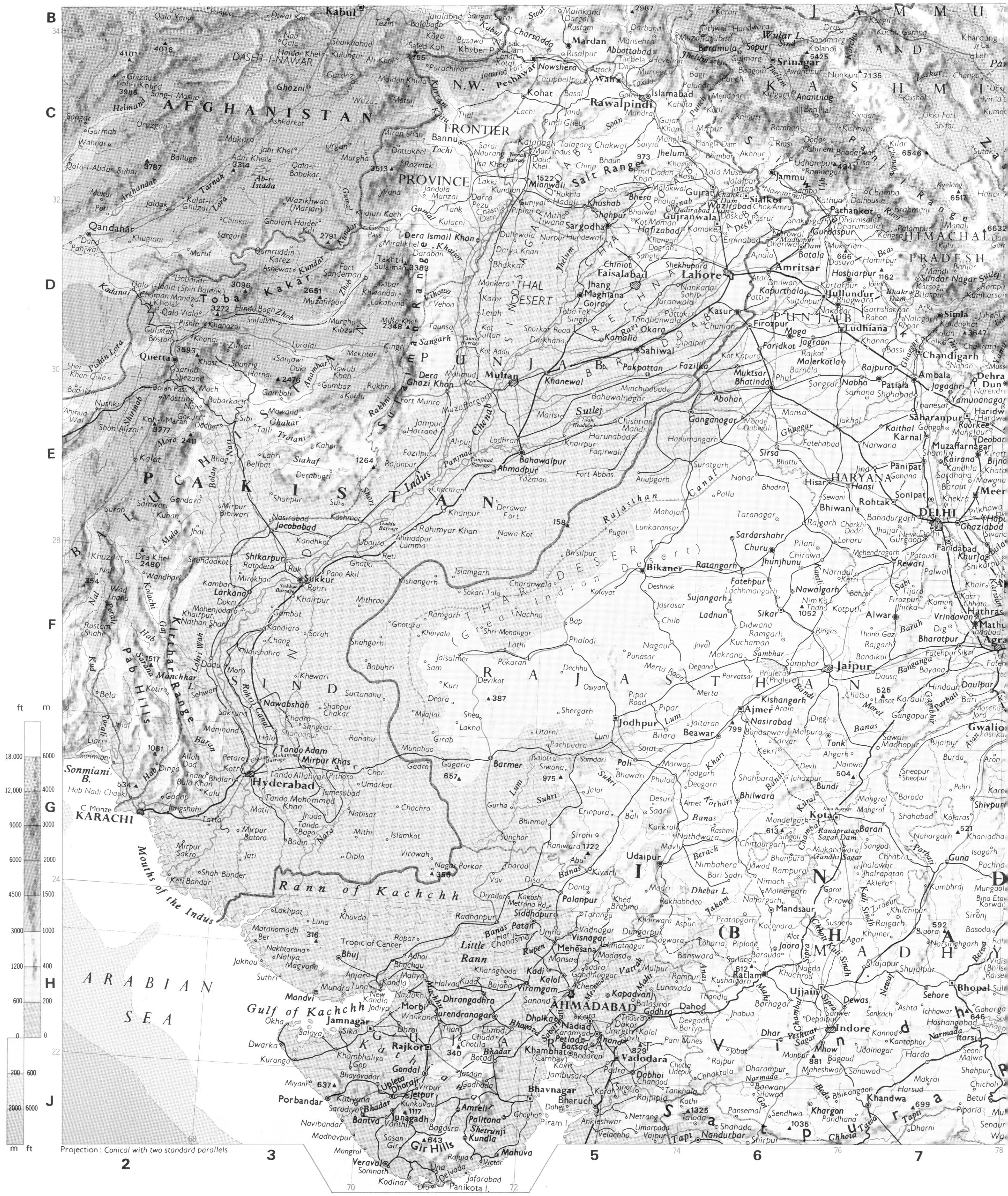

Countries and regions: AFGHANISTAN · PAKISTAN · BALUCHISTAN · SIND · PUNJAB · HARYANA · RAJASTHAN · HIMACHAL PRADESH · JAMMU AND KASHMIR · MADHYA · GUJARAT · N.W. FRONTIER PROVINCE · THAR DESERT (Great Indian Desert) · THAL DESERT · Rann of Kachchh · Little Rann

Water bodies: ARABIAN SEA · Gulf of Kachchh · Mouths of the Indus · Wular L. · Dhebar L.

Major cities: Kabul · Kandahar · Quetta · Karachi · Hyderabad · Sukkur · Larkana · Multan · Lahore · Amritsar · Srinagar · Rawalpindi · Islamabad · Peshawar · Faisalabad · Gujranwala · Sialkot · Jammu · Simla · Chandigarh · Ludhiana · Ambala · Dehra Dun · Delhi · Ghaziabad · Faridabad · Jaipur · Ajmer · Jodhpur · Udaipur · Bikaner · Agra · Mathura · Gwalior · Kota · Ahmadabad · Vadodara · Rajkot · Jamnagar · Bhavnagar · Porbandar · Bhuj · Indore · Ujjain · Bhopal · Khandwa

Rivers: Indus · Chenab · Jhelum · Ravi · Sutlej · Beas · Luni · Banas · Chambal · Narmada · Tapi · Mahi · Sabarmati

Spot heights (selected): 4101 · 4018 · 3995 · 3787 · 3593 · 3383 · 3096 · 2987 · 7135 · 6546 · 6632 · 6517 · 1522 · 973 · 1264 · 158 · 387 · 657 · 975 · 1722 · 637 · 643 · 1117 · 1325 · 1035 · 699 · 881 · 592 · 846 · 504 · 521 · 525 · 799 · 340 · 356 · 316 · 534 · 1061 · 1517 · 2480 · 2411 · 3277 · 3272 · 2651 · 2476 · 2791 · 2348 · 3513 · 3314 · 4755 · 4755

Elevation scale (ft / m):
18,000 / 6000
12,000 / 4000
9,000 / 3000
6,000 / 2000
4,500 / 1500
3,000 / 1000
1,200 / 400
600 / 200
0 / 0
200 / 600
2000 / 6000

Tropic of Cancer

Projection: Conical with two standard parallels

1 : 4 800 000

miles
km
East from Greenwich

JAMMU AND KASHMIR
On same scale as Main Map

CHINA

Kunlun Shan

N.W. FRONTIER PROVINCE

PUNJAB

Karakoram Range

Rawalpindi
Islamabad
Srinagar
Jammu
Gujrat
Sialkot

Nanga Parbat 8126
Masherbrum 7821
Gasherbrum 8068
Karakoram Pass 5575
Aksai Chin

KASHMIR

Zaskar Range

HIMACHAL PRADESH

Leh
Anantnag

Ngangong Kangri

CHINA

Gangdisê Shan

Mapam Yumco

La'nga Co

XIZANG
(TIBET)

Yarlung Zangbo Jiang (Brahmaputra)

Xigazê
Lhazê

Mt. Everest 8848
Makalu 8481
Kanchenjunga 8598

NEPAL

Katmandu
Pokhara

SIKKIM
Gangtok

BHUTAN

Darjiling
Kalimpang

ASSAM

Nanda Devi 7817

Moradabad
Rampur
Bareilly

Lucknow
KANPUR
Faizabad
Gorakhpur

Allahabad
VARANASI
(Banaras, Benares)
Mirzapur

Patna
Bhagalpur
Munger

BIHAR

UTTAR PRADESH

MADHYA PRADESH

Jabalpur
Bilaspur
Raipur
Raurkela
Sambalpur

Ranchi
Jamshedpur

Durgapur
Asansol
Dhanbad

BANGLADESH
DHAKA

Khulna
Jessore
Barisal

Haora
CALCUTTA
Kharagpur

BENGAL

Mouths of the Ganga
The Sandheads

Himalaya

Maikala Range

Vindhya Range

COPYRIGHT. GEORGE PHILIP & SON. LTD.

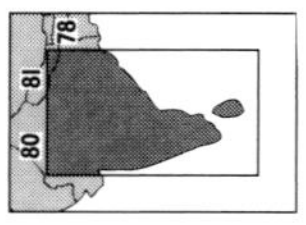

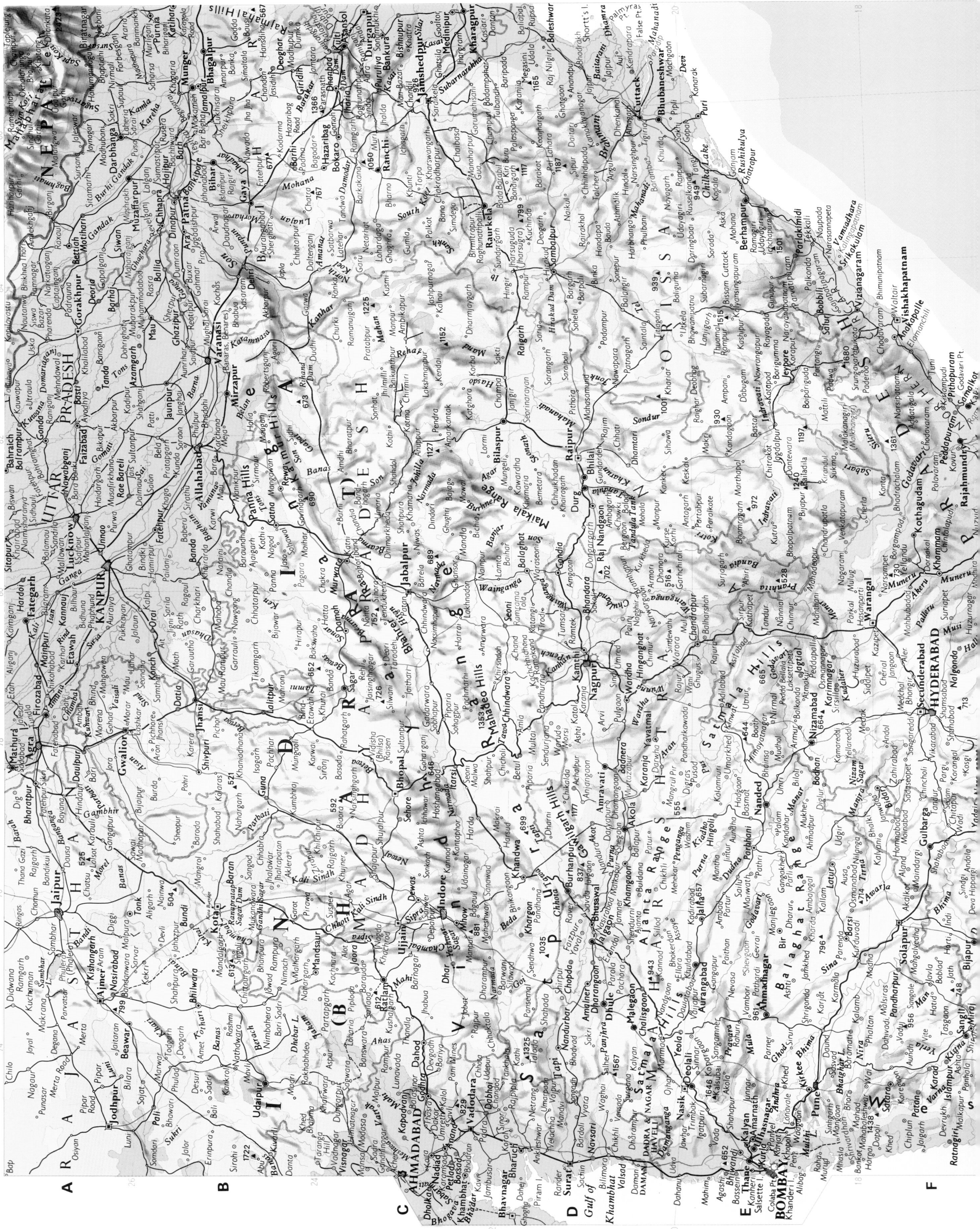
UTTAR PRADESH
MADHYA PRADESH
NEPAL
ORISSA
Gulf of Khambhat
Maikala Range
Mahadeo Hills
KANPUR
Lucknow
Allahabad
Varanasi
Patna
Gaya
Bhagalpur
Jamshedpur
Ranchi
Raurkela
Cuttack
Bhubaneshwar
Puri
Berhampur
Vishakhapatnam
Rajahmundry
HYDERABAD
Secunderabad
Warangal
Nagpur
Jabalpur
Raipur
Bilaspur
Durg
Bhilai
Indore
Ujjain
Bhopal
Gwalior
Jhansi
Agra
Jaipur
Jodhpur
Udaipur
AHMADABAD
Vadodara
Surat
BOMBAY
Pune
Solapur
Aurangabad

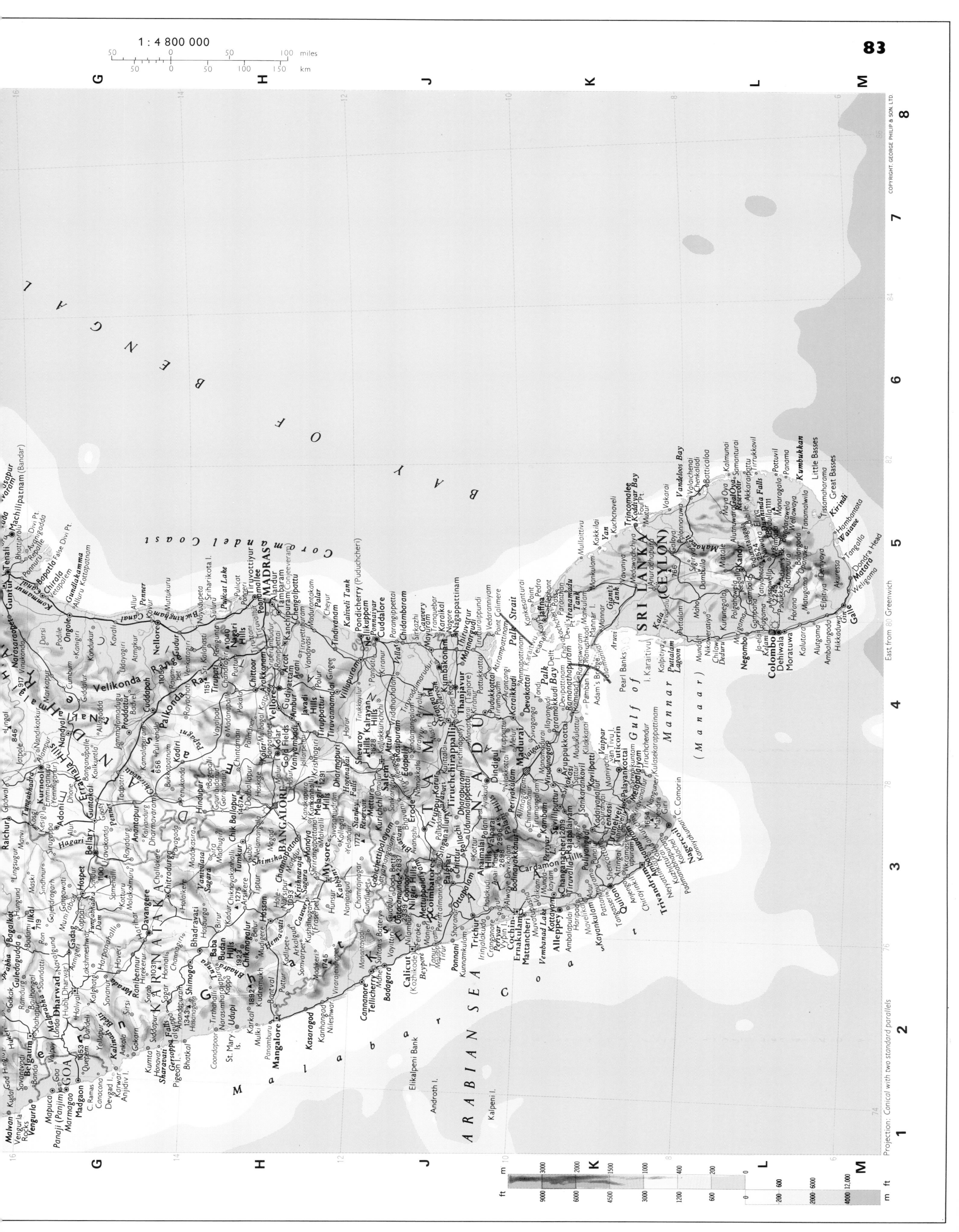

1 : 4 800 000

50   0   50   100 miles

50   0   50   100   150 km

Projection: Conical with two standard parallels

East from 80 Greenwich

BAY OF BENGAL

ARABIAN SEA

Coromandel Coast

Malabar Coast

Gulf of Mannar

Palk Strait

KARNATAKA

GOA

TAMIL NADU

KERALA

SRI LANKA (CEYLON)

MADRAS

BANGALORE

Mysore

Coimbatore

Calicut (Kozhikode)

Cochin

Madurai

Trivandrum

C. Comorin

Colombo

Trincomalee

Mangalore

Belgaum

Dharwad

Hubli

Kurnool

Bellary

Nellore

Guntur

Dondra Head

ft m
9000   3000
6000   2000
4500   1500
3000   1000
1200   400
600   200
0   0
600   200
6000   2000
12,000   4000
ft m

Projection: Conical with two standard parallels

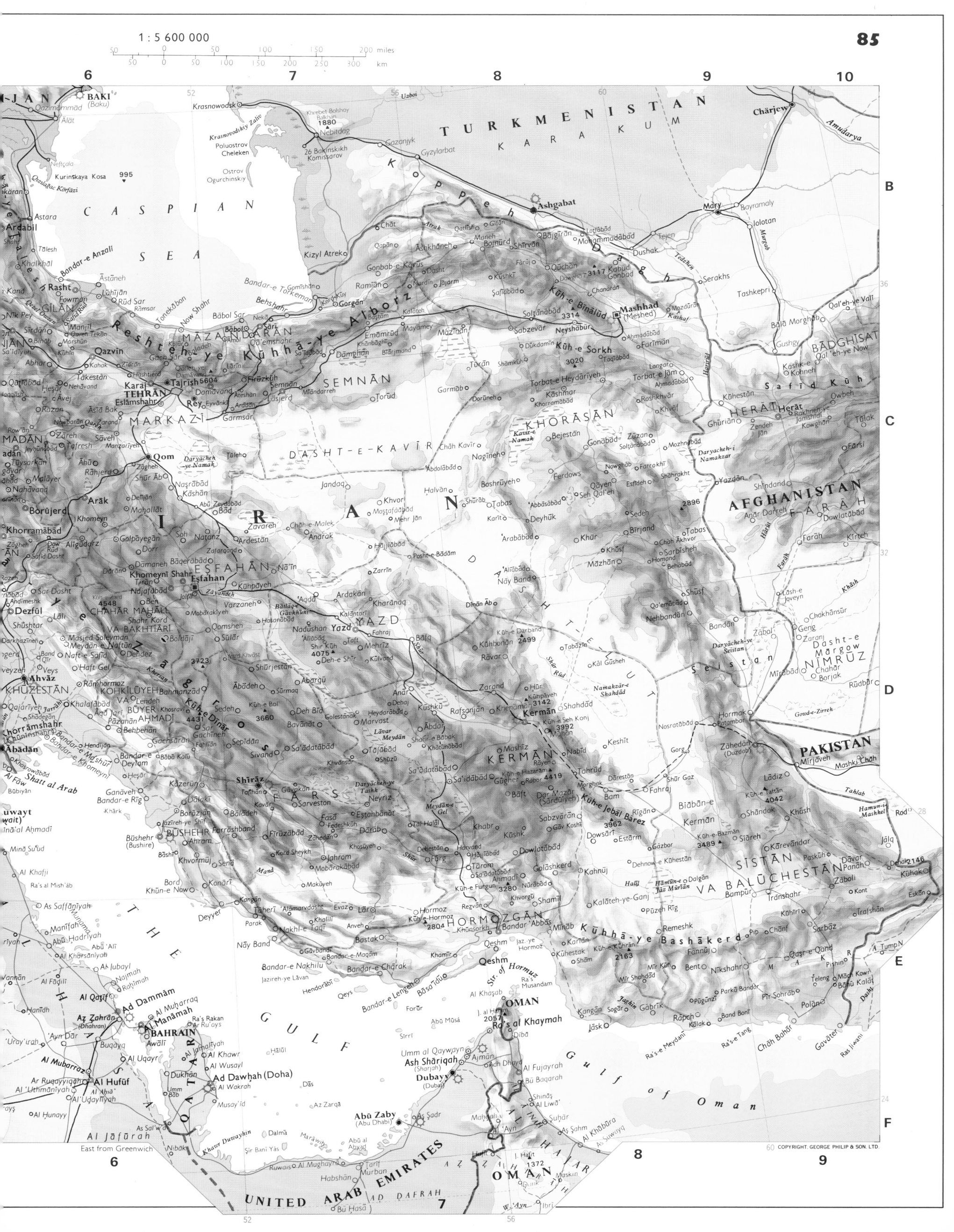

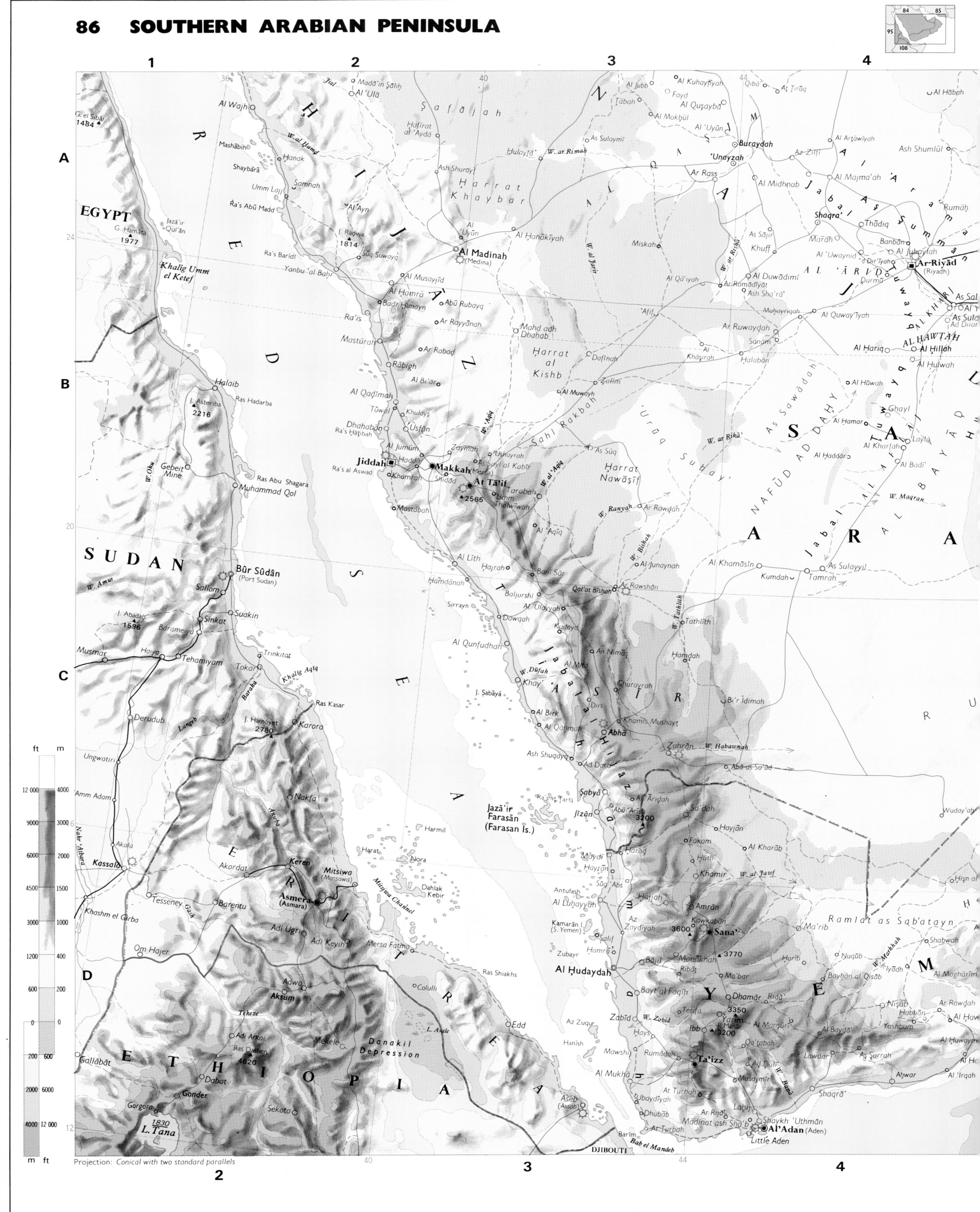

Projection: Conical with two standard parallels

1 : 5 600 000

50 0 50 100 150 200 miles

50 0 50 100 150 200 250 300 km

5 **6** **7** **8**

Abū Hadriyah
Abū 'Alī
Al Khārsaniyah
Al Jubayl
Najmah
Vannān
Al Fādilī
Ra'smah
Al Qatīf
Hanīdh
Ad Dammām
Al Muharraq
Al Manāmah
Uray'irah
Az Zahrān
(Dhahran)
BAHRAIN
'Ayn Dār
Awālī
Al Mubarraz
Buqayq
Al Uqayr
Ar Ruqayyiqah
Al Hufūf
Al Jamaliyah
Umm
Al Abā'
Al 'Uthmāniyah
Dukhān
Al Khawr
Al 'Udayliyah
'arys
As Sal'wā
Al Wusayl
Ad Dawhah
Al Wakrah
Musay'id

Harad
Al Jāfūrah
Nibāk
Khawr Duwayhin
Ruwais Al Mughaytā'
Al Khunn
Jirwān
Bunayyān
Habshān

D Al 'Ubaylah
Aţ Ţuwayrifah

B 'Aziz

I

A Al Qurayni

BIA

L

A

Az Zarqā'
Dās
Abū al
Dalmā
Abyaḍ
Şīr Banī Yās
Marāwih
Tarif
Murban
Abū Zaby
(Abu Dhabi)
Aş Şadr
UNITED ARAB EMIRATES
AD DAFRAH
Bū Hasā
Istaihah
Arādah
JIWA

Nāy Band
Bastak
Bandar-e Nakhilu
Bandar-e Moqām
Jazireh-ye Lāvan
Bandar-e Chārak
Khamīr
Qeshm
Hendorābi
Bandar-e Lengeh
Būso Tdū
Qeshm
Qeys
Forūr
Al Khasab
Str. of Hormuz
Ra's
Musandam
Sirrī
Abū Mūsā
J. al Hartm
2057
Umm al Qaywayn
Ra's al Khaymah
Dibā
Ash Shāriqah
(Sharjah)
Ajmān
Adh Dhayd
Dubayy
(Dubai)
Al Fujayrah
Bū Baqarah
Shinās
Al Liwā'
Maḥdah
Mahūlah
Al 'Ayn
Al Wāhāt al Buraymī
Ḥafit
Al Ayn
Hafit
1372
Dānk
Maskin
Ibrī
Bahlah
Nazwā
3019
Wadim
Izki
Ibrā
2151
Adam
Al Muḍaybī
Şulaym

Gāvbandi
Jaz-ye
Karān
Hormoz
Kūhestak
Jāsk
Qeshm
Shām
Gulf
Ra's-e Meydanī
Ra's-e Tang
Ḥaqat
Shinās
Khāmir
Şuḥār
Al Khābūra
As Suwayq
Wuthm
Alwā
Burkā'
Masqat (Muscat)
Matrah
Maskin
Al Mulāddah
R
Samā'il
Quryyāt
Tiwi
Şūr
Ra's al Hadd
Al Hadd
Tuwi
As Suwayh
Al Kāmil
Al Ashkharah

IRAN
2163
Mīr Kūh
Bent
Nīkshahr
Qasr-e Qand
Pishin
Tashin
Sogar
Jaghin
Gābrīk
Rāpch
Kālak
Band Boni
Pūr Sohrāb
Parkā Bandar
Pīc Sohrāb
Polān
Teleng
Mach Kowr
Bahū Kalāt
Dasht
Gavāter
Ras Jiwani

Gulf of Oman

AL HAJAR
AL HAJAR

Tropic of Cancer

OMAN
'Uwayfi
'Uwayfi
Ghalat
Hayy
Fīlim
Khalūl
Kalbān
Hukkān
Ra's Abū Rasās
Duqm
Khalīj Maşīrah

Ghubbat Sawqirah
Şawqirah

C

Ghalat
Maşīrah
Dawwah
Tūr al-Maşīrah

Jiddat al Harāsīs

Haymā'

W. Sabāh
DAHNĀ
W. Ainak
W. Qitbī
W. Muqshin
Ma'mul

20

24

Z
U
F
Ā
R
Sanāw
Thamarīt
Anzawr
Jabal Samhān
Hadbaram
Ḥāsik
W. Rakhtawt
W. Shu'ayt
W. Arabāh
W. Khudrah
W. Qinab
Thumūd
Bi'r Tamīs
Thamūd
W. Makhrūb
W. Aynak
J. al Qarā'
1678
Ḥabarūt
J. al Qamar
Mirbāt
Şadh
Rakhyūt
Salālah
Al Qurh
Al Faydamī
Al Fatk
Damqawt
W. Jiz'
Al Ghaydah

N

E
Fughmah
Qabr Hūd
Tarīm
Shibām
Şaywūn
Aynāt
Al Qatn
Al Hajarayn
Khuraydah
Hişn al Qarn
Al Ghaydah
2469
Qunfudh
Ghayl Bā Wazīr
Maşna'ah
Burūm
Shuhayr
Al Mukallā
'Ali
Al Hasy
Al Ghayl
'Itāb
Qishn
Sayhūt

Ghubbat al Qamar
Khalfūt
Ra's Fartak

A R A B I A N

S E A

Al Hallānīyah
Al Hāsikīyah
Ra's Nawş
Jazā'ir Khurīyā Murīyā
(Kuria Muria Is.)
(Oman)
Kuria Muria Bay
Al Qibliyah
Ra's ash Sharbatāt

Socotra
(Yemen)
Ra's Layht
Qalansīyah
Ra's Khalaf
Timareh
Ra's Māmī
Ra's Shu'b
Qādib
Sigra
Fahr

'Abd al Kūri
The Brothers
Ra's Qaţānan

A

B

D

12

16

COPYRIGHT GEORGE PHILIP & SON LTD

56

52

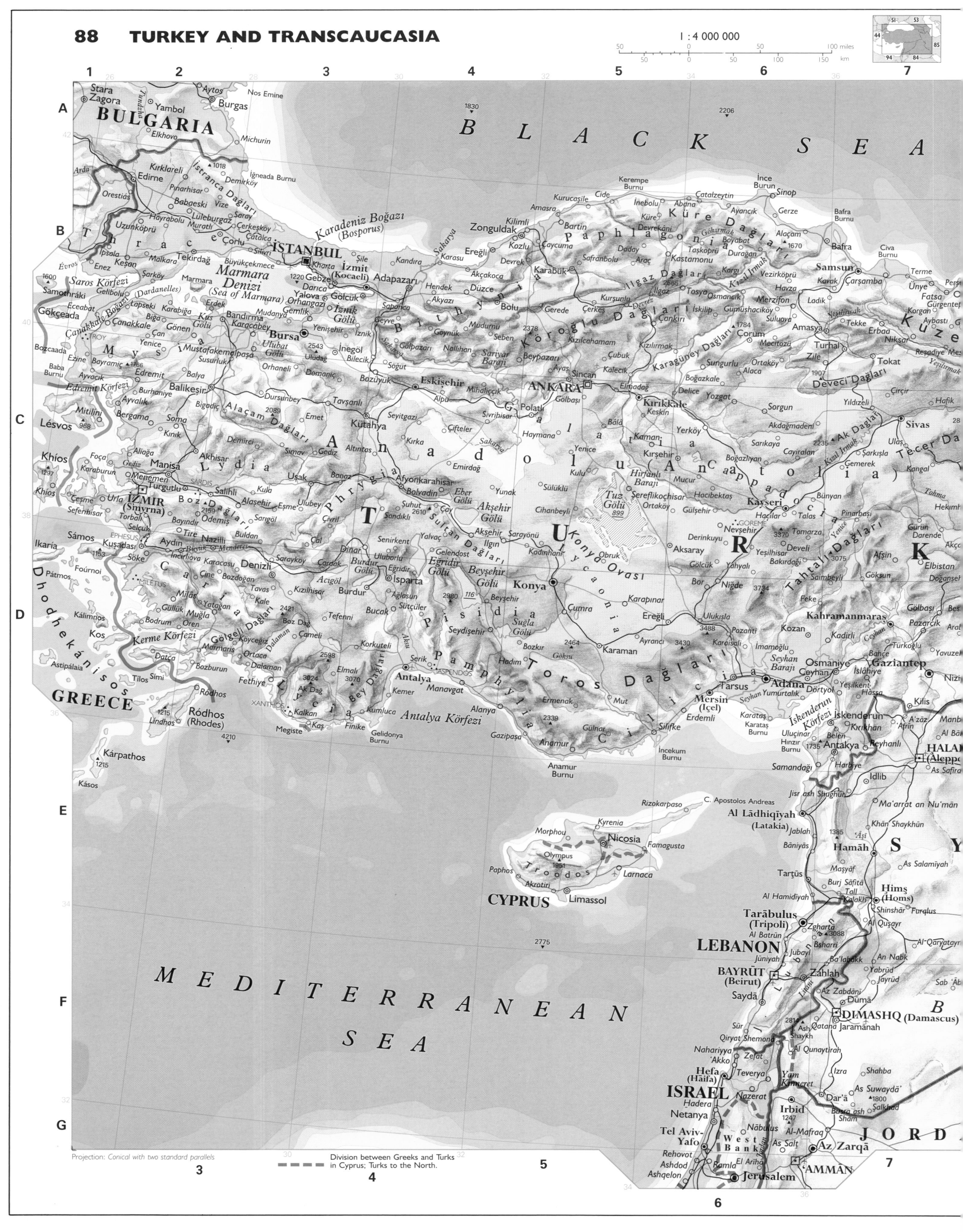

1 : 4 000 000

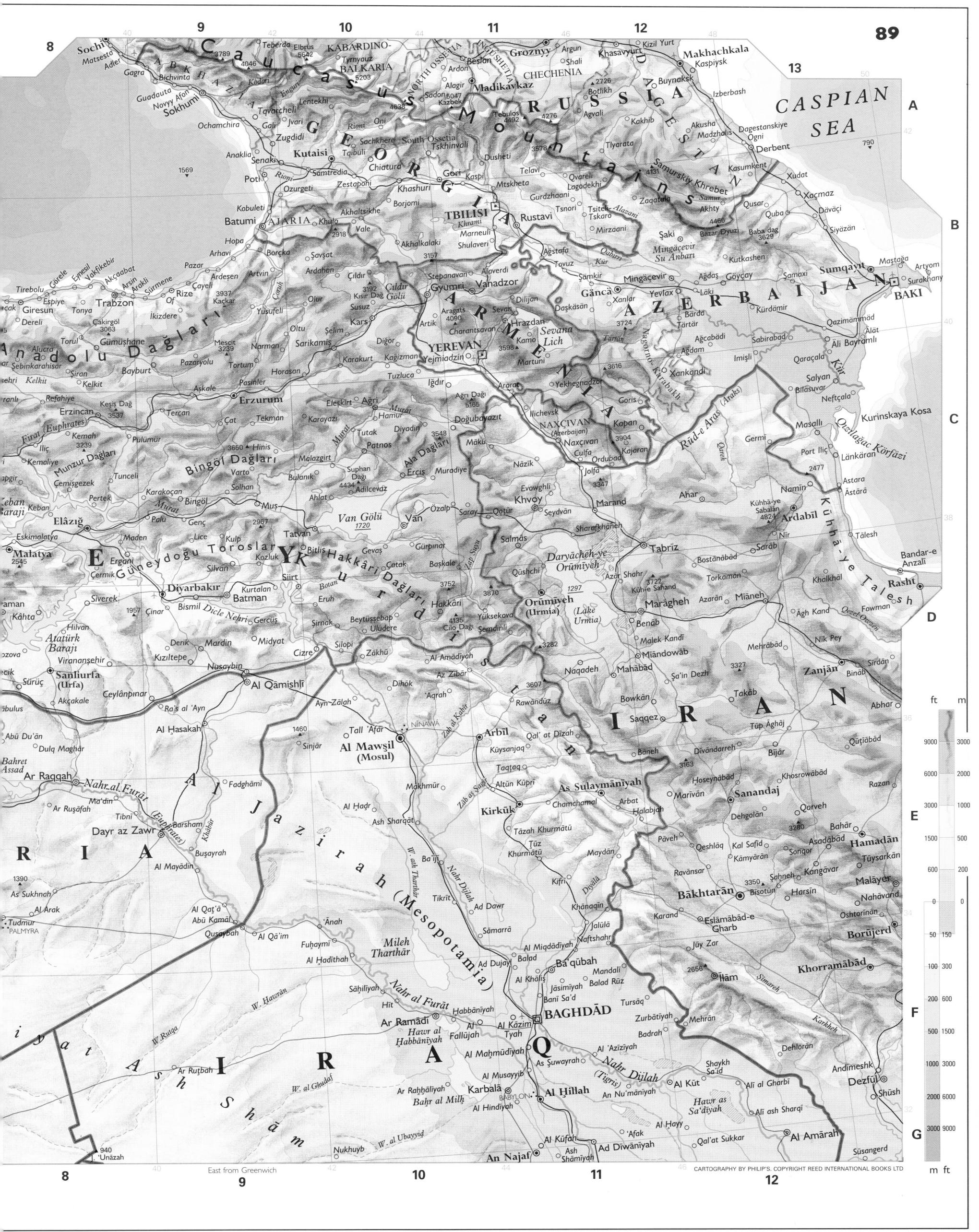

1 : 12 000 000

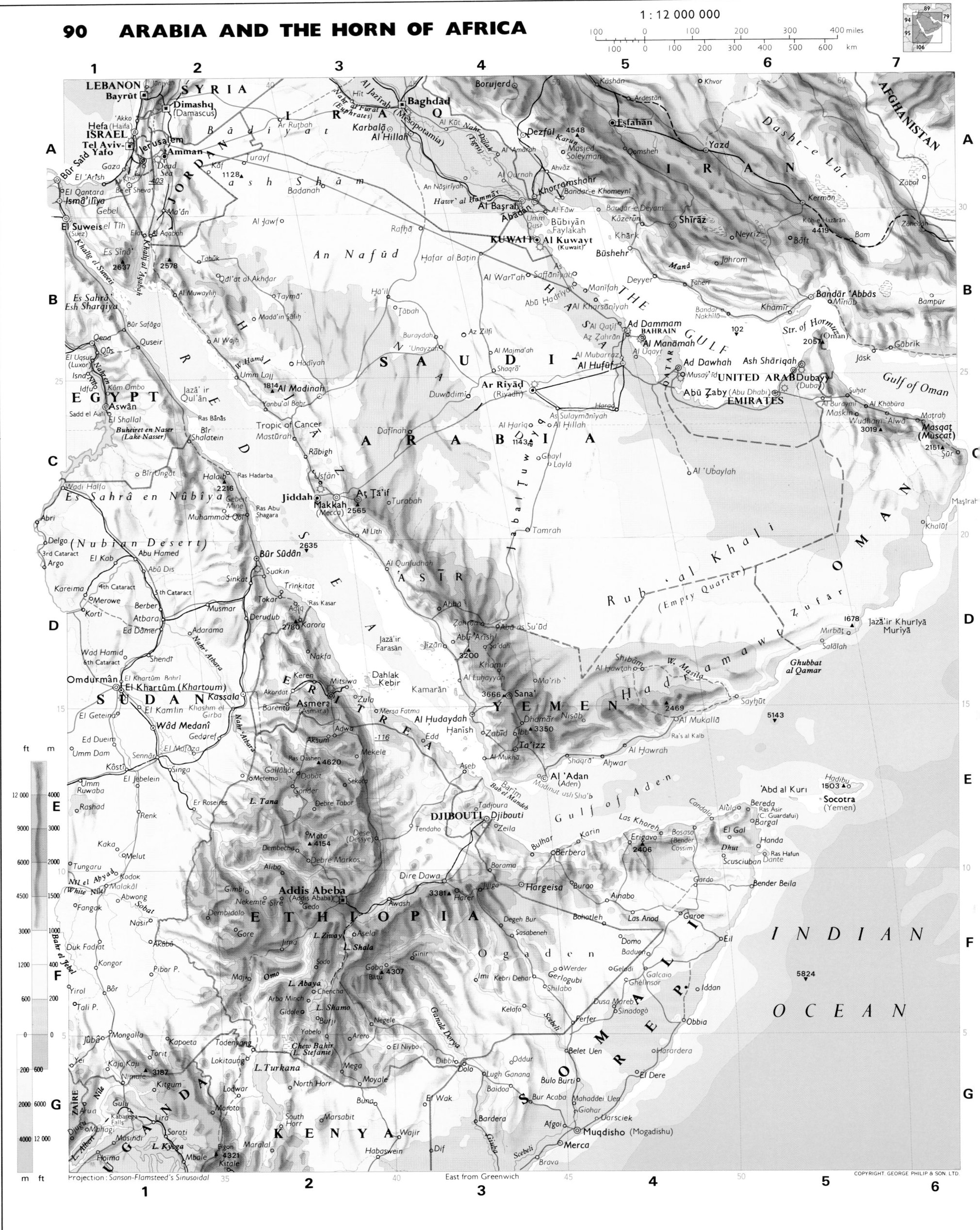

Projection : Sanson-Flamsteed's Sinusoidal

East from Greenwich

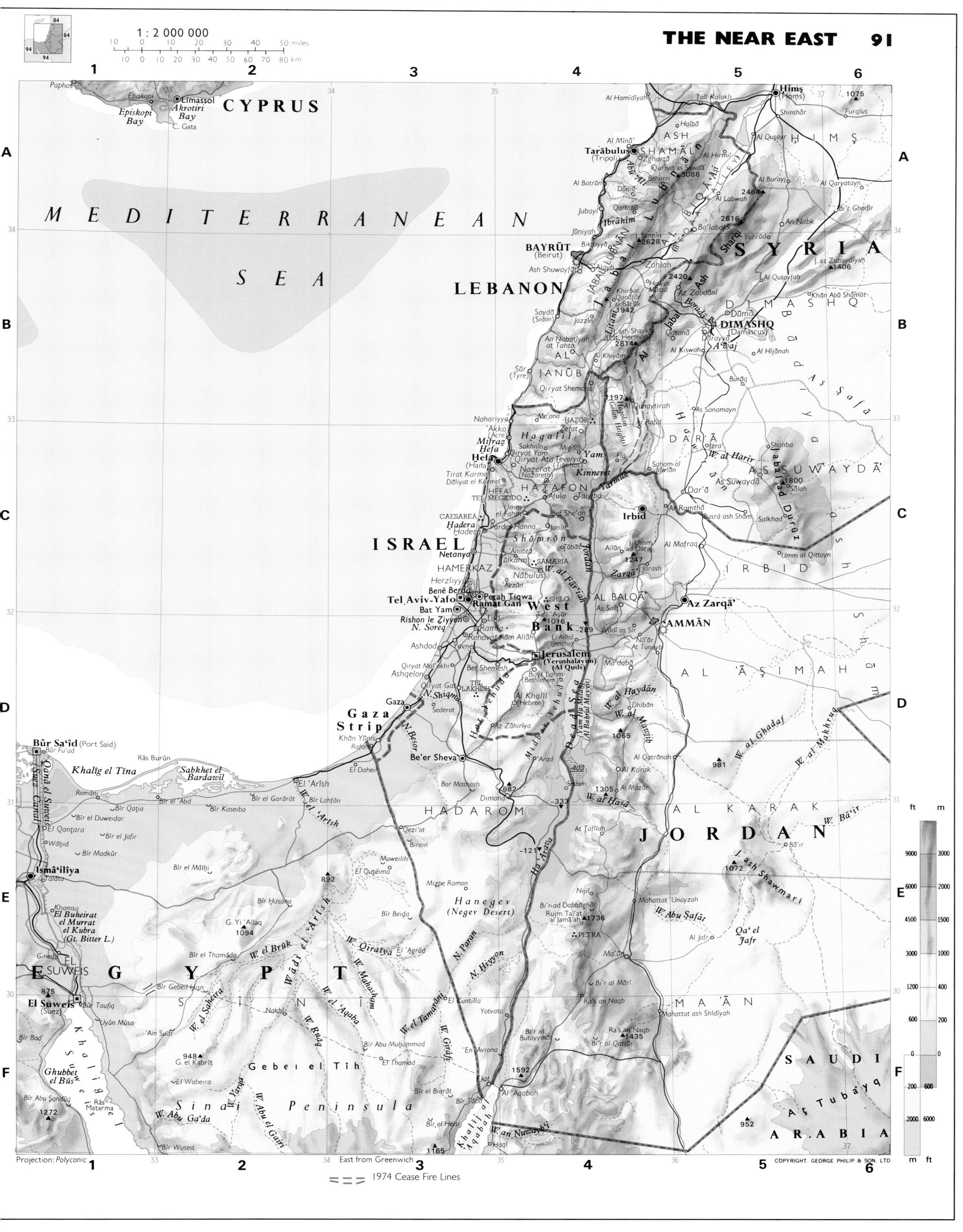

1 : 2 000 000

10   0   10   20   30   40   50 miles
10 0 10 20 30 40 50 60 70 80 km

1   **2**   **3**   **4**   **5**   **6**

CYPRUS

Paphos
Limassol
Episkopi
Bay
Akrotiri
Bay
C. Gata

M E D I T E R R A N E A N

S E A

Al Hamidiyah   Tall Kalakh   Hims (Homs)   1075   Furqlus
Halbā   Shinshār
Al Minā'   ASH   Al Qusayr   H I M Ş
Tarābulus (Tripoli)   SHAMĀL   Al Hirmil   Al Qaryatayn
Qurnat as Sawdā' 3088   Al Burayj   Bi'r Ghadīr
Al Batrūn   Dūmā   2464
Jubayl   Qartabā   Al Labwah
Ibrāhim   Ba'labakk   An Nabk
Jūniya   2616   Yabrūd
BAYRŪT (Beirut)   2628   Sannin   Shargī   SYRIA
Biklayyā   2420   J. az Zubaydiyah 1406
Ash Shuwayfāt   Zahlah   Az Zabdānī   Al Qutayfah
Alqylā   Khirbat   Hawsh   Khān Abū Shāmat
LEBANON   Qanāfar   Mūssā   Dūmā   Dārayyā   Damascus
Saydā (Sidon)   Jazzīn   al Barūk 1942   DIMASHQ (Damascus)
Qaṭanā   A'wāj   Al 'Asārī
An Nabatīyah   Ash Shaykh   Al Hijānah
at Tahtā   (Mt. Hermon) 2814   Al Kiswah
AL   Al Khiyām   Bi'r Qaṭṭa
Sūr (Tyre)   JANŪB   Buraq   S S a f ā
Qiryat Shemona   1197   Al Quṇayṭirah   As Sanamayn
Golan   DARĀ
Nahariyya   Me'ona   HAZOR   Heights   W. al Harīr   Shahba
Akko (Acre)   Zefat   Ar Rafid   AS SUWAYDĀ'
Hagalil   Sakhnīn   Miġdal   Fīq   Saham al   As Suwaydā
Hefa   Qiryat Yam   Qiryat Ata Teverya   Yam   Jawlān   1800
Mifraz   (Tiberias)   Kinneret   Yarmūk   Dar'ā   Sālah
Hefa (Haifa)   Nazerat (Nazareth)   As Suwaydā'
Tirat Karmel   HAZAFON   Afula   Tayibu   Busrā ash Shām   Salkhad
Dāliyat el Karmel   TEL MEGIDDO   Bet She'an   An Ramthā   IRBID
CAESAREA   Umm   Jantn   Irbid   Ailūn   Umm al Qittayn
Hadera   el Fahm   Shōmrōn   'Abābū   1247   Al Mafraq   Umm al Qittayn
Hadera   Pardes Hanna   W. al Fār'ah   Ajlūn   Jarash
ISRAEL   'Anabta   Zarqā'
Netanya   Tulkarm   SAMARIA   Jordan
Herzliyya   Nābulus   AL BALQĀ   Az Zarqā'
HAMERKAZ   As Salt   AMMAN
Benē Beraq   Tall 'Āṣūr   Na'ūr
Tel Aviv-Yafo   Petah Tiqwa   1016   Wādī as Sīr
Bat Yam   Ramat Gan   SHILO   289   Mādabā
Rishon le Ziyyon   West   Wādī al Mawjib
N. Soreq   Lod   Bank
Rehovot   Ramla   Rām Allāh   Al 'Arēd   W. al Haydān
Ashdod   Avne   (Jericho)   Dhības
Jerusalem (Yerushalayim)   W. al
Qiryat Mal'akhi   (Al Quds)   Bayt Lahm   Mūjib
Ashqelon   Bet Shemesh   (Bethlehem)   1065   W. al Ghadaf   W. al Makhruq
TEL   Al Karak
Qiryat Gat   LAKHISH   Al Khalīl (Hebron)   981
Gaza   N. Shiqma   Al Qaṭrānah
Gaza   Pāz Zāhiriya   1305   Al Mazār
Strip   Sederot   AL KARAK
Khān Yūnis   N. Besor   Dimona   Sdom   W. al Hasa   1072   'ash Shawmari
Rafah   El Daheri   682   -403   Bā'ir   W. Bā'ir
Be'er Sheva   Bor Mashash   JORDAN
Bûr Sa'îd (Port Said)   Arad   At Tafīlah   Bā'ir
Bûr Fu'ad   Rās Burûn   Sabkhet el   El 'Arîsh   HADAROM   -333   MA'ĀN
Khalîg el Tîna   Bardawil   Bīr el Garārāt   Bīr Lahfān   Dimona   At Tafīlah
Romāni   Bīr el 'Abd   -121   AL
Bir Qaṭia   Bīr Kaseiba   Ha Arava
El Qanṭara   Bīr el Duweidar   892   Mizpe Ramon   Nijil
El Qantara   Bīr el Jafir   El Quseima   Ma'ān
Ismâ'ilîya   Wāḥid   Bīr Madkūr   Hanegev   Bi'r ad Dabbāghāt
Khamsa   (Negev Desert)   Ruim Tal'at   W. Abu Safāt
El Buheirat   Bīr Hasana   Bīr Beiḍa   al Jamā'in 1736   Qa' el
el Murrat   G. Yi 'Allaq   PETRA   Jafr
el Kubra   1094   W. Qiratya   El 'Agrūd
(Gt. Bitter L.)   Bīr el Thamāda   N. Paran   N. Hiyyon
EL SUWEIS   El Suweis (Suez)   875   W. el Brūk   Ma'ān
Gineifa   S   W. el Ḥasana   Rā's an Naqb
El Suweis (Suez)   'Ain Sudr   Nakhl   El Kuntilla   Yotvata   MA'ĀN
Bīr Bad'   Uyūn Mūsa   W. el 'Aqaba   Mahattat ash Shidiyah
Ghubbet   Bīr Gebel Husn   1435
el Būs   G. el Kabrît   Gebel el Tîh   Ra's an Naqb
948   Bīr Abu Muḥammad   'En 'Avrona   Bi'r al Mārī
Bīr Abu Sandûq   1272   El Wabeira   El Thamad   952
Rās Matarma   W. Varqa   Sinai Peninsula   Bi'r al   SAUDI
Bīr Wuseit   W. Abu el Gaṭ   Bīr el Biarāt   Qaṭṭār   Aṭ Tubāyq
W. Abu Ga'da   1592   A R A B I A
Bīr el Hesi   Elat
1165   Al 'Aqabah   W. an Nuweybi
Khalīj al 'Aqaba   W. Haql

Projection: Polyconic   East from Greenwich   COPYRIGHT. GEORGE PHILIP & SON. LTD.

= = = 1974 Cease Fire Lines

ft   m
9000   3000
6000   2000
4500   1500
3000   1000
1200   400
600   200
0
200   600
2000   6000
m   ft

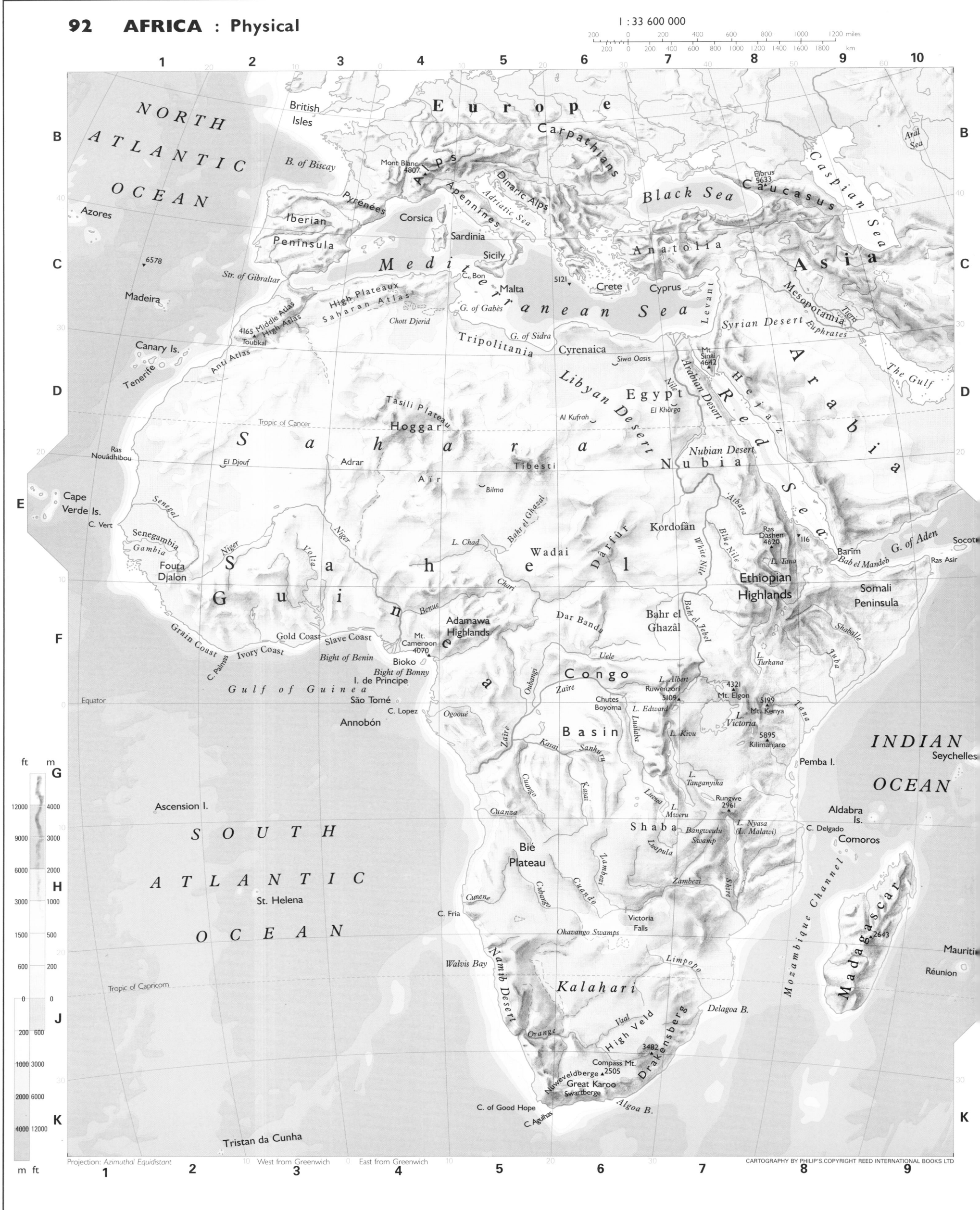

1 : 33 600 000

NORTH
ATLANTIC
OCEAN

Europe

British Isles

B. of Biscay

Carpathians

Alps

Mont Blanc
4807

Pyrénées

Apennines

Dinaric Alps

Adriatic Sea

Black Sea

Caucasus

Elbrus
5633

Caspian Sea

Aral Sea

Azores

Iberian
Peninsula

Corsica

Sardinia

6578

Madeira

Str. of Gibraltar

High Plateaux

Saharan Atlas

Sicily

C. Bon

Malta

Crete

Cyprus

5121

Anatolia

Asia

Mesopotamia

Tigris

Canary Is.

Tenerife

4165 Middle Atlas
High Atlas
Toubkal

Anti Atlas

Chott Djerid

Mediterranean Sea

G. of Gabès

Tripolitania

G. of Sidra

Cyrenaica

Siwa Oasis

Mt. Sinai
4642

Syrian Desert

Euphrates

Levant

Hejaz

The Gulf

Tropic of Cancer

Tasili Plateau

Hoggar

Sahara

Libyan Desert

Egypt

Al Kufrah

El Khârga

Nile

Arabian Desert

Red Sea

Arabia

Ras Nouâdhibou

El Djouf

Adrar

Aïr

Tibesti

Nubian Desert

Nubia

Ethiopian
Highlands

Cape
Verde Is.

C. Vert

Senegal

Bilma

L. Chad

Bahr el Ghazal

Wadai

Darfûr

Kordofân

Ras Dashen
4620

116

L. Tana

Barîm
Bab el Mandeb

G. of Aden

Ras Asir

Socotra

Senegambia

Gambia

Niger

Volta

Niger

Fouta Djalon

S   u   d   a   n

Benue

Chari

White Nile

Blue Nile

'Atbara

Somali
Peninsula

Grain Coast

Ivory Coast

C. Palmas

Gold Coast

Slave Coast

Bight of Benin

Mt. Cameroon
4070

Bioko

Adamawa
Highlands

Dar Banda

Uele

Bahr el Jebel

Bahr el
Ghazâl

Shabelle

Juba

Bight of Bonny

I. de Principe

São Tomé

C. Lopez

Ogooué

Gulf of Guinea

Equator

Annobón

Zaïre

Oubangi

Congo

Basin

Zaïre

Chutes Boyoma

L. Albert

Ruwenzori
5109

L. Edward

L. Kivu

Lualaba

L. Turkana

4321

Mt. Elgon

5199

Mt. Kenya

L. Victoria

5895
Kilimanjaro

Tana

INDIAN

OCEAN

Seychelles

Ascension I.

SOUTH

ATLANTIC

St. Helena

OCEAN

Kasai

Sankuru

Kasai

Cuango

Cuanza

L. Tanganyika

Luena

L. Mweru

Luapula

Lukuga

Pemba I.

Aldabra
Is.

Comoros

C. Delgado

Shaba

Bangweulu
Swamp

L. Nyasa
(L. Malawi)

Rungwe
2961

Bié
Plateau

Cuando

Cubango

Zambezi

Zambezi

Shire

Mozambique Channel

Madagascar

2643

Mauritius

Réunion

C. Fria

Cunene

Victoria
Falls

Tropic of Capricorn

Walvis Bay

Namib Desert

Okavango Swamps

Kalahari

Limpopo

Delagoa B.

Orange

Vaal

High Veld

Drakensberg

Compass Mt.
3482

C. of Good Hope

Niuweveldberge
Swartberge

Great Karoo
2505

Algoa B.

C. Agulhas

Tristan da Cunha

ft   m

12000   4000

9000   3000

6000   2000

3000   1000

1500   500

600   200

0   0

200   600

1000   3000

2000   6000

4000   12000

m ft

Projection: *Azimuthal Equidistant*

West from Greenwich

East from Greenwich

CARTOGRAPHY BY PHILIP'S. COPYRIGHT REED INTERNATIONAL BOOKS LTD

1 : 33 600 000

Projection: Azimuthal Equidistant

West from Greenwich   East from Greenwich

CARTOGRAPHY BY PHILIP'S.COPYRIGHT REED INTERNATIONAL BOOKS LTD

● Dakar Capital Cities

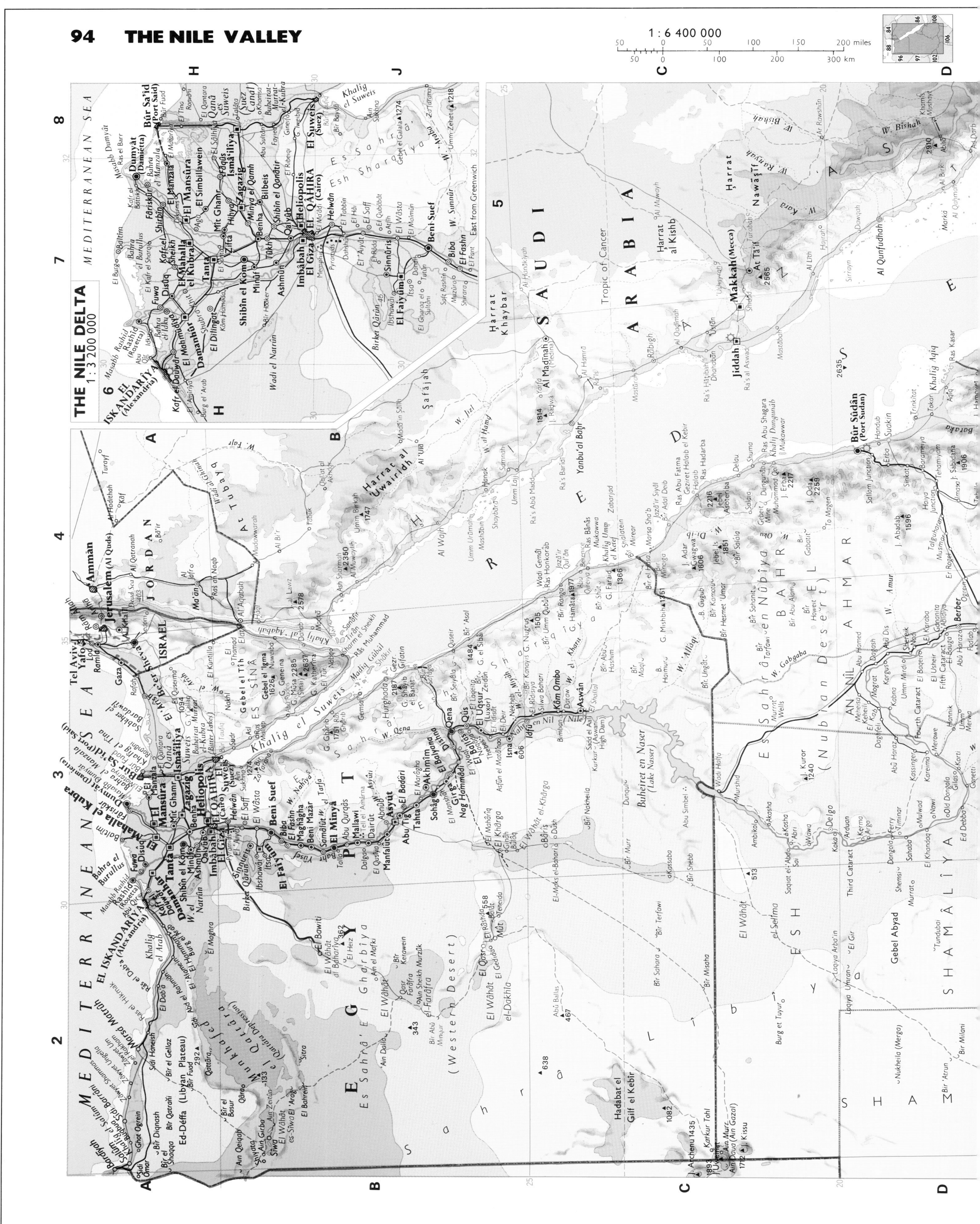

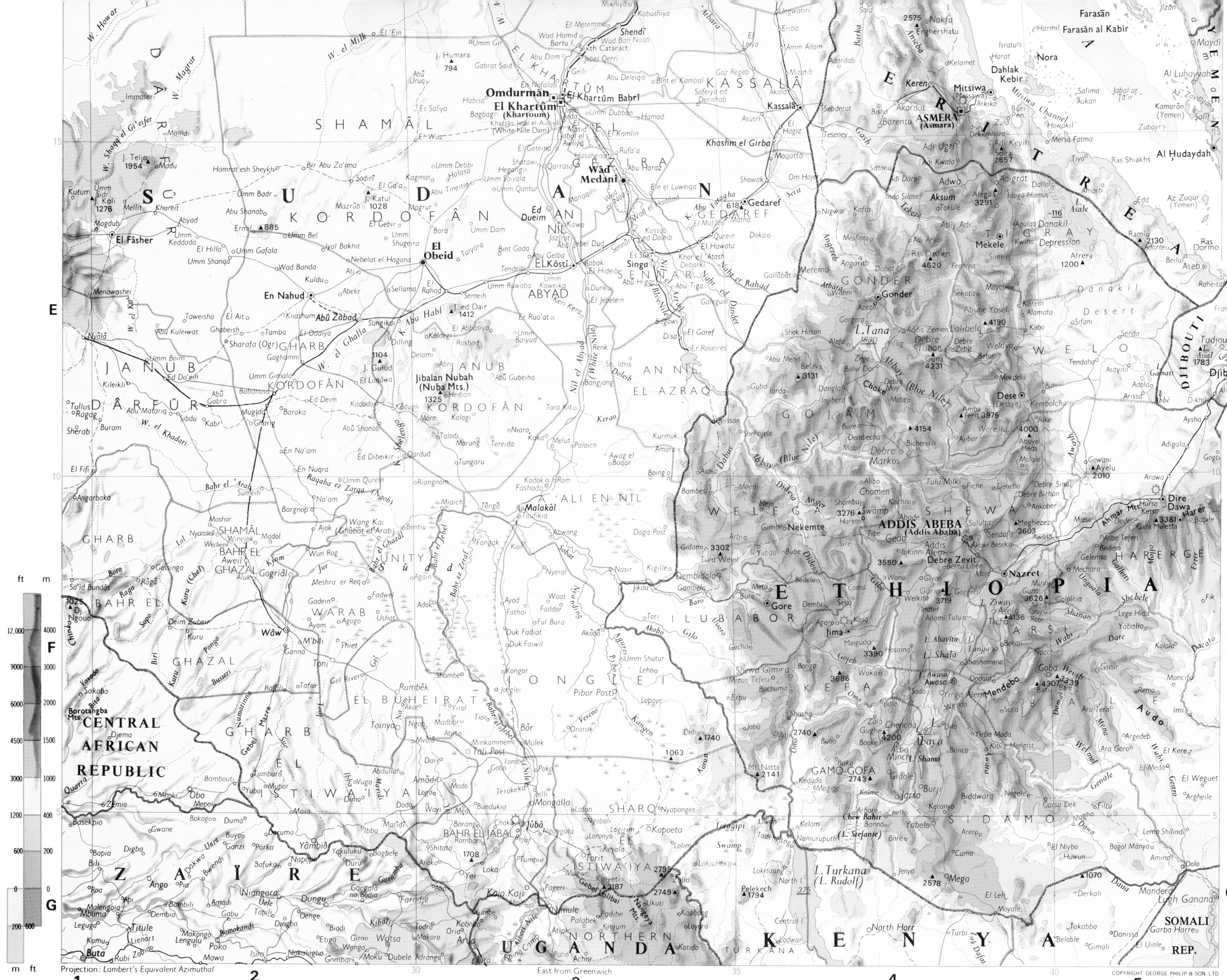

YEMEN

ERITREA

ETHIOPIA

DJIBOUTI

SOMALI REP.

SUDAN

ASH SHAMÂL

DÂR FÛR

KORDOFAN

KASSALA

GEDAREF

EN NIL

EL AZRAQ

AN NIL

KHARTÛM

EL KHARTÛM (Khartoum)

Omdurmân

El Obeid

El Fâsher

En Nahud

Kassala

Gedaref

Wâd Medani

Shendi

ADDIS ABEBA (Addis Ababa)

ASMERA (Asmara)

Mitsiwa

Keren

Gonder

Mekele

Aksum

Dese

Nazret

Dire Dawa

Dijbouti

L. Tana

Chóke

Abbay (Blue Nile)

Bahr el Jebel

Bahr el Zeraf

Bahr el Ghazal

L. Turkana (L. Rudolf)

KENYA

UGANDA

ZAÏRE

CENTRAL AFRICAN REPUBLIC

GAMO GOFA

SIDAMO

KEFFA

ILUBABOR

WELEGA

SHEWA

WELO

TIGRAY

GONDER

GOJAM

BÂHR EL GHAZAL

EL BUHEIRAT

WARAB

UNITY

SHARQ

JÛNGOLÎ

'ALÎ EN-NÎL

AL WÂHÂT

East from Greenwich

Projection: Lambert's Equivalent Azimuthal

COPYRIGHT GEORGE PHILIP & SON LTD

ft
12,000
9000
6000
4500
3000
1500
1000
400
200
0

m
4000
3000
2000
1500
1000
600
400
200
0
200-600

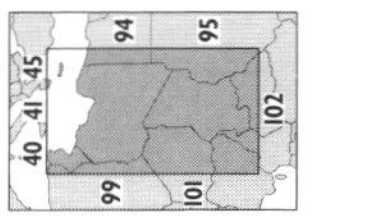

EGYPT

LIBYA

ALGERIA

TUNISIA

MEDITERRANEAN SEA

Sicilia

MALTA

CATÁNIA

Siracusa

Tripoli (It.)

Pantelleria

Lampedusa

Kríti

Iráklion

Dodecaneso

Kikládhes

Cyrenaica

Tripolitania

Tarābulus (Tripoli)

Banghāzī (Benghazi)

DARNAH

AL JABAL AL AKHDAR

BANGHĀZĪ

AJDĀBIYĀ

TUBRUQ

AL KUFRAH

Khalij Surt (Gulf of Sidra)

Sarir Calanscio

Sarir Nerastro

AL HARUJ AL ASWAD

Sahara

SURT

MISRĀTAH

AL KHUMS

SAWFAJJIN

YAFRAN

GHARYĀN

GHUDĀMIS

Ghudāmis

Al Hammādah al Hamrā

AWBĀRĪ

SĀBHĀ

Sābhāh (Sebha)

ASH SHĀTI

MARZŪQ

Marzūq

Idehan Marzūq

Idehan Awbārī

MURZUQ

TARHŪNAH

ZLĪTAN

Mesach Mellet

Plateau du Tinrhert

Grand Erg Oriental

Tropic of Cancer

TUNIS

Bizerte

Sfax

Sousse

Monastir

Kairouan

Gabès

G. de Gabès

Djerba

Zarzis

Médenine

Tatahouine

Nālūt

CONSTANTINE

Skikda

'Annaba

El Oued

Chott Melrhir

Chott el Djerid

Tozeur

Nefta

Gafsa

2328

1338

2254

Adrar 2254

1428

LIBYA
1. An Nuqāt Al Khams
2. Az Zāwiyah
3. Al 'Azīzīyah
4. Tarābulus
5. Al Khums
6. Tarhūnah
7. Zlītan
8. Misrātah

Siwa

El Wāhāt

Ed-Deffa

Libyan Plateau

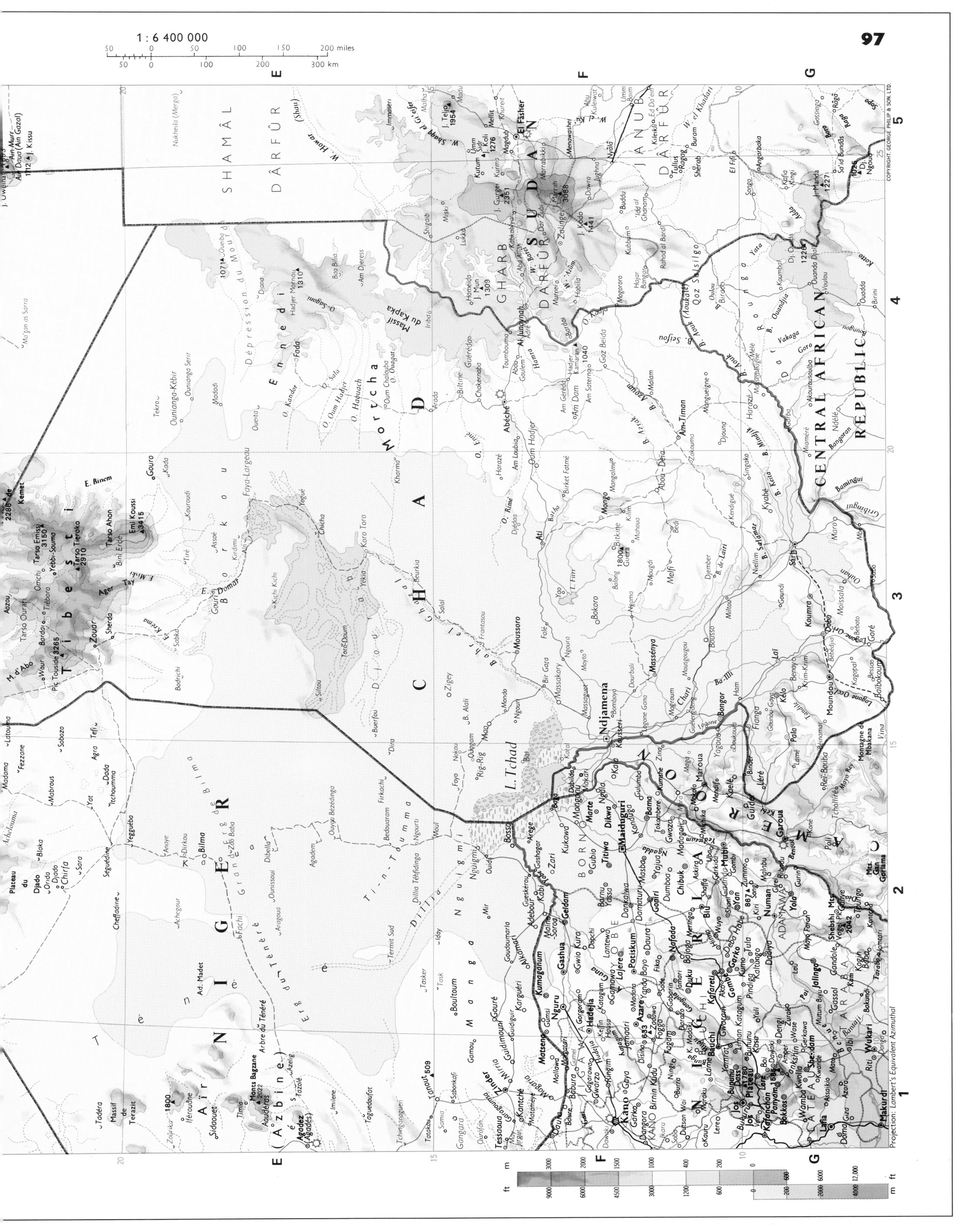

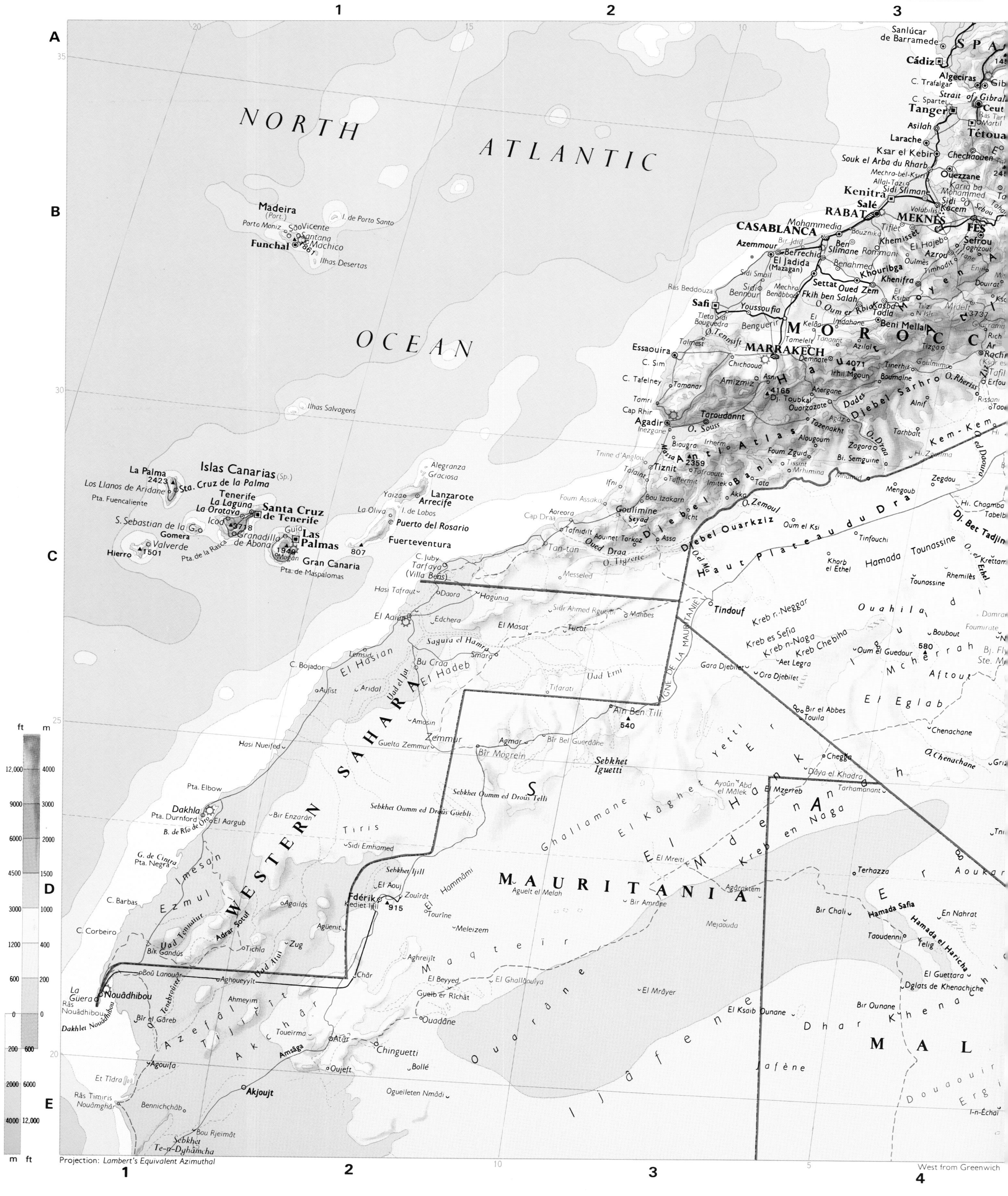

A

20
15
35

N O R T H

A T L A N T I C

O C E A N

Madeira
(Port.)
Porto Moniz I. de Porto Santo
São Vicente Santana
Machico
Funchal Faial

Ilhas Desertas

B

30

Ilhas Salvagens

La Palma
2423
Los Llanos de Aridane Sta. Cruz de la Palma
Pta. Fuencaliente
Islas Canarias (Sp.)
Tenerife
La Laguna **Santa Cruz**
La Orotava **de Tenerife**
Icod 3718
S. Sebastian de la G. Granadilla
Gomera de Abona
Valverde 1940
Hierro 1501 Pta. de la Rasca
Pta. de Maspalomas
Gran Canaria

Alegranza
Graciosa
Yaizao **Lanzarote**
Arrecife
La Oliva I. de Lobos
Puerto del Rosario
Fuerteventura
807

C

Cap Draa
Aoreora
Tafnidilt

Hasi Tafraut
oDaora Hagunia
El Aaiún
Edchera
Lemsian
C. Bojador El Hasian
Aridal Uad el Jut
Aufist
Zemmur Amasin
Hasi Nueifed
Guelta Zemmur

Saguia el Hamra
Smara
El Masat
Sidi Ahmed Rguetbi o Mahbes
Tucef
Uad Erni
Tifarati

C. Juby
Tarfaya
(Villa Bens)
Messeled

WESTERN SAHARA

25

D

Pta. Elbow
Dakhla
Pta. Durnford El Aargub
B. de Río de Oro
Bir Enzarân
G. de Cintra
Pta. Negra Sidi Emhamed
Tiris
Imesain
C. Barbas
Ezmul
Adrar Sotuf Agailas
Tichla Zug
C. Corbeiro Aguenit
Uad Tenualur
Bir Gandús
Aghoueyyît Châr
La Güera Bou Lanouâr
Nouâdhibou
Râs Ténérbrunil.
Nouâdhibou Bîr el Gâreb
Dakhlet Nouâdhibou Ahmeyim
Azefâl oOuadâne
Tijirît Tueirma
Akchâr Atâr
Ras Timiris Amsâga
Nouâmghâr Agouifa oOujeft
Bennichchâb Oguellen Nmâdi

Sebkhet Oumm ed Droûs Telli
Sebkhet Oumm ed Droûs Guebli
Fdérik
Kediet Ijill Zouîrât
915 Tourine
Meleizem
Aguelt el Melah Bîr Amrâne
Mejaûda
Maqteîr
Aghreijît
El Beyyed El Ghallâouîya
Guelb er Richât
Ouadâne
Chinguetti Bollé

MAURITANIA

Ghallamane

Sebkhet
Iguetti
Sebkhet Ijill
El Aouj
Hammâmi

Tindouf
Kreb r. Neggar
Kreb es Sefia
Kreb n-Naga
Gara Djebilet Kreb Chebiha
Ora Djebilet
Aïn Ben Tili
540
Bîr Bel Guerdâne
Bîr Mogrein
Agmar

Ouahila
Oum el Guedour 580
Mcherrah Bj. Fl.
Ste. M.
Aftout
El Eglab
Bîr el Abbes
Touila
Chenachane
o Chenachane
Chegga
Dâya el Khadra

Ayoûn’ Abd
el Mâlek Tarhamanant
Mzerreb
Kreb en Naga
Agâraktem
Terhazza
Aoukar
Bîr Chali
Hamada Safia
En Nahrat
Taoudenni
Telig **Hamada el Haricha**
El Guettara
Dglats de Khenachiche
Bîr Ounane
El Ksaib Ounane

Dhar Khena

E

20
La Güera

Et Tidra
Sebkhet
Te-n-Dghâmcha

Akjoujt

Ijâfène

Ouarâne

Ijâfène

Jafène

MAL

Douaouir
Ergî
I-n-Échaî

MOROCCO

Sanlúcar
de Barramede
Cádiz **SPA**
Algeciras
C. Trafalgar Strait of Gibral
C. Spartel **Ceut**
Tanger Râs Tarf
Asilah **Tétoua**
Larache Chechaouen
Ksar el Kebir
Souk el Arba du Rharb
Mechra-bel-Ksiri
Allal-Tazi Sidi
Karia ba
Sidi Slimane Mohammed
Kenitra Sidi Ta
Salé Kacem
RABAT **MEKNES** **FES**
Mohammedia Volubilis Sefrou
CASABLANCA Tiflet El Hajeb Azrou
Azemmour Bir Jdid Ben Taghzout
Berrechid Slimane Rommani El Hajeb
El Jadida Benahmed Oulmès Timhadit
(Mazagan) Khouribga Khenifra
Settat Oued Zem Douirat
Sidi Smaïl Kasba Beni Mellal 3737
Safi Mechra Fkih ben Salah Ksiba
Bennour Benâbbou Midelt
Youssoufia El Kelâa Imdahane
Tleta Sidi Qum er Rbia Tamelelt Rich
Bouguedra Benguerir Azilal Rissani
Essaouira Talmest Demnate 4071
C. Sim Chichaoua **MARRAKECH** Tinerhir Rachi
Amizmiz Asni Tafil
C. Tafelney Tamanar 4165 Irhil Mgoun
Dj. Toubkal Anergane Boumalne
Tamri Taroudannt Ouarzazate
Cap Rhir Inezgane Tazenakht Zagora O. Draa
Agadir O. Souss Tarhbalt
Biougra Irherm Alougoum
Oulad Foum Zguid Kem-Kem
Tnine d'Anglou 2359 Bi. Semguine
Tiznit Talaïne Tissint
Ifni Tafraoute Mrhimina
Foum Assaka Imitek Tata
Bou Izakarn Akka
Goulimine O. Zemoul
Seyad
Oued Draa Oum el Ksi
O. Tigzerte Assa
Tan-tan Aounet Torkoz
Tafnidilt Mengoub
Khorb Hi. Chagmba
el Ethel Tounassine O. Krettam
Rhemilès Tounassine
Mcherrah

Djebel Ouarkziz
Haut Plateau du Dra
Hamada Tounassine
Zegdou
Di. Bet Tadjm

ft m
12,000 4000
9000 3000
6000 2000
4500 1500
3000 1000
1200 400
600 200
0 0
200 600
2000 6000
4000 12,000
m ft

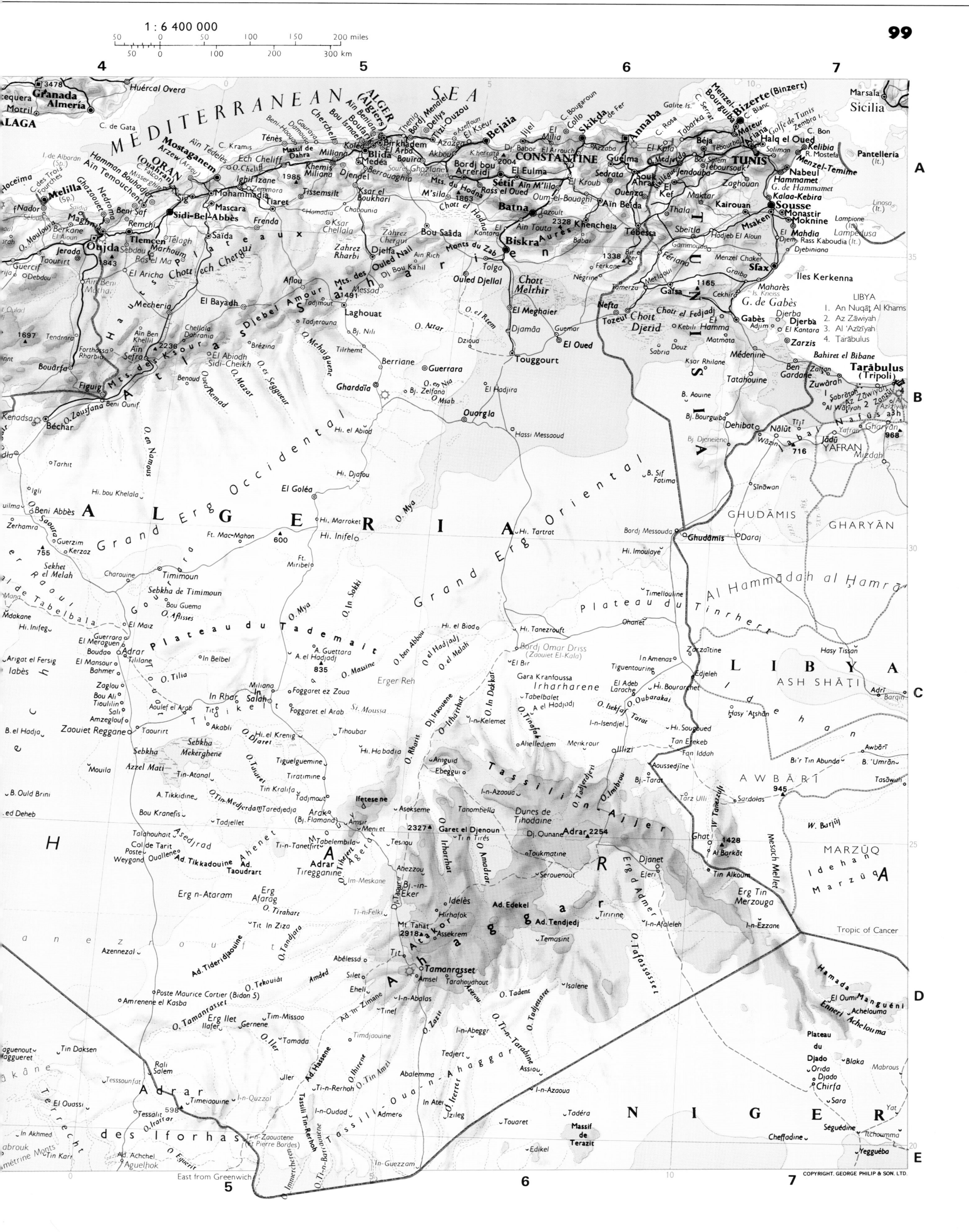

1 : 6 400 000

50   0   50   100   150   200 miles

50   0   100   200   300 km

MEDITERRANEAN SEA

LIBYA
1. An Nuqāt Al Khams
2. Az Zāwiyah
3. Al 'Azīzīyah
4. Tarābulus

ALGERIA

Grand Erg Occidental

Grand Erg Oriental

Plateau du Tademaït

Plateau du Tinrhert

Al Ḥammādah al Ḥamrā'

LIBYA

ASH SHĀṬI

GHUDĀMIS   GHARYĀN

Tassili-n-Ajjer

AWBĀRĪ

MARZŪQ

Idehan Marzūq

Ahaggar

Tassili Oua-n-Ahaggar

Tropic of Cancer

Adrar des Iforhas

NIGER

Plateau du Djado

East from Greenwich

COPYRIGHT. GEORGE PHILIP & SON. LTD.

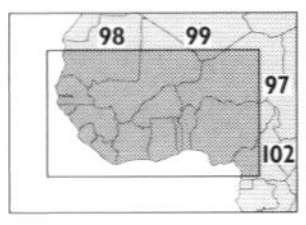

MAURITANIA

Nouakchott
Akjoujt
Oujeft
Oguelet en Nmâdi
Araouane
Sidi Moktar
Bennichâb
Boû Rjeimât
Sebkhet Te-n-Dghâmcha
Râs Timirist
Nouâmghâr
Et Tidra
Idîni
Rachid
Tidjikja
Gâneb
Tîchît
Aratâne
Akreijît
420
522
Moudjéria
Letfotar
Aleg
Mâl
Boûmdeïd
Togba
In Ahmer
Tâmchekket
Tagourâret
Oualâta
Aoukâr
Bourago
Ioûik
El Mreyyé

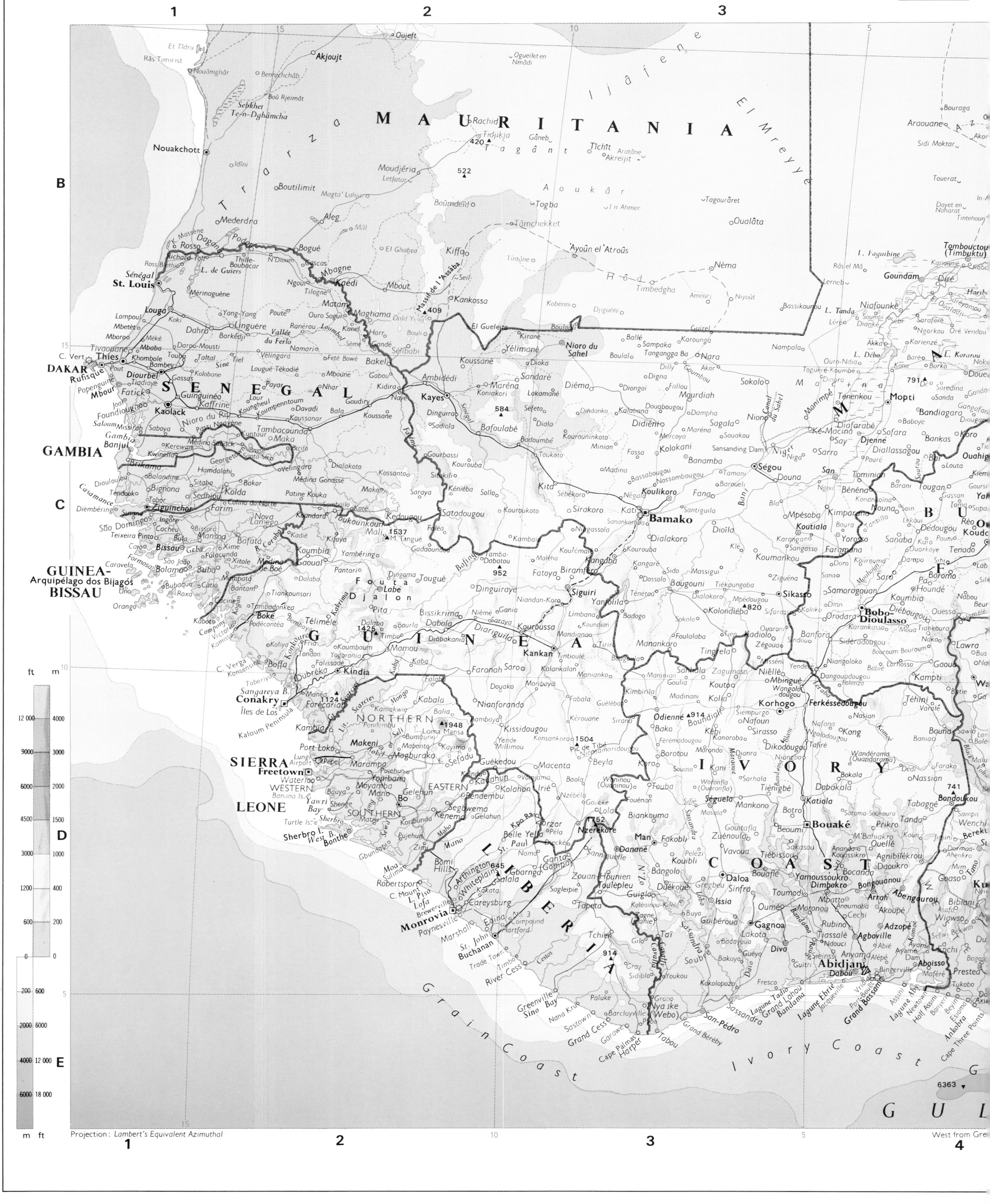

Mederdra
Boutilimit
Magta' Lahjar
El Ghabra
Kiffa
Kobenni
Djiguéni
Néma
Bassikounou
L. Tanda
Léro
Niafounké
Goundam
L. Faguibine
Tombouctou (Timbuktu)
Diré

Rosso
Massène
Dagan
Podor
Bogué
Kaédi
Mbout
Ouâlâta
Mbagne
Seïf
Timbedgha
'Ayoûn el 'Atroûs
Amoûri
Niyoût
Ré-ui
Guirel
Ras el Mâ
Goundam

Richard-Toll
Thillé-Boubacar
N'Dioum
Cascas
Ngoui
Tilogné
Matam
Maghama
Ould Yenjé
409
El Guelejta
Boulal
Kirane
Nioro du Sahel
Ballé
Sampaka
Tangagga Ba
Nara
Karounga
Nampala
Akka
Korienzé
Bambara-Maoundé

St. Louis Sénégal
L. de Guiers
Mérinaguène
Louga
Ouro Sogui
Kanel
Sémé
Gandé
Séribabi
Bakel
Koussané
Dioka
Digna
Dilly
Koumbou
Sokolo
Niono
Ténenkou
Mopti
791
Sofara
Bandiagara

DAKAR
C. Vert
Thiès
Khombole
Bambey
Diourbel
Rufisque
Popenguine
Mbour
Fatick
Gossas
Kaolack
SENEGAL
Guinguinéo
Kaffrine
Koungheul
Koumpentoum
Tambacounda
Maka
Goudiry
Nayé
Kayes
Maréna
Koniakari
Lakamané
Diéma
Sandaré
Diongoï
Fallou
Mourdiah
Doubabougou
Dampha
Sagala
Banamba
Sansanding Dam
Nioro
Ségou
Diafarabé
Say
Djenné
Diallassagou
Bankas
Korô
Ouahi

Banjul
GAMBIA
Brikama
Kaffrine
Passi
Nganey
Nioro du Rip
Tambacounda
Goumbou
Dinguiraye
Bafoulabé
Boundoumé
Toukoto
Médina
Bassabougou
Tamani
Baraouéli
Bla
Mpésoba
Kimparana
Nouna
Yorosso
Sanaba
Tenado

Ziguinchor
GUINEA-BISSAU
Arquipélago dos Bijagós
São Domingos
Farim
Bignona
Sédhiou
Kolda
Vélingara
Médina Gonasse
Kossanto
Saraya
Kéniéba
Sollo
Sébékoro
Kita
Négala
Koulikoro
Fana
Koutiala
Karangana
Sangasso
Faramana
BOBO
Boromo
Houndé
Kouka
Dédougou

Bissau
Bafatá
Gabú
Mali
1537
M. Tangué
Kédougou
Saraya
Koïna
Kambaï
Kourémalé
Kangaba
Siguiri
Kolondiéba
820
Sikasso
Dogo
Kadiolo
Niélé
Orodara
Bobo-Dioulasso
Banfora

GUINEA
Télimélé
Dalaba
1425
Pita
Labé
Fouta Djalon
Mamou
952
Dabola
Diaguiraya
Kouroussa
Kankan
1504
Siguiri
Kolondiéba
Sikasso
Korhogo
Ferkéssédougou
Niélé
Téhini

Conakry
Îles de Los
Forécariah
Dubréka
Kindia
Mamou
Kissidougou
Pic de Tibé
Beyla
IVORY COAST
Odienné
4914
Boundiali
Séguéla
Katiola
Bouaké
Yamoussoukro
Dimbokro
Bongouanou

SIERRA LEONE
Freetown
Waterloo
WESTERN
Port Loko
Makeni
Magburaka
Loma Mansa
1948
NORTHERN
Kabala
Bumbuna
EASTERN
Kenema
Bo
SOUTHERN
Pendembu
Koidu
Macenta
Nzérékoré
Man
Danané
Biankouma
Touba
Daloa
Gagnoa

Bonthe
Sherbro I.
Turtle Is.
Mano
LIBERIA
Voinjama
Gbarnga
Sanniquellie
Zouan-Hounien
Toulépleu
Guiglo
Tapeta
Tabou
Issia
Gagnoa
Divo
ABIDJAN
Dabou
Grand-Bassam

Monrovia
Paynesville
Marshall
Buchanan
Bomi Hills
Robertsport
Careysburg
Harbel
914
Greenville
Sino Bay
Harper
Cape Palmas
San-Pédro
Sassandra
Fresco

Grain Coast
Ivory Coast
Cape Three Points
6363
GULF

Projection: Lambert's Equivalent Azimuthal
West from Greenwich

ft m
12 000 4000
9000 3000
6000 2000
4500 1500
3000 1000
1200 400
600 200
0 0
200 600
2000 6000
4000 12 000
6000 18 000
m ft

1 : 6 400 000

50 0 50 100 150 200 miles
50 0 100 200 300 km

East from Greenwich

Inset:
NIGER
CHAD
N. E. NIGERIA
on same scale as general map

Lac Tchad
Bosso · Arege · Gashagar · Yobe · Zari · Kukawa · Baga · Dikwa · Dabilda · Mongonu · Ngala · Marte · Titiwa · Gubio · BORNO · Kala · Masbe · Yajua · Ngalda · Bama · Kumshe · Maiduguri · Konduga · Gulumba · Dumboa · Tokombere · Gwoza · Chibuk · Madagali · Mokolo · Maroua · Askira · Mishika · Mendif · Mubi · Gombi · Garkida · Malabu · Kaélé · Song · Zummo · Guider · Binder · Léré · Garoua · Poli · CAMEROON · ADAMAOUA · Benue

Main map labels:

ALGERIA

Adrar des Iforhas

NIGER

Aïr (Azbine)
Agadès (Agadez)
Monts Bagzane 2022

Ténéré

Tessalit · Kidal · Gao · Ansongo · Ménaka · Niamey · Dosso · Birni Ngaouré · Filingué · Tahoua · Tamaské · Keita · Madaoua · Birni Nkonni · Zinder · Mirria · Gouré · Nguigmi · Manga

BURKINA · Dori · Tillabéri · Ouallam · Tera · Maradi · Tessaoua · Aguié · Katsina · Daura · Kumaganam · Gashua · Nguru · Geidam

Ouagadougou (implied) · Fada N'Gourma · Diapaga · Kandi · Gusau · Sokoto · Argungu · Birnin Kebbi · KEBBI · SOKOTO · KATSINA · Kano · KANO · JIGAWA · Hadejia · Katagum · Azare · Potiskum · Gumel

Bolgatanga · Bawku · Mango · Nikki · Kaiama · Kontagora · Zaria · KADUNA · Kaduna · Bauchi · Gombe · Biu · Yola · ADAMAWA

GHANA · Tamale · Yendi · Bassari · Sokodé · BENIN · Parakou · Djougou · Jos · Jos Plateau · PLATEAU · Shendam · Jalingo · Numan

Kpalimé · TOGO · Abomey · Savalou · Save · Ilorin · OYO · Ogbomosho · Oyo · Offa · Osogbo · Osun · Ede · Iwo · Ife · Ilesha · Ikare-Ekiti · Akure · ONDO · Lokoja · Makurdi · Wukari · TARABA · Mts Gan Goriama

ACCRA · Tema · Winneba · Cape Coast · Lomé · Cotonou · Porto-Novo · LAGOS · Ikeja · Abeokuta · Ijebu-Ode · IBADAN · OGUN · Ondo · EDO · Benin City · Asaba · Onitsha · Enugu · ENUGU · Abakaliki · CROSS RIVER · Bamenda · Bafut · Bafoussam · Foumban · Dschang

Slave Coast · BIGHT of BENIN · Niger Delta · Warri · DELTA · Sapele · Ughelli · IMO · ABIA · Aba · Owerri · Umuahia · Port-Harcourt · Okrika · Uyo · AKWA IBOM · Eket · Calabar · Kumba · CAMEROON · DOUALA · Tiko · Buea · Limbe

BIGHT of BONNY
Bioko (Fernando Poo) · Pico Santa Isabel 2850 · Malabo · EQUATORIAL GUINEA

Cameroun 4070

GULF OF GUINEA

YAOUNDÉ · Edéa · Eséka · Ebolowa · Sangmélima · Mbalmayo

COPYRIGHT. GEORGE PHILIP & SON. LTD.

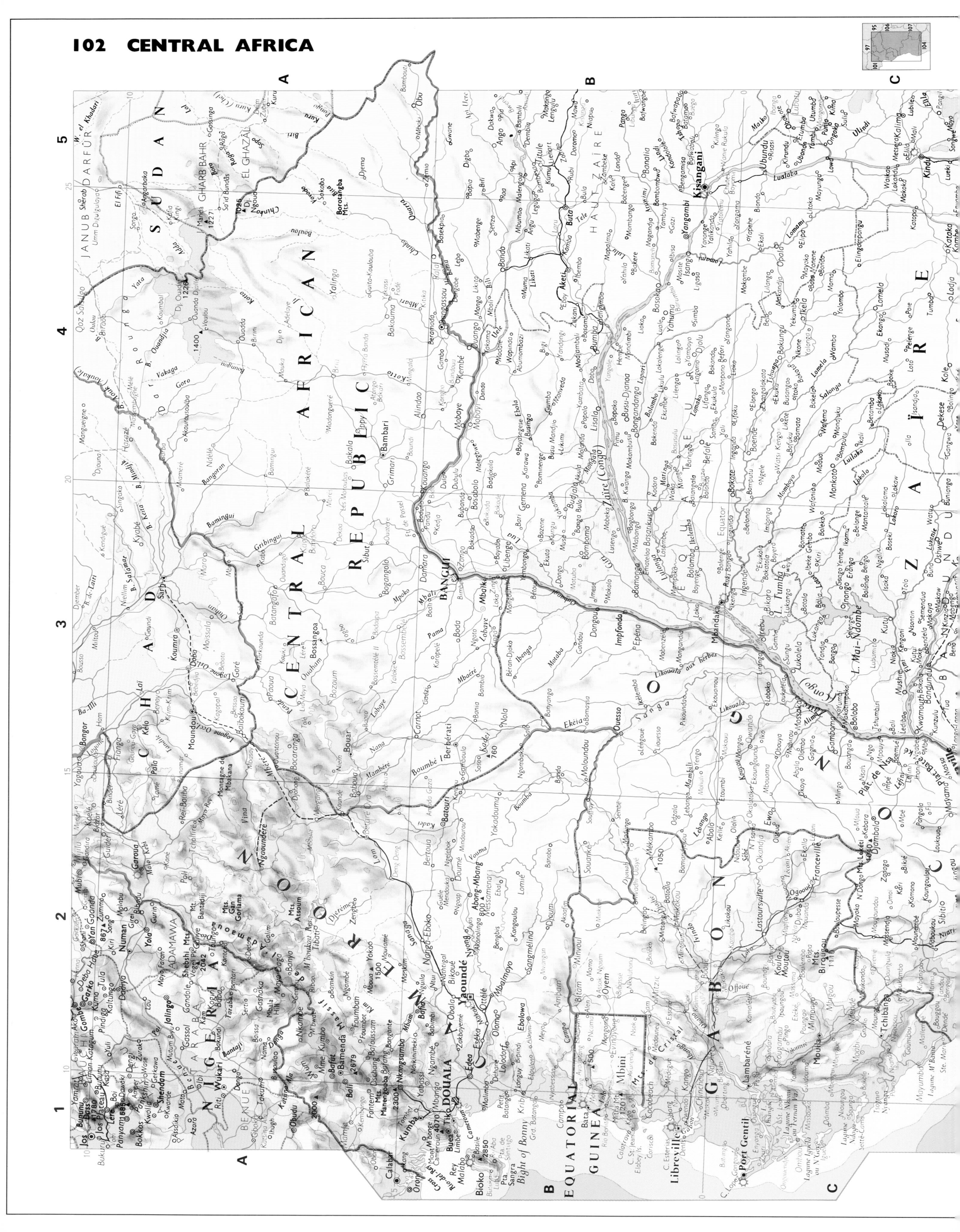

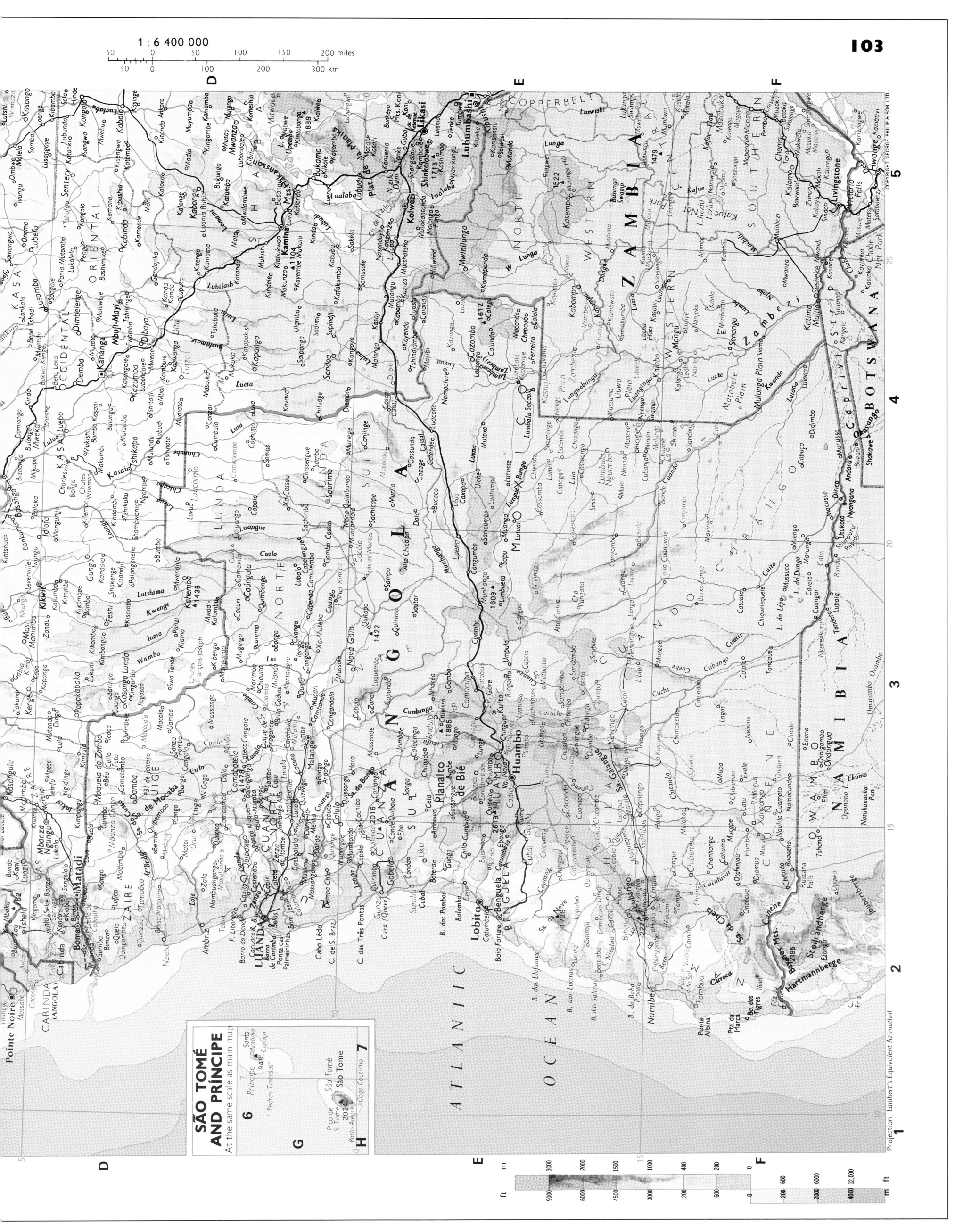

1 2 3

A N G O L A

NAMIBIA

BOTSWANA

SOUTH AFRICA

ZAMBIA

ATLANTIC OCEAN

NORTHERN CAPE

WESTERN CAPE

EASTERN CAPE

NORTH WEST

ORANGE FREE STATE

Tropic of Capricorn

Kalahari

Namib Desert

Kaokoveld

Damaraland

Ovamboland

Etosha Pan

Okavango Swamps

Chobe Nat. Park

Caprivi Strip

CUANDO CUBANGO

Windhoek

Walvisbaai (Walvis Bay)

Swakopmund

Lüderitz

Keetmanshoop

Mariental

Rehoboth

Okahandja

Otjiwarongo

Grootfontein

Tsumeb

Gobabis

Upington

Kimberley

Bloemfontein

CAPE TOWN (Kaapstad)

PORT ELIZABETH

Stellenbosch

Paarl

Worcester

George

Mosselbaai

Oudtshoorn

Beaufort West

De Aar

Great Karoo

Little Karoo

Victoria Falls

Livingstone

Maun

Ghanzi

Serowe

Gaborone

Lobatse

Molepolole

Kanye

Mafikeng

Vryburg

Kuruman

Springbok

Namaqualand

Port Nolloth

Alexander Bay

Oranjemund

C. Agulhas

Cape of Good Hope (Kaap die Goeie Hoop)

Projection: Lambert's Equivalent Azimuthal

ft m
9000 3000
6000 2000
4500 1500
3000 1000
1200 400
600 200
0 0
200 600
2000 6000
4000 12000
m ft

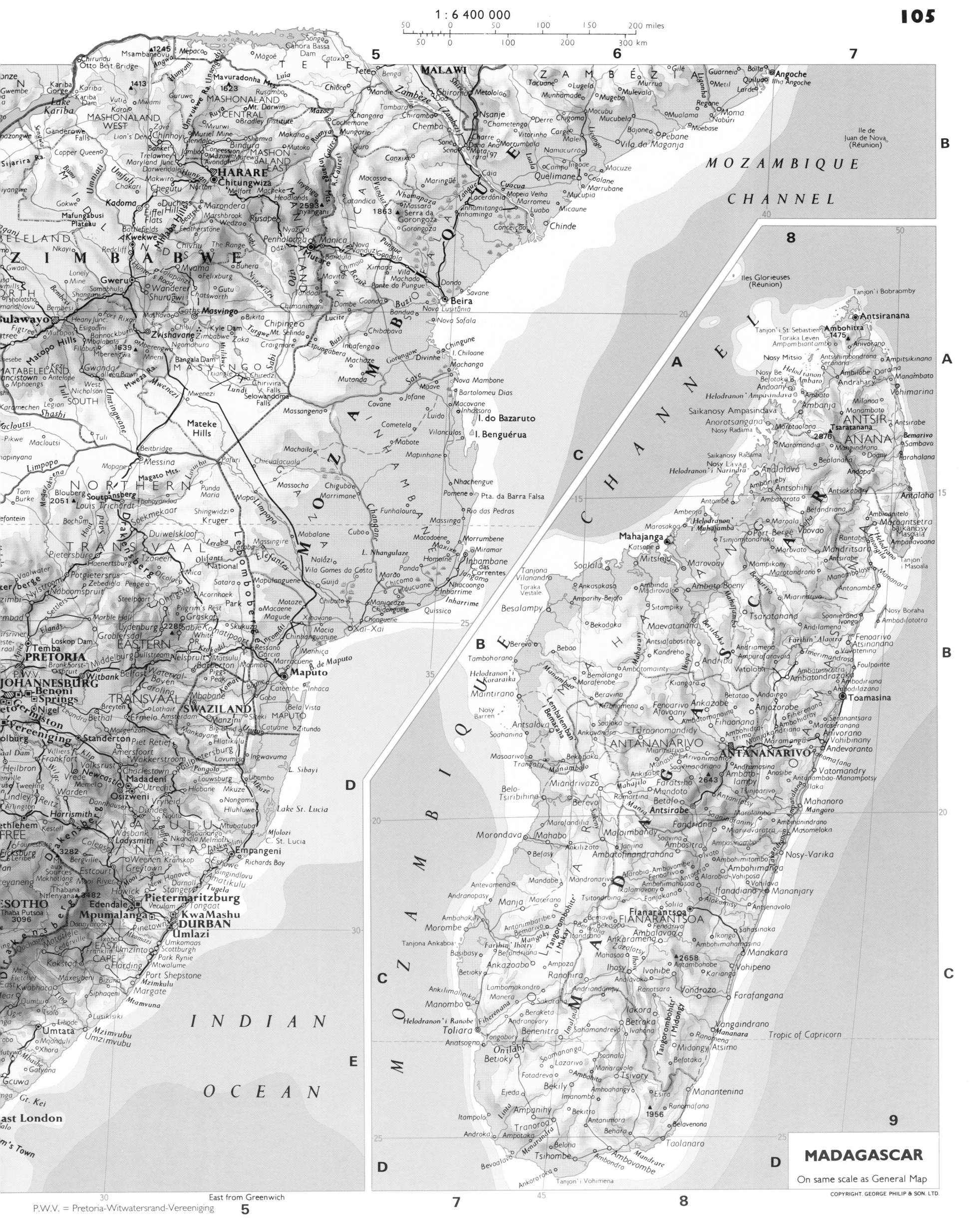

1 : 6 400 000

50   0   50   100   150   200 miles
50   0   100   200   300 km

MOZAMBIQUE CHANNEL

INDIAN OCEAN

ZIMBABWE

MOZAMBIQUE

MADAGASCAR

HARARE

Chitungwiza

Bulawayo

Beira

Maputo

PRETORIA

JOHANNESBURG

Benoni
Springs

Vereeniging

DURBAN

PIETERMARITZBURG

SWAZILAND

TRANSVAAL

NORTHERN

EASTERN

KWAZULU NATAL

LESOTHO

East London

ANTANANARIVO

Mahajanga

Antsiranana

ANTSIRANANA

Toamasina

ANTSIRABE

Fianarantsoa

FIANARANTSOA

Toliara

Tropic of Capricorn

P.W.V. = Pretoria-Witwatersrand-Vereeniging

East from Greenwich

MADAGASCAR

On same scale as General Map

COPYRIGHT. GEORGE PHILIP & SON. LTD.

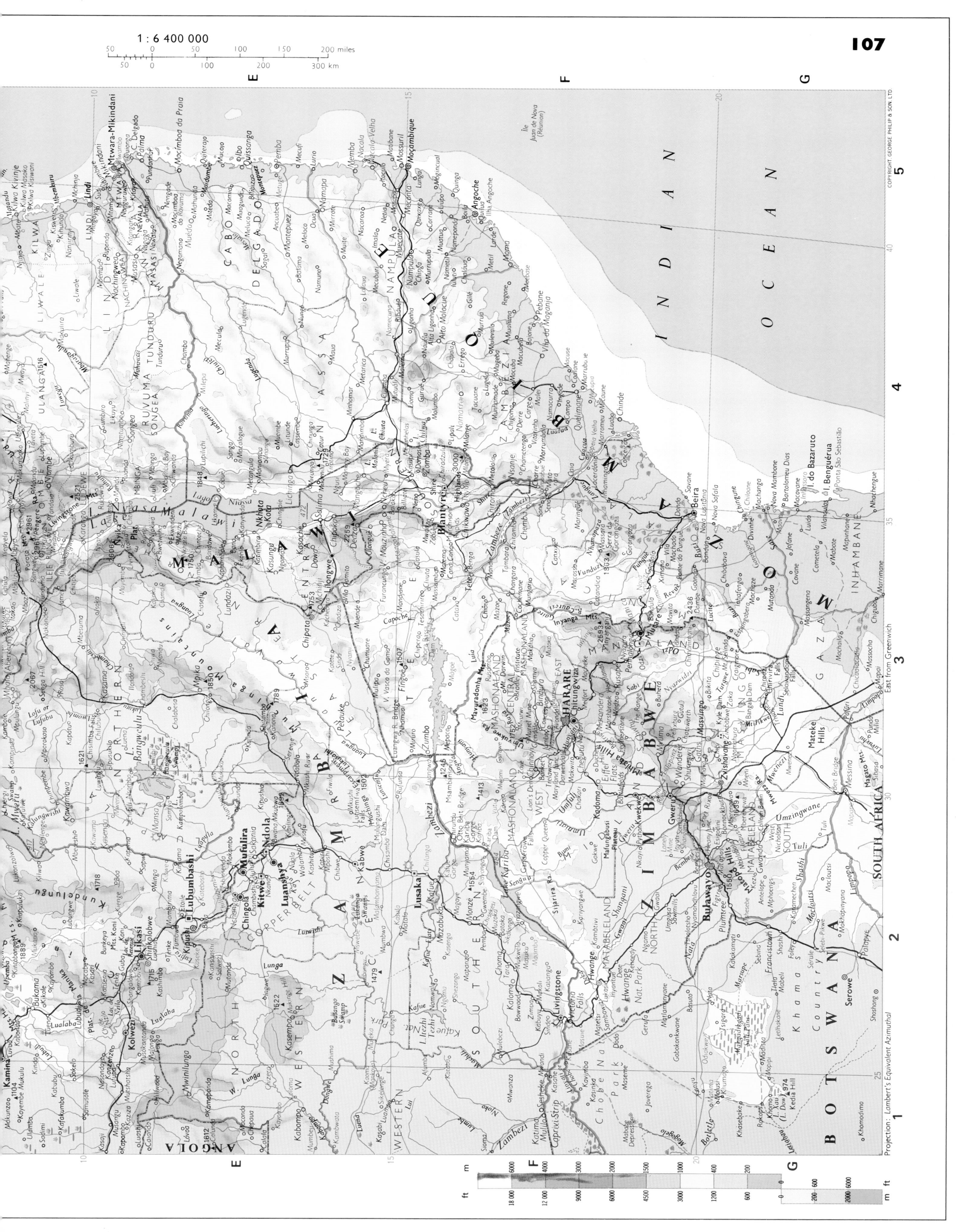

1 : 6 400 000

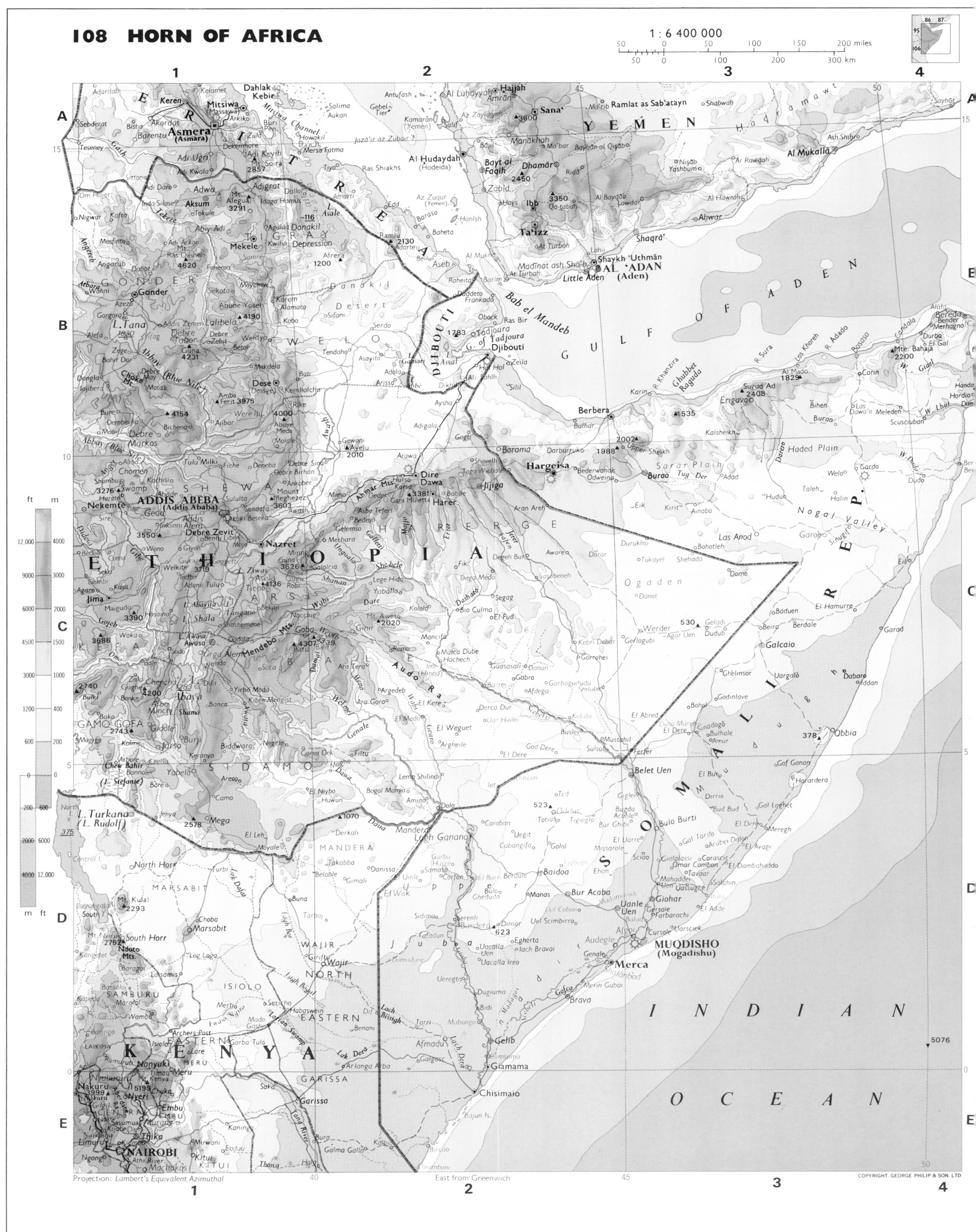

1 2 3 4 5 6 7 8 9 10

A
Mediterranean Sea • Bayrût • SYRIA • Baghdad • IRAN • Kâbul • Rawalpindi • Xi'an • CHINA
El Iskandariya • Tel Aviv Yafo • ISRAEL • Dimashq • Karbalâ • AFGHANISTAN • Lahore • Chengdu • Nanjing • Shanghai
Banghâzi • Jerusalem • IRAQ • Esfahân • Qandahâr • Multan • Mt. Everest 8848 • Wuhan • Hangzhou
El Qâhira • El Suweis • Al Basrah • Âbâdân • Quetta • Delhi • Chongqing • Changsha • Guiyang • Nanchang

B
LIBYA • EGYPT • SAUDI • BAHRAIN • QATAR • UNITED ARAB EMIRATES • Karachi • INDIA • Kanpur • Varanasi • NEPAL • BHUTAN • Brahmaputra • Kunming • Fuzhou
Aswân • Al Madînah • Ar Riyâd • G. of Oman • Ahmadabad • Calcutta • Dhaka • Chittagong • Guangzhou • TAIWAN
L. Nasser • Tropic of Cancer • ARABIA • OMAN • G. of Kutch • Narmada • BURMA • Mandalay • Hong Kong

C
Omdurmân • El Khartûm • YEMEN • Socotra (Yemen) • Bombay • Pune • Hyderabad • G. of Tonkin • Hainan
CHAD • SUDAN • Mitsiwa • Asmera • Arabian • Godavari • Cuttack • Andaman Is. (India) • Mergui Arch. • THAILAND • Bangkok • Paracel Is.
4620 • Gulf of Aden • Sea • Bangalore • Madras • CAMBODIA • Phnom Penh • South
CENTRAL AFRICA • ETHIOPIA • DJIBOUTI • Ras Asir (C. Guardafui) • Arabian • Lakshadweep Is. (India) • Isthmus of Kra • Phanh Bho • Ho Chi Minh • China

D
Addis Abeba • 4307 • Berbera • Basin • 5875 • Madurai • SRI LANKA (CEYLON) • Nicobar Is. (India) • Gulf of Thailand • Sea
Wâw • Mongalla • 5824 • MALDIVES • Colombo • Pidurutalagala 2524 • George Town • BRUNEI • SARAH BORNEO
L. Turkana • SOMALI REP. • Carlesberg • Kuala Lumpur • MALAYSIA • Kuching • SARAWAK
UGANDA • L. Albert • Kampala • KENYA • Muqdisho • Somali • Ridge • Nias • Singapore

E
ZAIRE • L. Edward • L. Kivu • RWANDA • Nairobi 5895 • Equator • Basin • Chagos Archipelago (Br.) • Diego Garcia • Mentawei Is. • Palembang • INDONESIA • Borneo
BURUNDI • Mt. Kenya 5199 • Mwanza • Victoria • SEYCHELLES • Mahe • Sumatera • Jakarta • Java Sea
TANZANIA • Kilimanjaro • Mombasa • Amirante Is. • Des Roches • Coetivy Is. • Sunda Strait • Bandung • Semarang • Surabaya
Tabora • Pemba • Zanzibar • Alphonse • Java • Flores Sea

F
ANGOLA • ZAMBIA • L. Mweru • Dar es Salaam • Aldabra Is. • St. Pierre • Providence • Cocos or Keeling Is. (Austral.) • Christmas I. (Austral.) • 7450
Lubumbashi • L. Bangweulu • Farquhar Is. • Agalega I. • 4819 • 6327
Lusaka • Zambezi • MALAWI • COMOROS • Mayotte (Fr.) • Tromelin I. • Cargados Garajos
ZIMBABWE • Blantyre • MADAGASCAR • Moçambique • Mahajanga • Rodriquez

G
NAMIBIA • BOTSWANA • Harare • Antananarivo 2643 • 5322 • Port St. Louis • MAURITIUS • Mascarene Islands • Tropic of Capricorn • N.W. Cape • Onslow
Bulawayo • MOZAMBIQUE • Toamasina • Denis • Réunion (Fr.) • Shark Bay • WESTERN AUSTRALIA
Gaborone • Beira • Toliara • Mascarene Basin • Geraldton • AUSTRALIA
Pretoria • Quelimane • 6400 • Kalgoorlie

H
Johannesburg • Bloemfontein • Durban • Madagascar • 1491 • 1104 • Perth • Fremantle
SOUTH AFRICA • Kimberley • Basin • Geographe Bay
Cape Town • East London • Port Elizabeth • Amsterdam I. (Fr.) • Albany
Kaap die Goeie Hoop • St. Paul I. (Fr.)

J
Agulhas • 5778 • Crozet • Pr. Edward Is. (S.A.) • Basin • 2899 • Southeast Indian Rise
Basin • Marion I. • Hog I. • Crozet Is. (Fr.) • Possession I.
Atlantic Indian Ridge • Kerguelen (Fr.)

K
Mc Donald Is. • Heard I. (Austral.) • 5141 • 5202
5848 • 4691
4850

L
Antarctic Circle • Enderby Land • Wilkes Land

M
Queen Maud Land • **Antarctica** • Adélie Land

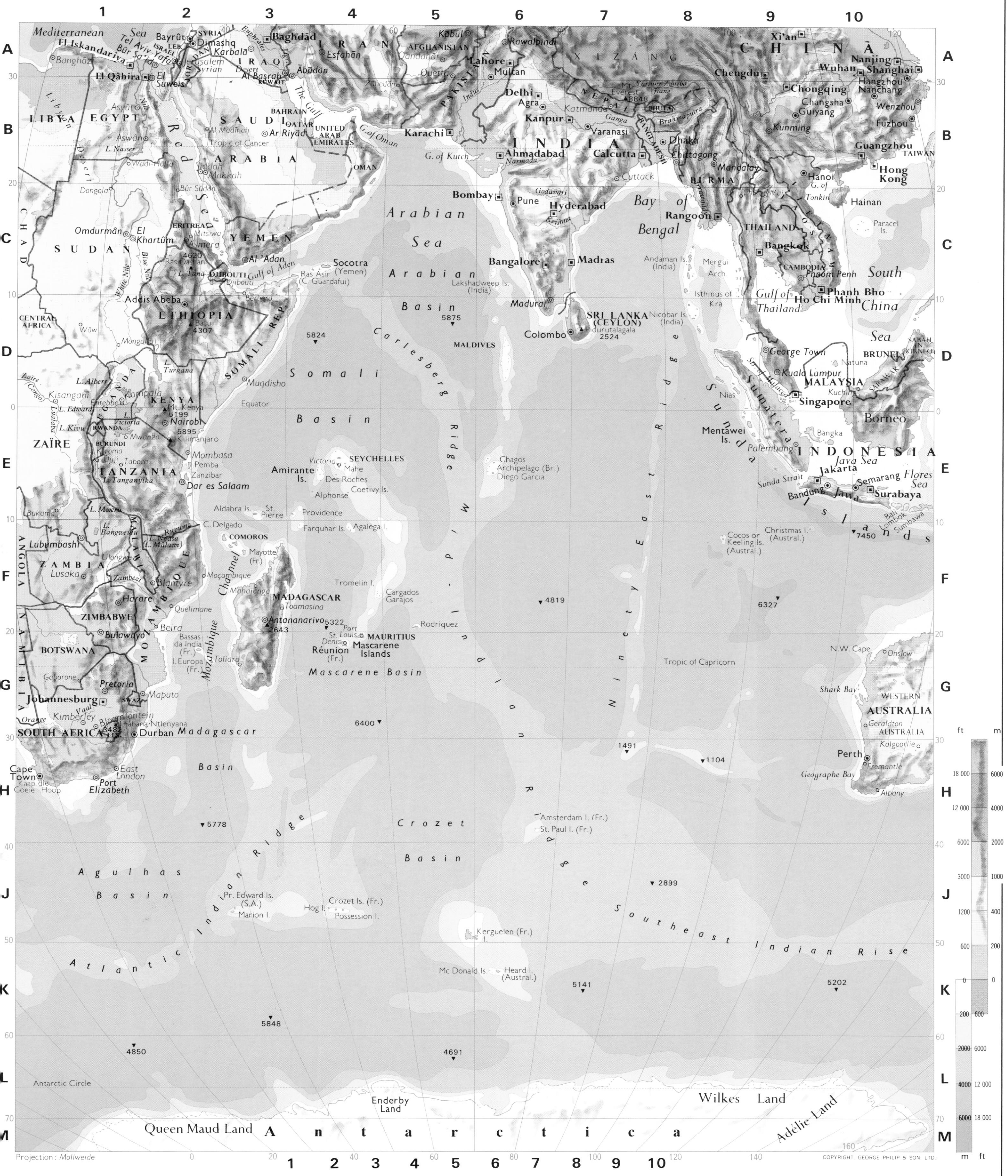

Projection: Mollweide

0 20 40 60 80 100 120 140 160

1 2 3 4 5 6 7 8 9 10

COPYRIGHT GEORGE PHILIP & SON LTD.

ft m
18 000 — 6000
12 000 — 4000
— 3000
6000 — 2000
— 1000
0 — 0
200 — 600
2000 — 6000
4000 — 12 000
6000 — 18 000
m ft

Projection : Lambert's Equivalent Azimuthal
East from Greenwich

1 : 16 000 000

100   0   100   200   300   400 miles

100   0   100   200   300   400   500   600 km

10   160   **11**   165   **12**   170   **13**   175   **14**   180   **15**   **16**

M   *e*   *l*   *a*   *n*   *e*   *s*   *i*   *a*

NAURU

Tamana

K I R I B A T I

Baker   175   Equator

6195   170

Bougainville

Choiseul

SOLOMON

Santa Isabel

New Georgia

Abariringa   A

ISLANDS

Malaita

Honiara   2331

Namumea

Phoenix Is.

Guadalcanal

San Cristóbal

TUVALU

Carondelet   5

Rennell   7223

Santa Cruz Is.

(Ellice Is.) Funafuti   Funafuti

Tokelau   B

Nukulaelae

Is. (N.Z.)

Fataka

Rotuma

10

Banks Is.

Mata-Utu   Uvea

WESTERN

Espíritu Santo   1880

VANUATU

Wallis & Futuna

SAMOA

Malakula

(New Hebrides)

Horn   (Fr.)

Savai'i   Apia   C

Upolu

Îs. D'Entrecasteaux

Port-Vila   Efate

Vanua Levu

Niuafo'ou

Tutuila   **AMERICAN**

s. Chesterfield

Viti Levu   1324   **FIJI**

SAMOA

15

Suva

Îs. Loyauté   7569

Lau Is.

1628

Vavau Is.

New

Ceve-i-Ra

Ha'apai Is.   **TONGA**

Niue   D

Caledonia

(N.Z.)

(Fr.)   Nouméa

Matthew

Nuku'Alofa

Tongatapu Is.

P   **A**   **C**   **I**   **F**   **I**   **C**

20

5303

10 882   Tonga Trench

Cook Is.   (N.Z.)

O   **C**   **E**   **A**   **N**   Tropic of Capricorn   E

Norfolk   (Austr.)

25

Lord Howe   (Austr.)

Raoul

Kermadec Is.

Kermadec Trench

(N.Z.)   F

734

10 047

T   *a*   *s*   *m*   *a*   *n*   *Sea*

North C.

30

Kaitaia

Whangarei

5267

Auckland

NORTH ISLAND

G

Hamilton

Bay of Plenty

New Plymouth

Rotorua

International Date Line

Gisborne

NEW

Raupehu   2797   Napier

Wanganui

ZEALAND

Palmerston

North

35

Nelson   Cook Strait   **Wellington**

Greymouth   Blenheim

SOUTH ISLAND

H

Mt.Cook   Southern Alps   **Christchurch**

3753

Timaru

40

Wakatipu   Chatham

(N.Z.)

Invercargill   **Dunedin**

Stewart   J

10   160   **11**   165   **12**   170   **13**   175   **14**   180   **15**   175 West from Greenwich 170   **16**   **17**   165   **18**   160

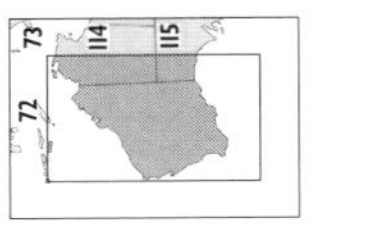

NORTHERN TERRITORY

Tanami Desert

Great Sandy Desert

TIMOR SEA

INDONESIA

Timor

INDIAN OCEAN

King Leopold Ranges

Hamersley Range

Darwin

Port Hedland

Broome

Melville I.

Bathurst I.

Cobourg Pen.

Katherine

Victoria River

Kimberley

Fitzroy

Wyndham

Derby

Karratha

Newman

Exmouth Gulf

Joseph Bonaparte Gulf

Macdonnell Ranges

Mt. Leisler 901

Lake Mackay

Mt. Zeil 1510

Mt. Liebig 1524

Tropic of Capricorn

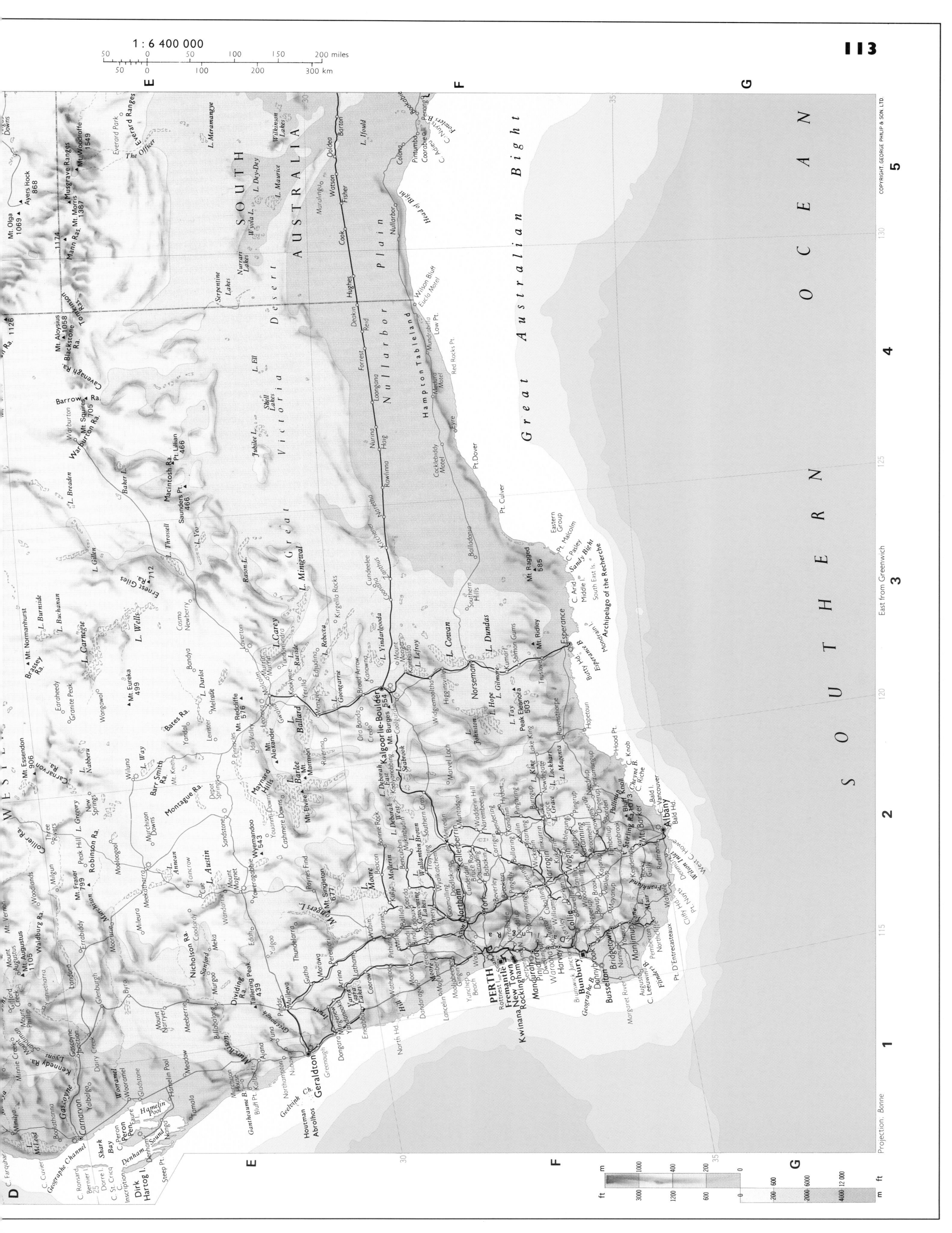

1 : 6 400 000

50   0   50   100   150   200 miles
50   0   100   200   300 km

E   F   G

S O U T H

A U S T R A L I A

W E S T E R N

Great Victoria Desert

Nullarbor Plain

Hampton Tableland

G r e a t   A u s t r a l i a n   B i g h t

S O U T H E R N   O C E A N

Shark Bay

Gascoyne

Geraldton

PERTH
Fremantle
New Town
Rockingham
Kwinana

Bunbury
Busselton

Albany

Esperance

Kalgoorlie-Boulder

Norseman

COPYRIGHT GEORGE PHILIP & SON LTD.

Projection. Bonne

East from Greenwich

ft
3000
1200
600

m
1000
400
200
0

200
600
2000
4000
6000
12 000

ft

m

Ayers Rock 868
Mt. Olga 1069
Mt. Woodroffe 1549
Musgrave Ranges
The Everard Ranges
Everard Park

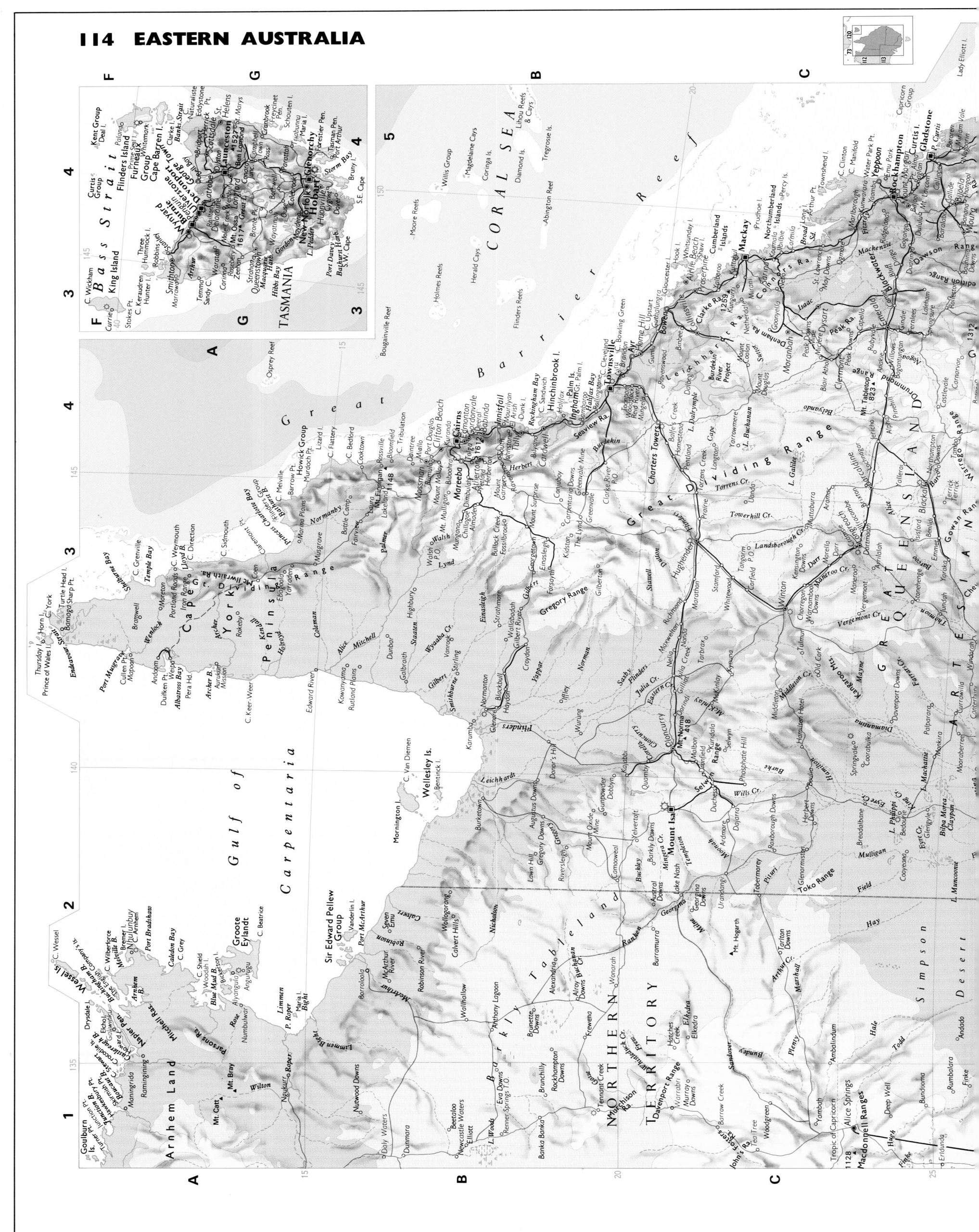

TASMANIA

Bass Strait

King Island

Flinders Island

Launceston

Hobart

CORAL SEA

Great Barrier Reef

Willis Group

Gulf of Carpentaria

Cape York Peninsula

Thursday I.

Wellesley Is.

Mornington I.

Arnhem Land

Groote Eylandt

Sir Edward Pellew Group

Wessel Is.

Cairns

Townsville

Mackay

Rockhampton

Gladstone

Mount Isa

NORTHERN TERRITORY

QUEENSLAND

Great Dividing Range

Charters Towers

Winton

Barkly Tableland

Simpson Desert

Alice Springs

Macdonnell Ranges

Tropic of Capricorn

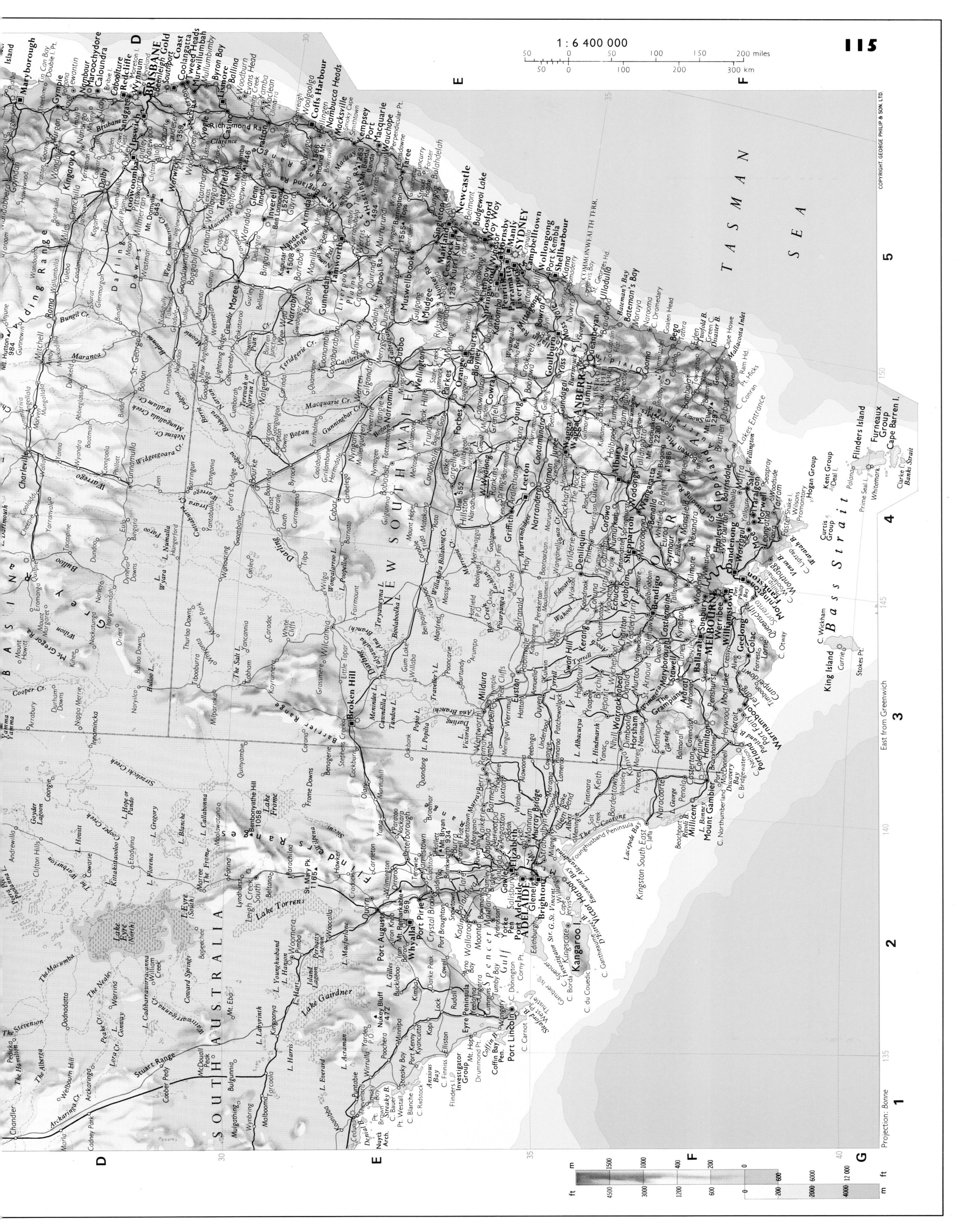

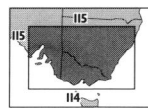

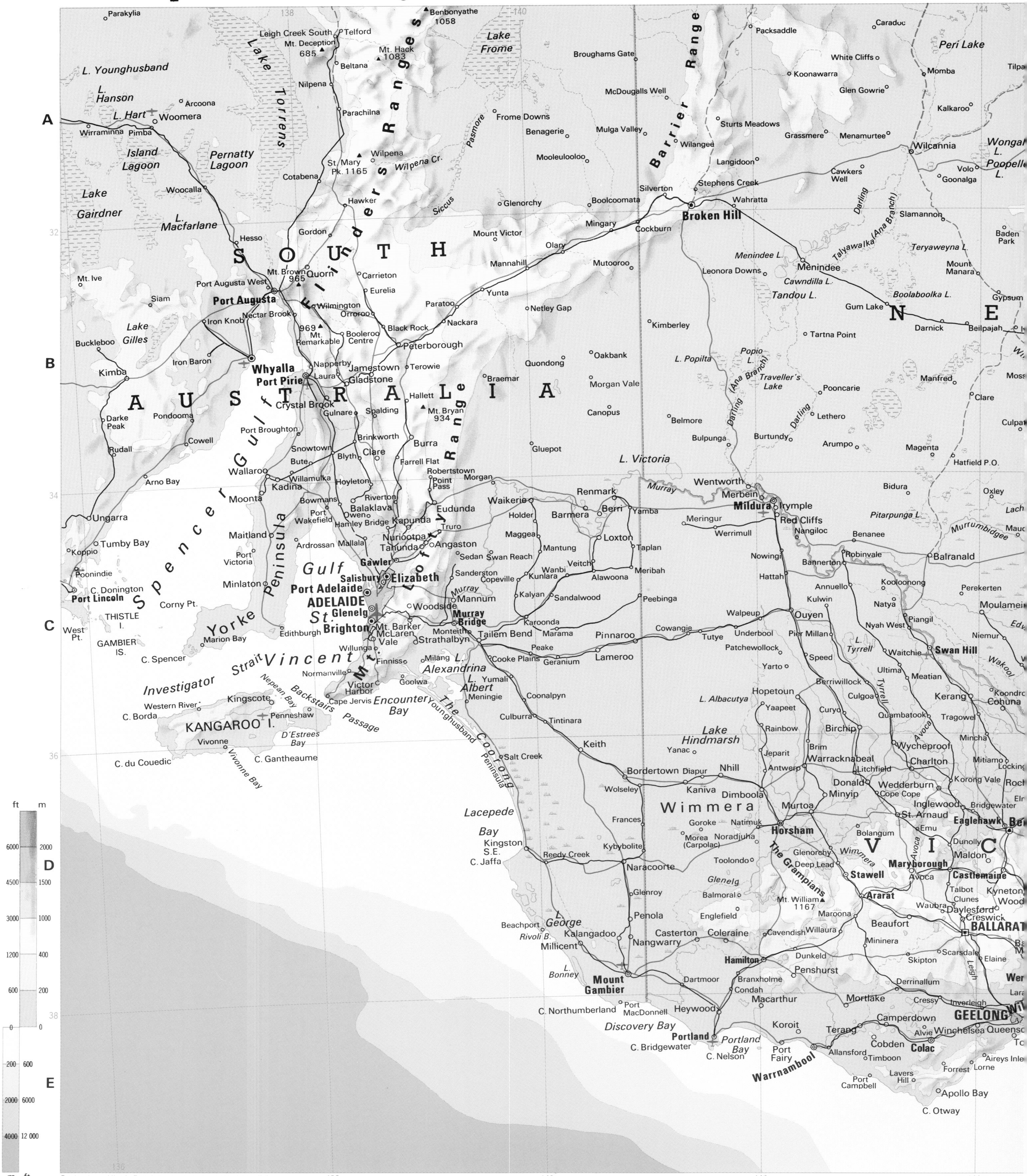

1 : 2 800 000

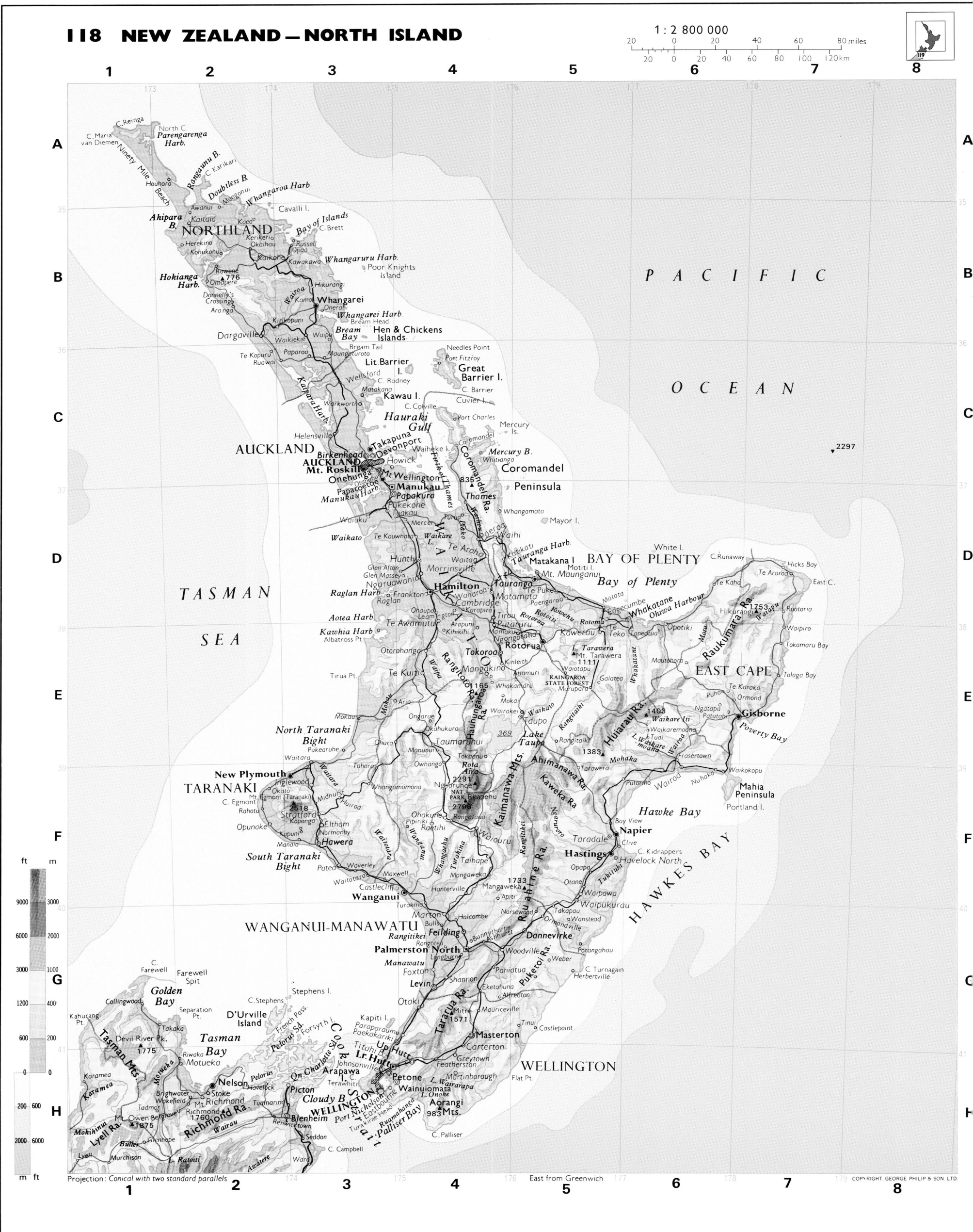

NORTHLAND

PACIFIC

OCEAN

AUCKLAND

Hauraki Gulf

Coromandel Peninsula

TASMAN

SEA

BAY OF PLENTY

Bay of Plenty

Hamilton

Rotorua

EAST CAPE

Gisborne

TARANAKI

New Plymouth

North Taranaki Bight

South Taranaki Bight

Lake Taupo

Napier

Hastings

Hawke Bay

Mahia Peninsula

H A W K E S   B A Y

Wanganui

WANGANUI-MANAWATU

Palmerston North

Golden Bay

Tasman Bay

Nelson

WELLINGTON

Blenheim

Cook Strait

Projection: Conical with two standard parallels

East from Greenwich

COPYRIGHT GEORGE PHILIP & SON LTD.

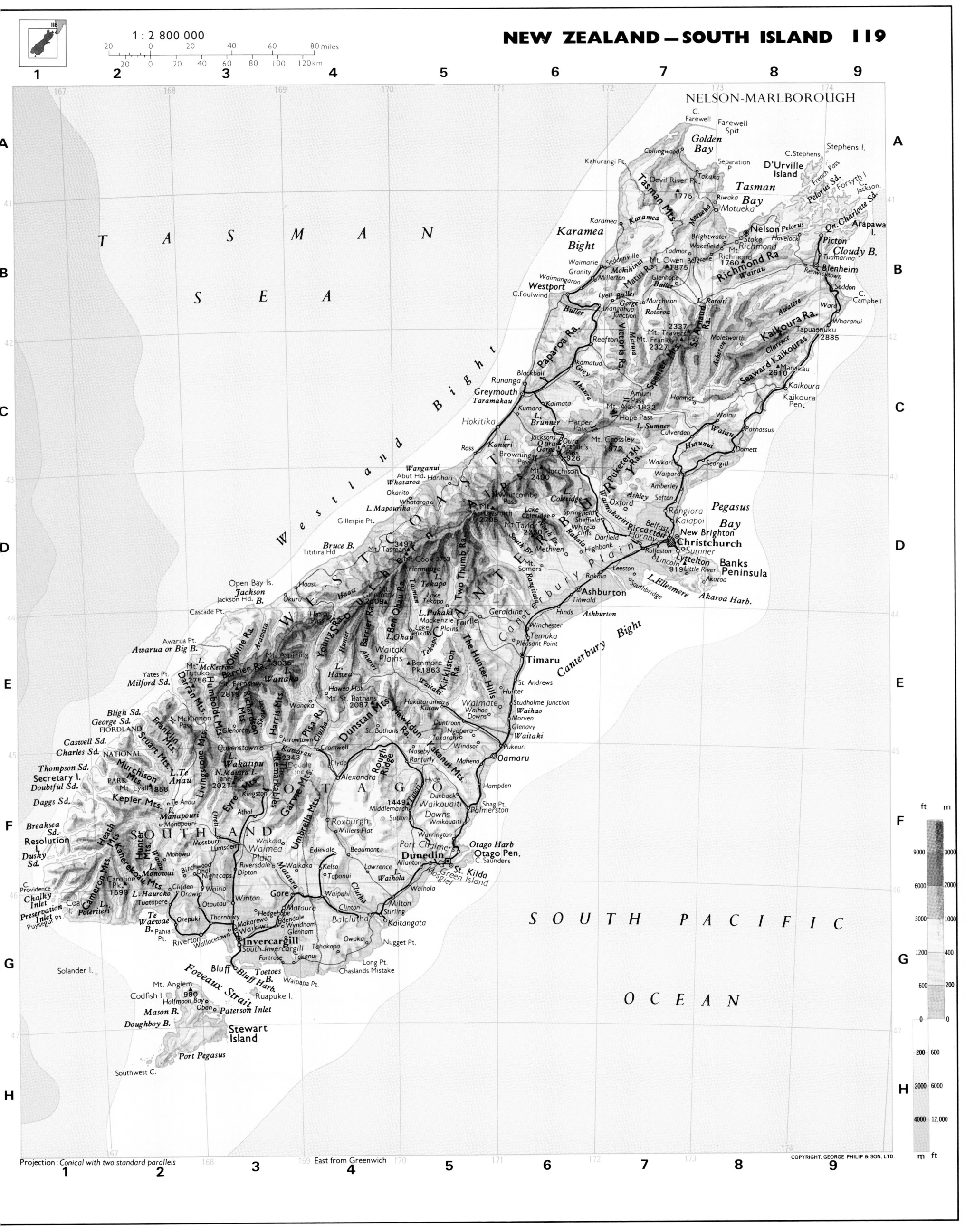

1 : 5 200 000

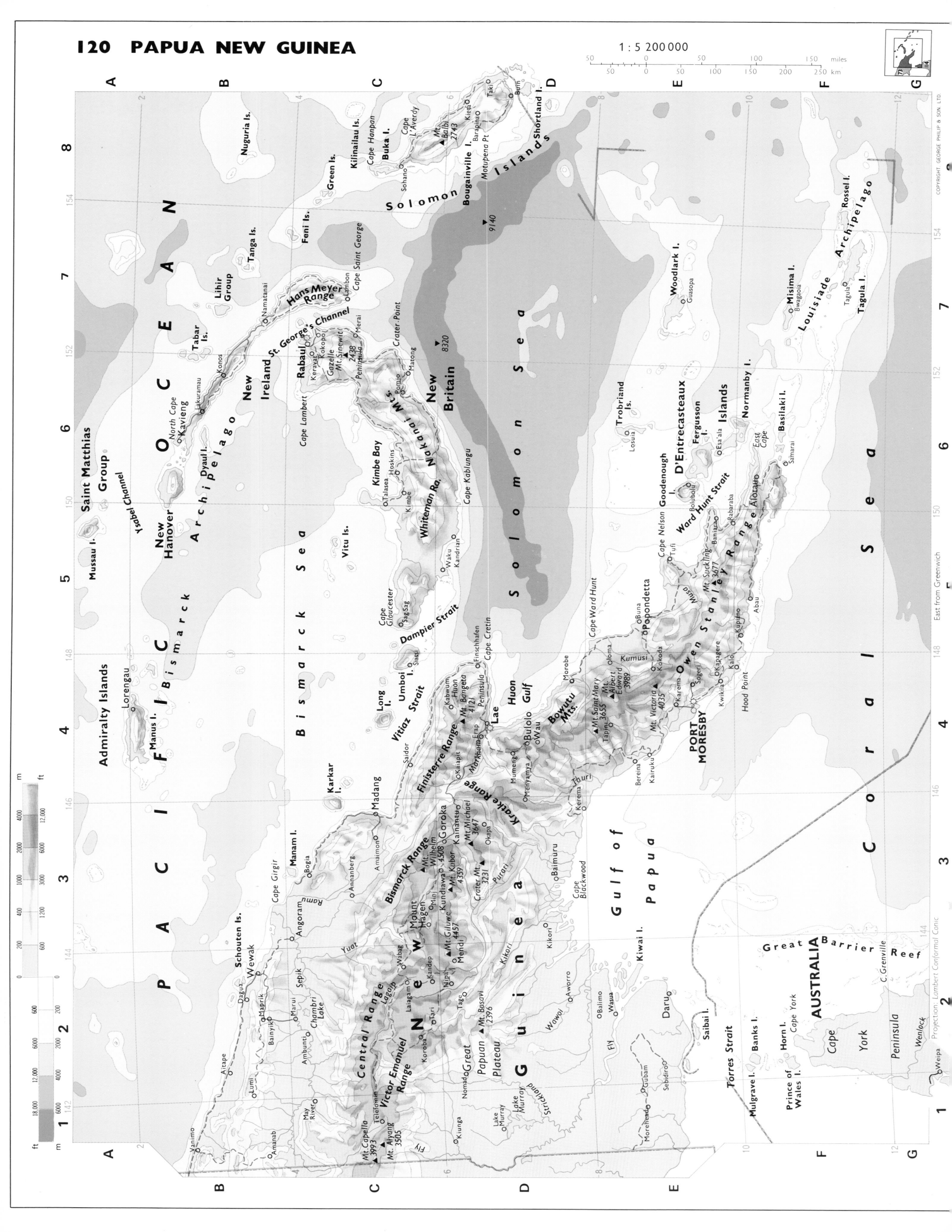

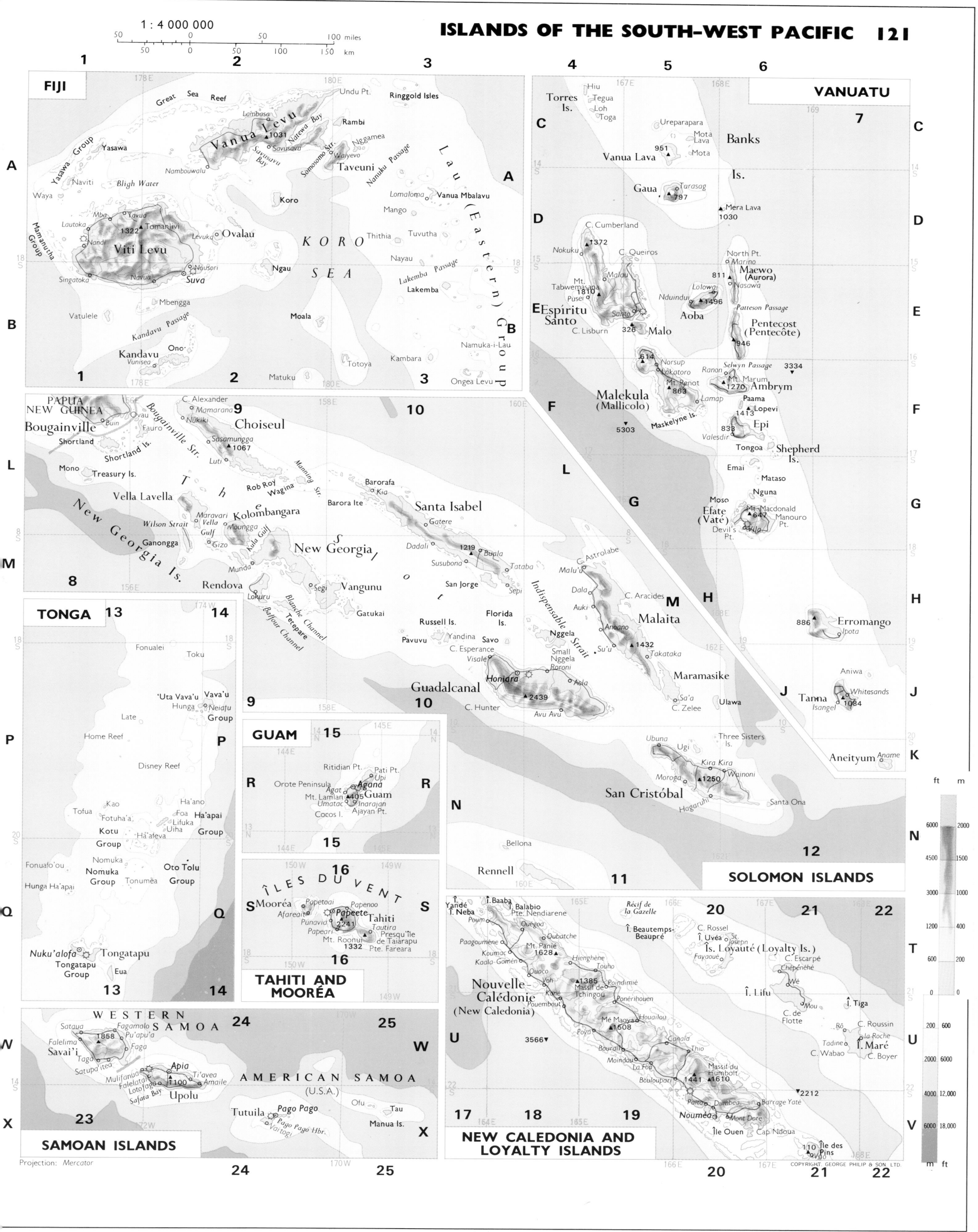

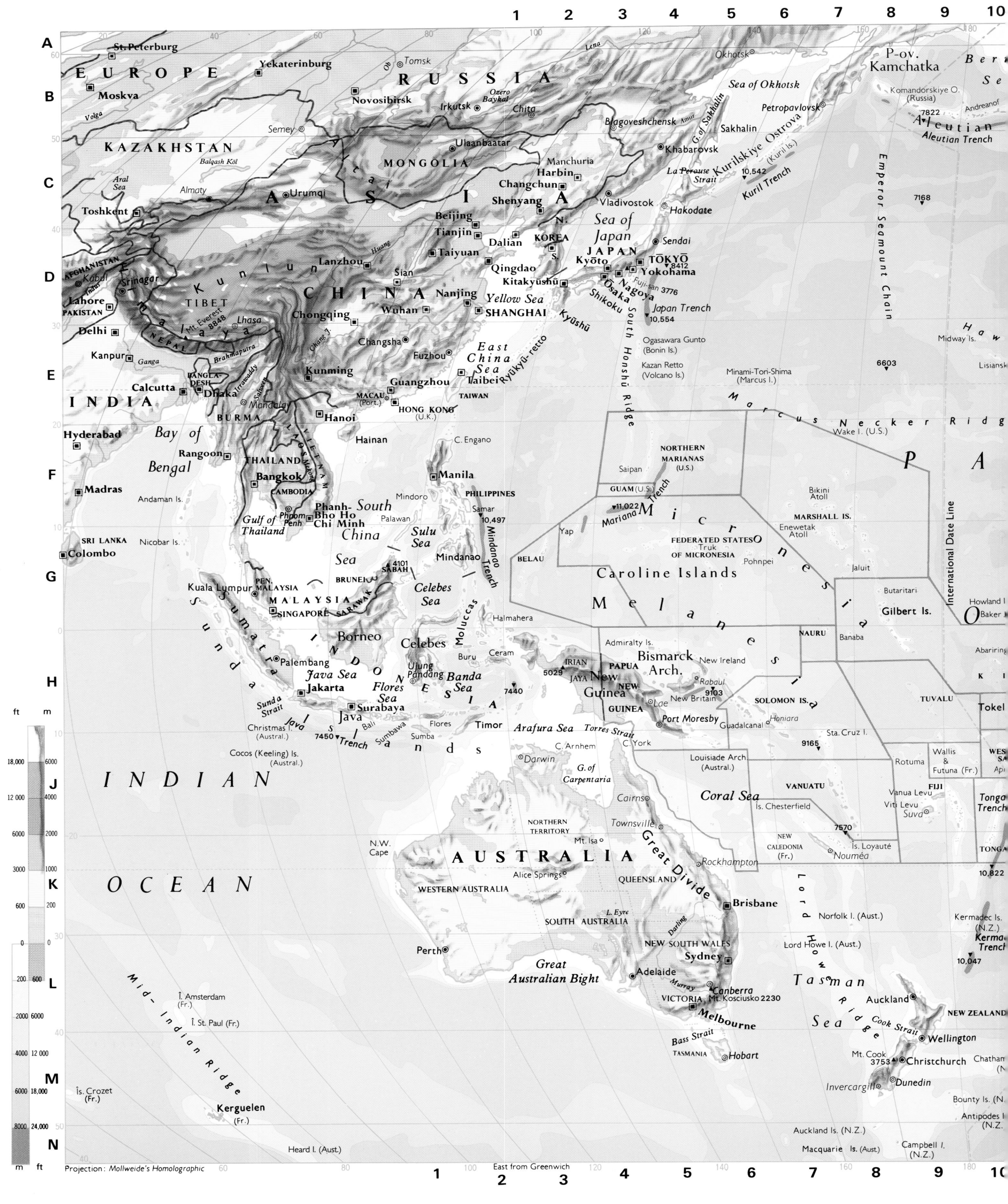

Projection: Mollweide's Homolographic

East from Greenwich

1 : 28 000 000

200   0   200   400   600   800 miles
400   0   400   800   1200 km

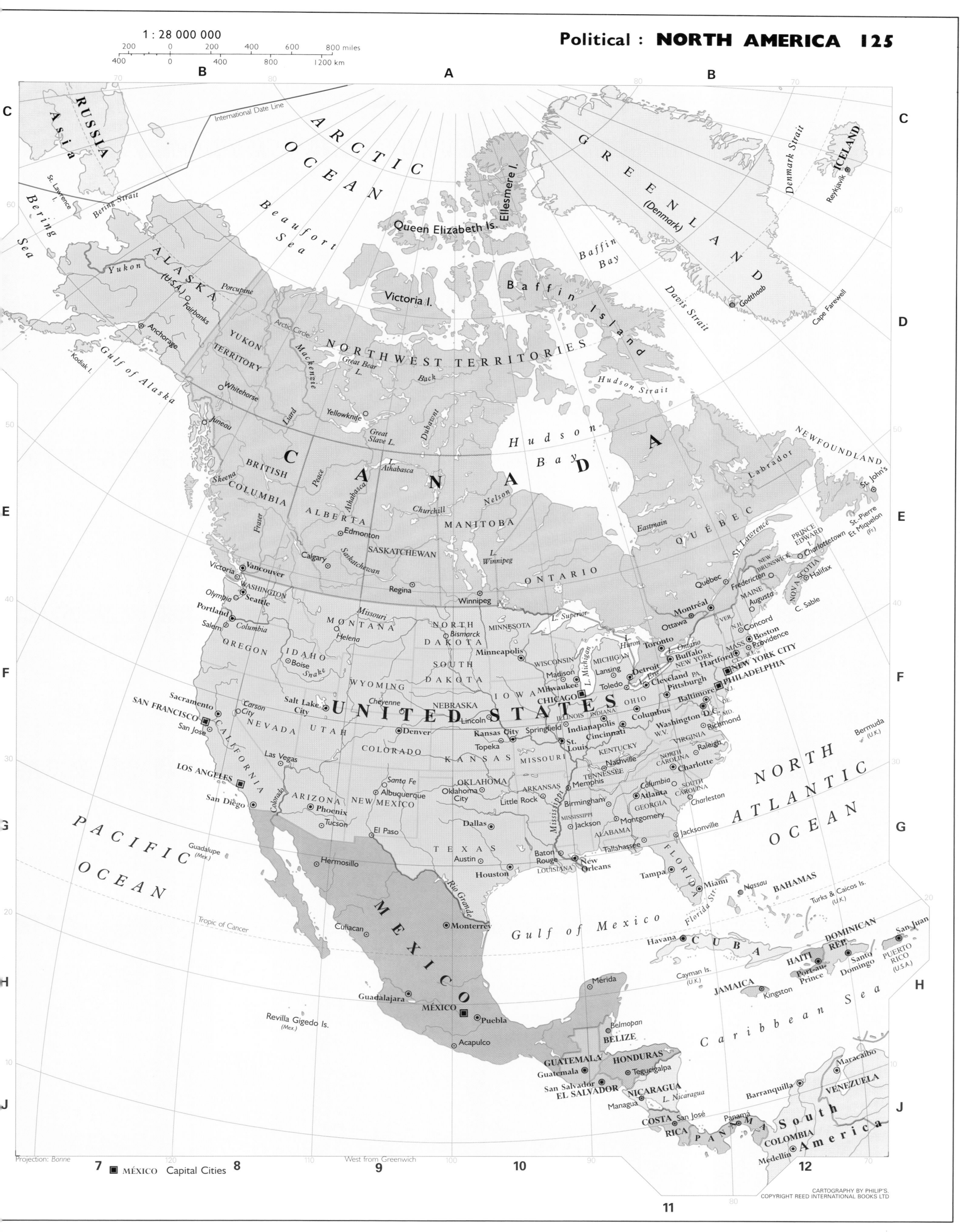

1 : 28 000 000

200 0 200 400 600 800 miles
400 0 400 800 1200 km

C **RUSSIA**
Asia
St. Lawrence I.
Bering
Sea
Bering Strait

ARCTIC
OCEAN

International Date Line

Beaufort
Sea

Queen Elizabeth Is.

Ellesmere I.

GREENLAND

Denmark Strait

ICELAND
Reykjavik

Baffin
Bay

Davis Strait

Godthåb

Cape Farewell

Yukon
ALASKA
(U.S.A.)
Porcupine
Fairbanks
Anchorage
Kodiak I.
Gulf of Alaska

Victoria I.

Baffin Island

YUKON
TERRITORY
Whitehorse
Mackenzie
Liard
Arctic Circle

Great Bear L.

NORTHWEST TERRITORIES

Hudson Strait

NEWFOUNDLAND

Juneau
BRITISH
COLUMBIA
Skeena
Fraser

C A N A D A
Yellowknife
Great Slave L.
Back
Dubawnt

Hudson
Bay

Labrador

St. John's

Peace
Athabasca
L. Athabasca
Churchill
Nelson

QUÉBEC
St. Lawrence

PRINCE EDWARD I.
St-Pierre Et Miquelon (Fr.)

ALBERTA
Edmonton
Calgary
SASKATCHEWAN
Saskatchewan
MANITOBA
L. Winnipeg

ONTARIO
Eastmain

Québec
NEW BRUNSWICK
Fredericton
Charlottetown
NOVA SCOTIA
Halifax
C. Sable

Victoria
Vancouver
WASHINGTON
Seattle
Olympia
Regina
Winnipeg

Portland
Columbia
Salem
OREGON

Montréal
Ottawa
MAINE
Augusta
VER.
N.H.
Concord
MASS. **Boston**
Hartford *Providence*
NEW YORK CITY
PHILADELPHIA
N.J.
Baltimore
DE.
Washington D.C.
MD.

MONTANA
Helena
Bismarck
NORTH DAKOTA
MINNESOTA
L. Superior
Minneapolis
WISCONSIN
Madison

L. Huron
Toronto
L. Ontario
Buffalo
L. Erie
Cleveland
PA.
Pittsburgh
IDAHO
Boise
Snake
WYOMING
Cheyenne
SOUTH DAKOTA

L. Michigan
Milwaukee
Lansing
MICHIGAN
Detroit
OHIO
Columbus
W.V.
Richmond
VIRGINIA

Sacramento
SAN FRANCISCO
Carson City
San Jose
NEVADA
Salt Lake City
UTAH
CALIFORNIA

Las Vegas

U N I T E D   S T A T E S
NEBRASKA
Lincoln
IOWA
ILLINOIS
CHICAGO
INDIANA
Indianapolis
Cincinnati
KENTUCKY
Nashville
TENNESSEE

Denver
COLORADO
Kansas City
Topeka
KANSAS
Springfield
St. Louis
MISSOURI

Raleigh
NORTH CAROLINA
Charlotte
SOUTH CAROLINA

Bermuda (U.K.)

NORTH
ATLANTIC
OCEAN

LOS ANGELES
San Diego
Santa Fe
Albuquerque
ARIZONA
Phoenix
Tucson
NEW MEXICO
Colorado

Oklahoma City
OKLAHOMA
ARKANSAS
Little Rock
Memphis
MISSISSIPPI

Columbia
Atlanta
GEORGIA
Birmingham
ALABAMA
Montgomery
Charleston

Guadalupe
(Mex.)

P A C I F I C

O C E A N

El Paso
Dallas
T E X A S
Austin
Houston
Baton Rouge
LOUISIANA
New Orleans
Jackson
Tallahassee
Jacksonville
FLORIDA
Tampa

Tropic of Cancer

Hermosillo
Culiacan
M E X I C O
Monterrey

Rio Grande

Gulf of Mexico

Miami
Nassau
BAHAMAS
Florida Str.
Turks & Caicos Is. (U.K.)

Havana
CUBA
Cayman Is. (U.K.)
JAMAICA
Kingston

DOMINICAN REP.
HAITI
Port-au-Prince
Santo Domingo
PUERTO RICO (U.S.A.)
San Juan

Caribbean   *Sea*

Revilla Gigedo Is. (Mex.)

Guadalajara
MÉXICO
Puebla
Mérida
Acapulco

Belmopan
BELIZE
GUATEMALA
Guatemala
HONDURAS
Tegucigalpa
EL SALVADOR
San Salvador
NICARAGUA
Managua
L. Nicaragua
COSTA RICA
San José
PANAMA
Panamá

Maracaibo
Barranquilla
VENEZUELA

South
America
COLOMBIA
Medellín

Projection: Bonne

7   ■ **MÉXICO** Capital Cities   8   9   *West from Greenwich*   10   11   12

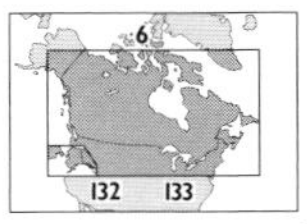

PACIFIC OCEAN

ALASKA

YUKON TERRITORY

BRITISH COLUMBIA

ALBERTA

SASKATCHEWAN

MANITOBA

NORTHWEST TERRITORIES

KITIKMEOT

KEEWATIN

Banks Island

Victoria Island

Prince of Wales Island

Boothia Peninsula

Amundsen Gulf

Coronation Gulf

Queen Maud Gulf

Melville Sound

Viscount Sound

Rocky Mountains

Mackenzie Mountains

Gr. Bear Lake

Gr. Slave L.

Lake Athabasca

Great Slave L.

WASHINGTON
MONTANA
NORTH DAKOTA
SOUTH DAKOTA
WYOMING
NEBRASKA
MINNESOTA

UNITED STATES

Vancouver
Victoria
Seattle
Tacoma
Spokane
Edmonton
Calgary
Lethbridge
Medicine Hat
Saskatoon
Regina
Moose Jaw
Winnipeg
St. Boniface
Brandon
Yellowknife
Whitehorse
Anchorage
Fairbanks
Minneapolis
St. Paul
Omaha
Sioux City
Des Moines

Queen Charlotte Is.

Vancouver I.

Projection: Bonne

ALASKA

1 : 24 000 000

100   0   100   200   300 miles
100   0   200   400 km

RUSSIA

Chukotskoye More (Chukchi Sea)

Brooks Range

Seward Pen.

BERING SEA

Bristol Bay

Kodiak I.

Aleutian Is.

Alaska Peninsula

GULF OF ALASKA

PACIFIC OCEAN

Barrow
Prudhoe Bay
Nome
Fairbanks
Anchorage
Valdez
Cordova
Juneau
Ketchikan
Whitehorse

Mt. McKinley 6194

West from Greenwich

1 : 12 000 000

West from Greenwich

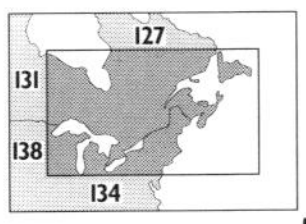

MANITOBA

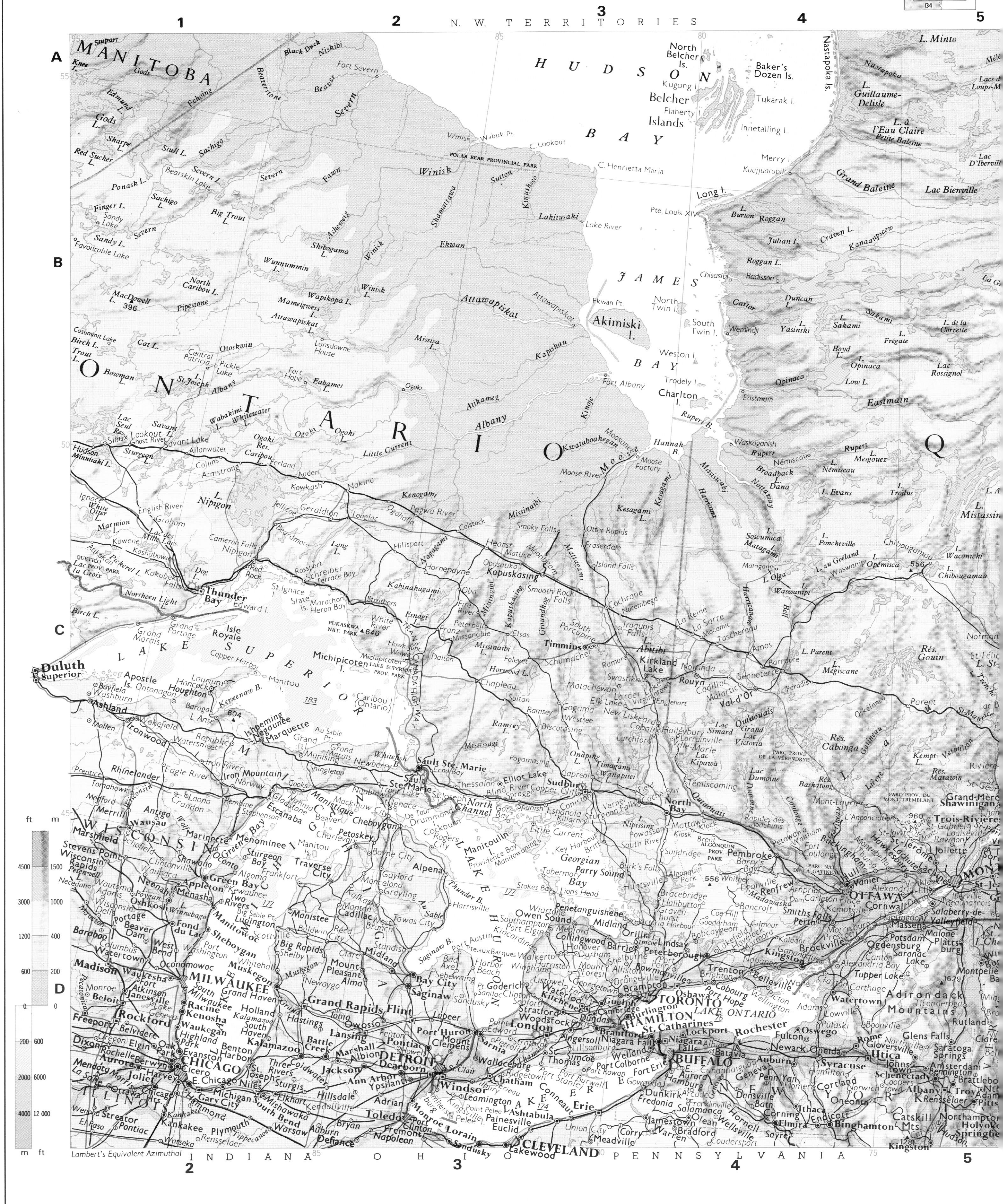

ONTARIO

QUEBEC

N.W. TERRITORIES

HUDSON BAY

JAMES BAY

LAKE SUPERIOR

LAKE HURON

LAKE ONTARIO

LAKE ERIE

LAKE MICHIGAN

WISCONSIN

MICHIGAN

ILLINOIS

INDIANA

OHIO

PENNSYLVANIA

NEW YORK

Belcher Islands

Akimiski I.

North Belcher Is.

Baker's Dozen Is.

Kugong I.

Tukarak I.

Flaherty I.

Innetalling I.

Georgian Bay

Manitoulin I.

Isle Royale

Apostle Is.

Polar Bear Provincial Park

Pukaskwa Nat. Park

Lake Superior Prov. Park

Algonquin Prov. Park

Quetico Prov. Park

Adirondack Mountains

Parc Nat. de la Gatineau

Parc Prov. de la Vérendrye

Parc Prov. du Mont-Tremblant

Cities and towns:

Thunder Bay, Duluth, Superior, Ashland, Ironwood, Marquette, Houghton, Sault Ste. Marie, Sudbury, North Bay, Timmins, Kirkland Lake, Kapuskasing, Hearst, Cochrane, Noranda, Rouyn, Val-d'Or, New Liskeard, Haileybury, Cobalt, Elliot Lake, Espanola, Parry Sound, Owen Sound, Collingwood, Barrie, Orillia, Midland, Peterborough, Lindsay, Belleville, Trenton, Cobourg, Oshawa, TORONTO, HAMILTON, St. Catharines, Niagara Falls, Welland, Brantford, Guelph, Kitchener, Cambridge, Woodstock, London, Stratford, Sarnia, Windsor, Chatham, Leamington, DETROIT, Dearborn, Ann Arbor, Toledo, CLEVELAND, Lakewood, Sandusky, Lorain, Elyria, Ashtabula, Erie, CHICAGO, Cicero, Aurora, Joliet, Gary, East Chicago, Hammond, MILWAUKEE, Racine, Kenosha, Waukesha, Madison, Beloit, Rockford, Freeport, Janesville, Green Bay, Appleton, Oshkosh, Fond du Lac, Sheboygan, Manitowoc, Wausau, Marshfield, Stevens Point, Wisconsin Rapids, Rhinelander, Merrill, Antigo, Marinette, Menominee, Escanaba, Iron Mountain, Grand Rapids, Muskegon, Lansing, Flint, Saginaw, Bay City, Pontiac, Battle Creek, Kalamazoo, Jackson, Port Huron, Mount Clemens, Holland, Traverse City, Petoskey, Cheboygan, Cadillac, Big Rapids, Mount Pleasant, Midland, Alma, Alpena, South Bend, Elkhart, Gary, Michigan City, OTTAWA, Hull, Vanier, Cornwall, Brockville, Kingston, Smiths Falls, Perth, Pembroke, Renfrew, Arnprior, Trois-Rivières, Shawinigan, Grand-Mère, Joliette, Montréal, Rochester, Buffalo, Niagara Falls, Lockport, Batavia, Tonawanda, Auburn, Syracuse, Utica, Rome, Oneida, Watertown, Ogdensburg, Massena, Potsdam, Plattsburg, Saranac Lake, Ticonderoga, Glens Falls, Saratoga Springs, Schenectady, Albany, Troy, Binghamton, Elmira, Corning, Ithaca, Oneonta, Amsterdam, Gloversville

Lake Nipigon, L. Seul, Lac Seul, Lake St. Joseph, Lake of the Woods

Trans-Canada Highway

Keweenaw B.

Saginaw B.

Thunder B.

Lambert's Equivalent Azimuthal

Map index: ft / m elevation legend: 4500 1500, 3000 1000, 1200 400, 600 200, 0 0, 600 200, 6000 2000, 12000 4000

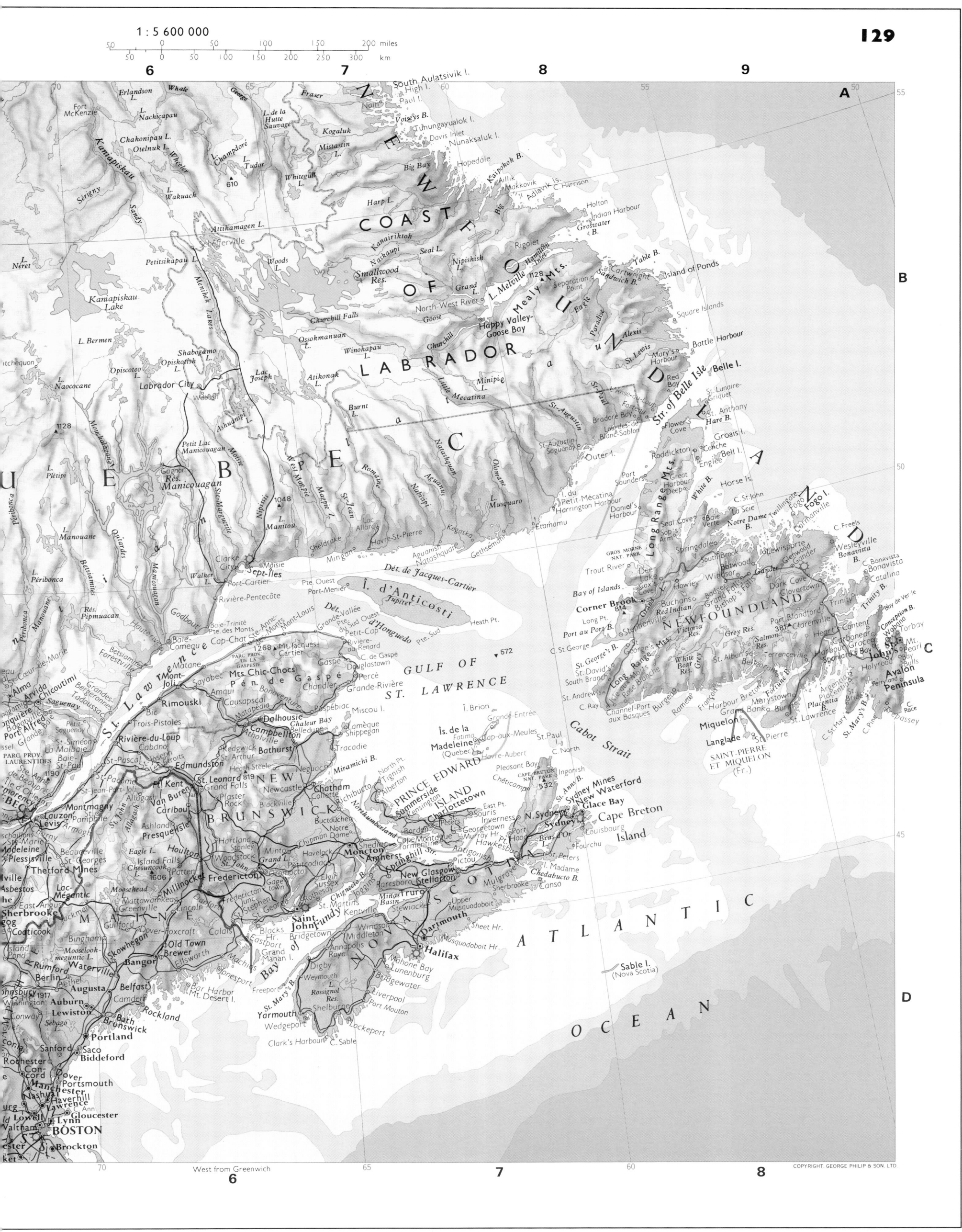

1 : 5 600 000

50 0 50 100 150 200 miles
50 0 50 100 150 200 250 300 km

COAST OF

LABRADOR

Q U E B E C

NEWFOUNDLAND

GULF OF
ST. LAWRENCE

Î. d'Anticosti

NEW
BRUNSWICK

PRINCE EDWARD
ISLAND

NOVA SCOTIA

Cape Breton
Island

Cabot Strait

SAINT-PIERRE
ET MIQUELON
(Fr.)

Avalon
Peninsula

Halifax

MAINE

BOSTON

ATLANTIC

OCEAN

Sable I.
(Nova Scotia)

West from Greenwich

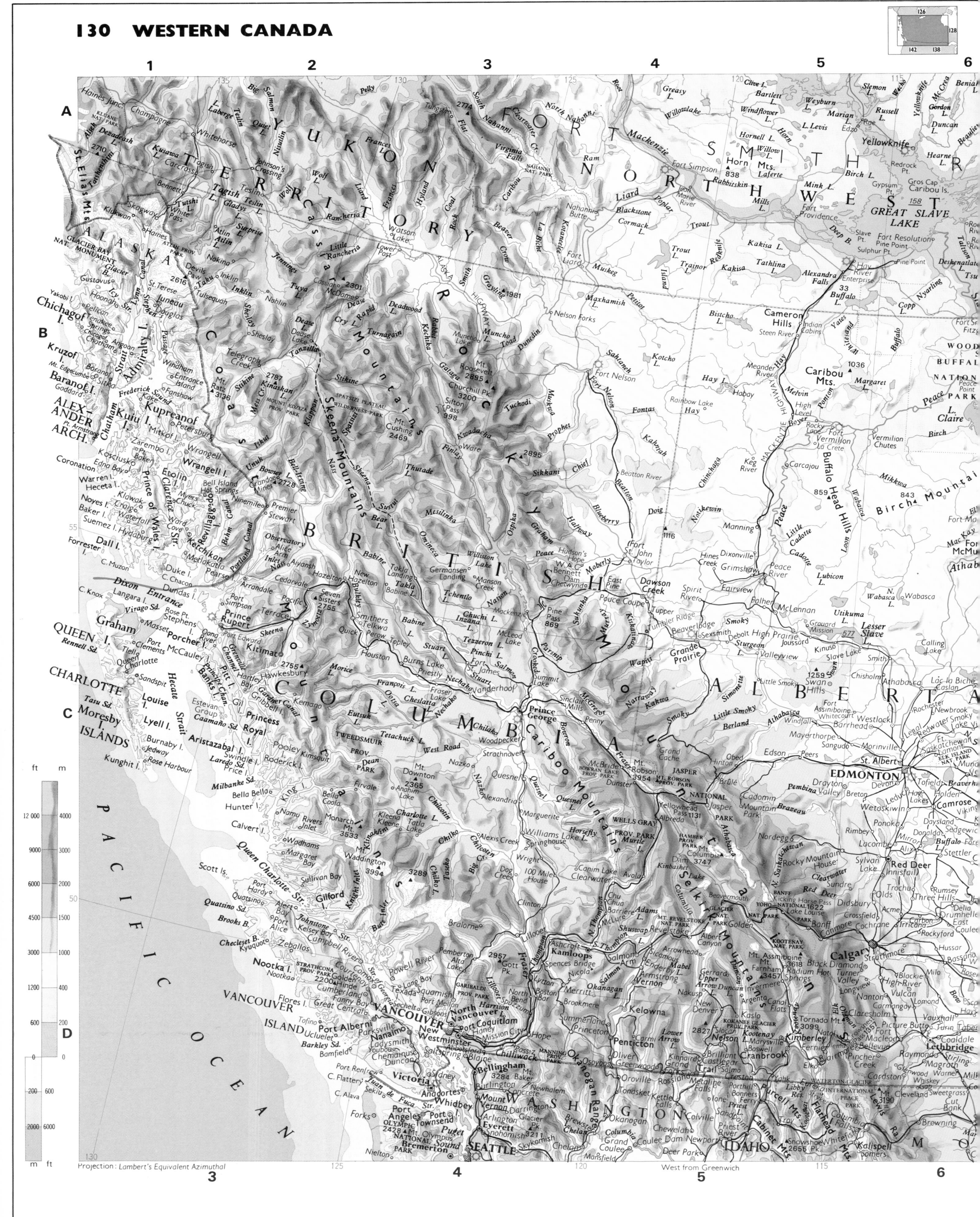

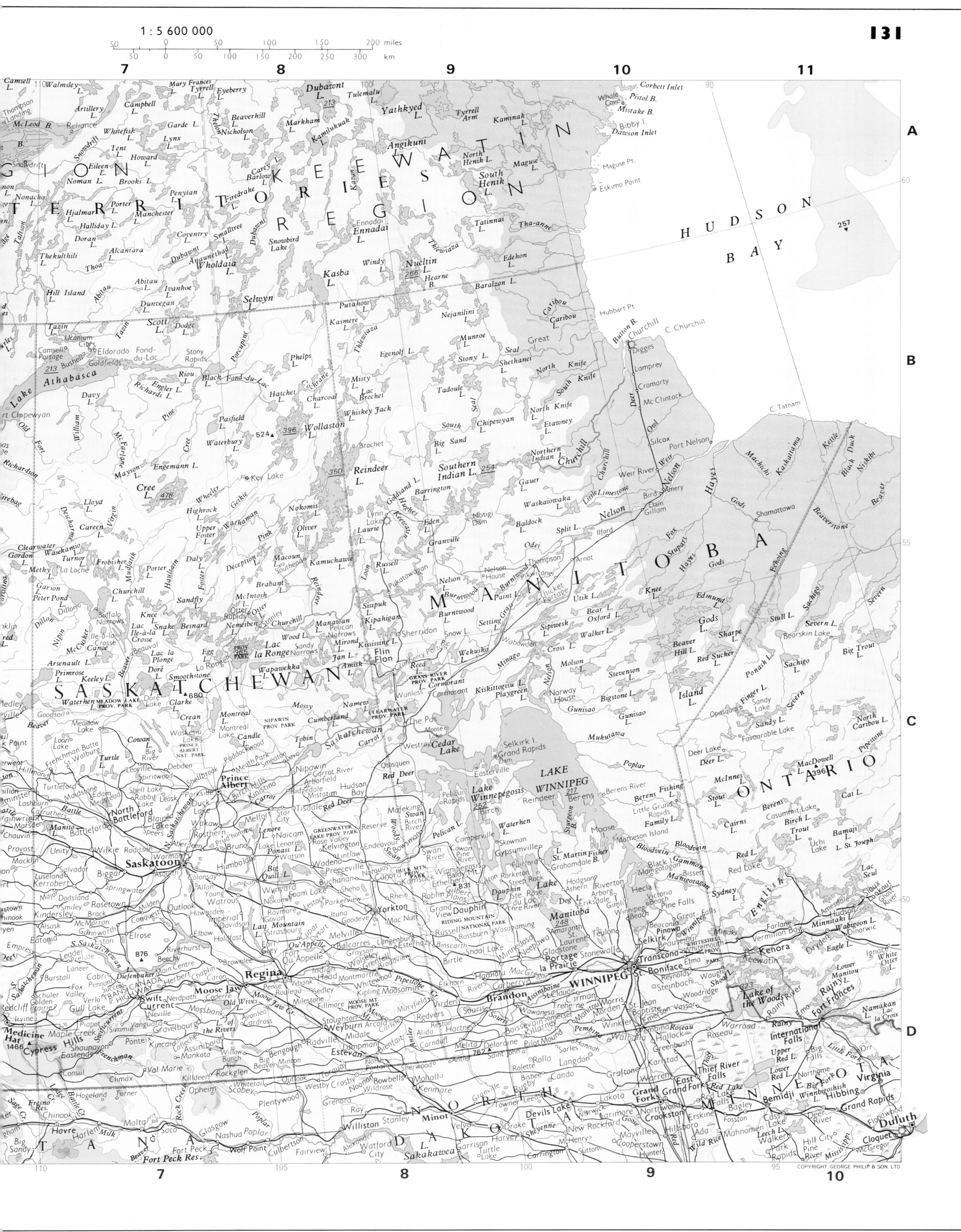

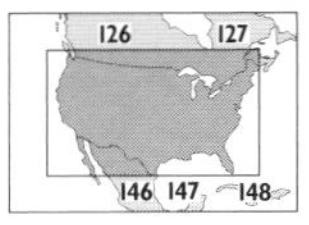

UNITED STATES

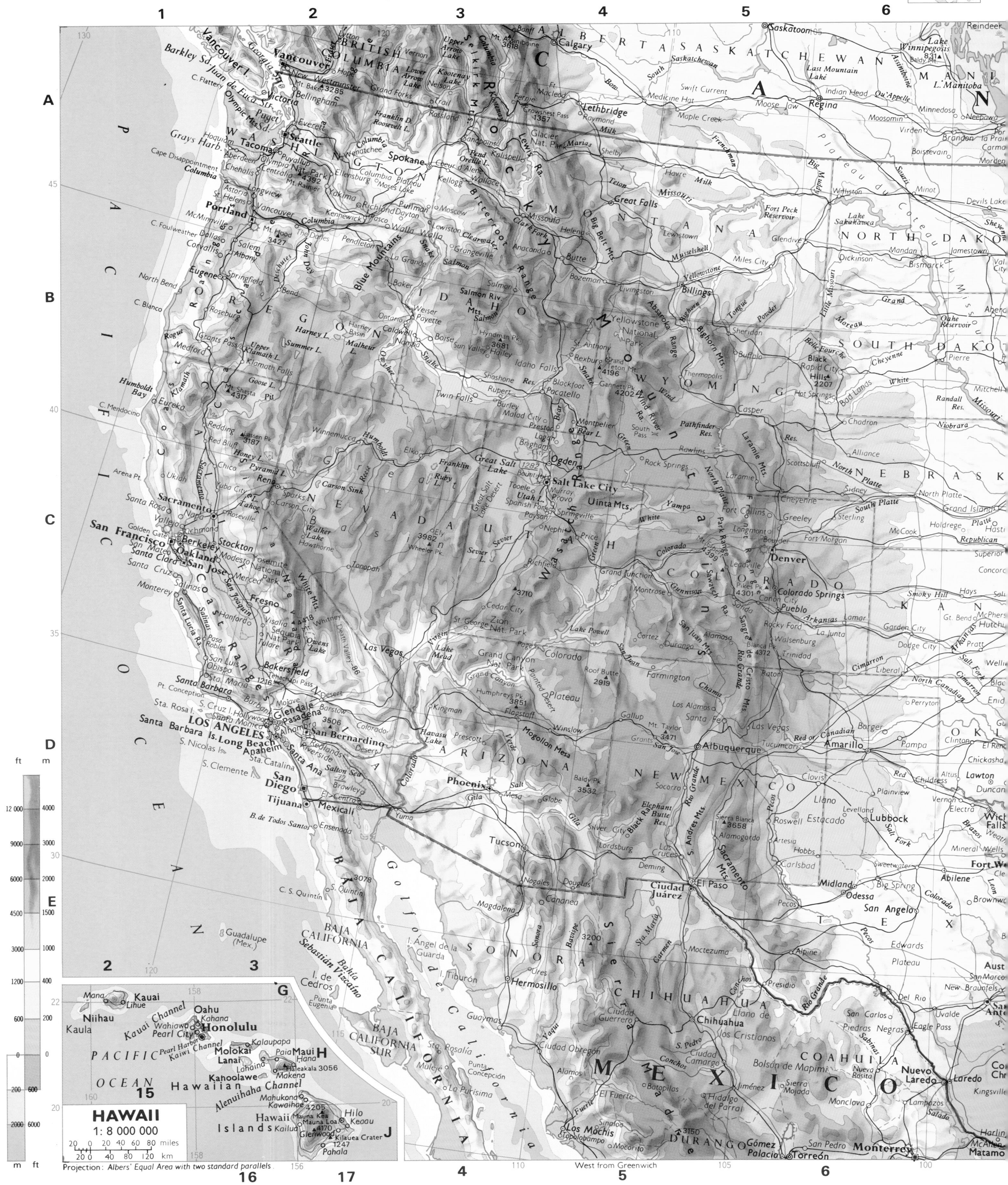

Map labels (selected):

BRITISH COLUMBIA · ALBERTA · SASKATCHEWAN · MANITOBA

Vancouver I. · Victoria · Vancouver · New Westminster · Bellingham · Seattle · Tacoma · Olympia · Everett · Spokane

WASHINGTON · OREGON · IDAHO · MONTANA · WYOMING · NEVADA · UTAH · COLORADO · ARIZONA · NEW MEXICO · CALIFORNIA · NORTH DAKOTA · SOUTH DAKOTA · NEBRASKA · KANSAS · OKLAHOMA · TEXAS

Calgary · Lethbridge · Medicine Hat · Moose Jaw · Regina · Saskatoon

Portland · Salem · Eugene · Corvallis · Albany · Medford · Roseburg

Great Falls · Helena · Butte · Bozeman · Billings · Missoula

Boise · Idaho Falls · Pocatello · Twin Falls

Yellowstone National Park · Casper · Cheyenne · Laramie

Salt Lake City · Ogden · Provo · Great Salt Lake

Denver · Colorado Springs · Pueblo · Grand Junction · Durango

Sacramento · San Francisco · Oakland · Berkeley · San Jose · Stockton · Modesto · Fresno · Bakersfield

LOS ANGELES · Long Beach · Pasadena · San Bernardino · Anaheim · Santa Ana · Riverside · San Diego · Tijuana · Mexicali

Las Vegas · Reno · Carson City

Phoenix · Tucson · Yuma · Flagstaff

Albuquerque · Santa Fe · Las Cruces · El Paso · Roswell · Lubbock · Amarillo

Ciudad Juárez · Chihuahua · Hermosillo · Ciudad Obregón · Los Mochis

BAJA CALIFORNIA · BAJA CALIFORNIA SUR · SONORA · CHIHUAHUA · COAHUILA · DURANGO · MEXICO

Nuevo Laredo · Laredo · Monterrey · Torreón · Gómez Palacio

PACIFIC OCEAN · Golfo de California

Hawaii inset:

HAWAII · 1:8 000 000

Kauai · Niihau · Oahu · Honolulu · Pearl City · Molokai · Lanai · Maui · Kahoolawe · Hawaii

Mauna Kea 4205 · Mauna Loa 4170 · Hilo · Kailua

PACIFIC OCEAN · Hawaiian Islands · Kaiwi Channel · Kauai Channel · Alenuihaha Channel

Projection: Albers' Equal Area with two standard parallels

West from Greenwich

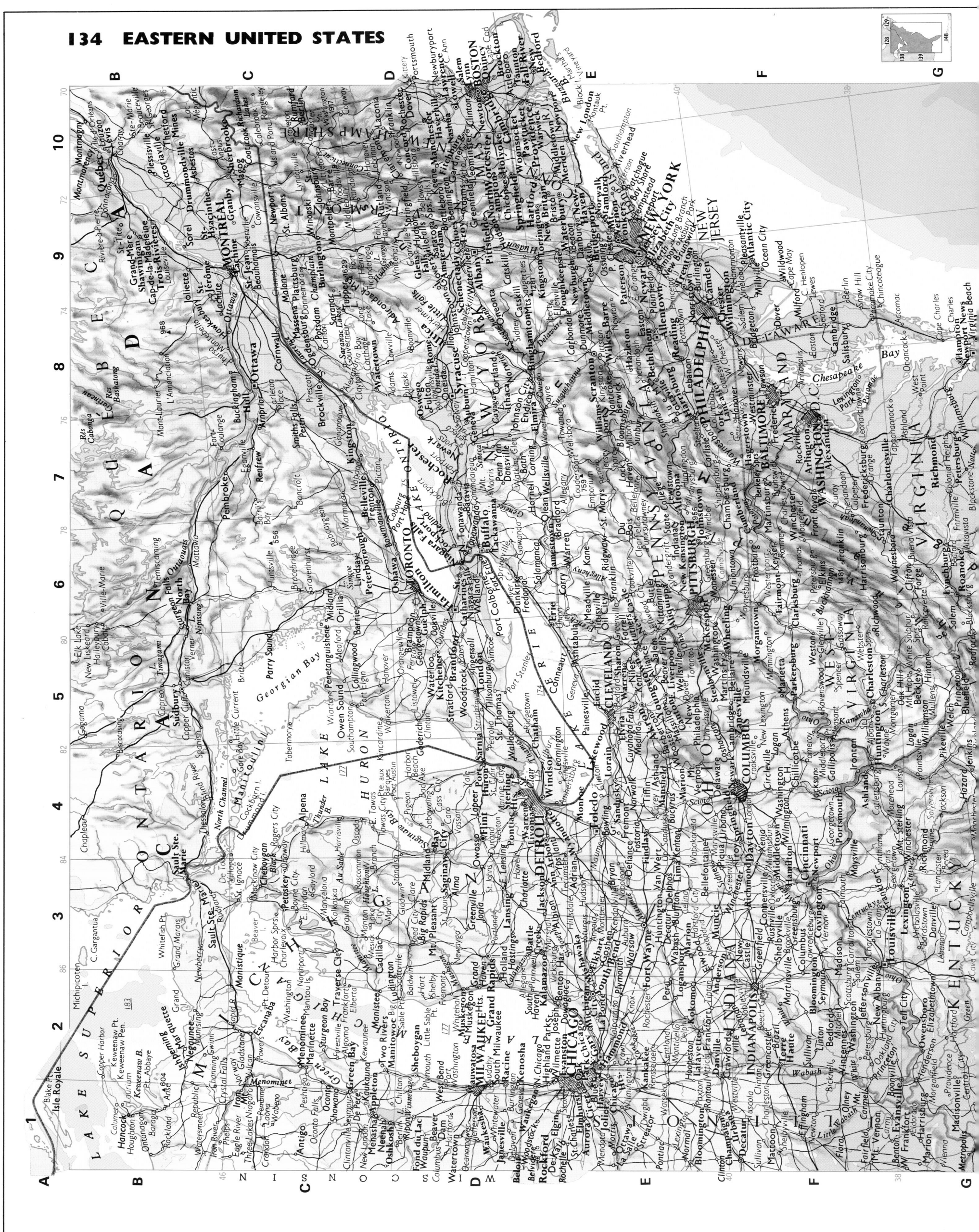

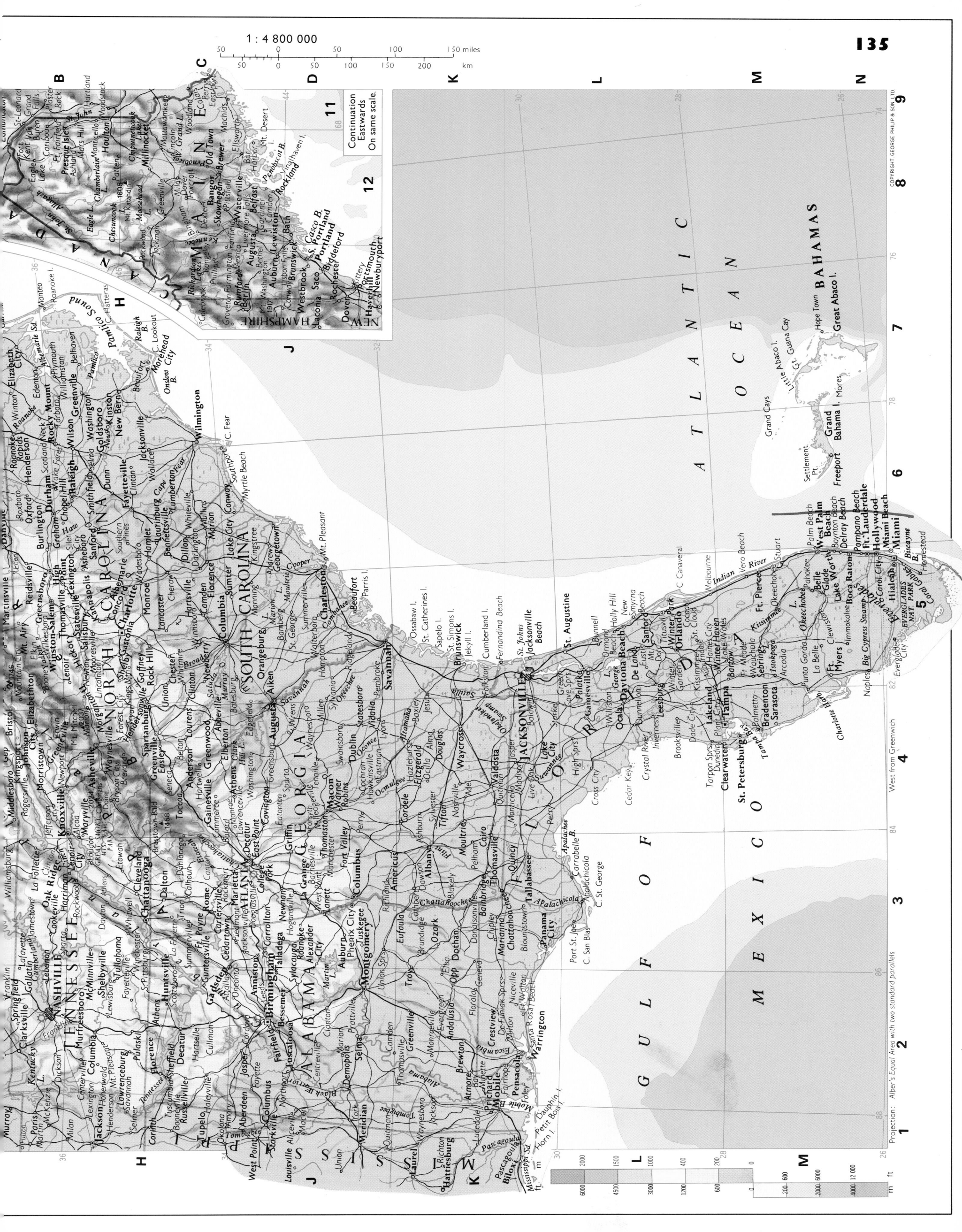

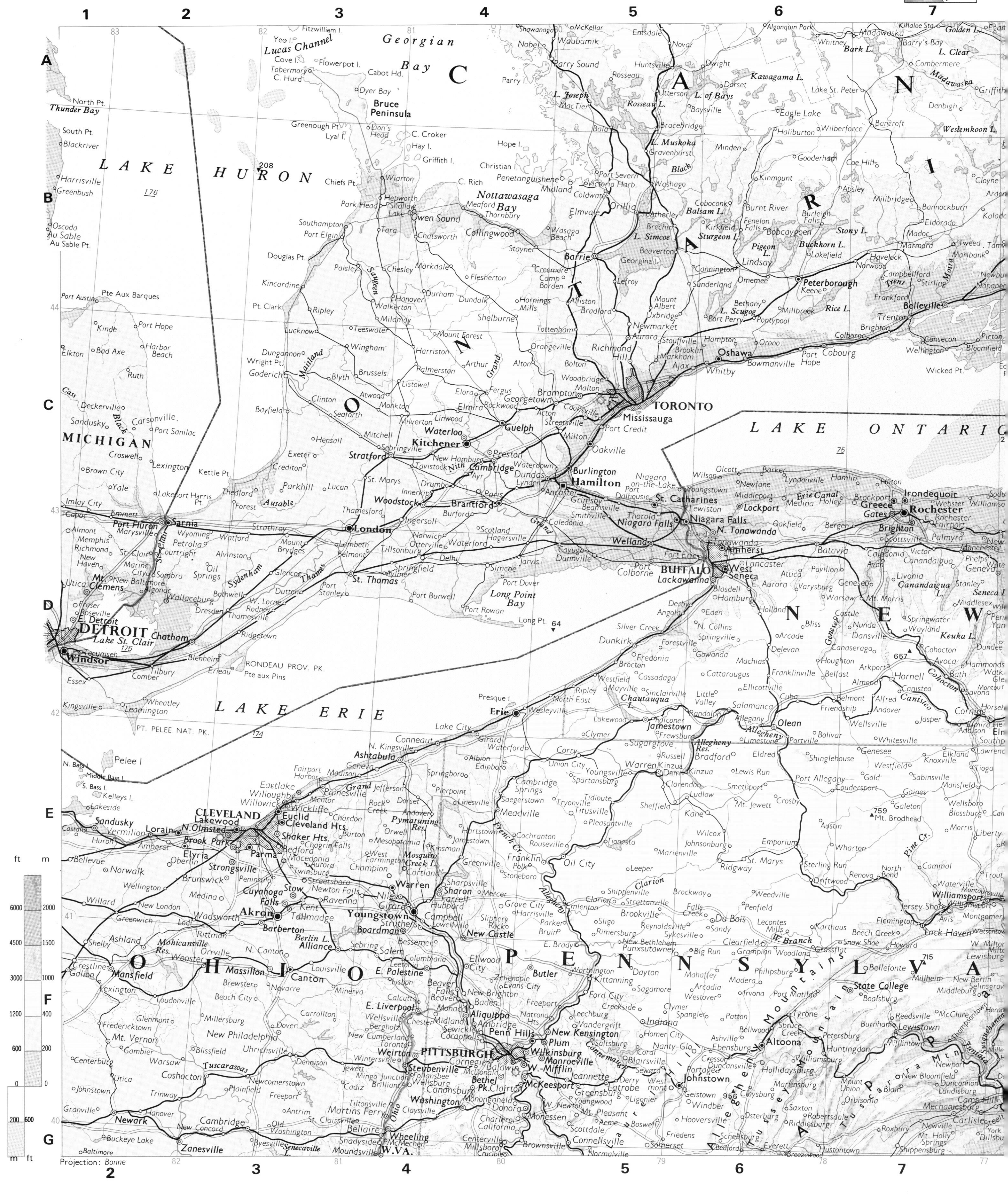

LAKE HURON

LAKE ONTARIO

LAKE ERIE

Georgian Bay

Bruce Peninsula

Nottawasaga Bay

Meaford Bay

MICHIGAN

ONTARIO

CANADA

TORONTO

Mississauga

Hamilton

Buffalo

Lackawanna

West Seneca

NEW YORK

DETROIT

Windsor

Lake St. Clair

CLEVELAND

Lakewood

Euclid

Shaker Hts.

Parma

Akron

Youngstown

Warren

OHIO

PENNSYLVANIA

PITTSBURGH

Penn Hills

McKeesport

Williamsport

State College

Altoona

W.VA.

Wheeling

Niagara Falls

St. Catharines

Rochester

Irondequoit

Greece

Gates

Brighton

Erie

Thunder Bay

Parry Sound

Huntsville

Peterborough

Belleville

Oshawa

Barrie

L. Simcoe

Owen Sound

Collingwood

Sarnia

Port Huron

London

St. Thomas

Long Point Bay

Ashtabula

Jamestown

Olean

Bradford

Johnstown

Allegheny Mts.

Projection: Bonne

ft	m
6000	2000
4500	1500
3000	1000
1200	400
600	200
0	0
200	600
m	ft

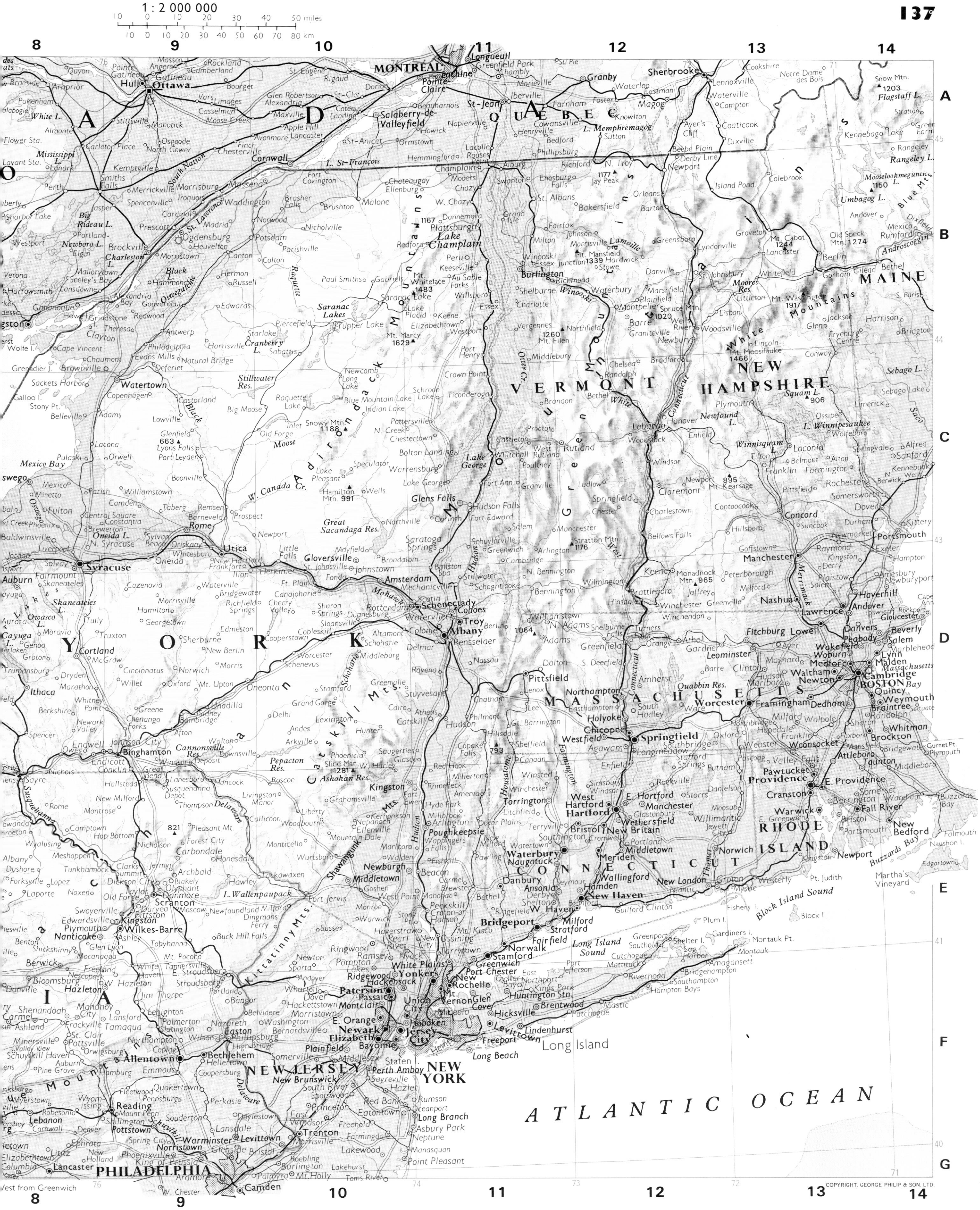

1 : 2 000 000

10 0 10 20 30 40 50 miles
10 0 10 20 30 40 50 60 70 80 km

8   9   10   11   12   13   14

MONTREAL Longueuil Greenfield Park Sherbrooke
Lachine

QUEBEC

ONTARIO

Ottawa Hull

Cornwall

Lake Champlain

Plattsburgh

Adirondack Mountains

Mt. Marcy 1629

Watertown

VERMONT

NEW HAMPSHIRE

Burlington Winooski

Montpelier Barre

Mt. Washington 1917 White Mountains

MAINE

Rutland

Lake George

Glens Falls

Saratoga Springs

Syracuse

Utica Rome

Gloversville

Amsterdam Schenectady Troy Albany Rensselaer

Cohoes

MASSACHUSETTS

Pittsfield

Springfield Chicopee Holyoke Northampton

Worcester

Framingham BOSTON Cambridge Quincy

Concord

Manchester Nashua

Lawrence Lowell Fitchburg

Portsmouth

NEW YORK

Binghamton Endicott Endwell

Scranton Wilkes-Barre Kingston

Ithaca Cortland

Oneonta

Kingston Poughkeepsie

Newburgh Middletown

CONNECTICUT

Hartford West Hartford E. Hartford New Britain

Waterbury Naugatuck Meriden Middletown

New Haven W. Haven Hamden

Bridgeport Stratford Milford Fairfield Norwalk

Danbury

New London Groton

RHODE ISLAND

Providence Pawtucket Cranston Warwick E. Providence

Fall River New Bedford

Block Island

Long Island Sound

Montauk Pt.

Stamford Greenwich Port Chester New Rochelle Yonkers White Plains

NEW JERSEY

Newark Jersey City Elizabeth Bayonne Paterson Passaic Montclair E. Orange

NEW YORK

Long Island

Hempstead Levittown Freeport Long Beach Lindenhurst

Trenton

PHILADELPHIA Camden

Reading Allentown Bethlehem Easton

Lancaster

ATLANTIC OCEAN

Asbury Park Long Branch Red Bank

West from Greenwich

8   9   10   11   12   13   14

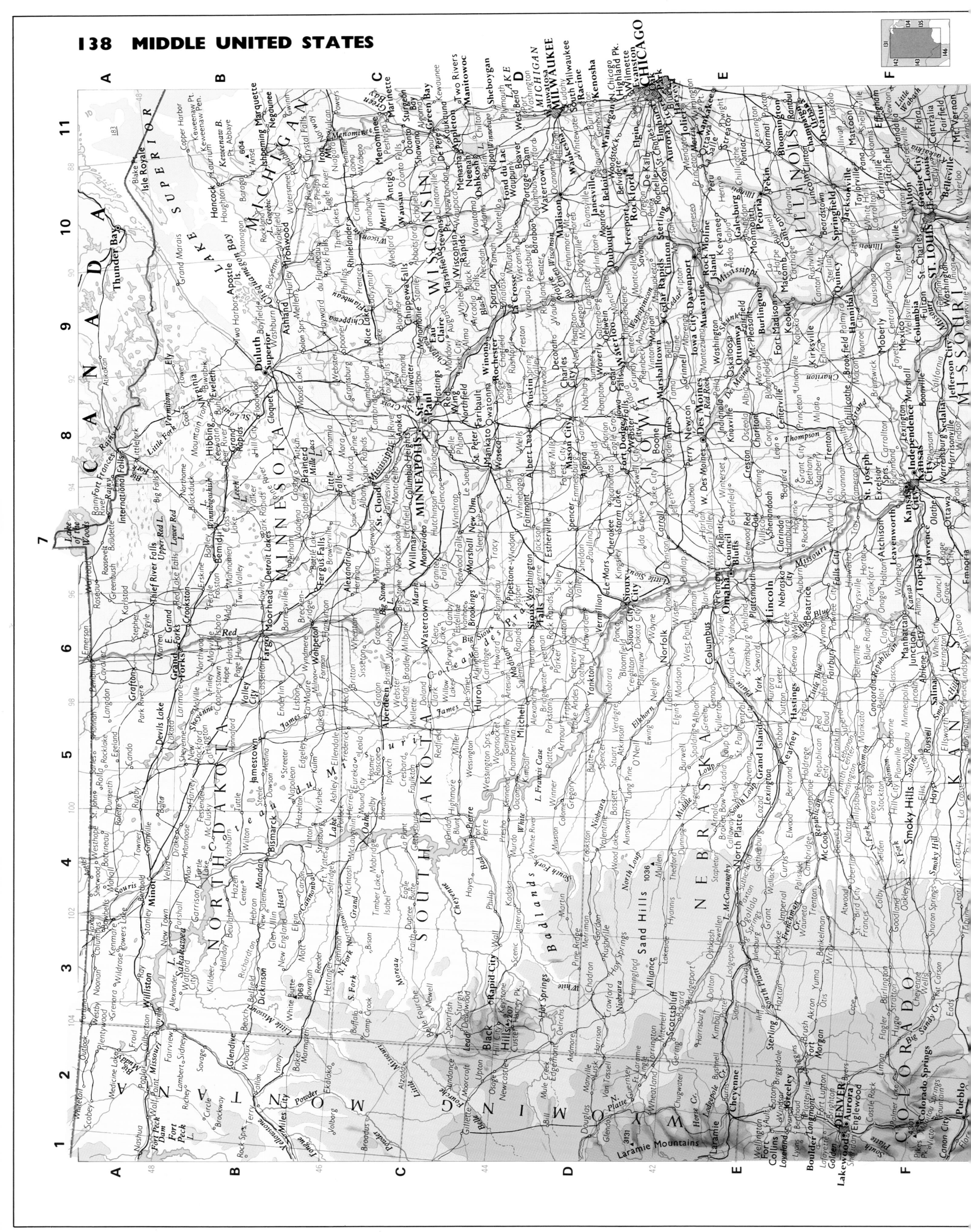

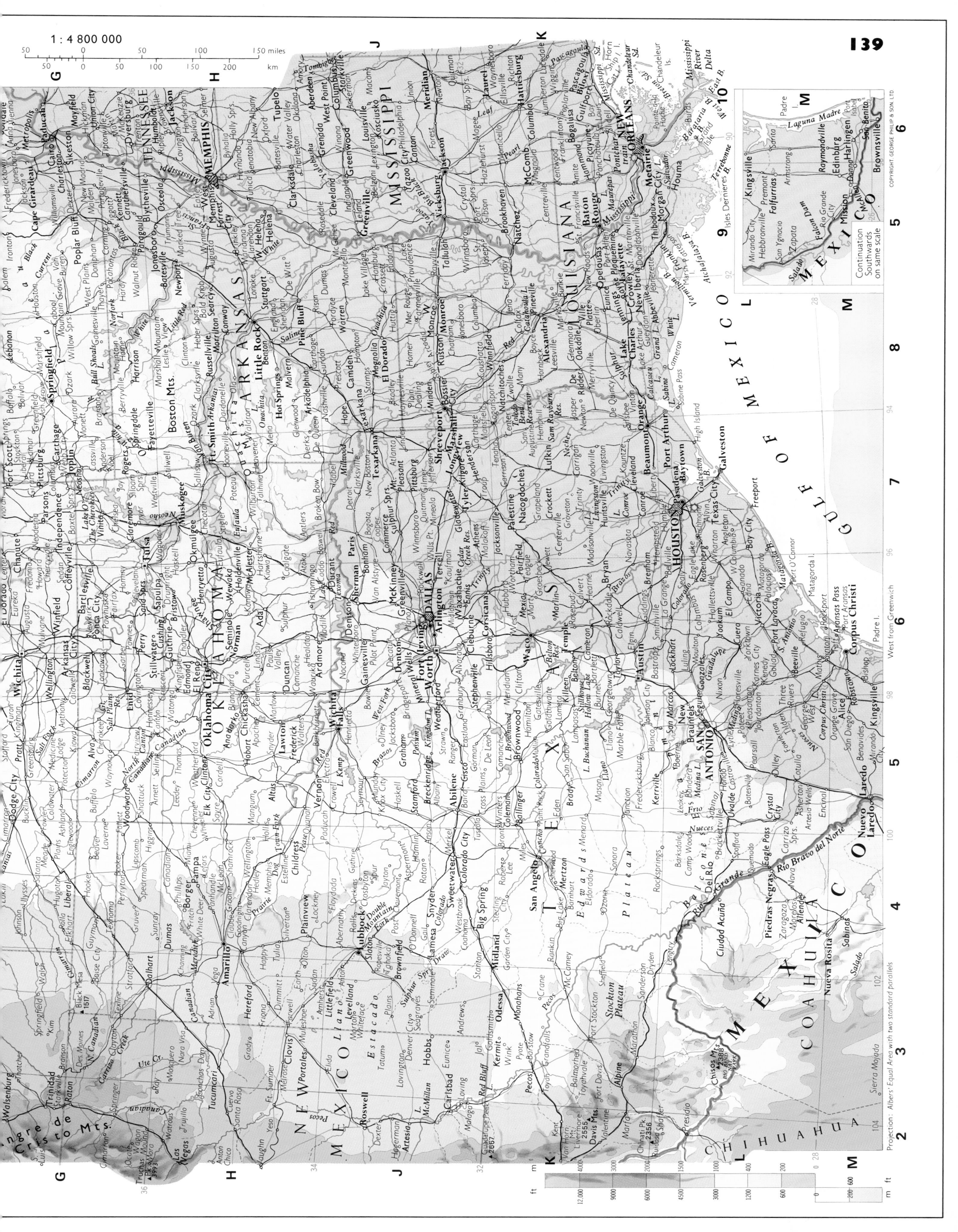

1 : 4 800 000

COPYRIGHT GEORGE PHILIP & SON, LTD.

Projection: Albers' Equal Area with two standard parallels

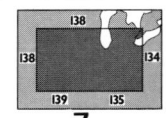

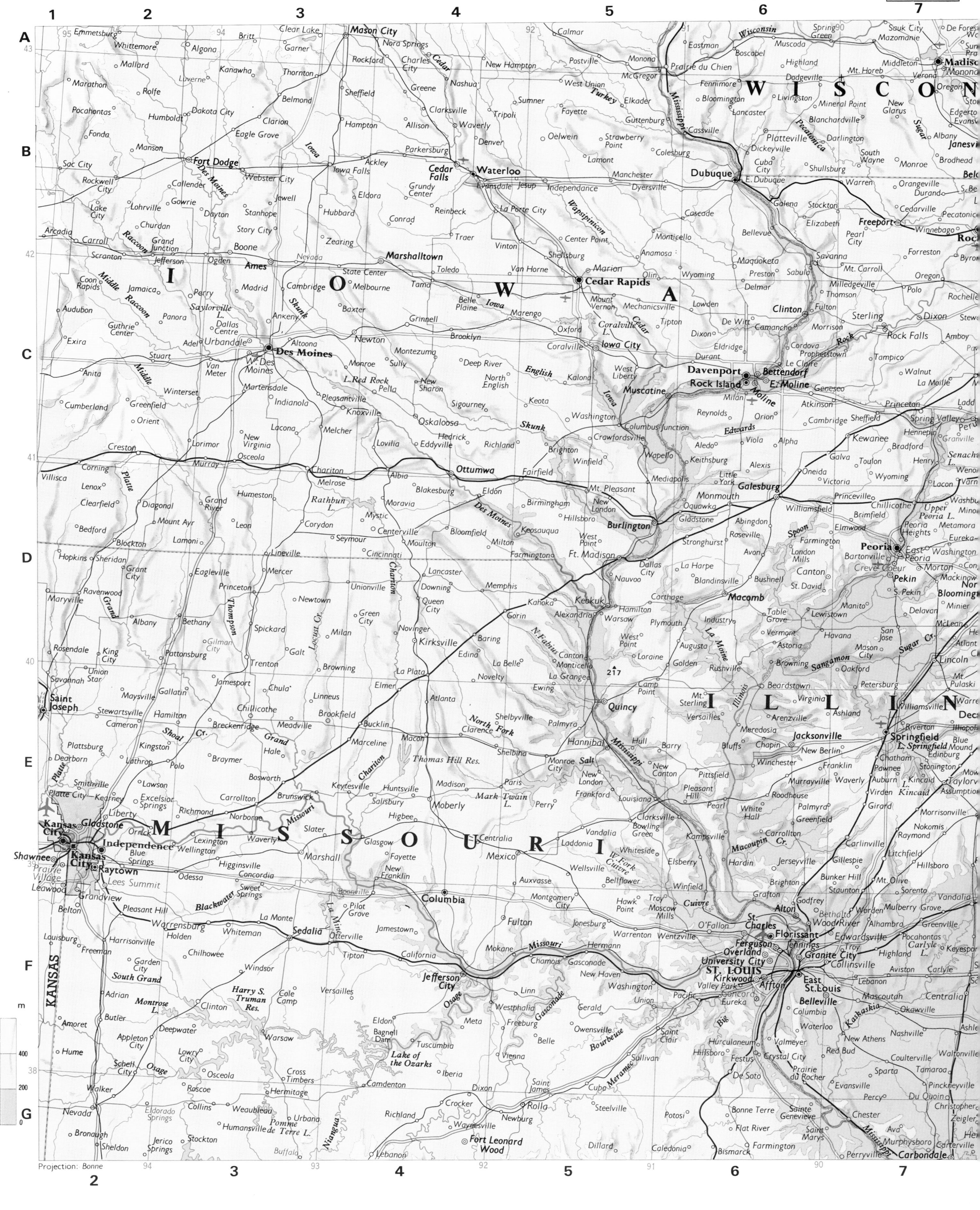

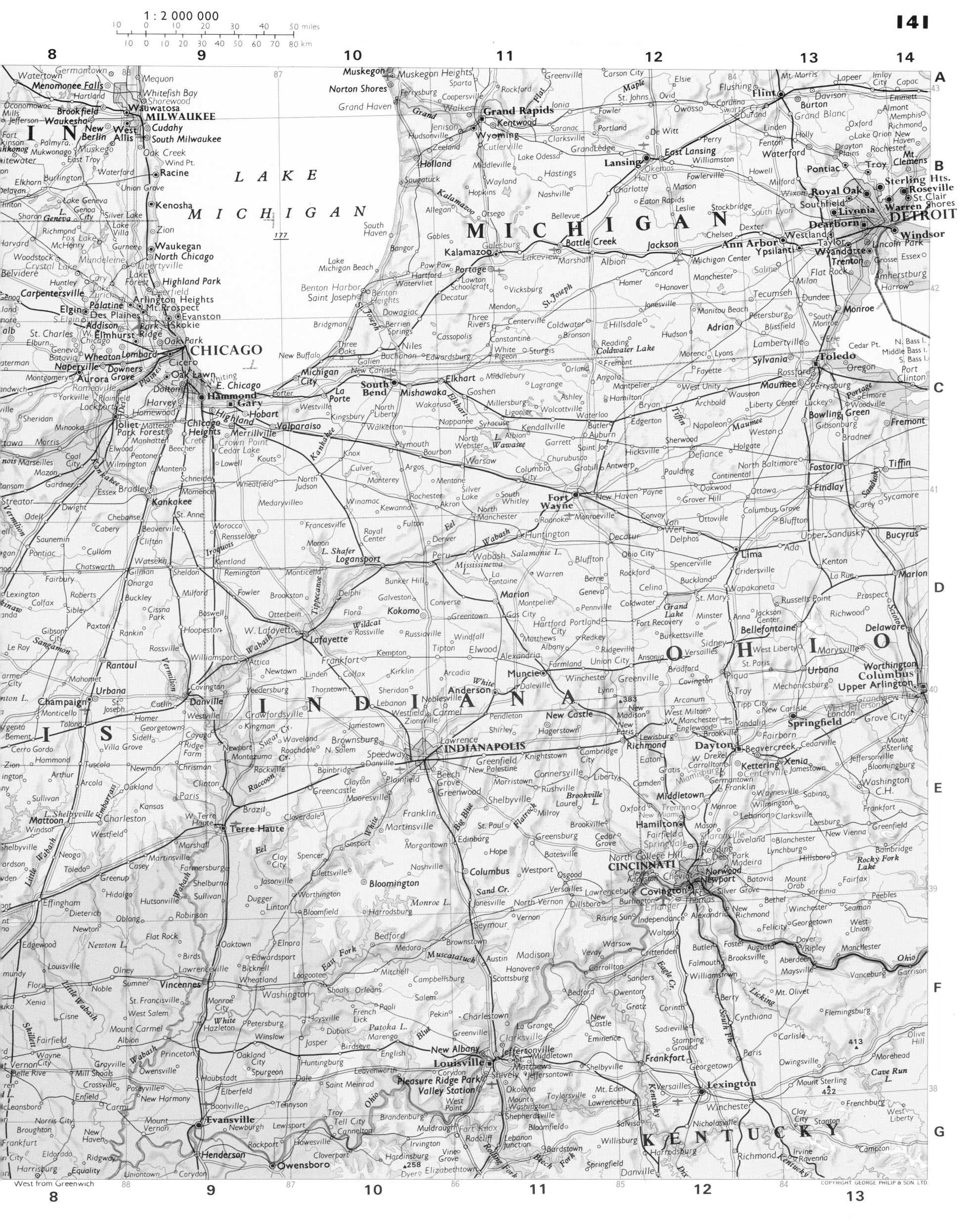

1 : 4 800 000

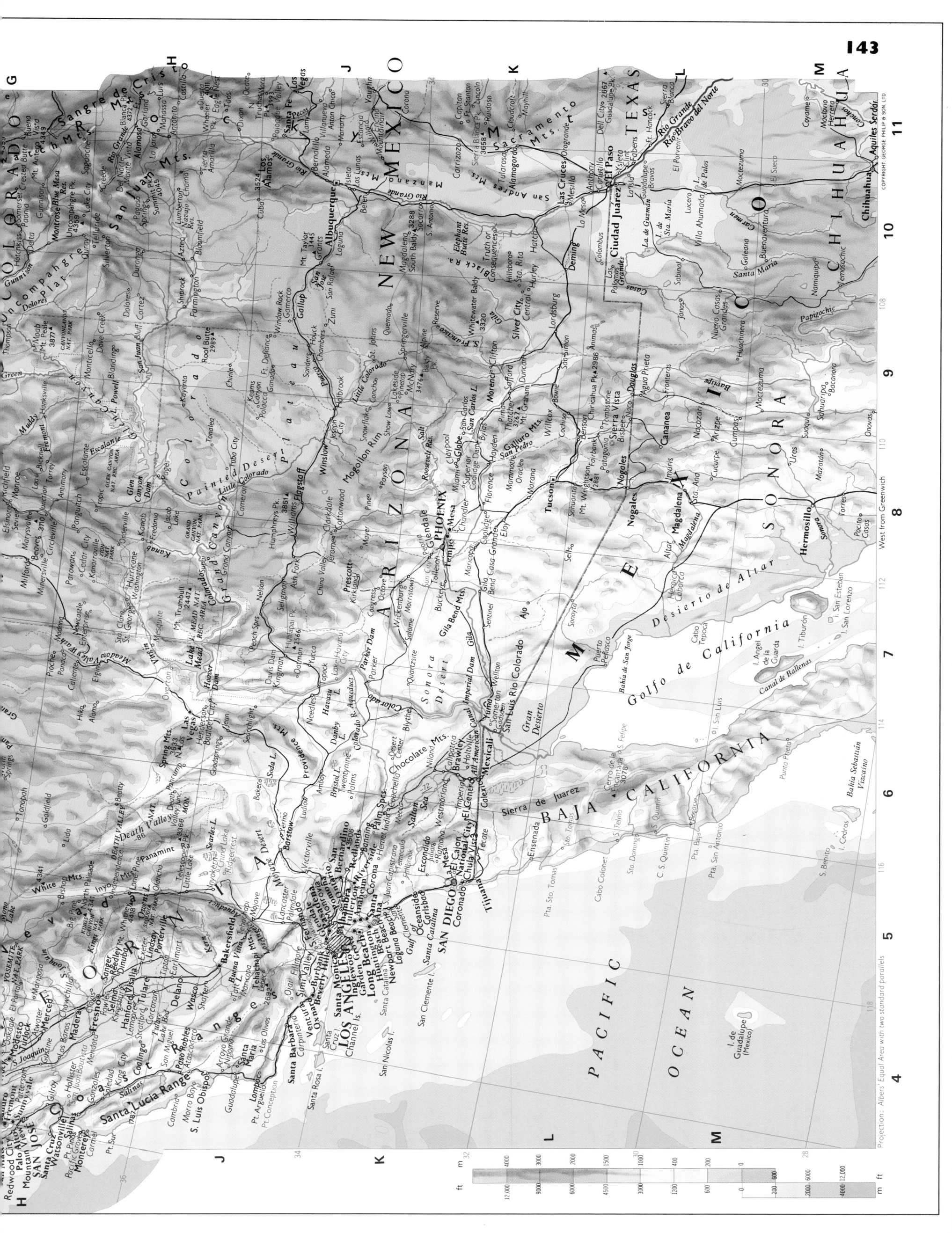

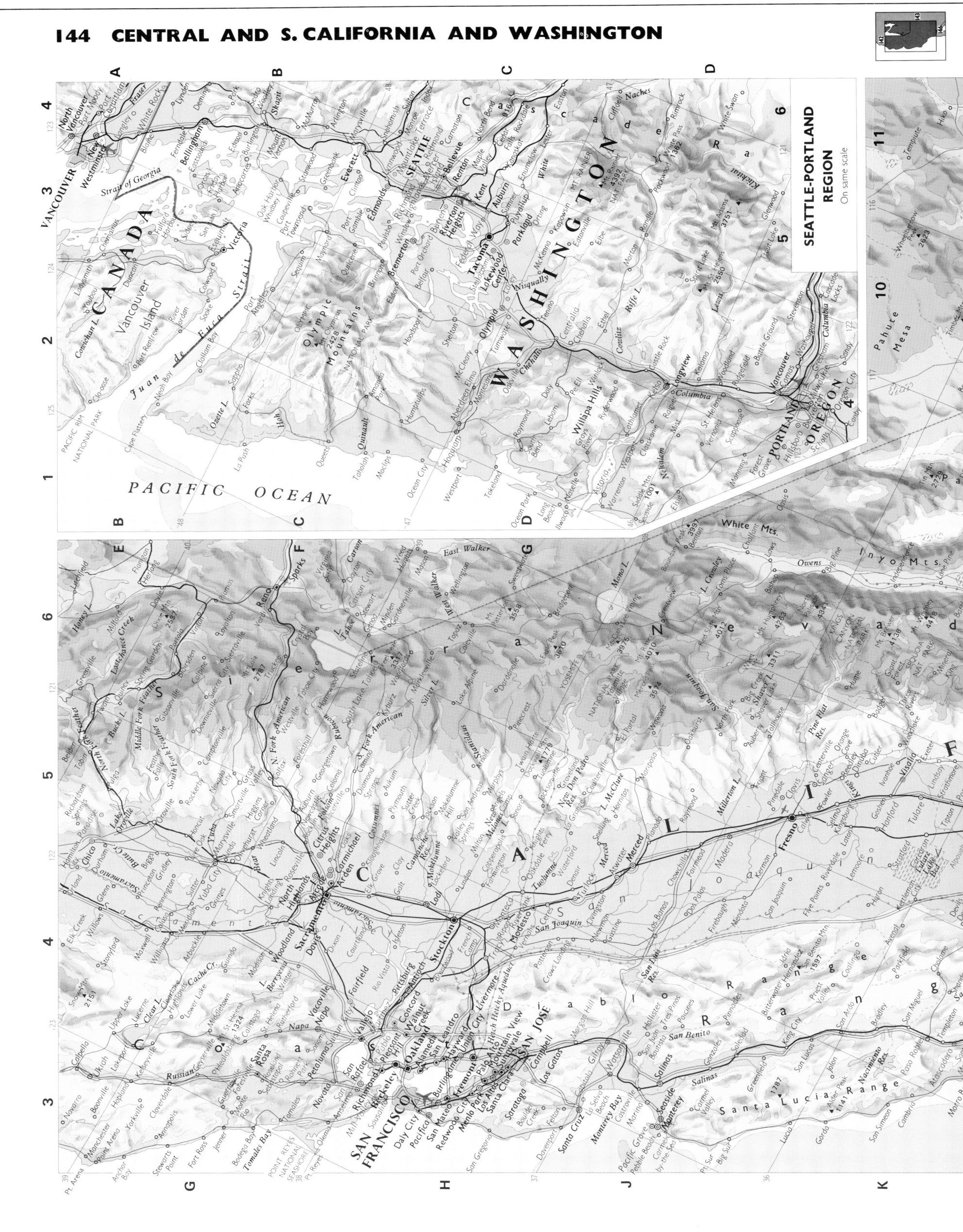

SEATTLE-PORTLAND REGION On same scale

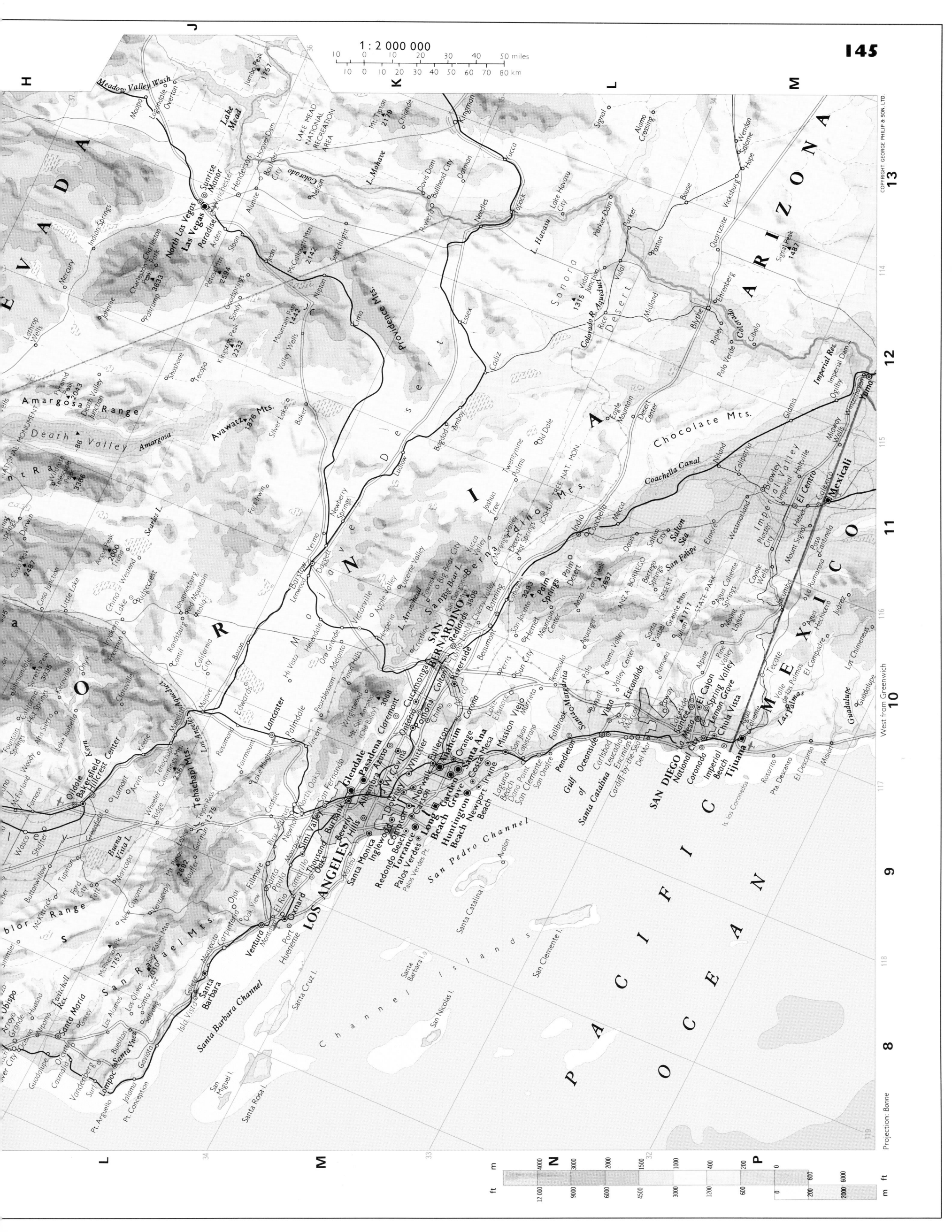

1 : 2 000 000

10   0   10   20   30   40   50 miles
10   0   10   20   30   40   50   60   70   80 km

N E V A D A

A R I Z O N A

C A L I F O R N I A

M E X I C O

P A C I F I C   O C E A N

LOS ANGELES

SAN BERNARDINO

SAN DIEGO

Death Valley

Lake Mead

Salton Sea

Mojave Desert

Colorado Desert

Chocolate Mts.

Santa Barbara Channel

Channel Islands

San Pedro Channel

Las Vegas

North Las Vegas

Santa Barbara

Santa Maria

Bakersfield

Lancaster

Palmdale

Pasadena

Long Beach

Santa Ana

Tijuana

Mexicali

El Centro

Yuma

Kingman

Needles

Palm Springs

Projection: Bonne

West from Greenwich

m   ft
4000 — 12 000
3000 — 9000
2000 — 6000
1500 — 4500
1000 — 3000
600 — 1200
400 — 600
200 —
0 —

ft   m
6000 — 2000
2000 — 600
500 — 200
0 — 0

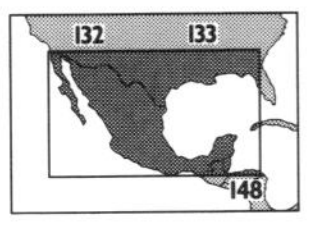

132 133
148

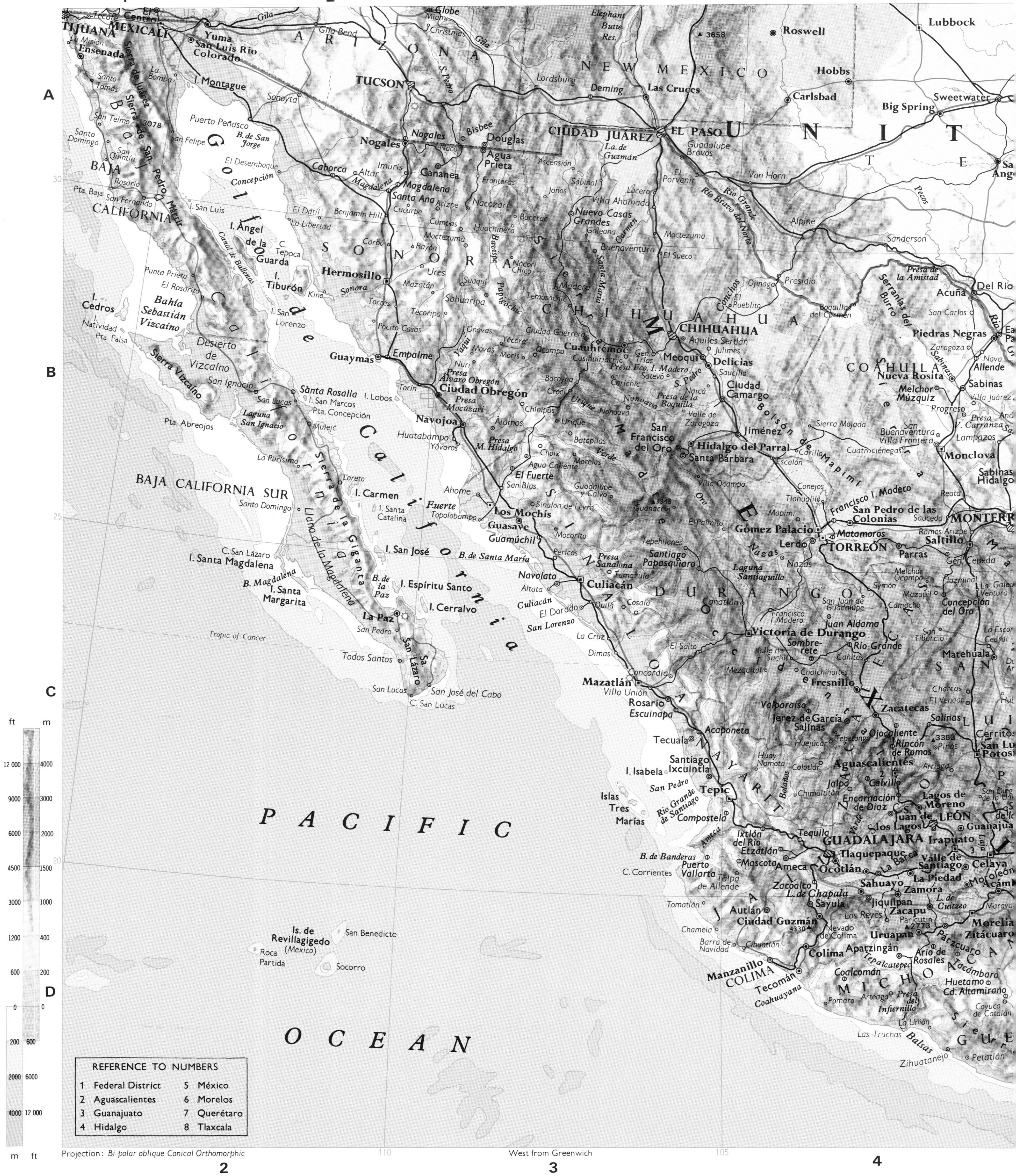

Grid references
1 2 3 4

A B C D

Map labels (selected)

United States / border region
Roswell, Lubbock, Hobbs, Carlsbad, Big Spring, Sweetwater, San Angelo, ARIZONA, NEW MEXICO, UNITED STATES, Tucson, Globe, Miami, Christmas, Gila Bend, Elephant Butte Res., 3658, Lordsburg, Deming, Las Cruces, Bisbee, Douglas, CIUDAD JUAREZ, EL PASO, Van Horn, Alpine, Sanderson, Del Rio, Presidio, Ojinaga, Pecos

Baja California
TIJUANA, MEXICALI, El Centro, Tecate, Ensenada, Misión, La Misión, Santo Tomás, San Telmo, 3078, Sierra de San Pedro Mártir, Sierra de San Juárez, Santo Domingo, San Quintín, San Fernando, Rosario, Pta. Baja, BAJA CALIFORNIA, Pta. Falsa, Natividad, I. Cedros, Punta Prieta, El Rosarito, Bahía Sebastián Vizcaíno, Desierto de Vizcaíno, Sierra Vizcaíno, Pta. Abreojos, BAJA CALIFORNIA SUR, Santo Domingo, San Ignacio, Laguna San Ignacio, La Purísima, San Lucas, Santa Rosalía, I. San Marcos, Pta. Concepción, Mulegé, Loreto, I. Carmen, I. Santa Catalina, C. San Lázaro, I. Santa Magdalena, B. Magdalena, I. Santa Margarita, I. San José, B. de la Paz, I. Espíritu Santo, La Paz, San Pedro, I. Cerralvo, Todos Santos, San Lázaro, San Lucas, San José del Cabo, C. San Lucas, Sierra de la Giganta, Llano de la Magdalena

Sonora / Gulf of California
Yuma, San Luis Río Colorado, Puerto Peñasco, B. de San Jorge, San Felipe, El Desemboque, I. San Luis, La Libertad, I. Ángel de la Guarda, C. Tepoca, I. Tiburón, El Datíl, Benjamín Hill, Nogales, Sonoyta, Caborca, Altar, Magdalena, Santa Ana, Cananea, Imuris, Arizpe, Cucurpe, Moctezuma, Hermosillo, Kino, I. San Esteban, I. Lobos, Guaymas, Empalme, Ures, Mazatán, Suaqui, Sahuaripa, Tecoripa, Pocito Casas, Onavas, Torres, Sonora, GOLFO DE CALIFORNIA, Yaqui, Nuri, Presa Álvaro Obregón, Ciudad Obregón, Torín, Huatabampo, Yávaros, Presa M. Hidalgo, Navojoa, Álamos, Presa Mocúzari, San Blas, Ahome, Los Mochis, Topolobampo, Fuerte, Guasave, Guamúchil

Chihuahua
Ascensión, La. de Guzmán, Janos, Buenaventura, El Sueco, Villa Ahumada, Nuevo Casas Grandes, Galeana, Carmen, Madera, Temósachic, Ciudad Guerrero, Cuauhtémoc, Cusihuiriáchic, Carichí, Creel, Bocoyna, Ocampo, Yécora, Maicoba, Moris, Uruáchic, Batopilas, Chínipas, Urique, Nonoava, CHIHUAHUA, Aquiles Serdán, Meoqui, Delicias, Saucillo, Santa Isabel, Presa de la Boquilla, Ciudad Camargo, Jiménez, Valle de Zaragoza, San Francisco del Oro, Hidalgo del Parral, Santa Bárbara, Villa Ocampo, Escalón, Conejos, Sierra Mojada, Tlahualilo, SIERRA MADRE OCCIDENTAL, Río Conchos, Río Bravo del Norte

Coahuila / Nuevo León
Acuña, Piedras Negras, Zaragoza, Nava, Allende, Sabinas, Villa Unión, Nueva Rosita, Melchor Múzquiz, Monclova, Sabinas Hidalgo, MONTERREY, Saltillo, Parras, Ramos Arizpe, General Cepeda, San Pedro de las Colonias, Matamoros, Gómez Palacio, Lerdo, TORREÓN, Viesca, Serranías del Burro, Presa de la Amistad, San Carlos, Boquillas del Carmen, BOLSÓN DE MAPIMÍ, Francisco I. Madero, Mapimí, Cuatrociénegas, Villa Frontera, San Buenaventura, Presa V. Carranza, Lampazos

Durango / Zacatecas / San Luis Potosí
DURANGO, Victoria de Durango, Canatlán, Santiago Papasquiaro, Tepehuanes, Guanaceví, El Palmito, Presa Lázaro Cárdenas, Nazas, Santiaguillo, Laguna, Cuencamé, Símon, Yerbanís, Pedriceña, Sombrerete, Río Grande, Cañitas, Fresnillo, Valparaíso, Jerez de García Salinas, Zacatecas, ZACATECAS, Ojocaliente, Salinas, Charcas, El Venado, Cerritos, San Luis Potosí, Matehuala, Catorce, El Salto, Concordia, Dimas, Mazatlán, Villa Unión, Rosario, Escuinapa, Acaponeta, Tecuala, Santiago Ixcuintla

Nayarit / Jalisco / Michoacán / Colima
NAYARIT, Tepic, I. Isabela, Islas Tres Marías, San Pedro, Río Grande de Santiago, Compostela, B. de Banderas, Puerto Vallarta, C. Corrientes, Talpa de Allende, Mascota, Ameca, Ixtlán del Río, Etzatlán, Tequila, GUADALAJARA, Tlaquepaque, Ocotlán, La Barca, Sahuayo, Zamora, La Piedad, Lagos de Moreno, Encarnación de Díaz, Aguascalientes, JALISCO, L. de Chapala, Sayula, Zacoalco, Autlán, Tomatlán, Chamela, Barra de Navidad, Cihuatlán, Manzanillo, COLIMA, Tecomán, Coahuayana, Pomaro, Colima, Nevado de Colima, 4330, Ciudad Guzmán, Los Reyes, Jiquilpan, Zacapu, L. de Cuitzeo, MICHOACÁN, Uruapan, Paricutín, Apatzingán, Coalcomán, Arteaga, Ario de Rosales, Tepalcatepec, Huetamo, Cd. Altamirano, La Unión, Las Truchas, Zihuatanejo, Petatlán, Morelia, Pátzcuaro, Zitácuaro, Acámbaro, Balsas

Central states
León, Guanajuato, Irapuato, Celaya, Valle de Santiago, Salamanca, Silao

Map geographic labels
GULF OF CALIFORNIA, PACIFIC OCEAN, Tropic of Cancer, Is. de Revillagigedo (Mexico), San Benedicto, Roca Partida, Socorro

Scale
ft / m
12 000 — 4000
9000 — 3000
6000 — 2000
4500 — 1500
3000 — 1000
1200 — 400
600 — 200
0 — 0
200 — 600
2000 — 6000
4000 — 12 000
m ft

REFERENCE TO NUMBERS
1 Federal District 5 México
2 Aguascalientes 6 Morelos
3 Guanajuato 7 Querétaro
4 Hidalgo 8 Tlaxcala

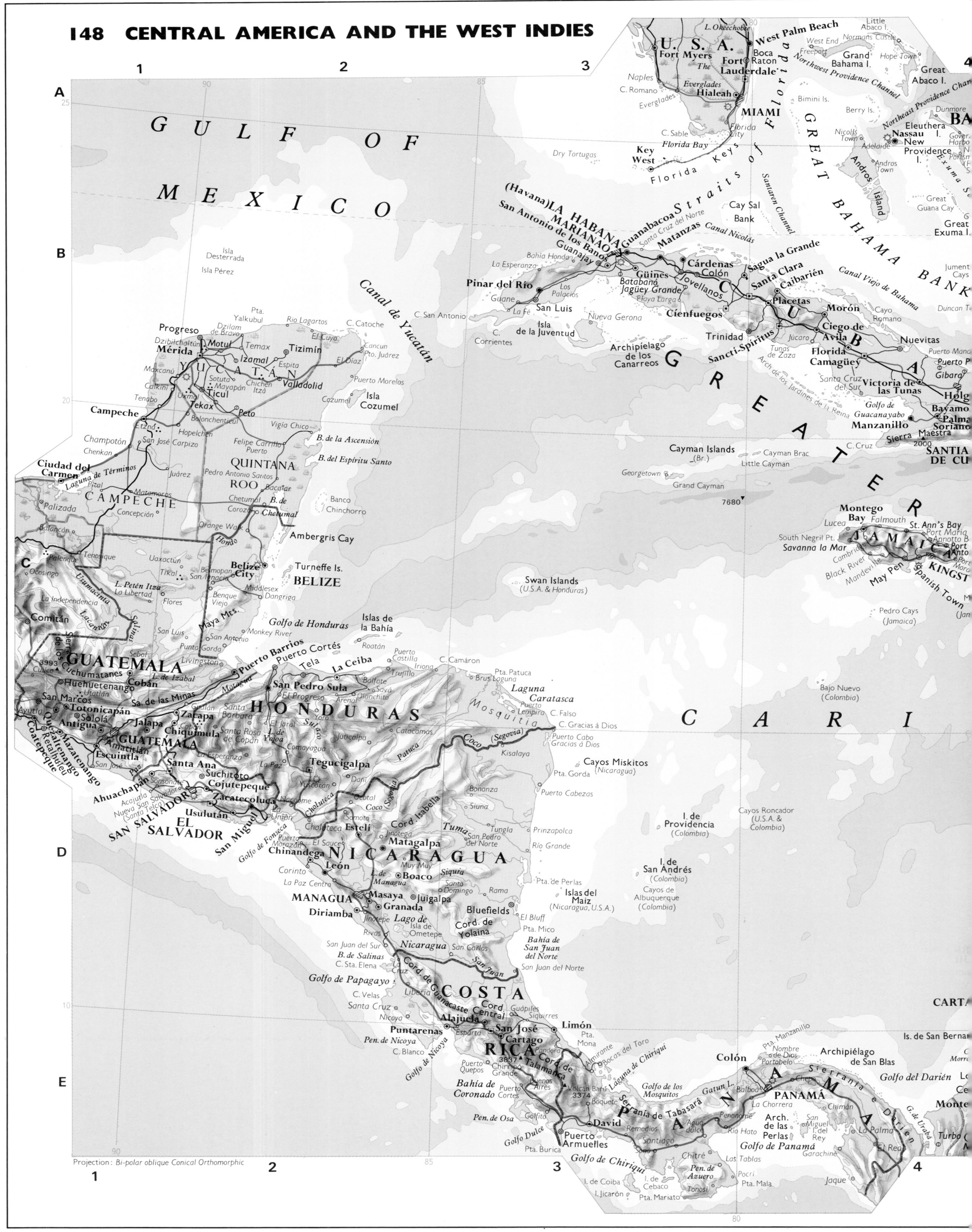

GULF OF MEXICO

U.S.A.
Fort Myers
Naples
C. Romano
Everglades
Hialeah
MIAMI
Key West
Dry Tortugas
Florida Bay
C. Sable
West Palm Beach
Boca Raton
Fort Lauderdale
Little Abaco I.
Great Abaco I.
Grand Bahama I.
West End
Freeport
Northwest Providence Channel
Bimini Is.
Berry Is.
Nassau
New Providence
Eleuthera I.
Governor's Harbour
Northeast Providence Channel
Great Guana Cay
Adelaide
Exuma Sound
Great Exuma I.

L. Okeechobee

Straits of Florida
(Havana) LA HABANA
MARIANAO
San Antonio de los Baños
Guanabacoa
Guanajay
Pinar del Río
Guane
La Fé
San Luis
Los Palacios
C. San Antonio
Bahía Honda
La Esperanza
Guïnes
Batabanó
Jagüey Grande
Cienfuegos
Nueva Gerona
Isla de la Juventud
C. Corrientes
Archipiélago de los Canarreos
Cárdenas
Colón
Jovellanos
Matanzas
Santa Clara
Caibarién
Placetas
Trinidad
Sancti-Spíritus
Santa Cruz del Norte
Sagua la Grande
Cay Sal Bank
Santaren Channel
Morón
Ciego de Ávila
Júcaro
Tunas de Zaza
Arch. de Zaza
Santa Cruz del Sur
Golfo de Guacanayabo
Manzanillo
Bayamo
Palma Soriano
Sierra Maestra
C. Cruz
Nuevitas
Camagüey
Victoria de las Tunas
Holguín
Gibara
Puerto Padre
Puerto P
2000
Canal Nicolás
Canal Viejo de Bahama
Cayo Romano
Duncan Te
Jumento Cays

GREATER
CUBA

GREAT BAHAMA BANK
Andros I.
Andros Town
Nicolls Town

Cayman Islands (Br.)
Georgetown
Grand Cayman
Cayman Brac
Little Cayman
7680

Swan Islands (U.S.A. & Honduras)

Montego Bay
Lucea
Falmouth
St. Ann's Bay
South Negril Pt.
Savanna la Mar
Black River
Mandeville
Spanish Town
KINGST
JAMAICA
May Pen
Port Maria
Annotto B
Port Antonio
Morant
Cambi

Pedro Cays (Jamaica)

CARI
CARI

Progreso
Dzilam de Bravo
Yalkubul
Pta.
Motul
Izamal
Temax
Tizimín
MÉRIDA
Dzibilchaltún
Maxcanú
Ticul
Tekax
Chichén Itzá
Mayapán
Valladolid
Sotuta
Peto
Bolonchenticul
C. Cuyo
Río Lagartos
C. Catoche
Cancún
Pto. Juárez
El Díaz
Isla Cozumel
Cozumel
Puerto Morelos
Vigía Chico
Campeche
Tenabo
Calkini
Umán
Hopelchén
Champotón
Chenkan
Felipe Carrillo Puerto
B. de la Ascensión
B. del Espíritu Santo
Ciudad del Carmen
Palizada
Laguna de Términos
Pital
Matamoros
Concepción
Balancán
Tenosique
Juárez
Pedro Antonio Santos
Bacalar
Chetumal
B. de Chetumal
Corozal
Banco Chinchorro
Orange Walk
Hondo
Ambergris Cay
Turneffe Is.
Middlesex
Dangriga
Palenque
Ocosingo
Tikal
Uaxactún
Flores
L. Petén Itzá
La Libertad
San Benito
Benque Viejo
Belmopan
BELIZE
Belize City
Comitán
La Independencia
Usumacinta
Sierra de los Cuchumatanes
3993
Culec
GUATEMALA
Huehuetenango
Cobán
San Marcos
Totonicapán
Sololá
Quezaltenango
Retalhuleu
Mazatenango
Antigua
Amatitlán
GUATEMALA
Escuintla
Santa Ana
Ahuachapán
Nueva San Salvador (Santa Tecla)
Acajutla
Sonsonate
SAN SALVADOR
Cojutepeque
Zacatecoluca
Usulután
San Miguel
EL SALVADOR
Golfo de Fonseca
La Unión
Chinandega
Corinto
León
La Paz Central
MANAGUA
Diriamba
Jinotepe
Masaya
Granada
Rivas
San Juan del Sur
B. de Salinas
C. Sta. Elena
Golfo de Papagayo
C. Velas
Santa Cruz
Nicoya
Pen. de Nicoya
Puntarenas
Esparta
Alajuela
San José
Cartago
COSTA RICA
Liberia
Cañas
Cord. de Guanacaste
Lago de Nicaragua
Isla de Ometepe
Cord. de Yolaina
Bluefields
El Bluff
Pta. Mico
Bahía de San Juan del Norte
San Juan del Norte
San Carlos
San Juan
Rama
Santo Domingo
Juigalpa
Boaco
Muy Muy
Siquia
Matagalpa
Jinotega
Estelí
Somoto
Ocotal
NICARAGUA
Chontales
Tuma
Río Grande
Prinzapolca
Siuna
Bonanza
Puerto Cabezas
Pta. Gorda
Cayos Miskitos (Nicaragua)
Kisalaya
Puerto Cabo Gracias á Dios
C. Gracias á Dios
(Segovia)
Coco
C. Falso
Puerto Castilla
Trujillo
Brus Laguna
Laguna Caratasca
C. Camarón
Iriona
Pta. Patuca
Mosquitia
La Ceiba
Tela
Balfate
Savá
Olanchito
Roatán
Islas de la Bahía
Puerto Cortés
San Pedro Sula
El Progreso
Santa Bárbara
Yoro
El Paraíso
Catacamas
Juticalpa
HONDURAS
Comayagua
TEGUCIGALPA
Danlí
Choluteca
Yuscarán
Nacaome
Chalchuapa
Santa Rosa de Copán
La Esperanza
Marcala
La Paz
Gracias
Puerto Barrios
Livingston
Golfo de Honduras
Monkey River
Punta Gorda
San Antonio
San Luis
Sebol
I. de Izabal
L. de Izabal
Motagua
Sa. de las Minas
Chiquimula
Zacapa
Jalapa
Ayutla
Coatepeque
Chimaltenango

Isla Desterrada
Isla Pérez
Canal de Yucatán

Pedro Cays

I. de Providencia (Colombia)
I. de San Andrés (Colombia)
Cayos Roncador (U.S.A. & Colombia)
Cayos de Albuquerque (Colombia)
Islas del Maíz (Nicaragua, U.S.A.)
Bajo Nuevo (Colombia)

Limón
Pta. Mona
Is. de San Bernar
CARTA
Colón
Pta. Manzanillo
Nombre de Dios
Portobelo
Gatun L.
Balboa
La Chorrera
PANAMÁ
Archipiélago de San Blas
Golfo del Darién
Serranía de San Blas
Chepo
San Miguel
Arch. de las Perlas
Golfo de Panamá
I. del Rey
Chimán
La Palma
El Real
Turbo
G. de Urabá
Garachiné
Serranía del Darién
Morti
Cord. Central
Siquirres
Guápiles
Turrialba
Pandora
Sixaola
Bocas del Toro
Almirante
Changuinola
Chiriquí Grande
Laguna de Chiriquí
Golfo de los Mosquitos
Penonomé
Aguadulce
Natá
Santiago
Chitré
Los Santos
Las Tablas
Pen. de Azuero
Pocrí
Pta. Mala
Tonosí
Pta. Mariato
David
Boquete
Volcán Barú
3374
Puerto Armuelles
Pta. Burica
Golfo de Chiriquí
Remedios
Río Hato
Serranía de Tabasará
Golfito
Golfo Dulce
Puerto Cortés
Pta. Burica
Buenos Aires
Pen. de Osa
Bahía de Coronado
Palmar
Cortés
Quepos
Puerto Quepos
Golfo de Nicoya
Chirripó Grande
3837
Cord. de Talamanca
Cartago

I. de Coiba
I. de Cebaco
I. Jicarón

Projection: Bi-polar oblique Conical Orthomorphic

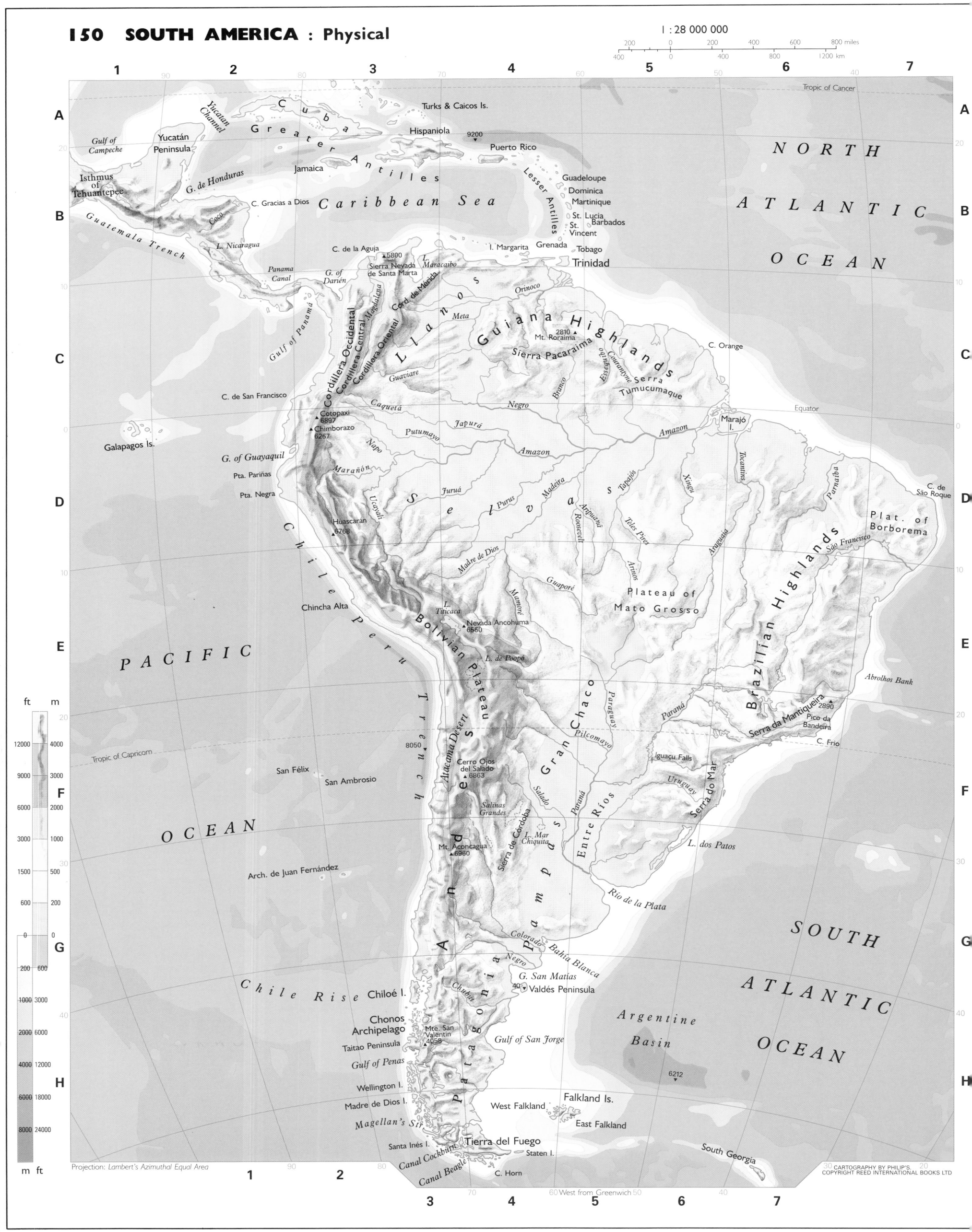

1 : 28 000 000

400 200 0 200 400 600 800 miles
400 0 400 800 1200 km

1 90 **2** 80 **3** 70 **4** 60 **5** 50 **6** 40 **7**

Tropic of Cancer

A — 20

Yucatán Channel

Cuba

Gulf of Campeche

Yucatán Peninsula

Greater Antilles

Turks & Caicos Is.

Hispaniola
9200

Puerto Rico

NORTH

ATLANTIC

Isthmus of Tehuantepec

G. de Honduras

Jamaica

Lesser Antilles

Guadeloupe
Dominica
Martinique

St. Lucia
St. Vincent
Barbados

B

Guatemala Trench

Coco

C. Gracias a Dios

Caribbean Sea

Grenada
Tobago

OCEAN

L. Nicaragua

Panama Canal

C. de la Aguja
5800

I. Margarita

Trinidad

Gulf of Darién

Sierra Nevada de Santa Marta

L. Maracaibo

Orinoco

Llanos

Guiana Highlands

C. Orange

C

C. de San Francisco

Cord. de Mérida

Meta

Mt. Roraima 2810

Sierra Pacaraima

Serra Tumucumaque

Gulf of Panama

Cordillera Occidental
Cordillera Central
Cordillera Oriental

Guaviare

Branco

Caureapanare

Cotopaxi 5897
Chimborazo 6267

Caquetá

Negro

Equator

Amazon

Marajó I.

D — 0

G. of Guayaquil

Pta. Pariñas

Pta. Negra

Napo

Putumayo

Japurá

Amazon

Amazon

Tocantins

C. de São Roque

Marañón

S

Juruá

Purus

Madeira

Tapajós

Xingu

Araguaia

Parnaíba

Plat. of Borborema

Ucayali

e

Juruá

Roosevelt

Teles Pires

São Francisco

Huascarán 6768

l

Madre de Dios

Arinos

Brazilian Highlands

E — 10

Chincha Alta

L. Titicaca

v

Guaporé

Mamoré

Plateau of Mato Grosso

Bolivian Plateau

Nevada Ancohuma 6550

a

Abrolhos Bank

PACIFIC

Chile

L. de Poopó

Gran Chaco

Paraguay

Paraná

Serra da Mantiqueira 2890
Pico da Bandeira

Peru

Trench

C. Frio

Tropic of Capricorn

Atacama Desert

8050

Pilcomayo

Iguaçu Falls

Serra do Mar

F

San Félix

San Ambrosio

Cerro Ojos del Salado 6863

Salado

Paraná

Uruguay

OCEAN

Salinas Grandes

Entre Ríos

L. dos Patos

Mt. Aconcagua 6960

Sierra de Córdoba

L. Mar Chiquita

P

Arch. de Juan Fernández

a

Río de la Plata

G — 30

Colorado

m

Bahía Blanca

SOUTH

A

Negro

p

G. San Matías

n

a

40 Valdés Peninsula

ATLANTIC

Chile Rise Chiloé I.

d

Chubut

Argentine

Chonos Archipelago

Mte. San Valentín 4058

Gulf of San Jorge

Basin

OCEAN

Taitao Peninsula

s

H

Gulf of Penas

6212

Wellington I.

Madre de Dios I.

West Falkland

Falkland Is.

Magellan's Str.

East Falkland

Santa Inés I.

Tierra del Fuego

Staten I.

South Georgia

Canal Cockburn
Canal Beagle

C. Horn

60 West from Greenwich 50

3 70 **4** 60 **5** 50 **6** 40 **7**

ft   m

12000 4000

9000   3000

6000   2000

3000   1000

1500   500

600   200

0   0

200   600

1000   3000

2000   6000

4000   12000

6000   18000

8000   24000

m ft

Projection: Lambert's Azimuthal Equal Area

1 : 28 000 000

200   0   200   400   600   800 miles
400   800   1200 km

1   2   3   4   5   6   7

90   80   70   60   50   40

Tropic of Cancer

A

Havana BAHAMAS
CUBA
Turks & Caicos Is.
(U.K.)

NORTH

B

HAITI DOMINICAN REP.
Port-au-Prince San Juan
JAMAICA Kingston PUERTO RICO (U.S.A.) Virgin Is. (U.K.)
MEXICO BELIZE ANTIGUA & BARBUDA
GUATEMALA HONDURAS ST. KITTS-NEVIS GUADELOUPE (Fr.)
Guatemala Tegucigalpa Basse-Terre DOMINICA
San Salvador NICARAGUA Fort-de-France MARTINIQUE (Fr.)
EL SALVADOR Managua Caribbean Sea Castries ST. LUCIA
COSTA San José ST. VINCENT BARBADOS
RICA PANAMA Panamá Kingstown Bridgetown
  GRENADA St. George's

ATLANTIC

OCEAN

Aruba Curaçao
Barranquilla C. de la Aguja Port of Spain TRINIDAD & TOBAGO
Cartagena Maracaibo Caracas
G. of Darién Barquisimeto Valencia
Gulf of Panamá Cúcuta Orinoco Ciudad Guayana
Medellín San Cristóbal VENEZUELA Georgetown
Bucaramanga Paramaribo
Bogotá GUYANA SURINAM Cayenne
Cali Magdalena FRENCH C. Orange
  GUIANA
COLOMBIA RORAIMA Esseguibo
  Branco AMAPÁ
Galapagos Is. Equator
(Ecuador) Quito Japurá
ECUADOR Napo Putumayo Marajó I.
Guayaquil Iquitos Manaus Belém
G. of Guayaquil Marañón Amazon Santarém São Luís
  Juruá Amazon Fortaleza
Chiclayo AMAZONAS Madeira PARÁ MARANHÃO Teresina C. de São Roque
Trujillo Purus Tapajós Xingu Tocantins CEARÁ RIO G. DO NORTE Natal
Chimbote ACRE Pôrto Velho PIAUÍ Parnaíba PARAÍBA Campina Grande
PERU RONDÔNIA BRAZIL PERNAMBUCO Recife
Callao LIMA Madre de Dios Araguaia ALAGOAS Maceió
  Cuzco Mamoré MATO GROSSO TOCANTINS SERGIPE Aracaju
  L. Titicaca BOLIVIA Cuiabá GOIÁS BAHÍA Salvador
Arequipa La Paz DIS. FED. Brasília
  Cochabamba Santa Cruz Goiânia São Francisco
Iquique Sucre MATO GROSSO MINAS GERAIS
  Paraguay DO SUL Belo ESPÍRITO SANTO
Tropic of Capricorn Ribeirão Horizonte Vitória
Antofagasta Salta PARAGUAY Prêto SÃO PAULO Campos
San Félix Pilcomayo Paraná Juiz R. DE J.
(Chile) San Ambrosio Asunción PARANÁ de Fora RIO DE
  (Chile) San Miguel SÃO Campinas JANEIRO
  de Tucumán Curitiba PAULO Niterói
  Resistencia SANTA CATARINA
  Corrientes Uruguay
Córdoba Santa Fe RIO GRANDE
Arch. de Juan Fernández San Juan Paraná DO SUL Pôrto Alegre
(Chile) Viña del Mar Mendoza Rosario URUGUAY Pelotas
  Valparaíso SANTIAGO BUENOS AIRES Montevideo
  Talca La Plata Río de la Plata
  Concepción ARGENTINA Bahía Mar del Plata
  Colorado Blanca
Valdivia Negro Viedma

PACIFIC

OCEAN

SOUTH

ATLANTIC

OCEAN

Comodoro Rivadavia
Gulf of San Jorge

Gulf of Penas

West Falkland FALKLAND IS. (U.K.)
Stanley
East Falkland

Magellan's Str.
Punta Arenas Tierra del Fuego
  South Georgia (U.K.)
C. Horn

Projection: Lambert's Azimuthal Equal Area

■ LIMA Capital Cities

West from Greenwich

CARTOGRAPHY BY PHILIP'S,
COPYRIGHT REED INTERNATIONAL BOOKS LTD

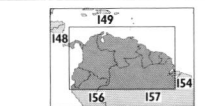

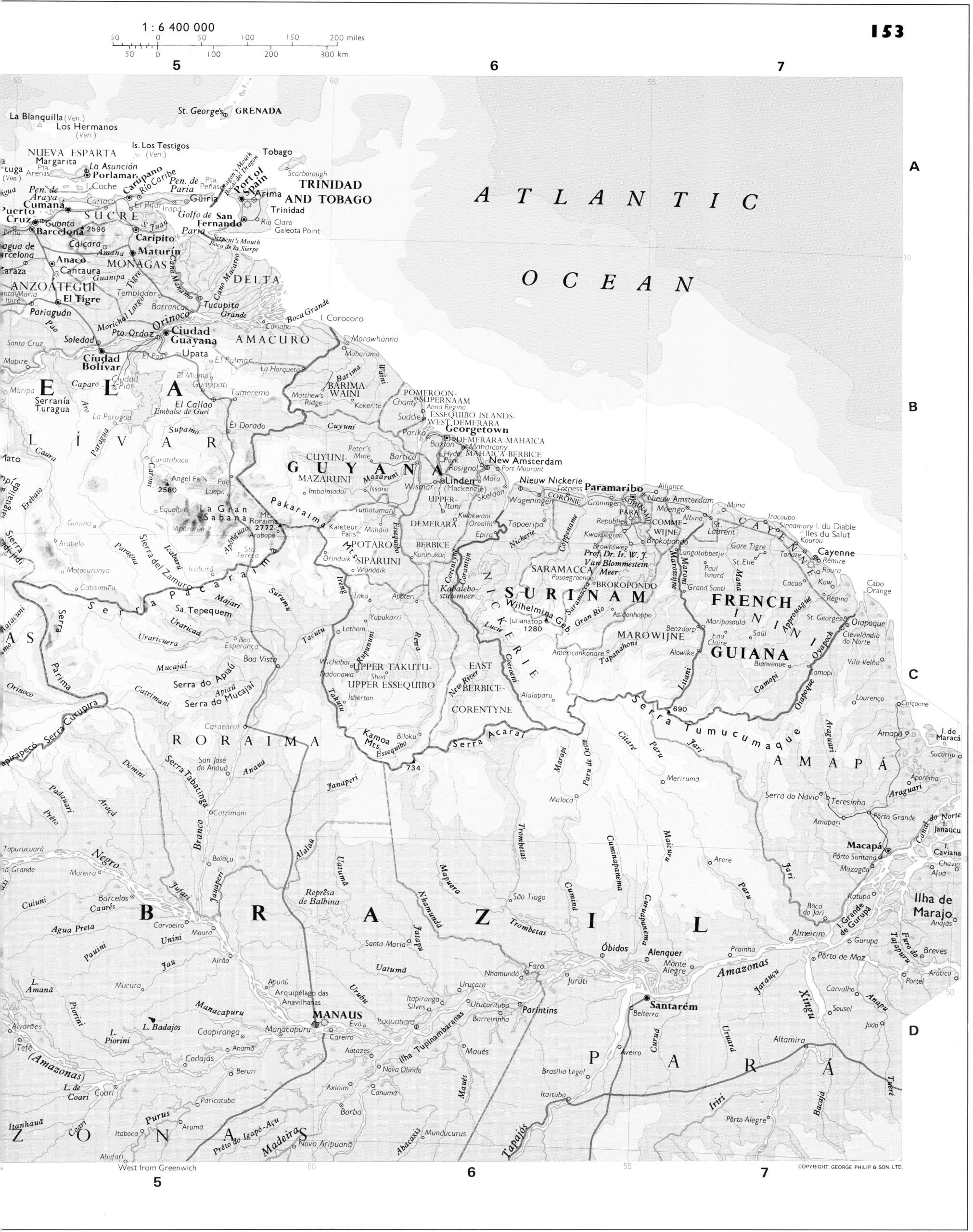

1 : 6 400 000

50   0   50   100   150   200 miles
50   0   100   200   300 km

West from Greenwich

COPYRIGHT. GEORGE PHILIP & SON. LTD.

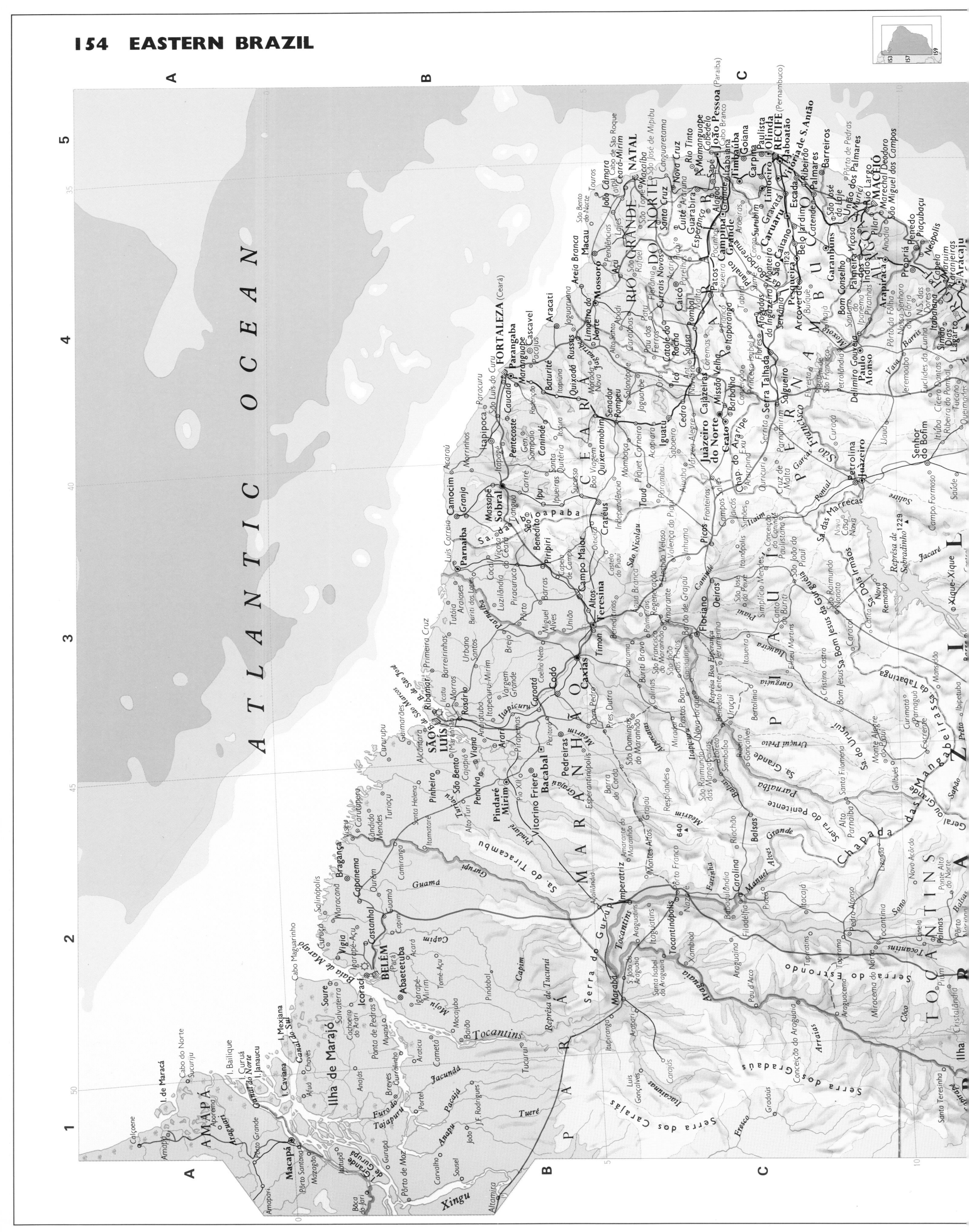

A T L A N T I C   O C E A N

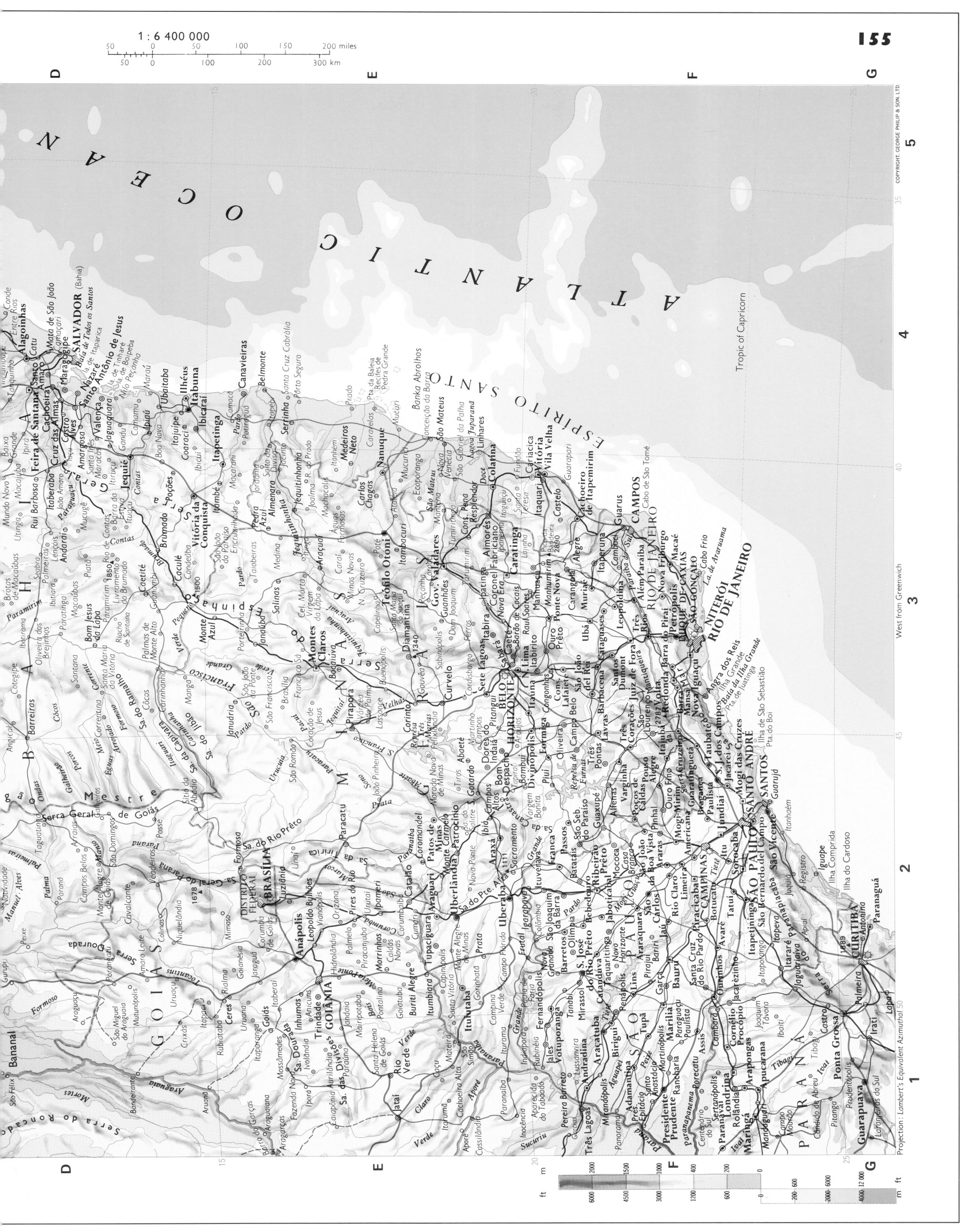

ATLANTIC OCEAN

1 : 6 400 000

50   0   50   100   150   200 miles
50   0   100   200   300 km

COPYRIGHT GEORGE PHILIP & SON, LTD.

Tropic of Capricorn

West from Greenwich

Projection: Lambert's Equivalent Azimuthal

ESPÍRITO SANTO

B A H I A

M I N A S   G E R A I S

G O I A S

S Ã O   P A U L O

P A R A N Á

D I S T R I T O   F E D E R A L

SALVADOR (Bahia)

RIO DE JANEIRO

SÃO PAULO

BELO HORIZONTE

BRASÍLIA

GOIÂNIA

CURITIBA

NITERÓI

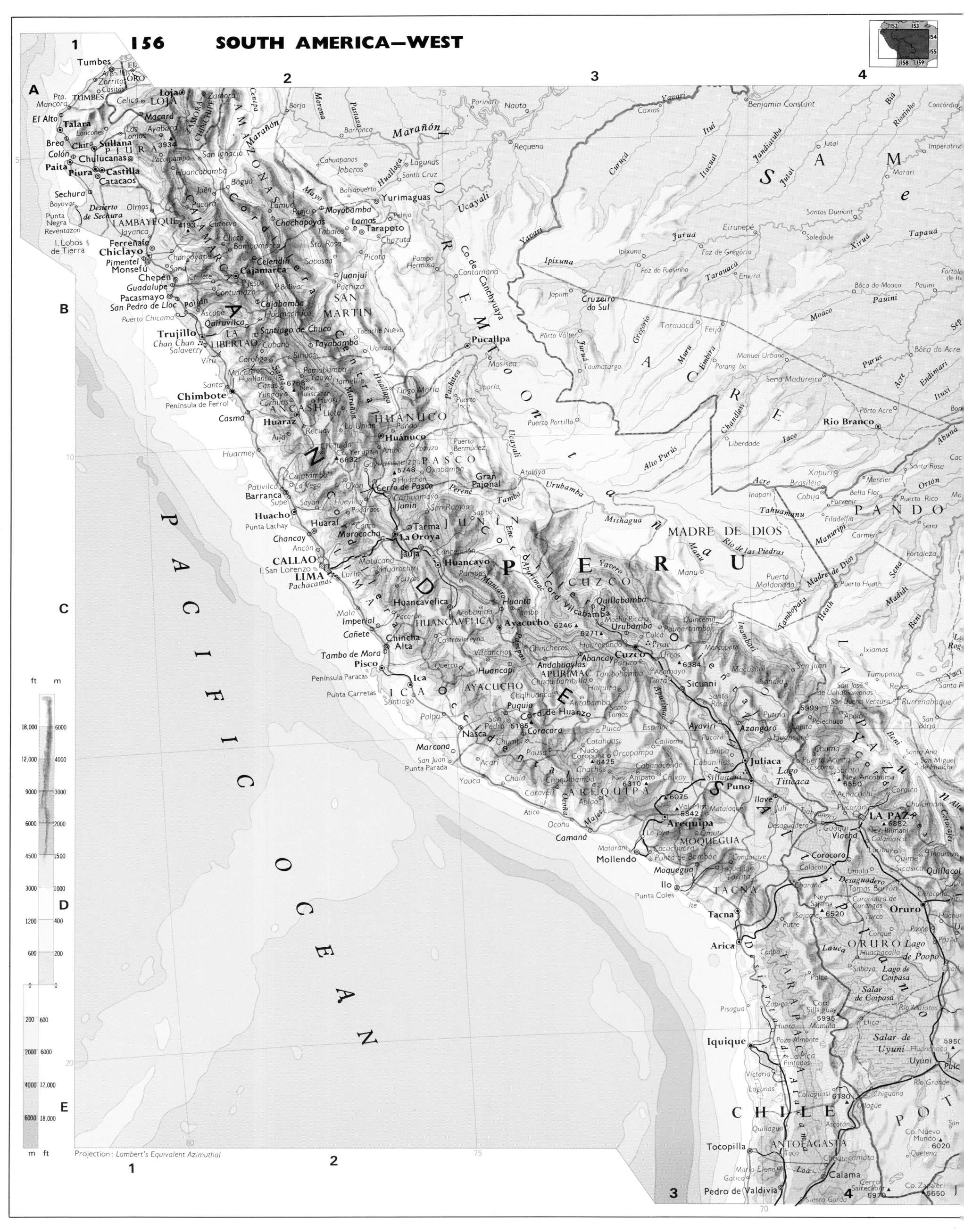

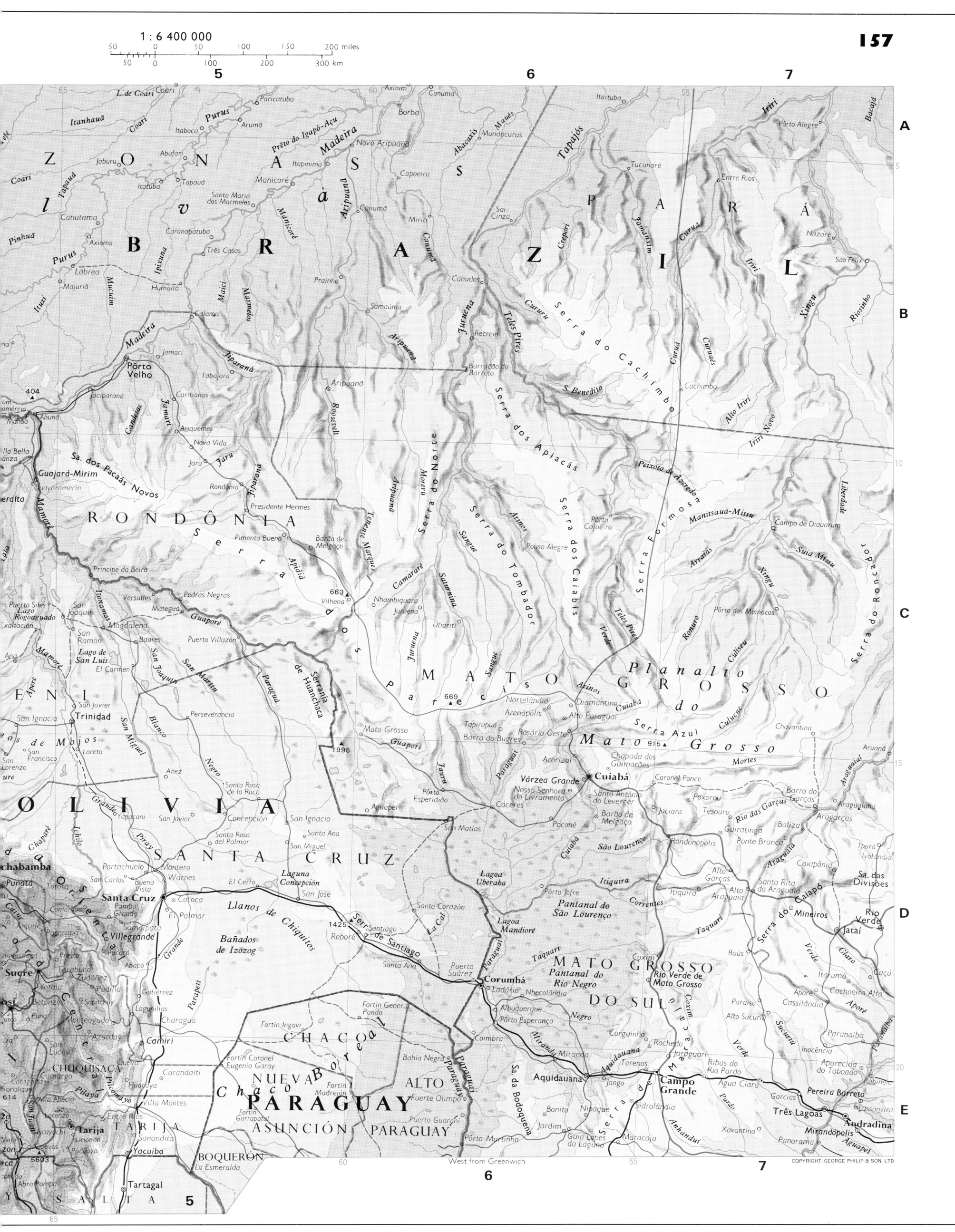

1 : 6 400 000

50   0   50   100   150   200 miles

50   0   100   200   300 km

5

6

7

A

B

C

D

E

Z O N A   B R A S

B R A Z I L

P A R Á

Pôrto Velho

R O N D Ô N I A

Sa. dos Pacaãs Novos

Serra do Cachimbo

Serra dos Apiacás

Serra Formosa

Serra dos Caiabis

Serra do Tombador

M A T O   G R O S S O

Planalto

do

Mato Grosso

Cuiabá

Várzea Grande

B O L I V I A

S A N T A   C R U Z

Santa Cruz

Trinidad

Llanos de Chiquitos

Bañados de Izozog

Corumbá

Pantanal do Rio Negro

M A T O   G R O S S O

D O   S U L

Rio Verde de Mato Grosso

Campo Grande

Aquidauana

Três Lagoas

Andradina

Rio Verde

Jataí

Caiapó

Sa. das Divisões

C H A C O

N U E V A

P A R A G U A Y

ASUNCIÓN   **PARAGUAY**

ALTO

Asunción

BOQUERÓN

La Esmeralda

Tarija

TARIJA

SALTA

Tartagal

West from Greenwich

COPYRIGHT. GEORGE PHILIP & SON. LTD.

5

6

7

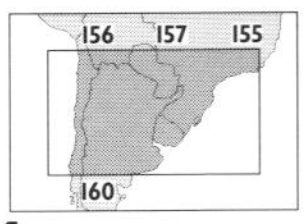

156 157 155
160

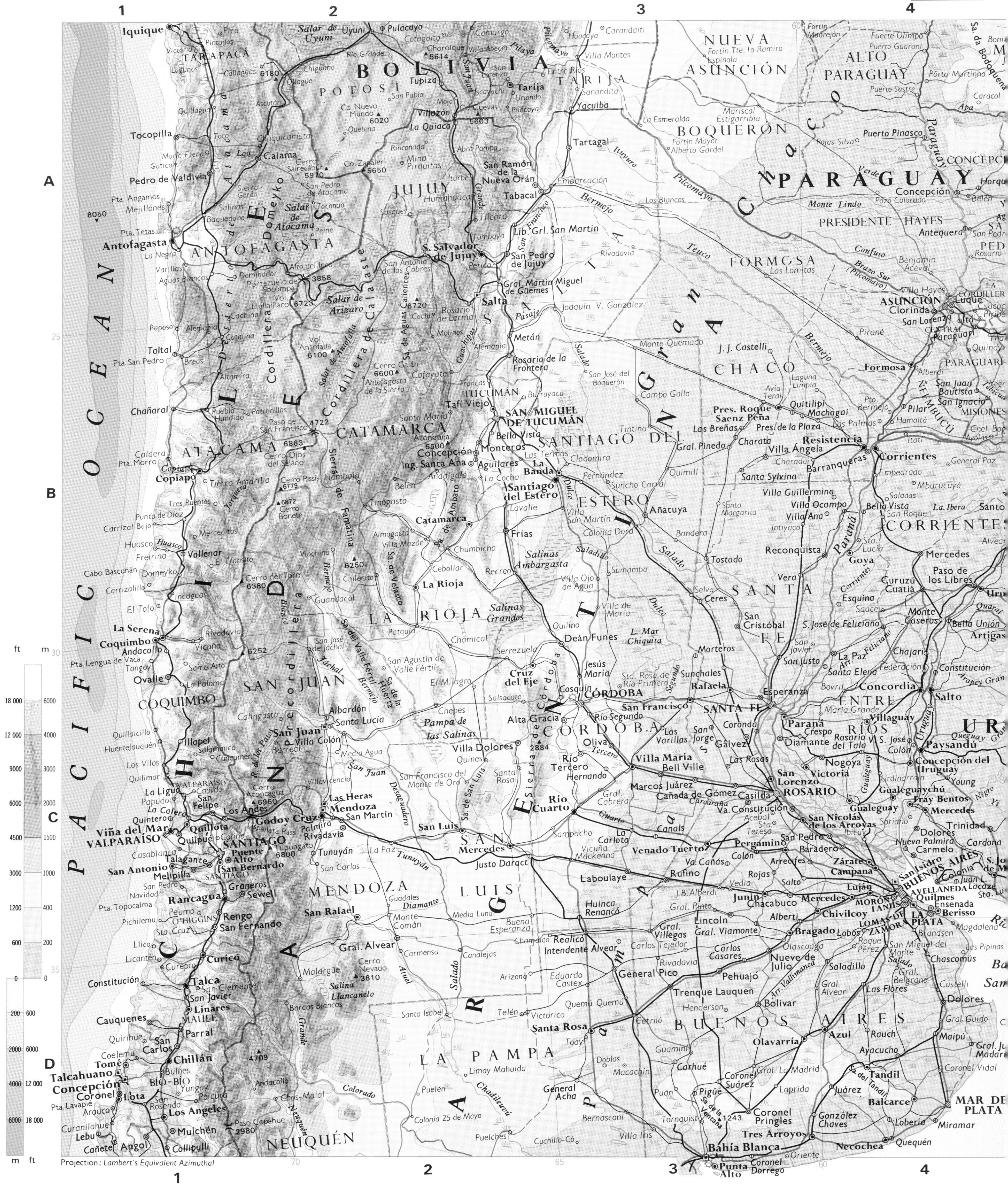

1 2 3 4

Iquique
Pica
Pintados
Victoria
Lagunas
TARAPACÁ
Salar de Uyuni
Pulacayo
Camargo
Carandaití
NUEVA
Fortín Tte. 1o Ramiro
Estigarribia
Fuerte Olimpo
Puerto Guaraní
Pôrto Murtinho
ALTO
PARAGUAY

Chiguana
POTOSÍ
Cotagaita
Chorolque
Villa Abecia
Pilaya
Pilcomayo
ASUNCIÓN
Fortín Mayor
Mariscal Estigarribia
Puerto Pinasco
Puerto Sastre

BOLIVIA
Río Grande
Tupiza
San Juan
Villa Montes
BOQUERÓN
Fortín Alberto Gardel
Rojas Silva

Tocopilla
Calama
Co. Nuevo Mundo 6020
Quetena
Villazón
La Quiaca
5803
TARIJA
Tarija
Uriondo
Yacuiba
Tartagal
La Esmeralda
PARAGUAY
Concepción
Belén
Horqu

Pedro de Valdivia
JUJUY
Iturbe
Humahuaca
San Ramón de la Nueva Orán
Embarcación
Los Blancos
Bermejo
PRESIDENTE HAYES
Antequera

ANTOFAGASTA
S. Salvador de Jujuy
San Pedro de Jujuy
Tabacal
FORMOSA
Confuso
Benjamín Aceval
ASUNCIÓN
Clorinda
San Lorenzo

Antofagasta
La Negra
Salta
Gral. Martín Miguel de Güemes
Joaquín V. González
Pirané
Formosa
MISIONES

Pacific Ocean

Salar de Atacama
SALTA
Rosario de la Frontera
Monte Quemado
Campo Gallo
CHACO
J. J. Castelli
San José del Boquerón
Resistencia
Corrientes

ATACAMA
Copiapó
Catamarca
SANTIAGO DEL ESTERO
Santiago del Estero
CORRIENTES
Goya

La Serena
Coquimbo
LA RIOJA
La Rioja
SANTA FE
Santa Fe
Paraná
ENTRE RÍOS
Concordia
Salto

COQUIMBO
SAN JUAN
San Juan
CÓRDOBA
Córdoba
Rosario
URUGUAY
Paysandú

VALPARAÍSO
SANTIAGO
MENDOZA
Mendoza
SAN LUIS
San Luis
Río Cuarto
BUENOS AIRES
Buenos Aires
La Plata

San Rafael
General Pico
Santa Rosa
LA PAMPA
Bahía Blanca
Mar del Plata

NEUQUÉN

Concepción
Talcahuano
Los Ángeles
BÍO-BÍO

Projection: Lambert's Equivalent Azimuthal

ft m
18 000 6000
12 000 4000
9000 3000
6000 2000
4500 1500
3000 1000
1200 400
600 200
0 0
600 200
2000 600
4000 12 000
6000 18 000
m ft

1 : 6 400 000

50 0 50 100 150 200 miles

50 0 100 200 300 km

5 **6** **7**

BELO
HORIZONTE
Lima
Itabirito
Vitória
Itaquari
Vila
Velha
Guarapari

Três Lagoas
Andradina
Mirassol
Olímpia
Passos
Oliveira
Cons.
Lafaiete
Ouro
Prêto
Ponte Nova
Pico da
Bandeira
2890
Castelo

Araçatuba
S. José
do Rio Prêto
Bebedouro
Catanduva
Ribeirão
Prêto
São Seb
do Paraíso
Campo Belo
São João
del Rei
Ubá
Muriaé
Cachoeiro
de Itapemirim

Xavantina
Panorama
Mirandópolis
Birigui
Taquaritinga
Jaboticabal
Guaxupé
Casa
Branca
Tres
Pontas
Lavras
Barbacena
Cataguases
Itaperuna

Adamantina
Penápolis
Lins
Mococa
Alfenas
Varginha
Tres
Dumont
Leopoldina
Cambuci
Guarus

Santo Anastácio
Pres.
Epitácio
Tupã
Novo
Horizonte
Poços de
Caldas
Pouso
Alegre
São
Lourenço
Juiz de Fora
Além Paraíba
CAMPOS

Presidente
Prudente
Martinópolis
Marília
Garça
Jaú
Rio Claro
Ouro Fino
Mogi-Mirim
Americana
Serra
Volta
da
Barra do Pirai
Nova Friburgo
Macaé
Cabo de
São Tomé

Paranavaí
Assis
Bauru
Piracicaba
CAMPINAS
Itajubá 2787
Barra
Mansa
Petrópolis
RIO DE JANEIRO

Nova
Esperança
Rolândia
Cornélio
Procópio
Ourinhos
Botucatu
Itu
Jundiaí
Bragança
Paulista
Redonda
Barra do Pirai
DUQUE DE CAXIAS
NITERÓI

Umuarama
Maringá
Apucarana
Avaré
Tatui
Sorocaba
SÃO PAULO
Taubaté
Jacareí
S. J. dos Campos
Angra dos Reis
NOVA IGUAÇU
SÃO GONÇALO
RIO DE JANEIRO

Cruzeiro
do Oeste
Mandaguari
Itapetininga
SANTO ANDRÉ
Ilha
Grande
Baía da Ilha Grande
Tropic of Capricorn

Guaíra
Goio
Erê
Itapeva
Itararé
Piraju
Jaguariaíva
São Bernardo
del Campo
São Vicente
SANTOS
Guarujá
Ilha de São Sebastião
Pta. do Boi

Foz do Iguaçu
GUARAPUAVA
Ibaití
Castro
Apiaí
Itanhaém
Juquiá
Registro

Ciudad
del Este
Ponta Grossa
Palmeira
CURITIBA
Antonina
Iguape

BRAZIL
Prudentópolis
Irati
Lapa
Paranaguá
Ilha Comprida
Ilha do Cardoso

União da
Vitória
Rio Negro
Guaratuba

PARANÁ
Mafra
São Francisco do Sul

Pto. União
Joinvile

Caçador
Blumenau
Itajaí

SANTA
CATARINA
Brusque

Chapecó
Joaçaba
Campos Novos
Rio do Sul
Itajaí

Erechim
Lajes
Ilha de Santa Catarina
Florianópolis

Caràzinho
Passo Fundo
Vacaria

Santo Ângelo
Cruz Alta
Tubarão
Laguna
Cabo Santa Marta Grande

Santa Maria
Santa Cruz
do Sul
Nôvo Hamburgo
Taquara
Criciúma
Araranguá

RIO GRANDE
Guaporé
Bento Gonçalves
Caxias do Sul

Alegrete
Cachoeira do Sul
Rio Pardo
Canoas
São
Leopoldo
Viamão
PÔRTO ALEGRE
Osorio

Santana do
Livramento
São
Gabriel
DO SUL
Camaquã

Rivera
Dom Pedrito
Bagé
Camaquã
Mostardas
Lagoa dos Patos

Pelotas
Rio Grande

Melo
Jaguarão
Lagoa
Mirim
Lagoa Mangueira

Treinta y Tres
Santa Vitória do Palmar

ATLANTIC

OCEAN

5304

COPYRIGHT. GEORGE PHILIP & SON. LTD

West from Greenwich

5 **6** **7**

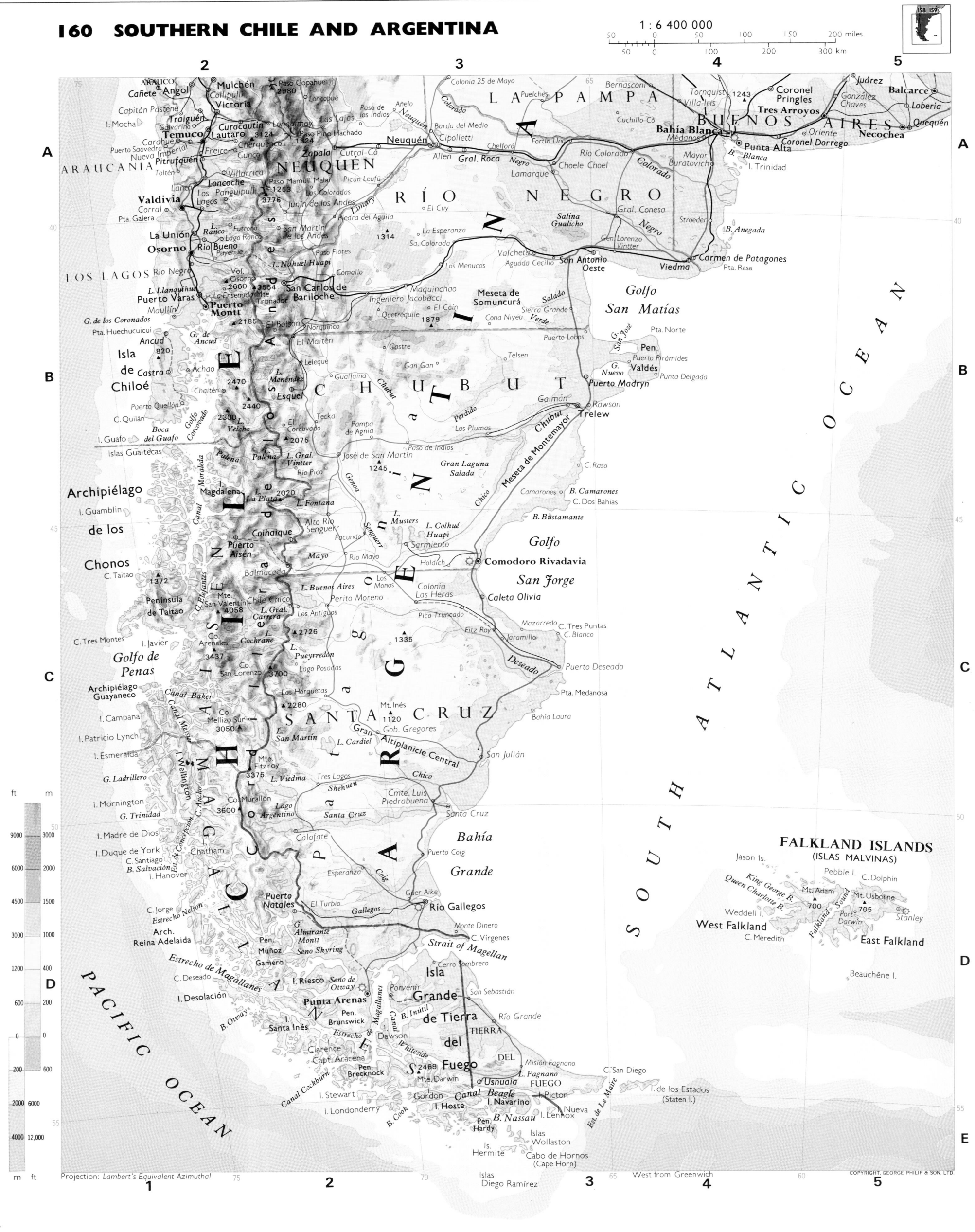

1 : 6 400 000

158 159

PACIFIC OCEAN

SOUTH ATLANTIC OCEAN

LA PAMPA

BUENOS AIRES

RÍO NEGRO

NEUQUÉN

ARAUCANIA

LOS LAGOS

CHUBUT

SANTA CRUZ

FALKLAND ISLANDS
(ISLAS MALVINAS)

West Falkland

East Falkland

Strait of Magellan

Isla Grande de Tierra del Fuego

TIERRA DEL FUEGO

Cabo de Hornos (Cape Horn)

Projection: Lambert's Equivalent Azimuthal

West from Greenwich

COPYRIGHT. GEORGE PHILIP & SON. LTD.

INDEX

The index contains the names of all the principal places and features shown on the World Maps. Each name is followed by an additional entry in italics giving the country or region within which it is located. The alphabetical order of names composed of two or more words is governed primarily by the first word and then by the second. This is an example of the rule:

Mīr Kūh, *Iran*	**85 E8**	26 22 N	58 55 E
Mīr Shahdād, *Iran*	**85 E8**	26 15 N	58 29 E
Miraj, *India*	**82 F2**	16 50 N	74 45 E
Miram Shah, *Pakistan*	**79 B3**	33 0 N	70 2 E
Miramar, *Mozam.*	**105 C6**	23 50 S	35 35 E

Physical features composed of a proper name (Erie) and a description (Lake) are positioned alphabetically by the proper name. The description is positioned after the proper name and is usually abbreviated:

Erie, L., *N. Amer.*	**136 D3**	42 15 N	81 0 W

Where a description forms part of a settlement or administrative name however, it is always written in full and put in its true alphabetic position:

Mount Morris, *U.S.A.*	**136 D7**	42 44 N	77 52 W

Names beginning with M' and Mc are indexed as if they were spelt Mac. Names beginning St. are alphabetised under Saint, but Sankt, Sint, Sant', Santa and San are all spelt in full and are alphabetised accordingly. If the same place name occurs two or more times in the index and all are in the same country, each is followed by the name of the administrative subdivision in which it is located. The names are placed in the alphabetical order of the subdivisions. For example:

Jackson, *Ky., U.S.A.*	**134 G4**	37 33 N	83 23 W
Jackson, *Mich., U.S.A.*	**141 B12**	42 15 N	84 24 W
Jackson, *Minn., U.S.A.*	**138 D7**	43 37 N	95 1 W

The number in bold type which follows each name in the index refers to the number of the map page where that feature or place will be found. This is usually the largest scale at which the place or feature appears.

The letter and figure which are in bold type immediately after the page number give the grid square on the map page, within which the feature is situated. The letter represents the latitude and the figure the longitude.

In some cases the feature itself may fall within the specified square, while the name is outside. This is usually the case only with features which are larger than a grid square.

For a more precise location the geographical coordinates which follow the letter/figure references give the latitude and the longitude of each place. The first set of figures represent the latitude which is the distance north or south of the Equator measured as an angle at the centre of the earth. The Equator is latitude 0°, the North Pole is 90°N, and the South Pole 90°S.

The second set of figures represent the longitude, which is the distance East or West of the prime meridian, which runs through Greenwich, England. Longitude is also measured as an angle at the centre of the earth and is given East or West of the prime meridian, from 0° to 180° in either direction.

The unit of measurement for latitude and longitude is the degree, which is subdivided into 60 minutes. Each index entry states the position of a place in degrees and minutes, a space being left between the degrees and the minutes.

The latitude is followed by N(orth) or S(outh) and the longitude by E(ast) or W(est).

Rivers are indexed to their mouths or confluences, and carry the symbol ➝ after their names. A solid square ■ follows the name of a country while, an open square □ refers to a first order administrative area.

ABBREVIATIONS USED IN THE INDEX

A.C.T. — Australian Capital Territory
Afghan. — Afghanistan
Ala. — Alabama
Alta. — Alberta
Amer. — America(n)
Arch. — Archipelago
Ariz. — Arizona
Ark. — Arkansas
Atl. Oc. — Atlantic Ocean
B. — Baie, Bahía, Bay, Bucht, Bugt
B.C. — British Columbia
Bangla. — Bangladesh
Barr. — Barrage
Bos.-H. — Bosnia-Herzegovina
C. — Cabo, Cap, Cape, Coast
C.A.R. — Central African Republic
C. Prov. — Cape Province
Calif. — California
Cent. — Central
Chan. — Channel
Colo. — Colorado
Conn. — Connecticut
Cord. — Cordillera
Cr. — Creek
Czech. — Czech Republic
D.C. — District of Columbia
Del. — Delaware
Dep. — Dependency
Des. — Desert
Dist. — District
Dj. — Djebel
Domin. — Dominica
Dom. Rep. — Dominican Republic
E. — East
El Salv. — El Salvador

Eq. Guin. — Equatorial Guinea
Fla. — Florida
Falk. Is. — Falkland Is.
G. — Golfe, Golfo, Gulf, Guba, Gebel
Ga. — Georgia
Gt. — Great, Greater
Guinea-Biss. — Guinea-Bissau
H.K. — Hong Kong
H.P. — Himachal Pradesh
Hants. — Hampshire
Harb. — Harbor, Harbour
Hd. — Head
Hts. — Heights
I.(s). — Île, Ilha, Insel, Isla, Island, Isle
Ill. — Illinois
Ind. — Indiana
Ind. Oc. — Indian Ocean
Ivory C. — Ivory Coast
J. — Jabal, Jebel, Jazira
Junc. — Junction
K. — Kap, Kapp
Kans. — Kansas
Kep. — Kepulauan
Ky. — Kentucky
L. — Lac, Lacul, Lago, Lagoa, Lake, Limni, Loch, Lough
La. — Louisiana
Liech. — Liechtenstein
Lux. — Luxembourg
Mad. P. — Madhya Pradesh
Madag. — Madagascar
Man. — Manitoba
Mass. — Massachusetts
Md. — Maryland

Me. — Maine
Medit. S. — Mediterranean Sea
Mich. — Michigan
Minn. — Minnesota
Miss. — Mississippi
Mo. — Missouri
Mont. — Montana
Mozam. — Mozambique
Mt.(e). — Mont, Monte, Monti, Montaña, Mountain
N. — Nord, Norte, North, Northern, Nouveau
N.B. — New Brunswick
N.C. — North Carolina
N. Cal. — New Caledonia
N. Dak. — North Dakota
N.H. — New Hampshire
N.I. — North Island
N.J. — New Jersey
N. Mex. — New Mexico
N.S. — Nova Scotia
N.S.W. — New South Wales
N.W.T. — North West Territory
N.Y. — New York
N.Z. — New Zealand
Nebr. — Nebraska
Neths. — Netherlands
Nev. — Nevada
Nfld. — Newfoundland
Nic. — Nicaragua
O. — Oued, Ouadi
Occ. — Occidentale
O.F.S. — Orange Free State
Okla. — Oklahoma
Ont. — Ontario
Or. — Orientale

Oreg. — Oregon
Os. — Ostrov
Oz. — Ozero
P. — Pass, Passo, Pasul, Pulau
P.E.I. — Prince Edward Island
Pa. — Pennsylvania
Pac. Oc. — Pacific Ocean
Papua N.G. — Papua New Guinea
Pass. — Passage
Pen. — Peninsula, Péninsule
Phil. — Philippines
Pk. — Park, Peak
Plat. — Plateau
P-ov. — Poluostrov
Prov. — Province, Provincial
Pt. — Point
Pta. — Ponta, Punta
Pte. — Pointe
Qué. — Québec
Queens. — Queensland
R. — Rio, River
R.I. — Rhode Island
Ra.(s). — Range(s)
Raj. — Rajasthan
Reg. — Region
Rep. — Republic
Res. — Reserve, Reservoir
S. — San, South, Sea
Si. Arabia — Saudi Arabia
S.C. — South Carolina
S. Dak. — South Dakota
S.I. — South Island
S. Leone — Sierra Leone
Sa. — Serra, Sierra
Sask. — Saskatchewan
Scot. — Scotland

Sd. — Sound
Sev. — Severnaya
Sib. — Siberia
Sprs. — Springs
St. — Saint, Sankt, Sint
Sta. — Santa, Station
Ste. — Sainte
Sto. — Santo
Str. — Strait, Stretto
Switz. — Switzerland
Tas. — Tasmania
Tenn. — Tennessee
Tex. — Texas
Tg. — Tanjung
Trin. & Tob. — Trinidad & Tobago
U.A.E. — United Arab Emirates
U.K. — United Kingdom
U.S.A. — United States of America
Ut. P. — Uttar Pradesh
Va. — Virginia
Vdkhr. — Vodokhranilishche
Vf. — Vîrful
Vic. — Victoria
Vol. — Volcano
Vt. — Vermont
W. — Wadi, West
W. Va. — West Virginia
Wash. — Washington
Wis. — Wisconsin
Wlkp. — Wielkopolski
Wyo. — Wyoming
Yorks. — Yorkshire
Yug. — Yugoslavia

162 A Coruña

A

A Coruña = La Coruña, Spain 36 B2 43 20N 8 25W
Aachen, Germany 26 E2 50 45N 6 6 E
Aadorf, Switz. 29 B7 47 30N 8 55 E
Aalborg = Ålborg, Denmark 15 G3 57 2N 9 54 E
Aalen, Germany 27 G6 48 51N 10 6 E
A'âli en Nîl □, Sudan 95 F3 9 30N 33 0 E
Aalsmeer, Neths. 20 D5 52 17N 4 43 E
Aalst, Belgium 21 G4 50 56N 4 2 E
Aalst, Neths. 21 F6 51 23N 5 29 E
Aalten, Neths. 20 E9 51 56N 6 35 E
Aalter, Belgium 21 F2 51 5N 3 28 E
Äänekoski, Finland 13 E21 62 36N 25 44 E
Aarau, Switz. 28 B6 47 23N 8 4 E
Aarberg, Switz. 28 B4 47 2N 7 16 E
Aardenburg, Belgium 21 F2 51 16N 3 28 E
Aare →, Switz. 28 A6 47 33N 8 14 E
Aargau □, Switz. 28 B6 47 26N 8 10 E
Aarhus = Århus, Denmark 15 H4 56 8N 10 11 E
Aarle, Neths. 21 E7 51 30N 5 38 E
Aarschot, Belgium 21 G5 50 59N 4 49 E
Aarsele, Belgium 21 G2 51 0N 3 26 E
Aartrijke, Belgium 21 F2 51 7N 3 6 E
Aarwangen, Switz. 28 B5 47 15N 7 46 E
Aba, China 68 A3 32 59N 101 42 E
Aba, Nigeria 101 D6 5 10N 7 19 E
Aba, Zaïre 106 B3 3 58N 30 17 E
Âbâ, Jazîrat, Sudan 95 E3 13 30N 32 31 E
Abacaxis →, Brazil 153 D6 3 54 S 58 47W
Ābādān, Iran 85 D6 30 22N 48 20 E
Abade, Ethiopia 95 F4 9 22N 38 3 E
Ābādeh, Iran 85 D7 31 8N 52 40 E
Abadin, Spain 36 B3 43 21N 7 29W
Abadla, Algeria 99 B4 31 2N 2 45W
Abaeté, Brazil 155 E2 19 9 S 45 27W
Abaeté →, Brazil 155 E2 18 2 S 45 12W
Abaetetuba, Brazil 154 B2 1 40 S 48 50W
Abagnar Qi, China 66 C9 43 52N 116 2 E
Abai, Paraguay 159 B4 25 58 S 55 54W
Abak, Nigeria 101 E6 4 58N 7 50 E
Abakaliki, Nigeria 101 D6 6 22N 8 2 E
Abakan, Russia 57 D10 53 40N 91 10 E
Abalemma, Niger 101 B6 16 12N 7 50 E
Abana, Turkey 88 B6 41 59N 34 1 E
Abancay, Peru 156 C3 13 35 S 72 55W
Abanilla, Spain 35 G3 38 12N 1 3W
Abano Terme, Italy 39 C8 45 22N 11 46 E
Abapó, Bolivia 157 D5 18 48 S 63 25W
Abarán, Spain 35 G3 38 12N 1 23W
Abariringa, Kiribati 122 H10 2 50 S 171 40W
Abarqū, Iran 85 D7 31 10N 53 20 E
Abashiri, Japan 60 B12 44 0N 144 15 E
Abashiri-Wan, Japan 60 B12 44 0N 144 30 E
Abau, Papua N. G. 120 F5 10 11 S 148 46 E
Abaújszántó, Hungary 31 C14 48 16N 21 12 E
Abay, Kazakhstan 56 E8 49 38N 72 53 E
Abaya, L., Ethiopia 95 F4 6 30N 37 50 E
Abaza, Russia 56 D10 52 39N 90 6 E
Abbadia San Salvatore, Italy 39 F8 42 53N 11 41 E
'Abbāsābād, Iran 85 C8 33 34N 58 23 E
Abbay = Nîl el Azraq →, Sudan 95 D3 15 38N 32 31 E
Abbaye, Pt., U.S.A. 134 B1 46 58N 88 8W
Abbé, L., Ethiopia 95 E5 11 8N 41 47 E
Abbeville, France 23 B8 50 6N 1 49 E
Abbeville, La., U.S.A. 139 K8 29 58N 92 8W
Abbeville, S.C., U.S.A. 135 H4 34 11N 82 23W
Abbiategrasso, Italy 38 C5 45 24N 8 54 E
Abbieglassie, Australia 115 D4 27 15 S 147 28 E
Abbot Ice Shelf, Antarctica 7 D16 73 0 S 92 0W
Abbotsford, Canada 130 D4 49 5N 122 20W
Abbotsford, U.S.A. 138 C9 44 57N 90 19W
Abbottabad, Pakistan 80 B5 34 10N 73 15 E
Abcoude, Neths. 20 D5 52 17N 4 59 E
Abd al Kūrī, Ind. Oc. 87 D6 12 5N 52 20 E
Ābdar, Iran 85 D7 30 16N 55 19 E
'Abdolābād, Iran 85 C8 34 12N 56 30 E
Abdulino, Russia 54 E4 53 42N 53 40 E
Abéché, Chad 97 F4 13 50N 20 35 E
Abejar, Spain 34 D2 41 48N 2 47W
Abekr, Sudan 95 E2 12 45N 28 50 E
Abêlessa, Algeria 99 D5 22 58N 4 47 E
Abengourou, Ivory C. 100 D4 6 42N 3 27W
Åbenrå, Denmark 15 J3 55 3N 9 25 E
Abensberg, Germany 27 G7 48 48N 11 51 E
Abeokuta, Nigeria 101 D5 7 3N 3 19 E
Aber, Uganda 106 B3 2 12N 32 25 E
Aberaeron, U.K. 17 E3 52 15N 4 15W
Aberayron = Aberaeron, U.K. 17 E3 52 15N 4 15W
Abercorn = Mbala, Zambia 107 D3 8 46 S 31 24 E
Abercorn, Australia 115 D5 25 12 S 151 5 E
Aberdare, U.K. 17 F4 51 43N 3 27W
Aberdare Ra., Kenya 106 C4 0 15 S 36 50 E
Aberdeen, Australia 117 B9 32 9 S 150 56 E
Aberdeen, Canada 131 C7 52 20N 106 8W
Aberdeen, S. Africa 104 E3 32 28 S 24 2 E
Aberdeen, U.K. 18 D6 57 9N 2 5W
Aberdeen, Ala., U.S.A. 135 J1 33 49N 88 33W
Aberdeen, Idaho, U.S.A. 142 E7 42 57N 112 50W
Aberdeen, Ohio, U.S.A. 141 F13 38 39N 83 46W
Aberdeen, S. Dak., U.S.A. 138 C5 45 28N 98 29W
Aberdeen, Wash., U.S.A. 144 D3 46 59N 123 50W
Aberdovey = Aberdyfi, U.K. 17 E3 52 33N 4 3W
Aberdyfi, U.K. 17 E3 52 33N 4 3W
Aberfeldy, Australia 117 D7 37 42 S 146 22 E
Aberfeldy, U.K. 18 E5 56 37N 3 51W
Abergaria-a-Velha, Portugal 36 E2 40 41N 8 32W
Abergavenny, U.K. 17 F4 51 49N 3 1W
Abernathy, U.S.A. 139 J4 33 50N 101 51W
Abert, L., U.S.A. 142 E3 42 38N 120 14W
Aberystwyth, U.K. 17 E3 52 25N 4 5W
Abha, Si. Arabia 94 D5 18 0N 42 34 E
Abhar, Iran 89 D13 36 9N 49 13 E
Abhayapuri, India 78 B3 26 24N 90 38 E
Abia □, Nigeria 101 D6 5 30N 7 35 E
Abidiya, Sudan 94 D3 18 18N 34 3 E
Abidjan, Ivory C. 100 D4 5 26N 3 58W
Abilene, Kans., U.S.A. 138 F6 38 55N 97 13W
Abilene, Tex., U.S.A. 139 J5 32 28N 99 43W
Abingdon, U.K. 17 F6 51 40N 1 17W
Abingdon, Ill., U.S.A. 140 D6 40 48N 90 24W
Abingdon, Va., U.S.A. 135 G5 36 43N 81 59W
Abington Reef, Australia 114 B4 18 0 S 149 35 E
Abitau →, Canada 131 B7 59 53N 109 3W
Abitau L., Canada 131 A7 60 27N 107 15W
Abitibi L., Canada 128 C4 48 40N 79 40W
Abiy Adi, Ethiopia 95 E4 13 39N 39 3 E
Abkhaz Republic = Abkhazia □, Georgia 53 J5 43 12N 41 5 E
Abkhazia □, Georgia 53 J5 43 12N 41 5 E
Abkit, Russia 57 C16 64 10N 157 10 E
Abminga, Australia 115 D1 26 8 S 134 51 E
Abnûb, Egypt 94 B3 27 18N 31 4 E
Åbo = Turku, Finland 13 F20 60 30N 22 19 E
Abo, Massif d', Chad 97 D3 21 41N 16 8 E
Abocho, Nigeria 101 D6 7 35N 6 56 E
Abohar, India 80 D6 30 10N 74 10 E
Aboisso, Ivory C. 100 D4 5 30N 3 5W
Abolo, Congo 102 B2 0 8N 14 16 E
Aboméy, Benin 101 D5 7 10N 2 5 E
Abondance, France 25 B10 46 18N 6 43 E
Abong-Mbang, Cameroon 102 B2 4 0N 13 8 E
Abongabong, Indonesia 74 B1 4 15N 96 48 E
Abonnema, Nigeria 101 E6 4 41N 6 49 E
Abony, Hungary 31 D13 47 12N 20 3 E
Aboso, Ghana 100 D4 5 23N 1 57W
Abou-Deïa, Chad 97 F3 11 20N 19 20 E
Abou Goulem, Chad 97 F4 13 37N 21 38 E
Aboyne, U.K. 18 D6 57 4N 2 47W
Abra □, Phil. 70 C3 17 35N 120 45 E
Abra de Ilog, Phil. 70 E3 13 27N 120 44 E
Abra Pampa, Argentina 158 A2 22 43 S 65 42W
Abrantes, Portugal 37 F2 39 24N 8 7W
Abraveses, Portugal 36 E3 40 41N 7 55W
Abreojos, Pta., Mexico 146 B2 26 50N 113 40W
Abreschviller, France 23 D14 48 39N 7 6 E
Abri, Esh Shamâliya, Sudan 94 C3 20 50N 30 27 E
Abri, Janub Kordofân, Sudan 95 E3 11 40N 30 21 E
Abrolhos, Banka, Brazil 155 E4 18 0 S 38 0W
Abrud, Romania 46 C4 46 19N 23 5 E
Abruzzi □, Italy 39 F10 42 15N 14 0 E
Absaroka Range, U.S.A. 142 D9 44 45N 109 50W
Abū al Khaşīb, Iraq 85 D6 30 25N 48 0 E
Abū 'Alī, Si. Arabia 85 E6 27 20N 49 27 E
Abū 'Alī →, Lebanon 91 A4 34 25N 35 50 E
Abu 'Arīsh, Si. Arabia 86 C3 16 53N 42 48 E
Abû Ballas, Egypt 94 C2 24 26N 27 36 E
Abu Deleiq, Sudan 95 D3 15 57N 33 48 E
Abu Dhabi = Abū Ẓāby, U.A.E. 85 E7 24 28N 54 22 E
Abû Dis, Sudan 94 D3 19 12N 33 38 E
Abû Dom, Sudan 95 D3 16 18N 32 25 E
Abû Du'ān, Syria 89 D8 36 25N 38 15 E
Abu el Gairi, W. →, Egypt 91 F2 29 35N 33 30 E
Abû Gabra, Sudan 95 E2 11 2N 26 50 E
Abu Ga'da, W. →, Egypt 91 F1 29 15N 32 53 E
Abû Gubeiha, Sudan 95 E3 11 30N 31 15 E
Abu Habl, Khawr →, Sudan 95 E3 12 37N 31 0 E
Abū Ḩadrīyah, Si. Arabia 85 E6 27 20N 48 58 E
Abu Hamed, Sudan 94 D3 19 32N 33 13 E
Abu Haraz, An Nîl el Azraq, Sudan 95 E3 14 35N 33 30 E
Abû Haraz, Esh Shamâliya, Sudan 94 D3 19 8N 32 18 E
Abû Higar, Sudan 95 E3 12 50N 33 59 E
Abū Kamāl, Syria 89 E9 34 30N 41 0 E
Abū Madd, Ra's, Si. Arabia 84 E3 24 50N 37 7 E
Abu Matariq, Sudan 95 E2 10 59N 26 9 E
Abu Qir, Egypt 94 H7 31 18N 30 0 E
Abu Qireiya, Egypt 94 C4 24 5N 35 28 E
Abu Qurqâs, Egypt 94 J7 28 1N 30 44 E
Abū Raşāş, Ra's, Oman 87 B7 20 10N 58 38 E
Abū Rubayq, Si. Arabia 86 B2 23 44N 39 42 E
Abu Şafāt, W. →, Jordan 91 E5 30 24N 36 7 E
Abû Simbel, Egypt 94 C3 22 18N 31 40 E
Abū Şukhayr, Iraq 89 G11 31 54N 44 30 E
Abu Tig, Egypt 94 B3 27 4N 31 15 E
Abu Tiga, Sudan 95 E3 12 47N 34 12 E
Abû Zabad, Sudan 95 E2 12 25N 29 10 E
Abū Ẓāby, U.A.E. 85 E7 24 28N 54 22 E
Abū Zeydābād, Iran 85 C6 33 54N 51 45 E
Abufari, Brazil 157 B5 5 25 S 62 59W
Abuja, Nigeria 101 D6 9 16N 7 2 E
Abukuma-Gawa →, Japan 60 E10 38 6N 140 52 E
Abukuma-Sammyaku, Japan 60 F10 37 30N 140 45 E
Abulug, Phil. 70 B3 18 27N 121 27 E
Abumombazi, Zaïre 102 B4 3 42N 22 10 E
Abunã, Brazil 157 B4 9 40 S 65 20W
Abunã →, Brazil 157 B4 9 41 S 65 20W
Abung, Phil. 70 E3 13 46N 121 26 E
Aburatsu, Japan 62 F3 31 34N 131 24 E
Aburo, Zaïre 106 B3 2 4N 30 53 E
Abut Hd., N.Z. 119 D5 43 7 S 170 15 E
Abwong, Sudan 95 F3 9 2N 32 14 E
Åby, Sweden 15 F10 58 40N 16 10 E
Aby, Lagune, Ivory C. 100 D4 5 15N 3 14W
Acacías, Colombia 152 C3 3 59N 73 46W
Acajutla, El Salv. 148 D2 13 36N 89 50W
Açailândia, Brazil 154 C2 5 0 S 47 30W
Acámbaro, Mexico 146 C4 20 0N 100 40W
Acanthus, Greece 44 D5 40 27N 23 47 E
Acaponeta, Mexico 146 C3 22 30N 105 20W
Acapulco, Mexico 147 D5 16 51N 99 56W
Acarai, Serra, Brazil 153 C6 1 50N 57 50W
Acaraú, Brazil 154 B3 2 53 S 40 7W
Acari, Brazil 154 C4 6 31 S 36 38W
Acarí, Peru 156 D3 15 25 S 74 36W
Acarigua, Venezuela 152 B4 9 33N 69 12W
Acatlán, Mexico 147 D5 18 10N 98 3W
Acayucan, Mexico 147 D6 17 59N 94 58W
Accéglio, Italy 38 D3 44 28N 7 0 E
Accomac, U.S.A. 134 G8 37 43N 75 40W
Accous, France 24 E3 43 0N 0 36W
Accra, Ghana 101 D4 5 35N 0 6W
Accrington, U.K. 16 D5 53 45N 2 22W
Acebal, Argentina 158 C3 33 20 S 60 50W
Aceh □, Indonesia 74 B1 4 15N 97 30 E
Acerenza, Italy 41 B8 40 48N 15 56 E
Acerra, Italy 41 B7 40 57N 14 22 E
Aceuchal, Spain 37 G4 38 39N 6 30W
Achacachi, Bolivia 156 D4 16 3 S 68 43W
Achaguas, Venezuela 152 B4 7 46N 68 14W
Achalpur, India 82 D3 21 22N 77 32 E
Achao, Chile 160 B2 42 28 S 73 30W
Achel, Belgium 21 F6 51 15N 5 29 E
Acheng, China 67 B14 45 30N 126 58 E
Achenkirch, Austria 30 D4 47 32N 11 45 E
Achensee, Austria 30 D4 47 26N 11 45 E
Acher, India 80 H5 23 10N 72 32 E
Achern, Germany 27 G4 48 37N 8 4 E
Acheron →, N.Z. 119 C8 42 16 S 173 4 E
Achill, Ireland 19 C2 53 56N 9 55W
Achill Hd., Ireland 19 C1 53 58N 10 15W
Achill I., Ireland 19 C1 53 58N 10 1W
Achill Sd., Ireland 19 C2 53 54N 9 56W
Achim, Germany 26 B5 53 1N 9 3 E
Achinsk, Russia 57 D10 56 20N 90 20 E
Achisay = Ashchysay, Kazakhstan 55 B4 43 35N 68 53 E
Achit, Russia 54 C6 56 48N 57 54 E
Achol, Sudan 95 F3 6 35N 31 32 E
Acıgöl, Turkey 88 D3 37 50N 29 50 E
Acireale, Italy 41 E8 37 37N 15 10 E
Ackerman, U.S.A. 139 J10 33 19N 89 11W
Ackley, U.S.A. 140 B3 42 33N 93 3W
Acklins I., Bahamas 149 B5 22 30N 74 0W
Acme, Canada 130 C6 51 33N 113 30W
Acobamba, Peru 156 C3 12 52 S 74 35W
Acomayo, Peru 156 C3 13 55 S 71 38W
Aconcagua, Cerro, Argentina 158 C2 32 39 S 70 0W
Aconquija, Mt., Argentina 158 B2 27 0 S 66 0W
Acopiara, Brazil 154 C4 6 6 S 39 27W
Açores, Is. dos = Azores, Atl. Oc. 8 E6 38 44N 29 0W
Acorizal, Brazil 157 D6 15 12 S 56 22W
Acquapendente, Italy 39 F8 42 44N 11 52 E
Acquasanta Terme, Italy 39 F10 42 46N 13 24 E
Acquaviva delle Fonti, Italy 41 B9 40 54N 16 50 E
Acqui Terme, Italy 38 D5 44 41N 8 28 E
Acraman, L., Australia 115 E2 32 2 S 135 23 E
Acre = 'Akko, Israel 91 C4 32 55N 35 4 E
Acre □, Brazil 156 B3 9 1 S 71 0W
Acre →, Brazil 156 B4 8 45 S 67 22W
Acri, Italy 41 C9 39 29N 16 23 E
Acs, Hungary 31 D11 47 42N 18 2 E
Actium, Greece 45 F2 38 57N 20 45 E
Acton, Canada 136 C4 43 38N 80 3W
Açu, Brazil 154 C4 5 34 S 36 54W
Ad Dahnā, Si. Arabia 87 A5 24 30N 48 10 E
Ad Ḍāli', Yemen 86 D4 13 42N 44 44 E
Ad Dammām, Si. Arabia 85 E6 26 20N 50 5 E
Ad Darb, Si. Arabia 86 C3 18 2N 43 7 E
Ad Dawhah, Qatar 85 E6 25 15N 51 35 E
Ad Dawr, Iraq 89 E10 34 27N 43 47 E
Ad Ḍiffah, Libya 96 B4 30 30N 24 30 E
Ad Dilam, Si. Arabia 86 B4 23 55N 47 10 E
Ad Dir'īyah, Si. Arabia 84 E5 24 44N 46 35 E
Ad Dīwānīyah, Iraq 89 F11 32 0N 45 0 E
Ad Dujayl, Iraq 89 F11 33 51N 44 14 E
Ad Durūz, J., Jordan 91 C5 32 35N 36 40 E
Ada, Ghana 101 D5 5 44N 0 40 E
Ada, Serbia, Yug. 42 B5 45 49N 20 9 E
Ada, Minn., U.S.A. 138 B6 47 18N 96 31W
Ada, Ohio, U.S.A. 141 D13 40 46N 83 49W
Ada, Okla., U.S.A. 139 H6 34 46N 96 41W
Adad, Somali Rep. 108 C3 9 27N 46 49 E
Adaja →, Spain 36 D6 41 32N 4 52W
Ådalslinden, Sweden 14 A10 63 27N 16 55 E
Adam, Oman 87 B7 22 15N 57 28 E
Adam, Mt., Falk. Is. 160 D4 51 34 S 60 4W
Adamantina, Brazil 155 F1 21 42 S 51 4W
Adamaoua, Massif de l', Cameroon 101 D7 7 20N 12 20 E
Adamawa □, Nigeria 101 D7 9 20N 12 30 E
Adamawa Highlands = Adamaoua, Massif de l', Cameroon 101 D7 7 20N 12 20 E
Adamello, Mte., Italy 38 B7 46 9N 10 30 E
Adami Tulu, Ethiopia 95 F4 7 53N 38 41 E
Adaminaby, Australia 117 D8 36 0 S 148 45 E
Adamovka, Russia 54 F7 51 32N 59 56 E
Adams, Phil. 70 B3 18 28N 120 54 E
Adams, Mass., U.S.A. 137 D11 42 38N 73 7W
Adams, N.Y., U.S.A. 137 C8 43 49N 76 1W
Adams, Wis., U.S.A. 138 D10 43 57N 89 49W
Adam's Bridge, Sri Lanka 83 K4 9 15N 79 40 E
Adams L., Canada 130 C5 51 10N 119 40W
Adams Mt., U.S.A. 144 D5 46 12N 121 30W
Adam's Peak, Sri Lanka 83 L5 6 48N 80 30 E
Adamuz, Spain 37 G6 38 2N 4 32W
Adana, Turkey 88 D6 37 0N 35 16 E
Adanero, Spain 36 E6 40 56N 4 36W
Adapazarı, Turkey 88 B4 40 48N 30 25 E
Adarama, Sudan 95 D3 17 10N 34 52 E
Adare, C., Antarctica 7 D11 71 0 S 171 0 E
Adaut, Indonesia 73 C4 8 8 S 131 7 E
Adavale, Australia 115 D3 25 52 S 144 32 E
Adda →, Italy 38 C6 45 8N 9 53 E
Addis Ababa = Addis Abeba, Ethiopia 95 F4 9 2N 38 42 E
Addis Abeba, Ethiopia 95 F4 9 2N 38 42 E
Addis Alem, Ethiopia 95 F4 9 0N 38 17 E
Addison, Ill., U.S.A. 141 C8 41 55N 88 0W
Addison, N.Y., U.S.A. 136 D7 42 1N 77 14W
Addo, S. Africa 104 E4 33 32 S 25 45 E
Addyston, U.S.A. 141 E12 39 8N 84 43W
Adebour, Niger 97 F2 13 17N 11 50 E
Ādeh, Iran 84 B5 37 42N 45 11 E
Adel, Ga., U.S.A. 135 K4 31 8N 83 25W
Adel, Iowa, U.S.A. 140 C2 41 37N 94 1W
Adelaide, Australia 116 C3 34 52 S 138 30 E
Adelaide, Bahamas 148 A4 25 4N 77 31W
Adelaide, S. Africa 104 E4 32 42 S 26 20 E
Adelaide I., Antarctica 7 C17 67 15 S 68 30W
Adelaide Pen., Canada 126 B10 68 15N 97 30W
Adelaide River, Australia 112 B5 13 15 S 131 7 E
Adelanto, U.S.A. 145 L9 34 35N 117 22W
Adelboden, Switz. 28 D5 46 29N 7 33 E
Adele I., Australia 112 C3 15 32 S 123 9 E
Adélie, Terre, Antarctica 7 C10 68 0 S 140 0 E
Adélie Land = Adélie, Terre, Antarctica 7 C10 68 0 S 140 0 E
Ademuz, Spain 34 E3 40 5N 1 13W
Aden = Al 'Adan, Yemen 86 D4 12 45N 45 0 E
Aden, G. of, Asia 90 E4 12 30N 47 30 E
Adendorp, S. Africa 104 E3 32 15 S 24 30 E
Adh Dhayd, U.A.E. 85 E7 25 17N 55 53 E
Adhoi, India 80 H4 23 26N 70 32 E
Adi, Indonesia 73 B4 4 15 S 133 30 E
Adi Daro, Ethiopia 95 E4 14 20N 38 14 E
Adi Keyih, Eritrea 95 E4 14 51N 39 22 E
Adi Kwala, Eritrea 95 E4 14 38N 38 48 E
Adi Ugri, Eritrea 95 E4 14 58N 38 48 E
Adieu, C., Australia 113 F5 32 0 S 132 10 E
Adieu Pt., Australia 112 C3 15 14 S 124 35 E
Adigala, Ethiopia 95 E5 10 24N 42 15 E
Adige →, Italy 39 C9 45 9N 12 20 E
Adigrat, Ethiopia 95 E4 14 20N 39 26 E
Adilabad, India 82 E4 19 33N 78 20 E
Adilcevaz, Turkey 89 C10 38 48N 42 43 E
Adin, U.S.A. 142 F3 41 12N 120 57W
Adinkerke, Belgium 21 F1 51 5N 2 36 E
Adirondack Mts., U.S.A. 137 C10 44 0N 74 0W
Adıyaman, Turkey 89 D8 37 45N 38 16 E
Adjim, Tunisia 96 B2 33 47N 10 50 E
Adjohon, Benin 101 D5 6 41N 2 32 E
Adjud, Romania 46 C8 46 7N 27 10 E
Adjumani, Uganda 106 B3 3 20N 31 50 E
Adlavik Is., Canada 129 B8 55 2N 57 45W
Adler, Russia 53 J4 43 28N 39 52 E
Adliswil, Switz. 29 B7 47 19N 8 32 E
Admer, Algeria 99 D6 20 21N 5 27 E
Admer, Erg d', Algeria 99 D6 24 0N 9 5 E
Admiralty G., Australia 112 B4 14 20 S 125 55 E
Admiralty I., U.S.A. 126 C6 57 30N 134 30W
Admiralty Inlet, U.S.A. 142 C2 48 8N 122 58W
Admiralty Is., Papua N. G. 120 B4 2 0 S 147 0 E
Ado, Nigeria 101 D5 6 36N 2 56 E
Ado-Ekiti, Nigeria 101 D6 7 38N 5 12 E
Adok, Sudan 95 F3 8 10N 30 20 E
Adola, Ethiopia 95 E5 11 14N 41 44 E
Adonara, Indonesia 72 C2 8 15 S 123 5 E
Adoni, India 83 G3 15 33N 77 18 E
Adony, Hungary 31 D11 47 6N 18 52 E
Adour →, France 24 E2 43 32N 1 32W
Adra, India 81 H12 23 30N 86 42 E
Adra, Spain 35 J1 36 43N 3 3W
Adrano, Italy 41 E7 37 40N 14 50 E
Adrar, Algeria 99 C4 27 51N 0 11W
Adrasman, Tajikistan 55 C4 40 38N 69 58 E
Adré, Chad 97 F4 13 40N 22 20 E
Adrī, Libya 96 C2 27 32N 13 2 E
Ádria, Italy 39 C9 45 4N 12 3 E
Adrian, Mich., U.S.A. 141 C12 41 54N 84 2W
Adrian, Mo., U.S.A. 140 F2 38 24N 94 21W
Adrian, Tex., U.S.A. 139 H3 35 16N 102 40W
Adriatic Sea, Medit. S. 10 G9 43 0N 16 0 E
Adua, Indonesia 73 B3 1 45 S 129 50 E
Adula, Switz. 29 D8 46 30N 9 3 E
Adung Long, Burma 78 A6 28 7N 97 42 E
Adur, India 83 K3 9 8N 76 40 E
Adwa, Ethiopia 95 E4 14 15N 38 52 E
Adygea □, Russia 53 H5 45 0N 40 0 E
Adzhar Republic = Ajaria □, Georgia 53 K6 41 30N 42 0 E
Adzopé, Ivory C. 100 D4 6 7N 3 49W
Ægean Sea, Medit. S. 45 F7 38 30N 25 0 E
Æolian Is. = Eólie, Ís., Italy 41 D7 38 30N 14 57 E
Aerhtai Shan, Mongolia 64 B4 46 40N 92 45 E
Ærø, Denmark 15 K4 54 52N 10 25 E
Ærøskøbing, Denmark 15 K4 54 53N 10 24 E
Aesch, Switz. 28 B5 47 28N 7 36 E
Aëtós, Greece 45 G3 37 15N 21 50 E
Afafi, Massif d', Niger 97 D3 22 11N 15 10 E
'Afak, Iraq 89 F11 32 4N 45 15 E
Afándou, Greece 32 C10 36 18N 28 12 E
Afarag, Erg, Algeria 99 D5 23 50N 2 47 E
Afars & Issas, Terr. of = Djibouti ■, Africa 90 E3 12 0N 43 0 E
Afdega, Ethiopia 108 C2 6 4N 43 30 E
Affreville = Khemis Miliana, Algeria 99 A5 36 11N 2 14 E
Affton, U.S.A. 140 F6 38 33N 90 20W
Afghanistan ■, Asia 79 B2 33 0N 65 0 E
Afgoi, Somali Rep. 108 D2 2 7N 44 59 E
'Afīf, Si. Arabia 86 B3 23 53N 42 56 E
Afikpo, Nigeria 101 D6 5 53N 7 54 E
Aflou, Algeria 99 B5 34 7N 2 3 E
Afmadu, Somali Rep. 108 D2 0 31N 42 4 E
Afogados da Ingàzeira, Brazil 154 C4 7 45 S 37 39W
Afognak I., U.S.A. 126 C4 58 15N 152 30W
Afragola, Italy 41 B7 40 55N 14 18 E
Afrera, Ethiopia 95 E5 13 16N 41 5 E
'Afrīn, Syria 88 D7 36 32N 36 50 E
Afşin, Turkey 88 C7 38 14N 36 55 E
Afton, U.S.A. 137 D9 42 14N 75 32W
Aftout, Algeria 98 C4 26 50N 3 45W
Afuá, Brazil 153 D7 0 15 S 50 20W
Afula, Israel 91 C4 32 37N 35 17 E
Afyonkarahisar, Turkey 88 C4 38 45N 30 33 E
Aga, Egypt 94 H7 30 55N 31 10 E
Agadès = Agadez, Niger 97 E1 16 58N 7 59 E
Agadez, Niger 97 E1 16 58N 7 59 E
Agadir, Morocco 98 B3 30 28N 9 55W
Agaete, Canary Is. 33 F4 28 6N 15 43W
Agailás, Mauritania 98 D2 22 37N 14 22W
Agana, Guam 121 R15 13 28N 144 45 E
Agapa, Russia 57 B9 71 27N 89 15 E
Agar, India 80 H7 23 40N 76 2 E
Agaro, Ethiopia 95 F4 7 50N 36 38 E
Agartala, India 78 D3 23 50N 91 23 E
Agăş, Romania 46 C7 46 28N 26 15 E
Agassiz, Canada 130 D4 49 14N 121 46W
Agats, Indonesia 73 C5 5 33 S 138 0 E
Agbélouvé, Togo 101 D5 6 35N 1 14 E
Agboville, Ivory C. 100 D4 5 55N 4 15W
Ağcabädi, Azerbaijan 53 K8 40 5N 47 27 E
Agcogan, Phil. 70 E3 12 4N 121 57 E
Ağdam, Azerbaijan 53 L8 40 0N 46 58 E
Ağdaş, Azerbaijan 53 K8 40 44N 47 22 E
Agde, France 24 E7 43 19N 3 28 E
Agde, C. d', France 24 E7 43 16N 3 28 E
Agdz, Morocco 98 B3 30 47N 6 30W

Al Khābūra 163

Agdzhabedi = Ağcabädi, *Azerbaijan* **53 K8** 40 5N 47 27 E
Agen, *France* **24 D4** 44 12N 0 38 E
Ageo, *Japan* **63 B11** 35 58N 139 36 E
Ager Tay, *Chad* **97 E3** 20 0N 17 41 E
Agersø, *Denmark* **15 J5** 55 13N 11 12 E
Ageyevo, *Russia* **52 C3** 54 10N 36 27 E
Agger, *Denmark* **15 H2** 56 47N 8 13 E
Āggius, *Italy* **40 B2** 40 56N 9 4 E
Āgh Kand, *Iran* **89 D13** 37 15N 48 4 E
Aghoueyyît, *Mauritania* **98 D1** 21 10N 15 6W
Aginskoye, *Russia* **57 D12** 51 6N 114 32 E
Agira, *Italy* **41 E7** 37 39N 14 31 E
Ağlasun, *Turkey* **88 D4** 37 39N 30 31 E
Agly →, *France* **24 F7** 42 46N 3 3 E
Agnibilékrou, *Ivory C.* **100 D4** 7 10N 3 11W
Agnita, *Romania* **46 D5** 45 59N 24 40 E
Agnone, *Italy* **41 A7** 41 48N 14 22 E
Ago, *Japan* **63 C8** 34 20N 136 51 E
Agofie, *Ghana* **101 D5** 8 27N 0 15 E
Agogna →, *Italy* **38 C5** 45 4N 8 54 E
Agogo, *Sudan* **95 F2** 7 50N 28 45 E
Agon, *France* **22 C5** 49 2N 1 34W
Agön, *Sweden* **14 C11** 61 34N 17 23 E
Agoo, *Phil.* **70 C3** 16 20N 120 22 E
Āgordo, *Italy* **39 B9** 46 18N 12 2 E
Agout →, *France* **24 E5** 43 47N 1 41 E
Agra, *India* **80 F7** 27 17N 77 58 E
Agrakhanskiuy Poluostrov, *Russia* **53 J8** 43 42N 47 36 E
Agramunt, *Spain* **34 D6** 41 48N 1 6 E
Agreda, *Spain* **34 D3** 41 51N 1 55W
Ağri, *Turkey* **89 C10** 39 44N 43 4 E
Agri →, *Italy* **41 B9** 40 13N 16 44 E
Ağrı Dağı, *Turkey* **89 C11** 39 50N 44 15 E
Ağrı Karakose, *Turkey* **49 G7** 39 44N 43 3 E
Agrigento, *Italy* **40 E6** 37 19N 13 34 E
Agrinion, *Greece* **45 F3** 38 37N 21 27 E
Agrópoli, *Italy* **41 B7** 40 21N 14 59 E
Ağstafa, *Azerbaijan* **53 K7** 41 7N 45 27 E
Água Branca, *Brazil* **154 C3** 5 50 S 42 40W
Agua Caliente, *Baja Calif. N., Mexico* **145 N10** 32 29N 116 59W
Agua Caliente, *Sinaloa, Mexico* **146 B3** 26 30N 108 20W
Agua Caliente Springs, *U.S.A.* **145 N10** 32 56N 116 19W
Água Clara, *Brazil* **157 E7** 20 25 S 52 45W
Agua Hechicero, *Mexico* **145 N10** 32 26N 116 14W
Agua Preta →, *Brazil* **153 D5** 1 41 S 63 48W
Agua Prieta, *Mexico* **146 A3** 31 20N 109 32W
Aguachica, *Colombia* **152 B3** 8 19N 73 38W
Aguada Cecilio, *Argentina* **160 B3** 40 51 S 65 51W
Aguadas, *Colombia* **152 B2** 5 40N 75 38W
Aguadilla, *Puerto Rico* **149 C6** 18 26N 67 10W
Aguadulce, *Panama* **148 E3** 8 15N 80 32W
Aguanga, *U.S.A.* **145 M10** 33 27N 116 51W
Aguanish, *Canada* **129 B7** 50 14N 62 2W
Aguanus →, *Canada* **129 B7** 50 13N 62 5W
Aguapeí, *Brazil* **157 D6** 16 12 S 59 43W
Aguapeí →, *Brazil* **155 F1** 21 0 S 51 0W
Aguapey →, *Argentina* **158 B4** 29 7 S 56 36W
Aguaray Guazú →, *Paraguay* **158 A4** 24 47 S 57 19W
Aguarico →, *Ecuador* **152 D2** 0 59 S 75 11W
Aguas →, *Spain* **34 D4** 41 20N 0 30W
Aguas Blancas, *Chile* **158 A2** 24 15 S 69 55W
Aguas Calientes, Sierra de, *Argentina* **158 B2** 25 26 S 66 40W
Águas Formosas, *Brazil* **155 E3** 17 5 S 40 57W
Aguascalientes, *Mexico* **146 C4** 21 53N 102 12W
Aguascalientes □, *Mexico* **146 C4** 22 0N 102 20W
Agudo, *Spain* **37 G6** 38 59N 4 52W
Águeda, *Portugal* **36 E2** 40 34N 8 27W
Agueda →, *Spain* **36 D4** 41 2N 6 56W
Aguié, *Niger* **101 C6** 13 31N 7 46 E
Aguilafuente, *Spain* **36 D6** 41 13N 4 7W
Aguilar, *Spain* **37 H6** 37 31N 4 40W
Aguilar de Campóo, *Spain* **36 C6** 42 47N 4 15W
Aguilares, *Argentina* **158 B2** 27 26 S 65 35W
Aguilas, *Spain* **35 H3** 37 23N 1 35W
Agüimes, *Canary Is.* **33 G4** 27 58N 15 27W
Aguja, C. de la, *Colombia* **152 A3** 11 18N 74 12W
Agulaa, *Ethiopia* **95 E4** 13 40N 39 40 E
Agulhas, C., *S. Africa* **104 E3** 34 52 S 20 0 E
Agulo, *Canary Is.* **33 F2** 28 11N 17 12W
Agung, *Indonesia* **75 D5** 8 20 S 115 28 E
Agur, *Uganda* **106 B3** 2 28N 32 55 E
Agusan →, *Phil.* **71 G5** 9 0N 125 30 E
Agusan del Norte □, *Phil.* **71 G5** 9 20N 125 10 E
Agusan del Sur □, *Phil.* **71 G5** 8 30N 125 30 E
Agustín Codazzi, *Colombia* **152 A3** 10 2N 73 14W
Agutaya I., *Phil.* **71 F3** 11 9N 120 58 E
Agvali, *Russia* **53 J8** 42 36N 46 8 E
Aha Mts., *Botswana* **104 B3** 19 45 S 21 0 E
Ahaggar, *Algeria* **99 D6** 23 0N 6 30 E
Ahamansu, *Ghana* **101 D5** 7 38N 0 35 E
Ahar, *Iran* **89 C12** 38 35N 47 0 E
Ahaura →, *N.Z.* **119 C6** 42 21 S 171 34 E
Ahaus, *Germany* **26 C3** 52 4N 7 0 E
Ahelledjem, *Algeria* **99 C6** 26 37N 6 58 E
Ahimanawa Ra., *N.Z.* **118 F5** 39 3 S 176 30 E
Ahipara B., *N.Z.* **118 B2** 35 5 S 173 5 E
Ahiri, *India* **82 E5** 19 30N 80 0 E
Ahlat, *Turkey* **89 C10** 38 45N 42 29 E
Ahlen, *Germany* **26 D3** 51 45N 7 53 E
Ahmad Wal, *Pakistan* **80 E1** 29 18N 65 58 E
Ahmadabad, *India* **80 H5** 23 0N 72 40 E
Aḩmadābād, *Khorāsān, Iran* **85 C9** 35 3N 60 50 E
Aḩmadābād, *Khorāsān, Iran* **85 C8** 35 49N 59 42 E
Aḩmadī, *Iran* **85 E8** 27 56N 56 42 E
Ahmadnagar, *India* **82 E2** 19 7N 74 46 E
Ahmadpur, *Pakistan* **80 E4** 29 12N 71 10 E
Ahmar, *Ethiopia* **95 F5** 9 20N 41 15 E
Ahmedabad = Ahmadabad, *India* **80 H5** 23 0N 72 40 E
Ahmednagar = Ahmadnagar, *India* **82 E2** 19 7N 74 46 E
Ahoada, *Nigeria* **101 D6** 5 8N 6 36 E
Ahome, *Mexico* **146 B3** 25 55N 109 11W
Ahr →, *Germany* **26 E3** 50 32N 7 16 E
Ahram, *Iran* **85 D6** 28 52N 51 16 E
Ahrax Pt., *Malta* **32 D1** 35 59N 14 22 E
Ahrensbök, *Germany* **26 A6** 54 8N 10 19 E
Ahrweiler, *Germany* **26 E3** 50 31N 7 3 E
Āhū, *Iran* **85 C6** 34 33N 50 2 E
Ahuachapán, *El Salv.* **148 D2** 13 54N 89 52W
Ahuriri →, *N.Z.* **119 E5** 44 31 S 170 12 E
Ahvāz, *Iran* **85 D6** 31 20N 48 40 E
Ahvenanmaa = Åland, *Finland* **13 F19** 60 15N 20 0 E
Aḩwar, *Yemen* **86 D4** 13 30N 46 40 E
Ahzar, *Mali* **101 B5** 15 30N 3 20 E
Aiari →, *Brazil* **152 C4** 1 22N 68 36W
Aichach, *Germany* **27 G7** 48 27N 11 8 E
Aichi □, *Japan* **63 B9** 35 0N 137 15 E
Aidone, *Italy* **41 E7** 37 25N 14 27 E
Aiello Cálabro, *Italy* **41 C9** 39 1N 16 10 E
Aigle, *Switz.* **28 D3** 46 18N 6 58 E
Aignay-le-Duc, *France* **23 E11** 47 40N 4 43 E
Aigoual, Mt., *France* **24 D7** 44 8N 3 35 E
Aigre, *France* **24 C4** 45 54N 0 1 E
Aigua, *Uruguay* **159 C5** 34 13 S 54 46W
Aigueperse, *France* **24 B7** 46 3N 3 13 E
Aigues →, *France* **25 D8** 44 7N 4 43 E
Aigues-Mortes, *France* **25 E8** 43 35N 4 12 E
Aigues-Mortes, G. d', *France* **25 E8** 43 31N 4 3 E
Aiguilles, *France* **25 D10** 44 47N 6 51 E
Aiguillon, *France* **24 D4** 44 18N 0 21 E
Aigurande, *France* **24 B5** 46 27N 1 49 E
Aihui, *China* **65 A7** 50 10N 127 30 E
Aija, *Peru* **156 B2** 9 50 S 77 45W
Aikawa, *Japan* **60 E9** 38 2N 138 15 E
Aiken, *U.S.A.* **135 J5** 33 34N 81 43W
Ailao Shan, *China* **68 F3** 24 0N 101 20 E
Aillant-sur-Tholon, *France* **23 E10** 47 52N 3 20 E
Aillik, *Canada* **129 A8** 55 11N 59 18W
Ailly-sur-Noye, *France* **23 C9** 49 45N 2 20 E
Ailsa Craig, *U.K.* **18 F3** 55 15N 5 6W
'Ailūn, *Jordan* **91 C4** 32 18N 35 47 E
Aim, *Russia* **57 D14** 59 0N 133 55 E
Aimere, *Indonesia* **72 C2** 8 45 S 121 3 E
Aimogasta, *Argentina* **158 B2** 28 33 S 66 50W
Aimorés, *Brazil* **155 E3** 19 30 S 41 4W
Ain □, *France* **25 B9** 46 5N 5 20 E
Ain →, *France* **25 C9** 45 45N 5 11 E
Aïn Beïda, *Algeria* **99 A6** 35 50N 7 29 E
Ain Ben Khellil, *Algeria* **99 B4** 33 15N 0 49W
Aïn Ben Tili, *Mauritania* **98 C3** 25 59N 9 27W
Aïn Beni Mathar, *Morocco* **99 B4** 34 1N 2 0W
Aïn Benian, *Algeria* **99 A5** 36 48N 2 55 E
Ain Dalla, *Egypt* **94 B2** 27 20N 27 23 E
Ain el Mafki, *Egypt* **94 B2** 27 30N 28 15 E
Ain Girba, *Egypt* **94 B2** 29 20N 25 14 E
Aïn M'lila, *Algeria* **99 A6** 36 2N 6 35 E
Ain Qeiqab, *Egypt* **94 B1** 29 42N 24 55 E
Aïn-Sefra, *Algeria* **99 B4** 32 47N 0 37W
Ain Sheikh Murzûk, *Egypt* **94 B2** 26 47N 27 45 E
'Ain Sudr, *Egypt* **91 F2** 29 50N 33 6 E
Ain Sukhna, *Egypt* **94 J8** 29 32N 32 20 E
Aïn Tédelès, *Algeria* **99 A5** 36 0N 0 21 E
Aïn-Témouchent, *Algeria* **99 A4** 35 16N 1 8W
Aïn Touta, *Algeria* **99 A6** 35 26N 5 54 E
Ain Zeitûn, *Egypt* **94 B2** 29 10N 25 48 E
Aïn Zorah, *Morocco* **99 B4** 34 37N 3 32W
Ainabo, *Somali Rep.* **108 C3** 9 0N 46 25 E
Ainaži, *Latvia* **13 H21** 57 50N 24 24 E
Aínos Óros, *Greece* **45 F2** 38 10N 20 35 E
Ainsworth, *U.S.A.* **138 D5** 42 33N 99 52W
Aioi, *Japan* **62 C6** 34 48N 134 28 E
Aipe, *Colombia* **152 C2** 3 13N 75 15W
Aiquile, *Bolivia* **157 D4** 18 10 S 65 10W
Aïr, *Niger* **97 E1** 18 30N 8 0 E
Air Hitam, *Malaysia* **77 M4** 1 55N 103 11 E
Airaines, *France* **23 C8** 49 58N 1 55 E
Airão, *Brazil* **153 D5** 1 56 S 61 22W
Airdrie, *U.K.* **18 F5** 55 52N 3 57W
Aire →, *France* **23 C11** 49 18N 4 49 E
Aire →, *U.K.* **16 D7** 53 43N 0 55W
Aire, I. del, *Spain* **33 B11** 39 48N 4 16 E
Aire-sur-la-Lys, *France* **23 B9** 50 37N 2 22 E
Aire-sur-l'Adour, *France* **24 E3** 43 42N 0 15W
Aireys Inlet, *Australia* **116 E6** 38 29 S 144 5 E
Airlie Beach, *Australia* **114 C4** 20 16 S 148 43 E
Airolo, *Switz.* **29 C7** 46 32N 8 37 E
Airvault, *France* **22 F6** 46 50N 0 8W
Aisch →, *Germany* **27 F7** 49 49N 10 58 E
Aisen □, *Chile* **160 C2** 46 30 S 73 0W
Aisne □, *France* **23 C10** 49 42N 3 40 E
Aisne →, *France* **23 C9** 49 26N 2 50 E
Aitana, Sierra de, *Spain* **35 G4** 38 35N 0 24W
Aitape, *Papua N. G.* **120 B2** 3 11 S 142 22 E
Aitkin, *U.S.A.* **138 B8** 46 32N 93 42W
Aitolía Kai Akarnanía □, *Greece* **45 F3** 38 45N 21 18 E
Aitolikón, *Greece* **45 F3** 38 26N 21 21 E
Aiuaba, *Brazil* **154 C3** 6 38 S 40 7W
Aiud, *Romania* **46 C4** 46 19N 23 44 E
Aix-en-Provence, *France* **25 E9** 43 32N 5 27 E
Aix-la-Chapelle = Aachen, *Germany* **26 E2** 50 45N 6 6 E
Aix-les-Bains, *France* **25 C9** 45 41N 5 53 E
Aixe-sur-Vienne, *France* **24 C5** 45 47N 1 9 E
Aiyang, Mt., *Papua N. G.* **120 C1** 5 10 S 141 20 E
Aiyansh, *Canada* **130 B3** 55 17N 129 2W
Áíyina, *Greece* **45 G5** 37 45N 23 26 E
Aiyínion, *Greece* **44 D4** 40 28N 22 28 E
Aiyion, *Greece* **45 F4** 38 15N 22 5 E
Aizawl, *India* **78 D4** 23 40N 92 44 E
Aizenay, *France* **22 F5** 46 44N 1 38W
Aizkraukle, *Latvia* **13 H21** 56 36N 25 11 E
Aizpute, *Latvia* **13 H19** 56 43N 21 40 E
Aizuwakamatsu, *Japan* **60 F9** 37 30N 139 56 E
Ajaccio, *France* **25 G12** 41 55N 8 40 E
Ajaccio, G. d', *France* **25 G12** 41 52N 8 40 E
Ajaju →, *Colombia* **152 C3** 0 59N 72 20W
Ajalpan, *Mexico* **147 D5** 18 22N 97 15W
Ajanta Ra., *India* **82 D2** 20 28N 75 50 E
Ajari Rep. = Ajaria □, *Georgia* **53 K6** 41 30N 42 0 E
Ajaria □, *Georgia* **53 K6** 41 30N 42 0 E
Ajax, *Canada* **136 C5** 43 50N 79 1W
Ajax, Mt., *N.Z.* **119 C7** 42 35 S 172 5 E
Ajayan Pt., *Guam* **121 R15** 13 15N 144 43 E
Ajdâbiyah, *Libya* **96 B4** 30 54N 20 4 E
Ajdovščina, *Slovenia* **39 C10** 45 54N 13 54 E
Ajibar, *Ethiopia* **95 E4** 10 35N 38 36 E
Ajka, *Hungary* **31 D10** 47 4N 17 31 E
'Ajmān, *U.A.E.* **85 E7** 25 25N 55 30 E
Ajmer, *India* **80 F6** 26 28N 74 37 E
Ajo, *U.S.A.* **143 K7** 32 22N 112 52W
Ajoie, *Switz.* **28 B4** 47 22N 7 0 E
Ajok, *Sudan* **95 F2** 9 15N 28 28 E
Ajuy, *Phil.* **71 F4** 11 10N 123 1 E
Ak Dağ, *Turkey* **88 D3** 36 30N 29 45 E
Ak Daglař, *Turkey* **88 C7** 39 32N 36 12 E
Akaba, *Togo* **101 D5** 8 10N 1 2 E
Akabira, *Japan* **60 C11** 43 33N 142 5 E
Akabli, *Algeria* **99 C5** 26 49N 1 31 E
Akaishi-Dake, *Japan* **63 B10** 35 27N 138 9 E
Akaishi-Sammyaku, *Japan* **63 B10** 35 25N 138 10 E
Akaki Beseka, *Ethiopia* **95 F4** 8 55N 38 45 E
Akala, *Sudan* **95 D4** 15 39N 36 13 E
Akamas □, *Cyprus* **32 D11** 35 3N 32 18 E
Akanthou, *Cyprus* **32 D12** 35 22N 33 45 E
Akaroa, *N.Z.* **119 D7** 43 49 S 172 59 E
Akaroa Harbour, *N.Z.* **119 D7** 43 50 S 172 55 E
Akasha, *Sudan* **94 C3** 21 10N 30 32 E
Akashi, *Japan* **62 C6** 34 45N 134 58 E
Akbou, *Algeria* **99 A5** 36 31N 4 31 E
Akbulak, *Russia* **54 F5** 51 1N 55 37 E
Akçaabat, *Turkey* **89 B8** 41 1N 39 34 E
Akçadağ, *Turkey* **88 C7** 38 27N 37 43 E
Akçakale, *Turkey* **89 D8** 36 41N 38 56 E
Akçakoca, *Turkey* **88 B4** 41 5N 31 8 E
Akchâr, *Mauritania* **98 D2** 20 20N 14 28W
Akdağmadeni, *Turkey* **88 C6** 39 39N 35 53 E
Akdala, *Kazakhstan* **55 A7** 45 2N 74 35 E
Akechi, *Japan* **63 B9** 35 18N 137 23 E
Akelamo, *Indonesia* **72 A3** 1 35N 129 40 E
Akershus fylke □, *Norway* **14 E5** 60 0N 11 10 E
Akeru →, *India* **82 F5** 17 25N 80 5 E
Aketi, *Zaïre* **102 B4** 2 38N 23 47 E
Akhaïa □, *Greece* **45 F3** 38 5N 21 45 E
Akhalkalaki, *Georgia* **53 K6** 41 27N 43 25 E
Akhaltsikhe, *Georgia* **53 K6** 41 40N 43 0 E
Akharnaí, *Greece* **45 F5** 38 5N 23 44 E
Akhelóös →, *Greece* **45 F3** 38 19N 21 7 E
Akhendria, *Greece* **45 K7** 34 58N 25 16 E
Akhéron →, *Greece* **44 E2** 39 20N 20 29 E
Akhisar, *Turkey* **88 C2** 38 56N 27 48 E
Akhladhókambos, *Greece* **45 G4** 37 31N 22 35 E
Akhmîm, *Egypt* **94 B3** 26 31N 31 47 E
Akhnur, *India* **81 C6** 32 52N 74 45 E
Akhtopol, *Bulgaria* **43 E12** 42 6N 27 56 E
Akhtuba →, *Russia* **53 G8** 47 41N 46 55 E
Akhtubinsk, *Russia* **53 F8** 48 13N 46 7 E
Akhty, *Russia* **53 K8** 41 30N 47 45 E
Akhtyrka = Okhtyrka, *Ukraine* **51 G8** 50 25N 35 0 E
Aki, *Japan* **62 D5** 33 30N 133 54 E
Aki-Nada, *Japan* **62 C4** 34 5N 132 40 E
Akiéni, *Gabon* **102 C2** 1 11 S 13 53 E
Akimiski I., *Canada* **128 B3** 52 50N 81 30W
Akimovka, *Ukraine* **51 J8** 46 44N 35 0 E
Akita, *Japan* **60 E10** 39 45N 140 7 E
Akita □, *Japan* **60 E10** 39 40N 140 30 E
Akjoujt, *Mauritania* **100 B2** 19 45N 14 15W
Akka, *Morocco* **98 C3** 29 22N 8 9W
Akkeshi, *Japan* **60 C12** 43 2N 144 51 E
'Akko, *Israel* **91 C4** 32 55N 35 4 E
Akkol, *Kazakhstan* **55 B5** 43 36N 70 45 E
Akkol, *Kazakhstan* **56 E8** 45 0N 75 39 E
Akköy, *Turkey* **45 G9** 37 30N 27 18 E
Akkrum, *Neths.* **20 B7** 53 3N 5 50 E
Aklampa, *Benin* **101 D5** 8 15N 2 10 E
Aklan □, *Phil.* **71 F4** 11 50N 122 30 E
Aklavik, *Canada* **126 B6** 68 12N 135 0W
Akmolinsk = Aqmola, *Kazakhstan* **56 D8** 51 10N 71 30 E
Akmonte, *Spain* **37 H4** 37 13N 6 38W
Akmuz, *Kyrgyzstan* **55 C8** 41 15N 76 10 E
Aknoul, *Morocco* **99 B4** 34 40N 3 55W
Akō, *Japan* **62 C6** 34 45N 134 24 E
Ako, *Nigeria* **101 C7** 10 19N 10 48 E
Akobo →, *Ethiopia* **95 F3** 7 48N 33 3 E
Akola, *India* **82 D3** 20 42N 77 2 E
Akonolinga, *Cameroon* **101 E7** 3 50N 12 18 E
Akordat, *Eritrea* **95 D4** 15 30N 37 40 E
Akosombo Dam, *Ghana* **101 D5** 6 20N 0 5 E
Akot, *India* **82 D3** 21 10N 77 10 E
Akot, *Sudan* **95 F3** 6 31N 30 9 E
Akpatok I., *Canada* **127 B13** 60 25N 68 8W
Åkrahamn, *Norway* **13 G11** 59 15N 5 10 E
Akranes, *Iceland* **12 D2** 64 19N 22 5W
Akreïjit, *Mauritania* **100 B3** 18 19N 9 11W
Akrítas Venétiko, Ákra, *Greece* **45 H3** 36 43N 21 54 E
Akron, *Colo., U.S.A.* **138 E3** 40 10N 103 13W
Akron, *Ind., U.S.A.* **141 C10** 41 2N 86 1W
Akron, *Ohio, U.S.A.* **136 E3** 41 5N 81 31W
Akrotiri, *Cyprus* **32 E11** 34 36N 32 57 E
Akrotíri, Ákra, *Greece* **44 D7** 40 26N 25 27 E
Akrotiri Bay, *Cyprus* **32 E12** 34 35N 33 10 E
Aksai Chin, *India* **81 B8** 35 15N 79 55 E
Aksaray, *Turkey* **88 C6** 38 25N 34 2 E
Aksarka, *Russia* **56 C7** 66 31N 67 50 E
Aksay, *Kazakhstan* **54 F4** 51 11N 53 0 E
Akşehir, *Turkey* **88 C4** 38 18N 31 30 E
Akşehir Gölü, *Turkey* **88 C4** 38 30N 31 25 E
Aksenovo Zilovskoye, *Russia* **57 D12** 53 20N 117 40 E
Akstafa = Ağstafa, *Azerbaijan* **53 K7** 41 7N 45 27 E
Aksu, *China* **64 B3** 41 5N 80 10 E
Aksu →, *Turkey* **88 D4** 36 52N 30 57 E
Aksum, *Ethiopia* **95 E4** 14 5N 38 40 E
Aktash, *Russia* **52 C11** 55 2N 52 0 E
Aktash, *Uzbekistan* **55 D2** 39 55N 65 55 E
Aktasty, *Kazakhstan* **54 F8** 50 42N 61 42 E
Aktogay, *Kazakhstan* **55 A8** 44 25N 76 44 E
Aktogay, *Kazakhstan* **56 E8** 46 57N 79 40 E
Aktsyabrski, *Belarus* **51 F5** 52 38N 28 53 E
Aktyubinsk = Aqtöbe, *Kazakhstan* **54 F6** 50 17N 57 10 E
Aktyuz, *Kyrgyzstan* **55 B8** 42 54N 76 7 E
Aku, *Nigeria* **101 D6** 6 40N 7 18 E
Akula, *Zaïre* **102 B4** 2 22N 20 12 E
Akune, *Japan* **62 E2** 32 1N 130 12 E
Akure, *Nigeria* **101 D6** 7 15N 5 5 E
Akureyri, *Iceland* **12 D4** 65 40N 18 6W
Akuseki-Shima, *Japan* **61 K4** 29 27N 129 37 E
Akusha, *Russia* **53 J8** 42 18N 47 30 E
Akwa-Ibom □, *Nigeria* **101 E6** 4 30N 7 30 E
Akyab = Sittwe, *Burma* **78 E4** 20 18N 92 45 E
Akyazı, *Turkey* **88 B4** 40 40N 30 38 E
Akzhar, *Kazakhstan* **55 B5** 43 8N 71 37 E
Al Abyār, *Libya* **96 B4** 32 9N 20 29 E
Al 'Adan, *Yemen* **86 D4** 12 45N 45 0 E
Al Aḩsā, *Si. Arabia* **85 E6** 25 50N 49 0 E
Al Ajfar, *Si. Arabia* **84 E4** 27 26N 43 0 E
Al Amādīyah, *Iraq* **89 D10** 37 5N 43 30 E
Al Amārah, *Iraq* **89 G12** 31 55N 47 15 E
Al 'Aqabah, *Jordan* **91 F4** 29 31N 35 0 E
Al' Aqīq, *Si. Arabia* **86 B3** 20 39N 41 25 E
Al Arak, *Syria* **89 E8** 34 38N 38 35 E
Al 'Aramah, *Si. Arabia* **84 E5** 25 30N 46 0 E
Al 'Ariḑah, *Si. Arabia* **86 C3** 17 3N 43 5 E
Al Arţāwīyah, *Si. Arabia* **84 E5** 26 31N 45 20 E
Al Ashkhara, *Oman* **87 B7** 21 50N 59 30 E
Al 'Āşimah □, *Jordan* **91 D5** 31 40N 36 30 E
Al' Assāfīyah, *Si. Arabia* **84 D3** 28 17N 38 59 E
Al 'Ayn, *Oman* **85 E7** 24 15N 55 45 E
Al 'Ayn, *Si. Arabia* **84 E3** 25 4N 38 6 E
Al A'ẓamīyah, *Iraq* **84 C5** 33 22N 44 22 E
Al 'Azīzīyah, *Iraq* **89 F11** 32 54N 45 4 E
Al 'Azīzīyah, *Libya* **96 B2** 32 30N 13 1 E
Al Bāb, *Syria* **88 D7** 36 23N 37 29 E
Al Bad', *Si. Arabia* **84 D2** 28 28N 35 1 E
Al Bādī, *Iraq* **84 C4** 35 56N 41 32 E
Al Badī', *Si. Arabia* **86 B4** 22 0N 46 35 E
Al Baḩrah, *Kuwait* **84 D5** 29 40N 47 52 E
Al Balqā' □, *Jordan* **91 C4** 32 5N 35 45 E
Al Barkāt, *Libya* **96 D2** 24 56N 10 14 E
Al Bārūk, J., *Lebanon* **91 B4** 33 39N 35 40 E
Al Başrah, *Iraq* **84 D5** 30 30N 47 50 E
Al Baţḩā, *Iraq* **84 D5** 31 6N 45 53 E
Al Batrūn, *Lebanon* **91 A4** 34 15N 35 40 E
Al Bayāḑ, *Si. Arabia* **86 B4** 22 0N 47 0 E
Al Bayḑā, *Libya* **96 B4** 32 50N 21 44 E
Al Bayḑā', *Yemen* **86 D4** 14 5N 45 42 E
Al Bayḑā □, *Libya* **96 B4** 32 0N 21 30 E
Al Bi'ār, *Si. Arabia* **86 B2** 22 39N 39 40 E
Al Biqā □, *Lebanon* **91 A5** 34 10N 36 10 E
Al Bi'r, *Si. Arabia* **84 D3** 28 51N 36 16 E
Al Birk, *Si. Arabia* **86 C3** 18 13N 41 33 E
Al Bu'ayrāt al Ḩasūn, *Libya* **96 B3** 31 24N 15 44 E
Al Burayj, *Syria* **91 A5** 34 15N 36 46 E
Al Fallūjah, *Iraq* **89 F10** 33 20N 43 55 E
Al Fatk, *Yemen* **87 C6** 16 31N 52 41 E
Al Fāw, *Iraq* **85 D6** 30 0N 48 30 E
Al Faydamī, *Yemen* **87 C6** 16 25N 52 26 E
Al Fujayrah, *U.A.E.* **85 E8** 25 7N 56 18 E
Al Ghadaf, W. →, *Jordan* **91 D5** 31 26N 36 43 E
Al Ghammās, *Iraq* **84 D5** 31 45N 44 37 E
Al Gharīb, *Libya* **96 B4** 32 35N 21 11 E
Al Ghayḑah, *Yemen* **87 C6** 16 13N 52 11 E
Al Ghayḑah, *Yemen* **87 D5** 14 55N 50 0 E
Al Ghayl, *Yemen* **87 D5** 15 30N 50 54 E
Al Hābah, *Si. Arabia* **84 E5** 27 10N 47 0 E
Al Ḩadd, *Oman* **87 B7** 22 32N 59 48 E
Al Ḩaddār, *Si. Arabia* **86 B4** 21 58N 45 57 E
Al Ḩadīthah, *Iraq* **89 E10** 34 0N 41 13 E
Al Ḩadīthah, *Si. Arabia* **84 D3** 31 28N 37 8 E
Al Ḩaḑr, *Iraq* **89 E10** 35 35N 42 44 E
Al Hājānah, *Syria* **91 B5** 33 20N 36 33 E
Al Hajarayn, *Yemen* **87 D5** 15 29N 48 20 E
Al Hallānīyah, *Oman* **87 C7** 17 30N 56 1 E
Al Hāmad, *Si. Arabia* **84 D3** 31 30N 39 30 E
Al Ḩamar, *Si. Arabia* **86 B4** 22 26N 46 12 E
Al Hamdāniyah, *Syria* **84 C3** 35 25N 36 50 E
Al Hamidīyah, *Syria* **91 A4** 34 42N 35 57 E
Al Hammādah al Ḩamrā', *Libya* **96 C2** 29 30N 12 0 E
Al Ḩammār, *Iraq* **84 D5** 30 57N 46 51 E
Al Ḩamrā', *Si. Arabia* **86 A2** 24 2N 38 55 E
Al Ḩarīq, *Si. Arabia* **86 B4** 23 29N 46 27 E
Al Harīr, W. →, *Syria* **91 C4** 32 44N 35 59 E
Al Harūj al Aswad, *Libya* **96 C3** 27 0N 17 10 E
Al Ḩasā, W. →, *Jordan* **91 D4** 31 4N 35 29 E
Al Ḩasakah, *Syria* **89 D9** 36 35N 40 45 E
Al Hāsikīyah, *Oman* **87 C6** 17 28N 55 36 E
Al Ḩasy, *Yemen* **87 D5** 14 3N 48 40 E
Al Ḩawrah, *Yemen* **86 D4** 13 50N 47 35 E
Al Ḩawţah, *Yemen* **86 D4** 14 23N 47 24 E
Al Ḩawţah □, *Si. Arabia* **86 B4** 23 30N 47 0 E
Al Haydān, W. →, *Jordan* **91 D4** 31 29N 35 34 E
Al Ḩayy, *Iraq* **89 F12** 32 5N 46 5 E
Al Ḩijāz, *Si. Arabia* **86 A2** 26 0N 37 30 E
Al Ḩillah, *Iraq* **89 F11** 32 30N 44 25 E
Al Ḩillah, *Si. Arabia* **86 B4** 23 35N 46 50 E
Al Hindīyah, *Iraq* **89 F11** 32 30N 44 10 E
Al Hirmil, *Lebanon* **91 A5** 34 26N 36 24 E
Al Hoceïma, *Morocco* **98 A4** 35 8N 3 58W
Al Ḩudaydah, *Yemen* **86 D3** 14 50N 43 0 E
Al Ḩufrah, *Libya* **96 C2** 25 32N 14 1 E
Al Hufūf, *Si. Arabia* **85 E6** 25 25N 49 45 E
Al Ḩulwah, *Si. Arabia* **86 B4** 23 24N 46 48 E
Al Ḩumayḑah, *Si. Arabia* **84 D2** 29 14N 34 56 E
Al Ḩunayy, *Si. Arabia* **85 E6** 25 58N 48 45 E
Al Ḩurayḑah, *Yemen* **87 D5** 15 36N 48 12 E
Al Husayyāt, *Libya* **96 B4** 30 24N 20 37 E
Al Hūwah, *Si. Arabia* **86 B4** 23 2N 45 48 E
Al Ḩuwaymī, *Yemen* **86 D4** 13 23N 44 28 E
Al 'Irqah, *Yemen* **86 D4** 13 39N 47 22 E
Al Īsāwīyah, *Si. Arabia* **84 D3** 30 43N 37 59 E
Al Ittihad = Madīnat ash Sha'b, *Yemen* **86 D4** 12 50N 45 0 E
Al Jafr, *Jordan* **91 E5** 30 18N 36 14 E
Al Jaghbūb, *Libya* **96 C4** 29 42N 24 38 E
Al Jahrah, *Kuwait* **84 D5** 29 25N 47 40 E
Al Jalāmīd, *Si. Arabia* **84 D3** 31 20N 39 45 E
Al Jamalīyah, *Qatar* **85 E6** 25 37N 51 5 E
Al Janūb □, *Lebanon* **91 B4** 33 20N 35 20 E
Al Jawf, *Libya* **96 D4** 24 10N 23 24 E
Al Jawf, *Si. Arabia* **84 D3** 29 55N 39 40 E
Al Jazirah, *Iraq* **89 E10** 33 30N 44 0 E
Al Jazirah, *Libya* **96 C4** 26 10N 21 20 E
Al Jithāmīyah, *Si. Arabia* **84 E4** 27 41N 41 43 E
Al Jubayl, *Si. Arabia* **85 E6** 27 0N 49 50 E
Al Jubaylah, *Si. Arabia* **84 E5** 24 55N 46 25 E
Al Jubb, *Si. Arabia* **84 E4** 27 11N 42 17 E
Al Jumūm, *Si. Arabia* **86 B2** 21 37N 39 42 E
Al Junaynah, *Sudan* **97 F4** 13 27N 22 45 E
Al Kabā'ish, *Iraq* **84 D5** 30 58N 47 0 E
Al Kāmil, *Oman* **87 B7** 22 13N 59 12 E
Al Karak, *Jordan* **91 D4** 31 11N 35 42 E
Al Karak □, *Jordan* **91 E5** 31 0N 36 0 E
Al Kāzim Tyah, *Iraq* **89 F11** 33 22N 44 12 E
Al Khābūra, *Oman* **87 B7** 23 57N 57 5 E

164 Al Khalīl

Al Khalīl, *West Bank* 91 D4 31 32N 35 6 E
Al Khāliş, *Iraq* 89 F11 33 49N 44 32 E
Al Khamāsīn, *Si. Arabia* 86 B4 20 29N 44 46 E
Al Kharāb, *Yemen* 86 C4 16 29N 44 18 E
Al Kharfah, *Si. Arabia* 86 B4 22 0N 46 35 E
Al Kharj, *Si. Arabia* 86 B4 24 0N 47 0 E
Al Khāşirah, *Si. Arabia* 86 B3 23 30N 43 47 E
Al Khawr, *Qatar* 85 E6 25 41N 51 30 E
Al Khiḍr, *Iraq* 84 D5 31 12N 45 33 E
Al Khiyām, *Lebanon* 91 B4 33 20N 35 36 E
Al Khums, *Libya* 96 B2 32 40N 14 17 E
Al Khums □, *Libya* 96 B2 31 20N 14 10 E
Al Kiswah, *Syria* 91 B5 33 23N 36 14 E
Al Kūfah, *Iraq* 89 F11 32 2N 44 24 E
Al Kufrah, *Libya* 96 D4 24 17N 23 15 E
Al Kuhayfīyah, *Si. Arabia* 84 E4 27 12N 43 3 E
Al Kūt, *Iraq* 89 F11 32 30N 46 0 E
Al Kuwayt, *Kuwait* 84 D5 29 30N 48 0 E
Al Labwah, *Lebanon* 91 A5 34 11N 36 20 E
Al Lādhiqīyah, *Syria* 88 E6 35 30N 35 45 E
Al Līth, *Si. Arabia* 86 B3 20 9N 40 15 E
Al Liwā', *Oman* 85 E8 24 31N 56 36 E
Al Luḥayyah, *Yemen* 86 D3 15 45N 42 40 E
Al Madīnah, *Iraq* 84 D5 30 57N 47 16 E
Al Madīnah, *Si. Arabia* 84 E3 24 35N 39 52 E
Al-Mafraq, *Jordan* 91 C5 32 17N 36 14 E
Al Maghārīm, *Yemen* 86 D4 15 1N 47 49 E
Al Maḥmūdīyah, *Iraq* 89 F11 33 3N 44 21 E
Al Majma'ah, *Si. Arabia* 84 E5 25 57N 45 22 E
Al Makhruq, W. →, *Jordan* 91 D6 31 28N 37 0 E
Al Makḥūl, *Si. Arabia* 84 E4 26 37N 42 39 E
Al Makīlī, *Libya* 96 B4 32 10N 22 17 E
Al Manā'if, *Si. Arabia* 87 B5 23 49N 51 20 E
Al Manāmah, *Bahrain* 85 E6 26 10N 50 30 E
Al Manşūrī, *Yemen* 86 D4 14 17N 45 16 E
Al Maqwa', *Kuwait* 84 D5 29 10N 47 59 E
Al Marj, *Libya* 96 B4 32 25N 20 30 E
Al Maţlā, *Kuwait* 84 D5 29 24N 47 40 E
Al Mawjib, W. →, *Jordan* 91 D4 31 28N 35 36 E
Al Mawşil, *Iraq* 89 D10 36 15N 43 5 E
Al Mayādin, *Syria* 89 E9 35 1N 40 27 E
Al Mazār, *Jordan* 91 D4 31 4N 35 41 E
Al Midhnab, *Si. Arabia* 84 E5 25 50N 44 18 E
Al Mīfá, *Si. Arabia* 86 C3 18 54N 41 57 E
Al Minā', *Lebanon* 91 A4 34 24N 35 49 E
Al Miqdādīyah, *Iraq* 89 E11 34 0N 45 0 E
Al Mubarraz, *Si. Arabia* 85 E6 25 30N 49 40 E
Al Muḍaybī, *Oman* 87 B7 22 34N 58 7 E
Al Mughayrā', *U.A.E.* 85 E7 24 5N 53 32 E
Al Muḥarraq, *Bahrain* 85 E6 26 15N 50 40 E
Al Mukallā, *Yemen* 87 D5 14 33N 49 2 E
Al Mukhā, *Yemen* 86 D3 13 18N 43 15 E
Al Muladdah, *Oman* 87 B7 23 45N 57 34 E
Al Musayjīd, *Si. Arabia* 84 E3 24 5N 39 5 E
Al Musayyib, *Iraq* 89 F11 32 49N 44 20 E
Al Muwayliḥ, *Si. Arabia* 84 E2 27 40N 35 30 E
Al Owuho = Otukpa, *Nigeria* 101 D6 7 9N 7 41 E
Al Qaddāḥīyah, *Libya* 96 B3 31 15N 15 9 E
Al Qaḍīmah, *Si. Arabia* 86 B2 22 20N 39 13 E
Al Qaḥmah, *Si. Arabia* 86 C3 18 0N 41 41 E
Al Qā'im, *Iraq* 89 E9 34 21N 41 7 E
Al Qalībah, *Si. Arabia* 84 D3 28 24N 37 42 E
Al Qāmishlī, *Syria* 89 D9 37 2N 41 14 E
Al Qaryah ash Sharqīyah, *Libya* 96 B2 30 28N 13 40 E
Al Qaryatayn, *Syria* 91 A6 34 12N 37 13 E
Al Qaşabát, *Libya* 96 B2 32 39N 14 1 E
Al Qaţ'ā, *Syria* 89 E9 34 40N 40 48 E
Al Qaţīf, *Si. Arabia* 85 E6 26 35N 50 0 E
Al Qaţn, *Yemen* 87 D5 15 51N 48 26 E
Al Qatrānah, *Jordan* 91 D5 31 12N 36 6 E
Al Qaţrūn, *Libya* 96 D3 24 56N 15 3 E
Al Qayşūmah, *Si. Arabia* 84 D5 28 20N 46 7 E
Al Qiblīyah, *Oman* 87 C7 17 30N 56 20 E
Al Quds = Jerusalem, *Israel* 91 D4 31 47N 35 10 E
Al Qunaytirah, *Syria* 91 C4 32 55N 35 45 E
Al Qunfudhah, *Si. Arabia* 86 C3 19 3N 41 4 E
Al Qurḥ, *Yemen* 87 C5 16 44N 51 29 E
Al Qurnah, *Iraq* 84 D5 31 1N 47 25 E
Al Quşayr, *Iraq* 84 D5 30 39N 45 50 E
Al Quşayr, *Syria* 91 A5 34 31N 36 34 E
Al Qutayfah, *Syria* 91 B5 33 44N 36 36 E
Al Quway'īyah, *Si. Arabia* 86 A4 24 3N 45 15 E
Al 'Ubaylah, *Si. Arabia* 87 B5 21 59N 50 57 E
Al' Uḍaylīyah, *Si. Arabia* 85 E6 25 8N 49 18 E
Al 'Ulā, *Si. Arabia* 84 E3 26 35N 38 0 E
Al 'Ulayyah, *Si. Arabia* 86 C3 19 39N 41 54 E
Al Uqaylah ash Sharqīgah, *Libya* 96 B3 30 12N 19 10 E
Al Uqayr, *Si. Arabia* 85 E6 25 40N 50 15 E
Al 'Uwaynid, *Si. Arabia* 84 E5 24 50N 46 0 E
Al 'Uwayqīlah, *Si. Arabia* 84 D4 30 30N 42 10 E
Al 'Uyūn, *Si. Arabia* 84 E4 26 30N 43 50 E
Al 'Uyūn, *Si. Arabia* 84 E3 24 33N 39 35 E
Al Wajh, *Si. Arabia* 84 E3 26 10N 36 30 E
Al Wakrah, *Qatar* 85 E6 25 10N 51 40 E
Al Wannān, *Si. Arabia* 85 E6 26 55N 48 24 E
Al Waqbah, *Si. Arabia* 84 D5 28 48N 45 33 E
Al Wari'āh, *Si. Arabia* 84 E5 27 51N 47 25 E
Al Wātīyah, *Libya* 96 B2 32 28N 11 57 E
Al Wusayl, *Qatar* 85 E6 25 29N 51 29 E
Ala, *Italy* 38 C8 45 45N 11 0 E
Ala Dağları, *Turkey* 89 C10 39 15N 43 33 E
Ala Tau Shankou = Dzhungarskiye Vorota, *Kazakhstan* 64 B3 45 0N 82 0 E
Alabama □, *U.S.A.* 135 J2 33 0N 87 0W
Alabama →, *U.S.A.* 135 K2 31 8N 87 57W
Alaca, *Turkey* 88 B6 40 10N 34 51 E
Alaçam, *Turkey* 88 B6 41 36N 35 36 E
Alaçam Dağları, *Turkey* 88 C3 39 18N 28 49 E
Alaçati, *Turkey* 45 F8 38 16N 26 23 E
Alaejos, *Spain* 36 D5 41 18N 5 13W
Alaérma, *Greece* 32 C9 36 9N 27 57 E
Alagir, *Russia* 53 J7 43 3N 44 14 E
Alagna Valsésia, *Italy* 38 C4 45 51N 7 56 E
Alagoa Grande, *Brazil* 154 C4 7 3 S 35 35W
Alagoas □, *Brazil* 154 C4 9 0 S 36 0W
Alagoinhas, *Brazil* 155 D4 12 7 S 38 20W
Alagón, *Spain* 34 D3 41 46N 1 12W
Alagón →, *Spain* 37 F4 39 44N 6 53W
Alajero, *Canary Is.* 33 F2 28 3N 17 13W
Alajuela, *Costa Rica* 148 D3 10 2N 84 8W
Alakamisy, *Madag.* 105 C8 21 19 S 47 14 E
Alakurtti, *Russia* 48 A5 67 0N 30 30 E
Alalapura, *Surinam* 153 C6 2 20N 56 25W
Alalaú →, *Brazil* 153 D5 0 30 S 61 9W
Alameda, *Spain* 37 H6 37 12N 4 39W
Alameda, *Calif., U.S.A.* 144 H4 37 46N 122 15W
Alameda, *N. Mex., U.S.A.* 143 J10 35 11N 106 37W
Alaminos, *Phil.* 70 C2 16 10N 119 59 E
Alamo, *U.S.A.* 145 J11 36 21N 115 10W
Alamo Crossing, *U.S.A.* 145 L13 34 16N 113 33W
Alamogordo, *U.S.A.* 143 K11 32 54N 105 57W
Alamos, *Mexico* 146 B3 27 0N 109 0W
Alamosa, *U.S.A.* 143 H11 37 28N 105 52W
Åland, *Finland* 13 F19 60 15N 20 0 E
Aland, *India* 82 F3 17 36N 76 35 E
Alandroal, *Portugal* 37 G3 38 41N 7 24W
Ålands hav, *Sweden* 13 F18 60 0N 19 30 E
Alandur, *India* 83 H5 13 0N 80 15 E
Alange, Presa de, *Spain* 37 G4 38 45N 6 18W
Alania = North Ossetia □, *Russia* 53 J7 43 30N 44 30 E
Alanis, *Spain* 37 G5 38 3N 5 43W
Alanya, *Turkey* 88 D5 36 38N 32 0 E
Alaotra, Farihin', *Madag.* 105 B8 17 30 S 48 30 E
Alapayevsk, *Russia* 54 C8 57 52N 61 42 E
Alar del Rey, *Spain* 36 C6 42 38N 4 20W
Alaraz, *Spain* 36 E5 40 45N 5 17W
Alaşehir, *Turkey* 88 C3 38 23N 28 30 E
Alaska □, *U.S.A.* 126 B5 64 0N 154 0W
Alaska, G. of, *Pac. Oc.* 126 C5 58 0N 145 0W
Alaska Highway, *Canada* 130 B3 60 0N 130 0W
Alaska Peninsula, *U.S.A.* 126 C4 56 0N 159 0W
Alaska Range, *U.S.A.* 126 B4 62 50N 151 0W
Alássio, *Italy* 38 D5 44 0N 8 10 E
Älät, *Azerbaijan* 53 L9 39 58N 49 25 E
Alatri, *Italy* 40 A6 41 43N 13 21 E
Alatyr, *Russia* 52 C8 54 55N 46 35 E
Alatyr →, *Russia* 52 C8 54 52N 46 36 E
Alausi, *Ecuador* 152 D2 2 0 S 78 50W
Álava □, *Spain* 34 C2 42 48N 2 28W
Alava, C., *U.S.A.* 142 B1 48 10N 124 44W
Alaverdi, *Armenia* 53 K7 41 15N 44 37 E
Alavus, *Finland* 13 E20 62 35N 23 36 E
Alawoona, *Australia* 116 C4 34 45 S 140 30 E
'Alayh, *Lebanon* 91 B4 33 46N 35 33 E
Alaykel, *Kyrgyzstan* 55 C7 40 15N 74 25 E
Alayor, *Spain* 33 B11 39 57N 4 8 E
Alayskiy Khrebet, *Kyrgyzstan* 55 D6 39 45N 72 0 E
Alazani →, *Azerbaijan* 53 K8 41 5N 46 40 E
Alba, *Italy* 38 D5 44 42N 8 2 E
Alba □, *Romania* 46 C4 46 10N 23 30 E
Alba de Tormes, *Spain* 36 E5 40 50N 5 30W
Alba-Iulia, *Romania* 46 C4 46 8N 23 39 E
Albac, *Romania* 46 C4 46 28N 23 1 E
Albacete, *Spain* 35 G3 39 0N 1 50W
Albacete □, *Spain* 35 G3 38 50N 2 0W
Albacutya, L., *Australia* 116 C4 35 45 S 141 58 E
Ålbæk, *Denmark* 15 G4 57 36N 10 25 E
Ålbæk Bugt, *Denmark* 15 G4 57 35N 10 40 E
Albaida, *Spain* 35 G4 38 51N 0 31W
Albalate de las Nogueras, *Spain* 34 E2 40 22N 2 18W
Albalate del Arzobispo, *Spain* 34 D4 41 6N 0 31W
Albania ■, *Europe* 44 C2 41 0N 20 0 E
Albano Laziale, *Italy* 40 A5 41 44N 12 39 E
Albany, *Australia* 113 G2 35 1 S 117 58 E
Albany, *Ga., U.S.A.* 135 K3 31 35N 84 10W
Albany, *Ind., U.S.A.* 141 D11 40 18N 85 14W
Albany, *Minn., U.S.A.* 138 C7 45 38N 94 34W
Albany, *Mo., U.S.A.* 140 D2 40 15N 94 20W
Albany, *N.Y., U.S.A.* 137 D11 42 39N 73 45W
Albany, *Oreg., U.S.A.* 142 D2 44 38N 123 6W
Albany, *Tex., U.S.A.* 139 J5 32 44N 99 18W
Albany, *Wis., U.S.A.* 140 B7 42 43N 89 26W
Albany →, *Canada* 128 B3 52 17N 81 31W
Albardón, *Argentina* 158 C2 31 20 S 68 30W
Albarracín, *Spain* 34 E3 40 25N 1 26W
Albarracín, Sierra de, *Spain* 34 E3 40 30N 1 30W
Albatross B., *Australia* 114 A3 12 45 S 141 30 E
Albatross Pt., *N.Z.* 118 E3 38 7 S 174 44 E
Albay □, *Phil.* 70 E4 13 13N 123 33 E
Albegna →, *Italy* 39 F8 42 30N 11 11 E
Albemarle, *U.S.A.* 135 H5 35 21N 80 11W
Albemarle Sd., *U.S.A.* 135 H7 36 5N 76 0W
Albenga, *Italy* 38 D5 44 3N 8 13 E
Alberche →, *Spain* 36 F6 39 58N 4 46W
Alberdi, *Paraguay* 158 B4 26 14 S 58 20W
Alberes, Mts., *Spain* 34 C7 42 28N 2 56 E
Alberique, *Spain* 35 F4 39 7N 0 31W
Albersdorf, *Germany* 26 A5 54 8N 9 17 E
Albert, *Australia* 117 B7 32 22 S 147 30 E
Albert, *France* 23 B9 50 0N 2 38 E
Albert, L., *Australia* 116 C3 35 30 S 139 10 E
Albert Canyon, *Canada* 130 C5 51 8N 117 41W
Albert Edward, Mt., *Papua N. G.* 120 E4 8 20 S 147 24 E
Albert Edward Ra., *Australia* 112 C4 18 17 S 127 57 E
Albert L., *Africa* 106 B3 1 30N 31 0 E
Albert Lea, *U.S.A.* 138 D8 43 39N 93 22W
Albert Nile →, *Uganda* 106 B3 3 36N 32 2 E
Albert Town, *Bahamas* 149 B5 22 37N 74 33W
Alberta □, *Canada* 130 C6 54 40N 115 0W
Alberti, *Argentina* 158 D3 35 1 S 60 16W
Albertinia, *S. Africa* 104 E3 34 11 S 21 34 E
Albertirsa, *Hungary* 31 D12 47 14N 19 37 E
Albertkanaal →, *Belgium* 21 F4 51 14N 4 26 E
Alberton, *Canada* 129 C7 46 50N 64 0W
Albertville = Kalemie, *Zaïre* 106 D2 5 55 S 29 9 E
Albertville, *France* 25 C10 45 40N 6 22 E
Albi, *France* 24 E6 43 56N 2 9 E
Albia, *U.S.A.* 140 C4 41 2N 92 48W
Albina, *Surinam* 153 B7 5 37N 54 15W
Albina, Ponta, *Angola* 103 F2 15 52 S 11 44 E
Albino, *Italy* 38 C6 45 46N 9 47 E
Albion, *Idaho, U.S.A.* 142 E7 42 25N 113 35W
Albion, *Ill., U.S.A.* 141 F8 38 23N 88 4W
Albion, *Ind., U.S.A.* 141 C11 41 24N 85 25W
Albion, *Mich., U.S.A.* 141 B12 42 15N 84 45W
Albion, *Nebr., U.S.A.* 138 E5 41 42N 98 0W
Albion, *Pa., U.S.A.* 136 E4 41 53N 80 22W
Alblasserdam, *Neths.* 20 E5 51 52N 4 40 E
Albocácer, *Spain* 34 E5 40 21N 0 1 E
Alborán, *Medit. S.* 37 K7 35 57N 3 0W
Alborea, *Spain* 35 F3 39 17N 1 24W
Ålborg, *Denmark* 15 G3 57 2N 9 54 E
Ålborg Bugt, *Denmark* 15 H4 56 50N 10 35 E
Alborz, Reshteh-ye Kūhhā-ye, *Iran* 85 C7 36 0N 52 0 E
Albox, *Spain* 35 H2 37 23N 2 8W
Albreda, *Canada* 130 C5 52 35N 119 10W
Albufeira, *Portugal* 37 H2 37 5N 8 15W
Albula →, *Switz.* 29 C8 46 38N 9 28 E
Albuñol, *Spain* 35 J1 36 48N 3 11W
Albuquerque, *Brazil* 157 D6 19 23 S 57 26W
Albuquerque, *U.S.A.* 143 J10 35 5N 106 39W
Albuquerque, Cayos de, *Caribbean* 148 D3 12 10N 81 50W
Alburg, *U.S.A.* 137 B11 44 59N 73 18W
Alburno, Mte., *Italy* 41 B8 40 33N 15 17 E
Alburquerque, *Spain* 37 F4 39 15N 6 59W
Albury, *Australia* 117 D7 36 3 S 146 56 E
Alby, *Sweden* 14 B9 62 30N 15 28 E
Alcácer do Sal, *Portugal* 37 G2 38 22N 8 33W
Alcaçovas, *Portugal* 37 G2 38 23N 8 9W
Alcala, *Phil.* 70 C3 17 54N 121 39 E
Alcalá de Chisvert, *Spain* 34 E5 40 19N 0 13 E
Alcalá de Guadaira, *Spain* 37 H5 37 20N 5 50W
Alcalá de Henares, *Spain* 34 E1 40 28N 3 22W
Alcalá de los Gazules, *Spain* 37 J5 36 29N 5 43W
Alcalá la Real, *Spain* 37 H7 37 27N 3 57W
Alcamo, *Italy* 40 E5 37 59N 12 55 E
Alcanadre, *Spain* 34 C2 42 24N 2 7W
Alcanadre →, *Spain* 34 D4 41 43N 0 12W
Alcanar, *Spain* 34 E5 40 33N 0 28 E
Alcanede, *Portugal* 37 F2 39 25N 8 49W
Alcanena, *Portugal* 37 F2 39 27N 8 40W
Alcañices, *Spain* 36 D4 41 41N 6 21W
Alcaníz, *Spain* 34 D4 41 2N 0 8W
Alcântara, *Brazil* 154 B3 2 20 S 44 30W
Alcántara, *Spain* 37 F4 39 41N 6 57W
Alcantara L., *Canada* 131 A7 60 57N 108 9W
Alcantarilla, *Spain* 35 H3 37 59N 1 12W
Alcaracejos, *Spain* 37 G6 38 24N 4 58W
Alcaraz, *Spain* 35 G2 38 40N 2 29W
Alcaraz, Sierra de, *Spain* 35 G2 38 40N 2 20W
Alcaudete, *Spain* 37 H6 37 35N 4 5W
Alcázar de San Juan, *Spain* 35 F1 39 24N 3 12W
Alchevsk, *Ukraine* 51 H10 48 30N 38 45 E
Alcira, *Spain* 35 F4 39 9N 0 30W
Alcoa, *U.S.A.* 135 H4 35 48N 83 59W
Alcobaça, *Portugal* 37 F2 39 32N 8 58W
Alcobendas, *Spain* 34 E1 40 32N 3 38W
Alcolea del Pinar, *Spain* 34 D2 41 2N 2 28W
Alcora, *Spain* 34 E4 40 5N 0 14W
Alcorcón, *Spain* 36 E7 40 20N 3 50W
Alcoutim, *Portugal* 37 H3 37 25N 7 28W
Alcova, *U.S.A.* 142 E10 42 34N 106 43W
Alcoy, *Spain* 35 G4 38 43N 0 30W
Alcubierre, Sierra de, *Spain* 34 D4 41 45N 0 22W
Alcublas, *Spain* 34 F4 39 48N 0 43W
Alcudia, *Spain* 33 B10 39 51N 3 7 E
Alcudia, B. de, *Spain* 33 B10 39 47N 3 15 E
Alcudia, Sierra de la, *Spain* 37 G6 38 34N 4 30W
Aldabra Is., *Seychelles* 93 G8 9 22 S 46 28 E
Aldama, *Mexico* 147 C5 23 0N 98 4W
Aldan, *Russia* 57 D13 58 40N 125 30 E
Aldan →, *Russia* 57 C13 63 28N 129 35 E
Aldea, Pta. de la, *Canary Is.* 33 G4 28 0N 15 50W
Aldeburgh, *U.K.* 17 E9 52 10N 1 37 E
Aldeia Nova, *Portugal* 37 H3 37 55N 7 24W
Alder, *U.S.A.* 142 D7 45 19N 112 6W
Alder Pk., *U.S.A.* 144 K5 35 53N 121 22W
Alderney, *U.K.* 17 H5 49 42N 2 11W
Aldershot, *U.K.* 17 F7 51 15N 0 44W
Aledo, *U.S.A.* 140 C6 41 12N 90 45W
Alefa, *Ethiopia* 95 E4 11 55N 36 55 E
Aleg, *Mauritania* 100 B2 17 3N 13 55W
Alegranza, *Canary Is.* 33 E6 29 23N 13 32W
Alegranza, I., *Canary Is.* 33 E6 29 23N 13 32W
Alegre, *Brazil* 155 F3 20 50 S 41 30W
Alegrete, *Brazil* 159 B4 29 40 S 56 0W
Alegria, *Phil.* 71 F5 11 47N 124 3 E
Aleisk, *Russia* 56 D9 52 40N 83 0 E
Aleksandriya = Oleksandriya, *Ukraine* 51 H7 48 42N 33 3 E
Aleksandriya = Oleksandriya, *Ukraine* 51 G4 50 37N 26 19 E
Aleksandriyskaya, *Russia* 53 J8 43 58N 47 14 E
Aleksandrov, *Russia* 52 B4 56 23N 38 44 E
Aleksandrov Gay, *Russia* 52 E9 50 9N 48 34 E
Aleksandrovac, *Serbia, Yug.* 42 D6 43 28N 21 3 E
Aleksandrovac, *Serbia, Yug.* 42 C6 44 28N 21 13 E
Aleksandrovka = Oleksandrovka, *Ukraine* 51 H7 48 55N 32 20 E
Aleksandrovo, *Bulgaria* 43 D9 43 14N 24 51 E
Aleksandrovsk, *Russia* 54 B6 59 9N 57 33 E
Aleksandrovsk-Sakhalinskiy, *Russia* 57 D15 50 50N 142 20 E
Aleksandrovskiy Zavod, *Russia* 57 D12 50 40N 117 50 E
Aleksandrovskoye, *Russia* 56 C8 60 35N 77 50 E
Aleksandrów Kujawski, *Poland* 47 C5 52 53N 18 43 E
Aleksandrów Łódźki, *Poland* 47 D6 51 49N 19 17 E
Alekseyevka, *Russia* 52 D10 52 35N 51 17 E
Alekseyevka, *Russia* 52 E4 50 43N 38 40 E
Aleksin, *Russia* 52 C3 54 31N 37 9 E
Aleksinac, *Serbia, Yug.* 42 D6 43 31N 21 42 E
Além Paraíba, *Brazil* 155 F3 21 52 S 42 41W
Alemania, *Argentina* 158 B2 25 40 S 65 30W
Alemania, *Chile* 158 B2 25 10 S 69 55W
Ålen, *Norway* 14 B5 62 51N 11 17 E
Alençon, *France* 22 D7 48 27N 0 4 E
Alenuihaha Channel, *U.S.A.* 132 H17 20 30N 156 0W
Aleppo = Ḥalab, *Syria* 88 D7 36 10N 37 15 E
Aléria, *France* 25 F13 42 5N 9 26 E
Alert Bay, *Canada* 130 C3 50 30N 126 55W
Alès, *France* 25 D8 44 9N 4 5 E
Aleşd, *Romania* 46 B3 47 3N 22 22 E
Alessándria, *Italy* 38 D5 44 54N 8 37 E
Ålestrup, *Denmark* 15 H3 56 42N 9 29 E
Ålesund, *Norway* 13 E12 62 28N 6 12 E
Alet-les-Bains, *France* 24 F6 42 59N 2 14 E
Aletschhorn, *Switz.* 28 D6 46 28N 8 0 E
Aleutian Is., *Pac. Oc.* 126 C2 52 0N 175 0W
Aleutian Trench, *Pac. Oc.* 122 B10 48 0N 180 0 E
Alexander, *U.S.A.* 138 B3 47 51N 103 39W
Alexander, Mt., *Australia* 113 E3 28 58 S 120 16 E
Alexander Arch., *U.S.A.* 130 B2 56 0N 136 0W
Alexander Bay, *S. Africa* 104 D2 28 40 S 16 30 E
Alexander City, *U.S.A.* 135 J3 32 56N 85 58W
Alexander I., *Antarctica* 7 C17 69 0 S 70 0W
Alexandra, *Australia* 117 D6 37 8 S 145 40 E
Alexandra, *N.Z.* 119 F4 45 14 S 169 25 E
Alexandra Falls, *Canada* 130 A5 60 29N 116 18W
Alexandretta = İskenderun, *Turkey* 88 D7 36 32N 36 10 E
Alexandria = El Iskandarîya, *Egypt* 94 H6 31 13N 29 58 E
Alexandria, *Australia* 114 B2 19 5 S 136 40 E
Alexandria, *B.C., Canada* 130 C4 52 35N 122 27W
Alexandria, *Ont., Canada* 128 C5 45 19N 74 38W
Alexandria, *Romania* 46 F6 43 57N 25 24 E
Alexandria, *S. Africa* 104 E4 33 38 S 26 28 E
Alexandria, *Ind., U.S.A.* 141 D11 40 16N 85 41W
Alexandria, *Ky., U.S.A.* 141 F12 38 58N 84 23W
Alexandria, *La., U.S.A.* 139 K8 31 18N 92 27W
Alexandria, *Minn., U.S.A.* 138 C7 45 53N 95 22W
Alexandria, *Mo., U.S.A.* 140 D5 40 27N 91 28W
Alexandria, *S. Dak., U.S.A.* 138 D6 43 39N 97 47W
Alexandria, *Va., U.S.A.* 134 F7 38 48N 77 3W
Alexandria Bay, *U.S.A.* 137 B9 44 20N 75 55W
Alexandrina, L., *Australia* 116 C3 35 25 S 139 10 E
Alexandroúpolis, *Greece* 44 D7 40 50N 25 54 E
Alexis, *U.S.A.* 140 C6 41 4N 90 33W
Alexis →, *Canada* 129 B8 52 33N 56 8W
Alexis Creek, *Canada* 130 C4 52 10N 123 20W
Alfabia, *Spain* 33 B9 39 44N 2 44 E
Alfambra, *Spain* 34 E3 40 33N 1 5W
Alfândega da Fé, *Portugal* 36 D4 41 20N 6 59W
Alfaro, *Spain* 34 C3 42 10N 1 50W
Alfatar, *Bulgaria* 43 D12 43 59N 27 13 E
Alfeld, *Germany* 26 D5 51 59N 9 50 E
Alfenas, *Brazil* 159 A6 21 20 S 46 10W
Alfiós →, *Greece* 45 G3 37 40N 21 33 E
Alfonsine, *Italy* 39 D9 44 30N 12 3 E
Alfonso XIII, *Phil.* 71 G1 9 15N 117 59 E
Alford, *U.K.* 18 D6 57 14N 2 41W
Alfred, *Maine, U.S.A.* 137 C14 43 29N 70 43W
Alfred, *N.Y., U.S.A.* 136 D7 42 16N 77 48W
Alfred Town, *Australia* 117 C7 35 8 S 147 30 E
Alfredton, *N.Z.* 118 G4 40 41 S 175 54 E
Alfreton, *U.K.* 16 D6 53 6N 1 24W
Alga, *Kazakhstan* 54 G6 49 53N 57 20 E
Algaida, *Spain* 33 B9 39 33N 2 53 E
Algar, *Spain* 37 J5 36 40N 5 39W
Algård, *Norway* 13 G11 58 46N 5 53 E
Algarinejo, *Spain* 37 H6 37 19N 4 9W
Algarve, *Portugal* 37 J2 36 58N 8 20W
Algeciras, *Spain* 37 J5 36 9N 5 28W
Algemesí, *Spain* 35 F4 39 11N 0 27W
Alger, *Algeria* 99 A5 36 42N 3 8 E
Algeria ■, *Africa* 99 C5 28 30N 2 0 E
Alghero, *Italy* 40 B1 40 33N 8 19 E
Algiers = Alger, *Algeria* 99 A5 36 42N 3 8 E
Algoa B., *S. Africa* 104 E4 33 50 S 25 45 E
Algodonales, *Spain* 37 J5 36 54N 5 24W
Algodor →, *Spain* 36 F7 39 55N 3 53W
Algoma, *U.S.A.* 134 C2 44 36N 87 26W
Algona, *U.S.A.* 140 A2 43 4N 94 14W
Algonac, *U.S.A.* 136 D2 42 37N 82 32W
Alhama de Almería, *Spain* 35 J2 36 57N 2 34W
Alhama de Aragón, *Spain* 34 D3 41 18N 1 54W
Alhama de Granada, *Spain* 37 J7 37 0N 3 59W
Alhama de Murcia, *Spain* 35 H3 37 51N 1 25W
Alhambra, *Spain* 35 G1 38 54N 3 4W
Alhambra, *Calif., U.S.A.* 145 L8 34 8N 118 6W
Alhambra, *Ill., U.S.A.* 140 F7 38 52N 89 45W
Alhaurín el Grande, *Spain* 37 J6 36 39N 4 41W
Alhucemas = Al Hoceïma, *Morocco* 98 A4 35 8N 3 58W
'Alī al Gharbī, *Iraq* 89 F12 32 30N 46 45 E
Alī ash Sharqī, *Iraq* 89 F12 32 7N 46 44 E
Äli Bayramlı, *Azerbaijan* 53 L9 39 59N 48 52 E
'Alī Khēl, *Afghan.* 79 B3 33 57N 69 43 E
Ali Sahîh, *Djibouti* 95 E5 11 10N 42 44 E
Alī Shāh, *Iran* 84 B5 38 9N 45 50 E
Ália, *Italy* 40 E6 37 47N 13 43 E
'Alīābād, *Khorāsān, Iran* 85 C8 32 30N 57 30 E
'Alīābād, *Kordestān, Iran* 84 C5 35 4N 46 58 E
'Alīābād, *Yazd, Iran* 85 D7 31 41N 53 49 E
Aliaga, *Spain* 34 E4 40 40N 0 42W
Aliağa, *Turkey* 88 C2 38 47N 26 59 E
Aliákmon →, *Greece* 44 D4 40 30N 22 36 E
Alibag, *India* 82 E1 18 38N 72 56 E
Alibo, *Ethiopia* 95 F4 9 52N 37 5 E
Alibunar, *Serbia, Yug.* 42 B5 45 5N 20 57 E
Alicante, *Spain* 35 G4 38 23N 0 30W
Alicante □, *Spain* 35 G4 38 30N 0 37W
Alice, *S. Africa* 104 E4 32 48 S 26 55 E
Alice, *U.S.A.* 139 M5 27 45N 98 5W
Alice →, *Queens., Australia* 114 C3 24 2 S 144 50 E
Alice →, *Queens., Australia* 114 B3 15 35 S 142 20 E
Alice, Punta dell', *Italy* 41 C10 39 24N 17 9 E
Alice Arm, *Canada* 130 B3 55 29N 129 31W
Alice Downs, *Australia* 112 C4 17 45 S 127 56 E
Alice Springs, *Australia* 114 C1 23 40 S 133 50 E
Alicedale, *S. Africa* 104 E4 33 15 S 26 4 E
Aliceville, *U.S.A.* 135 J1 33 8N 88 9W
Alicia, *Phil.* 71 G5 9 54N 124 26 E
Alick Cr. →, *Australia* 114 C3 20 55 S 142 20 E
Alicudi, I., *Italy* 41 D7 38 33N 14 20 E
Alida, *Canada* 131 D8 49 25N 101 55W
Aligarh, *Raj., India* 80 G7 25 55N 76 15 E
Aligarh, *Ut. P., India* 80 F8 27 55N 78 10 E
Alīgūdarz, *Iran* 85 C6 33 25N 49 45 E
Alijó, *Portugal* 36 D3 41 16N 7 27W
Alimena, *Italy* 41 E7 37 42N 14 7 E
Alimnía, *Greece* 32 C9 36 16N 27 43 E
Alimodian, *Phil.* 71 F4 10 49N 122 26 E
Alindao, *C.A.R.* 102 A4 5 2N 21 13 E
Alingsås, *Sweden* 15 G6 57 56N 12 31 E

Ampasindava, Helodranon' 165

Alipur, *Pakistan* 80 E4 29 25N 70 55 E
Alipur Duar, *India* 78 B2 26 30N 89 35 E
Aliquippa, *U.S.A.* 136 F4 40 37N 80 15W
Aliste →, *Spain* 36 D5 41 34N 5 58W
Alitus = Alytus, *Lithuania* 13 J21 54 24N 24 3 E
Alivérion, *Greece* 45 F6 38 24N 24 2 E
Aliwal North, *S. Africa* 104 E4 30 45 S 26 45 E
Alix, *Canada* 130 C6 52 24N 113 11W
Aljezur, *Portugal* 37 H2 37 18N 8 49W
Aljustrel, *Portugal* 37 H2 37 55N 8 10W
Alkamari, *Niger* 97 F2 13 27N 11 10 E
Alken, *Belgium* 21 G6 50 53N 5 18 E
Alkmaar, *Neths.* 20 C5 52 37N 4 45 E
All American Canal, *U.S.A.* 143 K6 32 45N 115 15W
Allacapan, *Phil.* 70 B3 18 15N 121 35 E
Allada, *Benin* 101 D5 6 41N 2 9 E
Allah Dad, *Pakistan* 80 G2 25 38N 67 34 E
Allahabad, *India* 81 G9 25 25N 81 58 E
Allakh-Yun, *Russia* 57 C14 60 50N 137 5 E
Allal Tazi, *Morocco* 98 B3 34 30N 6 20W
Allan, *Canada* 131 C7 51 53N 106 4W
Allanche, *France* 24 C6 45 14N 2 57 E
Allanmyo, *Burma* 78 F5 19 30N 95 17 E
Allanridge, *S. Africa* 104 D4 27 45 S 26 40 E
Allansford, *Australia* 116 E5 38 26 S 142 39 E
Allanton, *N.Z.* 119 F5 45 55 S 170 15 E
Allanwater, *Canada* 128 B1 50 14N 90 10W
Allaqi, Wadi →, *Egypt* 94 C3 23 7N 32 47 E
Allariz, *Spain* 36 C3 42 11N 7 50W
Allassac, *France* 24 C5 45 15N 1 29 E
Alle, *Belgium* 21 J5 49 51N 4 58 E
Allegan, *U.S.A.* 141 B11 42 32N 85 51W
Allegany, *U.S.A.* 136 D6 42 6N 78 30W
Allegheny →, *U.S.A.* 136 F5 40 27N 80 1W
Allegheny Plateau, *U.S.A.* 134 G6 38 0N 80 0W
Allegheny Reservoir, *U.S.A.* 136 E6 41 50N 79 0W
Allègre, *France* 24 C7 45 12N 3 41 E
Allen, *Argentina* 160 A3 38 58 S 67 50W
Allen, *Phil.* 70 E5 12 30N 124 17 E
Allen, Bog of, *Ireland* 19 C4 53 15N 7 0W
Allen, L., *Ireland* 19 B3 54 8N 8 4W
Allende, *Mexico* 146 B4 28 20N 100 50W
Allentown, *U.S.A.* 137 F9 40 37N 75 29W
Allentsteig, *Austria* 30 C8 48 41N 15 20 E
Alleppey, *India* 83 K3 9 30N 76 28 E
Aller →, *Germany* 26 C5 52 56N 9 12 E
Alleur, *Belgium* 21 G7 50 39N 5 31 E
Allevard, *France* 25 C10 45 24N 6 5 E
Alliance, *Surinam* 153 B7 5 50N 54 50W
Alliance, *Nebr., U.S.A.* 138 D3 42 6N 102 52W
Alliance, *Ohio, U.S.A.* 136 F3 40 55N 81 6W
Allier □, *France* 24 B6 46 25N 2 40 E
Allier →, *France* 23 F10 46 57N 3 4 E
Allingåbro, *Denmark* 15 H4 56 28N 10 20 E
Allison, *U.S.A.* 140 B4 42 45N 92 48W
Alliston, *Canada* 128 D4 44 9N 79 52W
Alloa, *U.K.* 18 E5 56 7N 3 47W
Allora, *Australia* 115 D5 28 2 S 152 0 E
Allos, *France* 25 D10 44 15N 6 38 E
Alluitsup Paa = Sydprøven, *Greenland* 6 C5 60 30N 45 35W
Alma, *Canada* 129 C5 48 35N 71 40W
Alma, *Ga., U.S.A.* 135 K4 31 33N 82 28W
Alma, *Kans., U.S.A.* 138 F6 39 1N 96 17W
Alma, *Mich., U.S.A.* 134 D3 43 23N 84 39W
Alma, *Nebr., U.S.A.* 138 E5 40 6N 99 22W
Alma, *Wis., U.S.A.* 138 C9 44 20N 91 55W
Alma Ata = Almaty, *Kazakhstan* 55 B8 43 15N 76 57 E
Almada, *Portugal* 37 G1 38 40N 9 9W
Almaden, *Australia* 114 B3 17 22 S 144 40 E
Almadén, *Spain* 37 G6 38 49N 4 52W
Almagro, *Spain* 37 G7 38 50N 3 45W
Almagro I., *Phil.* 71 F5 11 56N 124 18 E
Almalyk = Olmaliq, *Uzbekistan* 55 C4 40 50N 69 35 E
Almanor, L., *U.S.A.* 142 F3 40 14N 121 9W
Almansa, *Spain* 35 G3 38 51N 1 5W
Almanza, *Spain* 36 C5 42 39N 5 3W
Almanzor, Pico del Moro, *Spain* 36 E5 40 15N 5 18W
Almanzora →, *Spain* 35 H3 37 14N 1 46W
Almas, *Brazil* 155 D2 11 33 S 47 9W
Almaş, Mţii., *Romania* 46 E3 44 49N 22 12 E
Almaty, *Kazakhstan* 55 B8 43 15N 76 57 E
Almazán, *Spain* 34 D2 41 30N 2 30W
Almazora, *Spain* 34 F4 39 57N 0 3W
Almeirim, *Brazil* 153 D7 1 30 S 52 34W
Almeirim, *Portugal* 37 F2 39 12N 8 37W
Almelo, *Neths.* 20 D9 52 22N 6 42 E
Almenar, *Spain* 34 D2 41 43N 2 12W
Almenara, *Brazil* 155 E3 16 11 S 40 42W
Almenara, *Spain* 34 F4 39 46N 0 14W
Almenara, Sierra de, *Spain* 35 H3 37 34N 1 32W
Almendralejo, *Spain* 37 G4 38 41N 6 26W
Almería, *Spain* 35 J2 36 52N 2 27W
Almería □, *Spain* 35 H2 37 20N 2 20W
Almería, G. de, *Spain* 35 J2 36 41N 2 28W
Almetyevsk, *Russia* 52 C11 54 53N 52 20 E
Almirante, *Panama* 148 E3 9 10N 82 30W
Almirante Montt, G., *Chile* 160 D2 51 52 S 72 50W
Almiropótamos, *Greece* 45 F6 38 16N 24 11 E
Almirós, *Greece* 45 E4 39 11N 22 45 E
Almiroú, Kólpos, *Greece* 32 D6 35 23N 24 20 E
Almodôvar, *Portugal* 37 H2 37 31N 8 2W
Almodóvar del Campo, *Spain* 37 G6 38 43N 4 10W
Almogia, *Spain* 37 J6 36 50N 4 32W
Almonaster la Real, *Spain* 37 H4 37 52N 6 48W
Almont, *U.S.A.* 136 D1 42 55N 83 3W
Almonte, *Canada* 137 A8 45 14N 76 12W
Almonte →, *Spain* 37 F4 39 41N 6 28W
Almora, *India* 81 E8 29 38N 79 40 E
Almoradi, *Spain* 35 G4 38 7N 0 46W
Almorox, *Spain* 36 E6 40 14N 4 24W
Almoustarat, *Mali* 101 B5 17 35N 0 8 E
Almuñécar, *Spain* 37 J7 36 43N 3 41W
Alnif, *Morocco* 98 B3 31 10N 5 8W
Alnwick, *U.K.* 16 B6 55 24N 1 42W
Aloi, *Uganda* 106 B3 2 16N 33 10 E
Alon, *Burma* 78 D5 22 12N 95 5 E
Alor, *Indonesia* 72 C2 8 15 S 124 30 E
Alor Setar, *Malaysia* 77 J3 6 7N 100 22 E
Alora, *Spain* 37 J6 36 49N 4 46W
Alosno, *Spain* 37 H3 37 33N 7 7W
Alotau, *Papua N. G.* 120 F6 10 16 S 150 30 E
Alougoum, *Morocco* 98 B3 30 17N 6 56W
Aloysius, Mt., *Australia* 113 E4 26 0 S 128 38 E
Alpaugh, *U.S.A.* 144 K7 35 53N 119 29W
Alpedrinha, *Portugal* 36 E3 40 6N 7 27W
Alpena, *U.S.A.* 134 C4 45 4N 83 27W
Alpercatas →, *Brazil* 154 C3 6 2 S 44 19W
Alpes-de-Haute-Provence □, *France* 25 D10 44 8N 6 10 E
Alpes-Maritimes □, *France* 25 E11 43 55N 7 10 E
Alpha, *Australia* 114 C4 23 39 S 146 37 E
Alpha, *U.S.A.* 140 C6 41 12N 90 23W
Alphen, *Neths.* 21 F5 51 29N 4 58 E
Alphen aan den Rijn, *Neths.* 20 D5 52 7N 4 40 E
Alphonse, *Seychelles* 109 E4 7 0 S 52 45 E
Alpiarça, *Portugal* 37 F2 39 15N 8 35W
Alpine, *Ariz., U.S.A.* 143 K9 33 51N 109 9W
Alpine, *Calif., U.S.A.* 145 N10 32 50N 116 46W
Alpine, *Tex., U.S.A.* 139 K3 30 22N 103 40W
Alpnach, *Switz.* 29 C6 46 57N 8 17 E
Alps, *Europe* 10 F7 46 30N 9 30 E
Alpu, *Turkey* 88 C4 39 46N 30 58 E
Alrø, *Denmark* 15 J4 55 52N 10 5 E
Alroy Downs, *Australia* 114 B2 19 20 S 136 5 E
Alsace, *France* 23 D14 48 15N 7 25 E
Alsask, *Canada* 131 C7 51 21N 109 59W
Alsásua, *Spain* 34 C2 42 54N 2 10W
Alsdorf, *Germany* 26 E2 50 52N 6 10 E
Alsen, *Sweden* 14 A7 63 23N 13 56 E
Alsfeld, *Germany* 26 E5 50 44N 9 16 E
Alsónémedi, *Hungary* 31 D12 47 20N 19 15 E
Alsten, *Norway* 12 D15 65 58N 12 40 E
Alta, *Norway* 12 B20 69 57N 23 10 E
Alta, Sierra, *Spain* 34 E3 40 31N 1 30W
Alta Gracia, *Argentina* 158 C3 31 40 S 64 30W
Alta Lake, *Canada* 130 C4 50 10N 123 0W
Alta Sierra, *U.S.A.* 145 K8 35 42N 118 33W
Altaelva →, *Norway* 12 B20 69 54N 23 17 E
Altafjorden, *Norway* 12 A20 70 5N 23 5 E
Altagracia, *Venezuela* 152 A3 10 45N 71 30W
Altagracia de Orituco, *Venezuela* 152 B4 9 52N 66 23W
Altai = Aerhtai Shan, *Mongolia* 64 B4 46 40N 92 45 E
Altamachi →, *Bolivia* 156 D4 16 8 S 66 50W
Altamaha →, *U.S.A.* 135 K5 31 20N 81 20W
Altamira, *Brazil* 153 D7 3 12 S 52 10W
Altamira, *Chile* 158 B2 25 47 S 69 51W
Altamira, *Colombia* 152 C2 2 3N 75 47W
Altamira, *Mexico* 147 C5 22 24N 97 55W
Altamira, Cuevas de, *Spain* 36 B6 43 20N 4 5W
Altamont, *Ill., U.S.A.* 141 E8 39 4N 88 45W
Altamont, *N.Y., U.S.A.* 137 D10 42 43N 74 3W
Altamura, *Italy* 41 B9 40 49N 16 33 E
Altanbulag, *Mongolia* 64 A5 50 16N 106 30 E
Altar, *Mexico* 146 A2 30 40N 111 50W
Altata, *Mexico* 146 C3 24 30N 108 0W
Altavas, *Phil.* 71 F4 11 32N 122 29 E
Altavista, *U.S.A.* 134 G6 37 6N 79 17W
Altay, *China* 64 B3 47 48N 88 10 E
Altdorf, *Switz.* 29 C7 46 52N 8 36 E
Alte Mellum, *Germany* 26 B4 53 43N 8 10 E
Altea, *Spain* 35 G4 38 38N 0 2W
Altenberg, *Germany* 26 E9 50 45N 13 45 E
Altenbruch, *Germany* 26 B4 53 49N 8 46 E
Altenburg, *Germany* 26 E8 50 59N 12 25 E
Altenkirchen, *Mecklenburg-Vorpommern, Germany* 26 A9 54 38N 13 22 E
Altenkirchen, *Rhld-Pfz., Germany* 26 E3 50 41N 7 39 E
Altenmarkt, *Austria* 30 D7 47 43N 14 39 E
Altentreptow, *Germany* 26 B9 53 41N 13 14 E
Alter do Chão, *Portugal* 37 F3 39 12N 7 40W
Altıntas, *Turkey* 88 C4 39 4N 30 10 E
Altiplano, *Bolivia* 156 D4 17 0 S 68 0W
Altkirch, *France* 23 E14 47 37N 7 15 E
Altmühl →, *Germany* 27 G7 48 54N 11 52 E
Alto Adige = Trentino-Alto Adige □, *Italy* 38 B8 46 30N 11 20 E
Alto Araguaia, *Brazil* 157 D7 17 15 S 53 20W
Alto Cuchumatanes = Cuchumatanes, Sierra de los, *Guatemala* 148 C1 15 35N 91 25W
Alto Cuito, *Angola* 103 E3 13 27 S 18 49 E
Alto del Inca, *Chile* 158 A2 24 10 S 68 10W
Alto Garças, *Brazil* 157 D7 16 56 S 53 32W
Alto Iriri →, *Brazil* 157 B7 8 50 S 53 25W
Alto Ligonha, *Mozam.* 107 F4 15 30 S 38 11 E
Alto Molocue, *Mozam.* 107 F4 15 50 S 37 35 E
Alto Paraguai, *Brazil* 157 C6 14 30 S 56 31W
Alto Paraguay □, *Paraguay* 158 A4 21 0 S 58 30W
Alto Paraná □, *Paraguay* 159 B5 25 30 S 54 50W
Alto Parnaíba, *Brazil* 154 C2 9 6 S 45 57W
Alto Purús →, *Peru* 156 B3 9 12 S 70 28W
Alto Río Senguerr, *Argentina* 160 C2 45 2 S 70 50W
Alto Santo, *Brazil* 154 C4 5 31 S 38 15W
Alto Sucuriú, *Brazil* 157 D7 19 19 S 52 47W
Alto Turi, *Brazil* 154 B2 2 54 S 45 38W
Alton, *Canada* 136 C4 43 54N 80 5W
Alton, *U.S.A.* 140 F6 38 53N 90 11W
Alton Downs, *Australia* 115 D2 26 7 S 138 57 E
Altoona, *Iowa, U.S.A.* 140 C3 41 39N 93 28W
Altoona, *Pa., U.S.A.* 136 F6 40 31N 78 24W
Altopascio, *Italy* 38 E7 43 49N 10 40 E
Altos, *Brazil* 154 C3 5 3 S 42 28W
Altötting, *Germany* 27 G8 48 12N 12 39 E
Altstätten, *Switz.* 29 B9 47 22N 9 33 E
Altūn Kūprī, *Iraq* 89 E11 35 45N 44 9 E
Altun Shan, *China* 64 C3 38 30N 88 0 E
Alturas, *U.S.A.* 142 F3 41 29N 120 32W
Altus, *U.S.A.* 139 H5 34 38N 99 20W
Alubijid, *Phil.* 71 G5 8 35N 124 29 E
Alucra, *Turkey* 89 B8 40 22N 38 47 E
Alūksne, *Latvia* 13 H22 57 24N 27 3 E
Alùla, *Somali Rep.* 108 B4 11 50N 50 45 E
Alunite, *U.S.A.* 145 K12 35 59N 114 55W
Alupka, *Ukraine* 51 K8 44 23N 34 2 E
Alur Gajah, *Malaysia* 74 B2 2 23N 102 13 E
Alushta, *Ukraine* 51 K8 44 40N 34 25 E
Alusi, *Indonesia* 73 C4 7 35 S 131 40 E
Alustante, *Spain* 34 E3 40 36N 1 40W
Al'Uzayr, *Iraq* 84 D5 31 19N 47 25 E
Alva, *U.S.A.* 139 G5 36 48N 98 40W
Alvaiázere, *Portugal* 36 F2 39 49N 8 23W
Älvängen, *Sweden* 15 G6 57 58N 12 8 E
Alvarado, *Mexico* 147 D5 18 40N 95 50W
Alvarado, *U.S.A.* 139 J6 32 24N 97 13W
Alvarães, *Brazil* 153 D5 3 12 S 64 50W
Alvaro Obregón, Presa, *Mexico* 146 B3 27 55N 109 52W
Alvdal, *Norway* 14 B4 62 6N 10 37 E
Alvear, *Argentina* 158 B4 29 5 S 56 30W
Alverca, *Portugal* 37 G1 38 56N 9 1W
Alveringen, *Belgium* 21 F1 51 1N 2 43 E
Alvesta, *Sweden* 13 H16 56 54N 14 35 E
Alvie, *Australia* 116 E5 38 14 S 143 30 E
Alvin, *U.S.A.* 139 L7 29 26N 95 15W
Alvinston, *Canada* 136 D3 42 49N 81 52W
Alvito, *Portugal* 37 G3 38 15N 8 0W
Älvkarleby, *Sweden* 13 F17 60 34N 17 26 E
Älvros, *Sweden* 14 B8 62 3N 14 38 E
Älvsborgs län □, *Sweden* 15 F6 58 30N 12 30 E
Älvsbyn, *Sweden* 12 D19 65 40N 21 0 E
Älvsered, *Sweden* 15 G6 57 14N 12 51 E
Alwar, *India* 80 F7 27 38N 76 34 E
Alwaye, *India* 83 J3 10 8N 76 24 E
Alxa Zuoqi, *China* 66 E3 38 50N 105 40 E
Alyaskitovyy, *Russia* 57 C15 64 45N 141 30 E
Alyata = Älät, *Azerbaijan* 53 L9 39 58N 49 25 E
Alyth, *U.K.* 18 E5 56 38N 3 13W
Alytus, *Lithuania* 13 J21 54 24N 24 3 E
Alzada, *U.S.A.* 138 C2 45 2N 104 25W
Alzano Lombardo, *Italy* 38 C6 45 44N 9 43 E
Alzette →, *Lux.* 21 J8 49 45N 6 6 E
Alzey, *Germany* 27 F4 49 45N 8 7 E
Am Dam, *Chad* 97 F4 12 40N 20 35 E
Am Géréda, *Chad* 97 F4 12 53N 21 14 E
Am-Timan, *Chad* 97 F4 11 0N 20 10 E
Amacuro □, *Venezuela* 153 B5 8 50N 61 5W
Amadeus, L., *Australia* 113 D5 24 54 S 131 0 E
Amâdi, *Sudan* 95 F3 5 29N 30 25 E
Amadi, *Zaïre* 106 B2 3 40N 26 40 E
Amadjuak, *Canada* 127 B12 64 0N 72 39W
Amadjuak L., *Canada* 127 B12 65 0N 71 8W
Amadora, *Portugal* 37 G1 38 45N 9 13W
Amagasaki, *Japan* 63 C7 34 42N 135 20 E
Amager, *Denmark* 15 J6 55 37N 12 35 E
Amagi, *Japan* 62 D2 33 25N 130 39 E
Amahai, *Indonesia* 73 B3 3 20 S 128 55 E
Amaimon, *Papua N. G.* 120 C3 5 12 S 145 30 E
Amakusa-Nada, *Japan* 62 E2 32 35N 130 5 E
Amakusa-Shotō, *Japan* 62 E2 32 15N 130 10 E
Åmål, *Sweden* 13 G15 59 3N 12 42 E
Amalapuram, *India* 83 F5 16 35N 81 55 E
Amalfi, *Colombia* 152 B2 6 55N 75 4W
Amalfi, *Italy* 41 B7 40 38N 14 36 E
Amaliás, *Greece* 45 G3 37 47N 21 22 E
Amalner, *India* 82 D2 21 5N 75 5 E
Amambaí, *Brazil* 159 A4 23 5 S 55 13W
Amambaí →, *Brazil* 159 A5 23 22 S 53 56W
Amambay □, *Paraguay* 159 A4 23 0 S 56 0W
Amambay, Cordillera de, *S. Amer.* 159 A4 23 0 S 55 45W
Amami-Guntō, *Japan* 61 L4 27 16N 129 21 E
Amami-Ō-Shima, *Japan* 61 L4 28 0N 129 0 E
Amana →, *Venezuela* 153 B5 9 45N 62 39W
Amaná, L., *Brazil* 153 D5 2 35 S 64 40W
Amanab, *Papua N. G.* 120 B1 3 40 S 141 14 E
Amanda Park, *U.S.A.* 144 C3 47 28N 123 55W
Amándola, *Italy* 39 F10 42 59N 13 21 E
Amangeldy, *Kazakhstan* 56 D7 50 10N 65 10 E
Amantea, *Italy* 41 C9 39 8N 16 4 E
Amapá, *Brazil* 153 C7 2 5N 50 50W
Amapá □, *Brazil* 153 C7 1 40N 52 0W
Amapari, *Brazil* 153 C7 0 37N 51 39W
Amara, *Sudan* 95 E3 10 25N 34 10 E
Amarante, *Brazil* 154 C3 6 14 S 42 50W
Amarante, *Portugal* 36 D2 41 16N 8 5W
Amarante do Maranhão, *Brazil* 154 C2 5 36 S 46 45W
Amaranth, *Canada* 131 C9 50 36N 98 43W
Amarapura, *Burma* 78 E6 21 54N 96 3 E
Amaravati →, *India* 83 J4 11 0N 78 15 E
Amareleja, *Portugal* 37 G3 38 12N 7 13W
Amargosa, *Brazil* 155 D4 13 2 S 39 36W
Amargosa →, *U.S.A.* 145 J10 36 14N 116 51W
Amargosa Range, *U.S.A.* 145 J10 36 20N 116 45W
Amári, *Greece* 32 D6 35 13N 24 40 E
Amarillo, *U.S.A.* 139 H4 35 13N 101 50W
Amarnath, *India* 82 E1 19 12N 73 22 E
Amaro, Mte., *Italy* 39 F11 42 5N 14 5 E
Amaro Leite, *Brazil* 155 D2 13 58 S 49 9W
Amarpur, *India* 81 G12 25 5N 87 0 E
Amasra, *Turkey* 88 B5 41 45N 32 23 E
Amassama, *Nigeria* 101 D6 5 1N 6 2 E
Amasya, *Turkey* 88 B6 40 40N 35 50 E
Amatauá, *Brazil* 152 D4 3 29 S 68 6W
Amatikulu, *S. Africa* 105 D5 29 3 S 31 33 E
Amatitlán, *Guatemala* 148 D1 14 29N 90 38W
Amatrice, *Italy* 39 F10 42 38N 13 17 E
Amay, *Belgium* 21 G6 50 33N 5 19 E
Amazon = Amazonas →, *S. Amer.* 153 D7 0 5 S 50 0W
Amazonas □, *Brazil* 157 B5 5 0 S 65 0W
Amazonas □, *Peru* 156 B2 5 0 S 78 0W
Amazonas □, *Venezuela* 152 C4 3 30N 66 0W
Amazonas →, *S. Amer.* 153 D7 0 5 S 50 0W
Ambad, *India* 82 E2 19 38N 75 50 E
Ambahakily, *Madag.* 105 C7 21 36 S 43 41 E
Ambala, *India* 80 D7 30 23N 76 56 E
Ambalangoda, *Sri Lanka* 83 L5 6 15N 80 5 E
Ambalapulai, *India* 83 K3 9 25N 76 25 E
Ambalavao, *Madag.* 105 C8 21 50 S 46 56 E
Ambalindum, *Australia* 114 C2 23 23 S 135 0 E
Ambam, *Cameroon* 102 B2 2 20N 11 15 E
Ambanja, *Madag.* 105 A8 13 40 S 48 27 E
Ambarchik, *Russia* 57 C17 69 40N 162 20 E
Ambarijeby, *Madag.* 105 A8 14 56 S 47 41 E
Ambaro, Helodranon', *Madag.* 105 A8 13 23 S 48 38 E
Ambartsevo, *Russia* 56 D9 57 30N 83 52 E
Ambasamudram, *India* 83 K3 8 43N 77 25 E
Ambato, *Ecuador* 152 D2 1 5 S 78 42W
Ambato, Sierra de, *Argentina* 158 B2 28 25 S 66 10W
Ambato Boeny, *Madag.* 105 B8 16 28 S 46 43 E
Ambatofinandrahana, *Madag.* 105 C8 20 33 S 46 48 E
Ambatolampy, *Madag.* 105 B8 19 20 S 47 35 E
Ambatondrazaka, *Madag.* 105 B8 17 55 S 48 28 E
Ambatosoratra, *Madag.* 105 B8 17 37 S 48 31 E
Ambenja, *Madag.* 105 B8 15 17 S 46 58 E
Amberg, *Germany* 27 F7 49 26N 11 52 E
Ambergris Cay, *Belize* 147 D7 18 0N 88 0W
Ambérieu-en-Bugey, *France* 25 C9 45 57N 5 20 E
Amberley, *N.Z.* 119 D7 43 9 S 172 44 E
Ambert, *France* 24 C7 45 33N 3 44 E
Ambidédi, *Mali* 100 C2 14 35N 11 47W
Ambikapur, *India* 81 H10 23 15N 83 15 E
Ambikol, *Sudan* 94 C3 21 20N 30 50 E
Ambilobé, *Madag.* 105 A8 13 10 S 49 3 E
Ambinanindrano, *Madag.* 105 C8 20 5 S 48 23 E
Ambjörnarp, *Sweden* 15 G7 57 25N 13 17 E
Ambleside, *U.K.* 16 C5 54 26N 2 58W
Amblève, *Belgium* 21 H8 50 21N 6 10 E
Amblève →, *Belgium* 21 H7 50 25N 5 45 E
Ambo, *Ethiopia* 95 E4 12 20N 37 30 E
Ambo, *Peru* 156 C2 10 5 S 76 10W
Ambodifototra, *Madag.* 105 B8 16 59 S 49 52 E
Ambodilazana, *Madag.* 105 B8 18 6 S 49 10 E
Ambohimahasoa, *Madag.* 105 C8 21 7 S 47 13 E
Ambohimanga, *Madag.* 105 C8 20 52 S 47 36 E
Ambohitra, *Madag.* 105 A8 12 30 S 49 10 E
Amboise, *France* 22 E8 47 24N 1 2 E
Ambon, *Indonesia* 72 B3 3 43 S 128 12 E
Ambon, *Indonesia* 72 B3 3 35 S 128 20 E
Amboseli L., *Kenya* 106 C4 2 40 S 37 10 E
Ambositra, *Madag.* 105 C8 20 31 S 47 25 E
Ambovombé, *Madag.* 105 D8 25 11 S 46 5 E
Amboy, *Calif., U.S.A.* 145 L11 34 33N 115 45W
Amboy, *Ill., U.S.A.* 140 C7 41 44N 89 20W
Ambridge, *U.S.A.* 136 F4 40 36N 80 14W
Ambriz, *Angola* 103 D2 7 48 S 13 8 E
Ambrym, *Vanuatu* 121 F6 16 15 S 168 10 E
Ambunti, *Papua N. G.* 120 C2 4 13 S 142 52 E
Ambur, *India* 83 H4 12 48N 78 43 E
Amby, *Australia* 115 D4 26 30 S 148 11 E
Amchitka I., *U.S.A.* 126 C1 51 32N 179 0 E
Amderma, *Russia* 56 C7 69 45N 61 30 E
Ameca, *Mexico* 146 C4 20 30N 104 0W
Ameca →, *Mexico* 146 C3 20 40N 105 15W
Amecameca, *Mexico* 147 D5 19 7N 98 46W
Ameland, *Neths.* 20 B7 53 27N 5 45 E
Amélia, *Italy* 39 F9 42 33N 12 25 E
Amélie-les-Bains-Palalda, *France* 24 F6 42 29N 2 41 E
Amen, *Russia* 57 C18 68 45N 180 0 E
Amendolara, *Italy* 41 C9 39 57N 16 35 E
America, *Neths.* 21 F7 51 27N 5 59 E
American Falls, *U.S.A.* 142 E7 42 47N 112 51W
American Falls Reservoir, *U.S.A.* 142 E7 42 47N 112 52W
American Highland, *Antarctica* 7 D6 73 0 S 75 0 E
American Samoa ■, *Pac. Oc.* 121 X24 14 20 S 170 40W
Americana, *Brazil* 159 A6 22 45 S 47 20W
Americus, *U.S.A.* 135 J3 32 4N 84 14W
Amersfoort, *Neths.* 20 D6 52 9N 5 23 E
Amersfoort, *S. Africa* 105 D4 26 59 S 29 53 E
Amery, *Australia* 113 F2 31 9 S 117 5 E
Amery, *Canada* 131 B10 56 34N 94 3W
Amery Ice Shelf, *Antarctica* 7 C6 69 30 S 72 0 E
Ames, *U.S.A.* 140 B3 42 2N 93 37W
Amesbury, *U.S.A.* 137 D14 42 51N 70 56W
Amfíklia, *Greece* 45 F4 38 38N 22 35 E
Amfilokhía, *Greece* 45 F3 38 52N 21 9 E
Amfípolis, *Greece* 44 D5 40 48N 23 52 E
Ámfissa, *Greece* 45 F4 38 32N 22 22 E
Amga, *Russia* 57 C14 60 50N 132 0 E
Amga →, *Russia* 57 C14 62 38N 134 32 E
Amgu, *Russia* 57 E14 45 45N 137 15 E
Amgun →, *Russia* 57 D14 52 56N 139 38 E
Amherst, *Canada* 129 C7 45 48N 64 8W
Amherst, *Mass., U.S.A.* 137 D12 42 23N 72 31W
Amherst, *N.Y., U.S.A.* 136 D6 42 59N 78 48W
Amherst, *Ohio, U.S.A.* 136 E2 41 24N 82 14W
Amherst, *Tex., U.S.A.* 139 H3 34 1N 102 25W
Amherst I., *Canada* 137 B8 44 8N 76 43W
Amherstburg, *Canada* 128 D3 42 6N 83 6W
Amiata, Mte., *Italy* 39 F8 42 53N 11 37 E
Amiens, *France* 23 C9 49 54N 2 16 E
Amigdhalokefáli, *Greece* 45 J5 35 23N 23 30 E
Amili, *India* 78 A5 28 25N 95 52 E
Amindaion, *Greece* 44 D3 40 42N 21 42 E
Amīrābād, *Iran* 84 C5 33 20N 46 16 E
Amirante Is., *Seychelles* 109 E4 6 0 S 53 0 E
Amisk L., *Canada* 131 C8 54 35N 102 15W
Amistad, Presa de la, *Mexico* 146 B4 29 24N 101 0W
Amite, *U.S.A.* 139 K9 30 44N 90 30W
Amizmiz, *Morocco* 98 B3 31 12N 8 15W
Åmli, *Norway* 15 F2 58 45N 8 32 E
Amlwch, *U.K.* 16 D3 53 24N 4 20W
Amm Adam, *Sudan* 95 D4 16 20N 36 1 E
'Ammān, *Jordan* 91 D4 31 57N 35 52 E
Ammanford, *U.K.* 17 F3 51 48N 3 59W
Ammassalik = Angmagssalik, *Greenland* 6 C6 65 40N 37 20W
Ammerån, *Sweden* 14 A10 63 9N 16 13 E
Ammerån →, *Sweden* 14 A10 63 9N 16 13 E
Ammersee, *Germany* 27 G7 48 0N 11 7 E
Ammerzoden, *Neths.* 20 E6 51 45N 5 13 E
Amnat Charoen, *Thailand* 76 E5 15 51N 104 38 E
Amo Jiang →, *China* 68 F3 23 0N 101 50 E
Āmol, *Iran* 85 B7 36 23N 52 20 E
Amorebieta, *Spain* 34 B2 43 13N 2 44W
Amoret, *U.S.A.* 140 F2 38 15N 94 35W
Amorgós, *Greece* 45 H7 36 50N 25 57 E
Amory, *U.S.A.* 135 J1 33 59N 88 29W
Amos, *Canada* 128 C4 48 35N 78 5W
Åmot, *Buskerud, Norway* 14 E3 59 54N 9 54 E
Åmot, *Telemark, Norway* 14 E2 59 34N 8 0 E
Åmotsdal, *Norway* 14 E2 59 37N 8 26 E
Amour, Djebel, *Algeria* 99 B5 33 42N 1 37 E
Amoy = Xiamen, *China* 69 E12 24 25N 118 4 E
Ampang, *Malaysia* 77 L3 3 8N 101 45 E
Ampanihy, *Madag.* 105 C7 24 40 S 44 45 E
Ampasindava, Helodranon', *Madag.* 105 A8 13 40 S 48 15 E

166 Ampasindava, Saikanosy

Ampasindava, Saikanosy, *Madag.* **105 A8** 13 42 S 47 55 E
Ampato, Nevado, *Peru* **156 D3** 15 40 S 71 56W
Ampenan, *Indonesia* **75 D5** 8 35 S 116 13 E
Amper, *Nigeria* **101 D6** 9 25N 9 40 E
Amper →, *Germany* **27 G7** 48 29N 11 35 E
Ampezzo, *Italy* **39 B9** 46 25N 12 48 E
Amposta, *Spain* **34 E5** 40 43N 0 34 E
Ampotaka, *Madag.* **105 D7** 25 3 S 44 41 E
Ampoza, *Madag.* **105 C7** 22 20 S 44 44 E
Amqui, *Canada* **129 C6** 48 28N 67 27W
'Amrān, *Yemen* **86 D3** 15 41N 43 55 E
Amravati, *India* **82 D3** 20 55N 77 45 E
Amreli, *India* **80 J4** 21 35N 71 17 E
Amrenene el Kasba, *Algeria* **99 D5** 22 10N 0 30 E
Amriswil, *Switz.* **29 A8** 47 33N 9 18 E
Amritsar, *India* **80 D6** 31 35N 74 57 E
Amroha, *India* **81 E8** 28 53N 78 30 E
Amrum, *Germany* **26 A4** 54 38N 8 22 E
Amsel, *Algeria* **99 D6** 22 47N 5 29 E
Amsterdam, *Neths.* **20 D5** 52 23N 4 54 E
Amsterdam, *U.S.A.* **137 D10** 42 56N 74 11W
Amsterdam, I., *Ind. Oc.* **109 H6** 38 30 S 77 30 E
Amstetten, *Austria* **30 C7** 48 7N 14 51 E
Amudarya →, *Uzbekistan* **56 E6** 43 58N 59 34 E
Amulung, *Phil.* **70 C3** 17 50N 121 43 E
Amundsen Gulf, *Canada* **126 A7** 71 0N 124 0W
Amundsen Sea, *Antarctica* **7 D15** 72 0 S 115 0W
Amuntai, *Indonesia* **75 C5** 2 28 S 115 25 E
Amur, *Somali Rep.* **108 C3** 5 16N 46 30 E
Amur →, *Russia* **57 D15** 52 56N 141 10 E
Amurang, *Indonesia* **72 A2** 1 5N 124 40 E
Amuri Pass, *N.Z.* **119 C7** 42 31 S 172 11 E
Amurrio, *Spain* **34 B1** 43 3N 3 0W
Amursk, *Russia* **57 D14** 50 14N 136 54 E
Amurzet, *Russia* **57 E14** 47 50N 131 5 E
Amusco, *Spain* **36 C6** 42 10N 4 28W
Amutag, *Phil.* **70 E4** 12 23N 123 16 E
Amvrakikós Kólpos, *Greece* **45 F2** 39 0N 20 55 E
Amvrosiyivka, *Ukraine* **51 J10** 47 43N 38 30 E
Amyderya = Amudarya →, *Uzbekistan* **56 E6** 43 58N 59 34 E
Amzeglouf, *Algeria* **99 C5** 26 50N 0 1 E
An, *Burma* **78 F5** 19 48N 94 0 E
An Bien, *Vietnam* **77 H5** 9 45N 105 0 E
An Hoa, *Vietnam* **76 E7** 15 40N 108 5 E
An Khe, *Vietnam* **76 F7** 13 57N 108 39 E
An Nabatīyah at Tahta, *Lebanon* **91 B4** 33 23N 35 27 E
An Nabk, *Si. Arabia* **84 D3** 31 20N 37 20 E
An Nabk, *Syria* **91 A5** 34 2N 36 44 E
An Nabk Abū Qaşr, *Si. Arabia* **84 D3** 30 21N 38 34 E
An Nafūd, *Si. Arabia* **84 D4** 28 15N 41 0 E
An Najaf, *Iraq* **89 G11** 32 3N 44 15 E
An Nāşirīyah, *Iraq* **84 D5** 31 0N 46 15 E
An Nawfaliyah, *Libya* **96 B3** 30 54N 17 58 E
An Nhon, *Vietnam* **76 F7** 13 55N 109 7 E
An Nîl □, *Sudan* **94 D3** 19 30N 33 0 E
An Nîl el Abyad □, *Sudan* **95 E3** 14 0N 32 15 E
An Nîl el Azraq □, *Sudan* **95 E3** 11 30N 34 30 E
An Nimāş, *Si. Arabia* **86 C3** 19 7N 42 8 E
An Nu'ayrīyah, *Si. Arabia* **85 E6** 27 30N 48 30 E
An Nu'mānīyah, *Iraq* **89 F11** 32 32N 45 25 E
An Nuwayb'ī, W. →, *Si. Arabia* **91 F3** 29 18N 34 57 E
An Thoi, Dao, *Vietnam* **77 H4** 9 58N 104 0 E
An Uaimh, *Ireland* **19 C5** 53 39N 6 41W
Anabar →, *Russia* **57 B12** 73 8N 113 36 E
'Anabtā, *West Bank* **91 C4** 32 19N 35 7 E
Anabuki, *Japan* **62 C6** 34 2N 134 11 E
Anaco, *Venezuela* **153 B5** 9 27N 64 28W
Anaconda, *U.S.A.* **142 C7** 46 8N 112 57W
Anacortes, *U.S.A.* **144 B4** 48 30N 122 37W
Anacuao, Mt., *Phil.* **70 C3** 16 16N 121 53 E
Anadarko, *U.S.A.* **139 H5** 35 4N 98 15W
Anadia, *Brazil* **154 C4** 9 42 S 36 18W
Anadia, *Portugal* **36 E2** 40 26N 8 27W
Anadolu, *Turkey* **88 C5** 39 0N 30 0 E
Anadyr, *Russia* **57 C18** 64 35N 177 20 E
Anadyr →, *Russia* **57 C18** 64 55N 176 5 E
Anadyrskiy Zaliv, *Russia* **57 C19** 64 0N 180 0 E
Anáfi, *Greece* **45 H7** 36 22N 25 48 E
Anafópoulo, *Greece* **45 H7** 36 17N 25 50 E
Anaga, Pta. de, *Canary Is.* **33 F3** 28 34N 16 9W
Anagni, *Italy* **40 A6** 41 44N 13 9 E
'Ānah, *Iraq* **89 E10** 34 25N 42 0 E
Anaheim, *U.S.A.* **145 M9** 33 50N 117 55W
Anahim Lake, *Canada* **130 C3** 52 28N 125 18W
Anáhuac, *Mexico* **146 B4** 27 14N 100 9W
Anai Mudi, Mt., *India* **83 J3** 10 12N 77 4 E
Anaimalai Hills, *India* **83 J3** 10 20N 76 40 E
Anajás, *Brazil* **154 B2** 0 59 S 49 57W
Anajatuba, *Brazil* **154 B3** 3 16 S 44 37W
Anakapalle, *India* **82 F6** 17 42N 83 6 E
Anakie, *Australia* **114 C4** 23 32 S 147 45 E
Anaklia, *Georgia* **53 J5** 42 22N 41 35 E
Analalava, *Madag.* **105 A8** 14 35 S 48 0 E
Análipsis, *Greece* **32 A3** 39 36N 19 55 E
Anamã, *Brazil* **153 D5** 3 35 S 61 22W
Anambar →, *Pakistan* **80 D3** 30 15N 68 50 E
Anambas, Kepulauan, *Indonesia* **74 B3** 3 20N 106 30 E
Anambas Is. = Anambas, Kepulauan, *Indonesia* **74 B3** 3 20N 106 30 E
Anambra □, *Nigeria* **101 D6** 6 20N 7 0 E
Aname, *Vanuatu* **121 K7** 20 8 S 169 47 E
Anamoose, *U.S.A.* **138 B4** 47 53N 100 15W
Anamosa, *U.S.A.* **140 B5** 42 7N 91 17W
Anamur, *Turkey* **88 D5** 36 8N 32 58 E
Anamur Burnu, *Turkey* **88 D5** 36 2N 32 47 E
Anan, *Japan* **62 D6** 33 54N 134 40 E
Anand, *India* **80 H5** 22 32N 72 59 E
Anandpur, *India* **82 D8** 21 16N 86 13 E
Anánes, *Greece* **45 H6** 36 33N 24 9 E
Anantapur, *India* **83 G3** 14 39N 77 42 E
Anantnag, *India* **81 C6** 33 45N 75 10 E
Ananyiv, *Ukraine* **51 J5** 47 44N 29 58 E
Anao-aon, *Phil.* **71 G5** 9 47N 125 25 E
Anapa, *Russia* **53 H3** 44 55N 37 25 E
Anapodháris →, *Greece* **32 E7** 34 59N 25 20 E
Anápolis, *Brazil* **155 E2** 16 15 S 48 50W
Anapu →, *Brazil* **153 D7** 1 53 S 50 53W
Anār, *Iran* **85 D7** 30 55N 55 13 E
Anār Darreh, *Afghan.* **79 B1** 32 46N 61 39 E
Anārak, *Iran* **85 C7** 33 25N 53 40 E
Anatolia = Anadolu, *Turkey* **88 C5** 39 0N 30 0 E
Anatone, *U.S.A.* **142 C5** 46 8N 117 8W
Anatsogno, *Madag.* **105 C7** 23 33 S 43 46 E
Añatuya, *Argentina* **158 B3** 28 20 S 62 50W
Anauá →, *Brazil* **153 C5** 0 58N 61 21W
Anaunethad L., *Canada* **131 A8** 60 55N 104 25W
Anavilhanas, Arquipélago das, *Brazil* **153 D5** 2 42 S 60 45W
Anaye, *Niger* **97 E2** 19 15N 12 50 E
Anbyŏn, *N. Korea* **67 E14** 39 1N 127 35 E
Ancash □, *Peru* **156 B2** 9 30 S 77 45W
Ancenis, *France* **22 E5** 47 21N 1 10W
Ancho, Canal, *Chile* **160 D2** 50 0 S 74 20W
Anchor Bay, *U.S.A.* **144 G3** 38 48N 123 34W
Anchorage, *U.S.A.* **126 B5** 61 13N 149 54W
Anci, *China* **66 E9** 39 20N 116 40 E
Ancohuma, Nevada, *Bolivia* **156 D4** 16 0 S 68 50W
Ancón, *Peru* **156 C2** 11 50 S 77 10W
Ancona, *Italy* **39 E10** 43 38N 13 30 E
Ancud, *Chile* **160 B2** 42 0 S 73 50W
Ancud, G. de, *Chile* **160 B2** 42 0 S 73 0W
Anda, *China* **65 B7** 46 24N 125 19 E
Anda, *Phil.* **70 C2** 16 17N 119 57 E
Andacollo, *Argentina* **158 D1** 37 10 S 70 42W
Andacollo, *Chile* **158 C1** 30 5 S 71 10W
Andado, *Australia* **114 D2** 25 25 S 135 15 E
Andahuaylas, *Peru* **156 C3** 13 40 S 73 25W
Andalgalá, *Argentina* **158 B2** 27 40 S 66 30W
Åndalsnes, *Norway* **14 B1** 62 35N 7 43 E
Andalucía □, *Spain* **37 H6** 37 35N 5 0W
Andalusia, *U.S.A.* **135 K2** 31 18N 86 29W
Andalusia = Andalucía □, *Spain* **37 H6** 37 35N 5 0W
Andaman Is., *Ind. Oc.* **58 H13** 12 30N 92 30 E
Andara, *Namibia* **104 B3** 18 2 S 21 9 E
Andaraí, *Brazil* **155 D3** 12 48 S 41 20W
Andeer, *Switz.* **29 C8** 46 36N 9 26 E
Andelfingen, *Switz.* **29 A7** 47 36N 8 41 E
Andelot, *France* **23 D12** 48 15N 5 18 E
Andenes, *Norway* **12 B17** 69 19N 16 18 E
Andenne, *Belgium* **21 H6** 50 28N 5 5 E
Andéranboukane, *Mali* **101 B5** 15 26N 3 2 E
Anderlecht, *Belgium* **21 G4** 50 50N 4 19 E
Anderlues, *Belgium* **21 H4** 50 25N 4 16 E
Andermatt, *Switz.* **29 C7** 46 38N 8 35 E
Andernach, *Germany* **26 E3** 50 26N 7 24 E
Andernos-les-Bains, *France* **24 D2** 44 44N 1 6W
Anderslöv, *Sweden* **15 J7** 55 26N 13 19 E
Anderson, *Calif., U.S.A.* **142 F2** 40 27N 122 18W
Anderson, *Ind., U.S.A.* **141 D11** 40 10N 85 41W
Anderson, *Mo., U.S.A.* **139 G7** 36 39N 94 27W
Anderson, *S.C., U.S.A.* **135 H4** 34 31N 82 39W
Anderson →, *Canada* **126 B7** 69 42N 129 0W
Andes = Andes, Cord. de los, *S. Amer.* **156 C3** 20 0 S 68 0W
Andes, *S. Amer.* **156 C2** 10 0 S 75 53W
Andes, Cord. de los, *S. Amer.* **156 C3** 20 0 S 68 0W
Andfjorden, *Norway* **12 B17** 69 10N 16 20 E
Andhra, L., *India* **82 E1** 18 54N 73 32 E
Andhra Pradesh □, *India* **83 F4** 18 0N 79 0 E
Andijon, *Uzbekistan* **55 C6** 41 10N 72 15 E
Andikíthira, *Greece* **45 J5** 35 52N 23 15 E
Andīmeshk, *Iran* **89 F13** 32 27N 48 21 E
Andímilos, *Greece* **45 H6** 36 47N 24 12 E
Andíparos, *Greece* **45 H7** 37 0N 25 3 E
Andípaxoi, *Greece* **45 E2** 39 9N 20 13 E
Andípsara, *Greece* **45 F7** 38 30N 25 29 E
Andírrion, *Greece* **45 F3** 38 24N 21 46 E
Andizhan = Andijon, *Uzbekistan* **55 C6** 41 10N 72 15 E
Andkhvoy, *Afghan.* **79 A2** 36 52N 65 8 E
Andoany, *Madag.* **105 A8** 13 25 S 48 16 E
Andoas, *Peru* **152 D2** 2 55 S 76 25W
Andol, *India* **82 F4** 17 51N 78 4 E
Andong, *S. Korea* **67 F15** 36 40N 128 43 E
Andongwei, *China* **67 G10** 35 6N 119 20 E
Andorra ■, *Europe* **34 C6** 42 30N 1 30 E
Andorra La Vella, *Andorra* **34 C6** 42 31N 1 32 E
Andover, *U.K.* **17 F6** 51 12N 1 29W
Andover, *Mass., U.S.A.* **137 D13** 42 40N 71 8W
Andover, *N.Y., U.S.A.* **136 D7** 42 10N 77 48W
Andover, *Ohio, U.S.A.* **136 E4** 41 36N 80 34W
Andøya, *Norway* **12 B16** 69 10N 15 50 E
Andradina, *Brazil* **155 F1** 20 54 S 51 23W
Andrahary, Mt., *Madag.* **105 A8** 13 37 S 49 17 E
Andraitx, *Spain* **33 B9** 39 39N 2 25 E
Andramasina, *Madag.* **105 B8** 19 11 S 47 35 E
Andranopasy, *Madag.* **105 C7** 21 17 S 43 44 E
Andreanof Is., *U.S.A.* **126 C2** 52 0N 178 0W
Andreapol, *Russia* **50 D7** 56 40N 32 17 E
Andrespol, *Poland* **47 D6** 51 45N 19 34 E
Andrewilla, *Australia* **115 D2** 26 31 S 139 17 E
Andrews, *S.C., U.S.A.* **135 J6** 33 27N 79 34W
Andrews, *Tex., U.S.A.* **139 J3** 32 19N 102 33W
Andreyevka, *Russia* **52 D10** 52 19N 51 55 E
Ándria, *Italy* **41 A9** 41 13N 16 17 E
Andriba, *Madag.* **105 B8** 17 30 S 46 58 E
Andrijevica, *Montenegro, Yug.* **42 E4** 42 45N 19 48 E
Andrítsaina, *Greece* **45 G3** 37 29N 21 52 E
Androka, *Madag.* **105 C7** 24 58 S 44 2 E
Andropov = Rybinsk, *Russia* **52 A4** 58 5N 38 50 E
Ándros, *Greece* **45 G6** 37 50N 24 57 E
Andros I., *Bahamas* **148 B4** 24 30N 78 0W
Andros Town, *Bahamas* **148 B4** 24 43N 77 47W
Andrychów, *Poland* **31 B12** 49 51N 19 18 E
Andselv, *Norway* **12 B18** 69 4N 18 34 E
Andújar, *Spain* **37 G6** 38 3N 4 5W
Andulo, *Angola* **103 E3** 11 25 S 16 45 E
Anegada, B., *Argentina* **160 B4** 40 20 S 62 20W
Anegada I., *Virgin Is.* **149 C7** 18 45N 64 20W
Anegada Passage, *W. Indies* **149 C7** 18 15N 63 45W
Anécho, *Togo* **101 D5** 6 12N 1 34 E
Aneityum, *Vanuatu* **121 K7** 20 12 S 169 45 E
Añelo, *Argentina* **160 A3** 38 20 S 68 45W
Anergane, *Morocco* **98 B3** 31 4N 7 14W
Aneto, Pico de, *Spain* **34 C5** 42 37N 0 40 E
Añez, *Bolivia* **157 D5** 15 40 S 63 10W
Anfu, *China* **69 D10** 27 21N 114 40 E
Ang Thong, *Thailand* **76 E3** 14 35N 100 31 E
Angadanan, *Phil.* **70 C3** 16 45N 121 45 E
Angamos, Punta, *Chile* **158 A1** 23 1 S 70 32W
Angara →, *Russia* **57 D10** 58 5N 94 20 E
Angarab, *Ethiopia* **95 E4** 13 11N 37 7 E
Angarsk, *Russia* **57 D11** 52 30N 104 0 E
Angas Downs, *Australia* **113 E5** 25 2 S 132 14 E
Angas Hills, *Australia* **112 D4** 23 0 S 127 50 E
Angaston, *Australia* **116 C3** 34 30 S 139 8 E
Angat, *Phil.* **70 D3** 14 56N 121 2 E
Ånge, *Sweden* **14 B9** 62 31N 15 35 E
Ángel, Salto = Angel Falls, *Venezuela* **153 B5** 5 57N 62 30W
Ángel de la Guarda, I., *Mexico* **146 B2** 29 30N 113 30W
Angel Falls, *Venezuela* **153 B5** 5 57N 62 30W
Angeles, *Phil.* **70 D3** 15 9N 120 33 E
Ängelholm, *Sweden* **15 H6** 56 15N 12 58 E
Angellala, *Australia* **115 D4** 26 24 S 146 54 E
Angels Camp, *U.S.A.* **144 G6** 38 4N 120 32W
Anger →, *Ethiopia* **95 F4** 9 37N 36 6 E
Angereb →, *Ethiopia* **95 E4** 13 45N 36 40 E
Ångermanälven →, *Sweden* **14 B12** 62 40N 18 0 E
Ångermanland, *Sweden* **12 E18** 63 36N 17 45 E
Angermünde, *Germany* **26 B10** 53 0N 14 0 E
Angers, *Canada* **137 A9** 45 31N 75 29W
Angers, *France* **22 E6** 47 30N 0 35W
Angerville, *France* **23 D8** 48 19N 2 0 E
Ängesån →, *Sweden* **12 C20** 66 16N 22 47 E
Anghiari, *Italy* **39 E9** 43 32N 12 3 E
Angical, *Brazil* **155 D3** 12 0 S 44 42W
Angikuni L., *Canada* **131 A9** 62 0N 100 0W
Angkor, *Cambodia* **76 F4** 13 22N 103 50 E
Anglem Mt., *N.Z.* **119 G2** 46 45 S 167 53 E
Anglés, *Spain* **34 D7** 41 57N 2 38 E
Anglesey, *U.K.* **16 D3** 53 17N 4 20W
Anglet, *France* **24 E2** 43 29N 1 31W
Angleton, *U.S.A.* **139 L7** 29 10N 95 26W
Angleur, *Belgium* **21 G7** 50 36N 5 35 E
Anglin →, *France* **24 B4** 46 42N 0 52 E
Anglisidhes, *Cyprus* **32 E12** 34 51N 33 27 E
Anglure, *France* **23 D10** 48 35N 3 50 E
Angmagssalik, *Greenland* **6 C6** 65 40N 37 20W
Ango, *Zaïre* **106 B2** 4 10N 26 5 E
Angoche, *Mozam.* **107 F4** 16 8 S 39 55 E
Angoche, I., *Mozam.* **107 F4** 16 20 S 39 50 E
Angol, *Chile* **158 D1** 37 56 S 72 45W
Angola, *Ind., U.S.A.* **141 C12** 41 38N 85 0W
Angola, *N.Y., U.S.A.* **136 D5** 42 38N 79 2W
Angola ■, *Africa* **103 E3** 12 0 S 18 0 E
Angoon, *U.S.A.* **130 B2** 57 30N 134 35W
Angoram, *Papua N. G.* **120 C3** 4 4 S 144 4 E
Angoulême, *France* **24 C4** 45 39N 0 10 E
Angoumois, *France* **24 C4** 45 50N 0 25 E
Angra dos Reis, *Brazil* **159 A7** 23 0 S 44 10W
Angren, *Uzbekistan* **55 C5** 41 1N 70 12 E
Angtassom, *Cambodia* **77 G5** 11 1N 104 41 E
Angu, *Zaïre* **106 B1** 3 25N 24 28 E
Anguang, *China* **67 B12** 45 15N 123 45 E
Anguilla ■, *W. Indies* **149 C7** 18 14N 63 5W
Anguo, *China* **66 E8** 38 28N 115 15 E
Angurugu, *Australia* **114 A2** 14 0 S 136 25 E
Angus, Braes of, *U.K.* **18 E5** 56 51N 3 10W
Anhanduí →, *Brazil* **159 A5** 21 46 S 52 9W
Anhée, *Belgium* **21 H5** 50 18N 4 53 E
Anholt, *Denmark* **15 H5** 56 42N 11 33 E
Anhua, *China* **69 C8** 28 23N 111 12 E
Anhui □, *China* **69 B11** 32 0N 117 0 E
Anhwei = Anhui □, *China* **69 B11** 32 0N 117 0 E
Anichab, *Namibia* **104 C1** 21 0 S 14 46 E
Anicuns, *Brazil* **155 E2** 16 28 S 49 58W
Ánidhros, *Greece* **45 H7** 36 38N 25 43 E
Anie, *Togo* **101 D5** 7 42N 1 8 E
Animas, *U.S.A.* **143 L9** 31 57N 108 48W
Ånimskog, *Sweden* **15 F6** 58 53N 12 35 E
Anina, *Romania* **42 B6** 45 6N 21 51 E
Anini-y, *Phil.* **71 F3** 10 25N 121 55 E
Anita, *U.S.A.* **140 C2** 41 27N 94 46W
Anivorano, *Madag.* **105 B8** 18 44 S 48 58 E
Aniwa, *Vanuatu* **121 J7** 19 17 S 169 35 E
Anjalankoski, *Finland* **13 F22** 60 45N 26 51 E
Anjangaon, *India* **82 D3** 21 10N 77 20 E
Anjar, *India* **80 H4** 23 6N 70 10 E
Anjidiv I., *India* **83 G2** 14 40N 74 10 E
Anjō, *Japan* **63 C9** 34 57N 137 5 E
Anjou, *France* **22 E6** 47 20N 0 15W
Anjozorobe, *Madag.* **105 B8** 18 22 S 47 52 E
Anju, *N. Korea* **67 E13** 39 36N 125 40 E
Anka, *Nigeria* **101 C6** 12 13N 5 58 E
Ankaboa, Tanjona, *Madag.* **105 C7** 21 58 S 43 20 E
Ankang, *China* **66 H5** 32 40N 109 1 E
Ankara, *Turkey* **88 C5** 39 57N 32 54 E
Ankaramena, *Madag.* **105 C8** 21 57 S 46 39 E
Ankazoabo, *Madag.* **105 C7** 22 18 S 44 31 E
Ankazobe, *Madag.* **105 B8** 18 20 S 47 10 E
Ankeny, *U.S.A.* **140 C3** 41 44N 93 36W
Ankisabe, *Madag.* **105 B8** 19 17 S 46 29 E
Anklam, *Germany* **26 B9** 53 51N 13 41 E
Ankleshwar, *India* **82 D1** 21 38N 73 3 E
Ankober, *Ethiopia* **95 F4** 9 35N 39 40 E
Ankoro, *Zaïre* **106 D2** 6 45 S 26 55 E
Anlong, *China* **68 E5** 25 2N 105 27 E
Anlu, *China* **69 B9** 31 15N 113 45 E
Anmyŏn-do, *S. Korea* **67 F14** 36 25N 126 25 E
Ånn, *Sweden* **14 A6** 63 19N 12 34 E
Ann, C., *U.S.A.* **137 D14** 42 38N 70 35W
Ann Arbor, *U.S.A.* **141 B13** 42 17N 83 45W
Anna, *Russia* **52 E5** 51 28N 40 23 E
Anna, *Ill., U.S.A.* **139 G10** 37 28N 89 15W
Anna, *Ohio, U.S.A.* **141 D12** 40 24N 84 11W
Anna Plains, *Australia* **112 C3** 19 17 S 121 37 E
Anna Regina, *Guyana* **153 B6** 7 10N 58 30W
Annaba, *Algeria* **99 A6** 36 50N 7 46 E
Annaberg-Buchholz, *Germany* **26 E8** 50 34N 13 0 E
Annaka, *Japan* **63 A10** 36 19N 138 54 E
Annalee →, *Ireland* **19 B4** 54 2N 7 24W
Annam = Trung-Phan, *Vietnam* **76 E7** 16 0N 108 0 E
Annamitique, Chaîne, *Asia* **76 D6** 17 0N 106 0 E
Annan, *U.K.* **18 G5** 54 59N 3 16W
Annan →, *U.K.* **18 G5** 54 58N 3 16W
Annanberg, *Papua N. G.* **120 C3** 4 52 S 144 42 E
Annapolis, *U.S.A.* **134 F7** 38 59N 76 30W
Annapolis Royal, *Canada* **129 D6** 44 44N 65 32W
Annapurna, *Nepal* **81 E10** 28 34N 83 50 E
Annean, L., *Australia* **113 E2** 26 54 S 118 14 E
Anneberg, *Sweden* **15 G6** 57 32N 12 6 E
Annecy, *France* **25 C10** 45 55N 6 8 E
Annecy, L. d', *France* **25 C10** 45 52N 6 10 E
Annemasse, *France* **25 B10** 46 12N 6 16 E
Annenskiy Most, *Russia* **50 B9** 60 45N 37 10 E
Anning, *China* **68 E4** 24 55N 102 26 E
Anningie, *Australia* **112 D5** 21 50 S 133 7 E
Anniston, *U.S.A.* **135 J3** 33 39N 85 50W
Annobón, *Atl. Oc.* **93 G4** 1 25 S 5 36 E
Annonay, *France* **25 C8** 45 15N 4 40 E
Annot, *France* **25 E10** 43 58N 6 38 E
Annotto Bay, *Jamaica* **148 C4** 18 17N 76 45W
Annuello, *Australia* **116 C5** 34 53 S 142 55 E
Annville, *U.S.A.* **137 F8** 40 20N 76 31W
Annweiler, *Germany* **27 F3** 49 12N 7 57 E
Áno Arkhánai, *Greece* **45 J7** 35 16N 25 11 E
Áno Porróia, *Greece* **44 C5** 41 17N 23 2 E
Áno Viánnos, *Greece* **32 D7** 35 2N 25 21 E
Anoano, *Solomon Is.* **121 M11** 8 59 S 160 46 E
Anoka, *U.S.A.* **138 C8** 45 12N 93 23W
Anorotsangana, *Madag.* **105 A8** 13 56 S 47 55 E
Anóyia, *Greece* **32 D6** 35 16N 24 52 E
Anping, *Hebei, China* **66 E8** 38 15N 115 30 E
Anping, *Liaoning, China* **67 D12** 41 5N 123 30 E
Anpu Gang, *China* **68 G7** 21 25N 109 50 E
Anqing, *China* **69 B11** 30 30N 117 3 E
Anqiu, *China* **67 F10** 36 25N 119 10 E
Anren, *China* **69 D9** 26 43N 113 18 E
Ans, *Belgium* **21 G7** 50 39N 5 32 E
Ansai, *China* **66 F5** 36 50N 109 20 E
Ansbach, *Germany* **27 F6** 49 28N 10 34 E
Anseba →, *Eritrea* **95 D4** 16 0N 38 30 E
Anserma, *Colombia* **152 B2** 5 13N 75 48W
Anseroeul, *Belgium* **21 G3** 50 43N 3 32 E
Anshan, *China* **67 D12** 41 5N 122 58 E
Anshun, *China* **68 D5** 26 18N 105 57 E
Ansião, *Portugal* **36 F2** 39 56N 8 27W
Ansirabe, *Madag.* **105 B8** 19 55 S 47 2 E
Ansley, *U.S.A.* **138 E5** 41 18N 99 23W
Ansó, *Spain* **34 C4** 42 51N 0 48W
Anson, *U.S.A.* **139 J5** 32 45N 99 54W
Anson B., *Australia* **112 B5** 13 20 S 130 6 E
Ansongo, *Mali* **101 B5** 15 25N 0 35 E
Ansonia, *Conn., U.S.A.* **137 E11** 41 21N 73 5W
Ansonia, *Ohio, U.S.A.* **141 D12** 40 13N 84 38W
Anstruther, *U.K.* **18 E6** 56 14N 2 41W
Ansudu, *Indonesia* **73 B5** 2 11 S 139 22 E
Antabamba, *Peru* **156 C3** 14 40 S 73 0W
Antakya, *Turkey* **88 D7** 36 14N 36 10 E
Antalaha, *Madag.* **105 A9** 14 57 S 50 20 E
Antalya, *Turkey* **88 D4** 36 52N 30 45 E
Antalya Körfezi, *Turkey* **88 D4** 36 15N 31 30 E
Antananarivo, *Madag.* **105 B8** 18 55 S 47 31 E
Antananarivo □, *Madag.* **105 B8** 19 0 S 47 0 E
Antanimbaribe, *Madag.* **105 C7** 21 30 S 44 48 E
Antarctic Pen., *Antarctica* **7 C18** 67 0 S 60 0W
Antarctica **7 E3** 90 0 S 0 0 E
Antelope, *Zimbabwe* **107 G2** 21 2 S 28 31 E
Antenor Navarro, *Brazil* **154 C4** 6 44 S 38 27W
Antequera, *Paraguay* **158 A4** 24 8 S 57 7W
Antequera, *Spain* **37 H6** 37 5N 4 33W
Antero, Mt., *U.S.A.* **143 G10** 38 41N 106 15W
Anthemoús, *Greece* **44 D5** 40 31N 23 15 E
Anthony, *Kans., U.S.A.* **139 G5** 37 9N 98 2W
Anthony, *N. Mex., U.S.A.* **143 K10** 32 0N 106 36W
Anthony Lagoon, *Australia* **114 B2** 18 0 S 135 30 E
Anti Atlas, *Morocco* **98 C2** 30 0N 8 30W
Anti-Lebanon = Ash Sharqi, Al Jabal, *Lebanon* **91 B5** 33 40N 36 10 E
Antibes, *France* **25 E11** 43 34N 7 6 E
Antibes, C. d', *France* **25 E11** 43 31N 7 7 E
Anticosti, I. d', *Canada* **129 C7** 49 30N 63 0W
Antifer, C. d', *France* **22 C7** 49 41N 0 10 E
Antigo, *U.S.A.* **138 C10** 45 9N 89 9W
Antigonish, *Canada* **129 C7** 45 38N 61 58W
Antigua, *Canary Is.* **33 F5** 28 24N 14 1W
Antigua, *Guatemala* **148 D1** 14 34N 90 41W
Antigua, *W. Indies* **149 C7** 17 0N 61 50W
Antigua & Barbuda ■, *W. Indies* **149 C7** 17 20N 61 48W
Antilla, *Cuba* **148 B4** 20 40N 75 50W
Antimony, *U.S.A.* **143 G8** 38 7N 112 0W
Antioch, *U.S.A.* **144 G5** 38 1N 121 48W
Antioche, Pertuis d', *France* **24 B2** 46 6N 1 20W
Antioquia, *Colombia* **152 B2** 6 40N 75 55W
Antioquia □, *Colombia* **152 B2** 7 0N 75 30W
Antipodes Is., *Pac. Oc.* **122 M9** 49 45 S 178 40 E
Antique □, *Phil.* **71 F4** 11 10N 122 5 E
Antler, *U.S.A.* **138 A4** 48 59N 101 17W
Antler →, *Canada* **131 D8** 49 8N 101 0W
Antlers, *U.S.A.* **139 H7** 34 14N 95 37W
Antofagasta, *Chile* **158 A1** 23 50 S 70 30W
Antofagasta □, *Chile* **158 A2** 24 0 S 69 0W
Antofagasta de la Sierra, *Argentina* **158 B2** 26 5 S 67 20W
Antofalla, *Argentina* **158 B2** 25 30 S 68 5W
Antofalla, Salar de, *Argentina* **158 B2** 25 40 S 67 45W
Antoing, *Belgium* **21 G2** 50 34N 3 27 E
Anton, *U.S.A.* **139 J3** 33 49N 102 10W
Anton Chico, *U.S.A.* **143 J11** 35 12N 105 9W
Antongila, Helodrano, *Madag.* **105 B8** 15 30 S 49 50 E
Antonibé, *Madag.* **105 B8** 15 7 S 47 24 E
Antonibé, Presqu'île d', *Madag.* **105 A8** 14 55 S 47 20 E
Antonina, *Brazil* **159 B6** 25 26 S 48 42W
Antonito, *U.S.A.* **143 H10** 37 5N 106 0W
Antrain, *France* **22 D5** 48 28N 1 30W
Antrim, *U.K.* **19 B5** 54 43N 6 14W
Antrim □, *U.K.* **19 B5** 54 56N 6 25W
Antrim, Mts. of, *U.K.* **19 B5** 55 3N 6 14W
Antrim Plateau, *Australia* **112 C4** 18 8 S 128 20 E
Antrodoco, *Italy* **39 F10** 42 25N 13 5 E
Antropovo, *Russia* **52 A6** 58 24N 43 6 E
Antsalova, *Madag.* **105 B7** 18 40 S 44 37 E
Antsiranana, *Madag.* **105 A8** 12 25 S 49 20 E
Antsohihy, *Madag.* **105 A8** 14 50 S 47 59 E

Azeffoun 169

At Tāj, Libya 96 D4 24 13N 23 18 E
At Tamīmī, Libya 96 B4 32 20N 23 4 E
Aţ Ţirāq, Si. Arabia 84 E5 27 19N 44 33 E
At Turbah, Yemen 86 D4 13 13N 44 7 E
At Turbah, Yemen 86 D3 12 40N 43 30 E
Aţ Ţuwayrifah, Si. Arabia 87 B5 21 30N 49 35 E
Atacama □, Chile 158 B2 27 30 S 70 0W
Atacama, Desierto de, Chile 158 A2 24 0 S 69 20W
Atacama, Salar de, Chile 158 A2 23 30 S 68 20W
Ataco, Colombia 152 C2 3 35N 75 23W
Atakor, Algeria 99 D6 23 27N 5 31 E
Atakpamé, Togo 101 D5 7 31N 1 13 E
Atalándi, Greece 45 F4 38 39N 22 58 E
Atalaya, Peru 156 C3 10 45 S 73 50W
Atalaya de Femes, Canary Is. 33 F6 28 56N 13 47W
Ataléia, Brazil 155 E3 18 3 S 41 6W
Atambua, Indonesia 72 C2 9 7 S 124 54 E
Atami, Japan 63 B11 35 5N 139 4 E
Atankawng, Burma 78 C6 25 50N 97 47 E
Atapupu, Indonesia 72 C2 9 0 S 124 51 E
Atâr, Mauritania 98 D2 20 30N 13 5W
Ataram, Erg n-, Algeria 99 D5 23 57N 2 0 E
Atarfe, Spain 37 H7 37 13N 3 40W
Atascadero, U.S.A. 144 K6 35 29N 120 40W
Atasu, Kazakhstan 56 E8 48 30N 71 0 E
Atatürk Barajı, Turkey 89 D8 37 28N 38 30 E
Atauro, Indonesia 72 C3 8 10 S 125 30 E
Atbara, Sudan 94 D3 17 42N 33 59 E
'Atbara →, Sudan 94 D3 17 40N 33 56 E
Atbasar, Kazakhstan 56 D7 51 48N 68 20 E
Atbashi, Kyrgyzstan 55 C7 41 10N 75 48 E
Atbashi, Khrebet, Kyrgyzstan 55 C7 40 50N 75 30 E
Atchafalaya B., U.S.A. 139 L9 29 25N 91 25W
Atchison, U.S.A. 138 F7 39 34N 95 7W
Atebubu, Ghana 101 D4 7 47N 1 0W
Ateca, Spain 34 D3 41 20N 1 49W
Aterno →, Italy 39 F10 42 11N 13 51 E
Atesine, Alpi, Italy 38 B8 46 55N 11 30 E
Atessa, Italy 39 F11 42 4N 14 27 E
Ath, Belgium 21 G3 50 38N 3 47 E
Athabasca, Canada 130 C6 54 45N 113 20W
Athabasca →, Canada 131 B6 58 40N 110 50W
Athabasca, L., Canada 131 B7 59 15N 109 15W
Athboy, Ireland 19 C5 53 37N 6 56W
Athenry, Ireland 19 C3 53 18N 8 44W
Athens = Athínai, Greece 45 G5 37 58N 23 46 E
Athens, Ala., U.S.A. 135 H2 34 48N 86 58W
Athens, Ga., U.S.A. 135 J4 33 57N 83 23W
Athens, N.Y., U.S.A. 137 D11 42 16N 73 49W
Athens, Ohio, U.S.A. 134 F4 39 20N 82 6W
Athens, Pa., U.S.A. 137 E8 41 57N 76 31W
Athens, Tenn., U.S.A. 135 H3 35 27N 84 36W
Athens, Tex., U.S.A. 139 J7 32 12N 95 51W
Atherley, Canada 136 B5 44 37N 79 20W
Atherton, Australia 114 B4 17 17 S 145 30 E
Athiéme, Benin 101 D5 6 37N 1 40 E
Athienou, Cyprus 32 D12 35 3N 33 32 E
Athínai, Greece 45 G5 37 58N 23 46 E
Athlone, Ireland 19 C4 53 25N 7 56W
Athna, Cyprus 32 D12 35 3N 33 47 E
Athni, India 82 F2 16 44N 75 6 E
Athol, N.Z. 119 F3 45 30 S 168 35 E
Atholl, Forest of, U.K. 18 E5 56 51N 3 50W
Atholville, Canada 129 C6 47 59N 66 43W
Áthos, Greece 44 D6 40 9N 24 22 E
Athus, Belgium 21 J7 49 34N 5 50 E
Athy, Ireland 19 D5 53 0N 7 0W
Ati, Chad 97 F3 13 13N 18 20 E
Ati, Sudan 95 E2 13 5N 29 2 E
Atiak, Uganda 106 B3 3 12N 32 2 E
Atiamuri, N.Z. 118 E5 38 24 S 176 5 E
Atico, Peru 156 D3 16 14 S 73 40W
Atienza, Spain 34 D2 41 12N 2 52W
Atikokan, Canada 128 C1 48 45N 91 37W
Atikonak L., Canada 129 B7 52 40N 64 32W
Atimonan, Phil. 70 D3 14 0N 121 55 E
'Âtinah, W. →, Oman 87 C6 18 23N 53 28 E
Atirampattinam, India 83 J4 10 28N 79 20 E
Atka, Russia 57 C16 60 50N 151 48 E
Atkarsk, Russia 52 E7 51 55N 45 2 E
Atkinson, Ill., U.S.A. 140 C6 41 25N 90 1W
Atkinson, Nebr., U.S.A. 138 D5 42 32N 98 59W
Atlanta, Ga., U.S.A. 135 J3 33 45N 84 23W
Atlanta, Ill., U.S.A. 140 D7 40 16N 89 14W
Atlanta, Mo., U.S.A. 140 E4 39 54N 92 29W
Atlanta, Tex., U.S.A. 139 J7 33 7N 94 10W
Atlantic, U.S.A. 138 E7 41 24N 95 1W
Atlantic City, U.S.A. 134 F8 39 21N 74 27W
Atlantic Ocean 8 H7 0 0 20 0W
Atlántico □, Colombia 152 A2 10 45N 75 0W
Atlas Mts. = Haut Atlas, Morocco 98 B3 32 30N 5 0W
Atlin, Canada 130 B2 59 31N 133 41W
Atlin, L., Canada 130 B2 59 26N 133 45W
Atmakur, India 83 G4 14 37N 79 40 E
Atmore, U.S.A. 135 K2 31 2N 87 29W
Atō, Japan 62 C3 34 25N 131 40 E
Atok, Phil. 70 C3 16 35N 120 41 E
Atoka, U.S.A. 139 H6 34 23N 96 8W
Átokos, Greece 45 F2 38 28N 20 49 E
Atolia, U.S.A. 145 K9 35 19N 117 37W
Atouguia, Portugal 37 F1 39 20N 9 20W
Atoyac →, Mexico 147 D5 16 30N 97 31W
Atrak = Atrek →, Turkmenistan 85 B8 37 35N 53 58 E
Ätran, Sweden 15 G6 57 7N 12 57 E
Atrato →, Colombia 152 B2 8 17N 76 58W
Atrauli, India 80 E8 28 2N 78 20 E
Atrek →, Turkmenistan 85 B8 37 35N 53 58 E
Atri, Italy 39 F10 42 35N 13 58 E
Atsbi, Ethiopia 95 E4 13 52N 39 50 E
Atsoum, Mts., Cameroon 101 D7 6 41N 12 57 E
Atsugi, Japan 63 B11 35 25N 139 21 E
Atsumi, Japan 63 C9 34 35N 137 4 E
Atsumi-Wan, Japan 63 C9 34 44N 137 13 E
Atsuta, Japan 60 C10 43 24N 141 26 E
Attalla, U.S.A. 135 H2 34 1N 86 6W
Attáviros, Greece 32 C9 36 12N 27 50 E
Attawapiskat, Canada 128 B3 52 56N 82 24W
Attawapiskat →, Canada 128 B3 52 57N 82 18W
Attawapiskat, L., Canada 128 B2 52 18N 87 54W
Attendorn, Germany 26 D3 51 8N 7 56 E
Attersee, Austria 30 D6 47 55N 13 32 E
Attert, Belgium 21 J7 49 45N 5 47 E
Attica, U.S.A. 141 D9 40 18N 87 15W
Attichy, France 23 C10 49 25N 3 3 E
Attigny, France 23 C11 49 28N 4 35 E
Attikamagen L., Canada 129 A6 55 0N 66 30W
Attikí □, Greece 45 F5 38 10N 23 40 E
Attleboro, U.S.A. 137 E13 41 57N 71 17W
Attock, Pakistan 80 C5 33 52N 72 20 E
Attopeu, Laos 76 E6 14 48N 106 50 E
Attunga, Australia 117 A9 30 55 S 150 50 E
Attur, India 83 J4 11 35N 78 30 E
'Atūd, Yemen 87 D5 14 53N 48 10 E
Atuel →, Argentina 158 D2 36 17 S 66 50W
Åtvidaberg, Sweden 15 F10 58 12N 16 0 E
Atwater, U.S.A. 144 H6 37 21N 120 37W
Atwood, Canada 136 C3 43 40N 81 1W
Atwood, U.S.A. 138 F4 39 48N 101 3W
Atyraū, Kazakhstan 49 E9 47 5N 52 0 E
Au Sable →, U.S.A. 134 C4 44 25N 83 20W
Au Sable Pt., U.S.A. 128 C2 46 40N 86 10W
Aubagne, France 25 E9 43 17N 5 37 E
Aubange, Belgium 21 J7 49 34N 5 48 E
Aubarca, C., Spain 33 B7 39 4N 1 22 E
Aube □, France 23 D11 48 15N 4 10 E
Aube →, France 23 D10 48 34N 3 43 E
Aubel, Belgium 21 G7 50 42N 5 51 E
Aubenas, France 25 D8 44 37N 4 24 E
Aubenton, France 23 C11 49 50N 4 12 E
Auberry, U.S.A. 144 H7 37 7N 119 29W
Aubigny-sur-Nère, France 23 E9 47 30N 2 24 E
Aubin, France 24 D6 44 33N 2 15 E
Aubrac, Mts. d', France 24 D7 44 40N 3 2 E
Auburn, Ala., U.S.A. 135 J3 32 36N 85 29W
Auburn, Calif., U.S.A. 144 G5 38 54N 121 4W
Auburn, Ill., U.S.A. 140 E7 39 36N 89 45W
Auburn, Ind., U.S.A. 141 C11 41 22N 85 4W
Auburn, N.Y., U.S.A. 137 D8 42 56N 76 34W
Auburn, Nebr., U.S.A. 138 E7 40 23N 95 51W
Auburn, Wash., U.S.A. 144 C4 47 18N 122 14W
Auburn Ra., Australia 115 D5 25 15 S 150 30 E
Auburndale, U.S.A. 135 L5 28 4N 81 48W
Aubusson, France 24 C6 45 57N 2 11 E
Auch, France 24 E4 43 39N 0 36 E
Auchel, France 23 B9 50 30N 2 29 E
Auchi, Nigeria 101 D6 7 6N 6 13 E
Auckland, N.Z. 118 C3 36 52 S 174 46 E
Auckland □, N.Z. 118 E6 36 50 S 175 0 E
Auckland Is., Pac. Oc. 122 N8 50 40 S 166 5 E
Aude □, France 24 E6 43 8N 2 28 E
Aude →, France 24 E7 43 13N 3 14 E
Audegle, Somali Rep. 108 D2 1 59N 44 50 E
Auden, Canada 128 B2 50 14N 87 53W
Auderghem, Belgium 21 G4 50 49N 4 26 E
Auderville, France 22 C5 49 43N 1 57W
Audierne, France 22 D2 48 1N 4 34W
Audincourt, France 23 E13 47 30N 6 50 E
Audo, Ethiopia 95 F5 6 20N 41 50 E
Audubon, U.S.A. 140 C2 41 43N 94 56W
Aue, Germany 26 E8 50 35N 12 41 E
Auer = Ora, Italy 39 B8 46 21N 11 18 E
Auerbach, Germany 26 E8 50 30N 12 24 E
Aueti Paraná →, Brazil 152 D4 1 51 S 65 37W
Aufist, W. Sahara 98 C2 25 44N 14 39W
Augathella, Australia 115 D4 25 48 S 146 35 E
Augrabies Falls, S. Africa 104 D3 28 35 S 20 20 E
Augsburg, Germany 27 G6 48 25N 10 52 E
Augusta, Italy 41 E8 37 13N 15 13 E
Augusta, Ark., U.S.A. 139 H9 35 17N 91 22W
Augusta, Ga., U.S.A. 135 J5 33 28N 81 58W
Augusta, Ill., U.S.A. 140 D6 40 14N 90 57W
Augusta, Kans., U.S.A. 139 G6 37 41N 96 59W
Augusta, Ky., U.S.A. 141 F12 38 47N 84 0W
Augusta, Maine, U.S.A. 129 D6 44 19N 69 47W
Augusta, Mont., U.S.A. 142 C7 47 30N 112 24W
Augusta, Wis., U.S.A. 138 C9 44 41N 91 7W
Augustenborg, Denmark 15 K3 54 57N 9 53 E
Augustów, Poland 47 B9 53 51N 23 0 E
Augustus, Mt., Australia 112 D2 24 20 S 116 50 E
Augustus Downs, Australia 114 B2 18 35 S 139 55 E
Augustus I., Australia 112 C3 15 20 S 124 30 E
Aukan, Eritrea 95 D5 15 29N 40 50 E
Auki, Solomon Is. 121 M11 8 45 S 160 42 E
Aukum, U.S.A. 144 G6 38 34N 120 43W
Auld, L., Australia 112 D3 22 25 S 123 50 E
Aulla, Italy 38 D6 44 12N 9 58 E
Aulnay, France 24 B3 46 2N 0 22W
Aulne →, France 22 D2 48 17N 4 16W
Aulnoye-Aymeries, France 23 B10 50 12N 3 50 E
Ault, France 22 B8 50 8N 1 26 E
Ault, U.S.A. 138 E2 40 35N 104 44W
Aulus-les-Bains, France 24 F5 42 49N 1 19 E
Aumale, France 23 C8 49 46N 1 46 E
Aumont-Aubrac, France 24 D7 44 43N 3 17 E
Auna, Nigeria 101 C5 10 9N 4 42 E
Aundh, India 82 F2 17 33N 74 23 E
Aunis, France 24 B3 46 5N 0 50W
Auponhia, Indonesia 72 B3 1 58 S 125 27 E
Aups, France 25 E10 43 37N 6 15 E
Aur, P., Malaysia 77 L5 2 35N 104 10 E
Aura, Burma 78 B6 26 59N 97 57 E
Auraiya, India 81 F8 26 28N 79 33 E
Aurangabad, Bihar, India 81 G11 24 45N 84 18 E
Aurangabad, Maharashtra, India 82 E2 19 50N 75 23 E
Auray, France 22 E4 47 40N 2 59W
Aurès, Algeria 99 A6 35 8N 6 30 E
Aurich, Germany 26 B3 53 28N 7 28 E
Aurilândia, Brazil 155 E1 16 44 S 50 28W
Aurillac, France 24 D6 44 55N 2 26 E
Auronzo di Cadore, Italy 39 B9 46 33N 12 26 E
Aurora = Maewo, Vanuatu 121 E6 15 10 S 168 10 E
Aurora, Canada 136 C5 44 0N 79 28W
Aurora, Isabela, Phil. 70 C3 16 59N 121 38 E
Aurora, Quezon, Phil. 70 E4 13 21N 122 31 E
Aurora, S. Africa 104 E2 32 40 S 18 29 E
Aurora, Colo., U.S.A. 138 F2 39 44N 104 52W
Aurora, Ill., U.S.A. 141 C8 41 45N 88 19W
Aurora, Mo., U.S.A. 139 G8 36 58N 93 43W
Aurora, Nebr., U.S.A. 138 E6 40 52N 98 0W
Aurora, Ohio, U.S.A. 136 E3 41 21N 81 20W
Aursmoen, Norway 14 E5 59 55N 11 26 E
Aurukun Mission, Australia 114 A3 13 20 S 141 45 E
Aus, Namibia 104 D2 26 35 S 16 12 E
Auschwitz = Oświęcim, Poland 31 A12 50 2N 19 11 E
Austerlitz = Slavkov, Czech. 31 B9 49 10N 16 52 E
Austin, Ind., U.S.A. 141 F11 38 45N 85 49W
Austin, Minn., U.S.A. 138 D8 43 40N 92 58W
Austin, Nev., U.S.A. 142 G5 39 30N 117 4W
Austin, Pa., U.S.A. 136 E6 41 38N 78 6W
Austin, Tex., U.S.A. 139 K6 30 17N 97 45W
Austin, L., Australia 113 E2 27 40 S 118 0 E
Austra, Norway 12 D14 65 8N 11 55 E
Austral Downs, Australia 114 C2 20 30 S 137 45 E
Austral Is. = Tubuai Is., Pac. Oc. 123 K12 25 0 S 150 0W
Austral Seamount Chain, Pac. Oc. 123 K13 24 0 S 150 0W
Australia ■, Oceania 122 K5 23 0 S 135 0 E
Australian Alps, Australia 117 D8 36 30 S 148 30 E
Australian Capital Territory □, Australia 115 F4 35 30 S 149 0 E
Austria ■, Europe 30 E7 47 0N 14 0 E
Austvågøy, Norway 12 B16 68 20N 14 40 E
Autazes, Brazil 153 D6 3 35 S 59 8W
Autelbas, Belgium 21 J7 49 39N 5 52 E
Auterive, France 24 E5 43 21N 1 29 E
Authie →, France 23 B8 50 22N 1 38 E
Authon-du-Perche, France 22 D7 48 12N 0 54 E
Autlán, Mexico 146 D4 19 40N 104 30W
Autun, France 23 F11 46 58N 4 17 E
Auvelais, Belgium 21 H5 50 27N 4 38 E
Auvergne, Australia 112 C5 15 39 S 130 1 E
Auvergne, France 24 C7 45 20N 3 15 E
Auvergne, Mts. d', France 24 C6 45 20N 2 55 E
Auvézére →, France 24 C4 45 12N 0 50 E
Auxerre, France 23 E10 47 48N 3 32 E
Auxi-le-Château, France 23 B9 50 15N 2 8 E
Auxonne, France 23 E12 47 10N 5 20 E
Auxvasse, U.S.A. 140 E5 39 1N 91 54W
Auzances, France 24 B6 46 2N 2 30 E
Auzat-sur-Allier, France 24 C7 45 27N 3 19 E
Ava, U.S.A. 140 G7 37 53N 89 30W
Avallon, France 23 E10 47 30N 3 53 E
Avalon, U.S.A. 145 M8 33 21N 118 20W
Avalon Pen., Canada 129 C9 47 30N 53 20W
Avanigadda, India 83 G5 16 0N 80 56 E
Avaré, Brazil 159 A6 23 4 S 48 58W
Ávas, Greece 44 D7 40 57N 25 56 E
Avawatz Mts., U.S.A. 145 K10 35 40N 116 30W
Aveiro, Brazil 153 D6 3 10 S 55 5W
Aveiro, Portugal 36 E2 40 37N 8 38W
Aveiro □, Portugal 36 E2 40 40N 8 35W
Āvej, Iran 85 C6 35 40N 49 15 E
Avelgem, Belgium 21 G2 50 47N 3 27 E
Avellaneda, Argentina 158 C4 34 50 S 58 10W
Avellino, Italy 41 B7 40 54N 14 47 E
Avenal, U.S.A. 144 K6 36 0N 120 8W
Avenches, Switz. 28 C4 46 53N 7 2 E
Averøya, Norway 14 A1 63 0N 7 35 E
Aversa, Italy 41 B7 40 58N 14 12 E
Avery, U.S.A. 142 C6 47 15N 115 49W
Aves, I. de, W. Indies 149 C7 15 45N 63 55W
Aves, Is. de, Venezuela 149 D6 12 0N 67 30W
Avesnes-sur-Helpe, France 23 B10 50 8N 3 55 E
Avesta, Sweden 13 F17 60 9N 16 10 E
Aveyron □, France 24 D6 44 22N 2 45 E
Aveyron →, France 24 D5 44 5N 1 16 E
Avezzano, Italy 39 F10 42 2N 13 25 E
Avgó, Greece 45 J7 35 33N 25 37 E
Aviá Terai, Argentina 158 B3 26 45 S 60 50W
Aviano, Italy 39 B9 46 4N 12 36 E
Avigliana, Italy 38 C4 45 5N 7 23 E
Avigliano, Italy 41 B8 40 44N 15 43 E
Avignon, France 25 E8 43 57N 4 50 E
Ávila, Spain 36 E6 40 39N 4 43W
Ávila □, Spain 36 E6 40 30N 5 0W
Ávila, Sierra de, Spain 36 E5 40 40N 5 15W
Avila Beach, U.S.A. 145 K6 35 11N 120 44W
Avilés, Spain 36 B5 43 35N 5 57W
Avionárion, Greece 45 F6 38 31N 24 8 E
Avísio →, Italy 39 B8 46 7N 11 5 E
Aviston, U.S.A. 140 F7 38 36N 89 36W
Aviz, Portugal 37 F3 39 4N 7 53W
Avize, France 23 D11 48 59N 4 1 E
Avoca, Ireland 19 D5 52 51N 6 13W
Avoca, U.S.A. 136 D7 42 25N 77 25W
Avoca →, Australia 116 C5 35 40 S 143 43 E
Avola, Canada 130 C5 51 45N 119 19W
Avola, Italy 41 F8 36 56N 15 7 E
Avon, Ill., U.S.A. 140 D6 40 40N 90 26W
Avon, N.Y., U.S.A. 136 D7 42 55N 77 45W
Avon, S. Dak., U.S.A. 138 D5 43 0N 98 4W
Avon □, U.K. 17 F5 51 30N 2 40W
Avon →, Australia 113 F2 31 40 S 116 7 E
Avon →, Avon, U.K. 17 F5 51 29N 2 41W
Avon →, Hants., U.K. 17 G6 50 44N 1 46W
Avon →, Warks., U.K. 17 F5 52 10N 1 51W
Avondale, Zimbabwe 107 F3 17 43 S 30 58 E
Avonlea, Canada 131 D7 50 0N 105 0W
Avonmore, Canada 137 A10 45 10N 74 58W
Avonmouth, U.K. 17 F5 51 30N 2 42W
Avramov, Bulgaria 43 E11 42 45N 26 38 E
Avranches, France 22 D5 48 40N 1 20W
Avre →, France 22 D8 48 47N 1 22 E
Avrig, Romania 46 D5 45 43N 24 21 E
Avtovac, Bos.-H. 42 D3 43 9N 18 35 E
Avu Avu, Solomon Is. 121 M11 9 50 S 160 22 E
Awag el Baqar, Sudan 95 E3 10 10N 33 10 E
A'waj →, Syria 91 B5 33 23N 36 20 E
Awaji, Japan 63 C7 34 32N 135 1 E
Awaji-Shima, Japan 62 C6 34 30N 134 50 E
'Awālī, Bahrain 85 E6 26 0N 50 30 E
Awantipur, India 81 C6 33 55N 75 3 E
Awanui, N.Z. 118 B2 35 4 S 173 17 E
Awarja →, India 82 F3 17 5N 76 15 E
Awarua B., N.Z. 119 E3 44 28 S 168 4 E
Awarua Pt., N.Z. 119 E3 44 15 S 168 5 E
Awasa, L., Ethiopia 95 F4 7 0N 38 30 E
Awash, Ethiopia 90 F3 9 1N 40 10 E
Awash →, Ethiopia 95 E5 11 45N 41 5 E
Awaso, Ghana 100 D4 6 15N 2 22W
Awatere →, N.Z. 119 B9 41 37 S 174 10 E
Awbārī, Libya 96 C2 26 46N 12 57 E
Awbārī □, Libya 96 C2 26 35N 12 46 E
Awe, L., U.K. 18 E3 56 17N 5 16W
Aweil, Sudan 95 F2 8 42N 27 20 E
Awgu, Nigeria 101 D6 6 4N 7 24 E
Awjilah, Libya 96 C4 29 8N 21 7 E
Aworro, Papua N. G. 120 D2 7 43 S 143 11 E
Ax-les-Thermes, France 24 F5 42 44N 1 50 E
Axarfjörður, Iceland 12 C5 66 15N 16 45W
Axel, Neths. 21 F3 51 16N 3 55 E
Axel Heiberg I., Canada 6 B3 80 0N 90 0W
Axim, Ghana 100 E4 4 51N 2 15W
Axinim, Brazil 153 D6 4 2 S 59 22W
Axintele, Romania 46 E7 44 37N 26 47 E
Axioma, Brazil 157 B5 6 45 S 64 31W
Axiós →, Greece 44 D4 40 57N 22 35 E
Axminster, U.K. 17 G4 50 46N 3 0W
Axvall, Sweden 15 F7 58 23N 13 34 E
Aÿ, France 23 C11 49 3N 4 1 E
Ay →, Russia 54 C6 56 8N 57 40 E
Ayaantang, Eq. Guin. 102 B2 1 58N 10 24 E
Ayabaca, Peru 156 A2 4 40 S 79 53W
Ayabe, Japan 63 B7 35 20N 135 20 E
Ayacucho, Argentina 158 D4 37 5 S 58 20W
Ayacucho, Peru 156 C3 13 0 S 74 0W
Ayaguz, Kazakhstan 56 E9 48 10N 80 10 E
Ayakkuduk, Uzbekistan 55 C2 41 12N 65 12 E
Ayakudi, India 83 J3 10 28N 77 56 E
Ayala, Phil. 71 H3 6 57N 121 57 E
Ayamonte, Spain 37 H3 37 12N 7 24W
Ayan, Russia 57 D14 56 30N 138 16 E
Ayancık, Turkey 88 B6 41 57N 34 35 E
Ayapel, Colombia 152 B2 8 19N 75 9W
Ayas, Turkey 88 B5 40 2N 32 21 E
Ayaviri, Peru 156 C3 14 50 S 70 35W
Āybak, Afghan. 79 A3 36 15N 68 5 E
Aybastı, Turkey 88 B7 40 41N 37 23 E
Aydim, W. →, Oman 87 C6 18 8N 53 8 E
Aydın, Turkey 88 D2 37 51N 27 51 E
Aydyrlinskiy, Russia 54 E7 52 3N 59 50 E
Aye, Belgium 21 H6 50 14N 5 18 E
Ayenngré, Togo 101 D5 8 40N 1 1 E
Ayer's Cliff, Canada 137 A12 45 10N 72 3W
Ayers Rock, Australia 113 E5 25 23 S 131 5 E
Ayiá, Greece 44 E4 39 43N 22 45 E
Ayia Aikateríni, Ákra, Greece 32 A3 39 50N 19 50 E
Ayía Ánna, Greece 45 F5 38 52N 23 24 E
Ayia Dhéka, Greece 32 D6 35 3N 24 58 E
Ayia Gálini, Greece 32 D6 35 6N 24 41 E
Ayía Marína, Kásos, Greece 45 J8 35 27N 26 53 E
Ayía Marína, Leros, Greece 45 G8 37 11N 26 48 E
Ayia Napa, Cyprus 32 E13 34 59N 34 0 E
Ayía Paraskeví, Greece 44 E8 39 14N 26 16 E
Ayia Phyla, Cyprus 32 E12 34 43N 33 1 E
Ayía Rouméli, Greece 45 J5 35 14N 23 58 E
Ayia Varvára, Greece 32 D7 35 8N 25 1 E
Ayiássos, Greece 45 E8 39 5N 26 23 E
Áyion Óros, Greece 44 D6 40 25N 24 6 E
Áyios Amvrósios, Cyprus 32 D12 35 20N 33 35 E
Áyios Andréas, Greece 45 G4 37 21N 22 45 E
Áyios Evstrátios, Greece 44 E6 39 34N 24 58 E
Áyios Ioánnis, Ákra, Greece 32 D7 35 20N 25 40 E
Áyios Isidhoros, Greece 32 C9 36 9N 27 51 E
Áyios Kiríkos, Greece 45 G8 37 34N 26 17 E
Áyios Matthaíos, Greece 32 B3 39 30N 19 47 E
Áyios Mírono, Greece 45 J7 35 15N 25 1 E
Áyios Nikólaos, Greece 32 D7 35 11N 25 41 E
Áyios Pétros, Greece 45 F2 38 38N 20 33 E
Áyios Seryios, Cyprus 32 D12 35 12N 33 53 E
Áyios Theodhoros, Cyprus 32 D13 35 22N 34 1 E
Áyios Yeóryios, Greece 45 G5 37 28N 23 57 E
Aykathonisi, Greece 45 G8 37 28N 27 0 E
Aykino, Russia 48 B8 62 15N 49 56 E
Aylesbury, U.K. 17 F7 51 49N 0 49W
Aylmer, Canada 136 D4 42 46N 80 59W
Aylmer, L., Canada 126 B8 64 0N 110 8W
'Ayn al Ghazālah, Libya 96 B4 32 10N 23 20 E
Ayn Zālah, Iraq 89 D10 36 45N 42 35 E
Ayna, Spain 35 G2 38 34N 2 3W
Aynāt, Yemen 87 C5 16 4N 49 9 E
Ayni, Tajikistan 55 D4 39 23N 68 32 E
Ayolas, Paraguay 158 B4 27 10 S 56 59W
Ayom, Sudan 95 F2 7 49N 28 23 E
Ayon, Ostrov, Russia 57 C17 69 50N 169 0 E
Ayora, Spain 35 F3 39 3N 1 3W
Ayr, Australia 114 B4 19 35 S 147 25 E
Ayr, U.K. 18 F4 55 28N 4 38W
Ayr →, U.K. 18 F4 55 28N 4 38W
Ayrancı, Turkey 88 D5 37 21N 33 41 E
Ayre, Pt. of, U.K. 16 C3 54 25N 4 21W
Aysha, Ethiopia 95 E5 10 50N 42 23 E
Aytos, Bulgaria 43 E12 42 42N 27 16 E
Aytoska Planina, Bulgaria 43 E12 42 45N 27 30 E
Ayu, Kepulauan, Indonesia 73 A4 0 35N 131 5 E
Ayutla, Guatemala 148 D1 14 40N 92 10W
Ayutla, Mexico 147 D5 16 58N 99 17W
Ayvacık, Turkey 88 C2 39 36N 26 24 E
Ayvalık, Turkey 88 C2 39 20N 26 46 E
Aywaille, Belgium 21 H7 50 28N 5 40 E
Az Zabdānī, Syria 91 B5 33 43N 36 5 E
Az̧ Z̧āhirīyah, West Bank 91 D3 31 25N 34 58 E
Az̧ Z̧ahrān, Si. Arabia 85 E6 26 10N 50 7 E
Az Zarqā, Jordan 91 C5 32 5N 36 4 E
Az Zāwiyah, Libya 96 B2 32 52N 12 56 E
Az Zaydīyah, Yemen 86 D3 15 20N 43 1 E
Az Zībār, Iraq 89 D11 36 52N 44 4 E
Az-Zilfī, Si. Arabia 84 E5 26 12N 44 52 E
Az Zubayr, Iraq 84 D5 30 26N 47 40 E
Az Zuqur, Yemen 86 D3 14 0N 42 45 E
Az Zuwaytīnah, Libya 96 B4 30 58N 20 7 E
Azambuja, Portugal 37 F2 39 4N 8 51W
Azamgarh, India 81 F10 26 5N 83 13 E
Azangaro, Peru 156 C3 14 55 S 70 13W
Azaouak, Vallée de l', Mali 101 B5 15 50N 3 20 E
Āzār Shahr, Iran 89 D11 37 45N 45 59 E
Azarān, Iran 89 D12 37 25N 47 16 E
Azärbayjan = Azerbaijan ■, Asia 53 K9 40 20N 48 0 E
Āzarbāyjān-e Gharbī □, Iran 84 B5 37 0N 44 30 E
Āzarbāyjān-e Sharqī □, Iran 84 B5 37 20N 47 0 E
Azare, Nigeria 101 C7 11 55N 10 10 E
Azay-le-Rideau, France 22 E7 47 16N 0 30 E
A'zāz, Syria 88 D7 36 36N 37 4 E
Azazga, Algeria 99 A5 36 48N 4 22 E
Azbine = Aïr, Niger 97 E1 18 30N 8 0 E
Azefal, Mauritania 98 D2 21 0N 14 45W
Azeffoun, Algeria 99 A5 36 51N 4 26 E

170 Azemmour

Azemmour, Morocco 98 B3 33 20N 9 20W
Azerbaijan ■, Asia 53 K9 40 20N 48 0 E
Azerbaijchan = Azerbaijan ■, Asia 53 K9 40 20N 48 0 E
Azezo, Ethiopia 95 E4 12 28N 37 15 E
Azimganj, India 81 G13 24 14N 88 16 E
Aznalcóllar, Spain 37 H4 37 32N 6 17W
Azogues, Ecuador 152 D2 2 35 S 78 0W
Azores, Atl. Oc. 8 E6 38 44N 29 0W
Azov, Russia 53 G4 47 3N 39 25 E
Azov, Sea of, Europe 53 H3 46 0N 36 30 E
Azovskoye More = Azov, Sea of, Europe 53 H3 46 0N 36 30 E
Azovy, Russia 56 C7 64 55N 65 1 E
Azpeitia, Spain 34 B2 43 12N 2 19W
Azrou, Morocco 98 B3 33 28N 5 19W
Aztec, U.S.A. 143 H10 36 49N 107 59W
Azúa, Dom. Rep. 149 C5 18 25N 70 44W
Azuaga, Spain 37 G5 38 16N 5 39W
Azuara, Spain 34 D4 41 15N 0 53W
Azuay □, Ecuador 152 D2 2 55 S 79 0W
Azuer →, Spain 37 F7 39 8N 3 36W
Azuero, Pen. de, Panama 148 E3 7 30N 80 30W
Azul, Argentina 158 D4 36 42 S 59 43W
Azul, Serra, Brazil 157 C7 14 50 S 54 50W
Azurduy, Bolivia 157 D5 19 59 S 64 29W
Azusa, U.S.A. 145 L9 34 8N 117 52W
Azzaba, Algeria 99 A6 36 48N 7 6 E
Azzano Décimo, Italy 39 C9 45 52N 12 56 E
'Azzūn, West Bank 91 C4 32 10N 35 2 E

B

Ba Don, Vietnam 76 D6 17 45N 106 26 E
Ba Dong, Vietnam 77 H6 9 40N 106 33 E
Ba Ngoi = Cam Lam, Vietnam 77 G7 11 54N 109 10 E
Ba Ria, Vietnam 77 G6 10 30N 107 10 E
Ba Tri, Vietnam 77 G6 10 2N 106 36 E
Ba Xian, China 66 E9 39 8N 116 22 E
Baa, Indonesia 72 D2 10 50 S 123 0 E
Baaba, I., N. Cal. 121 T18 20 3 S 164 59 E
Baamonde, Spain 36 B3 43 7N 7 44W
Baao, Phil. 70 E4 13 27N 123 22 E
Baar, Switz. 29 B7 47 12N 8 32 E
Baarle Nassau, Belgium 21 F5 51 27N 4 56 E
Baarlo, Neths. 21 F8 51 20N 6 6 E
Baarn, Neths. 20 D6 52 12N 5 17 E
Bab el Mandeb, Red Sea 86 D3 12 35N 43 25 E
Baba, Bulgaria 43 E8 42 44N 23 59 E
Baba Burnu, Turkey 88 C2 39 29N 26 2 E
Baba dag, Azerbaijan 53 K9 41 0N 48 19 E
Bābā Kalū, Iran 85 D6 30 7N 50 49 E
Babaçulândia, Brazil 154 C2 7 13 S 47 46W
Babadag, Romania 46 E9 44 53N 28 44 E
Babadayhan, Turkmenistan 56 F7 37 42N 60 23 E
Babaeski, Turkey 88 B2 41 26N 27 6 E
Babahoyo, Ecuador 152 D2 1 40 S 79 30W
Babak, Phil. 71 H5 7 8N 125 41 E
Babakin, Australia 113 F2 32 7 S 118 1 E
Babana, Nigeria 101 C5 10 31N 3 46 E
Babar, Algeria 99 A6 35 10N 7 6 E
Babar, Indonesia 73 C3 8 0 S 129 30 E
Babar, Pakistan 80 D3 31 7N 69 32 E
Babarkach, Pakistan 80 E2 29 45N 68 0 E
Babayevo, Russia 50 C8 59 24N 35 55 E
Babb, U.S.A. 142 B7 48 51N 113 27W
Babenhausen, Germany 27 F4 49 57N 8 57 E
Babi Besar, P., Malaysia 77 L4 2 25N 103 59 E
Babia Gora, Europe 31 B12 49 38N 19 38 E
Babian Jiang →, China 68 F3 22 55N 101 47 E
Babile, Ethiopia 95 F5 9 16N 42 11 E
Babinda, Australia 114 B4 17 20 S 145 56 E
Babine, Canada 130 B3 55 22N 126 37W
Babine →, Canada 130 B3 55 45N 127 44W
Babine L., Canada 130 C3 54 48N 126 0W
Babo, Indonesia 73 B4 2 30 S 133 30 E
Babócsa, Hungary 31 E10 46 2N 17 21 E
Bābol, Iran 85 B7 36 40N 52 50 E
Bābol Sar, Iran 85 B7 36 45N 52 45 E
Baborów, Poland 31 A11 50 7N 18 1 E
Baboua, C.A.R. 102 A2 5 49N 14 58 E
Babruysk, Belarus 51 F5 53 10N 29 15 E
Babuna, Macedonia 42 F6 41 30N 21 40 E
Babura, Nigeria 101 C6 12 51N 8 59 E
Babusar Pass, Pakistan 81 B5 35 12N 73 59 E
Babušnica, Serbia, Yug. 42 D7 43 7N 22 27 E
Babuyan Chan., Phil. 70 B3 18 40N 121 30 E
Babuyan I., Phil. 70 B3 19 32N 121 57 E
Babuyan Is., Phil. 70 B3 19 15N 121 40 E
Babylon, Iraq 89 F11 32 34N 44 22 E
Bač, Serbia, Yug. 42 B4 45 29N 19 17 E
Bac Can, Vietnam 76 A5 22 8N 105 49 E
Bac Giang, Vietnam 76 B6 21 16N 106 11 E
Bac Ninh, Vietnam 76 B6 21 13N 106 4 E
Bac Phan, Vietnam 76 B5 22 0N 105 0 E
Bac Quang, Vietnam 76 A5 22 30N 104 48 E
Bacabal, Brazil 154 B3 4 15 S 44 45W
Bacacay, Phil. 70 E4 13 18N 123 47 E
Bacajá →, Brazil 153 D7 3 25 S 51 50W
Bacalar, Mexico 147 D7 18 50N 87 27W
Bacan, Kepulauan, Indonesia 72 B3 0 35 S 127 30 E
Bacan, Pulau, Indonesia 72 B3 0 50 S 127 30 E
Bacarra, Phil. 70 B3 18 15N 120 37 E
Bacău, Romania 46 C7 46 35N 26 55 E
Bacău □, Romania 46 C7 46 30N 26 45 E
Baccarat, France 23 D13 48 28N 6 42 E
Bacchus Marsh, Australia 116 D6 37 43 S 144 27 E
Bacerac, Mexico 146 A3 30 18N 108 50W
Băceşti, Romania 46 C8 46 50N 27 11 E
Bach Long Vi, Dao, Vietnam 76 B6 20 10N 107 40 E
Bachaquero, Venezuela 152 B3 9 56N 71 8W
Bacharach, Germany 27 E3 50 3N 7 44 E
Bachelina, Russia 56 D7 57 45N 67 20 E
Bachuma, Ethiopia 95 F4 6 48N 35 53 E
Bačina, Serbia, Yug. 42 D6 43 42N 21 23 E
Back →, Canada 126 B9 65 10N 104 0W
Bačka Palanka, Serbia, Yug. 42 B4 45 17N 19 27 E
Bačka Topola, Serbia, Yug. 42 B4 45 49N 19 39 E
Bäckefors, Sweden 15 F6 58 48N 12 9 E
Bački Petrovac, Serbia, Yug. 42 B4 45 29N 19 32 E
Backnang, Germany 27 G5 48 56N 9 26 E
Backstairs Passage, Australia 116 C3 35 40 S 138 5 E
Baco, Mt., Phil. 70 E3 12 49N 121 10 E
Bacolod, Phil. 71 F4 10 40N 122 57 E
Bacon, Phil. 70 E5 13 3N 124 3 E
Bacoor, Phil. 70 D3 14 28N 120 56 E
Bacqueville-en-Caux, France 22 C8 49 47N 1 0 E
Bacs-Kiskun □, Hungary 31 E12 46 43N 19 30 E
Bácsalmás, Hungary 31 E12 46 8N 19 17 E
Bacuag, Phil. 71 G5 9 36N 125 38 E
Bacuk, Malaysia 77 J4 6 4N 102 25 E
Baculin, Phil. 71 H6 7 27N 126 35 E
Bād, Iran 85 C7 33 41N 52 1 E
Bad →, U.S.A. 138 C4 44 21N 100 22W
Bad Aussee, Austria 30 D6 47 43N 13 45 E
Bad Axe, U.S.A. 136 C2 43 48N 83 0W
Bad Bergzabern, Germany 27 F4 49 6N 7 59 E
Bad Berleburg, Germany 26 D4 51 2N 8 26 E
Bad Bevensen, Germany 26 B6 53 5N 10 35 E
Bad Bramstedt, Germany 26 B5 53 55N 9 53 E
Bad Brückenau, Germany 27 E5 50 18N 9 47 E
Bad Doberan, Germany 26 A7 54 6N 11 53 E
Bad Driburg, Germany 26 D5 51 43N 9 1 E
Bad Ems, Germany 27 E3 50 20N 7 43 E
Bad Frankenhausen, Germany 26 D7 51 21N 11 5 E
Bad Freienwalde, Germany 26 C10 52 46N 14 1 E
Bad Godesberg, Germany 26 E3 50 41N 7 9 E
Bad Hersfeld, Germany 26 E5 50 52N 9 42 E
Bad Hofgastein, Austria 30 D6 47 17N 13 6 E
Bad Homburg, Germany 27 E4 50 13N 8 38 E
Bad Honnef, Germany 26 E3 50 38N 7 13 E
Bad Ischl, Austria 30 D6 47 44N 13 38 E
Bad Kissingen, Germany 27 E6 50 11N 10 4 E
Bad Königshofen, Germany 27 E6 50 17N 10 28 E
Bad Kreuznach, Germany 27 F3 49 50N 7 51 E
Bad Laasphe, Germany 26 E4 50 56N 8 25 E
Bad Lands, U.S.A. 138 D3 43 40N 102 10W
Bad Langensalza, Germany 26 D6 51 5N 10 38 E
Bad Lauterberg, Germany 26 D6 51 38N 10 28 E
Bad Leonfelden, Austria 30 C7 48 31N 14 18 E
Bad Lippspringe, Germany 26 D4 51 47N 8 49 E
Bad Mergentheim, Germany 27 F5 49 28N 9 42 E
Bad Münstereifel, Germany 26 E2 50 33N 6 46 E
Bad Muskau, Germany 26 D10 51 32N 14 35 E
Bad Nauheim, Germany 27 E4 50 21N 8 43 E
Bad Oeynhausen, Germany 26 C4 52 12N 8 46 E
Bad Oldesloe, Germany 26 B6 53 48N 10 22 E
Bad Orb, Germany 27 E5 50 12N 9 22 E
Bad Pyrmont, Germany 26 D5 51 59N 9 16 E
Bad Ragaz, Switz. 29 B9 47 0N 9 30 E
Bad Reichenhall, Germany 27 H8 47 43N 12 54 E
Bad Säckingen, Germany 27 H3 47 33N 7 56 E
Bad Salzuflen, Germany 26 C4 52 5N 8 45 E
Bad Segeberg, Germany 26 B6 53 56N 10 17 E
Bad Tölz, Germany 27 H7 47 45N 11 34 E
Bad Urach, Germany 27 G5 48 29N 9 23 E
Bad Waldsee, Germany 27 H5 47 55N 9 45 E
Bad Wildungen, Germany 26 D5 51 6N 9 7 E
Bad Wimpfen, Germany 27 F5 49 13N 9 11 E
Bad Windsheim, Germany 27 F6 49 36N 10 25 E
Badagara, India 83 J2 11 35N 75 40 E
Badagri, Nigeria 101 D5 6 25N 2 55 E
Badajós, L., Brazil 153 D5 3 15 S 62 50W
Badajoz, Spain 37 G4 38 50N 6 59W
Badajoz □, Spain 37 G4 38 40N 6 30W
Badakhshān □, Afghan. 79 A3 36 30N 71 0 E
Badalona, Spain 34 D7 41 26N 2 15 E
Badalzai, Afghan. 80 E1 29 50N 65 35 E
Badampahar, India 82 C8 22 10N 86 10 E
Badanah, Si. Arabia 84 D4 30 58N 41 30 E
Badarinath, India 81 D8 30 45N 79 30 E
Badas, Kepulauan, Indonesia 74 B3 0 45N 107 5 E
Baddo →, Pakistan 79 D2 28 0N 64 20 E
Bade, Indonesia 73 C5 7 10 S 139 35 E
Baden, Austria 31 C9 48 1N 16 13 E
Baden, Switz. 29 B6 47 28N 8 18 E
Baden-Baden, Germany 27 G4 48 44N 8 13 E
Baden Park, Australia 116 B6 32 8 S 144 12 E
Baden-Württemberg □, Germany 27 G5 48 20N 8 40 E
Badenoch, U.K. 18 E4 56 59N 4 15W
Badgastein, Austria 30 D6 47 7N 13 9 E
Badger, Canada 129 C8 49 0N 56 4W
Badger, U.S.A. 144 J7 36 38N 119 1W
Bādghīsāt □, Afghan. 79 B1 35 0N 63 0 E
Badgom, India 81 B6 34 1N 74 45 E
Badhoevedorp, Neths. 20 D5 52 20N 4 47 E
Badia Polésine, Italy 39 C8 45 5N 11 29 E
Badian, Phil. 71 G4 9 55N 123 24 E
Badin, Pakistan 79 D3 24 38N 68 54 E
Badnera, India 82 D3 20 48N 77 44 E
Badoc, Phil. 70 C3 17 56N 120 28 E
Badogo, Mali 100 C3 11 2N 8 13W
Badong, China 69 B8 31 1N 110 23 E
Badr Ḥunayn, Si. Arabia 86 B2 23 44N 38 46 E
Badrah, Iraq 89 F11 33 6N 45 58 E
Baduen, Somali Rep. 108 C3 7 15N 47 40 E
Badulla, Sri Lanka 83 L5 7 1N 81 7 E
Badupi, Burma 78 E4 21 36N 93 27 E
Baena, Spain 37 H6 37 37N 4 20W
Baerami Creek, Australia 117 B9 32 27 S 150 27 E
Baexem, Neths. 21 F7 51 13N 5 53 E
Baeza, Ecuador 152 D2 0 25 S 77 53W
Baeza, Spain 35 H1 37 57N 3 25W
Bafa Gölü, Turkey 45 G9 37 30N 27 29 E
Bafang, Cameroon 101 D7 5 9N 10 11 E
Bafatá, Guinea-Biss. 100 C2 12 8N 14 40W
Baffin B., Canada 124 B13 72 0N 64 0W
Baffin I., Canada 127 B12 68 0N 75 0W
Bafia, Cameroon 101 E7 4 40N 11 10 E
Bafilo, Togo 101 D5 9 22N 1 22 E
Bafing →, Mali 100 C2 13 49N 10 50W
Bafliyūn, Syria 84 B3 36 37N 36 59 E
Baflo, Neths. 20 B9 53 22N 6 31 E
Bafoulabé, Mali 100 C2 13 50N 10 55W
Bafoussam, Cameroon 101 D7 5 28N 10 25 E
Bāfq, Iran 85 D7 31 40N 55 25 E
Bafra, Turkey 88 B6 41 34N 35 54 E
Bafra Burnu, Turkey 88 B7 41 45N 36 2 E
Bāft, Iran 85 D8 29 15N 56 38 E
Bafut, Cameroon 101 D7 6 6N 10 2 E
Bafwasende, Zaïre 106 B2 1 3N 27 5 E
Bagabag, Phil. 70 C3 16 30N 121 15 E
Bagac, Phil. 70 D3 14 36N 120 23 E
Bagac Bay, Phil. 70 D3 14 36N 120 20 E
Bagalkot, India 83 F2 16 10N 75 40 E
Bagamoyo, Tanzania 106 D4 6 28 S 38 55 E
Bagamoyo □, Tanzania 106 D4 6 20 S 38 30 E
Bagan Datoh, Malaysia 77 L3 3 59N 100 47 E
Bagan Serai, Malaysia 77 K3 5 1N 100 32 E
Baganga, Phil. 71 H6 7 34N 126 33 E
Bagani, Namibia 104 B3 18 7 S 21 41 E
Bagansiapiapi, Indonesia 74 B2 2 12N 100 50 E
Bagasra, India 80 J4 21 30N 71 0 E
Bagata, Zaïre 102 C3 3 44 S 17 57 E
Bagawi, Sudan 95 E3 12 20N 34 18 E
Bagdad, U.S.A. 145 L11 34 35N 115 53W
Bagdarin, Russia 57 D12 54 26N 113 36 E
Bagé, Brazil 159 C5 31 20 S 54 15W
Bagenalstown = Muine Bheag, Ireland 19 D5 52 42N 6 58W
Baggs, U.S.A. 142 F10 41 2N 107 39W
Bagh, Pakistan 81 C5 33 59N 73 45 E
Baghdād, Iraq 89 F11 33 20N 44 30 E
Bagherhat, Bangla. 78 D2 22 40N 89 47 E
Bagheria, Italy 40 D6 38 5N 13 30 E
Baghlān, Afghan. 79 A3 36 12N 69 0 E
Baghlān □, Afghan. 79 B3 36 0N 68 30 E
Bagley, U.S.A. 138 B7 47 32N 95 24W
Bagnacavallo, Italy 39 D8 44 25N 12 0 E
Bagnara Cálabra, Italy 41 D8 38 17N 15 48 E
Bagnell Dam, U.S.A. 140 F4 38 14N 92 36W
Bagnères-de-Bigorre, France 24 E4 43 5N 0 9 E
Bagnères-de-Luchon, France 24 F4 42 47N 0 38 E
Bagni di Lucca, Italy 38 D7 44 1N 10 35 E
Bagno di Romagna, Italy 39 E8 43 50N 11 57 E
Bagnoles-de-l'Orne, France 22 D6 48 32N 0 25W
Bagnoli di Sopra, Italy 39 C8 45 11N 11 53 E
Bagnolo Mella, Italy 38 C7 45 26N 10 10 E
Bagnols-sur-Cèze, France 25 D8 44 10N 4 36 E
Bagnorégio, Italy 39 F9 42 37N 12 5 E
Bago, Phil. 71 F4 10 32N 122 50 E
Bagolino, Italy 38 C7 45 49N 10 28 E
Bagotville, Canada 129 C5 48 22N 70 54W
Bagrationovsk, Russia 13 J19 54 23N 20 39 E
Bagrdan, Serbia, Yug. 42 C6 44 5N 21 11 E
Bagua, Peru 156 B2 5 35 S 78 22W
Baguio, Phil. 70 C3 16 26N 120 34 E
Bahabón de Esgueva, Spain 34 D1 41 52N 3 43W
Bahadurabad Ghat, Bangla. 78 C2 25 11N 89 44 E
Bahadurgarh, India 80 E7 28 40N 76 57 E
Bahama, Canal Viejo de, W. Indies 148 B4 22 10N 77 30W
Bahamas ■, N. Amer. 149 B5 24 0N 75 0W
Bahār, Iran 89 E13 34 54N 48 26 E
Baharampur, India 81 G13 24 2N 88 27 E
Baharîya, El Wâhât al, Egypt 94 J6 28 0N 28 50 E
Bahau, Malaysia 77 L4 2 48N 102 26 E
Bahawalnagar, Pakistan 79 C4 30 0N 73 15 E
Bahawalpur, Pakistan 79 C3 29 24N 71 40 E
Bahçe, Turkey 88 D7 37 13N 36 34 E
Baheri, India 81 E8 28 45N 79 34 E
Bahi, Tanzania 106 D4 5 58 S 35 21 E
Bahi Swamp, Tanzania 106 D4 6 10 S 35 0 E
Bahía = Salvador, Brazil 155 D4 13 0 S 38 30W
Bahía □, Brazil 155 D3 12 0 S 42 0W
Bahía, Is. de la, Honduras 148 C2 16 45N 86 15W
Bahía Blanca, Argentina 158 D3 38 35 S 62 13W
Bahía de Caráquez, Ecuador 152 D1 0 40 S 80 27W
Bahía Honda, Cuba 148 B3 22 54N 83 10W
Bahía Laura, Argentina 160 C3 48 10 S 66 30W
Bahía Negra, Paraguay 157 E6 20 5 S 58 5W
Bahir Dar, Ethiopia 95 E4 11 37N 37 10 E
Bahlah, Oman 87 B7 22 58N 57 18 E
Bahmanzād, Iran 85 D6 31 15N 51 47 E
Bahmer, Algeria 99 C4 27 32N 0 10W
Bahönye, Hungary 31 E10 46 25N 17 28 E
Bahr Aouk →, C.A.R. 102 A3 8 40N 19 0 E
Bahr el Ahmar □, Sudan 94 C3 20 0N 35 0 E
Bahr el Ghazâl □, Sudan 95 F2 7 0N 28 0 E
Bahr Salamat →, Chad 97 G3 9 20N 18 0 E
Bahr Yûsef →, Egypt 94 J7 28 25N 30 35 E
Bahra el Burullus, Egypt 94 H7 31 28N 30 48 E
Bahraich, India 81 F9 27 38N 81 37 E
Bahrain ■, Asia 85 E6 26 0N 50 35 E
Bahror, India 80 F7 27 51N 76 20 E
Bāhū Kalāt, Iran 85 E9 25 43N 61 25 E
Bai, Mali 100 C4 13 35N 3 28W
Bai Bung, Mui, Vietnam 77 H5 8 38N 104 44 E
Bai Duc, Vietnam 76 C5 18 3N 105 49 E
Bai Thuong, Vietnam 76 C5 19 54N 105 23 E
Baia Farta, Angola 103 E2 12 40 S 13 11 E
Baia Mare, Romania 46 B4 47 40N 23 35 E
Baia-Sprie, Romania 46 B4 47 41N 23 43 E
Baião, Brazil 154 B2 2 40 S 49 40W
Baïbokoum, Chad 97 G3 7 46N 15 43 E
Baicheng, China 67 B12 45 38N 122 42 E
Băicoi, Romania 46 D6 45 3N 25 52 E
Baidoa, Somali Rep. 108 D2 3 8N 43 30 E
Baie Comeau, Canada 129 C6 49 12N 68 10W
Baie-St-Paul, Canada 129 C5 47 28N 70 32W
Baie Trinité, Canada 129 C6 49 25N 67 20W
Baie Verte, Canada 129 C8 49 55N 56 12W
Baignes-Ste.-Radegonde, France 24 C3 45 23N 0 25W
Baigneux-les-Juifs, France 23 E11 47 31N 4 39 E
Baihe, China 66 H6 32 50N 110 5 E
Baihe, Taiwan 69 F13 23 24N 120 24 E
Ba'ījī, Iraq 89 E10 35 0N 43 30 E
Baikal, L. = Baykal, Oz., Russia 57 D11 53 0N 108 0 E
Bailadila, Mt., India 82 E5 18 43N 81 15 E
Baile Atha Cliath = Dublin, Ireland 19 C5 53 21N 6 15W
Bailei, Ethiopia 95 F5 6 44N 40 18 E
Bailén, Spain 37 G7 38 8N 3 48W
Băileşti, Romania 46 E4 44 1N 23 20 E
Baileux, Belgium 21 H4 50 2N 4 23 E
Bailhongal, India 83 G2 15 55N 74 53 E
Bailique, Ilha, Brazil 154 A2 1 2N 49 58W
Bailleul, France 23 B9 50 44N 2 41 E
Bailundo, Angola 103 E3 12 10 S 15 50 E
Baima, China 68 A3 33 0N 100 26 E
Baimuru, Papua N. G. 120 D3 7 35 S 144 51 E
Bain-de-Bretagne, France 22 E5 47 50N 1 40W
Bainbridge, Ga., U.S.A. 135 K3 30 55N 84 35W
Bainbridge, Ind., U.S.A. 141 E10 39 46N 86 49W
Bainbridge, N.Y., U.S.A. 137 D9 42 18N 75 29W
Bainbridge, Ohio, U.S.A. 141 E13 39 14N 83 16W
Baing, Indonesia 72 D2 10 14 S 120 34 E
Bainiu, China 66 H7 32 50N 112 15 E
Bainville, U.S.A. 138 A2 48 8N 104 13W
Bainyik, Papua N. G. 120 B2 3 40 S 143 4 E
Bā'ir, Jordan 91 E5 30 45N 36 55 E
Baird, U.S.A. 139 J5 32 24N 99 24W
Baird Mts., U.S.A. 126 B3 67 0N 160 0W
Bairin Youqi, China 67 C10 43 30N 118 35 E
Bairin Zuoqi, China 67 C10 43 58N 119 15 E
Bairnsdale, Australia 117 D7 37 48 S 147 36 E
Bais, Phil. 71 G4 9 35N 123 7 E
Baisha, China 66 G7 34 20N 112 32 E
Baïsole →, France 24 E4 43 26N 0 25 E
Baissa, Nigeria 101 D7 7 14N 10 38 E
Baitadi, Nepal 81 E9 29 35N 80 25 E
Baixa Grande, Brazil 155 D3 11 57 S 40 11W
Baiyin, China 66 F3 36 45N 104 14 E
Baiyü, China 68 B2 31 16N 98 50 E
Baiyu Shan, China 66 F4 37 15N 107 30 E
Baiyuda, Sudan 94 D3 17 35N 32 7 E
Baj Baj, India 81 H13 22 30N 88 5 E
Baja, Hungary 31 E11 46 12N 18 59 E
Baja, Pta., Mexico 146 B1 29 50N 116 0W
Baja California, Mexico 146 A1 31 10N 115 12W
Baja California □, Mexico 146 B2 30 0N 115 0W
Baja California Sur □, Mexico 146 B2 25 50N 111 50W
Bajamar, Canary Is. 33 F3 28 33N 16 20W
Bajana, India 80 H4 23 7N 71 49 E
Bājgīrān, Iran 85 B8 37 36N 58 24 E
Bājil, Yemen 86 D3 15 4N 43 17 E
Bajimba, Mt., Australia 115 D5 29 17 S 152 6 E
Bajina Bašta, Serbia, Yug. 42 D4 43 58N 19 35 E
Bajmok, Serbia, Yug. 42 B4 45 57N 19 24 E
Bajo Nuevo, Caribbean 148 C4 15 40N 78 50W
Bajoga, Nigeria 101 C7 10 57N 11 20 E
Bajool, Australia 114 C5 23 40 S 150 35 E
Bak, Hungary 31 E9 46 43N 16 51 E
Bakal, Russia 54 D7 54 56N 58 48 E
Bakala, C.A.R. 102 A4 6 15N 20 20 E
Bakanas, Kazakhstan 55 A8 44 50N 76 15 E
Bakar, Croatia 39 C11 45 18N 14 32 E
Bakbakty, Kazakhstan 55 A8 44 36N 76 40 E
Bakchar, Russia 56 D9 57 1N 82 5 E
Bakel, Neths. 21 E7 51 30N 5 45 E
Bakel, Senegal 100 C2 14 56N 12 20W
Baker, Calif., U.S.A. 145 K10 35 16N 116 4W
Baker, Mont., U.S.A. 138 B2 46 22N 104 17W
Baker, Oreg., U.S.A. 142 D5 44 47N 117 50W
Baker, Canal, Chile 160 C2 47 45 S 74 45W
Baker, L., Canada 126 B10 64 0N 96 0W
Baker I., Pac. Oc. 122 G10 0 10N 176 35W
Baker L., Australia 113 E4 26 54 S 126 5 E
Baker Lake, Canada 126 B10 64 20N 96 3W
Baker Mt., U.S.A. 142 B3 48 50N 121 49W
Bakers Creek, Australia 114 C4 21 13 S 149 7 E
Baker's Dozen Is., Canada 128 A4 56 45N 78 45W
Bakersfield, Calif., U.S.A. 145 K7 35 23N 119 1W
Bakersfield, Vt., U.S.A. 137 B12 44 45N 72 48W
Bakhchysaray, Ukraine 51 K7 44 40N 33 45 E
Bakhmach, Ukraine 51 G7 51 10N 32 45 E
Bākhtarān, Iran 89 E12 34 23N 47 0 E
Bākhtarān □, Iran 84 C5 34 0N 46 30 E
Bakı, Azerbaijan 53 K9 40 29N 49 56 E
Bakırdaği, Turkey 88 C6 38 13N 35 46 E
Bakırköy, Turkey 43 F13 41 2N 28 53 E
Bakkafjörður, Iceland 12 C6 66 2N 14 48W
Bakkagerði, Iceland 12 D7 65 31N 13 49W
Bakony →, Hungary 31 D10 47 35N 17 54 E
Bakony Forest = Bakony Hegyseg, Hungary 31 D10 47 10N 17 30 E
Bakony Hegyseg, Hungary 31 D10 47 10N 17 30 E
Bakori, Nigeria 101 C6 11 34N 7 25 E
Bakouma, C.A.R. 102 A4 5 40N 22 56 E
Bakov, Czech. 30 A7 50 27N 14 55 E
Bakpakty = Bakbakty, Kazakhstan 55 A8 44 35N 76 40 E
Baksan, Russia 53 J6 43 42N 43 32 E
Baku = Bakı, Azerbaijan 53 K9 40 29N 49 56 E
Bakun, Phil. 70 C3 16 48N 120 40 E
Bakutis Coast, Antarctica 7 D15 74 0 S 120 0W
Bakwa-Kenge, Zaïre 103 C4 4 51 S 22 4 E
Baky = Bakı, Azerbaijan 53 K9 40 29N 49 56 E
Bala, Canada 136 A5 45 1N 79 37W
Bâlâ, Turkey 88 C5 39 32N 33 6 E
Bala, L., U.K. 16 E4 52 53N 3 37W
Bālā Morghāb, Afghan. 79 B1 35 35N 63 20 E
Balabac I., Phil. 71 H1 8 0N 117 0 E
Balabagh, Afghan. 80 B4 34 25N 70 12 E
Ba'labakk, Lebanon 91 A5 34 0N 36 10 E
Balabalangan, Kepulauan, Indonesia 75 C5 2 20 S 117 30 E
Balabio, I., N. Cal. 121 T18 20 7 S 164 11 E
Bălăciţa, Romania 46 E4 44 23N 23 8 E
Balad, Iraq 89 F11 34 1N 44 9 E
Balad Rūz, Iraq 89 F11 33 42N 45 5 E
Bālādeh, Fārs, Iran 85 D6 29 17N 51 56 E
Bālādeh, Māzandaran, Iran 85 B6 36 12N 51 48 E
Balaghat, India 82 D5 21 49N 80 12 E
Balaghat Ra., India 82 E3 18 50N 76 30 E
Balaguer, Spain 34 D5 41 50N 0 50 E
Balakété, C.A.R. 102 A3 6 56N 19 54 E
Balakhna, Russia 52 B6 56 25N 43 32 E
Balaklava, Australia 116 C3 34 7 S 138 22 E
Balaklava, Ukraine 51 K7 44 30N 33 30 E
Balakliya, Ukraine 51 H9 49 28N 36 55 E
Balakovo, Russia 52 D8 52 4N 47 55 E
Balamban, Phil. 71 F4 10 30N 123 43 E
Balambangan, Malaysia 75 A5 7 17N 116 55 E
Balancán, Mexico 147 D6 17 48N 91 32W
Balangiga, Phil. 71 F5 11 7N 125 23 E

Barapasi 171

Balangir, *India* **82 D6** 20 43N 83 35 E
Balapur, *India* **82 D3** 20 40N 76 45 E
Balashov, *Russia* **52 E6** 51 30N 43 10 E
Balasinor, *India* **80 H5** 22 57N 73 23 E
Balasore = Baleshwar, *India* **82 D8** 21 35N 87 3 E
Balassagyarmat, *Hungary* **31 C12** 48 4N 19 15 E
Balât, *Egypt* **94 B2** 25 36N 29 19 E
Balaton, *Hungary* **31 E10** 46 50N 17 40 E
Balatonfüred, *Hungary* **31 E10** 46 58N 17 54 E
Balatonszentgyörgy, *Hungary* **31 E10** 46 41N 17 19 E
Balayan, *Phil.* **70 E3** 13 57N 120 44 E
Balazote, *Spain* **35 G2** 38 54N 2 9W
Balbalan, *Phil.* **70 C3** 17 27N 121 12 E
Balbi, Mt., *Papua N. G.* **120 C8** 5 55 S 154 58 E
Balbina, Représa de, *Brazil* **153 D6** 2 0 S 59 30W
Balboa, *Panama* **148 E4** 8 57N 79 34W
Balbriggan, *Ireland* **19 C5** 53 37N 6 11W
Balcarce, *Argentina* **158 D4** 38 0 S 58 10W
Balcarres, *Canada* **131 C8** 50 50N 103 35W
Balchik, *Bulgaria* **43 D13** 43 28N 28 11 E
Balclutha, *N.Z.* **119 G4** 46 15 S 169 45 E
Bald Hd., *Australia* **113 G2** 35 6 S 118 1 E
Bald I., *Australia* **113 F2** 34 57 S 118 27 E
Bald Knob, *U.S.A.* **139 H9** 35 19N 91 34W
Baldock L., *Canada* **131 B9** 56 33N 97 57W
Baldwin, *Fla., U.S.A.* **135 K4** 30 18N 81 59W
Baldwin, *Mich., U.S.A.* **134 D3** 43 54N 85 51W
Baldwinsville, *U.S.A.* **137 C8** 43 10N 76 20W
Baldy Peak, *U.S.A.* **143 K9** 33 54N 109 34W
Bale, *Croatia* **39 C10** 45 4N 13 46 E
Bale □, *Ethiopia* **95 F5** 6 20N 41 30 E
Baleares □, *Spain* **34 F7** 39 30N 3 0 E
Baleares, Is., *Spain* **33 B10** 39 30N 3 0 E
Balearic Is. = Baleares, Is., *Spain* **33 B10** 39 30N 3 0 E
Baleia, Pta. da, *Brazil* **155 E4** 17 40 S 39 7W
Balen, *Belgium* **21 F6** 51 10N 5 10 E
Băleni, *Romania* **46 D8** 45 48N 27 51 E
Baler, *Phil.* **70 D3** 15 46N 121 34 E
Baler Bay, *Phil.* **70 D3** 15 50N 121 35 E
Balerna, *Switz.* **29 E8** 45 52N 9 0 E
Baleshwar, *India* **82 D8** 21 35N 87 3 E
Balezino, *Russia* **52 B11** 58 2N 53 6 E
Balfate, *Honduras* **148 C2** 15 48N 86 25W
Balfe's Creek, *Australia* **114 C4** 20 12 S 145 55 E
Balfour Channel, *Solomon Is.* **121 M9** 8 43 S 157 27 E
Balharshah, *India* **82 E4** 19 50N 79 23 E
Bali, *Cameroon* **101 D6** 5 54N 10 0 E
Balí, *Greece* **32 D6** 35 25N 24 47 E
Bali, *Indonesia* **75 D4** 8 20 S 115 0 E
Bali □, *Indonesia* **75 D4** 8 20 S 115 0 E
Bali, Selat, *Indonesia* **75 D4** 8 18 S 114 25 E
Balicuatro Is., *Phil.* **70 E5** 12 39N 124 24 E
Baligród, *Poland* **31 B15** 49 20N 22 17 E
Balıkeşir, *Turkey* **88 C2** 39 35N 27 58 E
Balikpapan, *Indonesia* **75 C5** 1 10 S 116 55 E
Balimbing, *Phil.* **71 J2** 5 5N 119 58 E
Balimo, *Papua N. G.* **120 E2** 8 6 S 142 57 E
Baling, *Malaysia* **77 K3** 5 41N 100 55 E
Balintang Channel, *Phil.* **70 B3** 19 49N 121 40 E
Balintang Is., *Phil.* **70 B4** 19 58N 122 9 E
Baliton, *Phil.* **71 J5** 5 44N 125 14 E
Baliza, *Brazil* **157 D7** 16 0 S 52 20W
Baljurshi, *Si. Arabia* **86 C3** 19 51N 41 33 E
Balk, *Neths.* **20 C7** 52 54N 5 35 E
Balkan Mts. = Stara Planina, *Bulgaria* **43 D8** 43 15N 23 0 E
Balkh □, *Afghan.* **79 A2** 36 50N 67 0 E
Balkhash = Balqash, *Kazakhstan* **56 E8** 46 50N 74 50 E
Balkhash, Ozero = Balqash Köl, *Kazakhstan* **56 E8** 46 0N 74 50 E
Ballachulish, *U.K.* **18 E3** 56 41N 5 8W
Balladonia, *Australia* **113 F3** 32 27 S 123 51 E
Ballara, *Australia* **116 B4** 32 19 S 140 45 E
Ballarat, *Australia* **115 F3** 37 33 S 143 50 E
Ballard, L., *Australia* **113 E3** 29 20 S 120 40 E
Ballater, *U.K.* **18 D5** 57 3N 3 3W
Balldale, *Australia* **117 C7** 35 50 S 146 33 E
Ballenas, Canal de, *Mexico* **146 B2** 29 10N 113 45W
Balleny Is., *Antarctica* **7 C11** 66 30 S 163 0 E
Ballesteros, *Phil.* **70 B3** 18 25N 121 31 E
Ballia, *India* **81 G11** 25 46N 84 12 E
Ballidu, *Australia* **113 F2** 30 35 S 116 45 E
Ballina, *Australia* **115 D5** 28 50 S 153 31 E
Ballina, *Mayo, Ireland* **19 B2** 54 7N 9 9W
Ballina, *Tipp., Ireland* **19 D3** 52 49N 8 26W
Ballinasloe, *Ireland* **19 C3** 53 20N 8 13W
Ballinger, *U.S.A.* **139 K5** 31 45N 99 57W
Ballinrobe, *Ireland* **19 C2** 53 38N 9 13W
Ballinskelligs B., *Ireland* **19 E1** 51 48N 10 13W
Ballon, *France* **22 D7** 48 10N 0 14 E
Ballycastle, *U.K.* **19 A5** 55 12N 6 15W
Ballymena, *U.K.* **19 B5** 54 52N 6 17W
Ballymena □, *U.K.* **19 B5** 54 53N 6 18W
Ballymoney, *U.K.* **19 A5** 55 5N 6 31W
Ballymoney □, *U.K.* **19 A5** 55 5N 6 23W
Ballyshannon, *Ireland* **19 B3** 54 30N 8 11W
Balmaceda, *Chile* **160 C2** 46 0 S 71 50W
Balmazújváros, *Hungary* **31 D14** 47 37N 21 21 E
Balmhorn, *Switz.* **28 D5** 46 26N 7 42 E
Balmoral, *Australia* **116 D4** 37 15 S 141 48 E
Balmoral, *U.K.* **18 D5** 57 3N 3 13W
Balmorhea, *U.S.A.* **139 K3** 30 59N 103 45W
Balombo, *Angola* **103 E2** 12 21 S 14 46 E
Balonne →, *Australia* **115 D4** 28 47 S 147 56 E
Balqash, *Kazakhstan* **56 E8** 46 50N 74 50 E
Balqash Köl, *Kazakhstan* **56 E8** 46 0N 74 50 E
Balrampur, *India* **81 F10** 27 30N 82 20 E
Balranald, *Australia* **116 C5** 34 38 S 143 33 E
Balş, *Romania* **46 E5** 44 22N 24 5 E
Balsapuerto, *Peru* **156 B2** 5 48 S 76 33W
Balsas, *Mexico* **147 D5** 18 0N 99 40W
Balsas →, *Maranhão, Brazil* **154 C3** 7 15 S 44 35W
Balsas →, *Tocantins, Brazil* **154 C2** 9 58 S 47 52W
Balsas →, *Mexico* **146 D4** 17 55N 102 10W
Bålsta, *Sweden* **14 E11** 59 35N 17 30 E
Balsthal, *Switz.* **28 B5** 47 19N 7 41 E
Balston Spa, *U.S.A.* **137 D11** 43 0N 73 52W
Balta, *Romania* **46 E3** 44 54N 22 38 E
Balta, *Ukraine* **51 H5** 48 2N 29 45 E
Balta, *U.S.A.* **138 A4** 48 10N 100 2W
Baltanás, *Spain* **36 D6** 41 56N 4 15W
Bălţi, *Moldova* **51 J4** 47 48N 28 0 E
Baltic Sea, *Europe* **13 H18** 57 0N 19 0 E
Baltîm, *Egypt* **94 H7** 31 35N 31 10 E
Baltimore, *Ireland* **19 E2** 51 29N 9 22W
Baltimore, *U.S.A.* **134 F7** 39 17N 76 37W
Baltit, *Pakistan* **81 A6** 36 15N 74 40 E
Baltiysk, *Russia* **13 J18** 54 41N 19 58 E
Baltrum, *Germany* **26 B3** 53 43N 7 24 E
Baluchistan □, *Pakistan* **79 D2** 27 30N 65 0 E
Balud, *Phil.* **70 E4** 12 2N 123 12 E
Balurghat, *India* **81 G13** 25 15N 88 44 E
Balvi, *Latvia* **13 H22** 57 8N 27 15 E
Balya, *Turkey* **88 C2** 39 44N 27 35 E
Balygychan, *Russia* **57 C16** 63 56N 154 12 E
Balzar, *Ecuador* **152 D2** 2 2 S 79 54W
Bam, *Iran* **85 D8** 29 7N 58 14 E
Bama, *China* **68 E6** 24 8N 107 12 E
Bama, *Nigeria* **101 C7** 11 33N 13 41 E
Bamako, *Mali* **100 C3** 12 34N 7 55W
Bamba, *Mali* **101 B4** 17 5N 1 24W
Bamba, *Zaïre* **103 D3** 5 45 S 18 23 E
Bambam, *Phil.* **70 D3** 15 40N 120 20 E
Bambamarca, *Peru* **156 B2** 6 36 S 78 32W
Bambang, *Phil.* **70 C3** 16 23N 121 6 E
Bambari, *C.A.R.* **102 A4** 5 40N 20 35 E
Bambaroo, *Australia* **114 B4** 18 50 S 146 10 E
Bamberg, *Germany* **27 F6** 49 54N 10 54 E
Bamberg, *U.S.A.* **135 J5** 33 18N 81 2W
Bambesi, *Ethiopia* **95 F3** 9 45N 34 40 E
Bambey, *Senegal* **100 C1** 14 42N 16 28W
Bambili, *Zaïre* **106 B2** 3 40N 26 0 E
Bambuí, *Brazil* **155 F2** 20 1 S 45 58W
Bamenda, *Cameroon* **101 D7** 5 57N 10 11 E
Bamfield, *Canada* **130 D3** 48 45N 125 10W
Bāmīān □, *Afghan.* **79 B2** 35 0N 67 0 E
Bamiancheng, *China* **67 C13** 43 15N 124 2 E
Bamingui, *C.A.R.* **102 A4** 7 34N 20 11 E
Bamkin, *Cameroon* **101 D7** 6 3N 11 27 E
Bampūr, *Iran* **85 E9** 27 15N 60 21 E
Ban Aranyaprathet, *Thailand* **76 F4** 13 41N 102 30 E
Ban Ban, *Laos* **76 C4** 19 31N 103 30 E
Ban Bang Hin, *Thailand* **77 H2** 9 32N 98 35 E
Ban Chiang Klang, *Thailand* **76 C3** 19 25N 100 55 E
Ban Chik, *Laos* **76 D4** 17 15N 102 22 E
Ban Choho, *Thailand* **76 E4** 15 2N 102 9 E
Ban Dan Lan Hoi, *Thailand* **76 D2** 17 0N 99 35 E
Ban Don = Surat Thani, *Thailand* **77 H2** 9 6N 99 20 E
Ban Don, *Vietnam* **76 F6** 12 53N 107 48 E
Ban Don, Ao, *Thailand* **77 H2** 9 20N 99 25 E
Ban Dong, *Thailand* **76 C3** 19 30N 100 59 E
Ban Hong, *Thailand* **76 C2** 18 18N 98 50 E
Ban Kaeng, *Thailand* **76 D3** 17 29N 100 7 E
Ban Keun, *Laos* **76 C4** 18 22N 102 35 E
Ban Khai, *Thailand* **76 F3** 12 46N 101 18 E
Ban Kheun, *Laos* **76 B3** 20 13N 101 7 E
Ban Khlong Kua, *Thailand* **77 J3** 6 57N 100 8 E
Ban Khuan Mao, *Thailand* **77 J2** 7 50N 99 37 E
Ban Khun Yuam, *Thailand* **76 C1** 18 49N 97 57 E
Ban Ko Yai Chim, *Thailand* **77 G2** 11 17N 99 26 E
Ban Kok, *Thailand* **76 D4** 16 40N 103 40 E
Ban Laem, *Thailand* **76 F2** 13 13N 99 59 E
Ban Lao Ngam, *Laos* **76 E6** 15 28N 106 10 E
Ban Le Kathe, *Thailand* **76 E2** 15 49N 98 53 E
Ban Mae Chedi, *Thailand* **76 C2** 19 11N 99 31 E
Ban Mae Laeng, *Thailand* **76 B2** 20 1N 99 17 E
Ban Mae Sariang, *Thailand* **76 C1** 18 10N 97 56 E
Ban Mê Thuôt = Buon Me Thuot, *Vietnam* **76 F7** 12 40N 108 3 E
Ban Mi, *Thailand* **76 E3** 15 3N 100 32 E
Ban Muong Mo, *Laos* **76 C4** 19 4N 103 58 E
Ban Na Mo, *Laos* **76 D5** 17 7N 105 40 E
Ban Na San, *Thailand* **77 H2** 8 53N 99 52 E
Ban Na Tong, *Laos* **76 B3** 20 56N 101 47 E
Ban Nam Bac, *Laos* **76 B4** 20 38N 102 20 E
Ban Nam Ma, *Laos* **76 A3** 22 2N 101 37 E
Ban Ngang, *Laos* **76 E6** 15 59N 106 11 E
Ban Nong Bok, *Laos* **76 D5** 17 5N 104 48 E
Ban Nong Boua, *Laos* **76 E6** 15 40N 106 33 E
Ban Nong Pling, *Thailand* **76 E3** 15 40N 100 10 E
Ban Pak Chan, *Thailand* **77 G2** 10 32N 98 51 E
Ban Phai, *Thailand* **76 D4** 16 4N 102 44 E
Ban Pong, *Thailand* **76 F2** 13 50N 99 55 E
Ban Ron Phibun, *Thailand* **77 H2** 8 9N 99 51 E
Ban Sanam Chai, *Thailand* **77 J3** 7 33N 100 25 E
Ban Sangkha, *Thailand* **76 E4** 14 37N 103 52 E
Ban Tak, *Thailand* **76 D2** 17 2N 99 4 E
Ban Tako, *Thailand* **76 E4** 14 5N 102 40 E
Ban Tha Dua, *Thailand* **76 D2** 17 59N 98 39 E
Ban Tha Li, *Thailand* **76 D3** 17 37N 101 25 E
Ban Tha Nun, *Thailand* **77 H2** 8 12N 98 18 E
Ban Thahine, *Laos* **76 E5** 14 12N 105 33 E
Ban Xien Kok, *Laos* **76 B3** 20 54N 100 39 E
Ban Yen Nhan, *Vietnam* **76 B6** 20 57N 106 2 E
Baña, Punta de la, *Spain* **34 E5** 40 33N 0 40 E
Banā, W. →, *Yemen* **86 D4** 13 3N 45 24 E
Banaba, *Kiribati* **122 H8** 0 45 S 169 50 E
Bañalbufar, *Spain* **33 B9** 39 42N 2 31 E
Banalia, *Zaïre* **106 B2** 1 32N 25 5 E
Banam, *Cambodia* **77 G5** 11 20N 105 17 E
Banamba, *Mali* **100 C3** 13 29N 7 22W
Banana, *Australia* **114 C5** 24 28 S 150 8 E
Bananal, I. do, *Brazil* **155 D1** 11 30 S 50 30W
Banaras = Varanasi, *India* **81 G10** 25 22N 83 0 E
Banas →, *Gujarat, India* **80 H4** 23 45N 71 25 E
Banas →, *Mad. P., India* **81 G9** 24 15N 81 30 E
Bânâs, Ras, *Egypt* **94 C4** 23 57N 35 50 E
Banaz, *Turkey* **88 C3** 38 44N 29 46 E
Banbān, *Si. Arabia* **84 E5** 25 1N 46 35 E
Banbridge, *U.K.* **19 B5** 54 22N 6 16W
Banbridge □, *U.K.* **19 B5** 54 21N 6 16W
Banbury, *U.K.* **17 E6** 52 4N 1 20W
Banchory, *U.K.* **18 D6** 57 3N 2 29W
Bancroft, *Canada* **128 C4** 45 3N 77 51W
Band, *Romania* **46 C5** 46 30N 24 25 E
Band Bonī, *Iran* **85 E8** 25 30N 59 33 E
Band-e Torkestān, *Afghan.* **79 B2** 35 30N 64 0 E
Band Qīr, *Iran* **85 D6** 31 39N 48 53 E
Banda, *Cameroon* **102 B2** 3 58N 14 32 E
Banda, *India* **81 G9** 25 30N 80 26 E
Banda, Kepulauan, *Indonesia* **73 B3** 4 37 S 129 50 E
Banda Aceh, *Indonesia* **74 A1** 5 35N 95 20 E
Banda Banda, Mt., *Australia* **117 A10** 31 10 S 152 28 E
Banda Elat, *Indonesia* **73 C4** 5 40 S 133 5 E
Banda Is. = Banda, Kepulauan, *Indonesia* **73 B3** 4 37 S 129 50 E
Banda Sea, *Indonesia* **72 C3** 6 0 S 130 0 E
Bandai-San, *Japan* **60 F10** 37 36N 140 4 E
Bandama →, *Ivory C.* **100 D3** 6 32N 5 30W
Bandān, *Iran* **85 D9** 31 23N 60 44 E
Bandanaira, *Indonesia* **73 B3** 4 32 S 129 54 E
Bandanwara, *India* **80 F6** 26 9N 74 38 E
Bandar = Machilipatnam, *India* **83 F5** 16 12N 81 8 E
Bandār 'Abbās, *Iran* **85 E8** 27 15N 56 15 E
Bandar-e Anzalī, *Iran* **89 D13** 37 30N 49 30 E
Bandar-e Bushehr = Būshehr, *Iran* **85 D6** 28 55N 50 55 E
Bandar-e Chārak, *Iran* **85 E7** 26 45N 54 20 E
Bandar-e Deylam, *Iran* **85 D6** 30 5N 50 10 E
Bandar-e Khomeynı, *Iran* **85 D6** 30 30N 49 5 E
Bandar-e Lengeh, *Iran* **85 E7** 26 35N 54 58 E
Bandar-e Maqām, *Iran* **85 E7** 26 56N 53 29 E
Bandar-e Ma'shur, *Iran* **85 D6** 30 35N 49 10 E
Bandar-e Nakhīlū, *Iran* **85 E7** 26 58N 53 30 E
Bandar-e Rīg, *Iran* **85 D6** 29 29N 50 38 E
Bandar-e Torkeman, *Iran* **85 B7** 37 0N 54 10 E
Bandar Maharani = Muar, *Malaysia* **77 L4** 2 3N 102 34 E
Bandar Penggaram = Batu Pahat, *Malaysia* **77 M4** 1 50N 102 56 E
Bandar Seri Begawan, *Brunei* **75 B4** 4 52N 115 0 E
Bandawe, *Malawi* **107 E3** 11 58 S 34 5 E
Bande, *Belgium* **21 H6** 50 10N 5 25 E
Bande, *Spain* **36 C3** 42 3N 7 58W
Bandeira, Pico da, *Brazil* **155 F3** 20 26 S 41 47W
Bandeirante, *Brazil* **155 D1** 13 41 S 50 48W
Bandera, *Argentina* **158 B3** 28 55 S 62 20W
Bandera, *U.S.A.* **139 L5** 29 44N 99 5W
Banderas, B. de, *Mexico* **146 C3** 20 40N 105 30W
Bandia →, *India* **82 E5** 19 2N 80 28 E
Bandiagara, *Mali* **100 C4** 14 12N 3 29W
Bandırma, *Turkey* **88 B3** 40 20N 28 0 E
Bandon, *Ireland* **19 E3** 51 44N 8 44W
Bandon →, *Ireland* **19 E3** 51 43N 8 37W
Bandoua, *C.A.R.* **102 B4** 4 39N 21 42 E
Bandula, *Mozam.* **107 F3** 19 0 S 33 7 E
Bandundu, *Zaïre* **102 C3** 3 15 S 17 22 E
Bandung, *Indonesia* **75 D3** 6 54 S 107 36 E
Bandya, *Australia* **113 E3** 27 40 S 122 5 E
Băneasa, *Romania* **46 D8** 45 56N 27 55 E
Bāneh, *Iran* **89 E11** 35 59N 45 53 E
Bañeres, *Spain* **35 G4** 38 44N 0 38W
Banes, *Cuba* **149 B4** 21 0N 75 42W
Banff, *Canada* **130 C5** 51 10N 115 34W
Banff, *U.K.* **18 D6** 57 40N 2 33W
Banff Nat. Park, *Canada* **130 C5** 51 30N 116 15W
Banfora, *Burkina Faso* **100 C4** 10 40N 4 40W
Bang Fai →, *Laos* **76 D5** 16 57N 104 45 E
Bang Hieng →, *Laos* **76 D5** 16 10N 105 10 E
Bang Krathum, *Thailand* **76 D3** 16 34N 100 18 E
Bang Lamung, *Thailand* **76 F3** 13 3N 100 56 E
Bang Mun Nak, *Thailand* **76 D3** 16 2N 100 23 E
Bang Pa In, *Thailand* **76 E3** 14 14N 100 35 E
Bang Rakam, *Thailand* **76 D3** 16 45N 100 7 E
Bang Saphan, *Thailand* **77 G2** 11 14N 99 28 E
Bangala Dam, *Zimbabwe* **107 G3** 21 7 S 31 25 E
Bangalore, *India* **83 H3** 12 59N 77 40 E
Bangante, *Cameroon* **101 D7** 5 8N 10 32 E
Bangaon, *India* **81 H13** 23 0N 88 47 E
Bangassou, *C.A.R.* **102 B4** 4 55N 23 7 E
Bangeta, Mt., *Papua N. G.* **120 D4** 6 21 S 147 3 E
Banggai, Kepulauan, *Indonesia* **72 B2** 1 40 S 123 30 E
Banggai Arch., *Indonesia* **72 B2** 2 0 S 123 15 E
Banggi, P., *Malaysia* **75 A5** 7 17N 117 12 E
Banghāzī, *Libya* **96 B4** 32 11N 20 3 E
Banghāzī □, *Libya* **96 B4** 32 7N 20 4 E
Bangil, *Indonesia* **75 D4** 7 36 S 112 50 E
Bangjang, *Sudan* **95 E3** 11 23N 32 41 E
Bangka, P., *Sulawesi, Indonesia* **72 A3** 1 50N 125 5 E
Bangka, P., *Sumatera, Indonesia* **74 C3** 2 0 S 105 50 E
Bangka, Selat, *Indonesia* **74 C3** 2 30 S 105 30 E
Bangkalan, *Indonesia* **75 D4** 7 2 S 112 46 E
Bangkinang, *Indonesia* **74 B2** 0 18N 101 5 E
Bangko, *Indonesia* **74 C2** 2 5 S 102 9 E
Bangkok, *Thailand* **76 F3** 13 45N 100 35 E
Bangladesh ■, *Asia* **78 C3** 24 0N 90 0 E
Bangolo, *Ivory C.* **100 D3** 7 1N 7 29W
Bangong Co, *India* **81 B8** 35 50N 79 20 E
Bangor, *Down, U.K.* **19 B6** 54 40N 5 40W
Bangor, *Gwynedd, U.K.* **16 D3** 53 14N 4 8W
Bangor, *Maine, U.S.A.* **129 D6** 44 48N 68 46W
Bangor, *Mich., U.S.A.* **141 B10** 42 18N 86 7W
Bangor, *Pa., U.S.A.* **137 F9** 40 52N 75 13W
Bangu, *Zaïre* **102 C3** 0 3 S 19 12 E
Bangued, *Phil.* **70 C3** 17 40N 120 37 E
Bangui, *C.A.R.* **102 B3** 4 23N 18 35 E
Bangui, *Phil.* **70 B3** 18 32N 120 46 E
Banguru, *Zaïre* **106 B2** 0 30N 27 10 E
Bangweulu, L., *Zambia* **107 E3** 11 0 S 30 0 E
Bangweulu Swamp, *Zambia* **107 E3** 11 20 S 30 15 E
Bani, *Dom. Rep.* **149 C5** 18 16N 70 22W
Bani, *Phil.* **70 C2** 16 11N 119 52 E
Bani →, *Mali* **100 C4** 14 30N 4 12W
Bani, Djebel, *Morocco* **98 C3** 29 16N 8 0W
Bani Bangou, *Niger* **101 B5** 15 3N 2 42 E
Banī Sa'd, *Iraq* **89 F11** 33 34N 44 32 E
Banī Sār, *Si. Arabia* **86 B3** 20 6N 41 27 E
Banī Walīd, *Libya* **96 B2** 31 36N 13 53 E
Bania, *Ivory C.* **100 D4** 9 4N 3 6W
Baniara, *Papua N. G.* **120 E5** 9 44 S 149 54 E
Banihal Pass, *India* **81 C6** 33 30N 75 12 E
Banīnah, *Libya* **96 B4** 32 0N 20 12 E
Bāniyās, *Syria* **88 E6** 35 10N 36 0 E
Banja Luka, *Bos.-H.* **42 C2** 44 49N 17 11 E
Banjar, *Indonesia* **75 D3** 7 24 S 108 30 E
Banjarmasin, *Indonesia* **75 C4** 3 20 S 114 35 E
Banjarnegara, *Indonesia* **75 D3** 7 24 S 109 42 E
Banjul, *Gambia* **100 C1** 13 28N 16 40W
Banka Banka, *Australia* **114 B1** 18 50 S 134 0 E
Banket, *Zimbabwe* **107 F3** 17 27 S 30 19 E
Bankilaré, *Niger* **101 C5** 14 35N 0 44 E
Bankipore, *India* **81 G11** 25 35N 85 10 E
Banks, Is., *Vanuatu* **121 C5** 13 50 S 167 30 E
Banks I., *B.C., Canada* **130 C3** 53 20N 130 0W
Banks I., *N.W.T., Canada* **126 A7** 73 15N 121 30W
Banks I., *Papua N. G.* **120 F2** 10 10 S 142 15 E
Banks Pen., *N.Z.* **119 D8** 43 45 S 173 15 E
Banks Str., *Australia* **114 G4** 40 40 S 148 10 E
Bankura, *India* **81 H12** 23 11N 87 18 E
Bankya, *Bulgaria* **42 E8** 42 43N 23 8 E
Bann →, *L'derry., U.K.* **19 A5** 55 8N 6 41W
Bann →, *Tyrone, U.K.* **19 B5** 54 30N 6 31W
Banna, *Phil.* **70 C3** 17 59N 120 39 E
Bannalec, *France* **22 E3** 47 57N 3 42W
Bannang Sata, *Thailand* **77 J3** 6 16N 101 16 E
Bannerton, *Australia* **116 C5** 34 42 S 142 47 E
Banning, *U.S.A.* **145 M10** 33 56N 116 53W
Banningville = Bandundu, *Zaïre* **102 C3** 3 15 S 17 22 E
Bannockburn, *Canada* **136 B7** 44 39N 77 33W
Bannockburn, *U.K.* **18 E5** 56 5N 3 55W
Bannockburn, *Zimbabwe* **107 G2** 20 17 S 29 48 E
Bannu, *Pakistan* **79 B3** 33 0N 70 18 E
Bañolas, *Spain* **34 C7** 42 16N 2 44 E
Banon, *France* **25 D9** 44 2N 5 38 E
Baños de la Encina, *Spain* **37 G7** 38 10N 3 46W
Baños de Molgas, *Spain* **36 C3** 42 15N 7 40W
Bánovce, *Slovak Rep.* **31 C11** 48 44N 18 16 E
Bansilan □, *Phil.* **71 H3** 6 40N 121 40 E
Banská Bystrica, *Slovak Rep.* **31 C12** 48 46N 19 14 E
Banská Štiavnica, *Slovak Rep.* **31 C11** 48 25N 18 55 E
Bansko, *Bulgaria* **43 F8** 41 52N 23 28 E
Banswara, *India* **80 H6** 23 32N 74 24 E
Bantayan, *Phil.* **71 F4** 11 10N 123 43 E
Bantayan I., *Phil.* **71 F4** 11 13N 123 44 E
Banten, *Indonesia* **74 D3** 6 5 S 106 8 E
Banton I., *Phil.* **70 E4** 12 56N 122 4 E
Bantry, *Ireland* **19 E2** 51 41N 9 27W
Bantry B., *Ireland* **19 E2** 51 37N 9 44W
Bantul, *Indonesia* **75 D4** 7 55 S 110 19 E
Bantva, *India* **80 J4** 21 29N 70 12 E
Bantval, *India* **83 H2** 12 55N 75 0 E
Banya, *Bulgaria* **43 E9** 42 33N 24 50 E
Banyak, Kepulauan, *Indonesia* **74 B1** 2 10N 97 10 E
Banyo, *Cameroon* **101 D7** 6 52N 11 45 E
Banyuls-sur-Mer, *France* **24 F7** 42 28N 3 8 E
Banyumas, *Indonesia* **75 D3** 7 32 S 109 18 E
Banyuwangi, *Indonesia* **75 D4** 8 13 S 114 21 E
Banzare Coast, *Antarctica* **7 C9** 68 0 S 125 0 E
Banzyville = Mobayi, *Zaïre* **102 B4** 4 15N 21 8 E
Bao Ha, *Vietnam* **76 A5** 22 11N 104 21 E
Bao Lac, *Vietnam* **76 A5** 22 57N 105 40 E
Bao Loc, *Vietnam* **77 G6** 11 32N 107 48 E
Bao'an = Shenzhen, *China* **69 F10** 22 27N 114 10 E
Baocheng, *China* **66 H4** 33 12N 106 56 E
Baode, *China* **66 E6** 39 1N 111 5 E
Baodi, *China* **67 E9** 39 38N 117 20 E
Baoding, *China* **66 E8** 38 50N 115 28 E
Baoji, *China* **66 G4** 34 20N 107 5 E
Baojing, *China* **68 C7** 28 45N 109 41 E
Baokang, *China* **69 B8** 31 54N 111 12 E
Baoro, *C.A.R.* **102 A3** 5 40N 15 58 E
Baoshan, *Shanghai, China* **69 B13** 31 27N 121 26 E
Baoshan, *Yunnan, China* **68 E2** 25 10N 99 5 E
Baotou, *China* **66 D6** 40 32N 110 2 E
Baoying, *China* **67 H10** 33 17N 119 20 E
Bap, *India* **80 F5** 27 23N 72 18 E
Bapatla, *India* **83 G5** 15 55N 80 30 E
Bapaume, *France* **23 B9** 50 7N 2 50 E
Bāqerābād, *Iran* **85 C6** 33 2N 51 58 E
Ba'qūbah, *Iraq* **89 F11** 33 45N 44 50 E
Baquedano, *Chile* **158 A2** 23 20 S 69 52W
Bar, *Montenegro, Yug.* **42 E4** 42 8N 19 8 E
Bar, *Ukraine* **51 H4** 49 4N 27 40 E
Bar Bigha, *India* **81 G11** 25 21N 85 47 E
Bar Harbor, *U.S.A.* **129 D6** 44 23N 68 13W
Bar-le-Duc, *France* **23 D12** 48 47N 5 10 E
Bar-sur-Aube, *France* **23 D11** 48 14N 4 40 E
Bar-sur-Seine, *France* **23 D11** 48 7N 4 20 E
Barabai, *Indonesia* **75 C5** 2 32 S 115 34 E
Barabinsk, *Russia* **56 D8** 55 20N 78 20 E
Baraboo, *U.S.A.* **138 D10** 43 28N 89 45W
Baracaldo, *Spain* **34 B2** 43 18N 2 59W
Baracoa, *Cuba* **149 B5** 20 20N 74 30W
Baradero, *Argentina* **158 C4** 33 52 S 59 29W
Baradine, *Australia* **117 A8** 30 56 S 149 4 E
Baraga, *U.S.A.* **138 B10** 46 47N 88 30W
Barahona, *Dom. Rep.* **149 C5** 18 13N 71 7W
Barahona, *Spain* **34 D2** 41 17N 2 39W
Baraka →, *Sudan* **94 D4** 18 13N 37 35 E
Barakot, *India* **81 J11** 21 33N 84 59 E
Barakpur, *India* **81 H13** 22 44N 88 30 E
Barakula, *Australia* **115 D5** 26 30 S 150 33 E
Baralaba, *Australia* **114 C4** 24 13 S 149 50 E
Baralzon L., *Canada* **131 B9** 60 0N 98 3W
Baram →, *Malaysia* **75 B4** 4 35N 113 58 E
Baramati, *India* **82 E2** 18 11N 74 33 E
Baramba, *India* **82 D7** 20 25N 85 23 E
Barameiya, *Sudan* **94 D4** 18 32N 36 38 E
Baramula, *India* **81 B6** 34 15N 74 20 E
Baran, *India* **80 G7** 25 9N 76 40 E
Baranavichy, *Belarus* **51 F4** 53 10N 26 0 E
Baranoa, *Colombia* **152 A3** 10 48N 74 55W
Baranof I., *U.S.A.* **130 B1** 57 0N 135 0W
Baranów Sandomierski, *Poland* **47 E8** 50 29N 21 30 E
Baranya □, *Hungary* **31 F11** 46 0N 18 15 E
Barão de Cocais, *Brazil* **155 E3** 19 56 S 43 28W
Barão de Grajaú, *Brazil* **154 C3** 6 45 S 43 1W
Barão de Melgaço, *Mato Grosso, Brazil* **157 D6** 16 14 S 55 52W
Barão de Melgaço, *Rondônia, Brazil* **157 C5** 11 50 S 60 45W
Baraolt, *Romania* **46 C6** 46 5N 25 34 E
Barapasi, *Indonesia* **73 B5** 2 15 S 137 5 E

174 Ben Bullen

Ben Bullen, Australia ... 117 B9 33 12 S 150 2 E
Ben Cruachan, U.K. 18 E3 56 26N 5 8W
Ben Dearg, U.K. 18 D4 57 47N 4 56W
Ben Gardane, Tunisia ... 96 B2 33 11N 11 11 E
Ben Hope, U.K. 18 C4 58 25N 4 36W
Ben Lawers, U.K. 18 E4 56 32N 4 14W
Ben Lomond, N.S.W., Australia .... 115 E5 30 1 S 151 43 E
Ben Lomond, Tas., Australia .... 114 G4 41 38 S 147 42 E
Ben Lomond, U.K. 18 E4 56 11N 4 38W
Ben Luc, Vietnam .... 77 G6 10 39N 106 29 E
Ben Macdhui, U.K. 18 D5 57 4N 3 40W
Ben Mhor, U.K. 18 D1 57 15N 7 18W
Ben More, Central, U.K. 18 E4 56 23N 4 32W
Ben More, Strath., U.K. . 18 E2 56 26N 6 1W
Ben More Assynt, U.K. . 18 C4 58 8N 4 52W
Ben Nevis, U.K. 18 E4 56 48N 5 1W
Ben Ohau Ra., N.Z. .... 119 E5 44 1 S 170 4 E
Ben Quang, Vietnam .... 76 D6 17 3N 106 55 E
Ben Slimane, Morocco .. 98 B3 33 38N 7 7W
Ben Tre, Vietnam .... 77 G6 10 3N 106 36 E
Ben Vorlich, U.K. 18 E4 56 21N 4 14W
Ben Wyvis, U.K. 18 D4 57 40N 4 35W
Bena, Nigeria .... 101 C6 11 20N 5 50 E
Bena Dibele, Zaïre ...... 103 C4 4 4 S 22 50 E
Bena-Leka, Zaïre .... 103 D4 5 8 S 22 10 E
Bena-Tshadi, Zaïre ..... 103 C4 4 40 S 22 49 E
Benāb, Iran .... 89 D12 37 20N 46 4 E
Benadir, Somali Rep. .... 108 D2 1 30N 44 30 E
Benagalbón, Spain 37 J6 36 45N 4 15W
Benagerie, Australia ..... 116 A4 31 25 S 140 22 E
Benahmed, Morocco 98 B3 33 4N 7 9W
Benalla, Australia ...... 117 D7 36 30 S 146 0 E
Benambra, Mt., Australia 117 D7 36 31 S 147 34 E
Benamejí, Spain 37 H6 37 16N 4 33W
Benares = Varanasi, India 81 G10 25 22N 83 0 E
Bénat, C., France ...... 25 E10 43 5N 6 22 E
Benavente, Portugal 37 G2 38 59N 8 49W
Benavente, Spain 36 C5 42 2N 5 43W
Benavides, Spain 36 C5 42 30N 5 54W
Benavides, U.S.A. 139 M5 27 36N 98 25W
Benbecula, U.K. 18 D1 57 26N 7 21W
Benbonyathe, Australia . 116 A3 30 25 S 139 11 E
Bencubbin, Australia ... 113 F2 30 48 S 117 52 E
Bend, U.S.A. 142 D3 44 4N 121 19W
Bendela, Zaïre .... 102 C3 3 18 S 17 36 E
Bender Beila, Somali Rep. 108 C4 9 30N 50 48 E
Bender Merchagno, Somali Rep. .... 108 B4 11 41N 50 34 E
Bendering, Australia .... 113 F2 32 23 S 118 18 E
Bendery = Tighina, Moldova .... 51 J5 46 50N 29 30 E
Bendigo, Australia ...... 116 D6 36 40 S 144 15 E
Bendorf, Germany ...... 26 E3 50 25N 7 35 E
Benē Beraq, Israel ...... 91 C3 32 6N 34 51 E
Beneden Knijpe, Neths. . 20 C7 52 58N 5 59 E
Beneditinos, Brazil ..... 154 C3 5 27 S 42 22W
Benedito Leite, Brazil ... 154 C3 7 13 S 44 34W
Bénéna, Mali 100 C4 13 9N 4 17W
Benenitra, Madag. ...... 105 C8 23 27 S 45 5 E
Benešŏv, Czech. .... 30 B7 49 46N 14 41 E
Bénestroff, France ...... 23 D13 48 54N 6 45 E
Benet, France 24 B3 46 22N 0 35W
Benevento, Italy .... 41 A7 41 8N 14 45 E
Benfeld, France .... 23 D14 48 22N 7 34 E
Benga, Mozam. .... 107 F3 16 11 S 33 40 E
Bengal, Bay of, Ind. Oc. 58 H12 15 0N 90 0 E
Bengbu, China .... 67 H9 32 58N 117 20 E
Benghazi = Banghāzī, Libya .... 96 B4 32 11N 20 3 E
Bengkalis, Indonesia .... 74 B2 1 30N 102 10 E
Bengkulu, Indonesia .... 74 C2 3 50 S 102 12 E
Bengkulu □, Indonesia .. 74 C2 3 48 S 102 16 E
Bengough, Canada 131 D7 49 25N 105 10W
Benguela, Angola .... 103 E2 12 37 S 13 25 E
Benguela □, Angola ..... 103 E2 13 0 S 13 30 E
Benguerir, Morocco 98 B3 32 16N 7 56W
Benguérua, I., Mozam. . 105 C6 21 58 S 35 28 E
Benguet □, Phil. .... 70 C3 16 30N 120 40 E
Benha, Egypt .... 94 H7 30 26N 31 8 E
Beni, Zaïre .... 106 B2 0 30N 29 27 E
Beni □, Bolivia .... 157 C4 14 0 S 65 0W
Beni →, Bolivia .... 157 C4 10 23 S 65 24W
Beni Abbès, Algeria 99 B4 30 5N 2 5W
Beni-Haoua, Algeria .... 99 A5 36 30N 1 30 E
Beni Mazâr, Egypt ...... 94 J7 28 32N 30 44 E
Beni Mellal, Morocco ... 98 B3 32 21N 6 21W
Beni Ounif, Algeria 99 B4 32 0N 1 10W
Beni Saf, Algeria 99 A4 35 17N 1 15W
Beni Suef, Egypt .... 94 J7 29 5N 31 6 E
Beniah L., Canada 130 A6 63 23N 112 17W
Benicarló, Spain .... 34 E5 40 23N 0 23 E
Benicia, U.S.A. 144 G4 38 3N 122 9W
Benidorm, Spain 35 G4 38 33N 0 9W
Benidorm, Islote de, Spain 35 G4 38 31N 0 9W
Benin ■, Africa .... 101 D5 10 0N 2 0 E
Benin, Bight of, W. Afr. . 101 D5 5 0N 3 0 E
Benin City, Nigeria ..... 101 D6 6 20N 5 31 E
Benisa, Spain .... 35 G5 38 43N 0 3 E
Benitses, Greece .... 32 A3 39 32N 19 55 E
Benjamin Aceval, Paraguay .... 158 A4 24 58 S 57 34W
Benjamin Constant, Brazil 152 D3 4 40 S 70 15W
Benjamin Hill, Mexico ... 146 A2 30 10N 111 10W
Benkelman, U.S.A. 138 E4 40 3N 101 32W
Benkovac, Croatia ...... 39 D12 44 2N 15 37 E
Benlidi, Australia ...... 114 C3 24 35 S 144 50 E
Benmore Pk., N.Z. ..... 119 E5 44 25 S 170 8 E
Bennebroek, Neths. ..... 20 D5 52 19N 4 36 E
Bennekom, Neths. ...... 20 D7 52 0N 5 41 E
Bennett, Canada 130 B2 59 56N 134 53W
Bennett, L., Australia ... 112 D5 22 50 S 131 2 E
Bennett, Ostrov, Russia . 57 B15 76 21N 148 56 E
Bennettsville, U.S.A. ... 135 H6 34 37N 79 41W
Bennington, U.S.A. 137 D11 43 0N 71 55W
Bénodet, France 22 E2 47 53N 4 7W
Benoni, S. Africa ...... 105 D4 26 11 S 28 18 E
Benoud, Algeria .... 99 B5 32 20N 0 16 E
Benoy, Chad .... 97 G3 8 59N 16 19 E
Benque Viejo, Belize 147 D7 17 5N 89 8W
Bensheim, Germany ..... 27 F4 49 40N 8 38 E
Benson, U.S.A. 143 L8 31 58N 110 18W
Bent, Iran .... 85 E8 26 20N 59 31 E
Benteng, Indonesia ..... 72 C2 6 10 S 120 30 E
Bentinck I., Australia ... 114 B2 17 3 S 139 35 E
Bentiu, Sudan .... 95 F2 9 10N 29 55 E
Bento Gonçalves, Brazil . 159 B5 29 10 S 51 31W
Benton, Ark., U.S.A. ... 139 H8 34 34N 92 35W
Benton, Calif., U.S.A. .. 144 H8 37 48N 118 32W
Benton, Ill., U.S.A. 140 G8 38 0N 88 55W
Benton Harbor, U.S.A. .. 141 B10 42 6N 86 27W
Bentu Liben, Ethiopia ... 95 F4 8 32N 38 21 E
Bentung, Malaysia ...... 77 L3 3 31N 101 55 E
Benue □, Nigeria .... 101 D6 7 20N 8 45 E
Benue →, Nigeria ...... 101 D6 7 48N 6 46 E
Benxi, China .... 67 D12 41 20N 123 48 E
Benzdorp, Surinam 153 C7 3 44N 54 5W
Beo, Indonesia .... 72 A3 4 25N 126 50 E
Beograd, Serbia, Yug. ... 42 C5 44 50N 20 37 E
Beowawe, U.S.A. 142 F5 40 35N 116 29W
Bepan Jiang →, China .. 68 E6 24 55N 106 5 E
Beppu, Japan .... 62 D3 33 15N 131 30 E
Beppu-Wan, Japan ..... 62 D3 33 18N 131 34 E
Beqaa Valley = Al Biqā □, Lebanon ..... 91 A5 34 10N 36 10 E
Bera, Bangla. .... 78 C2 24 5N 89 37 E
Berati, Albania .... 44 D1 40 43N 19 59 E
Berau →, Indonesia .... 75 B5 2 10N 117 42 E
Berau, Teluk, Indonesia . 73 B4 2 30 S 132 30 E
Berber, Sudan .... 94 D3 18 0N 34 0 E
Berbera, Somali Rep. .... 108 B3 10 30N 45 2 E
Berbérati, C.A.R. ...... 102 B3 4 15N 15 40 E
Berberia, C. del, Spain .. 33 C7 38 39N 1 24 E
Berbice →, Guyana 153 B6 6 20N 57 32W
Berceto, Italy .... 38 D7 44 31N 9 51 E
Berchtesgaden, Germany 27 H8 47 38N 13 0 E
Berck-Plage, France .... 23 B8 50 25N 1 36 E
Berdale, Somali Rep. ... 108 C3 7 4N 47 51 E
Berdichev = Berdychiv, Ukraine .... 51 H5 49 57N 28 30 E
Berdsk, Russia .... 56 D9 54 47N 83 2 E
Berdyansk, Ukraine .... 51 J9 46 45N 36 50 E
Berdyaush, Russia ..... 54 D7 55 9N 59 9 E
Berdychiv, Ukraine ..... 51 H5 49 57N 28 30 E
Berea, U.S.A. 134 G3 37 34N 84 17W
Berebere, Indonesia .... 72 A3 2 25N 128 45 E
Bereda, Somali Rep. .... 108 B4 11 45N 51 0 E
Berehove, Ukraine ..... 51 H2 48 15N 22 35 E
Bereina, Papua N. G. ... 120 E4 8 39 S 146 30 E
Berekum, Ghana 100 D4 7 29N 2 34W
Berenice, Egypt .... 94 C4 24 2N 35 25 E
Berens →, Canada 131 C9 52 25N 97 2W
Berens I., Canada 131 C9 52 18N 97 18W
Berens River, Canada ... 131 C9 52 25N 97 0W
Berestechko, Ukraine ... 51 G3 50 22N 25 5 E
Bereşti, Romania ...... 46 C8 46 6N 27 50 E
Beretău →, Romania .... 46 B2 47 10N 21 50 E
Berettyo →, Hungary ... 31 E14 46 59N 21 7 E
Berettyóújfalu, Hungary . 31 D14 47 13N 21 33 E
Berevo, Mahajanga, Madag. .... 105 B7 17 14 S 44 17 E
Berevo, Toliara, Madag. . 105 B7 19 44 S 44 58 E
Bereza, Belarus .... 51 F3 52 31N 24 51 E
Berezhany, Ukraine .... 51 H3 49 26N 24 58 E
Berezina = Byarezina →, Belarus .... 51 F6 52 33N 30 14 E
Berezivka, Ukraine .... 51 J6 47 14N 30 55 E
Berezna, Ukraine ...... 51 G6 51 35N 31 46 E
Berezniki, Russia ..... 54 B6 59 24N 56 46 E
Berezovo, Russia ..... 48 B11 64 0N 65 0 E
Berga, Spain .... 34 C6 42 6N 1 48 E
Bergama, Turkey ...... 88 C2 39 8N 27 15 E
Bergambacht, Neths. ... 20 E5 51 56N 4 48 E
Bérgamo, Italy .... 38 C6 45 41N 9 43 E
Bergantiños, Spain 36 B2 43 20N 8 40W
Bergara, Spain 34 B2 43 9N 2 28W
Bergedorf, Germany ... 26 B6 53 28N 10 6 E
Bergeijk, Neths. ...... 21 F6 51 19N 5 21 E
Bergen, Germany ..... 26 A9 54 25N 13 25 E
Bergen, Neths. .... 20 C5 52 40N 4 43 E
Bergen, Norway ...... 13 F11 60 20N 5 20 E
Bergen, U.S.A. 136 C7 43 5N 77 57W
Bergen-op-Zoom, Neths. . 21 F4 51 28N 4 18 E
Bergerac, France ..... 24 D4 44 51N 0 30 E
Bergheim, Germany .... 26 E2 50 57N 6 38 E
Berghem, Neths. ...... 20 E7 51 46N 5 33 E
Bergisch Gladbach, Germany .... 26 E3 50 59N 7 8 E
Bergschenhoek, Neths. . 20 E5 51 59N 4 30 E
Bergsjö, Sweden ...... 14 C11 61 59N 17 3 E
Bergues, France ...... 23 B9 50 58N 2 24 E
Bergum, Neths. .... 20 B7 53 13N 5 59 E
Bergville, S. Africa .... 105 D4 28 52 S 29 18 E
Berhala, Selat, Indonesia 74 C2 1 0 S 104 15 E
Berhampore = Baharampur, India .... 81 G13 24 2N 88 27 E
Berhampur, India ..... 82 E7 19 15N 84 54 E
Berheci →, Romania ... 46 C8 46 7N 27 19 E
Bering Sea, Pac. Oc. ... 126 C1 58 0N 171 0 E
Bering Strait, U.S.A. ... 126 B3 65 30N 169 0W
Beringen, Belgium ..... 21 F6 51 3N 5 14 E
Beringen, Switz. ...... 29 A7 47 38N 8 34 E
Beringovskiy, Russia ... 57 C18 63 3N 179 19 E
Berisso, Argentina .... 158 C4 34 56 S 57 50W
Berja, Spain 35 J2 36 50N 2 56W
Berkane, Morocco 99 B4 34 52N 2 20W
Berkel →, Neths. ..... 20 D8 52 8N 6 12 E
Berkeley, U.K. 17 F5 51 41N 2 27W
Berkeley, U.S.A. 144 H4 37 52N 122 16W
Berkeley Springs, U.S.A. 134 F6 39 38N 78 14W
Berkhout, Neths. ..... 20 C5 52 38N 4 59 E
Berkner I., Antarctica .. 7 D18 79 30 S 50 0W
Berkovitsa, Bulgaria ... 43 D8 43 16N 23 8 E
Berkshire □, U.K. 17 F6 51 25N 1 17W
Berlaar, Belgium ..... 21 F5 51 7N 4 39 E
Berland →, Canada ... 130 C5 54 0N 116 50W
Berlanga, Spain 37 G5 38 17N 5 50W
Berlare, Belgium ..... 21 F4 51 2N 4 2 E
Berlenga, I., Portugal .. 37 F1 39 25N 9 30W
Berlin, Germany ...... 26 C9 52 30N 13 25 E
Berlin, Md., U.S.A. ... 134 F8 38 20N 75 13W
Berlin, N.H., U.S.A. .. 137 B13 44 28N 71 11W
Berlin, Wis., U.S.A. ... 134 D1 43 58N 88 57W
Bermeja, Sierra, Spain . 37 J5 36 30N 5 11W
Bermejo →, Formosa, Argentina .... 158 B4 26 51 S 58 23W
Bermejo →, San Juan, Argentina .... 158 C2 32 30 S 67 30W
Bermeo, Spain 34 B2 43 25N 2 47W
Bermillo de Sayago, Spain 36 D4 41 22N 6 8W
Bermuda ■, Atl. Oc. ... 8 E2 32 45N 65 0W
Bern, Switz. .... 28 C4 46 57N 7 28 E
Bern □, Switz. .... 28 C5 46 45N 7 40 E
Bernado, U.S.A. 143 J10 34 30N 106 53W
Bernalda, Italy .... 41 B9 40 24N 16 41 E
Bernalillo, U.S.A. 143 J10 35 18N 106 33W
Bernam →, Malaysia ... 74 B2 3 45N 101 5 E
Bernardo de Irigoyen, Argentina .... 159 B5 26 15 S 53 40W
Bernardo O'Higgins □, Chile .... 158 C1 34 15 S 70 45W
Bernasconi, Argentina ... 158 D3 37 55 S 63 44W
Bernau, Bayern, Germany 27 H8 47 47N 12 22 E
Bernau, Brandenburg, Germany .... 26 C9 52 40N 13 35 E
Bernay, France .... 22 C7 49 5N 0 35 E
Bernburg, Germany ..... 26 D7 51 47N 11 44 E
Berndorf, Austria ...... 30 D9 47 59N 16 1 E
Berne = Bern, Switz. ... 28 C4 46 57N 7 28 E
Berne = Bern □, Switz. . 28 C5 46 45N 7 40 E
Berne, U.S.A. 141 D12 40 39N 84 57W
Berner Alpen, Switz. .... 28 D5 46 27N 7 35 E
Bernese Oberland = Oberland, Switz. ...... 28 C5 46 35N 7 38 E
Bernier I., Australia .... 113 D1 24 50 S 113 12 E
Bernina, Piz, Switz. .... 29 D9 46 20N 9 54 E
Bernina, Pizzo, Switz. ... 29 D9 46 22N 9 54 E
Bernissart, Belgium ..... 21 H3 50 28N 3 39 E
Bernkastel-Kues, Germany 27 F3 49 55N 7 3 E
Beroroha, Madag. ...... 105 C8 21 40 S 45 10 E
Béroubouay, Benin ..... 101 C5 10 34N 2 46 E
Beroun, Czech. .... 30 B7 49 57N 14 5 E
Berounka →, Czech. .... 30 B7 50 0N 14 22 E
Berovo, Macedonia ..... 42 F7 41 38N 22 51 E
Berrahal, Algeria ...... 99 A6 36 54N 7 33 E
Berre, Étang de, France . 25 E9 43 27N 5 5 E
Berrechid, Morocco 98 B3 33 18N 7 36W
Berri, Australia .... 116 C4 34 14 S 140 35 E
Berriane, Algeria ...... 99 B5 32 50N 3 46 E
Berrien Springs, U.S.A. . 141 C10 41 57N 86 20W
Berrigan, Australia ..... 117 C6 35 38 S 145 49 E
Berriwillock, Australia .. 116 C5 35 36 S 142 59 E
Berrouaghia, Algeria ... 99 A5 36 10N 2 53 E
Berry, Australia .... 117 C9 34 46 S 150 43 E
Berry, France .... 23 F8 46 50N 2 0 E
Berry, U.S.A. 141 F12 38 31N 84 23W
Berry Is., Bahamas 148 A4 25 40N 77 50W
Berryessa L., U.S.A. ... 144 G4 38 31N 122 6W
Berryville, U.S.A. 139 G8 36 22N 93 34W
Bersenbrück, Germany .. 26 C3 52 34N 7 56 E
Bershad, Ukraine ...... 51 H5 48 22N 29 31 E
Berthold, U.S.A. 138 A4 48 19N 101 44W
Berthoud, U.S.A. 138 E2 40 19N 105 5W
Bertincourt, France .... 23 B9 50 5N 2 58 E
Bertoua, Cameroon .... 102 B2 4 30N 13 45 E
Bertrand, U.S.A. 138 E5 40 32N 99 38W
Bertrange, Lux. .... 21 J8 49 37N 6 3 E
Bertrix, Belgium ...... 21 J6 49 51N 5 15 E
Beruri, Brazil .... 153 D5 3 54 S 61 22W
Berwick, U.S.A. 137 E8 41 3N 76 14W
Berwick-upon-Tweed, U.K. 16 B5 55 46N 2 0W
Berwyn Mts., U.K. 16 E4 52 54N 3 26W
Beryslav, Ukraine ..... 51 J7 46 50N 33 30 E
Berzasca, Romania .... 42 C6 44 39N 21 58 E
Berzence, Hungary .... 31 E10 46 12N 17 11 E
Besal, Pakistan .... 81 B5 35 4N 73 56 E
Besalampy, Madag. .... 105 B7 16 43 S 44 29 E
Besançon, France ..... 23 E13 47 15N 6 2 E
Besar, Indonesia ...... 75 C5 2 40 S 116 0 E
Besar, Gunong, Malaysia 74 A2 5 10N 101 18 E
Beshenkovichi, Belarus .. 50 E5 55 2N 29 29 E
Beška, Serbia, Yug. .... 42 B5 45 8N 20 6 E
Beskids = Beskydy, Europe .... 31 B11 49 35N 18 40 E
Beskydy, Europe ...... 31 B11 49 35N 18 40 E
Beslan, Russia .... 53 J7 43 15N 44 28 E
Besna Kobila, Serbia, Yug. .... 42 E7 42 31N 22 10 E
Besnard L., Canada 131 B7 55 25N 106 0W
Besni, Turkey .... 88 D7 37 41N 37 52 E
Besor, N. →, Egypt .... 91 D3 31 28N 34 22 E
Beşparmak Dağı, Turkey 45 G9 37 32N 27 30 E
Bessa Monteiro, Angola . 103 D2 7 7 S 13 44 E
Bessarabiya, Moldova ... 51 J5 47 0N 28 10 E
Bessarabka = Basarabeasca, Moldova 51 J5 46 21N 28 58 E
Bessèges, France ...... 25 D8 44 18N 4 8 E
Bessemer, Ala., U.S.A. .. 135 J2 33 24N 86 58W
Bessemer, Mich., U.S.A. 138 B9 46 29N 90 3W
Bessin, France 22 C5 49 18N 1 0W
Bessines-sur-Gartempe, France .... 24 B5 46 6N 1 22 E
Best, Neths. .... 21 E6 51 31N 5 23 E
Bet She'an, Israel ..... 91 C4 32 30N 35 30 E
Bet Shemesh, Israel .... 91 D3 31 44N 35 0 E
Bet Tadjine, Djebel, Algeria 98 C4 29 0N 3 30W
Betafo, Madag. .... 105 B8 19 50 S 46 51 E
Betancuria, Canary Is. .. 33 F5 28 25N 14 3W
Betanzos, Bolivia ..... 157 D4 19 34 S 65 27W
Betanzos, Spain 36 B2 43 15N 8 12W
Bétaré Oya, Cameroon .. 102 A2 5 40N 14 5 E
Bétera, Spain 34 F4 39 35N 0 28W
Bethal, S. Africa ...... 105 D4 26 27 S 29 28 E
Bethanien, Namibia .... 104 D2 26 31 S 17 8 E
Bethany, Ill., U.S.A. ... 141 E8 39 39N 88 45W
Bethany, Mo., U.S.A. .. 140 D2 40 16N 94 2W
Bethel, Alaska, U.S.A. .. 126 B3 60 48N 161 45W
Bethel, Ohio, U.S.A. ... 141 F12 38 58N 84 5W
Bethel, Vt., U.S.A. 137 C12 43 50N 72 38W
Bethel Park, U.S.A. ... 136 F4 40 20N 80 1W
Bethlehem = Bayt Lahm, West Bank .... 91 D4 31 43N 35 12 E
Bethlehem, S. Africa ... 105 D4 28 14 S 28 18 E
Bethlehem, U.S.A. 137 F9 40 37N 75 23W
Bethulie, S. Africa .... 104 E4 30 30 S 25 59 E
Béthune, France ...... 23 B9 50 30N 2 38 E
Béthune →, France .... 22 C8 49 53N 1 9 E
Bethungra, Australia ... 117 C7 34 45 S 147 51 E
Betijoque, Venezuela ... 152 B3 9 23N 70 44W
Betim, Brazil .... 155 E3 19 58 S 44 7W
Betioky, Madag. ...... 105 C7 23 48 S 44 20 E
Beton-Bazoches, France . 23 D10 48 42N 3 15 E
Betong, Thailand ...... 77 K3 5 45N 101 5 E
Betoota, Australia ..... 114 D3 25 45 S 140 42 E
Betroka, Madag. ...... 105 C8 23 16 S 46 0 E
Betsiamites, Canada ... 129 C6 48 56N 68 40W
Betsiamites →, Canada . 129 C6 48 56N 68 38W
Betsiboka →, Madag. ... 105 B8 16 3 S 46 36 E
Bettembourg, Lux. ...... 21 J8 49 31N 6 6 E
Bettendorf, U.S.A. 140 C6 41 32N 90 30W
Bettiah, India .... 81 F11 26 48N 84 33 E
Bèttola, Italy .... 38 D6 44 49N 9 36 E
Betul, India .... 82 D3 21 58N 77 59 E
Betung, Malaysia ...... 75 B4 1 24N 111 31 E
Betzdorf, Germany ..... 26 E3 50 46N 7 52 E
Beuca, Romania .... 46 E5 44 14N 24 56 E
Beuil, France .... 25 D10 44 6N 6 59 E
Beulah, U.S.A. 138 B4 47 16N 101 47W
Beuvron →, France .... 22 E8 47 29N 1 15 E
Beveren, Belgium ..... 21 F4 51 12N 4 16 E
Beverley, Australia .... 113 F2 32 9 S 116 56 E
Beverley, U.K. 16 D7 53 51N 0 26W
Beverlo, Belgium ..... 21 F6 51 6N 5 13 E
Beverly, Mass., U.S.A. . 137 D14 42 33N 70 53W
Beverly, Wash., U.S.A. . 142 C4 46 50N 119 56W
Beverly Hills, U.S.A. ... 145 L8 34 4N 118 25W
Beverwijk, Neths. ..... 20 D5 52 28N 4 38 E
Bex, Switz. .... 28 D4 46 15N 7 0 E
Bey Dağları, Turkey ... 88 D4 36 38N 30 29 E
Beya, Russia .... 57 D10 52 40N 92 30 E
Beyānlū, Iran .... 84 C5 36 0N 47 51 E
Beyin, Ghana 100 D4 5 1N 2 41W
Beykoz, Turkey .... 43 F14 41 8N 29 7 E
Beyla, Guinea 100 D3 8 30N 8 38W
Beynat, France .... 24 C5 45 8N 1 44 E
Beyneu, Kazakhstan ... 49 E10 45 18N 55 9 E
Beypazarı, Turkey ..... 88 B4 40 10N 31 56 E
Beypore →, India ..... 83 J2 11 10N 75 47 E
Beyşehir, Turkey ...... 88 D4 37 41N 31 43 E
Beyşehir Gölü, Turkey .. 88 D4 37 41N 31 33 E
Beytüşşebap, Turkey ... 89 D10 37 35N 43 10 E
Bezdan, Serbia, Yug. ... 42 B3 45 50N 18 57 E
Bezhetsk, Russia ..... 52 B3 57 47N 36 39 E
Bezhitsa, Russia ..... 48 D5 53 19N 34 17 E
Béziers, France ...... 24 E7 43 20N 3 12 E
Bezwada = Vijayawada, India .... 83 F5 16 31N 80 39 E
Bhadarwah, India ..... 81 C6 32 58N 75 46 E
Bhadra →, India ...... 83 H2 14 0N 75 20 E
Bhadrakh, India ...... 82 D8 21 10N 86 30 E
Bhadravati, India ..... 83 H2 13 49N 75 40 E
Bhagalpur, India ..... 81 G12 25 10N 87 0 E
Bhainsa, India .... 82 E3 19 10N 77 58 E
Bhairab →, Bangla. ... 78 D2 22 51N 89 34 E
Bhairab Bazar, Bangla. . 78 C3 24 4N 90 58 E
Bhakkar, Pakistan .... 79 C3 31 40N 71 5 E
Bhakra Dam, India .... 80 D7 31 30N 76 45 E
Bhamo, Burma .... 78 C6 24 15N 97 15 E
Bhamragarh, India .... 82 E5 19 30N 80 40 E
Bhandara, India ...... 82 D4 21 5N 79 42 E
Bhanrer Ra., India .... 80 H8 23 40N 79 45 E
Bharat = India ■, Asia . 59 G11 20 0N 78 0 E
Bharatpur, India ..... 80 F7 27 15N 77 30 E
Bharuch, India ...... 82 D1 21 47N 73 0 E
Bhatghar L., India .... 82 E1 18 10N 73 48 E
Bhatiapara Ghat, Bangla. 78 D2 23 13N 89 42 E
Bhatinda, India ...... 80 D6 30 15N 74 57 E
Bhatkal, India .... 83 H2 13 58N 74 35 E
Bhatpara, India ...... 81 H13 22 50N 88 25 E
Bhattiprolu, India .... 83 F5 16 7N 80 45 E
Bhaun, Pakistan ..... 80 C5 32 55N 72 40 E
Bhaunagar = Bhavnagar, India .... 80 J5 21 45N 72 10 E
Bhavani, India ...... 83 J3 11 27N 77 43 E
Bhavani →, India .... 83 J4 11 0N 78 15 E
Bhavnagar, India .... 80 J5 21 45N 72 10 E
Bhawanipatna, India .. 82 E5 19 55N 80 10 E
Bhera, Pakistan ..... 80 C5 32 29N 72 57 E
Bhilsa = Vidisha, India . 80 H7 23 28N 77 53 E
Bhilwara, India ..... 80 G6 25 25N 74 38 E
Bhima →, India ..... 82 F3 16 25N 77 17 E
Bhimavaram, India ... 83 F5 16 30N 81 30 E
Bhimbar, Pakistan ... 81 C6 32 59N 74 3 E
Bhind, India .... 81 F8 26 30N 78 46 E
Bhiwandi, India .... 82 E1 19 20N 73 0 E
Bhiwani, India ..... 80 E7 28 50N 76 9 E
Bhola, Bangla. ..... 78 D3 22 45N 90 35 E
Bhongir, India ..... 82 F4 17 30N 78 56 E
Bhopal, India ..... 80 H7 23 20N 77 30 E
Bhor, India .... 82 E1 18 12N 73 53 E
Bhubaneshwar, India . 82 D7 20 15N 85 50 E
Bhuj, India ...... 80 H3 23 15N 69 49 E
Bhumiphol Dam = Phumiphon, Khuan, Thailand .... 76 D2 17 15N 98 58 E
Bhusaval, India .... 82 D2 21 3N 75 46 E
Bhutan ■, Asia .... 78 B3 27 25N 90 30 E
Biá →, Brazil ..... 152 D4 3 28 S 67 23W
Biafra, B. of = Bonny, Bight of, Africa ..... 101 E6 3 30N 9 20 E
Biak, Indonesia .... 73 B5 1 10 S 136 6 E
Biała, Poland ..... 47 E4 50 24N 17 40 E
Biała →, Białystok, Poland .... 47 B10 53 11N 23 4 E
Biała →, Tarnów, Poland 31 A13 50 3N 20 55 E
Biała Piska, Poland ... 47 B9 53 37N 22 5 E
Biała Podlaska, Poland . 47 C10 52 4N 23 6 E
Biała Podlaska □, Poland 47 D10 52 0N 23 0 E
Biała Rawska, Poland .. 47 D7 51 48N 20 29 E
Białobrzegi, Poland ... 47 C8 52 27N 21 3 E
Białogard, Poland .... 47 A2 54 2N 15 58 E
Białowieża, Poland ... 47 C10 52 41N 23 49 E
Biały Bór, Poland .... 47 B3 53 53N 16 51 E
Białystok, Poland .... 47 B10 53 10N 23 10 E
Białystok □, Poland .. 47 B10 53 9N 23 10 E
Biancavilla, Italy .... 41 E7 37 38N 14 52 E
Bīārjmand, Iran ..... 85 B7 36 6N 55 53 E
Biaro, Indonesia .... 72 A3 2 5N 125 26 E
Biarritz, France 24 E2 43 29N 1 33W
Biasca, Switz. ..... 29 D7 46 22N 8 58 E
Biba, Egypt .... 94 J7 28 55N 31 0 E
Bibai, Japan ...... 60 C10 43 19N 141 52 E
Bibala, Angola .... 103 E2 14 44 S 13 24 E
Bibane, Bahiret el, Tunisia 96 B2 33 16N 11 13 E
Bibassé, Gabon .... 102 B2 1 27N 11 37 E
Bibbiena, Italy .... 39 E8 43 42N 11 49 E
Bibby I., Canada ... 131 A10 61 55N 93 0W
Biberach, Germany .. 27 G5 48 5N 9 47 E
Biberist, Switz. .... 28 B5 47 11N 7 34 E
Bibey →, Spain 36 C3 42 24N 7 13W
Bibiani, Ghana 100 D4 6 30N 2 8W
Bibile, Sri Lanka ... 83 L5 7 10N 81 25 E
Biboohra, Australia .. 114 B4 16 56 S 145 25 E
Bibungwa, Zaïre ... 106 C2 2 40 S 28 15 E
Bic, Canada 129 C6 48 20N 68 41W

Blackbull 175

Bicaj, Albania 44 C2 41 58N 20 25 E
Bicaz, Romania 46 C7 46 53N 26 5 E
Biccari, Italy 41 A8 41 23N 15 12 E
Bichena, Ethiopia 95 E4 10 28N 38 10 E
Bichvinta, Georgia 53 J5 43 9N 40 21 E
Bickerton I., Australia 114 A2 13 45 S 136 10 E
Bicknell, Ind., U.S.A. 141 F9 38 47N 87 19W
Bicknell, Utah, U.S.A. 143 G8 38 20N 111 33W
Bida, Nigeria 101 D6 9 3N 5 58 E
Bida, Zaïre 102 B3 4 55N 19 56 E
Bidar, India 82 F3 17 55N 77 35 E
Biddeford, U.S.A. 129 D5 43 30N 70 28W
Biddwara, Ethiopia 95 F4 5 11N 38 34 E
Bideford, U.K. 17 F3 51 1N 4 13W
Bidon 5 = Poste Maurice Cortier, Algeria 99 D5 22 14N 1 2 E
Bidor, Malaysia 77 K3 4 6N 101 15 E
Bidura, Australia 116 C5 34 10 S 143 21 E
Bié □, Angola 103 E3 12 30 S 17 0 E
Bié, Planalto de, Angola 103 E3 12 0 S 16 0 E
Bieber, U.S.A. 142 F3 41 7N 121 8W
Biebrza →, Poland 47 B9 53 13N 22 25 E
Biecz, Poland 31 B14 49 44N 21 15 E
Biel, Switz. 28 B4 47 8N 7 14 E
Bielawa, Poland 47 E3 50 43N 16 37 E
Bielé Karpaty, Europe 31 B10 49 5N 18 0 E
Bielefeld, Germany 26 C4 52 1N 8 33 E
Bielersee, Switz. 28 B4 47 6N 7 5 E
Biella, Italy 38 C5 45 34N 8 3 E
Bielsk Podlaski, Poland 47 C10 52 47N 23 12 E
Bielsko-Biała, Poland 31 B12 49 50N 19 2 E
Bielsko-Biała □, Poland 31 B12 49 45N 19 15 E
Bien Hoa, Vietnam 77 G6 10 57N 106 49 E
Bienfait, Canada 131 D8 49 10N 102 50W
Bienne = Biel, Switz. 28 B4 47 8N 7 14 E
Bienvenida, Spain 37 G4 38 18N 6 12W
Bienvenue, Fr. Guiana 153 C7 3 0N 52 30W
Bienville, L., Canada 128 A5 55 5N 72 40W
Biescas, Spain 34 C4 42 37N 0 20W
Biese →, Germany 26 C7 52 53N 11 46 E
Biesiesfontein, S. Africa 104 E2 30 57 S 17 58 E
Bietigheim, Germany 27 G5 48 58N 9 8 E
Bievre, Belgium 21 J6 49 57N 5 1 E
Biferno →, Italy 41 A8 41 59N 15 2 E
Bifoum, Gabon 102 C2 0 20 S 10 23 E
Big →, Canada 129 B8 54 50N 58 55W
Big →, U.S.A. 140 F6 38 28N 90 37W
Big B., Canada 129 A7 55 43N 60 35W
Big B., N.Z. 119 E3 44 28 S 168 4 E
Big Bear City, U.S.A. 145 L10 34 16N 116 51W
Big Bear Lake, U.S.A. 145 L10 34 15N 116 56W
Big Beaver, Canada 131 D7 49 10N 105 10W
Big Belt Mts., U.S.A. 142 C8 46 30N 111 25W
Big Bend, Swaziland 105 D5 26 50 S 31 58 E
Big Bend National Park, U.S.A. 139 L3 29 20N 103 5W
Big Black →, U.S.A. 139 J9 32 3N 91 4W
Big Blue →, Ind., U.S.A. 141 E11 39 12N 85 56W
Big Blue →, Kans., U.S.A. 138 F6 39 35N 96 34W
Big Cr. →, Canada 130 C4 51 42N 122 41W
Big Creek, U.S.A. 144 H7 37 11N 119 14W
Big Cypress Swamp, U.S.A. 135 M5 26 12N 81 10W
Big Falls, U.S.A. 138 A8 48 12N 93 48W
Big Fork →, U.S.A. 138 A8 48 31N 93 43W
Big Horn Mts. = Bighorn Mts., U.S.A. 142 D10 44 30N 107 30W
Big Lake, U.S.A. 139 K4 31 12N 101 28W
Big Moose, U.S.A. 137 C10 43 49N 74 58W
Big Muddy →, U.S.A. 140 G8 38 0N 89 0W
Big Muddy Cr. →, U.S.A. 138 A2 48 8N 104 36W
Big Pine, U.S.A. 144 H8 37 10N 118 17W
Big Piney, U.S.A. 142 E8 42 32N 110 7W
Big Quill L., Canada 131 C8 51 55N 104 50W
Big Rapids, U.S.A. 134 D3 43 42N 85 29W
Big River, Canada 131 C7 53 50N 107 0W
Big Run, U.S.A. 136 F6 40 57N 78 55W
Big Sable Pt., U.S.A. 134 C2 44 3N 86 1W
Big Sand L., Canada 131 B9 57 45N 99 45W
Big Sandy, U.S.A. 142 B8 48 11N 110 7W
Big Sandy Cr. →, U.S.A. 138 F3 38 7N 102 29W
Big Sioux →, U.S.A. 138 D6 42 29N 96 27W
Big Spring, U.S.A. 139 J4 32 15N 101 28W
Big Springs, U.S.A. 138 E3 41 4N 102 5W
Big Stone City, U.S.A. 138 C6 45 18N 96 28W
Big Stone Gap, U.S.A. 135 G4 36 52N 82 47W
Big Stone L., U.S.A. 138 C6 45 30N 96 35W
Big Sur, U.S.A. 144 J5 36 15N 121 48W
Big Timber, U.S.A. 142 D9 45 50N 109 57W
Big Trout L., Canada 128 B1 53 40N 90 0W
Biğa, Turkey 88 B2 40 13N 27 14 E
Bigadiç, Turkey 88 C3 39 22N 28 7 E
Biganos, France 24 D3 44 39N 0 59W
Bigfork, U.S.A. 142 B6 48 4N 114 4W
Biggar, Canada 131 C7 52 4N 108 0W
Biggar, U.K. 18 F5 55 38N 3 32W
Bigge I., Australia 112 B4 14 35 S 125 10 E
Biggenden, Australia 115 D5 25 31 S 152 4 E
Biggs, U.S.A. 144 F5 39 25N 121 43W
Bighorn, U.S.A. 142 C10 46 10N 107 27W
Bighorn →, U.S.A. 142 C10 46 10N 107 28W
Bighorn Mts., U.S.A. 142 D10 44 30N 107 30W
Bignona, Senegal 100 C1 12 52N 16 14W
Bigorre, France 24 E4 43 10N 0 5 E
Bigstone L., Canada 131 C9 53 42N 95 44W
Biguglia, Étang de, France 25 F13 42 36N 9 29 E
Bigwa, Tanzania 106 D4 7 10 S 39 10 E
Bihać, Bos.-H. 39 D12 44 49N 15 57 E
Bihar, India 81 G11 25 5N 85 40 E
Bihar □, India 81 G11 25 0N 86 0 E
Biharamulo, Tanzania 106 C3 2 25 S 31 25 E
Biharamulo □, Tanzania 106 C3 2 30 S 31 20 E
Biharkeresztes, Hungary 31 D14 47 8N 21 44 E
Bihor □, Romania 46 C3 47 0N 22 10 E
Bihor, Munţii, Romania 46 C3 46 29N 22 47 E
Bijagós, Arquipélago dos, Guinea-Biss. 100 C1 11 15N 16 10W
Bijaipur, India 80 F7 26 2N 77 20 E
Bijapur, Karnataka, India 82 F2 16 50N 75 55 E
Bijapur, Mad. P., India 82 E5 18 50N 80 50 E
Bījār, Iran 89 E12 35 52N 47 35 E
Bijeljina, Bos.-H. 42 C4 44 46N 19 17 E
Bijelo Polje, Montenegro, Yug. 42 D4 43 1N 19 45 E
Bijie, China 68 D5 27 20N 105 16 E
Bijni, India 78 B3 26 30N 90 40 E
Bijnor, India 80 E8 29 27N 78 11 E
Bikaner, India 80 E5 28 2N 73 18 E
Bikapur, India 81 F10 26 30N 82 7 E
Bikeqi, China 66 D6 40 43N 111 20 E
Bikfayyā, Lebanon 91 B4 33 55N 35 41 E
Bikin, Russia 57 E14 46 50N 134 20 E
Bikin →, Russia 60 A7 46 51N 134 2 E
Bikini Atoll, Pac. Oc. 122 F8 12 0N 167 30 E
Bikoro, Zaïre 102 C3 0 48 S 18 15 E
Bikoué, Cameroon 101 E7 3 55N 11 50 E
Bila Tserkva, Ukraine 51 H6 49 45N 30 10 E
Bilara, India 80 F5 26 14N 73 53 E
Bilaspara, India 78 B3 26 13N 90 14 E
Bilaspur, Mad. P., India 81 H10 22 2N 82 15 E
Bilaspur, Punjab, India 80 D7 31 19N 76 50 E
Biläsuvar, Azerbaijan 89 C13 39 27N 48 32 E
Bilauk Taungdan, Thailand 76 F2 13 0N 99 0 E
Bilbao, Spain 34 B2 43 16N 2 56W
Bilbeis, Egypt 94 H7 30 25N 31 34 E
Bilbo = Bilbao, Spain 34 B2 43 16N 2 56W
Bilbor, Romania 46 B6 47 6N 25 30 E
Bíldudalur, Iceland 12 D2 65 41N 23 36W
Bileća, Bos.-H. 42 E3 42 53N 18 27 E
Bilecik, Turkey 88 B4 40 5N 30 5 E
Biłgoraj, Poland 47 E9 50 33N 22 42 E
Bilhorod-Dnivstrovskyy, Ukraine 51 J6 46 11N 30 23 E
Bilibino, Russia 57 C17 68 3N 166 20 E
Bilibiza, Mozam. 107 E5 12 30 S 40 20 E
Bilin, Burma 78 G6 17 14N 97 15 E
Bilir, Russia 57 C14 65 40N 131 20 E
Biliran I., Phil. 71 F5 11 35N 124 28 E
Bilishti, Albania 44 D3 40 37N 21 2 E
Bill, U.S.A. 138 D2 43 14N 105 16W
Billabalong, Australia 113 E2 27 25 S 115 49 E
Billiluna, Australia 112 C4 19 37 S 127 41 E
Billingham, U.K. 16 C6 54 36N 1 17W
Billings, U.S.A. 142 D9 45 47N 108 30W
Billiton Is. = Belitung, Indonesia 75 C3 3 10 S 107 50 E
Billom, France 24 C7 45 43N 3 20 E
Bilma, Niger 97 E2 18 50N 13 30 E
Bilo Gora, Croatia 42 B2 45 53N 17 15 E
Biloela, Australia 114 C5 24 24 S 150 31 E
Bilohirsk, Ukraine 51 K8 45 3N 34 35 E
Biloku, Guyana 153 C6 1 50N 58 25W
Bilopillya, Ukraine 51 G8 51 14N 34 20 E
Biloxi, U.S.A. 139 K10 30 24N 88 53W
Bilpa Morea Claypan, Australia 114 D2 25 0 S 140 0 E
Bilthoven, Neths. 20 D6 52 8N 5 12 E
Biltine, Chad 97 F4 14 40N 20 50 E
Bilugyun, Burma 78 G6 16 24N 97 32 E
Bilyana, Australia 114 B4 18 5 S 145 50 E
Bilyarsk, Russia 52 C10 54 58N 50 22 E
Bilzen, Belgium 21 G7 50 52N 5 31 E
Bima, Indonesia 75 D5 8 22 S 118 49 E
Bimban, Egypt 94 C3 24 24N 32 54 E
Bimberi Pk., Australia 117 C8 35 44 S 148 51 E
Bimbila, Ghana 101 D5 8 54N 0 5 E
Bimbo, C.A.R. 102 B3 4 15N 18 33 E
Bimini Is., Bahamas 148 A4 25 42N 79 25W
Bin Xian, Heilongjiang, China 67 B14 45 42N 127 32 E
Bin Xian, Shaanxi, China 66 G5 35 2N 108 4 E
Bina-Etawah, India 80 G8 24 13N 78 14 E
Bināb, Iran 89 D13 36 35N 48 41 E
Binaiya, Indonesia 73 B3 3 11 S 129 26 E
Binalbagan, Phil. 71 F4 10 12N 122 50 E
Binalong, Australia 117 C8 34 40 S 148 39 E
Bīnālūd, Kūh-e, Iran 85 B8 36 30N 58 30 E
Binatang, Malaysia 75 B4 2 10N 111 40 E
Binbee, Australia 114 C4 20 19 S 147 56 E
Binche, Belgium 21 H4 50 26N 4 10 E
Binchuan, China 68 E3 25 42N 100 38 E
Binda, Australia 115 D4 27 52 S 147 21 E
Binda, Zaïre 103 D2 5 52 S 13 14 E
Bindle, Australia 115 D4 27 40 S 148 45 E
Bindoy, Phil. 71 G4 9 48N 123 5 E
Bindura, Zimbabwe 107 F3 17 18 S 31 18 E
Bingara, N.S.W., Australia 115 D5 29 52 S 150 36 E
Bingara, Queens., Australia 115 D3 28 10 S 144 37 E
Bingen, Germany 27 F3 49 57N 7 55 E
Bingerville, Ivory C. 100 D4 5 18N 3 49W
Bingham, U.S.A. 129 C6 45 3N 69 53W
Bingham Canyon, U.S.A. 142 F7 40 32N 112 9W
Binghamton, U.S.A. 137 D9 42 6N 75 55W
Bingöl, Turkey 89 C9 38 53N 40 29 E
Bingöl Dağları, Turkey 89 C9 39 16N 41 9 E
Binh Dinh = An Nhon, Vietnam 76 F7 13 55N 109 7 E
Binh Khe, Vietnam 76 F7 13 57N 108 51 E
Binh Son, Vietnam 76 E7 15 20N 108 40 E
Binhai, China 67 G10 34 2N 119 49 E
Binisatua, Spain 33 B11 39 50N 4 11 E
Binjai, Indonesia 74 B1 3 20N 98 30 E
Binnaway, Australia 117 A8 31 28 S 149 24 E
Binongko, Indonesia 72 C2 5 55 S 123 55 E
Binscarth, Canada 131 C8 50 37N 101 17W
Bintan, Indonesia 74 B2 1 0N 104 0 E
Bintuni, Indonesia 73 B4 2 7 S 133 32 E
Binyang, China 68 F7 23 12N 108 47 E
Binz, Germany 26 A9 54 24N 13 35 E
Binza, Zaïre 103 C3 4 21 S 15 14 E
Binzert = Bizerte, Tunisia 96 A1 37 15N 9 50 E
Bío Bío □, Chile 158 D1 37 35 S 72 0W
Biograd, Croatia 39 E12 43 56N 15 29 E
Bioko, Eq. Guin. 101 E6 3 30N 8 40 E
Biokovo, Croatia 42 D2 43 23N 17 0 E
Biougra, Morocco 98 B3 30 15N 9 14W
Bir, India 82 E2 19 4N 75 46 E
Bir, Ras, Djibouti 95 E5 12 0N 43 20 E
Bîr Abu Hashim, Egypt 94 C3 23 42N 34 6 E
Bîr Abu M'nqar, Egypt 94 B2 26 33N 27 33 E
Bîr Abu Muḥammad, Egypt 91 F3 29 44N 34 14 E
Bi'r ad Dabbāghāt, Jordan 91 E4 30 26N 35 32 E
Bîr Adal Deib, Sudan 94 C4 22 35N 36 10 E
Bi'r al Butayyihāt, Jordan 91 F4 29 47N 35 20 E
Bi'r al Mārī, Jordan 91 E4 30 4N 35 33 E
Bi'r al Qattār, Jordan 91 F4 29 47N 35 32 E
Bir 'Ali, Yemen 87 D5 14 1N 48 20 E
Bir Aouine, Tunisia 96 B1 32 25N 9 18 E
Bîr 'Asal, Egypt 94 B3 25 55N 34 20 E
Bir Autrun, Sudan 94 D2 18 15N 26 40 E
Bîr Beiḍa, Egypt 91 E3 30 25N 34 29 E
Bi'r Dhu'fān, Libya 96 B2 31 59N 14 32 E
Bîr Diqnash, Egypt 94 A2 31 3N 25 23 E
Bir el 'Abbes, Algeria 98 C3 26 7N 6 9W
Bîr el 'Abd, Egypt 91 D2 31 2N 33 0 E
Bir el Ater, Algeria 99 B6 34 46N 8 3 E
Bîr el Basur, Egypt 94 B2 29 51N 25 49 E
Bîr el Biarât, Egypt 91 F3 29 30N 34 43 E
Bîr el Duweidar, Egypt 91 E1 30 56N 32 32 E
Bîr el Garârât, Egypt 91 D2 31 3N 33 34 E
Bîr el Gellaz, Egypt 94 A2 30 50N 26 40 E
Bîr el Heisi, Egypt 91 F3 29 22N 34 36 E
Bîr el Jafir, Egypt 91 E1 30 50N 32 41 E
Bîr el Mâlḥi, Egypt 91 E2 30 38N 33 19 E
Bîr el Shaqqa, Egypt 94 A2 30 54N 25 1 E
Bîr el Thamâda, Egypt 91 E2 30 12N 33 27 E
Bîr Fuad, Egypt 94 A2 30 35N 26 28 E
Bîr Gara, Chad 97 F3 13 11N 15 58 E
Bîr Gebeil Ḥiṣn, Egypt 91 E2 30 2N 33 18 E
Bi'r Ghadīr, Syria 91 A6 34 6N 37 3 E
Bîr Haimur, Egypt 94 C3 22 45N 33 40 E
Bîr Ḥasana, Egypt 91 E2 30 29N 33 46 E
Bi'r Īdimah, Si. Arabia 86 C4 18 31N 44 12 E
Bi'r Jadīd, Iraq 84 C4 34 1N 42 54 E
Bir Jdid, Morocco 98 B3 33 26N 8 0W
Bîr Kanayis, Egypt 94 C3 24 59N 33 15 E
Bîr Kaseiba, Egypt 91 E2 31 0N 33 17 E
Bîr Kerawein, Egypt 94 B2 27 10N 28 25 E
Bîr Lahfân, Egypt 91 D2 31 0N 33 51 E
Bir Lahrache, Algeria 99 B6 32 1N 8 12 E
Bîr Madkûr, Egypt 91 E1 30 44N 32 33 E
Bîr Maql, Egypt 94 C3 23 7N 33 40 E
Bîr Misaha, Egypt 94 C2 22 13N 27 59 E
Bir Mogrein, Mauritania 98 C2 25 10N 11 25W
Bîr Murr, Egypt 94 C3 23 28N 30 10 E
Bi'r Muṭribah, Kuwait 84 D5 29 54N 47 17 E
Bîr Nakheila, Egypt 94 C3 24 1N 30 50 E
Bîr Qatia, Egypt 91 E1 30 58N 32 45 E
Bîr Qatrani, Egypt 94 A2 30 55N 26 10 E
Bîr Ranga, Egypt 94 C4 24 25N 35 15 E
Bîr Sahara, Egypt 94 C2 22 54N 28 40 E
Bîr Seiyâla, Egypt 94 B3 26 10N 33 50 E
Bir Semguine, Morocco 98 B3 30 1N 5 39W
Bîr Shalatein, Egypt 94 C4 23 5N 35 25 E
Bîr Shebb, Egypt 94 C2 22 25N 29 40 E
Bîr Shût, Egypt 94 C4 23 50N 35 15 E
Bi'r Tamis, Yemen 87 C5 16 45N 48 48 E
Bîr Terfawi, Egypt 94 C2 22 57N 28 55 E
Bîr Umm Qubûr, Egypt 94 C3 24 35N 34 2 E
Bîr Ungât, Egypt 94 C3 22 8N 33 48 E
Bîr Za'farâna, Egypt 94 J8 29 10N 32 40 E
Bîr Zāmūs, Libya 96 D3 24 16N 15 6 E
Bîr Zeidûn, Egypt 94 B3 25 45N 33 40 E
Bira, Indonesia 73 B4 2 3 S 132 2 E
Bîra, Romania 46 B8 47 2N 27 3 E
Biramféro, Guinea 100 C3 11 40N 9 10W
Birao, C.A.R. 102 A4 10 20N 22 47 E
Birawa, Zaïre 106 C2 2 20 S 28 48 E
Bîrca, Romania 46 F4 43 59N 23 36 E
Birch Hills, Canada 131 C7 52 59N 105 25W
Birch I., Canada 131 C9 52 26N 99 54W
Birch L., N.W.T., Canada 130 A5 62 4N 116 33W
Birch L., Ont., Canada 128 B1 51 23N 92 18W
Birch L., U.S.A. 128 C1 47 45N 91 51W
Birch Mts., Canada 130 B6 57 30N 113 10W
Birch River, Canada 131 C8 52 24N 101 6W
Birchip, Australia 116 C5 35 56 S 142 55 E
Birchiş, Romania 46 D3 45 58N 22 9 E
Birchwood, N.Z. 119 F2 45 55 S 167 53 E
Bird, Canada 131 B10 56 30N 94 13W
Bird City, U.S.A. 138 F4 39 45N 101 32W
Bird I. = Aves, I. de, W. Indies 149 C7 15 45N 63 55W
Birdaard, Neths. 20 B7 53 18N 5 53 E
Birdlip, U.K. 17 F5 51 50N 2 5W
Birds, U.S.A. 141 F9 38 50N 87 40W
Birdseye, U.S.A. 141 F10 38 19N 86 42W
Birdsville, Australia 114 D2 25 51 S 139 20 E
Birdum, Australia 112 C5 15 39 S 133 13 E
Birecik, Turkey 89 D8 37 2N 38 0 E
Birein, Israel 91 E3 30 50N 34 28 E
Bireuen, Indonesia 74 A1 5 14N 96 39 E
Birifo, Gambia 100 C2 13 30N 14 0W
Birigui, Brazil 159 A5 21 18 S 50 16W
Birini, C.A.R. 102 A4 7 51N 22 24 E
Birkenfeld, Germany 27 F3 49 38N 7 9 E
Birkenhead, N.Z. 118 C3 36 49 S 174 46 E
Birkenhead, U.K. 16 D4 53 23N 3 2W
Birket Qârûn, Egypt 94 J7 29 30N 30 40 E
Birkfeld, Austria 30 D8 47 21N 15 45 E
Birkhadem, Algeria 99 A5 36 43N 3 3 E
Bîrlad, Romania 46 C8 46 15N 27 38 E
Birlik, Kazakhstan 55 A6 44 5N 73 31 E
Birmingham, U.K. 17 E6 52 29N 1 52W
Birmingham, Ala., U.S.A. 135 J2 33 31N 86 48W
Birmingham, Iowa, U.S.A. 140 D5 40 53N 91 57W
Birmitrapur, India 82 C7 22 24N 84 46 E
Birni Ngaouré, Niger 101 C5 13 5N 2 51 E
Birni Nkonni, Niger 101 C6 13 55N 5 15 E
Birnin Gwari, Nigeria 101 C6 11 0N 6 45 E
Birnin Kebbi, Nigeria 101 C5 12 32N 4 12 E
Birnin Kudu, Nigeria 101 C6 11 30N 9 29 E
Birobidzhan, Russia 57 E14 48 50N 132 50 E
Birougou, Mts., Gabon 102 C2 1 51 S 12 20 E
Birr, Ireland 19 C4 53 7N 7 54W
Birrie →, Australia 115 D4 29 43 S 146 37 E
Birs →, Switz. 28 B5 47 24N 7 32 E
Birsilpur, India 80 E5 28 11N 72 15 E
Birsk, Russia 54 D5 55 25N 55 30 E
Birtin, Romania 46 C3 46 59N 22 31 E
Birtle, Canada 131 C8 50 30N 101 5W
Biryuchiy, Ukraine 51 J8 46 10N 35 0 E
Biržai, Lithuania 13 H21 56 11N 24 45 E
Bîrzava, Romania 46 C2 46 7N 21 59 E
Birzebugga, Malta 32 D2 35 49N 14 32 E
Bisa, Indonesia 72 B3 1 15 S 127 28 E
Bisáccia, Italy 41 A8 41 1N 15 22 E
Bisacquino, Italy 40 E6 37 42N 13 15 E
Bisai, Japan 63 B8 35 16N 136 44 E
Bisalpur, India 81 E8 28 14N 79 48 E
Bisbee, U.S.A. 143 L9 31 27N 109 55W
Biscarrosse et de Parentis, Étang de, France 24 D2 44 21N 1 10W
Biscay, B. of, Atl. Oc. 8 D8 45 0N 2 0W
Biscayne B., U.S.A. 135 N5 25 40N 80 12W
Biscéglie, Italy 41 A9 41 14N 16 30 E
Bischofshofen, Austria 30 D6 47 26N 13 14 E
Bischofswerda, Germany 26 D10 51 7N 14 10 E
Bischofszell, Switz. 29 B8 47 29N 9 15 E
Bischwiller, France 23 D14 48 46N 7 50 E
Biscoe Bay, Antarctica 7 D13 77 0 S 152 0W
Biscoe Is., Antarctica 7 C17 66 0 S 67 0W
Biscostasing, Canada 128 C3 47 18N 82 9W
Biscucuy, Venezuela 152 B4 9 22N 69 59W
Biševo, Croatia 39 F13 42 57N 16 3 E
Bisha, Eritrea 95 D4 15 30N 37 31 E
Bishah, W. →, Si. Arabia 86 B3 21 24N 43 26 E
Bishan, China 68 C6 29 33N 106 12 E
Bishkek, Kyrgyzstan 55 B7 42 54N 74 46 E
Bishnupur, India 81 H12 23 8N 87 20 E
Bisho, S. Africa 105 E4 32 50 S 27 23 E
Bishop, Calif., U.S.A. 144 H8 37 22N 118 24W
Bishop, Tex., U.S.A. 139 M6 27 35N 97 48W
Bishop Auckland, U.K. 16 C6 54 39N 1 40W
Bishop's Falls, Canada 129 C8 49 2N 55 30W
Bishop's Stortford, U.K. 17 F8 51 52N 0 10 E
Bisignano, Italy 41 C9 39 31N 16 17 E
Bisina, L., Uganda 106 B3 1 38N 33 56 E
Biskra, Algeria 99 B6 34 50N 5 44 E
Biskupiec, Poland 47 B7 53 53N 20 58 E
Bismarck, Mo., U.S.A. 140 G6 37 46N 90 38W
Bismarck, N. Dak., U.S.A. 138 B4 46 48N 100 47W
Bismarck Arch., Papua N. G. 120 B5 2 30 S 150 0 E
Bismarck Ra., Papua N. G. 120 C3 5 35 S 145 0 E
Bismarck Sea, Papua N. G. 120 C4 4 10 S 146 50 E
Bismark, Germany 26 C7 52 40N 11 33 E
Bismil, Turkey 89 D9 37 51N 40 40 E
Biso, Uganda 106 B3 1 44N 31 26 E
Bison, U.S.A. 138 C3 45 31N 102 28W
Bīsotūn, Iran 89 E12 34 23N 47 26 E
Bispgården, Sweden 14 A10 63 2N 16 40 E
Bissagos = Bijagós, Arquipélago dos, Guinea-Biss. 100 C1 11 15N 16 10W
Bissau, Guinea-Biss. 100 C1 11 45N 15 45W
Bissett, Canada 131 C9 51 2N 95 41W
Bissikrima, Guinea 100 C2 10 50N 10 58W
Bistcho L., Canada 130 B5 59 45N 118 50W
Bistreţu, Romania 46 F4 43 54N 23 23 E
Bistrica = Ilirska-Bistrica, Slovenia 39 C11 45 34N 14 14 E
Bistriţa, Romania 46 B5 47 9N 24 35 E
Bistriţa →, Romania 46 C7 46 30N 26 57 E
Bistriţa Năsăud □, Romania 46 B5 47 15N 24 30 E
Bistriţei, Munţii, Romania 46 B6 47 15N 25 40 E
Biswan, India 81 F9 27 29N 81 2 E
Bisztynek, Poland 47 A7 54 8N 20 53 E
Bitam, Gabon 102 B2 2 5N 11 25 E
Bitburg, Germany 27 F2 49 58N 6 31 E
Bitche, France 23 C14 49 2N 7 25 E
Bithynia, Turkey 88 B4 40 40N 31 0 E
Bitkine, Chad 97 F3 11 59N 18 13 E
Bitlis, Turkey 89 C10 38 20N 42 3 E
Bitola, Macedonia 42 F6 41 5N 21 10 E
Bitolj = Bitola, Macedonia 42 F6 41 5N 21 10 E
Bitonto, Italy 41 A9 41 6N 16 41 E
Bitter Creek, U.S.A. 142 F9 41 33N 108 33W
Bitter L. = Buheirat-Murrat-el-Kubra, Egypt 94 H8 30 18N 32 26 E
Bitterfeld, Germany 26 D8 51 37N 12 20 E
Bitterfontein, S. Africa 104 E2 31 1 S 18 32 E
Bitterroot →, U.S.A. 142 C6 46 52N 114 7W
Bitterroot Range, U.S.A. 142 D6 46 0N 114 20W
Bitterwater, U.S.A. 144 J6 36 23N 121 0W
Bitti, Italy 40 B2 40 29N 9 23 E
Bittou, Burkina Faso 101 C4 11 17N 0 18W
Biu, Nigeria 101 C7 10 40N 12 3 E
Bivolari, Romania 46 B8 47 31N 27 27 E
Bivolu, Romania 46 B6 47 16N 25 58 E
Biwa-Ko, Japan 63 B8 35 15N 136 10 E
Biwabik, U.S.A. 138 B8 47 32N 92 21W
Bixad, Romania 46 B4 47 56N 23 28 E
Biyang, China 66 H7 32 38N 113 21 E
Biylikol, Ozero, Kazakhstan 55 B5 43 5N 70 45 E
Biysk, Russia 56 D9 52 40N 85 0 E
Bizana, S. Africa 105 E4 30 50 S 29 52 E
Bizen, Japan 62 C6 34 43N 134 8 E
Bizerte, Tunisia 96 A1 37 15N 9 50 E
Bjargtangar, Iceland 12 D1 65 30N 24 30W
Bjelasica, Montenegro, Yug. 42 E4 42 50N 19 40 E
Bjelašnica, Bos.-H. 42 D3 43 43N 18 9 E
Bjelovar, Croatia 42 B1 45 56N 16 49 E
Bjerringbro, Denmark 15 H3 56 23N 9 39 E
Bjervamoen, Norway 14 E3 59 17N 9 5 E
Bjørnevatn, Norway 12 B23 69 40N 30 0 E
Bjørnøya, Arctic 6 B8 74 30N 19 0 E
Bjuv, Sweden 15 H6 56 5N 12 55 E
Blace, Serbia, Yug. 42 D6 43 18N 21 17 E
Blachownia, Poland 47 E5 50 49N 18 56 E
Black = Da →, Vietnam 76 B5 21 15N 105 20 E
Black →, Canada 136 B5 44 42N 79 19W
Black →, Ark., U.S.A. 139 H9 35 38N 91 20W
Black →, N.Y., U.S.A. 137 C8 43 59N 76 4W
Black →, Wis., U.S.A. 138 D9 43 57N 91 22W
Black Diamond, Canada 130 C6 50 45N 114 14W
Black Forest = Schwarzwald, Germany 27 H4 48 30N 8 20 E
Black Hills, U.S.A. 138 C3 44 0N 103 45W
Black I., Canada 131 C9 51 12N 96 30W
Black L., Canada 131 B7 59 12N 105 15W
Black L., U.S.A. 134 C3 45 28N 84 16W
Black Mesa, U.S.A. 139 G3 36 58N 102 58W
Black Mountain, Australia 117 A9 30 18 S 151 39 E
Black Mt. = Mynydd Du, U.K. 17 F4 51 52N 3 50W
Black Mts., U.K. 17 F4 51 55N 3 7W
Black Range, U.S.A. 143 K10 33 15N 107 50W
Black River, Jamaica 148 C4 18 0N 77 50W
Black River Falls, U.S.A. 138 C9 44 18N 90 51W
Black Rock, Australia 116 B3 32 50 S 138 44 E
Black Sea, Eurasia 49 F6 43 30N 35 0 E
Black Volta →, Africa 100 D4 8 41N 1 33W
Black Warrior →, U.S.A. 135 J2 32 32N 87 51W
Blackall, Australia 114 C4 24 25 S 145 45 E
Blackball, N.Z. 119 C6 42 22 S 171 26 E
Blackbull, Australia 114 B3 17 55 S 141 45 E

178 Brædstrup

Brædstrup, Denmark 15 J3 55 58N 9 37 E
Braemar, Australia 116 B3 33 12 S 139 35 E
Braeside, Canada 137 A8 45 28N 76 24W
Braga, Portugal 36 D2 41 35N 8 25W
Braga □, Portugal 36 D2 41 30N 8 30W
Bragado, Argentina 158 D3 35 2 S 60 27W
Bragança, Brazil 154 B2 1 0 S 47 2W
Bragança, Portugal 36 D4 41 48N 6 50W
Bragança □, Portugal 36 D4 41 30N 6 45W
Bragança Paulista, Brazil 159 A6 22 55 S 46 32W
Brahmanbaria, Bangla. 78 D3 23 58N 91 15 E
Brahmani →, India 82 D8 20 39N 86 46 E
Brahmaputra →, India 78 D2 23 58N 89 50 E
Braich-y-pwll, U.K. 16 E3 52 47N 4 46W
Braidwood, Australia 117 C8 35 27 S 149 49 E
Brăila, Romania 46 D8 45 19N 27 59 E
Brăila □, Romania 46 D8 45 5N 27 30 E
Braine-l'Alleud, Belgium 21 G4 50 42N 4 23 E
Braine-le-Comte, Belgium 21 G4 50 37N 4 8 E
Brainerd, U.S.A. 138 B7 46 22N 94 12W
Braintree, U.K. 17 F8 51 53N 0 34 E
Braintree, U.S.A. 137 D14 42 13N 71 0W
Brak →, S. Africa 104 D3 29 35 S 22 55 E
Brake, Germany 26 B4 53 20N 8 28 E
Brakel, Germany 26 D5 51 42N 9 11 E
Brakel, Neths. 20 E6 51 49N 5 5 E
Brakwater, Namibia 104 C2 22 28 S 17 3 E
Brålanda, Sweden 15 F6 58 34N 12 21 E
Bralorne, Canada 130 C4 50 50N 122 50W
Bramberg, Germany 27 E6 50 6N 10 40 E
Bramminge, Denmark 15 J2 55 28N 8 42 E
Brämön, Sweden 14 B11 62 14N 17 40 E
Brampton, Canada 128 D4 43 45N 79 45W
Bramsche, Germany 26 C3 52 24N 7 59 E
Bramwell, Australia 114 A3 12 8 S 142 37 E
Branco →, Brazil 153 D5 1 20 S 61 50W
Branco, C., Brazil 154 C5 7 9 S 34 47W
Brande, Denmark 15 J3 55 57N 9 8 E
Brandenburg = Neubrandenburg, Germany 26 B9 53 33N 13 15 E
Brandenburg, Germany 26 C8 52 25N 12 33 E
Brandenburg, U.S.A. 141 G10 38 0N 86 10W
Brandenburg □, Germany 26 C9 52 50N 13 0 E
Brandfort, S. Africa 104 D4 28 40 S 26 30 E
Brandon, Canada 131 D9 49 50N 99 57W
Brandon, U.S.A. 137 C11 43 48N 73 4W
Brandon B., Ireland 19 D1 52 17N 10 8W
Brandon Mt., Ireland 19 D1 52 15N 10 15W
Brandsen, Argentina 158 D4 35 10 S 58 15W
Brandval, Norway 14 D6 60 19N 12 1 E
Brandvlei, S. Africa 104 E3 30 25 S 20 30 E
Brandýs, Czech. 30 A7 50 10N 14 40 E
Branford, U.S.A. 137 E12 41 17N 72 49W
Braniewo, Poland 47 A6 54 25N 19 50 E
Bransfield Str., Antarctica 7 C18 63 0 S 59 0W
Brańsk, Poland 47 C9 52 45N 22 50 E
Branson, Colo., U.S.A. 139 G3 37 1N 103 53W
Branson, Mo., U.S.A. 139 G8 36 39N 93 13W
Brantford, Canada 128 D3 43 10N 80 15W
Brantôme, France 24 C4 45 22N 0 39 E
Branxholme, Australia 116 D4 37 52 S 141 49 E
Branxton, Australia 117 B9 32 38 S 151 21 E
Branzi, Italy 38 B6 46 0N 9 46 E
Bras d'Or, L., Canada 129 C7 45 50N 60 50W
Brasiléia, Brazil 156 C4 11 0 S 68 45W
Brasília, Brazil 155 E2 15 47 S 47 55W
Brasília Legal, Brazil 153 D6 3 49 S 55 36W
Braslaw, Belarus 13 J22 55 38N 27 0 E
Braslovce, Slovenia 39 B12 46 21N 15 3 E
Braşov, Romania 46 D6 45 38N 25 35 E
Braşov □, Romania 46 D6 45 45N 25 15 E
Brass, Nigeria 101 E6 4 35N 6 14 E
Brass →, Nigeria 101 E6 4 15N 6 13 E
Brassac-les-Mines, France 24 C7 45 24N 3 20 E
Brasschaat, Belgium 21 F4 51 19N 4 27 E
Brassey, Banjaran, Malaysia 75 B5 5 0N 117 15 E
Brassey Ra., Australia 113 E3 25 8 S 122 15 E
Brasstown Bald, U.S.A. 135 H4 34 53N 83 49W
Brastad, Sweden 13 G14 58 23N 11 30 E
Bratan = Morozov, Bulgaria 43 E10 42 30N 25 10 E
Bratislava, Slovak Rep. 31 C10 48 10N 17 7 E
Bratsigovo, Bulgaria 43 E9 42 1N 24 22 E
Bratsk, Russia 57 D11 56 10N 101 30 E
Brattleboro, U.S.A. 137 D12 42 51N 72 34W
Bratunac, Bos.-H. 42 C4 44 13N 19 21 E
Braunau, Austria 30 C6 48 15N 13 3 E
Braunschweig, Germany 26 C6 52 15N 10 31 E
Braunton, U.K. 17 F3 51 7N 4 10W
Brava, Somali Rep. 108 D2 1 20N 44 8 E
Bråviken, Sweden 14 F10 58 38N 16 32 E
Bravo del Norte →, Mexico 146 B5 25 57N 97 9W
Bravo del Norte, R. = Grande, Rio →, U.S.A. 139 N6 25 58N 97 9W
Brawley, U.S.A. 145 N11 32 59N 115 31W
Bray, Ireland 19 C5 53 13N 6 7W
Bray, Mt., Australia 114 A1 14 0 S 134 30 E
Bray, Pays de, France 23 C8 49 46N 1 26 E
Bray-sur-Seine, France 23 D10 48 25N 3 14 E
Braymer, U.S.A. 140 E3 39 35N 93 48W
Brazeau →, Canada 130 C5 52 55N 115 14W
Brazil, U.S.A. 141 E9 39 32N 87 8W
Brazil ■, S. Amer. 155 D2 12 0 S 50 0W
Brazo Sur →, S. Amer. 158 B4 25 21 S 57 42W
Brazos →, U.S.A. 139 L7 28 53N 95 23W
Brazzaville, Congo 103 C3 4 9 S 15 12 E
Brčko, Bos.-H. 42 C3 44 54N 18 46 E
Brda →, Poland 47 B5 53 8N 18 8 E
Brea, Peru 156 A1 4 40 S 81 7W
Breadalbane, Australia 114 C2 23 50 S 139 35 E
Breadalbane, U.K. 18 E4 56 30N 4 15W
Breaden, L., Australia 113 E4 25 51 S 125 28 E
Breaksea Sd., N.Z. 119 F1 45 35 S 166 35 E
Bream B., N.Z. 118 B3 35 56 S 174 28 E
Bream Hd., N.Z. 118 B3 35 51 S 174 36 E
Bream Tail, N.Z. 118 C3 36 3 S 174 36 E
Breas, Chile 158 B1 25 29 S 70 24W
Brebes, Indonesia 75 D3 6 52 S 109 3 E
Brechin, Canada 136 B5 44 32N 79 10W
Brechin, U.K. 18 E6 56 44N 2 39W
Brecht, Belgium 21 F5 51 21N 4 38 E
Breckenridge, Colo., U.S.A. 142 G10 39 29N 106 3W
Breckenridge, Minn., U.S.A. 138 B6 46 16N 96 35W
Breckenridge, Mo., U.S.A. 140 E3 39 46N 93 48W
Breckenridge, Tex., U.S.A. 139 J5 32 45N 98 54W
Breckland, U.K. 17 E8 52 30N 0 40 E
Brecknock, Pen., Chile 160 D2 54 35 S 71 30W
Břeclav, Czech. 31 C9 48 46N 16 53 E
Brecon, U.K. 17 F4 51 57N 3 23W
Brecon Beacons, U.K. 17 F4 51 53N 3 26W
Breda, Neths. 21 E5 51 35N 4 45 E
Bredasdorp, S. Africa 104 E3 34 33 S 20 2 E
Bredbo, Australia 117 C8 35 58 S 149 10 E
Bredene, Belgium 21 F1 51 14N 2 59 E
Bredstedt, Germany 26 A4 54 37N 8 55 E
Bredy, Russia 54 E8 52 26N 60 21 E
Bree, Belgium 21 F7 51 8N 5 35 E
Breezand, Neths. 20 C5 52 53N 4 49 E
Bregalnica →, Macedonia 42 F7 41 43N 22 9 E
Bregenz, Austria 30 D2 47 30N 9 45 E
Bregovo, Bulgaria 42 C7 44 9N 22 39 E
Bréhal, France 22 D5 48 53N 1 30W
Bréhat, I. de, France 22 D4 48 51N 3 0W
Breiðafjörður, Iceland 12 D2 65 15N 23 15W
Breil-sur-Roya, France 25 E11 43 56N 7 31 E
Breisach, Germany 27 G3 48 1N 7 36 E
Brejinho de Nazaré, Brazil 154 D2 11 1 S 48 34W
Brejo, Brazil 154 B3 3 41 S 42 47W
Bremen, Germany 26 B4 53 4N 8 47 E
Bremen □, Germany 26 B4 53 4N 8 50 E
Bremer I., Australia 114 A2 12 5 S 136 45 E
Bremerhaven, Germany 26 B4 53 33N 8 36 E
Bremerton, U.S.A. 144 C4 47 34N 122 38W
Bremervörde, Germany 26 B5 53 29N 9 8 E
Bremsnes, Norway 14 A1 63 6N 7 40 E
Brenes, Spain 37 H5 37 32N 5 54W
Brenham, U.S.A. 139 K6 30 10N 96 24W
Brenner P., Austria 30 D4 47 2N 11 30 E
Breno, Italy 38 C7 45 57N 10 18 E
Brent, Canada 128 C4 46 2N 78 29W
Brent, U.K. 17 F7 51 33N 0 16W
Brenta →, Italy 39 C9 45 11N 12 18 E
Brentwood, U.K. 17 F8 51 37N 0 19 E
Brentwood, U.S.A. 137 F11 40 47N 73 15W
Bréscia, Italy 38 C7 45 33N 10 15 E
Breskens, Neths. 21 F3 51 23N 3 33 E
Breslau = Wrocław, Poland 47 D4 51 5N 17 5 E
Bresle →, France 22 B8 50 4N 1 22 E
Bresles, France 23 C9 49 25N 2 13 E
Bressanone, Italy 39 B8 46 43N 11 39 E
Bressay, U.K. 18 A7 60 9N 1 6W
Bresse, France 23 F12 46 50N 5 10 E
Bressuire, France 22 F6 46 51N 0 30W
Brest, Belarus 51 F2 52 10N 23 40 E
Brest, France 22 D2 48 24N 4 31W
Brest-Litovsk = Brest, Belarus 51 F2 52 10N 23 40 E
Bretagne, France 22 D4 48 10N 3 0W
Bretçu, Romania 46 C7 46 7N 26 18 E
Breteuil, Eure, France 22 D7 48 50N 0 53 E
Breteuil, Oise, France 23 C9 49 38N 2 18 E
Breton, Canada 130 C6 53 7N 114 28W
Breton, Pertuis, France 24 B2 46 17N 1 25W
Breton Sd., U.S.A. 139 L10 29 35N 89 15W
Brett, C., N.Z. 118 B3 35 10 S 174 20 E
Bretten, Germany 27 F4 49 2N 8 42 E
Breukelen, Neths. 20 D6 52 10N 5 0 E
Brevard, U.S.A. 135 H4 35 14N 82 44W
Breves, Brazil 154 B1 1 40 S 50 29W
Brevik, Norway 14 E3 59 4N 9 42 E
Brewarrina, Australia 115 D4 30 0 S 146 51 E
Brewer, U.S.A. 129 D6 44 48N 68 46W
Brewer, Mt., U.S.A. 144 J8 36 44N 118 28W
Brewster, N.Y., U.S.A. 137 E11 41 23N 73 37W
Brewster, Wash., U.S.A. 142 B4 48 6N 119 47W
Brewster, Kap, Greenland 6 B6 70 7N 22 0W
Brewton, U.S.A. 135 K2 31 7N 87 4W
Breyten, S. Africa 105 D4 26 16 S 30 0 E
Brezhnev = Naberezhnyye Chelny, Russia 52 C11 55 42N 52 19 E
Brežice, Slovenia 39 C12 45 54N 15 35 E
Brézina, Algeria 99 B5 33 4N 1 14 E
Březnice, Czech. 30 B6 49 32N 13 57 E
Breznik, Bulgaria 42 E7 42 44N 22 50 E
Brezno, Slovak Rep. 31 C12 48 50N 19 40 E
Brezovo, Bulgaria 43 E10 42 21N 25 5 E
Bria, C.A.R. 102 A4 6 30N 21 58 E
Briançon, France 25 D10 44 54N 6 39 E
Briare, France 23 E9 47 38N 2 45 E
Bribbaree, Australia 117 C7 34 10 S 147 51 E
Bribie I., Australia 115 D5 27 0 S 153 10 E
Bricquebec, France 22 C5 49 28N 1 38W
Bridgehampton, U.S.A. 137 F12 40 56N 72 19W
Bridgend, U.K. 17 F4 51 30N 3 34W
Bridgeport, Calif., U.S.A. 144 G7 38 15N 119 14W
Bridgeport, Conn., U.S.A. 137 E11 41 11N 73 12W
Bridgeport, Nebr., U.S.A. 138 E3 41 40N 103 6W
Bridgeport, Tex., U.S.A. 139 J6 33 13N 97 45W
Bridger, U.S.A. 142 D9 45 18N 108 55W
Bridgeton, U.S.A. 134 F8 39 26N 75 14W
Bridgetown, Australia 113 F2 33 58 S 116 7 E
Bridgetown, Barbados 149 D8 13 5N 59 30W
Bridgetown, Canada 129 D6 44 55N 65 18W
Bridgewater, Australia 116 D5 36 36 S 143 59 E
Bridgewater, Canada 129 D7 44 25N 64 31W
Bridgewater, Mass., U.S.A. 137 E14 41 59N 70 58W
Bridgewater, S. Dak., U.S.A. 138 D6 43 33N 97 30W
Bridgewater, C., Australia 116 E4 38 23 S 141 23 E
Bridgman, U.S.A. 141 C10 41 57N 86 33W
Bridgnorth, U.K. 17 E5 52 32N 2 25W
Bridgton, U.S.A. 137 B14 44 3N 70 42W
Bridgwater, U.K. 17 F4 51 8N 2 59W
Bridlington, U.K. 16 C7 54 5N 0 12W
Bridport, Australia 114 G4 40 59 S 147 23 E
Bridport, U.K. 17 G5 50 44N 2 45W
Brie, Plaine de la, France 23 D10 48 35N 3 10 E
Brie-Comte-Robert, France 23 D9 48 40N 2 35 E
Briec, France 22 D2 48 6N 4 0W
Brielle, Neths. 20 E4 51 54N 4 10 E
Brienne-le-Château, France 23 D11 48 24N 4 30 E
Brienon-sur-Armançon, France 23 E10 47 59N 3 38 E
Brienz, Switz. 28 C6 46 46N 8 2 E
Brienzersee, Switz. 28 C5 46 44N 7 53 E
Brig, Switz. 28 D5 46 18N 7 59 E
Brigg, U.K. 16 D7 53 34N 0 28W
Briggsdale, U.S.A. 138 E2 40 38N 104 20W
Brigham City, U.S.A. 142 F7 41 31N 112 1W
Bright, Australia 117 D7 36 42 S 146 56 E
Brighton, Australia 116 C3 35 5 S 138 30 E
Brighton, Canada 128 D4 44 2N 77 44W
Brighton, U.K. 17 G7 50 49N 0 7W
Brighton, Colo., U.S.A. 138 F2 39 59N 104 49W
Brighton, Ill., U.S.A. 140 E6 39 2N 90 8W
Brighton, Iowa, U.S.A. 140 C5 41 10N 91 49W
Brightwater, N.Z. 119 B8 41 22 S 173 9 E
Brignogan-Plage, France 22 D2 48 40N 4 20W
Brignoles, France 25 E10 43 25N 6 5 E
Brihuega, Spain 34 E2 40 45N 2 52W
Brikama, Gambia 100 C1 13 15N 16 45W
Brilliant, Canada 130 D5 49 19N 117 38W
Brilliant, U.S.A. 136 F4 40 15N 80 39W
Brilon, Germany 26 D4 51 23N 8 25 E
Brim, Australia 116 D5 36 3 S 142 27 E
Brimfield, U.S.A. 140 D7 40 50N 89 53W
Bríndisi, Italy 41 B10 40 39N 17 55 E
Brinje, Croatia 39 D12 45 0N 15 9 E
Brinkley, U.S.A. 139 H9 34 53N 91 12W
Brinkworth, Australia 116 B3 33 42 S 138 26 E
Brinnon, U.S.A. 144 C4 47 41N 122 54W
Brion, I., Canada 129 C7 47 46N 61 26W
Brionne, France 22 C7 49 11N 0 43 E
Brionski, Croatia 39 D10 44 55N 13 45 E
Brioude, France 24 C7 45 18N 3 24 E
Briouze, France 22 D6 48 42N 0 23W
Brisbane, Australia 115 D5 27 25 S 153 2 E
Brisbane →, Australia 115 D5 27 24 S 153 9 E
Brisighella, Italy 39 D8 44 13N 11 46 E
Bristol, U.K. 17 F5 51 26N 2 35W
Bristol, Conn., U.S.A. 137 E12 41 40N 72 57W
Bristol, Pa., U.S.A. 137 F10 40 6N 74 51W
Bristol, R.I., U.S.A. 137 E13 41 40N 71 16W
Bristol, S. Dak., U.S.A. 138 C6 45 21N 97 45W
Bristol, Tenn., U.S.A. 135 G4 36 36N 82 11W
Bristol B., U.S.A. 126 C4 58 0N 160 0W
Bristol Channel, U.K. 17 F3 51 18N 4 30W
Bristol I., Antarctica 7 B1 58 45 S 28 0W
Bristol L., U.S.A. 143 J5 34 23N 116 50W
Bristow, U.S.A. 139 H6 35 50N 96 23W
British Columbia □, Canada 130 C3 55 0N 125 15W
British Isles, Europe 10 E5 54 0N 4 0W
Brits, S. Africa 105 D4 25 37 S 27 48 E
Britstown, S. Africa 104 E3 30 37 S 23 30 E
Britt, Canada 128 C3 45 46N 80 34W
Britt, U.S.A. 140 A3 43 6N 93 48W
Brittany = Bretagne, France 22 D4 48 10N 3 0W
Britton, U.S.A. 138 C6 45 48N 97 45W
Brive-la-Gaillarde, France 24 C5 45 10N 1 32 E
Briviesca, Spain 34 C1 42 32N 3 19W
Brixen = Bressanone, Italy 39 B8 46 43N 11 39 E
Brixton, Australia 114 C3 23 32 S 144 57 E
Brlik = Birlik, Kazakhstan 55 A6 44 5N 73 31 E
Brlik, Kazakhstan 55 B6 43 40N 73 49 E
Brno, Czech. 31 B9 49 10N 16 35 E
Bro, Sweden 14 E11 59 31N 17 38 E
Broach = Bharuch, India 82 D1 21 47N 73 0 E
Broad →, U.S.A. 135 J5 34 1N 81 4W
Broad Arrow, Australia 113 F3 30 23 S 121 15 E
Broad B., U.K. 18 C2 58 14N 6 18W
Broad Haven, Ireland 19 B2 54 20N 9 55W
Broad Law, U.K. 18 F5 55 30N 3 21W
Broad Sd., Australia 114 C4 22 0 S 149 45 E
Broadford, Australia 117 D6 37 14 S 145 4 E
Broadhurst Ra., Australia 112 D3 22 30 S 122 30 E
Broads, The, U.K. 16 E9 52 45N 1 30 E
Broadus, U.S.A. 138 C2 45 27N 105 25W
Broadview, Canada 131 C8 50 22N 102 35W
Broager, Denmark 15 K3 54 53N 9 40 E
Broaryd, Sweden 15 G7 57 7N 13 15 E
Brochet, Canada 131 B8 57 53N 101 40W
Brochet, L., Canada 131 B8 58 36N 101 35W
Brock, Canada 131 C7 51 26N 108 43W
Brocken, Germany 26 D6 51 47N 10 37 E
Brocklehurst, Australia 117 B8 32 9 S 148 38 E
Brockport, U.S.A. 136 C7 43 13N 77 56W
Brockton, U.S.A. 137 D13 42 5N 71 1W
Brockville, Canada 128 D4 44 35N 75 41W
Brockway, Mont., U.S.A. 138 B2 47 18N 105 45W
Brockway, Pa., U.S.A. 136 E6 41 15N 78 47W
Brocton, U.S.A. 136 D5 42 23N 79 26W
Brod, Macedonia 42 F6 41 35N 21 17 E
Brodarevo, Serbia, Yug. 42 D4 43 14N 19 44 E
Brodeur Pen., Canada 127 A11 72 30N 88 10W
Brodhead, U.S.A. 140 B7 42 37N 89 22W
Brodick, U.K. 18 F3 55 35N 5 9W
Brodnica, Poland 47 B6 53 15N 19 25 E
Brody, Ukraine 51 G3 50 5N 25 10 E
Broechem, Belgium 21 F5 51 11N 4 38 E
Broek, Neths. 20 D6 52 26N 5 0 E
Broek op Langedijk, Neths. 20 C5 52 41N 4 49 E
Brogan, U.S.A. 142 D5 44 15N 117 31W
Broglie, France 22 C7 49 2N 0 30 E
Brok, Poland 47 C8 52 43N 21 52 E
Broken Bow, Nebr., U.S.A. 138 E5 41 24N 99 38W
Broken Bow, Okla., U.S.A. 139 H7 34 2N 94 44W
Broken Hill = Kabwe, Zambia 107 E2 14 30 S 28 29 E
Broken Hill, Australia 116 A4 31 58 S 141 29 E
Brokind, Sweden 15 F9 58 13N 15 42 E
Brokopondo, Surinam 153 B7 5 3N 54 59W
Brokopondo □, Surinam 153 C6 4 30N 55 30W
Bromfield, U.K. 17 E5 52 24N 2 45W
Bromley, U.K. 17 F8 51 24N 0 2 E
Bronaugh, U.S.A. 140 G2 37 41N 94 28W
Brønderslev, Denmark 15 G3 57 16N 9 57 E
Brong-Ahafo □, Ghana 100 D4 7 50N 2 0W
Bronkhorstspruit, S. Africa 105 D4 25 46 S 28 45 E
Brønnøysund, Norway 12 D15 65 28N 12 14 E
Bronson, U.S.A. 141 C11 41 52N 85 12W
Bronte, Italy 41 E7 37 47N 14 50 E
Bronte, U.S.A. 139 K4 31 53N 100 18W
Bronte Park, Australia 114 G4 42 8 S 146 30 E
Brook Park, U.S.A. 136 E4 41 24N 80 51W
Brookes Point, Phil. 71 G1 8 47N 117 50 E
Brookfield, U.S.A. 140 E3 39 47N 93 4W
Brookhaven, U.S.A. 139 K9 31 35N 90 26W
Brookings, Oreg., U.S.A. 142 E1 42 3N 124 17W
Brookings, S. Dak., U.S.A. 138 C6 44 19N 96 48W
Brooklin, Canada 136 C6 43 55N 78 55W
Brooklyn, U.S.A. 140 C4 41 44N 92 27W
Brookmere, Canada 130 D4 49 52N 120 53W
Brooks, Canada 130 C6 50 35N 111 55W
Brooks B., Canada 130 C3 50 15N 127 55W
Brooks L., Canada 131 A7 61 55N 106 35W
Brooks Ra., U.S.A. 126 B5 68 40N 147 0W
Brookston, U.S.A. 141 D10 40 36N 86 52W
Brooksville, Fla., U.S.A. 135 L4 28 33N 82 23W
Brooksville, Ky., U.S.A. 141 F12 38 41N 84 4W
Brookville, U.S.A. 141 E11 39 25N 85 1W
Brooloo, Australia 115 D5 26 30 S 152 43 E
Broom, L., U.K. 18 D3 57 55N 5 15W
Broome, Australia 112 C3 18 0 S 122 15 E
Broomehill, Australia 113 F2 33 51 S 117 39 E
Broons, France 22 D4 48 20N 2 16W
Brora, U.K. 18 C5 58 0N 3 52W
Brora →, U.K. 18 C5 58 0N 3 51W
Brosna →, Ireland 19 C4 53 14N 7 58W
Broşteni, Romania 46 B6 47 14N 25 43 E
Brotas de Macaúbas, Brazil 155 D3 12 0 S 42 38W
Brothers, U.S.A. 142 E3 43 49N 120 36W
Brøttum, Norway 14 C4 61 2N 10 34 E
Brou, France 22 D8 48 13N 1 11 E
Brouage, Ramparts de, France 24 C2 45 52N 1 4W
Brough, U.K. 16 C5 54 32N 2 18W
Broughams Gate, Australia 116 A4 30 51 S 140 59 E
Broughton, U.S.A. 141 G8 37 56N 88 27W
Broughton Island, Canada 127 B13 67 33N 63 0W
Broughty Ferry, U.K. 18 E6 56 29N 2 51W
Broumov, Czech. 31 A9 50 35N 16 20 E
Brouwershaven, Neths. 20 E3 51 45N 3 55 E
Brouwershavensche Gat, Neths. 20 E3 51 46N 3 50 E
Brovary, Ukraine 51 G6 50 34N 30 48 E
Brovst, Denmark 15 G3 57 6N 9 31 E
Browerville, U.S.A. 138 B7 46 5N 94 52W
Brown, Mt., Australia 116 B2 32 30 S 138 0 E
Brown, Pt., Australia 115 E1 32 32 S 133 50 E
Brown Willy, U.K. 17 G3 50 35N 4 37W
Brownfield, U.S.A. 139 J3 33 11N 102 17W
Browning, Ill., U.S.A. 140 D6 40 8N 90 22W
Browning, Mo., U.S.A. 140 D3 40 3N 93 12W
Browning, Mont., U.S.A. 142 B7 48 34N 113 1W
Browning Pass, N.Z. 119 C6 42 55 S 171 22 E
Brownlee, Canada 131 C7 50 43N 106 1W
Brownsburg, U.S.A. 141 E10 39 51N 86 24W
Brownstown, U.S.A. 141 F10 38 53N 86 3W
Brownsville, Oreg., U.S.A. 142 D2 44 24N 122 59W
Brownsville, Tenn., U.S.A. 139 H10 35 36N 89 16W
Brownsville, Tex., U.S.A. 139 N6 25 54N 97 30W
Brownsweg, Surinam 153 B6 5 5N 55 15W
Brownwood, U.S.A. 139 K5 31 43N 98 59W
Brownwood, L., U.S.A. 139 K5 31 51N 98 35W
Browse I., Australia 112 B3 14 7 S 123 33 E
Broye →, Switz. 28 C3 46 52N 6 58 E
Brozas, Spain 37 F4 39 37N 6 47W
Bruas, Malaysia 77 K3 4 30N 100 47 E
Bruay-en-Artois, France 23 B9 50 29N 2 33 E
Bruce, Mt., Australia 112 D2 22 37 S 118 8 E
Bruce B., N.Z. 119 D4 43 35 S 169 42 E
Bruce Pen., Canada 136 A3 45 0N 81 30W
Bruce Rock, Australia 113 F2 31 52 S 118 8 E
Bruche →, France 23 D14 48 34N 7 43 E
Bruchsal, Germany 27 F4 49 7N 8 35 E
Bruck an der Leitha, Austria 31 C9 48 1N 16 47 E
Bruck an der Mur, Austria 30 D8 47 24N 15 16 E
Brue →, U.K. 17 F5 51 10N 2 59W
Brugelette, Belgium 21 G3 50 35N 3 52 E
Bruges = Brugge, Belgium 21 F2 51 13N 3 13 E
Brugg, Switz. 28 B6 47 29N 8 11 E
Brugge, Belgium 21 F2 51 13N 3 13 E
Brühl, Germany 26 E2 50 49N 6 54 E
Bruinisse, Neths. 21 E4 51 40N 4 5 E
Brûlé, Canada 130 C5 53 15N 117 58W
Brûlon, France 22 E6 47 58N 0 15W
Brûly, Belgium 21 J5 49 58N 4 32 E
Brumado, Brazil 155 D3 14 14 S 41 40W
Brumado →, Brazil 155 D3 14 13 S 41 40W
Brumath, France 23 D14 48 43N 7 40 E
Brummen, Neths. 20 D8 52 5N 6 10 E
Brumunddal, Norway 14 D4 60 53N 10 56 E
Brunchilly, Australia 114 B1 18 50 S 134 30 E
Brundidge, U.S.A. 135 K3 31 43N 85 49W
Bruneau, U.S.A. 142 E6 42 53N 115 48W
Bruneau →, U.S.A. 142 E6 42 56N 115 57W
Bruneck = Brunico, Italy 39 B8 46 48N 11 56 E
Brunei = Bandar Seri Begawan, Brunei 75 B4 4 52N 115 0 E
Brunei ■, Asia 75 B4 4 50N 115 0 E
Brunette Downs, Australia 114 B2 18 40 S 135 55 E
Brunflo, Sweden 14 A8 63 5N 14 50 E
Brunico, Italy 39 B8 46 48N 11 56 E
Brünig, P., Switz. 28 C6 46 46N 8 8 E
Brunkeberg, Norway 14 E2 59 26N 8 28 E
Brunna, Sweden 14 E11 59 52N 17 25 E
Brunnen, Switz. 29 C7 46 59N 8 37 E
Brunner, L., N.Z. 119 C6 42 37 S 171 27 E
Bruno, Canada 131 C7 52 20N 105 30W
Brunsbüttel, Germany 26 B5 53 53N 9 6 E
Brunssum, Neths. 21 G7 50 57N 5 59 E
Brunswick = Braunschweig, Germany 26 C6 52 15N 10 31 E
Brunswick, Ga., U.S.A. 135 K5 31 10N 81 30W
Brunswick, Maine, U.S.A. 129 D6 43 55N 69 58W
Brunswick, Md., U.S.A. 134 F7 39 19N 77 38W
Brunswick, Mo., U.S.A. 140 E3 39 26N 93 8W
Brunswick, Ohio, U.S.A. 136 E3 41 14N 81 51W
Brunswick, Pen. de, Chile 160 D2 53 30 S 71 30W
Brunswick B., Australia 112 C3 15 15 S 124 50 E
Brunswick Junction, Australia 113 F2 33 15 S 115 50 E

Būshehr 179

Bruntál, Czech. 31 B10 50 0N 17 27 E
Bruny I., Australia 114 G4 43 20 S 147 15 E
Brus Laguna, Honduras 148 C3 15 47N 84 35W
Brusartsi, Bulgaria 42 D8 43 40N 23 5 E
Brush, U.S.A. 138 E3 40 15N 103 37W
Brushton, U.S.A. 137 B10 44 50N 74 31W
Brusio, Switz. 29 D10 46 14N 10 8 E
Brusque, Brazil 159 B6 27 5 S 49 0W
Brussel, Belgium 21 G4 50 51N 4 21 E
Brussels = Brussel, Belgium 21 G4 50 51N 4 21 E
Brussels, Canada 136 C3 43 44N 81 15W
Brustem, Belgium 21 G6 50 48N 5 14 E
Bruthen, Australia 117 D7 37 42 S 147 50 E
Bruxelles = Brussel, Belgium 21 G4 50 51N 4 21 E
Bruyères, France 23 D13 48 10N 6 40 E
Brwinów, Poland 47 C7 52 9N 20 40 E
Bryagovo, Bulgaria 43 F10 41 58N 25 8 E
Bryan, Ohio, U.S.A. 141 C12 41 28N 84 33W
Bryan, Tex., U.S.A. 139 K6 30 40N 96 22W
Bryan, Mt., Australia 116 B3 33 30 S 139 0 E
Bryanka, Ukraine 51 H10 48 32N 38 45 E
Bryansk, Russia 52 D2 53 13N 34 25 E
Bryansk, Russia 53 H8 44 20N 47 10 E
Bryanskoye = Bryansk, Russia 53 H8 44 20N 47 10 E
Bryant, U.S.A. 138 C6 44 35N 97 28W
Bryne, Norway 13 G11 58 44N 5 38 E
Bryson City, U.S.A. 135 H4 35 26N 83 27W
Bryukhovetskaya, Russia 53 H4 45 48N 39 0 E
Brza Palanka, Serbia, Yug. 42 C7 44 28N 22 27 E
Brzava →, Serbia, Yug. 42 B5 45 21N 20 45 E
Brzeg, Poland 47 E4 50 52N 17 30 E
Brzeg Din, Poland 47 D3 51 16N 16 41 E
Brześć Kujawski, Poland 47 C5 52 36N 18 55 E
Brzesko, Poland 31 B13 49 59N 20 34 E
Brzeszcze, Poland 31 B12 49 59N 19 10 E
Brzeziny, Poland 47 D6 51 49N 19 42 E
Brzozów, Poland 31 B15 49 41N 22 3 E
Bsharri, Lebanon 91 A5 34 15N 36 0 E
Bū Athlah, Libya 96 B3 30 9N 15 39 E
Bū Baqarah, U.A.E. 85 E8 25 35N 56 25 E
Bu Craa, W. Sahara 98 C2 26 45N 12 50W
Bū Ḩasā, U.A.E. 85 F7 23 30N 53 20 E
Bua Yai, Thailand 76 E4 15 33N 102 26 E
Buad I., Phil. 71 F5 11 40N 124 51 E
Buala, Solomon Is. 121 M10 8 10 S 159 35 E
Buapinang, Indonesia 72 B2 4 40 S 121 30 E
Buba, Guinea-Biss. 100 C2 11 40N 14 59W
Bubanda, Zaïre 102 B3 4 14N 19 38 E
Bubanza, Burundi 106 C2 3 6 S 29 23 E
Būbiyān, Kuwait 85 D6 29 45N 48 15 E
Bucak, Turkey 88 D4 37 28N 30 36 E
Bucaramanga, Colombia 152 B3 7 0N 73 0W
Bucas Grande I., Phil. 71 G5 9 40N 125 57 E
Buccaneer Arch., Australia 112 C3 16 7 S 123 20 E
Bucchiánico, Italy 39 F11 42 18N 14 11 E
Bucecea, Romania 46 B7 47 47N 26 28 E
Buchach, Ukraine 51 H3 49 5N 25 25 E
Buchan, Australia 117 D8 37 30 S 148 12 E
Buchan, U.K. 18 D6 57 32N 2 21W
Buchan Ness, U.K. 18 D7 57 29N 1 46W
Buchanan, Canada 131 C8 51 40N 102 45W
Buchanan, Liberia 100 D2 5 57N 10 2W
Buchanan, U.S.A. 141 C10 41 50N 86 22W
Buchanan, L., Queens., Australia 114 C4 21 35 S 145 52 E
Buchanan, L., W. Austral., Australia 113 E3 25 33 S 123 2 E
Buchanan, L., U.S.A. 139 K5 30 45N 98 25W
Buchanan Cr. →, Australia 114 B2 19 13 S 136 33 E
Buchans, Canada 129 C8 48 50N 56 52W
Bucharest = Bucureşti, Romania 46 E7 44 27N 26 10 E
Buchholz, Germany 26 B5 53 19N 9 52 E
Buchloe, Germany 27 G6 48 1N 10 44 E
Buchon, Pt., U.S.A. 144 K6 35 15N 120 54W
Buchs, Switz. 29 B8 47 10N 9 28 E
Bückeburg, Germany 26 C5 52 16N 9 7 E
Buckeye, U.S.A. 143 K7 33 22N 112 35W
Buckhannon, U.S.A. 134 F5 39 0N 80 8W
Buckhaven, U.K. 18 E5 56 11N 3 3W
Buckie, U.K. 18 D6 57 41N 2 58W
Buckingham, Canada 128 C4 45 37N 75 24W
Buckingham, U.K. 17 F7 51 59N 0 57W
Buckingham B., Australia 114 A2 12 10 S 135 40 E
Buckingham Canal, India 83 G5 14 0N 80 5 E
Buckinghamshire □, U.K. 17 F7 51 53N 0 55W
Buckland, U.S.A. 141 D12 40 37N 84 16W
Buckle Hd., Australia 112 B4 14 26 S 127 52 E
Buckleboo, Australia 116 B2 32 54 S 136 12 E
Buckley, Ill., U.S.A. 141 D8 40 36N 88 2W
Buckley, Wash., U.S.A. 142 C2 47 10N 122 2W
Buckley →, Australia 114 C2 20 10 S 138 49 E
Bucklin, Kans., U.S.A. 139 G5 37 33N 99 38W
Bucklin, Mo., U.S.A. 140 E4 39 47N 92 53W
Bucks L., U.S.A. 144 F5 39 54N 121 12W
Buco Zau, Angola 103 C2 4 46 S 12 33 E
Bucquoy, France 23 B9 50 9N 2 43 E
Buctouche, Canada 129 C7 46 30N 64 45W
Bucureşti, Romania 46 E7 44 27N 26 10 E
Bucyrus, U.S.A. 141 D14 40 48N 82 59W
Budafok, Hungary 31 D12 47 26N 19 2 E
Budalin, Burma 78 D5 22 20N 95 10 E
Budapest, Hungary 31 D12 47 29N 19 5 E
Budaun, India 81 E8 28 5N 79 10 E
Budd Coast, Antarctica 7 C8 68 0 S 112 0 E
Buddabadah, Australia 117 A7 31 56 S 147 14 E
Buddusò, Italy 40 B2 40 35N 9 15 E
Bude, U.K. 17 G3 50 49N 4 34W
Budel, Neths. 21 F7 51 17N 5 34 E
Budennovsk, Russia 53 H7 44 50N 44 10 E
Budeşti, Romania 46 E7 44 13N 26 30 E
Budge Budge = Baj Baj, India 81 H13 22 30N 88 5 E
Budgewoi, Australia 117 B9 33 13 S 151 34 E
Búðareyri, Iceland 12 D6 65 2N 14 13W
Budia, Spain 34 E2 40 38N 2 46W
Budjala, Zaïre 102 B3 2 50N 19 40 E
Búdrio, Italy 39 D8 44 32N 11 32 E
Budva, Montenegro, Yug. 42 E3 42 17N 18 50 E
Budzyń, Poland 47 C3 52 54N 16 59 E
Buea, Cameroon 101 E6 4 10N 9 9 E
Buellton, U.S.A. 145 L6 34 37N 120 12W
Buena Vista, Bolivia 157 D5 17 27 S 63 40W
Buena Vista, Colo., U.S.A. 143 G10 38 51N 106 8W
Buena Vista, Va., U.S.A. 134 G6 37 44N 79 21W
Buena Vista L., U.S.A. 145 K7 35 12N 119 18W
Buenaventura, Colombia 152 C2 3 53N 77 4W
Buenaventura, Mexico 146 B3 29 50N 107 30W
Buenaventura, B. de, Colombia 152 C2 3 48N 77 17W
Buenavista, Luzon, Phil. 70 E4 13 35N 122 34 E
Buenavista, Mindanao, Phil. 71 G5 8 59N 125 24 E
Buenavista, Zamboanga del S., Phil. 71 H4 7 15N 122 16 E
Buendía, Pantano de, Spain 34 E2 40 25N 2 43W
Buenópolis, Brazil 155 E3 17 54 S 44 11W
Buenos Aires, Argentina 158 C4 34 30 S 58 20W
Buenos Aires, Colombia 152 C3 1 36N 73 18W
Buenos Aires, Costa Rica 148 E3 9 10N 83 20W
Buenos Aires □, Argentina 158 D4 36 30 S 60 0W
Buenos Aires, L., Chile 160 C2 46 35 S 72 30W
Buesaco, Colombia 152 C2 1 23N 77 9W
Buffalo, Mo., U.S.A. 139 G8 37 39N 93 6W
Buffalo, N.Y., U.S.A. 136 D6 42 53N 78 53W
Buffalo, Okla., U.S.A. 139 G5 36 50N 99 38W
Buffalo, S. Dak., U.S.A. 138 C3 45 35N 103 33W
Buffalo, Wyo., U.S.A. 142 D10 44 21N 106 42W
Buffalo →, Canada 130 A5 60 5N 115 5W
Buffalo Head Hills, Canada 130 B5 57 25N 115 55W
Buffalo L., Canada 130 C6 52 27N 112 54W
Buffalo Narrows, Canada 131 B7 55 51N 108 29W
Buffels →, S. Africa 104 D2 29 36 S 17 3 E
Buford, U.S.A. 135 H4 34 10N 84 0W
Bug = Buh →, Ukraine 51 J6 46 59N 31 58 E
Bug →, Poland 47 C8 52 31N 21 5 E
Buga, Colombia 152 C2 4 0N 76 15W
Buganda, Uganda 106 C3 0 0 31 30 E
Buganga, Uganda 106 C3 0 3 S 32 0 E
Bugasan, Phil. 71 H5 7 27N 124 14 E
Bugasong, Phil. 71 F4 11 3N 122 4 E
Bugeat, France 24 C5 45 36N 1 55 E
Bugel, Tanjung, Indonesia 75 D4 6 26 S 111 3 E
Buggenhout, Belgium 21 F4 51 1N 4 12 E
Bugibba, Malta 32 D1 35 57N 14 25 E
Bugojno, Bos.-H. 42 C2 44 2N 17 25 E
Bugsuk, Phil. 71 G1 8 15N 117 15 E
Buguey, Phil. 70 B3 18 17N 121 50 E
Bugulma, Russia 54 D4 54 33N 52 48 E
Buguma, Nigeria 101 E6 4 42N 6 55 E
Bugun Shara, Mongolia 64 B5 49 0N 104 0 E
Buguruslan, Russia 54 E4 53 39N 52 26 E
Buh →, Ukraine 51 J6 46 59N 31 58 E
Buhăeşti, Romania 46 C8 46 47N 27 32 E
Buheirat-Murrat-el-Kubra, Egypt 94 H8 30 18N 32 26 E
Buhl, Idaho, U.S.A. 142 E6 42 36N 114 46W
Buhl, Minn., U.S.A. 138 B8 47 30N 92 46W
Buhuşi, Romania 46 C7 46 41N 26 45 E
Buick, U.S.A. 139 G9 37 38N 91 2W
Builth Wells, U.K. 17 E4 52 9N 3 25W
Buin, Papua N. G. 121 L8 6 48 S 155 42 E
Buinsk, Russia 52 C9 55 0N 48 18 E
Buíque, Brazil 154 C4 8 37 S 37 9W
Buir Nur, Mongolia 65 B6 47 50N 117 42 E
Buis-les-Baronnies, France 25 D9 44 17N 5 16 E
Buitenpost, Neths. 20 B8 53 15N 6 9 E
Buitrago, Spain 36 E7 40 58N 3 38W
Bujalance, Spain 37 H6 37 54N 4 23W
Buján, Spain 36 C2 42 59N 8 36W
Bujanovac, Serbia, Yug. 42 E6 42 28N 21 44 E
Bujaraloz, Spain 34 D4 41 29N 0 10W
Buje, Croatia 39 C10 45 24N 13 39 E
Bujumbura, Burundi 106 C2 3 16 S 29 18 E
Bük, Hungary 31 D9 47 22N 16 45 E
Buk, Poland 47 C3 52 21N 16 30 E
Buka I., Papua N. G. 120 C8 5 10 S 154 35 E
Bukachacha, Russia 57 D12 52 55N 116 50 E
Bukama, Zaïre 107 D2 9 10 S 25 50 E
Bukavu, Zaïre 106 C2 2 20 S 28 52 E
Bukene, Tanzania 106 C3 4 15 S 32 48 E
Bukhara = Bukhoro, Uzbekistan 55 D2 39 48N 64 25 E
Bukhoro, Uzbekistan 55 D2 39 48N 64 25 E
Bukidnon □, Phil. 71 G5 8 0N 125 0 E
Bukima, Tanzania 106 C3 1 50 S 33 25 E
Bukit Mertajam, Malaysia 77 K3 5 22N 100 28 E
Bukittinggi, Indonesia 74 C2 0 20 S 100 20 E
Bukkapatnam, India 83 G3 14 14N 77 46 E
Bukoba, Tanzania 106 C3 1 20 S 31 49 E
Bukoba □, Tanzania 106 C3 1 30 S 32 0 E
Bukowno, Poland 31 A12 50 17N 19 35 E
Bukuru, Nigeria 101 D6 9 42N 8 48 E
Bukuya, Uganda 106 B3 0 40N 31 52 E
Bula, Guinea-Biss. 100 C1 12 7N 15 43W
Bula, Indonesia 73 B4 3 6 S 130 30 E
Bulacan, Phil. 70 E3 13 40N 120 21 E
Bulacan □, Phil. 70 D3 15 0N 121 5 E
Bülach, Switz. 29 A7 47 31N 8 32 E
Bulahdelah, Australia 117 B10 32 23 S 152 13 E
Bulalacao, Phil. 70 E3 12 31N 121 26 E
Bulan, Phil. 70 E4 12 40N 123 52 E
Bulanash, Russia 54 C9 57 16N 62 0 E
Bulancak, Turkey 89 B8 40 56N 38 14 E
Bulandshahr, India 80 E7 28 28N 77 51 E
Bulanık, Turkey 89 C10 39 4N 42 14 E
Bulanovo, Russia 54 E5 52 27N 55 10 E
Bûlâq, Egypt 94 B3 25 10N 30 38 E
Bulawayo, Zimbabwe 107 G2 20 7 S 28 32 E
Buldan, Turkey 88 C3 38 2N 28 50 E
Buldana, India 82 D3 20 30N 76 18 E
Buldon, Phil. 71 H5 7 33N 124 25 E
Bulgar, Russia 52 C9 54 57N 49 4 E
Bulgaria ■, Europe 43 E10 42 35N 25 30 E
Bulgroo, Australia 115 D3 25 47 S 143 58 E
Bulgunnia, Australia 115 E1 30 10 S 134 53 E
Bulhale, Somali Rep. 108 C3 5 20N 46 29 E
Bulhar, Somali Rep. 108 B2 10 25N 44 30 E
Buli, Teluk, Indonesia 72 A3 1 5N 128 25 E
Buliluyan, C., Phil. 71 G1 8 20N 117 15 E
Bulki, Ethiopia 95 F4 6 11N 36 31 E
Bulkley →, Canada 130 B3 55 15N 127 40W
Bull Shoals L., U.S.A. 139 G8 36 22N 92 35W
Bullange, Belgium 21 H8 50 24N 6 15 E
Bullaque →, Spain 37 G6 38 59N 4 17W
Bullara, Australia 112 D1 22 40 S 114 3 E
Bullaring, Australia 113 F2 32 30 S 117 45 E
Bullas, Spain 35 G3 38 2N 1 40W
Bulle, Switz. 28 C4 46 37N 7 3 E
Buller →, N.Z. 119 B6 41 44 S 171 36 E
Buller, Mt., Australia 117 D7 37 10 S 146 28 E
Buller Gorge, N.Z. 119 B7 41 40 S 172 10 E
Bulli, Australia 117 C9 34 15 S 150 57 E
Bullock Creek, Australia 114 B3 17 43 S 144 31 E
Bulloo →, Australia 115 D3 28 43 S 142 30 E
Bulloo Downs, Queens., Australia 115 D3 28 31 S 142 57 E
Bulloo Downs, W. Austral., Australia 112 D2 24 0 S 119 32 E
Bulloo L., Australia 115 D3 28 43 S 142 25 E
Bulls, N.Z. 118 G4 40 10 S 175 24 E
Bully-les-Mines, France 23 B9 50 27N 2 44 E
Bulnes, Chile 158 D1 36 42 S 72 19W
Bulo Burti, Somali Rep. 108 D3 3 50N 45 33 E
Bulo Ghedudo, Somali Rep. 108 D2 2 52N 43 1 E
Bulolo, Papua N. G. 120 D4 7 10 S 146 40 E
Bulongo, Zaïre 103 C4 4 45 S 21 30 E
Bulpunga, Australia 116 B4 33 47 S 141 45 E
Bulqiza, Albania 44 C2 41 30N 20 21 E
Bulsar = Valsad, India 82 D1 20 40N 72 58 E
Bultfontein, S. Africa 104 D4 28 18 S 26 10 E
Buluan, L., Phil. 71 H5 6 40N 124 49 E
Buluangan, Phil. 71 F4 10 24N 123 20 E
Bulukumba, Indonesia 72 C2 5 33 S 120 11 E
Bulun, Russia 57 B13 70 37N 127 30 E
Bulunghur, Uzbekistan 55 D3 39 46N 67 16 E
Bulungu, Zaïre 103 D4 6 4 S 21 54 E
Bulus, Russia 57 C13 63 10N 129 10 E
Bulusan, Phil. 70 E5 12 45N 124 8 E
Bumba, Zaïre 102 B4 2 13N 22 30 E
Bumbiri I., Tanzania 106 C3 1 40 S 31 55 E
Bumhkang, Burma 78 B6 26 51N 97 40 E
Bumhpa Bum, Burma 78 B6 26 51N 97 14 E
Bumi →, Zimbabwe 107 F2 17 0 S 28 20 E
Bumtang →, Bhutan 78 B3 26 56N 90 53 E
Buna, Kenya 106 B4 2 58N 39 30 E
Buna, Papua N. G. 120 E5 8 42 S 148 27 E
Bunawan, Agusan del S., Phil. 71 G5 8 12N 125 57 E
Bunawan, Davao del S., Phil. 71 H5 7 14N 125 38 E
Bunazi, Tanzania 106 C3 1 3 S 31 23 E
Bunbah, Khalīj, Libya 96 B4 32 20N 23 15 E
Bunbury, Australia 113 F2 33 20 S 115 35 E
Buncrana, Ireland 19 A4 55 8N 7 27W
Bundaberg, Australia 115 C5 24 54 S 152 22 E
Bünde, Germany 26 C4 52 11N 8 35 E
Bundey →, Australia 114 C2 21 46 S 135 37 E
Bundi, India 80 G6 25 30N 75 35 E
Bundooma, Australia 114 C1 24 54 S 134 16 E
Bundoran, Ireland 19 B3 54 28N 8 16W
Bundukia, Sudan 95 F3 5 14N 30 55 E
Bundure, Australia 117 C7 35 10 S 146 1 E
Bung Kan, Thailand 76 C4 18 23N 103 37 E
Bungatakada, Japan 62 D3 33 35N 131 25 E
Bungendore, Australia 117 C8 35 14 S 149 30 E
Bungil Cr. →, Australia 114 D4 27 5 S 149 5 E
Bungo, Gunong, Malaysia 75 B4 1 16N 110 9 E
Bungo-Suidō, Japan 62 E4 33 0N 132 15 E
Bungoma, Kenya 106 B3 0 34N 34 34 E
Bungu, Tanzania 106 D4 7 35 S 39 0 E
Bunia, Zaïre 106 B3 1 35N 30 20 E
Bunji, Pakistan 81 B6 35 45N 74 40 E
Bunker Hill, Ill., U.S.A. 140 E7 39 3N 89 57W
Bunker Hill, Ind., U.S.A. 141 D10 40 40N 86 6W
Bunkie, U.S.A. 139 K8 30 57N 92 11W
Bunnell, U.S.A. 135 L5 29 28N 81 16W
Bunnik, Neths. 20 D6 52 4N 5 12 E
Bunnythorpe, N.Z. 118 G4 40 16 S 175 39 E
Buñol, Spain 35 F4 39 25N 0 47W
Bunsbeek, Belgium 21 G5 50 50N 4 56 E
Bunschoten, Neths. 20 D6 52 14N 5 22 E
Buntok, Indonesia 75 C4 1 40 S 114 58 E
Bununu, Nigeria 101 D6 9 51N 9 32 E
Bununu Dass, Nigeria 101 C6 10 5N 9 31 E
Bünyan, Turkey 88 C6 38 51N 35 51 E
Bunyu, Indonesia 75 B5 3 35N 117 50 E
Bunza, Nigeria 101 C5 12 8N 4 0 E
Buol, Indonesia 72 A2 1 15N 121 32 E
Buon Brieng, Vietnam 76 F7 13 9N 108 12 E
Buon Me Thuot, Vietnam 76 F7 12 40N 108 3 E
Buong Long, Cambodia 76 F6 13 44N 106 59 E
Buorkhaya, Mys, Russia 57 B14 71 50N 132 40 E
Buqayq, Si. Arabia 85 E6 26 0N 49 45 E
Buqbua, Egypt 94 A2 31 29N 25 29 E
Bur Acaba, Somali Rep. 108 D2 3 12N 44 20 E
Bûr Fuad, Egypt 94 H8 31 15N 32 20 E
Bur Ghibi, Somali Rep. 108 D3 3 56N 45 7 E
Bûr Safâga, Egypt 94 B3 26 43N 33 57 E
Bûr Sa'îd, Egypt 94 H8 31 16N 32 18 E
Bûr Sûdân, Sudan 94 D4 19 32N 37 9 E
Bûr Taufiq, Egypt 94 J8 29 54N 32 32 E
Bura, Kenya 106 C4 1 4 S 39 58 E
Buran, Somali Rep. 108 B3 10 14N 48 44 E
Burao, Somali Rep. 108 C3 9 32N 45 32 E
Burāq, Syria 91 B5 33 11N 36 29 E
Buras, U.S.A. 139 L10 29 22N 89 32W
Burauen, Phil. 71 F5 10 58N 124 53 E
Buraydah, Si. Arabia 84 E5 26 20N 44 8 E
Burbank, U.S.A. 145 L8 34 11N 118 19W
Burcher, Australia 117 B7 33 30 S 147 16 E
Burdekin →, Australia 114 B4 19 38 S 147 25 E
Burdeos Bay, Phil. 70 D4 14 44N 122 6 E
Burdett, Canada 130 D6 49 50N 111 32W
Burdur, Turkey 88 D4 37 45N 30 17 E
Burdur Gölü, Turkey 88 D4 37 44N 30 10 E
Burdwan = Barddhaman, India 81 H12 23 14N 87 39 E
Bure, Ethiopia 95 E4 10 40N 37 4 E
Bure →, U.K. 16 E9 52 38N 1 43 E
Büren, Germany 26 D4 51 33N 8 35 E
Buren, Neths. 20 E6 51 55N 5 20 E
Bureya →, Russia 57 E13 49 27N 129 30 E
Burford, Canada 136 C4 43 7N 80 27W
Burg, Germany 26 C7 52 16N 11 51 E
Burg auf Fehmarn, Germany 26 A7 54 28N 11 9 E
Burg el Arab, Egypt 94 H6 30 54N 29 32 E
Burg et Tuyur, Sudan 94 C2 20 55N 27 56 E
Burg Stargard, Germany 26 B9 53 29N 13 18 E
Burgas, Bulgaria 43 E12 42 33N 27 29 E
Burgaski Zaliv, Bulgaria 43 E12 42 30N 27 39 E
Burgdorf, Germany 26 C5 52 27N 10 1 E
Burgdorf, Switz. 28 B5 47 3N 7 37 E
Burgenland □, Austria 31 D9 47 20N 16 20 E
Burgeo, Canada 129 C8 47 37N 57 38W
Burgersdorp, S. Africa 104 E4 31 0 S 26 20 E
Burges, Mt., Australia 113 F3 30 50 S 121 5 E
Burghausen, Germany 27 G8 48 9N 12 49 E
Búrgio, Italy 40 E6 37 36N 13 17 E
Bürglen, Switz. 29 C7 46 53N 8 40 E
Burglengenfeld, Germany 27 F8 49 12N 12 2 E
Burgo de Osma, Spain 34 D1 41 35N 3 4W
Burgohondo, Spain 36 E6 40 26N 4 47W
Burgos, Ilocos N., Phil. 70 B3 18 31N 120 39 E
Burgos, Pangasinan, Phil. 70 C2 16 4N 119 52 E
Burgos, Spain 34 C1 42 21N 3 41W
Burgos □, Spain 34 C1 42 21N 3 42W
Burgstädt, Germany 26 E8 50 54N 12 49 E
Burgsvik, Sweden 13 H18 57 3N 18 19 E
Burguillos del Cerro, Spain 37 G4 38 23N 6 35W
Burgundy = Bourgogne, France 23 F11 47 0N 4 50 E
Burhaniye, Turkey 88 C2 39 30N 26 58 E
Burhanpur, India 82 D3 21 18N 76 14 E
Burhou, U.K. 22 C4 49 45N 2 15W
Buri Pen., Eritrea 95 D4 15 25N 39 55 E
Burias, Phil. 70 E4 12 55N 123 5 E
Burias Pass, Phil. 70 E4 13 0N 123 15 E
Burica, Pta., Costa Rica 148 E3 8 3N 82 51W
Burigi, L., Tanzania 106 C3 2 2 S 31 22 E
Burin, Canada 129 C8 47 1N 55 14W
Buriram, Thailand 76 E4 15 0N 103 0 E
Buriti Alegre, Brazil 155 E2 18 9 S 49 3W
Buriti Bravo, Brazil 154 C3 5 50 S 43 50W
Buriti dos Lopes, Brazil 154 B3 3 10 S 41 52W
Burj Sāfitā, Syria 88 E7 34 48N 36 7 E
Burji, Ethiopia 95 F4 5 29N 37 51 E
Burkburnett, U.S.A. 139 H5 34 6N 98 34W
Burke, U.S.A. 142 C6 47 31N 115 49W
Burke →, Australia 114 C2 23 12 S 139 33 E
Burketown, Australia 114 B2 17 45 S 139 33 E
Burkettsville, U.S.A. 141 D12 40 21N 84 39W
Burkina Faso ■, Africa 100 C4 12 0N 1 0W
Burk's Falls, Canada 128 C4 45 37N 79 24W
Burley, U.S.A. 142 E7 42 32N 113 48W
Burli, Kazakhstan 54 F4 51 25N 52 40 E
Burlingame, U.S.A. 144 H4 37 35N 122 21W
Burlington, Canada 136 C5 43 18N 79 45W
Burlington, Colo., U.S.A. 138 F3 39 18N 102 16W
Burlington, Ill., U.S.A. 141 B8 42 3N 88 33W
Burlington, Iowa, U.S.A. 140 D5 40 49N 91 14W
Burlington, Kans., U.S.A. 138 F7 38 12N 95 45W
Burlington, Ky., U.S.A. 141 E12 39 2N 84 43W
Burlington, N.C., U.S.A. 135 G6 36 6N 79 26W
Burlington, N.J., U.S.A. 137 F10 40 4N 74 51W
Burlington, Vt., U.S.A. 137 B11 44 29N 73 12W
Burlington, Wash., U.S.A. 144 B4 48 28N 122 20W
Burlington, Wis., U.S.A. 134 D1 42 41N 88 17W
Burlyu-Tyube, Kazakhstan 56 E8 46 30N 79 10 E
Burma ■, Asia 78 E6 21 0N 96 30 E
Burnaby I., Canada 130 C2 52 25N 131 19W
Burnamwood, Australia 117 A6 31 7 S 144 53 E
Burnet, U.S.A. 139 K5 30 45N 98 14W
Burney, U.S.A. 142 F3 40 53N 121 40W
Burngup, Australia 113 F2 33 2 S 118 42 E
Burnham, U.S.A. 136 F7 40 38N 77 34W
Burnie, Australia 114 G4 41 4 S 145 56 E
Burnley, U.K. 16 D5 53 47N 2 14W
Burnoye, Kazakhstan 55 B5 42 36N 70 47 E
Burns, Oreg., U.S.A. 142 E4 43 35N 119 3W
Burns, Wyo., U.S.A. 138 E2 41 12N 104 21W
Burns Lake, Canada 130 C3 54 20N 125 45W
Burnside →, Canada 126 B9 66 51N 108 4W
Burnside, L., Australia 113 E3 25 22 S 123 0 E
Burnt River, Canada 136 B6 44 41N 78 42W
Burntwood →, Canada 131 B9 56 8N 96 34W
Burntwood L., Canada 131 B8 55 22N 100 26W
Burqān, Kuwait 84 D5 29 0N 47 57 E
Burra, Australia 116 B3 33 40 S 138 55 E
Burragorang, L., Australia 117 B9 33 52 S 150 37 E
Burramurra, Australia 114 C2 20 25 S 137 15 E
Burreli, Albania 44 C2 41 36N 20 1 E
Burren Junction, Australia 115 E4 30 7 S 148 59 E
Burrendong, L., Australia 117 B8 32 45 S 149 10 E
Burrendong Dam, Australia 115 E4 32 39 S 149 6 E
Burriana, Spain 34 F4 39 50N 0 4W
Burrinjuck Res., Australia 117 C8 35 0 S 148 36 E
Burro, Serranías del, Mexico 146 B4 29 0N 102 0W
Burruyacú, Argentina 158 B3 26 30 S 64 40W
Burry Port, U.K. 17 F3 51 41N 4 15W
Bursa, Turkey 88 B3 40 15N 29 5 E
Burseryd, Sweden 15 G7 57 12N 13 17 E
Burstall, Canada 131 C7 50 39N 109 54W
Burton, U.S.A. 141 B13 43 0N 83 40W
Burton L., Canada 128 B4 54 45N 78 20W
Burton upon Trent, U.K. 16 E6 52 48N 1 38W
Burtundy, Australia 116 B5 33 45 S 142 15 E
Buru, Indonesia 72 B3 3 30 S 126 30 E
Buruanga, Phil. 71 F3 11 51N 121 53 E
Burullus, Bahra el, Egypt 94 H7 31 25N 31 0 E
Burûm, Yemen 87 D5 14 22N 48 59 E
Burûn, Râs, Egypt 91 D2 31 14N 33 7 E
Burunday, Kazakhstan 55 B8 43 20N 76 51 E
Burundi ■, Africa 106 C3 3 15 S 30 0 E
Bururi, Burundi 106 C2 3 57 S 29 37 E
Burutu, Nigeria 101 D6 5 20N 5 29 E
Burwell, U.S.A. 138 E5 41 47N 99 8W
Bury, U.K. 16 D5 53 35N 2 17W
Bury St. Edmunds, U.K. 17 E8 52 15N 0 43 E
Buryatia □, Russia 57 D11 53 0N 110 0 E
Buryn, Ukraine 51 G7 51 13N 33 50 E
Burzenin, Poland 47 D5 51 28N 18 47 E
Busalla, Italy 38 D5 44 34N 8 57 E
Busango Swamp, Zambia 107 E2 14 15 S 25 45 E
Buşayrah, Syria 89 E9 35 9N 40 26 E
Buşayyah, Iraq 84 D5 30 0N 46 10 E
Busca, Italy 38 D4 44 31N 7 29 E
Bushati, Albania 44 C1 41 58N 19 34 E
Būshehr, Iran 85 D6 28 55N 50 55 E
Būshehr □, Iran 85 D6 28 20N 51 45 E

180 Bushell

Bushell, *Canada* 131 **B7** 59 31N 108 45W
Bushenyi, *Uganda* 106 **C3** 0 35 S 30 10 E
Bushire = Būshehr, *Iran* 85 **D6** 28 55N 50 55 E
Bushnell, *Ill., U.S.A.* 138 **E9** 40 33N 90 31W
Bushnell, *Nebr., U.S.A.* 138 **E3** 41 14N 103 54W
Busia □, *Kenya* 106 **B3** 0 25N 34 6 E
Busie, *Ghana* 100 **C4** 10 29N 2 22W
Businga, *Zaïre* 102 **B4** 3 16N 20 59 E
Buskerud fylke □, *Norway* 14 **D3** 60 13N 9 0 E
Busko Zdrój, *Poland* 47 **E7** 50 28N 20 42 E
Buskul, *Kazakhstan* 54 **E8** 53 45N 61 12 E
Buslei, *Ethiopia* 108 **C2** 5 28N 44 25 E
Busovača, *Bos.-H.* 42 **C2** 44 6N 17 53 E
Busra ash Shām, *Syria* 91 **C5** 32 30N 36 25 E
Bussang, *France* 23 **E13** 47 50N 6 50 E
Busselton, *Australia* 113 **F2** 33 42 S 115 15 E
Busseto, *Italy* 38 **D7** 44 59N 10 2 E
Bussigny, *Switz.* 28 **C3** 46 33N 6 33 E
Bussum, *Neths.* 20 **D6** 52 16N 5 10 E
Bustamante, B., *Argentina* 160 **C3** 45 5 S 66 18W
Busto, C., *Spain* 36 **B4** 43 34N 6 28W
Busto Arsízio, *Italy* 38 **C5** 45 37N 8 51 E
Busu-Djanoa, *Zaïre* 102 **B4** 1 43N 21 23 E
Busuanga, *Phil.* 70 **E2** 12 10N 120 0 E
Busuanga, *Phil.* 70 **E2** 12 14N 119 52 E
Büsum, *Germany* 26 **A4** 54 7N 8 51 E
Buta, *Zaïre* 106 **B1** 2 50N 24 53 E
Butare, *Rwanda* 106 **C2** 2 31 S 29 52 E
Butaritari, *Kiribati* 122 **G9** 3 30N 174 0 E
Bute, *Australia* 116 **B3** 33 51 S 138 2 E
Bute, *U.K.* 18 **F3** 55 48N 5 2W
Bute Inlet, *Canada* 130 **C4** 50 40N 124 53W
Butemba, *Uganda* 106 **B3** 1 9N 31 37 E
Butembo, *Zaïre* 106 **B2** 0 9N 29 18 E
Butera, *Italy* 41 **E7** 37 11N 14 11 E
Bütgenbach, *Belgium* 21 **H8** 50 26N 6 12 E
Butha Qi, *China* 65 **B7** 48 0N 122 32 E
Buthidaung, *Burma* 78 **E4** 20 52N 92 32 E
Butiaba, *Uganda* 106 **B3** 1 50N 31 20 E
Butkhāk, *Afghan.* 79 **B3** 34 30N 69 22 E
Butler, *Ind., U.S.A.* 141 **C12** 41 26N 84 52W
Butler, *Ky., U.S.A.* 141 **F12** 38 47N 84 22W
Butler, *Mo., U.S.A.* 140 **F2** 38 16N 94 20W
Butler, *Pa., U.S.A.* 136 **F5** 40 52N 79 54W
Butom Odrzánski, *Poland* 47 **D2** 51 44N 15 48 E
Buton, *Indonesia* 72 **C2** 5 0 S 122 45 E
Bütschwil, *Switz.* 29 **B8** 47 23N 9 5 E
Butte, *Mont., U.S.A.* 142 **C7** 46 0N 112 32W
Butte, *Nebr., U.S.A.* 138 **D5** 42 58N 98 51W
Butte Creek →, *U.S.A.* 144 **F5** 39 12N 121 56W
Butterworth = Gcuwa, *S. Africa* 105 **E4** 32 20 S 28 11 E
Butterworth, *Malaysia* 77 **K3** 5 24N 100 23 E
Buttfield, Mt., *Australia* 113 **D4** 24 45 S 128 9 E
Button B., *Canada* 131 **B10** 58 45N 94 23W
Buttonwillow, *U.S.A.* 145 **K7** 35 24N 119 28W
Butty Hd., *Australia* 113 **F3** 33 54 S 121 39 E
Butuan, *Phil.* 71 **G5** 8 57N 125 33 E
Butuku-Luba, *Eq. Guin.* 101 **E6** 3 29N 8 33 E
Butulan, *Phil.* 71 **J5** 5 38N 125 26 E
Butung = Buton, *Indonesia* 72 **C2** 5 0 S 122 45 E
Buturlinovka, *Russia* 52 **E5** 50 50N 40 35 E
Butzbach, *Germany* 26 **E4** 50 25N 8 40 E
Bützow, *Germany* 26 **B7** 53 50N 11 58 E
Buug, *Phil.* 71 **H4** 7 40N 123 2 E
Buxar, *India* 81 **G10** 25 34N 83 58 E
Buxton, *Guyana* 153 **B6** 6 48N 58 2W
Buxton, *U.K.* 16 **D6** 53 16N 1 54W
Buxy, *France* 23 **F11** 46 44N 4 40 E
Buy, *Russia* 52 **A5** 58 28N 41 28 E
Buynaksk, *Russia* 53 **J8** 42 48N 47 7 E
Büyük Kemikli Burun, *Turkey* 44 **D8** 40 20N 26 15 E
Büyük Menderes →, *Turkey* 88 **D2** 37 28N 27 11 E
Büyükçekmece, *Turkey* 88 **B3** 41 2N 28 35 E
Buzançais, *France* 22 **F8** 46 54N 1 25 E
Buzău, *Romania* 46 **D7** 45 10N 26 50 E
Buzău □, *Romania* 46 **D7** 45 20N 26 30 E
Buzău →, *Romania* 46 **D8** 45 26N 27 44 E
Buzău, Pasul, *Romania* 46 **D7** 45 35N 26 12 E
Buzen, *Japan* 62 **D3** 33 35N 131 5 E
Buzet, *Croatia* 39 **C10** 45 24N 13 58 E
Buzi →, *Mozam.* 107 **F3** 19 50 S 34 43 E
Buziaş, *Romania* 46 **D2** 45 38N 21 36 E
Buzuluk, *Russia* 54 **E4** 52 48N 52 12 E
Buzuluk →, *Russia* 52 **E6** 50 15N 42 7 E
Buzzards Bay, *U.S.A.* 137 **E14** 41 45N 70 37W
Bwagaoia, *Papua N. G.* 120 **F7** 10 40 S 152 52 E
Bwana Mkubwe, *Zaïre* 107 **E2** 13 8 S 28 38 E
Byala, *Ruse, Bulgaria* 43 **D10** 43 28N 25 44 E
Byala, *Varna, Bulgaria* 43 **E12** 42 53N 27 55 E
Byala Slatina, *Bulgaria* 43 **D8** 43 26N 23 55 E
Byarezina →, *Belarus* 51 **F6** 52 33N 30 14 E
Bychawa, *Poland* 47 **D9** 51 1N 22 36 E
Byczyna, *Poland* 47 **D5** 51 7N 18 12 E
Bydgoszcz, *Poland* 47 **B5** 53 10N 18 0 E
Bydgoszcz □, *Poland* 47 **B4** 53 16N 17 33 E
Byelarus = Belarus ■, *Europe* 50 **F4** 53 30N 27 0 E
Byelorussia = Belarus ■, *Europe* 50 **F4** 53 30N 27 0 E
Byers, *U.S.A.* 138 **F2** 39 43N 104 14W
Byesville, *U.S.A.* 136 **G3** 39 58N 81 32W
Byhalia, *U.S.A.* 139 **H10** 34 52N 89 41W
Bykhaw, *Belarus* 50 **F6** 53 31N 30 14 E
Bykhov = Bykhaw, *Belarus* 50 **F6** 53 31N 30 14 E
Bykovo, *Russia* 52 **F7** 49 50N 45 25 E
Bylas, *U.S.A.* 143 **K8** 33 8N 110 7W
Bylderup, *Denmark* 15 **K3** 54 57N 9 6 E
Bylot I., *Canada* 127 **A12** 73 13N 78 34W
Byrd, C., *Antarctica* 7 **C17** 69 38 S 76 7W
Byro, *Australia* 113 **E2** 26 5 S 116 11 E
Byrock, *Australia* 117 **A7** 30 40 S 146 27 E
Byron, *U.S.A.* 140 **B7** 42 8N 89 15W
Byron Bay, *Australia* 115 **D5** 28 43 S 153 37 E
Byrranga, Gory, *Russia* 57 **B11** 75 0N 100 0 E
Byrranga Mts. = Byrranga, Gory, *Russia* 57 **B11** 75 0N 100 0 E
Byrum, *Denmark* 15 **G5** 57 16N 11 0 E
Byske, *Sweden* 12 **D19** 64 57N 21 11 E
Byske älv →, *Sweden* 12 **D19** 64 57N 21 13 E
Bystrovka, *Kyrgyzstan* 55 **B7** 42 47N 75 42 E
Bystrzyca →, *Lublin, Poland* 47 **D9** 51 21N 22 46 E
Bystrzyca →, *Wrocław, Poland* 47 **D3** 51 12N 16 55 E
Bystrzyca Kłodzka, *Poland* 47 **E3** 50 19N 16 39 E
Bytom, *Poland* 47 **E5** 50 25N 18 54 E
Bytów, *Poland* 47 **A4** 54 10N 17 30 E
Byumba, *Rwanda* 106 **C3** 1 35 S 30 4 E
Bzenec, *Czech.* 31 **C10** 48 58N 17 18 E
Bzura →, *Poland* 47 **C7** 52 25N 20 15 E

C

Ca →, *Vietnam* 76 **C5** 18 45N 105 45 E
Ca Mau = Quan Long, *Vietnam* 77 **H5** 9 7N 105 8 E
Ca Mau, Mui = Bai Bung, Mui, *Vietnam* 77 **H5** 8 38N 104 44 E
Ca Na, *Vietnam* 77 **G7** 11 20N 108 54 E
Caacupé, *Paraguay* 158 **B4** 25 23 S 57 5W
Caála, *Angola* 103 **E3** 12 46 S 15 30 E
Caamano Sd., *Canada* 130 **C3** 52 55N 129 25W
Caapiranga, *Brazil* 153 **D5** 3 18 S 61 13W
Caazapá, *Paraguay* 158 **B4** 26 8 S 56 19W
Caazapá □, *Paraguay* 159 **B4** 26 10 S 56 0W
Cabadbaran, *Phil.* 71 **G5** 9 10N 125 38 E
Cabagan, *Phil.* 70 **C3** 17 26N 121 46 E
Cabalian, *Phil.* 71 **F5** 10 16N 125 10 E
Caballeria, C. de, *Spain* 33 **A11** 40 5N 4 5 E
Cabana, *Peru* 156 **B2** 8 25 S 78 5W
Cabanaconde, *Peru* 156 **D3** 15 38 S 71 58W
Cabañaquinta, *Spain* 36 **B5** 43 10N 5 38W
Cabanatuan, *Phil.* 70 **D3** 15 30N 120 58 E
Cabanes, *Spain* 34 **E5** 40 9N 0 2 E
Cabangon, *Phil.* 70 **D3** 15 10N 120 3 E
Cabanillas, *Peru* 156 **D3** 15 36 S 70 28W
Cabano, *Canada* 129 **C6** 47 40N 68 56W
Čabar, *Croatia* 39 **C11** 45 36N 14 39 E
Cabarroquis, *Phil.* 70 **C3** 16 50N 121 30 E
Cabarruyan I., *Phil.* 70 **C2** 16 18N 119 59 E
Cabazon, *U.S.A.* 145 **M10** 33 55N 116 47W
Cabcaben, *Phil.* 70 **D3** 14 27N 120 35 E
Cabedelo, *Brazil* 154 **C5** 7 0S 34 50W
Cabery, *U.S.A.* 141 **D8** 41 0N 88 12W
Cabeza del Buey, *Spain* 37 **G5** 38 44N 5 13W
Cabildo, *Chile* 158 **C1** 32 30 S 71 5W
Cabimas, *Venezuela* 152 **A3** 10 23N 71 25W
Cabinda, *Angola* 103 **D2** 5 33 S 12 11 E
Cabinda □, *Angola* 103 **D2** 5 0 S 12 30 E
Cabinet Mts., *U.S.A.* 142 **C6** 48 0N 115 30W
Cabiri, *Angola* 103 **D2** 8 52 S 13 39 E
Cabo Blanco, *Argentina* 160 **C3** 47 15 S 65 47W
Cabo Frio, *Brazil* 155 **F3** 22 51 S 42 3W
Cabo Pantoja, *Peru* 152 **D2** 1 0 S 75 10W
Cabo Raso, *Argentina* 160 **B3** 44 20 S 65 15W
Cabonga, Réservoir, *Canada* 128 **C4** 47 20N 76 40W
Cabool, *U.S.A.* 139 **G8** 37 7N 92 6W
Caboolture, *Australia* 115 **D5** 27 5 S 152 58 E
Cabora Bassa Dam = Cahora Bassa Dam, *Mozam.* 107 **F3** 15 20 S 32 50 E
Caborca, *Mexico* 146 **A2** 30 40N 112 10W
Cabot, Mt., *U.S.A.* 137 **B13** 44 30N 71 25W
Cabot Str., *Canada* 129 **C8** 47 15N 59 40W
Cabra, *Spain* 37 **H6** 37 30N 4 28W
Cabra del Santo Cristo, *Spain* 35 **H1** 37 42N 3 16W
Cábras, *Italy* 40 **C1** 39 56N 8 32 E
Cabrera, *Spain* 33 **B9** 39 8N 2 57 E
Cabrera, Sierra, *Spain* 36 **C4** 42 12N 6 40W
Cabri, *Canada* 131 **C7** 50 35N 108 25W
Cabriel →, *Spain* 35 **F3** 39 14N 1 3W
Cabruta, *Venezuela* 152 **B4** 7 50N 66 10W
Cabucgayan, *Phil.* 71 **F5** 11 29N 124 34 E
Cabugao, *Phil.* 70 **C3** 17 48N 120 27 E
Cabulauan Is., *Phil.* 71 **F3** 11 25N 120 8 E
Caburan = Jose Abad Santos, *Phil.* 71 **J5** 5 55N 125 39 E
Cabuyaro, *Colombia* 152 **C3** 4 18N 72 49W
Cacabelos, *Spain* 36 **C4** 42 36N 6 44W
Čačak, *Serbia, Yug.* 42 **D5** 43 54N 20 20 E
Cacao, *Fr. Guiana* 153 **C7** 4 33N 52 26W
Cáceres, *Brazil* 157 **D6** 16 5 S 57 40W
Cáceres, *Colombia* 152 **B2** 7 35N 75 20W
Cáceres, *Spain* 37 **F4** 39 26N 6 23W
Cáceres □, *Spain* 37 **F4** 39 45N 6 0W
Cache Bay, *Canada* 128 **C4** 46 22N 80 0W
Cache Cr. →, *U.S.A.* 144 **G5** 38 42N 121 42W
Cachepo, *Portugal* 37 **H3** 37 20N 7 49W
Cachéu, *Guinea-Biss.* 100 **C1** 12 14N 16 8W
Cachi, *Argentina* 158 **B2** 25 5 S 66 10W
Cachimbo, *Brazil* 157 **B7** 8 57 S 54 54W
Cachimbo, Serra do, *Brazil* 157 **B6** 9 30 S 55 30W
Cachingues, *Angola* 103 **E3** 13 5 S 16 43 E
Cachoeira, *Brazil* 155 **D4** 12 30 S 39 0W
Cachoeira Alta, *Brazil* 155 **E1** 18 48 S 50 58W
Cachoeira de Itapemirim, *Brazil* 155 **F3** 20 51 S 41 7W
Cachoeira do Sul, *Brazil* 159 **C5** 30 3 S 52 53W
Cachoeiro do Arari, *Brazil* 154 **B2** 1 1 S 48 58W
Cachopo, *Portugal* 37 **H3** 37 20N 7 49W
Cachuela Esperanza, *Bolivia* 157 **C4** 10 32 S 65 38W
Cacólo, *Angola* 103 **E3** 10 9 S 19 21 E
Caconda, *Angola* 103 **E3** 13 48 S 15 8 E
Cacongo, *Angola* 103 **D2** 5 11 S 12 5 E
Caçu, *Brazil* 155 **E1** 18 37 S 51 4W
Cacula, *Angola* 103 **E2** 14 29 S 14 10 E
Caculé, *Brazil* 155 **D3** 14 30 S 42 13W
Cacuso, *Angola* 103 **D3** 9 25 S 15 45 E
Cadarache, *France* 25 **E9** 43 41N 5 43 E
Čadca, *Slovak Rep.* 31 **B11** 49 26N 18 45 E
Caddo, *U.S.A.* 139 **H6** 34 7N 96 16W
Cadell Cr. →, *Australia* 114 **C3** 22 35 S 141 51 E
Cadenazzo, *Switz.* 29 **D7** 46 9N 8 57 E
Cader Idris, *U.K.* 16 **E4** 52 42N 3 53W
Cadí, Sierra del, *Spain* 34 **C6** 42 17N 1 42 E
Cadibarrawirracanna, L., *Australia* 115 **D2** 28 52 S 135 27 E
Cadillac, *Canada* 128 **C4** 48 14N 78 23W
Cadillac, *France* 24 **D3** 44 38N 0 20W
Cadillac, *U.S.A.* 134 **C3** 44 15N 85 24W
Cadiz, *Phil.* 71 **F4** 10 57N 123 15 E
Cádiz, *Spain* 37 **J4** 36 30N 6 20W
Cadiz, *U.S.A.* 136 **F4** 40 22N 81 0W
Cádiz □, *Spain* 37 **J5** 36 36N 5 45W
Cádiz, G. de, *Spain* 37 **J4** 36 40N 7 0W
Cadney Park, *Australia* 115 **D1** 27 55 S 134 3 E
Cadomin, *Canada* 130 **C5** 53 2N 117 20W
Cadotte →, *Canada* 130 **B5** 56 43N 117 10W
Cadours, *France* 24 **E5** 43 44N 1 2 E
Cadoux, *Australia* 113 **F2** 30 46 S 117 7 E
Caen, *France* 22 **C6** 49 10N 0 22W
Caernarfon, *U.K.* 16 **D3** 53 8N 4 16W
Caernarfon B., *U.K.* 16 **D3** 53 4N 4 40W
Caernarvon = Caernarfon, *U.K.* 16 **D3** 53 8N 4 16W
Caerphilly, *U.K.* 17 **F4** 51 35N 3 13W
Caesarea, *Israel* 91 **C3** 32 30N 34 53 E
Caeté, *Brazil* 155 **E3** 19 55 S 43 40W
Caetité, *Brazil* 155 **D3** 13 50 S 42 32W
Cafayate, *Argentina* 158 **B2** 26 2 S 66 0W
Cafifi, *Colombia* 152 **B3** 5 13N 71 4W
Cafu, *Angola* 103 **F3** 16 30 S 15 8 E
Cagayan □, *Phil.* 70 **B3** 18 0N 121 50 E
Cagayan →, *Phil.* 70 **B3** 18 25N 121 42 E
Cagayan de Oro, *Phil.* 71 **G5** 8 30N 124 40 E
Cagayan Is., *Phil.* 71 **G3** 9 40N 121 16 E
Cagayan Sulu I., *Phil.* 71 **H2** 7 1N 118 30 E
Cagli, *Italy* 39 **E9** 43 33N 12 39 E
Cágliari, *Italy* 40 **C2** 39 13N 9 7 E
Cágliari, G. di, *Italy* 40 **C2** 39 8N 9 11 E
Cagnano Varano, *Italy* 41 **A8** 41 49N 15 47 E
Cagnes-sur-Mer, *France* 25 **E11** 43 40N 7 9 E
Caguán →, *Colombia* 152 **D3** 0 8 S 74 18W
Caguas, *Puerto Rico* 149 **C6** 18 14N 66 2W
Caha Mts., *Ireland* 19 **E2** 51 45N 9 40W
Cahama, *Angola* 103 **F2** 16 17 S 14 19 E
Caher, *Ireland* 19 **D4** 52 22N 7 56W
Caherciveen, *Ireland* 19 **E1** 51 56N 10 14W
Cahora Bassa Dam, *Mozam.* 107 **F3** 15 20 S 32 50 E
Cahore Pt., *Ireland* 19 **D5** 52 33N 6 12W
Cahors, *France* 24 **D5** 44 27N 1 27 E
Cahuapanas, *Peru* 156 **B2** 5 15 S 77 0W
Cahuinari →, *Colombia* 152 **D3** 1 21 S 70 44W
Cahul, *Moldova* 51 **K5** 45 50N 28 15 E
Cai Bau, Dao, *Vietnam* 76 **B6** 21 10N 107 27 E
Cai Nuoc, *Vietnam* 77 **H5** 8 56N 105 1 E
Caia, *Mozam.* 107 **F4** 17 51 S 35 24 E
Caiabis, Serra dos, *Brazil* 157 **C6** 11 30 S 56 30W
Caianda, *Angola* 103 **E4** 11 2 S 23 31 E
Caiapó, Serra do, *Brazil* 157 **D7** 17 0 S 52 0W
Caiapônia, *Brazil* 157 **D7** 16 57 S 51 49W
Caibarién, *Cuba* 148 **B4** 22 30N 79 30W
Caibiran, *Phil.* 71 **F5** 11 34N 124 35 E
Caicara, *Bolívar, Venezuela* 152 **B4** 7 38N 66 10W
Caicara, *Monagas, Venezuela* 153 **B5** 9 52N 63 38W
Caicó, *Brazil* 154 **C4** 6 20 S 37 0W
Caicos Is., *W. Indies* 149 **B5** 21 40N 71 40W
Caicos Passage, *W. Indies* 149 **B5** 22 45N 72 45W
Cailloma, *Peru* 156 **D3** 15 9 S 71 45W
Caine →, *Bolivia* 157 **D4** 18 23 S 65 21W
Caird Coast, *Antarctica* 7 **D1** 75 0 S 25 0W
Cairn Gorm, *U.K.* 18 **D5** 57 7N 3 39W
Cairn Toul, *U.K.* 18 **D5** 57 3N 3 44W
Cairngorm Mts., *U.K.* 18 **D5** 57 6N 3 42W
Cairns, *Australia* 114 **B4** 16 57 S 145 45 E
Cairo = El Qâhira, *Egypt* 94 **H7** 30 1N 31 14 E
Cairo, *Ga., U.S.A.* 135 **K3** 30 52N 84 13W
Cairo, *Ill., U.S.A.* 139 **G10** 37 0N 89 11W
Cairo Montenotte, *Italy* 38 **D5** 44 23N 8 16 E
Caithness, Ord of, *U.K.* 18 **C5** 58 8N 3 36W
Caiundo, *Angola* 103 **F3** 15 50 S 17 28 E
Caiza, *Bolivia* 157 **E4** 20 2 S 65 40W
Cajabamba, *Peru* 156 **B2** 7 38 S 78 4W
Cajamarca, *Peru* 156 **B2** 7 5 S 78 28W
Cajamarca □, *Peru* 156 **B2** 6 15 S 78 50W
Cajapió, *Brazil* 154 **B3** 2 58 S 44 48W
Cajarc, *France* 24 **D5** 44 29N 1 50 E
Cajatambo, *Peru* 156 **C2** 10 30 S 77 2W
Cajàzeiras, *Brazil* 154 **C4** 6 52 S 38 30W
Čajetina, *Serbia, Yug.* 42 **D4** 43 47N 19 42 E
Cajidiocan, *Phil.* 70 **E4** 12 22N 122 41 E
Čajniče, *Bos.-H.* 42 **D4** 43 34N 19 5 E
Çakirgol, *Turkey* 89 **B8** 40 33N 39 40 E
Čakovec, *Croatia* 39 **B13** 46 23N 16 26 E
Çal, *Turkey* 88 **C3** 38 4N 29 23 E
Cala, *Spain* 37 **H4** 37 59N 6 21W
Cala →, *Spain* 37 **H4** 37 38N 6 5W
Cala Cadolar, Punta de, *Spain* 35 **G6** 38 38N 1 35 E
Cala d'Or, *Spain* 33 **B10** 39 23N 3 14 E
Cala Figuera, C., *Spain* 33 **B9** 39 27N 2 31 E
Cala Forcat, *Spain* 33 **A10** 40 0N 3 47 E
Cala Mayor, *Spain* 33 **B9** 39 33N 2 37 E
Cala Mezquida, *Spain* 33 **B11** 39 55N 4 16 E
Cala Millor, *Spain* 33 **B10** 39 35N 3 22 E
Cala Ratjada, *Spain* 33 **B10** 39 43N 3 27 E
Calabanga, *Phil.* 70 **E4** 13 42N 123 17 E
Calabar, *Nigeria* 101 **E6** 4 57N 8 20 E
Calabozo, *Venezuela* 152 **B4** 9 0N 67 28W
Calábria □, *Italy* 41 **C9** 39 0N 16 30 E
Calaburras, Pta. de, *Spain* 37 **J6** 36 30N 4 38W
Calaceite, *Spain* 34 **D5** 41 1N 0 11 E
Calacota, *Bolivia* 156 **D4** 17 16 S 68 38W
Calafat, *Romania* 46 **F3** 43 58N 22 59 E
Calafate, *Argentina* 160 **D2** 50 19 S 72 15W
Calahorra, *Spain* 34 **C3** 42 18N 1 59W
Calais, *France* 23 **B8** 50 57N 1 56 E
Calais, *U.S.A.* 129 **C6** 45 11N 67 17W
Calais, Pas de, *France* 23 **B8** 50 30N 1 20 E
Calalaste, Cord. de, *Argentina* 158 **B2** 25 0 S 67 0W
Calalayan, *Phil.* 71 **F2** 11 30N 119 38 E
Calama, *Brazil* 157 **B5** 8 0 S 62 50W
Calama, *Chile* 158 **A2** 22 30 S 68 55W
Calamar, *Bolívar, Colombia* 152 **A3** 10 15N 74 55W
Calamar, *Vaupés, Colombia* 152 **C3** 1 58N 72 32W
Calamarca, *Bolivia* 156 **D4** 16 55 S 68 9W
Calamba, *Cavite, Phil.* 71 **G4** 8 35N 123 39 E
Calamba, *Misamis, Phil.* 71 **F4** 10 11N 123 17 E
Calamba, *Negros, Phil.* 70 **D3** 14 13N 121 10 E
Calamian Group, *Phil.* 71 **F2** 11 50N 119 55 E
Calamocha, *Spain* 34 **E3** 40 50N 1 17W
Calán Porter, *Spain* 33 **B11** 39 52N 4 8 E
Calañas, *Spain* 37 **H4** 37 40N 6 53W
Calanda, *Spain* 34 **E4** 40 56N 0 15W
Calandagan I., *Phil.* 71 **F3** 10 39N 120 15 E
Calandula, *Angola* 103 **D3** 9 6 S 15 57 E
Calang, *Indonesia* 74 **B1** 4 37N 95 37 E
Calangiánus, *Italy* 40 **B2** 40 56N 9 11 E
Calanscio, Sarīr, *Libya* 96 **C4** 27 0N 21 30 E
Calapan, *Phil.* 70 **E3** 13 25N 121 7 E
Calape, *Phil.* 71 **G4** 9 54N 123 52 E
Călăraşi, *Romania* 46 **E8** 44 12N 27 20 E
Călăraşi □, *Romania* 46 **E8** 44 10N 27 0 E
Calasparra, *Spain* 35 **G3** 38 14N 1 41W
Calatafimi, *Italy* 40 **E5** 37 55N 12 52 E
Calatagan, *Phil.* 70 **E3** 13 50N 120 38 E
Calatayud, *Spain* 34 **D3** 41 20N 1 40W
Calato = Kálathos, *Greece* 45 **H10** 36 9N 28 8 E
Calauag, *Phil.* 70 **E4** 13 55N 122 15 E
Calavà, C., *Italy* 41 **D7** 38 11N 14 55 E
Calavite, C., *Phil.* 70 **E3** 13 26N 120 20 E
Calavite Pass, *Phil.* 70 **E3** 13 36N 120 25 E
Calayan, *Phil.* 70 **B3** 19 16N 121 28 E
Calayan I., *Phil.* 70 **B3** 19 20N 121 27 E
Calbayog, *Phil.* 70 **E5** 12 4N 124 38 E
Calbe, *Germany* 26 **D7** 51 54N 11 46 E
Calca, *Peru* 156 **C3** 13 22 S 72 0W
Calcasieu L., *U.S.A.* 139 **L8** 29 55N 93 18W
Calci, *Italy* 38 **E7** 43 44N 10 31 E
Calcutta, *India* 81 **H13** 22 36N 88 24 E
Caldaro, *Italy* 39 **B8** 46 25N 11 14 E
Caldas □, *Colombia* 152 **B2** 5 15N 75 30W
Caldas da Rainha, *Portugal* 37 **F1** 39 24N 9 8W
Caldas de Reyes, *Spain* 36 **C2** 42 36N 8 39W
Caldas Novas, *Brazil* 155 **E2** 17 45 S 48 38W
Calder →, *U.K.* 16 **D6** 53 44N 1 22W
Caldera, *Chile* 158 **B1** 27 5 S 70 55W
Caldwell, *Idaho, U.S.A.* 142 **E5** 43 40N 116 41W
Caldwell, *Kans., U.S.A.* 139 **G6** 37 2N 97 37W
Caldwell, *Tex., U.S.A.* 139 **K6** 30 32N 96 42W
Caledon, *S. Africa* 104 **E2** 34 14 S 19 26 E
Caledon →, *S. Africa* 104 **E4** 30 31 S 26 5 E
Caledon B., *Australia* 114 **A2** 12 45 S 137 0 E
Caledonia, *Canada* 136 **C5** 43 7N 79 58W
Caledonia, *Mo., U.S.A.* 140 **G6** 37 45N 90 46W
Caledonia, *N.Y., U.S.A.* 136 **D7** 42 58N 77 51W
Calella, *Spain* 34 **D7** 41 37N 2 40 E
Calemba, *Angola* 104 **B2** 16 0 S 15 44 E
Calenzana, *France* 25 **F12** 42 31N 8 51 E
Caleta Olivia, *Argentina* 160 **C3** 46 25 S 67 25W
Calexico, *U.S.A.* 145 **N11** 32 40N 115 30W
Calf of Man, *U.K.* 16 **C3** 54 3N 4 48W
Calgary, *Canada* 130 **C6** 51 0N 114 10W
Calheta, *Madeira* 33 **D2** 32 44N 17 11W
Calhoun, *U.S.A.* 135 **H3** 34 30N 84 57W
Cali, *Colombia* 152 **C2** 3 25N 76 35W
Calicut, *India* 83 **J2** 11 15N 75 43 E
Caliente, *U.S.A.* 143 **H6** 37 37N 114 31W
California, *Mo., U.S.A.* 140 **F4** 38 38N 92 34W
California, *Pa., U.S.A.* 136 **F5** 40 4N 79 54W
California □, *U.S.A.* 143 **H4** 37 30N 119 30W
California, Baja, *Mexico* 146 **A1** 32 10N 115 12W
California, Baja, T.N. = Baja California □, *Mexico* 146 **B2** 30 0N 115 0W
California, Baja, T.S. = Baja California Sur □, *Mexico* 146 **B2** 25 50N 111 50W
California, G. de, *Mexico* 146 **B2** 27 0N 111 0W
California City, *U.S.A.* 145 **K9** 35 10N 117 55W
California Hot Springs, *U.S.A.* 145 **K8** 35 51N 118 41W
Călimăneşti, *Romania* 46 **D5** 45 14N 24 20 E
Călimani, Munţii, *Romania* 46 **B5** 47 12N 25 0 E
Călineşti, *Romania* 46 **D5** 45 21N 24 18 E
Calingasta, *Argentina* 158 **C2** 31 15 S 69 30W
Calinog, *Phil.* 71 **F4** 11 7N 122 32 E
Calintaan, *Phil.* 70 **E3** 12 35N 120 57 E
Calipatria, *U.S.A.* 145 **M11** 33 8N 115 31W
Calistoga, *U.S.A.* 144 **G4** 38 35N 122 35W
Calitri, *Italy* 41 **B8** 40 54N 15 26 E
Calitzdorp, *S. Africa* 104 **E3** 33 33 S 21 42 E
Callabonna, L., *Australia* 115 **D3** 29 40 S 140 5 E
Callac, *France* 22 **D3** 48 25N 3 27W
Callan, *Ireland* 19 **D4** 52 32N 7 24W
Callander, *U.K.* 18 **E4** 56 15N 4 13W
Callang, *Phil.* 70 **C3** 17 2N 121 38 E
Callantsoog, *Neths.* 20 **C5** 52 50N 4 42 E
Callao, *Peru* 156 **C2** 12 0 S 77 0W
Callaway, *U.S.A.* 138 **E5** 41 18N 99 56W
Callender, *U.S.A.* 140 **B2** 42 22N 94 17W
Calles, *Mexico* 147 **C5** 23 2N 98 42W
Callide, *Australia* 114 **C5** 24 18 S 150 28 E
Calling Lake, *Canada* 130 **B6** 55 15N 113 12W
Calliope, *Australia* 114 **C5** 24 0 S 151 16 E
Callosa de Ensarriá, *Spain* 35 **G4** 38 40N 0 8W
Callosa de Segura, *Spain* 35 **G4** 38 7N 0 53W
Calmar, *U.S.A.* 140 **A5** 43 11N 91 52W
Calola, *Angola* 103 **F3** 16 25 S 17 48 E
Calolbon, *Phil.* 70 **E5** 13 36N 124 6 E
Caloocan, *Phil.* 70 **D3** 14 39N 120 58 E
Calore →, *Italy* 41 **A7** 41 11N 14 28 E
Caloundra, *Australia* 115 **D5** 26 45 S 153 10 E
Calpe, *Spain* 35 **G5** 38 39N 0 3 E
Calpella, *U.S.A.* 144 **F3** 39 14N 123 12W
Calpine, *U.S.A.* 144 **F6** 39 40N 120 27W
Calstock, *Canada* 128 **C3** 49 47N 84 9W
Caltabellotta, *Italy* 40 **E6** 37 34N 13 13 E
Caltagirone, *Italy* 41 **E7** 37 14N 14 31 E
Caltanissetta, *Italy* 41 **E7** 37 29N 14 4 E
Calucinga, *Angola* 103 **E3** 11 18 S 16 12 E
Calulo, *Angola* 103 **E2** 10 1 S 14 56 E
Calumet, *U.S.A.* 134 **B1** 47 14N 88 27W
Calunda, *Angola* 103 **E4** 12 7 S 23 36 E
Caluquembe, *Angola* 103 **E2** 13 47 S 14 44 E
Caluso, *Italy* 38 **C4** 45 18N 7 53 E
Caluya I., *Phil.* 71 **F3** 11 55N 121 34 E
Calvados □, *France* 22 **C6** 49 5N 0 15W
Calvert, *U.S.A.* 139 **K6** 30 59N 96 40W
Calvert →, *Australia* 114 **B2** 16 17 S 137 44 E
Calvert Hills, *Australia* 114 **B2** 17 15 S 137 20 E
Calvert I., *Canada* 130 **C3** 51 30N 128 0W
Calvert Ra., *Australia* 112 **D3** 24 0 S 122 30 E
Calvi, *France* 25 **F12** 42 34N 8 45 E
Calvillo, *Mexico* 146 **C4** 21 51N 102 43W
Calvinia, *S. Africa* 104 **E2** 31 28 S 19 45 E
Calw, *Germany* 27 **G4** 48 42N 8 44 E
Calwa, *U.S.A.* 144 **J7** 36 42N 119 46W

Caraveli 181

Calzada Almuradiel, *Spain* **35 G1** 38 32N 3 28W
Calzada de Calatrava, *Spain* **37 G7** 38 42N 3 46W
Cam →, *U.K.* **17 E8** 52 21N 0 16 E
Cam Lam, *Vietnam* **77 G7** 11 54N 109 10 E
Cam Pha, *Vietnam* **76 B6** 21 7N 107 18 E
Cam Ranh, *Vietnam* **77 G7** 11 54N 109 12 E
Cam Xuyen, *Vietnam* **76 C6** 18 15N 106 0 E
Camabatela, *Angola* **103 D3** 8 20 S 15 26 E
Camacã, *Brazil* **155 E4** 15 24 S 39 30W
Camaçari, *Brazil* **155 D4** 12 41 S 38 18W
Camacha, *Madeira* **33 D3** 32 41N 16 49W
Camacho, *Mexico* **146 C4** 24 25N 102 18W
Camacupa, *Angola* **103 E3** 11 58 S 17 22 E
Camaguán, *Venezuela* **152 B4** 8 6N 67 36W
Camagüey, *Cuba* **148 B4** 21 20N 78 0W
Camaiore, *Italy* **38 E7** 43 56N 10 18 E
Camamu, *Brazil* **155 D4** 13 57 S 39 7W
Camaná, *Peru* **156 D3** 16 30 S 72 50W
Camanche, *U.S.A.* **140 C6** 41 47N 90 15W
Camanche Reservoir, *U.S.A.* **144 G6** 38 14N 121 1W
Camanongue, *Angola* **103 E4** 11 24 S 20 17 E
Camaquã →, *Brazil* **159 C5** 31 17 S 51 47W
Câmara de Lobos, *Madeira* **33 D3** 32 39N 16 59W
Camararé →, *Brazil* **157 C6** 12 15 S 58 55W
Camarat, C., *France* **25 E10** 43 12N 6 41 E
Camaret, *France* **22 D2** 48 16N 4 37W
Camargo, *Bolivia* **157 E4** 20 38 S 65 15W
Camargue, *France* **25 E8** 43 34N 4 34 E
Camarillo, *U.S.A.* **145 L7** 34 13N 119 2W
Camariñas, *Spain* **36 B1** 43 8N 9 12W
Camarines Norte □, *Phil.* **70 D4** 14 10N 122 45 E
Camarines Sur □, *Phil.* **70 E4** 13 40N 123 20 E
Camarón, C., *Honduras* **148 C2** 16 0N 85 5W
Camarones, *Argentina* **160 B3** 44 50 S 65 40W
Camarones, B., *Argentina* **160 B3** 44 45 S 65 35W
Camas, *U.S.A.* **144 E4** 45 35N 122 24W
Camas Valley, *U.S.A.* **142 E2** 43 2N 123 40W
Camaxilo, *Angola* **103 D3** 8 21 S 18 56 E
Cambados, *Spain* **36 C2** 42 31N 8 49W
Cambamba, *Angola* **103 D2** 8 53 S 14 44 E
Cambará, *Brazil* **159 A5** 23 2 S 50 5W
Cambay = Khambhat, *India* **80 H5** 22 23N 72 33 E
Cambay, G. of = Khambat, G. of, *India* **80 J5** 20 45N 72 30 E
Cambil, *Spain* **35 H1** 37 40N 3 33W
Cambo-les-Bains, *France* **24 E2** 43 22N 1 23W
Cambodia ■, *Asia* **76 F5** 12 15N 105 0 E
Camborne, *U.K.* **17 G2** 50 12N 5 19W
Cambrai, *France* **23 B10** 50 11N 3 14 E
Cambria, *U.S.A.* **144 K5** 35 34N 121 5W
Cambrian Mts., *U.K.* **17 E4** 52 3N 3 57W
Cambridge, *Canada* **128 D3** 43 23N 80 15W
Cambridge, *Jamaica* **148 C4** 18 18N 77 54W
Cambridge, *N.Z.* **118 D4** 37 54 S 175 29 E
Cambridge, *U.K.* **17 E8** 52 12N 0 8 E
Cambridge, *Idaho, U.S.A.* **142 D5** 44 34N 116 41W
Cambridge, *Ill., U.S.A.* **140 C6** 41 18N 90 12W
Cambridge, *Iowa, U.S.A.* **140 C3** 41 54N 93 32W
Cambridge, *Mass., U.S.A.* **137 D13** 42 22N 71 6W
Cambridge, *Md., U.S.A.* **134 F7** 38 34N 76 5W
Cambridge, *Minn., U.S.A.* **138 C8** 45 34N 93 13W
Cambridge, *N.Y., U.S.A.* **137 C11** 43 2N 73 22W
Cambridge, *Nebr., U.S.A.* **138 E4** 40 17N 100 10W
Cambridge, *Ohio, U.S.A.* **136 F3** 40 2N 81 35W
Cambridge Bay, *Canada* **126 B9** 69 10N 105 0W
Cambridge City, *U.S.A.* **141 E11** 39 49N 85 10W
Cambridge G., *Australia* **112 B4** 14 55 S 128 15 E
Cambridge Springs, *U.S.A.* **136 E4** 41 48N 80 4W
Cambridgeshire □, *U.K.* **17 E8** 52 25N 0 7W
Cambrils, *Spain* **34 D6** 41 8N 1 3 E
Cambuci, *Brazil* **155 F3** 21 35 S 41 55W
Cambundi-Catembo, *Angola* **103 E3** 10 10 S 17 35 E
Camden, *Australia* **117 C9** 34 1 S 150 43 E
Camden, *Ala., U.S.A.* **135 K2** 31 59N 87 17W
Camden, *Ark., U.S.A.* **139 J8** 33 35N 92 50W
Camden, *Maine, U.S.A.* **129 D6** 44 13N 69 4W
Camden, *N.J., U.S.A.* **137 G9** 39 56N 75 7W
Camden, *Ohio, U.S.A.* **141 E12** 39 38N 84 39W
Camden, *S.C., U.S.A.* **135 H5** 34 16N 80 36W
Camden Sd., *Australia* **112 C3** 15 27 S 124 25 E
Camdenton, *U.S.A.* **139 F8** 38 1N 92 45W
Çameli, *Turkey* **88 D3** 37 5N 29 24 E
Camembert, *France* **22 D7** 48 53N 0 10 E
Cámeri, *Italy* **38 C5** 45 30N 8 40 E
Camerino, *Italy* **39 E10** 43 8N 13 4 E
Cameron, *Ariz., U.S.A.* **143 J8** 35 53N 111 25W
Cameron, *La., U.S.A.* **139 L8** 29 48N 93 20W
Cameron, *Mo., U.S.A.* **140 E2** 39 44N 94 14W
Cameron, *Tex., U.S.A.* **139 K6** 30 51N 96 59W
Cameron Falls, *Canada* **128 C2** 49 8N 88 19W
Cameron Highlands, *Malaysia* **77 K3** 4 27N 101 22 E
Cameron Hills, *Canada* **130 B5** 59 48N 118 0W
Cameron Mts., *N.Z.* **119 G1** 46 1 S 167 0 E
Cameroon ■, *Africa* **102 A2** 6 0N 12 30 E
Camerota, *Italy* **41 B8** 40 2N 15 22 E
Cameroun →, *Cameroon* **101 E6** 4 0N 9 35 E
Cameroun, Mt., *Cameroon* **101 E6** 4 13N 9 10 E
Cametá, *Brazil* **154 B2** 2 12 S 49 30W
Camiguin □, *Phil.* **71 G5** 9 11N 124 42 E
Camiguin I., *Phil.* **70 B3** 18 56N 121 55 E
Camiling, *Phil.* **70 D3** 15 42N 120 24 E
Caminha, *Portugal* **36 D2** 41 50N 8 50W
Camino, *U.S.A.* **144 G6** 38 44N 120 41W
Camira Creek, *Australia* **115 D5** 29 15 S 152 58 E
Camiranga, *Brazil* **154 B2** 1 48 S 46 17W
Camiri, *Bolivia* **157 E5** 20 3 S 63 31W
Camissombo, *Angola* **103 D4** 8 7 S 20 38 E
Cammal, *U.S.A.* **136 E7** 41 24N 77 28W
Camocim, *Brazil* **154 B3** 2 55 S 40 50W
Camogli, *Italy* **38 D6** 44 21N 9 9 E
Camooweal, *Australia* **114 B2** 19 56 S 138 7 E
Camopi, *Fr. Guiana* **153 C7** 3 12N 52 17W
Camopi →, *Fr. Guiana* **153 C7** 3 10N 52 20W
Camotes Is., *Phil.* **71 F5** 10 40N 124 24 E
Camotes Sea, *Phil.* **71 F5** 10 30N 124 15 E
Camp Crook, *U.S.A.* **138 C3** 45 33N 103 59W
Camp Nelson, *U.S.A.* **145 J8** 36 8N 118 39W
Camp Point, *U.S.A.* **140 D5** 40 3N 91 4W
Camp Wood, *U.S.A.* **139 L4** 29 40N 100 1W
Campagna, *Italy* **41 B8** 40 40N 15 6 E
Campana, *Argentina* **158 C4** 34 10 S 58 55W
Campana, I., *Chile* **160 C1** 48 20 S 75 20W
Campanário, *Madeira* **33 D2** 32 39N 17 2W
Campanario, *Spain* **37 G5** 38 52N 5 36W
Campània □, *Italy* **41 B7** 41 0N 14 30 E
Campbell, *S. Africa* **104 D3** 28 48 S 23 44 E
Campbell, *Calif., U.S.A.* **144 H5** 37 17N 121 57W
Campbell, *Ohio, U.S.A.* **136 E4** 41 5N 80 37W
Campbell, C., *N.Z.* **119 B9** 41 47 S 174 18 E
Campbell I., *Pac. Oc.* **122 N8** 52 30 S 169 0 E
Campbell L., *Canada* **131 A7** 63 14N 106 55W
Campbell River, *Canada* **130 C3** 50 5N 125 20W
Campbell Town, *Australia* **114 G4** 41 52 S 147 30 E
Campbellford, *Canada* **136 B7** 44 18N 77 48W
Campbellpur, *Pakistan* **80 C5** 33 46N 72 26 E
Campbellsburg, *U.S.A.* **141 F10** 38 39N 86 16W
Campbellsville, *U.S.A.* **134 G3** 37 21N 85 20W
Campbellton, *Canada* **129 C6** 47 57N 66 43W
Campbelltown, *Australia* **117 C9** 34 4 S 150 49 E
Campbeltown, *U.K.* **18 F3** 55 26N 5 36W
Campeche, *Mexico* **147 D6** 19 50N 90 32W
Campeche □, *Mexico* **147 D6** 19 50N 90 32W
Campeche, B. de, *Mexico* **147 D6** 19 30N 93 0W
Camperdown, *Australia* **116 E5** 38 14 S 143 9 E
Camperville, *Canada* **131 C8** 51 59N 100 9W
Campi Salentina, *Italy* **41 B11** 40 24N 18 1 E
Campidano, *Italy* **40 C1** 39 30N 8 47 E
Campíglia Maríttima, *Italy* **38 E7** 43 4N 10 37 E
Campillo de Altobuey, *Spain* **34 F3** 39 36N 1 49W
Campillo de Llerena, *Spain* **37 G5** 38 30N 5 50W
Campillos, *Spain* **37 H6** 37 4N 4 51W
Campina Grande, *Brazil* **154 C4** 7 20 S 35 47W
Campina Verde, *Brazil* **155 E2** 19 31 S 49 28W
Campinas, *Brazil* **159 A6** 22 50 S 47 0W
Campine, *Belgium* **21 F6** 51 8N 5 20 E
Campli, *Italy* **39 F10** 42 43N 13 41 E
Campo, *Cameroon* **102 B1** 2 22N 9 50 E
Campo, *Spain* **34 C5** 42 25N 0 24 E
Campo Belo, *Brazil* **155 F2** 20 52 S 45 16W
Campo de Criptana, *Spain* **35 F1** 39 24N 3 7W
Campo de Diauarum, *Brazil* **157 C7** 11 12 S 53 14W
Campo de Gibraltar, *Spain* **37 J5** 36 15N 5 25W
Campo Flórido, *Brazil* **155 E2** 19 47 S 48 35W
Campo Formoso, *Brazil* **154 D3** 10 30 S 40 20W
Campo Grande, *Brazil* **157 E7** 20 25 S 54 40W
Campo Maíor, *Brazil* **154 B3** 4 50 S 42 12W
Campo Maior, *Portugal* **37 G3** 38 59N 7 7W
Campo Mourão, *Brazil* **159 A5** 24 3 S 52 22W
Campo Tencia, *Switz.* **29 D7** 46 26N 8 43 E
Campo Túres, *Italy* **39 B8** 46 53N 11 55 E
Campoalegre, *Colombia* **152 C2** 2 41N 75 20W
Campobasso, *Italy* **41 A7** 41 34N 14 39 E
Campobello di Licata, *Italy* **40 E6** 37 15N 13 55 E
Campobello di Mazara, *Italy* **40 E5** 37 38N 12 45 E
Campofelice di Roccella, *Italy* **40 E6** 37 59N 13 53 E
Camporeale, *Italy* **40 E6** 37 54N 13 6 E
Campos, *Brazil* **155 F3** 21 50 S 41 20W
Campos Altos, *Brazil* **155 E2** 19 47 S 46 10W
Campos Belos, *Brazil* **155 D2** 13 10 S 47 3W
Campos del Puerto, *Spain* **33 B10** 39 26N 3 1 E
Campos Novos, *Brazil* **159 B5** 27 21 S 51 50W
Campos Sales, *Brazil* **154 C3** 7 4 S 40 23W
Camprodón, *Spain* **34 C7** 42 19N 2 23 E
Campton, *U.S.A.* **141 G13** 37 44N 83 33W
Camptonville, *U.S.A.* **144 F5** 39 27N 121 3W
Campuya →, *Peru* **152 D3** 1 40 S 73 30W
Camrose, *Canada* **130 C6** 53 0N 112 50W
Camsell Portage, *Canada* **131 B7** 59 37N 109 15W
Çan, *Turkey* **88 B2** 40 2N 27 3 E
Can Clavo, *Spain* **33 C7** 38 57N 1 27 E
Can Creu, *Spain* **33 C7** 38 58N 1 28 E
Can Gio, *Vietnam* **77 G6** 10 25N 106 58 E
Can Tho, *Vietnam* **77 G5** 10 2N 105 46 E
Canaan, *U.S.A.* **137 D11** 42 2N 73 20W
Canada ■, *N. Amer.* **126 C10** 60 0N 100 0W
Cañada de Gómez, *Argentina* **158 C3** 32 40 S 61 30W
Canadian, *U.S.A.* **139 H4** 35 55N 100 23W
Canadian →, *U.S.A.* **139 H7** 35 28N 95 3W
Canadian Shield, *Canada* **127 C10** 53 0N 75 0W
Çanakkale, *Turkey* **88 B2** 40 8N 26 24 E
Çanakkale Boğazı, *Turkey* **88 B2** 40 17N 26 32 E
Canal Flats, *Canada* **130 C5** 50 10N 115 48W
Canala, *N. Cal.* **121 U19** 21 32 S 165 57 E
Canalejas, *Argentina* **158 D2** 35 15 S 66 34W
Canals, *Argentina* **158 C3** 33 35 S 62 53W
Canals, *Spain* **35 G4** 38 58N 0 35W
Canandaigua, *U.S.A.* **136 D7** 42 54N 77 17W
Cananea, *Mexico* **146 A2** 31 0N 110 20W
Cañar, *Ecuador* **152 D2** 2 33 S 78 56W
Cañar □, *Ecuador* **152 D2** 2 30 S 79 0W
Canarias, Is., *Atl. Oc.* **33 F4** 28 30N 16 0W
Canarreos, Arch. de los, *Cuba* **148 B3** 21 35N 81 40W
Canary Is. = Canarias, Is., *Atl. Oc.* **33 F4** 28 30N 16 0W
Canastra, Serra da, *Brazil* **155 F2** 20 0 S 46 20W
Canatlán, *Mexico* **146 C4** 24 31N 104 47W
Canaveral, C., *U.S.A.* **135 L5** 28 27N 80 32W
Cañaveras, *Spain* **34 E2** 40 27N 2 24W
Canavieiras, *Brazil* **155 E4** 15 39 S 39 0W
Canbelego, *Australia* **117 A7** 31 32 S 146 18 E
Canberra, *Australia* **117 C8** 35 15 S 149 8 E
Canby, *Calif., U.S.A.* **142 F3** 41 27N 120 52W
Canby, *Minn., U.S.A.* **138 C6** 44 43N 96 16W
Canby, *Oreg., U.S.A.* **144 E4** 45 16N 122 42W
Cancale, *France* **22 D5** 48 40N 1 50W
Canche →, *France* **23 B8** 50 31N 1 39 E
Canchyuaya, Cordillera de, *Peru* **156 B3** 7 30 S 74 0W
Cancún, *Mexico* **147 C7** 21 8N 86 44W
Candala, *Somali Rep.* **108 B3** 11 30N 49 58 E
Candarave, *Peru* **156 D3** 17 15 S 70 13W
Candas, *Spain* **36 B5** 43 35N 5 45W
Candé, *France* **22 E5** 47 34N 1 2W
Candeias →, *Brazil* **157 B5** 8 39 S 63 31W
Candela, *Italy* **41 A8** 41 8N 15 31 E
Candelaria, *Argentina* **159 B4** 27 29 S 55 44W
Candelaria, *Canary Is.* **33 F3** 28 22N 16 22W
Candelaria, *Phil.* **70 E3** 13 56N 121 25 E
Candelaria, Pta. de la, *Spain* **36 B2** 43 45N 8 0W
Candeleda, *Spain* **36 E5** 40 10N 5 14W
Candelo, *Australia* **117 D8** 36 47 S 149 43 E
Candia = Iráklion, *Greece* **32 D7** 35 20N 25 12 E
Candia, Sea of = Crete, Sea of, *Greece* **45 H6** 36 0N 25 0 E
Cândido de Abreu, *Brazil* **155 F1** 24 35 S 51 20W
Cândido Mendes, *Brazil* **154 B2** 1 27 S 45 43W
Candle L., *Canada* **131 C7** 53 50N 105 18W
Candlemas I., *Antarctica* **7 B1** 57 3 S 26 40W
Cando, *U.S.A.* **138 A5** 48 32N 99 12W
Candon, *Phil.* **70 C3** 17 12N 120 27 E
Candoni, *Phil.* **71 G4** 9 48N 122 30 E
Canea = Khaniá, *Greece* **32 D6** 35 30N 24 4 E
Canela, *Brazil* **154 D2** 10 15 S 48 25W
Canelli, *Italy* **38 D5** 44 43N 8 17 E
Canelones, *Uruguay* **159 C4** 34 32 S 56 17W
Canet-Plage, *France* **24 F7** 42 41N 3 2 E
Cañete, *Chile* **158 D1** 37 50 S 73 30W
Cañete, *Peru* **156 C2** 13 8 S 76 30W
Cañete, *Spain* **34 E3** 40 3N 1 54W
Cañete de las Torres, *Spain* **37 H6** 37 53N 4 19W
Canfranc, *Spain* **34 C4** 42 42N 0 31W
Cangamba, *Angola* **103 E3** 13 40 S 19 54 E
Cangandala, *Angola* **103 D3** 9 45 S 16 33 E
Cangas, *Spain* **36 C2** 42 16N 8 47W
Cangas de Narcea, *Spain* **36 B4** 43 10N 6 32W
Cangas de Onís, *Spain* **36 B5** 43 21N 5 8W
Cangoa, *Angola* **103 E3** 13 8 S 18 30 E
Cangombo, *Angola* **103 E3** 14 24 S 19 59 E
Cangongo, *Angola* **103 D3** 9 24 S 17 30 E
Canguaretama, *Brazil* **154 C4** 6 20 S 35 5W
Canguçu, *Brazil* **159 C5** 31 22 S 52 43W
Cangxi, *China* **68 B5** 31 47N 105 59 E
Cangyuan, *China* **68 F2** 23 12N 99 14 E
Cangzhou, *China* **66 E9** 38 19N 116 52 E
Canhoca, *Angola* **103 D2** 9 15 S 14 41 E
Canicattì, *Italy* **40 E6** 37 21N 13 51 E
Canicattini Bagni, *Italy* **41 E8** 37 2N 15 4 E
Canigao Channel, *Phil.* **71 F5** 10 15N 124 42 E
Canim Lake, *Canada* **130 C4** 51 47N 120 54W
Canindé, *Brazil* **154 B4** 4 22 S 39 19W
Canindé →, *Brazil* **154 C3** 6 15 S 42 52W
Canindeyu □, *Paraguay* **159 A4** 24 10 S 55 0W
Canisteo, *U.S.A.* **136 D7** 42 16N 77 36W
Canisteo →, *U.S.A.* **136 D7** 42 7N 77 8W
Cañitas, *Mexico* **146 C4** 23 36N 102 43W
Cañizal, *Spain* **36 D5** 41 12N 5 22W
Canjáyar, *Spain* **35 H2** 37 1N 2 44W
Canjinge, *Angola* **103 E4** 10 12 S 21 17 E
Çankırı, *Turkey* **88 B5** 40 40N 33 37 E
Cankuzo, *Burundi* **106 C3** 3 10 S 30 31 E
Canmore, *Canada* **130 C5** 51 7N 115 18W
Cann River, *Australia* **117 D8** 37 35 S 149 7 E
Canna, *U.K.* **18 D2** 57 3N 6 33W
Cannanore, *India* **83 J2** 11 53N 75 27 E
Cannelton, *U.S.A.* **141 G10** 37 55N 86 45W
Cannes, *France* **25 E11** 43 32N 7 1 E
Canning Town = Port Canning, *India* **81 H13** 22 23N 88 40 E
Cannington, *Canada* **136 B5** 44 20N 79 2W
Cannock, *U.K.* **16 E5** 52 41N 2 1W
Cannon Ball →, *U.S.A.* **138 B4** 46 20N 100 38W
Cannondale Mt., *Australia* **114 D4** 25 13 S 148 57 E
Caño Colorado, *Colombia* **152 C4** 2 18N 68 22W
Canoas, *Brazil* **159 B5** 29 56 S 51 11W
Canoe L., *Canada* **131 B7** 55 10N 108 15W
Canon City, *U.S.A.* **138 F2** 38 27N 105 14W
Canopus, *Australia* **116 B4** 33 29 S 140 42 E
Canora, *Canada* **131 C8** 51 40N 102 30W
Canosa di Púglia, *Italy* **41 A9** 41 13N 16 4 E
Canowindra, *Australia* **117 B8** 33 35 S 148 38 E
Canso, *Canada* **129 C7** 45 20N 61 0W
Canta, *Peru* **156 C2** 11 29 S 76 37W
Cantabria □, *Spain* **36 B6** 43 10N 4 0W
Cantabria, Sierra de, *Spain* **34 C2** 42 40N 2 30W
Cantabrian Mts. = Cantábrica, Cordillera, *Spain* **36 C5** 43 0N 5 10W
Cantábrica, Cordillera, *Spain* **36 C5** 43 0N 5 10W
Cantal □, *France* **24 C6** 45 5N 2 45 E
Cantal, Plomb du, *France* **24 C6** 45 3N 2 45 E
Cantanhede, *Portugal* **36 E2** 40 20N 8 36W
Cantaura, *Venezuela* **153 B5** 9 19N 64 21W
Cantavieja, *Spain* **34 E4** 40 31N 0 25W
Čantavir, *Serbia, Yug.* **42 B4** 45 55N 19 46 E
Canterbury, *Australia* **114 D3** 25 23 S 141 53 E
Canterbury, *U.K.* **17 F9** 51 16N 1 6 E
Canterbury □, *N.Z.* **119 D6** 43 45 S 171 19 E
Canterbury Bight, *N.Z.* **119 E6** 44 16 S 171 55 E
Canterbury Plains, *N.Z.* **119 D6** 43 55 S 171 22 E
Cantil, *U.S.A.* **145 K9** 35 18N 117 58W
Cantilan, *Phil.* **71 G5** 9 20N 125 58 E
Cantillana, *Spain* **37 H5** 37 36N 5 50W
Canto do Buriti, *Brazil* **154 C3** 8 7 S 42 58W
Canton = Guangzhou, *China* **69 F9** 23 5N 113 10 E
Canton, *Ga., U.S.A.* **135 H3** 34 14N 84 29W
Canton, *Ill., U.S.A.* **140 D6** 40 33N 90 2W
Canton, *Miss., U.S.A.* **139 J9** 32 37N 90 2W
Canton, *Mo., U.S.A.* **140 D5** 40 8N 91 32W
Canton, *N.Y., U.S.A.* **137 B9** 44 36N 75 10W
Canton, *Ohio, U.S.A.* **136 F3** 40 48N 81 23W
Canton, *Okla., U.S.A.* **139 G5** 36 3N 98 35W
Canton, *S. Dak., U.S.A.* **138 D6** 43 18N 96 35W
Canton L., *U.S.A.* **139 G5** 36 6N 98 35W
Cantù, *Italy* **38 C6** 45 44N 9 8 E
Canudos, *Brazil* **157 B6** 7 13 S 58 5W
Canumã, *Amazonas, Brazil* **153 D6** 4 2 S 59 4W
Canumã, *Amazonas, Brazil* **157 B5** 6 8 S 60 10W
Canumã →, *Brazil* **157 A6** 3 55 S 59 10W
Canutama, *Brazil* **157 B5** 6 30 S 64 20W
Canutillo, *U.S.A.* **143 L10** 31 55N 106 36W
Canyon, *Tex., U.S.A.* **139 H4** 34 59N 101 55W
Canyon, *Wyo., U.S.A.* **142 D8** 44 43N 110 36W
Canyonlands National Park, *U.S.A.* **143 G9** 38 15N 110 0W
Canyonville, *U.S.A.* **142 E2** 42 56N 123 17W
Canzo, *Italy* **38 C6** 45 54N 9 18 E
Cao Bang, *Vietnam* **76 A6** 22 40N 106 15 E
Cao He →, *China* **67 D13** 40 10N 124 32 E
Cao Lanh, *Vietnam* **77 G5** 10 27N 105 38 E
Cao Xian, *China* **66 G8** 34 50N 115 35 E
Caoayan, *Phil.* **70 C3** 17 37N 120 23 E
Cáorle, *Italy* **39 C9** 45 36N 12 53 E
Cap-aux-Meules, *Canada* **129 C7** 47 23N 61 52W
Cap-Chat, *Canada* **129 C6** 49 6N 66 40W
Cap-de-la-Madeleine, *Canada* **128 C5** 46 22N 72 31W
Cap-Haïtien, *Haiti* **149 C5** 19 40N 72 20W
Cap St.-Jacques = Vung Tau, *Vietnam* **77 G6** 10 21N 107 4 E
Capa, *Vietnam* **76 A4** 22 21N 103 50 E
Capa Stilo, *Italy* **41 D9** 38 25N 16 35 E
Capáccio, *Italy* **41 B8** 40 25N 15 5 E
Capaia, *Angola* **103 D4** 8 27 S 20 13 E
Capalonga, *Phil.* **70 D4** 14 20N 122 30 E
Capanaparo →, *Venezuela* **152 B4** 7 1N 67 7W
Capanema, *Brazil* **154 B2** 1 12 S 47 11W
Caparo →, *Barinas, Venezuela* **152 B3** 7 46N 70 23W
Caparo →, *Bolívar, Venezuela* **153 B5** 7 30N 64 0W
Capatárida, *Venezuela* **152 A3** 11 11N 70 37W
Capayas, *Phil.* **71 F2** 10 28N 119 39 E
Capbreton, *France* **24 E2** 43 39N 1 26W
Capdenac, *France* **24 D6** 44 34N 2 5 E
Cape →, *Australia* **114 C4** 20 59 S 146 51 E
Cape Barren I., *Australia* **114 G4** 40 25 S 148 15 E
Cape Breton Highlands Nat. Park, *Canada* **129 C7** 46 50N 60 40W
Cape Breton I., *Canada* **129 C7** 46 0N 60 30W
Cape Charles, *U.S.A.* **134 G8** 37 16N 76 1W
Cape Coast, *Ghana* **101 D4** 5 5N 1 15W
Cape Dorset, *Canada* **127 B12** 64 14N 76 32W
Cape Dyer, *Canada* **127 B13** 66 30N 61 22W
Cape Fear →, *U.S.A.* **135 H6** 33 53N 78 1W
Cape Girardeau, *U.S.A.* **139 G10** 37 19N 89 32W
Cape Jervis, *Australia* **116 C3** 35 40 S 138 5 E
Cape May, *U.S.A.* **134 F8** 38 56N 74 56W
Cape May Point, *U.S.A.* **133 C12** 38 56N 74 58W
Cape Palmas, *Liberia* **100 E3** 4 25N 7 49W
Cape Tormentine, *Canada* **129 C7** 46 8N 63 47W
Cape Town, *S. Africa* **104 E2** 33 55 S 18 22 E
Cape Verde Is. ■, *Atl. Oc.* **8 G6** 17 10N 25 20W
Cape Vincent, *U.S.A.* **137 B8** 44 8N 76 20W
Cape York Peninsula, *Australia* **114 A3** 12 0 S 142 30 E
Capela, *Brazil* **154 D4** 10 30 S 37 0W
Capela de Campo, *Brazil* **154 B3** 4 40 S 41 55W
Capele, *Angola* **103 E2** 13 39 S 14 53 E
Capelinha, *Brazil* **155 E3** 17 42 S 42 31W
Capella, *Australia* **114 C4** 23 2 S 148 1 E
Capella, Mt., *Papua N. G.* **120 C1** 5 4 S 141 8 E
Capenda Camulemba, *Angola* **103 D3** 9 24 S 18 27 E
Capendu, *France* **24 E6** 43 11N 2 31 E
Capestang, *France* **24 E7** 43 20N 3 2 E
Capim, *Brazil* **154 B2** 1 41 S 47 47W
Capim →, *Brazil* **154 B2** 1 40 S 47 47W
Capinópolis, *Brazil* **155 E2** 18 41 S 49 35W
Capinota, *Bolivia* **156 D4** 17 43 S 66 14W
Capitan, *U.S.A.* **143 K11** 33 35N 105 35W
Capitán Aracena, I., *Chile* **160 D2** 54 10 S 71 20W
Capitán Pastene, *Chile* **160 A2** 38 13 S 73 1W
Capitola, *U.S.A.* **144 J5** 36 59N 121 57W
Capivara, Serra da, *Brazil* **155 D3** 14 35 S 45 0W
Capiz □, *Phil.* **71 F4** 11 35N 122 30 E
Capizzi, *Italy* **41 E7** 37 51N 14 29 E
Čapljina, *Bos.-H.* **42 D2** 43 10N 17 43 E
Capoche →, *Mozam.* **107 F3** 15 35 S 33 0 E
Capoeira, *Brazil* **157 B6** 5 37 S 59 33W
Capolo, *Angola* **103 E2** 10 22 S 14 7 E
Cappadocia, *Turkey* **88 C6** 39 0N 35 0 E
Capraia, *Italy* **38 E6** 43 2N 9 50 E
Caprarola, *Italy* **39 F9** 42 19N 12 14 E
Capreol, *Canada* **128 C3** 46 43N 80 56W
Caprera, *Italy* **40 A2** 41 12N 9 28 E
Capri, *Italy* **41 B7** 40 33N 14 14 E
Capricorn Group, *Australia* **114 C5** 23 30 S 151 55 E
Capricorn Ra., *Australia* **112 D2** 23 20 S 116 50 E
Caprino Veronese, *Italy* **38 C7** 45 36N 10 47 E
Caprivi Strip, *Namibia* **104 B3** 18 0 S 23 0 E
Captainganj, *India* **81 F10** 26 55N 83 45 E
Captain's Flat, *Australia* **117 C8** 35 35 S 149 27 E
Captieux, *France* **24 D3** 44 18N 0 16W
Capu-Lapu, *Phil.* **71 F4** 10 20N 123 55 E
Cápua, *Italy* **41 A7** 41 6N 14 12 E
Capul I., *Phil.* **70 E5** 12 26N 124 10 E
Caquetá □, *Colombia* **152 C3** 1 0N 74 0W
Caquetá →, *Colombia* **152 D4** 1 15 S 69 15W
Carabalan, *Phil.* **71 F4** 10 6N 122 57 E
Carabao I., *Phil.* **70 E3** 12 4N 121 56 E
Carabobo, *Venezuela* **152 A4** 10 2N 68 5W
Carabobo □, *Venezuela* **152 A4** 10 10N 68 5W
Caracal, *Romania* **46 E5** 44 8N 24 22 E
Caracaraí, *Brazil* **153 C5** 1 50N 61 8W
Caracas, *Venezuela* **152 A4** 10 30N 66 55W
Caracol, *Brazil* **154 C3** 9 15 S 43 22W
Caracollo, *Bolivia* **156 D4** 17 39 S 67 10W
Caradoc, *Australia* **116 A5** 30 35 S 143 5 E
Caragabal, *Australia* **117 B7** 33 49 S 147 45 E
Caráglio, *Italy* **38 D4** 44 25N 7 26 E
Carahue, *Chile* **160 A2** 38 43 S 73 12W
Caraí, *Brazil* **155 E3** 17 12 S 41 42W
Carajás, Serra dos, *Brazil* **154 C1** 6 0 S 51 30W
Caramoan, *Phil.* **70 E4** 13 46N 123 52 E
Caranapatuba, *Brazil* **157 B5** 6 38 S 62 34W
Carandaiti, *Bolivia* **157 E5** 20 45 S 63 4W
Carangola, *Brazil* **155 F3** 20 44 S 42 5W
Carani, *Australia* **113 F2** 30 57 S 116 28 E
Caransebeş, *Romania* **46 D3** 45 28N 22 18 E
Carantec, *France* **22 D3** 48 40N 3 55W
Caraparaná →, *Colombia* **152 D3** 1 45 S 73 13W
Carapelle →, *Italy* **41 A8** 41 30N 15 55 E
Caras, *Peru* **156 B2** 9 3 S 77 47W
Caraş Severin □, *Romania* **42 B7** 45 10N 22 10 E
Caraşova, *Romania* **42 B6** 45 11N 21 51 E
Caratasca, L., *Honduras* **148 C3** 15 20N 83 40W
Caratinga, *Brazil* **155 E3** 19 50 S 42 10W
Caraúbas, *Brazil* **154 C4** 5 43 S 37 33W
Caravaca, *Spain* **35 G3** 38 8N 1 52W
Caravággio, *Italy* **38 C6** 45 30N 9 38 E
Caravelas, *Brazil* **155 E4** 17 45 S 39 15W
Caraveli, *Peru* **156 D3** 15 45 S 73 25W

182 Caràzinho

Caràzinho, *Brazil* **159 B5** 28 16 S 52 46W
Carballino, *Spain* **36 C2** 42 26N 8 5W
Carballo, *Spain* **36 B2** 43 13N 8 41W
Carberry, *Canada* **131 D9** 49 50N 99 25W
Carbia, *Spain* **36 C2** 42 48N 8 14W
Carbó, *Mexico* **146 B2** 29 42N 110 58W
Carbon, *Canada* **130 C6** 51 30N 113 9W
Carbonara, C., *Italy* **40 C2** 39 6N 9 31 E
Carbondale, *Colo., U.S.A.* **142 G10** 39 24N 107 13W
Carbondale, *Ill., U.S.A.* **139 G10** 37 44N 89 13W
Carbondale, *Pa., U.S.A.* **137 E9** 41 35N 75 30W
Carbonear, *Canada* **129 C9** 47 42N 53 13W
Carboneras, *Spain* **35 J3** 37 0N 1 53W
Carboneras de Guadazaón, *Spain* **34 F3** 39 54N 1 50W
Carbonia, *Italy* **40 C1** 39 10N 8 30 E
Carcabuey, *Spain* **37 H6** 37 27N 4 17W
Carcagente, *Spain* **35 F4** 39 8N 0 28W
Carcajou, *Canada* **130 B5** 57 47N 117 6W
Carcar, *Phil.* **71 F4** 10 6N 123 38 E
Carcasse, C., *Haiti* **149 C5** 18 30N 74 28W
Carcassonne, *France* **24 E6** 43 13N 2 20 E
Carche, *Spain* **35 G3** 38 26N 1 9W
Carchi □, *Ecuador* **152 C2** 0 45N 78 0W
Carcoar, *Australia* **117 B8** 33 36 S 149 8 E
Carcross, *Canada* **126 B6** 60 13N 134 45W
Cardabia, *Australia* **112 D1** 23 2 S 113 48 E
Çardak, *Turkey* **88 D3** 37 49N 29 39 E
Cardamon Hills, *India* **83 K3** 9 30N 77 15 E
Cárdenas, *Cuba* **148 B3** 23 0N 81 30W
Cárdenas, *San Luis Potosí, Mexico* **147 C5** 22 0N 99 41W
Cárdenas, *Tabasco, Mexico* **147 D6** 17 59N 93 21W
Cardenete, *Spain* **34 F3** 39 46N 1 41W
Cardiel, L., *Argentina* **160 C2** 48 55 S 71 10W
Cardiff, *U.K.* **17 F4** 51 29N 3 10W
Cardiff-by-the-Sea, *U.S.A.* **145 M9** 33 1N 117 17W
Cardigan, *U.K.* **17 E3** 52 5N 4 40W
Cardigan B., *U.K.* **17 E3** 52 30N 4 30W
Cardinal, *Canada* **137 B9** 44 47N 75 23W
Cardón, Punta, *Venezuela* **152 A3** 11 37N 70 14W
Cardona, *Spain* **34 D6** 41 56N 1 40 E
Cardona, *Uruguay* **158 C4** 33 53 S 57 18W
Cardoner →, *Spain* **34 D6** 41 41N 1 51 E
Cardross, *Canada* **131 D7** 49 50N 105 40W
Cardston, *Canada* **130 D6** 49 15N 113 20W
Cardwell, *Australia* **114 B4** 18 14 S 146 2 E
Careen L., *Canada* **131 B7** 57 0N 108 11W
Carei, *Romania* **46 B3** 47 40N 22 29 E
Careiro, *Brazil* **153 D6** 3 12 S 59 45W
Careme, *Indonesia* **75 D3** 6 55 S 108 27 E
Carentan, *France* **22 C5** 49 19N 1 15W
Carey, *Idaho, U.S.A.* **142 E7** 43 19N 113 57W
Carey, *Ohio, U.S.A.* **141 D13** 40 57N 83 23W
Carey, L., *Australia* **113 E3** 29 0 S 122 15 E
Carey L., *Canada* **131 A8** 62 12N 102 55W
Careysburg, *Liberia* **100 D2** 6 34N 10 30W
Cargados Garajos, *Ind. Oc.* **109 F4** 17 0 S 59 0 E
Cargèse, *France* **25 F12** 42 7N 8 35 E
Carhaix-Plouguer, *France* **22 D3** 48 18N 3 36W
Carhuamayo, *Peru* **156 C2** 10 51 S 76 4W
Carhuas, *Peru* **156 B2** 9 15 S 77 39W
Carhué, *Argentina* **158 D3** 37 10 S 62 50W
Caria, *Turkey* **88 D3** 37 20N 28 10 E
Cariacica, *Brazil* **155 F3** 20 16 S 40 25W
Cariango, *Angola* **103 E3** 10 37 S 15 20 E
Caribbean Sea, *W. Indies* **149 C5** 15 0N 75 0W
Cariboo Mts., *Canada* **130 C4** 53 0N 121 0W
Caribou, *U.S.A.* **129 C6** 46 52N 68 1W
Caribou →, *Man., Canada* **131 B10** 59 20N 94 44W
Caribou →, *N.W.T., Canada* **130 A3** 61 27N 125 45W
Caribou I., *Canada* **128 C2** 47 22N 85 49W
Caribou Is., *Canada* **130 A6** 61 55N 113 15W
Caribou L., *Man., Canada* **131 B9** 59 21N 96 10W
Caribou L., *Ont., Canada* **128 B2** 50 25N 89 5W
Caribou Mts., *Canada* **130 B5** 59 12N 115 40W
Carichic, *Mexico* **146 B3** 27 56N 107 3W
Carigara, *Phil.* **71 F5** 11 18N 124 41 E
Carignan, *France* **23 C12** 49 38N 5 10 E
Carignano, *Italy* **38 D4** 44 55N 7 40 E
Carillo, *Mexico* **146 B4** 26 50N 103 55W
Carin, *Somali Rep.* **108 B3** 10 59N 49 13 E
Carinda, *Australia* **117 A7** 30 28 S 147 41 E
Cariñena, *Spain* **34 D3** 41 20N 1 13W
Carinhanha, *Brazil* **155 D3** 14 15 S 44 46W
Carinhanha →, *Brazil* **155 D3** 14 20 S 43 47W
Carini, *Italy* **40 D6** 38 8N 13 11 E
Carinola, *Italy* **40 A6** 41 11N 13 58 E
Carinthia = Kärnten □, *Austria* **30 E6** 46 52N 13 30 E
Caripito, *Venezuela* **153 A5** 10 8N 63 6W
Caritianas, *Brazil* **157 B5** 9 20 S 63 6W
Carlbrod = Dimitrovgrad, *Serbia, Yug.* **42 D7** 43 2N 22 48 E
Carlentini, *Italy* **41 E8** 37 16N 15 1 E
Carles, *Phil.* **71 F4** 11 34N 123 8 E
Carleton Place, *Canada* **128 C4** 45 8N 76 9W
Carletonville, *S. Africa* **104 D4** 26 23 S 27 22 E
Carlin, *U.S.A.* **142 F5** 40 43N 116 7W
Carlingford L., *U.K.* **19 B5** 54 3N 6 9W
Carlinville, *U.S.A.* **140 E7** 39 17N 89 53W
Carlisle, *U.K.* **16 C5** 54 54N 2 56W
Carlisle, *Ky., U.S.A.* **141 F12** 38 19N 84 1W
Carlisle, *Pa., U.S.A.* **136 F7** 40 12N 77 12W
Carlit, Pic, *France* **24 F5** 42 35N 1 55 E
Carloforte, *Italy* **40 C1** 39 8N 8 18 E
Carlos Casares, *Argentina* **158 D3** 35 32 S 61 20W
Carlos Chagas, *Brazil* **155 E3** 17 43 S 40 45W
Carlos Tejedor, *Argentina* **158 D3** 35 25 S 62 25W
Carlow, *Ireland* **19 D5** 52 50N 6 56W
Carlow □, *Ireland* **19 D5** 52 43N 6 50W
Carlsbad, *Calif., U.S.A.* **145 M9** 33 10N 117 21W
Carlsbad, *N. Mex., U.S.A.* **139 J2** 32 25N 104 14W
Carlyle, *Canada* **131 D8** 49 40N 102 20W
Carlyle, *U.S.A.* **138 F10** 38 37N 89 22W
Carlyle Res., *U.S.A.* **140 F7** 38 37N 89 21W
Carmacks, *Canada* **126 B6** 62 5N 136 16W
Carmagnola, *Italy* **38 D4** 44 51N 7 43 E
Carman, *Canada* **131 D9** 49 30N 98 0W
Carmangay, *Canada* **130 C6** 50 10N 113 10W
Carmanville, *Canada* **129 C9** 49 23N 54 19W
Carmarthen, *U.K.* **17 F3** 51 52N 4 19W
Carmarthen B., *U.K.* **17 F3** 51 40N 4 30W
Carmaux, *France* **24 D6** 44 3N 2 10 E
Carmel, *Ind., U.S.A.* **141 E10** 39 59N 86 8W
Carmel, *N.Y., U.S.A.* **137 E11** 41 26N 73 41W
Carmel-by-the-Sea, *U.S.A.* **144 J5** 36 33N 121 55W
Carmel Valley, *U.S.A.* **144 J5** 36 29N 121 43W
Carmelo, *Uruguay* **158 C4** 34 0 S 58 20W
Carmen, *Bolivia* **156 C4** 11 40 S 67 51W
Carmen, *Colombia* **152 B2** 9 43N 75 8W
Carmen, *Paraguay* **159 B4** 27 13 S 56 12W
Carmen, *Bohol, Phil.* **71 G5** 9 50N 124 12 E
Carmen, *Cebu, Phil.* **71 F5** 10 35N 124 1 E
Carmen, *Mindanao, Phil.* **71 H5** 7 13N 124 45 E
Carmen →, *Mexico* **146 A3** 30 42N 106 29W
Carmen, I., *Mexico* **146 B2** 26 0N 111 20W
Carmen de Patagones, *Argentina* **160 B4** 40 50 S 63 0W
Cármenes, *Spain* **36 C5** 42 58N 5 34W
Carmensa, *Argentina* **158 D2** 35 15 S 67 40W
Carmi, *U.S.A.* **141 F8** 38 5N 88 10W
Carmichael, *U.S.A.* **144 G5** 38 38N 121 19W
Carmila, *Australia* **114 C4** 21 55 S 149 24 E
Carmona, *Spain* **37 H5** 37 28N 5 42W
Carnac, *France* **22 E3** 47 35N 3 6W
Carnarvon, *Queens., Australia* **114 C4** 24 48 S 147 45 E
Carnarvon, *W. Austral., Australia* **113 D1** 24 51 S 113 42 E
Carnarvon, *S. Africa* **104 E3** 30 56 S 22 8 E
Carnarvon Ra., *Queens., Australia* **114 D4** 25 15 S 148 30 E
Carnarvon Ra., *W. Austral., Australia* **113 E3** 25 20 S 120 45 E
Carnation, *U.S.A.* **144 C5** 47 39N 121 55W
Carnaxide, *Portugal* **37 G1** 38 43N 9 14W
Carndonagh, *Ireland* **19 A4** 55 16N 7 15W
Carnduff, *Canada* **131 D8** 49 10N 101 50W
Carnegie, *U.S.A.* **136 F4** 40 24N 80 5W
Carnegie, L., *Australia* **113 E3** 26 5 S 122 30 E
Carnic Alps = Karnische Alpen, *Europe* **30 E6** 46 36N 13 0 E
Carniche Alpi = Karnische Alpen, *Europe* **30 E6** 46 36N 13 0 E
Carnot, *C.A.R.* **102 B3** 4 59N 15 56 E
Carnot, C., *Australia* **115 E2** 34 57 S 135 38 E
Carnot B., *Australia* **112 C3** 17 20 S 122 15 E
Carnsore Pt., *Ireland* **19 D5** 52 10N 6 22W
Caro, *U.S.A.* **134 D4** 43 29N 83 24W
Carol City, *U.S.A.* **135 N5** 25 56N 80 16W
Carolina, *Brazil* **154 C2** 7 10 S 47 30W
Carolina, *Puerto Rico* **149 C6** 18 23N 65 58W
Carolina, *S. Africa* **105 D5** 26 5 S 30 6 E
Caroline I., *Kiribati* **123 H12** 9 15 S 150 3W
Caroline Is., *Pac. Oc.* **122 G6** 8 0N 150 0 E
Caroline Pk., *N.Z.* **119 F2** 45 57 S 167 15 E
Caron, *Canada* **131 C7** 50 30N 105 50W
Caroni →, *Venezuela* **153 B5** 8 21N 62 43W
Caronie = Nébrodi, Monti, *Italy* **41 E7** 37 54N 14 35 E
Caroona, *Australia* **115 E5** 31 24 S 150 26 E
Carora, *Venezuela* **152 A3** 10 11N 70 5W
Carovigno, *Italy* **41 B10** 40 42N 17 39 E
Carpathians, *Europe* **10 F10** 49 30N 21 0 E
Carpaţii Meridionali, *Romania* **46 D5** 45 30N 25 0 E
Carpenédolo, *Italy* **38 C7** 45 22N 10 26 E
Carpentaria, G. of, *Australia* **114 A2** 14 0 S 139 0 E
Carpentaria Downs, *Australia* **114 B3** 18 44 S 144 20 E
Carpentersville, *U.S.A.* **141 B8** 42 6N 88 17W
Carpentras, *France* **25 D9** 44 3N 5 2 E
Carpi, *Italy* **38 D7** 44 47N 10 53 E
Carpina, *Brazil* **154 C4** 7 51 S 35 15W
Carpino, *Italy* **41 A8** 41 51N 15 51 E
Carpinteria, *U.S.A.* **145 L7** 34 24N 119 31W
Carpio, *Spain* **36 D5** 41 13N 5 7W
Carpolac = Morea, *Australia* **116 D4** 36 45 S 141 18 E
Carr Boyd Ra., *Australia* **112 C4** 16 15 S 128 35 E
Carrabelle, *U.S.A.* **135 L3** 29 51N 84 40W
Carranglan, *Phil.* **70 D3** 15 58N 121 4 E
Carranya, *Australia* **112 C4** 19 14 S 127 46 E
Carrara, *Italy* **38 D7** 44 5N 10 6 E
Carrascal, *Phil.* **71 G5** 9 22N 125 56 E
Carrascosa del Campo, *Spain* **34 E2** 40 2N 2 45W
Carrauntoohill, *Ireland* **19 E2** 52 0N 9 45W
Carretas, Punta, *Peru* **156 C2** 14 12 S 76 17W
Carrick-on-Shannon, *Ireland* **19 C3** 53 57N 8 5W
Carrick-on-Suir, *Ireland* **19 D4** 52 21N 7 24W
Carrickfergus, *U.K.* **19 B6** 54 43N 5 49W
Carrickfergus □, *U.K.* **19 B6** 54 43N 5 49W
Carrickmacross, *Ireland* **19 C5** 53 59N 6 43W
Carrieton, *Australia* **116 B3** 32 25 S 138 31 E
Carrington, *U.S.A.* **138 B5** 47 27N 99 8W
Carrión →, *Spain* **36 D6** 41 53N 4 32W
Carrión de los Condes, *Spain* **36 C6** 42 20N 4 37W
Carrizal Bajo, *Chile* **158 B1** 28 5 S 71 20W
Carrizalillo, *Chile* **158 B1** 29 5 S 71 30W
Carrizo Cr. →, *U.S.A.* **139 G3** 36 55N 103 55W
Carrizo Springs, *U.S.A.* **139 L5** 28 31N 99 52W
Carrizozo, *U.S.A.* **143 K11** 33 38N 105 53W
Carroll, *U.S.A.* **140 B2** 42 4N 94 52W
Carrollton, *Ga., U.S.A.* **135 J3** 33 35N 85 5W
Carrollton, *Ill., U.S.A.* **140 E6** 39 18N 90 24W
Carrollton, *Ky., U.S.A.* **141 F11** 38 41N 85 11W
Carrollton, *Mo., U.S.A.* **140 E3** 39 22N 93 30W
Carrollton, *Ohio, U.S.A.* **136 F3** 40 34N 81 5W
Carron →, *U.K.* **18 D3** 57 19N 5 26W
Carron, L., *U.K.* **18 D3** 57 22N 5 35W
Carrot →, *Canada* **131 C8** 53 50N 101 17W
Carrot River, *Canada* **131 C8** 53 17N 103 35W
Carrouges, *France* **22 D6** 48 34N 0 10W
Carruthers, *Canada* **131 C7** 52 52N 109 16W
Çarşamba, *Turkey* **88 B7** 41 11N 36 44 E
Carse of Gowrie, *U.K.* **18 E5** 56 30N 3 10W
Carsóli, *Italy* **39 F10** 42 6N 13 5 E
Carson, *Calif., U.S.A.* **145 M8** 33 48N 118 17W
Carson, *N. Dak., U.S.A.* **138 B4** 46 25N 101 34W
Carson →, *U.S.A.* **144 F8** 39 45N 118 40W
Carson City, *Mich., U.S.A.* **141 A12** 43 11N 84 51W
Carson City, *Nev., U.S.A.* **144 F7** 39 10N 119 46W
Carson Sink, *U.S.A.* **142 G4** 39 50N 118 25W
Carstairs, *U.K.* **18 F5** 55 42N 3 41W
Cartagena, *Colombia* **152 A2** 10 25N 75 33W
Cartagena, *Spain* **35 H4** 37 38N 0 59W
Cartago, *Colombia* **152 C2** 4 45N 75 55W
Cartago, *Costa Rica* **148 E3** 9 50N 83 55W
Cartaxo, *Portugal* **37 F2** 39 10N 8 47W
Cartaya, *Spain* **37 H3** 37 16N 7 9W
Carteret, *France* **22 C5** 49 23N 1 47W
Cartersville, *U.S.A.* **135 H3** 34 10N 84 48W
Carterton, *N.Z.* **118 H4** 41 2 S 175 31 E
Carterville, *U.S.A.* **140 G7** 37 46N 89 5W
Carthage, *Ark., U.S.A.* **139 H8** 34 4N 92 33W
Carthage, *Ill., U.S.A.* **140 D5** 40 25N 91 8W
Carthage, *Mo., U.S.A.* **139 G7** 37 11N 94 19W
Carthage, *S. Dak., U.S.A.* **138 C6** 44 10N 97 43W
Carthage, *Tex., U.S.A.* **139 J7** 32 9N 94 20W
Cartier I., *Australia* **112 B3** 12 31 S 123 29 E
Cartwright, *Canada* **129 B8** 53 41N 56 58W
Caruaru, *Brazil* **154 C4** 8 15 S 35 55W
Carubig, *Phil.* **70 E5** 12 24N 125 3 E
Carúpano, *Venezuela* **153 A5** 10 39N 63 15W
Caruray, *Phil.* **71 F2** 10 20N 119 0 E
Carutapera, *Brazil* **154 B2** 1 13 S 46 1W
Caruthersville, *U.S.A.* **139 G10** 36 11N 89 39W
Carvalho, *Brazil* **153 D7** 2 16 S 51 29W
Carvin, *France* **23 B9** 50 30N 2 57 E
Carvoeiro, *Brazil* **153 D5** 1 30 S 61 59W
Carvoeiro, C., *Portugal* **37 F1** 39 21N 9 24W
Casa Branca, *Brazil* **155 F2** 21 46 S 47 4W
Casa Branca, *Portugal* **37 G2** 38 29N 8 12W
Casa Grande, *U.S.A.* **143 K8** 32 53N 111 45W
Casablanca, *Chile* **158 C1** 33 20 S 71 25W
Casablanca, *Morocco* **98 B3** 33 36N 7 36W
Casacalenda, *Italy* **41 A7** 41 44N 14 51 E
Casal di Principe, *Italy* **41 B7** 41 0N 14 8 E
Casalbordino, *Italy* **39 F11** 42 10N 14 35 E
Casale Monferrato, *Italy* **38 C5** 45 8N 8 27 E
Casalmaggiore, *Italy* **38 D7** 44 59N 10 26 E
Casalpusterlengo, *Italy* **38 C6** 45 11N 9 39 E
Casamance →, *Senegal* **100 C1** 12 33N 16 46W
Casamássima, *Italy* **41 B9** 40 57N 16 55 E
Casanare □, *Colombia* **152 B3** 5 30N 72 0W
Casanare →, *Colombia* **152 B4** 6 2N 69 51W
Casarano, *Italy* **41 B11** 40 0N 18 10 E
Casares, *Spain* **37 J5** 36 27N 5 16W
Casas Grandes, *Mexico* **146 A3** 30 22N 108 0W
Casas Ibañez, *Spain* **35 F3** 39 17N 1 30W
Casasimarro, *Spain* **35 F2** 39 22N 2 3W
Casatejada, *Spain* **36 F5** 39 54N 5 40W
Casavieja, *Spain* **36 E6** 40 17N 4 46W
Cascade, *Idaho, U.S.A.* **142 D5** 44 31N 116 2W
Cascade, *Iowa, U.S.A.* **140 B6** 42 18N 91 1W
Cascade, *Mont., U.S.A.* **142 C8** 47 16N 111 42W
Cascade Locks, *U.S.A.* **144 E5** 45 40N 121 54W
Cascade Pt., *N.Z.* **119 E3** 44 1 S 168 20 E
Cascade Ra., *U.S.A.* **144 D5** 47 0N 121 30W
Cascais, *Portugal* **37 G1** 38 41N 9 25W
Cascavel, *Brazil* **159 A5** 24 57 S 53 28W
Cáscina, *Italy* **38 E7** 43 41N 10 33 E
Caselle Torinese, *Italy* **38 C4** 45 10N 7 39 E
Caserta, *Italy* **41 A7** 41 4N 14 20 E
Cashel, *Ireland* **19 D4** 52 31N 7 53W
Cashmere, *U.S.A.* **142 C3** 47 31N 120 28W
Cashmere Downs, *Australia* **113 E2** 28 57 S 119 35 E
Casibare →, *Colombia* **152 C3** 3 48N 72 18W
Casiguran, *Phil.* **70 C4** 16 22N 122 7 E
Casiguran Sound, *Phil.* **70 C3** 16 6N 121 58 E
Casilda, *Argentina* **158 C3** 33 10 S 61 10W
Casimcea, *Romania* **46 E9** 44 45N 28 23 E
Casino, *Australia* **115 D5** 28 52 S 153 3 E
Casiquiare →, *Venezuela* **152 C4** 2 1N 67 7W
Casitas, *Peru* **156 A1** 3 54 S 80 39W
Caslan, *Canada* **130 C6** 54 38N 112 31W
Čáslav, *Czech.* **30 B8** 49 54N 15 22 E
Casma, *Peru* **156 B2** 9 30 S 78 20W
Casmalia, *U.S.A.* **145 L6** 34 50N 120 32W
Casola Valsenio, *Italy* **39 D8** 44 12N 11 40 E
Cásoli, *Italy* **39 F11** 42 7N 14 18 E
Caspe, *Spain* **34 D4** 41 14N 0 1W
Casper, *U.S.A.* **142 E10** 42 51N 106 19W
Caspian Depression, *Eurasia* **53 G9** 47 0N 48 0 E
Caspian Sea, *Eurasia* **53 J10** 43 0N 50 0 E
Casquets, *U.K.* **22 C4** 49 46N 2 15W
Cass City, *U.S.A.* **134 D4** 43 36N 83 11W
Cass Lake, *U.S.A.* **138 B7** 47 23N 94 37W
Cassá de la Selva, *Spain* **34 D7** 41 53N 2 52 E
Cassai, *Angola* **103 E4** 10 33 S 21 59 E
Cassamba, *Angola* **103 E4** 13 6 S 20 18 E
Cassano Iónio, *Italy* **41 C9** 39 47N 16 20 E
Cassel, *France* **23 B9** 50 48N 2 30 E
Casselman, *Canada* **137 A9** 45 19N 75 5W
Casselton, *U.S.A.* **138 B6** 46 54N 97 13W
Cassiar, *Canada* **130 B3** 59 16N 129 40W
Cassiar Mts., *Canada* **130 B2** 59 30N 130 30W
Cassilândia, *Brazil* **157 D7** 19 9 S 51 45W
Cassinga, *Angola* **103 F3** 15 5 S 16 4 E
Cassino, *Italy* **40 A6** 41 30N 13 49 E
Cassis, *France* **25 E9** 43 14N 5 32 E
Cassoalala, *Angola* **103 D2** 9 30 S 14 22 E
Cassoango, *Angola* **103 E4** 13 42 S 20 56 E
Cassopolis, *U.S.A.* **141 C10** 41 55N 86 1W
Cassunda, *Angola* **103 E4** 10 57 S 21 3 E
Cassville, *Mo., U.S.A.* **139 G8** 36 41N 93 52W
Cassville, *Wis., U.S.A.* **140 B6** 42 43N 90 59W
Cástagneto Carducci, *Italy* **38 E7** 43 9N 10 36 E
Castaic, *U.S.A.* **145 L8** 34 30N 118 38W
Castanhal, *Brazil* **154 B2** 1 18 S 47 55W
Casteau, *Belgium* **21 G4** 50 32N 4 2 E
Castéggio, *Italy* **38 C6** 45 0N 9 7 E
Castejón de Monegros, *Spain* **34 D4** 41 37N 0 15W
Castel di Sangro, *Italy* **39 G11** 41 47N 14 6 E
Castel San Giovanni, *Italy* **38 C6** 45 4N 9 26 E
Castel San Pietro Terme, *Italy* **39 D8** 44 24N 11 35 E
Castelbuono, *Italy* **41 E7** 37 56N 14 5 E
Casteldelfino, *Italy* **38 D4** 44 35N 7 4 E
Castelfiorentino, *Italy* **38 E7** 43 36N 10 58 E
Castelfranco Emília, *Italy* **38 D8** 44 37N 11 3 E
Castelfranco Véneto, *Italy* **39 C8** 45 40N 11 55 E
Casteljaloux, *France* **24 D4** 44 19N 0 6 E
Castellabate, *Italy* **41 B7** 40 17N 14 57 E
Castellammare, G. di, *Italy* **40 D5** 38 8N 12 54 E
Castellammare del Golfo, *Italy* **40 D5** 38 1N 12 53 E
Castellammare di Stábia, *Italy* **41 B7** 40 42N 14 29 E
Castellamonte, *Italy* **38 C4** 45 23N 7 42 E
Castellana Grotte, *Italy* **41 B10** 40 53N 17 10 E
Castellane, *France* **25 E10** 43 50N 6 31 E
Castellaneta, *Italy* **41 B9** 40 38N 16 56 E
Castellar de Santisteban, *Spain* **35 G1** 38 16N 3 8W
Castelleone, *Italy* **38 C6** 45 18N 9 46 E
Castelli, *Argentina* **158 D4** 36 7 S 57 47W
Castelló de Ampurias, *Spain* **34 C8** 42 15N 3 4 E
Castellón □, *Spain* **34 E4** 40 15N 0 5W
Castellón de la Plana, *Spain* **34 F4** 39 58N 0 3W
Castellote, *Spain* **34 E4** 40 48N 0 15W
Castelltersol, *Spain* **34 D7** 41 45N 2 8 E
Castelmáuro, *Italy* **41 A7** 41 50N 14 43 E
Castelnau-de-Médoc, *France* **24 C3** 45 2N 0 48W
Castelnaudary, *France* **24 E5** 43 20N 1 58 E
Castelnovo ne' Monti, *Italy* **38 D7** 44 26N 10 24 E
Castelnuovo di Val di Cécina, *Italy* **38 E7** 43 12N 10 59 E
Castelo, *Brazil* **155 F3** 20 33 S 41 14W
Castelo Branco, *Portugal* **36 F3** 39 50N 7 31W
Castelo Branco □, *Portugal* **36 F3** 39 52N 7 45W
Castelo de Paiva, *Portugal* **36 D2** 41 2N 8 16W
Castelo de Vide, *Portugal* **37 F3** 39 25N 7 27W
Castelo do Piauí, *Brazil* **154 C3** 5 20 S 41 33W
Castelsarrasin, *France* **24 D5** 44 2N 1 7 E
Casteltérmini, *Italy* **40 E6** 37 32N 13 39 E
Castelvetrano, *Italy* **40 E5** 37 41N 12 47 E
Castendo, *Angola* **103 D2** 8 39 S 14 10 E
Casterton, *Australia* **116 D4** 37 30 S 141 30 E
Castets, *France* **24 E2** 43 52N 1 6W
Castiglione del Lago, *Italy* **39 E9** 43 7N 12 3 E
Castiglione della Pescáia, *Italy* **38 F7** 42 46N 10 53 E
Castiglione delle Stiviere, *Italy* **38 C7** 45 23N 10 29 E
Castiglione Fiorentino, *Italy* **39 E8** 43 20N 11 55 E
Castilblanco, *Spain* **37 F5** 39 17N 5 5W
Castilla, *Peru* **156 B1** 5 12 S 80 38W
Castilla, Playa de, *Spain* **37 H4** 37 0N 6 33W
Castilla La Mancha □, *Spain* **37 F7** 39 30N 3 30W
Castilla y Leon □, *Spain* **36 D6** 42 0N 5 0W
Castillon, Barr. de, *France* **25 E10** 43 53N 6 33 E
Castillon-en-Couserans, *France* **24 F5** 42 56N 1 1 E
Castillon-la-Bataille, *France* **24 D3** 44 51N 0 2W
Castillonès, *France* **24 D4** 44 39N 0 37 E
Castillos, *Uruguay* **159 C5** 34 12 S 53 52W
Castle Dale, *U.S.A.* **142 G8** 39 13N 111 1W
Castle Douglas, *U.K.* **18 G5** 54 56N 3 56W
Castle Point, *N.Z.* **118 G5** 40 54 S 176 15 E
Castle Rock, *Colo., U.S.A.* **138 F2** 39 22N 104 51W
Castle Rock, *Wash., U.S.A.* **144 D4** 46 17N 122 54W
Castlebar, *Ireland* **19 C2** 53 52N 9 18W
Castleblaney, *Ireland* **19 B5** 54 7N 6 44W
Castlecliff, *N.Z.* **118 F3** 39 57 S 174 59 E
Castlegar, *Canada* **130 D5** 49 20N 117 40W
Castlemaine, *Australia* **116 D6** 37 2 S 144 12 E
Castlerea, *Ireland* **19 C3** 53 46N 8 29W
Castlereagh □, *U.K.* **19 B6** 54 33N 5 53W
Castlereagh →, *Australia* **117 A7** 30 12 S 147 32 E
Castlereagh B., *Australia* **114 A2** 12 10 S 135 10 E
Castletown, *U.K.* **16 C3** 54 5N 4 38W
Castletown Bearhaven, *Ireland* **19 E2** 51 39N 9 55W
Castlevale, *Australia* **114 C4** 24 30 S 146 48 E
Castor, *Canada* **130 C6** 52 15N 111 50W
Castres, *France* **24 E6** 43 37N 2 13 E
Castricum, *Neths.* **20 C5** 52 33N 4 40 E
Castries, *St. Lucia* **149 D7** 14 2N 60 58W
Castril, *Spain* **35 H2** 37 48N 2 46W
Castro, *Brazil* **159 A5** 24 45 S 50 0W
Castro, *Chile* **160 B2** 42 30 S 73 50W
Castro Alves, *Brazil* **155 D4** 12 46 S 39 33W
Castro del Río, *Spain* **37 H6** 37 41N 4 29W
Castro Marim, *Portugal* **37 H3** 37 13N 7 26W
Castro Urdiales, *Spain* **34 B1** 43 23N 3 11W
Castro Verde, *Portugal* **37 H2** 37 41N 8 4W
Castrojeriz, *Spain* **36 C6** 42 17N 4 9W
Castropol, *Spain* **36 B3** 43 32N 7 0W
Castroreale, *Italy* **41 D8** 38 6N 15 12 E
Castrovillari, *Italy* **41 C9** 39 49N 16 12 E
Castroville, *Calif., U.S.A.* **144 J5** 36 46N 121 45W
Castroville, *Tex., U.S.A.* **139 L5** 29 21N 98 53W
Castrovirreyna, *Peru* **156 C2** 13 20 S 75 18W
Castuera, *Spain* **37 G5** 38 43N 5 37W
Casummit Lake, *Canada* **128 B1** 51 29N 92 22W
Caswell Sound, *N.Z.* **119 E2** 44 59 S 167 8 E
Çat, *Turkey* **89 C9** 39 40N 41 3 E
Cat Ba, Dao, *Vietnam* **76 B6** 20 50N 107 0 E
Cat I., *Bahamas* **149 B4** 24 30N 75 30W
Cat I., *U.S.A.* **139 K10** 30 14N 89 6W
Cat L., *Canada* **128 B1** 51 40N 91 50W
Čata, *Slovak Rep.* **31 D11** 47 58N 18 38 E
Catabola, *Angola* **103 E3** 12 9 S 17 16 E
Catacamas, *Honduras* **148 D2** 14 54N 85 56W
Catacáos, *Peru* **156 B1** 5 20 S 80 45W
Cataguases, *Brazil* **155 F3** 21 23 S 42 39W
Catagupan, *Phil.* **71 G1** 8 1N 116 58 E
Catahoula L., *U.S.A.* **139 K8** 31 31N 92 7W
Çatak, *Turkey* **89 C10** 38 1N 43 8 E
Catalão, *Brazil* **155 E2** 18 10 S 47 57W
Çatalca, *Turkey* **88 B3** 41 8N 28 27 E
Catalina, *Canada* **129 C9** 48 31N 53 4W
Catalonia = Cataluña □, *Spain* **34 D6** 41 40N 1 15 E
Cataluña □, *Spain* **34 D6** 41 40N 1 15 E
Çatalzeytin, *Turkey* **88 B6** 41 57N 34 12 E
Catamarca, *Argentina* **158 B2** 28 30 S 65 50W
Catamarca □, *Argentina* **158 B2** 27 0 S 65 50W
Catanauan, *Phil.* **70 E4** 13 36N 122 19 E
Catanduanes, *Phil.* **70 E5** 13 50N 124 20 E
Catanduva, *Brazil* **159 A6** 21 5 S 48 58W
Catània, *Italy* **41 E8** 37 31N 15 4 E
Catània, G. di, *Italy* **41 E8** 37 24N 15 9 E

Changhai 183

Catanzaro, *Italy* **41 D9** 38 54N 16 35 E
Catarman, *Camiguin, Phil.* **71 G5** 9 8N 124 40 E
Catarman, *N. Samar, Phil.* **70 E5** 12 28N 124 35 E
Catbalogan, *Phil.* **71 F5** 11 46N 124 53 E
Cateel, *Phil.* **71 H6** 7 47N 126 24 E
Cateel Bay, *Phil.* **71 H6** 7 54N 126 25 E
Catende, *Angola* **103 E4** 11 14 S 21 30 E
Catende, *Brazil* **154 C4** 8 40 S 35 43W
Catete, *Angola* **103 D2** 9 6 S 13 43 E
Cathcart, *Australia* **117 D8** 36 52 S 149 24 E
Cathcart, *S. Africa* **104 E4** 32 18 S 27 10 E
Cathlamet, *U.S.A.* **144 D3** 46 12N 123 23W
Catio, *Guinea-Biss.* **100 C1** 11 17N 15 15W
Catismiña, *Venezuela* **153 C5** 4 5N 63 40W
Catita, *Brazil* **154 C3** 9 31 S 43 1W
Catlettsburg, *U.S.A.* **134 F4** 38 25N 82 36W
Catlin, *U.S.A.* **141 D9** 40 4N 87 42W
Catmon, *Phil.* **71 F5** 10 43N 124 1 E
Catoche, C., *Mexico* **147 C7** 21 40N 87 8W
Catolé do Rocha, *Brazil* **154 C4** 6 21 S 37 45W
Catral, *Spain* **35 G4** 38 10N 0 47W
Catria, Mt., *Italy* **39 E9** 43 28N 12 42 E
Catrimani, *Brazil* **153 C5** 0 27N 61 41W
Catrimani →, *Brazil* **153 C5** 0 28N 61 44W
Catskill, *U.S.A.* **137 D11** 42 14N 73 52W
Catskill Mts., *U.S.A.* **137 D10** 42 10N 74 25W
Catt, Mt., *Australia* **114 A1** 13 49 S 134 23 E
Cattaraugus, *U.S.A.* **136 D6** 42 22N 78 52W
Cattólica, *Italy* **39 E9** 43 58N 12 44 E
Cattólica Eraclea, *Italy* **40 E6** 37 26N 13 24 E
Catu, *Brazil* **155 D4** 12 21 S 38 23W
Catuala, *Angola* **103 F3** 16 25 S 19 2 E
Catumbela, *Angola* **103 E2** 12 25 S 13 34 E
Catur, *Mozam.* **107 E4** 13 45 S 35 30 E
Catwick Is., *Vietnam* **77 G7** 10 0N 109 0 E
Cauayan, *Phil.* **70 C3** 16 56N 121 46 E
Cauca □, *Colombia* **152 C2** 2 30N 76 50W
Cauca →, *Colombia* **152 B3** 8 54N 74 28W
Caucaia, *Brazil* **154 B4** 3 40 S 38 35W
Caucasia, *Colombia* **152 B2** 8 0N 75 12W
Caucasus Mountains, *Eurasia* **53 J7** 42 50N 44 0 E
Caudebec-en-Caux, *France* **22 C7** 49 30N 0 42 E
Caudebec-lès-Elbeuf, *France* **22 C8** 49 17N 1 2 E
Caudete, *Spain* **35 G3** 38 42N 1 2W
Caudry, *France* **23 B10** 50 7N 3 22 E
Caulnes, *France* **22 D4** 48 18N 2 10W
Caulónia, *Italy* **41 D9** 38 23N 16 24 E
Caúngula, *Angola* **103 D3** 8 26 S 18 38 E
Cauquenes, *Chile* **158 D1** 36 0 S 72 22W
Caura →, *Venezuela* **153 B5** 7 38N 64 53W
Caurés →, *Brazil* **153 D5** 1 21 S 62 20W
Cauresi →, *Mozam.* **107 F3** 17 8 S 33 0 E
Causapscal, *Canada* **129 C6** 48 19N 67 12W
Caussade, *France* **24 D5** 44 10N 1 33 E
Causse-Méjean, *France* **24 D7** 44 18N 3 42 E
Cauterets, *France* **24 F3** 42 52N 0 8W
Cautín □, *Chile* **160 A2** 39 0 S 72 30W
Caux, Pays de, *France* **22 C7** 49 38N 0 35 E
Cava dei Tirreni, *Italy* **41 B7** 40 42N 14 42 E
Cávado →, *Portugal* **36 D2** 41 32N 8 48W
Cavaillon, *France* **25 E9** 43 50N 5 2 E
Cavalaire-sur-Mer, *France* **25 E10** 43 10N 6 33 E
Cavalcante, *Brazil* **155 D2** 13 48 S 47 30W
Cavalese, *Italy* **39 B8** 46 17N 11 27 E
Cavalier, *U.S.A.* **138 A6** 48 48N 97 37W
Cavalla = Cavally →, *Africa* **100 E3** 4 22N 7 32W
Cavalli Is., *N.Z.* **118 B2** 35 0 S 173 58 E
Cavallo, I. de, *France* **25 G13** 41 22N 9 16 E
Cavally →, *Africa* **100 E3** 4 22N 7 32W
Cavan, *Ireland* **19 C4** 54 0N 7 22W
Cavan □, *Ireland* **19 C4** 54 1N 7 16W
Cavárzere, *Italy* **39 C9** 45 8N 12 5 E
Cave City, *U.S.A.* **134 G3** 37 8N 85 58W
Cavenagh Ra., *Australia* **113 E4** 26 12 S 127 55 E
Cavendish, *Australia* **116 D5** 37 31 S 142 2 E
Caviana, I., *Brazil* **153 C7** 0 10N 50 10W
Cavite, *Phil.* **70 D3** 14 29N 120 55 E
Cavite □, *Phil.* **70 D3** 14 15N 120 50 E
Cavour, *Italy* **38 D4** 44 47N 7 22 E
Cavtat, *Croatia* **42 E3** 42 35N 18 13 E
Cawkers Well, *Australia* **116 A5** 31 41 S 142 57 E
Cawndilla L., *Australia* **116 B5** 32 30 S 142 15 E
Cawnpore = Kanpur, *India* **81 F9** 26 28N 80 20 E
Caxias, *Brazil* **154 B3** 4 55 S 43 20W
Caxias do Sul, *Brazil* **159 B5** 29 10 S 51 10W
Caxito, *Angola* **103 D2** 8 30 S 13 30 E
Caxopa, *Angola* **103 E4** 11 52 S 20 52 E
Çay, *Turkey* **88 C4** 38 35N 31 1 E
Cay Sal Bank, *Bahamas* **148 B3** 23 45N 80 0W
Cayambe, *Napo, Ecuador* **152 C2** 0 2N 77 59W
Cayambe, *Quito, Ecuador* **152 C2** 0 3N 78 8W
Çaycuma, *Turkey* **88 B5** 41 25N 32 4 E
Çayeli, *Turkey* **89 B9** 41 5N 40 45 E
Cayenne, *Fr. Guiana* **153 B7** 5 5N 52 18W
Cayenne □, *Fr. Guiana* **153 C7** 5 0N 53 0W
Cayeux-sur-Mer, *France* **23 B8** 50 10N 1 30 E
Cayiralan, *Turkey* **88 C6** 39 17N 35 38 E
Caylus, *France* **24 D5** 44 15N 1 47 E
Cayman Brac, *Cayman Is.* **148 C4** 19 43N 79 49W
Cayman Is. ■, *W. Indies* **148 C3** 19 40N 80 30W
Cayo Romano, *Cuba* **149 B4** 22 0N 78 0W
Cayuga, *Canada* **136 D5** 42 59N 79 50W
Cayuga, *Ind., U.S.A.* **141 E9** 39 57N 87 28W
Cayuga, *N.Y., U.S.A.* **137 D8** 42 54N 76 44W
Cayuga L., *U.S.A.* **137 D8** 42 41N 76 41W
Cazaje, *Angola* **103 E4** 11 2 S 20 45 E
Cazalla de la Sierra, *Spain* **37 H5** 37 56N 5 45W
Căzănești, *Romania* **46 E8** 44 36N 27 3 E
Cazaux et de Sanguinet, Étang de, *France* **24 D2** 44 29N 1 10W
Cazères, *France* **24 E5** 43 13N 1 5 E
Cazin, *Bos.-H.* **39 D12** 44 57N 15 57 E
Čazma, *Croatia* **39 C13** 45 45N 16 39 E
Čazma →, *Croatia* **39 C13** 45 35N 16 29 E
Cazombo, *Angola* **103 E4** 11 54 S 22 56 E
Cazorla, *Spain* **35 H1** 37 55N 3 2W
Cazorla, *Venezuela* **152 B4** 8 1N 67 0W
Cazorla, Sierra de, *Spain* **35 G2** 38 5N 2 55W
Cea →, *Spain* **36 C5** 42 0N 5 36W
Ceadâr-Lunga, *Moldova* **51 J5** 46 3N 28 51 E
Ceamurlia de Jos, *Romania* **46 E9** 44 43N 28 47 E
Ceanannus Mor, *Ireland* **19 C5** 53 44N 6 53W
Ceará = Fortaleza, *Brazil* **154 B4** 3 45 S 38 35W
Ceará □, *Brazil* **154 C4** 5 0 S 40 0W
Ceará Mirim, *Brazil* **154 C4** 5 38 S 35 25W
Ceauru, L., *Romania* **46 E4** 44 58N 23 11 E
Cebaco, I. de, *Panama* **148 E3** 7 33N 81 9W
Cebollar, *Argentina* **158 B2** 29 10 S 66 35W
Cebollera, Sierra de, *Spain* **34 D2** 42 0N 2 30W
Cebreros, *Spain* **36 E6** 40 27N 4 28W
Cebu, *Phil.* **71 F4** 10 18N 123 54 E
Ceccano, *Italy* **40 A6** 41 34N 13 20 E
Cece, *Hungary* **31 E11** 46 46N 18 39 E
Cechi, *Ivory C.* **100 D4** 6 15N 4 25W
Čechy, *Czech.* **30 B6** 49 58N 13 30 E
Cecil Plains, *Australia* **115 D5** 27 30 S 151 11 E
Cécina, *Italy* **38 E7** 43 19N 10 31 E
Cécina →, *Italy* **38 E7** 43 18N 10 29 E
Ceclavin, *Spain* **36 F4** 39 50N 6 45W
Cedar →, *U.S.A.* **140 C5** 41 17N 91 21W
Cedar City, *U.S.A.* **143 H7** 37 41N 113 4W
Cedar Creek Reservoir, *U.S.A.* **139 J6** 32 11N 96 4W
Cedar Falls, *Iowa, U.S.A.* **140 B4** 42 32N 92 27W
Cedar Falls, *Wash., U.S.A.* **144 C5** 47 25N 121 45W
Cedar Grove, *U.S.A.* **141 E12** 39 22N 84 56W
Cedar Key, *U.S.A.* **135 L4** 29 8N 83 2W
Cedar L., *Canada* **131 C8** 53 10N 100 0W
Cedar Lake, *U.S.A.* **141 C9** 41 22N 87 26W
Cedar Point, *U.S.A.* **141 C13** 41 44N 83 21W
Cedar Rapids, *U.S.A.* **140 C5** 41 59N 91 40W
Cedartown, *U.S.A.* **135 H3** 34 1N 85 15W
Cedarvale, *Canada* **130 B3** 55 1N 128 22W
Cedarville, *S. Africa* **105 E4** 30 23 S 29 3 E
Cedarville, *Calif., U.S.A.* **142 F3** 41 32N 120 10W
Cedarville, *Ill., U.S.A.* **140 B7** 42 23N 89 38W
Cedarville, *Ohio, U.S.A.* **141 E13** 39 44N 83 49W
Cedeira, *Spain* **36 B2** 43 39N 8 2W
Cedral, *Mexico* **146 C4** 23 50N 100 42W
Cedrino →, *Italy* **40 B2** 40 11N 9 24 E
Cedro, *Brazil* **154 C4** 6 34 S 39 3W
Cedros, I. de, *Mexico* **146 B1** 28 10N 115 20W
Ceduna, *Australia* **115 E1** 32 7 S 133 46 E
Cedynia, *Poland* **47 C1** 52 53N 14 12 E
Cefalù, *Italy* **41 D7** 38 2N 14 1 E
Cega →, *Spain* **36 D6** 41 33N 4 46W
Cegléd, *Hungary* **31 D12** 47 11N 19 47 E
Céglie Messápico, *Italy* **41 B10** 40 39N 17 31 E
Cehegín, *Spain* **35 G3** 38 6N 1 48W
Ceheng, *China* **68 E5** 24 58N 105 48 E
Cehu-Silvaniei, *Romania* **46 B4** 47 24N 23 9 E
Ceica, *Romania* **46 C3** 46 53N 22 10 E
Ceira →, *Portugal* **36 E2** 40 13N 8 16W
Cekhira, *Tunisia* **96 B2** 34 20N 10 5 E
Cela, *Angola* **103 E3** 11 25 S 15 7 E
Celano, *Italy* **39 F10** 42 5N 13 33 E
Celanova, *Spain* **36 C3** 42 9N 7 58W
Celaya, *Mexico* **146 C4** 20 31N 100 37W
Celbridge, *Ireland* **19 C5** 53 20N 6 32W
Celebes = Sulawesi □, *Indonesia* **72 B2** 2 0 S 120 0 E
Celebes Sea, *Indonesia* **72 A2** 3 0N 123 0 E
Čelendín, *Peru* **156 B2** 6 52 S 78 10W
Čelić, *Bos.-H.* **42 C3** 44 43N 18 47 E
Celica, *Ecuador* **152 D2** 4 7 S 79 59W
Celina, *U.S.A.* **141 D12** 40 33N 84 35W
Celje, *Slovenia* **39 B12** 46 16N 15 18 E
Celldömölk, *Hungary* **31 D10** 47 16N 17 10 E
Celle, *Germany* **26 C6** 52 37N 10 4 E
Celles, *Belgium* **21 G2** 50 42N 3 28 E
Celorico da Beira, *Portugal* **36 E3** 40 38N 7 24W
Cement, *U.S.A.* **139 H5** 34 56N 98 8W
Çemişgezek, *Turkey* **89 C8** 39 3N 38 56 E
Cenepa →, *Peru* **152 D2** 4 40 S 78 10W
Cengong, *China* **68 D7** 27 13N 108 44 E
Ceno →, *Italy* **38 D7** 44 43N 10 5 E
Cenon, *France* **24 D3** 44 51N 0 31 E
Centallo, *Italy* **38 D4** 44 30N 7 35 E
Centenário do Sul, *Brazil* **155 F1** 22 48 S 51 36W
Center, *N. Dak., U.S.A.* **138 B4** 47 7N 101 18W
Center, *Tex., U.S.A.* **139 K7** 31 48N 94 11W
Center Point, *U.S.A.* **140 B5** 42 12N 91 46W
Centerfield, *U.S.A.* **143 G8** 39 8N 111 49W
Centerville, *Calif., U.S.A.* **144 J7** 36 44N 119 30W
Centerville, *Iowa, U.S.A.* **140 D4** 40 44N 92 52W
Centerville, *Mich., U.S.A.* **141 C11** 41 55N 85 32W
Centerville, *Pa., U.S.A.* **136 F5** 40 3N 79 59W
Centerville, *S. Dak., U.S.A.* **138 D6** 43 7N 96 58W
Centerville, *Tenn., U.S.A.* **135 H2** 35 47N 87 28W
Centerville, *Tex., U.S.A.* **139 K7** 31 16N 95 59W
Cento, *Italy* **39 D8** 44 43N 11 17 E
Central, *Brazil* **154 D3** 11 8 S 42 8W
Central, *U.S.A.* **143 K9** 32 47N 108 9W
Central □, *Kenya* **106 C4** 0 30 S 37 30 E
Central □, *Malawi* **107 E3** 13 30 S 33 30 E
Central □, *U.K.* **18 E4** 56 10N 4 30W
Central □, *Zambia* **107 E2** 14 25 S 28 50 E
Central, Cordillera, *Bolivia* **157 D5** 18 30 S 64 55W
Central, Cordillera, *Colombia* **152 C2** 5 0N 75 0W
Central, Cordillera, *Costa Rica* **148 D3** 10 10N 84 5W
Central, Cordillera, *Dom. Rep.* **149 C5** 19 15N 71 0W
Central, Cordillera, *Peru* **156 B2** 7 0 S 77 30W
Central, Cordillera, *Phil.* **70 C3** 17 20N 120 57 E
Central, Sistema, *Spain* **36 E5** 40 40N 5 55W
Central African Rep. ■, *Africa* **102 A4** 7 0N 20 0 E
Central City, *Ky., U.S.A.* **134 G2** 37 18N 87 7W
Central City, *Nebr., U.S.A.* **138 E5** 41 7N 98 0W
Central I., *Kenya* **106 B4** 3 30N 36 0 E
Central Makran Range, *Pakistan* **79 D2** 26 30N 64 15 E
Central Patricia, *Canada* **128 B1** 51 30N 90 9W
Central Ra., *Papua N. G.* **120 C2** 5 0 S 143 0 E
Central Russian Uplands, *Europe* **10 E13** 54 0N 36 0 E
Central Siberian Plateau, *Russia* **58 C14** 65 0N 105 0 E
Centralia, *Ill., U.S.A.* **140 F7** 38 32N 89 8W
Centralia, *Mo., U.S.A.* **140 E4** 39 13N 92 8W
Centralia, *Wash., U.S.A.* **144 D4** 46 43N 122 58W
Centreville, *Ala., U.S.A.* **135 J2** 32 57N 87 8W
Centreville, *Miss., U.S.A.* **139 K9** 31 5N 91 4W
Centúripe, *Italy* **41 E7** 37 37N 14 44 E
Cephalonia = Kefallinía, *Greece* **45 F2** 38 20N 20 30 E
Čepin, *Croatia* **42 B3** 45 32N 18 34 E
Ceprano, *Italy* **40 A6** 41 33N 13 31 E
Ceptura, *Romania* **46 D7** 45 1N 26 21 E
Cepu, *Indonesia* **75 D4** 7 9 S 111 35 E
Ceram = Seram, *Indonesia* **73 B3** 3 10 S 129 0 E
Ceram Sea = Seram Sea, *Indonesia* **72 B3** 2 30 S 128 30 E
Cerbère, *France* **24 F7** 42 26N 3 10 E
Cerbicales, Is., *France* **25 G13** 41 33N 9 22 E
Cerbu, *Romania* **46 E5** 44 46N 24 46 E
Cercal, *Portugal* **37 H2** 37 48N 8 40W
Cercemaggiore, *Italy* **41 A7** 41 28N 14 43 E
Cerdaña, *Spain* **34 C6** 42 22N 1 35 E
Cerdedo, *Spain* **36 C2** 42 33N 8 23W
Cère →, *France* **24 D5** 44 55N 1 49 E
Cerea, *Italy* **39 C8** 45 12N 11 13 E
Ceres, *Argentina* **158 B3** 29 55 S 61 55W
Ceres, *Brazil* **155 E2** 15 17 S 49 35W
Ceres, *Italy* **38 C4** 45 19N 7 22 E
Ceres, *S. Africa* **104 E2** 33 21 S 19 18 E
Ceres, *U.S.A.* **144 H6** 37 35N 120 57W
Céret, *France* **24 F6** 42 30N 2 42 E
Cereté, *Colombia* **152 B2** 8 53N 75 48W
Cerfontaine, *Belgium* **21 H4** 50 11N 4 26 E
Cerignola, *Italy* **41 A8** 41 17N 15 53 E
Cerigo = Kíthira, *Greece* **45 H5** 36 8N 23 0 E
Cérilly, *France* **24 B6** 46 37N 2 50 E
Cerisiers, *France* **23 D10** 48 8N 3 30 E
Cerizay, *France* **22 F6** 46 50N 0 40W
Çerkeş, *Turkey* **88 B5** 40 49N 32 52 E
Çerkeşköy, *Turkey* **88 B2** 41 17N 27 59 E
Cerknica, *Slovenia* **39 C11** 45 48N 14 21 E
Cermerno, *Serbia, Yug.* **42 D5** 43 35N 20 25 E
Çermik, *Turkey* **89 C8** 38 8N 39 26 E
Cerna, *Romania* **46 D9** 45 4N 28 17 E
Cerna →, *Romania* **46 E5** 44 15N 24 25 E
Cernavodă, *Romania* **46 E9** 44 22N 28 3 E
Cernay, *France* **23 E14** 47 44N 7 10 E
Cernik, *Croatia* **42 B2** 45 17N 17 22 E
Cerralvo, I., *Mexico* **146 C3** 24 20N 109 45W
Cerreto Sannita, *Italy* **41 A7** 41 17N 14 33 E
Cerritos, *Mexico* **146 C4** 22 27N 100 20W
Cerro Gordo, *U.S.A.* **141 E8** 39 53N 88 44W
Cerro Sombrero, *Chile* **160 D3** 52 45 S 69 15W
Certaldo, *Italy* **38 E8** 43 33N 11 2 E
Cervaro →, *Italy* **41 A8** 41 30N 15 52 E
Cervera, *Spain* **34 D6** 41 40N 1 16 E
Cervera de Pisuerga, *Spain* **36 C6** 42 51N 4 30W
Cervera del Río Alhama, *Spain* **34 C3** 42 2N 1 58W
Cérvia, *Italy* **39 D9** 44 15N 12 22 E
Cervignano del Friuli, *Italy* **39 C10** 45 49N 13 20 E
Cervinara, *Italy* **41 A7** 41 2N 14 37 E
Cervione, *France* **25 F13** 42 20N 9 29 E
Cervo, *Spain* **36 B3** 43 40N 7 24W
César □, *Colombia* **152 B3** 9 0N 73 30W
Cesaro, *Italy* **41 E7** 37 50N 14 38 E
Cesena, *Italy* **39 D9** 44 8N 12 15 E
Cesenático, *Italy* **39 D9** 44 12N 12 24 E
Cēsis, *Latvia* **13 H21** 57 18N 25 15 E
Česká Lípa, *Czech.* **30 A7** 50 45N 14 30 E
Česka Republika = Czech Rep. ■, *Europe* **30 B7** 50 0N 15 0 E
Česká Třebová, *Czech.* **31 B9** 49 54N 16 27 E
České Budějovice, *Czech.* **30 C7** 48 55N 14 25 E
České Velenice, *Czech.* **30 C8** 48 45N 15 1 E
Ceskomoravská Vrchovina, *Czech.* **30 B8** 49 30N 15 40 E
Český Brod, *Czech.* **30 A7** 50 4N 14 52 E
Český Krumlov, *Czech.* **30 C7** 48 43N 14 21 E
Český Těšin, *Czech.* **31 B11** 49 45N 18 39 E
Çeşme, *Turkey* **88 C2** 38 20N 26 23 E
Cessnock, *Australia* **117 B9** 32 50 S 151 21 E
Cestos →, *Liberia* **100 D3** 5 40N 9 10W
Cetate, *Romania* **46 E4** 44 7N 23 2 E
Cétin Grad, *Croatia* **39 C12** 45 9N 15 45 E
Cetina →, *Croatia* **39 E13** 43 26N 16 42 E
Cetinje, *Montenegro, Yug.* **42 E3** 42 23N 18 59 E
Cetraro, *Italy* **41 C8** 39 31N 15 55 E
Ceuta, *N. Afr.* **98 A3** 35 52N 5 18W
Ceva, *Italy* **38 D5** 44 23N 8 2 E
Cévennes, *France* **24 D7** 44 10N 3 50 E
Ceyhan, *Turkey* **88 D6** 37 4N 35 47 E
Ceyhan →, *Turkey* **88 D6** 36 38N 35 40 E
Ceylânpınar, *Turkey* **89 D9** 36 50N 40 2 E
Ceylon = Sri Lanka ■, *Asia* **83 L5** 7 30N 80 50 E
Cèze →, *France* **25 D8** 44 6N 4 43 E
Cha-am, *Thailand* **76 F2** 12 48N 99 58 E
Chá Pungana, *Angola* **103 E3** 13 44 S 18 39 E
Chaam, *Neths.* **21 E5** 51 30N 4 52 E
Chabeuil, *France* **25 D9** 44 54N 5 3 E
Chablais, *France* **25 B10** 46 20N 6 36 E
Chablis, *France* **23 E10** 47 47N 3 48 E
Chabounia, *Algeria* **99 A5** 35 30N 2 38 E
Chacabuco, *Argentina* **158 C3** 34 40 S 60 27W
Chachapoyas, *Peru* **156 B2** 6 15 S 77 50W
Chachasp, *Peru* **156 D3** 15 30 S 72 15W
Chachoengsao, *Thailand* **76 F3** 13 42N 101 5 E
Chachro, *Pakistan* **80 G4** 25 5N 70 15 E
Chaco □, *Argentina* **158 B3** 26 30 S 61 0W
Chaco □, *Paraguay* **158 B3** 26 0 S 60 0W
Chad ■, *Africa* **97 F3** 15 0N 17 15 E
Chad, L. = Tchad, L., *Chad* **97 F2** 13 30N 14 30 E
Chadan, *Russia* **57 D10** 51 17N 91 35 E
Chadileuvú →, *Argentina* **158 D2** 37 46 S 66 0W
Chadiza, *Zambia* **107 E3** 14 45 S 32 27 E
Chadron, *U.S.A.* **138 D3** 42 50N 103 0W
Chadyr-Lunga = Ceadâr-Lunga, *Moldova* **51 J5** 46 3N 28 51 E
Chae Hom, *Thailand* **76 C2** 18 43N 99 35 E
Chaem →, *Thailand* **76 C2** 18 11N 98 38 E
Chaeryŏng, *N. Korea* **67 E13** 38 24N 125 36 E
Chagda, *Russia* **57 D14** 58 45N 130 38 E
Chagny, *France* **23 F11** 46 57N 4 45 E
Chagoda, *Russia* **50 C8** 59 10N 35 15 E
Chagos Arch., *Ind. Oc.* **58 K11** 6 0 S 72 0 E
Chāh Ākhvor, *Iran* **85 C8** 32 41N 59 40 E
Chāh Bahār, *Iran* **85 E9** 25 20N 60 40 E
Chāh-e-Malek, *Iran* **85 D8** 28 35N 59 7 E
Chāh Gay Hills, *Afghan.* **79 C1** 29 30N 64 0 E
Chāh Kavīr, *Iran* **85 D7** 31 45N 54 52 E
Chahār Borjak, *Afghan.* **79 C1** 30 17N 62 3 E
Chahtung, *Burma* **78 B7** 26 41N 98 10 E
Chaillé-les-Marais, *France* **24 B2** 46 25N 1 2W
Chainat, *Thailand* **76 E3** 15 11N 100 8 E
Chaitén, *Chile* **160 B2** 42 55 S 72 43W
Chaiya, *Thailand* **77 H2** 9 23N 99 14 E
Chaj Doab, *Pakistan* **80 C5** 32 15N 73 0 E
Chajari, *Argentina* **158 C4** 30 42 S 58 0W
Chakaria, *Bangla.* **78 E4** 21 45N 92 5 E
Chake Chake, *Tanzania* **106 D4** 5 15 S 39 45 E
Chakhānsūr, *Afghan.* **79 C1** 31 10N 62 0 E
Chakonipau, L., *Canada* **129 A6** 56 18N 68 30W
Chakradharpur, *India* **81 H11** 22 45N 85 40 E
Chakwadam, *Burma* **78 B7** 27 29N 98 31 E
Chakwal, *Pakistan* **79 B4** 32 56N 72 53 E
Chala, *Peru* **156 D3** 15 48 S 74 20W
Chalais, *France* **24 C4** 45 16N 0 3 E
Chalakudi, *India* **83 J3** 10 18N 76 20 E
Chalchihuites, *Mexico* **146 C4** 23 29N 103 53W
Chalcis = Khalkís, *Greece* **45 F5** 38 27N 23 42 E
Chaleur B., *Canada* **129 C6** 47 55N 65 30W
Chalfant, *U.S.A.* **144 H8** 37 32N 118 21W
Chalhuanca, *Peru* **156 C3** 14 15 S 73 15W
Chalindrey, *France* **23 E12** 47 43N 5 26 E
Chaling, *China* **69 D9** 26 58N 113 30 E
Chalisgaon, *India* **82 D2** 20 30N 75 10 E
Chalkar = Shalkar, *Kazakhstan* **54 F3** 50 40N 51 53 E
Chalkar, Ozero = Shalkar, Ozero, *Kazakhstan* **54 F3** 50 35N 51 47 E
Chalky Inlet, *N.Z.* **119 G1** 46 3 S 166 31 E
Challans, *France* **22 F5** 46 50N 1 52W
Challapata, *Bolivia* **156 D4** 18 53 S 66 50W
Challis, *U.S.A.* **142 D6** 44 30N 114 14W
Chalna, *India* **81 H13** 22 36N 89 35 E
Chalon-sur-Saône, *France* **23 F11** 46 48N 4 50 E
Chalonnes-sur-Loire, *France* **22 E6** 47 20N 0 45W
Châlons-sur-Marne, *France* **23 D11** 48 58N 4 20 E
Châlus, *France* **24 C4** 45 39N 0 58 E
Chalyaphum, *Thailand* **76 E4** 15 48N 102 2 E
Cham, *Germany* **27 F8** 49 13N 12 39 E
Cham, *Switz.* **29 B6** 47 11N 8 28 E
Cham, Cu Lao, *Vietnam* **76 E7** 15 57N 108 30 E
Chama, *U.S.A.* **143 H10** 36 54N 106 35W
Chamah, Gunong, *Malaysia* **74 A2** 5 13N 101 35 E
Chaman, *Pakistan* **79 C2** 30 58N 66 25 E
Chamba, *India* **80 C7** 32 35N 76 10 E
Chamba, *Tanzania* **107 E4** 11 37 S 37 0 E
Chambal →, *India* **81 F8** 26 29N 79 15 E
Chamberlain, *U.S.A.* **138 D5** 43 49N 99 20W
Chamberlain →, *Australia* **112 C4** 15 30 S 127 54 E
Chambers, *U.S.A.* **143 J9** 35 11N 109 26W
Chambersburg, *U.S.A.* **134 F7** 39 56N 77 40W
Chambéry, *France* **25 C9** 45 34N 5 55 E
Chambly, *Canada* **137 A11** 45 27N 73 17W
Chambord, *Canada* **129 C5** 48 25N 72 6W
Chambri L., *Papua N. G.* **120 C2** 4 15 S 143 10 E
Chamchamal, *Iraq* **89 E11** 35 32N 44 50 E
Chamela, *Mexico* **146 D3** 19 32N 105 5W
Chamical, *Argentina* **158 C2** 30 22 S 66 27W
Chamkar Luong, *Cambodia* **77 G4** 11 0N 103 45 E
Chamois, *U.S.A.* **140 F5** 38 41N 91 46W
Chamonix-Mont Blanc, *France* **25 C10** 45 55N 6 51 E
Champa, *India* **81 H10** 22 2N 82 43 E
Champagne, *Canada* **130 A1** 60 49N 136 30W
Champagne, *France* **23 D11** 48 40N 4 20 E
Champagne, Plaine de, *France* **23 D11** 49 0N 4 30 E
Champagnole, *France* **23 F12** 46 45N 5 55 E
Champaign, *U.S.A.* **141 D8** 40 7N 88 15W
Champassak, *Laos* **76 E5** 14 53N 105 52 E
Champaubert, *France* **23 D10** 48 50N 3 45 E
Champdeniers, *France* **24 B3** 46 29N 0 25W
Champeix, *France* **24 C7** 45 37N 3 8 E
Champlain, *Canada* **134 B9** 46 27N 72 24W
Champlain, *U.S.A.* **137 B11** 44 59N 73 27W
Champlain, L., *U.S.A.* **137 B11** 44 40N 73 20W
Champotón, *Mexico* **147 D6** 19 20N 90 50W
Chamrajnagar, *India* **83 J3** 11 52N 76 52 E
Chamusca, *Portugal* **37 F2** 39 21N 8 29W
Chan Chan, *Peru* **156 B2** 8 7S 79 0W
Chana, *Thailand* **77 J3** 6 55N 100 44 E
Chañaral, *Chile* **158 B1** 26 23 S 70 40W
Chanārān, *Iran* **85 B8** 36 39N 59 6 E
Chanasma, *India* **80 H5** 23 44N 72 5 E
Chancay, *Peru* **156 C2** 11 32 S 77 25W
Chancy, *Switz.* **28 D1** 46 8N 5 58 E
Chandannagar, *India* **81 H13** 22 52N 88 24 E
Chandausi, *India* **81 E8** 28 27N 78 49 E
Chandeleur Is., *U.S.A.* **139 L10** 29 55N 88 57W
Chandeleur Sd., *U.S.A.* **139 L10** 29 55N 89 0W
Chandigarh, *India* **80 D7** 30 43N 76 47 E
Chandler, *Australia* **115 D1** 27 0 S 133 19 E
Chandler, *Canada* **129 C7** 48 18N 64 46W
Chandler, *Ariz., U.S.A.* **143 K8** 33 18N 111 50W
Chandler, *Okla., U.S.A.* **139 H6** 35 42N 96 53W
Chandlers Pk., *Australia* **117 A9** 30 15 S 151 48 E
Chandless →, *Brazil* **156 B4** 9 8 S 69 51W
Chandpur, *Bangla.* **78 D3** 23 8N 90 45 E
Chandpur, *India* **80 E8** 29 8N 78 19 E
Chandrapur, *India* **82 E4** 19 57N 79 25 E
Chānf, *Iran* **85 E9** 26 38N 60 29 E
Chang, *Pakistan* **80 F3** 26 59N 68 30 E
Chang, Ko, *Thailand* **77 F4** 12 0N 102 23 E
Ch'ang Chiang = Chang Jiang →, *China* **69 B13** 31 48N 121 10 E
Chang Jiang →, *China* **69 B13** 31 48N 121 10 E
Changa, *India* **81 C7** 33 53N 77 35 E
Changanacheri, *India* **83 K3** 9 25N 76 31 E
Changane →, *Mozam.* **105 C5** 24 30 S 33 30 E
Changbai, *China* **67 D15** 41 25N 128 5 E
Changbai Shan, *China* **67 C15** 42 20N 129 0 E
Changchiak'ou = Zhangjiakou, *China* **66 D8** 40 48N 114 55 E
Ch'angchou = Changzhou, *China* **69 B12** 31 47N 119 58 E
Changchun, *China* **67 C13** 43 57N 125 17 E
Changchunling, *China* **67 B13** 45 18N 125 27 E
Changde, *China* **69 C8** 29 4N 111 35 E
Changdo-ri, *N. Korea* **67 E14** 38 30N 127 40 E
Changfeng, *China* **69 A11** 32 28N 117 10 E
Changhai = Shanghai, *China* **69 B13** 31 15N 121 26 E

Cieszyn 185

Chiange, *Angola* **103 F2** 15 35 S 13 40 E
Chiapa →, *Mexico* **147 D6** 16 42N 93 0W
Chiapa de Corzo, *Mexico* **147 D6** 16 42N 93 0W
Chiapas □, *Mexico* **147 D6** 17 0N 92 45W
Chiaramonte Gulfi, *Italy* **41 E7** 37 2N 14 42 E
Chiaravalle, *Italy* **39 E10** 43 36N 13 19 E
Chiaravalle Centrale, *Italy* **41 D9** 38 41N 16 25 E
Chiari, *Italy* **38 C6** 45 32N 9 56 E
Chiasso, *Switz.* **29 E8** 45 50N 9 2 E
Chiatura, *Georgia* **53 J6** 42 15N 43 17 E
Chiautla, *Mexico* **147 D5** 18 18N 98 34W
Chiávari, *Italy* **38 D6** 44 19N 9 19 E
Chiavenna, *Italy* **38 B6** 46 19N 9 24 E
Chiba, *Japan* **63 B12** 35 30N 140 7 E
Chiba □, *Japan* **63 B12** 35 30N 140 20 E
Chibabava, *Mozam.* **105 C5** 20 17 S 33 35 E
Chibemba, *Cunene, Angola* **103 F2** 15 48 S 14 8 E
Chibemba, *Huila, Angola* **103 F3** 16 20 S 15 20 E
Chibia, *Angola* **103 F2** 15 10 S 13 42 E
Chibougamau, *Canada* **128 C5** 49 56N 74 24W
Chibougamau L., *Canada* **128 C5** 49 50N 74 20W
Chibuk, *Nigeria* **101 C7** 10 52N 12 50 E
Chic-Chocs, Mts., *Canada* **129 C6** 48 55N 66 0W
Chicacole = Srikakulam, *India* **82 E6** 18 14N 83 58 E
Chicago, *U.S.A.* **141 C9** 41 53N 87 38W
Chicago Heights, *U.S.A.* **141 C9** 41 30N 87 38W
Chichagof I., *U.S.A.* **130 B1** 57 30N 135 30W
Chichaoua, *Morocco* **98 B3** 31 32N 8 44W
Chicheng, *China* **66 D8** 40 55N 115 55 E
Chichester, *U.K.* **17 G7** 50 50N 0 47W
Chichibu, *Japan* **63 A11** 36 5N 139 10 E
Ch'ich'ihaerh = Qiqihar, *China* **57 E13** 47 26N 124 0 E
Chickasha, *U.S.A.* **139 H5** 35 3N 97 58W
Chiclana de la Frontera, *Spain* **37 J4** 36 26N 6 9W
Chiclayo, *Peru* **156 B2** 6 42 S 79 50W
Chico, *U.S.A.* **144 F5** 39 44N 121 50W
Chico →, *Chubut, Argentina* **160 B3** 44 0 S 67 0W
Chico →, *Santa Cruz, Argentina* **160 C3** 50 0 S 68 30W
Chicomo, *Mozam.* **105 C5** 24 31 S 34 6 E
Chicontepec, *Mexico* **147 C5** 20 58N 98 10W
Chicopee, *U.S.A.* **137 D12** 42 9N 72 37W
Chicoutimi, *Canada* **129 C5** 48 28N 71 5W
Chicualacuala, *Mozam.* **105 C5** 22 6 S 31 42 E
Chidambaram, *India* **83 J4** 11 20N 79 45 E
Chidenguele, *Mozam.* **105 C5** 24 55 S 34 11 E
Chidley, C., *Canada* **127 B13** 60 23N 64 26W
Chiede, *Angola* **103 F3** 17 15 S 16 22 E
Chiefs Pt., *Canada* **136 B3** 44 41N 81 18W
Chiem Hoa, *Vietnam* **76 A5** 22 12N 105 17 E
Chiemsee, *Germany* **27 H8** 47 53N 12 28 E
Chiengi, *Zambia* **107 D2** 8 45 S 29 10 E
Chiengmai = Chiang Mai, *Thailand* **76 C2** 18 47N 98 59 E
Chiengo, *Angola* **103 E4** 13 20 S 21 55 E
Chienti →, *Italy* **39 E10** 43 18N 13 45 E
Chieri, *Italy* **38 D4** 45 1N 7 49 E
Chiers →, *France* **23 C11** 49 39N 4 59 E
Chiese →, *Italy* **38 C7** 45 8N 10 25 E
Chieti, *Italy* **39 F11** 42 21N 14 10 E
Chièvres, *Belgium* **21 G3** 50 35N 3 48 E
Chifeng, *China* **67 C10** 42 18N 118 58 E
Chigasaki, *Japan* **63 B11** 35 19N 139 24 E
Chigirin, *Ukraine* **51 H7** 49 4N 32 38 E
Chignecto B., *Canada* **129 C7** 45 30N 64 40W
Chigorodó, *Colombia* **152 B2** 7 41N 76 42W
Chiguana, *Bolivia* **158 A2** 21 0 S 67 58W
Chiha-ri, *N. Korea* **67 E14** 38 40N 126 30 E
Chihli, G. of = Bo Hai, *China* **67 E10** 39 0N 119 0 E
Chihuahua, *Mexico* **146 B3** 28 40N 106 3W
Chihuahua □, *Mexico* **146 B3** 28 40N 106 3W
Chiili, *Kazakhstan* **55 A3** 44 20N 66 15 E
Chik Bollapur, *India* **83 H3** 13 25N 77 45 E
Chikhli, *India* **82 D3** 20 20N 76 18 E
Chikmagalur, *India* **83 H2** 13 15N 75 45 E
Chikodi, *India* **83 F2** 16 26N 74 38 E
Chikugo, *Japan* **62 D2** 33 14N 130 28 E
Chikuma-Gawa →, *Japan* **63 A10** 36 59N 138 35 E
Chikwawa, *Malawi* **107 F3** 16 2 S 34 50 E
Chilac, *Mexico* **147 D5** 18 20N 97 24W
Chilako →, *Canada* **130 C4** 53 53N 122 57W
Chilam Chavki, *Pakistan* **81 B6** 35 5N 75 5 E
Chilanga, *Zambia* **107 F2** 15 33 S 28 16 E
Chilapa, *Mexico* **147 D5** 17 40N 99 11W
Chilas, *Pakistan* **81 B6** 35 25N 74 5 E
Chilcotin →, *Canada* **130 C4** 51 44N 122 23W
Childers, *Australia* **115 D5** 25 15 S 152 17 E
Childress, *U.S.A.* **139 H4** 34 25N 100 13W
Chile ■, *S. Amer.* **160 B2** 35 0 S 72 0W
Chile Chico, *Chile* **160 C2** 46 33 S 71 44W
Chile Rise, *Pac. Oc.* **123 L18** 38 0 S 92 0W
Chilecito, *Argentina* **158 B2** 29 10 S 67 30W
Chilete, *Peru* **156 B2** 7 10 S 78 50W
Chilhowee, *U.S.A.* **140 F3** 38 36N 93 51W
Chilia, Braţul →, *Romania* **46 D10** 45 25N 29 20 E
Chilik, *Kazakhstan* **55 B9** 43 33N 78 17 E
Chililabombwe, *Zambia* **107 E2** 12 18 S 27 43 E
Chilin = Jilin, *China* **67 C14** 43 44N 126 30 E
Chilka L., *India* **82 E7** 19 40N 85 25 E
Chilko →, *Canada* **130 C4** 52 0N 123 40W
Chilko, L., *Canada* **130 C4** 51 20N 124 10W
Chillagoe, *Australia* **114 B3** 17 7 S 144 33 E
Chillán, *Chile* **158 D1** 36 40 S 72 10W
Chillicothe, *Ill., U.S.A.* **140 D7** 40 55N 89 29W
Chillicothe, *Mo., U.S.A.* **140 E3** 39 48N 93 33W
Chillicothe, *Ohio, U.S.A.* **134 F4** 39 20N 82 59W
Chilliwack, *Canada* **130 D4** 49 10N 121 54W
Chilo, *India* **80 F5** 27 25N 73 32 E
Chiloane, I., *Mozam.* **105 C5** 20 40 S 34 55 E
Chiloé □, *Chile* **160 B2** 43 0 S 73 0W
Chiloé, I. de, *Chile* **160 B2** 42 30 S 73 50W
Chilonda, *Angola* **103 E3** 11 19 S 16 12 E
Chilpancingo, *Mexico* **147 D5** 17 30N 99 30W
Chiltern, *Australia* **117 D7** 36 10 S 146 36 E
Chiltern Hills, *U.K.* **17 F7** 51 40N 0 53W
Chilton, *U.S.A.* **134 C1** 44 2N 88 10W
Chiluage, *Angola* **103 D4** 9 30 S 21 50 E
Chilubi, *Zambia* **107 E2** 11 5 S 29 58 E
Chilubula, *Zambia* **107 E3** 10 14 S 30 51 E
Chilumba, *Malawi* **107 E3** 10 28 S 34 12 E
Chilung, *Taiwan* **65 D7** 25 3N 121 45 E
Chilwa, L., *Malawi* **107 F4** 15 15 S 35 40 E
Chimaltitán, *Mexico* **146 C4** 21 46N 103 50W
Chimán, *Panama* **148 E4** 8 45N 78 40W
Chimay, *Belgium* **21 H4** 50 3N 4 20 E
Chimbay, *Uzbekistan* **56 E6** 42 57N 59 47 E
Chimborazo, *Ecuador* **152 D2** 1 29 S 78 55W
Chimborazo □, *Ecuador* **152 D2** 1 0 S 78 40W
Chimbote, *Peru* **156 B2** 9 0 S 78 35W
Chimion, *Uzbekistan* **55 C5** 40 15N 71 32 E
Chimkent = Shymkent, *Kazakhstan* **55 B4** 42 18N 69 36 E
Chimoio, *Mozam.* **107 F3** 19 4 S 33 30 E
Chimpembe, *Zambia* **107 D2** 9 31 S 29 33 E
Chin □, *Burma* **78 D4** 22 0N 93 0 E
Chin Hills, *Burma* **78 D4** 22 30N 93 30 E
Chin Ling Shan = Qinling Shandi, *China* **66 H5** 33 50N 108 10 E
China, *Mexico* **147 B5** 25 40N 99 20W
China ■, *Asia* **66 E3** 30 0N 110 0 E
China Lake, *U.S.A.* **145 K9** 35 44N 117 37W
Chinacota, *Colombia* **152 B3** 7 37N 72 36W
Chinan = Jinan, *China* **66 F9** 36 38N 117 1 E
Chinandega, *Nic.* **148 D2** 12 35N 87 12W
Chinati Peak, *U.S.A.* **139 K2** 29 57N 104 29W
Chincha Alta, *Peru* **156 C2** 13 25 S 76 7W
Chinchilla, *Australia* **115 D5** 26 45 S 150 38 E
Chinchilla de Monte Aragón, *Spain* **35 G3** 38 53N 1 40W
Chinchón, *Spain* **34 E1** 40 9N 3 26W
Chinchorro, Banco, *Mexico* **147 D7** 18 35N 87 20W
Chinchou = Jinzhou, *China* **67 D11** 41 5N 121 3 E
Chinchoua, *Gabon* **102 B1** 0 1N 9 48 E
Chincoteague, *U.S.A.* **134 G8** 37 56N 75 23W
Chinde, *Mozam.* **107 F4** 18 35 S 36 30 E
Chindo, *S. Korea* **67 G14** 34 28N 126 15 E
Chindwin →, *Burma* **78 E5** 21 26N 95 15 E
Chineni, *India* **81 C6** 33 2N 75 15 E
Chinga, *Mozam.* **107 F4** 15 13 S 38 35 E
Chingirlau, *Kazakhstan* **54 F5** 51 7N 54 7 E
Chingola, *Zambia* **107 E2** 12 31 S 27 53 E
Chingole, *Malawi* **107 E3** 13 4 S 34 17 E
Chingoroi, *Angola* **103 E2** 13 37 S 14 1 E
Ch'ingtao = Qingdao, *China* **67 F11** 36 5N 120 20 E
Chinguar, *Angola* **103 E3** 12 25 S 16 45 E
Chinguetti, *Mauritania* **98 D2** 20 25N 12 24W
Chingune, *Mozam.* **105 C5** 20 33 S 34 58 E
Chinhae, *S. Korea* **67 G15** 35 9N 128 47 E
Chinhanguanine, *Mozam.* **105 D5** 25 21 S 32 30 E
Chinhoyi, *Zimbabwe* **107 F3** 17 20 S 30 8 E
Chiniot, *Pakistan* **79 C4** 31 45N 73 0 E
Chínipas, *Mexico* **146 B3** 27 22N 108 32W
Chinju, *S. Korea* **67 G15** 35 12N 128 2 E
Chinle, *U.S.A.* **143 H9** 36 9N 109 33W
Chinnamanur, *India* **83 K3** 9 50N 77 24 E
Chinnampo, *N. Korea* **67 E13** 38 52N 125 10 E
Chinnur, *India* **82 E4** 18 57N 79 49 E
Chino, *Japan* **63 B10** 35 59N 138 9 E
Chino, *U.S.A.* **145 L9** 34 1N 117 41W
Chino Valley, *U.S.A.* **143 J7** 34 45N 112 27W
Chinon, *France* **22 E7** 47 10N 0 15 E
Chinook, *Canada* **131 C6** 51 28N 110 59W
Chinook, *U.S.A.* **142 B9** 48 35N 109 14W
Chinsali, *Zambia* **107 E3** 10 30 S 32 2 E
Chintamani, *India* **83 H4** 13 26N 78 3 E
Chióggia, *Italy* **39 C9** 45 13N 12 17 E
Chíos = Khíos, *Greece* **45 F8** 38 27N 26 9 E
Chipata, *Zambia* **107 E3** 13 38 S 32 28 E
Chipewyan L., *Canada* **131 B9** 58 0N 98 27W
Chipinge, *Zimbabwe* **107 G3** 20 13 S 32 28 E
Chipiona, *Spain* **37 J4** 36 44N 6 26W
Chipley, *U.S.A.* **135 K3** 30 47N 85 32W
Chiplun, *India* **82 F1** 17 31N 73 34 E
Chipman, *Canada* **129 C6** 46 6N 65 53W
Chipoka, *Malawi* **107 E3** 13 57 S 34 28 E
Chippenham, *U.K.* **17 F5** 51 27N 2 6W
Chippewa →, *U.S.A.* **138 C8** 44 25N 92 5W
Chippewa Falls, *U.S.A.* **138 C9** 44 56N 91 24W
Chiprovtsi, *Bulgaria* **42 D7** 43 24N 22 52 E
Chiquián, *Peru* **156 C2** 10 10 S 77 0W
Chiquimula, *Guatemala* **148 D2** 14 51N 89 37W
Chiquinquira, *Colombia* **152 B3** 5 37N 73 50W
Chiquitos, Llanos de, *Bolivia* **157 D5** 18 0 S 61 30W
Chir →, *Russia* **53 F6** 48 30N 43 0 E
Chirala, *India* **83 G5** 15 50N 80 26 E
Chiramba, *Mozam.* **107 F3** 16 55 S 34 39 E
Chiran, *Japan* **62 F2** 31 22N 130 27 E
Chirawa, *India* **80 E6** 28 14N 75 42 E
Chirayinkil, *India* **83 K3** 8 41N 76 49 E
Chirchiq, *Uzbekistan* **55 C4** 41 29N 69 35 E
Chirfa, *Niger* **97 D2** 20 55N 12 22 E
Chirgua →, *Venezuela* **152 B4** 8 54N 67 58W
Chiricahua Peak, *U.S.A.* **143 L9** 31 51N 109 18W
Chiriquí, G. de, *Panama* **148 E3** 8 0N 82 10W
Chiriquí, L. de, *Panama* **148 E3** 9 10N 82 0W
Chirivira Falls, *Zimbabwe* **107 G3** 21 10 S 32 12 E
Chirnogi, *Romania* **46 E7** 44 7N 26 32 E
Chirpan, *Bulgaria* **43 E10** 42 10N 25 19 E
Chirripó Grande, Cerro, *Costa Rica* **148 E3** 9 29N 83 29W
Chisamba, *Zambia* **107 E2** 14 55 S 28 20 E
Chisasibi, *Canada* **128 B4** 53 50N 79 0W
Chishmy, *Russia* **54 D5** 54 35N 55 23 E
Chisholm, *Canada* **130 C6** 54 55N 114 10W
Chishtian Mandi, *Pakistan* **80 E5** 29 50N 72 55 E
Chishui, *China* **68 C5** 28 30N 105 42 E
Chishui He →, *China* **68 C5** 28 49N 105 50 E
Chisimaio, *Somali Rep.* **108 E2** 0 22 S 42 32 E
Chisimba Falls, *Zambia* **107 E3** 10 12 S 30 56 E
Chişinău, *Moldova* **51 J5** 47 0N 28 50 E
Chisineu Criş, *Romania* **46 C2** 46 32N 21 37 E
Chisone →, *Italy* **38 D4** 44 49N 7 25 E
Chisos Mts., *U.S.A.* **139 L3** 29 5N 103 15W
Chistopol, *Russia* **52 C10** 55 25N 50 38 E
Chita, *Colombia* **152 B3** 6 11N 72 28W
Chita, *Russia* **57 D12** 52 0N 113 35 E
Chitado, *Angola* **103 F2** 17 10 S 14 8 E
Chitapur, *India* **82 F3** 17 10N 77 5 E
Chitembo, *Angola* **103 E3** 13 30 S 16 50 E
Chitipa, *Malawi* **107 D3** 9 41 S 33 19 E
Chitose, *Japan* **60 C10** 42 49N 141 39 E
Chitrakot, *India* **82 E5** 19 10N 81 40 E
Chitral, *Pakistan* **79 B3** 35 50N 71 56 E
Chitravati →, *India* **83 G4** 14 45N 78 15 E
Chitré, *Panama* **148 E3** 7 59N 80 27W
Chittagong, *Bangla.* **78 D3** 22 19N 91 48 E
Chittagong □, *Bangla.* **78 C3** 24 5N 91 0 E
Chittaurgarh, *India* **80 G6** 24 52N 74 38 E
Chittoor, *India* **83 H4** 13 15N 79 5 E
Chittur, *India* **83 J3** 10 40N 76 45 E
Chitungwiza, *Zimbabwe* **107 F3** 18 0 S 31 6 E
Chiumba, *Angola* **103 E3** 12 29 S 16 8 E
Chiume, *Angola* **103 F4** 15 3 S 21 14 E
Chiusa, *Italy* **39 B8** 46 38N 11 34 E
Chiusi, *Italy* **39 E8** 43 1N 11 57 E
Chiva, *Spain* **35 F4** 39 27N 0 41W
Chivacoa, *Venezuela* **152 A4** 10 10N 68 54W
Chivasso, *Italy* **38 C4** 45 11N 7 53 E
Chivay, *Peru* **156 D3** 15 40 S 71 35W
Chivhu, *Zimbabwe* **107 F3** 19 2 S 30 52 E
Chivilcoy, *Argentina* **158 C4** 34 55 S 60 0W
Chiwanda, *Tanzania* **107 E3** 11 23 S 34 55 E
Chixi, *China* **69 G9** 22 0N 112 58 E
Chizera, *Zambia* **107 E1** 13 10 S 25 0 E
Chkalov = Orenburg, *Russia* **54 F5** 51 45N 55 6 E
Chkolovsk, *Russia* **52 B6** 56 50N 43 10 E
Chloride, *U.S.A.* **145 K12** 35 25N 114 12W
Chlumec, *Czech.* **30 A8** 50 9N 15 29 E
Chmielnik, *Poland* **47 E7** 50 37N 20 43 E
Cho Bo, *Vietnam* **76 B5** 20 46N 105 10 E
Cho-do, *N. Korea* **67 E13** 38 30N 124 40 E
Cho Phuoc Hai, *Vietnam* **77 G6** 10 26N 107 18 E
Choa Chukang, *Malaysia* **74 B2** 1 22N 103 41 E
Choba, *Kenya* **106 B4** 2 30N 38 5 E
Chobe National Park, *Botswana* **104 B3** 18 0 S 25 0 E
Chochiwŏn, *S. Korea* **67 F14** 36 37N 127 18 E
Chocianów, *Poland* **47 D2** 51 27N 15 55 E
Chociwel, *Poland* **47 B2** 53 29N 15 21 E
Chocó □, *Colombia* **152 B2** 6 0N 77 0W
Chocontá, *Colombia* **152 B3** 5 9N 73 41W
Choctawhatchee B., *U.S.A.* **133 D9** 30 20N 86 20W
Chodaków, *Poland* **47 C7** 52 16N 20 18 E
Chodavaram, *India* **82 F6** 17 50N 82 57 E
Chodecz, *Poland* **47 C6** 52 24N 19 2 E
Chodziez, *Poland* **47 C3** 52 58N 16 58 E
Choele Choel, *Argentina* **160 A3** 39 11 S 65 40W
Chōfu, *Japan* **63 B11** 35 39N 139 33 E
Choiseul, *Solomon Is.* **121 L9** 7 0 S 156 40 E
Choisy-le-Roi, *France* **23 D9** 48 45N 2 24 E
Choix, *Mexico* **146 B3** 26 40N 108 23W
Chojna, *Poland* **47 C1** 52 58N 14 25 E
Chojnice, *Poland* **47 B4** 53 42N 17 32 E
Chojnów, *Poland* **47 D2** 51 18N 15 58 E
Chōkai-San, *Japan* **60 E10** 39 6N 140 3 E
Choke, *Ethiopia* **95 E4** 11 18N 37 15 E
Chokurdakh, *Russia* **57 B15** 70 38N 147 55 E
Cholame, *U.S.A.* **144 K6** 35 44N 120 18W
Cholet, *France* **22 E6** 47 4N 0 52W
Cholpon-Ata, *Kyrgyzstan* **55 B8** 42 40N 77 6 E
Choluteca, *Honduras* **148 D2** 13 20N 87 14W
Choluteca →, *Honduras* **148 D2** 13 0N 87 20W
Chom Bung, *Thailand* **76 F2** 13 37N 99 36 E
Chom Thong, *Thailand* **76 C2** 18 25N 98 41 E
Choma, *Zambia* **107 F2** 16 48 S 26 59 E
Chomen Swamp, *Ethiopia* **95 F4** 9 20N 37 10 E
Chomun, *India* **80 F6** 27 15N 75 40 E
Chomutov, *Czech.* **30 A6** 50 28N 13 23 E
Chon Buri, *Thailand* **76 F3** 13 21N 101 1 E
Chon Thanh, *Vietnam* **77 G6** 11 24N 106 36 E
Chonan, *S. Korea* **67 F14** 36 48N 127 9 E
Chone, *Ecuador* **152 D2** 0 40 S 80 0W
Chong Kai, *Cambodia* **76 F4** 13 57N 103 35 E
Chong Mek, *Thailand* **76 E5** 15 10N 105 27 E
Chong'an, *China* **69 D12** 27 45N 118 0 E
Chongde, *China* **69 B13** 30 32N 120 26 E
Chŏngdo, *S. Korea* **67 G15** 35 38N 128 42 E
Chŏngha, *S. Korea* **67 F15** 36 12N 129 21 E
Chŏngjin, *N. Korea* **67 D15** 41 47N 129 50 E
Chŏngju, *N. Korea* **67 E13** 39 40N 125 5 E
Chŏngju, *S. Korea* **67 F14** 36 39N 127 27 E
Chongli, *China* **66 D8** 40 58N 115 15 E
Chongming, *China* **69 B13** 31 38N 121 23 E
Chongming Dao, *China* **69 B13** 31 40N 121 30 E
Chongoyape, *Peru* **156 B2** 6 35 S 79 25W
Chongqing, *Sichuan, China* **68 C6** 29 35N 106 25 E
Chongqing, *Sichuan, China* **68 B4** 30 38N 103 40 E
Chongren, *China* **69 D11** 27 46N 116 3 E
Chŏngŭp, *S. Korea* **67 G14** 35 35N 126 50 E
Chongzuo, *China* **68 F6** 22 23N 107 20 E
Chŏnju, *S. Korea* **67 G14** 35 50N 127 4 E
Chonos, Arch. de los, *Chile* **160 C2** 45 0 S 75 0W
Chop, *Ukraine* **51 H2** 48 26N 22 12 E
Chopda, *India* **82 D2** 21 20N 75 15 E
Chopim →, *Brazil* **159 B5** 25 35 S 53 5W
Chorbat La, *India* **81 B7** 34 42N 76 37 E
Chorley, *U.K.* **16 D5** 53 39N 2 38W
Chornobyl, *Ukraine* **51 G6** 51 20N 30 15 E
Chornomorske, *Ukraine* **51 K7** 45 31N 32 40 E
Chorolque, Cerro, *Bolivia* **158 A2** 20 59 S 66 5W
Choroszcz, *Poland* **47 B9** 53 10N 22 59 E
Chorregon, *Australia* **114 C3** 22 40 S 143 32 E
Chortkiv, *Ukraine* **51 H3** 49 2N 25 46 E
Chŏrwŏn, *S. Korea* **67 E14** 38 15N 127 10 E
Chorzele, *Poland* **47 B7** 53 15N 20 55 E
Chorzów, *Poland* **47 E5** 50 18N 18 57 E
Chos-Malal, *Argentina* **158 D1** 37 20 S 70 15W
Chosan, *N. Korea* **67 D13** 40 50N 125 47 E
Chōshi, *Japan* **63 B12** 35 45N 140 51 E
Choszczno, *Poland* **47 B2** 53 7N 15 25 E
Chota, *Peru* **156 B2** 6 33 S 78 39W
Choteau, *U.S.A.* **142 C7** 47 49N 112 11W
Chotila, *India* **80 H4** 22 23N 71 15 E
Chowchilla, *U.S.A.* **144 H6** 37 7N 120 16W
Chowkham, *Burma* **78 E6** 20 52N 97 28 E
Choybalsan, *Mongolia* **65 B6** 48 4N 114 30 E
Chrisman, *U.S.A.* **141 E9** 39 48N 87 41W
Christchurch, *N.Z.* **119 D7** 43 33 S 172 47 E
Christchurch, *U.K.* **17 G6** 50 44N 1 47W
Christian I., *Canada* **136 B4** 44 50N 80 12W
Christiana, *S. Africa* **104 D4** 27 52 S 25 8 E
Christiansfeld, *Denmark* **15 J3** 55 21N 9 29 E
Christiansted, *Virgin Is.* **149 C7** 17 45N 64 42W
Christie B., *Canada* **131 A6** 62 32N 111 10W
Christina →, *Canada* **131 B6** 56 40N 111 3W
Christmas Cr. →, *Australia* **112 C4** 18 29 S 125 23 E
Christmas Creek, *Australia* **112 C4** 18 29 S 125 23 E
Christmas I. = Kiritimati, *Kiribati* **123 G12** 1 58N 157 27W
Christmas I., *Ind. Oc.* **109 F9** 10 30 S 105 40 E
Christopher L., *Australia* **113 D4** 24 49 S 127 42 E
Chrudim, *Czech.* **30 B8** 49 58N 15 43 E
Chrzanów, *Poland* **31 A12** 50 10N 19 21 E
Chtimba, *Malawi* **107 E3** 10 35 S 34 13 E
Chu = Shu, *Kazakhstan* **55 B6** 43 36N 73 42 E
Chu = Shu →, *Kazakhstan* **55 A3** 45 0N 67 44 E
Chu →, *Vietnam* **76 C5** 19 53N 105 45 E
Chu Chua, *Canada* **130 C4** 51 22N 120 10W
Chu Lai, *Vietnam* **76 E7** 15 28N 108 45 E
Chu Xian, *China* **69 A12** 32 19N 118 20 E
Chuadanga, *Bangla.* **78 D2** 23 38N 88 51 E
Ch'uanchou = Quanzhou, *China* **69 E12** 24 55N 118 34 E
Chuankou, *China* **66 G6** 34 20N 110 59 E
Chūbu □, *Japan* **63 A9** 36 45N 137 30 E
Chubut □, *Argentina* **160 B3** 43 30 S 69 0W
Chubut →, *Argentina* **160 B3** 43 20 S 65 5W
Chuchi L., *Canada* **130 B4** 55 12N 124 30W
Chudovo, *Russia* **50 C6** 59 10N 31 41 E
Chudskoye, Oz., *Russia* **13 G22** 58 13N 27 30 E
Chūgoku □, *Japan* **62 C4** 35 0N 133 0 E
Chūgoku-Sanchi, *Japan* **62 C4** 35 0N 133 0 E
Chuguyev = Chuhuyiv, *Ukraine* **51 H9** 49 55N 36 45 E
Chugwater, *U.S.A.* **138 E2** 41 46N 104 50W
Chuhuyiv, *Ukraine* **51 H9** 49 55N 36 45 E
Chukchi Sea, *Russia* **57 C19** 68 0N 175 0W
Chukotskoye Nagorye, *Russia* **57 C18** 68 0N 175 0 E
Chula, *U.S.A.* **140 E3** 39 55N 93 29W
Chula Vista, *U.S.A.* **145 N9** 32 39N 117 5W
Chulakkurgan, *Kazakhstan* **55 B4** 43 46N 69 9 E
Chulman, *Russia* **57 D13** 56 52N 124 52 E
Chulucanas, *Peru* **156 B1** 5 8 S 80 10W
Chulumani, *Bolivia* **156 D4** 16 24 S 67 31W
Chulym →, *Russia* **56 D9** 57 43N 83 51 E
Chum Phae, *Thailand* **76 D4** 16 40N 102 6 E
Chum Saeng, *Thailand* **76 E3** 15 55N 100 15 E
Chuma, *Bolivia* **156 D4** 15 24 S 68 56W
Chumar, *India* **81 C8** 32 40N 78 35 E
Chumbicha, *Argentina* **158 B2** 29 0 S 66 10W
Chumerna, *Bulgaria* **43 E10** 42 45N 25 55 E
Chumikan, *Russia* **57 D14** 54 40N 135 10 E
Chumphon, *Thailand* **77 G2** 10 35N 99 14 E
Chumpi, *Peru* **156 D3** 15 4 S 73 46W
Chumuare, *Mozam.* **107 E3** 14 31 S 31 50 E
Chumunjin, *S. Korea* **67 F15** 37 55N 128 54 E
Chuna →, *Russia* **57 D10** 57 47N 94 37 E
Chun'an, *China* **69 C12** 29 35N 119 3 E
Chunchŏn, *S. Korea* **67 F14** 37 58N 127 44 E
Chunchura, *India* **81 H13** 22 53N 88 27 E
Chunga, *Zambia* **107 F2** 15 0 S 26 2 E
Chunggang-ŭp, *N. Korea* **67 D14** 41 48N 126 48 E
Chunghwa, *N. Korea* **67 E13** 38 52N 125 47 E
Chungju, *S. Korea* **67 F14** 36 58N 127 58 E
Chungking = Chongqing, *China* **68 C6** 29 35N 106 25 E
Chungmu, *S. Korea* **67 G15** 34 50N 128 20 E
Chungt'iaoshan = Zhongtiao Shan, *China* **66 G6** 35 0N 111 10 E
Chunian, *Pakistan* **80 D6** 30 57N 74 0 E
Chunya, *Tanzania* **107 D3** 8 30 S 33 27 E
Chunya □, *Tanzania* **106 D3** 7 48 S 33 0 E
Chunyang, *China* **67 C15** 43 38N 129 23 E
Chuquibamba, *Peru* **156 D3** 15 47 S 72 44W
Chuquibambilla, *Peru* **156 C3** 14 7 S 72 41W
Chuquicamata, *Chile* **158 A2** 22 15 S 69 0W
Chuquisaca □, *Bolivia* **157 E5** 20 30 S 63 30W
Chur, *Switz.* **29 C9** 46 52N 9 32 E
Churachandpur, *India* **78 C4** 24 20N 93 40 E
Churchill, *Canada* **131 B10** 58 47N 94 11W
Churchill →, *Man., Canada* **131 B10** 58 47N 94 12W
Churchill →, *Nfld., Canada* **129 B7** 53 19N 60 10W
Churchill, C., *Canada* **131 B10** 58 46N 93 12W
Churchill Falls, *Canada* **129 B7** 53 36N 64 19W
Churchill L., *Canada* **131 B7** 55 55N 108 20W
Churchill Pk., *Canada* **130 B3** 58 10N 125 10W
Churdan, *U.S.A.* **140 B2** 42 9N 94 29W
Churfisten, *Switz.* **29 B8** 47 8N 9 17 E
Churu, *India* **80 E6** 28 20N 74 50 E
Churubusco, *U.S.A.* **141 C11** 41 14N 85 19W
Churún Merú = Angel Falls, *Venezuela* **153 B5** 5 57N 62 30W
Churwalden, *Switz.* **29 C9** 46 47N 9 33 E
Chushal, *India* **81 C8** 33 40N 78 40 E
Chusovaya →, *Russia* **54 B6** 58 12N 56 54 E
Chusovoy, *Russia* **54 B6** 58 22N 57 50 E
Chust, *Uzbekistan* **55 C5** 41 0N 71 13 E
Chuuronjang, *N. Korea* **67 D15** 41 35N 129 40 E
Chuvash Republic = Chuvashia □, *Russia* **52 C8** 55 30N 47 0 E
Chuvashia □, *Russia* **52 C8** 55 30N 47 0 E
Chuwārtah, *Iraq* **84 C5** 35 43N 45 34 E
Chuxiong, *China* **68 E3** 25 2N 101 28 E
Ci Xian, *China* **66 F8** 36 20N 114 25 E
Ciacova, *Romania* **46 D2** 45 35N 21 10 E
Ciamis, *Indonesia* **75 D3** 7 20 S 108 21 E
Cianjur, *Indonesia* **74 D3** 6 49 S 107 8 E
Cibola, *U.S.A.* **145 M12** 33 17N 114 42W
Cicero, *U.S.A.* **134 E2** 41 48N 87 48W
Cicero, *Ill., U.S.A.* **141 C9** 41 51N 87 45W
Cícero Dantas, *Brazil* **154 D4** 10 36 S 38 23W
Cidacos →, *Spain* **34 C3** 42 21N 1 38W
Cide, *Turkey* **88 B5** 41 53N 33 1 E
Ciechanów, *Poland* **47 C7** 52 52N 20 38 E
Ciechanów □, *Poland* **47 B7** 53 0N 20 30 E
Ciechanowiec, *Poland* **47 C9** 52 40N 22 31 E
Ciechocinek, *Poland* **47 C5** 52 53N 18 45 E
Ciego de Avila, *Cuba* **148 B4** 21 50N 78 50W
Ciénaga, *Colombia* **152 A3** 11 1N 74 15W
Ciénaga de Oro, *Colombia* **152 B2** 8 53N 75 37W
Cienfuegos, *Cuba* **148 B3** 22 10N 80 30W
Cieplice Sląskie Zdrój, *Poland* **47 E2** 50 50N 15 40 E
Cierp, *France* **24 F4** 42 55N 0 40 E
Cíes, Is., *Spain* **36 C2** 42 12N 8 55W
Cieszanów, *Poland* **47 E10** 50 14N 23 8 E
Cieszyn, *Poland* **31 B11** 49 45N 18 35 E

Corinth 187

Colfax, *Ill., U.S.A.* **141 D8** 40 34N 88 37W
Colfax, *Ind., U.S.A.* **141 D10** 40 12N 86 40W
Colfax, *La., U.S.A.* **139 K8** 31 31N 92 42W
Colfax, *Wash., U.S.A.* **142 C5** 46 53N 117 22W
Colhué Huapi, L., *Argentina* **160 C3** 45 30 S 69 0W
Cólico, *Italy* **38 B6** 46 8N 9 22 E
Coligny, *France* **25 B9** 46 23N 5 21 E
Coligny, *S. Africa* **105 D4** 26 17 S 26 15 E
Colima, *Mexico* **146 D4** 19 14N 103 43W
Colima □, *Mexico* **146 D4** 19 10N 103 40W
Colima, Nevado de, *Mexico* **146 D4** 19 35N 103 45W
Colina, *Chile* **158 C1** 33 13 S 70 45W
Colina do Norte, *Guinea-Biss.* **100 C2** 12 28N 15 0W
Colinas, *Goiás, Brazil* **155 D2** 14 15 S 48 2W
Colinas, *Maranhão, Brazil* **154 C3** 6 0 S 44 10W
Colinton, *Australia* **117 C8** 35 50 S 149 10 E
Coll, *U.K.* **18 E2** 56 39N 6 34W
Collaguasi, *Chile* **158 A2** 21 5 S 68 45W
Collarada, Peña, *Spain* **34 C4** 42 43N 0 29W
Collarenebri, *Australia* **115 D4** 29 33 S 148 34 E
Collbran, *U.S.A.* **143 G10** 39 14N 107 58W
Colle di Val d'Elsa, *Italy* **39 E8** 43 25N 11 7 E
Colle Salvetti, *Italy* **38 E7** 43 34N 10 27 E
Colle Sannita, *Italy* **41 A7** 41 22N 14 50 E
Collécchio, *Italy* **38 D7** 44 45N 10 13 E
Colleen Bawn, *Zimbabwe* **107 G2** 21 0 S 29 12 E
College Park, *U.S.A.* **135 J3** 33 40N 84 27W
Collette, *Canada* **129 C6** 46 40N 65 30W
Collie, *N.S.W., Australia* **117 A8** 31 41 S 148 18 E
Collie, *W. Austral., Australia* **113 F2** 33 22 S 116 8 E
Collier B., *Australia* **112 C3** 16 10 S 124 15 E
Collier Ra., *Australia* **112 D2** 24 45 S 119 10 E
Colline Metallifere, *Italy* **38 E7** 43 10N 11 0 E
Collingwood, *Canada* **128 D3** 44 29N 80 13W
Collingwood, *N.Z.* **119 A7** 40 41 S 172 40 E
Collins, *Canada* **128 B2** 50 17N 89 27W
Collins, *U.S.A.* **140 G3** 37 54N 93 37W
Collinsville, *Australia* **114 C4** 20 30 S 147 56 E
Collinsville, *U.S.A.* **140 F7** 38 40N 89 59W
Collipulli, *Chile* **158 D1** 37 55 S 72 30W
Collo, *Algeria* **99 A6** 36 58N 6 37 E
Collonges, *France* **25 B9** 46 9N 5 52 E
Collooney, *Ireland* **19 B3** 54 11N 8 29W
Colmar, *France* **23 D14** 48 5N 7 20 E
Colmars, *France* **25 D10** 44 11N 6 39 E
Colmenar, *Spain* **37 J6** 36 54N 4 20W
Colmenar de Oreja, *Spain* **34 E1** 40 6N 3 25W
Colmenar Viejo, *Spain* **36 E7** 40 39N 3 47W
Colne, *U.K.* **16 D5** 53 51N 2 9W
Colo →, *Australia* **117 B9** 33 25 S 150 52 E
Cologna Véneta, *Italy* **39 C8** 45 18N 11 23 E
Cologne = Köln, *Germany* **26 E2** 50 56N 6 57 E
Colom, I., *Spain* **33 B11** 39 58N 4 16 E
Coloma, *U.S.A.* **144 G6** 38 48N 120 53W
Colomb-Béchar = Béchar, *Algeria* **99 B4** 31 38N 2 18W
Colombey-les-Belles, *France* **23 D12** 48 32N 5 54 E
Colombey-les-Deux-Églises, *France* **23 D11** 48 13N 4 50 E
Colômbia, *Brazil* **155 F2** 20 10 S 48 40W
Colombia ■, *S. Amer.* **152 C3** 3 45N 73 0W
Colombier, *Switz.* **28 C3** 46 58N 6 53 E
Colombo, *Sri Lanka* **83 L4** 6 56N 79 58 E
Colome, *U.S.A.* **138 D5** 43 16N 99 43W
Colón, *Argentina* **158 C4** 32 12 S 58 10W
Colón, *Cuba* **148 B3** 22 42N 80 54W
Colón, *Panama* **148 E4** 9 20N 79 54W
Colón, *Peru* **156 A1** 5 0 S 81 0W
Colona, *Australia* **113 F5** 31 38 S 132 4 E
Colonella, *Italy* **39 F10** 42 52N 13 50 E
Colonia, *Uruguay* **158 C4** 34 25 S 57 50W
Colonia de San Jordi, *Spain* **33 B9** 39 19N 2 59 E
Colonia Dora, *Argentina* **158 B3** 28 34 S 62 59W
Colonial Heights, *U.S.A.* **134 G7** 37 15N 77 25W
Colonne, C. delle, *Italy* **41 C10** 39 2N 17 12 E
Colonsay, *Canada* **131 C7** 51 59N 105 52W
Colonsay, *U.K.* **18 E2** 56 5N 6 12W
Colorado □, *U.S.A.* **143 G10** 39 30N 105 30W
Colorado →, *Argentina* **160 A4** 39 50 S 62 8W
Colorado →, *N. Amer.* **143 L6** 31 45N 114 40W
Colorado →, *U.S.A.* **139 L7** 28 36N 95 59W
Colorado City, *U.S.A.* **139 J4** 32 24N 100 52W
Colorado Desert, *U.S.A.* **132 D3** 34 20N 116 0W
Colorado Plateau, *U.S.A.* **143 H8** 37 0N 111 0W
Colorado River Aqueduct, *U.S.A.* **145 L12** 34 17N 114 10W
Colorado Springs, *U.S.A.* **138 F2** 38 50N 104 49W
Colorno, *Italy* **38 D7** 44 56N 10 23 E
Colotlán, *Mexico* **146 C4** 22 6N 103 16W
Colquechaca, *Bolivia* **157 D4** 18 40 S 66 1W
Colton, *Calif., U.S.A.* **145 L9** 34 4N 117 20W
Colton, *N.Y., U.S.A.* **137 B10** 44 33N 74 56W
Colton, *Wash., U.S.A.* **142 C5** 46 34N 117 8W
Columbia, *Ill., U.S.A.* **140 F6** 38 27N 90 12W
Columbia, *La., U.S.A.* **139 J8** 32 6N 92 5W
Columbia, *Miss., U.S.A.* **139 K10** 31 15N 89 50W
Columbia, *Mo., U.S.A.* **140 F4** 38 57N 92 20W
Columbia, *Pa., U.S.A.* **137 F8** 40 2N 76 30W
Columbia, *S.C., U.S.A.* **135 H5** 34 0N 81 2W
Columbia, *Tenn., U.S.A.* **135 H2** 35 37N 87 2W
Columbia →, *U.S.A.* **142 C1** 46 15N 124 5W
Columbia, C., *Canada* **6 A4** 83 0N 70 0W
Columbia, District of □, *U.S.A.* **134 F7** 38 55N 77 0W
Columbia, Mt., *Canada* **130 C5** 52 8N 117 20W
Columbia Basin, *U.S.A.* **142 C4** 46 45N 119 5W
Columbia Falls, *U.S.A.* **142 B6** 48 23N 114 11W
Columbia Heights, *U.S.A.* **138 C8** 45 3N 93 15W
Columbiana, *U.S.A.* **136 F4** 40 53N 80 42W
Columbretes, Is., *Spain* **34 F5** 39 50N 0 50 E
Columbus, *Ga., U.S.A.* **135 J3** 32 28N 84 59W
Columbus, *Ind., U.S.A.* **141 E11** 39 13N 85 55W
Columbus, *Kans., U.S.A.* **139 G7** 37 10N 94 50W
Columbus, *Miss., U.S.A.* **135 J1** 33 30N 88 25W
Columbus, *Mont., U.S.A.* **142 D9** 45 38N 109 15W
Columbus, *N. Dak., U.S.A.* **138 A3** 48 54N 102 47W
Columbus, *N. Mex., U.S.A.* **143 L10** 31 50N 107 38W
Columbus, *Nebr., U.S.A.* **138 E6** 41 26N 97 22W
Columbus, *Ohio, U.S.A.* **141 E13** 39 58N 83 0W
Columbus, *Tex., U.S.A.* **139 L6** 29 42N 96 33W
Columbus, *Wis., U.S.A.* **138 D10** 43 21N 89 1W
Columbus Grove, *U.S.A.* **141 D12** 40 55N 84 4W
Columbus Junction, *U.S.A.* **140 C5** 41 17N 91 22W
Colunga, *Spain* **36 B5** 43 29N 5 16W
Colusa, *U.S.A.* **144 F4** 39 13N 122 1W
Colville, *U.S.A.* **142 B5** 48 33N 117 54W
Colville →, *U.S.A.* **126 A4** 70 25N 150 30W
Colville, C., *N.Z.* **118 C4** 36 29 S 175 21 E
Colwyn Bay, *U.K.* **16 D4** 53 18N 3 44W
Coma, *Ethiopia* **95 F4** 8 29N 36 53 E
Comácchio, *Italy* **39 D9** 44 42N 12 11 E
Comalcalco, *Mexico* **147 D6** 18 16N 93 13W
Comallo, *Argentina* **160 B2** 41 0 S 70 5W
Comana, *Romania* **46 E7** 44 10N 26 10 E
Comanche, *Okla., U.S.A.* **139 H6** 34 22N 97 58W
Comanche, *Tex., U.S.A.* **139 K5** 31 54N 98 36W
Comandante Luis Piedrabuena, *Argentina* **160 C3** 49 59 S 68 54W
Comăneşti, *Romania* **46 C7** 46 25N 26 26 E
Comarapa, *Bolivia* **157 D5** 17 54 S 64 29W
Comayagua, *Honduras* **148 D2** 14 25N 87 37W
Combahee →, *U.S.A.* **135 J5** 32 30N 80 31W
Combara, *Australia* **117 A8** 31 10 S 148 22 E
Combeaufontaine, *France* **23 E12** 47 38N 5 54 E
Comber, *Canada* **136 D2** 42 14N 82 33W
Combermere Bay, *Burma* **78 F4** 19 37N 93 34 E
Comblain-au-Pont, *Belgium* **21 H7** 50 29N 5 35 E
Combles, *France* **23 B9** 50 2N 2 50 E
Combourg, *France* **22 D5** 48 25N 1 46W
Comboyne, *Australia* **117 A10** 31 34 S 152 27 E
Combronde, *France* **24 C7** 45 58N 3 5 E
Comeragh Mts., *Ireland* **19 D4** 52 18N 7 34W
Comet, *Australia* **114 C4** 23 36 S 148 38 E
Comilla, *Bangla.* **78 D3** 23 28N 91 10 E
Comines, *Belgium* **21 G1** 50 46N 3 0 E
Comino, *Malta* **32 C1** 36 2N 14 20 E
Comino, C., *Italy* **40 B2** 40 32N 9 49 E
Cómiso, *Italy* **41 F7** 36 56N 14 36 E
Comitán, *Mexico* **147 D6** 16 18N 92 9W
Commentry, *France* **24 B6** 46 20N 2 46 E
Commerce, *Ga., U.S.A.* **135 H4** 34 12N 83 28W
Commerce, *Tex., U.S.A.* **139 J7** 33 15N 95 54W
Commercy, *France* **23 D12** 48 43N 5 34 E
Commewijne □, *Surinam* **153 B7** 5 25N 54 45W
Committee B., *Canada* **127 B11** 68 30N 86 30W
Commonwealth B., *Antarctica* **7 C10** 67 0 S 144 0 E
Commoron Cr. →, *Australia* **115 D5** 28 22 S 150 8 E
Communism Pk. = Kommunizma, Pik, *Tajikistan* **55 D6** 39 0N 72 2 E
Como, *Italy* **38 C6** 45 47N 9 5 E
Como, L. di, *Italy* **38 B6** 46 0N 9 11 E
Comodoro Rivadavia, *Argentina* **160 C3** 45 50 S 67 40W
Comorin, C., *India* **83 K3** 8 3N 77 40 E
Comorişte, *Romania* **46 D2** 45 10N 21 35 E
Comoro Is. = Comoros ■, *Ind. Oc.* **93 H8** 12 10 S 44 15 E
Comoros ■, *Ind. Oc.* **93 H8** 12 10 S 44 15 E
Comox, *Canada* **130 D4** 49 42N 124 55W
Compiègne, *France* **23 C9** 49 24N 2 50 E
Comporta, *Portugal* **37 G2** 38 22N 8 46W
Compostela, *Mexico* **146 C4** 21 15N 104 53W
Compostela, *Phil.* **71 H6** 7 40N 126 2 E
Comprida, I., *Brazil* **159 A6** 24 50 S 47 42W
Compton, *U.S.A.* **145 M8** 33 54N 118 13W
Compton Downs, *Australia* **115 E4** 30 28 S 146 30 E
Comrat, *Moldova* **51 J5** 46 18N 28 40 E
Con Cuong, *Vietnam* **76 C5** 19 2N 104 54 E
Con Son, *Vietnam* **77 H6** 8 41N 106 37 E
Cona Niyeu, *Argentina* **160 B3** 41 58 S 67 0W
Conakry, *Guinea* **100 D2** 9 29N 13 49W
Conara Junction, *Australia* **114 G4** 41 50 S 147 26 E
Conargo, *Australia* **117 C6** 35 16 S 145 10 E
Concarneau, *France* **22 E3** 47 52N 3 56W
Conceição, *Brazil* **154 C4** 7 33 S 38 31W
Conceição, *Mozam.* **107 F4** 18 47 S 36 7 E
Conceição da Barra, *Brazil* **155 E4** 18 35 S 39 45W
Conceição do Araguaia, *Brazil* **154 C2** 8 0 S 49 2W
Conceição do Canindé, *Brazil* **154 C3** 7 54 S 41 34W
Concepción, *Argentina* **158 B2** 27 20 S 65 35W
Concepción, *Bolivia* **157 D5** 16 15 S 62 8W
Concepción, *Chile* **158 D1** 36 50 S 73 0W
Concepción, *Mexico* **147 D6** 18 15N 90 5W
Concepción, *Paraguay* **158 A4** 23 22 S 57 26W
Concepción, *Peru* **156 C2** 11 54 S 75 19W
Concepción □, *Chile* **158 D1** 37 0 S 72 30W
Concepción →, *Mexico* **146 A2** 30 32N 113 2W
Concepción, Est. de, *Chile* **160 D2** 50 30 S 74 55W
Concepción, L., *Bolivia* **157 D5** 17 20 S 61 20W
Concepción, Punta, *Mexico* **146 B2** 26 55N 111 59W
Concepción del Oro, *Mexico* **146 C4** 24 40N 101 30W
Concepción del Uruguay, *Argentina* **158 C4** 32 35 S 58 20W
Conception, Pt., *U.S.A.* **145 L6** 34 27N 120 28W
Conception B., *Namibia* **104 C1** 23 55 S 14 22 E
Conception I., *Bahamas* **149 B4** 23 52N 75 9W
Concession, *Zimbabwe* **107 F3** 17 27 S 30 56 E
Conchas Dam, *U.S.A.* **139 H2** 35 22N 104 11W
Conche, *Canada* **129 B8** 50 55N 55 58W
Conches-en-Ouche, *France* **22 D7** 48 58N 0 56 E
Concho, *U.S.A.* **143 J9** 34 28N 109 36W
Concho →, *U.S.A.* **139 K5** 31 34N 99 43W
Conchos →, *Chihuahua, Mexico* **146 B4** 29 32N 105 0W
Conchos →, *Tamaulipas, Mexico* **147 B5** 25 9N 98 35W
Concord, *Calif., U.S.A.* **144 H4** 37 59N 122 2W
Concord, *Mich., U.S.A.* **141 B12** 42 11N 84 38W
Concord, *N.C., U.S.A.* **135 H5** 35 25N 80 35W
Concord, *N.H., U.S.A.* **137 C13** 43 12N 71 32W
Concordia, *Argentina* **158 C4** 31 20 S 58 2W
Concórdia, *Brazil* **152 D4** 4 36 S 66 36W
Concordia, *Mexico* **146 C3** 23 18N 106 2W
Concordia, *Kans., U.S.A.* **138 F6** 39 34N 97 40W
Concordia, *Mo., U.S.A.* **140 F3** 38 59N 93 34W
Concots, *France* **24 D5** 44 26N 1 40 E
Concrete, *U.S.A.* **142 B3** 48 32N 121 45W
Condah, *Australia* **116 D4** 37 57 S 141 44 E
Condamine, *Australia* **115 D5** 26 56 S 150 9 E
Condat, *France* **24 C6** 45 21N 2 46 E
Condé, *Angola* **103 E2** 10 50 S 14 37 E
Conde, *Brazil* **155 D4** 11 49 S 37 37W
Conde, *U.S.A.* **138 C5** 45 9N 98 6W
Condé-sur-l'Escaut, *France* **23 B10** 50 26N 3 34 E
Condé-sur-Noireau, *France* **22 D6** 48 51N 0 33W
Condeúba, *Brazil* **155 D3** 14 52 S 42 0W
Condobolin, *Australia* **115 E4** 33 4 S 147 6 E
Condom, *France* **24 E4** 43 57N 0 22 E
Condon, *U.S.A.* **142 D3** 45 14N 120 11W
Condove, *Italy* **38 C4** 45 7N 7 18 E
Conegliano, *Italy* **39 C9** 45 53N 12 18 E
Conejera, I., *Spain* **33 B9** 39 11N 2 58 E
Conejos, *Mexico* **146 B4** 26 14N 103 53W
Conflans-en-Jarnisy, *France* **23 C12** 49 10N 5 52 E
Confolens, *France* **24 B4** 46 2N 0 40 E
Confuso →, *Paraguay* **158 B4** 25 9 S 57 34W
Congjiang, *China* **68 E7** 25 43N 108 52 E
Congleton, *U.K.* **16 D5** 53 10N 2 13W
Congo = Zaïre →, *Africa* **103 D2** 6 4 S 12 24 E
Congo, *Brazil* **154 C4** 7 48 S 36 40W
Congo (Kinshasa) = Zaïre ■, *Africa* **103 C4** 3 0 S 23 0 E
Congo ■, *Africa* **102 C3** 1 0 S 16 0 E
Congo Basin, *Africa* **92 G6** 0 10 S 24 30 E
Congonhas, *Brazil* **155 F3** 20 30 S 43 52W
Congress, *U.S.A.* **143 J7** 34 9N 112 51W
Conil, *Spain* **37 J4** 36 17N 6 10W
Coniston, *Canada* **128 C3** 46 29N 80 51W
Conjeeveram = Kanchipuram, *India* **83 H4** 12 52N 79 45 E
Conjuboy, *Australia* **114 B3** 18 35 S 144 35 E
Conklin, *Canada* **131 B6** 55 38N 111 5W
Conlea, *Australia* **115 E3** 30 7 S 144 35 E
Conn, L., *Ireland* **19 B2** 54 3N 9 15W
Connacht □, *Ireland* **19 C3** 53 43N 9 12W
Conneaut, *U.S.A.* **136 E4** 41 57N 80 34W
Connecticut □, *U.S.A.* **137 E12** 41 30N 72 45W
Connecticut →, *U.S.A.* **137 E12** 41 16N 72 20W
Connell, *U.S.A.* **142 C4** 46 40N 118 52W
Connellsville, *U.S.A.* **136 F5** 40 1N 79 35W
Connemara, *Ireland* **19 C2** 53 29N 9 45W
Connemaugh →, *U.S.A.* **136 F5** 40 28N 79 19W
Conner, *Phil.* **70 C3** 17 48N 121 19 E
Connerré, *France* **22 D7** 48 3N 0 30 E
Connersville, *U.S.A.* **141 E11** 39 39N 85 8W
Connors Ra., *Australia* **114 C4** 21 40 S 149 10 E
Conoble, *Australia* **117 B6** 32 55 S 144 33 E
Cononaco →, *Ecuador* **152 D2** 1 32 S 75 35W
Cononbridge, *U.K.* **18 D4** 57 34N 4 27W
Conquest, *Canada* **131 C7** 51 32N 107 14W
Conrad, *Iowa, U.S.A.* **140 B4** 42 14N 92 52W
Conrad, *Mont., U.S.A.* **142 B8** 48 10N 111 57W
Conran, C., *Australia* **117 D8** 37 49 S 148 44 E
Conroe, *U.S.A.* **139 K7** 30 19N 95 27W
Conselheiro Lafaiete, *Brazil* **155 F3** 20 40 S 43 48W
Conselheiro Pena, *Brazil* **155 E3** 19 10 S 41 30W
Consort, *Canada* **131 C6** 52 1N 110 46W
Constance = Konstanz, *Germany* **27 H5** 47 40N 9 10 E
Constance, L. = Bodensee, *Europe* **29 A8** 47 35N 9 25 E
Constanţa, *Romania* **46 E9** 44 14N 28 38 E
Constanţa □, *Romania* **46 E9** 44 15N 28 15 E
Constantina, *Spain* **37 H5** 37 51N 5 40W
Constantine, *Algeria* **99 A6** 36 25N 6 42 E
Constantine, *U.S.A.* **141 C11** 41 50N 85 40W
Constitución, *Chile* **158 D1** 35 20 S 72 30W
Constitución, *Uruguay* **158 C4** 31 0 S 57 50W
Consuegra, *Spain* **37 F7** 39 28N 3 36W
Consul, *Canada* **131 D7** 49 20N 109 30W
Contact, *U.S.A.* **142 F6** 41 46N 114 45W
Contai, *India* **81 J12** 21 54N 87 46 E
Contamana, *Peru* **156 B3** 7 19 S 74 55W
Contarina, *Italy* **39 C9** 45 2N 12 13 E
Contas →, *Brazil* **155 D4** 14 17 S 39 1W
Contes, *France* **25 E11** 43 49N 7 19 E
Conthey, *Switz.* **28 D4** 46 14N 7 18 E
Continental, *U.S.A.* **141 C12** 41 6N 84 16W
Contoocook, *U.S.A.* **137 C13** 43 13N 71 45W
Contra Costa, *Mozam.* **105 D5** 25 9 S 33 30 E
Contres, *France* **22 E8** 47 24N 1 26 E
Contrexéville, *France* **23 D12** 48 10N 5 53 E
Contumaza, *Peru* **156 B2** 7 23 S 78 57W
Convención, *Colombia* **152 B3** 8 28N 73 21W
Conversano, *Italy* **41 B10** 40 58N 17 7 E
Converse, *U.S.A.* **141 D11** 40 35N 85 52W
Convoy, *U.S.A.* **141 D12** 40 55N 84 43W
Conway = Conwy, *U.K.* **16 D4** 53 17N 3 50W
Conway = Conwy →, *U.K.* **16 D4** 53 17N 3 50W
Conway, *Ark., U.S.A.* **139 H8** 35 5N 92 26W
Conway, *N.H., U.S.A.* **137 C13** 43 59N 71 7W
Conway, *S.C., U.S.A.* **135 J6** 33 51N 79 3W
Conway, L., *Australia* **115 D2** 28 17 S 135 35 E
Conwy, *U.K.* **16 D4** 53 17N 3 50W
Conwy →, *U.K.* **16 D4** 53 17N 3 50W
Coober Pedy, *Australia* **115 D1** 29 1 S 134 43 E
Cooch Behar = Koch Bihar, *India* **78 B2** 26 22N 89 29 E
Coodardy, *Australia* **113 E2** 27 15 S 117 39 E
Cook, *Australia* **113 F5** 30 37 S 130 25 E
Cook, *U.S.A.* **138 B8** 47 49N 92 39W
Cook, B., *Chile* **160 E2** 55 10 S 70 0W
Cook, Mt., *N.Z.* **119 D5** 43 36 S 170 9 E
Cook Inlet, *U.S.A.* **126 C4** 60 0N 152 0W
Cook Is., *Pac. Oc.* **123 J11** 17 0 S 160 0W
Cook Strait, *N.Z.* **118 H3** 41 15 S 174 29 E
Cooke Plains, *Australia* **116 C3** 35 23 S 139 34 E
Cookeville, *U.S.A.* **135 G3** 36 10N 85 30W
Cookhouse, *S. Africa* **104 E4** 32 44 S 25 47 E
Cookshire, *Canada* **137 A13** 45 25N 71 38W
Cookstown, *U.K.* **19 B5** 54 39N 6 45W
Cookstown □, *U.K.* **19 B5** 54 40N 6 43W
Cooksville, *Canada* **136 C5** 43 36N 79 35W
Cooktown, *Australia* **114 B4** 15 30 S 145 16 E
Coolabah, *Australia* **117 A7** 31 1 S 146 43 E
Cooladdi, *Australia* **115 D4** 26 37 S 145 23 E
Coolah, *Australia* **117 A8** 31 48 S 149 41 E
Coolamon, *Australia* **115 E4** 34 46 S 147 8 E
Coolangatta, *Australia* **115 D5** 28 11 S 153 29 E
Coolgardie, *Australia* **113 F3** 30 55 S 121 8 E
Coolibah, *Australia* **112 C5** 15 33 S 130 56 E
Coolidge, *U.S.A.* **143 K8** 32 59N 111 31W
Coolidge Dam, *U.S.A.* **143 K8** 33 0N 110 20W
Cooma, *Australia* **117 D8** 36 12 S 149 8 E
Coon Rapids, *U.S.A.* **140 C2** 41 53N 94 41W
Coonabarabran, *Australia* **117 A8** 31 14 S 149 18 E
Coonalpyn, *Australia* **116 C3** 35 43 S 139 52 E
Coonamble, *Australia* **117 A8** 30 56 S 148 27 E
Coonana, *Australia* **113 F3** 31 0 S 123 0 E
Coondapoor, *India* **83 H2** 13 42N 74 40 E
Coongie, *Australia* **115 D3** 27 9 S 140 8 E
Coongoola, *Australia* **115 D4** 27 43 S 145 51 E
Cooninie, L., *Australia* **115 D2** 26 4 S 139 59 E
Coonoor, *India* **83 J3** 11 21N 76 45 E
Cooper, *U.S.A.* **139 J7** 33 23N 95 42W
Cooper →, *U.S.A.* **135 J6** 32 50N 79 56W
Cooper Cr. →, *N. Terr., Australia* **110 C5** 12 7 S 132 41 E
Cooper Cr. →, *S. Austral., Australia* **115 D2** 28 29 S 137 46 E
Cooperstown, *N. Dak., U.S.A.* **138 B5** 47 27N 98 8W
Cooperstown, *N.Y., U.S.A.* **137 D10** 42 42N 74 56W
Coopersville, *U.S.A.* **141 A11** 43 4N 85 57W
Coorabie, *Australia* **113 F5** 31 54 S 132 18 E
Coorabulka, *Australia* **114 C3** 23 41 S 140 20 E
Coorow, *Australia* **113 E2** 29 53 S 116 2 E
Cooroy, *Australia* **115 D5** 26 22 S 152 54 E
Coos Bay, *U.S.A.* **142 E1** 43 22N 124 13W
Cootamundra, *Australia* **117 C8** 34 36 S 148 1 E
Cootehill, *Ireland* **19 B4** 54 4N 7 5W
Cooyar, *Australia* **115 D5** 26 59 S 151 51 E
Cooyeana, *Australia* **114 C2** 24 29 S 138 45 E
Copahue Paso, *Argentina* **158 D1** 37 49 S 71 8W
Copainalá, *Mexico* **147 D6** 17 8N 93 11W
Copán, *Honduras* **148 D2** 14 50N 89 9W
Copatana, *Brazil* **152 D4** 2 48 S 67 4W
Cope, *U.S.A.* **138 F3** 39 40N 102 51W
Cope, C., *Spain* **35 H3** 37 26N 1 28W
Cope Cope, *Australia* **116 D5** 36 27 S 143 5 E
Copenhagen = København, *Denmark* **15 J6** 55 41N 12 34 E
Copertino, *Italy* **41 B11** 40 16N 18 3 E
Copeville, *Australia* **116 C3** 34 47 S 139 51 E
Copiapó, *Chile* **158 B1** 27 30 S 70 20W
Copiapó →, *Chile* **158 B1** 27 19 S 70 56W
Copley, *Australia* **116 A3** 30 36 S 138 26 E
Copp L., *Canada* **130 A6** 60 14N 114 40W
Copparo, *Italy* **39 D8** 44 54N 11 49 E
Coppename →, *Surinam* **153 B6** 5 48N 55 55W
Copper Center, *U.S.A.* **126 B5** 61 58N 145 18W
Copper Cliff, *Canada* **128 C3** 46 28N 81 4W
Copper Harbor, *U.S.A.* **134 B2** 47 28N 87 53W
Copper Queen, *Zimbabwe* **107 F2** 17 29 S 29 18 E
Copperbelt □, *Zambia* **107 E2** 13 15 S 27 30 E
Coppermine, *Canada* **126 B8** 67 50N 115 5W
Coppermine →, *Canada* **126 B8** 67 49N 116 4W
Copperopolis, *U.S.A.* **144 H6** 37 58N 120 38W
Coquet →, *U.K.* **16 B6** 55 20N 1 32W
Coquilhatville = Mbandaka, *Zaïre* **102 B3** 0 1N 18 18 E
Coquille, *U.S.A.* **142 E1** 43 11N 124 11W
Coquimbo, *Chile* **158 B1** 30 0 S 71 20W
Coquimbo □, *Chile* **158 C1** 31 0 S 71 0W
Corabia, *Romania* **46 F5** 43 48N 24 30 E
Coração de Jesus, *Brazil* **155 E3** 16 43 S 44 22W
Coracora, *Peru* **156 D3** 15 5 S 73 45W
Coradi, Is., *Italy* **41 B10** 40 27N 17 10 E
Coral Bay, *Phil.* **71 G1** 8 25N 117 20 E
Coral Gables, *U.S.A.* **135 N5** 25 45N 80 16W
Coral Harbour, *Canada* **127 B11** 64 8N 83 10W
Coral Sea, *Pac. Oc.* **122 J7** 15 0 S 150 0 E
Coralville, *U.S.A.* **140 C5** 41 40N 91 35W
Coralville Res., *U.S.A.* **140 C5** 41 50N 91 40W
Corantijn →, *Surinam* **153 B6** 5 50N 57 8W
Coraopolis, *U.S.A.* **136 F4** 40 31N 80 10W
Corato, *Italy* **41 A9** 41 9N 16 25 E
Corbeil-Essonnes, *France* **23 D9** 48 36N 2 26 E
Corbie, *France* **23 C9** 49 54N 2 30 E
Corbières, *France* **24 F6** 42 55N 2 35 E
Corbigny, *France* **23 E10** 47 16N 3 40 E
Corbin, *U.S.A.* **134 G3** 36 57N 84 6W
Corbion, *Belgium* **21 J6** 49 48N 5 0 E
Corbones →, *Spain* **37 H5** 37 36N 5 39W
Corby, *U.K.* **17 E7** 52 30N 0 41W
Corby Glen, *U.K.* **17 E7** 52 49N 0 30W
Corcaigh = Cork, *Ireland* **19 E3** 51 54N 8 29W
Corcoles →, *Spain* **35 F1** 39 40N 3 18W
Corcoran, *U.S.A.* **144 J7** 36 6N 119 33W
Corcubión, *Spain* **36 C1** 42 56N 9 12W
Cordele, *U.S.A.* **135 K4** 31 58N 83 47W
Cordell, *U.S.A.* **139 H5** 35 17N 98 59W
Cordenons, *Italy* **39 C9** 45 59N 12 42 E
Cordes, *France* **24 D5** 44 5N 1 57 E
Cordisburgo, *Brazil* **155 E3** 19 7 S 44 21W
Córdoba, *Argentina* **158 C3** 31 20 S 64 10W
Córdoba, *Mexico* **147 D5** 18 50N 97 0W
Córdoba, *Spain* **37 H6** 37 50N 4 50W
Córdoba □, *Argentina* **158 C3** 31 22 S 64 15W
Córdoba □, *Colombia* **152 B2** 8 20N 75 40W
Córdoba □, *Spain* **37 G6** 38 5N 5 0W
Córdoba, Sierra de, *Argentina* **158 C3** 31 10 S 64 25W
Cordon, *Phil.* **70 C3** 16 42N 121 32 E
Cordova, *Ala., U.S.A.* **135 J2** 33 46N 87 11W
Cordova, *Alaska, U.S.A.* **126 B5** 60 33N 145 45W
Cordova, *Ill., U.S.A.* **140 C6** 41 41N 90 19W
Corella, *Spain* **34 C3** 42 7N 1 48W
Corella →, *Australia* **114 B3** 19 34 S 140 47 E
Coremas, *Brazil* **154 C4** 7 1 S 37 58W
Corentyne →, *Guyana* **153 B6** 5 50N 57 8W
Corfield, *Australia* **114 C3** 21 40 S 143 21 E
Corfu = Kérkira, *Greece* **32 A3** 39 38N 19 50 E
Corfu, Str of, *Greece* **32 A4** 39 34N 20 0 E
Corgo, *Spain* **36 C3** 42 56N 7 25W
Corguinho, *Brazil* **157 D7** 19 53 S 54 52W
Cori, *Italy* **40 A5** 41 39N 12 55 E
Coria, *Spain* **36 F4** 39 58N 6 33W
Coricudgy, *Australia* **117 B9** 32 51 S 150 24 E
Corigliano Cálabro, *Italy* **41 C9** 39 36N 16 31 E
Coringa Is., *Australia* **114 B4** 16 58 S 149 58 E
Corinna, *Australia* **114 G4** 41 35 S 145 10 E
Corinth = Kórinthos, *Greece* **45 G4** 37 56N 22 55 E
Corinth, *Ky., U.S.A.* **141 F12** 38 30N 84 34W

Dammarie 189

Csurgo, *Hungary* **31 E10** 46 16N 17 9 E
Cu Lao Hon, *Vietnam* **77 G7** 10 54N 108 18 E
Cua Rao, *Vietnam* **76 C5** 19 16N 104 27 E
Cuácua →, *Mozam.* **107 F4** 17 54 S 37 0 E
Cuamato, *Angola* **103 F3** 17 2 S 15 7 E
Cuamba, *Mozam.* **107 E4** 14 45 S 36 22 E
Cuando →, *Angola* **103 F4** 17 30 S 23 15 E
Cuando Cubango □, *Angola* **103 F3** 16 25 S 20 0 E
Cuangar, *Angola* **103 F3** 17 36 S 18 39 E
Cuango, *Angola* **103 D3** 6 15 S 16 42 E
Cuanza →, *Angola* **103 D2** 9 2 S 13 30 E
Cuanza Norte □, *Angola* **103 D2** 8 50 S 14 30 E
Cuanza Sul □, *Angola* **103 E2** 10 50 S 14 50 E
Cuarto →, *Argentina* **158 C3** 33 25 S 63 2W
Cuatrociénegas, *Mexico* **146 B4** 26 59N 102 5W
Cuauhtémoc, *Mexico* **146 B3** 28 25N 106 52W
Cuba, *Portugal* **37 G3** 38 10N 7 54W
Cuba, *Mo., U.S.A.* **140 F5** 38 4N 91 24W
Cuba, *N. Mex., U.S.A.* **143 J10** 36 1N 107 4W
Cuba, *N.Y., U.S.A.* **136 D6** 42 13N 78 17W
Cuba ■, *W. Indies* **148 B4** 22 0N 79 0W
Cuba City, *U.S.A.* **140 B6** 42 36N 90 26W
Cubal, *Angola* **103 E2** 12 26 S 14 3 E
Cuballing, *Australia* **113 F2** 32 50 S 117 10 E
Cubango →, *Africa* **103 F4** 18 50 S 22 25 E
Cubanja, *Angola* **103 E4** 14 49 S 21 20 E
Cubia, *Angola* **103 F4** 15 58 S 21 42 E
Çubuk, *Turkey* **88 B5** 40 14N 33 3 E
Cucamonga, *U.S.A.* **145 L9** 34 10N 117 30W
Cuchi, *Angola* **103 E3** 14 37 S 16 58 E
Cuchillo-Có, *Argentina* **160 A4** 38 20 S 64 37W
Cuchivero →, *Venezuela* **152 B4** 7 40N 65 57W
Cuchumatanes, Sierra de los, *Guatemala* **148 C1** 15 35N 91 25W
Cucuí, *Brazil* **152 C4** 1 12N 66 50W
Cucurpe, *Mexico* **146 A2** 30 20N 110 43W
Cucurrupí, *Colombia* **152 C2** 4 23N 76 56W
Cúcuta, *Colombia* **152 B3** 7 54N 72 31W
Cudahy, *U.S.A.* **141 B9** 42 58N 87 52W
Cudalbi, *Romania* **46 D8** 45 46N 27 41 E
Cuddalore, *India* **83 J4** 11 46N 79 45 E
Cuddapah, *India* **83 G4** 14 30N 78 47 E
Cuddapan, L., *Australia* **114 D3** 25 45 S 141 26 E
Cudgewa, *Australia* **117 D7** 36 10 S 147 42 E
Cudillero, *Spain* **36 B4** 43 33N 6 9W
Cue, *Australia* **113 E2** 27 25 S 117 54 E
Cuéllar, *Spain* **36 D6** 41 23N 4 21W
Cuemba, *Angola* **103 E3** 11 50 S 17 42 E
Cuenca, *Ecuador* **152 D2** 2 50 S 79 9W
Cuenca, *Spain* **34 E2** 40 5N 2 10W
Cuenca □, *Spain* **34 F2** 40 0N 2 0W
Cuenca, Serranía de, *Spain* **34 F3** 39 55N 1 50W
Cuerdo del Pozo, Pantano de la, *Spain* **34 D2** 41 51N 2 44W
Cuernavaca, *Mexico* **147 D5** 18 55N 99 15W
Cuero, *U.S.A.* **139 L6** 29 6N 97 17W
Cuers, *France* **25 E10** 43 14N 6 5 E
Cuervo, *U.S.A.* **139 H2** 35 2N 104 25W
Cuesmes, *Belgium* **21 H3** 50 26N 3 56 E
Cuevas, Cerro, *Bolivia* **157 E4** 22 0 S 65 12W
Cuevas del Almanzora, *Spain* **35 H3** 37 18N 1 58W
Cuevo, *Bolivia* **157 E5** 20 15 S 63 30W
Cugir, *Romania* **46 D4** 45 48N 23 25 E
Cuiabá, *Brazil* **157 D6** 15 30 S 56 0W
Cuiabá →, *Brazil* **157 D6** 17 5 S 56 36W
Cuilco, *Guatemala* **148 C1** 15 24N 91 58W
Cuillin Hills, *U.K.* **18 D2** 57 13N 6 15W
Cuillin Sd., *U.K.* **18 D2** 57 4N 6 20W
Cuima, *Angola* **103 E3** 13 25 S 15 45 E
Cuiseaux, *France* **25 B9** 46 30N 5 22 E
Cuité, *Brazil* **154 C4** 6 29 S 36 9W
Cuito →, *Angola* **103 F4** 18 1 S 20 48 E
Cuito Cuanavale, *Angola* **103 F3** 15 10 S 19 10 E
Cuitzeo, L. de, *Mexico* **146 D4** 19 55N 101 5W
Cuiuni →, *Brazil* **153 D5** 0 45 S 63 7W
Cuivre →, *U.S.A.* **140 F6** 38 55N 90 44W
Cuivre, West Fork →, *U.S.A.* **140 E6** 39 2N 90 58W
Cujmir, *Romania* **46 E3** 44 13N 22 57 E
Cukai, *Malaysia* **77 K4** 4 13N 103 25 E
Culaba, *Phil.* **71 F5** 11 40N 124 32 E
Culan, *France* **24 B6** 46 34N 2 20 E
Culasi, *Phil.* **71 F4** 11 26N 122 3 E
Culauan, *Phil.* **71 J5** 5 58N 125 40 E
Culbertson, *U.S.A.* **138 A2** 48 9N 104 31W
Culburra, *Australia* **116 C3** 35 50 S 139 58 E
Culcairn, *Australia* **117 C7** 35 41 S 147 3 E
Culebra, Sierra de la, *Spain* **36 D4** 41 55N 6 20W
Culemborg, *Neths.* **20 E6** 51 58N 5 14 E
Culfa, *Azerbaijan* **89 C11** 38 57N 45 38 E
Culgoa, *Australia* **116 C5** 35 44 S 143 6 E
Culgoa →, *Australia* **115 D4** 29 56 S 146 20 E
Culiacán, *Mexico* **146 C3** 24 50N 107 23W
Culiacán →, *Mexico* **146 C3** 24 30N 107 42W
Culion, *Phil.* **71 F3** 11 54N 120 1 E
Culiseu →, *Brazil* **157 C7** 12 14 S 53 17W
Cúllar de Baza, *Spain* **35 H2** 37 35N 2 34W
Cullarin Ra., *Australia* **117 C8** 34 30 S 149 30 E
Cullen, *U.K.* **18 D6** 57 42N 2 49W
Cullen Pt., *Australia* **114 A3** 11 57 S 141 54 E
Cullera, *Spain* **35 F4** 39 9N 0 17W
Cullman, *U.S.A.* **135 H2** 34 11N 86 51W
Culloden, *U.K.* **18 D4** 57 30N 4 9W
Cullom, *U.S.A.* **141 D8** 40 53N 88 16W
Culoz, *France* **25 C9** 45 47N 5 46 E
Culpataro, *Australia* **116 B6** 33 40 S 144 22 E
Culpeper, *U.S.A.* **134 F7** 38 30N 78 0W
Culuene →, *Brazil* **157 C7** 12 56 S 52 51W
Culver, *U.S.A.* **141 C10** 41 13N 86 25W
Culver, Pt., *Australia* **113 F3** 32 54 S 124 43 E
Culverden, *N.Z.* **119 C7** 42 47 S 172 49 E
Cuma, *Angola* **103 E3** 12 52 S 15 5 E
Cumali, *Turkey* **45 H9** 36 42N 27 28 E
Cumaná, *Venezuela* **153 A5** 10 30N 64 5W
Cumare, *Colombia* **152 C3** 0 49N 72 32W
Cumari, *Brazil* **155 E2** 18 16 S 48 11W
Cumberland, *Canada* **130 D3** 49 40N 125 0W
Cumberland, *Iowa, U.S.A.* **140 C2** 41 16N 94 52W
Cumberland, *Md., U.S.A.* **134 F6** 39 39N 78 46W
Cumberland, *Wis., U.S.A.* **138 C8** 45 32N 92 1W
Cumberland →, *U.S.A.* **135 G2** 36 15N 87 0W
Cumberland, C., *Vanuatu* **121 D4** 14 39 S 166 37 E
Cumberland I., *U.S.A.* **135 K5** 30 50N 81 25W
Cumberland Is., *Australia* **114 C4** 20 35 S 149 10 E
Cumberland L., *Canada* **131 C8** 54 3N 102 18W
Cumberland Pen., *Canada* **127 B13** 67 0N 64 0W
Cumberland Plateau, *U.S.A.* **135 H3** 36 0N 85 0W
Cumberland Sd., *Canada* **127 B13** 65 30N 66 0W
Cumborah, *Australia* **115 D4** 29 40 S 147 45 E
Cumbres Mayores, *Spain* **37 G4** 38 4N 6 39W
Cumbria □, *U.K.* **16 C5** 54 42N 2 52W
Cumbrian Mts., *U.K.* **16 C4** 54 30N 3 0W
Cumbum, *India* **83 G4** 15 40N 79 10 E
Cuminá →, *Brazil* **153 D6** 1 30 S 56 0W
Cuminapanema →, *Brazil* **153 D7** 1 9 S 54 54W
Cummings Mt., *U.S.A.* **145 K8** 35 2N 118 34W
Cummins, *Australia* **115 E2** 34 16 S 135 43 E
Cumnock, *Australia* **117 B8** 32 59 S 148 46 E
Cumnock, *U.K.* **18 F4** 55 28N 4 17W
Cumpas, *Mexico* **146 A3** 30 0N 109 48W
Cumplida, Pta., *Canary Is.* **33 F2** 28 50N 17 48W
Çumra, *Turkey* **88 D5** 37 34N 32 45 E
Cuncumén, *Chile* **158 C1** 31 53 S 70 38W
Cundeelee, *Australia* **113 F3** 30 43 S 123 26 E
Cunderdin, *Australia* **113 F2** 31 37 S 117 12 E
Cundinamarca □, *Colombia* **152 C3** 5 0N 74 0W
Cunene □, *Angola* **103 F3** 16 30 S 15 0 E
Cunene →, *Angola* **103 F2** 17 20 S 11 50 E
Cúneo, *Italy* **38 D4** 44 23N 7 32 E
Cunhinga, *Angola* **103 E3** 12 11 S 16 47 E
Cunillera, I., *Spain* **33 C7** 38 59N 1 13 E
Cunjamba, *Angola* **103 F4** 15 27 S 20 10 E
Cunlhat, *France* **24 C7** 45 38N 3 32 E
Cunnamulla, *Australia* **115 D4** 28 2 S 145 38 E
Cuorgnè, *Italy* **38 C4** 45 23N 7 39 E
Cupar, *Canada* **131 C8** 50 57N 104 10W
Cupar, *U.K.* **18 E5** 56 19N 3 1W
Cupica, G. de, *Colombia* **152 B2** 6 25N 77 30W
Ćuprija, *Serbia, Yug.* **42 D6** 43 57N 21 26 E
Curaçá, *Brazil* **154 C4** 8 59 S 39 54W
Curaçao, *Neth. Ant.* **149 D6** 12 10N 69 0W
Curacautín, *Chile* **160 A2** 38 26 S 71 53W
Curahuara de Carangas, *Bolivia* **156 D4** 17 52 S 68 26W
Curanilahue, *Chile* **158 D1** 37 29 S 73 28W
Curaray →, *Peru* **152 D3** 2 20 S 74 5W
Curatabaca, *Venezuela* **153 B5** 6 19N 62 51W
Cure →, *France* **23 E10** 47 40N 3 41 E
Curepto, *Chile* **158 D1** 35 8 S 72 1W
Curiapo, *Venezuela* **153 B5** 8 33N 61 5W
Curicó, *Chile* **158 C1** 34 55 S 71 20W
Curicó □, *Chile* **158 C1** 34 50 S 71 15W
Curicuriari →, *Brazil* **152 D4** 0 14 S 66 48W
Curimatá, *Brazil* **154 D3** 10 2 S 44 17W
Curiplaya, *Colombia* **152 C3** 0 16N 74 52W
Curitiba, *Brazil* **159 B6** 25 20 S 49 10W
Currabubula, *Australia* **117 A9** 31 16 S 150 44 E
Currais Novos, *Brazil* **154 C4** 6 13 S 36 30W
Curralinho, *Brazil* **154 B2** 1 45 S 49 46W
Currant, *U.S.A.* **142 G6** 38 51N 115 32W
Curranyalpa, *Australia* **117 A6** 30 53 S 144 39 E
Curraweena, *Australia* **117 A6** 30 47 S 145 54 E
Currawilla, *Australia* **114 D3** 25 10 S 141 20 E
Current →, *U.S.A.* **139 G9** 36 15N 90 55W
Currie, *Australia* **114 F3** 39 56 S 143 53 E
Currie, *U.S.A.* **142 F6** 40 16N 114 45W
Currituck Sd., *U.S.A.* **135 G8** 36 20N 75 52W
Cursole, *Somali Rep.* **108 D3** 2 14N 45 25 E
Curtea de Argeş, *Romania* **46 D5** 45 12N 24 42 E
Curtis, *Spain* **36 B2** 43 7N 8 4W
Curtis, *U.S.A.* **138 E4** 40 38N 100 31W
Curtis Group, *Australia* **114 F4** 39 30 S 146 37 E
Curtis I., *Australia* **114 C5** 23 35 S 151 10 E
Curuá →, *Pará, Brazil* **153 D7** 2 24 S 54 5W
Curuá →, *Pará, Brazil* **157 B7** 5 23 S 54 22W
Curuá, I., *Brazil* **154 A1** 0 48N 50 10W
Curuaés →, *Brazil* **157 B7** 7 30 S 54 45W
Curuápanema →, *Brazil* **153 D6** 2 25 S 55 2W
Curuçá, *Brazil* **154 B2** 0 43 S 47 50W
Curuguaty, *Paraguay* **159 A4** 24 31 S 55 42W
Çürüksu Çayi →, *Turkey* **49 G4** 37 27N 27 11 E
Curup, *Indonesia* **74 C2** 4 26 S 102 13 E
Curupira, Serra, *S. Amer.* **153 C5** 1 25N 64 30W
Cururu →, *Brazil* **157 B6** 7 12 S 58 3W
Cururupu, *Brazil* **154 B3** 1 50 S 44 50W
Curuzú Cuatiá, *Argentina* **158 B4** 29 50 S 58 5W
Curvelo, *Brazil* **155 E3** 18 45 S 44 27W
Curyo, *Australia* **116 C5** 35 50 S 142 47 E
Cushing, *U.S.A.* **139 H6** 35 59N 96 46W
Cushing, Mt., *Canada* **130 B3** 57 35N 126 57W
Cusihuiriáchic, *Mexico* **146 B3** 28 10N 106 50W
Cusna, Mte., *Italy* **38 D7** 44 17N 10 23 E
Cusset, *France* **24 B7** 46 8N 3 28 E
Custer, *U.S.A.* **138 D3** 43 46N 103 36W
Cut Bank, *U.S.A.* **142 B7** 48 38N 112 20W
Cutervo, *Peru* **156 B2** 6 25 S 78 55W
Cuthbert, *U.S.A.* **135 K3** 31 46N 84 48W
Cutler, *U.S.A.* **144 J7** 36 31N 119 17W
Cutral-Có, *Argentina* **160 A3** 38 58 S 69 15W
Cutro, *Italy* **41 C9** 39 2N 16 59 E
Cuttaburra →, *Australia* **115 D3** 29 43 S 144 22 E
Cuttack, *India* **82 D7** 20 25N 85 57 E
Cuvelai, *Angola* **103 F3** 15 44 S 15 50 E
Cuvier, C., *Australia* **113 D1** 23 14 S 113 22 E
Cuvier I., *N.Z.* **118 C4** 36 27 S 175 50 E
Cuxhaven, *Germany* **26 B4** 53 51N 8 41 E
Cuyabeno, *Ecuador* **152 D2** 0 16 S 75 53W
Cuyahoga Falls, *U.S.A.* **136 E3** 41 8N 81 29W
Cuyapo, *Phil.* **70 D3** 15 46N 120 40 E
Cuyo, *Phil.* **71 F3** 10 50N 121 5 E
Cuyo East Pass, *Phil.* **71 F3** 11 0N 121 28 E
Cuyo I., *Phil.* **71 F3** 10 51N 121 2 E
Cuyo West Pass, *Phil.* **71 F3** 11 0N 120 30 E
Cuyuni →, *Guyana* **153 B6** 6 23N 58 41W
Cuzco, *Bolivia* **156 E4** 20 0 S 66 50W
Cuzco, *Peru* **156 C3** 13 32 S 72 0W
Cuzco □, *Peru* **156 C3** 13 31 S 71 59W
Čvrsnica, *Bos.-H.* **42 D2** 43 36N 17 35 E
Cwmbran, *U.K.* **17 F4** 51 39N 3 2W
Cyangugu, *Rwanda* **106 C2** 2 29 S 28 54 E
Cybinka, *Poland* **47 C1** 52 12N 14 46 E
Cyclades = Kikládhes, *Greece* **45 G6** 37 20N 24 30 E
Cygnet, *Australia* **114 G4** 43 8 S 147 1 E
Cynthiana, *U.S.A.* **141 F12** 38 23N 84 18W
Cypress Hills, *Canada* **131 D7** 49 40N 109 30W
Cyprus ■, *Asia* **32 E12** 35 0N 33 0 E
Cyrenaica, *Libya* **96 B4** 27 0N 23 0 E
Cyrene = Shaḥḥāt, *Libya* **96 B4** 32 48N 21 54 E
Czaplinek, *Poland* **47 B3** 53 34N 16 14 E
Czar, *Canada* **131 C6** 52 27N 110 50W
Czarna →, *Piotrkow Trybunalski, Poland* **47 D6** 51 18N 19 55 E
Czarna →, *Tarnobrzeg, Poland* **47 E8** 50 3N 21 21 E
Czarna Woda, *Poland* **47 B5** 53 51N 18 6 E
Czarne, *Poland* **47 B3** 53 42N 16 58 E
Czarnków, *Poland* **47 C3** 52 55N 16 38 E
Czech Rep. ■, *Europe* **30 B7** 50 0N 15 0 E
Czechowice-Dziedzice, *Poland* **31 B11** 49 54N 18 59 E
Czeladz, *Poland* **47 E6** 50 16N 19 2 E
Czempiń, *Poland* **47 C3** 52 9N 16 33 E
Czeremcha, *Poland* **47 C10** 52 31N 23 21 E
Czersk, *Poland* **47 B4** 53 46N 17 58 E
Czerwieńsk, *Poland* **47 C2** 52 1N 15 13 E
Czerwionka, *Poland* **31 A11** 50 7N 18 37 E
Częstochowa, *Poland* **47 E6** 50 49N 19 7 E
Częstochowa □, *Poland* **47 E6** 50 45N 19 0 E
Człopa, *Poland* **47 B3** 53 6N 16 6 E
Człuchów, *Poland* **47 B4** 53 41N 17 22 E
Czyzew, *Poland* **47 C9** 52 48N 22 19 E

D

Da →, *Vietnam* **76 B5** 21 15N 105 20 E
Da Hinggan Ling, *China* **65 B7** 48 0N 121 0 E
Da Lat, *Vietnam* **77 G7** 11 56N 108 25 E
Da Nang, *Vietnam* **76 D7** 16 4N 108 13 E
Da Qaidam, *China* **64 C4** 37 50N 95 15 E
Da Yunhe →, *China* **67 G11** 34 25N 120 5 E
Da'an, *China* **67 B13** 45 30N 124 7 E
Daap, *Phil.* **71 H4** 7 4N 122 12 E
Daarlerveen, *Neths.* **20 D9** 52 26N 6 34 E
Dab'a, Râs el, *Egypt* **94 H6** 31 3N 28 31 E
Daba Shan, *China* **68 B7** 32 0N 109 0 E
Dabai, *Nigeria* **101 C6** 11 25N 5 15 E
Dabajuro, *Venezuela* **152 A3** 11 2N 70 40W
Dabakala, *Ivory C.* **100 D4** 8 15N 4 20W
Dabaro, *Somali Rep.* **108 C3** 6 21N 48 43 E
Dabeiba, *Colombia* **152 B2** 7 1N 76 16W
Dabhoi, *India* **80 H5** 22 10N 73 20 E
Dąbie, *Konin, Poland* **47 C5** 52 5N 18 50 E
Dąbie, *Szczecin, Poland* **47 B1** 53 27N 14 45 E
Dabie Shan, *China* **69 B10** 31 20N 115 20 E
Dabo, *Indonesia* **74 C2** 0 30 S 104 33 E
Dabola, *Guinea* **100 C2** 10 50N 11 5W
Dabou, *Ivory C.* **100 D4** 5 20N 4 23W
Daboya, *Ghana* **101 D4** 9 30N 1 20W
Dabrowa Górnicza, *Poland* **47 E6** 50 15N 19 10 E
Dabrowa Tarnówska, *Poland* **31 A13** 50 10N 20 59 E
Dąbrówno, *Poland* **47 B7** 53 27N 20 2 E
Dabu, *China* **69 E11** 24 22N 116 41 E
Dabung, *Malaysia* **77 K4** 5 23N 102 1 E
Dabus →, *Ethiopia* **95 E4** 10 48N 35 10 E
Dacato →, *Ethiopia* **95 F5** 7 25N 42 40 E
Dacca = Dhaka, *Bangla.* **78 D3** 23 43N 90 26 E
Dacca = Dhaka □, *Bangla.* **78 C3** 24 25N 90 25 E
Dachau, *Germany* **27 G7** 48 15N 11 26 E
Dadale, *Solomon Is.* **121 M10** 8 7 S 159 6 E
Dadanawa, *Guyana* **153 C6** 2 50N 59 30W
Daday, *Turkey* **88 B5** 41 28N 33 27 E
Dade City, *U.S.A.* **135 L4** 28 22N 82 11W
Dades, Oued →, *Morocco* **98 B3** 30 58N 6 44W
Dadiya, *Nigeria* **101 D7** 9 35N 11 24 E
Dadra and Nagar Haveli □, *India* **82 D1** 20 5N 73 0 E
Dadri = Charkhi Dadri, *India* **80 E7** 28 37N 76 17 E
Dadu, *Pakistan* **79 D2** 26 45N 67 45 E
Dadu He →, *China* **68 C4** 29 31N 103 46 E
Dăeni, *Romania* **46 E9** 44 51N 28 10 E
Daet, *Phil.* **70 D4** 14 2N 122 55 E
Dafang, *China* **68 D5** 27 9N 105 39 E
Dagana, *Senegal* **100 B1** 16 30N 15 35W
Dagash, *Sudan* **94 D3** 19 19N 33 25 E
Dagestan □, *Russia* **53 J8** 42 30N 47 0 E
Dagestanskiye Ogni, *Russia* **53 J9** 42 6N 48 12 E
Dagg Sd., *N.Z.* **119 F1** 45 23 S 166 45 E
Daggett, *U.S.A.* **145 L10** 34 52N 116 52W
Daghestan Republic = Dagestan □, *Russia* **53 J8** 42 30N 47 0 E
Daghfeli, *Sudan* **94 D3** 19 18N 32 40 E
Dağlıq Qarabağ = Nagorno-Karabakh, *Azerbaijan* **89 C12** 39 55N 46 45 E
Dagö = Hiiumaa, *Estonia* **13 G20** 58 50N 22 45 E
Dagu, *China* **67 E9** 38 59N 117 40 E
Dagua, *Papua N. G.* **120 B2** 3 27 S 143 20 E
Daguan, *China* **68 D4** 27 43N 103 56 E
Dagupan, *Phil.* **70 C3** 16 3N 120 20 E
Dahab, *Egypt* **94 B3** 28 31N 34 31 E
Dahlak Kebir, *Eritrea* **90 D3** 15 50N 40 10 E
Dahlenburg, *Germany* **26 B6** 53 11N 10 44 E
Dahlgren, *U.S.A.* **141 F8** 38 12N 88 41W
Dahlonega, *U.S.A.* **135 H4** 34 32N 83 59W
Dahme, *Germany* **26 D9** 51 52N 13 25 E
Dahod, *India* **80 H6** 22 50N 74 15 E
Dahomey = Benin ■, *Africa* **101 D5** 10 0N 2 0 E
Dahong Shan, *China* **69 B9** 31 25N 113 0 E
Dahra, *Senegal* **100 B1** 15 22N 15 30W
Dahra, Massif de, *Algeria* **99 A5** 36 7N 1 21 E
Dahy, Nafūd ad, *Si. Arabia* **86 B4** 22 0N 45 25 E
Dai Hao, *Vietnam* **76 C6** 18 1N 106 25 E
Dai-Sen, *Japan* **62 B5** 35 22N 133 32 E
Dai Shan, *China* **69 B14** 30 25N 122 10 E
Dai Xian, *China* **66 E7** 39 4N 112 58 E
Daicheng, *China* **66 E9** 38 42N 116 38 E
Daigo, *Japan* **63 A12** 36 46N 140 21 E
Daimanji-San, *Japan* **62 A5** 36 14N 133 20 E
Daimiel, *Spain* **35 F1** 39 5N 3 35W
Daingean, *Ireland* **19 C4** 53 18N 7 17W
Dainkog, *China* **68 A1** 32 30N 97 58 E
Daintree, *Australia* **114 B4** 16 20 S 145 20 E
Daiō-Misaki, *Japan* **63 C8** 34 15N 136 45 E
Dairût, *Egypt* **94 B3** 27 34N 30 43 E
Daisetsu-Zan, *Japan* **60 C11** 43 30N 142 57 E
Daitari, *India* **82 D7** 21 10N 85 46 E
Daito, *Japan* **62 B4** 35 19N 132 58 E
Dajarra, *Australia* **114 C2** 21 42 S 139 30 E
Dajia, *Taiwan* **69 E13** 24 22N 120 37 E
Dajin Chuan →, *China* **68 B3** 31 16N 101 59 E
Dak Dam, *Cambodia* **76 F6** 12 20N 107 21 E
Dak Nhe, *Vietnam* **76 E6** 15 28N 107 48 E
Dak Pek, *Vietnam* **76 E6** 15 4N 107 44 E
Dak Song, *Vietnam* **77 F6** 12 19N 107 35 E
Dak Sui, *Vietnam* **76 E6** 14 55N 107 43 E
Dakar, *Senegal* **100 C1** 14 34N 17 29W
Dakhla, *W. Sahara* **98 D1** 23 50N 15 53W
Dakhla, El Wâhât el-, *Egypt* **94 B2** 25 30N 28 50 E
Dakhovskaya, *Russia* **49 F7** 44 13N 40 13 E
Dakingari, *Nigeria* **101 C5** 11 37N 4 1 E
Dakor, *India* **80 H5** 22 45N 73 11 E
Dakoro, *Niger* **101 C6** 14 31N 6 46 E
Dakota City, *Iowa, U.S.A.* **140 B2** 42 43N 94 12W
Dakota City, *Nebr., U.S.A.* **138 D6** 42 25N 96 25W
Đakovica, *Serbia, Yug.* **42 E5** 42 22N 20 26 E
Đakovo, *Croatia* **42 B3** 45 19N 18 24 E
Dala, *Angola* **103 E4** 11 3 S 20 17 E
Dala, *Solomon Is.* **121 M11** 8 30 S 160 41 E
Dalaba, *Guinea* **100 C2** 10 42N 12 15W
Dalachi, *China* **66 F3** 36 48N 105 0 E
Dalaguete, *Phil.* **71 G4** 9 46N 123 32 E
Dalai Nur, *China* **66 C9** 43 20N 116 45 E
Dālakī, *Iran* **85 D6** 29 26N 51 17 E
Dalälven, *Sweden* **13 F17** 60 12N 16 43 E
Dalaman, *Turkey* **88 D3** 36 48N 28 47 E
Dalaman →, *Turkey* **88 D3** 36 41N 28 43 E
Dalandzadgad, *Mongolia* **66 C3** 43 27N 104 30 E
Dalanganem Is., *Phil.* **71 F3** 10 40N 120 17 E
Dalarna, *Sweden* **13 F16** 61 0N 14 0 E
Dalat, *Malaysia* **75 B4** 2 44N 111 56 E
Dālbandīn, *Pakistan* **79 C2** 29 0N 64 23 E
Dalbeattie, *U.K.* **18 G5** 54 56N 3 50W
Dalbosjön, *Sweden* **15 F6** 58 40N 12 45 E
Dalby, *Australia* **115 D5** 27 10 S 151 17 E
Dalby, *Sweden* **15 J7** 55 40N 13 22 E
Dale, *U.S.A.* **141 F10** 38 10N 86 59W
Dalen, *Neths.* **20 C9** 52 42N 6 46 E
Dalen, *Norway* **14 E2** 59 26N 8 0 E
Dalet, *Burma* **78 F4** 19 59N 93 51 E
Daletme, *Burma* **78 E4** 21 36N 92 46 E
Daleville, *U.S.A.* **141 D11** 40 7N 85 33W
Dalfsen, *Neths.* **20 C8** 52 31N 6 16 E
Dalga, *Egypt* **94 B3** 27 39N 30 41 E
Dalgān, *Iran* **85 E8** 27 31N 59 19 E
Dalhart, *U.S.A.* **139 G3** 36 4N 102 31W
Dalhousie, *Canada* **129 C6** 48 5N 66 26W
Dalhousie, *India* **80 C6** 32 38N 75 58 E
Dali, *Shaanxi, China* **66 G5** 34 48N 109 58 E
Dali, *Yunnan, China* **68 E3** 25 40N 100 10 E
Dalian, *China* **67 E11** 38 50N 121 40 E
Daliang Shan, *China* **68 D4** 28 0N 102 45 E
Dalias, *Spain* **35 J2** 36 49N 2 52W
Daling He →, *China* **67 D11** 40 55N 121 40 E
Dāliyat el Karmel, *Israel* **91 C4** 32 43N 35 2 E
Dalj, *Croatia* **42 B3** 45 29N 18 59 E
Dalkeith, *U.K.* **18 F5** 55 54N 3 4W
Dall I., *U.S.A.* **130 C2** 54 59N 133 25W
Dallarnil, *Australia* **115 D5** 25 19 S 152 2 E
Dallas, *Oreg., U.S.A.* **142 D2** 44 55N 123 19W
Dallas, *Tex., U.S.A.* **139 J6** 32 47N 96 49W
Dallas Center, *U.S.A.* **140 C3** 41 41N 93 58W
Dallas City, *U.S.A.* **140 D5** 40 38N 91 10W
Dallol, *Ethiopia* **95 E5** 14 14N 40 17 E
Dalmacija, *Croatia* **42 D2** 43 20N 17 0 E
Dalmatia = Dalmacija, *Croatia* **42 D2** 43 20N 17 0 E
Dalmatovo, *Russia* **54 C9** 56 16N 62 56 E
Dalmellington, *U.K.* **18 F4** 55 19N 4 23W
Dalnegorsk, *Russia* **57 E14** 44 32N 135 33 E
Dalnerechensk, *Russia* **57 E14** 45 50N 133 40 E
Daloa, *Ivory C.* **100 D3** 7 0N 6 30W
Dalou Shan, *China* **68 C6** 28 15N 107 0 E
Dalsjöfors, *Sweden* **15 G7** 57 46N 13 5 E
Dalskog, *Sweden* **15 F6** 58 44N 12 18 E
Dalsland, *Sweden* **13 G14** 58 50N 12 15 E
Daltenganj, *India* **81 G11** 24 0N 84 4 E
Dalton, *Canada* **128 C3** 48 11N 84 1W
Dalton, *Ga., U.S.A.* **135 H3** 34 46N 84 58W
Dalton, *Mass., U.S.A.* **137 D11** 42 28N 73 11W
Dalton, *Nebr., U.S.A.* **138 E3** 41 25N 102 58W
Dalton Iceberg Tongue, *Antarctica* **7 C9** 66 15 S 121 30 E
Dalupiri I., *Cagayan, Phil.* **70 B3** 19 5N 121 12 E
Dalupiri I., *N. Samar, Phil.* **70 E5** 12 25N 124 16 E
Dalvík, *Iceland* **12 D4** 65 58N 18 32W
Daly →, *Australia* **112 B5** 13 35 S 130 19 E
Daly City, *U.S.A.* **144 H4** 37 42N 122 28W
Daly L., *Canada* **131 B7** 56 32N 105 39W
Daly Waters, *Australia* **114 B1** 16 15 S 133 24 E
Dam Doi, *Vietnam* **77 H5** 8 50N 105 12 E
Dam Ha, *Vietnam* **76 B6** 21 21N 107 36 E
Daman, *India* **82 D1** 20 25N 72 57 E
Daman □, *India* **82 D1** 20 25N 72 58 E
Dāmaneh, *Iran* **85 C6** 33 1N 50 29 E
Damanhûr, *Egypt* **94 H7** 31 0N 30 30 E
Damanzhuang, *China* **66 E9** 38 5N 116 35 E
Damar, *Indonesia* **72 C3** 7 7 S 128 40 E
Damara, *C.A.R.* **102 B3** 4 58N 18 42 E
Damaraland, *Namibia* **104 C2** 21 0 S 17 0 E
Damascus = Dimashq, *Syria* **91 B5** 33 30N 36 18 E
Damaturu, *Nigeria* **101 C7** 11 45N 11 55 E
Damāvand, *Iran* **85 C7** 35 47N 52 0 E
Damāvand, Qolleh-ye, *Iran* **85 C7** 35 56N 52 10 E
Damba, *Angola* **103 D3** 6 44 S 15 20 E
Dame Marie, *Haiti* **149 C5** 18 36N 74 26W
Dāmghān, *Iran* **85 B7** 36 10N 54 17 E
Dămienesti, *Romania* **46 C8** 46 44N 27 1 E
Damietta = Dumyât, *Egypt* **94 H7** 31 24N 31 48 E
Daming, *China* **66 F8** 36 15N 115 6 E
Damīr Qābū, *Syria* **84 B4** 36 58N 41 51 E
Dammam = Ad Dammām, *Si. Arabia* **85 E6** 26 20N 50 5 E
Dammarie, *France* **22 D8** 48 20N 1 30 E

Delvinákion, Greece 44 E2 39 57N 20 32 E
Delvinë, Albania 44 E2 39 59N 20 4 E
Demak, Indonesia 75 D4 6 53 S 110 38 E
Demanda, Sierra de la, Spain 34 C1 42 15N 3 0W
Demavand = Damāvand, Iran 85 C7 35 47N 52 0 E
Demba, Zaïre 103 D4 5 28 S 22 15 E
Demba Chio, Angola 103 D2 9 41 S 13 41 E
Dembecha, Ethiopia 95 E4 10 32N 37 30 E
Dembi, Ethiopia 95 F4 8 5N 36 25 E
Dembia, Zaïre 106 B2 3 33N 25 48 E
Dembidolo, Ethiopia 95 F3 8 34N 34 50 E
Demer →, Belgium 21 G5 50 57N 4 42 E
Demetrias, Greece 44 E5 39 22N 23 1 E
Demidov, Russia 50 E6 55 16N 31 30 E
Deming, N. Mex., U.S.A. 143 K10 32 16N 107 46W
Deming, Wash., U.S.A. 144 B4 48 50N 122 13W
Demini →, Brazil 153 D5 0 46 S 62 56W
Demirci, Turkey 88 C3 39 2N 28 38 E
Demirköy, Turkey 88 B2 41 49N 27 45 E
Demmin, Germany 26 B9 53 54N 13 2 E
Demnate, Morocco 98 B3 31 44N 6 59W
Demonte, Italy 38 D4 44 19N 7 17 E
Demopolis, U.S.A. 135 J2 32 31N 87 50W
Dempo, Indonesia 74 C2 4 2 S 103 15 E
Demyansk, Russia 50 D7 57 40N 32 27 E
Den Burg, Neths. 20 B5 53 3N 4 47 E
Den Chai, Thailand 76 D3 17 59N 100 4 E
Den Dungen, Neths. 21 E6 51 41N 5 22 E
Den Haag = 's-Gravenhage, Neths. 20 D4 52 7N 4 17 E
Den Ham, Neths. 20 D8 52 28N 6 30 E
Den Helder, Neths. 20 C5 52 57N 4 45 E
Den Hulst, Neths. 20 C8 52 36N 6 16 E
Den Oever, Neths. 20 C6 52 56N 5 2 E
Denain, France 23 B10 50 20N 3 22 E
Denair, U.S.A. 144 H6 37 32N 120 48W
Denau, Uzbekistan 55 D3 38 16N 67 54 E
Denbigh, U.K. 16 D4 53 12N 3 25W
Dendang, Indonesia 75 C3 3 7 S 107 56 E
Dender →, Belgium 21 F4 51 2N 4 6 E
Denderhoutem, Belgium 21 G4 50 53N 4 2 E
Denderleeuw, Belgium 21 G4 50 54N 4 5 E
Dendermonde, Belgium 21 F4 51 2N 4 5 E
Deneba, Ethiopia 95 F4 9 47N 39 10 E
Denekamp, Neths. 20 D10 52 22N 7 1 E
Denezhkin Kamen, Gora, Russia 54 A7 60 25N 59 32 E
Deng Deng, Cameroon 102 A2 5 12N 13 31 E
Deng Xian, China 69 A9 32 34N 112 4 E
Dengchuan, China 68 E3 25 59N 100 3 E
Denge, Nigeria 101 C6 12 52N 5 21 E
Dengfeng, China 66 G7 34 25N 113 2 E
Dengi, Nigeria 101 D6 9 25N 9 55 E
Dengkou, China 66 D4 40 18N 106 55 E
Denham, Australia 113 E1 25 56 S 113 31 E
Denham Ra., Australia 114 C4 21 55 S 147 46 E
Denham Sd., Australia 113 E1 25 45 S 113 15 E
Denia, Spain 35 G5 38 49N 0 8 E
Denial B., Australia 115 E1 32 14 S 133 32 E
Deniliquin, Australia 115 F3 35 30 S 144 58 E
Denison, Iowa, U.S.A. 138 D7 42 1N 95 21W
Denison, Tex., U.S.A. 139 J6 33 45N 96 33W
Denison Plains, Australia 112 C4 18 35 S 128 0 E
Denizli, Turkey 88 D3 37 42N 29 2 E
Denman, Australia 117 B9 32 24 S 150 42 E
Denman Glacier, Antarctica 7 C7 66 45 S 99 25 E
Denmark, Australia 113 F2 34 59 S 117 25 E
Denmark ■, Europe 15 J3 55 30N 9 0 E
Denmark Str., Atl. Oc. 8 B6 66 0N 30 0W
Dennison, U.S.A. 136 F3 40 24N 81 19W
Denpasar, Indonesia 75 D5 8 45 S 115 14 E
Denton, Mont., U.S.A. 142 C9 47 19N 109 57W
Denton, Tex., U.S.A. 139 J6 33 13N 97 8W
D'Entrecasteaux, Pt., Australia 113 F2 34 50 S 115 57 E
D'Entrecasteaux Is., Papua N. G. 120 E6 9 0 S 151 0 E
Dents du Midi, Switz. 28 D3 46 10N 6 56 E
Denu, Ghana 101 D5 6 4N 1 8 E
Denver, Colo., U.S.A. 138 F2 39 44N 104 59W
Denver, Ind., U.S.A. 141 D10 40 52N 86 5W
Denver, Iowa, U.S.A. 140 B4 42 40N 92 20W
Denver City, U.S.A. 139 J3 32 58N 102 50W
Deoband, India 80 E7 29 42N 77 43 E
Deobhog, India 82 E6 19 53N 82 44 E
Deogarh, India 82 D7 21 32N 84 45 E
Deoghar, India 81 G12 24 30N 86 42 E
Deolali, India 82 E1 19 58N 73 50 E
Deoli = Devli, India 80 G6 25 50N 75 20 E
Deoria, India 81 F10 26 31N 83 48 E
Deosai Mts., Pakistan 81 B6 35 40N 75 0 E
Deping, China 67 F9 37 25N 116 58 E
Deposit, U.S.A. 137 D9 42 5N 75 25W
Depot Springs, Australia 113 E3 27 55 S 120 3 E
Deputatskiy, Russia 57 C14 69 18N 139 54 E
Dêqên, China 68 C2 28 34N 98 51 E
Deqing, China 69 F8 23 8N 111 42 E
Dera Ghazi Khan, Pakistan 79 C3 30 5N 70 43 E
Dera Ismail Khan, Pakistan 79 C3 31 50N 70 50 E
Derbent, Russia 53 J9 42 5N 48 15 E
Derby, Australia 112 C3 17 18 S 123 38 E
Derby, U.K. 16 E6 52 56N 1 28W
Derby, Conn., U.S.A. 137 E11 41 19N 73 5W
Derby, Kans., U.S.A. 139 G6 37 33N 97 16W
Derby, N.Y., U.S.A. 136 D6 42 41N 78 58W
Derbyshire □, U.K. 16 E6 53 11N 1 38W
Derecske, Hungary 31 D14 47 20N 21 33 E
Dereli, Turkey 89 B8 40 44N 38 26 E
Derg →, U.K. 19 B4 54 44N 7 26W
Derg, L., Ireland 19 D3 53 0N 8 20W
Dergachi = Derhaci, Ukraine 51 G9 50 9N 36 11 E
Derhaci, Ukraine 51 G9 50 9N 36 11 E
Derik, Turkey 89 D9 37 21N 40 18 E
Derinkuyu, Turkey 88 C6 38 22N 34 45 E
Dermantsi, Bulgaria 43 D9 43 8N 24 17 E
Dernieres, Isles, U.S.A. 139 L9 29 2N 90 50W
Dêrong, China 68 C2 28 44N 99 9 E
Derrinallum, Australia 116 D5 37 57 S 143 15 E
Derry = Londonderry, U.K. 19 B4 55 0N 7 20W
Derryveagh Mts., Ireland 19 B3 54 56N 8 11W
Derudub, Sudan 94 D4 17 31N 36 7 E
Derval, France 22 E5 47 40N 1 41W
Dervéni, Greece 45 F4 38 8N 22 25 E
Derventa, Bos.-H. 42 C2 44 59N 17 55 E
Derwent, Canada 131 C6 53 41N 110 58W
Derwent →, Derby, U.K. 16 E6 52 57N 1 28W
Derwent →, N. Yorks., U.K. 16 D7 53 45N 0 58W
Derwent Water, U.K. 16 C4 54 35N 3 9W
Des Moines, Iowa, U.S.A. 140 C3 41 35N 93 37W
Des Moines, N. Mex., U.S.A. 139 G3 36 46N 103 50W
Des Moines →, U.S.A. 138 E9 40 23N 91 25W
Des Plaines, U.S.A. 141 B9 42 3N 87 52W
Des Plaines →, U.S.A. 141 C8 41 23N 88 15W
Desaguadero →, Argentina 158 C2 34 30 S 66 46W
Desaguadero →, Bolivia 156 D4 16 35 S 69 5W
Descanso, Pta., Mexico 145 N9 32 21N 117 3W
Descartes, France 24 B4 46 59N 0 42 E
Deschaillons, Canada 129 C5 46 32N 72 7W
Descharme →, Canada 131 B7 56 51N 109 13W
Deschutes →, U.S.A. 142 D3 45 38N 120 55W
Dese, Ethiopia 90 E2 11 5N 39 40 E
Deseado, C., Chile 160 D2 52 45 S 74 42W
Desenzano del Garda, Italy 38 C7 45 28N 10 32 E
Desert Center, U.S.A. 145 M11 33 43N 115 24W
Desert Hot Springs, U.S.A. 145 M10 33 58N 116 30W
Désirade, I., Guadeloupe 149 C7 16 18N 61 3W
Deskenatlata L., Canada 130 A6 60 55N 112 3W
Desna →, Ukraine 51 G6 50 33N 30 32 E
Desnătui →, Romania 46 E4 44 15N 23 27 E
Desolación, I., Chile 160 D2 53 0 S 74 0W
Despeñaperros, Paso, Spain 35 G1 38 24N 3 30W
Despotovac, Serbia, Yug. 42 C6 44 6N 21 30 E
Dessau, Germany 26 D8 51 51N 12 14 E
Dessel, Belgium 21 F6 51 15N 5 7 E
Dessye = Dese, Ethiopia 90 E2 11 5N 39 40 E
D'Estrees B., Australia 116 C2 35 55 S 137 45 E
Desuri, India 80 G5 25 18N 73 35 E
Desvres, France 23 B8 50 40N 1 48 E
Det Udom, Thailand 76 E5 14 54N 105 5 E
Deta, Romania 42 B6 45 24N 21 13 E
Dete, Zimbabwe 107 F2 18 38 S 26 50 E
Detinja →, Serbia, Yug. 42 D4 43 51N 20 5 E
Detmold, Germany 26 D4 51 56N 8 52 E
Detour, Pt., U.S.A. 134 C2 45 40N 86 40W
Detroit, Mich., U.S.A. 128 D3 42 20N 83 3W
Detroit, Tex., U.S.A. 139 J7 33 40N 95 16W
Detroit Lakes, U.S.A. 138 B7 46 49N 95 51W
Deurne, Belgium 21 F4 51 12N 4 24 E
Deurne, Neths. 21 F7 51 27N 5 49 E
Deutsche Bucht, Germany 26 A4 54 15N 8 0 E
Deutschlandsberg, Austria 30 E8 46 49N 15 14 E
Deux-Sèvres □, France 22 F6 46 35N 0 20W
Deva, Romania 46 D3 45 53N 22 55 E
Devakottai, India 83 K4 9 55N 78 45 E
Devaprayag, India 81 D8 30 13N 78 35 E
Dévaványa, Hungary 31 D13 47 2N 20 59 E
Deveci Dağları, Turkey 88 B7 40 6N 36 15 E
Devecser, Hungary 31 D10 47 6N 17 26 E
Develi, Turkey 88 C6 38 23N 35 29 E
Deventer, Neths. 20 D8 52 15N 6 10 E
Deveron →, U.K. 18 D6 57 41N 2 32W
Devesel, Romania 46 E3 44 28N 22 41 E
Devgad I., India 83 G2 14 48N 74 5 E
Devgadh Bariya, India 80 H5 22 40N 73 55 E
Devil River Pk., N.Z. 119 A7 40 56 S 172 37 E
Devils Den, U.S.A. 144 K7 35 46N 119 58W
Devils Lake, U.S.A. 138 A5 48 7N 98 52W
Devils Paw, Canada 130 B2 58 47N 134 0W
Devil's Pt., Sri Lanka 83 K5 9 26N 80 6 E
Devil's Pt., Vanuatu 121 G6 17 44 S 168 11 E
Devin, Bulgaria 43 F9 41 44N 24 24 E
Devizes, U.K. 17 F6 51 22N 1 58W
Devli, India 80 G6 25 50N 75 20 E
Devnya, Bulgaria 43 D12 43 13N 27 33 E
Devolii →, Albania 44 D2 40 57N 20 15 E
Devon, Canada 130 C6 53 24N 113 44W
Devon □, U.K. 17 G4 50 50N 3 40W
Devon I., Canada 6 B3 75 10N 85 0W
Devonport, Australia 114 G4 41 10 S 146 22 E
Devonport, N.Z. 118 C3 36 49 S 174 49 E
Devonport, U.K. 17 G3 50 22N 4 11W
Devrek, Turkey 88 B4 41 13N 31 57 E
Devrekâni, Turkey 88 B5 41 36N 33 50 E
Devrez →, Turkey 88 B6 41 6N 34 25 E
Dewas, India 80 H7 22 59N 76 3 E
Dewetsdorp, S. Africa 104 D4 29 33 S 26 39 E
Dewsbury, U.K. 16 D6 53 42N 1 37W
Dexing, China 69 C11 28 46N 117 30 E
Dexter, Mich., U.S.A. 141 B13 42 20N 83 53W
Dexter, Mo., U.S.A. 139 G9 36 48N 89 57W
Dexter, N. Mex., U.S.A. 139 J2 33 12N 104 22W
Dey-Dey, L., Australia 113 E5 29 12 S 131 4 E
Deyang, China 68 B5 31 3N 104 27 E
Deyhūk, Iran 85 C8 33 15N 57 30 E
Deyyer, Iran 85 E6 27 55N 51 55 E
Dezadeash L., Canada 130 A1 60 28N 136 58W
Dezfūl, Iran 89 F13 32 20N 48 30 E
Dezhneva, Mys, Russia 57 C19 66 5N 169 40W
Dezhou, China 66 F9 37 26N 116 18 E
Dháfni, Kríti, Greece 32 D7 35 13N 25 3 E
Dháfni, Pelopónnisos, Greece 45 G4 37 48N 22 1 E
Dhahaban, Si. Arabia 86 B2 21 58N 39 3 E
Dhahiriya = Az Zāhirīyah, West Bank 91 D3 31 25N 34 58 E
Dhahran = Az Zahrān, Si. Arabia 85 E6 26 10N 50 7 E
Dhaka, Bangla. 78 D3 23 43N 90 26 E
Dhaka □, Bangla. 78 C3 24 25N 90 25 E
Dhali, Cyprus 32 D12 35 1N 33 25 E
Dhamangaon, India 82 D4 20 45N 78 15 E
Dhamar, Yemen 86 D4 14 30N 44 20 E
Dhamási, Greece 44 E4 39 43N 22 11 E
Dhampur, India 81 E8 29 19N 78 33 E
Dhamtari, India 82 D5 20 42N 81 35 E
Dhanbad, India 81 H12 23 50N 86 30 E
Dhankuta, Nepal 81 F12 26 55N 87 40 E
Dhanora, India 82 D5 20 20N 80 22 E
Dhar, India 80 H6 22 35N 75 26 E
Dharampur, Gujarat, India 82 D1 20 32N 73 17 E
Dharampur, Mad. P., India 80 H6 22 13N 75 18 E
Dharamsala = Dharmsala, India 80 C7 32 16N 76 23 E
Dharapuram, India 83 J3 10 45N 77 34 E
Dharmapuri, India 83 H4 12 10N 78 10 E
Dharmavaram, India 83 G3 14 29N 77 44 E
Dharmsala, India 80 C7 32 16N 76 23 E
Dharwad, India 83 G2 15 22N 75 15 E
Dhaulagiri, Nepal 81 E10 28 39N 83 28 E
Dhebar, L., India 80 G6 24 10N 74 0 E
Dheftera, Cyprus 32 D12 35 5N 33 16 E
Dhenkanal, India 82 D7 20 45N 85 35 E
Dhenoúsa, Greece 45 G7 37 8N 25 48 E
Dherinia, Cyprus 32 D12 35 3N 33 57 E
Dheskáti, Greece 44 E3 39 55N 21 49 E
Dhespotikó, Greece 45 H6 36 57N 24 58 E
Dhestina, Greece 45 F4 38 25N 22 31 E
Dhiarrizos →, Cyprus 32 E11 34 41N 32 34 E
Dhībān, Jordan 91 D4 31 30N 35 46 E
Dhidhimótikhon, Greece 44 C8 41 22N 26 29 E
Dhíkti Óros, Greece 32 D7 35 8N 25 22 E
Dhilianáta, Greece 45 F2 38 15N 20 34 E
Dhílos, Greece 45 G7 37 23N 25 15 E
Dhimitsána, Greece 45 G4 37 36N 22 3 E
Dhírfis, Greece 45 F5 38 40N 23 54 E
Dhodhekánisos, Greece 45 H8 36 35N 27 0 E
Dhokós, Greece 45 G5 37 20N 23 20 E
Dholiana, Greece 44 E2 39 54N 20 32 E
Dholka, India 80 H5 22 44N 72 29 E
Dhomokós, Greece 45 E4 39 10N 22 18 E
Dhoraji, India 80 J4 21 45N 70 37 E
Dhoxáton, Greece 44 C6 41 9N 24 16 E
Dhragonísi, Greece 45 G7 37 27N 25 29 E
Dhráhstis, Ákra, Greece 32 A3 39 48N 19 40 E
Dhrangadhra, India 80 H4 22 59N 71 31 E
Dhrápanon, Ákra, Greece 32 D6 35 28N 24 14 E
Dhriopís, Greece 45 G6 37 25N 24 35 E
Dhrol, India 80 H4 22 33N 70 25 E
Dhubāb, Yemen 86 D3 12 56N 43 25 E
Dhuburi, India 78 B2 26 2N 89 59 E
Dhulasar, Bangla. 78 E3 21 52N 90 14 E
Dhule, India 82 D2 20 58N 74 50 E
Dhupdhara, India 78 B3 26 10N 91 4 E
Dhut →, Somali Rep. 90 E5 10 30N 50 0 E
Di Linh, Vietnam 77 G7 11 35N 108 4 E
Di Linh, Cao Nguyen, Vietnam 77 G7 11 30N 108 0 E
Día, Greece 32 D7 35 28N 25 14 E
Diablo, Mt., U.S.A. 144 H5 37 53N 121 56W
Diablo Range, U.S.A. 144 J5 37 20N 121 25W
Diafarabé, Mali 100 C4 14 9N 4 57W
Diagonal, U.S.A. 140 D2 40 49N 94 20W
Diala, Mali 100 C3 14 10N 9 58W
Dialakoro, Mali 100 C3 12 18N 7 54W
Diallassagou, Mali 100 C4 13 47N 3 41W
Diamante, Argentina 158 C3 32 5 S 60 40W
Diamante →, Argentina 158 C2 34 30 S 66 46W
Diamantina, Brazil 155 E3 18 17 S 43 40W
Diamantina →, Australia 115 D2 26 45 S 139 10 E
Diamantino, Brazil 157 C6 14 30 S 56 30W
Diamond Harbour, India 81 H13 22 11N 88 14 E
Diamond Is., Australia 114 B5 17 25 S 151 5 E
Diamond Mts., U.S.A. 142 G6 39 50N 115 30W
Diamond Springs, U.S.A. 144 G6 38 42N 120 49W
Diamondville, U.S.A. 142 F8 41 47N 110 32W
Dianbai, China 69 G8 21 33N 111 0 E
Diancheng, China 69 G8 21 30N 111 4 E
Diano Marina, Italy 38 E5 43 54N 8 5 E
Dianópolis, Brazil 155 D2 11 38 S 46 50W
Dianra, Ivory C. 100 D3 8 45N 6 14W
Diapaga, Burkina Faso 101 C5 12 5N 1 46 E
Diapangou, Burkina Faso 101 C5 12 5N 0 10 E
Diapur, Australia 116 D4 36 19 S 141 29 E
Diariguila, Guinea 100 C2 10 35N 10 2W
Dībā, Oman 85 E8 25 45N 56 16 E
Dibaya, Zaïre 103 D4 6 30 S 22 57 E
Dibaya-Lubue, Zaïre 103 C3 4 12 S 19 54 E
Dibbi, Ethiopia 90 G3 4 10N 41 52 E
Dibete, Botswana 104 C4 23 45 S 26 32 E
Dibrugarh, India 78 B5 27 29N 94 55 E
Dickeyville, U.S.A. 140 B6 42 38N 90 36W
Dickinson, U.S.A. 138 B3 46 53N 102 47W
Dickson, Russia 56 B9 73 40N 80 5 E
Dickson, U.S.A. 135 G2 36 5N 87 23W
Dickson City, U.S.A. 137 E9 41 29N 75 40W
Dicle Nehri →, Turkey 89 D9 37 44N 41 10 E
Dicomano, Italy 39 E8 43 53N 11 31 E
Didam, Neths. 20 E8 51 57N 6 8 E
Didesa, W. →, Ethiopia 95 E4 10 2N 35 32 E
Didiéni, Mali 100 C3 13 53N 8 6W
Didsbury, Canada 130 C6 51 35N 114 10W
Didwana, India 80 F6 27 23N 74 36 E
Die, France 25 D9 44 47N 5 22 E
Diébougou, Burkina Faso 100 C4 11 0N 3 15W
Diefenbaker L., Canada 131 C7 51 0N 106 55W
Diego Garcia, Ind. Oc. 109 E6 7 50 S 72 50 E
Diekirch, Lux. 21 J8 49 52N 6 10 E
Diélette, France 22 C5 49 33N 1 52W
Diéma, Mali 100 C3 14 32N 9 12W
Diémbéring, Senegal 100 C1 12 29N 16 47W
Diemen, Neths. 20 D5 52 21N 4 58 E
Dien Ban, Vietnam 76 E7 15 53N 108 16 E
Dien Bien, Vietnam 76 B4 21 20N 103 0 E
Dien Khanh, Vietnam 77 F7 12 15N 109 6 E
Diepenbeek, Belgium 21 G6 50 54N 5 25 E
Diepenheim, Neths. 20 D9 52 12N 6 33 E
Diepenveen, Neths. 20 D8 52 18N 6 9 E
Diepholz, Germany 26 C4 52 37N 8 22 E
Diepoldsau, Switz. 29 B9 47 23N 9 40 E
Dieppe, France 22 C8 49 54N 1 4 E
Dieren, Neths. 20 D8 52 3N 6 6 E
Dierks, U.S.A. 139 H7 34 7N 94 1W
Diessen, Neths. 21 F6 51 29N 5 10 E
Diessenhofen, Switz. 29 A7 47 42N 8 46 E
Diest, Belgium 21 G6 50 58N 5 4 E
Dieterich, U.S.A. 141 E8 39 4N 88 23W
Dietikon, Switz. 29 B6 47 24N 8 24 E
Dieulefit, France 25 D9 44 32N 5 4 E
Dieuze, France 23 D13 48 49N 6 43 E
Diever, Neths. 20 C8 52 51N 6 19 E
Differdange, Lux. 21 J7 49 31N 5 54 E
Diffun, Phil. 70 C3 16 36N 121 33 E
Dig, India 80 F7 27 28N 77 20 E
Digba, Zaïre 106 B2 4 25N 25 48 E
Digboi, India 78 B5 27 23N 95 38 E
Digby, Canada 129 D6 44 38N 65 50W
Digges, Canada 131 B10 58 40N 94 0W
Digges Is., Canada 127 B12 62 40N 77 50W
Dighinala, Bangla. 78 D4 23 15N 92 5 E
Dighton, U.S.A. 138 F4 38 29N 100 28W
Diglur, India 82 E3 18 34N 77 33 E
Digne-les-Bains, France 25 D10 44 5N 6 12 E
Digoin, France 24 B7 46 29N 3 58 E
Digor, Turkey 89 B10 40 22N 43 25 E
Digos, Phil. 71 H5 6 45N 125 20 E
Digranes, Iceland 12 C6 66 4N 14 44W
Digras, India 82 D3 20 6N 77 45 E
Digul →, Indonesia 73 C5 7 7 S 138 42 E
Dīhōk, Iraq 89 D10 36 50N 43 1 E
Dijlah, Nahr →, Asia 84 D5 31 0N 47 25 E
Dijle →, Belgium 21 G5 50 58N 4 41 E
Dijon, France 23 E12 47 20N 5 3 E
Dikala, Sudan 95 G3 4 45N 31 28 E
Dikimdya, Russia 57 D13 59 1N 121 47 E
Dikkil, Djibouti 95 E5 11 8N 42 20 E
Dikomu di Kai, Botswana 104 C3 24 58 S 24 36 E
Diksmuide, Belgium 21 F1 51 2N 2 52 E
Dikson = Dickson, Russia 56 B9 73 40N 80 5 E
Dikwa, Nigeria 101 C7 12 4N 13 30 E
Dila, Ethiopia 95 F4 6 21N 38 22 E
Dilbeek, Belgium 21 G4 50 51N 4 17 E
Dili, Indonesia 72 C3 8 39 S 125 34 E
Dilijan, Armenia 53 K7 40 46N 44 57 E
Dilizhan = Dilijan, Armenia 53 K7 40 46N 44 57 E
Dilj, Croatia 42 B3 45 29N 18 1 E
Dillard, U.S.A. 140 G5 37 44N 91 13W
Dillenburg, Germany 26 E4 50 43N 8 17 E
Dilley, U.S.A. 139 L5 28 40N 99 10W
Dilling, Sudan 95 E2 12 3N 29 35 E
Dillingen, Germany 27 G6 48 21N 10 30 E
Dillingham, U.S.A. 126 C4 59 3N 158 28W
Dillon, Canada 131 B7 55 56N 108 35W
Dillon, Mont., U.S.A. 142 D7 45 13N 112 38W
Dillon, S.C., U.S.A. 135 H6 34 25N 79 22W
Dillon →, Canada 131 B7 55 56N 108 56W
Dillsboro, U.S.A. 141 E11 39 1N 85 4W
Dilolo, Zaïre 103 E4 10 28 S 22 18 E
Dilsen, Belgium 21 F7 51 2N 5 44 E
Dilston, Australia 114 G4 41 22 S 147 10 E
Dimapur, India 78 C4 25 54N 93 45 E
Dimas, Mexico 146 C3 23 43N 106 47W
Dimasalang, Phil. 70 E4 12 12N 123 51 E
Dimashq, Syria 91 B5 33 30N 36 18 E
Dimashq □, Syria 91 B5 33 30N 36 30 E
Dimbaza, S. Africa 105 E4 32 50 S 27 14 E
Dimbelenge, Zaïre 103 D4 5 33 S 23 7 E
Dimbokro, Ivory C. 100 D4 6 45N 4 46W
Dimboola, Australia 116 D5 36 28 S 142 7 E
Dîmboviţa □, Romania 46 E6 45 0N 25 30 E
Dîmboviţa →, Romania 46 E7 44 5N 26 35 E
Dîmbovnic →, Romania 46 E6 44 28N 25 18 E
Dimbulah, Australia 114 B4 17 8 S 145 4 E
Dimitrovgrad, Bulgaria 43 E10 42 5N 25 35 E
Dimitrovgrad, Russia 52 C9 54 14N 49 39 E
Dimitrovgrad, Serbia, Yug. 42 D7 43 2N 22 48 E
Dimitrovo = Pernik, Bulgaria 42 E8 42 35N 23 2 E
Dimmitt, U.S.A. 139 H3 34 33N 102 19W
Dimo, Sudan 95 F2 5 19N 29 10 E
Dimona, Israel 91 D4 31 2N 35 1 E
Dimovo, Bulgaria 42 D7 43 43N 22 50 E
Dinagat, Phil. 71 F5 10 10N 125 40 E
Dinaig, Phil. 71 H5 7 11N 124 10 E
Dinajpur, Bangla. 78 C2 25 33N 88 43 E
Dinalupihan, Phil. 70 D3 14 52N 120 28 E
Dinan, France 22 D4 48 28N 2 2W
Dīnān Āb, Iran 85 C8 32 4N 56 49 E
Dinant, Belgium 21 H5 50 16N 4 55 E
Dinapur, India 81 G11 25 38N 85 5 E
Dinar, Turkey 88 C4 38 5N 30 10 E
Dīnār, Kūh-e, Iran 85 D6 30 42N 51 46 E
Dinara Planina, Croatia 39 E13 44 0N 16 30 E
Dinard, France 22 D4 48 38N 2 6W
Dinaric Alps = Dinara Planina, Croatia 39 E13 44 0N 16 30 E
Dinas, Phil. 71 H4 7 38N 123 20 E
Dinder, Nahr ed →, Sudan 95 E3 14 6N 33 40 E
Dindi →, India 83 F4 16 24N 78 15 E
Dindigul, India 83 J3 10 25N 78 0 E
Ding Xian, China 66 E8 38 30N 114 59 E
Dingalan, Phil. 70 D3 15 18N 121 25 E
Dingalan Bay, Phil. 70 D3 15 18N 121 25 E
Dingbian, China 66 F4 37 35N 107 32 E
Dingelstädt, Germany 26 D6 51 18N 10 19 E
Dinghai, China 69 B14 30 1N 122 6 E
Dingle, Ireland 19 D1 52 9N 10 17W
Dingle B., Ireland 19 D1 52 3N 10 20W
Dingmans Ferry, U.S.A. 137 E10 41 13N 74 55W
Dingnan, China 69 E10 24 45N 115 0 E
Dingo, Australia 114 C4 23 38 S 149 19 E
Dingolfing, Germany 27 G8 48 37N 12 30 E
Dingras, Phil. 70 B3 18 6N 120 42 E
Dingtao, China 66 G8 35 5N 115 35 E
Dinguiraye, Guinea 100 C2 11 18N 10 49W
Dingwall, U.K. 18 D4 57 36N 4 26W
Dingxi, China 66 G3 35 30N 104 33 E
Dingxiang, China 66 E7 38 30N 112 58 E
Dingyuan, China 69 A11 32 32N 117 41 E
Dinh, Mui, Vietnam 77 G7 11 22N 109 1 E
Dinh Lap, Vietnam 76 B6 21 33N 107 6 E
Dinhata, India 78 B2 26 8N 89 27 E
Dinkel →, Neths. 20 D9 52 30N 6 58 E
Dinokwe, Botswana 104 C4 23 29 S 26 37 E
Dinosaur National Monument, U.S.A. 142 F9 40 30N 108 45W
Dinslaken, Germany 21 E9 51 33N 6 44 E
Dinsor, Somali Rep. 108 D2 2 24N 42 59 E
Dintel →, Neths. 21 E4 51 39N 4 22 E
Dinteloord, Neths. 21 E4 51 38N 4 22 E
Dinuba, U.S.A. 144 J7 36 32N 119 23W
Dinxperlo, Neths. 20 E9 51 52N 6 30 E
Diósgyör, Hungary 31 C13 48 7N 20 43 E
Diosig, Romania 46 B3 47 18N 22 2 E
Diourbel, Senegal 100 C1 14 39N 16 12W
Dipaculao, Phil. 70 D3 15 51N 121 32 E
Diphu Pass, India 78 A6 28 9N 97 20 E
Diplo, Pakistan 80 G3 24 35N 69 35 E
Dipolog, Phil. 71 G4 8 36N 123 20 E
Dipşa, Romania 46 C5 46 58N 24 27 E
Dipton, N.Z. 119 F3 45 54 S 168 22 E
Dir, Pakistan 79 B3 35 8N 71 59 E
Diré, Mali 100 B4 16 20N 3 25W

Dzhetym, Khrebet 193

Dover, Ohio, U.S.A. 136 F3 40 32N 81 29W
Dover, Pt., Australia 113 F4 32 32 S 125 32 E
Dover, Str. of, Europe 22 B8 51 0N 1 30 E
Dover-Foxcroft, U.S.A. 129 C6 45 11N 69 13W
Dover Plains, U.S.A. 137 E11 41 43N 73 35W
Dovey → = Dyfi →, U.K. 17 E4 52 32N 4 3W
Dovrefjell, Norway 14 B3 62 15N 9 33 E
Dow Rūd, Iran 85 C6 33 28N 49 4 E
Dowa, Malawi 107 E3 13 38 S 33 58 E
Dowagiac, U.S.A. 141 C10 41 59N 86 6W
Dowgha'i, Iran 85 B8 36 54N 58 32 E
Dowlat Yār, Afghan. 79 B2 34 30N 65 45 E
Dowlatābād, Farāh, Afghan. 79 B1 32 47N 62 40 E
Dowlatābād, Fāryāb, Afghan. 79 A2 36 26N 64 55 E
Dowlatābād, Iran 85 D8 28 20N 56 40 E
Down □, U.K. 19 B6 54 23N 6 2W
Downers Grove, U.S.A. 141 C8 41 48N 88 1W
Downey, Calif., U.S.A. 145 M8 33 56N 118 7W
Downey, Idaho, U.S.A. 142 E7 42 26N 112 7W
Downham Market, U.K. 17 E8 52 37N 0 23 E
Downieville, U.S.A. 144 F6 39 34N 120 50W
Downing, U.S.A. 140 D4 40 29N 92 22W
Downpatrick, U.K. 19 B6 54 20N 5 43W
Downpatrick Hd., Ireland 19 B2 54 20N 9 21W
Dowsārī, Iran 85 D8 28 25N 57 59 E
Dowshī, Afghan. 79 B3 35 35N 68 43 E
Doyle, U.S.A. 144 E6 40 2N 120 6W
Doylestown, U.S.A. 137 F9 40 21N 75 10W
Draa, C., Morocco 98 C2 28 47N 11 0W
Draa, Oued →, Morocco 98 C2 28 40N 11 10W
Drac →, France 25 C9 45 12N 5 42 E
Drachten, Neths. 20 B8 53 7N 6 5 E
Drăgăneşti, Romania 46 E5 44 9N 24 32 E
Drăgăneşti-Vlaşca, Romania 46 E6 44 5N 25 33 E
Dragaš, Serbia, Yug. 42 E5 42 5N 20 35 E
Drăgăşani, Romania 46 E5 44 39N 24 17 E
Dragichyn, Belarus 51 F3 52 15N 25 8 E
Dragina, Serbia, Yug. 42 C4 44 30N 19 25 E
Dragocvet, Serbia, Yug. 42 D6 43 58N 21 15 E
Dragoman, Prokhod, Bulgaria 42 E7 42 58N 22 53 E
Dragonera, I., Spain 33 B9 39 35N 2 19 E
Dragovishtitsa, Bulgaria 42 E7 42 22N 22 39 E
Draguignan, France 25 E10 43 32N 6 27 E
Drain, U.S.A. 142 E2 43 40N 123 19W
Drake, Australia 115 D5 28 55 S 152 25 E
Drake, U.S.A. 138 B4 47 55N 100 23W
Drake Passage, S. Ocean 7 B17 58 0 S 68 0W
Drakensberg, S. Africa 105 E4 31 0 S 28 0 E
Dráma, Greece 44 C6 41 9N 24 10 E
Dráma □, Greece 44 C6 41 20N 24 0 E
Drammen, Norway 14 E4 59 42N 10 12 E
Drangajökull, Iceland 12 C2 66 9N 22 15W
Drangedal, Norway 14 E3 59 6N 9 3 E
Dranov, Ostrov, Romania 46 E10 44 55N 29 30 E
Dras, India 81 B6 34 25N 75 48 E
Drau = Drava →, Croatia 31 F11 45 33N 18 55 E
Drava →, Croatia 31 F11 45 33N 18 55 E
Draveil, France 23 D9 48 41N 2 25 E
Dravograd, Slovenia 39 B12 46 36N 15 5 E
Drawa →, Poland 47 C2 52 52N 15 59 E
Drawno, Poland 47 B2 53 13N 15 46 E
Drawsko Pomorskie, Poland 47 B2 53 35N 15 50 E
Drayton Plains, U.S.A. 141 B13 42 42N 83 23W
Drayton Valley, Canada 130 C6 53 12N 114 58W
Dreibergen, Neths. 20 D6 52 3N 5 17 E
Dren, Serbia, Yug. 42 D5 43 8N 20 44 E
Drenthe □, Neths. 20 C9 52 52N 6 40 E
Drentsche Hoofdvaart, Neths. 20 C8 52 39N 6 4 E
Drepanum, C., Cyprus 32 E11 34 54N 32 19 E
Dresden, Canada 136 D2 42 35N 82 11W
Dresden, Germany 26 D9 51 3N 13 44 E
Dreux, France 22 D8 48 44N 1 23 E
Drexel, U.S.A. 141 E12 39 45N 84 18W
Drezdenko, Poland 47 C2 52 50N 15 49 E
Driel, Neths. 20 E7 51 57N 5 49 E
Driffield, U.K. 16 C7 54 0N 0 26W
Driftwood, U.S.A. 136 E6 41 20N 78 8W
Driggs, U.S.A. 142 E8 43 44N 111 6W
Drin i zi →, Albania 44 C2 41 37N 20 28 E
Drina →, Bos.-H. 42 C4 44 53N 19 21 E
Drincea →, Romania 46 E3 44 20N 22 55 E
Drînceni, Romania 46 C9 46 49N 28 10 E
Drini →, Albania 44 B2 42 1N 19 38 E
Drinjača →, Bos.-H. 42 C4 44 15N 19 8 E
Drissa = Vyerkhnyadzvinsk, Belarus 50 E4 55 45N 27 58 E
Drivstua, Norway 14 B3 62 26N 9 47 E
Drniš, Croatia 39 E13 43 51N 16 10 E
Drøbak, Norway 14 E4 59 39N 10 39 E
Drobin, Poland 47 C6 52 42N 19 58 E
Drochia, Moldova 51 H4 48 2N 27 48 E
Drogheda, Ireland 19 C5 53 43N 6 22W
Drogichin = Dragichyn, Belarus 51 F3 52 15N 25 8 E
Drogobych = Drohobych, Ukraine 51 H2 49 20N 23 30 E
Drohiczyn, Poland 47 C9 52 24N 22 39 E
Drohobych, Ukraine 51 H2 49 20N 23 30 E
Droichead Atha = Drogheda, Ireland 19 C5 53 43N 6 22W
Droichead Nua, Ireland 19 C5 53 11N 6 48W
Droitwich, U.K. 17 E5 52 16N 2 8W
Drôme □, France 25 D9 44 38N 5 15 E
Drôme →, France 25 D8 44 46N 4 46 E
Dromedary, C., Australia 117 D9 36 17 S 150 10 E
Dronero, Italy 38 D4 44 28N 7 22 E
Dronfield, Australia 114 C3 21 12 S 140 3 E
Dronne →, France 24 C3 45 2N 0 9W
Dronninglund, Denmark 15 G4 57 10N 10 19 E
Dronrijp, Neths. 20 B7 53 11N 5 39 E
Dropt →, France 24 D3 44 35N 0 6W
Drosendorf, Austria 30 C8 48 52N 15 37 E
Drouin, Australia 117 E6 38 10 S 145 53 E
Drouzhba, Bulgaria 43 D13 43 15N 28 0 E
Drumbo, Canada 136 C4 43 16N 80 35W
Drumheller, Canada 130 C6 51 25N 112 40W
Drummond, U.S.A. 142 C7 46 40N 113 9W
Drummond I., U.S.A. 128 C3 46 1N 83 39W
Drummond Pt., Australia 115 E2 34 9 S 135 16 E
Drummond Ra., Australia 114 C4 23 45 S 147 10 E
Drummondville, Canada 128 C5 45 55N 72 25W
Drumright, U.S.A. 139 H6 35 59N 96 36W
Drunen, Neths. 21 E6 51 41N 5 8 E
Druskininkai, Lithuania 13 J20 54 3N 23 58 E
Drut →, Belarus 51 F6 53 8N 30 5 E
Druten, Neths. 20 E7 51 53N 5 36 E
Druya, Belarus 50 E4 55 45N 27 28 E
Druzhina, Russia 57 C15 68 14N 145 18 E
Drvar, Bos.-H. 39 D13 44 21N 16 23 E
Drvenik, Croatia 39 E13 43 27N 16 3 E
Drwęca →, Poland 47 C5 53 0N 18 42 E
Dry Tortugas, U.S.A. 148 B3 24 38N 82 55W
Dryanovo, Bulgaria 43 E10 42 59N 25 28 E
Dryden, Canada 131 D10 49 47N 92 50W
Dryden, U.S.A. 139 K3 30 3N 102 7W
Drygalski I., Antarctica 7 C7 66 0 S 92 0 E
Drysdale →, Australia 112 B4 13 59 S 126 51 E
Drysdale I., Australia 114 A2 11 41 S 136 0 E
Drzewiczka →, Poland 47 D7 51 36N 20 36 E
Dschang, Cameroon 101 D7 5 32N 10 3 E
Du Bois, U.S.A. 136 E6 41 8N 78 46W
Du Quoin, U.S.A. 140 F7 38 1N 89 14W
Duanesburg, U.S.A. 137 D10 42 45N 74 11W
Duaringa, Australia 114 C4 23 42 S 149 42 E
Ḑubā, Si. Arabia 84 E2 27 10N 35 40 E
Dubai = Dubayy, U.A.E. 85 E7 25 18N 55 20 E
Dubăsari, Moldova 51 J5 47 15N 29 10 E
Dubăsari Vdkhr., Moldova 51 J5 47 30N 29 0 E
Dubawnt →, Canada 131 A8 64 33N 100 6W
Dubawnt, L., Canada 131 A8 63 4N 101 42W
Dubayy, U.A.E. 85 E7 25 18N 55 20 E
Dubbeldam, Neths. 20 E5 51 47N 4 43 E
Dubbo, Australia 117 B8 32 11 S 148 35 E
Dubele, Zaïre 106 B2 2 56N 29 35 E
Dübendorf, Switz. 29 B7 47 24N 8 37 E
Dubica, Croatia 39 C13 45 11N 16 48 E
Dublin, Ireland 19 C5 53 20N 6 15W
Dublin, Ga., U.S.A. 135 J4 32 32N 82 54W
Dublin, Tex., U.S.A. 139 J5 32 5N 98 21W
Dublin □, Ireland 19 C5 53 24N 6 20W
Dublin B., Ireland 19 C5 53 18N 6 5W
Dubna, Russia 52 B3 56 44N 37 10 E
Dubno, Ukraine 51 G3 50 25N 25 45 E
Dubois, Idaho, U.S.A. 142 D7 44 10N 112 14W
Dubois, Ind., U.S.A. 141 F10 38 27N 86 48W
Dubossary = Dubăsari, Moldova 51 J5 47 15N 29 10 E
Dubossary Vdkhr. = Dubăsari Vdkhr., Moldova 51 J5 47 30N 29 0 E
Dubovka, Russia 53 F7 49 5N 44 50 E
Dubovskoye, Russia 53 G6 47 28N 42 46 E
Dubrajpur, India 81 H12 23 48N 87 25 E
Dubréka, Guinea 100 D2 9 46N 13 31W
Dubrovitsa = Dubrovytsya, Ukraine 51 G4 51 31N 26 35 E
Dubrovnik, Croatia 42 E3 42 39N 18 6 E
Dubrovskoye, Russia 57 D12 58 55N 111 10 E
Dubrovytsya, Ukraine 51 G4 51 31N 26 35 E
Dubulu, Zaïre 102 B4 4 18N 20 16 E
Dubuque, U.S.A. 140 B6 42 30N 90 41W
Duchang, China 69 C11 29 18N 116 12 E
Duchesne, U.S.A. 142 F8 40 10N 110 24W
Duchess, Australia 114 C2 21 20 S 139 50 E
Ducie I., Pac. Oc. 123 K15 24 40 S 124 48W
Duck Cr. →, Australia 112 D2 22 37 S 116 53 E
Duck Lake, Canada 131 C7 52 50N 106 16W
Duck Mountain Prov. Park, Canada 131 C8 51 45N 101 0W
Duckwall, Mt., U.S.A. 144 H6 37 58N 120 7W
Düdelange, Lux. 21 K8 49 29N 6 5 E
Duderstadt, Germany 26 D6 51 31N 10 15 E
Dudhnai, India 78 C3 25 59N 90 47 E
Düdingen, Switz. 28 C4 46 52N 7 12 E
Dudinka, Russia 57 C9 69 30N 86 13 E
Dudley, U.K. 17 E5 52 31N 2 5W
Dudna →, India 82 E3 19 17N 76 54 E
Dudo, Somali Rep. 108 C4 9 20N 50 12 E
Dudub, Ethiopia 108 C3 6 55N 46 43 E
Duenas, Phil. 71 F4 11 4N 122 37 E
Dueñas, Spain 36 D6 41 52N 4 33W
Dueré, Brazil 155 D2 11 20 S 49 17W
Duero = Douro →, Europe 36 D2 41 8N 8 40W
Dūfah, W. →, Si. Arabia 86 C3 18 45N 41 49 E
Duffel, Belgium 21 F5 51 6N 4 30 E
Dufftown, U.K. 18 D5 57 27N 3 8W
Dufourspitz, Switz. 28 E5 45 56N 7 52 E
Dugger, U.S.A. 141 E9 39 4N 87 18W
Dugi Otok, Croatia 39 E12 44 0N 15 3 E
Dugiuma, Somali Rep. 108 D2 1 15N 42 34 E
Dugo Selo, Croatia 39 C13 45 51N 16 18 E
Duifken Pt., Australia 114 A3 12 33 S 141 38 E
Duisburg, Germany 26 D2 51 26N 6 45 E
Duitama, Colombia 152 B3 5 50N 73 2W
Duiveland, Neths. 21 E4 51 38N 4 0 E
Duiwelskloof, S. Africa 105 C5 23 42 S 30 10 E
Dukati, Albania 44 D1 40 16N 19 32 E
Dūkdamīn, Iran 85 C8 35 59N 57 43 E
Duke I., U.S.A. 130 C2 54 50N 131 20W
Dukelský Průsmyk, Slovak Rep. 31 B14 49 25N 21 42 E
Dukhān, Qatar 85 E6 25 25N 50 50 E
Dukhovshchina, Russia 50 E7 55 15N 32 27 E
Duki, Pakistan 79 C3 30 14N 68 25 E
Dukla, Poland 31 B14 49 30N 21 35 E
Duku, Bauchi, Nigeria 101 C7 10 43N 10 43 E
Duku, Sokoto, Nigeria 101 C5 11 11N 4 55 E
Dulag, Phil. 71 F5 10 57N 125 2 E
Dulce →, Argentina 158 C3 30 32 S 62 33W
Dulce, G., Costa Rica 148 E3 8 40N 83 20W
Dulf, Iraq 84 C5 35 7N 45 51 E
Dŭlgopol, Bulgaria 43 D12 43 3N 27 22 E
Duliu, China 66 E9 39 2N 116 55 E
Dullewala, Pakistan 80 D4 31 50N 71 25 E
Dülmen, Germany 26 D3 51 49N 7 17 E
Dulovo, Bulgaria 43 D12 43 48N 27 9 E
Dulq Maghār, Syria 89 D8 36 22N 38 39 E
Dululu, Australia 114 C5 23 48 S 150 15 E
Duluth, U.S.A. 138 B8 46 47N 92 6W
Dum Dum, India 81 H13 22 39N 88 33 E
Dum Hadjer, Chad 97 F3 13 18N 19 41 E
Dūmā, Lebanon 91 A4 34 12N 35 50 E
Dūmā, Syria 91 B5 33 34N 36 24 E
Dumaguete, Phil. 71 G4 9 17N 123 15 E
Dumai, Indonesia 74 B2 1 35N 101 28 E
Dumalinao, Phil. 71 H4 7 49N 123 23 E
Dumanguilas Bay, Phil. 71 H4 7 34N 123 4 E
Dumaran, Phil. 71 F2 10 33N 119 50 E
Dumas, Ark., U.S.A. 139 J9 33 53N 91 29W
Dumas, Tex., U.S.A. 139 H4 35 52N 101 58W
Dumbarton, U.K. 18 F4 55 57N 4 33W
Dumbea, N. Cal. 121 V20 22 10 S 166 27 E
Dumbleyung, Australia 113 F2 33 17 S 117 42 E
Dumbo, Angola 103 E3 14 6 S 17 24 E
Dumbrăveni, Romania 46 C5 46 14N 24 34 E
Dumfries, U.K. 18 F5 55 4N 3 37W
Dumfries & Galloway □, U.K. 18 F5 55 9N 3 58W
Dumingag, Phil. 71 G4 8 20N 123 20 E
Dumka, India 81 G12 24 12N 87 15 E
Dümmer See, Germany 26 C4 52 31N 8 20 E
Dumoine →, Canada 128 C4 46 13N 77 51W
Dumoine L., Canada 128 C4 46 55N 77 55W
Dumraon, India 81 G11 25 33N 84 8 E
Dumyât, Egypt 94 H7 31 24N 31 48 E
Dumyât, Masabb, Egypt 94 H7 31 28N 31 51 E
Dún Dealgan = Dundalk, Ireland 19 B5 54 1N 6 24W
Dun Laoghaire, Ireland 19 C5 53 17N 6 8W
Dun-le-Palestel, France 24 B5 46 18N 1 39 E
Dun-sur-Auron, France 23 F9 46 53N 2 33 E
Duna = Dunărea →, Europe 51 K5 45 20N 29 40 E
Duna →, Hungary 31 F11 45 51N 18 48 E
Dunaföldvár, Hungary 31 E11 46 50N 18 57 E
Dunaj = Dunărea →, Europe 51 K5 45 20N 29 40 E
Dunaj →, Slovak Rep. 31 D11 47 50N 18 50 E
Dunajec →, Poland 31 A13 50 15N 20 44 E
Dunajská Streda, Slovak Rep. 31 D10 48 0N 17 37 E
Dunapatai, Hungary 31 E12 46 39N 19 4 E
Dunărea →, Europe 51 K5 45 20N 29 40 E
Dunaszekcsö, Hungary 31 E11 46 6N 18 45 E
Dunaújváros, Hungary 31 E11 47 0N 18 57 E
Dunav = Dunărea →, Europe 51 K5 45 20N 29 40 E
Dunav →, Serbia, Yug. 42 C6 44 47N 20 20 E
Dunavtsi, Bulgaria 42 D7 43 57N 22 53 E
Dunay, Russia 60 C6 42 52N 132 22 E
Dunback, N.Z. 119 F5 45 23 S 170 36 E
Dunbar, Australia 114 B3 16 0 S 142 22 E
Dunbar, U.K. 18 E6 56 0N 2 31W
Dunblane, U.K. 18 E5 56 11N 3 58W
Duncan, Canada 130 D4 48 45N 123 40W
Duncan, Ariz., U.S.A. 143 K9 32 43N 109 6W
Duncan, Okla., U.S.A. 139 H6 34 30N 97 57W
Duncan, L., Canada 128 B4 53 29N 77 58W
Duncan L., Canada 130 A6 62 51N 113 58W
Duncan Town, Bahamas 148 B4 22 15N 75 45W
Duncannon, U.S.A. 136 F7 40 23N 77 2W
Dundalk, Canada 136 B4 44 10N 80 24W
Dundalk, Ireland 19 B5 54 1N 6 24W
Dundalk Bay, Ireland 19 C5 53 55N 6 15W
Dundas, Canada 128 D4 43 17N 79 59W
Dundas, L., Australia 113 F3 32 35 S 121 50 E
Dundas I., Canada 130 C2 54 30N 130 50W
Dundas Str., Australia 112 B5 11 15 S 131 35 E
Dundee, S. Africa 105 D5 28 11 S 30 15 E
Dundee, U.K. 18 E6 56 28N 2 59W
Dundee, U.S.A. 141 C13 41 57N 83 40W
Dundgovĭ □, Mongolia 66 B4 45 10N 106 0 E
Dundoo, Australia 115 D3 27 40 S 144 37 E
Dundrum, U.K. 19 B6 54 16N 5 52W
Dundrum B., U.K. 19 B6 54 13N 5 47W
Dundwara, India 81 F8 27 48N 79 9 E
Dunedin, N.Z. 119 F5 45 50 S 170 33 E
Dunedin, U.S.A. 135 L4 28 1N 82 47W
Dunedin →, Canada 130 B4 59 30N 124 5W
Dunfermline, U.K. 18 E5 56 5N 3 27W
Dungannon, Canada 136 C3 43 51N 81 36W
Dungannon, U.K. 19 B5 54 31N 6 46W
Dungannon □, U.K. 19 B5 54 30N 6 55W
Dungarpur, India 80 H5 23 52N 73 45 E
Dungarvan, Ireland 19 D4 52 5N 7 37W
Dungarvan Harbour, Ireland 19 D4 52 4N 7 35W
Dungeness, U.K. 17 G8 50 54N 0 59 E
Dungo, L. do, Angola 103 F3 17 15 S 19 0 E
Dungog, Australia 117 B9 32 22 S 151 46 E
Dungu, Zaïre 106 B2 3 40N 28 32 E
Dungunâb, Sudan 94 C4 21 10N 37 9 E
Dungunâb, Khalij, Sudan 94 C4 21 5N 37 12 E
Dunhinda Falls, Sri Lanka 83 L5 7 5N 81 6 E
Dunhua, China 67 C15 43 20N 128 14 E
Dunhuang, China 64 B4 40 8N 94 36 E
Dunières, France 25 C8 45 13N 4 20 E
Dunk I., Australia 114 B4 17 59 S 146 29 E
Dunkeld, Australia 116 D5 37 40 S 142 22 E
Dunkeld, U.K. 18 E5 56 34N 3 35W
Dunkerque, France 23 A9 51 2N 2 20 E
Dunkery Beacon, U.K. 17 F4 51 9N 3 36W
Dunkirk = Dunkerque, France 23 A9 51 2N 2 20 E
Dunkirk, U.S.A. 136 D5 42 29N 79 20W
Dunkuj, Sudan 95 E3 12 50N 32 49 E
Dunkwa, Central, Ghana 100 D4 6 0N 1 47W
Dunkwa, Central, Ghana 101 D4 5 30N 1 0W
Dunlap, U.S.A. 138 E7 41 51N 95 36W
Dúnleary = Dun Laoghaire, Ireland 19 C5 53 17N 6 8W
Dunmanus B., Ireland 19 E2 51 31N 9 50W
Dunmara, Australia 114 B1 16 42 S 133 25 E
Dunmore, U.S.A. 137 E9 41 25N 75 38W
Dunmore Hd., Ireland 19 D1 52 10N 10 35W
Dunmore Town, Bahamas 148 A4 25 30N 76 39W
Dunn, U.S.A. 135 H6 35 19N 78 37W
Dunnellon, U.S.A. 135 L4 29 3N 82 28W
Dunnet Hd., U.K. 18 C5 58 40N 3 21W
Dunning, U.S.A. 138 E4 41 50N 100 6W
Dunnville, Canada 136 D5 42 54N 79 36W
Dunolly, Australia 116 D5 36 51 S 143 44 E
Dunoon, U.K. 18 F4 55 57N 4 56W
Dunqul, Egypt 94 C3 23 26N 31 37 E
Duns, U.K. 18 F6 55 47N 2 20W
Dunseith, U.S.A. 138 A4 48 50N 100 3W
Dunsmuir, U.S.A. 142 F2 41 13N 122 16W
Dunstable, U.K. 17 F7 51 53N 0 32W
Dunstan Mts., N.Z. 119 E4 44 53 S 169 35 E
Dunster, Canada 130 C5 53 8N 119 50W
Duntroon, N.Z. 119 E5 44 51 S 170 40 E
Dunvegan L., Canada 131 A7 60 8N 107 10W
Duolun, China 66 C9 42 12N 116 28 E
Duong Dong, Vietnam 77 G4 10 13N 103 58 E
Dupax, Phil. 70 C3 16 17N 121 5 E
Dupree, U.S.A. 138 C4 45 4N 101 35W
Dupuyer, U.S.A. 142 B7 48 13N 112 30W
Duqm, Oman 87 C7 19 39N 57 42 E
Duque de Caxias, Brazil 155 F3 22 45 S 43 19W
Duque de York, I., Chile 160 D1 50 37 S 75 25W
Durack →, Australia 112 C4 15 33 S 127 52 E
Durack Ra., Australia 112 C4 16 50 S 127 40 E
Durağan, Turkey 88 B6 41 25N 35 3 E
Durance →, France 25 E8 43 55N 4 45 E
Durand, Ill., U.S.A. 140 B7 42 26N 89 20W
Durand, Mich., U.S.A. 141 B13 42 55N 83 59W
Durango = Victoria de Durango, Mexico 146 C4 24 3N 104 39W
Durango, Spain 34 B2 43 13N 2 40W
Durango, U.S.A. 143 H10 37 16N 107 53W
Durango □, Mexico 146 C4 25 0N 105 0W
Duranillin, Australia 113 F2 33 30 S 116 45 E
Durant, Iowa, U.S.A. 140 C6 41 36N 90 54W
Durant, Okla., U.S.A. 139 J6 33 59N 96 25W
Duratón →, Spain 36 D6 41 37N 4 7W
Durazno, Uruguay 158 C4 33 25 S 56 31W
Durazzo = Durrësi, Albania 44 C1 41 19N 19 28 E
Durban, France 24 F6 42 59N 2 49 E
Durban, S. Africa 105 D5 29 49 S 31 1 E
Durbo, Somali Rep. 108 B4 11 37N 50 20 E
Dúrcal, Spain 37 J7 37 0N 3 34W
Đurđevac, Croatia 42 A2 46 2N 17 3 E
Düren, Germany 26 E2 50 48N 6 29 E
Durg, India 82 D5 21 15N 81 22 E
Durgapur, India 81 H12 23 30N 87 20 E
Durham, Canada 128 D3 44 10N 80 49W
Durham, U.K. 16 C6 54 47N 1 34W
Durham, Calif., U.S.A. 144 F5 39 39N 121 48W
Durham, N.C., U.S.A. 135 G6 35 59N 78 54W
Durham □, U.K. 16 C6 54 42N 1 45W
Durham Downs, Australia 115 D4 26 6 S 149 5 E
Ḑurmā, Si. Arabia 86 A4 24 37N 46 8 E
Durmitor, Montenegro, Yug. 42 D4 43 10N 19 0 E
Durness, U.K. 18 C4 58 34N 4 45W
Durrësi, Albania 44 C1 41 19N 19 28 E
Durrie, Australia 114 D3 25 40 S 140 15 E
Dursunbey, Turkey 88 C3 39 35N 28 37 E
Durtal, France 22 E6 47 40N 0 18W
Duru, Zaïre 106 B2 4 14N 28 50 E
D'Urville, Tanjung, Indonesia 73 B5 1 28 S 137 54 E
D'Urville I., N.Z. 119 A8 40 50 S 173 55 E
Duryea, U.S.A. 137 E9 41 20N 75 45W
Dusa Mareb, Somali Rep. 108 C3 5 30N 46 15 E
Dûsh, Egypt 94 C3 24 35N 30 41 E
Dushak, Turkmenistan 56 F7 37 13N 60 1 E
Dushan, China 68 E6 25 48N 107 30 E
Dushanbe, Tajikistan 55 D4 38 33N 68 48 E
Dusheti, Georgia 53 J7 42 10N 44 42 E
Dusky Sd., N.Z. 119 F1 45 47 S 166 30 E
Dussejour, C., Australia 112 B4 14 45 S 128 13 E
Düsseldorf, Germany 26 D2 51 14N 6 47 E
Dussen, Neths. 20 E5 51 44N 4 59 E
Duszniki-Zdrój, Poland 47 E3 50 24N 16 24 E
Dutch Harbor, U.S.A. 126 C3 53 53N 166 32W
Dutlwe, Botswana 104 C3 23 58 S 23 46 E
Dutsan Wai, Nigeria 101 C6 10 50N 8 10 E
Dutton, Canada 136 D3 42 39N 81 30W
Dutton →, Australia 114 C3 20 44 S 143 10 E
Duved, Sweden 14 A6 63 24N 12 55 E
Duvno, Bos.-H. 42 D2 43 42N 17 13 E
Duyun, China 68 D6 26 18N 107 29 E
Düzce, Turkey 88 B4 40 50N 31 10 E
Duzdab = Zāhedān, Iran 85 D9 29 30N 60 50 E
Dve Mogili, Bulgaria 43 D10 43 35N 25 55 E
Dvina, Severnaya →, Russia 48 B7 64 32N 40 30 E
Dvinsk = Daugavpils, Latvia 13 J22 55 53N 26 32 E
Dvinskaya Guba, Russia 48 B6 65 0N 39 0 E
Dvor, Croatia 39 C13 45 4N 16 22 E
Dvorce, Czech. 31 B10 49 50N 17 34 E
Dvur Králové, Czech. 30 A8 50 27N 15 50 E
Dwarka, India 80 H3 22 18N 69 8 E
Dwellingup, Australia 113 F2 32 43 S 116 4 E
Dwight, Canada 136 A5 45 20N 79 1W
Dwight, U.S.A. 141 C8 41 5N 88 26W
Dyatkovo, Russia 52 D2 53 40N 34 27 E
Dyatlovo = Dzyatlava, Belarus 50 F3 53 28N 25 28 E
Dyer, U.S.A. 141 G10 37 24N 86 13W
Dyer, C., Canada 127 B13 66 40N 61 0W
Dyer Plateau, Antarctica 7 D17 70 45 S 65 30W
Dyerbeldzhin, Kyrgyzstan 55 C7 41 13N 74 54 E
Dyersburg, U.S.A. 139 G10 36 3N 89 23W
Dyersville, U.S.A. 140 B5 42 29N 91 8W
Dyfed □, U.K. 17 E3 52 0N 4 30W
Dyfi →, U.K. 17 E4 52 32N 4 3W
Dyje →, Czech. 31 C9 48 37N 16 56 E
Dyle →, Belgium 21 G5 50 58N 4 41 E
Dymer, Ukraine 51 G6 50 47N 30 18 E
Dynevor Downs, Australia 115 D3 28 10 S 144 20 E
Dynów, Poland 31 B15 49 50N 22 11 E
Dysart, Canada 131 C8 50 57N 104 2W
Dyurtyuli, Russia 54 D5 55 9N 54 40 E
Dzamin Üüd, Mongolia 66 C6 43 50N 111 58 E
Dzerzhinsk, Russia 52 B6 56 14N 43 30 E
Dzhalal-Abad = Jalal-Abad, Kyrgyzstan 55 C6 40 56N 73 0 E
Dzhalinda, Russia 57 D13 53 26N 124 0 E
Dzhambeyty, Kazakhstan 54 F4 50 16N 52 35 E
Dzhambul = Zhambyl, Kazakhstan 55 B5 42 54N 71 22 E
Dzhambul, Gora = Zhambyl, Gora, Kazakhstan 55 A6 44 54N 73 0 E
Dzhankoy, Ukraine 51 K8 45 40N 34 20 E
Dzhanybek, Kazakhstan 52 F8 49 25N 46 50 E
Dzhardzhan, Russia 57 C13 68 10N 124 10 E
Dzharkurgan = Jarqürghon, Uzbekistan 55 E3 37 31N 67 25 E
Dzharylhach, Ostriv, Ukraine 51 J7 46 2N 32 55 E
Dzhetygara = Zhetiqara, Kazakhstan 56 D7 52 11N 61 12 E
Dzhetym, Khrebet, Kyrgyzstan 55 C8 41 30N 77 0 E

El Gezira □, *Sudan* 95 E3 15 0N 33 0 E
El Gîza, *Egypt* 94 H7 30 0N 31 10 E
El Goléa, *Algeria* 99 B5 30 30N 2 50 E
El Hadeb, *W. Sahara* 98 C2 25 51N 13 0W
El Hadjira, *Algeria* 99 B6 32 36N 5 30 E
El Hagiz, *Sudan* 95 D4 15 15N 35 50 E
El Hajeb, *Morocco* 98 B3 33 43N 5 13W
El Hammam, *Egypt* 94 H6 30 52N 29 25 E
El Hammâmi, *Mauritania* 98 D2 23 3N 11 30W
El Hamurre, *Somali Rep.* 108 C3 7 13N 48 54 E
El Hank, *Mauritania* 98 D3 24 30N 7 0W
El Hasian, *W. Sahara* 98 C2 26 20N 14 0W
El Hawata, *Sudan* 95 E3 13 25N 34 42 E
El Heiz, *Egypt* 94 B2 27 50N 28 40 E
El 'Idisât, *Egypt* 94 B3 25 30N 32 35 E
El Iskandarîya, *Egypt* 94 H6 31 13N 29 58 E
El Jadida, *Morocco* 98 B3 33 11N 8 17W
El Jebelein, *Sudan* 95 E3 12 40N 32 55 E
El Kab, *Sudan* 94 D3 19 27N 32 46 E
El Kabrît, G., *Egypt* 91 F2 29 42N 33 16 E
El Kala, *Algeria* 99 A6 36 50N 8 30 E
El Kalâa, *Morocco* 98 B3 32 4N 7 27W
El Kamlin, *Sudan* 95 D3 15 3N 33 11 E
El Kantara, *Algeria* 99 A6 35 14N 5 45 E
El Kantara, *Tunisia* 96 B2 33 45N 10 58 E
El Karaba, *Sudan* 94 D3 18 32N 33 41 E
El Kef, *Tunisia* 96 A1 36 12N 8 47 E
El Khandaq, *Sudan* 94 D3 18 30N 30 30 E
El Khârga, *Egypt* 94 B3 25 30N 30 33 E
El Khartûm, *Sudan* 95 D3 15 31N 32 35 E
El Khartûm □, *Sudan* 95 D3 16 0N 33 0 E
El Khartûm Bahrî, *Sudan* 95 D3 15 40N 32 31 E
El Khroub, *Algeria* 99 A6 36 10N 6 55 E
El Kseur, *Algeria* 99 A5 36 46N 4 49 E
El Ksiba, *Morocco* 98 B3 32 45N 6 1W
El Kuntilla, *Egypt* 91 E3 30 1N 34 45 E
El Laqâwa, *Sudan* 95 E2 11 25N 29 1 E
El Laqeita, *Egypt* 94 B3 25 50N 33 15 E
El Leiya, *Sudan* 95 D4 16 15N 35 28 E
El Mafâza, *Sudan* 95 E3 13 38N 34 30 E
El Mahalla el Kubra,
Egypt 94 H7 31 0N 31 0 E
El Mahârîq, *Egypt* 94 B3 25 35N 30 35 E
El Mahmûdîya, *Egypt* 94 H7 31 10N 30 32 E
El Maitén, *Argentina* 160 B2 42 3 S 71 10W
El Maiz, *Algeria* 99 C4 28 19N 0 9W
El-Maks el-Bahari, *Egypt* 94 C3 24 30N 30 40 E
El Manshâh, *Egypt* 94 B3 26 26N 31 50 E
El Mansour, *Algeria* 99 C4 27 47N 0 14W
El Mansûra, *Egypt* 94 H7 31 0N 31 19 E
El Mantico, *Venezuela* 153 B5 7 38N 62 45W
El Manzala, *Egypt* 94 H7 31 10N 31 50 E
El Marâgha, *Egypt* 94 B3 26 35N 31 10 E
El Masid, *Sudan* 95 D3 15 15N 33 0 E
El Matariya, *Egypt* 94 H8 31 15N 32 0 E
El Medano, *Canary Is.* 33 F3 28 3N 16 32W
El Meghaier, *Algeria* 99 B6 33 55N 5 58 E
El Meraguen, *Algeria* 99 C4 28 0N 0 7W
El Metemma, *Sudan* 95 D3 16 50N 33 10 E
El Miamo, *Venezuela* 153 B5 7 39N 61 46W
El Milagro, *Argentina* 158 C2 30 59 S 65 59W
El Milia, *Algeria* 99 A6 36 51N 6 13 E
El Minyâ, *Egypt* 94 J7 28 7N 30 33 E
El Molar, *Spain* 34 E1 40 42N 3 45W
El Mreyye, *Mauritania* 100 B3 18 0N 6 0W
El Nido, *Phil.* 71 F2 11 10N 119 25 E
El Obeid, *Sudan* 95 E3 13 8N 30 10 E
El Odaiya, *Sudan* 95 E2 12 8N 28 12 E
El Oro, *Mexico* 147 D4 19 48N 100 8W
El Oro □, *Ecuador* 152 D2 3 30 S 79 50W
El Oued, *Algeria* 99 B6 33 20N 6 58 E
El Palmar, *Bolivia* 157 D5 17 50 S 63 9W
El Palmar, *Venezuela* 153 B5 7 58N 61 53W
El Palmito, Presa, *Mexico* 146 B3 25 40N 105 30W
El Panadés, *Spain* 34 D6 41 10N 1 30 E
El Pardo, *Spain* 36 E7 40 31N 3 47W
El Paso, *Ill., U.S.A.* 140 D7 40 44N 89 1W
El Paso, *Tex., U.S.A.* 143 L10 31 45N 106 29W
El Paso Robles, *U.S.A.* 144 K6 35 38N 120 41W
El Pedernoso, *Spain* 35 F2 39 29N 2 45W
El Pedroso, *Spain* 37 H5 37 51N 5 45W
El Pobo de Dueñas, *Spain* 34 E3 40 46N 1 39W
El Portal, *U.S.A.* 144 H7 37 41N 119 47W
El Porvenir, *Mexico* 146 A3 31 15N 105 51W
El Prat de Llobregat,
Spain 34 D7 41 18N 2 3 E
El Progreso, *Honduras* 148 C2 15 26N 87 51W
El Provencío, *Spain* 35 F2 39 23N 2 35W
El Pueblito, *Mexico* 146 B3 29 3N 105 4W
El Pueblo, *Canary Is.* 33 F2 28 36N 17 47W
El Puerto de Santa María,
Spain 37 J4 36 36N 6 13W
El Qâhira, *Egypt* 94 H7 30 1N 31 14 E
El Qantara, *Egypt* 91 E1 30 51N 32 20 E
El Qasr, *Egypt* 94 B2 25 44N 28 42 E
El Quseima, *Egypt* 91 E3 30 40N 34 15 E
El Qusîya, *Egypt* 94 B3 27 29N 30 44 E
El Râshda, *Egypt* 94 B2 25 36N 28 57 E
El Reno, *U.S.A.* 139 H6 35 32N 97 57W
El Ribero, *Spain* 36 C2 42 30N 8 30W
El Rîdisiya, *Egypt* 94 C3 24 56N 32 51 E
El Rio, *U.S.A.* 145 L7 34 14N 119 10W
El Ronquillo, *Spain* 37 H4 37 44N 6 10W
El Roque, Pta., *Canary Is.* 33 F4 28 10N 15 25W
El Rosarito, *Mexico* 146 B2 28 38N 114 4W
El Rubio, *Spain* 37 H5 37 22N 5 0W
El Saff, *Egypt* 94 J7 29 34N 31 16 E
El Saheira, W. —►, *Egypt* 91 E2 30 5N 33 25 E
El Salto, *Mexico* 146 C3 23 47N 105 22W
El Salvador ■,
Cent. Amer. 148 D2 13 50N 89 0W
El Sancejo, *Spain* 37 H5 37 4N 5 6W
El Sauce, *Nic.* 148 D2 13 0N 86 40W
El Shallal, *Egypt* 94 C3 24 0N 32 53 E
El Simbillawein, *Egypt* 94 H7 30 48N 31 13 E
El Sombrero, *Venezuela* 152 B4 9 23N 67 3W
El Suweis, *Egypt* 94 J8 29 58N 32 31 E
El Tamarâni, W. —►,
Egypt 91 E3 30 7N 34 43 E
El Thamad, *Egypt* 91 F3 29 40N 34 28 E
El Tigre, *Venezuela* 153 B5 8 44N 64 15W
El Tîh, G., *Egypt* 91 F2 29 40N 33 50 E
El Tîna, Khalîg, *Egypt* 91 D1 31 10N 32 40 E
El Tocuyo, *Venezuela* 152 B4 9 47N 69 48W
El Tofo, *Chile* 158 B1 29 22 S 71 18W
El Tránsito, *Chile* 158 B1 28 52 S 70 17W
El Tûr, *Egypt* 94 J8 28 14N 33 36 E
El Turbio, *Argentina* 160 D2 51 45 S 72 5W
El Uinle, *Somali Rep.* 108 D2 3 4N 41 42 E
El Uqsur, *Egypt* 94 B3 25 41N 32 38 E
El Vado, *Spain* 34 D1 41 2N 3 18W
El Vallés, *Spain* 34 D7 41 35N 2 20 E
El Venado, *Mexico* 146 C4 22 56N 101 10W
El Vigía, *Venezuela* 152 B3 8 38N 71 39W
El Wabeira, *Egypt* 91 F2 29 34N 33 6 E
El Wak, *Kenya* 106 B5 2 49N 40 56 E
El Wak, *Somali Rep.* 108 D2 2 44N 41 1 E
El Waqf, *Egypt* 94 B3 25 45N 32 15 E
El Wâsta, *Egypt* 94 J7 29 19N 31 12 E
El Weguet, *Ethiopia* 95 F5 5 28N 42 17 E
El Wuz, *Sudan* 95 D3 15 5N 30 7 E
Elafónisos, *Greece* 45 H4 36 29N 22 56 E
Elaine, *Australia* 116 D6 37 44 S 144 2 E
Elamanchili, *India* 82 F6 17 33N 82 50 E
Elands, *Australia* 117 A10 31 37 S 152 20 E
Élassa, *Greece* 45 J8 35 18N 26 21 E
Elassón, *Greece* 44 E4 39 53N 22 12 E
Elat, *Israel* 91 F3 29 30N 34 56 E
Eláthia, *Greece* 45 F4 38 37N 22 46 E
Elâzığ, *Turkey* 89 C8 38 37N 39 14 E
Elba, *Italy* 38 F7 42 46N 10 17 E
Elba, *U.S.A.* 135 K2 31 25N 86 4W
Elbasani, *Albania* 44 C2 41 9N 20 9 E
Elbasani-Berati □,
Albania 44 D2 40 58N 20 0 E
Elbe, *U.S.A.* 144 D4 46 45N 122 10W
Elbe —►, *Europe* 26 B4 53 50N 9 0 E
Elbe-Seiten Kanal,
Germany 26 C6 52 45N 10 32 E
Elberfeld, *U.S.A.* 141 F9 38 10N 87 27W
Elbert, Mt., *U.S.A.* 143 G10 39 7N 106 27W
Elberta, *U.S.A.* 134 C2 44 37N 86 14W
Elberton, *U.S.A.* 135 H4 34 7N 82 52W
Elbeuf, *France* 22 C8 49 17N 1 2 E
Elbing = Elbląg, *Poland* 47 A6 54 10N 19 25 E
Elbistan, *Turkey* 88 C7 38 13N 37 15 E
Elbląg, *Poland* 47 A6 54 10N 19 25 E
Elbląg □, *Poland* 47 A6 54 15N 19 30 E
Elbow, *Canada* 131 C7 51 7N 106 35W
Elbrus, *Asia* 53 J6 43 21N 42 30 E
Elburg, *Neths.* 20 D7 52 26N 5 50 E
Elburn, *U.S.A.* 141 C8 41 54N 88 28W
Elburz Mts. = Alborz,
Reshteh-ye Kūhhā-ye,
Iran 85 C7 36 0N 52 0 E
Elche, *Spain* 35 G4 38 15N 0 42W
Elche de la Sierra, *Spain* 35 G2 38 27N 2 3W
Elcho I., *Australia* 114 A2 11 55 S 135 45 E
Elda, *Spain* 35 G4 38 29N 0 47W
Eldon, *Mo., U.S.A.* 140 F4 38 21N 92 35W
Eldon, *Wash., U.S.A.* 144 C3 47 33N 123 3W
Eldora, *U.S.A.* 140 B3 42 22N 93 5W
Eldorado, *Argentina* 159 B5 26 28 S 54 43W
Eldorado, *Canada* 131 B7 59 35N 108 30W
Eldorado, *Mexico* 146 C3 24 20N 107 22W
Eldorado, *Ill., U.S.A.* 141 G8 37 49N 88 26W
Eldorado, *Tex., U.S.A.* 139 K4 30 52N 100 36W
Eldorado Springs, *U.S.A.* 139 G8 37 52N 94 1W
Eldoret, *Kenya* 106 B4 0 30N 35 17 E
Eldred, *U.S.A.* 136 E6 41 58N 78 23W
Eldridge, *U.S.A.* 140 C6 41 39N 90 35W
Elea, C., *Cyprus* 32 D13 35 19N 34 4 E
Electra, *U.S.A.* 139 H5 34 2N 98 55W
Elefantes —►, *Mozam.* 105 C5 24 10 S 32 40 E
Elefantes, G., *Chile* 160 C2 46 28 S 73 49W
Elektrogorsk, *Russia* 52 C4 55 56N 38 50 E
Elektrostal, *Russia* 52 C4 55 41N 38 32 E
Elele, *Nigeria* 101 D6 5 5N 6 50 E
Elena, *Bulgaria* 43 E10 42 55N 25 53 E
Elephant Butte Reservoir,
U.S.A. 143 K10 33 9N 107 11W
Elephant I., *Antarctica* 7 C18 61 0 S 55 0W
Elephant Pass, *Sri Lanka* 83 K5 9 35N 80 25 E
Elesbão Veloso, *Brazil* 154 C3 6 13 S 42 8W
Eleshnitsa, *Bulgaria* 43 F8 41 52N 23 36 E
Eleşkirt, *Turkey* 89 C10 39 50N 42 50 E
Eleuthera, *Bahamas* 148 A4 25 0N 76 20W
Elevsís, *Greece* 45 F5 38 4N 23 26 E
Elevtheroúpolis, *Greece* 44 D6 40 52N 24 20 E
Elgepiggen, *Norway* 14 B5 62 10N 11 21 E
Elgeyo-Marakwet □,
Kenya 106 B4 0 45N 35 30 E
Elgg, *Switz.* 29 B7 47 29N 8 52 E
Elgin, *N.B., Canada* 129 C6 45 48N 65 10W
Elgin, *Ont., Canada* 137 B8 44 36N 76 13W
Elgin, *U.K.* 18 D5 57 39N 3 19W
Elgin, *Ill., U.S.A.* 141 B8 42 2N 88 17W
Elgin, *N. Dak., U.S.A.* 138 B4 46 24N 101 51W
Elgin, *Nebr., U.S.A.* 138 E5 41 59N 98 5W
Elgin, *Nev., U.S.A.* 143 H6 37 21N 114 32W
Elgin, *Oreg., U.S.A.* 142 D5 45 34N 117 55W
Elgin, *Tex., U.S.A.* 139 K6 30 21N 97 22W
Elgon, Mt., *Africa* 106 B3 1 10N 34 30 E
Eliase, *Indonesia* 73 C4 8 21 S 130 48 E
Elida, *U.S.A.* 139 J3 33 57N 103 39W
Elikón, *Greece* 45 F4 38 18N 22 45 E
Elim, *S. Africa* 104 E2 34 35 S 19 45 E
Elin Pelin, *Bulgaria* 43 E8 42 40N 23 36 E
Elisabethville =
Lubumbashi, *Zaïre* 107 E2 11 40 S 27 28 E
Eliseu Martins, *Brazil* 154 C3 8 13 S 43 42W
Elista, *Russia* 53 G7 46 16N 44 14 E
Elizabeth, *Australia* 116 C3 34 42 S 138 41 E
Elizabeth, *Ill., U.S.A.* 140 B6 42 19N 90 13W
Elizabeth, *N.J., U.S.A.* 137 F10 40 40N 74 13W
Elizabeth City, *U.S.A.* 135 G7 36 18N 76 14W
Elizabethton, *U.S.A.* 135 G4 36 21N 82 13W
Elizabethtown, *Ky.,*
U.S.A. 134 G3 37 42N 85 52W
Elizabethtown, *N.Y.,*
U.S.A. 137 B11 44 13N 73 36W
Elizabethtown, *Pa.,*
U.S.A. 137 F8 40 9N 76 36W
Elizondo, *Spain* 34 B3 43 12N 1 30W
Ełk, *Poland* 47 B9 53 50N 22 21 E
Ełk —►, *Poland* 47 B9 53 41N 22 28 E
Elk City, *U.S.A.* 139 H5 35 25N 99 25W
Elk Creek, *U.S.A.* 144 F4 39 36N 122 32W
Elk Grove, *U.S.A.* 144 G5 38 25N 121 22W
Elk Island Nat. Park,
Canada 130 C6 53 35N 112 59W
Elk Lake, *Canada* 128 C3 47 40N 80 25W
Elk Point, *Canada* 131 C6 53 54N 110 55W
Elk River, *Idaho, U.S.A.* 142 C5 46 47N 116 11W
Elk River, *Minn., U.S.A.* 138 C8 45 18N 93 35W
Elkader, *U.S.A.* 140 B5 42 51N 91 24W
Elkedra, *Australia* 114 C2 21 9 S 135 33 E
Elkedra —►, *Australia* 114 C2 21 8 S 136 22 E
Elkhart, *Ind., U.S.A.* 141 C11 41 41N 85 58W
Elkhart, *Kans., U.S.A.* 139 G4 37 0N 101 54W
Elkhart —►, *U.S.A.* 141 C11 41 41N 85 58W
Elkhorn, *Canada* 131 D8 49 59N 101 14W
Elkhorn, *U.S.A.* 141 B8 42 40N 88 33W
Elkhorn —►, *U.S.A.* 138 E6 41 8N 96 19W
Elkhovo, *Bulgaria* 43 E11 42 10N 26 40 E
Elkin, *U.S.A.* 135 G5 36 15N 80 51W
Elkins, *U.S.A.* 134 F6 38 55N 79 51W
Elko, *Canada* 130 D5 49 20N 115 10W
Elko, *U.S.A.* 142 F6 40 50N 115 46W
Ell, L., *Australia* 113 E4 29 13 S 127 46 E
Ellecom, *Neths.* 20 D8 52 2N 6 6 E
Ellef Ringnes I., *Canada* 6 B2 78 30N 102 2W
Ellendale, *Australia* 112 C3 17 56 S 124 48 E
Ellendale, *U.S.A.* 138 B5 46 0N 98 32W
Ellensburg, *U.S.A.* 142 C3 46 59N 120 34W
Ellenville, *U.S.A.* 137 E10 41 43N 74 24W
Ellerston, *Australia* 117 A9 31 49 S 151 20 E
Ellery, Mt., *Australia* 117 D8 37 28 S 148 47 E
Ellesmere, L., *N.Z.* 119 H7 47 47 S 172 28 E
Ellesmere I., *Canada* 6 B4 79 30N 80 0W
Ellesmere Port, *U.K.* 16 D5 53 17N 2 54W
Ellettsville, *U.S.A.* 141 E10 39 14N 86 38W
Ellezelles, *Belgium* 21 G3 50 44N 3 42 E
Ellice Is. = Tuvalu ■,
Pac. Oc. 122 H9 8 0 S 178 0 E
Ellinwood, *U.S.A.* 138 F5 38 21N 98 35W
Elliot, *Australia* 114 B1 17 33 S 133 32 E
Elliot, *S. Africa* 105 E4 31 22 S 27 48 E
Elliot Lake, *Canada* 128 C3 46 25N 82 35W
Elliotdale = Xhora,
S. Africa 105 E4 31 55 S 28 38 E
Ellis, *U.S.A.* 138 F5 38 56N 99 34W
Elliston, *Australia* 115 E1 33 39 S 134 53 E
Ellisville, *U.S.A.* 139 K10 31 36N 89 12W
Ellon, *U.K.* 18 D6 57 22N 2 4W
Ellore = Eluru, *India* 82 F5 16 48N 81 8 E
Ells —►, *Canada* 130 B6 57 18N 111 40W
Ellsworth, *U.S.A.* 138 F5 38 44N 98 14W
Ellsworth Land, *Antarctica* 7 D16 76 0 S 89 0W
Ellsworth Mts., *Antarctica* 7 D16 78 30 S 85 0W
Ellwangen, *Germany* 27 G6 48 57N 10 8 E
Ellwood City, *U.S.A.* 136 F4 40 52N 80 17W
Elm, *Switz.* 29 C8 46 54N 9 10 E
Elma, *Canada* 131 D9 49 52N 95 55W
Elma, *U.S.A.* 144 D3 47 0N 123 25W
Elmadağ, *Turkey* 88 C5 39 55N 33 14 E
Elmalı, *Turkey* 88 D3 36 44N 29 56 E
Elmer, *U.S.A.* 140 E4 39 57N 92 39W
Elmhurst, *U.S.A.* 141 C9 41 53N 87 56W
Elmina, *Ghana* 101 D4 5 5N 1 21W
Elmira, *Canada* 136 C4 43 36N 80 33W
Elmira, *U.S.A.* 136 D8 42 6N 76 48W
Elmore, *Australia* 116 D6 36 30 S 144 37 E
Elmore, *Calif., U.S.A.* 145 M11 33 7N 115 49W
Elmore, *Minn., U.S.A.* 141 C13 41 29N 83 18W
Elmshorn, *Germany* 26 B5 53 43N 9 40 E
Elmvale, *Canada* 136 B5 44 35N 79 52W
Elmwood, *U.S.A.* 140 D6 40 47N 89 58W
Elne, *France* 24 F6 42 36N 2 58 E
Elnora, *U.S.A.* 141 F9 38 53N 87 5W
Elora, *Canada* 136 C4 43 41N 80 26W
Elorza, *Venezuela* 152 B4 7 3N 69 31W
Elos, *Greece* 45 H4 36 46N 22 43 E
Eloúnda, *Greece* 32 D7 35 16N 25 42 E
Eloy, *U.S.A.* 143 K8 32 45N 111 33W
Éloyes, *France* 23 D13 48 6N 6 36 E
Elrose, *Canada* 131 C7 51 12N 108 0W
Elsas, *Canada* 128 C3 48 32N 82 55W
Elsie, *U.S.A.* 144 E3 45 52N 123 36W
Elsinore = Helsingør,
Denmark 15 H6 56 2N 12 35 E
Elsinore, *Australia* 117 A6 31 35 S 145 11 E
Elsinore, *U.S.A.* 143 G7 38 41N 112 9W
Elspe, *Germany* 26 D4 51 10N 8 1 E
Elspeet, *Neths.* 20 D7 52 17N 5 48 E
Elst, *Neths.* 20 E7 51 55N 5 51 E
Elster —►, *Germany* 26 D7 51 25N 11 57 E
Elsterwerda, *Germany* 26 D9 51 27N 13 31 E
Elten, *Neths.* 20 E8 51 52N 6 9 E
Eltham, *Australia* 117 D6 37 43 S 145 12 E
Eltham, *N.Z.* 118 F3 39 26 S 174 19 E
Elton, *Russia* 53 F8 49 5N 46 52 E
Elton, Ozero, *Russia* 53 F8 49 5N 46 42 E
Eluanbi, *Taiwan* 69 G13 21 51N 120 50 E
Eluru, *India* 82 F5 16 48N 81 8 E
Elvas, *Portugal* 37 G3 38 50N 7 10W
Elven, *France* 22 E4 47 44N 2 36W
Elverum, *Norway* 14 D5 60 53N 11 34 E
Elvire —►, *Australia* 112 C4 17 51 S 128 11 E
Elvo —►, *Italy* 38 C5 45 23N 8 21 E
Elvran, *Norway* 14 A5 63 24N 11 3 E
Elwood, *Ill., U.S.A.* 141 C8 41 24N 88 7W
Elwood, *Ind., U.S.A.* 141 D11 40 17N 85 50W
Elwood, *Nebr., U.S.A.* 138 E5 40 36N 99 52W
Elx = Elche, *Spain* 35 G4 38 15N 0 42W
Ely, *U.K.* 17 E8 52 24N 0 16 E
Ely, *Minn., U.S.A.* 138 B9 47 55N 91 51W
Ely, *Nev., U.S.A.* 142 G6 39 15N 114 54W
Elyria, *U.S.A.* 136 E2 41 22N 82 7W
Elyrus, *Greece* 32 D5 35 15N 23 45 E
Elz —►, *Germany* 27 G3 48 18N 7 44 E
Emai, *Vanuatu* 121 G6 17 4 S 168 24 E
Emāmrūd, *Iran* 85 B7 36 30N 55 0 E
Emba = Embi,
Kazakhstan 56 E6 48 50N 58 8 E
Emba = Embi —►,
Kazakhstan 49 E9 46 55N 53 28 E
Embarcación, *Argentina* 158 A3 23 10 S 64 0W
Embarras —►, *U.S.A.* 141 F9 38 39N 87 37W
Embarras Portage, *Canada* 131 B6 58 27N 111 28W
Embetsu, *Japan* 60 B10 44 44N 141 47 E
Embi, *Kazakhstan* 56 E6 48 50N 58 8 E
Embi —►, *Kazakhstan* 49 E9 46 55N 53 28 E
Embira —►, *Brazil* 156 B3 7 19 S 70 15W
Embóna, *Greece* 32 C9 36 13N 27 51 E
Embrach, *Switz.* 29 A7 47 30N 8 36 E
Embrun, *France* 25 D10 44 34N 6 30 E
Embu, *Kenya* 106 C4 0 32 S 37 38 E
Embu □, *Kenya* 106 C4 0 30 S 37 35 E
Emden, *Germany* 26 B3 53 21N 7 12 E
Emerald, *Australia* 114 C4 23 32 S 148 10 E
Emerson, *Canada* 131 D9 49 0N 97 10W
Emery, *U.S.A.* 143 G8 38 55N 111 15W
Emet, *Turkey* 88 C3 39 20N 29 15 E
Emi Koussi, *Chad* 97 E3 19 45N 18 55 E
Emília-Romagna □, *Italy* 38 D7 44 45N 11 0 E
Emilius, Mte., *Italy* 38 C4 45 45N 7 20 E
Eminabad, *Pakistan* 80 C6 32 2N 74 8 E
Emine, Nos, *Bulgaria* 43 E12 42 40N 27 56 E
Eminence, *U.S.A.* 141 F11 38 22N 85 11W
Emirdağ, *Turkey* 88 C4 39 2N 31 8 E
Emlenton, *U.S.A.* 136 E5 41 11N 79 43W
Emlichheim, *Germany* 26 C2 52 37N 6 51 E
Emme —►, *Switz.* 28 B5 47 14N 7 32 E
Emmeloord, *Neths.* 20 C7 52 44N 5 46 E
Emmen, *Neths.* 20 C9 52 48N 6 57 E
Emmendingen, *Germany* 27 G3 48 6N 7 51 E
Emmental, *Switz.* 28 C4 46 55N 7 20 E
Emmer-Compascuum,
Neths. 20 C10 52 49N 7 2 E
Emmerich, *Germany* 26 D2 51 50N 6 14 E
Emmet, *Australia* 114 C3 24 45 S 144 30 E
Emmetsburg, *U.S.A.* 140 A2 43 7N 94 41W
Emmett, *U.S.A.* 142 E5 43 52N 116 30W
Emöd, *Hungary* 31 D13 47 57N 20 47 E
Emona, *Bulgaria* 43 E12 42 43N 27 53 E
Empalme, *Mexico* 146 B2 28 1N 110 49W
Empangeni, *S. Africa* 105 D5 28 50 S 31 52 E
Empedrado, *Argentina* 158 B4 28 0 S 58 46W
Emperor Seamount Chain,
Pac. Oc. 122 D9 40 0N 170 0 E
Empoli, *Italy* 38 E7 43 43N 10 57 E
Emporia, *Kans., U.S.A.* 138 F6 38 25N 96 11W
Emporia, *Va., U.S.A.* 135 G7 36 42N 77 32W
Emporium, *U.S.A.* 136 E6 41 31N 78 14W
Empress, *Canada* 131 C6 50 57N 110 0W
Emptinne, *Belgium* 21 H6 50 19N 5 8 E
Empty Quarter = Rub' al
Khali, *Si. Arabia* 87 C5 18 0N 48 0 E
Ems —►, *Germany* 26 B3 53 20N 7 12 E
Emsdale, *Canada* 136 A5 45 32N 79 19W
Emsdetten, *Germany* 26 C3 52 10N 7 32 E
Emu, *Australia* 116 D5 36 44 S 143 26 E
Emu, *China* 67 C15 43 40N 128 6 E
Emu Park, *Australia* 114 C5 23 13 S 150 50 E
'En 'Avrona, *Israel* 91 F3 29 43N 35 0 E
En Nahud, *Sudan* 95 E2 12 45N 28 25 E
Ena, *Japan* 63 B9 35 25N 137 25 E
Ena-San, *Japan* 63 B9 35 26N 137 36 E
Enafors, *Sweden* 14 A6 63 17N 12 20 E
Enambú, *Colombia* 152 C3 1 1N 70 17W
Enana, *Namibia* 104 B2 17 30 S 16 23 E
Enånger, *Sweden* 14 C11 61 30N 17 9 E
Enaratoli, *Indonesia* 73 B5 3 55 S 136 21 E
Enard B., *U.K.* 18 C3 58 5N 5 20W
Enare = Inarijärvi,
Finland 12 B22 69 0N 28 0 E
Encantadas, Serra, *Brazil* 159 C5 30 40 S 53 0W
Encarnación, *Paraguay* 159 B4 27 15 S 55 50W
Encarnación de Diaz,
Mexico 146 C4 21 30N 102 13W
Enchi, *Ghana* 100 D4 5 53N 2 48W
Encinal, *U.S.A.* 139 L5 28 2N 99 21W
Encinitas, *U.S.A.* 145 M9 33 3N 117 17W
Encino, *U.S.A.* 143 J11 34 39N 105 28W
Encontrados, *Venezuela* 152 B3 9 3N 72 14W
Encounter B., *Australia* 116 C3 35 45 S 138 45 E
Encruzilhada, *Brazil* 155 E3 15 31 S 40 54W
Ende, *Indonesia* 72 C2 8 45 S 121 40 E
Endeavour, *Canada* 131 C8 52 10N 102 39W
Endeavour Str., *Australia* 114 A3 10 45 S 142 0 E
Endelave, *Denmark* 15 J4 55 46N 10 18 E
Enderbury I., *Kiribati* 122 H10 3 8 S 171 5W
Enderby, *Canada* 130 C5 50 35N 119 10W
Enderby I., *Australia* 112 D2 20 35 S 116 30 E
Enderby Land, *Antarctica* 7 C5 66 0 S 53 0 E
Enderlin, *U.S.A.* 138 B6 46 38N 97 36W
Endicott, *N.Y., U.S.A.* 137 D8 42 6N 76 4W
Endicott, *Wash., U.S.A.* 142 C5 46 56N 117 41W
Endimari —►, *Brazil* 156 B4 8 46 S 66 7W
Endröd, *Hungary* 31 E13 46 55N 20 47 E
Endyalgout I., *Australia* 112 B5 11 40 S 132 35 E
Ene —►, *Peru* 156 C3 11 10 S 74 18W
Energetik, *Russia* 54 F7 51 45N 58 45 E
Enewetak Atoll, *Pac. Oc.* 122 F8 11 30N 162 15 E
Enez, *Turkey* 88 B2 40 45N 26 5 E
Enfield, *U.K.* 17 F7 51 38N 0 5W
Enfield, *U.S.A.* 141 F8 38 6N 88 20W
Engadin, *Switz.* 27 J6 46 45N 10 10 E
Engaño, C., *Dom. Rep.* 149 C6 18 30N 68 20W
Engaño, C., *Phil.* 70 B4 18 35N 122 23 E
Engcobo, *S. Africa* 105 E4 31 37 S 28 0 E
Engelberg, *Switz.* 29 C6 46 48N 8 26 E
Engels, *Russia* 53 E8 51 28N 46 6 E
Engemann L., *Canada* 131 B7 58 0N 106 55W
Enger, *Norway* 14 D4 60 35N 10 20 E
Enggano, *Indonesia* 74 D2 5 20 S 102 40 E
Enghien, *Belgium* 21 G4 50 37N 4 2 E
Engil, *Morocco* 98 B4 33 12N 4 32W
Engkilili, *Malaysia* 75 B4 1 3N 111 42 E
England, *U.S.A.* 139 H9 34 33N 91 58W
England □, *U.K.* 11 E5 53 0N 2 0W
Englee, *Canada* 129 B8 50 45N 56 5W
Englefield, *Australia* 116 D4 37 21 S 141 48 E
Englehart, *Canada* 128 C4 47 49N 79 52W
Engler L., *Canada* 131 B7 59 8N 106 52W
Englewood, *Colo., U.S.A.* 138 F2 39 39N 104 59W
Englewood, *Kans., U.S.A.* 139 G5 37 2N 99 59W
Englewood, *Ohio, U.S.A.* 141 E12 39 53N 84 18W
English, *U.S.A.* 141 F10 38 20N 86 28W
English —►, *Canada* 131 C10 50 35N 93 30W
English —►, *U.S.A.* 140 C5 41 29N 91 32W
English Bazar = Ingraj
Bazar, *India* 81 G13 24 58N 88 10 E
English Channel, *Europe* 17 G6 50 0N 2 0W
English River, *Canada* 128 C1 49 14N 91 0W
Enguri —►, *Georgia* 53 J5 42 27N 41 38 E
Enid, *U.S.A.* 139 G6 36 24N 97 53W
Enipévs —►, *Greece* 44 E4 39 22N 22 17 E
Enkhuizen, *Neths.* 20 C6 52 42N 5 17 E
Enköping, *Sweden* 14 E11 59 37N 17 4 E
Enle, *China* 68 E3 24 0N 101 9 E
Enna, *Italy* 41 E7 37 34N 14 16 E
Ennadai, *Canada* 131 A8 61 8N 100 53W
Ennadai L., *Canada* 131 A8 61 0N 101 0W
Ennedi, *Chad* 97 E4 17 15N 22 0 E
Enngonia, *Australia* 115 D4 29 21 S 145 50 E
Ennis, *Ireland* 19 D3 52 51N 8 59W

Feshi 197

Eyasi, L., *Tanzania* 106 C4 3 30 S 35 0 E
Eyeberry L., *Canada* 131 A8 63 8N 104 43W
Eyemouth, *U.K.* 18 F6 55 52N 2 5W
Eygurande, *France* 24 C6 45 40N 2 26 E
Eyjafjörður, *Iceland* 12 C4 66 15N 18 30W
Eymet, *France* 24 D4 44 40N 0 25 E
Eymoutiers, *France* 24 C5 45 40N 1 45 E
Eynesil, *Turkey* 89 B8 41 4N 39 9 E
Eyre, *Australia* 113 F4 32 15 S 126 18 E
Eyre (North), L., *Australia* 115 D2 28 30 S 137 20 E
Eyre (South), L., *Australia* 115 D2 29 18 S 137 25 E
Eyre, L., *Australia* 110 F6 29 30 S 137 26 E
Eyre Cr. →, *Australia* 115 D2 26 40 S 139 0 E
Eyre Mts., *N.Z.* 119 F3 45 25 S 168 25 E
Eyre Pen., *Australia* 115 E2 33 30 S 136 17 E
Eysturoy, *Færoe Is.* 12 E9 62 13N 6 54W
Eyvānkī, *Iran* 85 C6 35 24N 51 56 E
Ez Zeidab, *Sudan* 94 D3 17 25N 33 55 E
Ezcaray, *Spain* 34 C2 42 19N 3 0W
Ezine, *Turkey* 88 C2 39 48N 26 20 E
Ezmul, *Mauritania* 98 D1 22 15N 15 40W
Ezouza →, *Cyprus* 32 E11 34 44N 32 27 E

F

F.Y.R.O.M. = Macedonia ■, *Europe* 42 F6 41 53N 21 40 E
Fabens, *U.S.A.* 143 L10 31 30N 106 10W
Fåborg, *Denmark* 15 J4 55 6N 10 15 E
Fabriano, *Italy* 39 E9 43 20N 12 54 E
Făcăeni, *Romania* 46 E8 44 32N 27 53 E
Facatativá, *Colombia* 152 C3 4 49N 74 22W
Fachi, *Niger* 97 E2 18 6N 11 34 E
Facture, *France* 24 D3 44 39N 0 58W
Fada, *Chad* 97 E4 17 13N 21 34 E
Fada-n-Gourma, *Burkina Faso* 101 C5 12 10N 0 30 E
Fadd, *Hungary* 31 E11 46 28N 18 49 E
Faddeyevskiy, Ostrov, *Russia* 57 B15 76 0N 144 0 E
Fadghāmī, *Syria* 89 E9 35 53N 40 52 E
Fadlab, *Sudan* 94 D3 17 42N 34 2 E
Faenza, *Italy* 39 D8 44 17N 11 53 E
Færoe Is. = Føroyar, *Atl. Oc.* 12 F9 62 0N 7 0W
Fafa, *Mali* 101 B5 15 22N 0 48 E
Fafe, *Portugal* 36 D2 41 27N 8 11W
Faga, *W. Samoa* 121 W23 13 39 S 172 31W
Fagam, *Nigeria* 101 C7 11 1N 10 1 E
Fagamalo, *W. Samoa* 121 W23 13 25 S 172 21W
Făgăras, *Romania* 46 D5 45 48N 24 58 E
Făgăras, Munţii, *Romania* 46 D5 45 40N 24 40 E
Fågelsjö, *Sweden* 14 C8 61 50N 14 35 E
Fagersta, *Sweden* 13 F16 60 1N 15 46 E
Făget, *Romania* 46 D3 45 52N 22 10 E
Făget, Munţii, *Romania* 46 B4 47 40N 23 10 E
Fagnano, L., *Argentina* 160 D3 54 30 S 68 0W
Fagnano Castello, *Italy* 41 C9 39 34N 16 3 E
Fagnières, *France* 23 D11 48 58N 4 20 E
Fahlīān, *Iran* 85 D6 30 11N 51 28 E
Fahr, *Yemen* 87 D6 12 26N 54 8 E
Fahraj, *Kermān, Iran* 85 D8 29 0N 59 0 E
Fahraj, *Yazd, Iran* 85 D7 31 46N 54 36 E
Faial, *Madeira* 33 D3 32 47N 16 53W
Faido, *Switz.* 29 D7 46 29N 8 48 E
Fair Hd., *U.K.* 19 A5 55 14N 6 9W
Fair Oaks, *U.S.A.* 144 G5 38 39N 121 16W
Fairbank, *U.S.A.* 143 L8 31 43N 110 11W
Fairbanks, *U.S.A.* 126 B5 64 51N 147 43W
Fairborn, *U.S.A.* 141 E12 39 49N 84 2W
Fairbury, *Ill., U.S.A.* 141 D8 40 45N 88 31W
Fairbury, *Nebr., U.S.A.* 138 E6 40 8N 97 11W
Faire, *Phil.* 70 C3 17 53N 121 34 E
Fairfax, *Ohio, U.S.A.* 141 E13 39 5N 83 37W
Fairfax, *Okla., U.S.A.* 139 G6 36 34N 96 42W
Fairfield, *Australia* 117 B9 33 53 S 150 57 E
Fairfield, *Ala., U.S.A.* 135 J2 33 29N 86 55W
Fairfield, *Calif., U.S.A.* 144 G4 38 15N 122 3W
Fairfield, *Conn., U.S.A.* 137 E11 41 9N 73 16W
Fairfield, *Idaho, U.S.A.* 142 E6 43 21N 114 44W
Fairfield, *Ill., U.S.A.* 141 F8 38 23N 88 22W
Fairfield, *Iowa, U.S.A.* 140 C5 40 56N 91 57W
Fairfield, *Mont., U.S.A.* 142 C8 47 37N 111 59W
Fairfield, *Ohio, U.S.A.* 141 E12 39 21N 84 34W
Fairfield, *Tex., U.S.A.* 139 K7 31 44N 96 10W
Fairford, *Canada* 131 C9 51 37N 98 38W
Fairhope, *U.S.A.* 135 K2 30 31N 87 54W
Fairlie, *N.Z.* 119 E5 44 5 S 170 49 E
Fairmead, *U.S.A.* 144 H6 37 5N 120 10W
Fairmont, *Minn., U.S.A.* 138 D7 43 39N 94 28W
Fairmont, *W. Va., U.S.A.* 134 F5 39 29N 80 9W
Fairmount, *U.S.A.* 145 L8 34 45N 118 26W
Fairplay, *U.S.A.* 143 G11 39 15N 106 2W
Fairport, *U.S.A.* 136 C7 43 6N 77 27W
Fairport Harbor, *U.S.A.* 136 E3 41 45N 81 17W
Fairview, *Australia* 114 B3 15 31 S 144 17 E
Fairview, *Canada* 130 B5 56 5N 118 25W
Fairview, *Mont., U.S.A.* 138 B2 47 51N 104 3W
Fairview, *Okla., U.S.A.* 139 G5 36 16N 98 29W
Fairview, *Utah, U.S.A.* 142 G8 39 50N 111 0W
Fairweather, Mt., *U.S.A.* 126 C6 58 55N 137 32W
Faisalabad, *Pakistan* 79 C4 31 30N 73 5 E
Faith, *U.S.A.* 138 C3 45 2N 102 2W
Faizabad, *India* 81 F10 26 45N 82 10 E
Faizpur, *India* 82 D2 21 14N 75 49 E
Fajardo, *Puerto Rico* 149 C6 18 20N 65 39W
Fakam, *Yemen* 86 C3 16 38N 43 49 E
Fakfak, *Indonesia* 73 B4 3 0 S 132 15 E
Fakiya, *Bulgaria* 43 E12 42 10N 27 6 E
Fakobli, *Ivory C.* 100 D3 7 23N 7 23W
Fakse, *Denmark* 15 J6 55 15N 12 8 E
Fakse B., *Denmark* 15 J6 55 11N 12 15 E
Fakse Ladeplads, *Denmark* 15 J6 55 11N 12 9 E
Faku, *China* 67 C12 42 32N 123 21 E
Falaise, *France* 22 D6 48 54N 0 12W
Falaise, Mui, *Vietnam* 76 C5 19 6N 105 45 E
Falakrón Óros, *Greece* 44 C5 41 15N 23 58 E
Falam, *Burma* 78 D4 23 0N 93 45 E
Falces, *Spain* 34 C3 42 24N 1 48W
Fălciu, *Romania* 46 C9 46 17N 28 7 E
Falcón □, *Venezuela* 152 A4 11 0N 69 50W
Falcon, C., *Algeria* 99 A4 35 50N 0 50W
Falcón, C., *Spain* 33 C7 38 50N 1 23 E
Falcon Dam, *U.S.A.* 139 M5 26 50N 99 20W
Falconara Maríttima, *Italy* 39 E10 43 37N 13 24 E
Falconer, *U.S.A.* 136 D5 42 7N 79 13W
Faléa, *Mali* 100 C2 12 16N 11 17W
Falelatai, *W. Samoa* 121 W24 13 55 S 171 59W
Falelima, *W. Samoa* 121 W23 13 32 S 172 41W
Falenki, *Russia* 54 B3 58 22N 51 35 E
Faleshty = Făleşti, *Moldova* 51 J4 47 32N 27 44 E
Făleşti, *Moldova* 51 J4 47 32N 27 44 E
Falfurrias, *U.S.A.* 139 M5 27 14N 98 9W
Falher, *Canada* 130 B5 55 44N 117 15W
Falirakí, *Greece* 32 C10 36 22N 28 12 E
Falkenberg, *Germany* 26 D9 51 35N 13 14 E
Falkenberg, *Sweden* 15 H6 56 54N 12 30 E
Falkensee, *Germany* 26 C9 52 34N 13 4 E
Falkenstein, *Germany* 26 E8 50 28N 12 20 E
Falkirk, *U.K.* 18 F5 56 0N 3 47W
Falkland, East, I., *Falk. Is.* 160 D5 51 40 S 58 30W
Falkland, West, I., *Falk. Is.* 160 D4 51 40 S 60 0W
Falkland Is. □, *Atl. Oc.* 160 D5 51 30 S 59 0W
Falkland Sd., *Falk. Is.* 160 D5 52 0 S 60 0W
Falkonéra, *Greece* 45 H5 36 50N 23 52 E
Falköping, *Sweden* 15 F7 58 12N 13 33 E
Fall River, *U.S.A.* 137 E13 41 43N 71 10W
Fall River Mills, *U.S.A.* 142 F3 41 3N 121 26W
Fallbrook, *U.S.A.* 143 K5 33 25N 117 12W
Fallbrook, *Calif., U.S.A.* 145 M9 33 23N 117 15W
Fallon, *Mont., U.S.A.* 138 B2 46 50N 105 8W
Fallon, *Nev., U.S.A.* 142 G4 39 28N 118 47W
Falls City, *Nebr., U.S.A.* 138 E7 40 3N 95 36W
Falls City, *Oreg., U.S.A.* 142 D2 44 52N 123 26W
Falls Creek, *U.S.A.* 136 E6 41 9N 78 48W
Falmouth, *Jamaica* 148 C4 18 30N 77 40W
Falmouth, *U.K.* 17 G2 50 9N 5 5W
Falmouth, *U.S.A.* 141 F12 38 41N 84 20W
False B., *S. Africa* 104 E2 34 15 S 18 40 E
False Divi Pt., *India* 83 G5 15 43N 80 50 E
Falset, *Spain* 34 D5 41 7N 0 50 E
Falso, C., *Honduras* 148 C3 15 12N 83 21W
Falster, *Denmark* 15 K5 54 45N 11 55 E
Falsterbo, *Sweden* 15 J6 55 23N 12 50 E
Fălticeni, *Romania* 46 B7 47 21N 26 20 E
Falun, *Sweden* 13 F16 60 37N 15 37 E
Famagusta, *Cyprus* 32 D12 35 8N 33 55 E
Famagusta Bay, *Cyprus* 32 D13 35 15N 34 0 E
Famatina, Sierra de, *Argentina* 158 B2 27 30 S 68 0W
Family L., *Canada* 131 C9 51 54N 95 27W
Famoso, *U.S.A.* 145 K7 35 37N 119 12W
Fan Xian, *China* 66 G8 35 55N 115 38 E
Fana, *Mali* 100 C3 13 0N 6 56W
Fanárion, *Greece* 44 E3 39 24N 21 47 E
Fandriana, *Madag.* 105 C8 20 14 S 47 21 E
Fang, *Thailand* 76 C2 19 55N 99 13 E
Fang Xian, *China* 69 A8 32 3N 110 40 E
Fangchang, *China* 69 B12 31 5N 118 4 E
Fangcheng, *Guangxi Zhuangzu, China* 68 G7 21 42N 108 21 E
Fangcheng, *Henan, China* 66 H7 33 18N 112 59 E
Fangliao, *Taiwan* 69 F13 22 22N 120 38 E
Fangshan, *China* 66 E6 38 3N 111 25 E
Fangzi, *China* 67 F10 36 33N 119 10 E
Fani i Madh →, *Albania* 44 C2 41 56N 20 16 E
Fanjiatun, *China* 67 C13 43 40N 125 15 E
Fannich, L., *U.K.* 18 D4 57 38N 4 59W
Fannūj, *Iran* 85 E8 26 35N 59 38 E
Fanny Bay, *Canada* 130 D4 49 37N 124 48W
Fanø, *Denmark* 15 J2 55 25N 8 25 E
Fano, *Italy* 39 E10 43 50N 13 1 E
Fanshaw, *U.S.A.* 130 B2 57 11N 133 30W
Fanshi, *China* 66 E7 39 12N 113 20 E
Fao = Al Fāw, *Iraq* 85 D6 30 0N 48 30 E
Faqirwali, *Pakistan* 80 E5 29 27N 73 0 E
Fara in Sabina, *Italy* 39 F9 42 12N 12 43 E
Farab, *Turkmenistan* 55 D1 39 9N 63 36 E
Faradje, *Zaïre* 106 B2 3 50N 29 45 E
Farafangana, *Madag.* 105 C8 22 49 S 47 50 E
Farâfra, El Wâhât el-, *Egypt* 94 B2 27 15N 28 20 E
Farāh, *Afghan.* 79 B1 32 20N 62 7 E
Farāh □, *Afghan.* 79 B1 32 25N 62 10 E
Farahalana, *Madag.* 105 A9 14 26 S 50 10 E
Faraid, Gebel, *Egypt* 94 C4 23 33N 35 19 E
Faramana, *Burkina Faso* 100 C4 11 56N 4 45W
Faranah, *Guinea* 100 C2 10 3N 10 45W
Farasān, Jazā'ir, *Si. Arabia* 86 C3 16 45N 41 55 E
Farasan Is. = Farasān, Jazā'ir, *Si. Arabia* 86 C3 16 45N 41 55 E
Faratsiho, *Madag.* 105 B8 19 24 S 46 57 E
Farbarachi, *Somali Rep.* 108 D3 2 30N 45 30 E
Fardes →, *Spain* 35 H1 37 35N 3 0W
Fareham, *U.K.* 17 G6 50 51N 1 11W
Farewell, C., *N.Z.* 119 A7 40 29 S 172 43 E
Farewell C. = Farvel, Kap, *Greenland* 124 D15 59 48N 43 55W
Farewell Spit, *N.Z.* 119 A8 40 35 S 173 0 E
Farghona, *Uzbekistan* 55 C5 40 23N 71 19 E
Farghonskaya Dolina, *Uzbekistan* 55 C5 40 50N 71 30 E
Fargo, *U.S.A.* 138 B6 46 53N 96 48W
Fār'iah, W. al →, *West Bank* 91 C4 32 12N 35 27 E
Faribault, *U.S.A.* 138 C8 44 18N 93 16W
Faridkot, *India* 80 D6 30 44N 74 45 E
Faridpur, *Bangla.* 78 D2 23 15N 89 55 E
Färila, *Sweden* 14 C9 61 48N 15 50 E
Farim, *Guinea-Biss.* 100 C1 12 27N 15 9W
Farīmān, *Iran* 85 C8 35 40N 59 49 E
Farina, *Australia* 115 E2 30 3 S 138 15 E
Farinha →, *Brazil* 154 C2 6 51 S 47 30W
Fariones, Pta., *Canary Is.* 33 E6 29 13N 13 28W
Fâriskûr, *Egypt* 94 H7 31 20N 31 43 E
Farmakonisi, *Greece* 45 G9 37 17N 27 8 E
Farmer City, *U.S.A.* 141 D8 40 15N 88 39W
Farmersburg, *U.S.A.* 141 E9 39 15N 87 23W
Farmerville, *U.S.A.* 139 J8 32 47N 92 24W
Farmington, *Calif., U.S.A.* 144 H6 37 55N 120 59W
Farmington, *Ill., U.S.A.* 140 D6 40 42N 90 0W
Farmington, *Iowa, U.S.A.* 140 D5 40 38N 91 44W
Farmington, *Mo., U.S.A.* 140 G6 37 47N 90 25W
Farmington, *N.H., U.S.A.* 137 C13 43 24N 71 4W
Farmington, *N. Mex., U.S.A.* 143 H9 36 44N 108 12W
Farmington, *Utah, U.S.A.* 142 F8 41 0N 111 12W
Farmington →, *U.S.A.* 137 E12 41 51N 72 38W
Farmland, *U.S.A.* 141 D11 40 15N 85 5W
Farmville, *U.S.A.* 134 G6 37 18N 78 24W
Farnborough, *U.K.* 17 F7 51 16N 0 45W
Farne Is., *U.K.* 16 B6 55 38N 1 37W
Farnham, *Canada* 137 A12 45 17N 72 59W
Faro, *Brazil* 153 D6 2 10 S 56 39W
Faro, *Portugal* 37 H3 37 2N 7 55W
Fårö, *Sweden* 13 H18 57 55N 19 5 E
Faro □, *Portugal* 37 H2 37 12N 8 10W
Farquhar, C., *Australia* 113 D1 23 50 S 113 36 E
Farquhar Is., *Seychelles* 109 F4 11 0 S 52 0 E
Farrars Cr. →, *Australia* 114 D3 25 35 S 140 43 E
Farrāshband, *Iran* 85 D7 28 57N 52 5 E
Farrell, *U.S.A.* 136 E4 41 13N 80 30W
Farrell Flat, *Australia* 116 B3 33 48 S 138 48 E
Farrokhī, *Iran* 85 C8 33 50N 59 31 E
Farruch, C., *Spain* 33 B10 39 47N 3 21 E
Farrukhabad-cum-Fatehgarh, *India* 81 F8 27 30N 79 32 E
Fārs □, *Iran* 85 D7 29 30N 55 0 E
Fársala, *Greece* 44 E4 39 17N 22 23 E
Fārsī, *Afghan.* 79 B1 33 47N 63 15 E
Farsø, *Denmark* 15 H3 56 46N 9 19 E
Farsund, *Norway* 13 G12 58 5N 6 55 E
Fartak, Râs, *Si. Arabia* 84 D2 28 5N 34 34 E
Fartak, Ra's, *Yemen* 87 D6 15 38N 52 15 E
Fartura, Serra da, *Brazil* 159 B5 26 21 S 52 52W
Faru, *Nigeria* 101 C6 12 48N 6 12 E
Fārūj, *Iran* 85 B8 37 14N 58 14 E
Farum, *Denmark* 15 J6 55 49N 12 21 E
Farvel, Kap, *Greenland* 124 D15 59 48N 43 55W
Farwell, *U.S.A.* 139 H3 34 23N 103 2W
Fāryāb □, *Afghan.* 79 B2 36 0N 65 0 E
Fasā, *Iran* 85 D7 29 0N 53 39 E
Fasano, *Italy* 41 B10 40 50N 17 22 E
Fashoda, *Sudan* 95 F3 9 50N 32 2 E
Fastiv, *Ukraine* 51 G5 50 7N 29 57 E
Fastnet Rock, *Ireland* 19 E2 51 22N 9 37W
Fastov = Fastiv, *Ukraine* 51 G5 50 7N 29 57 E
Fatagar, Tanjung, *Indonesia* 73 B4 2 46 S 131 57 E
Fatehgarh, *India* 81 F8 27 25N 79 35 E
Fatehpur, *Raj., India* 80 F6 28 0N 74 40 E
Fatehpur, *Ut. P., India* 81 G9 25 56N 81 13 E
Fatesh, *Russia* 52 D2 52 8N 35 57 E
Fatick, *Senegal* 100 C1 14 19N 16 27W
Fatima, *Canada* 129 C7 47 24N 61 53W
Fátima, *Portugal* 37 F2 39 37N 8 39W
Fatoya, *Guinea* 100 C3 11 37N 9 10W
Fatsa, *Turkey* 88 B7 41 2N 37 31 E
Faucille, Col de la, *France* 25 B10 46 22N 6 2 E
Faulkton, *U.S.A.* 138 C5 45 2N 99 8W
Faulquemont, *France* 23 C13 49 3N 6 36 E
Fauquembergues, *France* 23 B9 50 36N 2 5 E
Faure I., *Australia* 113 E1 25 52 S 113 50 E
Făurei, *Romania* 46 D8 45 6N 27 19 E
Fauresmith, *S. Africa* 104 D4 29 44 S 25 17 E
Fauro, *Solomon Is.* 121 L9 6 55 S 156 7 E
Fauske, *Norway* 12 C16 67 17N 15 25 E
Fauvillers, *Belgium* 21 J7 49 51N 5 40 E
Favara, *Italy* 40 E6 37 19N 13 39 E
Favaritx, C., *Spain* 33 A11 40 0N 4 15 E
Favignana, *Italy* 40 E5 37 56N 12 20 E
Favignana, I., *Italy* 40 E5 37 56N 12 19 E
Favourable Lake, *Canada* 128 B1 52 50N 93 39W
Fawn →, *Canada* 128 A2 55 20N 87 35W
Fawnskin, *U.S.A.* 145 L10 34 16N 116 56W
Faxaflói, *Iceland* 12 D2 64 29N 23 0W
Faya-Largeau, *Chad* 97 E3 17 58N 19 6 E
Fayaoué, *Vanuatu* 121 K4 20 38 S 166 33 E
Fayd, *Si. Arabia* 84 E4 27 1N 42 52 E
Fayence, *France* 25 E10 43 38N 6 42 E
Fayette, *Ala., U.S.A.* 135 J2 33 41N 87 50W
Fayette, *Iowa, U.S.A.* 140 B5 42 51N 91 48W
Fayette, *Mo., U.S.A.* 140 E4 39 9N 92 41W
Fayette, *Ohio, U.S.A.* 141 C12 41 40N 84 20W
Fayetteville, *Ark., U.S.A.* 139 G7 36 4N 94 10W
Fayetteville, *N.C., U.S.A.* 135 H6 35 3N 78 53W
Fayetteville, *Tenn., U.S.A.* 135 H2 35 9N 86 34W
Fayón, *Spain* 34 D5 41 15N 0 20 E
Fazenda Libongo, *Angola* 103 D2 8 24 S 13 24 E
Fazenda Nova, *Brazil* 155 E1 16 11 S 50 48W
Fazilka, *India* 80 D6 30 27N 74 2 E
Fazilpur, *Pakistan* 80 E4 29 18N 70 29 E
Fdérik, *Mauritania* 98 D2 22 40N 12 45W
Feale →, *Ireland* 19 D2 52 27N 9 37W
Fear, C., *U.S.A.* 135 J7 33 50N 77 58W
Feather →, *U.S.A.* 142 G3 38 47N 121 36W
Feather Falls, *U.S.A.* 144 F5 39 36N 121 16W
Featherston, *N.Z.* 118 H4 41 6 S 175 20 E
Featherstone, *Zimbabwe* 107 F3 18 42 S 30 55 E
Fécamp, *France* 22 C7 49 45N 0 22 E
Fedala = Mohammedia, *Morocco* 98 B3 33 44N 7 21W
Federación, *Argentina* 158 C4 31 0 S 57 55W
Fedeshkūh, *Iran* 85 D7 28 49N 53 50 E
Fedjadj, Chott el, *Tunisia* 96 B1 33 52N 9 14 E
Fehérgyarmat, *Hungary* 31 D15 48 0N 22 30 E
Fehmarn, *Germany* 26 A7 54 27N 11 7 E
Fehmarn Bælt, *Denmark* 13 J14 54 35N 11 20 E
Fei Xian, *China* 67 G9 35 18N 117 59 E
Feijó, *Brazil* 156 B3 8 9 S 70 21W
Feilding, *N.Z.* 118 G4 40 13 S 175 35 E
Feira de Santana, *Brazil* 155 D4 12 15 S 38 57W
Feixiang, *China* 66 F8 36 30N 114 45 E
Fejér □, *Hungary* 31 D11 47 9N 18 30 E
Fejø, *Denmark* 15 K5 54 55N 11 30 E
Feke, *Turkey* 88 D6 37 48N 35 56 E
Fekete →, *Hungary* 31 F11 45 47N 18 15 E
Felanitx, *Spain* 33 B10 39 28N 3 9 E
Feldbach, *Austria* 30 E8 46 57N 15 52 E
Feldberg, *Baden-W., Germany* 27 H3 47 52N 8 0 E
Feldberg, *Mecklenburg-Vorpommern, Germany* 26 B9 53 20N 13 25 E
Feldkirch, *Austria* 30 D2 47 15N 9 37 E
Feldkirchen, *Austria* 30 E7 46 44N 14 6 E
Felicity, *U.S.A.* 141 F12 38 51N 84 6W
Felipe Carrillo Puerto, *Mexico* 147 D7 19 38N 88 3W
Felixlândia, *Brazil* 155 E3 18 47 S 44 55W
Felixstowe, *U.K.* 17 F9 51 58N 1 23 E
Felletin, *France* 24 C6 45 53N 2 11 E
Felton, *U.K.* 16 B6 55 18N 1 42W
Felton, *U.S.A.* 144 H4 37 3N 122 4W
Feltre, *Italy* 39 B8 46 1N 11 54 E
Femø, *Denmark* 15 K5 54 58N 11 53 E
Femunden, *Norway* 14 B5 62 10N 11 53 E
Fen He →, *China* 66 G6 35 36N 110 42 E
Fenelon Falls, *Canada* 136 B6 44 32N 78 45W
Feneroa, *Ethiopia* 95 E4 13 5N 39 3 E
Feng Xian, *Jiangsu, China* 66 G9 34 43N 116 35 E
Feng Xian, *Shaanxi, China* 66 H4 33 54N 106 40 E
Fengári, *Greece* 44 D7 40 25N 25 32 E
Fengcheng, *Jiangxi, China* 69 C10 28 12N 115 48 E
Fengcheng, *Liaoning, China* 67 D13 40 28N 124 5 E
Fengdu, *China* 68 C6 29 55N 107 41 E
Fengfeng, *China* 66 F8 36 28N 114 8 E
Fenggang, *China* 68 D6 27 57N 107 47 E
Fenghua, *China* 69 C13 29 40N 121 25 E
Fenghuang, *China* 68 D7 27 57N 109 29 E
Fenghuangzui, *China* 68 A7 33 30N 109 23 E
Fengjie, *China* 68 B7 31 5N 109 36 E
Fengkai, *China* 69 F8 23 24N 111 30 E
Fengle, *China* 69 B9 31 29N 112 29 E
Fengning, *China* 66 D9 41 10N 116 33 E
Fengqing, *China* 68 E2 24 38N 99 55 E
Fengqiu, *China* 66 G8 35 2N 114 25 E
Fengrun, *China* 67 E10 39 48N 118 8 E
Fengshan, *Guangxi Zhuangzu, China* 68 E7 24 39N 109 15 E
Fengshan, *Guangxi Zhuangzu, China* 68 E6 24 31N 107 3 E
Fengtai, *Anhui, China* 69 A11 32 50N 116 40 E
Fengtai, *Beijing, China* 66 E9 39 50N 116 18 E
Fengxian, *China* 69 B13 30 55N 121 26 E
Fengxiang, *China* 66 G4 34 29N 107 25 E
Fengxin, *China* 69 C10 28 41N 115 18 E
Fengyang, *China* 67 H9 32 51N 117 29 E
Fengyi, *China* 68 E3 25 37N 100 20 E
Fengzhen, *China* 66 D7 40 25N 113 2 E
Feni Is., *Papua N. G.* 120 C7 4 0 S 153 40 E
Fenit, *Ireland* 19 D2 52 17N 9 51W
Fennimore, *U.S.A.* 140 B6 42 59N 90 39W
Fenny, *Bangla.* 78 D3 22 55N 91 32 E
Feno, C. de, *France* 25 G12 41 58N 8 33 E
Fenoarivo Afovoany, *Madag.* 105 B8 18 26 S 46 34 E
Fenoarivo Atsinanana, *Madag.* 105 B8 17 22 S 49 25 E
Fens, The, *U.K.* 16 E8 52 38N 0 2W
Fenton, *U.S.A.* 141 B13 42 48N 83 42W
Fenxi, *China* 66 F6 36 40N 111 31 E
Fenyang, *China* 66 F6 37 18N 111 48 E
Fenyi, *China* 69 D10 27 45N 114 47 E
Feodosiya, *Ukraine* 51 K8 45 2N 35 16 E
Fer, C. de, *Algeria* 99 A6 37 3N 7 10 E
Ferdows, *Iran* 85 C8 33 58N 58 2 E
Fère-Champenoise, *France* 23 D10 48 45N 3 59 E
Fère-en-Tardenois, *France* 23 C10 49 10N 3 30 E
Ferentino, *Italy* 40 A6 41 42N 13 15 E
Ferfer, *Somali Rep.* 108 C3 5 4N 45 9 E
Fergana = Farghona, *Uzbekistan* 55 C5 40 23N 71 19 E
Ferganskaya Dolina = Farghonskaya Dolina, *Uzbekistan* 55 C5 40 50N 71 30 E
Ferganskiy Khrebet, *Kyrgyzstan* 55 C6 41 0N 73 50 E
Fergus, *Canada* 128 D3 43 43N 80 24W
Fergus Falls, *U.S.A.* 138 B6 46 17N 96 4W
Ferguson, *U.S.A.* 140 F6 38 45N 90 18W
Fergusson I., *Papua N. G.* 120 E6 9 30 S 150 45 E
Fériana, *Tunisia* 96 B1 34 59N 8 33 E
Feričanci, *Croatia* 42 B3 45 32N 18 0 E
Ferkane, *Algeria* 99 B6 34 37N 7 26 E
Ferkéssédougou, *Ivory C.* 100 D3 9 35N 5 6W
Ferlach, *Austria* 30 E7 46 32N 14 18 E
Ferland, *Canada* 128 B2 50 19N 88 27W
Ferlo, Vallée du, *Senegal* 100 B2 15 15N 14 15W
Fermanagh □, *U.K.* 19 B4 54 21N 7 40W
Fermo, *Italy* 39 E10 43 9N 13 43 E
Fermoselle, *Spain* 36 D4 41 19N 6 27W
Fermoy, *Ireland* 19 D3 52 9N 8 16W
Fernán Nuñéz, *Spain* 37 H6 37 40N 4 44W
Fernández, *Argentina* 158 B3 27 55 S 63 50W
Fernandina Beach, *U.S.A.* 135 K5 30 40N 81 27W
Fernando de Noronha, *Brazil* 154 B5 4 0 S 33 10W
Fernando Póo = Bioko, *Eq. Guin.* 101 E6 3 30N 8 40 E
Fernandópolis, *Brazil* 155 F1 20 16 S 50 14W
Ferndale, *Calif., U.S.A.* 142 F1 40 35N 124 16W
Ferndale, *Wash., U.S.A.* 144 B4 48 51N 122 36W
Fernie, *Canada* 130 D5 49 30N 115 5W
Fernlees, *Australia* 114 C4 23 51 S 148 7 E
Fernley, *U.S.A.* 142 G4 39 36N 119 15W
Feroke, *India* 83 J2 11 9N 75 46 E
Ferozepore = Firozpur, *India* 80 D6 30 55N 74 40 E
Férrai, *Greece* 44 D8 40 53N 26 10 E
Ferrandina, *Italy* 41 B9 40 29N 16 27 E
Ferrara, *Italy* 39 D8 44 50N 11 35 E
Ferrato, C., *Italy* 40 C2 39 18N 9 38 E
Ferreira do Alentejo, *Portugal* 37 G2 38 4N 8 6W
Ferreñafe, *Peru* 156 B2 6 42 S 79 50W
Ferrerías, *Spain* 33 B11 39 59N 4 1 E
Ferret, C., *France* 24 D2 44 38N 1 15W
Ferrette, *France* 23 E14 47 30N 7 20 E
Ferriday, *U.S.A.* 139 K9 31 38N 91 33W
Ferrières, *France* 23 D9 48 5N 2 48 E
Ferriete, *Italy* 38 D6 44 40N 9 30 E
Ferrol = El Ferrol, *Spain* 36 B2 43 29N 8 15W
Ferrol, Pen. de, *Peru* 156 B2 9 10 S 78 35W
Ferron, *U.S.A.* 143 G8 39 5N 111 8W
Ferros, *Brazil* 155 E3 19 14 S 43 2W
Ferryland, *Canada* 129 C9 47 2N 52 53W
Ferrysburg, *U.S.A.* 141 A10 43 5N 86 13W
Fertile, *U.S.A.* 138 B6 47 32N 96 17W
Fertília, *Italy* 40 B1 40 36N 8 17 E
Fertöszentmiklós, *Hungary* 31 D9 47 35N 16 53 E
Fès, *Morocco* 98 B4 34 0N 5 0W
Feschaux, *Belgium* 21 H5 50 9N 4 54 E
Feshi, *Zaïre* 103 D3 6 8 S 18 10 E

Girdle Ness 201

Gayny, *Russia* .... **54 A5** 60 18N 54 19 E
Gaysin = Haysyn, *Ukraine* **51 H5** 48 57N 29 25 E
Gayvoron = Hayvoron,
Ukraine .... **51 H5** 48 22N 29 52 E
Gaza, *Gaza Strip* .... **91 D3** 31 30N 34 28 E
Gaza □, *Mozam.* .... **105 C5** 23 10 S 32 45 E
Gaza Strip □, *Asia* ..... **91 D3** 31 29N 34 25 E
Gazaoua, *Niger* .... **97 F1** 13 32N 7 55 E
Gāzbor, *Iran* .... **85 D8** 28 5N 58 51 E
Gazelle Pen., *Papua N. G.* **120 C6** 4 40 S 152 0 E
Gazi, *Zaïre* .... **106 B1** 1 3N 24 30 E
Gaziantep, *Turkey* ...... **88 D7** 37 6N 37 23 E
Gazipaşa, *Turkey* .... **88 D5** 36 16N 32 18 E
Gazli, *Uzbekistan* .... **56 E7** 40 14N 63 24 E
Gbarnga, *Liberia* **100 D3** 7 19N 9 13W
Gbekebo, *Nigeria* .... **101 D5** 6 20N 4 56 E
Gboko, *Nigeria* .... **101 D6** 7 17N 9 4 E
Gbongan, *Nigeria* .... **101 D5** 7 28N 4 20 E
Gcuwa, *S. Africa* .... **105 E4** 32 20 S 28 11 E
Gdańsk, *Poland* .... **47 A5** 54 22N 18 40 E
Gdańsk □, *Poland* ...... **47 A5** 54 10N 18 30 E
Gdańska, Zatoka, *Poland* **47 A6** 54 30N 19 20 E
Gdov, *Russia* .... **13 G22** 58 48N 27 55 E
Gdynia, *Poland* .... **47 A5** 54 35N 18 33 E
Gebe, *Indonesia* .... **73 A3** 0 5N 129 25 E
Gebeit Mine, *Sudan* ..... **94 C4** 21 3N 36 29 E
Gebel Mûsa, *Egypt* ..... **94 J8** 28 32N 33 59 E
Gebze, *Turkey* .... **88 B3** 40 47N 29 25 E
Gecha, *Ethiopia* .... **95 F4** 7 30N 35 18 E
Gedaref, *Sudan* .... **95 E4** 14 2N 35 28 E
Gede, Tanjung, *Indonesia* **74 D3** 6 46 S 105 12 E
Gedinne, *Belgium* .... **21 J5** 49 59N 4 56 E
Gediz, *Turkey* .... **88 C3** 39 1N 29 24 E
Gediz →, *Turkey* .... **88 C2** 38 35N 26 48 E
Gedo, *Ethiopia* .... **95 F4** 9 2N 37 25 E
Gèdre, *France* .... **24 F4** 42 47N 0 2 E
Gedser, *Denmark* .... **15 K5** 54 35N 11 55 E
Gedser Odde, *Denmark* . **15 K5** 54 30N 11 58 E
Geegully Cr. →, *Australia* **112 C3** 18 32 S 123 41 E
Geel, *Belgium* .... **21 F5** 51 10N 4 59 E
Geelong, *Australia* ...... **116 E6** 38 10 S 144 22 E
Geelvink Chan., *Australia* **113 E1** 28 30 S 114 0 E
Geer →, *Belgium* ...... **21 G7** 50 51N 5 42 E
Geesthacht, *Germany* ... **26 B6** 53 26N 10 22 E
Geffen, *Neths.* .... **20 E6** 51 44N 5 28 E
Geidam, *Nigeria* .... **101 C7** 12 57N 11 57 E
Geikie →, *Canada* **131 B8** 57 45N 103 52W
Geili, *Sudan* .... **95 D3** 16 1N 32 37 E
Geilo, *Norway* .... **14 D2** 60 32N 8 14 E
Geinica, *Slovak Rep.* .... **31 C13** 48 51N 20 55 E
Geisingen, *Germany* .... **27 H4** 47 54N 8 38 E
Geislingen, *Germany* ... **27 G5** 48 37N 9 50 E
Geita, *Tanzania* .... **106 C3** 2 48 S 32 12 E
Geita □, *Tanzania* ...... **106 C3** 2 50 S 32 10 E
Gejiu, *China* .... **68 F4** 23 20N 103 10 E
Gel →, *Sudan* .... **95 F2** 7 5N 29 10 E
Gel River, *Sudan* .... **95 F2** 7 5N 29 10 E
Gela, *Italy* .... **41 E7** 37 4N 14 15 E
Gela, G. di, *Italy* .... **41 F7** 37 0N 14 20 E
Geladi, *Ethiopia* .... **108 C3** 6 59N 46 30 E
Gelderland □, *Neths.* ... **20 D8** 52 5N 6 10 E
Geldermalsen, *Neths.* ... **20 E6** 51 53N 5 17 E
Geldern, *Germany* ...... **26 D2** 51 31N 6 20 E
Geldrop, *Neths.* .... **21 F7** 51 25N 5 32 E
Geleen, *Neths.* .... **21 G7** 50 57N 5 49 E
Gelehun, *S. Leone* **100 D2** 8 20N 11 40W
Gelendost, *Turkey* ...... **88 C4** 38 7N 31 1 E
Gelendzhik, *Russia* ..... **53 H4** 44 33N 38 10 E
Gelib, *Somali Rep.* ...... **108 D2** 0 29N 42 46 E
Gelibolu, *Turkey* .... **88 B2** 40 28N 26 43 E
Gelidonya Burnu, *Turkey* **88 D4** 36 12N 30 24 E
Gelnhausen, *Germany* ... **27 E5** 50 11N 9 11 E
Gelsenkirchen, *Germany* . **26 D3** 51 32N 7 1 E
Gelting, *Germany* .... **26 A5** 54 45N 9 53 E
Gemas, *Malaysia* .... **77 L4** 2 37N 102 36 E
Gembloux, *Belgium* ..... **21 G5** 50 34N 4 43 E
Gemena, *Zaïre* .... **102 B3** 3 13N 19 48 E
Gemerek, *Turkey* .... **88 C7** 39 15N 36 10 E
Gemert, *Neths.* .... **21 E7** 51 33N 5 41 E
Gemlik, *Turkey* .... **88 B3** 40 26N 29 9 E
Gemona del Friuli, *Italy* . **39 B10** 46 16N 13 9 E
Gemsa, *Egypt* .... **94 B3** 27 39N 33 35 E
Gemünden, *Germany* .... **27 E5** 50 3N 9 42 E
Genale, *Ethiopia* .... **95 F4** 6 0N 39 30 E
Genale, *Somali Rep.* ..... **108 D2** 1 48N 44 42 E
Genappe, *Belgium* ...... **21 G4** 50 37N 4 30 E
Genç, *Turkey* .... **89 C9** 38 44N 40 34 E
Gençay, *France* .... **24 B4** 46 23N 0 23 E
Gendringen, *Neths.* ..... **20 E8** 51 52N 6 21 E
Gendt, *Neths.* .... **20 E7** 51 53N 5 59 E
Geneina, Gebel, *Egypt* .. **94 J8** 29 2N 33 55 E
Genemuiden, *Neths.* .... **20 C8** 52 38N 6 2 E
General Acha, *Argentina* **158 D3** 37 20 S 64 38W
General Alvear,
Buenos Aires, Argentina **158 D3** 36 0 S 60 0W
General Alvear, *Mendoza,*
Argentina .... **158 D2** 35 0 S 67 40W
General Artigas, *Paraguay* **158 B4** 26 52 S 56 16W
General Belgrano,
Argentina .... **158 D4** 36 35 S 58 47W
General Cabrera,
Argentina .... **158 C3** 32 53 S 63 52W
General Carrera, L., *Chile* **160 C2** 46 35 S 72 0W
General Cepeda, *Mexico* . **146 B4** 25 23N 101 27W
General Conesa, *Argentina* **160 B4** 40 6 S 64 25W
General Guido, *Argentina* **158 D4** 36 40 S 57 50W
General Juan Madariaga,
Argentina .... **158 D4** 37 0 S 57 0W
General La Madrid,
Argentina .... **158 D3** 37 17 S 61 20W
General Lorenzo Vintter,
Argentina .... **160 B4** 40 45 S 64 26W
General Luna, *Phil.* ..... **70 E4** 13 41N 122 10 E
General MacArthur, *Phil.* **71 F5** 11 18N 125 28 E
General Martin Miguel de
Güemes, *Argentina* .... **158 A3** 24 50 S 65 0W
General Paz, *Argentina* . **158 B4** 27 45 S 57 36W
General Pico, *Argentina* . **158 D3** 35 45 S 63 50W
General Pinedo, *Argentina* **158 B3** 27 15 S 61 20W
General Pinto, *Argentina* **158 C3** 34 45 S 61 50W
General Sampaio, *Brazil* . **154 B4** 4 2 S 39 29W
General Santos, *Phil.* .... **71 H5** 6 5N 125 14 E
General Tinio, *Phil.* ..... **70 D3** 15 39N 121 10 E
General Toshevo, *Bulgaria* **43 D13** 43 42N 28 6 E
General Trevino, *Mexico* **147 B5** 26 14N 99 29W
General Trías, *Mexico* ... **146 B3** 28 21N 106 22W
General Viamonte,
Argentina .... **158 D3** 35 1 S 61 3W
General Villegas,
Argentina .... **158 D3** 35 5 S 63 0W
General Vintter, L.,
Argentina .... **160 B2** 43 55 S 71 40W
Generoso, Mte., *Switz.* .. **29 E8** 45 56N 9 2 E
Genesee, *Idaho, U.S.A.* . **142 C5** 46 33N 116 56W
Genesee, *Pa., U.S.A.* ... **136 E7** 41 59N 77 54W
Genesee →, *U.S.A.* **136 C7** 43 16N 77 36W
Geneseo, *Ill., U.S.A.* **140 C6** 41 27N 90 9W
Geneseo, *Kans., U.S.A.* . **138 F5** 38 31N 98 10W
Geneseo, *N.Y., U.S.A.* .. **136 D7** 42 48N 77 49W
Geneva = Genève, *Switz.* **28 D2** 46 12N 6 9 E
Geneva, *Ala., U.S.A.* **135 K3** 31 2N 85 52W
Geneva, *Ill., U.S.A.* **141 C8** 41 53N 88 18W
Geneva, *Ind., U.S.A.* **141 D12** 40 36N 84 58W
Geneva, *N.Y., U.S.A.* ... **136 D7** 42 52N 76 59W
Geneva, *Nebr., U.S.A.* .. **138 E6** 40 32N 97 36W
Geneva, *Ohio, U.S.A.* ... **136 E4** 41 48N 80 57W
Geneva, L. = Léman, L.,
Europe .... **28 D3** 46 26N 6 30 E
Geneva, L., *U.S.A.* **141 B8** 42 38N 88 30W
Genève, *Switz.* .... **28 D2** 46 12N 6 9 E
Genève □, *Switz.* .... **28 D2** 46 10N 6 10 E
Geng, *Afghan.* .... **79 C1** 31 22N 61 28 E
Gengenbach, *Germany* .. **27 G4** 48 24N 8 0 E
Gengma, *China* .... **68 F2** 23 32N 99 20 E
Genichesk = Henichesk,
Ukraine .... **51 J8** 46 12N 34 50 E
Genil →, *Spain* **37 H5** 37 42N 5 19W
Génissiat, Barr. de, *France* **25 B9** 46 1N 5 48 E
Genk, *Belgium* .... **21 G7** 50 58N 5 32 E
Genkai-Nada, *Japan* .... **62 D2** 34 0N 130 0 E
Genlis, *France* .... **23 E12** 47 11N 5 12 E
Gennargentu, Mti. del,
Italy .... **40 C2** 40 1N 9 19 E
Gennep, *Neths.* .... **21 E7** 51 41N 5 59 E
Gennes, *France* **22 E6** 47 20N 0 17W
Genoa = Génova, *Italy* .. **38 D5** 44 25N 8 57 E
Genoa, *Australia* .... **117 D8** 37 29 S 149 35 E
Genoa, *Ill., U.S.A.* **141 B8** 42 6N 88 42W
Genoa, *N.Y., U.S.A.* **137 D8** 42 40N 76 32W
Genoa, *Nebr., U.S.A.* ... **138 E6** 41 27N 97 44W
Genoa, *Nev., U.S.A.* **144 F7** 39 2N 119 50W
Genoa →, *Argentina* .... **160 B2** 44 55 S 70 5W
Genoa City, *U.S.A.* **141 B8** 42 30N 88 20W
Génova, *Italy* .... **38 D5** 44 25N 8 57 E
Génova, G. di, *Italy* ..... **38 E6** 44 0N 9 0 E
Gent, *Belgium* .... **21 F3** 51 2N 3 42 E
Gentbrugge, *Belgium* .... **21 F3** 51 3N 3 47 E
Genthin, *Germany* ...... **26 C8** 52 25N 12 9 E
Gentio do Ouro, *Brazil* .. **154 D3** 11 25 S 42 30W
Geographe B., *Australia* . **113 F2** 33 30 S 115 15 E
Geographe Chan.,
Australia .... **113 D1** 24 30 S 113 0 E
Geokchay = Göyçay,
Azerbaijan .... **53 K8** 40 42N 47 43 E
Georga, Zemlya, *Russia* . **56 A5** 80 30N 49 0 E
George, *S. Africa* .... **104 E3** 33 58 S 22 29 E
George →, *Canada* **129 A6** 58 49N 66 10W
George, L., *N.S.W.,*
Australia .... **117 C8** 35 10 S 149 25 E
George, L., *S. Austral.,*
Australia .... **116 D4** 37 25 S 140 0 E
George, L., *W. Austral.,*
Australia .... **112 D3** 22 45 S 123 40 E
George, L., *Uganda* ..... **106 B3** 0 5N 30 10 E
George, L., *Fla., U.S.A.* . **135 L5** 29 17N 81 36W
George, L., *N.Y., U.S.A.* **137 C11** 43 37N 73 33W
George Gill Ra., *Australia* **112 D5** 24 22 S 131 45 E
George River =
Kangiqsualujjuaq,
Canada **127 C13** 58 30N 65 59W
George Sound, *N.Z.* .... **119 E2** 44 52 S 167 25 E
George Town, *Bahamas* . **148 B4** 23 33N 75 47W
George Town, *Malaysia* . **77 K3** 5 25N 100 15 E
George V Land,
Antarctica .... **7 C10** 69 0 S 148 0 E
George VI Sound,
Antarctica .... **7 D17** 71 0 S 68 0W
George West, *U.S.A.* **139 L5** 28 20N 98 7W
Georgetown, *Australia* ... **114 B3** 18 17 S 143 33 E
Georgetown, *Ont.,*
Canada **128 D4** 43 40N 79 56W
Georgetown, *P.E.I.,*
Canada **129 C7** 46 13N 62 24W
Georgetown, *Cayman Is.* **148 C3** 19 20N 81 24W
Georgetown, *Gambia* **100 C2** 13 30N 14 47W
Georgetown, *Guyana* **153 B6** 6 50N 58 12W
Georgetown, *Calif.,*
U.S.A. **144 G6** 38 54N 120 50W
Georgetown, *Colo.,*
U.S.A. **142 G11** 39 42N 105 42W
Georgetown, *Ill., U.S.A.* . **141 E9** 39 59N 87 38W
Georgetown, *Ky., U.S.A.* **134 F3** 38 13N 84 33W
Georgetown, *Ohio,*
U.S.A. **141 F13** 38 52N 83 54W
Georgetown, *S.C., U.S.A.* **135 J6** 33 23N 79 17W
Georgetown, *Tex., U.S.A.* **139 K6** 30 38N 97 41W
Georgi Dimitrov, *Bulgaria* **43 E8** 42 15N 23 54 E
Georgi Dimitrov, Yazovir,
Bulgaria .... **43 E10** 42 37N 25 18 E
Georgia □, *U.S.A.* **135 J4** 32 50N 83 15W
Georgia ■, *Asia* .... **53 J6** 42 0N 43 0 E
Georgia, Str. of, *Canada* . **130 D4** 49 25N 124 0W
Georgian B., *Canada* **128 C3** 45 15N 81 0W
Georgina →, *Australia* .. **114 C2** 23 30 S 139 47 E
Georgina Downs,
Australia .... **114 C2** 21 10 S 137 40 E
Georgiu-Dezh = Liski,
Russia .... **52 E4** 51 3N 39 30 E
Georgiyevka, *Kazakhstan* **55 B7** 43 3N 74 43 E
Georgiyevsk, *Russia* .... **53 H6** 44 12N 43 28 E
Gera, *Germany* .... **26 E8** 50 53N 12 4 E
Geraardsbergen, *Belgium* **21 G3** 50 45N 3 53 E
Geral, Serra, *Bahia, Brazil* **155 D3** 14 0 S 41 0W
Geral, Serra, *Goiás, Brazil* **154 D2** 11 15 S 46 30W
Geral, Serra,
Sta. Catarina, Brazil ... **159 B6** 26 25 S 50 0W
Geral de Goiás, Serra,
Brazil .... **155 D2** 12 0 S 46 0W
Geral do Paranã Serra,
Brazil .... **155 E2** 15 0 S 47 30W
Gerald, *U.S.A.* **140 F5** 38 24N 91 20W
Geraldine, *N.Z.* .... **119 E6** 44 5 S 171 15 E
Geraldine, *U.S.A.* **142 C8** 47 36N 110 16W
Geraldton, *Australia* .... **113 E1** 28 48 S 114 32 E
Geraldton, *Canada* **128 C2** 49 44N 86 59W
Geranium, *Australia* ..... **116 C4** 35 23 S 140 11 E
Gérardmer, *France* ...... **23 D13** 48 3N 6 50 E
Gercüş, *Turkey* .... **89 D9** 37 34N 41 23 E
Gerede, *Turkey* .... **88 B5** 40 45N 32 10 E
Gereshk, *Afghan.* .... **79 C2** 31 47N 64 35 E
Gérgal, *Spain* **35 H2** 37 7N 2 31W
Gerik, *Malaysia* .... **77 K3** 5 50N 101 15 E
Gering, *U.S.A.* **138 E3** 41 50N 103 40W
Gerlach, *U.S.A.* **142 F4** 40 39N 119 21W
Gerlachovka, *Slovak Rep.* **31 B13** 49 11N 20 7 E
Gerlogubi, *Ethiopia* ..... **108 C3** 6 53N 45 3 E
German Planina,
Macedonia .... **42 E7** 42 20N 22 0 E
Germansen Landing,
Canada **130 B4** 55 43N 124 40W
Germantown, *U.S.A.* **141 E12** 39 38N 84 22W
Germany ■, *Europe* ..... **26 E6** 51 0N 10 0 E
Germersheim, *Germany* . **27 F4** 49 12N 8 22 E
Germī, *Iran* .... **89 C13** 39 1N 48 3 E
Germiston, *S. Africa* .... **105 D4** 26 15 S 28 10 E
Gernsheim, *Germany* .... **27 F4** 49 45N 8 30 E
Gero, *Japan* .... **63 B9** 35 48N 137 14 E
Gerolstein, *Germany* .... **27 E2** 50 13N 6 39 E
Gerolzhofen, *Germany* .. **27 F6** 49 54N 10 21 E
Gerona, *Spain* .... **34 D7** 41 58N 2 46 E
Gerona □, *Spain* .... **34 C7** 42 11N 2 30 E
Gérouville, *Belgium* ..... **21 J6** 49 37N 5 26 E
Gerrard, *Canada* **130 C5** 50 30N 117 17W
Gerringong, *Australia* ... **117 C9** 34 46 S 150 47 E
Gers □, *France* .... **24 E4** 43 35N 0 30 E
Gers →, *France* .... **24 D4** 44 9N 0 39 E
Gersfeld, *Germany* ...... **26 E5** 50 27N 9 56 E
Gersoppa Falls, *India* ... **83 G2** 14 12N 74 46 E
Gerze, *Turkey* .... **88 B6** 41 48N 35 12 E
Geseke, *Germany* .... **26 D4** 51 38N 8 31 E
Geser, *Indonesia* .... **73 B4** 3 50 S 130 54 E
Gesso →, *Italy* .... **38 D4** 44 24N 7 33 E
Gestro, Wabi →,
Ethiopia .... **95 G5** 4 12N 42 2 E
Gesves, *Belgium* .... **21 H6** 50 24N 5 4 E
Getafe, *Spain* **36 E7** 40 18N 3 44W
Gethsémani, *Canada* **129 B7** 50 13N 60 40W
Gettysburg, *Pa., U.S.A.* . **134 F7** 39 50N 77 14W
Gettysburg, *S. Dak.,*
U.S.A. **138 C5** 45 1N 99 57W
Getz Ice Shelf, *Antarctica* **7 D14** 75 0 S 130 0W
Geul →, *Neths.* .... **21 G7** 50 53N 5 43 E
Geureudong, Mt.,
Indonesia .... **74 B1** 4 13N 96 42 E
Geurie, *Australia* .... **117 B8** 32 22 S 148 50 E
Gevaş, *Turkey* .... **89 C10** 38 15N 43 6 E
Gévaudan, *France* ...... **24 D7** 44 40N 3 40 E
Gevgelija, *Macedonia* .... **42 F7** 41 9N 22 30 E
Gévora →, *Spain* **37 G4** 38 53N 6 57W
Gex, *France* .... **25 B10** 46 21N 6 3 E
Geyikli, *Turkey* .... **44 E8** 39 50N 26 12 E
Geyser, *U.S.A.* **142 C8** 47 16N 110 30W
Geyserville, *U.S.A.* **144 G4** 38 42N 122 54W
Geyve, *Turkey* .... **88 B4** 40 30N 30 18 E
Ghâbat el Arab = Wang
Kai, *Sudan* .... **95 F2** 9 3N 29 23 E
Ghaghara →, *India* ..... **81 G11** 25 45N 84 40 E
Ghalat, *Oman* .... **87 B7** 21 6N 58 53 E
Ghalla, Wadi el →,
Sudan .... **95 E2** 10 25N 27 32 E
Ghallamane, *Mauritania* . **98 D3** 23 15N 10 0W
Ghana ■, *W. Afr.* **101 D4** 8 0N 1 0W
Ghansor, *India* .... **81 H9** 22 39N 80 1 E
Ghanzi, *Botswana* .... **104 C3** 21 50 S 21 34 E
Ghanzi □, *Botswana* .... **104 C3** 21 50 S 21 45 E
Gharb el Istiwa'iya □,
Sudan .... **95 F2** 5 0N 30 0 E
Gharbîya, Es Sahrâ el,
Egypt .... **94 B2** 27 40N 26 30 E
Ghard Abû Muharik,
Egypt .... **94 B2** 26 50N 30 0 E
Ghardaïa, *Algeria* .... **99 B5** 32 20N 3 37 E
Ghârib, G., *Egypt* ...... **94 J8** 28 6N 32 54 E
Gharm, W. →, *Oman* ... **87 C7** 19 57N 57 38 E
Gharyān, *Libya* .... **96 B2** 32 10N 13 0 E
Gharyān □, *Libya* ...... **96 B2** 30 35N 13 0 E
Ghat, *Libya* .... **96 D2** 24 59N 10 11 E
Ghatal, *India* .... **81 H12** 22 40N 87 46 E
Ghatampur, *India* .... **81 F9** 26 8N 80 13 E
Ghatere, *Solomon Is.* .... **121 L10** 7 55 S 159 0 E
Ghatprabha →, *India* ... **83 F2** 16 15N 75 20 E
Ghaṭṭī, *Si. Arabia* .... **84 D3** 31 16N 37 31 E
Ghawdex = Gozo, *Malta* **32 C1** 36 3N 14 13 E
Ghayl, *Si. Arabia* .... **86 B4** 21 40N 46 20 E
Ghayl Bā Wazīr, *Yemen* . **87 D5** 14 47N 49 22 E
Ghazal, Bahr el →, *Chad* **97 F3** 13 0N 15 47 E
Ghazâl, Bahr el →,
Sudan .... **95 F3** 9 31N 30 25 E
Ghazaouet, *Algeria* **99 A4** 35 8N 1 50W
Ghaziabad, *India* .... **80 E7** 28 42N 77 26 E
Ghazipur, *India* .... **81 G10** 25 38N 83 35 E
Ghaznī, *Afghan.* .... **79 B3** 33 30N 68 28 E
Ghaznī □, *Afghan.* ..... **79 B3** 32 10N 68 20 E
Ghedi, *Italy* .... **38 C7** 45 24N 10 16 E
Ghelari, *Romania* .... **46 D3** 45 38N 22 45 E
Ghèlinsor, *Somali Rep.* .. **108 C3** 6 28N 46 39 E
Ghent = Gent, *Belgium* . **21 F3** 51 2N 3 42 E
Gheorghe Gheorghiu-
Dej = Oneşti, *Romania* **46 C7** 46 17N 26 47 E
Gheorgheni, *Romania* ... **46 C6** 46 43N 25 41 E
Ghergani, *Romania* ..... **46 E6** 44 37N 25 37 E
Gherla, *Romania* .... **46 B4** 47 2N 23 57 E
Ghilarza, *Italy* .... **40 B1** 40 7N 8 50 E
Ghisonaccia, *France* .... **25 F13** 42 1N 9 26 E
Ghisoni, *France* .... **25 F13** 42 7N 9 12 E
Ghizao, *Afghan.* .... **80 C1** 33 20N 65 44 E
Ghizar →, *Pakistan* .... **81 A5** 36 15N 73 43 E
Ghod →, *India* .... **82 E2** 18 30N 74 35 E
Ghogha, *India* .... **80 J5** 21 40N 72 20 E
Ghot Ogrein, *Egypt* .... **94 A2** 31 10N 25 20 E
Ghotaru, *India* .... **80 F4** 27 20N 70 1 E
Ghotki, *Pakistan* .... **80 E3** 28 5N 69 21 E
Ghowr □, *Afghan.* ...... **79 B2** 34 0N 64 20 E
Ghudaf, W. al →, *Iraq* . **89 F10** 32 56N 43 30 E
Ghudāmis, *Libya* .... **96 B1** 30 11N 9 29 E
Ghughri, *India* .... **81 H9** 22 39N 80 41 E
Ghugus, *India* .... **82 E4** 19 58N 79 12 E
Ghulam Mohammad
Barrage, *Pakistan* ..... **80 G3** 25 30N 68 20 E
Ghūrayrah, *Si. Arabia* ... **86 C3** 18 37N 42 41 E
Ghūrīān, *Afghan.* ...... **79 B1** 34 17N 61 25 E
Gia Dinh, *Vietnam* ...... **77 G6** 10 49N 106 42 E
Gia Lai = Pleiku, *Vietnam* **76 F7** 13 57N 108 0 E
Gia Nghia, *Vietnam* ..... **77 G6** 11 58N 107 42 E
Gia Ngoc, *Vietnam* ..... **76 E7** 14 50N 108 58 E
Gia Vuc, *Vietnam* ...... **76 E7** 14 42N 108 34 E
Giamama, *Somali Rep.* .. **108 D2** 0 4N 42 44 E
Giannutri, *Italy* .... **38 F8** 42 15N 11 6 E
Giant Forest, *U.S.A.* **144 J8** 36 36N 118 43W
Giant Mts. = Krkonoše,
Czech. .... **30 A8** 50 50N 15 35 E
Giants Causeway, *U.K.* .. **19 A5** 55 16N 6 29W
Giarabub = Al Jaghbūb,
Libya .... **96 C4** 29 42N 24 38 E
Giarre, *Italy* .... **41 E8** 37 43N 15 11 E
Giaveno, *Italy* .... **38 C4** 45 2N 7 21 E
Gibara, *Cuba* **148 B4** 21 9N 76 11W
Gibb River, *Australia* ... **112 C4** 16 26 S 126 26 E
Gibbon, *U.S.A.* **138 E5** 40 45N 98 51W
Gibe →, *Ethiopia* ...... **95 F4** 7 20N 37 36 E
Gibellina, *Italy* .... **40 E6** 37 47N 12 58 E
Gibraléon, *Spain* **37 H4** 37 23N 6 58W
Gibraltar ■, *Europe* **37 J5** 36 7N 5 22W
Gibraltar, Str. of,
Medit. S. **37 K5** 35 55N 5 40W
Gibson City, *U.S.A.* **141 D8** 40 28N 88 22W
Gibson Desert, *Australia* . **112 D4** 24 0 S 126 0 E
Gibsonburg, *U.S.A.* **141 C13** 41 23N 83 19W
Gibsons, *Canada* **130 D4** 49 24N 123 32W
Gibsonville, *U.S.A.* **144 F6** 39 46N 120 54W
Giddalur, *India* .... **83 G4** 15 20N 78 57 E
Giddings, *U.S.A.* **139 K6** 30 11N 96 56W
Gidole, *Ethiopia* .... **95 F4** 5 40N 37 25 E
Gien, *France* .... **23 E9** 47 40N 2 36 E
Giessen, *Germany* ...... **26 E4** 50 34N 8 41 E
Gieten, *Neths.* .... **20 B9** 53 1N 6 46 E
Gīfān, *Iran* .... **85 B8** 37 54N 57 28 E
Gifatin, Geziret, *Egypt* .. **94 B3** 27 10N 33 50 E
Gifford Creek, *Australia* . **112 D2** 24 3 S 116 16 E
Gifhorn, *Germany* ...... **26 C6** 52 30N 10 33 E
Gifu, *Japan* .... **63 B8** 35 30N 136 45 E
Gifu □, *Japan* .... **63 B8** 35 40N 137 0 E
Gigant, *Russia* .... **53 G5** 46 28N 41 20 E
Giganta, Sa. de la, *Mexico* **146 B2** 25 30N 111 30W
Gigen, *Bulgaria* .... **43 D9** 43 40N 24 28 E
Gigha, *U.K.* **18 F3** 55 42N 5 44W
Giglei, *Somali Rep.* ..... **108 C3** 5 25N 45 20 E
Giglio, *Italy* .... **38 F7** 42 20N 10 52 E
Gigmoto, *Phil.* .... **70 E5** 13 47N 124 23 E
Gignac, *France* .... **24 E7** 43 39N 3 32 E
Gigüela →, *Spain* **35 F1** 39 8N 3 44W
Gijón, *Spain* **36 B5** 43 32N 5 42W
Gil I., *Canada* **130 C3** 53 12N 129 15W
Gila →, *U.S.A.* **143 K6** 32 43N 114 33W
Gila Bend, *U.S.A.* **143 K7** 32 57N 112 43W
Gila Bend Mts., *U.S.A.* . **143 K7** 33 10N 113 0W
Gīlān □, *Iran* .... **85 B6** 37 0N 50 0 E
Gilău, *Romania* .... **46 C4** 46 45N 23 23 E
Gilbert →, *Australia* .... **114 B3** 16 35 S 141 15 E
Gilbert Is., *Kiribati* ..... **122 G9** 1 0N 172 0 E
Gilbert Plains, *Canada* .. **131 C8** 51 9N 100 28W
Gilbert River, *Australia* . **114 B3** 18 9 S 142 52 E
Gilberton, *Australia* ..... **114 B3** 19 16 S 143 35 E
Gilbués, *Brazil* .... **154 C2** 9 50 S 45 21W
Gilf el Kebîr, Hadabat el,
Egypt .... **94 C2** 23 50N 25 50 E
Gilford I., *Canada* **130 C3** 50 40N 126 30W
Gilgandra, *Australia* .... **117 A8** 31 43 S 148 39 E
Gilgil, *Kenya* .... **106 C4** 0 30 S 36 20 E
Gilgit, *India* .... **81 B6** 35 50N 74 15 E
Gilgit →, *Pakistan* ..... **81 B6** 35 44N 74 37 E
Gilgunnia, *Australia* .... **117 B7** 32 26 S 146 2 E
Giljeva Planina,
Serbia, Yug. .... **42 D4** 43 9N 20 0 E
Gillam, *Canada* **131 B10** 56 20N 94 40W
Gilleleje, *Denmark* ..... **15 H6** 56 8N 12 19 E
Gillen, L., *Australia* .... **113 E3** 26 11 S 124 38 E
Gilles, L., *Australia* .... **116 B2** 32 50 S 136 45 E
Gillespie, *U.S.A.* **140 E7** 39 8N 89 49W
Gillespies Pt., *N.Z.* .... **119 D4** 43 24 S 169 49 E
Gillette, *U.S.A.* **138 C2** 44 18N 105 30W
Gilliat, *Australia* .... **114 C3** 20 40 S 141 28 E
Gillingham, *U.K.* .... **17 F8** 51 23N 0 33 E
Gilly, *Belgium* .... **21 H4** 50 25N 4 29 E
Gilman, *U.S.A.* **140 D9** 40 46N 88 0W
Gilman City, *U.S.A.* **140 D3** 40 8N 93 53W
Gilmer, *U.S.A.* **139 J7** 32 44N 94 57W
Gilmore, *Australia* ...... **117 C8** 35 20 S 148 12 E
Gilmore, L., *Australia* ... **113 F3** 32 29 S 121 37 E
Gilmour, *Canada* **128 D4** 44 48N 77 37W
Gilo →, *Ethiopia* .... **95 F3** 8 10N 33 15 E
Gilort →, *Romania* ..... **46 E4** 44 38N 23 32 E
Gilroy, *U.S.A.* **144 H5** 37 1N 121 34W
Giluwe, Mt., *Papua N. G.* **120 D2** 6 8 S 143 52 E
Gilze, *Neths.* .... **21 E5** 51 32N 4 57 E
Gimbi, *Ethiopia* .... **95 F4** 9 3N 35 42 E
Gimigliano, *Italy* .... **41 D9** 38 58N 16 32 E
Gimli, *Canada* **131 C9** 50 40N 97 0W
Gimone →, *France* ..... **24 E5** 44 0N 1 6 E
Gimont, *France* .... **24 E4** 43 38N 0 52 E
Gin →, *Sri Lanka* ...... **83 L5** 6 5N 80 7 E
Gin Gin, *Australia* ...... **115 D5** 25 0 S 151 58 E
Ginâh, *Egypt* .... **94 B3** 25 21N 30 30 E
Ginatilan, *Phil.* .... **71 G4** 9 34N 123 19 E
Gindie, *Australia* .... **114 C4** 23 44 S 148 8 E
Gingin, *Australia* .... **113 F2** 31 22 S 115 54 E
Gîngiova, *Romania* ..... **46 F4** 43 54N 23 50 E
Gingoog, *Phil.* .... **71 G5** 8 50N 125 7 E
Ginir, *Ethiopia* .... **90 F3** 7 6N 40 40 E
Ginosa, *Italy* .... **41 B9** 40 35N 16 45 E
Ginzo de Limia, *Spain* ... **36 C3** 42 3N 7 47W
Giohar, *Somali Rep.* .... **108 D3** 2 48N 45 30 E
Gióia, G. di, *Italy* ...... **41 D8** 38 30N 15 45 E
Gióia del Colle, *Italy* ... **41 B9** 40 48N 16 55 E
Gióia Táuro, *Italy* ...... **41 D8** 38 25N 15 54 E
Gioiosa Iónica, *Italy* .... **41 D9** 38 20N 16 18 E
Gióna, Óros, *Greece* .... **45 F4** 38 38N 22 14 E
Giovi, Passo dei, *Italy* .. **38 D5** 44 33N 8 57 E
Giovinazzo, *Italy* .... **41 A9** 41 11N 16 40 E
Gir Hills, *India* .... **80 J4** 21 0N 71 0 E
Girab, *India* .... **80 F4** 26 2N 70 38 E
Girâfi, W. →, *Egypt* ... **91 F3** 29 58N 34 39 E
Giraltovce, *Slovak Rep.* . **31 B14** 49 7N 21 32 E
Girard, *Ill., U.S.A.* **140 E7** 39 27N 89 47W
Girard, *Kans., U.S.A.* ... **139 G7** 37 31N 94 51W
Girard, *Ohio, U.S.A.* **136 E4** 41 9N 80 42W
Girard, *Pa., U.S.A.* **136 D4** 42 0N 80 19W
Girardot, *Colombia* **152 C3** 4 18N 74 48W
Girdle Ness, *U.K.* **18 D6** 57 9N 2 3W

Hamélé 205

Guiping, *China* 69 F8 23 21N 110 2 E
Guipúzcoa □, *Spain* 34 B2 43 12N 2 15W
Guir, O. →, *Algeria* 99 B4 31 29N 2 17W
Guiratinga, *Brazil* 157 D7 16 21 S 53 45W
Güiria, *Venezuela* 153 A5 10 32N 62 18W
Guiscard, *France* 23 C10 49 40N 3 1 E
Guise, *France* 23 C10 49 52N 3 35 E
Guitiriz, *Spain* 36 B3 43 11N 7 50W
Guiuan, *Phil.* 71 F5 11 5N 125 55 E
Guixi, *China* 69 C11 28 16N 117 15 E
Guiyang, *Guizhou, China* 68 D6 26 32N 106 40 E
Guiyang, *Hunan, China* 69 E9 25 46N 112 42 E
Guizhou □, *China* 68 D6 27 0N 107 0 E
Gujan-Mestras, *France* 24 D2 44 38N 1 4W
Gujarat □, *India* 80 H4 23 20N 71 0 E
Gujiang, *China* 69 D10 27 11N 114 47 E
Gujranwala, *Pakistan* 79 B4 32 10N 74 12 E
Gujrat, *Pakistan* 79 B4 32 40N 74 2 E
Gukovo, *Russia* 53 F5 48 1N 39 58 E
Gulargambone, *Australia* 117 A8 31 20 S 148 30 E
Gulbarga, *India* 82 F3 17 20N 76 50 E
Gulbene, *Latvia* 13 H22 57 8N 26 52 E
Gulcha, *Kyrgyzstan* 55 C6 40 19N 73 26 E
Guledagudda, *India* 83 F2 16 3N 75 48 E
Gulf, The, *Asia* 85 E6 27 0N 50 0 E
Gulfport, *U.S.A.* 139 K10 30 22N 89 6W
Gulgong, *Australia* 117 B8 32 20 S 149 49 E
Gulin, *China* 68 C5 28 1N 105 50 E
Gulistan, *Pakistan* 80 D2 30 30N 66 35 E
Guliston, *Uzbekistan* 55 C4 40 29N 68 46 E
Gull Lake, *Canada* 131 C7 50 10N 108 29W
Gullegem, *Belgium* 21 G2 50 51N 3 13 E
Güllük, *Turkey* 88 D2 37 14N 27 35 E
Gulma, *Nigeria* 101 C5 12 40N 4 23 E
Gulmarg, *India* 81 B6 34 3N 74 25 E
Gülnar, *Turkey* 88 D5 36 19N 33 24 E
Gulnare, *Australia* 116 B3 33 27 S 138 27 E
Gulpen, *Neths.* 21 G7 50 49N 5 53 E
Gülpinar, *Turkey* 44 E8 39 32N 26 10 E
Gülşehir, *Turkey* 88 C6 38 44N 34 37 E
Gulshad, *Kazakhstan* 56 E8 46 45N 74 25 E
Gulsvik, *Norway* 14 D3 60 24N 9 38 E
Gulu, *Uganda* 106 B3 2 48N 32 17 E
Gŭlŭbovo, *Bulgaria* 43 E10 42 8N 25 55 E
Gulwe, *Tanzania* 106 D4 6 30 S 36 25 E
Gulyaypole = Hulyaypole, *Ukraine* 51 J9 47 45N 36 21 E
Gum Lake, *Australia* 116 B5 32 42 S 143 9 E
Gumaca, *Phil.* 70 E4 13 55N 122 6 E
Gumal →, *Pakistan* 80 D4 31 40N 71 50 E
Gumbaz, *Pakistan* 80 D3 30 2N 69 0 E
Gumel, *Nigeria* 101 C6 12 39N 9 22 E
Gumiel de Hizán, *Spain* 34 D1 41 46N 3 41W
Gumlu, *Australia* 114 B4 19 53 S 147 41 E
Gumma □, *Japan* 63 A10 36 30N 138 20 E
Gummersbach, *Germany* 26 D3 51 1N 7 34 E
Gummi, *Nigeria* 101 C6 12 4N 5 9 E
Gümüşhaciköy, *Turkey* 88 B6 40 50N 35 18 E
Gümüşhane, *Turkey* 89 B8 40 30N 39 30 E
Gumzai, *Indonesia* 73 C4 5 28 S 134 42 E
Guna, *Ethiopia* 95 E4 11 50N 37 40 E
Guna, *India* 80 G7 24 40N 77 19 E
Gundagai, *Australia* 117 C8 35 3 S 148 6 E
Gundelfingen, *Germany* 27 G6 48 19N 10 22 E
Gundih, *Indonesia* 75 D4 7 10 S 110 56 E
Gundlakamma →, *India* 83 G5 15 30N 80 15 E
Gunebang, *Australia* 117 B7 33 1 S 146 38 E
Güneydoğu Toroslar, *Turkey* 89 C9 38 20N 40 30 E
Gungal, *Australia* 117 B9 32 17 S 150 32 E
Gungu, *Zaïre* 103 D3 5 43 S 19 20 E
Gunisao →, *Canada* 131 C9 53 56N 97 53W
Gunisao L., *Canada* 131 C9 53 33N 96 15W
Gunnbjørn Fjeld, *Greenland* 6 C6 68 55N 29 47W
Gunnedah, *Australia* 117 A9 30 59 S 150 15 E
Gunniguldrie, *Australia* 117 B7 33 12 S 146 8 E
Gunningbar Cr. →, *Australia* 117 A7 31 14 S 147 6 E
Gunnison, *Colo., U.S.A.* 143 G10 38 33N 106 56W
Gunnison, *Utah, U.S.A.* 142 G8 39 9N 111 49W
Gunnison →, *U.S.A.* 143 G9 39 4N 108 35W
Gunpowder, *Australia* 114 B2 19 42 S 139 22 E
Guntakal, *India* 83 G3 15 11N 77 27 E
Guntersville, *U.S.A.* 135 H2 34 21N 86 18W
Guntong, *Malaysia* 77 K3 4 36N 101 3 E
Guntur, *India* 83 F5 16 23N 80 30 E
Gunungapi, *Indonesia* 72 C3 6 45 S 126 30 E
Gunungsitoli, *Indonesia* 74 B1 1 15N 97 30 E
Gunupur, *India* 82 E6 19 5N 83 50 E
Günz →, *Germany* 27 G6 48 27N 10 16 E
Gunza, *Angola* 103 E2 10 50 S 13 50 E
Günzburg, *Germany* 27 G6 48 26N 10 17 E
Gunzenhausen, *Germany* 27 F6 49 7N 10 44 E
Guo He →, *China* 67 H9 32 59N 117 10 E
Guoyang, *China* 66 H9 33 32N 116 12 E
Gupis, *Pakistan* 81 A5 36 15N 73 20 E
Gura Humorului, *Romania* 46 B6 47 35N 25 53 E
Gura-Teghii, *Romania* 46 D7 45 30N 26 25 E
Gurag, *Ethiopia* 95 F4 8 20N 38 20 E
Gurdaspur, *India* 80 C6 32 5N 75 31 E
Gurdon, *U.S.A.* 139 J8 33 55N 93 9W
Gurdzhaani, *Georgia* 53 K7 41 43N 45 52 E
Gurgaon, *India* 80 E7 28 27N 77 1 E
Gürgentepe, *Turkey* 88 B7 40 51N 37 50 E
Gurghiu, Munţii, *Romania* 46 C6 46 41N 25 15 E
Gurgueia →, *Brazil* 154 C3 6 50 S 43 24W
Gurha, *India* 80 G4 25 12N 71 39 E
Guri, Embalse de, *Venezuela* 153 B5 7 50N 62 52W
Gurk →, *Austria* 30 E7 46 35N 14 31 E
Gurkha, *Nepal* 81 E11 28 5N 84 40 E
Gurley, *Australia* 115 D4 29 45 S 149 48 E
Gurnee, *U.S.A.* 141 B9 42 22N 87 55W
Gürpınar, *Turkey* 89 C10 38 18N 43 25 E
Gurué, *Mozam.* 107 F4 15 25 S 36 58 E
Gurun, *Malaysia* 77 K3 5 49N 100 27 E
Gürün, *Turkey* 88 C7 38 43N 37 15 E
Gurupá, *Brazil* 153 D7 1 25 S 51 35W
Gurupá, I. Grande de, *Brazil* 153 D7 1 25 S 51 45W
Gurupi, *Brazil* 155 D2 11 43 S 49 4W
Gurupi →, *Brazil* 154 B2 1 13 S 46 6W
Gurupi, Serra do, *Brazil* 154 C2 5 0 S 47 50W
Guryev = Atyraū, *Kazakhstan* 49 E9 47 5N 52 0 E
Gus-Khrustalnyy, *Russia* 52 C5 55 42N 40 44 E
Gusau, *Nigeria* 101 C6 12 12N 6 40 E
Gusev, *Russia* 13 J20 54 35N 22 10 E
Gushan, *China* 67 E12 39 50N 123 35 E
Gushgy, *Turkmenistan* 56 F7 35 20N 62 18 E
Gushi, *China* 69 A10 32 11N 115 41 E
Gushiago, *Ghana* 101 D4 9 55N 0 15W
Gusinje, *Montenegro, Yug.* 42 E4 42 35N 19 50 E
Gusinoozersk, *Russia* 57 D11 51 16N 106 27 E
Gúspini, *Italy* 40 C1 39 32N 8 37 E
Güssing, *Austria* 31 D9 47 3N 16 20 E
Gustanj, *Slovenia* 39 B11 46 36N 14 59 E
Gustine, *U.S.A.* 144 H6 37 16N 121 0W
Güstrow, *Germany* 26 B8 53 47N 12 10 E
Gusum, *Sweden* 15 F10 58 16N 16 30 E
Guta = Kalárovo, *Slovak Rep.* 31 D11 47 54N 18 0 E
Gütersloh, *Germany* 26 D4 51 54N 8 24 E
Gutha, *Australia* 113 E2 28 58 S 115 55 E
Guthalongra, *Australia* 114 B4 19 52 S 147 50 E
Guthrie, *U.S.A.* 139 H6 35 53N 97 25W
Guthrie Center, *U.S.A.* 140 C2 41 41N 94 30W
Gutian, *China* 69 D12 26 32N 118 43 E
Gutiérrez, *Bolivia* 157 D5 19 25 S 63 34W
Guttannen, *Switz.* 29 C6 46 38N 8 18 E
Guttenberg, *U.S.A.* 140 B5 42 47N 91 6W
Guyana ■, *S. Amer.* 153 B6 5 0N 59 0W
Guyane française = French Guiana ■, *S. Amer.* 153 C7 4 0N 53 0W
Guyang, *China* 66 D6 41 0N 110 5 E
Guyenne, *France* 24 D4 44 30N 0 40 E
Guymon, *U.S.A.* 139 G4 36 41N 101 29W
Guyra, *Australia* 115 E5 30 15 S 151 40 E
Guyuan, *Hebei, China* 66 D8 41 37N 115 40 E
Guyuan, *Ningxia Huizu, China* 66 F4 36 0N 106 20 E
Guzar, *Uzbekistan* 55 D3 38 36N 66 15 E
Guzhang, *China* 68 C7 28 42N 109 58 E
Guzhen, *China* 67 H9 33 22N 117 18 E
Guzmán, L. de, *Mexico* 146 A3 31 25N 107 25W
Gvardeysk, *Russia* 13 J19 54 39N 21 5 E
Gvardeyskoye, *Ukraine* 51 K8 45 7N 34 1 E
Gwa, *Burma* 78 G5 17 36N 94 34 E
Gwaai, *Zimbabwe* 107 F2 19 15 S 27 45 E
Gwabegar, *Australia* 117 A8 30 31 S 149 0 E
Gwadabawa, *Nigeria* 101 C6 13 28N 5 15 E
Gwādar, *Pakistan* 79 D1 25 10N 62 18 E
Gwagwada, *Nigeria* 101 C6 10 15N 7 15 E
Gwalia, *Australia* 113 E3 28 54 S 121 20 E
Gwalior, *India* 80 F8 26 12N 78 10 E
Gwanda, *Zimbabwe* 107 G2 20 55 S 29 0 E
Gwandu, *Nigeria* 101 C5 12 30N 4 41 E
Gwane, *Zaïre* 106 B2 4 45N 25 48 E
Gwaram, *Nigeria* 101 C7 10 15N 10 25 E
Gwarzo, *Nigeria* 101 C6 12 20N 8 55 E
Gwda →, *Poland* 47 B3 53 3N 16 44 E
Gweebarra B., *Ireland* 19 B3 54 51N 8 23W
Gweedore, *Ireland* 19 A3 55 3N 8 13W
Gwent □, *U.K.* 17 F5 51 40N 2 57W
Gweru, *Zimbabwe* 107 F2 19 28 S 29 45 E
Gwi, *Nigeria* 101 D6 9 0N 7 10 E
Gwinn, *U.S.A.* 134 B2 46 19N 87 27W
Gwio Kura, *Nigeria* 101 C7 12 40N 11 2 E
Gwol, *Ghana* 100 C4 10 58N 1 59W
Gwoza, *Nigeria* 101 C7 11 5N 13 40 E
Gwydir →, *Australia* 115 D4 29 27 S 149 48 E
Gwynedd □, *U.K.* 16 E4 53 0N 4 0W
Gyandzha = Gäncä, *Azerbaijan* 53 K8 40 45N 46 20 E
Gyaring Hu, *China* 64 C4 34 50N 97 40 E
Gydanskiy P-ov., *Russia* 56 C8 70 0N 78 0 E
Gympie, *Australia* 115 D5 26 11 S 152 38 E
Gyobingauk, *Burma* 78 F5 18 13N 95 39 E
Gyoda, *Japan* 63 A11 36 10N 139 30 E
Gyoma, *Hungary* 31 E13 46 56N 20 50 E
Gyöngyös, *Hungary* 31 D12 47 48N 19 56 E
Györ, *Hungary* 31 D10 47 41N 17 40 E
Györ-Sopron □, *Hungary* 31 D10 47 40N 17 20 E
Gypsum Palace, *Australia* 116 B6 32 37 S 144 9 E
Gypsum Pt., *Canada* 130 A6 61 53N 114 35W
Gypsumville, *Canada* 131 C9 51 45N 98 40W
Gyula, *Hungary* 31 E14 46 38N 21 17 E
Gyumri, *Armenia* 53 K6 40 47N 43 50 E
Gyzylarbat, *Turkmenistan* 56 F6 39 4N 56 23 E
Gzhatsk = Gagarin, *Russia* 52 C2 55 38N 35 0 E

H

Ha 'Arava →, *Israel* 91 E4 30 50N 35 20 E
Ha Coi, *Vietnam* 76 B6 21 26N 107 46 E
Ha Dong, *Vietnam* 76 B5 20 58N 105 46 E
Ha Giang, *Vietnam* 76 A5 22 50N 104 59 E
Ha Tien, *Vietnam* 77 G5 10 23N 104 29 E
Ha Tinh, *Vietnam* 76 C5 18 20N 105 54 E
Ha Trung, *Vietnam* 76 C5 19 58N 105 50 E
Haacht, *Belgium* 21 G5 50 59N 4 37 E
Ha'afeva, *Tonga* 121 P13 19 57 S 174 43W
Haag, *Germany* 27 G8 48 9N 12 11 E
Haaksbergen, *Neths.* 20 D9 52 9N 6 45 E
Haaltert, *Belgium* 21 G4 50 55N 4 1 E
Haamstede, *Neths.* 21 E3 51 42N 3 45 E
Ha'ano, *Tonga* 121 P13 19 41 S 174 18W
Ha'apai Group, *Tonga* 121 P13 19 47 S 174 27W
Haapsalu, *Estonia* 13 G20 58 56N 23 30 E
Haarlem, *Neths.* 20 D5 52 23N 4 39 E
Haast, *N.Z.* 119 D4 43 51 S 169 1 E
Haast →, *N.Z.* 119 D4 43 50 S 169 2 E
Haast Bluff, *Australia* 112 D5 23 22 S 132 0 E
Haast Pass, *N.Z.* 119 E4 44 6 S 169 21 E
Haastrecht, *Neths.* 20 E5 52 0N 4 47 E
Hab Nadi Chauki, *Pakistan* 80 G2 25 0N 66 50 E
Ḩabarūt, *Yemen* 87 C6 17 18N 52 44 E
Habaswein, *Kenya* 106 B4 1 2N 39 30 E
Ḩabawnah, W. →, *Si. Arabia* 86 C4 17 57N 44 58 E
Habay, *Canada* 130 B5 58 50N 118 44W
Habay-la-Neuve, *Belgium* 21 J7 49 44N 5 38 E
Ḩabbān, *Yemen* 86 D4 14 21N 47 5 E
Ḩabbānīyah, *Iraq* 89 F10 33 17N 43 29 E
Ḩabbānīyah, Hawr al, *Iraq* 89 F10 33 17N 43 29 E
Habiganj, *Bangla.* 78 C3 24 24N 91 30 E
Haboro, *Japan* 60 B10 44 22N 141 42 E
Haccourt, *Belgium* 21 G7 50 44N 5 40 E
Hachenburg, *Germany* 26 E3 50 40N 7 49 E
Hachijō-Jima, *Japan* 63 D11 33 5N 139 45 E
Hachinohe, *Japan* 60 D10 40 30N 141 29 E
Hachiōji, *Japan* 63 B11 35 40N 139 20 E
Hachŏn, *N. Korea* 67 D15 41 29N 129 2 E
Hachy, *Belgium* 21 J7 49 42N 5 41 E
Hacıbektaş, *Turkey* 88 C6 38 56N 34 33 E
Hacılar, *Turkey* 88 C6 38 38N 35 26 E
Hackensack, *U.S.A.* 137 F10 40 53N 74 3W
Hadali, *Pakistan* 80 C5 32 16N 72 11 E
Hadarba, Ras, *Sudan* 94 C4 22 4N 36 51 E
Hadarom □, *Israel* 91 E3 31 0N 35 0 E
Ḩadbaram, *Oman* 87 C6 17 27N 55 15 E
Hadd, Ras al, *Oman* 87 B7 22 35N 59 50 E
Ḩaddā, *Si. Arabia* 86 B2 21 27N 39 34 E
Haddington, *U.K.* 18 F6 55 57N 2 47W
Haddon Rig, *Australia* 117 A7 31 27 S 147 52 E
Haded Plain, *Somali Rep.* 108 C3 9 46N 48 2 E
Hadejia, *Nigeria* 101 C7 12 30N 10 5 E
Hadejia →, *Nigeria* 101 C7 12 50N 10 51 E
Haden, *Australia* 115 D5 27 13 S 151 54 E
Ḩadera, *Israel* 91 C3 32 27N 34 55 E
Ḩadera, N. →, *Israel* 91 C3 32 28N 34 52 E
Haderslev, *Denmark* 15 J3 55 15N 9 30 E
Hadházteglas, *Hungary* 31 D14 47 40N 21 40 E
Hadhramaut = Ḩaḑramawt, *Yemen* 87 D5 15 30N 49 30 E
Hadım, *Turkey* 88 D5 36 58N 32 26 E
Hadjeb El Aïoun, *Tunisia* 96 A1 35 21N 9 32 E
Hadong, *S. Korea* 67 G14 35 5N 127 44 E
Ḩaḑramawt, *Yemen* 87 D5 15 30N 49 30 E
Ḩaḑramawt, W. →, *Yemen* 87 D5 16 0N 48 53 E
Ḩadrānīyah, *Iraq* 84 C4 35 38N 43 14 E
Hadrian's Wall, *U.K.* 16 C5 55 0N 2 30W
Hadsten, *Denmark* 15 H4 56 19N 10 3 E
Hadsund, *Denmark* 15 H4 56 44N 10 8 E
Hadyach, *Ukraine* 51 G8 50 21N 34 0 E
Haeju, *N. Korea* 67 E13 38 3N 125 45 E
Haenam, *S. Korea* 67 G14 34 34N 126 35 E
Haerhpin = Harbin, *China* 67 B14 45 48N 126 40 E
Hafar al Bāţin, *Si. Arabia* 84 D5 28 32N 45 52 E
Hafik, *Turkey* 88 C7 39 51N 37 23 E
Ḩafīrat al 'Aydā, *Si. Arabia* 84 E3 26 26N 39 12 E
Ḩafit, *Oman* 87 B6 23 59N 55 49 E
Hafizabad, *Pakistan* 80 C5 32 5N 73 40 E
Haflong, *India* 78 C4 25 10N 93 5 E
Hafnarfjörður, *Iceland* 12 D3 64 4N 21 57W
Hafun, Ras, *Somali Rep.* 90 E5 10 29N 51 30 E
Hagalil, *Israel* 91 C4 32 53N 35 18 E
Hagari →, *India* 83 G3 15 40N 77 0 E
Hagdan, *Phil.* 71 F4 11 20N 123 54 E
Hagen, *Germany* 26 D3 51 21N 7 27 E
Hagenow, *Germany* 26 B7 53 26N 11 12 E
Hagerman, *U.S.A.* 139 J2 33 7N 104 20W
Hagerstown, *Ind., U.S.A.* 141 E11 39 55N 85 10W
Hagerstown, *Md., U.S.A.* 134 F7 39 39N 77 43W
Hagetmau, *France* 24 E3 43 39N 0 37W
Hagfors, *Sweden* 13 F15 60 3N 13 45 E
Häggenås, *Sweden* 14 A8 63 24N 14 55 E
Hagi, *Japan* 62 C3 34 30N 131 22 E
Hagolan, *Syria* 91 B4 33 0N 35 45 E
Hagondange-Briey, *France* 23 C13 49 16N 6 11 E
Hagonoy, *Phil.* 70 D3 14 50N 120 44 E
Hags Hd., *Ireland* 19 D2 52 57N 9 28W
Hague, C. de la, *France* 22 C5 49 44N 1 56W
Hague, The = 's-Gravenhage, *Neths.* 20 D4 52 7N 4 17 E
Haguenau, *France* 23 D14 48 49N 7 47 E
Hai □, *Tanzania* 106 C4 3 10 S 37 10 E
Hai Duong, *Vietnam* 76 B6 20 56N 106 19 E
Hai'an, *Guangdong, China* 69 G8 20 18N 110 11 E
Hai'an, *Jiangsu, China* 69 A13 32 37N 120 27 E
Haicheng, *Fujian, China* 69 E11 24 23N 117 48 E
Haicheng, *Liaoning, China* 67 D12 40 50N 122 45 E
Haidar Khel, *Afghan.* 80 C3 33 58N 68 38 E
Haifa = Ḩefa, *Israel* 91 C3 32 46N 35 0 E
Haifeng, *China* 69 F10 22 58N 115 10 E
Haig, *Australia* 113 F4 30 55 S 126 10 E
Haiger, *Germany* 26 E4 50 43N 8 12 E
Haikang, *China* 69 G8 20 52N 110 8 E
Haikou, *China* 65 D6 20 1N 110 16 E
Ḩā'il, *Si. Arabia* 84 E4 27 28N 41 45 E
Hailakandi, *India* 78 C4 24 42N 92 34 E
Hailar, *China* 65 B6 49 10N 119 38 E
Hailey, *U.S.A.* 142 E6 43 31N 114 19W
Haileybury, *Canada* 128 C4 47 30N 79 38W
Hailin, *China* 67 B15 44 37N 129 30 E
Hailing Dao, *China* 69 G8 21 35N 111 47 E
Hailong, *China* 67 C13 42 32N 125 40 E
Hailuoto, *Finland* 12 D21 65 3N 24 45 E
Haimen, *Guangdong, China* 69 F11 23 15N 116 38 E
Haimen, *Jiangsu, China* 69 B13 31 52N 121 10 E
Haimen, *Zhejiang, China* 69 C13 28 40N 121 24 E
Hainan □, *China* 65 E5 19 0N 109 30 E
Hainaut □, *Belgium* 21 H4 50 30N 4 0 E
Hainburg, *Austria* 31 C9 48 9N 16 56 E
Haines, *U.S.A.* 142 D5 44 55N 117 56W
Haines City, *U.S.A.* 135 L5 28 7N 81 38W
Haines Junction, *Canada* 130 A1 60 45N 137 30W
Hainfeld, *Austria* 30 C8 48 3N 15 48 E
Haining, *China* 69 B13 30 28N 120 40 E
Haiphong, *Vietnam* 64 D5 20 47N 106 41 E
Haiti ■, *W. Indies* 149 C5 19 0N 72 30W
Haiya Junction, *Sudan* 94 D4 18 20N 36 21 E
Haiyan, *China* 69 B13 30 28N 120 58 E
Haiyang, *China* 67 F11 36 47N 121 9 E
Haiyuan, *Guangxi Zhuangzu, China* 68 F6 22 8N 107 35 E
Haiyuan, *Ningxia Huizu, China* 66 F3 36 35N 105 52 E
Haizhou, *China* 67 G10 34 37N 119 7 E
Haizhou Wan, *China* 67 G10 34 50N 119 20 E
Hajar Bangar, *Sudan* 97 F4 10 40N 22 45 E
Hajdú-Bihar □, *Hungary* 31 D14 47 30N 21 30 E
Hajdúböszörmény, *Hungary* 31 D14 47 40N 21 30 E
Hajdúdorog, *Hungary* 31 D14 47 48N 21 30 E
Hajdúnánás, *Hungary* 31 D14 47 50N 21 26 E
Hajdúsámson, *Hungary* 31 D14 47 37N 21 42 E
Hajdúszoboszló, *Hungary* 31 D14 47 27N 21 22 E
Hajiganj, *Bangla.* 78 D3 23 15N 90 50 E
Hajipur, *India* 81 G11 25 45N 85 13 E
Ḩajjah, *Yemen* 86 D3 15 42N 43 36 E
Ḩājjī Muḩsin, *Iraq* 84 C5 32 35N 45 29 E
Ḩājjīābād, *Eşfahan, Iran* 85 C7 33 41N 54 50 E
Ḩājjīābād, *Hormozgān, Iran* 85 D7 28 19N 55 55 E
Hajnówka, *Poland* 47 C10 52 47N 23 35 E
Ḩajrah, *Si. Arabia* 86 B3 20 14N 41 3 E
Haka, *Burma* 78 D4 22 39N 93 37 E
Hakansson, Mts., *Zaïre* 103 D5 8 40 S 25 45 E
Håkantorp, *Sweden* 15 F6 58 18N 12 55 E
Hakataramea, *N.Z.* 119 E5 44 43 S 170 30 E
Ḩakkan, *Oman* 87 B7 20 22N 58 47 E
Hakkâri, *Turkey* 89 D10 37 34N 43 44 E
Hakkâri Dağları, *Turkey* 89 C10 38 2N 42 58 E
Hakken-Zan, *Japan* 63 C7 34 10N 135 54 E
Hakodate, *Japan* 60 D10 41 45N 140 44 E
Hakota, *Japan* 63 A12 36 5N 140 30 E
Haku-San, *Japan* 63 A8 36 9N 136 46 E
Hakui, *Japan* 61 F8 36 53N 136 47 E
Hakun, *Burma* 78 B5 26 46N 95 42 E
Hala, *Pakistan* 79 D3 25 43N 68 20 E
Ḩalab, *Syria* 88 D7 36 10N 37 15 E
Ḩalaban, *Si. Arabia* 86 B4 23 29N 44 23 E
Ḩalabjah, *Iraq* 89 E11 35 10N 45 58 E
Halaib, *Sudan* 94 C4 22 12N 36 30 E
Halanzy, *Belgium* 21 J7 49 33N 5 44 E
Hālat 'Ammār, *Si. Arabia* 84 D3 29 10N 36 4 E
Halbā, *Lebanon* 91 A5 34 34N 36 6 E
Halberstadt, *Germany* 26 D7 51 54N 11 3 E
Halcombe, *N.Z.* 118 G4 40 8 S 175 30 E
Halcon, Mt., *Phil.* 70 E3 13 0N 121 30 E
Halden, *Norway* 14 E5 59 9N 11 23 E
Haldensleben, *Germany* 26 C7 52 17N 11 24 E
Haldwani, *India* 81 E8 29 31N 79 30 E
Hale, *U.S.A.* 140 E3 39 36N 93 20W
Hale →, *Australia* 114 C2 24 56 S 135 53 E
Haleakala Crater, *U.S.A.* 132 H16 20 43N 156 16W
Halen, *Belgium* 21 G6 50 57N 5 6 E
Haleyville, *U.S.A.* 135 H2 34 14N 87 37W
Half Assini, *Ghana* 100 D4 5 1N 2 50W
Halfmoon Bay, *N.Z.* 119 G3 46 50 S 168 5 E
Halfway →, *Canada* 130 B4 56 12N 121 32W
Haliburton, *Canada* 128 C4 45 3N 78 30W
Halicarnassus, *Turkey* 45 G9 37 3N 27 30 E
Halifax, *Australia* 114 B4 18 32 S 146 22 E
Halifax, *Canada* 129 D7 44 38N 63 35W
Halifax, *U.K.* 16 D6 53 43N 1 52W
Halifax B., *Australia* 114 B4 18 50 S 147 0 E
Halifax I., *Namibia* 104 D2 26 38 S 15 4 E
Halīl →, *Iran* 85 E8 27 40N 58 30 E
Halin, *Somali Rep.* 108 C3 9 6N 48 37 E
Hall, *Austria* 30 D4 47 17N 11 30 E
Hall Beach, *Canada* 127 B11 68 46N 81 12W
Hall Pt., *Australia* 112 C3 15 40 S 124 23 E
Halland, *Sweden* 13 H15 57 8N 12 47 E
Hallands län □, *Sweden* 15 H6 56 50N 12 50 E
Hallands Väderö, *Sweden* 15 H6 56 27N 12 34 E
Hallandsås, *Sweden* 15 H7 56 22N 13 0 E
Halle, *Belgium* 21 G4 50 44N 4 13 E
Halle, *Nordrhein-Westfalen, Germany* 26 C4 52 3N 8 22 E
Halle, *Sachsen-Anhalt, Germany* 26 D7 51 30N 11 56 E
Hällefors, *Sweden* 13 G16 59 47N 14 31 E
Hallein, *Austria* 30 D6 47 40N 13 5 E
Hällekis, *Sweden* 15 F7 58 38N 13 27 E
Hallett, *Australia* 116 B3 33 25 S 138 55 E
Hallettsville, *U.S.A.* 139 L6 29 27N 96 57W
Hällevadsholm, *Sweden* 15 F5 58 35N 11 33 E
Hallia →, *India* 82 F4 16 55N 79 20 E
Halliday, *U.S.A.* 138 B3 47 21N 102 20W
Halliday L., *Canada* 131 A7 61 21N 108 56W
Hallim, *S. Korea* 67 H14 33 24N 126 15 E
Hallingdalselva →, *Norway* 13 F13 60 40N 8 50 E
Hallock, *U.S.A.* 131 D9 48 47N 96 57W
Halls Creek, *Australia* 112 C4 18 16 S 127 38 E
Hallsberg, *Sweden* 13 G16 59 5N 15 7 E
Hallstahammar, *Sweden* 14 E10 59 38N 16 15 E
Hallstatt, *Austria* 30 D6 47 33N 13 38 E
Hallstead, *U.S.A.* 137 E9 41 58N 75 45W
Halmahera, *Indonesia* 72 A3 0 40N 128 0 E
Halmeu, *Romania* 46 B4 47 57N 23 2 E
Halmstad, *Sweden* 15 H6 56 41N 12 52 E
Halq el Oued, *Tunisia* 96 A2 36 53N 10 18 E
Hals, *Denmark* 15 H4 56 59N 10 18 E
Halsafjorden, *Norway* 14 A2 63 5N 8 10 E
Hälsingborg = Helsingborg, *Sweden* 15 H6 56 3N 12 42 E
Hälsingland, *Sweden* 13 F16 61 40N 16 5 E
Halstad, *U.S.A.* 138 B6 47 21N 96 50W
Haltdalen, *Norway* 14 B5 62 56N 11 8 E
Haltern, *Germany* 26 D3 51 44N 7 11 E
Halti, *Finland* 12 B19 69 17N 21 18 E
Halul, *Qatar* 85 E7 25 40N 52 40 E
Ḩalvān, *Iran* 85 C8 33 57N 56 15 E
Ham, *France* 23 C10 49 45N 3 4 E
Ham Tan, *Vietnam* 77 G6 10 40N 107 45 E
Ham Yen, *Vietnam* 76 A5 22 4N 105 3 E
Hamab, *Namibia* 104 D2 28 7 S 19 16 E
Hamad, *Sudan* 95 D3 15 20N 33 32 E
Hamada, *Japan* 62 C4 34 56N 132 4 E
Hamadān, *Iran* 89 E13 34 52N 48 32 E
Hamadān □, *Iran* 85 C6 35 0N 49 0 E
Hamadia, *Algeria* 99 A5 35 28N 1 57 E
Hamāh, *Syria* 88 E7 35 5N 36 40 E
Hamakita, *Japan* 63 C9 34 45N 137 47 E
Hamamatsu, *Japan* 63 C9 34 45N 137 45 E
Hamar, *Norway* 14 D5 60 48N 11 7 E
Hamâta, Gebel, *Egypt* 94 C3 24 17N 35 0 E
Hamber Prov. Park, *Canada* 130 C5 52 20N 118 0W
Hamburg, *Germany* 26 B5 53 33N 9 59 E
Hamburg, *Ark., U.S.A.* 139 J9 33 14N 91 48W
Hamburg, *Iowa, U.S.A.* 138 E7 40 36N 95 39W
Hamburg, *N.Y., U.S.A.* 136 D6 42 43N 78 50W
Hamburg, *Pa., U.S.A.* 137 F9 40 33N 75 59W
Hamburg □, *Germany* 26 B6 53 30N 10 0 E
Ḩamḑ, W. al →, *Si. Arabia* 84 E3 24 55N 36 20 E
Ḩamdah, *Si. Arabia* 86 C3 19 2N 43 36 E
Ḩamdānah, *Si. Arabia* 86 C3 19 59N 40 34 E
Hamden, *U.S.A.* 137 E12 41 23N 72 54W
Häme, *Finland* 13 F20 61 38N 25 10 E
Hämeenlinna, *Finland* 13 F21 61 0N 24 28 E
Hamélé, *Ghana* 100 C4 10 56N 2 45W

208 Hinckley

Hinckley, *U.S.A.* **142 G7** 39 20N 112 40W
Hindås, *Sweden* **15 G6** 57 42N 12 27 E
Hindaun, *India* **80 F7** 26 44N 77 5 E
Hindmarsh, L., *Australia* **116 D4** 36 5 S 141 55 E
Hindol, *India* **82 D7** 20 40N 85 10 E
Hinds, *N.Z.* **119 D6** 43 59 S 171 36 E
Hindsholm, *Denmark* **15 J4** 55 30N 10 40 E
Hindu Bagh, *Pakistan* **79 C2** 30 56N 67 50 E
Hindu Kush, *Asia* **79 B3** 36 0N 71 0 E
Hindupur, *India* **83 H3** 13 49N 77 32 E
Hines Creek, *Canada* **130 B5** 56 20N 118 40W
Hinganghat, *India* **82 D4** 20 30N 78 52 E
Hingeon, *Belgium* **21 G5** 50 32N 4 59 E
Hingham, *U.S.A.* **142 B8** 48 33N 110 25W
Hingoli, *India* **82 E3** 19 41N 77 15 E
Hinigaran, *Phil.* **71 F4** 10 16N 122 50 E
Hinis, *Turkey* **89 C9** 39 22N 41 43 E
Hinna = Imi, *Ethiopia* **90 F3** 6 28N 42 10 E
Hinna, *Nigeria* **101 C7** 10 25N 11 35 E
Hinnøya, *Norway* **12 B16** 68 35N 15 50 E
Hino, *Japan* **63 C8** 35 0N 136 15 E
Hinojosa del Duque, *Spain* **37 G5** 38 30N 5 9W
Hinokage, *Japan* **62 E3** 32 39N 131 24 E
Hinsdale, *U.S.A.* **142 B10** 48 24N 107 5W
Hinterrhein →, *Switz.* **29 C8** 46 40N 9 25 E
Hinton, *Canada* **130 C5** 53 26N 117 34W
Hinton, *U.S.A.* **134 G5** 37 40N 80 54W
Hinuangan, *Phil.* **71 F5** 10 25N 125 12 E
Hinwil, *Switz.* **29 B7** 47 18N 8 51 E
Hınzır Burnu, *Turkey* **88 D6** 36 19N 35 46 E
Hippolytushoef, *Neths.* **20 C5** 52 54N 4 58 E
Hirado, *Japan* **62 D1** 33 22N 129 33 E
Hirado-Shima, *Japan* **62 D1** 33 20N 129 30 E
Hirakarta, *Japan* **63 C7** 34 48N 135 40 E
Hirakud, *India* **82 D6** 21 32N 83 51 E
Hirakud Dam, *India* **82 D6** 21 32N 83 45 E
Hirata, *Japan* **62 B4** 35 24N 132 49 E
Hiratsuka, *Japan* **63 B11** 35 19N 139 21 E
Hirfanlı Baraji, *Turkey* **88 C5** 39 18N 33 31 E
Hirhafok, *Algeria* **99 D6** 23 49N 5 45 E
Hîrlău, *Romania* **46 B7** 47 23N 26 55 E
Hiromi, *Japan* **62 D4** 33 13N 132 36 E
Hiroo, *Japan* **60 C11** 42 17N 143 19 E
Hirosaki, *Japan* **60 D10** 40 34N 140 28 E
Hiroshima, *Japan* **62 C4** 34 24N 132 30 E
Hiroshima □, *Japan* **62 C4** 34 50N 133 0 E
Hiroshima-Wan, *Japan* **62 C4** 34 5N 132 20 E
Hirsholmene, *Denmark* **15 G4** 57 30N 10 36 E
Hirson, *France* **23 C11** 49 55N 4 4 E
Hîrşova, *Romania* **46 E8** 44 40N 27 59 E
Hirtshals, *Denmark* **15 G3** 57 36N 9 57 E
Hisar, *India* **80 E6** 29 12N 75 45 E
Hisb →, *Iraq* **84 D5** 31 45N 44 17 E
Ḥismá, *Si. Arabia* **84 D3** 28 30N 36 0 E
Ḥisn al 'Abr, *Yemen* **86 C4** 16 8N 47 14 E
Ḥiṣn al Qarn, *Yemen* **87 D5** 15 8N 49 7 E
Hispaniola, *W. Indies* **149 C5** 19 0N 71 0W
Hīt, *Iraq* **89 F10** 33 38N 42 49 E
Hita, *Japan* **62 D2** 33 20N 130 58 E
Hitachi, *Japan* **63 A12** 36 36N 140 39 E
Hitachiota, *Japan* **63 A12** 36 30N 140 30 E
Hitchin, *U.K.* **17 F7** 51 58N 0 16W
Hitoyoshi, *Japan* **62 E2** 32 13N 130 45 E
Hitra, *Norway* **12 E13** 63 30N 8 45 E
Hitzacker, *Germany* **26 B7** 53 9N 11 2 E
Hiu, *Vanuatu* **121 C4** 13 10 S 166 35 E
Hiuchi-Nada, *Japan* **62 C5** 34 5N 133 20 E
Ḥiyyon, N. →, *Israel* **91 E4** 30 25N 35 10 E
Hjalmar L., *Canada* **131 A7** 61 33N 109 25W
Hjälmare kanal, *Sweden* **14 E9** 59 20N 15 59 E
Hjälmaren, *Sweden* **14 E9** 59 18N 15 40 E
Hjartdal, *Norway* **14 E2** 59 37N 8 41 E
Hjerkinn, *Norway* **14 B3** 62 13N 9 33 E
Hjørring, *Denmark* **15 G3** 57 29N 9 59 E
Hjortkvarn, *Sweden* **15 F9** 58 54N 15 26 E
Hkakabo Razi, *Burma* **78 B6** 28 17N 97 46 E
Hko-ut, *Burma* **78 E7** 20 58N 98 2 E
Hkyenhpa, *Burma* **78 B6** 27 43N 97 25 E
Hlaingbwe, *Burma* **78 G6** 17 8N 97 50 E
Hlinsko, *Czech.* **30 B8** 49 45N 15 54 E
Hlohovec, *Slovak Rep.* **31 C10** 48 26N 17 49 E
Hluhluwe, *S. Africa* **105 D5** 28 1 S 32 15 E
Hlukhiv, *Ukraine* **51 G7** 51 40N 33 58 E
Hlwaze, *Burma* **78 F6** 18 54N 96 37 E
Hlyboka, *Ukraine* **51 H3** 48 5N 25 56 E
Hlybokaye, *Belarus* **50 E4** 55 10N 27 45 E
Ho, *Ghana* **101 D5** 6 37N 0 27 E
Ho Chi Minh City = Phanh Bho Ho Chi Minh, *Vietnam* **77 G6** 10 58N 106 40 E
Ho Thuong, *Vietnam* **76 C5** 19 32N 105 48 E
Hoa Binh, *Vietnam* **76 B5** 20 50N 105 20 E
Hoa Da, *Vietnam* **77 G7** 11 16N 108 40 E
Hoa Hiep, *Vietnam* **77 G5** 11 34N 105 51 E
Hoai Nhon, *Vietnam* **76 E7** 14 28N 109 1 E
Hoang Lien Son, *Vietnam* **76 A4** 22 0N 104 0 E
Hoare B., *Canada* **127 B13** 65 17N 62 30W
Hobart, *Australia* **114 G4** 42 50 S 147 21 E
Hobart, *Ind., U.S.A.* **141 C9** 41 32N 87 15W
Hobart, *Okla., U.S.A.* **139 H5** 35 1N 99 6W
Hobbs, *U.S.A.* **139 J3** 32 42N 103 8W
Hobbs Coast, *Antarctica* **7 D14** 74 50 S 131 0W
Hobo, *Colombia* **152 C2** 2 35N 75 30W
Hoboken, *Belgium* **21 F4** 51 11N 4 21 E
Hoboken, *U.S.A.* **137 F10** 40 45N 74 4W
Hobro, *Denmark* **15 H3** 56 39N 9 46 E
Hobscheid, *Lux.* **21 J7** 49 42N 5 57 E
Hoburgen, *Sweden* **13 H18** 56 55N 18 7 E
Hochdorf, *Switz.* **29 B6** 47 10N 8 17 E
Hochschwab, *Austria* **30 D8** 47 35N 15 0 E
Höchstadt, *Germany* **27 F6** 49 42N 10 47 E
Hockenheim, *Germany* **27 F4** 49 19N 8 32 E
Hodaka-Dake, *Japan* **63 A9** 36 17N 137 39 E
Hodgson, *Canada* **131 C9** 51 13N 97 36W
Hódmezövásárhely, *Hungary* **31 E13** 46 28N 20 22 E
Hodna, Chott el, *Algeria* **99 A5** 35 26N 4 43 E
Hodna, Monts du, *Algeria* **99 A5** 35 52N 4 42 E
Hodonín, *Czech.* **31 C10** 48 50N 17 10 E
Hoeamdong, *N. Korea* **67 C16** 42 30N 130 16 E
Hœdic, I. de, *France* **22 E4** 47 20N 2 53W
Hoegaarden, *Belgium* **21 G5** 50 47N 4 53 E
Hoek van Holland, *Neths.* **20 E4** 52 0N 4 7 E
Hoeksche Waard, *Neths.* **20 E4** 51 46N 4 25 E
Hoenderloo, *Neths.* **20 D7** 52 7N 5 52 E
Hoengsŏng, *S. Korea* **67 F14** 37 29N 127 59 E
Hoensbroek, *Neths.* **21 G7** 50 55N 5 55 E
Hoeryong, *N. Korea* **67 C15** 42 30N 129 45 E
Hoeselt, *Belgium* **21 G6** 50 51N 5 29 E
Hoeven, *Neths.* **21 E5** 51 35N 4 35 E
Hoeyang, *N. Korea* **67 E14** 38 43N 127 36 E
Hof, *Germany* **27 E7** 50 19N 11 55 E
Höfðakaupstaður, *Iceland* **12 D3** 65 50N 20 19W
Hofgeismar, *Germany* **26 D5** 51 29N 9 23 E
Hofmeyr, *S. Africa* **104 E4** 31 39 S 25 50 E
Höfn, *Iceland* **12 D6** 64 15N 15 13W
Hofors, *Sweden* **13 F17** 60 31N 16 15 E
Hofsjökull, *Iceland* **12 D4** 64 49N 18 48W
Hōfu, *Japan* **62 C3** 34 3N 131 34 E
Hogan Group, *Australia* **114 F4** 39 13 S 147 1 E
Hogansville, *U.S.A.* **135 J3** 33 10N 84 55W
Hogeland, *U.S.A.* **142 B9** 48 51N 108 40W
Hogenakai Falls, *India* **83 H3** 12 6N 77 50 E
Hoggar = Ahaggar, *Algeria* **99 D6** 23 0N 6 30 E
Hōgo-Kaikyō, *Japan* **62 D3** 33 20N 131 58 E
Högsäter, *Sweden* **15 F6** 58 38N 12 5 E
Hogsty Reef, *Bahamas* **149 B5** 21 41N 73 48W
Hoh →, *U.S.A.* **144 C2** 47 45N 124 29W
Hohe Rhön, *Germany* **27 E5** 50 24N 9 58 E
Hohe Tauern, *Austria* **30 D5** 47 11N 12 40 E
Hohe Venn, *Belgium* **21 H8** 50 30N 6 5 E
Hohenau, *Austria* **31 C9** 48 36N 16 55 E
Hohenems, *Austria* **30 D2** 47 22N 9 42 E
Hohenstein-Ernstthal, *Germany* **26 E8** 50 48N 12 42 E
Hohenwald, *U.S.A.* **135 H2** 35 33N 87 33W
Hohenwestedt, *Germany* **26 A5** 54 5N 9 40 E
Hohhot, *China* **66 D6** 40 52N 111 40 E
Hóhlakas, *Greece* **32 D9** 35 57N 27 53 E
Hohoe, *Ghana* **101 D5** 7 8N 0 32 E
Hoi An, *Vietnam* **76 E7** 15 30N 108 19 E
Hoi Xuan, *Vietnam* **76 B5** 20 25N 105 9 E
Hoisington, *U.S.A.* **138 F5** 38 31N 98 47W
Højer, *Denmark* **15 K2** 54 58N 8 42 E
Hōjō, *Japan* **62 D4** 33 58N 132 46 E
Hökerum, *Sweden* **15 G7** 57 51N 13 16 E
Hokianga Harbour, *N.Z.* **118 B2** 35 31 S 173 22 E
Hokitika, *N.Z.* **119 C5** 42 42 S 171 0 E
Hokkaidō □, *Japan* **60 C11** 43 30N 143 0 E
Hokksund, *Norway* **14 E3** 59 48N 9 54 E
Hol-Hol, *Djibouti* **95 E5** 11 20N 42 50 E
Hola Pristan, *Ukraine* **51 J7** 46 29N 32 32 E
Holbæk, *Denmark* **15 J5** 55 43N 11 43 E
Holbrook, *Australia* **117 C7** 35 42 S 147 18 E
Holbrook, *U.S.A.* **143 J8** 34 54N 110 10W
Holden, *Canada* **130 C6** 53 13N 112 11W
Holden, *Mo., U.S.A.* **140 F3** 38 43N 94 1W
Holden, *Utah, U.S.A.* **142 G7** 39 6N 112 16W
Holdenville, *U.S.A.* **139 H6** 35 5N 96 24W
Holder, *Australia* **116 C3** 34 21 S 140 0 E
Holderness, *U.K.* **16 D7** 53 45N 0 5W
Holdfast, *Canada* **131 C7** 50 58N 105 25W
Holdich, *Argentina* **160 C3** 45 57 S 68 13W
Holdrege, *U.S.A.* **138 E5** 40 26N 99 23W
Hole-Narsipur, *India* **83 H3** 12 48N 76 16 E
Holešov, *Czech.* **31 B10** 49 20N 17 35 E
Holgate, *U.S.A.* **141 C12** 41 15N 84 8W
Holguín, *Cuba* **148 B4** 20 50N 76 20W
Holíč, *Slovak Rep.* **31 C10** 48 49N 17 10 E
Hollabrunn, *Austria* **30 C9** 48 34N 16 5 E
Hollams Bird I., *Namibia* **104 C1** 24 40 S 14 30 E
Holland, *U.S.A.* **141 B10** 42 47N 86 7W
Hollandia = Jayapura, *Indonesia* **73 B6** 2 28 S 140 38 E
Hollandsch Diep, *Neths.* **21 E5** 51 41N 4 30 E
Hollandsch IJssel →, *Neths.* **20 E5** 51 55N 4 34 E
Hollfeld, *Germany* **27 F7** 49 56N 11 18 E
Hollidaysburg, *U.S.A.* **136 F6** 40 26N 78 24W
Hollis, *U.S.A.* **139 H5** 34 41N 99 55W
Hollister, *Calif., U.S.A.* **144 J5** 36 51N 121 24W
Hollister, *Idaho, U.S.A.* **142 E6** 42 21N 114 35W
Hollum, *Neths.* **20 B7** 53 26N 5 38 E
Holly, *Colo., U.S.A.* **138 F3** 38 3N 102 7W
Holly, *Mich., U.S.A.* **141 B13** 42 48N 83 38W
Holly Hill, *U.S.A.* **135 L5** 29 16N 81 3W
Holly Springs, *U.S.A.* **139 H10** 34 46N 89 27W
Hollywood, *Calif., U.S.A.* **143 J4** 34 7N 118 25W
Hollywood, *Fla., U.S.A.* **135 N5** 26 1N 80 9W
Holm, *Sweden* **14 B10** 62 40N 16 40 E
Holman Island, *Canada* **126 A8** 70 42N 117 41W
Hólmavík, *Iceland* **12 D3** 65 42N 21 40W
Holmes Reefs, *Australia* **114 B4** 16 27 S 148 0 E
Holmestrand, *Norway* **14 E4** 59 31N 10 14 E
Holmsbu, *Norway* **14 E4** 59 32N 10 27 E
Holmsjön, *Sweden* **14 B9** 62 26N 15 20 E
Holmsland Klit, *Denmark* **15 J2** 56 0N 8 5 E
Holmsund, *Sweden* **12 E19** 63 41N 20 20 E
Holod, *Romania* **46 C3** 46 49N 22 8 E
Holroyd →, *Australia* **114 A3** 14 10 S 141 36 E
Holstebro, *Denmark* **15 H2** 56 22N 8 37 E
Holsworthy, *U.K.* **17 G3** 50 48N 4 22W
Holte, *Denmark* **15 J6** 55 50N 12 29 E
Holten, *Neths.* **20 D8** 52 17N 6 26 E
Holton, *Canada* **129 B8** 54 31N 57 12W
Holton, *U.S.A.* **138 F7** 39 28N 95 44W
Holtville, *U.S.A.* **145 N11** 32 49N 115 23W
Holwerd, *Neths.* **20 B7** 53 22N 5 54 E
Holy Cross, *U.S.A.* **126 B4** 62 12N 159 46W
Holy I., *Gwynedd, U.K.* **16 D3** 53 17N 4 37W
Holy I., *Northumb., U.K.* **16 B6** 55 40N 1 47W
Holyhead, *U.K.* **16 D3** 53 18N 4 38W
Holyoke, *Colo., U.S.A.* **138 E3** 40 35N 102 18W
Holyoke, *Mass., U.S.A.* **137 D12** 42 12N 72 37W
Holyrood, *Canada* **129 C9** 47 27N 53 8W
Holzkirchen, *Germany* **27 H7** 47 52N 11 42 E
Holzminden, *Germany* **26 D5** 51 50N 9 28 E
Homa Bay, *Kenya* **106 C3** 0 36 S 34 30 E
Homa Bay □, *Kenya* **106 C3** 0 50 S 34 30 E
Homalin, *Burma* **78 C5** 24 55N 95 0 E
Homand, *Iran* **85 C8** 32 28N 59 37 E
Homberg, *Germany* **26 D5** 51 2N 9 25 E
Hombori, *Mali* **101 B4** 15 20N 1 38W
Homburg, *Germany* **27 F3** 49 28N 7 18 E
Home B., *Canada* **127 B13** 68 40N 67 10W
Home Hill, *Australia* **114 B4** 19 43 S 147 25 E
Home Reef, *Tonga* **121 P13** 18 59 S 174 47W
Homedale, *U.S.A.* **142 E5** 43 37N 116 56W
Homer, *Alaska, U.S.A.* **126 C4** 59 39N 151 33W
Homer, *Ill., U.S.A.* **141 D9** 40 4N 87 57W
Homer, *La., U.S.A.* **139 J8** 32 48N 93 4W
Homer, *Mich., U.S.A.* **141 B12** 42 9N 84 49W
Homestead, *Australia* **114 C4** 20 20 S 145 40 E
Homestead, *Fla., U.S.A.* **135 N5** 25 28N 80 29W
Homestead, *Oreg., U.S.A.* **142 D5** 45 2N 116 51W
Homewood, *Calif., U.S.A.* **144 F6** 39 4N 120 8W
Homewood, *Ill., U.S.A.* **141 C9** 41 34N 87 40W
Hominy, *U.S.A.* **139 G6** 36 25N 96 24W
Homnabad, *India* **82 F3** 17 45N 77 11 E
Homoine, *Mozam.* **105 C6** 23 55 S 35 8 E
Homoljske Planina, *Serbia, Yug.* **42 C6** 44 10N 21 45 E
Homonhan I., *Phil.* **71 F5** 10 44N 125 43 E
Homorod, *Romania* **46 C6** 46 5N 25 15 E
Homs = Ḥimṣ, *Syria* **91 A5** 34 40N 36 45 E
Homyel, *Belarus* **51 F6** 52 28N 31 0 E
Hon Chong, *Vietnam* **77 G5** 10 25N 104 30 E
Hon Me, *Vietnam* **76 C5** 19 23N 105 56 E
Hon Quan, *Vietnam* **77 G6** 11 40N 106 50 E
Honan = Henan □, *China* **66 G8** 34 0N 114 0 E
Honbetsu, *Japan* **60 C11** 43 7N 143 37 E
Honcut, *U.S.A.* **144 F5** 39 20N 121 32W
Honda, *Colombia* **152 B3** 5 12N 74 45W
Honda Bay, *Phil.* **71 G2** 9 53N 118 49 E
Hondeklipbaai, *S. Africa* **104 E2** 30 19 S 17 17 E
Hondo, *Japan* **62 E2** 32 27N 130 12 E
Hondo, *U.S.A.* **139 L5** 29 21N 99 9W
Hondo →, *Belize* **147 D7** 18 25N 88 21W
Honduras ■, *Cent. Amer.* **148 D2** 14 40N 86 30W
Honduras, G. de, *Caribbean* **148 C2** 16 50N 87 0W
Hønefoss, *Norway* **13 F14** 60 10N 10 18 E
Honesdale, *U.S.A.* **137 E9** 41 34N 75 16W
Honey L., *U.S.A.* **144 E6** 40 15N 120 19W
Honfleur, *France* **22 C7** 49 25N 0 13 E
Hong Gai, *Vietnam* **76 B6** 20 57N 107 5 E
Hong He →, *China* **66 H8** 32 25N 115 35 E
Hong Kong ■, *Asia* **69 F10** 22 11N 114 14 E
Hong'an, *China* **69 B10** 31 20N 114 40 E
Hongchŏn, *S. Korea* **67 F14** 37 44N 127 53 E
Hongha →, *Vietnam* **64 D5** 22 0N 104 0 E
Honghai Wan, *China* **69 F10** 22 40N 115 0 E
Honghu, *China* **69 C9** 29 50N 113 30 E
Hongjiang, *China* **68 D7** 27 7N 109 59 E
Hongliu He →, *China* **66 F5** 38 0N 109 50 E
Hongor, *Mongolia* **66 B7** 45 45N 112 50 E
Hongsa, *Laos* **76 C3** 19 43N 101 20 E
Hongshui He →, *China* **68 F7** 23 48N 109 30 E
Hongsŏng, *S. Korea* **67 F14** 36 37N 126 38 E
Hongtong, *China* **66 F6** 36 16N 111 40 E
Honguedo, Détroit d', *Canada* **129 C7** 49 15N 64 0W
Hongwon, *N. Korea* **67 E14** 40 0N 127 56 E
Hongya, *China* **68 C4** 29 57N 103 22 E
Hongyuan, *China* **68 A4** 32 51N 102 40 E
Hongze Hu, *China* **67 H10** 33 15N 118 35 E
Honiara, *Solomon Is.* **121 M10** 9 27 S 159 57 E
Honiton, *U.K.* **17 G4** 50 47N 3 11W
Honjō, *Akita, Japan* **60 E10** 39 23N 140 3 E
Honjō, *Gumma, Japan* **63 A11** 36 14N 139 11 E
Honkawane, *Japan* **63 B10** 35 5N 138 5 E
Honkorâb, Ras, *Egypt* **94 C4** 24 35N 35 10 E
Honningsvåg, *Norway* **12 A21** 70 59N 25 59 E
Honolulu, *U.S.A.* **132 H16** 21 19N 157 52W
Honshū, *Japan* **61 G9** 36 0N 138 0 E
Hontoria del Pinar, *Spain* **34 D1** 41 50N 3 10W
Hood, Mt., *U.S.A.* **142 D3** 45 23N 121 42W
Hood, Pt., *Australia* **113 F2** 34 23 S 119 34 E
Hood Pt., *Papua N. G.* **120 F4** 10 4 S 147 45 E
Hood River, *U.S.A.* **142 D3** 45 43N 121 31W
Hoodsport, *U.S.A.* **144 C3** 47 24N 123 9W
Hooge, *Germany* **26 A4** 54 34N 8 33 E
Hoogerheide, *Neths.* **21 F4** 51 26N 4 20 E
Hoogeveen, *Neths.* **20 C8** 52 44N 6 28 E
Hoogeveensche Vaart, *Neths.* **20 C8** 52 42N 6 12 E
Hoogezand, *Neths.* **20 B9** 53 11N 6 45 E
Hooghly = Hughli →, *India* **81 J13** 21 56N 88 4 E
Hooghly-Chinsura = Chunchura, *India* **81 H13** 22 53N 88 27 E
Hoogkerk, *Neths.* **20 B9** 53 13N 6 30 E
Hooglede, *Belgium* **21 G2** 50 59N 3 5 E
Hoogstraten, *Belgium* **21 F5** 51 24N 4 46 E
Hoogvliet, *Neths.* **20 E4** 51 52N 4 21 E
Hook Hd., *Ireland* **19 D5** 52 7N 6 56W
Hook I., *Australia* **114 C4** 20 4 S 149 0 E
Hook of Holland = Hoek van Holland, *Neths.* **20 E4** 52 0N 4 7 E
Hooker, *U.S.A.* **139 G4** 36 52N 101 13W
Hooker Creek, *Australia* **112 C5** 18 23 S 130 38 E
Hoopeston, *U.S.A.* **141 D9** 40 28N 87 40W
Hoopstad, *S. Africa* **104 D4** 27 50 S 25 55 E
Hoorn, *Neths.* **20 C6** 52 38N 5 4 E
Hoover Dam, *U.S.A.* **145 K12** 36 1N 114 44W
Hooversville, *U.S.A.* **136 F6** 40 9N 78 55W
Hop Bottom, *U.S.A.* **137 E9** 41 42N 75 46W
Hopa, *Turkey* **89 B9** 41 28N 41 30 E
Hope, *Canada* **130 D4** 49 25N 121 25W
Hope, *Ariz., U.S.A.* **145 M13** 33 43N 113 42W
Hope, *Ark., U.S.A.* **139 J8** 33 40N 93 36W
Hope, *Ind., U.S.A.* **141 E11** 39 18N 85 46W
Hope, *N. Dak., U.S.A.* **138 B6** 47 19N 97 43W
Hope, L., *Australia* **115 D2** 28 24 S 139 18 E
Hope, Pt., *U.S.A.* **126 B3** 68 20N 166 50W
Hope Pass, *N.Z.* **119 C7** 42 36 S 172 6 E
Hope Town, *Bahamas* **148 A4** 26 35N 76 57W
Hopedale, *Canada* **129 A7** 55 28N 60 13W
Hopefield, *S. Africa* **104 E2** 33 3 S 18 22 E
Hopei = Hebei □, *China* **66 E9** 39 0N 116 0 E
Hopelchén, *Mexico* **147 D7** 19 46N 89 50W
Hopetoun, *Vic., Australia* **116 C5** 35 42 S 142 22 E
Hopetoun, *W. Austral., Australia* **113 F3** 33 57 S 120 7 E
Hopetown, *S. Africa* **104 D3** 29 34 S 24 3 E
Hopin, *Burma* **78 C6** 24 58N 96 30 E
Hopkins, *Mich., U.S.A.* **141 B11** 42 37N 85 46W
Hopkins, *Mo., U.S.A.* **140 D2** 40 33N 94 49W
Hopkins, L., *Australia* **112 D4** 24 15 S 128 35 E
Hopkinsville, *U.S.A.* **135 G2** 36 52N 87 29W
Hopland, *U.S.A.* **144 G3** 38 58N 123 7W
Hoptrup, *Denmark* **15 J3** 55 11N 9 28 E
Hoquiam, *U.S.A.* **144 D3** 46 59N 123 53W
Hōrai, *Japan* **63 C9** 34 58N 137 32 E
Horasan, *Turkey* **89 B10** 40 3N 42 11 E
Horazdovice, *Czech.* **30 B6** 49 19N 13 42 E
Horcajo de Santiago, *Spain* **34 F1** 39 50N 3 1W
Horden Hills, *Australia* **112 D5** 20 15 S 130 0 E
Hordio, *Somali Rep.* **108 B4** 10 33N 51 6 E
Horezu, *Romania* **46 D5** 45 6N 24 0 E
Horgen, *Switz.* **29 B7** 47 15N 8 35 E
Horgoš, *Serbia, Yug.* **42 A5** 46 10N 20 0 E
Horice, *Czech.* **30 A8** 50 21N 15 39 E
Horinger, *China* **66 D6** 40 28N 111 48 E
Horki, *Belarus* **50 E6** 54 17N 30 59 E
Horlick Mts., *Antarctica* **7 E15** 84 0 S 102 0W
Horlivka, *Ukraine* **51 H10** 48 19N 38 5 E
Hormoz, *Iran* **85 E7** 27 35N 55 0 E
Hormoz, Jaz. ye, *Iran* **85 E8** 27 8N 56 28 E
Hormuz, Str. of, *The Gulf* **85 E8** 26 30N 56 30 E
Horn, *Austria* **30 C8** 48 39N 15 40 E
Horn, *Iceland* **12 C2** 66 28N 22 28W
Horn, *Neths.* **21 F7** 51 12N 5 57 E
Horn →, *Canada* **130 A5** 61 30N 118 1W
Horn, Cape = Hornos, C. de, *Chile* **160 E3** 55 50 S 67 30W
Horn, Is., *Wall. & F. Is.* **111 C15** 14 16 S 178 6W
Horn Head, *Ireland* **19 A3** 55 14N 8 0W
Horn I., *Australia* **114 A3** 10 37 S 142 17 E
Horn I., *U.S.A.* **135 K1** 30 14N 88 39W
Horn Mts., *Canada* **130 A5** 62 15N 119 15W
Hornachuelos, *Spain* **37 H5** 37 50N 5 14W
Hornavan, *Sweden* **12 C17** 66 15N 17 30 E
Hornbæk, *Denmark* **15 H6** 56 5N 12 26 E
Hornbeck, *U.S.A.* **139 K8** 31 20N 93 24W
Hornbrook, *U.S.A.* **142 F2** 41 55N 122 33W
Hornburg, *Germany* **26 C6** 52 2N 10 36 E
Hornby, *N.Z.* **119 D7** 43 33 S 172 33 E
Horncastle, *U.K.* **16 D7** 53 13N 0 7W
Hornell, *U.S.A.* **136 D7** 42 20N 77 40W
Hornell L., *Canada* **130 A5** 62 20N 119 25W
Hornepayne, *Canada* **128 C3** 49 14N 84 48W
Hornitos, *U.S.A.* **144 H6** 37 30N 120 14W
Hornos, C. de, *Chile* **160 E3** 55 50 S 67 30W
Hornoy, *France* **23 C8** 49 50N 1 54 E
Hornsby, *Australia* **117 B9** 33 42 S 151 2 E
Hornsea, *U.K.* **16 D7** 53 55N 0 11W
Hornslandet, *Sweden* **14 C11** 61 35N 17 37 E
Hornslet, *Denmark* **15 H4** 56 18N 10 19 E
Hornu, *Belgium* **21 H3** 50 26N 3 50 E
Hörnum, *Germany* **26 A4** 54 45N 8 17 E
Horobetsu, *Japan* **60 C10** 42 24N 141 6 E
Horodenka, *Ukraine* **51 H3** 48 41N 25 29 E
Horodnya, *Ukraine* **51 G6** 51 55N 31 33 E
Horodok, *Ukraine* **51 H2** 49 46N 23 32 E
Horodok, *Ukraine* **51 H4** 49 10N 26 34 E
Horodyshche, *Ukraine* **51 H6** 49 17N 31 27 E
Horokhiv, *Ukraine* **51 G3** 50 30N 24 45 E
Horovice, *Czech.* **30 B6** 49 48N 13 53 E
Horqin Youyi Qianqi, *China* **67 A12** 46 5N 122 3 E
Horqueta, *Paraguay* **158 A4** 23 15 S 56 55W
Horred, *Sweden* **15 G6** 57 22N 12 28 E
Horse Creek, *U.S.A.* **138 E3** 41 57N 105 10W
Horse Is., *Canada* **129 B8** 50 15N 55 50W
Horsefly L., *Canada* **130 C4** 52 25N 121 0W
Horsens, *Denmark* **15 J3** 55 52N 9 51 E
Horsens Fjord, *Denmark* **15 J4** 55 50N 10 0 E
Horsham, *Australia* **116 D5** 36 44 S 142 13 E
Horsham, *U.K.* **17 F7** 51 4N 0 20W
Horšovský Týn, *Czech.* **30 B5** 49 31N 12 58 E
Horst, *Neths.* **21 F8** 51 27N 6 3 E
Horten, *Norway* **14 E4** 59 25N 10 32 E
Hortobágy →, *Hungary* **31 D14** 47 30N 21 6 E
Horton, *U.S.A.* **138 F7** 39 40N 95 32W
Horton →, *Canada* **126 B7** 69 56N 126 52W
Horw, *Switz.* **29 B6** 47 1N 8 19 E
Horwood, L., *Canada* **128 C3** 48 5N 82 20W
Hosaina, *Ethiopia* **95 F4** 7 30N 37 47 E
Hosdurga, *India* **83 H3** 13 49N 76 17 E
Ḥoseynābād, *Khuzestān, Iran* **85 C6** 32 45N 48 20 E
Ḥoseynābād, *Kordestān, Iran* **89 E12** 35 33N 47 8 E
Hoshangabad, *India* **80 H7** 22 45N 77 45 E
Hoshiarpur, *India* **80 D6** 31 30N 75 58 E
Hosingen, *Lux.* **21 H8** 50 1N 6 6 E
Hoskins, *Papua N. G.* **120 C6** 5 29 S 150 27 E
Hosmer, *U.S.A.* **138 C5** 45 34N 99 28W
Hososhima, *Japan* **62 E3** 32 26N 131 40 E
Hospental, *Switz.* **29 C7** 46 37N 8 34 E
Hospet, *India* **83 G3** 15 15N 76 20 E
Hospitalet de Llobregat, *Spain* **34 D7** 41 21N 2 6 E
Hoste, I., *Chile* **160 E3** 55 0 S 69 0W
Hostens, *France* **24 D3** 44 30N 0 40W
Hot, *Thailand* **76 C2** 18 8N 98 29 E
Hot Creek Range, *U.S.A.* **142 G5** 38 40N 116 20W
Hot Springs, *Ark., U.S.A.* **139 H8** 34 31N 93 3W
Hot Springs, *S. Dak., U.S.A.* **138 D3** 43 26N 103 29W
Hotagen, *Sweden* **12 E16** 63 50N 14 30 E
Hotan, *China* **64 C2** 37 25N 79 55 E
Hotazel, *S. Africa* **104 D3** 27 17 S 22 58 E
Hotchkiss, *U.S.A.* **143 G10** 38 48N 107 43W
Hotham, C., *Australia* **112 B5** 12 2 S 131 18 E
Hoting, *Sweden* **12 D17** 64 8N 16 15 E
Hotolishti, *Albania* **44 C2** 41 10N 20 25 E
Hotte, Massif de la, *Haiti* **149 C5** 18 30N 73 45W
Hottentotsbaai, *Namibia* **104 D1** 26 8 S 14 59 E
Hotton, *Belgium* **21 H6** 50 16N 5 26 E
Houailou, *N. Cal.* **121 U19** 21 17 S 165 38 E
Houat, I. de, *France* **22 E4** 47 24N 2 58W
Houck, *U.S.A.* **143 J9** 35 20N 109 10W
Houdan, *France* **23 D8** 48 48N 1 35 E
Houdeng-Goegnies, *Belgium* **21 H4** 50 29N 4 10 E
Houei Sai, *Laos* **76 B3** 20 18N 100 26 E
Houffalize, *Belgium* **21 H7** 50 8N 5 48 E
Houghton, *U.S.A.* **138 B10** 47 7N 88 34W
Houghton L., *U.S.A.* **134 C3** 44 21N 84 44W
Houghton-le-Spring, *U.K.* **16 C6** 54 51N 1 28W
Houhora Heads, *N.Z.* **118 A2** 34 49 S 173 9 E
Houille →, *Belgium* **21 H5** 50 8N 4 50 E
Houlton, *U.S.A.* **129 C6** 46 8N 67 51W
Houma, *U.S.A.* **139 L9** 29 36N 90 43W
Houndé, *Burkina Faso* **100 C4** 11 34N 3 31W
Hourtin, *France* **24 C2** 45 11N 1 4W
Hourtin-Carcans, Étang d', *France* **24 C2** 45 10N 1 6W
Houston, *Canada* **130 C3** 54 25N 126 39W
Houston, *Mo., U.S.A.* **139 G9** 37 22N 91 58W
Houston, *Tex., U.S.A.* **139 L7** 29 46N 95 22W
Houten, *Neths.* **20 D6** 52 2N 5 10 E
Houthalen, *Belgium* **21 F6** 51 2N 5 23 E
Houthem, *Belgium* **21 G1** 50 48N 2 57 E

212 J.F. Rodrigues

J

J.F. Rodrigues, *Brazil* . . . **154 B1** 2 55 S 50 20W
Jaba, *Ethiopia* . . . . **95 F4** 6 20N 35 7 E
Jabal el Awlîya, *Sudan* . . **95 D3** 15 10N 32 31 E
Jabal Lubnān, *Lebanon* . . **91 B4** 33 45N 35 40 E
Jabalón →, *Spain* **37 G6** 38 53N 4 5W
Jabalpur, *India* . . . . **81 H8** 23 9N 79 58 E
Jabbūl, *Syria* . . . . **84 B3** 36 4N 37 30 E
Jablah, *Syria* . . . . **88 E6** 35 20N 36 0 E
Jablanac, *Croatia* . . . . . . **39 D11** 44 42N 14 56 E
Jablonec, *Czech.* . . . . **30 A8** 50 43N 15 10 E
Jablonica, *Slovak Rep.* . . . **31 C10** 48 37N 17 26 E
Jabłonowo, *Poland* . . . . . . **47 B6** 53 23N 19 10 E
Jaboatão, *Brazil* . . . . **154 C4** 8 7 S 35 1W
Jabonga, *Phil.* . . . . **71 G5** 9 20N 125 32 E
Jaboticabal, *Brazil* . . . . . . **159 A6** 21 15 S 48 17W
Jabukovac, *Serbia, Yug.* . **42 C7** 44 22N 22 21 E
Jaburu, *Brazil* . . . . **157 B5** 5 30 S 64 0W
Jaca, *Spain* **34 C4** 42 35N 0 33W
Jacaré →, *Brazil* . . . . **154 D3** 10 3 S 42 13W
Jacareí, *Brazil* . . . . **159 A6** 23 20 S 46 0W
Jacarèzinho, *Brazil* . . . . . **159 A6** 23 5 S 49 58W
Jáchymov, *Czech.* . . . . **30 A5** 50 22N 12 55 E
Jaciara, *Brazil* . . . . **157 D7** 15 59 S 54 57W
Jacinto, *Brazil* . . . . **155 E3** 16 10 S 40 17W
Jaciparaná, *Brazil* . . . . . . **157 B5** 9 15 S 64 23W
Jackman, *U.S.A.* **129 C5** 45 35N 70 17W
Jacksboro, *U.S.A.* **139 J5** 33 14N 98 15W
Jackson, *Australia* . . . . . . **115 D4** 26 39 S 149 39 E
Jackson, *Ala., U.S.A.* . . . **135 K2** 31 31N 87 53W
Jackson, *Calif., U.S.A.* . . **144 G6** 38 21N 120 46W
Jackson, *Ky., U.S.A.* **134 G4** 37 33N 83 23W
Jackson, *Mich., U.S.A.* . . **141 B12** 42 15N 84 24W
Jackson, *Minn., U.S.A.* . . **138 D7** 43 37N 95 1W
Jackson, *Miss., U.S.A.* . . **139 J9** 32 18N 90 12W
Jackson, *Mo., U.S.A.* . . . **139 G10** 37 23N 89 40W
Jackson, *Ohio, U.S.A.* . . . **134 F4** 39 3N 82 39W
Jackson, *Tenn., U.S.A.* . . **135 H1** 35 37N 88 49W
Jackson, *Wyo., U.S.A.* . . . **142 E8** 43 29N 110 46W
Jackson, C., *N.Z.* . . . . **119 A9** 40 59 S 174 20 E
Jackson B., *N.Z.* . . . . **119 D3** 43 58 S 168 42 E
Jackson Center, *U.S.A.* . . **141 D12** 40 27N 84 4W
Jackson Hd., *N.Z.* . . . . **119 D3** 43 58 S 168 37 E
Jackson L., *U.S.A.* **142 E8** 43 52N 110 36W
Jacksons, *N.Z.* . . . . **119 C6** 42 46 S 171 32 E
Jacksonville, *Ala., U.S.A.* **135 J3** 33 49N 85 46W
Jacksonville, *Calif.,*
U.S.A. **144 H6** 37 52N 120 24W
Jacksonville, *Fla., U.S.A.* **135 K5** 30 20N 81 39W
Jacksonville, *Ill., U.S.A.* . **140 E6** 39 44N 90 14W
Jacksonville, *N.C., U.S.A.* **135 H7** 34 45N 77 26W
Jacksonville, *Oreg.,*
U.S.A. **142 E2** 42 19N 122 57W
Jacksonville, *Tex., U.S.A.* **139 K7** 31 58N 95 17W
Jacksonville Beach,
U.S.A. **135 K5** 30 17N 81 24W
Jacmel, *Haiti* **149 C5** 18 14N 72 32W
Jacob Lake, *U.S.A.* **143 H7** 36 43N 112 13W
Jacobabad, *Pakistan* . . . . . **79 C3** 28 20N 68 29 E
Jacobina, *Brazil* . . . . **154 D3** 11 11 S 40 30W
Jacques-Cartier, Mt.,
Canada **129 C6** 48 57N 66 0W
Jacqueville, *Ivory C.* **100 D4** 5 12N 4 25W
Jacuí →, *Brazil* . . . . **159 C5** 30 2 S 51 15W
Jacumba, *U.S.A.* **145 N10** 32 37N 116 11W
Jacundá →, *Brazil* . . . . . **154 B1** 1 57 S 50 26W
Jade, *Germany* . . . . **26 B4** 53 20N 8 14 E
Jadebusen, *Germany* . . . . **26 B4** 53 29N 8 12 E
Jadoigne, *Belgium* . . . . . . **21 G5** 50 43N 4 52 E
Jadotville = Likasi, *Zaïre* **107 E2** 10 55 S 26 48 E
Jadovnik, *Serbia, Yug.* . . **42 D4** 43 20N 19 45 E
Jadów, *Poland* . . . . **47 C8** 52 28N 21 38 E
Jadraque, *Spain* **34 E2** 40 55N 2 55W
Jādū, *Libya* . . . . **96 B2** 32 0N 12 0 E
Jaén, *Peru* . . . . **156 B2** 5 25 S 78 40W
Jaén, *Spain* **37 H7** 37 44N 3 43W
Jaén □, *Spain* **37 H7** 37 50N 3 30W
Jafène, *Africa* **98 D3** 20 35N 5 30W
Jaffa = Tel Aviv-Yafo,
Israel . . . . **91 C3** 32 4N 34 48 E
Jaffa, C., *Australia* . . . . . **116 D3** 36 58 S 139 40 E
Jaffna, *Sri Lanka* . . . . **83 K5** 9 45N 80 2 E
Jagadhri, *India* . . . . **80 D7** 30 10N 77 20 E
Jagadishpur, *India* . . . . . . **81 G11** 25 30N 84 21 E
Jagdalpur, *India* . . . . **82 E5** 19 3N 82 0 E
Jagersfontein, *S. Africa* . . **104 D4** 29 44 S 25 27 E
Jagst →, *Germany* . . . . . **27 F5** 49 14N 9 10 E
Jagtial, *India* . . . . **82 E4** 18 50N 79 0 E
Jaguaquara, *Brazil* . . . . . **155 D4** 13 32 S 39 58W
Jaguariaíva, *Brazil* . . . . . **159 A6** 24 10 S 49 50W
Jaguaribe, *Brazil* . . . . **154 C4** 5 53 S 38 37W
Jaguaribe →, *Brazil* . . . . **154 B4** 4 25 S 37 45W
Jaguaruana, *Brazil* . . . . . . **154 B4** 4 50 S 37 47W
Jagüey Grande, *Cuba* . . . **148 B3** 22 35N 81 7W
Jagungal, Mt., *Australia* . **117 D8** 36 8 S 148 22 E
Jahangirabad, *India* . . . . . **80 E8** 28 19N 78 4 E
Jahrom, *Iran* . . . . **85 D7** 28 30N 53 31 E
Jaicós, *Brazil* . . . . **154 C3** 7 21 S 41 8W
Jailolo, *Indonesia* . . . . **72 A3** 1 5N 127 30 E
Jailolo, Selat, *Indonesia* . **73 A3** 0 5N 129 5 E
Jaintiapur, *Bangla.* . . . . . . **78 C4** 25 8N 92 7 E
Jaipur, *India* . . . . **80 F6** 27 0N 75 50 E
Jājarm, *Iran* . . . . **85 B8** 36 58N 56 27 E
Jajce, *Bos.-H.* . . . . **42 C2** 44 19N 17 17 E
Jajpur, *India* . . . . **82 D8** 20 53N 86 22 E
Jakarta, *Indonesia* . . . . . . **74 D3** 6 9 S 106 49 E
Jakobstad = Pietarsaari,
Finland . . . . **12 E20** 63 40N 22 43 E
Jakupica, *Macedonia* . . . . **42 F6** 41 45N 21 22 E
Jal, *U.S.A.* **139 J3** 32 7N 103 12W
Jalal-Abad, *Kyrgyzstan* . . **55 C6** 40 56N 73 0 E
Jalalabad, *Afghan.* . . . . . . **79 B3** 34 30N 70 29 E
Jalalabad, *India* . . . . **81 F8** 27 41N 79 42 E
Jalalpur Jattan, *Pakistan* . **80 C6** 32 38N 74 11 E
Jalama, *U.S.A.* **145 L6** 34 29N 120 29W
Jalapa, *Guatemala* **148 D2** 14 39N 89 59W
Jalapa Enríquez, *Mexico* . **147 D5** 19 32N 96 55W
Jalasjärvi, *Finland* . . . . . . **13 E20** 62 29N 22 47 E
Jalaun, *India* . . . . **81 F8** 26 8N 79 25 E
Jaldak, *Afghan.* . . . . **79 C2** 31 58N 66 43 E
Jales, *Brazil* . . . . **155 F1** 20 10 S 50 33W
Jaleswar, *Nepal* . . . . **81 F11** 26 38N 85 48 E
Jalgaon, *Maharashtra,*
India . . . . **82 D3** 21 2N 76 31 E
Jalgaon, *Maharashtra,*
India . . . . **82 D2** 21 0N 75 42 E
Jalhay, *Belgium* . . . . **21 G7** 50 33N 5 58 E
Jalībah, *Iraq* . . . . **84 D5** 30 35N 46 32 E
Jalingo, *Nigeria* . . . . **101 D7** 8 55N 11 25 E
Jalisco □, *Mexico* **146 C4** 20 0N 104 0W
Jalkot, *Pakistan* . . . . **81 B5** 35 14N 73 24 E
Jallas →, *Spain* **36 C1** 42 54N 9 8W
Jalna, *India* . . . . **82 E2** 19 48N 75 38 E
Jalón →, *Spain* **34 D3** 41 47N 1 4W
Jalpa, *Mexico* **146 C4** 21 38N 102 58W
Jalpaiguri, *India* . . . . **78 B2** 26 32N 88 46 E
Jalq, *Iran* . . . . **79 D1** 27 35N 62 46 E
Jaluit I., *Pac. Oc.* . . . . **122 G8** 6 0N 169 30 E
Jalūlā, *Iraq* . . . . **89 E11** 34 16N 45 10 E
Jamaari, *Nigeria* . . . . **101 C6** 11 44N 9 53 E
Jamaica, *U.S.A.* **140 C2** 41 51N 94 18W
Jamaica ■, *W. Indies* **148 C4** 18 10N 77 30W
Jamalpur, *Bangla.* . . . . . . **78 C2** 24 52N 89 56 E
Jamalpur, *India* . . . . **81 G12** 25 18N 86 28 E
Jamalpurganj, *India* . . . . . **81 H13** 23 2N 88 1 E
Jamanxim →, *Brazil* . . . . **157 A6** 4 43 S 56 18W
Jamari, *Brazil* . . . . **157 B5** 8 45 S 63 27W
Jamari →, *Brazil* . . . . **157 B5** 8 27 S 63 30W
Jambe, *Indonesia* . . . . **73 B4** 1 15 S 132 10 E
Jambes, *Belgium* . . . . **21 H5** 50 27N 4 52 E
Jambi, *Indonesia* . . . . **74 C2** 1 38 S 103 30 E
Jambi □, *Indonesia* . . . . . **74 C2** 1 30 S 102 30 E
Jambusar, *India* . . . . **80 H5** 22 3N 72 51 E
James →, *U.S.A.* **138 D6** 42 52N 97 18W
James B., *Canada* **127 C11** 51 30N 80 0W
James Ras., *Australia* . . . **112 D5** 24 10 S 132 30 E
James Ross I., *Antarctica* **7 C18** 63 58 S 57 50W
Jamesport, *U.S.A.* **140 E3** 39 58N 93 48W
Jamestown, *Australia* . . . . **116 B3** 33 10 S 138 32 E
Jamestown, *S. Africa* . . . . **104 E4** 31 6 S 26 45 E
Jamestown, *Ind., U.S.A.* . **141 E10** 39 56N 86 38W
Jamestown, *Ky., U.S.A.* . **134 G3** 36 59N 85 4W
Jamestown, *Mo., U.S.A.* . **140 F4** 38 48N 92 30W
Jamestown, *N. Dak.,*
U.S.A. **138 B5** 46 54N 98 42W
Jamestown, *N.Y., U.S.A.* . **136 D5** 42 6N 79 14W
Jamestown, *Ohio, U.S.A.* . **141 E13** 39 39N 83 33W
Jamestown, *Pa., U.S.A.* . **136 E4** 41 29N 80 27W
Jamestown, *Tenn., U.S.A.* **135 G3** 36 26N 84 56W
Jamīlābād, *Iran* . . . . **85 C6** 34 24N 48 28 E
Jamiltepec, *Mexico* **147 D5** 16 17N 97 49W
Jamkhandi, *India* . . . . **82 F2** 16 30N 75 15 E
Jammalamadugu, *India* . . **83 G4** 14 51N 78 25 E
Jammerbugt, *Denmark* . . **15 G3** 57 15N 9 20 E
Jammu, *India* . . . . **80 C6** 32 43N 74 54 E
Jammu & Kashmir □,
India . . . . **81 B7** 34 25N 77 0 E
Jamnagar, *India* . . . . **80 H4** 22 30N 70 6 E
Jamner, *India* . . . . **82 D2** 20 45N 75 52 E
Jamoigne, *Belgium* . . . . . . **21 J6** 49 41N 5 24 E
Jampur, *Pakistan* . . . . **79 C3** 29 39N 70 40 E
Jamrud, *Pakistan* . . . . **79 B3** 33 59N 71 24 E
Jämsä, *Finland* . . . . **13 F21** 61 53N 25 10 E
Jamshedpur, *India* . . . . . . **81 H12** 22 44N 86 12 E
Jamtara, *India* . . . . **81 H12** 23 59N 86 49 E
Jämtland, *Sweden* . . . . . . **12 E15** 63 31N 14 0 E
Jämtlands län □, *Sweden* **14 B7** 62 40N 13 50 E
Jamuna →, *Bangla.* . . . . **78 D2** 23 51N 89 45 E
Jamurki, *Bangla.* . . . . **78 C3** 24 9N 90 2 E
Jan L., *Canada* **131 C8** 54 56N 102 55W
Jan Mayen, *Arctic* **6 B7** 71 0N 9 0W
Janakkala, *Finland* . . . . . . **13 F21** 60 54N 24 36 E
Janaúba, *Brazil* . . . . **155 E3** 15 48 S 43 19W
Janaucu, I., *Brazil* **154 A1** 0 30N 50 10W
Jand, *Pakistan* . . . . **80 C5** 33 30N 72 6 E
Janda, L. de la, *Spain* . . . **37 J5** 36 15N 5 45W
Jandaia, *Brazil* . . . . **155 E1** 17 6 S 50 7W
Jandaq, *Iran* . . . . **85 C7** 34 3N 54 22 E
Jandia, *Canary Is.* **33 F5** 28 6N 14 21W
Jandia, Pta. de, *Canary Is.* **33 F5** 28 3N 14 31W
Jandiatuba →, *Brazil* . . . **152 D4** 3 28 S 68 42W
Jandola, *Pakistan* . . . . **80 C4** 32 20N 70 9 E
Jandowae, *Australia* . . . . **115 D5** 26 45 S 151 7 E
Jandrain-Jandrenouilles,
Belgium . . . . **21 G5** 50 40N 4 58 E
Jándula →, *Spain* **37 G6** 38 3N 4 6W
Jane Pk., *N.Z.* . . . . **119 F3** 45 15 S 168 20 E
Janesville, *U.S.A.* **140 B7** 42 41N 89 1W
Janga, *Ghana* **101 C4** 10 5N 1 0W
Jango, *Brazil* . . . . **157 E6** 20 27 S 55 29W
Jangoon, *India* . . . . **82 F4** 17 44N 79 5 E
Janhtang Ga, *Burma* . . . . **78 B6** 26 32N 96 38 E
Jānī Kheyl, *Afghan.* . . . . . **79 B3** 32 46N 68 24 E
Janikowo, *Poland* . . . . **47 C5** 52 45N 18 7 E
Janīn, *West Bank* . . . . **91 C4** 32 28N 35 18 E
Janinà = Ioánnina □,
Greece . . . . **44 E2** 39 39N 20 57 E
Janiuay, *Phil.* . . . . **71 F4** 10 58N 122 30 E
Janja, *Bos.-H.* . . . . **42 C4** 44 40N 19 17 E
Janjevo, *Serbia, Yug.* . . . . **42 E6** 42 35N 21 19 E
Janjina, *Croatia* . . . . **42 E2** 42 58N 17 25 E
Janos, *Mexico* **146 A3** 30 45N 108 10W
Jánoshalma, *Hungary* . . . **31 E12** 46 18N 19 21 E
Jánosháza, *Hungary* . . . . **31 D10** 47 8N 17 12 E
Jánossomorja, *Hungary* . . **31 D10** 47 47N 17 11 E
Janów, *Poland* . . . . **47 E6** 50 44N 19 27 E
Janów Lubelski, *Poland* . . **47 E9** 50 48N 22 23 E
Janów Podlaski, *Poland* . . **47 C10** 52 11N 23 11 E
Janowiec Wielkopolski,
Poland . . . . **47 C4** 52 45N 17 30 E
Januária, *Brazil* . . . . **155 E3** 15 25 S 44 25W
Janub Dârfûr □, *Sudan* . . **95 E2** 11 0N 25 0 E
Janub Kordofân □, *Sudan* **95 E3** 12 0N 30 0 E
Janubio, *Canary Is.* **33 F6** 28 56N 13 50W
Janville, *France* . . . . **23 D8** 48 10N 1 50 E
Janzé, *France* **22 E5** 47 55N 1 28W
Jaora, *India* . . . . **80 H6** 23 40N 75 10 E
Japan ■, *Asia* . . . . **61 G8** 36 0N 136 0 E
Japan, Sea of, *Asia* . . . . . **60 E7** 40 0N 135 0 E
Japan Trench, *Pac. Oc.* . . **122 D6** 32 0N 142 0 E
Japen = Yapen, *Indonesia* **73 B5** 1 50 S 136 0 E
Japurá →, *Brazil* . . . . . . **152 D4** 3 8 S 65 46W
Jaque, *Panama* **152 B2** 7 27N 78 8W
Jarābulus, *Syria* . . . . **89 D8** 36 49N 38 1 E
Jaraguá, *Brazil* . . . . **155 E2** 15 45 S 49 20W
Jaraguari, *Brazil* . . . . **157 E7** 20 9 S 54 35W
Jaraicejo, *Spain* **37 F5** 39 40N 5 49W
Jaraiz, *Spain* **36 E5** 40 4N 5 45W
Jarama →, *Spain* **34 E1** 40 24N 3 32W
Jaramānah, *Syria* . . . . **88 F7** 33 29N 36 21 E
Jaramillo, *Argentina* . . . . . **160 C3** 47 10 S 67 7W
Jarandilla, *Spain* **36 E5** 40 8N 5 39W
Jaranwala, *Pakistan* . . . . . **79 C4** 31 15N 73 26 E
Jarash, *Jordan* . . . . **91 C4** 32 17N 35 54 E
Jarauçu →, *Brazil* . . . . . . **153 D7** 1 48 S 52 22W
Jardas al 'Abīd, *Libya* . . . **96 B4** 32 18N 20 59 E
Jardim, *Brazil* . . . . **158 A4** 21 28 S 56 2W
Jardín →, *Spain* **35 G2** 38 50N 2 10W
Jardines de la Reina, Is.,
Cuba **148 B4** 20 50N 78 50W
Jargalang, *China* . . . . **67 C12** 43 5N 122 55 E
Jargalant = Hovd,
Mongolia . . . . **64 B4** 48 2N 91 37 E
Jargeau, *France* . . . . **23 E9** 47 50N 2 1 E
Jari →, *Brazil* . . . . **153 D7** 1 9 S 51 54W
Jarīr, W. al →, *Si. Arabia* **84 E4** 25 38N 42 30 E
Jarmen, *Germany* . . . . **26 B9** 53 54N 13 20 E
Jarnac, *France* **24 C3** 45 40N 0 11W
Jarny, *France* . . . . **23 C12** 49 9N 5 53 E
Jaro, *Phil.* . . . . **71 F5** 11 11N 124 47 E
Jarocin, *Poland* . . . . **47 D4** 51 59N 17 29 E
Jaroměr, *Czech.* . . . . **30 A8** 50 22N 15 52 E
Jarosław, *Poland* . . . . **31 A15** 50 2N 22 42 E
Järpås, *Sweden* . . . . **15 F6** 58 23N 12 57 E
Järpen, *Sweden* . . . . **14 A7** 63 21N 13 26 E
Jarqŭrghon, *Uzbekistan* . . **55 E3** 37 31N 67 25 E
Jarrahdale, *Australia* . . . . **113 F2** 32 24 S 116 5 E
Jarres, Plaine des, *Laos* . **76 C4** 19 27N 103 10 E
Jarso, *Ethiopia* . . . . **95 F4** 5 15N 37 30 E
Jartai, *China* . . . . **66 E3** 39 45N 105 48 E
Jaru, *Brazil* . . . . **157 C5** 10 26 S 62 27W
Jaru →, *Brazil* . . . . **157 C5** 10 5 S 61 59W
Jarud Qi, *China* . . . . **67 B11** 44 28N 120 50 E
Järvenpää, *Finland* . . . . . **13 F21** 60 29N 25 5 E
Jarvis, *Canada* **136 D4** 42 53N 80 6W
Jarvis I., *Pac. Oc.* . . . . . . **123 H12** 0 15 S 159 55W
Jarvornik, *Czech.* . . . . **31 A10** 50 23N 17 2 E
Jarwa, *India* . . . . **81 F10** 27 38N 82 30 E
Jaša Tomić, *Serbia, Yug.* . **42 B5** 45 26N 20 50 E
Jasaan, *Phil.* . . . . **71 G5** 8 39N 124 45 E
Jasien, *Poland* . . . . **47 D2** 51 46N 15 0 E
Jāsimīyah, *Iraq* . . . . **89 F11** 33 45N 44 41 E
Jasin, *Malaysia* . . . . **77 L4** 2 20N 102 26 E
Jāsk, *Iran* . . . . **85 E8** 25 38N 57 45 E
Jasło, *Poland* . . . . **31 B14** 49 45N 21 30 E
Jason, Is., *Falk. Is.* . . . . . **160 D4** 51 0 S 61 0W
Jasonville, *U.S.A.* **141 E9** 39 10N 87 12W
Jasper, *Alta., Canada* . . . **130 C5** 52 55N 118 5W
Jasper, *Ont., Canada* **137 B9** 44 52N 75 57W
Jasper, *Ala., U.S.A.* **135 J2** 33 50N 87 17W
Jasper, *Fla., U.S.A.* **135 K4** 30 31N 82 57W
Jasper, *Ind., U.S.A.* **141 F10** 38 24N 86 56W
Jasper, *Minn., U.S.A.* . . . **138 D6** 43 51N 96 24W
Jasper, *Tex., U.S.A.* **139 K8** 30 56N 94 1W
Jasper Nat. Park, *Canada* **130 C5** 52 50N 118 8W
Jassy = Iaşi, *Romania* . . . **46 B8** 47 10N 27 40 E
Jastrebarsko, *Croatia* . . . **39 C12** 45 41N 15 39 E
Jastrowie, *Poland* . . . . **47 B3** 53 26N 16 49 E
Jastrzębie Zdrój, *Poland* . **31 B11** 49 57N 18 35 E
Jászapáti, *Hungary* . . . . . **31 D13** 47 32N 20 10 E
Jászárokszállás, *Hungary* . **31 D13** 47 39N 20 1 E
Jászberény, *Hungary* . . . . **31 D12** 47 30N 19 55 E
Jászkiser, *Hungary* . . . . . **31 D13** 47 27N 20 20 E
Jászladány, *Hungary* . . . . **31 D13** 47 23N 20 10 E
Jataí, *Brazil* . . . . **155 E1** 17 58 S 51 48W
Jatapu →, *Brazil* . . . . **153 D6** 2 13 S 58 17W
Jati, *Pakistan* . . . . **80 G3** 24 20N 68 19 E
Jatibarang, *Indonesia* . . . . **75 D3** 6 28 S 108 18 E
Jatinegara, *Indonesia* . . . . **74 D3** 6 13 S 106 52 E
Játiva, *Spain* **35 G4** 39 0N 0 32W
Jatobal, *Brazil* . . . . **154 B2** 4 35 S 49 33W
Jáu, *Angola* . . . . **103 F2** 15 12 S 13 31 E
Jaú, *Brazil* . . . . **159 A6** 22 10 S 48 30W
Jaú →, *Brazil* . . . . **153 D5** 1 54 S 61 26W
Jauaperí →, *Brazil* . . . . . **153 D5** 1 26 S 61 35W
Jauche, *Belgium* . . . . **21 G5** 50 41N 4 57 E
Jauja, *Peru* . . . . **156 C2** 11 45 S 75 15W
Jaunpur, *India* . . . . **81 G10** 25 46N 82 44 E
Jauru →, *Brazil* . . . . **157 D6** 16 22 S 57 46W
Java = Jawa, *Indonesia* . . **75 D4** 7 0 S 110 0 E
Java Sea, *Indonesia* . . . . . **75 C3** 4 35 S 107 15 E
Java Trench, *Ind. Oc.* . . . **74 D2** 9 0 S 105 0 E
Javadi Hills, *India* . . . . . . **83 H4** 12 40N 78 40 E
Jávea, *Spain* . . . . **35 G5** 38 48N 0 10 E
Javhlant = Ulyasutay,
Mongolia . . . . **64 B4** 47 56N 97 28 E
Javier, I., *Chile* . . . . **160 C2** 47 5 S 74 25W
Javla, *India* . . . . **82 F2** 17 18N 75 9 E
Javron, *France* **22 D6** 48 25N 0 25W
Jawa, *Indonesia* . . . . **75 D4** 7 0 S 110 0 E
Jawf, W. al →, *Yemen* . . **86 D4** 15 50N 45 30 E
Jawor, *Poland* . . . . **47 D3** 51 4N 16 11 E
Jaworzno, *Poland* . . . . **31 A12** 50 13N 19 11 E
Jay, *U.S.A.* **139 G7** 36 25N 94 48W
Jaya, Puncak, *Indonesia* . **73 B5** 3 57 S 137 17 E
Jayanca, *Peru* . . . . **156 B2** 6 24 S 79 50W
Jayanti, *India* . . . . **78 B2** 26 45N 89 40 E
Jayapura, *Indonesia* . . . . . **73 B6** 2 28 S 140 38 E
Jayawijaya, Pegunungan,
Indonesia . . . . **73 B5** 5 0 S 139 0 E
Jayrūd, *Syria* . . . . **88 F7** 33 49N 36 44 E
Jayton, *U.S.A.* **139 J4** 33 15N 100 34W
Jazīreh-ye Shīf, *Iran* . . . . **85 D6** 29 4N 50 54 E
Jazminal, *Mexico* **146 C4** 24 56N 101 25W
Jazzīn, *Lebanon* . . . . **91 B4** 33 31N 35 35 E
Jean, *U.S.A.* **145 K11** 35 47N 115 20W
Jean Marie River, *Canada* **130 A4** 61 32N 120 38W
Jean Rabel, *Haiti* **149 C5** 19 50N 73 5W
Jeanerette, *U.S.A.* **139 L9** 29 55N 91 40W
Jeanette, Ostrov, *Russia* . **57 B16** 76 43N 158 0 E
Jeannette, *U.S.A.* **136 F5** 40 20N 79 36W
Jebba, *Morocco* **98 A4** 35 11N 4 43W
Jebba, *Nigeria* . . . . **101 D5** 9 9N 4 48 E
Jebel, Bahr el →, *Sudan* . **95 F3** 9 30N 30 25 E
Jebel Qerri, *Sudan* . . . . . . **95 D3** 16 16N 32 50 E
Jeberos, *Peru* . . . . **156 B2** 5 15 S 76 10W
Jedburgh, *U.K.* **18 F6** 55 28N 2 33W
Jedda = Jiddah,
Si. Arabia . . . . **86 B2** 21 29N 39 10 E
Jedlicze, *Poland* . . . . **31 B14** 49 43N 21 40 E
Jedlnia-Letnisko, *Poland* . **47 D8** 51 25N 21 19 E
Jędrzejów, *Poland* . . . . . . **47 E7** 50 35N 20 15 E
Jedwabne, *Poland* . . . . . . **47 B9** 53 17N 22 18 E
Jedway, *Canada* **130 C2** 52 17N 131 14W
Jeetzel →, *Germany* . . . . **26 B7** 53 9N 11 3 E
Jefferson, *Iowa, U.S.A.* . . **140 B2** 42 1N 94 23W
Jefferson, *Ohio, U.S.A.* . . **136 E4** 41 44N 80 46W
Jefferson, *Tex., U.S.A.* . . **139 J7** 32 46N 94 21W
Jefferson, *Wis., U.S.A.* . . **138 D10** 43 0N 88 48W
Jefferson, Mt., *Nev.,*
U.S.A. **142 G5** 38 51N 117 0W
Jefferson, Mt., *Oreg.,*
U.S.A. **142 D3** 44 41N 121 48W
Jefferson City, *Mo.,*
U.S.A. **140 F4** 38 34N 92 10W
Jefferson City, *Tenn.,*
U.S.A. **135 G4** 36 7N 83 30W
Jeffersontown, *U.S.A.* . . . **141 F11** 38 12N 85 35W
Jeffersonville, *Ind.,*
U.S.A. **141 F11** 38 17N 85 44W
Jeffersonville, *Ohio,*
U.S.A. **141 E13** 39 39N 83 34W
Jega, *Nigeria* . . . . **101 C5** 12 15N 4 23 E
Jēkabpils, *Latvia* . . . . **13 H21** 56 29N 25 57 E
Jelenia Góra, *Poland* . . . . **47 E2** 50 50N 15 45 E
Jelenia Góra □, *Poland* . . **47 D2** 51 0N 15 30 E
Jelgava, *Latvia* . . . . **13 H20** 56 41N 23 49 E
Jelica, *Serbia, Yug.* . . . . . **42 D5** 43 50N 20 17 E
Jelli, *Sudan* . . . . **95 F3** 5 25N 31 45 E
Jellicoe, *Canada* **128 C2** 49 40N 87 30W
Jelšava, *Slovak Rep.* . . . . **31 C13** 48 37N 20 15 E
Jemaja, *Indonesia* . . . . **74 B3** 3 5N 105 45 E
Jemaluang, *Malaysia* . . . . **77 L4** 2 16N 103 52 E
Jemappes, *Belgium* . . . . . **21 H3** 50 27N 3 54 E
Jember, *Indonesia* . . . . . . **75 D4** 8 11 S 113 41 E
Jemeppe, *Belgium* . . . . . . **21 G7** 50 37N 5 30 E
Jemnice, *Czech.* . . . . **30 B8** 49 1N 15 34 E
Jena, *Germany* . . . . **26 E7** 50 54N 11 35 E
Jena, *U.S.A.* **139 K8** 31 41N 92 8W
Jenbach, *Austria* . . . . **30 D4** 47 24N 11 47 E
Jendouba, *Tunisia* . . . . . . **96 A1** 36 29N 8 47 E
Jenkins, *U.S.A.* **134 G4** 37 10N 82 38W
Jenner, *U.S.A.* **144 G3** 38 27N 123 7W
Jennings, *La., U.S.A.* . . . **139 K8** 30 13N 92 40W
Jennings, *Mo., U.S.A.* . . . **140 F6** 38 43N 90 16W
Jennings →, *Canada* **130 B2** 59 38N 132 5W
Jepara, *Indonesia* . . . . **75 D3** 7 40 S 109 14 E
Jeparit, *Australia* . . . . **116 D5** 36 8 S 142 1 E
Jequié, *Brazil* . . . . **155 D3** 13 51 S 40 5W
Jequitaí →, *Brazil* . . . . . . **155 E3** 17 4 S 44 50W
Jequitinhonha, *Brazil* . . . . **155 E3** 16 30 S 41 0W
Jequitinhonha →, *Brazil* **155 E4** 15 51 S 38 53W
Jerada, *Morocco* **99 B4** 34 17N 2 10W
Jerantut, *Malaysia* . . . . . . **77 L4** 3 56N 102 22 E
Jérémie, *Haiti* **149 C5** 18 40N 74 10W
Jeremoabo, *Brazil* . . . . **154 D4** 10 4 S 38 21W
Jerez, Punta, *Mexico* **147 C5** 22 58N 97 40W
Jerez de García Salinas,
Mexico **146 C4** 22 39N 103 0W
Jerez de la Frontera, *Spain* **37 J4** 36 41N 6 7W
Jerez de los Caballeros,
Spain **37 G4** 38 20N 6 45W
Jericho = Arīḩā, *Syria* . . **84 C3** 35 49N 36 35 E
Jericho = El Arīḩā,
West Bank . . . . **91 D4** 31 52N 35 27 E
Jericho, *Australia* . . . . **114 C4** 23 38 S 146 6 E
Jerichow, *Germany* . . . . . **26 C8** 52 30N 12 1 E
Jerico Springs, *U.S.A.* . . . **140 G2** 37 37N 94 1W
Jerilderie, *Australia* . . . . . **117 C6** 35 20 S 145 41 E
Jermyn, *U.S.A.* **137 E9** 41 31N 75 31W
Jerome, *U.S.A.* **143 J8** 34 45N 112 7W
Jersey, *U.K.* **17 H5** 49 11N 2 7W
Jersey City, *U.S.A.* **137 F10** 40 44N 74 4W
Jersey Shore, *U.S.A.* **136 E7** 41 12N 77 15W
Jerseyville, *U.S.A.* **140 E6** 39 7N 90 20W
Jerusalem, *Israel* . . . . **91 D4** 31 47N 35 10 E
Jervis B., *Australia* . . . . . **117 C9** 35 8 S 150 46 E
Jesenice, *Slovenia* . . . . . . **39 B11** 46 28N 14 3 E
Jeseník, *Czech.* . . . . **31 B10** 50 0N 17 8 E
Jesenké, *Slovak Rep.* . . . **31 C13** 48 20N 20 10 E
Jesselton = Kota
Kinabalu, *Malaysia* . . . . **75 A5** 6 0N 116 4 E
Jessnitz, *Germany* . . . . . . **26 D8** 51 40N 12 18 E
Jessore, *Bangla.* . . . . **78 D2** 23 10N 89 10 E
Jesup, *Ga., U.S.A.* **135 K5** 31 36N 81 53W
Jesup, *Iowa, U.S.A.* **140 B4** 42 29N 92 4W
Jesús, *Peru* . . . . **156 B2** 7 15 S 78 25W
Jesús Carranza, *Mexico* . . **147 D5** 17 28N 95 1W
Jesús María, *Argentina* . . **158 C3** 30 59 S 64 5W
Jetafe, *Phil.* . . . . **71 F5** 10 9N 124 9 E
Jetmore, *U.S.A.* **139 F5** 38 4N 99 54W
Jetpur, *India* . . . . **80 J4** 21 45N 70 10 E
Jette, *Belgium* . . . . **21 G4** 50 53N 4 20 E
Jevnaker, *Norway* . . . . **14 D4** 60 15N 10 26 E
Jewell, *U.S.A.* **140 B3** 42 20N 93 39W
Jewett, *Ohio, U.S.A.* **136 F3** 40 22N 81 2W
Jewett, *Tex., U.S.A.* **139 K6** 31 22N 96 9W
Jewett City, *U.S.A.* **137 E13** 41 36N 72 0W
Jeyḩūnābād, *Iran* . . . . **85 C6** 34 58N 48 59 E
Jeypore, *India* . . . . **82 E6** 18 50N 82 38 E
Jeziorak, Jezioro, *Poland* . **47 B6** 53 40N 19 35 E
Jeziorany, *Poland* . . . . **47 B7** 53 58N 20 46 E
Jeziorka →, *Poland* . . . . **47 D7** 51 59N 20 57 E
Jhajjar, *India* . . . . **80 E7** 28 37N 76 42 E
Jhal Jhao, *Pakistan* . . . . . **79 D2** 26 20N 65 35 E
Jhalakati, *Bangla.* . . . . . . **78 D3** 22 39N 90 12 E
Jhalawar, *India* . . . . **80 G7** 24 40N 76 10 E
Jhang Maghiana, *Pakistan* **79 C4** 31 15N 72 22 E
Jhansi, *India* . . . . **81 G8** 25 30N 78 36 E
Jharia, *India* . . . . **81 H12** 23 45N 86 26 E
Jharsuguda, *India* . . . . **82 D7** 21 56N 84 5 E
Jhelum, *Pakistan* . . . . **79 B4** 33 0N 73 45 E
Jhelum →, *Pakistan* . . . . **80 D5** 31 20N 72 10 E
Jhunjhunu, *India* . . . . **80 E6** 28 10N 75 30 E
Ji Xian, *Hebei, China* . . . . **66 F8** 37 35N 115 30 E
Ji Xian, *Henan, China* . . . **66 G8** 35 22N 114 5 E
Ji Xian, *Shanxi, China* . . . **66 F6** 36 7N 110 40 E
Jia Xian, *Henan, China* . . **66 H7** 33 59N 113 12 E
Jia Xian, *Shaanxi, China* . **66 E6** 38 12N 110 28 E
Jiading, *China* . . . . **69 B13** 31 22N 121 15 E
Jiahe, *China* . . . . **69 E9** 25 38N 112 19 E
Jiali, *Taiwan* . . . . **69 F13** 23 12N 120 10 E
Jialing Jiang →, *China* . . **68 C6** 29 30N 106 20 E
Jiamusi, *China* . . . . **65 B8** 46 40N 130 26 E
Ji'an, *Jiangxi, China* . . . . . **69 D10** 27 6N 114 59 E
Ji'an, *Jilin, China* . . . . **67 D14** 41 5N 126 10 E
Jianchang, *China* . . . . **67 D11** 40 55N 120 35 E
Jianchangying, *China* . . . . **67 D10** 40 10N 118 50 E
Jianchuan, *China* . . . . **68 D2** 26 38N 99 55 E
Jiande, *China* . . . . **69 C12** 29 23N 119 15 E
Jiangbei, *China* . . . . **68 C6** 29 40N 106 34 E
Jiangcheng, *China* . . . . . . **68 F3** 22 36N 101 52 E
Jiangdi, *China* . . . . **68 D4** 26 57N 103 37 E
Jiange, *China* . . . . **68 A5** 32 4N 105 32 E
Jiangjin, *China* . . . . **68 C6** 29 14N 106 14 E

Kachira, L. 213

Jiangkou, China 68 D7 27 40N 108 49 E
Jiangle, China 69 D11 26 42N 117 23 E
Jiangling, China 69 B9 30 25N 112 12 E
Jiangmen, China 69 F9 22 32N 113 0 E
Jiangshan, China 69 C12 28 40N 118 37 E
Jiangsu □, China 67 H10 33 0N 120 0 E
Jiangxi □, China 69 D10 27 30N 116 0 E
Jiangyin, China 69 B13 31 54N 120 17 E
Jiangyong, China 69 E8 25 20N 111 22 E
Jiangyou, China 68 B5 31 44N 104 43 E
Jianhe, China 68 D7 26 37N 108 31 E
Jianli, China 69 C9 29 46N 112 56 E
Jianning, China 69 D11 26 50N 116 50 E
Jian'ou, China 69 D12 27 3N 118 17 E
Jianshi, China 68 B7 30 37N 109 38 E
Jianshui, China 68 F4 23 36N 102 43 E
Jianyang, Fujian, China 69 D12 27 20N 118 5 E
Jianyang, Sichuan, China 68 B5 30 24N 104 33 E
Jiao Xian, China 67 F11 36 18N 120 1 E
Jiaohe, Hebei, China 66 E9 38 2N 116 20 E
Jiaohe, Jilin, China 67 C14 43 40N 127 22 E
Jiaoling, China 69 E11 24 41N 116 12 E
Jiaozhou Wan, China 67 F11 36 5N 120 10 E
Jiaozuo, China 66 G7 35 16N 113 12 E
Jiashan, China 69 A11 32 46N 117 59 E
Jiawang, China 67 G9 34 28N 117 26 E
Jiaxiang, China 66 G9 35 25N 116 20 E
Jiaxing, China 69 B13 30 49N 120 45 E
Jiayi, Taiwan 69 F13 23 30N 120 24 E
Jiayu, China 69 C9 29 55N 113 55 E
Jibão, Serra do, Brazil 155 D3 14 48 S 45 0W
Jibiya, Nigeria 101 C6 13 5N 7 12 E
Jibou, Romania 46 B4 47 15N 23 17 E
Jibuti = Djibouti ■, Africa 90 E3 12 0N 43 0 E
Jicarón, I., Panama 148 E3 7 10N 81 50W
Jičín, Czech. 30 A8 50 25N 15 28 E
Jiddah, Si. Arabia 86 B2 21 29N 39 10 E
Jieshou, China 66 H8 33 18N 115 22 E
Jiexiu, China 66 F6 37 2N 111 55 E
Jieyang, China 69 F11 23 35N 116 21 E
Jigawa □, Nigeria 101 C6 12 0N 9 45 E
Jiggalong, Australia 112 D3 23 21 S 120 47 E
Jihlava, Czech. 30 B8 49 28N 15 35 E
Jihlava →, Czech. 30 C9 48 55N 16 36 E
Jihočeský □, Czech. 30 B7 49 8N 14 35 E
Jihomoravský □, Czech. 31 B9 49 5N 16 30 E
Jijel, Algeria 99 A6 36 52N 5 50 E
Jijiga, Ethiopia 108 C2 9 20N 42 50 E
Jijona, Spain 35 G4 38 34N 0 30W
Jikamshi, Nigeria 101 C6 12 12N 7 45 E
Jilin, China 67 C14 43 44N 126 30 E
Jilin □, China 67 C13 44 0N 127 0 E
Jiloca →, Spain 34 D3 41 21N 1 39W
Jilong, Taiwan 69 E13 25 8N 121 42 E
Jílové, Czech. 30 B7 49 52N 14 29 E
Jima, Ethiopia 95 F4 7 40N 36 47 E
Jimbolia, Romania 46 D1 45 47N 20 43 E
Jimena de la Frontera, Spain 37 J5 36 27N 5 24W
Jimenbuen, Australia 117 D8 36 42 S 148 53 E
Jiménez, Mexico 146 B4 27 10N 104 54W
Jimenez, Phil. 71 G4 8 20N 123 50 E
Jimo, China 67 F11 36 23N 120 30 E
Jin Jiang →, China 69 C10 28 24N 115 48 E
Jin Xian, Hebei, China 66 E8 38 2N 115 2 E
Jin Xian, Liaoning, China 67 E11 38 55N 121 42 E
Jinan, China 66 F9 36 38N 117 1 E
Jincheng, China 66 G7 35 29N 112 50 E
Jinchuan, China 68 B4 31 30N 102 3 E
Jind, India 80 E7 29 19N 76 22 E
Jindabyne, Australia 117 D8 36 25 S 148 35 E
Jindrichuv Hradeç, Czech. 30 B8 49 10N 15 2 E
Jing He →, China 66 G5 34 27N 109 4 E
Jing Shan, China 69 B8 31 20N 111 35 E
Jing Xian, Anhui, China 69 B12 30 38N 118 25 E
Jing Xian, Hunan, China 68 D7 26 33N 109 40 E
Jing'an, China 69 C10 28 50N 115 17 E
Jingbian, China 66 F5 37 20N 108 30 E
Jingchuan, China 66 G4 35 20N 107 20 E
Jingde, China 69 B12 30 15N 118 27 E
Jingdezhen, China 69 C11 29 20N 117 11 E
Jingdong, China 68 E3 24 23N 100 47 E
Jinggu, China 68 F3 23 35N 100 41 E
Jinghai, China 66 E9 38 55N 116 55 E
Jinghong, China 68 F3 22 0N 100 45 E
Jingjiang, China 69 A13 32 2N 120 16 E
Jingle, China 66 E6 38 20N 111 55 E
Jingmen, China 69 B9 31 0N 112 10 E
Jingning, China 66 G3 35 30N 105 43 E
Jingpo Hu, China 67 C15 43 55N 128 55 E
Jingshan, China 69 B9 31 1N 113 7 E
Jingtai, China 66 F3 37 10N 104 6 E
Jingxi, China 68 F6 23 8N 106 27 E
Jingxing, China 66 E8 38 2N 114 8 E
Jingyang, China 66 G5 34 30N 108 50 E
Jingyu, China 67 C14 42 25N 126 45 E
Jingyuan, China 66 F3 36 30N 104 40 E
Jingziguan, China 66 H6 33 15N 111 0 E
Jinhua, China 69 C12 29 8N 119 38 E
Jining, Nei Mongol Zizhiqu, China 66 D7 41 5N 113 0 E
Jining, Shandong, China 66 G9 35 22N 116 34 E
Jinja, Uganda 106 B3 0 25N 33 12 E
Jinjang, Malaysia 77 L3 3 13N 101 39 E
Jinji, China 66 F4 37 58N 106 8 E
Jinjiang, Fujian, China 69 E12 24 43N 118 33 E
Jinjiang, Yunnan, China 68 D3 26 14N 100 34 E
Jinjie, China 68 F6 23 15N 107 18 E
Jinjini, Ghana 100 D4 7 26N 3 42W
Jinkou, China 69 B10 30 20N 114 8 E
Jinmen Dao, China 69 E12 24 25N 118 25 E
Jinning, China 68 E4 24 38N 102 38 E
Jinotega, Nic. 148 D2 13 6N 85 59W
Jinotepe, Nic. 148 D2 11 50N 86 10W
Jinping, Guizhou, China 68 D7 26 41N 109 10 E
Jinping, Yunnan, China 68 F4 22 45N 103 18 E
Jinsha, China 68 D6 27 29N 106 12 E
Jinsha Jiang →, China 68 C5 28 50N 104 36 E
Jinshan, China 69 B13 30 54N 121 8 E
Jinshi, China 69 C8 29 40N 111 50 E
Jintan, China 69 B12 31 42N 119 36 E
Jintotolo Channel, Phil. 71 F4 11 48N 123 5 E
Jinxi, Jiangxi, China 69 D11 27 56N 116 45 E
Jinxi, Liaoning, China 67 D11 40 52N 120 50 E
Jinxian, China 69 C11 28 26N 116 17 E
Jinxiang, China 66 G9 35 5N 116 22 E
Jinyun, China 69 C13 28 35N 120 5 E
Jinzhai, China 69 B10 31 40N 115 53 E
Jinzhou, China 67 D11 41 5N 121 3 E
Jiparaná →, Brazil 157 B5 8 3 S 62 52W
Jipijapa, Ecuador 152 D1 1 0 S 80 40W
Jiquilpan, Mexico 146 D4 19 57N 102 42W
Jirwān, Si. Arabia 87 B5 23 27N 50 53 E
Jishan, China 66 G6 35 34N 110 58 E
Jishou, China 68 C7 28 21N 109 43 E
Jishui, China 69 D10 27 12N 115 8 E
Jisr ash Shughūr, Syria 88 E7 35 49N 36 18 E
Jitarning, Australia 113 F2 32 48 S 117 57 E
Jitra, Malaysia 77 J3 6 16N 100 25 E
Jiu →, Romania 46 F4 43 47N 23 48 E
Jiudengkou, China 66 E4 39 56N 106 40 E
Jiujiang, Guangdong, China 69 F9 22 50N 113 0 E
Jiujiang, Jiangxi, China 69 C10 29 42N 115 58 E
Jiuling Shan, China 69 C10 28 40N 114 40 E
Jiulong, China 68 C3 28 57N 101 31 E
Jiutai, China 67 B13 44 10N 125 50 E
Jiuxiangcheng, China 66 H8 33 12N 114 50 E
Jiuxincheng, China 66 E8 39 17N 115 59 E
Jiuyuhang, China 69 B12 30 18N 119 56 E
Jixi, Anhui, China 69 B12 30 5N 118 34 E
Jixi, Heilongjiang, China 67 B16 45 20N 130 50 E
Jiyang, China 67 F9 37 0N 117 12 E
Jiz', W. →, Yemen 87 C6 16 12N 52 14 E
Jīzān, Si. Arabia 86 C3 17 0N 42 20 E
Jize, China 66 F8 36 54N 114 56 E
Jizera →, Czech. 30 A7 50 10N 14 43 E
Jizō-Zaki, Japan 62 B5 35 34N 133 20 E
Jizzakh, Uzbekistan 55 C3 40 6N 67 50 E
Joaçaba, Brazil 159 B5 27 5 S 51 31W
Joaíma, Brazil 155 E3 16 39 S 41 2W
João, Brazil 154 B1 2 46 S 50 59W
João Amaro, Brazil 155 D3 12 46 S 40 22W
João Câmara, Brazil 154 C4 5 32 S 35 48W
João Pessoa, Brazil 154 C5 7 10 S 34 52W
João Pinheiro, Brazil 155 E2 17 45 S 46 10W
Joaquim Távora, Brazil 155 F2 23 30 S 49 58W
Joaquín V. González, Argentina 158 B3 25 10 S 64 0W
Jobourg, Nez de, France 22 C5 49 41N 1 57W
Jódar, Spain 35 H1 37 50N 3 21W
Jodhpur, India 80 F5 26 23N 73 8 E
Joensuu, Finland 48 B4 62 37N 29 49 E
Jœuf, France 23 C13 49 12N 6 0 E
Jofane, Mozam. 105 C5 21 15 S 34 18 E
Jõgeva, Estonia 13 G22 58 45N 26 24 E
Joggins, Canada 129 C7 45 42N 64 27W
Jogjakarta = Yogyakarta, Indonesia 75 D4 7 49 S 110 22 E
Jõhana, Japan 63 A8 36 30N 136 57 E
Johannesburg, S. Africa 105 D4 26 10 S 28 2 E
Johannesburg, U.S.A. 145 K9 35 22N 117 38W
Jōhen, Japan 62 E4 32 58N 132 32 E
John Day, U.S.A. 142 D4 44 25N 118 57W
John Day →, U.S.A. 142 D3 45 44N 120 39W
John H. Kerr Reservoir, U.S.A. 135 G6 36 36N 78 18W
John o' Groats, U.K. 18 C5 58 38N 3 4W
Johnnie, U.S.A. 145 J10 36 25N 116 5W
John's Ra., Australia 114 C1 21 55 S 133 23 E
Johnson, U.S.A. 139 G4 37 34N 101 45W
Johnson City, Ill., U.S.A. 140 G8 37 49N 88 56W
Johnson City, N.Y., U.S.A. 137 D9 42 7N 75 58W
Johnson City, Tenn., U.S.A. 135 G4 36 19N 82 21W
Johnson City, Tex., U.S.A. 139 K5 30 17N 98 25W
Johnsonburg, U.S.A. 136 E6 41 29N 78 41W
Johnsondale, U.S.A. 145 K8 35 58N 118 32W
Johnson's Crossing, Canada 130 A2 60 29N 133 18W
Johnsonville, N.Z. 118 H3 41 13 S 174 48 E
Johnston, L., Australia 113 F3 32 25 S 120 30 E
Johnston Falls = Mambilima Falls, Zambia 107 E2 10 31 S 28 45 E
Johnston I., Pac. Oc. 123 F11 17 10N 169 8W
Johnstone Str., Canada 130 C3 50 28N 126 0W
Johnstown, N.Y., U.S.A. 137 C10 43 0N 74 22W
Johnstown, Pa., U.S.A. 136 F6 40 20N 78 55W
Johor □, Malaysia 74 B2 2 5N 103 20 E
Johor Baharu, Malaysia 77 M4 1 28N 103 46 E
Jõhvi, Estonia 13 G22 59 22N 27 27 E
Joigny, France 23 E10 47 58N 3 20 E
Joinvile, Brazil 159 B6 26 15 S 48 55W
Joinville, France 23 D12 48 27N 5 10 E
Joinville I., Antarctica 7 C18 65 0 S 55 30W
Jojutla, Mexico 147 D5 18 37N 99 11W
Jokkmokk, Sweden 12 C18 66 35N 19 50 E
Jökulsá á Bru →, Iceland 12 D6 65 40N 14 16W
Jökulsá á Fjöllum →, Iceland 12 C5 66 10N 16 30W
Jolfā, Āzarbājān-e Sharqī, Iran 89 C11 38 57N 45 38 E
Jolfā, Eşfahan, Iran 85 C6 32 58N 51 37 E
Joliet, U.S.A. 141 C8 41 32N 88 5W
Joliette, Canada 128 C5 46 3N 73 24W
Jolo, Phil. 71 H3 6 0N 121 0 E
Jolo Group, Phil. 71 J3 6 0N 121 9 E
Jolon, U.S.A. 144 K5 35 58N 121 9W
Jomalig, Phil. 70 D4 14 42N 122 22 E
Jombang, Indonesia 75 D4 7 33 S 112 14 E
Jomda, China 68 B2 31 28N 98 12 E
Jome, Indonesia 72 B3 1 16 S 127 30 E
Jomfruland, Norway 15 F3 58 52N 9 36 E
Jönåker, Sweden 15 F10 58 44N 16 40 E
Jonava, Lithuania 13 J21 55 8N 24 12 E
Jones, Phil. 70 C3 16 33N 121 42 E
Jones Sound, Canada 6 B3 76 0N 85 0W
Jonesboro, Ark., U.S.A. 139 H9 35 50N 90 42W
Jonesboro, Ill., U.S.A. 139 G10 37 27N 89 16W
Jonesboro, La., U.S.A. 139 J8 32 15N 92 43W
Jonesburg, U.S.A. 140 F5 38 51N 91 18W
Jonesport, U.S.A. 129 D6 44 32N 67 37W
Jonesville, Ind., U.S.A. 141 E11 39 5N 85 54W
Jonesville, Mich., U.S.A. 141 C12 41 59N 84 40W
Jonglei, Sudan 95 F3 6 25N 30 50 E
Jonglei □, Sudan 95 F3 7 30N 32 30 E
Joniškis, Lithuania 13 H20 56 13N 23 35 E
Jönköping, Sweden 13 H16 57 45N 14 10 E
Jonquière, Canada 129 C5 48 27N 71 14W
Jonsberg, Sweden 15 F10 58 30N 16 48 E
Jonsered, Sweden 15 G6 57 45N 12 10 E
Jonzac, France 24 C3 45 27N 0 28W
Joplin, U.S.A. 139 G7 37 6N 94 31W
Jordan, U.S.A. 142 C10 47 19N 106 55W
Jordan ■, Asia 91 E5 31 0N 36 0 E
Jordan →, Asia 91 D4 31 48N 35 32 E
Jordan Valley, U.S.A. 142 E5 42 59N 117 3W
Jordânia, Brazil 155 E3 15 55 S 40 11W
Jordanów, Poland 31 B12 49 41N 19 49 E
Jorge, C., Chile 160 D1 51 40 S 75 35W
Jorhat, India 78 B5 26 45N 94 12 E
Jorm, Afghan. 79 A3 36 50N 70 52 E
Jörn, Sweden 12 D19 65 4N 20 1 E
Jorong, Indonesia 75 C4 3 58 S 114 56 E
Jørpeland, Norway 13 G11 59 3N 6 1 E
Jorquera →, Chile 158 B2 28 3 S 69 58W
Jos, Nigeria 101 D6 9 53N 8 51 E
Jošanička Banja, Serbia, Yug. 42 D5 43 24N 20 47 E
Jose Abad Santos, Phil. 71 J5 5 55N 125 39 E
José Batlle y Ordóñez, Uruguay 159 C4 33 20 S 55 10W
José de San Martín, Argentina 160 B2 44 4 S 70 26W
Jose Panganiban, Phil. 70 D4 14 17N 122 41 E
Joseni, Romania 46 C6 46 42N 25 29 E
Joseph, U.S.A. 142 D5 45 21N 117 14W
Joseph, L., Nfld., Canada 129 B6 52 45N 65 18W
Joseph, L., Ont., Canada 136 A5 45 10N 79 44W
Joseph Bonaparte G., Australia 112 B4 14 35 S 128 50 E
Joseph City, U.S.A. 143 J8 34 57N 110 20W
Joshua Tree, U.S.A. 145 L10 34 8N 116 19W
Joshua Tree National Monument, U.S.A. 145 M10 33 55N 116 0W
Josselin, France 22 E4 47 57N 2 33W
Jostedalsbreen, Norway 13 F12 61 40N 6 59 E
Jotunheimen, Norway 14 C2 61 35N 8 25 E
Jourdanton, U.S.A. 139 L5 28 55N 98 33W
Joure, Neths. 20 C7 52 58N 5 48 E
Joussard, Canada 130 B5 55 22N 115 50W
Joutseno, Finland 50 B5 61 7N 28 31 E
Jovellanos, Cuba 148 B3 22 40N 81 10W
Jovellar, Phil. 70 E4 13 4N 123 36 E
Jowai, India 78 C4 25 26N 92 12 E
Jowzjān □, Afghan. 79 A2 36 10N 66 0 E
Joyeuse, France 25 D8 44 29N 4 16 E
Józefów, Poland 47 C8 52 10N 21 11 E
Ju Xian, China 67 F10 36 35N 118 20 E
Juan Aldama, Mexico 146 C4 24 20N 103 23W
Juan Bautista Alberdi, Argentina 158 C3 34 26 S 61 48W
Juan de Fuca Str., Canada 144 B2 48 15N 124 0W
Juan de Nova, Ind. Oc. 105 B7 17 3 S 43 45 E
Juan Fernández, Arch. de, Pac. Oc. 123 L20 33 50 S 80 0W
Juan José Castelli, Argentina 158 B3 25 27 S 60 57W
Juan L. Lacaze, Uruguay 158 C4 34 26 S 57 25W
Juanjuí, Peru 156 B2 7 10 S 76 45W
Juankoski, Finland 12 E23 63 3N 28 19 E
Juárez, Argentina 158 D4 37 40 S 59 43W
Juárez, Mexico 145 N11 32 20N 115 57W
Juárez, Sierra de, Mexico 146 A1 32 0N 116 0W
Juatinga, Ponta de, Brazil 155 F3 23 17 S 44 30W
Juàzeiro, Brazil 154 C3 9 30 S 40 30W
Juàzeiro do Norte, Brazil 154 C4 7 10 S 39 18W
Jubay, Phil. 71 F5 11 33N 124 18 E
Jubayl, Lebanon 91 A4 34 5N 35 39 E
Jubbah, Si. Arabia 84 D4 28 2N 40 56 E
Jubbulpore = Jabalpur, India 81 H8 23 9N 79 58 E
Jübek, Germany 26 A5 54 33N 9 22 E
Jubga, Russia 53 H4 44 19N 38 48 E
Jubilee L., Australia 113 E4 29 0 S 126 50 E
Juby, C., Morocco 98 C2 28 0N 12 59W
Júcar →, Spain 35 F4 39 5N 0 10W
Júcaro, Cuba 148 B4 21 37N 78 51W
Juchitán, Mexico 147 D5 16 27N 95 5W
Judaea = Har Yehuda, Israel 91 D3 31 35N 34 57 E
Judenburg, Austria 30 D7 47 12N 14 38 E
Judith →, U.S.A. 142 C9 47 44N 109 39W
Judith, Pt., U.S.A. 137 E13 41 22N 71 29W
Judith Gap, U.S.A. 142 C9 46 41N 109 45W
Jufari →, Brazil 153 D5 1 13 S 62 0W
Jugoslavia = Yugoslavia ■, Europe 42 D5 44 0N 20 0 E
Juigalpa, Nic. 148 D2 12 6N 85 26W
Juillac, France 24 C5 45 20N 1 19 E
Juist, Germany 26 B2 53 40N 6 59 E
Juiz de Fora, Brazil 155 F3 21 43 S 43 19W
Jujuy □, Argentina 158 A2 23 20 S 65 40W
Julesburg, U.S.A. 138 E3 40 59N 102 16W
Juli, Peru 156 D4 16 10 S 69 25W
Julia Cr. →, Australia 114 C3 20 0 S 141 11 E
Julia Creek, Australia 114 C3 20 39 S 141 44 E
Juliaca, Peru 156 D3 15 25 S 70 10W
Julian, U.S.A. 145 M10 33 4N 116 38W
Julian Alps = Julijske Alpe, Slovenia 39 B11 46 15N 14 1 E
Julianakanaal, Neths. 21 F7 51 6N 5 52 E
Julianatop, Surinam 153 C6 3 40N 56 30W
Julianehåb, Greenland 6 C5 60 43N 46 0W
Jülich, Germany 26 E2 50 55N 6 22 E
Julierpass, Switz. 29 D9 46 28N 9 32 E
Julijske Alpe, Slovenia 39 B11 46 15N 14 1 E
Julimes, Mexico 146 B3 28 25N 105 27W
Jullundur, India 80 D6 31 20N 75 40 E
Julu, China 66 F8 37 15N 115 2 E
Jumbo, Zimbabwe 107 F3 17 30 S 30 58 E
Jumbo Pk., U.S.A. 145 J12 36 12N 114 11W
Jumentos Cays, Bahamas 149 B4 23 0N 75 40W
Jumet, Belgium 21 H4 50 27N 4 25 E
Jumilla, Spain 35 G3 38 28N 1 19W
Jumla, Nepal 81 E10 29 15N 82 13 E
Jumna = Yamuna →, India 81 G9 25 30N 81 53 E
Junagadh, India 80 J4 21 30N 70 30 E
Junaynah, Si. Arabia 86 B4 22 33N 46 18 E
Junction, Tex., U.S.A. 139 K5 30 29N 99 46W
Junction, Utah, U.S.A. 143 G7 38 14N 112 13W
Junction B., Australia 114 A1 11 52 S 133 55 E
Junction City, Kans., U.S.A. 138 F6 39 2N 96 50W
Junction City, Oreg., U.S.A. 142 D2 44 13N 123 12W
Junction Pt., Australia 114 A1 11 45 S 133 50 E
Jundah, Australia 114 C3 24 46 S 143 2 E
Jundiaí, Brazil 159 A6 24 30 S 47 0W
Juneau, U.S.A. 126 C6 58 18N 134 25W
Junee, Australia 117 C7 34 53 S 147 35 E
Jungfrau, Switz. 28 C5 46 32N 7 58 E
Junggar Pendi, China 64 B3 44 30N 86 0 E
Junglinster, Lux. 21 J8 49 43N 6 15 E
Jungshahi, Pakistan 80 G2 24 52N 67 44 E
Juniata →, U.S.A. 136 F7 40 30N 77 40W
Junín, Argentina 158 C3 34 33 S 60 57W
Junín, Peru 156 C2 11 12 S 76 0W
Junín □, Peru 156 C3 11 30 S 75 0W
Junín de los Andes, Argentina 160 A2 39 45 S 71 0W
Jūniyah, Lebanon 91 B4 33 59N 35 38 E
Junnar, India 82 E1 19 12N 73 58 E
Juntura, U.S.A. 142 E4 43 45N 118 5W
Juparanã, L., Brazil 155 E3 19 16 S 40 8W
Jupiter →, Canada 129 C7 49 29N 63 37W
Juquiá, Brazil 155 F2 24 19 S 47 38W
Jur, Nahr el →, Sudan 95 F2 8 45N 29 15 E
Jura = Jura, Mts. du, Europe 25 B10 46 40N 6 5 E
Jura = Schwäbische Alb, Germany 27 G5 48 20N 9 30 E
Jura, Europe 23 F13 46 35N 6 5 E
Jura, U.K. 18 F3 56 0N 5 50W
Jura □, France 23 F12 46 47N 5 45 E
Jura, Mts. du, Europe 25 B10 46 40N 6 5 E
Jura, Sd. of, U.K. 18 F3 55 57N 5 45W
Jura Suisse, Switz. 28 B3 47 10N 7 0 E
Jurado, Colombia 152 B2 7 7N 77 46W
Jurbarkas, Lithuania 13 J20 55 6N 22 47 E
Jurilovca, Romania 46 E9 44 46N 28 52 E
Jūrmala, Latvia 13 H20 56 58N 23 34 E
Jurong, China 69 B12 31 57N 119 9 E
Juruá →, Brazil 152 D4 2 37 S 65 44W
Juruena, Brazil 157 C6 13 0 S 58 10W
Juruena →, Brazil 157 B6 7 20 S 58 3W
Juruti, Brazil 153 D6 2 9 S 56 4W
Jussey, France 23 E12 47 50N 5 55 E
Justo Daract, Argentina 158 C2 33 52 S 65 12W
Jutaí, Brazil 156 B4 5 11 S 68 54W
Jutaí →, Brazil 152 D4 2 43 S 66 57W
Jüterbog, Germany 26 D9 51 59N 13 5 E
Juticalpa, Honduras 148 D2 14 40N 86 12W
Jutland = Jylland, Denmark 15 H3 56 25N 9 30 E
Jutphaas, Neths. 20 D6 52 2N 5 6 E
Juventud, I. de la, Cuba 148 B3 21 40N 82 40W
Juvigny-sous-Andaine, France 22 D6 48 32N 0 30W
Juvisy-sur-Orge, France 23 D9 48 42N 2 22 E
Jūy Zar, Iran 89 F12 33 50N 46 18 E
Juye, China 66 G9 35 22N 116 5 E
Juzennecourt, France 23 D11 48 10N 4 58 E
Jvari, Georgia 53 J6 42 42N 42 4 E
Jylland, Denmark 15 H3 56 25N 9 30 E
Jyväskylä, Finland 13 E21 62 14N 25 50 E

K

K2, Mt., Pakistan 81 B7 35 58N 76 32 E
Kaala-Gomén, N. Cal. 121 T18 20 40 S 164 25 E
Kaap Plateau, S. Africa 104 D3 28 30 S 24 0 E
Kaapkruis, Namibia 104 C1 21 55 S 13 57 E
Kaapstad = Cape Town, S. Africa 104 E2 33 55 S 18 22 E
Kaatsheuvel, Neths. 21 E6 51 39N 5 2 E
Kabacan, Phil. 71 H5 7 8N 124 49 E
Kabaena, Indonesia 72 C2 5 15 S 122 0 E
Kabala, S. Leone 100 D2 9 38N 11 37W
Kabale, Uganda 106 C3 1 15 S 30 0 E
Kabalo, Zaïre 103 D5 6 0 S 27 0 E
Kabambare, Zaïre 106 C2 4 41 S 27 39 E
Kabango, Zaïre 107 D2 8 35 S 28 30 E
Kabanjahe, Indonesia 74 B1 3 6N 98 30 E
Kabankalan, Phil. 71 G4 9 59N 122 49 E
Kabara, Mali 100 B4 16 40N 2 50W
Kabardinka, Russia 53 H4 44 40N 37 57 E
Kabardino-Balkar Republic = Kabardino Balkaria □, Russia 53 J6 43 30N 43 30 E
Kabardino Balkaria □, Russia 53 J6 43 30N 43 30 E
Kabare, Indonesia 73 B4 0 4 S 130 58 E
Kabarega Falls, Uganda 106 B3 2 15N 31 30 E
Kabasalan, Phil. 71 H4 7 47N 122 44 E
Kabba, Nigeria 101 D6 7 50N 6 3 E
Kabe, Japan 62 C4 34 31N 132 31 E
Kabi, Niger 97 F2 13 30N 12 35 E
Kabin Buri, Thailand 76 F3 13 57N 101 43 E
Kabinakagami L., Canada 128 C3 48 54N 84 25W
Kabīr, Zab al →, Iraq 89 D10 36 1N 43 24 E
Kabkabīyah, Sudan 97 F4 13 50N 24 0 E
Kablungu, C., Papua N. G. 120 D6 6 20 S 150 1 E
Kabna, Sudan 94 D3 19 6N 32 40 E
Kabo, C.A.R. 102 A3 7 35N 18 38 E
Kabompo, Zambia 107 E1 13 36 S 24 14 E
Kabondo, Zaïre 103 D5 8 58 S 25 40 E
Kabongo, Zaïre 103 D5 7 22 S 25 33 E
Kabou, Togo 101 D5 9 28N 0 55 E
Kaboudia, Rass, Tunisia 96 A2 35 13N 11 10 E
Kabra, Australia 114 C5 23 25 S 150 25 E
Kabūd Gonbad, Iran 85 B8 37 5N 59 45 E
Kabugao, Phil. 70 B3 18 2N 121 11 E
Kābul, Afghan. 79 B3 34 28N 69 11 E
Kābul □, Afghan. 79 B3 34 30N 69 0 E
Kabul →, Pakistan 79 B4 33 55N 72 14 E
Kabunga, Zaïre 106 C2 1 38 S 28 3 E
Kaburuang, Indonesia 72 A3 3 50N 126 30 E
Kabushiya, Sudan 95 D3 16 54N 33 41 E
Kabwe, Zambia 107 E2 14 30 S 28 29 E
Kabwum, Papua N. G. 120 D4 6 11 S 147 15 E
Kačanik, Serbia, Yug. 42 E6 42 13N 21 12 E
Kachchh, Gulf of, India 80 H3 22 50N 69 15 E
Kachchh, Rann of, India 80 G4 24 0N 70 0 E
Kachebera, Zambia 107 E3 13 50 S 32 50 E
Kachin □, Burma 78 B6 26 0N 97 30 E
Kachira, L., Uganda 106 C3 0 40 S 31 7 E

Katipunan 215

Kanggye, N. Korea 67 D14 41 0N 126 35 E
Kanggyŏng, S. Korea 67 F14 36 10N 127 0 E
Kanghwa, S. Korea 67 F14 37 45N 126 30 E
Kangiqsualujjuaq, Canada 127 C13 58 30N 65 59W
Kangiqsujuaq, Canada 127 B12 61 30N 72 0W
Kangirsuk, Canada 127 B13 60 0N 70 0W
Kangnŭng, S. Korea 67 F15 37 45N 128 54 E
Kango, Gabon 102 B2 0 11N 10 5 E
Kangoya, Zaïre 103 D4 9 55 S 22 48 E
Kangping, China 67 C12 42 43N 123 18 E
Kangpokpi, India 78 C4 25 8N 93 58 E
Kangyidaung, Burma 78 G5 16 56N 94 54 E
Kanhangad, India 83 H2 12 21N 74 58 E
Kanheri, India 82 E1 19 13N 72 50 E
Kani, Ivory C. 100 D3 8 29N 6 36W
Kaniama, Zaïre 103 D4 7 30 S 24 12 E
Kaniapiskau →, Canada 129 A6 56 40N 69 30W
Kaniapiskau L., Canada 129 B6 54 10N 69 55W
Kanibadam, Tajikistan 55 C5 40 17N 70 25 E
Kaniere, L., N.Z. 119 C6 42 50 S 171 10 E
Kanin, Poluostrov, Russia 48 A8 68 0N 45 0 E
Kanin Nos, Mys, Russia 48 A7 68 39N 43 32 E
Kanin Pen. = Kanin, Poluostrov, Russia 48 A8 68 0N 45 0 E
Kanina, Albania 44 D1 40 23N 19 30 E
Kaniva, Australia 116 D4 36 22 S 141 18 E
Kanjiža, Serbia, Yug. 42 A5 46 3N 20 4 E
Kanjut Sar, Pakistan 81 A6 36 7N 75 25 E
Kankaanpää, Finland 13 F20 61 44N 22 50 E
Kankakee, U.S.A. 141 C9 41 7N 87 52W
Kankakee →, U.S.A. 141 C8 41 23N 88 15W
Kankan, Guinea 100 C3 10 23N 9 15W
Kankendy = Xankändi, Azerbaijan 89 C12 39 52N 46 49 E
Kanker, India 82 D5 20 10N 81 40 E
Kankunskiy, Russia 57 D13 57 37N 126 8 E
Kanmuri-Yama, Japan 62 C4 34 30N 132 4 E
Kannabe, Japan 62 C5 34 32N 133 23 E
Kannapolis, U.S.A. 135 H5 35 30N 80 37W
Kannauj, India 81 F8 27 3N 79 56 E
Kano, Nigeria 101 C6 12 2N 8 30 E
Kano □, Nigeria 101 C6 11 30N 8 30 E
Kan'onji, Japan 62 C5 34 7N 133 39 E
Kanoroba, Ivory C. 100 D3 9 7N 6 8W
Kanowha, U.S.A. 140 B3 42 57N 93 47W
Kanowit, Malaysia 75 B4 2 14N 112 20 E
Kanowna, Australia 113 F3 30 32 S 121 31 E
Kanoya, Japan 62 F2 31 25N 130 50 E
Kanpetlet, Burma 78 E4 21 10N 93 59 E
Kanpur, India 81 F9 26 28N 80 20 E
Kansas, U.S.A. 141 E9 39 33N 87 56W
Kansas □, U.S.A. 138 F6 38 30N 99 0W
Kansas →, U.S.A. 138 F7 39 7N 94 37W
Kansas City, Kans., U.S.A. 140 E2 39 7N 94 38W
Kansas City, Mo., U.S.A. 140 E2 39 6N 94 35W
Kansenia, Zaïre 107 E2 10 20 S 26 0 E
Kansk, Russia 57 D10 56 20N 95 37 E
Kansŏng, S. Korea 67 E15 38 24N 128 30 E
Kansu = Gansu □, China 66 G3 36 0N 104 0 E
Kant, Kyrgyzstan 55 B7 42 53N 74 51 E
Kantang, Thailand 77 J2 7 25N 99 31 E
Kantché, Niger 97 F1 13 31N 8 30 E
Kanté, Togo 101 D5 9 57N 1 3 E
Kantemirovka, Russia 52 F4 49 43N 39 55 E
Kantharalak, Thailand 76 E5 14 39N 104 39 E
Kantō □, Japan 63 A11 36 15N 139 30 E
Kantō-Heiya, Japan 63 A11 36 0N 139 30 E
Kantō-Sanchi, Japan 63 B10 35 59N 138 50 E
Kantu-long, Burma 78 F6 19 57N 97 36 E
Kanturk, Ireland 19 D3 52 11N 8 54W
Kanuma, Japan 63 A11 36 34N 139 42 E
Kanus, Namibia 104 D2 27 50 S 18 39 E
Kanye, Botswana 104 C4 24 55 S 25 28 E
Kanzenze, Zaïre 103 E5 10 30 S 25 12 E
Kanzi, Ras, Tanzania 106 D4 7 1 S 39 33 E
Kao, Fiji 121 P13 19 40 S 175 1W
Kaohsiung = Gaoxiong, Taiwan 69 F13 22 38N 120 18 E
Kaohsiung, Taiwan 65 D7 22 35N 120 16 E
Kaokoveld, Namibia 104 B1 19 15 S 14 30 E
Kaolack, Senegal 100 C1 14 5N 16 8W
Kaoshan, China 67 B13 44 38N 124 50 E
Kaouar, Niger 97 E2 19 5N 12 52 E
Kapadvanj, India 80 H5 23 5N 73 0 E
Kapagere, Papua N. G. 120 E4 9 46 S 147 42 E
Kapan, Armenia 89 C12 39 18N 46 27 E
Kapanga, Zaïre 103 D4 8 30 S 22 40 E
Kapatagan, Phil. 71 H4 7 52N 123 44 E
Kapchagai = Qapshaghay, Kazakhstan 55 B8 43 51N 77 14 E
Kapchagaiskoye Vdkhr. = Qapshaghay Bögeni, Kazakhstan 55 B8 43 45N 77 50 E
Kapellen, Belgium 21 F4 51 19N 4 25 E
Kapéllo, Ákra, Greece 45 H5 36 9N 23 3 E
Kapema, Zaïre 107 E2 10 45 S 28 22 E
Kapfenberg, Austria 30 D8 47 26N 15 18 E
Kapia, Zaïre 103 C3 4 17 S 19 46 E
Kapiri Mposhi, Zambia 107 E2 13 59 S 28 43 E
Kāpīsā □, Afghan. 79 B3 35 0N 69 20 E
Kapiskau →, Canada 128 B3 52 47N 81 55W
Kapit, Malaysia 75 B4 2 0N 112 55 E
Kapiti I., N.Z. 118 G3 40 50 S 174 56 E
Kapka, Massif du, Chad 97 E4 15 7N 21 45 E
Kaplice, Czech. 30 C7 48 42N 14 30 E
Kapoe, Thailand 77 H2 9 34N 98 32 E
Kapoeta, Sudan 95 G3 4 50N 33 35 E
Kápolnásnyék, Hungary 31 D11 47 16N 18 41 E
Kaponga, N.Z. 118 F3 39 29 S 174 9 E
Kapos →, Hungary 31 E11 46 44N 18 30 E
Kaposvár, Hungary 31 E10 46 25N 17 47 E
Kapowsin, U.S.A. 144 D4 46 59N 122 13W
Kappeln, Germany 26 A5 54 40N 9 55 E
Kapps, Namibia 104 C2 22 32 S 17 18 E
Kaprije, Croatia 39 E12 43 42N 15 43 E
Kaprijke, Belgium 21 F3 51 13N 3 38 E
Kapsan, N. Korea 67 D15 41 4N 128 19 E
Kapsukas = Mariyampole, Lithuania 13 J20 54 33N 23 19 E
Kapuas, Indonesia 75 C4 3 10 S 114 5 E
Kapuas →, Indonesia 75 C3 0 25 S 109 20 E
Kapuas Hulu, Pegunungan, Malaysia 75 B4 1 30N 113 30 E
Kapuas Hulu Ra. = Kapuas Hulu, Pegunungan, Malaysia 75 B4 1 30N 113 30 E
Kapulo, Zaïre 107 D2 8 18 S 29 15 E
Kapunda, Australia 116 C3 34 20 S 138 56 E
Kapuni, N.Z. 118 F3 39 29 S 174 8 E
Kapurthala, India 80 D6 31 23N 75 25 E
Kapuskasing, Canada 128 C3 49 25N 82 30W
Kapuskasing →, Canada 128 C3 49 49N 82 0W
Kapustin Yar, Russia 53 F7 48 37N 45 40 E
Kaputar, Australia 115 E5 30 15 S 150 10 E
Kaputir, Kenya 106 B4 2 5N 35 28 E
Kapuvár, Hungary 31 D10 47 36N 17 1 E
Kara, Russia 56 C7 69 10N 65 0 E
Kara, Turkey 45 H9 36 58N 27 30 E
Kara Balta, Kyrgyzstan 55 B6 42 50N 73 49 E
Kara Bogaz Gol, Zaliv = Garabogazköl Aylagy, Turkmenistan 49 F9 41 0N 53 30 E
Kara Kalpak Republic = Karakalpakstan □, Uzbekistan 56 E6 43 0N 58 0 E
Kara Kum, Turkmenistan 56 F6 39 30N 60 0 E
Kara-Saki, Japan 62 C1 34 41N 129 30 E
Kara Sea, Russia 56 B7 75 0N 70 0 E
Kara Su, Kyrgyzstan 55 C6 40 44N 72 53 E
Karabash, Russia 54 D8 55 29N 60 14 E
Karabekaul, Turkmenistan 55 D2 38 30N 64 8 E
Karabiğa, Turkey 88 B2 40 24N 27 18 E
Karabük, Turkey 88 B5 41 12N 32 37 E
Karabulak, Kazakhstan 55 A9 44 54N 78 30 E
Karaburun, Turkey 88 C2 38 41N 26 28 E
Karaburuni, Albania 44 D1 40 25N 19 20 E
Karabutak = Qarabutaq, Kazakhstan 54 G8 49 59N 60 14 E
Karacabey, Turkey 88 B3 40 12N 28 21 E
Karacasu, Turkey 88 D3 37 43N 28 35 E
Karachala = Qaraçala, Azerbaijan 53 L9 39 45N 48 53 E
Karachayevsk, Russia 53 J5 43 50N 41 55 E
Karachev, Russia 52 D2 53 10N 35 5 E
Karachey-Cherkessia □, Russia 53 J5 43 40N 41 30 E
Karachi, Pakistan 79 D2 24 53N 67 0 E
Karád, Hungary 31 E10 46 41N 17 51 E
Karad, India 82 F2 17 15N 74 10 E
Karadeniz Boğazı, Turkey 88 B3 41 10N 29 10 E
Karaga, Ghana 101 D4 9 58N 0 28W
Karaganda = Qaraghandy, Kazakhstan 56 E8 49 50N 73 10 E
Karagayly, Kazakhstan 56 E8 49 26N 76 0 E
Karaginskiy, Ostrov, Russia 57 D17 58 45N 164 0 E
Karagiye, Vpadina, Kazakhstan 49 F9 43 27N 51 45 E
Karagiye Depression = Karagiye, Vpadina, Kazakhstan 49 F9 43 27N 51 45 E
Karagüney Dağları, Turkey 88 B6 40 30N 34 40 E
Karagwe □, Tanzania 106 C3 2 0 S 31 0 E
Karaikal, India 83 J4 10 59N 79 50 E
Karaikkudi, India 83 J4 10 5N 78 45 E
Karaisali, Turkey 88 D6 37 16N 35 2 E
Karaitivu I., Sri Lanka 83 K4 9 45N 79 52 E
Karaj, Iran 85 C6 35 48N 51 0 E
Karak, Malaysia 77 L4 3 25N 102 2 E
Karakalpakstan □, Uzbekistan 56 E6 43 0N 58 0 E
Karakas, Kazakhstan 56 E9 48 20N 83 30 E
Karakitang, Indonesia 72 A3 3 14N 125 28 E
Karaklis = Vanadzor, Armenia 53 K7 40 48N 44 30 E
Karakoçan, Turkey 89 C9 38 57N 40 2 E
Karakoram Pass, Pakistan 81 B7 35 33N 77 50 E
Karakoram Ra., Pakistan 81 B7 35 30N 77 0 E
Karakul, Tajikistan 55 D6 39 2N 73 33 E
Karakul, Uzbekistan 55 D1 39 22N 63 50 E
Karakuldzha, Kyrgyzstan 55 C6 40 39N 73 26 E
Karakurt, Turkey 89 B10 40 10N 42 37 E
Karal, Chad 97 F2 12 50N 14 46 E
Karalon, Russia 57 D12 57 5N 115 50 E
Karaman, Turkey 88 D5 37 14N 33 13 E
Karamay, China 64 B3 45 30N 84 58 E
Karambu, Indonesia 75 C5 3 53 S 116 6 E
Karamea, N.Z. 119 B7 41 14 S 172 6 E
Karamea →, N.Z. 119 B7 41 13 S 172 26 E
Karamea Bight, N.Z. 119 B6 41 22 S 171 40 E
Karamet Niyaz, Turkmenistan 55 E2 37 45N 64 34 E
Karamsad, India 80 H5 22 35N 72 50 E
Karand, Iran 89 E12 34 16N 46 15 E
Karanganyar, Indonesia 75 D3 7 38 S 109 37 E
Karanja, India 82 D3 20 29N 77 31 E
Karapınar, Turkey 88 D5 37 41N 33 30 E
Karapiro, N.Z. 118 D4 37 53 S 175 32 E
Karasburg, Namibia 104 D2 28 0 S 18 44 E
Karasino, Russia 56 C9 66 50N 86 50 E
Karasjok, Norway 12 B21 69 27N 25 30 E
Karasu, Turkey 88 B4 41 4N 30 46 E
Karasuk, Russia 56 D8 53 44N 78 2 E
Karasuyama, Japan 63 A12 36 39N 140 9 E
Karataş, Turkey 88 D6 36 36N 35 21 E
Karataş Burnu, Turkey 88 D6 36 31N 35 24 E
Karatau = Qarataū, Kazakhstan 55 B5 43 10N 70 28 E
Karatau, Khrebet, Kazakhstan 55 B4 43 30N 69 30 E
Karativu, Sri Lanka 83 K4 8 22N 79 47 E
Karatobe, Kazakhstan 54 G4 49 44N 53 30 E
Karatoya →, India 78 C2 24 7N 89 36 E
Karaturuk, Kazakhstan 55 B8 43 35N 77 50 E
Karaul-Bazar, Uzbekistan 55 D2 39 30N 64 48 E
Karauli, India 80 F7 26 30N 77 4 E
Karávi, Greece 45 H5 36 49N 23 37 E
Karavostasi, Cyprus 32 D11 35 8N 32 50 E
Karawa, Zaïre 102 B4 3 18N 20 17 E
Karawang, Indonesia 75 D3 6 30 S 107 15 E
Karawanken, Europe 30 E7 46 30N 14 40 E
Karayazı, Turkey 89 C10 39 41N 42 9 E
Karazhal, Kazakhstan 56 E8 48 2N 70 49 E
Karbalā, Iraq 89 F11 32 36N 44 3 E
Kårböle, Sweden 14 C9 61 59N 15 22 E
Karcag, Hungary 31 D13 47 19N 20 57 E
Karcha →, Pakistan 81 B7 34 45N 76 10 E
Karda, Russia 57 D11 55 0N 103 16 E
Kardeljovo, Croatia 42 D2 43 4N 17 26 E
Kardhámila, Greece 45 F8 38 35N 26 5 E
Kardhítsa, Greece 44 E3 39 23N 21 54 E
Kardhítsa □, Greece 44 E3 39 15N 21 50 E
Kärdla, Estonia 13 G20 58 50N 22 40 E
Kareeberge, S. Africa 104 E3 30 59 S 21 50 E
Kareima, Sudan 94 D3 18 30N 31 49 E
Karelia □, Russia 48 A5 65 30N 32 30 E
Karelian Republic = Karelia □, Russia 48 A5 65 30N 32 30 E
Karema, Papua N. G. 120 E4 9 12 S 147 18 E
Karevāndar, Iran 85 E9 27 53N 60 44 E
Kargasok, Russia 56 D9 59 3N 80 53 E
Kargat, Russia 56 D9 55 10N 80 15 E
Kargı, Turkey 88 B6 41 11N 34 30 E
Kargil, India 81 B7 34 32N 76 12 E
Kargopol, Russia 50 B10 61 30N 38 58 E
Kargowa, Poland 47 C2 52 5N 15 51 E
Karguéri, Niger 97 F2 13 27N 10 30 E
Karia ba Mohammed, Morocco 98 B3 34 22N 5 12W
Kariaí, Greece 44 D6 40 14N 24 19 E
Karīān, Iran 85 E8 26 57N 57 14 E
Kariba, Zimbabwe 107 F2 16 28 S 28 50 E
Kariba, L., Zimbabwe 107 F2 16 40 S 28 25 E
Kariba Dam, Zimbabwe 107 F2 16 30 S 28 35 E
Kariba Gorge, Zambia 107 F2 16 30 S 28 50 E
Karibib, Namibia 104 C2 22 0 S 15 56 E
Karikari, C., N.Z. 118 A2 34 46 S 173 24 E
Karimata, Kepulauan, Indonesia 75 C3 1 25 S 109 0 E
Karimata, Selat, Indonesia 75 C3 2 0 S 108 40 E
Karimata Is. = Karimata, Kepulauan, Indonesia 75 C3 1 25 S 109 0 E
Karimnagar, India 82 E4 18 26N 79 10 E
Karimunjawa, Kepulauan, Indonesia 75 D4 5 50 S 110 30 E
Karin, Somali Rep. 108 B3 10 50N 45 52 E
Káristos, Greece 45 F6 38 1N 24 29 E
Karīt, Iran 85 C8 33 29N 56 55 E
Kariya, Japan 63 C9 34 58N 137 1 E
Karjala, Finland 50 A5 62 0N 30 25 E
Karkal, India 83 H2 13 15N 74 56 E
Karkar I., Papua N. G. 120 C4 4 40 S 146 0 E
Karkaralinsk = Qarqaraly, Kazakhstan 56 E8 49 26N 75 30 E
Karkinitska Zatoka, Ukraine 51 K7 45 56N 33 0 E
Karkinitskiy Zaliv = Karkinitska Zatoka, Ukraine 51 K7 45 56N 33 0 E
Karkur Tohl, Egypt 94 C2 22 5N 25 5 E
Karl Liebknecht, Russia 52 E2 51 40N 35 35 E
Karl-Marx-Stadt = Chemnitz, Germany 26 E8 50 51N 12 54 E
Karla, L. = Voiviïs Límni, Greece 44 E4 39 30N 22 45 E
Karlino, Poland 47 A2 54 3N 15 53 E
Karlivka, Ukraine 51 H8 49 29N 35 8 E
Karlobag, Croatia 39 D12 44 32N 15 5 E
Karlovac, Croatia 39 C12 45 31N 15 36 E
Karlovka = Karlivka, Ukraine 51 H8 49 29N 35 8 E
Karlovo, Bulgaria 43 E9 42 38N 24 47 E
Karlovy Vary, Czech. 30 A5 50 13N 12 51 E
Karlsbad = Karlovy Vary, Czech. 30 A5 50 13N 12 51 E
Karlsborg, Sweden 15 F8 58 33N 14 33 E
Karlshamn, Sweden 13 H16 56 10N 14 51 E
Karlskoga, Sweden 13 G16 59 22N 14 33 E
Karlskrona, Sweden 13 H16 56 10N 15 35 E
Karlsruhe, Germany 27 F4 49 0N 8 23 E
Karlstad, Sweden 13 G15 59 23N 13 30 E
Karlstad, U.S.A. 138 A6 48 35N 96 31W
Karlstadt, Germany 27 F5 49 57N 10 47 E
Karnal, India 80 E7 29 42N 77 2 E
Karnali →, Nepal 81 E9 28 45N 81 16 E
Karnaphuli Res., Bangla. 78 D4 22 40N 92 20 E
Karnataka □, India 83 H3 13 15N 77 0 E
Karnes City, U.S.A. 139 L6 28 53N 97 54W
Karnische Alpen, Europe 30 E6 46 36N 13 0 E
Kärnten □, Austria 30 E6 46 52N 13 30 E
Karo, Mali 100 C4 12 16N 3 18W
Karoi, Zimbabwe 107 F2 16 48 S 29 45 E
Karomatan, Phil. 71 H4 7 55N 123 44 E
Karonga, Malawi 107 D3 9 57 S 33 55 E
Karoonda, Australia 116 C3 35 1 S 139 59 E
Karora, Sudan 94 D4 17 44N 38 15 E
Káros, Greece 45 H7 36 54N 25 40 E
Karousádhes, Greece 44 E1 39 47N 19 45 E
Karpasia □, Cyprus 32 D13 35 32N 34 15 E
Kárpathos, Greece 45 J9 35 37N 27 10 E
Kárpathos, Stenón, Greece 45 J9 36 0N 27 30 E
Karpinsk, Russia 54 B8 59 45N 60 1 E
Karpogory, Russia 48 B7 64 0N 44 27 E
Karpuz Burnu = Apostolos Andreas, C., Cyprus 32 D13 35 42N 34 35 E
Karrebæk, Denmark 15 J5 55 12N 11 39 E
Kars, Turkey 89 B10 40 40N 43 5 E
Karsakpay, Kazakhstan 56 E7 47 55N 66 40 E
Karsha, Kazakhstan 52 F10 49 45N 51 35 E
Karshi = Qarshi, Uzbekistan 55 D2 38 53N 65 48 E
Karsiyang, India 81 F13 26 56N 88 18 E
Karst, Croatia 39 C11 45 35N 14 0 E
Karsun, Russia 48 D8 54 14N 46 57 E
Kartál Óros, Greece 44 C7 41 15N 25 13 E
Kartaly, Russia 54 E8 53 3N 60 40 E
Kartapur, India 80 D6 31 27N 75 32 E
Karthaus, U.S.A. 136 E6 41 8N 78 9W
Kartuzy, Poland 47 A5 54 22N 18 10 E
Karuah, Australia 117 B9 32 37 S 151 56 E
Karufa, Indonesia 73 B4 3 50 S 133 20 E
Karumba, Australia 114 B3 17 31 S 140 50 E
Karumo, Tanzania 106 C3 2 25 S 32 50 E
Karumwa, Tanzania 106 C3 3 12 S 32 38 E
Karungu, Kenya 106 C3 0 50 S 34 10 E
Karup, Denmark 15 H3 56 19N 9 10 E
Karur, India 83 J4 10 59N 78 2 E
Karviná, Czech. 31 B11 49 53N 18 25 E
Karwi, India 81 G9 25 12N 80 57 E
Kaş, Turkey 88 D3 36 11N 29 37 E
Kasache, Malawi 107 E3 13 25 S 34 20 E
Kasai, Japan 62 C6 34 55N 134 52 E
Kasai →, Zaïre 103 C3 3 30 S 16 10 E
Kasai Occidental □, Zaïre 103 D4 6 0 S 22 0 E
Kasai Oriental □, Zaïre 103 D4 5 0 S 24 30 E
Kasaji, Zaïre 103 E4 10 25 S 23 27 E
Kasama, Japan 63 A12 36 23N 140 16 E
Kasama, Zambia 107 E3 10 16 S 31 9 E
Kasan, Uzbekistan 55 D2 39 2N 65 35 E
Kasan-dong, N. Korea 67 D14 41 18N 126 55 E
Kasane, Namibia 104 B3 17 34 S 24 50 E
Kasanga, Tanzania 107 D3 8 30 S 31 10 E
Kasangulu, Zaïre 103 C3 4 33 S 15 15 E
Kasaoka, Japan 62 C5 34 30N 133 30 E
Kasaragod, India 83 H2 12 30N 74 58 E
Kasat, Burma 78 G7 15 56N 98 13 E
Kasba, Bangla. 78 D3 23 45N 91 2 E
Kasba L., Canada 131 A8 60 20N 102 10W
Kasba Tadla, Morocco 98 B3 32 36N 6 17W
Kaseda, Japan 62 F2 31 25N 130 19 E
Kāseh Garān, Iran 84 C5 34 5N 46 2 E
Kasempa, Zambia 107 E2 13 30 S 25 44 E
Kasenga, Zaïre 107 E2 10 20 S 28 45 E
Kasese, Uganda 106 B3 0 13N 30 3 E
Kasewa, Zambia 107 E2 14 28 S 28 53 E
Kasganj, India 81 F8 27 48N 78 42 E
Kashabowie, Canada 128 C1 48 40N 90 26W
Kāshān, Iran 85 C6 34 5N 51 30 E
Kashi, China 64 C2 39 30N 76 2 E
Kashihara, Japan 63 C7 34 27N 135 46 E
Kashima, Ibaraki, Japan 63 B12 35 58N 140 38 E
Kashima, Saga, Japan 62 D2 33 7N 130 6 E
Kashima-Nada, Japan 63 B12 36 0N 140 45 E
Kashimbo, Zaïre 107 E2 11 12 S 26 19 E
Kashin, Russia 52 B3 57 20N 37 36 E
Kashipur, Orissa, India 82 E6 19 16N 83 3 E
Kashipur, Ut. P., India 81 E8 29 15N 79 0 E
Kashira, Russia 52 C4 54 45N 38 10 E
Kashiwa, Japan 63 B11 35 52N 139 59 E
Kashiwazaki, Japan 61 F9 37 22N 138 33 E
Kashk-e Kohneh, Afghan. 79 B1 34 55N 62 30 E
Kashkasu, Kyrgyzstan 55 D6 39 54N 72 44 E
Kāshmar, Iran 85 C8 35 16N 58 26 E
Kashmir, Asia 81 C7 34 0N 76 0 E
Kashmor, Pakistan 79 C3 28 28N 69 32 E
Kashpirovka, Russia 52 D9 53 0N 48 30 E
Kashun Noerh = Gaxun Nur, China 64 B5 42 22N 100 30 E
Kasimov, Russia 52 C5 54 55N 41 20 E
Kasinge, Zaïre 106 D2 6 15 S 26 58 E
Kasiruta, Indonesia 72 B3 0 25 S 127 12 E
Kaskaskia →, U.S.A. 140 G7 37 58N 89 57W
Kaskattama →, Canada 131 B10 57 3N 90 4W
Kaskelen = Qaskeleng, Kazakhstan 55 B8 43 20N 76 35 E
Kaskinen, Finland 13 E19 62 22N 21 15 E
Kasli, Russia 54 D8 55 53N 60 46 E
Kaslo, Canada 130 D5 49 55N 116 55W
Kasmere L., Canada 131 B8 59 34N 101 10W
Kasongan, Indonesia 75 C4 2 0 S 113 23 E
Kasongo, Zaïre 106 C2 4 30 S 26 33 E
Kasongo Lunda, Zaïre 103 D3 6 35 S 16 49 E
Kásos, Greece 45 J8 35 20N 26 55 E
Kásos, Stenón, Greece 45 J8 35 30N 26 30 E
Kaspi, Georgia 53 K7 41 59N 44 26 E
Kaspichan, Bulgaria 43 D12 43 18N 27 11 E
Kaspiysk, Russia 53 J8 42 52N 47 40 E
Kaspiyskiy, Russia 53 H8 45 22N 47 23 E
Kassab ed Doleib, Sudan 95 E3 13 30N 33 35 E
Kassaba, Egypt 94 C2 22 40N 29 55 E
Kassalâ, Sudan 95 D4 15 30N 36 0 E
Kassalâ □, Sudan 95 D4 15 20N 36 26 E
Kassándra, Greece 44 D5 40 0N 23 30 E
Kassansay, Uzbekistan 55 C5 41 15N 71 31 E
Kassel, Germany 26 D5 51 18N 9 26 E
Kassinger, Sudan 94 D3 18 46N 31 51 E
Kassiópi, Greece 32 A3 39 48N 19 53 E
Kassue, Indonesia 73 C5 6 58 S 139 21 E
Kastamonu, Turkey 88 B5 41 25N 33 43 E
Kastav, Croatia 39 C11 45 22N 14 20 E
Kastélli, Greece 32 D5 35 29N 23 38 E
Kastéllion, Greece 32 D7 35 12N 25 20 E
Kastellou, Ákra, Greece 45 J9 35 30N 27 15 E
Kasterlee, Belgium 21 F5 51 15N 4 59 E
Kastóri, Greece 45 G4 37 10N 22 17 E
Kastoría, Greece 44 D3 40 30N 21 19 E
Kastoría □, Greece 44 D3 40 30N 21 15 E
Kastorías, L., Greece 44 D3 40 30N 21 20 E
Kastornoye, Russia 52 E4 51 55N 38 2 E
Kastós, Greece 45 F2 38 35N 20 55 E
Kástron, Greece 44 E7 39 50N 25 2 E
Kastrosikiá, Greece 45 E2 39 6N 20 36 E
Kastsyukovichy, Belarus 50 F7 53 20N 32 4 E
Kasugai, Japan 63 B8 35 12N 136 59 E
Kasukabe, Japan 63 B11 35 58N 139 49 E
Kasulu, Tanzania 106 C3 4 37 S 30 5 E
Kasulu □, Tanzania 106 C3 4 37 S 30 5 E
Kasumi, Japan 62 B6 35 38N 134 38 E
Kasumiga-Ura, Japan 63 B12 36 0N 140 25 E
Kasumkent, Russia 53 K9 41 47N 48 15 E
Kasungu, Malawi 107 E3 13 0 S 33 29 E
Kasur, Pakistan 79 C4 31 5N 74 25 E
Kata, Russia 57 D11 58 46N 102 40 E
Kataba, Zambia 107 F2 16 5 S 25 10 E
Katako Kombe, Zaïre 102 C4 3 25 S 24 20 E
Katákolon, Greece 45 G3 37 38N 21 19 E
Katale, Tanzania 106 C3 4 52 S 31 7 E
Katamatite, Australia 117 D6 36 6 S 145 41 E
Katanda, Kivu, Zaïre 106 C2 0 55 S 29 21 E
Katanda, Shaba, Zaïre 103 D4 7 52 S 24 13 E
Katanga = Shaba □, Zaïre 106 D2 8 0 S 25 0 E
Katangi, India 82 D4 21 56N 79 50 E
Katanglad Mts., Phil. 71 G5 8 6N 124 54 E
Katangli, Russia 57 D15 51 42N 143 14 E
Katapakishi, Zaïre 103 D4 8 15 S 22 49 E
Katastári, Greece 45 G2 37 50N 20 45 E
Katav Ivanovsk, Russia 54 D7 54 45N 58 12 E
Katavi Swamp, Tanzania 106 D3 6 50 S 31 10 E
Kataysk, Russia 54 C9 56 20N 62 30 E
Katchiungo, Angola 103 E3 12 35 S 16 13 E
Kateríni, Greece 44 D4 40 18N 22 37 E
Katherîna, Gebel, Egypt 94 J8 28 30N 33 57 E
Katherine, Australia 112 B5 14 27 S 132 20 E
Kathiawar, India 80 H4 22 20N 71 0 E
Kathikas, Cyprus 32 E11 34 55N 32 25 E
Kati, Mali 100 C3 12 41N 8 4W
Katihar, India 81 G12 25 34N 87 36 E
Katikati, N.Z. 118 D4 37 32 S 175 57 E
Katima Mulilo, Zambia 104 B3 17 28 S 24 13 E
Katimbira, Malawi 107 E3 12 40 S 34 0 E
Katingan = Mendawai →, Indonesia 75 C4 3 30 S 113 0 E
Katiola, Ivory C. 100 D3 8 10N 5 10W
Katipunan, Phil. 71 G4 8 31N 123 17 E

Katlanovo, *Macedonia* **42 F6** 41 52N 21 40 E
Katmandu, *Nepal* **81 F11** 27 45N 85 20 E
Kato Akhaïa, *Greece* **45 F3** 38 8N 21 33 E
Káto Arkhánai, *Greece* **32 D7** 35 15N 25 10 E
Káto Khorió, *Greece* **32 D7** 35 3N 25 47 E
Kato Pyrgos, *Cyprus* **32 D11** 35 11N 32 41 E
Káto Stavros, *Greece* **44 D5** 40 39N 23 43 E
Katol, *India* **82 D4** 21 17N 78 38 E
Katompe, *Zaïre* **103 D5** 6 2 S 26 23 E
Katonga —►, *Uganda* **106 B3** 0 34N 31 50 E
Katoomba, *Australia* **117 B9** 33 41 S 150 19 E
Katowice, *Poland* **47 E6** 50 17N 19 5 E
Katowice □, *Poland* **47 E6** 50 10N 19 0 E
Katrine, L., *U.K.* **18 E4** 56 15N 4 30W
Katrineholm, *Sweden* **14 E10** 59 9N 16 12 E
Katsepe, *Madag.* **105 B8** 15 45 S 46 15 E
Katsina, *Nigeria* **101 C6** 13 0N 7 32 E
Katsina □, *Nigeria* **101 C6** 12 30N 7 30 E
Katsina Ala —►, *Nigeria* **101 D6** 7 10N 9 20 E
Katsumoto, *Japan* **62 D1** 33 51N 129 42 E
Katsuta, *Japan* **63 A12** 36 25N 140 31 E
Katsuura, *Japan* **63 B12** 35 10N 140 20 E
Katsuyama, *Japan* **63 A8** 36 3N 136 30 E
Kattakurgan, *Uzbekistan* **55 D3** 39 55N 66 15 E
Kattaviá, *Greece* **32 D9** 35 57N 27 46 E
Kattegat, *Denmark* **15 H5** 57 0N 11 20 E
Katumba, *Zaïre* **103 D5** 7 40 S 25 17 E
Katungu, *Kenya* **106 C5** 2 55 S 40 3 E
Katwa, *India* **81 H13** 23 30N 88 5 E
Katwijk-aan-Zee, *Neths.* **20 D4** 52 12N 4 24 E
Katy, *Poland* **47 D3** 51 2N 16 45 E
Kauai, *U.S.A.* **132 H15** 22 3N 159 30W
Kauai Channel, *U.S.A.* **132 H15** 21 45N 158 50W
Kaub, *Germany* **27 E3** 50 5N 7 46 E
Kaufbeuren, *Germany* **27 H6** 47 53N 10 57 E
Kaufman, *U.S.A.* **139 J6** 32 35N 96 19W
Kauhajoki, *Finland* **13 E20** 62 25N 22 10 E
Kaukauna, *U.S.A.* **134 C1** 44 17N 88 17W
Kaukauveld, *Namibia* **104 C3** 20 0 S 20 15 E
Kaunas, *Lithuania* **13 J20** 54 54N 23 54 E
Kaunghein, *Burma* **78 C5** 25 41N 95 26 E
Kaupalatmada, Mt., *Indonesia* **72 B3** 3 30 S 126 10 E
Kaura Namoda, *Nigeria* **101 C6** 12 37N 6 33 E
Kautokeino, *Norway* **12 B20** 69 0N 23 4 E
Kavacha, *Russia* **57 C17** 60 16N 169 51 E
Kavadarci, *Macedonia* **42 F7** 41 26N 22 3 E
Kavaja, *Albania* **44 C1** 41 11N 19 33 E
Kavak, *Turkey* **88 B7** 41 4N 36 3 E
Kavalerovo, *Russia* **60 B7** 44 15N 135 4 E
Kavali, *India* **83 G5** 14 55N 80 1 E
Kaválla, *Greece* **44 D6** 40 57N 24 28 E
Kaválla □, *Greece* **44 C6** 41 5N 24 30 E
Kaválla Kólpos, *Greece* **44 D6** 40 50N 24 25 E
Kavār, *Iran* **85 D7** 29 11N 52 44 E
Kavarna, *Bulgaria* **43 D13** 43 26N 28 22 E
Kavieng, *Papua N. G.* **120 B6** 2 36 S 150 51 E
Kavkaz, *Russia* **53 H3** 45 20N 36 40 E
Kavos, *Greece* **32 B4** 39 23N 20 3 E
Kavoúsi, *Greece* **45 J7** 35 7N 25 51 E
Kaw, *Fr. Guiana* **153 C7** 4 30N 52 15W
Kawa, *Sudan* **95 E3** 13 42N 32 34 E
Kawachi-Nagano, *Japan* **63 C7** 34 28N 135 31 E
Kawagama L., *Canada* **136 A6** 45 18N 78 45W
Kawagoe, *Japan* **63 B11** 35 55N 139 29 E
Kawaguchi, *Japan* **63 B11** 35 52N 139 45 E
Kawaihae, *U.S.A.* **132 H17** 20 3N 155 50W
Kawakawa, *N.Z.* **118 B3** 35 23 S 174 6 E
Kawambwa, *Zambia* **107 D2** 9 48 S 29 3 E
Kawanoe, *Japan* **62 C5** 34 1N 133 34 E
Kawarau, *N.Z.* **119 F3** 45 3 S 168 45 E
Kawardha, *India* **81 J9** 22 0N 81 17 E
Kawasaki, *Japan* **63 B11** 35 35N 139 42 E
Kawau I., *N.Z.* **118 C3** 36 25 S 174 52 E
Kaweka Ra., *N.Z.* **118 F5** 39 17 S 176 19 E
Kawene, *Canada* **128 C1** 48 45N 91 15W
Kawerau, *N.Z.* **118 E5** 38 7 S 176 42 E
Kawhia Harbour, *N.Z.* **118 E3** 38 5 S 174 51 E
Kawio, Kepulauan, *Indonesia* **72 A3** 4 30N 125 30 E
Kawit, *Phil.* **71 H3** 6 57N 121 58 E
Kawkabān, *Yemen* **86 D3** 15 30N 43 54 E
Kawkareik, *Burma* **78 G7** 16 33N 98 14 E
Kawlin, *Burma* **78 D5** 23 47N 95 41 E
Kawthoolei = Kawthule □, *Burma* **78 G6** 18 0N 97 30 E
Kawthule □, *Burma* **78 G6** 18 0N 97 30 E
Kawya, *Burma* **78 C5** 24 50N 94 58 E
Kay, *Russia* **54 B4** 59 57N 52 59 E
Kaya, *Burkina Faso* **101 C4** 13 4N 1 10W
Kayah □, *Burma* **78 F6** 19 15N 97 15 E
Kayan, *Burma* **78 G6** 16 54N 96 34 E
Kayan —►, *Indonesia* **75 B5** 2 55N 117 35 E
Kayankulam, *India* **83 K3** 9 10N 76 33 E
Kayapa, *Phil.* **70 C3** 16 22N 120 53 E
Kaycee, *U.S.A.* **142 E10** 43 43N 106 38W
Kayeli, *Indonesia* **72 B3** 3 20 S 127 10 E
Kayenta, *U.S.A.* **143 H8** 36 44N 110 15W
Kayes, *Congo* **103 C2** 4 25 S 11 41 E
Kayes, *Mali* **100 C2** 14 25N 11 30W
Kayima, *S. Leone* **100 D2** 8 54N 11 15W
Kayl, *Lux.* **21 K8** 49 29N 6 2 E
Kayoa, *Indonesia* **72 A3** 0 1N 127 28 E
Kayomba, *Zambia* **107 E1** 13 11 S 24 2 E
Kayoro, *Ghana* **101 C4** 11 0N 1 28W
Kayrakkumskoye Vdkhr., *Tajikistan* **55 C4** 40 20N 70 0 E
Kayrunnera, *Australia* **115 E3** 30 40 S 142 30 E
Kaysatskoye, *Russia* **52 F8** 49 47N 46 49 E
Kayseri, *Turkey* **88 C6** 38 45N 35 30 E
Kaysville, *U.S.A.* **142 F8** 41 2N 111 56W
Kayuagung, *Indonesia* **74 C2** 3 24 S 104 50 E
Kazachye, *Russia* **57 B14** 70 52N 135 58 E
Kazakhstan ■, *Asia* **56 E7** 50 0N 70 0 E
Kazan, *Russia* **54 C1** 55 50N 49 10 E
Kazan-Rettō, *Pac. Oc.* **122 E6** 25 0N 141 0 E
Kazanlŭk, *Bulgaria* **43 E10** 42 38N 25 20 E
Kazanskaya, *Russia* **52 F5** 49 50N 41 10 E
Kazarman, *Kyrgyzstan* **55 C6** 41 24N 73 59 E
Kazatin = Kozyatyn, *Ukraine* **51 H5** 49 45N 28 50 E
Kazbek, *Russia* **53 J7** 42 42N 44 30 E
Kāzerūn, *Iran* **85 D6** 29 38N 51 40 E
Kazhim, *Russia* **54 A3** 60 21N 51 33 E
Kazi Magomed = Qazimämmäd, *Azerbaijan* **53 K9** 40 3N 49 0 E
Kazimierz Dolny, *Poland* **47 D8** 51 19N 21 57 E
Kazimierza Wielka, *Poland* **47 E7** 50 15N 20 30 E
Kazincbarcika, *Hungary* **31 C13** 48 17N 20 36 E
Kazo, *Japan* **63 A11** 36 7N 139 36 E
Kaztalovka, *Kazakhstan* **52 F9** 49 47N 48 43 E
Kazu, *Burma* **78 C6** 25 27N 97 46 E
Kazumba, *Zaïre* **103 D4** 6 25 S 22 5 E
Kazuno, *Japan* **60 D10** 40 10N 140 45 E
Kazym —►, *Russia* **56 C7** 63 54N 65 50 E
Kcynia, *Poland* **47 C4** 53 0N 17 30 E
Ke-hsi Mansam, *Burma* **78 E6** 21 56N 97 50 E
Ké-Macina, *Mali* **100 C3** 13 58N 5 22W
Kéa, *Greece* **45 G6** 37 35N 24 22 E
Keams Canyon, *U.S.A.* **143 J8** 35 49N 110 12W
Kearney, *Mo., U.S.A.* **140 E2** 39 22N 94 22W
Kearney, *Nebr., U.S.A.* **138 E5** 40 42N 99 5W
Keban, *Turkey* **89 C8** 38 50N 38 50 E
Keban Baraji, *Turkey* **89 C8** 38 41N 38 33 E
Kebbi □, *Nigeria* **101 C6** 11 30N 8 30 E
Kébi, *Ivory C.* **100 D3** 9 18N 6 37W
Kebili, *Tunisia* **96 B1** 33 47N 9 0 E
Kebnekaise, *Sweden* **12 C18** 67 53N 18 33 E
Kebri Dehar, *Ethiopia* **108 C2** 6 45N 44 17 E
Kebumen, *Indonesia* **75 D3** 7 42 S 109 40 E
Kecel, *Hungary* **31 E12** 46 31N 19 16 E
Kechika —►, *Canada* **130 B3** 59 41N 127 12W
Kecskemét, *Hungary* **31 E12** 46 57N 19 42 E
Kedada, *Ethiopia* **95 F4** 5 25N 35 58 E
Kedah □, *Malaysia* **74 A2** 5 50N 100 40 E
Kedgwick, *Canada* **129 C6** 47 40N 67 20W
Kédhros Óros, *Greece* **32 D6** 35 11N 24 37 E
Kedia Hill, *Botswana* **104 C3** 21 28 S 24 37 E
Kediniai, *Lithuania* **13 J21** 55 15N 24 2 E
Kediri, *Indonesia* **75 D4** 7 51 S 112 1 E
Kédougou, *Senegal* **100 C2** 12 35N 12 10W
Kedzierzyn, *Poland* **47 E5** 50 20N 18 12 E
Keeler, *U.S.A.* **144 J9** 36 29N 117 52W
Keeley L., *Canada* **131 C7** 54 54N 108 8W
Keeling Is. = Cocos Is., *Ind. Oc.* **109 F8** 12 10 S 96 55 E
Keelung = Chilung, *Taiwan* **65 D7** 25 3N 121 45 E
Keene, *Calif., U.S.A.* **145 K8** 35 13N 118 33W
Keene, *N.H., U.S.A.* **137 D12** 42 56N 72 17W
Keeper Hill, *Ireland* **19 D3** 52 45N 8 16W
Keer-Weer, C., *Australia* **114 A3** 14 0 S 141 32 E
Keerbergen, *Belgium* **21 F5** 51 1N 4 38 E
Keeseville, *U.S.A.* **137 B11** 44 29N 73 30W
Keeten Mastgat, *Neths.* **21 E4** 51 36N 4 0 E
Keetmanshoop, *Namibia* **104 D2** 26 35 S 18 8 E
Keewatin, *U.S.A.* **138 B8** 47 24N 93 5W
Keewatin □, *Canada* **131 A9** 63 20N 95 0W
Keewatin —►, *Canada* **131 B8** 56 29N 100 46W
Kefa □, *Ethiopia* **95 F4** 6 55N 36 30 E
Kefallinía, *Greece* **45 F2** 38 20N 20 30 E
Kefamenanu, *Indonesia* **72 C2** 9 28 S 124 29 E
Keffi, *Nigeria* **101 D6** 8 55N 7 43 E
Keflavík, *Iceland* **12 D2** 64 2N 22 35W
Keg River, *Canada* **130 B5** 57 54N 117 55W
Kegalla, *Sri Lanka* **83 L5** 7 15N 80 21 E
Kegaska, *Canada* **129 B7** 50 9N 61 18W
Kehl, *Germany* **27 G3** 48 34N 7 50 E
Keighley, *U.K.* **16 D6** 53 52N 1 54W
Keila, *Estonia* **13 G21** 59 18N 24 25 E
Keimoes, *S. Africa* **104 D3** 28 41 S 20 59 E
Keita, *Niger* **101 C6** 14 46N 5 56 E
Keitele, *Finland* **12 E22** 63 10N 26 20 E
Keith, *Australia* **116 D4** 36 6 S 140 20 E
Keith, *U.K.* **18 D6** 57 32N 2 57W
Keith Arm, *Canada* **126 B7** 64 20N 122 15W
Keithsburg, *U.S.A.* **140 C6** 41 6N 90 56W
Kejser Franz Joseph Fjord = Kong Franz Joseph Fd., *Greenland* **6 B6** 73 30N 24 30W
Kekaygyr, *Kyrgyzstan* **55 C7** 40 42N 75 32 E
Kekri, *India* **80 G6** 26 0N 75 10 E
Këlʼ, *Russia* **57 C13** 69 30N 124 10 E
Kelamet, *Eritrea* **95 D4** 16 0N 38 30 E
Kelan, *China* **66 E6** 38 43N 111 31 E
Kelang, *Malaysia* **77 L3** 3 2N 101 26 E
Kelani Ganga —►, *Sri Lanka* **83 L4** 6 58N 79 50 E
Kelantan □, *Malaysia* **74 A2** 5 10N 102 0 E
Kelantan —►, *Malaysia* **77 J4** 6 13N 102 14 E
Këlcyra, *Albania* **44 D2** 40 22N 20 12 E
Keles —►, *Kazakhstan* **55 C4** 41 1N 68 37 E
Kelheim, *Germany* **27 G7** 48 54N 11 52 E
Kelibia, *Tunisia* **96 A2** 36 50N 11 3 E
Kelkit, *Turkey* **89 B8** 40 7N 39 16 E
Kelkit —►, *Turkey* **88 B7** 40 45N 36 32 E
Kellé, *Congo* **102 C2** 0 8 S 14 38 E
Keller, *U.S.A.* **142 B4** 48 5N 118 41W
Kellerberrin, *Australia* **113 F2** 31 36 S 117 38 E
Kellett, C., *Canada* **6 B1** 72 0N 126 0W
Kelleys I., *U.S.A.* **136 E2** 41 36N 82 42W
Kellogg, *U.S.A.* **142 C5** 47 32N 116 7W
Kells = Ceanannus Mor, *Ireland* **19 C5** 53 44N 6 53W
Kélo, *Chad* **97 G3** 9 10N 15 45 E
Kelokedhara, *Cyprus* **32 E11** 34 48N 32 39 E
Kelowna, *Canada* **130 D5** 49 50N 119 25W
Kelsey Bay, *Canada* **130 C3** 50 25N 126 0W
Kelseyville, *U.S.A.* **144 G4** 38 59N 122 50W
Kelso, *N.Z.* **119 F4** 45 54 S 169 15 E
Kelso, *U.K.* **18 F6** 55 36N 2 26W
Kelso, *U.S.A.* **144 D4** 46 9N 122 54W
Keltemashat, *Kazakhstan* **55 B5** 42 25N 70 8 E
Keluang, *Malaysia* **77 L4** 2 3N 103 18 E
Kelvington, *Canada* **131 C8** 52 10N 103 30W
Kem, *Russia* **48 B5** 65 0N 34 38 E
Kem —►, *Russia* **48 B5** 64 57N 34 41 E
Kem-Kem, *Morocco* **98 B4** 30 40N 4 30W
Kema, *Indonesia* **72 A3** 1 22N 125 8 E
Kemah, *Turkey* **89 C8** 39 32N 39 5 E
Kemaliye, *Turkey* **89 C8** 39 16N 38 29 E
Kemano, *Canada* **130 C3** 53 35N 128 0W
Kemapyu, *Burma* **78 F6** 18 49N 97 19 E
Kemasik, *Malaysia* **77 K4** 4 25N 103 27 E
Kembé, *C.A.R.* **102 B4** 4 36N 21 54 E
Kembolcha, *Ethiopia* **95 E4** 11 2N 39 42 E
Kemer, *Turkey* **88 D4** 36 35N 30 33 E
Kemerovo, *Russia* **56 D9** 55 20N 86 5 E
Kemi, *Finland* **12 D21** 65 44N 24 34 E
Kemi älv = Kemijoki —►, *Finland* **12 D21** 65 47N 24 32 E
Kemijärvi, *Finland* **12 C22** 66 43N 27 22 E
Kemijoki —►, *Finland* **12 D21** 65 47N 24 32 E
Kemmel, *Belgium* **21 G1** 50 47N 2 50 E
Kemmerer, *U.S.A.* **142 F8** 41 48N 110 32W
Kemmuna = Comino, *Malta* **32 C1** 36 2N 14 20 E
Kemp, L., *U.S.A.* **139 J5** 33 46N 99 9W
Kemp Land, *Antarctica* **7 C5** 69 0 S 55 0 E
Kempsey, *Australia* **117 A10** 31 1 S 152 50 E
Kempt, L., *Canada* **128 C5** 47 25N 74 22W
Kempten, *Germany* **27 H6** 47 45N 10 17 E
Kempton, *U.S.A.* **141 D10** 40 17N 86 14W
Kemptville, *Canada* **128 C4** 45 0N 75 38W
Kenadsa, *Algeria* **99 B4** 31 48N 2 26W
Kendal, *Indonesia* **75 D4** 6 56 S 110 14 E
Kendal, *U.K.* **16 C5** 54 20N 2 44W
Kendall, *Australia* **117 A10** 31 35 S 152 44 E
Kendall —►, *Australia* **114 A3** 14 4 S 141 35 E
Kendallville, *U.S.A.* **141 C11** 41 27N 85 16W
Kendari, *Indonesia* **72 B2** 3 50 S 122 30 E
Kendawangan, *Indonesia* **75 C4** 2 32 S 110 17 E
Kende, *Nigeria* **101 C5** 11 30N 4 12 E
Kendenup, *Australia* **113 F2** 34 30 S 117 38 E
Kendervicës, Mal e., *Albania* **44 D1** 40 15N 19 52 E
Kendrapara, *India* **82 D8** 20 35N 86 30 E
Kendrew, *S. Africa* **104 E3** 32 32 S 24 30 E
Kendrick, *U.S.A.* **142 C5** 46 37N 116 39W
Kene Thao, *Laos* **76 D3** 17 44N 101 10 E
Kenedy, *U.S.A.* **139 L6** 28 49N 97 51W
Kenema, *S. Leone* **100 D2** 7 50N 11 14W
Keng Kok, *Laos* **76 D5** 16 26N 105 12 E
Keng Tawng, *Burma* **78 E7** 20 45N 98 18 E
Kengani, *Zaïre* **102 C3** 2 59 S 17 36 E
Kenge, *Zaïre* **103 C3** 4 50 S 17 4 E
Kengeja, *Tanzania* **106 D4** 5 26 S 39 45 E
Kenhardt, *S. Africa* **104 D3** 29 19 S 21 12 E
Kenimekh, *Uzbekistan* **55 C2** 40 16N 65 7 E
Kenitra, *Morocco* **98 B3** 34 15N 6 40W
Kenli, *China* **67 F10** 37 30N 118 20 E
Kenmare, *Ireland* **19 E2** 51 53N 9 36W
Kenmare, *U.S.A.* **138 A3** 48 41N 102 5W
Kenmare —►, *Ireland* **19 E2** 51 48N 9 51W
Kennebec, *U.S.A.* **138 D5** 43 54N 99 52W
Kennedy, *Zimbabwe* **107 F2** 18 52 S 27 10 E
Kennedy Ra., *Australia* **113 D2** 24 45 S 115 10 E
Kennet —►, *U.K.* **17 F7** 51 27N 0 57W
Kenneth Ra., *Australia* **112 D2** 23 50 S 117 8 E
Kennett, *U.S.A.* **139 G9** 36 14N 90 3W
Kennewick, *U.S.A.* **142 C4** 46 12N 119 7W
Kénogami, *Canada* **129 C5** 48 25N 71 15W
Kenogami —►, *Canada* **128 B3** 51 6N 84 28W
Kenora, *Canada* **131 D10** 49 47N 94 29W
Kenosha, *U.S.A.* **141 B9** 42 35N 87 49W
Kensington, *Canada* **129 C7** 46 28N 63 34W
Kensington, *U.S.A.* **138 F5** 39 46N 99 2W
Kensington Downs, *Australia* **114 C3** 22 31 S 144 19 E
Kent, *Ohio, U.S.A.* **136 E3** 41 9N 81 22W
Kent, *Oreg., U.S.A.* **142 D3** 45 12N 120 42W
Kent, *Tex., U.S.A.* **139 K2** 31 4N 104 13W
Kent, *Wash., U.S.A.* **144 C4** 47 23N 122 14W
Kent □, *U.K.* **17 F8** 51 12N 0 40 E
Kent Group, *Australia* **114 F4** 39 30 S 147 20 E
Kent Pen., *Canada* **126 B9** 68 30N 107 0W
Kentau, *Kazakhstan* **55 B4** 43 32N 68 36 E
Kentland, *U.S.A.* **141 D9** 40 46N 87 27W
Kenton, *U.S.A.* **141 D13** 40 39N 83 37W
Kentucky, *Australia* **117 A9** 30 45 S 151 28 E
Kentucky □, *U.S.A.* **134 G3** 37 0N 84 0W
Kentucky —►, *U.S.A.* **141 F11** 38 41N 85 11W
Kentucky L., *U.S.A.* **135 G2** 37 1N 88 16W
Kentville, *Canada* **129 C7** 45 6N 64 29W
Kentwood, *U.S.A.* **139 K9** 30 56N 90 31W
Kenya ■, *Africa* **106 B4** 1 0N 38 0 E
Kenya, Mt., *Kenya* **106 C4** 0 10 S 37 18 E
Kenzou, *Cameroon* **102 B3** 4 10N 15 2 E
Keo Neua, Deo, *Vietnam* **76 C5** 18 23N 105 10 E
Keokuk, *U.S.A.* **140 D5** 40 24N 91 24W
Keosauqua, *U.S.A.* **140 D5** 40 44N 91 58W
Keota, *U.S.A.* **140 C5** 41 22N 91 57W
Kep, *Cambodia* **77 G5** 10 29N 104 19 E
Kep, *Vietnam* **76 B6** 21 24N 106 16 E
Kepi, *Indonesia* **73 C5** 6 32 S 139 19 E
Kepice, *Poland* **47 A3** 54 16N 16 51 E
Kepler Mts., *N.Z.* **119 F2** 45 25 S 167 20 E
Kępno, *Poland* **47 D4** 51 18N 17 58 E
Kerala □, *India* **83 J3** 11 0N 76 15 E
Kerama-Rettō, *Japan* **61 L3** 26 5N 127 15 E
Keran, *Pakistan* **81 B5** 34 35N 73 59 E
Kerang, *Australia* **116 C5** 35 40 S 143 55 E
Keratéa, *Greece* **45 G5** 37 48N 23 58 E
Keraudren, C., *Australia* **112 C2** 19 58 S 119 45 E
Kerava, *Finland* **13 F21** 60 25N 25 5 E
Keravat, *Papua N. G.* **120 C7** 4 17 S 152 2 E
Kerch, *Ukraine* **51 K9** 45 20N 36 20 E
Kerchenskiy Proliv, *Black Sea* **51 K9** 45 10N 36 30 E
Kerchoual, *Mali* **101 B5** 17 12N 0 20 E
Kerema, *Papua N. G.* **120 D3** 7 58 S 145 50 E
Kerempe Burnu, *Turkey* **88 A5** 42 2N 33 20 E
Keren, *Eritrea* **95 D4** 15 45N 38 28 E
Kerewan, *Gambia* **100 C1** 13 29N 16 10W
Kerguelen, *Ind. Oc.* **109 J5** 49 15 S 69 10 E
Keri, *Greece* **45 G2** 37 40N 20 49 E
Keri Kera, *Sudan* **95 E3** 12 21N 32 42 E
Kericho, *Kenya* **106 C4** 0 22 S 35 15 E
Kericho □, *Kenya* **106 C4** 0 30 S 35 15 E
Kerikeri, *N.Z.* **118 B2** 35 12 S 173 59 E
Kerinci, *Indonesia* **74 C2** 1 40 S 101 15 E
Kerkdriel, *Neths.* **20 E6** 51 47N 5 20 E
Kerkenna, Is., *Tunisia* **96 B2** 34 48N 11 11 E
Kerki, *Turkmenistan* **55 E2** 37 50N 65 12 E
Kerkinítis, Límni, *Greece* **44 C5** 41 12N 23 10 E
Kérkira, *Greece* **32 A3** 39 38N 19 50 E
Kerkrade, *Neths.* **21 G8** 50 53N 6 4 E
Kerma, *Sudan* **94 D3** 19 33N 30 32 E
Kermadec Is., *Pac. Oc.* **122 K10** 30 0 S 178 15W
Kermadec Trench, *Pac. Oc.* **122 L10** 30 30 S 176 0W
Kermān, *Iran* **85 D8** 30 15N 57 1 E
Kerman, *U.S.A.* **144 J6** 36 43N 120 4W
Kermān □, *Iran* **85 D8** 30 0N 57 0 E
Kermānshāh = Bākhtarān, *Iran* **89 E12** 34 23N 47 0 E
Kerme Körfezi, *Turkey* **88 D2** 36 55N 27 50 E
Kermen, *Bulgaria* **43 E11** 42 30N 26 16 E
Kermit, *U.S.A.* **139 K3** 31 52N 103 6W
Kern —►, *U.S.A.* **145 K7** 35 16N 119 18W
Kerns, *Switz.* **29 C6** 46 54N 8 17 E
Kernville, *U.S.A.* **145 K8** 35 45N 118 26W
Keroh, *Malaysia* **77 K3** 5 43N 101 1 E
Kerrobert, *Canada* **131 C7** 51 56N 109 8W
Kerrville, *U.S.A.* **139 K5** 30 3N 99 8W
Kerry □, *Ireland* **19 D2** 52 7N 9 35W
Kerry Hd., *Ireland* **19 D2** 52 25N 9 56W
Kersa, *Ethiopia* **95 F5** 9 28N 41 48 E
Kerteminde, *Denmark* **15 J4** 55 28N 10 39 E
Kertosono, *Indonesia* **75 D4** 7 38 S 112 9 E
Kerulen —►, *Asia* **65 B6** 48 48N 117 0 E
Kerzaz, *Algeria* **99 C4** 29 29N 1 37W
Kerzers, *Switz.* **28 C4** 46 59N 7 12 E
Kesagami —►, *Canada* **128 B4** 51 40N 79 45W
Kesagami L., *Canada* **128 B3** 50 23N 80 15W
Keşan, *Turkey* **88 B2** 40 49N 26 38 E
Kesch, Piz, *Switz.* **29 C9** 46 38N 9 53 E
Kesennuma, *Japan* **60 E10** 38 54N 141 35 E
Keshit, *Iran* **85 D8** 29 43N 58 17 E
Keşiş Dağ, *Turkey* **89 C8** 39 47N 39 46 E
Keskin, *Turkey* **88 C5** 39 40N 33 36 E
Kessel, *Belgium* **21 F5** 51 8N 4 38 E
Kessel, *Neths.* **21 F8** 51 17N 6 3 E
Kessel-Lo, *Belgium* **21 G5** 50 53N 4 43 E
Kestell, *S. Africa* **105 D4** 28 17 S 28 42 E
Kestenga, *Russia* **48 A5** 65 50N 31 45 E
Kesteren, *Neths.* **20 E7** 51 56N 5 34 E
Keswick, *U.K.* **16 C4** 54 36N 3 8W
Keszthely, *Hungary* **31 E10** 46 50N 17 15 E
Ket —►, *Russia* **56 D9** 58 55N 81 32 E
Keta, *Ghana* **101 D5** 5 49N 1 0 E
Ketapang, *Indonesia* **75 C4** 1 55 S 110 0 E
Ketchikan, *U.S.A.* **126 C6** 55 21N 131 39W
Ketchum, *U.S.A.* **142 E6** 43 41N 114 22W
Kete Krachi, *Ghana* **101 D4** 7 46N 0 1W
Ketef, Khalîg Umm el, *Egypt* **94 C4** 23 40N 35 35 E
Ketelmeer, *Neths.* **20 C7** 52 36N 5 46 E
Keti Bandar, *Pakistan* **80 G2** 24 8N 67 27 E
Ketri, *India* **80 E6** 28 1N 75 50 E
Kętrzyn, *Poland* **47 A8** 54 7N 21 22 E
Kettering, *U.K.* **17 E7** 52 24N 0 43W
Kettering, *U.S.A.* **141 E12** 39 41N 84 10W
Kettle —►, *Canada* **131 B11** 56 40N 89 34W
Kettle Falls, *U.S.A.* **142 B4** 48 37N 118 3W
Kettleman City, *U.S.A.* **144 J7** 36 1N 119 58W
Kety, *Poland* **31 B12** 49 51N 19 16 E
Keuruu, *Finland* **13 E21** 62 16N 24 41 E
Kevin, *U.S.A.* **142 B8** 48 45N 111 58W
Kewanee, *U.S.A.* **140 C7** 41 14N 89 56W
Kewanna, *U.S.A.* **141 C10** 41 1N 86 25W
Kewaunee, *U.S.A.* **134 C2** 44 27N 87 31W
Keweenaw B., *U.S.A.* **134 B1** 47 0N 88 15W
Keweenaw Pen., *U.S.A.* **134 B2** 47 30N 88 0W
Keweenaw Pt., *U.S.A.* **134 B2** 47 25N 87 43W
Key Harbour, *Canada* **128 C3** 45 50N 80 45W
Key West, *U.S.A.* **133 F10** 24 33N 81 48W
Keyesport, *U.S.A.* **140 F7** 38 45N 89 17W
Keyser, *U.S.A.* **134 F6** 39 26N 78 59W
Keystone, *U.S.A.* **138 D3** 43 54N 103 25W
Keytesville, *U.S.A.* **140 E4** 39 26N 92 56W
Kez, *Russia* **54 C4** 57 55N 53 46 E
Kezhma, *Russia* **57 D11** 58 59N 101 9 E
Kežmarok, *Slovak Rep.* **31 B13** 49 10N 20 28 E
Khabarovo, *Russia* **56 C7** 69 30N 60 30 E
Khabarovsk, *Russia* **57 E14** 48 30N 135 5 E
Khabr, *Iran* **85 D8** 28 51N 56 22 E
Khābūr —►, *Syria* **89 E9** 35 17N 40 35 E
Khachmas = Xaçmaz, *Azerbaijan* **53 K9** 41 31N 48 42 E
Khachrod, *India* **80 H6** 23 25N 75 20 E
Khadari, W. el —►, *Sudan* **95 E2** 10 29N 27 15 E
Khadro, *Pakistan* **80 F3** 26 11N 68 50 E
Khadyzhensk, *Russia* **53 H4** 44 26N 39 32 E
Khadzhilyangar, *India* **81 B8** 35 45N 79 20 E
Khagaria, *India* **81 G12** 25 30N 86 32 E
Khaipur, *Bahawalpur, Pakistan* **80 E5** 29 34N 72 17 E
Khaipur, *Hyderabad, Pakistan* **80 F3** 27 32N 68 49 E
Khair, *India* **80 F7** 27 57N 77 46 E
Khairabad, *India* **81 F9** 27 33N 80 47 E
Khairagarh, *India* **81 J9** 21 27N 81 2 E
Khairpur, *Pakistan* **79 D3** 27 32N 68 49 E
Khāk Dow, *Afghan.* **79 B2** 34 57N 67 16 E
Khakassia □, *Russia* **56 D9** 53 0N 90 0 E
Khakhea, *Botswana* **104 C3** 24 48 S 23 22 E
Khalach, *Turkmenistan* **55 D2** 38 4N 64 52 E
Khalafābād, *Iran* **85 D6** 30 54N 49 24 E
Khalfallah, *Algeria* **99 B5** 34 20N 0 16 E
Khalfūt, *Yemen* **87 D6** 15 52N 52 10 E
Khalilabad, *India* **81 F10** 26 48N 83 5 E
Khalīlī, *Iran* **85 E7** 27 38N 53 17 E
Khalkhāl, *Iran* **89 D13** 37 37N 48 32 E
Khálki, *Greece* **44 E4** 39 36N 22 30 E
Khalkidhikí □, *Greece* **44 D5** 40 25N 23 20 E
Khalkís, *Greece* **45 F5** 38 27N 23 42 E
Khalmer-Sede = Tazovskiy, *Russia* **56 C8** 67 30N 78 44 E
Khalmer Yu, *Russia* **48 A12** 67 58N 65 1 E
Khalturin, *Russia* **54 B2** 58 40N 48 50 E
Khalūf, *Oman* **90 C6** 20 30N 58 13 E
Kham Keut, *Laos* **76 C5** 18 15N 104 43 E
Khamaria, *India* **82 C5** 23 10N 80 52 E
Khamas Country, *Botswana* **104 C4** 21 45 S 26 30 E
Khambat, G. of, *India* **80 J5** 20 45N 72 30 E
Khambhaliya, *India* **80 H3** 22 14N 69 41 E
Khambhat, *India* **80 H5** 22 23N 72 33 E
Khamgaon, *India* **82 D3** 20 42N 76 37 E
Khamilonísion, *Greece* **45 J8** 35 50N 26 15 E
Khamīr, *Iran* **85 E7** 26 57N 55 36 E
Khamir, *Yemen* **86 C3** 16 2N 44 0 E
Khamīs Mushayţ, *Si. Arabia* **86 C3** 18 18N 42 44 E
Khammam, *India* **82 F5** 17 11N 80 6 E
Khamsa, *Egypt* **91 E1** 30 27N 32 23 E
Khān Abū Shāmat, *Syria* **91 B5** 33 39N 36 53 E
Khān Azād, *Iraq* **84 C5** 33 7N 44 22 E
Khān Mujiddah, *Iraq* **84 C4** 32 21N 43 48 E
Khān Shaykhūn, *Syria* **88 E7** 35 26N 36 38 E
Khān Yūnis, *Gaza Strip* **91 D3** 31 21N 34 18 E
Khānābād, *Afghan.* **79 A3** 36 45N 69 5 E
Khanabad, *Uzbekistan* **55 C5** 40 59N 70 38 E
Khānaqīn, *Iraq* **89 E11** 34 23N 45 25 E
Khānbāghī, *Iran* **85 B7** 36 10N 55 25 E

Kirgiziya Steppe 217

Khandrá, Greece 45 J8 35 3N 26 8 E
Khandwa, India 82 D3 21 49N 76 22 E
Khandyga, Russia 57 C14 62 42N 135 35 E
Khāneh, Iran 84 B5 36 41N 45 8 E
Khanewal, Pakistan 79 C3 30 20N 71 55 E
Khanh Duong, Vietnam 76 F7 12 44N 108 44 E
Khaniá, Greece 32 D6 35 30N 24 4 E
Khaniá □, Greece 32 D6 35 30N 24 0 E
Khaníon, Kólpos, Greece 32 D5 35 33N 23 55 E
Khanka, Ozero, Asia 57 E14 45 0N 132 24 E
Khankendy = Xankändi, Azerbaijan 89 C12 39 52N 46 49 E
Khanna, India 80 D7 30 42N 76 16 E
Khanpur, Pakistan 79 C3 28 42N 70 35 E
Khantau, Kazakhstan 55 A6 44 13N 73 48 E
Khanty-Mansiysk, Russia 56 C7 61 0N 69 0 E
Khapalu, Pakistan 81 B7 35 10N 76 20 E
Khapcheranga, Russia 57 E12 49 42N 112 24 E
Kharabali, Russia 53 G8 47 25N 47 15 E
Kharagpur, India 81 H12 22 20N 87 25 E
Khárakas, Greece 32 D7 35 1N 25 7 E
Kharan Kalat, Pakistan 79 C2 28 34N 65 21 E
Kharānaq, Iran 85 C7 32 20N 54 45 E
Kharda, India 82 E2 18 40N 75 34 E
Khardung La, India 81 B7 34 20N 77 43 E
Khârga, El Wâhât el, Egypt 94 B3 25 10N 30 35 E
Khargon, India 82 D2 21 45N 75 40 E
Kharit, Wadi el →, Egypt 94 C3 24 26N 33 3 E
Khārk, Jazireh, Iran 85 D6 29 15N 50 28 E
Kharkiv, Ukraine 51 H9 49 58N 36 20 E
Kharkov = Kharkiv, Ukraine 51 H9 49 58N 36 20 E
Kharmanli, Bulgaria 43 F10 41 55N 25 55 E
Kharovsk, Russia 50 C11 59 56N 40 13 E
Kharta, Turkey 88 B3 40 55N 29 7 E
Khartoum = El Khartûm, Sudan 95 D3 15 31N 32 35 E
Khasan, Russia 60 C5 42 25N 130 40 E
Khasavyurt, Russia 53 J8 43 16N 46 40 E
Khāsh, Iran 85 D9 28 15N 61 15 E
Khashm el Girba, Sudan 95 E4 14 59N 35 58 E
Khashuri, Georgia 53 J6 42 3N 43 35 E
Khasi Hills, India 78 C3 25 30N 91 30 E
Khaskovo, Bulgaria 43 F10 41 56N 25 30 E
Khatanga, Russia 57 B11 72 0N 102 20 E
Khatanga →, Russia 57 B11 72 55N 106 0 E
Khatauli, India 80 E7 29 17N 77 43 E
Khātūnābād, Iran 85 C6 35 30N 51 40 E
Khatyrchi, Uzbekistan 55 C2 40 2N 65 58 E
Khatyrka, Russia 57 C18 62 3N 175 15 E
Khavast, Uzbekistan 55 C4 40 10N 68 49 E
Khawlaf, Ra's, Yemen 87 D6 12 40N 54 7 E
Khay', Si. Arabia 86 C3 18 45N 41 24 E
Khaybar, Harrat, Si. Arabia 84 E4 25 45N 40 0 E
Khaydarken, Kyrgyzstan 55 D5 39 57N 71 20 E
Khāzimiyah, Iraq 84 C4 34 46N 43 37 E
Khazzân Jabal el Awliyâ, Sudan 95 D3 15 24N 32 20 E
Khe Bo, Vietnam 76 C5 19 8N 104 41 E
Khe Long, Vietnam 76 B5 21 29N 104 46 E
Khed, Maharashtra, India 82 F1 17 43N 73 27 E
Khed, Maharashtra, India 82 E1 18 51N 73 56 E
Khekra, India 80 E7 28 52N 77 20 E
Khemarak Phouminville, Cambodia 77 G4 11 37N 102 59 E
Khemelnik, Ukraine 51 H4 49 33N 27 58 E
Khemis Miliana, Algeria 99 A5 36 11N 2 14 E
Khemissèt, Morocco 98 B3 33 50N 6 1W
Khemmarat, Thailand 76 D5 16 10N 105 15 E
Khenāmān, Iran 85 D8 30 27N 56 29 E
Khenchela, Algeria 99 A6 35 28N 7 11 E
Khenifra, Morocco 98 B3 32 58N 5 46W
Kherrata, Algeria 99 A6 36 27N 5 13 E
Khérson, Greece 44 C4 41 5N 22 47 E
Kherson, Ukraine 51 J7 46 35N 32 35 E
Khersónisos Akrotíri, Greece 32 D6 35 30N 24 10 E
Kheta →, Russia 57 B11 71 54N 102 6 E
Khiliomódhion, Greece 45 G4 37 48N 22 51 E
Khilok, Russia 57 D12 51 30N 110 45 E
Khimki, Russia 52 C3 55 50N 37 20 E
Khíos, Greece 45 F8 38 27N 26 9 E
Khirbat Qanāfār, Lebanon 91 B4 33 39N 35 43 E
Khisar-Momina Banya, Bulgaria 43 E9 42 30N 24 44 E
Khiuma = Hiiumaa, Estonia 13 G20 58 50N 22 45 E
Khiva, Uzbekistan 56 E7 41 30N 60 18 E
Khīyāv, Iran 84 B5 38 30N 47 45 E
Khlebarovo, Bulgaria 43 D11 43 37N 26 15 E
Khlong Khlung, Thailand 76 D2 16 12N 99 43 E
Khmelnitskiy = Khmelnytskyy, Ukraine 51 H4 49 23N 27 0 E
Khmelnytskyy, Ukraine 51 H4 49 23N 27 0 E
Khmer Rep. = Cambodia ■, Asia 76 F5 12 15N 105 0 E
Khoai, Hon, Vietnam 77 H5 8 26N 104 50 E
Khodoriv, Ukraine 51 H3 49 24N 24 19 E
Khodzent = Khudzhand, Tajikistan 55 C4 40 17N 69 37 E
Khojak P., Afghan. 79 C2 30 55N 66 30 E
Khok Kloi, Thailand 77 H2 8 17N 98 19 E
Khok Pho, Thailand 77 J3 6 43N 101 6 E
Kholm, Afghan. 79 A2 36 45N 67 40 E
Kholm, Russia 50 D6 57 10N 31 15 E
Kholmsk, Russia 57 E15 47 40N 142 5 E
Khomas Hochland, Namibia 104 C2 22 40 S 16 0 E
Khomeyn, Iran 85 C6 33 40N 50 7 E
Khon Kaen, Thailand 76 D4 16 30N 102 47 E
Khong, Laos 76 E5 14 7N 105 51 E
Khong Sedone, Laos 76 E5 15 34N 105 49 E
Khonuu, Russia 57 C15 66 30N 143 12 E
Khoper →, Russia 52 F6 49 30N 42 20 E
Khor el 'Atash, Sudan 95 E3 13 20N 34 15 E
Khóra, Greece 45 G3 37 3N 21 42 E
Khóra Sfakíon, Greece 32 D6 35 15N 24 9 E
Khorāsān □, Iran 85 C8 34 0N 58 0 E
Khorat = Nakhon Ratchasima, Thailand 76 E4 14 59N 102 12 E
Khorat, Cao Nguyen, Thailand 76 E4 15 30N 102 50 E
Khorb el Ethel, Algeria 98 C3 28 30N 6 17W
Khorixas, Namibia 104 C1 20 16 S 14 59 E
Khorol, Ukraine 51 H7 49 48N 33 15 E
Khorramābād, Khorāsān, Iran 85 C8 35 6N 57 57 E
Khorramābād, Lorestān, Iran 89 F13 33 30N 48 25 E
Khorrāmshahr, Iran 85 D6 30 29N 48 15 E
Khorugh, Tajikistan 55 E5 37 30N 71 36 E
Khosravī, Iran 85 D6 30 48N 51 28 E
Khosrowābād, Khuzestān, Iran 85 D6 30 10N 48 25 E
Khosrowābād, Kordestān, Iran 89 E12 35 31N 47 38 E
Khosūyeh, Iran 85 D7 28 32N 54 26 E
Khotyn, Ukraine 51 H4 48 31N 26 27 E
Khouribga, Morocco 98 B3 32 58N 6 57W
Khowai, Bangla. 78 C3 24 5N 91 40 E
Khoyniki, Belarus 51 G5 51 54N 29 55 E
Khrami →, Georgia 53 K7 41 25N 45 0 E
Khrenovoye, Russia 52 E5 51 4N 40 16 E
Khristianá, Greece 45 H7 36 14N 25 13 E
Khromtau, Kazakhstan 54 F7 50 17N 58 27 E
Khrysokhou B., Cyprus 32 D11 35 6N 32 25 E
Khtapodhiá, Greece 45 G7 37 24N 25 34 E
Khu Khan, Thailand 76 E5 14 42N 104 12 E
Khudrah, W. →, Yemen 87 C5 18 10N 50 20 E
Khudzhand, Tajikistan 55 C4 40 17N 69 37 E
Khuff, Si. Arabia 84 E5 24 55N 44 53 E
Khūgīānī, Qandahar, Afghan. 79 C2 31 34N 66 32 E
Khūgīānī, Qandahar, Afghan. 79 C2 31 28N 65 14 E
Khulayş, Si. Arabia 86 B2 22 9N 39 19 E
Khulna, Bangla. 78 D2 22 45N 89 34 E
Khulna □, Bangla. 78 D2 22 25N 89 35 E
Khulo, Georgia 53 K6 41 33N 42 19 E
Khumago, Botswana 104 C3 20 26 S 24 32 E
Khumrah, Si. Arabia 86 B2 21 22N 39 13 E
Khūnsorkh, Iran 85 E8 27 9N 56 7 E
Khūr, Iran 85 C8 32 55N 58 18 E
Khurai, India 80 G8 24 3N 78 23 E
Khuraydah, Yemen 87 D5 15 33N 48 18 E
Khurayş, Si. Arabia 85 E6 25 6N 48 2 E
Khūrīyā Mūrīyā, Jazā 'ir, Oman 87 C6 17 30N 55 58 E
Khurja, India 80 E7 28 15N 77 58 E
Khūsf, Iran 85 C8 32 46N 58 53 E
Khushab, Pakistan 79 B4 32 20N 72 20 E
Khust, Ukraine 51 H2 48 10N 23 18 E
Khuzdar, Pakistan 79 D2 27 52N 66 30 E
Khūzestān □, Iran 85 D6 31 0N 49 0 E
Khvājeh, Iran 84 B5 38 9N 46 35 E
Khvājeh Moḩammad, Kūh-e, Afghan. 79 A3 36 22N 70 17 E
Khvalynsk, Russia 52 D9 52 30N 48 2 E
Khvānsār, Iran 85 D7 29 56N 54 8 E
Khvatovka, Russia 52 D8 52 24N 46 32 E
Khvor, Iran 85 C7 33 45N 55 0 E
Khvorgū, Iran 85 E8 27 34N 56 27 E
Khvormūj, Iran 85 D6 28 40N 51 30 E
Khvoy, Iran 89 C11 38 35N 45 0 E
Khvoynaya, Russia 50 C8 58 58N 34 28 E
Khyber Pass, Afghan. 79 B3 34 10N 71 8 E
Kia, Solomon Is. 121 L10 7 32 S 158 26 E
Kiabukwa, Zaïre 103 D4 8 40 S 24 48 E
Kiadho →, India 82 E3 19 37N 77 40 E
Kiama, Australia 117 C9 34 40 S 150 50 E
Kiamba, Phil. 71 H5 6 2N 124 46 E
Kiambi, Zaïre 106 D2 7 15 S 28 0 E
Kiambu, Kenya 106 C4 1 8 S 36 50 E
Kiangsi = Jiangxi □, China 69 D10 27 30N 116 0 E
Kiangsu = Jiangsu □, China 67 H10 33 0N 120 0 E
Kiáton, Greece 45 F4 38 2N 22 43 E
Kibæk, Denmark 15 H2 56 2N 8 51 E
Kibanga Port, Uganda 106 B3 0 10N 32 58 E
Kibangou, Congo 102 C2 3 26 S 12 22 E
Kibara, Tanzania 106 C3 2 8 S 33 30 E
Kibare, Mts., Zaïre 106 D2 8 25 S 27 10 E
Kibawe, Phil. 71 H5 7 34N 125 0 E
Kibombo, Zaïre 103 C5 3 57 S 25 53 E
Kibondo, Tanzania 106 C3 3 35 S 30 45 E
Kibondo □, Tanzania 106 C3 4 0 S 30 55 E
Kibumbu, Burundi 106 C2 3 32 S 29 45 E
Kibungu, Rwanda 106 C3 2 10 S 30 32 E
Kibuye, Burundi 106 C2 3 39 S 29 59 E
Kibuye, Rwanda 106 C2 2 3 S 29 21 E
Kibwesa, Tanzania 106 D2 6 30 S 29 58 E
Kibwezi, Kenya 106 C4 2 27 S 37 57 E
Kičevo, Macedonia 42 F5 41 34N 20 59 E
Kichiga, Russia 57 D17 59 50N 163 5 E
Kicking Horse Pass, Canada 130 C5 51 28N 116 16W
Kidal, Mali 101 B5 18 26N 1 22 E
Kidapawan, Phil. 71 H5 7 1N 125 3 E
Kidderminster, U.K. 17 E5 52 24N 2 15W
Kidete, Tanzania 106 D4 6 25 S 37 17 E
Kidira, Senegal 100 C2 14 28N 12 13W
Kidnappers, C., N.Z. 118 F6 39 38 S 177 5 E
Kidston, Australia 114 B3 18 52 S 144 8 E
Kidugallo, Tanzania 106 D4 6 49 S 38 15 E
Kidurong, Tanjong, Malaysia 75 B4 3 16N 113 3 E
Kiel, Germany 26 A6 54 19N 10 8 E
Kiel Kanal = Nord-Ostsee-Kanal →, Germany 26 A5 54 12N 9 32 E
Kielce, Poland 47 E7 50 52N 20 42 E
Kielce □, Poland 47 E7 50 40N 20 40 E
Kieldrecht, Belgium 21 F4 51 17N 4 11 E
Kieler Bucht, Germany 26 A6 54 35N 10 25 E
Kien Binh, Vietnam 77 H5 9 55N 105 19 E
Kien Tan, Vietnam 77 G5 10 7N 105 17 E
Kienge, Zaïre 107 E2 10 30 S 27 30 E
Kiessé, Niger 101 C5 13 29N 4 1 E
Kieta, Papua N. G. 120 D8 6 12 S 155 36 E
Kiev = Kyyiv, Ukraine 51 G6 50 30N 30 28 E
Kiffa, Mauritania 100 B2 16 37N 11 24W
Kifisiá, Greece 45 F5 38 4N 23 49 E
Kifissós →, Greece 45 F5 38 35N 23 20 E
Kifrī, Iraq 89 E11 34 45N 45 0 E
Kigali, Rwanda 106 C3 1 59 S 30 4 E
Kigarama, Tanzania 106 C3 1 1 S 31 50 E
Kigoma □, Tanzania 106 D2 5 0 S 30 0 E
Kigoma-Ujiji, Tanzania 106 C2 4 55 S 29 36 E
Kigomasha, Ras, Tanzania 106 C4 4 58 S 38 58 E
Kihee, Australia 115 D3 27 23 S 142 37 E
Kihikihi, N.Z. 118 E4 38 2 S 175 22 E
Kihnu, Estonia 13 G21 58 9N 24 1 E
Kii-Hantō, Japan 63 D7 34 0N 135 45 E
Kii-Sanchi, Japan 63 C8 34 20N 136 0 E
Kii-Suidō, Japan 62 D6 33 40N 134 45 E
Kikaiga-Shima, Japan 61 K4 28 19N 129 59 E
Kikinda, Serbia, Yug. 42 B5 45 50N 20 30 E
Kikládhes, Greece 45 G6 37 20N 24 30 E
Kikládhes □, Greece 45 G6 37 0N 25 0 E
Kikoira, Australia 117 B7 33 39 S 146 40 E
Kikori, Papua N. G. 120 D3 7 25 S 144 15 E
Kikori →, Papua N. G. 120 D3 7 38 S 144 20 E
Kikuchi, Japan 62 E2 32 59N 130 47 E
Kikwit, Zaïre 103 D3 5 0 S 18 45 E
Kila' Drosh, Pakistan 79 B3 35 33N 71 52 E
Kilakkarai, India 83 K4 9 12N 78 47 E
Kílalki, Greece 45 H9 36 15N 27 35 E
Kilauea Crater, U.S.A. 132 J17 19 25N 155 17W
Kilchberg, Switz. 29 B7 47 18N 8 33 E
Kilcoy, Australia 115 D5 26 59 S 152 30 E
Kildare, Ireland 19 C5 53 9N 6 55W
Kildare □, Ireland 19 C5 53 10N 6 50W
Kilembe, Zaïre 103 D3 5 42 S 19 55 E
Kilgore, U.S.A. 139 J7 32 23N 94 53W
Kilifi, Kenya 106 C4 3 40 S 39 48 E
Kilifi □, Kenya 106 C4 3 30 S 39 40 E
Kilimanjaro, Tanzania 106 C4 3 7 S 37 20 E
Kilimanjaro □, Tanzania 106 C4 4 0 S 38 0 E
Kilimli, Turkey 88 B4 41 28N 31 50 E
Kilinailau Is., Papua N. G. 120 C8 4 45 S 155 20 E
Kilindini, Kenya 106 C4 4 4 S 39 40 E
Kilis, Turkey 88 D7 36 42N 37 6 E
Kiliya, Ukraine 51 K5 45 28N 29 16 E
Kilju, N. Korea 67 D15 40 57N 129 25 E
Kilkee, Ireland 19 D2 52 41N 9 39W
Kilkenny, Ireland 19 D4 52 39N 7 15W
Kilkenny □, Ireland 19 D4 52 35N 7 15W
Kilkieran B., Ireland 19 C2 53 20N 9 41W
Kilkís, Greece 44 D4 40 58N 22 57 E
Kilkís □, Greece 44 C4 41 5N 22 50 E
Killala, Ireland 19 B2 54 13N 9 12W
Killala B., Ireland 19 B2 54 16N 9 8W
Killaloe, Ireland 19 D3 52 48N 8 28W
Killaloe Sta., Canada 136 A7 45 33N 77 25W
Killam, Canada 130 C6 52 47N 111 51W
Killarney, Australia 115 D5 28 20 S 152 18 E
Killarney, Canada 128 C3 45 55N 81 30W
Killarney, Ireland 19 D2 52 4N 9 30W
Killarney, Lakes of, Ireland 19 E2 52 0N 9 30W
Killary Harbour, Ireland 19 C2 53 38N 9 52W
Killdeer, Canada 131 D7 49 6N 106 22W
Killdeer, U.S.A. 138 B3 47 26N 102 48W
Killeen, U.S.A. 139 K6 31 7N 97 44W
Killiecrankie, Pass of, U.K. 18 E5 56 44N 3 46W
Killin, U.K. 18 E4 56 28N 4 19W
Killíni, Ilía, Greece 45 G3 37 55N 21 8 E
Killíni, Korinthía, Greece 45 G4 37 54N 22 25 E
Killybegs, Ireland 19 B3 54 38N 8 26W
Kilmarnock, U.K. 18 F4 55 37N 4 29W
Kilmez, Russia 52 B10 56 58N 50 55 E
Kilmez →, Russia 52 B10 56 58N 50 28 E
Kilmore, Australia 117 D6 37 25 S 144 53 E
Kilondo, Tanzania 107 D3 9 45 S 34 20 E
Kilosa, Tanzania 106 D4 6 48 S 37 0 E
Kilosa □, Tanzania 106 D4 6 48 S 37 0 E
Kilrush, Ireland 19 D2 52 38N 9 29W
Kilwa □, Tanzania 107 D4 9 0 S 39 0 E
Kilwa Kisiwani, Tanzania 107 D4 8 58 S 39 32 E
Kilwa Kivinje, Tanzania 107 D4 8 45 S 39 25 E
Kilwa Masoko, Tanzania 107 D4 8 55 S 39 30 E
Kim, U.S.A. 139 G3 37 15N 103 21W
Kimaam, Indonesia 73 C5 7 58 S 138 53 E
Kimamba, Tanzania 106 D4 6 45 S 37 10 E
Kimba, Australia 116 B2 33 8 S 136 23 E
Kimball, Nebr., U.S.A. 138 E3 41 14N 103 40W
Kimball, S. Dak., U.S.A. 138 D5 43 45N 98 57W
Kimbe, Papua N. G. 120 C6 5 33 S 150 11 E
Kimbe B., Papua N. G. 120 C6 5 15 S 150 30 E
Kimberley, Australia 116 B4 32 50 S 141 4 E
Kimberley, Canada 130 D5 49 40N 115 59W
Kimberley, S. Africa 104 D3 28 43 S 24 46 E
Kimberley Downs, Australia 112 C3 17 24 S 124 22 E
Kimberley Plateau, Australia 110 D4 16 20 S 127 0 E
Kimberly, U.S.A. 142 E6 42 32N 114 22W
Kimchaek, N. Korea 67 D15 40 40N 129 10 E
Kimchŏn, S. Korea 67 F15 36 11N 128 4 E
Kími, Greece 45 F6 38 38N 24 6 E
Kimje, S. Korea 67 G14 35 48N 126 45 E
Kímolos, Greece 45 H6 36 48N 24 37 E
Kimovsk, Russia 52 C3 54 0N 38 29 E
Kimparana, Mali 100 C4 12 48N 5 0W
Kimry, Russia 52 B3 56 55N 37 15 E
Kimsquit, Canada 130 C3 52 45N 126 57W
Kimstad, Sweden 15 F9 58 35N 15 58 E
Kimvula, Zaïre 103 D3 5 44 S 15 58 E
Kinabalu, Gunong, Malaysia 75 A5 6 3N 116 14 E
Kínaros, Greece 45 H8 36 59N 26 15 E
Kinaskan L., Canada 130 B2 57 38N 130 8W
Kinbasket L., Canada 130 C5 52 0N 118 10W
Kincaid, Canada 131 D7 49 40N 107 0W
Kincaid, U.S.A. 140 E7 39 35N 89 25W
Kincardine, Canada 128 D3 44 10N 81 40W
Kinda, Kasai Or., Zaïre 103 D5 9 18 S 25 4 E
Kinda, Shaba, Zaïre 103 C4 4 47 S 21 48 E
Kinder Scout, U.K. 16 D6 53 24N 1 52W
Kindersley, Canada 131 C7 51 30N 109 10W
Kindia, Guinea 100 C2 10 0N 12 52W
Kindu, Zaïre 102 C5 2 55 S 25 50 E
Kinel, Russia 52 D10 53 15N 50 40 E
Kineshma, Russia 52 B6 57 30N 42 5 E
Kinesi, Tanzania 106 C3 1 25 S 33 50 E
King, L., Australia 113 F2 33 10 S 119 35 E
King, Mt., Australia 114 D4 25 10 S 147 30 E
King City, Calif., U.S.A. 144 J5 36 13N 121 8W
King City, Mo., U.S.A. 140 D2 40 3N 94 31W
King Cr. →, Australia 114 C2 24 35 S 139 30 E
King Edward →, Australia 112 B4 14 14 S 126 35 E
King Frederick VI Land = Kong Frederik VI.s Kyst, Greenland 6 C5 63 0N 43 0W
King George B., Falk. Is. 160 D4 51 30 S 60 30W
King George I., Antarctica 7 C18 60 0 S 60 0W
King George Is., Canada 127 C11 57 20N 80 30W
King I., Australia 114 F3 39 50 S 144 0 E
King I., Canada 130 C3 52 10N 127 40W
King Leopold Ras., Australia 112 C4 17 30 S 125 45 E
King Sd., Australia 112 C3 16 50 S 123 20 E
King William I., Canada 126 B10 69 10N 97 25W
King William's Town, S. Africa 104 E4 32 51 S 27 22 E
Kingaroy, Australia 115 D5 26 32 S 151 51 E
Kingfisher, U.S.A. 139 H6 35 52N 97 56W
Kingirbān, Iraq 84 C5 34 40N 44 54 E
Kingisepp = Kuressaare, Estonia 13 G20 58 15N 22 30 E
Kingisepp = Kuressaare, Russia 50 C5 59 25N 28 40 E
Kingking, Phil. 71 H5 7 9N 125 54 E
Kingman, Ariz., U.S.A. 145 K12 35 12N 114 4W
Kingman, Ind., U.S.A. 141 E9 39 58N 87 18W
Kingman, Kans., U.S.A. 139 G5 37 39N 98 7W
Kingoonya, Australia 115 E2 30 55 S 135 19 E
Kings →, U.S.A. 144 J7 36 3N 119 50W
Kings Canyon National Park, U.S.A. 144 J8 36 50N 118 40W
King's Lynn, U.K. 16 E8 52 45N 0 24 E
Kings Mountain, U.S.A. 135 H5 35 15N 81 20W
King's Peak, U.S.A. 142 F8 40 46N 110 27W
Kingsbridge, U.K. 17 G4 50 17N 3 47W
Kingsburg, U.S.A. 144 J7 36 31N 119 33W
Kingsbury, U.S.A. 141 C10 41 31N 86 42W
Kingscote, Australia 116 C2 35 40 S 137 38 E
Kingscourt, Ireland 19 C5 53 55N 6 48W
Kingsley, U.S.A. 138 D7 42 35N 95 58W
Kingsport, U.S.A. 135 G4 36 33N 82 33W
Kingston, Canada 128 D4 44 14N 76 30W
Kingston, Jamaica 148 C4 18 0N 76 50W
Kingston, N.Z. 119 F3 45 20 S 168 43 E
Kingston, Mo., U.S.A. 140 E2 39 39N 94 2W
Kingston, N.Y., U.S.A. 137 E10 41 56N 73 59W
Kingston, Pa., U.S.A. 137 E9 41 16N 75 54W
Kingston, R.I., U.S.A. 137 E13 41 29N 71 30W
Kingston Pk., U.S.A. 145 K11 35 45N 115 54W
Kingston South East, Australia 116 D3 36 51 S 139 55 E
Kingston upon Hull, U.K. 16 D7 53 45N 0 21W
Kingston-upon-Thames, U.K. 17 F7 51 24N 0 17W
Kingstown, Australia 117 A9 30 29 S 151 6 E
Kingstown, St. Vincent 149 D7 13 10N 61 10W
Kingstree, U.S.A. 135 J6 33 40N 79 50W
Kingsville, Canada 128 D3 42 2N 82 45W
Kingsville, U.S.A. 139 M6 27 31N 97 52W
Kingussie, U.K. 18 D4 57 6N 4 2W
Kınık, Turkey 88 C2 39 5N 27 23 E
Kinistino, Canada 131 C7 52 57N 105 2W
Kinkala, Congo 103 C2 4 18 S 14 49 E
Kinki □, Japan 63 D8 33 45N 136 0 E
Kinleith, N.Z. 118 E4 38 20 S 175 56 E
Kinmount, Canada 136 B6 44 48N 78 45W
Kinmundy, U.S.A. 141 F8 38 46N 88 51W
Kinna, Sweden 15 G6 57 32N 12 42 E
Kinnaird, Canada 130 D5 49 17N 117 39W
Kinnairds Hd., U.K. 18 D7 57 43N 2 1W
Kinnared, Sweden 15 G7 57 2N 13 7 E
Kinnarodden, Norway 10 A11 71 8N 27 40 E
Kino, Mexico 146 B2 28 45N 111 59W
Kinogitan, Phil. 71 G5 9 0N 124 48 E
Kinoje →, Canada 128 B3 52 8N 81 25W
Kinomoto, Japan 63 B8 35 30N 136 13 E
Kinoni, Uganda 106 C3 0 41 S 30 28 E
Kinrooi, Belgium 21 F7 51 9N 5 45 E
Kinross, U.K. 18 E5 56 13N 3 25W
Kinsale, Ireland 19 E3 51 42N 8 31W
Kinsale, Old Hd. of, Ireland 19 E3 51 37N 8 33W
Kinsha = Chang Jiang →, China 69 B13 31 48N 121 10 E
Kinshasa, Zaïre 103 C3 4 20 S 15 15 E
Kinsley, U.S.A. 139 G5 37 55N 99 25W
Kinston, U.S.A. 135 H7 35 16N 77 35W
Kintampo, Ghana 101 D4 8 5N 1 41W
Kintap, Indonesia 75 C5 3 51 S 115 13 E
Kintore Ra., Australia 112 D4 23 15 S 128 47 E
Kintyre, U.K. 18 F3 55 30N 5 35W
Kintyre, Mull of, U.K. 18 F3 55 17N 5 47W
Kinu, Burma 78 D5 22 46N 95 37 E
Kinu-Gawa →, Japan 63 B11 35 36N 139 57 E
Kinushseo →, Canada 128 A3 55 15N 83 45W
Kinuso, Canada 130 B5 55 20N 115 25W
Kinyangiri, Tanzania 106 C3 4 25 S 34 37 E
Kinzig →, Germany 27 G3 48 36N 7 49 E
Kinzua, U.S.A. 136 E6 41 52N 78 58W
Kinzua Dam, U.S.A. 136 E5 41 53N 79 0W
Kióni, Greece 45 F2 38 27N 20 41 E
Kiosk, Canada 128 C4 46 6N 78 53W
Kiowa, Kans., U.S.A. 139 G5 37 1N 98 29W
Kiowa, Okla., U.S.A. 139 H7 34 43N 95 54W
Kipahigan L., Canada 131 B8 55 20N 101 55W
Kipanga, Tanzania 106 D4 6 15 S 35 20 E
Kiparissía, Greece 45 G3 37 15N 21 40 E
Kiparissiakós Kólpos, Greece 45 G3 37 25N 21 25 E
Kipembawe, Tanzania 106 D3 7 38 S 33 27 E
Kipengere Ra., Tanzania 107 D3 9 12 S 34 15 E
Kipili, Tanzania 106 D3 7 28 S 30 32 E
Kipini, Kenya 106 C5 2 30 S 40 32 E
Kipling, Canada 131 C8 50 6N 102 38W
Kippure, Ireland 19 C5 53 11N 6 21W
Kipungot, Phil. 71 H5 6 24N 124 4 E
Kipushi, Zaïre 107 E2 11 48 S 27 12 E
Kira Kira, Solomon Is. 121 N11 10 27 S 161 56 E
Kirandul, India 82 E5 18 33N 81 10 E
Kiratpur, India 80 E8 29 32N 78 12 E
Kirchberg, Switz. 28 B5 47 5N 7 35 E
Kirchhain, Germany 26 E4 50 47N 8 56 E
Kirchheim, Germany 27 G5 48 39N 9 27 E
Kirchheim-Bolanden, Germany 27 F4 49 40N 8 0 E
Kirchschlag, Austria 31 D9 47 30N 16 19 E
Kirensk, Russia 57 D11 57 50N 107 55 E
Kirgella Rocks, Australia 113 F3 30 5 S 122 50 E
Kirghizia = Kyrgyzstan ■, Asia 55 C7 42 0N 75 0 E
Kirghizstan = Kyrgyzstan ■, Asia 55 C7 42 0N 75 0 E
Kirgiziya Steppe, Eurasia 54 G4 50 0N 55 0 E

Kraśnik Fabryczny 219

Komoran, Pulau, Indonesia 73 C5 8 18 S 138 45 E
Komoro, Japan 63 A10 36 19N 138 26 E
Komotini, Greece 44 C7 41 9N 25 26 E
Komovi, Montenegro, Yug. 42 E4 42 41N 19 39 E
Kompasberg, S. Africa 104 E3 31 45 S 24 32 E
Kompong Bang, Cambodia 77 F5 12 24N 104 40 E
Kompong Cham, Cambodia 77 F5 12 0N 105 30 E
Kompong Chhnang, Cambodia 77 F5 12 20N 104 35 E
Kompong Chikreng, Cambodia 76 F5 13 5N 104 18 E
Kompong Kleang, Cambodia 76 F5 13 6N 104 8 E
Kompong Luong, Cambodia 77 G5 11 49N 104 48 E
Kompong Pranak, Cambodia 76 F5 13 35N 104 55 E
Kompong Som, Cambodia 77 G4 10 38N 103 30 E
Kompong Som, Chhung, Cambodia 77 G4 10 50N 103 32 E
Kompong Speu, Cambodia 77 G5 11 26N 104 32 E
Kompong Sralao, Cambodia 76 E5 14 5N 105 46 E
Kompong Thom, Cambodia 76 F5 12 35N 104 51 E
Kompong Trabeck, Cambodia 76 F5 13 6N 105 14 E
Kompong Trabeck, Cambodia 77 G5 11 9N 105 28 E
Kompong Trach, Cambodia 77 G5 11 25N 105 48 E
Kompong Tralach, Cambodia 77 G5 11 54N 104 47 E
Komrat = Comrat, Moldova 51 J5 46 18N 28 40 E
Komsberg, S. Africa 104 E3 32 40 S 20 45 E
Komsomolabad, Tajikistan 55 D4 38 50N 69 55 E
Komsomolets, Kazakhstan 54 E9 53 45N 62 2 E
Komsomolets, Ostrov, Russia 57 A10 80 30N 95 0 E
Komsomolsk, Russia 52 B5 57 2N 40 20 E
Komsomolsk, Russia 57 D14 50 30N 137 0 E
Komsomolsk, Turkmenistan 55 D1 39 2N 63 36 E
Komsomolskiy, Russia 52 C7 54 27N 45 33 E
Konakovo, Russia 52 B3 56 40N 36 51 E
Konarhá □, Afghan. 79 B3 35 30N 71 3 E
Konārī, Iran 85 D6 28 13N 51 36 E
Konawa, U.S.A. 139 H6 34 58N 96 45W
Konch, India 81 G8 26 0N 79 10 E
Kondagaon, India 82 E5 19 35N 81 35 E
Kondakovo, Russia 57 C16 69 36N 152 0 E
Konde, Tanzania 106 C4 4 57 S 39 45 E
Kondiá, Greece 44 E7 39 49N 25 10 E
Kondinin, Australia 113 F2 32 34 S 118 8 E
Kondo, Zaïre 103 D2 5 35 S 13 0 E
Kondoa, Tanzania 106 C4 4 55 S 35 50 E
Kondoa □, Tanzania 106 D4 5 0S 36 0 E
Kondókali, Greece 32 A3 39 38N 19 51 E
Kondopaga, Russia 50 A8 62 12N 34 17 E
Kondratyevo, Russia 57 D10 57 22N 98 15 E
Kondrovo, Russia 52 C2 54 48N 35 56 E
Konduga, Nigeria 101 C7 11 35N 13 26 E
Kondukur, India 83 G4 15 12N 79 57 E
Koné, N. Cal. 121 U18 21 4 S 164 52 E
Köneürgench, Turkmenistan 56 E6 42 19N 59 10 E
Konevo, Russia 50 A10 62 8N 39 20 E
Kong, Ivory C. 100 D4 8 54N 4 36W
Kong →, Cambodia 76 F5 13 32N 105 58 E
Kong, Koh, Cambodia 77 G4 11 20N 103 0 E
Kong Christian IX.s Land, Greenland 6 C6 68 0N 36 0W
Kong Christian X.s Land, Greenland 6 B6 74 0N 29 0W
Kong Franz Joseph Fd., Greenland 6 B6 73 30N 24 30W
Kong Frederik IX.s Land, Greenland 6 C5 67 0N 52 0W
Kong Frederik VI.s Kyst, Greenland 6 C5 63 0N 43 0W
Kong Frederik VIII.s Land, Greenland 6 B6 78 30N 26 0W
Kong Oscar Fjord, Greenland 6 B6 72 20N 24 0W
Kongbo, C.A.R. 102 B4 4 44N 21 23 E
Kongeå →, Denmark 15 J3 55 23N 8 39 E
Kongju, S. Korea 67 F14 36 30N 127 0 E
Kongkemul, Indonesia 75 B4 1 52N 112 11 E
Konglu, Burma 78 B6 27 13N 97 57 E
Kongolo, Kasai Or., Zaïre 103 D4 5 26 S 24 49 E
Kongolo, Shaba, Zaïre 106 D2 5 22 S 27 0 E
Kongor, Sudan 95 F3 7 1N 31 27 E
Kongoussi, Burkina Faso 101 C4 13 19N 1 32W
Kongsberg, Norway 14 E3 59 39N 9 39 E
Kongsvinger, Norway 14 D6 60 12N 12 2 E
Kongwa, Tanzania 106 D4 6 11 S 36 26 E
Koni, Zaïre 107 E2 10 40 S 27 11 E
Koni, Mts., Zaïre 107 E2 10 36 S 27 10 E
Koniecpol, Poland 47 E6 50 46N 19 40 E
Königsberg = Kaliningrad, Russia 13 J19 54 42N 20 32 E
Königslutter, Germany 26 C6 52 15N 10 49 E
Königswusterhausen, Germany 26 C9 52 19N 13 38 E
Konin, Poland 47 C5 52 12N 18 15 E
Konin □, Poland 47 C5 52 15N 18 30 E
Konispoli, Albania 44 E2 39 42N 20 10 E
Kónitsa, Greece 44 D2 40 5N 20 48 E
Köniz, Switz. 28 C4 46 56N 7 25 E
Konjic, Bos.-H. 42 D2 43 42N 17 58 E
Konjice, Slovenia 39 B12 46 20N 15 28 E
Konkiep, Namibia 104 D2 26 49 S 17 15 E
Konkouré →, Guinea 100 C2 9 50N 13 42W
Könnern, Germany 26 D7 51 41N 11 47 E
Konnur, India 83 F2 16 14N 74 49 E
Kono, S. Leone 100 D2 8 30N 11 5W
Konolfingen, Switz. 28 C5 46 54N 7 38 E
Konongo, Ghana 101 D4 6 40N 1 15W
Konos, Papua N. G. 120 B6 3 10 S 151 44 E
Konosha, Russia 50 B11 61 0N 40 5 E
Kōnosu, Japan 63 A11 36 3N 139 31 E
Konotop, Ukraine 51 G7 51 12N 33 7 E
Końskie, Poland 47 D7 51 15N 20 23 E
Konstantinovka = Kostyantynivka, Ukraine 51 H9 48 32N 37 39 E
Konstantinovsk, Russia 53 G5 47 33N 41 10 E
Konstantynów Łódźki, Poland 47 D6 51 45N 19 20 E
Konstanz, Germany 27 H5 47 40N 9 10 E
Kont, Iran 85 E9 26 55N 61 50 E
Kontagora, Nigeria 101 C6 10 23N 5 27 E
Kontich, Belgium 21 F4 51 8N 4 26 E
Kontum, Vietnam 76 E7 14 24N 108 0 E
Kontum, Plateau du, Vietnam 76 E7 14 30N 108 30 E
Konya, Turkey 88 D5 37 52N 32 35 E
Konya Ovası, Turkey 88 C5 38 9N 33 5 E
Konyin, Burma 78 D5 22 58N 94 42 E
Konz, Germany 27 F2 49 42N 6 34 E
Konza, Kenya 106 C4 1 45 S 37 7 E
Konzhakovskiy Kamen, Gora, Russia 54 B7 59 38N 59 8 E
Kookynie, Australia 113 E3 29 17 S 121 22 E
Kooline, Australia 112 D2 22 57 S 116 20 E
Kooloonong, Australia 116 C5 34 48 S 143 10 E
Koolyanobbing, Australia 113 F2 30 48 S 119 36 E
Koondrook, Australia 116 C6 35 33 S 144 8 E
Koonibba, Australia 115 E1 31 54 S 133 25 E
Koorawatha, Australia 117 C8 34 2 S 148 33 E
Koorda, Australia 113 F2 30 48 S 117 35 E
Kooskia, U.S.A. 142 C6 46 9N 115 59W
Kootenai →, Canada 142 B5 49 15N 117 39W
Kootenay L., Canada 130 D5 49 45N 116 50W
Kootenay Nat. Park, Canada 130 C5 51 0N 116 0W
Kootjieskolk, S. Africa 104 E3 31 15 S 20 21 E
Kopa, Kazakhstan 55 B7 43 31N 75 50 E
Kopanovka, Russia 53 G8 47 28N 46 50 E
Kopaonik, Serbia, Yug. 42 D6 43 10N 20 50 E
Kopargaon, India 82 E2 19 51N 74 28 E
Kópavogur, Iceland 12 D3 64 6N 21 55W
Koper, Slovenia 39 C10 45 31N 13 44 E
Kopervik, Norway 13 G11 59 17N 5 17 E
Kopeysk, Russia 54 D8 55 7N 61 37 E
Kopi, Australia 115 E2 33 24 S 135 40 E
Köping, Sweden 14 E10 59 31N 16 3 E
Kopiste, Croatia 39 F13 42 48N 16 42 E
Kopliku, Albania 44 B1 42 15N 19 25 E
Köpmanholmen, Sweden 14 A12 63 10N 18 35 E
Koppal, India 83 G3 15 23N 76 5 E
Koppang, Norway 14 C5 61 34N 11 3 E
Kopperå, Norway 14 A5 63 24N 11 50 E
Koppies, S. Africa 105 D4 27 20 S 27 30 E
Koppio, Australia 116 C1 34 26 S 135 51 E
Koprivlen, Bulgaria 43 F8 41 36N 23 53 E
Koprivnica, Croatia 39 B13 46 12N 16 45 E
Koprivshtitsa, Bulgaria 43 E9 42 40N 24 19 E
Kopychyntsi, Ukraine 51 H3 49 7N 25 58 E
Korab, Macedonia 42 F5 41 44N 20 40 E
Korakiána, Greece 32 A3 39 42N 19 45 E
Koraput, India 82 E6 18 50N 82 40 E
Korba, India 81 H10 22 20N 82 45 E
Korbach, Germany 26 D4 51 16N 8 52 E
Korbu, G., Malaysia 77 K3 4 41N 101 18 E
Korça, Albania 44 D2 40 37N 20 50 E
Korça □, Albania 44 D2 40 40N 20 50 E
Korce = Korça, Albania 44 D2 40 37N 20 50 E
Korčula, Croatia 39 F14 42 56N 16 57 E
Korčulanski Kanal, Croatia 39 E13 43 3N 16 40 E
Kord Kūy, Iran 85 B7 36 48N 54 7 E
Kord Sheykh, Iran 85 D7 28 31N 52 53 E
Kordestān □, Iran 84 C5 36 0N 47 0 E
Korea, North ■, Asia 67 E14 40 0N 127 0 E
Korea, South ■, Asia 67 F15 36 0N 128 0 E
Korea Bay, Korea 67 E13 39 0N 124 0 E
Korea Strait, Asia 67 G15 34 0N 129 30 E
Koregaon, India 82 F2 17 40N 74 10 E
Korenevo, Russia 52 E2 51 27N 34 55 E
Korenovsk, Russia 53 H4 45 30N 39 22 E
Korets, Ukraine 51 G4 50 40N 27 5 E
Korgan, Turkey 88 B7 40 44N 37 13 E
Korgus, Sudan 94 D3 19 16N 33 29 E
Korhogo, Ivory C. 100 D3 9 29N 5 28W
Koribundu, S. Leone 100 D2 7 41N 11 46W
Korim, Indonesia 73 B5 0 58 S 136 10 E
Korinthía □, Greece 45 G4 37 50N 22 35 E
Korinthiakós Kólpos, Greece 45 F4 38 16N 22 30 E
Kórinthos, Greece 45 G4 37 56N 22 55 E
Korioumé, Mali 100 B4 16 35N 3 0W
Koríssa, Límni, Greece 32 B3 39 27N 19 53 E
Kōriyama, Japan 60 F10 37 24N 140 23 E
Korkino, Russia 54 D8 54 54N 61 23 E
Korkuteli, Turkey 88 D4 37 2N 30 11 E
Korla, China 64 B3 41 45N 86 4 E
Kormakiti, C., Cyprus 32 D11 35 23N 32 56 E
Körmend, Hungary 31 D9 47 5N 16 35 E
Kornat, Croatia 39 E12 43 50N 15 20 E
Korneshty = Cornești, Moldova 51 J5 47 21N 28 1 E
Korneuburg, Austria 31 C9 48 20N 16 20 E
Kornsjø, Norway 14 F5 58 57N 11 39 E
Kornstad, Norway 14 B1 62 59N 7 27 E
Koro, Fiji 121 A2 17 19 S 179 23 E
Koro, Ivory C. 100 D3 8 32N 7 30W
Koro, Mali 100 C4 14 1N 2 58W
Koro Sea, Fiji 121 A3 17 30 S 179 45W
Koro Toro, Chad 97 E3 16 5N 18 30 E
Koroba, Papua N. G. 120 C2 5 44 S 142 47 E
Korocha, Russia 52 E3 50 54N 37 19 E
Köroğlu Dağları, Turkey 88 B5 40 38N 33 0 E
Korogwe, Tanzania 106 D4 5 5S 38 25 E
Korogwe □, Tanzania 106 D4 5 0S 38 20 E
Koroit, Australia 116 E5 38 18 S 142 24 E
Koronadal, Phil. 71 H5 6 12N 125 1 E
Korong Vale, Australia 116 D5 36 22 S 143 45 E
Koróni, Greece 45 H3 36 48N 21 57 E
Korónia, Limni, Greece 44 D5 40 47N 23 37 E
Koronís, Greece 45 G7 37 12N 25 35 E
Koronowo, Poland 47 B4 53 19N 17 55 E
Körös →, Hungary 31 E13 46 43N 20 12 E
Körösterssa, Hungary 31 E14 46 53N 21 3 E
Korosten, Ukraine 51 G5 50 54N 28 36 E
Korostyshev, Ukraine 51 G5 50 19N 29 4 E
Korotoyak, Russia 52 E4 51 1N 39 2 E
Korraraika, Helodranon' i, Madag. 105 B7 17 45 S 43 57 E
Korsakov, Russia 57 E15 46 36N 142 42 E
Korshunovo, Russia 57 D12 58 37N 110 10 E
Korsør, Denmark 13 J14 55 20N 11 9 E
Korsun Shevchenkovskiy, Ukraine 51 H6 49 26N 31 16 E
Korsze, Poland 47 A8 54 11N 21 9 E
Kortemark, Belgium 21 F2 51 2N 3 3 E
Kortessem, Belgium 21 G6 50 52N 5 23 E
Korti, Sudan 94 D3 18 6N 31 33 E
Kortrijk, Belgium 21 G2 50 50N 3 17 E
Korumburra, Australia 117 E6 38 26 S 145 50 E
Korwai, India 80 G8 24 7N 78 5 E
Koryakskoye Nagorye, Russia 57 C18 61 0N 171 0 E
Koryŏng, S. Korea 67 G15 35 44N 128 15 E
Koryukovka, Ukraine 51 G7 51 46N 32 16 E
Kos, Greece 45 H9 36 50N 27 15 E
Kosa, Ethiopia 95 F4 7 50N 36 50 E
Kosa, Russia 54 B5 59 56N 55 0 E
Kosa →, Russia 54 A5 60 11N 55 10 E
Kosaya Gora, Russia 52 C3 54 10N 37 30 E
Koschagyl, Kazakhstan 49 E9 46 40N 54 0 E
Kościan, Poland 47 C3 52 5N 16 40 E
Kościerzyna, Poland 47 A4 54 8N 17 59 E
Kosciusko, U.S.A. 139 J10 33 4N 89 35W
Kosciusko, Mt., Australia 117 D8 36 27 S 148 16 E
Kosciusko I., U.S.A. 130 B2 56 0N 133 40W
Kösély →, Hungary 31 D14 47 25N 21 5 E
Kosgi, India 82 F3 16 58N 77 43 E
Kosha, Sudan 94 C3 20 50N 30 30 E
Koshigaya, Japan 63 B11 35 54N 139 48 E
K'oshih = Kashi, China 64 C2 39 30N 76 2 E
Koshiki-Rettō, Japan 62 F1 31 45N 129 49 E
Koshkonong L., U.S.A. 141 B8 42 52N 88 58W
Kōshoku, Japan 63 A10 36 38N 138 6 E
Koshtëbë, Kyrgyzstan 55 C7 41 5N 74 15 E
Kosi, India 80 F7 27 48N 77 29 E
Košice, Slovak Rep. 31 C14 48 42N 21 15 E
Kosjerić, Serbia, Yug. 42 D4 44 0N 19 55 E
Koskhinoú, Greece 32 C10 36 23N 28 13 E
Koslan, Russia 48 B8 63 34N 49 14 E
Kosŏng, N. Korea 67 E15 38 40N 128 22 E
Kosovo □, Serbia, Yug. 42 E5 42 30N 21 0 E
Kosovska-Mitrovica = Titova-Mitrovica, Serbia, Yug. 42 E5 42 54N 20 52 E
Kostajnica, Croatia 39 C13 45 17N 16 30 E
Kostamuksa, Russia 48 B5 62 34N 32 44 E
Kostanjevica, Slovenia 39 C12 45 51N 15 27 E
Kostelec, Czech. 31 A9 50 14N 16 35 E
Kostenets, Bulgaria 43 E8 42 15N 23 52 E
Koster, S. Africa 104 D4 25 52 S 26 54 E
Kôstī, Sudan 95 E3 13 8N 32 43 E
Kostolac, Serbia, Yug. 42 C6 44 37N 21 15 E
Kostopil, Ukraine 51 G4 50 51N 26 22 E
Kostroma, Russia 52 B5 57 50N 40 58 E
Kostromskoye Vdkhr., Russia 52 B5 57 52N 40 49 E
Kostrzyn, Gorzow Wlkp., Poland 47 C1 52 35N 14 39 E
Kostrzyn, Poznań, Poland 47 C4 52 24N 17 14 E
Kostyantynivka, Ukraine 51 H9 48 32N 37 39 E
Kostyukovichi = Kastsyukovichy, Belarus 50 F7 53 20N 32 4 E
Koszalin, Poland 47 A3 54 11N 16 8 E
Koszalin □, Poland 47 B3 53 40N 16 10 E
Köszeg, Hungary 31 D9 47 23N 16 33 E
Kot Addu, Pakistan 79 C3 30 30N 71 0 E
Kot Moman, Pakistan 80 C5 32 13N 73 0 E
Kota, India 80 G6 25 14N 75 49 E
Kota Baharu, Malaysia 77 J4 6 7N 102 14 E
Kota Belud, Malaysia 75 A5 6 21N 116 26 E
Kota Kinabalu, Malaysia 75 A5 6 0N 116 4 E
Kota Tinggi, Malaysia 77 M4 1 44N 103 53 E
Kotaagung, Indonesia 74 D2 5 38 S 104 29 E
Kotabaru, Indonesia 75 C5 3 20 S 116 20 E
Kotabumi, Indonesia 74 C2 4 49 S 104 54 E
Kotagede, Indonesia 75 D4 7 54 S 110 26 E
Kotamobagu, Indonesia 72 A2 0 57N 124 31 E
Kotaneelee →, Canada 130 A4 60 11N 123 42W
Kotawaringin, Indonesia 75 C4 2 28 S 111 27 E
Kotchandpur, Bangla. 78 D2 23 24N 89 1 E
Kotcho L., Canada 130 B4 59 7N 121 12W
Kotel, Bulgaria 43 E11 42 52N 26 26 E
Kotelnich, Russia 52 A9 58 22N 48 24 E
Kotelnikovo, Russia 53 G6 47 38N 43 8 E
Kotelnyy, Ostrov, Russia 57 B14 75 10N 139 0 E
Kothagudam, India 82 F5 17 30N 80 40 E
Kothapet, India 82 E4 19 21N 79 28 E
Köthen, Germany 26 D7 51 45N 11 59 E
Kothi, India 81 G9 24 45N 80 40 E
Kotiro, Pakistan 80 F2 26 17N 67 13 E
Kotka, Finland 13 F22 60 28N 26 58 E
Kotlas, Russia 48 B8 61 17N 46 43 E
Kotlenska Planina, Bulgaria 43 E11 42 56N 26 30 E
Kotli, Pakistan 80 C5 33 30N 73 55 E
Kotmul, Pakistan 81 B6 35 32N 75 10 E
Kotohira, Japan 62 C5 34 11N 133 49 E
Kotonkoro, Nigeria 101 C6 11 3N 5 58 E
Kotor, Montenegro, Yug. 42 E3 42 25N 18 47 E
Kotor Varoš, Bos.-H. 42 C2 44 38N 17 22 E
Kotoriba, Croatia 39 B13 46 23N 16 48 E
Kotovo, Russia 52 E7 50 22N 44 45 E
Kotovsk, Russia 52 D5 52 36N 41 32 E
Kotovsk, Ukraine 51 J5 47 45N 29 35 E
Kotputli, India 80 F7 27 43N 76 12 E
Kotri, Pakistan 79 D3 25 22N 68 22 E
Kotri →, India 82 E5 19 15N 80 35 E
Kótronas, Greece 45 H4 36 38N 22 29 E
Kötschach-Mauthen, Austria 30 E6 46 41N 13 1 E
Kottayam, India 83 K3 9 35N 76 33 E
Kottur, India 83 J3 10 34N 76 56 E
Kotuy →, Russia 57 B11 71 54N 102 6 E
Kotzebue, U.S.A. 126 B3 66 53N 162 39W
Kouango, C.A.R. 102 B4 5 0N 20 10 E
Koudekerke, Neths. 21 F3 51 29N 3 33 E
Koudougou, Burkina Faso 100 C4 12 10N 2 20W
Koufonisi, Greece 32 E8 34 56N 26 8 E
Koufonísia, Greece 45 H7 36 57N 25 35 E
Kougaberge, S. Africa 104 E3 33 48 S 23 50 E
Kouibli, Ivory C. 100 D3 7 15N 7 14W
Kouilou →, Congo 103 C2 4 10 S 12 5 E
Kouki, C.A.R. 102 A3 7 22N 17 3 E
Koula Moutou, Gabon 102 C2 1 15 S 12 25 E
Koulen, Cambodia 76 F5 13 50N 104 40 E
Koulikoro, Mali 100 C3 12 40N 7 50W
Kouloúra, Greece 32 A3 39 42N 19 54 E
Koúm-bournoú, Ákra, Greece 32 C10 36 15N 28 11 E
Koumac, N. Cal. 121 T18 20 33 S 164 17 E
Koumala, Australia 114 C4 21 38 S 149 15 E
Koumankou, Mali 100 C3 11 58N 6 6W
Koumbia, Burkina Faso 100 C4 11 10N 3 50W
Koumbia, Guinea 100 C2 11 48N 13 29W
Koumboum, Guinea 100 C2 10 25N 13 0W
Koumpenntoum, Senegal 100 C2 13 59N 14 34W
Koumra, Chad 97 G3 8 50N 17 35 E
Koundara, Guinea 100 C2 12 29N 13 18W
Koundé, C.A.R. 102 A2 6 7N 14 38 E
Kounradskiy, Kazakhstan 56 E8 46 59N 75 0 E
Kountze, U.S.A. 139 K7 30 22N 94 19W
Koupéla, Burkina Faso 101 C4 12 11N 0 21W
Kouris →, Cyprus 32 E11 34 38N 32 54 E
Kourizo, Passe de, Chad 96 D3 22 28N 15 27 E
Kourou, Fr. Guiana 153 B7 5 9N 52 39W
Kouroussa, Guinea 100 C3 10 45N 9 45W
Koussané, Mali 100 C2 14 53N 11 14W
Kousseri, Cameroon 97 F2 12 0N 14 55 E
Koutiala, Mali 100 C3 12 25N 5 23W
Kouto, Ivory C. 100 D3 9 53N 6 25W
Kouts, U.S.A. 141 C9 41 19N 87 2W
Kouvé, Togo 101 D5 6 25N 1 25 E
Kouvola, Finland 13 F22 60 52N 26 43 E
Kovačica, Serbia, Yug. 42 B5 45 5N 20 38 E
Kovdor, Russia 48 A5 67 34N 30 24 E
Kovel, Ukraine 51 G3 51 11N 24 38 E
Kovilpatti, India 83 K3 9 10N 77 50 E
Kovin, Serbia, Yug. 42 C5 44 44N 20 59 E
Kovrov, Russia 52 B5 56 25N 41 25 E
Kovur, Andhra Pradesh, India 82 F5 17 3N 81 39 E
Kovur, Andhra Pradesh, India 83 G5 14 30N 80 1 E
Kowal, Poland 47 C6 52 32N 19 7 E
Kowalewo Pomorskie, Poland 47 B5 53 10N 18 52 E
Kowanyama, Australia 114 B3 15 29 S 141 44 E
Kowghān, Afghan. 79 B1 34 12N 63 2 E
Kowkash, Canada 128 B2 50 20N 87 12W
Kowloon, H.K. 69 F10 22 20N 114 15 E
Kowŏn, N. Korea 67 E14 39 26N 127 14 E
Kōyama, Japan 62 F2 31 20N 130 56 E
Köyceğiz, Turkey 88 D3 36 57N 28 40 E
Koytash, Uzbekistan 55 C3 40 11N 67 19 E
Koyuk, U.S.A. 126 B3 64 56N 161 9W
Koyukuk →, U.S.A. 126 B4 64 55N 157 32W
Koyulhisar, Turkey 88 B7 40 20N 37 52 E
Koza, Japan 61 L3 26 19N 127 46 E
Kozan, Turkey 88 D6 37 26N 35 50 E
Kozáni, Greece 44 D3 40 19N 21 47 E
Kozáni □, Greece 44 D3 40 18N 21 45 E
Kozara, Bos.-H. 39 D14 45 0N 17 0 E
Kozarac, Bos.-H. 39 D13 44 58N 16 48 E
Kozelets, Ukraine 51 G6 50 55N 31 7 E
Kozelsk, Russia 52 C2 54 2N 35 48 E
Kozhikode = Calicut, India 83 J2 11 15N 75 43 E
Kozhva, Russia 48 A10 65 10N 57 0 E
Koziegłowy, Poland 47 E6 50 37N 19 8 E
Kozienice, Poland 47 D8 51 35N 21 34 E
Kozje, Slovenia 39 B12 46 5N 15 35 E
Kozle, Poland 47 E5 50 20N 18 8 E
Kozloduy, Bulgaria 43 D8 43 45N 23 42 E
Kozlovets, Bulgaria 43 D10 43 30N 25 20 E
Kozlovka, Russia 52 C9 55 52N 48 14 E
Kozlu, Turkey 88 B4 41 26N 31 45 E
Kozluk, Turkey 89 C9 38 11N 41 31 E
Koźmin, Poland 47 D4 51 48N 17 27 E
Kozmodemyansk, Russia 52 B8 56 20N 46 36 E
Kōzu-Shima, Japan 63 C11 34 13N 139 10 E
Kozuchów, Poland 47 D2 51 45N 15 31 E
Kozyatyn, Ukraine 51 H5 49 45N 28 50 E
Kpabia, Ghana 101 D4 9 10N 0 20W
Kpalimé, Togo 101 D5 6 57N 0 44 E
Kpandae, Ghana 101 D4 8 30N 0 2W
Kpessi, Togo 101 D5 8 4N 1 16 E
Kra, Isthmus of = Kra, Kho Khot, Thailand 77 G2 10 15N 99 30 E
Kra, Kho Khot, Thailand 77 G2 10 15N 99 30 E
Kra Buri, Thailand 77 G2 10 22N 98 46 E
Krabbendijke, Neths. 21 F4 51 26N 4 7 E
Krabi, Thailand 77 H2 8 4N 98 55 E
Kragan, Indonesia 75 D4 6 43 S 111 38 E
Kragerø, Norway 14 F3 58 52N 9 25 E
Kragujevac, Serbia, Yug. 42 C5 44 2N 20 56 E
Krajenka, Poland 47 B3 53 18N 16 59 E
Krajina, Bos.-H. 39 D13 44 45N 16 35 E
Krakatau = Rakata, Pulau, Indonesia 74 D3 6 10 S 105 20 E
Krakor, Cambodia 76 F5 12 32N 104 12 E
Kraków, Poland 31 A12 50 4N 19 57 E
Kraków □, Poland 31 A13 50 0N 20 0 E
Kraksaan, Indonesia 75 D4 7 43 S 113 23 E
Kråkstad, Norway 14 E4 59 39N 10 55 E
Kralanh, Cambodia 76 F4 13 35N 103 25 E
Králiky, Czech. 31 A9 50 6N 16 45 E
Kraljevo, Serbia, Yug. 42 D5 43 44N 20 41 E
Kralovice, Czech. 30 B6 49 59N 13 29 E
Královský Chlmec, Slovak Rep. 31 C14 48 27N 22 0 E
Kralupy, Czech. 30 A7 50 13N 14 20 E
Kramatorsk, Ukraine 51 H9 48 50N 37 30 E
Kramfors, Sweden 14 B11 62 55N 17 48 E
Kramis, C., Algeria 99 A5 36 26N 0 45 E
Krångede, Sweden 14 A10 63 9N 16 10 E
Kraniá, Greece 44 E3 39 53N 21 18 E
Kranídhion, Greece 45 G5 37 20N 23 10 E
Kranj, Slovenia 39 B11 46 16N 14 22 E
Kranjska Gora, Slovenia 39 B10 46 29N 13 48 E
Krankskop, S. Africa 105 D5 28 0S 30 47 E
Krapina, Croatia 39 B12 46 10N 15 52 E
Krapina →, Croatia 39 C12 45 50N 15 50 E
Krapkowice, Poland 47 E4 50 29N 17 56 E
Krasavino, Russia 48 B8 60 58N 46 29 E
Krashyy Klyuch, Russia 54 D6 55 23N 56 39 E
Kraskino, Russia 57 E14 42 44N 130 48 E
Krāslava, Latvia 13 J22 55 54N 27 10 E
Kraslice, Czech. 30 A5 50 19N 12 31 E
Krasnaya Gorbatka, Russia 52 C5 55 52N 41 45 E
Krasnaya Polyana, Russia 53 J5 43 40N 40 13 E
Kraśnik, Poland 47 E9 50 55N 22 5 E
Kraśnik Fabryczny, Poland 47 E9 50 58N 22 11 E

Labytnangi 221

Kushka = Gushgy, *Turkmenistan* **56 F7** 35 20N 62 18 E
Kūshkī, *Īlām, Iran* **84 C5** 33 31N 47 13 E
Kūshkī, *Khorāsān, Iran* **85 B8** 37 2N 57 26 E
Kūshkū, *Iran* **85 E7** 27 19N 53 28 E
Kushnarenkovo, *Russia* **54 D5** 55 6N 55 22 E
Kushol, *India* **81 C7** 33 40N 76 36 E
Kushrabat, *Uzbekistan* **55 C3** 40 18N 66 32 E
Kushtia, *Bangla.* **78 D2** 23 55N 89 5 E
Kushum →, *Kazakhstan* **52 F10** 49 20N 50 30 E
Kushva, *Russia* **54 B7** 58 18N 59 45 E
Kuskokwim →, *U.S.A.* **126 B3** 60 5N 162 25W
Kuskokwim B., *U.S.A.* **126 C3** 59 45N 162 25W
Küsnacht, *Switz.* **29 B7** 47 19N 8 35 E
Kussharo-Ko, *Japan* **60 C12** 43 38N 144 21 E
Küssnacht, *Switz.* **29 B6** 47 5N 8 26 E
Kustanay = Qostanay, *Kazakhstan* **56 D7** 53 10N 63 35 E
Kusu, *Japan* **62 D3** 33 16N 131 9 E
Kut, Ko, *Thailand* **77 G4** 11 40N 102 35 E
Kutacane, *Indonesia* **74 B1** 3 50N 97 50 E
Kütahya, *Turkey* **88 C4** 39 30N 30 2 E
Kutaisi, *Georgia* **53 J6** 42 19N 42 40 E
Kutaraja = Banda Aceh, *Indonesia* **74 A1** 5 35N 95 20 E
Kutch, Gulf of = Kachchh, Gulf of, *India* **80 H3** 22 50N 69 15 E
Kutch, Rann of = Kachchh, Rann of, *India* **80 G4** 24 0N 70 0 E
Kutina, *Croatia* **39 C13** 45 29N 16 48 E
Kutiyana, *India* **80 J4** 21 36N 70 2 E
Kutjevo, *Croatia* **42 B2** 45 23N 17 55 E
Kutkai, *Burma* **78 D6** 23 27N 97 56 E
Kutkashen, *Azerbaijan* **53 K8** 40 58N 47 47 E
Kutná Hora, *Czech.* **30 B8** 49 57N 15 16 E
Kutno, *Poland* **47 C6** 52 15N 19 23 E
Kuttabul, *Australia* **114 C4** 21 5 S 148 48 E
Kutu, *Zaïre* **102 C3** 2 40 S 18 11 E
Kutum, *Sudan* **95 E1** 14 10N 24 40 E
Kúty, *Slovak Rep.* **31 C10** 48 40N 17 3 E
Kuujjuaq, *Canada* **127 C13** 58 6N 68 15W
Kuŭp-tong, *N. Korea* **67 D14** 40 45N 126 1 E
Kuurne, *Belgium* **21 G2** 50 51N 3 18 E
Kuusamo, *Finland* **12 D23** 65 57N 29 8 E
Kuusankoski, *Finland* **13 F22** 60 55N 26 38 E
Kuvandyk, *Russia* **54 F6** 51 28N 57 21 E
Kuvango, *Angola* **103 E3** 14 28 S 16 20 E
Kuvasay, *Uzbekistan* **55 C5** 40 18N 71 59 E
Kuvshinovo, *Russia* **52 B2** 57 2N 34 11 E
Kuwait = Al Kuwayt, *Kuwait* **84 D5** 29 30N 48 0 E
Kuwait ■, *Asia* **84 D5** 29 30N 47 30 E
Kuwana, *Japan* **63 B8** 35 5N 136 43 E
Kuybyshev = Samara, *Russia* **52 D10** 53 8N 50 6 E
Kuybyshev, *Russia* **56 D8** 55 27N 78 19 E
Kuybyshevo, *Ukraine* **51 J9** 47 25N 36 40 E
Kuybyshevo, *Uzbekistan* **55 C5** 40 20N 71 15 E
Kuybyshevskiy, *Tajikistan* **55 E4** 37 52N 68 44 E
Kuybyshevskoye Vdkhr., *Russia* **52 C9** 55 2N 49 30 E
Kuye He →, *China* **66 E6** 38 23N 110 46 E
Kūyeh, *Iran* **84 B5** 38 45N 47 57 E
Kuylyuk, *Uzbekistan* **55 C4** 41 14N 69 17 E
Kūysanjaq, *Iraq* **89 D11** 36 5N 44 38 E
Kuyto, Ozero, *Russia* **48 B5** 65 6N 31 20 E
Kuyumba, *Russia* **57 C10** 60 58N 96 59 E
Kuzey Anadolu Dağları, *Turkey* **88 B7** 41 30N 35 0 E
Kuzhitturai, *India* **83 K3** 8 18N 77 11 E
Kuzino, *Russia* **54 C7** 57 1N 59 27 E
Kuzmin, *Serbia, Yug.* **42 B4** 45 2N 19 25 E
Kuznetsk, *Russia* **52 D8** 53 12N 46 40 E
Kuzomen, *Russia* **48 A6** 66 22N 36 50 E
Kvænangen, *Norway* **12 A19** 70 5N 21 15 E
Kvaløy, *Norway* **12 B18** 69 40N 18 30 E
Kvam, *Norway* **14 C3** 61 40N 9 42 E
Kvareli = Qvareli, *Georgia* **53 K7** 41 57N 45 47 E
Kvarner, *Croatia* **39 D11** 44 50N 14 10 E
Kvarnerič, *Croatia* **39 D11** 44 43N 14 37 E
Kviteseid, *Norway* **14 E2** 59 24N 8 29 E
Kwabhaca, *S. Africa* **105 E4** 30 51 S 29 0 E
Kwadacha →, *Canada* **130 B3** 57 28N 125 38W
Kwakhanai, *Botswana* **104 C3** 21 39 S 21 16 E
Kwakoegron, *Surinam* **153 B6** 5 12N 55 25W
Kwale, *Kenya* **106 C4** 4 15 S 39 31 E
Kwale, *Nigeria* **101 D6** 5 46N 6 26 E
Kwale □, *Kenya* **106 C4** 4 15 S 39 10 E
KwaMashu, *S. Africa* **105 D5** 29 45 S 30 58 E
Kwamouth, *Zaïre* **102 C3** 3 9 S 16 12 E
Kwando →, *Africa* **103 F4** 18 27 S 23 32 E
Kwangdaeri, *N. Korea* **67 D14** 40 31N 127 32 E
Kwangju, *S. Korea* **67 G14** 35 9N 126 54 E
Kwango →, *Zaïre* **102 C3** 3 14 S 17 22 E
Kwangsi-Chuang = Guangxi Zhuangzu Zizhiqu □, *China* **68 E7** 24 0N 109 0 E
Kwangtung = Guangdong □, *China* **69 F9** 23 0N 113 0 E
Kwara □, *Nigeria* **101 D5** 8 45N 4 30 E
Kwataboahegan →, *Canada* **128 B3** 51 9N 80 50W
Kwatisore, *Indonesia* **73 B4** 3 18 S 134 50 E
KwaZulu Natal □, *S. Africa* **105 D5** 29 0 S 30 0 E
Kweichow = Guizhou □, *China* **68 D6** 27 0N 107 0 E
Kwekwe, *Zimbabwe* **107 F2** 18 58 S 29 48 E
Kwidzyn, *Poland* **47 B5** 53 44N 18 55 E
Kwikila, *Papua N. G.* **120 E4** 9 49 S 147 38 E
Kwimba □, *Tanzania* **106 C3** 3 0 S 33 0 E
Kwinana New Town, *Australia* **113 F2** 32 15 S 115 47 E
Kwisa →, *Poland* **47 D2** 51 34N 15 24 E
Kwoka, *Indonesia* **73 B4** 0 31 S 132 27 E
Kya-in-Seikkyi, *Burma* **78 G7** 16 2N 98 8 E
Kyabé, *Chad* **97 G3** 9 30N 19 0 E
Kyabra Cr. →, *Australia* **115 D3** 25 36 S 142 55 E
Kyabram, *Australia* **115 F4** 36 19 S 145 4 E
Kyaiklat, *Burma* **78 G5** 16 25N 95 40 E
Kyaikmaraw, *Burma* **78 G6** 16 23N 97 44 E
Kyaikthin, *Burma* **78 D5** 23 32N 95 40 E
Kyaikto, *Burma* **76 D1** 17 20N 97 3 E
Kyakhta, *Russia* **57 D11** 50 30N 106 25 E
Kyancutta, *Australia* **115 E2** 33 8 S 135 33 E
Kyangin, *Burma* **78 F5** 18 20N 95 20 E
Kyaukhnyat, *Burma* **78 F6** 18 15N 97 31 E
Kyaukse, *Burma* **78 E6** 21 36N 96 10 E
Kyauktaw, *Burma* **78 E4** 20 51N 92 59 E
Kyawkku, *Burma* **78 E6** 21 48N 96 56 E
Kyburz, *U.S.A.* **144 G6** 38 47N 120 18W
Kybybolite, *Australia* **116 D4** 36 53 S 140 55 E
Kyeintali, *Burma* **78 G5** 18 0N 94 29 E
Kyenjojo, *Uganda* **106 B3** 0 40N 30 37 E
Kyidaunggan, *Burma* **78 F6** 19 53N 96 12 E
Kyle Dam, *Zimbabwe* **107 G3** 20 15 S 31 0 E
Kyle of Lochalsh, *U.K.* **18 D3** 57 17N 5 44W
Kyll →, *Germany* **27 F2** 49 48N 6 41 E
Kyllburg, *Germany* **27 E2** 50 2N 6 34 E
Kymijoki →, *Finland* **13 F22** 60 30N 26 55 E
Kyneton, *Australia* **116 D6** 37 10 S 144 29 E
Kynuna, *Australia* **114 C3** 21 37 S 141 55 E
Kyō-ga-Saki, *Japan* **63 B7** 35 45N 135 15 E
Kyoga, L., *Uganda* **106 B3** 1 35N 33 0 E
Kyogle, *Australia* **115 D5** 28 40 S 153 0 E
Kyongju, *S. Korea* **67 G15** 35 51N 129 14 E
Kyŏngsŏng, *N. Korea* **67 D15** 41 35N 129 36 E
Kyōto, *Japan* **63 B7** 35 0N 135 45 E
Kyōto □, *Japan* **63 B7** 35 15N 135 45 E
Kyparissovouno, *Cyprus* **32 D12** 35 19N 33 10 E
Kyperounda, *Cyprus* **32 E11** 34 56N 32 58 E
Kyren, *Russia* **57 D11** 51 45N 101 45 E
Kyrenia, *Cyprus* **32 D12** 35 20N 33 20 E
Kyrgyzstan ■, *Asia* **55 C7** 42 0N 75 0 E
Kyritz, *Germany* **26 C8** 52 56N 12 24 E
Kyrönjoki →, *Finland* **12 E19** 63 14N 21 45 E
Kyshtym, *Russia* **54 D8** 55 42N 60 34 E
Kystatyam, *Russia* **57 C13** 67 20N 123 10 E
Kytal Ktakh, *Russia* **57 C13** 65 30N 123 40 E
Kythréa, *Cyprus* **32 D12** 35 15N 33 29 E
Kytlym, *Russia* **54 B7** 59 30N 59 12 E
Kyu-hkok, *Burma* **78 C7** 24 4N 98 4 E
Kyulyunken, *Russia* **57 C14** 64 10N 137 5 E
Kyunhla, *Burma* **78 D5** 23 25N 95 15 E
Kyuquot, *Canada* **130 C3** 50 3N 127 25W
Kyurdamir = Kürdämir, *Azerbaijan* **53 K9** 40 25N 48 3 E
Kyūshū, *Japan* **62 E3** 33 0N 131 0 E
Kyūshū □, *Japan* **62 E3** 33 0N 131 0 E
Kyūshū-Sanchi, *Japan* **62 E3** 32 35N 131 17 E
Kyustendil, *Bulgaria* **42 E7** 42 16N 22 41 E
Kyusyur, *Russia* **57 B13** 70 19N 127 30 E
Kywong, *Australia* **117 C7** 34 58 S 146 44 E
Kyyiv, *Ukraine* **51 G6** 50 30N 30 28 E
Kyyivske Vdskh., *Ukraine* **51 G6** 51 0N 30 25 E
Kyzyl, *Russia* **57 D10** 51 50N 94 30 E
Kyzyl-Kyya, *Kyrgyzstan* **55 C6** 40 16N 72 8 E
Kyzylkum, Peski, *Uzbekistan* **55 B2** 42 30N 65 0 E
Kyzylsu →, *Kyrgyzstan* **55 D6** 38 50N 70 0 E
Kzyl-Orda = Qyzylorda, *Kazakhstan* **55 A2** 44 48N 65 28 E

L

La Albuera, *Spain* **37 G4** 38 45N 6 49W
La Albufera, *Spain* **35 F4** 39 20N 0 27W
La Alcarria, *Spain* **34 E2** 40 31N 2 45W
La Algaba, *Spain* **37 H4** 37 27N 6 1W
La Almarcha, *Spain* **34 F2** 39 41N 2 24W
La Almunia de Doña Godina, *Spain* **34 D3** 41 29N 1 23W
La Asunción, *Venezuela* **153 A5** 11 2N 63 53W
La Banda, *Argentina* **158 B3** 27 45 S 64 10W
La Bañeza, *Spain* **36 C5** 42 17N 5 54W
La Barca, *Mexico* **146 C4** 20 20N 102 40W
La Barge, *U.S.A.* **142 E8** 42 16N 110 12W
La Bassée, *France* **23 B9** 50 31N 2 49 E
La Bastide-Puylaurent, *France* **24 D7** 44 35N 3 55 E
La Baule, *France* **22 E4** 47 17N 2 24W
La Belle, *Fla., U.S.A.* **135 M5** 26 46N 81 26W
La Belle, *Mo., U.S.A.* **140 D5** 40 7N 91 55W
La Biche →, *Canada* **130 B4** 59 57N 123 50W
La Bisbal, *Spain* **34 D8** 41 58N 3 2 E
La Blanquilla, *Venezuela* **153 A5** 11 51N 64 37W
La Bomba, *Mexico* **146 A1** 31 53N 115 2W
La Bresse, *France* **23 D13** 48 2N 6 53 E
La Bureba, *Spain* **34 C1** 42 36N 3 24W
La Cal →, *Bolivia* **157 D6** 17 27 S 58 15W
La Calera, *Chile* **158 C1** 32 50 S 71 10W
La Campiña, *Spain* **37 H6** 37 45N 4 45W
La Canal, *Spain* **33 C7** 38 51N 1 23 E
La Cañiza, *Spain* **36 C2** 42 13N 8 16W
La Capelle, *France* **23 C10** 49 59N 3 50 E
La Carlota, *Argentina* **158 C3** 33 30 S 63 20W
La Carlota, *Phil.* **71 F4** 10 25N 122 55 E
La Carolina, *Spain* **37 G7** 38 17N 3 38W
La Castellana, *Phil.* **71 F4** 10 20N 123 3 E
La Cavalerie, *France* **24 D7** 44 1N 3 10 E
La Ceiba, *Honduras* **148 C2** 15 40N 86 50W
La Chaise-Dieu, *France* **24 C7** 45 18N 3 42 E
La Chaize-le-Vicomte, *France* **22 F5** 46 40N 1 18W
La Chapelle d'Angillon, *France* **23 E9** 47 21N 2 25 E
La Chapelle-Glain, *France* **22 E5** 47 38N 1 11W
La Charité-sur-Loire, *France* **23 E10** 47 10N 3 1 E
La Chartre-sur-le-Loir, *France* **22 E7** 47 44N 0 34 E
La Châtaigneraie, *France* **24 B3** 46 39N 0 44W
La Châtre, *France* **24 B5** 46 35N 2 0 E
La Chaux de Fonds, *Switz.* **28 B3** 47 7N 6 50 E
La Chorrera, *Colombia* **152 D3** 0 44 S 73 1W
La Ciotat, *France* **25 E9** 43 10N 5 37 E
La Clayette, *France* **25 B8** 46 17N 4 19 E
La Cocha, *Argentina* **158 B2** 27 50 S 65 40W
La Concepción = Ri-Aba, *Eq. Guin.* **101 E6** 3 28N 8 40 E
La Concepción, *Venezuela* **152 A3** 10 30N 71 50W
La Concordia, *Mexico* **147 D6** 16 8N 92 38W
La Conner, *U.S.A.* **142 B2** 48 23N 122 30W
La Coruña, *Spain* **36 B2** 43 20N 8 25W
La Coruña □, *Spain* **36 B2** 43 10N 8 30W
La Côte, *Switz.* **28 D2** 46 25N 6 15 E
La Côte-St.-André, *France* **25 C9** 45 24N 5 15 E
La Courtine-le-Trucq, *France* **24 C6** 45 41N 2 15 E
La Crau, *France* **25 E8** 43 32N 4 40 E
La Crete, *Canada* **130 B5** 58 11N 116 24W
La Crosse, *Kans., U.S.A.* **138 F5** 38 32N 99 18W
La Crosse, *Wis., U.S.A.* **138 D9** 43 48N 91 15W
La Cruz, *Costa Rica* **148 D2** 11 4N 85 39W
La Cruz, *Mexico* **146 C3** 23 55N 106 54W
La Dorada, *Colombia* **152 B3** 5 30N 74 40W
La Ensenada, *Chile* **160 B2** 41 12 S 72 33W
La Escondida, *Mexico* **146 C5** 24 6N 99 55W
La Esmeralda, *Paraguay* **158 A3** 22 16 S 62 33W
La Esperanza, *Argentina* **160 B3** 40 26 S 68 32W
La Esperanza, *Cuba* **148 B3** 22 46N 83 44W
La Esperanza, *Honduras* **148 D2** 14 15N 88 10W
La Estrada, *Spain* **36 C2** 42 43N 8 27W
La Fayette, *U.S.A.* **135 H3** 34 42N 85 17W
La Fé, *Cuba* **148 B3** 22 2N 84 15W
La Fère, *France* **23 C10** 49 39N 3 21 E
La Ferté-Bernard, *France* **22 D7** 48 10N 0 40 E
La Ferté-Macé, *France* **22 D6** 48 35N 0 22W
La Ferté-St.-Aubin, *France* **23 E8** 47 42N 1 57 E
La Ferté-sous-Jouarre, *France* **23 D10** 48 56N 3 8 E
La Ferté-Vidame, *France* **22 D7** 48 37N 0 53 E
La Flèche, *France* **22 E6** 47 42N 0 4W
La Foa, *N. Cal.* **121 U19** 21 43 S 165 50 E
La Follette, *U.S.A.* **135 G3** 36 23N 84 7W
La Fontaine, *U.S.A.* **141 D11** 40 40N 85 43W
La Fregeneda, *Spain* **36 E4** 40 58N 6 54W
La Fría, *Venezuela* **152 B3** 8 13N 72 15W
La Fuente de San Esteban, *Spain* **36 E4** 40 49N 6 15W
La Gineta, *Spain* **35 F2** 39 8N 2 1W
La Gloria, *Colombia* **152 B3** 8 37N 73 48W
La Gran Sabana, *Venezuela* **153 B5** 5 30N 61 30W
La Grand-Combe, *France* **25 D8** 44 13N 4 2 E
La Grande, *U.S.A.* **142 D4** 45 20N 118 5W
La Grande-Motte, *France* **25 E8** 43 23N 4 5 E
La Grange, *Calif., U.S.A.* **144 H6** 37 42N 120 27W
La Grange, *Ga., U.S.A.* **135 J3** 33 2N 85 2W
La Grange, *Ky., U.S.A.* **134 F3** 38 25N 85 23W
La Grange, *Mo., U.S.A.* **140 D5** 40 3N 91 35W
La Grange, *Tex., U.S.A.* **139 L6** 29 54N 96 52W
La Grita, *Venezuela* **152 B3** 8 8N 71 59W
La Guaira, *Venezuela* **152 A4** 10 36N 66 56W
La Guardia, *Spain* **36 D2** 41 56N 8 52W
La Gudiña, *Spain* **36 C3** 42 4N 7 8W
La Güera, *Mauritania* **98 D1** 20 51N 17 0W
La Guerche-de-Bretagne, *France* **22 E5** 47 57N 1 16W
La Guerche-sur-l'Aubois, *France* **23 F9** 46 58N 2 56 E
La Habana, *Cuba* **148 B3** 23 8N 82 22W
La Harpe, *U.S.A.* **140 D6** 40 35N 90 58W
La Haye-du-Puits, *France* **22 C5** 49 17N 1 33W
La Horqueta, *Venezuela* **153 B5** 7 55N 60 20W
La Horra, *Spain* **36 D7** 41 44N 3 53W
La Independencia, *Mexico* **147 D6** 16 31N 91 47W
La Isabela, *Dom. Rep.* **149 C5** 19 58N 71 2W
La Jara, *U.S.A.* **143 H11** 37 16N 105 58W
La Joya, *Peru* **156 D3** 16 43 S 71 52W
La Junquera, *Spain* **34 C7** 42 25N 2 53 E
La Junta, *U.S.A.* **139 F3** 37 59N 103 33W
La Laguna, *Canary Is.* **33 F3** 28 28N 16 18W
La Libertad, *Guatemala* **148 C1** 16 47N 90 7W
La Libertad, *Mexico* **146 B2** 29 55N 112 41W
La Libertad □, *Peru* **156 B2** 8 0 S 78 30W
La Ligua, *Chile* **158 C1** 32 30 S 71 16W
La Línea de la Concepción, *Spain* **37 J5** 36 15N 5 23W
La Loche, *Canada* **131 B7** 56 29N 109 26W
La Londe-les-Maures, *France* **25 E10** 43 8N 6 14 E
La Lora, *Spain* **36 C7** 42 45N 4 0W
La Loupe, *France* **22 D8** 48 29N 1 1 E
La Louvière, *Belgium* **21 H4** 50 27N 4 10 E
La Machine, *France* **23 F10** 46 54N 3 27 E
La Maddalena, *Italy* **40 A2** 41 13N 9 24 E
La Malbaie, *Canada* **129 C5** 47 40N 70 10W
La Mancha, *Spain* **35 F2** 39 10N 2 54W
La Mariña, *Spain* **36 B3** 43 30N 7 40W
La Mesa, *Calif., U.S.A.* **145 N9** 32 46N 117 3W
La Mesa, *N. Mex., U.S.A.* **143 K10** 32 7N 106 42W
La Misión, *Mexico* **146 A1** 32 5N 116 50W
La Moille, *U.S.A.* **140 C7** 41 32N 89 17W
La Moine →, *U.S.A.* **140 E6** 39 59N 90 31W
La Monte, *U.S.A.* **140 F3** 38 46N 93 26W
La Mothe-Achard, *France* **22 F5** 46 37N 1 40W
La Motte, *France* **25 D10** 44 20N 6 3 E
La Motte-Chalançon, *France* **25 D9** 44 30N 5 21 E
La Moure, *U.S.A.* **138 B5** 46 21N 98 18W
La Muela, *Spain* **34 D3** 41 36N 1 7W
La Mure, *France* **25 D9** 44 55N 5 48 E
La Negra, *Chile* **158 A1** 23 46 S 70 18W
La Neuveville, *Switz.* **28 B4** 47 4N 7 6 E
La Oliva, *Canary Is.* **33 F6** 28 36N 13 57W
La Oraya, *Peru* **156 C2** 11 32 S 75 54W
La Orotava, *Canary Is.* **33 F3** 28 22N 16 31W
La Pacaudière, *France* **24 B7** 46 11N 3 52 E
La Palma, *Canary Is.* **33 F2** 28 40N 17 50W
La Palma, *Panama* **148 E4** 8 15N 78 0W
La Palma del Condado, *Spain* **37 H4** 37 21N 6 38W
La Paloma, *Chile* **158 C1** 30 35 S 71 0W
La Pampa □, *Argentina* **158 D2** 36 50 S 66 0W
La Paragua, *Venezuela* **153 B5** 6 50N 63 20W
La Paz, *Entre Ríos, Argentina* **158 C4** 30 50 S 59 45W
La Paz, *San Luis, Argentina* **158 C2** 33 30 S 67 20W
La Paz, *Bolivia* **156 D4** 16 20 S 68 10W
La Paz, *Honduras* **148 D2** 14 20N 87 47W
La Paz, *Mexico* **146 C2** 24 10N 110 20W
La Paz, *Phil.* **70 D3** 15 26N 120 45 E
La Paz □, *Bolivia* **156 D4** 15 30 S 68 0W
La Paz Centro, *Nic.* **148 D2** 12 20N 86 41W
La Pedrera, *Colombia* **152 D4** 1 18 S 69 43W
La Perouse Str., *Asia* **60 B11** 45 40N 142 0 E
La Pesca, *Mexico* **147 C5** 23 46N 97 47W
La Piedad, *Mexico* **146 C4** 20 20N 102 1W
La Pine, *U.S.A.* **142 E3** 43 40N 121 30W
La Plant, *U.S.A.* **138 C4** 45 9N 100 39W
La Plata, *Argentina* **158 D4** 35 0 S 57 55W
La Plata, *Colombia* **152 C2** 2 23N 75 53W
La Plata, *U.S.A.* **140 D4** 40 2N 92 29W
La Plata, L., *Argentina* **160 B2** 44 55 S 71 50W
La Pobla de Lillet, *Spain* **34 C6** 42 16N 1 59 E
La Pola de Gordón, *Spain* **36 C5** 42 51N 5 41W
La Porte, *U.S.A.* **141 C10** 41 36N 86 43W
La Porte City, *U.S.A.* **140 B4** 42 19N 92 12W
La Puebla, *Spain* **34 F8** 39 46N 3 1 E
La Puebla de Cazalla, *Spain* **37 H5** 37 10N 5 20W
La Puebla de los Infantes, *Spain* **37 H5** 37 47N 5 24W
La Puebla de Montalbán, *Spain* **36 F6** 39 52N 4 22W
La Puerta, *Spain* **35 G2** 38 22N 2 45W
La Punt, *Switz.* **29 C9** 46 35N 9 56 E
La Purísima, *Mexico* **146 B2** 26 10N 112 4W
La Push, *U.S.A.* **144 C2** 47 55N 124 38W
La Quiaca, *Argentina* **158 A2** 22 5 S 65 35W
La Rambla, *Spain* **37 H6** 37 37N 4 45W
La Reine, *Canada* **128 C4** 48 50N 79 30W
La Réole, *France* **24 D3** 44 35N 0 1W
La Restinga, *Canary Is.* **33 G2** 27 38N 17 59W
La Rioja, *Argentina* **158 B2** 29 20 S 67 0W
La Rioja □, *Argentina* **158 B2** 29 30 S 67 0W
La Rioja □, *Spain* **34 C2** 42 20N 2 20W
La Robla, *Spain* **36 C5** 42 50N 5 41W
La Roche, *Switz.* **28 C4** 46 42N 7 7 E
La Roche-Bernard, *France* **22 E4** 47 31N 2 19W
La Roche-Canillac, *France* **24 C5** 45 12N 1 57 E
La Roche-en-Ardenne, *Belgium* **21 H7** 50 11N 5 35 E
La Roche-sur-Yon, *France* **22 F5** 46 40N 1 25W
La Rochefoucauld, *France* **24 C4** 45 44N 0 24 E
La Rochelle, *France* **24 B2** 46 10N 1 9W
La Roda, *Albacete, Spain* **35 F2** 39 13N 2 15W
La Roda, *Sevilla, Spain* **37 H6** 37 12N 4 46W
La Romana, *Dom. Rep.* **149 C6** 18 27N 68 57W
La Ronge, *Canada* **131 B7** 55 5N 105 20W
La Rue, *U.S.A.* **141 D13** 40 35N 83 23W
La Rumorosa, *Mexico* **145 N10** 32 33N 116 4W
La Sabina, *Spain* **33 C7** 38 44N 1 25 E
La Sagra, *Spain* **35 H2** 37 57N 2 35W
La Salle, *U.S.A.* **140 C7** 41 20N 89 6W
La Sanabria, *Spain* **36 C4** 42 0N 6 30W
La Santa, *Canary Is.* **33 E6** 29 5N 13 40W
La Sarraz, *Switz.* **28 C3** 46 38N 6 32 E
La Sarre, *Canada* **128 C4** 48 45N 79 15W
La Scie, *Canada* **129 C8** 49 57N 55 36W
La Selva, *Spain* **34 D7** 42 0N 2 45 E
La Selva Beach, *U.S.A.* **144 J5** 36 56N 121 51W
La Serena, *Chile* **158 B1** 29 55 S 71 10W
La Serena, *Spain* **37 G5** 38 45N 5 40W
La Seyne-sur-Mer, *France* **25 E9** 43 7N 5 52 E
La Sila, *Italy* **41 C9** 39 15N 16 35 E
La Solana, *Spain* **35 G1** 38 59N 3 14W
La Souterraine, *France* **24 B5** 46 15N 1 30 E
La Spézia, *Italy* **38 D6** 44 7N 9 50 E
La Suze-sur-Sarthe, *France* **22 E7** 47 53N 0 2 E
La Tagua, *Colombia* **152 C3** 0 3N 74 40W
La Teste, *France* **24 D2** 44 37N 1 8W
La Tortuga, *Venezuela* **149 D6** 11 0N 65 22W
La Tour-du-Pin, *France* **25 C9** 45 33N 5 27 E
La Tranche-sur-Mer, *France* **22 F5** 46 20N 1 27W
La Tremblade, *France* **24 C2** 45 46N 1 8W
La Trinidad, *Phil.* **70 C3** 16 28N 120 35 E
La Tuque, *Canada* **128 C5** 47 30N 72 50W
La Unión, *Chile* **160 B2** 40 10 S 73 0W
La Unión, *Colombia* **152 C2** 1 35N 77 5W
La Unión, *El Salv.* **148 D2** 13 20N 87 50W
La Unión, *Mexico* **146 D4** 17 58N 101 49W
La Unión, *Peru* **156 B2** 9 43 S 76 45W
La Unión, *Spain* **35 H4** 37 38N 0 53W
La Union □, *Phil.* **70 C3** 16 30N 120 25 E
La Urbana, *Venezuela* **152 B4** 7 8N 66 56W
La Vecilla, *Spain* **36 C5** 42 51N 5 27W
La Vega, *Dom. Rep.* **149 C5** 19 20N 70 30W
La Vega, *Peru* **156 C2** 10 41 S 77 44W
La Vela, *Venezuela* **152 A4** 11 27N 69 34W
La Veleta, *Spain* **37 H7** 37 1N 3 22W
La Venta, *Mexico* **147 D6** 18 8N 94 3W
La Ventura, *Mexico* **146 C4** 24 38N 100 54W
La Venturosa, *Colombia* **152 B4** 6 8N 68 48W
La Victoria, *Venezuela* **152 A4** 10 14N 67 20W
La Voulte-sur-Rhône, *France* **25 D8** 44 48N 4 46 E
La Zarza, *Spain* **37 H4** 37 42N 6 51W
Laa, *Austria* **31 C9** 48 43N 16 23 E
Laaber →, *Germany* **27 G8** 48 55N 12 30 E
Laage, *Germany* **26 B8** 53 55N 12 21 E
Laba →, *Russia* **53 H4** 45 11N 39 42 E
Laban, *Burma* **78 C6** 25 52N 96 40 E
Labason, *Phil.* **71 G4** 8 4N 122 31 E
Labastide-Murat, *France* **24 D5** 44 39N 1 33 E
Labastide-Rouairoux, *France* **24 E6** 43 28N 2 39 E
Labbézenga, *Mali* **101 B5** 15 2N 0 48 E
Labdah = Leptis Magna, *Libya* **96 B2** 32 40N 14 12 E
Labe = Elbe →, *Europe* **26 B4** 53 50N 9 0 E
Labé, *Guinea* **100 C2** 11 24N 12 16W
Laberec →, *Slovak Rep.* **31 C14** 48 37N 21 58 E
Laberge, L., *Canada* **130 A1** 61 11N 135 12W
Labian, Tanjong, *Malaysia* **75 A5** 5 9N 119 13 E
Labig Pt., *Phil.* **70 B4** 18 25N 122 25 E
Labin, *Croatia* **39 C11** 45 5N 14 8 E
Labinsk, *Russia* **53 H5** 44 40N 40 48 E
Labis, *Malaysia* **77 L4** 2 22N 103 2 E
Labiszyn, *Poland* **47 C4** 52 57N 17 54 E
Labo, *Phil.* **70 D4** 14 9N 122 51 E
Laboe, *Germany* **26 A6** 54 24N 10 13 E
Laboka, *Gabon* **102 B2** 0 19N 11 32 E
Labouheyre, *France* **24 D3** 44 13N 0 55W
Laboulaye, *Argentina* **158 C3** 34 10 S 63 30W
Labra, Peña, *Spain* **36 B6** 43 3N 4 26W
Labrador, Coast of □, *Canada* **129 B7** 53 20N 61 0W
Labrador City, *Canada* **129 B6** 52 57N 66 55W
Lábrea, *Brazil* **157 B5** 7 15 S 64 51W
Labrède, *France* **24 D3** 44 41N 0 32W
Labuan, Pulau, *Malaysia* **75 A5** 5 21N 115 13 E
Labuha, *Indonesia* **72 B3** 0 30 S 127 30 E
Labuhan, *Indonesia* **74 D3** 6 22 S 105 50 E
Labuhanbajo, *Indonesia* **72 C2** 8 28 S 120 1 E
Labuissière, *Belgium* **21 H4** 50 19N 4 11 E
Labuk, Telok, *Malaysia* **75 A5** 6 10N 117 50 E
Labutta, *Burma* **78 G5** 16 9N 94 46 E
Labyrinth, L., *Australia* **115 E2** 30 40 S 135 11 E
Labytnangi, *Russia* **48 A12** 66 39N 66 21 E

Łabźenica, Poland 47 B4 53 18N 17 15 E
Lac Allard, Canada 129 B7 50 33N 63 24W
Lac Bouchette, Canada 129 C5 48 16N 72 11W
Lac du Flambeau, U.S.A. 138 B10 45 58N 89 53W
Lac Édouard, Canada 128 C5 47 40N 72 16W
Lac La Biche, Canada 130 C6 54 45N 111 58W
Lac la Martre, Canada 126 B8 63 8N 117 16W
Lac-Mégantic, Canada 129 C5 45 35N 70 53W
Lac Seul, Res., Canada 128 B1 50 25N 92 30W
Lac Thien, Vietnam 76 F7 12 25N 108 11 E
Lacanau, France 24 D2 44 58N 1 5W
Lacanau, Étang de, France 24 D2 44 58N 1 7W
Lacantúm →, Mexico 147 D6 16 36N 90 40W
Lacara →, Spain 37 G4 38 55N 6 25W
Lacaune, France 24 E6 43 43N 2 40 E
Lacaune, Mts. de, France 24 E6 43 43N 2 50 E
Laccadive Is. = Lakshadweep Is., Ind. Oc. 58 H11 10 0N 72 30 E
Lacepede B., Australia 116 D3 36 40 S 139 40 E
Lacepede Is., Australia 112 C3 16 55 S 122 0 E
Lacerdónia, Mozam. 107 F4 18 3 S 35 35 E
Lacey, U.S.A. 144 C4 47 7N 122 49W
Lachay, Pta., Peru 156 C2 11 17 S 77 44W
Lachen, India 78 B2 27 46N 88 36 E
Lachen, Switz. 29 B7 47 12N 8 51 E
Lachhmangarh, India 80 F6 27 50N 75 4 E
Lachi, Pakistan 80 C4 33 25N 71 20 E
Lachine, Canada 128 C5 45 30N 73 40W
Lachlan →, Australia 116 C5 34 22 S 143 55 E
Lachute, Canada 128 C5 45 39N 74 21W
Lackawanna, U.S.A. 136 D6 42 50N 78 50W
Lacolle, Canada 137 A11 45 5N 73 22W
Lacombe, Canada 130 C6 52 30N 113 44W
Lacon, U.S.A. 140 C7 41 2N 89 24W
Lacona, Iowa, U.S.A. 140 C3 41 12N 93 23W
Lacona, N.Y., U.S.A. 137 C8 43 39N 76 10W
Láconi, Italy 40 C2 39 54N 9 4 E
Laconia, U.S.A. 137 C13 43 32N 71 28W
Lacq, France 24 E3 43 25N 0 35W
Lacrosse, U.S.A. 142 C5 46 51N 117 58W
Lacub, Phil. 70 C3 17 40N 120 53 E
Ladakh Ra., India 81 B8 34 0N 78 0 E
Ladário, Brazil 157 D6 19 1 S 57 35W
Ladd, U.S.A. 140 C7 41 23N 89 13W
Laddonia, U.S.A. 140 E5 39 15N 91 39W
Lądekzdrój, Poland 47 E3 50 21N 16 53 E
Ládhon →, Greece 45 G3 37 40N 21 50 E
Ladik, Turkey 88 B6 40 57N 35 58 E
Ladismith, S. Africa 104 E3 33 28 S 21 15 E
Lādīz, Iran 85 D9 28 55N 61 15 E
Ladnun, India 80 F6 27 38N 74 25 E
Ladoga, L. = Ladozhskoye Ozero, Russia 50 B6 61 15N 30 30 E
Ladon, France 23 D9 48 0N 2 30 E
Ladozhskoye Ozero, Russia 50 B6 61 15N 30 30 E
Ladrillero, G., Chile 160 C1 49 20 S 75 35W
Lady Grey, S. Africa 104 E4 30 43 S 27 13 E
Ladybrand, S. Africa 104 D4 29 9 S 27 29 E
Ladysmith, Canada 130 D4 49 0N 123 49W
Ladysmith, S. Africa 105 D4 28 32 S 29 46 E
Ladysmith, U.S.A. 138 C9 45 28N 91 12W
Lae, Papua N. G. 120 D4 6 40 S 147 2 E
Laem Ngop, Thailand 77 F4 12 10N 102 26 E
Laem Pho, Thailand 77 J3 6 55N 101 19 E
Læsø, Denmark 15 G4 57 15N 10 53 E
Læsø Rende, Denmark 15 G4 57 20N 10 45 E
Lafayette, Colo., U.S.A. 138 F2 39 58N 105 12W
Lafayette, Ind., U.S.A. 141 D10 40 25N 86 54W
Lafayette, La., U.S.A. 139 K9 30 14N 92 1W
Lafayette, Tenn., U.S.A. 135 G3 36 31N 86 2W
Laferte →, Canada 130 A5 61 53N 117 44W
Lafia, Nigeria 101 D6 8 30N 8 34 E
Lafiagi, Nigeria 101 D6 8 52N 5 20 E
Lafleche, Canada 131 D7 49 45N 106 40W
Lafon, Sudan 95 F3 5 5N 32 29 E
Laforsen, Sweden 14 C9 61 56N 15 3 E
Lagaip →, Papua N. G. 120 C2 5 4 S 142 52 E
Lagan →, Sweden 15 H7 56 56N 13 58 E
Lagan →, U.K. 19 B6 54 36N 5 55W
Lagangilang, Phil. 70 C3 17 37N 120 44 E
Lagarfljót →, Iceland 12 D6 65 40N 14 18W
Lagarto, Brazil 154 D4 10 54 S 37 41W
Lagawe, Phil. 70 C3 16 49N 121 6 E
Lage, Germany 26 D4 51 59N 8 48 E
Lage, Spain 36 B2 43 13N 9 0W
Lage-Mierde, Neths. 21 F6 51 25N 5 9 E
Lågen →, Oppland, Norway 13 F14 61 8N 10 25 E
Lågen →, Vestfold, Norway 13 G14 59 3N 10 3 E
Lägerdorf, Germany 26 B5 53 53N 9 34 E
Laghmān □, Afghan. 79 B3 34 20N 70 0 E
Laghouat, Algeria 99 B5 33 50N 2 59 E
Lagnieu, France 25 C9 45 55N 5 20 E
Lagny, France 23 D9 48 52N 2 44 E
Lago, Italy 41 C9 39 10N 16 9 E
Lago Posadas, Argentina 160 C2 47 30 S 71 40W
Lago Ranco, Chile 160 B2 40 19 S 72 30W
Lagôa, Portugal 37 H2 37 8N 8 27W
Lagoaça, Portugal 36 D4 41 11N 6 44W
Lagodekhi, Georgia 53 K8 41 50N 46 22 E
Lagónegro, Italy 41 B8 40 8N 15 45 E
Lagonoy Gulf, Phil. 70 E4 13 50N 123 50 E
Lagos, Nigeria 101 D5 6 25N 3 27 E
Lagos, Portugal 37 H2 37 5N 8 41W
Lagos de Moreno, Mexico 146 C4 21 21N 101 55W
Lagrange, Australia 112 C3 18 45 S 121 43 E
Lagrange, U.S.A. 141 C11 41 39N 85 25W
Lagrange B., Australia 112 C3 18 38 S 121 42 E
Laguardia, Spain 34 C2 42 33N 2 35W
Laguépie, France 24 D5 44 8N 1 57 E
Laguna, Brazil 159 B6 28 30 S 48 50W
Laguna, U.S.A. 143 J10 35 2N 107 25W
Laguna □, Phil. 70 D3 14 10N 121 20 E
Laguna Beach, U.S.A. 145 M9 33 33N 117 47W
Laguna de la Janda, Spain 37 J5 36 15N 5 45W
Laguna Limpia, Argentina 158 B4 26 32 S 59 45W
Laguna Madre, U.S.A. 147 B5 27 0N 97 20W
Lagunas, Chile 158 A2 21 0 S 69 45W
Lagunas, Peru 156 B2 5 10 S 75 35W
Lagunillas, Bolivia 157 D5 19 38 S 63 43W
Lahad Datu, Malaysia 75 A5 5 0N 118 20 E
Lahan Sai, Thailand 76 E4 14 25N 102 52 E
Lahanam, Laos 76 D5 16 16N 105 16 E
Laharpur, India 81 F9 27 43N 80 56 E
Lahat, Indonesia 74 C2 3 45 S 103 30 E
Lahe, Burma 78 B5 26 20N 95 26 E
Lahewa, Indonesia 74 B1 1 22N 97 12 E
Lahiang Lahiang, Phil. 71 H3 6 1N 121 23 E
Laḥij, Yemen 86 D4 13 4N 44 53 E
Lāhījān, Iran 85 B6 37 10N 50 6 E
Lahn →, Germany 27 E3 50 19N 7 37 E
Laholm, Sweden 15 H7 56 30N 13 2 E
Laholmsbukten, Sweden 15 H6 56 30N 12 45 E
Lahontan Reservoir, U.S.A. 142 G4 39 28N 119 4W
Lahore, Pakistan 79 C4 31 32N 74 22 E
Lahpongsel, Burma 78 B7 27 7N 98 25 E
Lahr, Germany 27 G3 48 20N 7 53 E
Lahti, Finland 13 F21 60 58N 25 40 E
Lahtis = Lahti, Finland 13 F21 60 58N 25 40 E
Laï, Chad 97 G3 9 25N 16 18 E
Lai Chau, Vietnam 76 A4 22 5N 103 3 E
Lai-hka, Burma 78 E6 21 16N 97 40 E
Laiagam, Papua N. G. 120 C2 5 33 S 143 30 E
Lai'an, China 69 A12 32 28N 118 30 E
Laibin, China 68 F7 23 42N 109 14 E
Laidley, Australia 115 D5 27 39 S 152 20 E
Laifeng, China 68 C7 29 27N 109 20 E
L'Aigle, France 22 D7 48 46N 0 38 E
Laignes, France 23 E11 47 50N 4 20 E
L'Aiguillon-sur-Mer, France 24 B2 46 20N 1 18W
Laikipia □, Kenya 106 B4 0 30N 36 30 E
Laila = Laylá, Si. Arabia 86 B4 22 10N 46 40 E
Laingsburg, S. Africa 104 E3 33 9 S 20 52 E
Lainio älv →, Sweden 12 C20 67 35N 22 40 E
Lairg, U.K. 18 C4 58 2N 4 24W
Lais, Phil. 71 H5 6 20N 125 39 E
Laishui, China 66 E8 39 23N 115 45 E
Laiwu, China 67 F9 36 15N 117 40 E
Laixi, China 67 F11 36 50N 120 31 E
Laiyang, China 67 F11 36 59N 120 45 E
Laiyuan, China 66 E8 39 20N 114 40 E
Laizhou Wan, China 67 F10 37 30N 119 30 E
Laja →, Mexico 146 C4 20 55N 100 46W
Lajere, Nigeria 101 C7 12 10N 11 25 E
Lajes, Rio Grande do N., Brazil 154 C4 5 41 S 36 14W
Lajes, Sta. Catarina, Brazil 159 B5 27 48 S 50 20W
Lajinha, Brazil 155 F3 20 9 S 41 37W
Lajkovac, Serbia, Yug. 42 C5 44 27N 20 14 E
Lajosmizse, Hungary 31 D12 47 3N 19 32 E
Lak Sao, Laos 76 C5 18 11N 104 59 E
Lakaband, Pakistan 80 D3 31 2N 69 15 E
Lakatoro, Vanuatu 121 F5 16 0 S 167 0 E
Lake Alpine, U.S.A. 144 G7 38 29N 120 0W
Lake Andes, U.S.A. 138 D5 43 9N 98 32W
Lake Anse, U.S.A. 134 B1 46 42N 88 25W
Lake Arthur, U.S.A. 139 K8 30 5N 92 41W
Lake Cargelligo, Australia 117 B7 33 15 S 146 22 E
Lake Charles, U.S.A. 139 K8 30 14N 93 13W
Lake City, Colo., U.S.A. 143 G10 38 2N 107 19W
Lake City, Fla., U.S.A. 135 K4 30 11N 82 38W
Lake City, Iowa, U.S.A. 140 B2 42 16N 94 44W
Lake City, Mich., U.S.A. 134 C3 44 20N 85 13W
Lake City, Minn., U.S.A. 138 C8 44 27N 92 16W
Lake City, Pa., U.S.A. 136 D4 42 1N 80 21W
Lake City, S.C., U.S.A. 135 J6 33 52N 79 45W
Lake Coleridge, N.Z. 119 D6 43 17 S 171 30 E
Lake Forest, U.S.A. 141 B9 42 15N 87 50W
Lake Geneva, U.S.A. 141 B8 42 36N 88 26W
Lake George, U.S.A. 137 C11 43 26N 73 43W
Lake Grace, Australia 113 F2 33 7 S 118 28 E
Lake Harbour, Canada 127 B13 62 50N 69 50W
Lake Havasu City, U.S.A. 145 L12 34 27N 114 22W
Lake Hughes, U.S.A. 145 L8 34 41N 118 26W
Lake Isabella, U.S.A. 145 K8 35 38N 118 28W
Lake King, Australia 113 F2 33 5 S 119 45 E
Lake Lenore, Canada 131 C8 52 24N 104 59W
Lake Louise, Canada 130 C5 51 30N 116 10W
Lake Mead National Recreation Area, U.S.A. 145 K12 36 15N 114 30W
Lake Michigan Beach, U.S.A. 141 B10 42 13N 86 25W
Lake Mills, Iowa, U.S.A. 138 D8 43 25N 93 32W
Lake Mills, Wis., U.S.A. 141 A8 43 5N 88 55W
Lake Murray, Papua N. G. 120 D1 6 48 S 141 29 E
Lake Nash, Australia 114 C2 20 57 S 138 0 E
Lake Odessa, U.S.A. 141 B11 42 47N 85 8W
Lake Orion, U.S.A. 141 B13 42 47N 83 14W
Lake Providence, U.S.A. 139 J9 32 48N 91 10W
Lake Pukaki, N.Z. 119 E5 44 11 S 170 8 E
Lake River, Canada 128 B3 54 30N 82 31W
Lake Superior Prov. Park, Canada 128 C3 47 45N 84 45W
Lake Tekapo, N.Z. 119 D5 44 0 S 170 30 E
Lake Villa, U.S.A. 141 B8 42 25N 88 5W
Lake Village, U.S.A. 139 J9 33 20N 91 17W
Lake Wales, U.S.A. 135 M5 27 54N 81 35W
Lake Worth, U.S.A. 135 M5 26 37N 80 3W
Lakefield, Canada 128 D4 44 25N 78 16W
Lakeland, Australia 114 B3 15 49 S 144 57 E
Lakeland, U.S.A. 135 L5 28 3N 81 57W
Lakemba, Fiji 121 B3 18 13 S 178 47W
Lakeport, U.S.A. 144 F4 39 3N 122 55W
Lakes Entrance, Australia 117 D8 37 50 S 148 0 E
Lakeside, Ariz., U.S.A. 143 J9 34 9N 109 58W
Lakeside, Calif., U.S.A. 145 N10 32 52N 116 55W
Lakeside, Nebr., U.S.A. 138 D3 42 3N 102 26W
Lakeview, U.S.A. 142 E3 42 11N 120 21W
Lakewood, Colo., U.S.A. 138 F2 39 44N 105 5W
Lakewood, N.J., U.S.A. 137 F10 40 6N 74 13W
Lakewood, Ohio, U.S.A. 136 E3 41 29N 81 48W
Lakewood Center, U.S.A. 144 C4 47 11N 122 32W
Lakhaniá, Greece 32 D9 35 58N 27 54 E
Lakhipur, Assam, India 78 C4 24 48N 93 0 E
Lakhipur, Assam, India 78 B3 26 2N 90 18 E
Lakhonpheng, Laos 76 E5 15 54N 105 34 E
Lakhpat, India 80 H3 23 48N 68 47 E
Läki, Azerbaijan 53 K8 40 34N 47 22 E
Lakin, U.S.A. 139 G4 37 57N 101 15W
Lakitusaki →, Canada 128 B3 54 21N 82 25W
Lákkoi, Greece 32 D5 35 24N 23 57 E
Lakonía □, Greece 45 H4 36 55N 22 30 E
Lakonikós Kólpos, Greece 45 H4 36 40N 22 40 E
Lakor, Indonesia 72 C3 8 15 S 128 17 E
Lakota, Ivory C. 100 D3 5 50N 5 30W
Lakota, U.S.A. 138 A5 48 2N 98 21W
Laksefjorden, Norway 12 A22 70 45N 26 50 E
Lakselv, Norway 12 A21 70 2N 25 0 E
Lakshadweep Is., Ind. Oc. 58 H11 10 0N 72 30 E
Laksham, Bangla. 78 D3 23 14N 91 8 E
Lakshmeshwar, India 83 G2 15 9N 75 28 E
Lakshmikantapur, India 81 H13 22 5N 88 20 E
Lakshmipur, Bangla. 78 D3 22 58N 90 50 E
Lakuramau, Papua N. G. 120 B6 2 54 S 151 15 E
Lal-lo, Phil. 70 B3 18 12N 121 40 E
Lala, Phil. 71 H4 7 59N 123 46 E
Lala Musa, Pakistan 80 C5 32 40N 73 57 E
Lalago, Tanzania 106 C3 3 28 S 33 58 E
Lalapanzi, Zimbabwe 107 F3 19 20 S 30 15 E
Lalganj, India 81 G11 25 52N 85 13 E
Lalibela, Ethiopia 95 E4 12 2N 39 2 E
Lalin, China 67 B14 45 12N 127 0 E
Lalín, Spain 36 C2 42 40N 8 5W
Lalin He →, China 67 B13 45 32N 125 40 E
Lalinde, France 24 D4 44 50N 0 44 E
Lalitpur, India 81 G8 24 42N 78 28 E
Lam, Vietnam 76 B6 21 21N 106 31 E
Lam Pao Res., Thailand 76 D4 16 50N 103 15 E
Lama Kara, Togo 101 D5 9 30N 1 15 E
Lamag, Malaysia 75 A5 5 29N 117 49 E
Lamaipum, Burma 78 C6 25 40N 97 57 E
Lamap, Vanuatu 121 F5 16 26 S 167 43 E
Lamar, Colo., U.S.A. 138 F3 38 5N 102 37W
Lamar, Mo., U.S.A. 139 G7 37 30N 94 16W
Lamarque, Argentina 160 A3 39 24 S 65 40W
Lamas, Peru 156 B2 6 28 S 76 31W
Lamastre, France 25 D8 44 59N 4 35 E
Lambach, Austria 30 C6 48 6N 13 51 E
Lamballe, France 22 D4 48 29N 2 31W
Lambaréné, Gabon 102 C2 0 41 S 10 12 E
Lambasa, Fiji 121 A2 16 30 S 179 10 E
Lambay I., Ireland 19 C5 53 29N 6 1W
Lambayeque □, Peru 156 B2 6 45 S 80 0W
Lambert, U.S.A. 138 B2 47 41N 104 37W
Lambert, C., Papua N. G. 120 C6 4 11 S 151 31 E
Lambert Glacier, Antarctica 7 D6 71 0 S 70 0 E
Lamberts Bay, S. Africa 104 E2 32 5 S 18 17 E
Lambesc, France 25 E9 43 39N 5 16 E
Lámbia, Greece 45 G3 37 52N 21 53 E
Lambon, Papua N. G. 120 C7 4 45 S 152 48 E
Lambro →, Italy 38 C6 45 8N 9 32 E
Lambunao, Phil. 71 F4 11 3N 122 29 E
Lame, Nigeria 101 C6 10 30N 9 20 E
Lame Deer, U.S.A. 142 D10 45 37N 106 40W
Lamego, Portugal 36 D3 41 5N 7 52W
Lamèque, Canada 129 C7 47 45N 64 38W
Lameroo, Australia 116 C4 35 19 S 140 33 E
Lamesa, U.S.A. 139 J4 32 44N 101 58W
Lamía, Greece 45 F4 38 55N 22 26 E
Lamitan, Phil. 71 H4 6 39N 122 8 E
Lammermuir Hills, U.K. 18 F6 55 50N 2 40W
Lamon Bay, Phil. 70 D4 14 30N 122 20 E
Lamongan, Indonesia 75 D4 7 7 S 112 25 E
Lamoni, U.S.A. 140 D3 40 37N 93 56W
Lamont, Canada 130 C6 53 46N 112 50W
Lamont, Calif., U.S.A. 145 K8 35 15N 118 55W
Lamont, Iowa, U.S.A. 140 B5 42 35N 91 40W
Lampa, Peru 156 D3 15 22 S 70 22W
Lampang, Thailand 76 C2 18 16N 99 32 E
Lampasas, U.S.A. 139 K5 31 4N 98 11W
Lampaul, France 22 D1 48 28N 5 7W
Lampazos de Naranjo, Mexico 146 B4 27 2N 100 32W
Lampeter, U.K. 17 E3 52 7N 4 4W
Lampione, Medit. S. 96 A2 35 33N 12 20 E
Lampman, Canada 131 D8 49 25N 102 50W
Lamprechtshausen, Austria 30 D5 48 0N 12 58 E
Lamprey, Canada 131 B10 58 33N 94 8W
Lampung □, Indonesia 74 D2 5 30 S 104 30 E
Lamu, Burma 78 F5 19 14N 94 10 E
Lamu, Kenya 106 C5 2 16 S 40 55 E
Lamu □, Kenya 106 C5 2 0 S 40 45 E
Lamud, Peru 156 B2 6 10 S 77 57W
Lamut, Phil. 70 C3 16 39N 121 14 E
Lamy, U.S.A. 143 J11 35 29N 105 53W
Lan Xian, China 66 E6 38 15N 111 35 E
Lan Yu, Taiwan 69 F13 22 5N 121 35 E
Lanai I., U.S.A. 132 H16 20 50N 156 55W
Lanak La, India 81 B8 34 27N 79 32 E
Lanaken, Belgium 21 G7 50 53N 5 39 E
Lanak'o Shank'ou = Lanak La, India 81 B8 34 27N 79 32 E
Lanao, L., Phil. 71 H5 7 52N 124 15 E
Lanao del Norte □, Phil. 71 H5 8 0N 124 10 E
Lanao del Sur □, Phil. 71 H5 7 40N 124 15 E
Lanark, Canada 137 A8 45 1N 76 22W
Lanark, U.K. 18 F5 55 40N 3 47W
Lancang, China 68 F2 22 36N 99 58 E
Lancang Jiang →, China 68 G3 21 40N 101 10 E
Lancashire □, U.K. 16 D5 53 50N 2 48W
Lancaster, Canada 137 A10 45 10N 74 30W
Lancaster, U.K. 16 C5 54 3N 2 48W
Lancaster, Calif., U.S.A. 145 L8 34 42N 118 8W
Lancaster, Ky., U.S.A. 134 G3 37 37N 84 35W
Lancaster, Mo., U.S.A. 140 D4 40 31N 92 32W
Lancaster, N.H., U.S.A. 137 B13 44 29N 71 34W
Lancaster, N.Y., U.S.A. 136 D6 42 54N 78 40W
Lancaster, Pa., U.S.A. 137 F8 40 2N 76 19W
Lancaster, S.C., U.S.A. 135 H5 34 43N 80 46W
Lancaster, Wis., U.S.A. 140 B6 42 51N 90 43W
Lancaster Sd., Canada 127 A11 74 13N 84 0W
Lancer, Canada 131 C7 50 48N 108 53W
Lanchow = Lanzhou, China 66 F2 36 1N 103 52 E
Lanciano, Italy 39 F11 42 14N 14 23 E
Lanco, Chile 160 A2 39 24 S 72 46W
Lancones, Peru 156 A1 4 30 S 80 30W
Lancun, China 67 F11 36 25N 120 10 E
Łańcut, Poland 31 A15 50 10N 22 13 E
Lancy, Switz. 28 D2 46 12N 6 8 E
Landau, Bayern, Germany 27 G8 48 40N 12 41 E
Landau, Rhld-Pfz., Germany 27 F4 49 12N 8 6 E
Landay, Afghan. 79 C1 30 31N 63 47 E
Landeck, Austria 30 D3 47 9N 10 34 E
Landen, Belgium 21 G6 50 45N 5 3 E
Lander, U.S.A. 142 E9 42 50N 108 44W
Lander →, Australia 112 D5 22 0 S 132 0 E
Landerneau, France 22 D2 48 28N 4 17W
Landeryd, Sweden 15 G7 57 7N 13 15 E
Landes, France 24 D3 44 0N 1 0W
Landes □, France 24 E3 43 57N 0 48W
Landete, Spain 34 F3 39 56N 1 25W
Landi Kotal, Pakistan 79 B3 34 7N 71 6 E
Landivisiau, France 22 D2 48 31N 4 6W
Landor, Australia 113 E2 25 10 S 116 54 E
Landquart, Switz. 29 C9 46 58N 9 32 E
Landquart →, Switz. 29 C9 46 50N 9 47 E
Landrecies, France 23 B10 50 7N 3 40 E
Land's End, U.K. 17 G2 50 4N 5 44W
Landsberg, Germany 27 G6 48 2N 10 53 E
Landsborough Cr. →, Australia 114 C3 22 28 S 144 35 E
Landshut, Germany 27 G8 48 34N 12 8 E
Landskrona, Sweden 15 J6 55 53N 12 50 E
Landstuhl, Germany 27 F3 49 24N 7 33 E
Landvetter, Sweden 15 G6 57 41N 12 17 E
Laneffe, Belgium 21 H5 50 17N 4 30 E
Lanesboro, U.S.A. 137 E9 41 57N 75 34W
Lanett, U.S.A. 135 J3 32 52N 85 12W
Lang Bay, Canada 130 D4 49 45N 124 21W
Lang Qua, Vietnam 76 A5 22 16N 104 27 E
Lang Shan, China 66 D4 41 0N 106 30 E
Lang Son, Vietnam 76 B6 21 52N 106 42 E
Lang Suan, Thailand 77 H2 9 57N 99 4 E
Lángadhás, Greece 44 D5 40 46N 23 2 E
Langádhia, Greece 45 G4 37 43N 22 1 E
Långan →, Sweden 14 A8 63 19N 14 44 E
Langar, Iran 85 C9 35 23N 60 25 E
Langara I., Canada 130 C2 54 14N 133 1W
Langatabbetje, Surinam 153 C7 4 59N 54 28W
Langdai, China 68 D5 26 6N 105 21 E
Langdon, U.S.A. 138 A5 48 45N 98 22W
Langdorp, Belgium 21 G5 50 59N 4 52 E
Langeac, France 24 C7 45 7N 3 29 E
Langeais, France 22 E7 47 20N 0 24 E
Langeb Baraka →, Sudan 94 D4 17 28N 36 50 E
Langeberg, S. Africa 104 E3 33 55 S 21 0 E
Langeberge, S. Africa 104 D3 28 15 S 22 33 E
Langeland, Denmark 15 K4 54 56N 10 48 E
Langemark, Belgium 21 G1 50 55N 2 55 E
Langen, Germany 27 F4 49 59N 8 40 E
Langenburg, Canada 131 C8 50 51N 101 43W
Langeneß, Germany 26 A4 54 38N 8 36 E
Langenlois, Austria 30 C8 48 29N 15 40 E
Langenthal, Switz. 28 B5 47 13N 7 47 E
Langeoog, Germany 26 B3 53 45N 7 32 E
Langeskov, Denmark 15 J4 55 22N 10 35 E
Langesund, Norway 14 F3 59 0N 9 45 E
Länghem, Sweden 15 G7 57 36N 13 14 E
Langhirano, Italy 38 D7 44 37N 10 16 E
Langholm, U.K. 18 F6 55 9N 3 0W
Langhus, Norway 14 E4 59 45N 10 50 E
Langidoon, Australia 116 A5 31 36 S 142 2 E
Langjökull, Iceland 12 D3 64 39N 20 12W
Langkawi, P., Malaysia 77 J2 6 25N 99 45 E
Langklip, S. Africa 104 D3 28 12 S 20 20 E
Langkon, Malaysia 75 A5 6 30N 116 40 E
Langlade, St- P. & M. 129 C8 46 50N 56 20W
Langlois, U.S.A. 142 E1 42 56N 124 27W
Langnau, Switz. 28 C5 46 56N 7 47 E
Langogne, France 24 D7 44 43N 3 50 E
Langon, France 24 D3 44 33N 0 16W
Langøya, Norway 12 B16 68 45N 14 50 E
Langres, France 23 E12 47 52N 5 20 E
Langres, Plateau de, France 23 E12 47 45N 5 3 E
Langsa, Indonesia 74 B1 4 30N 97 57 E
Långsele, Sweden 14 A11 63 12N 17 4 E
Langtao, Burma 78 B6 27 15N 97 34 E
Langting, India 78 C4 25 31N 93 7 E
Langtry, U.S.A. 139 L4 29 49N 101 34W
Langu, Thailand 77 J2 6 53N 99 47 E
Languedoc, France 24 E7 43 58N 3 55 E
Langwies, Switz. 29 C9 46 50N 9 44 E
Langxi, China 69 B12 31 10N 119 12 E
Langxiangzhen, China 66 E9 39 43N 116 8 E
Langzhong, China 68 B5 31 38N 105 58 E
Lanigan, Canada 131 C7 51 51N 105 2W
Lankao, China 66 G8 34 48N 114 50 E
Länkäran, Azerbaijan 89 C13 38 48N 48 52 E
Lannemezan, France 24 E4 43 8N 0 23 E
Lannilis, France 22 D2 48 35N 4 32W
Lannion, France 22 D3 48 46N 3 29W
L'Annonciation, Canada 128 C5 46 25N 74 55W
Lanouaille, France 24 C5 45 24N 1 9 E
Lanping, China 68 D2 26 28N 99 15 E
Lansdale, U.S.A. 137 F9 40 14N 75 17W
Lansdowne, Australia 117 A10 31 48 S 152 30 E
Lansdowne, Canada 137 B8 44 24N 76 1W
Lansdowne House, Canada 128 B2 52 14N 87 53W
L'Anse, U.S.A. 128 C2 46 45N 88 27W
L'Anse au Loup, Canada 129 B8 51 32N 56 50W
Lansford, U.S.A. 137 F9 40 50N 75 53W
Lanshan, China 69 E9 25 24N 112 10 E
Lansing, U.S.A. 141 B12 42 44N 84 33W
Lanslebourg-Mont-Cenis, France 25 C10 45 17N 6 52 E
Lanta Yai, Ko, Thailand 77 J2 7 35N 99 3 E
Lantian, China 66 G5 34 11N 109 20 E
Lanus, Argentina 158 C4 34 44 S 58 27W
Lanusei, Italy 40 C2 39 52N 9 34 E
Lanuza, Phil. 71 G6 9 14N 126 4 E
Lanxi, China 69 C12 29 13N 119 28 E
Lanzarote, Canary Is. 33 E6 29 0N 13 40W
Lanzhou, China 66 F2 36 1N 103 52 E
Lanzo Torinese, Italy 38 C4 45 16N 7 28 E
Lao →, Italy 41 C8 39 47N 15 48 E
Lao Bao, Laos 76 D6 16 35N 106 30 E
Lao Cai, Vietnam 76 A4 22 30N 103 57 E
Laoag, Phil. 70 B3 18 7N 120 34 E
Laoang, Phil. 70 E5 12 32N 125 8 E
Laoha He →, China 67 C11 43 25N 120 35 E
Laois □, Ireland 19 D4 52 57N 7 36W
Laon, France 23 C10 49 33N 3 35 E
Laona, U.S.A. 134 C1 45 34N 88 40W
Laos ■, Asia 76 D5 17 45N 105 0 E
Lapa, Brazil 159 B6 25 46 S 49 44W
Lapalisse, France 24 B7 46 15N 3 38 E
Lapeer, U.S.A. 141 A13 43 3N 83 19W
Lapithos, Cyprus 32 D12 35 21N 33 11 E
Lapland = Lappland, Europe 12 B21 68 7N 24 0 E
Lapog, Phil. 70 C3 17 45N 120 27 E
Laporte, U.S.A. 137 E8 41 25N 76 30W
Lapovo, Serbia, Yug. 42 C6 44 10N 21 2 E
Lappeenranta, Finland 13 F23 61 3N 28 12 E
Lappland, Europe 12 B21 68 7N 24 0 E

Mackenzie 227

Luchuan, *China* 69 F8 22 21N 110 12 E
Lucie →, *Surinam* 153 C6 3 35N 57 38W
Lucira, *Angola* 103 E2 14 0 S 12 35 E
Luckau, *Germany* 26 D9 51 50N 13 42 E
Luckenwalde, *Germany* 26 C9 52 5N 13 10 E
Luckey, *U.S.A.* 141 C13 41 27N 83 29W
Lucknow, *India* 81 F9 26 50N 81 0 E
Luçon, *France* 24 B2 46 28N 1 10W
Lucusse, *Angola* 103 E4 12 32 S 20 48 E
Lüda = Dalian, *China* 67 E11 38 50N 121 40 E
Luda Kamchiya →, *Bulgaria* 43 D12 43 3N 27 29 E
Ludbreg, *Croatia* 39 B13 46 15N 16 38 E
Lüdenscheid, *Germany* 26 D3 51 13N 7 37 E
Lüderitz, *Namibia* 104 D2 26 41 S 15 8 E
Ludewe □, *Tanzania* 107 D3 10 0 S 34 50 E
Ludhiana, *India* 80 D6 30 57N 75 56 E
Ludian, *China* 68 D4 27 10N 103 33 E
Luding Qiao, *China* 68 C4 29 53N 102 12 E
Lüdinghausen, *Germany* 26 D3 51 46N 7 27 E
Ludington, *U.S.A.* 134 D2 43 57N 86 27W
Ludlow, *U.K.* 17 E5 52 22N 2 42W
Ludlow, *Calif., U.S.A.* 145 L10 34 43N 116 10W
Ludlow, *Vt., U.S.A.* 137 C12 43 24N 72 42W
Ludus, *Romania* 46 C5 46 29N 24 5 E
Ludvika, *Sweden* 13 F16 60 8N 15 14 E
Ludwigsburg, *Germany* 27 G5 48 53N 9 11 E
Ludwigshafen, *Germany* 27 F4 49 29N 8 26 E
Ludwigslust, *Germany* 26 B7 53 19N 11 30 E
Ludza, *Latvia* 50 D4 56 32N 27 43 E
Lue, *Australia* 117 B8 32 38 S 149 50 E
Luebo, *Zaïre* 103 D4 5 21 S 21 23 E
Lueki, *Zaïre* 102 C5 3 20 S 25 48 E
Luena, *Angola* 103 E3 12 13 S 19 51 E
Luena, *Zaïre* 107 D2 9 28 S 25 43 E
Luena, *Zambia* 107 E3 10 40 S 30 25 E
Luepa, *Venezuela* 153 B5 5 43N 61 31W
Lüeyang, *China* 66 H4 33 22N 106 10 E
Lufeng, *Guangdong, China* 69 F10 22 57N 115 38 E
Lufeng, *Yunnan, China* 68 E4 25 0N 102 5 E
Lufico, *Angola* 103 D2 6 24 S 13 23 E
Lufira →, *Zaïre* 107 D2 9 30 S 27 0 E
Lufkin, *U.S.A.* 139 K7 31 21N 94 44W
Lufupa, *Zaïre* 103 E4 10 37 S 24 56 E
Luga, *Russia* 50 C5 58 40N 29 55 E
Luga →, *Russia* 50 C5 59 40N 28 18 E
Lugang, *Taiwan* 69 E13 24 4N 120 23 E
Lugano, *Switz.* 29 D7 46 0N 8 57 E
Lugano, L. di, *Switz.* 29 E8 46 0N 9 0 E
Lugansk = Luhansk, *Ukraine* 51 H10 48 38N 39 15 E
Lugard's Falls, *Kenya* 106 C4 3 6 S 38 41 E
Lugela, *Mozam.* 107 F4 16 25 S 36 43 E
Lugenda →, *Mozam.* 107 E4 11 25 S 38 33 E
Lugh Ganana, *Somali Rep.* 108 D2 3 48N 42 34 E
Lugnaquilla, *Ireland* 19 D5 52 58N 6 28W
Lugnvik, *Sweden* 14 B11 62 56N 17 55 E
Lugo, *Italy* 39 D8 44 25N 11 54 E
Lugo, *Spain* 36 B3 43 2N 7 35W
Lugo □, *Spain* 36 C3 43 0N 7 30W
Lugoj, *Romania* 42 B6 45 42N 21 57 E
Lugones, *Spain* 36 B5 43 26N 5 50W
Lugovoy, *Kazakhstan* 55 B6 42 55N 72 43 E
Luhansk, *Ukraine* 51 H10 48 38N 39 15 E
Luhe, *China* 69 A12 32 19N 118 50 E
Luhe →, *Germany* 26 B6 53 23N 10 13 E
Luhuo, *China* 68 B3 31 21N 100 48 E
Luiana, *Angola* 103 F4 17 25 S 22 59 E
Luimneach = Limerick, *Ireland* 19 D3 52 40N 8 37W
Luino, *Italy* 38 C5 46 0N 8 44 E
Luís Correia, *Brazil* 154 B3 3 0 S 41 35W
Luís Gonçalves, *Brazil* 154 C1 5 37 S 50 25W
Luitpold Coast, *Antarctica* 7 D1 78 30 S 32 0W
Luiza, *Zaïre* 103 D4 7 40 S 22 30 E
Luizi, *Zaïre* 106 D2 6 0 S 27 25 E
Luján, *Argentina* 158 C4 34 45 S 59 5W
Lujiang, *China* 69 B11 31 20N 117 15 E
Lukala, *Zaïre* 103 D2 5 31 S 14 32 E
Lukanga Swamp, *Zambia* 107 E2 14 30 S 27 40 E
Lukenie →, *Zaïre* 102 C3 3 0 S 18 50 E
Lukhisaral, *India* 81 G12 25 11N 86 5 E
Lŭki, *Bulgaria* 43 F9 41 50N 24 43 E
Lukk, *Libya* 96 B4 32 1N 24 46 E
Lukolela, *Equateur, Zaïre* 102 C3 1 10 S 17 12 E
Lukolela, *Kasai Or., Zaïre* 103 D4 5 23 S 24 32 E
Lukosi, *Zimbabwe* 107 F2 18 30 S 26 30 E
Lukovit, *Bulgaria* 43 D9 43 13N 24 11 E
Łuków, *Poland* 47 D9 51 55N 22 23 E
Lukoyanov, *Russia* 52 C7 55 2N 44 29 E
Lule älv →, *Sweden* 12 D19 65 35N 22 10 E
Luleå, *Sweden* 12 D20 65 35N 22 10 E
Lüleburgaz, *Turkey* 88 B2 41 23N 27 22 E
Luliang, *China* 68 E4 25 0N 103 40 E
Luling, *U.S.A.* 139 L6 29 41N 97 39W
Lulong, *China* 67 E10 39 53N 118 51 E
Lulonga →, *Zaïre* 102 B3 1 0N 18 10 E
Lulua →, *Zaïre* 103 C4 4 30 S 20 30 E
Luluabourg = Kananga, *Zaïre* 103 D4 5 55 S 22 18 E
Lumai, *Angola* 103 E4 13 13 S 21 25 E
Lumajang, *Indonesia* 75 D4 8 8 S 113 13 E
Lumaku, Gunong, *Malaysia* 75 B5 4 52N 115 38 E
Lumbala Kaquengue, *Angola* 103 E4 12 39 S 22 34 E
Lumbala N'guimbo, *Angola* 103 E4 14 18 S 21 18 E
Lumberton, *Miss., U.S.A.* 139 K10 31 0N 89 27W
Lumberton, *N.C., U.S.A.* 135 H6 34 37N 79 0W
Lumberton, *N. Mex., U.S.A.* 143 H10 36 56N 106 56W
Lumbres, *France* 23 B9 50 40N 2 5 E
Lumbwa, *Kenya* 106 C4 0 12 S 35 28 E
Lumding, *India* 78 C4 25 46N 93 10 E
Lumi, *Papua N. G.* 120 B2 3 30 S 142 2 E
Lummen, *Belgium* 21 G6 50 59N 5 12 E
Lumsden, *N.Z.* 119 F3 45 44 S 168 27 E
Lumut, *Malaysia* 77 K3 4 13N 100 37 E
Lumut, Tg., *Indonesia* 74 C3 3 50 S 105 58 E
Luna, *Luzon, Phil.* 70 B3 18 18N 121 21 E
Luna, *Luzon, Phil.* 70 C3 16 51N 120 23 E
Lunan, *China* 68 E4 24 40N 103 18 E
Lunavada, *India* 80 H5 23 8N 73 37 E
Lunca, *Romania* 46 B6 47 22N 25 1 E
Lund, *Sweden* 15 J7 55 44N 13 12 E
Lund, *U.S.A.* 142 G6 38 52N 115 0W
Lunda Norte □, *Angola* 103 D3 8 0 S 20 0 E
Lunda Sul □, *Angola* 103 D4 10 0 S 20 0 E
Lundazi, *Zambia* 107 E3 12 20 S 33 7 E
Lunderskov, *Denmark* 15 J3 55 29N 9 19 E
Lundi →, *Zimbabwe* 107 G3 21 43 S 32 34 E
Lundu, *Malaysia* 75 B3 1 40N 109 50 E
Lundy, *U.K.* 17 F3 51 10N 4 41W
Lune →, *U.K.* 16 C5 54 0N 2 51W
Lüneburg, *Germany* 26 B6 53 15N 10 24 E
Lüneburg Heath = Lüneburger Heide, *Germany* 26 C6 53 10N 10 12 E
Lüneburger Heide, *Germany* 26 C6 53 10N 10 12 E
Lunel, *France* 25 E8 43 39N 4 9 E
Lünen, *Germany* 26 D3 51 37N 7 32 E
Lunenburg, *Canada* 129 D7 44 22N 64 18W
Lunéville, *France* 23 D13 48 36N 6 30 E
Lunga →, *Zambia* 107 E2 14 34 S 26 25 E
Lungern, *Switz.* 28 C6 46 48N 8 10 E
Lungi Airport, *S. Leone* 100 D2 8 40N 13 17W
Lunglei, *India* 78 D4 22 55N 92 45 E
Lungngo, *Burma* 78 E4 21 57N 93 36 E
Luni, *India* 80 F5 26 0N 73 6 E
Luni →, *India* 80 G4 24 41N 71 14 E
Luninets = Luninyets, *Belarus* 51 F4 52 15N 26 50 E
Luning, *U.S.A.* 142 G4 38 30N 118 11W
Lunino, *Russia* 52 D7 53 38N 45 18 E
Luninyets, *Belarus* 51 F4 52 15N 26 50 E
Lunner, *Norway* 14 D4 60 19N 10 35 E
Lunsemfwa →, *Zambia* 107 E3 14 54 S 30 12 E
Lunsemfwa Falls, *Zambia* 107 E2 14 30 S 29 6 E
Lunteren, *Neths.* 20 D7 52 5N 5 38 E
Luo He →, *China* 66 G6 34 35N 110 20 E
Luocheng, *China* 68 E7 24 48N 108 53 E
Luochuan, *China* 66 G5 35 45N 109 26 E
Luoci, *China* 68 E4 25 19N 102 18 E
Luodian, *China* 68 E6 25 24N 106 43 E
Luoding, *China* 69 F8 22 45N 111 40 E
Luodong, *Taiwan* 69 E13 24 41N 121 46 E
Luofu, *Zaïre* 106 C2 0 10 S 29 15 E
Luohe, *China* 66 H8 33 32N 114 2 E
Luojiang, *China* 68 B5 31 18N 104 33 E
Luonan, *China* 66 G6 34 5N 110 10 E
Luoning, *China* 66 G6 34 35N 111 40 E
Luoshan, *China* 69 A10 32 13N 114 30 E
Luotian, *China* 69 B10 30 46N 115 22 E
Luoyang, *China* 66 G7 34 40N 112 26 E
Luoyuan, *China* 69 D12 26 28N 119 30 E
Luozi, *Zaïre* 103 C2 4 54 S 14 0 E
Luozigou, *China* 67 C16 43 42N 130 18 E
Lupanshui, *China* 68 D5 26 38N 104 48 E
Lupeni, *Romania* 46 D4 45 21N 23 13 E
Lupilichi, *Mozam.* 107 E4 11 47 S 35 13 E
Lupire, *Angola* 103 E3 14 36 S 19 29 E
Łupków, *Poland* 31 B15 49 15N 22 4 E
Lupoing, *China* 68 E5 24 53N 104 21 E
Lupon, *Phil.* 71 H5 6 54N 126 0 E
Luquan, *China* 68 E4 25 35N 102 25 E
Luque, *Paraguay* 158 B4 25 19 S 57 25W
Luque, *Spain* 37 H6 37 35N 4 16W
Luray, *U.S.A.* 134 F6 38 40N 78 28W
Lure, *France* 23 E13 47 40N 6 30 E
Luremo, *Angola* 103 D3 8 30 S 17 50 E
Lurgan, *U.K.* 19 B5 54 28N 6 19W
Luri, *France* 25 F13 42 53N 9 23 E
Luribay, *Bolivia* 156 D4 17 6 S 67 39W
Lurin, *Peru* 156 C2 12 17 S 76 52W
Lusaka, *Zambia* 107 F2 15 28 S 28 16 E
Lusambo, *Zaïre* 103 C4 4 58 S 23 28 E
Lusangaye, *Zaïre* 106 C2 4 54 S 26 0 E
Luseland, *Canada* 131 C7 52 5N 109 24W
Lushan, *Henan, China* 66 H7 33 45N 112 55 E
Lushan, *Sichuan, China* 68 B4 30 12N 102 52 E
Lushi, *China* 66 G6 34 3N 111 3 E
Lushnja, *Albania* 44 D1 40 55N 19 41 E
Lushoto, *Tanzania* 106 C4 4 47 S 38 20 E
Lushoto □, *Tanzania* 106 C4 4 45 S 38 20 E
Lushui, *China* 68 E2 25 58N 98 44 E
Lüshun, *China* 67 E11 38 45N 121 15 E
Lusignan, *France* 24 B4 46 26N 0 8 E
Lusigny-sur-Barse, *France* 23 D11 48 16N 4 15 E
Lusk, *U.S.A.* 138 D2 42 46N 104 27W
Lussac-les-Châteaux, *France* 24 B4 46 24N 0 43 E
Lussanvira, *Brazil* 155 F1 20 42 S 51 7W
Luta = Dalian, *China* 67 E11 38 50N 121 40 E
Lutembo, *Angola* 103 E4 13 26 S 21 16 E
Luti, *Solomon Is.* 121 L9 7 14 S 157 0 E
Luton, *U.K.* 17 F7 51 53N 0 24W
Lutong, *Malaysia* 75 B4 4 28N 114 0 E
Lutry, *Switz.* 28 C3 46 31N 6 42 E
Lutsk, *Ukraine* 51 G3 50 50N 25 15 E
Lutuai, *Angola* 103 E4 12 41 S 20 7 E
Lützow Holmbukta, *Antarctica* 7 C4 69 10 S 37 30 E
Lutzputs, *S. Africa* 104 D3 28 3 S 20 40 E
Luverne, *U.S.A.* 138 D6 43 39N 96 13W
Luvo, *Angola* 103 D2 5 51 S 14 5 E
Luvua, *Zaïre* 103 D5 8 48 S 25 17 E
Luvua →, *Zaïre* 106 D2 6 50 S 27 30 E
Luwegu →, *Tanzania* 107 D4 8 31 S 37 23 E
Luwuk, *Indonesia* 72 B2 0 56 S 122 47 E
Luxembourg, *Lux.* 21 J8 49 37N 6 9 E
Luxembourg □, *Belgium* 21 J7 49 58N 5 30 E
Luxembourg ■, *Europe* 21 J8 49 45N 6 0 E
Luxeuil-les-Bains, *France* 23 E13 47 49N 6 24 E
Luxi, *Hunan, China* 69 C8 28 20N 110 7 E
Luxi, *Yunnan, China* 68 E4 24 40N 103 55 E
Luxi, *Yunnan, China* 68 E2 24 27N 98 36 E
Luxor = El Uqsur, *Egypt* 94 B3 25 41N 32 38 E
Luy →, *France* 24 E2 43 39N 1 9W
Luy-de-Béarn →, *France* 24 E3 43 39N 0 48W
Luy-de-France →, *France* 24 E3 43 39N 0 48W
Luyi, *China* 66 H8 33 50N 115 35 E
Luyksgestel, *Neths.* 21 F6 51 17N 5 20 E
Luz-St.-Sauveur, *France* 24 F4 42 53N 0 1 E
Luza, *Russia* 48 B8 60 39N 47 10 E
Luzern, *Switz.* 29 B6 47 3N 8 18 E
Luzern □, *Switz.* 28 B5 47 2N 7 55 E
Luzhai, *China* 68 E7 24 29N 109 42 E
Luzhou, *China* 68 C5 28 52N 105 20 E
Luziânia, *Brazil* 155 E2 16 20 S 48 0W
Luzilândia, *Brazil* 154 B3 3 28 S 42 22W
Luzon, *Phil.* 70 C3 16 0N 121 0 E
Luzy, *France* 23 F10 46 47N 3 58 E
Luzzi, *Italy* 41 C9 39 27N 16 17 E
Lviv, *Ukraine* 51 H3 49 50N 24 0 E
Lvov = Lviv, *Ukraine* 51 H3 49 50N 24 0 E
Lwówek, *Poland* 47 C3 52 28N 16 10 E
Lwówek Śląski, *Poland* 47 D2 51 7N 15 38 E
Lyakhavichy, *Belarus* 51 F4 53 2N 26 32 E
Lyakhovskiye, Ostrova, *Russia* 57 B15 73 40N 141 0 E
Lyaki = Läki, *Azerbaijan* 53 K8 40 34N 47 22 E
Lyall Mt., *N.Z.* 119 F2 45 16 S 167 32 E
Lyallpur = Faisalabad, *Pakistan* 79 C4 31 30N 73 5 E
Lyalya →, *Russia* 54 B8 59 9N 61 29 E
Lyaskovets, *Bulgaria* 43 D10 43 6N 25 44 E
Lycaonia, *Turkey* 88 D5 38 0N 33 0 E
Lychen, *Germany* 26 B9 53 12N 13 18 E
Lychkova, *Russia* 50 D7 57 55N 32 24 E
Lycia, *Turkey* 88 D3 36 30N 29 30 E
Lycksele, *Sweden* 12 D18 64 38N 18 40 E
Lycosura, *Greece* 45 G4 37 20N 22 3 E
Lydda = Lod, *Israel* 91 D3 31 57N 34 54 E
Lydenburg, *S. Africa* 105 D5 25 10 S 30 29 E
Lydia, *Turkey* 88 C3 38 48N 28 19 E
Lyell, *N.Z.* 119 B7 41 48 S 172 4 E
Lyell I., *Canada* 130 C2 52 40N 131 35W
Lyepyel, *Belarus* 50 E5 54 50N 28 40 E
Lygnern, *Sweden* 15 G6 57 30N 12 15 E
Lyman, *U.S.A.* 142 F8 41 20N 110 18W
Lyme Regis, *U.K.* 17 G5 50 43N 2 57W
Lymington, *U.K.* 17 G6 50 45N 1 32W
Łyna →, *Poland* 13 J19 54 37N 21 14 E
Lynchburg, *Ohio, U.S.A.* 141 E13 39 15N 83 48W
Lynchburg, *Va., U.S.A.* 134 G6 37 25N 79 9W
Lynd →, *Australia* 114 B3 16 28 S 143 18 E
Lynd Ra., *Australia* 115 D4 25 30 S 149 20 E
Lynden, *Canada* 136 C4 43 14N 80 9W
Lynden, *U.S.A.* 144 B4 48 57N 122 27W
Lyndhurst, *Queens., Australia* 114 B3 19 12 S 144 20 E
Lyndhurst, *S. Austral., Australia* 115 E2 30 15 S 138 18 E
Lyndon →, *Australia* 113 D1 23 29 S 114 6 E
Lyndonville, *N.Y., U.S.A.* 136 C6 43 20N 78 23W
Lyndonville, *Vt., U.S.A.* 137 B12 44 31N 72 1W
Lyngdal, *Norway* 14 E3 59 54N 9 32 E
Lyngen, *Norway* 12 B19 69 45N 20 30 E
Lynher Reef, *Australia* 112 C3 15 27 S 121 55 E
Lynn, *Ind., U.S.A.* 141 D12 40 3N 84 56W
Lynn, *Mass., U.S.A.* 137 D14 42 28N 70 57W
Lynn Canal, *U.S.A.* 130 B1 58 50N 135 15W
Lynn Lake, *Canada* 131 B8 56 51N 101 3W
Lynnwood, *U.S.A.* 144 C4 47 49N 122 19W
Lynton, *U.K.* 17 F4 51 13N 3 50W
Lyntupy, *Belarus* 13 J22 55 4N 26 23 E
Lynx L., *Canada* 131 A7 62 25N 106 15W
Lyø, *Denmark* 15 J4 55 3N 10 9 E
Lyon, *France* 25 C8 45 46N 4 50 E
Lyonnais, *France* 25 C8 45 45N 4 15 E
Lyons = Lyon, *France* 25 C8 45 46N 4 50 E
Lyons, *Colo., U.S.A.* 138 E2 40 14N 105 16W
Lyons, *Ga., U.S.A.* 135 J4 32 12N 82 19W
Lyons, *Kans., U.S.A.* 138 F5 38 21N 98 12W
Lyons, *N.Y., U.S.A.* 136 C8 43 5N 77 0W
Lyozna, *Belarus* 50 E6 55 0N 30 50 E
Lyrestad, *Sweden* 15 F8 58 48N 14 4 E
Lys = Leie →, *Belgium* 23 A10 51 2N 3 45 E
Lysá, *Czech.* 30 A7 50 11N 14 51 E
Lysekil, *Sweden* 15 F5 58 17N 11 26 E
Lyskovo, *Russia* 52 B7 56 0N 45 3 E
Lyss, *Switz.* 28 B4 47 4N 7 19 E
Lysva, *Russia* 54 B6 58 7N 57 49 E
Lysychansk, *Ukraine* 51 H10 48 55N 38 30 E
Lytle, *U.S.A.* 139 L5 29 14N 98 48W
Lyttelton, *N.Z.* 119 D7 43 35 S 172 44 E
Lytton, *Canada* 130 C4 50 13N 121 31W
Lyuban, *Russia* 50 C6 59 16N 31 18 E
Lyubertsy, *Russia* 52 C3 55 39N 37 50 E
Lyubim, *Russia* 52 A5 58 20N 40 39 E
Lyubimets, *Bulgaria* 43 F11 41 50N 26 5 E
Lyuboml, *Ukraine* 51 G3 51 11N 24 4 E
Lyubotyn, *Ukraine* 51 H8 50 0N 36 0 E
Lyubytino, *Russia* 50 C7 58 50N 33 16 E
Lyudinovo, *Russia* 52 D2 53 52N 34 28 E

M

Ma →, *Vietnam* 76 C5 19 47N 105 56 E
Ma'adaba, *Jordan* 91 E4 30 43N 35 47 E
Maamba, *Zambia* 104 B4 17 17 S 26 28 E
Ma'ān, *Jordan* 91 E4 30 12N 35 44 E
Ma'ān □, *Jordan* 91 F5 30 0N 36 0 E
Maanselkä, *Finland* 12 C23 63 52N 28 32 E
Ma'anshan, *China* 69 B12 31 44N 118 29 E
Maarheeze, *Neths.* 21 F7 51 19N 5 36 E
Maarianhamina, *Finland* 13 F18 60 5N 19 55 E
Maarn, *Neths.* 20 D6 52 3N 5 22 E
Ma'arrat an Nu'mān, *Syria* 88 E7 35 43N 36 43 E
Maarssen, *Neths.* 20 D6 52 9N 5 2 E
Maartensdijk, *Neths.* 20 D6 52 9N 5 10 E
Maas →, *Neths.* 20 E5 51 45N 4 32 E
Maasbracht, *Belgium* 21 F7 51 9N 5 54 E
Maasbree, *Neths.* 21 F8 51 22N 6 3 E
Maasdam, *Neths.* 20 E5 51 48N 4 34 E
Maasdijk, *Neths.* 20 E4 51 58N 4 13 E
Maaseik, *Belgium* 21 F7 51 6N 5 45 E
Maasland, *Neths.* 20 E4 51 57N 4 16 E
Maasniel, *Neths.* 21 F8 51 12N 6 1 E
Maassluis, *Neths.* 20 E4 51 56N 4 16 E
Maastricht, *Neths.* 21 G7 50 50N 5 40 E
Maave, *Mozam.* 105 C5 21 4 S 34 47 E
Ma'bar, *Yemen* 86 D4 14 48N 44 17 E
Mabaruma, *Guyana* 153 B6 8 10N 59 50W
Mabein, *Burma* 78 D6 23 29N 96 37 E
Mabel L., *Canada* 130 C5 50 35N 118 43W
Mabenge, *Zaïre* 106 B1 4 15N 24 12 E
Mabian, *China* 68 C4 28 47N 103 37 E
Mablethorpe, *U.K.* 16 D8 53 20N 0 15 E
Maboma, *Zaïre* 106 B2 2 30N 28 10 E
Maboukou, *Congo* 102 C2 3 3 S 12 31 E
Mabrouk, *Mali* 101 B4 19 29N 1 15W
Mabton, *U.S.A.* 142 C3 46 13N 120 0W
Mabungo, *Somali Rep.* 108 D2 0 49N 42 35 E
Mac Bac, *Vietnam* 77 H6 9 46N 106 7 E
Macachín, *Argentina* 158 D3 37 10 S 63 43W
Macaé, *Brazil* 155 F3 22 20 S 41 43W
Macaíba, *Brazil* 154 C4 5 51 S 35 21W
Macajuba, *Brazil* 155 D3 12 9 S 40 22W
Macalelon, *Phil.* 70 E4 13 45N 122 8 E
McAlester, *U.S.A.* 139 H7 34 56N 95 46W
McAllen, *U.S.A.* 139 M5 26 12N 98 14W
Macamic, *Canada* 128 C4 48 45N 79 0W
Macao = Macau ■, *China* 69 F9 22 16N 113 35 E
Macão, *Portugal* 37 F3 39 35N 7 59W
Macapá, *Brazil* 153 C7 0 5N 51 4W
Macará, *Ecuador* 152 D2 4 23 S 79 57W
Macarani, *Brazil* 155 E3 15 33 S 40 24W
Macarena, Serranía de la, *Colombia* 152 C3 2 45N 73 55W
Macarthur, *Australia* 116 E5 38 5 S 142 0 E
McArthur →, *Australia* 114 B2 15 54 S 136 40 E
McArthur, Port, *Australia* 114 B2 16 4 S 136 23 E
McArthur River, *Australia* 114 B2 16 27 S 136 7 E
Macas, *Ecuador* 152 D2 2 19 S 78 7W
Macate, *Peru* 156 B2 8 48 S 78 7W
Macau, *Brazil* 154 C4 5 15 S 36 40W
Macau ■, *China* 69 F9 22 16N 113 35 E
Macaúbas, *Brazil* 155 D3 13 2 S 42 42W
Macaya →, *Colombia* 152 C3 0 59N 72 20W
McBride, *Canada* 130 C4 53 20N 120 19W
McCall, *U.S.A.* 142 D5 44 55N 116 6W
McCamey, *U.S.A.* 139 K3 31 8N 102 14W
McCammon, *U.S.A.* 142 E7 42 39N 112 12W
McCauley I., *Canada* 130 C2 53 40N 130 15W
McCleary, *U.S.A.* 144 C3 47 3N 123 16W
Macclesfield, *U.K.* 16 D5 53 15N 2 8W
McClintock, *Canada* 131 B10 57 50N 94 10W
M'Clintock Chan., *Canada* 126 A9 72 0N 102 0W
McClintock Ra., *Australia* 112 C4 18 44 S 127 38 E
McCloud, *U.S.A.* 142 F2 41 15N 122 8W
McCluer I., *Australia* 112 B5 11 5 S 133 0 E
McClure, *U.S.A.* 136 F7 40 42N 77 19W
McClure, L., *U.S.A.* 144 H6 37 35N 120 16W
M'Clure Str., *Canada* 124 B8 75 0N 119 0W
McClusky, *U.S.A.* 138 B4 47 29N 100 27W
McComb, *U.S.A.* 139 K9 31 15N 90 27W
McConaughy, L., *U.S.A.* 138 E4 41 14N 101 40W
McCook, *U.S.A.* 138 E4 40 12N 100 38W
McCullough Mt., *U.S.A.* 145 K11 35 35N 115 13W
McCusker →, *Canada* 131 B7 55 32N 108 39W
McDame, *Canada* 130 B3 59 44N 128 59W
McDermitt, *U.S.A.* 142 F5 41 59N 117 43W
Macdonald, L., *Australia* 112 D4 23 30 S 129 0 E
Macdonald, Mt., *Vanuatu* 121 G6 17 36 S 168 23 E
McDonald Is., *Ind. Oc.* 109 K6 53 0 S 73 0 E
Macdonnell Ras., *Australia* 112 D5 23 40 S 133 0 E
McDouall Peak, *Australia* 115 D1 29 51 S 134 55 E
Macdougall L., *Canada* 126 B10 66 0N 98 27W
McDougalls Well, *Australia* 116 A4 31 8 S 141 15 E
MacDowell L., *Canada* 128 B1 52 15N 92 45W
Macduff, *U.K.* 18 D6 57 40N 2 31W
Maceda, *Spain* 36 C3 42 16N 7 39W
Macedonia = Makedhonía □, *Greece* 44 D3 40 39N 22 0 E
Macedonia ■, *Europe* 42 F6 41 53N 21 40 E
Maceió, *Brazil* 154 C4 9 40 S 35 41W
Maceira, *Portugal* 37 F2 39 41N 8 55W
Macenta, *Guinea* 100 D3 8 35N 9 32W
Macerata, *Italy* 39 E10 43 18N 13 27 E
McFarland, *U.S.A.* 145 K7 35 41N 119 14W
McFarlane →, *Canada* 131 B7 59 12N 107 58W
Macfarlane, L., *Australia* 116 B2 32 0 S 136 40 E
McGehee, *U.S.A.* 139 J9 33 38N 91 24W
McGill, *U.S.A.* 142 G6 39 23N 114 47W
Macgillycuddy's Reeks, *Ireland* 19 D2 51 58N 9 45W
MacGregor, *Canada* 131 D9 49 57N 98 48W
McGregor, *U.S.A.* 140 A5 43 1N 91 11W
McGregor →, *Canada* 130 B4 55 10N 122 0W
McGregor Ra., *Australia* 115 D3 27 0 S 142 45 E
Māch Kowr, *Iran* 85 E9 25 48N 61 28 E
Machacalis, *Brazil* 155 E3 17 5 S 40 45W
Machado = Jiparaná →, *Brazil* 157 B5 8 3 S 62 52W
Machagai, *Argentina* 158 B3 26 56 S 60 2W
Machakos, *Kenya* 106 C4 1 30 S 37 15 E
Machakos □, *Kenya* 106 C4 1 30 S 37 15 E
Machala, *Ecuador* 152 D2 3 20 S 79 57W
Machanga, *Mozam.* 105 C6 20 59 S 35 0 E
Machattie, L., *Australia* 114 C2 24 50 S 139 48 E
Machava, *Mozam.* 105 D5 25 54 S 32 28 E
Machece, *Mozam.* 107 F4 19 15 S 35 32 E
Machecoul, *France* 22 F5 47 0N 1 49W
Machelen, *Belgium* 21 G4 50 55N 4 26 E
Macheng, *China* 69 B10 31 12N 115 2 E
McHenry, *U.S.A.* 141 B8 42 21N 88 16W
Machevna, *Russia* 57 C18 61 20N 172 20 E
Machezo, *Spain* 37 F6 39 21N 4 20W
Machias, *U.S.A.* 129 D6 44 43N 67 28W
Machichaco, C., *Spain* 34 B2 43 28N 2 47W
Machichi →, *Canada* 131 B10 57 3N 92 6W
Machico, *Madeira* 33 D3 32 43N 16 44W
Machida, *Japan* 63 B11 35 28N 139 23 E
Machilipatnam, *India* 83 F5 16 12N 81 8 E
Machiques, *Venezuela* 152 A3 10 4N 72 34W
Machupicchu, *Peru* 156 C3 13 8 S 72 30W
Machynlleth, *U.K.* 17 E4 52 35N 3 50W
Maciejowice, *Poland* 47 D8 51 36N 21 26 E
McIlwraith Ra., *Australia* 114 A3 13 50 S 143 20 E
Măcin, *Romania* 46 D9 45 16N 28 8 E
Macina, *Mali* 100 C4 14 50N 5 0W
McIntosh, *U.S.A.* 138 C4 45 55N 101 21W
McIntosh L., *Canada* 131 B8 55 45N 105 0W
Macintosh Ra., *Australia* 113 E4 27 39 S 125 32 E
Macintyre →, *Australia* 115 D5 28 37 S 150 47 E
Macizo Galaico, *Spain* 36 C3 42 30N 7 30W
Mackay, *Australia* 114 C4 21 8 S 149 11 E
Mackay, *U.S.A.* 142 E7 43 55N 113 37W
MacKay →, *Canada* 130 B6 57 10N 111 38W
Mackay, L., *Australia* 112 D4 22 30 S 129 0 E
McKay Ra., *Australia* 112 D3 23 0 S 122 30 E
McKeesport, *U.S.A.* 136 F5 40 21N 79 52W
McKenna, *U.S.A.* 144 D4 46 56N 122 33W
Mackenzie, *Canada* 130 B4 55 20N 123 5W
Mackenzie, *Guyana* 153 B6 6 0N 58 17W
McKenzie, *U.S.A.* 135 G1 36 8N 88 31W
Mackenzie →, *Australia* 114 C4 23 38 S 149 46 E
Mackenzie →, *Canada* 126 B6 69 10N 134 20W

228 McKenzie

McKenzie →, *U.S.A.* . . . **142 D2** 44 7N 123 6W
McKenzie Bay, *Canada* . **6 B1** 69 0N 137 30W
Mackenzie City = Linden, *Guyana* **153 B6** 6 0N 58 10W
Mackenzie Highway, *Canada* **130 B5** 58 0N 117 15W
Mackenzie Mts., *Canada* . **126 B6** 64 0N 130 0W
Mackenzie Plains, *N.Z.* . . **119 E5** 44 10 S 170 25 E
McKerrow L., *N.Z.* . . . . **119 E3** 44 25 S 168 5 E
Mackinaw, *U.S.A.* **140 D7** 40 32N 89 21W
Mackinaw →, *U.S.A.* . . . **140 D7** 40 33N 89 44W
Mackinaw City, *U.S.A.* . . **134 C3** 45 47N 84 44W
McKinlay, *Australia* . . . . . **114 C3** 21 16 S 141 18 E
McKinlay →, *Australia* . **114 C3** 20 50 S 141 28 E
McKinley, Mt., *U.S.A.* . . . **126 B4** 63 4N 151 0W
McKinley Sea, *Arctic* . . . . **6 A7** 82 0N 0 0 E
McKinney, *U.S.A.* **139 J6** 33 12N 96 37W
McKinnon Pass, *N.Z.* . . . . **119 E3** 44 52 S 168 12 E
Mackinnon Road, *Kenya* **106 C4** 3 40 S 39 1 E
Macksville, *Australia* . . . **117 A10** 30 40 S 152 56 E
McLaren Vale, *Australia* . **116 C3** 35 13 S 138 31 E
McLaughlin, *U.S.A.* **138 C4** 45 49N 100 49W
Maclean, *Australia* . . . . . . **115 D5** 29 26 S 153 16 E
McLean, *Ill., U.S.A.* **140 D7** 40 19N 89 10W
McLean, *Tex., U.S.A.* **139 H4** 35 14N 100 36W
McLeansboro, *U.S.A.* **138 F10** 38 6N 88 32W
Maclear, *S. Africa* . . . . . . **105 E4** 31 2 S 28 23 E
Macleay →, *Australia* . . . **117 A10** 30 56 S 153 0 E
McLennan, *Canada* **130 B5** 55 42N 116 50W
MacLeod, B., *Canada* . . . **131 A7** 62 53N 110 0W
McLeod, L., *Australia* . . . **113 D1** 24 9 S 113 47 E
MacLeod Lake, *Canada* . . **130 C4** 54 58N 123 0W
McLoughlin, Mt., *U.S.A.* . **142 E2** 42 27N 122 19W
McLure, *Canada* **130 C4** 51 2N 120 13W
McMechen, *U.S.A.* **136 G4** 39 57N 80 44W
McMillan, L., *U.S.A.* **139 J2** 32 36N 104 21W
McMinnville, *Oreg., U.S.A.* **142 D2** 45 13N 123 12W
McMinnville, *Tenn., U.S.A.* **135 H3** 35 41N 85 46W
McMorran, *Canada* **131 C7** 51 19N 108 42W
McMurdo Sd., *Antarctica* **7 D11** 77 0 S 170 0 E
McMurray = Fort McMurray, *Canada* . . . **130 B6** 56 44N 111 7W
McMurray, *U.S.A.* **144 B4** 48 19N 122 14W
McNary, *U.S.A.* **143 J9** 34 4N 109 51W
MacNutt, *Canada* **131 C8** 51 5N 101 36W
Maco, *Phil.* . . . . **71 H5** 7 20N 125 50 E
Macocolo, *Angola* . . . . . . **103 D3** 6 47 S 16 8 E
Macodoene, *Mozam.* . . . . . **105 C6** 23 32 S 35 5 E
Macomb, *U.S.A.* **140 D6** 40 27N 90 40W
Macomer, *Italy* . . . . **40 B1** 40 16N 8 47 E
Mâcon, *France* . . . . **25 B8** 46 19N 4 50 E
Macon, *Ga., U.S.A.* **135 J4** 32 51N 83 38W
Macon, *Ill., U.S.A.* **140 E7** 39 43N 89 0W
Macon, *Miss., U.S.A.* **135 J1** 33 7N 88 34W
Macon, *Mo., U.S.A.* **140 E4** 39 44N 92 28W
Macondo, *Angola* . . . . **103 E4** 12 37 S 23 46 E
Macossa, *Mozam.* . . . . **107 F3** 17 55 S 33 56 E
Macoun L., *Canada* **131 B8** 56 32N 103 40W
Macoupin Cr. →, *U.S.A.* . **140 E6** 39 11N 90 38W
Macovane, *Mozam.* . . . . . **105 C6** 21 30 S 35 2 E
McPherson, *U.S.A.* **138 F6** 38 22N 97 40W
McPherson Pk., *U.S.A.* . . **145 L7** 34 53N 119 53W
McPherson Ra., *Australia* **115 D5** 28 15 S 153 15 E
Macquarie Harbour, *Australia* . . . . **114 G4** 42 15 S 145 23 E
Macquarie Is., *Pac. Oc.* . **122 N7** 54 36 S 158 55 E
MacRobertson Land, *Antarctica* . . . . **7 D6** 71 0 S 64 0 E
Macroom, *Ireland* **19 E3** 51 54N 8 57W
Macroy, *Australia* . . . . **112 D2** 20 53 S 118 2 E
MacTier, *Canada* **136 A5** 45 9N 79 46W
Macubela, *Mozam.* . . . . . . **107 F4** 16 53 S 37 49 E
Macugnaga, *Italy* . . . . **38 C4** 45 58N 7 58 E
Macuiza, *Mozam.* . . . . **107 F3** 18 7 S 34 29 E
Macujer, *Colombia* **152 C3** 0 24N 73 10W
Macusani, *Peru* . . . . **156 C3** 14 4 S 70 29W
Macuse, *Mozam.* . . . . **107 F4** 17 45 S 37 10 E
Macuspana, *Mexico* **147 D6** 17 46N 92 36W
Macusse, *Angola* . . . . **103 F4** 17 48 S 20 23 E
McVille, *U.S.A.* **138 B5** 47 46N 98 11W
Madadeni, *S. Africa* . . . . . **105 D5** 27 43 S 30 3 E
Madadi, *Chad* . . . . **97 E4** 18 28N 20 45 E
Madagali, *Nigeria* . . . . . . **101 C7** 10 56N 13 33 E
Madagascar ■, *Africa* . . . **105 C8** 20 0 S 47 0 E
Madā'in Sālih, *Si. Arabia* **84 E3** 26 46N 37 57 E
Madalag, *Phil.* . . . . **71 F4** 11 32N 122 18 E
Madama, *Niger* . . . . **97 D2** 22 0N 13 40 E
Madame I., *Canada* **129 C7** 45 30N 60 58W
Madan, *Bulgaria* . . . . **43 F9** 41 30N 24 57 E
Madanapalle, *India* . . . . . **83 H4** 13 33N 78 28 E
Madang, *Papua N. G.* . . . . **120 C3** 5 12 S 145 49 E
Madaoua, *Niger* . . . . **101 C6** 14 5N 6 27 E
Madara, *Nigeria* . . . . **101 C7** 11 45N 10 35 E
Madaripur, *Bangla.* . . . . . **78 D3** 23 19N 90 15 E
Madauk, *Burma* . . . . **78 G6** 17 56N 96 52 E
Madawaska, *Canada* **136 A7** 45 30N 78 0W
Madawaska →, *Canada* . **128 C4** 45 27N 76 21W
Madaya, *Burma* . . . . **78 D6** 22 12N 96 10 E
Madbar, *Sudan* . . . . **95 F3** 6 17N 30 45 E
Maddaloni, *Italy* . . . . **41 A7** 41 2N 14 23 E
Made, *Neths.* . . . . **21 E5** 51 41N 4 49 E
Madeira, *Atl. Oc.* **33 D3** 32 50N 17 0W
Madeira, *U.S.A.* **141 E12** 39 11N 84 22W
Madeira →, *Brazil* . . . . . **153 D6** 3 22 S 58 45W
Madeleine, Is. de la, *Canada* **129 C7** 47 30N 61 40W
Maden, *Turkey* . . . . **89 C8** 38 23N 39 40 E
Madera, *U.S.A.* **144 J6** 36 57N 120 3W
Madgaon, *India* . . . . **83 G1** 15 12N 73 58 E
Madha, *India* . . . . **82 F2** 18 0N 75 30 E
Madhubani, *India* . . . . **81 F12** 26 21N 86 7 E
Madhumati →, *Bangla.* . **78 D2** 22 53N 89 52 E
Madhya Pradesh □, *India* **80 J7** 21 50N 78 0 E
Madian, *China* . . . . **69 A11** 33 0N 116 6 E
Madidi →, *Bolivia* . . . . . **156 C4** 12 32 S 66 52W
Madikeri, *India* . . . . **83 H2** 12 30N 75 45 E
Madill, *U.S.A.* **139 H6** 34 6N 96 46W
Madimba, *Angola* . . . . **103 D2** 6 36 S 14 23 E
Madimba, *Zaïre* . . . . **103 C3** 4 58 S 15 5 E
Ma'din, *Syria* . . . . **89 E8** 35 45N 39 36 E
Madīnat ash Sha'b, *Yemen* **86 D4** 12 50N 45 0 E
Madingou, *Congo* . . . . **102 C2** 4 10 S 13 33 E
Madirovalo, *Madag.* . . . . . **105 B8** 16 26 S 46 32 E
Madison, *Calif., U.S.A.* . . **144 G5** 38 41N 121 59W
Madison, *Fla., U.S.A.* **135 K4** 30 28N 83 25W
Madison, *Ind., U.S.A.* . . . **141 F11** 38 44N 85 23W
Madison, *Mo., U.S.A.* **140 E4** 39 28N 92 13W
Madison, *Nebr., U.S.A.* . . **138 E6** 41 50N 97 27W
Madison, *Ohio, U.S.A.* . . . **136 E3** 41 46N 81 3W
Madison, *S. Dak., U.S.A.* **138 D6** 44 0N 97 7W
Madison, *Wis., U.S.A.* . . . **140 A7** 43 4N 89 24W
Madison →, *U.S.A.* **142 D8** 45 56N 111 31W
Madisonville, *Ky., U.S.A.* **134 G2** 37 20N 87 30W
Madisonville, *Tex., U.S.A.* **139 K7** 30 57N 95 55W
Madista, *Botswana* . . . . . . **104 C4** 21 15 S 25 6 E
Madiun, *Indonesia* . . . . . . **75 D4** 7 38 S 111 32 E
Madley, *U.K.* **17 E5** 52 2N 2 51W
Madol, *Sudan* . . . . **95 F2** 9 3N 27 45 E
Madon →, *France* . . . . . . **23 D13** 48 36N 6 6 E
Madona, *Latvia* . . . . **13 H22** 56 53N 26 5 E
Madrakah, Ra's al, *Oman* **87 C7** 19 0N 57 50 E
Madras = Tamil Nadu □, *India* . . . . **83 J3** 11 0N 77 0 E
Madras, *India* . . . . **83 H5** 13 8N 80 19 E
Madras, *U.S.A.* **142 D3** 44 38N 121 8W
Madre, L., *Mexico* **147 B5** 25 0N 97 30W
Madre, Laguna, *U.S.A.* . . **139 M6** 27 0N 97 30W
Madre, Sierra, *Phil.* . . . . . **70 C4** 17 0N 122 0 E
Madre de Dios □, *Peru* . . **156 C3** 12 0 S 70 15W
Madre de Dios →, *Bolivia* . . . . **156 C4** 10 59 S 66 8W
Madre de Dios, I., *Chile* . **160 D1** 50 20 S 75 10W
Madre del Sur, Sierra, *Mexico* **147 D5** 17 30N 100 0W
Madre Occidental, Sierra, *Mexico* **146 B3** 27 0N 107 0W
Madre Oriental, Sierra, *Mexico* **146 C4** 25 0N 100 0W
Madri, *India* . . . . **80 G5** 24 16N 73 32 E
Madrid, *Spain* **36 E7** 40 25N 3 45W
Madrid, *U.S.A.* **140 C3** 41 53N 93 49W
Madrid □, *Spain* **36 E7** 40 30N 3 45W
Madridejos, *Spain* **37 F7** 39 28N 3 33W
Madrigal de las Altas Torres, *Spain* **36 D6** 41 5N 5 0W
Madrona, Sierra, *Spain* . . **37 G6** 38 27N 4 16W
Madroñera, *Spain* **37 F5** 39 26N 5 42W
Madu, *Sudan* . . . . **95 E2** 14 37N 26 4 E
Madura, Selat, *Indonesia* **75 D4** 7 30 S 113 20 E
Madura Motel, *Australia* . **113 F4** 31 55 S 127 0 E
Madurai, *India* . . . . **83 K4** 9 55N 78 10 E
Madurantakam, *India* . . . . **83 H4** 12 30N 79 50 E
Madzhalis, *Russia* . . . . **53 J8** 42 9N 47 47 E
Mae Chan, *Thailand* . . . . . **76 B2** 20 9N 99 52 E
Mae Hong Son, *Thailand* . **76 C2** 19 16N 98 1 E
Mae Khlong →, *Thailand* **76 F3** 13 24N 100 0 E
Mae Phrik, *Thailand* . . . . **76 D2** 17 27N 99 7 E
Mae Ramat, *Thailand* . . . **76 D2** 16 58N 98 31 E
Mae Rim, *Thailand* . . . . . **76 C2** 18 54N 98 57 E
Mae Sot, *Thailand* . . . . . . **76 D2** 16 43N 98 34 E
Mae Suai, *Thailand* . . . . . **76 C2** 19 39N 99 33 E
Mae Tha, *Thailand* . . . . . . **76 C2** 18 28N 99 8 E
Maebaru, *Japan* . . . . **62 D2** 33 33N 130 12 E
Maebashi, *Japan* . . . . **63 A11** 36 24N 139 4 E
Maella, *Spain* . . . . **34 D5** 41 8N 0 7 E
Măeruş, *Romania* . . . . **46 D6** 45 53N 25 31 E
Maesteg, *U.K.* **17 F4** 51 36N 3 40W
Maestra, Sierra, *Cuba* . . . **148 B4** 20 15N 77 0W
Maestrazgo, Mts. del, *Spain* **34 E4** 40 30N 0 25W
Maestre de Campo I., *Phil.* . . . . **70 E3** 12 56N 121 42 E
Maevatanana, *Madag.* . . . **105 B8** 16 56 S 46 49 E
Maewo, *Vanuatu* . . . . **121 E6** 15 10 S 168 10 E
Ma'fan, *Libya* . . . . **96 C2** 25 56N 14 29 E
Mafeking = Mafikeng, *S. Africa* . . . . **104 D4** 25 50 S 25 38 E
Mafeking, *Canada* **131 C8** 52 40N 101 10W
Maféré, *Ivory C.* **100 D4** 5 30N 3 2W
Mafeteng, *Lesotho* . . . . . . **104 D4** 29 51 S 27 15 E
Maffe, *Belgium* . . . . **21 H6** 50 21N 5 19 E
Maffra, *Australia* . . . . **117 D7** 37 53 S 146 58 E
Mafia I., *Tanzania* . . . . . . **106 D4** 7 45 S 39 50 E
Mafikeng, *S. Africa* . . . . . **104 D4** 25 50 S 25 38 E
Mafra, *Brazil* . . . . **159 B6** 26 10 S 49 55W
Mafra, *Portugal* **37 G1** 38 55N 9 20W
Mafungabusi Plateau, *Zimbabwe* . . . . **107 F2** 18 30 S 29 8 E
Magadan, *Russia* . . . . **57 D16** 59 38N 150 50 E
Magadi, *Kenya* . . . . **106 C4** 1 54 S 36 19 E
Magadi, L., *Kenya* . . . . . . **106 C4** 1 54 S 36 19 E
Magaliesburg, *S. Africa* . **105 D4** 26 0 S 27 32 E
Magallanes, *Phil.* . . . . **70 E4** 12 50N 123 50 E
Magallanes □, *Chile* . . . . **160 D2** 52 0 S 72 0W
Magallanes, Estrecho de, *Chile* . . . . **160 D2** 52 30 S 75 0W
Magangué, *Colombia* **152 B3** 9 14N 74 45W
Maganoy, *Phil.* . . . . **71 H5** 6 51N 124 31 E
Magaria, *Niger* . . . . **97 F1** 13 4N 9 5 E
Magburaka, *S. Leone* . . . **100 D2** 8 47N 12 0W
Magdalen Is. = Madeleine, Is. de la, *Canada* **129 C7** 47 30N 61 40W
Magdalena, *Argentina* . . . **158 D4** 35 5 S 57 30W
Magdalena, *Bolivia* . . . . . **157 C5** 13 13 S 63 57W
Magdalena, *Mexico* **146 A2** 30 50N 112 0W
Magdalena, *U.S.A.* **143 J10** 34 7N 107 15W
Magdalena □, *Colombia* . **152 A3** 10 0N 74 0W
Magdalena →, *Colombia* **152 A3** 11 6N 74 51W
Magdalena →, *Mexico* . . **146 A2** 30 40N 112 25W
Magdalena, B., *Mexico* . . **146 C2** 24 30N 112 10W
Magdalena, I., *Chile* . . . . **160 B2** 44 40 S 73 0W
Magdalena, Llano de la, *Mexico* **146 C2** 25 0N 111 30W
Magdeburg, *Germany* . . . **26 C7** 52 7N 11 38 E
Magdelaine Cays, *Australia* . . . . **114 B5** 16 33 S 150 18 E
Magdub, *Sudan* . . . . **95 E2** 13 42N 25 5 E
Magee, *U.S.A.* **139 K10** 31 52N 89 44W
Magee, I., *U.K.* **19 B6** 54 48N 5 43W
Magelang, *Indonesia* . . . . **75 D4** 7 29 S 110 13 E
Magellan's Str. = Magallanes, Estrecho de, *Chile* . . . . **160 D2** 52 30 S 75 0W
Magenta, *Australia* . . . . . **116 B5** 33 51 S 143 34 E
Magenta, *Italy* . . . . **38 C5** 45 28N 8 53 E
Magenta, L., *Australia* . . **113 F2** 33 30 S 119 2 E
Magerøya, *Norway* . . . . . . **12 A21** 71 3N 25 40 E
Maggea, *Australia* . . . . . . **116 C4** 34 28 S 140 2 E
Maggia, *Switz.* . . . . **29 D7** 46 15N 8 42 E
Maggia →, *Switz.* . . . . . . **29 D7** 46 18N 8 36 E
Maggiorasca, Mte., *Italy* . **38 D6** 44 33N 9 29 E
Maggiore, L., *Italy* . . . . . . **38 C5** 45 57N 8 39 E
Maghama, *Mauritania* . . . **100 B2** 15 32N 12 57W
Magherafelt, *U.K.* **19 B5** 54 45N 6 37W
Maghnia, *Algeria* **99 B4** 34 50N 1 43W
Magione, *Italy* . . . . **39 E9** 43 8N 12 12 E
Magistralnyy, *Russia* . . . . **57 D11** 56 16N 107 36 E
Maglaj, *Bos.-H.* . . . . **42 C3** 44 33N 18 7 E
Magliano in Toscana, *Italy* **39 F8** 42 36N 11 17 E
Máglie, *Italy* . . . . **41 B11** 40 7N 18 18 E
Magnac-Laval, *France* . . . **24 B5** 46 13N 1 11 E
Magnetic Pole (North) = North Magnetic Pole, *Canada* **6 B2** 77 58N 102 8W
Magnetic Pole (South) = South Magnetic Pole, *Antarctica* . . . . **7 C9** 64 8 S 138 8 E
Magnisía □, *Greece* . . . . . **44 E4** 39 15N 22 45 E
Magnitogorsk, *Russia* . . . . **54 E7** 53 27N 59 4 E
Magnolia, *Ark., U.S.A.* . . **139 J8** 33 16N 93 14W
Magnolia, *Miss., U.S.A.* . **139 K9** 31 9N 90 28W
Magnor, *Norway* . . . . **14 E6** 59 56N 12 15 E
Magny-en-Vexin, *France* . **23 C8** 49 9N 1 47 E
Magog, *Canada* **129 C5** 45 18N 72 9W
Magoro, *Uganda* . . . . **106 B3** 1 45N 34 12 E
Magosa = Famagusta, *Cyprus* . . . . **32 D12** 35 8N 33 55 E
Magouládhes, *Greece* . . . . **32 A3** 39 45N 19 42 E
Magoye, *Zambia* . . . . **107 F2** 16 1 S 27 30 E
Magpie L., *Canada* **129 B7** 51 0N 64 41W
Magrath, *Canada* **130 D6** 49 25N 112 50W
Magro →, *Spain* **35 F4** 39 11N 0 25W
Magrur, Wadi →, *Sudan* **95 D2** 16 5N 26 30 E
Magsingal, *Phil.* . . . . **70 C3** 17 41N 120 25 E
Magu □, *Tanzania* . . . . . . **106 C3** 2 31 S 33 28 E
Maguan, *China* . . . . **68 F5** 23 0N 104 21 E
Maguarinho, C., *Brazil* . . **154 B2** 0 15 S 48 30W
Maguindanao □, *Phil.* . . . **71 H5** 7 5N 124 0 E
Mağusa = Famagusta, *Cyprus* . . . . **32 D12** 35 8N 33 55 E
Maguse L., *Canada* **131 A9** 61 40N 95 10W
Maguse Pt., *Canada* **131 A10** 61 20N 93 50W
Magwe, *Burma* . . . . **78 E5** 20 10N 95 0 E
Maha Sarakham, *Thailand* **76 D4** 16 12N 103 16 E
Mahābād, *Iran* . . . . **89 D11** 36 50N 45 45 E
Mahabaleshwar, *India* . . . **82 F1** 17 58N 73 43 E
Mahabharat Lekh, *Nepal* **81 E9** 28 30N 82 0 E
Mahabo, *Madag.* . . . . **105 C7** 20 23 S 44 40 E
Mahad, *India* . . . . **82 E1** 18 6N 73 29 E
Mahaddei Uen, *Somali Rep.* . . . . **108 D3** 2 58N 45 32 E
Mahadeo Hills, *India* . . . . **80 H8** 22 20N 78 30 E
Mahadeopur, *India* . . . . . . **82 E5** 18 48N 80 0 E
Mahagi, *Zaïre* . . . . **106 B3** 2 20N 31 0 E
Mahaicony, *Guyana* **153 B6** 6 36N 57 48W
Mahajamba →, *Madag.* . . **105 B8** 15 33 S 47 8 E
Mahajamba, Helodranon' i, *Madag.* . . . . **105 B8** 15 24 S 47 5 E
Mahajan, *India* . . . . **80 E5** 28 48N 73 56 E
Mahajanga, *Madag.* . . . . . . **105 B8** 15 40 S 46 25 E
Mahajanga □, *Madag.* . . . **105 B8** 17 0 S 47 0 E
Mahajilo →, *Madag.* . . . . **105 B8** 19 42 S 45 22 E
Mahakam →, *Indonesia* . . **75 C5** 0 35 S 117 17 E
Mahalapye, *Botswana* . . . **104 C4** 23 1 S 26 51 E
Maḩallāt, *Iran* . . . . **85 C6** 33 55N 50 30 E
Māhān, *Iran* . . . . **85 D8** 30 5N 57 18 E
Mahanadi →, *India* . . . . . **82 D8** 20 20N 86 25 E
Mahanoro, *Madag.* . . . . . . **105 B8** 19 54 S 48 48 E
Mahanoy City, *U.S.A.* . . . **137 F8** 40 49N 76 9W
Mahaplag, *Phil.* . . . . **71 F5** 10 35N 124 57 E
Maharashtra □, *India* . . . **82 D2** 20 30N 75 30 E
Maharès, *Tunisia* . . . . **96 B2** 34 32N 10 29 E
Mahari Mts., *Tanzania* . . **106 D2** 6 20 S 30 0 E
Mahasham, W. →, *Egypt* **91 E3** 30 15N 34 10 E
Mahasolo, *Madag.* . . . . **105 B8** 19 7 S 46 22 E
Mahattat ash Shīdīyah, *Jordan* . . . . **91 F4** 29 55N 35 55 E
Mahattat 'Unayzah, *Jordan* . . . . **91 E4** 30 30N 35 47 E
Mahaweli →, *Sri Lanka* . **83 K5** 8 30N 81 15 E
Mahaxay, *Laos* . . . . **76 D5** 17 22N 105 12 E
Mahbes, *W. Sahara* **98 C3** 27 10N 9 50W
Mahbubabad, *India* . . . . . **82 F5** 17 42N 80 2 E
Mahbubnagar, *India* . . . . . **82 F3** 16 45N 77 59 E
Maḩḑah, *Oman* . . . . **85 E7** 24 24N 55 59 E
Mahdia, *Guyana* **153 B6** 5 13N 59 8W
Mahdia, *Tunisia* . . . . **96 A2** 35 28N 11 0 E
Mahe, *Jammu & Kashmir, India* . . . . **81 C8** 33 10N 78 32 E
Mahé, *Pondicherry, India* **83 J2** 11 42N 75 34 E
Mahé, *Seychelles* . . . . **109 E4** 5 0 S 55 30 E
Mahendra Giri, *India* . . . **83 K3** 8 20N 77 30 E
Mahendraganj, *India* . . . . **78 C2** 25 20N 89 45 E
Mahenge, *Tanzania* . . . . . **107 D4** 8 45 S 36 41 E
Maheno, *N.Z.* . . . . **119 F5** 45 10 S 170 50 E
Mahesana, *India* . . . . **80 H5** 23 39N 72 26 E
Mahia Pen., *N.Z.* . . . . **118 F6** 39 9 S 177 55 E
Mahilyow, *Belarus* . . . . . . **50 F6** 53 55N 30 18 E
Mahirija, *Morocco* **99 B4** 34 0N 3 16W
Mahlaing, *Burma* . . . . **78 E5** 21 6N 95 39 E
Mahmiya, *Sudan* . . . . **95 D3** 17 12N 33 43 E
Mahmud Kot, *Pakistan* . . **80 D4** 30 16N 71 0 E
Mahmudia, *Romania* . . . . . **46 D10** 45 5N 29 5 E
Mahnomen, *U.S.A.* **138 B7** 47 19N 95 58W
Mahoba, *India* . . . . **81 G8** 25 15N 79 55 E
Mahomet, *U.S.A.* **141 D8** 40 12N 88 24W
Mahón, *Spain* . . . . **33 B11** 39 53N 4 16 E
Mahone Bay, *Canada* **129 D7** 44 30N 64 20W
Mahuta, *Nigeria* . . . . **101 C5** 11 32N 4 58 E
Mai-Ndombe, L., *Zaïre* . . **102 C3** 2 0 S 18 20 E
Mai-Sai, *Thailand* . . . . **76 B2** 20 20N 99 55 E
Maibara, *Japan* . . . . **63 B8** 35 19N 136 17 E
Maicao, *Colombia* **152 A3** 11 23N 72 13W
Maîche, *France* . . . . **23 E13** 47 16N 6 48 E
Maici →, *Brazil* . . . . **157 B5** 6 30 S 61 43W
Maicurú →, *Brazil* . . . . . **153 D7** 2 14 S 54 17W
Máida, *Italy* . . . . **41 D9** 38 51N 16 22 E
Maidan Khula, *Afghan.* . . **80 C3** 33 36N 69 50 E
Maidenhead, *U.K.* **17 F7** 51 31N 0 42W
Maidi, *Yemen* . . . . **95 D5** 16 20N 42 45 E
Maidstone, *Canada* **131 C7** 53 5N 109 20W
Maidstone, *U.K.* . . . . **17 F8** 51 16N 0 32 E
Maiduguri, *Nigeria* . . . . . . **101 C7** 12 0N 13 20 E
Maignelay, *France* . . . . **23 C9** 49 32N 2 30 E
Maigo, *Phil.* . . . . **71 G4** 8 10N 123 57 E
Maigualida, Sierra, *Venezuela* **153 B4** 5 30N 65 10W
Maigudo, *Ethiopia* . . . . . . **95 F4** 7 30N 37 8 E
Maijdi, *Bangla.* . . . . **78 D3** 22 48N 91 10 E
Maikala Ra., *India* . . . . . . **82 D5** 22 0N 81 0 E
Maikoor, *Indonesia* . . . . . **73 C4** 6 8 S 134 6 E
Mailly-le-Camp, *France* . . **23 D11** 48 41N 4 12 E
Mailsi, *Pakistan* . . . . **80 E5** 29 48N 72 15 E
Maimbung, *Phil.* . . . . **71 J3** 5 56N 121 2 E
Main →, *Germany* . . . . . . **27 F4** 50 0N 8 18 E
Main →, *U.K.* **19 B5** 54 48N 6 18W
Main Centre, *Canada* **131 C7** 50 35N 107 21W
Mainburg, *Germany* . . . . . **27 G7** 48 38N 11 47 E
Maine, *France* **22 E6** 47 55N 0 25W
Maine □, *U.S.A.* **129 C6** 45 20N 69 0W
Maine →, *Ireland* **19 D2** 52 9N 9 45W
Maine-et-Loire □, *France* **22 E6** 47 31N 0 30W
Maïne-Soroa, *Niger* . . . . . **101 C7** 13 13N 12 2 E
Maingkwan, *Burma* . . . . . . **78 B6** 26 15N 96 37 E
Mainit, *Phil.* . . . . **71 G5** 9 32N 125 32 E
Mainit, L., *Phil.* . . . . **71 G5** 9 31N 125 30 E
Mainkaing, *Burma* . . . . . . **78 C5** 24 48N 95 16 E
Mainland, *Orkney, U.K.* . . **18 C5** 58 59N 3 8W
Mainland, *Shet., U.K.* . . . **18 A7** 60 15N 1 22W
Mainpuri, *India* . . . . **81 F8** 27 18N 79 4 E
Maintenon, *France* . . . . . . **23 D8** 48 35N 1 35 E
Maintirano, *Madag.* . . . . . **105 B7** 18 3 S 44 1 E
Mainvault, *Belgium* . . . . . **21 G3** 50 39N 3 43 E
Mainz, *Germany* . . . . **27 F4** 50 1N 8 14 E
Maipú, *Argentina* . . . . **158 D4** 36 52 S 57 50W
Maiquetía, *Venezuela* **152 A4** 10 36N 66 57W
Maira →, *Italy* . . . . **38 D4** 44 49N 7 38 E
Mairabari, *India* . . . . **78 B4** 26 30N 92 22 E
Mairipotaba, *Brazil* . . . . . **155 E2** 17 18 S 49 28W
Maisí, *Cuba* **149 B5** 20 17N 74 9W
Maisi, Pta. de, *Cuba* **149 B5** 20 10N 74 10W
Maisse, *France* . . . . **23 D9** 48 24N 2 21 E
Maissin, *Belgium* . . . . **21 J6** 49 58N 5 10 E
Maitland, *N.S.W., Australia* . . . . **117 B9** 32 33 S 151 36 E
Maitland, *S. Austral., Australia* . . . . **116 C2** 34 23 S 137 40 E
Maitland →, *Canada* . . . **136 C3** 43 45N 81 43W
Maitland, Banjaran, *Malaysia* . . . . **75 B5** 4 55N 116 37 E
Maiyema, *Nigeria* . . . . **101 C5** 12 5N 4 25 E
Maiyuan, *China* . . . . **69 E11** 25 34N 117 28 E
Maiz, Is. del, *Nic.* **148 D3** 12 15N 83 4W
Maizuru, *Japan* . . . . **63 B7** 35 25N 135 22 E
Majagual, *Colombia* **152 B3** 8 33N 74 38W
Majalengka, *Indonesia* . . . **75 D3** 6 50 S 108 13 E
Majari →, *Brazil* **153 C5** 3 29N 60 58W
Majene, *Indonesia* . . . . . . **72 B1** 3 38 S 118 57 E
Majes →, *Peru* . . . . **156 D3** 16 40 S 72 44W
Majevica, *Bos.-H.* . . . . **42 C3** 44 45N 18 50 E
Maji, *Ethiopia* . . . . **95 F4** 6 12N 35 30 E
Majiang, *China* . . . . **68 D6** 26 28N 107 32 E
Major, *Canada* **131 C7** 51 52N 109 37W
Majorca = Mallorca, *Spain* . . . . **33 B10** 39 30N 3 0 E
Majors Creek, *Australia* . **117 C8** 35 33 S 149 45 E
Majuriã, *Brazil* . . . . **157 B5** 7 30 S 64 55W
Maka, *Senegal* **100 C2** 13 40N 14 10W
Makak, *Cameroon* . . . . **101 E7** 3 36N 11 0 E
Makakou, *Gabon* . . . . **102 C2** 0 11 S 12 12 E
Makale, *Indonesia* . . . . **72 B1** 3 6 S 119 51 E
Makamba, *Burundi* . . . . . . **106 C2** 4 8 S 29 49 E
Makarewa, *N.Z.* . . . . **119 G3** 46 20 S 168 21 E
Makarikari = Makgadikgadi Salt Pans, *Botswana* . . . . **104 C4** 20 40 S 25 45 E
Makarovo, *Russia* . . . . **57 D11** 57 40N 107 45 E
Makarska, *Croatia* . . . . . . **42 D2** 43 20N 17 2 E
Makaryev, *Russia* . . . . **52 B6** 57 52N 43 50 E
Makasar = Ujung Pandang, *Indonesia* . . . **72 C1** 5 10 S 119 20 E
Makasar, Selat, *Indonesia* **72 B1** 1 0 S 118 20 E
Makasar, Str. of = Makasar, Selat, *Indonesia* . . . . **72 B1** 1 0 S 118 20 E
Makat, *Kazakhstan* . . . . . . **49 E9** 47 39N 53 19 E
Makedhonía □, *Greece* . . . **44 D3** 40 39N 22 0 E
Makedonija = Macedonia ■, *Europe* . . **42 F6** 41 53N 21 40 E
Makena, *U.S.A.* **132 H16** 20 39N 156 27W
Makeni, *S. Leone* **100 D2** 8 55N 12 5W
Makeyevka = Makiyivka, *Ukraine* . . . . **51 H9** 48 0N 38 0 E
Makgadikgadi Salt Pans, *Botswana* . . . . **104 C4** 20 40 S 25 45 E
Makhachkala, *Russia* . . . . **53 J8** 43 0N 47 30 E
Makharadze = Ozurgeti, *Georgia* . . . . **53 K5** 41 55N 42 2 E
Makhmūr, *Iraq* . . . . **89 E10** 35 46N 43 35 E
Makhyah, W. →, *Yemen* **87 C5** 17 40N 49 1 E
Makian, *Indonesia* . . . . . . **72 A3** 0 20N 127 20 E
Makindu, *Kenya* . . . . **106 C4** 2 18 S 37 50 E
Makinsk, *Kazakhstan* . . . . **56 D8** 52 37N 70 26 E
Makiyivka, *Ukraine* . . . . . **51 H9** 48 0N 38 0 E
Makkah, *Si. Arabia* . . . . . **86 B2** 21 30N 39 54 E
Makkovik, *Canada* **129 A8** 55 10N 59 10W
Makkum, *Neths.* . . . . **20 B6** 53 3N 5 25 E
Makó, *Hungary* . . . . **31 E13** 46 14N 20 33 E
Makok, *Gabon* . . . . **102 C1** 0 1 S 9 35 E
Makokou, *Gabon* . . . . **102 B2** 0 40N 12 50 E
Makongo, *Zaïre* . . . . **106 B2** 3 25N 26 17 E
Makoro, *Zaïre* . . . . **106 B2** 3 10N 29 59 E
Makoua, *Congo* . . . . **102 C3** 0 5 S 15 50 E
Maków Mazowiecki, *Poland* . . . . **47 C8** 52 52N 21 6 E
Maków Podhal, *Poland* . . **31 B12** 49 43N 19 45 E
Makrá, *Greece* . . . . **45 H7** 36 15N 25 54 E
Makran, *Asia* . . . . **79 D1** 26 13N 61 30 E
Makran Coast Range, *Pakistan* . . . . **79 D2** 25 40N 64 0 E
Makrana, *India* . . . . **80 F6** 27 2N 74 46 E
Mákri, *Greece* . . . . **44 D7** 40 52N 25 40 E
Makriyialos, *Greece* . . . . . **32 D7** 35 2N 25 59 E
Maksimkin Yar, *Russia* . . **56 D9** 58 42N 86 50 E
Maktar, *Tunisia* . . . . **96 A1** 35 48N 9 12 E
Mākū, *Iran* . . . . **84 B5** 39 15N 44 31 E
Makum, *India* . . . . **78 B5** 27 30N 95 23 E
Makumbi, *Zaïre* . . . . **103 D4** 5 50 S 20 43 E
Makunda, *Botswana* . . . . . **104 C3** 22 30 S 20 7 E
Makurazaki, *Japan* . . . . . . **62 F2** 31 15N 130 20 E
Makurdi, *Nigeria* . . . . **101 D6** 7 43N 8 35 E
Makūyeh, *Iran* . . . . **85 D7** 28 7N 53 9 E
Makwassie, *S. Africa* . . . . **104 D4** 27 17 S 26 0 E
Mal, *India* . . . . **78 B2** 26 51N 88 45 E
Mal B., *Ireland* **19 D2** 52 50N 9 30W

Mannum 229

Mal i Gjalicës së Lumës, Albania 44 B2 42 2N 20 25 E
Mal i Gribës, Albania 44 D1 40 17N 19 45 E
Mal i Nemërçkës, Albania 44 D2 40 15N 20 15 E
Mala, Peru 156 C2 12 40 S 76 38W
Mala, Pta., Panama 148 E3 7 28N 80 2W
Mala Belozërka, Ukraine 51 J8 47 12N 34 56 E
Mala Kapela, Croatia 39 D12 44 45N 15 30 E
Mala Vyska, Ukraine 51 H6 48 39N 31 36 E
Malabang, Phil. 71 H5 7 36N 124 3 E
Malabar Coast, India 83 J2 11 0N 75 0 E
Malabo = Rey Malabo, Eq. Guin. 101 E6 3 45N 8 50 E
Malabon, Phil. 70 D3 14 21N 121 0 E
Malabrigo Pt., Phil. 70 E3 13 36N 121 15 E
Malabungan, Phil. 71 G1 9 3N 117 38 E
Malacca, Str. of, Indonesia 77 L3 3 0N 101 0 E
Malacky, Slovak Rep. 31 C10 48 27N 17 0 E
Malad City, U.S.A. 142 E7 42 12N 112 15W
Maladzyechna, Belarus 50 E4 54 20N 26 50 E
Málaga, Colombia 152 B3 6 42N 72 44W
Málaga, Spain 37 J6 36 43N 4 23W
Malaga, U.S.A. 139 J2 32 14N 104 4W
Málaga □, Spain 37 J6 36 38N 4 58W
Malagarasi, Tanzania 106 D3 5 5 S 30 50 E
Malagarasi →, Tanzania 106 D2 5 12 S 29 47 E
Malagón, Spain 37 F7 39 11N 3 52W
Malagón →, Spain 37 H3 37 35N 7 29W
Malaimbandy, Madag. 105 C8 20 20 S 45 36 E
Malaita, Pac. Oc. 121 M11 9 0 S 161 0 E
Malakâl, Sudan 95 F3 9 33N 31 40 E
Malakand, Pakistan 79 B3 34 40N 71 55 E
Malakoff, U.S.A. 139 J7 32 10N 96 1W
Malalag, Phil. 71 H5 6 36N 125 24 E
Malam, Chad 97 F4 11 27N 20 59 E
Malamyzh, Russia 57 E14 49 50N 136 50 E
Malang, Indonesia 75 D4 7 59 S 112 45 E
Malangas, Phil. 71 H4 7 37N 123 1 E
Malange □, Angola 103 D3 9 30 S 16 0 E
Malangen, Norway 12 B18 69 24N 18 37 E
Malanje, Angola 103 D3 9 36 S 16 17 E
Mälaren, Sweden 14 E11 59 30N 17 10 E
Malargüe, Argentina 158 D2 35 32 S 69 30W
Malartic, Canada 128 C4 48 9N 78 9W
Malaryta, Belarus 51 G3 51 50N 24 3 E
Malatya, Turkey 89 C8 38 25N 38 20 E
Malawali, Malaysia 75 A5 7 3N 117 18 E
Malawi ■, Africa 107 E3 11 55 S 34 0 E
Malawi, L., Africa 107 E3 12 30 S 34 30 E
Malay, Phil. 71 F3 11 54N 121 55 E
Malay Pen., Asia 77 J3 7 25N 100 0 E
Malaya Belozërka = Mala Belozërka, Ukraine 51 J8 47 12N 34 56 E
Malaya Vishera, Russia 50 C7 58 55N 32 25 E
Malaya Viska = Mala Vyska, Ukraine 51 H6 48 39N 31 36 E
Malaybalay, Phil. 71 G5 8 5N 125 7 E
Malāyer, Iran 89 E13 34 19N 48 51 E
Malaysia ■, Asia 74 B4 5 0N 110 0 E
Malazgirt, Turkey 89 C10 39 10N 42 33 E
Malbon, Australia 114 C3 21 5 S 140 17 E
Malbooma, Australia 115 E1 30 41 S 134 11 E
Malbork, Poland 47 A6 54 3N 19 1 E
Malca Dube, Ethiopia 108 C2 6 47N 42 4 E
Malcésine, Italy 38 C7 45 46N 10 48 E
Malchin, Germany 26 B8 53 44N 12 44 E
Malchow, Germany 26 B8 53 28N 12 25 E
Malcolm, Australia 113 E3 28 51 S 121 25 E
Malcolm, Pt., Australia 113 F3 33 48 S 123 45 E
Malczyce, Poland 47 D3 51 14N 16 29 E
Maldegem, Belgium 21 F2 51 14N 3 26 E
Malden, Mass., U.S.A. 137 D13 42 26N 71 4W
Malden, Mo., U.S.A. 139 G10 36 34N 89 57W
Malden I., Kiribati 123 H12 4 3 S 155 1W
Maldives ■, Ind. Oc. 59 J11 5 0N 73 0 E
Maldon, Australia 116 D6 37 0 S 144 6 E
Maldonado, Uruguay 159 C5 34 59 S 55 0W
Maldonado, Punta, Mexico 147 D5 16 19N 98 35W
Malè, Italy 38 B7 46 21N 10 55 E
Malé Karpaty, Slovak Rep. 31 C10 48 30N 17 20 E
Maléa, Ákra, Greece 45 H5 36 28N 23 7 E
Malebo, Pool, Africa 103 C3 4 17 S 15 20 E
Malegaon, India 82 D2 20 30N 74 38 E
Malei, Mozam. 107 F4 17 12 S 36 58 E
Malek Kandī, Iran 89 D12 37 9N 46 6 E
Malekula, Vanuatu 121 F5 16 15 S 167 30 E
Malela, Bas Zaïre, Zaïre 103 D2 5 59 S 12 37 E
Malela, Kivu, Zaïre 103 C5 4 22 S 26 8 E
Malema, Mozam. 107 E4 14 57 S 37 20 E
Máleme, Greece 32 D5 35 31N 23 49 E
Malerkotla, India 80 D6 30 32N 75 58 E
Máles, Greece 32 D7 35 6N 25 35 E
Malesherbes, France 23 D9 48 15N 2 24 E
Maleshevska Planina, Europe 42 F8 41 38N 23 7 E
Malestroit, France 22 E4 47 49N 2 25W
Malfa, Italy 41 D7 38 35N 14 50 E
Malgobek, Russia 53 J7 43 30N 44 34 E
Malgomaj, Sweden 12 D17 64 40N 16 30 E
Malgrat, Spain 34 D7 41 39N 2 46 E
Malha, Sudan 95 D2 15 8N 25 10 E
Malheur →, U.S.A. 142 D5 44 4N 116 59W
Malheur L., U.S.A. 142 E4 43 20N 118 48W
Mali, Guinea 100 C2 12 10N 12 20W
Mali ■, Africa 100 B4 17 0N 3 0W
Mali Hka →, Burma 78 C6 25 42N 97 30 E
Mali Kanal, Serbia, Yug. 42 B4 45 36N 19 24 E
Malibu, U.S.A. 145 L8 34 2N 118 41W
Maligaya, Phil. 70 E3 12 59N 121 30 E
Malik, Indonesia 72 B2 0 39 S 123 16 E
Malili, Indonesia 72 B2 2 42 S 121 6 E
Malimba, Mts., Zaïre 106 D2 7 30 S 29 30 E
Malin Hd., Ireland 19 A4 55 23N 7 23W
Malindang, Mt., Phil. 71 G4 8 13N 123 38 E
Malindi, Kenya 106 C5 3 12 S 40 5 E
Malines = Mechelen, Belgium 21 F4 51 2N 4 29 E
Malino, Indonesia 72 A2 1 0N 121 0 E
Malinyi, Tanzania 107 D4 8 56 S 36 0 E
Malipo, China 68 F5 23 7N 104 42 E
Maliqi, Albania 44 D2 40 45N 20 48 E
Malita, Phil. 71 H5 6 19N 125 39 E
Maljenik, Serbia, Yug. 42 D6 43 59N 21 55 E
Malkapur, Maharashtra, India 82 F3 16 57N 76 17 E
Malkapur, Maharashtra, India 82 D1 20 53N 73 58 E
Malkara, Turkey 88 B2 40 53N 26 53 E
Małkinia Górna, Poland 47 C9 52 42N 22 5 E
Malko Tŭrnovo, Bulgaria 43 F12 41 59N 27 31 E
Mallacoota, Australia 117 D8 37 40 S 149 40 E
Mallacoota Inlet, Australia 117 D8 37 34 S 149 40 E
Mallaig, U.K. 18 E3 57 0N 5 50W
Mallala, Australia 116 C3 34 26 S 138 30 E
Mallard, U.S.A. 140 B2 42 56N 94 41W
Mallawan, India 81 F9 27 4N 80 12 E
Mallawi, Egypt 94 B3 27 44N 30 44 E
Malleco □, Chile 160 A2 38 10 S 72 20W
Mallemort, France 25 E9 43 43N 5 11 E
Málles Venosta, Italy 38 B7 46 41N 10 32 E
Mállia, Greece 32 D7 35 17N 25 27 E
Mallicolo = Malekula, Vanuatu 121 F5 16 15 S 167 30 E
Mallig, Phil. 70 C3 17 8N 121 42 E
Mallión, Kólpos, Greece 32 D7 35 19N 25 27 E
Mallorca, Spain 33 B10 39 30N 3 0 E
Mallorytown, Canada 137 B9 44 29N 75 53W
Mallow, Ireland 19 D3 52 8N 8 39W
Malmberget, Sweden 12 C19 67 11N 20 40 E
Malmédy, Belgium 21 H8 50 25N 6 2 E
Malmesbury, S. Africa 104 E2 33 28 S 18 41 E
Malmö, Sweden 15 J6 55 36N 12 59 E
Malmöhus län □, Sweden 15 J7 55 45N 13 30 E
Malmslätt, Sweden 15 F9 58 27N 15 33 E
Malmyzh, Russia 52 B10 56 31N 50 41 E
Malnaş, Romania 46 C6 46 2N 25 49 E
Malo, Vanuatu 121 E5 15 40 S 167 11 E
Malo Konare, Bulgaria 43 E9 42 12N 24 24 E
Maloarkhangelsk, Russia 52 D3 52 28N 36 30 E
Maloca, Brazil 153 C6 0 43N 55 57W
Maloja, Switz. 29 D9 46 25N 9 35 E
Maloja, P., Switz. 29 D9 46 23N 9 42 E
Malolos, Phil. 70 D3 14 50N 120 49 E
Malombe L., Malawi 107 E4 14 40 S 35 15 E
Malomir, Bulgaria 43 E11 42 16N 26 30 E
Malone, U.S.A. 137 B10 44 51N 74 18W
Malong, China 68 E4 25 24N 103 34 E
Malonga, Zaïre 103 E4 10 24 S 23 10 E
Malorad, Bulgaria 43 D8 43 28N 23 41 E
Måløy, Norway 13 F11 61 57N 5 6 E
Maloyaroslovets, Russia 52 C3 55 2N 36 20 E
Malozemelskaya Tundra, Russia 48 A9 67 0N 50 0 E
Malpartida, Spain 37 F4 39 26N 6 30W
Malpaso, Canary Is. 33 G1 27 43N 18 3W
Malpica, Spain 36 B2 43 19N 8 50W
Malprabha →, India 83 F3 16 20N 76 5 E
Mals = Málles Venosta, Italy 38 B7 46 41N 10 32 E
Malta, Brazil 154 C4 6 54 S 37 31W
Malta, Idaho, U.S.A. 142 E7 42 18N 113 22W
Malta, Mont., U.S.A. 142 B10 48 21N 107 52W
Malta ■, Europe 32 D1 35 50N 14 30 E
Malta Channel, Medit. S. 40 F6 36 40N 14 0 E
Maltahöhe, Namibia 104 C2 24 55 S 17 0 E
Malters, Switz. 28 B6 47 3N 8 11 E
Malton, Canada 136 C5 43 42N 79 38W
Malton, U.K. 16 C7 54 8N 0 49W
Malu'a, Solomon Is. 121 M11 8 0 S 160 0 E
Maluku, Indonesia 72 B3 1 0 S 127 0 E
Maluku □, Indonesia 72 B3 3 0 S 128 0 E
Maluku Sea = Molucca Sea, Indonesia 72 A3 2 0 S 124 0 E
Malumfashi, Nigeria 101 C6 11 48N 7 39 E
Malungun, Phil. 71 H5 6 16N 125 14 E
Maluso, Phil. 71 H3 6 33N 121 53 E
Malvalli, India 83 H3 12 28N 77 8 E
Malvan, India 83 F1 16 2N 73 30 E
Malvern, U.S.A. 139 H8 34 22N 92 49W
Malvern Hills, U.K. 17 E5 52 0N 2 19W
Malvik, Norway 14 A4 63 25N 10 40 E
Malvinas, Is. = Falkland Is. □, Atl. Oc. 160 D5 51 30 S 59 0W
Malya, Tanzania 106 C3 3 5 S 33 38 E
Malybay, Kazakhstan 55 B9 43 30N 78 25 E
Malyn, Ukraine 51 G5 50 46N 29 3 E
Malyy Lyakhovskiy, Ostrov, Russia 57 B15 74 7N 140 36 E
Malyy Nimnyr, Russia 57 D13 57 50N 125 10 E
Mama, Russia 57 D12 58 18N 112 54 E
Mamadysh, Russia 52 C10 55 44N 51 23 E
Mamaia, Romania 46 E9 44 18N 28 37 E
Mamaku, N.Z. 118 E5 38 5 S 176 8 E
Mamanguape, Brazil 154 C4 6 50 S 35 4W
Mamasa, Indonesia 72 B1 2 55 S 119 20 E
Mambajao, Phil. 71 G5 9 15N 124 43 E
Mambasa, Zaïre 106 B2 1 22N 29 3 E
Mamberamo →, Indonesia 73 B5 2 0 S 137 50 E
Mambilima Falls, Zambia 107 E2 10 31 S 28 45 E
Mambirima, Zaïre 107 E2 11 25 S 27 33 E
Mambo, Tanzania 106 C4 4 52 S 38 22 E
Mambrui, Kenya 106 C5 3 5 S 40 5 E
Mamburao, Phil. 70 E3 13 13N 120 39 E
Mameigwess L., Canada 128 B2 52 35N 87 50W
Mamer, Lux. 21 J8 49 38N 6 2 E
Mamers, France 22 D7 48 21N 0 22 E
Mamfe, Cameroon 101 D6 5 50N 9 15 E
Māmī, Ra's, Yemen 87 D6 12 32N 54 30 E
Mamiña, Chile 156 E4 20 5 S 69 14W
Mámmola, Italy 41 D9 38 22N 16 14 E
Mammoth, U.S.A. 143 K8 32 43N 110 39W
Mamoré →, Bolivia 157 C4 10 23 S 65 53W
Mamou, Guinea 100 C2 10 15N 12 0W
Mamparang Mts., Phil. 70 C3 16 21N 121 28 E
Mampatá, Guinea-Biss. 100 C2 11 54N 14 53W
Mampong, Ghana 101 D4 7 6N 1 26W
Mamry, Jezioro, Poland 47 A8 54 5N 21 50 E
Mamuil Malal, Paso, S. Amer. 160 A2 39 35 S 71 28W
Mamuju, Indonesia 72 B1 2 41 S 118 50 E
Ma'mūl, Oman 87 C6 18 8N 55 16 E
Man, Ivory C. 100 D3 7 30N 7 40W
Man →, India 82 F2 17 31N 75 32 E
Man, I. of, U.K. 16 C3 54 15N 4 30W
Man Na, Burma 78 D6 23 27N 97 19 E
Man Tun, Burma 78 D7 23 52N 98 38 E
Mana, Fr. Guiana 153 B7 5 45N 53 55W
Mana →, Fr. Guiana 153 B7 5 45N 53 55W
Måna →, Norway 14 E2 59 55N 8 48 E
Manaar, G. of = Mannar, G. of, Asia 83 K4 8 30N 79 0 E
Manabí □, Ecuador 152 D1 0 40 S 80 5W
Manacacías →, Colombia 152 C3 4 23N 72 4W
Manacapuru, Brazil 153 D5 3 16 S 60 37W
Manacapuru →, Brazil 153 D5 3 18 S 60 37W
Manacor, Spain 33 B10 39 34N 3 13 E
Manado, Indonesia 72 A2 1 29N 124 51 E
Manage, Belgium 21 G4 50 31N 4 15 E
Managua, Nic. 148 D2 12 6N 86 20W
Managua, L., Nic. 148 D2 12 20N 86 30W
Manaia, N.Z. 118 F3 39 33 S 174 8 E
Manakara, Madag. 105 C8 22 8 S 48 1 E
Manakau Mt., N.Z. 119 C8 42 15 S 173 42 E
Manākhah, Yemen 86 D3 15 5N 43 44 E
Manakino, N.Z. 118 E4 38 22 S 175 47 E
Manam I., Papua N. G. 120 C3 4 5 S 145 0 E
Manama = Al Manāmah, Bahrain 85 E6 26 10N 50 30 E
Manambao →, Madag. 105 B7 17 35 S 44 0 E
Manambato, Madag. 105 A8 13 43 S 49 7 E
Manambolo →, Madag. 105 B7 19 18 S 44 22 E
Manambolosy, Madag. 105 B8 16 2 S 49 40 E
Mananara, Madag. 105 B8 16 10 S 49 46 E
Mananara →, Madag. 105 C8 23 21 S 47 42 E
Mananjary, Madag. 105 C8 21 13 S 48 20 E
Manantenina, Madag. 105 C8 24 17 S 47 19 E
Manaos = Manaus, Brazil 153 D6 3 0 S 60 0W
Manapala, Phil. 71 F4 10 58N 123 5 E
Manapire →, Venezuela 152 B4 7 42N 66 7W
Manapouri, N.Z. 119 F2 45 34 S 167 39 E
Manapouri, L., N.Z. 119 F2 45 32 S 167 32 E
Manar →, India 82 E3 18 50N 77 20 E
Manār, Jabal, Yemen 86 D4 14 2N 44 17 E
Manas, China 64 B3 44 17N 85 56 E
Manas, Somali Rep. 108 D2 2 57N 43 28 E
Manas, Gora, Kyrgyzstan 55 B5 42 22N 71 2 E
Manaslu, Nepal 81 E11 28 33N 84 33 E
Manasquan, U.S.A. 137 F10 40 8N 74 3W
Manassa, U.S.A. 143 H11 37 11N 105 56W
Manatuto, Indonesia 72 C3 8 30 S 126 1 E
Manaung, Burma 78 F4 18 45N 93 40 E
Manaus, Brazil 153 D6 3 0 S 60 0W
Manavgat, Turkey 88 D4 36 47N 31 26 E
Manawan L., Canada 131 B8 55 24N 103 14W
Manawatu →, N.Z. 118 G4 40 28 S 175 12 E
Manay, Phil. 71 H6 7 17N 126 33 E
Manbij, Syria 88 D7 36 31N 37 57 E
Mancelona, U.S.A. 134 C3 44 54N 85 4W
Mancha Real, Spain 37 H7 37 48N 3 39W
Manche □, France 22 C5 49 10N 1 20W
Manchester, U.K. 16 D5 53 29N 2 12W
Manchester, Calif., U.S.A. 144 G3 38 58N 123 41W
Manchester, Conn., U.S.A. 137 E12 41 47N 72 31W
Manchester, Ga., U.S.A. 135 J3 32 51N 84 37W
Manchester, Iowa, U.S.A. 140 B5 42 29N 91 27W
Manchester, Ky., U.S.A. 134 G4 37 9N 83 46W
Manchester, Mich., U.S.A. 141 B12 42 9N 84 2W
Manchester, N.H., U.S.A. 137 D13 42 59N 71 28W
Manchester, N.Y., U.S.A. 136 D7 42 56N 77 16W
Manchester, Vt., U.S.A. 137 C11 43 10N 73 5W
Manchester L., Canada 131 A7 61 28N 107 29W
Manchuria = Dongbei, China 67 D13 42 0N 125 0 E
Manciano, Italy 39 F8 42 35N 11 31 E
Mancifa, Ethiopia 95 F5 6 53N 41 50 E
Mancora, Pta., Peru 156 A1 4 9 S 81 1W
Mand →, Iran 85 D7 28 20N 52 30 E
Manda, Chunya, Tanzania 106 D3 6 51 S 32 29 E
Manda, Ludewe, Tanzania 107 E3 10 30 S 34 40 E
Mandabé, Madag. 105 C7 21 0 S 44 55 E
Mandaguari, Brazil 159 A5 23 32 S 51 42W
Mandah, Mongolia 66 B5 44 27N 108 2 E
Mandal, Norway 13 G12 58 2N 7 25 E
Mandalay, Burma 78 D6 22 0N 96 4 E
Mandale = Mandalay, Burma 78 D6 22 0N 96 4 E
Mandalgovi, Mongolia 66 B4 45 45N 106 10 E
Mandalī, Iraq 89 F11 33 43N 45 28 E
Mandalya Körfezi, Turkey 45 G9 37 15N 27 20 E
Mandan, U.S.A. 138 B4 46 50N 100 54W
Mandaon, Phil. 70 E4 12 13N 123 17 E
Mandapeta, India 82 F5 16 47N 81 56 E
Mandar, Teluk, Indonesia 72 B1 3 35 S 119 15 E
Mandas, Italy 40 C2 39 40N 9 8 E
Mandasor = Mandsaur, India 80 G6 24 3N 75 8 E
Mandaue, Phil. 71 F4 10 20N 123 56 E
Mandayar, Phil. 71 H6 7 34N 126 14 E
Mandelieu-la-Napoule, France 25 E10 43 34N 6 57 E
Mandera, Kenya 106 B5 3 55N 41 53 E
Mandera □, Kenya 106 B5 3 30N 41 0 E
Manderfeld, Belgium 21 H8 50 20N 6 20 E
Mandi, India 80 D7 31 39N 76 58 E
Mandimba, Mozam. 107 E4 14 20 S 35 40 E
Mandioli, Indonesia 72 B3 0 40 S 127 20 E
Mandioré, L., S. Amer. 157 D6 18 8 S 57 33W
Mandji I. = Lopez I., Gabon 102 C1 0 50 S 8 47 E
Mandla, India 81 H9 22 39N 80 30 E
Mandø, Denmark 15 J2 55 18N 8 33 E
Mandoto, Madag. 105 B8 19 34 S 46 17 E
Mandoúdhion, Greece 45 F5 38 48N 23 29 E
Mandra, Pakistan 80 C5 33 23N 73 12 E
Mandráki, Greece 45 H9 36 36N 27 11 E
Mandrare →, Madag. 105 D8 25 10 S 46 30 E
Mandritsara, Madag. 105 B8 15 50 S 48 49 E
Mandsaur, India 80 G6 24 3N 75 8 E
Mandurah, Australia 113 F2 32 36 S 115 48 E
Mandúria, Italy 41 B10 40 24N 17 38 E
Mandvi, India 80 H3 22 51N 69 22 E
Mandya, India 83 H3 12 30N 77 0 E
Mandzai, Pakistan 80 D2 30 55N 67 6 E
Mané, Burkina Faso 101 C4 12 59N 1 21W
Maneh, Iran 85 B8 37 39N 57 7 E
Manengouba, Mts., Cameroon 101 D6 5 0N 9 50 E
Maner →, India 82 E4 18 30N 79 40 E
Maneroo, Australia 114 C3 23 22 S 143 53 E
Maneroo Cr. →, Australia 114 C3 23 21 S 143 53 E
Manfalût, Egypt 94 B3 27 20N 30 52 E
Manfred, Australia 116 B5 33 19 S 143 45 E
Manfredónia, Italy 41 A8 41 38N 15 55 E
Manfredónia, G. di, Italy 41 A9 41 35N 16 5 E
Manga, Brazil 155 D3 14 46 S 43 56W
Manga, Burkina Faso 101 C4 11 40N 1 4W
Manga, Niger 97 F2 15 0N 14 0 E
Mangabeiras, Chapada das, Brazil 154 D2 10 0 S 46 30W
Mangal, Phil. 71 H3 6 25N 121 58 E
Mangalagiri, India 83 F5 16 26N 80 36 E
Mangaldai, India 78 B4 26 26N 92 2 E
Mangalia, Romania 46 F9 43 50N 28 35 E
Mangalore, Australia 117 D6 36 56 S 145 10 E
Mangalore, India 83 H2 12 55N 74 47 E
Manganeses, Spain 36 D5 41 45N 5 43W
Mangaon, India 82 E1 18 15N 73 20 E
Mangaweka, N.Z. 118 F4 39 48 S 175 47 E
Mangaweka, Mt., N.Z. 118 F5 39 49 S 176 5 E
Mange, Zaïre 102 B4 0 54N 20 30 E
Manggar, Indonesia 75 C3 2 50 S 108 10 E
Manggawitu, Indonesia 73 B4 4 8 S 133 32 E
Mangin Range, Burma 78 C5 24 15N 95 45 E
Mangkalihat, Tanjung, Indonesia 75 B5 1 2N 118 59 E
Mangla Dam, Pakistan 81 C5 33 9N 73 44 E
Manglares, C., Colombia 152 C2 1 36N 79 2W
Manglaur, India 80 E7 29 44N 77 49 E
Mangnai, China 64 C4 37 52N 91 43 E
Mango, Togo 101 C5 10 20N 0 30 E
Mangoche, Malawi 107 E4 14 25 S 35 16 E
Mangoky →, Madag. 105 C7 21 29 S 43 41 E
Mangole, Indonesia 72 B3 1 50 S 125 55 E
Mangombe, Zaïre 106 C2 1 20 S 26 48 E
Mangonui, N.Z. 118 B2 35 1 S 173 32 E
Mangualde, Portugal 36 E3 40 38N 7 48W
Mangueigne, Chad 97 F4 10 30N 21 15 E
Mangueira, L. da, Brazil 159 C5 33 0 S 52 50W
Manguéni, Hamada, Niger 96 D2 22 35N 12 40 E
Mangum, U.S.A. 139 H5 34 53N 99 30W
Mangyshlak Poluostrov, Kazakhstan 56 E6 44 30N 52 30 E
Manhattan, U.S.A. 138 F6 39 11N 96 35W
Manhatten, U.S.A. 141 C9 41 26N 87 59W
Manhiça, Mozam. 105 D5 25 23 S 32 49 E
Manhuaçu, Brazil 155 F3 20 15 S 42 2W
Manhumirim, Brazil 155 F3 20 22 S 41 57W
Maní, Colombia 152 C3 4 49N 72 17W
Mania →, Madag. 105 B8 19 42 S 45 22 E
Maniago, Italy 39 B9 46 10N 12 43 E
Manica, Mozam. 105 B5 18 58 S 32 59 E
Manica e Sofala □, Mozam. 105 B5 19 10 S 33 45 E
Manicaland □, Zimbabwe 107 F3 19 0 S 32 30 E
Manicoré, Brazil 157 B5 5 48 S 61 16W
Manicoré →, Brazil 157 B5 5 51 S 61 19W
Manicouagan →, Canada 129 C6 49 30N 68 30W
Manīfah, Si. Arabia 85 E6 27 44N 49 0 E
Manifold, Australia 114 C5 22 41 S 150 40 E
Manifold, C., Australia 114 C5 22 41 S 150 50 E
Maniganggo, China 68 B2 31 56N 99 10 E
Manigotagan, Canada 131 C9 51 6N 96 18W
Manihiki, Cook Is. 123 J11 10 24 S 161 1W
Manika, Plateau de la, Zaïre 107 E2 10 0 S 25 5 E
Manikganj, Bangla. 78 D3 23 52N 90 0 E
Manila, Phil. 70 D3 14 40N 121 3 E
Manila, U.S.A. 142 F9 40 59N 109 43W
Manila B., Phil. 70 D3 14 40N 120 35 E
Manilla, Australia 117 A9 30 45 S 150 43 E
Manimpé, Mali 100 C3 14 11N 5 28W
Maningrida, Australia 114 A1 12 3 S 134 13 E
Manipur □, India 78 C4 25 0N 94 0 E
Manipur →, Burma 78 D5 23 45N 94 20 E
Manisa, Turkey 88 C2 38 38N 27 30 E
Manistee, U.S.A. 134 C2 44 15N 86 19W
Manistee →, U.S.A. 134 C2 44 15N 86 21W
Manistique, U.S.A. 134 C2 45 57N 86 15W
Manito, U.S.A. 140 D7 40 26N 89 47W
Manito L., Canada 131 C7 52 43N 109 43W
Manitoba □, Canada 131 B9 55 30N 97 0W
Manitoba, L., Canada 131 C9 51 0N 98 45W
Manitou, Canada 131 D9 49 15N 98 32W
Manitou Beach, U.S.A. 141 C12 41 58N 84 19W
Manitou I., U.S.A. 128 C2 47 25N 87 37W
Manitou Is., U.S.A. 134 C2 45 8N 86 0W
Manitou L., Canada 129 B6 50 55N 65 17W
Manitou Springs, U.S.A. 138 F2 38 52N 104 55W
Manitoulin I., Canada 128 C3 45 40N 82 30W
Manitowaning, Canada 128 C3 45 46N 81 49W
Manitowoc, U.S.A. 134 C2 44 5N 87 40W
Manitsauá-Missu →, Brazil 157 C7 10 58 S 53 20W
Manizales, Colombia 152 B2 5 5N 75 32W
Manja, Madag. 105 C7 21 26 S 44 20 E
Manjacaze, Mozam. 105 C5 24 45 S 34 0 E
Manjakandriana, Madag. 105 B8 18 55 S 47 47 E
Manjeri, India 83 J3 11 7N 76 11 E
Manjhand, Pakistan 79 D3 25 50N 68 10 E
Manjil, Iran 85 B6 36 46N 49 30 E
Manjimup, Australia 113 F2 34 15 S 116 6 E
Manjra →, India 82 E3 18 49N 77 52 E
Mankato, Kans., U.S.A. 138 F5 39 47N 98 13W
Mankato, Minn., U.S.A. 138 C8 44 10N 94 0W
Mankayan, Phil. 70 C3 16 52N 120 47 E
Mankayane, Swaziland 105 D5 26 40 S 31 4 E
Mankono, Ivory C. 100 D3 8 1N 6 10W
Mankota, Canada 131 D7 49 25N 107 5W
Manlay, Mongolia 66 B4 44 9N 107 0 E
Manlleu, Spain 34 C7 42 2N 2 17 E
Manly, Australia 117 B9 33 48 S 151 17 E
Manmad, India 82 D2 20 18N 74 28 E
Mann Ras., Australia 113 E5 26 6 S 130 5 E
Manna, Indonesia 74 C2 4 25 S 102 55 E
Mannahill, Australia 116 B3 32 25 S 140 0 E
Mannar, Sri Lanka 83 K4 9 1N 79 54 E
Mannar, G. of, Asia 83 K4 8 30N 79 0 E
Mannar I., Sri Lanka 83 K4 9 5N 79 45 E
Mannargudi, India 83 J4 10 45N 79 51 E
Mannduque □, Phil. 70 E4 13 18N 122 0 E
Männedorf, Switz. 29 B7 47 15N 8 43 E
Mannheim, Germany 27 F4 49 29N 8 29 E
Manning, Canada 130 B5 56 53N 117 39W
Manning, Oreg., U.S.A. 144 E3 45 45N 123 13W
Manning, S.C., U.S.A. 135 J5 33 42N 80 13W
Manning Prov. Park, Canada 130 D4 49 5N 120 45W
Manning Str., Solomon Is. 121 L10 7 30 S 158 0 E
Mannington, U.S.A. 134 F5 39 32N 80 21W
Mannu →, Italy 40 C2 39 16N 9 0 E
Mannu, C., Italy 40 B1 40 3N 8 21 E
Mannum, Australia 116 C3 34 50 S 139 20 E

Mano, S. Leone 100 D2 8 3N 12 2W
Manoa, Bolivia 157 B4 9 40 S 65 27W
Manokwari, Indonesia 73 B4 0 54 S 134 0 E
Manolás, Greece 45 F3 38 4N 21 21 E
Manolo Fortich, Phil. 71 G5 8 28N 124 50 E
Manombo, Madag. 105 C7 22 57 S 43 28 E
Manono, Zaïre 106 D2 7 15 S 27 25 E
Manosque, France 25 E9 43 49N 5 47 E
Manouane, L., Canada 129 B5 50 45N 70 45W
Manouro, Pt., Vanuatu 121 G6 17 41 S 168 36 E
Manpojin, N. Korea 67 D14 41 6N 126 24 E
Manresa, Spain 34 D6 41 48N 1 50 E
Mansa, Gujarat, India 80 H5 23 27N 72 45 E
Mansa, Punjab, India 80 E6 30 0N 75 27 E
Mansa, Zambia 107 E2 11 13 S 28 55 E
Mansalay, Phil. 70 E3 12 31N 121 26 E
Mansehra, Pakistan 80 B5 34 20N 73 15 E
Mansel I., Canada 127 B11 62 0N 80 0W
Mansfield, Australia 117 D7 37 4 S 146 6 E
Mansfield, U.K. 16 D6 53 9N 1 11W
Mansfield, La., U.S.A. 139 J8 32 2N 93 43W
Mansfield, Mass., U.S.A. 137 D13 42 2N 71 13W
Mansfield, Ohio, U.S.A. 136 F2 40 45N 82 31W
Mansfield, Pa., U.S.A. 136 E7 41 48N 77 5W
Mansfield, Wash., U.S.A. 142 C4 47 49N 119 38W
Mansi, Burma 78 C5 24 48N 95 52 E
Mansidão, Brazil 154 D3 10 43 S 44 2W
Mansilla de las Mulas, Spain 36 C5 42 30N 5 25W
Mansle, France 24 C4 45 52N 0 12 E
Manso →, Brazil 155 D2 13 50 S 47 0W
Mansoa, Guinea-Biss. 100 C1 12 0N 15 20W
Manson, U.S.A. 140 B2 42 32N 94 32W
Manson Creek, Canada 130 B4 55 37N 124 32W
Mansoura, Algeria 99 A5 36 1N 4 31 E
Manta, Ecuador 152 D1 1 0 S 80 40W
Manta, B. de, Ecuador 152 D1 0 54 S 80 44W
Mantalingajan, Mt., Phil. 71 G1 8 55N 117 45 E
Mantare, Tanzania 106 C3 2 42 S 33 13 E
Manteca, U.S.A. 144 H5 37 48N 121 13W
Mantecal, Venezuela 152 B4 7 34N 69 17W
Mantena, Brazil 155 E3 18 47 S 40 59W
Manteno, U.S.A. 141 C9 41 15N 87 50W
Manteo, U.S.A. 135 H8 35 55N 75 40W
Mantes-la-Jolie, France 23 D8 48 58N 1 41 E
Manthani, India 82 E4 18 40N 79 35 E
Manthelan, France 22 E7 47 9N 0 47 E
Manti, U.S.A. 142 G8 39 16N 111 38W
Mantiqueira, Serra da, Brazil 155 F3 22 0 S 44 0W
Manton, U.S.A. 134 C3 44 25N 85 24W
Mantorp, Sweden 15 F9 58 21N 15 20 E
Mántova, Italy 38 C7 45 9N 10 48 E
Mänttä, Finland 13 E21 62 0N 24 40 E
Mantua = Mántova, Italy 38 C7 45 9N 10 48 E
Mantung, Australia 116 C4 34 35 S 140 3 E
Manturovo, Russia 52 A7 58 23N 44 45 E
Manu, Peru 156 C3 12 10 S 70 51W
Manu →, Peru 156 C3 12 16 S 70 55W
Manua Is., Amer. Samoa 121 X25 14 13 S 169 35W
Manuae, Cook Is. 123 J12 19 30 S 159 0W
Manuel Alves →, Brazil 155 D2 11 19 S 48 28W
Manuel Alves Grande →, Brazil 154 C2 7 27 S 47 35W
Manuel Urbano, Brazil 156 B4 8 53 S 69 18W
Manui, Indonesia 72 B2 3 35 S 123 5 E
Manukau Harbour, N.Z. 118 D3 37 3 S 174 45 E
Manunui, N.Z. 118 E4 38 54 S 175 21 E
Manurewa, N.Z. 118 D3 37 1 S 174 54 E
Manuripi →, Bolivia 156 C4 11 6 S 67 36W
Manus I., Papua N. G. 120 B4 2 0 S 147 0 E
Manvi, India 83 G3 15 57N 76 59 E
Manville, U.S.A. 138 D2 42 47N 104 37W
Manwath, India 82 E3 19 19N 76 32 E
Many, U.S.A. 139 K8 31 34N 93 29W
Manyara, L., Tanzania 106 C4 3 40 S 35 50 E
Manych →, Russia 53 G5 47 13N 40 40 E
Manych-Gudilo, Ozero, Russia 53 G6 46 24N 42 38 E
Manyonga →, Tanzania 106 C3 4 10 S 34 15 E
Manyoni, Tanzania 106 D3 5 45 S 34 55 E
Manyoni □, Tanzania 106 D3 6 30 S 34 30 E
Manzai, Pakistan 79 B3 32 12N 70 15 E
Manzala, Bahra el, Egypt 94 H7 31 10N 31 56 E
Manzanares, Spain 35 F1 39 2N 3 22W
Manzaneda, Cabeza de, Spain 36 C3 42 12N 7 15W
Manzanillo, Cuba 148 B4 20 20N 77 31W
Manzanillo, Mexico 146 D4 19 0N 104 20W
Manzanillo, Pta., Panama 148 E4 9 30N 79 40W
Manzano Mts., U.S.A. 143 J10 34 40N 106 20W
Manzarīyeh, Iran 85 C6 34 53N 50 50 E
Manzhouli, China 65 B6 49 35N 117 25 E
Manzini, Swaziland 105 D5 26 30 S 31 25 E
Mao, Chad 97 F3 14 4N 15 19 E
Maoke, Pegunungan, Indonesia 73 B5 3 40 S 137 30 E
Maolin, China 67 C12 43 58N 123 30 E
Maoming, China 69 G8 21 50N 110 54 E
Maowen, China 68 B4 31 41N 103 49 E
Maoxing, China 67 B13 45 28N 124 40 E
Mapam Yumco, China 64 C3 30 45N 81 28 E
Mapastepec, Mexico 147 D6 15 26N 92 54W
Mapia, Kepulauan, Indonesia 73 A4 0 50N 134 20 E
Mapimí, Mexico 146 B4 25 50N 103 50W
Mapimí, Bolsón de, Mexico 146 B4 27 30N 104 15W
Maping, China 69 B9 31 34N 113 32 E
Mapinga, Tanzania 106 D4 6 40 S 39 12 E
Mapinhane, Mozam. 105 C6 22 20 S 35 0 E
Mapire, Venezuela 153 B5 7 45N 64 42W
Maple →, U.S.A. 141 B12 42 59N 84 57W
Maple Creek, Canada 131 D7 49 55N 109 29W
Maple Valley, U.S.A. 144 C4 47 25N 122 3W
Mapleton, U.S.A. 142 D2 44 2N 123 52W
Mapourika, L., N.Z. 119 D5 43 16 S 170 12 E
Maprik, Papua N. G. 120 B2 3 44 S 143 3 E
Mapuca, India 83 G1 15 36N 73 46 E
Mapuera →, Brazil 153 D6 1 5 S 57 2W
Maputing Baybay, Phil. 70 E4 12 45N 123 20 E
Maputo, Mozam. 105 D5 25 58 S 32 32 E
Maputo, B. de, Mozam. 105 D5 25 50 S 32 45 E
Maqiaohe, China 67 B16 44 40N 130 30 E
Maqnā, Si. Arabia 84 D2 28 25N 34 50 E
Maqran, W. →, Si. Arabia 86 B4 20 55N 47 12 E
Maqteïr, Mauritania 98 D2 21 50N 11 40W
Maqueda Channel, Phil. 70 E5 13 42N 124 1 E
Maquela do Zombo, Angola 103 D3 6 0 S 15 15 E
Maquinchao, Argentina 160 B3 41 15 S 68 50W
Maquoketa, U.S.A. 140 B6 42 4N 90 40W
Mar, Serra do, Brazil 159 B6 25 30 S 49 0W
Mar Chiquita, L., Argentina 158 C3 30 40 S 62 50W
Mar del Plata, Argentina 158 D4 38 0 S 57 30W
Mar Menor, Spain 35 H4 37 40N 0 45W
Mara, Guyana 153 B6 6 0N 57 36W
Mara, India 78 A5 28 11N 94 14 E
Mara, Tanzania 106 C3 1 30 S 34 32 E
Mara □, Tanzania 106 C3 1 45 S 34 20 E
Maraã, Brazil 152 D4 1 52 S 65 25W
Marabá, Brazil 154 C2 5 20 S 49 5W
Maracá, I. de, Brazil 153 C7 2 10N 50 30W
Maracaibo, Venezuela 152 A3 10 40N 71 37W
Maracaibo, L. de, Venezuela 152 B3 9 40N 71 30W
Maracaju, Brazil 159 A4 21 38 S 55 9W
Maracajú, Serra de, Brazil 157 E6 23 57 S 55 1W
Maracanã, Brazil 154 B2 0 46 S 47 27W
Maracás, Brazil 155 D3 13 26 S 40 18W
Maracay, Venezuela 152 A4 10 15N 67 28W
Marādah, Libya 96 C3 29 15N 19 15 E
Maradi, Niger 101 C6 13 29N 7 20 E
Maradun, Nigeria 101 C6 12 35N 6 18 E
Marāgheh, Iran 89 D12 37 30N 46 12 E
Maragogipe, Brazil 155 D4 12 46 S 38 55W
Maragondon, Phil. 70 D3 14 16N 120 44 E
Marāh, Si. Arabia 84 E5 25 0N 45 35 E
Marajó, B. de, Brazil 154 B2 1 0 S 48 30W
Marajó, I. de, Brazil 154 B2 1 0 S 49 30W
Marākand, Iran 84 B5 38 51N 45 16 E
Maralal, Kenya 106 B4 1 0N 36 38 E
Maralinga, Australia 113 F5 30 13 S 131 32 E
Marama, Australia 116 C4 35 10 S 140 10 E
Maramasike, Solomon Is. 121 M11 9 30 S 161 25 E
Marampa, S. Leone 100 D2 8 45N 12 28W
Maramureş □, Romania 46 B4 47 45N 24 0 E
Maran, Malaysia 77 L4 3 35N 102 45 E
Marana, U.S.A. 143 K8 32 27N 111 13W
Maranboy, Australia 112 B5 14 40 S 132 39 E
Maranchón, Spain 34 D2 41 6N 2 15W
Marand, Iran 89 C11 38 30N 45 45 E
Marang, Malaysia 77 K4 5 12N 103 13 E
Maranguape, Brazil 154 B4 3 55 S 38 50W
Maranhão = São Luís, Brazil 154 B3 2 39 S 44 15W
Maranhão □, Brazil 154 B2 5 0 S 46 0W
Marano, L. di, Italy 39 C10 45 44N 13 10 E
Maranoa →, Australia 115 D4 27 50 S 148 37 E
Marañón →, Peru 156 A3 4 30 S 73 35W
Marão, Mozam. 105 C5 24 18 S 34 2 E
Marapi →, Brazil 153 C6 0 37N 55 58W
Marari, Brazil 156 B4 5 43 S 67 47W
Maraş = Kahramanmaraş, Turkey 88 D7 37 37N 36 53 E
Mărăşeşti, Romania 46 D8 45 52N 27 14 E
Maratea, Italy 41 C8 39 59N 15 43 E
Marateca, Portugal 37 G2 38 34N 8 40W
Marathasa □, Cyprus 32 E11 34 59N 32 51 E
Marathókambos, Greece 45 G8 37 43N 26 42 E
Marathon, Australia 114 C3 20 51 S 143 32 E
Marathon, Canada 128 C2 48 44N 86 23W
Marathón, Greece 45 F5 38 11N 23 58 E
Marathon, Iowa, U.S.A. 140 B2 42 52N 94 59W
Marathon, N.Y., U.S.A. 137 D8 42 27N 76 2W
Marathon, Tex., U.S.A. 139 K3 30 12N 103 15W
Marathóvouno, Cyprus 32 D12 35 13N 33 37 E
Maratua, Indonesia 75 B5 2 10N 118 35 E
Maraú, Brazil 155 D4 14 6 S 39 0W
Maravatío, Mexico 146 D4 19 51N 100 25W
Marawi City, Phil. 71 G5 8 0N 124 21 E
Marāwih, U.A.E. 85 E7 24 18N 53 18 E
Marbella, Spain 37 J6 36 30N 4 57W
Marble Bar, Australia 112 D2 21 9 S 119 44 E
Marble Falls, U.S.A. 139 K5 30 35N 98 16W
Marblehead, U.S.A. 137 D14 42 30N 70 51W
Marburg, Germany 26 E4 50 47N 8 46 E
Marby, Sweden 14 A8 63 7N 14 18 E
Marcal →, Hungary 31 D10 47 41N 17 32 E
Marcali, Hungary 31 E10 46 35N 17 25 E
Marcapata, Peru 156 C3 13 31 S 70 52W
Marcaria, Italy 38 C7 45 7N 10 34 E
Marceline, U.S.A. 140 E4 39 43N 92 57W
March, U.K. 17 E8 52 33N 0 5 E
Marchal, Zaïre 103 D2 5 16 S 14 58 E
Marchand = Rommani, Morocco 98 B3 33 31N 6 40W
Marche, France 24 B5 46 5N 1 20 E
Marche □, Italy 39 E10 43 30N 13 15 E
Marche-en-Famenne, Belgium 21 H6 50 14N 5 19 E
Marchena, Spain 37 H5 37 18N 5 23W
Marches = Marche □, Italy 39 E10 43 30N 13 15 E
Marciana Marina, Italy 38 F7 42 48N 10 12 E
Marcianise, Italy 41 A7 41 2N 14 17 E
Marcigny, France 25 B8 46 17N 4 2 E
Marcillat-en-Combraille, France 24 B6 46 12N 2 38 E
Marcinelle, Belgium 21 H4 50 24N 4 26 E
Marck, France 23 B8 50 57N 1 57 E
Marckolsheim, France 23 D14 48 10N 7 30 E
Marcona, Peru 156 D2 15 10 S 75 0W
Marcos Juárez, Argentina 158 C3 32 42 S 62 5W
Marcus I. = Minami-Tori-Shima, Pac. Oc. 122 E7 24 0N 153 45 E
Marcus Necker Ridge, Pac. Oc. 122 F9 20 0N 175 0 E
Marcy, Mt., U.S.A. 137 B11 44 7N 73 56W
Mardan, Pakistan 79 B4 34 20N 72 0 E
Mardie, Australia 112 D2 21 12 S 115 59 E
Mardin, Turkey 89 D9 37 20N 40 43 E
Maré, I., N. Cal. 121 U22 21 30 S 168 0 E
Marechal Deodoro, Brazil 154 C4 9 43 S 35 54W
Maree, L., U.K. 18 D3 57 40N 5 26W
Mareeba, Australia 114 B4 16 59 S 145 28 E
Marek = Stanke Dimitrov, Bulgaria 42 E8 42 17N 23 9 E
Marek, Indonesia 72 B2 4 41 S 120 24 E
Maremma, Italy 38 F8 42 30N 11 30 E
Maréna, Mali 100 C3 14 0N 7 20W
Marenberg, Slovenia 39 B12 46 38N 15 13 E
Marengo, U.S.A. 140 C4 41 48N 92 4W
Marennes, France 24 C2 45 49N 1 7W
Marenyi, Kenya 106 C4 4 22 S 39 8 E
Marerano, Madag. 105 C7 21 23 S 44 52 E
Maréttimo, Italy 40 E5 37 58N 12 4 E
Mareuil-sur-Lay, France 24 B2 46 32N 1 14W
Marfa, U.S.A. 139 K2 30 19N 104 1W
Marfa Pt., Malta 32 D1 35 59N 14 19 E
Marganets = Marhanets, Ukraine 51 J8 47 40N 34 40 E
Margaret →, Australia 112 C4 18 9 S 125 41 E
Margaret Bay, Canada 130 C3 51 20N 127 35W
Margaret L., Canada 130 B5 58 56N 115 25W
Margaret River, Australia 112 C4 18 38 S 126 52 E
Margarita, I. de, Venezuela 153 A5 11 0N 64 0W
Margarítion, Greece 44 E2 39 22N 20 26 E
Margaritovo, Russia 60 C7 43 25N 134 45 E
Margate, S. Africa 105 E5 30 50 S 30 20 E
Margate, U.K. 17 F9 51 23N 1 23 E
Margelan = Marghilon, Uzbekistan 55 C5 40 27N 71 42 E
Margeride, Mts. de la, France 24 D7 44 43N 3 38 E
Margherita, India 78 B5 27 16N 95 40 E
Margherita di Savoia, Italy 41 A9 41 22N 16 9 E
Marghilon, Uzbekistan 55 C5 40 27N 71 42 E
Marghita, Romania 46 B3 47 22N 22 22 E
Margonin, Poland 47 C4 52 58N 17 5 E
Margosatubig, Phil. 71 H4 7 34N 123 10 E
Marguerite, Canada 130 C4 52 30N 122 25W
Marhanets, Ukraine 51 J8 47 40N 34 40 E
Marhoum, Algeria 99 B4 34 27N 0 11W
Mari El □, Russia 52 B8 56 30N 48 0 E
Mari Republic = Mari El □, Russia 52 B8 56 30N 48 0 E
María Elena, Chile 158 A2 22 18 S 69 40W
María Grande, Argentina 158 C4 31 45 S 59 55W
Maria I., N. Terr., Australia 114 A2 14 52 S 135 45 E
Maria I., Tas., Australia 114 G4 42 35 S 148 0 E
Maria van Diemen, C., N.Z. 118 A1 34 29 S 172 40 E
Mariager, Denmark 15 H4 56 40N 10 0 E
Mariager Fjord, Denmark 15 H4 56 42N 10 19 E
Mariakani, Kenya 106 C4 3 50 S 39 27 E
Marian L., Canada 130 A5 63 0N 116 15W
Mariana Trench, Pac. Oc. 122 F6 13 0N 145 0 E
Marianao, Cuba 148 B3 23 8N 82 24W
Mariani, India 78 B5 26 39N 94 19 E
Marianna, Ark., U.S.A. 139 H9 34 46N 90 46W
Marianna, Fla., U.S.A. 135 K3 30 46N 85 14W
Mariánské Lázně, Czech. 30 B5 49 48N 12 41 E
Marias →, U.S.A. 142 C8 47 56N 110 30W
Mariato, Punta, Panama 148 E3 7 12N 80 52W
Mariazell, Austria 30 D8 47 47N 15 19 E
Ma'rib, Yemen 86 D4 15 25N 45 21 E
Maribo, Denmark 15 K5 54 48N 11 30 E
Maribor, Slovenia 39 B12 46 36N 15 40 E
Maricaban I., Phil. 70 E3 13 39N 120 53 E
Maricalom, Phil. 71 G4 9 42N 122 25 E
Marico →, Africa 104 C4 23 35 S 26 57 E
Maricopa, Ariz., U.S.A. 143 K7 33 4N 112 3W
Maricopa, Calif., U.S.A. 145 K7 35 4N 119 24W
Maricourt, Canada 127 C12 56 34N 70 49W
Marīdī, Sudan 95 G2 4 55N 29 25 E
Maridi, Wadi →, Sudan 95 F2 6 15N 29 21 E
Marié →, Brazil 152 D4 0 27 S 66 26W
Marie Byrd Land, Antarctica 7 D14 79 30 S 125 0W
Marie-Galante, Guadeloupe 149 C7 15 56N 61 16W
Mariecourt = Kangiqsujuaq, Canada 127 B12 61 30N 72 0W
Mariefred, Sweden 14 E11 59 15N 17 12 E
Marienbad = Mariánské Lázně, Czech. 30 B5 49 48N 12 41 E
Marienberg, Germany 26 E9 50 39N 13 9 E
Marienberg, Neths. 20 D9 52 2N 6 35 E
Marienbourg, Belgium 21 H5 50 6N 4 31 E
Mariental, Namibia 104 C2 24 36 S 18 0 E
Marienville, U.S.A. 136 E5 41 28N 79 8W
Mariestad, Sweden 15 F7 58 43N 13 50 E
Marietta, Ga., U.S.A. 135 J3 33 57N 84 33W
Marietta, Ohio, U.S.A. 134 F5 39 25N 81 27W
Marieville, Canada 137 A11 45 26N 73 10W
Marignane, France 25 E9 43 25N 5 13 E
Marihatag, Phil. 71 G6 8 48N 126 18 E
Mariinsk, Russia 56 D9 56 10N 87 20 E
Mariinskiy Posad, Russia 52 B8 56 10N 47 45 E
Marijampolė, Lithuania 13 J20 54 33N 23 19 E
Marília, Brazil 159 A5 22 13 S 50 0W
Marillana, Australia 112 D2 22 37 S 119 16 E
Marimba, Angola 103 D3 8 28 S 17 8 E
Marín, Spain 36 C2 42 23N 8 42W
Marina, U.S.A. 144 J5 36 41N 121 48W
Marina di Cirò, Italy 41 C10 39 22N 17 8 E
Marina Plains, Australia 114 A3 14 37 S 143 57 E
Marinduque, Phil. 70 E3 13 25N 122 0 E
Marine City, U.S.A. 134 D4 42 43N 82 30W
Marineo, Italy 40 E6 37 57N 13 25 E
Marinette, U.S.A. 134 C2 45 6N 87 38W
Maringá, Brazil 159 A5 23 26 S 52 2W
Marinha Grande, Portugal 37 F2 39 45N 8 56W
Marion, Ala., U.S.A. 135 J2 32 38N 87 19W
Marion, Ill., U.S.A. 139 G10 37 44N 88 56W
Marion, Ind., U.S.A. 141 D11 40 32N 85 40W
Marion, Iowa, U.S.A. 140 B5 42 2N 91 36W
Marion, Kans., U.S.A. 138 F6 38 21N 97 1W
Marion, Mich., U.S.A. 134 C3 44 6N 85 9W
Marion, N.C., U.S.A. 135 H4 35 41N 82 1W
Marion, Ohio, U.S.A. 141 D13 40 35N 83 8W
Marion, S.C., U.S.A. 135 H6 34 11N 79 24W
Marion, Va., U.S.A. 135 G5 36 50N 81 31W
Marion, L., U.S.A. 135 J5 33 28N 80 10W
Marion Bay, Australia 116 C2 35 12 S 136 59 E
Marion I., Ind. Oc. 109 J2 47 0 S 38 0 E
Maripa, Venezuela 153 B4 7 26N 65 9W
Maripasoula, Fr. Guiana 153 C7 3 40N 54 4W
Maripipi I., Phil. 71 F5 11 47N 124 19 E
Mariposa, U.S.A. 144 H7 37 29N 119 58W
Mariscal Estigarribia, Paraguay 158 A3 22 3 S 60 40W
Maritime Alps = Maritimes, Alpes, Europe 25 D11 44 10N 7 10 E
Maritimes, Alpes, Europe 25 D11 44 10N 7 10 E
Maritsa = Évros →, Bulgaria 88 B2 41 40N 26 34 E
Maritsa, Bulgaria 43 E10 42 1N 25 50 E
Maritsa, Greece 32 C10 36 22N 28 10 E
Mariupol, Ukraine 51 J9 47 5N 37 31 E
Marīvān, Iran 89 E12 35 30N 46 25 E
Markah, W. →, Yemen 86 D4 14 59N 46 36 E
Markam, China 68 C2 29 42N 98 38 E
Markapur, India 83 G4 15 44N 79 19 E
Markazī □, Iran 85 C6 35 0N 49 30 E
Markdale, Canada 136 B4 44 19N 80 39W
Marke, Belgium 21 G2 50 48N 3 14 E
Marked Tree, U.S.A. 139 H9 35 32N 90 25W
Markelsdorfer Huk, Germany 26 A6 54 33N 11 0 E
Marken, Neths. 20 D6 52 26N 5 12 E
Markermeer, Neths. 20 C6 52 33N 5 15 E
Market Drayton, U.K. 16 E5 52 54N 2 29W
Market Harborough, U.K. 17 E7 52 29N 0 55W
Markham, Canada 136 C5 43 52N 79 16W
Markham →, Papua N. G. 120 D4 6 41 S 147 2 E
Markham, Mt., Antarctica 7 E11 83 0 S 164 0 E
Markham L., Canada 131 A8 62 30N 102 35W
Marki, Poland 47 C8 52 20N 21 2 E
Markleeville, U.S.A. 144 G7 38 42N 119 47W
Markoupoulon, Greece 45 G5 37 53N 23 57 E
Markovac, Serbia, Yug. 42 C6 44 14N 21 7 E
Markovo, Russia 57 C17 64 40N 169 40 E
Markoye, Burkina Faso 101 C5 14 39N 0 2 E
Marks, Russia 52 E8 51 45N 46 50 E
Marksville, U.S.A. 139 K8 31 8N 92 4W
Markt Schwaben, Germany 27 G7 48 11N 11 52 E
Marktredwitz, Germany 27 E8 50 1N 12 6 E
Marla, Australia 115 D1 27 19 S 133 33 E
Marlboro, U.S.A. 137 D13 42 19N 71 33W
Marlborough, Australia 114 C4 22 46 S 149 52 E
Marlborough Downs, U.K. 17 F6 51 27N 1 53W
Marle, France 23 C10 49 43N 3 47 E
Marlin, U.S.A. 139 K6 31 18N 96 54W
Marlow, Germany 26 A8 54 9N 12 33 E
Marlow, U.S.A. 139 H6 34 39N 97 58W
Marly-le-Grand, Switz. 28 C4 46 47N 7 10 E
Marmagao, India 83 G1 15 25N 73 56 E
Marmande, France 24 D4 44 30N 0 10 E
Marmara, Turkey 88 B2 40 35N 27 38 E
Marmara, Sea of = Marmara Denizi, Turkey 88 B3 40 45N 28 15 E
Marmara Denizi, Turkey 88 B3 40 45N 28 15 E
Marmaris, Turkey 88 D3 36 50N 28 14 E
Marmarth, U.S.A. 138 B3 46 18N 103 54W
Marmelos →, Brazil 157 B5 6 6 S 61 46W
Marmion, Mt., Australia 113 E2 29 16 S 119 50 E
Marmion L., Canada 128 C1 48 55N 91 20W
Marmolada, Mte., Italy 39 B8 46 26N 11 51 E
Marmolejo, Spain 37 G6 38 3N 4 13W
Marmora, Canada 128 D4 44 28N 77 41W
Marnay, France 23 E12 47 16N 5 48 E
Marne, Germany 26 B5 53 56N 9 0 E
Marne □, France 23 D11 48 50N 4 10 E
Marne →, France 23 D9 48 48N 2 24 E
Marneuli, Georgia 53 K7 41 30N 44 48 E
Maro, Chad 97 G3 8 30N 19 0 E
Maroa, Venezuela 152 C4 2 43N 67 33W
Maroala, Madag. 105 B8 15 23 S 47 59 E
Maroantsetra, Madag. 105 B8 15 26 S 49 44 E
Maromandia, Madag. 105 A8 14 13 S 48 5 E
Marondera, Zimbabwe 107 F3 18 5 S 31 42 E
Maroni →, Fr. Guiana 153 B7 5 30N 54 0W
Marónia, Greece 44 D7 40 53N 25 24 E
Maronne →, France 24 C5 45 5N 1 56 E
Maroochydore, Australia 115 D5 26 29 S 153 5 E
Maroona, Australia 116 D5 37 27 S 142 54 E
Maros, Indonesia 72 C1 5 0 S 119 34 E
Maros →, Hungary 31 E13 46 15N 20 13 E
Marosakoa, Madag. 105 B8 15 26 S 46 38 E
Marostica, Italy 39 C8 45 44N 11 40 E
Maroua, Cameroon 101 C7 10 40N 14 20 E
Marovoay, Madag. 105 B8 16 6 S 46 39 E
Marowijne □, Surinam 153 C7 4 0N 55 0W
Marowijne →, Surinam 153 B7 5 45N 53 58W
Marquard, S. Africa 104 D4 28 40 S 27 28 E
Marquesa, Portugal 37 G1 38 41N 9 9W
Marquesas Is. = Marquises, Is., Pac. Oc. 123 H14 9 30 S 140 0W
Marquette, U.S.A. 134 B2 46 33N 87 24W
Marquise, France 23 B8 50 50N 1 40 E
Marquises, Is., Pac. Oc. 123 H14 9 30 S 140 0W
Marra, Gebel, Sudan 95 F2 7 20N 27 35 E
Marracuene, Mozam. 105 D5 25 45 S 32 35 E
Marradi, Italy 39 D8 44 4N 11 37 E
Marrakech, Morocco 98 B3 31 9N 8 0W
Marrawah, Australia 114 G3 40 55 S 144 42 E
Marrecas, Serra das, Brazil 154 C3 9 0 S 41 0W
Marree, Australia 115 D2 29 39 S 138 1 E
Marrilla, Australia 112 D1 22 31 S 114 25 E
Marrimane, Mozam. 105 C5 22 58 S 33 34 E
Marromeu, Mozam. 105 B6 18 15 S 36 25 E
Marroquí, Punta, Spain 37 K5 36 0N 5 37W
Marrowie Cr. →, Australia 117 B6 33 23 S 145 40 E
Marrubane, Mozam. 107 F4 18 0 S 37 0 E
Marrum, Neths. 20 B7 53 19N 5 48 E
Marrupa, Mozam. 107 E4 13 8 S 37 30 E
Marsa el Brega, Libya 96 B3 30 24N 19 37 E
Marsá Matrûh, Egypt 94 A2 31 19N 27 9 E
Marsá Susah, Libya 96 B4 32 52N 21 59 E
Marsabit, Kenya 106 B4 2 18N 38 0 E
Marsabit □, Kenya 106 B4 2 45N 37 45 E
Marsala, Italy 40 E5 37 48N 12 26 E
Marsalforn, Malta 32 C1 36 4N 14 15 E
Marsberg, Germany 26 D4 51 28N 8 52 E
Marsciano, Italy 39 F9 42 54N 12 20 E
Marsden, Australia 117 B7 33 47 S 147 32 E
Marsdiep, Neths. 20 C5 52 58N 4 46 E
Marseillan, France 24 E7 43 23N 3 31 E
Marseille, France 25 E9 43 18N 5 23 E
Marseilles = Marseille, France 25 E9 43 18N 5 23 E
Marseilles, U.S.A. 141 C8 41 20N 88 43W
Marsh I., U.S.A. 139 L9 29 34N 91 53W
Marsh L., U.S.A. 138 C6 45 5N 96 0W
Marshall, Liberia 100 D2 6 8N 10 22W
Marshall, Ark., U.S.A. 139 H8 35 55N 92 38W

Mindoro 233

Merrill, Oreg., U.S.A. ... 142 E3 42 1N 121 36W
Merrill, Wis., U.S.A. 138 C10 45 11N 89 41W
Merrillville, U.S.A. 141 C9 41 29N 87 20W
Merriman, U.S.A. 138 D4 42 55N 101 42W
Merritt, Canada 130 C4 50 10N 120 45W
Merriwa, Australia ...... 117 B9 32 6 S 150 22 E
Merriwagga, Australia ... 117 B6 33 47 S 145 43 E
Merry I., Canada 128 A4 55 29N 77 31W
Merrygoen, Australia ..... 117 A8 31 51 S 149 12 E
Merryville, U.S.A. 139 K8 30 45N 93 33W
Mersa Fatma, Eritrea .... 90 E3 14 57N 40 17 E
Mersch, Lux. .... 21 J8 49 44N 6 7 E
Merseburg, Germany ..... 26 D7 51 22N 11 59 E
Mersey →, U.K. 16 D5 53 25N 3 1W
Merseyside □, U.K. 16 D5 53 31N 3 2W
Mersin, Turkey .... 88 D6 36 51N 34 36 E
Mersing, Malaysia ...... 77 L4 2 25N 103 50 E
Merta, India .... 80 F6 26 39N 74 4 E
Mertert, Lux. .... 21 J8 49 43N 6 29 E
Merthyr Tydfil, U.K. 17 F4 51 45N 3 22W
Mértola, Portugal 37 H3 37 40N 7 40W
Mertzig, Lux. .... 21 J8 49 51N 6 1 E
Mertzon, U.S.A. 139 K4 31 16N 100 49W
Méru, France .... 23 C9 49 13N 2 8 E
Meru, Kenya .... 106 B4 0 3N 37 40 E
Meru, Tanzania .... 106 C4 3 15 S 36 46 E
Meru □, Kenya .... 106 B4 0 3N 37 46 E
Merville, France .... 23 B9 50 38N 2 38 E
Méry-sur-Seine, France .. 23 D10 48 31N 3 54 E
Merzifon, Turkey .... 88 B6 40 53N 35 32 E
Merzig, Germany .... 27 F2 49 26N 6 38 E
Merzouga, Erg Tin,
Algeria .... 99 D7 24 0N 11 4 E
Mesa, U.S.A. 143 K8 33 25N 111 50W
Mesach Mellet, Libya ... 96 D2 24 30N 11 30 E
Mesagne, Italy .... 41 B10 40 34N 17 48 E
Mesanagrós, Greece ..... 32 C9 36 1N 27 49 E
Mesaoría □, Cyprus ..... 32 D12 35 12N 33 14 E
Mesarás, Kólpos, Greece . 32 D6 35 6N 24 47 E
Meschede, Germany ..... 26 D4 51 20N 8 18 E
Mescit, Turkey .... 89 B9 40 21N 41 11 E
Mesfinto, Ethiopia ...... 95 E4 13 20N 37 22 E
Mesgouez, L., Canada ... 128 B4 51 20N 75 0W
Meshchovsk, Russia ..... 52 C2 54 22N 35 17 E
Meshed = Mashhad, Iran 85 B8 36 20N 59 35 E
Meshoppen, U.S.A. 137 E8 41 36N 76 3W
Meshra er Req, Sudan ... 95 F2 8 25N 29 18 E
Mesick, U.S.A. 134 C3 44 24N 85 43W
Mesilinka →, Canada ... 130 B4 56 6N 124 30W
Mesilla, U.S.A. 143 K10 32 16N 106 48W
Meslay-du-Maine, France 22 E6 47 58N 0 33W
Mesocco, Switz. .... 29 D8 46 23N 9 12 E
Mesolóngion, Greece .... 45 F3 38 21N 21 28 E
Mesopotamia = Al
Jazirah, Iraq .... 89 E10 33 30N 44 0 E
Mesoraca, Italy .... 41 C9 39 5N 16 48 E
Mésou Volímais, Greece . 45 G2 37 53N 20 35 E
Mesquite, U.S.A. 143 H6 36 47N 114 6W
Mess Cr. →, Canada ... 130 B2 57 55N 131 14W
Messac, France 22 E5 47 49N 1 50W
Messad, Algeria .... 99 B5 34 8N 3 30 E
Messalo →, Mozam. .... 107 E4 12 25 S 39 15 E
Méssaména, Cameroon ... 101 E7 3 48N 12 49 E
Messancy, Belgium ...... 21 J7 49 36N 5 49 E
Messeue, Greece .... 45 G3 37 12N 21 58 E
Messier, Canal, Chile .... 160 C2 48 20 S 74 33W
Messina, Italy .... 41 D8 38 11N 15 34 E
Messina, S. Africa ...... 105 C5 22 20 S 30 5 E
Messina, Str. di, Italy ... 41 D8 38 15N 15 35 E
Messíni, Greece .... 45 G4 37 4N 22 1 E
Messínia □, Greece ..... 45 G3 37 10N 22 0 E
Messiniakós Kólpos,
Greece .... 45 H4 36 45N 22 5 E
Messkirch, Germany .... 27 H5 47 59N 9 7 E
Messonghi, Greece ..... 32 B3 39 29N 19 56 E
Mesta →, Bulgaria ..... 43 F9 40 54N 24 49 E
Mestà, Ákra, Greece .... 45 F7 38 16N 25 53 E
Mestanza, Spain 37 G6 38 35N 4 4W
Město Teplá, Czech. .... 30 B5 49 59N 12 52 E
Mestre, Italy .... 39 C9 45 29N 12 15 E
Mestre, Espigão, Brazil .. 155 D2 12 30 S 46 10W
Městys Zelezná Ruda,
Czech. .... 30 B6 49 8N 13 15 E
Mesudiye, Turkey .... 88 B7 40 28N 37 46 E
Meta, U.S.A. 140 F4 38 19N 92 10W
Meta □, Colombia 152 C3 3 30N 73 0W
Meta →, S. Amer. 152 B4 6 12N 67 28W
Metairie, U.S.A. 139 L9 29 58N 90 10W
Metalici, Munţii, Romania 46 C3 46 15N 22 50 E
Metaline Falls, U.S.A. ... 142 B5 48 52N 117 22W
Metamora, U.S.A. 140 D7 40 47N 89 22W
Metán, Argentina .... 158 B3 25 30 S 65 0W
Metangula, Mozam. ..... 107 E3 12 40 S 34 50 E
Metáuro →, Italy ...... 39 E10 43 50N 13 3 E
Metema, Ethiopia .... 95 E4 12 56N 36 13 E
Metengobalame, Mozam. 107 E3 14 49 S 34 30 E
Méthana, Greece .... 45 G5 37 35N 23 23 E
Methóni, Greece .... 45 H3 36 49N 21 42 E
Methven, N.Z. .... 119 D6 43 38 S 171 40 E
Methy L., Canada 131 B7 56 28N 109 30W
Metil, Mozam. .... 107 F4 16 24 S 39 0 E
Metkovets, Bulgaria ..... 43 D8 43 37N 23 10 E
Metković, Croatia ...... 42 D2 43 6N 17 39 E
Metlakatla, U.S.A. 130 B2 55 8N 131 35W
Metlaoui, Tunisia ...... 96 B1 34 24N 8 24 E
Metlika, Slovenia ...... 39 C12 45 40N 15 20 E
Metro, Indonesia .... 74 D3 5 5 S 105 20 E
Metropolis, U.S.A. 139 G10 37 9N 88 44W
Métsovon, Greece ...... 44 E3 39 48N 21 12 E
Mettet, Belgium .... 21 H5 50 19N 4 41 E
Mettuppalaiyam, India ... 83 J3 11 18N 76 59 E
Mettur, India .... 83 J3 11 48N 77 47 E
Metz, France .... 23 C13 49 8N 6 10 E
Meulaboh, Indonesia .... 74 B1 4 11N 96 3 E
Meulan, France .... 23 C8 49 3N 1 55 E
Meung-sur-Loire, France . 23 E8 47 50N 1 40 E
Meureudu, Indonesia .... 74 A1 5 19N 96 10 E
Meurthe →, France ..... 23 D13 48 47N 6 9 E
Meurthe-et-Moselle □,
France .... 23 D13 48 52N 6 0 E
Meuse □, France .... 23 C12 49 8N 5 25 E
Meuse →, Europe ...... 21 G7 50 45N 5 41 E
Meuselwitz, Germany .... 26 D8 51 2N 12 17 E
Meutapok, Mt., Malaysia 75 A5 5 40N 117 0 E
Mexborough, U.K. 16 D6 53 30N 1 15W
Mexia, U.S.A. 139 K6 31 41N 96 29W
Mexiana, I., Brazil 154 A2 0 0 49 30W
Mexicali, Mexico 146 A1 32 40N 115 30W
México, Mexico 147 D5 19 20N 99 10W
Mexico, Maine, U.S.A. .. 137 B14 44 34N 70 33W
Mexico, Mo., U.S.A. 140 E5 39 10N 91 53W
México □, Mexico 146 D5 19 20N 99 10W
Mexico ■, Cent. Amer. .. 146 C4 25 0N 105 0W
Mexico, G. of,
Cent. Amer. 147 C7 25 0N 90 0W
Meyenburg, Germany ... 26 B8 53 18N 12 14 E
Meymac, France .... 24 C6 45 32N 2 10 E
Meymaneh, Afghan. .... 79 B2 35 53N 64 38 E
Meyrargues, France .... 25 E9 43 38N 5 32 E
Meyrueis, France ...... 24 D7 44 12N 3 27 E
Meyssac, France .... 24 C5 45 3N 1 40 E
Mezdra, Bulgaria .... 43 D8 43 12N 23 42 E
Mèze, France .... 24 E7 43 27N 3 36 E
Mezen, Russia .... 48 A7 65 50N 44 20 E
Mezen →, Russia ...... 48 A7 65 44N 44 22 E
Mézene, France .... 25 D8 44 54N 4 11 E
Mezeş, Munţii, Romania . 46 B4 47 5N 23 5 E
Mezha →, Russia ...... 50 E6 55 44N 31 33 E
Mézidon, France 22 C6 49 5N 0 1W
Mézilhac, France .... 25 D8 44 49N 4 21 E
Mézin, France .... 24 D4 44 4N 0 16 E
Mezöberény, Hungary ... 31 E14 46 49N 21 3 E
Mezöfalva, Hungary ..... 31 E11 46 55N 18 49 E
Mezöhegyes, Hungary ... 31 E13 46 19N 20 49 E
Mezökövácsháza, Hungary 31 E13 46 25N 20 57 E
Mezökövesd, Hungary ... 31 D13 47 49N 20 35 E
Mézos, France 24 D2 44 5N 1 10W
Mezötúr, Hungary ...... 31 E13 46 58N 20 41 E
Mezquital, Mexico 146 C4 23 29N 104 23W
Mezzolombardo, Italy ... 38 B8 46 13N 11 5 E
Mgeta, Tanzania .... 107 D4 8 22 S 36 6 E
Mglin, Russia .... 51 F7 53 2N 32 50 E
Mhlaba Hills, Zimbabwe . 107 F3 18 30 S 30 30 E
Mhow, India .... 80 H6 22 33N 75 50 E
Mi-Shima, Japan .... 62 C3 34 46N 131 9 E
Miahuatlán, Mexico 147 D5 16 21N 96 36W
Miajadas, Spain 37 F5 39 9N 5 54W
Miallo, Australia .... 114 B4 16 28 S 145 22 E
Miami, Ariz., U.S.A. 143 K8 33 24N 110 52W
Miami, Fla., U.S.A. 135 N5 25 47N 80 11W
Miami, Tex., U.S.A. 139 H4 35 42N 100 38W
Miami →, U.S.A. 134 F3 39 20N 84 40W
Miami Beach, U.S.A. ... 135 N5 25 47N 80 8W
Miamisburg, U.S.A. 141 E12 39 38N 84 17W
Mian Xian, China ...... 66 H4 33 10N 106 32 E
Mianchi, China .... 66 G6 34 48N 111 48 E
Miāndowāb, Iran .... 89 D12 37 0N 46 5 E
Miandrivazo, Madag. .... 105 B8 19 31 S 45 29 E
Mīāneh, Iran .... 89 D12 37 30N 47 40 E
Mianning, China .... 68 C4 28 32N 102 9 E
Mianwali, Pakistan ..... 79 B3 32 38N 71 28 E
Mianyang, Hubei, China . 69 B9 30 25N 113 25 E
Mianyang, Sichuan, China 68 B5 31 22N 104 47 E
Mianzhu, China .... 68 B5 31 22N 104 7 E
Miaoli, Taiwan .... 69 E13 24 37N 120 49 E
Miarinarivo, Madag. .... 105 B8 18 57 S 46 55 E
Miass, Russia .... 54 D8 54 59N 60 6 E
Miasteczko Kraj, Poland . 47 B4 53 7N 17 1 E
Miastko, Poland .... 47 B3 54 0N 16 58 E
Micăsasa, Romania ..... 46 C5 46 7N 24 7 E
Michael, Mt.,
Papua N. G. .... 120 D3 6 27 S 145 22 E
Michalovce, Slovak Rep. . 31 C14 48 47N 21 58 E
Michelstadt, Germany ... 27 F5 49 40N 8 2 E
Michigan □, U.S.A. 134 C3 44 0N 85 0W
Michigan, L., U.S.A. 134 C2 44 0N 87 0W
Michigan Center, U.S.A. 141 B12 42 14N 84 20W
Michigan City, U.S.A. ... 141 C10 41 43N 86 54W
Michikamau L., Canada . 129 B7 54 20N 63 10W
Michipicoten, Canada ... 128 C3 47 55N 84 55W
Michipicoten I., Canada . 128 C2 47 40N 85 40W
Michoacan □, Mexico ... 146 D4 19 0N 102 0W
Michurin, Bulgaria ..... 43 E12 42 9N 27 51 E
Michurinsk, Russia ..... 52 D5 52 58N 40 27 E
Miclere, Australia ...... 114 C4 22 34 S 147 32 E
Mico, Pta., Nic. 148 D3 12 0N 83 30W
Micronesia, Federated
States of ■, Pac. Oc. .. 122 G7 9 0N 150 0 E
Mid Glamorgan □, U.K. 17 F4 51 38N 3 26W
Mid-Indian Ridge,
Ind. Oc. .... 109 H6 30 0 S 75 0 E
Midai, P., Indonesia .... 75 B3 3 0N 107 47 E
Midale, Canada 131 D8 49 25N 103 20W
Middagsfjället, Sweden .. 14 A6 63 27N 12 19 E
Middelbeers, Neths. .... 21 F6 51 28N 5 15 E
Middelburg, Neths. ..... 21 F3 51 30N 3 36 E
Middelburg, Eastern Cape,
S. Africa .... 104 E3 31 30 S 25 0 E
Middelburg,
Eastern Trans., S. Africa 105 D4 25 49 S 29 28 E
Middelfart, Denmark .... 15 J3 55 30N 9 43 E
Middelharnis, Neths. .... 20 E4 51 46N 4 10 E
Middelkerke, Belgium ... 21 F1 51 11N 2 49 E
Middelrode, Neths. ..... 21 E6 51 41N 5 26 E
Middelwit, S. Africa .... 104 C4 24 51 S 27 3 E
Middle →, U.S.A. 140 C3 41 26N 93 30W
Middle Alkali L., U.S.A. 142 F3 41 27N 120 5W
Middle Fork Feather →,
U.S.A. 144 F5 38 33N 121 30W
Middle I., Australia .... 113 F3 34 6 S 123 11 E
Middle Loup →, U.S.A. 138 E5 41 17N 98 24W
Middle Raccoon →,
U.S.A. 140 C3 41 35N 93 35W
Middleboro, U.S.A. 137 E14 41 54N 70 55W
Middleburg, N.Y., U.S.A. 137 D10 42 36N 74 20W
Middleburg, Pa., U.S.A. . 136 F7 40 47N 77 3W
Middlebury, Ind., U.S.A. 141 C11 41 41N 85 42W
Middlebury, Vt., U.S.A. . 137 B11 44 1N 73 10W
Middlemarch, N.Z. ..... 119 F5 45 30 S 170 9 E
Middleport, U.S.A. 134 F4 39 0N 82 3W
Middlesboro, Ky., U.S.A. 133 C10 36 36N 83 43W
Middlesboro, Ky., U.S.A. 135 G4 36 36N 83 43W
Middlesbrough, U.K. 16 C6 54 35N 1 13W
Middlesex, Belize 148 C2 17 2N 88 31W
Middlesex, U.S.A. 137 F10 40 36N 74 30W
Middleton, Australia .... 114 C3 22 22 S 141 32 E
Middleton, Canada 129 D6 44 57N 65 4W
Middleton, U.S.A. 140 A7 43 6N 89 30W
Middletown, Calif.,
U.S.A. 144 G4 38 45N 122 37W
Middletown, Conn.,
U.S.A. 137 E12 41 34N 72 39W
Middletown, N.Y., U.S.A. 137 E10 41 27N 74 25W
Middletown, Ohio, U.S.A. 141 E12 39 31N 84 24W
Middletown, Pa., U.S.A. 137 F8 40 12N 76 44W
Middleville, U.S.A. 141 B11 42 43N 85 28W
Midelt, Morocco 98 B4 32 46N 4 44W
Midhirst, N.Z. .... 118 F3 39 17 S 174 18 E
Midi, Canal du →, France 24 E5 43 45N 1 21 E
Midi d'Ossau, Pic du,
France 24 F3 42 50N 0 26W
Midland, Canada 128 D4 44 45N 79 50W
Midland, Calif., U.S.A. .. 145 M12 33 52N 114 48W
Midland, Mich., U.S.A. . 134 D3 43 37N 84 14W
Midland, Pa., U.S.A. ... 136 F4 40 39N 80 27W
Midland, Tex., U.S.A. ... 139 K3 32 0N 102 3W
Midlands □, Zimbabwe .. 107 F2 19 40 S 29 0 E
Midleton, Ireland 19 E3 51 55N 8 10W
Midlothian, U.S.A. 139 J6 32 30N 97 0W
Midongy, Tangorombohitr'
i, Madag. .... 105 C8 23 30 S 47 0 E
Midongy Atsimo, Madag. 105 C8 23 35 S 47 1 E
Midou □, France 24 E3 43 54N 0 30W
Midouze →, France 24 E3 43 48N 0 51W
Midsayap, Phil. .... 71 H5 7 12N 124 32 E
Midu, China .... 68 E3 25 18N 100 30 E
Midway Is., Pac. Oc. ... 122 E10 28 13N 177 22W
Midway Wells, U.S.A. ... 145 N11 32 41N 115 7W
Midwest, U.S.A. 133 B9 42 0N 90 0W
Midwest, Wyo., U.S.A. .. 142 E10 43 25N 106 16W
Midwolda, Neths. ...... 20 B9 53 12N 6 52 E
Midyat, Turkey .... 89 D9 37 25N 41 23 E
Midzŏr, Bulgaria .... 42 D7 43 24N 22 40 E
Mie □, Japan .... 63 C8 34 30N 136 10 E
Miechów, Poland ...... 47 E7 50 21N 20 5 E
Miedwie, Jezioro, Poland 47 B1 53 17N 14 54 E
Międzybód, Poland .... 47 D4 51 25N 17 34 E
Międzychód, Poland .... 47 C2 52 35N 15 53 E
Międzylesie, Poland .... 47 E3 50 8N 16 40 E
Międzyrzec Podlaski,
Poland .... 47 D9 51 58N 22 45 E
Międzyrzecz, Poland .... 47 C2 52 26N 15 35 E
Międzyzdroje, Poland ... 47 B1 53 56N 14 26 E
Miejska, Poland .... 47 D3 51 39N 16 58 E
Miélan, France .... 24 E4 43 27N 0 19 E
Mielec, Poland .... 47 E8 50 15N 21 25 E
Mienga, Angola .... 103 F3 17 12 S 19 48 E
Miercurea Ciuc, Romania 46 C6 46 21N 25 48 E
Mieres, Spain 36 B5 43 18N 5 48W
Mierlo, Neths. .... 21 F7 51 27N 5 37 E
Mieroszów, Poland ..... 47 E3 50 40N 16 11 E
Mieso, Ethiopia .... 95 F5 9 15N 40 43 E
Mieszkowice, Poland ... 47 C1 52 47N 14 30 E
Mifflintown, U.S.A. 136 F7 40 34N 77 24W
Mifraz Hefa, Israel ..... 91 C4 32 52N 35 0 E
Migdāl, Israel .... 91 C4 32 51N 35 30 E
Migennes, France ...... 23 E10 47 58N 3 31 E
Migliarino, Italy ...... 39 D8 44 46N 11 56 E
Miguel Alemán, Presa,
Mexico 147 D5 18 15N 96 40W
Miguel Alves, Brazil ... 154 B3 4 11 S 42 55W
Miguel Calmon, Brazil .. 154 D3 11 26 S 40 36W
Mihaliççık, Turkey ..... 88 C4 39 53N 31 30 E
Mihara, Japan .... 62 C5 34 24N 133 5 E
Mihara-Yama, Japan .... 63 C11 34 43N 139 23 E
Mijares →, Spain 34 F4 39 55N 0 1W
Mijas, Spain 37 J6 36 36N 4 40W
Mikese, Tanzania ...... 106 D4 6 48 S 37 55 E
Mikha-Tskhakaya =
Senaki, Georgia ...... 53 J6 42 15N 42 7 E
Mikhailovka =
Mykhaylivka, Ukraine . 51 J8 47 12N 35 15 E
Mikhaylov, Russia ..... 52 C4 54 14N 39 0 E
Mikhaylovgrad, Bulgaria . 43 D8 43 27N 23 16 E
Mikhaylovka, Russia ... 52 E6 50 3N 43 5 E
Mikhaylovski, Russia ... 54 C7 56 27N 59 7 E
Mikhnevo, Russia ..... 52 C3 55 4N 37 59 E
Miki, Hyōgo, Japan .... 62 C6 34 48N 134 59 E
Miki, Kagawa, Japan .... 62 C6 34 12N 134 7 E
Mikínai, Greece .... 45 G4 37 43N 22 46 E
Mikkeli, Finland ...... 13 F22 61 43N 27 15 E
Mikkwa →, Canada 130 B6 58 25N 114 46W
Mikniya, Sudan .... 95 D3 17 0N 33 45 E
Mikołajki, Poland ..... 47 B8 53 49N 21 37 E
Mikołów, Poland ...... 31 A11 50 10N 18 50 E
Míkonos, Greece ...... 45 G7 37 30N 25 25 E
Mikrí Préspa, Límni,
Greece .... 44 D3 40 47N 21 3 E
Mikrón Dhérion, Greece . 44 C8 41 19N 26 6 E
Mikstat, Poland ...... 47 D4 51 32N 17 59 E
Mikulov, Czech. ...... 31 C9 48 48N 16 39 E
Mikumi, Tanzania ..... 106 D4 7 26 S 37 0 E
Mikun, Russia .... 48 B9 62 20N 50 0 E
Mikuni, Japan .... 63 A8 36 13N 136 9 E
Mikuni-Tōge, Japan .... 63 A10 36 50N 138 50 E
Mikura-Jima, Japan .... 63 D11 33 52N 139 36 E
Milaca, U.S.A. 138 C8 45 45N 93 39W
Milagro, Ecuador ..... 152 D2 2 11 S 79 36W
Milagros, Phil. .... 70 E4 12 13N 123 30 E
Milan = Milano, Italy ... 38 C6 45 28N 9 12 E
Milan, Ill., U.S.A. 140 C6 41 27N 90 34W
Milan, Mich., U.S.A. ... 141 B13 42 5N 83 41W
Milan, Mo., U.S.A. 140 D3 40 12N 93 7W
Milan, Tenn., U.S.A. ... 135 H1 35 55N 88 46W
Milang, S. Austral.,
Australia .... 115 E2 32 2 S 139 10 E
Milang, S. Austral.,
Australia .... 116 C3 35 24 S 138 58 E
Milange, Mozam. ..... 107 F4 16 3 S 35 45 E
Milano, Italy .... 38 C6 45 28N 9 12 E
Milâs, Turkey .... 88 D2 37 20N 27 50 E
Mílatos, Greece ...... 32 D7 35 18N 25 34 E
Milazzo, Italy .... 41 D8 38 13N 15 13 E
Milbank, U.S.A. 138 C6 45 13N 96 38W
Milden, Canada 131 C7 51 29N 107 32W
Mildmay, Canada 136 B3 44 3N 81 7W
Mildura, Australia ..... 116 C5 34 13 S 142 9 E
Mile, China .... 68 E4 24 28N 103 20 E
Miléai, Greece .... 44 E5 39 20N 23 9 E
Miles, Australia ...... 115 D5 26 40 S 150 9 E
Miles, U.S.A. 139 K4 31 36N 100 11W
Miles City, U.S.A. 138 B2 46 25N 105 51W
Milestone, Canada 131 D8 49 59N 104 31W
Mileto, Italy .... 41 D9 38 36N 16 4 E
Miletto, Mte., Italy .... 41 A7 41 27N 14 22 E
Miletus, Turkey ...... 88 D2 37 30N 27 18 E
Mileura, Australia ..... 113 E2 26 22 S 117 20 E
Milevsko, Czech. ..... 30 B7 49 27N 14 21 E
Milford, Calif., U.S.A. .. 144 E6 40 10N 120 22W
Milford, Conn., U.S.A. .. 137 E11 41 14N 73 3W
Milford, Del., U.S.A. ... 134 F8 38 55N 75 26W
Milford, Ill., U.S.A. 141 D9 40 38N 87 42W
Milford, Mass., U.S.A. .. 137 D13 42 8N 71 31W
Milford, Mich., U.S.A. .. 141 B13 42 35N 83 36W
Milford, Pa., U.S.A. ... 137 E10 41 19N 74 48W
Milford, Utah, U.S.A. ... 143 G7 38 24N 113 1W
Milford Haven, U.K. ... 17 F2 51 42N 5 7W
Milford Sd., N.Z. ..... 119 E2 44 41 S 167 47 E
Milgun, Australia ..... 113 D2 24 56 S 118 18 E
Milḥ, Baḥr al, Iraq .... 89 F10 32 40N 43 35 E
Miliana, Aïn Salah,
Algeria .... 99 C5 27 20N 2 32 E
Miliana, Médéa, Algeria . 99 A5 36 20N 2 15 E
Milicz, Poland .... 47 D4 51 31N 17 19 E
Miling, Australia ..... 113 F2 30 30 S 116 17 E
Militello in Val di Catánia,
Italy .... 41 E7 37 16N 14 48 E
Milk →, U.S.A. 142 B10 48 4N 106 19W
Milk, Wadi el →, Sudan 94 D3 17 55N 30 20 E
Milk River, Canada 130 D6 49 10N 112 5W
Mill, Neths. .... 21 E7 51 41N 5 48 E
Mill City, U.S.A. 142 D2 44 45N 122 29W
Mill I., Antarctica .... 7 C8 66 0 S 101 30 E
Mill Shoals, U.S.A. ... 141 F8 38 15N 88 21W
Mill Valley, U.S.A. ... 144 H4 37 54N 122 32W
Millau, France .... 24 D7 44 8N 3 4 E
Millbridge, Canada ... 136 B7 44 41N 77 36W
Millbrook, Canada 136 B6 44 10N 78 29W
Mille Lacs, L. des, Canada 128 C1 48 45N 90 35W
Mille Lacs L., U.S.A. .. 138 B8 46 15N 93 39W
Milledgeville, Ga., U.S.A. 135 J4 33 5N 83 14W
Milledgeville, Ill., U.S.A. 140 C7 41 58N 89 46W
Millen, U.S.A. 135 J5 32 48N 81 57W
Miller, U.S.A. 138 C5 44 31N 98 59W
Millerovo, Russia .... 53 F5 48 57N 40 28 E
Miller's Flat, N.Z. .... 119 F4 45 39 S 169 23 E
Millersburg, Ind., U.S.A. 141 C11 41 32N 85 42W
Millersburg, Ohio, U.S.A. 136 F3 40 33N 81 55W
Millersburg, Pa., U.S.A. 136 F8 40 32N 76 58W
Millerton, N.Z. ...... 119 B6 41 39 S 171 54 E
Millerton, U.S.A. 137 E11 41 57N 73 31W
Millerton L., U.S.A. .. 144 J7 37 1N 119 41W
Millevaches, Plateau de,
France .... 24 C6 45 45N 2 0 E
Millicent, Australia ... 116 D4 37 34 S 140 21 E
Millingen, Neths. .... 20 E8 51 52N 6 2 E
Millinocket, U.S.A. ... 129 C6 45 39N 68 43W
Millmerran, Australia .. 115 D5 27 53 S 151 16 E
Mills L., Canada 130 A5 61 30N 118 20W
Millsboro, U.S.A. 136 G4 40 0N 80 0W
Milltown Malbay, Ireland 19 D2 52 52N 9 24W
Millville, U.S.A. 134 F8 39 24N 75 2W
Millwood L., U.S.A. ... 139 J8 33 42N 93 58W
Milly-la-Forêt, France .. 23 D9 48 24N 2 28 E
Milna, Croatia .... 39 E13 43 20N 16 28 E
Milne →, Australia ... 114 C2 21 10 S 137 33 E
Milne Inlet, Canada ... 127 A11 72 30N 80 0W
Milnor, U.S.A. 138 B6 46 16N 97 27W
Milo, Canada 130 C6 50 34N 112 53W
Mílos, Greece .... 45 H6 36 44N 24 25 E
Miloševo, Serbia, Yug. .. 42 B5 45 42N 20 20 E
Miłosław, Poland ..... 47 C4 52 12N 17 32 E
Milparinka P.O., Australia 115 D3 29 46 S 141 57 E
Milroy, U.S.A. 141 E11 39 30N 85 28W
Miltenberg, Germany .. 27 F5 49 41N 10 16 E
Milton, Canada 136 C5 43 31N 79 53W
Milton, N.Z. .... 119 G4 46 7 S 169 59 E
Milton, U.K. 18 D4 57 18N 4 32W
Milton, Calif., U.S.A. .. 144 G6 38 3N 120 51W
Milton, Fla., U.S.A. ... 135 K2 30 38N 87 3W
Milton, Iowa, U.S.A. .. 140 D4 40 41N 92 10W
Milton, Pa., U.S.A. ... 136 F8 41 1N 76 51W
Milton, Wis., U.S.A. .. 141 B8 42 47N 88 56W
Milton-Freewater, U.S.A. 142 D4 45 56N 118 23W
Milton Keynes, U.K. .. 17 E7 52 1N 0 44W
Miltou, Chad .... 97 F3 10 14N 17 26 E
Milverton, Canada 136 C4 43 34N 80 55W
Milwaukee, U.S.A. ... 141 A9 43 2N 87 55W
Milwaukee Deep, Atl. Oc. 8 G2 19 50N 68 0W
Milwaukie, U.S.A. ... 144 E4 45 27N 122 38W
Mim, Ghana 100 D4 6 57N 2 33W
Mimizan, France 24 D2 44 12N 1 13W
Mimon, Czech. ...... 30 A7 50 38N 14 43 E
Mimongo, Gabon ..... 102 C2 1 11 S 11 36 E
Mimoso, Brazil ...... 155 E2 15 10 S 48 5W
Min Chiang →, China .. 69 E12 26 0N 119 35 E
Min Jiang →, China ... 68 C5 28 45N 104 40 E
Min-Kush, Kyrgyzstan .. 55 C7 41 40N 74 28 E
Min Xian, China ..... 66 G3 34 25N 104 5 E
Mina, U.S.A. 143 G4 38 24N 118 7W
Mina Pirquitas, Argentina 158 A2 22 40 S 66 30W
Mīnā Su'ud, Si. Arabia .. 85 D6 28 45N 48 28 E
Mīnā'al Aḥmadī, Kuwait . 85 D6 29 5N 48 10 E
Mīnāb, Iran .... 85 E8 27 10N 57 1 E
Minago →, Canada ... 131 C9 54 33N 98 59W
Minakami, Japan ..... 63 A10 36 49N 138 59 E
Minaki, Canada 131 D10 49 59N 94 40W
Minakuchi, Japan .... 63 C8 34 58N 136 10 E
Minamata, Japan .... 62 E2 32 10N 130 30 E
Minami-Tori-Shima,
Pac. Oc. .... 122 E7 24 0N 153 45 E
Minas, Uruguay ..... 159 C4 34 20 S 55 10W
Minas, Sierra de las,
Guatemala 148 C2 15 9N 89 31W
Minas Basin, Canada .. 129 C7 45 20N 64 12W
Minas de Rio Tinto, Spain 37 H4 37 42N 6 35W
Minas de San Quintín,
Spain 37 G6 38 49N 4 23W
Minas Gerais □, Brazil .. 155 E2 18 50 S 46 0W
Minas Novas, Brazil ... 155 E3 17 15 S 42 36W
Minatitlán, Mexico 147 D6 17 59N 94 31W
Minbu, Burma .... 78 E5 20 10N 94 52 E
Minbya, Burma ...... 78 E4 20 22N 93 16 E
Mincio →, Italy ..... 38 C7 45 4N 10 59 E
Mindanao, Phil. ..... 71 H5 8 0N 125 0 E
Mindanao Sea = Bohol
Sea, Phil. .... 71 G5 9 0N 124 0 E
Mindanao Trench,
Pac. Oc. .... 70 E5 12 0N 126 6 E
Mindel →, Germany .. 27 G6 48 31N 10 23 E
Mindelheim, Germany .. 27 G6 48 2N 10 29 E
Minden, Canada 136 B6 44 55N 78 43W
Minden, Germany .... 26 C4 52 17N 8 55 E
Minden, La., U.S.A. .. 139 J8 32 37N 93 17W
Minden, Nev., U.S.A. .. 144 G7 38 57N 119 46W
Mindiptana, Indonesia .. 73 C6 5 55 S 140 22 E
Mindon, Burma ...... 78 F5 19 21N 94 44 E
Mindoro, Phil. ...... 70 E3 13 0N 121 0 E

234 Mindoro Occidental

Mindoro Occidental □, Phil. 70 E3 13 0N 120 55 E
Mindoro Oriental □, Phil. 70 E3 13 0N 121 5 E
Mindoro Str., Phil. 70 E3 12 30N 120 30 E
Mindouli, Congo 103 C2 4 12 S 14 28 E
Mine, Japan 62 C3 34 12N 131 7 E
Minehead, U.K. 17 F4 51 12N 3 29W
Mineiros, Brazil 157 D7 17 34 S 52 34W
Mineola, U.S.A. 139 J7 32 40N 95 29W
Mineral King, U.S.A. 144 J8 36 27N 118 36W
Mineral Point, U.S.A. 140 B6 42 52N 90 11W
Mineral Wells, U.S.A. 139 J5 32 48N 98 7W
Mineralnyye Vody, Russia 53 H6 44 15N 43 8 E
Minersville, Pa., U.S.A. 137 F8 40 41N 76 16W
Minersville, Utah, U.S.A. 143 G7 38 13N 112 56W
Minerva, U.S.A. 136 F3 40 44N 81 6W
Minervino Murge, Italy 41 A9 41 5N 16 5 E
Minetto, U.S.A. 137 C8 43 24N 76 28W
Mingäçevir, Azerbaijan 53 K8 40 45N 47 0 E
Mingäçevir Su Anbarı, Azerbaijan 53 K8 40 57N 46 50 E
Mingan, Canada 129 B7 50 20N 64 0W
Mingary, Australia 116 B4 32 8 S 140 45 E
Mingechaur = Mingäçevir, Azerbaijan 53 K8 40 45N 47 0 E
Mingechaurskoye Vdkhr. = Mingäçevir Su Anbarı, Azerbaijan 53 K8 40 57N 46 50 E
Mingela, Australia 114 B4 19 52 S 146 38 E
Mingenew, Australia 113 E2 29 12 S 115 21 E
Mingera Cr. →, Australia 114 C2 20 38 S 137 45 E
Minggang, China 69 A10 32 24N 114 3 E
Mingin, Burma 78 D5 22 50N 94 30 E
Minglanilla, Spain 34 F3 39 34N 1 38W
Minglun, China 68 E7 25 10N 108 21 E
Mingorria, Spain 36 E6 40 45N 4 40W
Mingt'iehkaitafan = Mintaka Pass, Pakistan 81 A6 37 0N 74 58 E
Mingxi, China 69 D11 26 18N 117 12 E
Mingyuegue, China 67 C15 43 2N 128 50 E
Minho = Miño →, Spain 36 D2 41 52N 8 40W
Minhou, China 69 E12 26 0N 119 15 E
Minićevo, Serbia, Yug. 42 D7 43 42N 22 18 E
Minidoka, U.S.A. 142 E7 42 45N 113 29W
Minier, U.S.A. 140 D7 40 26N 89 19W
Minigwal, L., Australia 113 E3 29 31 S 123 14 E
Minilya, Australia 113 D1 23 55 S 114 0 E
Minilya →, Australia 113 D1 23 45 S 114 0 E
Mininera, Australia 116 D5 37 37 S 142 58 E
Minipi, L., Canada 129 B7 52 25N 60 45W
Minj, Papua N. G. 120 C3 5 54 S 144 37 E
Mink L., Canada 130 A5 61 54N 117 40W
Minlaton, Australia 116 C2 34 45 S 137 35 E
Minna, Nigeria 101 D6 9 37N 6 30 E
Minneapolis, Kans., U.S.A. 138 F6 39 8N 97 42W
Minneapolis, Minn., U.S.A. 138 C8 44 59N 93 16W
Minnedosa, Canada 131 C9 50 14N 99 50W
Minnesota □, U.S.A. 138 B7 46 0N 94 15W
Minnesund, Norway 14 D5 60 23N 11 14 E
Minnie Creek, Australia 113 D2 24 3 S 115 42 E
Minnipa, Australia 115 E2 32 51 S 135 9 E
Minnitaki L., Canada 128 C1 49 57N 92 10W
Mino, Japan 63 B8 35 32N 136 55 E
Miño →, Spain 36 D2 41 52N 8 40W
Mino-Kamo, Japan 63 B9 35 23N 137 2 E
Mino-Mikawa-Kōgen, Japan 63 B9 35 10N 137 23 E
Minoa, Greece 45 J7 35 6N 25 45 E
Minobu, Japan 63 B10 35 22N 138 26 E
Minobu-Sanchi, Japan 63 B10 35 14N 138 20 E
Minonk, U.S.A. 140 D7 40 54N 89 2W
Minooka, U.S.A. 141 C8 41 27N 88 16W
Minorca = Menorca, Spain 33 B11 40 0N 4 0 E
Minore, Australia 117 B8 32 14 S 148 27 E
Minot, U.S.A. 138 A4 48 14N 101 18W
Minqin, China 66 E2 38 38N 103 20 E
Minqing, China 69 D12 26 15N 118 50 E
Minsen, Germany 26 B3 53 41N 7 58 E
Minsk, Belarus 50 F4 53 52N 27 30 E
Mińsk Mazowiecki, Poland 47 C8 52 10N 21 33 E
Minster, U.S.A. 141 D12 40 24N 84 23W
Mintaka Pass, Pakistan 81 A6 37 0N 74 58 E
Minthami, Burma 78 D5 23 55N 94 16 E
Minto, U.S.A. 126 B5 64 53N 149 11W
Minton, Canada 131 D8 49 10N 104 35W
Mintoum, Gabon 102 B2 0 27N 12 16 E
Minturn, U.S.A. 142 G10 39 35N 106 26W
Minturno, Italy 40 A6 41 15N 13 45 E
Minûf, Egypt 94 H7 30 26N 30 52 E
Minusinsk, Russia 57 D10 53 50N 91 20 E
Minutang, India 78 A6 28 15N 96 30 E
Minvoul, Gabon 102 B2 2 9N 12 8 E
Minwakh, Yemen 87 C5 16 48N 48 6 E
Minya el Qamh, Egypt 94 H7 30 31N 31 21 E
Minyar, Russia 54 D6 55 4N 57 33 E
Minyip, Australia 116 D5 36 29 S 142 36 E
Mionica, Serbia, Yug. 42 C5 44 14N 20 6 E
Mir, Niger 97 F2 14 5N 11 59 E
Mīr Kūh, Iran 85 E8 26 22N 58 55 E
Mīr Shahdād, Iran 85 E8 26 15N 58 29 E
Mira, Italy 39 C9 45 26N 12 8 E
Mira, Portugal 36 E2 40 26N 8 44W
Mira →, Colombia 152 C2 1 36N 79 1W
Mira →, Portugal 37 H2 37 43N 8 47W
Mira por vos Cay, Bahamas 149 B5 22 9N 74 30W
Mīrābād, Afghan. 79 C1 30 25N 61 50 E
Mirabella Eclano, Italy 41 A7 41 2N 14 59 E
Miracema do Norte, Brazil 154 C2 9 33 S 48 24W
Mirador, Brazil 154 C3 6 22 S 44 22W
Miraflores, Colombia 152 C3 1 25N 72 13W
Miraj, India 82 F2 16 50N 74 45 E
Miram Shah, Pakistan 79 B3 33 0N 70 2 E
Miramar, Argentina 158 D4 38 15 S 57 50W
Miramar, Mozam. 105 C6 23 50 S 35 35 E
Miramas, France 25 E8 43 33N 4 59 E
Mirambeau, France 24 C3 45 23N 0 35W
Miramichi B., Canada 129 C7 47 15N 65 0W
Miramont-de-Guyenne, France 24 D4 44 37N 0 21 E
Miranda, Brazil 157 E6 20 10 S 56 15W
Miranda □, Venezuela 152 A4 10 15N 66 25W
Miranda →, Brazil 157 D6 19 25 S 57 20W
Miranda de Ebro, Spain 34 C2 42 41N 2 57W
Miranda do Corvo, Spain 36 E2 40 6N 8 20W
Miranda do Douro, Portugal 36 D4 41 30N 6 16W
Mirande, France 24 E4 43 31N 0 25 E
Mirandela, Portugal 36 D3 41 32N 7 10W
Mirando City, U.S.A. 139 M5 27 26N 99 0W
Mirándola, Italy 38 D8 44 53N 11 4 E
Mirandópolis, Brazil 159 A5 21 9 S 51 6W
Mirango, Malawi 107 E3 13 32 S 34 58 E
Mirani, Australia 114 C4 21 9 S 148 53 E
Mirano, Italy 39 C9 45 30N 12 7 E
Mirassol, Brazil 159 A6 20 46 S 49 28W
Mirbāţ, Oman 87 C6 17 0N 54 45 E
Mirboo North, Australia 117 E7 38 24 S 146 10 E
Mirear, Egypt 94 C4 23 15N 35 41 E
Mirebeau, Côte-d'Or, France 23 E12 47 25N 5 20 E
Mirebeau, Vienne, France 22 F7 46 49N 0 10 E
Mirecourt, France 23 D13 48 20N 6 10 E
Mirgorod = Myrhorod, Ukraine 51 H7 49 58N 33 37 E
Miri, Malaysia 75 B4 4 23N 113 59 E
Miriam Vale, Australia 114 C5 24 20 S 151 33 E
Mirim, L., S. Amer. 159 C5 32 45 S 52 50W
Mirimire, Venezuela 152 A4 11 10N 68 43W
Miriti, Brazil 157 B6 6 15 S 59 0W
Mirnyy, Russia 57 C12 62 33N 113 53 E
Miroč, Serbia, Yug. 42 C7 44 32N 22 16 E
Mirond L., Canada 131 B8 55 6N 102 47W
Mirosławiec, Poland 47 B3 53 20N 16 5 E
Mirpur, Pakistan 79 B4 33 32N 73 56 E
Mirpur Bibiwari, Pakistan 80 E2 28 33N 67 44 E
Mirpur Khas, Pakistan 79 D3 25 30N 69 0 E
Mirpur Sakro, Pakistan 80 G2 24 33N 67 41 E
Mirria, Niger 97 F1 13 43N 9 7 E
Mirror, Canada 130 C6 52 30N 113 7W
Mîrşani, Romania 46 E4 44 1N 23 59 E
Mirsk, Poland 47 E2 50 58N 15 23 E
Miryang, S. Korea 67 G15 35 31N 128 44 E
Mirzaani, Georgia 53 K8 41 24N 46 5 E
Mirzapur, India 81 G10 25 10N 82 34 E
Mirzapur-cum-Vindhyachal = Mirzapur, India 81 G10 25 10N 82 34 E
Misamis Occidental □, Phil. 71 G4 8 20N 123 42 E
Misamis Oriental □, Phil. 71 G5 8 45N 125 0 E
Misantla, Mexico 147 D5 19 56N 96 50W
Misawa, Japan 60 D10 40 41N 141 24 E
Miscou I., Canada 129 C7 47 57N 64 31W
Mish'āb, Ra'as al, Si. Arabia 85 D6 28 15N 48 43 E
Mishagua →, Peru 156 C3 11 12 S 72 58W
Mishan, China 65 B8 45 37N 131 48 E
Mishawaka, U.S.A. 141 C10 41 40N 86 11W
Mishbih, Gebel, Egypt 94 C3 22 38N 34 44 E
Mishima, Japan 63 B10 35 10N 138 52 E
Mishmi Hills, India 78 A5 29 0N 96 0 E
Misilmeri, Italy 40 D6 38 2N 13 27 E
Misima I., Papua N. G. 120 F7 10 40 S 152 45 E
Misión, Mexico 145 N10 32 6N 116 53W
Misión Fagnano, Argentina 160 D3 54 32 S 67 17W
Misiones □, Argentina 159 B5 27 0 S 55 0W
Misiones □, Paraguay 158 B4 27 0 S 56 0W
Miskah, Si. Arabia 84 E4 24 49N 42 56 E
Miskitos, Cayos, Nic. 148 D3 14 26N 82 50W
Miskolc, Hungary 31 C13 48 7N 20 50 E
Misoke, Zaïre 106 C2 0 42 S 28 2 E
Misool, Indonesia 73 B4 1 52 S 130 10 E
Misrātah, Libya 96 B3 32 24N 15 3 E
Misrātah □, Libya 96 C3 33 30N 15 0 E
Missanabie, Canada 128 C3 48 20N 84 6W
Missão Velha, Brazil 154 C4 7 15 S 39 10W
Missinaibi →, Canada 128 B3 50 43N 81 29W
Missinaibi L., Canada 128 C3 48 23N 83 40W
Mission, S. Dak., U.S.A. 138 D4 43 18N 100 39W
Mission, Tex., U.S.A. 139 M5 26 13N 98 20W
Mission City, Canada 130 D4 49 10N 122 15W
Mission Viejo, U.S.A. 145 M9 33 36N 117 40W
Missisa L., Canada 128 B2 52 20N 85 7W
Mississagi →, Canada 128 C3 46 15N 83 9W
Mississinewa Res., U.S.A. 141 D10 40 46N 86 3W
Mississippi □, U.S.A. 139 J10 33 0N 90 0W
Mississippi →, U.S.A. 139 L10 29 9N 89 15W
Mississippi L., Canada 137 A8 45 5N 76 10W
Mississippi River Delta, U.S.A. 139 L9 29 10N 89 15W
Mississippi Sd., U.S.A. 139 K10 30 20N 89 0W
Missoula, U.S.A. 142 C6 46 52N 114 1W
Missour, Morocco 98 B4 33 3N 4 0W
Missouri □, U.S.A. 138 F8 38 25N 92 30W
Missouri →, U.S.A. 138 F9 38 49N 90 7W
Missouri Valley, U.S.A. 138 E7 41 34N 95 53W
Mist, U.S.A. 144 E3 45 59N 123 15W
Mistake B., Canada 131 A10 62 8N 93 0W
Mistassini →, Canada 129 C5 48 42N 72 20W
Mistassini L., Canada 128 B5 51 0N 73 30W
Mistastin L., Canada 129 A7 55 57N 63 20W
Mistatim, Canada 131 C8 52 52N 103 22W
Mistelbach, Austria 31 C9 48 34N 16 34 E
Misterbianco, Italy 41 E8 37 31N 15 0 E
Mistretta, Italy 41 E7 37 56N 14 22 E
Misty L., Canada 131 B8 58 53N 101 40W
Misugi, Japan 63 C8 34 31N 136 16 E
Misumi, Japan 62 E2 32 37N 130 27 E
Misurata = Misrātah, Libya 96 B3 32 24N 15 3 E
Mît Ghamr, Egypt 94 H7 30 42N 31 12 E
Mitaka, Japan 63 B11 35 40N 139 33 E
Mitan, Uzbekistan 55 C3 40 5N 66 35 E
Mitatib, Sudan 95 D4 15 59N 36 12 E
Mitchell, Australia 115 D4 26 29 S 147 58 E
Mitchell, Canada 136 C3 43 28N 81 12W
Mitchell, Ind., U.S.A. 141 F10 38 44N 86 28W
Mitchell, Nebr., U.S.A. 138 E3 41 57N 103 49W
Mitchell, Oreg., U.S.A. 142 D3 44 34N 120 9W
Mitchell, S. Dak., U.S.A. 138 D5 43 43N 98 2W
Mitchell →, Australia 114 B3 15 12 S 141 35 E
Mitchell, Mt., U.S.A. 135 H4 35 46N 82 16W
Mitchell Ras., Australia 114 A2 12 49 S 135 36 E
Mitchelstown, Ireland 19 D3 52 15N 8 16W
Mitha Tiwana, Pakistan 80 C5 32 13N 72 6 E
Míthimna, Greece 44 E8 39 20N 26 12 E
Mitiamo, Australia 116 D6 36 12 S 144 15 E
Mitilíni, Greece 45 E8 39 6N 26 35 E
Mitilinoí, Greece 45 G8 37 42N 26 56 E
Mito, Japan 63 A12 36 20N 140 30 E
Mitre, Mt., N.Z. 118 G4 40 50 S 175 30 E
Mitrofanovka, Russia 52 F4 49 58N 39 42 E
Mitrovica = Titova-Mitrovica, Serbia, Yug. 42 E5 42 54N 20 52 E
Mitsinjo, Madag. 105 B8 16 1 S 45 52 E
Mitsiwa, Eritrea 95 D4 15 35N 39 25 E
Mitsiwa Channel, Eritrea 95 D5 15 30N 40 0 E
Mitsukaidō, Japan 63 A11 36 1N 139 59 E
Mittagong, Australia 117 C9 34 28 S 150 29 E
Mittelland, Switz. 28 C4 46 50N 7 23 E
Mittellandkanal →, Germany 26 C3 52 20N 8 28 E
Mittenwalde, Germany 26 C9 52 15N 13 31 E
Mitterteich, Germany 27 F8 49 57N 12 14 E
Mittweida, Germany 26 E8 50 59N 12 59 E
Mitú, Colombia 152 C3 1 8N 70 3W
Mituas, Colombia 152 C4 3 52N 68 49W
Mitumba, Tanzania 106 D3 7 8 S 31 2 E
Mitumba, Chaîne des, Zaïre 106 D2 7 0 S 27 30 E
Mitumba Mts. = Mitumba, Chaîne des, Zaïre 106 D2 7 0 S 27 30 E
Mitwaba, Zaïre 107 D2 8 2 S 27 17 E
Mityana, Uganda 106 B3 0 23N 32 2 E
Mitzic, Gabon 102 B2 0 45N 11 40 E
Miura, Japan 63 B11 35 12N 139 40 E
Mixteco →, Mexico 147 D5 18 11N 98 30W
Miyagi □, Japan 60 E10 38 15N 140 45 E
Miyâh, W. el →, Egypt 94 B3 25 0N 33 23 E
Miyah, W. el →, Syria 84 C3 34 44N 39 57 E
Miyake-Jima, Japan 63 C11 34 5N 139 30 E
Miyako, Japan 60 E10 39 40N 141 59 E
Miyako-Jima, Japan 61 M2 24 45N 125 20 E
Miyako-Rettō, Japan 61 M2 24 24N 125 0 E
Miyakonojō, Japan 62 F3 31 40N 131 5 E
Miyanojō, Japan 62 F2 31 54N 130 27 E
Miyanoura-Dake, Japan 61 J5 30 20N 130 31 E
Miyata, Japan 62 D2 33 49N 130 42 E
Miyazaki, Japan 62 F3 31 56N 131 30 E
Miyazaki □, Japan 62 E3 32 30N 131 30 E
Miyazu, Japan 63 B7 35 35N 135 10 E
Miyet, Bahr el = Dead Sea, Asia 91 D4 31 30N 35 30 E
Miyi, China 68 D4 26 47N 102 9 E
Miyoshi, Japan 62 C4 34 48N 132 51 E
Miyun, China 66 D9 40 28N 116 50 E
Miyun Shuiku, China 67 D9 40 30N 117 0 E
Mizamis = Ozamiz, Phil. 71 G4 8 15N 123 50 E
Mizdah, Libya 96 B2 31 30N 13 0 E
Mizen Hd., Cork, Ireland 19 E2 51 27N 9 50W
Mizen Hd., Wick., Ireland 19 D5 52 52N 6 4W
Mizhi, China 66 F6 37 47N 110 12 E
Mizil, Romania 46 E7 44 59N 26 29 E
Mizoram □, India 78 D4 23 30N 92 40 E
Mizpe Ramon, Israel 91 E3 30 34N 34 49 E
Mizuho, Japan 63 B7 35 6N 135 17 E
Mizunami, Japan 63 B9 35 22N 137 15 E
Mizusawa, Japan 60 E10 39 8N 141 8 E
Mjöbäck, Sweden 15 G6 57 28N 12 53 E
Mjölby, Sweden 15 F9 58 20N 15 10 E
Mjörn, Sweden 15 G6 57 55N 12 25 E
Mjøsa, Norway 14 D5 60 40N 11 0 E
Mkata, Tanzania 106 D4 5 45 S 38 20 E
Mkokotoni, Tanzania 106 D4 5 55 S 39 15 E
Mkomazi, Tanzania 106 C4 4 40 S 38 7 E
Mkomazi →, S. Africa 105 E5 30 12 S 30 50 E
Mkulwe, Tanzania 107 D3 8 37 S 32 20 E
Mkumbi, Ras, Tanzania 106 D4 7 38 S 39 55 E
Mkushi, Zambia 107 E2 14 25 S 29 15 E
Mkushi River, Zambia 107 E2 13 32 S 29 45 E
Mkuze, S. Africa 105 D5 27 10 S 32 0 E
Mladá Boleslav, Czech. 30 A7 50 27N 14 53 E
Mladenovac, Serbia, Yug. 42 C5 44 28N 20 44 E
Mlala Hills, Tanzania 106 D3 6 50 S 31 40 E
Mlange, Malawi 107 F4 16 2 S 35 33 E
Mlava →, Serbia, Yug. 42 C6 44 45N 21 13 E
Mława, Poland 47 B7 53 9N 20 25 E
Mliniště, Bos.-H. 39 D13 44 15N 16 50 E
Mljet, Croatia 42 E2 42 43N 17 30 E
Mljetski Kanal, Croatia 42 E2 42 48N 17 35 E
Młynąry, Poland 47 A6 54 12N 19 46 E
Mmabatho, S. Africa 104 D4 25 49 S 25 30 E
Mme, Cameroon 101 D7 6 18N 10 14 E
Mo i Rana, Norway 12 C16 66 20N 14 7 E
Moa, Indonesia 72 C3 8 0 S 128 0 E
Moa →, S. Leone 100 D2 6 59N 11 36W
Moab, U.S.A. 143 G9 38 35N 109 33W
Moabi, Gabon 102 C2 2 24 S 10 59 E
Moaco →, Brazil 156 B4 7 41 S 68 18W
Moala, Fiji 121 B2 18 36 S 179 53 E
Moalie Park, Australia 115 D3 29 42 S 143 3 E
Moaña, Spain 36 C2 42 18N 8 43W
Moba, Zaïre 106 D2 7 0 S 29 48 E
Mobara, Japan 63 B12 35 25N 140 18 E
Mobārakābād, Iran 85 D7 28 24N 53 20 E
Mobārakīyeh, Iran 85 C6 32 23N 51 37 E
Mobaye, C.A.R. 102 B4 4 25N 21 5 E
Mobayi, Zaïre 102 B4 4 15N 21 8 E
Moberly, U.S.A. 140 E4 39 25N 92 26W
Moberly →, Canada 130 B4 56 12N 120 55W
Mobile, U.S.A. 135 K1 30 41N 88 3W
Mobile B., U.S.A. 135 K2 30 30N 88 0W
Mobridge, U.S.A. 138 C4 45 32N 100 26W
Mobutu Sese Seko, L. = Albert L., Africa 106 B3 1 30N 31 0 E
Moc Chau, Vietnam 76 B5 20 50N 104 38 E
Moc Hoa, Vietnam 77 G5 10 46N 105 56 E
Mocabe, Sa. de, Angola 103 D3 7 12 S 15 0 E
Mocabe Kasari, Zaïre 107 D2 9 58 S 26 12 E
Moçajuba, Brazil 154 B2 2 35 S 49 30W
Moçambique, Mozam. 107 F5 15 3 S 40 42 E
Moçâmedes = Namibe, Angola 103 F2 15 7 S 12 11 E
Mocapra →, Venezuela 152 B4 7 56N 66 46W
Mocha, I., Chile 160 A2 38 22 S 73 56W
Mochudi, Botswana 104 C4 24 27 S 26 7 E
Mocimboa da Praia, Mozam. 107 E5 11 25 S 40 20 E
Mociu, Romania 46 C5 46 46N 24 3 E
Moclips, U.S.A. 144 C2 47 14N 124 13W
Mocoa, Colombia 152 C2 1 7N 76 35W
Mococa, Brazil 159 A6 21 28 S 47 0W
Mocorito, Mexico 146 B3 25 30N 107 53W
Moctezuma, Mexico 146 B3 29 50N 109 0W
Moctezuma →, Mexico 147 C5 21 59N 98 34W
Mocuba, Mozam. 107 F4 16 54 S 36 57 E
Mocúzari, Presa, Mexico 146 B3 27 10N 109 10W
Moda, Burma 78 C6 24 22N 96 29 E
Modane, France 25 C10 45 12N 6 40 E
Modasa, India 80 H5 23 30N 73 21 E
Modave, Belgium 21 H6 50 27N 5 18 E
Modder →, S. Africa 104 D3 29 2 S 24 37 E
Modderrivier, S. Africa 104 D3 29 2 S 24 38 E
Módena, Italy 38 D7 44 40N 10 55 E
Modena, U.S.A. 143 H7 37 48N 113 56W
Modesto, U.S.A. 144 H6 37 39N 121 0W
Módica, Italy 41 F7 36 52N 14 46 E
Modigliana, Italy 39 D8 44 9N 11 47 E
Modjamboli, Zaïre 102 B4 2 28N 22 6 E
Modlin, Poland 47 C7 52 24N 20 41 E
Mödling, Austria 31 C9 48 5N 16 17 E
Modo, Sudan 95 F3 5 31N 30 33 E
Modra, Slovak Rep. 31 C10 48 19N 17 20 E
Modriča, Bos.-H. 42 C3 44 57N 18 17 E
Moe, Australia 117 E7 38 12 S 146 19 E
Moebase, Mozam. 107 F4 17 3 S 38 41 E
Moëlan-sur-Mer, France 22 E3 47 49N 3 38W
Moengo, Surinam 153 B7 5 45N 54 20W
Moergestel, Neths. 21 E6 51 33N 5 11 E
Moers, Germany 21 F9 51 27N 6 36 E
Moësa →, Switz. 29 D8 46 12N 9 10 E
Moffat, U.K. 18 F5 55 20N 3 27W
Moga, India 80 D6 30 48N 75 8 E
Mogadishu = Muqdisho, Somali Rep. 108 D3 2 2N 45 25 E
Mogador = Essaouira, Morocco 98 B3 31 32N 9 42W
Mogadouro, Portugal 36 D4 41 22N 6 47W
Mogalakwena →, S. Africa 105 C4 22 38 S 28 40 E
Mogami →, Japan 60 E10 38 45N 140 0 E
Mogán, Canary Is. 33 G4 27 53N 15 43W
Mogaung, Burma 78 C6 25 20N 97 0 E
Møgeltønder, Denmark 15 K2 54 57N 8 48 E
Mogente, Spain 35 G4 38 52N 0 45W
Mogho, Ethiopia 95 G5 4 54N 40 16 E
Mogi das Cruzes, Brazil 159 A6 23 31 S 46 11W
Mogi-Guaçu →, Brazil 159 A6 20 53 S 48 10W
Mogi-Mirim, Brazil 159 A6 22 29 S 47 0W
Mogielnica, Poland 47 D7 51 42N 20 41 E
Mogilev = Mahilyow, Belarus 50 F6 53 55N 30 18 E
Mogilev-Podolskiy = Mohyliv-Podilskyy, Ukraine 51 H4 48 26N 27 48 E
Mogilno, Poland 47 C4 52 39N 17 55 E
Mogincual, Mozam. 107 F5 15 35 S 40 25 E
Mogliano Véneto, Italy 39 C9 45 33N 12 14 E
Mogocha, Russia 57 D12 53 40N 119 50 E
Mogoi, Indonesia 73 B4 1 55 S 133 10 E
Mogok, Burma 78 D6 23 0N 96 40 E
Mogriguy, Australia 117 B8 32 3 S 148 40 E
Moguer, Spain 37 H4 37 15N 6 52W
Mogumber, Australia 113 F2 31 2 S 116 3 E
Mohács, Hungary 31 F11 45 58N 18 41 E
Mohaka →, N.Z. 118 F6 39 7 S 177 12 E
Mohales Hoek, Lesotho 104 E4 30 7 S 27 26 E
Mohall, U.S.A. 138 A4 48 46N 101 31W
Moḩammadābād, Iran 85 B8 37 52N 59 5 E
Mohammadia, Algeria 99 A5 35 33N 0 3 E
Mohammedia, Morocco 98 B3 33 44N 7 21W
Mohave, L., U.S.A. 145 K12 35 12N 114 34W
Mohawk →, U.S.A. 137 D11 42 47N 73 41W
Möhne →, Germany 26 D3 51 29N 7 57 E
Mohnyin, Burma 78 C6 24 47N 96 22 E
Moholm, Sweden 15 F8 58 37N 14 5 E
Mohoro, Tanzania 106 D4 8 6 S 39 8 E
Mohyliv-Podilskyy, Ukraine 51 H4 48 26N 27 48 E
Moia, Sudan 95 F2 5 3N 28 2 E
Moidart, L., U.K. 18 E3 56 47N 5 52W
Moinabad, India 82 F3 17 44N 77 16 E
Moindou, N. Cal. 121 U19 21 42 S 165 41 E
Moinești, Romania 46 C7 46 28N 26 31 E
Mointy, Kazakhstan 56 E8 47 10N 73 18 E
Moirans, France 25 C9 45 20N 5 33 E
Moirans-en-Montagne, France 25 B9 46 26N 5 43 E
Moíres, Greece 32 D6 35 4N 24 56 E
Moisaküla, Estonia 13 G21 58 3N 25 12 E
Moisie, Canada 129 B6 50 12N 66 1W
Moisie →, Canada 129 B6 50 14N 66 5W
Moissac, France 24 D5 44 7N 1 5 E
Moïssala, Chad 97 G3 8 21N 17 46 E
Moita, Portugal 37 G2 38 38N 8 58W
Mojácar, Spain 35 H3 37 6N 1 55W
Mojados, Spain 36 D6 41 26N 4 40W
Mojave, U.S.A. 145 K8 35 3N 118 10W
Mojave Desert, U.S.A. 145 L10 35 0N 116 30W
Mojiang, China 68 F3 23 37N 101 35 E
Mojo, Bolivia 158 A2 21 48 S 65 33W
Mojo, Ethiopia 95 F4 8 35N 39 5 E
Mojokerto, Indonesia 75 D4 7 28 S 112 26 E
Mojos, Llanos de, Bolivia 157 D5 15 10 S 65 0W
Moju →, Brazil 154 B2 1 40 S 48 25W
Mokai, N.Z. 118 E4 38 32 S 175 56 E
Mokambo, Zaïre 107 E2 12 25 S 28 20 E
Mokameh, India 81 G11 25 24N 85 55 E
Mokane, U.S.A. 140 F5 38 41N 91 53W
Mokau, N.Z. 118 E3 38 42 S 174 39 E
Mokau →, N.Z. 118 E3 38 35 S 174 35 E
Mokelumne →, U.S.A. 144 G5 38 13N 121 28W
Mokelumne Hill, U.S.A. 144 G6 38 18N 120 43W
Mokhós, Greece 32 D7 35 16N 25 27 E
Mokhotlong, Lesotho 105 D4 29 22 S 29 2 E
Mokihinui →, N.Z. 119 B6 41 33 S 171 58 E
Moknine, Tunisia 96 A2 35 35N 10 58 E
Mokpalin, Burma 78 G6 17 26N 96 53 E
Mokra Gora, Serbia, Yug. 42 E5 42 50N 20 30 E
Mokronog, Slovenia 39 C12 45 57N 15 9 E
Moksha →, Russia 52 C6 54 45N 41 53 E
Mokshan, Russia 52 D7 53 25N 44 35 E
Mol, Belgium 21 F6 51 11N 5 5 E
Mola, C. de la, Spain 34 F9 39 40N 4 20 E
Mola di Bari, Italy 41 A10 41 3N 17 5 E
Moláoi, Greece 45 H4 36 49N 22 56 E
Molat, Croatia 39 D11 44 15N 14 50 E
Molave, Phil. 71 G4 8 5N 123 30 E
Molchanovo, Russia 56 D9 57 40N 83 50 E
Mold, U.K. 16 D4 53 9N 3 8W
Moldava nad Bodvou, Slovak Rep. 31 C14 48 38N 21 0 E
Moldavia, Romania 46 C8 46 30N 27 0 E

Moquegua 235

Moldavia = Moldova ■, *Europe* **51 J5** 47 0N 28 0 E
Molde, *Norway* **12 E12** 62 45N 7 9 E
Moldotau, Khrebet, *Kyrgyzstan* **55 C7** 41 35N 75 0 E
Moldova ■, *Europe* **51 J5** 47 0N 28 0 E
Moldova Nouă, *Romania* **46 E2** 44 45N 21 41 E
Moldoveana, *Romania* **46 D5** 45 36N 24 45 E
Molepolole, *Botswana* **104 C4** 24 28 S 25 28 E
Moléson, *Switz.* **28 C4** 46 33N 7 1 E
Molesworth, *N.Z.* **119 C8** 42 5 S 173 16 E
Molfetta, *Italy* **41 A9** 41 12N 16 36 E
Molina de Aragón, *Spain* **34 E3** 40 46N 1 52W
Moline, *U.S.A.* **140 C6** 41 30N 90 31W
Molinella, *Italy* **39 D8** 44 37N 11 40 E
Molinos, *Argentina* **158 B2** 25 28 S 66 15W
Moliro, *Zaïre* **106 D3** 8 12 S 30 30 E
Molise □, *Italy* **39 G11** 41 38N 14 29 E
Moliterno, *Italy* **41 B8** 40 14N 15 52 E
Mollahat, *Bangla.* **81 H13** 22 56N 89 48 E
Mölle, *Sweden* **15 H6** 56 17N 12 31 E
Molledo, *Spain* **36 B6** 43 8N 4 6W
Mollendo, *Peru* **156 D3** 17 0 S 72 0W
Mollerin, L., *Australia* **113 F2** 30 30 S 117 35 E
Mollerusa, *Spain* **34 D5** 41 37N 0 54 E
Mollina, *Spain* **37 H6** 37 8N 4 38W
Mölln, *Germany* **26 B6** 53 39N 10 32 E
Mölltorp, *Sweden* **15 F8** 58 30N 14 26 E
Mölndal, *Sweden* **15 G6** 57 40N 12 3 E
Molo, *Burma* **78 D6** 23 22N 96 53 E
Molochansk, *Ukraine* **51 J8** 47 15N 35 35 E
Molochnoye, Ozero, *Ukraine* **51 J8** 46 30N 35 20 E
Molodechno = Maladzyechna, *Belarus* **50 E4** 54 20N 26 50 E
Molokai, *U.S.A.* **132 H16** 21 8N 157 0W
Moloma →, *Russia* **54 B2** 58 29N 48 33 E
Molong, *Australia* **117 B8** 33 5 S 148 54 E
Molopo →, *Africa* **104 D3** 27 30 S 20 13 E
Mólos, *Greece* **45 F4** 38 47N 22 37 E
Molotov = Perm, *Russia* **54 C6** 58 0N 56 10 E
Moloundou, *Cameroon* **102 B3** 2 8N 15 15 E
Molsheim, *France* **23 D14** 48 33N 7 29 E
Molson L., *Canada* **131 C9** 54 22N 96 40W
Molteno, *S. Africa* **104 E4** 31 22 S 26 22 E
Molu, *Indonesia* **73 C4** 6 45 S 131 40 E
Molucca Sea, *Indonesia* **72 A3** 2 0 S 124 0 E
Moluccas = Maluku, *Indonesia* **72 B3** 1 0 S 127 0 E
Molundo, *Phil.* **71 H5** 7 57N 124 23 E
Moma, *Mozam.* **107 F4** 16 47 S 39 4 E
Moma, *Zaïre* **102 C4** 1 35 S 23 52 E
Momba, *Australia* **116 A5** 30 58 S 143 30 E
Mombaça, *Brazil* **154 C4** 5 43 S 39 45W
Mombasa, *Kenya* **106 C4** 4 2 S 39 43 E
Mombetsu, *Japan* **60 B11** 44 21N 143 22 E
Mombil, *Burma* **78 B7** 27 46N 98 6 E
Mombuey, *Spain* **36 C4** 42 3N 6 20W
Momchilgrad, *Bulgaria* **43 F10** 41 33N 25 23 E
Momence, *U.S.A.* **141 C9** 41 10N 87 40W
Momi, *Zaïre* **106 C2** 1 42 S 27 0 E
Momignies, *Belgium* **21 H4** 50 2N 4 10 E
Mompog Pass, *Phil.* **70 E4** 13 34N 122 13 E
Mompós, *Colombia* **152 B3** 9 14N 74 26W
Møn, *Denmark* **15 K6** 54 57N 12 15 E
Mona, Canal de la, *W. Indies* **149 C6** 18 30N 67 45W
Mona, Isla, *Puerto Rico* **149 C6** 18 5N 67 54W
Mona, Pta., *Costa Rica* **148 E3** 9 37N 82 36W
Mona, Pta., *Spain* **37 J7** 36 43N 3 45W
Mona Quimbundo, *Angola* **103 D3** 9 55 S 19 58 E
Monach Is., *U.K.* **18 D1** 57 32N 7 40W
Monaco ■, *Europe* **25 E11** 43 46N 7 23 E
Monadhliath Mts., *U.K.* **18 D4** 57 10N 4 4W
Monagas □, *Venezuela* **153 B5** 9 20N 63 0W
Monaghan, *Ireland* **19 B5** 54 15N 6 57W
Monaghan □, *Ireland* **19 B5** 54 11N 6 56W
Monahans, *U.S.A.* **139 K3** 31 36N 102 54W
Monapo, *Mozam.* **107 E5** 14 56 S 40 19 E
Monarch Mt., *Canada* **130 C3** 51 55N 125 57W
Monastir = Bitola, *Macedonia* **42 F6** 41 5N 21 10 E
Monastir, *Tunisia* **96 A2** 35 50N 10 49 E
Moncada, *Phil.* **70 D3** 15 44N 120 34 E
Moncada, *Spain* **34 F4** 39 30N 0 24W
Moncalieri, *Italy* **38 D4** 45 0N 7 41 E
Moncalvo, *Italy* **38 C5** 45 3N 8 16 E
Moncão, *Portugal* **36 C2** 42 4N 8 27W
Moncarapacho, *Portugal* **37 H3** 37 5N 7 46W
Moncayo, Sierra del, *Spain* **34 D3** 41 48N 1 50W
Monchegorsk, *Russia* **48 A5** 67 54N 32 58 E
Mönchengladbach, *Germany* **26 D2** 51 11N 6 27 E
Monchique, *Portugal* **37 H2** 37 19N 8 38W
Monclova, *Mexico* **146 B4** 26 50N 101 30W
Moncontour, *France* **22 D4** 48 22N 2 38W
Moncoutant, *France* **24 B3** 46 43N 0 35W
Moncton, *Canada* **129 C7** 46 7N 64 51W
Mondego →, *Portugal* **36 E2** 40 9N 8 52W
Mondego, C., *Portugal* **36 E2** 40 11N 8 54W
Mondeodo, *Indonesia* **72 B2** 3 34 S 122 9 E
Mondo, *Chad* **97 F3** 13 47N 15 32 E
Mondolfo, *Italy* **39 E10** 43 45N 13 6 E
Mondoñedo, *Spain* **36 B3** 43 25N 7 23W
Mondovì, *Italy* **38 D4** 44 23N 7 49 E
Mondovi, *U.S.A.* **138 C9** 44 34N 91 40W
Mondragon, *France* **25 D8** 44 13N 4 44 E
Mondragon, *Phil.* **70 E5** 12 31N 124 45 E
Mondragone, *Italy* **40 A6** 41 7N 13 53 E
Mondrain I., *Australia* **113 F3** 34 9 S 122 14 E
Monduli □, *Tanzania* **106 C4** 3 0 S 36 0 E
Monemvasía, *Greece* **45 H5** 36 41N 23 3 E
Monessen, *U.S.A.* **136 F5** 40 9N 79 54W
Monesterio, *Spain* **37 G4** 38 6N 6 15W
Monestier-de-Clermont, *France* **25 D9** 44 55N 5 38 E
Monett, *U.S.A.* **139 G8** 36 55N 93 55W
Monfalcone, *Italy* **39 C10** 45 49N 13 32 E
Monflanquin, *France* **24 D4** 44 32N 0 47 E
Monforte, *Portugal* **37 F3** 39 6N 7 25W
Monforte de Lemos, *Spain* **36 C3** 42 31N 7 33W
Mong Hta, *Burma* **78 F7** 19 50N 98 35 E
Mong Ket, *Burma* **78 D7** 23 8N 98 22 E
Mong Kung, *Burma* **78 E6** 21 35N 97 35 E
Mong Kyawt, *Burma* **78 F7** 19 56N 98 45 E
Mong Nai, *Burma* **78 E6** 20 32N 97 46 E
Mong Ping, *Burma* **78 E7** 21 22N 99 2 E
Mong Pu, *Burma* **78 E7** 20 55N 98 44 E
Mong Ton, *Burma* **78 E7** 20 17N 98 45 E
Mong Tung, *Burma* **78 D6** 22 2N 97 41 E
Mong Yai, *Burma* **78 D7** 22 21N 98 3 E
Monga, *Zaïre* **102 B4** 4 12N 22 49 E
Mongalla, *Sudan* **95 F3** 5 8N 31 42 E
Mongers, L., *Australia* **113 E2** 29 25 S 117 5 E
Monghyr = Munger, *India* **81 G12** 25 23N 86 30 E
Mongibello = Etna, *Italy* **41 E8** 37 50N 14 55 E
Mongla, *Bangla.* **78 D2** 22 8N 89 35 E
Mongngaw, *Burma* **78 D6** 22 47N 96 59 E
Mongo, *Chad* **97 F3** 12 14N 18 43 E
Mongó, *Eq. Guin.* **102 B2** 1 52N 10 10 E
Mongolia ■, *Asia* **57 E10** 47 0N 103 0 E
Mongomo, *Eq. Guin.* **102 B2** 1 38N 11 19 E
Mongonu, *Nigeria* **101 C7** 12 40N 13 32 E
Mongororo, *Chad* **97 F4** 12 3N 22 26 E
Mongu, *Zambia* **103 F4** 15 16 S 23 12 E
Môngua, *Angola* **103 F3** 16 43 S 15 20 E
Monistrol-d'Allier, *France* **24 D7** 44 58N 3 38 E
Monistrol-sur-Loire, *France* **25 C8** 45 17N 4 11 E
Monkayo, *Phil.* **71 H6** 7 50N 126 5 E
Monkey Bay, *Malawi* **107 E4** 14 7 S 35 1 E
Monkey River, *Belize* **147 D7** 16 22N 88 29W
Mońki, *Poland* **47 B9** 53 23N 22 48 E
Monkira, *Australia* **114 C3** 24 46 S 140 30 E
Monkoto, *Zaïre* **102 C4** 1 38 S 20 35 E
Monmouth, *U.K.* **17 F5** 51 48N 2 42W
Monmouth, *U.S.A.* **140 D6** 40 55N 90 39W
Mono, *Solomon Is.* **121 L8** 7 20 S 155 35 E
Mono L., *U.S.A.* **144 H7** 38 1N 119 1W
Monolith, *U.S.A.* **145 K8** 35 7N 118 22W
Monólithos, *Greece* **32 C9** 36 7N 27 45 E
Monon, *U.S.A.* **141 D10** 40 52N 86 53W
Monona, *Iowa, U.S.A.* **140 A5** 43 3N 91 23W
Monona, *Wis., U.S.A.* **140 A7** 43 4N 89 20W
Monongahela, *U.S.A.* **136 F5** 40 12N 79 56W
Monópoli, *Italy* **41 B10** 40 57N 17 18 E
Monor, *Hungary* **31 D12** 47 21N 19 27 E
Monóvar, *Spain* **35 G4** 38 28N 0 53W
Monowai, *N.Z.* **119 F2** 45 53 S 167 31 E
Monowai, L., *N.Z.* **119 F2** 45 53 S 167 25 E
Monqoumba, *C.A.R.* **102 B3** 3 33N 18 40 E
Monreal del Campo, *Spain* **34 E3** 40 47N 1 20W
Monreale, *Italy* **40 D6** 38 5N 13 17 E
Monroe, *Ga., U.S.A.* **135 J4** 33 47N 83 43W
Monroe, *Iowa, U.S.A.* **140 C3** 41 31N 93 6W
Monroe, *La., U.S.A.* **139 J8** 32 30N 92 7W
Monroe, *Mich., U.S.A.* **141 C13** 41 55N 83 24W
Monroe, *N.C., U.S.A.* **135 H5** 34 59N 80 33W
Monroe, *N.Y., U.S.A.* **137 E10** 41 20N 74 11W
Monroe, *Ohio, U.S.A.* **141 E12** 39 27N 84 22W
Monroe, *Utah, U.S.A.* **143 G7** 38 38N 112 7W
Monroe, *Wash., U.S.A.* **144 C5** 47 51N 121 58W
Monroe, *Wis., U.S.A.* **140 B7** 42 36N 89 38W
Monroe City, *U.S.A.* **140 E5** 39 39N 91 44W
Monroe Res., *U.S.A.* **141 E10** 39 1N 86 31W
Monroeville, *Ala., U.S.A.* **135 K2** 31 31N 87 20W
Monroeville, *Ind., U.S.A.* **141 D12** 40 59N 84 52W
Monroeville, *Pa., U.S.A.* **136 F5** 40 26N 79 45W
Monrovia, *Liberia* **100 D2** 6 18N 10 47W
Monrovia, *U.S.A.* **143 J4** 34 7N 118 1W
Mons, *Belgium* **21 H3** 50 27N 3 58 E
Monsaraz, *Portugal* **37 G3** 38 28N 7 22W
Monse, *Indonesia* **72 B2** 4 0 S 123 10 E
Monsefú, *Peru* **156 B2** 6 52 S 79 52W
Monségur, *France* **24 D4** 44 38N 0 4 E
Monsélice, *Italy* **39 C8** 45 14N 11 45 E
Monster, *Neths.* **20 D4** 52 1N 4 10 E
Mont Cenis, Col du, *France* **25 C10** 45 15N 6 55 E
Mont-de-Marsan, *France* **24 E3** 43 54N 0 31W
Mont-Joli, *Canada* **129 C6** 48 37N 68 10W
Mont-Laurier, *Canada* **128 C4** 46 35N 75 30W
Mont-St.-Michel, Le = Le Mont-St.-Michel, *France* **22 D5** 48 40N 1 30W
Mont-sous-Vaudrey, *France* **23 F12** 46 58N 5 36 E
Mont-sur-Marchienne, *Belgium* **21 H4** 50 23N 4 24 E
Mont Tremblant Prov. Park, *Canada* **128 C5** 46 30N 74 30W
Montabaur, *Germany* **26 E3** 50 25N 7 50 E
Montagnac, *France* **24 E7** 43 29N 3 28 E
Montagnana, *Italy* **39 C8** 45 14N 11 28 E
Montagu, *S. Africa* **104 E3** 33 45 S 20 8 E
Montagu I., *Antarctica* **7 B1** 58 25 S 26 20W
Montague, *Canada* **129 C7** 46 10N 62 39W
Montague, *U.S.A.* **142 F2** 41 44N 122 32W
Montague, I., *Mexico* **146 A2** 31 40N 114 56W
Montague Ra., *Australia* **113 E2** 27 15 S 119 30 E
Montague Sd., *Australia* **112 B4** 14 28 S 125 20 E
Montaigu, *France* **22 F5** 46 59N 1 18W
Montalbán, *Spain* **34 E4** 40 50N 0 45W
Montalbano di Elicona, *Italy* **41 D8** 38 1N 15 0 E
Montalbano Iónico, *Italy* **41 B9** 40 17N 16 34 E
Montalbo, *Spain* **34 F2** 39 53N 2 42W
Montalcino, *Italy* **39 E8** 43 3N 11 29 E
Montalegre, *Portugal* **36 D3** 41 49N 7 47W
Montalto di Castro, *Italy* **39 F8** 42 21N 11 37 E
Montalto Uffugo, *Italy* **41 C9** 39 24N 16 9 E
Montalvo, *U.S.A.* **145 L7** 34 15N 119 12W
Montamarta, *Spain* **36 D5** 41 39N 5 49W
Montaña, *Peru* **156 B3** 6 0 S 73 0W
Montana, *Switz.* **28 D4** 46 19N 7 29 E
Montana □, *U.S.A.* **142 C9** 47 0N 110 0W
Montaña Clara, I., *Canary Is.* **33 E6** 29 17N 13 33W
Montánchez, *Spain* **37 F4** 39 15N 6 8W
Montañita, *Colombia* **152 C2** 1 22N 75 28W
Montargis, *France* **23 E9** 47 59N 2 43 E
Montauban, *France* **24 D5** 44 2N 1 21 E
Montauk, *U.S.A.* **137 E13** 41 3N 71 57W
Montauk Pt., *U.S.A.* **137 E13** 41 4N 71 52W
Montbard, *France* **23 E11** 47 38N 4 20 E
Montbéliard, *France* **23 E13** 47 31N 6 48 E
Montblanch, *Spain* **34 D6** 41 23N 1 4 E
Montbrison, *France* **25 C8** 45 36N 4 3 E
Montcalm, Pic de, *France* **24 F5** 42 40N 1 25 E
Montceau-les-Mines, *France* **23 F11** 46 40N 4 23 E
Montchanin, *France* **25 B8** 46 47N 4 30 E
Montclair, *U.S.A.* **137 F10** 40 49N 74 13W
Montcornet, *France* **23 C11** 49 40N 4 1 E
Montcuq, *France* **24 D5** 44 21N 1 13 E
Montdidier, *France* **23 C9** 49 38N 2 35 E
Monte Albán, *Mexico* **147 D5** 17 2N 96 45W
Monte Alegre, *Brazil* **153 D7** 2 0 S 54 0W
Monte Alegre de Goiás, *Brazil* **155 D2** 13 14 S 47 10W
Monte Alegre de Minas, *Brazil* **155 E2** 18 52 S 48 52W
Monte Azul, *Brazil* **155 E3** 15 9 S 42 53W
Monte Bello Is., *Australia* **112 D2** 20 30 S 115 45 E
Monte-Carlo, *Monaco* **25 E11** 43 46N 7 23 E
Monte Carmelo, *Brazil* **155 E2** 18 43 S 47 29W
Monte Caseros, *Argentina* **158 C4** 30 10 S 57 50W
Monte Comán, *Argentina* **158 C2** 34 40 S 67 53W
Monte Cristi, *Dom. Rep.* **149 C5** 19 52N 71 39W
Monte Dinero, *Argentina* **160 D3** 52 18 S 68 33W
Monte Lindo →, *Paraguay* **158 A4** 23 56 S 57 12W
Monte Quemado, *Argentina* **158 B3** 25 53 S 62 41W
Monte Redondo, *Portugal* **36 F2** 39 53N 8 50W
Monte Rio, *U.S.A.* **144 G4** 38 28N 123 0W
Monte San Giovanni Campano, *Italy* **40 A6** 41 38N 13 31 E
Monte San Savino, *Italy* **39 E8** 43 20N 11 43 E
Monte Sant' Ángelo, *Italy* **41 A8** 41 42N 15 59 E
Monte Santu, C. di, *Italy* **40 B2** 40 5N 9 44 E
Monte Vista, *U.S.A.* **143 H10** 37 35N 106 9W
Monteagudo, *Argentina* **159 B5** 27 14 S 54 8W
Monteagudo, *Bolivia* **157 D5** 19 49 S 63 59W
Montealegre, *Spain* **35 G3** 38 48N 1 17W
Montebello, *Canada* **128 C5** 45 40N 74 55W
Montebelluna, *Italy* **39 C9** 45 47N 12 3 E
Montebourg, *France* **22 C5** 49 30N 1 20W
Montecastrilli, *Italy* **39 F9** 42 39N 12 29 E
Montecatini Terme, *Italy* **38 E7** 43 53N 10 46 E
Montecito, *U.S.A.* **145 L7** 34 26N 119 40W
Montecristi, *Ecuador* **152 D1** 1 0 S 80 40W
Montecristo, *Italy* **38 F7** 42 20N 10 19 E
Montefalco, *Italy* **39 F9** 42 54N 12 39 E
Montefiascone, *Italy* **39 F9** 42 32N 12 2 E
Montefrío, *Spain* **37 H6** 37 20N 4 0W
Montegnée, *Belgium* **21 G7** 50 38N 5 31 E
Montego Bay, *Jamaica* **148 C4** 18 30N 78 0W
Montegranaro, *Italy* **39 E10** 43 14N 13 38 E
Monteiro, *Brazil* **154 C4** 7 48 S 37 2W
Monteith, *Australia* **116 C3** 35 11 S 139 23 E
Montejicar, *Spain* **35 H1** 37 33N 3 30W
Montejinnie, *Australia* **112 C5** 16 40 S 131 38 E
Montelíbano, *Colombia* **152 B2** 8 5N 75 29W
Montélimar, *France* **25 D8** 44 33N 4 45 E
Montella, *Italy* **41 B8** 40 51N 15 1 E
Montellano, *Spain* **37 J5** 36 59N 5 36W
Montello, *U.S.A.* **138 D10** 43 48N 89 20W
Montelupo Fiorentino, *Italy* **38 E8** 43 44N 11 1 E
Montemor-o-Novo, *Portugal* **37 G2** 38 40N 8 12W
Montemor-o-Velho, *Portugal* **36 E2** 40 11N 8 40W
Montemorelos, *Mexico* **147 B5** 25 11N 99 42W
Montendre, *France* **24 C3** 45 16N 0 26W
Montenegro, *Brazil* **159 B5** 29 39 S 51 29W
Montenegro □, *Yugoslavia* **42 E4** 42 40N 19 20 E
Montenero di Bisáccia, *Italy* **39 G11** 41 57N 14 47 E
Montepuez, *Mozam.* **107 E4** 13 8 S 38 59 E
Montepuez →, *Mozam.* **107 E5** 12 32 S 40 27 E
Montepulciano, *Italy* **39 E8** 43 5N 11 47 E
Montereale, *Italy* **39 F10** 42 31N 13 15 E
Montereau-Fault-Yonne, *France* **23 D9** 48 22N 2 57 E
Monterey, *Calif., U.S.A.* **144 J5** 36 37N 121 55W
Monterey, *Ind., U.S.A.* **141 C10** 41 11N 86 30W
Monterey B., *U.S.A.* **144 J5** 36 45N 122 0W
Montería, *Colombia* **152 B2** 8 46N 75 53W
Montero, *Bolivia* **157 D5** 17 20 S 63 15W
Monteros, *Argentina* **158 B2** 27 11 S 65 30W
Monterotondo, *Italy* **39 F9** 42 3N 12 37 E
Monterrey, *Mexico* **146 B4** 25 40N 100 30W
Montes Altos, *Brazil* **154 C2** 5 50 S 47 4W
Montes Claros, *Brazil* **155 E3** 16 30 S 43 50W
Montesano, *U.S.A.* **144 D3** 46 59N 123 36W
Montesárchio, *Italy* **41 A7** 41 4N 14 38 E
Montescaglioso, *Italy* **41 B9** 40 33N 16 40 E
Montesilvano Marina, *Italy* **39 F11** 42 29N 14 8 E
Montevarchi, *Italy* **39 E8** 43 31N 11 34 E
Montevideo, *Uruguay* **159 C4** 34 50 S 56 11W
Montevideo, *U.S.A.* **138 C7** 44 57N 95 43W
Montezuma, *Ind., U.S.A.* **141 E9** 39 48N 87 22W
Montezuma, *Iowa, U.S.A.* **140 C4** 41 35N 92 32W
Montfaucon, *France* **23 C12** 49 16N 5 8 E
Montfaucon-en-Velay, *France* **25 C8** 45 11N 4 20 E
Montfort, *France* **22 D5** 48 9N 1 58W
Montfort, *Neths.* **21 F7** 51 7N 5 58 E
Montfort-l'Amaury, *France* **23 D8** 48 47N 1 49 E
Montgenèvre, *France* **25 D10** 44 56N 6 43 E
Montgomery = Sahiwal, *Pakistan* **79 C4** 30 45N 73 8 E
Montgomery, *U.K.* **17 E4** 52 34N 3 8W
Montgomery, *Ala., U.S.A.* **135 J2** 32 23N 86 19W
Montgomery, *Ill., U.S.A.* **141 C8** 41 44N 88 21W
Montgomery, *W. Va., U.S.A.* **134 F5** 38 11N 81 19W
Montgomery City, *U.S.A.* **140 F5** 38 59N 91 30W
Montguyon, *France* **24 C3** 45 12N 0 12W
Monthey, *Switz.* **28 D3** 46 15N 6 56 E
Monticelli d'Ongina, *Italy* **38 C6** 45 5N 9 56 E
Monticello, *Ark., U.S.A.* **139 J9** 33 38N 91 47W
Monticello, *Fla., U.S.A.* **135 K4** 30 33N 83 52W
Monticello, *Ill., U.S.A.* **141 D8** 40 1N 88 34W
Monticello, *Ind., U.S.A.* **141 D10** 40 45N 86 46W
Monticello, *Iowa, U.S.A.* **140 B5** 42 15N 91 12W
Monticello, *Ky., U.S.A.* **135 G3** 36 50N 84 51W
Monticello, *Minn., U.S.A.* **138 C8** 45 18N 93 48W
Monticello, *Miss., U.S.A.* **139 K9** 31 33N 90 7W
Monticello, *Mo., U.S.A.* **140 D5** 40 7N 91 43W
Monticello, *N.Y., U.S.A.* **137 E10** 41 39N 74 42W
Monticello, *Utah, U.S.A.* **143 H9** 37 52N 109 21W
Montichiari, *Italy* **38 C7** 45 25N 10 23 E
Montier-en-Der, *France* **23 D11** 48 30N 4 45 E
Montignac, *France* **24 C5** 45 4N 1 10 E
Montignies-sur-Sambre, *Belgium* **21 H4** 50 24N 4 29 E
Montigny, *France* **23 C13** 49 7N 6 10 E
Montigny-sur-Aube, *France* **23 E11** 47 57N 4 45 E
Montijo, *Spain* **37 G4** 38 52N 6 39W
Montijo, Presa de, *Spain* **37 G4** 38 52N 6 26W
Montilla, *Spain* **37 H6** 37 36N 4 40W
Montlhéry, *France* **23 D9** 48 39N 2 15 E
Montluçon, *France* **24 B6** 46 22N 2 36 E
Montmagny, *Canada* **129 C5** 46 58N 70 34W
Montmarault, *France* **24 B6** 46 19N 2 57 E
Montmartre, *Canada* **131 C8** 50 14N 103 27W
Montmédy, *France* **23 C12** 49 30N 5 20 E
Montmélian, *France* **25 C10** 45 30N 6 4 E
Montmirail, *France* **23 D10** 48 51N 3 30 E
Montmoreau-St.-Cybard, *France* **24 C4** 45 23N 0 8 E
Montmorency, *Canada* **129 C5** 46 53N 71 11W
Montmorillon, *France* **24 B4** 46 26N 0 50 E
Montmort, *France* **23 D10** 48 55N 3 49 E
Monto, *Australia* **114 C5** 24 52 S 151 6 E
Montoir-sur-le-Loir, *France* **22 E7** 47 45N 0 52 E
Montório al Vomano, *Italy* **39 F10** 42 35N 13 38 E
Montoro, *Spain* **37 G6** 38 1N 4 27W
Montour Falls, *U.S.A.* **136 D8** 42 21N 76 51W
Montpelier, *Idaho, U.S.A.* **142 E8** 42 19N 111 18W
Montpelier, *Ind., U.S.A.* **141 D11** 40 33N 85 17W
Montpelier, *Ohio, U.S.A.* **141 C12** 41 35N 84 37W
Montpelier, *Vt., U.S.A.* **137 B12** 44 16N 72 35W
Montpellier, *France* **24 E7** 43 37N 3 52 E
Montpezat-de-Quercy, *France* **24 D5** 44 15N 1 30 E
Montpon-Ménestérol, *France* **24 D4** 45 0N 0 11 E
Montréal, *Canada* **128 C5** 45 31N 73 34W
Montréal, *France* **24 E6** 43 13N 2 8 E
Montreal L., *Canada* **131 C7** 54 20N 105 45W
Montreal Lake, *Canada* **131 C7** 54 3N 105 46W
Montredon-Labessonnié, *France* **24 E6** 43 45N 2 18 E
Montréjeau, *France* **24 E4** 43 6N 0 35 E
Montrésor, *France* **22 E8** 47 10N 1 10 E
Montreuil, *France* **23 B8** 50 27N 1 45 E
Montreuil-Bellay, *France* **22 E6** 47 8N 0 9W
Montreux, *Switz.* **28 D3** 46 26N 6 55 E
Montrevault, *France* **22 E5** 47 17N 1 2W
Montrevel-en-Bresse, *France* **25 B9** 46 21N 5 8 E
Montrichard, *France* **22 E8** 47 20N 1 10 E
Montrose, *U.K.* **18 E6** 56 44N 2 27W
Montrose, *Colo., U.S.A.* **143 G10** 38 29N 107 53W
Montrose, *Pa., U.S.A.* **137 E9** 41 50N 75 53W
Montrose, L., *U.S.A.* **140 F3** 38 18N 93 50W
Monts, Pte. des, *Canada* **129 C6** 49 20N 67 12W
Monts-sur-Guesnes, *France* **22 F7** 46 55N 0 13 E
Montsalvy, *France* **24 D6** 44 41N 2 30 E
Montsant, Sierra de, *Spain* **34 D5** 41 17N 1 0 E
Montsauche, *France* **23 E11** 47 13N 4 2 E
Montsech, Sierra del, *Spain* **34 C5** 42 0N 0 45 E
Montseny, *Spain* **34 D2** 41 55N 2 25W
Montserrat, *Spain* **34 D6** 41 36N 1 49 E
Montserrat ■, *W. Indies* **149 C7** 16 40N 62 10W
Montuenga, *Spain* **36 D6** 41 3N 4 38W
Montuiri, *Spain* **33 B9** 39 34N 2 59 E
Monveda, *Zaïre* **102 B4** 2 52N 21 30 E
Monyo, *Burma* **78 G5** 17 59N 95 30 E
Monywa, *Burma* **78 D5** 22 7N 95 11 E
Monza, *Italy* **38 C6** 45 35N 9 16 E
Monze, *Zambia* **107 F2** 16 17 S 27 29 E
Monze, C., *Pakistan* **79 D2** 24 47N 66 37 E
Monzón, *Spain* **34 D5** 41 52N 0 10 E
Mooi River, *S. Africa* **105 D4** 29 13 S 29 50 E
Mook, *Neths.* **20 E7** 51 46N 5 54 E
Mo'oka, *Japan* **63 A12** 36 26N 140 1 E
Moolawatana, *Australia* **115 D2** 29 55 S 139 45 E
Mooleulooloo, *Australia* **116 A4** 31 36 S 140 32 E
Mooliabeenee, *Australia* **113 F2** 31 20 S 116 2 E
Mooloogool, *Australia* **113 E2** 26 2 S 119 5 E
Moomin Cr. →, *Australia* **115 D4** 29 44 S 149 20 E
Moonah →, *Australia* **114 C2** 22 3 S 138 33 E
Moonbeam, *Canada* **128 C3** 49 20N 82 10W
Moonda, L., *Australia* **114 D3** 25 52 S 140 25 E
Moonie, *Australia* **115 D5** 27 46 S 150 20 E
Moonie →, *Australia* **115 D4** 29 19 S 148 43 E
Moonta, *Australia* **116 C2** 34 6 S 137 32 E
Moora, *Australia* **113 F2** 30 37 S 115 58 E
Mooraberree, *Australia* **114 D3** 25 13 S 140 54 E
Moorarie, *Australia* **113 E2** 25 56 S 117 35 E
Moorcroft, *U.S.A.* **138 C2** 44 16N 104 57W
Moore →, *Australia* **113 F2** 31 22 S 115 30 E
Moore, L., *Australia* **113 E2** 29 50 S 117 35 E
Moore Reefs, *Australia* **114 B4** 16 0 S 149 5 E
Moorefield, *U.S.A.* **134 F6** 39 5N 78 59W
Moores Res., *U.S.A.* **137 B13** 44 45N 71 50W
Mooresville, *Ind., U.S.A.* **141 E10** 39 37N 86 22W
Mooresville, *N.C., U.S.A.* **135 H5** 35 35N 80 48W
Moorfoot Hills, *U.K.* **18 F5** 55 44N 3 8W
Moorhead, *U.S.A.* **138 B6** 46 53N 96 45W
Moorland, *Australia* **117 A10** 31 46 S 152 38 E
Mooroopna, *Australia* **117 D6** 36 25 S 145 22 E
Moorpark, *U.S.A.* **145 L8** 34 17N 118 53W
Moorreesburg, *S. Africa* **104 E2** 33 6 S 18 38 E
Moorslede, *Belgium* **21 G2** 50 54N 3 4 E
Moosburg, *Germany* **27 G7** 48 27N 11 56 E
Moose →, *Canada* **128 B3** 51 20N 80 25W
Moose Factory, *Canada* **128 B3** 51 16N 80 32W
Moose I., *Canada* **131 C9** 51 42N 97 10W
Moose Jaw, *Canada* **131 C7** 50 24N 105 30W
Moose Jaw →, *Canada* **131 C7** 50 34N 105 18W
Moose Lake, *Canada* **131 C8** 53 43N 100 20W
Moose Lake, *U.S.A.* **138 B8** 46 27N 92 46W
Moose Mountain Cr. →, *Canada* **131 D8** 49 13N 102 12W
Moose Mountain Prov. Park, *Canada* **131 D8** 49 48N 102 25W
Moose River, *Canada* **128 B3** 50 48N 81 17W
Moosehead L., *U.S.A.* **129 C6** 45 38N 69 40W
Moosomin, *Canada* **131 C8** 50 9N 101 40W
Moosonee, *Canada* **128 B3** 51 17N 80 39W
Moosup, *U.S.A.* **137 E13** 41 43N 71 53W
Mopeia Velha, *Mozam.* **107 F4** 17 30 S 35 40 E
Mopipi, *Botswana* **104 C3** 21 6 S 24 55 E
Mopoi, *C.A.R.* **102 A5** 5 6N 26 54 E
Mopti, *Mali* **100 C4** 14 30N 4 0W
Moqatta, *Sudan* **95 E4** 14 38N 35 50 E
Moquegua, *Peru* **156 D3** 17 15 S 70 46W

236 Moquegua

Moquegua □, *Peru* **156 D3** 16 50 S 70 55W
Mór, *Hungary* **31 D11** 47 25N 18 12 E
Móra, *Portugal* **37 G2** 38 55N 8 10W
Mora, *Sweden* **13 F16** 61 2N 14 38 E
Mora, *Minn., U.S.A.* **138 C8** 45 53N 93 18W
Mora, *N. Mex., U.S.A.* **143 J11** 35 58N 105 20W
Mora de Ebro, *Spain* **34 D5** 41 6N 0 38 E
Mora de Rubielos, *Spain* **34 E4** 40 15N 0 45W
Mora la Nueva, *Spain* **34 D5** 41 7N 0 39 E
Morača →, *Montenegro, Yug.* **42 E4** 42 20N 19 9 E
Morada Nova, *Brazil* **154 C4** 5 7 S 38 23W
Morada Nova de Minas, *Brazil* **155 E2** 18 37 S 45 22W
Moradabad, *India* **81 E8** 28 50N 78 50 E
Morafenobe, *Madag.* **105 B7** 17 50 S 44 53 E
Morąg, *Poland* **47 B6** 53 55N 19 56 E
Moral de Calatrava, *Spain* **35 G1** 38 51N 3 33W
Moraleja, *Spain* **36 E4** 40 6N 6 43W
Morales, *Colombia* **152 C2** 2 45N 76 38W
Moramanga, *Madag.* **105 B8** 18 56 S 48 12 E
Moran, *Kans., U.S.A.* **139 G7** 37 55N 95 10W
Moran, *Wyo., U.S.A.* **142 E8** 43 53N 110 37W
Moranbah, *Australia* **114 C4** 22 1 S 148 6 E
Morano Cálabro, *Italy* **41 C9** 39 50N 16 8 E
Morant Cays, *Jamaica* **148 C4** 17 22N 76 0W
Morant Pt., *Jamaica* **148 C4** 17 55N 76 12W
Morar, L., *U.K.* **18 E3** 56 57N 5 40W
Moratalla, *Spain* **35 G3** 38 14N 1 49W
Moratuwa, *Sri Lanka* **83 L4** 6 45N 79 55 E
Morava →, *Europe* **31 C9** 48 10N 16 59 E
Moravia, *U.S.A.* **140 D4** 40 53N 92 49W
Moravian Hts. = Ceskomoravská Vrchovina, *Czech.* **30 B8** 49 30N 15 40 E
Moravica →, *Serbia, Yug.* **42 D5** 43 52N 20 8 E
Moravice →, *Czech.* **31 B10** 49 50N 17 43 E
Moraviţa, *Romania* **42 B6** 45 17N 21 14 E
Moravská Třebová, *Czech.* **31 B9** 49 45N 16 40 E
Moravské Budějovice, *Czech.* **30 B8** 49 4N 15 49 E
Morawa, *Australia* **113 E2** 29 13 S 116 0 E
Morawhanna, *Guyana* **153 B6** 8 30N 59 40W
Moray Firth, *U.K.* **18 D5** 57 40N 3 52W
Morbach, *Germany* **27 F3** 49 48N 7 6 E
Morbegno, *Italy* **38 B6** 46 8N 9 34 E
Morbi, *India* **80 H4** 22 50N 70 42 E
Morbihan □, *France* **22 E4** 47 55N 2 50W
Morcenx, *France* **24 D3** 44 3N 0 55W
Mordelles, *France* **22 D5** 48 5N 1 52W
Morden, *Canada* **131 D9** 49 15N 98 10W
Mordovian Republic = Mordvinia □, *Russia* **52 C7** 54 20N 44 30 E
Mordovo, *Russia* **52 D5** 52 6N 40 50 E
Mordvinia □, *Russia* **52 C7** 54 20N 44 30 E
Mordy, *Poland* **47 C9** 52 13N 22 31 E
Møre og Romsdal fylke □, *Norway* **14 B2** 62 30N 8 0 E
Morea, *Australia* **116 D4** 36 45 S 141 18 E
Morea, *Greece* **10 H10** 37 45N 22 10 E
Moreau →, *U.S.A.* **138 C4** 45 18N 100 43W
Morecambe, *U.K.* **16 C5** 54 5N 2 52W
Morecambe B., *U.K.* **16 C5** 54 7N 3 0W
Moree, *Australia* **115 D4** 29 28 S 149 54 E
Morehead, *Papua N. G.* **120 E1** 8 41 S 141 41 E
Morehead, *U.S.A.* **141 F13** 38 11N 83 26W
Morehead City, *U.S.A.* **135 H7** 34 43N 76 43W
Morelia, *Mexico* **146 D4** 19 42N 101 7W
Morella, *Australia* **114 C3** 23 0 S 143 52 E
Morella, *Spain* **34 E4** 40 35N 0 5W
Morelos, *Mexico* **146 B3** 26 42N 107 40W
Morelos □, *Mexico* **147 D5** 18 40N 99 10W
Morena, Sierra, *Spain* **37 G7** 38 20N 4 0W
Morenci, *Ariz., U.S.A.* **143 K9** 33 5N 109 22W
Morenci, *Mich., U.S.A.* **141 C12** 41 43N 84 13W
Moreni, *Romania* **46 E6** 44 59N 25 36 E
Morero, *Bolivia* **157 C4** 11 9 S 66 15W
Moreru →, *Brazil* **157 C6** 10 10 S 59 15W
Moresby I., *Canada* **130 C2** 52 30N 131 40W
Morestel, *France* **25 C9** 45 40N 5 28 E
Moret-sur-Loing, *France* **23 D9** 48 22N 2 58 E
Moreton, *Australia* **114 A3** 12 22 S 142 40 E
Moreton I., *Australia* **115 D5** 27 10 S 153 25 E
Moreuil, *France* **23 C9** 49 46N 2 30 E
Morey, *Spain* **33 B10** 39 44N 3 20 E
Morez, *France* **25 B10** 46 31N 6 2 E
Morgan, *Australia* **116 C3** 34 2 S 139 35 E
Morgan, *U.S.A.* **142 F8** 41 2N 111 41W
Morgan City, *U.S.A.* **139 L9** 29 42N 91 12W
Morgan Hill, *U.S.A.* **144 H5** 37 8N 121 39W
Morgan Vale, *Australia* **116 B4** 33 10 S 140 32 E
Morganfield, *U.S.A.* **134 G2** 37 41N 87 55W
Morganton, *U.S.A.* **135 H5** 35 45N 81 41W
Morgantown, *Ind., U.S.A.* **141 E10** 39 22N 86 16W
Morgantown, *W. Va., U.S.A.* **134 F6** 39 38N 79 57W
Morgat, *France* **22 D2** 48 15N 4 32W
Morgenzon, *S. Africa* **105 D4** 26 45 S 29 36 E
Morges, *Switz.* **28 C2** 46 31N 6 29 E
Morghak, *Iran* **85 D8** 29 7N 57 54 E
Morhange, *France* **23 D13** 48 55N 6 38 E
Mori, *Italy* **38 C7** 45 51N 10 59 E
Morialmée, *Belgium* **21 H5** 50 17N 4 35 E
Morice L., *Canada* **130 C3** 53 50N 127 40W
Morichal, *Colombia* **152 C3** 2 10N 70 34W
Morichal Largo →, *Venezuela* **153 B5** 9 27N 62 25W
Moriguchi, *Japan* **63 C7** 34 44N 135 34 E
Moriki, *Nigeria* **101 C6** 12 52N 6 30 E
Morinville, *Canada* **130 C6** 53 49N 113 41W
Morioka, *Japan* **60 E10** 39 45N 141 8 E
Moris, *Mexico* **146 B3** 28 8N 108 32W
Morisset, *Australia* **117 B9** 33 6 S 151 30 E
Morlaàs, *France* **24 E3** 43 21N 0 18W
Morlaix, *France* **22 D3** 48 36N 3 52W
Morlanwelz, *Belgium* **21 H4** 50 28N 4 15 E
Mormanno, *Italy* **41 C8** 39 53N 15 59 E
Mormant, *France* **23 D9** 48 37N 2 52 E
Mornington, *Vic., Australia* **117 E6** 38 15 S 145 5 E
Mornington, *W. Austral., Australia* **112 C4** 17 31 S 126 6 E
Mornington, I., *Chile* **160 C1** 49 50 S 75 30W
Mornington I., *Australia* **114 B2** 16 30 S 139 30 E
Mórnos →, *Greece* **45 F3** 38 25N 21 50 E
Moro, *Sudan* **95 E3** 10 50N 30 9 E
Moro G., *Phil.* **71 H4** 6 30N 123 0 E
Morobe, *Papua N. G.* **120 D4** 7 49 S 147 38 E
Morocco, *U.S.A.* **141 D9** 40 57N 87 27W
Morocco ■, *N. Afr.* **98 B3** 32 0N 5 50W
Morococha, *Peru* **156 C2** 11 40 S 76 5W
Morogoro, *Tanzania* **106 D4** 6 50 S 37 40 E
Morogoro □, *Tanzania* **106 D4** 8 0 S 37 0 E
Moroleón, *Mexico* **146 C4** 20 8N 101 32W
Morombe, *Madag.* **105 C7** 21 45 S 43 22 E
Moron, *Argentina* **158 C4** 34 39 S 58 37W
Morón, *Cuba* **148 B4** 22 8N 78 39W
Morón de Almazán, *Spain* **34 D2** 41 29N 2 27W
Morón de la Frontera, *Spain* **37 H5** 37 6N 5 28W
Morona →, *Peru* **152 D2** 4 40 S 77 10W
Morona-Santiago □, *Ecuador* **152 D2** 2 30 S 78 0W
Morondava, *Madag.* **105 C7** 20 17 S 44 17 E
Morondo, *Ivory C.* **100 D3** 8 57N 6 47W
Morong, *Phil.* **70 D3** 14 41N 120 16 E
Morongo Valley, *U.S.A.* **145 L10** 34 3N 116 37W
Moronou, *Ivory C.* **100 D4** 6 16N 4 59W
Morotai, *Indonesia* **72 A3** 2 10N 128 30 E
Moroto, *Uganda* **106 B3** 2 28N 34 42 E
Moroto Summit, *Kenya* **106 B3** 2 30N 34 43 E
Morozov, *Bulgaria* **43 E10** 42 30N 25 10 E
Morozovsk, *Russia* **53 F5** 48 25N 41 50 E
Morpeth, *U.K.* **16 B6** 55 10N 1 41W
Morphou, *Cyprus* **32 D11** 35 12N 32 59 E
Morphou Bay, *Cyprus* **32 D11** 35 15N 32 50 E
Morrelganj, *Bangla.* **78 D2** 22 28N 89 51 E
Morrilton, *U.S.A.* **139 H8** 35 9N 92 44W
Morrinhos, *Ceará, Brazil* **154 B3** 3 14 S 40 7W
Morrinhos, *Minas Gerais, Brazil* **155 E2** 17 45 S 49 10W
Morrinsville, *N.Z.* **118 D4** 37 40 S 175 32 E
Morris, *Canada* **131 D9** 49 25N 97 22W
Morris, *Ill., U.S.A.* **141 C8** 41 22N 88 26W
Morris, *Minn., U.S.A.* **138 C7** 45 35N 95 55W
Morris, Mt., *Australia* **113 E5** 26 9 S 131 4 E
Morrisburg, *Canada* **128 D4** 44 55N 75 7W
Morrison, *U.S.A.* **140 C7** 41 49N 89 58W
Morrisonville, *U.S.A.* **140 E7** 39 25N 89 27W
Morristown, *Ariz., U.S.A.* **143 K7** 33 51N 112 37W
Morristown, *Ind., U.S.A.* **141 E11** 39 40N 85 42W
Morristown, *N.J., U.S.A.* **137 F10** 40 48N 74 29W
Morristown, *S. Dak., U.S.A.* **138 C4** 45 56N 101 43W
Morristown, *Tenn., U.S.A.* **135 G4** 36 13N 83 18W
Morro, Pta., *Chile* **158 B1** 27 6 S 71 0W
Morro Bay, *U.S.A.* **144 K6** 35 22N 120 51W
Morro del Jable, *Canary Is.* **33 F5** 28 3N 14 23W
Morro do Chapéu, *Brazil* **155 D3** 11 33 S 41 9W
Morro Jable, Pta. de, *Canary Is.* **33 F5** 28 2N 14 20W
Morros, *Brazil* **154 B3** 2 52 S 44 3W
Morrosquillo, G. de, *Colombia* **148 E4** 9 35N 75 40W
Morrumbene, *Mozam.* **105 C6** 23 31 S 35 16 E
Mors Nykøbing, *Denmark* **15 H2** 56 50N 8 45 E
Morshansk, *Russia* **52 D5** 53 28N 41 50 E
Mörsil, *Sweden* **14 A7** 63 19N 13 40 E
Mortagne →, *France* **23 D13** 48 33N 6 27 E
Mortagne-au-Perche, *France* **22 D7** 48 31N 0 33 E
Mortagne-sur-Gironde, *France* **24 C3** 45 28N 0 47W
Mortagne-sur-Sèvre, *France* **22 F6** 46 59N 0 57W
Mortain, *France* **22 D6** 48 40N 0 57W
Mortara, *Italy* **38 C5** 45 15N 8 44 E
Mortcha, *Chad* **97 E4** 16 0N 21 10 E
Morteau, *France* **23 E13** 47 3N 6 35 E
Morteros, *Argentina* **158 C3** 30 50 S 62 0W
Mortes, R. das →, *Brazil* **155 D1** 11 45 S 50 44W
Mortlake, *Australia* **116 E5** 38 5 S 142 50 E
Morton, *Ill., U.S.A.* **140 D7** 40 37N 89 28W
Morton, *Tex., U.S.A.* **139 J3** 33 44N 102 46W
Morton, *Wash., U.S.A.* **144 D4** 46 34N 122 17W
Mortsel, *Belgium* **21 F4** 51 11N 4 27 E
Morundah, *Australia* **117 C7** 34 57 S 146 19 E
Moruya, *Australia* **117 C9** 35 58 S 150 3 E
Morvan, *France* **23 E11** 47 5N 4 3 E
Morven, *Australia* **115 D4** 26 22 S 147 5 E
Morven, *N.Z.* **119 E6** 44 50 S 171 6 E
Morvern, *U.K.* **18 E3** 56 38N 5 44W
Morwell, *Australia* **117 E7** 38 10 S 146 22 E
Moryn, *Poland* **47 C1** 52 51N 14 22 E
Morzhovets, Ostrov, *Russia* **48 A7** 66 44N 42 35 E
Mosalsk, *Russia* **52 C2** 54 30N 34 55 E
Mosbach, *Germany* **27 F5** 49 21N 9 9 E
Mošćenice, *Croatia* **39 C11** 45 17N 14 16 E
Mosciano Sant' Ángelo, *Italy* **39 F10** 42 42N 13 52 E
Moscos Is., *Burma* **76 E1** 14 0N 97 30 E
Moscow = Moskva, *Russia* **52 C3** 55 45N 37 35 E
Moscow, *U.S.A.* **142 C5** 46 44N 117 0W
Mosel →, *Europe* **27 E3** 50 22N 7 36 E
Moselle = Mosel →, *Europe* **27 E3** 50 22N 7 36 E
Moselle □, *France* **23 D13** 48 59N 6 33 E
Moses Lake, *U.S.A.* **142 C4** 47 8N 119 17W
Mosgiel, *N.Z.* **119 F5** 45 53 S 170 21 E
Moshi, *Tanzania* **106 C4** 3 22 S 37 18 E
Moshi □, *Tanzania* **106 C4** 3 22 S 37 18 E
Moshupa, *Botswana* **104 C4** 24 46 S 25 29 E
Mośina, *Poland* **47 C3** 52 15N 16 50 E
Mosjøen, *Norway* **12 D15** 65 51N 13 12 E
Moskenesøya, *Norway* **12 C15** 67 58N 13 0 E
Moskenstraumen, *Norway* **12 C15** 67 47N 12 45 E
Moskva, *Russia* **52 C3** 55 45N 37 35 E
Moskva →, *Russia* **52 C4** 55 5N 38 51 E
Moslavačka Gora, *Croatia* **39 C13** 45 40N 16 37 E
Moso, *Vanuatu* **121 G6** 17 30 S 168 15 E
Mosomane, *Botswana* **104 C4** 24 2 S 26 19 E
Moson-magyaróvar, *Hungary* **31 D10** 47 52N 17 18 E
Mošorin, *Serbia, Yug.* **42 B5** 45 19N 20 4 E
Mospino, *Ukraine* **51 J9** 47 52N 38 0 E
Mosquera, *Colombia* **152 C2** 2 35N 78 24W
Mosquero, *U.S.A.* **139 H3** 35 47N 103 58W
Mosqueruela, *Spain* **34 E4** 40 21N 0 27W
Mosquitia, *Honduras* **148 C3** 15 20N 84 10W
Mosquitos, G. de los, *Panama* **148 E3** 9 15N 81 10W
Moss, *Norway* **14 E4** 59 27N 10 40 E
Moss Vale, *Australia* **117 C9** 34 32 S 150 25 E
Mossaka, *Congo* **102 C3** 1 15 S 16 45 E
Mossâmedes, *Brazil* **155 E1** 16 7 S 50 11W
Mossbank, *Canada* **131 D7** 49 56N 105 56W
Mossburn, *N.Z.* **119 F3** 45 41 S 168 15 E
Mosselbaai, *S. Africa* **104 E3** 34 11 S 22 8 E
Mossendjo, *Congo* **102 C2** 2 55 S 12 42 E
Mosses, Col des, *Switz.* **28 D4** 46 25N 7 7 E
Mossgiel, *Australia* **116 B6** 33 15 S 144 5 E
Mossman, *Australia* **114 B4** 16 21 S 145 15 E
Mossoró, *Brazil* **154 C4** 5 10 S 37 15W
Mossuril, *Mozam.* **107 E5** 14 58 S 40 42 E
Mossy →, *Canada* **131 C8** 54 5N 102 58W
Most, *Czech.* **30 A6** 50 31N 13 38 E
Mosta, *Malta* **32 D1** 35 54N 14 24 E
Moşţafáābād, *Iran* **85 C7** 33 39N 54 53 E
Mostaganem, *Algeria* **99 A5** 35 54N 0 5 E
Mostar, *Bos.-H.* **42 D2** 43 22N 17 50 E
Mostardas, *Brazil* **159 C5** 31 2 S 50 51W
Mostefa, Rass, *Tunisia* **96 A2** 36 55N 11 3 E
Mostiska = Mostyska, *Ukraine* **51 H2** 49 48N 23 4 E
Móstoles, *Spain* **36 E7** 40 19N 3 53W
Mosty = Masty, *Belarus* **50 F3** 53 27N 24 38 E
Mostyska, *Ukraine* **51 H2** 49 48N 23 4 E
Mosul = Al Mawşil, *Iraq* **89 D10** 36 15N 43 5 E
Mosulpo, *S. Korea* **67 H14** 33 20N 126 17 E
Mota, *Vanuatu* **121 C5** 13 49 S 167 42 E
Mota del Cuervo, *Spain* **34 F2** 39 30N 2 52W
Mota del Marqués, *Spain* **36 D5** 41 38N 5 11W
Mota Lava, *Vanuatu* **121 C5** 13 40 S 167 40 E
Motagua →, *Guatemala* **148 C2** 15 44N 88 14W
Motala, *Sweden* **15 F9** 58 32N 15 1 E
Motegi, *Japan* **63 A12** 36 32N 140 11 E
Motherwell, *U.K.* **18 F5** 55 47N 3 58W
Motihari, *India* **81 F11** 26 30N 84 55 E
Motilla del Palancar, *Spain* **34 F3** 39 34N 1 55W
Motiti I., *N.Z.* **118 D5** 37 38 S 176 25 E
Motnik, *Slovenia* **39 B11** 46 14N 14 54 E
Motocurunya, *Venezuela* **153 C5** 4 24N 64 5W
Motovun, *Croatia* **39 C10** 45 20N 13 50 E
Motozintla de Mendoza, *Mexico* **147 D6** 15 21N 92 14W
Motril, *Spain* **35 J1** 36 31N 3 37W
Motru →, *Romania* **46 E4** 44 32N 23 31 E
Mott, *U.S.A.* **138 B3** 46 23N 102 20W
Móttola, *Italy* **41 B10** 40 38N 17 2 E
Motu, *N.Z.* **118 E6** 38 18 S 177 40 E
Motu →, *N.Z.* **118 D6** 37 51 S 177 35 E
Motueka, *N.Z.* **119 B8** 41 7 S 173 1 E
Motueka →, *N.Z.* **119 B8** 41 5 S 173 1 E
Motul, *Mexico* **147 C7** 21 0N 89 20W
Motupena Pt., *Papua N. G.* **120 D8** 6 30 S 155 10 E
Mouanda, *Gabon* **102 C2** 1 28 S 13 7 E
Mouchalagane →, *Canada* **129 B6** 50 56N 68 41W
Moúdhros, *Greece* **44 E7** 39 50N 25 18 E
Mouding, *China* **68 E3** 25 20N 101 28 E
Moudjeria, *Mauritania* **100 B2** 17 50N 12 28W
Moudon, *Switz.* **28 C3** 46 40N 6 49 E
Mougoundou, *Congo* **102 C2** 2 40 S 12 41 E
Mouila, *Gabon* **102 C2** 1 50 S 11 0 E
Mouka, *C.A.R.* **102 A4** 7 16N 21 52 E
Moulamein, *Australia* **116 C6** 35 3 S 144 1 E
Moulianá, *Greece* **32 D7** 35 10N 25 59 E
Moulins, *France* **24 B7** 46 35N 3 19 E
Moulmein, *Burma* **78 G6** 16 30N 97 40 E
Moulmeingyun, *Burma* **78 G5** 16 23N 95 16 E
Moulouya, O. →, *Morocco* **99 A4** 35 5N 2 25W
Moulton, *Iowa, U.S.A.* **140 D4** 40 41N 92 41W
Moulton, *Tex., U.S.A.* **139 L6** 29 35N 97 9W
Moultrie, *U.S.A.* **135 K4** 31 11N 83 47W
Moultrie, L., *U.S.A.* **135 J5** 33 20N 80 5W
Mound City, *Mo., U.S.A.* **138 E7** 40 7N 95 14W
Mound City, *S. Dak., U.S.A.* **138 C4** 45 44N 100 4W
Moúnda, Ákra, *Greece* **45 F2** 38 5N 20 45 E
Moundou, *Chad* **97 G3** 8 40N 16 10 E
Moundsville, *U.S.A.* **136 G4** 39 55N 80 44W
Mounembé, *Congo* **102 C2** 3 20 S 12 32 E
Moung, *Cambodia* **76 F4** 12 46N 103 27 E
Moungoudi, *Congo* **102 C2** 2 45 S 11 46 E
Mount Airy, *U.S.A.* **135 G5** 36 31N 80 37W
Mount Albert, *Canada* **136 B5** 44 8N 79 19W
Mount Amherst, *Australia* **112 C4** 18 24 S 126 58 E
Mount Angel, *U.S.A.* **142 D2** 45 4N 122 48W
Mount Augustus, *Australia* **112 D2** 24 20 S 116 56 E
Mount Ayr, *U.S.A.* **140 D2** 40 43N 94 14W
Mount Barker, *S. Austral., Australia* **116 C3** 35 5 S 138 52 E
Mount Barker, *W. Austral., Australia* **113 F2** 34 38 S 117 40 E
Mount Beauty, *Australia* **117 D7** 36 47 S 147 10 E
Mount Carmel, *U.S.A.* **141 F9** 38 25N 87 46W
Mount Carroll, *U.S.A.* **140 B7** 42 6N 89 59W
Mount Clemens, *U.S.A.* **128 D3** 42 35N 82 53W
Mount Coolon, *Australia* **114 C4** 21 25 S 147 25 E
Mount Darwin, *Zimbabwe* **107 F3** 16 47 S 31 38 E
Mount Desert I., *U.S.A.* **129 D6** 44 21N 68 20W
Mount Dora, *U.S.A.* **135 L5** 28 48N 81 38W
Mount Douglas, *Australia* **114 C4** 21 35 S 146 50 E
Mount Eba, *Australia* **115 E2** 30 11 S 135 40 E
Mount Eden, *U.S.A.* **141 F11** 38 3N 85 9W
Mount Edgecumbe, *U.S.A.* **130 B1** 57 3N 135 21W
Mount Elizabeth, *Australia* **112 C4** 16 0 S 125 50 E
Mount Fletcher, *S. Africa* **105 E4** 30 40 S 28 30 E
Mount Forest, *Canada* **128 D3** 43 59N 80 43W
Mount Gambier, *Australia* **116 D4** 37 50 S 140 46 E
Mount Garnet, *Australia* **114 B4** 17 37 S 145 6 E
Mount Hagen, *Papua N. G.* **120 C3** 5 52 S 144 16 E
Mount Hope, *N.S.W., Australia* **117 B6** 32 51 S 145 51 E
Mount Hope, *S. Austral., Australia* **115 E2** 34 7 S 135 23 E
Mount Hope, *U.S.A.* **134 G5** 37 54N 81 10W
Mount Horeb, *U.S.A.* **140 A7** 43 1N 89 44W
Mount Howitt, *Australia* **115 D3** 26 31 S 142 16 E
Mount Isa, *Australia* **114 C2** 20 42 S 139 26 E
Mount Keith, *Australia* **113 E3** 27 15 S 120 30 E
Mount Laguna, *U.S.A.* **145 N10** 32 52N 116 25W
Mount Larcom, *Australia* **114 C5** 23 48 S 150 59 E
Mount Lofty Ra., *Australia* **116 C3** 34 35 S 139 5 E
Mount McKinley National Park, *U.S.A.* **126 B5** 63 30N 150 0W
Mount Magnet, *Australia* **113 E2** 28 2 S 117 47 E
Mount Manara, *Australia* **116 B5** 32 29 S 143 58 E
Mount Margaret, *Australia* **115 D3** 26 54 S 143 21 E
Mount Maunganui, *N.Z.* **118 D5** 37 40 S 176 14 E
Mount Molloy, *Australia* **114 B4** 16 42 S 145 20 E
Mount Monger, *Australia* **113 F3** 31 0 S 122 0 E
Mount Morgan, *Australia* **114 C5** 23 40 S 150 25 E
Mount Morris, *U.S.A.* **136 D7** 42 44N 77 52W
Mount Mulligan, *Australia* **114 B3** 16 45 S 144 47 E
Mount Narryer, *Australia* **113 E2** 26 30 S 115 55 E
Mount Olive, *U.S.A.* **140 E7** 39 4N 89 44W
Mount Olivet, *U.S.A.* **141 F12** 38 32N 84 2W
Mount Olympus = Uludağ, *Turkey* **88 B3** 40 4N 29 13 E
Mount Orab, *U.S.A.* **141 E13** 39 2N 83 55W
Mount Oxide Mine, *Australia* **114 B2** 19 30 S 139 29 E
Mount Pearl, *Canada* **129 C9** 47 31N 52 47W
Mount Perry, *Australia* **115 D5** 25 13 S 151 42 E
Mount Phillips, *Australia* **112 D2** 24 25 S 116 15 E
Mount Pleasant, *Iowa, U.S.A.* **140 D5** 40 58N 91 33W
Mount Pleasant, *Mich., U.S.A.* **134 D3** 43 36N 84 46W
Mount Pleasant, *Pa., U.S.A.* **136 F5** 40 9N 79 33W
Mount Pleasant, *S.C., U.S.A.* **135 J6** 32 47N 79 52W
Mount Pleasant, *Tenn., U.S.A.* **135 H2** 35 32N 87 12W
Mount Pleasant, *Tex., U.S.A.* **139 J7** 33 9N 94 58W
Mount Pleasant, *Utah, U.S.A.* **142 G8** 39 33N 111 27W
Mount Pocono, *U.S.A.* **137 E9** 41 7N 75 22W
Mount Pulaski, *U.S.A.* **140 D7** 40 1N 89 17W
Mount Rainier National Park, *U.S.A.* **144 D5** 46 55N 121 50W
Mount Revelstoke Nat. Park, *Canada* **130 C5** 51 5N 118 30W
Mount Robson Prov. Park, *Canada* **130 C5** 53 0N 119 0W
Mount Roskill, *N.Z.* **118 C3** 36 55 S 174 45 E
Mount Sandiman, *Australia* **113 D2** 24 25 S 115 30 E
Mount Shasta, *U.S.A.* **142 F2** 41 19N 122 19W
Mount Signal, *U.S.A.* **145 N11** 32 39N 115 37W
Mount Somers, *N.Z.* **119 D6** 43 45 S 171 27 E
Mount Sterling, *Ill., U.S.A.* **140 E6** 39 59N 90 45W
Mount Sterling, *Ky., U.S.A.* **141 F13** 38 4N 83 56W
Mount Sterling, *Ohio, U.S.A.* **141 E13** 39 43N 83 16W
Mount Surprise, *Australia* **114 B3** 18 10 S 144 17 E
Mount Union, *U.S.A.* **136 F7** 40 23N 77 53W
Mount Vernon, *Australia* **112 D2** 24 9 S 118 2 E
Mount Vernon, *Ind., U.S.A.* **138 F10** 38 17N 88 57W
Mount Vernon, *Ind., U.S.A.* **141 F8** 37 56N 87 54W
Mount Vernon, *Iowa, U.S.A.* **140 C5** 41 55N 91 23W
Mount Vernon, *N.Y., U.S.A.* **137 F11** 40 55N 73 50W
Mount Vernon, *Ohio, U.S.A.* **136 F2** 40 23N 82 29W
Mount Vernon, *Wash., U.S.A.* **144 B4** 48 25N 122 20W
Mount Victor, *Australia* **116 B3** 32 11 S 139 44 E
Mount Washington, *U.S.A.* **141 F11** 38 3N 85 33W
Mount Wellington, *N.Z.* **118 C3** 36 55 S 174 52 E
Mount Zion, *U.S.A.* **141 E8** 39 46N 88 53W
Mountain □, *Phil.* **70 C3** 17 20N 121 10 E
Mountain Center, *U.S.A.* **145 M10** 33 42N 116 44W
Mountain City, *Nev., U.S.A.* **142 F6** 41 50N 115 58W
Mountain City, *Tenn., U.S.A.* **135 G5** 36 29N 81 48W
Mountain Grove, *U.S.A.* **139 G8** 37 8N 92 16W
Mountain Home, *Ark., U.S.A.* **139 G8** 36 20N 92 23W
Mountain Home, *Idaho, U.S.A.* **142 E6** 43 8N 115 41W
Mountain Iron, *U.S.A.* **138 B8** 47 32N 92 37W
Mountain Park, *Canada* **130 C5** 52 50N 117 15W
Mountain Pass, *U.S.A.* **145 K11** 35 29N 115 35W
Mountain View, *Ark., U.S.A.* **139 H8** 35 52N 92 7W
Mountain View, *Calif., U.S.A.* **144 H4** 37 23N 122 5W
Mountainair, *U.S.A.* **143 J10** 34 31N 106 15W
Mountmellick, *Ireland* **19 C4** 53 7N 7 20W
Moura, *Australia* **114 C4** 24 35 S 149 58 E
Moura, *Brazil* **153 D5** 1 32 S 61 38W
Moura, *Portugal* **37 G3** 38 7N 7 30W
Mourão, *Portugal* **37 G3** 38 22N 7 22W
Mourdi, Dépression du, *Chad* **97 E4** 18 10N 23 0 E
Mourdiah, *Mali* **100 C3** 14 35N 7 25W
Mourenx-Ville-Nouvelle, *France* **24 E3** 43 22N 0 38W
Mouri, *Ghana* **101 D4** 5 6N 1 14W
Mourilyan, *Australia* **114 B4** 17 35 S 146 3 E
Mourmelon-le-Grand, *France* **23 C11** 49 8N 4 22 E
Mourne →, *U.K.* **19 B4** 54 52N 7 26W
Mourne Mts., *U.K.* **19 B5** 54 10N 6 0W
Mournies, *Greece* **32 D6** 35 29N 24 1 E
Mouscron, *Belgium* **21 G2** 50 45N 3 12 E
Moussoro, *Chad* **97 F3** 13 41N 16 35 E
Mouthe, *France* **23 F13** 46 44N 6 12 E
Moutier, *Switz.* **28 B4** 47 16N 7 21 E
Moûtiers, *France* **25 C10** 45 29N 6 32 E
Moutohara, *N.Z.* **118 E6** 38 27 S 177 32 E
Moutong, *Indonesia* **72 A2** 0 28N 121 13 E
Mouy, *France* **23 C9** 49 18N 2 20 E
Mouzáki, *Greece* **44 E3** 39 25N 21 37 E
Movas, *Mexico* **146 B3** 28 10N 109 25W
Moville, *Ireland* **19 A4** 55 11N 7 3W
Moweaqua, *U.S.A.* **140 E7** 39 38N 89 1W
Moxhe, *Belgium* **21 G6** 50 38N 5 5 E

Mutanda 237

Moxico □, *Angola* ...... **103 E4** 12 0 S 20 30 E
Moxotó →, *Brazil* ...... **154 C4** 9 19 S 38 14W
Moy →, *Ireland* **19 B3** 54 8N 9 8W
Moyale, *Kenya* .... **90 G2** 3 30N 39 0 E
Moyamba, *S. Leone* **100 D2** 8 4N 12 30W
Moyen Atlas, *Morocco* .. **98 B3** 33 0N 5 0W
Moyle □, *U.K.* **19 A5** 55 10N 6 15W
Moyo, *Indonesia* .... **72 C1** 8 10 S 117 40 E
Moyobamba, *Peru* ...... **156 B2** 6 0 S 77 0W
Moyyero →, *Russia* .... **57 C11** 68 44N 103 42 E
Mozambique =
Moçambique, *Mozam.* . **107 F5** 15 3 S 40 42 E
Mozambique ■, *Africa* .. **107 F4** 19 0 S 35 0 E
Mozambique Chan.,
Africa .... **105 B7** 17 30 S 42 30 E
Mozdok, *Russia* .... **53 J7** 43 45N 44 48 E
Mozdūrān, *Iran* .... **85 B9** 36 9N 60 35 E
Mozhaysk, *Russia* .... **52 C3** 55 30N 36 2 E
Mozhga, *Russia* .... **52 B11** 56 26N 52 15 E
Mozhnābād, *Iran* .... **85 C9** 34 7N 60 6 E
Mozirje, *Slovenia* ...... **39 B11** 46 22N 14 58 E
Mozyr = Mazyr, *Belarus* . **51 F5** 51 59N 29 15 E
Mpanda, *Tanzania* ...... **106 D3** 6 23 S 31 1 E
Mpanda □, *Tanzania* .... **106 D3** 6 23 S 31 40 E
Mpésoba, *Mali* **100 C3** 12 31N 5 39W
Mpika, *Zambia* .... **107 E3** 11 51 S 31 25 E
Mpulungu, *Zambia* ...... **107 D3** 8 51 S 31 5 E
Mpumalanga, *S. Africa* .. **105 D5** 29 50 S 30 33 E
Mpwapwa, *Tanzania* .... **106 D4** 6 23 S 36 30 E
Mpwapwa □, *Tanzania* .. **106 D4** 6 30 S 36 20 E
Mqinvartsveri = Kazbek,
Russia .... **53 J7** 42 42N 44 30 E
Mrągowo, *Poland* .... **47 B8** 53 52N 21 18 E
Mramor, *Serbia, Yug.* ... **42 D6** 43 20N 21 45 E
Mrimina, *Morocco* **98 C3** 29 50N 7 9W
Mrkonjić Grad, *Bos.-H.* . **42 C2** 44 26N 17 4 E
Mrkopalj, *Croatia* ...... **39 C11** 45 21N 14 52 E
Mrocza, *Poland* .... **47 B4** 53 16N 17 35 E
Msab, Oued en →,
Algeria .... **99 B6** 32 15N 5 0 E
Msaken, *Tunisia* .... **96 A2** 35 49N 10 33 E
Msambansovu, *Zimbabwe* **107 F3** 15 50 S 30 3 E
M'sila, *Algeria* .... **99 A5** 35 46N 4 30 E
Msoro, *Zambia* .... **107 E3** 13 35 S 31 50 E
Msta →, *Russia* .... **50 C6** 58 25N 31 20 E
Mstislavl = Mstsislaw,
Belarus .... **50 E6** 54 0N 31 50 E
Mstsislaw, *Belarus* ...... **50 E6** 54 0N 31 50 E
Mszana Dolna, *Poland* .. **31 B13** 49 41N 20 5 E
Mszczonów, *Poland* ..... **47 D7** 51 58N 20 33 E
Mtama, *Tanzania* .... **107 E4** 10 17 S 39 21 E
Mtilikwe →, *Zimbabwe* . **107 G3** 21 9 S 31 30 E
Mtsensk, *Russia* .... **52 D3** 53 17N 36 36 E
Mtskheta, *Georgia* ...... **53 K7** 41 52N 44 45 E
Mtubatuba, *S. Africa* ... **105 D5** 28 30 S 32 8 E
Mtwara-Mikindani,
Tanzania .... **107 E5** 10 20 S 40 20 E
Mu →, *Burma* .... **78 E5** 21 56N 95 38 E
Mu Gia, Deo, *Vietnam* .. **76 D5** 17 40N 105 47 E
Mu Us Shamo, *China* ... **66 E5** 39 0N 109 0 E
Muacandalo, *Angola* .... **103 E3** 10 2 S 19 40 E
Muaná, *Brazil* .... **154 B2** 1 25 S 49 15W
Muanda, *Zaïre* .... **103 D2** 6 0 S 12 20 E
Muang Chiang Rai,
Thailand .... **76 C2** 19 52N 99 50 E
Muang Lamphun,
Thailand .... **76 C2** 18 40N 99 2 E
Muang Pak Beng, *Laos* .. **76 C3** 19 54N 101 8 E
Muar, *Malaysia* .... **77 L4** 2 3N 102 34 E
Muarabungo, *Indonesia* .. **74 C2** 1 28 S 102 52 E
Muaraenim, *Indonesia* ... **74 C2** 3 40 S 103 50 E
Muarajuloi, *Indonesia* ... **75 C4** 0 12 S 114 3 E
Muarakaman, *Indonesia* . **75 C5** 0 2 S 116 45 E
Muaratebo, *Indonesia* ... **74 C2** 1 30 S 102 26 E
Muaratembesi, *Indonesia* **74 C2** 1 42 S 103 8 E
Muaratewe, *Indonesia* ... **75 C4** 0 58 S 114 52 E
Mubarakpur, *India* ...... **81 F10** 26 6N 83 18 E
Mubarraz = Al Mubarraz,
Si. Arabia .... **85 E6** 25 30N 49 40 E
Mubende, *Uganda* ...... **106 B3** 0 33N 31 22 E
Mubi, *Nigeria* .... **101 C7** 10 18N 13 16 E
Mubur, P., *Indonesia* .... **77 L6** 3 20N 106 12 E
Mucajaí →, *Brazil* **153 C5** 2 25N 60 52W
Mucajaí, Serra do, *Brazil* **153 C5** 2 23N 61 10W
Mucari, *Angola* .... **103 D3** 9 30 S 16 54 E
Muchachos, Roque de los,
Canary Is. **33 F2** 28 44N 17 52W
Mücheln, *Germany* ...... **26 D7** 51 17N 11 47 E
Muchinga Mts., *Zambia* . **107 E3** 11 30 S 31 30 E
Muchkapskiy, *Russia* .... **52 E6** 51 52N 42 28 E
Muck, *U.K.* **18 E2** 56 50N 6 15W
Muckadilla, *Australia* ... **115 D4** 26 35 S 148 23 E
Muco →, *Colombia* **152 C3** 4 15N 70 21W
Mucoma, *Angola* .... **103 F2** 15 18 S 13 39 E
Muconda, *Angola* .... **103 E4** 10 31 S 21 15 E
Mucuim →, *Brazil* ...... **157 B5** 6 33 S 64 18W
Mucur, *Turkey* .... **88 C6** 39 3N 34 22 E
Mucura, *Brazil* .... **153 D5** 2 31 S 62 43W
Mucuri, *Brazil* .... **155 E4** 18 0 S 39 36W
Mucurici, *Brazil* .... **155 E3** 18 6 S 40 31W
Mucusso, *Angola* .... **103 F4** 18 1 S 21 25 E
Muda, *Canary Is.* **33 F6** 28 34N 13 57W
Mudan Jiang →, *China* . **67 A15** 46 20N 129 30 E
Mudanjiang, *China* ..... **67 B15** 44 38N 129 30 E
Mudanya, *Turkey* .... **88 B3** 40 25N 28 50 E
Muddy Cr. →, *U.S.A.* .. **143 H8** 38 24N 110 42W
Mudgee, *Australia* ...... **117 B8** 32 32 S 149 31 E
Mudjatik →, *Canada* **131 B7** 56 1N 107 36W
Mudon, *Burma* .... **78 G6** 16 15N 97 44 E
Mudugh, *Somali Rep.* ... **108 C3** 7 0N 47 30 E
Mudurnu, *Turkey* .... **88 B4** 40 27N 31 12 E
Muecate, *Mozam.* .... **107 E4** 14 55 S 39 40 E
Muêda, *Mozam.* .... **107 E4** 11 36 S 39 28 E
Mueller Ra., *Australia* ... **112 C4** 18 18 S 126 46 E
Muende, *Mozam.* .... **107 E3** 14 28 S 33 0 E
Muerto, Mar, *Mexico* ... **147 D6** 16 10N 94 10W
Muertos, Punta de los,
Spain **35 J3** 36 57N 1 54W
Mufindi □, *Tanzania* .... **107 D4** 8 30 S 35 20 E
Mufu Shan, *China* ...... **69 C10** 29 20N 114 30 E
Mufulira, *Zambia* .... **107 E2** 12 32 S 28 15 E
Mufumbiro Range, *Africa* **106 C2** 1 25 S 29 30 E
Mugardos, *Spain* **36 B2** 43 27N 8 15W
Muge, *Portugal* **37 F2** 39 3N 8 40W
Muge →, *Portugal* **37 F2** 39 8N 8 44W
Múggia, *Italy* .... **39 C10** 45 36N 13 46 E
Mughayrā', *Si. Arabia* ... **84 D3** 29 17N 37 41 E
Mugi, *Japan* .... **62 D6** 33 40N 134 25 E
Mugia, *Spain* **36 B1** 43 3N 9 10W
Mugila, Mts., *Zaïre* ..... **106 D2** 7 0 S 28 50 E
Muğla, *Turkey* .... **88 D3** 37 15N 28 22 E
Müglizh, *Bulgaria* .... **43 E10** 42 37N 25 32 E
Mugu, *Nepal* .... **81 E10** 29 45N 82 30 E
Muhammad, Râs, *Egypt* . **94 B3** 27 44N 34 16 E
Muhammad Qol, *Sudan* . **94 C4** 20 53N 37 9 E
Muhammadabad, *India* .. **81 F10** 26 4N 83 25 E
Muḩayriqah, *Si. Arabia* .. **86 B4** 23 59N 45 4 E
Muhesi →, *Tanzania* ... **106 D4** 7 0 S 35 20 E
Muheza □, *Tanzania* .... **106 C4** 5 0 S 39 0 E
Mühldorf, *Germany* ..... **27 G8** 48 14N 12 32 E
Mühlhausen, *Germany* .. **26 D6** 51 12N 10 27 E
Mühlig Hofmann fjella,
Antarctica .... **7 D3** 72 30 S 5 0 E
Muhos, *Finland* .... **12 D22** 64 47N 25 59 E
Muhu, *Estonia* .... **13 G20** 58 36N 23 11 E
Muhutwe, *Tanzania* ..... **106 C3** 1 35 S 31 45 E
Muiden, *Neths.* .... **20 D6** 52 20N 5 4 E
Muikamachi, *Japan* ..... **61 F9** 37 15N 138 50 E
Muine Bheag, *Ireland* ... **19 D5** 52 42N 6 58W
Muiños, *Spain* **36 D3** 41 58N 7 59W
Muir, L., *Australia* ...... **113 F2** 34 30 S 116 40 E
Mukacheve, *Ukraine* .... **51 H2** 48 27N 22 45 E
Mukachevo = Mukacheve,
Ukraine .... **51 H2** 48 27N 22 45 E
Mukah, *Malaysia* .... **75 B4** 2 55N 112 5 E
Mukawwa, Geziret, *Egypt* **94 C4** 23 55N 35 53 E
Mukdahan, *Thailand* .... **76 D5** 16 32N 104 43 E
Mukden = Shenyang,
China .... **67 D12** 41 48N 123 27 E
Mukhtolovo, *Russia* ..... **52 C6** 55 29N 43 15 E
Mukhtuya = Lensk,
Russia .... **57 C12** 60 48N 114 55 E
Mukinbudin, *Australia* .. **113 F2** 30 55 S 118 5 E
Mukishi, *Zaïre* .... **103 D4** 8 30 S 24 44 E
Mukomuko, *Indonesia* ... **74 C2** 2 30 S 101 10 E
Mukomwenze, *Zaïre* .... **106 D2** 6 49 S 27 15 E
Mukry, *Turkmenistan* ... **55 E2** 37 54N 65 12 E
Muktsar, *India* .... **80 D6** 30 30N 74 30 E
Mukur, *Afghan.* .... **80 C2** 32 50N 67 42 E
Mukutawa →, *Canada* .. **131 C9** 53 10N 97 24W
Mukwela, *Zambia* ...... **107 F2** 17 0 S 26 40 E
Mukwonago, *U.S.A.* **141 B8** 42 52N 88 20W
Mula, *Spain* **35 G3** 38 3N 1 33W
Mula →, *India* .... **82 E2** 18 34N 74 21 E
Mulanay, *Phil.* .... **70 E4** 13 31N 122 24 E
Mulange, *Zaïre* .... **106 C2** 3 40 S 27 10 E
Mulberry Grove, *U.S.A.* . **140 F7** 38 56N 89 16W
Mulchén, *Chile* .... **158 D1** 37 45 S 72 20W
Mulde →, *Germany* .... **26 D8** 51 53N 12 15 E
Muldraugh, *U.S.A.* **141 G11** 37 56N 85 59W
Mule Creek, *U.S.A.* **138 D2** 43 19N 104 8W
Muleba, *Tanzania* ...... **106 C3** 1 50 S 31 37 E
Muleba □, *Tanzania* .... **106 C3** 2 0 S 31 30 E
Mulegns, *Switz.* .... **29 C9** 46 32N 9 38 E
Muleshoe, *U.S.A.* **139 H3** 34 13N 102 43W
Mulga Valley, *Australia* .. **116 A4** 31 8 S 141 3 E
Mulgathing, *Australia* ... **115 E1** 30 15 S 134 8 E
Mulgrave, *Canada* **129 C7** 45 38N 61 31W
Mulgrave I., *Papua N. G.* **120 F2** 10 5 S 142 10 E
Mulhacén, *Spain* **35 H1** 37 4N 3 20W
Mülheim, *Germany* ..... **26 D2** 51 25N 6 54 E
Mulhouse, *France* .... **23 E14** 47 40N 7 20 E
Muli, *China* .... **68 D3** 27 52N 101 8 E
Mulifanua, *W. Samoa* ... **121 W24** 13 50 S 171 59W
Muling, *China* .... **67 B16** 44 35N 130 10 E
Mull, *U.K.* **18 E3** 56 25N 5 56W
Mullaittvu, *Sri Lanka* ... **83 K5** 9 15N 80 49 E
Mullen, *U.S.A.* **138 D4** 42 3N 101 1W
Mullengudgery, *Australia* **117 A7** 31 43 S 147 23 E
Mullens, *U.S.A.* **134 G5** 37 35N 81 23W
Muller, Pegunungan,
Indonesia .... **75 B4** 0 30N 113 30 E
Mullet Pen., *Ireland* **19 B1** 54 13N 10 2W
Mullewa, *Australia* ..... **113 E2** 28 29 S 115 30 E
Müllheim, *Germany* ..... **27 H3** 47 47N 7 36 E
Mulligan →, *Australia* .. **114 C2** 25 0 S 139 0 E
Mullin, *U.S.A.* **139 K5** 31 33N 98 40W
Mullingar, *Ireland* **19 C4** 53 31N 7 21W
Mullins, *U.S.A.* **135 H6** 34 12N 79 15W
Mullumbimby, *Australia* . **115 D5** 28 30 S 153 30 E
Mulobezi, *Zambia* ...... **107 F2** 16 45 S 25 7 E
Mulshi L., *India* .... **82 E1** 18 30N 73 48 E
Multai, *India* .... **82 D4** 21 50N 78 21 E
Multan, *Pakistan* .... **79 C3** 30 15N 71 36 E
Multrå, *Sweden* .... **14 A11** 63 10N 17 24 E
Mulu, Gunong, *Malaysia* **75 B4** 4 3N 114 56 E
Mulumbe, Mts., *Zaïre* ... **107 D2** 8 40 S 27 30 E
Mulungushi Dam, *Zambia* **107 E2** 14 48 S 28 48 E
Mulvane, *U.S.A.* **139 G6** 37 29N 97 15W
Mulwad, *Sudan* .... **94 D3** 18 45N 30 39 E
Mulwala, *Australia* ..... **117 C7** 35 59 S 146 0 E
Mumbaī = Bombay, *India* **82 E1** 18 55N 72 50 E
Mumbondo, *Angola* ..... **103 E2** 10 9 S 14 15 E
Mumbwa, *Zambia* ...... **107 E2** 15 0 S 27 0 E
Mumeng, *Papua N. G.* .. **120 D4** 7 1 S 146 37 E
Mumra, *Russia* .... **53 H8** 45 45N 47 41 E
Mun →, *Thailand* ...... **76 E5** 15 19N 105 30 E
Muna, *Indonesia* .... **72 B2** 5 0 S 122 30 E
Munamagi, *Estonia* ..... **13 H22** 57 43N 27 4 E
Münchberg, *Germany* ... **27 E7** 50 11N 11 47 E
Müncheberg, *Germany* .. **26 C10** 52 30N 14 9 E
München, *Germany* .... **27 G7** 48 8N 11 34 E
Munchen-Gladbach =
Mönchengladbach,
Germany .... **26 D2** 51 11N 6 27 E
Muncho Lake, *Canada* .. **130 B3** 59 0N 125 50W
Munchŏn, *N. Korea* ..... **67 E14** 39 14N 127 19 E
Münchwilen, *Switz.* ..... **29 B7** 47 28N 8 59 E
Muncie, *U.S.A.* **141 D11** 40 12N 85 23W
Muncoonie, L., *Australia* **114 D2** 25 12 S 138 40 E
Munda, *Solomon Is.* .... **121 M9** 8 20 S 157 16 E
Mundakayam, *India* .... **83 K3** 9 30N 76 50 E
Mundala, *Indonesia* .... **73 B6** 4 30 S 141 0 E
Mundare, *Canada* **130 C6** 53 35N 112 20W
Munday, *U.S.A.* **139 J5** 33 27N 99 38W
Münden, *Germany* ..... **26 D5** 51 25N 9 38 E
Mundiwindi, *Australia* .. **112 D3** 23 47 S 120 9 E
Mundo →, *Spain* **35 G2** 38 30N 2 15W
Mundo Novo, *Brazil* .... **155 D3** 11 50 S 40 29W
Mundra, *India* .... **80 H3** 22 54N 69 48 E
Mundrabilla, *Australia* .. **113 F4** 31 52 S 127 51 E
Munducurus, *Brazil* .... **153 D6** 4 47 S 58 16W
Munenga, *Angola* ...... **103 E2** 10 2 S 14 41 E
Munera, *Spain* **35 F2** 39 2N 2 29W
Muneru →, *India* ...... **83 F5** 16 45N 80 3 E
Mungallala, *Australia* ... **115 D4** 26 28 S 147 34 E
Mungallala Cr. →,
Australia .... **115 D4** 28 53 S 147 5 E
Mungana, *Australia* ..... **114 B3** 17 8 S 144 27 E
Mungaoli, *India* .... **80 G8** 24 24N 78 7 E
Mungari, *Mozam.* .... **107 F3** 17 12 S 33 30 E
Mungbere, *Zaïre* .... **106 B2** 2 36N 28 28 E
Munger, *India* .... **81 G12** 25 23N 86 30 E
Mungindi, *Australia* ..... **115 D4** 28 58 S 149 1 E
Munhango, *Angola* ...... **103 E3** 12 10 S 18 38 E
Munich = München,
Germany .... **27 G7** 48 8N 11 34 E
Munising, *U.S.A.* **134 B2** 46 25N 86 40W
Munka-Ljungby, *Sweden* . **15 H6** 56 16N 12 58 E
Munkedal, *Sweden* ...... **15 F5** 58 28N 11 40 E
Munku-Sardyk, *Russia* ... **57 D11** 51 45N 100 20 E
Münnerstadt, *Germany* .. **27 E6** 50 14N 10 12 E
Munoz, *Phil.* .... **70 D3** 15 43N 120 54 E
Muñoz Gamero, Pen.,
Chile .... **160 D2** 52 30 S 73 5W
Munro, *Australia* .... **117 D7** 37 56 S 147 11 E
Munroe L., *Canada* **131 B9** 59 13N 98 35W
Munsan, *S. Korea* ...... **67 F14** 37 51N 126 48 E
Munshiganj, *Bangla.* .... **78 D3** 23 33N 90 32 E
Münsingen, *Switz.* ...... **28 C5** 46 52N 7 32 E
Munster, *France* .... **23 D14** 48 2N 7 8 E
Munster, *Niedersachsen*,
Germany .... **26 C6** 52 58N 10 5 E
Münster,
Nordrhein-Westfalen,
Germany .... **26 D3** 51 58N 7 37 E
Münster, *Switz.* .... **29 D6** 46 29N 8 17 E
Munster □, *Ireland* **19 D3** 52 18N 8 44W
Muntadgin, *Australia* .... **113 F2** 31 45 S 118 33 E
Muntele Mare, *Romania* . **46 C4** 46 30N 23 12 E
Muntok, *Indonesia* ...... **74 C3** 2 5 S 105 10 E
Munyama, *Zambia* ...... **107 F2** 16 5 S 28 31 E
Munzur Dağları, *Turkey* . **89 C8** 39 30N 39 10 E
Muong Beng, *Laos* ...... **76 B3** 20 23N 101 46 E
Muong Boum, *Vietnam* .. **76 A4** 22 24N 102 49 E
Muong Et, *Laos* .... **76 B5** 20 49N 104 1 E
Muong Hai, *Laos* .... **76 B3** 21 3N 101 49 E
Muong Hiem, *Laos* ..... **76 B4** 20 5N 103 22 E
Muong Houn, *Laos* ..... **76 B3** 20 8N 101 23 E
Muong Hung, *Vietnam* .. **76 B4** 20 56N 103 53 E
Muong Kau, *Laos* ...... **76 E5** 15 6N 105 47 E
Muong Khao, *Laos* ..... **76 C4** 19 38N 103 32 E
Muong Khoua, *Laos* .... **76 B4** 21 5N 102 31 E
Muong Liep, *Laos* ...... **76 C3** 18 29N 101 40 E
Muong May, *Laos* ...... **76 E6** 14 49N 106 56 E
Muong Ngeun, *Laos* .... **76 B3** 20 36N 101 3 E
Muong Ngoi, *Laos* ..... **76 B4** 20 43N 102 41 E
Muong Nhie, *Vietnam* .. **76 A4** 22 12N 102 28 E
Muong Nong, *Laos* ..... **76 D6** 16 22N 106 30 E
Muong Ou Tay, *Laos* ... **76 A3** 22 7N 101 48 E
Muong Oua, *Laos* ...... **76 C3** 18 18N 101 20 E
Muong Peun, *Laos* ..... **76 B4** 20 13N 103 52 E
Muong Phalane, *Laos* ... **76 D5** 16 39N 105 34 E
Muong Phieng, *Laos* .... **76 C3** 19 6N 101 32 E
Muong Phine, *Laos* ..... **76 D6** 16 32N 106 2 E
Muong Sai, *Laos* .... **76 B3** 20 42N 101 59 E
Muong Saiapoun, *Laos* .. **76 C3** 18 24N 101 31 E
Muong Sen, *Vietnam* ... **76 C5** 19 24N 104 8 E
Muong Sing, *Laos* ..... **76 B3** 21 11N 101 9 E
Muong Son, *Laos* ...... **76 B4** 20 27N 103 19 E
Muong Soui, *Laos* ..... **76 C4** 19 33N 102 52 E
Muong Va, *Laos* .... **76 B4** 21 53N 102 19 E
Muong Xia, *Vietnam* ... **76 B5** 20 19N 104 50 E
Muonio, *Finland* .... **12 C20** 67 57N 23 40 E
Muonionjoki →, *Finland* **12 C20** 67 11N 23 34 E
Muotathal, *Switz.* ...... **29 C7** 46 58N 8 46 E
Mupa, *Angola* .... **103 F3** 16 5 S 15 50 E
Muping, *China* .... **67 F11** 37 22N 121 36 E
Muqaddam, Wadi →,
Sudan .... **94 D3** 18 4N 31 30 E
Muqdisho, *Somali Rep.* .. **108 D3** 2 2N 45 25 E
Muqshin, W. →, *Oman* . **87 C6** 19 44N 55 14 E
Muquequete, *Angola* .... **103 E2** 14 50 S 14 16 E
Mur →, *Austria* .... **30 E9** 46 18N 16 52 E
Mur-de-Bretagne, *France* **22 D4** 48 12N 3 0W
Mura →, *Slovenia* ...... **39 B13** 46 30N 16 33 E
Muradiye, *Turkey* ...... **89 C10** 39 0N 43 44 E
Murakami, *Japan* .... **60 E9** 38 14N 139 29 E
Murallón, Cuerro, *Chile* . **160 C2** 49 48 S 73 30W
Muralto, *Switz.* .... **29 D7** 46 11N 8 49 E
Muranda, *Rwanda* ...... **106 C2** 1 52 S 29 20 E
Murang'a, *Kenya* .... **106 C4** 0 45 S 37 9 E
Murashi, *Russia* .... **54 B2** 59 30N 49 0 E
Murat, *France* .... **24 C6** 45 7N 2 53 E
Murat →, *Turkey* .... **89 C9** 38 46N 40 0 E
Muratlı, *Turkey* .... **88 B2** 41 10N 27 29 E
Murau, *Austria* .... **30 D7** 47 6N 14 10 E
Muravera, *Italy* .... **40 C2** 39 25N 9 34 E
Murayama, *Japan* ...... **60 E10** 38 30N 140 25 E
Murban, *U.A.E.* .... **85 F7** 23 50N 53 45 E
Murça, *Portugal* **36 D3** 41 24N 7 28W
Murchison, *N.Z.* .... **119 B7** 41 49 S 172 21 E
Murchison →, *Australia* . **113 E1** 27 45 S 114 0 E
Murchison, Mt., *Antarctica* **7 D11** 73 0 S 168 0 E
Murchison Falls =
Kabarega Falls, *Uganda* **106 B3** 2 15N 31 30 E
Murchison House,
Australia .... **113 E1** 27 39 S 114 14 E
Murchison Mt., *N.Z.* .... **119 D6** 43 0 S 171 22 E
Murchison Mts., *N.Z.* ... **119 F2** 45 13 S 167 23 E
Murchison Ra., *Australia* **114 C1** 20 0 S 134 10 E
Murchison Rapids, *Malawi* **107 F3** 15 55 S 34 35 E
Murcia, *Spain* **35 G3** 38 5N 1 10W
Murcia □, *Spain* **35 H3** 37 50N 1 30W
Murdo, *U.S.A.* **138 D4** 43 53N 100 43W
Murdoch Pt., *Australia* .. **114 A3** 14 37 S 144 55 E
Mureş □, *Romania* ..... **46 C5** 46 45N 24 40 E
Mureş →, *Romania* .... **46 C1** 46 15N 20 13 E
Mureşul = Mureş →,
Romania .... **46 C1** 46 15N 20 13 E
Muret, *France* .... **24 E5** 43 30N 1 20 E
Murfatlar, *Romania* ..... **46 E9** 44 10N 28 26 E
Murfreesboro, *U.S.A.* ... **135 H2** 35 51N 86 24W
Murg, *Switz.* .... **29 B8** 47 6N 9 13 E
Murg →, *Germany* ...... **27 G4** 48 55N 8 10 E
Murgab = Murghob,
Tajikistan .... **55 D7** 38 10N 74 2 E
Murgeni, *Romania* ...... **46 C9** 46 12N 28 1 E
Murgenthal, *Switz.* ..... **28 B5** 47 16N 7 50 E
Murghob, *Tajikistan* .... **55 D7** 38 10N 74 2 E
Murgon, *Australia* ...... **115 D5** 26 15 S 151 54 E
Murgoo, *Australia* ...... **113 E2** 27 24 S 116 28 E
Muri, *Switz.* .... **29 B6** 47 17N 8 21 E
Muria, *Indonesia* .... **75 D4** 6 36 S 110 53 E
Muriaé, *Brazil* .... **155 F3** 21 8 S 42 23W
Murias de Paredes, *Spain* **36 C4** 42 52N 6 11W
Murici, *Brazil* .... **154 C4** 9 19 S 35 56W
Muriége, *Angola* .... **103 D4** 9 58 S 21 11 E
Muriel Mine, *Zimbabwe* . **107 F3** 17 14 S 30 40 E
Murila, *Angola* .... **103 E4** 10 44 S 20 20 E
Müritz-see, *Germany* .... **26 B8** 53 25N 12 42 E
Murka, *Kenya* .... **106 C4** 3 27 S 38 0 E
Murmansk, *Russia* ...... **48 A5** 68 57N 33 10 E
Murmerwoude, *Neths.* ... **20 B8** 53 18N 6 0 E
Murnau, *Germany* ...... **27 H7** 47 40N 11 12 E
Muro, *France* .... **25 F12** 42 34N 8 54 E
Muro, *Spain* .... **33 B10** 39 44N 3 3 E
Muro, C. de, *France* .... **25 G12** 41 44N 8 37 E
Muro Lucano, *Italy* ..... **41 B8** 40 45N 15 29 E
Murom, *Russia* .... **52 C6** 55 35N 42 3 E
Muroran, *Japan* .... **60 C10** 42 25N 141 0 E
Muros, *Spain* **36 C1** 42 45N 9 5W
Muros y de Noya, Ría de,
Spain **36 C2** 42 45N 9 0W
Muroto, *Japan* .... **62 D6** 33 18N 134 9 E
Muroto-Misaki, *Japan* ... **62 D6** 33 15N 134 10 E
Murowana Goślina,
Poland .... **47 C4** 52 35N 17 0 E
Murphy, *U.S.A.* **142 E5** 43 13N 116 33W
Murphys, *U.S.A.* **144 G6** 38 8N 120 28W
Murphysboro, *U.S.A.* ... **139 G10** 37 46N 89 20W
Murrat, *Sudan* .... **94 D2** 18 51N 29 33 E
Murray, *Iowa, U.S.A.* ... **140 C3** 41 3N 93 57W
Murray, *Ky., U.S.A.* **135 G1** 36 37N 88 19W
Murray, *Utah, U.S.A.* ... **142 F8** 40 40N 111 53W
Murray →, *Australia* ... **116 C3** 35 20 S 139 22 E
Murray →, *Canada* **130 B4** 56 11N 120 45W
Murray, L., *Papua N. G.* **120 D1** 7 0 S 141 35 E
Murray, L., *U.S.A.* **135 H5** 34 3N 81 13W
Murray Bridge, *Australia* **116 C3** 35 6 S 139 14 E
Murray Downs, *Australia* **114 C1** 21 4 S 134 40 E
Murray Harbour, *Canada* **129 C7** 46 0N 62 28W
Murraysburg, *S. Africa* .. **104 E3** 31 58 S 23 47 E
Murrayville, *U.S.A.* **140 E6** 39 35N 90 15W
Murree, *Pakistan* .... **80 C5** 33 56N 73 28 E
Murrieta, *U.S.A.* **145 M9** 33 33N 117 13W
Murrin Murrin, *Australia* **113 E3** 28 58 S 121 33 E
Murrumbidgee →,
Australia .... **116 C5** 34 43 S 143 12 E
Murrumburrah, *Australia* **117 C8** 34 32 S 148 22 E
Murrurundi, *Australia* ... **117 A9** 31 42 S 150 51 E
Murshid, *Sudan* .... **94 C3** 21 40N 31 10 E
Murshidabad, *India* ..... **81 G13** 24 11N 88 19 E
Murska Sobota, *Slovenia* . **39 B13** 46 39N 16 12 E
Murtazapur, *India* ...... **82 D3** 20 40N 77 25 E
Murten, *Switz.* .... **28 C4** 46 56N 7 4 E
Murtensee, *Switz.* ...... **28 C4** 46 56N 7 7 E
Murtle L., *Canada* **130 C5** 52 8N 119 38W
Murtoa, *Australia* ...... **116 D5** 36 35 S 142 28 E
Murtosa, *Portugal* **36 E2** 40 44N 8 40W
Muru →, *Brazil* .... **156 B3** 8 9 S 70 45W
Murungu, *Tanzania* .... **106 C3** 4 12 S 31 10 E
Murupara, *N.Z.* .... **118 E5** 38 28 S 176 42 E
Murwara, *India* .... **81 H9** 23 46N 80 28 E
Murwillumbah, *Australia* . **115 D5** 28 18 S 153 27 E
Mürz →, *Austria* .... **30 D8** 47 30N 15 25 E
Mürzzuschlag, *Austria* .. **30 D8** 47 36N 15 41 E
Muş, *Turkey* .... **89 C9** 38 45N 41 30 E
Musa, *Zaïre* .... **102 B3** 2 40N 19 18 E
Musa →, *Papua N. G.* .. **120 E5** 9 3 S 148 55 E
Mûsa, G., *Egypt* .... **94 J8** 28 33N 33 59 E
Musa Khel, *Pakistan* .... **79 C3** 30 59N 69 52 E
Mūsá Qal'eh, *Afghan.* ... **79 B2** 32 20N 64 50 E
Musala, *Bulgaria* .... **43 E8** 42 13N 23 37 E
Musala, *Indonesia* ...... **74 B1** 1 41N 98 28 E
Musan, *N. Korea* .... **67 C15** 42 12N 129 12 E
Musangu, *Zaïre* .... **103 E4** 10 28 S 23 55 E
Musasa, *Tanzania* ...... **106 C3** 3 25 S 31 30 E
Musashino, *Japan* ...... **63 B11** 35 42N 139 34 E
Musay'īd, *Qatar* .... **85 E6** 25 0N 51 33 E
Musaymīr, *Yemen* ...... **86 D4** 13 27N 44 37 E
Muscat = Masqat, *Oman* **87 B7** 23 37N 58 36 E
Muscat & Oman =
Oman ■, *Asia* .... **87 B7** 23 0N 58 0 E
Muscatine, *U.S.A.* **140 C5** 41 25N 91 3W
Muscoda, *U.S.A.* **140 A6** 43 11N 90 27W
Musel, *Spain* **36 B5** 43 34N 5 42W
Musgrave, *Australia* .... **114 A3** 14 47 S 143 30 E
Musgrave Ras., *Australia* **113 E5** 26 0 S 132 0 E
Mushie, *Zaïre* .... **102 C3** 2 56 S 16 55 E
Mushin, *Nigeria* .... **101 D5** 6 32N 3 21 E
Musi →, *India* .... **82 F4** 16 41N 79 40 E
Musi →, *Indonesia* ..... **74 C2** 2 20 S 104 56 E
Muskeg →, *Canada* **130 A4** 60 20N 123 20W
Muskegon, *U.S.A.* **141 A10** 43 14N 86 16W
Muskegon →, *U.S.A.* .. **134 D2** 43 14N 86 21W
Muskegon Heights, *U.S.A.* **141 A10** 43 12N 86 16W
Muskogee, *U.S.A.* **139 H7** 35 45N 95 22W
Muskwa →, *Canada* **130 B4** 58 47N 122 48W
Muslīmiyah, *Syria* ...... **84 B3** 36 19N 37 12 E
Musmar, *Sudan* .... **94 D4** 18 13N 35 40 E
Musofu, *Zambia* .... **107 E2** 13 30 S 29 0 E
Musoma, *Tanzania* ..... **106 C3** 1 30 S 33 48 E
Musoma □, *Tanzania* ... **106 C3** 1 50 S 34 30 E
Musquaro, L., *Canada* .. **129 B7** 50 38N 61 5W
Musquodoboit Harbour,
Canada **129 D7** 44 50N 63 9W
Mussau I., *Papua N. G.* . **120 A5** 1 30 S 149 40 E
Musselburgh, *U.K.* **18 F5** 55 57N 3 2W
Musselkanaal, *Neths.* .... **20 C10** 52 57N 7 0 E
Musselshell →, *U.S.A.* .. **142 C10** 47 21N 107 57W
Mussende, *Angola* ...... **103 E3** 10 32 S 16 5 E
Mussidan, *France* .... **24 C4** 45 2N 0 22 E
Mussolo, *Angola* .... **103 D3** 9 59 S 17 19 E
Mussomeli, *Italy* .... **40 E6** 37 35N 13 45 E
Musson, *Belgium* .... **21 J7** 49 33N 5 42 E
Mussoorie, *India* .... **80 D8** 30 27N 78 6 E
Mussuco, *Angola* .... **103 F3** 17 2 S 19 3 E
Mustafakemalpaşa, *Turkey* **88 B3** 40 2N 28 24 E
Mustahil, *Ethiopia* ...... **108 C2** 5 16N 44 45 E
Mustang, *Nepal* .... **81 E10** 29 10N 83 55 E
Musters, L., *Argentina* .. **160 C3** 45 20 S 69 25W
Musudan, *N. Korea* ..... **67 D15** 40 50N 129 43 E
Muswellbrook, *Australia* . **117 B9** 32 16 S 150 56 E
Muszyna, *Poland* .... **31 B13** 49 22N 20 55 E
Mût, *Egypt* .... **94 B2** 25 28N 28 58 E
Mut, *Turkey* .... **88 D5** 36 40N 33 28 E
Mutanda, *Mozam.* ...... **105 C5** 21 0 S 33 34 E

240 Neusiedl

Neusiedl, *Austria* **31 D9** 47 57N 16 50 E
Neusiedler See, *Austria* **31 D9** 47 50N 16 47 E
Neuss, *Germany* **21 F9** 51 11N 6 42 E
Neussargues-Moissac, *France* **24 C7** 45 9N 3 0 E
Neustadt, *Baden-W., Germany* **27 H4** 47 54N 8 12 E
Neustadt, *Bayern, Germany* **27 F8** 49 44N 12 10 E
Neustadt, *Bayern, Germany* **27 G7** 48 48N 11 46 E
Neustadt, *Bayern, Germany* **27 F6** 49 34N 10 37 E
Neustadt, *Bayern, Germany* **27 E7** 50 19N 11 7 E
Neustadt, *Brandenburg, Germany* **26 C8** 52 50N 12 27 E
Neustadt, *Hessen, Germany* **26 E5** 50 51N 9 9 E
Neustadt, *Niedersachsen, Germany* **26 C5** 52 30N 9 30 E
Neustadt, *Rhld-Pfz., Germany* **27 F4** 49 21N 8 10 E
Neustadt, *Schleswig-Holstein, Germany* **26 A6** 54 6N 10 49 E
Neustadt, *Thüringen, Germany* **26 E7** 50 45N 11 43 E
Neustrelitz, *Germany* **26 B9** 53 21N 13 4 E
Neuvic, *France* **24 C6** 45 23N 2 16 E
Neuville, *Belgium* **21 H5** 50 11N 4 32 E
Neuville-aux-Bois, *France* **23 D9** 48 4N 2 3 E
Neuville-de-Poitou, *France* **24 B4** 46 41N 0 15 E
Neuville-sur-Saône, *France* **25 C8** 45 52N 4 51 E
Neuvy-le-Roi, *France* **22 E7** 47 36N 0 36 E
Neuvy-St.-Sépulchre, *France* **24 B5** 46 35N 1 48 E
Neuvy-sur-Barangeon, *France* **23 E9** 47 20N 2 15 E
Neuwerk, *Germany* **26 B4** 53 55N 8 30 E
Neuwied, *Germany* **26 E3** 50 26N 7 29 E
Neva →, *Russia* **50 C6** 59 50N 30 30 E
Nevada, *Iowa, U.S.A.* **140 B3** 42 1N 93 27W
Nevada, *Mo., U.S.A.* **139 G7** 37 51N 94 22W
Nevada □, *U.S.A.* **142 G5** 39 0N 117 0W
Nevada, Sierra, *Spain* **35 H1** 37 3N 3 15W
Nevada, Sierra, *U.S.A.* **142 G3** 39 0N 120 30W
Nevada City, *U.S.A.* **144 F6** 39 16N 121 1W
Nevado, Cerro, *Argentina* **158 D2** 35 30 S 68 32W
Nevanka, *Russia* **57 D10** 56 31N 98 55 E
Nevasa, *India* **82 E2** 19 34N 75 0 E
Nevel, *Russia* **50 D5** 56 0N 29 55 E
Nevele, *Belgium* **21 F3** 51 3N 3 33 E
Nevers, *France* **23 F10** 47 0N 3 9 E
Nevertire, *Australia* **117 A7** 31 50 S 147 44 E
Nevesinje, *Bos.-H.* **42 D3** 43 14N 18 6 E
Neville, *Canada* **131 D7** 49 58N 107 39W
Nevinnomyssk, *Russia* **53 H6** 44 40N 42 0 E
Nevis, *W. Indies* **149 C7** 17 0N 62 30W
Nevrokop = Gotse Delchev, *Bulgaria* **43 F8** 41 43N 23 46 E
Nevşehir, *Turkey* **88 C6** 38 33N 34 40 E
Nevyansk, *Russia* **54 C8** 57 30N 60 13 E
New →, *Guyana* **153 C6** 3 20N 57 37W
New Albany, *Ind., U.S.A.* **141 F11** 38 18N 85 49W
New Albany, *Miss., U.S.A.* **139 H10** 34 29N 89 0W
New Albany, *Pa., U.S.A.* **137 E8** 41 36N 76 27W
New Amsterdam, *Guyana* **153 B6** 6 15N 57 36W
New Angledool, *Australia* **115 D4** 29 5 S 147 55 E
New Athens, *U.S.A.* **140 F7** 38 19N 89 53W
New Bedford, *U.S.A.* **137 E14** 41 38N 70 56W
New Berlin, *Ill., U.S.A.* **140 E7** 39 44N 89 55W
New Berlin, *Wis., U.S.A.* **141 B8** 42 59N 88 6W
New Bern, *U.S.A.* **135 H7** 35 7N 77 3W
New Bethlehem, *U.S.A.* **136 E5** 41 0N 79 20W
New Bloomfield, *U.S.A.* **136 F7** 40 25N 77 11W
New Boston, *U.S.A.* **139 J7** 33 28N 94 25W
New Braunfels, *U.S.A.* **139 L5** 29 42N 98 8W
New Brighton, *N.Z.* **119 D7** 43 29 S 172 43 E
New Brighton, *U.S.A.* **136 F4** 40 42N 80 19W
New Britain, *Papua N. G.* **120 C6** 5 50 S 150 20 E
New Britain, *U.S.A.* **137 E12** 41 40N 72 47W
New Brunswick, *U.S.A.* **137 F10** 40 30N 74 27W
New Brunswick □, *Canada* **129 C6** 46 50N 66 30W
New Buffalo, *U.S.A.* **141 C10** 41 47N 86 45W
New Bussa, *Nigeria* **101 D5** 9 53N 4 31 E
New Caledonia ■, *Pac. Oc.* **121 U19** 21 0 S 165 0 E
New Canton, *U.S.A.* **140 E5** 39 37N 91 8W
New Carlisle, *Ind., U.S.A.* **141 C10** 41 45N 86 32W
New Carlisle, *Ohio, U.S.A.* **141 E12** 39 56N 84 2W
New Castle, *Ind., U.S.A.* **141 E11** 39 55N 85 22W
New Castle, *Ky., U.S.A.* **141 F11** 38 26N 85 10W
New Castle, *Pa., U.S.A.* **136 E4** 41 0N 80 21W
New City, *U.S.A.* **137 E11** 41 9N 73 59W
New Cumberland, *U.S.A.* **136 F4** 40 30N 80 36W
New Cuyama, *U.S.A.* **145 L7** 34 57N 119 38W
New Delhi, *India* **80 E7** 28 37N 77 13 E
New Denver, *Canada* **130 D5** 50 0N 117 25W
New Don Pedro Reservoir, *U.S.A.* **144 H6** 37 43N 120 24W
New England, *U.S.A.* **138 B3** 46 32N 102 52W
New England Ra., *Australia* **115 E5** 30 20 S 151 45 E
New Forest, *U.K.* **17 G6** 50 53N 1 34W
New Franklin, *U.S.A.* **140 E4** 39 1N 92 44W
New Georgia Is., *Solomon Is.* **121 M9** 8 15 S 157 30 E
New Glarus, *U.S.A.* **140 B7** 42 49N 89 38W
New Glasgow, *Canada* **129 C7** 45 35N 62 36W
New Guinea, *Oceania* **122 H5** 4 0 S 136 0 E
New Hamburg, *Canada* **136 C4** 43 23N 80 42W
New Hampshire □, *U.S.A.* **137 C13** 44 0N 71 30W
New Hampton, *U.S.A.* **140 A4** 43 3N 92 19W
New Hanover, *Papua N. G.* **120 B6** 2 30 S 150 10 E
New Hanover, *S. Africa* **105 D5** 29 22 S 30 31 E
New Harmony, *U.S.A.* **141 F9** 38 8N 87 56W
New Haven, *Conn., U.S.A.* **137 E12** 41 18N 72 55W
New Haven, *Ill., U.S.A.* **141 G8** 37 55N 88 8W
New Haven, *Ind., U.S.A.* **141 C11** 41 4N 85 1W
New Haven, *Mich., U.S.A.* **136 D2** 42 44N 82 48W
New Haven, *Mo., U.S.A.* **140 F5** 38 37N 91 13W
New Hazelton, *Canada* **130 B3** 55 20N 127 30W
New Hebrides = Vanuatu ■, *Pac. Oc.* **121 E6** 15 0 S 168 0 E
New Iberia, *U.S.A.* **139 K9** 30 1N 91 49W
New Ireland, *Papua N. G.* **120 B6** 3 20 S 151 50 E
New Jersey □, *U.S.A.* **137 F10** 40 0N 74 30W
New Kensington, *U.S.A.* **136 F5** 40 34N 79 46W
New Lexington, *U.S.A.* **134 F4** 39 43N 82 13W
New Liskeard, *Canada* **128 C4** 47 31N 79 41W
New London, *Conn., U.S.A.* **137 E12** 41 22N 72 6W
New London, *Iowa, U.S.A.* **140 D5** 40 55N 91 24W
New London, *Minn., U.S.A.* **138 C7** 45 18N 94 56W
New London, *Mo., U.S.A.* **140 E5** 39 35N 91 24W
New London, *Ohio, U.S.A.* **136 E2** 41 5N 82 24W
New London, *Wis., U.S.A.* **138 C10** 44 23N 88 45W
New Madison, *U.S.A.* **141 E12** 39 58N 84 43W
New Madrid, *U.S.A.* **139 G10** 36 36N 89 32W
New Meadows, *U.S.A.* **142 D5** 44 58N 116 18W
New Melones L., *U.S.A.* **144 H6** 37 57N 120 31W
New Mexico □, *U.S.A.* **143 J10** 34 30N 106 0W
New Miami, *U.S.A.* **141 E12** 39 26N 84 32W
New Milford, *Conn., U.S.A.* **137 E11** 41 35N 73 25W
New Milford, *Pa., U.S.A.* **137 E9** 41 52N 75 44W
New Norcia, *Australia* **113 F2** 30 57 S 116 13 E
New Norfolk, *Australia* **114 G4** 42 46 S 147 2 E
New Orleans, *U.S.A.* **139 K9** 29 58N 90 4W
New Palestine, *U.S.A.* **141 E11** 39 45N 85 52W
New Paris, *U.S.A.* **141 E12** 39 51N 84 48W
New Pekin, *U.S.A.* **141 F10** 38 31N 86 2W
New Philadelphia, *U.S.A.* **136 F3** 40 30N 81 27W
New Plymouth, *N.Z.* **118 F3** 39 4 S 174 5 E
New Plymouth, *U.S.A.* **142 E5** 43 58N 116 49W
New Providence, *Bahamas* **148 A4** 25 25N 78 35W
New Radnor, *U.K.* **17 E4** 52 15N 3 9W
New Richmond, *Ohio, U.S.A.* **141 F12** 38 57N 84 17W
New Richmond, *Wis., U.S.A.* **138 C8** 45 7N 92 32W
New Roads, *U.S.A.* **139 K9** 30 42N 91 26W
New Rochelle, *U.S.A.* **137 F11** 40 55N 73 47W
New Rockford, *U.S.A.* **138 B5** 47 41N 99 8W
New Ross, *Ireland* **19 D5** 52 23N 6 57W
New Salem, *U.S.A.* **138 B4** 46 51N 101 25W
New Scone, *U.K.* **18 E5** 56 25N 3 24W
New Sharon, *U.S.A.* **140 C4** 41 28N 92 39W
New Siberian Is. = Novosibirskiye Ostrova, *Russia* **57 B15** 75 0N 142 0 E
New Smyrna Beach, *U.S.A.* **135 L5** 29 1N 80 56W
New South Wales □, *Australia* **115 E4** 33 0 S 146 0 E
New Springs, *Australia* **113 E3** 25 49 S 120 1 E
New Town, *U.S.A.* **138 A3** 47 59N 102 30W
New Ulm, *U.S.A.* **138 C7** 44 19N 94 28W
New Vienna, *U.S.A.* **141 E13** 39 19N 83 42W
New Virginia, *U.S.A.* **140 C3** 41 11N 93 44W
New Washington, *Phil.* **71 F4** 11 39N 122 26 E
New Waterford, *Canada* **129 C7** 46 13N 60 4W
New Westminster, *Canada* **130 D4** 49 13N 122 55W
New York □, *U.S.A.* **137 D9** 43 0N 75 0W
New York City, *U.S.A.* **137 F11** 40 45N 74 0W
New Zealand ■, *Oceania* **118 G5** 40 0 S 176 0 E
Newala, *Tanzania* **107 E4** 10 58 S 39 18 E
Newala □, *Tanzania* **107 E4** 10 46 S 39 20 E
Newark, *Del., U.S.A.* **134 F8** 39 41N 75 46W
Newark, *N.J., U.S.A.* **137 F10** 40 44N 74 10W
Newark, *N.Y., U.S.A.* **136 C7** 43 3N 77 6W
Newark, *Ohio, U.S.A.* **136 F2** 40 3N 82 24W
Newark-on-Trent, *U.K.* **16 D7** 53 5N 0 48W
Newaygo, *U.S.A.* **134 D3** 43 25N 85 48W
Newberg, *Mo., U.S.A.* **140 G5** 37 55N 91 54W
Newberg, *Oreg., U.S.A.* **142 D2** 45 18N 122 58W
Newberry, *Mich., U.S.A.* **134 B3** 46 21N 85 30W
Newberry, *S.C., U.S.A.* **135 H5** 34 17N 81 37W
Newberry Springs, *U.S.A.* **145 L10** 34 50N 116 41W
Newbridge = Droichead Nua, *Ireland* **19 C5** 53 11N 6 48W
Newbrook, *Canada* **130 C6** 54 24N 112 57W
Newburgh, *Ind., U.S.A.* **141 G9** 37 57N 87 24W
Newburgh, *N.Y., U.S.A.* **137 E10** 41 30N 74 1W
Newbury, *U.K.* **17 F6** 51 24N 1 20W
Newbury, *U.S.A.* **137 B12** 43 19N 72 3W
Newburyport, *U.S.A.* **137 D14** 42 49N 70 53W
Newcastle, *Australia* **117 B9** 33 0 S 151 46 E
Newcastle, *Canada* **129 C6** 47 1N 65 38W
Newcastle, *S. Africa* **105 D4** 27 45 S 29 58 E
Newcastle, *U.K.* **19 B6** 54 13N 5 54W
Newcastle, *Calif., U.S.A.* **144 G5** 38 53N 121 8W
Newcastle, *Wyo., U.S.A.* **138 D2** 43 50N 104 11W
Newcastle Emlyn, *U.K.* **17 E3** 52 2N 4 28W
Newcastle Ra., *Australia* **112 C5** 15 45 S 130 15 E
Newcastle-under-Lyme, *U.K.* **16 D5** 53 1N 2 14W
Newcastle-upon-Tyne, *U.K.* **16 C6** 54 58N 1 36W
Newcastle Waters, *Australia* **114 B1** 17 30 S 133 28 E
Newdegate, *Australia* **113 F2** 33 6 S 119 0 E
Newell, *U.S.A.* **138 C3** 44 43N 103 25W
Newfoundland □, *Canada* **129 B8** 53 0N 58 0W
Newhalem, *U.S.A.* **130 D4** 48 40N 121 15W
Newhall, *U.S.A.* **145 L8** 34 23N 118 32W
Newham, *U.K.* **17 F8** 51 31N 0 3 E
Newhaven, *U.K.* **17 G8** 50 47N 0 3 E
Newkirk, *U.S.A.* **139 G6** 36 53N 97 3W
Newman, *Australia* **112 D2** 23 18 S 119 45 E
Newman, *Calif., U.S.A.* **144 H5** 37 19N 121 1W
Newman, *Ill., U.S.A.* **141 E9** 39 48N 87 59W
Newmarket, *Canada* **136 B5** 44 3N 79 28W
Newmarket, *Ireland* **19 D3** 52 13N 9 0W
Newmarket, *U.K.* **17 E8** 52 15N 0 25 E
Newmarket, *U.S.A.* **137 C14** 43 5N 70 56W
Newnan, *U.S.A.* **135 J3** 33 23N 84 48W
Newport, *Gwent, U.K.* **17 F5** 51 35N 3 0W
Newport, *I. of W., U.K.* **17 G6** 50 42N 1 17W
Newport, *Ark., U.S.A.* **139 H9** 35 37N 91 16W
Newport, *Ind., U.S.A.* **141 E9** 39 53N 87 25W
Newport, *Ky., U.S.A.* **141 E12** 39 5N 84 30W
Newport, *N.H., U.S.A.* **137 C12** 43 22N 72 10W
Newport, *Oreg., U.S.A.* **142 D1** 44 39N 124 3W
Newport, *Pa., U.S.A.* **136 F7** 40 29N 77 8W
Newport, *R.I., U.S.A.* **137 E13** 41 29N 71 19W
Newport, *Tenn., U.S.A.* **135 H4** 35 58N 83 11W
Newport, *Vt., U.S.A.* **137 B12** 44 56N 72 13W
Newport, *Wash., U.S.A.* **142 B5** 48 11N 117 3W
Newport Beach, *U.S.A.* **145 M9** 33 37N 117 56W
Newport News, *U.S.A.* **134 G7** 36 59N 76 25W
Newquay, *U.K.* **17 G2** 50 25N 5 6W
Newry, *U.K.* **19 B5** 54 11N 6 21W
Newry & Mourne □, *U.K.* **19 B5** 54 10N 6 15W
Newton, *Ill., U.S.A.* **141 F8** 38 59N 88 10W
Newton, *Iowa, U.S.A.* **140 C3** 41 42N 93 3W
Newton, *Mass., U.S.A.* **137 D13** 42 21N 71 12W
Newton, *Miss., U.S.A.* **139 J10** 32 19N 89 10W
Newton, *N.C., U.S.A.* **135 H5** 35 40N 81 13W
Newton, *N.J., U.S.A.* **137 E10** 41 3N 74 45W
Newton, *Tex., U.S.A.* **139 K8** 30 51N 93 46W
Newton Abbot, *U.K.* **17 G4** 50 32N 3 37W
Newton Boyd, *Australia* **115 D5** 29 45 S 152 16 E
Newton Stewart, *U.K.* **18 G4** 54 57N 4 30W
Newtonmore, *U.K.* **18 D4** 57 4N 4 8W
Newtown, *U.K.* **17 E4** 52 31N 3 19W
Newtown, *U.S.A.* **140 D3** 40 22N 93 20W
Newtownabbey □, *U.K.* **19 B6** 54 45N 6 0W
Newtownards, *U.K.* **19 B6** 54 36N 5 42W
Newville, *U.S.A.* **136 F7** 40 10N 77 24W
Nexon, *France* **24 C5** 45 41N 1 11 E
Neya, *Russia* **52 A6** 58 21N 43 49 E
Neyrīz, *Iran* **85 D7** 29 15N 54 19 E
Neyshābūr, *Iran* **85 B8** 36 10N 58 50 E
Neyyattinkara, *India* **83 K3** 8 26N 77 5 E
Nezhin = Nizhyn, *Ukraine* **51 G6** 51 5N 31 55 E
Nezperce, *U.S.A.* **142 C5** 46 14N 116 14W
Ngabang, *Indonesia* **75 B3** 0 23N 109 55 E
Ngabordamlu, Tanjung, *Indonesia* **73 C4** 6 56 S 134 11 E
N'Gage, *Angola* **103 D3** 7 46 S 15 15 E
Ngaiphaipi, *Burma* **78 D4** 22 14N 93 15 E
Ngambé, *Cameroon* **101 D7** 5 48N 11 29 E
Ngami Depression, *Botswana* **104 C3** 20 30 S 22 46 E
Ngamo, *Zimbabwe* **107 F2** 19 3 S 27 32 E
Nganjuk, *Indonesia* **75 D4** 7 32 S 111 55 E
Ngao, *Thailand* **76 C2** 18 46N 99 59 E
Ngaoundéré, *Cameroon* **102 A2** 7 15N 13 35 E
Ngapara, *N.Z.* **119 E5** 44 57 S 170 46 E
Ngara, *Tanzania* **106 C3** 2 29 S 30 40 E
Ngara □, *Tanzania* **106 C3** 2 29 S 30 40 E
Ngaruawahia, *N.Z.* **118 D4** 37 42 S 175 11 E
Ngaruroro →, *N.Z.* **118 F5** 39 34 S 176 55 E
Ngatapa, *N.Z.* **118 E6** 38 32 S 177 45 E
Ngathainggyaung, *Burma* **78 G5** 17 24N 95 5 E
Ngauruhoe, Mt., *N.Z.* **118 F4** 39 13 S 175 45 E
Ngawi, *Indonesia* **75 D4** 7 24 S 111 26 E
Nggela, *Solomon Is.* **121 M11** 9 5 S 160 15 E
Nghia Lo, *Vietnam* **76 B5** 21 33N 104 28 E
Ngidinga, *Zaïre* **103 D3** 5 37 S 15 17 E
Ngo, *Congo* **102 C3** 2 29 S 15 45 E
N'Gola, *Angola* **103 E2** 14 10 S 14 30 E
Ngoma, *Malawi* **107 E3** 13 8 S 33 45 E
Ngomahura, *Zimbabwe* **107 G3** 20 26 S 30 43 E
Ngomba, *Tanzania* **107 D3** 8 20 S 32 53 E
Ngongotaha, *N.Z.* **118 E5** 38 5 S 176 12 E
Ngop, *Sudan* **95 F3** 6 17N 30 9 E
Ngoring Hu, *China* **64 C4** 34 55N 97 5 E
Ngorkou, *Mali* **100 B4** 15 40N 3 41W
Ngorongoro, *Tanzania* **106 C4** 3 11 S 35 32 E
Ngouri, *Chad* **97 F3** 13 38N 15 22 E
Ngourti, *Niger* **97 E2** 15 19N 13 12 E
Ngozi, *Burundi* **106 C2** 2 54 S 29 50 E
Ngudu, *Tanzania* **106 C3** 2 58 S 33 25 E
Nguigmi, *Niger* **97 F2** 14 20N 13 20 E
Ngukurr, *Australia* **114 A1** 14 44 S 134 44 E
Ngunga, *Tanzania* **106 C3** 3 37 S 33 37 E
Nguru, *Nigeria* **101 C7** 12 56N 10 29 E
Nguru Mts., *Tanzania* **106 D4** 6 0 S 37 30 E
Nguyen Binh, *Vietnam* **76 A5** 22 39N 105 56 E
Nha Trang, *Vietnam* **77 F7** 12 16N 109 10 E
Nhacoongo, *Mozam.* **105 C6** 24 18 S 35 14 E
Nhamaabué, *Mozam.* **107 F4** 17 25 S 35 5 E
Nhambiquara, *Brazil* **157 C6** 12 50 S 59 49W
Nhamundá, *Brazil* **153 D6** 2 14 S 56 43W
Nhamundá →, *Brazil* **153 D6** 2 12 S 56 41W
Nhangutazi, L., *Mozam.* **105 C5** 24 0 S 34 30 E
Nhecolândia, *Brazil* **157 D6** 19 17 S 56 58W
Nhill, *Australia* **116 D4** 36 18 S 141 40 E
Nho Quan, *Vietnam* **76 B5** 20 18N 105 45 E
Nhulunbuy, *Australia* **114 A2** 12 10 S 137 20 E
Nhundo, *Angola* **103 E4** 14 25 S 21 23 E
Nia-nia, *Zaïre* **106 B2** 1 30N 27 40 E
Niafounké, *Mali* **100 B4** 16 0N 4 5W
Niagara, *U.S.A.* **134 C1** 45 45N 88 0W
Niagara Falls, *Canada* **128 D4** 43 7N 79 5W
Niagara Falls, *U.S.A.* **136 C6** 43 5N 79 4W
Niagara-on-the-Lake, *Canada* **136 C5** 43 15N 79 4W
Niah, *Malaysia* **75 B4** 3 58N 113 46 E
Niamey, *Niger* **101 C5** 13 27N 2 6 E
Nianforando, *Guinea* **100 D2** 9 37N 10 36W
Nianfors, *Sweden* **14 C10** 61 36N 16 46 E
Niangara, *Zaïre* **106 B2** 3 42N 27 50 E
Niangua →, *U.S.A.* **140 G4** 38 0N 92 48W
Nias, *Indonesia* **74 B1** 1 0N 97 30 E
Niassa □, *Mozam.* **107 E4** 13 30 S 36 0 E
Nibāk, *Si. Arabia* **87 A5** 24 25N 50 50 E
Nibbiano, *Italy* **38 D6** 44 54N 9 19 E
Nibe, *Denmark* **15 H3** 56 59N 9 38 E
Nicaragua ■, *Cent. Amer.* **148 D2** 11 40N 85 30W
Nicaragua, L. de, *Nic.* **148 D2** 12 0N 85 30W
Nicastro, *Italy* **41 D9** 38 59N 16 19 E
Nice, *France* **25 E11** 43 42N 7 14 E
Niceville, *U.S.A.* **135 K2** 30 31N 86 30W
Nichinan, *Japan* **62 F3** 31 38N 131 23 E
Nicholás, Canal, *W. Indies* **148 B3** 23 30N 80 5W
Nicholasville, *U.S.A.* **141 G12** 37 53N 84 34W
Nichols, *U.S.A.* **137 D8** 42 1N 76 22W
Nicholson, *Australia* **112 C4** 18 2 S 128 54 E
Nicholson, *U.S.A.* **137 E9** 41 37N 75 47W
Nicholson →, *Australia* **114 B2** 17 31 S 139 36 E
Nicholson Ra., *Australia* **113 E2** 27 15 S 116 45 E
Nickerie □, *Surinam* **153 C6** 4 0N 57 0W
Nickerie →, *Surinam* **153 B6** 5 58N 57 0W
Nicobar Is., *Ind. Oc.* **58 J13** 9 0N 93 0 E
Nicoclí, *Colombia* **152 B2** 8 26N 76 48W
Nicola, *Canada* **130 C4** 50 12N 120 40W
Nicolet, *Canada* **128 C5** 46 17N 72 35W
Nicolls Town, *Bahamas* **148 A4** 25 8N 78 0W
Nicopolis, *Greece* **45 E2** 39 2N 20 37 E
Nicosia, *Cyprus* **32 D12** 35 10N 33 25 E
Nicosia, *Italy* **41 E7** 37 45N 14 24 E
Nicótera, *Italy* **41 D8** 38 33N 15 56 E
Nicoya, *Costa Rica* **148 D2** 10 9N 85 27W
Nicoya, G. de, *Costa Rica* **148 E3** 10 0N 85 0W
Nicoya, Pen. de, *Costa Rica* **148 E2** 9 45N 85 40W
Nidau, *Switz.* **28 B4** 47 7N 7 15 E
Nidd →, *U.K.* **16 C6** 53 59N 1 23W
Nidda, *Germany* **26 E5** 50 23N 9 1 E
Nidda →, *Germany* **27 E4** 50 17N 8 48 E
Nidwalden □, *Switz.* **29 C6** 46 50N 8 25 E
Nidzica, *Poland* **47 B7** 53 25N 20 28 E
Niebüll, *Germany* **26 A4** 54 46N 8 48 E
Nied →, *Germany* **23 C13** 49 23N 6 40 E
Niederaula, *Germany* **26 E5** 50 47N 9 36 E
Niederbipp, *Switz.* **28 B5** 47 16N 7 42 E
Niederbronn-les-Bains, *France* **23 D14** 48 57N 7 39 E
Niedere Tauern, *Austria* **30 D7** 47 20N 14 0 E
Niederösterreich □, *Austria* **30 C8** 48 25N 15 40 E
Niedersachsen □, *Germany* **26 C5** 53 8N 9 0 E
Niefang, *Eq. Guin.* **102 B2** 1 50N 10 14 E
Niekerkshoop, *S. Africa* **104 D3** 29 19 S 22 51 E
Niel, *Belgium* **21 F4** 51 7N 4 20 E
Niellé, *Ivory C.* **100 C3** 10 5N 5 38W
Niem, *C.A.R.* **102 A3** 6 12N 15 14 E
Niemba, *Zaïre* **106 D2** 5 58 S 28 24 E
Niemcza, *Poland* **47 E3** 50 42N 16 47 E
Niemen = Neman →, *Lithuania* **13 J19** 55 25N 21 10 E
Niemodlin, *Poland* **47 E4** 50 38N 17 38 E
Niemur, *Australia* **116 C6** 35 17 S 144 9 E
Nienburg, *Germany* **26 C5** 52 39N 9 13 E
Niepołomice, *Poland* **31 A13** 50 3N 20 13 E
Niers →, *Germany* **26 D2** 51 43N 5 57 E
Niesen, *Switz.* **28 C5** 46 38N 7 39 E
Niesky, *Germany* **26 D10** 51 17N 14 49 E
Nieszawa, *Poland* **47 C5** 52 52N 18 50 E
Nieu Bethesda, *S. Africa* **104 E3** 31 51 S 24 34 E
Nieuw-Amsterdam, *Neths.* **20 C9** 52 43N 6 52 E
Nieuw Amsterdam, *Surinam* **153 B6** 5 53N 55 5W
Nieuw Beijerland, *Neths.* **20 E4** 51 49N 4 20 E
Nieuw-Dordrecht, *Neths.* **20 C9** 52 45N 6 59 E
Nieuw Loosdrecht, *Neths.* **20 D6** 52 12N 5 8 E
Nieuw Nickerie, *Surinam* **153 B6** 6 0N 56 59W
Nieuw-Schoonebeek, *Neths.* **20 C10** 52 39N 7 0 E
Nieuw-Vennep, *Neths.* **20 D5** 52 16N 4 38 E
Nieuw-Vossemeer, *Neths.* **21 E4** 51 34N 4 12 E
Nieuwe-Niedorp, *Neths.* **20 C5** 52 44N 4 54 E
Nieuwe-Pekela, *Neths.* **20 B9** 53 5N 6 58 E
Nieuwe-Schans, *Neths.* **20 B10** 53 11N 7 12 E
Nieuwendijk, *Neths.* **20 E5** 51 46N 4 55 E
Nieuwerkerken, *Belgium* **21 G6** 50 52N 5 12 E
Nieuwkoop, *Neths.* **20 D5** 52 9N 4 48 E
Nieuwleusen, *Neths.* **20 C8** 52 34N 6 17 E
Nieuwnamen, *Neths.* **21 F4** 51 18N 4 9 E
Nieuwolda, *Neths.* **20 B9** 53 15N 6 58 E
Nieuwoudtville, *S. Africa* **104 E2** 31 23 S 19 7 E
Nieuwpoort, *Belgium* **21 F1** 51 8N 2 45 E
Nieuwveen, *Neths.* **20 D5** 52 12N 4 46 E
Nieves, *Spain* **36 C2** 42 7N 8 26W
Nieves, Pico de las, *Canary Is.* **33 G4** 27 57N 15 35W
Nièvre □, *France* **23 E10** 47 10N 3 40 E
Nigata, *Japan* **62 C4** 34 13N 132 39 E
Niğde, *Turkey* **88 D6** 37 58N 34 40 E
Nigel, *S. Africa* **105 D4** 26 27 S 28 25 E
Niger □, *Nigeria* **101 C6** 10 0N 5 30 E
Niger ■, *W. Afr.* **97 E2** 17 30N 10 0 E
Niger →, *W. Afr.* **101 D6** 5 33N 6 33 E
Nigeria ■, *W. Afr.* **101 D6** 8 30N 8 0 E
Nightcaps, *N.Z.* **119 F3** 45 57 S 168 2 E
Nigríta, *Greece* **44 D5** 40 56N 23 29 E
Nihtaur, *India* **81 E8** 29 20N 78 23 E
Nii-Jima, *Japan* **63 C11** 34 20N 139 15 E
Niigata, *Japan* **60 F9** 37 58N 139 0 E
Niigata □, *Japan* **61 F9** 37 15N 138 45 E
Niihama, *Japan* **62 D5** 33 55N 133 16 E
Niihau, *U.S.A.* **132 H14** 21 54N 160 9W
Niimi, *Japan* **62 C5** 34 59N 133 28 E
Niitsu, *Japan* **60 F9** 37 48N 139 7 E
Níjar, *Spain* **35 J2** 36 53N 2 15W
Nijil, *Jordan* **91 E4** 30 32N 35 33 E
Nijkerk, *Neths.* **20 D7** 52 13N 5 30 E
Nijlen, *Belgium* **21 F5** 51 10N 4 40 E
Nijmegen, *Neths.* **20 E7** 51 50N 5 52 E
Nijverdal, *Neths.* **20 D8** 52 22N 6 28 E
Nīk Pey, *Iran* **89 D13** 36 50N 48 10 E
Nike, *Nigeria* **101 D6** 6 26N 7 29 E
Nikiniki, *Indonesia* **72 C2** 9 49 S 124 30 E
Nikítas, *Greece* **44 D5** 40 13N 23 34 E
Nikki, *Benin* **101 D5** 9 58N 3 12 E
Nikkō, *Japan* **63 A11** 36 45N 139 35 E
Nikolayev = Mykolayiv, *Ukraine* **51 J7** 46 58N 32 0 E
Nikolayevsk, *Russia* **52 E7** 50 0N 45 35 E
Nikolayevsk-na-Amur, *Russia* **57 D15** 53 8N 140 44 E
Nikolsk, *Russia* **52 D8** 53 49N 46 4 E
Nikolskoye, *Russia* **57 D17** 55 12N 166 0 E
Nikopol, *Bulgaria* **43 D9** 43 43N 24 54 E
Nikopol, *Ukraine* **51 J8** 47 35N 34 25 E
Niksar, *Turkey* **88 B7** 40 31N 37 2 E
Nīkshahr, *Iran* **85 E9** 26 15N 60 10 E
Nikšić, *Montenegro, Yug.* **42 E3** 42 50N 18 57 E
Nîl, Nahr en →, *Africa* **94 H7** 30 10N 31 6 E
Nîl el Abyad →, *Sudan* **95 D3** 15 38N 32 31 E
Nîl el Azraq →, *Sudan* **95 D3** 15 38N 32 31 E
Niland, *U.S.A.* **145 M11** 33 14N 115 31W
Nile = Nîl, Nahr en →, *Africa* **94 H7** 30 10N 31 6 E
Nile Delta, *Egypt* **94 H7** 31 40N 31 0 E
Niles, *U.S.A.* **136 E4** 41 11N 80 46W
Nilgiri Hills, *India* **83 J3** 11 30N 76 30 E
Nilo Peçanha, *Brazil* **155 D4** 13 37 S 39 6W
Nilpena, *Australia* **116 A3** 30 58 S 138 20 E
Nimach, *India* **80 G6** 24 30N 74 56 E
Nimbahera, *India* **80 G6** 24 37N 74 45 E
Nîmes, *France* **25 E8** 43 50N 4 23 E

Northcliffe 241

Nimfaíon, Ákra = Pínnes, Ákra, Greece 44 D6 40 5N 24 20 E
Nimmitabel, Australia 117 D8 36 29 S 149 15 E
Nimule, Sudan 95 G3 3 32N 32 3 E
Nin, Croatia 39 D12 44 16N 15 12 E
Nīnawá, Iraq 89 D10 36 25N 43 10 E
Ninda, Angola 103 E4 14 47 S 21 24 E
Nindigully, Australia 115 D4 28 21 S 148 50 E
Ninemile, U.S.A. 130 B2 56 0N 130 7W
Ninety Mile Beach, N.Z. 118 A1 34 48 S 173 0 E
Ninety Mile Beach, The, Australia 117 E7 38 15 S 147 24 E
Nineveh = Nīnawá, Iraq 89 D10 36 25N 43 10 E
Ning Xian, China 66 G4 35 30N 107 58 E
Ningaloo, Australia 112 D1 22 41 S 113 41 E
Ning'an, China 67 B15 44 22N 129 20 E
Ningbo, China 69 C13 29 51N 121 28 E
Ningcheng, China 67 D10 41 32N 119 53 E
Ningde, China 69 D12 26 38N 119 23 E
Ningdu, China 69 D10 26 25N 115 59 E
Ninggang, China 69 D9 26 42N 113 55 E
Ningguo, China 69 B12 30 35N 119 0 E
Ninghai, China 69 C13 29 15N 121 27 E
Ninghua, China 69 D11 26 14N 116 45 E
Ningjin, China 66 F8 37 35N 114 57 E
Ningjing Shan, China 68 B2 30 0N 98 20 E
Ninglang, China 68 D3 27 20N 100 55 E
Ningling, China 66 G8 34 25N 115 22 E
Ningming, China 68 F6 22 8N 107 4 E
Ningnan, China 68 D4 27 5N 102 36 E
Ningpo = Ningbo, China 69 C13 29 51N 121 28 E
Ningqiang, China 66 H4 32 47N 106 15 E
Ningshan, China 66 H5 33 21N 108 21 E
Ningsia Hui A.R. = Ningxia Huizu Zizhiqu □, China 66 E3 38 0N 106 0 E
Ningwu, China 66 E7 39 0N 112 18 E
Ningxia Huizu Zizhiqu □, China 66 E3 38 0N 106 0 E
Ningxiang, China 69 C9 28 15N 112 30 E
Ningyang, China 66 G9 35 47N 116 45 E
Ningyuan, China 69 E8 25 37N 111 57 E
Ninh Binh, Vietnam 76 B5 20 15N 105 55 E
Ninh Giang, Vietnam 76 B6 20 44N 106 24 E
Ninh Hoa, Vietnam 76 F7 12 30N 109 7 E
Ninh Ma, Vietnam 76 F7 12 48N 109 21 E
Ninove, Belgium 21 G4 50 51N 4 2 E
Nioaque, Brazil 159 A4 21 5 S 55 50W
Niobrara, U.S.A. 138 D6 42 45N 98 2W
Niobrara →, U.S.A. 138 D6 42 46N 98 3W
Nioki, Zaïre 102 C3 2 47 S 17 40 E
Niono, Mali 100 C3 14 15N 6 0W
Nioro du Rip, Senegal 100 C1 13 40N 15 50W
Nioro du Sahel, Mali 100 B3 15 15N 9 30W
Niort, France 24 B3 46 19N 0 29W
Nipa, Papua N. G. 120 D2 6 9 S 143 29 E
Nipani, India 83 F2 16 20N 74 25 E
Nipawin, Canada 131 C8 53 20N 104 0W
Nipawin Prov. Park, Canada 131 C8 54 0N 104 37W
Nipigon, Canada 128 C2 49 0N 88 17W
Nipigon, L., Canada 128 C2 49 50N 88 30W
Nipin →, Canada 131 B7 55 46N 108 35W
Nipishish L., Canada 129 B7 54 12N 60 45W
Nipissing L., Canada 128 C4 46 20N 80 0W
Nipomo, U.S.A. 145 K6 35 3N 120 29W
Nipton, U.S.A. 145 K11 35 28N 115 16W
Niquelândia, Brazil 155 D2 14 33 S 48 23W
Nīr, Iran 89 C12 38 2N 47 59 E
Nira →, India 82 F2 17 58N 75 8 E
Nirasaki, Japan 63 B10 35 42N 138 27 E
Nirmal, India 82 E4 19 3N 78 20 E
Nirmali, India 81 F12 26 20N 86 35 E
Niš, Serbia, Yug. 42 D6 43 19N 21 58 E
Nisa, Portugal 37 F3 39 30N 7 41W
Nişāb, Si. Arabia 84 D5 29 11N 44 43 E
Nişāb, Yemen 86 D4 14 25N 46 29 E
Nišava →, Serbia, Yug. 42 D6 43 20N 21 46 E
Niscemi, Italy 41 E7 37 9N 14 23 E
Nishi-Sonogi-Hantō, Japan 62 E1 32 55N 129 45 E
Nishinomiya, Japan 63 C7 34 45N 135 20 E
Nishin'omote, Japan 61 J5 30 43N 130 59 E
Nishio, Japan 63 C9 34 52N 137 3 E
Nishiwaki, Japan 62 C6 34 59N 134 58 E
Nísíros, Greece 45 H9 36 35N 27 12 E
Niskibi →, Canada 128 A2 56 29N 88 9W
Nisko, Poland 47 E9 50 35N 22 7 E
Nispen, Neths. 21 F4 51 29N 4 28 E
Nisporeni, Moldova 46 B9 47 4N 28 10 E
Nisqually →, U.S.A. 144 C4 47 6N 122 42W
Nissáki, Greece 32 A3 39 43N 19 52 E
Nissan →, Sweden 15 H6 56 40N 12 51 E
Nissedal, Norway 14 E2 59 10N 8 30 E
Nisser, Norway 14 E2 59 7N 8 28 E
Nissum Bredning, Denmark 13 H13 56 40N 8 20 E
Nissum Fjord, Denmark 15 H2 56 20N 8 11 E
Nistelrode, Neths. 21 E7 51 42N 5 34 E
Nistru = Dnister →, Europe 51 J6 46 18N 30 17 E
Nisutlin →, Canada 130 A2 60 14N 132 34W
Nitchequon, Canada 129 B5 53 10N 70 58W
Niterói, Brazil 155 F3 22 52 S 43 0W
Nith →, U.K. 18 F5 55 14N 3 33W
Nitra, Slovak Rep. 31 C11 48 19N 18 4 E
Nitra →, Slovak Rep. 31 D11 47 46N 18 10 E
Nittedal, Norway 14 D4 60 1N 10 57 E
Nittenau, Germany 27 F8 49 12N 12 16 E
Niuafo'ou, Tonga 111 D15 15 30 S 175 58W
Niue, Cook Is. 123 J11 19 2 S 169 54W
Niulan Jiang →, China 68 D4 27 30N 103 5 E
Niut, Indonesia 75 B4 0 55N 110 6 E
Niutou Shan, China 69 C13 29 5N 121 59 E
Niuzhuang, China 67 D12 40 58N 122 28 E
Nivala, Finland 12 E21 63 56N 24 57 E
Nivelles, Belgium 21 G4 50 35N 4 20 E
Nivernais, France 23 E10 47 15N 3 30 E
Nixon, U.S.A. 139 L6 29 16N 97 46W
Nizam Sagar, India 82 E3 18 10N 77 58 E
Nizamabad, India 82 E4 18 45N 78 7 E
Nizamghat, India 78 A5 28 20N 95 45 E
Nizhne Kolymsk, Russia 57 C17 68 34N 160 55 E
Nizhne-Vartovsk, Russia 56 C8 60 56N 76 38 E
Nizhneangarsk, Russia 57 D11 55 47N 109 30 E
Nizhnegorskiy = Nyzhnohirskyy, Ukraine 51 K8 45 27N 34 38 E
Nizhnekamsk, Russia 52 C10 55 38N 51 49 E
Nizhnekamskoye Vdkhr., Russia 54 D4 55 56N 52 56 E
Nizhneudinsk, Russia 57 D10 54 54N 99 3 E
Nizhneyansk, Russia 57 B14 71 26N 136 4 E
Nizhniy Chir, Russia 53 F6 48 22N 43 5 E
Nizhniy Lomov, Russia 52 D6 53 34N 43 38 E
Nizhniy Novgorod, Russia 52 B7 56 20N 44 0 E
Nizhniy Pyandzh, Tajikistan 55 E4 37 12N 68 35 E
Nizhniy Tagil, Russia 54 C7 57 55N 59 57 E
Nizhniye Sergi, Russia 54 C7 56 40N 59 18 E
Nizhnyaya Salda, Russia 54 B8 58 8N 60 42 E
Nizhyn, Ukraine 51 G6 51 5N 31 55 E
Nizip, Turkey 88 D7 37 5N 37 50 E
Nízké Tatry, Slovak Rep. 31 C12 48 55N 19 30 E
Nizza Monferrato, Italy 38 D5 44 46N 8 21 E
Njakwa, Malawi 107 E3 11 1 S 33 56 E
Njanji, Zambia 107 E3 14 25 S 31 46 E
Njinjo, Tanzania 107 D4 8 48 S 38 54 E
Njombe, Tanzania 107 D3 9 20 S 34 50 E
Njombe □, Tanzania 107 D3 9 20 S 34 49 E
Njombe →, Tanzania 106 D4 6 56 S 35 6 E
Nkambe, Cameroon 101 D7 6 35N 10 40 E
Nkana, Zambia 107 E2 12 50 S 28 8 E
Nkawkaw, Ghana 101 D4 6 36N 0 49W
Nkayi, Zimbabwe 107 F2 19 41 S 29 20 E
Nkhota Kota, Malawi 107 E3 12 56 S 34 15 E
Nkolabona, Gabon 102 B2 1 14N 11 43 E
Nkone, Zaïre 102 C4 1 2 S 22 20 E
Nkongsamba, Cameroon 101 E6 4 55N 9 55 E
Nkunga, Zaïre 103 C3 4 41 S 18 34 E
Nkurenkuru, Namibia 104 B2 17 42 S 18 32 E
Nkwanta, Ghana 100 D4 6 10N 2 10W
Noakhali = Maijdi, Bangla. 78 D3 22 48N 91 10 E
Noatak, U.S.A. 126 B3 67 34N 162 58W
Nobel, Canada 136 A4 45 25N 80 6W
Nobeoka, Japan 62 E3 32 36N 131 41 E
Nōbi-Heiya, Japan 63 B8 35 15N 136 45 E
Noble, U.S.A. 141 F8 38 42N 88 14W
Noblejas, Spain 34 F1 39 58N 3 26W
Noblesville, U.S.A. 141 D11 40 3N 86 1W
Noce →, Italy 38 B8 46 9N 11 4 E
Nocera Inferiore, Italy 41 B7 40 44N 14 38 E
Nocera Terinese, Italy 41 C9 39 2N 16 9 E
Nocera Umbra, Italy 39 E9 43 5N 12 47 E
Noci, Italy 41 B10 40 48N 17 7 E
Nockatunga, Australia 115 D3 27 42 S 142 42 E
Nocona, U.S.A. 139 J6 33 47N 97 44W
Nocrich, Romania 46 D5 45 55N 24 26 E
Noda, Japan 63 B11 35 56N 139 52 E
Noel, U.S.A. 139 G7 36 33N 94 29W
Nogal Valley, Somali Rep. 108 C3 8 35N 48 35 E
Nogales, Mexico 146 A2 31 20N 110 56W
Nogales, U.S.A. 143 L8 31 20N 110 56W
Nogat →, Poland 47 A6 54 17N 19 17 E
Nōgata, Japan 62 D2 33 48N 130 44 E
Nogent-en-Bassigny, France 23 D12 48 1N 5 20 E
Nogent-le-Rotrou, France 22 D7 48 20N 0 50 E
Nogent-sur-Seine, France 23 D10 48 30N 3 30 E
Noggerup, Australia 113 F2 33 32 S 116 5 E
Noginsk, Moskva, Russia 52 C4 55 50N 38 25 E
Noginsk, Sib., Russia 57 C10 64 30N 90 50 E
Nogoa →, Australia 114 C4 23 40 S 147 55 E
Nogoyá, Argentina 158 C4 32 24 S 59 48W
Nógrád □, Hungary 31 D12 48 0N 19 30 E
Noguera de Ramuin, Spain 36 C3 42 21N 7 43W
Noguera Pallaresa →, Spain 34 D5 41 55N 0 55 E
Noguera Ribagorzana →, Spain 34 D5 41 40N 0 43 E
Nohar, India 80 E6 29 11N 74 49 E
Noing, Phil. 71 J5 5 40N 125 28 E
Noire, Montagne, France 24 E6 43 28N 2 18 E
Noire, Mt., France 22 D3 48 11N 3 40W
Noirétable, France 24 C7 45 48N 3 46 E
Noirmoutier, I. de, France 22 F4 46 58N 2 10W
Noirmoutier-en-l'Ile, France 22 F4 47 0N 2 14W
Nojane, Botswana 104 C3 23 15 S 20 14 E
Nojima-Zaki, Japan 63 C11 34 54N 139 53 E
Nok Kundi, Pakistan 79 C1 28 50N 62 45 E
Nokaneng, Botswana 104 B3 19 40 S 22 17 E
Nokhtuysk, Russia 57 C12 60 0N 117 45 E
Nokia, Finland 13 F20 61 30N 23 30 E
Nokomis, Canada 131 C8 51 35N 105 0W
Nokomis, U.S.A. 140 E7 39 18N 89 18W
Nokomis L., Canada 131 B8 57 0N 103 0W
Nokou, Chad 97 F2 14 35N 14 47 E
Nol, Sweden 15 G6 57 56N 12 5 E
Nola, C.A.R. 102 B3 3 35N 16 4 E
Nola, Italy 41 B7 40 55N 14 33 E
Nolay, France 23 F11 46 58N 4 35 E
Noli, C. di, Italy 38 D5 44 12N 8 25 E
Nolinsk, Russia 52 B9 57 28N 49 57 E
Noma Omuramba →, Namibia 104 B3 18 52 S 20 53 E
Noma-Saki, Japan 62 F2 31 25N 130 7 E
Nomad, Papua N. G. 120 D2 6 19 S 142 13 E
Noman L., Canada 131 A7 62 15N 108 55W
Nombre de Dios, Panama 148 E4 9 34N 79 28W
Nome, U.S.A. 126 B3 64 30N 165 25W
Nomo-Zaki, Japan 62 E1 32 35N 129 44 E
Nomuka, Tonga 121 Q13 20 17 S 174 48W
Nomuka Group, Tonga 121 Q13 20 20 S 174 48W
Nonacho L., Canada 131 A7 61 42N 109 40W
Nonancourt, France 22 D8 48 47N 1 11 E
Nonant-le-Pin, France 22 D7 48 42N 0 12 E
Nonda, Australia 114 C3 20 40 S 142 28 E
Nong Chang, Thailand 76 E2 15 23N 99 51 E
Nong Het, Laos 76 C4 19 29N 103 59 E
Nong Khai, Thailand 76 D4 17 50N 102 46 E
Nong'an, China 67 B13 44 25N 125 5 E
Nongoma, S. Africa 105 D5 27 58 S 31 35 E
Nonoava, Mexico 146 B3 27 28N 106 44W
Nonoc I., Phil. 71 G5 9 51N 125 37 E
Nonthaburi, Thailand 76 F3 13 51N 100 34 E
Nontron, France 24 C4 45 31N 0 40 E
Nonza, France 25 F13 42 47N 9 21 E
Noonamah, Australia 112 B5 12 40 S 131 4 E
Noonan, U.S.A. 138 A3 48 54N 103 1W
Noondoo, Australia 115 D4 28 35 S 148 30 E
Noonkanbah, Australia 112 C3 18 30 S 124 50 E
Noord-Bergum, Neths. 20 B8 53 14N 6 1 E
Noord Brabant □, Neths. 21 E6 51 40N 5 0 E
Noord Holland □, Neths. 20 D5 52 30N 4 45 E
Noordbeveland, Neths. 21 E3 51 35N 3 50 E
Noordeloos, Neths. 20 E5 51 55N 4 56 E
Noordhollandsch Kanaal, Neths. 20 C5 52 55N 4 48 E
Noordhorn, Neths. 20 B8 53 16N 6 24 E
Noordoostpolder, Neths. 20 C7 52 45N 5 45 E
Noordwijk aan Zee, Neths. 20 D4 52 14N 4 26 E
Noordwijk-Binnen, Neths. 20 D4 52 14N 4 27 E
Noordwijkerhout, Neths. 20 D5 52 16N 4 30 E
Noordzee Kanaal, Neths. 20 D5 52 28N 4 35 E
Noorwolde, Neths. 20 C8 52 54N 6 8 E
Nootka, Canada 130 D3 49 38N 126 38W
Nootka I., Canada 130 D3 49 32N 126 42W
Nóqui, Angola 103 D2 5 55 S 13 30 E
Nora, Eritrea 95 D5 16 6N 40 4 E
Nora Springs, U.S.A. 140 A4 43 9N 93 1W
Noranda, Canada 128 C4 48 20N 79 0W
Norborne, U.S.A. 140 E3 39 18N 93 40W
Nórcia, Italy 39 F10 42 48N 13 5 E
Norco, U.S.A. 145 M9 33 56N 117 33W
Nord □, France 23 B10 50 15N 3 30 E
Nord-Ostsee-Kanal →, Germany 26 A5 54 12N 9 32 E
Nordagutu, Norway 14 E3 59 25N 9 20 E
Nordaustlandet, Svalbard 6 B9 79 14N 23 0 E
Nordborg, Denmark 15 J3 55 5N 9 50 E
Nordby, Århus, Denmark 15 J4 55 58N 10 32 E
Nordby, Ribe, Denmark 15 J2 55 27N 8 24 E
Norddeich, Germany 26 B3 53 36N 7 9 E
Nordegg, Canada 130 C5 52 29N 116 5W
Norden, Germany 26 B3 53 35N 7 12 E
Nordenham, Germany 26 B4 53 30N 8 30 E
Norderhov, Norway 14 D4 60 7N 10 17 E
Norderney, Germany 26 B3 53 42N 7 9 E
Nordfjord, Norway 13 F11 61 55N 5 30 E
Nordfriesische Inseln, Germany 26 A4 54 40N 8 20 E
Nordhausen, Germany 26 D6 51 30N 10 47 E
Nordhorn, Germany 26 C3 52 26N 7 4 E
Norðoyar, Færoe Is. 12 E9 62 17N 6 35W
Nordjyllands Amtskommune □, Denmark 15 H4 57 0N 10 0 E
Nordkapp, Norway 12 A21 71 10N 25 50 E
Nordkapp, Svalbard 6 A9 80 31N 20 0 E
Nordkinn = Kinnarodden, Norway 10 A11 71 8N 27 40 E
Nordkinn-halvøya, Norway 12 A22 70 55N 27 40 E
Nördlingen, Germany 27 G6 48 48N 10 30 E
Nordrhein-Westfalen □, Germany 26 D3 51 45N 7 30 E
Nordstrand, Germany 26 A4 54 30N 8 52 E
Nordvik, Russia 57 B12 74 2N 111 32 E
Nore, Norway 14 D3 60 10N 9 0 E
Norefjell, Norway 14 D3 60 16N 9 29 E
Norembega, Canada 128 C3 48 59N 80 43W
Noresund, Norway 14 D3 60 11N 9 37 E
Norfolk, Nebr., U.S.A. 138 D6 42 2N 97 25W
Norfolk, Va., U.S.A. 134 G7 36 51N 76 17W
Norfolk □, U.K. 16 E9 52 39N 0 54 E
Norfolk Broads, U.K. 16 E9 52 30N 1 15 E
Norfolk I., Pac. Oc. 122 K8 28 58 S 168 3 E
Norfork Res., U.S.A. 139 G8 36 13N 92 15W
Norg, Neths. 20 B8 53 4N 6 28 E
Norilsk, Russia 57 C9 69 20N 88 6 E
Norley, Australia 115 D3 27 45 S 143 48 E
Norma, Mt., Australia 114 C3 20 55 S 140 42 E
Normal, U.S.A. 140 D8 40 31N 88 59W
Norman, U.S.A. 139 H6 35 13N 97 26W
Norman →, Australia 114 B3 19 18 S 141 51 E
Norman Wells, Canada 126 B7 65 17N 126 51W
Normanby, N.Z. 118 F3 39 32 S 174 18 E
Normanby →, Australia 114 A3 14 23 S 144 10 E
Normanby I., Papua N. G. 120 F6 10 5 S 151 5 E
Normandie, France 22 D7 48 45N 0 10 E
Normandie, Collines de, France 22 D6 48 45N 0 45W
Normandin, Canada 128 C5 48 49N 72 31W
Normandy = Normandie, France 22 D7 48 45N 0 10 E
Normanhurst, Mt., Australia 113 E3 25 4 S 122 30 E
Normanton, Australia 114 B3 17 40 S 141 10 E
Normanville, Australia 116 C3 35 27 S 138 18 E
Norquay, Canada 131 C8 51 53N 102 5W
Norquinco, Argentina 160 B2 41 51 S 70 55W
Norrbotten □, Sweden 12 C19 66 30N 22 30 E
Nørre Åby, Denmark 15 J3 55 27N 9 52 E
Nørre Nebel, Denmark 15 J2 55 47N 8 17 E
Nørresundby, Denmark 15 G3 57 5N 9 52 E
Norris, U.S.A. 142 D8 45 34N 111 41W
Norris City, U.S.A. 141 G8 37 59N 88 20W
Norristown, U.S.A. 137 F9 40 7N 75 21W
Norrköping, Sweden 15 F10 58 37N 16 11 E
Norrland, Sweden 13 E16 62 15N 15 45 E
Norrtälje, Sweden 14 E12 59 46N 18 42 E
Norseman, Australia 113 F3 32 8 S 121 43 E
Norsewood, N.Z. 118 G5 40 3 S 176 13 E
Norsholm, Sweden 15 F9 58 31N 15 59 E
Norsk, Russia 57 D14 52 30N 130 5 E
Norsup, Vanuatu 121 F5 16 3 S 167 24 E
Norte, Pta., Argentina 160 B4 42 5 S 63 46W
Norte, Pta. del, Canary Is. 33 G2 27 51N 17 57W
Norte de Santander □, Colombia 152 B3 8 0N 73 0W
Nortelândia, Brazil 157 C6 14 25 S 56 48W
North Adams, U.S.A. 137 D11 42 42N 73 7W
North Atlantic Ocean, Atl. Oc. 8 F4 30 0N 50 0W
North Baltimore, U.S.A. 141 C13 41 11N 83 41W
North Battleford, Canada 131 C7 52 50N 108 17W
North Bay, Canada 128 C4 46 20N 79 30W
North Belcher Is., Canada 128 A4 56 50N 79 50W
North Bend, Canada 130 D4 49 50N 121 27W
North Bend, Oreg., U.S.A. 142 E1 43 24N 124 14W
North Bend, Pa., U.S.A. 136 E7 41 20N 77 42W
North Bend, Wash., U.S.A. 144 C5 47 30N 121 47W
North Berwick, U.K. 18 E6 56 4N 2 42W
North Berwick, U.S.A. 137 C14 43 18N 70 44W
North Buganda □, Uganda 106 B3 1 0N 32 0 E
North C., Canada 129 C7 47 2N 60 20W
North C., N.Z. 118 A2 34 23 S 173 4 E
North C., Papua N. G. 120 B6 2 32 S 150 50 E
North Canadian →, U.S.A. 139 H7 35 16N 95 31W
North Cape = Nordkapp, Norway 12 A21 71 10N 25 50 E
North Cape = Nordkapp, Svalbard 6 A9 80 31N 20 0 E
North Caribou L., Canada 128 B1 52 50N 90 40W
North Carolina □, U.S.A. 135 H5 35 30N 80 0W
North Channel, Canada 128 C3 46 0N 83 0W
North Channel, U.K. 18 G3 55 13N 5 52W
North Chicago, U.S.A. 141 B9 42 19N 87 51W
North College Hill, U.S.A. 141 E12 39 13N 84 33W
North Cotabato □, Phil. 71 H5 7 10N 125 0 E
North Dakota □, U.S.A. 138 B5 47 30N 100 15W
North Dandalup, Australia 113 F2 32 30 S 115 57 E
North Down □, U.K. 19 B6 54 40N 5 45W
North Downs, U.K. 17 F8 51 19N 0 21 E
North East, U.S.A. 136 D5 42 13N 79 50W
North East Frontier Agency = Arunachal Pradesh □, India 78 A5 28 0N 95 0 E
North East Providence Chan., W. Indies 148 A4 26 0N 76 0W
North Eastern □, Kenya 106 B5 1 30N 40 0 E
North English, U.S.A. 140 C4 41 31N 92 5W
North Esk →, U.K. 18 E6 56 46N 2 24W
North European Plain, Europe 10 D11 55 0N 25 0 E
North Fabius →, U.S.A. 140 E5 39 54N 91 30W
North Foreland, U.K. 17 F9 51 22N 1 28 E
North Fork, U.S.A. 144 H7 37 14N 119 21W
North Fork, Salt →, U.S.A. 140 E5 39 26N 91 53W
North Fork American →, U.S.A. 144 G5 38 57N 120 59W
North Fork Feather →, U.S.A. 144 F5 38 33N 121 30W
North Frisian Is. = Nordfriesische Inseln, Germany 26 A4 54 40N 8 20 E
North Henik L., Canada 131 A9 61 45N 97 40W
North Highlands, U.S.A. 144 G5 38 40N 121 23W
North Horr, Kenya 106 B4 3 20N 37 8 E
North I., Kenya 106 B4 4 5N 36 5 E
North I., N.Z. 118 E4 38 0 S 175 0 E
North Judson, U.S.A. 141 C10 41 13N 86 46W
North Kingsville, U.S.A. 136 E4 41 54N 80 42W
North Knife →, Canada 131 B10 58 53N 94 45W
North Koel →, India 81 G10 24 45N 83 50 E
North Korea ■, Asia 67 E14 40 0N 127 0 E
North Lakhimpur, India 78 B5 27 14N 94 7 E
North Las Vegas, U.S.A. 145 J11 36 12N 115 7W
North Liberty, U.S.A. 141 C10 41 32N 86 26W
North Loup →, U.S.A. 138 E5 41 17N 98 24W
North Magnetic Pole, Canada 6 B2 77 58N 102 8W
North Manchester, U.S.A. 141 D11 41 0N 85 46W
North Minch, U.K. 18 C3 58 5N 5 55W
North Nahanni →, Canada 130 A4 62 15N 123 20W
North Olmsted, U.S.A. 136 E3 41 25N 81 56W
North Ossetia □, Russia 53 J7 43 30N 44 30 E
North Pagai, I. = Pagai Utara, Indonesia 74 C2 2 35 S 100 0 E
North Palisade, U.S.A. 144 H8 37 6N 118 31W
North Platte, U.S.A. 138 E4 41 8N 100 46W
North Platte →, U.S.A. 138 E4 41 7N 100 42W
North Pole, Arctic 6 A 90 0N 0 0 E
North Portal, Canada 131 D8 49 0N 102 33W
North Powder, U.S.A. 142 D5 45 2N 117 55W
North Pt., Canada 129 C7 47 5N 64 0W
North Pt., Vanuatu 121 D6 14 56 S 168 6 E
North Rhine Westphalia = Nordrhein-Westfalen □, Germany 26 D3 51 45N 7 30 E
North Ronaldsay, U.K. 18 B6 59 22N 2 26W
North Saskatchewan →, Canada 131 C7 53 15N 105 5W
North Sea, Europe 10 D6 56 0N 4 0 E
North Sporades = Voríai Sporádhes, Greece 45 E5 39 15N 23 30 E
North Sydney, Canada 129 C7 46 12N 60 15W
North Taranaki Bight, N.Z. 118 E3 38 50 S 174 15 E
North Thompson →, Canada 130 C4 50 40N 120 20W
North Tonawanda, U.S.A. 136 C6 43 2N 78 53W
North Troy, U.S.A. 137 B12 45 0N 72 24W
North Truchas Pk., U.S.A. 143 J11 36 0N 105 30W
North Twin I., Canada 128 B3 53 20N 80 0W
North Tyne →, U.K. 16 C5 55 0N 2 8W
North Uist, U.K. 18 D1 57 40N 7 15W
North Vancouver, Canada 130 D4 49 25N 123 3W
North Vernon, U.S.A. 141 E11 39 0N 85 38W
North Wabasca L., Canada 130 B6 56 0N 113 55W
North Walsham, U.K. 16 E9 52 50N 1 22 E
North Webster, U.S.A. 141 C11 41 25N 85 48W
North-West □, S. Africa 104 D4 27 0 S 25 0 E
North West C., Australia 112 D1 21 45 S 114 9 E
North West Christmas I. Ridge, Pac. Oc. 123 G11 6 30N 165 0W
North West Frontier □, Pakistan 79 B3 34 0N 71 0 E
North West Highlands, U.K. 18 D3 57 33N 4 58W
North West Providence Channel, W. Indies 148 A4 26 0N 78 0W
North West River, Canada 129 B7 53 30N 60 10W
North West Territories □, Canada 126 B9 67 0N 110 0W
North Western □, Zambia 107 E2 13 30 S 25 30 E
North York Moors, U.K. 16 C7 54 23N 0 53W
North Yorkshire □, U.K. 16 C6 54 15N 1 25W
Northallerton, U.K. 16 C6 54 20N 1 26W
Northam, S. Africa 104 C4 24 56 S 27 18 E
Northampton, Australia 113 E1 28 27 S 114 33 E
Northampton, U.K. 17 E7 52 15N 0 53W
Northampton, Mass., U.S.A. 137 D12 42 19N 72 38W
Northampton, Pa., U.S.A. 137 F9 40 41N 75 30W
Northampton Downs, Australia 114 C4 24 35 S 145 48 E
Northamptonshire □, U.K. 17 E7 52 16N 0 55W
Northbridge, U.S.A. 137 D13 42 9N 71 39W
Northcliffe, Australia 113 F2 34 39 S 116 7 E

O

Pacific 245

Oryakhovo, *Bulgaria* **43 D8** 43 40N 23 57 E
Orzinuovi, *Italy* **38 C6** 45 24N 9 55 E
Orzyc →, *Poland* **47 C8** 52 46N 21 14 E
Orzysz, *Poland* **47 B8** 53 50N 21 58 E
Osa, *Russia* **54 C5** 57 17N 55 26 E
Osa →, *Poland* **47 B5** 53 33N 18 46 E
Osa, Pen. de, *Costa Rica* **148 E3** 8 0N 84 0W
Osage, *Iowa, U.S.A.* **138 D8** 43 17N 92 49W
Osage, *Wyo., U.S.A.* **138 D2** 43 59N 104 25W
Osage →, *U.S.A.* **140 F5** 38 35N 91 57W
Osage City, *U.S.A.* **138 F7** 38 38N 95 50W
Ōsaka, *Japan* **63 C7** 34 40N 135 30 E
Ōsaka □, *Japan* **63 C7** 34 30N 135 30 E
Ōsaka-Wan, *Japan* **63 C7** 34 30N 135 18 E
Osan, *S. Korea* **67 F14** 37 11N 127 4 E
Osawatomie, *U.S.A.* **138 F7** 38 31N 94 57W
Osborne, *U.S.A.* **138 F5** 39 26N 98 42W
Osceola, *Ark., U.S.A.* **139 H10** 35 42N 89 58W
Osceola, *Iowa, U.S.A.* **140 C3** 41 2N 93 46W
Osceola, *Mo., U.S.A.* **140 F3** 38 3N 93 42W
Oschatz, *Germany* **26 D9** 51 17N 13 6 E
Oschersleben, *Germany* **26 C7** 52 2N 11 14 E
Óschiri, *Italy* **40 B2** 40 43N 9 6 E
Oscoda, *U.S.A.* **136 B1** 44 26N 83 20W
Osečina, *Serbia, Yug.* **42 C4** 44 23N 19 34 E
Ösel = Saaremaa, *Estonia* **13 G20** 58 30N 22 30 E
Osery, *Russia* **52 C4** 54 52N 38 28 E
Osgood, *U.S.A.* **141 E11** 39 8N 85 18W
Osh, *Kyrgyzstan* **55 C6** 40 37N 72 49 E
Oshawa, *Canada* **128 D4** 43 50N 78 50W
Oshima, *Japan* **62 D4** 33 55N 132 14 E
Oshkosh, *Nebr., U.S.A.* **138 E3** 41 24N 102 21W
Oshkosh, *Wis., U.S.A.* **138 C10** 44 1N 88 33W
Oshmyany = Ashmyany, *Belarus* **13 J21** 54 26N 25 52 E
Oshnovīyeh, *Iran* **84 B5** 37 2N 45 6 E
Oshogbo, *Nigeria* **101 D5** 7 48N 4 37 E
Oshtorīnān, *Iran* **89 E13** 34 1N 48 38 E
Oshwe, *Zaïre* **102 C3** 3 25 S 19 28 E
Osica de Jos, *Romania* **46 E5** 44 14N 24 20 E
Osieczna, *Poland* **47 D3** 51 55N 16 40 E
Osijek, *Croatia* **42 B3** 45 34N 18 41 E
Osilo, *Italy* **40 B1** 40 45N 8 40 E
Osimo, *Italy* **39 E10** 43 28N 13 30 E
Osintorf, *Belarus* **50 E6** 54 40N 30 39 E
Osipenko = Berdyansk, *Ukraine* **51 J9** 46 45N 36 50 E
Osipovichi = Asipovichy, *Belarus* **50 F5** 53 19N 28 33 E
Osizweni, *S. Africa* **105 D5** 27 49 S 30 7 E
Oskaloosa, *U.S.A.* **140 C4** 41 18N 92 39W
Oskarshamn, *Sweden* **13 H17** 57 15N 16 27 E
Oskélanéo, *Canada* **128 C4** 48 5N 75 15W
Öskemen, *Kazakhstan* **56 E9** 50 0N 82 36 E
Oskol →, *Ukraine* **51 H9** 49 6N 37 25 E
Oslo, *Norway* **14 E4** 59 55N 10 45 E
Oslob, *Phil.* **71 G4** 9 31N 123 26 E
Oslofjorden, *Norway* **14 E4** 59 20N 10 35 E
Osmanabad, *India* **82 E3** 18 5N 76 10 E
Osmancık, *Turkey* **88 B6** 40 58N 34 47 E
Osmaniye, *Turkey* **88 D7** 37 5N 36 10 E
Ösmo, *Sweden* **14 F11** 58 58N 17 55 E
Osnabrück, *Germany* **26 C4** 52 17N 8 3 E
Ośno Lubuskie, *Poland* **47 C1** 52 28N 14 51 E
Osobláha, *Czech.* **31 A10** 50 17N 17 44 E
Osogovska Planina, *Macedonia* **42 E7** 42 10N 22 30 E
Osor, *Italy* **39 D11** 44 42N 14 24 E
Osorio, *Brazil* **159 B5** 29 53 S 50 17W
Osorno, *Chile* **160 B2** 40 25 S 73 0W
Osorno, *Spain* **36 C6** 42 24N 4 22W
Osorno □, *Chile* **160 B2** 40 34 S 73 9W
Osorno, Vol., *Chile* **160 B2** 41 0 S 72 30W
Osoyoos, *Canada* **130 D5** 49 0N 119 30W
Osøyri, *Norway* **13 F11** 60 9N 5 30 E
Ospika →, *Canada* **130 B4** 56 20N 124 0W
Osprey Reef, *Australia* **114 A4** 13 52 S 146 36 E
Oss, *Neths.* **20 E7** 51 46N 5 32 E
Ossa, Mt., *Australia* **114 G4** 41 52 S 146 3 E
Óssa, Oros, *Greece* **44 E4** 39 47N 22 42 E
Ossa de Montiel, *Spain* **35 G2** 38 58N 2 45W
Ossabaw I., *U.S.A.* **135 K5** 31 50N 81 5W
Osse →, *France* **24 D4** 44 7N 0 17 E
Ossendrecht, *Neths.* **21 F4** 51 24N 4 20 E
Ossining, *U.S.A.* **137 E11** 41 10N 73 55W
Ossipee, *U.S.A.* **137 C13** 43 41N 71 7W
Ossokmanuan L., *Canada* **129 B7** 53 25N 65 0W
Ossora, *Russia* **57 D17** 59 20N 163 13 E
Ostashkov, *Russia* **52 B1** 57 4N 33 2 E
Oste →, *Germany* **26 B5** 53 49N 9 2 E
Ostend = Oostende, *Belgium* **21 F1** 51 15N 2 54 E
Oster, *Ukraine* **51 G6** 50 57N 30 53 E
Osterburg, *Germany* **26 C7** 52 47N 11 45 E
Osterburken, *Germany* **27 F5** 49 25N 9 26 E
Österdalälven, *Sweden* **13 F16** 61 30N 13 45 E
Østerdalen, *Norway* **13 F14** 61 40N 10 50 E
Östergötlands län □, *Sweden* **15 F9** 58 35N 15 45 E
Osterholz-Scharmbeck, *Germany* **26 B4** 53 13N 8 47 E
Østerild, *Denmark* **15 G2** 57 2N 8 51 E
Ostermundigen, *Switz.* **28 C4** 46 58N 7 27 E
Östersund, *Sweden* **14 A8** 63 10N 14 38 E
Østfold fylke □, *Norway* **14 E5** 59 25N 11 25 E
Ostfriesische Inseln, *Germany* **26 B3** 53 42N 7 0 E
Ostfriesland, *Germany* **26 B3** 53 20N 7 30 E
Óstia, Lido di, *Italy* **40 A5** 41 43N 12 17 E
Ostíglia, *Italy* **39 C8** 45 4N 11 8 E
Ostra, *Italy* **39 E10** 43 37N 13 9 E
Ostrava, *Czech.* **31 B11** 49 51N 18 18 E
Ostróda, *Poland* **47 B6** 53 42N 19 58 E
Ostrogozhsk, *Russia* **52 E4** 50 55N 39 7 E
Ostrogróg Szamotuły, *Poland* **47 C3** 52 37N 16 33 E
Ostroh, *Ukraine* **51 G4** 50 20N 26 30 E
Ostrołęka, *Poland* **47 B8** 53 4N 21 32 E
Ostrołeka □, *Poland* **47 C8** 53 0N 21 30 E
Ostrov, *Bulgaria* **43 D9** 43 40N 24 9 E
Ostrov, *Romania* **46 E8** 44 6N 27 24 E
Ostrov, *Russia* **50 D5** 57 25N 28 20 E
Ostrów Lubelski, *Poland* **47 D9** 51 29N 22 51 E
Ostrów Mazowiecka, *Poland* **47 C8** 52 50N 21 51 E
Ostrów Wielkopolski, *Poland* **47 D4** 51 36N 17 44 E
Ostrowiec-Świętokrzyski, *Poland* **47 E8** 50 55N 21 22 E
Ostrozac, *Bos.-H.* **42 D2** 43 43N 17 49 E
Ostrzeszów, *Poland* **47 D4** 51 25N 17 52 E
Ostseebad Kühlungsborn, *Germany* **26 A7** 54 8N 11 44 E
Osttirol □, *Austria* **27 J8** 46 50N 12 30 E
Ostuni, *Italy* **41 B10** 40 44N 17 35 E
Osum →, *Bulgaria* **43 D9** 43 40N 24 50 E
Osumi →, *Albania* **44 D2** 40 40N 20 10 E
Ōsumi-Hantō, *Japan* **62 F2** 31 20N 130 55 E
Ōsumi-Kaikyō, *Japan* **61 J5** 30 55N 131 0 E
Ōsumi-Shotō, *Japan* **61 J5** 30 30N 130 0 E
Osun □, *Nigeria* **101 D5** 7 30N 4 30 E
Osuna, *Spain* **37 H5** 37 14N 5 8W
Oswego, *U.S.A.* **137 C8** 43 27N 76 31W
Oswestry, *U.K.* **16 E4** 52 52N 3 3W
Oświęcim, *Poland* **31 A12** 50 2N 19 11 E
Ōta, *Japan* **63 A11** 36 18N 139 22 E
Ota-Gawa →, *Japan* **62 C4** 34 21N 132 18 E
Otago □, *N.Z.* **119 E4** 45 15 S 170 0 E
Otago Harbour, *N.Z.* **119 F5** 45 47 S 170 42 E
Otago Pen., *N.Z.* **119 F5** 45 48 S 170 39 E
Otahuhu, *N.Z.* **118 C3** 36 56 S 174 51 E
Ōtake, *Japan* **62 C4** 34 12N 132 13 E
Ōtaki, *Japan* **63 B12** 35 17N 140 15 E
Otaki, *N.Z.* **118 G4** 40 45 S 175 10 E
Otane, *N.Z.* **118 F5** 39 54 S 176 39 E
Otar, *Kazakhstan* **55 B7** 43 32N 75 12 E
Otaru, *Japan* **60 C10** 43 10N 141 0 E
Otaru-Wan = Ishikari-Wan, *Japan* **60 C10** 43 25N 141 1 E
Otautau, *N.Z.* **119 G3** 46 9 S 168 1 E
Otava →, *Czech.* **30 B7** 49 26N 14 12 E
Otavalo, *Ecuador* **152 C2** 0 13N 78 20W
Otavi, *Namibia* **104 B2** 19 40 S 17 24 E
Otchinjau, *Angola* **103 F2** 16 30 S 13 56 E
Otelec, *Romania* **46 D1** 45 36N 20 50 E
Otero de Rey, *Spain* **36 B3** 43 6N 7 36W
Othello, *U.S.A.* **142 C4** 46 50N 119 10W
Othonoí, *Greece* **44 E1** 39 52N 19 22 E
Óthris, Óros, *Greece* **45 E4** 39 4N 22 42 E
Otira, *N.Z.* **119 C6** 42 49 S 171 35 E
Otira Gorge, *N.Z.* **119 C6** 42 53 S 171 33 E
Otis, *U.S.A.* **138 E3** 40 9N 102 58W
Otjiwarongo, *Namibia* **104 C2** 20 30 S 16 33 E
Otmuchów, *Poland* **47 E4** 50 28N 17 10 E
Oto Tolu Group, *Tonga* **121 Q13** 20 21 S 174 32W
Otočac, *Croatia* **39 D12** 44 53N 15 12 E
Otoineppu, *Japan* **60 B11** 44 44N 142 16 E
Oton, *Phil.* **71 F4** 10 42N 122 29 E
Otorohanga, *N.Z.* **118 E4** 38 12 S 175 14 E
Otoskwin →, *Canada* **128 B2** 52 13N 88 6W
Otosquen, *Canada* **131 C8** 53 17N 102 1W
Ōtoyo, *Japan* **62 D5** 33 43N 133 45 E
Otra →, *Norway* **13 G13** 58 9N 8 1 E
Otradnyy, *Russia* **52 D10** 53 22N 51 21 E
Otranto, *Italy* **41 B11** 40 9N 18 28 E
Otranto, C. d', *Italy* **41 B11** 40 7N 18 30 E
Otranto, Str. of, *Italy* **41 B11** 40 15N 18 40 E
Otse, *S. Africa* **104 D4** 25 2 S 25 45 E
Otsego, *U.S.A.* **141 B11** 42 27N 85 42W
Ōtsu, *Japan* **63 B7** 35 0N 135 50 E
Ōtsuki, *Japan* **63 B10** 35 36N 138 57 E
Otta, *Norway* **14 C3** 61 46N 9 32 E
Ottapalam, *India* **83 J3** 10 46N 76 23 E
Ottawa = Outaouais →, *Canada* **128 C5** 45 27N 74 8W
Ottawa, *Canada* **128 C4** 45 27N 75 42W
Ottawa, *Ill., U.S.A.* **138 E10** 41 21N 88 51W
Ottawa, *Kans., U.S.A.* **138 F7** 38 37N 95 16W
Ottawa, *Ohio, U.S.A.* **141 C12** 41 1N 84 3W
Ottawa Is., *Canada* **127 C11** 59 35N 80 10W
Ottélé, *Cameroon* **101 E7** 3 38N 11 19 E
Ottensheim, *Austria* **30 C7** 48 21N 14 12 E
Otter L., *Canada* **131 B8** 55 35N 104 39W
Otter Rapids, *Ont., Canada* **128 B3** 50 11N 81 39W
Otter Rapids, *Sask., Canada* **131 B8** 55 38N 104 44W
Otterbein, *U.S.A.* **141 D9** 40 29N 87 6W
Otterndorf, *Germany* **26 B4** 53 48N 8 53 E
Otterup, *Denmark* **15 J4** 55 30N 10 22 E
Otterville, *Canada* **136 D4** 42 55N 80 36W
Otterville, *U.S.A.* **140 F4** 38 42N 93 0W
Ottignies, *Belgium* **21 G5** 50 40N 4 33 E
Otto Beit Bridge, *Zimbabwe* **107 F2** 15 59 S 28 56 E
Ottosdal, *S. Africa* **104 D4** 26 46 S 25 59 E
Ottoville, *U.S.A.* **141 D12** 40 57N 84 22W
Ottsjön, *Sweden* **14 A7** 63 13N 13 2 E
Ottumwa, *U.S.A.* **140 C4** 41 1N 92 25W
Otu, *Nigeria* **101 D5** 8 14N 3 22 E
Otukpa, *Nigeria* **101 D6** 7 9N 7 41 E
Oturkpo, *Nigeria* **101 D6** 7 16N 8 8 E
Otway, B., *Chile* **160 D2** 53 30 S 74 0W
Otway, C., *Australia* **116 E5** 38 52 S 143 30 E
Otwock, *Poland* **47 C8** 52 5N 21 20 E
Ötz, *Austria* **30 D3** 47 13N 10 53 E
Ötz →, *Austria* **30 D3** 47 14N 10 50 E
Ötztaler Alpen, *Austria* **30 E4** 46 56N 11 0 E
Ou →, *Laos* **76 B4** 20 4N 102 13 E
Ou Neua, *Laos* **76 A3** 22 18N 101 48 E
Ou-Sammyaku, *Japan* **60 E10** 39 20N 140 35 E
Ouachita →, *U.S.A.* **139 K9** 31 38N 91 49W
Ouachita, L., *U.S.A.* **139 H8** 34 34N 93 12W
Ouachita Mts., *U.S.A.* **139 H7** 34 40N 94 25W
Ouaco, *N. Cal.* **121 T18** 20 50 S 164 29 E
Ouadâne, *Mauritania* **98 D2** 20 50N 11 40W
Ouadda, *C.A.R.* **102 A4** 8 15N 22 20 E
Ouagadougou, *Burkina Faso* **101 C4** 12 25N 1 30W
Ouagam, *Chad* **97 F2** 14 22N 14 42 E
Ouahigouya, *Burkina Faso* **100 C4** 13 31N 2 25W
Ouahila, *Algeria* **98 C3** 27 50N 5 0W
Ouahran = Oran, *Algeria* **99 A4** 35 45N 0 39W
Oualâta, *Mauritania* **100 B3** 17 20N 6 55W
Ouallene, *Algeria* **99 D5** 24 41N 1 11 E
Ouanda Djallé, *C.A.R.* **102 A4** 8 55N 22 53 E
Ouandago, *C.A.R.* **102 A3** 7 13N 18 50 E
Ouango, *C.A.R.* **102 B4** 4 19N 22 30 E
Ouarâne, *Mauritania* **98 D2** 21 0N 10 30W
Ouargla, *Algeria* **99 B6** 31 59N 5 16 E
Ouarkziz, Djebel, *Algeria* **98 C3** 28 50N 8 0W
Ouarzazate, *Morocco* **98 B3** 30 55N 6 50W
Ouatagouna, *Mali* **101 B5** 15 11N 0 43 E
Ouatere, *C.A.R.* **102 A3** 5 30N 19 8 E
Oubangi →, *Zaïre* **102 C3** 0 30 S 17 50 E
Oubarakai, O. →, *Algeria* **99 C6** 27 20N 9 0 E
Oubatche, *N. Cal.* **121 T18** 20 26 S 164 39 E
Ouche →, *France* **23 E12** 47 6N 5 16 E
Oud-Beijerland, *Neths.* **20 E4** 51 50N 4 25 E
Oud-Gastel, *Neths.* **21 E4** 51 35N 4 28 E
Oud Turnhout, *Belgium* **21 F6** 51 19N 5 0 E
Ouddorp, *Neths.* **20 E3** 51 50N 3 57 E
Oude-Pekela, *Neths.* **20 B10** 53 6N 7 0 E
Oude Rijn →, *Neths.* **20 D4** 52 12N 4 24 E
Oudega, *Neths.* **20 B8** 53 8N 6 0 E
Oudenaarde, *Belgium* **21 G3** 50 50N 3 37 E
Oudenbosch, *Neths.* **21 E5** 51 35N 4 32 E
Oudenburg, *Belgium* **21 F2** 51 11N 3 1 E
Ouderkerk, *Utrecht, Neths.* **20 D5** 52 18N 4 55 E
Ouderkerk, *Zuid-Holland, Neths.* **20 E5** 51 56N 4 38 E
Oudeschild, *Neths.* **20 B5** 53 2N 4 50 E
Oudewater, *Neths.* **20 D5** 52 2N 4 52 E
Oudkarspel, *Neths.* **20 C5** 52 43N 4 49 E
Oudon, *France* **22 E5** 47 22N 1 19W
Oudon →, *France* **22 E6** 47 41N 0 53W
Oudtshoorn, *S. Africa* **104 E3** 33 35 S 22 14 E
Oued Zem, *Morocco* **98 B3** 32 52N 6 34W
Ouégoa, *N. Cal.* **121 T18** 20 20 S 164 26 E
Ouellé, *Ivory C.* **100 D4** 7 26N 4 1W
Ouen, I., *N. Cal.* **121 V20** 22 26 S 166 49 E
Ouenza, *Algeria* **99 A6** 35 57N 8 4 E
Ouessa, *Burkina Faso* **100 C4** 11 4N 2 47W
Ouessant, I. d', *France* **22 D1** 48 28N 5 6W
Ouesso, *Congo* **102 B3** 1 37N 16 5 E
Ouest, Pte., *Canada* **129 C7** 49 52N 64 40W
Ouezzane, *Morocco* **98 B3** 34 51N 5 35W
Ouffet, *Belgium* **21 H6** 50 26N 5 28 E
Ouidah, *Benin* **101 D5** 6 25N 2 0 E
Ouistreham, *France* **22 C6** 49 17N 0 18W
Oujda, *Morocco* **99 B4** 34 41N 1 55W
Oujeft, *Mauritania* **98 D2** 20 2N 13 0W
Oulainen, *Finland* **12 D21** 64 17N 24 47 E
Ould Yenjé, *Mauritania* **100 B2** 15 38N 12 16W
Ouled Djellal, *Algeria* **99 B6** 34 28N 5 2 E
Ouled Naïl, Mts. des, *Algeria* **99 B5** 34 30N 3 30 E
Oulmès, *Morocco* **98 B3** 33 17N 6 0W
Oulu, *Finland* **12 D21** 65 1N 25 29 E
Oulujärvi, *Finland* **12 D22** 64 25N 27 15 E
Oulujoki →, *Finland* **12 D21** 65 1N 25 30 E
Oulx, *Italy* **38 C3** 45 2N 6 50 E
Oum Chalouba, *Chad* **97 E4** 15 48N 20 46 E
Oum-el-Bouaghi, *Algeria* **99 A6** 35 55N 7 6 E
Oum el Ksi, *Algeria* **98 C3** 29 4N 6 59W
Oum-er-Rbia, O. →, *Morocco* **98 B3** 33 19N 8 21W
Oumè, *Ivory C.* **100 D3** 6 21N 5 27W
Ounane, Dj., *Algeria* **99 C6** 25 4N 7 19 E
Ounasjoki →, *Finland* **12 C21** 66 31N 25 40 E
Ounguati, *Namibia* **104 C2** 22 0 S 15 46 E
Ounianga-Kébir, *Chad* **97 E4** 19 4N 20 29 E
Ounianga Sérir, *Chad* **97 E4** 18 54N 20 51 E
Our →, *Lux.* **21 J8** 49 55N 6 5 E
Ouray, *U.S.A.* **143 G10** 38 1N 107 40W
Ourcq →, *France* **23 C10** 49 1N 3 1 E
Ourém, *Brazil* **154 B2** 1 33 S 47 6W
Ourense = Orense, *Spain* **36 C3** 42 19N 7 55W
Ouricuri, *Brazil* **154 C3** 7 53 S 40 5W
Ourinhos, *Brazil* **159 A6** 23 0 S 49 54W
Ourique, *Portugal* **37 H2** 37 38N 8 16W
Ouro Fino, *Brazil* **159 A6** 22 16 S 46 25W
Ouro Prêto, *Brazil* **155 F3** 20 20 S 43 30W
Ouro Sogui, *Senegal* **100 B2** 15 36N 13 19W
Oursi, *Burkina Faso* **101 C4** 14 41N 0 27W
Ourthe →, *Belgium* **21 H7** 50 29N 5 35 E
Ouse, *Australia* **114 G4** 42 38 S 146 42 E
Ouse →, *E. Susx., U.K.* **17 G8** 50 47N 0 4 E
Ouse →, *N. Yorks., U.K.* **16 C8** 53 44N 0 55W
Oust, *France* **24 F5** 42 52N 1 13 E
Oust →, *France* **22 E4** 47 35N 2 6W
Outaouais →, *Canada* **128 C5** 45 27N 74 8W
Outardes →, *Canada* **129 C6** 49 24N 69 30W
Outat Oulad el Haj, *Morocco* **99 B4** 33 22N 3 42W
Outer Hebrides, *U.K.* **18 D1** 57 30N 7 40W
Outer I., *Canada* **129 B8** 51 10N 58 35W
Outes, *Spain* **36 C2** 42 52N 8 55W
Outjo, *Namibia* **104 C2** 20 5 S 16 7 E
Outlook, *Canada* **131 C7** 51 30N 107 0W
Outlook, *U.S.A.* **138 A2** 48 53N 104 47W
Outokumpu, *Finland* **12 E23** 62 43N 29 1 E
Outreau, *France* **23 B8** 50 40N 1 36 E
Ouvèze →, *France* **25 E8** 43 59N 4 51 E
Ouyen, *Australia* **116 C5** 35 1 S 142 22 E
Ouzouer-le-Marché, *France* **23 E8** 47 54N 1 32 E
Ovada, *Italy* **38 D5** 44 38N 8 38 E
Ovalau, *Fiji* **121 A2** 17 40 S 178 48 E
Ovalle, *Chile* **158 C1** 30 33 S 71 18W
Ovamboland, *Namibia* **104 B2** 18 30 S 16 0 E
Ovar, *Portugal* **36 E2** 40 51N 8 40W
Ovejas, *Colombia* **152 B2** 9 32N 75 14W
Ovens, *Australia* **117 D7** 36 35 S 146 46 E
Overdinkel, *Neths.* **20 D10** 52 14N 7 2 E
Overflakkee, *Neths.* **20 E4** 51 44N 4 10 E
Overijse, *Belgium* **21 G5** 50 47N 4 32 E
Overijssel □, *Neths.* **20 D9** 52 25N 6 35 E
Overijsselsch Kanaal →, *Neths.* **20 C8** 52 31N 6 6 E
Overland, *U.S.A.* **140 F6** 38 41N 90 22W
Overpelt, *Belgium* **21 F6** 51 12N 5 20 E
Overton, *U.S.A.* **145 J12** 36 33N 114 27W
Övertorneå, *Sweden* **12 C20** 66 23N 23 38 E
Ovid, *Colo., U.S.A.* **138 E3** 40 58N 102 23W
Ovid, *Mich., U.S.A.* **141 A12** 43 1N 84 22W
Ovidiopol, *Ukraine* **51 J6** 46 15N 30 30 E
Oviedo, *Spain* **36 B5** 43 25N 5 50W
Oviedo □, *Spain* **36 B5** 43 20N 6 0W
Oviken, *Sweden* **14 A8** 63 0N 14 23 E
Oviksfjällen, *Sweden* **14 B7** 63 0N 13 49 E
Oviši, *Latvia* **13 H19** 57 33N 21 44 E
Övör Hangay □, *Mongolia* **66 B2** 45 0N 103 0 E
Ovoro, *Nigeria* **101 D6** 5 26N 7 16 E
Øvre Årdal, *Norway* **13 F12** 61 19N 7 48 E
Ovruch, *Ukraine* **51 G5** 51 25N 28 45 E
Owaka, *N.Z.* **119 G4** 46 27 S 169 40 E
Owambo = Ovamboland, *Namibia* **104 B2** 18 30 S 16 0 E
Owando, *Congo* **102 C3** 0 29 S 15 55 E
Owase, *Japan* **63 C8** 34 7N 136 12 E
Owatonna, *U.S.A.* **138 C8** 44 5N 93 14W
Owbeh, *Afghan.* **79 B1** 34 28N 63 10 E
Owego, *U.S.A.* **137 D8** 42 6N 76 16W
Owen, *Australia* **116 C3** 34 15 S 138 32 E
Owen Falls Dam, *Uganda* **106 B3** 0 30N 33 5 E
Owen Mt., *N.Z.* **119 B7** 41 35 S 172 33 E
Owen Sound, *Canada* **128 D3** 44 35N 80 55W
Owen Stanley Ra., *Papua N. G.* **120 E4** 8 30 S 147 0 E
Owendo, *Gabon* **102 B1** 0 17N 9 30 E
Owens →, *U.S.A.* **144 J9** 36 32N 117 59W
Owens L., *U.S.A.* **145 J9** 36 26N 117 57W
Owensboro, *U.S.A.* **141 G9** 37 46N 87 7W
Owensville, *Ind., U.S.A.* **141 F9** 38 16N 87 41W
Owensville, *Mo., U.S.A.* **140 F5** 38 21N 91 30W
Owenton, *U.S.A.* **141 F12** 38 32N 84 50W
Owerri, *Nigeria* **101 D6** 5 29N 7 0 E
Owhango, *N.Z.* **118 F4** 39 0 S 175 23 E
Owingsville, *U.S.A.* **141 F13** 38 9N 83 46W
Owl →, *Canada* **131 B10** 57 51N 92 44W
Owo, *Nigeria* **101 D6** 7 10N 5 39 E
Owosso, *U.S.A.* **141 B12** 43 0N 84 10W
Owyhee, *U.S.A.* **142 F5** 41 57N 116 6W
Owyhee →, *U.S.A.* **142 E5** 43 49N 117 2W
Owyhee, L., *U.S.A.* **142 E5** 43 38N 117 14W
Oxapampa, *Peru* **156 C2** 10 33 S 75 26W
Oxelösund, *Sweden* **15 F11** 58 43N 17 15 E
Oxford, *N.Z.* **119 D7** 43 18 S 172 11 E
Oxford, *U.K.* **17 F6** 51 46N 1 15W
Oxford, *Iowa, U.S.A.* **140 C5** 41 43N 91 47W
Oxford, *Mich., U.S.A.* **141 B13** 42 49N 83 16W
Oxford, *Miss., U.S.A.* **139 H10** 34 22N 89 31W
Oxford, *N.C., U.S.A.* **135 G6** 36 19N 78 35W
Oxford, *Ohio, U.S.A.* **141 E12** 39 31N 84 45W
Oxford L., *Canada* **131 C9** 54 51N 95 37W
Oxfordshire □, *U.K.* **17 F6** 51 48N 1 16W
Oxía, *Greece* **45 F3** 38 16N 21 5 E
Oxílithos, *Greece* **45 F6** 38 35N 24 7 E
Oxley, *Australia* **116 C6** 34 11 S 144 6 E
Oxnard, *U.S.A.* **145 L7** 34 12N 119 11W
Oxus = Amudarya →, *Uzbekistan* **56 E6** 43 58N 59 34 E
Oya, *Malaysia* **75 B4** 2 55N 111 55 E
Oyabe, *Japan* **63 A8** 36 47N 136 56 E
Oyama, *Japan* **63 A11** 36 18N 139 48 E
Oyana, *Japan* **62 E2** 32 32N 130 30 E
Oyapock →, *Fr. Guiana* **153 C7** 4 8N 51 40W
Oyem, *Gabon* **102 B2** 1 34N 11 31 E
Oyen, *Canada* **131 C6** 51 22N 110 28W
Øyeren, *Norway* **14 E5** 59 50N 11 15 E
Oykel →, *U.K.* **18 D4** 57 56N 4 26W
Oymyakon, *Russia* **57 C15** 63 25N 142 44 E
Oyo, *Nigeria* **101 D5** 7 46N 3 56 E
Oyo □, *Nigeria* **101 D5** 8 15N 3 30 E
Oyón, *Peru* **156 C2** 10 37 S 76 47W
Oyonnax, *France* **25 B9** 46 16N 5 40 E
Oyster Bay, *U.S.A.* **137 F11** 40 52N 73 32W
Oytal, *Kazakhstan* **55 B6** 42 54N 73 17 E
Ōyūbari, *Japan* **60 C11** 43 1N 142 5 E
Özalp, *Turkey* **89 C10** 38 39N 43 59 E
Ozamiz, *Phil.* **71 G4** 8 15N 123 50 E
Ozark, *Ala., U.S.A.* **135 K3** 31 28N 85 39W
Ozark, *Ark., U.S.A.* **139 H8** 35 29N 93 50W
Ozark, *Mo., U.S.A.* **139 G8** 37 1N 93 12W
Ozark Plateau, *U.S.A.* **139 G9** 37 20N 91 40W
Ozarks, L. of the, *U.S.A.* **140 F4** 38 12N 92 38W
Ózd, *Hungary* **31 C13** 48 14N 20 15 E
Ozernoye, *Russia* **52 E10** 51 46N 51 28 E
Ozërnyy, *Russia* **54 F8** 51 8N 60 50 E
Ozette L., *U.S.A.* **144 B2** 48 6N 124 38W
Özgön, *Kyrgyzstan* **55 C6** 40 46N 73 18 E
Ozieri, *Italy* **40 B2** 40 35N 9 0 E
Ozimek, *Poland* **47 E5** 50 41N 18 11 E
Ozinki, *Russia* **52 E9** 51 12N 49 44 E
Ozona, *U.S.A.* **139 K4** 30 43N 101 12W
Ozorków, *Poland* **47 D6** 51 57N 19 16 E
Ozren, *Bos.-H.* **42 D3** 43 55N 18 29 E
Ozu, *Ehime, Japan* **62 D4** 33 30N 132 33 E
Ozu, *Kumamoto, Japan* **62 E2** 32 52N 130 52 E
Ozuluama, *Mexico* **147 C5** 21 40N 97 50W
Ozun, *Romania* **46 D6** 45 47N 25 50 E
Ozurgeti, *Georgia* **53 K5** 41 55N 42 2 E

P

P.W.V. □, *S. Africa* **105 D4** 26 0 S 28 0 E
Pa, *Burkina Faso* **100 C4** 11 33N 3 19W
Pa-an, *Burma* **78 G6** 16 51N 97 40 E
Pa Mong Dam, *Thailand* **76 D4** 18 0N 102 22 E
Paagoumène, *N. Cal.* **121 T18** 20 29 S 164 11 E
Paal, *Belgium* **21 F6** 51 2N 5 10 E
Paama, *Vanuatu* **121 F6** 16 28 S 168 14 E
Paamiut = Frederikshåb, *Greenland* **6 C5** 62 0N 49 43W
Paar →, *Germany* **27 G6** 48 46N 11 36 E
Paarl, *S. Africa* **104 E2** 33 45 S 18 56 E
Paauilo, *U.S.A.* **132 H17** 20 2N 155 22W
Pab Hills, *Pakistan* **79 D2** 26 30N 66 45 E
Pabianice, *Poland* **47 D6** 51 40N 19 20 E
Pabna, *Bangla.* **78 C2** 24 1N 89 18 E
Pabo, *Uganda* **106 B3** 3 1N 32 10 E
Pacaás Novos, Serra dos, *Brazil* **157 C5** 10 45 S 64 15W
Pacaipampa, *Peru* **156 B2** 5 35 S 79 39W
Pacaja →, *Brazil* **154 B1** 1 56 S 50 50W
Pacajus, *Brazil* **154 B4** 4 10 S 38 31W
Pacaraima, Sierra, *Venezuela* **153 C5** 4 0N 62 30W
Pacarán, *Peru* **156 C2** 12 50 S 76 3W
Pacaraos, *Peru* **156 C2** 11 12 S 76 42W
Pacasmayo, *Peru* **156 B2** 7 20 S 79 35W
Paceco, *Italy* **40 E5** 37 59N 12 33 E
Pachacamac, *Peru* **156 C2** 12 14 S 77 53W
Pachhar, *India* **80 G7** 24 40N 77 42 E
Pachino, *Italy* **41 F8** 36 43N 15 5 E
Pachitea →, *Peru* **156 B3** 8 46 S 74 33W
Pachiza, *Peru* **156 B2** 7 16 S 76 46W
Pacho, *Colombia* **152 B3** 5 8N 74 10W
Pachora, *India* **82 D2** 20 38N 75 29 E
Pachuca, *Mexico* **147 C5** 20 10N 98 40W
Pacific, *Canada* **130 C3** 54 48N 128 28W
Pacific, *U.S.A.* **140 F6** 38 29N 90 45W

Peking 247

Paraíba do Sul →, Brazil 155 F3 21 37 S 41 3W
Parainen, Finland 13 F20 60 18N 22 18 E
Paraiso, Mexico 147 D6 18 24N 93 14W
Parak, Iran 85 E7 27 38N 52 25 E
Parakhino Paddubye, Russia 50 C7 58 26N 33 10 E
Parakou, Benin 101 D5 9 25N 2 40 E
Parakylia, Australia 116 A2 30 24 S 136 25 E
Paralimni, Cyprus 32 D12 35 2N 33 58 E
Parálion-Astrous, Greece 45 G4 37 25N 22 45 E
Paramakkudi, India 83 K4 9 31N 78 39 E
Paramaribo, Surinam 153 B6 5 50N 55 10W
Parambu, Brazil 154 C3 6 13 S 40 43W
Paramillo, Nudo del, Colombia 152 B2 7 4N 75 55W
Paramirim, Brazil 155 D3 13 26 S 42 15W
Paramirim →, Brazil 155 D3 11 34 S 43 18W
Paramithiá, Greece 44 E2 39 30N 20 35 E
Paramushir, Ostrov, Russia 57 D16 50 24N 156 0 E
Paran →, Israel 91 E4 30 20N 35 10 E
Paraná, Argentina 158 C3 31 45 S 60 30W
Paranã, Brazil 155 D2 12 30 S 47 48W
Paraná □, Brazil 159 A5 24 30 S 51 0W
Paraná →, Argentina 158 C4 33 43 S 59 15W
Paranaguá, Brazil 159 B6 25 30 S 48 30W
Paranaíba →, Brazil 155 F1 20 6 S 51 4W
Paranapanema →, Brazil 159 A5 22 40 S 53 9W
Paranapiacaba, Serra do, Brazil 159 A6 24 31 S 48 35W
Paranavaí, Brazil 159 A5 23 4 S 52 56W
Parang, Jolo, Phil. 71 J3 5 55N 120 54 E
Parang, Mindanao, Phil. 71 H5 7 23N 124 16 E
Parangaba, Brazil 154 B4 3 45 S 38 33W
Parangippettai, India 83 J4 11 30N 79 48 E
Paraparaumu, N.Z. 118 G4 40 57 S 175 3 E
Parapóla, Greece 45 H5 36 55N 23 27 E
Paraspóri, Ákra, Greece 45 J9 35 55N 27 15 E
Paratinga, Brazil 155 D3 12 40 S 43 10W
Paratoo, Australia 116 B3 32 42 S 139 20 E
Parattah, Australia 114 G4 42 22 S 147 23 E
Paraúna, Brazil 155 E1 16 55 S 50 26W
Paray-le-Monial, France 25 B8 46 27N 4 7 E
Parbati →, India 80 G7 25 50N 76 30 E
Parbatipur, Bangla. 78 C2 25 39N 88 55 E
Parbhani, India 82 E3 19 8N 76 52 E
Parchim, Germany 26 B7 53 26N 11 52 E
Parczew, Poland 47 D9 51 40N 22 52 E
Pardes Hanna, Israel 91 C3 32 28N 34 57 E
Pardilla, Spain 36 D7 41 33N 3 43W
Pardo →, Bahia, Brazil 155 E4 15 40 S 39 0W
Pardo →, Mato Grosso, Brazil 159 A5 21 46 S 52 9W
Pardo →, Minas Gerais, Brazil 155 E3 15 48 S 44 48W
Pardo →, São Paulo, Brazil 155 F2 20 10 S 48 38W
Pardubice, Czech. 30 A8 50 3N 15 45 E
Pare, Indonesia 75 D4 7 43 S 112 12 E
Pare □, Tanzania 106 C4 4 10 S 38 0 E
Pare Mts., Tanzania 106 C4 4 0 S 37 45 E
Parecis, Serra dos, Brazil 157 C6 13 0 S 60 0W
Paredes de Nava, Spain 36 C6 42 9N 4 42W
Pareh, Iran 84 B5 38 52N 45 42 E
Parelhas, Brazil 154 C4 6 41 S 36 39W
Paren, Russia 57 C17 62 30N 163 15 E
Parengarenga Harbour, N.Z. 118 A1 34 31 S 173 0 E
Parent, Canada 128 C5 47 55N 74 35W
Parent, L., Canada 128 C4 48 31N 77 1W
Parentis-en-Born, France 24 D2 44 21N 1 4W
Parepare, Indonesia 72 B1 4 0 S 119 40 E
Parfino, Russia 50 D6 57 59N 31 34 E
Pargo, Pta. do, Madeira 33 D2 32 49N 17 17W
Parguba, Russia 48 B5 62 20N 34 27 E
Paria, G. de, Venezuela 153 A5 10 20N 62 0W
Paria, Pen. de, Venezuela 153 A5 10 50N 62 30W
Pariaguán, Venezuela 153 B5 8 51N 64 34W
Pariaman, Indonesia 74 C2 0 47 S 100 11 E
Paricatuba, Brazil 153 D5 4 26 S 61 53W
Paricutín, Cerro, Mexico 146 D4 19 28N 102 15W
Parigi, Java, Indonesia 75 D3 7 42 S 108 29 E
Parigi, Sulawesi, Indonesia 72 B2 0 50 S 120 5 E
Parika, Guyana 153 B6 6 50N 58 20W
Parikkala, Finland 50 B5 61 33N 29 31 E
Parima, Serra, Brazil 153 C5 2 30N 64 0W
Parinari, Peru 156 A3 4 35 S 74 25W
Parincea, Romania 46 C8 46 27N 27 9 E
Parîngul Mare, Romania 46 D4 45 20N 23 37 E
Parintins, Brazil 153 D6 2 40 S 56 50W
Paris, Canada 128 D3 43 12N 80 25W
Paris, France 23 D9 48 50N 2 20 E
Paris, Idaho, U.S.A. 142 E8 42 14N 111 24W
Paris, Ill., U.S.A. 141 E9 39 36N 87 42W
Paris, Ky., U.S.A. 141 F12 38 13N 84 15W
Paris, Mo., U.S.A. 140 E5 39 29N 92 0W
Paris, Tenn., U.S.A. 135 G1 36 18N 88 19W
Paris, Tex., U.S.A. 139 J7 33 40N 95 33W
Paris, Ville de □, France 23 D9 48 50N 2 20 E
Parish, U.S.A. 137 C8 43 25N 76 8W
Pariti, Indonesia 72 D2 10 15 S 123 45 E
Park, U.S.A. 144 B4 48 45N 122 18W
Park City, U.S.A. 142 F8 40 39N 111 30W
Park Falls, U.S.A. 138 C9 45 56N 90 27W
Park Forest, U.S.A. 141 C9 41 29N 87 40W
Park Range, U.S.A. 142 G10 40 0N 106 30W
Park Rapids, U.S.A. 138 B7 46 55N 95 4W
Park Ridge, U.S.A. 141 B9 42 2N 87 51W
Park River, U.S.A. 138 A6 48 24N 97 45W
Park Rynie, S. Africa 105 E5 30 25 S 30 45 E
Parkā Bandar, Iran 85 E8 25 55N 59 35 E
Parkano, Finland 13 E20 62 1N 23 0 E
Parkent, Uzbekistan 55 C4 41 18N 69 40 E
Parker, Ariz., U.S.A. 145 L12 34 9N 114 17W
Parker, S. Dak., U.S.A. 138 D6 43 24N 97 8W
Parker Dam, U.S.A. 145 L12 34 18N 114 8W
Parkersburg, Iowa, U.S.A. 140 B4 42 35N 92 47W
Parkersburg, W. Va., U.S.A. 134 F5 39 16N 81 34W
Parkerview, Canada 131 C8 51 21N 103 18W
Parkes, Australia 117 B8 33 9 S 148 11 E
Parkfield, U.S.A. 144 K6 35 54N 120 26W
Parkhar, Tajikistan 55 E4 37 30N 69 34 E
Parkland, U.S.A. 144 C4 47 9N 122 26W
Parkside, Canada 131 C7 53 10N 106 33W
Parkston, U.S.A. 138 D5 43 24N 97 59W
Parksville, Canada 130 D4 49 20N 124 21W
Parlakimidi, India 82 E7 18 45N 84 5 E
Parli, India 82 E3 18 50N 76 35 E
Parma, Italy 38 D7 44 48N 10 20 E
Parma, Idaho, U.S.A. 142 E5 43 47N 116 57W
Parma, Ohio, U.S.A. 136 E3 41 23N 81 43W
Parma →, Italy 38 D7 44 56N 10 26 E
Parnaguá, Brazil 154 D3 10 10 S 44 38W
Parnaíba, Piauí, Brazil 154 B3 2 54 S 41 47W
Parnaíba, São Paulo, Brazil 157 D7 19 34 S 51 14W
Parnaíba →, Brazil 154 B3 3 0 S 41 50W
Parnamirim, Brazil 154 C4 8 5 S 39 34W
Parnarama, Brazil 154 C3 5 31 S 43 6W
Parnassós, Greece 45 F4 38 35N 22 30 E
Parnassus, N.Z. 119 C8 42 42 S 173 23 E
Párnis, Greece 45 F5 38 14N 23 45 E
Párnon Óros, Greece 45 G4 37 15N 22 45 E
Pärnu, Estonia 13 G21 58 28N 24 33 E
Parola, India 82 D2 20 47N 75 7 E
Paroo →, Australia 115 E3 31 28 S 143 32 E
Páros, Greece 45 G7 37 5N 25 12 E
Parowan, U.S.A. 143 H7 37 51N 112 50W
Parpaillon, France 25 D10 44 30N 6 40 E
Parral, Chile 158 D1 36 10 S 71 52W
Parramatta, Australia 117 B9 33 48 S 151 1 E
Parras, Mexico 146 B4 25 30N 102 20W
Parrett →, U.K. 17 F5 51 12N 3 1W
Parris I., U.S.A. 135 J5 32 20N 80 41W
Parrsboro, Canada 129 C7 45 30N 64 25W
Parry Is., Canada 6 B2 77 0N 110 0W
Parry Sound, Canada 128 C3 45 20N 80 0W
Parsberg, Germany 27 F7 49 10N 11 43 E
Parseta →, Poland 47 A2 54 11N 15 34 E
Parshall, U.S.A. 138 B3 47 57N 102 8W
Parsnip →, Canada 130 B4 55 10N 123 2W
Parsons, U.S.A. 139 G7 37 20N 95 16W
Parsons Ra., Australia 114 A2 13 30 S 135 15 E
Partabpur, India 82 E5 20 0N 80 42 E
Partanna, Italy 40 E5 37 43N 12 53 E
Parthenay, France 22 F6 46 38N 0 16W
Partinico, Italy 40 D6 38 3N 13 7 E
Partur, India 82 E3 19 40N 76 14 E
Paru →, Brazil 153 D7 1 33 S 52 38W
Parú →, Venezuela 152 C4 4 20N 66 27W
Paru de Oeste →, Brazil 153 C6 1 30N 56 0W
Parubcan, Phil. 70 E4 13 43N 123 45 E
Parucito →, Venezuela 152 B4 5 18N 65 59W
Parur, India 83 J3 10 13N 76 14 E
Paruro, Peru 156 C3 13 45 S 71 50W
Parvān □, Afghan. 79 B3 35 0N 69 0 E
Parvatipuram, India 82 E6 18 50N 83 25 E
Parys, S. Africa 104 D4 26 52 S 27 29 E
Pas-de-Calais □, France 23 B9 50 30N 2 10 E
Pasadena, Calif., U.S.A. 145 L8 34 9N 118 9W
Pasadena, Tex., U.S.A. 139 L7 29 43N 95 13W
Pasaje, Ecuador 152 D2 3 23 S 79 50W
Pasaje →, Argentina 158 B3 25 39 S 63 56W
Pasay, Phil. 70 D3 14 33N 121 0 E
Pascagoula, U.S.A. 139 K10 30 21N 88 33W
Pascagoula →, U.S.A. 139 K10 30 23N 88 37W
Paşcani, Romania 46 B7 47 14N 26 45 E
Pasco, U.S.A. 142 C4 46 14N 119 6W
Pasco □, Peru 156 C2 10 40 S 75 0W
Pasco, Cerro de, Peru 156 C2 10 45 S 76 10W
Pascua, I. de, Pac. Oc. 123 K17 27 0 S 109 0W
Pasewalk, Germany 26 B10 53 30N 13 58 E
Pasfield L., Canada 131 B7 58 24N 105 20W
Pasha →, Russia 50 B7 60 29N 32 55 E
Pashiwari, Pakistan 81 B6 34 40N 75 10 E
Pashiya, Russia 54 B7 58 33N 58 26 E
Pashmakli = Smolyan, Bulgaria 43 F9 41 36N 24 38 E
Pasighat, India 78 A5 28 4N 95 21 E
Pasinler, Turkey 89 C9 39 59N 41 41 E
Pasir Mas, Malaysia 74 A2 6 2N 102 8 E
Pasirian, Indonesia 75 D4 8 13 S 113 8 E
Paskūh, Iran 85 E9 27 34N 61 39 E
Pasłeka →, Poland 47 A6 54 26N 19 46 E
Pasley, C., Australia 113 F3 33 52 S 123 35 E
Pašman, Croatia 39 E12 43 58N 15 20 E
Pasmore →, Australia 116 A3 31 5 S 139 49 E
Pasni, Pakistan 79 D1 25 15N 63 27 E
Paso Cantinela, Mexico 145 N11 32 33N 115 47W
Paso de Indios, Argentina 160 B3 43 55 S 69 0W
Paso de los Libres, Argentina 158 B4 29 44 S 57 10W
Paso de los Toros, Uruguay 158 C4 32 45 S 56 30W
Paso Flores, Argentina 160 B2 40 35 S 70 38W
Paso Robles, U.S.A. 143 J3 35 38N 120 41W
Pasorapa, Bolivia 157 D5 18 16 S 64 37W
Paspébiac, Canada 129 C6 48 3N 65 17W
Pasrur, Pakistan 80 C6 32 16N 74 43 E
Passage West, Ireland 19 E3 51 52N 8 21W
Passaic, U.S.A. 137 F10 40 51N 74 7W
Passau, Germany 27 G9 48 34N 13 28 E
Passendale, Belgium 21 G2 50 54N 3 2 E
Passero, C., Italy 41 F8 36 41N 15 10 E
Passi, Phil. 71 F4 11 6N 122 38 E
Passo Fundo, Brazil 159 B5 28 10 S 52 20W
Passos, Brazil 155 F2 20 45 S 46 37W
Passow, Germany 26 B10 53 8N 14 6 E
Passwang, Switz. 28 B5 47 22N 7 41 E
Passy, France 25 C10 45 55N 6 41 E
Pastavy, Belarus 13 J22 55 4N 26 50 E
Pastaza □, Ecuador 152 D2 2 0 S 77 0W
Pastaza →, Peru 152 D2 4 50 S 76 52W
Pastęk, Poland 47 A6 54 3N 19 41 E
Pasto, Colombia 152 C2 1 13N 77 17W
Pastos Bons, Brazil 154 C3 6 36 S 44 5W
Pastrana, Spain 34 E2 40 27N 2 53W
Pasuquin, Phil. 70 B3 18 20N 120 37 E
Pasuruan, Indonesia 75 D4 7 40 S 112 44 E
Pasym, Poland 47 B7 53 48N 20 49 E
Pásztó, Hungary 31 D12 47 52N 19 43 E
Patagonia, Argentina 160 C2 45 0 S 69 0W
Patagonia, U.S.A. 143 L8 31 33N 110 45W
Patambar, Iran 85 D9 29 45N 60 17 E
Patan, Gujarat, India 82 F1 17 22N 73 57 E
Patan, Maharashtra, India 80 H5 23 54N 72 14 E
Patani, Indonesia 72 A3 0 20N 128 50 E
Pataudi, India 80 E7 28 18N 76 48 E
Patay, France 23 D8 48 2N 1 40 E
Patchewollock, Australia 116 C5 35 22 S 142 12 E
Patchogue, U.S.A. 137 F11 40 46N 73 1W
Patea, N.Z. 118 F3 39 45 S 174 30 E
Pategi, Nigeria 101 D6 8 50N 5 45 E
Patensie, S. Africa 104 E3 33 46 S 24 49 E
Paternò, Italy 41 E7 37 34N 14 53 E
Pateros, U.S.A. 142 B4 48 3N 119 54W
Paterson, Australia 117 B9 32 35 S 151 36 E
Paterson, U.S.A. 137 F10 40 55N 74 11W
Paterson Inlet, N.Z. 119 G3 46 56 S 168 12 E
Paterson Ra., Australia 112 D3 21 45 S 122 10 E
Paterswolde, Neths. 20 B9 53 9N 6 34 E
Pathankot, India 80 C6 32 18N 75 45 E
Patharghata, Bangla. 78 D2 22 2N 89 58 E
Pathfinder Reservoir, U.S.A. 142 E10 42 28N 106 51W
Pathiu, Thailand 77 G2 10 42N 99 19 E
Pathum Thani, Thailand 76 E3 14 1N 100 32 E
Pati, Indonesia 75 D4 6 45 S 111 1 E
Pati Pt., Guam 121 R15 13 40N 144 50 E
Patía, Colombia 152 C2 2 4N 77 4W
Patía →, Colombia 152 C2 2 13N 78 40W
Patiala, India 80 D7 30 23N 76 26 E
Patine Kouka, Senegal 100 C2 12 45N 13 45W
Pativilca, Peru 156 C2 10 42 S 77 48W
Patkai Bum, India 78 B5 27 0N 95 30 E
Pátmos, Greece 45 G8 37 21N 26 36 E
Patna, India 81 G11 25 35N 85 12 E
Patnongon, Phil. 71 F3 10 55N 122 0 E
Patnos, Turkey 89 C10 39 13N 42 51 E
Patonga, Uganda 106 B3 2 45N 33 15 E
Patos, Brazil 154 C4 6 55 S 37 16W
Patos, L. dos, Brazil 159 C5 31 20 S 51 0W
Patos de Minas, Brazil 155 E2 18 35 S 46 32W
Patosi, Albania 44 D1 40 42N 19 38 E
Patquía, Argentina 158 C2 30 2 S 66 55W
Pátrai, Greece 45 F3 38 14N 21 47 E
Pátraikós Kólpos, Greece 45 F3 38 17N 21 30 E
Patras = Pátrai, Greece 45 F3 38 14N 21 47 E
Patricio Lynch, I., Chile 160 C1 48 35 S 75 30W
Patrocínio, Brazil 155 E2 18 57 S 47 0W
Patta, Kenya 106 C5 2 10 S 41 0 E
Pattada, Italy 40 B2 40 35N 9 6 E
Pattanapuram, India 83 K3 9 6N 76 50 E
Pattani, Thailand 77 J3 6 48N 101 15 E
Patten, U.S.A. 129 C6 46 0N 68 38W
Patterson, Calif., U.S.A. 144 H5 37 28N 121 8W
Patterson, La., U.S.A. 139 L9 29 42N 91 18W
Patterson, Mt., U.S.A. 144 G7 38 29N 119 20W
Patteson, Passage, Vanuatu 121 E6 15 26 S 168 12 E
Patti, India 80 D6 31 17N 74 54 E
Patti, Italy 41 D7 38 8N 14 58 E
Pattoki, Pakistan 80 D5 31 5N 73 52 E
Patton, U.S.A. 136 F6 40 38N 78 39W
Pattonsburg, U.S.A. 140 D2 40 3N 94 8W
Pattukkattai, India 83 J4 10 25N 79 20 E
Patu, Brazil 154 C4 6 6 S 37 38W
Patuakhali, Bangla. 78 D3 22 20N 90 25 E
Patuca →, Honduras 148 C3 15 50N 84 18W
Patuca, Punta, Honduras 148 C3 15 49N 84 14W
Pâturages, Belgium 21 H3 50 25N 3 52 E
Pátzcuaro, Mexico 146 D4 19 30N 101 40W
Pau, France 24 E3 43 19N 0 25W
Pau, Gave de →, France 24 E2 43 33N 1 12W
Pau d' Arco, Brazil 154 C2 7 30 S 49 22W
Pau dos Ferros, Brazil 154 C4 6 7 S 38 10W
Paucartambo, Peru 156 C3 13 19 S 71 35W
Pauillac, France 24 C3 45 11N 0 46W
Pauini, Brazil 156 B4 7 40 S 66 58W
Pauini →, Brazil 153 D5 1 42 S 62 50W
Pauk, Burma 78 E5 21 27N 94 30 E
Paul I., Canada 129 A7 56 30N 61 20W
Paul Isnard, Fr. Guiana 153 C7 4 47N 54 1W
Paulding, U.S.A. 141 C12 41 8N 84 35W
Paulhan, France 24 E7 43 33N 3 28 E
Paulis = Isiro, Zaïre 106 B2 2 53N 27 40 E
Paulista, Brazil 154 C5 7 57 S 34 53W
Paulistana, Brazil 154 C3 8 9 S 41 9W
Paullina, U.S.A. 138 D7 42 59N 95 41W
Paulo Afonso, Brazil 154 C4 9 21 S 38 15W
Paulo de Faria, Brazil 155 F2 20 2 S 49 24W
Paulpietersburg, S. Africa 105 D5 27 23 S 30 50 E
Pauls Valley, U.S.A. 139 H6 34 44N 97 13W
Pauma Valley, U.S.A. 145 M10 33 16N 116 58W
Paungde, Burma 78 F5 18 29N 95 30 E
Pauni, India 82 D4 20 48N 79 40 E
Pausa, Peru 156 D3 15 16 S 73 22W
Pauto →, Colombia 152 B3 5 9N 70 55W
Pāveh, Iran 89 E12 35 3N 46 22 E
Pavelets, Russia 52 D4 53 49N 39 14 E
Pavia, Italy 38 C6 45 7N 9 8 E
Pāvilosta, Latvia 13 H19 56 53N 21 14 E
Pavlikeni, Bulgaria 43 D10 43 14N 25 20 E
Pavlodar, Kazakhstan 56 D8 52 33N 77 0 E
Pavlograd = Pavlohrad, Ukraine 51 H8 48 30N 35 52 E
Pavlohrad, Ukraine 51 H8 48 30N 35 52 E
Pavlovo, Russia 52 C6 55 58N 43 5 E
Pavlovo, Russia 57 C12 63 5N 115 25 E
Pavlovsk, Russia 52 E5 50 26N 40 5 E
Pavlovskaya, Russia 53 G4 46 17N 39 47 E
Pavlovskiy-Posad, Russia 52 C4 55 47N 38 42 E
Pavullo nel Frignano, Italy 38 D7 44 20N 10 50 E
Pavuvu, Solomon Is. 121 M10 9 4 S 159 8 E
Paw Paw, U.S.A. 141 B11 42 13N 85 53W
Pawahku, Burma 78 B7 26 11N 98 40 E
Pawan →, Indonesia 75 C4 1 55 S 110 0 E
Pawhuska, U.S.A. 139 G6 36 40N 96 20W
Pawling, U.S.A. 137 E11 41 34N 73 36W
Pawnee, Ill., U.S.A. 140 E7 39 36N 89 35W
Pawnee, Okla., U.S.A. 139 G6 36 20N 96 48W
Pawnee City, U.S.A. 138 E6 40 7N 96 9W
Pawpaw, U.S.A. 140 C8 41 41N 88 59W
Pawtucket, U.S.A. 137 E13 41 53N 71 23W
Paximádhia, Greece 32 D6 35 0N 24 35 E
Paxoí, Greece 44 E2 39 14N 20 12 E
Paxton, Ill., U.S.A. 141 D8 40 27N 88 6W
Paxton, Nebr., U.S.A. 138 E4 41 7N 101 21W
Payakumbuh, Indonesia 74 C2 0 20 S 100 35 E
Payerne, Switz. 28 C3 46 49N 6 56 E
Payette, U.S.A. 142 D5 44 5N 116 56W
Paymogo, Spain 37 H3 37 44N 7 21W
Payne, U.S.A. 141 C12 41 5N 84 44W
Payne Bay = Kangirsuk, Canada 127 B13 60 0N 70 0W
Paynes Find, Australia 113 E2 29 15 S 117 42 E
Paynesville, Liberia 100 D2 6 20N 10 45W
Paynesville, U.S.A. 138 C7 45 23N 94 43W
Paysandú, Uruguay 158 C4 32 19 S 58 8W
Payson, Ariz., U.S.A. 143 J8 34 14N 111 20W
Payson, Utah, U.S.A. 142 F8 40 3N 111 44W
Paz →, Guatemala 148 D1 13 44N 90 10W
Paz, B. la, Mexico 146 C2 24 15N 110 25W
Pāzanān, Iran 85 D6 30 35N 49 59 E
Pazar, Turkey 89 B9 41 10N 40 50 E
Pazarcık, Turkey 88 D7 37 30N 37 17 E
Pazardzhik, Bulgaria 43 E9 42 12N 24 20 E
Pazaryolu, Turkey 89 B9 40 21N 40 47 E
Pazin, Croatia 39 C10 45 14N 13 56 E
Pazña, Bolivia 156 D4 18 36 S 66 55W
Pčinja →, Macedonia 42 F6 41 50N 21 45 E
Pe Ell, U.S.A. 144 D3 46 34N 123 18W
Peabody, U.S.A. 137 D14 42 31N 70 56W
Peace →, Canada 130 B6 59 0N 111 25W
Peace Point, Canada 130 B6 59 7N 112 27W
Peace River, Canada 130 B5 56 15N 117 18W
Peach Springs, U.S.A. 143 J7 35 32N 113 25W
Peak, The = Kinder Scout, U.K. 16 D6 53 24N 1 52W
Peak Downs, Australia 114 C4 22 55 S 148 5 E
Peak Downs Mine, Australia 114 C4 22 17 S 148 11 E
Peak Hill, N.S.W., Australia 117 B8 32 47 S 148 11 E
Peak Hill, W. Austral., Australia 113 E2 25 35 S 118 43 E
Peak Ra., Australia 114 C4 22 50 S 148 20 E
Peake, Australia 116 C3 35 25 S 139 55 E
Peake Cr. →, Australia 115 D2 28 2 S 136 7 E
Peale, Mt., U.S.A. 143 G9 38 26N 109 14W
Pearblossom, U.S.A. 145 L9 34 30N 117 55W
Pearl, U.S.A. 140 E6 39 28N 90 38W
Pearl →, U.S.A. 139 K10 30 11N 89 32W
Pearl Banks, Sri Lanka 83 K4 8 45N 79 45 E
Pearl City, Hawaii, U.S.A. 132 H16 21 24N 157 59W
Pearl City, Ill., U.S.A. 140 B7 42 16N 89 50W
Pearsall, U.S.A. 139 L5 28 54N 99 6W
Pearse I., Canada 130 C2 54 52N 130 14W
Peary Land, Greenland 6 A6 82 40N 33 0W
Pease →, U.S.A. 139 H5 34 12N 99 2W
Pebane, Mozam. 107 F4 17 10 S 38 8 E
Pebas, Peru 152 D3 3 10 S 71 46W
Pebble, I., Falk. Is. 160 D5 51 20 S 59 40W
Pebble Beach, U.S.A. 144 J5 36 34N 121 57W
Peć, Serbia, Yug. 42 E5 42 40N 20 17 E
Peçanha, Brazil 155 E3 18 33 S 42 34W
Pecatonica, U.S.A. 140 B7 42 19N 89 22W
Pecatonica →, U.S.A. 140 B7 42 26N 89 12W
Péccioli, Italy 38 E7 43 33N 10 43 E
Pechea, Romania 46 D8 45 36N 27 49 E
Pechenga, Russia 48 A5 69 29N 31 4 E
Pechenizhyn, Ukraine 51 H3 48 30N 24 48 E
Pechiguera, Pta., Canary Is. 33 F6 28 51N 13 53W
Pechnezhskoye Vdkhr., Ukraine 51 G9 50 5N 36 54 E
Pechora →, Russia 48 A9 68 13N 54 15 E
Pechorskaya Guba, Russia 48 A9 68 40N 54 0 E
Pecica, Romania 46 C2 46 10N 21 3 E
Pečka, Serbia, Yug. 42 C4 44 18N 19 33 E
Pécora, C., Italy 40 C1 39 27N 8 23 E
Pečory, Russia 13 H22 57 48N 27 40 E
Pecos, U.S.A. 139 K3 31 26N 103 30W
Pecos →, U.S.A. 139 L3 29 42N 101 22W
Pécs, Hungary 31 E11 46 5N 18 15 E
Peddapalli, India 82 E4 18 40N 79 24 E
Peddapuram, India 82 F6 17 6N 82 8 E
Pedder, L., Australia 114 G4 42 55 S 146 10 E
Peddie, S. Africa 105 E4 33 14 S 27 7 E
Pédernales, Dom. Rep. 149 C5 18 2N 71 44W
Pedieos →, Cyprus 32 D12 35 10N 33 54 E
Pedirka, Australia 115 D2 26 40 S 135 14 E
Pedra Azul, Brazil 155 E3 16 2 S 41 17W
Pedra Grande, Recifes de, Brazil 155 E4 17 45 S 38 58W
Pedras Negras, Brazil 157 C5 12 51 S 62 54W
Pedreiras, Brazil 154 B3 4 32 S 44 40W
Pedro Afonso, Brazil 154 C2 9 0 S 48 10W
Pedro Cays, Jamaica 148 C4 17 5N 77 48W
Pedro Chico, Colombia 152 C3 1 4N 70 25W
Pedro de Valdivia, Chile 158 A2 22 55 S 69 38W
Pedro Juan Caballero, Paraguay 159 A4 22 30 S 55 40W
Pedro Muñoz, Spain 35 F2 39 25N 2 56W
Pedrógão Grande, Portugal 36 F2 39 55N 8 9W
Peebinga, Australia 116 C4 34 52 S 140 57 E
Peebles, U.K. 18 F5 55 40N 3 11W
Peebles, U.S.A. 141 F13 38 57N 83 24W
Peekskill, U.S.A. 137 E11 41 17N 73 55W
Peel, U.K. 16 C3 54 13N 4 40W
Peel →, Australia 117 A9 30 50 S 150 29 E
Peel →, Canada 126 B6 67 0N 135 0W
Peelwood, Australia 117 C8 34 7 S 149 27 E
Peene →, Germany 26 A9 54 9N 13 46 E
Peera Peera Poolanna L., Australia 115 D2 26 30 S 138 0 E
Peers, Canada 130 C5 53 40N 116 0W
Pegasus Bay, N.Z. 119 D8 43 20 S 173 10 E
Peggau, Austria 30 D8 47 12N 15 21 E
Pegnitz, Germany 27 F7 49 44N 11 31 E
Pegnitz →, Germany 27 F6 49 30N 10 59 E
Pego, Spain 35 G4 38 51N 0 8W
Pegu, Burma 78 G6 17 20N 96 29 E
Pegu Yoma, Burma 78 F5 19 0N 96 0 E
Pehčevo, Macedonia 42 F7 41 41N 22 55 E
Pehuajó, Argentina 158 D3 35 45 S 62 0W
Pei Xian, China 66 G9 34 44N 116 55 E
Peine, Chile 158 A2 23 45 S 68 8W
Peine, Germany 26 C6 52 19N 10 14 E
Peip'ing = Beijing, China 66 E9 39 55N 116 20 E
Peipus, L. = Chudskoye, Oz., Russia 13 G22 58 13N 27 30 E
Peissenberg, Germany 27 H7 47 48N 11 4 E
Peitz, Germany 26 D10 51 51N 14 24 E
Peixe, Brazil 155 D2 12 0 S 48 40W
Peixe →, Brazil 155 F1 21 31 S 51 58W
Peixoto de Azeredo →, Brazil 157 C6 10 6 S 55 31W
Peize, Neths. 20 B8 53 9N 6 30 E
Pek →, Serbia, Yug. 42 C6 44 45N 21 29 E
Pekalongan, Indonesia 75 D3 6 53 S 109 40 E
Pekan, Malaysia 77 L4 3 30N 103 25 E
Pekanbaru, Indonesia 74 B2 0 30N 101 15 E
Pekin, U.S.A. 140 D7 40 35N 89 40W
Peking = Beijing, China 66 E9 39 55N 116 20 E

250 Pleasant Bay

Pleasant Bay, *Canada* **129 C7** 46 51N 60 48W
Pleasant Hill, *Calif., U.S.A.* **144 H4** 37 57N 122 4W
Pleasant Hill, *Ill., U.S.A.* **140 E6** 39 27N 90 52W
Pleasant Hill, *Mo., U.S.A.* **140 F2** 38 47N 94 16W
Pleasant Hills, *Australia* **117 C7** 35 28 S 146 50 E
Pleasant Pt., *N.Z.* **119 E6** 44 16 S 171 9 E
Pleasanton, *U.S.A.* **139 L5** 28 58N 98 29W
Pleasantville, *Iowa, U.S.A.* **140 C3** 41 23N 93 18W
Pleasantville, *N.J., U.S.A.* **134 F8** 39 24N 74 32W
Pleasure Ridge Park, *U.S.A.* **141 F11** 38 9N 85 50W
Pléaux, *France* **24 C6** 45 8N 2 13 E
Pleiku, *Vietnam* **76 F7** 13 57N 108 0 E
Plélan-le-Grand, *France* **22 D4** 48 0N 2 7W
Plémet-la-Pierre, *France* **22 D4** 48 11N 2 36W
Pléneuf-Val-André, *France* **22 D4** 48 35N 2 32W
Plenița, *Romania* **46 E4** 44 14N 23 10 E
Plenty →, *Australia* **114 C2** 23 25 S 136 31 E
Plenty, B. of, *N.Z.* **118 D6** 37 45 S 177 0 E
Plentywood, *U.S.A.* **138 A2** 48 47N 104 34W
Plesetsk, *Russia* **48 B7** 62 43N 40 20 E
Plessisville, *Canada* **129 C5** 46 14N 71 47W
Plestin-les-Grèves, *France* **22 D3** 48 40N 3 39W
Pleszew, *Poland* **47 D4** 51 53N 17 47 E
Pleternica, *Croatia* **42 B2** 45 17N 17 48 E
Pletipi L., *Canada* **129 B5** 51 44N 70 6W
Pleven, *Bulgaria* **43 D9** 43 26N 24 37 E
Plevlja, *Montenegro, Yug.* **42 D4** 43 21N 19 21 E
Ploče = Kardeljovo, *Croatia* **42 D2** 43 4N 17 26 E
Płock, *Poland* **47 C6** 52 32N 19 40 E
Płock □, *Poland* **47 C6** 52 30N 19 45 E
Plöcken Passo, *Italy* **39 B9** 46 37N 12 57 E
Ploegsteert, *Belgium* **21 G1** 50 44N 2 53 E
Ploemeur, *France* **22 E3** 47 44N 3 26W
Ploërmel, *France* **22 E4** 47 55N 2 26W
Ploiești, *Romania* **46 E7** 44 57N 26 5 E
Plomárion, *Greece* **45 F8** 38 58N 26 24 E
Plombières-les-Bains, *France* **23 E13** 47 58N 6 27 E
Plomin, *Croatia* **39 C11** 45 8N 14 10 E
Plön, *Germany* **26 A6** 54 9N 10 24 E
Plöner See, *Germany* **26 A6** 54 10N 10 22 E
Plonge, Lac la, *Canada* **131 B7** 55 8N 107 20W
Płońsk, *Poland* **47 C7** 52 37N 20 21 E
Płoty, *Poland* **47 B2** 53 48N 15 18 E
Plouaret, *France* **22 D3** 48 37N 3 28W
Plouay, *France* **22 E3** 47 55N 3 21W
Ploučnice →, *Czech.* **30 A7** 50 46N 14 13 E
Ploudalmézeau, *France* **22 D2** 48 34N 4 41W
Plougasnou, *France* **22 D3** 48 42N 3 49W
Plouha, *France* **22 D4** 48 41N 2 57W
Plouhinec, *France* **22 E2** 48 0N 4 29W
Plovdiv, *Bulgaria* **43 E9** 42 8N 24 44 E
Plum, *U.S.A.* **136 F5** 40 29N 79 47W
Plum I., *U.S.A.* **137 E12** 41 11N 72 12W
Plumas, *U.S.A.* **144 F7** 39 45N 119 4W
Plummer, *U.S.A.* **142 C5** 47 20N 116 53W
Plumtree, *Zimbabwe* **107 G2** 20 27 S 27 55 E
Plunge, *Lithuania* **13 J19** 55 53N 21 59 E
Pluvigner, *France* **22 E3** 47 46N 3 1W
Plymouth, *U.K.* **17 G3** 50 22N 4 10W
Plymouth, *Calif., U.S.A.* **144 G6** 38 29N 120 51W
Plymouth, *Ill., U.S.A.* **140 D6** 40 18N 90 58W
Plymouth, *Ind., U.S.A.* **141 C10** 41 21N 86 19W
Plymouth, *Mass., U.S.A.* **137 E14** 41 57N 70 40W
Plymouth, *N.C., U.S.A.* **135 H7** 35 52N 76 43W
Plymouth, *N.H., U.S.A.* **137 C13** 43 46N 71 41W
Plymouth, *Pa., U.S.A.* **137 E9** 41 14N 75 57W
Plymouth, *Wis., U.S.A.* **134 D2** 43 45N 87 59W
Plynlimon = Pumlumon Fawr, *U.K.* **17 E4** 52 28N 3 46W
Plyusa, *Russia* **50 C5** 58 28N 29 27 E
Plyusa →, *Russia* **50 C5** 59 4N 28 6 E
Plyussa = Plyusa, *Russia* **50 C5** 58 28N 29 27 E
Plyussa = Plyusa →, *Russia* **50 C5** 59 4N 28 6 E
Plzeň, *Czech.* **30 B6** 49 45N 13 22 E
Pniewy, *Poland* **47 C3** 52 31N 16 16 E
Pô, *Burkina Faso* **101 C4** 11 14N 1 5W
Po →, *Italy* **39 D9** 44 57N 12 4 E
Po, Foci del, *Italy* **39 D9** 44 55N 12 30 E
Po Hai = Bo Hai, *China* **67 E10** 39 0N 119 0 E
Pobé, *Benin* **101 D5** 7 0N 2 56 E
Pobeda, *Russia* **57 C15** 65 12N 146 12 E
Pobedino, *Russia* **57 E15** 49 51N 142 49 E
Pobedy Pik, *Kyrgyzstan* **56 E8** 40 45N 79 58 E
Pobiedziska, *Poland* **47 C4** 52 29N 17 11 E
Pobla de Segur, *Spain* **34 C5** 42 15N 0 58 E
Pobladura de Valle, *Spain* **36 C5** 42 6N 5 44W
Pocahontas, *Ark., U.S.A.* **139 G9** 36 16N 90 58W
Pocahontas, *Ill., U.S.A.* **140 F7** 38 50N 89 33W
Pocahontas, *Iowa, U.S.A.* **140 B2** 42 44N 94 40W
Pocatello, *U.S.A.* **142 E7** 42 52N 112 27W
Počátky, *Czech.* **30 B8** 49 15N 15 14 E
Pochep, *Russia* **51 F7** 52 58N 33 29 E
Pochinki, *Russia* **52 C7** 54 41N 44 59 E
Pochinok, *Russia* **50 E7** 54 28N 32 29 E
Pöchlarn, *Austria* **30 C8** 48 12N 15 12 E
Pochutla, *Mexico* **147 D5** 15 50N 96 31W
Poci, *Venezuela* **153 B5** 5 57N 61 29W
Pocinhos, *Brazil* **154 C4** 7 4 S 36 3W
Pocito Casas, *Mexico* **146 B2** 28 32N 111 6W
Poções, *Brazil* **155 D3** 14 31 S 40 21W
Pocomoke City, *U.S.A.* **134 F8** 38 5N 75 34W
Poconé, *Brazil* **157 D6** 16 15 S 56 37W
Poços de Caldas, *Brazil* **159 A6** 21 50 S 46 33W
Poddebice, *Poland* **47 D5** 51 54N 18 58 E
Poděbrady, *Czech.* **30 A8** 50 9N 15 8 E
Podensac, *France* **24 D3** 44 40N 0 22W
Podgorač, *Croatia* **42 B3** 45 27N 18 13 E
Podgorica, *Montenegro, Yug.* **42 E4** 42 30N 19 19 E
Podilska Vysochyna, *Ukraine* **51 H4** 49 0N 28 0 E
Podkamennaya Tunguska →, *Russia* **57 C10** 61 50N 90 13 E
Podlapac, *Croatia* **39 D12** 44 37N 15 47 E
Podmokly, *Czech.* **30 A7** 50 48N 14 10 E
Podoleni, *Romania* **46 C7** 46 46N 26 39 E
Podolínec, *Slovak Rep.* **31 B13** 49 16N 20 31 E
Podolsk, *Russia* **52 C3** 55 25N 37 30 E
Podor, *Senegal* **100 B1** 16 40N 15 2W
Podporozhy, *Russia* **50 B8** 60 55N 34 2 E
Podravska Slatina, *Croatia* **42 B2** 45 42N 17 45 E
Podu Turcului, *Romania* **46 C8** 46 11N 27 25 E
Podujevo, *Serbia, Yug.* **42 E6** 42 54N 21 10 E
Poel, *Germany* **26 B7** 54 0N 11 25 E
Pofadder, *S. Africa* **104 D2** 29 10 S 19 22 E
Pogamasing, *Canada* **128 C3** 46 55N 81 50W
Poggiardo, *Italy* **41 B11** 40 3N 18 23 E
Poggibonsi, *Italy* **39 E8** 43 28N 11 9 E
Pogoanele, *Romania* **46 E8** 44 55N 27 0 E
Pogorzela, *Poland* **47 D4** 51 50N 17 12 E
Pogoso, *Zaïre* **103 D3** 6 46 S 17 12 E
Pogradeci, *Albania* **44 D2** 40 57N 20 37 E
Pogranitšnyi, *Russia* **60 B5** 44 25N 131 24 E
Poh, *Indonesia* **72 B2** 0 46 S 122 51 E
Pohang, *S. Korea* **67 F15** 36 1N 129 23 E
Pohjanmaa, *Finland* **12 E20** 62 58N 22 50 E
Pohnpei, *Pac. Oc.* **122 G7** 6 55N 158 10 E
Pohorelá, *Slovak Rep.* **31 C13** 48 50N 20 2 E
Pohořelice, *Czech.* **31 C9** 48 59N 16 31 E
Pohorje, *Slovenia* **39 B12** 46 30N 15 20 E
Poiana Mare, *Romania* **46 F4** 43 57N 23 5 E
Poiana Ruscăi, Munții, *Romania* **46 D3** 45 45N 22 25 E
Poindimié, *N. Cal.* **121 T19** 20 56 S 165 20 E
Poinsett, C., *Antarctica* **7 C8** 65 42 S 113 18 E
Point Edward, *Canada* **128 D3** 43 0N 82 30W
Point Pass, *Australia* **116 C3** 34 5 S 139 5 E
Point Pedro, *Sri Lanka* **83 K5** 9 50N 80 15 E
Point Pleasant, *N.J., U.S.A.* **137 F10** 40 5N 74 4W
Point Pleasant, *W. Va., U.S.A.* **134 F4** 38 51N 82 8W
Pointe-à-la Hache, *U.S.A.* **139 L10** 29 35N 89 55W
Pointe-à-Pitre, *Guadeloupe* **149 C7** 16 10N 61 30W
Pointe Noire, *Congo* **103 C2** 4 48 S 11 53 E
Poirino, *Italy* **38 D4** 44 56N 7 48 E
Poisonbush Ra., *Australia* **112 D3** 22 30 S 121 30 E
Poissy, *France* **23 D9** 48 55N 2 2 E
Poitiers, *France* **22 F7** 46 35N 0 20 E
Poitou, *France* **24 B3** 46 40N 0 10W
Poitou, Seuil du, *France* **24 B4** 46 20N 0 10 E
Poix de Picardie, *France* **23 C8** 49 47N 1 58 E
Poix-Terron, *France* **23 C11** 49 38N 4 38 E
Pojoaque Valley, *U.S.A.* **143 J11** 35 54N 106 1W
Pokataroo, *Australia* **115 D4** 29 30 S 148 36 E
Pokhvistnevo, *Russia* **52 D11** 53 36N 52 0 E
Poko, *Sudan* **95 F3** 5 41N 31 55 E
Poko, *Zaïre* **106 B2** 3 7N 26 52 E
Pokrov, *Russia* **52 C4** 55 55N 39 7 E
Pokrovka, *Kyrgyzstan* **55 B9** 42 20N 78 0 E
Pokrovsk = Engels, *Russia* **52 E8** 51 28N 46 6 E
Pokrovsk, *Russia* **57 C13** 61 29N 129 0 E
Pokrovsk-Uralskiy, *Russia* **54 A7** 60 10N 59 49 E
Pokrovskoye, *Russia* **53 G4** 47 25N 38 54 E
Pol, *Spain* **36 B3** 43 9N 7 20W
Pola = Pula, *Croatia* **39 D10** 44 54N 13 57 E
Pola, *Russia* **50 D7** 57 55N 32 0 E
Pola de Allande, *Spain* **36 B4** 43 16N 6 37W
Pola de Lena, *Spain* **36 B5** 43 10N 5 49W
Pola de Siero, *Spain* **36 B5** 43 24N 5 39W
Pola de Somiedo, *Spain* **36 B4** 43 5N 6 15W
Polacca, *U.S.A.* **143 J8** 35 50N 110 23W
Polan, *Iran* **85 E9** 25 30N 61 10 E
Poland ■, *Europe* **47 D7** 52 0N 20 0 E
Polanów, *Poland* **47 A3** 54 7N 16 41 E
Polatlı, *Turkey* **88 C5** 39 36N 32 9 E
Polatsk, *Belarus* **50 E5** 55 30N 28 50 E
Polcura, *Chile* **158 D1** 37 17 S 71 43W
Połczyn Zdrój, *Poland* **47 B3** 53 47N 16 5 E
Polden Hills, *U.K.* **17 F5** 51 7N 2 50W
Polesk, *Russia* **13 J19** 54 50N 21 8 E
Polesye = Pripet Marshes, *Europe* **51 F5** 52 10N 28 10 E
Polevskoy, *Russia* **54 C8** 56 26N 60 11 E
Polgar, *Hungary* **31 D14** 47 54N 21 6 E
Pŏlgyo-ri, *S. Korea* **67 G14** 34 51N 127 21 E
Poli, *Cameroon* **102 A2** 8 34N 13 15 E
Políaigos, *Greece* **45 H6** 36 45N 24 38 E
Policastro, G. di, *Italy* **41 C8** 40 0N 15 35 E
Police, *Poland* **47 B1** 53 33N 14 33 E
Polička, *Czech.* **31 B9** 49 43N 16 15 E
Polignano a Mare, *Italy* **41 B10** 41 0N 17 13 E
Poligny, *France* **23 F12** 46 50N 5 42 E
Políkhnitas, *Greece* **45 E8** 39 4N 26 10 E
Polillo, *Phil.* **70 D3** 14 43N 121 56 E
Polillo Is., *Phil.* **70 D4** 14 56N 122 0 E
Polillo Strait, *Phil.* **70 D3** 14 44N 121 51 E
Polis, *Cyprus* **32 D11** 35 2N 32 26 E
Polístena, *Italy* **41 D9** 38 24N 16 4 E
Políyiros, *Greece* **44 D5** 40 23N 23 25 E
Polk, *U.S.A.* **136 E5** 41 22N 79 56W
Polkowice, *Poland* **47 D3** 51 29N 16 3 E
Polla, *Italy* **41 B8** 40 31N 15 29 E
Pollachi, *India* **83 J3** 10 35N 77 0 E
Pollensa, *Spain* **33 B10** 39 54N 3 1 E
Pollensa, B. de, *Spain* **33 B10** 39 53N 3 8 E
Póllica, *Italy* **41 B8** 40 11N 15 3 E
Pollino, Mte., *Italy* **41 C9** 39 55N 16 11 E
Pollock, *U.S.A.* **138 C4** 45 55N 100 17W
Polna, *Russia* **50 C5** 58 31N 28 5 E
Polnovat, *Russia* **56 C7** 63 50N 65 54 E
Polo, *Ill., U.S.A.* **140 C7** 41 59N 89 35W
Polo, *Mo., U.S.A.* **140 E2** 39 33N 94 3W
Pology, *Ukraine* **51 J9** 47 29N 36 15 E
Polonne, *Ukraine* **51 G4** 50 6N 27 30 E
Polonnoye = Polonne, *Ukraine* **51 G4** 50 6N 27 30 E
Polski Trŭmbesh, *Bulgaria* **43 D10** 43 20N 25 38 E
Polsko Kosovo, *Bulgaria* **43 D10** 43 23N 25 38 E
Polson, *U.S.A.* **142 C6** 47 41N 114 9W
Poltava, *Ukraine* **51 H8** 49 35N 34 35 E
Põltsamaa, *Estonia* **13 G21** 58 41N 25 58 E
Polunochnoye, *Russia* **48 B11** 60 52N 60 25 E
Polur, *India* **83 H4** 12 32N 79 11 E
Põlva, *Estonia* **13 G22** 58 3N 27 3 E
Polyanovgrad, *Bulgaria* **43 E11** 42 39N 26 59 E
Polyarny, *Russia* **48 A5** 69 8N 33 20 E
Polynesia, *Pac. Oc.* **123 H11** 10 0 S 162 0W
Polynésie française = French Polynesia ■, *Pac. Oc.* **123 J13** 20 0 S 145 0W
Pomarance, *Italy* **38 E7** 43 18N 10 52 E
Pomarico, *Italy* **41 B9** 40 31N 16 33 E
Pomaro, *Mexico* **146 D4** 18 20N 103 18W
Pombal, *Brazil* **154 C4** 6 45 S 37 50W
Pombal, *Portugal* **36 F2** 39 55N 8 40W
Pómbia, *Greece* **32 D6** 35 0N 24 51 E
Pomeroy, *Ohio, U.S.A.* **134 F4** 39 2N 82 2W
Pomeroy, *Wash., U.S.A.* **142 C5** 46 28N 117 36W
Pomichna, *Ukraine* **51 H6** 48 13N 31 36 E
Pomio, *Papua N. G.* **120 C6** 5 32 S 151 33 E
Pomme de Terre L., *U.S.A.* **140 G3** 37 54N 93 19W
Pomona, *U.S.A.* **145 L9** 34 4N 117 45W
Pomorie, *Bulgaria* **43 E12** 42 32N 27 41 E
Pomos, *Cyprus* **32 D11** 35 9N 32 33 E
Pomos, C., *Cyprus* **32 D11** 35 10N 32 33 E
Pompano Beach, *U.S.A.* **135 M5** 26 14N 80 8W
Pompei, *Italy* **41 B7** 40 45N 14 30 E
Pompey, *France* **23 D13** 48 46N 6 6 E
Pompeys Pillar, *U.S.A.* **142 D10** 45 59N 107 57W
Ponape = Pohnpei, *Pac. Oc.* **122 G7** 6 55N 158 10 E
Ponask, L., *Canada* **128 B1** 54 0N 92 41W
Ponass L., *Canada* **131 C8** 52 16N 103 58W
Ponca, *U.S.A.* **138 D6** 42 34N 96 43W
Ponca City, *U.S.A.* **139 G6** 36 42N 97 5W
Ponce, *Puerto Rico* **149 C6** 18 1N 66 37W
Ponchatoula, *U.S.A.* **139 K9** 30 26N 90 26W
Poncheville, L., *Canada* **128 B4** 50 10N 76 55W
Poncin, *France* **25 B9** 46 6N 5 25 E
Pond, *U.S.A.* **145 K7** 35 43N 119 20W
Pond Inlet, *Canada* **127 A12** 72 40N 77 0W
Pondicherry, *India* **83 J4** 11 59N 79 50 E
Pondooma, *Australia* **116 B2** 33 29 S 136 59 E
Pondrôme, *Belgium* **21 H6** 50 6N 5 2 E
Ponds, I. of, *Canada* **129 B8** 53 27N 55 52W
Ponérihouen, *N. Cal.* **121 U19** 21 5 S 165 24 E
Ponferrada, *Spain* **36 C4** 42 32N 6 35W
Pongo, Wadi →, *Sudan* **95 F2** 8 42N 27 40 E
Poniatowa, *Poland* **47 D9** 51 11N 22 3 E
Poniec, *Poland* **47 D3** 51 48N 16 50 E
Ponikva, *Slovenia* **39 B12** 46 16N 15 26 E
Ponnaiyar →, *India* **83 J4** 11 50N 79 45 E
Ponnani, *India* **83 J2** 10 45N 75 59 E
Ponneri, *India* **83 H5** 13 20N 80 15 E
Ponnuru, *India* **83 F5** 16 5N 80 34 E
Ponoka, *Canada* **130 C6** 52 42N 113 40W
Ponomarevka, *Russia* **54 E5** 53 19N 54 8 E
Ponorogo, *Indonesia* **75 D4** 7 52 S 111 27 E
Ponot, *Phil.* **71 G4** 8 25N 123 0 E
Ponoy, *Russia* **48 A7** 67 0N 41 13 E
Ponoy →, *Russia* **48 A7** 66 59N 41 17 E
Pons, *France* **24 C3** 45 35N 0 34W
Pons, *Spain* **34 D6** 41 55N 1 12 E
Ponsul →, *Portugal* **37 F3** 39 40N 7 31W
Pont-à-Celles, *Belgium* **21 G4** 50 30N 4 22 E
Pont-à-Mousson, *France* **23 D13** 48 54N 6 1 E
Pont-Audemer, *France* **22 C7** 49 21N 0 30 E
Pont-Aven, *France* **22 E3** 47 51N 3 47W
Pont Canavese, *Italy* **38 C4** 45 25N 7 36 E
Pont-de-Roide, *France* **23 E13** 47 23N 6 45 E
Pont-de-Salars, *France* **24 D6** 44 18N 2 44 E
Pont-de-Vaux, *France* **23 F11** 46 26N 4 56 E
Pont-de-Veyle, *France* **25 B8** 46 17N 4 53 E
Pont-l'Abbé, *France* **22 E2** 47 52N 4 15W
Pont-l'Évêque, *France* **22 C7** 49 18N 0 11 E
Pont-St.-Esprit, *France* **25 D8** 44 16N 4 40 E
Pont-Saint-Martin, *Italy* **38 C4** 45 36N 7 48 E
Pont-sur-Yonne, *France* **23 D10** 48 18N 3 10 E
Ponta de Pedras, *Brazil* **154 B2** 1 23 S 48 52W
Ponta do Sol, *Madeira* **33 D2** 32 42N 17 7W
Ponta Grossa, *Brazil* **159 B5** 25 7 S 50 10W
Ponta Pora, *Brazil* **159 A4** 22 20 S 55 35W
Pontacq, *France* **24 E3** 43 11N 0 8W
Pontailler-sur-Saône, *France* **23 E12** 47 13N 5 25 E
Pontal →, *Brazil* **154 C3** 9 8 S 40 12W
Pontalina, *Brazil* **155 E2** 17 31 S 49 27W
Pontarlier, *France* **23 F13** 46 54N 6 20 E
Pontassieve, *Italy* **39 E8** 43 46N 11 26 E
Pontaubault, *France* **22 D5** 48 40N 1 20W
Pontaumur, *France* **24 C6** 45 52N 2 40 E
Pontcharra, *France* **25 C10** 45 26N 6 1 E
Pontchartrain L., *U.S.A.* **139 K9** 30 5N 90 5W
Pontchâteau, *France* **22 E4** 47 25N 2 5W
Ponte Alta, Serra do, *Brazil* **155 E2** 19 42 S 47 40W
Ponte Alta do Norte, *Brazil* **154 D2** 10 45 S 47 34W
Ponte Branca, *Brazil* **157 D7** 16 27 S 52 40W
Ponte da Barca, *Portugal* **36 D2** 41 48N 8 25W
Ponte de Sor, *Portugal* **37 F3** 39 17N 7 57W
Ponte dell'Ólio, *Italy* **38 D6** 44 52N 9 39 E
Ponte di Legno, *Italy* **38 B7** 46 16N 10 31 E
Ponte do Lima, *Portugal* **36 D2** 41 46N 8 35W
Ponte do Pungué, *Mozam.* **107 F3** 19 30 S 34 33 E
Ponte-Leccia, *France* **25 F13** 42 28N 9 13 E
Ponte nelle Alpi, *Italy* **39 B9** 46 11N 12 16 E
Ponte Nova, *Brazil* **155 F3** 20 25 S 42 54W
Ponte San Pietro, *Italy* **38 C6** 45 42N 9 35 E
Pontebba, *Italy* **39 B10** 46 30N 13 18 E
Pontecorvo, *Italy* **40 A6** 41 27N 13 40 E
Pontedera, *Italy* **38 E7** 43 40N 10 38 E
Pontefract, *U.K.* **16 D6** 53 42N 1 18W
Ponteix, *Canada* **131 D7** 49 46N 107 29W
Pontelandolfo, *Italy* **41 A7** 41 17N 14 41 E
Pontevedra, *Negros, Phil.* **71 F4** 10 22N 122 52 E
Pontevedra, *Panay, Phil.* **71 F4** 11 29N 122 50 E
Pontevedra, *Spain* **36 C2** 42 26N 8 40W
Pontevedra □, *Spain* **36 C2** 42 25N 8 39W
Pontevedra, R. de →, *Spain* **36 C2** 42 22N 8 45W
Pontevico, *Italy* **38 C7** 45 16N 10 5 E
Pontiac, *Ill., U.S.A.* **138 E10** 40 53N 88 38W
Pontiac, *Mich., U.S.A.* **141 B13** 42 38N 83 18W
Pontian Kecil, *Malaysia* **77 M4** 1 29N 103 23 E
Pontianak, *Indonesia* **75 C3** 0 3 S 109 15 E
Pontic Mts. = Kuzey Anadolu Dağları, *Turkey* **88 B7** 41 30N 35 0 E
Pontine Is. = Ponziane, Ísole, *Italy* **40 B5** 40 55N 12 57 E
Pontine Mts. = Kuzey Anadolu Dağları, *Turkey* **88 B7** 41 30N 35 0 E
Pontínia, *Italy* **40 A6** 41 25N 12 55 E
Pontivy, *France* **22 D4** 48 5N 2 58W
Pontoise, *France* **23 C9** 49 3N 2 5 E
Ponton →, *Canada* **130 B5** 58 27N 116 11W
Pontorson, *France* **22 D5** 48 34N 1 30W
Pontrémoli, *Italy* **38 D6** 44 22N 9 53 E
Pontresina, *Switz.* **29 D9** 46 29N 9 48 E
Pontrieux, *France* **22 D3** 48 42N 3 10W
Pontypool, *Canada* **136 B6** 44 6N 78 38W
Pontypool, *U.K.* **17 F4** 51 42N 3 2W
Pontypridd, *U.K.* **17 F4** 51 36N 3 20W
Ponza, *Italy* **40 B5** 40 54N 12 58 E
Ponziane, Ísole, *Italy* **40 B5** 40 55N 12 57 E
Poochera, *Australia* **115 E1** 32 43 S 134 51 E
Poole, *U.K.* **17 G6** 50 43N 1 59W
Pooley I., *Canada* **130 C3** 52 45N 128 15W
Poona = Pune, *India* **82 E1** 18 29N 73 57 E
Poona Bayabao, *Phil.* **71 H5** 7 56N 124 17 E
Poonamallee, *India* **83 H5** 13 3N 80 10 E
Pooncarie, *Australia* **116 B5** 33 22 S 142 31 E
Poonindie, *Australia* **116 C1** 34 34 S 135 54 E
Poopelloe L., *Australia* **116 A6** 31 40 S 144 0 E
Poopó, *Bolivia* **156 D4** 18 23 S 66 59W
Poopó, L. de, *Bolivia* **156 D4** 18 30 S 67 35W
Poor Knights Is., *N.Z.* **118 B3** 35 29 S 174 43 E
Popanyinning, *Australia* **113 F2** 32 40 S 117 2 E
Popayán, *Colombia* **152 C2** 2 27N 76 36W
Poperinge, *Belgium* **21 G1** 50 51N 2 42 E
Popigay, *Russia* **57 B12** 72 1N 110 39 E
Popilta, L., *Australia* **116 B4** 33 10 S 141 42 E
Popina, *Bulgaria* **43 C11** 44 7N 26 57 E
Popio L., *Australia* **116 B4** 33 10 S 141 52 E
Poplar, *U.S.A.* **138 A2** 48 7N 105 12W
Poplar →, *Man., Canada* **131 C9** 53 0N 97 19W
Poplar →, *N.W.T., Canada* **130 A4** 61 22N 121 52W
Poplar Bluff, *U.S.A.* **139 G9** 36 46N 90 24W
Poplarville, *U.S.A.* **139 K10** 30 51N 89 32W
Popocatépetl, Volcán, *Mexico* **147 D5** 19 2N 98 38W
Popokabaka, *Zaïre* **103 D3** 5 41 S 16 40 E
Pópoli, *Italy* **39 F10** 42 10N 13 50 E
Popondetta, *Papua N. G.* **120 E5** 8 48 S 148 17 E
Popovača, *Croatia* **39 C13** 45 30N 16 41 E
Popovo, *Bulgaria* **43 D11** 43 21N 26 18 E
Poppel, *Belgium* **21 F6** 51 27N 5 2 E
Poprád, *Slovak Rep.* **31 B13** 49 3N 20 18 E
Poprád →, *Slovak Rep.* **31 B13** 49 38N 20 42 E
Poradaha, *Bangla.* **78 D2** 23 51N 89 1 E
Porali →, *Pakistan* **79 D2** 25 35N 66 26 E
Porangaba, *Brazil* **156 B3** 8 48 S 70 36W
Porangahau, *N.Z.* **118 G5** 40 17 S 176 37 E
Porangatu, *Brazil* **155 D2** 13 26 S 49 10W
Porbandar, *India* **80 J3** 21 44N 69 43 E
Porce →, *Colombia* **152 B3** 7 28N 74 53W
Porcher I., *Canada* **130 C2** 53 50N 130 30W
Porco, *Bolivia* **157 D4** 19 50 S 65 59W
Porcos →, *Brazil* **155 D2** 12 42 S 45 7W
Porcuna, *Spain* **37 H6** 37 52N 4 11W
Porcupine →, *Canada* **131 B8** 59 11N 104 46W
Porcupine →, *U.S.A.* **126 B5** 66 34N 145 19W
Pordenone, *Italy* **39 C9** 45 57N 12 39 E
Pordim, *Bulgaria* **43 D9** 43 23N 24 51 E
Poreč, *Croatia* **39 C10** 45 14N 13 36 E
Porecatu, *Brazil* **155 F1** 22 43 S 51 24W
Poretskoye, *Russia* **52 C8** 55 9N 46 21 E
Pori, *Finland* **13 F19** 61 29N 21 48 E
Porí, *Greece* **45 J5** 35 58N 23 13 E
Porkhov, *Russia* **50 D5** 57 45N 29 38 E
Porlamar, *Venezuela* **153 A5** 10 57N 63 51W
Porlezza, *Italy* **38 B6** 46 2N 9 7 E
Porma →, *Spain* **36 C5** 42 49N 5 28W
Pornic, *France* **22 E4** 47 7N 2 5W
Poronaysk, *Russia* **57 E15** 49 13N 143 0 E
Póros, *Greece* **45 G5** 37 30N 23 30 E
Poroshiri-Dake, *Japan* **60 C11** 42 41N 142 52 E
Poroszló, *Hungary* **31 D13** 47 39N 20 40 E
Poroto Mts., *Tanzania* **107 D3** 9 0 S 33 30 E
Porpoise B., *Antarctica* **7 C9** 66 0 S 127 0 E
Porquerolles, I. de, *France* **25 F10** 43 0N 6 13 E
Porrentruy, *Switz.* **28 B4** 47 25N 7 6 E
Porreras, *Spain* **33 B10** 39 31N 3 2 E
Porretta, Passo di, *Italy* **38 D7** 44 2N 10 56 E
Porsangen, *Norway* **12 A21** 70 40N 25 40 E
Porsgrunn, *Norway* **14 E3** 59 10N 9 40 E
Port Adelaide, *Australia* **116 C3** 34 46 S 138 30 E
Port Alberni, *Canada* **130 D4** 49 14N 124 50W
Port Albert, *Australia* **117 E7** 38 42 S 146 42 E
Port Alfred, *Canada* **129 C5** 48 18N 70 53W
Port Alfred, *S. Africa* **104 E4** 33 36 S 26 55 E
Port Alice, *Canada* **130 C3** 50 20N 127 25W
Port Allegany, *U.S.A.* **136 E6** 41 48N 78 17W
Port Allen, *U.S.A.* **139 K9** 30 27N 91 12W
Port Alma, *Australia* **114 C5** 23 38 S 150 53 E
Port Angeles, *U.S.A.* **144 B3** 48 7N 123 27W
Port Antonio, *Jamaica* **148 C4** 18 10N 76 30W
Port Aransas, *U.S.A.* **139 M6** 27 50N 97 4W
Port Arthur = Lüshun, *China* **67 E11** 38 45N 121 15 E
Port Arthur, *Australia* **114 G4** 43 7 S 147 50 E
Port Arthur, *U.S.A.* **139 L8** 29 54N 93 56W
Port au Port B., *Canada* **129 C8** 48 40N 58 50W
Port-au-Prince, *Haiti* **149 C5** 18 40N 72 20W
Port Augusta, *Australia* **116 B2** 32 30 S 137 50 E
Port Augusta West, *Australia* **116 B2** 32 29 S 137 29 E
Port Austin, *U.S.A.* **128 D3** 44 3N 83 1W
Port Bell, *Uganda* **106 B3** 0 18N 32 35 E
Port Bergé Vaovao, *Madag.* **105 B8** 15 33 S 47 40 E
Port Blandford, *Canada* **129 C9** 48 20N 54 10W
Port Bou, *Spain* **34 C8** 42 25N 3 9 E
Port Bouët, *Ivory C.* **100 D4** 5 16N 3 57W
Port Bradshaw, *Australia* **114 A2** 12 30 S 137 20 E
Port Broughton, *Australia* **116 B2** 33 37 S 137 56 E
Port Burwell, *Canada* **128 D3** 42 40N 80 48W
Port Campbell, *Australia* **116 E5** 38 37 S 143 1 E
Port Canning, *India* **81 H13** 22 23N 88 40 E
Port-Cartier, *Canada* **129 B6** 50 2N 66 50W
Port Chalmers, *N.Z.* **119 F5** 45 49 S 170 30 E
Port Charles, *N.Z.* **118 C4** 36 33 S 175 30 E
Port Chester, *U.S.A.* **137 F11** 41 0N 73 40W
Port Clements, *Canada* **130 C2** 53 40N 132 10W
Port Clinton, *U.S.A.* **141 C14** 41 31N 82 56W
Port Colborne, *Canada* **128 D4** 42 50N 79 10W
Port Coquitlam, *Canada* **130 D4** 49 15N 122 45W
Port Credit, *Canada* **136 C5** 43 33N 79 35W
Port Curtis, *Australia* **114 C5** 23 57 S 151 20 E
Port Dalhousie, *Canada* **136 C5** 43 13N 79 16W
Port Darwin, *Australia* **112 B5** 12 24 S 130 45 E
Port Darwin, *Falk. Is.* **160 D5** 51 50 S 59 0W
Port Davey, *Australia* **114 G4** 43 16 S 145 55 E
Port-de-Bouc, *France* **25 E8** 43 24N 4 59 E
Port-de-Paix, *Haiti* **149 C5** 19 50N 72 50W
Port Dickson, *Malaysia* **77 L3** 2 30N 101 49 E
Port Douglas, *Australia* **114 B4** 16 30 S 145 30 E

Prikaspiyskaya Nizmennost 251

Port Dover, Canada **136 D4** 42 47N 80 12W
Port Edward, Canada **130 C2** 54 12N 130 10W
Port Elgin, Canada **128 D3** 44 25N 81 25W
Port Elizabeth, S. Africa **104 E4** 33 58 S 25 40 E
Port Ellen, U.K. **18 F2** 55 38N 6 11W
Port-en-Bessin, France **22 C6** 49 21N 0 45W
Port Erin, U.K. **16 C3** 54 5N 4 45W
Port Essington, Australia **112 B5** 11 15 S 132 10 E
Port Etienne = Nouâdhibou, Mauritania **98 D1** 20 54N 17 0W
Port Fairy, Australia **116 E5** 38 22 S 142 12 E
Port Fitzroy, N.Z. **118 C4** 36 8 S 175 20 E
Port Fouâd = Bûr Fuad, Egypt **94 H8** 31 15N 32 20 E
Port Gamble, U.S.A. **144 C4** 47 51N 122 35W
Port-Gentil, Gabon **102 C1** 0 40 S 8 50 E
Port Gibson, U.S.A. **139 K9** 31 58N 90 59W
Port Glasgow, U.K. **18 F4** 55 56N 4 41W
Port Harcourt, Nigeria **101 E6** 4 40N 7 10 E
Port Hardy, Canada **130 C3** 50 41N 127 30W
Port Harrison = Inukjuak, Canada **127 C12** 58 25N 78 15W
Port Hawkesbury, Canada **129 C7** 45 36N 61 22W
Port Hedland, Australia **112 D2** 20 25 S 118 35 E
Port Henry, U.S.A. **137 B11** 44 3N 73 28W
Port Hood, Canada **129 C7** 46 0N 61 32W
Port Hope, Canada **128 D4** 43 56N 78 20W
Port Hueneme, U.S.A. **145 L7** 34 7N 119 12W
Port Huron, U.S.A. **134 D4** 42 58N 82 26W
Port Iliç, Azerbaijan **89 C13** 38 53N 48 47 E
Port Isabel, U.S.A. **139 M6** 26 5N 97 12W
Port Jefferson, U.S.A. **137 F11** 40 57N 73 3W
Port Jervis, U.S.A. **137 E10** 41 22N 74 41W
Port-Joinville, France **22 F4** 46 45N 2 23W
Port Katon, Russia **53 G4** 46 52N 38 46 E
Port Kelang = Pelabuhan Kelang, Malaysia **77 L3** 3 0N 101 23 E
Port Kembla, Australia **117 C9** 34 52 S 150 49 E
Port Kenny, Australia **115 E1** 33 10 S 134 41 E
Port-la-Nouvelle, France **24 E7** 43 1N 3 3 E
Port Lairge = Waterford, Ireland **19 D4** 52 15N 7 8W
Port Laoise, Ireland **19 C4** 53 2N 7 18W
Port Lavaca, U.S.A. **139 L6** 28 37N 96 38W
Port-Leucate, France **24 F7** 42 53N 3 3 E
Port Lincoln, Australia **116 C1** 34 42 S 135 52 E
Port Loko, S. Leone **100 D2** 8 48N 12 46W
Port Louis, France **22 E3** 47 42N 3 22W
Port Louis, Mauritius **109 G4** 20 10 S 57 30 E
Port Lyautey = Kenitra, Morocco **98 B3** 34 15N 6 40W
Port MacDonnell, Australia **116 E4** 38 5 S 140 48 E
Port Macquarie, Australia **117 A10** 31 25 S 152 25 E
Port Maria, Jamaica **148 C4** 18 25N 76 55W
Port Mellon, Canada **130 D4** 49 32N 123 31W
Port-Menier, Canada **129 C7** 49 51N 64 15W
Port Morant, Jamaica **148 C4** 17 54N 76 19W
Port Moresby, Papua N. G. **120 E4** 9 24 S 147 8 E
Port Mourant, Guyana **153 B6** 6 15N 57 20W
Port Mouton, Canada **129 D7** 43 58N 64 50W
Port Musgrave, Australia **114 A3** 11 55 S 141 50 E
Port-Navalo, France **22 E4** 47 34N 2 54W
Port Nelson, Canada **131 B10** 57 3N 92 36W
Port Nicholson, N.Z. **118 H3** 41 20 S 174 52 E
Port Nolloth, S. Africa **104 D2** 29 17 S 16 52 E
Port Nouveau-Québec = Kangiqsualujjuaq, Canada **127 C13** 58 30N 65 59W
Port O'Connor, U.S.A. **139 L6** 28 26N 96 24W
Port of Spain, Trin. & Tob. **149 D7** 10 40N 61 31W
Port Orchard, U.S.A. **144 C4** 47 32N 122 38W
Port Orford, U.S.A. **142 E1** 42 45N 124 30W
Port Pegasus, N.Z. **119 H2** 47 12 S 167 41 E
Port Perry, Canada **128 D4** 44 6N 78 56W
Port Phillip B., Australia **115 F3** 38 10 S 144 50 E
Port Pirie, Australia **116 B3** 33 10 S 138 1 E
Port Pólnocny, Poland **47 A5** 54 25N 18 42 E
Port Radium = Echo Bay, Canada **126 B8** 66 5N 117 55W
Port Renfrew, Canada **130 D4** 48 30N 124 20W
Port Roper, Australia **114 A2** 14 45 S 135 25 E
Port Rowan, Canada **128 D3** 42 40N 80 30W
Port Safaga = Bûr Safâga, Egypt **94 B3** 26 43N 33 57 E
Port Said = Bûr Sa'îd, Egypt **94 H8** 31 16N 32 18 E
Port St. Joe, U.S.A. **135 L3** 29 49N 85 18W
Port St. Johns, S. Africa **105 E4** 31 38 S 29 33 E
Port-St.-Louis-du-Rhône, France **25 E8** 43 23N 4 49 E
Port San Vicente, Phil. **70 B4** 18 30N 122 8 E
Port Sanilac, U.S.A. **128 D3** 43 26N 82 33W
Port Saunders, Canada **129 B8** 50 40N 57 18W
Port Severn, Canada **136 B5** 44 48N 79 43W
Port Shepstone, S. Africa **105 E5** 30 44 S 30 28 E
Port Simpson, Canada **130 C2** 54 30N 130 20W
Port Stanley = Stanley, Falk. Is. **160 D5** 51 40 S 59 51W
Port Stanley, Canada **128 D3** 42 40N 81 10W
Port Sudan = Bûr Sûdân, Sudan **94 D4** 19 32N 37 9 E
Port-sur-Saône, France **23 E13** 47 42N 6 2 E
Port Talbot, U.K. **17 F4** 51 35N 3 47W
Port Taufiq = Bûr Taufiq, Egypt **94 J8** 29 54N 32 32 E
Port Townsend, U.S.A. **144 B4** 48 7N 122 45W
Port-Vendres, France **24 F7** 42 32N 3 8 E
Port Victoria, Australia **116 C2** 34 30 S 137 29 E
Port Vila, Pac. Oc. **111 D12** 17 45 S 168 18 E
Port Vladimir, Russia **48 A5** 69 25N 33 6 E
Port Wakefield, Australia **116 C3** 34 12 S 138 10 E
Port Washington, U.S.A. **134 D2** 43 23N 87 53W
Port Weld, Malaysia **77 K3** 4 50N 100 38 E
Portachuelo, Bolivia **157 D5** 17 10 S 63 20W
Portadown, U.K. **19 B5** 54 25N 6 27W
Portage, Mich., U.S.A. **141 B11** 42 12N 85 35W
Portage, Wis., U.S.A. **138 D10** 43 33N 89 28W
Portage →, U.S.A. **141 C14** 41 31N 83 5W
Portage La Prairie, Canada **131 D9** 49 58N 98 18W
Portageville, U.S.A. **139 G10** 36 26N 89 42W
Portalegre, Portugal **37 F3** 39 19N 7 25W
Portalegre □, Portugal **37 F3** 39 20N 7 40W
Portales, U.S.A. **139 H3** 34 11N 103 20W
Portarlington, Ireland **19 C4** 53 9N 7 14W
Porteirinha, Brazil **155 E3** 15 44 S 43 2W
Portel, Brazil **154 B1** 1 57 S 50 49W
Portel, Portugal **37 G3** 38 19N 7 41W
Porter, U.S.A. **141 C9** 41 36N 87 4W
Porter L., N.W.T., Canada **131 A7** 61 41N 108 5W
Porter L., Sask., Canada **131 B7** 56 20N 107 20W
Porterville, S. Africa **104 E2** 33 0 S 19 0 E
Porterville, U.S.A. **144 J8** 36 4N 119 1W
Porthcawl, U.K. **17 F4** 51 29N 3 42W
Porthill, U.S.A. **142 B5** 48 59N 116 30W
Portile de Fier, Europe **46 E3** 44 42N 22 30 E
Portimão, Portugal **37 H2** 37 8N 8 32W
Portland, N.S.W., Australia **117 B8** 33 20 S 150 0 E
Portland, Vic., Australia **116 E4** 38 20 S 141 35 E
Portland, Canada **137 B8** 44 42N 76 12W
Portland, Conn., U.S.A. **137 E12** 41 34N 72 38W
Portland, Ind., U.S.A. **141 D12** 40 26N 84 59W
Portland, Maine, U.S.A. **129 D5** 43 39N 70 16W
Portland, Mich., U.S.A. **141 B12** 42 52N 84 54W
Portland, Oreg., U.S.A. **144 E4** 45 32N 122 37W
Portland, I. of, U.K. **17 G5** 50 33N 2 26W
Portland B., Australia **116 E4** 38 15 S 141 45 E
Portland Bill, U.K. **17 G5** 50 31N 2 28W
Portland I., N.Z. **118 F6** 39 20 S 177 51 E
Portland Prom., Canada **127 C12** 58 40N 78 33W
Portlands Roads, Australia **114 A3** 12 36 S 143 25 E
Portneuf, Canada **129 C5** 46 43N 71 55W
Pôrto, Brazil **154 B3** 3 54 S 42 42W
Porto, France **25 F12** 42 16N 8 42 E
Porto, Portugal **36 D2** 41 8N 8 40W
Porto □, Portugal **36 D2** 41 8N 8 20W
Porto, G. de, France **25 F12** 42 17N 8 34 E
Pôrto Acre, Brazil **156 B4** 9 34 S 67 31W
Pôrto Alegre, Pará, Brazil **153 D7** 4 22 S 52 44W
Pôrto Alegre, Rio Grande do S., Brazil **159 C5** 30 5 S 51 10W
Porto Amboim = Gunza, Angola **103 E2** 10 50 S 13 50 E
Porto Argentera, Italy **38 D4** 44 15N 7 27 E
Porto Azzurro, Italy **38 F7** 42 46N 10 24 E
Porto Botte, Italy **40 C1** 39 3N 8 33 E
Pôrto Cajueiro, Brazil **157 C6** 11 3 S 55 53W
Porto Civitanova, Italy **39 E10** 43 19N 13 44 E
Porto Cristo, Spain **33 B10** 39 33N 3 20 E
Pôrto da Fôlha, Brazil **154 C4** 9 55 S 37 17W
Pôrto de Móz, Brazil **153 D7** 1 41 S 52 13W
Pôrto de Pedras, Brazil **154 C4** 9 10 S 35 17W
Pôrto des Meinacos, Brazil **157 C7** 12 33 S 53 7W
Porto Empédocle, Italy **40 E6** 37 17N 13 32 E
Pôrto Esperança, Brazil **157 D6** 19 37 S 57 29W
Pôrto Esperidão, Brazil **157 D6** 15 51 S 58 28W
Pôrto Franco, Brazil **154 C2** 6 20 S 47 24W
Porto Garibaldi, Italy **39 D9** 44 41N 12 14 E
Pôrto Grande, Brazil **153 C7** 0 42N 51 24W
Pôrto Jofre, Brazil **157 D6** 17 20 S 56 48W
Pórto Lágo, Greece **44 D7** 40 58N 25 6 E
Porto Mendes, Brazil **159 A5** 24 30 S 54 15W
Porto Moniz, Madeira **33 D2** 32 52N 17 11W
Pôrto Murtinho, Brazil **157 E6** 21 45 S 57 55W
Pôrto Nacional, Brazil **154 D2** 10 40 S 48 30W
Porto Novo, Benin **101 D5** 6 23N 2 42 E
Porto Petro, Spain **33 B10** 39 22N 3 13 E
Porto Recanati, Italy **39 E10** 43 26N 13 40 E
Porto San Giórgio, Italy **39 E10** 43 11N 13 48 E
Pôrto Santana, Brazil **153 D7** 0 3 S 51 11W
Porto Santo Stéfano, Italy **38 F8** 42 26N 11 7 E
Pôrto São José, Brazil **159 A5** 22 43 S 53 10W
Pôrto Seguro, Brazil **155 E4** 16 26 S 39 5W
Porto Tolle, Italy **39 D9** 44 56N 12 22 E
Porto Tórres, Italy **40 B1** 40 50N 8 24 E
Pôrto União, Brazil **159 B5** 26 10 S 51 10W
Pôrto Válter, Brazil **156 B3** 8 15 S 72 40W
Porto-Vecchio, France **25 G13** 41 35N 9 16 E
Pôrto Velho, Brazil **157 B5** 8 46 S 63 54W
Portobelo, Panama **148 E4** 9 35N 79 42W
Portoferráio, Italy **38 F7** 42 48N 10 20 E
Portogruaro, Italy **39 C9** 45 47N 12 50 E
Portola, U.S.A. **144 F6** 39 49N 120 28W
Portomaggiore, Italy **39 D8** 44 42N 11 48 E
Portoscuso, Italy **40 C1** 39 12N 8 24 E
Portovénere, Italy **38 D6** 44 3N 9 51 E
Portoviejo, Ecuador **152 D1** 1 7 S 80 28W
Portpatrick, U.K. **18 G3** 54 51N 5 7W
Portree, U.K. **18 D2** 57 25N 6 12W
Portrush, U.K. **19 A5** 55 12N 6 40W
Portsall, France **22 D2** 48 37N 4 45W
Portsmouth, Domin. **149 C7** 15 34N 61 27W
Portsmouth, U.K. **17 G6** 50 48N 1 6W
Portsmouth, N.H., U.S.A. **137 C14** 43 5N 70 45W
Portsmouth, Ohio, U.S.A. **134 F4** 38 44N 82 57W
Portsmouth, R.I., U.S.A. **137 E13** 41 36N 71 15W
Portsmouth, Va., U.S.A. **134 G7** 36 50N 76 18W
Portsoy, U.K. **18 D6** 57 41N 2 41W
Porttipahtan tekojärvi, Finland **12 B22** 68 5N 26 40 E
Portugal ■, Europe **36 F3** 40 0N 8 0W
Portugalete, Spain **34 B1** 43 19N 3 4W
Portuguesa □, Venezuela **152 B4** 9 10N 69 15W
Portumna, Ireland **19 C3** 53 6N 8 14W
Portville, U.S.A. **136 D6** 42 3N 78 20W
Porvenir, Bolivia **156 C4** 11 10 S 68 50W
Porvenir, Chile **160 D2** 53 10 S 70 16W
Porvoo, Finland **13 F21** 60 24N 25 40 E
Porzuna, Spain **37 F6** 39 9N 4 9W
Posada →, Italy **40 B2** 40 39N 9 45 E
Posadas, Argentina **159 B4** 27 30 S 55 50W
Posadas, Spain **37 H5** 37 47N 5 11W
Poschiavo, Switz. **29 D10** 46 19N 10 4 E
Posets, Spain **34 C5** 42 39N 0 25 E
Poseyville, U.S.A. **141 F9** 38 10N 87 47W
Poshan = Boshan, China **67 F9** 36 28N 117 49 E
Posht-e-Badam, Iran **85 C7** 33 2N 55 23 E
Posídhion, Ákra, Greece **44 E5** 39 57N 23 30 E
Posidium, Greece **45 J9** 35 30N 27 10 E
Poso, Indonesia **72 B2** 1 20 S 120 55 E
Poso, Danau, Indonesia **72 B2** 1 52 S 120 35 E
Posoegroenoe, Surinam **153 C6** 4 23N 55 43W
Posong, S. Korea **67 G14** 34 46N 127 5 E
Posse, Brazil **155 D2** 14 4 S 46 18W
Possel, C.A.R. **102 A3** 5 5N 19 10 E
Possession I., Antarctica **7 D11** 72 4 S 172 0 E
Pössneck, Germany **26 E7** 50 42N 11 34 E
Post, U.S.A. **139 J4** 33 12N 101 23W
Post Falls, U.S.A. **142 C5** 47 43N 116 57W
Postavy = Pastavy, Belarus **13 J22** 55 4N 26 50 E
Poste Maurice Cortier, Algeria **99 D5** 22 14N 1 2 E
Postmasburg, S. Africa **104 D3** 28 18 S 23 5 E
Postojna, Slovenia **39 C11** 45 46N 14 12 E
Poston, U.S.A. **145 M12** 34 0N 114 24W
Postville, U.S.A. **140 A5** 43 5N 91 34W
Potamós, Andikíthira, Greece **45 H4** 35 52N 23 15 E
Potamós, Kíthira, Greece **45 H4** 36 15N 22 58 E
Potchefstroom, S. Africa **104 D4** 26 41 S 27 7 E
Potcoava, Romania **46 E5** 44 30N 24 39 E
Poté, Brazil **155 E3** 17 49 S 41 49W
Poteau, U.S.A. **139 H7** 35 3N 94 37W
Poteet, U.S.A. **139 L5** 29 2N 98 35W
Potelu, Lacul, Romania **46 F5** 43 44N 24 20 E
Potenza, Italy **41 B8** 40 38N 15 48 E
Potenza →, Italy **39 E10** 43 25N 13 40 E
Potenza Picena, Italy **39 E10** 43 22N 13 37 E
Poteriteri, L., N.Z. **119 G2** 46 5 S 167 10 E
Potes, Spain **36 B6** 43 15N 4 42W
Potgietersrus, S. Africa **105 C4** 24 10 S 28 55 E
Poti, Georgia **53 J5** 42 10N 41 38 E
Potiraguá, Brazil **155 E4** 15 36 S 39 53W
Potiskum, Nigeria **101 C7** 11 39N 11 2 E
Potlogi, Romania **46 E6** 44 34N 25 34 E
Potomac →, U.S.A. **134 F7** 38 0N 76 23W
Potosí, Bolivia **157 D4** 19 38 S 65 50W
Potosi, U.S.A. **140 G6** 37 56N 90 47W
Potosí □, Bolivia **156 E4** 20 31 S 67 0W
Potosi Mt., U.S.A. **145 K11** 35 57N 115 29W
Pototan, Phil. **71 F4** 10 54N 122 38 E
Potrerillos, Chile **158 B2** 26 30 S 69 30W
Potsdam, Germany **26 C9** 52 25N 13 4 E
Potsdam, U.S.A. **137 B10** 44 40N 74 59W
Pottenstein, Germany **27 F7** 49 46N 11 24 E
Potter, U.S.A. **138 E3** 41 13N 103 19W
Pottery Hill = Abû Ballas, Egypt **94 C2** 24 26N 27 36 E
Pottstown, U.S.A. **137 F9** 40 15N 75 39W
Pottsville, U.S.A. **137 F8** 40 41N 76 12W
Pouancé, France **22 E5** 47 44N 1 10W
Pouce Coupé, Canada **130 B4** 55 40N 120 10W
Pouembout, N. Cal. **121 U18** 21 8 S 164 53 E
Poughkeepsie, U.S.A. **137 E11** 41 42N 73 56W
Pouilly-sur-Loire, France **23 E9** 47 17N 2 57 E
Poulaphouca Res., Ireland **19 C5** 53 8N 6 30W
Poulsbo, U.S.A. **144 C4** 47 44N 122 39W
Poum, N. Cal. **121 T18** 20 14 S 164 2 E
Poumadji, C.A.R. **102 A4** 5 56N 22 10 E
Pounga-Nganda, Gabon **102 C2** 2 58 S 10 51 E
Pourri, Mt., France **25 C10** 45 32N 6 52 E
Pouso Alegre, Mato Grosso, Brazil **157 C6** 11 46 S 57 16W
Pouso Alegre, Minas Gerais, Brazil **159 A6** 22 14 S 45 57W
Pouzauges, France **22 F6** 46 47N 0 50W
Povenets, Russia **48 B5** 62 50N 34 50 E
Poverty B., N.Z. **118 E7** 38 43 S 178 2 E
Povlen, Serbia, Yug. **42 C4** 44 9N 19 44 E
Póvoa de Lanhosa, Portugal **36 D2** 41 33N 8 15W
Póvoa de Varzim, Portugal **36 D2** 41 25N 8 46W
Povorino, Russia **52 E6** 51 12N 42 5 E
Powassan, Canada **128 C4** 46 5N 79 25W
Poway, U.S.A. **145 N9** 32 58N 117 2W
Powder →, U.S.A. **138 B2** 46 45N 105 26W
Powder River, U.S.A. **142 E10** 43 2N 106 59W
Powell, U.S.A. **142 D9** 44 45N 108 46W
Powell L., U.S.A. **143 H8** 36 57N 111 29W
Powell River, Canada **130 D4** 49 50N 124 35W
Powers, Mich., U.S.A. **134 C2** 45 41N 87 32W
Powers, Oreg., U.S.A. **142 E1** 42 53N 124 4W
Powers Lake, U.S.A. **138 A3** 48 34N 102 39W
Powys □, U.K. **17 E4** 52 20N 3 20W
Poxoreu, Brazil **157 D7** 15 50 S 54 23W
Poya, N. Cal. **121 U19** 21 19 S 165 7 E
Poyang Hu, China **69 C11** 29 5N 116 20 E
Poyarkovo, Russia **57 E13** 49 36N 128 41 E
Poysdorf, Austria **31 C9** 48 40N 16 37 E
Poza de la Sal, Spain **34 C1** 42 35N 3 31W
Poza Rica, Mexico **147 C5** 20 33N 97 27W
Pozanti, Turkey **88 D6** 37 25N 34 50 E
Požarevac, Serbia, Yug. **42 C6** 44 35N 21 18 E
Požega, Serbia, Yug. **42 D5** 43 53N 20 2 E
Pozhva, Russia **54 B6** 59 5N 56 5 E
Pozi, Taiwan **69 F13** 23 30N 120 13 E
Poznań, Poland **47 C3** 52 25N 16 55 E
Poznań □, Poland **47 C4** 52 30N 17 0 E
Pozo, U.S.A. **145 K6** 35 20N 120 24W
Pozo Alcón, Spain **35 H2** 37 42N 2 56W
Pozo Almonte, Chile **156 E4** 20 10 S 69 50W
Pozo Colorado, Paraguay **158 A4** 23 30 S 58 45W
Pozo del Dátil, Mexico **146 B2** 30 0N 112 15W
Pozoblanco, Spain **37 G6** 38 23N 4 51W
Pozorrubio, Phil. **70 C3** 16 7N 120 33 E
Pozuzo, Peru **156 C2** 10 5 S 75 35W
Pozzallo, Italy **41 F7** 36 43N 14 51 E
Pozzuoli, Italy **41 B7** 40 49N 14 7 E
Pra →, Ghana **101 D4** 5 1N 1 37W
Prabuty, Poland **47 B6** 53 47N 19 15 E
Prača, Bos.-H. **42 D3** 43 47N 18 43 E
Prachatice, Czech. **30 B7** 49 1N 14 0 E
Prachin Buri, Thailand **76 E3** 14 0N 101 25 E
Prachuap Khiri Khan, Thailand **77 G2** 11 49N 99 48 E
Pradelles, France **24 D7** 44 46N 3 52 E
Pradera, Colombia **152 C2** 3 25N 76 15W
Prades, France **24 F6** 42 38N 2 23 E
Prado, Brazil **155 E4** 17 20 S 39 13W
Prado del Rey, Spain **37 J5** 36 48N 5 33W
Præstø, Denmark **15 J6** 55 8N 12 2 E
Pragersko, Slovenia **39 B12** 46 27N 15 42 E
Prague = Praha, Czech. **30 A7** 50 5N 14 22 E
Praha, Czech. **30 A7** 50 5N 14 22 E
Prahecq, France **24 B3** 46 19N 0 26W
Prahita →, India **82 E4** 19 0N 79 55 E
Prahova □, Romania **46 D6** 45 10N 26 0 E
Prahova →, Romania **46 E6** 44 50N 25 50 E
Prahovo, Serbia, Yug. **42 C7** 44 18N 22 39 E
Praia, C. Verde Is. **8 G6** 14 55N 23 30W
Praid, Romania **46 C6** 46 32N 25 10 E
Prainha, Amazonas, Brazil **157 B5** 7 10 S 60 30W
Prainha, Pará, Brazil **153 D7** 1 45 S 53 30W
Prairie, Australia **114 C3** 20 50 S 144 35 E
Prairie →, U.S.A. **139 H5** 34 30N 99 23W
Prairie City, U.S.A. **142 D4** 44 28N 118 43W
Prairie du Chien, U.S.A. **140 A5** 43 3N 91 9W
Prairie du Rocher, U.S.A. **140 F6** 38 5N 90 6W
Prairies, Canada **126 C9** 52 0N 108 0W
Pramánda, Greece **44 E3** 39 32N 21 8 E
Pran Buri, Thailand **76 F2** 12 23N 99 55 E
Prang, Ghana **101 D4** 8 1N 0 56W
Prasonísi, Ákra, Greece **32 D9** 35 42N 27 46 E
Praszka, Poland **47 D5** 51 5N 18 31 E
Prata, Brazil **155 E2** 19 25 S 48 54W
Pratapgarh, India **80 G6** 24 2N 74 40 E
Prática di Mare, Italy **40 A5** 41 40N 12 26 E
Prätigau, Switz. **29 C9** 46 56N 9 44 E
Prato, Italy **38 E8** 43 53N 11 6 E
Prátola Peligna, Italy **39 F10** 42 6N 13 52 E
Pratovécchio, Italy **39 E8** 43 44N 11 43 E
Prats-de-Mollo-la-Preste, France **24 F6** 42 25N 2 27 E
Pratt, U.S.A. **139 G5** 37 39N 98 44W
Pratteln, Switz. **28 A5** 47 31N 7 41 E
Prattville, U.S.A. **135 J2** 32 28N 86 29W
Pravara →, India **82 E2** 19 35N 74 45 E
Pravdinsk, Russia **52 B6** 56 29N 43 28 E
Pravia, Spain **36 B4** 43 30N 6 12W
Praya, Indonesia **75 D5** 8 39 S 116 17 E
Pré-en-Pail, France **22 D6** 48 28N 0 12W
Pré-Saint-Didier, Italy **38 C4** 45 46N 6 59 E
Precordillera, Argentina **158 C2** 30 0 S 69 1W
Predáppio, Italy **39 D8** 44 6N 11 59 E
Predazzo, Italy **39 B8** 46 19N 11 36 E
Predejane, Serbia, Yug. **42 E7** 42 51N 22 9 E
Preeceville, Canada **131 C8** 51 57N 102 40W
Préfailles, France **22 E4** 47 9N 2 11W
Pregonero, Venezuela **152 B3** 8 1N 71 46W
Pregrada, Croatia **39 B12** 46 11N 15 45 E
Preiļi, Latvia **13 H22** 56 18N 26 43 E
Preko, Croatia **39 D12** 44 7N 15 14 E
Prelate, Canada **131 C7** 50 51N 109 24W
Prelog, Croatia **39 B13** 46 18N 16 32 E
Premier, Canada **130 B3** 56 4N 129 56W
Premont, U.S.A. **139 M5** 27 22N 98 7W
Premuda, Croatia **39 D11** 44 20N 14 36 E
Prenj, Bos.-H. **42 D2** 43 33N 17 53 E
Prenjasi, Albania **44 C2** 41 6N 20 32 E
Prentice, U.S.A. **138 C9** 45 33N 90 17W
Prenzlau, Germany **26 B9** 53 19N 13 50 E
Preobrazheniye, Russia **60 C6** 42 54N 133 54 E
Přerov, Czech. **31 B10** 49 28N 17 27 E
Presanella, Italy **38 B7** 46 13N 10 40 E
Prescott, Canada **128 D4** 44 45N 75 30W
Prescott, Ariz., U.S.A. **143 J7** 34 33N 112 28W
Prescott, Ark., U.S.A. **139 J8** 33 48N 93 23W
Preservation Inlet, N.Z. **119 G1** 46 8 S 166 35 E
Preševo, Serbia, Yug. **42 E6** 42 19N 21 39 E
Presho, U.S.A. **138 D4** 43 54N 100 3W
Presicce, Italy **41 C11** 39 54N 18 16 E
Presidencia de la Plaza, Argentina **158 B4** 27 0 S 59 50W
Presidencia Roque Saenz Peña, Argentina **158 B3** 26 45 S 60 30W
Presidente Epitácio, Brazil **155 F1** 21 56 S 52 6W
Presidente Hayes □, Paraguay **158 A4** 24 0 S 59 0W
Presidente Hermes, Brazil **157 C5** 11 17 S 61 55W
Presidente Prudente, Brazil **159 A5** 22 5 S 51 25W
Presidente Roxas, Phil. **71 F4** 11 26N 122 56 E
Presidio, Mexico **146 B4** 29 29N 104 23W
Presidio, U.S.A. **139 L2** 29 34N 104 22W
Preslav, Bulgaria **43 D11** 43 10N 26 52 E
Preslavska Planina, Bulgaria **43 D11** 43 10N 26 45 E
Prešov, Slovak Rep. **31 C14** 49 0N 21 15 E
Prespa, Bulgaria **43 F9** 41 44N 24 55 E
Prespa, L. = Prespansko Jezero, Macedonia **44 D3** 40 55N 21 0 E
Prespansko Jezero, Macedonia **44 D3** 40 55N 21 0 E
Presque Isle, U.S.A. **129 C6** 46 41N 68 1W
Presseger See, Austria **30 E6** 46 37N 13 26 E
Prestbury, U.K. **17 F5** 51 54N 2 2W
Prestea, Ghana **100 D4** 5 22N 2 7W
Presteigne, U.K. **17 E5** 52 17N 3 0W
Přeštice, Czech. **30 B6** 49 34N 13 20 E
Presto, Bolivia **157 D5** 18 55 S 64 56W
Preston, Canada **136 C4** 43 23N 80 21W
Preston, U.K. **16 D5** 53 46N 2 42W
Preston, Idaho, U.S.A. **142 E8** 42 6N 111 53W
Preston, Iowa, U.S.A. **140 B6** 42 3N 90 24W
Preston, Minn., U.S.A. **138 D8** 43 40N 92 5W
Preston, Nev., U.S.A. **142 G6** 38 55N 115 4W
Preston, C., Australia **112 D2** 20 51 S 116 12 E
Prestonpans, U.K. **18 F6** 55 58N 2 58W
Prestwick, U.K. **18 F4** 55 29N 4 37W
Prêto →, Amazonas, Brazil **153 D5** 0 8 S 64 6W
Prêto →, Bahia, Brazil **154 D3** 11 21 S 43 52W
Prêto do Igapó-Açu →, Brazil **153 D6** 4 26 S 59 48W
Pretoria, S. Africa **105 D4** 25 44 S 28 12 E
Preuilly-sur-Claise, France **22 F7** 46 51N 0 56 E
Préveza, Greece **45 F2** 38 57N 20 47 E
Préveza □, Greece **44 E2** 39 20N 20 40 E
Priazovskoye, Ukraine **51 J8** 46 44N 35 40 E
Pribilof Is., Bering S. **6 D17** 56 0N 170 0W
Priboj, Serbia, Yug. **42 D4** 43 35N 19 32 E
Příbram, Czech. **30 B7** 49 41N 14 2 E
Price, U.S.A. **142 G8** 39 36N 110 49W
Price I., Canada **130 C3** 52 23N 128 41W
Prichard, U.S.A. **135 K1** 30 44N 88 5W
Priego, Spain **34 E2** 40 26N 2 21W
Priego de Córdoba, Spain **37 H6** 37 27N 4 12W
Priekule, Latvia **13 H19** 56 26N 21 35 E
Prien, Germany **27 H8** 47 51N 12 20 E
Prienai, Lithuania **13 J20** 54 40N 23 55 E
Prieska, S. Africa **104 D3** 29 40 S 22 42 E
Priest →, U.S.A. **142 B5** 48 12N 116 54W
Priest L., U.S.A. **142 B5** 48 35N 116 52W
Priest Valley, U.S.A. **144 J6** 36 10N 120 39W
Priestly, Canada **130 C3** 54 8N 125 20W
Prieto Diaz, Phil. **70 E5** 13 2N 124 12 E
Prievidza, Slovak Rep. **31 C11** 48 46N 18 36 E
Prijedor, Bos.-H. **39 D13** 44 58N 16 41 E
Prijepolje, Serbia, Yug. **42 D4** 43 27N 19 40 E
Prikaspiyskaya Nizmennost = Caspian Depression, Eurasia **53 G9** 47 0N 48 0 E

252 Prikubanskaya Nizmennost

Prikubanskaya Nizmennost, *Russia* **53 H4** 45 39N 38 33 E
Prilep, *Macedonia* **42 F6** 41 21N 21 37 E
Priluki = Pryluky, *Ukraine* **51 G7** 50 30N 32 24 E
Prime Seal I., *Australia* **114 G4** 40 3 S 147 43 E
Primeira Cruz, *Brazil* **154 B3** 2 30 S 43 26W
Primorsk, *Russia* **50 B5** 60 22N 28 37 E
Primorsko, *Bulgaria* **43 E12** 42 15N 27 44 E
Primorsko-Akhtarsk, *Russia* **53 G4** 46 2N 38 10 E
Primorskoye, *Ukraine* **51 J9** 46 48N 36 20 E
Primrose L., *Canada* **131 C7** 54 55N 109 45W
Prince Albert, *Canada* **131 C7** 53 15N 105 50W
Prince Albert, *S. Africa* **104 E3** 33 12 S 22 2 E
Prince Albert Mts., *Antarctica* **7 D11** 76 0 S 161 30 E
Prince Albert Nat. Park, *Canada* **131 C7** 54 0N 106 25W
Prince Albert Pen., *Canada* **126 A8** 72 30N 116 0W
Prince Albert Sd., *Canada* **126 A8** 70 25N 115 0W
Prince Alfred, C., *Canada* **6 B1** 74 20N 124 40W
Prince Charles I., *Canada* **127 B12** 67 47N 76 12W
Prince Charles Mts., *Antarctica* **7 D6** 72 0 S 67 0 E
Prince Edward I. □, *Canada* **129 C7** 46 20N 63 20W
Prince Edward Is., *Ind. Oc.* **109 J2** 46 35 S 38 0 E
Prince George, *Canada* **130 C4** 53 55N 122 50W
Prince of Wales, C., *U.S.A.* **124 C3** 65 36N 168 5W
Prince of Wales I., *Australia* **114 A3** 10 40 S 142 10 E
Prince of Wales I., *Canada* **126 A10** 73 0N 99 0W
Prince of Wales I., *U.S.A.* **130 B2** 55 47N 132 50W
Prince Patrick I., *Canada* **6 B2** 77 0N 120 0W
Prince Regent Inlet, *Canada* **6 B3** 73 0N 90 0W
Prince Rupert, *Canada* **130 C2** 54 20N 130 20W
Princenhage, *Neths.* **21 F5** 51 9N 4 45 E
Princesa Isabel, *Brazil* **154 C4** 7 44 S 38 0W
Princess Charlotte B., *Australia* **114 A3** 14 25 S 144 0 E
Princess May Ras., *Australia* **112 C4** 15 30 S 125 30 E
Princess Royal I., *Canada* **130 C3** 53 0N 128 40W
Princeton, *Canada* **130 D4** 49 27N 120 30W
Princeton, *Calif., U.S.A.* **144 F4** 39 24N 122 1W
Princeton, *Ill., U.S.A.* **140 C7** 41 23N 89 28W
Princeton, *Ind., U.S.A.* **141 F9** 38 21N 87 34W
Princeton, *Ky., U.S.A.* **134 G2** 37 7N 87 53W
Princeton, *Mo., U.S.A.* **140 D3** 40 24N 93 35W
Princeton, *N.J., U.S.A.* **137 F10** 40 21N 74 39W
Princeton, *W. Va., U.S.A.* **134 G5** 37 22N 81 6W
Princeville, *U.S.A.* **140 D7** 40 56N 89 46W
Principe, I. de, *Atl. Oc.* **102 B1** 1 37N 7 27 E
Principe Chan., *Canada* **130 C2** 53 28N 130 0W
Principe da Beira, *Brazil* **157 C5** 12 20 S 64 30W
Prineville, *U.S.A.* **142 D3** 44 18N 120 51W
Prins Harald Kyst, *Antarctica* **7 D4** 70 0 S 35 1 E
Prinsesse Astrid Kyst, *Antarctica* **7 D3** 70 45 S 12 30 E
Prinsesse Ragnhild Kyst, *Antarctica* **7 D4** 70 15 S 27 30 E
Prinzapolca, *Nic.* **148 D3** 13 20N 83 35W
Prior, C., *Spain* **36 B2** 43 34N 8 17W
Priozersk, *Russia* **50 B6** 61 2N 30 7 E
Pripet = Prypyat →, *Europe* **51 G6** 51 20N 30 15 E
Pripet Marshes, *Europe* **51 F5** 52 10N 28 10 E
Pripyat Marshes = Pripet Marshes, *Europe* **51 F5** 52 10N 28 10 E
Pripyats = Prypyat →, *Europe* **51 G6** 51 20N 30 15 E
Prislop, Pasul, *Romania* **46 B6** 47 37N 25 15 E
Pristen, *Russia* **52 E3** 51 15N 36 44 E
Priština, *Serbia, Yug.* **42 E6** 42 40N 21 13 E
Pritzwalk, *Germany* **26 B8** 53 9N 12 10 E
Privas, *France* **25 D8** 44 45N 4 37 E
Priverno, *Italy* **40 A6** 41 28N 13 11 E
Privolzhsk, *Russia* **52 B5** 57 23N 41 16 E
Privolzhskaya Vozvyshennost, *Russia* **52 E7** 51 0N 46 0 E
Privolzhskiy, *Russia* **52 E8** 51 25N 46 3 E
Privolzhye, *Russia* **52 D9** 52 52N 48 33 E
Priyutnoye, *Russia* **53 G6** 46 12N 43 40 E
Priyutovo, *Russia* **54 E4** 53 55N 53 59 E
Prizren, *Serbia, Yug.* **42 E5** 42 13N 20 45 E
Prizzi, *Italy* **40 E6** 37 43N 13 26 E
Prnjavor, *Bos.-H.* **42 C2** 44 52N 17 43 E
Probolinggo, *Indonesia* **75 D4** 7 46 S 113 13 E
Prochowice, *Poland* **47 D3** 51 17N 16 20 E
Procida, *Italy* **41 B7** 40 46N 14 2 E
Proddatur, *India* **83 G4** 14 45N 78 30 E
Prodhromos, *Cyprus* **32 E11** 34 57N 32 50 E
Proença-a-Nova, *Portugal* **37 F3** 39 45N 7 54W
Prof. Van Blommestein Meer, *Surinam* **153 C6** 4 45N 55 5W
Profítis Ilías, *Greece* **32 C9** 36 17N 27 56 E
Profondeville, *Belgium* **21 H5** 50 23N 4 52 E
Progreso, *Mexico* **147 C7** 21 20N 89 40W
Prokhladnyy, *Russia* **53 J7** 43 50N 44 2 E
Prokletije, *Albania* **44 B1** 42 30N 19 45 E
Prokopyevsk, *Russia* **56 D9** 54 0N 86 45 E
Prokuplje, *Serbia, Yug.* **42 D6** 43 16N 21 36 E
Proletarsk, *Russia* **53 G5** 46 42N 41 50 E
Proletarskaya = Proletarsk, *Russia* **53 G5** 46 42N 41 50 E
Prome = Pyè, *Burma* **78 F5** 18 49N 95 13 E
Prophet →, *Canada* **130 B4** 58 48N 122 40W
Prophetstown, *U.S.A.* **140 C7** 41 40N 89 56W
Propriá, *Brazil* **154 D4** 10 13 S 36 51W
Propriano, *France* **25 G12** 41 41N 8 52 E
Proserpine, *Australia* **114 C4** 20 21 S 148 36 E
Prosna, *Poland* **47 D5** 51 1N 18 30 E
Prosperidad, *Phil.* **71 G5** 8 34N 125 52 E
Prosser, *U.S.A.* **142 C4** 46 12N 119 46W
Prostějov, *Czech.* **31 B10** 49 30N 17 9 E
Prostki, *Poland* **47 B9** 53 42N 22 25 E
Proston, *Australia* **115 D5** 26 8 S 151 32 E
Proszowice, *Poland* **31 A13** 50 13N 20 16 E
Protection, *U.S.A.* **139 G5** 37 12N 99 29W
Próti, *Greece* **45 G3** 37 5N 21 32 E
Provadiya, *Bulgaria* **43 D12** 43 12N 27 30 E
Proven, *Belgium* **21 G1** 50 54N 2 40 E
Provence, *France* **25 E9** 43 40N 5 46 E
Providence, *Ky., U.S.A.* **134 G2** 37 24N 87 46W
Providence, *R.I., U.S.A.* **137 E13** 41 49N 71 24W
Providence Bay, *Canada* **128 C3** 45 41N 82 15W
Providence C., *N.Z.* **119 F1** 45 59 S 166 29 E
Providence Mts., *U.S.A.* **143 J6** 35 10N 115 15W
Providencia, *Ecuador* **152 D2** 0 28 S 76 28W
Providencia, I. de, *Colombia* **148 D3** 13 25N 81 26W
Provideniya, *Russia* **57 C19** 64 23N 173 18W
Provins, *France* **23 D10** 48 33N 3 15 E
Provo, *U.S.A.* **142 F8** 40 14N 111 39W
Provost, *Canada* **131 C6** 52 25N 110 20W
Prozor, *Bos.-H.* **42 D2** 43 50N 17 34 E
Prudentópolis, *Brazil* **155 G1** 25 12 S 50 57W
Prud'homme, *Canada* **131 C7** 52 20N 105 54W
Prudnik, *Poland* **47 E4** 50 20N 17 38 E
Prüm, *Germany* **27 E2** 50 12N 6 25 E
Pruszcz Gdański, *Poland* **47 A5** 54 17N 18 40 E
Pruszków, *Poland* **47 C7** 52 9N 20 49 E
Prut →, *Romania* **46 D9** 45 28N 28 10 E
Pruzhany, *Belarus* **51 F3** 52 33N 24 28 E
Prvič, *Croatia* **39 D11** 44 55N 14 47 E
Prydz B., *Antarctica* **7 C6** 69 0 S 74 0 E
Pryluky, *Ukraine* **51 G7** 50 30N 32 24 E
Pryor, *U.S.A.* **139 G7** 36 19N 95 19W
Prypyat →, *Europe* **51 G6** 51 20N 30 15 E
Przasnysz, *Poland* **47 B7** 53 2N 20 45 E
Przedbórz, *Poland* **47 D6** 51 6N 19 53 E
Przedecz, *Poland* **47 C5** 52 20N 18 53 E
Przemyśl, *Poland* **31 B15** 49 50N 22 45 E
Przeworsk, *Poland* **31 A15** 50 6N 22 32 E
Przewóz, *Poland* **47 D1** 51 28N 14 57 E
Przhevalsk, *Kyrgyzstan* **55 B9** 42 30N 78 20 E
Przysuchla, *Poland* **47 D7** 51 22N 20 38 E
Psakhná, *Greece* **45 F5** 38 34N 23 35 E
Psará, *Greece* **45 F7** 38 37N 25 38 E
Psathoúra, *Greece* **44 E6** 39 30N 24 12 E
Psel →, *Ukraine* **51 H7** 49 10N 33 37 E
Pserimos, *Greece* **45 H9** 36 56N 27 12 E
Psíra, *Greece* **32 D7** 35 12N 25 52 E
Pskem →, *Uzbekistan* **55 C5** 41 38N 70 1 E
Pskemskiy Khrebet, *Uzbekistan* **55 C5** 42 0N 70 45 E
Pskent, *Uzbekistan* **55 C4** 40 54N 69 20 E
Pskov, *Russia* **50 D5** 57 50N 28 25 E
Pskovskoye, Ozero, *Russia* **13 H22** 58 0N 27 58 E
Psunj, *Croatia* **42 B2** 45 25N 17 19 E
Pszczyna, *Poland* **31 B11** 49 59N 18 58 E
Pteleón, *Greece* **45 E4** 39 3N 22 57 E
Ptich = Ptsich →, *Belarus* **51 F5** 52 9N 28 52 E
Ptolemaís, *Greece* **44 D3** 40 30N 21 43 E
Ptsich →, *Belarus* **51 F5** 52 9N 28 52 E
Ptuj, *Slovenia* **39 B12** 46 28N 15 50 E
Ptujska Gora, *Slovenia* **39 B12** 46 23N 15 47 E
Pu Xian, *China* **66 F6** 36 24N 111 6 E
Pua, *Thailand* **76 C3** 19 11N 100 55 E
Puán, *Argentina* **158 D3** 37 30 S 62 45W
Pu'an, *China* **68 E5** 25 46N 104 57 E
Puan, *S. Korea* **67 G14** 35 44N 126 44 E
Pu'apu'a, *W. Samoa* **121 W23** 13 34 S 172 9W
Pubei, *China* **68 F7** 22 16N 109 31 E
Pucacuro →, *Peru* **152 D3** 3 20 S 74 58W
Pucallpa, *Peru* **156 B3** 8 25 S 74 30W
Pucará, *Bolivia* **157 D5** 18 43 S 64 11W
Pucará, *Peru* **156 D3** 15 5 S 70 24W
Pucarani, *Bolivia* **156 D4** 16 23 S 68 30W
Pucheng, *China* **69 D12** 27 59N 118 31 E
Pucheni, *Romania* **46 D6** 45 12N 25 17 E
Pucio Pt., *Phil.* **71 F3** 11 46N 121 51 E
Pučišće, *Croatia* **39 E13** 43 22N 16 43 E
Puck, *Poland* **47 A5** 54 45N 18 23 E
Pucka, Zatoka, *Poland* **47 A5** 54 30N 18 40 E
Pudasjärvi, *Finland* **12 D22** 65 23N 26 53 E
Puding, *China* **68 D5** 26 18N 105 44 E
Pudozh, *Russia* **50 B9** 61 48N 36 32 E
Pudtol, *Phil.* **70 B3** 18 13N 121 22 E
Pudukkottai, *India* **83 J4** 10 28N 78 47 E
Puebla, *Mexico* **147 D5** 19 3N 98 12W
Puebla □, *Mexico* **147 D5** 18 30N 98 0W
Puebla de Alcocer, *Spain* **37 G5** 38 59N 5 14W
Puebla de Don Fadrique, *Spain* **35 H2** 37 58N 2 25W
Puebla de Don Rodrigo, *Spain* **37 F6** 39 5N 4 37W
Puebla de Guzmán, *Spain* **37 H3** 37 37N 7 15W
Puebla de Sanabria, *Spain* **36 C4** 42 4N 6 38W
Puebla de Trives, *Spain* **36 C3** 42 20N 7 10W
Puebla del Caramiñal, *Spain* **36 C2** 42 37N 8 56W
Pueblo, *U.S.A.* **138 F2** 38 16N 104 37W
Pueblo Hundido, *Chile* **158 B1** 26 20 S 70 5W
Pueblo Nuevo, *Venezuela* **152 B3** 8 26N 71 26W
Puelches, *Argentina* **158 D2** 38 5 S 65 51W
Puelén, *Argentina* **158 D2** 37 32 S 67 38W
Puente Alto, *Chile* **158 C1** 33 32 S 70 35W
Puente del Arzobispo, *Spain* **36 F5** 39 48N 5 10W
Puente-Genil, *Spain* **37 H6** 37 22N 4 47W
Puente la Reina, *Spain* **34 C3** 42 40N 1 49W
Puenteareas, *Spain* **36 C2** 42 10N 8 28W
Puentedeume, *Spain* **36 B2** 43 24N 8 10W
Puentes de Garcia Rodriguez, *Spain* **36 B3** 43 27N 7 50W
Pu'er, *China* **68 F3** 23 0N 101 15 E
Puerco →, *U.S.A.* **143 J10** 34 22N 107 50W
Puerta Galera, *Phil.* **70 E3** 13 30N 120 57 E
Puerto, *Canary Is.* **33 F2** 28 5N 17 20W
Puerto Acosta, *Bolivia* **156 D4** 15 32 S 69 15W
Puerto Aisén, *Chile* **160 C2** 45 27 S 73 0W
Puerto Ángel, *Mexico* **147 D5** 15 40N 96 29W
Puerto Arista, *Mexico* **147 D6** 15 56N 93 48W
Puerto Armuelles, *Panama* **148 E3** 8 20N 82 51W
Puerto Ayacucho, *Venezuela* **152 B4** 5 40N 67 35W
Puerto Barrios, *Guatemala* **148 C2** 15 40N 88 32W
Puerto Bermejo, *Argentina* **158 B4** 26 55 S 58 34W
Puerto Bermúdez, *Peru* **156 C3** 10 20 S 74 58W
Puerto Bolívar, *Ecuador* **152 D2** 3 19 S 79 55W
Puerto Cabello, *Venezuela* **152 A4** 10 28N 68 1W
Puerto Cabezas, *Nic.* **148 D3** 14 0N 83 30W
Puerto Cabo Gracias á Dios, *Nic.* **148 D3** 15 0N 83 10W
Puerto Capaz = Jebba, *Morocco* **98 A4** 35 11N 4 43W
Puerto Carreño, *Colombia* **152 B4** 6 12N 67 22W
Puerto Castilla, *Honduras* **148 C2** 16 0N 86 0W
Puerto Chicama, *Peru* **156 B2** 7 45 S 79 20W
Puerto Coig, *Argentina* **160 D3** 50 54 S 69 15W
Puerto Cortes, *Costa Rica* **148 E3** 8 55N 84 0W
Puerto Cortés, *Honduras* **148 C2** 15 51N 88 0W
Puerto Cumarebo, *Venezuela* **152 A4** 11 29N 69 30W
Puerto de Alcudia, *Spain* **33 B10** 39 50N 3 7 E
Puerto de Andraitx, *Spain* **33 B9** 39 32N 2 23 E
Puerto de Cabrera, *Spain* **33 B9** 39 8N 2 56 E
Puerto de Gran Tarajal, *Canary Is.* **33 F5** 28 13N 14 1W
Puerto de la Cruz, *Canary Is.* **33 F3** 28 24N 16 32W
Puerto de Pozo Negro, *Canary Is.* **33 F6** 28 19N 13 55W
Puerto de Sóller, *Spain* **33 B9** 39 48N 2 42 E
Puerto del Carmen, *Canary Is.* **33 F6** 28 55N 13 38W
Puerto del Rosario, *Canary Is.* **33 F6** 28 30N 13 52W
Puerto Deseado, *Argentina* **160 C3** 47 55 S 66 0W
Puerto Guaraní, *Paraguay* **157 E6** 21 18 S 57 55W
Puerto Heath, *Bolivia* **156 C4** 12 34 S 68 39W
Puerto Huitoto, *Colombia* **152 C3** 0 18N 74 3W
Puerto Inca, *Peru* **156 B3** 9 22 S 74 54W
Puerto Juárez, *Mexico* **147 C7** 21 11N 86 49W
Puerto La Cruz, *Venezuela* **153 A5** 10 13N 64 38W
Puerto Leguízamo, *Colombia* **152 D3** 0 12 S 74 46W
Puerto Limón, *Colombia* **152 C3** 3 23N 73 30W
Puerto Lobos, *Argentina* **160 B3** 42 0 S 65 3W
Puerto López, *Colombia* **152 C3** 4 5N 72 58W
Puerto Lumbreras, *Spain* **35 H3** 37 34N 1 48W
Puerto Madryn, *Argentina* **160 B3** 42 48 S 65 4W
Puerto Maldonado, *Peru* **156 C4** 12 30 S 69 10W
Puerto Manotí, *Cuba* **148 B4** 21 22N 76 50W
Puerto Mazarrón, *Spain* **35 H3** 37 34N 1 15W
Puerto Mercedes, *Colombia* **152 C3** 1 11N 72 53W
Puerto Miraña, *Colombia* **152 D3** 1 20 S 70 19W
Puerto Montt, *Chile* **160 B2** 41 28 S 73 0W
Puerto Morelos, *Mexico* **147 C7** 20 49N 86 52W
Puerto Nariño, *Colombia* **152 C4** 4 56N 67 48W
Puerto Natales, *Chile* **160 D2** 51 45 S 72 15W
Puerto Nuevo, *Colombia* **152 B4** 5 53N 69 56W
Puerto Nutrias, *Venezuela* **152 B4** 8 5N 69 18W
Puerto Ordaz, *Venezuela* **153 B5** 8 16N 62 44W
Puerto Padre, *Cuba* **148 B4** 21 13N 76 35W
Puerto Páez, *Venezuela* **152 B4** 6 13N 67 28W
Puerto Peñasco, *Mexico* **146 A2** 31 20N 113 33W
Puerto Pinasco, *Paraguay* **158 A4** 22 36 S 57 50W
Puerto Pirámides, *Argentina* **160 B4** 42 35 S 64 20W
Puerto Plata, *Dom. Rep.* **149 C5** 19 48N 70 45W
Puerto Pollensa, *Spain* **33 B10** 39 54N 3 4 E
Puerto Portillo, *Peru* **156 B3** 9 45 S 72 42W
Puerto Princesa, *Phil.* **71 G2** 9 46N 118 45 E
Puerto Quellón, *Chile* **160 B2** 43 7 S 73 37W
Puerto Quepos, *Costa Rica* **148 E3** 9 29N 84 6W
Puerto Real, *Spain* **37 J4** 36 33N 6 12W
Puerto Rico, *Bolivia* **156 C4** 11 5 S 67 38W
Puerto Rico, *Canary Is.* **33 G4** 27 47N 15 42W
Puerto Rico ■, *W. Indies* **149 C6** 18 15N 66 45W
Puerto Rico Trench, *Atl. Oc.* **149 C6** 19 50N 66 0W
Puerto Saavedra, *Chile* **160 A2** 38 47 S 73 24W
Puerto Sastre, *Paraguay* **158 A4** 22 2 S 57 55W
Puerto Siles, *Bolivia* **157 C4** 12 48 S 65 5W
Puerto Suárez, *Bolivia* **157 D6** 18 58 S 57 52W
Puerto Tejada, *Colombia* **152 C2** 3 14N 76 24W
Puerto Umbría, *Colombia* **152 C2** 0 52N 76 33W
Puerto Vallarta, *Mexico* **146 C3** 20 36N 105 15W
Puerto Varas, *Chile* **160 B2** 41 19 S 72 59W
Puerto Villazón, *Bolivia* **157 C5** 13 32 S 61 57W
Puerto Wilches, *Colombia* **152 B3** 7 21N 73 54W
Puertollano, *Spain* **37 G6** 38 43N 4 7W
Puertomarin, *Spain* **36 C3** 42 48N 7 36W
Puesto Cunambo, *Peru* **152 D2** 2 10 S 76 0W
Pueyrredón, L., *Argentina* **160 C2** 47 20 S 72 0W
Pugachev, *Russia* **52 D9** 52 0N 48 49 E
Puge, *China* **68 D4** 27 20N 102 31 E
Puge, *Tanzania* **106 C3** 4 45 S 33 11 E
Puget Sound, *U.S.A.* **142 C2** 47 50N 122 30W
Puget-Théniers, *France* **25 E10** 43 58N 6 53 E
Púglia □, *Italy* **41 B9** 41 15N 16 15 E
Pugo, *Phil.* **70 C3** 16 8N 121 31 E
Pugŏdong, *N. Korea* **67 C16** 42 5N 130 0 E
Pugu, *Tanzania* **106 D4** 6 55 S 39 4 E
Pūgūnzī, *Iran* **85 E8** 25 49N 59 10 E
Puha, *N.Z.* **118 E6** 38 30 S 177 50 E
Pui, *Romania* **46 D4** 45 30N 23 4 E
Puica, *Peru* **156 C3** 15 0 S 72 33W
Puieşti, *Romania* **46 C8** 46 25N 27 33 E
Puig Mayor, *Spain* **33 B9** 39 48N 2 47 E
Puigcerdá, *Spain* **34 C6** 42 24N 1 50 E
Puigmal, *Spain* **34 C7** 42 23N 2 7 E
Puigpuñent, *Spain* **33 B9** 39 38N 2 32 E
Puisaye, Collines de la, *France* **23 E10** 47 37N 3 20 E
Puiseaux, *France* **23 D9** 48 11N 2 30 E
Pujilí, *Ecuador* **152 D2** 0 57 S 78 41W
Pujon-chosuji, *N. Korea* **67 D14** 40 35N 127 35 E
Puka, *Albania* **44 B1** 42 2N 19 53 E
Pukaki L., *N.Z.* **119 E5** 44 4 S 170 1 E
Pukapuka, *Cook Is.* **123 J11** 10 53 S 165 49W
Pukatawagan, *Canada* **131 B8** 55 45N 101 20W
Pukchin, *N. Korea* **67 D13** 40 12N 125 45 E
Pukchŏng, *N. Korea* **67 D15** 40 14N 128 10 E
Pukearuhe, *N.Z.* **118 E3** 38 55 S 174 31 E
Pukekohe, *N.Z.* **118 D3** 37 12 S 174 55 E
Puketeraki Ra., *N.Z.* **119 C7** 42 58 S 172 13 E
Puketoi Ra., *N.Z.* **118 G5** 40 30 S 176 5 E
Pukeuri, *N.Z.* **119 F6** 45 4 S 171 2 E
Pukou, *China* **69 A12** 32 7N 118 38 E
Pula, *Croatia* **39 D10** 44 54N 13 57 E
Pula, *Italy* **40 D2** 39 1N 9 0 E
Pulacayo, *Bolivia* **156 E4** 20 25 S 66 41W
Pulaski, *N.Y., U.S.A.* **137 C8** 43 34N 76 8W
Pulaski, *Tenn., U.S.A.* **135 H2** 35 12N 87 2W
Pulaski, *Va., U.S.A.* **134 G5** 37 3N 80 47W
Puławy, *Poland* **47 D8** 51 23N 21 59 E
Pulga, *U.S.A.* **144 F5** 39 48N 121 29W
Pulgaon, *India* **82 D4** 20 44N 78 21 E
Pulicat, L., *India* **83 H5** 13 40N 80 15 E
Puliyangudi, *India* **83 K3** 9 11N 77 24 E
Pullabooka, *Australia* **117 B7** 33 44 S 147 46 E
Pullman, *U.S.A.* **142 C5** 46 44N 117 10W
Pulog, *Phil.* **70 C3** 16 40N 120 50 E
Púlpito do Sul, *Angola* **103 F2** 15 46 S 12 0 E
Pułtusk, *Poland* **47 C8** 52 43N 21 6 E
Pülümür, *Turkey* **89 C8** 39 30N 39 51 E
Pulupandan, *Phil.* **71 F4** 10 31N 122 48 E
Pumlumon Fawr, *U.K.* **17 E4** 52 28N 3 46W
Puna, *Bolivia* **157 D4** 19 45 S 65 28W
Puná, I., *Ecuador* **152 D1** 2 55 S 80 5W
Punakha, *Bhutan* **78 B2** 27 42N 89 52 E
Punalur, *India* **83 K3** 9 0N 76 56 E
Punasar, *India* **80 F5** 27 6N 73 6 E
Punata, *Bolivia* **157 D4** 17 32 S 65 50W
Punch, *India* **81 C6** 33 48N 74 4 E
Pune, *India* **82 E1** 18 29N 73 57 E
Pungsan, *N. Korea* **67 D15** 40 50N 128 9 E
Pungue, Ponte de, *Mozam.* **107 F3** 19 0 S 34 0 E
Puning, *China* **69 F11** 23 20N 116 12 E
Punjab □, *India* **80 D6** 31 0N 76 0 E
Punjab □, *Pakistan* **79 C4** 32 0N 74 30 E
Puno, *Peru* **156 D3** 15 55 S 70 3W
Punta Alta, *Argentina* **160 A4** 38 53 S 62 4W
Punta Arenas, *Chile* **160 D2** 53 10 S 71 0W
Punta Cardón, *Venezuela* **152 A3** 11 38N 70 14W
Punta Coles, *Peru* **156 D3** 17 43 S 71 23W
Punta de Bombón, *Peru* **156 D3** 17 10 S 71 48W
Punta de Díaz, *Chile* **158 B1** 28 0 S 70 45W
Punta Delgado, *Argentina* **160 B4** 42 43 S 63 38W
Punta Gorda, *Belize* **147 D7** 16 10N 88 45W
Punta Gorda, *U.S.A.* **135 M5** 26 56N 82 3W
Punta Prieta, *Mexico* **146 B2** 28 58N 114 17W
Punta Prima, *Spain* **33 B11** 39 48N 4 16 E
Puntabie, *Australia* **115 E1** 32 12 S 134 13 E
Puntarenas, *Costa Rica* **148 E3** 10 0N 84 50W
Punto Fijo, *Venezuela* **152 A3** 11 50N 70 13W
Punxsatawney, *U.S.A.* **136 F5** 40 57N 78 59W
Puqi, *China* **69 C9** 29 40N 113 50 E
Puquio, *Peru* **156 C3** 14 45 S 74 10W
Pur →, *Russia* **56 C8** 67 31N 77 55 E
Purace, Vol., *Colombia* **152 C2** 2 21N 76 23W
Puračić, *Bos.-H.* **42 C3** 44 33N 18 28 E
Puralia = Puruliya, *India* **81 H12** 23 17N 86 24 E
Purari →, *Papua N. G.* **120 D3** 7 49 S 145 0 E
Purbeck, Isle of, *U.K.* **17 G5** 50 39N 1 59W
Purcell, *U.S.A.* **139 H6** 35 1N 97 22W
Purchena Tetica, *Spain* **35 H2** 37 21N 2 21W
Puri, *India* **82 E7** 19 50N 85 58 E
Purificación, *Colombia* **152 C3** 3 51N 74 55W
Purmerend, *Neths.* **20 C5** 52 32N 4 58 E
Purna →, *India* **82 E3** 19 6N 77 2 E
Purnia, *India* **81 G12** 25 45N 87 31 E
Purukcahu, *Indonesia* **75 C4** 0 35 S 114 35 E
Puruliya, *India* **81 H12** 23 17N 86 24 E
Purus →, *Brazil* **153 D5** 3 42 S 61 28W
Puruvesi, *Finland* **50 B5** 61 50N 29 30 E
Pŭrvomay, *Bulgaria* **43 E10** 42 8N 25 17 E
Purwakarta, *Indonesia* **75 D3** 6 35 S 107 29 E
Purwodadi, *Jawa, Indonesia* **75 D4** 7 7 S 110 55 E
Purwodadi, *Jawa, Indonesia* **75 D3** 7 51 S 110 0 E
Purwokerto, *Indonesia* **75 D3** 7 25 S 109 14 E
Purworejo, *Indonesia* **75 D4** 7 43 S 110 2 E
Puryŏng, *N. Korea* **67 C15** 42 5N 129 43 E
Pus →, *India* **82 E3** 19 55N 77 55 E
Pusad, *India* **82 E3** 19 56N 77 36 E
Pusan, *S. Korea* **67 G15** 35 5N 129 0 E
Pushchino, *Russia* **57 D16** 54 10N 158 0 E
Pushkin, *Russia* **50 C6** 59 45N 30 25 E
Pushkino, *Russia* **52 E8** 51 16N 47 0 E
Pushkino, *Russia* **52 B3** 56 2N 37 49 E
Püspökladány, *Hungary* **31 D14** 47 19N 21 6 E
Pustoshka, *Russia* **50 D5** 56 20N 29 30 E
Puszczykowo, *Poland* **47 C3** 52 18N 16 49 E
Putahow L., *Canada* **131 B8** 59 54N 100 40W
Putao, *Burma* **78 B6** 27 28N 97 30 E
Putaruru, *N.Z.* **118 E4** 38 2 S 175 50 E
Putbus, *Germany* **26 A9** 54 22N 13 28 E
Puţeni, *Romania* **46 D8** 45 49N 27 42 E
Putian, *China* **69 E12** 25 23N 119 0 E
Putignano, *Italy* **41 B10** 40 51N 17 7 E
Putina, *Peru* **156 C4** 14 55 S 69 55W
Puting, Tanjung, *Indonesia* **75 C4** 3 31 S 111 46 E
Putlitz, *Germany* **26 B8** 53 15N 12 2 E
Putna, *Romania* **46 B6** 47 50N 25 33 E
Putna →, *Romania* **46 D8** 45 42N 27 26 E
Putnam, *U.S.A.* **137 E13** 41 55N 71 55W
Putnok, *Hungary* **31 C13** 48 18N 20 26 E
Putorana, Gory, *Russia* **57 C10** 69 0N 95 0 E
Putorino, *N.Z.* **118 F5** 39 4 S 176 58 E
Putre, *Chile* **156 D4** 18 12 S 69 35W
Puttalam Lagoon, *Sri Lanka* **83 K4** 8 15N 79 45 E
Putte, *Neths.* **21 F4** 51 22N 4 24 E
Putten, *Neths.* **20 D7** 52 16N 5 36 E
Puttgarden, *Germany* **26 A7** 54 30N 11 10 E
Puttur, *India* **83 H2** 12 46N 75 12 E
Putty, *Australia* **117 B9** 32 57 S 150 42 E
Putumayo →, *S. Amer.* **152 D4** 3 7 S 67 58W
Putuo, *China* **69 C14** 29 56N 122 20 E
Putussibau, *Indonesia* **75 B4** 0 50N 112 56 E
Pututahi, *N.Z.* **118 E6** 38 39 S 177 53 E
Puurs, *Belgium* **21 F4** 51 5N 4 17 E
Puy-de-Dôme, *France* **24 C6** 45 46N 2 57 E
Puy-de-Dôme □, *France* **24 C7** 45 40N 3 5 E
Puy-Guillaume, *France* **24 C7** 45 57N 3 29 E
Puy-l'Évêque, *France* **24 D5** 44 31N 1 9 E
Puyallup, *U.S.A.* **144 C4** 47 12N 122 18W
Puyang, *China* **66 G8** 35 40N 115 1 E
Puyehue, *Chile* **160 B2** 40 40 S 72 37W
Puylaurens, *France* **24 E6** 43 35N 2 0 E
Puyo, *Ecuador* **152 D2** 1 28 S 77 59W
Puysegur Pt., *N.Z.* **119 G1** 46 9 S 166 37 E
Pūzeh Rīg, *Iran* **85 E8** 27 20N 58 40 E
Pwani □, *Tanzania* **106 D4** 7 0 S 39 0 E
Pweto, *Zaïre* **107 D2** 8 25 S 28 51 E
Pwinbyu, *Burma* **78 E5** 20 23N 94 40 E
Pwllheli, *U.K.* **16 E3** 52 53N 4 25W
Pya-ozero, *Russia* **48 A5** 66 5N 30 58 E
Pyana →, *Russia* **52 C8** 55 43N 46 1 E
Pyandzh, *Tajikistan* **55 E4** 37 14N 69 6 E
Pyandzh →, *Afghan.* **79 A2** 37 15N 67 15 E
Pyandzh →, *Tajikistan* **55 E4** 37 6N 68 20 E
Pyapon, *Burma* **78 G5** 16 20N 95 40 E
Pyasina →, *Russia* **57 B9** 73 30N 87 0 E
Pyatigorsk, *Russia* **53 H6** 44 2N 43 6 E
Pyatykhatky, *Ukraine* **51 H7** 48 28N 33 38 E

Ripatransone 255

Reedsport, *U.S.A.* **142 E1** 43 42N 124 6W
Reedy Creek, *Australia* **116 D4** 36 58 S 140 2 E
Reefton, *Australia* **117 C7** 34 15 S 147 27 E
Reefton, *N.Z.* **119 C6** 42 6 S 171 51 E
Refahiye, *Turkey* **89 C8** 39 54N 38 47 E
Refugio, *U.S.A.* **139 L6** 28 18N 97 17W
Rega →, *Poland* **47 A2** 54 10N 15 18 E
Regalbuto, *Italy* **41 E7** 37 39N 14 38 E
Regen, *Germany* **27 G9** 48 58N 13 8 E
Regen →, *Germany* **27 F8** 49 1N 12 6 E
Regeneração, *Brazil* **154 C3** 6 15 S 42 41W
Regensburg, *Germany* **27 F8** 49 1N 12 6 E
Regensdorf, *Switz.* **29 B6** 47 26N 8 28 E
Réggio di Calábria, *Italy* **41 D8** 38 6N 15 39 E
Réggio nell'Emília, *Italy* **38 D7** 44 43N 10 36 E
Regina, *Canada* **131 C8** 50 27N 104 35W
Régina, *Fr. Guiana* **153 C7** 4 19N 52 8W
Registro, *Brazil* **159 A6** 24 29 S 47 49W
Reguengos de Monsaraz, *Portugal* **37 G3** 38 25N 7 32W
Rehar →, *India* **81 H10** 23 55N 82 40 E
Rehoboth, *Namibia* **104 C2** 23 15 S 17 4 E
Rehovot, *Israel* **91 D3** 31 54N 34 48 E
Rei-Bouba, *Cameroon* **102 A2** 8 40N 14 15 E
Reichenbach, *Germany* **26 E8** 50 37N 12 17 E
Reichenbach, *Switz.* **28 C5** 46 38N 7 42 E
Reid, *Australia* **113 F4** 30 49 S 128 26 E
Reid River, *Australia* **114 B4** 19 40 S 146 48 E
Reiden, *Switz.* **28 B5** 47 14N 7 59 E
Reidsville, *U.S.A.* **135 G6** 36 21N 79 40W
Reigate, *U.K.* **17 F7** 51 14N 0 12W
Reillo, *Spain* **34 F3** 39 54N 1 53W
Reims, *France* **23 C11** 49 15N 4 1 E
Reina Adelaida, Arch., *Chile* **160 D2** 52 20 S 74 0W
Reinach, *Aargau, Switz.* **28 B6** 47 14N 8 11 E
Reinach, *Basel, Switz.* **28 B5** 47 29N 7 35 E
Reinbeck, *U.S.A.* **140 B4** 42 19N 92 36W
Reindeer →, *Canada* **131 B8** 55 36N 103 11W
Reindeer I., *Canada* **131 C9** 52 30N 98 0W
Reindeer L., *Canada* **131 B8** 57 15N 102 15W
Reinga, C., *N.Z.* **118 A1** 34 25 S 172 43 E
Reinosa, *Spain* **36 B6** 43 2N 4 15W
Reinosa, Paso, *Spain* **36 C6** 42 56N 4 10W
Reitdiep, *Neths.* **20 B8** 53 20N 6 20 E
Reitz, *S. Africa* **105 D4** 27 48 S 28 29 E
Reivilo, *S. Africa* **104 D3** 27 36 S 24 8 E
Rejmyre, *Sweden* **15 F9** 58 50N 15 55 E
Rejowiec Fabryczny, *Poland* **47 D10** 51 5N 23 17 E
Reka →, *Slovenia* **39 C11** 45 40N 14 0 E
Rekinniki, *Russia* **57 C17** 60 51N 163 40 E
Rekovac, *Serbia, Yug.* **42 D6** 43 51N 21 3 E
Reliance, *Canada* **131 A7** 63 0N 109 20W
Remad, Oued →, *Algeria* **99 B4** 33 28N 1 20W
Rémalard, *France* **22 D7** 48 26N 0 47 E
Remarkable, Mt., *Australia* **116 B3** 32 48 S 138 10 E
Rembang, *Indonesia* **75 D4** 6 42 S 111 21 E
Rembau, *Malaysia* **74 B2** 2 35N 102 6 E
Remchi, *Algeria* **99 A4** 35 2N 1 26W
Remedios, *Colombia* **152 B3** 7 2N 74 41W
Remedios, *Panama* **148 E3** 8 15N 81 50W
Remeshk, *Iran* **85 E8** 26 55N 58 50 E
Remetea, *Romania* **46 C6** 46 45N 25 29 E
Remich, *Lux.* **21 J8** 49 32N 6 22 E
Remington, *U.S.A.* **141 D9** 40 46N 87 9W
Rémire, *Fr. Guiana* **153 C7** 4 53N 52 17W
Remiremont, *France* **23 D13** 48 2N 6 36 E
Remo, *Ethiopia* **95 F5** 6 48N 41 20 E
Remontnoye, *Russia* **53 G6** 46 34N 43 37 E
Remoulins, *France* **25 E8** 43 55N 4 35 E
Remscheid, *Germany* **26 D3** 51 11N 7 12 E
Ren Xian, *China* **66 F8** 37 8N 114 40 E
Renascença, *Brazil* **152 D4** 3 50 S 66 21W
Rend Lake, *U.S.A.* **140 F8** 38 2N 88 58W
Rende, *Italy* **41 C9** 39 20N 16 11 E
Rendeux, *Belgium* **21 H7** 50 14N 5 30 E
Rendína, *Greece* **45 E3** 39 4N 21 58 E
Rendova, *Solomon Is.* **121 M9** 8 33 S 157 17 E
Rendsburg, *Germany* **26 A5** 54 17N 9 39 E
Rene, *Russia* **57 C19** 66 2N 179 25W
Renfrew, *Canada* **128 C4** 45 30N 76 40W
Renfrew, *U.K.* **18 F4** 55 52N 4 24W
Rengat, *Indonesia* **74 C2** 0 30 S 102 45 E
Rengo, *Chile* **158 C1** 34 24 S 70 50W
Renhua, *China* **69 E9** 25 5N 113 40 E
Renhuai, *China* **68 D6** 27 48N 106 24 E
Reni, *Ukraine* **51 K5** 45 28N 28 15 E
Renigunta, *India* **83 H4** 13 38N 79 30 E
Renk, *Sudan* **95 E3** 11 50N 32 50 E
Renkum, *Neths.* **20 E7** 51 58N 5 43 E
Renmark, *Australia* **116 C4** 34 11 S 140 43 E
Rennell, *Solomon Is.* **121 N11** 11 40 S 160 10 E
Rennell Sd., *Canada* **130 C2** 53 23N 132 35W
Renner Springs T.O., *Australia* **114 B1** 18 20 S 133 47 E
Rennes, *France* **22 D5** 48 7N 1 41W
Rennes, Bassin de, *France* **22 E5** 48 0N 1 30W
Reno, *U.S.A.* **144 F7** 39 31N 119 48W
Reno →, *Italy* **39 D9** 44 38N 12 16 E
Renovo, *U.S.A.* **136 E7** 41 20N 77 45W
Renqiu, *China* **66 E9** 38 43N 116 5 E
Rensselaer, *Ind., U.S.A.* **141 D9** 40 57N 87 9W
Rensselaer, *N.Y., U.S.A.* **137 D11** 42 38N 73 45W
Rentería, *Spain* **34 B3** 43 19N 1 54W
Renton, *U.S.A.* **144 C4** 47 29N 122 12W
Renwick, *N.Z.* **119 B8** 41 30 S 173 51 E
Réo, *Burkina Faso* **100 C4** 12 28N 2 35W
Reotipur, *India* **81 G10** 25 33N 83 45 E
Repalle, *India* **83 F5** 16 2N 80 45 E
Répcelak, *Hungary* **31 D10** 47 24N 17 1 E
Republic, *Mich., U.S.A.* **134 B2** 46 25N 87 59W
Republic, *Wash., U.S.A.* **142 B4** 48 39N 118 44W
Republican →, *U.S.A.* **138 F6** 39 4N 96 48W
Republican City, *U.S.A.* **138 E5** 40 6N 99 13W
Republiek, *Surinam* **153 B6** 5 30N 55 13W
Repulse Bay, *Canada* **127 B11** 66 30N 86 30W
Requena, *Peru* **156 B3** 5 5 S 73 52W
Requena, *Spain* **35 F3** 39 30N 1 4W
Resadiye = Datça, *Turkey* **88 D2** 36 46N 27 40 E
Reşadiye, *Turkey* **88 B7** 40 23N 37 20 E
Resele, *Sweden* **14 A11** 63 20N 17 5 E
Resen, *Macedonia* **42 F6** 41 5N 21 0 E
Reserve, *Canada* **131 C8** 52 28N 102 39W
Reserve, *U.S.A.* **143 K9** 33 43N 108 45W
Resht = Rasht, *Iran* **89 D13** 37 20N 49 40 E
Resistencia, *Argentina* **158 B4** 27 30 S 59 0W
Reşiţa, *Romania* **42 B6** 45 18N 21 53 E
Resko, *Poland* **47 B2** 53 47N 15 25 E
Resolution I., *Canada* **127 B13** 61 30N 65 0W
Resolution I., *N.Z.* **119 F1** 45 40 S 166 40 E
Resplandes, *Brazil* **154 C2** 6 17 S 45 13W
Resplendor, *Brazil* **155 E3** 19 20 S 41 15W
Ressano Garcia, *Mozam.* **105 D5** 25 25 S 32 0 E
Reston, *Canada* **131 D8** 49 33N 101 6W
Reszel, *Poland* **47 A8** 54 4N 21 10 E
Retalhuleu, *Guatemala* **148 D1** 14 33N 91 46W
Reteag, *Romania* **46 B5** 47 10N 24 0 E
Retenue, L. de, *Zaïre* **107 E2** 11 0 S 27 0 E
Retford, *U.K.* **16 D7** 53 19N 0 56W
Rethel, *France* **23 C11** 49 30N 4 20 E
Rethem, *Germany* **26 C5** 52 47N 9 22 E
Réthímnon, *Greece* **32 D6** 35 18N 24 30 E
Réthímnon □, *Greece* **32 D6** 35 23N 24 28 E
Retiche, Alpi, *Switz.* **29 D10** 46 30N 10 0 E
Retie, *Belgium* **21 F6** 51 16N 5 5 E
Retiers, *France* **22 E5** 47 55N 1 23W
Retortillo, *Spain* **36 E4** 40 48N 6 21W
Rétság, *Hungary* **31 D12** 47 58N 19 10 E
Reuland, *Belgium* **21 H8** 50 12N 6 8 E
Réunion ■, *Ind. Oc.* **93 J9** 21 0 S 56 0 E
Reus, *Spain* **34 D6** 41 10N 1 5 E
Reusel, *Neths.* **21 F6** 51 21N 5 9 E
Reuss →, *Switz.* **29 B6** 47 16N 8 24 E
Reutlingen, *Germany* **27 G5** 48 29N 9 12 E
Reutte, *Austria* **30 D3** 47 29N 10 42 E
Reuver, *Neths.* **21 F8** 51 17N 6 5 E
Reval = Tallinn, *Estonia* **13 G21** 59 22N 24 48 E
Revda, *Russia* **54 C7** 56 48N 59 57 E
Revel, *France* **24 E6** 43 28N 2 0 E
Revelganj, *India* **81 G11** 25 50N 84 40 E
Revelstoke, *Canada* **130 C5** 51 0N 118 10W
Reventazón, *Peru* **156 B1** 6 10 S 80 58W
Revigny-sur-Ornain, *France* **23 D11** 48 49N 4 59 E
Revilla Gigedo, Is., *Pac. Oc.* **123 F16** 18 40N 112 0W
Revillagigedo I., *U.S.A.* **130 B2** 55 50N 131 20W
Revin, *France* **23 C11** 49 55N 4 39 E
Revolyutsii, Pik, *Tajikistan* **55 D6** 38 31N 72 21 E
Revuè →, *Mozam.* **107 F3** 19 50 S 34 0 E
Rewa, *India* **81 G9** 24 33N 81 25 E
Rewa →, *Guyana* **153 C6** 3 19N 58 42W
Rewari, *India* **80 E7** 28 15N 76 40 E
Rexburg, *U.S.A.* **142 E8** 43 49N 111 47W
Rey, *Iran* **85 C6** 35 35N 51 25 E
Rey, Rio del →, *Nigeria* **101 E6** 4 30N 8 48 E
Rey Malabo, *Eq. Guin.* **101 E6** 3 45N 8 50 E
Reyes, *Bolivia* **156 C4** 14 19 S 67 23W
Reyes, Pt., *U.S.A.* **144 H3** 38 0N 123 0W
Reyhanlı, *Turkey* **88 D7** 36 16N 36 35 E
Reykjahlið, *Iceland* **12 D5** 65 40N 16 55W
Reykjanes, *Iceland* **12 E2** 63 48N 22 40W
Reykjavík, *Iceland* **12 D3** 64 10N 21 57W
Reynolds, *Canada* **131 D9** 49 40N 95 55W
Reynolds, *U.S.A.* **140 C6** 41 20N 90 40W
Reynolds Ra., *Australia* **112 D5** 22 30 S 133 0 E
Reynoldsville, *U.S.A.* **136 E6** 41 5N 78 58W
Reynosa, *Mexico* **147 B5** 26 5N 98 18W
Rēzekne, *Latvia* **13 H22** 56 30N 27 17 E
Rezh, *Russia* **54 C8** 57 23N 61 24 E
Rezovo, *Bulgaria* **43 E13** 42 0N 28 0 E
Rezvān, *Iran* **85 E8** 27 34N 56 6 E
Rgotina, *Serbia, Yug.* **42 C7** 44 1N 22 17 E
Rhamnus, *Greece* **45 F6** 38 12N 24 3 E
Rharis, O. →, *Algeria* **99 C6** 26 0N 5 4 E
Rhayader, *U.K.* **17 E4** 52 18N 3 29W
Rheden, *Neths.* **20 D8** 52 3N 6 3 E
Rhein, *Canada* **131 C8** 51 25N 102 15W
Rhein →, *Europe* **20 E8** 51 52N 6 2 E
Rhein-Main-Donau-Kanal, *Germany* **27 F7** 49 1N 11 27 E
Rheinbach, *Germany* **26 E2** 50 38N 6 57 E
Rheine, *Germany* **26 C3** 52 17N 7 26 E
Rheineck, *Switz.* **29 B9** 47 28N 9 31 E
Rheinfelden, *Switz.* **28 A5** 47 32N 7 47 E
Rheinland-Pfalz □, *Germany* **27 E2** 50 0N 7 0 E
Rheinsberg, *Germany* **26 B8** 53 6N 12 54 E
Rheinwaldhorn, *Switz.* **29 D8** 46 30N 9 3 E
Rhenen, *Neths.* **20 E7** 51 58N 5 33 E
Rheriss, Oued →, *Morocco* **98 B4** 30 50N 4 34W
Rheydt, *Germany* **26 D2** 51 9N 6 26 E
Rhin = Rhein →, *Europe* **20 E8** 51 52N 6 2 E
Rhinau, *France* **23 D14** 48 19N 7 43 E
Rhine = Rhein →, *Europe* **20 E8** 51 52N 6 2 E
Rhineland-Palatinate = Rheinland-Pfalz □, *Germany* **27 E2** 50 0N 7 0 E
Rhinelander, *U.S.A.* **138 C10** 45 38N 89 25W
Rhino Camp, *Uganda* **106 B3** 3 0N 31 22 E
Rhir, Cap, *Morocco* **98 B3** 30 38N 9 54W
Rhisnes, *Belgium* **21 G5** 50 31N 4 48 E
Rho, *Italy* **38 C6** 45 32N 9 2 E
Rhode Island □, *U.S.A.* **137 E13** 41 40N 71 30W
Rhodes = Ródhos, *Greece* **32 C10** 36 15N 28 10 E
Rhodesia = Zimbabwe ■, *Africa* **107 F2** 19 0 S 30 0 E
Rhodope Mts. = Rhodopi Planina, *Bulgaria* **43 F9** 41 40N 24 20 E
Rhodopi Planina, *Bulgaria* **43 F9** 41 40N 24 20 E
Rhön = Hohe Rhön, *Germany* **27 E5** 50 24N 9 58 E
Rhondda, *U.K.* **17 F4** 51 39N 3 31W
Rhône □, *France* **25 C8** 45 54N 4 35 E
Rhône →, *France* **25 E8** 43 28N 4 42 E
Rhum, *U.K.* **18 E2** 57 0N 6 20W
Rhyl, *U.K.* **16 D4** 53 20N 3 29W
Rhymney, *U.K.* **17 F4** 51 46N 3 17W
Ri-Aba, *Eq. Guin.* **101 E6** 3 28N 8 40 E
Riachão, *Brazil* **154 C2** 7 20 S 46 37W
Riacho de Santana, *Brazil* **155 D3** 13 37 S 42 57W
Rialma, *Brazil* **155 E2** 15 18 S 49 34W
Riang, *India* **78 B4** 27 31N 92 56 E
Riaño, *Spain* **36 C5** 42 59N 5 0W
Rians, *France* **25 E9** 43 37N 5 44 E
Riansares →, *Spain* **34 F1** 39 32N 3 18W
Riasi, *India* **81 C6** 33 10N 74 50 E
Riau □, *Indonesia* **74 B2** 0 0 102 35 E
Riau, Kepulauan, *Indonesia* **74 B2** 0 30N 104 20 E
Riau Arch. = Riau, Kepulauan, *Indonesia* **74 B2** 0 30N 104 20 E
Riaza, *Spain* **34 D1** 41 18N 3 30W
Riaza →, *Spain* **34 D1** 41 42N 3 55W
Riba de Saelices, *Spain* **34 E2** 40 55N 2 17W
Ribadavia, *Spain* **36 C2** 42 17N 8 8W
Ribadeo, *Spain* **36 B3** 43 35N 7 5W
Ribadesella, *Spain* **36 B5** 43 30N 5 7W
Ribamar, *Brazil* **154 B3** 2 33 S 44 3W
Ribas, *Spain* **34 C7** 42 19N 2 15 E
Ribas do Rio Pardo, *Brazil* **157 E7** 20 27 S 53 46W
Ribāţ, *Yemen* **86 D4** 14 18N 44 15 E
Ribble →, *U.K.* **16 C5** 53 52N 2 25W
Ribe, *Denmark* **15 J2** 55 19N 8 44 E
Ribeauvillé, *France* **23 D14** 48 10N 7 20 E
Ribécourt, *France* **23 C9** 49 30N 2 55 E
Ribeira, *Spain* **36 C2** 42 36N 8 58W
Ribeira Brava, *Madeira* **33 D2** 32 41N 17 4W
Ribeira do Pombal, *Brazil* **154 D4** 10 50 S 38 32W
Ribeirão Prêto, *Brazil* **159 A6** 21 10 S 47 50W
Ribeiro Gonçalves, *Brazil* **154 C2** 7 32 S 45 14W
Ribemont, *France* **23 C10** 49 47N 3 27 E
Ribera, *Italy* **40 E6** 37 30N 13 16 E
Ribérac, *France* **24 C4** 45 15N 0 20 E
Riberalta, *Bolivia* **157 C4** 11 0 S 66 0W
Ribnica, *Slovenia* **39 C11** 45 45N 14 45 E
Rîbniţa, *Moldova* **51 J5** 47 45N 29 0 E
Ribnitz-Damgarten, *Germany* **26 A8** 54 15N 12 27 E
Ričany, *Czech.* **30 B7** 50 0N 14 40 E
Riccarton, *N.Z.* **119 D7** 43 32 S 172 37 E
Riccia, *Italy* **41 A7** 41 30N 14 50 E
Riccione, *Italy* **39 D9** 43 59N 12 39 E
Rice, *U.S.A.* **145 L12** 34 5N 114 51W
Rice L., *Canada* **136 B6** 44 12N 78 10W
Rice Lake, *U.S.A.* **138 C9** 45 30N 91 44W
Rich, *Morocco* **98 B4** 32 16N 4 30W
Rich Hill, *U.S.A.* **139 F7** 38 6N 94 22W
Richards Bay, *S. Africa* **105 D5** 28 48 S 32 6 E
Richards L., *Canada* **131 B7** 59 10N 107 10W
Richardson →, *Canada* **131 B6** 58 25N 111 14W
Richardson Mts., *N.Z.* **119 E3** 44 49 S 168 34 E
Richardson Springs, *U.S.A.* **144 F5** 39 51N 121 46W
Richardton, *U.S.A.* **138 B3** 46 53N 102 19W
Riche, C., *Australia* **113 F2** 34 36 S 118 47 E
Richelieu, *France* **22 E7** 47 0N 0 20 E
Richey, *U.S.A.* **138 B2** 47 39N 105 4W
Richfield, *Idaho, U.S.A.* **142 E6** 43 3N 114 9W
Richfield, *Utah, U.S.A.* **143 G8** 38 46N 112 5W
Richford, *U.S.A.* **137 B12** 45 0N 72 40W
Richibucto, *Canada* **129 C7** 46 42N 64 54W
Richland, *Ga., U.S.A.* **135 J3** 32 5N 84 40W
Richland, *Iowa, U.S.A.* **140 C4** 41 13N 92 0W
Richland, *Mo., U.S.A.* **140 G4** 37 51N 92 26W
Richland, *Oreg., U.S.A.* **142 D5** 44 46N 117 10W
Richland, *Wash., U.S.A.* **142 C4** 46 17N 119 18W
Richland Center, *U.S.A.* **138 D9** 43 21N 90 23W
Richlands, *U.S.A.* **134 G5** 37 6N 81 48W
Richmond, *N.S.W., Australia* **117 B9** 33 35 S 150 42 E
Richmond, *Queens., Australia* **114 C3** 20 43 S 143 8 E
Richmond, *N.Z.* **119 B8** 41 20 S 173 12 E
Richmond, *U.K.* **16 C6** 54 25N 1 43W
Richmond, *Calif., U.S.A.* **144 H4** 37 56N 122 21W
Richmond, *Ind., U.S.A.* **141 E12** 39 50N 84 53W
Richmond, *Ky., U.S.A.* **141 G12** 37 45N 84 18W
Richmond, *Mich., U.S.A.* **136 D2** 42 49N 82 45W
Richmond, *Mo., U.S.A.* **138 F8** 39 17N 93 58W
Richmond, *Tex., U.S.A.* **139 L7** 29 35N 95 46W
Richmond, *Utah, U.S.A.* **142 F8** 41 56N 111 48W
Richmond, *Va., U.S.A.* **134 G7** 37 33N 77 27W
Richmond, Mt., *N.Z.* **119 B8** 41 32 S 173 22 E
Richmond Ra., *Australia* **115 D5** 29 0 S 152 45 E
Richmond Ra., *N.Z.* **119 B8** 41 32 S 173 22 E
Richmond-upon-Thames, *U.K.* **17 F7** 51 27N 0 17W
Richterswil, *Switz.* **29 B7** 47 13N 8 43 E
Richton, *U.S.A.* **135 K1** 31 16N 88 56W
Richwood, *Ohio, U.S.A.* **141 D13** 40 26N 83 18W
Richwood, *W. Va., U.S.A.* **134 F5** 38 14N 80 32W
Ricla, *Spain* **34 D3** 41 31N 1 24W
Ricupe, *Angola* **103 E4** 14 37 S 21 25 E
Ridā', *Yemen* **86 D4** 14 25N 44 50 E
Ridder, *Kazakhstan* **56 D9** 50 20N 83 30 E
Ridderkerk, *Neths.* **20 E5** 51 52N 4 35 E
Riddes, *Switz.* **28 D4** 46 11N 7 14 E
Ridge Farm, *U.S.A.* **141 E9** 39 54N 87 39W
Ridgecrest, *U.S.A.* **145 K9** 35 38N 117 40W
Ridgedale, *Canada* **131 C8** 53 0N 104 10W
Ridgefield, *U.S.A.* **144 E4** 45 49N 122 45W
Ridgeland, *U.S.A.* **135 J5** 32 29N 80 59W
Ridgelands, *Australia* **114 C5** 23 16 S 150 17 E
Ridgetown, *Canada* **128 D3** 42 26N 81 52W
Ridgeville, *U.S.A.* **141 D11** 40 18N 85 2W
Ridgewood, *U.S.A.* **137 F10** 40 59N 74 7W
Ridgway, *Ill., U.S.A.* **141 G8** 37 48N 88 16W
Ridgway, *Pa., U.S.A.* **136 E6** 41 25N 78 44W
Riding Mountain Nat. Park, *Canada* **131 C8** 50 50N 100 0W
Ridley, Mt., *Australia* **113 F3** 33 12 S 122 7 E
Ried, *Austria* **30 C6** 48 14N 13 30 E
Riedlingen, *Germany* **27 G5** 48 9N 9 28 E
Riel, *Neths.* **21 E6** 51 31N 5 1 E
Rienza →, *Italy* **39 B8** 46 49N 11 47 E
Riesa, *Germany* **26 D9** 51 19N 13 17 E
Riesco, I., *Chile* **160 D2** 52 55 S 72 40W
Riesi, *Italy* **41 E7** 37 17N 14 5 E
Riet →, *S. Africa* **104 D3** 29 0 S 23 54 E
Rieti, *Italy* **39 F9** 42 24N 12 51 E
Rieupeyroux, *France* **24 D6** 44 19N 2 12 E
Riez, *France* **25 E10** 43 49N 6 6 E
Riffe L., *U.S.A.* **144 D4** 46 32N 122 26W
Rifle, *U.S.A.* **142 G10** 39 32N 107 47W
Rift Valley □, *Kenya* **106 B4** 0 20N 36 0 E
Rig Rig, *Chad* **97 F2** 14 13N 14 25 E
Rīga, *Latvia* **13 H21** 56 53N 24 8 E
Riga, G. of, *Latvia* **13 H20** 57 40N 23 45 E
Rīgān, *Iran* **85 D8** 28 37N 58 58 E
Rīgas Jūras Līcis = Riga, G. of, *Latvia* **13 H20** 57 40N 23 45 E
Rigaud, *Canada* **137 A10** 45 29N 74 18W
Rigby, *U.S.A.* **142 E8** 43 40N 111 55W
Rīgestān □, *Afghan.* **79 C2** 30 15N 65 0 E
Riggins, *U.S.A.* **142 D5** 45 25N 116 19W
Rignac, *France* **24 D6** 44 25N 2 16 E
Rigolet, *Canada* **129 B8** 54 10N 58 23W
Riihimäki, *Finland* **13 F21** 60 45N 24 48 E
Riiser-Larsen-halvøya, *Antarctica* **7 C4** 68 0 S 35 0 E
Rijau, *Nigeria* **101 C6** 11 8N 5 17 E
Rijeka, *Croatia* **39 C11** 45 20N 14 21 E
Rijeka Crnojevica, *Montenegro, Yug.* **42 E4** 42 24N 19 1 E
Rijen, *Neths.* **21 E5** 51 35N 4 55 E
Rijkevorsel, *Belgium* **21 F5** 51 21N 4 46 E
Rijn →, *Neths.* **20 D4** 52 12N 4 21 E
Rijnsberg, *Neths.* **20 D4** 52 11N 4 27 E
Rijsbergen, *Neths.* **21 E5** 51 31N 4 41 E
Rijssen, *Neths.* **20 D9** 52 19N 6 31 E
Rijswijk, *Neths.* **20 D4** 52 4N 4 22 E
Rikā', W. ar →, *Si. Arabia* **86 B4** 22 25N 44 50 E
Rike, *Ethiopia* **95 E4** 10 50N 39 53 E
Rikuzentakada, *Japan* **60 E10** 39 0N 141 40 E
Rila, *Bulgaria* **43 E8** 42 7N 23 7 E
Rila Planina, *Bulgaria* **42 E8** 42 10N 23 0 E
Riley, *U.S.A.* **142 E4** 43 32N 119 28W
Rima →, *Nigeria* **101 C6** 13 4N 5 10 E
Rimah, Wadi ar →, *Si. Arabia* **84 E4** 26 5N 41 30 E
Rimavská Sobota, *Slovak Rep.* **31 C13** 48 22N 20 2 E
Rimbey, *Canada* **130 C6** 52 35N 114 15W
Rimbo, *Sweden* **14 E12** 59 44N 18 21 E
Rimi, *Nigeria* **101 C6** 12 58N 7 43 E
Rímini, *Italy* **39 D9** 44 3N 12 33 E
Rîmna →, *Romania* **46 D8** 45 36N 27 3 E
Rîmnicu Sărat, *Romania* **46 D8** 45 26N 27 3 E
Rîmnicu Vîlcea, *Romania* **46 D5** 45 9N 24 21 E
Rimouski, *Canada* **129 C6** 48 27N 68 30W
Rimrock, *U.S.A.* **144 D5** 46 38N 121 10W
Rinca, *Indonesia* **72 C1** 8 45 S 119 35 E
Rincón de Romos, *Mexico* **146 C4** 22 14N 102 18W
Rinconada, *Argentina* **158 A2** 22 26 S 66 10W
Ringarum, *Sweden* **15 F10** 58 21N 16 26 E
Ringe, *Denmark* **15 J4** 55 13N 10 28 E
Ringgold Is., *Fiji* **121 A3** 16 15 S 179 25W
Ringim, *Nigeria* **101 C6** 12 13N 9 10 E
Ringkøbing, *Denmark* **15 H2** 56 5N 8 15 E
Ringling, *U.S.A.* **142 C8** 46 16N 110 49W
Ringsaker, *Norway* **14 D4** 60 54N 10 45 E
Ringsted, *Denmark* **15 J5** 55 25N 11 46 E
Ringvassøy, *Norway* **12 B18** 69 56N 19 15 E
Rinía, *Greece* **45 G7** 37 23N 25 13 E
Rinjani, *Indonesia* **75 D5** 8 24 S 116 28 E
Rinteln, *Germany* **26 C5** 52 10N 9 8 E
Río, Punta del, *Spain* **35 J2** 36 49N 2 24W
Rio Branco, *Brazil* **156 B4** 9 58 S 67 49W
Río Branco, *Uruguay* **159 C5** 32 40 S 53 40W
Rio Brilhante, *Brazil* **159 A5** 21 48 S 54 33W
Río Bueno, *Chile* **160 B2** 40 19 S 72 58W
Río Chico, *Venezuela* **152 A4** 10 19N 65 59W
Rio Claro, *Brazil* **159 A6** 22 19 S 47 35W
Rio Claro, *Trin. & Tob.* **149 D7** 10 20N 61 25W
Río Colorado, *Argentina* **160 A4** 39 0 S 64 0W
Río Cuarto, *Argentina* **158 C3** 33 10 S 64 25W
Rio das Pedras, *Mozam.* **105 C6** 23 8 S 35 28 E
Rio de Contas, *Brazil* **155 D3** 13 36 S 41 48W
Rio de Janeiro, *Brazil* **155 F3** 23 0 S 43 12W
Rio de Janeiro □, *Brazil* **155 F3** 22 50 S 43 0W
Rio do Prado, *Brazil* **155 E3** 16 35 S 40 34W
Rio do Sul, *Brazil* **159 B6** 27 13 S 49 37W
Río Gallegos, *Argentina* **160 D3** 51 35 S 69 15W
Río Grande, *Argentina* **160 D3** 53 50 S 67 45W
Río Grande, *Bolivia* **156 E4** 20 51 S 67 17W
Rio Grande, *Brazil* **159 C5** 32 0 S 52 20W
Río Grande, *Mexico* **146 C4** 23 50N 103 2W
Río Grande, *Nic.* **148 D3** 12 54N 83 33W
Rio Grande →, *U.S.A.* **139 N6** 25 57N 97 9W
Rio Grande City, *U.S.A.* **139 M5** 26 23N 98 49W
Río Grande del Norte →, *N. Amer.* **133 E7** 26 0N 97 0W
Rio Grande do Norte □, *Brazil* **154 C4** 5 40 S 36 0W
Rio Grande do Sul □, *Brazil* **159 C5** 30 0 S 53 0W
Río Hato, *Panama* **148 E3** 8 22N 80 10W
Río Lagartos, *Mexico* **147 C7** 21 36N 88 10W
Rio Largo, *Brazil* **154 C4** 9 28 S 35 50W
Rio Maior, *Portugal* **37 F2** 39 19N 8 57W
Rio Marina, *Italy* **38 F7** 42 49N 10 25 E
Río Mayo, *Argentina* **160 C2** 45 40 S 70 15W
Río Mulatos, *Bolivia* **156 D4** 19 40 S 66 50W
Río Muni = Mbini □, *Eq. Guin.* **102 B2** 1 30N 10 0 E
Rio Negro, *Brazil* **159 B6** 26 0 S 49 55W
Río Negro, *Chile* **160 B2** 40 47 S 73 14W
Rio Negro, Pantanal do, *Brazil* **157 D6** 19 0 S 56 0W
Rio Pardo, *Brazil* **159 C5** 30 0 S 52 30W
Río Pico, *Argentina* **160 B2** 44 0 S 70 22W
Rio Real, *Brazil* **155 D4** 11 28 S 37 56W
Río Segundo, *Argentina* **158 C3** 31 40 S 63 59W
Río Tercero, *Argentina* **158 C3** 32 15 S 64 8W
Rio Tinto, *Brazil* **154 C4** 6 48 S 35 5W
Rio Tinto, *Portugal* **36 D2** 41 11N 8 34W
Rio Verde, *Brazil* **155 E1** 17 50 S 51 0W
Río Verde, *Mexico* **147 C5** 21 56N 99 59W
Rio Verde de Mato Grosso, *Brazil* **157 D7** 18 56 S 54 52W
Rio Vista, *U.S.A.* **144 G5** 38 10N 121 42W
Ríobamba, *Ecuador* **152 D2** 1 50 S 78 45W
Ríohacha, *Colombia* **152 A3** 11 33N 72 55W
Rioja, *Peru* **156 B2** 6 11 S 77 5W
Riom, *France* **24 C7** 45 54N 3 7 E
Riom-ès-Montagnes, *France* **24 C6** 45 17N 2 39 E
Rion-des-Landes, *France* **24 E3** 43 55N 0 56W
Rionegro, *Colombia* **152 B2** 6 9N 75 22W
Rionero in Vúlture, *Italy* **41 B8** 40 55N 15 40 E
Rioni →, *Georgia* **53 J5** 42 14N 41 44 E
Rios, *Spain* **36 D3** 41 58N 7 16W
Riosucio, *Caldas, Colombia* **152 B2** 5 30N 75 40W
Ríosucio, *Choco, Colombia* **152 B2** 7 27N 77 7W
Riou L., *Canada* **131 B7** 59 7N 106 25W
Rioz, *France* **23 E13** 47 26N 6 5 E
Riozinho →, *Brazil* **152 D4** 2 55 S 67 7W
Riparia, Dora →, *Italy* **38 C4** 45 7N 7 24 E
Ripatransone, *Italy* **39 F10** 43 0N 13 46 E

Ross-on-Wye, U.K. 17 F5 51 54N 2 34W
Ross Sea, Antarctica 7 D11 74 0 S 178 0 E
Rossa, Switz. 29 D8 46 23N 9 8 E
Rossan Pt., Ireland 19 B3 54 42N 8 47W
Rossano Cálabro, Italy 41 C9 39 36N 16 39 E
Rossburn, Canada 131 C8 50 40N 100 49W
Rosseau, Canada 136 A5 45 16N 79 39W
Rossel, C., Vanuatu 121 K4 20 23 S 166 36 E
Rossford, U.S.A. 141 C13 41 36N 83 34W
Rossignol, L., Canada 128 B5 52 43N 73 40W
Rossignol Res., Canada 129 D6 44 12N 65 10W
Rossland, Canada 130 D5 49 6N 117 50W
Rosslare, Ireland 19 D5 52 17N 6 24W
Rosslau, Germany 26 D8 51 52N 12 15 E
Rosso, Mauritania 100 B1 16 40N 15 45W
Rosso, C., France 25 F12 42 13N 8 32 E
Rossosh, Russia 52 E4 50 15N 39 28 E
Rossport, Canada 128 C2 48 50N 87 30W
Rossum, Neths. 20 E6 51 48N 5 20 E
Røssvatnet, Norway 12 D16 65 45N 14 5 E
Rossville, Australia 114 B4 15 48 S 145 15 E
Rossville, U.S.A. 141 D10 40 25N 86 36W
Røst, Norway 12 C15 67 32N 12 0 E
Rostāq, Afghan. 79 A3 37 7N 69 49 E
Rosthern, Canada 131 C7 52 40N 106 20W
Rostock, Germany 26 A8 54 5N 12 8 E
Rostov, Russia 52 B4 57 14N 39 25 E
Rostov, Russia 53 G4 47 15N 39 45 E
Rostrenen, France 22 D3 48 14N 3 21W
Roswell, U.S.A. 139 J2 33 24N 104 32W
Rosyth, U.K. 18 E5 56 2N 3 25W
Rota, Spain 37 J4 36 37N 6 20W
Rotälven →, Sweden 14 C8 61 15N 14 3 E
Rotan, U.S.A. 139 J4 32 51N 100 28W
Rotem, Belgium 21 F7 51 3N 5 45 E
Rotenburg, Germany 26 B5 53 6N 9 25 E
Roth, Germany 27 F7 49 15N 11 5 E
Rothaargebirge, Germany 26 E4 51 2N 8 13 E
Rothenburg, Switz. 29 B6 47 6N 8 16 E
Rothenburg ob der
Tauber, Germany 27 F6 49 23N 10 11 E
Rother →, U.K. 17 G8 50 59N 0 45 E
Rotherham, U.K. 16 D6 53 26N 1 20W
Rothes, U.K. 18 D5 57 32N 3 13W
Rothesay, Canada 129 C6 45 23N 66 0W
Rothesay, U.K. 18 F3 55 50N 5 3W
Rothrist, Switz. 28 B5 47 18N 7 54 E
Roti, Indonesia 72 D2 10 50 S 123 0 E
Roto, Australia 117 B6 33 0 S 145 30 E
Roto Aira L., N.Z. 118 E5 38 5 S 175 45 E
Rotoehu L., N.Z. 118 E5 38 1 S 176 32 E
Rotoiti, L., N.Z. 118 E5 38 2 S 176 26 E
Rotoiti L., N.Z. 119 B7 41 51 S 172 49 E
Rotoma L., N.Z. 118 E5 38 2 S 176 35 E
Rotondella, Italy 41 B9 40 10N 16 31 E
Rotoroa, L., N.Z. 119 B7 41 55 S 172 39 E
Rotorua, N.Z. 118 E5 38 9 S 176 16 E
Rotorua, L., N.Z. 118 E5 38 5 S 176 18 E
Rotselaar, Belgium 21 G5 50 57N 4 42 E
Rott →, Germany 27 G9 48 27N 13 25 E
Rotten →, Switz. 28 D5 46 18N 7 36 E
Rottenburg, Germany 27 G4 48 28N 8 55 E
Rottenmann, Austria 30 D7 47 31N 14 22 E
Rotterdam, Neths. 20 E5 51 55N 4 30 E
Rottnest I., Australia 113 F2 32 0 S 115 27 E
Rottumeroog, Neths. 20 A9 53 33N 6 34 E
Rottweil, Germany 27 G4 48 9N 8 37 E
Rotuma, Fiji 122 J9 12 25 S 177 5 E
Roubaix, France 23 B10 50 40N 3 10 E
Roudnice, Czech. 30 A7 50 25N 14 15 E
Rouen, France 22 C8 49 27N 1 4 E
Rouergue, France 24 D5 44 15N 2 30 E
Rough Ridge, N.Z. 119 F4 45 10 S 169 55 E
Rouillac, France 24 C3 45 47N 0 4W
Rouleau, Canada 131 C8 50 10N 104 56W
Round Mountain, U.S.A. 142 G5 38 43N 117 4W
Round Mt., Australia 115 E5 30 26 S 152 16 E
Roundup, U.S.A. 142 C9 46 27N 108 33W
Roura, Fr. Guiana 153 C7 4 44N 52 20W
Rousay, U.K. 18 B5 59 10N 3 2W
Rouses Point, U.S.A. 137 B11 44 59N 73 22W
Roussillon, Isère, France 25 C8 45 24N 4 49 E
Roussillon, Pyrénées-Or.,
France 24 F6 42 30N 2 35 E
Roussin, C., N. Cal. 121 U21 21 20 S 167 59 E
Rouveen, Neths. 20 C8 52 37N 6 11 E
Rouxville, S. Africa 104 E4 30 25 S 26 50 E
Rouyn, Canada 128 C4 48 20N 79 0W
Rovaniemi, Finland 12 C21 66 29N 25 41 E
Rovato, Italy 38 C7 45 34N 10 0 E
Rovenki, Ukraine 51 H10 48 5N 39 21 E
Rovereto, Italy 38 C8 45 53N 11 3 E
Rovigo, Italy 39 C8 45 4N 11 47 E
Rovinari, Romania 46 E4 44 56N 23 10 E
Rovinj, Croatia 39 C10 45 5N 13 40 E
Rovira, Colombia 152 C2 4 15N 75 20W
Rovno = Rivne, Ukraine 51 G4 50 40N 26 10 E
Rovnoye, Russia 52 E8 50 52N 46 3 E
Rovuma →, Tanzania 107 E5 10 29 S 40 28 E
Row'ān, Iran 85 C6 35 8N 48 51 E
Rowena, Australia 115 D4 29 48 S 148 55 E
Rowes, Australia 117 D8 37 0 S 149 6 E
Rowley Shoals, Australia 112 C2 17 30 S 119 0 E
Roxa, Guinea-Biss. 100 C1 11 15N 15 45W
Roxas = Barbacan, Phil. 71 F2 10 20N 119 21 E
Roxas, Capiz, Phil. 71 F4 11 36N 122 49 E
Roxas, Isabela, Phil. 70 C3 17 8N 121 36 E
Roxas, Mindoro, Phil. 70 E3 12 35N 121 31 E
Roxboro, U.S.A. 135 G6 36 24N 78 59W
Roxborough Downs,
Australia 114 C2 22 30 S 138 45 E
Roxburgh, N.Z. 119 F4 45 33 S 169 19 E
Roxen, Sweden 15 F9 58 30N 15 40 E
Roy, Mont., U.S.A. 142 C9 47 20N 108 58W
Roy, N. Mex., U.S.A. 139 H2 35 57N 104 12W
Roy Hill, Australia 112 D2 22 37 S 119 58 E
Roya, Peña, Spain 34 E4 40 25N 0 40W
Royal Center, U.S.A. 141 D10 40 52N 86 30W
Royal Leamington Spa,
U.K. 17 E6 52 18N 1 31W
Royal Tunbridge Wells,
U.K. 17 F8 51 7N 0 16 E
Royalla, Australia 117 C8 35 30 S 149 9 E
Royan, France 24 C2 45 37N 1 2W
Roye, France 23 C9 49 42N 2 48 E
Røyken, Norway 14 E4 59 45N 10 23 E
Rožaj, Montenegro, Yug. 42 E5 42 50N 20 15 E
Rózan, Poland 47 C8 52 52N 21 25 E
Rozay-en-Brie, France 23 D9 48 41N 2 58 E
Rozdilna, Ukraine 51 J6 46 50N 30 2 E
Rozhyshche, Ukraine 51 G3 50 54N 25 15 E
Rožňava, Slovak Rep. 31 C13 48 37N 20 35 E
Rozogi, Poland 47 B8 53 48N 21 9 E
Rozoy-sur-Serre, France 23 C11 49 40N 4 8 E
Rozwadów, Poland 47 E9 50 37N 22 2 E
Rrësheni, Albania 44 C1 41 47N 19 49 E
Rrogozhino, Albania 44 C1 41 2N 19 50 E
Rtanj, Serbia, Yug. 42 D6 43 45N 21 50 E
Rtishchevo, Russia 52 D6 52 18N 43 46 E
Rúa, Spain 36 C3 42 24N 7 6W
Ruacaná, Angola 103 F2 17 20 S 14 12 E
Ruahine Ra., N.Z. 118 F5 39 55 S 176 2 E
Ruamahanga →, N.Z. 118 H4 41 24 S 175 8 E
Ruapehu, N.Z. 118 F4 39 17 S 175 35 E
Ruapuke I., N.Z. 119 G3 46 46 S 168 31 E
Ruâq, W. →, Egypt 91 F2 30 0N 33 49 E
Ruatoria, N.Z. 118 D7 37 55 S 178 20 E
Ruawai, N.Z. 118 C2 36 8 S 173 59 E
Rub' al Khali, Si. Arabia 87 C5 18 0N 48 0 E
Rubeho Mts., Tanzania 106 D4 6 50 S 36 25 E
Rubezhnoye = Rubizhne,
Ukraine 51 H10 49 6N 38 25 E
Rubh a' Mhail, U.K. 18 F2 55 56N 6 8W
Rubha Hunish, U.K. 18 D2 57 42N 6 20W
Rubha Robhanais =
Lewis, Butt of, U.K. 18 C2 58 31N 6 16W
Rubiataba, Brazil 155 E2 15 8 S 49 48W
Rubicon →, U.S.A. 144 G5 38 53N 121 4W
Rubicone →, Italy 39 D9 44 8N 12 28 E
Rubinéia, Brazil 155 F1 20 13 S 51 2W
Rubino, Ivory C. 100 D4 6 4N 4 18W
Rubio, Venezuela 152 B3 7 43N 72 22W
Rubizhne, Ukraine 51 H10 49 6N 38 25 E
Rubtsovsk, Russia 56 D9 51 30N 81 10 E
Ruby L., U.S.A. 142 F6 40 10N 115 28W
Ruby Mts., U.S.A. 142 F6 40 30N 115 20W
Rucheng, China 69 E9 25 33N 113 38 E
Ruciane-Nida, Poland 47 B8 53 40N 21 32 E
Rud, Norway 14 D4 60 1N 10 1 E
Rūd Sar, Iran 85 B6 37 8N 50 18 E
Ruda Śląska, Poland 47 E5 50 16N 18 50 E
Rudall, Australia 116 B2 33 43 S 136 17 E
Rudall →, Australia 112 D3 22 34 S 122 13 E
Rūdbār, Afghan. 79 C1 30 9N 62 36 E
Ruden, Germany 26 A9 54 13N 13 47 E
Rüdersdorf, Germany 26 C9 52 27N 13 47 E
Rudewa, Tanzania 107 E3 10 7 S 34 40 E
Rudkøbing, Denmark 15 K4 54 56N 10 41 E
Rudna, Poland 47 D3 51 30N 16 17 E
Rudnichnyy, Russia 48 C9 59 38N 52 26 E
Rudnik, Bulgaria 43 E12 42 36N 27 30 E
Rudnik, Poland 47 E9 50 26N 22 15 E
Rudnik, Serbia, Yug. 42 C5 44 7N 20 35 E
Rudnogorsk, Russia 57 D11 57 15N 103 42 E
Rudnya, Russia 50 E6 54 55N 31 7 E
Rudnyy, Kazakhstan 54 E9 52 57N 63 7 E
Rudo, Bos.-H. 42 D4 43 41N 19 23 E
Rudolf, Ostrov, Russia 56 A6 81 45N 58 30 E
Rudolstadt, Germany 26 E7 50 44N 11 19 E
Rudong, China 69 A13 32 20N 121 12 E
Rudozem, Bulgaria 43 F9 41 29N 24 51 E
Rudyard, U.S.A. 134 B3 46 14N 84 36W
Rue, France 23 B8 50 15N 1 40 E
Ruelle, France 24 C4 45 41N 0 14 E
Rufa'a, Sudan 95 E3 14 44N 33 22 E
Ruffec, France 24 B4 46 2N 0 12 E
Rufiji □, Tanzania 106 D4 8 0 S 38 30 E
Rufiji →, Tanzania 106 D4 7 50 S 39 15 E
Rufino, Argentina 158 C3 34 20 S 62 50W
Rufisque, Senegal 100 C1 14 40N 17 15W
Rufunsa, Zambia 107 F2 15 4 S 29 34 E
Rugao, China 69 A13 32 23N 120 31 E
Rugby, U.K. 17 E6 52 23N 1 16W
Rugby, U.S.A. 138 A5 48 22N 100 0W
Rügen, Germany 26 A9 54 22N 13 24 E
Rugles, France 22 D7 48 50N 0 40 E
Ruhea, Bangla. 78 B2 26 10N 88 25 E
Ruhengeri, Rwanda 106 C2 1 30 S 29 36 E
Ruhla, Germany 26 E6 50 54N 10 23 E
Ruhland, Germany 26 D9 51 27N 13 51 E
Ruhnu saar, Estonia 13 H20 57 48N 23 15 E
Ruhr →, Germany 26 D2 51 27N 6 43 E
Ruhuhu →, Tanzania 107 E3 10 31 S 34 34 E
Rui Barbosa, Brazil 155 D3 12 18 S 40 27W
Rui'an, China 69 D13 27 47N 120 40 E
Ruichang, China 69 C10 29 40N 115 39 E
Ruidosa, U.S.A. 139 L2 29 59N 104 41W
Ruidoso, U.S.A. 143 K11 33 20N 105 41W
Ruili, China 68 E1 24 1N 97 43 E
Ruinen, Neths. 20 C8 52 46N 6 21 E
Ruinerwold, Neths. 20 C8 52 44N 6 15 E
Ruiten A Kanaal →,
Neths. 20 C10 52 54N 7 8 E
Ruivo, Pico, Madeira 33 D3 32 45N 16 56W
Ruj, Bulgaria 42 E7 42 52N 22 42 E
Rujen, Macedonia 42 E7 42 9N 22 30 E
Rujm Tal'at al Jamā'ah,
Jordan 91 E4 30 24N 35 30 E
Ruk, Pakistan 80 F3 27 50N 68 42 E
Rukwa □, Tanzania 106 D3 7 0 S 31 30 E
Rukwa L., Tanzania 106 D3 8 0 S 32 20 E
Rulhieres, C., Australia 112 B4 13 56 S 127 22 E
Rulles, Belgium 21 J7 49 43N 5 32 E
Rum Cay, Bahamas 149 B5 23 40N 74 58W
Rum Jungle, Australia 112 B5 13 0 S 130 59 E
Ruma, Serbia, Yug. 42 B4 45 0N 19 50 E
Rumādah, Yemen 86 D3 13 34N 43 52 E
Rumāḩ, Si. Arabia 84 E5 25 29N 47 10 E
Rumania = Romania ■,
Europe 46 C5 46 0N 25 0 E
Rumaylah, Iraq 84 D5 30 47N 47 37 E
Rumaylah, 'Urūq ar,
Si. Arabia 86 B4 21 0N 47 30 E
Rumbalara, Australia 114 D1 25 20 S 134 29 E
Rumbêk, Sudan 95 F2 6 54N 29 37 E
Rumbeke, Belgium 21 G2 50 56N 3 10 E
Rumburk, Czech. 30 A7 50 57N 14 32 E
Rumelange, Lux. 21 K8 49 27N 6 2 E
Rumford, U.S.A. 137 B14 44 33N 70 33W
Rumia, Poland 47 A5 54 37N 18 25 E
Rumilly, France 25 C9 45 53N 5 56 E
Rumoi, Japan 60 C10 43 56N 141 39 E
Rumonge, Burundi 106 C2 3 59 S 29 26 E
Rumsey, Canada 130 C6 51 51N 112 48W
Rumula, Australia 114 B4 16 35 S 145 20 E
Rumuruti, Kenya 106 B4 0 17N 36 32 E
Runan, China 66 H8 33 0N 114 30 E
Runanga, N.Z. 119 C6 42 25 S 171 15 E
Runaway, C., N.Z. 118 D6 37 32 S 177 59 E
Runcorn, U.K. 16 D5 53 21N 2 44W
Rungwa, Tanzania 106 D3 6 55 S 33 32 E
Rungwa →, Tanzania 106 D3 7 36 S 31 50 E
Rungwe, Tanzania 107 D3 9 11 S 33 32 E
Rungwe □, Tanzania 107 D3 9 25 S 33 32 E
Runka, Nigeria 101 C6 12 28N 7 20 E
Runton Ra., Australia 112 D3 23 31 S 123 6 E
Ruokolahti, Finland 50 B5 61 17N 28 50 E
Ruoqiang, China 64 C3 38 55N 88 10 E
Rupa, India 78 B4 27 15N 92 21 E
Rupar, India 80 D7 31 2N 76 38 E
Rupat, Indonesia 74 B2 1 45N 101 40 E
Rupea, Romania 46 C6 46 2N 25 13 E
Rupert →, Canada 128 B4 51 29N 78 45W
Rupert House =
Waskaganish, Canada 128 B4 51 30N 78 40W
Rupsa, Bangla. 78 E2 21 44N 89 30 E
Rupununi →, Guyana 153 C6 4 3N 58 35W
Rur →, Germany 26 E2 51 11N 5 59 E
Rurrenabaque, Bolivia 156 C4 14 30 S 67 32W
Rus →, Spain 35 F2 39 30N 2 30W
Rusambo, Zimbabwe 107 F3 16 30 S 32 4 E
Rusape, Zimbabwe 107 F3 18 35 S 32 8 E
Ruschuk = Ruse, Bulgaria 43 D10 43 48N 25 59 E
Ruse, Bulgaria 43 D10 43 48N 25 59 E
Ruşeţu, Romania 46 E8 44 57N 27 14 E
Rushan, China 67 F11 36 56N 121 30 E
Rushden, U.K. 17 E7 52 18N 0 35W
Rushford, U.S.A. 138 D9 43 49N 91 46W
Rushville, Ill., U.S.A. 140 D6 40 7N 90 34W
Rushville, Ind., U.S.A. 141 E11 39 37N 85 27W
Rushville, Nebr., U.S.A. 138 D3 42 43N 102 28W
Rushworth, Australia 117 D6 36 32 S 145 1 E
Russas, Brazil 154 B4 4 55 S 37 50W
Russell, Canada 131 C8 50 50N 101 20W
Russell, N.Z. 118 B3 35 16 S 174 10 E
Russell, U.S.A. 138 F5 38 54N 98 52W
Russell Is., Solomon Is. 121 M10 9 4 S 159 12 E
Russell L., Man., Canada 131 B8 56 15N 101 30W
Russell L., N.W.T.,
Canada 130 A5 63 5N 115 44W
Russellkonda, India 82 E7 19 57N 84 42 E
Russellville, Ala., U.S.A. 135 H2 34 30N 87 44W
Russellville, Ark., U.S.A. 139 H8 35 17N 93 8W
Russellville, Ky., U.S.A. 135 G2 36 51N 86 53W
Russi, Italy 39 D9 44 22N 12 2 E
Russia ■, Eurasia 57 C11 62 0N 105 0 E
Russian →, U.S.A. 144 G3 38 27N 123 8W
Russiaville, U.S.A. 141 D10 40 25N 86 16W
Russkaya Polyana,
Kazakhstan 56 D8 53 47N 73 53 E
Russkoye Ustie, Russia 6 B15 71 0N 149 0 E
Rust, Austria 31 D9 47 49N 16 42 E
Rustam, Pakistan 80 B5 34 25N 72 13 E
Rustam Shahr, Pakistan 80 F2 26 58N 66 6 E
Rustavi, Georgia 53 K7 41 30N 45 0 E
Rustenburg, S. Africa 104 D4 25 41 S 27 14 E
Ruston, U.S.A. 139 J8 32 32N 92 38W
Ruswil, Switz. 28 B6 47 5N 8 8 E
Rutana, Burundi 106 C2 3 55 S 30 0 E
Rute, Spain 37 H6 37 19N 4 23W
Ruteng, Indonesia 72 C2 8 35 S 120 30 E
Ruth, Mich., U.S.A. 136 C2 43 42N 82 45W
Ruth, Nev., U.S.A. 142 G6 39 17N 114 59W
Rutherford, U.S.A. 144 G4 38 26N 122 24W
Rutherglen, Australia 117 D7 36 5 S 146 29 E
Rutherglen, U.K. 18 F4 55 49N 4 13W
Rüti, Switz. 29 B7 47 16N 8 51 E
Rutigliano, Italy 41 A10 41 1N 17 0 E
Rutland Plains, Australia 114 B3 15 38 S 141 43 E
Rutledge →, Canada 131 A6 61 4N 112 0W
Rutledge L., Canada 131 A6 61 33N 110 47W
Rutqa, W. →, Syria 89 E9 34 30N 41 3 E
Rutshuru, Zaïre 106 C2 1 13 S 29 25 E
Ruurlo, Neths. 20 D8 52 5N 6 24 E
Ruvo di Púglia, Italy 41 A9 41 7N 16 29 E
Ruvu, Tanzania 106 D4 6 49 S 38 43 E
Ruvu →, Tanzania 106 D4 6 23 S 38 52 E
Ruvuma □, Tanzania 107 E4 10 20 S 36 0 E
Ruwais, U.A.E. 85 E7 24 5N 52 50 E
Ruwenzori, Africa 106 B2 0 30N 29 55 E
Ruyigi, Burundi 106 C3 3 29 S 30 15 E
Ruyuan, China 69 E9 24 46N 113 16 E
Ruzayevka, Russia 52 C7 54 4N 45 0 E
Rûzhevo Konare, Bulgaria 43 E9 42 23N 24 46 E
Ružomberok, Slovak Rep. 31 B12 49 3N 19 17 E
Rwanda ■, Africa 106 C3 2 0 S 30 0 E
Ry, Denmark 15 H3 56 5N 9 45 E
Ryakhovo, Bulgaria 43 D11 43 58N 26 18 E
Ryan, L., U.K. 18 G3 55 0N 5 2W
Ryazan, Russia 52 C4 54 40N 39 40 E
Ryazhsk, Russia 52 D5 53 45N 40 3 E
Rybache = Rybachye,
Kazakhstan 56 E9 46 40N 81 20 E
Rybachiy Poluostrov,
Russia 48 A5 69 43N 32 0 E
Rybachye = Ysyk-Köl,
Kyrgyzstan 55 B8 42 26N 76 12 E
Rybachye, Kazakhstan 56 E9 46 40N 81 20 E
Rybinsk, Russia 52 A4 58 5N 38 50 E
Rybinskoye Vdkhr.,
Russia 50 C10 58 30N 38 25 E
Rybnik, Poland 31 A11 50 6N 18 32 E
Rybnitsa = Rîbniţa,
Moldova 51 J5 47 45N 29 0 E
Rybnoye, Russia 52 C4 54 45N 39 30 E
Rychwał, Poland 47 C5 52 4N 18 10 E
Ryde, U.K. 17 G6 50 43N 1 9W
Ryderwood, U.S.A. 144 D3 46 23N 123 3W
Rydöbruk, Sweden 15 H7 56 58N 13 7 E
Rydułtowy, Poland 31 A11 50 4N 18 23 E
Rydzyna, Poland 47 D3 51 47N 16 39 E
Rye, U.K. 17 G8 50 57N 0 45 E
Rye →, U.K. 16 C7 54 11N 0 44W
Rye Patch Reservoir,
U.S.A. 142 F4 40 28N 118 19W
Ryegate, U.S.A. 142 C9 46 18N 109 15W
Ryki, Poland 47 D8 51 38N 21 56 E
Rylsk, Russia 52 E2 51 36N 34 43 E
Rylstone, Australia 117 B8 32 46 S 149 58 E
Rymanów, Poland 31 B14 49 35N 21 51 E
Ryn, Poland 47 B8 53 57N 21 34 E
Ryn Peski, Kazakhstan 53 G9 47 30N 49 0 E
Ryōhaku-Sanchi, Japan 63 A8 36 9N 136 49 E
Ryōthu, Japan 60 E9 38 5N 138 26 E
Rypin, Poland 47 B6 53 3N 19 25 E
Ryūgasaki, Japan 63 B12 35 54N 140 11 E
Ryūkyū Is. = Ryūkyū-
rettō, Japan 61 M2 26 0N 126 0 E
Ryūkyū-rettō, Japan 61 M2 26 0N 126 0 E
Rzepin, Poland 47 C1 52 20N 14 49 E
Rzeszów, Poland 31 A14 50 5N 21 58 E
Rzeszów □, Poland 31 B15 50 0N 22 0 E
Rzhev, Russia 52 B2 56 20N 34 20 E

S

Sa, Thailand 76 C3 18 34N 100 45 E
Sa Dec, Vietnam 77 G5 10 20N 105 46 E
Sa-koi, Burma 78 F6 19 54N 97 3 E
Sa'ādatābād, Fārs, Iran 85 D7 30 10N 53 5 E
Sa'ādatābād, Kermān, Iran 85 D7 28 3N 55 53 E
Saale →, Germany 26 D7 51 56N 11 54 E
Saaler Bodden, Germany 26 A8 54 20N 12 27 E
Saalfeld, Germany 26 E7 50 38N 11 21 E
Saalfelden, Austria 30 D5 47 25N 12 51 E
Saane →, Switz. 28 B4 47 8N 7 10 E
Saar →, Europe 23 C13 49 41N 6 32 E
Saarbrücken, Germany 27 F2 49 14N 6 59 E
Saarburg, Germany 27 F2 49 36N 6 32 E
Saaremaa, Estonia 13 G20 58 30N 22 30 E
Saarijärvi, Finland 13 E21 62 43N 25 16 E
Saariselkä, Finland 12 B23 68 16N 28 15 E
Saarland □, Germany 23 C13 49 20N 7 0 E
Saarlouis, Germany 27 F2 49 18N 6 45 E
Saas Fee, Switz. 28 D5 46 7N 7 56 E
Sab 'Ābar, Syria 88 F7 33 46N 37 41 E
Saba, W. Indies 149 C7 17 42N 63 26W
Šabac, Serbia, Yug. 42 C4 44 48N 19 42 E
Sabadell, Spain 34 D7 41 28N 2 7 E
Sabae, Japan 63 B8 35 57N 136 11 E
Sabah □, Malaysia 75 A5 6 0N 117 0 E
Sabak Bernam, Malaysia 77 L3 3 46N 100 58 E
Sabalān, Kūhhā-ye, Iran 89 C12 38 15N 47 45 E
Sábana de la Mar,
Dom. Rep. 149 C6 19 7N 69 24W
Sábanalarga, Colombia 152 A3 10 38N 74 55W
Sabang, Indonesia 74 A1 5 50N 95 15 E
Sabangan, Phil. 70 C3 17 0N 120 55 E
Sabará, Brazil 155 E3 19 55 S 43 46W
Sabari →, India 82 F5 17 35N 81 16 E
Sab'atayn, Ramlat as,
Yemen 86 D4 15 30N 46 10 E
Sabattis, U.S.A. 137 B10 44 6N 74 40W
Sabáudia, Italy 40 A6 41 18N 13 1 E
Sabaya, Bolivia 156 D4 19 1 S 68 23W
Sabāyā, Jaz., Si. Arabia 86 C3 18 35N 41 3 E
Saberania, Indonesia 73 B5 2 5 S 138 18 E
Sabhah, Libya 96 C2 27 9N 14 29 E
Sabhah □, Libya 96 C2 26 0N 14 0 E
Sabie, S. Africa 105 D5 25 10 S 30 48 E
Sabina, U.S.A. 141 E13 39 29N 83 38W
Sabinal, Mexico 146 A3 30 58N 107 25W
Sabinal, U.S.A. 139 L5 29 19N 99 28W
Sabinal, Punta del, Spain 35 J2 36 43N 2 44W
Sabinas, Mexico 146 B4 27 50N 101 10W
Sabinas →, Mexico 146 B4 27 37N 100 42W
Sabinas Hidalgo, Mexico 146 B4 26 33N 100 10W
Sabine →, U.S.A. 139 L8 29 59N 93 47W
Sabine L., U.S.A. 139 L8 29 53N 93 51W
Sabine Pass, U.S.A. 139 L8 29 44N 93 54W
Sabinópolis, Brazil 155 E3 18 40 S 43 6W
Sabinov, Slovak Rep. 31 B14 49 6N 21 5 E
Sabirabad, Azerbaijan 53 K9 40 5N 48 30 E
Sabkhat Tāwurghā', Libya 96 B3 31 51N 15 15 E
Sabkhet el Bardawîl,
Egypt 91 D2 31 10N 33 15 E
Sablayan, Phil. 70 E3 12 50N 120 50 E
Sable, C., Canada 129 D6 43 29N 65 38W
Sable, C., U.S.A. 133 E10 25 9N 81 8W
Sable I., Canada 129 D8 44 0N 60 0W
Sablé-sur-Sarthe, France 22 E6 47 50N 0 20W
Saboeiro, Brazil 154 C4 6 32 S 39 54W
Sabor →, Portugal 36 D3 41 10N 7 7W
Sabou, Burkina Faso 100 C4 12 1N 2 15W
Sabrātah, Libya 96 B2 32 47N 12 29 E
Sabria, Tunisia 96 B1 33 22N 8 45 E
Sabrina Coast, Antarctica 7 C9 68 0 S 120 0 E
Sabtang I., Phil. 70 A3 20 19N 121 52 E
Sabugal, Portugal 36 E3 40 20N 7 5W
Sabula, U.S.A. 140 B6 42 4N 90 10W
Sabulubek, Indonesia 74 C1 1 36 S 98 40 E
Şabyā, Si. Arabia 86 C3 17 9N 42 37 E
Sabzevār, Iran 85 B8 36 15N 57 40 E
Sabzvārān, Iran 85 D8 28 45N 57 50 E
Sac City, U.S.A. 140 B2 42 25N 95 0W
Sacedón, Spain 34 E2 40 29N 2 41W
Sachigo →, Canada 128 A2 55 6N 88 58W
Sachigo, L., Canada 128 B1 53 50N 92 12W
Sachkhere, Georgia 53 J6 42 25N 43 28 E
Sachseln, Switz. 29 C6 46 52N 8 15 E
Sachsen □, Germany 26 E9 50 55N 13 10 E
Sachsen-Anhalt □,
Germany 26 D8 52 0N 12 0 E
Sacile, Italy 39 C9 45 57N 12 30 E
Sackets Harbor, U.S.A. 137 C8 43 57N 76 7W
Saco, Maine, U.S.A. 135 D10 43 30N 70 27W
Saco, Mont., U.S.A. 142 B10 48 28N 107 21W
Sacramento, Brazil 155 E2 19 53 S 47 27W
Sacramento, U.S.A. 144 G5 38 35N 121 29W
Sacramento →, U.S.A. 144 G5 38 3N 121 56W
Sacramento Mts., U.S.A. 143 K11 32 30N 105 30W
Sacramento Valley,
U.S.A. 144 G5 39 30N 122 0W
Sacratif, C., Spain 35 J1 36 42N 3 28W
Săcueni, Romania 46 B3 47 20N 22 5 E
Sada, Spain 36 B2 43 22N 8 15W
Sada-Misaki-Hantō, Japan 62 D4 33 22N 132 1 E
Sádaba, Spain 34 C3 42 19N 1 12W
Sadani, Tanzania 106 D4 5 58 S 38 35 E
Sadao, Thailand 77 J3 6 38N 100 26 E
Sadaseopet, India 82 F3 17 38N 77 59 E
Sadd el Aali, Egypt 94 C3 23 54N 32 54 E
Saddle Mt., U.S.A. 144 E3 45 58N 123 41W
Sade, Nigeria 101 C7 11 22N 10 45 E
Şadḩ, Oman 87 C6 17 3N 55 4 E

San Antonio, C. 259

St.-Pierre-d'Oléron, France 24 C2 45 57N 1 19W
St.-Pierre-Église, France 22 C5 49 40N 1 24W
St.-Pierre-en-Port, France 22 C7 49 48N 0 30 E
St.-Pierre et Miquelon □, St- P. & M. 129 C8 46 55N 56 10W
St.-Pierre-le-Moûtier, France 23 F10 46 47N 3 7 E
St.-Pierre-sur-Dives, France 22 C6 49 2N 0 1W
St.-Pieters Leew, Belgium 21 G4 50 47N 4 16 E
St.-Pol-de-Léon, France 22 D2 48 41N 4 0W
St.-Pol-sur-Mer, France 23 A9 51 1N 2 20 E
St.-Pol-sur-Ternoise, France 23 B9 50 23N 2 20 E
St.-Pons, France 24 E6 43 30N 2 45 E
St.-Pourçain-sur-Sioule, France 24 B7 46 18N 3 18 E
St.-Quay-Portrieux, France 22 D4 48 39N 2 51W
St.-Quentin, France 23 C10 49 50N 3 16 E
St.-Rambert-d'Albon, France 25 C8 45 17N 4 49 E
St.-Raphaël, France 25 E10 43 25N 6 46 E
St. Regis, U.S.A. 142 C6 47 18N 115 6W
St.-Rémy-de-Provence, France 25 E8 43 48N 4 50 E
St.-Renan, France 22 D2 48 26N 4 37W
St.-Saëns, France 22 C8 49 41N 1 16 E
St.-Sauveur-en-Puisaye, France 23 E10 47 37N 3 12 E
St.-Sauveur-le-Vicomte, France 22 C5 49 23N 1 32W
St.-Savin, France 24 B4 46 34N 0 53 E
St.-Savinien, France 24 C3 45 53N 0 42W
St. Sebastien, Tanjon' i, Madag. 105 A8 12 26 S 48 44 E
St.-Seine-l'Abbaye, France 23 E11 47 26N 4 47 E
St.-Sernin-sur-Rance, France 24 E6 43 54N 2 35 E
St.-Servan-sur-Mer, France 22 D4 48 38N 2 2W
St.-Sever, France 24 E3 43 45N 0 35W
St.-Sever-Calvados, France 22 D5 48 50N 1 3W
St-Siméon, Canada 129 C6 47 51N 69 54W
St. Stephen, Canada 129 C6 45 16N 67 17W
St.-Sulpice, France 24 E5 43 46N 1 41 E
St.-Sulpice-Laurière, France 24 B5 46 3N 1 29 E
St.-Syprien, France 24 F7 42 37N 3 2 E
St.-Thégonnec, France 22 D3 48 31N 3 57W
St. Thomas, Canada 128 D3 42 45N 81 10W
St. Thomas I., Virgin Is. 149 C7 18 20N 64 55W
St-Tite, Canada 128 C5 46 45N 72 34W
St.-Tropez, France 25 E10 43 17N 6 38 E
St. Troud = Sint Truiden, Belgium 21 G6 50 48N 5 10 E
St.-Vaast-la-Hougue, France 22 C5 49 35N 1 17W
St.-Valéry-en-Caux, France 22 C7 49 52N 0 43 E
St.-Valéry-sur-Somme, France 23 B8 50 11N 1 38 E
St.-Vallier, France 25 C8 45 11N 4 50 E
St.-Vallier-de-Thiey, France 25 E10 43 42N 6 51 E
St.-Varent, France 22 F6 46 53N 0 13W
St. Vincent = São Vicente, C. Verde Is. 8 G6 18 0N 26 1W
St. Vincent, W. Indies 149 D7 13 10N 61 10W
St. Vincent, G., Australia 116 C3 35 0 S 138 0 E
St. Vincent & the Grenadines ■, W. Indies 149 D7 13 0N 61 10W
St.-Vincent-de-Tyrosse, France 24 E2 43 39N 1 19W
St. Vincent Passage, W. Indies 149 D7 13 30N 61 0W
St-Vith, Belgium 21 H8 50 17N 6 9 E
St.-Yrieix-la-Perche, France 24 C5 45 31N 1 12 E
Ste.-Adresse, France 22 C7 49 31N 0 5 E
Ste-Agathe-des-Monts, Canada 128 C5 46 3N 74 17W
Ste-Anne de Beaupré, Canada 129 C5 47 2N 70 58W
Ste-Anne-des-Monts, Canada 129 C6 49 8N 66 30W
Ste-Croix, Switz. 28 C3 46 49N 6 34 E
Ste.-Enimie, France 24 D7 44 22N 3 26 E
Ste.-Foy-la-Grande, France 24 D4 44 50N 0 13 E
Ste. Genevieve, U.S.A. 140 G6 37 59N 90 2W
Ste.-Hermine, France 24 B2 46 32N 1 4W
Ste.-Livrade-sur-Lot, France 24 D4 44 24N 0 36 E
Ste-Marguerite →, Canada 129 B6 50 9N 66 36W
Ste.-Marie, Martinique 149 D7 14 48N 61 1W
Ste.-Marie-aux-Mines, France 23 D14 48 15N 7 12 E
Ste-Marie de la Madeleine, Canada 129 C5 46 26N 71 0W
Ste.-Maure-de-Touraine, France 22 E7 47 7N 0 37 E
Ste.-Maxime, France 25 E10 43 19N 6 39 E
Ste.-Menehould, France 23 C11 49 5N 4 54 E
Ste.-Mère-Église, France 22 C5 49 24N 1 19W
Ste.-Rose, Guadeloupe 149 C7 16 20N 61 45W
Ste. Rose du Lac, Canada 131 C9 51 4N 99 30W
Saintes, France 24 C3 45 45N 0 37W
Saintes, I. des, Guadeloupe 149 C7 15 50N 61 35W
Stes.-Maries-de-la-Mer, France 25 E8 43 26N 4 26 E
Saintonge, France 24 C3 45 40N 0 50W
Saipan, Pac. Oc. 122 F6 15 12N 145 45 E
Sairecábur, Cerro, Bolivia 158 A2 22 43 S 67 54W
Saitama □, Japan 63 A11 36 25N 139 30 E
Saito, Japan 62 E3 32 3N 131 24 E
Sajama, Bolivia 156 D4 18 7 S 69 0W
Sajan, Serbia, Yug. 42 B5 45 50N 20 20 E
Sajó, Hungary 31 C13 48 12N 20 44 E
Sajum, India 81 C8 33 20N 79 0 E
Sak →, S. Africa 104 E3 30 52 S 20 25 E
Sakai, Japan 63 C7 34 30N 135 30 E
Sakaide, Japan 62 C5 34 15N 133 50 E
Sakaiminato, Japan 62 B5 35 38N 133 11 E
Sakākah, Si. Arabia 84 D4 30 0N 40 8 E
Sakakawea, L., U.S.A. 138 B3 47 30N 101 25W
Sakami, L., Canada 128 B4 53 15N 77 0W
Sâkâne, 'Erg i-n, Mali 98 D4 20 30N 1 30W
Sakania, Zaïre 107 E2 12 43 S 28 30 E
Sakarya = Adapazarı, Turkey 88 B4 40 48N 30 25 E
Sakarya →, Turkey 88 B4 41 7N 30 39 E
Sakashima-Guntō, Japan 61 M2 24 46N 124 0 E
Sakata, Japan 60 E9 38 55N 139 50 E
Sakchu, N. Korea 67 D13 40 23N 125 2 E
Sakeny →, Madag. 105 C8 20 0 S 45 25 E
Sakété, Benin 101 D5 6 40N 2 45 E
Sakha □, Russia 57 C13 62 0N 130 0 E
Sakhalin, Russia 57 D15 51 0N 143 0 E
Sakhalinskiy Zaliv, Russia 57 D15 54 0N 141 0 E
Sakhi Gopal, India 82 E7 19 58N 85 50 E
Şaki, Azerbaijan 53 K8 41 10N 47 5 E
Šakiai, Lithuania 13 J20 54 59N 23 0 E
Sakmara →, Russia 54 F5 51 46N 55 1 E
Sakon Nakhon, Thailand 76 D5 17 10N 104 9 E
Sakrand, Pakistan 80 F3 26 10N 68 15 E
Sakri, India 82 D2 21 2N 74 20 E
Sakrivier, S. Africa 104 E3 30 54 S 20 28 E
Sakskøbing, Denmark 15 K5 54 49N 11 39 E
Saku, Japan 63 A10 36 17N 138 31 E
Sakuma, Japan 63 B9 35 3N 137 49 E
Sakura, Japan 63 B12 35 43N 140 14 E
Sakurai, Japan 63 C7 34 30N 135 51 E
Saky, Ukraine 51 K7 45 9N 33 34 E
Sal →, Russia 53 G5 47 31N 40 45 E
Šal'a, Slovak Rep. 31 C10 48 10N 17 50 E
Sala, Sweden 13 G17 59 58N 16 35 E
Sala Consilina, Italy 41 B8 40 23N 15 36 E
Sala-y-Gómez, Pac. Oc. 123 K17 26 28 S 105 28W
Salaberry-de-Valleyfield, Canada 128 C5 45 15N 74 8W
Saladas, Argentina 158 B4 28 15 S 58 40W
Saladillo, Argentina 158 D4 35 40 S 59 55W
Salado →, Buenos Aires, Argentina 158 D4 35 44 S 57 22W
Salado →, La Pampa, Argentina 160 A3 37 30 S 67 0W
Salado →, Río Negro, Argentina 160 B3 41 34 S 65 3W
Salado →, Santa Fe, Argentina 158 C3 31 40 S 60 41W
Salado →, Mexico 146 B5 26 52N 99 19W
Salaga, Ghana 101 D4 8 31N 0 31W
Sālah, Syria 91 C5 32 40N 36 45 E
Sălaj □, Romania 46 B4 47 15N 23 0 E
Sálakhos, Greece 32 C9 36 17N 27 57 E
Salala, Liberia 100 D2 6 42N 10 7W
Salala, Sudan 94 C4 21 17N 36 16 E
Salālah, Oman 87 C6 16 56N 53 59 E
Salamanca, Chile 158 C1 31 46 S 70 59W
Salamanca, Spain 36 E5 40 58N 5 39W
Salamanca, U.S.A. 136 D6 42 10N 78 43W
Salamanca □, Spain 36 E5 40 57N 5 40W
Salāmatābād, Iran 84 C5 35 39N 47 50 E
Salamina, Colombia 152 B2 5 25N 75 29W
Salamis, Cyprus 32 D12 35 11N 33 54 E
Salamís, Greece 45 G5 37 56N 23 30 E
Salamonie L., U.S.A. 141 D11 40 46N 85 37W
Salar de Atacama, Chile 158 A2 23 30 S 68 25W
Salar de Uyuni, Bolivia 156 E4 20 30 S 67 45W
Sălard, Romania 46 B3 47 12N 22 3 E
Salas, Spain 36 B4 43 25N 6 15W
Salas de los Infantes, Spain 34 C1 42 2N 3 17W
Salatiga, Indonesia 75 D4 7 19 S 110 30 E
Salavat, Russia 54 E5 53 21N 55 55 E
Salaverry, Peru 156 B2 8 15 S 79 0W
Salawati, Indonesia 73 B4 1 7 S 130 52 E
Salay, Phil. 71 G5 8 52N 124 47 E
Salayar, Indonesia 72 C2 6 7 S 120 30 E
Salazar →, Spain 34 C3 42 40N 1 20W
Salbris, France 23 E9 47 25N 2 3 E
Salcia, Romania 46 F5 43 56N 24 55 E
Salcombe, U.K. 17 G4 50 14N 3 47W
Saldaña, Spain 36 C6 42 32N 4 48W
Saldanha, S. Africa 104 E2 33 0 S 17 58 E
Saldanha B., S. Africa 104 E2 33 6 S 18 0 E
Saldus, Latvia 13 H20 56 38N 22 30 E
Sale, Australia 117 E7 38 6 S 147 6 E
Salé, Morocco 98 B3 34 3N 6 48W
Sale, U.K. 16 D5 53 26N 2 19W
Salekhard, Russia 48 A12 66 30N 66 35 E
Salem, India 83 J4 11 40N 78 11 E
Salem, Ill., U.S.A. 140 F8 38 38N 88 57W
Salem, Ind., U.S.A. 141 F10 38 36N 86 6W
Salem, Mass., U.S.A. 137 D14 42 31N 70 53W
Salem, Mo., U.S.A. 139 G9 37 39N 91 32W
Salem, N.J., U.S.A. 134 F8 39 34N 75 28W
Salem, Ohio, U.S.A. 136 F4 40 54N 80 52W
Salem, Oreg., U.S.A. 142 D2 44 56N 123 2W
Salem, S. Dak., U.S.A. 138 D6 43 44N 97 23W
Salem, Va., U.S.A. 134 G5 37 18N 80 3W
Salemi, Italy 40 E5 37 49N 12 48 E
Salernes, France 25 E10 43 34N 6 15 E
Salerno, Italy 41 B7 40 41N 14 47 E
Salerno, G. di, Italy 41 B7 40 32N 14 42 E
Salford, U.K. 16 D5 53 30N 2 18W
Salgir →, Ukraine 51 K8 45 38N 35 1 E
Salgótarján, Hungary 31 C12 48 5N 19 47 E
Salgueiro, Brazil 154 C4 8 4 S 39 6W
Salida, U.S.A. 132 C5 38 32N 106 0W
Salies-de-Béarn, France 24 E3 43 28N 0 56W
Şalif, Yemen 86 D3 15 18N 42 41 E
Salihli, Turkey 88 C3 38 28N 28 8 E
Salihorsk, Belarus 51 F4 52 51N 27 27 E
Salin, Burma 78 E5 20 35N 94 40 E
Salina, Italy 41 D7 38 34N 14 50 E
Salina, U.S.A. 138 F6 38 50N 97 37W
Salina Cruz, Mexico 147 D5 16 10N 95 10W
Salinas, Brazil 155 E3 16 10 S 42 10W
Salinas, Chile 158 A2 23 31 S 69 29W
Salinas, Ecuador 152 D1 2 10 S 80 58W
Salinas, U.S.A. 144 J5 36 40N 121 39W
Salinas →, Guatemala 147 D6 16 28N 90 31W
Salinas →, U.S.A. 144 J5 36 45N 121 48W
Salinas, B. de, Nic. 148 D2 11 4N 85 45W
Salinas, C. de, Spain 33 B10 39 16N 3 4 E
Salinas, Pampa de las, Argentina 158 C2 31 58 S 66 42W
Salinas Ambargasta, Argentina 158 B3 29 0 S 65 0W
Salinas de Hidalgo, Mexico 146 C4 22 30N 101 40W
Salinas Grandes, Argentina 158 B2 30 0 S 65 0W
Saline →, Ark., U.S.A. 139 J8 33 10N 92 8W
Saline →, Kans., U.S.A. 138 F6 38 52N 97 30W
Salines, Spain 33 B10 39 21N 3 3 E
Salinópolis, Brazil 154 B2 0 40 S 47 20W
Salins-les-Bains, France 23 F12 46 58N 5 52 E
Salir, Portugal 37 H2 37 14N 8 2W
Salisbury = Harare, Zimbabwe 107 F3 17 43 S 31 2 E
Salisbury, Australia 116 C3 34 46 S 138 40 E
Salisbury, U.K. 17 F6 51 4N 1 47W
Salisbury, Md., U.S.A. 134 F8 38 22N 75 36W
Salisbury, Mo., U.S.A. 140 E4 39 25N 92 48W
Salisbury, N.C., U.S.A. 135 H5 35 40N 80 29W
Salisbury Plain, U.K. 17 F6 51 14N 1 55W
Săliște, Romania 46 D4 45 45N 23 56 E
Salitre →, Brazil 154 C3 9 29 S 40 39W
Salka, Nigeria 101 C5 10 20N 4 58 E
Şalkhad, Syria 91 C5 32 29N 36 43 E
Salla, Finland 12 C23 66 50N 28 49 E
Sallent, Spain 34 D6 41 49N 1 54 E
Salles-Curan, France 24 D6 44 11N 2 48 E
Salling, Denmark 15 H2 56 40N 8 55 E
Sallisaw, U.S.A. 139 H7 35 28N 94 47W
Sallom Junction, Sudan 94 D4 19 17N 37 6 E
Salluit, Canada 127 B12 62 14N 75 38W
Salmās, Iran 89 C11 38 11N 44 47 E
Salmerón, Spain 34 E2 40 33N 2 29W
Salmo, Canada 130 D5 49 10N 117 20W
Salmon, U.S.A. 142 D7 45 11N 113 54W
Salmon →, Canada 130 C4 54 3N 122 40W
Salmon →, U.S.A. 142 D5 45 51N 116 47W
Salmon Arm, Canada 130 C5 50 40N 119 15W
Salmon Falls, U.S.A. 142 E6 42 48N 114 59W
Salmon Gums, Australia 113 F3 32 59 S 121 38 E
Salmon Res., Canada 129 C8 48 5N 56 0W
Salmon River Mts., U.S.A. 142 D6 45 0N 114 30W
Salo, Finland 13 F20 60 22N 23 10 E
Salò, Italy 38 C7 45 36N 10 31 E
Salobreña, Spain 37 J7 36 44N 3 35W
Salome, U.S.A. 145 M13 33 47N 113 37W
Salon-de-Provence, France 25 E9 43 39N 5 6 E
Salonica = Thessaloníki, Greece 44 D4 40 38N 22 58 E
Salonta, Romania 46 C2 46 49N 21 42 E
Salor →, Spain 37 F3 39 39N 7 3W
Salou, C., Spain 34 D6 41 3N 1 10 E
Salpausselkä, Finland 13 F22 61 0N 27 0 E
Salsacate, Argentina 158 C2 31 20 S 65 5W
Salses, France 24 F6 42 50N 2 55 E
Salsette I., India 82 E1 19 5N 72 50 E
Salsk, Russia 53 G5 46 28N 41 30 E
Salso →, Italy 40 E6 37 6N 13 57 E
Salsomaggiore Terme, Italy 38 D6 44 49N 9 59 E
Salt →, Canada 130 B6 60 0N 112 25W
Salt →, Ariz., U.S.A. 143 K7 33 23N 112 19W
Salt →, Mo., U.S.A. 140 E5 39 28N 91 4W
Salt Creek, Australia 116 D3 36 8 S 139 38 E
Salt Fork Arkansas →, U.S.A. 139 G6 36 36N 97 3W
Salt Lake City, U.S.A. 142 F8 40 45N 111 53W
Salt Range, Pakistan 80 C5 32 30N 72 25 E
Salta, Argentina 158 A2 24 57 S 65 25W
Salta □, Argentina 158 A2 24 48 S 65 30W
Saltcoats, U.K. 18 F4 55 38N 4 47W
Saltee Is., Ireland 19 D5 52 7N 6 37W
Saltfjellet, Norway 12 C16 66 40N 15 15 E
Saltfjorden, Norway 12 C16 67 15N 14 10 E
Saltholm, Denmark 15 J6 55 38N 12 43 E
Saltillo, Mexico 146 B4 25 25N 101 0W
Salto, Argentina 158 C3 34 20 S 60 15W
Salto, Uruguay 158 C4 31 27 S 57 50W
Salto da Divisa, Brazil 155 E4 16 0 S 39 57W
Salton City, U.S.A. 145 M11 33 29N 115 51W
Salton Sea, U.S.A. 145 M11 33 15N 115 45W
Saltpond, Ghana 101 D4 5 15N 1 3W
Saltsjöbaden, Sweden 14 E12 59 15N 18 20 E
Saltville, U.S.A. 134 G5 36 53N 81 46W
Saluda →, U.S.A. 135 H5 34 1N 81 4W
Salûm, Egypt 94 A2 31 31N 25 7 E
Salûm, Khâlig el, Egypt 94 A2 31 35N 25 24 E
Salur, India 82 E6 18 27N 83 18 E
Salut, Is. du, Fr. Guiana 153 B7 5 15N 52 35W
Saluzzo, Italy 38 D4 44 39N 7 29 E
Salvacion, Phil. 71 G2 9 56N 118 47 E
Salvación, B., Chile 160 D1 50 50 S 75 10W
Salvador, Brazil 155 D4 13 0 S 38 30W
Salvador, Canada 131 C7 52 10N 109 32W
Salvador, L., U.S.A. 139 L9 29 43N 90 15W
Salvaterra, Brazil 154 B2 0 46 S 48 31W
Salvaterra de Magos, Portugal 37 F2 39 1N 8 47W
Salvisa, U.S.A. 141 G12 37 54N 84 51W
Sálvora, I., Spain 36 C2 42 30N 8 58W
Salween →, Burma 78 G6 16 31N 97 37 E
Salyan, Azerbaijan 89 C13 39 33N 48 59 E
Salyersville, U.S.A. 134 G4 37 45N 83 4W
Salza →, Austria 30 D7 47 40N 14 43 E
Salzach →, Austria 30 C5 48 12N 12 56 E
Salzburg, Austria 30 D6 47 48N 13 2 E
Salzburg □, Austria 30 D6 47 15N 13 0 E
Salzgitter, Germany 26 C6 52 9N 10 19 E
Salzwedel, Germany 26 C7 52 52N 11 10 E
Sam, Gabon 102 B2 0 58N 11 16 E
Sam Neua, Laos 76 B5 20 29N 104 5 E
Sam Ngao, Thailand 76 D2 17 18N 99 0 E
Sam Rayburn Reservoir, U.S.A. 139 K7 31 4N 94 5W
Sam Son, Vietnam 76 C5 19 44N 105 54 E
Sam Teu, Laos 76 C5 19 59N 104 38 E
Sama, Russia 56 C7 60 12N 60 22 E
Sama de Langreo, Spain 36 B5 43 18N 5 40W
Samacimbo, Angola 103 E3 13 33 S 16 59 E
Samagaltay, Russia 57 D10 50 36N 95 3 E
Samā'il, Oman 87 B7 23 40N 57 50 E
Samaipata, Bolivia 157 D5 18 9 S 63 52W
Samal, Phil. 71 H5 7 5N 125 42 E
Samal I., Phil. 71 H5 7 3N 125 44 E
Samales Group, Phil. 71 J3 6 0N 122 0 E
Samalkot, India 82 F6 17 3N 82 13 E
Samâlût, Egypt 94 J7 28 20N 30 42 E
Samana, India 80 D7 30 10N 76 13 E
Samana Cay, Bahamas 149 B5 23 3N 73 45W
Samandağı, Turkey 88 D6 36 5N 35 59 E
Samanga, Tanzania 107 D4 8 20 S 39 13 E
Samangān □, Afghan. 79 A3 36 15N 68 3 E
Samangwa, Zaïre 103 C4 4 23 S 24 10 E
Samani, Japan 60 C11 42 7N 142 56 E
Samar, Phil. 71 E5 12 0N 125 0 E
Samar □, Phil. 71 F5 11 50N 125 0 E
Samar Sea, Phil. 70 E5 12 0N 124 15 E
Samara, Russia 52 D10 53 8N 50 6 E
Samara →, Russia 52 D10 53 10N 50 4 E
Samara →, Ukraine 51 H8 48 28N 35 7 E
Samarai, Papua N. G. 120 F6 10 39 S 150 41 E
Samaria = Shōmrōn, West Bank 91 C4 32 15N 35 13 E
Samariá, Greece 32 D5 35 17N 23 58 E
Samarinda, Indonesia 75 C5 0 30 S 117 9 E
Samarkand = Samarqand, Uzbekistan 55 D3 39 40N 66 55 E
Samarqand, Uzbekistan 55 D3 39 40N 66 55 E
Sāmarrā, Iraq 89 E10 34 12N 43 52 E
Samastipur, India 81 G11 25 50N 85 50 E
Samatan, France 24 E4 43 29N 0 55 E
Samaúma, Brazil 157 B5 7 50 S 60 2W
Şamaxi, Azerbaijan 53 K9 40 38N 48 37 E
Samba, India 81 C6 32 32N 75 10 E
Samba, Zaïre 103 C5 4 38 S 26 22 E
Samba Caju, Angola 103 D3 8 46 S 15 24 E
Sambaíba, Brazil 154 C2 7 8 S 45 21W
Sambalpur, India 82 D7 21 28N 84 4 E
Sambar, Tanjung, Indonesia 75 C4 2 59 S 110 19 E
Sambas, Indonesia 75 B3 1 20N 109 20 E
Sambava, Madag. 105 A9 14 16 S 50 10 E
Sambawizi, Zimbabwe 107 F2 18 24 S 26 13 E
Sambhal, India 81 E8 28 35N 78 37 E
Sambhar, India 80 F6 26 52N 75 6 E
Sambiase, Italy 41 D9 38 58N 16 17 E
Sambir, Ukraine 51 H2 49 30N 23 10 E
Sambonifacio, Italy 38 C8 45 24N 11 16 E
Sambor, Cambodia 76 F6 12 46N 106 0 E
Sambre →, Europe 21 H5 50 27N 4 52 E
Sambuca di Sicilia, Italy 40 E6 37 39N 13 7 E
Samburu □, Kenya 106 B4 1 10N 37 0 E
Samchŏk, S. Korea 67 F15 37 30N 129 10 E
Samchonpo, S. Korea 67 G15 35 0N 128 6 E
Same, Tanzania 106 C4 4 2 S 37 38 E
Samedan, Switz. 29 C9 46 32N 9 52 E
Samer, France 23 B8 50 38N 1 44 E
Samfya, Zambia 107 E2 11 22 S 29 31 E
Samḩān, Jabal, Oman 87 C6 17 12N 54 55 E
Sámi, Greece 45 F2 38 15N 20 39 E
Şämkir, Azerbaijan 53 K8 40 50N 46 0 E
Samnah, Si. Arabia 84 E3 25 10N 37 15 E
Samnaun, Switz. 29 C10 46 57N 10 22 E
Samnū, Libya 96 C2 27 15N 14 55 E
Samo Alto, Chile 158 C1 30 22 S 71 0W
Samoan Is., Pac. Oc. 121 X24 14 0 S 171 0W
Samobor, Croatia 39 C12 45 47N 15 44 E
Samoëns, France 25 B10 46 5N 6 45 E
Samokov, Bulgaria 43 E8 42 18N 23 35 E
Samoorombón, B., Argentina 158 D4 36 5 S 57 20W
Samorogouan, Burkina Faso 100 C4 11 21N 4 57W
Sámos, Greece 45 G8 37 45N 26 50 E
Samoš, Serbia, Yug. 42 B5 45 13N 20 49 E
Samos, Spain 36 C3 42 44N 7 20W
Samosir, Indonesia 74 B1 2 55N 98 50 E
Samothráki, Évros, Greece 44 D7 40 28N 25 28 E
Samothráki, Ionioi Nísoi, Greece 44 E1 39 48N 19 31 E
Samothráki, Kérkira, Greece 32 A3 39 48N 19 31 E
Samoylovka, Russia 52 E6 51 12N 43 43 E
Sampa, Ghana 100 D4 8 0N 2 36W
Sampacho, Argentina 158 C3 33 20 S 64 50W
Sampang, Indonesia 75 D4 7 11 S 113 13 E
Samper de Calanda, Spain 34 D4 41 11N 0 28W
Sampit, Indonesia 75 C4 2 34 S 113 0 E
Sampit →, Indonesia 75 C4 2 44 S 112 54 E
Sampit, Teluk, Indonesia 75 C4 3 5 S 113 3 E
Samrée, Belgium 21 H7 50 13N 5 39 E
Samrong, Cambodia 76 E4 14 15N 103 30 E
Samrong, Thailand 76 E3 15 10N 100 40 E
Samsø, Denmark 15 J4 55 50N 10 35 E
Samsø Bælt, Denmark 15 J4 55 45N 10 45 E
Samsonovo, Turkmenistan 55 E2 37 53N 65 15 E
Samsun, Turkey 88 B7 41 15N 36 22 E
Samsun Dağı, Turkey 45 G9 37 45N 27 10 E
Samtredia, Georgia 53 J6 42 7N 42 24 E
Samui, Ko, Thailand 77 H3 9 30N 100 0 E
Samur →, Russia 53 K9 41 53N 48 32 E
Samurskiy Khrebet, Russia 53 K8 41 55N 47 11 E
Samusole, Zaïre 103 E4 10 2 S 24 0 E
Samut Prakan, Thailand 76 F3 13 32N 100 40 E
Samut Sakhon, Thailand 76 F3 13 31N 100 13 E
Samut Songkhram →, Thailand 76 F3 13 24N 100 1 E
Samwari, Pakistan 80 E2 28 30N 66 46 E
San, Mali 100 C4 13 15N 4 57W
San →, Cambodia 76 F5 13 32N 105 57 E
San →, Poland 31 A14 50 45N 21 51 E
San Adrián, C. de, Spain 36 B2 43 21N 8 50W
San Agustín, Colombia 152 C2 1 53N 76 16W
San Agustin, C., Phil. 71 H6 6 20N 126 13 E
San Agustín de Valle Fértil, Argentina 158 C2 30 35 S 67 30W
San Ambrosio, Pac. Oc. 123 K20 26 28 S 79 53W
San Andreas, U.S.A. 144 G6 38 12N 120 41W
San Andres, Phil. 70 E4 13 19N 122 41 E
San Andrés, I. de, Caribbean 148 D3 12 42N 81 46W
San Andres Mts., U.S.A. 143 K10 33 0N 106 30W
San Andrés Tuxtla, Mexico 147 D5 18 30N 95 20W
San Angelo, U.S.A. 139 K4 31 28N 100 26W
San Anselmo, U.S.A. 144 H4 37 59N 122 34W
San Antonio, Belize 147 D7 16 15N 89 2W
San Antonio, Chile 158 C1 33 40 S 71 40W
San Antonio, Phil. 70 D3 14 57N 120 5 E
San Antonio, Spain 33 C7 38 59N 1 19 E
San Antonio, N. Mex., U.S.A. 143 K10 33 55N 106 52W
San Antonio, Tex., U.S.A. 139 L5 29 25N 98 30W
San Antonio, Venezuela 152 C4 3 30N 66 44W
San Antonio →, U.S.A. 139 L6 28 30N 96 54W
San Antonio, C., Argentina 158 D4 36 15 S 56 40W
San Antonio, C., Cuba 148 B3 21 50N 84 57W

260 San Antonio, C. de

San Antonio, C. de, *Spain* **35 G5** 38 48N 0 12 E
San Antonio, Mt., *U.S.A.* **145 L9** 34 17N 117 38W
San Antonio Bay, *Phil.* **71 G1** 8 38N 117 35 E
San Antonio de los Baños, *Cuba* **148 B3** 22 54N 82 31W
San Antonio de los Cobres, *Argentina* **158 A2** 24 10 S 66 17W
San Antonio Oeste, *Argentina* **160 B4** 40 40 S 65 0W
San Arcángelo, *Italy* **41 B9** 40 14N 16 14 E
San Ardo, *U.S.A.* **144 J6** 36 1N 120 54W
San Augustín, *Canary Is.* **33 G4** 27 47N 15 32W
San Augustine, *U.S.A.* **139 K7** 31 30N 94 7W
San Bartolomé, *Canary Is.* **33 F6** 28 59N 13 37W
San Bartolomé de Tirajana, *Canary Is.* **33 G4** 27 54N 15 34W
San Bartolomeo in Galdo, *Italy* **41 A8** 41 24N 15 1 E
San Benedetto del Tronto, *Italy* **39 F10** 42 57N 13 53 E
San Benedetto Po, *Italy* **38 C7** 45 2N 10 55 E
San Benedicto, I., *Mexico* **146 D2** 19 18N 110 49W
San Benito, *U.S.A.* **139 M6** 26 8N 97 38W
San Benito →, *U.S.A.* **144 J5** 36 53N 121 34W
San Benito Mt., *U.S.A.* **144 J6** 36 22N 120 37W
San Bernardino, *U.S.A.* **145 L9** 34 7N 117 19W
San Bernardino, Paso del, *Switz.* **29 D8** 46 28N 9 11 E
San Bernardino Mts., *U.S.A.* **145 L10** 34 10N 116 45W
San Bernardino Str., *Phil.* **70 E5** 13 0N 125 0 E
San Bernardo, *Chile* **158 C1** 33 40 S 70 50W
San Bernardo, I. de, *Colombia* **152 B2** 9 45N 75 50W
San Blas, *Mexico* **146 B3** 26 4N 108 46W
San Blas, Arch. de, *Panama* **148 E4** 9 50N 78 31W
San Blas, C., *U.S.A.* **135 L3** 29 40N 85 21W
San Borja, *Bolivia* **156 C4** 14 50 S 66 52W
San Buenaventura, *Bolivia* **156 C4** 14 28 S 67 35W
San Buenaventura, *Mexico* **146 B4** 27 5N 101 32W
San Carlos = Butuku-Luba, *Eq. Guin.* **101 E6** 3 29N 8 33 E
San Carlos, *Argentina* **158 C2** 33 50 S 69 0W
San Carlos, *Bolivia* **157 D5** 17 24 S 63 45W
San Carlos, *Chile* **158 D1** 36 10 S 72 0W
San Carlos, *Mexico* **146 B4** 29 0N 100 54W
San Carlos, *Nic.* **148 D3** 11 12N 84 50W
San Carlos, *Negros, Phil.* **71 F4** 10 29N 123 25 E
San Carlos, *Pangasinan, Phil.* **70 D3** 15 55N 120 20 E
San Carlos, *Spain* **33 B8** 39 3N 1 34 E
San Carlos, *Uruguay* **159 C5** 34 46 S 54 58W
San Carlos, *U.S.A.* **143 K8** 33 21N 110 27W
San Carlos, *Amazonas, Venezuela* **152 C4** 1 55N 67 4W
San Carlos, *Cojedes, Venezuela* **152 B4** 9 40N 68 36W
San Carlos de Bariloche, *Argentina* **160 B2** 41 10 S 71 25W
San Carlos de la Rápita, *Spain* **34 E5** 40 37N 0 35 E
San Carlos del Zulia, *Venezuela* **152 B3** 9 1N 71 55W
San Carlos L., *U.S.A.* **143 K8** 33 11N 110 32W
San Cataldo, *Italy* **40 E6** 37 29N 13 59 E
San Celoni, *Spain* **34 D7** 41 42N 2 30 E
San Clemente, *Chile* **158 D1** 35 30 S 71 29W
San Clemente, *Spain* **35 F2** 39 24N 2 25W
San Clemente, *U.S.A.* **145 M9** 33 26N 117 37W
San Clemente I., *U.S.A.* **145 N8** 32 53N 118 29W
San Costanzo, *Italy* **39 E10** 43 46N 13 4 E
San Cristóbal, *Argentina* **158 C3** 30 20 S 61 10W
San Cristóbal, *Colombia* **152 D3** 2 18 S 73 2W
San Cristóbal, *Dom. Rep.* **149 C5** 18 25N 70 6W
San Cristóbal, *Mexico* **147 D6** 16 50N 92 33W
San Cristóbal, *Solomon Is.* **121 N11** 10 30 S 161 0 E
San Cristóbal, *Spain* **33 B11** 39 57N 4 3 E
San Cristóbal, *Venezuela* **152 B3** 7 46N 72 14W
San Damiano d'Asti, *Italy* **38 D5** 44 50N 8 4 E
San Daniele del Friuli, *Italy* **39 B10** 46 9N 13 0 E
San Demétrio Corone, *Italy* **41 C9** 39 34N 16 22 E
San Diego, *Calif., U.S.A.* **145 N9** 32 43N 117 9W
San Diego, *Tex., U.S.A.* **139 M5** 27 46N 98 14W
San Diego, C., *Argentina* **160 D3** 54 40 S 65 10W
San Diego de la Unión, *Mexico* **146 C4** 21 28N 100 52W
San Dimitri, Ras, *Malta* **32 C1** 36 4N 14 11 E
San Dionisio, *Phil.* **71 F4** 11 16N 123 6 E
San Doná di Piave, *Italy* **39 C9** 45 38N 12 34 E
San Elpídio a Mare, *Italy* **39 E10** 43 16N 13 41 E
San Emilio, *Phil.* **70 C3** 17 14N 120 37 E
San Estanislao, *Paraguay* **158 A4** 24 39 S 56 26W
San Esteban de Gormaz, *Spain* **34 D1** 41 34N 3 13W
San Fabian, *Phil.* **70 C3** 16 5N 120 25 E
San Felice sul Panaro, *Italy* **38 D8** 44 50N 11 8 E
San Felipe, *Chile* **158 C1** 32 43 S 70 42W
San Felipe, *Colombia* **152 C4** 1 55N 67 6W
San Felipe, *Mexico* **146 A2** 31 0N 114 52W
San Felipe, *Phil.* **70 D3** 15 4N 120 4 E
San Felipe, *Venezuela* **152 A4** 10 20N 68 44W
San Felipe →, *U.S.A.* **145 M11** 33 12N 115 49W
San Felíu de Guíxols, *Spain* **34 D8** 41 45N 3 1 E
San Felíu de Llobregat, *Spain* **34 D7** 41 23N 2 2 E
San Félix, *Pac. Oc.* **123 K20** 26 23 S 80 0W
San Fernando, *Chile* **158 C1** 34 30 S 71 0W
San Fernando, *Mexico* **146 B1** 29 55N 115 10W
San Fernando, *Cebu, Phil.* **71 F4** 10 10N 123 42 E
San Fernando, *La Union, Phil.* **70 C3** 16 40N 120 23 E
San Fernando, *Pampanga, Phil.* **70 D3** 15 5N 120 37 E
San Fernando, *Tablas, Phil.* **70 E4** 12 18N 122 36 E
San Fernando, *Baleares, Spain* **33 C7** 38 42N 1 28 E
San Fernando, *Cádiz, Spain* **37 J4** 36 28N 6 17W
San Fernando, *Trin. & Tob.* **149 D7** 10 20N 61 30W
San Fernando, *U.S.A.* **145 L8** 34 17N 118 26W
San Fernando →, *Mexico* **146 C5** 24 55N 98 10W
San Fernando de Apure, *Venezuela* **152 B4** 7 54N 67 15W
San Fernando de Atabapo, *Venezuela* **152 C4** 4 3N 67 42W
San Fernando di Púglia, *Italy* **41 A9** 41 18N 16 5 E
San Francisco, *Argentina* **158 C3** 31 30 S 62 5W
San Francisco, *Bolivia* **157 D4** 15 16 S 65 31W
San Francisco, *Cebu, Phil.* **71 F5** 10 39N 124 23 E
San Francisco, *Leyte, Phil.* **71 F5** 10 4N 125 9 E
San Francisco, *Mindanao, Phil.* **71 G5** 8 30N 125 56 E
San Francisco, *U.S.A.* **144 H4** 37 47N 122 25W
San Francisco →, *U.S.A.* **143 K9** 32 59N 109 22W
San Francisco, Paso de, *S. Amer.* **158 B2** 27 0 S 68 0W
San Francisco de Macorîs, *Dom. Rep.* **149 C5** 19 19N 70 15W
San Francisco del Monte de Oro, *Argentina* **158 C2** 32 36 S 66 8W
San Francisco del Oro, *Mexico* **146 B3** 26 52N 105 50W
San Francisco Javier, *Spain* **33 C7** 38 42N 1 26 E
San Francisco Solano, Pta., *Colombia* **152 B2** 6 18N 77 29W
San Fratello, *Italy* **41 D7** 38 1N 14 36 E
San Gabriel, *Ecuador* **152 C2** 0 36N 77 49W
San Gavino Monreale, *Italy* **40 C1** 39 33N 8 47 E
San Gil, *Colombia* **152 B3** 6 33N 73 8W
San Gimignano, *Italy* **38 E8** 43 28N 11 2 E
San Giórgio di Nogaro, *Italy* **39 C10** 45 50N 13 13 E
San Giórgio Iónico, *Italy* **41 B10** 40 27N 17 23 E
San Giovanni Bianco, *Italy* **38 C6** 45 52N 9 39 E
San Giovanni in Fiore, *Italy* **41 C9** 39 15N 16 42 E
San Giovanni in Persiceto, *Italy* **39 D8** 44 38N 11 11 E
San Giovanni Rotondo, *Italy* **41 A8** 41 42N 15 44 E
San Giovanni Valdarno, *Italy* **39 E8** 43 34N 11 32 E
San Giuliano Terme, *Italy* **38 E7** 43 45N 10 26 E
San Gorgonio Mt., *U.S.A.* **145 L10** 34 7N 116 51W
San Gottardo, P. del, *Switz.* **29 C7** 46 33N 8 33 E
San Gregorio, *Uruguay* **159 C4** 32 37 S 55 40W
San Gregorio, *U.S.A.* **144 H4** 37 20N 122 23W
San Guiseppe Iato, *Italy* **40 E6** 37 57N 13 11 E
San Ignacio, *Belize* **147 D7** 17 10N 89 0W
San Ignacio, *Bolivia* **157 D5** 16 20 S 60 55W
San Ignacio, *Mexico* **146 B2** 27 27N 113 0W
San Ignacio, *Paraguay* **158 B4** 26 52 S 57 3W
San Ignacio, L., *Mexico* **146 B2** 26 50N 113 11W
San Ildefonso, *Phil.* **70 D3** 15 5N 120 56 E
San Ildefonso, C., *Phil.* **70 C4** 16 0N 122 1 E
San Isidro, *Argentina* **158 C4** 34 29 S 58 31W
San Jacinto, *Colombia* **152 B2** 9 50N 75 8W
San Jacinto, *Phil.* **70 E4** 12 34N 123 44 E
San Jacinto, *U.S.A.* **145 M10** 33 47N 116 57W
San Jaime, *Spain* **33 B11** 39 54N 4 4 E
San Javier, *Misiones, Argentina* **159 B4** 27 55 S 55 5W
San Javier, *Santa Fe, Argentina* **158 C4** 30 40 S 59 55W
San Javier, *Beni, Bolivia* **157 C5** 14 34 S 64 42W
San Javier, *Santa Cruz, Bolivia* **157 D5** 16 18 S 62 30W
San Javier, *Chile* **158 D1** 35 40 S 71 45W
San Javier, *Spain* **35 H4** 37 49N 0 50W
San Jerónimo, Sa. de, *Colombia* **152 B2** 8 0N 75 50W
San Jeronimo Taviche, *Mexico* **147 D5** 16 38N 96 32W
San Joaquín, *Bolivia* **157 C5** 13 4 S 64 49W
San Joaquin, *Phil.* **71 F4** 10 35N 122 8 E
San Joaquin, *U.S.A.* **144 J6** 36 36N 120 11W
San Joaquín, *Venezuela* **152 A4** 10 16N 67 47W
San Joaquín →, *Bolivia* **157 C5** 13 8 S 63 41W
San Joaquin →, *U.S.A.* **144 G5** 38 4N 121 51W
San Joaquin Valley, *U.S.A.* **144 J6** 37 20N 121 0W
San Jordi, *Spain* **33 B9** 39 33N 2 46 E
San Jorge, *Argentina* **158 C3** 31 54 S 61 50W
San Jorge, *Spain* **33 C7** 38 54N 1 24 E
San Jorge, B. de, *Mexico* **146 A2** 31 20N 113 20W
San Jorge, G., *Argentina* **160 C3** 46 0 S 66 0W
San Jorge, G. de, *Spain* **34 E4** 40 53N 1 2 E
San José, *Bolivia* **157 D5** 17 53 S 60 50W
San José, *Costa Rica* **148 E3** 9 55N 84 2W
San José, *Guatemala* **148 D1** 14 0N 90 50W
San José, *Mexico* **146 C2** 25 0N 110 50W
San Jose, *Phil.* **70 D3** 15 45N 120 55 E
San José, *Spain* **33 C7** 38 55N 1 18 E
San Jose, *Calif., U.S.A.* **144 H5** 37 20N 121 53W
San Jose, *Ill., U.S.A.* **140 D7** 40 18N 89 36W
San Jose →, *U.S.A.* **143 J10** 34 25N 106 45W
San Jose de Buenovista, *Phil.* **70 E3** 12 27N 121 4 E
San José de Feliciano, *Argentina* **158 C4** 30 26 S 58 46W
San José de Jáchal, *Argentina* **158 C2** 30 15 S 68 46W
San José de Mayo, *Uruguay* **158 C4** 34 27 S 56 40W
San José de Ocune, *Colombia* **152 C3** 4 15N 70 20W
San José de Uchapiamonas, *Bolivia* **156 C4** 14 13 S 68 5W
San José del Cabo, *Mexico* **146 C3** 23 0N 109 40W
San José del Guaviare, *Colombia* **152 C3** 2 35N 72 38W
San José do Anauá, *Brazil* **153 C5** 0 58N 61 22W
San Juan, *Argentina* **158 C2** 31 30 S 68 30W
San Juan, *Colombia* **152 B2** 8 46N 76 32W
San Juan, *Mexico* **146 C4** 21 20N 102 50W
San Juan, *Ica, Peru* **156 D2** 15 22 S 75 7W
San Juan, *Puno, Peru* **156 C4** 14 2 S 69 19W
San Juan, *Luzon, Phil.* **70 C3** 16 40N 120 20 E
San Juan, *Mindanao, Phil.* **71 G6** 8 25N 126 20 E
San Juan, *Puerto Rico* **149 C6** 18 28N 66 7W
San Juan □, *Argentina* **158 C2** 31 9 S 69 0W
San Juan →, *Argentina* **158 C2** 32 20 S 67 25W
San Juan →, *Bolivia* **157 E4** 21 2 S 65 19W
San Juan →, *Colombia* **152 C2** 4 3N 77 27W
San Juan →, *Nic.* **148 D3** 10 56N 83 42W
San Juan →, *U.S.A.* **143 H8** 37 16N 110 26W
San Juan →, *Venezuela* **153 A5** 10 14N 62 38W
San Juan, C., *Eq. Guin.* **102 B1** 1 5N 9 20 E
San Juan Bautista, *Paraguay* **158 B4** 26 37 S 57 6W
San Juan Bautista, *Spain* **33 B8** 39 5N 1 31 E
San Juan Bautista, *U.S.A.* **144 J5** 36 51N 121 32W
San Juan Bautista Valle Nacional, *Mexico* **147 D5** 17 47N 96 19W
San Juan Capistrano, *U.S.A.* **145 M9** 33 30N 117 40W
San Juan Cr. →, *U.S.A.* **144 J5** 35 40N 120 22W
San Juan de Guadalupe, *Mexico* **146 C4** 24 38N 102 44W
San Juan de los Morros, *Venezuela* **152 B4** 9 55N 67 21W
San Juan del César, *Colombia* **152 A3** 10 46N 73 1W
San Juan del Norte, *Nic.* **148 D3** 10 58N 83 40W
San Juan del Norte, B. de, *Nic.* **148 D3** 11 0N 83 40W
San Juan del Puerto, *Spain* **37 H4** 37 20N 6 50W
San Juan del Río, *Mexico* **147 C5** 20 25N 100 0W
San Juan del Sur, *Nic.* **148 D2** 11 20N 85 51W
San Juan I., *U.S.A.* **144 B3** 48 32N 123 5W
San Juan Mts., *U.S.A.* **143 H10** 37 30N 107 0W
San Julián, *Argentina* **160 C3** 49 15 S 67 45W
San Julian, *Phil.* **71 F5** 11 45N 125 27 E
San Just, Sierra de, *Spain* **34 E4** 40 45N 0 49W
San Justo, *Argentina* **158 C3** 30 47 S 60 30W
San Kamphaeng, *Thailand* **76 C2** 18 45N 99 8 E
San Lázaro, C., *Mexico* **146 C2** 24 50N 112 18W
San Lázaro, Sa., *Mexico* **146 C3** 23 25N 110 0W
San Leandro, *U.S.A.* **144 H4** 37 44N 122 9W
San Leonardo, *Spain* **34 D1** 41 51N 3 5W
San Lorenzo, *Argentina* **158 C3** 32 45 S 60 45W
San Lorenzo, *Beni, Bolivia* **157 D4** 15 22 S 65 48W
San Lorenzo, *Tarija, Bolivia* **157 E5** 21 26 S 64 47W
San Lorenzo, *Ecuador* **152 C2** 1 15N 78 50W
San Lorenzo, *Paraguay* **158 B4** 25 20 S 57 32W
San Lorenzo, *Spain* **33 B10** 39 37N 3 17 E
San Lorenzo, *Venezuela* **152 B3** 9 47N 71 4W
San Lorenzo →, *Mexico* **146 C3** 24 15N 107 24W
San Lorenzo, I., *Mexico* **146 B2** 28 35N 112 50W
San Lorenzo, I., *Peru* **156 C2** 12 7 S 77 15W
San Lorenzo, Mt., *Argentina* **160 C2** 47 40 S 72 20W
San Lorenzo de la Parrilla, *Spain* **34 F2** 39 51N 2 22W
San Lorenzo de Morunys, *Spain* **34 C6** 42 8N 1 35 E
San Lucas, *Bolivia* **157 E4** 20 5 S 65 7W
San Lucas, *Baja Calif. S., Mexico* **146 C3** 22 53N 109 54W
San Lucas, *Baja Calif. S., Mexico* **146 B2** 27 10N 112 14W
San Lucas, *U.S.A.* **144 J5** 36 8N 121 1W
San Lucas, C., *Mexico* **146 C3** 22 50N 110 0W
San Lúcido, *Italy* **41 C9** 39 18N 16 3 E
San Luis, *Argentina* **158 C2** 33 20 S 66 20W
San Luis, *Cuba* **148 B3** 22 17N 83 46W
San Luis, *Guatemala* **148 C2** 16 14N 89 27W
San Luis, *U.S.A.* **143 H11** 37 12N 105 25W
San Luis □, *Argentina* **158 C2** 34 0 S 66 0W
San Luis, I., *Mexico* **146 B2** 29 58N 114 26W
San Luis, L. de, *Bolivia* **157 C5** 13 45 S 64 0W
San Luis, Sierra de, *Argentina* **158 C2** 32 30 S 66 10W
San Luis de la Paz, *Mexico* **146 C4** 21 19N 100 32W
San Luis Obispo, *U.S.A.* **145 K6** 35 17N 120 40W
San Luis Potosí, *Mexico* **146 C4** 22 9N 100 59W
San Luis Potosí □, *Mexico* **146 C4** 22 10N 101 0W
San Luis Reservoir, *U.S.A.* **144 H5** 37 4N 121 5W
San Luis Río Colorado, *Mexico* **146 A2** 32 29N 114 58W
San Manuel, *Phil.* **70 C3** 16 4N 120 40 E
San Marco Argentano, *Italy* **41 C9** 39 33N 16 7 E
San Marco dei Cavoti, *Italy* **41 A7** 41 18N 14 53 E
San Marco in Lámis, *Italy* **41 A8** 41 43N 15 38 E
San Marcos, *Colombia* **152 B2** 8 39N 75 8W
San Marcos, *Guatemala* **148 D1** 14 59N 91 52W
San Marcos, *Mexico* **146 B2** 27 13N 112 6W
San Marcos, *U.S.A.* **139 L6** 29 53N 97 56W
San Marino ■, *Europe* **39 E9** 43 56N 12 25 E
San Martín, *Argentina* **158 C2** 33 5 S 68 28W
San Martín, *Colombia* **152 C3** 3 42N 73 42W
San Martín →, *Bolivia* **157 C5** 13 8 S 63 43W
San Martín, L., *Argentina* **160 C2** 48 50 S 72 50W
San Martin de los Andes, *Argentina* **160 B2** 40 10 S 71 20W
San Martín de Valdeiglesias, *Spain* **36 E6** 40 21N 4 24W
San Martino de Calvi, *Italy* **38 C6** 45 57N 9 41 E
San Mateo, *Agusan del N., Phil.* **71 G5** 8 48N 125 33 E
San Mateo, *Isabela, Phil.* **70 C3** 16 54N 121 33 E
San Mateo, *Baleares, Spain* **33 B7** 39 3N 1 23 E
San Mateo, *Valencia, Spain* **34 E5** 40 28N 0 10 E
San Mateo, *U.S.A.* **144 H4** 37 34N 122 19W
San Matías, *Bolivia* **157 D6** 16 25 S 58 20W
San Matías, G., *Argentina* **160 B4** 41 30 S 64 0W
San Miguel = Linapacan, *Phil.* **71 F2** 11 30N 119 52 E
San Miguel, *El Salv.* **148 D2** 13 30N 88 12W
San Miguel, *Panama* **148 E4** 8 27N 78 55W
San Miguel, *Lanao del N., Phil.* **71 G5** 9 3N 125 59 E
San Miguel, *Lanao del S., Phil.* **71 G5** 8 13N 124 14 E
San Miguel, *Spain* **33 B7** 39 3N 1 26 E
San Miguel, *U.S.A.* **144 K6** 35 45N 120 42W
San Miguel, *Venezuela* **152 B4** 9 40N 65 11W
San Miguel →, *Bolivia* **157 C5** 13 52 S 63 56W
San Miguel →, *S. Amer.* **152 C2** 0 25N 76 30W
San Miguel de Huachi, *Bolivia* **156 D4** 15 40 S 67 15W
San Miguel de Salinas, *Spain* **35 H4** 37 59N 0 47W
San Miguel de Tucumán, *Argentina* **158 B2** 26 50 S 65 20W
San Miguel del Monte, *Argentina* **158 D4** 35 23 S 58 50W
San Miguel I., *U.S.A.* **145 L6** 34 2N 120 23W
San Miguel Is., *Phil.* **71 H2** 7 45N 118 28 E
San Miniato, *Italy* **38 E7** 43 41N 10 51 E
San Narciso, *Quezon, Phil.* **70 E4** 13 34N 122 34 E
San Narciso, *Zambales, Phil.* **70 D3** 15 2N 120 3 E
San Nicolás, *Canary Is.* **33 G4** 27 58N 15 47W
San Nicolas, *Phil.* **70 B3** 18 10N 120 36 E
San Nicolás de los Arroyas, *Argentina* **158 C3** 33 25 S 60 10W
San Nicolas I., *U.S.A.* **145 M7** 33 15N 119 30W
San Onofre, *Colombia* **152 B2** 9 44N 75 32W
San Onofre, *U.S.A.* **145 M9** 33 22N 117 34W
San Pablo, *Bolivia* **158 A2** 21 43 S 66 38W
San Pablo, *Isabela, Phil.* **70 C3** 17 27N 121 48 E
San Pablo, *Laguna, Phil.* **70 D3** 14 11N 121 31 E
San Páolo di Civitate, *Italy* **41 A8** 41 44N 15 15 E
San Pascual, *Phil.* **70 E4** 13 8N 122 59 E
San Pedro, *Buenos Aires, Argentina* **159 B5** 26 30 S 54 10W
San Pedro, *Jujuy, Argentina* **158 A3** 24 12 S 64 55W
San Pedro, *Colombia* **152 C3** 4 56N 71 53W
San-Pédro, *Ivory C.* **100 E3** 4 50N 6 33W
San Pedro, *Mexico* **146 C2** 23 55N 110 17W
San Pedro, *Peru* **156 C3** 14 49 S 74 5W
San Pedro □, *Paraguay* **158 A4** 24 0 S 57 0W
San Pedro →, *Chihuahua, Mexico* **146 B3** 28 20N 106 10W
San Pedro →, *Michoacan, Mexico* **146 D4** 19 23N 103 51W
San Pedro →, *Nayarit, Mexico* **146 C3** 21 45N 105 30W
San Pedro →, *U.S.A.* **143 K8** 32 59N 110 47W
San Pedro, Pta., *Chile* **158 B1** 25 30 S 70 38W
San Pedro, Sierra de, *Spain* **37 F4** 39 18N 6 40W
San Pedro Channel, *U.S.A.* **145 M8** 33 30N 118 25W
San Pedro de Arimena, *Colombia* **152 C3** 4 37N 71 42W
San Pedro de Atacama, *Chile* **158 A2** 22 55 S 68 15W
San Pedro de Jujuy, *Argentina* **158 A3** 24 12 S 64 55W
San Pedro de las Colonias, *Mexico* **146 B4** 25 50N 102 59W
San Pedro de Lloc, *Peru* **156 B2** 7 15 S 79 28W
San Pedro de Macorís, *Dom. Rep.* **149 C6** 18 30N 69 18W
San Pedro del Norte, *Nic.* **148 D3** 13 4N 84 33W
San Pedro del Paraná, *Paraguay* **158 B4** 26 43 S 56 13W
San Pedro del Pinatar, *Spain* **35 H4** 37 50N 0 50W
San Pedro Mártir, Sierra, *Mexico* **146 A1** 31 0N 115 30W
San Pedro Mixtepec, *Mexico* **147 D5** 16 2N 97 7W
San Pedro Ocampo = Melchor Ocampo, *Mexico* **146 C4** 24 52N 101 40W
San Pedro Sula, *Honduras* **148 C2** 15 30N 88 0W
San Pieto, *Italy* **40 C1** 39 8N 8 17 E
San Pietro Vernótico, *Italy* **41 B11** 40 29N 18 0 E
San Quintín, *Mexico* **146 A1** 30 29N 115 57W
San Rafael, *Argentina* **158 C2** 34 40 S 68 21W
San Rafael, *Calif., U.S.A.* **144 H4** 37 58N 122 32W
San Rafael, *N. Mex., U.S.A.* **143 J10** 35 7N 107 53W
San Rafael, *Venezuela* **152 A3** 10 58N 71 46W
San Rafael Mt., *U.S.A.* **145 L7** 34 41N 119 52W
San Rafael Mts., *U.S.A.* **145 L7** 34 40N 119 50W
San Ramón, *Bolivia* **157 C5** 13 17 S 64 43W
San Ramón, *Peru* **156 C2** 11 8 S 75 20W
San Ramón de la Nueva Orán, *Argentina* **158 A3** 23 10 S 64 20W
San Remo, *Italy* **38 E4** 43 49N 7 46 E
San Román, C., *Venezuela* **152 A3** 12 12N 70 0W
San Roque, *Argentina* **158 B4** 28 25 S 58 45W
San Roque, *Phil.* **70 E5** 12 37N 124 52 E
San Roque, *Spain* **37 J5** 36 17N 5 21W
San Rosendo, *Chile* **158 D1** 37 16 S 72 43W
San Saba, *U.S.A.* **139 K5** 31 12N 98 43W
San Salvador, *Bahamas* **149 B5** 24 0N 74 40W
San Salvador, *El Salv.* **148 D2** 13 40N 89 10W
San Salvador, *Spain* **33 B10** 39 27N 3 11 E
San Salvador de Jujuy, *Argentina* **158 A3** 24 10 S 64 48W
San Salvador I., *Bahamas* **149 B5** 24 0N 74 32W
San Sebastián, *Argentina* **160 D3** 53 10 S 68 30W
San Sebastián, *Spain* **34 B3** 43 17N 1 58W
San Sebastián, *Venezuela* **152 B4** 9 57N 67 11W
San Sebastian de la Gomera, *Canary Is.* **33 F2** 28 5N 17 7W
San Serra, *Spain* **33 B10** 39 43N 3 13 E
San Serverino Marche, *Italy* **39 E10** 43 13N 13 10 E
San Simeon, *U.S.A.* **144 K5** 35 39N 121 11W
San Simon, *U.S.A.* **143 K9** 32 16N 109 14W
San Stéfano di Cadore, *Italy* **39 B9** 46 34N 12 33 E
San Telmo, *Mexico* **146 A1** 30 58N 116 6W
San Telmo, *Spain* **33 B9** 39 35N 2 21 E
San Teodoro, *Phil.* **70 E3** 13 26N 121 1 E
San Tiburcio, *Mexico* **146 C4** 24 8N 101 32W
San Valentin, Mte., *Chile* **160 C2** 46 30 S 73 30W
San Vicente de Alcántara, *Spain* **37 F3** 39 22N 7 8W
San Vicente de la Barquera, *Spain* **36 B6** 43 23N 4 29W
San Vicente del Caguán, *Colombia* **152 C3** 2 7N 74 46W
San Vincenzo, *Italy* **38 E7** 43 6N 10 32 E
San Vito, *Italy* **40 C2** 39 26N 9 32 E
San Vito, C., *Italy* **40 D5** 38 11N 12 41 E
San Vito al Tagliamento, *Italy* **39 C9** 45 54N 12 52 E
San Vito Chietino, *Italy* **39 F11** 42 18N 14 27 E
San Vito dei Normanni, *Italy* **41 B10** 40 39N 17 42 E
San Yanaro, *Colombia* **152 C4** 2 47N 69 42W

Sepīdān 263

Schagen, *Neths.* 20 C5 52 49N 4 48 E
Schaijk, *Neths.* 20 E7 51 44N 5 38 E
Schalkhaar, *Neths.* 20 D8 52 17N 6 12 E
Schalkwijk, *Neths.* 20 E6 52 0N 5 11 E
Schangnau, *Switz.* 28 C5 46 50N 7 47 E
Schänis, *Switz.* 29 B8 47 10N 9 3 E
Schärding, *Austria* 30 C6 48 27N 13 27 E
Scharhörn, *Germany* 26 B4 53 57N 8 24 E
Scharnitz, *Austria* 30 D4 47 23N 11 15 E
Scheessel, *Germany* 26 B5 53 10N 9 29 E
Schefferville, *Canada* 129 B6 54 48N 66 50W
Scheibbs, *Austria* 30 C8 48 1N 15 9 E
Schelde →, *Belgium* 21 F4 51 15N 4 16 E
Schell City, *U.S.A.* 140 F2 38 1N 94 7W
Schell Creek Ra., *U.S.A.* 142 G6 39 15N 114 30W
Schenectady, *U.S.A.* 137 D11 42 49N 73 57W
Scherfede, *Germany* 26 D5 51 32N 9 2 E
Scherpenheuvel, *Belgium* 21 G5 50 58N 4 58 E
Scherpenisse, *Neths.* 21 E4 51 33N 4 6 E
Scherpenzeel, *Neths.* 20 D7 52 5N 5 30 E
Schesaplana, *Switz.* 29 B9 47 5N 9 43 E
Schesslitz, *Germany* 27 F7 49 58N 11 1 E
Scheveningen, *Neths.* 20 D4 52 6N 4 16 E
Schiedam, *Neths.* 20 E4 51 55N 4 25 E
Schiermonnikoog, *Neths.* 20 B8 53 30N 6 15 E
Schiers, *Switz.* 29 C9 46 58N 9 41 E
Schifferstadt, *Germany* 27 F4 49 23N 8 22 E
Schifflange, *Lux.* 21 K8 49 30N 6 1 E
Schijndel, *Neths.* 21 E6 51 37N 5 27 E
Schiltigheim, *France* 23 D14 48 35N 7 45 E
Schio, *Italy* 39 C8 45 43N 11 21 E
Schipbeek, *Neths.* 20 D8 52 14N 6 10 E
Schipluiden, *Neths.* 20 E4 51 59N 4 19 E
Schirmeck, *France* 23 D14 48 29N 7 12 E
Schladming, *Austria* 30 D6 47 23N 13 41 E
Schlanders = Silandro, *Italy* 38 B7 46 38N 10 46 E
Schlei →, *Germany* 26 A5 54 40N 10 0 E
Schleiden, *Germany* 26 E2 50 31N 6 28 E
Schleiz, *Germany* 26 E7 50 35N 11 49 E
Schleswig, *Germany* 26 A5 54 31N 9 34 E
Schleswig-Holstein □, *Germany* 26 A5 54 30N 9 30 E
Schlieren, *Switz.* 29 B6 47 26N 8 27 E
Schlüchtern, *Germany* 27 E5 50 20N 9 32 E
Schmalkalden, *Germany* 26 E6 50 44N 10 26 E
Schmölln, *Brandenburg, Germany* 26 B10 53 17N 14 5 E
Schmölln, *Thüringen, Germany* 26 E8 50 54N 12 19 E
Schneeberg, *Austria* 30 D8 47 47N 15 48 E
Schneeberg, *Germany* 26 E8 50 35N 12 38 E
Schneider, *U.S.A.* 141 C9 41 13N 87 28W
Schoenberg, *Belgium* 21 H8 50 17N 6 16 E
Schofield, *U.S.A.* 138 C10 44 54N 89 36W
Scholls, *U.S.A.* 144 E4 45 24N 122 56W
Schönberg, *Mecklenburg-Vorpommern, Germany* 26 B6 53 52N 10 56 E
Schönberg, *Schleswig-Holstein, Germany* 26 A6 54 23N 10 21 E
Schönebeck, *Germany* 26 C7 52 2N 11 44 E
Schönenwerd, *Switz.* 28 B5 47 23N 8 0 E
Schongau, *Germany* 27 H6 47 47N 10 53 E
Schöningen, *Germany* 26 C6 52 8N 10 56 E
Schoolcraft, *U.S.A.* 141 B11 42 7N 85 38W
Schoondijke, *Neths.* 21 F3 51 21N 3 33 E
Schoonebeek, *Neths.* 20 C9 52 39N 6 52 E
Schoonhoven, *Neths.* 20 E5 51 57N 4 51 E
Schoorl, *Neths.* 20 C5 52 42N 4 42 E
Schortens, *Germany* 26 B3 53 31N 7 56 E
Schoten, *Belgium* 21 F5 51 16N 4 30 E
Schouten I., *Australia* 114 G4 42 20 S 148 20 E
Schouten Is. = Supiori, *Indonesia* 73 B5 1 0 S 136 0 E
Schouwen, *Neths.* 21 E3 51 43N 3 45 E
Schramberg, *Germany* 27 G4 48 13N 8 22 E
Schrankogl, *Austria* 30 D4 47 3N 11 7 E
Schreckhorn, *Switz.* 28 C6 46 36N 8 7 E
Schreiber, *Canada* 128 C2 48 45N 87 20W
Schrobenhausen, *Germany* 27 G7 48 34N 11 16 E
Schruns, *Austria* 30 D2 47 5N 9 56 E
Schuler, *Canada* 131 C6 50 20N 110 6W
Schuls, *Switz.* 29 C10 46 48N 10 18 E
Schumacher, *Canada* 128 C3 48 30N 81 16W
Schüpfen, *Switz.* 28 B4 47 2N 7 24 E
Schüpfheim, *Switz.* 28 C6 46 57N 8 2 E
Schurz, *U.S.A.* 142 G4 38 57N 118 49W
Schuyler, *U.S.A.* 138 E6 41 27N 97 4W
Schuylkill Haven, *U.S.A.* 137 F8 40 37N 76 11W
Schwabach, *Germany* 27 F7 49 19N 11 2 E
Schwäbisch Gmünd, *Germany* 27 G5 48 48N 9 47 E
Schwäbisch Hall, *Germany* 27 F5 49 6N 9 44 E
Schwäbische Alb, *Germany* 27 G5 48 20N 9 30 E
Schwabmünchen, *Germany* 27 G6 48 10N 10 46 E
Schwanden, *Switz.* 29 C8 46 58N 9 5 E
Schwandorf, *Germany* 27 F8 49 20N 12 7 E
Schwaner, Pegunungan, *Indonesia* 75 C4 1 0 S 112 30 E
Schwarmstedt, *Germany* 26 C5 52 31N 9 38 E
Schwarzach →, *Austria* 30 E5 46 56N 12 35 E
Schwärze, *Germany* 26 C9 52 50N 13 49 E
Schwarze Elster →, *Germany* 26 D8 51 48N 12 50 E
Schwarzenberg, *Germany* 26 E8 50 32N 12 47 E
Schwarzenburg, *Switz.* 28 C4 46 49N 7 20 E
Schwarzwald, *Germany* 27 H4 48 30N 8 20 E
Schwaz, *Austria* 30 D4 47 20N 11 44 E
Schwedt, *Germany* 26 B10 53 3N 14 16 E
Schweinfurt, *Germany* 27 E6 50 3N 10 14 E
Schweizer Mittelland, *Switz.* 28 C4 47 0N 7 15 E
Schweizer-Reneke, *S. Africa* 104 D4 27 11 S 25 18 E
Schwenningen = Villingen-Schwenningen, *Germany* 27 G4 48 3N 8 26 E
Schwerin, *Germany* 26 B7 53 36N 11 22 E
Schweriner See, *Germany* 26 B7 53 43N 11 28 E
Schwetzingen, *Germany* 27 F4 49 23N 8 35 E
Schwyz, *Switz.* 29 B7 47 2N 8 39 E
Schwyz □, *Switz.* 29 B7 47 2N 8 39 E
Sciacca, *Italy* 40 E6 37 30N 13 3 E
Sciao, *Somali Rep.* 108 D3 3 26N 45 21 E
Scicli, *Italy* 41 F7 36 47N 14 42 E
Scilla, *Italy* 41 D8 38 15N 15 43 E
Scilly, Isles of, *U.K.* 17 H1 49 56N 6 22W
Ścinawa, *Poland* 47 D3 51 25N 16 26 E
Scione, *Greece* 44 E5 39 57N 23 36 E
Scioto →, *U.S.A.* 134 F4 38 44N 83 1W
Scobey, *U.S.A.* 138 A2 48 47N 105 25W
Scone, *Australia* 117 B9 32 5 S 150 52 E
Scórdia, *Italy* 41 E7 37 18N 14 51 E
Scoresbysund, *Greenland* 6 B6 70 20N 23 0W
Scorno, Punta dello, *Italy* 40 A1 41 7N 8 19 E
Scotia, *Calif., U.S.A.* 142 F1 40 29N 124 6W
Scotia, *N.Y., U.S.A.* 137 D11 42 50N 73 58W
Scotia Sea, *Antarctica* 7 B18 56 5 S 56 0W
Scotland, *U.S.A.* 138 D6 43 9N 97 43W
Scotland □, *U.K.* 18 E5 57 0N 4 0W
Scotland Neck, *U.S.A.* 135 G7 36 8N 77 25W
Scott, C., *Australia* 112 B4 13 30 S 129 49 E
Scott City, *U.S.A.* 138 F4 38 29N 100 54W
Scott Glacier, *Antarctica* 7 C8 66 15 S 100 5 E
Scott I., *Antarctica* 7 C11 67 0 S 179 0 E
Scott Inlet, *Canada* 127 A12 71 0N 71 0W
Scott Is., *Canada* 130 C3 50 48N 128 40W
Scott L., *Canada* 131 B7 59 55N 106 18W
Scott Reef, *Australia* 112 B3 14 0 S 121 50 E
Scottburgh, *S. Africa* 105 E5 30 15 S 30 47 E
Scottdale, *U.S.A.* 136 F5 40 6N 79 35W
Scottsbluff, *U.S.A.* 138 E3 41 52N 103 40W
Scottsboro, *U.S.A.* 135 H2 34 40N 86 2W
Scottsburg, *U.S.A.* 141 F11 38 41N 85 47W
Scottsdale, *Australia* 114 G4 41 9 S 147 31 E
Scottsville, *Ky., U.S.A.* 135 G2 36 45N 86 11W
Scottsville, *N.Y., U.S.A.* 136 C7 43 2N 77 47W
Scottville, *U.S.A.* 134 D2 43 58N 86 17W
Scranton, *Iowa, U.S.A.* 140 B2 42 1N 94 33W
Scranton, *Pa., U.S.A.* 137 E9 41 25N 75 40W
Scugog, L., *Canada* 136 B6 44 10N 78 55W
Scunthorpe, *U.K.* 16 D7 53 36N 0 39W
Scuol, *Switz.* 29 C10 46 48N 10 17 E
Scusciuban, *Somali Rep.* 108 B4 10 18N 50 12 E
Scutari = Üsküdar, *Turkey* 49 F4 41 0N 29 5 E
Seabra, *Brazil* 155 D3 12 25 S 41 46W
Seabrook, L., *Australia* 113 F2 30 55 S 119 40 E
Seaford, *U.S.A.* 134 F8 38 39N 75 37W
Seaforth, *Canada* 128 D3 43 35N 81 25W
Seagraves, *U.S.A.* 139 J3 32 57N 102 34W
Seal →, *Canada* 131 B10 59 4N 94 48W
Seal Cove, *Canada* 129 C8 49 57N 56 22W
Seal L., *Canada* 129 B7 54 20N 61 30W
Sealy, *U.S.A.* 139 L6 29 47N 96 9W
Seaman, *U.S.A.* 141 F13 38 57N 83 34W
Searchlight, *U.S.A.* 145 K12 35 28N 114 55W
Searcy, *U.S.A.* 139 H9 35 15N 91 44W
Searles L., *U.S.A.* 145 K9 35 44N 117 21W
Seaside, *Calif., U.S.A.* 144 J5 36 37N 121 50W
Seaside, *Oreg., U.S.A.* 144 E3 46 0N 123 56W
Seaspray, *Australia* 117 E7 38 25 S 147 15 E
Seattle, *U.S.A.* 144 C4 47 36N 122 20W
Seaview Ra., *Australia* 114 B4 18 40 S 145 45 E
Seaward Kaikouras, Mts., *N.Z.* 119 C8 42 10 S 173 44 E
Sebangka, *Indonesia* 74 B2 0 7N 104 36 E
Sebastián Vizcaíno, B., *Mexico* 146 B2 28 0N 114 30W
Sebastopol = Sevastopol, *Ukraine* 51 K7 44 35N 33 30 E
Sebastopol, *U.S.A.* 144 G4 38 24N 122 49W
Sebderat, *Eritrea* 95 D4 15 26N 36 42 E
Sebdou, *Algeria* 99 B4 34 38N 1 19W
Seben, *Turkey* 88 B4 40 24N 31 34 E
Sebeş, *Romania* 46 D4 45 58N 23 34 E
Sebeşului, Munţii, *Romania* 46 D4 45 36N 23 40 E
Sebewaing, *U.S.A.* 134 D4 43 44N 83 27W
Sebezh, *Russia* 50 D5 56 14N 28 22 E
Sebha = Sabhah, *Libya* 96 C2 27 9N 14 29 E
Sébi, *Mali* 100 B4 15 50N 4 12W
Şebinkarahisar, *Turkey* 89 B8 40 22N 38 28 E
Sebiş, *Romania* 46 C3 46 23N 22 13 E
Sebkhet Te-n-Dghâmcha, *Mauritania* 100 B1 18 30N 15 55W
Sebkra Azzel Mati, *Algeria* 99 C5 26 10N 0 43 E
Sebkra Mekerghene, *Algeria* 99 C5 26 21N 1 30 E
Seblat, *Indonesia* 74 C2 3 14 S 101 38 E
Sebnitz, *Germany* 26 E10 50 58N 14 15 E
Sebou, Oued →, *Morocco* 98 B3 34 16N 6 40W
Sebring, *Fla., U.S.A.* 135 M5 27 30N 81 27W
Sebring, *Ohio, U.S.A.* 136 F3 40 55N 81 2W
Sebringville, *Canada* 136 C3 43 24N 81 4W
Sebta = Ceuta, *N. Afr.* 98 A3 35 52N 5 18W
Sebuku, *Indonesia* 75 C5 3 30 S 116 25 E
Sebuku, Teluk, *Malaysia* 75 B5 4 0N 118 10 E
Sečanj, *Serbia, Yug.* 42 B5 45 25N 20 47 E
Secchia →, *Italy* 38 C8 45 4N 11 0 E
Sechelt, *Canada* 130 D4 49 25N 123 42W
Sechura, *Peru* 156 B1 5 39 S 80 50W
Sechura, Desierto de, *Peru* 156 B1 6 0 S 80 30W
Seclin, *France* 23 B10 50 33N 3 2 E
Secondigny, *France* 22 F6 46 37N 0 26W
Sečovce, *Slovak Rep.* 31 C14 48 42N 21 40 E
Secretary I., *N.Z.* 119 F1 45 15 S 166 56 E
Secunderabad, *India* 82 F4 17 28N 78 30 E
Sécure →, *Bolivia* 157 D5 15 10 S 64 52W
Sedalia, *U.S.A.* 140 F3 38 42N 93 14W
Sedan, *Australia* 116 C3 34 34 S 139 19 E
Sedan, *France* 23 C11 49 43N 4 57 E
Sedan, *U.S.A.* 139 G6 37 8N 96 11W
Sedano, *Spain* 34 C1 42 43N 3 49W
Seddon, *N.Z.* 119 B9 41 40 S 174 7 E
Seddonville, *N.Z.* 119 B7 41 33 S 172 1 E
Sedeh, *Fārs, Iran* 85 D7 30 45N 52 11 E
Sedeh, *Khorāsān, Iran* 85 C8 33 20N 59 14 E
Sederot, *Israel* 91 D3 31 32N 34 37 E
Sedgewick, *Canada* 130 C6 52 48N 111 41W
Sedhiou, *Senegal* 100 C1 12 44N 15 30W
Sedičany, *Czech.* 30 B7 49 40N 14 25 E
Sedico, *Italy* 39 B9 46 8N 12 6 E
Sedienie, *Bulgaria* 43 E9 42 16N 24 33 E
Sedley, *Canada* 131 C8 50 10N 104 0W
Sedova, Pik, *Russia* 56 B6 73 29N 54 58 E
Sedrata, *Algeria* 99 A6 36 7N 7 31 E
Sedro-Woolley, *U.S.A.* 144 B4 48 30N 122 14W
Sedrun, *Switz.* 29 C7 46 36N 8 47 E
Sedziszów Małapolski, *Poland* 31 A14 50 5N 21 45 E
Seebad Ahlbeck, *Germany* 26 B10 53 56N 14 10 E
Seefeld, *Austria* 30 D4 47 19N 11 13 E
Seehausen, *Germany* 26 C7 52 54N 11 45 E
Seeheim, *Namibia* 104 D2 26 50 S 17 45 E
Seekoei →, *S. Africa* 104 E4 30 18 S 25 1 E
Seelow, *Germany* 26 C10 52 32N 14 23 E
Sées, *France* 22 D7 48 38N 0 10 E
Seesen, *Germany* 26 D6 51 54N 10 10 E
Sefadu, *S. Leone* 100 D2 8 35N 10 58W
Seferihisar, *Turkey* 88 C2 38 10N 26 50 E
Séfeto, *Mali* 100 C3 14 8N 9 49W
Sefrou, *Morocco* 98 B4 33 52N 4 52W
Sefton, *N.Z.* 119 D7 43 15 S 172 41 E
Sefuri-San, *Japan* 62 D2 33 28N 130 18 E
Sefwi Bekwai, *Ghana* 100 D4 6 10N 2 25W
Seg-ozero, *Russia* 48 B5 63 20N 33 46 E
Segag, *Ethiopia* 108 C2 7 39N 42 50 E
Segamat, *Malaysia* 77 L4 2 30N 102 50 E
Segarcea, *Romania* 46 E4 44 6N 23 43 E
Segbwema, *S. Leone* 100 D2 8 0N 11 0W
Seget, *Indonesia* 73 B4 1 24 S 130 58 E
Segezha, *Russia* 48 B5 63 44N 34 19 E
Seggueur, O. →, *Algeria* 99 B5 32 14N 1 48 E
Segonzac, *France* 24 C3 45 36N 0 14W
Segorbe, *Spain* 34 F4 39 50N 0 30W
Ségou, *Mali* 100 C3 13 30N 6 16W
Segovia = Coco →, *Cent. Amer.* 148 D3 15 0N 83 8W
Segovia, *Colombia* 152 B3 7 7N 74 42W
Segovia, *Spain* 36 E6 40 57N 4 10W
Segovia □, *Spain* 36 E6 40 55N 4 10W
Segré, *France* 22 E6 47 40N 0 52W
Segre →, *Spain* 34 D5 41 40N 0 43 E
Séguéla, *Ivory C.* 100 D3 7 55N 6 40W
Seguin, *U.S.A.* 139 L6 29 34N 97 58W
Segundo →, *Argentina* 158 C3 30 53 S 62 44W
Segura →, *Spain* 35 G4 38 3N 0 44W
Segura, Sierra de, *Spain* 35 G2 38 5N 2 45W
Seh Qal'eh, *Iran* 85 C8 33 40N 58 24 E
Sehitwa, *Botswana* 104 C3 20 30 S 22 30 E
Sehore, *India* 80 H7 23 10N 77 5 E
Sehwan, *Pakistan* 79 D2 26 28N 67 53 E
Şeica Mare, *Romania* 46 C5 46 1N 24 7 E
Seikpyu, *Burma* 78 E5 20 54N 94 48 E
Seiland, *Norway* 12 A20 70 25N 23 15 E
Seiling, *U.S.A.* 139 G5 36 9N 98 56W
Seille →, *Moselle, France* 23 C13 49 7N 6 11 E
Seille →, *Saône-et-Loire, France* 25 B8 46 31N 4 57 E
Seilles, *Belgium* 21 G6 50 30N 5 6 E
Sein, I. de, *France* 22 D2 48 2N 4 52W
Seinäjoki, *Finland* 13 E20 62 40N 22 51 E
Seine →, *France* 22 C7 49 26N 0 26 E
Seine, B. de la, *France* 22 C6 49 40N 0 40W
Seine-et-Marne □, *France* 23 D9 48 45N 3 0 E
Seine-Maritime □, *France* 22 C7 49 40N 1 0 E
Seine-St.-Denis □, *France* 23 D9 48 58N 2 24 E
Seini, *Romania* 46 B4 47 44N 23 21 E
Seistan, *Iran* 85 D9 30 50N 61 0 E
Sejerø, *Denmark* 15 J5 55 54N 11 9 E
Sejerø Bugt, *Denmark* 15 J5 55 53N 11 15 E
Sejny, *Poland* 47 A10 54 6N 23 21 E
Seka, *Ethiopia* 95 F4 8 10N 36 52 E
Sekayu, *Indonesia* 74 C2 2 51 S 103 51 E
Seke, *Tanzania* 106 C3 3 20 S 33 31 E
Seke-Banza, *Zaïre* 103 D2 5 20 S 13 16 E
Sekenke, *Tanzania* 106 C3 4 18 S 34 11 E
Seki, *Japan* 63 B8 35 29N 136 55 E
Sekigahara, *Japan* 63 B8 35 22N 136 28 E
Sekondi-Takoradi, *Ghana* 100 E4 4 58N 1 45W
Seksna, *Russia* 50 C10 59 13N 38 30 E
Sekuma, *Botswana* 104 C3 24 36 S 23 50 E
Selah, *U.S.A.* 142 C3 46 39N 120 32W
Selama, *Malaysia* 77 K3 5 12N 100 42 E
Selangor □, *Malaysia* 74 B2 3 10N 101 30 E
Selárgius, *Italy* 40 C2 39 16N 9 10 E
Selaru, *Indonesia* 73 C4 8 9 S 131 0 E
Selb, *Germany* 27 E8 50 10N 12 7 E
Selby, *U.K.* 16 D6 53 47N 1 5W
Selby, *U.S.A.* 138 C4 45 31N 100 2W
Selca, *Croatia* 39 E13 43 20N 16 50 E
Selçuk, *Turkey* 88 D2 37 56N 27 22 E
Selden, *U.S.A.* 138 F4 39 33N 100 34W
Sele →, *Italy* 41 B7 40 29N 14 56 E
Selemdzha →, *Russia* 57 D13 51 42N 128 53 E
Selenga = Selenge Mörön →, *Asia* 64 A5 52 16N 106 16 E
Selenge, *Zaïre* 102 C3 1 58 S 18 11 E
Selenge Mörön →, *Asia* 64 A5 52 16N 106 16 E
Selenica, *Albania* 44 D1 40 33N 19 39 E
Selenter See, *Germany* 26 A6 54 18N 10 26 E
Sélestat, *France* 23 D14 48 16N 7 26 E
Seletan, Tg., *Indonesia* 75 C4 4 10 S 114 40 E
Selevac, *Serbia, Yug.* 42 C5 44 28N 20 52 E
Selfridge, *U.S.A.* 138 B4 46 2N 100 56W
Sélibabi, *Mauritania* 100 B2 15 10N 12 15W
Seliger, Ozero, *Russia* 52 B1 57 15N 33 0 E
Seligman, *U.S.A.* 143 J7 35 20N 112 53W
Şelim, *Turkey* 89 B10 40 30N 42 46 E
Selîma, El Wâhât el, *Sudan* 94 C2 21 22N 29 19 E
Selinda Spillway, *Botswana* 104 B3 18 35 S 23 10 E
Selinoús, *Greece* 45 G3 37 35N 21 37 E
Selizharovo, *Russia* 52 B1 56 51N 33 27 E
Seljord, *Norway* 14 E2 59 30N 8 40 E
Selkirk, *Canada* 131 C9 50 10N 96 55W
Selkirk, *U.K.* 18 F6 55 33N 2 50W
Selkirk I., *Canada* 131 C9 53 20N 99 6W
Selkirk Mts., *Canada* 130 C5 51 15N 117 40W
Selles-sur-Cher, *France* 23 E8 47 16N 1 33 E
Selliá, *Greece* 32 D6 35 12N 24 23 E
Sellières, *France* 23 F12 46 50N 5 32 E
Sells, *U.S.A.* 143 L8 31 55N 111 53W
Sellye, *Hungary* 31 F10 45 52N 17 51 E
Selma, *Ala., U.S.A.* 135 J2 32 25N 87 1W
Selma, *Calif., U.S.A.* 144 J7 36 34N 119 37W
Selma, *N.C., U.S.A.* 135 H6 35 32N 78 17W
Selmer, *U.S.A.* 135 H1 35 10N 88 36W
Selo, *Greece* 44 C7 41 10N 25 53 E
Selong, *Indonesia* 75 D5 8 39 S 116 32 E
Selongey, *France* 23 E12 47 36N 5 10 E
Selowandoma Falls, *Zimbabwe* 107 G3 21 15 S 31 50 E
Selpele, *Indonesia* 73 B4 0 1 S 130 5 E
Selsey Bill, *U.K.* 17 G7 50 43N 0 47W
Seltso, *Russia* 52 D2 53 22N 34 4 E
Seltz, *France* 23 D15 48 54N 8 4 E
Selu, *Indonesia* 73 C4 7 32 S 130 55 E
Sélune →, *France* 22 D5 48 38N 1 22W
Selva, *Argentina* 158 B3 29 50 S 62 0W
Selva, *Italy* 39 B8 46 33N 11 46 E
Selva, *Spain* 34 D6 41 13N 1 8 E
Selvas, *Brazil* 156 B4 6 30 S 67 0W
Selwyn, *Australia* 114 C3 21 32 S 140 30 E
Selwyn L., *Canada* 131 A8 60 0N 104 30W
Selwyn Passage, *Vanuatu* 121 F6 16 3 S 168 12 E
Selwyn Ra., *Australia* 114 C3 21 10 S 140 0 E
Selyatyn, *Ukraine* 46 B6 47 50N 25 12 E
Semani →, *Albania* 44 D1 40 47N 19 30 E
Semara, *W. Sahara* 98 C2 26 48N 11 41W
Semarang, *Indonesia* 75 D4 7 0 S 110 26 E
Sematan, *Malaysia* 75 B3 1 48N 109 46 E
Semau, *Indonesia* 72 D2 10 13 S 123 22 E
Sembabule, *Uganda* 106 C3 0 4 S 31 25 E
Sembé, *Congo* 102 B2 1 39N 14 36 E
Şemdinli, *Turkey* 89 D11 37 18N 44 35 E
Sémé, *Senegal* 100 B2 15 4N 13 41W
Semeih, *Sudan* 95 E3 12 43N 30 53 E
Semenov, *Russia* 52 B7 56 43N 44 30 E
Semenovka, *Ukraine* 51 H7 49 37N 33 10 E
Semenovka, *Ukraine* 51 F7 52 8N 32 36 E
Semeru, *Indonesia* 75 D4 8 4 S 112 55 E
Semey, *Kazakhstan* 56 D9 50 30N 80 10 E
Semikarakorskiy, *Russia* 53 G5 47 31N 40 48 E
Semiluki, *Russia* 52 E4 51 41N 39 2 E
Seminoe Reservoir, *U.S.A.* 142 E10 42 9N 106 55W
Seminole, *Okla., U.S.A.* 139 H6 35 14N 96 41W
Seminole, *Tex., U.S.A.* 139 J3 32 43N 102 39W
Semiozernoye, *Kazakhstan* 56 D7 52 22N 64 8 E
Semipalatinsk = Semey, *Kazakhstan* 56 D9 50 30N 80 10 E
Semirara I., *Phil.* 70 E3 12 4N 121 23 E
Semirara Is., *Phil.* 71 F3 12 0N 121 20 E
Semisopochnoi I., *U.S.A.* 126 C2 51 55N 179 36 E
Semitau, *Indonesia* 75 B4 0 29N 111 57 E
Semiyarka, *Kazakhstan* 56 D8 50 55N 78 23 E
Semiyarskoye = Semiyarka, *Kazakhstan* 56 D8 50 55N 78 23 E
Semmering P., *Austria* 30 D8 47 41N 15 45 E
Semnān, *Iran* 85 C7 35 40N 53 23 E
Semnān □, *Iran* 85 C7 36 0N 54 0 E
Semois →, *Europe* 21 J5 49 53N 4 44 E
Sempang Mengayou, Tanjong, *Malaysia* 75 A5 7 0N 116 40 E
Semporna, *Malaysia* 75 B5 4 30N 118 33 E
Semuda, *Indonesia* 75 C4 2 51 S 112 58 E
Semur-en-Auxois, *France* 23 E11 47 30N 4 20 E
Sena, *Bolivia* 156 C4 11 32 S 67 11W
Senā, *Iran* 85 D6 28 27N 51 36 E
Sena, *Mozam.* 107 F3 17 25 S 35 0 E
Sena →, *Bolivia* 156 C4 11 31 S 67 11W
Sena Madureira, *Brazil* 156 B4 9 5 S 68 45W
Senador Pompeu, *Brazil* 154 C4 5 40 S 39 20W
Senaja, *Malaysia* 75 A5 6 45N 117 3 E
Senaki, *Georgia* 53 J6 42 15N 42 7 E
Senanga, *Zambia* 104 B3 16 2 S 23 14 E
Senatobia, *U.S.A.* 139 H10 34 37N 89 58W
Sendafa, *Ethiopia* 95 F4 9 11N 39 3 E
Sendai, *Kagoshima, Japan* 62 F2 31 50N 130 20 E
Sendai, *Miyagi, Japan* 60 E10 38 15N 140 53 E
Sendai-Wan, *Japan* 60 E10 38 15N 141 0 E
Sendamangalam, *India* 83 J4 11 17N 78 17 E
Sendenhorst, *Germany* 26 D3 51 50N 7 49 E
Sendurjana, *India* 82 D4 21 32N 78 17 E
Senec, *Slovak Rep.* 31 C10 48 12N 17 23 E
Seneca, *Oreg., U.S.A.* 142 D4 44 8N 118 58W
Seneca, *S.C., U.S.A.* 135 H4 34 41N 82 57W
Seneca Falls, *U.S.A.* 137 D8 42 55N 76 48W
Seneca L., *U.S.A.* 136 D8 42 40N 76 54W
Seneffe, *Belgium* 21 G4 50 32N 4 16 E
Senegal ■, *W. Afr.* 100 C2 14 30N 14 30W
Senegal →, *W. Afr.* 100 B1 15 48N 16 32W
Senegambia, *Africa* 92 E2 12 45N 12 0W
Senekal, *S. Africa* 105 D4 28 20 S 27 36 E
Senftenberg, *Germany* 26 D10 51 32N 14 0 E
Senga Hill, *Zambia* 107 D3 9 19 S 31 11 E
Senge Khambab = Indus →, *Pakistan* 79 D2 24 20N 67 47 E
Sengerema □, *Tanzania* 106 C3 2 10 S 32 20 E
Sengiley, *Russia* 52 D9 53 58N 48 46 E
Sengkang, *Indonesia* 72 B2 4 8 S 120 1 E
Sengua →, *Zimbabwe* 107 F2 17 7 S 28 5 E
Senguerr →, *Argentina* 160 C3 45 35 S 68 50W
Senhor-do-Bonfim, *Brazil* 154 D3 10 30 S 40 10W
Senica, *Slovak Rep.* 31 C10 48 41N 17 25 E
Senigállia, *Italy* 39 E10 43 42N 13 13 E
Seniku, *Burma* 78 C6 25 32N 97 48 E
Senio →, *Italy* 39 D9 44 35N 12 15 E
Senirkent, *Turkey* 88 C4 38 6N 30 33 E
Senise, *Italy* 41 B9 40 9N 16 17 E
Senj, *Croatia* 39 D11 45 0N 14 58 E
Senja, *Norway* 12 B17 69 25N 17 30 E
Senlis, *France* 23 C9 49 13N 2 35 E
Senmonorom, *Cambodia* 76 F6 12 27N 107 12 E
Sennâr, *Sudan* 95 E3 13 30N 33 35 E
Senne →, *Belgium* 21 G4 50 42N 4 13 E
Senneterre, *Canada* 128 C4 48 25N 77 15W
Senniquelle, *Liberia* 100 D3 7 19N 8 38W
Senno, *Belarus* 50 E5 54 45N 29 43 E
Sennori, *Italy* 40 B1 40 47N 8 35 E
Seno, *Laos* 76 D5 16 35N 104 50 E
Senonches, *France* 22 D8 48 34N 1 2 E
Senorbì, *Italy* 40 C2 39 32N 9 8 E
Senožeče, *Slovenia* 39 C11 45 43N 14 3 E
Sens, *France* 23 D10 48 11N 3 15 E
Senta, *Serbia, Yug.* 42 B5 45 55N 20 3 E
Sentani, *Indonesia* 73 B6 2 36 S 140 37 E
Sentery, *Zaïre* 103 D5 5 17 S 25 42 E
Sentinel, *U.S.A.* 143 K7 32 52N 113 13W
Sentolo, *Indonesia* 75 D4 7 55 S 110 13 E
Senya Beraku, *Ghana* 101 D4 5 28N 0 31W
Senye, *Eq. Guin.* 102 B1 1 34N 9 50 E
Seo de Urgel, *Spain* 34 C6 42 22N 1 23 E
Seohara, *India* 81 E8 29 15N 78 33 E
Seoni, *India* 81 H8 22 5N 79 30 E
Seoriuarayan, *India* 82 D6 21 45N 82 34 E
Seoul = Sŏul, *S. Korea* 67 F14 37 31N 126 58 E
Separation Point, *Canada* 129 B8 53 37N 57 25W
Separation Pt., *N.Z.* 119 A7 40 47 S 172 59 E
Sepīdān, *Iran* 85 D7 30 20N 52 5 E

Silverton 265

Sherbro I., S. Leone 100 D2 7 30N 12 40W
Sherbrooke, Canada 129 C5 45 28N 71 57W
Sherda, Chad 97 D3 20 7N 16 46 E
Shereik, Sudan 94 D3 18 44N 33 47 E
Sheridan, Ark., U.S.A. 139 H8 34 19N 92 24W
Sheridan, Ill., U.S.A. 141 C8 41 32N 88 41W
Sheridan, Ind., U.S.A. 141 D10 40 8N 86 13W
Sheridan, Mo., U.S.A. 140 D2 40 31N 94 37W
Sheridan, Wyo., U.S.A. 142 D10 44 48N 106 58W
Sherkot, India 81 E8 29 22N 78 35 E
Sherman, U.S.A. 139 J6 33 40N 96 35W
Shērpūr, Afghan. 79 B3 34 32N 69 10 E
Sherpur, Bangla. 78 C3 25 0N 90 0 E
Sherridon, Canada 131 B8 55 8N 101 5W
Sherwood, N. Dak., U.S.A. 138 A4 48 57N 101 38W
Sherwood, Ohio, U.S.A. 141 C12 41 17N 84 33W
Sherwood, Tex., U.S.A. 139 K4 31 18N 100 45W
Sherwood Forest, U.K. 16 D6 53 6N 1 7W
Sheslay, Canada 130 B2 58 17N 131 52W
Sheslay →, Canada 130 B2 58 48N 132 5W
Shethanei L., Canada 131 B9 58 48N 97 50W
Shetland □, U.K. 18 A7 60 30N 1 30W
Shetland Is., U.K. 18 A7 60 30N 1 30W
Shevaroy Hills, India 83 J4 11 58N 78 12 E
Shewa □, Ethiopia 95 F4 9 33N 38 10 E
Shewa Gimira, Ethiopia 95 F4 7 4N 35 51 E
Sheyenne, U.S.A. 138 B5 47 50N 99 7W
Sheyenne →, U.S.A. 138 B6 47 2N 96 50W
Shibām, Yemen 87 D5 16 0N 48 36 E
Shibata, Japan 60 F9 37 57N 139 20 E
Shibecha, Japan 60 C12 43 17N 144 36 E
Shibetsu, Japan 60 B11 44 10N 142 23 E
Shibîn el Kôm, Egypt 94 H7 30 31N 30 55 E
Shibîn el Qanâtir, Egypt 94 H7 30 19N 31 19 E
Shibing, China 68 D7 27 2N 108 7 E
Shibogama L., Canada 128 B2 53 35N 88 15W
Shibukawa, Japan 63 A10 36 29N 139 0 E
Shibushi, Japan 62 F3 31 25N 131 8 E
Shibushi-Wan, Japan 62 F3 31 24N 131 8 E
Shicheng, China 69 D11 26 22N 116 20 E
Shickshock Mts. = Chic-Chocs, Mts., Canada 129 C6 48 55N 66 0W
Shidād, Si. Arabia 86 B3 21 19N 40 3 E
Shidao, China 67 F12 36 50N 122 25 E
Shidian, China 68 E2 24 40N 99 5 E
Shido, Japan 62 C6 34 19N 134 10 E
Shiel, L., U.K. 18 E3 56 48N 5 34W
Shield, C., Australia 114 A2 13 20 S 136 20 E
Shiga □, Japan 63 B8 35 20N 136 0 E
Shigaib, Sudan 97 E4 15 5N 23 35 E
Shigaraki, Japan 63 C8 34 57N 136 2 E
Shigu, China 68 D2 26 51N 99 56 E
Shiguaigou, China 66 D6 40 52N 110 15 E
Shihan, W. →, Yemen 87 C5 17 24N 51 26 E
Shihchiachuangi = Shijiazhuang, China 66 E8 38 2N 114 28 E
Shiiba, Japan 62 E3 32 29N 131 4 E
Shijaku, Albania 44 C1 41 21N 19 33 E
Shijiazhuang, China 66 E8 38 2N 114 28 E
Shijiu Hu, China 69 B12 31 25N 118 50 E
Shikarpur, India 80 E8 28 17N 78 7 E
Shikarpur, Pakistan 79 D3 27 57N 68 39 E
Shikine-Jima, Japan 63 C11 34 19N 139 13 E
Shikoku, Japan 62 D5 33 30N 133 30 E
Shikoku □, Japan 62 D5 33 30N 133 30 E
Shikoku-Sanchi, Japan 62 D5 33 30N 133 30 E
Shilabo, Ethiopia 90 F3 6 22N 44 32 E
Shiliguri, India 78 B2 26 45N 88 25 E
Shilka, Russia 57 D12 52 0N 115 55 E
Shilka →, Russia 57 D13 53 20N 121 26 E
Shillelagh, Ireland 19 D5 52 45N 6 32W
Shillong, India 78 C3 25 35N 91 53 E
Shilo, West Bank 91 C4 32 4N 35 18 E
Shilong, China 69 F9 23 5N 113 52 E
Shilou, China 66 F6 37 0N 110 48 E
Shilovo, Russia 52 C5 54 25N 40 57 E
Shima-Hantō, Japan 63 C8 34 22N 136 45 E
Shimabara, Japan 62 E2 32 48N 130 20 E
Shimada, Japan 63 C10 34 49N 138 10 E
Shimane □, Japan 62 C4 35 0N 132 30 E
Shimane-Hantō, Japan 62 B5 35 30N 133 0 E
Shimanovsk, Russia 57 D13 52 15N 127 30 E
Shimen, China 69 C8 29 35N 111 20 E
Shimenjie, China 69 C11 29 29N 116 48 E
Shimian, China 68 C4 29 17N 102 23 E
Shimizu, Japan 63 C10 35 0N 138 30 E
Shimo-Jima, Japan 62 E2 32 15N 130 7 E
Shimo-Koshiki-Jima, Japan 62 F1 31 40N 129 43 E
Shimoda, Japan 63 C10 34 40N 138 57 E
Shimodate, Japan 63 A11 36 20N 139 55 E
Shimoga, India 83 H2 13 57N 75 32 E
Shimoni, Kenya 106 C4 4 38 S 39 20 E
Shimonita, Japan 63 A10 36 13N 138 47 E
Shimonoseki, Japan 62 D2 33 58N 130 55 E
Shimotsuma, Japan 63 A11 36 11N 139 58 E
Shimpuru Rapids, Angola 103 F3 17 45 S 19 55 E
Shimsha →, India 83 H3 13 15N 77 10 E
Shimsk, Russia 50 C6 58 15N 30 50 E
Shin, L., U.K. 18 C4 58 5N 4 30W
Shin-Tone →, Japan 63 B12 35 44N 140 51 E
Shinan, China 68 F7 22 44N 109 53 E
Shinano →, Japan 61 F9 36 50N 138 30 E
Shīndand, Afghan. 79 B1 33 12N 62 8 E
Shingbwiyang, Burma 78 B6 26 41N 96 13 E
Shingleton, U.S.A. 128 C2 46 21N 86 28W
Shingū, Japan 63 D7 33 40N 135 55 E
Shinji, Japan 62 B4 35 24N 132 54 E
Shinji Ko, Japan 62 B4 35 26N 132 57 E
Shinjō, Japan 60 E10 38 46N 140 18 E
Shinkafe, Nigeria 101 C6 13 8N 6 29 E
Shīnkay, Afghan. 79 C2 31 57N 67 26 E
Shinminato, Japan 63 A9 36 47N 137 4 E
Shinonoi, Japan 63 A10 36 35N 138 9 E
Shinshār, Syria 91 A5 34 36N 36 43 E
Shinshiro, Japan 63 C9 34 54N 137 30 E
Shinyanga, Tanzania 106 C3 3 45 S 33 27 E
Shinyanga □, Tanzania 106 C3 3 50 S 34 0 E
Shio-no-Misaki, Japan 63 D7 33 25N 135 45 E
Shiogama, Japan 60 E10 38 19N 141 1 E
Shiojiri, Japan 63 A9 36 6N 137 58 E
Ship I., U.S.A. 139 K10 30 13N 88 55W
Shipehenski Prokhod, Bulgaria 43 E10 42 45N 25 15 E
Shiping, China 68 F4 23 45N 102 23 E
Shippegan, Canada 129 C7 47 45N 64 45W
Shippensburg, U.S.A. 136 F7 40 3N 77 31W
Shiprock, U.S.A. 143 H9 36 47N 108 41W
Shiqian, China 68 D7 27 32N 108 13 E
Shiqma, N. →, Israel 91 D3 31 37N 34 30 E
Shiquan, China 66 H5 33 5N 108 15 E
Shīr Kūh, Iran 85 D7 31 39N 54 3 E
Shirabad = Sherabad, Uzbekistan 55 E3 37 40N 67 1 E
Shiragami-Misaki, Japan 60 D10 41 24N 140 12 E
Shirahama, Japan 63 D7 33 41N 135 20 E
Shirakawa, Fukushima, Japan 61 F10 37 7N 140 13 E
Shirakawa, Gifu, Japan 63 A8 36 17N 136 56 E
Shirane-San, Gumma, Japan 63 A11 36 48N 139 22 E
Shirane-San, Yamanashi, Japan 63 B10 35 42N 138 9 E
Shiraoi, Japan 60 C10 42 33N 141 21 E
Shīrāz, Iran 85 D7 29 42N 52 30 E
Shirbīn, Egypt 94 H7 31 11N 31 32 E
Shire →, Africa 107 F4 17 42 S 35 19 E
Shiretoko-Misaki, Japan 60 B12 44 21N 145 20 E
Shirinab →, Pakistan 80 D2 30 15N 66 28 E
Shiriya-Zaki, Japan 60 D10 41 25N 141 30 E
Shirley, U.S.A. 141 E11 39 53N 85 35W
Shiroishi, Japan 60 E10 38 0N 140 37 E
Shirol, India 82 F2 16 47N 74 41 E
Shirpur, India 82 D2 21 21N 74 57 E
Shīrvān, Iran 85 B8 37 30N 57 50 E
Shirwa, L. = Chilwa, L., Malawi 107 F4 15 15 S 35 40 E
Shishmanova, Bulgaria 43 E8 42 58N 23 12 E
Shishou, China 69 C9 29 38N 112 22 E
Shitai, China 69 B11 30 12N 117 25 E
Shively, U.S.A. 141 F11 38 12N 85 49W
Shivpuri, India 80 G7 25 26N 77 42 E
Shixian, China 67 C15 43 5N 129 50 E
Shixing, China 69 E10 24 46N 114 5 E
Shiyan, China 69 A8 32 35N 110 45 E
Shiyata, Egypt 94 B2 29 25N 25 7 E
Shizhu, China 68 C7 29 58N 108 7 E
Shizong, China 68 E5 24 50N 104 0 E
Shizuishan, China 66 E4 39 15N 106 50 E
Shizuoka, Japan 63 C10 34 57N 138 24 E
Shizuoka □, Japan 63 B10 35 15N 138 40 E
Shklov = Shklow, Belarus 50 E6 54 16N 30 15 E
Shklow, Belarus 50 E6 54 16N 30 15 E
Shkoder = Shkodra, Albania 44 B1 42 4N 19 32 E
Shkodra, Albania 44 B1 42 4N 19 32 E
Shkodra □, Albania 44 B1 42 25N 19 20 E
Shkumbini →, Albania 44 C1 41 2N 19 31 E
Shmidta, O., Russia 57 A10 81 0N 91 0 E
Shō-Gawa →, Japan 63 A9 36 47N 137 4 E
Shoal Cr. →, U.S.A. 140 E3 39 44N 93 32W
Shoal Lake, Canada 131 C8 50 30N 100 35W
Shoals, U.S.A. 141 F10 38 40N 86 47W
Shōbara, Japan 62 C5 34 51N 133 1 E
Shōdo-Shima, Japan 62 C6 34 30N 134 15 E
Shoeburyness, U.K. 17 F8 51 32N 0 49 E
Shokpar, Kazakhstan 55 B7 43 49N 74 21 E
Sholapur = Solapur, India 82 F2 17 43N 75 56 E
Shologontsy, Russia 57 C12 66 13N 114 0 E
Shōmrōn, West Bank 91 C4 32 15N 35 13 E
Shoranur, India 83 J3 10 46N 76 19 E
Shorapur, India 83 F3 16 31N 76 48 E
Shortland I., Solomon Is. 121 L8 7 0 S 155 45 E
Shoshone, Calif., U.S.A. 145 K10 35 58N 116 16W
Shoshone, Idaho, U.S.A. 142 E6 42 56N 114 25W
Shoshone L., U.S.A. 142 D8 44 22N 110 43W
Shoshone Mts., U.S.A. 142 G5 39 20N 117 25W
Shoshong, Botswana 104 C4 22 56 S 26 31 E
Shoshoni, U.S.A. 142 E9 43 14N 108 7W
Shostka, Ukraine 51 G7 51 57N 33 32 E
Shou Xian, China 69 A11 32 37N 116 42 E
Shouchang, China 69 C12 29 18N 119 12 E
Shouguang, China 67 F10 37 52N 118 45 E
Shouning, China 69 D12 27 27N 119 31 E
Shouyang, China 66 F7 37 54N 113 8 E
Show Low, U.S.A. 143 J9 34 15N 110 2W
Shpola, Ukraine 51 H6 49 1N 31 30 E
Shreveport, U.S.A. 139 J8 32 31N 93 45W
Shrewsbury, U.K. 16 E5 52 43N 2 45W
Shrirampur, India 81 H13 22 44N 88 21 E
Shrirangapattana, India 83 H3 12 26N 76 43 E
Shropshire □, U.K. 17 E5 52 36N 2 45W
Shu, Kazakhstan 55 B6 43 36N 73 42 E
Shu →, Kazakhstan 55 A3 45 0N 67 44 E
Shuangbai, China 68 E3 24 42N 101 38 E
Shuangcheng, China 67 B14 45 20N 126 15 E
Shuangfeng, China 69 D9 27 29N 112 11 E
Shuanggou, China 67 G9 34 2N 117 30 E
Shuangjiang, China 68 F2 23 26N 99 58 E
Shuangliao, China 67 C12 43 29N 123 30 E
Shuangshanzi, China 67 D10 40 20N 119 8 E
Shuangyang, China 67 C13 43 28N 125 40 E
Shuangyashan, China 65 B8 46 28N 131 5 E
Shu'b, Ra's, Yemen 87 D6 12 30N 53 25 E
Shucheng, China 69 B11 31 28N 116 57 E
Shugozero, Russia 50 C8 59 54N 34 10 E
Shuguri Falls, Tanzania 107 D4 8 33 S 37 22 E
Shuḩayr, Yemen 87 D5 14 41N 49 23 E
Shuiji, China 69 D12 27 13N 118 20 E
Shuiye, China 66 F8 36 7N 114 8 E
Shujalpur, India 80 H7 23 18N 76 46 E
Shukpa Kunzang, India 81 B8 34 22N 78 22 E
Shulan, China 67 B14 44 28N 127 0 E
Shulaveri, Georgia 53 K7 41 22N 44 45 E
Shule, China 64 C2 39 25N 76 3 E
Shullsburg, U.S.A. 140 B6 42 35N 90 13W
Shumagin Is., U.S.A. 126 C4 55 7N 159 45W
Shumerlya, Russia 52 C8 55 30N 46 25 E
Shumikha, Russia 54 D9 55 10N 63 15 E
Shunchang, China 69 D11 26 54N 117 48 E
Shunde, China 69 F9 22 42N 113 14 E
Shungay, Kazakhstan 53 F8 48 30N 46 45 E
Shungnak, U.S.A. 126 B4 66 52N 157 9W
Shuo Xian, China 66 E7 39 20N 112 33 E
Shūr →, Iran 85 D7 28 30N 55 0 E
Shūr Āb, Iran 85 C6 34 23N 51 11 E
Shūr Gaz, Iran 85 D8 29 10N 59 20 E
Shūrāb, Iran 85 C8 33 43N 56 29 E
Shurab, Tajikistan 55 C5 40 3N 70 33 E
Shurchi, Uzbekistan 55 E3 37 59N 67 47 E
Shūrjestān, Iran 85 D7 31 24N 52 25 E
Shurkhua, Burma 78 D4 22 15N 93 38 E
Shurugwi, Zimbabwe 107 F3 19 40 S 30 0 E
Shūsf, Iran 85 D9 31 50N 60 5 E
Shūsh, Iran 89 F13 32 11N 48 15 E
Shūshtar, Iran 85 D6 32 0N 48 50 E
Shuswap L., Canada 130 C5 50 55N 119 3W
Shuya, Russia 52 B5 56 50N 41 28 E
Shuyang, China 67 G10 34 10N 118 42 E
Shuzenji, Japan 63 C10 34 58N 138 56 E
Shūzū, Iran 85 D7 29 52N 54 30 E
Shwebo, Burma 78 D5 22 30N 95 45 E
Shwegu, Burma 78 C6 24 15N 96 26 E
Shwegun, Burma 78 G6 17 9N 97 39 E
Shwenyaung, Burma 78 E6 20 46N 96 57 E
Shymkent, Kazakhstan 55 B4 42 18N 69 36 E
Shyok, India 81 B8 34 15N 78 12 E
Shyok →, Pakistan 81 B6 35 13N 75 53 E
Si Chon, Thailand 77 H2 9 0N 99 54 E
Si Kiang = Xi Jiang →, China 69 F9 22 5N 113 20 E
Si-ngan = Xi'an, China 66 G5 34 15N 109 0 E
Si Prachan, Thailand 76 E3 14 37N 100 9 E
Si Racha, Thailand 76 F3 13 10N 100 48 E
Si Xian, China 67 H9 33 30N 117 50 E
Siahan Range, Pakistan 79 D2 27 30N 64 40 E
Siak →, Indonesia 74 B2 1 13N 102 9 E
Siaksriindrapura, Indonesia 74 B2 0 51N 102 0 E
Sialkot, Pakistan 79 B4 32 32N 74 30 E
Sialsuk, India 78 D4 23 24N 92 45 E
Siam = Thailand ■, Asia 76 E4 16 0N 102 0 E
Siam, Australia 116 B2 32 35 S 136 41 E
Siantan, P., Indonesia 77 L6 3 10N 106 15 E
Siàpo →, Venezuela 152 C4 2 7N 66 28W
Sīāreh, Iran 85 D9 28 5N 60 14 E
Siargao, Phil. 71 G6 9 52N 126 3 E
Siari, Pakistan 81 B7 34 55N 76 40 E
Siari, Phil. 71 G4 8 19N 122 58 E
Siasi I., Phil. 71 J3 5 33N 120 51 E
Siassi, Papua N. G. 120 C4 5 40 S 147 51 E
Siátista, Greece 44 D3 40 15N 21 33 E
Siaton, Phil. 71 G4 9 4N 123 2 E
Siau, Indonesia 72 A3 2 50N 125 25 E
Šiauliai, Lithuania 13 J20 55 56N 23 15 E
Siaya □, Kenya 106 B3 0 0 34 20 E
Siazan = Siyäzän, Azerbaijan 53 K9 41 3N 49 10 E
Sibâi, Gebel el, Egypt 94 B3 25 45N 34 10 E
Sibang, Gabon 102 B1 0 25N 9 31 E
Sibari, Italy 41 C9 39 47N 16 27 E
Sibay, Russia 54 E7 52 42N 58 39 E
Sibay I., Phil. 71 F3 11 51N 121 29 E
Sibayi, L., S. Africa 105 D5 27 20 S 32 45 E
Šibenik, Croatia 39 E12 43 48N 15 54 E
Siberia, Russia 58 C11 60 0N 100 0 E
Siberut, Indonesia 74 C1 1 30 S 99 0 E
Sibi, Pakistan 79 C2 29 30N 67 54 E
Sibil, Indonesia 73 B6 4 59 S 140 35 E
Sibiti, Congo 102 C2 3 38 S 13 19 E
Sibiu, Romania 46 D5 45 45N 24 9 E
Sibiu □, Romania 46 D5 45 50N 24 15 E
Sibley, Ill., U.S.A. 141 D8 40 35N 88 23W
Sibley, Iowa, U.S.A. 138 D7 43 24N 95 45W
Sibley, La., U.S.A. 139 J8 32 33N 93 18W
Sibolga, Indonesia 74 B1 1 42N 98 45 E
Sibret, Belgium 21 J7 49 58N 5 38 E
Sibsagar, India 78 B5 27 0N 94 36 E
Sibu, Malaysia 75 B4 2 18N 111 49 E
Sibuco, Phil. 71 H4 7 20N 122 10 E
Sibuguey B., Phil. 71 H4 7 50N 122 45 E
Sibut, C.A.R. 102 A3 5 46N 19 10 E
Sibutu, Phil. 71 J2 4 45N 119 30 E
Sibutu Group, Phil. 71 J2 4 45N 119 20 E
Sibutu Passage, E. Indies 71 J2 4 50N 120 0 E
Sibuyan, Phil. 70 E4 12 25N 122 40 E
Sibuyan Sea, Phil. 70 E4 12 30N 122 20 E
Sicamous, Canada 130 C5 50 49N 119 0W
Sicapoo, Mt., Phil. 70 B3 18 1N 120 56 E
Sichuan □, China 68 B5 31 0N 104 0 E
Sicilia, Italy 41 E7 37 30N 14 30 E
Sicilia □, Italy 41 E7 37 45N 14 15 E
Sicilia, Canale di, Italy 40 E5 37 25N 12 30 E
Sicilian Channel = Sicilia, Canale di, Italy 40 E5 37 25N 12 30 E
Sicily = Sicilia, Italy 41 E7 37 30N 14 30 E
Sicuani, Peru 156 C3 14 21 S 71 10W
Şiculiana, Italy 40 E6 37 20N 13 25 E
Šid, Serbia, Yug. 42 B4 45 8N 19 14 E
Sidamo □, Ethiopia 95 G4 5 0N 37 50 E
Sidaouet, Niger 97 E1 18 34N 8 3 E
Sidári, Greece 32 A3 39 47N 19 41 E
Siddeburen, Neths. 20 B9 53 15N 6 52 E
Siddhapur, India 80 H5 23 56N 72 25 E
Siddipet, India 82 E4 18 5N 78 51 E
Sidell, U.S.A. 141 E9 39 55N 87 49W
Sidéradougou, Burkina Faso 100 C4 10 42N 4 12W
Siderno, Italy 41 D9 38 16N 16 18 E
Sídheros, Ákra, Greece 32 D8 35 19N 26 19 E
Sidhirókastron, Greece 44 C5 41 13N 23 24 E
Sîdi Abd el Rahmân, Egypt 94 H6 30 55N 29 44 E
Sîdi Barrâni, Egypt 94 A2 31 38N 25 58 E
Sidi-bel-Abbès, Algeria 99 A4 35 13N 0 39W
Sidi Bennour, Morocco 98 B3 32 40N 8 25W
Sidi Haneish, Egypt 94 A2 31 10N 27 35 E
Sidi Kacem, Morocco 98 B3 34 11N 5 49W
Sidi Omar, Egypt 94 A1 31 24N 24 57 E
Sidi Slimane, Morocco 98 B3 34 16N 5 56W
Sidi Smaïl, Morocco 98 B3 32 50N 8 31W
Sidi 'Uzayz, Libya 96 B4 31 41N 24 55 E
Sidlaw Hills, U.K. 18 E5 56 32N 3 2W
Sidley, Mt., Antarctica 7 D14 77 2 S 126 2W
Sidmouth, U.K. 17 G4 50 40N 3 15W
Sidmouth, C., Australia 114 A3 13 25 S 143 36 E
Sidney, Canada 130 D4 48 39N 123 24W
Sidney, Mont., U.S.A. 138 B2 47 43N 104 9W
Sidney, N.Y., U.S.A. 137 D9 42 19N 75 24W
Sidney, Nebr., U.S.A. 138 E3 41 8N 102 59W
Sidney, Ohio, U.S.A. 141 D12 40 17N 84 9W
Sidoarjo, Indonesia 75 D4 7 27 S 112 43 E
Sidoktaya, Burma 78 E5 20 27N 94 15 E
Sidon = Saydā, Lebanon 91 B4 33 35N 35 25 E
Sidra, G. of = Surt, Khalīj, Libya 96 B3 31 40N 18 30 E
Siedlce, Poland 47 C9 52 10N 22 20 E
Siedlce □, Poland 47 C9 52 0N 22 0 E
Sieg →, Germany 26 E3 50 46N 7 6 E
Siegburg, Germany 26 E3 50 47N 7 12 E
Siegen, Germany 26 E4 50 51N 8 0 E
Siem Pang, Cambodia 76 E6 14 7N 106 23 E
Siem Reap, Cambodia 76 F4 13 20N 103 52 E
Siena, Italy 39 E8 43 19N 11 21 E
Sieniawa, Poland 31 A15 50 11N 22 38 E
Sieradz, Poland 47 D5 51 37N 18 41 E
Sieraków, Poland 47 C3 52 39N 16 2 E
Sierck-les-Bains, France 23 C13 49 26N 6 20 E
Sierpc, Poland 47 C6 52 55N 19 43 E
Sierpe, Bocas de la, Venezuela 153 B5 10 0N 61 30W
Sierra Blanca, U.S.A. 143 L11 31 11N 105 22W
Sierra Blanca Peak, U.S.A. 143 K11 33 23N 105 49W
Sierra City, U.S.A. 144 F6 39 34N 120 38W
Sierra Colorada, Argentina 160 B3 40 35 S 67 50W
Sierra de Yeguas, Spain 37 H6 37 7N 4 52W
Sierra Gorda, Chile 158 A2 22 50 S 69 15W
Sierra Grande, Argentina 160 B3 41 36 S 65 22W
Sierra Leone ■, W. Afr. 100 D2 9 0N 12 0W
Sierra Madre, Mexico 147 D6 16 0N 93 0W
Sierra Mojada, Mexico 146 B4 27 19N 103 42W
Sierraville, U.S.A. 144 F6 39 36N 120 22W
Sierre, Switz. 28 D5 46 17N 7 31 E
Sífnos, Greece 45 H6 37 0N 24 45 E
Sifton, Canada 131 C8 51 21N 100 8W
Sifton Pass, Canada 130 B3 57 52N 126 15W
Sig, Algeria 99 A4 35 32N 0 12W
Sigaboy, Phil. 71 H6 6 39N 126 5 E
Sigdal, Norway 14 D3 60 3N 9 38 E
Sigean, France 24 E6 43 2N 2 58 E
Sighetu-Marmatiei, Romania 46 B4 47 57N 23 52 E
Sighişoara, Romania 46 C5 46 12N 24 50 E
Sigira, Yemen 87 D6 12 37N 54 20 E
Sigli, Indonesia 74 A1 5 25N 96 0 E
Siglufjörður, Iceland 12 C4 66 12N 18 55W
Sigmaringen, Germany 27 G5 48 5N 9 12 E
Signakhi = Tsnori, Georgia 53 K7 41 40N 45 57 E
Signal, U.S.A. 145 L13 34 30N 113 38W
Signal Pk., U.S.A. 145 M12 33 20N 114 2W
Signau, Switz. 28 C5 46 56N 7 45 E
Signy-l'Abbaye, France 23 C11 49 40N 4 25 E
Sigourney, U.S.A. 140 C4 41 20N 92 12W
Sigsig, Ecuador 152 D2 3 0 S 78 50W
Sigtuna, Sweden 14 E11 59 36N 17 44 E
Sigüenza, Spain 34 D2 41 3N 2 40W
Siguiri, Guinea 100 C3 11 31N 9 10W
Sigulda, Latvia 13 H21 57 10N 24 55 E
Sigurd, U.S.A. 143 G8 38 50N 111 58W
Sihanoukville = Kompong Som, Cambodia 77 G4 10 38N 103 30 E
Sihaus, Peru 156 B2 8 40 S 77 40W
Sihui, China 69 F9 23 20N 112 40 E
Siikajoki →, Finland 12 D21 64 50N 24 43 E
Siilinjärvi, Finland 12 E22 63 4N 27 39 E
Siirt, Turkey 89 D9 37 57N 41 55 E
Siit, Phil. 71 J5 5 59N 124 13 E
Sijarira Ra., Zimbabwe 107 F2 17 36 S 27 45 E
Sijungjung, Indonesia 74 C2 0 42 S 100 58 E
Sikao, Thailand 77 J2 7 34N 99 21 E
Sikar, India 80 F6 27 33N 75 10 E
Sikasso, Mali 100 C3 11 18N 5 35W
Sikeston, U.S.A. 139 G10 36 53N 89 35W
Sikhote Alin, Khrebet, Russia 57 E14 45 0N 136 0 E
Sikhote Alin Ra. = Sikhote Alin, Khrebet, Russia 57 E14 45 0N 136 0 E
Sikiá., Greece 44 D5 40 2N 23 56 E
Síkinos, Greece 45 H7 36 40N 25 8 E
Sikkani Chief →, Canada 130 B4 57 47N 122 15W
Sikkim □, India 78 B2 27 50N 88 30 E
Siklós, Hungary 31 F11 45 50N 18 19 E
Sikotu-Ko, Japan 60 C10 42 45N 141 25 E
Sil →, Spain 36 C3 42 27N 7 43W
Silacayoapan, Mexico 147 D5 17 30N 98 9W
Silandro, Italy 38 B7 46 38N 10 46 E
Silanga, Phil. 71 F2 11 1N 119 34 E
Silay, Phil. 71 F4 10 47N 123 14 E
Silba, Croatia 39 D11 44 24N 14 41 E
Silchar, India 78 C4 24 49N 92 48 E
Silcox, Canada 131 B10 57 12N 94 10W
Şile, Turkey 88 B3 41 10N 29 37 E
Silenrieux, Belgium 21 H4 50 14N 4 27 E
Siler City, U.S.A. 135 H6 35 44N 79 28W
Sileru →, India 82 F5 17 49N 81 24 E
Silet, Algeria 99 D5 22 44N 4 37 E
Silgarhi Doti, Nepal 81 E9 29 15N 81 0 E
Silghat, India 78 B4 26 35N 93 0 E
Silifke, Turkey 88 D5 36 22N 33 58 E
Siliguri = Shiliguri, India 78 B2 26 45N 88 25 E
Siling Co, China 64 C3 31 50N 89 20 E
Silíqua, Italy 40 C1 39 18N 8 48 E
Silistra, Bulgaria 43 C12 44 6N 27 19 E
Silivri, Turkey 88 B3 41 4N 28 14 E
Siljan, Sweden 13 F16 60 55N 14 45 E
Silkeborg, Denmark 15 H3 56 10N 9 32 E
Sillajhuay, Cordillera, Chile 156 D4 19 46 S 68 40W
Sillamäe, Estonia 13 G22 59 24N 27 45 E
Sillé-le-Guillaume, France 22 D6 48 10N 0 8W
Sillustani, Peru 156 D3 15 50 S 70 7W
Siloam Springs, U.S.A. 139 G7 36 11N 94 32W
Silopi, Turkey 89 D10 37 15N 42 27 E
Silsbee, U.S.A. 139 K7 30 21N 94 11W
Šilute, Lithuania 13 J19 55 21N 21 33 E
Silva Porto = Kuito, Angola 103 E3 12 22 S 16 55 E
Silvan, Turkey 89 C9 38 7N 41 2 E
Silvaplana, Switz. 29 D9 46 28N 9 48 E
Silver City, N. Mex., U.S.A. 143 K9 32 46N 108 17W
Silver City, Nev., U.S.A. 142 G4 39 15N 119 48W
Silver Cr. →, U.S.A. 142 E4 43 16N 119 13W
Silver Creek, U.S.A. 136 D5 42 33N 79 10W
Silver Grove, U.S.A. 141 E12 39 2N 84 24W
Silver L., Calif., U.S.A. 144 G6 38 39N 120 6W
Silver L., Calif., U.S.A. 145 K10 35 21N 116 7W
Silver Lake, Ind., U.S.A. 141 C11 41 4N 85 53W
Silver Lake, Oreg., U.S.A. 142 E3 43 8N 121 3W
Silver Lake, Wis., U.S.A. 141 B8 42 33N 88 13W
Silver Streams, S. Africa 104 D3 28 20 S 23 33 E
Silverton, Australia 116 A4 31 52 S 141 10 E
Silverton, Colo., U.S.A. 143 H10 37 49N 107 40W
Silverton, Tex., U.S.A. 139 H4 34 28N 101 19W

Sulingen 269

Steenwijk, Neths. 20 C8 52 47N 6 7 E
Steep Pt., Australia 113 E1 26 8 S 113 8 E
Steep Rock, Canada 131 C9 51 30N 98 48W
Ştefăneşti, Romania 46 B8 47 44N 27 15 E
Stefanie L. = Chew Bahir, Ethiopia 95 G4 4 40N 36 50 E
Stefansson Bay, Antarctica 7 C5 67 20 S 59 8 E
Steffisburg, Switz. 28 C5 46 47N 7 38 E
Stege, Denmark 15 K6 55 0N 12 18 E
Steiermark □, Austria 30 D8 47 26N 15 0 E
Steigerwald, Germany 27 F6 49 44N 10 26 E
Steilacoom, U.S.A. 144 C4 47 10N 122 36W
Stein, Neths. 21 G7 50 58N 5 45 E
Steinbach, Canada 131 D9 49 32N 96 40W
Steinfort, Lux. 21 J7 49 39N 5 55 E
Steinfurt, Germany 26 C3 52 9N 7 20 E
Steinheim, Germany 26 D5 51 51N 9 5 E
Steinhuder Meer, Germany 26 C5 52 29N 9 21 E
Steinkjer, Norway 12 D14 64 1N 11 31 E
Steinkopf, S. Africa 104 D2 29 18 S 17 43 E
Stekene, Belgium 21 F4 51 12N 4 2 E
Stellarton, Canada 129 C7 45 32N 62 30W
Stellenbosch, S. Africa 104 E2 33 58 S 18 50 E
Stellendam, Neths. 20 E4 51 49N 4 1 E
Stelvio, Paso dello, Italy 38 B7 46 32N 10 27 E
Stemshaug, Norway 14 A2 63 19N 8 44 E
Stendal, Germany 26 C7 52 36N 11 53 E
Stene, Belgium 21 F1 51 12N 2 56 E
Stenstorp, Sweden 15 F7 58 17N 13 45 E
Steornabhaigh = Stornoway, U.K. 18 C2 58 13N 6 23W
Stepanakert = Xankändi, Azerbaijan 89 C12 39 52N 46 49 E
Stepanavan, Armenia 53 K7 41 1N 44 23 E
Stephen, U.S.A. 138 A6 48 27N 96 53W
Stephens, C., N.Z. 119 A8 40 42 S 173 58 E
Stephens Creek, Australia 116 A4 31 50 S 141 30 E
Stephens I., Canada 130 C2 54 10N 130 45W
Stephens I., N.Z. 119 A9 40 40 S 174 1 E
Stephenville, Canada 129 C8 48 31N 58 35W
Stephenville, U.S.A. 139 J5 32 13N 98 12W
Stepnica, Poland 47 B1 53 38N 14 36 E
Stepnoi = Elista, Russia 53 G7 46 16N 44 14 E
Stepnoye, Russia 54 D8 54 4N 60 26 E
Stepnyak, Kazakhstan 56 D8 52 50N 70 50 E
Steppe, Asia 58 E7 50 0N 50 0 E
Stereá Ellas □, Greece 45 F4 38 50N 22 0 E
Sterkstroom, S. Africa 104 E4 31 32 S 26 32 E
Sterling, Colo., U.S.A. 138 E3 40 37N 103 13W
Sterling, Ill., U.S.A. 140 C7 41 48N 89 42W
Sterling, Kans., U.S.A. 138 F5 38 13N 98 12W
Sterling City, U.S.A. 139 K4 31 51N 101 0W
Sterling Heights, U.S.A. 141 B13 42 35N 83 0W
Sterling Run, U.S.A. 136 E6 41 25N 78 12W
Sterlitamak, Russia 54 E6 53 40N 56 0 E
Šternberg, Germany 26 B7 53 42N 11 50 E
Šternberk, Czech. 31 B10 49 45N 17 15 E
Stérnes, Greece 32 D6 35 30N 24 9 E
Sterzing = Vipiteno, Italy 39 B8 46 54N 11 26 E
Stettin = Szczecin, Poland 47 B1 53 27N 14 27 E
Stettiner Haff, Germany 26 B10 53 47N 14 15 E
Stettler, Canada 130 C6 52 19N 112 40W
Steubenville, U.S.A. 136 F4 40 22N 80 37W
Stevens Point, U.S.A. 138 C10 44 31N 89 34W
Stevenson, U.S.A. 144 E5 45 42N 121 53W
Stevenson L., Canada 131 C9 53 55N 96 0W
Stevns Klint, Denmark 15 J6 55 17N 12 28 E
Steward, U.S.A. 140 C7 41 51N 89 1W
Stewardson, U.S.A. 141 E8 39 16N 88 38W
Stewart, B.C., Canada 130 B3 55 56N 129 57W
Stewart, N.W.T., Canada 126 B6 63 19N 139 26W
Stewart, U.S.A. 144 F7 39 5N 119 46W
Stewart, C., Australia 114 A1 11 57 S 134 56 E
Stewart, I., Chile 160 D2 54 50 S 71 15W
Stewart I., N.Z. 119 G2 46 58 S 167 54 E
Stewarts Point, U.S.A. 144 G3 38 39N 123 24W
Stewartsville, U.S.A. 140 E2 39 45N 94 30W
Stewiacke, Canada 129 C7 45 9N 63 22W
Steynsburg, S. Africa 104 E4 31 15 S 25 49 E
Steyr, Austria 30 C7 48 3N 14 25 E
Steyr →, Austria 30 C7 48 17N 14 15 E
Steytlerville, S. Africa 104 E3 33 17 S 24 19 E
Stia, Italy 39 E8 43 48N 11 42 E
Stiens, Neths. 20 B7 53 16N 5 46 E
Stigler, U.S.A. 139 H7 35 15N 95 8W
Stigliano, Italy 41 B9 40 24N 16 14 E
Stigsnæs, Denmark 15 J5 55 13N 11 18 E
Stigtomta, Sweden 15 F10 58 47N 16 48 E
Stikine →, Canada 130 B2 56 40N 132 30W
Stilfontein, S. Africa 104 D4 26 51 S 26 50 E
Stilís, Greece 45 F4 38 55N 22 47 E
Stillwater, N.Z. 119 C6 42 27 S 171 20 E
Stillwater, Minn., U.S.A. 138 C8 45 3N 92 49W
Stillwater, N.Y., U.S.A. 137 D11 42 55N 73 41W
Stillwater, Okla., U.S.A. 139 G6 36 7N 97 4W
Stillwater Range, U.S.A. 142 G4 39 50N 118 5W
Stilwell, U.S.A. 139 H7 35 49N 94 38W
Ştimfalías, L., Greece 45 G4 37 51N 22 27 E
Štip, Macedonia 42 F7 41 42N 22 10 E
Stíra, Greece 45 F6 38 9N 24 14 E
Stirling, Australia 114 B3 17 12 S 141 35 E
Stirling, Canada 130 D6 49 30N 112 30W
Stirling, N.Z. 119 G4 46 14 S 169 49 E
Stirling, U.K. 18 E5 56 8N 3 57W
Stirling Ra., Australia 113 F2 34 23 S 118 0 E
Stittsville, Canada 137 A9 45 15N 75 55W
Stjernøya, Norway 12 A20 70 20N 22 40 E
Stjørdalshalsen, Norway 12 E14 63 29N 10 51 E
Stockach, Germany 27 H5 47 50N 9 1 E
Stockbridge, U.S.A. 141 B12 42 27N 84 11W
Stockerau, Austria 31 C9 48 24N 16 12 E
Stockett, U.S.A. 142 C8 47 21N 111 10W
Stockholm, Sweden 14 E12 59 20N 18 3 E
Stockholms län □, Sweden 14 E12 59 30N 18 20 E
Stockhorn, Switz. 28 C5 46 42N 7 33 E
Stockport, U.K. 16 D5 53 25N 2 9W
Stockton, Australia 117 B9 32 50 S 151 47 E
Stockton, Calif., U.S.A. 144 H5 37 58N 121 17W
Stockton, Ill., U.S.A. 140 B6 42 21N 90 1W
Stockton, Kans., U.S.A. 138 F5 39 26N 99 16W
Stockton, Mo., U.S.A. 139 G8 37 42N 93 48W
Stockton-on-Tees, U.K. 16 C6 54 35N 1 19W
Stockvik, Sweden 14 B11 62 17N 17 23 E
Stoczek Łukowski, Poland 47 D8 51 58N 21 58 E
Stöde, Sweden 14 B10 62 28N 16 35 E
Stogovo, Macedonia 42 F5 41 31N 20 38 E
Stoke, N.Z. 119 B8 41 19 S 173 14 E
Stoke on Trent, U.K. 16 D5 53 1N 2 11W
Stokes Bay, Canada 128 C3 45 0N 81 28W
Stokes Pt., Australia 114 G3 40 10 S 143 56 E
Stokes Ra., Australia 112 C5 15 50 S 130 50 E
Stokksnes, Iceland 12 D6 64 14N 14 58W
Stokmarknes, Norway 12 B16 68 34N 14 54 E
Stolac, Bos.-H. 42 D2 43 8N 17 59 E
Stolberg, Germany 26 E2 50 47N 6 13 E
Stolbovaya, Russia 57 C16 64 50N 153 50 E
Stolbovoy, Ostrov, Russia 57 D17 74 44N 135 14 E
Stolbtsy = Stowbtsy, Belarus 50 F4 53 30N 26 43 E
Stolin, Belarus 51 G4 51 53N 26 50 E
Stolnici, Romania 46 E5 44 31N 24 48 E
Stolwijk, Neths. 20 E5 51 59N 4 47 E
Stomíon, Greece 32 D5 35 21N 23 32 E
Ston, Croatia 42 E2 42 51N 17 43 E
Stonehaven, U.K. 18 E6 56 59N 2 12W
Stonehenge, Australia 114 C3 24 22 S 143 17 E
Stonewall, Canada 131 C9 50 10N 97 19W
Stonington, U.S.A. 140 E7 39 44N 89 12W
Stony L., Man., Canada 131 B9 58 51N 98 40W
Stony L., Ont., Canada 136 B6 44 30N 78 5W
Stony Rapids, Canada 131 B7 59 16N 105 50W
Stony Tunguska = Podkamennaya Tunguska →, Russia 57 C10 61 50N 90 13 E
Stonyford, U.S.A. 144 F4 39 23N 122 33W
Stopnica, Poland 47 E7 50 27N 20 57 E
Stora Lulevatten, Sweden 12 C18 67 10N 19 30 E
Storavan, Sweden 12 D18 65 45N 18 10 E
Stord, Norway 13 G11 59 52N 5 23 E
Store Bælt, Denmark 15 J5 55 20N 11 0 E
Store Creek, Australia 117 B8 32 54 S 149 6 E
Store Heddinge, Denmark 15 J6 55 18N 12 23 E
Støren, Norway 14 A4 63 3N 10 18 E
Storm B., Australia 114 G4 43 10 S 147 30 E
Storm Lake, U.S.A. 138 D7 42 39N 95 13W
Stormberge, S. Africa 104 E4 31 16 S 26 17 E
Stormsrivier, S. Africa 104 E3 33 59 S 23 52 E
Stornoway, U.K. 18 C2 58 13N 6 23W
Storozhinets = Storozhynets, Ukraine 51 H3 48 14N 25 45 E
Storozhynets, Ukraine 51 H3 48 14N 25 45 E
Storsjö, Sweden 14 B7 62 49N 13 5 E
Storsjøen, Hedmark, Norway 14 D5 60 20N 11 40 E
Storsjøen, Hedmark, Norway 14 C5 61 30N 11 14 E
Storsjön, Sweden 14 B7 63 9N 14 30 E
Storstrøms Amt. □, Denmark 15 K5 54 50N 11 45 E
Storuman, Sweden 12 D17 65 5N 17 10 E
Storuman, sjö, Sweden 12 D17 65 13N 16 50 E
Story City, U.S.A. 140 B3 42 11N 93 36W
Stoughton, Canada 131 D8 49 40N 103 0W
Stoughton, U.S.A. 140 B7 42 55N 89 13W
Stour →, Dorset, U.K. 17 G5 50 43N 1 47W
Stour →, Here. & Worcs., U.K. 17 E5 52 21N 2 17W
Stour →, Kent, U.K. 17 F9 51 18N 1 22 E
Stour →, Suffolk, U.K. 17 F9 51 57N 1 4 E
Stourbridge, U.K. 17 E5 52 28N 2 8W
Stout, L., Canada 131 C10 52 0N 94 40W
Stove Pipe Wells Village, U.S.A. 145 J9 36 35N 117 11W
Stowbtsy, Belarus 50 F4 53 30N 26 43 E
Stowmarket, U.K. 17 E9 52 12N 1 0 E
Strabane, U.K. 19 B4 54 50N 7 27W
Strabane □, U.K. 19 B4 54 45N 7 25W
Stracin, Macedonia 42 E7 42 13N 22 2 E
Stradella, Italy 38 C6 45 5N 9 18 E
Strahan, Australia 114 G4 42 9 S 145 20 E
Strakonice, Czech. 30 B6 49 15N 13 53 E
Straldzha, Bulgaria 43 E11 42 35N 26 40 E
Stralsund, Germany 26 A9 54 18N 13 4 E
Strand, S. Africa 104 E2 34 9 S 18 48 E
Stranda, Møre og Romsdal, Norway 13 E12 62 19N 6 58 E
Stranda, Nord-Trøndelag, Norway 12 E14 63 33N 10 14 E
Strangford L., U.K. 19 B6 54 30N 5 37W
Strängnäs, Sweden 14 E11 59 23N 17 2 E
Strangsville, U.S.A. 136 E3 41 19N 81 50W
Stranraer, U.K. 18 G3 54 54N 5 1W
Strasbourg, Canada 131 C8 51 4N 104 55W
Strasbourg, France 23 D14 48 35N 7 42 E
Strasburg, Germany 26 B9 53 30N 13 43 E
Strasburg, U.S.A. 138 B4 46 8N 100 10W
Strassen, Lux. 21 J8 49 37N 6 4 E
Stratford, N.S.W., Australia 117 B9 32 7 S 151 55 E
Stratford, Vic., Australia 117 D7 37 59 S 147 7 E
Stratford, Canada 128 D3 43 23N 81 0W
Stratford, N.Z. 118 F3 39 20 S 174 19 E
Stratford, Calif., U.S.A. 144 J7 36 11N 119 49W
Stratford, Conn., U.S.A. 137 E11 41 12N 73 8W
Stratford, Tex., U.S.A. 139 G3 36 20N 102 4W
Stratford-upon-Avon, U.K. 17 E6 52 12N 1 42W
Strath Spey, U.K. 18 D5 57 9N 3 49W
Strathalbyn, Australia 116 C3 35 13 S 138 53 E
Strathclyde □, U.K. 18 F4 56 0N 4 50W
Strathcona Prov. Park, Canada 130 D3 49 38N 125 40W
Strathmore, Australia 114 B3 17 50 S 142 35 E
Strathmore, Canada 130 C6 51 5N 113 18W
Strathmore, U.K. 18 E5 56 37N 3 7W
Strathmore, U.S.A. 144 J7 36 9N 119 4W
Strathnaver, Canada 130 C4 53 20N 122 33W
Strathpeffer, U.K. 18 D4 57 35N 4 32W
Strathroy, Canada 128 D3 42 58N 81 38W
Strathy Pt., U.K. 18 C4 58 36N 4 1W
Stratton, U.S.A. 138 F3 39 19N 102 36W
Straubing, Germany 27 G8 48 52N 12 34 E
Straumnes, Iceland 12 C2 66 26N 23 8W
Strausberg, Germany 26 C9 52 35N 13 54 E
Strawberry Point, U.S.A. 140 B5 42 41N 91 32W
Strawberry Reservoir, U.S.A. 142 F8 40 8N 111 9W
Strawn, U.S.A. 139 J5 32 33N 98 30W
Strážnice, Czech. 31 C10 48 54N 17 19 E
Streaky B., Australia 115 E1 32 48 S 134 13 E
Streaky Bay, Australia 115 E1 32 51 S 134 18 E
Streator, U.S.A. 138 E10 41 8N 88 50W
Středočeský □, Czech. 30 B7 49 55N 14 30 E
Středoslovenský □, Slovak Rep. 31 C12 48 30N 19 15 E
Streé, Belgium 21 H4 50 17N 4 18 E
Streeter, U.S.A. 138 B5 46 39N 99 21W
Streetsville, Canada 136 C5 43 35N 79 42W
Strehaia, Romania 46 E4 44 37N 23 10 E
Strelcha, Bulgaria 43 E9 42 25N 24 19 E
Strelka, Russia 57 D10 58 5N 93 3 E
Streng →, Cambodia 76 F4 13 12N 103 37 E
Strésa, Italy 38 C5 45 52N 8 28 E
Streymoy, Færoe Is. 12 E9 62 8N 7 5W
Strezhevoy, Russia 56 C8 60 42N 77 34 E
Stříbro, Czech. 30 B6 49 44N 13 2 E
Strickland →, Papua N. G. 120 D1 7 35 S 141 36 E
Strijen, Neths. 20 E5 51 45N 4 33 E
Strimón →, Greece 44 D5 40 46N 23 51 E
Strimonikós Kólpos, Greece 44 D5 40 33N 24 0 E
Stroeder, Argentina 160 B4 40 12 S 62 37W
Strofádhes, Greece 45 G3 37 15N 21 0 E
Strömbacka, Sweden 14 C10 61 58N 16 44 E
Strómboli, Italy 41 D8 38 47N 15 13 E
Stromeferry, U.K. 18 D3 57 21N 5 33W
Stromness, U.K. 18 C5 58 58N 3 17W
Stromsburg, U.S.A. 138 E6 41 7N 97 36W
Strömstad, Sweden 13 G14 58 56N 11 10 E
Strömsund, Sweden 12 E16 63 51N 15 33 E
Stronghurst, U.S.A. 140 D6 40 45N 90 55W
Stróngoli, Italy 41 C10 39 16N 17 3 E
Stronsay, U.K. 18 B6 59 7N 2 35W
Stropkov, Slovak Rep. 31 B14 49 13N 21 39 E
Stroud, U.K. 17 F5 51 45N 2 13W
Stroud Road, Australia 117 B9 32 18 S 151 57 E
Stroudsburg, U.S.A. 137 F9 40 59N 75 12W
Stroumbi, Cyprus 32 E11 34 53N 32 29 E
Struer, Denmark 15 H2 56 30N 8 35 E
Struga, Macedonia 42 F5 41 13N 20 44 E
Strugi Krasnyye, Russia 50 C5 58 21N 29 1 E
Strumica, Macedonia 42 F7 41 28N 22 41 E
Strumica →, Europe 42 F8 41 20N 23 22 E
Struthers, Canada 128 C2 48 41N 85 51W
Struthers, U.S.A. 136 E4 41 4N 80 39W
Stryama, Bulgaria 43 E9 42 16N 24 54 E
Stryker, U.S.A. 142 B6 48 41N 114 46W
Stryków, Poland 47 D6 51 55N 19 33 E
Stryy, Ukraine 51 H2 49 16N 23 48 E
Strzegom, Poland 47 E3 50 58N 16 20 E
Strzelce Krajeńskie, Poland 47 C2 52 52N 15 33 E
Strzelce Opolskie, Poland 47 E5 50 31N 18 18 E
Strzelecki Cr. →, Australia 115 D2 29 37 S 139 59 E
Strzelin, Poland 47 E4 50 46N 17 2 E
Strzelno, Poland 47 C5 52 35N 18 9 E
Strzybnica, Poland 47 E5 50 28N 18 48 E
Strzyzów, Poland 31 B14 49 52N 21 47 E
Stuart, Fla., U.S.A. 135 M5 27 12N 80 15W
Stuart, Iowa, U.S.A. 140 C2 41 30N 94 19W
Stuart, Nebr., U.S.A. 138 D5 42 36N 99 8W
Stuart →, Canada 130 C4 54 0N 123 35W
Stuart Bluff Ra., Australia 112 D5 22 50 S 131 52 E
Stuart L., Canada 130 C4 54 30N 124 30W
Stuart Mts., N.Z. 119 F2 45 2 S 167 39 E
Stuart Ra., Australia 115 D1 29 10 S 134 56 E
Stubbekøbing, Denmark 15 K6 54 53N 12 9 E
Stuben, Austria 30 D3 47 10N 10 8 E
Studen Kladenets, Yazovir, Bulgaria 43 F10 41 37N 25 30 E
Studholme, N.Z. 119 E6 44 42 S 171 9 E
Stugun, Sweden 14 A9 63 10N 15 40 E
Stull, L., Canada 128 B1 54 24N 92 34W
Stung Treng, Cambodia 76 F5 13 31N 105 58 E
Stupart →, Canada 131 B10 56 0N 93 25W
Stupino, Russia 52 C4 54 57N 38 2 E
Sturgeon B., Canada 131 C9 52 0N 97 50W
Sturgeon Bay, U.S.A. 134 C2 44 50N 87 23W
Sturgeon Falls, Canada 128 C4 46 25N 79 57W
Sturgeon L., Alta., Canada 130 B5 55 6N 117 32W
Sturgeon L., Ont., Canada 128 B1 50 0N 90 45W
Sturgeon L., Ont., Canada 136 B6 44 28N 78 43W
Sturgis, Mich., U.S.A. 141 C11 41 48N 85 25W
Sturgis, S. Dak., U.S.A. 138 C3 44 25N 103 31W
Štúrovo, Slovak Rep. 31 D11 47 48N 18 41 E
Sturt Cr. →, Australia 112 C4 19 8 S 127 50 E
Sturt Creek, Australia 112 C4 19 12 S 128 8 E
Sturts Meadows, Australia 116 A4 31 18 S 141 42 E
Stutterheim, S. Africa 104 E4 32 33 S 27 28 E
Stuttgart, Germany 27 G5 48 48N 9 11 E
Stuttgart, U.S.A. 139 H9 34 30N 91 33W
Stuyvesant, U.S.A. 137 D11 42 23N 73 45W
Stykkishólmur, Iceland 12 D2 65 2N 22 40W
Styria = Steiermark □, Austria 30 D8 47 26N 15 0 E
Su-no-Saki, Japan 63 C11 34 58N 139 45 E
Su Xian, China 66 H9 33 41N 116 59 E
Suai, Indonesia 72 C3 9 21 S 125 17 E
Suakin, Sudan 94 D4 19 8N 37 20 E
Sual, Phil. 70 C3 16 4N 120 5 E
Suan, N. Korea 67 E14 38 42N 126 22 E
Suapure →, Venezuela 152 B4 6 48N 67 1W
Suaqui, Mexico 146 B3 29 12N 109 41W
Suatá →, Venezuela 153 B4 7 52N 65 22W
Subang, Indonesia 75 D3 6 34 S 107 45 E
Subansiri →, India 78 B4 26 48N 93 50 E
Subayhah, Si. Arabia 84 D3 30 2N 38 50 E
Subi, Indonesia 75 B3 2 58N 108 50 E
Subiaco, Italy 39 G10 41 56N 13 5 E
Subotica, Serbia, Yug. 42 A4 46 6N 19 39 E
Success, Canada 131 C7 50 28N 108 6W
Suceava, Romania 46 B7 47 38N 26 16 E
Suceava □, Romania 46 B6 47 37N 25 40 E
Suceava →, Romania 46 B7 47 38N 26 16 E
Sucha-Beskidzka, Poland 31 B12 49 44N 19 35 E
Suchan, Poland 47 B2 53 18N 15 18 E
Suchan, Russia 60 C6 43 8N 133 9 E
Suchedniów, Poland 47 D7 51 3N 20 49 E
Suchitoto, El Salv. 148 D2 13 56N 89 0W
Suchou = Suzhou, China 69 B13 31 19N 120 38 E
Süchow = Xuzhou, China 67 G9 34 18N 117 10 E
Suchowola, Poland 47 B10 53 33N 23 3 E
Sucio →, Colombia 152 B2 7 27N 77 7W
Suck →, Ireland 19 C3 53 17N 8 3W
Suckling, Mt., Papua N. G. 120 E5 9 49 S 148 53 E
Sucre, Bolivia 157 D4 19 0 S 65 15W
Sucre, Colombia 152 B3 8 49N 74 44W
Sucre □, Colombia 152 B2 8 50N 75 40W
Sucre □, Venezuela 153 A5 10 25N 63 30W
Sucuaro, Colombia 152 C4 4 34N 68 50W
Sućuraj, Croatia 39 E14 43 10N 17 8 E
Sucuriju, Brazil 154 A2 1 39N 49 57W
Sucuriú →, Brazil 157 E7 20 47 S 51 38W
Sud, Pte., Canada 129 C7 49 3N 62 14W
Sud-Ouest, Pte. du, Canada 129 C7 49 23N 63 36W
Suda →, Russia 50 C9 59 0N 37 40 E
Sudak, Ukraine 51 K8 44 51N 34 57 E
Sudan, U.S.A. 139 H3 34 4N 102 32W
Sudan ■, Africa 95 E3 15 0N 30 0 E
Sudbury, Canada 128 C3 46 30N 81 0W
Sudbury, U.K. 17 E8 52 2N 0 45 E
Sûdd, Sudan 95 F2 8 20N 30 0 E
Suddie, Guyana 153 B6 7 8N 58 29W
Süderbrarup, Germany 26 A5 54 38N 9 45 E
Süderlügum, Germany 26 A4 54 52N 8 54 E
Süderoogsand, Germany 26 A4 54 27N 8 28 E
Sudeten Mts. = Sudety, Europe 31 A9 50 20N 16 45 E
Sudety, Europe 31 A9 50 20N 16 45 E
Suðuroy, Færoe Is. 12 F9 61 32N 6 50W
Sudi, Tanzania 107 E4 10 11 S 39 57 E
Sudirman, Pegunungan, Indonesia 73 B5 4 30 S 137 0 E
Sudiţi, Romania 46 E8 44 35N 27 38 E
Sudogda, Russia 52 C5 55 55N 40 50 E
Sudr, Egypt 94 J8 29 40N 32 42 E
Sudzha, Russia 52 E2 51 14N 35 17 E
Sueca, Spain 35 F4 39 12N 0 21W
Suedala, Sweden 15 J7 55 30N 13 15 E
Suez = El Suweis, Egypt 94 J8 29 58N 32 31 E
Suez, G. of = Suweis, Khalîg el, Egypt 94 J8 28 40N 33 0 E
Suez Canal = Suweis, Qanâ es, Egypt 94 H8 31 0N 32 20 E
Suffield, Canada 131 C6 50 12N 111 10W
Suffolk, U.S.A. 134 G7 36 44N 76 35W
Suffolk □, U.K. 17 E9 52 16N 1 0 E
Sufi-Kurgan, Kyrgyzstan 55 C6 40 2N 73 30 E
Suga no-Sen, Japan 62 B6 35 25N 134 25 E
Sugag, Romania 46 D4 45 47N 23 37 E
Sugar →, Ill., U.S.A. 140 B7 42 26N 89 12W
Sugar →, Ind., U.S.A. 141 E9 39 50N 87 23W
Sugar City, U.S.A. 138 F3 38 14N 103 40W
Sugar Cr. →, U.S.A. 140 D7 40 9N 89 38W
Sugbai Passage, Phil. 71 J3 5 22N 120 33 E
Suğla Gölü, Turkey 88 D5 37 20N 32 0 E
Sugluk = Salluit, Canada 127 B12 62 14N 75 38W
Sugny, Belgium 21 J5 49 49N 4 54 E
Suhaia, L., Romania 46 F6 43 45N 25 15 E
Suḥār, Oman 85 E8 24 20N 56 40 E
Sühbaatar □, Mongolia 66 B8 45 30N 114 0 E
Suhl, Germany 26 E6 50 36N 10 42 E
Suhr, Switz. 28 B6 47 22N 8 5 E
Şuhut, Turkey 88 C4 38 31N 30 32 E
Sui Xian, Henan, China 66 G8 34 25N 115 2 E
Sui Xian, Henan, China 69 B9 31 42N 113 24 E
Suiá Missu →, Brazil 157 C7 11 13 S 53 15W
Suichang, China 69 C12 28 29N 119 15 E
Suichuan, China 69 D10 26 20N 114 32 E
Suide, China 66 F6 37 30N 110 12 E
Suifenhe, China 67 B16 44 25N 131 10 E
Suihua, China 65 B7 46 32N 126 55 E
Suijiang, China 68 C4 28 40N 103 59 E
Suining, Hunan, China 69 D8 26 35N 110 10 E
Suining, Jiangsu, China 67 H9 33 56N 117 58 E
Suining, Sichuan, China 68 B5 30 26N 105 35 E
Suiping, China 66 H7 33 10N 113 59 E
Suippes, France 23 C11 49 8N 4 30 E
Suir →, Ireland 19 D4 52 16N 7 9W
Suita, Japan 63 C7 34 45N 135 32 E
Suixi, China 69 G8 21 19N 110 18 E
Suiyang, Guizhou, China 68 D6 27 58N 107 18 E
Suiyang, Heilongjiang, China 67 B16 44 30N 130 56 E
Suizhong, China 67 D11 40 21N 120 20 E
Sujangarh, India 80 F6 27 42N 74 31 E
Sujica, Bos.-H. 42 D2 43 52N 17 11 E
Sukabumi, Indonesia 74 D3 6 56 S 106 50 E
Sukadana, Kalimantan, Indonesia 75 C4 1 10 S 110 0 E
Sukadana, Sumatera, Indonesia 74 D3 5 5 S 105 33 E
Sukagawa, Japan 61 F10 37 17N 140 23 E
Sukaraja, Indonesia 75 C4 2 28 S 110 25 E
Sukarnapura = Jayapura, Indonesia 73 B6 2 28 S 140 38 E
Sukchŏn, N. Korea 67 E13 39 22N 125 35 E
Sukhindol, Bulgaria 43 D10 43 11N 25 10 E
Sukhinichi, Russia 52 C2 54 8N 35 10 E
Sukhona →, Russia 48 C6 61 15N 46 39 E
Sukhothai, Thailand 76 D2 17 1N 99 49 E
Sukhoy Log, Russia 54 C9 56 55N 62 1 E
Sukhumi = Sokhumi, Georgia 53 J5 43 0N 41 0 E
Sukkur, Pakistan 79 D3 27 42N 68 54 E
Sukkur Barrage, Pakistan 80 F3 27 40N 68 50 E
Sukma, India 82 E5 18 24N 81 45 E
Sukovo, Serbia, Yug. 42 D7 43 4N 22 37 E
Sukumo, Japan 62 E4 32 56N 132 44 E
Sukunka →, Canada 130 B4 55 45N 121 15W
Sul, Canal do, Brazil 154 B2 0 10 S 48 30W
Sula →, Ukraine 51 H7 49 40N 32 41 E
Sula, Kepulauan, Indonesia 72 B3 1 45 S 125 0 E
Sulaco →, Honduras 148 C2 15 2N 87 44W
Sulaiman Range, Pakistan 80 D3 30 30N 69 50 E
Sulak →, Russia 53 J8 43 20N 47 34 E
Sūlār, Iran 85 D6 31 53N 51 54 E
Sulawesi □, Indonesia 72 B2 2 0 S 120 0 E
Sulawesi Sea = Celebes Sea, Indonesia 72 A2 3 0N 123 0 E
Sulechów, Poland 47 C2 52 5N 15 40 E
Sulęcin, Poland 47 C2 52 26N 15 10 E
Sulejów, Poland 47 D6 51 26N 19 53 E
Sulejówek, Poland 47 C8 52 13N 21 17 E
Sulgen, Switz. 29 A8 47 33N 9 7 E
Sulima, S. Leone 100 D2 6 58N 11 32W
Sulina, Romania 46 D10 45 10N 29 40 E
Sulina, Braţul →, Romania 46 D10 45 10N 29 20 E
Sulingen, Germany 26 C4 52 41N 8 48 E

Tangxi 271

Tabar Is., *Papua N. G.* **120 B7** 2 50 S 152 0 E
Tabarca, I. de, *Spain* **35 G4** 38 17N 0 30W
Tabarka, *Tunisia* **96 A1** 36 56N 8 46 E
Ţabas, *Khorāsān, Iran* **85 C9** 32 48N 60 12 E
Ţabas, *Khorāsān, Iran* **85 C8** 33 35N 56 55 E
Tabasará, Serranía de, *Panama* **148 E3** 8 35N 81 40W
Tabasco □, *Mexico* **147 D6** 17 45N 93 30W
Tabatinga, Serra da, *Brazil* **154 D3** 10 30 S 44 0W
Tabayin, *Burma* **78 D5** 22 42N 95 20 E
Tabāzīn, *Iran* **85 D8** 31 12N 57 54 E
Tabelbala, Kahal de, *Algeria* **99 C4** 28 47N 2 0W
Taber, *Canada* **130 D6** 49 47N 112 8W
Tabernas, *Spain* **35 H2** 37 4N 2 26W
Tabernes de Valldigna, *Spain* **35 F4** 39 5N 0 13W
Tabi, *Angola* **103 D2** 8 10 S 13 18 E
Tabira, *Brazil* **154 C4** 7 35 S 37 33W
Tablas, *Phil.* **70 E4** 12 25N 122 2 E
Tablas Strait, *Phil.* **70 E3** 12 40N 121 48 E
Table B. = Tafelbaai, *S. Africa* **104 E2** 33 35 S 18 25 E
Table B., *Canada* **129 B8** 53 40N 56 25W
Table Grove, *U.S.A.* **140 D6** 40 20N 90 27W
Table Mt., *S. Africa* **104 E2** 34 0 S 18 22 E
Tableland, *Australia* **112 C4** 17 16 S 126 51 E
Tabletop, Mt., *Australia* **114 C4** 23 24 S 147 11 E
Tabogon, *Phil.* **71 F5** 10 57N 124 2 E
Tábor, *Czech.* **30 B7** 49 25N 14 39 E
Tabora, *Tanzania* **106 D3** 5 2 S 32 50 E
Tabora □, *Tanzania* **106 D3** 5 0 S 33 0 E
Tabou, *Ivory C.* **100 E3** 4 30N 7 20W
Tabrīz, *Iran* **89 C12** 38 7N 46 20 E
Tabuaeran, *Pac. Oc.* **123 G12** 3 51N 159 22W
Tabuelan, *Phil.* **71 F4** 10 49N 123 52 E
Tabuenca, *Spain* **34 D3** 41 42N 1 33W
Tabuk, *Phil.* **70 C3** 17 24N 121 25 E
Tabūk, *Si. Arabia* **84 D3** 28 23N 36 36 E
Tabwemasana, Mt., *Vanuatu* **121 E4** 15 20 S 166 44 E
Tacámbaro de Codallos, *Mexico* **146 D4** 19 14N 101 28W
Tacheng, *China* **64 B3** 46 40N 82 58 E
Tachibana-Wan, *Japan* **62 E2** 32 45N 130 7 E
Tachikawa, *Japan* **63 B11** 35 42N 139 25 E
Tach'ing Shan = Daqing Shan, *China* **66 D6** 40 40N 111 0 E
Táchira □, *Venezuela* **152 B3** 8 7N 72 15W
Tachov, *Czech.* **30 B5** 49 47N 12 39 E
Tácina →, *Italy* **41 D9** 38 57N 16 55 E
Tacloban, *Phil.* **71 F5** 11 15N 124 58 E
Tacna, *Peru* **156 D3** 18 0 S 70 20W
Tacna □, *Peru* **156 D3** 17 40 S 70 20W
Tacoma, *U.S.A.* **144 C4** 47 14N 122 26W
Tacuarembó, *Uruguay* **159 C4** 31 45 S 56 0W
Tacutu →, *Brazil* **153 C5** 3 1N 60 29W
Tademaït, Plateau du, *Algeria* **99 C5** 28 30N 2 30 E
Tadent, O. →, *Algeria* **99 D6** 22 25N 6 40 E
Tadjerdjeri, O. →, *Algeria* **99 C6** 26 0N 8 0 E
Tadjerouna, *Algeria* **99 B5** 33 31N 2 3 E
Tadjettaret, O. →, *Algeria* **99 D6** 21 20N 7 22 E
Tadjmout, *Oasis, Algeria* **99 B5** 33 52N 2 30 E
Tadjmout, *Saoura, Algeria* **99 C5** 25 37N 3 48 E
Tadjoura, *Djibouti* **90 E3** 11 50N 42 55 E
Tadjoura, Golfe de, *Djibouti* **95 E5** 11 50N 43 0 E
Tadmor, *N.Z.* **119 B7** 41 27 S 172 45 E
Tadotsu, *Japan* **62 C5** 34 16N 133 45 E
Tadoule, L., *Canada* **131 B9** 58 36N 98 20W
Tadoussac, *Canada* **129 C6** 48 11N 69 42W
Tadzhikistan = Tajikistan ■, *Asia* **55 D5** 38 30N 70 0 E
Taechŏn-ni, *S. Korea* **67 F14** 36 21N 126 36 E
Taegu, *S. Korea* **67 G15** 35 50N 128 37 E
Taegwan, *N. Korea* **67 D13** 40 13N 125 12 E
Taejŏn, *S. Korea* **67 F14** 36 20N 127 28 E
Tafalla, *Spain* **34 C3** 42 30N 1 41W
Tafar, *Sudan* **95 F2** 6 52N 28 15 E
Tafassasset, O. →, *Algeria* **99 D6** 22 0N 9 57 E
Tafelbaai, *S. Africa* **104 E2** 33 35 S 18 25 E
Tafelney, C., *Morocco* **98 B3** 31 3N 9 51W
Tafermaar, *Indonesia* **73 C4** 6 47 S 134 10 E
Taffermit, *Morocco* **98 C3** 29 37N 9 15W
Tafí Viejo, *Argentina* **158 B2** 26 43 S 65 17W
Tafīhān, *Iran* **85 D7** 29 25N 52 39 E
Tafiré, *Ivory C.* **100 D3** 9 4N 5 4W
Tafnidilt, *Morocco* **98 C2** 28 47N 10 58W
Tafraoute, *Morocco* **98 C3** 29 50N 8 58W
Taft, *Iran* **85 D7** 31 45N 54 14 E
Taft, *Phil.* **71 F5** 11 57N 125 30 E
Taft, *Calif., U.S.A.* **145 K7** 35 8N 119 28W
Taft, *Tex., U.S.A.* **139 M6** 27 59N 97 24W
Taga, *W. Samoa* **121 W23** 13 46 S 172 28W
Taga Dzong, *Bhutan* **78 B2** 27 5N 89 55 E
Tagana-an, *Phil.* **71 G5** 9 42N 125 35 E
Taganrog, *Russia* **53 G4** 47 12N 38 50 E
Taganrogskiy Zaliv, *Russia* **53 G4** 47 0N 38 30 E
Tagânt, *Mauritania* **100 B2** 18 20N 11 0W
Tagap Ga, *Burma* **78 B6** 26 56N 96 13 E
Tagapula I., *Phil.* **70 E5** 12 4N 124 12 E
Tagatay, *Phil.* **70 D3** 14 6N 120 56 E
Tagauayan I., *Phil.* **71 F3** 10 58N 121 13 E
Tagbilaran, *Phil.* **71 G4** 9 39N 123 51 E
Tage, *Papua N. G.* **120 D2** 6 19 S 143 20 E
Tággia, *Italy* **38 E4** 43 52N 7 51 E
Taghzout, *Morocco* **98 B4** 33 30N 4 49W
Tagish, *Canada* **130 A2** 60 19N 134 16W
Tagish L., *Canada* **130 A2** 60 10N 134 20W
Tagkawayan, *Phil.* **70 E4** 13 58N 122 32 E
Tagliacozzo, *Italy* **39 F10** 42 4N 13 14 E
Tagliamento →, *Italy* **39 C10** 45 38N 13 6 E
Táglio di Po, *Italy* **39 D9** 45 0N 12 12 E
Tagna, *Colombia* **152 D3** 2 24 S 70 37W
Tago, *Phil.* **71 G6** 9 2N 126 13 E
Tago, Mt., *Phil.* **71 G5** 8 23N 125 5 E
Tagomago, I. de, *Spain* **33 B8** 39 2N 1 39 E
Taguatinga, *Brazil* **155 D3** 12 16 S 42 26W
Tagudin, *Phil.* **70 C3** 16 56N 120 27 E
Tagula, *Papua N. G.* **120 F7** 11 22 S 153 15 E
Tagula I., *Papua N. G.* **120 F7** 11 30 S 153 30 E
Tagum, *Phil.* **71 H5** 7 33N 125 53 E
Tagus = Tejo →, *Europe* **37 G1** 38 40N 9 24W
Tahakopa, *N.Z.* **119 G4** 46 30 S 169 23 E
Tahala, *Morocco* **98 B4** 34 0N 4 28W
Tahan, Gunong, *Malaysia* **77 K4** 4 34N 102 17 E
Tahānah-ye sūr Gol, *Afghan.* **79 C2** 31 43N 67 53 E
Tahara, *Japan* **63 C9** 34 40N 137 16 E
Tahat, *Algeria* **99 D6** 23 18N 5 33 E
Tāherī, *Iran* **85 E7** 27 43N 52 20 E
Tahiti, *Pac. Oc.* **123 J13** 17 37 S 149 27W
Tahoe, L., *U.S.A.* **144 G6** 39 6N 120 2W
Tahoe City, *U.S.A.* **144 F6** 39 10N 120 9W
Taholah, *U.S.A.* **144 C2** 47 21N 124 17W
Tahora, *N.Z.* **118 F3** 39 2 S 174 49 E
Tahoua, *Niger* **101 C6** 14 57N 5 16 E
Tahta, *Egypt* **94 B3** 26 44N 31 32 E
Tahtalı Dağları, *Turkey* **88 C7** 38 20N 36 0 E
Tahuamanu →, *Bolivia* **156 C4** 11 6 S 67 36W
Tahulandang, *Indonesia* **72 A3** 2 27N 125 23 E
Tahuna, *Indonesia* **72 A3** 3 38N 125 30 E
Taï, *Ivory C.* **100 D3** 5 55N 7 30W
Tai Shan, *China* **67 F9** 36 25N 117 20 E
Tai Xian, *China* **69 A13** 32 30N 120 7 E
Tai'an, *China* **67 F9** 36 12N 117 8 E
Taibei, *Taiwan* **69 E13** 25 4N 121 29 E
Taibique, *Canary Is.* **33 G2** 27 42N 17 58W
Taibus Qi, *China* **66 D8** 41 54N 115 22 E
T'aichung = Taizhong, *Taiwan* **69 E13** 24 12N 120 35 E
Taidong, *Taiwan* **69 F13** 22 43N 121 9 E
Taieri →, *N.Z.* **119 G5** 46 3 S 170 12 E
Taiga Madema, *Libya* **96 D3** 23 46N 15 25 E
Taigu, *China* **66 F7** 37 28N 112 30 E
Taihang Shan, *China* **66 G7** 36 0N 113 30 E
Taihape, *N.Z.* **118 F4** 39 41 S 175 48 E
Taihe, *Anhui, China* **66 H8** 33 20N 115 42 E
Taihe, *Jiangxi, China* **69 D10** 26 47N 114 52 E
Taihu, *China* **69 B11** 30 22N 116 20 E
Taijiang, *China* **68 D7** 26 39N 108 21 E
Taikang, *China* **66 G8** 34 5N 114 50 E
Taikkyi, *Burma* **78 G6** 17 20N 96 0 E
Tailem Bend, *Australia* **116 C3** 35 12 S 139 29 E
Tailfingen, *Germany* **27 G5** 48 15N 9 1 E
Taimyr Peninsula = Taymyr, Poluostrov, *Russia* **57 B11** 75 0N 100 0 E
Tain, *U.K.* **18 D4** 57 49N 4 4W
Tainan, *Taiwan* **69 F13** 23 17N 120 18 E
Taínaron, Ákra, *Greece* **45 H4** 36 22N 22 27 E
Tainggyo, *Burma* **78 G5** 17 49N 94 29 E
Taining, *China* **69 D11** 26 54N 117 9 E
Taintignies, *Belgium* **21 G2** 50 33N 3 22 E
Taiobeiras, *Brazil* **155 E3** 15 49 S 42 14W
T'aipei = Taibei, *Taiwan* **69 E13** 25 4N 121 29 E
Taiping, *China* **69 B12** 30 15N 118 6 E
Taiping, *Malaysia* **77 K3** 4 51N 100 44 E
Taipingzhen, *China* **66 H6** 33 35N 111 42 E
Taipu, *Brazil* **154 C4** 5 37 S 35 36W
Tairbeart = Tarbert, *U.K.* **18 D2** 57 54N 6 49W
Taisha, *Japan* **62 B4** 35 24N 132 40 E
Taishan, *China* **69 F9** 22 14N 112 41 E
Taishun, *China* **69 D12** 27 30N 119 42 E
Taita □, *Kenya* **106 C4** 4 0 S 38 30 E
Taita Hills, *Kenya* **106 C4** 3 25 S 38 15 E
Taitao, C., *Chile* **160 C1** 45 53 S 75 5W
Taitao, Pen. de, *Chile* **160 C2** 46 30 S 75 0W
Taivalkoski, *Finland* **12 D23** 65 33N 28 12 E
Taiwan ■, *Asia* **69 F13** 23 30N 121 0 E
Taiwan Shan, *Taiwan* **69 F13** 23 40N 121 20 E
Taixing, *China* **69 A13** 32 11N 120 0 E
Taïyetos Óros, *Greece* **45 H4** 37 0N 22 23 E
Taiyiba, *Israel* **91 C4** 32 36N 35 27 E
Taiyuan, *China* **66 F7** 37 52N 112 33 E
Taizhong, *Taiwan* **69 E13** 24 12N 120 35 E
Taizhou, *China* **69 A12** 32 28N 119 55 E
Taizhou Liedao, *China* **69 C13** 28 30N 121 55 E
Ta'izz, *Yemen* **86 D4** 13 35N 44 2 E
Tājābād, *Iran* **85 D7** 30 2N 54 24 E
Tajapuru, Furo do, *Brazil* **154 B1** 1 50 S 50 25W
Tajarḩī, *Libya* **96 D2** 24 21N 14 28 E
Tajikistan ■, *Asia* **55 D5** 38 30N 70 0 E
Tajima, *Japan* **61 F9** 37 12N 139 46 E
Tajimi, *Japan* **63 B9** 35 19N 137 8 E
Tajo = Tejo →, *Europe* **37 G1** 38 40N 9 24W
Tajrīsh, *Iran* **85 C6** 35 48N 51 25 E
Tājūrā, *Libya* **96 B2** 32 51N 13 21 E
Tak, *Thailand* **76 D2** 16 52N 99 8 E
Takāb, *Iran* **89 D12** 36 24N 47 7 E
Takachiho, *Japan* **62 E3** 32 42N 131 18 E
Takada, *Japan* **61 F9** 37 7N 138 15 E
Takahagi, *Japan* **61 F10** 36 43N 140 45 E
Takahashi, *Japan* **62 C5** 34 51N 133 39 E
Takaka, *N.Z.* **119 A7** 40 51 S 172 50 E
Takamatsu, *Japan* **62 C6** 34 20N 134 5 E
Takanabe, *Japan* **62 E3** 32 8N 131 30 E
Takaoka, *Japan* **63 A8** 36 47N 137 0 E
Takapau, *N.Z.* **118 G5** 40 2 S 176 21 E
Takapuna, *N.Z.* **118 C3** 36 47 S 174 47 E
Takasago, *Japan* **62 C6** 34 45N 134 48 E
Takasaki, *Japan* **63 A10** 36 20N 139 0 E
Takase, *Japan* **62 C5** 34 7N 133 48 E
Takatsuki, *Japan* **63 C7** 34 51N 135 37 E
Takaungu, *Kenya* **106 C4** 3 38 S 39 52 E
Takawa, *Japan* **62 D2** 33 38N 130 51 E
Takayama, *Japan* **63 A9** 36 18N 137 11 E
Takayama-Bonchi, *Japan* **63 B9** 36 0N 137 18 E
Take-Shima, *Japan* **61 J5** 30 49N 130 26 E
Takefu, *Japan* **63 B8** 35 50N 136 10 E
Takehara, *Japan* **62 C4** 34 21N 132 55 E
Takengon, *Indonesia* **74 B1** 4 45N 96 50 E
Takeo, *Cambodia* **77 G5** 10 59N 104 47 E
Takeo, *Japan* **62 D2** 33 12N 130 1 E
Tåkern, *Sweden* **15 F8** 58 22N 14 45 E
Tākestān, *Iran* **85 C6** 36 0N 49 40 E
Taketa, *Japan* **62 E3** 32 58N 131 24 E
Takh, *India* **81 C7** 33 6N 77 32 E
Takhār □, *Afghan.* **79 A3** 36 40N 70 0 E
Takhman, *Cambodia* **77 G5** 11 29N 104 57 E
Taki, *Papua N. G.* **120 D8** 6 29 S 155 52 E
Takikawa, *Japan* **60 C10** 43 33N 141 54 E
Takla L., *Canada* **130 B3** 55 15N 125 45W
Takla Landing, *Canada* **130 B3** 55 30N 125 50W
Takla Makan = Taklamakan Shamo, *China* **58 F12** 38 0N 83 0 E
Taklamakan Shamo, *China* **58 F12** 38 0N 83 0 E
Taku, *Japan* **62 D2** 33 18N 130 3 E
Taku →, *Canada* **130 B2** 58 30N 133 50W
Takum, *Nigeria* **101 D6** 7 18N 9 36 E
Takuma, *Japan* **62 C5** 34 13N 133 40 E
Takundi, *Zaïre* **103 C3** 4 45 S 16 34 E
Takuran, *Phil.* **71 H4** 7 51N 123 34 E
Takutu →, *Guyana* **153 C5** 3 1N 60 29W
Tal Halāl, *Iran* **85 D7** 28 54N 55 1 E
Tala, *Uruguay* **159 C4** 34 21 S 55 46W
Talachyn, *Belarus* **50 E5** 54 25N 29 42 E
Talacogan, *Phil.* **71 G5** 8 32N 125 39 E
Talagante, *Chile* **158 C1** 33 40 S 70 50W
Talaïnt, *Morocco* **98 C3** 29 41N 9 40W
Talak, *Niger* **101 B6** 18 0N 5 0 E
Talakag, *Phil.* **71 G5** 8 16N 124 37 E
Talamanca, Cordillera de, *Cent. Amer.* **148 E3** 9 20N 83 20W
Talara, *Peru* **156 A1** 4 38 S 81 18W
Talas, *Kyrgyzstan* **55 B6** 42 30N 72 13 E
Talas, *Turkey* **88 C6** 38 41N 35 33 E
Talas →, *Kazakhstan* **55 B5** 44 0N 70 20 E
Talasea, *Papua N. G.* **120 C6** 5 20 S 150 2 E
Talasskiy Alatau, Khrebet, *Kyrgyzstan* **55 B6** 42 15N 72 0 E
Talâta, *Egypt* **91 E1** 30 36N 32 20 E
Talata Mafara, *Nigeria* **101 C6** 12 38N 6 4 E
Talaud, Kepulauan, *Indonesia* **72 A3** 4 30N 127 10 E
Talaud Is. = Talaud, Kepulauan, *Indonesia* **72 A3** 4 30N 127 10 E
Talavera de la Reina, *Spain* **36 F6** 39 55N 4 46W
Talawana, *Australia* **112 D3** 22 51 S 121 9 E
Talawgyi, *Burma* **78 C6** 25 4N 97 19 E
Talayan, *Phil.* **71 H5** 6 52N 124 24 E
Talbert, Sillon de, *France* **22 D3** 48 53N 3 5W
Talbot, C., *Australia* **112 B4** 13 48 S 126 43 E
Talbragar →, *Australia* **117 B8** 32 12 S 148 37 E
Talca, *Chile* **158 D1** 35 28 S 71 40W
Talca □, *Chile* **158 D1** 35 20 S 71 46W
Talcahuano, *Chile* **158 D1** 36 40 S 73 10W
Talcher, *India* **82 D7** 21 0N 85 18 E
Talcho, *Niger* **101 C5** 14 44N 3 28 E
Taldy Kurgan = Taldyqorghan, *Kazakhstan* **56 E8** 45 10N 78 45 E
Taldyqorghan, *Kazakhstan* **56 E8** 45 10N 78 45 E
Tālesh, *Iran* **89 D13** 37 58N 48 58 E
Ţalesh, Kūhhā-ye, *Iran* **89 D13** 37 42N 48 55 E
Talgar, *Kazakhstan* **55 B8** 43 19N 77 15 E
Talgar, Pik, *Kazakhstan* **55 B8** 43 5N 77 20 E
Talguharai, *Sudan* **94 D4** 18 19N 35 56 E
Tali Post, *Sudan* **95 F3** 5 55N 30 44 E
Talibon, *Phil.* **71 F5** 10 9N 124 20 E
Talibong, Ko, *Thailand* **77 J2** 7 15N 99 23 E
Talihina, *U.S.A.* **139 H7** 34 45N 95 3W
Talikota, *India* **83 F3** 16 29N 76 17 E
Talimardzhan, *Turkmenistan* **55 D2** 38 23N 65 37 E
Talisay, *Phil.* **71 F4** 10 44N 122 58 E
Talisayan, *Phil.* **71 G5** 9 0N 124 55 E
Talitsa, *Russia* **54 C9** 57 0N 63 43 E
Taliwang, *Indonesia* **72 C1** 8 50 S 116 55 E
Tall 'Afar, *Iraq* **89 D10** 36 22N 42 27 E
Tall 'Asūr, *West Bank* **91 D4** 31 59N 35 17 E
Tall Kalakh, *Syria* **91 A5** 34 41N 36 15 E
Talla, *Egypt* **94 J7** 28 5N 30 43 E
Talladega, *U.S.A.* **135 J2** 33 26N 86 6W
Tallahassee, *U.S.A.* **135 K3** 30 27N 84 17W
Tallangatta, *Australia* **117 D7** 36 15 S 147 19 E
Tallarook, *Australia* **117 D6** 37 5 S 145 6 E
Tallawang, *Australia* **117 B8** 32 12 S 149 28 E
Tallering Pk., *Australia* **113 E2** 28 6 S 115 37 E
Tallinn, *Estonia* **13 G21** 59 22N 24 48 E
Tallulah, *U.S.A.* **139 J9** 32 25N 91 11W
Tălmaciu, *Romania* **46 D5** 45 38N 24 19 E
Talmest, *Morocco* **98 B3** 31 48N 9 21W
Talmont, *France* **24 B2** 46 27N 1 37W
Talne, *Ukraine* **51 H6** 48 50N 30 44 E
Talnoye = Talne, *Ukraine* **51 H6** 48 50N 30 44 E
Taloda, *India* **82 D2** 21 34N 74 11 E
Talodi, *Sudan* **95 E3** 10 35N 30 22 E
Talomo, *Phil.* **71 H5** 7 3N 125 32 E
Talovaya, *Russia* **52 E5** 51 6N 40 45 E
Talpa de Allende, *Mexico* **146 C4** 20 23N 104 51W
Tālqān, *Afghan.* **79 A3** 36 44N 69 33 E
Talsi, *Latvia* **13 H20** 57 10N 22 30 E
Talsinnt, *Morocco* **99 B4** 32 33N 3 27W
Taltal, *Chile* **158 B1** 25 23 S 70 33W
Taltson →, *Canada* **130 A6** 61 24N 112 46W
Talwood, *Australia* **115 D4** 28 29 S 149 29 E
Talyawalka Cr. →, *Australia* **116 B5** 32 28 S 142 22 E
Tam Chau, *Vietnam* **77 G5** 10 48N 105 12 E
Tam Ky, *Vietnam* **76 E7** 15 34N 108 29 E
Tam Quan, *Vietnam* **76 E7** 14 35N 109 3 E
Tama, *U.S.A.* **140 C4** 41 58N 92 35W
Tamala, *Australia* **113 E1** 26 42 S 113 47 E
Tamalameque, *Colombia* **152 B3** 8 52N 73 49W
Tamale, *Ghana* **101 D4** 9 22N 0 50W
Taman, *Russia* **53 H3** 45 14N 36 41 E
Tamana, *Japan* **62 E2** 32 58N 130 32 E
Tamanar, *Morocco* **98 B3** 31 1N 9 46W
Tamano, *Japan* **62 C5** 34 29N 133 59 E
Tamanrasset, *Algeria* **99 D6** 22 50N 5 30 E
Tamanrasset, O. →, *Algeria* **99 D5** 22 0N 2 0 E
Tamanthi, *Burma* **78 C5** 25 19N 95 17 E
Tamaqua, *U.S.A.* **137 F9** 40 48N 75 58W
Tamar →, *U.K.* **17 G3** 50 27N 4 15W
Támara, *Colombia* **152 B3** 5 50N 72 10W
Tamarang, *Australia* **117 A9** 31 27 S 150 5 E
Tamarinda, *Spain* **33 B10** 39 55N 3 49 E
Tamarite de Litera, *Spain* **34 D5** 41 52N 0 25 E
Tamaroa, *U.S.A.* **140 F7** 38 8N 89 14W
Tamashima, *Japan* **62 C5** 34 32N 133 40 E
Tamási, *Hungary* **31 E11** 46 40N 18 18 E
Tamaské, *Niger* **101 C6** 14 49N 5 43 E
Tamaulipas □, *Mexico* **147 C5** 24 0N 99 0W
Tamaulipas, Sierra de, *Mexico* **147 C5** 23 30N 98 20W
Tamazula, *Mexico* **146 C3** 24 55N 106 58W
Tamazunchale, *Mexico* **147 C5** 21 16N 98 47W
Tamba-Dabatou, *Guinea* **100 C2** 11 50N 10 40W
Tambacounda, *Senegal* **100 C2** 13 45N 13 40W
Tambelan, Kepulauan, *Indonesia* **74 B3** 1 0N 107 30 E
Tambellup, *Australia* **113 F2** 34 4 S 117 37 E
Tambo, *Australia* **114 C4** 24 54 S 146 14 E
Tambo, *Peru* **156 C3** 12 57 S 74 1W
Tambo →, *Peru* **156 C3** 10 42 S 73 47W
Tambo de Mora, *Peru* **156 C2** 13 30 S 76 8W
Tambobamba, *Peru* **156 C3** 13 54 S 72 8W
Tambohorano, *Madag.* **105 B7** 17 30 S 43 58 E
Tambopata →, *Peru* **156 C4** 13 21 S 69 36W
Tambora, *Indonesia* **72 C1** 8 12 S 118 5 E
Tamboritha, Mt., *Australia* **117 D7** 37 31 S 146 40 E
Tambov, *Russia* **52 D5** 52 45N 41 28 E
Tambre →, *Spain* **36 C2** 42 49N 8 53W
Tambuku, *Indonesia* **75 D4** 7 8 S 113 40 E
Tambun Sigumbal, *Phil.* **71 H3** 6 5N 121 47 E
Tamburâ, *Sudan* **95 F2** 5 40N 27 25 E
Tambuyukan, Gunong, *Malaysia* **75 A5** 6 13N 116 39 E
Tâmchekket, *Mauritania* **100 B2** 17 25N 10 40W
Tamdybulak, *Uzbekistan* **55 C2** 41 46N 64 36 E
Tame, *Colombia* **152 B3** 6 28N 71 44W
Tamega →, *Portugal* **36 D2** 41 5N 8 21W
Tamelelt, *Morocco* **98 B3** 31 50N 7 32W
Tamenglong, *India* **78 C4** 25 0N 93 35 E
Tamerlanovka, *Kazakhstan* **55 B4** 42 36N 69 17 E
Tamerza, *Tunisia* **96 B1** 34 23N 7 58 E
Tamiahua, L. de, *Mexico* **147 C5** 21 30N 97 30W
Tamil Nadu □, *India* **83 J3** 11 0N 77 0 E
Tamines, *Belgium* **21 H5** 50 26N 4 36 E
Tamis →, *Serbia, Yug.* **46 E1** 44 51N 20 39 E
Tamluk, *India* **81 H12** 22 18N 87 58 E
Tammerfors = Tampere, *Finland* **13 F20** 61 30N 23 50 E
Tammisaari, *Finland* **13 F20** 60 0N 23 26 E
Tamo Abu, Pegunungan, *Malaysia* **75 B5** 3 10N 115 5 E
Tampa, *U.S.A.* **135 M4** 27 57N 82 27W
Tampa B., *U.S.A.* **135 M4** 27 50N 82 30W
Tampere, *Finland* **13 F20** 61 30N 23 50 E
Tampico, *Mexico* **147 C5** 22 20N 97 50W
Tampico, *U.S.A.* **140 C7** 41 38N 89 47W
Tampin, *Malaysia* **77 L4** 2 28N 102 13 E
Tamrah, *Si. Arabia* **86 B4** 20 24N 45 25 E
Tamri, *Morocco* **98 B3** 30 49N 9 50W
Tamrida = Qādib, *Yemen* **87 D6** 12 37N 53 57 E
Tamsweg, *Austria* **30 D6** 47 7N 13 49 E
Tamuja →, *Spain* **37 F4** 39 38N 6 29W
Tamworth, *Australia* **117 A9** 31 7 S 150 58 E
Tamworth, *U.K.* **17 E6** 52 39N 1 41W
Tamyang, *S. Korea* **67 G14** 35 19N 126 59 E
Tan An, *Vietnam* **77 G6** 10 32N 106 25 E
Tan-tan, *Morocco* **98 C2** 28 29N 11 1W
Tana →, *Kenya* **106 C5** 2 32 S 40 31 E
Tana →, *Norway* **12 A23** 70 30N 28 14 E
Tana, L., *Ethiopia* **95 E4** 13 5N 37 30 E
Tana River, *Kenya* **106 C4** 2 0 S 39 30 E
Tanabe, *Japan* **63 D7** 33 44N 135 22 E
Tanabi, *Brazil* **155 F2** 20 37 S 49 37W
Tanafjorden, *Norway* **12 A23** 70 45N 28 25 E
Tanaga, Pta., *Canary Is.* **33 G1** 27 42N 18 10W
Tanagro →, *Italy* **41 B8** 40 38N 15 14 E
Tanah Merah, *Malaysia* **74 A2** 5 48N 102 9 E
Tanahbala, *Indonesia* **74 C1** 0 30 S 98 30 E
Tanahgrogot, *Indonesia* **75 C5** 1 55 S 116 15 E
Tanahjampea, *Indonesia* **72 C2** 7 10 S 120 35 E
Tanahmasa, *Indonesia* **74 C1** 0 12 S 98 39 E
Tanahmerah, *Indonesia* **73 C6** 6 5 S 140 16 E
Tanakura, *Japan* **61 F10** 37 10N 140 20 E
Tanami, *Australia* **112 C4** 19 59 S 129 43 E
Tanami Desert, *Australia* **112 C5** 18 50 S 132 0 E
Tanana, *U.S.A.* **126 B4** 65 10N 152 4W
Tanana →, *U.S.A.* **126 B4** 65 10N 151 58W
Tananarive = Antananarivo, *Madag.* **105 B8** 18 55 S 47 31 E
Tanannt, *Morocco* **98 B3** 31 54N 6 56W
Tánaro →, *Italy* **38 C5** 44 55N 8 40 E
Tanauan, *Batangas, Phil.* **70 D3** 14 5N 121 10 E
Tanauan, *Leyte, Phil.* **71 F5** 11 7N 125 1 E
Tanaunella, *Italy* **40 B2** 40 42N 9 45 E
Tanay, *Phil.* **70 D3** 14 30N 121 17 E
Tanba-Sanchi, *Japan* **63 B7** 35 7N 135 48 E
Tanbar, *Australia* **114 D3** 25 51 S 141 55 E
Tancarville, *France* **22 C7** 49 29N 0 28 E
Tancheng, *China* **67 G10** 34 25N 118 20 E
Tanchŏn, *N. Korea* **67 D15** 40 27N 128 54 E
Tanda, *Ut. P., India* **81 F10** 26 33N 82 35 E
Tanda, *Ut. P., India* **81 E8** 28 57N 78 56 E
Tanda, *Ivory C.* **100 D4** 7 48N 3 10W
Tandag, *Phil.* **71 G6** 9 4N 126 9 E
Tandaia, *Tanzania* **107 D3** 9 25 S 34 15 E
Tăndărei, *Romania* **46 E8** 44 39N 27 40 E
Tandaué, *Angola* **103 F3** 16 58 S 18 5 E
Tandil, *Argentina* **158 D4** 37 15 S 59 6W
Tandil, Sa. del, *Argentina* **158 D4** 37 30 S 59 0W
Tandlianwala, *Pakistan* **80 D5** 31 3N 73 9 E
Tando Adam, *Pakistan* **79 D3** 25 45N 68 40 E
Tandou L., *Australia* **116 B5** 32 40 S 142 5 E
Tandsbyn, *Sweden* **14 A8** 63 3N 14 45 E
Tandur, *India* **82 E4** 19 11N 79 30 E
Tane-ga-Shima, *Japan* **61 J5** 30 30N 131 0 E
Taneatua, *N.Z.* **118 E6** 38 4 S 177 1 E
Tanen Tong Dan, *Burma* **76 D2** 16 30N 98 30 E
Tanew →, *Poland* **47 E9** 50 29N 22 16 E
Tanezrouft, *Algeria* **99 D5** 23 9N 0 11 E
Tang, Koh, *Cambodia* **77 G4** 10 16N 103 7 E
Tang Krasang, *Cambodia* **76 F5** 12 34N 105 3 E
Tanga, *Tanzania* **106 D4** 5 5 S 39 2 E
Tanga □, *Tanzania* **106 D4** 5 20 S 38 0 E
Tanga Is., *Papua N. G.* **120 B7** 3 20 S 153 15 E
Tangail, *Bangla.* **78 C2** 24 15N 89 55 E
Tanganyika, L., *Africa* **106 D2** 6 40 S 30 0 E
Tangawan, *Phil.* **71 H4** 7 30N 122 20 E
Tanger, *Morocco* **98 A3** 35 50N 5 49W
Tangerang, *Indonesia* **74 D3** 6 11 S 106 37 E
Tangerhütte, *Germany* **26 C7** 52 26N 11 48 E
Tangermünde, *Germany* **26 C7** 52 33N 11 58 E
Tanggu, *China* **67 E9** 39 2N 117 40 E
Tanggula Shan, *China* **64 C4** 32 40N 92 10 E
Tanghe, *China* **66 H7** 32 47N 112 50 E
Tangier = Tanger, *Morocco* **98 A3** 35 50N 5 49W
Tangkeleboke, *Indonesia* **72 B2** 3 10 S 121 30 E
Tangorin P.O., *Australia* **114 C3** 21 47 S 144 12 E
Tangshan, *China* **67 E10** 39 38N 118 10 E
Tangtou, *China* **67 G10** 35 28N 118 30 E
Tangub, *Phil.* **71 G4** 8 3N 123 44 E
Tanguiéta, *Benin* **101 C5** 10 35N 1 21 E
Tangxi, *China* **69 C12** 29 3N 119 25 E

Thmar Puok 273

Tekapo, L., *N.Z.* .... **119 D5** 43 53 S 170 33 E
Tekax, *Mexico* **147 C7** 20 11N 89 18W
Tekeli, *Kazakhstan* ...... **55 A9** 44 50N 79 0 E
Tekeze →, *Ethiopia* .... **95 E4** 14 20N 35 50 E
Tekija, *Serbia, Yug.* ..... **42 C7** 44 42N 22 26 E
Tekirdağ, *Turkey* .... **88 B2** 40 58N 27 30 E
Tekkali, *India* .... **82 E7** 18 37N 84 15 E
Tekke, *Turkey* .... **88 B7** 40 42N 36 12 E
Tekman, *Turkey* .... **89 C9** 39 38N 41 29 E
Tekoa, *U.S.A.* **142 C5** 47 14N 117 4W
Tekouiât, O. →, *Algeria* **99 D5** 22 25N 2 35 E
Tel Aviv-Yafo, *Israel* .... **91 C3** 32 4N 34 48 E
Tel Lakhish, *Israel* ...... **91 D3** 31 34N 34 51 E
Tel Megiddo, *Israel* ..... **91 C4** 32 35N 35 11 E
Tela, *Honduras* **148 C2** 15 40N 87 28W
Télagh, *Algeria* **99 B4** 34 51N 0 32W
Telanaipura = Jambi,
Indonesia .... **74 C2** 1 38 S 103 30 E
Telavi, *Georgia* .... **53 J7** 42 0N 45 30 E
Telciu, *Romania* .... **46 B5** 47 25N 24 24 E
Telde, *Canary Is.* **33 G4** 27 59N 15 25W
Telefomin, *Papua N. G.* . **120 C1** 5 10 S 141 31 E
Telegraph Creek, *Canada* **130 B2** 58 0N 131 10W
Telekhany =
Tsyelyakhany, *Belarus* . **51 F3** 52 30N 25 46 E
Telemark, *Norway* ...... **13 G12** 59 15N 7 40 E
Telemark fylke □, *Norway* **14 E2** 59 25N 8 30 E
Telén, *Argentina* .... **158 D2** 36 15 S 65 31W
Telen →, *Indonesia* .... **75 C5** 0 10 S 117 20 E
Teleng, *Iran* .... **85 E9** 25 47N 61 3 E
Teleño, *Spain* **36 C4** 42 23N 6 22W
Teleorman □, *Romania* .. **46 E6** 44 0N 25 0 E
Teleorman →, *Romania* **46 E6** 44 15N 25 20 E
Teles Pires →, *Brazil* ... **157 B6** 7 21 S 58 3W
Telescope Pk., *U.S.A.* **145 J9** 36 10N 117 5W
Teletaye, *Mali* .... **101 B5** 16 31N 1 30 E
Telford, *U.K.* **16 E5** 52 40N 2 27W
Telfs, *Austria* .... **30 D4** 47 19N 11 4 E
Télimélé, *Guinea* **100 C2** 10 54N 13 2W
Telkwa, *Canada* **130 C3** 54 41N 127 5W
Tell City, *U.S.A.* **141 G10** 37 57N 86 46W
Tellicherry, *India* .... **83 J2** 11 45N 75 30 E
Tellin, *Belgium* .... **21 H6** 50 5N 5 13 E
Telluride, *U.S.A.* **143 H10** 37 56N 107 49W
Telok Datok, *Malaysia* .. **74 B2** 2 49N 101 31 E
Teloloapán, *Mexico* **147 D5** 18 21N 99 51W
Telpos Iz, *Russia* .... **48 B10** 63 16N 59 13 E
Telsen, *Argentina* .... **160 B3** 42 30 S 66 50W
Telšiai, *Lithuania* .... **13 H20** 55 59N 22 14 E
Teltow, *Germany* .... **26 C9** 52 24N 13 15 E
Teluk Anson, *Malaysia* .. **77 K3** 4 3N 101 0 E
Teluk Betung =
Tanjungkarang
Telukbetung, *Indonesia* **74 D3** 5 20 S 105 10 E
Teluk Cenderawasih,
Indonesia .... **73 B5** 2 30 S 135 20 E
Teluk Intan = Teluk
Anson, *Malaysia* ...... **77 K3** 4 3N 101 0 E
Telukbutun, *Indonesia* ... **75 B3** 4 13N 108 12 E
Telukdalem, *Indonesia* .. **74 B1** 0 33N 97 50 E
Tema, *Ghana* .... **101 D5** 5 41N 0 0 E
Temanggung, *Indonesia* .. **75 D4** 7 18 S 110 10 E
Temapache, *Mexico* **147 C5** 21 4N 97 38W
Temax, *Mexico* **147 C7** 21 10N 88 50W
Temba, *S. Africa* .... **105 D4** 25 20 S 28 17 E
Tembe, *Zaïre* .... **106 C2** 0 16 S 28 14 E
Tembesi, *Indonesia* ..... **74 C2** 1 43 S 103 6 E
Tembilahan, *Indonesia* ... **74 C2** 0 19 S 103 9 E
Temblador, *Venezuela* ... **153 B5** 8 59N 62 44W
Tembleque, *Spain* **34 F1** 39 41N 3 30W
Temblor Range, *U.S.A.* . **145 K7** 35 20N 119 50W
Teme →, *U.K.* **17 E5** 52 11N 2 13W
Temecula, *U.S.A.* **145 M9** 33 30N 117 9W
Temerloh, *Malaysia* ..... **77 L4** 3 27N 102 25 E
Temir, *Kazakhstan* ...... **56 E6** 49 1N 57 14 E
Temirtau, *Kazakhstan* .... **56 D8** 50 5N 72 56 E
Temirtau, *Russia* .... **56 D9** 53 10N 87 30 E
Témiscaming, *Canada* ... **128 C4** 46 44N 79 5W
Temma, *Australia* .... **114 G3** 41 12 S 144 48 E
Temnikov, *Russia* .... **52 C6** 54 40N 43 11 E
Temo →, *Italy* .... **40 B1** 40 17N 8 28 E
Temora, *Australia* ...... **117 C7** 34 30 S 147 30 E
Temosachic, *Mexico* **146 B3** 28 58N 107 50W
Tempe, *U.S.A.* **143 K8** 33 25N 111 56W
Tempe Downs, *Australia* . **112 D5** 24 22 S 132 24 E
Tëmpio Pausânia, *Italy* .. **40 B2** 40 54N 9 6 E
Tempiute, *U.S.A.* **144 H11** 37 39N 115 38W
Temple, *U.S.A.* **139 K6** 31 6N 97 21W
Temple B., *Australia* .... **114 A3** 12 15 S 143 3 E
Templemore, *Ireland* **19 D4** 52 47N 7 51W
Templeton, *U.S.A.* **144 K6** 35 33N 120 42W
Templeton →, *Australia* **114 C2** 21 0 S 138 40 E
Templeuve, *Belgium* .... **21 G2** 50 39N 3 17 E
Templin, *Germany* ...... **26 B9** 53 7N 13 28 E
Tempoal, *Mexico* **147 C5** 21 31N 98 23W
Temryuk, *Russia* .... **53 H3** 45 15N 37 24 E
Temse, *Belgium* .... **21 F4** 51 7N 4 13 E
Temska →, *Serbia, Yug.* **42 D7** 43 17N 22 33 E
Temuco, *Chile* .... **160 A2** 38 45 S 72 40W
Temuka, *N.Z.* .... **119 E6** 44 14 S 171 17 E
Ten Boer, *Neths.* .... **20 B9** 53 16N 6 42 E
Tena, *Ecuador* .... **152 D2** 0 59 S 77 49W
Tenabo, *Mexico* **147 C6** 20 2N 90 12W
Tenaha, *U.S.A.* **139 K7** 31 57N 94 15W
Tenali, *India* .... **83 F5** 16 15N 80 35 E
Tenancingo, *Mexico* **147 D5** 19 0N 99 33W
Tenango, *Mexico* **147 D5** 19 7N 99 33W
Tenasserim, *Burma* ..... **77 F2** 12 6N 99 3 E
Tenasserim □, *Burma* ... **76 F2** 14 0N 98 30 E
Tenay, *France* .... **25 C9** 45 55N 5 31 E
Tenby, *U.K.* **17 F3** 51 40N 4 42W
Tenda, Col di, *France* ... **25 D11** 44 7N 7 36 E
Tendaho, *Ethiopia* ...... **90 E3** 11 48N 40 54 E
Tende, *France* .... **25 D11** 44 5N 7 35 E
Tendelti, *Sudan* .... **95 E3** 13 1N 31 55 E
Tendjedi, Adrar, *Algeria* . **99 D6** 23 41N 7 32 E
Tendrara, *Morocco* **99 B4** 33 3N 1 58W
Tendre, Mt., *Switz.* ..... **28 C2** 46 35N 6 18 E
Tendrovskaya Kosa,
Ukraine .... **51 J6** 46 16N 31 35 E
Teneida, *Egypt* .... **94 B2** 25 30N 29 19 E
Tenente Marques →,
Brazil .... **157 C6** 11 10 S 59 56W
Ténéré, *Niger* .... **97 E2** 19 0N 10 30 E
Ténéré, Erg du, *Niger* ... **97 E2** 17 35N 10 55 E
Tenerife, *Canary Is.* **33 F3** 28 15N 16 35W
Tenerife, Pico, *Canary Is.* **33 G1** 27 43N 18 1W
Ténès, *Algeria* .... **99 A5** 36 31N 1 14 E
Teng Xian,
Guangxi Zhuangzu,
China .... **69 F8** 23 21N 110 56 E
Teng Xian, *Shandong,*
China .... **67 G9** 35 5N 117 10 E
Tengah □, *Indonesia* .... **72 B2** 2 0 S 122 0 E
Tengah Kepulauan,
Indonesia .... **75 D5** 7 5 S 118 15 E
Tengchong, *China* ...... **68 E2** 25 0N 98 28 E
Tengchowfu = Penglai,
China .... **67 F11** 37 48N 120 42 E
Tenggara □, *Indonesia* .. **72 B2** 3 0 S 122 0 E
Tenggarong, *Indonesia* .. **75 C5** 0 24 S 116 58 E
Tenggol, P., *Malaysia* ... **77 K4** 4 48N 103 41 E
Tengiz, Ozero,
Kazakhstan .... **56 E7** 48 5N 63 7 E
Tengiz, Ozero,
Kazakhstan .... **56 D7** 50 30N 69 0 E
Tenigerbad, *Switz.* ...... **29 C7** 46 42N 8 57 E
Tenino, *U.S.A.* **144 D4** 46 51N 122 51W
Tenkasi, *India* .... **83 K3** 8 55N 77 20 E
Tenke, *Shaba, Zaïre* .... **107 E2** 11 22 S 26 40 E
Tenke, *Shaba, Zaïre* .... **107 E2** 10 32 S 26 7 E
Tenkodogo, *Burkina Faso* **101 C4** 11 54N 0 19W
Tenna →, *Italy* .... **39 E10** 43 14N 13 47 E
Tennant Creek, *Australia* **114 B1** 19 30 S 134 15 E
Tennessee □, *U.S.A.* **135 H2** 36 0N 86 30W
Tennessee →, *U.S.A.* ... **134 G1** 37 4N 88 34W
Tenneville, *Belgium* ..... **21 H7** 50 6N 5 32 E
Tennille, *U.S.A.* **135 J4** 32 56N 82 48W
Tennsift, Oued →,
Morocco **98 B3** 32 3N 9 28W
Tennyson, *U.S.A.* **141 F9** 38 5N 87 7W
Teno, Pta. de, *Canary Is.* **33 F3** 28 21N 16 55W
Tenom, *Malaysia* .... **75 A5** 5 4N 115 57 E
Tenosique, *Mexico* **147 D6** 17 30N 91 24W
Tenri, *Japan* .... **63 C7** 34 39N 135 49 E
Tenryū, *Japan* .... **63 C9** 34 52N 137 49 E
Tenryū-Gawa →, *Japan* . **63 B9** 35 39N 137 48 E
Tent L., *Canada* **131 A7** 62 25N 107 54W
Tentelomatinan, *Indonesia* **72 A2** 0 56N 121 48 E
Tenterfield, *Australia* .... **115 D5** 29 0 S 152 0 E
Teófilo Otoni, *Brazil* .... **155 E3** 17 50 S 41 30W
Teotihuacán, *Mexico* **147 D5** 19 44N 98 50W
Tepa, *Indonesia* .... **73 C3** 7 52 S 129 31 E
Tepalcatepec →, *Mexico* **146 D4** 18 35N 101 59W
Tepehuanes, *Mexico* **146 B3** 25 21N 105 44W
Tepelena, *Albania* ...... **44 D2** 40 17N 20 2 E
Tepequem, Serra, *Brazil* . **153 C5** 3 45N 61 45W
Tepetongo, *Mexico* **146 C4** 22 28N 103 9W
Tepic, *Mexico* **146 C4** 21 30N 104 54W
Teplice, *Czech.* .... **30 A6** 50 40N 13 48 E
Teploklyuchenka,
Kyrgyzstan .... **55 B9** 42 30N 78 30 E
Tepoca, C., *Mexico* **146 A2** 30 20N 112 25W
Tequila, *Mexico* **146 C4** 20 54N 103 47W
Ter →, *Spain* .... **34 C8** 42 2N 3 12 E
Ter Apel, *Neths.* .... **20 C10** 52 53N 7 5 E
Téra, *Niger* .... **101 C5** 14 0N 0 45 E
Tera →, *Spain* **36 D5** 41 54N 5 44W
Teraina, *Kiribati* **123 G11** 4 43N 160 25W
Téramo, *Italy* .... **39 F10** 42 39N 13 42 E
Terang, *Australia* .... **116 E5** 38 15 S 142 55 E
Terawhiti, C., *N.Z.* ...... **118 H3** 41 16 S 174 38 E
Terazit, Massif de, *Niger* . **97 D1** 20 2N 8 30 E
Terborg, *Neths.* .... **20 E8** 51 56N 6 22 E
Tercan, *Turkey* .... **89 C9** 39 47N 40 23 E
Tercero →, *Argentina* .. **158 C3** 32 58 S 61 47W
Terdal, *India* .... **82 F2** 16 33N 75 3 E
Terebovlya, *Ukraine* .... **51 H3** 49 18N 25 44 E
Teregova, *Romania* ..... **46 D3** 45 10N 22 16 E
Terek →, *Russia* .... **53 J8** 44 0N 47 30 E
Terek-Say, *Kyrgyzstan* ... **55 C5** 41 30N 71 11 E
Terengganu □, *Malaysia* . **74 B2** 4 55N 103 0 E
Terenos, *Brazil* .... **157 E7** 20 26 S 54 50W
Tereshka →, *Russia* .... **52 E8** 51 48N 46 26 E
Teresina, *Brazil* .... **154 C3** 5 9 S 42 45W
Teresinha, *Brazil* **153 C7** 0 58N 52 2W
Terespol, *Poland* .... **47 C10** 52 5N 23 37 E
Terewah, L., *Australia* .. **115 D4** 29 52 S 147 35 E
Terges →, *Portugal* **37 H3** 37 49N 7 41W
Tergnier, *France* .... **23 C10** 49 40N 3 17 E
Terhazza, *Mali* **98 D3** 23 38N 5 22W
Terheijden, *Neths.* ...... **21 E5** 51 38N 4 45 E
Teridgerie Cr. →,
Australia .... **115 E4** 30 25 S 148 50 E
Terifa, *Yemen* .... **86 D3** 14 24N 43 48 E
Terlizzi, *Italy* .... **41 A9** 41 8N 16 32 E
Terme, *Turkey* .... **88 B7** 41 11N 37 0 E
Termez = Termiz,
Uzbekistan .... **55 E3** 37 15N 67 15 E
Términi Imerese, *Italy* ... **40 E6** 37 59N 13 42 E
Términos, L. de, *Mexico* . **147 D6** 18 35N 91 30W
Termiz, *Uzbekistan* ..... **55 E3** 37 15N 67 15 E
Térmoli, *Italy* .... **39 F12** 42 0N 15 0 E
Ternate, *Indonesia* ...... **72 A3** 0 45N 127 25 E
Terneuzen, *Neths.* ...... **21 F3** 51 20N 3 50 E
Terney, *Russia* .... **57 E14** 45 3N 136 37 E
Terni, *Italy* .... **39 F9** 42 34N 12 38 E
Ternitz, *Austria* .... **30 D9** 47 43N 16 2 E
Ternopil, *Ukraine* ...... **51 H3** 49 30N 25 40 E
Ternopol = Ternopil,
Ukraine .... **51 H3** 49 30N 25 40 E
Terowie, *N.S.W.,*
Australia .... **115 E4** 32 27 S 147 52 E
Terowie, *S. Austral.,*
Australia .... **115 E2** 33 8 S 138 55 E
Terra Bella, *U.S.A.* **145 K7** 35 58N 119 3W
Terrace, *Canada* **130 C3** 54 30N 128 35W
Terrace Bay, *Canada* **128 C2** 48 47N 87 5W
Terracina, *Italy* .... **40 A6** 41 17N 13 15 E
Terralba, *Italy* .... **40 C1** 39 43N 8 39 E
Terranova = Ólbia, *Italy* **40 B2** 40 55N 9 31 E
Terranuova Bracciolini,
Italy .... **39 E8** 43 33N 11 35 E
Terrasini Favarotta, *Italy* **40 D6** 38 10N 13 4 E
Terrassa = Tarrasa, *Spain* **34 D7** 41 34N 2 1 E
Terrasson-la-Villedieu,
France .... **24 C5** 45 8N 1 18 E
Terre Haute, *U.S.A.* **141 E9** 39 28N 87 25W
Terrebonne B., *U.S.A.* ... **139 L9** 29 5N 90 35W
Terrecht, *Mali* **99 D4** 20 10N 0 10W
Terrell, *U.S.A.* **139 J6** 32 44N 96 17W
Terrenceville, *Canada* ... **129 C9** 47 40N 54 44W
Terrick Terrick, *Australia* **114 C4** 24 44 S 145 5 E
Terry, *U.S.A.* **138 B2** 46 47N 105 19W
Terschelling, *Neths.* ..... **20 B6** 53 25N 5 20 E
Terskey Alatau, Khrebet,
Kyrgyzstan .... **55 C8** 41 50N 77 0 E
Tersko-Kumskiy
Kanal →, *Russia* ..... **53 H7** 44 32N 44 38 E
Terter = Tärtär →,
Azerbaijan .... **53 K8** 40 26N 47 20 E
Teruel, *Spain* **34 E3** 40 22N 1 8W
Teruel □, *Spain* **34 E4** 40 48N 1 0W
Tervel, *Bulgaria* .... **43 D12** 43 45N 27 28 E
Tervola, *Finland* .... **12 C21** 66 6N 24 49 E
Teryaweyna L., *Australia* **116 B5** 32 18 S 143 22 E
Tešanj, *Bos.-H.* .... **42 C2** 44 38N 18 1 E
Teseney, *Eritrea* .... **95 D4** 15 5N 36 42 E
Tesha →, *Russia* .... **52 C6** 55 38N 42 9 E
Teshio, *Japan* .... **60 B10** 44 53N 141 44 E
Teshio-Gawa →, *Japan* . **60 B10** 44 53N 141 45 E
Tešica, *Serbia, Yug.* ..... **42 D6** 43 27N 21 45 E
Tesiyn Gol →, *Mongolia* **64 A4** 50 40N 93 20 E
Teslić, *Bos.-H.* .... **42 C2** 44 37N 17 54 E
Teslin, *Canada* **130 A2** 60 10N 132 43W
Teslin →, *Canada* **130 A2** 61 34N 134 35W
Teslin L., *Canada* **130 A2** 60 15N 132 57W
Tesouro, *Brazil* .... **157 D7** 16 4 S 53 34W
Tessalit, *Mali* .... **101 A5** 20 12N 1 0 E
Tessaoua, *Niger* .... **97 F1** 13 47N 7 56 E
Tessenderlo, *Belgium* .... **21 F6** 51 4N 5 5 E
Tessin, *Germany* .... **26 A8** 54 2N 12 26 E
Tessit, *Mali* .... **101 B5** 15 13N 0 18 E
Test →, *U.K.* **17 F6** 50 56N 1 29W
Testa del Gargano, *Italy* . **41 A9** 41 50N 16 10 E
Tét, *Hungary* .... **31 D10** 47 30N 17 33 E
Têt →, *France* .... **24 F7** 42 44N 3 2 E
Tetachuck L., *Canada* ... **130 C3** 53 18N 125 55W
Tetas, Pta., *Chile* .... **158 A1** 23 31 S 70 38W
Tete, *Mozam.* .... **107 F3** 16 13 S 33 33 E
Tete □, *Mozam.* .... **107 F3** 15 15 S 32 40 E
Teterev →, *Ukraine* .... **51 G6** 51 1N 30 5 E
Teteringen, *Neths.* ...... **21 E5** 51 37N 4 49 E
Teterow, *Germany* ...... **26 B8** 53 46N 12 34 E
Teteven, *Bulgaria* ...... **43 E9** 42 58N 24 17 E
Tethul →, *Canada* **130 A6** 60 35N 112 12W
Tetiyev, *Ukraine* .... **51 H5** 49 22N 29 38 E
Teton →, *U.S.A.* **142 C8** 47 56N 110 31W
Tétouan, *Morocco* **98 A3** 35 35N 5 21W
Tetovo, *Macedonia* ..... **42 E6** 42 1N 21 2 E
Tetuán = Tétouan,
Morocco **98 A3** 35 35N 5 21W
Tetyukhe Pristan, *Russia* . **60 B7** 44 22N 135 48 E
Tetyushi, *Russia* .... **52 C9** 54 55N 48 49 E
Teuco →, *Argentina* ... **158 B3** 25 35 S 60 11W
Teufen, *Switz.* .... **29 B8** 47 24N 9 23 E
Teulada, *Italy* .... **40 D1** 38 58N 8 46 E
Teulon, *Canada* **131 C9** 50 23N 97 16W
Teun, *Indonesia* .... **73 C3** 6 59 S 129 8 E
Teutoburger Wald,
Germany .... **26 C4** 52 5N 8 22 E
Tévere →, *Italy* .... **39 G9** 41 44N 12 14 E
Teverya, *Israel* .... **91 C4** 32 47N 35 32 E
Teviot →, *U.K.* **18 F6** 55 29N 2 38W
Tewantin, *Australia* ..... **115 D5** 26 27 S 153 3 E
Tewkesbury, *U.K.* **17 F5** 51 59N 2 9W
Texada I., *Canada* **130 D4** 49 40N 124 25W
Texarkana, *Ark., U.S.A.* . **139 J8** 33 26N 94 2W
Texarkana, *Tex., U.S.A.* . **139 J7** 33 26N 94 3W
Texas, *Australia* .... **115 D5** 28 49 S 151 9 E
Texas □, *U.S.A.* **139 K5** 31 40N 98 30W
Texas City, *U.S.A.* **139 L7** 29 24N 94 54W
Texel, *Neths.* .... **20 B5** 53 5N 4 50 E
Texhoma, *U.S.A.* **139 G4** 36 30N 101 47W
Texline, *U.S.A.* **139 G3** 36 23N 103 2W
Texoma, L., *U.S.A.* **139 J6** 33 50N 96 34W
Teykovo, *Russia* .... **52 B5** 56 55N 40 30 E
Teyvareh, *Afghan.* ...... **79 B2** 33 30N 64 24 E
Teza →, *Russia* .... **52 B5** 56 32N 41 53 E
Tezin, *Afghan.* .... **80 B3** 34 24N 69 30 E
Teziutlán, *Mexico* **147 D5** 19 50N 97 22W
Tezpur, *India* .... **78 B4** 26 40N 92 45 E
Tezzeron L., *Canada* **130 C4** 54 43N 124 30W
Tha-anne →, *Canada* ... **131 A10** 60 31N 94 37W
Tha Deua, *Laos* .... **76 D4** 17 57N 102 53 E
Tha Deua, *Laos* .... **76 C3** 19 26N 101 50 E
Tha Pla, *Thailand* ...... **76 D3** 17 48N 100 32 E
Tha Rua, *Thailand* ...... **76 E3** 14 34N 100 44 E
Tha Sala, *Thailand* ..... **77 H2** 8 40N 99 56 E
Tha Song Yang, *Thailand* **76 D1** 17 34N 97 55 E
Thaba Putsoa, *Lesotho* .. **105 D4** 29 45 S 28 0 E
Thabana Ntlenyana,
Lesotho .... **105 D4** 29 30 S 29 16 E
Thabazimbi, *S. Africa* ... **105 C4** 24 40 S 27 21 E
Thabeikkyin, *Burma* .... **78 D5** 22 53N 95 59 E
Thai Binh, *Vietnam* ..... **76 B6** 20 35N 106 1 E
Thai Hoa, *Vietnam* ..... **76 C5** 19 20N 105 20 E
Thai Muang, *Thailand* ... **77 H2** 8 24N 98 16 E
Thai Nguyen, *Vietnam* ... **76 B5** 21 35N 105 55 E
Thailand ■, *Asia* .... **76 E4** 16 0N 102 0 E
Thailand, G. of, *Asia* ... **77 G3** 11 30N 101 0 E
Thakhek, *Laos* .... **76 D5** 17 25N 104 45 E
Thakurgaon, *Bangla.* .... **78 B2** 26 2N 88 34 E
Thal, *Pakistan* .... **79 B3** 33 28N 70 33 E
Thal Desert, *Pakistan* ... **80 D4** 31 10N 71 30 E
Thala, *Tunisia* .... **96 A1** 35 35N 8 40 E
Thalabarivat, *Cambodia* . **76 F5** 13 33N 105 57 E
Thalkirch, *Switz.* .... **29 C8** 46 39N 9 17 E
Thallon, *Australia* ...... **115 D4** 28 39 S 148 49 E
Thalwil, *Switz.* .... **29 B7** 47 17N 8 35 E
Thamarīt, *Oman* .... **87 C6** 17 39N 54 2 E
Thame →, *U.K.* **17 F6** 51 39N 1 9W
Thames, *N.Z.* .... **118 D4** 37 7 S 175 34 E
Thames →, *Canada* **128 D3** 42 20N 82 25W
Thames →, *U.K.* .... **17 F8** 51 29N 0 34 E
Thames →, *U.S.A.* **137 E12** 41 18N 72 5W
Thames, Firth of, *N.Z.* .. **118 D4** 37 0 S 175 25 E
Thamesford, *Canada* **136 C3** 43 4N 81 0W
Thamesville, *Canada* **136 D3** 42 33N 81 59W
Thāmit, W. →, *Libya* .. **96 B3** 31 11N 16 8 E
Thamūd, *Yemen* .... **87 C5** 17 18N 49 55 E
Than Uyen, *Vietnam* .... **76 B4** 22 0N 103 54 E
Thanbyuzayat, *Burma* ... **78 G6** 15 58N 97 44 E
Thane, *India* .... **82 E1** 19 12N 72 59 E
Thanesar, *India* .... **80 D7** 30 1N 76 52 E
Thanet, I. of, *U.K.* ..... **17 F9** 51 21N 1 20 E
Thangoo, *Australia* ..... **112 C3** 18 10 S 122 22 E
Thangool, *Australia* ..... **114 C5** 24 38 S 150 42 E
Thanh Hoa, *Vietnam* .... **76 C5** 19 48N 105 46 E
Thanh Hung, *Vietnam* ... **77 H5** 9 55N 105 43 E
Thanh Pho Ho Chi
Minh = Phanh Bho Ho
Chi Minh, *Vietnam* .... **77 G6** 10 58N 106 40 E
Thanh Thuy, *Vietnam* ... **76 A5** 22 55N 104 51 E
Thanjavur, *India* .... **83 J4** 10 48N 79 12 E
Thann, *France* .... **23 E14** 47 48N 7 5 E
Thaon-les-Vosges, *France* **23 D13** 48 15N 6 24 E
Thap Sakae, *Thailand* ... **77 G2** 11 30N 99 37 E
Thap Than, *Thailand* .... **76 E2** 15 27N 99 54 E
Thar Desert, *India* ...... **80 F4** 28 0N 72 0 E
Tharad, *India* .... **80 G4** 24 30N 71 44 E
Thargomindah, *Australia* . **115 D3** 27 58 S 143 46 E
Tharrawaddy, *Burma* .... **78 G5** 17 38N 95 48 E
Tharrawaw, *Burma* ..... **78 G5** 17 41N 95 28 E
Tharthār, Mileh, *Iraq* ... **89 E10** 34 0N 43 15 E
Tharthār, W. ath →, *Iraq* **89 E10** 33 59N 43 12 E
Thasopoúla, *Greece* ..... **44 D6** 40 49N 24 45 E
Thásos, *Greece* .... **44 D6** 40 40N 24 40 E
That Khe, *Vietnam* ...... **76 A6** 22 16N 106 28 E
Thatcher, *Ariz., U.S.A.* . **143 K9** 32 51N 109 46W
Thatcher, *Colo., U.S.A.* . **139 G2** 37 33N 104 7W
Thaton, *Burma* .... **78 G6** 16 55N 97 22 E
Thau, Bassin de, *France* . **24 E7** 43 23N 3 36 E
Thaungdut, *Burma* ...... **78 C5** 24 30N 94 40 E
Thayer, *U.S.A.* **139 G9** 36 31N 91 33W
Thayetmyo, *Burma* ..... **78 F5** 19 20N 95 10 E
Thayngen, *Switz.* .... **29 A7** 47 49N 8 43 E
The Alberga →, *Australia* **115 D2** 27 6 S 135 33 E
The Bight, *Bahamas* **149 B4** 24 19N 75 24W
The Brothers, *Yemen* ... **87 D6** 12 8N 53 10 E
The Coorong, *Australia* .. **116 C3** 35 50 S 139 20 E
The Dalles, *U.S.A.* **142 D3** 45 36N 121 10W
The English Company's
Is., *Australia* .... **114 A2** 11 50 S 136 32 E
The Entrance, *Australia* . **117 B9** 33 21 S 151 30 E
The Frome →, *Australia* **115 D2** 29 8 S 137 54 E
The Grampians, *Australia* **116 D5** 37 0 S 142 20 E
The Great Divide = Great
Dividing Ra., *Australia* **114 C4** 23 0 S 146 0 E
The Hague = 's-
Gravenhage, *Neths.* ... **20 D4** 52 7N 4 17 E
The Hamilton →,
Australia .... **115 D2** 26 40 S 135 19 E
The Hunter Hills, *N.Z.* .. **119 E5** 44 26 S 170 46 E
The Macumba →,
Australia .... **115 D2** 27 52 S 137 12 E
The Neales →, *Australia* **115 D2** 28 8 S 136 47 E
The Oaks, *Australia* ..... **117 C9** 34 3 S 150 34 E
The Officer →, *Australia* **113 E5** 27 46 S 132 30 E
The Pas, *Canada* **131 C8** 53 45N 101 15W
The Range, *Zimbabwe* .. **107 F3** 19 2 S 31 2 E
The Remarkables, *N.Z.* .. **119 F3** 45 10 S 168 50 E
The Rock, *Australia* ..... **115 F4** 35 15 S 147 2 E
The Salt L., *Australia* ... **115 E3** 30 6 S 142 8 E
The Stevenson →,
Australia .... **115 D2** 27 6 S 135 33 E
The Warburton →,
Australia .... **115 D2** 28 4 S 137 28 E
Thebes = Thívai, *Greece* **45 F5** 38 19N 23 19 E
Thebes, *Egypt* .... **94 B3** 25 40N 32 35 E
Thedford, *Canada* **136 C3** 43 9N 81 51W
Thedford, *U.S.A.* **138 E4** 41 59N 100 35W
Theebine, *Australia* ..... **115 D5** 25 57 S 152 34 E
Thekulthili L., *Canada* .. **131 A7** 61 3N 110 0W
Thelon →, *Canada* **131 A8** 62 35N 104 3W
Thénezay, *France* **22 F6** 46 44N 0 2W
Thenia, *Algeria* .... **99 A5** 36 44N 3 33 E
Thenon, *France* .... **24 C5** 45 9N 1 4 E
Theodore, *Australia* ..... **114 C5** 24 55 S 150 3 E
Thepha, *Thailand* ...... **77 J3** 6 52N 100 58 E
Thérain →, *France* ..... **23 C9** 49 15N 2 27 E
Theresa, *U.S.A.* **137 B9** 44 13N 75 48W
Thermaïkós Kólpos,
Greece .... **44 D4** 40 15N 22 45 E
Thermopolis, *U.S.A.* **142 E9** 43 39N 108 13W
Thermopylae P., *Greece* . **45 F4** 38 48N 22 35 E
Thesprotía □, *Greece* ... **44 E2** 39 27N 20 22 E
Thessalía □, *Greece* .... **44 E3** 39 25N 21 50 E
Thessalon, *Canada* **128 C3** 46 20N 83 30W
Thessaloníki, *Greece* .... **44 D4** 40 38N 22 58 E
Thessaloníki □, *Greece* .. **44 D5** 40 45N 23 0 E
Thessaloniki, Gulf of =
Thermaïkós Kólpos,
Greece .... **44 D4** 40 15N 22 45 E
Thessaly = Thessalía □,
Greece .... **44 E3** 39 25N 21 50 E
Thetford, *U.K.* .... **17 E8** 52 25N 0 45 E
Thetford Mines, *Canada* . **129 C5** 46 8N 71 18W
Theun →, *Laos* .... **76 C5** 18 19N 104 0 E
Theunissen, *S. Africa* ... **104 D4** 28 26 S 26 43 E
Theux, *Belgium* .... **21 G7** 50 32N 5 49 E
Thevenard, *Australia* .... **115 E1** 32 9 S 133 38 E
Thiámis →, *Greece* ..... **44 E2** 39 15N 20 6 E
Thiberville, *France* ..... **22 C7** 49 8N 0 27 E
Thibodaux, *U.S.A.* **139 L9** 29 48N 90 49W
Thicket Portage, *Canada* . **131 B9** 55 19N 97 42W
Thief River Falls, *U.S.A.* . **138 A6** 48 7N 96 10W
Thiel Mts., *Antarctica* ... **7 E16** 85 15 S 91 0W
Thiene, *Italy* .... **39 C8** 45 42N 11 29 E
Thiérache, *France* ...... **23 C10** 49 51N 3 45 E
Thiers, *France* .... **24 C7** 45 52N 3 33 E
Thies, *Senegal* **100 C1** 14 50N 16 51W
Thiet, *Sudan* .... **95 F2** 7 37N 28 49 E
Thika, *Kenya* .... **106 C4** 1 1 S 37 5 E
Thille-Boubacar, *Senegal* . **100 B1** 16 31N 15 5W
Thimphu, *Bhutan* .... **78 B2** 27 31N 89 45 E
þingvallavatn, *Iceland* ... **12 D3** 64 11N 21 9W
Thio, *N. Cal.* .... **121 U20** 21 37 S 166 14 E
Thionville, *France* ...... **23 C13** 49 20N 6 10 E
Thíra, *Greece* .... **45 H7** 36 23N 25 27 E
Thirasía, *Greece* .... **45 H7** 36 26N 25 21 E
Thirsk, *U.K.* **16 C6** 54 14N 1 19W
Thiruvarur, *India* .... **83 J4** 10 46N 79 38 E
Thisted, *Denmark* ...... **13 H13** 56 58N 8 40 E
Thistle I., *Australia* ..... **116 C2** 35 0 S 136 8 E
Thitgy, *Burma* .... **78 F6** 18 15N 96 13 E
Thithia, *Fiji* .... **121 A3** 17 45 S 179 18W
Thitpokpin, *Burma* ..... **78 F5** 19 24N 95 58 E
Thívai, *Greece* .... **45 F5** 38 19N 23 19 E
Thiviers, *France* .... **24 C4** 45 25N 0 54 E
Thizy, *France* .... **25 B8** 46 2N 4 18 E
þjórsá →, *Iceland* **12 E3** 63 47N 20 48W
Thlewiaza →, *Man.,*
Canada **131 B8** 59 43N 100 5W
Thlewiaza →, *N.W.T.,*
Canada **131 A10** 60 29N 94 40W
Thmar Puok, *Cambodia* . **76 F4** 13 57N 103 4 E

Tower 275

Tobago, *W. Indies* **149 D7** 11 10N 60 30W
Tobarra, *Spain* **35 G3** 38 37N 1 44W
Tobelo, *Indonesia* **72 A3** 1 45N 127 56 E
Tobermorey, *Australia* **114 C2** 22 12 S 138 0 E
Tobermory, *Canada* **128 C3** 45 12N 81 40W
Tobermory, *U.K.* **18 E2** 56 38N 6 5W
Tobin, *U.S.A.* **144 F5** 39 55N 121 19W
Tobin, L., *Australia* **112 D4** 21 45 S 125 49 E
Tobin L., *Canada* **131 C8** 53 35N 103 30W
Toblach = Dobbiaco, *Italy* **39 B9** 46 44N 12 14 E
Toboali, *Indonesia* **74 C3** 3 0 S 106 25 E
Tobol, *Kazakhstan* **54 E9** 52 40N 62 39 E
Tobol →, *Russia* **56 D7** 58 10N 68 12 E
Toboli, *Indonesia* **72 B2** 0 38 S 120 5 E
Tobolsk, *Russia* **56 D7** 58 15N 68 10 E
Toboso, *Phil.* **71 F4** 10 43N 123 31 E
Tobruk = Tubruq, *Libya* **96 B4** 32 7N 23 55 E
Tobyhanna, *U.S.A.* **137 E9** 41 11N 75 25W
Tobyl = Tobol →, *Russia* **56 D7** 58 10N 68 12 E
Tocache Nuevo, *Peru* **156 B2** 8 9 S 76 26W
Tocantínia, *Brazil* **154 C2** 9 33 S 48 22W
Tocantinópolis, *Brazil* **154 C2** 6 20 S 47 25W
Tocantins □, *Brazil* **154 D2** 10 0 S 48 0W
Tocantins →, *Brazil* **154 B2** 1 45 S 49 10W
Toccoa, *U.S.A.* **135 H4** 34 35N 83 19W
Toce →, *Italy* **38 C5** 45 56N 8 29 E
Tochigi, *Japan* **63 A11** 36 25N 139 45 E
Tochigi □, *Japan* **63 A11** 36 45N 139 45 E
Tocina, *Spain* **37 H5** 37 37N 5 44W
Tocopilla, *Chile* **158 A1** 22 5 S 70 10W
Tocumwal, *Australia* **117 C6** 35 51 S 145 31 E
Tocuyo →, *Venezuela* **152 A4** 11 3N 68 23W
Tocuyo de la Costa, *Venezuela* **152 A4** 11 2N 68 23W
Todd →, *Australia* **114 C2** 24 52 S 135 48 E
Todeli, *Indonesia* **72 B2** 1 38 S 124 34 E
Todenyang, *Kenya* **106 B4** 4 35N 35 56 E
Todi, *Italy* **39 F9** 42 47N 12 24 E
Tödi, *Switz.* **29 C7** 46 48N 8 55 E
Todos os Santos, B. de, *Brazil* **155 D4** 12 48 S 38 38W
Todos Santos, *Mexico* **146 C2** 23 27N 110 13W
Todtnau, *Germany* **27 H3** 47 49N 7 56 E
Toecé, *Burkina Faso* **101 C4** 11 50N 1 16W
Toetoes B., *N.Z.* **119 G3** 46 42 S 168 41 E
Tofield, *Canada* **130 C6** 53 25N 112 40W
Tofino, *Canada* **130 D3** 49 11N 125 55W
Töfsingdalens nationalpark, *Sweden* **14 B6** 62 15N 12 44 E
Toftlund, *Denmark* **15 J3** 55 11N 9 2 E
Tofua, *Tonga* **121 P13** 19 45 S 175 5W
Toga, *Vanuatu* **121 C4** 13 26 S 166 42 E
Tōgane, *Japan* **63 B12** 35 33N 140 22 E
Togba, *Mauritania* **100 B2** 17 26N 10 12W
Togbo, *C.A.R.* **102 A3** 6 0N 17 27 E
Toggenburg, *Switz.* **29 B8** 47 16N 9 9 E
Togian, Kepulauan, *Indonesia* **72 B2** 0 20 S 121 50 E
Togliatti, *Russia* **52 D9** 53 32N 49 24 E
Togo ■, *W. Afr.* **101 D5** 8 30N 1 35 E
Togtoh, *China* **66 D6** 40 15N 111 10 E
Toguzak →, *Kazakhstan* **54 D9** 54 3N 62 44 E
Tohma →, *Turkey* **88 C7** 38 29N 38 23 E
Tōhoku □, *Japan* **60 E10** 39 50N 141 45 E
Toi, *Japan* **63 C10** 34 54N 138 47 E
Toinya, *Sudan* **95 F2** 6 17N 29 46 E
Tojikiston = Tajikistan ■, *Asia* **55 D5** 38 30N 70 0 E
Tojo, *Indonesia* **72 B2** 1 20 S 121 15 E
Tōjō, *Japan* **62 C5** 34 53N 133 16 E
Tok →, *Russia* **54 E4** 52 46N 52 22 E
Toka, *Guyana* **153 C6** 3 58N 59 17W
Tokaanu, *N.Z.* **118 E4** 38 58 S 175 46 E
Tokachi-Dake, *Japan* **60 C11** 43 17N 142 5 E
Tokachi-Gawa →, *Japan* **60 C11** 42 44N 143 42 E
Tokai, *Japan* **63 B8** 35 2N 136 55 E
Tokaj, *Hungary* **31 C14** 48 8N 21 27 E
Tokala, *Indonesia* **72 B2** 1 30 S 121 40 E
Tōkamachi, *Japan* **61 F9** 37 8N 138 43 E
Tokanui, *N.Z.* **119 G3** 46 34 S 168 56 E
Tokar, *Sudan* **94 D4** 18 27N 37 56 E
Tokara-Rettō, *Japan* **61 K4** 29 37N 129 43 E
Tokarahi, *N.Z.* **119 E5** 44 56 S 170 39 E
Tokashiki-Shima, *Japan* **61 L3** 26 11N 127 21 E
Tokat, *Turkey* **88 B7** 40 22N 36 35 E
Tŏkchŏn, *N. Korea* **67 E14** 39 45N 126 18 E
Tokeland, *U.S.A.* **144 D3** 46 42N 123 59W
Tokelau Is., *Pac. Oc.* **122 H10** 9 0 S 171 45W
Toki, *Japan* **63 B9** 35 18N 137 8 E
Tokmak, *Kyrgyzstan* **55 B7** 42 49N 75 15 E
Tokmak, *Ukraine* **51 J8** 47 16N 35 42 E
Toko Ra., *Australia* **114 C2** 23 5 S 138 20 E
Tokomaru Bay, *N.Z.* **118 E7** 38 8 S 178 22 E
Tokoname, *Japan* **63 C8** 34 53N 136 51 E
Tokoro-Gawa →, *Japan* **60 B12** 44 7N 144 5 E
Tokoroa, *N.Z.* **118 E4** 38 13 S 175 50 E
Tokorozawa, *Japan* **63 B11** 35 47N 139 28 E
Toktogul, *Kyrgyzstan* **55 C6** 41 50N 72 50 E
Toku, *Tonga* **121 P13** 18 10 S 174 11W
Tokuji, *Japan* **62 C3** 34 11N 131 42 E
Tokuno-Shima, *Japan* **61 L4** 27 56N 128 55 E
Tokushima, *Japan* **62 C6** 34 4N 134 34 E
Tokushima □, *Japan* **62 D6** 33 55N 134 0 E
Tokuyama, *Japan* **62 C3** 34 3N 131 50 E
Tōkyō, *Japan* **63 B11** 35 45N 139 45 E
Tōkyō □, *Japan* **63 B11** 35 40N 139 30 E
Tōkyō-Wan, *Japan* **63 B11** 35 25N 139 47 E
Tokzār, *Afghan.* **79 B2** 35 52N 66 26 E
Tolaga Bay, *N.Z.* **118 E7** 38 21 S 178 20 E
Tolbukhin = Dobrich, *Bulgaria* **43 D12** 43 37N 27 49 E
Toledo, *Phil.* **71 F4** 10 23N 123 38 E
Toledo, *Spain* **36 F6** 39 50N 4 2W
Toledo, *Ill., U.S.A.* **141 E8** 39 16N 88 15W
Toledo, *Iowa, U.S.A.* **140 B4** 42 0N 92 35W
Toledo, *Ohio, U.S.A.* **141 C13** 41 39N 83 33W
Toledo, *Oreg., U.S.A.* **142 D2** 44 37N 123 56W
Toledo, *Wash., U.S.A.* **142 C2** 46 26N 122 51W
Toledo, Montes de, *Spain* **37 F6** 39 33N 4 20W
Tolentino, *Italy* **39 E10** 43 12N 13 17 E
Tolga, *Algeria* **99 B6** 34 40N 5 22 E
Tolga, *Norway* **14 B5** 62 26N 11 1 E
Toliara, *Madag.* **105 C7** 23 21 S 43 40 E
Toliara □, *Madag.* **105 C8** 21 0 S 45 0 E
Tolima, *Colombia* **152 C2** 4 40N 75 19W
Tolima □, *Colombia* **152 C2** 3 45N 75 15W
Tolitoli, *Indonesia* **72 A2** 1 5N 120 50 E
Tolkamer, *Neths.* **20 E8** 51 52N 6 6 E
Tolkmicko, *Poland* **47 A6** 54 19N 19 31 E
Tolleson, *U.S.A.* **143 K7** 33 27N 112 16W
Tollhouse, *U.S.A.* **144 H7** 37 1N 119 24W
Tolmachevo, *Russia* **50 C5** 58 56N 29 51 E
Tolmezzo, *Italy* **39 B10** 46 24N 13 1 E
Tolmin, *Slovenia* **39 B10** 46 11N 13 45 E
Tolna, *Hungary* **31 E11** 46 25N 18 48 E
Tolna □, *Hungary* **31 E11** 46 30N 18 30 E
Tolo, *Zaïre* **102 C3** 2 55 S 18 34 E
Tolo, Teluk, *Indonesia* **72 B2** 2 20 S 122 10 E
Tolochin = Talachyn, *Belarus* **50 E5** 54 25N 29 42 E
Tolong Bay, *Phil.* **71 G4** 9 20N 122 49 E
Tolono, *U.S.A.* **141 E8** 39 59N 88 16W
Tolosa, *Spain* **34 B2** 43 8N 2 5W
Tolox, *Spain* **37 J6** 36 41N 4 54W
Toltén, *Chile* **160 A2** 39 13 S 74 14W
Toluca, *Mexico* **147 D5** 19 20N 99 40W
Tolybay, *Kazakhstan* **54 F9** 50 31N 62 19 E
Tom Burke, *S. Africa* **105 C4** 23 5 S 28 0 E
Tom Price, *Australia* **112 D2** 22 40 S 117 48 E
Tomah, *U.S.A.* **138 D9** 43 59N 90 30W
Tomahawk, *U.S.A.* **138 C10** 45 28N 89 44W
Tomakomai, *Japan* **60 C10** 42 38N 141 36 E
Tomales, *U.S.A.* **144 G4** 38 15N 122 53W
Tomales B., *U.S.A.* **144 G3** 38 15N 123 58W
Tomanlivi, *Fiji* **121 A2** 17 37 S 178 1 E
Tomar, *Portugal* **37 F2** 39 36N 8 25W
Tómaros Óros, *Greece* **44 E2** 39 29N 20 48 E
Tomarza, *Turkey* **88 C6** 38 27N 35 48 E
Tomás Barrón, *Bolivia* **156 D4** 17 35 S 67 31W
Tomaszów Mazowiecki, *Poland* **47 D6** 51 30N 19 57 E
Tomatlán, *Mexico* **146 D3** 19 56N 105 15W
Tombador, Serra do, *Brazil* **157 C6** 12 0 S 58 0W
Tombé, *Sudan* **95 F3** 5 53N 31 40 E
Tombigbee →, *U.S.A.* **135 K2** 31 8N 87 57W
Tombôco, *Angola* **103 D2** 6 48 S 13 18 E
Tombouctou, *Mali* **100 B4** 16 50N 3 0W
Tombstone, *U.S.A.* **143 L8** 31 43N 110 4W
Tombua, *Angola* **103 F2** 15 55 S 11 55 E
Tomé, *Chile* **158 D1** 36 36 S 72 57W
Tomé-Açu, *Brazil* **154 B2** 2 25 S 48 9W
Tomelilla, *Sweden* **15 J7** 55 33N 13 58 E
Tomelloso, *Spain* **35 F1** 39 10N 3 2W
Tomingley, *Australia* **117 B8** 32 26 S 148 16 E
Tomini, *Indonesia* **72 A2** 0 30N 120 30 E
Tomini, Teluk, *Indonesia* **72 B2** 0 10 S 122 0 E
Tominian, *Mali* **100 C4** 13 17N 4 35W
Tomiño, *Spain* **36 D2** 41 59N 8 46W
Tomkinson Ras., *Australia* **113 E4** 26 11 S 129 5 E
Tommot, *Russia* **57 D13** 59 4N 126 20 E
Tomnavoulin, *U.K.* **18 D5** 57 19N 3 19W
Tomnop Ta Suos, *Cambodia* **77 G5** 11 20N 104 15 E
Tomo, *Colombia* **152 C4** 2 38N 67 32W
Tomo, *Japan* **62 C5** 34 23N 133 23 E
Tomo →, *Colombia* **152 B4** 5 20N 67 48W
Tomobe, *Japan* **63 A12** 36 20N 140 20 E
Tomorit, *Albania* **44 D2** 40 42N 20 11 E
Toms Place, *U.S.A.* **144 H8** 37 34N 118 41W
Toms River, *U.S.A.* **137 G10** 39 58N 74 12W
Tomsk, *Russia* **56 D9** 56 30N 85 5 E
Tonalá, *Mexico* **147 D6** 16 8N 93 41W
Tonale, Passo del, *Italy* **38 B7** 46 16N 10 35 E
Tonalea, *U.S.A.* **143 H8** 36 19N 110 56W
Tonami, *Japan* **63 A8** 36 40N 136 58 E
Tonantins, *Brazil* **152 D4** 2 45 S 67 45W
Tonasket, *U.S.A.* **142 B4** 48 42N 119 26W
Tonate, *Fr. Guiana* **153 C7** 5 0N 52 28W
Tonawanda, *U.S.A.* **136 D6** 43 1N 78 53W
Tonbridge, *U.K.* **17 F8** 51 11N 0 17 E
Tondano, *Indonesia* **72 A2** 1 35N 124 54 E
Tondela, *Portugal* **36 E2** 40 31N 8 5W
Tønder, *Denmark* **15 K2** 54 58N 8 50 E
Tondi, *India* **83 K4** 9 45N 79 4 E
Tondi Kiwindi, *Niger* **101 C5** 14 28N 2 2 E
Tondibi, *Mali* **101 B4** 16 39N 0 14W
Tonekābon, *Iran* **85 B6** 36 45N 51 12 E
Tong Xian, *China* **66 E9** 39 55N 116 35 E
Tonga ■, *Pac. Oc.* **121 P13** 19 50 S 174 30W
Tonga Trench, *Pac. Oc.* **122 J10** 18 0 S 173 0W
Tongaat, *S. Africa* **105 D5** 29 33 S 31 9 E
Tongala, *Australia* **117 D6** 36 14 S 144 56 E
Tong'an, *China* **69 E12** 24 37N 118 8 E
Tongareva, *Cook Is.* **123 H12** 9 0 S 158 0W
Tongatapu, *Tonga* **121 Q14** 21 10 S 174 0W
Tongatapu Group, *Tonga* **121 Q13** 21 0 S 175 0W
Tongbai, *China* **69 A9** 32 20N 113 23 E
Tongcheng, *Anhui, China* **69 B11** 31 4N 116 56 E
Tongcheng, *Hubei, China* **69 C9** 29 15N 113 50 E
Tongchŏn-ni, *N. Korea* **67 E14** 39 50N 127 25 E
Tongchuan, *China* **66 G5** 35 6N 109 3 E
Tongdao, *China* **68 D7** 26 10N 109 42 E
Tongeren, *Belgium* **21 G6** 50 47N 5 28 E
Tonggu, *China* **69 C10** 28 31N 114 20 E
Tongguan, *China* **66 G6** 34 40N 110 25 E
Tonghai, *China* **68 E4** 24 10N 102 53 E
Tonghua, *China* **67 D13** 41 42N 125 58 E
Tongjiang, *China* **68 B6** 31 58N 107 11 E
Tongjosŏn Man, *N. Korea* **67 E14** 39 30N 128 0 E
Tongking, G. of = Tonkin, G. of, *Asia* **64 E5** 20 0N 108 0 E
Tongliang, *China* **68 C6** 29 50N 106 3 E
Tongliao, *China* **67 C12** 43 38N 122 18 E
Tongling, *China* **69 B11** 30 55N 117 48 E
Tonglu, *China* **69 C12** 29 45N 119 37 E
Tongnae, *S. Korea* **67 G15** 35 12N 129 5 E
Tongnan, *China* **68 B5** 30 9N 105 50 E
Tongoa, *Vanuatu* **121 F6** 16 54 S 168 34 E
Tongobory, *Madag.* **105 C7** 23 32 S 44 20 E
Tongoy, *Chile* **158 C1** 30 16 S 71 31W
Tongren, *China* **68 D7** 27 43N 109 11 E
Tongres = Tongeren, *Belgium* **21 G6** 50 47N 5 28 E
Tongsa Dzong, *Bhutan* **78 B3** 27 31N 90 31 E
Tongue, *U.K.* **18 C4** 58 29N 4 25W
Tongue →, *U.S.A.* **138 B2** 46 25N 105 52W
Tongwei, *China* **66 G3** 35 0N 105 5 E
Tongxin, *China* **66 F3** 36 59N 105 58 E
Tongyang, *N. Korea* **67 E14** 39 9N 126 53 E
Tongyu, *China* **67 B12** 44 45N 123 4 E
Tongzi, *China* **68 C6** 28 9N 106 49 E
Tonica, *U.S.A.* **140 C7** 41 13N 89 4W
Tonj, *Sudan* **95 F2** 7 20N 28 44 E
Tonk, *India* **80 F6** 26 6N 75 54 E
Tonkawa, *U.S.A.* **139 G6** 36 41N 97 18W
Tonkin = Bac Phan, *Vietnam* **76 B5** 22 0N 105 0 E
Tonkin, G. of, *Asia* **64 E5** 20 0N 108 0 E
Tonlé Sap, *Cambodia* **76 F4** 13 0N 104 0 E
Tonnay-Charente, *France* **24 C3** 45 56N 0 55W
Tonneins, *France* **24 D4** 44 23N 0 19 E
Tonnerre, *France* **23 E10** 47 51N 3 59 E
Tönning, *Germany* **26 A4** 54 19N 8 56 E
Tono, *Japan* **60 E10** 39 19N 141 32 E
Tonopah, *U.S.A.* **143 G5** 38 4N 117 14W
Tonoshō, *Japan* **62 C6** 34 29N 134 11 E
Tonosí, *Panama* **148 E3** 7 20N 80 20W
Tønsberg, *Norway* **14 E4** 59 19N 10 25 E
Tonumea, *Tonga* **121 Q13** 20 30 S 174 30W
Tonya, *Turkey* **89 B8** 40 53N 39 16 E
Tonzang, *Burma* **78 D4** 23 36N 93 42 E
Tonzi, *Burma* **78 C5** 24 39N 94 57 E
Tooele, *U.S.A.* **142 F7** 40 32N 112 18W
Toolondo, *Australia* **116 D4** 36 58 S 141 58 E
Toompine, *Australia* **115 D3** 27 15 S 144 19 E
Toongi, *Australia* **117 B8** 32 28 S 148 30 E
Toonpan, *Australia* **114 B4** 19 28 S 146 48 E
Toora, *Australia* **117 E7** 38 39 S 146 23 E
Toora-Khem, *Russia* **57 D10** 52 28N 96 17 E
Toowoomba, *Australia* **115 D5** 27 32 S 151 56 E
Top-ozero, *Russia* **48 A5** 65 35N 32 0 E
Topalu, *Romania* **46 E9** 44 31N 28 3 E
Topaz, *U.S.A.* **144 G7** 38 41N 119 30W
Topeka, *U.S.A.* **138 F7** 39 3N 95 40W
Topki, *Russia* **56 D9** 55 20N 85 35 E
Topl'a →, *Slovak Rep.* **31 C14** 48 45N 21 45 E
Topley, *Canada* **130 C3** 54 49N 126 18W
Toplica →, *Serbia, Yug.* **42 D6** 43 15N 21 49 E
Toplița, *Romania* **46 C6** 46 55N 25 20 E
Topocalma, Pta., *Chile* **158 C1** 34 10 S 72 2W
Topock, *U.S.A.* **145 L12** 34 46N 114 29W
Topola, *Serbia, Yug.* **42 C5** 44 17N 20 41 E
Topolčane, *Macedonia* **42 F6** 41 14N 21 56 E
Topol'čany, *Slovak Rep.* **31 C11** 48 35N 18 12 E
Topolnitsa →, *Bulgaria* **43 E9** 42 11N 24 18 E
Topolobampo, *Mexico* **146 B3** 25 40N 109 4W
Topolovgrad, *Bulgaria* **43 E11** 42 5N 26 20 E
Topolvătu Mare, *Romania* **46 D2** 45 46N 21 41 E
Toppenish, *U.S.A.* **142 C3** 46 23N 120 19W
Topusko, *Croatia* **39 C12** 45 18N 15 59 E
Toquepala, *Peru* **156 D3** 17 24 S 70 25W
Torá, *Spain* **34 D6** 41 49N 1 25 E
Tora Kit, *Sudan* **95 E3** 11 2N 32 36 E
Toraka Vestale, *Madag.* **105 B7** 16 20 S 43 58 E
Torata, *Peru* **156 D3** 17 23 S 70 1W
Torbalı, *Turkey* **88 C2** 38 10N 27 21 E
Torbay, *Canada* **129 C9** 47 40N 52 42W
Torbay, *U.K.* **17 G4** 50 26N 3 31W
Tørdal, *Norway* **14 E2** 59 10N 8 45 E
Tordesillas, *Spain* **36 D6** 41 30N 5 0W
Tordoya, *Spain* **36 B2** 43 6N 8 36W
Töreboda, *Sweden* **15 F8** 58 41N 14 7 E
Torgau, *Germany* **26 D8** 51 34N 13 0 E
Torgelow, *Germany* **26 B9** 53 37N 13 59 E
Torhout, *Belgium* **21 F2** 51 5N 3 7 E
Tori, *Ethiopia* **95 F3** 7 53N 33 35 E
Tori-Shima, *Japan* **61 J10** 30 29N 140 19 E
Torigni-sur-Vire, *France* **22 C6** 49 3N 0 58W
Torija, *Spain* **34 E1** 40 44N 3 2W
Torin, *Mexico* **146 B2** 27 33N 110 15W
Toriñana, C., *Spain* **36 B1** 43 3N 9 17W
Torino, *Italy* **38 C4** 45 3N 7 40 E
Torit, *Sudan* **95 G3** 4 27N 32 31 E
Torkamān, *Iran* **89 D12** 37 35N 47 23 E
Torkovichi, *Russia* **50 C6** 58 51N 30 21 E
Tormac, *Romania* **42 B6** 45 30N 21 30 E
Tormes →, *Spain* **36 D4** 41 18N 6 29W
Tornado Mt., *Canada* **130 D6** 49 55N 114 40W
Tornala, *Slovak Rep.* **31 C13** 48 25N 20 20 E
Torne älv →, *Sweden* **12 D21** 65 50N 24 12 E
Torneå = Tornio, *Finland* **12 D21** 65 50N 24 12 E
Torneträsk, *Sweden* **12 B18** 68 24N 19 15 E
Tornio, *Finland* **12 D21** 65 50N 24 12 E
Tornionjoki →, *Finland* **12 D21** 65 50N 24 12 E
Tornquist, *Argentina* **158 D3** 38 8 S 62 15W
Toro, *Baleares, Spain* **33 B11** 39 59N 4 8 E
Toro, *Zamora, Spain* **36 D5** 41 35N 5 24W
Torö, *Sweden* **15 F11** 58 48N 17 50 E
Toro, Cerro del, *Chile* **158 B2** 29 10 S 69 50W
Toro Pk., *U.S.A.* **145 M10** 33 34N 116 24W
Törökszentmiklós, *Hungary* **31 D13** 47 11N 20 27 E
Toronaíos Kólpos, *Greece* **44 D5** 40 5N 23 30 E
Toronto, *Australia* **117 B9** 33 0 S 151 30 E
Toronto, *Canada* **128 D4** 43 39N 79 20W
Toronto, *U.S.A.* **136 F4** 40 28N 80 36W
Toropets, *Russia* **50 D6** 56 30N 31 40 E
Tororo, *Uganda* **106 B3** 0 45N 34 12 E
Toros Dağları, *Turkey* **88 D5** 37 0N 32 30 E
Torotoro, *Bolivia* **157 D4** 18 7 S 65 46W
Torpshammar, *Sweden* **14 B10** 62 29N 16 20 E
Torquay, *Australia* **116 E6** 38 20 S 144 19 E
Torquay, *Canada* **131 D8** 49 9N 103 30W
Torquay, *U.K.* **17 G4** 50 27N 3 32W
Torquemada, *Spain* **36 C6** 42 2N 4 19W
Torralba de Calatrava, *Spain* **37 F7** 39 1N 3 44W
Torrance, *U.S.A.* **145 M8** 33 50N 118 19W
Torrão, *Portugal* **37 G2** 38 16N 8 11W
Torre Annunziata, *Italy* **41 B7** 40 45N 14 27 E
Tôrre de Moncorvo, *Portugal* **36 D3** 41 12N 7 8W
Torre del Greco, *Italy* **41 B7** 40 47N 14 22 E
Torre del Mar, *Spain* **37 J6** 36 44N 4 6W
Torre-Pacheco, *Spain* **35 H4** 37 44N 0 57W
Torre Péllice, *Italy* **38 D4** 44 49N 7 13 E
Torreblanca, *Spain* **34 E5** 40 14N 0 12 E
Torrecampo, *Spain* **37 G6** 38 29N 4 41W
Torrecilla en Cameros, *Spain* **34 C2** 42 15N 2 38W
Torredembarra, *Spain* **34 D6** 41 9N 1 24 E
Torredonjimeno, *Spain* **37 H7** 37 46N 3 57W
Torrejoncillo, *Spain* **36 F4** 39 54N 6 28W
Torrelaguna, *Spain* **34 E1** 40 50N 3 38W
Torrelavega, *Spain* **36 B6** 43 20N 4 5W
Torremaggiore, *Italy* **41 A8** 41 41N 15 17 E
Torremolinos, *Spain* **37 J6** 36 38N 4 30W
Torrens, L., *Australia* **116 A2** 31 0 S 137 50 E
Torrens Cr. →, *Australia* **114 C4** 22 23 S 145 9 E
Torrens Creek, *Australia* **114 C4** 20 48 S 145 3 E
Torrente, *Spain* **35 F4** 39 27N 0 28W
Torrenueva, *Spain* **35 G1** 38 38N 3 22W
Torreón, *Mexico* **146 B4** 25 33N 103 26W
Torreperogil, *Spain* **35 G1** 38 2N 3 17W
Torres, *Mexico* **146 B2** 28 46N 110 47W
Torres, Is., *Vanuatu* **121 C4** 13 15 S 166 37 E
Torres Novas, *Portugal* **37 F2** 39 27N 8 33W
Torres Strait, *Australia* **120 E2** 9 50 S 142 20 E
Torres Vedras, *Portugal* **37 F1** 39 5N 9 15W
Torrevieja, *Spain* **35 H4** 37 59N 0 42W
Torrey, *U.S.A.* **143 G8** 38 18N 111 25W
Torridge →, *U.K.* **17 G3** 51 0N 4 13W
Torridon, L., *U.K.* **18 D3** 57 35N 5 50W
Torrijos, *Phil.* **70 E4** 13 19N 122 5 E
Torrijos, *Spain* **36 F6** 39 59N 4 18W
Torrington, *Conn., U.S.A.* **137 E11** 41 48N 73 7W
Torrington, *Wyo., U.S.A.* **138 D2** 42 4N 104 11W
Torroella de Montgri, *Spain* **34 C8** 42 2N 3 8 E
Torrox, *Spain* **37 J7** 36 46N 3 57W
Tórshavn, *Færoe Is.* **12 E9** 62 5N 6 56W
Torsö, *Sweden* **15 F7** 58 48N 13 45 E
Tortola, *Virgin Is.* **149 C7** 18 19N 64 45W
Tórtoles de Esgueva, *Spain* **36 D6** 41 49N 4 2W
Tortona, *Italy* **38 D5** 44 54N 8 52 E
Tortoreto, *Italy* **39 F10** 42 48N 13 55 E
Tortorici, *Italy* **41 D7** 38 2N 14 49 E
Tortosa, *Spain* **34 E5** 40 49N 0 31 E
Tortosa, C. de, *Spain* **34 E5** 40 41N 0 52 E
Tortosendo, *Portugal* **36 E3** 40 15N 7 31W
Tortue, I. de la, *Haiti* **149 B5** 20 5N 72 57W
Tortum, *Turkey* **89 B9** 40 19N 41 35 E
Ţorūd, *Iran* **85 C7** 35 25N 55 5 E
Torugart, Pereval = Turugart, Pereval, *Kyrgyzstan* **55 C7** 40 32N 75 24 E
Torul, *Turkey* **89 B8** 40 34N 39 18 E
Toruń, *Poland* **47 B5** 53 2N 18 39 E
Toruń □, *Poland* **47 B6** 53 20N 19 0 E
Torup, *Denmark* **15 G3** 57 5N 9 5 E
Torup, *Sweden* **15 H7** 56 57N 13 5 E
Tory I., *Ireland* **19 A3** 55 16N 8 14W
Torysa →, *Slovak Rep.* **31 C14** 48 39N 21 21 E
Torzhok, *Russia* **52 B2** 57 5N 34 55 E
Tosa, *Japan* **62 D5** 33 24N 133 23 E
Tosa-Shimizu, *Japan* **62 E4** 32 52N 132 58 E
Tosa-Wan, *Japan* **62 D5** 33 15N 133 30 E
Tosa-yamada, *Japan* **62 D5** 33 36N 133 38 E
Toscana □, *Italy* **38 E8** 43 25N 11 0 E
Toscano, Arcipelago, *Italy* **38 F7** 42 30N 10 30 E
Toshkent, *Uzbekistan* **55 C4** 41 20N 69 10 E
Tosno, *Russia* **50 C6** 59 38N 30 46 E
Tossa, *Spain* **34 D7** 41 43N 2 56 E
Tostado, *Argentina* **158 B3** 29 15 S 61 50W
Tostedt, *Germany* **26 B5** 53 17N 9 42 E
Tostón, Pta. de, *Canary Is.* **33 F5** 28 42N 14 2W
Tosu, *Japan* **62 D2** 33 22N 130 31 E
Tosya, *Turkey* **88 B6** 41 1N 34 2 E
Toszek, *Poland* **47 E5** 50 27N 18 32 E
Totana, *Spain* **35 H3** 37 45N 1 30W
Toten, *Norway* **14 D4** 60 40N 10 40 E
Toteng, *Botswana* **104 C3** 20 22 S 22 58 E
Tôtes, *France* **22 C8** 49 41N 1 3 E
Tótkomlós, *Hungary* **31 E13** 46 24N 20 45 E
Totma, *Russia* **48 C7** 60 0N 42 40 E
Totnes, *U.K.* **17 G4** 50 26N 3 42W
Totness, *Surinam* **153 B6** 5 53N 56 19W
Totonicapán, *Guatemala* **148 D1** 14 58N 91 12W
Totora, *Bolivia* **157 D4** 17 42 S 65 9W
Totoya, I., *Fiji* **121 B3** 18 57 S 179 50W
Totskoye, *Russia* **54 E4** 52 32N 52 45 E
Totten Glacier, *Antarctica* **7 C8** 66 45 S 116 10 E
Tottenham, *Australia* **117 B7** 32 14 S 147 21 E
Tottenham, *Canada* **136 B5** 44 1N 79 49W
Tottori, *Japan* **62 B6** 35 30N 134 15 E
Tottori □, *Japan* **62 B6** 35 30N 134 12 E
Touat, *Algeria* **99 C5** 27 27N 0 30 E
Touba, *Ivory C.* **100 D3** 8 22N 7 40W
Toubkal, Djebel, *Morocco* **98 B3** 31 0N 8 0W
Toucy, *France* **23 E10** 47 44N 3 15 E
Tougan, *Burkina Faso* **100 C4** 13 11N 2 58W
Touggourt, *Algeria* **99 B6** 33 6N 6 4 E
Tougué, *Guinea* **100 C2** 11 25N 11 50W
Touho, *N. Cal.* **121 T19** 20 47 S 165 14 E
Toukmatine, *Algeria* **99 D6** 24 49N 7 11 E
Toul, *France* **23 D12** 48 40N 5 53 E
Toulepleu, *Ivory C.* **100 D3** 6 32N 8 24W
Toulon, *France* **25 E9** 43 10N 5 55 E
Toulon, *U.S.A.* **140 C7** 41 6N 89 52W
Toulouse, *France* **24 E5** 43 37N 1 27 E
Toummo, *Niger* **96 D2** 22 45N 14 8 E
Toummo Dhoba, *Niger* **96 D2** 22 30N 14 31 E
Toumodi, *Ivory C.* **100 D3** 6 32N 5 4W
Tounassine, Hamada, *Algeria* **98 C3** 28 48N 5 0W
Toungoo, *Burma* **78 F6** 19 0N 96 30 E
Touques →, *France* **22 C7** 49 22N 0 8 E
Touraine, *France* **22 E7** 47 20N 0 30 E
Tourane = Da Nang, *Vietnam* **76 D7** 16 4N 108 13 E
Tourcoing, *France* **23 B10** 50 42N 3 10 E
Tourine, *Mauritania* **98 D2** 22 23N 11 50W
Tournai, *Belgium* **21 G2** 50 35N 3 25 E
Tournan-en-Brie, *France* **23 D9** 48 44N 2 46 E
Tournay, *France* **24 E4** 43 13N 0 13 E
Tournon, *France* **25 C8** 45 4N 4 50 E
Tournon-St.-Martin, *France* **22 F7** 46 45N 0 58 E
Tournus, *France* **25 B8** 46 35N 4 54 E
Touros, *Brazil* **154 C4** 5 12 S 35 28W
Tours, *France* **22 E7** 47 22N 0 40 E
Touside, Pic, *Chad* **97 D3** 21 1N 16 29 E
Touwsrivier, *S. Africa* **104 E3** 33 20 S 20 2 E
Tovar, *Venezuela* **152 B3** 8 20N 71 46W
Tovarkovskiy, *Russia* **52 D4** 53 40N 38 14 E
Tovdal, *Norway* **15 F2** 58 47N 8 10 E
Tovdalselva →, *Norway* **15 F2** 58 11N 8 7 E
Tovuz, *Azerbaijan* **53 K7** 41 0N 45 40 E
Towada, *Japan* **60 D10** 40 37N 141 13 E
Towada-Ko, *Japan* **60 D10** 40 28N 140 55 E
Towamba, *Australia* **117 D8** 37 6 S 149 43 E
Towanda, *Ill., U.S.A.* **141 D8** 40 36N 88 53W
Towanda, *Pa., U.S.A.* **137 E8** 41 46N 76 27W
Tower, *U.S.A.* **138 B8** 47 48N 92 17W

Tula, *Hidalgo, Mexico* ... **147 C5** 20 5N 99 20W
Tula, *Tamaulipas, Mexico* **147 C5** 23 0N 99 40W
Tula, *Nigeria* .... **101 D7** 9 51N 11 27 E
Tula, *Russia* .... **52 C3** 54 13N 37 38 E
Tulak, *Afghan.* .... **79 B1** 33 55N 63 40 E
Tulancingo, *Mexico* **147 C5** 20 5N 99 22W
Tulangbawang →,
Indonesia .... **74 C3** 4 24 S 105 52 E
Tulare, *U.S.A.* **144 J7** 36 13N 119 21W
Tulare Lake Bed, *U.S.A.* **144 K7** 36 0N 119 48W
Tularosa, *U.S.A.* **143 K10** 33 5N 106 1W
Tulbagh, *S. Africa* ...... **104 E2** 33 16 S 19 6 E
Tulcán, *Ecuador* **152 C2** 0 48N 77 43W
Tulcea, *Romania* .... **46 D9** 45 13N 28 46 E
Tulcea □, *Romania* ..... **46 D9** 45 0N 28 30 E
Tulchyn, *Ukraine* .... **51 H5** 48 41N 28 49 E
Tūleh, *Iran* .... **85 C7** 34 35N 52 33 E
Tulemalu L., *Canada* **131 A9** 62 58N 99 25W
Tulgheş, *Romania* .... **46 C6** 46 58N 25 45 E
Tuli, *Indonesia* .... **72 B2** 1 24 S 122 26 E
Tuli, *Zimbabwe* .... **107 G2** 21 58 S 29 13 E
Tulia, *U.S.A.* **139 H4** 34 32N 101 46W
Ţulkarm, *West Bank* .... **91 C4** 32 19N 35 2 E
Tullahoma, *U.S.A.* **135 H2** 35 22N 86 13W
Tullamore, *Australia* .... **117 B7** 32 39 S 147 36 E
Tullamore, *Ireland* **19 C4** 53 16N 7 31W
Tulle, *France* .... **24 C5** 45 16N 1 46 E
Tullibigeal, *Australia* ... **117 B7** 33 25 S 146 44 E
Tullins, *France* .... **25 C9** 45 18N 5 29 E
Tulln, *Austria* .... **30 C9** 48 20N 16 4 E
Tullow, *Ireland* **19 D4** 52 49N 6 45W
Tullus, *Sudan* .... **95 E1** 11 7N 24 31 E
Tully, *Australia* .... **114 B4** 17 56 S 145 55 E
Ţulmaythah, *Libya* ...... **96 B4** 32 40N 20 55 E
Tulmur, *Australia* .... **114 C3** 22 40 S 142 20 E
Tulnici, *Romania* .... **46 D7** 45 51N 26 38 E
Tulovo, *Bulgaria* .... **43 E10** 42 33N 25 32 E
Tulsa, *U.S.A.* **139 G7** 36 10N 95 55W
Tulsequah, *Canada* **130 B2** 58 39N 133 35W
Tulu Milki, *Ethiopia* ..... **95 F4** 9 55N 38 20 E
Tulu Welel, *Ethiopia* .... **95 F3** 8 56N 34 47 E
Tulua, *Colombia* **152 C2** 4 6N 76 11W
Tulun, *Russia* .... **57 D11** 54 32N 100 35 E
Tulungagung, *Indonesia* . **75 D4** 8 5 S 111 54 E
Tuma, *Russia* .... **52 C5** 55 10N 40 30 E
Tuma →, *Nic.* **148 D3** 13 6N 84 35W
Tumaco, *Colombia* **152 C2** 1 50N 78 45W
Tumaco, Ensenada,
Colombia **152 C2** 1 55N 78 45W
Tumatumari, *Guyana* **153 B6** 5 20N 58 55W
Tumauini, *Phil.* .... **70 C3** 17 17N 121 49 E
Tumba, *Sweden* .... **14 E11** 59 12N 17 48 E
Tumba, L., *Zaïre* .... **102 C3** 0 50 S 18 0 E
Tumbarumba, *Australia* .. **117 C8** 35 44 S 148 0 E
Tumbaya, *Argentina* ..... **158 A2** 23 50 S 65 26W
Túmbes, *Peru* .... **156 A1** 3 37 S 80 27W
Tumbes □, *Peru* .... **156 A1** 3 50 S 80 30W
Tumbwe, *Zaïre* .... **107 E2** 11 25 S 27 15 E
Tumby Bay, *Australia* ... **116 C2** 34 21 S 136 8 E
Tumd Youqi, *China* ..... **66 D6** 40 30N 110 30 E
Tumen, *China* .... **67 C15** 43 0N 129 50 E
Tumen Jiang →, *China* . **67 C16** 42 20N 130 35 E
Tumeremo, *Venezuela* ... **153 B5** 7 18N 61 30W
Tumiritinga, *Brazil* ..... **155 E3** 18 58 S 41 38W
Tumkur, *India* .... **83 H3** 13 18N 77 6 E
Tummel, L., *U.K.* **18 E5** 56 43N 3 55W
Tump, *Pakistan* .... **79 D1** 26 7N 62 16 E
Tumpat, *Malaysia* .... **77 J4** 6 11N 102 10 E
Tumsar, *India* .... **82 D4** 21 26N 79 45 E
Tumu, *Ghana* **100 C4** 10 56N 1 56W
Tumucumaque, Serra,
Brazil **153 C7** 2 0N 55 0W
Tumupasa, *Bolivia* ...... **156 C4** 14 9 S 67 55W
Tumut, *Australia* .... **117 C8** 35 16 S 148 13 E
Tumwater, *U.S.A.* **142 C2** 47 1N 122 54W
Tunas de Zaza, *Cuba* **148 B4** 21 39N 79 34W
Tunbridge Wells = Royal
Tunbridge Wells, *U.K.* **17 F8** 51 7N 0 16 E
Tunceli, *Turkey* .... **89 C8** 39 6N 39 31 E
Tuncurry, *Australia* ..... **117 B10** 32 17 S 152 29 E
Tunduru, *Tanzania* ...... **107 E4** 11 8 S 37 25 E
Tunduru □, *Tanzania* ... **107 E4** 11 5 S 37 22 E
Tundzha →, *Bulgaria* ... **43 F11** 41 40N 26 35 E
Tunga →, *India* .... **83 G2** 15 0N 75 50 E
Tunga Pass, *India* ...... **78 A5** 29 0N 94 14 E
Tungabhadra →, *India* .. **83 G4** 15 57N 78 15 E
Tungabhadra Dam, *India* **83 G2** 15 0N 75 50 E
Tungaru, *Sudan* .... **95 E3** 10 9N 30 52 E
Tungi, *Bangla.* .... **78 D3** 23 53N 90 24 E
Tungla, *Nic.* **148 D3** 13 24N 84 21W
Tungsten, *Canada* **130 A3** 61 57N 128 16W
Tungurahua □, *Ecuador* . **152 D2** 1 15 S 78 35W
Tunguska, Nizhnyaya →,
Russia .... **57 C9** 65 48N 88 4 E
Tuni, *India* .... **82 F6** 17 22N 82 36 E
Tunia, *Colombia* **152 C2** 2 41N 76 31W
Tunica, *U.S.A.* **139 H9** 34 41N 90 23W
Tunis, *Tunisia* .... **96 A2** 36 50N 10 11 E
Tunis, Golfe de, *Tunisia* . **96 A2** 37 0N 10 30 E
Tunisia ■, *Africa* .... **96 B1** 33 30N 9 10 E
Tunja, *Colombia* **152 B3** 5 33N 73 25W
Tunkhannock, *U.S.A.* ... **137 E9** 41 32N 75 57W
Tunliu, *China* .... **66 F7** 36 13N 112 52 E
Tunnsjøen, *Norway* ..... **12 D15** 64 45N 13 25 E
Tunungayualok I., *Canada* **129 A7** 56 0N 61 0W
Tunuyán, *Argentina* ..... **158 C2** 33 35 S 69 0W
Tunuyán →, *Argentina* .. **158 C2** 33 33 S 67 30W
Tunxi, *China* .... **69 C12** 29 42N 118 25 E
Tuo Jiang →, *China* .... **68 C5** 28 50N 105 35 E
Tuolumne, *U.S.A.* **144 H6** 37 58N 120 15W
Tuolumne →, *U.S.A.* ... **144 H5** 37 36N 121 13W
Tuoy-Khaya, *Russia* .... **57 C12** 62 32N 111 25 E
Tūp Āghāj, *Iran* .... **89 D12** 36 3N 47 50 E
Tupã, *Brazil* .... **159 A5** 21 57 S 50 28W
Tupaciguara, *Brazil* ..... **155 E2** 18 35 S 48 42W
Tupelo, *U.S.A.* **135 H1** 34 16N 88 43W
Tupik, *Russia* .... **52 C1** 55 42N 33 22 E
Tupik, *Russia* .... **57 D12** 54 26N 119 57 E
Tupinambaranas, *Brazil* . **153 D6** 3 0 S 58 0W
Tupirama, *Brazil* .... **154 C2** 8 58 S 48 12W
Tupiratins, *Brazil* ...... **154 C2** 8 23 S 48 8W
Tupiza, *Bolivia* .... **158 A2** 21 30 S 65 40W
Tupižnica, *Serbia, Yug.* .. **42 D7** 43 43N 22 10 E
Tupman, *U.S.A.* **145 K7** 35 18N 119 21W
Tupper, *Canada* **130 B4** 55 32N 120 1W
Tupper Lake, *U.S.A.* **137 B10** 44 14N 74 28W
Tupungato, Cerro,
S. Amer. .... **158 C2** 33 15 S 69 50W
Tuquan, *China* .... **67 B11** 45 18N 121 38 E
Túquerres, *Colombia* **152 C2** 1 5N 77 37W
Tura, *India* .... **78 C3** 25 30N 90 16 E
Tura, *Russia* .... **57 C11** 64 20N 100 17 E
Turabah, *Si. Arabia* .... **84 D4** 28 20N 43 15 E
Turagua, Serranía,
Venezuela **153 B5** 7 20N 64 35W
Turaiyur, *India* .... **83 J4** 11 9N 78 38 E
Turakina, *N.Z.* .... **118 G4** 40 3 S 175 16 E
Turakina →, *N.Z.* ...... **118 G4** 40 5 S 175 8 E
Turakirae Hd., *N.Z.* .... **118 H3** 41 26 S 174 56 E
Tūrān, *Iran* .... **85 C8** 35 39N 56 42 E
Turan, *Russia* .... **57 D10** 51 55N 95 0 E
Turayf, *Si. Arabia* ...... **84 D3** 31 41N 38 39 E
Turbacz, *Poland* .... **31 B13** 49 30N 20 8 E
Turbe, *Bos.-H.* .... **42 C2** 44 15N 17 35 E
Turbenthal, *Switz.* ...... **29 B7** 47 27N 8 51 E
Turda, *Romania* .... **46 C4** 46 34N 23 47 E
Turégano, *Spain* **36 D6** 41 9N 4 1W
Turek, *Poland* .... **47 C5** 52 3N 18 30 E
Turen, *Venezuela* **152 B4** 9 17N 69 6W
Turfan = Turpan, *China* . **64 B3** 43 58N 89 10 E
Tŭrgovishte, *Bulgaria* ... **43 D11** 43 17N 26 38 E
Turgutlu, *Turkey* .... **88 C2** 38 30N 27 48 E
Turhal, *Turkey* .... **88 B7** 40 24N 36 5 E
Turia →, *Spain* **35 F4** 39 27N 0 19W
Turiaçu, *Brazil* .... **154 B2** 1 40 S 45 19W
Turiaçu →, *Brazil* ...... **154 B2** 1 36 S 45 19W
Turiec →, *Slovak Rep.* .. **31 B11** 49 7N 18 55 E
Turin = Torino, *Italy* .... **38 C4** 45 3N 7 40 E
Turin, *Canada* **130 D6** 49 58N 112 31W
Turinsk, *Russia* .... **54 B9** 58 3N 63 42 E
Turkana □, *Kenya* ...... **106 B4** 3 0N 35 30 E
Turkana, L., *Africa* ..... **106 B4** 3 30N 36 5 E
Turkestan = Türkistan,
Kazakhstan .... **55 B4** 43 17N 68 16 E
Turkestanskiy, Khrebet,
Tajikistan .... **55 D4** 39 35N 69 0 E
Túrkeve, *Hungary* ...... **31 D13** 47 6N 20 44 E
Turkey ■, *Eurasia* ...... **88 C7** 39 0N 36 0 E
Turkey →, *U.S.A.* **140 B5** 42 43N 91 2W
Turkey Creek, *Australia* . **112 C4** 17 2 S 128 12 E
Turki, *Russia* .... **52 D6** 52 0N 43 15 E
Türkistan, *Kazakhstan* ... **55 B4** 43 17N 68 16 E
Turkmenistan ■, *Asia* ... **56 F6** 39 0N 59 0 E
Türkoğlu, *Turkey* .... **88 D7** 37 23N 36 50 E
Turks & Caicos Is. ■,
W. Indies **149 B5** 21 20N 71 20W
Turks Island Passage,
W. Indies **149 B5** 21 30N 71 30W
Turku, *Finland* .... **13 F20** 60 30N 22 19 E
Turkwe →, *Kenya* ...... **106 B4** 3 6N 36 6 E
Turlock, *U.S.A.* **144 H6** 37 30N 120 51W
Turnagain →, *Canada* .. **130 B3** 59 12N 127 35W
Turnagain, C., *N.Z.* ..... **118 G5** 40 28 S 176 38 E
Turneffe Is., *Belize* **147 D7** 17 20N 87 50W
Turner, *Australia* .... **112 C4** 17 52 S 128 16 E
Turner, *U.S.A.* **142 B9** 48 51N 108 24W
Turner Pt., *Australia* ... **114 A1** 11 47 S 133 32 E
Turner Valley, *Canada* .. **130 C6** 50 40N 114 17W
Turners Falls, *U.S.A.* ... **137 D12** 42 36N 72 33W
Turnhout, *Belgium* ...... **21 F5** 51 19N 4 57 E
Türnitz, *Austria* .... **30 D8** 47 55N 15 29 E
Turnor L., *Canada* **131 B7** 56 35N 108 35W
Turnov, *Czech.* .... **30 A8** 50 34N 15 10 E
Tŭrnovo = Veliko
Tŭrnovo, *Bulgaria* .... **43 D10** 43 5N 25 41 E
Turnu Măgurele, *Romania* **46 F5** 43 46N 24 56 E
Turnu Roşu, P., *Romania* **46 D5** 45 33N 24 17 E
Turobin, *Poland* .... **47 E9** 50 50N 22 44 E
Turon, *U.S.A.* **139 G5** 37 48N 98 26W
Tuross Head, *Australia* .. **117 D9** 36 3 S 150 8 E
Turpan, *China* .... **64 B3** 43 58N 89 10 E
Turrës, Kalaja e, *Albania* **44 C1** 41 10N 19 28 E
Turriff, *U.K.* **18 D6** 57 32N 2 27W
Tursāq, *Iraq* .... **89 F11** 33 27N 45 47 E
Tursi, *Italy* .... **41 B9** 40 15N 16 28 E
Turtle Head I., *Australia* **114 A3** 10 56 S 142 37 E
Turtle Is., *Phil.* .... **71 H2** 6 7N 118 14 E
Turtle L., *Canada* **131 C7** 53 36N 108 38W
Turtle Lake, *N. Dak.,*
U.S.A. **138 B4** 47 31N 100 53W
Turtle Lake, *Wis., U.S.A.* **138 C8** 45 24N 92 8W
Turtleford, *Canada* **131 C7** 53 23N 108 57W
Turua, *N.Z.* .... **118 D4** 37 14 S 175 35 E
Turugart, Pereval,
Kyrgyzstan .... **55 C7** 40 32N 75 24 E
Turukhansk, *Russia* ..... **57 C9** 65 21N 88 5 E
Turzovka, *Slovak Rep.* .. **31 B11** 49 25N 18 35 E
Tuscaloosa, *U.S.A.* **135 J2** 33 12N 87 34W
Tuscánia, *Italy* .... **39 F8** 42 25N 11 52 E
Tuscany = Toscana □,
Italy .... **38 E8** 43 25N 11 0 E
Tuscola, *Ill., U.S.A.* **141 E8** 39 48N 88 17W
Tuscola, *Tex., U.S.A.* ... **139 J5** 32 12N 99 48W
Tuscumbia, *Ala., U.S.A.* **135 H2** 34 44N 87 42W
Tuscumbia, *Mo., U.S.A.* . **140 F4** 38 14N 92 28W
Tuskar Rock, *Ireland* **19 D5** 52 12N 6 10W
Tuskegee, *U.S.A.* **135 J3** 32 25N 85 42W
Tustna, *Norway* .... **14 A2** 63 10N 8 5 E
Tuszyn, *Poland* .... **47 D6** 51 36N 19 33 E
Tutak, *Turkey* .... **89 C10** 39 31N 42 46 E
Tutayev, *Russia* .... **52 B4** 57 53N 39 32 E
Tuticorin, *India* .... **83 K4** 8 50N 78 12 E
Tutin, *Serbia, Yug.* ..... **42 E5** 42 58N 20 20 E
Tutóia, *Brazil* .... **154 B3** 2 45 S 42 20W
Tutong, *Brunei* .... **75 B4** 4 47N 114 40 E
Tutova →, *Romania* .... **46 C8** 46 20N 27 30 E
Tutrakan, *Bulgaria* ...... **43 C11** 44 2N 26 40 E
Tutshi L., *Canada* **130 B2** 59 56N 134 30W
Tuttle, *U.S.A.* **138 B5** 47 9N 100 0W
Tuttlingen, *Germany* .... **27 H4** 47 58N 8 48 E
Tutuala, *Indonesia* ...... **72 C3** 8 25 S 127 15 E
Tutuila, *Amer. Samoa* ... **121 X24** 14 19 S 170 50W
Tutuko Mt., *N.Z.* .... **119 E3** 44 35 S 168 1 E
Tututepec, *Mexico* **147 D5** 16 9N 97 38W
Tutye, *Australia* .... **116 C4** 35 12 S 141 29 E
Tuva □, *Russia* .... **57 D10** 51 30N 95 0 E
Tuvalu ■, *Pac. Oc.* ..... **122 H9** 8 0 S 178 0 E
Tuwal, *Si. Arabia* ...... **86 B2** 22 17N 39 6 E
Tuxpan, *Mexico* **147 C5** 20 58N 97 23W
Tuxtla Gutiérrez, *Mexico* **147 D6** 16 45N 93 7W
Tuy, *Spain* **36 C2** 42 3N 8 39W
Tuy An, *Vietnam* .... **76 F7** 13 17N 109 16 E
Tuy Duc, *Vietnam* ...... **77 F6** 12 15N 107 27 E
Tuy Hoa, *Vietnam* ...... **76 F7** 13 5N 109 10 E
Tuy Phong, *Vietnam* .... **77 G7** 11 14N 108 43 E
Tuya L., *Canada* **130 B2** 59 7N 130 35W
Tuyen Hoa, *Vietnam* .... **76 D6** 17 50N 106 10 E
Tuyen Quang, *Vietnam* .. **76 B5** 21 50N 105 10 E
Tuymazy, *Russia* .... **54 D4** 54 36N 53 42 E
Tūysarkān, *Iran* .... **89 E13** 34 33N 48 27 E
Tuz Gölü, *Turkey* .... **88 C5** 38 42N 33 18 E
Ţūz Khurmātū, *Iraq* ..... **89 E11** 34 56N 44 38 E
Tuzkan, Ozero,
Uzbekistan .... **55 C3** 40 35N 67 28 E
Tuzla, *Bos.-H.* .... **42 C3** 44 34N 18 41 E
Tuzlov →, *Russia* ...... **53 G4** 47 17N 39 57 E
Tuzluca, *Turkey* .... **89 B10** 40 3N 43 39 E
Tvååker, *Sweden* .... **15 G6** 57 4N 12 25 E
Tvedestrand, *Norway* .... **15 F2** 58 38N 8 58 E
Tver, *Russia* .... **52 B2** 56 55N 35 55 E
Tvŭrditsa, *Bulgaria* ...... **43 E10** 42 42N 25 53 E
Twain, *U.S.A.* **144 E5** 40 1N 121 3W
Twain Harte, *U.S.A.* **144 G6** 38 2N 120 14W
Twardogóra, *Poland* .... **47 D4** 51 23N 17 28 E
Tweed, *Canada* **136 B7** 44 29N 77 19W
Tweed →, *U.K.* **18 F7** 55 45N 2 0W
Tweed Heads, *Australia* . **115 D5** 28 10 S 153 31 E
Tweedsmuir Prov. Park,
Canada **130 C3** 53 0N 126 20W
Twello, *Neths.* .... **20 D8** 52 14N 6 6 E
Twentynine Palms, *U.S.A.* **145 L10** 34 8N 116 3W
Twillingate, *Canada* **129 C9** 49 42N 54 45W
Twin Bridges, *U.S.A.* ... **142 D7** 45 33N 112 20W
Twin Falls, *U.S.A.* **142 E6** 42 34N 114 28W
Twin Valley, *U.S.A.* **138 B6** 47 16N 96 16W
Twinnge, *Burma* .... **78 D6** 23 10N 96 2 E
Twisp, *U.S.A.* **142 B3** 48 22N 120 7W
Twistringen, *Germany* ... **26 C4** 52 48N 8 37 E
Two Harbors, *U.S.A.* ... **138 B9** 47 2N 91 40W
Two Hills, *Canada* **130 C6** 53 43N 111 52W
Two Rivers, *U.S.A.* **134 C2** 44 9N 87 34W
Two Thumbs Ra., *N.Z.* . **119 D5** 43 45 S 170 44 E
Twofold B., *Australia* ... **117 D8** 37 8 S 149 59 E
Tyachiv, *Ukraine* .... **51 H2** 48 1N 23 35 E
Tychy, *Poland* .... **31 A11** 50 9N 18 59 E
Tyczyn, *Poland* .... **31 B15** 49 58N 22 2 E
Tykocin, *Poland* .... **47 B9** 53 13N 22 46 E
Tyler, *U.S.A.* **133 D7** 32 18N 95 17W
Tyler, *Minn., U.S.A.* **138 C6** 44 18N 96 8W
Tyler, *Tex., U.S.A.* **139 J7** 32 21N 95 18W
Tyligul →, *Ukraine* ..... **51 J6** 47 4N 30 57 E
Tylldal, *Norway* .... **14 B4** 62 8N 10 48 E
Týn nad Vltavou, *Czech.* **30 B7** 49 13N 14 26 E
Tynda, *Russia* .... **57 D13** 55 10N 124 43 E
Tyne →, *U.K.* **16 C6** 54 59N 1 32W
Tyne & Wear □, *U.K.* ... **16 C6** 55 6N 1 17W
Tynemouth, *U.K.* **16 B6** 55 1N 1 26W
Tynset, *Norway* .... **14 B4** 62 17N 10 47 E
Tyre = Sūr, *Lebanon* .... **91 B4** 33 19N 35 16 E
Tyrifjorden, *Norway* .... **14 D4** 60 2N 10 8 E
Tyringe, *Sweden* .... **15 H7** 56 9N 13 35 E
Tyristrand, *Norway* ..... **14 D4** 60 4N 10 6 E
Tyrnyauz, *Russia* .... **53 J6** 43 21N 42 45 E
Tyrol = Tirol □, *Austria* . **30 D3** 47 3N 10 43 E
Tyrone, *U.S.A.* **136 F6** 40 40N 78 14W
Tyrrell →, *Australia* .... **116 C5** 35 26 S 142 51 E
Tyrrell, L., *Australia* .... **116 C5** 35 20 S 142 50 E
Tyrrell Arm, *Canada* **131 A9** 62 27N 97 30W
Tyrrell L., *Canada* **131 A7** 63 7N 105 27W
Tyrrhenian Sea, *Medit. S.* **40 B5** 40 0N 12 30 E
Tysfjorden, *Norway* ..... **12 B17** 68 7N 16 25 E
Tystberga, *Sweden* ...... **15 F11** 58 51N 17 15 E
Tyub Karagan, Mys,
Kazakhstan .... **53 H10** 44 40N 50 19 E
Tyuleni, Ostrova,
Kazakhstan .... **53 H10** 45 2N 50 16 E
Tyuleniy, *Russia* .... **53 H8** 44 28N 47 30 E
Tyuleniy, Mys, *Azerbaijan* **53 K10** 40 12N 50 22 E
Tyulgan, *Russia* .... **54 E6** 52 22N 56 12 E
Tyumen, *Russia* .... **56 D7** 57 11N 65 29 E
Tyumen-Aryk, *Kazakhstan* **55 A3** 44 2N 67 1 E
Tyup, *Kyrgyzstan* ...... **55 B9** 42 45N 78 20 E
Tywi →, *U.K.* **17 F3** 51 48N 4 21W
Tywyn, *U.K.* **17 E3** 52 35N 4 5W
Tzaneen, *S. Africa* ...... **105 C5** 23 47 S 30 9 E
Tzermiádhes, *Greece* .... **32 D7** 35 12N 25 29 E
Tzermiádhes Neápolis,
Greece .... **45 J7** 35 11N 25 29 E
Tzoumérka, Óros, *Greece* **44 E3** 39 30N 21 26 E
Tzukong = Zigong, *China* **68 C5** 29 15N 104 48 E
Tzummarum, *Neths.* ..... **20 B7** 53 14N 5 32 E

U

U Taphao, *Thailand* ..... **76 F3** 12 35N 101 0 E
U.S.A. = United States of
America ■, *N. Amer.* . **132 C7** 37 0N 96 0W
Uacalla Iero, *Somali Rep.* **108 D2** 1 48N 42 38 E
Uachadi, Sierra, *Venezuela* **153 C4** 4 54N 65 18W
Uainambi, *Colombia* **152 C4** 1 43N 69 51W
Uanda, *Australia* .... **114 C3** 21 37 S 144 55 E
Uanle Uen, *Somali Rep.* . **108 D2** 2 37N 44 54 E
Uarsciek, *Somali Rep.* .. **108 D3** 2 28N 45 55 E
Uascen, *Somali Rep.* .... **108 D2** 4 11N 43 13 E
Uasin □, *Kenya* .... **106 B4** 0 30N 35 20 E
Uato-Udo, *Indonesia* .... **72 C3** 9 7 S 125 36 E
Uatumã →, *Brazil* ...... **153 D6** 2 26 S 57 37W
Uauá, *Brazil* .... **154 C4** 9 50 S 39 28W
Uaupés, *Brazil* .... **152 D4** 0 8 S 67 5W
Uaupés →, *Brazil* **152 C4** 0 2N 67 16W
Uaxactún, *Guatemala* ... **148 C2** 17 25N 89 29W
Ub, *Serbia, Yug.* .... **42 C5** 44 28N 20 6 E
Ubá, *Brazil* .... **155 F3** 21 8 S 43 0W
Ubaitaba, *Brazil* .... **155 D4** 14 18 S 39 20W
Ubangi = Oubangi →,
Zaïre .... **102 C3** 0 30 S 17 50 E
Ubaté, *Colombia* **152 B3** 5 19N 73 49W
Ubauro, *Pakistan* .... **80 E3** 28 15N 69 45 E
Ubay, *Phil.* .... **71 F5** 10 3N 124 28 E
'Ubaydīyah, *Yemen* ..... **86 D3** 13 7N 43 20 E
Ubaye →, *France* .... **25 D10** 44 28N 6 18 E
Ubayyiḍ, W. al →, *Iraq* . **89 F10** 32 34N 43 48 E
Ube, *Japan* .... **62 D3** 33 56N 131 15 E
Ubeda, *Spain* **35 G1** 38 3N 3 23W
Uberaba, *Brazil* .... **155 E2** 19 50 S 47 55W
Uberaba, L., *Brazil* ..... **157 D6** 17 30 S 57 50W
Uberlândia, *Brazil* ...... **155 E2** 19 0 S 48 20W
Überlingen, *Germany* ... **27 H5** 47 46N 9 10 E
Ubiaja, *Nigeria* .... **101 D6** 6 41N 6 22 E
Ubolratna Res., *Thailand* **76 D4** 16 45N 102 30 E
Ubombo, *S. Africa* ...... **105 D5** 27 31 S 32 4 E
Ubon Ratchathani,
Thailand .... **76 E5** 15 15N 104 50 E
Ubondo, *Zaïre* .... **106 C2** 0 55 S 25 42 E
Ubort →, *Belarus* ...... **51 F5** 52 6N 28 30 E
Ubrique, *Spain* **37 J5** 36 41N 5 27W
Ubundu, *Zaïre* .... **102 C5** 0 22 S 25 30 E
Ucayali →, *Peru* .... **156 A3** 4 30 S 73 30W
Uccle, *Belgium* .... **21 G4** 50 48N 4 22 E
Uchaly, *Russia* .... **54 D7** 54 19N 59 27 E
Uchi Lake, *Canada* **131 C10** 51 5N 92 35W
Uchiko, *Japan* .... **62 D4** 33 33N 132 39 E
Uchiura-Wan, *Japan* .... **60 C10** 42 25N 140 40 E
Uchiza, *Peru* .... **156 B2** 8 25 S 76 20W
Uchte, *Germany* .... **26 C4** 52 30N 8 54 E
Uchur →, *Russia* .... **57 D14** 58 48N 130 35 E
Ucluelet, *Canada* **130 D3** 48 57N 125 32W
Ucuriş, *Romania* .... **46 C2** 46 41N 21 58 E
Uda →, *Russia* .... **57 D14** 54 42N 135 14 E
Udaipur, *India* .... **80 G5** 24 36N 73 44 E
Udaipur Garhi, *Nepal* ... **81 F12** 27 0N 86 35 E
Udbina, *Croatia* .... **39 D12** 44 31N 15 47 E
Uddel, *Neths.* .... **20 D7** 52 15N 5 48 E
Uddevalla, *Sweden* ...... **15 F5** 58 21N 11 55 E
Uddjaur, *Sweden* .... **12 D17** 65 56N 17 49 E
Uden, *Neths.* .... **21 E7** 51 40N 5 37 E
Udgir, *India* .... **82 E3** 18 25N 77 5 E
Udhampur, *India* .... **81 C6** 33 0N 75 5 E
Udi, *Nigeria* .... **101 D6** 6 17N 7 21 E
Údine, *Italy* .... **39 B10** 46 3N 13 14 E
Udmurtia □, *Russia* ..... **54 C4** 57 30N 52 30 E
Udon Thani, *Thailand* ... **76 D4** 17 29N 102 46 E
Udumalaippettai, *India* .. **83 J3** 10 35N 77 15 E
Udupi, *India* .... **83 H2** 13 25N 74 42 E
Udvoy Balkan, *Bulgaria* . **43 E11** 42 50N 26 50 E
Udzungwa Range,
Tanzania .... **107 D4** 9 30 S 35 10 E
Ueckermünde, *Germany* . **26 B10** 53 44N 14 1 E
Ueda, *Japan* .... **63 A10** 36 24N 138 16 E
Uedineniya, Os., *Russia* . **6 B12** 78 0N 85 0 E
Uel Scimbirro,
Somali Rep. .... **108 D2** 2 23N 44 14 E
Uele →, *Zaïre* .... **102 B4** 3 45N 24 45 E
Uelen, *Russia* **57 C19** 66 10N 170 0W
Uelzen, *Germany* .... **26 C6** 52 57N 10 32 E
Ueno, *Japan* .... **63 C8** 34 45N 136 8 E
Uetendorf, *Switz.* .... **28 C5** 46 47N 7 34 E
Ufa, *Russia* .... **54 D6** 54 45N 55 55 E
Ufa →, *Russia* .... **54 D6** 54 40N 56 0 E
Uffenheim, *Germany* .... **27 F6** 49 33N 10 14 E
Ugab →, *Namibia* ...... **104 C1** 20 55 S 13 30 E
Ugalla →, *Tanzania* .... **106 D3** 5 8 S 30 42 E
Ugamskiy, Khrebet,
Kazakhstan .... **55 B5** 42 20N 70 30 E
Uganda ■, *Africa* .... **106 B3** 2 0N 32 0 E
Ugchelen, *Neths.* .... **20 D7** 52 11N 5 56 E
Ugento, *Italy* .... **41 C11** 39 56N 18 10 E
Ugep, *Nigeria* .... **101 D6** 5 53N 8 2 E
Ugie, *S. Africa* .... **105 E4** 31 10 S 28 13 E
Ugijar, *Spain* **35 J1** 36 58N 3 7W
Ugine, *France* .... **25 C10** 45 45N 6 25 E
Uglegorsk, *Russia* ...... **57 E15** 49 5N 142 2 E
Ugleuralskiy, *Russia* .... **54 B6** 59 8N 57 35 E
Uglich, *Russia* .... **52 B4** 57 33N 38 20 E
Ugljane, *Croatia* .... **39 E13** 43 35N 16 46 E
Ugolyak, *Russia* .... **57 C13** 64 33N 120 30 E
Ugra →, *Russia* .... **52 C3** 54 30N 36 7 E
Ugûn Mûsa, *Egypt* ..... **91 F1** 29 53N 32 40 E
Ugŭrchin, *Bulgaria* ..... **43 D9** 43 6N 24 26 E
Uh →, *Slovak Rep.* ..... **31 C14** 48 7N 21 25 E
Uherske Hradiště, *Czech.* **31 B10** 49 4N 17 30 E
Uhersky Brod, *Czech.* ... **31 B10** 49 1N 17 40 E
Úhlava →, *Czech.* ...... **30 B6** 49 45N 13 24 E
Uhrichsville, *U.S.A.* **136 F3** 40 24N 81 21W
Uibhist a Deas = South
Uist, *U.K.* **18 D1** 57 20N 7 15W
Uibhist a Tuath = North
Uist, *U.K.* **18 D1** 57 40N 7 15W
Uíge, *Angola* .... **103 D2** 7 30 S 14 40 E
Uige □, *Angola* .... **103 D3** 7 0 S 16 0 E
Uiha, *Tonga* .... **121 P13** 19 54 S 174 25W
Uijŏngbu, *S. Korea* ..... **67 F14** 37 48N 127 0 E
Ŭiju, *N. Korea* .... **67 D13** 40 15N 124 35 E
Uinta Mts., *U.S.A.* **142 F8** 40 45N 110 30W
Uitenhage, *S. Africa* .... **104 E4** 33 40 S 25 28 E
Uitgeest, *Neths.* .... **20 C5** 52 32N 4 43 E
Uithoorn, *Neths.* .... **20 D5** 52 14N 4 50 E
Uithuizen, *Neths.* .... **20 B9** 53 24N 6 41 E
Uitkerke, *Belgium* ...... **21 F2** 51 18N 3 9 E
Újfehértó, *Hungary* ..... **31 D14** 47 49N 21 41 E
Ujhani, *India* .... **81 F8** 28 0N 79 6 E
Uji, *Japan* .... **63 C7** 34 53N 135 48 E
Uji-guntō, *Japan* .... **61 J4** 31 15N 129 25 E
Ujjain, *India* .... **80 H6** 23 9N 75 43 E
Újpest, *Hungary* .... **31 D12** 47 32N 19 6 E
Újszász, *Hungary* ...... **31 D13** 47 19N 20 7 E
Ujung Pandang, *Indonesia* **72 C1** 5 10 S 119 20 E
Uka, *Russia* .... **57 D17** 57 50N 162 0 E
Ukara I., *Tanzania* ...... **106 C3** 1 50 S 33 0 E
Uke-Shima, *Japan* ...... **61 K4** 28 2N 129 14 E
Ukerewe □, *Tanzania* ... **106 C3** 2 0 S 32 30 E
Ukerewe I., *Tanzania* ... **106 C3** 2 0 S 33 0 E
Ukholovo, *Russia* .... **52 D5** 53 47N 40 30 E
Ukhrul, *India* .... **78 C5** 25 10N 94 25 E
Ukhta, *Russia* .... **48 B9** 63 34N 53 41 E
Ukiah, *U.S.A.* **144 F3** 39 9N 123 13W
Ukki Fort, *India* .... **81 C7** 33 28N 76 54 E
Ukmerge, *Lithuania* ..... **13 J21** 55 15N 24 45 E
Ukraine ■, *Europe* ...... **51 H7** 49 0N 32 0 E
Uku, *Angola* .... **103 E2** 11 24 S 14 22 E
Ukwi, *Botswana* .... **104 C3** 23 29 S 20 30 E
Ulaanbaatar, *Mongolia* .. **57 E11** 47 55N 106 53 E
Ulaangom, *Mongolia* ... **64 A4** 50 5N 92 10 E
Ulamambri, *Australia* ... **117 A8** 31 19 S 149 23 E
Ulamba, *Zaïre* .... **107 D4** 9 3 S 23 38 E
Ulan Bator =
Ulaanbaatar, *Mongolia* **57 E11** 47 55N 106 53 E
Ulan Erge, *Russia* ...... **53 G7** 46 19N 44 53 E
Ulan Khol, *Russia* ...... **53 H8** 45 27N 47 4 E
Ulan Ude, *Russia* ...... **57 D11** 51 45N 107 40 E
Ulanbel, *Kazakhstan* .... **55 A5** 44 50N 71 7 E
Ulanga □, *Tanzania* .... **107 D4** 8 40 S 36 50 E
Ulanów, *Poland* .... **47 E9** 50 30N 22 16 E
Ulaş, *Turkey* .... **88 C7** 39 26N 37 2 E

V

Vennesla 279

Værøy, *Norway* **12 C15** 67 40N 12 40 E
Vágar, *Færoe Is.* **12 E9** 62 5N 7 15W
Vagney, *France* **23 E13** 48 1N 6 43 E
Vagnhärad, *Sweden* **14 F11** 58 57N 17 33 E
Vagos, *Portugal* **36 E2** 40 33N 8 42W
Vågsfjorden, *Norway* **12 B17** 68 50N 16 50 E
Váh →, *Slovak Rep.* **31 D11** 47 43N 18 7 E
Vahsel B., *Antarctica* **7 D1** 75 0 S 35 0W
Vái, *Greece* **32 D8** 35 15N 26 18 E
Vaigach, *Russia* **56 B6** 70 10N 59 0 E
Vaigai →, *India* **83 K4** 9 15N 79 10 E
Vaiges, *France* **22 D6** 48 2N 0 30W
Vaihingen, *Germany* **27 G4** 48 54N 8 57 E
Vaijapur, *India* **82 E2** 19 58N 74 45 E
Vaikam, *India* **83 K3** 9 45N 76 25 E
Vailly-sur-Aisne, *France* **23 C10** 49 24N 3 31 E
Vaippar →, *India* **83 K4** 9 0N 78 25 E
Vaison-la-Romaine, *France* **25 D9** 44 14N 5 4 E
Vajpur, *India* **82 D1** 21 24N 73 17 E
Vakarel, *Bulgaria* **43 E8** 42 35N 23 40 E
Vakfikebir, *Turkey* **89 B8** 41 2N 39 17 E
Vakh →, *Russia* **56 C8** 60 45N 76 45 E
Vakhsh →, *Tajikistan* **55 E4** 37 6N 68 18 E
Vakhtan, *Russia* **52 B8** 57 53N 46 47 E
Vál, *Hungary* **31 D11** 47 22N 18 40 E
Val-de-Marne □, *France* **23 D9** 48 45N 2 28 E
Val-d'Oise □, *France* **23 C9** 49 5N 2 10 E
Val d'Or, *Canada* **128 C4** 48 7N 77 47W
Val Marie, *Canada* **131 D7** 49 15N 107 45W
Valaam, *Russia* **50 B6** 61 22N 30 57 E
Valadares, *Portugal* **36 D2** 41 5N 8 38W
Valahia, *Romania* **46 E5** 44 35N 25 0 E
Valais □, *Switz.* **28 D5** 46 12N 7 45 E
Valais, Alpes du, *Switz.* **28 D5** 46 5N 7 35 E
Valandovo, *Macedonia* **42 F7** 41 19N 22 34 E
Valašské Meziříčí, *Czech.* **31 B10** 49 29N 17 59 E
Valáxa, *Greece* **45 F6** 38 50N 24 29 E
Vâlcani, *Romania* **42 A5** 46 0N 20 26 E
Valcheta, *Argentina* **160 B3** 40 40 S 66 8W
Valdagno, *Italy* **39 C8** 45 39N 11 18 E
Valdahon, *France* **23 E13** 47 8N 6 21 E
Valday, *Russia* **52 B1** 57 58N 33 9 E
Valdayskaya Vozvyshennost, *Russia* **52 B1** 57 0N 33 30 E
Valdeazogues →, *Spain* **37 G6** 38 45N 4 55W
Valdemarsvik, *Sweden* **15 F10** 58 14N 16 40 E
Valdepeñas, *Ciudad Real, Spain* **37 G7** 38 43N 3 25W
Valdepeñas, *Jaén, Spain* **37 H7** 37 33N 3 47W
Valderaduey →, *Spain* **36 D5** 41 31N 5 42W
Valderrobres, *Spain* **34 E5** 40 53N 0 9 E
Valdés, Pen., *Argentina* **160 B4** 42 30 S 63 45W
Valdez, *Ecuador* **152 C2** 1 15N 79 0W
Valdez, *U.S.A.* **126 B5** 61 7N 146 16W
Valdivia, *Chile* **160 A2** 39 50 S 73 14W
Valdivia, *Colombia* **152 B2** 7 11N 75 27W
Valdivia □, *Chile* **160 B2** 40 0 S 73 0W
Valdobbiádene, *Italy* **39 C9** 45 17N 12 24 E
Valdosta, *U.S.A.* **135 K4** 30 50N 83 17W
Valdoviño, *Spain* **36 B2** 43 36N 8 8W
Valdres, *Norway* **14 D3** 61 5N 9 5 E
Vale, *Georgia* **53 K6** 41 30N 42 58 E
Vale, *U.S.A.* **142 E5** 43 59N 117 15W
Valea lui Mihai, *Romania* **46 B3** 47 32N 22 11 E
Valença, *Brazil* **155 D4** 13 20 S 39 5W
Valença, *Portugal* **36 C2** 42 1N 8 34W
Valença do Piauí, *Brazil* **154 C3** 6 20 S 41 45W
Valençay, *France* **23 E8** 47 9N 1 34 E
Valence, *Drôme, France* **25 D8** 44 57N 4 54 E
Valence, *Tarn-et-Garonne, France* **24 D4** 44 6N 0 53 E
Valencia, *Phil.* **71 H5** 7 57N 125 3 E
Valencia, *Spain* **35 F4** 39 27N 0 23W
Valencia, *Venezuela* **152 A4** 10 11N 68 0W
Valencia □, *Spain* **35 F4** 39 20N 0 40W
Valencia, G. de, *Spain* **35 F5** 39 30N 0 20 E
Valencia de Alcántara, *Spain* **37 F3** 39 25N 7 14W
Valencia de Don Juan, *Spain* **36 C5** 42 17N 5 31W
Valencia del Ventoso, *Spain* **37 G4** 38 15N 6 29W
Valencia Harbour, *Ireland* **19 E1** 51 56N 10 19W
Valencia I., *Ireland* **19 E1** 51 54N 10 22W
Valenciennes, *France* **23 B10** 50 20N 3 34 E
Văleni, *Romania* **46 E5** 44 15N 24 45 E
Valensole, *France* **25 E9** 43 50N 5 59 E
Valentim, Sa. do, *Brazil* **154 C3** 6 0 S 43 30W
Valentin, *Russia* **60 C7** 43 8N 134 17 E
Valentine, *Nebr., U.S.A.* **138 D4** 42 52N 100 33W
Valentine, *Tex., U.S.A.* **139 K2** 30 35N 104 30W
Valenza, *Italy* **38 C5** 45 1N 8 38 E
Valera, *Venezuela* **152 B3** 9 19N 70 37W
Valga, *Estonia* **13 H22** 57 47N 26 2 E
Valguarnera Caropepe, *Italy* **41 E7** 37 30N 14 23 E
Valier, *U.S.A.* **142 B7** 48 18N 112 16W
Valinco, G. de, *France* **25 G12** 41 40N 8 52 E
Valjevo, *Serbia, Yug.* **42 C4** 44 18N 19 53 E
Valka, *Latvia* **13 H21** 57 42N 25 57 E
Valkeakoski, *Finland* **13 F20** 61 16N 24 2 E
Valkenburg, *Neths.* **21 G7** 50 52N 5 50 E
Valkenswaard, *Neths.* **21 F6** 51 21N 5 29 E
Vall de Uxó, *Spain* **34 F4** 39 49N 0 15W
Valla, *Sweden* **14 E10** 59 2N 16 20 E
Valladolid, *Mexico* **147 C7** 20 40N 88 11W
Valladolid, *Spain* **36 D6** 41 38N 4 43W
Valladolid □, *Spain* **36 D6** 41 38N 4 43W
Vallata, *Italy* **41 A8** 41 2N 15 15 E
Valldemosa, *Spain* **33 B9** 39 43N 2 37 E
Valle d'Aosta □, *Italy* **38 C4** 45 45N 7 15 E
Valle de Arán, *Spain* **34 C5** 42 50N 0 55 E
Valle de Cabuérniga, *Spain* **36 B6** 43 14N 4 18W
Valle de la Pascua, *Venezuela* **152 B4** 9 13N 66 0W
Valle de las Palmas, *Mexico* **145 N10** 32 20N 116 43W
Valle de Santiago, *Mexico* **146 C4** 20 25N 101 15W
Valle de Suchil, *Mexico* **146 C4** 23 38N 103 55W
Valle de Zaragoza, *Mexico* **146 B3** 27 28N 105 49W
Valle del Cauca □, *Colombia* **152 C2** 3 45N 76 30W
Valle Fértil, Sierra del, *Argentina* **158 C2** 30 20 S 68 0W
Valle Hermoso, *Mexico* **147 B5** 25 35N 97 40W
Vallecas, *Spain* **36 E7** 40 23N 3 41W
Valledupar, *Colombia* **152 A3** 10 29N 73 15W
Vallehermoso, *Canary Is.* **33 F2** 28 10N 17 15W
Vallejo, *U.S.A.* **144 G4** 38 7N 122 14W
Vallenar, *Chile* **158 B1** 28 30 S 70 50W
Valleraugue, *France* **24 D7** 44 6N 3 39 E
Vallet, *France* **22 E5** 47 10N 1 15W
Valletta, *Malta* **32 D2** 35 54N 14 31 E
Valley Center, *U.S.A.* **145 M9** 33 13N 117 2W
Valley City, *U.S.A.* **138 B6** 46 55N 98 0W
Valley Falls, *U.S.A.* **142 E3** 42 29N 120 17W
Valley Park, *U.S.A.* **140 F6** 38 33N 90 29W
Valley Springs, *U.S.A.* **144 G6** 38 12N 120 50W
Valley Station, *U.S.A.* **141 F11** 38 6N 85 52W
Valley Wells, *U.S.A.* **145 K11** 35 27N 115 46W
Valleyview, *Canada* **130 B5** 55 5N 117 17W
Valli di Comácchio, *Italy* **39 D9** 44 40N 12 15 E
Vallimanca, Arroyo, *Argentina* **158 D4** 35 40 S 59 10W
Vallo della Lucánia, *Italy* **41 B8** 40 14N 15 16 E
Vallon-Pont-d'Arc, *France* **25 D8** 44 24N 4 24 E
Vallorbe, *Switz.* **28 C2** 46 42N 6 20 E
Valls, *Spain* **34 D6** 41 18N 1 15 E
Vallsta, *Sweden* **14 C10** 61 31N 16 22 E
Valmaseda, *Spain* **34 B1** 43 11N 3 12W
Valmeyer, *U.S.A.* **140 F6** 38 18N 90 19W
Valmiera, *Latvia* **13 H21** 57 37N 25 29 E
Valmont, *France* **22 C7** 49 45N 0 30 E
Valmontone, *Italy* **40 A5** 41 46N 12 57 E
Valmy, *France* **23 C11** 49 5N 4 45 E
Valnera, Mte., *Spain* **34 B1** 43 9N 3 40W
Valognes, *France* **22 C5** 49 30N 1 28W
Valona = Vlóra, *Albania* **44 D1** 40 32N 19 28 E
Valongo, *Portugal* **36 D2** 41 8N 8 30W
Valozhyn, *Belarus* **50 E4** 54 3N 26 30 E
Valpaços, *Portugal* **36 D3** 41 36N 7 17W
Valparaíso, *Chile* **158 C1** 33 2 S 71 40W
Valparaíso, *Mexico* **146 C4** 22 50N 103 32W
Valparaiso, *U.S.A.* **141 C9** 41 28N 87 4W
Valparaíso □, *Chile* **158 C1** 33 2 S 71 40W
Valpovo, *Croatia* **42 B3** 45 39N 18 25 E
Valréas, *France* **25 D8** 44 24N 5 0 E
Vals, *Switz.* **29 C8** 46 39N 9 11 E
Vals →, *S. Africa* **104 D4** 27 23 S 26 30 E
Vals, Tanjung, *Indonesia* **73 C5** 8 26 S 137 25 E
Vals-les-Bains, *France* **25 D8** 44 42N 4 24 E
Valsad, *India* **82 D1** 20 40N 72 58 E
Valskog, *Sweden* **14 E9** 59 27N 15 57 E
Válta, *Greece* **44 D5** 40 3N 23 25 E
Valtellina, *Italy* **38 B6** 46 11N 9 55 E
Valuyki, *Russia* **52 E4** 50 10N 38 5 E
Valverde, *Canary Is.* **33 G2** 27 48N 17 55W
Valverde del Camino, *Spain* **37 H4** 37 35N 6 47W
Valverde del Fresno, *Spain* **36 E4** 40 15N 6 51W
Vama, *Romania* **46 B6** 47 34N 25 42 E
Vammala, *Finland* **13 F20** 61 20N 22 54 E
Vámos, *Greece* **32 D6** 35 24N 24 13 E
Vamsadhara →, *India* **82 E7** 18 21N 84 8 E
Van, *Turkey* **89 C10** 38 30N 43 20 E
Van, L. = Van Gölü, *Turkey* **89 C10** 38 30N 43 0 E
Van Alstyne, *U.S.A.* **139 J6** 33 25N 96 35W
Van Bruyssel, *Canada* **129 C5** 47 56N 72 9W
Van Buren, *Canada* **129 C6** 47 10N 67 55W
Van Buren, *Ark., U.S.A.* **139 H7** 35 26N 94 21W
Van Buren, *Maine, U.S.A.* **135 B11** 47 10N 67 58W
Van Buren, *Mo., U.S.A.* **139 G9** 37 0N 91 1W
Van Canh, *Vietnam* **76 F7** 13 37N 109 0 E
Van Diemen, C., *N. Terr., Australia* **112 B5** 11 9 S 130 24 E
Van Diemen, C., *Queens., Australia* **114 B2** 16 30 S 139 46 E
Van Diemen G., *Australia* **112 B5** 11 45 S 132 0 E
Van Gölü, *Turkey* **89 C10** 38 30N 43 0 E
Van Horn, *U.S.A.* **139 K2** 31 3N 104 50W
Van Horne, *U.S.A.* **140 B4** 42 1N 92 4W
Van Ninh, *Vietnam* **76 F7** 12 42N 109 14 E
Van Rees, Pegunungan, *Indonesia* **73 B5** 2 35 S 138 15 E
Van Tassell, *U.S.A.* **138 D2** 42 40N 104 5W
Van Tivu, *India* **83 K4** 8 51N 78 15 E
Van Wert, *U.S.A.* **141 D12** 40 52N 84 35W
Van Yen, *Vietnam* **76 B5** 21 4N 104 42 E
Vanadzor, *Armenia* **53 K7** 40 48N 44 30 E
Vanavara, *Russia* **57 C11** 60 22N 102 16 E
Vanceburg, *U.S.A.* **141 F13** 38 36N 83 19W
Vancouver, *Canada* **130 D4** 49 15N 123 10W
Vancouver, *U.S.A.* **144 E4** 45 38N 122 40W
Vancouver, C., *Australia* **113 G2** 35 2 S 118 11 E
Vancouver I., *Canada* **130 D3** 49 50N 126 0W
Vandalia, *Ill., U.S.A.* **140 F7** 38 58N 89 6W
Vandalia, *Mo., U.S.A.* **140 E5** 39 19N 91 29W
Vandalia, *Ohio, U.S.A.* **141 E12** 39 54N 84 12W
Vandavasi, *India* **83 H4** 12 30N 79 30 E
Vandeloos B., *Sri Lanka* **83 K5** 8 0N 81 45 E
Vandenburg, *U.S.A.* **145 L6** 34 35N 120 33W
Vanderbijlpark, *S. Africa* **105 D4** 26 42 S 27 54 E
Vandergrift, *U.S.A.* **136 F5** 40 36N 79 34W
Vanderhoof, *Canada* **130 C4** 54 0N 124 0W
Vanderkloof Dam, *S. Africa* **104 E3** 30 4 S 24 40 E
Vanderlin I., *Australia* **114 B2** 15 44 S 137 2 E
Vandyke, *Australia* **114 C4** 24 10 S 147 51 E
Vänern, *Sweden* **15 F7** 58 47N 13 30 E
Vänersborg, *Sweden* **15 F6** 58 26N 12 19 E
Vang Vieng, *Laos* **76 C4** 18 58N 102 32 E
Vanga, *Kenya* **106 C4** 4 35 S 39 12 E
Vangaindrano, *Madag.* **105 C8** 23 21 S 47 36 E
Vanguard, *Canada* **131 D7** 49 55N 107 20W
Vangunu, *Solomon Is.* **121 M10** 8 40 S 158 5 E
Vanier, *Canada* **128 C4** 45 27N 75 40W
Vanimo, *Papua N. G.* **120 B1** 2 42 S 141 21 E
Vanivilasa Sagara, *India* **83 H3** 13 45N 76 30 E
Vaniyambadi, *India* **83 H4** 12 46N 78 44 E
Vankleek Hill, *Canada* **128 C5** 45 32N 74 40W
Vanna, *Norway* **12 A18** 70 6N 19 50 E
Vännäs, *Sweden* **12 E18** 63 58N 19 48 E
Vannes, *France* **22 E4** 47 40N 2 47W
Vanoise, Massif de la, *France* **25 C10** 45 25N 6 40 E
Vanrhynsdorp, *S. Africa* **104 E2** 31 36 S 18 44 E
Vanrook, *Australia* **114 B3** 16 57 S 141 57 E
Vansbro, *Sweden* **13 F16** 60 32N 14 15 E
Vansittart B., *Australia* **112 B4** 14 3 S 126 17 E
Vantaa, *Finland* **13 F21** 60 18N 24 58 E
Vanthli, *India* **80 J4** 21 28N 70 25 E
Vanua Levu, *Fiji* **121 A2** 16 33 S 179 15 E
Vanuatu ■, *Pac. Oc.* **121 E6** 15 0 S 168 0 E
Vanwyksvlei, *S. Africa* **104 E3** 30 18 S 21 49 E
Vanzylsrus, *S. Africa* **104 D3** 26 52 S 22 4 E
Vapnyarka, *Ukraine* **51 H5** 48 32N 28 45 E
Var □, *France* **25 E10** 43 27N 6 18 E
Var →, *France* **25 E11** 43 39N 7 12 E
Vara, *Sweden* **15 F6** 58 16N 12 55 E
Varada →, *India* **83 G2** 15 0N 75 40 E
Varades, *France* **22 E5** 47 25N 1 1W
Varáita →, *Italy* **38 D4** 44 9N 7 53 E
Varallo, *Italy* **38 C5** 45 49N 8 15 E
Varanasi, *India* **81 G10** 25 22N 83 0 E
Varanger-halvøya, *Norway* **12 A23** 70 25N 29 30 E
Varangerfjorden, *Norway* **12 A23** 70 3N 29 25 E
Varaždin, *Croatia* **39 B13** 46 20N 16 20 E
Varazze, *Italy* **38 D5** 44 22N 8 34 E
Varberg, *Sweden* **15 G6** 57 6N 12 20 E
Vardar = Axiós →, *Greece* **44 D4** 40 57N 22 35 E
Varde, *Denmark* **15 J2** 55 38N 8 29 E
Varde Å →, *Denmark* **15 J2** 55 35N 8 19 E
Vardø, *Norway* **12 A24** 70 23N 31 5 E
Varel, *Germany* **26 B4** 53 23N 8 8 E
Varella, Mui, *Vietnam* **76 F7** 12 54N 109 26 E
Varėna, *Lithuania* **13 J21** 54 12N 24 30 E
Varennes-sur-Allier, *France* **24 B7** 46 19N 3 24 E
Vareš, *Bos.-H.* **42 C3** 44 12N 18 23 E
Varese, *Italy* **38 C5** 45 48N 8 50 E
Varese Lígure, *Italy* **38 D6** 44 49N 9 45 E
Vårgårda, *Sweden* **15 F6** 58 2N 12 49 E
Vargem Bonita, *Brazil* **155 F2** 20 20 S 46 22W
Vargem Grande, *Brazil* **154 B3** 3 33 S 43 56W
Varginha, *Brazil* **159 A6** 21 33 S 45 25W
Vargön, *Sweden* **15 F6** 58 22N 12 20 E
Variadero, *U.S.A.* **139 H2** 35 43N 104 17W
Varillas, *Chile* **158 A1** 24 0 S 70 10W
Väring, *Sweden* **15 F8** 58 30N 14 0 E
Varkaus, *Finland* **13 E22** 62 19N 27 50 E
Varna, *Bulgaria* **43 D12** 43 13N 27 56 E
Varna, *Russia* **54 E8** 53 24N 60 58 E
Varna, *U.S.A.* **140 C7** 41 2N 89 14W
Varna →, *India* **82 F2** 16 48N 74 32 E
Värnamo, *Sweden* **13 H16** 57 10N 14 3 E
Varnsdorf, *Czech.* **30 A7** 50 55N 14 35 E
Värö, *Sweden* **15 G6** 57 16N 12 11 E
Vars, *Canada* **137 A9** 45 21N 75 21W
Varsseveld, *Neths.* **20 E8** 51 56N 6 29 E
Varto, *Turkey* **89 C9** 39 10N 41 27 E
Varvarin, *Serbia, Yug.* **42 D6** 43 43N 21 20 E
Varzaneh, *Iran* **85 C7** 32 25N 52 40 E
Várzea Alegre, *Brazil* **154 C4** 6 47 S 39 17W
Várzea da Palma, *Brazil* **155 E3** 17 36 S 44 44W
Várzea Grande, *Brazil* **157 D6** 15 39 S 56 8W
Varzi, *Italy* **38 D6** 44 49N 9 12 E
Varzo, *Italy* **38 B5** 46 12N 8 15 E
Varzy, *France* **23 E10** 47 22N 3 20 E
Vas □, *Hungary* **31 D9** 47 10N 16 55 E
Vasa Barris →, *Brazil* **154 D4** 11 10 S 37 10W
Vásárosnamény, *Hungary* **31 C15** 48 9N 22 19 E
Vascão →, *Portugal* **37 H3** 37 31N 7 31W
Vaşcău, *Romania* **46 C3** 46 28N 22 30 E
Vascongadas = País Vasco □, *Spain* **34 C2** 42 50N 2 45W
Vāshīr, *Afghan.* **79 B1** 32 16N 63 51 E
Vasht = Khāsh, *Iran* **85 D9** 28 15N 61 15 E
Vasilevichi, *Belarus* **51 F5** 52 15N 29 50 E
Vasilikón, *Greece* **45 F5** 38 25N 23 40 E
Vasilkov = Vasylkiv, *Ukraine* **51 G6** 50 7N 30 15 E
Vaslui, *Romania* **46 C8** 46 38N 27 42 E
Vaslui □, *Romania* **46 C8** 46 30N 27 45 E
Vassar, *Canada* **131 D9** 49 10N 95 55W
Vassar, *U.S.A.* **134 D4** 43 22N 83 35W
Västerås, *Sweden* **14 E10** 59 37N 16 38 E
Västerbotten, *Sweden* **12 D18** 64 36N 20 4 E
Västerdalälven →, *Sweden* **13 F16** 60 30N 14 7 E
Västernorrlands län □, *Sweden* **14 A11** 63 30N 17 30 E
Västervik, *Sweden* **13 H17** 57 43N 16 33 E
Västmanland, *Sweden* **13 G16** 59 45N 16 20 E
Vasto, *Italy* **39 F11** 42 8N 14 40 E
Vasvár, *Hungary* **31 D9** 47 3N 16 47 E
Vasylkiv, *Ukraine* **51 G6** 50 7N 30 15 E
Vatan, *France* **23 E8** 47 4N 1 50 E
Vaté = Efate, I., *Vanuatu* **121 G6** 17 40 S 168 25 E
Vathí, *Itháki, Greece* **45 F2** 38 18N 20 40 E
Vathí, *Sámos, Greece* **45 G9** 37 46N 27 1 E
Váthia, *Greece* **45 H4** 36 29N 22 29 E
Vatican City ■, *Europe* **39 G9** 41 54N 12 27 E
Vaticano, C., *Italy* **41 D8** 38 37N 15 50 E
Vatili, *Cyprus* **32 D12** 35 6N 33 40 E
Vatin, *Serbia, Yug.* **42 B6** 45 12N 21 20 E
Vatnajökull, *Iceland* **12 D5** 64 30N 16 48W
Vatnås, *Norway* **14 E3** 59 58N 9 37 E
Vatólakkos, *Greece* **32 D5** 35 27N 23 53 E
Vatoloha, *Madag.* **105 B8** 17 52 S 47 48 E
Vatomandry, *Madag.* **105 B8** 19 20 S 48 59 E
Vatra-Dornei, *Romania* **46 B6** 47 22N 25 22 E
Vättern, *Sweden* **15 F8** 58 25N 14 30 E
Vättis, *Switz.* **29 C8** 46 55N 9 27 E
Vatulele, *Fiji* **121 B1** 18 33 S 177 37 E
Vaucluse □, *France* **25 E9** 43 50N 5 20 E
Vaucouleurs, *France* **23 D12** 48 37N 5 40 E
Vaud □, *Switz.* **28 C2** 46 35N 6 30 E
Vaughn, *Mont., U.S.A.* **142 C8** 47 33N 111 33W
Vaughn, *N. Mex., U.S.A.* **143 J11** 34 36N 105 13W
Vaulruz, *Switz.* **28 C3** 46 38N 6 58 E
Vaupés = Uaupés →, *Brazil* **152 C4** 0 2N 67 16W
Vaupes □, *Colombia* **152 C3** 1 0N 71 0W
Vauvert, *France* **25 E8** 43 42N 4 17 E
Vauxhall, *Canada* **130 C6** 50 5N 112 9W
Vava'u, *Tonga* **121 P14** 18 36 S 174 0W
Vavoua, *Ivory C.* **100 D3** 7 23N 6 29W
Vawkavysk, *Belarus* **51 F3** 53 9N 24 30 E
Vaxholm, *Sweden* **14 E12** 59 25N 18 20 E
Växjö, *Sweden* **13 H16** 56 52N 14 50 E
Vaygach, Ostrov, *Russia* **56 C6** 70 0N 60 0 E
Váyia, Ákra, *Greece* **32 C10** 36 15N 28 11 E
Vazovgrad, *Bulgaria* **43 E9** 42 39N 24 45 E
Veadeiros, *Brazil* **155 D2** 14 7 S 47 31W
Vechta, *Germany* **26 C4** 52 44N 8 17 E
Vechte →, *Neths.* **20 C8** 52 34N 6 6 E
Vecsés, *Hungary* **31 D12** 47 26N 19 19 E
Vedaranniyam, *India* **83 J4** 10 25N 79 50 E
Veddige, *Sweden* **15 G6** 57 17N 12 20 E
Vedea →, *Romania* **46 F6** 43 53N 25 59 E
Vedia, *Argentina* **158 C3** 34 30 S 61 31W
Vedra, I. del, *Spain* **33 C7** 38 52N 1 12 E
Vedrin, *Belgium* **21 G5** 50 30N 4 52 E
Veendam, *Neths.* **20 B9** 53 5N 6 52 E
Veenendaal, *Neths.* **20 D7** 52 2N 5 34 E
Veerle, *Belgium* **21 F5** 51 4N 4 59 E
Vefsna →, *Norway* **12 D15** 65 48N 13 10 E
Vega, *Norway* **12 D14** 65 40N 11 55 E
Vega, *U.S.A.* **139 H3** 35 15N 102 26W
Vegadeo, *Spain* **36 B3** 43 27N 7 4W
Veghel, *Neths.* **21 E7** 51 37N 5 32 E
Vegorritis, Límni, *Greece* **44 D3** 40 45N 21 45 E
Vegreville, *Canada* **130 C6** 53 30N 112 5W
Vegusdal, *Norway* **15 F2** 58 32N 8 10 E
Veii, *Italy* **39 F9** 42 0N 12 24 E
Veitch, *Australia* **116 C4** 34 39 S 140 31 E
Vejen, *Denmark* **15 J3** 55 30N 9 9 E
Vejer de la Frontera, *Spain* **37 J5** 36 15N 5 59W
Vejle, *Denmark* **15 J3** 55 43N 9 30 E
Vejle Fjord, *Denmark* **15 J3** 55 40N 9 50 E
Vela Luka, *Croatia* **39 F13** 42 59N 16 44 E
Velanai I., *Sri Lanka* **83 K4** 9 45N 79 45 E
Velas, C., *Costa Rica* **148 D2** 10 21N 85 52W
Velasco, Sierra de, *Argentina* **158 B2** 29 20 S 67 10W
Velay, Mts. du, *France* **24 D7** 45 0N 3 40 E
Velddrif, *S. Africa* **104 E2** 32 42 S 18 11 E
Veldegem, *Belgium* **21 F2** 51 7N 3 10 E
Velden, *Neths.* **21 F8** 51 25N 6 10 E
Veldhoven, *Neths.* **21 F6** 51 24N 5 25 E
Velebit Planina, *Croatia* **39 D12** 44 50N 15 20 E
Velebitski Kanal, *Croatia* **39 D11** 44 45N 14 55 E
Veleka →, *Bulgaria* **43 E12** 42 4N 27 58 E
Velenje, *Slovenia* **39 B12** 46 23N 15 8 E
Velestínon, *Greece* **44 E4** 39 23N 22 43 E
Velež, *Bos.-H.* **42 D3** 43 19N 18 2 E
Vélez, *Colombia* **152 B3** 6 1N 73 41W
Vélez Blanco, *Spain* **35 H2** 37 41N 2 5W
Vélez Málaga, *Spain* **37 J6** 36 48N 4 5W
Vélez Rubio, *Spain* **35 H2** 37 41N 2 5W
Velhas →, *Brazil* **155 E3** 17 13 S 44 49W
Velik Jastrebac, *Serbia, Yug.* **42 D6** 43 25N 21 30 E
Velika, *Croatia* **42 B2** 45 27N 17 40 E
Velika Gorica, *Croatia* **39 C13** 45 44N 16 5 E
Velika Gradište, *Serbia, Yug.* **42 C6** 44 46N 21 29 E
Velika Kapela, *Croatia* **39 C12** 45 10N 15 5 E
Velika Kladuša, *Bos.-H.* **39 C12** 45 11N 15 48 E
Velika Morava →, *Serbia, Yug.* **42 C6** 44 43N 21 3 E
Velika Plana, *Serbia, Yug.* **42 C6** 44 20N 21 1 E
Velikaya →, *Russia* **50 D5** 57 48N 28 10 E
Velikaya Kema, *Russia* **60 B8** 45 30N 137 12 E
Velikaya Lepetikha, *Ukraine* **51 J7** 47 2N 33 58 E
Veliké Kapušany, *Slovak Rep.* **31 C15** 48 34N 22 5 E
Velike Lašče, *Slovenia* **39 C11** 45 49N 14 45 E
Veliki Backu Kanal, *Serbia, Yug.* **42 B4** 45 45N 19 15 E
Veliki Popović, *Serbia, Yug.* **42 C6** 44 8N 21 18 E
Veliki Ustyug, *Russia* **48 B8** 60 47N 46 20 E
Velikiye Luki, *Russia* **50 D6** 56 25N 30 32 E
Veliko Tŭrnovo, *Bulgaria* **43 D10** 43 5N 25 41 E
Velikonda Range, *India* **83 G4** 14 45N 79 10 E
Velingrad, *Bulgaria* **43 E8** 42 4N 23 58 E
Velino, Mte., *Italy* **39 F10** 42 9N 13 23 E
Velizh, *Russia* **50 E6** 55 36N 31 11 E
Velké Karlovice, *Czech.* **31 B11** 49 20N 18 17 E
Velke Meziřici, *Czech.* **30 B9** 49 21N 16 1 E
Veľký Žitný ostrov, *Slovak Rep.* **31 C10** 48 5N 17 20 E
Vella G., *Solomon Is.* **121 M9** 8 0 S 156 50 E
Vella Lavella, *Solomon Is.* **121 L9** 7 45 S 156 40 E
Vellar →, *India* **83 J4** 11 30N 79 36 E
Velletri, *Italy* **40 A5** 41 41N 12 47 E
Vellinge, *Sweden* **15 J7** 55 29N 13 0 E
Vellore, *India* **83 H4** 12 57N 79 10 E
Velp, *Neths.* **20 D7** 52 0N 5 59 E
Velsen-Noord, *Neths.* **20 D5** 52 27N 4 40 E
Velsk, *Russia* **50 B11** 61 10N 42 5 E
Velten, *Germany* **26 C9** 52 42N 13 10 E
Veluwe Meer, *Neths.* **20 D7** 52 24N 5 44 E
Velva, *U.S.A.* **138 A4** 48 4N 100 56W
Velvendós, *Greece* **44 D4** 40 15N 22 6 E
Vembanad L., *India* **83 K3** 9 36N 76 15 E
Veme, *Norway* **14 D4** 60 14N 10 7 E
Ven, *Sweden* **15 J6** 55 55N 12 45 E
Venaco, *France* **25 F13** 42 14N 9 11 E
Venado Tuerto, *Argentina* **158 C3** 33 50 S 62 0W
Venafro, *Italy* **41 A7** 41 29N 14 2 E
Venarey-les-Laumes, *France* **23 E11** 47 32N 4 26 E
Venaria, *Italy* **38 C4** 45 8N 7 38 E
Venčane, *Serbia, Yug.* **42 C5** 44 24N 20 28 E
Vence, *France* **25 E11** 43 43N 7 6 E
Vendas Novas, *Portugal* **37 G2** 38 39N 8 27W
Vendée □, *France* **22 F5** 46 50N 1 35W
Vendée →, *France* **22 F5** 46 20N 1 10W
Vendéen, Bocage, *France* **24 B2** 46 40N 1 20W
Vendeuvre-sur-Barse, *France* **23 D11** 48 14N 4 28 E
Vendôme, *France* **22 E8** 47 47N 1 3 E
Vendrell, *Spain* **34 D6** 41 10N 1 30 E
Vendsyssel, *Denmark* **15 G4** 57 22N 10 0 E
Véneta, L., *Italy* **39 C9** 45 23N 12 25 E
Véneto □, *Italy* **39 C8** 45 30N 12 0 E
Venev, *Russia* **52 C4** 54 22N 38 17 E
Venézia, *Italy* **39 C9** 45 27N 12 21 E
Venézia, G. di, *Italy* **39 C10** 45 15N 13 0 E
Venezuela ■, *S. Amer.* **152 B4** 8 0N 66 0W
Venezuela, G. de, *Venezuela* **152 A3** 11 30N 71 0W
Vengurla, *India* **83 G1** 15 53N 73 45 E
Vengurla Rocks, *India* **83 G1** 15 55N 73 22 E
Venice = Venézia, *Italy* **39 C9** 45 27N 12 21 E
Venkatagiri, *India* **83 G4** 14 0N 79 35 E
Venkatapuram, *India* **82 E5** 18 20N 80 30 E
Venlo, *Neths.* **21 F8** 51 22N 6 11 E
Vennesla, *Norway* **13 G12** 58 15N 8 0 E

Vukovar 281

Villanueva del Fresno, Spain 37 G3 38 23N 7 10W
Villanueva y Geltrú, Spain 34 D6 41 13N 1 40 E
Villaodrid, Spain 36 B3 43 20N 7 11W
Villaputzu, Italy 40 C2 39 26N 9 34 E
Villar del Arzobispo, Spain 34 F4 39 44N 0 50W
Villar del Rey, Spain 37 F4 39 7N 6 50W
Villarcayo, Spain 34 C1 42 56N 3 34W
Villard-Bonnot, France 25 C9 45 14N 5 53 E
Villard-de-Lans, France 25 C9 45 3N 5 33 E
Villarino de los Aires, Spain 36 D4 41 18N 6 23W
Villarosa, Italy 41 E7 37 35N 14 10 E
Villarramiel, Spain 36 C6 42 2N 4 55W
Villarreal, Spain 34 F4 39 55N 0 3W
Villarrica, Chile 160 A2 39 15 S 72 15W
Villarrica, Paraguay 158 B4 25 40 S 56 30W
Villarrobledo, Spain 35 F2 39 18N 2 36W
Villarroya de la Sierra, Spain 34 D3 41 27N 1 46W
Villarrubia de los Ojos, Spain 35 F1 39 14N 3 36W
Villars-les-Dombes, France 25 B9 46 0N 5 3 E
Villarta de San Juan, Spain 35 F1 39 15N 3 25W
Villasayas, Spain 34 D2 41 24N 2 39W
Villaseca de los Gamitos, Spain 36 D4 41 2N 6 7W
Villastar, Spain 34 E3 40 17N 1 9W
Villatobas, Spain 34 F1 39 54N 3 20W
Villavicencio, Argentina 158 C2 32 28 S 69 0W
Villavicencio, Colombia 152 C3 4 9N 73 37W
Villaviciosa, Spain 36 B5 43 32N 5 27W
Villazón, Bolivia 158 A2 22 0 S 65 35W
Ville-Marie, Canada 128 C4 47 20N 79 30W
Ville Platte, U.S.A. 139 K8 30 41N 92 17W
Villedieu-les-Poêlles, France 22 D5 48 50N 1 13W
Villefort, France 24 D7 44 28N 3 56 E
Villefranche-de-Lauragais, France 24 E5 43 25N 1 44 E
Villefranche-de-Rouergue, France 24 D6 44 21N 2 2 E
Villefranche-du-Périgord, France 24 D5 44 38N 1 5 E
Villefranche-sur-Cher, France 23 E8 47 18N 1 46 E
Villefranche-sur-Saône, France 25 C8 45 59N 4 43 E
Villegrande, Bolivia 157 D5 18 30 S 64 10W
Villel, Spain 34 E3 40 14N 1 12W
Villemaur-sur-Vanne, France 23 D10 48 15N 3 44 E
Villemur-sur-Tarn, France 24 E5 43 51N 1 31 E
Villena, Spain 35 G4 38 39N 0 52W
Villenauxe-la-Grande, France 23 D10 48 35N 3 33 E
Villenave-d'Ornon, France 24 D3 44 46N 0 33W
Villeneuve, Italy 38 C4 45 42N 7 14 E
Villeneuve, Switz. 28 D3 46 24N 6 56 E
Villeneuve-l'Archevêque, France 23 D10 48 14N 3 32 E
Villeneuve-lès-Avignon, France 25 E8 43 58N 4 49 E
Villeneuve-St.-Georges, France 23 D9 48 44N 2 28 E
Villeneuve-sur-Allier, France 24 B7 46 40N 3 13 E
Villeneuve-sur-Lot, France 24 D4 44 24N 0 42 E
Villeréal, France 24 D4 44 38N 0 45 E
Villers-Bocage, France 22 C6 49 3N 0 40W
Villers-Bretonneux, France 23 C9 49 50N 2 30 E
Villers-Cotterêts, France 23 C10 49 15N 3 4 E
Villers-le-Bouillet, Belgium 21 G6 50 34N 5 15 E
Villers-le-Gambon, Belgium 21 H5 50 11N 4 37 E
Villers-sur-Mer, France 22 C6 49 21N 0 2W
Villersexel, France 23 E13 47 33N 6 26 E
Villerupt, France 23 C12 49 28N 5 55 E
Villerville, France 22 C7 49 26N 0 5 E
Villiers, S. Africa 105 D4 27 2 S 28 36 E
Villingen-Schwenningen, Germany 27 G4 48 3N 8 26 E
Villisca, U.S.A. 140 D2 40 56N 94 59W
Villupuram, India 83 J4 11 59N 79 31 E
Vilna, Canada 130 C6 54 7N 111 55W
Vilnius, Lithuania 13 J21 54 38N 25 19 E
Vils →, Germany 27 G9 48 37N 13 11 E
Vilsbiburg, Germany 27 G8 48 26N 12 22 E
Vilshofen, Germany 27 G9 48 37N 13 11 E
Vilusi, Montenegro, Yug. 42 E3 42 44N 18 34 E
Vilvoorde, Belgium 21 G4 50 56N 4 26 E
Vilyuy →, Russia 57 C13 64 24N 126 26 E
Vilyuysk, Russia 57 C13 63 40N 121 35 E
Vimercate, Italy 38 C6 45 37N 9 22 E
Vimiosa, Portugal 36 D4 41 35N 6 31W
Vimoutiers, France 22 D7 48 57N 0 10 E
Vimperk, Czech. 30 B6 49 3N 13 46 E
Viña del Mar, Chile 158 C1 33 0 S 71 30W
Vinaroz, Spain 34 E5 40 30N 0 27 E
Vincennes, U.S.A. 141 F9 38 41N 87 32W
Vincent, U.S.A. 145 L8 34 33N 118 11W
Vinces, Ecuador 152 D2 1 32 S 79 45W
Vinchina, Argentina 158 B2 28 45 S 68 15W
Vindelälven →, Sweden 12 E18 63 55N 19 50 E
Vindeln, Sweden 12 D18 64 12N 19 43 E
Vinderup, Denmark 15 H2 56 29N 8 45 E
Vindhya Ra., India 80 H7 22 50N 77 0 E
Vine Grove, U.S.A. 141 G11 37 49N 85 59W
Vineland, U.S.A. 134 F8 39 29N 75 2W
Vinga, Romania 46 C2 46 0N 21 14 E
Vingnes, Norway 14 C4 61 7N 10 26 E
Vinh, Vietnam 76 C5 18 45N 105 38 E
Vinh Linh, Vietnam 76 D6 17 4N 107 2 E
Vinh Long, Vietnam 77 G5 10 16N 105 57 E
Vinh Yen, Vietnam 76 B5 21 21N 105 35 E
Vinhais, Portugal 36 D3 41 50N 7 5W
Vinica, Croatia 39 B13 46 20N 16 9 E
Vinica, Slovenia 39 C12 45 28N 15 16 E
Vinita, U.S.A. 139 G7 36 39N 95 9W
Vinkeveen, Neths. 20 D5 52 13N 4 56 E
Vinkovci, Croatia 42 B3 45 19N 18 48 E
Vinnitsa = Vinnytsya, Ukraine 51 H5 49 15N 28 30 E
Vinnytsya, Ukraine 51 H5 49 15N 28 30 E
Vinstra, Norway 14 C3 61 37N 9 44 E
Vintar, Phil. 70 B3 18 14N 120 39 E
Vinton, Calif., U.S.A. 144 F6 39 48N 120 10W
Vinton, Iowa, U.S.A. 140 B4 42 10N 92 1W
Vinton, La., U.S.A. 139 K8 30 11N 93 35W
Vinţu de Jos, Romania 46 D4 46 0N 23 30 E
Viöl, Germany 26 A5 54 34N 9 11 E
Viola, U.S.A. 140 C6 41 12N 90 35W
Violet Town, Australia 117 D6 36 38 S 145 42 E
Vipava, Slovenia 39 C10 45 51N 13 58 E
Vipiteno, Italy 39 B8 46 54N 11 26 E
Vir, Croatia 39 D12 44 17N 15 3 E
Vir, Tajikistan 55 E6 37 45N 72 5 E
Virac, Phil. 70 E5 13 30N 124 20 E
Virachei, Cambodia 76 F6 13 59N 106 49 E
Virago Sd., Canada 130 C2 54 0N 132 30W
Virajpet = Virarajendrapet, India 83 H2 12 10N 75 50 E
Viramgam, India 80 H5 23 5N 72 0 E
Viranşehir, Turkey 89 D8 37 13N 39 45 E
Virarajendrapet, India 83 H2 12 10N 75 50 E
Viravanallur, India 83 K3 8 40N 77 30 E
Virden, Canada 131 D8 49 50N 100 56W
Virden, U.S.A. 140 E7 39 30N 89 46W
Vire, France 22 D6 48 50N 0 53W
Vire →, France 22 C5 49 20N 1 7W
Virgem da Lapa, Brazil 155 E3 16 49 S 42 21W
Vírgenes, C., Argentina 160 D3 52 19 S 68 21W
Virgin →, Canada 131 B7 57 2N 108 17W
Virgin →, U.S.A. 143 H6 36 28N 114 21W
Virgin Gorda, Virgin Is. 149 C7 18 30N 64 26W
Virgin Is. (British) ■, W. Indies 149 C7 18 30N 64 30W
Virgin Is. (U.S.) ■, W. Indies 149 C7 18 20N 65 0W
Virginia, S. Africa 104 D4 28 8 S 26 55 E
Virginia, Ill., U.S.A. 140 E6 39 57N 90 13W
Virginia, Minn., U.S.A. 138 B8 47 31N 92 32W
Virginia □, U.S.A. 134 G7 37 30N 78 45W
Virginia Beach, U.S.A. 134 G8 36 51N 75 59W
Virginia City, Mont., U.S.A. 142 D8 45 18N 111 56W
Virginia City, Nev., U.S.A. 144 F7 39 19N 119 39W
Virginia Falls, Canada 130 A3 61 38N 125 42W
Virginiatown, Canada 128 C4 48 9N 79 36W
Virieu-le-Grand, France 25 C9 45 51N 5 39 E
Virje, Croatia 42 A1 46 4N 16 59 E
Viroqua, U.S.A. 138 D9 43 34N 90 53W
Virovitica, Croatia 42 B2 45 51N 17 21 E
Virpazar, Montenegro, Yug. 42 E4 42 14N 19 6 E
Virton, Belgium 21 J7 49 35N 5 32 E
Virú, Peru 156 B2 8 25 S 78 45W
Virudunagar, India 83 K3 9 30N 77 58 E
Vis, Croatia 39 E13 43 4N 16 10 E
Vis Kanal, Croatia 39 E13 43 4N 16 5 E
Visalia, U.S.A. 144 J7 36 20N 119 18W
Visayan Sea, Phil. 71 F4 11 30N 123 30 E
Visby, Sweden 13 H18 57 37N 18 18 E
Viscount Melville Sd., Canada 6 B2 74 10N 108 0W
Visé, Belgium 21 G7 50 44N 5 41 E
Višegrad, Bos.-H. 42 D4 43 47N 19 17 E
Viseu, Brazil 154 B2 1 10 S 46 5W
Viseu, Portugal 36 E3 40 40N 7 55W
Viseu □, Portugal 36 E3 40 40N 7 55W
Vişeu de Sus, Romania 46 B5 47 45N 24 25 E
Vishakhapatnam, India 82 F6 17 45N 83 20 E
Vishera →, Russia 54 A6 59 55N 56 25 E
Viskafors, Sweden 15 G6 57 37N 12 50 E
Visnagar, India 80 H5 23 45N 72 32 E
Višnja Gora, Slovenia 39 C11 45 58N 14 45 E
Viso, Mte., Italy 38 D4 44 38N 7 5 E
Viso del Marqués, Spain 35 G1 38 32N 3 34W
Visoko, Bos.-H. 42 D3 43 58N 18 10 E
Visokoi I., Antarctica 7 B1 56 43 S 27 15W
Visp, Switz. 28 D5 46 17N 7 52 E
Vispa →, Switz. 28 D5 46 9N 7 48 E
Visselhövede, Germany 26 C5 52 59N 9 34 E
Vissoie, Switz. 28 D5 46 13N 7 36 E
Vista, U.S.A. 145 M9 33 12N 117 14W
Vistonikos, Ormos, Greece 44 D7 41 0N 25 7 E
Vistula = Wisła →, Poland 47 A5 54 22N 18 55 E
Vit →, Bulgaria 43 D9 43 30N 24 30 E
Vitanje, Slovenia 39 B12 46 25N 15 18 E
Vitebsk = Vitsyebsk, Belarus 50 E6 55 10N 30 15 E
Viterbo, Italy 39 F9 42 25N 12 6 E
Viti Levu, Fiji 121 A1 17 30 S 177 30 E
Vitiaz Str., Papua N. G. 120 C4 5 40 S 147 10 E
Vitigudino, Spain 36 D4 41 1N 6 26W
Vitim, Russia 57 D12 59 28N 112 35 E
Vitim →, Russia 57 D12 59 26N 112 34 E
Vitina, Bos.-H. 42 D2 43 17N 17 29 E
Vitína, Greece 45 G4 37 40N 22 10 E
Vitória, Brazil 155 F3 20 20 S 40 22W
Vitoria, Spain 34 C2 42 50N 2 41W
Vitória da Conquista, Brazil 155 D3 14 51 S 40 51W
Vitória de São Antão, Brazil 154 C4 8 10 S 35 20W
Vitorino Friere, Brazil 154 B2 4 4 S 45 10W
Vitré, France 22 D5 48 8N 1 12W
Vitry-le-François, France 23 D11 48 43N 4 33 E
Vitsi, Óros, Greece 44 D3 40 40N 21 25 E
Vitsyebsk, Belarus 50 E6 55 10N 30 15 E
Vitteaux, France 23 E11 47 24N 4 30 E
Vittel, France 23 D12 48 12N 5 57 E
Vittória, Italy 41 F7 36 57N 14 32 E
Vittório Véneto, Italy 39 C9 45 59N 12 18 E
Vitu Is., Papua N. G. 120 C5 4 50 S 149 25 E
Vivario, France 25 F13 42 10N 9 11 E
Vivegnis, Belgium 21 G7 50 42N 5 39 E
Viver, Spain 34 F4 39 55N 0 36W
Vivero, Spain 36 B3 43 39N 7 38W
Viviers, France 25 D8 44 30N 4 40 E
Vivonne, Australia 116 C2 35 59 S 137 9 E
Vivonne, France 24 B4 46 25N 0 15 E
Vivonne B., Australia 116 C2 35 59 S 137 9 E
Vizcaíno, Desierto de, Mexico 146 B2 27 40N 113 50W
Vizcaíno, Sierra, Mexico 146 B2 27 30N 114 0W
Vizcaya □, Spain 34 B2 43 15N 2 45W
Vize, Turkey 88 B2 41 34N 27 45 E
Vizianagaram, India 82 E6 18 6N 83 30 E
Vizille, France 25 C9 45 5N 5 46 E
Viziñada, Croatia 39 C10 45 20N 13 46 E
Viziru, Romania 46 D8 45 0N 27 43 E
Vizovice, Czech. 31 B10 49 12N 17 56 E
Vizzini, Italy 41 E7 37 10N 14 45 E
Vjosa →, Albania 44 D1 40 37N 19 24 E
Vlaardingen, Neths. 20 E4 51 55N 4 21 E
Vlădeasa, Romania 46 C3 46 47N 22 50 E
Vladicin Han, Serbia, Yug. 42 E7 42 42N 22 1 E
Vladikavkaz, Russia 53 J7 43 0N 44 35 E
Vladimir, Russia 52 B5 56 15N 40 30 E
Vladimir Volynskiy = Volodymyr-Volynskyy, Ukraine 51 G3 50 50N 24 18 E
Vladimirci, Serbia, Yug. 42 C4 44 36N 19 45 E
Vladimirovac, Serbia, Yug. 42 B5 45 1N 20 53 E
Vladimirovka, Russia 53 F8 48 27N 46 10 E
Vladimirovo, Bulgaria 43 D8 43 32N 23 22 E
Vladimorvka, Kazakhstan 52 E10 50 51N 51 8 E
Vladislavovka, Ukraine 51 K8 45 12N 35 29 E
Vladivostok, Russia 57 E14 43 10N 131 53 E
Vlamertinge, Belgium 21 G1 50 51N 2 49 E
Vlasenica, Bos.-H. 42 C3 44 11N 18 59 E
Vlašić, Bos.-H. 42 C2 44 19N 17 37 E
Vlašim, Czech. 30 B7 49 40N 14 53 E
Vlasinsko Jezero, Serbia, Yug. 42 E7 42 44N 22 22 E
Vlasotinci, Serbia, Yug. 42 E7 42 59N 22 7 E
Vleuten, Neths. 20 D6 52 6N 5 1 E
Vlieland, Neths. 20 B5 53 16N 4 55 E
Vliestroom, Neths. 20 B6 53 19N 5 8 E
Vlijmen, Neths. 21 E6 51 42N 5 14 E
Vlissingen, Neths. 21 F3 51 26N 3 34 E
Vlóra, Albania 44 D1 40 32N 19 28 E
Vlóra □, Albania 44 D2 40 12N 20 0 E
Vlorës, Gjiri i, Albania 44 D1 40 29N 19 27 E
Vltava →, Czech. 30 A7 50 21N 14 30 E
Vo Dat, Vietnam 77 G6 11 9N 107 31 E
Vobarno, Italy 38 C7 45 39N 10 30 E
Voćin, Croatia 42 B2 45 37N 17 33 E
Vöcklabruck, Austria 30 C6 48 1N 13 39 E
Vodice, Croatia 39 E12 43 47N 15 47 E
Vodňany, Czech. 30 B7 49 9N 14 11 E
Vodnjan, Croatia 39 D10 44 59N 13 52 E
Vogelkop = Doberai, Jazirah, Indonesia 73 B4 1 25 S 133 0 E
Vogelsberg, Germany 26 E5 50 31N 9 12 E
Voghera, Italy 38 D6 44 59N 9 1 E
Voh, N. Cal. 121 T18 20 58 S 164 42 E
Vohibinany, Madag. 105 B8 18 49 S 49 4 E
Vohimarina, Madag. 105 A9 13 25 S 50 0 E
Vohimena, Tanjon' i, Madag. 105 D8 25 36 S 45 8 E
Vohipeno, Madag. 105 C8 22 22 S 47 51 E
Voi, Kenya 106 C4 3 25 S 38 32 E
Void, France 23 D12 48 40N 5 36 E
Voineşti, Iaşi, Romania 46 B8 47 5N 27 27 E
Voineşti, Prahova, Romania 46 D6 45 5N 25 14 E
Voíotía □, Greece 45 F5 38 20N 23 0 E
Voiron, France 25 C9 45 22N 5 35 E
Voisey B., Canada 129 A7 56 15N 61 50W
Voitsberg, Austria 30 D8 47 3N 15 9 E
Voiviïs Límni, Greece 44 E4 39 30N 22 45 E
Vojens, Denmark 15 J3 55 16N 9 18 E
Vojmsjön, Sweden 12 D17 64 55N 16 40 E
Vojnić, Croatia 39 C12 45 19N 15 43 E
Vojnik, Italy 39 B12 46 18N 15 19 E
Vojvodina □, Serbia, Yug. 42 B4 45 20N 20 0 E
Vokhtoga, Russia 50 C11 58 46N 41 8 E
Volary, Czech. 30 C6 48 54N 13 52 E
Volborg, U.S.A. 138 C2 45 51N 105 41W
Volcano Is. = Kazan-Rettō, Pac. Oc. 122 E6 25 0N 141 0 E
Volchansk = Vovchansk, Ukraine 51 G9 50 17N 36 58 E
Volchansk, Russia 54 B8 59 56N 60 4 E
Volchayevka, Russia 57 E14 48 40N 134 30 E
Volchya →, Ukraine 51 H8 48 32N 36 0 E
Volda, Norway 13 E12 62 9N 6 5 E
Volendam, Neths. 20 D6 52 30N 5 4 E
Volga, Russia 52 A4 57 58N 38 16 E
Volga →, Russia 53 G9 46 0N 48 30 E
Volga Hts. = Privolzhskaya Vozvyshennost, Russia 52 E7 51 0N 46 0 E
Volgo-Baltiyskiy Kanal, Russia 50 B9 60 0N 38 0 E
Volgo-Donskoy Kanal, Russia 53 F7 48 40N 43 37 E
Volgodonsk, Russia 53 G6 47 33N 42 5 E
Volgograd, Russia 53 F7 48 40N 44 25 E
Volgogradskoye Vdkhr., Russia 52 E6 50 0N 45 20 E
Volgorechensk, Russia 52 B5 57 28N 41 14 E
Volissós, Greece 45 F7 38 29N 25 54 E
Volkach, Germany 27 F6 49 52N 10 14 E
Volkerak, Neths. 21 E4 51 39N 4 18 E
Völkermarkt, Austria 30 E7 46 39N 14 39 E
Volkhov, Russia 50 C7 59 55N 32 15 E
Volkhov →, Russia 50 B7 60 8N 32 20 E
Völklingen, Germany 27 F2 49 15N 6 50 E
Volkovysk = Vawkavysk, Belarus 51 F3 53 9N 24 30 E
Volksrust, S. Africa 105 D4 27 24 S 29 53 E
Vollenhove, Neths. 20 C7 52 40N 5 58 E
Volnansk, Ukraine 51 H8 48 2N 35 29 E
Volnovakha, Ukraine 51 J9 47 35N 37 30 E
Volochanka, Russia 57 B10 71 0N 94 28 E
Volodarsk, Russia 52 B6 56 12N 43 15 E
Volodymyr-Volynskyy, Ukraine 51 G3 50 50N 24 18 E
Vologda, Russia 50 C10 59 10N 39 45 E
Volokolamsk, Russia 52 B2 56 5N 35 57 E
Volokonovka, Russia 52 E3 50 33N 37 52 E
Vólos, Greece 44 E4 39 24N 22 59 E
Volosovo, Russia 50 C5 59 27N 29 32 E
Volovets, Ukraine 51 H2 48 43N 23 11 E
Volovo, Russia 52 D4 53 35N 38 1 E
Volozhin = Valozhyn, Belarus 50 E4 54 3N 26 30 E
Volsk, Russia 52 D8 52 5N 47 22 E
Volta →, Ghana 101 D5 5 46N 0 41 E
Volta, L., Ghana 101 D5 7 30N 0 15 E
Volta Blanche = White Volta →, Ghana 101 D4 9 10N 1 15W
Volta Redonda, Brazil 155 F3 22 31 S 44 5W
Voltaire, C., Australia 112 B4 14 16 S 125 35 E
Volterra, Italy 38 E7 43 24N 10 51 E
Voltri, Italy 38 D5 44 26N 8 45 E
Volturara Appula, Italy 41 A8 41 30N 15 3 E
Volturno →, Italy 40 A6 41 1N 13 55 E
Volubilis, Morocco 98 B3 34 2N 5 33W
Volujak, Bos.-H. 42 D2 43 53N 17 47 E
Vólvi, L., Greece 44 D5 40 40N 23 34 E
Volvo, Australia 116 A5 31 41 S 143 57 E
Volzhsk, Russia 52 C9 55 57N 48 23 E
Volzhskiy, Russia 53 F7 48 56N 44 46 E
Vondrozo, Madag. 105 C8 22 49 S 47 20 E
Vónitsa, Greece 45 F2 38 53N 20 58 E
Voorburg, Neths. 20 D4 52 5N 4 24 E
Voorne Putten, Neths. 20 E4 51 52N 4 10 E
Voorst, Neths. 20 D8 52 10N 6 8 E
Voorthuizen, Neths. 20 D7 52 11N 5 36 E
Vopnafjörður, Iceland 12 D6 65 45N 14 50W
Vorarlberg □, Austria 30 D2 47 20N 10 0 E
Vóras Óros, Greece 44 D3 40 57N 21 45 E
Vorbasse, Denmark 15 J3 55 39N 9 6 E
Vorden, Neths. 20 D8 52 6N 6 19 E
Vorderrhein →, Switz. 29 C8 46 49N 9 25 E
Vordingborg, Denmark 15 K5 55 0N 11 54 E
Voreppe, France 25 C9 45 18N 5 39 E
Voríai Sporádhes, Greece 45 E5 39 15N 23 30 E
Vórios Evvoïkos Kólpos, Greece 45 F5 38 45N 23 15 E
Vorkuta, Russia 48 A11 67 48N 64 20 E
Vorma →, Norway 14 D5 60 9N 11 29 E
Vormsi, Estonia 13 G20 59 1N 23 13 E
Vorona →, Russia 52 E6 51 22N 42 3 E
Voronezh, Russia 52 E4 51 40N 39 10 E
Voronezh, Ukraine 51 G7 51 47N 33 28 E
Voronezh →, Russia 52 E4 51 32N 39 10 E
Vorontsovo-Aleksandrovskoye = Zelenokumsk, Russia 53 H6 44 24N 44 0 E
Voroshilovgrad = Luhansk, Ukraine 51 H10 48 38N 39 15 E
Voroshilovsk = Alchevsk, Ukraine 51 H10 48 30N 38 45 E
Vorovskoye, Russia 57 D16 54 30N 155 50 E
Vorselaar, Belgium 21 F5 51 12N 4 46 E
Vorskla →, Ukraine 51 H8 48 50N 34 10 E
Vörts Järv, Estonia 13 G22 58 16N 26 3 E
Võru, Estonia 13 H22 57 48N 26 54 E
Vorukh, Kyrgyzstan 55 D5 39 52N 70 35 E
Vorupør, Denmark 15 H2 56 58N 8 22 E
Vosges, France 23 D14 48 20N 7 10 E
Vosges □, France 23 D13 48 12N 6 20 E
Voskopoja, Albania 44 D2 40 40N 20 33 E
Voskresensk, Russia 52 C4 55 19N 38 43 E
Voskresenskoye, Russia 52 B7 56 51N 45 30 E
Voss, Norway 13 F12 60 38N 6 26 E
Vosselaar, Belgium 21 F5 51 19N 4 52 E
Vostok I., Kiribati 123 J12 10 5 S 152 23W
Votice, Czech. 30 B7 49 38N 14 39 E
Votkinsk, Russia 54 C4 57 0N 53 55 E
Votkinskoye Vdkhr., Russia 54 C5 57 22N 55 12 E
Vouga →, Portugal 36 E2 40 41N 8 40W
Vouillé, France 22 F7 46 38N 0 10 E
Voulou, C.A.R. 102 A4 8 33N 22 36 E
Vouvray, France 22 E7 47 25N 0 48 E
Vouvry, Switz. 28 D3 46 21N 6 51 E
Voúxa, Ákra, Greece 32 D5 35 37N 23 32 E
Vouzela, Portugal 36 E2 40 43N 8 7W
Vouziers, France 23 C11 49 22N 4 40 E
Vovchansk, Ukraine 51 G9 50 17N 36 58 E
Voves, France 23 D8 48 15N 1 38 E
Voxna, Sweden 14 C9 61 20N 15 40 E
Vozhe Ozero, Russia 50 B10 60 45N 39 0 E
Vozhega, Russia 50 B11 60 29N 40 12 E
Voznesenka, Russia 57 D10 56 40N 95 3 E
Voznesensk, Ukraine 51 J6 47 35N 31 21 E
Voznesenye, Russia 50 B8 61 0N 35 28 E
Vráble, Slovak Rep. 31 C11 48 15N 18 16 E
Vraćevšnica, Serbia, Yug. 42 C5 44 2N 20 34 E
Vrådal, Norway 14 E2 59 20N 8 25 E
Vraka, Albania 44 B1 42 8N 19 28 E
Vrakhnéïka, Greece 45 F3 38 10N 21 40 E
Vrancea □, Romania 46 D7 45 50N 26 45 E
Vrancei, Munţii, Romania 46 D7 46 0N 26 30 E
Vrangelya, Ostrov, Russia 57 B19 71 0N 180 0 E
Vranica, Bos.-H. 42 D2 43 55N 17 50 E
Vranje, Serbia, Yug. 42 E6 42 34N 21 54 E
Vranjska Banja, Serbia, Yug. 42 E7 42 34N 22 1 E
Vranov, Slovak Rep. 31 C14 48 53N 21 40 E
Vransko, Slovenia 39 B11 46 17N 14 58 E
Vratsa, Bulgaria 43 D8 43 13N 23 30 E
Vrbas, Serbia, Yug. 42 B4 45 40N 19 40 E
Vrbas →, Bos.-H. 42 B2 45 8N 17 29 E
Vrbnik, Croatia 39 C11 45 4N 14 40 E
Vrbovec, Croatia 39 C13 45 53N 16 28 E
Vrbovsko, Croatia 39 C12 45 24N 15 5 E
Vrchlabí, Czech. 30 A8 50 38N 15 37 E
Vrede, S. Africa 105 D4 27 24 S 29 6 E
Vredefort, S. Africa 104 D4 27 0 S 27 22 E
Vredenburg, S. Africa 104 E2 32 56 S 18 0 E
Vredendal, S. Africa 104 E2 31 41 S 18 35 E
Vreeswijk, Neths. 20 D6 52 1N 5 6 E
Vrena, Sweden 15 F10 58 54N 16 41 E
Vrgorac, Croatia 42 D2 43 12N 17 20 E
Vrhnika, Slovenia 39 C11 45 58N 14 15 E
Vriddhachalam, India 83 J4 11 30N 79 20 E
Vridi, Ivory C. 100 D4 5 15N 4 3W
Vries, Neths. 20 B9 53 5N 6 35 E
Vriezenveen, Neths. 20 D9 52 25N 6 38 E
Vrindavan, India 80 F7 27 37N 77 40 E
Vríses, Greece 32 D6 35 23N 24 13 E
Vrnograč, Bos.-H. 39 C12 45 10N 15 57 E
Vrondádhes, Greece 45 F8 38 25N 26 7 E
Vroomshoop, Neths. 20 D9 52 27N 6 34 E
Vrpolje, Croatia 42 B3 45 13N 18 24 E
Vršac, Serbia, Yug. 42 B6 45 8N 21 18 E
Vrsacki Kanal, Serbia, Yug. 42 B5 45 15N 21 0 E
Vryburg, S. Africa 104 D3 26 55 S 24 45 E
Vryheid, S. Africa 105 D5 27 45 S 30 47 E
Vsetín, Czech. 31 B11 49 20N 18 0 E
Vu Liet, Vietnam 76 C5 18 43N 105 23 E
Vucha →, Bulgaria 43 E9 42 10N 24 26 E
Vučitrn, Serbia, Yug. 42 E5 42 49N 20 59 E
Vught, Neths. 21 E6 51 38N 5 20 E
Vukovar, Croatia 42 B3 45 21N 18 59 E

284 Western Samoa

Western Samoa ■, *Pac. Oc.* **121 X24** 14 0 S 172 0W
Westernport, *U.S.A.* **134 F6** 39 29N 79 3W
Westerschelde →, *Neths.* **21 F2** 51 25N 3 25 E
Westerstede, *Germany* **26 B3** 53 15N 7 55 E
Westervoort, *Neths.* **20 E7** 51 58N 5 59 E
Westerwald, *Germany* **26 E4** 50 38N 7 56 E
Westfield, *Ill., U.S.A.* **141 E8** 39 27N 88 0W
Westfield, *Ind., U.S.A.* **141 D10** 40 2N 86 8W
Westfield, *Mass., U.S.A.* **137 D12** 42 7N 72 45W
Westfield, *N.Y., U.S.A.* **136 D5** 42 20N 79 35W
Westfield, *Pa., U.S.A.* **136 E7** 41 55N 77 32W
Westgat, *Neths.* **21 E3** 51 39N 3 44 E
Westhope, *U.S.A.* **138 A4** 48 55N 101 1W
Westkapelle, *Belgium* **21 F2** 51 19N 3 19 E
Westkapelle, *Neths.* **21 E2** 51 31N 3 28 E
Westland, *U.S.A.* **141 B13** 42 15N 83 20W
Westland Bight, *N.Z.* **119 C5** 42 55 S 170 5 E
Westlock, *Canada* **130 C6** 54 9N 113 55W
Westmalle, *Belgium* **21 F5** 51 18N 4 42 E
Westmeath □, *Ireland* **19 C4** 53 33N 7 34W
Westminster, *U.S.A.* **134 F7** 39 34N 76 59W
Westmorland, *U.S.A.* **143 K6** 33 2N 115 37W
Weston, *Malaysia* **75 A5** 5 10N 115 35 E
Weston, *Ohio, U.S.A.* **141 C13** 41 21N 83 47W
Weston, *Oreg., U.S.A.* **142 D4** 45 49N 118 26W
Weston, *W. Va., U.S.A.* **134 F5** 39 2N 80 28W
Weston I., *Canada* **128 B4** 52 33N 79 36W
Weston-super-Mare, *U.K.* **17 F5** 51 21N 2 58W
Westphalia, *U.S.A.* **140 F5** 38 26N 92 0W
Westport, *Canada* **137 B8** 44 40N 76 25W
Westport, *Ireland* **19 C2** 53 48N 9 31W
Westport, *N.Z.* **119 B6** 41 46 S 171 37 E
Westport, *Ind., U.S.A.* **141 E11** 39 11N 85 34W
Westport, *Oreg., U.S.A.* **144 D3** 46 8N 123 23W
Westport, *Wash., U.S.A.* **142 C1** 46 53N 124 6W
Westray, *Canada* **131 C8** 53 36N 101 24W
Westray, *U.K.* **18 B6** 59 18N 3 0W
Westree, *Canada* **128 C3** 47 26N 81 34W
Westville, *Calif., U.S.A.* **144 F6** 39 8N 120 42W
Westville, *Ill., U.S.A.* **141 D9** 40 2N 87 38W
Westville, *Ind., U.S.A.* **141 C10** 41 35N 86 55W
Westville, *Okla., U.S.A.* **139 G7** 35 58N 94 40W
Westwood, *U.S.A.* **142 F3** 40 18N 121 0W
Wetar, *Indonesia* **72 C3** 7 30 S 126 30 E
Wetaskiwin, *Canada* **130 C6** 52 55N 113 24W
Wethersfield, *U.S.A.* **137 E12** 41 42N 72 40W
Wetlet, *Burma* **78 D5** 22 20N 95 53 E
Wetteren, *Belgium* **21 G3** 51 0N 3 52 E
Wettingen, *Switz.* **29 B6** 47 28N 8 20 E
Wetzikon, *Switz.* **29 B7** 47 19N 8 48 E
Wetzlar, *Germany* **26 E4** 50 32N 8 31 E
Wevelgem, *Belgium* **21 G2** 50 49N 3 12 E
Wewak, *Papua N. G.* **120 B2** 3 38 S 143 41 E
Wewoka, *U.S.A.* **139 H6** 35 9N 96 30W
Wexford, *Ireland* **19 D5** 52 20N 6 28W
Wexford □, *Ireland* **19 D5** 52 20N 6 25W
Wexford Harbour, *Ireland* **19 D5** 52 20N 6 25W
Weyburn, *Canada* **131 D8** 49 40N 103 50W
Weyburn L., *Canada* **130 A5** 63 0N 117 59W
Weyer, *Austria* **30 D7** 47 51N 14 40 E
Weyib →, *Ethiopia* **95 F5** 7 15N 40 15 E
Weymouth, *Canada* **129 D6** 44 30N 66 1W
Weymouth, *U.K.* **17 G5** 50 37N 2 28W
Weymouth, *U.S.A.* **137 D14** 42 13N 70 58W
Weymouth, C., *Australia* **114 A3** 12 37 S 143 27 E
Wezemaal, *Belgium* **21 G5** 50 57N 4 45 E
Wezep, *Neths.* **20 D7** 52 28N 6 0 E
Whakamaru, *N.Z.* **118 E4** 38 23 S 175 50 E
Whakatane, *N.Z.* **118 D6** 37 57 S 177 1 E
Whakatane →, *N.Z.* **118 D6** 37 57 S 177 1 E
Whale →, *Canada* **129 A6** 58 15N 67 40W
Whale Cove, *Canada* **131 A10** 62 11N 92 36W
Whales, B. of, *Antarctica* **7 D12** 78 0 S 165 0W
Whalsay, *U.K.* **18 A7** 60 22N 0 59W
Whangaehu →, *N.Z.* **118 G4** 40 3 S 175 6 E
Whangamata, *N.Z.* **118 D4** 37 12 S 175 53 E
Whangamomona, *N.Z.* **118 F3** 39 8 S 174 44 E
Whangarei, *N.Z.* **118 B3** 35 43 S 174 21 E
Whangarei Harb., *N.Z.* **118 B3** 35 45 S 174 28 E
Whangaroa Harb., *N.Z.* **118 B2** 35 4 S 173 46 E
Whangaruru Harb., *N.Z.* **118 B3** 35 24 S 174 23 E
Wharanui, *N.Z.* **119 B9** 41 55 S 174 6 E
Wharfe →, *U.K.* **16 D6** 53 51N 1 9W
Wharfedale, *U.K.* **16 C5** 54 6N 2 1W
Wharton, *N.J., U.S.A.* **137 F10** 40 54N 74 35W
Wharton, *Pa., U.S.A.* **136 E6** 41 31N 78 1W
Wharton, *Tex., U.S.A.* **139 L6** 29 19N 96 6W
Whataroa, *N.Z.* **119 D5** 43 18 S 170 24 E
Whataroa →, *N.Z.* **119 D5** 43 7 S 170 16 E
Wheatfield, *U.S.A.* **141 C9** 41 13N 87 4W
Wheatland, *Calif., U.S.A.* **144 F5** 39 1N 121 25W
Wheatland, *Ind., U.S.A.* **141 F9** 38 40N 87 19W
Wheatland, *Wyo., U.S.A.* **138 D2** 42 3N 104 58W
Wheatley, *Canada* **136 D2** 42 6N 82 27W
Wheaton, *Ill., U.S.A.* **141 C8** 41 52N 88 6W
Wheaton, *Minn., U.S.A.* **138 C6** 45 48N 96 30W
Wheelbarrow Pk., *U.S.A.* **144 H10** 37 26N 116 5W
Wheeler, *Oreg., U.S.A.* **142 D2** 45 41N 123 53W
Wheeler, *Tex., U.S.A.* **139 H4** 35 27N 100 16W
Wheeler →, *Canada* **131 B7** 57 25N 105 30W
Wheeler Pk., *N. Mex., U.S.A.* **143 H11** 36 34N 105 25W
Wheeler Pk., *Nev., U.S.A.* **143 G6** 38 57N 114 15W
Wheeler Ridge, *U.S.A.* **145 L8** 35 0N 118 57W
Wheeling, *U.S.A.* **136 F4** 40 4N 80 43W
Whernside, *U.K.* **16 C5** 54 14N 2 24W
Whidbey I., *U.S.A.* **130 D4** 48 12N 122 17W
Whiskey Gap, *Canada* **130 D6** 49 0N 113 3W
Whiskey Jack L., *Canada* **131 B8** 58 23N 101 55W
Whistleduck Cr. →, *Australia* **114 C2** 20 15 S 135 18 E
Whitby, *Canada* **136 C6** 43 52N 78 56W
Whitby, *U.K.* **16 C7** 54 29N 0 37W
Whitcombe Pass, *N.Z.* **119 D5** 43 13 S 170 55 E
White →, *Ark., U.S.A.* **139 J9** 33 57N 91 5W
White →, *Ind., U.S.A.* **141 F9** 38 25N 87 45W
White →, *S. Dak., U.S.A.* **138 D5** 43 42N 99 27W
White →, *Utah, U.S.A.* **142 F9** 40 4N 109 41W
White →, *Wash., U.S.A.* **144 C4** 47 12N 122 15W
White, East Fork →, *U.S.A.* **141 F9** 38 33N 87 14W
White, L., *Australia* **112 D4** 21 9 S 128 56 E
White B., *Canada* **129 B8** 50 0N 56 35W
White Bear Res., *Canada* **129 C8** 48 10N 57 5W
White Bird, *U.S.A.* **142 D5** 45 46N 116 18W
White Butte, *U.S.A.* **138 B3** 46 23N 103 18W
White City, *U.S.A.* **138 F6** 38 48N 96 44W
White Cliffs, *Australia* **116 A5** 30 50 S 143 10 E
White Deer, *U.S.A.* **139 H4** 35 26N 101 10W
White Hall, *U.S.A.* **140 E6** 39 26N 90 24W
White Haven, *U.S.A.* **137 E9** 41 4N 75 47W
White Horse, Vale of, *U.K.* **17 F6** 51 37N 1 30W
White I., *N.Z.* **118 D6** 37 30 S 177 13 E
White L., *Canada* **137 A8** 45 18N 76 31W
White L., *U.S.A.* **139 L8** 29 44N 92 30W
White Mts., *Calif., U.S.A.* **144 H8** 37 30N 118 15W
White Mts., *N.H., U.S.A.* **137 B13** 44 15N 71 15W
White Nile = Nîl el Abyad →, *Sudan* **95 D3** 15 38N 32 31 E
White Nile Dam = Khazzân Jabal el Awliyâ, *Sudan* **95 D3** 15 24N 32 20 E
White Otter L., *Canada* **128 C1** 49 5N 91 55W
White Pass, *Canada* **130 B1** 59 40N 135 3W
White Pass, *U.S.A.* **144 D5** 46 38N 121 24W
White Pigeon, *U.S.A.* **141 C11** 41 48N 85 39W
White Plains, *U.S.A.* **137 E11** 41 2N 73 46W
White River, *Canada* **128 C2** 48 35N 85 20W
White River, *S. Africa* **105 D5** 25 20 S 31 0 E
White River, *U.S.A.* **138 D4** 43 34N 100 45W
White Russia = Belarus ■, *Europe* **50 F4** 53 30N 27 0 E
White Sea = Beloye More, *Russia* **48 A6** 66 30N 38 0 E
White Sulphur Springs, *Mont., U.S.A.* **142 C8** 46 33N 110 54W
White Sulphur Springs, *W. Va., U.S.A.* **134 G5** 37 48N 80 18W
White Swan, *U.S.A.* **144 D6** 46 23N 120 44W
White Volta →, *Ghana* **101 D4** 9 10N 1 15W
Whitecliffs, *N.Z.* **119 D6** 43 26 S 171 55 E
Whitecourt, *Canada* **130 C5** 54 10N 115 45W
Whiteface, *U.S.A.* **139 J3** 33 36N 102 37W
Whitefield, *U.S.A.* **137 B13** 44 23N 71 37W
Whitefish, *U.S.A.* **142 B6** 48 25N 114 20W
Whitefish Bay, *U.S.A.* **141 A9** 43 23N 87 54W
Whitefish L., *Canada* **131 A7** 62 41N 106 48W
Whitefish Point, *U.S.A.* **134 B3** 46 45N 84 59W
Whitegull, L., *Canada* **129 A7** 55 27N 64 17W
Whitehall, *Mich., U.S.A.* **134 D2** 43 24N 86 21W
Whitehall, *Mont., U.S.A.* **142 D7** 45 52N 112 6W
Whitehall, *N.Y., U.S.A.* **137 C11** 43 33N 73 24W
Whitehall, *Wis., U.S.A.* **138 C9** 44 22N 91 19W
Whitehaven, *U.K.* **16 C4** 54 33N 3 35W
Whitehorse, *Canada* **130 A1** 60 43N 135 3W
Whiteman, *U.S.A.* **140 F3** 38 45N 93 40W
Whiteman Ra., *Papua N. G.* **120 C5** 5 55 S 150 0 E
Whitemark, *Australia* **114 G4** 40 7 S 148 3 E
Whitemouth, *Canada* **131 D9** 49 57N 95 58W
Whiteplains, *Liberia* **100 D2** 6 28N 10 40W
Whitesboro, *N.Y., U.S.A.* **137 C9** 43 7N 75 18W
Whitesboro, *Tex., U.S.A.* **139 J6** 33 39N 96 54W
Whiteshell Prov. Park, *Canada* **131 C9** 50 0N 95 40W
Whiteside, *U.S.A.* **140 E5** 39 12N 91 2W
Whiteside, Canal, *Chile* **160 D2** 53 55 S 70 15W
Whitetail, *U.S.A.* **138 A2** 48 54N 105 10W
Whiteville, *U.S.A.* **135 H6** 34 20N 78 42W
Whitewater, *U.S.A.* **141 B8** 42 50N 88 44W
Whitewater Baldy, *U.S.A.* **143 K9** 33 20N 108 39W
Whitewater L., *Canada* **128 B2** 50 50N 89 10W
Whitewood, *Australia* **114 C3** 21 28 S 143 30 E
Whitewood, *Canada* **131 C8** 50 20N 102 20W
Whitfield, *Australia* **117 D7** 36 42 S 146 24 E
Whithorn, *U.K.* **18 G4** 54 44N 4 26W
Whitianga, *N.Z.* **118 D4** 36 47 S 175 41 E
Whiting, *U.S.A.* **141 C9** 41 41N 87 29W
Whitman, *U.S.A.* **137 D14** 42 5N 70 56W
Whitmire, *U.S.A.* **135 H5** 34 30N 81 37W
Whitney, *Canada* **128 C4** 45 31N 78 14W
Whitney, Mt., *U.S.A.* **144 J8** 36 35N 118 18W
Whitney Point, *U.S.A.* **137 D9** 42 20N 75 58W
Whitstable, *U.K.* **17 F9** 51 21N 1 3 E
Whitsunday I., *Australia* **114 C4** 20 15 S 149 4 E
Whittemore, *U.S.A.* **140 A2** 43 4N 94 26W
Whittier, *U.S.A.* **145 M8** 33 58N 118 3W
Whittlesea, *Australia* **117 D6** 37 27 S 145 9 E
Whitwell, *U.S.A.* **135 H3** 35 12N 85 31W
Wholdaia L., *Canada* **131 A8** 60 43N 104 20W
Whyalla, *Australia* **116 B2** 33 2 S 137 30 E
Whyjonta, *Australia* **115 D3** 29 41 S 142 28 E
Wiarton, *Canada* **128 D3** 44 40N 81 10W
Wiawso, *Ghana* **100 D4** 6 10N 2 25W
Wiazów, *Poland* **47 E4** 50 50N 17 10 E
Wibaux, *U.S.A.* **138 B2** 46 59N 104 11W
Wichabai, *Guyana* **153 C6** 2 57N 59 35W
Wichian Buri, *Thailand* **76 E3** 15 39N 101 7 E
Wichita, *U.S.A.* **139 G6** 37 42N 97 20W
Wichita Falls, *U.S.A.* **139 J5** 33 54N 98 30W
Wick, *U.K.* **18 C5** 58 26N 3 5W
Wickenburg, *U.S.A.* **143 K7** 33 58N 112 44W
Wickepin, *Australia* **113 F2** 32 50 S 117 30 E
Wickham, C., *Australia* **114 F3** 39 35 S 143 57 E
Wickliffe, *U.S.A.* **136 E3** 41 36N 81 28W
Wicklow, *Ireland* **19 D5** 52 59N 6 3W
Wicklow □, *Ireland* **19 D5** 52 57N 6 25W
Wicklow Hd., *Ireland* **19 D5** 52 58N 6 0W
Widawa, *Poland* **47 D5** 51 27N 18 51 E
Widawka, *Poland* **47 D6** 51 7N 19 36 E
Widgiemooltha, *Australia* **113 F3** 31 30 S 121 34 E
Widnes, *U.K.* **16 D5** 53 23N 2 45W
Więcbork, *Poland* **47 B4** 53 21N 17 30 E
Wiedenbrück, *Germany* **26 D4** 51 52N 8 15 E
Wiehl, *Germany* **26 E3** 50 56N 7 34 E
Wiek, *Germany* **26 A9** 54 37N 13 17 E
Wielbark, *Poland* **47 B7** 53 24N 20 55 E
Wieleń, *Poland* **47 C3** 52 53N 16 9 E
Wieliczka, *Poland* **31 B13** 50 0N 20 5 E
Wieluń, *Poland* **47 D5** 51 15N 18 34 E
Wien, *Austria* **31 C9** 48 12N 16 22 E
Wiener Neustadt, *Austria* **31 D9** 47 49N 16 16 E
Wieprz →, *Koszalin, Poland* **47 A3** 54 26N 16 35 E
Wieprz →, *Lublin, Poland* **47 D8** 51 34N 21 49 E
Wierden, *Neths.* **20 D9** 52 22N 6 35 E
Wiers, *Belgium* **21 H3** 50 30N 3 32 E
Wieruszów, *Poland* **47 D5** 51 19N 18 9 E
Wiesbaden, *Germany* **27 E4** 50 4N 8 14 E
Wiesental, *Germany* **27 F4** 49 13N 8 31 E
Wigan, *U.K.* **16 D5** 53 33N 2 38W
Wiggins, *Colo., U.S.A.* **138 E2** 40 14N 104 4W
Wiggins, *Miss., U.S.A.* **139 K10** 30 51N 89 8W
Wight, I. of □, *U.K.* **17 G6** 50 40N 1 20W
Wigry, Jezioro, *Poland* **47 A10** 54 2N 23 8 E
Wigtown, *U.K.* **18 G4** 54 53N 4 27W
Wigtown B., *U.K.* **18 G4** 54 46N 4 15W
Wijchen, *Neths.* **20 E7** 51 48N 5 44 E
Wijhe, *Neths.* **20 D8** 52 23N 6 8 E
Wijk bij Duurstede, *Neths.* **20 E6** 51 59N 5 21 E
Wil, *Switz.* **29 B8** 47 28N 9 3 E
Wilamowice, *Poland* **31 B12** 49 55N 19 9 E
Wilangee, *Australia* **116 A4** 31 28 S 141 20 E
Wilber, *U.S.A.* **138 E6** 40 29N 96 58W
Wilberforce, *Canada* **136 A6** 45 2N 78 13W
Wilberforce, C., *Australia* **114 A2** 11 54 S 136 35 E
Wilburton, *U.S.A.* **139 H7** 34 55N 95 19W
Wilcannia, *Australia* **116 A5** 31 30 S 143 26 E
Wilcox, *U.S.A.* **136 E6** 41 35N 78 41W
Wildbad, *Germany* **27 G4** 48 44N 8 33 E
Wildcat →, *U.S.A.* **141 D10** 40 28N 86 52W
Wildervank, *Neths.* **20 B9** 53 5N 6 52 E
Wildeshausen, *Germany* **26 C4** 52 54N 8 27 E
Wildhorn, *Switz.* **28 D4** 46 22N 7 21 E
Wildon, *Austria* **30 E8** 46 52N 15 31 E
Wildrose, *Calif., U.S.A.* **145 J9** 36 14N 117 11W
Wildrose, *N. Dak., U.S.A.* **138 A3** 48 38N 103 11W
Wildspitze, *Austria* **30 E3** 46 53N 10 53 E
Wildstrubel, *Switz.* **28 D5** 46 24N 7 32 E
Wildwood, *U.S.A.* **134 F8** 38 59N 74 50W
Wilga →, *Poland* **47 D8** 51 52N 21 18 E
Wilgaroon, *Australia* **117 A6** 30 52 S 145 42 E
Wilge →, *S. Africa* **105 D4** 27 3 S 28 20 E
Wilhelm, Mt., *Papua N. G.* **120 C3** 5 50 S 145 1 E
Wilhelm II Coast, *Antarctica* **7 C7** 68 0 S 90 0 E
Wilhelm-Pieck-Stadt-Guben, *Germany* **26 D10** 51 57N 14 42 E
Wilhelmina, Geb., *Surinam* **153 C6** 3 50N 56 30W
Wilhelmina Kanaal, *Neths.* **21 E6** 51 36N 5 6 E
Wilhelmsburg, *Austria* **30 C8** 48 6N 15 36 E
Wilhelmshaven, *Germany* **26 B4** 53 31N 8 7 E
Wilhelmstal, *Namibia* **104 C2** 21 58 S 16 21 E
Wilkes-Barre, *U.S.A.* **137 E9** 41 15N 75 53W
Wilkesboro, *U.S.A.* **135 G5** 36 9N 81 10W
Wilkie, *Canada* **131 C7** 52 27N 108 42W
Wilkinsburg, *U.S.A.* **136 F5** 40 26N 79 53W
Wilkinson Lakes, *Australia* **113 E5** 29 40 S 132 39 E
Willamina, *U.S.A.* **142 D2** 45 5N 123 29W
Willamulka, *Australia* **116 B2** 33 55 S 137 52 E
Willandra Billabong Creek →, *Australia* **116 B6** 33 22 S 145 52 E
Willapa B., *U.S.A.* **142 C2** 46 40N 124 0W
Willapa Hills, *U.S.A.* **144 D3** 46 35N 123 25W
Willard, *N. Mex., U.S.A.* **143 J10** 34 36N 106 2W
Willard, *Utah, U.S.A.* **142 F7** 41 25N 112 2W
Willaura, *Australia* **116 D5** 37 31 S 142 45 E
Willbriggie, *Australia* **117 C7** 34 28 S 146 2 E
Willcox, *U.S.A.* **143 K9** 32 15N 109 50W
Willebroek, *Belgium* **21 F4** 51 4N 4 22 E
Willemstad, *Neth. Ant.* **149 D6** 12 5N 69 0W
Willeroo, *Australia* **112 C5** 15 14 S 131 37 E
William →, *Canada* **131 B7** 59 8N 109 19W
William, Mt., *Australia* **116 D5** 37 17 S 142 35 E
William Creek, *Australia* **115 D2** 28 58 S 136 22 E
Williambury, *Australia* **113 D2** 23 45 S 115 12 E
Williams, *Australia* **113 F2** 33 2 S 116 52 E
Williams, *Ariz., U.S.A.* **143 J7** 35 15N 112 11W
Williams, *Calif., U.S.A.* **144 F4** 39 9N 122 9W
Williams Lake, *Canada* **130 C4** 52 10N 122 10W
Williamsburg, *Ky., U.S.A.* **135 G3** 36 44N 84 10W
Williamsburg, *Pa., U.S.A.* **136 F6** 40 28N 78 12W
Williamsburg, *Va., U.S.A.* **134 G7** 37 17N 76 44W
Williamsfield, *U.S.A.* **140 D6** 40 55N 90 1W
Williamson, *N.Y., U.S.A.* **136 C7** 43 14N 77 11W
Williamson, *W. Va., U.S.A.* **134 G4** 37 41N 82 17W
Williamsport, *Ind., U.S.A.* **141 D9** 40 17N 87 17W
Williamsport, *Pa., U.S.A.* **136 E7** 41 15N 77 0W
Williamston, *Mich., U.S.A.* **141 B12** 42 41N 84 17W
Williamston, *N.C., U.S.A.* **135 H7** 35 51N 77 4W
Williamstown, *Australia* **117 D6** 37 51 S 144 52 E
Williamstown, *Ky., U.S.A.* **141 F12** 38 38N 84 34W
Williamstown, *Mass., U.S.A.* **137 D11** 42 41N 73 12W
Williamstown, *N.Y., U.S.A.* **137 C9** 43 26N 75 53W
Williamsville, *Ill., U.S.A.* **140 E7** 39 57N 89 33W
Williamsville, *Mo., U.S.A.* **139 G9** 36 58N 90 33W
Willimantic, *U.S.A.* **137 E12** 41 43N 72 13W
Willis Group, *Australia* **114 B5** 16 18 S 150 0 E
Willisau, *Switz.* **28 B6** 47 7N 8 0 E
Willisburg, *U.S.A.* **141 G11** 37 49N 85 8W
Williston, *S. Africa* **104 E3** 31 20 S 20 53 E
Williston, *Fla., U.S.A.* **135 L4** 29 23N 82 27W
Williston, *N. Dak., U.S.A.* **138 A3** 48 9N 103 37W
Williston L., *Canada* **130 B4** 56 0N 124 0W
Willits, *U.S.A.* **142 G2** 39 25N 123 21W
Willmar, *U.S.A.* **138 C7** 45 7N 95 3W
Willoughby, *U.S.A.* **136 E3** 41 39N 81 24W
Willow Bunch, *Canada* **131 D7** 49 20N 105 35W
Willow L., *Canada* **130 A5** 62 10N 119 8W
Willow Lake, *U.S.A.* **138 C6** 44 38N 97 38W
Willow Springs, *U.S.A.* **139 G8** 37 0N 91 58W
Willow Tree, *Australia* **117 A9** 31 40 S 150 45 E
Willow Wall, The, *China* **67 C12** 42 10N 122 0 E
Willowlake →, *Canada* **130 A4** 62 42N 123 8W
Willowmore, *S. Africa* **104 E3** 33 15 S 23 30 E
Willows, *Australia* **114 C4** 23 39 S 147 25 E
Willows, *U.S.A.* **144 F4** 39 31N 122 10W
Willowvale = Gatyana, *S. Africa* **105 E4** 32 16 S 28 31 E
Wills, L., *Australia* **112 D4** 21 25 S 128 51 E
Wills Cr. →, *Australia* **114 C3** 22 43 S 140 2 E
Wills Point, *U.S.A.* **139 J7** 32 43N 96 1W
Willunga, *Australia* **116 C3** 35 15 S 138 30 E
Wilmette, *U.S.A.* **134 D2** 42 5N 87 42W
Wilmington, *Australia* **116 B3** 32 39 S 138 7 E
Wilmington, *Del., U.S.A.* **134 F8** 39 45N 75 33W
Wilmington, *Ill., U.S.A.* **141 C8** 41 18N 88 9W
Wilmington, *N.C., U.S.A.* **135 H7** 34 14N 77 55W
Wilmington, *Ohio, U.S.A.* **141 E13** 39 27N 83 50W
Wilpena Cr. →, *Australia* **116 A3** 31 25 S 139 29 E
Wilrijk, *Belgium* **21 F4** 51 9N 4 22 E
Wilsall, *U.S.A.* **142 D8** 45 59N 110 38W
Wilson, *U.S.A.* **135 H7** 35 44N 77 55W
Wilson →, *Queens., Australia* **115 D3** 27 38 S 141 24 E
Wilson →, *W. Austral., Australia* **112 C4** 16 48 S 128 16 E
Wilson Bluff, *Australia* **113 F4** 31 41 S 129 0 E
Wilson Str., *Solomon Is.* **121 M9** 8 0 S 156 39 E
Wilsons Promontory, *Australia* **117 E7** 38 55 S 146 25 E
Wilster, *Germany* **26 B5** 53 55N 9 23 E
Wilton, *U.K.* **17 F6** 51 5N 1 51W
Wilton, *U.S.A.* **138 B4** 47 10N 100 47W
Wilton →, *Australia* **114 A1** 14 45 S 134 33 E
Wiltshire □, *U.K.* **17 F6** 51 18N 1 53W
Wiltz, *Lux.* **21 J7** 49 57N 5 55 E
Wiluna, *Australia* **113 E3** 26 36 S 120 14 E
Wimereux, *France* **23 B8** 50 45N 1 37 E
Wimmera →, *Australia* **116 D4** 36 8 S 141 56 E
Winam G., *Kenya* **106 C3** 0 20 S 34 15 E
Winamac, *U.S.A.* **141 C10** 41 3N 86 36W
Winburg, *S. Africa* **104 D4** 28 30 S 27 2 E
Winchelsea, *Australia* **116 E6** 38 10 S 144 1 E
Winchendon, *U.S.A.* **137 D12** 42 41N 72 3W
Winchester, *N.Z.* **119 E6** 44 11 S 171 17 E
Winchester, *U.K.* **17 F6** 51 4N 1 18W
Winchester, *Conn., U.S.A.* **137 E11** 41 53N 73 9W
Winchester, *Idaho, U.S.A.* **142 C5** 46 14N 116 38W
Winchester, *Ill., U.S.A.* **140 E6** 39 38N 90 27W
Winchester, *Ind., U.S.A.* **141 D12** 40 10N 84 59W
Winchester, *Ky., U.S.A.* **141 G12** 38 0N 84 11W
Winchester, *N.H., U.S.A.* **137 D12** 42 46N 72 23W
Winchester, *Nev., U.S.A.* **145 J11** 36 6N 115 10W
Winchester, *Ohio, U.S.A.* **141 F13** 38 57N 83 40W
Winchester, *Tenn., U.S.A.* **135 H2** 35 11N 86 7W
Winchester, *Va., U.S.A.* **134 F6** 39 11N 78 10W
Wind →, *U.S.A.* **142 E9** 43 12N 108 12W
Wind Point, *U.S.A.* **141 B9** 42 47N 87 46W
Wind River Range, *U.S.A.* **142 E9** 43 0N 109 30W
Windau = Ventspils, *Latvia* **13 H19** 57 25N 21 32 E
Windber, *U.S.A.* **136 F6** 40 14N 78 50W
Windermere, L., *U.K.* **16 C5** 54 22N 2 56W
Windfall, *Canada* **130 C5** 54 12N 116 13W
Windfall, *U.S.A.* **141 D11** 40 22N 85 57W
Windflower L., *Canada* **130 A5** 62 52N 118 30W
Windhoek, *Namibia* **104 C2** 22 35 S 17 4 E
Windischgarsten, *Austria* **30 D7** 47 42N 14 21 E
Windom, *U.S.A.* **138 D7** 43 52N 95 7W
Windorah, *Australia* **114 D3** 25 24 S 142 36 E
Window Rock, *U.S.A.* **143 J9** 35 41N 109 3W
Windrush →, *U.K.* **17 F6** 51 43N 1 24W
Windsor, *Australia* **117 B9** 33 37 S 150 50 E
Windsor, *N.S., Canada* **129 D7** 44 59N 64 5W
Windsor, *Nfld., Canada* **129 C8** 48 57N 55 40W
Windsor, *Ont., Canada* **128 D3** 42 18N 83 0W
Windsor, *N.Z.* **119 E5** 44 59 S 170 49 E
Windsor, *U.K.* **17 F7** 51 29N 0 36W
Windsor, *Colo., U.S.A.* **138 E2** 40 29N 104 54W
Windsor, *Conn., U.S.A.* **137 E12** 41 50N 72 39W
Windsor, *Ill., U.S.A.* **141 E8** 39 26N 88 36W
Windsor, *Mo., U.S.A.* **140 F3** 38 32N 93 31W
Windsor, *N.Y., U.S.A.* **137 D9** 42 5N 75 37W
Windsor, *Vt., U.S.A.* **137 C12** 43 29N 72 24W
Windsorton, *S. Africa* **104 D3** 28 16 S 24 44 E
Windward Is., *W. Indies* **149 D7** 13 0N 61 0W
Windward Passage = Vientos, Paso de los, *Caribbean* **149 C5** 20 0N 74 0W
Windy L., *Canada* **131 A8** 60 20N 100 2W
Winefred L., *Canada* **131 B6** 55 30N 110 30W
Winejok, *Sudan* **95 F2** 9 1N 27 30 E
Winfield, *Iowa, U.S.A.* **140 C5** 41 7N 91 26W
Winfield, *Kans., U.S.A.* **139 G6** 37 15N 96 59W
Winfield, *Mo., U.S.A.* **140 E6** 39 0N 90 44W
Wingate Mts., *Australia* **112 B5** 14 25 S 130 40 E
Wingen, *Australia* **117 A9** 31 54 S 150 54 E
Wingene, *Belgium* **21 F2** 51 3N 3 17 E
Wingham, *Australia* **117 A10** 31 48 S 152 22 E
Wingham, *Canada* **128 D3** 43 55N 81 20W
Winifred, *U.S.A.* **142 C9** 47 34N 109 23W
Winisk, *Canada* **128 A2** 55 20N 85 15W
Winisk →, *Canada* **128 A2** 55 17N 85 5W
Winisk L., *Canada* **128 B2** 52 55N 87 22W
Wink, *U.S.A.* **139 K3** 31 45N 103 9W
Winkler, *Canada* **131 D9** 49 10N 97 56W
Winklern, *Austria* **30 E5** 46 52N 12 52 E
Winlock, *U.S.A.* **144 D4** 46 30N 122 56W
Winneba, *Ghana* **101 D4** 5 25N 0 36W
Winnebago, *Ill., U.S.A.* **140 B7** 42 16N 89 15W
Winnebago, *Minn., U.S.A.* **138 D7** 43 46N 94 10W
Winnebago, L., *U.S.A.* **134 D1** 44 0N 88 26W
Winnecke Cr. →, *Australia* **112 C5** 18 35 S 131 34 E
Winnemucca, *U.S.A.* **142 F5** 40 58N 117 44W
Winnemucca L., *U.S.A.* **142 F4** 40 7N 119 21W
Winner, *U.S.A.* **138 D5** 43 22N 99 52W
Winnett, *U.S.A.* **142 C9** 47 0N 108 21W
Winnfield, *U.S.A.* **139 K8** 31 56N 92 38W
Winnibigoshish, L., *U.S.A.* **138 B7** 47 27N 94 13W
Winning, *Australia* **112 D1** 23 9 S 114 30 E
Winnipeg, *Canada* **131 D9** 49 54N 97 9W
Winnipeg →, *Canada* **131 C9** 50 38N 96 19W
Winnipeg, L., *Canada* **131 C9** 52 0N 97 0W
Winnipeg Beach, *Canada* **131 C9** 50 30N 96 58W
Winnipegosis, *Canada* **131 C9** 51 39N 99 55W
Winnipegosis L., *Canada* **131 C9** 52 30N 100 0W
Winnipesaukee, L., *U.S.A.* **137 C13** 43 38N 71 21W
Winnsboro, *La., U.S.A.* **139 J9** 32 10N 91 43W
Winnsboro, *S.C., U.S.A.* **135 H5** 34 23N 81 5W
Winnsboro, *Tex., U.S.A.* **139 J7** 32 58N 95 17W
Winokapau, L., *Canada* **129 B7** 53 15N 62 50W
Winona, *Minn., U.S.A.* **138 C9** 44 3N 91 39W
Winona, *Miss., U.S.A.* **139 J10** 33 29N 89 44W
Winooski, *U.S.A.* **137 B11** 44 29N 73 11W
Winschoten, *Neths.* **20 B10** 53 9N 7 3 E
Winsen, *Germany* **26 B6** 53 22N 10 13 E
Winslow, *Ariz., U.S.A.* **143 J8** 35 2N 110 42W
Winslow, *Ind., U.S.A.* **141 F9** 38 23N 87 13W
Winslow, *Wash., U.S.A.* **144 C4** 47 38N 122 31W
Winsted, *U.S.A.* **137 E11** 41 55N 73 4W

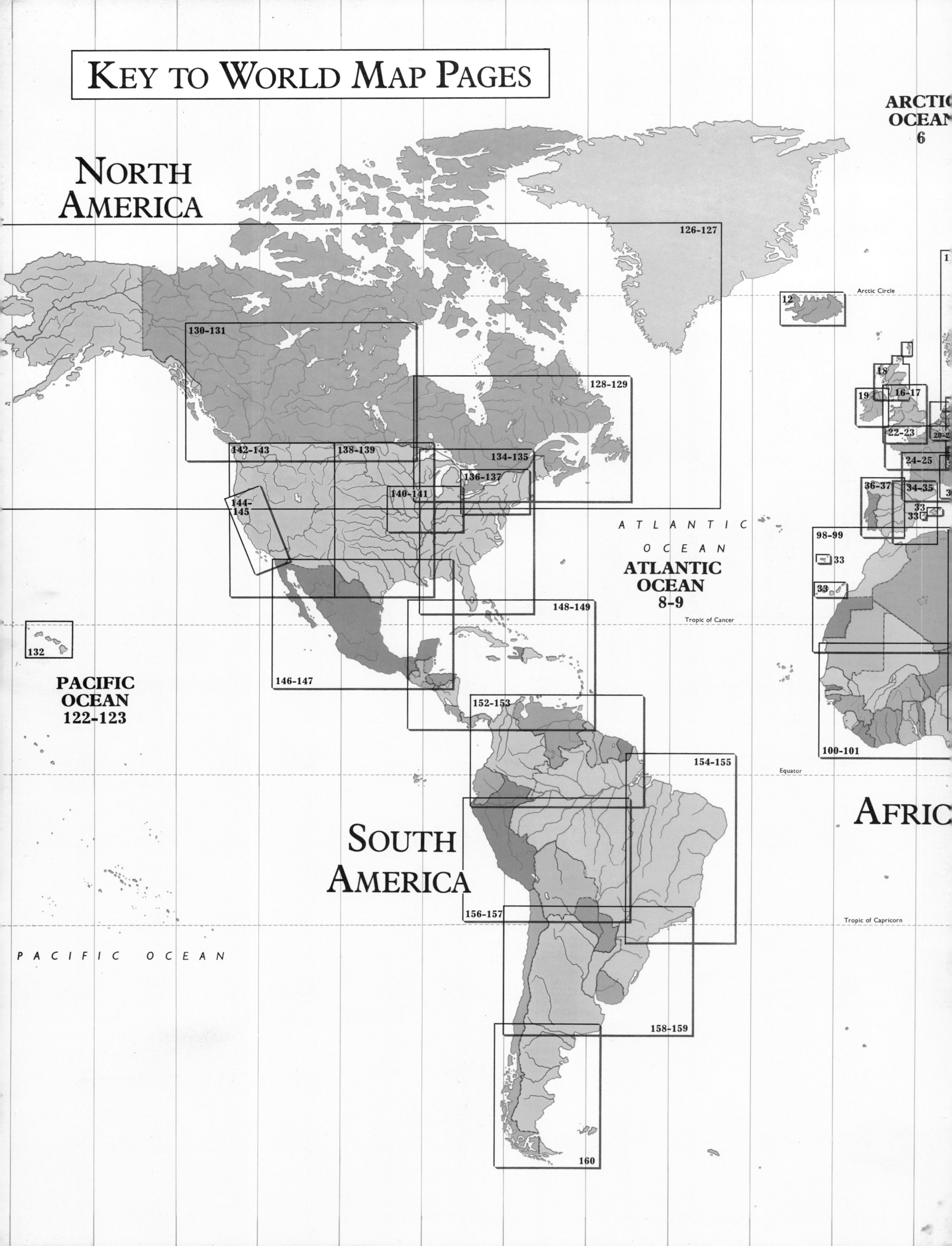

KEY TO WORLD MAP PAGES

NORTH AMERICA

ARCTIC OCEAN 6

Arctic Circle

126-127

130-131

128-129

142-143   138-139   134-135

136-137

140-141

144-145

ATLANTIC OCEAN

ATLANTIC OCEAN 8-9

Tropic of Cancer

148-149

132

146-147

PACIFIC OCEAN 122-123

152-153

SOUTH AMERICA

154-155

Equator

AFRICA

156-157

Tropic of Capricorn

PACIFIC OCEAN

158-159

160

12

18

19   16-17

22-23   20-2

24-25

36-37   34-35

33

33

98-99   33

33

100-101